Survey of

ORGANIC SYNTHESES

SURVEY OF

ORGANIC SYNTHESES

CALVIN A. BUEHLER

University of Tennessee

DONALD E. PEARSON

Vanderbilt University

Wiley-Interscience

A DIVISION OF JOHN WILEY & SONS, INC.

New York · London · Sydney · Toronto

Library of Congress Catalogue Card Number: 73-112590

SBN 471 11670 X

Printed in the United States of America

10 9 8 7 6 5 4 3 2

PREFACE

In this book, designed for the advanced undergraduate and graduate student, and the research chemist, we have attempted to bring together the principal methods for synthesizing the main types of organic compounds. To keep the work within the bounds of one volume, we deal largely with hydrocarbons and derivatives possessing functional groups that contain carbon, hydrogen, oxygen, nitrogen, or the halogens. What to include and what to omit has often been a difficult problem. We have discussed some rather uncommon methods because of their potential if studied more thoroughly.

The book consists of 20 chapters that discuss the different functional groups of interest. *How the functional group is created from other functional groups is the main concern of the work.* The method of classification to be used within each chapter has presented certain problems. An ideal one would permit a reaction to be covered at one place only. Such a goal could not be achieved because of the variable nature of some organic reactions and because many of the groups of interest are polyfunctional. The breakdown of each chapter is therefore often arbitrary. Typical chapter divisions are: (*a*) oxidation or reduction, each of which takes advantage of a difference in oxidation level;[1] (*b*) solvolysis or metathesis in which the oxidation level remains the same; (*c*) electrophilic reactions that involve the addition (or substitution) of some electron-deficient species to an unsaturated linkage; (*d*) nucleophilic reactions in which almost always a carbanion adds to an unsaturated linkage; (*e*) cycloaddition; and (*f*) free radical reactions. By the use of cross references duplication is kept to a minimum.

The pattern in which each reaction is covered follows a regular plan. First, there is a general equation illustrating the reaction. Second, the reaction is discussed with emphasis on its particular value and its limitations. Third, a

---

[1] The various oxidation levels with types of compound in each are the following. (*a*) Hydrocarbon level: lowest oxidation state which includes alkanes, organometallics, and alkylboranes. (*b*) Alcohol level: halides, alkenes, amines, ethers, thiols, and sulfides. (*c*) Carbonyl level: aldehydes, ketones, *gem* dihalides, vinyl halides, vinyl esters, acetylenes, allenes, diolefins, imines, acetals, ketals, epoxides, α-haloethers, and thiocarbonyl compounds. (*d*) Acid level: acid halides, acid anhydrides, esters, amides, carboxylates, trihalomethyl compounds, nitriles, ortho esters, and alkoxy- or chloroacetylenes. (*e*) Carbon dioxide level: highest oxidation state that includes derivatives of carbonic acid, carbon tetrahalides, and ortho carbonic esters.

limited amount of theory is sometimes included, not only to account for the path that the reaction takes, but also for its limitations and any by-products formed. Fourth, examples are given, some in sufficient detail to permit preparation in the laboratory and others to illustrate minor variations not necessarily presented in the discussion.

Any work of this sort must of necessity lean heavily on the chemical literature. Many references are included, those of particular interest being reviews on the various syntheses. Perhaps the most heavily used sources, exclusive of the original literature covered through 1968 with a few references of 1969, are the following:

1. W. Theilheimer, "Synthetic Methods of Organic Chemistry," Vols. 1–21, S. Barger, New York, 1946–1967 incl.

2. R. B. Wagner and Harry D. Zook, "Synthetic Organic Chemistry," John Wiley and Sons, New York, 1953.

3. *Org. Syn.*, Coll. Vols. I, II, III, and IV (1941, 1943, 1955, 1963); *Org, Syn.*, **40–48** (1960–1968).

4. *Org. Reactions*, Vol. 1–XVII (1942–1969).

An effort—difficult as it is in such a rapidly expanding area—has been made to include the most recent developments in the field.

We are greatly indebted to Dr. Henry E. Baumgarten, University of Nebraska, who examined thoroughly all chapters with the exception of Chapter 2, and to Dr. James G. Traynham of Louisiana State University who reviewed Chapter 2. The constructive comments of these two professors have been unusually helpful. Our special gratitude is due as well to Dr. David A. Shirley and Dr. Jerome F. Eastham, both of the University of Tennessee, and Dr. Thomas M. Harris, Vanderbilt University, for their advice and suggestions. Finally, we express thanks to our respective departments for the encouragement given during the writing and for the secretarial assistance made available in getting the manuscript into its final shape.

Perhaps the two most difficult tasks in the construction of a book of this type are (*a*) to include the most modern procedures at the expense of those that are passé and (*b*) to include the great detail involved in an errorless fashion. We shall be grateful if our attention is called to any lapses of these or of other sorts.

February 1970
Knoxville, Tennessee   Calvin A. Buehler
Nashville, Tennessee   Donald E. Pearson

CONTENTS

ABBREVIATIONS

Ac	Acetyl
Bu	Butyl
DBN	1,5-Diazabicyclo [3.4.0.] nonene-5
DCC	Dicyclohexylcarbodiimide
DDQ	2,3-Dichloro-5,6-dicyano-1,4-benzoquinone

$$C_6H_5COC\overset{\mid}{H}C_6H_5$$

Desyl	$C_6H_5COCHC_6H_5$
Diglyme	Diethylene glycol dimethyl ether
DMF	Dimethylformamide
DMSO	Dimethyl sulfoxide
Chloretone	α,α,α-Trichloro-t-butyl alcohol
Et	Ethyl
Glyme	1,2-Dimethoxyethane
Hexamine	Hexamethylenetetramine
Me	Methyl
Ms	Mesyl, CH_3SO_2—
NBS	N-Bromosuccinimide
NCS	N-Chlorosuccinimide
Ph or φ	Phenyl
PPA	Polyphosphoric acid
Pr	Propyl
PTAB	Phenyltrimethylammonium perbromide
Py	Pyridine
Sulfolane	Tetramethylene sulfone

$$\textbf{Thexyl} \quad (CH_3)_2CH-\overset{\overset{\textstyle CH_3}{\mid}}{\underset{\underset{\textstyle CH_3}{\mid}}{C}}-$$

THF	Tetrahydrofuran
Trityl	$(C_6H_5)_3C$—
Ts	Tosyl, p-$CH_3C_6H_4SO_2$—

Survey of

ORGANIC SYNTHESES

Chapter 1

ALKANES, CYCLOALKANES, AND ARENES

I

To eliminate as much repetition as possible, the alkanes, cycloalkanes, and arenes are discussed in the same chapter. Seemingly a simple chapter, it soon reveals itself as a Pandora's box of methods of synthesis, the most glamorous of which are the free radical methods (G) presently in an active state of development. One would suspect that methods of reduction (A) are the most important for production of hydrocarbons since they are in the lowest state of oxidation of any class of organic compounds. They are important methods, but Friedel-Crafts procedures (D) are the most numerous, and this does not seem surprising when one realizes the potentialities of harnessing carbonium ion behavior. The chapter is not all-inclusive. Very little is said about hydrocarbons from petroleum or from natural sources. Less is stated about polymerization processes which produce polyethylene and like polymers but are capable of producing lower molecular hydrocarbons by telomerization. It is hoped that the majority of general methods have been included. In each section an effort has been made to point out the best or the simplest method of synthesis, together with recent modifications which may be helpful in improving yields. The modifications usually are based on a better understanding of the events that are occurring or can occur in a reaction. Such an understanding leads to better choice of environment or technique, including even the simple choice of sequence of reagent addition.

Noteworthy improvements have been made in almost all types of classical syntheses of hydrocarbons. Conversion of phenols into arenes, once a most difficult synthesis, has now been accomplished in a number of ways (A.2). New reagents have been found for the reduction of halides of all types to hydrocarbons (A.3) and for the reduction of olefins (A.5). Wurtz-type syntheses have taken on a new look with the introduction of the versatile reagent, lithium copperdialkyl, LiCuR$_2$ (B.2). And of course small- and large-ring alicyclic and bicyclic hydrocarbons are no longer substances to be coveted, but rather substances which may be readily obtainable via methods described in G, H, and other sections.

A. Reduction

1. From Aldehydes or Ketones

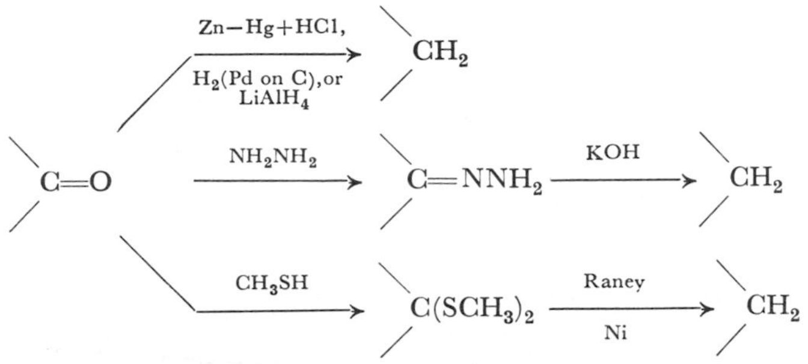

Aldehydes or ketones may be reduced directly to hydrocarbons by the use of zinc amalgam in hydrochloric acid (Clemmensen), by hydrogen in the presence of a catalyst such as palladium on carbon or Raney nickel, or by a metal hydride such as lithium aluminum hydride. Indirect methods have also been employed. By the Wolff–Kishner reduction the carbonyl compound is first converted into the hydrazone, which gives the hydrocarbon with alkali. It is also possible with ketones to form thioketals, which with Raney nickel in ethanol yield the hydrocarbon.

Of the direct methods, that of Clemmensen, which has been reviewed [1] appears to have been the most widely used. Although it has been utilized in synthesizing hydrocarbons from a great variety of aldehydes and ketones, it is the most satisfactory when applied to ketones, particularly of the aliphatic, alicyclic, and aliphatic-aromatic types. The method consists in refluxing the carbonyl compound with a large excess of amalgamated zinc and hydrochloric acid either alone, or with a miscible solvent such as ethanol, acetic acid, or dioxane, or with an immiscible solvent such as toluene. Yields vary greatly, although in many cases they are very satisfactory. Among the by-products to be found are olefins, pinacols, and traces of carbinols. Homogeneity favors pinacol formation. The addition of acetic acid to acetophenone and the Clemmensen reducing agent lowers the yield of ethylbenzene from 80 to 27% and increases the pinacol yield correspondingly. A decrease in the mineral acid concentration favors olefin formation; for example, decreasing the acid from 20 to 3% increases styrene formation from 2 to 26% [2].

The use of toluene introduced by Martin [3] has permitted the yield of γ-phenylbutyric acid from the γ-keto acid to be increased from 72–78% to 90%. Other modifications found to be helpful are concerned with the method of preparing the zinc and the rapidity of stirring [4]. For phenolic ketones, dilute ethanol is the preferred solvent (see Ex. b.1) [5, 6].

References for Section A are on pp. 19—22.

Some ketones, particularly those of polycyclic hydrocarbons, may be reduced satisfactorily with zinc in sodium hydroxide containing a trace of copper sulfate [7].

The mechanism of the Clemmensen reduction is complex and not completely understood but involves a zinc adduct [8].

$$\mathrm{C{=}O} + Zn \longrightarrow \begin{bmatrix} \ \ \ \ \ \overset{O}{\diagup\diagdown} \\ C \!-\!-\!-\! Zn \end{bmatrix} \xrightarrow[Zn]{H^{\oplus}} \begin{bmatrix} \ \ \ \ ZnOH \\ C \\ Zn^{\oplus} \end{bmatrix} \xrightarrow{H^{\oplus},H_2O} CH_2$$

A second direct method of reduction of carbonyl compounds to hydrocarbons is that of catalytic hydrogenation. It is restricted, however, to the reduction of carbonyl compounds which form benzylic alcohols (or other alcohols prone to hydrogenolysis) as intermediates. With catalysts such as palladium [9–11], or Raney nickel [12], satisfactory yields of the hydrocarbon have been obtained under mild conditions. Hydrogenation under these conditions offers an advantage over the Clemmensen reduction in that structural features such as the lactam ring and even some carbonyl groups [12] are not attacked.

The final method of direct reduction of carbonyl compounds involves the use of metal hydrides. Although such reducing agents ordinarily produce alcohols, over longer periods of time [13] or in the presence of aluminum chloride [14, 15], hydrocarbons are obtained. With lithium aluminum hydride and aluminum chloride in equimolecular amounts in ether at 35°, diaryl and alkyl aryl ketones reduce, usually in good yield, to the corresponding hydrocarbons [15]. This procedure offers the advantage of simplicity, and pure products are readily obtained. Suggestions about the mechanism of the reduction are available [13, 14].

The most widely used of the indirect methods is that of Wolff-Kishner, which has been reviewed [16]. This method serves well as a complementary one to the Clemmensen reduction. It may be employed for compounds such as pyrroles which are not sensitive to bases but are sensitive to the acidic conditions prevailing in the Clemmensen reduction. The Wolff-Kishner procedure has also been employed with greater success than the Clemmensen among compounds of high molecular weight. Fortunately the intermediate hydrazone need not be isolated, and recent modifications permit operation at atmospheric pressure. Although it was demonstrated that the Wolff-Kishner reduction can be carried out at room temperature in dimethyl sulfoxide solvent [17], it has since been shown that a somewhat higher temperature is beneficial for the reaction in this solvent and that a protic solvent in mixture with dimethyl sulfoxide is absolutely necessary [18]. Lower temperatures may also be employed by the use of potassium t-butoxide in boiling toluene [19] but, with keto esters, triethanolamine is more satisfactory than a glycol solvent [20]. The Huang-Minlon modification of the Wolff-Kishner [21, 22] is by far the most widely employed method for reducing ketones. By this method the carbonyl compound in anhydrous hydrazine and an alkali metal hydroxide or alkoxide is heated in a high-boiling polar solvent, such as ethylene glycol or diethylene glycol, in such a way that, to the degree possible, water is kept out or removed from the system. The mechanism involves decomposition of the anion of the hydrazone [18].

$$\diagdown C{=}NNH_2 + OH^\ominus \underset{\longleftarrow}{\longrightarrow} \diagdown C{=}N\overset{\ominus}{N}H \longleftrightarrow \overset{|}{\underset{|}{\ominus}}C{-}N{=}NH \overset{BH}{\longrightarrow}$$

$$\overset{|}{\underset{|}{H}}C N{=}NH \overset{OH^\ominus}{\underset{\longleftarrow}{\longrightarrow}} \overset{|}{\underset{|}{H}}C N{=}N^\ominus \longrightarrow \overset{|}{\underset{|}{H}}C^\ominus + N_2 \overset{BH}{\longrightarrow} {-}CH_2{-}$$

If only hydrazone is present (with traces of potassium hydroxide), the reduction presumably utilizes the hydrogen of the hydrazone via a series of anion and proton sources (see Ex. b.4). In other examples, platinum, silver oxide, and porous ground plate are combined with potassium hydroxide to facilitate decomposition [23], but their advantage has not been clearly demonstrated. The most important side reaction is azine formation, and occasionally, with alicyclic ketones, aromatization and alkyl group migration occur [23].

It is of interest to note that semicarbazones with alkali [24] or with hydrazine and alkali [25] may be reduced as well to hydrocarbons by the Wolff-Kishner reaction.

The final indirect method of reduction has been employed for carbonyl compounds which are sensitive to acids and bases or which for other reasons cannot be reduced satisfactorily by the usual methods. In this procedure the thioacetal or thioketal is formed with an alkyl thiol or an alkanedithiol, after which desulfurization is accomplished with hydrogen-saturated Raney nickel [26] (see Ex. b.7) or hydrazine and alkali [27].

(a) **Preparation of 1,2,3,4-tetrahydro-2-naphthaleneacetic acid.**   A mixture of 1-oxo-1,2,3,4-tetrahydro-2-naphthylideneacetic acid, 10 g., 67 g. of freshly amalgamated Zn, 80 ml. of concentrated HCl, 23 ml. of water, and 100

ml. of toluene while being stirred was refluxed for 24 hr. From the mixture 7.58 g. (81%) of the acid was isolated, m.p. 83.5–86.0°, by the usual methods including decolorization with carbon. The pure acid melts at 88.3–88.8° [28].

(b) **Other examples**

(1) *o-n*-**Heptylphenol** (81–86% by refluxing *o*-heptanoylphenol and amalgamated mossy zinc in a solution of HCl, water, and ethanol for at least 8 hr.) [5].

(2) β-**Benzyl-γ-butyrolactone** (98% from β-benzoyl-γ-butyrolactone by hydrogenation in methanol in the presence of palladium chloride at 50 p.s.i. for 2 hr.) [11].

(3) **Diphenylmethane** (92% from benzophenone, lithium aluminum hydride, and aluminum chloride in ether) [15].

(4) Lanostanyl acetate (69% from 7,11-dioxolanostanyl acetate by reduction with hydrazine and sodium in diethylene glycol under anhydrous conditions at 180°) [22].

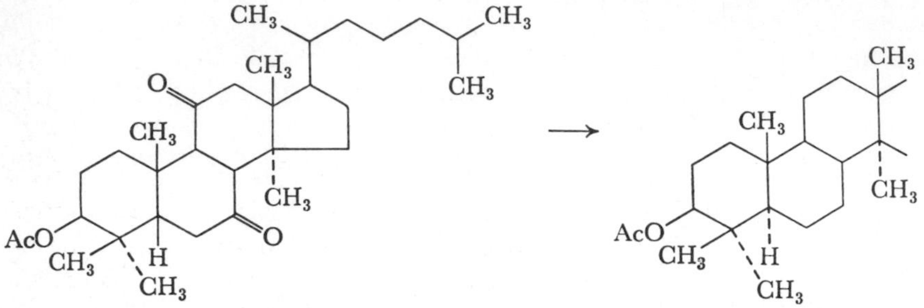

(5) n-Hexylbenzene.　Caprophenone hydrazone (137 g.) and 2 g. of powdered potassium hydroxide were heated cautiously with a free flame until nitrogen evolution began. (If overheated, the evolution of nitrogen is overly vigorous but can be controlled by ice-cooling.) After the nitrogen evolution had subsided, the hydrocarbon was removed by distillation at water aspirator pressure. The distillate was washed with dilute acid, dried, and redistilled (82%), b.p. 70° (2 mm.), $n^{19.3}$D 1.4860 [29].

(6) 1,2-Di-9-anthrylethane (60% from 9-anthraldehyde and lithium aluminum hydride in refluxing tetrahydrofuran, an abnormal reduction pathway) [30].

(7) 7,7,10-Trimethyl-$\Delta^{1(9)}$-octalin (74% overall from 7,7,10-trimethyl-$\Delta^{1(9)}$-octal-2-one by conversion first into the thioketal by the use of

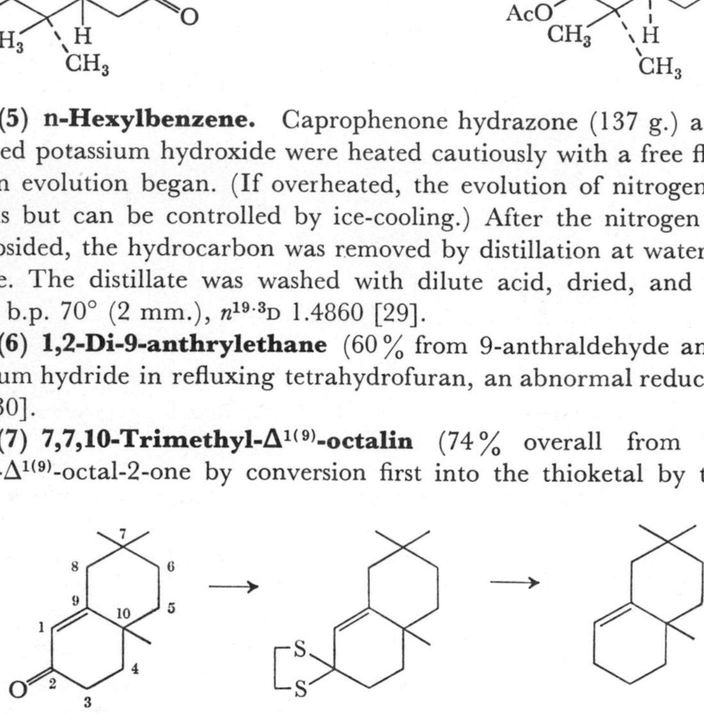

boron trifluoride etherate and ethanedithiol at 0° and then into the hydrocarbon by the use of W-2 Raney nickel and absolute ethanol at reflux temperature for 14 hr.) [26].

2. From Alcohols or Phenols

$$(R)ArOH \xrightarrow[HI]{P} (R)ArH$$

A great variety of reducing agents has been employed to convert alcohols or phenols into hydrocarbons. In catalytic reduction, cobalt-on-alumina, [31], copper chromite [32], copper chromite and graphite [33], nickel on kieselguhr with and without thiophene [34], tungsten disulfide [35], palladium-barium sulfate preferably activated with perchloric acid [36], and Raney nickel [37] have been employed as catalysts. Hydrogenolysis of this type eliminates possible rearrangements which may occur via carbonium ions [31]. Hydrogenolysis occurs best with benzylic-type alcohols, but rarely with other types.

The most general reducing agent for carbinols is a combination of phosphorus and hydriodic acid. The combination can be used under mild conditions, dilute hydriodic acid, to reduce hydroxyl groups [38] or under vigorous conditions, sealed tube and 190° temperature, to reduce phenolic groups [39] or even carboxyl groups to the hydrocarbon [40]. Among the miscellaneous reducing agents are hydriodic acid in acetic anhydride [41], zinc alone or with acetic acid [42], zinc with acetic and hydrochloric acids [43], or sodium in liquid ammonia [44]. It appears to be a broad generalization that any carbinol which can be converted into a carbonium ion in boiling formic acid will reduce to a hydrocarbon. The reduction succeeds best with triphenyl carbinols [45]. It is not clear which is the most effective coreagent with formic acid: sodium carbonate [46] or strong mineral acid (see Ex. c).

The conversion of phenols into hydrocarbons poses a difficult problem since the carbon-oxygen bond is considerably stronger than that in alcohols. The solution to the problem is to attach the oxygen atom of the phenol to another group or atom which forms a stronger bond. The best general procedure seems to be the formation of 1-phenyltetrazole ethers which can be cleaved under reducing conditions to the hydrocarbon [47]. Aryl diethyl phosphates with

$$
\text{ArO}-
\begin{array}{c}
\text{N}-\text{N} \\
\diagup \quad \parallel \\
\quad \quad \parallel \\
\diagdown \quad \parallel \\
\text{N}-\text{N} \\
\mid \\
\text{C}_6\text{H}_5
\end{array}
\xrightarrow[\text{H}_2, 35°, \text{Parr shaker, 15 hr.}]{\text{5\% Pd on C(15\% by wt.)}}
\begin{array}{c}
\text{ArH} \\
33–94\%
\end{array}
$$

sodium in liquid ammonia [48, 49] or aryl p-toluenesulfonates over Raney nickel [50] are also reduced to hydrocarbons. In one of the best general methods of reduction, alkyl sulfonate esters are reduced with lithium aluminum hydride [51]. The reduction appears to have been employed largely with sulfonates of primary alcohols of ring systems or with sugars, in which case methyl derivatives are produced. Sometimes it is desirable to form intermediates such as the benzyl thioether [52] or thiocyanate [53] before the reduction.

(a) **Preparation of 1,5-dimesitylpentane.** 1,5-Dimesitylpentanediol-1,2, 1 g. in a mixture of 5 ml. each of acetic anhydride and glacial acetic acid, was treated dropwise with 10 ml. of 48% hydriodic acid. After boiling had subsided, the mixture was refluxed for 30 min. Cooling and pouring in water followed by refrigeration and crystallization from ethanol gave 0.82 g. (81%) of 1,5-dimesitylpentane, m.p. 114.5–115° [41].

(b) **Preparation of naphthalene** (97% of β-naphthyl diethyl phosphate from β-naphthol, diethyl phosphate, and triethylamine; 95% of the hydrocarbon from reduction of the phosphate with sodium in liquid ammonia) [48].

(c) **Preparation of 2-phenyl-4-(diphenylmethyl)-quinoline** (60%
from 1 g. of the carbinol in 10 ml. of refluxing 90% formic acid to which 4 ml. of
concentrated sulfuric acid was added dropwise. The cooled mixture was made
basic, and the precipitated quinoline derivative recrystallized from aqueous
acetone to give cream-colored crystals, m.p. 200–201°. The product is not a
hydrocarbon, but the procedure is relevant to the production of hydrocarbons)
[29].

(d) **Diphenylacetic acid** (94–97% from benzilic acid, red phosphorus,
iodine, and water (or red phosphorus and dilute hydriodic acid) in acetic acid
refluxed for at least 2.5 hr.) [54].

3. From Halides

$$RX \xrightarrow{2\ (H)} RH + HX$$

Two new methods are noteworthy. Lithium in tetrahydrofuran and t-butyl
alcohol appears to be a reducing agent of wide applicability [55]. It displaces
chlorine from vinylic, allylic, geminal, aromatic (see Ex. a), and even bridgehead chlorides. The method, simple in operation, found applications later [56].
The second reducing reagent, magnesium and isopropyl alcohol, has been
shown to be effective in reducing alkyl chlorides, bromides, and iodides [57],

$$RX + Mg + (CH_3)_2CHOH \rightarrow RH + (CH_3)_2CHOMgX$$

some alkyl fluorides and aryl halides (see Ex. b). A trace of iodine initiates the
reaction, and, for difficultly reducible halides, some decalin is added. Groups
such as amino, phenolic ester, and ethylenic linkages do not interfere, but
nitro and carbonyl groups do.

So many other reagents have been used for the reduction of halides that a
sampling of the more promising ones is simply listed:

Zinc and acetic acid for aliphatic halides (see Ex. c.1).
Zinc and 10% pyridine in acetic acid for aromatic halides [58].
Zinc-cupric acetate in methanol and formamide for trichloromethyl groups
(see Ex. c.3).
Lithium aluminum hydride for aliphatic halides [59].
Sodium in wet methanol for *gem*-dihalides [60].
Sodium in amyl alcohol for aromatic halides [61].
Sodium borohydride with some water for reactive halides, *sec* or *tert* types
[62] (see Ex. c.4).
Tributylstannane for mono- or didebromination of *gem*-dibromocyclopro-
panes [63] or dechlorination of *gem*-chlorofluorocyclopropanes [64].
Triphenylstannane for aromatic halides (see Ex. c.2).
Tributyl and dibutyl stannanes for reductions of all types of halides [65].
Raney nickel in alkaline solution for aliphatic halides [66], for halo aromatic
acids or halophenols [67], for trifluoromethyl groups [68], for halophenols [69].
Stannous chloride in dioxane and hydrogen chloride for benzylic halides [70].
Palladium on charcoal and hydrazine for aromatic halides [71].

Dehalogenation of both aliphatic and aromatic halides with hydrogen and platinum metals has been studied extensively. Conditions are so important that the summarized literature [72, Chap. 24] should be consulted. But, in general, dehalogenation is best carried out in basic solution as illustrated by the rate of uptake of hydrogen (ml./min.) for chlorobenzene in the two best media studied:

	5% Pd/C	Pt/C	Rh/C
Acetic acid, sodium acetate	55	45	11
Ethanol, sodium hydroxide	100	8	4

Under the proper conditions the dehalogenations may be selective in the presence of vinyl or carbonyl groups as illustrated for the latter:

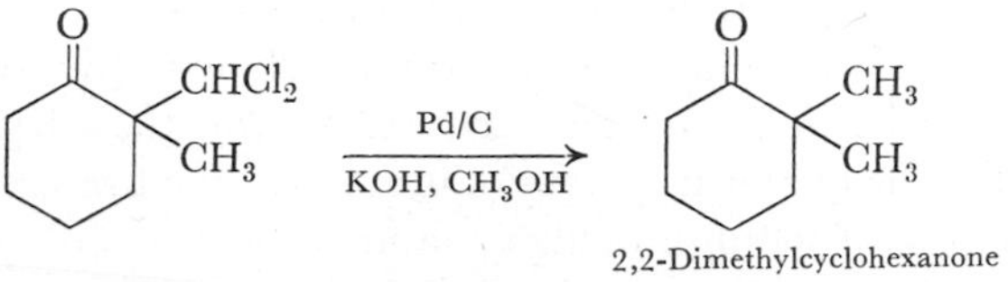

2,2-Dimethylcyclohexanone

(a) Preparation of benznorbornadiene [73].

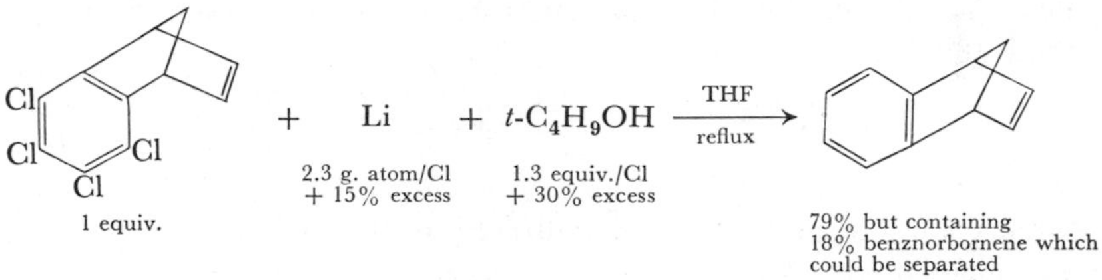

1 equiv.   2.3 g. atom/Cl + 15% excess   1.3 equiv./Cl + 30% excess

79% but containing 18% benznorbornene which could be separated

(b) Preparation of benzene (70–83% from 0.25 mole of magnesium powder, 50 ml. of decalin, and a crystal of iodine heated to reflux under N_2, to which 0.1 mole of chlorobenzene and 0.15 mole of 2-propanol were added dropwise; stirring was initiated as soon as exothermic reaction began) [74].

(c) Other examples

(1) n-Hexadecane (85% from cetyl iodide by reduction with zinc dust in glacial acetic acid, saturated with hydrogen chloride, on a steam bath for 25 hr.) [75].

(2) Benzene (61–72% from bromobenzene by refluxing with triphenyltin hydride [76].

(3) 2-Amino-4,6-dimethyltriazine (90% from 2,4,6-tris(trichloro-

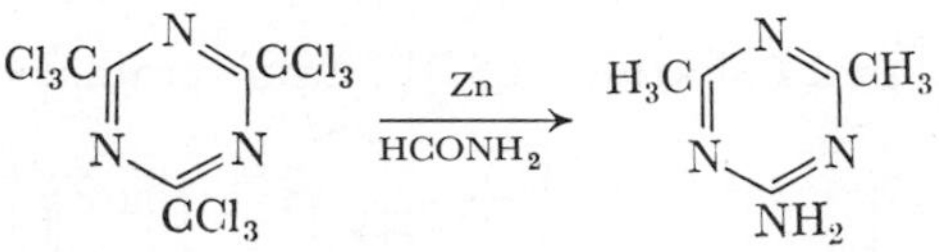

methyl)-1,3,5-triazine (simultaneous reduction and replacement of a trichloromethyl group by an amino group) and zinc dust in methanol-formamide containing cupric acetate) [77].

(4) **Triphenylmethane** (96% by gas chromatography from a solution of 65% aqueous diglyme at 50° containing 4.0 M NaBH$_4$ and 1.0 M NaOH and sufficient triphenylmethyl chloride to make the total composition, if homogeneous, 0.5 M) [78].

4. From Tertiary Amines or Quaternary Ammonium Salts

$$\text{RCH}_2\text{N(CH}_3)_2 \xrightarrow[\text{Raney Ni}]{\text{H}_2} \text{RCH}_3 \xleftarrow{\text{Na—Hg}} \text{RCH}_2\overset{\oplus}{\text{N}}(\text{CH}_3)_3\overset{\ominus}{\text{I}}$$

Both tertiary amines (usually of the benzylic type) and quaternary ammonium salts have been reduced to hydrocarbons with yields of 70–90%. Common reducing agents for the tertiary amines are hydrogen in the presence of palladium on carbon [79], palladized strontium carbonate [80], copper chromite [81], or Raney nickel [82], sodium methoxide in methanol [83], and benzyl mercaptan [84]. The quaternary ammonium salts have been reduced to hydrocarbons by the use of sodium amalgam and water [85, 86], hydrogen in the presence of palladium-carbon and sodium hydroxide [87], and lithium aluminum hydride [88]. Aromatic quaternary salts have been reduced by irradiation in alcoholic solvents, but the reaction is neither general nor well adapted to synthesis [89].

$$p\text{-CH}_3\text{C}_6\text{H}_4\text{N(CH}_3)_3\text{I} \xrightarrow[\text{CH}_3\text{OH}]{h\nu} \text{C}_6\text{H}_5\text{CH}_3$$
Toluene
78%

Primary amines can be reduced indirectly to the hydrocarbon in low yields via the sulfonamide [90]. The process succeeds to about the same degree with

$$n\text{-C}_6\text{H}_{13}\text{NHSO}_2\text{C}_6\text{H}_5 + \text{NH}_2\text{OSO}_3\text{H} \xrightarrow[\text{aq. NaOH}]{\text{Reflux in}} n\text{-C}_6\text{H}_{13}\overset{\overset{\displaystyle\text{NH}_2}{|}}{\text{N}}\text{SO}_2\text{C}_6\text{H}_5 \longrightarrow$$

$$\text{C}_6\text{H}_5\text{SO}_2{}^{\ominus} + [n\text{-C}_6\text{H}_{13}\text{N}{=}\text{NH}] \xrightarrow{-\text{N}_2} n\text{-C}_6\text{H}_{14}$$
Hexane
35–40%

β-naphthylamine, a fact which suggests a general application to aromatic primary amines.

(a) **Preparation of hemimellitene** (85–90% from 2,3-dimethyl-benzyltrimethylammonium iodide, 5% sodium amalgam, and water at steam-bath temperature for about 24 hr.) [86].

(b) **Preparation of skatole** (83.9% from gramine by hydrogenation

in the presence of palladium-carbon in ethanol for about 20 hr.) [79].

5. From Alkenes and Arenes

$$RCH{=}CH_2 + H_2 \xrightarrow[\text{C}]{\text{Pd on}} RCH_2CH_3$$

The hydrogenation of alkenes and arenes has been reviewed recently [91]. The most common hydrogenation catalysts are nickel, platinum, palladium, rhodium, and ruthenium. The use of copper-chromium oxide catalysts is on the decline.

Success has also been achieved in this reduction with sodium borohydride-boron trifluoride by refluxing the organoborane first formed with propionic acid [92]. For the preparation of hydrogenation catalysts see [93]. Of these

$$RCH{=}CH_2 \xrightarrow{NaBH_4,\ BF_3} (RCH_2CH_2)_3B \xrightarrow{C_2H_5COOH} 3\ RCH_2CH_3$$

catalysts, nickel, which is usually employed at high temperature and pressure, is the most common. Its activity may be increased by supporting it on a substance such as kieselguhr. As Raney nickel, it comes in a variety of forms. Platinum is satisfactory for the hydrogenation of most functional groups under rather mild conditions, such as temperatures below 70° and hydrogen pressure under 60 p.s.i. Its activity may be increased by preparation from chloroplatinic acid and sodium borohydride [94] or tribenzylsilane [95] or by the addition of perchloric acid to the Adams catalyst [96]. Platinum oxide on silicic acid is so active at room temperature that it may be used with advantage to determine the degree of unsaturation [97]. The popularity of palladium, usually supported on activated charcoal, barium sulfate, or strontium carbonate, is on the increase. The use of rhodium and ruthenium is on the increase because of their selectivity in reducing unsaturated alcohols, halides, and esters without hydrogenolysis [98].

It is of interest to note that homogeneous catalysts are now under investigation. These catalysts appear to offer three advantages over heterogeneous catalysts: they are capable of reducing compounds of high molecular weight such as polymers; they are capable of reducing unsaturated sulfur compounds; and they are less susceptible to poisoning effects.

Thus thiophenol in trace amounts has no effect on the reduction of ergosterol with hydrogen in the presence of chlororhodium tris(triphenylphosphine), $((C_6H_5)_3P)_3RhCl$ [99; see also 100]. Sulfides such as phenyl propyl sulfide have no effect in higher concentrations. Phenyl allyl sulfide is reduced satisfactorily with the rhodium catalyst to phenyl n-propyl sulfide.

Another useful homogeneous catalyst is prepared from ruthenium chloride [101].

$$6\ (C_6H_5)_3P + RuCl_3 \xrightarrow[\text{H}_2]{\text{CH}_3\text{OH}} ((C_6H_5)_3P)_3RuHCl$$

Alkenes

Isolated double bonds are hydrogenated more readily than any other functional group with the exception of acetylenes, allenes, and in some cases aromatic nitro groups. Of the catalysts employed, palladium is superior at low temperature and pressure, but platinum or rhodium is preferred in hydrogenating a double bond in a compound which also contains a moiety subject to hydrogenolysis. To eliminate undesirable double bond migration, Raney nickel and ruthenium are the catalysts preferred, and palladium is the catalyst to be avoided. It is interesting to note that activated alkenes (activated alkynes and carbonyl groups as well) may be smoothly reduced with W-2 Raney nickel and a hydrogen donor such as cyclohexanol [102].

In the addition of hydrogen to alkenes the addition occurs *cis* on the least hindered side [103]. At least, the higher the hydrogen pressure, the more *cis* hydrogenation product is obtained [104]. As an example of *cis* addition, *trans*-1,2-dimethylstyrene 1-I gives the *dl* (*threo*) product 1-II, while the *cis* isomer 1-III yields the *meso* (*erythro*) product 1-IV. As a rule such hydrogenations occur with good yields.

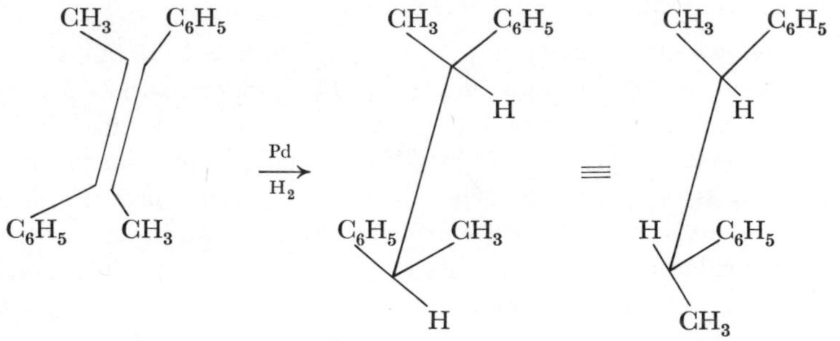

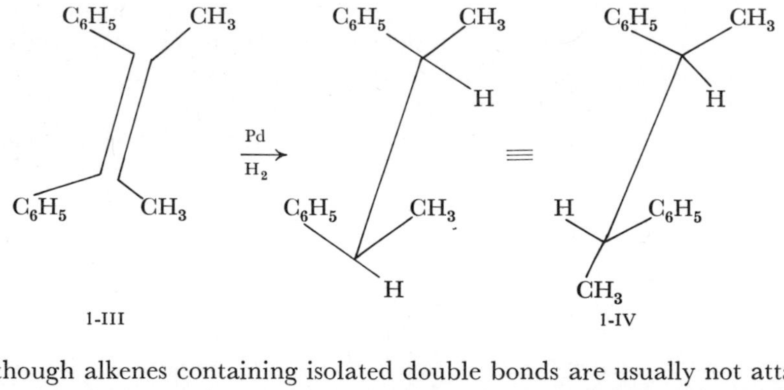

Although alkenes containing isolated double bonds are usually not attacked by metal hydrides, such as lithium aluminum hydride [105], other functional groups in conjugation with the olefinic group or other metal hydrides may lead to reduction [106]. As has already been indicated, alkenes may be reduced to the corresponding alkane by the use of sodium borohydride and boron trifluoride etherate in diglyme [92].

Addition of an alkane (RH) by an indirect method may be accomplished as shown in the accompanying formulation. In this manner highly reactive

$$R_3B + 3\ CH_2{=}CCH{=}O \xrightarrow{\ H_2O\ } 3\ RCH_2CHCHO + B(OH)_3$$
$$\qquad\qquad\quad |\qquad\qquad\qquad\qquad\qquad |$$
$$\qquad\qquad\quad Br\qquad\qquad\qquad\qquad\qquad Br$$

aldehydes, such as 2-bromo-3-cyclohexylpropanal, which must be used immediately or stored as an acetal, have been prepared in 65–97 % yields [107].

Two rather new catalysts in hydrogenation are nickel boride, prepared from nickel acetate and sodium borohydride in aqueous solution, called P-1 [108], and in alcoholic solution, called P-2 [109]. Both these catalysts exhibit a low tendency toward isomerization of olefins. The P-1 product is more active than Raney nickel, and the P-2 product exhibits remarkable selectivity. With the latter the half-life for the hydrogenation of 1-octene is 7 minutes, while that for cyclohexene is 200 minutes, facts which indicate sufficient selectivity to hydrogenate terminal double bonds in the presence of nonterminal ones.

A similar selectivity is shown by 5 % ruthenium on Norite [110]. In a water dispersion this catalyst promotes the hydrogenation of monosubstituted in preference to di- and trisubstituted olefins, i.e., 1-octene rather than 2-octene or cyclohexene. Likewise, palladium permits the hydrogenation of the carbon-carbon double bond in preference to the carbon-oxygen double bond, i.e., an α,β unsaturated aldehyde gives the saturated aldehyde [111].

Arenes

The nature of the substituent in the arene plays a part in determining the ease of catalytic hydrogenation. In fact, the order of reactivity varies with different catalysts. Thus for Raney nickel the order is $\phi OH > \phi H > \phi NH_2 > \phi COONa$ [112]; for platinum in acetic acid, $\phi OH > \phi NH_2 > \phi H > \phi COOH > \phi CH_3$ [113]; for rhodium in methanol or ruthenium in aqueous media, $\phi H > \phi CH_3 > \phi OH > \phi NH_2$ [113]. Alkyl benzenes in ethanol may usually be hydrogenated over 5 % palladium on charcoal at room temperature and low pressure (see Ex. b), while with W-2 Raney nickel a temperature of 100–150° and a pressure of 2000–2500 p.s.i. is required. In these cases the yields are good.

Although catalytic hydrogenation far outweighs other methods of reducing olefins, chemical means are available as well. Hydride addition to olefins conjugated with electron-withdrawing groups has been mentioned. In addition, isolated double bonds can be reduced by diimide (for reviews on reductions with the reagent produced from various sources see [114]); produced *in situ* [115]:

$$p\text{-}CH_3C_6H_4SO_2NHNH_2\ (100\%\ \text{excess}) + \text{cyclohexene} \xrightarrow[\text{diglyme under }N_2]{\text{Refluxing}}$$

$$\text{cyclohexane}\ (98\%) + p\text{-}CH_3C_6H_4SO_2H + N_2$$

And, in Friedel-Crafts reactions, hydride shifts can occur to produce hydrocarbons. Tertiary carbon hydrogen atoms (R_3CH) are the best source of hydride atoms, but they need not necessarily be present to bring about reduction. For example, in the alkylation of benzene with cetyl chloride in the presence of aluminum chloride, nearly 10% of the halide was converted into cetane [29]. And diphenylethylenes, including substituted ones, when stirred at room temperature with aluminum chloride in benzene, are converted into bibenzyl as the major product [116]. The first step apparently is complete in a short time [117].

$$Ar_2C{=}CH_2 \xrightarrow[C_6H_6,\ HCl]{AlCl_3} \quad (C_6H_5)_2C{=}CH_2 + 2\ ArH \xrightarrow{\hspace{2cm}} C_6H_5CH_2CH_2C_6H_5$$

Sodium hydrazide, especially in the presence of hydrazine, reduces non-isolated carbon-carbon double bonds [118] in yields varying from 43 to 96%.

(a) Hydrogenation of an alkene (general procedure) The alkene, 1 g. in 10–25 ml. of ethanol, was hydrogenated over 50–100 mg. of 5% palladium on charcoal at room temperature under an atmosphere of hydrogen to give a good yield of the alkane [119].

(b) Hydrogenation of an alkyl benzene (general procedure) The alkyl benzene, 10 g. in 50–100 ml. of ethanol, was hydrogenated over 2.0–2.5 g. of 5% rhodium on charcoal at room temperature and 60 p.s.i. of hydrogen to give a good yield of the alkyl cyclohexane. Sometimes it was necessary to increase the temperature to 50° to hasten the reaction [120].

(c) 2-Methylaminopropylcyclohexane (90% by the hydrogenation of 2-methylaminopropylbenzene in the presence of ruthenium at 90° and 70 atm. pressure for 1 hr.) [121].

6. From Quinones

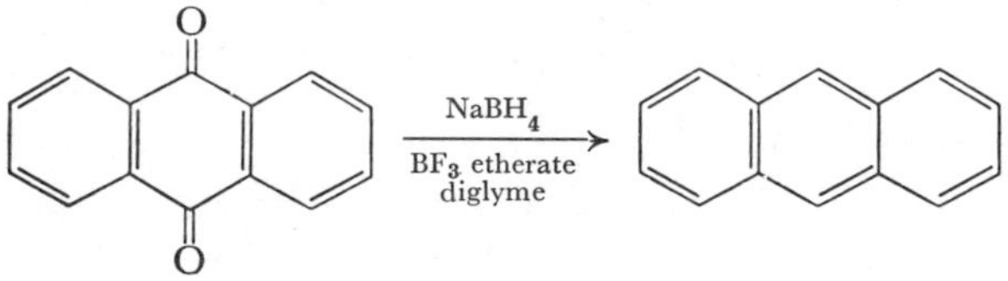

Although the milder reducing conditions convert quinones into dihydric phenols (Chapter 5, C.1), more strenuous conditions give arenes. To effect such a reduction, stannous chloride, concentrated hydrochloric acid, and acetic acid followed by zinc and sodium hydroxide [122], lithium aluminum hydride [123], aluminum and cyclohexanol [124] and a mixture of sodium borohydride and boron trifluoride-etherate in diglyme [125] have been employed. It is claimed that, for the one-stage reduction with lithium aluminum hydride to be successful, the quinone oxygens must be in an exposed ring [123]. Yields from a series of substituted anthraquinones vary from 60 to 70%.

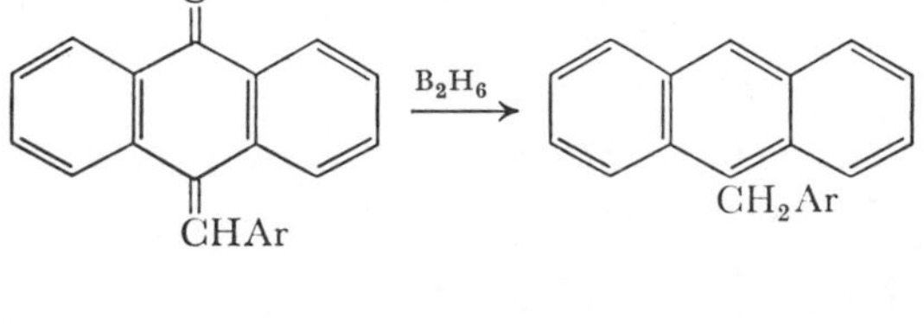

For 10-arylmethyleneanthraquinones, diborane (but not lithium aluminum hydride) effects the reduction to the hydrocarbons [126].

(a) Preparation of anthracene.   To 2 g. of anthraquinone in 10 ml. of diglyme, 10 ml. of a solution of sodium borohydride in diglyme was added with stirring and then, at 25°, 5 ml. of a 2 M solution of boron trifluoride-etherate in diglyme was added during 5 min. After stirring in a closed vessel for 2 hr. at 25–30°, the mixture was acidified and the solvent was removed under reduced pressure. The hexane extract of the residue was run through a short column of alumina to give a colorless, fluorescent percolate which yielded 1.25 g. (73 %) of anthracene [125].

(b) Preparation of chrysene (70 % from chrysene 3,6-quinone by reduction with lithium aluminum hydride in THF) [123].

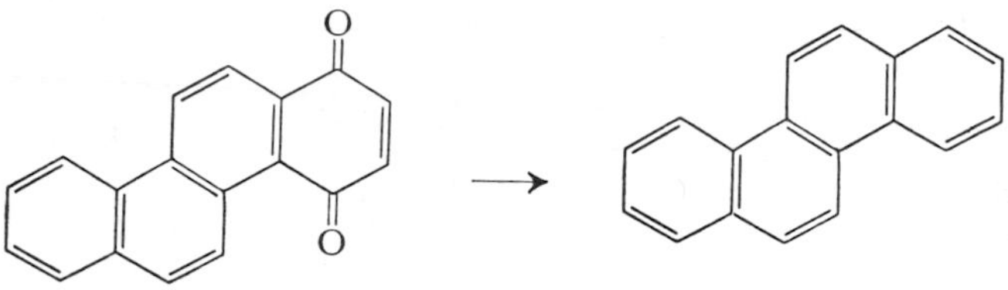

7. From Thiols and Sulfides

$$RSH(RSR') \xrightarrow{\text{Raney Ni}} RH$$

Desulfurization with Raney nickel saturated with hydrogen is used frequently to prepare acids, glycols, and other derivatives [127]. It can be used as well to prepare hydrocarbons [128]. The most important point to note is that the ratio of Raney nickel to substrate is very high, on the order of 10 to 1 or higher. Aromatic thiols tend to give sulfides rather than hydrocarbons [129]. To minimize the reduction of aromatic rings a special Raney nickel catalyst, which

$$2 \text{ ArSH} \xrightarrow{\text{Raney Ni}} \text{ArSAr}$$

effects clean desulfurization in 10 minutes in refluxing alcohol, has been developed [130].

The mechanism of Raney nickel desulfurization has been discussed [131].

Triethyl phosphite in a reaction catalyzed by irradiation also reduces mercaptans to alkanes in good yield [132].

$$C_8H_{17}SH + (C_2H_5O)_3P \xrightarrow{hv} \underset{88\%}{C_8H_{18}} + (C_2H_5O)_3PS$$

In some instances, sodium in liquid ammonia excels Raney nickel as a desulfurization agent [133]; see Ex. b.

(a) Preparation of camphane [134].

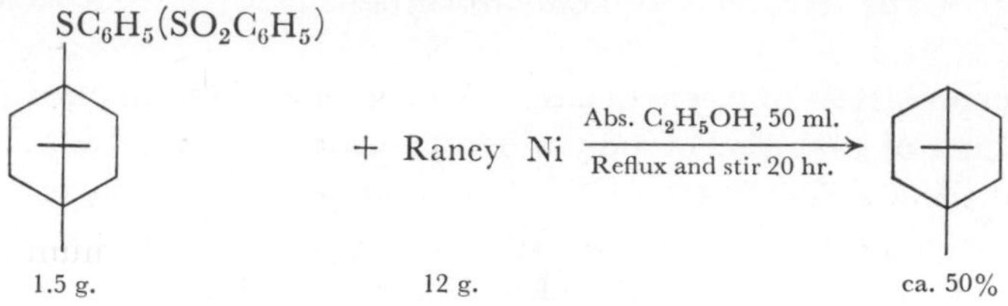

$SC_6H_5(SO_2C_6H_5)$

$+$ Raney Ni $\xrightarrow[\text{Reflux and stir 20 hr.}]{\text{Abs. } C_2H_5OH, 50 \text{ ml.}}$

1.5 g.   12 g.   ca. 50%

(b) Preparation of 2,7-dimethyloct-2,6-diene [135].

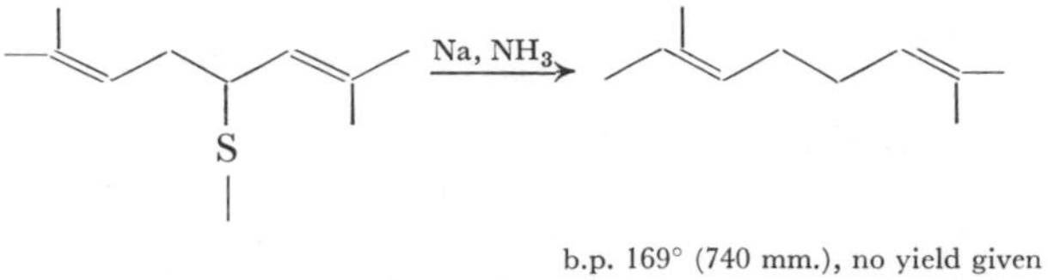

$\xrightarrow{\text{Na, NH}_3}$

b.p. 169° (740 mm.), no yield given

8. From Diazonium Salts (Deamination)

$$\overset{\oplus}{A}r\overset{\ominus}{N_2}X + H_3PO_2 + H_2O \rightarrow ArH + H_3PO_3 + HX + N_2$$

Arenes may be obtained from diazonium salts by reduction, a method which has been reviewed [136]. The method is not too common, although it serves well in preparing compounds not readily obtainable except when advantage is taken of the directing ability of the amino group. Since the amines are *o*- and *p*-directing while the amine salts are *m*-directing, considerable latitude exists in preparing derivatives in this manner. To illustrate the principle involved, the preparation of 2,4,6-tribromobenzoic acid in 70–80% yield from *m*-amino-benzoic is shown [137].

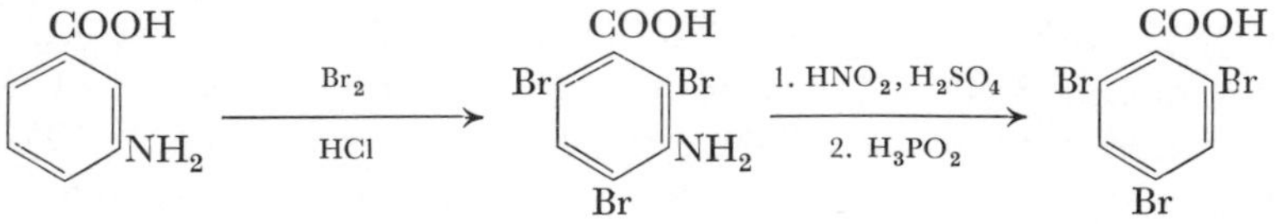

The original reducing agent employed was ethyl alcohol, which is not always satisfactory because ethyl aryl ethers may also be formed. The yield of hydro-carbons in methanol increases with the addition of sodium methoxide and reaches a maximum with 2 equivalents [138]. Hypophosphorous acid in fivefold excess, which gives yields with aqueous solutions of the diazonium salt at 0–5° of 60–85%, is the preferred reagent [139]; reduction also may be accomplished with dioxane [140]. If the diazonium salt decomposes at 0°, the diazotization itself may be carried out in hypophosphorous acid; for example, the conversion of 5-aminotetrazole into tetrazole is thus possible in 75–80% yield [141]. Other reducing agents which have been used are alkaline formaldehyde [142], sodium stannite [143], and dimethylformamide [144]. Such reagents are often unsatisfactory because of the hydrolytic cleavage brought on by alkaline media.

Recently sodium borohydride has been used successfully in reducing dia-zonium borofluorides in nonaqueous media [145]. Either the solid borohydride is added to the chilled methanolic solution or suspension or a chilled solution of the borohydride in dimethylformamide is added to a chilled solution of the diazonium salt in the same solvent. The yields with a limited number of borofluoride salts varied from 48 to 77%. Methoxy, carboxy, carboethoxy, and nitro groups were unaffected.

Another nonaqueous decomposition is carried out with hexafluorophosphates and tetramethylurea [146]. The hydrogen probably is abstracted from

$$ArN_2^{\oplus}PF_6^{\ominus} + (CH_3)_2N\overset{O}{\overset{\|}{C}}N(CH_3)_2 \xrightarrow[65°]{25° \text{ up to}} N_2 + ArO\overset{N_{\oplus}(CH_3)_2}{\underset{N(CH_3)_2}{C}} \longrightarrow ArH$$
$$\quad 50 \text{ g.} \qquad\qquad 200 \text{ g.}$$

one of the methyl groups attached to quaternary nitrogen. Yields are below that of hypophosphorous reduction with the exception of anilines containing electron-withdrawing groups. For example, anthranilic acid is converted into benzoic acid in 80–85% yield.

For the decomposition of aliphatic primary amines to hydrocarbons the best procedure appears to involve the formation of the diimide, $RN{=\!=}NH$, from the

$$RNH_2 \xrightarrow{ArSO_2Cl} \underset{1 \text{ g.}}{RNHSO_2Ar} \xrightarrow[\underset{\text{portionwise}}{10 \text{ g. } NH_2OSO_3H}]{100 \text{ ml. aq. NaOH}} RH + ArSO_2H + N_2$$

sulfonamide [147]. Yields are quite variable, however, owing in part to decomposition of the hydroxylamine-O-sulfonic acid in alkaline solution.

(a) Preparation of 3,3′-dimethylbiphenyl (76–82% from the tetrazotization of o-tolidine and then the addition of 30% hypophosphorous acid) [139].

(b) Preparation of biphenyl (75% from the addition of a chilled solution of sodium borohydride in dimethylformamide to the chilled diazonium salt of benzidine in the same solvent) [145].

9. From Alkoxides

$$2 \text{ RONa} + TiCl_4 \xrightarrow{C_6H_6} (RO)_2TiCl_2 \xrightarrow{K} (RO)_2Ti \xrightarrow{\Delta} RR + TiO_2$$

A clever utilization of a redox system enables one to produce hydrocarbons from alkoxides [148]. In this coupling the yield of bibenzyl from sodium benzyloxide was about 50%. In the few other cases investigated the yields were lower.

10. From Hydrocarbons

$$RR' + H_2 \rightarrow RH + R'H$$

The hydrogenolysis of the carbon-carbon bond is a high-energy reaction and takes place only under conditions of structural strain in the molecule. Even cyclopropane is not readily reduced unless further strain features are incorporated such as exist in methylenecyclopropane or phenylcyclopropane. For cyclobutanes even more strenuous conditions are needed for hydrogenolysis [149].

The Birch reduction conditions (discussed more appropriately under Alkenes, Chapter 2, B.3) can be adjusted to yield hydrocarbons from polynuclear compounds. To stop at the dihydro stage of anthracene reduction, ferric chloride is an excellent inhibitor. In general, ferric chloride inhibits all stages of the Birch reduction except that where stable dianions are formed. The accompanying equations illustrate two ways of obtaining hydrocarbons from benzanthracene [150].

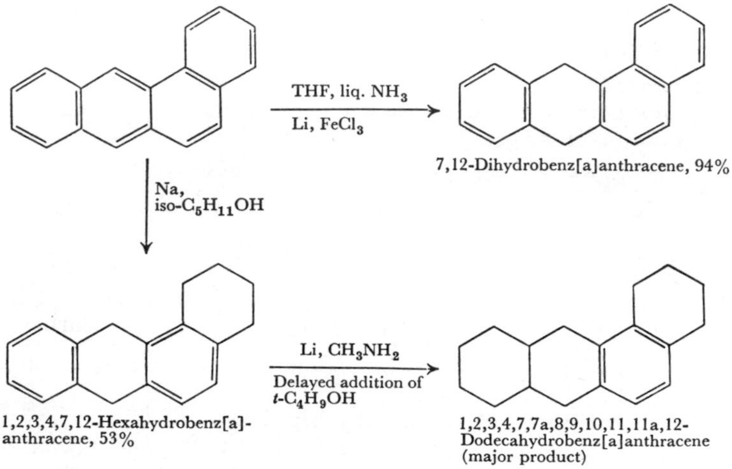

7,12-Dihydrobenz[a]anthracene, 94%

1,2,3,4,7,12-Hexahydrobenz[a]-anthracene, 53%

1,2,3,4,7,7a,8,9,10,11,11a,12-Dodecahydrobenz[a]anthracene (major product)

Examples of hydrogenolyses of strained cyclopropanes and cyclobutanes are given in Ex. a and b. But some phenylcyclopropanes may be cleaved by sodium in liquid ammonia [151]:

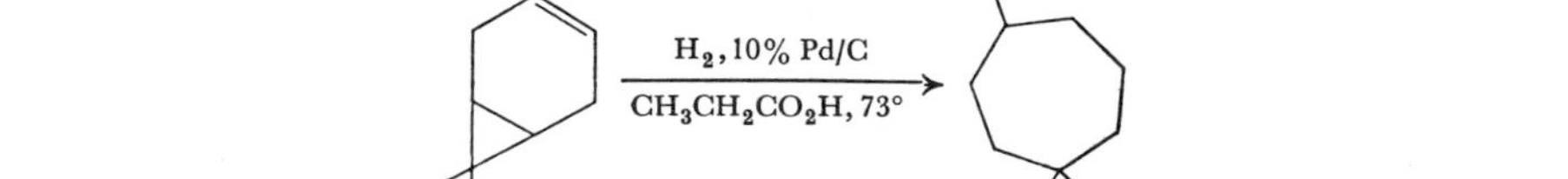

$(C_6H_5)_2CHCH_2CH_2CH_3 + (C_6H_5)_2CHCH(CH_3)_2$

1,1-Diphenylbutane, 80%   1,1-Diphenyl-2-methylpropane, 14%

(a) Preparation of 1,1,4-trimethylcycloheptane [152].

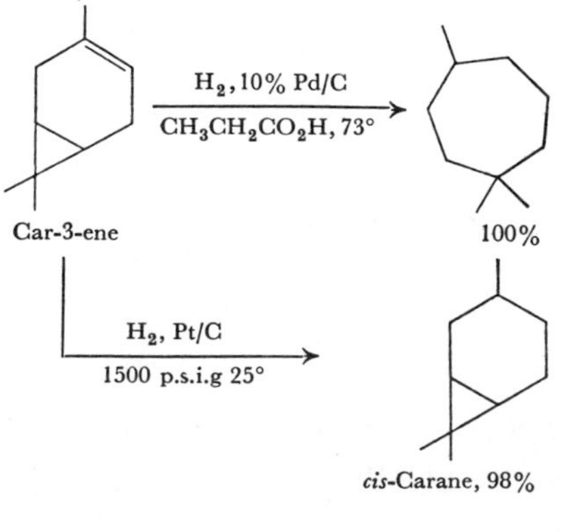

Car-3-ene

100%

cis-Carane, 98%

(b) Preparation of o-dibenzylbenzene [153]. The corresponding

benzocyclobutene (without phenyl groups) does not hydrogenolyze under comparable conditions [154].

1. E. L. Martin, *Org. Reactions*, **1**, 155 (1942); D. Staschewski, *Angew. Chem.*, **71**, 726 (1959).
2. G. E. Risinger *et al.*, *Chem. Ind.* (*London*), 679 (1965).
3. E. L. Martin, *J. Am. Chem. Soc.*, **58**, 1438 (1936).
4. L. F. Fieser *et al.*, *J. Am. Chem. Soc.*, **70**, 3197 (1948).
5. R. R. Read and J. Wood, Jr., *Org. Syn.*, Coll. Vol. **3**, 444 (1955).
6. R. Schwarz and H. Hering, *Org. Syn.*, Coll. Vol. **4**, 203 (1963).
7. L. F. Fieser and E. B. Hershberg, *J. Am. Chem. Soc.*, **62**, 49 (1940).
8. G. E. Risinger, unpublished results; T. Nakabayashi, *J. Am. Chem. Soc.*, **82**, 3906 (1960).
9. R. L. Letsinger and J. D. Jamison, *J. Am. Chem. Soc.*, **83**, 193 (1961).
10. T. P. C. Mulholland and G. Ward, *J. Chem. Soc.*, 4676 (1954).
11. J. Novák *et al.*, *Coll. Czech. Chem. Commun.*, **19**, 1264 (1954); J. Rothe and H. Zimmer, *J. Org. Chem.*, **24**, 586 (1959).
12. C. F. Koelsch and F. M. Robinson, *J. Org. Chem.*, **21**, 1211 (1956).
13. L. H. Conover and D. S. Tarbell, *J. Am. Chem. Soc.*, **72**, 3586 (1950).
14. B. R. Brown and A. M. S. White, *J. Chem. Soc.*, 3755 (1957).
15. R. F. Nystrom and C. R. A. Berger, *J. Am. Chem. Soc.*, **80**, 2896 (1958).
16. D. Todd, *Org. Reactions*, **4**, 378 (1948); H. H. Szmant, *Angew. Chem. Intern.*, *Ed. Engl.* **7**, 120 (1968).
17. D. J. Cram *et al.*, *J. Am. Chem. Soc.*, **84**, 1734 (1962).
18. H. H. Szmant and M. N. Román, *J. Am. Chem. Soc.*, **88**, 4034 (1966).
19. H. B. Henbest *et al.*, *J. Chem. Soc.*, 1855 (1963); M. F. Grundon and M. D. Scott, *J. Chem. Soc.*, 5674 (1964).
20. P. D. Gardner *et al.*, *J. Am. Chem. Soc.*, **78**, 3425 (1956).
21. Huang-Minlon, *J. Am. Chem. Soc.*, **71**, 3301 (1949).
22. D. H. R. Barton *et al.*, *J. Chem. Soc.*, 2056 (1955).
23. E. M. Tarasova and V. A. Tùlupova, *J. Gen. Chem. USSR* (*Eng. Transl.*), **31**, 1812 (1961).
24. P. A. Canton *et al.*, *J. Org. Chem.*, **21**, 918 (1956).
25. W. V. Ruyle *et al.*, *J. Org. Chem.*, **25**, 1260 (1960).
26. F. Sondheimer and S. Wolfe, *Can. J. Chem.*, **37**, 1870 (1959).
27. V. Georgian *et al.*, *J. Am. Chem. Soc.*, **81**, 5834 (1959).
28. M. S. Newman *et al.*, *J. Org. Chem.*, **23**, 1832 (1958).
29. D. E. Pearson, unpublished results.
30. K. C. Schreiber and W. Emerson, *J. Org. Chem.*, **31**, 95 (1966).
31. T. A. Ford *et al.*, *J. Am. Chem. Soc.*, **70**, 3793 (1948).
32. W. Reeve and J. D. Sterling, Jr., *J. Am. Chem. Soc.*, **71**, 3657 (1949).
33. D. G. Manly and A. P. Dunlop, *J. Org. Chem.*, **23**, 1093 (1958).
34. H. Pines *et al.*, *J. Am. Chem. Soc.*, **77**, 5099 (1955).
35. S. Landa and J. Mostecký, *Collection Czech. Chem. Commun.*, **20**, 430 (1955).
36. K. W. Rosenmund and E. Karg, *Chem. Ber.*, **75**, 1850 (1942).
37. M. Metayer, *Ann. Chim.* (*Paris*) **4**, 212 (1949).
38. K. N. F. Shaw *et al.*, *J. Org. Chem.*, **21**, 1149 (1956).
39. C. F. Koelsch and R. M. Lindquist, *J. Org. Chem.*, **21**, 657 (1956).
40. E. H. Rodd, *Chemistry of Carbon Compounds*, Elsevier Publishing Co., New York, 2nd Ed., 1964, Vol. 1, Pt. A, p. 361.
41. R. C. Fuson and H. P. Wallingford, *J. Am. Chem. Soc.*, **75**, 5950 (1953).
42. M. E. Gross and H. P. Lankelma, *J. Am. Chem. Soc.*, **73**, 3439 (1951).
43. A. C. Cope *et al.*, *Org. Syn.*, Coll. Vol. **4**, 218 (1963).

44. N. N. Shorygina *et al.*, *Zh. Obshch, Khim.*, **19,** 1558 (1949); *C.A.* **44,** 3919 (1950).

45. R. Grinter and S. F. Mason, *Trans. Faraday Soc.*, **60,** 889 (1964).

46. R. L. Shriner and C. N. Wolf, *J. Am. Chem. Soc.*, **73,** 891 (1951).

47. W. J. Musliner and J. W. Gates, Jr., *J. Am. Chem. Soc.*, **88,** 4271 (1966).

48. G. W. Kenner and N. R. Williams, *J. Chem. Soc.*, 522 (1955).

49. S. W. Pelletier and D. M. Locke, *J. Org. Chem.*, **23,** 131 (1958); J. Fishman and M. Tomasz, *ibid.*, **27,** 365 (1962); E. Caspi *et al.*, *J. Chem. Soc.*, 212 (1963).

50. G. W. Kenner and M. A. Murray, *J. Chem. Soc.*, S178 (1949).

51. H. Zinner *et al.*, *Chem. Ber.*, **92,** 1618 (1959); E. Hardegger *et al.*, *Helv. Chim. Acta*, **41,** 2401 (1958); H. Rapoport and R. M. Bonner, *J. Am. Chem. Soc.*, **73,** 2872 (1951); A. Eschenmoser and A. Frey, *Helv. Chim. Acta*, **35,** 1660 (1952).

52. A. S. Hussey *et al.*, *J. Am. Chem. Soc.*, **75,** 4727 (1953).

53. R. M. Hann *et al.*, *J. Am. Chem. Soc.*, **72,** 561 (1950).

54. C. S. Marvel *et al.*, *Org. Syn.*, Coll. Vol. **1,** 224 (1941).

55. S. Winstein *et al.*, *Chem. Ind.* (*London*), 405 (1960).

56. L. F. Fieser and D. H. Sachs, *J. Org. Chem.*, **29,** 1113 (1964); D. I. Schuster and F.-T. Lee, *Tetrahedron Letters*, 4119 (1965).

57. D. Bryce-Smith *et al.*, *Proc. Chem. Soc.*, 219 (1963).

58. W. Deuschel, *Helv. Chim. Acta.*, **34,** 2403 (1951).

59. W. G. Brown, *Org. Reactions*, **6,** 469 (1951).

60. S. Winstein and J. Sonnenberg, *J. Am. Chem. Soc.*, **83,** 3235 (1961).

61. C. A. Buehler *et al.*, *J. Org. Chem.*, **8,** 316 (1943).

62. H. C. Brown and H. M. Bell, *J. Org. Chem.*, **27,** 1928 (1962); *J. Am. Chem. Soc.*, **88,** 1473 (1966).

63. D. Seyferth *et al.*, *J. Org. Chem.*, **28,** 703 (1963).

64. J. P. Oliver *et al.*, *Tetrahedron Letters*, 3419 (1964).

65. H. G. Kuivila *et al.*, *J. Am. Chem. Soc.*, **84,** 3584 (1962).

66. H. Stetter and J. Mayer, *Chem. Ber.*, **92,** 2664 (1959).

67. N. P. Buu-Hoï *et al.*, *Bull. Soc. Chim. France*, 2442 (1963).

68. N. P. Buu-Hoï *et al.*, *Compt. Rend.*, **257,** 3182 (1963).

69. H. Hart, *J. Am. Chem. Soc.*, **71,** 1966 (1949).

70. R. B. Sandin and L. F. Fieser, *J. Am. Chem. Soc.*, **62,** 3098 (1940).

71. W. L. Mosby, *Chem. Ind.* (*London*), 1348 (1959); *J. Org. Chem.*, **24,** 421 (1959).

72. P. N. Rylander, *Catalytic Hydrogenation over Platinum Metals*, Academic Press, New York, 1967, Chap. 24.

73. P. Bruck, *Tetrahedron Letters*, 449 (1962).

74. D. Bryce-Smith and B. J. Wakefield, *Org. Syn.*, **47,** 103 (1967).

75. P. A. Levene, *Org. Syn.*, Coll. Vol. **2,** 320 (1943).

76. L. A. Rothmann and E. I. Becker, *J. Org. Chem.*, **24,** 294 (1959).

77. C. Grundmann and G. Weisse, *Chem. Ber.*, **84,** 684 (1951).

78. H. M. Bell and H. C. Brown, *J. Am. Chem. Soc.*, **88,** 1473 (1966).

79. B. Marchand, *Chem. Ber.*, **95,** 577 (1962).

80. J. W. Cornforth *et al.*, *J. Chem. Soc.*, 3348 (1955).

81. W. Reeve and A. Sadle, *J. Am. Chem. Soc.*, **72,** 3252 (1950).

82. E. M. Schultz and J. B. Bicking, *J. Am. Chem. Soc.*, **75,** 1128 (1953).

83. H. Rapoport *et al.*, *J. Am. Chem. Soc.*, **73,** 2718 (1951).

84. F. Poppelsdorf and S. J. Holt, *J. Chem. Soc.*, 1124 (1954).

85. S. W. Cantor and C. R. Hauser, *J. Am. Chem. Soc.*, **73,** 4122 (1951).

86. W. R. Brasen and C. R. Hauser, *Org. Syn.*, Coll. Vol. **4,** 508 (1963).

87. E. E. van Tamelen *et al.*, *Tetrahedron*, **14,** 8 (1961).

88. K. Hafner and J. Schneider, *Ann. Chem.*, **624,** 37 (1959).

89. T. D. Walsh and R. C. Long, *J. Am. Chem. Soc.*, **89,** 3943 (1967).

90. A. Nickon and A. S. Hill, *J. Am. Chem. Soc.*, **86,** 1152 (1964).

91. R. L. Augustine, *Catalytic Hydrogenation*, Marcel Dekker, New York, 1965, Chap. 4; Ref. 72, Chaps. 5, 18.

92. H. C. Brown and K. Murray, *J. Am. Chem. Soc.*, **81,** 4108 (1959).
93. Ref. 72, pp. 23–26; Ref. 91, pp. 147–153.
94. H. C. Brown and C. A. Brown, *J. Am. Chem. Soc.*, **84,** 1494 (1962).
95. R. W. Bott *et al.*, *Proc. Chem. Soc.*, 337 (1962).
96. E. B. Hershberg *et al.*, *J. Am. Chem. Soc.*, **73,** 1144 (1951).
97. F. A. Vandenheuvel, *Anal. Chem.*, **28,** 362 (1956).
98. Ref. 72, p. 84.
99. A. J. Birch and K. A. M. Walker, *Tetrahedron Letters*, 1935 (1967).
100. G. Wilkinson, *Chem. Commun.*, 131 (1965).
101. I. Jardine and F. J. McQuillan, *Tetrahedron Letters*, 4871 (1966).
102. E. C. Kleiderer and E. C. Kornfeld, *J. Org. Chem.*, **13,** 455 (1948).
103. F. v. Wessely and H. Welleba, *Chem. Ber.*, **74,** 777 (1941).
104. S. Siegel, *Advan. Catalysis*, **16,** 123 (1966).
105. N. G. Gaylord, *Reduction with Complex Metal Hydrides*, Interscience Publishers, New York, 1956, p. 925.
106. B. Franzus and E. I. Snyder, *J. Am. Chem. Soc.*, **87,** 3423 (1965).
107. H. C. Brown *et al.*, *J. Am. Chem. Soc.*, **90,** 4165 (1968).
108. C. A. Brown and H. C. Brown, *J. Am. Chem. Soc.*, **85,** 1003 (1963).
109. H. C. Brown and C. A. Brown, *J. Am. Chem. Soc.*, **85,** 1005 (1963).
110. L. M. Berkowitz and P. N. Rylander, *J. Org. Chem.*, **24,** 708 (1959).
111. Ref. 72, p. 107.
112. A. V. Lozovoĭ, *J. Gen. Chem.*, *USSR* (*Eng. Transl.*), **10,** 1855 (1940).
113. P. N. Rylander and D. R. Steele, *Engelhard Ind. Tech. Bull.*, **3,** 19 (1962).
114. C. E. Miller, *J. Chem. Ed.*, **42,** 254 (1965); S. Hünig *et al.*, *Angew. Chem. Intern. Ed. Engl.*, **4,** 271 (1965).
115. R. S. Dewey and E. E. van Tamelen, *J. Am. Chem. Soc.*, **83,** 3729 (1961).
116. R. C. Fuson *et al.*, *J. Am. Chem. Soc.*, **57,** 2208 (1935); **58,** 1745 (1936).
117. A. Streitwieser, Jr. and W. J. Downs, *J. Org. Chem.*, **27,** 625 (1962).
118. T. Kauffmann *et al.*, *Chem. Ber.*, **96,** 999 (1963).
119. Ref. 91, p. 57.
120. Ref. 91, p. 72.
121. M. Freifelder and G. R. Stone, *J. Am. Chem. Soc.*, **80,** 5270 (1958); *J. Org. Chem.*, **27,** 3568 (1962).
122. G. M. Badger and A. R. M. Gibb, *J. Chem. Soc.*, 799 (1949).
123. W. Davies and Q. N. Porter, *J. Chem. Soc.*, 4967 (1957).
124. V. Bruckner *et al.*, *Tetrahedron Letters*, No. 1, 5 (1960).
125. D. S. Bapat *et al.*, *Tetrahedron Letters* No. 5, 15 (1960).
126. M. Rabinovitz and G. Salemnik, *J. Org. Chem.*, **33,** 3935 (1968).
127. L. F. and M. Fieser, *Reagents for Organic Synthesis*, John Wiley and Sons, New York, 1967, p. 727.
128. E. E. Reid, *Organic Chemistry of Bivalent Sulfur*, Vol. 1, Chemical Publishing Co., New York, 1958, p. 115.
129. H. Hauptmann *et al.*, *Ann. Chem.*, **576,** 45 (1952).
130. Ref. 127, p. 729.
131. N. Kharasch and C. Y. Meyers, *The Chemistry of Organic Sulfur Compounds*, Vol. 2, Pergamon Press, New York, 1966, p. 35.
132. F. W. Hoffmann *et al.*, *J. Am. Chem. Soc.*, **78,** 6414 (1956).
133. R. C. Krug and S. Tocker, *J. Org. Chem.*, **20,** 1 (1955).
134. E. E. van Tamelen and E. A. Grant, *J. Am. Chem. Soc.*, **81,** 2160 (1959).
135. J. E. Baldwin *et al.*, *J. Am. Chem. Soc.*, **90,** 4758 (1968).
136. N. Kornblum, *Org. Reactions*, **2,** 262 (1944).
137. M. M. Robison and B. L. Robison, *Org. Syn.*, Coll. Vol. **4,** 947 (1963).
138. J. F. Bunnett and H. Takayama, *J. Org. Chem.*, **33,** 1924 (1968).
139. N. Kornblum, *Org. Syn.*, Coll. Vol. **3,** 295 (1955).
140. A. H. Lewin and T. Cohen, *J. Org. Chem.*, **32,** 3844 (1967).

141. R. A. Henry and W. G. Finnigan, *J. Am. Chem. Soc.*, **76**, 290 (1954).
142. R. Q. Brewster and J. A. Poje, *J. Am. Chem. Soc.*, **61**, 2418 (1939).
143. L. F. Fieser and H. Heymann, *J. Am. Chem. Soc.*, **64**, 376 (1942).
144. H. Zollinger, *Azo and Diazochemistry*, Interscience Publishers, New York, 1961, p. 168.
145. J. B. Hendrickson, *J. Am. Chem. Soc.*, **83**, 1251 (1961).
146. K. G. Rutherford and W. A. Redmond, *J. Org. Chem.*, **28**, 568 (1963).
147. A. Nickon and A. S. Hill, *J. Am. Chem. Soc.*, **86**, 1152 (1964).
148. E. E. van Tamelen and M. A. Schwartz, *J. Am. Chem. Soc.*, **87**, 3277 (1965).
149. Ref. 72, p. 468.
150. R. G. Harvey and K. Urberg, *J. Org. Chem.*, **33**, 2206 and 2570 (1968).
151. H. M. Walborsky and J. B. Pierce, *J. Org. Chem.*, **33**, 4102 (1968).
152. Ref. 72, p. 470.
153. F. R. Jensen and W. E. Coleman, *J. Am. Chem. Soc.*, **80**, 6149 (1958).
154. M. P. Cava and R. Pohlke, *J. Org. Chem.*, **28**, 1012 (1963).

B. Organometallic Methods

1. Hydrolysis

$$RMgX + HOH \rightarrow RH + MgXOH$$

This synthesis represents an indirect method of preparing hydrocarbons from alkyl, aryl, or cycloalkyl halides. The direct method, of course, is hydrogenolysis of the halide (see A.3). But the indirect method has its advantages both in ease of preparation and in the incorporation of deuterium into the hydrocarbon [1]:

$$2,5\text{-Dichlorophenylmagnesium iodide in ether} \xrightarrow{D_2O} 1,4\text{-dichlorobenzene-2d}$$

only 76.8 mole % purity

Evidently, some reduction of the Grignard reagent occurs from the hydrogen atoms of the ether solvent. The Grignard reagent is usually prepared in ether or tetrahydrofuran, or benzene or toluene medium may be employed if 1 mole of tetrahydrofuran per mole of halide is present [2] or if triethylamine is used as a complexing agent [3]. The latter procedure reduces the fire hazard because the tetrahydrofuran is bound to the Grignard reagent.

(a) **Preparation of pentane** (50–53% from 2-bromopentane which was converted into the Grignard reagent in *n*-butyl ether; the latter was hydrolyzed with aqueous sulfuric acid) [4].

(b) **Preparation of cyclobutane** (83% from cyclobutyl bromide via the Grignard reagent which was hydrolyzed with *n*-butyl alcohol [5].

2. Coupling with Halides

$$RMgX + R'X \rightarrow RR' + MgX_2$$

This synthesis has been of value particularly in the preparation of monoalkyl aromatic hydrocarbons and aliphatic hydrocarbons having quaternary carbon atoms. Not only the Grignard reagent, but also the alkyls of zinc, copper, lithium, and sodium have been the most widely used in the coupling. Sodium

References for Section B are on pp. 28–29.

is involved in the well-known Wurtz or Wurtz-Fittig reaction. The sodium may be employed as naphthalenesodium, a reagent which serves particularly well in the coupling of benzyl halides [6]. In addition to alkyl halides, alkyl sulfates [7] and alkyl esters of aryl sulfonic acids [8] may be employed to introduce the alkyl group.

Although Grignard reagents couple with tertiary alkyl halides in low yields [9], the method is suitable for the preparation of highly branched hydrocarbons such as neopentane [10], neohexane [11], and hexamethylethane [12]. It is of interest to note that coupling yields of Grignard reagents with t-alkyl halides are improved by shifting from ether to heptane as the solvent after the Grignard reagent has been prepared (see Ex. a). Incidentally, Grignard reagents can also be made in nonethereal solvents [13].

The Grignard reagent of benzyl chloride and other related active-halide types couple readily to yield alkylbenzenes. In this way n-amylbenzene [14], 50–59% yield, and n-propylbenzene [15], 70–75% yield, have been produced. This method is also applicable to more highly substituted chloromethylarenes. Alkyl fluorides are much less active toward displacement by the Grignard reagent. Indeed, they are unreactive enough to be contained in the Grignard reagent, as in $F(CH_2)nMgBr$ [16].

Although the coupling of Grignard reagents with active halides operates mainly via a displacement reaction, indications are that a free radical mechanism may encroach, the more so in the presence of the slightest amount of metallic impurities. To detect the free radical mechanism, cumene is often added to trap the free radical, as shown by the formation of the coupling product, 2,3-diphenyl-2,3-dimethylbutane. Butyl iodide and magnesium in cumene, for example, yield about 18% of the coupled cumene [17], an amount comparable to that obtained with lithium and butyl iodide.

In order to favor the free radical mechanism, catalytic amounts of cobalt (or cuprous chloride) are added to the Grignard reagent, probably to form intermediate cobalt alkyls RCoR′ [18]. The free radical behavior of Grignard reagents is quite complex and not well suited for syntheses. For example, ethylmagnesium bromide in the presence of cuprous chloride gives alkyl exchange with an alkyl halide: $C_2H_5MgBr + RBr \rightarrow RMgBr + C_2H_5Br$ and coupling with itself to form butane (12%) and disproportionation to give ethane (48%) and ethylene (40%) [19]; for alkyl exchange see also [20].

The Wurtz or Wurtz-Fittig reaction is similar in many respects to the Grignard coupling reaction, perhaps differing because of the more heterogeneous nature of the reaction components. The mechanism is predominantly

Wurtz: $\qquad RX + 2\,Na \xrightarrow{} RNa + NaX \xrightarrow{RX} RR$

Wurtz-Fittig: $\quad ArX + 2\,Na \xrightarrow{} ArNa \xrightarrow{RX} ArR \qquad$ (see Ex. c.3)

displacement by the sodium alkyl or aryl [21]. The most recent study of the Wurtz reaction shows that yields of 40–60% of hydrocarbons may be obtained by adding the alkyl chloride (C_6 to C_{18}) slowly to sodium while holding the mixture as near room temperature as possible [22]. High-speed stirring should have a favorable effect. Such stirring has been employed to advantage in the reaction of amyl chloride and sodium in toluene [23]. Curiously, the addition of

$$C_5H_{11}Cl + 2\ Na \longrightarrow C_5H_{11}Na \xrightarrow{C_6H_5CH_3} C_6H_5CH_2Na \xrightarrow[RONa]{C_5H_{11}Cl} C_6H_5CH_2C_5H_{11}$$

$$\text{Phenylhexane}$$
$$76\%$$

an alkoxide (RONa) has a pronounced effect on yield, perhaps by providing an ionic environment in the nonpolar medium suitable for coupling to take place. In addition pyrophoric lead has been recommended as a reagent for Wurtz coupling particularly with reactive halides of various types, such as acyl halides (coupling to diketones) or α-chlorocarboxylic acids (coupling to succinic acids) [24].

Recently it has been shown [25] that perfluoroalkyl iodides may be coupled with aromatic iodides by means of a copper complex. The reaction, though

$$CF_3CF_2CF_2I + Cu \xrightarrow[\text{solvent}]{\text{Aprotic}} \text{complex} \xrightarrow[100°]{C_6H_5I} CF_3CF_2CF_2C_6H_5$$

$$\text{Phenylheptafluoropropane}$$

specific, is particularly advantageous since reactive aromatic halides, such as 2-iodothiophene, do not couple with themselves as in the high-temperature Ullmann reaction (B.3) when brought together with the preformed copper complex.

In addition the oxidative coupling of copper(I)ate complexes has been accomplished under mild conditions [26]. As one example, n-octane has been produced from butyllithium as follows:

$$2\ C_4H_9Li + [CuI \cdot P(C_4H_9)_3]_4 \xrightarrow[-78°]{THF} (C_4H_9)_2CuLi \xrightarrow[-78°]{O_2} C_8H_{18}$$

The Wurtz reaction has been applied to dihalides in attempts to synthesize cycloalkanes. In this case, zinc is preferred to sodium, but unfortunately the yields are not good except in the synthesis of cyclopropane. By using 1 mole of 1,3-dichloropropane with 100% excess of zinc dust, 1 mole of sodium carbonate, and $\frac{1}{6}$ mole of sodium iodide in aqueous ethanol a 95% yield of crude cyclopropane was obtained [27]. Good yields were also obtained by employing anhydrous acetamide as the solvent. Cyclopropane may also be made by the coupling of trimethylene dibromide (or 3-bromopropyl tosylate) with chromous ion, Cr^{II}, complexed with ethylenediamine [28]. Cyclobutane is made by specific coupling with lithium amalgam [29] and cyclopropylbenzene in 75–85% yield from 1,3-dibromo-1-phenylpropane and a zinc-copper couple in DMF at 7–9° [30].

$$Br(CH_2)_4Br + Li(Hg) \xrightarrow[\text{3 hr.}]{\text{Dioxane, reflux}} \boxed{}$$

$$70\%$$

Coupling of alkyl- or aryllithiums with active alkyl halides is similar to the

Wurtz reaction, but the reaction differs in being carried out in discrete steps. The alkyllithiums can be obtained from the reaction of a halide with lithium metal or by halogen-metal interconversion [31]. The aryllithium then may be

$$C_4H_9Li \text{ (available commercially)} + ArX \rightarrow ArLi + C_4H_9X$$

coupled with an active alkyl halide. However, this reaction has not been used frequently for synthesis of hydrocarbons.

Furthermore, stable anions of certain relatively acidic hydrocarbons can be alkylated by the use of sodamide or lithium amide and an alkyl halide [32]. Much more di-n-butylfluorene is formed with sodamide as a catalyst.

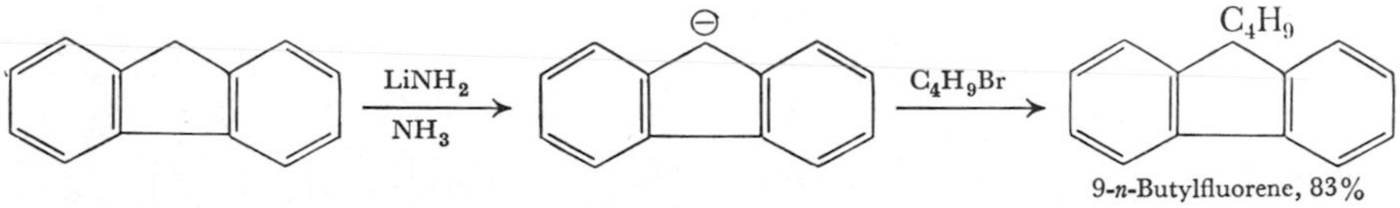

9-n-Butylfluorene, 83%

It must be borne in mind that very strong bases are liable to form, as an intermediate, benzyne, which then adds an aryllithium reagent to give a coupling product [33]:

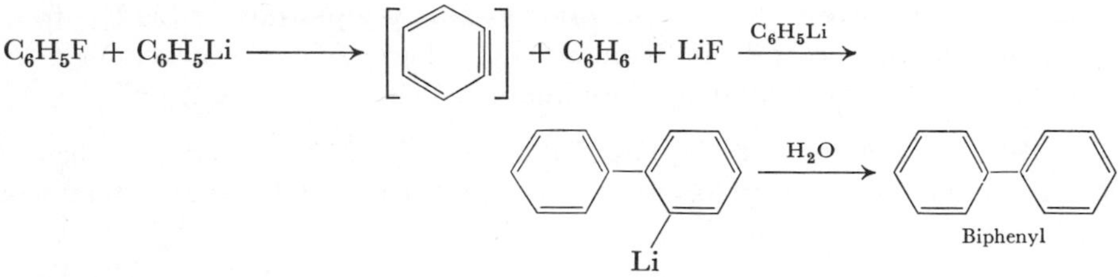

Biphenyl

Although in this case the product is identical whether simple displacement or benzyne formation takes place, the product is not identical if the benzyne is asymmetrical:

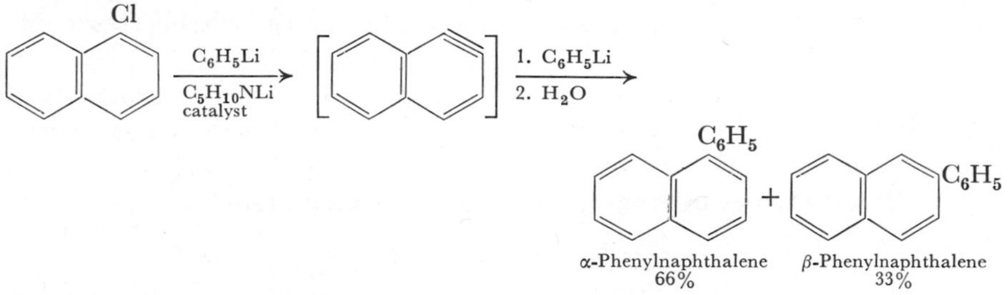

α-Phenylnaphthalene   β-Phenylnaphthalene
66%   33%

The lithium piperidide is an essential catalyst for good yields. Benzyne formation can be expected with very strong bases and fluoro- or chlorobenzenes, rarely with bromo- or iodobenzenes.

Lastly, a very promising general method of coupling has been demonstrated, not one that will replace simple Wurtz reactions, but one that should be considered for synthesis of inaccessible hydrocarbons. It involves the coupling of lithium dialkylcopper with alkyl, vinyl, or aryl halides. Syntheses have been carried out with lithium dimethyl-(diethyl-or dibutyl-)copper and with alkyl groups containing carboxyl or substituted amide groups [34]. Optimum temperature seems to increase within the range -95 to $25°$ as activity of the halide

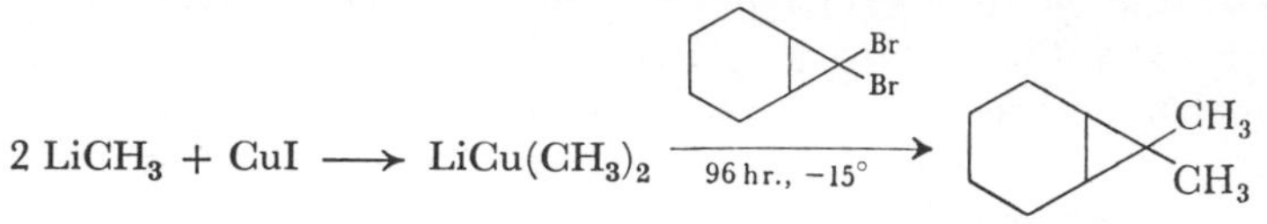

$$2\ LiCH_3 + CuI \longrightarrow LiCu(CH_3)_2 \xrightarrow[96\ hr.,\ -15°]{} $$

7,7-Dimethylnorcarane
65%

$$LiCu(CH_3)_2 \xrightarrow[25°]{C_6H_5I,\ 14\ hr.} C_6H_5CH_3$$

Toluene
90%

$$LiCu(CH_3)_2 \xrightarrow[2.5\ hr.,\ 0°]{trans\text{-}C_6H_5CH=CHBr} trans\text{-}C_6H_5CH=CHCH_3$$

β-Methylstyrene 81%

decreases. sec-Halides tend to dehydrohalogenate rather than couple with the copper reagent. An octaphenyl has been prepared from an iodotetraphenyl, butyllithium, and cupric chloride in a somewhat similar process [35].

(a) Preparation of 2,4-dimethyl-2-benzylpentane (59% from benzylmagnesium chloride in heptane, the ether having been removed by distillation, after which 2-chloro-2,4-dimethylpentane was added) [36].

(b) Preparation of n-propylbenzene (70–75% based on benzyl chloride which was converted into the Grignard reagent followed by treatment with diethyl sulfate) [15].

(c) Other examples.

    (1) 2,2′-Methylenedithiophene (66% from thiophene allowed to react first with n-butyllithium in ether under nitrogen and then with α-chloromethylthiophene) [37].

    (2) 1-n-Butylnaphthalene (87% from 1-bromonaphthalene treated first with n-butyllithium in ether and then with a 20% excess of n-butyl bromide) [38].

    (3) n-Butylbenzene (Wurtz-Fittig) (65–70% by the slow addition of a mixture of butyl bromide and bromobenzene to sodium shavings covered with ether at 20°) [39].

    (4) n-Octane, bibenzyl, and n-butylbenzene [40]

$$2\ C_4H_9Br + 2\ Na \rightarrow C_4H_9C_4H_9 + 2\ NaBr$$
24 g.   8 g.   n-Octane, 48%

$$2\ C_6H_5CH_2Cl + 2\ Na \rightarrow C_6H_5CH_2CH_2C_6H_5 + 2\ NaCl$$
25 g.   6 g.   Bibenzyl, 40%

$$C_6H_5Br + C_4H_9Br + 2\ Na \rightarrow C_6H_5C_4H_9 + 2\ NaBr$$
26 g.   24.6 g.   11.2 g.   n-Butylbenzene, 40%

(5) 1,3-Dimethylbicyclobutane. This reaction is an attractive alternative for Wurtz coupling to cyclic hydrocarbons [41].

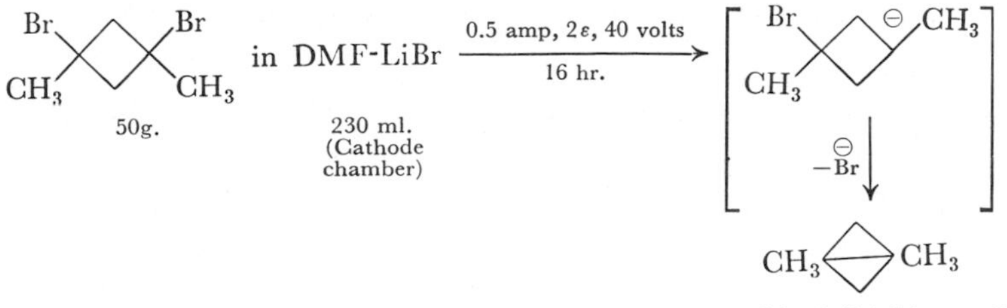

1,3-Dimethylbicyclo(1.1.0)butane, 55–94%

3. Coupling of Aryl Halides (Ullmann)

$$2 \, ArX + Cu \rightarrow ArAr + CuX_2$$

The Ullman reaction has been reviewed [42]. It is of particular value in the synthesis of biphenyl and its derivatives. The iodides are more reactive than bromides, which are in turn more reactive than chlorides. However, bromides and chlorides may be employed satisfactorily if there are electron-withdrawing substituents, such as the nitro group, in the o- or p-position. Yields of symmetrical biaryls rarely exceed 80% in this synthesis; for unsymmetrical biaryls the yields, as expected, are lower. But, interestingly good yields of mixed biaryls can be obtained by maintaining a temperature low enough that the least reactive aryl halide will not react with itself. The mechanism must be quite complicated, but the observation above indicates that a two-step process is involved, possibly formation of a copper salt deposited on a copper surface from the more reactive halide followed by homolytic or heterolytic displacement from the least active halide. Or perhaps a cuprous anion, $^{\ominus}(CuAr_2)$, is involved [34]. Cuprous oxide rather than copper has been implicated as the active agent [43]. Indeed the synthesis of biphenylene requires cuprous oxide [44]:

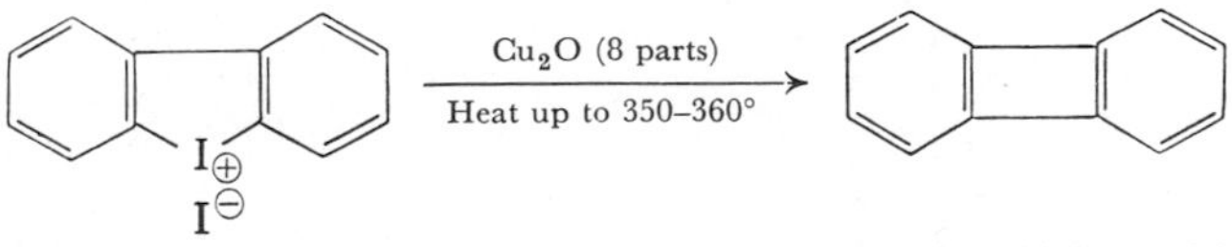

Biphenylene, 21% ave. yield

The reaction is simple to operate, the halide or halides being heated with copper bronze which has been activated by washing with iodine in acetone solution followed by a hydrochloric acid wash [45]. Diluents such as nitrobenzene, toluene, or naphthalene can be used but are not always necessary. Dimethylformamide has proved particularly efficacious with aryl halides of relatively high activity [46], but one must be aware that reduction rather than coupling may take place with this solvent [47]. In this case, tetramethylurea is recommended as the diluent [48]. Interfering groups such as amino, carboxyl, and hydroxyl groups may be carried in the substrate structure if they are properly protected.

An elegant new procedure for coupling may well supplant the Ullman synthesis in many instances [49]. Yields ranged from 50 to 99 % for a number of

$$p\text{-}CH_3C_6H_4MgBr \ + \ TlBr \ \xrightarrow[\text{N}_2,\ \text{reflux 4 hr.}]{\text{THF, C}_6\text{H}_6} \ p\text{-}CH_3C_6H_4C_6H_4CH_3\text{-}p$$

0.067 mole   0.135 mole   4,4′-Dimethylbiphenyl, 91%

aromatic and aliphatic halides, including the coupling of 2-bromopentane to 4,5-dimethyloctane (50 %). About the only unattractive feature is the toxicity of thallium salts.

(a) Preparation of 2,2′-dinitrobiphenyl (52–61 % from *o*-chloro-nitrobenzene, copper bronze, and sand at 215–225° for 2.7 hr.) [50]. The same compound can also be made in 99.6 % yield by heating *o*-nitrochlorobenzene at 60° for 60 hr. with a tenfold excess of activated copper bronze and no solvent. This is the lowest temperature and highest yield ever reported for an Ullmann reaction [47].

(b) Preparation of 2,4′-dimethyl-2′,4-dinitrobiphenyl (35.2 % of

impure product from 2-iodo-5-nitrotoluene, 3-nitro-4-bromotoluene, and copper powder at 200–320°) [51].

1. J. A. Zoltewicz and J. F. Bunnett, *J. Am. Chem. Soc.*, **87**, 2640 (1965).
2. T. Leigh, *Chem. Ind.* (*London*), 426 (1965).
3. E. C. Ashby and R. Reed, *J. Org. Chem.*, **31**, 971 (1966).
4. C. R. Noller, *Org. Syn.*, Coll. Vol. **2**, 478 (1943).
5. J. Cason and R. L. Way, *J. Org. Chem.*, **14**, 31 (1949).
6. H. Güsten and L. Horner, *Angew Chem. Intern. Ed. Engl.*, **1**, 455 (1962).
7. L. I. Smith, *Org. Syn.*, Coll. Vol. **1**, 471 (1941).
8. H. Gilman and J. Robinson, *Org. Syn.*, Coll. Vol. **2**, 47 (1943).
9. H. Soroos and H. B. Willis, *J. Am. Chem. Soc.*, **63**, 881 (1941).
10. F. C. Whitmore and G. H. Fleming, *J. Am. Chem. Soc.*, **55**, 3803 (1933).
11. F. C. Whitmore *et al.*, *J. Am. Chem. Soc.*, **60**, 2539 (1938).
12. G. Calingaert, *J. Am. Chem. Soc.*, **66**, 1389 (1944).
13. D. Bryce-Smith *et al.*, *J. Chem. Soc.*, 1175 (1961); *Org. Syn.* **47**, 113 (1967).
14. H. Gilman and J. Robinson, *Org. Syn.*, Coll. Vol. **2**, 47 (1943).
15. H. Gilman and W. E. Catlin, *Org. Syn.*, Coll. Vol. **1**, 471 (1941).
16. F. L. M. Pattison and W. C. Howell, *J. Org. Chem.*, **21**, 879 (1956).
17. D. Bryce-Smith and G. F. Cox, *J. Chem. Soc.*, 1050 (1958).
18. F. W. Frey, Jr., *J. Org. Chem.*, **26**, 5187 (1961).
19. V. D. Parker and C. R. Noller, *J. Am. Chem. Soc.*, **86**, 1112 (1964).
20. L. H. Slaugh, *J. Am. Chem. Soc.*, **83**, 2734 (1961).
21. S. E. Ulrich *et al.*, *J. Am. Chem. Soc.*, **80**, 622 (1958).
22. R. Ulrich, *Gannon Coll. Chem. J.*, 14 (1964); *C.A.*, **62**, 6381 (1965).
23. A. A. Morton and A. E. Brachman, *J. Am. Chem. Soc.*, **73**, 4363 (1951).
24. L. Mészáros, *Tetrahedron Letters*, 4951 (1967).
25. I. M. White, *Chem. Eng. News*, 40, August 14, 1967.
26. G. M. Whiteside *et al.*, *J. Am. Chem. Soc.*, **89**, 5302 (1967).
27. H. B. Hass *et al.*, *Ind. Eng. Chem.*, **28**, 1178 (1936).
28. J. K. Kochi and D. M. Singleton, *J. Org. Chem.*, **33**, 1027 (1968).

29. D. S. Connor and E. R. Wilson, *Tetrahedron Letters*, 4925 (1967).
30. H. Shechter *et al.*, *Org. Syn.*, **44,** 30 (1964).
31. R. G. Jones and H. Gilman, *Org. Reactions*, **6,** 339 (1951).
32. W. S. Murphy and C. R. Hauser, *J. Org. Chem.*, **31,** 85 (1966).
33. H. Heaney, *Chem. Rev.*, **62,** 81 (1962).
34. E. J. Corey and G. H. Posner, *J. Am. Chem. Soc.*, **89,** 3911 (1967); **90,** 5615 (1968).
35. W. Heitz and R. Ullrich, *Makromol. Chem.*, **98,** 29 (1966); *C.A.*, **67,** 1062 (1967).
36. A. D. Petrov *et al.*, *J. Gen. Chem. USSR (Eng. Transl.)*, **29,** 49 (1959).
37. N. Lofgren and C. Tegner, *Acta Chem. Scand.*, **6,** 1020 (1952).
38. H. Gilman *et al.*, *J. Org. Chem.*, **22,** 685 (1957).
39. R. R. Read *et al.*, *Org. Syn.*, Coll. Vol. **3,** 157 (1955).
40. R. J. W. Cremlyn and R. H. Still, *Named and Miscellaneous Reactions in Practical Organic Chemistry*, John Wiley and Sons, New York, 1967, pp. 151–153.
41. M. R. Rifi, *J. Am. Chem. Soc.*, **89,** 4442 (1967).
42. P. E. Fanta, *Chem. Rev.*, **38,** 139 (1946); **64,** 613 (1964).
43. R. G. R. Bacon *et al.*, *Tetrahedron Letters*, 2003 (1967).
44. W. Baker *et al.*, *J. Chem. Soc.*, 1476 (1954).
45. A. I. Vogel, *Practical Organic Chemistry*, Longmans, Green and Co., New York, 1951, p. 188.
46. N. Kornblum and D. L. Kendall, *J. Am. Chem. Soc.*, **74,** 5782 (1952).
47. M. D. Rausch, *J. Org. Chem.*, **26,** 1802 (1961).
48. A. Lüttringhaus and H. W. Dirksen, *Angew. Chem. Intern. Ed. Engl.*, **3,** 260 (1964).
49. E. C. Taylor *et al.*, *J. Am. Chem. Soc.*, **90,** 2423 (1968).
50. R. C. Fuson and E. A. Cleveland, *Org. Syn.*, Coll. Vol. **3,** 339 (1955).
51. R. B. Carlin and G. E. Folz, *J. Am. Chem. Soc.*, **78,** 1997 (1956).

C. Nucleophilic Reactions

1. From Alkenes

$$R_2C{=}CH_2 \xrightarrow[\text{ether}]{Na} R_2CHCH_2CH_2CHR_2$$

The conditions for this very old reaction have been improved recently so that it has the potentiality of a much wider scope. An intermediate sodio derivative is formed which can react in three ways:

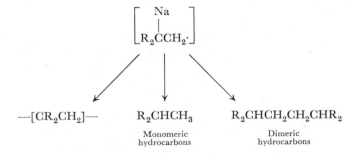

Polymerization occurs particularly with styrene and dienes. The monomeric hydrocarbon is favored if some protonic source such as *N*-ethylaniline is incorporated in the reaction mixture. The dimeric hydrocarbon is favored if finely divided sodium is used with a sodium carrier such as terphenyl and with a good complexing ether such as tetrahydrofuran [1]. Apparently terminal olefins

References for Section C are on p. 33.

of any type undergo reductive dimerization with the preceding modifications, but unsymmetrical diarylethylenes are so prone to form the dimeric hydrocarbon that simpler conditions are possible [2].

Alkylboranes, prepared from olefins and diborane, also can be dimerized by silver nitrate (see Ex. b). This reaction is more related in mechanism to the Wurtz (B.2) or Ullmann (B.3) reactions.

In a related reaction, paracyclophanes are prepared in superior yield (see Ex. c) from ditosylates probably via the *p*-xylylene derivative [3]:

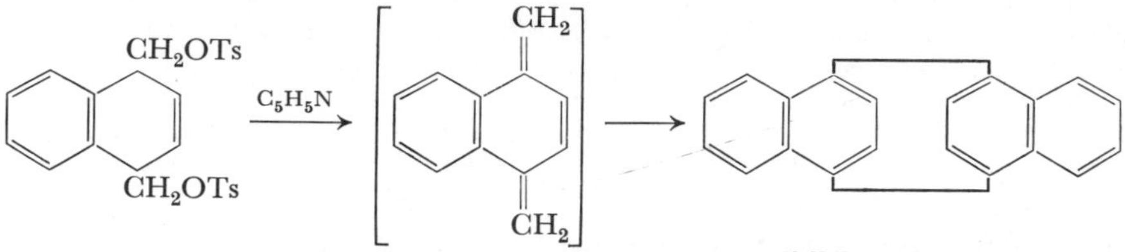

[2.2]-Paracyclonaphthane

(a) **Preparation of 1,1,4,4-tetraphenylbutane** ("exclusively" from 3 g. of 1,1-diphenylethylene and 10 g. of sodium powder in 250 ml. of dry ether) [4].

(b) **Preparation of dodecane** (79% from 1-hexene, sodium borohydride, and boron trifluoride etherate in diglyme, cooled to −20° and treated consecutively with water, to remove excess hydride, aqueous sodium hydroxide, and then aqueous silver nitrate; internal olefins gave yields of about 35–50%) [5].

(c) **Preparation of (2.2)-paracyclonaphthane** (90% from the above ditosylate by solvolysis with boiling pyridine for 5 hr. [3].

2. From Alkenes and Carbanions

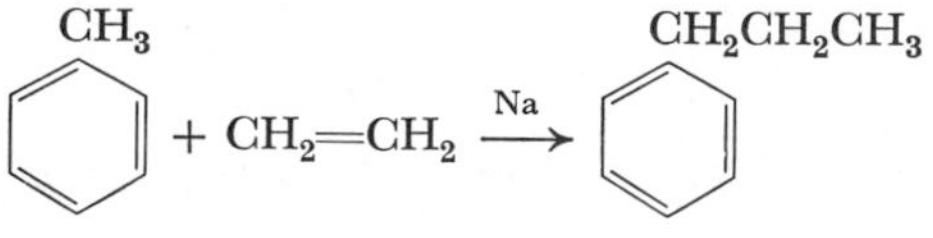

This synthesis has been reviewed [6]. It permits one to enlarge the alkyl group catalytically although the arylalkane must possess at least one benzylic hydrogen. Ethylene is the alkene which has been employed most widely in the reaction, but others such as propylene and styrene may be utilized. The catalysts used are organosodium compounds [7], alkali metals [8], benzylsodium [9], and potassium-graphite [10].

With ethylene, temperatures in the 150–200° range and pressures varying from 0–70 atms. are employed when sodium is the catalyst. The chief by-products are more highly ethyl-substituted arenes and ethane. More recently, more satisfactory yields were obtained with α-methylstyrene by operating at lower pressures and temperatures with potassium as catalyst [11]. For example, α-methylstyrene with toluene, ethylbenzene, and *n*-propylbenzene gave, respectively, 1,3-diphenylbutane in 73%, 2,4-diphenylpentane in 85%, and 2,4-diphenylhexane in 75% yields.

The mechanism proposed by Pines [12] is shown below. A catalyst such as sodium probably metalates the benzylic compound, 1-V, to form the benzylic carbanion, 1-VI, which reacts with the alkene to give the extended carbanion 1-VII. By a transmetalation reaction the latter reacts with 1-V to form the product 1-VIII and the benzyl anion, which is available for further reaction.

$$C_6H_5CH_3 + Na \longrightarrow C_6H_5\overset{\ominus}{C}H_2\overset{\oplus}{N}a \xrightarrow{CH_2=CH_2} C_6H_5CH_2CH_2\overset{\ominus}{C}H_2\overset{\oplus}{N}a \xrightarrow{C_6H_5CH_3}$$

1-V   1-VI   1-VII

$$C_6H_5CH_2CH_2CH_3 + C_6H_5\overset{\ominus}{C}H_2$$

1-VIII

(a) **Preparation of 2,4-diphenylpentane** Ethylbenzene, 30 g., was treated with 2 g. of freshly cut potassium, after which the mixture in a helium

$$C_6H_5\overset{\overset{\text{CH}_3}{|}}{C}H_2 + C_6H_5\overset{\overset{\text{CH}_3}{|}}{C}=CH_2 \longrightarrow C_6H_5\overset{\overset{\text{CH}_3}{|}}{C}HCH_2\overset{\overset{\text{CH}_3}{|}}{C}HC_6H_5$$

atmosphere was heated to 105° and stirred for 30 min. Then 30 g. more of ethylbenzene was added dropwise and stirring was continued for another 30 min. Freshly distilled α-methylstyrene, 24 g., dissolved in 46 g. of ethylbenzene was finally added to the reaction mixture in 45–60 min. Total time was 2 hr. By fractionation to remove ethylbenzene, the desired hydrocarbon, b.p. 118–119° (1 mm.), was obtained in 84.5% yield [11].

3. From Fluorene and Sodium Alkoxides

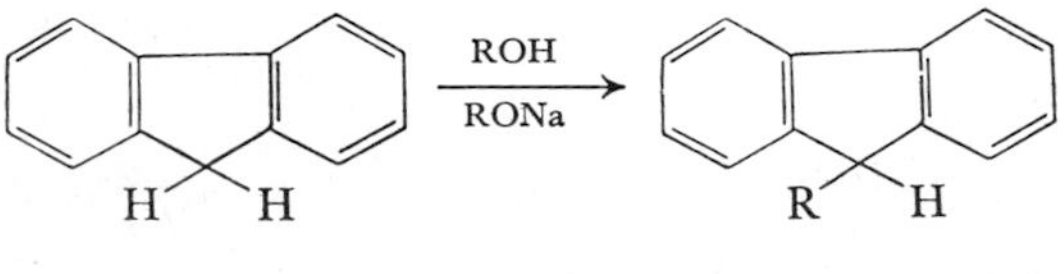

9-Alkylfluorenes are prepared from fluorene and the sodium *n*-alkoxide, pressure being used for the introduction of the lower alkyl groups [13, 14]. Yields vary from 58 to 92%. The reaction is similar to the Guerbet reaction (see Alcohols, Chapter 4, G.3):

$$CH_3ONa \xrightarrow{\Delta} H_2C=O + NaH \xrightarrow{R_2\overset{\ominus}{C}H} R_2C=CH_2 \xrightarrow{NaH} R_2\overset{\overset{\text{H}}{|}}{C}CH_3$$

(a) **Preparation of 9-methylfluorene** (76–86% from fluorene and sodium methoxide in methanol at 220° under pressure) [14].

4. From Aldols

Aldols or, more specifically, ketone β, β′-diols are potential sources of hydrocarbons. So many side reactions can take place in the ring closure, however, that yields are quite poor. Perhaps the driving force for the formation of any hydrocarbon at all from this reaction is the stabilization wrought by aromatization of the ring system. Although the ketone diols usually are made by base catalysis, they can be made by acid catalysis and ring closure in a single reaction.

(a) **Mesitylene** (13–15 % from acetone in sulfuric acid, held first at 10°, allowed to warm to room temperature, and then heated until an exothermic reaction occurs; the mixture finally was steam-distilled) [15]. Among by-products of this reaction were durene, pentamethylbenzene, 3,4,5-trimethyl-*t*-butybenzene, and an unknown crystalline compound, $C_{14}H_{20}$, m.p. 79° [16].

(b) **Preparation of dodecahydro-1,2,3,4,5,6,7,8,9,10,11,12-triphenylene.**

The unusual ketodiol

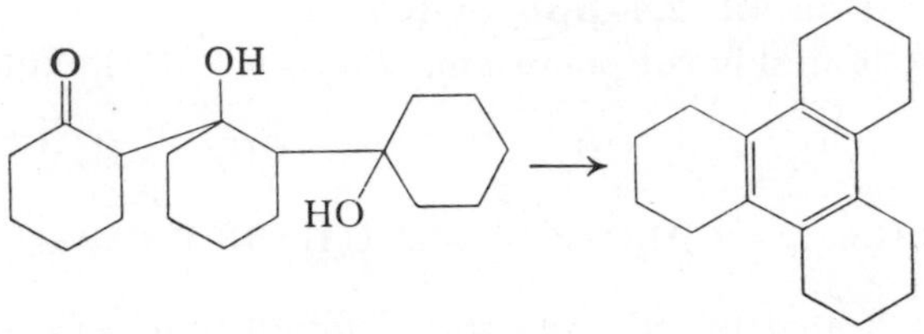

was made in 40 % yield by allowing a mixture of solid sodium hydroxide and cyclohexanone to stand at room temperature for a number of days. The ketodiol was then heated to 350° over a period of 4 hr. to yield 75.5 % dodecahydro-triphenylene [17]. The same hydrocarbon can be obtained from cyclohexanone and sulfuric acid in about 8 % yield [18].

5. From Arenes and Carbanions

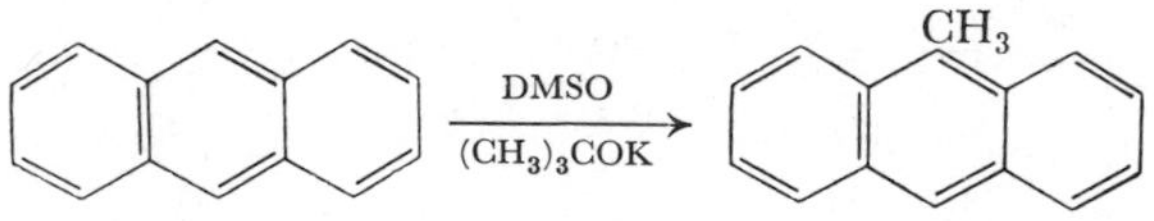

Although nucleophilic substitution in unsubstituted arenes is rare, such substitution is possible in condensed ring systems through attack of the methyl sulfinylmethide anion. Such ions may be produced from dimethyl sulfoxide by the use of potassium *t*-butoxide [19, 20] or sodium hydride [21, 22]. The reaction is a low-temperature one which sometimes gives mixtures but at times may be manipulated to give one product in considerable excess. Some poly-nuclear heterocycles have been methylated, often in high yield, by this procedure.

(a) **Preparation of 9,10-dimethylanthracene.** Anthracene, 10 g. (0.056 mole), in a mixture of 85 ml. of DMSO and 40 ml. of THF was added to a solution of 5.3 g. (0.22 mole) of sodium hydride in 75 ml. of DMSO at 25° during a 5-min. period with rapid stirring. After stirring for 8 hr. at 25°, the mixture was treated with aqueous acid. The precipitate resulting was dissolved in chloroform, and the washed and dried solution when analyzed by gas-liquid partition chromatography indicated a 96% yield [20].

1. C. E. Frank and W. E. Foster, *J. Org. Chem.*, **26**, 303 (1961).
2. K. N. and B. K. Campbell, *Chem. Rev.*, **31**, 77 (1942).
3. G. W. Brown and F. Sondheimer, *J. Am. Chem. Soc.*, **89**, 7116 (1967).
4. W. Schlenk *et al.*, *Chem. Ber.*, **47**, 473 (1914).
5. H. C. Brown and C. H. Snyder, *J. Am. Chem. Soc.*, **83**, 1002 (1961).
6. H. Pines and L. A. Schaap, *Advan. Catalysis*, **12**, 126 (1960).
7. H. Pines and V. Mark, *J. Am. Chem. Soc.*, **78**, 4316 (1956).
8. R. M. Schramm and G. E. Laaglois, Preprints, 136th *Meeting Am. Chem. Soc.* (*Petroleum Division*), *Atlantic City*, 1959, Paper B-53.
9. R. D. Closson *et al.*, *J. Org. Chem.*, **22**, 646 (1957).
10. H. Podall and W. E. Foster, *J. Org. Chem.*, **23**, 401 (1958).
11. J. Shabtai and II. Pines, *J. Org. Chem.*, **26**, 4225 (1961).
12. H. Pines *et al.*, *J. Am. Chem. Soc.*, **77**, 554 (1955).
13. K. L. Schoen and E. I. Becker, *J. Am. Chem. Soc.*, **77**, 6030 (1955).
14. K. L. Schoen and E. I. Becker, *Org. Syn.*, Coll. Vol. **4**, 623 (1963).
15. R. Adams and R. W. Hufferd, *Org. Syn.*, Coll. Vol. **1**, 341 (1941).
16. S. Landa and V. Sesulka, *Chem. Listy*, **51**, 1159 (1957); *C.A.*, **51**, 13814 (1957).
17. S. V. Svetozarskii *et al.*, *J. Gen. Chem.*, USSR (*Eng. Transl.*), **29**, 1428 (1959).
18. C. M. Buess and D. D. Lawson, *Chem. Rev.*, **60**, 313 (1960).
19. A. Schriesheim *et al.*, *J. Org. Chem.*, **30**, 3233 (1965).
20. G. A. Russell and S. A. Weiner, *J. Org. Chem.*, **31**, 248 (1966).
21. E. J. Corey and M. Chaykovsky, *J. Am. Chem. Soc.*, **84**, 866 (1962).

D. Friedel-Crafts Alkylations and Related Reactions

1. From Arenes and Alkylating Agents

$$ArH + RX \xrightarrow{AlCl_3} ArR + HX$$

Friedel-Crafts alkylation, which has been widely reviewed [1–3], is a highly diverse and complex subject. It is reviewed here with as much brevity as possible and with emphasis on the synthetic aspects.

General Characteristics

1. Alkylation with aluminum chloride and without moderating solvents is extremely fast, a matter of fractions of a minute [4].

2. The tendency for dialkylation (or disproportionation to dialkylbenzene) is great. This is a matter of the relative basicity of the alkylbenzenes, a fact which

References for Section D are on pp. 47–49.

determines the stability of the sigma complexes. The relative stabilities are in the order

$$C_6R_6H^{\oplus} > C_6R_5H_2^{\oplus} > C_6R_4H_3^{\oplus} > C_6R_3H_4^{\oplus} > C_6R_2H_5^{\oplus} > C_6RH_6^{\oplus}$$

In practice, then, attempts to produce monoalkyl benzenes will yield appreciable amounts of di- and polyalkylbenzenes. Polyalkylation is minimized by using large excesses of the aromatic hydrocarbon, low temperature, and moderating solvents.

3. Alkylation is reversible. The phenomenon lends itself to synthesis because

$$2\ C_6H_5R \rightarrow RC_6H_4R + C_6H_6$$

or

$$ArR + Ar'H \rightarrow Ar'R + ArH$$

alkylation can be run under partial kinetic control or under thermodynamic control, processes which yield different products. The products of kinetic control are those which are first formed in alkylation and tend to favor o- and p-substitution. For example, under the very mildest conditions (aluminum chloride in acetonitrile) the isopropylation of toluene yields 63% o-, 25% p-, and 12% m-cymene [5]. It is possible that this ratio is not a true kinetic control product ratio, but it is the closest that one can obtain under the conditions that bring about alkylation. The product of thermodynamic control is the meta-isomer free of o- and p-isomers, a product which is obtained by employing at least 1 equivalent of a strong Lewis acid with a cocatalyst, e.g., $BF_3 + H_2F_2$ or $AlCl_3 + HCl$. For example, any cymene (o-, m-, or p-methylisopropylbenzene) when dissolved in hydrofluoric acid with more than 1 equivalent of boron fluoride yields pure meta-substituted cymene within minutes [6]. This isomer forms because in the presence of a strong Lewis acid (HBF_4 or, better, $AlCl_3 + HCl$), the most stable sigma complex is favored:

The charge is dispersed on both alkyl-substituted carbon atoms. Similarly for trialkylbenzenes, the symmetrical 1,3,5-trialkylbenzene is the most stable as the sigma complex.

Moreover, a difference exists in the ease of isomerization of alkyl groups; it depends on their ability to carry a partial or complete positive charge. The methyl group is the most difficult and the t-butyl group is the easiest to isomerize, while ethyl and isopropyl groups occupy intermediate positions in the reactivity sequence.

The aromatic nucleus also cannot be neglected in contributing to ease of isomerization. The more active the nucleus toward alkylation, the more readily it is dialkylated. It is often said that tri-t-butylphenol is an excellent source of the t-butyl carbonium ion. Simply heating it in dilute acid or acetic acid, for example, generates isobutylene.

4. The alkylation agent can isomerize during a substitution process as a consequence of carbonium ion hydride shifts. Definitive studies of these isomerizations are available only since the advent of gas chromatography. The best of these studies indicates that primary halides (with no tertiary carbon atoms in the alkyl chain) yield a mixture of secondary and primary alkylbenzenes [7]; for a summary see [8]. n-Propyl chloride behaves similarly.

$$n\text{-}C_4H_9Cl + C_6H_6 \xrightarrow[\substack{AlCl_3, \\ \text{catalytic} \\ \text{amounts}}]{0°} n\text{-butylbenzene} + sec\text{-butylbenzene}$$
$$\phantom{n\text{-}C_4H_9Cl + C_6H_6 \longrightarrow} 32\text{--}36\% \qquad\qquad 64\text{--}68\%$$

No experiments are available to indicate whether moderating solvents or milder catalysts will overcome isomerization, although some experiments show that disproportionation takes place without isomerization [9]:

$$n\text{-}C_4H_9C_6H_5 \xrightarrow[100°, 3 \text{ hr.}]{AlCl_3} C_6H_6 + (n\text{-}C_4H_9)_2C_6H_4$$
$$\phantom{n\text{-}C_4H_9C_6H_5 \longrightarrow} 15\% \qquad 28\%$$

Thus pure primary alkylbenzenes, containing an alkyl group whose halide rearranges, are best prepared by means other than alkylation, preferably acylation followed by Wolff-Kishner reduction:

$$ArH + RCOCl \xrightarrow{\text{Catalyst}} ArCOR \xrightarrow{W\text{--}K} ArCH_2R$$

Secondary alkyl halides are extraordinarily prone to isomerization. More specifically, hydride shifts occur with great ease. For example, both 2- and 3-pentanols with benzene and boron trifluoride at 0° yield an identical mixture of 2.6 parts of 2-pentylbenzene to 1 part of 3-pentylbenzene [10]. sec-Butylchloride, benzene, and catalytic amounts of aluminum chloride give a 71 % yield of pure sec-butylbenzene at 0°, but at 80° a mixture of 45 % isobutylbenzene and 22 % sec-butylbenzene is obtained [7]. The appearance of isobutylbenzene may seem extraordinary, but it is a consequence of the fact that in reversible reactions of this type the primary alkylbenzene is more stable than the secondary. The only t-alkyl halide that behaves normally is t-butyl chloride. High yields of t-butylbenzenes can be obtained either from t-butyl chloride or from isobutyl chloride [7]. t-Amyl chloride, on the other hand, behaves as follows [11]:

$$(CH_3)_2\overset{\overset{\displaystyle C_2H_5}{\displaystyle |}}{C}Cl + C_6H_6 \xrightarrow{AlCl_3}{0°} (CH_3)_2CHCHCH_3 + (CH_3)_2\overset{\overset{\displaystyle C_2H_5}{\displaystyle |}}{C}C_6H_5$$
$$ \underset{C_6H_5}{|}$$
$$ 85\% \qquad\qquad 15\%$$

Reports are available that ferric chloride gives a higher yield of t-amylbenzene (60 %) [12], but the product was not examined in the most critical manner (GLC). Moderating solvents such as nitromethane, however, minimize isomerization in the preparation of t-amylbenzene (see Ex. b.3). The conclusion from these data is that the energy barrier of isomerization of the t-butyl cation

$$(CH_3)_3C^\oplus \rightarrow \overset{\oplus}{C}H_2CH(CH_3)_2$$

is too high to give abnormal products in the Friedel-Crafts reaction, but that it is sufficiently lower for the t-amyl cation

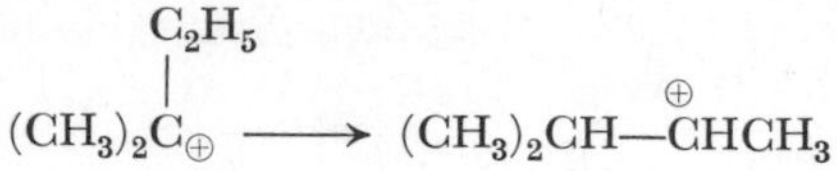

$$(CH_3)_2\overset{\overset{\displaystyle C_2H_5}{|}}{C}_\oplus \longrightarrow (CH_3)_2CH-\overset{\oplus}{C}HCH_3$$

to yield isomerization products. It is interesting that alkylation of benzene with neopentyl chloride yields pure t-amylbenzene [10]. Another phenomenon to be aware of in alkylation with t-alkyl halides is the steric hindrance problem. For instance, a t-alkyl group cannot be inserted adjacent to a methyl group in the Friedel-Crafts reaction. Thus p-xylene does not alkylate with t-butyl chloride and forms an alkylated product derived from a secondary halide with t-amyl chloride

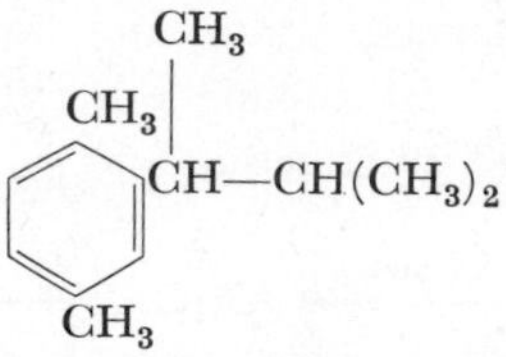

in 54% yield [13]. Intramolecular alkylations with ω-phenylalkyl halides to produce tetralins appear to proceed with less rearrangement than intermolecular ones [14]. But intramolecular alkylations to produce indanes do tend to give rearrangements with aluminum chloride, and also, but to a lesser extent, with ferric chloride and hydrogen chloride in carbon bisulfide.

5. The alkyl group can fragment. Fragmentation has been observed usually under slightly more drastic conditions than are needed for alkylation. sec-Butylbenzene, for example, can give rise to the formation of some ethylbenzene, and the pathway is complex [15]:

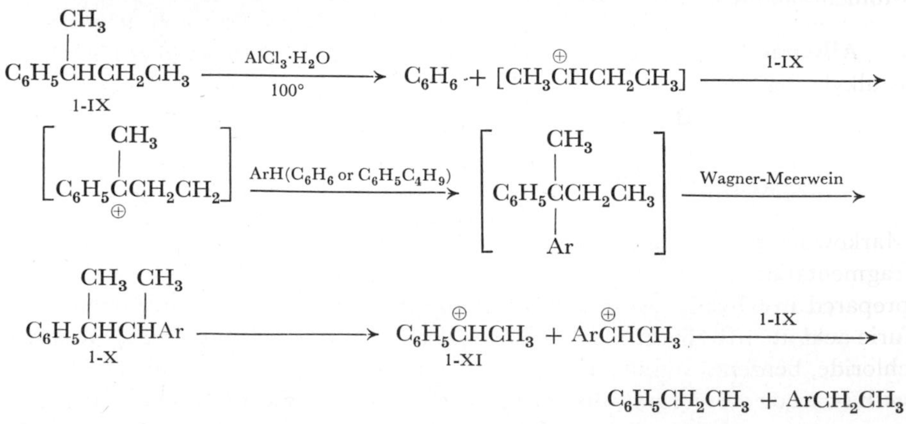

Hydrocarbon 1-X (Ar = C_6H_5) has been isolated from this reaction and has been shown to cleave readily to ethylbenzene. This fragmentation, which is by no means an isolated case [13], takes place readily because of the formation of a benzylic-type cation, 1-XI.

Alkylation Catalysts

See [16]. Such a wide variety of catalysts has been used in alkylation that it is difficult to assess their particular advantages. In general, one uses the mildest catalyst possible to minimize isomerization. With n-alkyl halides and fairly resistant aromatic substrates such as the halobenzenes, the stronger catalysts are needed. On the other hand, with benzylic or t-alkyl halides and reactive substrates such as polynuclear compounds, the weaker catalysts or occasionally no catalyst at all can be used. A series of catalysts has been studied in their ability to alkylate benzene. They fall in the following sequence [17]: $Al_2Br_6 >$ $Ga_2Br_6 > Ga_2Cl_6 > Fe_2Cl_6 > SbCl_5 > ZrCl_4 > BF_3 > BCl_3 > SnCl_4 > SbCl_3$. Catalysts that have been used are: titanium tetrachloride [18], zinc chloride [19], 83.5% aqueous zinc chloride [20], ferric chloride [21], boron trifluoride [22], proton acids (hydrogen fluoride, hydrogen chloride, and sulfuric acid), and even freshly cut aluminum or aluminum activated with gaseous hydrogen chloride or mercuric chloride [23].

Among the important alkylating agents are the following [16].

Alkyl halides. Their activity is in the sequence RF > RCl > RBr > RI. The alkyl fluorides are active enough in alkylation to be selective in competition with a chloride [24].

$$ClCH_2CH_2F + C_6H_6 \xrightarrow[-10°, 30min.]{BBr_3} \underset{50\%}{C_6H_5CH_2CH_2Cl}$$

Sulfonates or p-toluenesulfonates [25]. Eighty-one percent butylbenzenes of undetermined structure are obtained from n-butyl p-toluenesulfonate, benzene, and aluminum chloride [26]. The sulfonates can also be used without catalysts by heating in the hydrocarbon at 115–120° or lower; for example, cyclohexylbenzene is prepared in 60–65% yield from cyclohexyl p-toluenesulfonate in refluxing benzene [27].

Alkenes [28]. Alkenes are used more in industry than in the laboratory as alkylating agents. They, of course, give rise to hydrocarbons derived from

$$RCH{=}CH_2 \xrightarrow{H^{\oplus}, AlCl_3} [R\overset{\oplus}{C}HCH_3] \xrightarrow{ArH} Ar\overset{\overset{\displaystyle R}{|}}{C}HCH_3$$

Markownikoff addition and are subject to all the dangers of isomerization and fragmentation characteristic of alkyl halides. Cyclohexylbenzene has been prepared in 65–85% yield from benzene, cyclohexene, and concentrated sulfuric acid at 5–10° for $2\frac{1}{2}$ hr. [29], and t-butylbenzene in 89% yield from ferric chloride, benzene, and isobutylene [30]. Hydrocarbons, including cyclic types, with t-carbon atoms can intercede in the alkylation of an aromatic hydrocarbon with an olefin [31]:

$$CH_2=\overset{\underset{\displaystyle CH_3}{|}}{C}CH_2Cl + \underset{}{\bigcirc} \xrightarrow[2-4°]{AlCl_3} \left[CH_3\overset{\underset{\displaystyle CH_3}{|}}{C}HCH_2Cl \right] + \left[\overset{CH_3}{\underset{\oplus}{\bigcirc}} \right] \xrightarrow{C_6H_6}$$

$$\underset{39\%}{C_6H_5CH_2CH(CH_3)_2} + C_6H_5\underset{\substack{\text{1-Methyl-1-}\\\text{phenylcyclo-}\\\text{hexane or}\\\text{isomer 52\%}}}{\overset{CH_3}{\diagup\!\bigcirc}}$$

The appearance of isobutylbenzene may indicate another pathway, but an intermolecular hydride transfer occurs at some place in the course of events.

Alcohols.   Sulfuric acid or boron trifluoride [10] is used more often than other catalysts with alcohols. Benzylic alcohols are convenient alkylating agents since they are active enough to need little catalyst and nonvolatile enough to permit removal of the water of reaction by the Dean-Stark tube method [32].

$$p\text{-}CH_3C_6H_4CH_2OH + C_6H_5CH_3 \xrightarrow[\text{Reflux, remove H}_2O]{\text{Toluenesulfonic acid}} \underset{87\%}{(p\text{-}CH_3C_6H_4)_2CH_2}$$

Carbonyl compounds.   Theoretically, carbonyl compounds should lead to diarylalkanes, but the process is severely limited by side reactions [33].

$$R_2C=O + 2\ ArH \xrightarrow{AlCl_3} R_2CAr_2 + H_2O$$

Benzaldehyde, ferric chloride, and benzene mixed by shaking at 40–60° for 5 hr. yield 30 % triphenylmethane and 6 % anthracene [34]. Anthracene always seems to be a by-product when benzylic ions are formed. Paraldehyde and toluene in hydrofluoric acid at 0–5° for 1–3 hours give 95 % 1,1-ditolylethane [35]. Trioxymethylene and arenes yield diarylmethanes. Hydrogen fluoride seems to be the reaction solvent of choice for these condensations. Simple aliphatic ketones do not appear to give diphenylmethanes with aromatic hydrocarbons, although they form di(p-hydroxyphenyl)-methanes perfectly well with phenols.

Moderating Solvents

See [16]. Certain solvents moderate the catalytic action of Lewis acids by forming complexes, which are the alkylating catalysts. Nitromethane is one of the most effective solvents (see Ex. b.3). It solubilizes its own weight of aluminum chloride in an aromatic solvent, a desirable feature for an otherwise heterogeneous medium which eliminates the use of an inordinate amount of polar solvent to be removed at the end of the reaction. Acetonitrile and carbon bisulfide are other moderating solvents which give mild reaction conditions.

Practical Comments

The strong Lewis acids such as aluminum chloride are so reactive that they are not "anhydrous." Small amounts of water vapor have reacted to form some aluminum oxyhalides and entrapped hydrogen chloride. Purification can be brought about by sublimation, but in the majority of cases it is not worth the trouble. We are so partial to Baker's powdered resublimed aluminum chloride that little experience has been gained with other forms of aluminum chloride. The aluminum chloride is best weighed in the bottle; the amount is transferred rapidly to the reaction vessel; and the bottle is reweighed to obtain the weight employed. The procedure minimizes exposure to moisture. Aluminum bromide is so difficult to handle and store that we prefer to prepare it *in situ* from aluminum shavings and bromine. The only commercial aluminum bromide worth handling appears to be that sealed in ampoules or that supplied in cyclohexane solution (Michigan Chemical Corporation).

(a) **Preparation of *t*-butylbenzene** (80% from benzene, anhydrous ferric chloride, and *t*-butyl chloride at 10–25°) [36].

(b) **Other examples**

(1) **Dibenzyl** (97% from β-bromoethylbenzene, benzene, and aluminum bromide at 45° for 3 hr.) [37].

(2) **1,3,5-Tri-*t*-butylbenzene**   (Wagner-Meerwein isomerization) (75% from *p*-di-*tert*-butylbenzene, *t*-butyl chloride, and aluminum chloride at 0 to −7° for 50 min.) [38].

(3) ***t*-Amylbenzene** (69% from benzene, *t*-amyl chloride, and an aluminum chloride-nitromethane complex) [39].

(4) **Ethylbenzene** (96% from 8 moles of benzene, 1 mole of ethyl bromide, and 0.01 g. atom of aluminum shavings heated until reaction started; cooling to 5–10° for 10–15 min., after which heating to 70–80° for 35–40 min. was maintained; finally the mixture was kept at room temperature for 2 hr.; $C_2H_5AlBr_2$ may be the alkylating agent) [40].

(5) **1,1,2,3-Tetramethylindene** (80% from benzene, 1,1-dibromotetramethylcyclopropane, and aluminum chloride) [41].

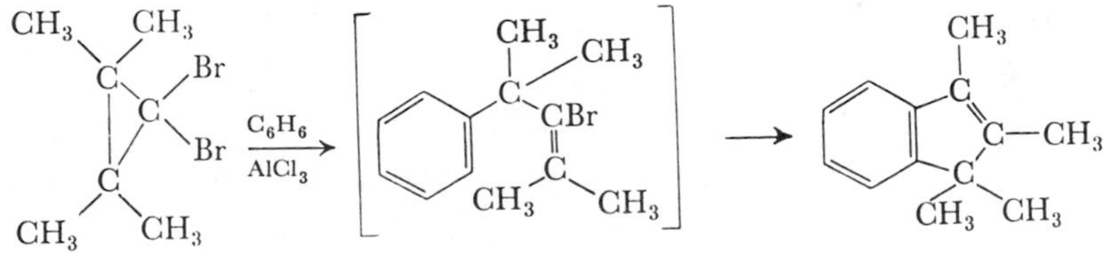

2. From Aromatic Ketones (Cyclization)

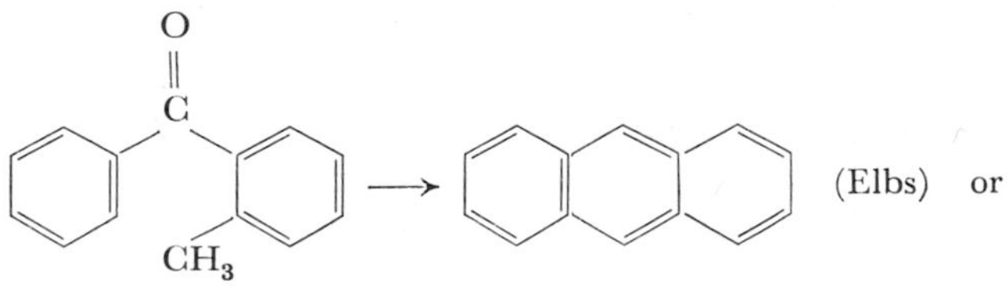

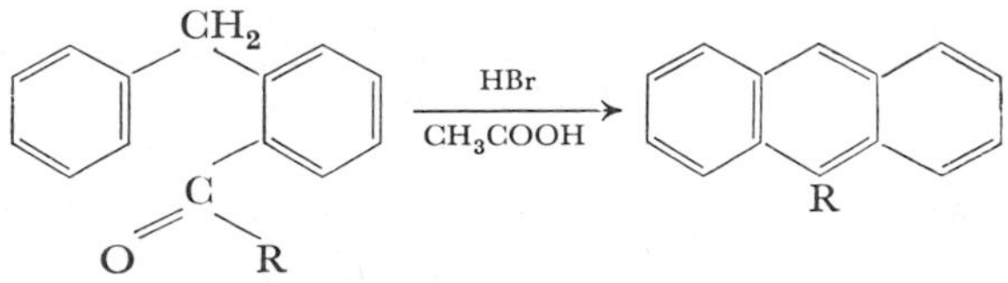

(Elbs)   or

The Elbs reaction, which has been reviewed [42], consists in heating a diaryl ketone with a methyl group *ortho* to the carbonyl group to produce a polynuclear aromatic hydrocarbon. Although the yields are not high, in some instances the method is the best available. For example, in the synthesis of 1,2,5,6-dibenzanthracene, widely used for the production of cancer in animals, the method is the most rapid and economical one known. Since no great success has been achieved in finding a catalyst, the ketone is simply heated (usually 400–450°) at the lowest temperature at which water is steadily eliminated. Side reactions such as aroyl migration, elimination of alkyl, halo, and methyl substitutents, degradation of isopropyl to methyl, hydrogenation, dehydrogenation, and intramolecular disproportionation sometimes occur. The mechanism of the reaction is not clear.

Polynuclear aromatic hydrocarbons may also be synthesized by the dehydration of *o*-acyldiphenylmethanes, a process called cyclodehydration [43, 44]. In this case the dehydrating agent used is either hydrobromic-acetic acid or sulfuric acid. By this synthesis a series of 9- and 10-alkyl- and 9- and 10-arylanthracenes, phenanthrenes, and benzanthracenes have been prepared in fair to good yields. The first step in the mechanism appears to be protonation to form the conjugate acid 1-XII which attacks the *ortho* position of the adjoining ring to form the complex 1-XIII, which in turn loses first a proton and then water to yield the polynuclear hydrocarbon 1-XIV [45].

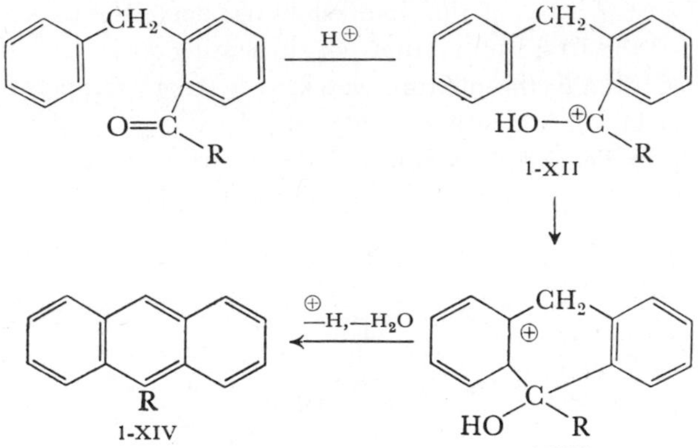

Many other types of cyclization to produce hydrocarbons are described in the Olah treatise [44]; see also Ex. c.

(a) Preparation of 1,2-benzanthracene (61 % from *o*-tolyl α-naphthyl

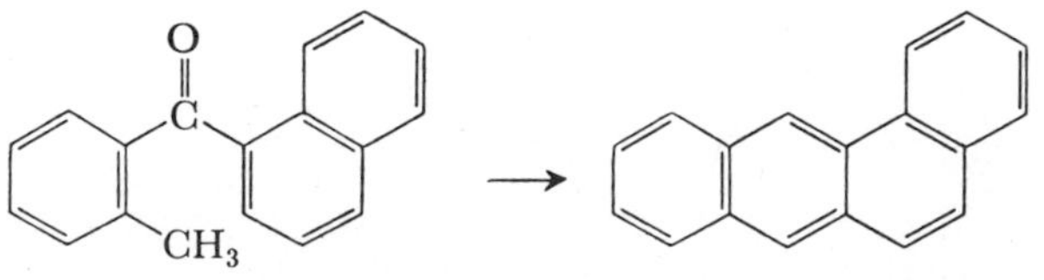

ketone and zinc dust heated in a metal bath at 410° for 3 hr.) [46].

(b) Preparation of 9-methylanthracene (80% by refluxing a mixture of *o*-benzylacetophenone, acetic acid, and 34% hydrobromic acid for 3 days) [47].

(c) Preparation of 1,3-dimethylnaphthalene [48]

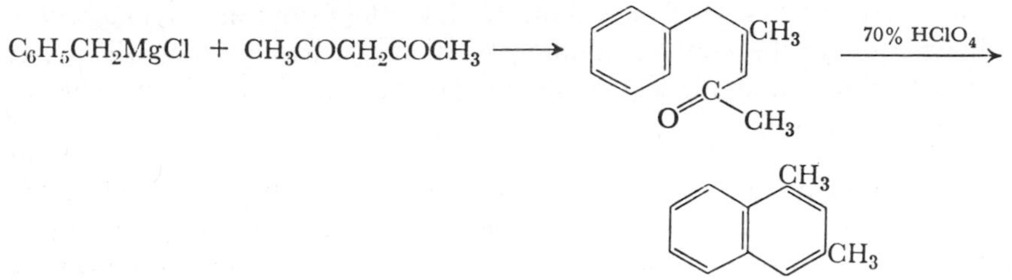
C6H5CH2MgCl + CH3COCH2COCH3
CH3
C
O
CH3
70% HClO4
CH3
CH3

40% yield for last step

3. Rearrangement of Hydrocarbons

Rearrangement of hydrocarbons [49] is not a good method of preparation, not because the hydrocarbons are inert but because mixtures, frequently of great complexity, are formed from exposure of a hydrocarbon to a strong Lewis acid. Such products are to be expected from carbonium ion rearrangements, hydride shifts, dimerization, and fragmentation.

The rearrangement of cyclohexane, via the cyclopentylmethyl carbonium ion, 1-XV, to an equilibrium mixture of methylcyclopentane and cyclohexane is one of the best examples studied. Since tertiary carbonium ions are estimated to

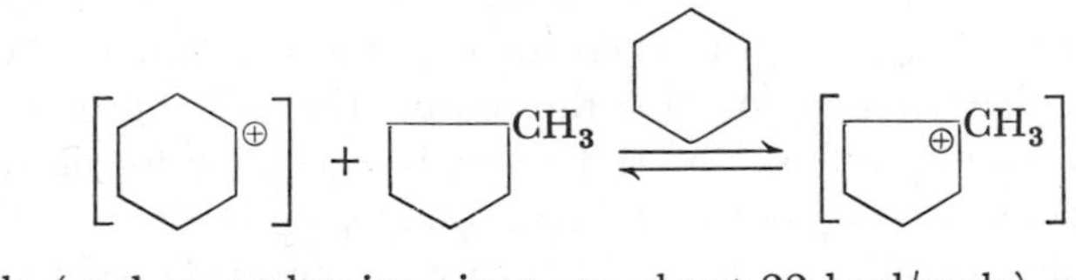
R⊕
(from an olefin)
−AlCl3
⊕
CH2⊕
1-XV
Fast
⊕
+
CH3
⊕
CH3

be 33 kcal/mole (and *sec*-carbonium ions are about 22 kcal/mole) more stable than primary carbonium ions, doubt has been cast on the intermediate formation of 1-XV [50]. It is more likely that dimerization to a C_{12} carbonium ion which is not of the primary type occurs. The C_{12} carbonium ion then fragments to the C_6 hydrocarbons. On the other hand, the carbon atoms in propane ($CH_3\overset{14}{C}H_2CH_3$) exposed to water-activated aluminum bromide are completely scrambled. Yet no propane with two labeled carbon atoms was detected, an observation which argues against the intermediate dimer [49].

Any butyl alcohol dissolved in sulfuric acid separates into a sulfuric acid-insoluble phase, which is a complex mixture of hydrocarbons, and a soluble phase which consists of stable cations derived from methylated cyclopentadienes [51]. Indeed, any alcohol, olefin, or hydrocarbon which can be converted into a carbonium ion when heated with polyphosphoric acid forms in part a complex mixture of hydrocarbons including small amounts of aromatic hydrocarbons [52]. Cyclononane with aluminum chloride is rearranged to propylcyclohexane at 20° and to trimethylcyclohexanes at 50° [53]. Cyclohexane and cyclodecane are unchanged under the same conditions. No doubt, rearrangement will occur if some carbonium ion source such as an olefin is added to the aluminum chloride-cycloalkane mixture.

(a) **Preparation of adamantane** (13.5–15% from *endo*-tetrahydro-dicyclopentadiene and aluminum chloride at 150–180°) [54]. (Apparently any tricyclic alkane is capable of rearrangement to the adamantane structure) [55].

(b) **Preparation of isoheptane** (mixture of 2- and 3-methylhexanes and dimethylpentanes, 70%, from *n*-heptane, isobutane (5 parts), aluminum chloride and hydrogen chloride in an autoclave at 30° for 15 min. The isobutane apparently acts as a hydride source suppressing fragmentation) [56]. Methyl-cyclopentane is a better hydride source [57].

4. From Alkanes and Alkenes

$$\underset{\underset{CH_3}{|}}{CH_3CHCH_3} + CH_2{=}CH_2 \xrightarrow{AlCl_3} \underset{\underset{CH}{/}\ \ \underset{CH_3}{\backslash}}{CH_3CHCHCH_3}$$

This synthesis is not a likely laboratory procedure but, because of its commercial importance in producing hydrocarbons with a high octane rating, it is discussed briefly. A detailed discussion may be found [58].

The alkylation of alkanes may be accomplished by thermal or catalytic conditions. The former requires temperatures around 500° and pressure varying from 150 to 300 atm., while for the latter lower temperature (-30 to $100°$) and pressures only high enough to keep some of the reactant in the liquid phase suffice. It is interesting to note that the composition of the products obtained by the two procedures differs. This difference is accounted for on the basis of the mechanism, the thermal process following a free radical path and the catalytic process a carbonium ion path. In either case, mixtures of products are obtained, a fact which detracts from the method if pure alkanes are desired.

Of the two methods of alkylation the catalytic is the more common one. It utilizes the usual Friedel-Crafts catalysts such as: anhydrous metal halides (aluminum chloride, aluminum bromide, or boron trifluoride), preferably with a hydrogen halide promoter; and protonic acids (hydrogen fluoride or sulfuric acid). For more details, [58] should be consulted. The mechanism of the reaction between isobutane and ethylene, for example, which gives largely 2,3-dimethylbutane is represented below [59]. The carbonium ion from iso-butane, 1-XVI, reacts with ethylene to give the carbonium ion 1-XVII which eventually rearranges to give the most stable carbonium ion 1-XVIII. In the last step this ion with isobutane produces 2,3-dimethylbutane and 1-XVI available for further reaction. Isobutane reacts similarly with other olefins such as isobutylene to yield isooctane.

$$(CH_3)_3\overset{\oplus}{C} + CH_2{=}CH_2 \longrightarrow [(CH_3)_3C{-}CH_2\overset{\oplus}{CH_2}] \longrightarrow [(CH_3)_3C{-}\overset{\oplus}{CH}CH_3] \longrightarrow$$

1-XVI   1-XVII

$$[(CH_3)_2\overset{\oplus}{C}{-}CH(CH_3)_2] \xrightarrow{(CH_3)_3CH} (CH_3)_2CHCH(CH_3)_2 + (CH_3)_3\overset{\oplus}{C}$$

1-XVIII   1-XVI

No discussion of alkylation with olefins is complete without some reference to the Ziegler-Natta catalysts as vehicles for polymerizing olefins [60]. Alkylaluminums (AlR_3) can polymerize to rather short-chain polymers by heating, but a combination of alkylaluminums and titanium chloride forms a different catalytic system which permits low-pressure polymerization of olefins to isotactic polymers of very high molecular weight. These facts indicate that the active catalyst contains both a titanium and an aluminum atom, the presence of which controls the building of a chain [61]. The Ti-Et bond probably is

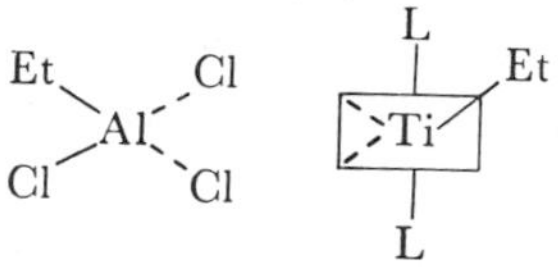

$$L = CH_2{=}CH_2$$

weakened to the extent that alkylation of one of the ligands takes place. Although the synthesis of hydrocarbons is limited by this process, it may be adaptable to telomerization processes, particularly with hydrogen or aromatics which will lead to useful hydrocarbons of low molecular weight.

(a) **Preparation of 2,3-dimethylbutane** (68–75 volume % of 270–300 weight % of alkylate, based on ethylene consumed in its reaction with isobutane, using aluminum chloride-hydrocarbon complex catalyst and ethyl chloride promoter; the original must be consulted for additional details) [62].

5. Rearrangement of Alkylbenzenes (Jacobsen)

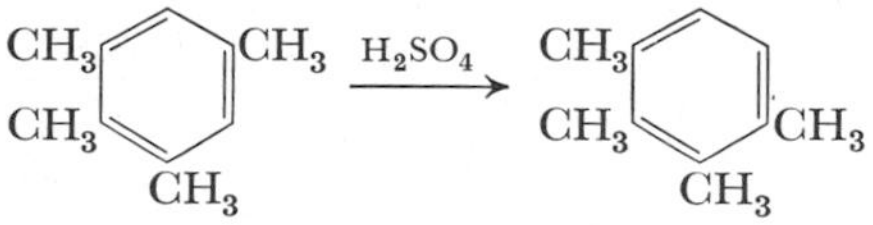

The migration of an alkyl group (or a halogen atom if present) occurs in sulfonic acids when allowed to remain in contact with sulfuric acid. The migration may be intramolecular or intermolecular, and the net result is that the groups end up in neighboring positions. Thus the synthesis, which has been reviewed [63], is of value in preparing vicinal substituted alkylbenzenes. The rearrangement occurs only with the tetra- and pentaalkyl derivatives, often in high yield.

Although it may seem odd that vicinal alkylbenzenes are the thermodynamically stable products, the fact is explicable in that the rearrangement occurs in the sulfonic acid [64]:

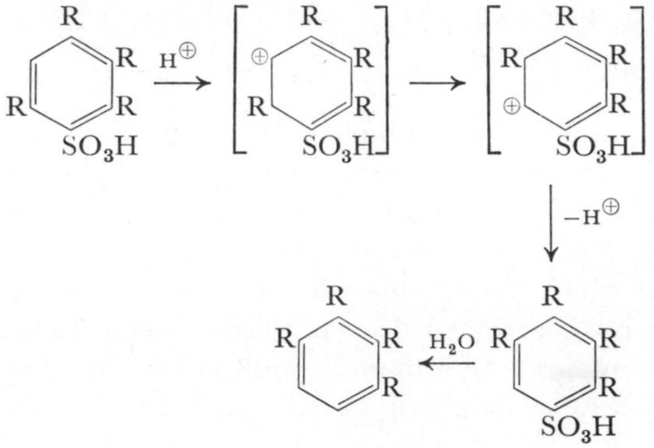

Steric requirements are such that as few R groups as possible should flank the sulfonic acid group in the stablest sigma complex. An analogous situation is found in the rearrangement of 2,4,6-trialkylacetophenones to 3,4,5-trialkylacetophenones [65].

(a) **Preparation of 1,2,3,4-tetraethylbenzene** (90.7 % from a mixture of the sulfonic acids of 1,2,4,5- and 1,2,3,5-tetraethylbenzenes and 50 % sulfuric acid heated and then hydrolyzed by the introduction of steam) [66].

6. From Sulfonic Acids [67]

$$ArSO_3H \xrightarrow[H_2O]{H^{\oplus}} ArH + H_2SO_4$$

Desulfonation of aromatic sulfonic acids is an electrophilic substitution by a protonic species. Or it is simply the reverse reaction of aromatic sulfonation. Thus desulfonation must occur via the sigma complex

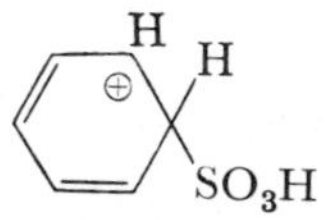

in which either an $\overset{\oplus}{S}O_3H$ is lost to form benzene or a proton is lost to form the sulfonic acid. Desulfonation has all the characteristics to be expected of an electrophilic reaction: the stronger the acid and the more concentrated, the faster the desulfonation occurs although a concentration of acid will be reached eventually where sulfonation will compete with desulfonation; and, the more readily the aromatic nucleus is substituted, the easier is its sulfonic acid de-sulfonated as illustrated in desulfonation with phosphoric acid:

Ar of Sulfonic acid	Approx. Desulfonation Temp.
C_6H_5-	$227°$
$4\text{-}CH_3C_6H_4-$	$186°$
$3,4\text{-}(CH_3)_2C_6H_3-$	$175°$
$2,4\text{-}(CH_3)_2C_6H_3-$	$137°$
$2,4,6\text{-}(CH_3)_3C_6H_2-$	$80°$ in HCl
$(CH_3)_5C_6-$	$25°$ in H_2SO_4

One notes that *o*-alkyl substituents increase the rate of desulfonation con-siderably. Phosphoric acid seems to be the acid of choice in the temperature range 190–220° or a 30% aqueous solution in an autoclave. Other mineral acids may be used, although sulfuric acid is the most common. One difficulty with this acid is that it has a tendency to oxidize as well as to furnish protons.

Desulfonation is not a very useful reaction, but it can be employed to separate mixtures of alkylbenzene isomers. For example, the results of distilling a mixture of ethyltoluenesulfonic acids with steam in about 50% sulfuric acid-water by volume were:

	Percentages of		
	o-	*m*-	*p*-
Initial composition	1.4	64	34.6
Fraction below 160°	0	99.4	0.6
Fraction boiling 160–180°	0	77	23
Fraction boiling 180–200°	0	8.4	91.6

Generally, the *meta* isomer sulfonates and desulfonates the most readily, as indicated above for desulfonation. Desulfonation is also employed when the sulfonic acid group has been introduced to block substitution in a particular position (see preparation of *o*-bromophenol) [68]. No specific example of large-scale desulfonation is available, but one could be designed from the following small-scale operation [69]:

$$ArSO_3H \xrightarrow[\substack{250°, \; 1 \; hr}]{\substack{20 \; ml. \; H_3PO_4, \; 2 \; g. \; Sn}} ArH$$

10–200 mg.   quantitative yield

The tin (or stannous chloride) prevents charring.

Alkanesulfonic acids cannot be desulfonated in the same manner as arene-sulfonic acids because no sigma complexes can be formed. However, they can be desulfonated by fusion with alkali [70]:

$$RSO_3Na + NaOH + KOH \xrightarrow{330°} RH$$

70–80%

7. From Hydrocarbons by Dehydrogenative Coupling

This reaction [71] is more applicable to polynuclear hydrocarbon synthesis [72], but it is capable of giving some product with any aromatic hydrocarbon. The mechanism is similar to that of any Friedel-Crafts reaction except that the electrophilic species is a sigma complex

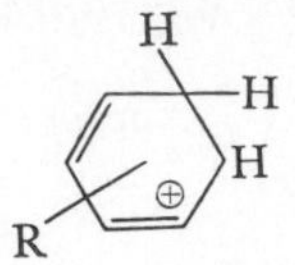

made from aluminum chloride and its cocatalyst, HCl, which then attacks an aromatic nucleus to form another sigma complex

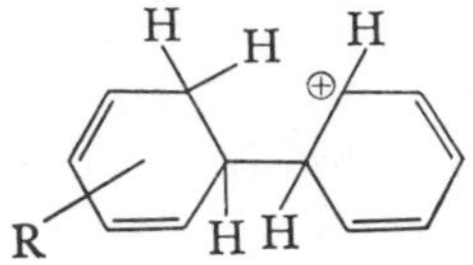

One can see that the latter is an unusual complex—one ring is rearomatized by simple loss of a proton to

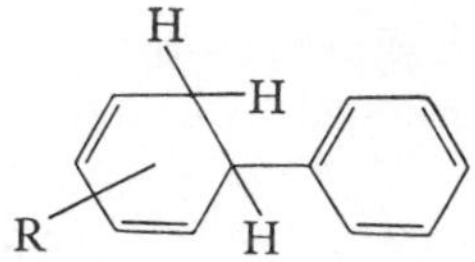

but the other ring must be rearomatized by dehydrogenation or oxidation. Too few examples in the literature recognize the dual nature of the rearomatization process. That it is important is shown by the smooth dehydrogenative coupling of benzene itself when carried out properly [73]. Hydrogen chloride serves to

$$C_6H_6 \xrightarrow[\text{CuCl}_2,\, 36°,\, 15 \text{ min.}]{\text{Water-activated AlCl}_3,} \text{p-polyphenyl (high m.w.)}$$
$$60\%$$

form the sigma complex, and cupric chloride serves to dehydrogenate the di-hydro compound. The product of this reaction suggests another problem: how to stop at the dimer stage since dimers usually are the objective of synthesis. In most cases this problem has not been solved very well, judging from the low yields of dimer and high yields of pot residues reported. In other cases structural features, such as proximity of appropriate carbon atoms, favor single coupling [74]:

$$\text{o-terphenyl} \xrightarrow[\text{NaCl, 4 g., 200°}]{\text{AlCl}_3,\, 16 \text{ g.}}$$
2 g.

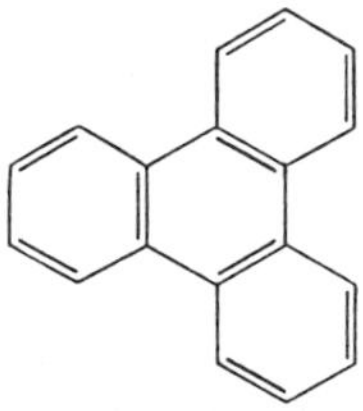

+ p-terphenyl

Triphenylene, 20% yield

In Friedel-Crafts reactions with naphthalene, 1,1′- (or 2,2′-)binaphthyl is often a by-product, but attempts to make either the major product have not been successful [75]. A better method (57%) is the Ullmann condensation with 1-bromonaphthalene [76].

Some polynuclear hydrocarbons with aluminum chloride undergo profound rearrangements. Although the yields are poor in all dehydrogenative couplings and rearrangements, the procedure is simple enough to have merit in certain cases.

(a) Preparation of 2'-methyl-1,2,4,5-dibenzopyrene [77]

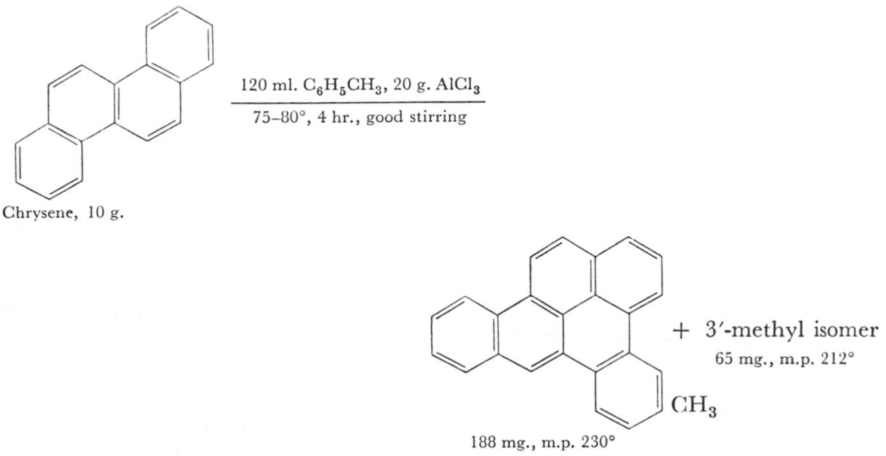

$$120 \text{ ml. } C_6H_5CH_3, 20 \text{ g. } AlCl_3$$
$$\overline{75-80°, 4 \text{ hr., good stirring}}$$

Chrysene, 10 g.

+ 3'-methyl isomer
65 mg., m.p. 212°

CH₃

188 mg., m.p. 230°

(b) Preparation of p-quaterphenyl.

(26% from 15.4 g. of molten biphenyl, 16 g. of aluminum chloride at 80°, to which 1 ml. of nitrobenzene was added in 15 min. and the mixture was then heated another 15 min. The brown product was extracted thoroughly with boiling water, then with ether, and the residue (from the ethereal solution) was sublimed at 100–300° (1 mm.)). The simplicity of the method compensates in part for the low yield [78].

1. C. C. Price, *Org. Reactions*, **3**, 1 (1946); Annual reviews on Unit Processes, *Ind. Eng. Chem.*
2. G. A. Olah *ed.*, *Friedel-Crafts and Related Reactions*, Vol. **2**, Interscience Publishers, New York, 1964, Pts. 1 and 2.
3. G. Baddeley, *Quart. Rev. (London)*, **8**, 355 (1954).
4. H. C. Brown and H. Jungk, *J. Am. Chem. Soc.*, **78**, 2182 (1956).
5. G. A. Olah *et al.*, *J. Am. Chem. Soc.*, **86**, 1046 (1964).
6. D. E. Pearson *et al.*, *Org. Syn.*, **47**, 40 (1967).
7. R. M. Roberts and D. Shiengthong, *J. Am. Chem. Soc.*, **82**, 732 (1960).
8. K. L. Marsi and S. H. Wilen, *J. Chem. Ed.*, **40**, 214 (1963).
9. R. E. Kinney and L. A. Hamilton, *J. Am. Chem. Soc.*, **76**, 786 (1954).
10. A. Streitwieser, Jr., *et al.*, *J. Am. Chem. Soc.*, **81**, 1110 (1959).
11. L. Schmerling and J. P. West, *J. Am. Chem. Soc.*, **76**, 1917 (1954).
12. C. E. Boord *et al.*, *J. Am. Chem. Soc.*, **74**, 292 (1952).
13. L. Schmerling *et al.*, *J. Am. Chem. Soc.*, **81**, 2718 (1959).
14. A. A. Khalaf and R. M. Roberts, *J. Org. Chem.*, **31**, 89 (1966).
15. R. M. Roberts *et al.*, *J. Am. Chem. Soc.*, **85**, 3454 (1963).
16. G. A. Olah, Ref. 2, Vol. 1, 1963, p. 201.
17. G. A. Russell, *J. Am. Chem. Soc.*, **81**, 4834 (1959).
18. N. M. Cullinane and D. M. Leyshon, *J. Chem. Soc.*, 2942 (1954); R. Pajeau, *Compt. Rend.*, 247, 935 (1958).
19. A. B. Kuchkarov and I. P. Tsukervanik, *J. Gen. Chem. USSR (Eng. Transl.)*, **20**, 458 (1950).

20. R. Jenny, *Compt. Rend.*, **246**, 3477 (1958).

21. M. Inatome *et al.*, *J. Am. Chem. Soc.*, **74**, 292 (1952); B. W. Larner and A. T. Peters, *J. Chem. Soc.*, 680 (1952).

22. S. P. Malchick and R. B. Hannan, *J. Am. Chem. Soc.*, **81**, 2119 (1959); V. V. Korshak and N. N. Lebedev, *J. Gen. Chem. USSR*, (*Eng. Transl.*), **20**, 266 (1950).

23. M. B. Turova-Polyak and I. R. Davydova, *Zh. Obshch. Khim.*, **26**, 2710 (1956); *C.A.*, **51**, 7317 (1957); M. B. Turova-Polyak and M. A. Maslova, *J. Gen. Chem. USSR*, (*Eng. Transl.*), **27**, 976 (1957).

24. G. A. Olah and J. J. Kuhn, *J. Org. Chem.*, **29**, 2317 (1964).

25. F. A. Drahowzal, Ref. 2, Vol. 2, Pt. 1, p. 641.

26. F. A. Drahowzal *et al.*, *Chem. Ann.*, **580**, 210 (1953).

27. W. J. Hickinbottom *et al.*, *Chem. Ind.* (*London*), 539 (1955); 1359 (1957).

28. S. H. Patinkin and B. S. Friedman, Ref. 25, p. 1.

29. B. B. Corson and V. N. Ipatieff, *Org. Syn.*, Coll. Vol. **2**, 151 (1943).

30. W. M. Potts and L. L. Carpenter, *J. Am. Chem. Soc.*, **61**, 663 (1939).

31. L. Schmerling *et al.*, *J. Am. Chem. Soc.*, **80**, 576 (1958).

32. E. F. Pratt and H. J. E. Segrave, *J. Am. Chem. Soc.*, **81**, 5369 (1959).

33. J. E. Hofmann and A. Schriesheim, Ref. 25, p. 597.

34. A. Schaarschmidt *et al.*, *Chem. Ber.*, **58B**, 1914 (1925).

35. J. E. Hofmann and A. Schriesheim, Ref. 25, p. 630.

36. D. Nightingale *et al.*, *Org. Reactions*, **3**, 17 (1946).

37. I. P. Tsukervanik and T. G. Garkovets, *Zh. Obshch. Khim.*, **26**, 1653 (1956); *C.A.*, **51**, 1909 (1957).

38. L. R. C. Barclay and E. E. Betts, *Can. J. Chem.*, **33**, 672 (1955).

39. B. S. Friedman *et al.*, *J. Am. Chem. Soc.*, **80**, 5867 (1958).

40. M. B. Turova-Polyak and M. A. Maslova, *Zh. Obshch. Khim.*, **26**, 2185 (1956); *C.A.*, **51** 4972 (1957).

41. L. Skattebøl and B. Boulette, *J. Org. Chem.*, **31**, 81 (1966).

42. L. F. Fieser, *Org. Reactions*, **1**, 129 (1942).

43. C. K. Bradsher, *Chem. Rev.*, **38**, 447 (1946).

44. L. R. C. Barclay, Ref. 2, Pt. 2, p. 785.

45. C. K. Bradsher and L. J. Wissow, *J. Am. Chem. Soc.*, **68**, 1094 (1946).

46. L. F. Fieser and E. B. Hershberg, *J. Am. Chem. Soc.*, **59**, 2502 (1937).

47. C. K. Bradsher, *J. Am. Chem. Soc.*, **62**, 1077 (1940).

48. A. T. Balaban and A. Barabas, *Chem. Ind.* (*London*), 404 (1967).

49. H. Pines and N. E. Hoffman, Ref. 2, Pt. 2, p. 1211.

50. G. J. Karabatsos *et al.*, *J. Am. Chem. Soc.*, **85**, 733 (1963).

51. N. C. Deno *et al.*, *J. Am. Chem. Soc.*, **86**, 1745 (1964).

52. D. E. Pearson, unpublished work.

53. M. B. Turova-Polyak *et al.*, *J. Gen. Chem., USSR*, (*Eng. Transl.*), **31**, 1849 (1961).

54. P. Von R. Schleyer *et al.*, *Org. Syn.*, **42**, 8 (1962).

55. D. J. Cash and P. Wilder, *Tetrahedron Letters*, 6445 (1966).

56. H. J. Hepp and L. E. Drehman, *Ind. Eng. Chem.*, **52**, 207 (1960).

57. J. E. Hofmann, *J. Org. Chem.*, **29**, 3039 (1964).

58. L. Schmerling, Ref. 2, Pt. 2, p. 1075.

59. P. D. Bartlett *et al.*, *J. Am. Chem. Soc.*, **66**, 1531 (1944); L. Schmerling, *ibid.*, **66**, 1422 (1944); **67**, 1778 (1945); **68**, 275 (1946).

60. A. V. Topchiev *et al.*, *Russ. Chem. Rev.*, (*Eng. Transl.*), **30**, 192 (1961).

61. G. Henrici-Olivé and S. Olivé, *Angew. Chem. Intern. Ed. Engl.*, **6**, 790 (1967).

62. R. B. Thompson, *Ind. Eng. Chem.*, **40**, 1265 (1948).

63. L. I. Smith, *Org. Reactions*, **1**, 370 (1942).

64. E. N. Marvell and B. M. Graybill, *J. Org. Chem.*, **30**, 4014 (1965).

65. P. J. McNulty and D. E. Pearson, *J. Am. Chem. Soc.*, **81**, 612 (1959).

66. L. I. Smith and C. O. Guss, *Org. Reactions*, **1**, 383 (1942).

67. E. E. Gilbert, *Sulfonation and Related Reactions*, Interscience Publishers, New York, 1965.

68. R. C. Huston and M. M. Ballard, *Org. Syn.*, Coll. Vol. **2**, 97 (1943).
69. S. Nishi, *Bunseki Kagaku*, **14**, 912 (1965); *C.A.*, **64**, 3858 (1965).
70. J. Pollerberg, *Fette, Seifen, Anstrichmittel*, **67**, 927 (1965); *C.A.*, **64**, 8503 (1966).
71. A. T. Balaban and C. D. Nenitzescu, Ref. 2, Pt. 2, p. 979.
72. E. Clar, *Polycyclic Hydrocarbons*, Academic Press, New York, 1964, Vols. 1 and 2.
73. P. Kovacic and A. Kyriakis, *J. Am. Chem. Soc.*, **85**, 454 (1963).
74. C. F. H. Allen and F. P. Pingert, *J. Am. Chem. Soc.*, **64**, 1365 (1942).
75. G. Baddeley *et al.*, *J. Chem. Soc.*, 100 (1952).
76. P. G. Copeland *et al.*, *J. Chem. Soc.*, 1689 (1960).
77. N. P. Buu-Hoi and D. Lavit-Lamy, *Bull. Soc. Chem. France*, 341 (1963).
78. S. K. Dayal and K. N. Rao, *Indian J. Chem.*, **5**, 122 (1967).

E. Dehydrogenation of Hydroaromatic Hydrocarbons

Although dehydrogenation has been of greater value in determining the structure of natural, highly hydrogenated carbocycles by converting them into more easily identifiable aromatic hydrocarbons, the method also serves at times as a preparatory one. It has been reviewed in some detail [1] and in a briefer form [2]. The most common dehydrogenating agents are sulfur, selenium, or a metal such as platinum or palladium, although other metals such as nickel or rhodium and the compounds chloranil, with or without irradiation [3,4] 2,3-dichloro-5,6-dicyano-1,4-benzoquinone [5], and trityl perchlorate [6] have also been employed. The latter appears to be the most effective reagent for converting perinaphthanones into perinaphthenones and chromanones into chromones [7]. If sulfur is used, the temperature is kept fairly low (230–250°); with selenium, high temperatures are necessary (300–330°). By the catalytic method (use of Pt or Pd) the compound in the vapor state may be passed over the heated catalyst at 300–350°, but it is more convenient to use the liquid-phase method. As a rule, dehydrogenation results satisfactorily by heating the substance with one-tenth part of 10 % palladium-on-charcoal catalyst to about 310–320°. At times, solvents such as naphthalene or quinoline are used when sulfur or selenium are the dehydrogenating agents. Bubbling carbon dioxide through the reaction mixture as well as vigorous boiling facilitates the removal of hydrogen or hydrogen acceptors, such as benzene [8] or oleic acid [9], may be used.

A novel catalyst which converts cyclic dienes as shown into aromatic hydrocarbons, sometimes at room temperature [10], is N-lithium ethylenediamine.

Potassium *t*-butoxide acts similarly since at 250–300° in a closed vessel over

References for Section E are on p. 52.

nitrogen it converts *d*-limonene into *p*-cymene [11]:

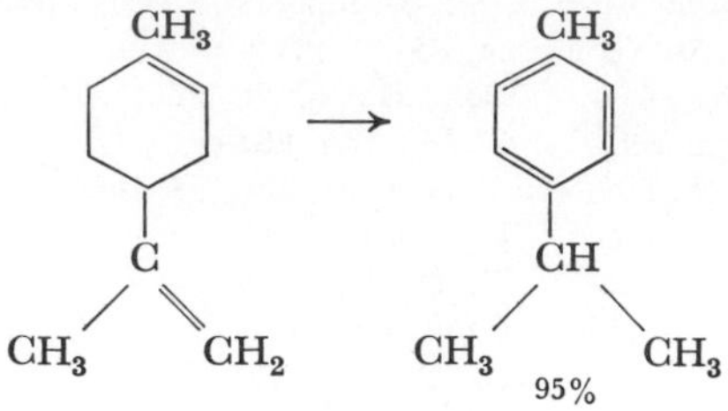

Some dihydro compounds disproportionate by heat alone under pressure [12], the driving force being the formation of the aromatic hydrocarbon. For example,

$$\textit{cis-}\text{Dihydronaphthalene} \xrightarrow[\substack{\text{cyclohexane} \\ \text{(sealed tube)}}]{95°, 20 \text{ hr.}} 710 \text{ mg. of product}$$

1 g.

consisting principally of three products, naphthalene, 1,2,9,10-tetrahydronaphthalene, and 1,2,5,6,9,10-hexahydronaphthalene in the ratios 1.00:0.56:0.23 respectively.

Temperature control is important in carrying out the reaction. Even so, a variety of effects other than simple dehydrogenation of existing rings sometimes occurs. For example, methyl or ethyl groups may be eliminated [13].

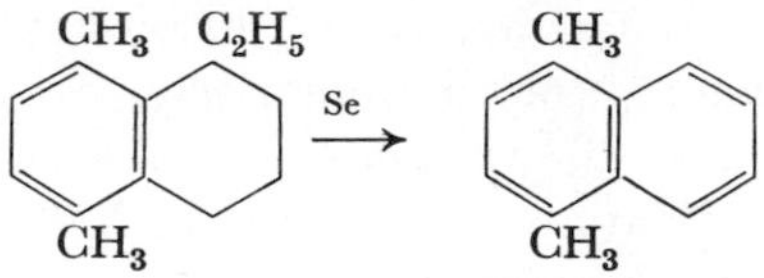

1,4-Dimethylnaphthalene

or new rings may be formed [14]. Indeed, the dehydrogenative coupling by this method usually is superior to that of the electrophilic method with aluminum chloride, discussed in D.7.

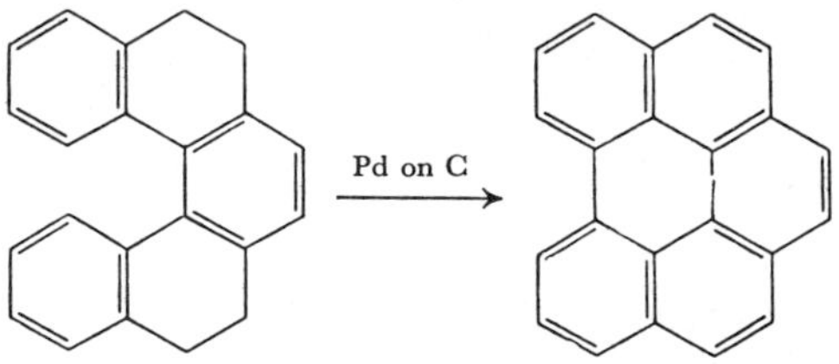

1,12-Benzoperylene, 95%

Particularly effective for a dehydrogenative coupling of this sort is a palladium-platinum charcoal catalyst [15].

Quaternary methyl groups are also almost always completely eliminated, although with the catalytic dehydrogenating agents some migration occurs [16]. In spite of these limitations the synthesis often gives superior yields.

Some modifications have been helpful in improving yields. When 1,1'-binaphthyl mixed with decalin is passed over a Pd-Pt-C catalyst at 490°, a 40%

yield of perylene is obtained [17]. The decalin prevents rearrangement to 1,2'-binaphthyl, perhaps acting simply as a heat exchange agent, but perhaps having some more direct effect such as saturating the catalyst with hydrogen. The naphthalene produced from decalin is removed easily by steam distillation. The saturation of the noble metal catalysts with hydrogen may slow down dehydrogenation. On the basis of this rationale, some tetrahydrobinaphthyls have been dehydrogenated with a Pd-S-C catalyst at 300°. Omission of the sulfur does not give dehydrogenation products [18]. The sulfur probably acts as a hydrogen acceptor in that it forms hydrogen sulfide.

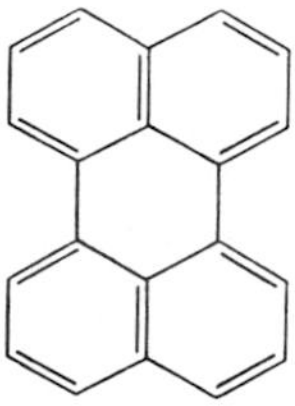

(a) **Preparation of 1-phenylnaphthalene** (91–94% from 1-phenyl-dialin and sulfur at 250–270° for 20 min.) [19].

(b) **Other examples**

(1) **Fluoranthene** (91% from 6b,7,10,10a-tetrahydrofluoranthene heated with palladium-on-barium sulfate at 270–290° for 30 min.) [20].

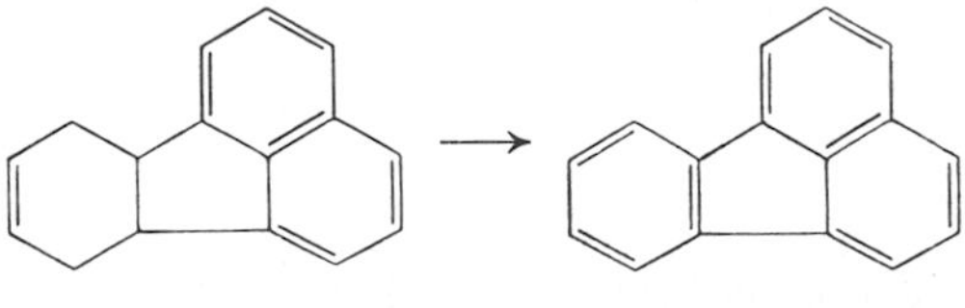

(2) **Pentaphenylbenzene** (99% from 1,2,3,4,5-pentaphenylcyclo-hexadiene-1,3 and chloranil in benzene by irradiation under nitrogen for $2\frac{1}{3}$ hr.) [4].

(3) **Azulene** (naphthalene was first reduced, by hydrogen in the presence of the specific catalyst, tungsten trisulfide, to decalin at 400° and 200

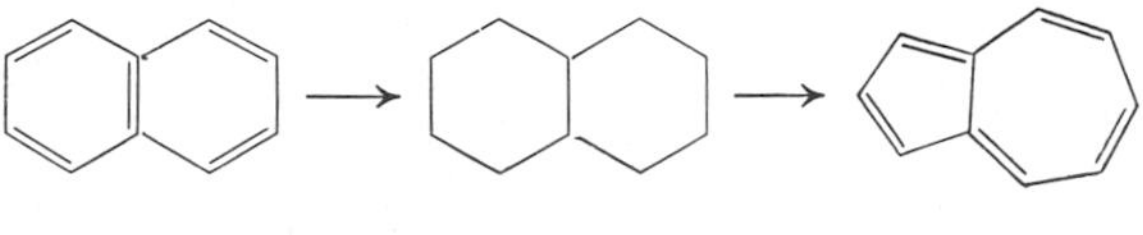

atm; the decalin containing some perhydroazulene was then dehydrogenated with Pd-Pt-C catalyst at 380°; the azulene was isolated by solution in concentrated hydrochloric acid followed by dilution with water, yield 0.2%) [2].

Azulene is best prepared, however, by dehydrogenation of cyclodecene over Pd-C at 300° [22].

(4) 1H-Benzonaphthene-1-one. (80 % by boiling perinaphthan-1-one and trityl perchlorate in acetic acid) [6].

1. P. A. Plattner, *Newer Methods of Preparative Organic Chemistry*, *Eng. Transl.* Interscience Publishers, New York, 1948, Vol. 1, p. 21.
2. L. F. and M. Fieser, *Advanced Organic Chemistry*, Reinhold Publishing Corp., New York, 1961, p. 646.
3. H. M. Crawford and H. B. Nelson, *J. Am. Chem. Soc.*, **68,** 134 (1946).
4. G. R. Evanega *et al.*, *J. Org. Chem.*, **27,** 13 (1962).
5. A. M. Creighton and L. M. Jackman, *J. Chem. Soc.*, 3138 (1960).
6. W. Bonthrone and D. H. Reid, *J. Chem. Soc.*, 2773 (1959).
7. A. Schönberg and G. Schütz, *Chem. Ber.*, **93,** 1466 (1960).
8. M. S. Newman and D. Lednicer, *J. Am. Chem. Soc.*, **78,** 4765 (1956).
9. H. A. Silverwood and M. Orchin, *J. Org. Chem.*, **27,** 3401 (1962).
10. I. Wender *et al.*, *J. Org. Chem.*, **23,** 1136 (1958).
11. H. Pines and L. A. Schaap, *J. Am. Chem. Soc.*, **79,** 2956 (1957).
12. W. von E. Doering and J. W. Rosenthal, *J. Am. Chem. Soc.*, **89,** 4534 (1967).
13. W. Cocker *et al.*, *J. Chem. Soc.*, 72 (1952).
14. Y. Altman and D. Ginsburg, *J. Chem. Soc.*, 466 (1959).
15. P. G. Copeland *et al.*, *J. Chem. Soc.*, 1687 (1960); 1689 (1960).
16. R. P. Linstead *et al.*, *J. Chem. Soc.*, 1146 (1937); R. P. Linstead and S. L. S. Thomas, *ibid.* 1127 (1940).
17. P. G. Copeland *et al.*, *J. Chem. Soc.*, 1689 (1960).
18. H. M. Crawford and V. R. Supanekar, *J. Chem. Soc.*, 2380 (1964).
19. R. Weiss, *Org. Syn.*, Coll. Vol. **3,** 729 (1955).
20. M. C. Kloetzel and H. E. Mertel, *J. Am. Chem. Soc.*, **72,** 4786 (1950).
21. W. Baker *et al.*, *J. Chem. Soc.*, 4149 (1953).
22. V. Prelog and K. Schenker, *Helv. Chim. Acta*, **36,** 1181 (1953).

F. Decarboxylation

1. From Acids

$$\text{ArCOOH} \xrightarrow[\text{Quinoline}]{\text{Cu-}} \text{ArH}$$

$$\text{ArCOOCH}_3 \xrightarrow[\text{HBr}]{48\%} \text{ArH}$$

Decarboxylation is much more satisfactory for aromatic and heterocyclic carboxylic acids, although good yields of methane may be obtained by heating sodium acetate with soda lime. Other aliphatic acid salts do not give good yields of the corresponding hydrocarbon [1]. Heating alone [2] or under alkaline conditions such as with dimethylaniline [3] with copper oxide [4], or with

References for Section F are on pp. 54–55.

copper and quinoline [5] is the usual procedure. Yields from cyclic carboxylic acids are often in the 70–90% range.

In strong basic media the reaction appears to proceed via the anion which is

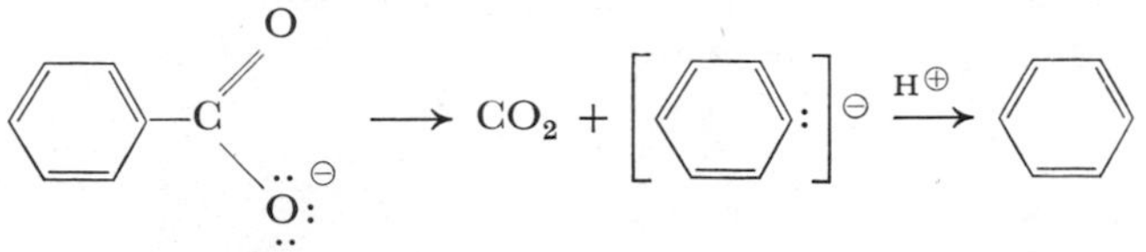

particularly susceptible to cleavage if there are strong electron-attracting substitutents in the aromatic ring. In copper-containing media the reaction appears to proceed by a free radical mechanism.

The esters corresponding to readily decarboxylated acids may be decarbalkoxylated as well under alkaline conditions provided by potassium hydroxide

$$YCH_2CO_2R \longrightarrow YCH_3 + CO_2 + ROH$$

$$Y = \overset{O}{\underset{}{\underset{}{C}}} \ . \ , \ -CO_2R, \ -NO_2, \text{ or others}$$

[6], alkali metal carbonates [7], or lithium iodide in collidine [8].

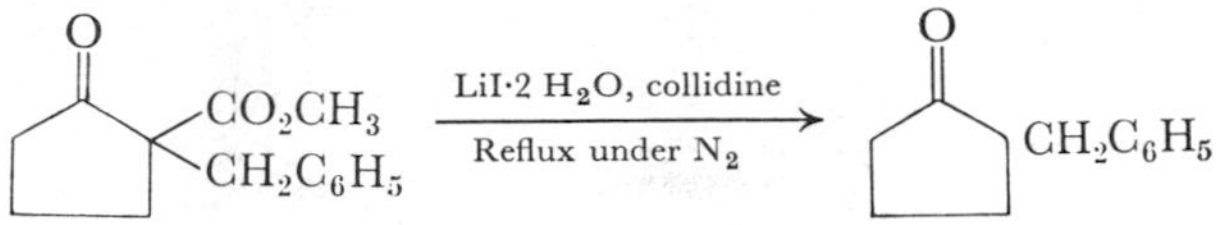

2-Benzylcyclopentanone,
72–76%

Ester (or acid) groups attached to aromatic rings (mainly phenols), susceptible to electrophilic attack, are decarboxylated readily with acid [9].

(a) Preparation of 3-methylfuran (83–89% by heating 3-methyl-2-furoic acid with copper powder and dry quinoline) [5].

(b) Other examples

(1) Imidazole (68–76% by heating imidazole-4,5-dicarboxylic acid with powdered copper oxide) [4].

(2) 2,6-Dimethylpyridine (63–65% from 2,6-dimethyl-3,5-dicarbethoxypyridine by treatment first with potassium hydroxide followed by heating the dry potassium salt with calcium oxide) [10].

2. From Maleic Anhydride Adducts

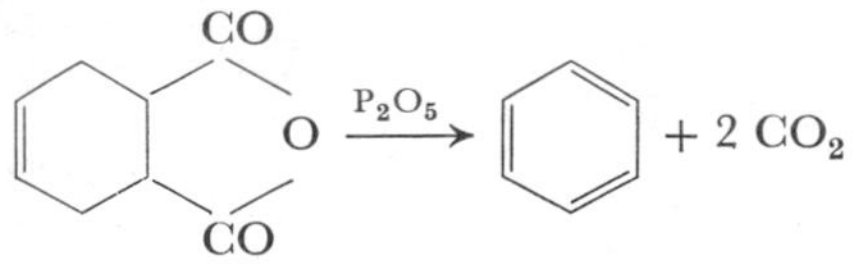

It has been noted that Δ^4-tetrahydrophthalic anhydride can be decarboxylated to benzene by heating with phosphorus pentoxide [11]. The tetrahydropthhalic anhydride, of course, is obtained from butadiene and maleic anhydride. The decarboxylation has considerable utility because substituted butadienes may be used to form maleic anhydride adducts, which on decarboxylation yield arenes. Since butadienes are not particularly accessible, the more available unsaturated carbinols, obtained from the addition of allylmagnesium chloride and ketones, have been used directly in the Diels-Alder reaction to form a butadiene *in situ*, which then reacts with maleic anhydride:

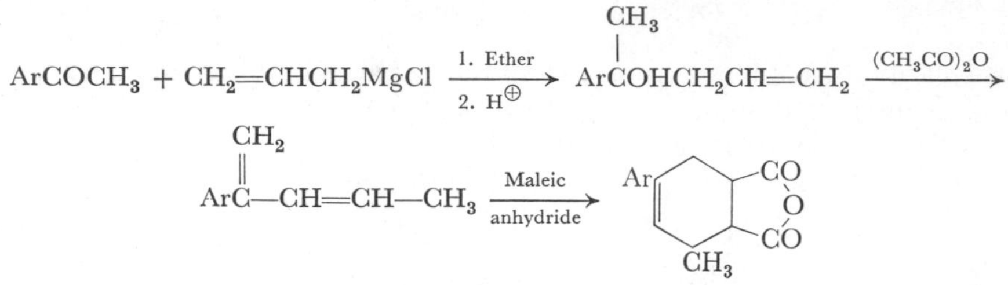

$$\text{ArCOCH}_3 + \text{CH}_2{=}\text{CHCH}_2\text{MgCl} \xrightarrow[\text{2. H}^{\oplus}]{\text{1. Ether}} \text{Ar}\overset{\overset{\displaystyle CH_3}{|}}{C}\text{OHCH}_2\text{CH}{=}\text{CH}_2 \xrightarrow{(CH_3CO)_2O}$$

$$\overset{\overset{\displaystyle CH_2}{\|}}{\text{ArC}}{-}\text{CH}{=}\text{CH}{-}\text{CH}_3 \xrightarrow[\text{anhydride}]{\text{Maleic}}$$

Some oxidized adducts of maleic anhydride have been decarboxylated with soda lime [12]:

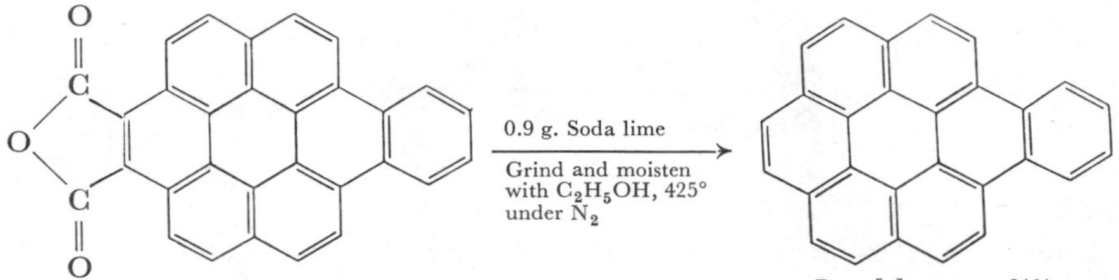

Benzo[α]coronene, 64%

(a) **Preparation of 3-methylbiphenyl.** 4-Phenyl-4-hydroxyl-1-pentene was made first from acetophenone and allylmagnesium chloride. The unsaturated carbinol, maleic anhydride, acetic anhydride, and catalytic quantities of potassium hydrogen sulfate (dehydrating agent) and phenothiazine (inhibitor?) were refluxed in xylene to yield 70% of Δ⁴-4-phenyl-6-methyltetrahydrophthalic anhydride. Equimolecular quantities of phosphorus pentoxide and the maleic anhydride adduct then were heated gradually to 260–300° until carbon dioxide evolution ceased to yield 70% of 3-methylbiphenyl [11].

1. T. S. Oakwood and M. R. Miller, *J. Am. Chem. Soc.*, **72,** 1849 (1950).
2. W. R. Boehme, *Org. Syn.*, Coll. Vol. **4,** 590 (1963); R. B. Sandin and R. A. McKee, *Org. Syn.*, Coll. Vol. **2,** 100 (1943).
3. D. S. Tarbell *et al.*, *Org. Syn.*, Coll. Vol. **3,** 267 (1955).
4. H. R. Snyder *et al.*, *Org. Syn.*, Coll. Vol. **3,** 471 (1955).
5. D. M. Burness, *Org. Syn.*, Coll. Vol. **4,** 628 (1963).
6. W. A. Orth and W. Riedl, *Ann. Chem.*, **663,** 83 (1963); M. S. Schechter *et al.*, *J. Am. Chem. Soc.*, **74,** 4902 (1952).
7. C. A. Grob and O. Weissbach, *Helv. Chim. Acta*, **44,** 1748 (1961).
8. F. Elsinger, *Org. Syn.*, **45,** 7 (1965).
9. E. Biekert and W. Schäfer, *Chem. Ber.*, **93,** 642 (1960); H. Muxfeldt *et al.*, *ibid.*, **95,** 2581 (1962).

10. A. Singer and S. M. McElvain, *Org. Syn.*, Coll. Vol. **2**, 214 (1943).
11. V. R. Skvarchenko *et al.*, *J. Gen. Chem. USSR* (*Eng. Transl.*), **32**, 211 (1962).
12. G. Stork and K. Matsuda, U.S. Patent 3, 364,274, Jan. 16, 1968; *C.A.*, **69**, 975 (1968).

G. Free Radical Reactions

A clear delineation of mechanism is not achieved in this section. One difficulty arises from the fact that coupling may be brought about either homolytically or heterolytically in some cases. In the main, however, these reactions are best carried out under homolytic conditions and in addition are discussed more uniformly in the light of homolytic theory.

1. From Amines or Diazo Compounds by Coupling with Aromatic Nuclei (Gomberg-Bachmann)

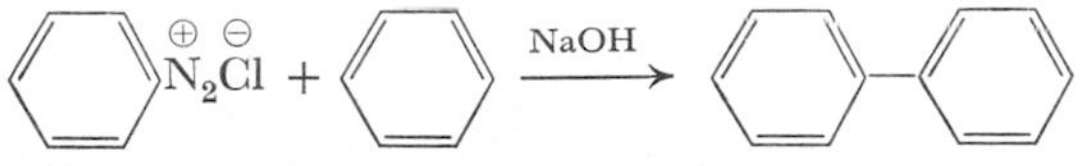

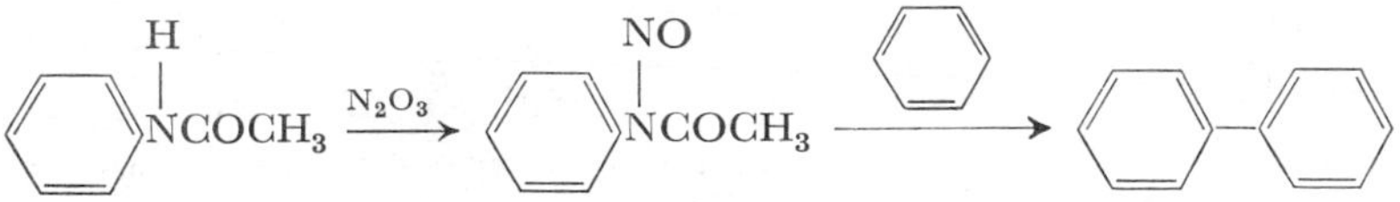

These two syntheses have been reviewed [1]. Although yields generally are not high, the reaction gives access to unsymmetrical biaryls. In both reactions the one necessary requirement is that the aromatic compound to be coupled with the diazo compound must be a liquid.

In the first method the cold diazonium salt produced by diazotization is converted by 15–40% aqueous sodium hydroxide into the diazo hydroxide which in the presence of the liquid aromatic compound forms the biphenyl. Thus isolation of the unstable, explosive diazohydroxide is avoided. Modifications have consisted in using sufficient alkali to form the sodium diazotate [2] or in using sodium acetate instead of sodium hydroxide [3]. Maximum yields starting with a series of amines were 44% by employing sodium hydroxide and 60% with sodium acetate [3]. With the diazonium tetrafluoroborate ring closures of the Pschorr type are possible by refluxing the salt in pyridine [4]:

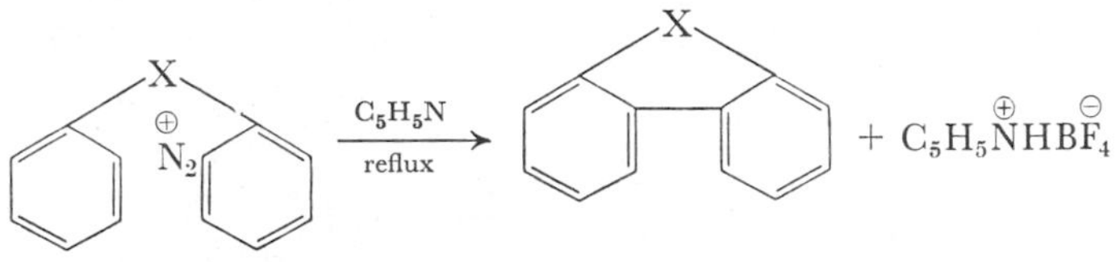

In the second method the acetylated amine in an ice-cold solution or suspension in acetic acid and acetic anhydride is treated with nitrogen trioxide, or nitrosyl chloride is added to the acetylamine and sodium acetate in acetic acid. The nitroso derivative formed is precipitated by pouring the solution into ice water, after which it is treated with the aromatic liquid. Yields by this procedure

References for Section G are on pp. 66–67.

are sometimes in the 60 % range. Thus the yields are as a rule better than in the first method, although the first one is less troublesome to carry out.

Examples of hydrocarbons prepared in this manner are α- and β-phenylnaphthalenes [5], o-, m-, and p-methylbiphenyls [2], and m- and p-terphenyls [6]. Arylation of the thiophene and pyridine nuclei has also been accomplished [7].

More recently three other methods of coupling have been employed. The

$$2 \; \langle \text{C}_6\text{H}_5 \rangle\text{NH}_2 \xrightarrow[\text{C}_6\text{H}_6]{\text{CH}_3(\text{CH}_2)_4\text{ONO}} \langle \text{C}_6\text{H}_5 \rangle{-}\langle \text{C}_6\text{H}_5 \rangle$$

$$\underset{\text{NO}_2}{\langle \rangle}\text{N}_2\text{Cl} \xrightarrow[\text{Na}_2\text{CO}_3]{(\text{CH}_3)_2\text{NH}} \underset{\text{NO}_2}{\langle \rangle}\text{N}{=}\text{NN(CH}_3)_2 \xrightarrow[\text{TsOH}]{\text{C}_6\text{H}_6} \underset{\text{NO}_2}{\langle \rangle}\text{C}_6\text{H}_5$$

$$\text{R}_2\text{NH} + \text{HNF}_2 \rightarrow \begin{bmatrix} \text{H} \\ | \\ \text{R}_2\text{NNF} \end{bmatrix} \rightarrow \text{RR} + \text{N}_2 + \text{HF}$$

first one is very simple in that it consists in heating a mixture of aromatic primary amine and pentyl nitrite in the hydrocarbon with which it couples until nitrogen gas evolution ceases (see Ex. a). Yields are in the range 50–60 %, and application seems general. The second method converts the diazonium salt into the 3,3-dimethyltriazine, which is separated and refluxed in a benzene solution while p-toluenesulfonic acid is added over a period of 4–4.5 hours [8]. The yield in the last step is 34–42 %.

The third method employs difluoramine, an unusual and versatile reagent. It reacts with secondary amines, as already shown, to form coupling products [9]. Dibenzylamine gives bibenzyl in 53 % yield and azetidine, ⌐NH, cyclopropane in 40 % yield. Primary amines are converted into hydrocarbons by

$$3 \, \text{RNH}_2 + \text{HNF}_2 \longrightarrow 2 \, \text{RNH}_3\text{F} + \text{N}_2 + \text{RH}$$

difluoramine. Yields with this reagent are rather low, but improvement with additional experimentation may be forthcoming.

Thermal decomposition of diazotized 2-amino-2′-methylbiphenyls yields a substituted fluorene as well as the corresponding phenol as a by-product [10]:

$$\text{CH}_3\langle \rangle\text{CH}_3 \; \text{N}_2^{\oplus}\langle \rangle\underset{\text{CH}_3}{\overset{\text{CH}_3}{}} \xrightarrow[\Delta]{2.5\% \text{ aq. H}_2\text{SO}_4} \text{CH}_3\langle \rangle\overset{\text{CH}_2}{\langle \rangle}\underset{\text{CH}_3}{\overset{\text{CH}_3}{}}$$

2,3,7-Trimethylfluorene, 36%

This conversion is known as the Mascarelli reaction.

The mechanism of all couplings has been characterized rather loosely as homolytic; obviously, the differences occurring depend on the environment; see [11] for a detailed discussion. The most convincing evidence in favor of the homolytic mechanism is the pattern of orientation when an aromatic solvent other than benzene is used [12]. Thus, as shown in the accompanying tabulation, similar isomer ratios are obtained in phenylation. The indiscriminate

	o	m	p
Nitrobenzene	58	10	32
Chlorobenzene	60	26	14
Toluene	66	19	14

coupling regardless of the electronic characteristics of the substituent suggests strongly a free radical mechanism. Moreover, the same couplings brought about by decomposition of different sources of free radicals, benzoyl peroxide, lead tetrabenzoate, or phenyl iodosobenzoate, produce almost indistinguishable patterns of orientation.

Since the diazonium couplings are for the most part free radical in nature, it is curious that very few preparations of merit are carried out with copper or cuprous oxide catalysis. The Pschorr coupling of o-diazostilbenes to phenanthrenes has benefited from the addition of copper [13], and other couplings may benefit as well.

(a) **Preparation of p-chlorobiphenyl** (73% from 7 g. of p-chloroaniline and 9 g. of pentyl nitrite warmed in 200 ml. of benzene until nitrogen gas evolution was substantial, followed by reflux for 2 hr.) [14].

(b) **Preparation of 4-methylbiphenyl** (22% from diazotized p-toluidine added to 10 N sodium hydroxide and benzene, all of which was kept at 5° by ice) [15].

(c) **Other examples**

(1) **p-Terphenyl** (69% from bis-nitrosoacetyl-1,4-phenylenediamine in 300 ml. of C_6H_6 at 35° until N_2 evolution ceased) [16].

(2) **9,10-Di-p-nitrophenylanthracene** (62% from anthracene and cupric chloride in aqueous acetone to which 2 equivalents of p-nitrobenzenediazonium salt in water were added slowly) [17].

2. From Aroyl Peroxides and Ketone Peroxides

$$(ArCO_2)_2 + ArH \longrightarrow ArCOOH + ArAr + CO_2$$

Normally aroyl peroxides and arenes give low yields of the aromatic acid and the biphenyl. By the introduction of aromatic nitro compounds the yields are increased substantially [18, 19]. For example, with m-dinitrobenzene as the additive the yields of biphenyls are increased from about 50 to 80–90%. The method is applicable to the synthesis of arylbenzenes in which the aryl group contains only halogen, alkyl, or aryl substituents. In some cases substituted benzenes (C_6H_5Br, $C_6H_5NO_2$) react with aroyl peroxides to give reasonable yields of the biaryl without the presence of an additive [20, 19].

Not only acyl peroxides but also ketone peroxides yield hydrocarbons. The procedure, involving photolysis or thermolysis only, is quite simple and offers promise in preparing ring hydrocarbons [21].

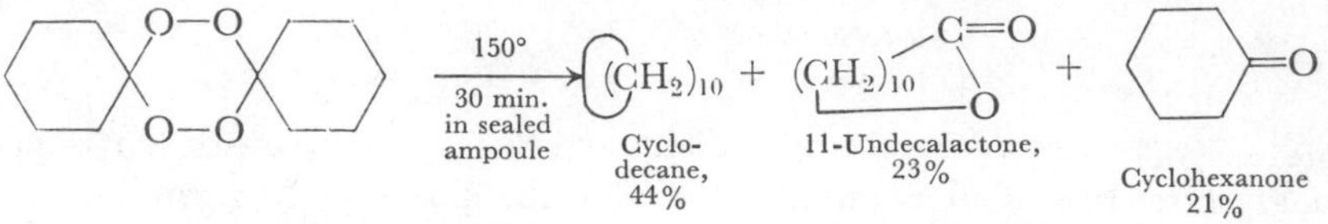

Although cyclodecane is a commercial product, prepared from butadiene and ethylene by oligomerization, the present process may be the simplest route to other cycloalkanes. The hydrocarbon apparently is formed via free radicals held in a solvent cage:

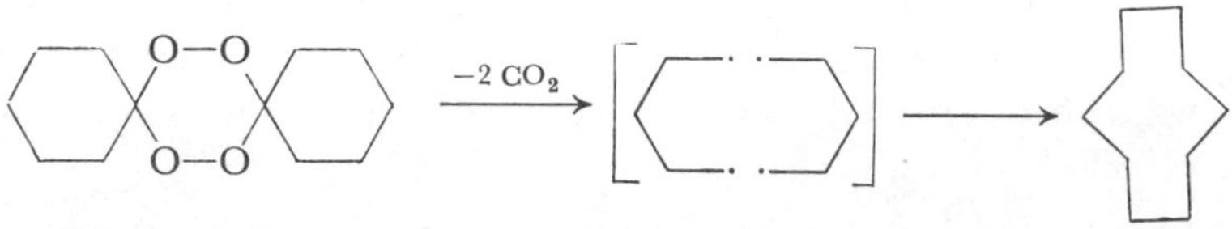

(a) **Preparation of 4-chlorobiphenyl** (90% from bis-p-chlorobenzoyl peroxide, 12 g., in a solution of m-dinitrobenzene, 2 g., in 400 ml. of benzene boiled for 26 hr. The p-chlorobenzoic acid was removed by cooling.

3. From Aryl Iodides and Hydrocarbons by Photochemical Coupling

In an elegant, versatile, and recent procedure, aryl iodides have been shown to photolyze to free radicals which couple with the aromatic solvent [22, 23].

$$p\text{-}CH_3C_6H_4I \xrightarrow[h\nu, \text{ 18 hr.}]{250\text{ml. } C_6H_6} p\text{-}CH_3C_6H_4C_6H_5$$
$$74\%$$

Traces of biphenyl from reduction of the iodide may contaminate the product. Although the scale of preparation is small, more concentrated solutions possibly may be used, or radiation equipment adapted to larger scale may be helpful.

The photochemical coupling reaction is not only applicable to iodo compounds but also to specific hydrocarbons in the presence of iodine [24]:

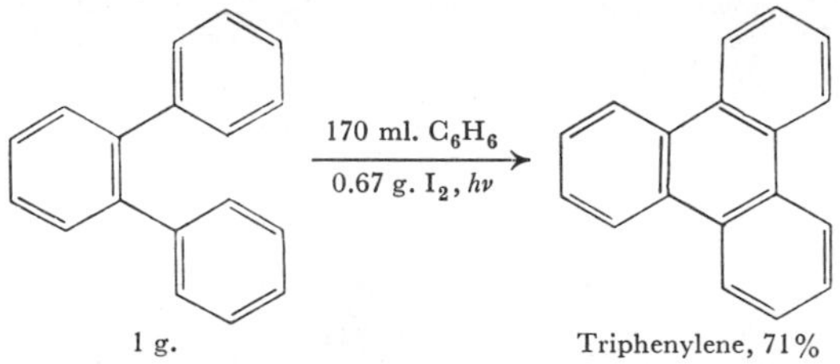

No doubt other hydrocarbons, such as stilbene (to phenanthrene), can be coupled, although benzene itself does not couple to any appreciable extent [22].

The initial step in the processes above brings to mind the possibilities that a free radical can be generated from any hydrocarbon [25]:

$$C_6H_{12} \text{ (cyclohexane)} \xrightarrow[h\nu]{C_6H_5COC_6H_5} [C_6H_{11}\cdot] + [(C_6H_5)_2CHO\cdot]$$

The hydrogen atom is abstracted from cyclohexane about one-fourth as readily as from the carbinol group of isopropyl alcohol, which fact places the abstraction in the realm of synthetic possibilities as shown [26]:

$$C_6H_5CH(CH_3)_2 + (C_6H_5)_2C{=}O \xrightarrow[\text{144 hr.}]{h\nu \text{ (high-pressure Hg lamp)}} \underset{\text{Dicumyl,57 g.}}{C_6H_5\overset{\overset{\displaystyle CH_3}{|}}{\underset{\underset{\displaystyle CH_3}{|}}{C}}\overline{\quad\quad}\overset{\overset{\displaystyle CH_3}{|}}{\underset{\underset{\displaystyle CH_3}{|}}{C}}C_6H_5}$$

600 g.   91 g.

Other sources of free radicals give coupling reactions. For example, recently nitrobenzene has been coupled with benzene in the gaseous phase [27]:

$$C_6H_5NO_2 \text{ and excess } C_6H_6 \xrightarrow[\substack{\text{9 sec. contact time} \\ \text{over Vycor chips}}]{\substack{600° \text{ in N}_2 \text{ stream} \\ \text{at 45 ml./min.}}} \text{biphenyl} + \text{terphenyl}$$

62%   14%

4. From Olefins or from Aromatic Hydrocarbons by Irradiation

The photochemistry field is expanding to such an extent that this section will be out of date by the time it is published. Nevertheless, although the course of photochemical reactions is the least predictable of any type of reaction, some generalizations can be made. Theoretically, any double bond can be excited to a singlet or triplet state which can undergo a number of reactions (which in this section are restricted to hydrocarbon production):

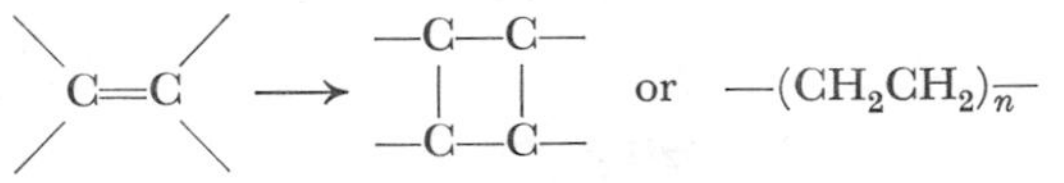

$$\begin{matrix} \diagdown \quad \diagup \\ C{=}C \\ \diagup \quad \diagdown \end{matrix} \longrightarrow \begin{matrix} -C-C- \\ |\quad | \\ -C-C- \end{matrix} \quad \text{or} \quad -(CH_2CH_2)_{\overline{n}}$$

The dimerization reaction to form a cyclobutane is the most desirable from the standpoint of synthesis, although one can visualize possibilities in the polymerization process if end groups can stop the polymerization at some discrete stage. Alicyclic enones are the most versatile compounds to use in formation of a cyclobutane or even a cubane [28]. The unsaturated ketone dimerizes to a mixture of the isomeric substituted cyclobutanes:

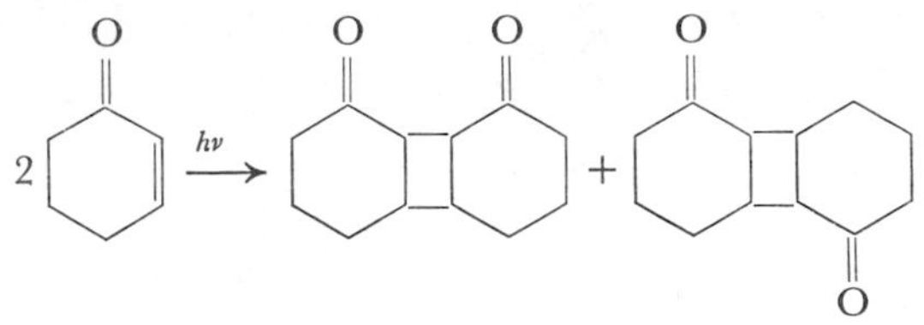

With olefins the addition has been shown to proceed stepwise:

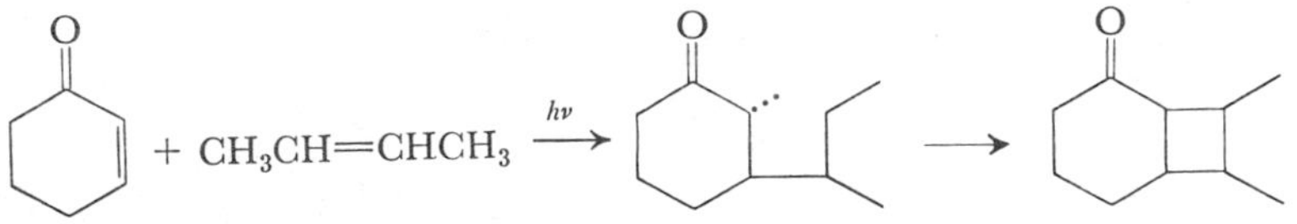

cis or trans

With either cis- or trans-2-butene the same ratio of isomers is obtained [29].

Another reaction that can occur is addition of some free radical:

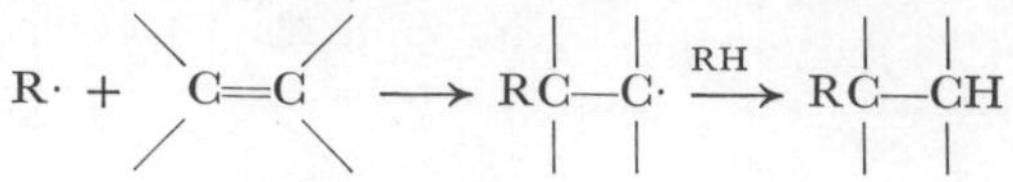

$$R\cdot + \quad \diagdown C\!=\!C \diagup \quad \longrightarrow \quad RC\!-\!C\cdot \quad \xrightarrow{RH} \quad RC\!-\!CH$$

Excitation of aromatic hydrocarbons leads to profound changes:

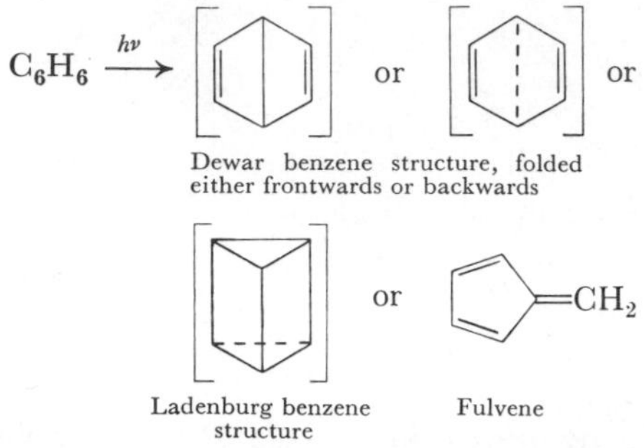

$$C_6H_6 \quad \xrightarrow{h\nu}$$

Dewar benzene structure, folded
either frontwards or backwards

or

Ladenburg benzene   Fulvene
structure

Reactions to illustrate the preceding mechanisms are listed below; *see also* [30]. It will be noted that the yields are low probably because so many side reactions, including those in which the product itself is subject to photoexcitation, can take place. The ideal way to conduct a photochemical reaction is to use monochromatic light (or very narrow banded polychromatic light) which will excite the substrate but not the product. Although the yields are low and the irradiated amounts small, some of the products are exotic and inaccessible by any other method.

(a) **Preparation of *trans*-tricyclo[5.3.0.0^{2,6}]decane** (55% from uv ir-

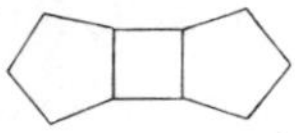

radiation for 100 hours of an equal mixture of acetone and cyclopentene; other products are 3.5% dicyclopentane, 10% dicyclopentene, and 30% dicyclopentadienes) [31].

(b) **Other examples**

(1) **Binorbornylene** (24% as two stereoisomers from the irradiation

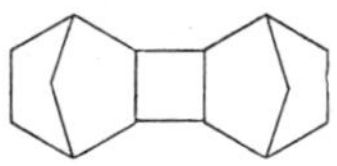

of norbornene in acetone; benzophenone as a sensitizer, rather than acetone, gave a mixture of products) [32].

(2) **Heptacyclene** (30% as a mixture of two stereoisomers from

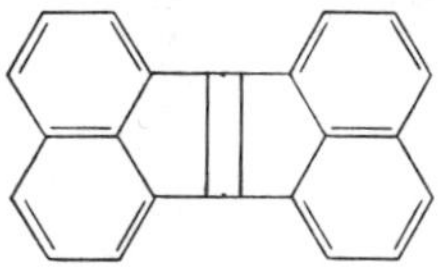

sunlight irradiation of acenaphthylene in benzene) [33].

(3) Naphthalene-tolane adduct (29% from the irradiation of

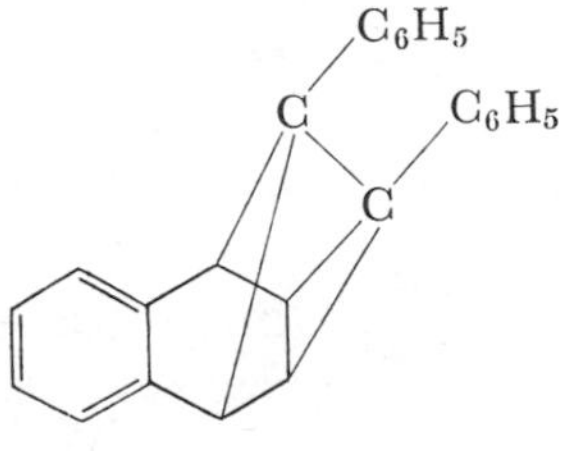

naphthalene and diphenylacetylene) [34].

(4) 1,2,5-Tri-*t*-butylbicyclo[2.2.0]hexa-2,5-diene ("significant"

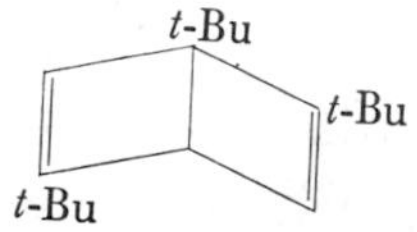

amounts by irradiation of 1,2,4-tri-*t*-butylbenzene) [35].

(5) 1,2,4-Trimethylbenzene-1,2,4-C_{14} (from mesitylene-1,3,5-C_{14} by irradiation in isohexane; this reaction illustrates the profound change which occurs in the benzenoid nucleus) [36].

(6) Tricyclo[3.3.0.0^{2,6}] octane (30% from the uv irradiation of 1,5-

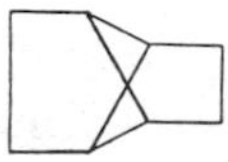

cyclooctadiene-cuprous chloride complex in ether; the modification of using complexes may be helpful in extending the scope of hydrocarbon synthesis from olefins) [37].

5. From Decarbonylation of Carbonyl Compounds [38].

The general reactions perceived are:

$$\mathrm{RCH_2CH_2\overset{\displaystyle H}{\overset{|}{C}}{=}O} \xrightarrow[\text{Sensitizer}]{h\nu} [\mathrm{RCH_2CH_2\dot{C}{=}O}] \xrightarrow{\quad -\mathrm{CO}\quad}$$

$$[\mathrm{RCH_2CH_2\cdot}] \xrightarrow{\mathrm{RCH_2CH_2\overset{H}{\overset{|}{C}}{=}O}} \mathrm{RCH_2CH_3} \quad \text{or} \quad \xrightarrow{\quad \mathrm{ArH}\quad} \mathrm{RCH_2CH_2Ar}$$

A series of tetralins, for example, has been made from δ-phenylvaleraldehydes [39]. Cyclic ketones under drastic photochemical conditions *or* with special structural features can extrude carbon monoxide to form various products as shown in the following example [40]:

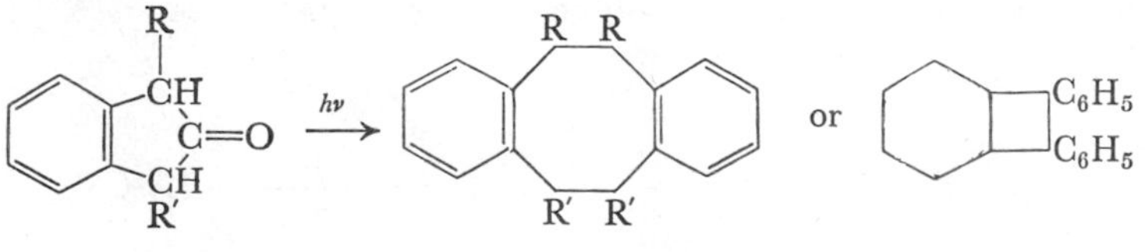

R = C_6H_5, R′ = H. 1,2,5,6-Dibenzo-3,4-diphenyl-cyclooctadiene-1,5 above 50%

R, R′ = C_6H_5 1,2-Diphenyl-3,4-benzocyclobutene-3 practically quantitative

These are unusual cases, however, and more often simple redox reactions or ring contraction can occur.

The aldehyde decompositions are performed more frequently with peroxide catalysts [39]. However, photochemical rearrangements can occur in decarbonylation of aryl-substituted aliphatic aldehydes [41]:

$$\overset{14}{(C_6H_5)_2CHCH(C_6H_5)CH=\!O} \xrightarrow[C_6H_5SH]{h\nu,\,180°} \overset{14}{(C_6H_5)_2CHCH_2C_6H_5} + \overset{14}{(C_6H_5)_2C=\!CHC_6H_5}$$

$$\begin{array}{cc} \text{1,1,2-Triphenylethane 29\%} & \text{Triphenylethylene 4\%} \\ \text{5\% phenyl migration} & \end{array}$$

In this reaction the thiophenol minimizes dimerization.

Two chemical reagents are available for the decarbonylation of aldehydes: palladium (see Ex. b.1) or chloro-tris(triphenylphosphine)rhodium (see Ex. b.2).

(a) **Preparation of ethylbenzene** (51–77% from 0.0186 mole of β-phenylpropionaldehyde and 0.028 mole of di-*t*-butylperoxide in *o*-dichlorobenzene to which a second portion of the peroxide was added later) [42].

(b) **Other examples**

(1) **Naphthalene** (80% from 1-naphthaldehyde and 1% Pd on C heated in a carbon dioxide atmosphere at 220° for 0.5 hr.) [43].

(2) **Styrene** [44].

$$C_6H_5CH=\!CHCH=\!O + RhCl[(C_6H_5)_3P]_3 \xrightarrow[\text{Reflux 15 min.}]{C_6H_6,\,10\ ml.}$$

$$\underset{77\%}{C_6H_5CH=\!CH_2} + \underset{93\%}{RhClCO[(C_6H_5)_3P]_2} + (C_6H_5)_3P$$

6. From Carbenes (or Organometallic Intermediates)

Carbenes [45] are most active agents which can be sources of unusual hydrocarbons. The most important means of generation is from the photolysis of diazoalkanes:

$$RCHN_2 \xrightarrow{h\nu} RCH: +\ N_2$$

But other means, of which the most common from a synthetic viewpoint is the reaction of an alkyl halide and a strong base, are available [46].

$$RCH_2Cl \xrightarrow{C_6H_5Li} RCH: \quad \text{or}$$

$$R_2CCl \xrightarrow{2\ RLi} R_2C: \quad \text{or}$$

$$ICH_2HgI \xrightarrow{(C_6H_5)_2Hg} CH_2: \qquad\qquad \text{[ref. 47]}$$

A fourth method consists of the decomposition of tosylhydrazones[48] (see Ex. b.3 also):

$$\underset{\substack{\text{NaOCH}_3 \\ C_5H_5N, \\ 55\text{–}80°}}{p\text{-}C_7H_7SO_2NHN=\!\overset{\displaystyle Ar}{\underset{\displaystyle }{C}}R} \longrightarrow \underset{\substack{| \\ R \\ 20\text{–}70\%}}{ArC=\!N_2} \xrightarrow{130°} [Ar\overset{..}{C}R]$$

The last, but most useful, method is the Simmons-Smith reaction which involves the reagent as shown [49]:

$$CH_2I_2 + Zn\text{—}Cu \text{ couple} \longrightarrow (ICH_2)ZnI \quad or \quad (ICH_2)_2Zn, ZnI_2$$

Although this reagent is not a carbene, the preparation bears some resemblance to that of a carbene and thus is considered here. A highly reactive Zn-Cu couple has been prepared recently (see Ex. a), but it has been demonstrated that zinc dust itself works well in dimethoxyethane solvent (see Ex. b.1). In addition a simulated but highly reactive Simmons-Smith reagent can be made from zinc

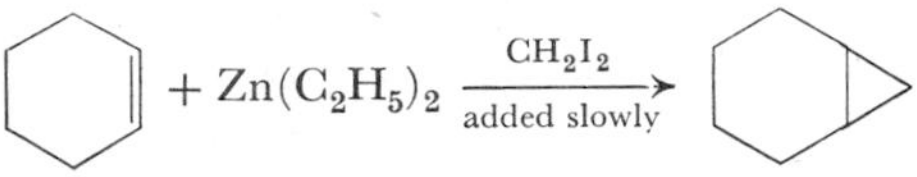

$$\text{(hexene ring)} + Zn(C_2H_5)_2 \xrightarrow[\text{added slowly}]{CH_2I_2} \text{(norcarane)}$$

Norcarane, 53%

diethyl [50]. The reaction must be conducted in an inert atmosphere because of the inflammable nature of zinc diethyl in air.

In the synthesis of spirocyclic hydrocarbons the method has been improved by removing the excess of methylene iodide before workup by adding 2-methylbutene-2 and more zinc-copper couple and then refluxing for 24 hours [51].

It is interesting to note that the Simmons-Smith reaction is stereospecific in that it gives *cis* addition. These two equations illustrate this fact:

$$\underset{H}{\overset{H\quad H}{CH_3CH_2C=C\text{—}CH_2CH_3}} \longrightarrow CH_3CH_2\underset{\underset{\underset{CH_2}{\diagdown\quad\diagup}}{}}{\overset{H\quad H}{C\text{———}C}}CH_2CH_3 \quad \text{[ref. 52]}$$

cis-1,2-Diethylcyclopropane

$$\text{(bicyclic structure)} \longrightarrow \text{(tricyclic structure)} \quad \text{[ref. 53]}$$

exo-Tricyclo[3.2.1.0²,⁴]octane

The outstanding reaction of the zinc haloalkyls is addition to an olefin to form

$$CH_2I_2 + Zn(Cu) + RCH=CHR \longrightarrow \underset{RCH\text{———}CHR}{\overset{CH_2}{\diagup\quad\diagdown}}$$

a cyclopropane. It will be noted that this reaction is quite different from the behavior shown by carbenes with olefins.

The most active carbene (singlet carbene) generated in a nonpolar medium is so indiscriminate in its attack on hydrocarbons that it gives spectacular statistical insertion [54] as shown:

$$CH_2N_2 \xrightarrow[\text{heptane}]{hv} \text{methylheptanes}$$

	% Calcd.	% Found
C_1 insertion	37.5	38
C_2 insertion	25	25
C_3 insertion	25	24
C_4 insertion	12.5	13

A second carbene, called a triplet carbene, is known; it is derived from singlet carbene by dissipation of some of its energy or by generation from a different source, such as cyclodiazomethane [55]:

Whatever the difference between the two species, an energy difference (and structure) or one of solvation (or coordination), it is agreed that two forms of carbene exist. One is nonselective, the other is more selective in insertion. Insertion in aliphatic hydrocarbons is obtained not only with carbene but also in aromatic hydrocarbons to yield cycloheptatrienes (see Ex. b.2). The examples will best illustrate the versatility of the type in synthesizing hydrocarbons.

(a) Preparation of norcarane. A zinc-copper couple was made by

treating either zinc dust or granules with a hot solution of cupric acetate in acetic acid, followed by washing with acetic acid and then ether. Cyclohexene and dibromomethane were added slowly to a stirred suspension of the Zn-Cu couple in ether. After distillation, 61 % of norcarane was obtained. By the same procedure, bicyclo[6.1.0]nonane was obtained in 82 % yield from cyclooctene and methylene iodide [56]; *see also* [57].

(b) Other examples

(1) 9,10-Dihydro-9,10-methylenephenanthrene (25 % from phenanthrene, methylene iodide, and zinc dust in refluxing dimethoxyethane) [58].

(2) Cycloheptatriene (32 % together with 9 % toluene from a benzene solution of diazomethane by irradiation for 18–24 hr. The diazomethane was prepared *in situ* from nitrosomethylurea in benzene. The separation of the two products required a highly efficient distillation column) [59].

(3) *cis*-Bicyclo[5.3.0]decane [60].

	Hexadecane		
	$\longrightarrow$		
NNHSO$_2$C$_6$H$_4$CH$_3$-p		62%	*cis*-Decalin, 18%

(4) 7,7'-Spirobi(bicyclo[4.1.0])heptane. This reaction illustrates the

Br
Br + CH$_3$Li / Ether → 10% + C=C 30%

utilization of the *gem*-dibromides so readily prepared from an olefin and dibromocarbene (Br$_2$C:) [61].

7. Electrolysis of Alkali Metal Carboxylates (Kolbe)

$$2\,RCOONa + 2H_2O \xrightarrow{\text{Electrolysis}} \underset{\text{Anode}}{RR + 2CO_2} + \underset{\text{Cathode}}{2NaOH + H_2}$$

This electrolysis has been reviewed [62]. The products of the reaction are dependent on the experimental conditions employed. To obtain the alkane in aqueous solution it is important to use a platinum (or iridium) anode, a high anode current density, an acidic medium at low temperature, and a high concentration of alkali carboxylate. Methanol with or without water has been utilized as a solvent also, and in this case the nature of the anode, the variation in current density, concentration, and temperature are not so critical. The side reactions produce alkenes, alcohols, and esters. Best yields of alkanes are obtained with straight-chain carboxylic acids containing six or more carbon atoms. Mixtures of two carboxylic acids give the expected one unsymmetrical and two symmetrical alkanes. α-Branched, α,β-unsaturated, and aromatic carboxylic acids react only with difficulty, if at all. The dibasic acids from malonic to sebacic do not give alkanes; however, half esters of these acids produce diesters successfully.

$$2\,C_2H_5OOC(CH_2)_nCOOK \xrightarrow{\text{Electrolysis}} C_2H_5OOC(CH_2)_{2n}COOC_2H_5$$

Although yields in the Kolbe reaction are ordinarily low, a recent procedure for the electrolysis of an acid ester leads to 68–74% yields (see Ex. b.1).

The reaction proceeds via free radicals:

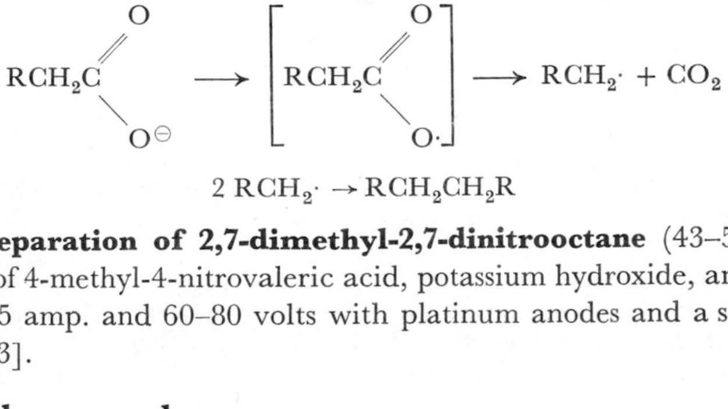

$$RCH_2C\overset{O}{\underset{O^\ominus}{\Big\backslash}}\ \longrightarrow\ \left[RCH_2C\overset{O}{\underset{O\cdot}{\Big\backslash}}\right]\ \longrightarrow\ RCH_2\cdot + CO_2$$

$$2\,RCH_2\cdot \rightarrow RCH_2CH_2R$$

(a) **Preparation of 2,7-dimethyl-2,7-dinitrooctane** (43–56% by the electrolysis of 4-methyl-4-nitrovaleric acid, potassium hydroxide, and methanol at about 3–5 amp. and 60–80 volts with platinum anodes and a stainless steel cathode) [63].

(b) **Other examples**

(1) **Dimethyl octadecanedioate** (68–74% by the electrolysis of methyl hydrogen sebecate in sodium methoxide-methanol solution with the use of platinum electrodes and a current flow of 1–2 amp. for 30–40 hr.) [64].

(2) *cis-?*-2,4-Dicarbomethoxybicyclobutane (13% by the electrolysis

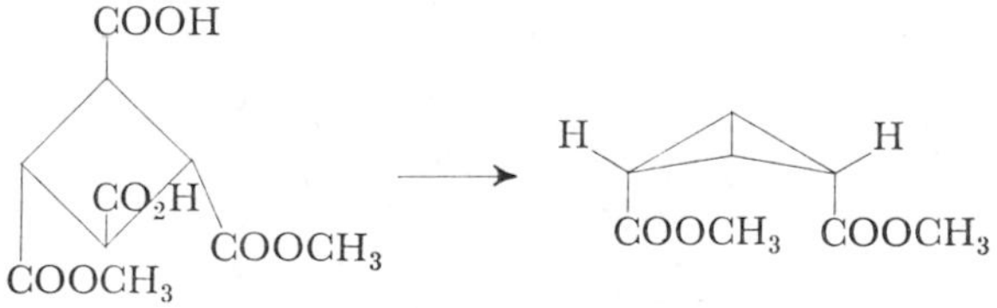

(80 volts, 0.8 amp.) of *trans,trans,trans*-1,3-dicarboxy-2,4-dicarbomethoxy-cyclobutane in anhydrous methanol containing a trace of sodium methoxide between Pt electrodes for 4 hr.; first example of ring closure by the Kolbe reaction) [65].

1. W. E. Bachmann and R. A. Hoffman, *Org. Reactions*, **2**, 224 (1944).
2. M. Gomberg and J. C. Pernert, *J. Am. Chem. Soc.*, **48**, 1372 (1926).
3. J. Elks *et al.*, *J. Chem. Soc.*, 1284 (1940).
4. R. A. Abramovitch and A. Robson, *J. Chem. Soc.*, (C) 1101 (1967).
5. H. H. Hodgson and E. Marsden, *J. Chem. Soc.*, 208 (1940); D. H. Hey and S. E. Lawton, *J. Chem. Soc.*, 374 (1940).
6. H. France *et al.*, *J. Chem. Soc.*, 1364 (1938); 1288 (1939).
7. W. E. Bachmann and R. A. Hoffman, *Org. Reactions*, **2**, 236 (1944); J. W. Haworth *et al.*, *J. Chem. Soc.*, 349 (1940).
8. C. E. Kaslow and R. M. Summers, *Org. Syn.*, Coll. Vol. **4**, 718 (1963).
9. C. L. Bumgardner *et al.*, *J. Am. Chem. Soc.*, **85**, 97 (1963).
10. I. Puskas and E. K. Fields, *J. Org. Chem.*, **33**, 4237 (1968).
11. H. Zollinger, *Azo and Diazo Chemistry*, Interscience Publishers, New York, 1961.
12. D. R. Augood and G. H. Williams, *Chem. Rev.*, **57**, 123 (1957).
13. D. F. DeTar, *Org. Reactions*, **9**, 409 (1957).
14. J. I. G. Cadogan, *J. Chem. Soc.*, 4257 (1962).
15. M. Gomberg and J. C. Pernert, *Org. Reactions*, **2**, 247 (1944).
16. H. France *et al.*, *Org. Reactions*, **2**, 252 (1944).
17. S. C. Dickerman *et al.*, *J. Org. Chem.*, **29**, 26 (1964).
18. D. H. Hey *et al.*, *Chem. Ind.* (*London*), 83 (1963).
19. D. H. Hey *et al.*, *J. Chem. Soc.*, (C) 1153 (1967).
20. G. B. Gill and G. H. Williams, *J. Chem. Soc.*, 7127 (1965).
21. P. R. Story *et al.*, *J. Am. Chem. Soc.*, **90**, 817 (1968).
22. W. Wolf and N. Kharasch, *J. Org. Chem.*, **30**, 2493 (1965).
23. N. Kharasch, *Photochemical Arylations*, Intra-Science Research Foundation, P.O. Box 430, Santa Monica, Calif., 90406, 1967; R. K. Sharma and N. Kharasch, *Angew. Chem. Intern Ed. Engl.*, **7**, 36 (1968).
24. N. Kharasch *et al.*, *Chem. Commun.*, 242 (1965).
25. C. Walling and M. J. Gibian, *J. Am. Chem. Soc.*, **86**, 3902 (1964).
26. J. H. McCracken *et al.*, U.S. Patent 3,384,658, May 21, 1968; *C.A.*, **69**, 7177 (1968).
27. E. K. Fields and S. Meyerson, *J. Am. Chem. Soc.*, **89**, 724 (1967).
28. P. E. Eaton, *Accounts Chem. Research*, **1**, 50 (1968).
29. E. J. Corey *et al.*, *J. Am. Chem. Soc.*, **86**, 5570 (1964).
30. R. O. Kan., *Organic Photochemistry*, McGraw-Hill Book Co., New York, 1966.
31. H.-D. Scharf and F. Korte, *Chem. Ber.*, **97**, 2425 (1964).
32. H.-D. Scharf and F. Korte, *Tetrahedron Letters*, 821 (1963).
33. K. Dziewoński and C. Paschalski, *Chem. Ber.* **46**, 1986 (1913).
34. W. H. F. Sasse *et al.*, *Tetrahedron Letters*, 3373 (1965).
35. E. E. Van Tamelen and S. P. Pappas, *J. Am. Chem. Soc.*, **84**, 3789 (1962); H. G. Viehe, *Angew. Chem.*, **77**, 768 (1965).
36. K. E. Wilzbach and L. Kaplan, *J. Am. Chem. Soc.*, **87**, 4004 (1965).
37. R. Srinivasan, *J. Am. Chem. Soc.*, **85**, 3048 (1963).
38. Ref. 30, p. 74.
39. M. Julia, *Record Chem. Progr.* (*Kresge-Hooker Sci. Lib.*), **25**, 3 (1964).
40. G. Quinkert, *Pure Appl. Chem.*, **9**, 607 (1964); G. Quinkert *et al.*, *Tetrahedron Letters*, 1863 (1963).
41. W. A. Bonner and F. D. Mango, *J. Org. Chem.*, **29**, 29 (1964); P. de Mayo and S. T. Reid, *Quart Rev.* (*London*), **15**, 393 (1961).
42. L. H. Slaugh, *J. Am. Chem. Soc.*, **81**, 2262 (1959).
43. J. O. Hawthorne and M. H. Wilt, *J. Org. Chem.*, **25**, 2215 (1960).
44. J. Tsuji and K. Ohno, *Tetrahedron Letters*, 3969 (1965).

45. J. Hine, *Divalent Carbon*, The Ronald Press Co., New York, 1964.
46. W. Kirmse, *Angew. Chem., Intern. Ed. Engl.*, **4**, 1 (1965).
47. D. Seyferth and M. A. Eisert, *J. Am. Chem. Soc.*, **86**, 121 (1964).
48. P. A. S. Smith, *The Chemistry of Open-Chain Organic Nitrogen Compounds*, W. A. Benjamin, New York, 1966, Vol. **2**, p. 187.
49. H. E. Simmons *et al.*, *J. Am. Chem. Soc.*, **86**, 1337, 1347 (1964).
50. J. Furukama *et al.*, *Tetrahedron Letters*, 3353 (1966).
51. R. J. Wineman *et al.*, *J. Org. Chem.*, **26**, 3122 (1961).
52. H. E. Simmons *et al.*, *J. Am. Chem. Soc.*, **81**, 4256 (1959).
53. H. E. Simmons *et al.*, *J. Am. Chem. Soc.*, **86**, 1347 (1964).
54. I. Dvoretzky *et al.*, *J. Am. Chem. Soc.*, **83**, 1934 (1961).
55. H. M. Frey and I. D. R. Stevens, *Proc. Chem. Soc.*, 79 (1962).
56. E. LeGoff, *J. Org. Chem.*, **29**, 2048 (1964).
57. R. D. Smith and H. E. Simmons, *Org. Syn.*, **41**, 72 (1961).
58. I. Dvoretzky *et al.*, *J. Am. Chem. Soc.*, **87**, 2763 (1965).
59. W. von E. Doering and L. H. Knox, *J. Am. Chem. Soc.*, **75**, 297 (1953).
60. L. Friedman and H. Shechter; Ref. 45, p. 111.
61. W. R. Moore and H. R. Ward, *J. Org. Chem.*, **25**, 2073 (1960).
62. B. C. L. Weedon, *Quart. Rev.* (*London*), **6**, 380 (1952); G. W. Thiessen, *Record Chem. Progr.* (*Kresge-Hooker Sci. Lib.*), **21**, 243 (1960).
63. W. H. Sharkey and C. M. Lang-Kammerer, *Org. Syn.*, **41**, 24 (1961).
64. S. Swann, Jr., and W. E. Garrison, Jr., *Org. Syn.*, **41**, 33 (1961).
65. A. F. Vellturo and G. W. Griffin, *J. Am. Chem. Soc.*, **87**, 3021 (1965).

H. Cycloaddition

The most characteristic products of the thermal polymerization of acetylene in the temperature range 400–600° are carbon and benzene, the latter being produced in yields of 20–90% depending on the flow rate and the contact surface [1]. The reaction is hazardous and is more amenable to industrial than to laboratory manipulation. On the other hand, the liquid-phase trimerization of substituted acetylenes with the appropriate catalyst is not difficult in the laboratory. For example, hexalkylbenzenes are produced from dialkylacetylenes in the presence of a Ziegler catalyst [2], and hexaphenylbenzene is produced in 80–85% yield from tolane ($C_6H_5C{\equiv}CC_6H_5$) by the use of palladium chloride-benzonitrile ($C_6H_5CN)_2PdCl_2$ complex [3]. Similarly triphenylbenzene in 80% yield is obtained from phenylacetylene in the presence of a titanium chloride-diethylaluminum chloride, $Al(C_2H_5)_2ClTiCl_4$, catalyst [4]. Acetylene can also be tetramized to cyclooctatetraene in about 70% yield by the use of nickel halide or cyanide catalysts in tetrahydrofuran under anhydrous conditions [5].

A number of organometallic compounds are capable of cyclizing nonterminal acetylenes to benzenoid structures [6]. The most important of them appears to be a nickel carbonyl, $Ni(CO)_2[(C_6H_5)_3P]_2$, which is capable of trimerizing almost any acetylene provided no bulky group is present [7]. Yields vary greatly, however. Other catalysts are $Co_2(CO)_8$ or $[Co(CO)_4]_2Hg$ in dioxane, the latter yielding 95% hexa-(p-chlorophenyl)-benzene from 1,2-di-(4-chlorophenyl)-acetylene. Diarylcobalt behaves like a true catalyst in that a small amount brings about a high conversion of 2-butyne into hexamethylbenzene [6]. Triphenylchromium in THF, on the other hand, converts 2-butyne into 1,2,3,4-tetramethylnaphthalene.

References for Section H are on p. 68.

Although the Diels-Alder reaction is discussed in Alkenes, Chapter 2, C.2, Ex. b.2 and b.3 which utilize benzyne intermediates are shown here.

(a) Preparation of hexaphenylbenzene (84%, m.p. 454–456°, by refluxing tetraphenylcyclopentadienone and tolane in benzophenone at 300° until color has lightened and evolution of carbon monoxide has ceased) [8].

(b) Other examples

(1) **1,2,3,4-Tetraphenylnaphthalene.** (82–90% from diphenyliodonium-2-carboxylate

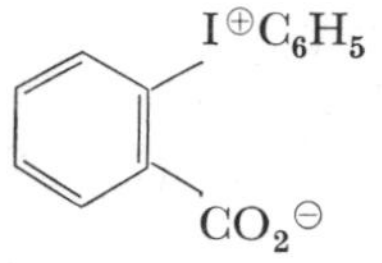

and tetraphenylcyclopentadienone in refluxing diethylbenzene) [9].

(2) **Triptycene** (59% from anthranilic acid added slowly to a refluxing solution of amyl nitrite and anthracene in methylene chloride. The excess anthracene was removed by forming its maleic anhydride adduct) [10].

(3) **1,2,3,4,5,8-Hexamethylnaphthalene** [11].

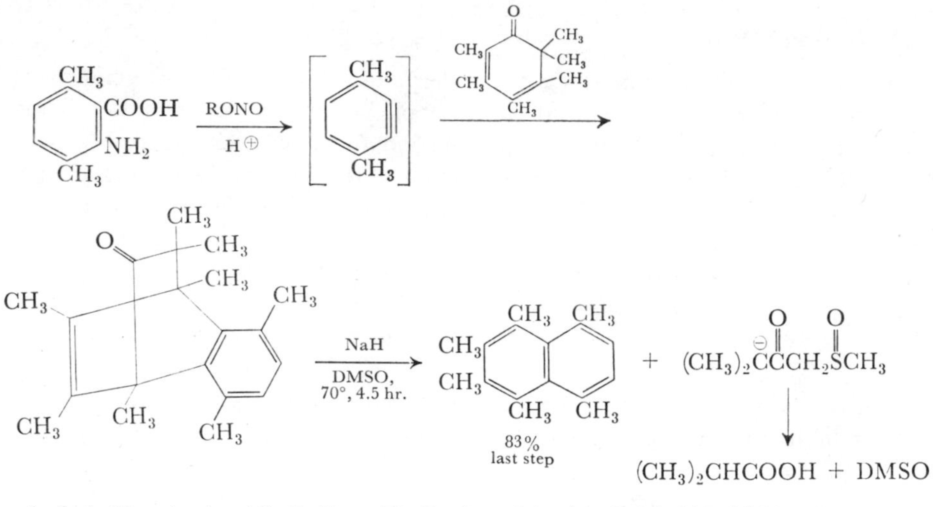

1. J. A. Nieuwland and R. R. Vogt, *The Chemistry of Acetylene*, Reinhold Publishing Corp., New York, 1945, Chap. 5.
2. H. Hopff and A. Gati, *Helv. Chim. Acta*, **48,** 509 (1965).
3. A. T. Blomquist and P. M. Maitlis, *J. Am. Chem. Soc.*, **84,** 2329 (1962).
4. A. Furlani *et al.*, *J. Polymer Sci.*, Part B, **5,** 523 (1967).
5. W. Reppe, *Acetylene Chemistry*, Charles A. Meyer & Co., 25 Vanderbilt Avenue, New York, 17, 1949, pp. 133, 182.
6. H. Zeiss, *Organometallic Chemistry*, Reinhold Publishing Corp., New York, 1960, p. 411.
7. V. O. Reikhsfel'd and K. L. Makovetskii, *Russ. Chem. Rev.*, **35,** 510 (1966).
8. L. F. Fieser, *Org. Syn.*, **46,** 44 (1966).
9. L. F. Fieser and M. J. Haddadin, *Org. Syn.*, **46,** 107 (1966).
10. L. Friedman and F. M. Logullo, *J. Am. Chem. Soc.*, **85,** 1549 (1963).
11. H. Hart *et al.*, *J. Am. Chem. Soc.*, **89,** 4554 (1967).

Chapter 2

ALKENES, CYCLOALKENES, AND DIENES

In this chapter will be found the principal methods of synthesizing compounds containing the double bond. Typical preparations are included as well as recent innovations and methods of a more limited scope.

For dehydration, the potassium acid sulfate method appears to be the most general and safest; however, vapor-phase dehydration over thorium oxide possesses interesting orientation control (see A.1). For dehydrohalogenation, alkoxide in nonpolar solvents offers much flexibility, but DBN, 1,5-diazabicyclo-[4.3.0]-non-5-ene, may soon challenge the position of the alkoxides (see A.2). The reduction of acetylenes is by far the most common method of preparing *cis*-olefins (see B.1). The preparation of *trans* isomers is less difficult because they are frequently the principal products of most eliminations and of other reactions, but the dehaloalkoxylation of α-alkyl-β-chlorotetrahydropyrans is a reliable method for preparing a *trans*-olefin as an intermediate for further transformations (see A.4). The phosphorane method is another procedure to consider for the preparation of either *cis*- or *trans*-olefins (see E.2). When the possibility exists of forming either 1- or 2-olefins from a single starting material, many procedures are available to control the orientation (see A.1, A.3, A.8, and particularly A.16). To synthesize nonterminal olefins, the old, reliable procedure of Boord still seems to be the best (see A.4). The greatest ferment in the development of new procedures for olefin synthesis is taking place in the Ziegler-type catalytic methods (see C.1) and in photochemical reactions (see

C.3). There is no doubt that these sections will be out of date by the time of publication. Many other clever methods of more limited scope are to be found in this chapter.

A word should be said about the olefins themselves. Unadorned olefins, i.e., those without other functional groups, are reactive substances. When it is considered that most of them are liquids, also, care should be taken in their isolation, purification, and storage. We have some skepticism about the purity of some of the olefins described in this chapter.

See the General References below; they deal with olefin preparation.

General References

F. Asinger, *Olefins*. Pergamon Press, New York, 1966.
W. H. Saunders, Jr., in S. Patai, ed., *The Chemistry of Alkenes*, Interscience Publishers, New York, 1964, Chap. 2.

A. Elimination

Elimination to produce olefins takes many forms not only in the species of molecule eliminated, but also in the variety of mechanisms by which the molecules are eliminated. This section is divided on the basis of the type of molecule eliminated, and the mechanism is discussed at appropriate places. A good general reference on mechanisms of elimination is that of Banthorpe [1].

1. From Alcohols (Dehydration)

$$RCH_2CHOHR \rightarrow RCH{=}CHR + H_2O$$

A great variety of dehydrating agents has been employed in the widely used synthesis of alkenes from alcohols. Among the most common are acids, such as sulfuric [2, 3], anhydrous or aqueous oxalic [4], or phosphoric [5, 6], acidic oxides such as phosphorus pentoxide [7, 8], bases such as potassium hydroxide [8–10], and salts such as sodium or potassium acid sulfate [8, 11], and iodine [12], dimethyl sulfoxide [13], phenyl isocyanate [14], N-bromosuccinimide in pyridine [15] or phosphorus oxychloride or thionyl chloride [16]. It is interesting to note that of the two chlorides the thionyl chloride is the more powerful, while the phosphorus oxychloride is the more specific. Of all the methods listed above, the potassium hydrogen sulfate dehydration is perhaps the most generally used, particularly in the preparation of styrenes [17]. The dehydration is

$$ArH \xrightarrow{\quad\quad} ArCOCH_3 \xrightarrow{\quad\quad} ArCHOHCH_3 \xrightarrow{KHSO_4} ArCH{=}CH_2$$

carried out usually under vacuum to remove the olefin as formed, but extraction of the reaction mixture can be performed if the olefin is relatively nonvolatile (see Ex. a).

Many alcohols may be dehydrated satisfactorily by passing their vapors over alumina at 350–400°. With 2-octanol, for example, the main product is 2-octene, the product anticipated by the Saytzeff rule (see A.2). Surprisingly, the

References for Section A are on pp. 102–106.

rare earth catalysts, such as thorium oxide, give the product expected by the Hofmann rule [18].

$$2\text{-Octanol} \xrightarrow[\text{ThO}_2]{350\text{--}400°, \ 0.1\text{--}0.5 \ \text{sec.}} \underset{95\text{--}97\%}{1\text{-octene}} + \underset{3\text{--}5\%}{2\text{-octene}}$$

Among the monohydric alcohols the ease of dehydration varies and is in the order: tertiary > secondary > primary. Thus, if sulfuric acid is used as the dehydrating agent, a lower concentration and/or a lower temperature would be required for tertiary than for secondary alcohols, and similarly milder conditions would be required for secondary as compared to primary alcohols. In fact, under the proper conditions it is possible to dehydrate only the alcoholic function which is more susceptible to attack:

$$(CH_3)_2\overset{\displaystyle |}{\underset{\displaystyle OH}{C}}CH_2CH_2CH_2OH \xrightarrow[-H_2O]{\overset{\oplus}{H}} (CH_3)_2C{=}CHCH_2CH_2OH$$

The formation of an alkene from a primary alcohol with the use of sulfuric acid probably involves SN_2 and E_2 reactions [19]. The SN_2 step leads to the oxonium ion 2-I which in the E_2 step gives the alkene 2-II.

$$CH_3CH_2OH + H_2SO_4 \rightarrow \underset{\text{2-I}}{CH_3CH_2\overset{\oplus}{O}H_2} + \overset{\ominus}{H}SO_4$$

$$\overset{\ominus}{H}SO_4 + \underset{\underset{\displaystyle H}{|}}{CH_2}{-}CH_2{-}\overset{\oplus}{O}H_2 \rightarrow \underset{\text{2-II}}{CH_2{=}CH_2} + H_2O + H_2SO_4$$

For tertiary alcohols and also to some extent for secondary alcohols, SN_1 and E_1 reactions occur. Thus for tertiary butyl alcohol the first-formed oxonium ion 2-III by the loss of water in an SN_1 step gives the carbonium ion 2-IV, which by an E_1 step yields the alkene 2-V.

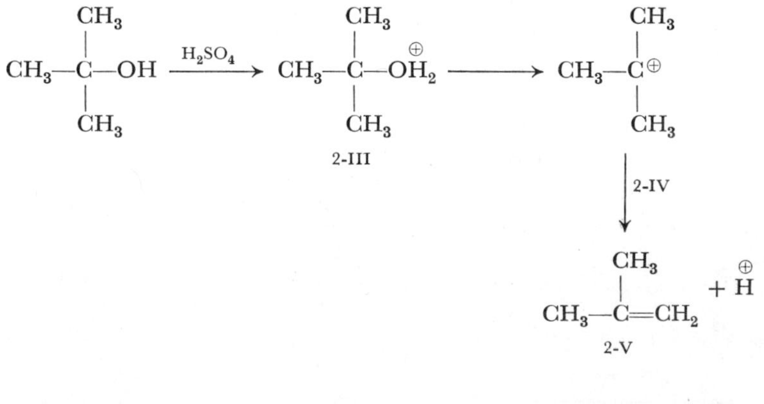

$$\underset{\underset{\displaystyle CH_3}{|}}{\overset{\overset{\displaystyle CH_3}{|}}{CH_3{-}C{-}OH}} \xrightarrow{H_2SO_4} \underset{\underset{\displaystyle CH_3}{|}}{\overset{\overset{\displaystyle CH_3}{|}}{CH_3{-}C{-}\overset{\oplus}{O}H_2}} \longrightarrow \underset{\underset{\displaystyle CH_3}{|}}{\overset{\overset{\displaystyle CH_3}{|}}{CH_3{-}\overset{\oplus}{C}}}$$

2-III

2-IV ↓

$$\underset{\text{2-V}}{\overset{\overset{\displaystyle CH_3}{|}}{CH_3{-}C{=}CH_2}} + \overset{\oplus}{H}$$

Acid dehydration of alcohols to give specific olefins is unsatisfactory even with the mildest catalysts. Three examples are given to illustrate isomerization possibilities.

1. 1-Octanol and phosphoric acid yield mainly 2-octene, a small amount of 3-octene, and no 1-octene [20].

2. 3,3-Dimethyl-2-butanol (and any one of various types of acid dehydration reagents) yields mainly tetramethylethylene, 2,3-dimethyl-1-butene, and traces of *t*-butylethylene [21].

3. The assertion that selective dehydration of α-terpineol is possible by proper choice of acidic catalysts has been refuted [22] as shown:

| α-Terpineol | Terpinolene | Dipentene | α- and γ-Terpinene | $\Delta^{2,4(8)}$-*p*-Menthadiene |

With oxalic acid in water, a very mild dehydrating medium, the yield of terpinolene with time reaches 25 % and then decreases to 10 %, and of dipentene reaches 12 % and decreases to 5 %; the last three products become predominant in the latter phases of the dehydration. The cleanest dehydration that could be found was with pyrolytic potassium acid sulfate (fused until fumes were evolved) in which a 70 % yield of olefin, containing 75 % terpinolene, was obtained although in our experience pyrolytic potassium hydrogen sulfate is prone to give charred products. The situation is not completely discouraging, however, because: the thermodynamically stable product (in acid) may be desired; other means of decomposition (see following sections) are available; and limited isomerization methods are possible (see D).

Furthermore, the thoria-catalyzed dehydration under flow methods previously mentioned is a uniquely *cis* elimination which gives rise to 1-olefins from 2-alkanols. This process must involve specific chemisorption of the alcohol on the thorium-oxygen lattice so that a concerted dehydration takes place, the smaller alkyl group being better accommodated than the larger alkyl group in the crystal lattice (see Ex. c).

For dehydration of aldols and related compounds to alkenes see Alcohols, Chapter 4, G.1, and Aldehydes, Chapter 10, F.3. Similar dehydrations via the Knoevenagel and Doebner reactions are found in Carboxylic Acid Esters, Chapter 14, C.4, and Carboxylic Acids, Chapter 13, D.2, respectively.

The diene, 2,3-dimethyl-1,3-butadiene, may be prepared from pinacol by the use of a variety of dehydrating agents such as hydrobromic acid, aluminum oxide, and calcium phosphate [23–25] in yields as high as 86 %. In a less direct manner, *trans*-1-phenyl-1,3-butadiene may be obtained in 72–75 % yield by the acid hydrolysis (30 % sulfuric) of the adduct formed between cinnamaldehyde and methylmagnesium bromide [26]. (A trace of phenyl-β-naphthylamine is added to inhibit polymerization.)

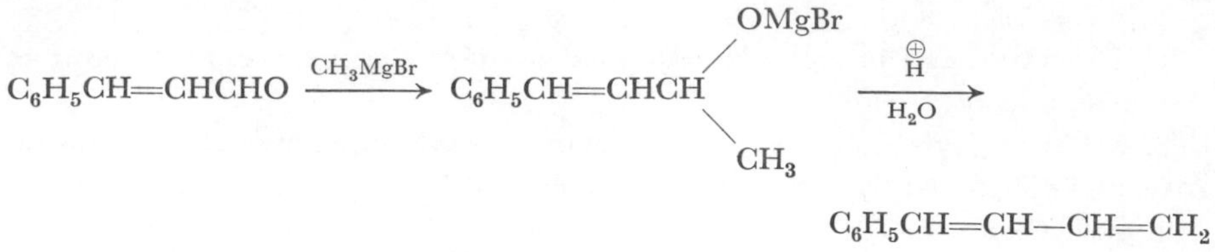

$$C_6H_5CH{=}CHCHO \xrightarrow{CH_3MgBr} C_6H_5CH{=}CHCH\begin{smallmatrix}OMgBr\\CH_3\end{smallmatrix} \xrightarrow[H_2O]{\overset{\oplus}{H}}$$

$$C_6H_5CH{=}CH{-}CH{=}CH_2$$

(a) Preparation of 2-phenyl-4-isopropenylquinoline.   Freshly fused

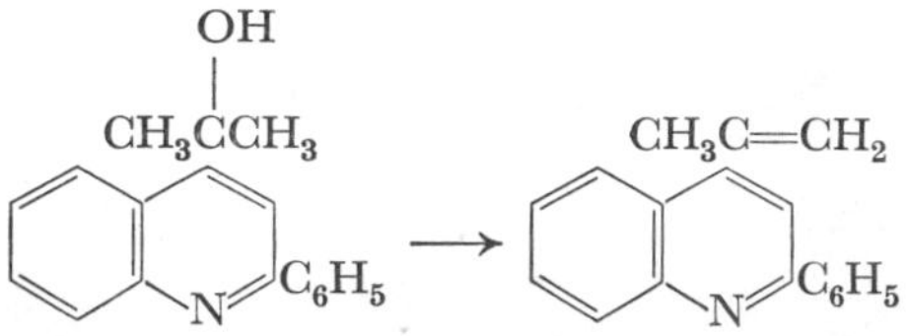

potassium hydrogen sulfate (38 g.) and 2-phenyl-4-(2-hydroxypropyl)-quinoline (15 g.) were mixed intimately and transferred rapidly to a flask which was kept at 180° for 3 hr. The cooled mixture was made basic with aqueous sodium hydroxide and extracted several times with a total volume of 400 ml. of ether. The ether extract was washed with water, dried, concentrated, to give 12 g., 86 % of a heavy oil which distilled at 167° (0.13 mm.). (Secondary alcohols of the structure shown above do not dehydrate so easily but tend to yield ethers [27].)

(b) Preparation of cyclohexene (79–84 % from cyclohexanol and 85 % phosphoric acid at 165–170°) [6; *see also* 28].

(c) Preparation of 4-methyl-1-pentene.   The dehydration was carried out in a Pyrex flow system consisting of a vaporizer, a catalyst bed, and a trap. The catalyst bed was filled with 50 g. of glass beads, 0.57 mm. diameter, coated with 3 g. of thorium oxide by wetting with 0.5 ml. of the alcohol to be dehydrated. 4-Methyl-2-pentanol (1 mole) was passed at 60 ml./hr. under 100 mm. vacuum over the catalyst bed held at 387°. The condensate was collected, dried, and distilled to give an 87 % conversion to the titled olefin (97 % 1-olefin, 3 % 2-olefin) [29].

(d) Other examples

(1) Stilbene (94 % from 2,2-diphenylethanol and phosphorus tribromide in benzene overnight; rearrangement also occurs with other acidic reagents) [30].

(2) Tetraphenylethylene (83 % from 1,1,2,2-tetraphenylethanol and thionyl chloride in chloroform for 1 hr.) [31].

(3) 2-Pentene (65–80 % by distilling a mixture of equal volumes of concentrated sulfuric acid and water while pentanol-2 was slowly added) [3].

(4) 2,3-Diphenyl-1,2-dihydronaphthalene

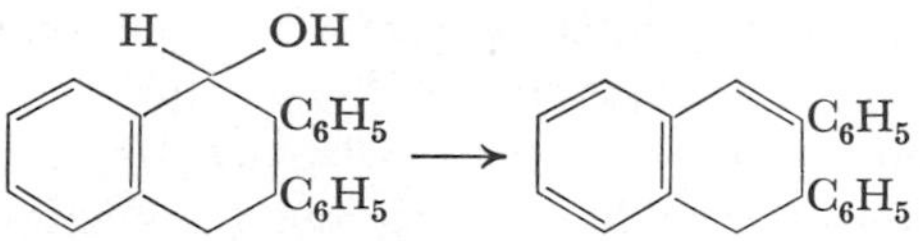

(96% from 1-hydroxy-2,3-diphenyl-1,2,3,4-tetrahydronaphthalene by refluxing with Lucas reagent for 2 hr.) [32].

(5) **2,3-Dimethyl-1,3-butadiene** (79–86% from pinacol by passing the vapor over activated alumina at 420–470° under reduced pressure [33].

2. From Alkyl Halides (Dehydrohalogenation)

$$RCH_2CHXR' \xrightarrow[C_2H_5OH]{KOH} RCH{=}CHR'$$

Dehydrohalogenation is a complex reaction because of problems of orientation, *cis* vs. *trans* elimination, and competing side reactions. For synthesis the process is carried out mainly via an E_2 elimination. In this transition complex

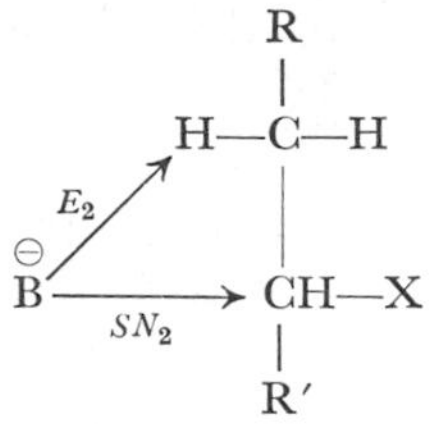

it is noted that the nucleophile, $\overset{\ominus}{B}$, can remove the β-proton to produce elimination or it can attack the α-carbon to give the SN_2 product. α-Elimination of a proton can also take place:

$$CH_3CH_2CH_2CH_2Cl + CH_3ONa \rightarrow CH_3CH_2CH_2CH\text{:} \rightarrow CH_3CH_2CH{=}CH_2$$

With sodium methoxide, 10% α-elimination and 90% β-elimination occurs as shown by deuterium labeling [34]. With stronger bases such as phenylsodium, α-elimination occurs to the extent of 94%. Normally α-elimination leads to the same product as β-elimination, but the former can lead to cyclopropane formation in the case of branched-chain alkyl halides.

Elimination and substitution usually accompany each other. However, a nucleophile, $\overset{\ominus}{B}$, of strong basic character, steric factors at the α-carbon atom or in the nucleophile, electron-withdrawing groups in R, and high temperature usually favor elimination. Differences in the halide-leaving group, with the exception of the fluoride, have little effect on the elimination/solvolysis product ratio, but the tosylate-leaving group under SN_2 conditions gives only traces of the elimination product [35].

$$n\text{-}C_{18}H_{37}Br \xrightarrow[(CH_3)_3COH, 40°]{(CH_3)_3COK} \text{olefin} + \text{ether}$$
$$ 85\% \qquad 12\%$$

$$n\text{-}C_{18}H_{37}OTs \xrightarrow[(CH_3)_3COH, 40°]{(CH_3)_3COK} \text{olefin} + \text{ether}$$
$$ 1\% \qquad 99\%$$

β-Phenylethyl tosylate ($C_6H_5CH_2CH_2OTs$), on the other hand, yields predominantly the elimination product because of the electron-withdrawing nature of the phenyl group and the conjugation of the olefinic and phenyl groups in the product. Indeed, the formation of styrenes by dehydrohalogenation is a facile process brought about by numerous strong and weak bases [17].

When relatively simple isomeric olefins can be formed, dehydrohalogenation under normal conditions (alcoholic bases) gives the Saytzeff olefin in predominant amounts [36]:

$$2\text{-}C_5H_{11}Br + KOC_2H_5 \rightarrow CH_3CH_2CH{=}CHCH_3 + CH_3CH_2CH_2CH{=}CH_2$$

Saytzeff olefin,   Hofmann olefin,
18% *cis*, 51% *trans*   31%

But steric effects in the alkyl group, the attacking nucleophile, or the leaving group will increase the tendency toward Hofmann elimination. For example, in the reaction of the tertiary halide

$$\begin{array}{c} CH_3 \\ | \\ RCH_2CBr \\ | \\ CH_3 \end{array}$$

with potassium ethoxide the 1-olefin yield varies from 30% when $R = CH_3$ to 86% when $R = t\text{-}C_4H_9$ [37]; in the reaction of t-pentyl bromide with the potassium alkoxide, KOR, the 1-olefin yields are as follows: when $R = C_2H_5$, 30%; $t\text{-}C_4H_9$, 72.5%; $t\text{-}C_5H_{11}$, 77.5%; $t\text{-}C_7H_{15}$, 89% [37].

It has been known for a long time that onium ions as leaving groups in E_2 reactions give a preponderance of the Hofmann olefin. For example, $t\text{-}C_5H_{11}\overset{\oplus}{N}(CH_3)_3$ with alcoholic potassium hydroxide gives an 84% yield of olefin, 93% of which is 2-methyl-1-butene [37]. On the other hand, it has been discovered recently that the fluoride ion as a leaving group gives considerable 1-olefin [38]. Saunders attributes the increasing amount of 1-olefin with decreasing size of the halogen to greater carbanion character of the transition state associated with the fluoride ion, but Brown and Klimisch [39] in a more comprehensive study attribute the cause to increased solvation as the size of the halogen decreases, an effect which introduces a steric factor into the leaving

$$2\text{-}C_5H_{11}X \xrightarrow[C_2H_5OH]{NaOC_2H_5} CH_3CH_2CH_2CH{=}CH_2$$

X = F,   82%
X = Cl,   35%
X = Br,   25%
X = I,   20%

group. Both explanations may be correct, the direction of orientation depending not only on steric effects in the leaving group and the nucleophile, but also on increasing carbanion character of the transition state with more electron-attracting leaving groups and with stronger bases. All these factors favor Hofmann-type elimination [40].

From the dehydrohalogenation of cyclic compounds it is apparent that *trans* elimination of the halide and hydrogen atoms (see transition representation at the beginning of A.2) is a dominating factor which takes precedence over Saytzeff orientation [41]. But *trans* elimination of the halide and hydrogen atoms does not mean that a *trans*-olefin is obtained. Indeed, dehydrohalogenation usually gives mixtures of *cis*- and *trans*-olefins; see preceding example, [36]. And the literature [42] suggests that no thermodynamic equilibrium in strongly basic solution favors exclusive formation of either the *cis*- or *trans*-olefin:

$$\text{1-Pentene} \xrightarrow[\text{DMSO,55°}]{t\text{-C}_4\text{H}_9\text{OK}} \underset{60.2\%}{\text{trans-2-pentene}} + \underset{36.9\%}{\text{cis-2-pentene}} + \underset{3.1\%}{\text{1-pentene}}$$

Thus one must turn ordinarily to other methods of olefin formation to obtain pure *cis* or *trans* isomers. However, an interesting and exceptional example of orientation control by choice of base is known [43]:

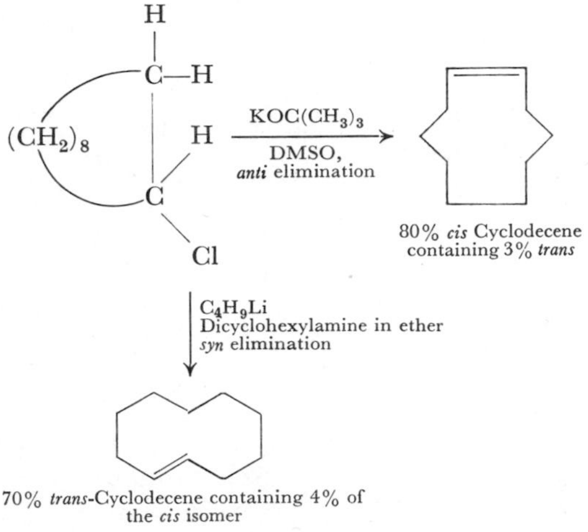

80% *cis* Cyclodecene containing 3% *trans*

70% *trans*-Cyclodecene containing 4% of the *cis* isomer

In dimethyl sulfoxide the base, being well-dissociated, favors the normal path of *anti* elimination. In ether the lithium base, behaving more like an ion pair, favors *syn* elimination. Other examples show the same trend [44]. Bond angles dictate that only *cis* forms of cycloolefins from C_3 to C_7 are isolable, that *cis* forms from C_8 to C_{11} are more stable thermodynamically than *trans* forms, and that *trans* are more stable than *cis* forms in C_{12} and higher cycloolefins. In the formation of acyclic olefins stereospecific dehydrohalogenation is not easily attained. On the assumption that linear (*trans*) dehydrohalogenation is the lowest energetic state of transformation and with the provision that conformation in the transition state is restricted, it is possible to obtain a *trans*-olefin as illustrated.

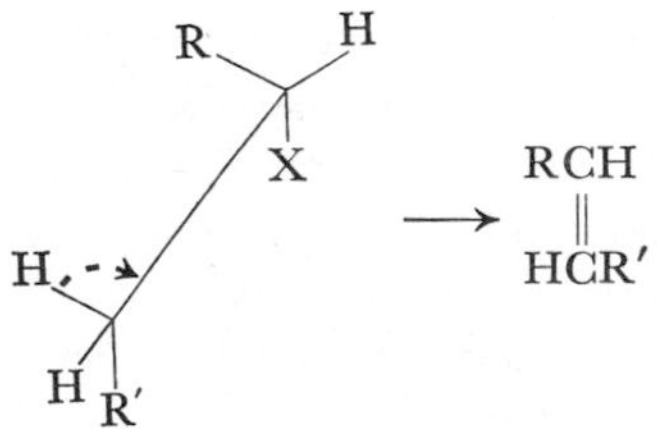

The greater the steric repulsion between R and R′, the greater is the amount of *trans*-isomer without encroachment of the product from other conformations as in

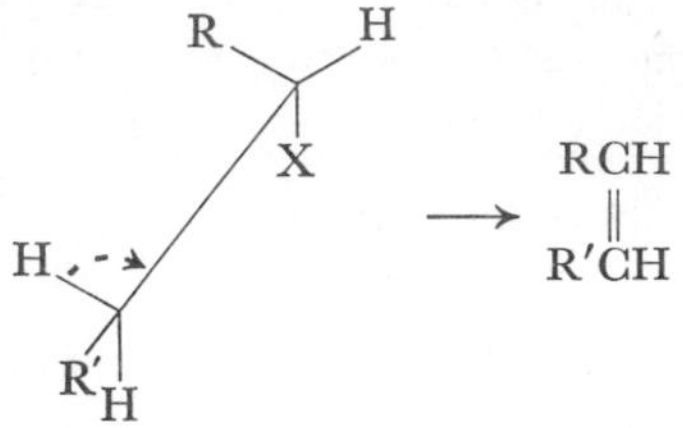

$$\text{(structure)} \longrightarrow \begin{array}{c} \text{RCH} \\ \| \\ \text{R'CH} \end{array}$$

Although the mechanism is not so simple as the above, a remarkable example of the application of the preceding principles is shown where R is represented by a cyclopropyl and R′ by a butenyl group [45]:

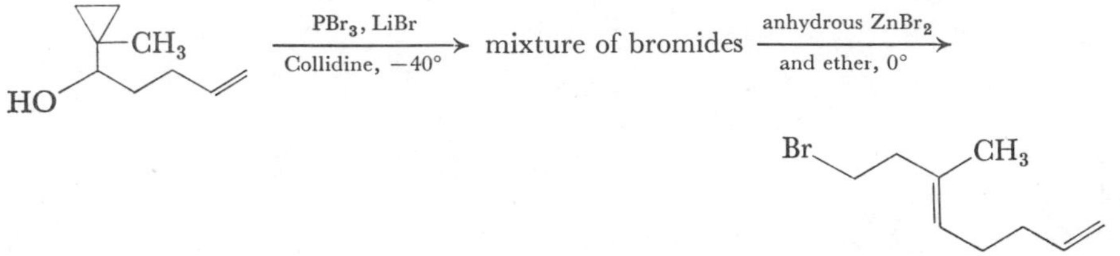

$$\text{(cyclopropyl structure with —CH}_3\text{, HO)} \xrightarrow[\text{Collidine, } -40°]{\text{PBr}_3, \text{LiBr}} \text{mixture of bromides} \xrightarrow[\text{and ether, } 0°]{\text{anhydrous ZnBr}_2}$$

trans-1-Bromo-3-methyl-3,7-octadiene, 71% overall, only a trace of *cis*

Although alcoholic potassium hydroxide is the most common reagent, dimethylaniline [46], dimethylformamide [47], triethylamine [48], aniline, [49], pyridine [50], collidine [51], quinoline [52], sodium acetate [53], and silver oxide [54] have also been used. Recently a new organic base, 1,5-diazabicyclo-[3.4.0]nonene-5, DBN, has been claimed to be the most versatile dehydrohalogenating agent known [55, 62]. DBN is made readily and is available commercially. It is a strong proton abstractor and its cation probably forms an

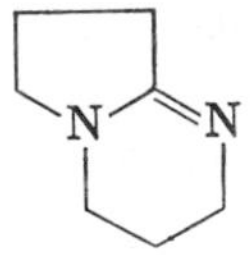

unusually stable ion pair with the leaving halide group. Manipulation is very simple, but its use is too new to recommend it unreservedly. In fact at least in a limited number of cases 1,5-diazabicyclo(4.5.0)undecene-5 is more effective than DBN (Ex. b). Potassium *t*-butoxide, in addition to its advantages already mentioned, is the most satisfactory reagent for the dehydrobromination of α-bromo acids to Δ²-alkenoic acids [56]; see Ex. c.8). Lithium bromide-lithium carbonate in dimethylformamide [57] may be used particularly in the steroid series where it usually offers an advantage over the lithium bromide alone [58] as well as over collidine [59]. For unreactive chlorides, such as bornyl chloride and 2,6-dichlorocamphane, sodium 2-*n*-butylcyclohexoxide is a superior dehydrohalogenation reagent [60].

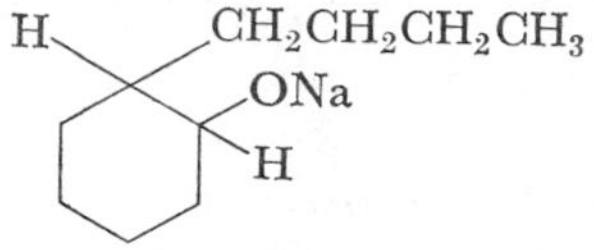

$$\text{H} \quad \text{CH}_2\text{CH}_2\text{CH}_2\text{CH}_3$$
$$\text{ONa}$$
$$\text{H}$$

Applied to allylic halides, the base-catalyzed elimination reaction gives 1,3-dienes. For example, 3-chlorocyclohexene with dimethylaniline gives cyclohexadiene (80%) [46], while 1-p-nitrophenyl-4-chloro-2-butene with methanolic potassium hydroxide yields 1-(p-nitrophenyl)-1,3-butadiene (57–62% based on the p-nitroaniline employed as a starting material) [61]; see Ex. c.4).

(a) **Preparation of octadecene-1** (85% by refluxing n-octadecyl bromide with 1 N solution of potassium t-butoxide in anhydrous t-butyl alcohol at 80° for 20 hr.) [35].

(b) **Preparation of 3-heptene.** Equimolecular quantities of 4-bromoheptane and DBN (1,5-diazabicyclo[3.4.0]non-5-ene) were warmed to 80–90° for a short while and the olefin was removed by distillation in 60% yield. Similarly 1,5-diazabicyclo[4.5.0]undecene-5 gave a 91% yield. Yields on other alkyl halides ranged from 78 to 91%. The reaction can be run in dimethyl sulfoxide solvent also [62].

(c) **Other examples**

(1) **1-(2-Naphthyl)-butene-1** (81% based on 2-n-butylnaphthalene which was first converted into 1-bromo-1-(2-naphthyl)-butane, the latter being dehydrobrominated by heating with quinoline at 150–160°), [63].

(2) **1-Cyclohexylcyclohexene** (90.5% from 1-bromo-1-cyclohexyl-cyclohexane refluxed 0.5 hr. with aniline in toluene) [49].

(3) **2-Vinylthiophene** (50–55% from thiophene, paraldehyde, and

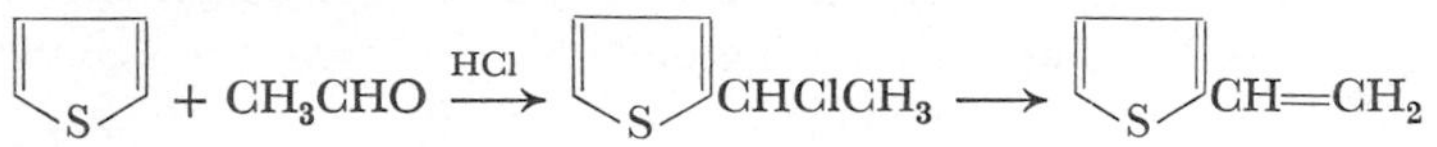

$$\text{S} + \text{CH}_3\text{CHO} \xrightarrow{\text{HCl}} \text{S}\text{-CHClCH}_3 \longrightarrow \text{S}\text{-CH}=\text{CH}_2$$

concentrated hydrochloric acid, in which mixture hydrogen chloride was introduced for 25 min at 10–13°; in the final step, pyridine and α-nitroso-β-naphthol (stabilizer) were added and distillation was accomplished under nitrogen at low pressure) [64].

(4) **1-(p-Nitrophenyl)-1,3-butadiene** (57–62%, based on the starting p-nitroaniline, from 1-(p-nitrophenyl)-4-chloro-2-butene and methanolic potassium hydroxide at 15–30° for about 30 min.) [61].

(5) **Methylcyclooctatetraene** (53% from 9,9-dibromobicyclo [6.1.0]-4-nonene

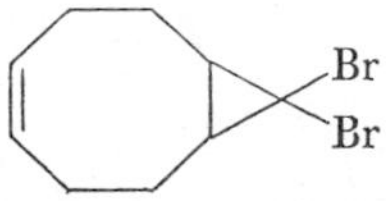

$$\text{Br}$$
$$\text{Br}$$

and potassium t-butoxide at 40–45°; this reaction demonstrates the use of the readily available dibromobicycloalkanes or -enes) [65].

(6) 1,3-Cyclohexadiene (54–55%, b.p. 80–83° by distillation from dropwise addition of 1,2-dibromocyclohexane to sodium isopropoxide in triethylene glycol dimethyl ether at 110°) [66].

(7) 2-Methyl-2-cyclohexenone (43–45% from 2-methyl-2-chlorocyclohexanone and lithium chloride in dimethylformide heated to 100°) [67]. α-Halo ketones also dehydrohalogenate spontaneously during the formation of the 2,4-dinitrophenylhydrazones [68].

(8) 2-Methylenedodecanoic acid (35–40% from 2-methyldode-

$$CH_3(CH_2)_9\overset{\overset{\displaystyle CH_3}{|}}{C}HCOOH \xrightarrow[PBr_3]{Br_2} CH_3(CH_2)_9\overset{\overset{\displaystyle CH_3}{|}}{\underset{\underset{\displaystyle Br}{|}}{C}}COBr \xrightarrow{(CH_3)_3COK}$$

$$CH_3(CH_2)_9\overset{\overset{\displaystyle CH_2}{||}}{C}COOC(CH_3)_3 \xrightarrow[2.\ H_2SO_4]{1.\ NaOH} CH_3(CH_2)_9\overset{\overset{\displaystyle CH_2}{||}}{C}COOH$$

canoic acid) [69].

3. From Dihalides (Dehalogenation)

$$RCHXCHXR \xrightarrow[95\%\ C_2H_5OH]{Zn} RCH{=}CHR$$

$$2\ R_2CX_2 \xrightarrow[Ether]{Mg} \overset{R}{\underset{R}{\diagdown}}C{=}C\overset{R}{\underset{R}{\diagup}}$$

(a) From 1,2-dihalides. This synthesis, a *trans* elimination, is of limited value since the 1,2-dihalides are often prepared from the alkene. It has found use, however, in the purification of alkenes by conversion into dihalides followed by dehalogenation. The dehalogenating agents which have been employed are zinc in methanol with a trace of zinc chloride [70] or in ethanol [71, 72], zinc and ether in the presence of acetic acid [73], sodium iodide in acetone [74], magnesium and magnesium iodide in dry ether [75], thiourea [76], trimethyl phosphite [77], triethyl phosphite [78], triphenylphosphine [79], sodium methyl sulfinylmethide [80], Cr(II) salts [81], and heat alone [82].

Heat applied to a mixture of quinoline and a dibromide with hydrogen atoms on adjacent carbon atoms gives a 1,3-diene [83]; see Ex. b.4.

$$C_6H_5CH_2CHBrCHBrCH_2C_6H_5 \rightarrow C_6H_5CH{=}CHCH{=}CHC_6H_5$$
trans-trans-1,4-Diphenylbutadiene

Chromium(II) salts in aqueous dimethylformamide [84] or complexed with ethylenediamine [81] are capable of converting *vicinal* dihalides to olefins by reductive elimination:

$$\underset{}{\overset{}{\bigcirc}}\overset{Br}{\underset{Br}{}} + 2\ Cr^{(2+)} \xrightarrow[ethylenediamine]{DMF\ or} \bigcirc\!\!| + 2\ (Cr^{(3+)},\ Br^{\ominus})$$

Cyclohexene, 99%

With Cr^{II} salts [81] or strong bases [85] as dehalogenation agents, the reaction apparently is stereospecific.

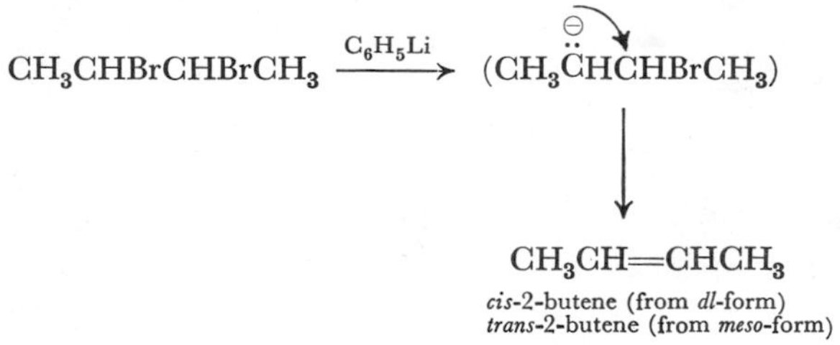

$$CH_3CHBrCHBrCH_3 \xrightarrow{C_6H_5Li} (CH_3\overset{..}{\overset{\ominus}{C}}HCHBrCH_3)$$

$$\downarrow$$

$$CH_3CH{=}CHCH_3$$

cis-2-butene (from dl-form)
trans-2-butene (from meso-form)

(b) From *gem* dihalides.   This elimination appears to involve a carbene as an intermediate [86], although other mechanisms are probably involved. Some of the reagents employed, such as sodium iodide in acetone or pyridine [87], magnesium in ether [88] (see Ex. b.5) and sodium methyl sulfinylmethide [86] are also capable of dehalogenating 1,2-dihalides satisfactorily. Others utilized for *gem*-dihalides are iron pentacarbonyl [89] and copper [90].

The *gem*-dihalocyclopropanes now so readily available from dihalocarbenes and olefins are convenient intermediates for increasing the chain length of the olefin by one carbon atom via a carbene (see Ex. b.5 and Alkanes, Chapter 1, G.6).

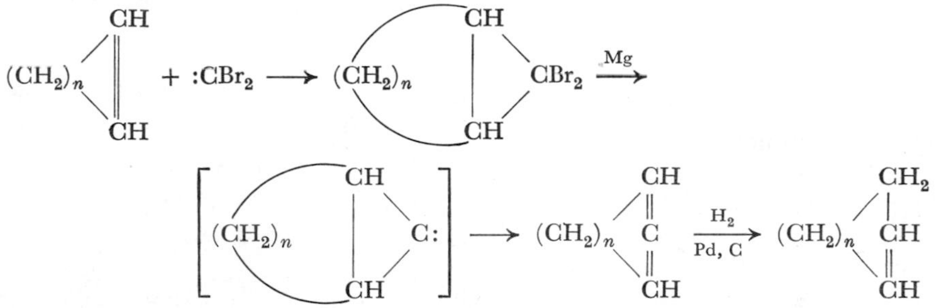

A remarkable one-step operation utilizes the ingredients of the dihalocyclopropane reaction beginning with the olefin and chloroform (see Ex. b.6).

It is interesting to note that some dibromocyclopropanes rearrange simply on heating as shown [91]:

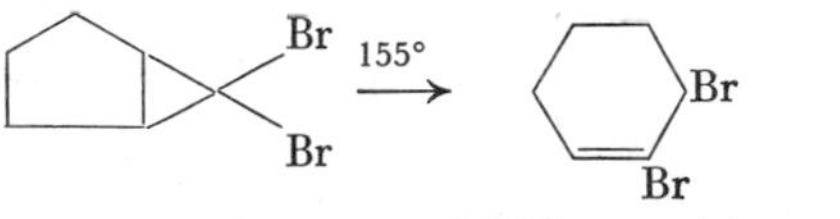

2,3-Dibromocyclohexene

(a) Preparation of 1,3-diphenylbutene-1 (89% by boiling 1,3-diphenyl-1,2-dibromobutane and zinc dust in absolute ethanol) [71].

(b) Other examples

(1) Tetraphenylethylene (55–70% by boiling diphenyldichloromethane in anhydrous benzene with copper) [90; *see also* 89].

(2) Perchlorofulvalene (85% from perchloro-9,10-dihydrofulvalene

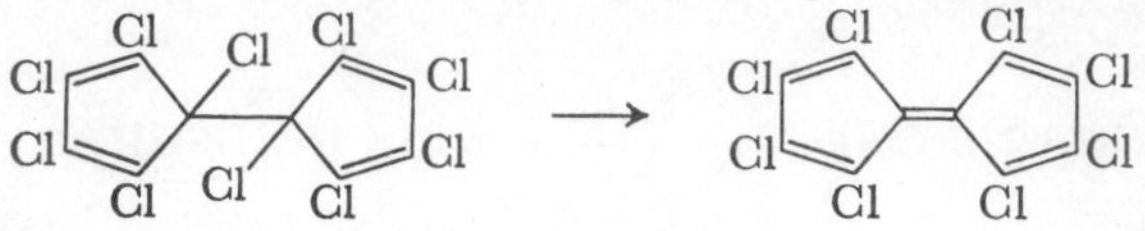

heated to 250°) [82; *see also* 78, 92].

(3) Allene (80% from 2,3-dichloropropene, zinc dust, and aqueous ethanol at reflux temperature) [93].

(4) *trans-trans*-1,4-diphenyl-1,3-butadiene (92% by heating 1,4-diphenyl-2,3-dibromobutane with quinoline 0.5 hr. at 160°) [83].

(5) 1,2-Cyclononadiene (59% by adding 9,9-dibromobicyclo[6.1.0]-nonane in ether dropwise to stirred magnesium turnings in ether. A latent period was noted followed by a rapid reaction. Perhaps better control could be achieved by activating the magnesium with ethylene bromide before the addition of the dibromide) [88].

(6) 2-Methyl-3-chloro-1,3-pentadiene (104 g. from heating 84 g.

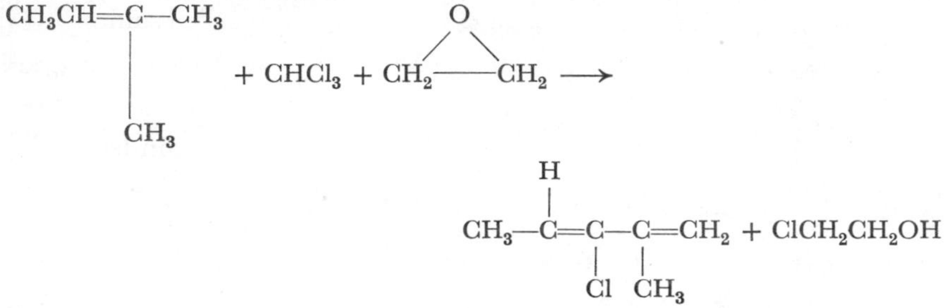

of 2-methyl-2-butene, 180 g. of chloroform, 132 g. of ethylene oxide, 2 g. of tetraethylammonium bromide, and 0.5 g. of hydroquinone for 10 hr. in an autoclave at 150°) [94].

(7) 1,2-Cyclotridecadiene. The intermediate organometallic com-

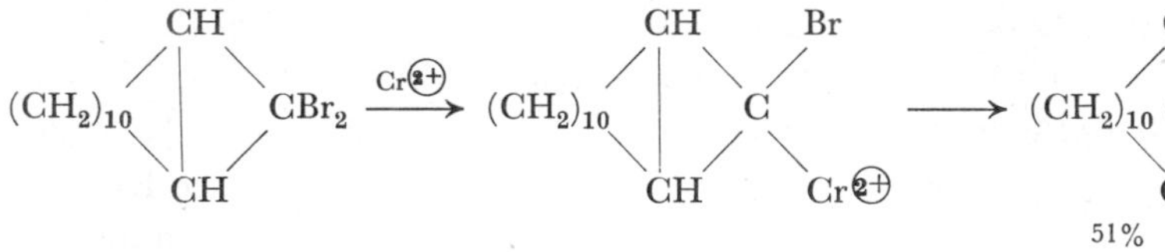

pound is indicated by the fact that, when optically active ligands such as the tartrate are used with the chromous salt, optically active allenes are formed. Butyllithium may also be used in this conversion [95].

4. From Halo Ethers (Boord)

$$RCHBrCHBrOC_2H_5 \xrightarrow{R'MgBr} RCHBrCHOC_2H_5 \xrightarrow{Zn} RCH=CHR'$$
$$\quad\quad\quad\quad\quad\quad\quad\quad\quad\quad | \quad\quad\quad\quad$$
$$\quad\quad\quad\quad\quad\quad\quad\quad\quad\quad R' \quad\quad\quad\quad$$

This synthesis [96] is of importance in synthesizing substituted alkenes. Although several steps are involved, it offers the advantage of giving products whose structure is established by the course of the reaction. The α,β-dibromo ether may be prepared as follows in yields of 70–90%:

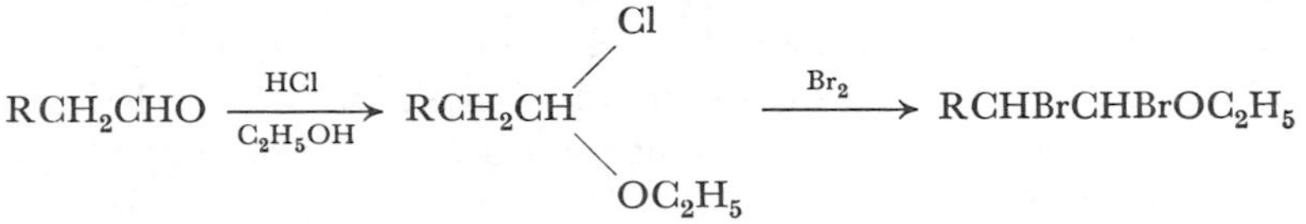

$$RCH_2CHO \xrightarrow[C_2H_5OH]{HCl} RCH_2CH\begin{smallmatrix}Cl\\|\\\\|\\OC_2H_5\end{smallmatrix} \xrightarrow{Br_2} RCHBrCHBrOC_2H_5$$

It is possible to prepare more highly substituted alkenes by starting with the β-bromo ether:

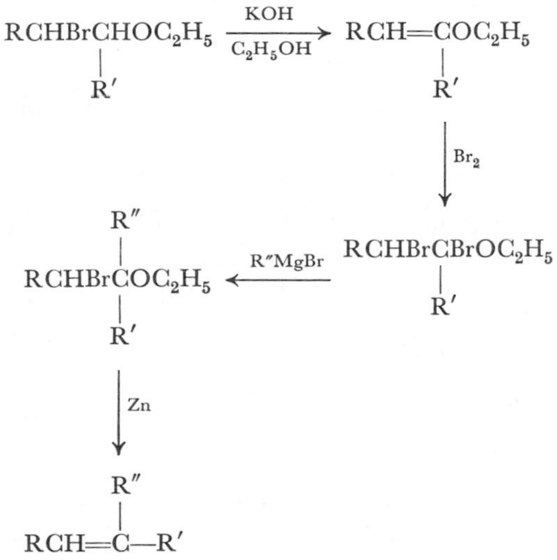

$$\underset{R'}{RCHBrCHOC_2H_5} \xrightarrow[C_2H_5OH]{KOH} \underset{R'}{RCH{=}COC_2H_5}$$

$\downarrow Br_2$

$$\underset{R'}{RCHBrCBrOC_2H_5}$$

$$\underset{R'}{\overset{R''}{RCHBrCOC_2H_5}} \xleftarrow{R''MgBr} \underset{R'}{RCHBrCBrOC_2H_5}$$

$\downarrow Zn$

$$\overset{R''}{RCH{=}C{-}R'}$$

The preparation of halo ethers is discussed in Chapter 6, B.7, B.8. This type and the α,β-dibromo ethers need not be isolated in the synthesis. It will be noted that the α-, but not the β-halogen, is affected by the Grignard reagent. For the elimination of the bromo and ethoxy groups in the final step, zinc [97, 98] is usually employed. In this step the yields vary greatly, but are sometimes as high as 75%. Sometimes a trace of zinc chloride improves the yield (see Ex. a). The stereochemistry of the reaction appears not to have been investigated, although the formation of *cis* and *trans* forms has been reported [99]. 1,4-Diolefins [97] have also been prepared through the use of allylmagnesium bromide in the reaction sequence.

The Crombie-Harper dehaloalkoxylation method is a powerful tool for obtaining pure *trans*-olefins which can be converted into other pure *trans* derivatives. The method is based on the observation that either *cis*- or *trans*-α-alkyl-β-chlorotetrahydropyrans are transformed to the open-chain *trans*-olefin by sodium in ether [100]. Pehaps an alkylsodium, which opens up to an alkoxy

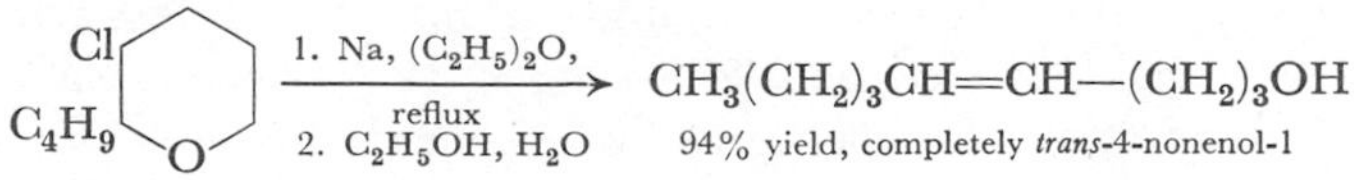

$$\underset{\substack{C_4H_9}}{\overset{Cl}{\bigcirc}}\text{(pyran ring)} \xrightarrow[\substack{reflux\\2.\ C_2H_5OH,\ H_2O}]{1.\ Na,\ (C_2H_5)_2O,} CH_3(CH_2)_3CH{=}CH{-}(CH_2)_3OH$$

Either *cis* or *trans*   94% yield, completely *trans*-4-nonenol-1 anion, forms. The *cis*-isomer may be obtained by reduction of the corresponding acetylene compound (see B.1). Considerable variation in the structure of the *trans* derivative is possible. The alkyl group in the α-position of the tetrahydropyran can be varied, the variation depending only on the coupling of the α,β-dichlorotetrahydropyran and the Grignard reagent, and the alcohol group in the product can be converted into halide and the halide subjected to various displacement reactions.

(a) Preparation of 1,4-pentadiene (72–76% from α-allyl-β-bromo-

$$BrCH_2CHOC_2H_5 \longrightarrow CH_2{=}CHCH_2CH{=}CH_2$$
$$|$$
$$CH_2CH{=}CH_2$$

ethyl ethyl ether, zinc dust, and zinc chloride in butyl alcohol gradually heated until the diene distills over) [97].

(b) Preparation of hexadecene-1 (62.5% from α-tetradecyl-β-

$$BrCH_2CHOC_2H_5 \longrightarrow CH_3(CH_2)_{12}CH_2CH{=}CH_2$$
$$|$$
$$CH_2(CH_2)_{12}CH_3$$

bromoethyl ethyl ether and zinc dust in butanol refluxed for 24 hr.) [98].

5. From Halohydrins

$$\begin{array}{c} R \quad OH \quad I \quad H \\ \diagdown \mid \quad \mid \diagup \\ C{-}C \longrightarrow \\ \diagup \quad \diagdown \\ R' \qquad R^2 \end{array} \qquad \begin{array}{c} R \\ \diagdown \\ C{=}CHR^2 \\ \diagup \\ R' \end{array}$$

Although olefins have been produced from chloro- and bromohydrins, the iodohydrins lead to better yields [101]. The present method of olefin formation is of interest in that, like the dehydrohalogenation of halides (A.2), the elimination is stereospecific. To avoid using the chlorohydrin (2-VII), available from the chloroketone (2-VI), directly, the British investigators proceeded to the epoxide (2-VIII) which was converted into the iodohydrin (2-IX). The latter with stannous chloride, phosphorus oxychloride, and pyridine gave a 80–85% yield of "3-methyl-*cis*-2-pentene" (2-X). In a similar manner, "3-methyl-*trans*-2-pentene" was obtained in 80–85% yield from 3-chloro-2-ketobutane, in which case the Grignard reagent in the first step was ethylmagnesium bromide. In this sequence of reactions it is assumed that the alkyl group of the Grignard reagent adds largely *anti* to the alkyl group on the adjoining carbon atom, in accordance with Cram's rule [102].

(a) Preparation of "3-methyl-*cis*-2-pentene." 2,3-Epoxy-3-methyl pentane (2-VIII), 22 g., was added to a mixture of 45 g. of sodium iodide, 4 g. of sodium acetate, 40 ml. of acetic acid, and 100 ml. of propionic acid at −30°. For ½ hr. the mixture was kept at −20 to −30°, after which it was allowed to warm to 0°, kept at that temperature for another ½ hr., and poured into a mixture of ether and aqueous sodium bicarbonate. From the ethereal layer, washed with a little sodium bisulfite solution and water, dried with magnesium sulfate, and then evaporated, 49 g. (97 %) of the iodohydrin (2-IX) was obtained. To this was added an ice-cooled solution of 70 g. of anhydrous stannous chloride in 250 ml. of pyridine followed by, over a 5-min. period, 18 ml. of phosphoryl chloride in 50 ml. of pyridine. Cooling for about 10–15 min. led to solidification and, after 2 hr., 300 ml. of water was added. Distillation gave 15 g. (83 %) of 3-methyl-2-pentene. Redistillation and infrared analysis indicated that the product contained 80–85 % of "3-methyl-*cis*-2-pentene" (2-X) [101].

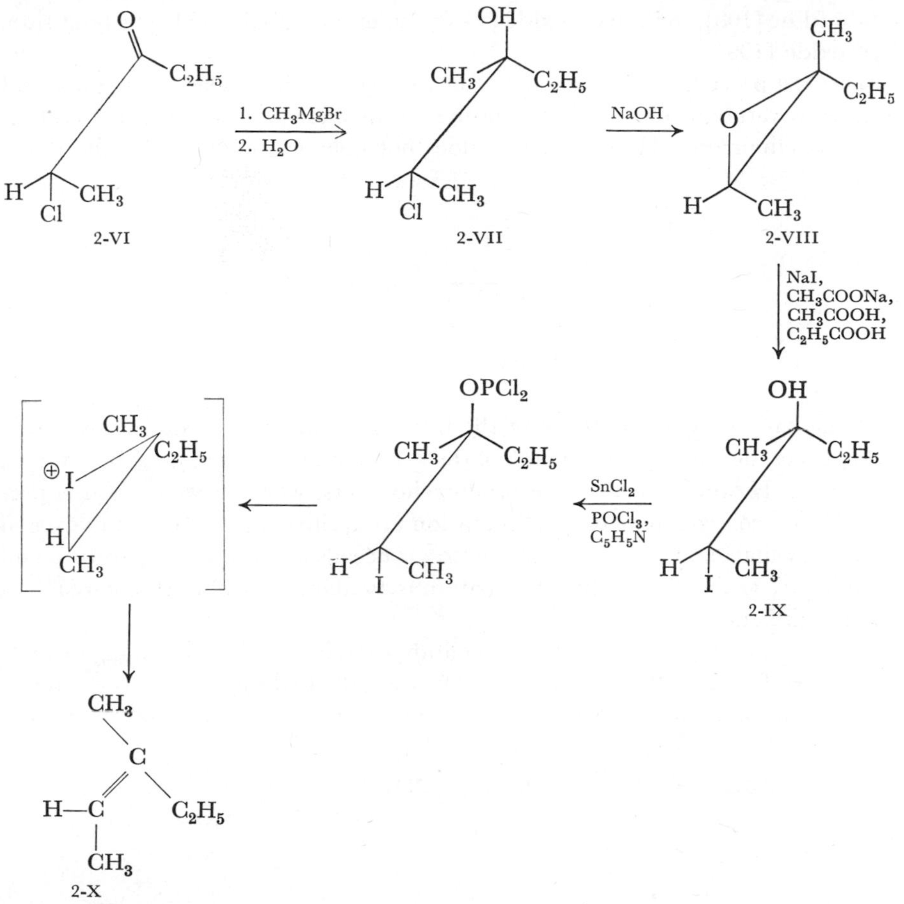

6. From Esters

$$
\underset{RCH_2CHCH_3}{\overset{\overset{\displaystyle OCOR}{|}}{}} \xrightarrow{\Delta} RCH{=}CHCH_3
$$

A variety of esters such as acetates [103–110], aryl sulfonates [111], stearates [112], carbonates and carbamates [113], and borates [114] have been employed to produce olefins. Elimination is accomplished simply by heating or by treatment with a base such as lithium amide in liquid ammonia [107], pyridine [112],

2,6-lutidine [108], sodium ethoxide [110], dimethyl sulfoxide [111] or potassium hydroxide [109].

The best procedure is to pass the ester through a tube packed with glass held at a temperature of 450–600°, depending on the ester (see Ex. a). The pyrolysis gives *cis* elimination [115], but this does not give a *cis*-olefin since the freely

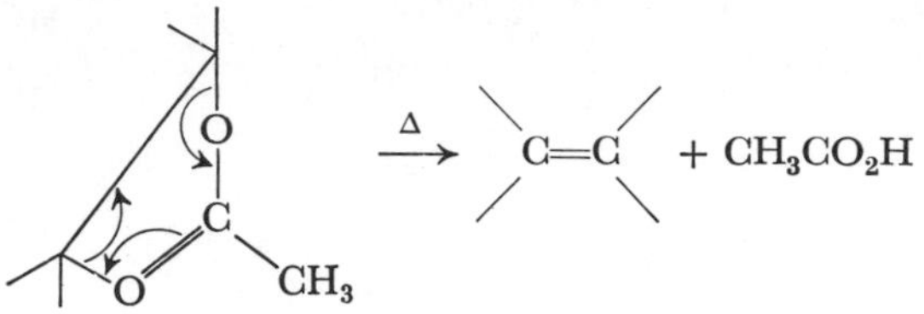

$$\overset{\Delta}{\longrightarrow} \quad \overset{\textstyle\diagdown}{\underset{\textstyle\diagup}{C}}{=}\overset{\textstyle\diagup}{\underset{\textstyle\diagdown}{C}} \quad + \quad CH_3CO_2H$$

rotating carbon allows a choice in the hydrogen atom extraction. Only subtle preferences, depending on the bulk of the groups on the α- and β-carbon atoms, are noted. However, in the case of alicyclic esters, where free rotation is prevented, the consequences of *cis* elimination are quite evident. If a *cis*-hydrogen atom is available, the elimination reaction proceeds at a lower temperature and with higher yield. If no *cis*-hydrogen atom is available, the elimination reaction is more difficult.

Despite earlier reports, pyrolytic *cis* elimination from esters does not give preferential Hofmann or Saytzeff products. Rather, the hydrogen atom is extracted almost in a statistical manner. Again, subtle differences from the statistical ratio

$$CH_3CHCH_2CH_3 \overset{\Delta}{\longrightarrow} \underset{\text{3 parts}}{CH_2{=}CHCH_2CH_3} + \underset{\text{2 parts}}{CH_3CH{=}CHCH_3}$$
$$\underset{\displaystyle \underset{O}{\overset{\|}{}}}{OCCH_3}$$

may be noted. Nevertheless, if the olefin is stable at this high temperature and if the exposure time is short, the synthesis is attractive in providing inaccessible olefins (see Ex. b.1). For olefins which are somewhat unstable, the pyrolysis can be conducted under lower conversion conditions (lower temperatures or faster flow rates).

A mixture of the alcohol and acetic anhydride, rather than the acetate ester, also can be converted into the olefin (see Ex. b.3).

A most unusual elimination of acid from an ester is illustrated and bears some resemblance to a retrograde Prins reaction (Alcohols, Chapter 4, B.4) [116].

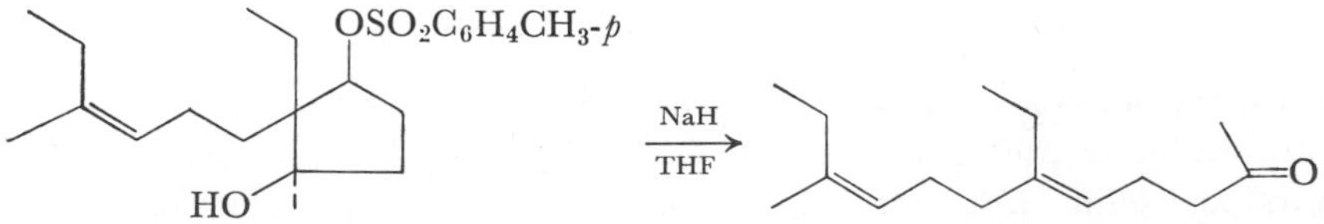

$$OSO_2C_6H_4CH_3\text{-}p$$

$$\xrightarrow[\text{THF}]{\text{NaH}}$$

cis

trans,cis-6-Ethyl-10-methyl-dodeca-5,9-dien-2-one, 80%

As with *vicinal* dihalides (A.3), *vicinal* ditosylates or dimesylates may be converted by sodium iodide in acetone into alkenes [117, 118]:

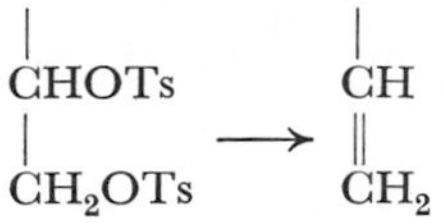

$$
\begin{array}{ccc}
\overset{|}{\underset{|}{\text{CHOTs}}} & & \overset{|}{\text{CH}} \\
\overset{|}{\text{CH}_2\text{OTs}} & \longrightarrow & \overset{||}{\text{CH}_2}
\end{array}
$$

The reaction gave surprising results when applied to the four 2,3-diols obtained from 5α-spirostane-12-one [119]:

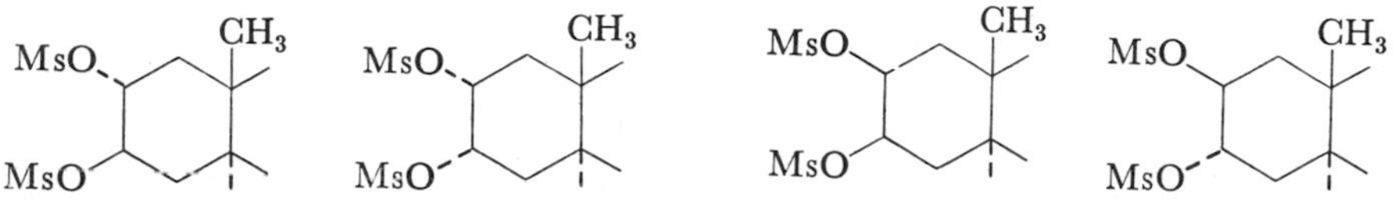

React with NaI   Do not react with NaI

On the basis of the elimination of bromine from the dibromocholestanes, it would be expected that the last compound shown, the *trans*-diaxial isomer, would lose the two mesyl groups the most readily. However, with sodium iodide in acetone the first two shown were converted smoothly into the olefin, while the last two were essentially unchanged.

(a) **Preparation of 1,4-pentadiene** (63–71% from 1,5-pentanediol diacetate by heating under nitrogen in a specially designed glass apparatus) [120].

(b) **Other examples**

(1) **1,3-Dimethylenecyclohexane** (61% conversion or 94% yield

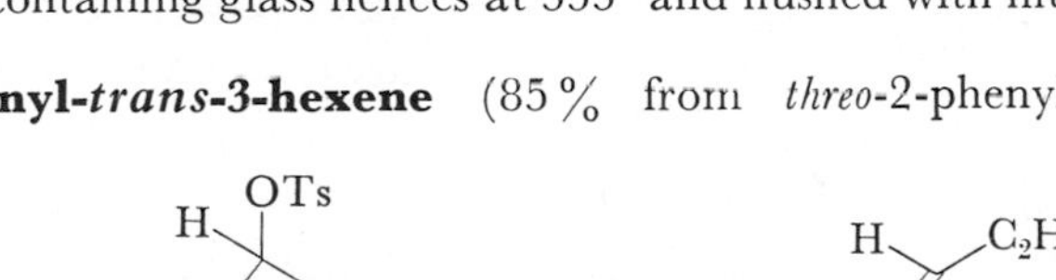

2-XI   2-XII

based on recovered 2-XI and 2-XII by dropping the 1,3-diacetate through a pyrolysis tube containing glass helices at 555° and flushed with nitrogen) [103].

(2) **3-Phenyl-*trans*-3-hexene** (85% from *threo*-2-phenyl-1-ethylbutyl

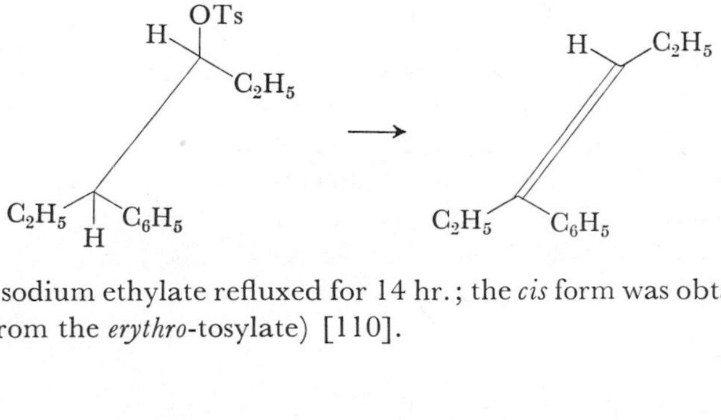

tosylate and sodium ethylate refluxed for 14 hr.; the *cis* form was obtained in the same yield from the *erythro*-tosylate) [110].

(3) **1-Octadecene** (68% from a solution of 1-octadecanol in 1.5 equivalents of acetic anhydride passed downward through a column heated to 600° and packed with silica wool, exposure time 2 min. The condensate was refluxed with alcoholic potassium hydroxide to remove unreacted ester) [121].

7. From Xanthates (Chugaev)

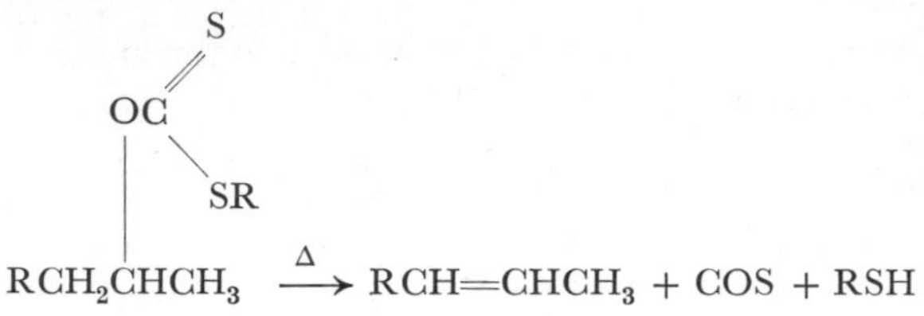

$$\underset{\underset{\displaystyle RCH_2\overset{|}{C}HCH_3}{}}{\overset{\displaystyle \overset{S}{\overset{\|}{OC}}}{\underset{\displaystyle SR}{}}} \xrightarrow{\Delta} RCH{=}CHCH_3 + COS + RSH$$

This synthesis was discussed in a recent review [122]. The reaction is similar to that of the pyrolysis of esters (A.6) but is carried out at lower temperatures and thus has less chance for rearrangement. For example, to cite one case, 3,3-dimethylpentanol-2 forms the S-methyl xanthate in 50% yield, and this ester in turn yields 3,3-dimethylpentene-1 in a 67% yield [123].

$$CH_3CHOHC(CH_3)_2CH_2CH_3 \rightarrow \text{S-methyl xanthate} \rightarrow CH_2{=}CHC(CH_3)_2CH_2CH_3$$

Dehydration of such branched-chain alcohols leads to mixtures of alkenes rather than to an unrearranged product. One advantage of the pyrolysis of xanthates is that all precursors are prepared in basic solutions, a condition which is often desirable. That the xanthates are sometimes difficult to prepare and purify and that the olefin produced may be contaminated with sulfur-containing impurities may be mentioned as disadvantages.

The method has been applied largely to secondary alcohols in the acyclic and alicyclic series. Its mechanism is similar to that given (A.6) for the pyrolysis of

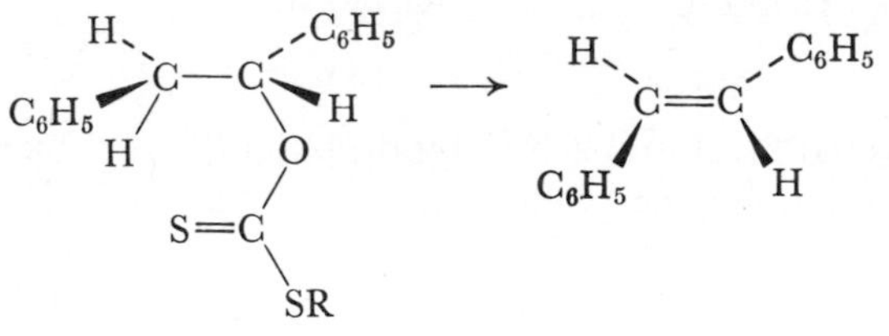

2-XIII

acetates. In the transition state of 2-XIII, a β-hydrogen atom and the xanthate group must be coplanar. Again, as with acetate pyrolyses, this does not mean that *cis*-olefins are obtained. But it does help to predict the olefin obtained from the Chugaev pyrolysis of xanthates of alicyclic alcohols (see Ex. b.3) and from diastereoisomeric alcohols. Examples are the conversion of the S-methyl xanthates of *erythro*- (2-XIV) and *threo*- (2-XVI) 1,2-diphenyl-1-propanols (see Ex. b.2) into *trans*- (2-XV) and *cis*- (2-XVII) α-methylstilbenes, respectively (see Ex. b.2) [102].

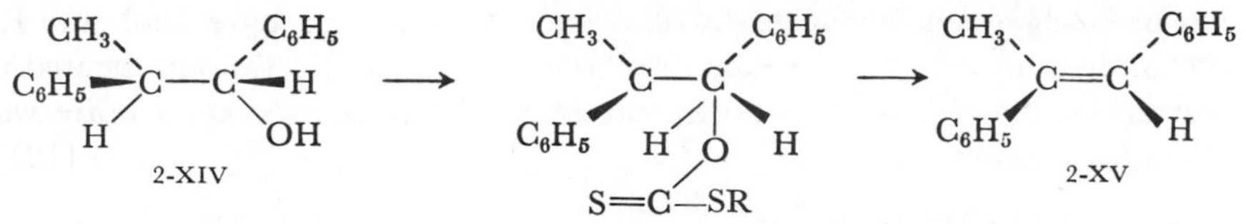

$$2\text{-XIV} \longrightarrow \underset{S=C-SR}{} \longrightarrow 2\text{-XV}$$

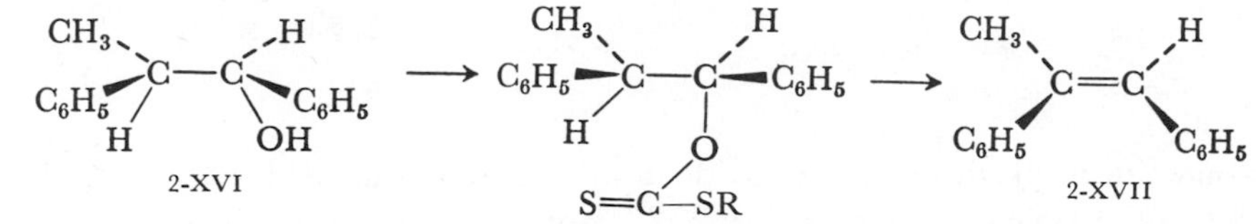

$$2\text{-XVI} \longrightarrow \underset{S=C-SR}{} \longrightarrow 2\text{-XVII}$$

(a) Preparation of cyclopentene (70 % as the dibromide by pyrolyzing the xanthate of cyclopentanol by dropping it in boiling biphenyl; the xanthate was prepared by refluxing a mixture of sodium hydride and cyclopentanol in ether for 3 hr., by then adding carbon disulfide and refluxing for 3 hr., and finally by adding methyl iodide and refluxing for 3 hr.) [124].

(b) Other examples

   (1) _t_-Butylethylene (58 % by distilling the xanthate of 3,3-dimethyl-2-butanol at atmospheric pressure) [123].

   (2) _trans_-α-Methylstilbene (77 % from the xanthate of _erythro_-1,2-diphenyl-1-propanol heated at 20–22 mm. pressure and at 130–195°; the _cis_-isomer, 65 %, was obtained similarly from the xanthate of _threo_-1,2-diphenyl-1-propanol) [102].

   (3) 3-Benzyl-1-cyclopentene (90 %, 95–98 % pure, from the pyrolysis of the methyl xanthate of _cis_-2-benzyl-1-cyclopentanol. The pyrolysis of the methyl sulfite ester of the alcohol above (Berti elimination) gave a mixture of roughly 48 % each of 1- and 3-benzylcyclopentene) [125].

8. From Quaternary Ammonium Hydroxides (Hofmann)

$$\underset{\overset{|}{CH_3}}{RCHN(\overset{\oplus}{C}H_3)_3\overset{\ominus}{O}H} \xrightarrow{\Delta} RCH=CH_2 + N(CH_3)_3 + H_2O$$

This elimination reaction has been the subject of a recent review [126]. Although it has perhaps been employed most widely in the structural determination of alkaloids and in studies of elimination reactions, it is used occasionally in synthesis. In contrast to the Saytzeff rule (see A.2) which, in cases in which the formation of 1- or 2-alkenes is possible, predicts the formation of 2-alkenes, the quaternary ammonium hydroxide route leads to 1-alkenes (Hofmann rule). It has now been shown that, by using the _t_-butoxide ion on the alkyl halide [127, 37] (see also A.2), 1-alkenes are produced in excess over 2-alkenes. In view of the steps required in going from the amine to the quaternary ammonium hydroxide:

$$ArCH_2NH_2 \xrightarrow[\text{HCHO}]{\text{HCOOH}} ArCH_2N(CH_3)_2 \xrightarrow{CH_3I}$$

$$ArCH_2\overset{\oplus}{N}(CH_3)_3\overset{\ominus}{I} \xrightarrow{Ag_2O} ArCH_2N(CH_3)_3OH$$

it has been suggested that it might be more desirable to prepare 1-alkenes via alkyl halides rather than via primary or tertiary amines [127, 37]; (see also A.2). Nevertheless, the yield of 1-olefin by elimination from quaternary ammonium hydroxides is usually high [128]:

$$CH_3CH_2CH_2CHCH_3 \xrightarrow[\Delta]{NaOC_2H_5} CH_3CH_2CH_2CH{=}CH_2$$

with the substituent $\overset{\oplus}{N}(CH_3)_3$ and $I^{\ominus}$ on the carbon, giving 63%, 94% pure.

If t-butoxide is substituted for ethoxide in this reaction, the yield is 84% and the purity of the 1-olefin is increased to 97%. However, some of the alcohol corresponding to the olefin is usually present [129]. In fact, the alcohol is formed exclusively in the case of derivatives of tetrahydroquinolines and pavine:

Contrary to statements in textbooks, ethylene is not preferentially eliminated in all cases [130]:

$$(CH_3)_3C\overset{\overset{\displaystyle Et}{|}\,\oplus}{N}Me_2 \longrightarrow 7.2\% \; CH_2{=}CH_2 + 92.8\% \; (CH_3)_2C{=}CH_2$$

The accompanying equations show the leveling effect of multiple substituents

$$Prop_2\overset{\oplus}{N}But_2 \rightarrow 63\% \; CH_3CH{=}CH_2 + 37\% \; CH_3CH_2CH{=}CH_2$$

$$Prop_3\overset{\oplus}{N}But \rightarrow 83\% \; CH_3CH{=}CH_2 + 17\% \; CH_3CH_2CH{=}CH_2$$

$$PropN\overset{\oplus}{N}But_3 \rightarrow 36\% \; CH_3CH{=}CH_2 + 64\% \; CH_3CH_2CH{=}CH_2$$

[131]. A more complete compilation of Hofmann degradation products is available. [132].

The most useful method for carrying out the reaction is to heat and concentrate an aqueous solution of the quaternary ammonium hydroxide under reduced pressure until decomposition occurs. The hydroxyl ion is usually preferred for converting the quaternary ammonium salt into its hydroxide, although Amberlite IRA-400(OH) [133], alkoxides, phenoxides, and carbonates [134, 135] have also been employed. As a rule the last-mentioned basic anions give less olefin and more alcohol as a side reaction product.

For success in the elimination it is necessary that the quaternary ammonium hydroxide contain a β-hydrogen atom. The mechanism usually, although not always, follows the E_2 pattern, as previously discussed under dehydrohalogenation of halides (A.2). Again, the consequences of *trans* elimination of the hydrogen atom and the amino group can be noted in certain diastereoisomeric quaternary salts [136]. It is interesting to note that the t-butoxide, in place of the ethoxide, gave the *trans*-olefin from both diastereoisomers.

The reaction has been applied to aliphatic, alicyclic, and heterocyclic amines. In alicyclic amines containing seven members or less, only the more stable *cis*-olefin is produced. However, in the case of eight-, nine-, and ten-membered rings both *cis*- and *trans*-olefins are produced, the latter predominating [137–139]. Yields are variable but often satisfactory.

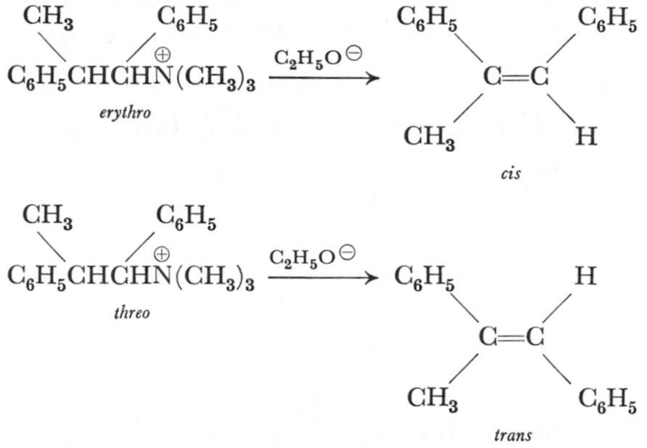

$$\underset{erythro}{\underset{C_6H_5}{CH_3}\overset{C_6H_5}{\underset{\oplus}{CHCHN(CH_3)_3}}} \xrightarrow{C_2H_5O^{\ominus}} \underset{cis}{\underset{CH_3}{\overset{C_6H_5}{C}}=\underset{H}{\overset{C_6H_5}{C}}}$$

$$\underset{threo}{\underset{C_6H_5}{CH_3}\overset{C_6H_5}{\underset{\oplus}{CHCHN(CH_3)_3}}} \xrightarrow{C_2H_5O^{\ominus}} \underset{trans}{\underset{CH_3}{\overset{C_6H_5}{C}}=\underset{C_6H_5}{\overset{H}{C}}}$$

In a modification of the Hofmann degradation the quaternary ammonium bromide may be substituted for the hydroxide [140]. With phenyllithium, for example, trimethylcyclooctylammonium bromide shaken in ether for 1 day gives cyclooctene, in 64 % yield, containing 81 % *cis*. With potassium amide in

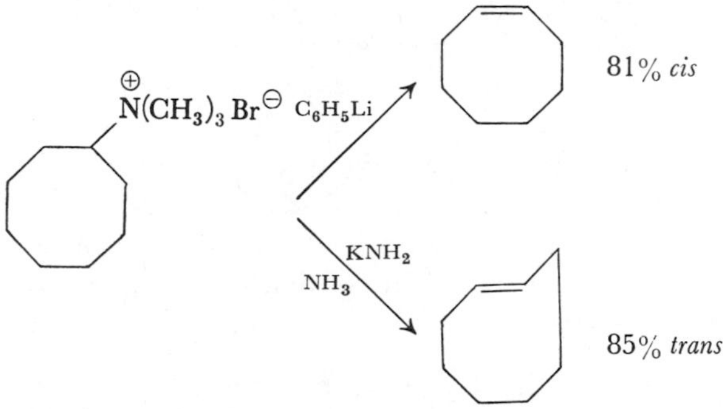

81% *cis*

85% *trans*

liquid ammonia at the boiling point for 4 hours the quaternary salt gives a 68 % yield containing 85 % of the *trans*-cyclene [141].

(a) Preparation of *cis*- and *trans*-cyclooctene.   N,N,N,-Trimethyl-cyclooctylammonium iodide, 29.2 g., in 150 ml. of water was stirred for 4 hr. at room temperature with moist neutral silver oxide freshly precipitated from 34 g. of silver nitrate. The solid, removed by filtration, was washed with 75 ml. of water, after which the filtrate was concentrated at 30 mm. (bath temperature 65°). The residual quaternary base was heated in a nitrogen atmosphere at 11 mm. (bath temperature 105–120°). The distillate, collected in a trap cooled with liquid nitrogen, was acidified with dilute sulfuric acid. After freezing the aqueous layer with dry ice, the mixture of olefins was pipetted off and distilled through a semimicro column to give 9.73 g. (89 %) of cyclooctene. Infrared analysis indicated that it consisted of 60 % *trans*- and 40 % *cis*-cyclooctene [142].

(b) Preparation of 1-hexene (60% from hexylamine methylated with dimethyl sulfate and sodium hydroxide, after which the quaternary salt was heated first with 20% sulfuric acid and then with a slight excess of barium hydroxide. The filtrate, after concentration, was heated with 50% potassium hydroxide solution and distilled to yield, after purification, 60% of the alkene [143].

9. From β-Dialkylamino Ketones and Related Compounds

Dialkylamino groups when attached β to an electron-withdrawing group are notorious for their ease of elimination to yield the unsaturated compound:

$$R_2NCH_2CH_2Y \overset{B\ominus}{\rightleftharpoons} R_2NCH_2\overset{\ominus}{C}HY \overset{BH}{\longrightarrow} CH_2{=}CHY + R_2NH$$

Y = electron-withdrawing group

The relatively strong acidity of the α-hydrogen atoms accounts for this decomposition. Common sources of olefins via this method are Mannich bases [144] and 2- or 4-(β dialkylaminoethyl) pyridines or quinolines (see Ex. b):

$$RCOCH_2CH_2NR_2' \longrightarrow RCOCH{=}CH_2$$

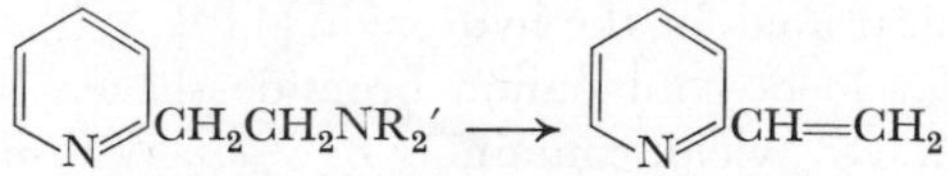

Mannich bases [145–147] are converted, with the loss of a secondary amine, by dimethylaniline, oxalic acid, or Dowtherm, into ketoalkenes. In other cases the β-dialkylamino ketone can be used *in situ* as a source of the unsaturated ketone [144].

As one can see, this reaction is related to the Hofmann degradation. The elimination from a tertiary amine is difficult and is restricted to compounds which have an activated hydrogen atom on the carbon attached to the carbon containing the amino group. The elimination from a quaternary ammonium salt is easy and nonrestricted:

$$CH_3CH_2 \overset{-\delta}{\cdots\cdots}NR_2 \quad \text{difficult}$$

$$CH_3CH_2 \cdots\cdots\overset{\oplus}{N}R_3 \quad \text{easy}$$

If electron-withdrawing groups are attached to the nitrogen atom, the leaving group should be more stable as an anion:

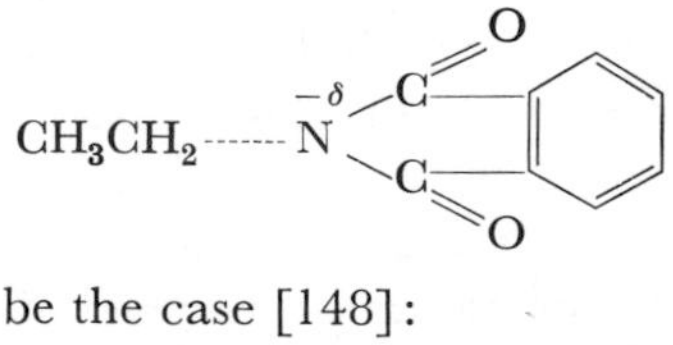

Such has been found to be the case [148]:

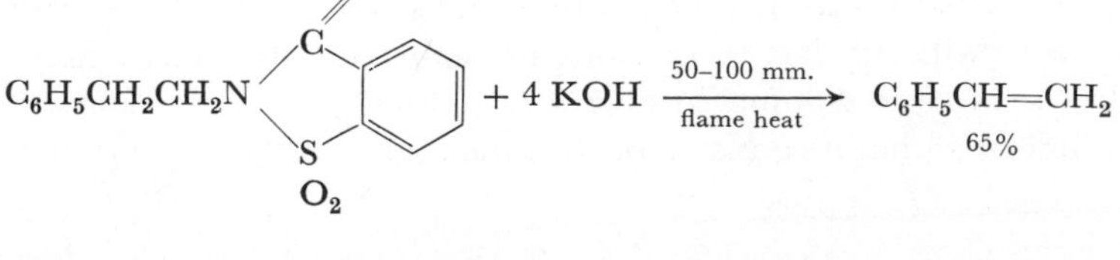

It would be interesting to see if amines with less acidic β-hydrogen atoms are amenable to this treatment.

(a) Preparation of α-phenoxyacrylophenone. A mixture of 50 g. (0.186 mole) of β-dimethylamino-α-phenoxypropiophenone and 50 ml. of dimethylaniline was refluxed for 1 hr. An ethereal solution of the mixture, after extraction twice with 400 ml. of 1 N hydrochloric acid, was dried and evaporated. Crystallization of the residue from ethanol gave 35.4 g. (85%) of the acrylophenone [146].

(b) Preparation of 1-vinylisoquinoline (70% by distilling 1-β-dimethylaminoethylisoquinoline with solid potassium hydroxide and a trace of N-phenyl-β-naphthylamine under reduced pressure) [149].

10. From Amine Oxides (Cope)

$$R_2CHCR_2 \longrightarrow R_2C{=}CR_2 + (CH_3)_2NOH$$
$$O \longleftarrow N(CH_3)_2$$

A review of the Cope elimination has been published [150]. Although this reaction has been used less than the Hofmann elimination (A.8), it sometimes offers an advantage in the ease of manipulation and lack of isomerization. Like the Hofmann reaction, it has been applied to acyclic, alicyclic, and heterocyclic amines. The elimination ordinarily proceeds by pyrolysis at 120–150°, but it may also be accomplished at 25° in an anhydrous mixture of dimethyl sulfoxide and tetrahydrofuran [151]. The mechanism appears to be a *cis* elimination, as was the case in the pyrolysis of xanthates (Chugaev) (A.7) and esters (A.6).

$$R_2C{-}CH_2$$
$$H \quad N(CH_3)_2 \longrightarrow R_2C{=}CH_2 + (CH_3)_2NOH$$
$$O$$

Thus β hydrogen atoms are again necessary, but it is interesting to note that there is no decided preference for attack at a β-methyl group as in the case of the Hofmann reaction [152]. Yields are often satisfactory.

(a) Preparation of methylenecyclohexane (79–88% from N,N-dimethylcyclohexylmethyl amine oxidized by 30% hydrogen peroxide in methanol, after which the amine oxide hydrate was heated under vacuum) [153].

11. From Sulfoxides, Sulfones, β-Hydroxysulfinamides, and β-Hydroxyphosphonamides

$$RCH_2CH_2SOR \xrightarrow{\Delta} RCH{=}CH_2 + RSOH$$

$$\underset{R}{\overset{R}{\diagdown}}C{-}C\underset{R}{\overset{R}{\diagup}} \xrightarrow{\Delta} \underset{R}{\overset{R}{\diagdown}}C{=}C\underset{R}{\overset{R}{\diagup}}$$
$$SO_2$$

$$(CH_3)_2CHSOCH(CH_3)_2 \xrightarrow[\text{butoxide}]{K\,t\text{-}} 2\ CH_3CH{=}CH_2$$

$$(CH_3)_2CHSO_2CH(CH_3)_2 \xrightarrow[\text{butoxide}]{K\,t\text{-}} 2\ CH_3CH{=}CH_2$$

Sulfoxides and sulfones may be converted into olefins by pyrolysis [154–156] or base-catalyzed β-elimination [157, 158]. The two recent detailed studies of the pyrolysis of sulfoxides to yield alkenes are those of 1,2-diphenylpropyl phenyl sulfoxide [154] and 3-phenylpropyl methyl sulfoxide [155]. At comparatively low temperature the reaction is stereospecific with the occurrence of *cis* elimination, as was the case in the Cope pyrolysis of amine oxides (A.10). Since

$$RCH_2CH_2SOR \longrightarrow RCH \!-\!\!-\! CH_2 \longrightarrow RCH{=}CH_2 + RSOH$$

dimethyl sulfoxide can be condensed with a variety of olefins, elimination from this product yields an olefin with one more carbon atom (see Ex. b). A series of cyclic olefins has been prepared recently by the elimination of phenyl sulfenic acid, C_6H_5SOH [159].

The formation of olefins from sulfones has been accomplished with sodium ethoxide in ethanol at temperatures of 200° or above [160]. More recently it has been found [157, 158] that potassium *t*-butoxide in dimethyl sulfoxide is the preferred reagent. With a limited number of sulfoxides and sulfones at 55° this reagent gave excellent yields in some cases. In fact, the reagent gives 1,3-butadiene in 80% yield from tetramethylene sulfoxide in 117 hours:

$$\xrightarrow{\;(CH_3)_3C\overset{\ominus}{O}\;} CH_2{=}CH{-}CH{=}CH_2 + H_2\overset{\ominus}{S}O$$

It has been found that the sulfones are degraded more readily than sulfoxides and that the decreasing order of reactivity among sulfoxides is $t\text{-}C_4H_9SOC_4H_9\text{-}t$ > $iso\text{-}C_3H_7SOC_3H_7\text{-}iso$ > $n\text{-}C_4H_9SOC_4H_9\text{-}n$. The mechanism, illustrated with diisopropyl sulfoxide, appears to involve the formation of the carbanion 2-XVIII by the elimination of a β-hydrogen. By the elimination of an alkyl sulfenate ion 2-XX, the carbanion gives the alkene 2-XIX. A second mole of the alkene 2-XIX is produced by the action of the reagent on the alkyl sulfenate ion:

$$\longrightarrow CH_3CH{=}CH_2 + (CH_3)_2CH\overset{\ominus}{S}O \xrightarrow{\;\;K^{\oplus}\;\;} CH_3CH{=}CH_2 + KSOH$$

$$2\text{-XIX} \qquad 2\text{-XX} \qquad 2\text{-XIX}$$

2-XVIII

Aryl-conjugated olefins, such as styrene and 1,1-diphenylethylene, add to dimethyl sulfoxide under basic conditions to give, in almost quantitative yield, the corresponding methyl 3-aryl sulfoxides from which phenyl-substituted alkenes may be produced [155].

Four adaptations which increase the versatility of sulfur dioxide extrusion are shown:

1. Via the chloromethyl sulfone (Romberg-Bäckland reaction) [161]:

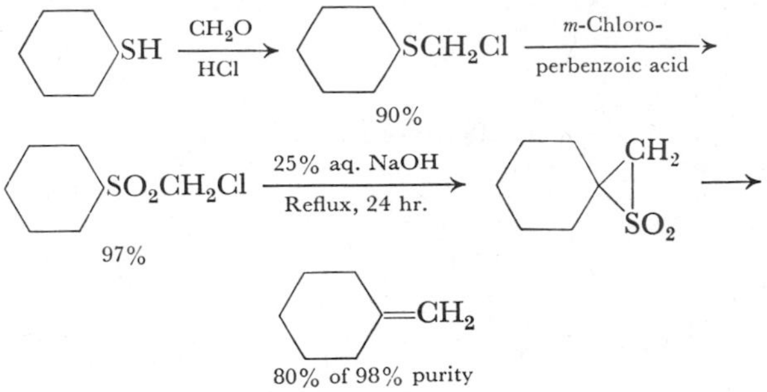

90%

97%

80% of 98% purity

1-Heptene was prepared in 54% overall yield from n-hexylmercaptan in a similar manner. 2-Butene, consisting of 78.8% *cis* and 21.2% *trans*, was obtained [161] in 75% yield from α-chlorodiethylsulfone. These results are typical of the Romberg-Bäcklund reaction.

2. From the sulfonyl chloride [162]:

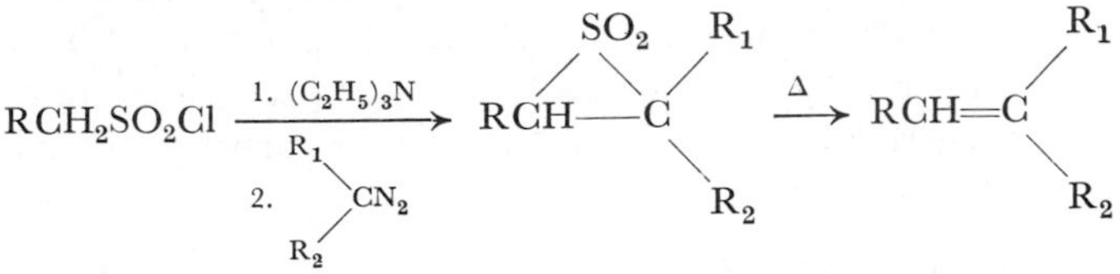

$$RCH_2SO_2Cl \xrightarrow[\substack{2. \underset{R_2}{\overset{R_1}{\diagup}}CN_2}]{1. (C_2H_5)_3N} RCH-\underset{R_2}{\overset{SO_2\ R_1}{C}} \xrightarrow{\Delta} RCH=\underset{R_2}{\overset{R_1}{C}}$$

Yields with a series of sulfonyl chlorides vary from 35–97% (see Ex. c).

3. The Staudinger-Pfenninger elimination. This is a rather complex elimination suitable for highly substituted ethylenes. It involves the simultaneous elimination of nitrogen and sulfur dioxide with coupling of the two fragments [163]:

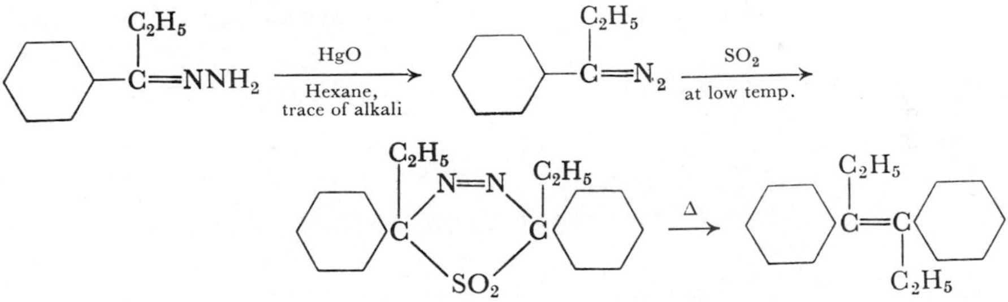

The overall yield starting from the ketone was 8%.

4. From β-hydroxy sulfinamides [164]. The following series of reactions produces olefins in high yield:

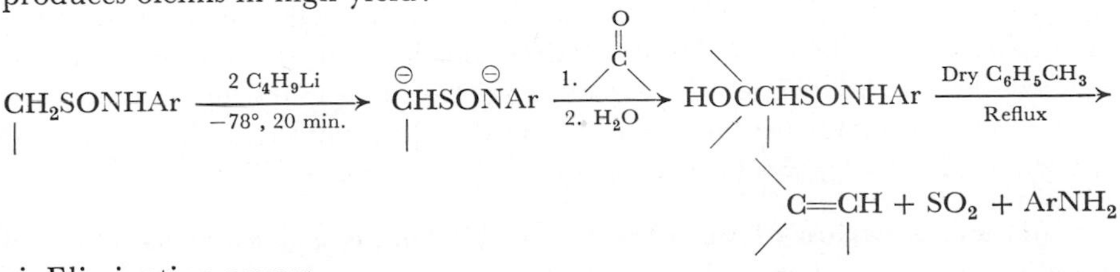

cis Elimination occurs.

Of greater versatility, however, is elimination from β-hydroxy phosphon-amides [165]:

$$\underset{\underset{R'}{|}}{RCHPO(N(CH_3)_2)_2} \xrightarrow{C_4H_9Li} \underset{\underset{R'}{|}}{R\overset{\ominus}{C}PO(N(CH_3)_2)_2} \xrightarrow[2.\ H_2O]{1.\ \overset{\overset{O}{\parallel}}{\underset{}{C}}\triangle}$$

$$\underset{\underset{R'}{|}}{\overset{\overset{OH}{|}}{-C}\underset{\underset{R'}{|}}{\overset{\overset{R}{|}}{-C}PO(N(CH_3)_2)_2}} \xrightarrow[C_6H_6,\ \text{reflux 3–12 hr.}]{\text{Silica gel (to remove amide)}} \underset{R'}{\overset{R}{>C=C<}} + HOPO(N(CH_3)_2)_2$$

$$\underset{74\text{–}98\%}{} \qquad\qquad\qquad\qquad\qquad\qquad\qquad\qquad \underset{53\text{–}93\%}{}$$

Furthermore the properly substituted β-hydroxy phosphonamides may be separated by crystallization. Since each diastereoisomer eliminates in a *cis* manner, the *cis*-olefin should be obtained from one isomer and the *trans*-olefin from the other. *cis*-1-Phenylpropene was obtained in 90% yield by heating the corresponding β-hydroxy phosphonamide diastereoisomer melting at 80.5–82°.

Elimination occurring in the intermediate from a carbonyl compound and a phosphorane is discussed in E.2.

(a) Preparation of propylene. Diisopropyl sulfoxide, 2.24 mmoles, was treated with 7 ml. of a 0.6 M solution of potassium t-butoxide in dimethyl sulfoxide at 55°. After 17 hr., gas chromatographic-mass spectral analysis indicated a 95 mole % yield; diisopropyl sulfone gave a quantitative yield of propylene by a similar treatment for 24 hr. [157].

(b) Preparation of 3-phenyl-2-methylpropene (92% from allyl-

$$C_6H_5CH_2CH{=}CH_2 + CH_3SOCH_3 \xrightarrow[1\ \text{equiv.}]{NaH} [C_6H_5CH{=}CHCH_3] \xrightarrow{\ominus CH_2SOCH_3}$$

$$\underset{\underset{}{}}{C_6H_5CH_2\overset{\overset{CH_3}{|}}{C}HCH_2SOCH_3} \xrightarrow{\Delta} C_6H_5CH_2\overset{\overset{CH_3}{|}}{C}{=}CH_2$$

benzene and dimethyl sulfoxide in the presence of sodium hydride at 25° followed by pyrolysis of the sulfoxide formed at 165°) [155].

(c) Preparation of benzylethylene. An ethereal solution of diazo-methane and triethylamine was added slowly with cooling to β-phenylethyl sulfonyl chloride. The triethylamine hydrochloride separated and nitrogen was evolved. The cyclic sulfone was recovered (99%) on evaporating the filtrate at room temperature. On heating this compound to 80°, sulfur dioxide was evolved and the olefin (97%) was formed [162].

(d) Preparation of butadiene. 3-Sulfolene is a convenient source of butadiene for use without isolation in the Diels-Alder reaction, as shown [166].

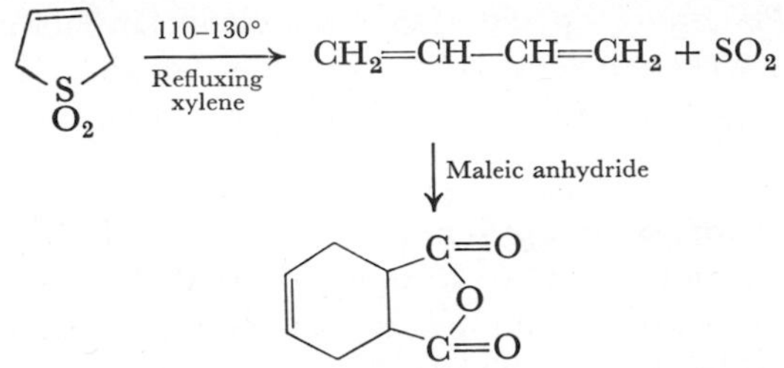

$$CH_2=CH-CH=CH_2 + SO_2$$

at 110–130° Refluxing xylene

↓ Maleic anhydride

82–90%

12. From 1,2-Epithioalkanes or Epoxides

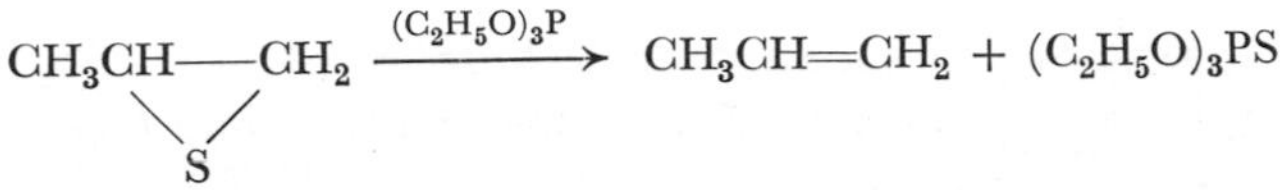

$$CH_3CH\text{——}CH_2 \xrightarrow{(C_2H_5O)_3P} CH_3CH=CH_2 + (C_2H_5O)_3PS$$

(with S bridging)

Sulfur has been removed from 1,2-epithioalkanes by the use of Raney nickel or copper bronze [167], by triethyl phosphite [168], by tributylphosphine [169], by triphenylphosphine [170], and by alkyl- or aryllithium [171]. The yields by the first two methods are excellent. The last-mentioned method appears to be of more value in preparing thiols, particularly thiophenols. Discussions of the mechanism of the elimination with triethyl phosphite and the alkyl- or aryllithium reagents are available [168, 171].

With triethyl phosphite, tributylphosphine, or phenyllithium, the elimination is stereospecific in that the *cis*-episulfide yields the *cis*- while the *trans*-isomer yields the *trans*-olefin. With the same reagents the epoxides give almost the opposite results [169; *see also* 172]:

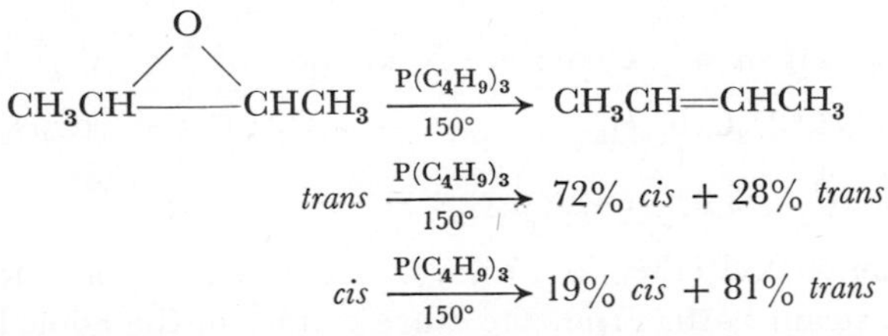

$$CH_3CH\text{——}CHCH_3 \xrightarrow[150°]{P(C_4H_9)_3} CH_3CH=CHCH_3$$

(with O bridging)

$$trans \xrightarrow[150°]{P(C_4H_9)_3} 72\%\ cis + 28\%\ trans$$

$$cis \xrightarrow[150°]{P(C_4H_9)_3} 19\%\ cis + 81\%\ trans$$

Obviously, mechanisms of elimination from episulfides and epoxides must be different; that from the epoxide may proceed as follows:

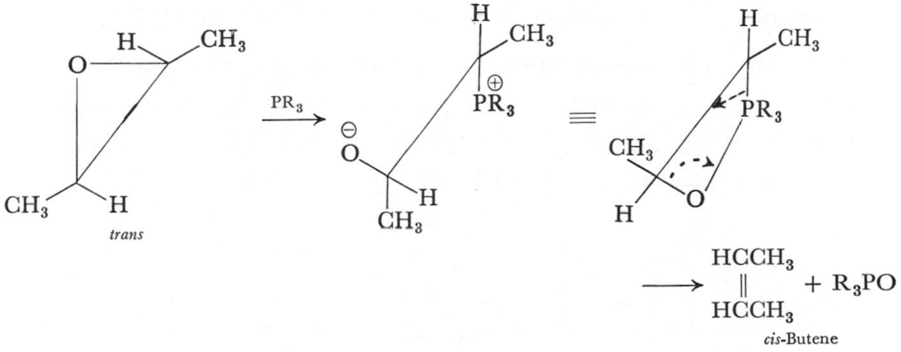

$$\longrightarrow \begin{array}{c} HCCH_3 \\ \| \\ HCCH_3 \end{array} + R_3PO$$

cis-Butene

Reduction elimination of epoxides to form olefins in high yield has been carried out with Cr(2+) ethylenediamine complex at 25° [173]. Actually the desulfurization of episulfides proceeds more readily than the deoxygenation of epoxides [170].

(a) Preparation of propylene. A mixture of 14.8 g. (0.20 mole) of 1,2-epithiopropane and 33.2 g. (0.20 mole) of triethyl phosphite was distilled through a 30-cm. Fenske column over a 3 hr. period, at which time the temperature of the residue had reached 180°. The propylene (97%) was recovered during the distillation as 1,2-dibromopropane in traps included in the apparatus [168].

(b) Preparation of tetraphenylethylene (Almost quantitative by heating tetraphenylethylene sulfide in ethanol with Raney nickel for about 5 min.) [167].

(c) Preparation of ethyl cinnamate (82% by heating ethyl β-phenylglycidate with triphenylphosphine at 125–178° in the presence of hydroquinone) [174].

13. From Ethers

$$RCH_2CH_2OR' \rightarrow RCH{=}CH_2 + R'OH$$

In a limited number of cases, alcohol has been removed from ethers to yield alkenes. Usually other functional groups such as the amino [175], carbalkoxy [176], or hydroxy [177] are present in the starting material. The reagents which have been used to effect the elimination are 20% hydrochloric acid [175], metaphosphoric acid [177], and phosphorus pentoxide [176]. Yields are good.

(a) Preparation of stilbene. A solution of 5 g. of β-dimethylamino-

$$\underset{\underset{OCH_2CH_2N(CH_3)_2}{|}}{C_6H_5CH_2CHC_6H_5} \longrightarrow C_6H_5CH{=}CHC_6H_5$$

ethyl-1,2-diphenylethyl ether in 25 ml. of 20% hydrochloric acid was heated for 1 hr. on a steam bath. From the ether extract of the cooled mixture, previously washed with water, 3.34 g. (95%) of crude stilbene was obtained [175].

(b) Preparation of ethyl methacrylate (90% by heating ethyl α-ethoxyisobutyrate with phosphorus pentoxide at 80° for 2 hr.) [176].

14. From Diazo Compounds, Tosylhydrazones, and the Like

Decomposition of diazo compounds leads to olefins. The intermediate is a carbene which can isomerize to an olefin or a cyclopropane:

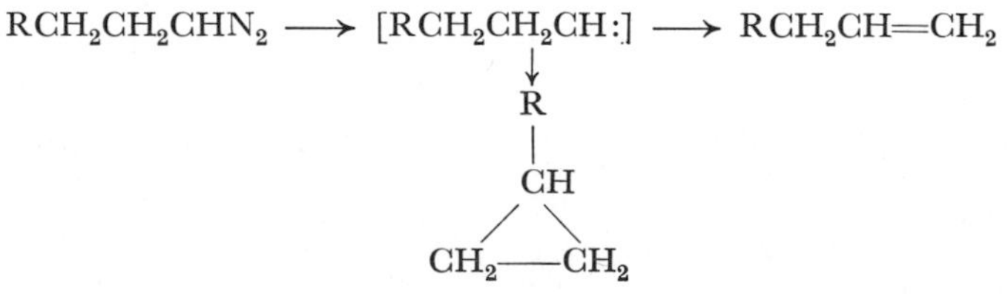

$$RCH_2CH_2CHN_2 \longrightarrow [RCH_2CH_2CH{:}] \longrightarrow RCH_2CH{=}CH_2$$

$$\underset{\underset{\underset{CH_2{-\!-}CH_2}{\diagup\ \diagdown}}{CH}}{\overset{|}{R}}$$

Branched chains tend to increase the yield of cyclopropane. The decomposition can be brought about by heat, acid, or photochemical excitation. Silver chloride is a good catalytic agent for the thermal decomposition. A comparison of the various means of effecting the reaction has been made recently [178]:

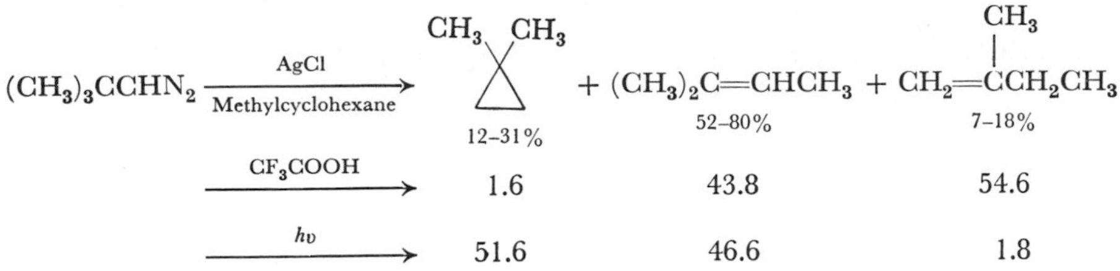

$$(CH_3)_3CCHN_2 \xrightarrow[\text{Methylcyclohexane}]{AgCl} \triangle + (CH_3)_2C=CHCH_3 + CH_2=\overset{\overset{\displaystyle CH_3}{|}}{C}CH_2CH_3$$

	$\triangle$	$(CH_3)_2C=CHCH_3$	$CH_2=\overset{CH_3}{C}CH_2CH_3$
	12–31%	52–80%	7–18%
$\xrightarrow{CF_3COOH}$	1.6	43.8	54.6
$\xrightarrow{h\upsilon}$	51.6	46.6	1.8

Overall yields with a series of diazo compounds and silver salts were 60–98%.

More useful in the preparation of olefins is the reaction of tosylhydrazones with methyllithium as shown [179–182]: 169980

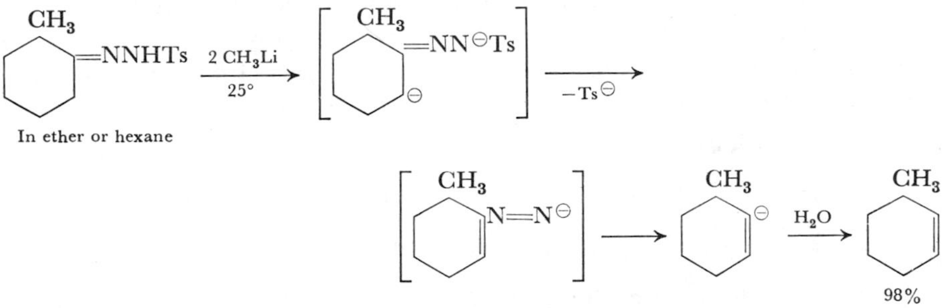

In ether or hexane

2-Bornene was prepared from camphor in quantitative yield.

In an unrelated reaction but one utilizing the formation of nitrogen gas as a driving force, azines have been converted in poor yield into olefins (see Ex. a).

Diazomethane and aryl-substituted diazomethanes in the presence of trityl perchlorate as a catalyst result in dimerization to form ethylenes [183] (see Ex. b):

$$(C_6H_5)_2C\overset{\oplus}{=}N\overset{\ominus}{=}N \xrightarrow{(C_6H_5)_3CClO_4} (C_6H_5)_2C=C(C_6H_5)_2 + N_2$$

It is suggested that the steps are addition, rearrangement, and elimination.

(a) Preparation of 1,2-di-α-naphthylethylene. α-Naphthaldehyde-azine (3 g.) and activated copper (0.15 g.) were heated to 270° and then gradually to 300° as controlled by the rate of nitrogen evolution. The crude mixture was recrystallized four times from benzene and other solvents and then chromatographed on alumina to yield 0.3 g. of the desired product, m.p. 161° [184].

(b) Preparation of tetraphenylethylene (97% from 64.5 mmoles of diphenyldiazomethane in 100 ml. of dry ether added to 3.35 mmoles of trityl perchlorate in 100 ml. of ether at 0°) [183].

15. From Lithium Epoxides (Elimination of Li$_2$O) and from Some Anions

An unusual elimination occurs in the addition of alkyllithiums to epoxides [185]:

$$(CH_3)_3CCH\overset{O}{\underset{}{\diagup\!\diagdown}}CH_2 + LiC(CH_3)_3 \xrightarrow[\text{in pentane}]{\text{Reflux 24 hr.}} \left[(CH_3)_3CCH\overset{O}{\underset{}{\diagup\!\diagdown}}CHLi\right] \longrightarrow$$

3 equiv.

$$\left[(CH_3)_3C\overset{OLi}{\underset{|}{-}}CH-CH:\right] \longrightarrow (CH_3)_3C\overset{\overset{Li}{|}}{\underset{\underset{H}{|}}{-}}C\overset{O}{\underset{\diagdown}{\diagup}}CH\underset{C(CH_3)_3}{} \xrightarrow{-Li_2O} (CH_3)_3C\diagdown\qquad H$$

$$C=C$$

$$H\quad C(CH_3)_3$$

trans-1,2-Di-*t*-butylethylene,

Yields range from 39 to 67 % in other cases, but it was very poor in the addition of C$_2$H$_5$Li to 1-butene oxide.

Another unusual reaction is the elimination of ethylene from organometallic derivatives which form stable anions [186]:

$$Ar_2\overset{\overset{CH_3}{|}}{C}-CH_2CH_2Li \xrightarrow[\text{Very rapid}]{\text{THF, } -40°} Ar_2\overset{\overset{CH_3}{|}}{C}-Li + CH_2=CH_2$$

Triarylmethyl and 9-methyl-9-fluorenyl anions were formed similarly.

16. From Alkylboron Compounds

$$\left[R\overset{\overset{CH_3}{|}}{CH}\right]_3B \rightleftharpoons 3\,RCH=CH_3 + BH_3 \rightleftharpoons (RCH_2CH_2)_3B \xrightarrow{R'CH=CH_2}$$

$$3\,RCH=CH_2 + (R'CH_2CH_2)_3B$$

The fruitful research of H. C. Brown and co-workers with boron alkyls encompasses the synthesis of certain terminal olefins which ordinarily would be difficult to obtain. Advantage is taken of the dissociation of the trialkylborons to olefins and recombination to give the more stable trialkylborons. The more stable alkylborons are those in which the boron is attached to a terminal carbon, a fact which suggests that steric compression in the trialkylboron determines its thermal stability. Essentially, then, this process is a procedure for preparing 1-alkenes from isomeric alkenes having the olefinic group more centrally located:

$$RCH=CHCH_3 \underset{}{\overset{R_2'BH}{\rightleftharpoons}} RCH_2\overset{\overset{BR_2'}{|}}{CH}-CH_3 \rightleftharpoons RCH_2CH=CH_2$$

In addition a high-boiling olefin, R″CH=CH$_2$, can displace a lower-boiling one, RCH$_2$CH=CH$_2$, from the boron alkyl (see also Ex. a):

$$RCH_2CH_2CH_2BR_2' \xrightarrow{R''CH=CH_2} RCH_2CH=CH_2 + R''CH_2CH_2BR_2'$$

A further modification is the conversion of acetylenic compounds into vinyl bromides with some stereospecific selectivity [187]:

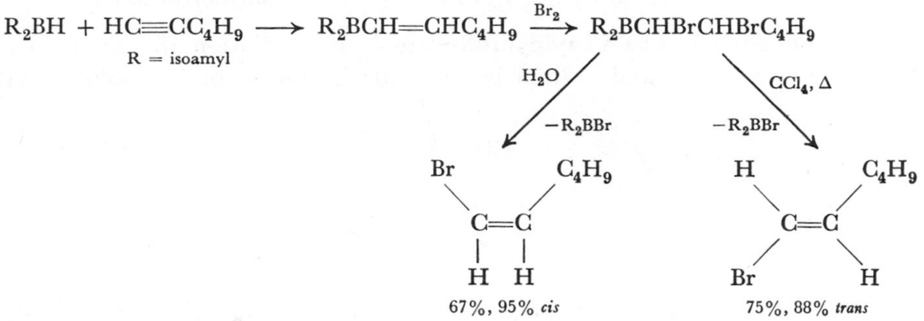
R₂BH + HC≡CC₄H₉ → R₂BCH=CHC₄H₉ → R₂BCHBrCHBrC₄H₉
R = isoamyl
Br₂
H₂O
−R₂BBr
CCl₄, Δ
−R₂BBr
Br
C₄H₉
C=C
H H
67%, 95% cis
H
C₄H₉
C=C
Br
H
75%, 88% trans

An even more unusual conversion of an acetylenic compound into an olefin has been discovered by Zweifel [188]:

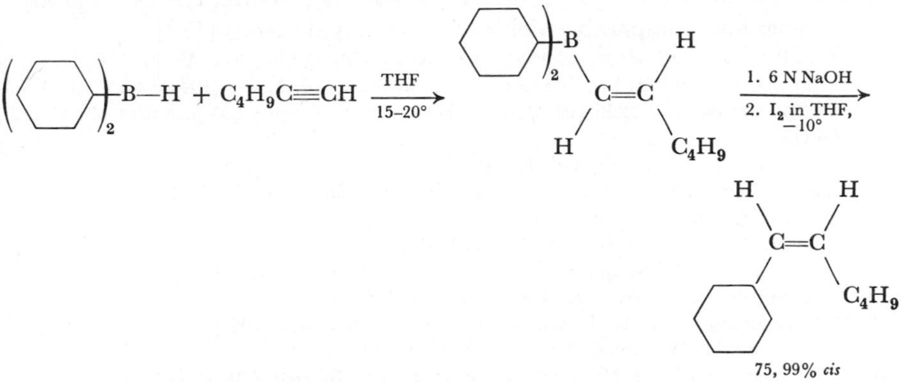
B—H + C₄H₉C≡CH
THF
15–20°
B
H
C=C
H
C₄H₉
1. 6 N NaOH
2. I₂ in THF, −10°
H
H
C=C
C₄H₉
75, 99% cis

In a later publication [189] a selective oxidizing agent which permits the synthesis of *cis,trans*-butadienes by iodine coupling has been found:

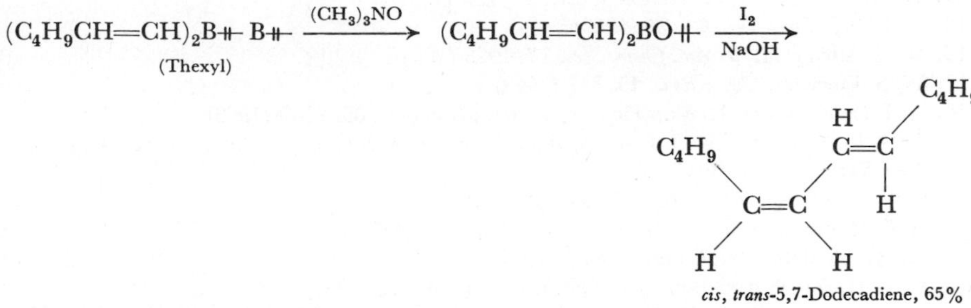
(C₄H₉CH=CH)₂B (Thexyl)
(CH₃)₃NO
(C₄H₉CH=CH)₂BO
I₂
NaOH
C₄H₉
H
C=C
H
C₄H₉
C=C
H
H
cis, trans-5,7-Dodecadiene, 65%

cis, cis-Butadienes are obtained from the addition of disobutylaluminum hydride to dialkylacetylenes [190].

Alkenes may also be obtained from triorganoboranes by treatment with phenyl (bromodichloromethyl) mercury [191]:

$$(RCH_2CH_2)_3B \xrightarrow{C_6H_5HgCCl_2Br} RCH_2CH_2CH{=}CHCH_2R$$

Tri-*n*-hexylborane, for example, gives 6-tridecene (52 % *cis*, 48 % *trans*) in 58 % yield. It is thought that the mechanism involves a nucleophilic attack by dichlorocarbene on boron followed by alkyl group migration from boron to carbon.

(a) Preparation of Vinylcyclohexane.   To a solution of 1-ethylcyclohexene (100 mmoles) and sodium borohydride in 30 ml. of 1 *M* solution in

diglyme was added boron trifluoride (11 ml. of a 3.65 *M* solution in diglyme), and the mixture was refluxed 4 hr. 1-Decene (200 mmoles) was then added, and after refluxing for 6 hr. more the mixture was fractionated through a small Vigreux column to yield 62 % of $CH_2{=}CHC_6H_{11}$. 1-Methyl- or isopropylcyclohexenes give comparable yields of the terminal olefins [192].

1. D. V. Banthorpe, *Elimination Reactions*, Elsevier Publishing Co., New York, (1963).
2. G. H. Coleman and H. F. Johnstone, *Org. Syn.*, Coll. Vol. **1**, 183 (1941); C. F. H. Allen and S. Converse, *ibid.*, 226 (1941); H. Adkins and W. Zartman, *Org. Syn.*, Coll. Vol. **2**, 606 (1943).
3. J. F. Norris, *Org. Syn.* Coll. Vol. **1**, 430 (1941).
4. R. B. Carlin and D. A. Constantine, *J. Am. Chem. Soc.*, **69** 50 (1947); R. E. Miller and F. F. Nord, *J. Org. Chem.*, **15**, 89 (1950).
5. E. Levas, *Ann. Chim. (Paris)*, (12) **3**, 145 (1948).
6. B. B. Corson and V. N. Ipatieff, *Org. Syn.*, Coll. Vol. **2**, 151 (1943).
7. N. Campbell and D. Kidd, *J. Chem. Soc.*, 2154 (1954).
8. G. B. Bachman and L. L. Lewis, *J. Am. Chem. Soc.*, **69**, 2022 (1947).
9. E. Profft and H.-W. Linke, *Chem. Ber.*, **93**, 2591 (1960).
10. J. W. Schick and H. D. Hartough, *J. Am. Chem. Soc.*, **70**, 1646 (1948).
11. J. R. Dice *et al.*, *J. Am. Chem. Soc.*, **72**, 1738 (1950).
12. B. B. Elsner and H. E. Strauss, *J. Chem. Soc.*, 588 (1957).
13. V. J. Traynelis *et al.*, *J. Org. Chem.*, **27**, 2377 (1962); **29**, 123, 221 (1964).
14. W. Oroshnik *et al.*, *J. Am. Chem. Soc.*, **74**, 295 (1952).
15. R. Filler, *Chem. Rev.*, **63**, 21 (1963).
16. W. S. Allen *et al.*, *J. Am. Chem. Soc.*, **77**, 1028 (1955).
17. W. S. Emerson, *Chem. Rev.*, **45**, 347 (1949).
18. A. J. Lundeen and R. Van Hoozer, *J. Am. Chem. Soc.*, **85**, 2180 (1963).
19. J. D. Roberts and M. C. Caserio, *Basic Principles of Organic Chemistry*, W. A. Benjamin, New York, 1964, p. 395.
20. E. W. Abel *et al.*, *Chem. Ind. (London)*, 158 (1958).
21. F. C. Whitmore *et al.*, *J. Am. Chem. Soc.*, **56**, 1395 (1934).
22. E. von Rudloff, *Can. J. Chem.*, **39**, 1 (1961).
23. C. F. H. Allen ad A. Bell, *Org. Syn.*, Coll. Vol. **3**, 312 (1955).
24. L. W. Newton and E. R. Coburn, *Org. Syn.*, Coll. Vol. **3**, 313 (1955).
25. L. K. Friedlin and V. Z. Sharf, *Bull. Acad. Sci. USSR, Div. Chem. Sci. (English Transl.)*, 646 (1962).
26. O. Grummitt and E. I. Becker, *Org. Syn.*, Coll. Vol. **4**, 771 (1963).
27. J. B. Wommack and D. E. Pearson, unpublished results.
28. G. H. Coleman and H. F. Johnstone, *Org. Syn.*, Coll. Vol. **1**, 183 (1941); H. Waldmann and F. Petru, *Chem. Ber.*, **83**, 287 (1950).
29. A. J. Lundeen and R. Van Hoozer, *J. Org. Chem.*, **32**, 3386 (1967).

30. P. J. Hamrick, Jr. and C. R. Hauser, *J. Org. Chem.*, **26,** 4199 (1961).
31. C. R. Hauser *et al.*, *J. Am. Chem. Soc.*, **78,** 1653 (1956).
32. H. M. Crawford and H. B. Nelson, *J. Am. Chem. Soc.*, **68,** 134 (1946).
33. L. W. Newton and E. R. Coburn, *Org. Syn.*, Coll. Vol. **3,** 313 (1955).
34. L. Friedman and J. G. Berger, *J. Am. Chem. Soc.*, **83,** 492 (1961).
35. R. T. Arnold *et al.*, *J. Am. Chem. Soc.*, **86,** 3072 (1964).
36. H. C. Brown and O. H. Wheeler, *J. Am. Chem. Soc.*, **78,** 2199 (1956).
37. H. C. Brown and I. Moritani, *J. Am. Chem. Soc.*, **78,** 2203 (1956).
38. W. H. Saunders *et al.*, *J. Am. Chem. Soc.*, **87,** 2401 (1965); R. A. Bartsch and J. F. Bunnett, *J. Am. Chem. Soc.*, **90,** 408 (1968).
39. H. C. Brown and R. L. Klimisch, *J. Am. Chem. Soc.*, **88,** 1425 (1966).
40. D. H. Froemsdorf and M. D. Robbins, *J. Am. Chem. Soc.*, **89,** 1737 (1967).
41. S. J. Cristol *et al.*, *J. Am. Chem. Soc.*, **73,** 674 (1951).
42. A. Schriesheim *et al.*, *J. Am. Chem. Soc.*, **85,** 2115 (1963); W. B. Smith and W. H. Watson, *J. Am. Chem. Soc.*, **84,** 3174 (1962).
43. J. G. Traynham *et al.*, *J. Org. Chem.*, **32,** 510 (1967).
44. J. Sicher *et al.*, *Collection Czech. Chem. Commun.*, **31,** 4273 (1966); **32,** 2122 (1967); *Chem. Commun.*, 66 (1967); 394 (1967); *Tetrahedron Letters*, 4269 (1968).
45. W. S. Johnson *et al.*, *J. Am. Chem. Soc.*, **90,** 2882 (1968).
46. C. A. Grob *et al.*, *Helv. Chim. Acta.*, **40,** 130 (1957).
47. S. Bernstein *et al.*, *J. Am. Chem. Soc.*, **86,** 2309 (1964).
48. C. C. Price and J. M. Judge, *Org. Syn.*, **45,** 22 (1965).
49. A. N. Kost and I. I. Grandberg, *Zh. Obshch. Khim.*, **25,** 2064 (1955); *C.A.*, **50,** 8609 (1956).
50. W. S. Emerson and T. M. Patrick, Jr., *Org. Syn.*, Coll. Vol. **4,** 980 (1963).
51. H. H. Inhoffen *et al.*, *Chem. Ann.*, **585,** 132 (1954).
52. G. B. Pickering and J. C. Smith, *Rec. Trav. Chim.*, **69,** 535 (1950).
53. N. H. Cromwell *et al.*, *Org. Syn.*, Coll. Vol. **3,** 125 (1955).
54. U. Steiner and H. Schinz, *Helv. Chim. Acta*, **34,** 1176 (1951).
55. K. Eiter *et al.*, *Chem. Ber.*, **99,** 2012 (1966).
56. J. Cason *et al.*, *J. Org. Chem.*, **18,** 850 (1953).
57. R. Joly *et al.*, *Bull. Soc. Chim. France*, 366 (1958).
58. R. P. Holysz, *J. Am. Chem. Soc.*, **75,** 4432 (1953).
59. H. Plieninger *et al.*, *Chem. Ber.*, **94,** 2115 (1961); E. J. Corey and A. G. Hortmann, *J. Am. Chem. Soc.*, **87,** 5736 (1965).
60. M. Hanack *et al.*, *Chem. Ber.*, **95,** 191 (1962); *Ann. Chem.*, **652,** 96 (1962).
61. G. A. Ropp and E. C. Coyner, *Org. Syn.*, Coll. Vol. **4,** 727 (1963).
62. H. Oediger and F. Möller, *Angew. Chem.*, **79,** 53 (1967).
63. G. B. Pickering and J. C. Smith, *Rec. Trav. Chim.*, **69,** 535 (1950).
64. W. S. Emerson and T. M. Patrick, Jr., *Org. Syn.*, Coll. Vol. **4,** 980 (1963).
65. P. D. Gardner *et al.*, *J. Am. Chem. Soc.*, **87,** 3158 (1965).
66. J. P. Schaefer and L. Endres, *Org. Syn.*, **47,** 31 (1967).
67. W. S. Johnson *et al.*, *Org. Syn.*, Coll. Vol. **4,** 162 (1963).
68. W. W. Rinne *et al.*, *J. Am. Chem. Soc.*, **72,** 5759 (1950).
69. C. F. Allen and M. J. Kalm, *Org. Syn.*, Coll. Vol. **4,** 616 (1963).
70. J. C. Sauer, *Org. Syn.*, Vol. **4,** 268 (1963); *see also* T. Alfrey, Jr., *et al.*, *J. Am. Chem. Soc.*, **74,** 2097 (1952).
71. P. E. Spoerri and M. J. Rosen, *J. Am. Chem. Soc.*, **72,** 4918 (1950).
72. R. N. Haszeldine *et al.*, *J. Chem. Soc.*, 2040 (1954).
73. L. F. Fieser, *Org. Syn.*, Coll. Vol. **4,** 195 (1963).
74. L. F. and M. Fieser, *Reagents for Organic Synthesis*, John Wiley and Sons, New York, 1967, p. 1089.
75. R. K. Summerbell and R. R. Umhoefer, *J. Am. Chem. Soc.*, **61,** 3016 (1939).
76. K. M. Ibne-Rasa *et al.*, *Chem. Ind.* (*London*), 1418 (1966).
77. S. Dershowitz and S. Proskauer, *J. Org. Chem.*, **26,** 3595 (1961).
78. V. Mark, *Tetrahedron Letters*, 333 (1961).

79. C. C. Tung and A. J. Speziale, *J. Org. Chem.*, **28,** 1521 (1963).
80. P. D. Gardner *et al.*, *Chem. Ind.* (*London*), 345 (1965).
81. D. M. Singleton and J. K. Kochi, *J. Am. Chem. Soc.*, **89,** 6547 (1967); **90,** 1582 (1968).
82. E. T. McBee *et al.*, *J. Am. Chem. Soc.*, **77,** 4942 (1955).
83. A. V. Dombrovskiĭ and A. P. Terent'ev, *Zh. Obshch. Khim.*, **26,** 2776 (1956); *C.A.*, **51,** 7337 (1957).
84. C. E. Castro and W. C. Kray, *J. Am. Chem. Soc.*, **85,** 2768 (1963).
85. H. J. S. and H. Winkler, *Ann. Chem.*, **705,** 76 (1967).
86. P. D. Gardner *et al.*, *Chem. Ind.* (*London*), 766 (1965).
87. J. H. Gorvin, *J. Chem. Soc.*, 678 (1959).
88. P. D. Gardner and M. Narayana, *J. Org. Chem.*, **26,** 3518 (1961).
89. C. E. Coffey, *J. Am. Chem. Soc.*, **83,** 1623 (1961).
90. R. E. Buckles and G. M. Matlack, *Org. Syn.*, Coll. Vol. **4,** 914 (1963).
91. J. Sonnenberg and S. Winstein, *J. Org. Chem.*, **27,** 748 (1962).
92. V. Mark, *Org. Syn.*, **46,** 93 (1966).
93. H. N. Cripps and E. F. Kiefer, *Org. Syn.*, **42,** 12 (1962).
94. C. Finger *et al.*, German Patent 1,235,293, March 2, 1967; *C.A.*, **67,** 9353 (1967).
95. H. Nozaki *et al.*, *Tetrahedron Letters*, 2087 (1968).
96. C. E. Boord *et al.*, *J. Am. Chem. Soc.*, **52,** 3396 (1930); C. G. Schmitt and C. E. Boord, *ibid.*, **54,** 751 (1932); F. J. Soday and C. E. Boord, *ibid.*, **55,** 3293 (1933).
97. O. Grummitt *et al.*, *Org. Syn.*, Coll. Vol. **4,** 748 (1963).
98. C. Niemann and C. D. Wagner, *J. Org. Chem.*, **7,** 227 (1942).
99. L. Crombie, *Quart. Rev.* (*London*), **6,** 131 (1952).
100. N. Green *et al.*, *J. Med. Chem.*, **10,** 533 (1967).
101. J. W. Cornforth *et al.*, *J. Chem. Soc.*, 112 (1959).
102. D. J. Cram and F. A. A. Elhafez, *J. Am. Chem. Soc.*, **74,** 5828 (1952).
103. W. J. Bailey and J. Economy, *J. Org. Chem.*, **23,** 1002 (1958).
104. J. M. Stewart *et al.*, *J. Org. Chem.*, **25,** 913 (1960).
105. E. R. Alexander and A. Mudrak, *J. Am. Chem. Soc.*, **72,** 1810 (1950).
106. W. S. Emerson, *Chem. Rev.*, **45,** 347 (1949).
107. P. J. Hamrick, Jr., and C. R. Hauser, *J. Org. Chem.*, **26,** 4199 (1961).
108. R. T. Blickenstaff and F. C. Chang, *J. Am. Chem. Soc.*, **80,** 2726 (1958).
109. G. Eglinton and M. C. Whiting, *J. Chem. Soc.*, 3650 (1950).
110. D. J. Cram *et al.*, *J. Am. Chem. Soc.*, **75,** 2293 (1953).
111. H. R. Nace, *J. Am. Chem. Soc.*, **81,** 5428 (1959).
112. M. A. Davis and W. J. Hickinbottom, *J. Chem. Soc.*, 1998 (1957).
113. G. L. O'Connor and H. R. Nace, *J. Am. Chem. Soc.*, **75,** 2118 (1953).
114. G. L. O'Connor and H. R. Nace, *J. Am. Chem. Soc.*, **77,** 1578 (1955).
115. C. H. DePuy and R. W. King, *Chem. Rev.*, **60,** 431 (1960).
116. J. B. Siddall *et al.*, *J. Am. Chem. Soc.*, **90,** 6224 (1968).
117. P. Bladon and L. N. Owen, *J. Chem. Soc.*, 598 (1950).
118. A. B. Foster and W. G. Overend, *ibid.*, 3452 (1951).
119. H. L. Slates and N. L. Wender, *J. Am. Chem. Soc.*, **78,** 3749 (1956).
120. R. E. Benson and B. C. McKusick, *Org. Syn.*, Coll. Vol. **4,** 746 (1963).
121. D. W. Aubrey *et al.*, *Chem. Ind.* (*London*), 681 (1965).
122. H. R. Nace, *Org. Reactions*, **12,** 57 (1962).
123. I. Schurman and C. E. Boord, *J. Am. Chem. Soc.*, **55,** 4930 (1933).
124. J. D. Roberts and C. W. Sauer, *J. Am. Chem. Soc.*, **71,** 3925 (1949).
125. L. S. McNamara and C. C. Price, *J. Org. Chem.*, **27,** 1230 (1962).
126. A. C. Cope and E. R. Trumbell, *Org. Reactions*, **11,** 317 (1960).
127. H. C. Brown and I. Moritani, *J. Am. Chem. Soc.*, **75,** 4112 (1953).
128. I. N. Feit and W. H. Saunders, Jr., *Chem. Commun.*, 610 (1967).
129. R. J. Baumgarten, *J. Chem. Ed.*, **45,** 122 (1968).
130. L. D. Freedman, *J. Chem. Ed.*, **43,** 662 (1966).
131. P. A. S. Smith and S. Frank, *J. Am. Chem. Soc.*, **74,** 509 (1952).

132. D. V. Banthorpe *et al.*, *J. Chem. Soc.*, 4054 (1960).
133. J. Weinstock and V. Boekelheide, *J. Am. Chem. Soc.*, **75,** 2546 (1953).
134. W. Hanhart and C. K. Ingold, *J. Chem. Soc.*, 997 (1927).
135. C. K. Ingold and C. S. Patel, *J. Chem. Soc.*, 68 (1933).
136. D. J. Cram *et al.*, *J. Am. Chem. Soc.*, **78,** 790 (1956).
137. K. Ziegler and H. Wilms, *Ann. Chem.*, **567,** 1 (1950).
138. A. T. Blomquist *et al.*, *J. Am. Chem. Soc.*, **74,** 3643 (1952).
139. A. C. Cope *et al.*, *J. Am. Chem. Soc.*, **77,** 1628 (1955); **81,** 3153 (1959); **82,** 1744 (1960).
140. J. Rabiant and G. Wittig, *Bull. Soc. Chim. France*, 798 (1957).
141. G. Wittig and R. Polster, *Ann. Chem.*, **612,** 102 (1958).
142. A. C. Cope *et al.*, *J. Am. Soc.*, **75,** 3212 (1953).
143. J. von Braun and E. Anton, *Chem. Ber.*, **64,** 2865 (1931).
144. F. F. Blicke, *Org. Reactions*, **1,** 303 (1942).
145. H. J. Hagemeyer, Jr., *J. Am. Chem. Soc.*, **71,** 1119 (1949).
146. J. B. Wright, *J. Org. Chem.*, **25,** 1867 (1960).
147. H. M. E. Cardwell, *J. Chem. Soc.*, 1056 (1950).
148. R. J. Baumgarten and P. L. DeChristopher, *Tetrahedron Letters*, 3027 (1967).
149. V. Boekelheide and A. L. Sieg, *J. Org. Chem.*, **19,** 587 (1954).
150. A. C. Cope and E. R. Trumbell, *Org. Reactions*, **11,** 361 (1960).
151. D. J. Cram *et al.*, *J. Am. Chem. Soc.*, **84,** 1734 (1962).
152. A. C. Cope *et al.*, *J. Am. Chem. Soc.*, **79,** 4720 (1957).
153. A. C. Cope and E. Ciganek, *Org. Syn.*, Coll. Vol. **4,** 612 (1963).
154. C. A. Kingsbury and D. J. Cram, *J. Am. Chem. Soc.*, **82,** 1810 (1960).
155. C. Walling and L. Bollyky, *J. Org. Chem.*, **29,** 2699 (1964).
156. L. V. Vargha and E. Kovács, *Chem. Ber.*, **75,** 794 (1942).
157. J. E. Hofmann *et al.*, *Chem. Ind.* (*London*), 1243 (1963).
158. T. J. Wallace *et al.*, *J. Am. Chem. Soc.*, **85,** 2739 (1963).
159. J. L. Kice and J. D. Campbell, *J. Org. Chem.*, **32,** 1631 (1967).
160. G. W. Fenton and C. K. Ingold, *J. Chem. Soc.*, 705 (1930).
161. L. A. Paquette, *Accounts Chem. Research*, **1,** 209 (1968); *J. Am. Chem. Soc.*, **86,** 4383 (1964).
162. G. Opitz and K. Fischer, *Angew. Chem.*, **77,** 41 (1965).
163. H. H. Inhoffen *et al.*, *Ann. Chem.*, **694,** 19 (1966).
164. E. J. Corey and T. Durst, *J. Am. Chem. Soc.*, **90,** 5548 and 5553 (1968).
165. E. J. Corey and G. T. Kwiatkowski, *J. Am. Chem. Soc.*, **90,** 6816 (1968).
166. T. E. Sample and L. F. Hatch, *J. Chem. Ed.*, **45,** 55 (1968).
167. A. Schönberg and E. Frese, *Chem. Ber.*, **95,** 2810 (1962).
168. R. D. Schuetz and R. L. Jacobs, *J. Org. Chem.*, **23,** 1799 (1958); **26,** 3467 (1961).
169. M. J. Boskin and D. B. Denney, *Chem. Ind.* (*London*), 330 (1959).
170. R. E. Davis, *J. Org. Chem.*, **23,** 1767 (1958).
171. F. G. Bordwell *et al.*, *J. Am. Chem. Soc.*, **76,** 1082 (1954).
172. N. P. Neureiter and F. G. Bordwell, *J. Am. Chem. Soc.*, **81,** 578 (1959).
173. J. K. Kochi *et al.*, *Tetrahedron*, **24,** 3503 (1968).
174. G. Wittig and W. Haag, *Chem. Ber.*, **88,** 1654 (1955).
175. N. Sperber *et al.*, *J. Am. Chem. Soc.*, **72,** 3068 (1950).
176. Ch. Weizmann *et al.*, *J. Am. Chem. Soc.*, **70,** 1153 (1948).
177. M. Viscontini and H. Köhler, *Helv. Chim. Acta*, **37,** 41 (1954).
178. W. Kirmse and K. Horn, *Chem. Ber.*, **100,** 2698 (1967).
179. R. H. Shapiro and M. J. Heath, *J. Am. Chem. Soc.*, **89,** 5734 (1967).
180. H. Shechter *et al.*, *J. Am. Chem. Soc.*, **89,** 5736, 7112 (1967).
181. R. K. Bartlett and T. S. Stevens, *J. Chem. Soc.*, (C), 1964 (1967).
182. J. W. Wilt and P. J. Chenier, *J. Am. Chem. Soc.*, **90,** 7366 (1968).
183. H. W. Whitlock, Jr., *J. Am. Chem. Soc.*, **84,** 2807 (1962).
184. L. Y. Malkes and L. V. Shubina, *J. Gen. Chem. USSR.* (*Eng. Transl.*), **32,** 281 (1962).
185. J. K. Crandall and L.-H. C. Lin, *J. Am. Chem. Soc.*, **89,** 4527 (1967).
186. H. P. Fischer *et al.*, *Chimia*, **22,** 338 (1968).

187. H. C. Brown *et al.*, *J. Am. Chem. Soc.*, **89**, 4531 (1967); **81**, 1512 (1959).
188. G. Zweifel *et al.*, *J. Am. Chem. Soc.*, **89**, 3652 (1967).
189. G. Zweifel *et al.*, *J. Am. Chem. Soc.*, **90**, 6243 (1968).
190. G. Wilke and H. Müller, *Ann. Chem.*, **629**, 222 (1960).
191. D. Seyferth and B. Prokai, *J. Am. Chem. Soc.*, **88**, 1834 (1966).
192. H. C. Brown *et al.*, *J. Am. Chem. Soc.*, **89**, 567 (1967).

B. Reduction

1. From Acetylenes

$$RC{\equiv}CR + H_2 \xrightarrow{\text{Pd-BaSO}_4} RCH{=}CHR \qquad cis$$

$$RC{\equiv}CR \xrightarrow{\text{LiAlH}_4} RCH{=}CHR \qquad trans$$

Recent discussions of the catalytic hydrogenation of acetylenes to olefins are available [1, 2]. Deactivated palladium at room temperature and atmospheric pressure is the preferred catalyst for the hydrogenation. This noble metal on calcium carbonate deactivated by lead acetate and quinoline (Lindlar's catalyst) [3] has been rather widely used, although the metal on barium sulfate deactivated by an equal weight of pure, synthetic quinoline is superior both in ease of preparation and in reproducibility of results [4]. In the case of the partial reduction of acetylenic carbinols, palladium on barium carbonate deactivated with powdered potassium hydroxide is the preferred catalyst [5]. Recently a new catalyst, nickel boride, designated P-2, formed by the reaction of sodium borohydride with nickel acetate in ethanol, has shown remarkable promise [6]. It is selective in its action and exhibits a low tendency toward producing isomerization. The hydrogenation of hexyne-3, for example, in the presence of this catalyst produces 98–99 % of *cis*-hexene-3. Various forms of Raney nickel [7–9] at room temperature and atmospheric pressure may be employed in the hydrogenation, but as a rule such catalysts are inferior to deactivated palladium. Metals such as platinum, rhodium, and ruthenium are nonselective in the reduction.

The acetylenic group is hydrogenated over palladium or nickel more readily than any other functional group. The absorption of the group on the catalyst appears to be so strong that with the proper amounts of catalyst and substrate all other functional groups will be displaced. Usually an amount of catalyst equal to 1–2 % of the weight of substrate [1] is employed. In the case of a disubstituted acetylene the product of the selective hydrogenation is almost exclusively *cis*. Under mild conditions no double bond migration or isomerization occurs.

The synthesis has been utilized as well in the preparation of ethylenic derivatives. For example, olefinic acids [9], olefinic alcohols [8, 10], olefinic esters [10, 4], cyclic olefinic acetals [11], and olefinic enol ethers [12] have been synthesized from the corresponding alkynes.

A second method for converting alkynes into *cis*- alkenes in good yield is via the addition product formed with a dialkylborane [13] having bulky alkyl groups. On treatment with acetic acid at 0°, the unsaturated boron derivative

References for Section B are on p. 112.

gives the *cis*-olefin. Alkynes may also be converted into *cis*- and *trans*-alkenes via the dibromide of the borane addition product (see A.16).

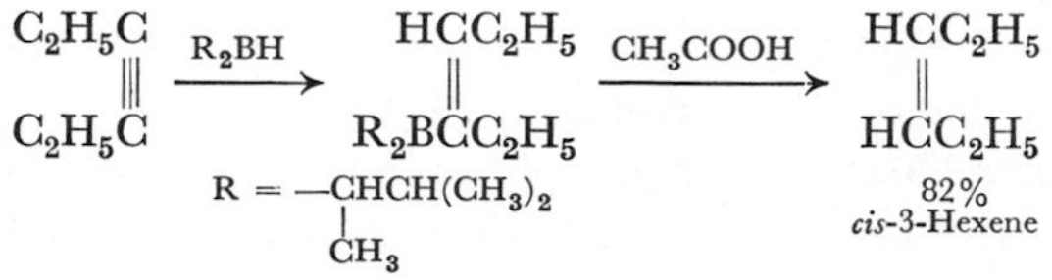

$$\underset{\substack{\|\\ C_2H_5C}}{C_2H_5C} \xrightarrow{R_2BH} \underset{\substack{\|\\ R_2BCC_2H_5}}{HCC_2H_5} \xrightarrow{CH_3COOH} \underset{\substack{\|\\ HCC_2H_5}}{HCC_2H_5}$$

$$R = -CHCH(CH_3)_2$$
$$\qquad\quad |$$
$$\qquad\quad CH_3$$

82%
cis-3-Hexene

The synthesis of pure *trans*-olefins from acetylenes has also been accomplished by reduction with sodium in liquid ammonia [7], lithium in liquid ammonia [14], or with lithium aluminum hydride in tetrahydrofuran [15]. Recent work favors higher temperatures for the latter reductions [16] which occurs stepwise as follows:

$$LiAlD_4 + HC \equiv CH \rightarrow LiAlD_3CH = CHD \xrightarrow{D_2O} DCH = CHD$$

It is interesting to note that partial catalytic hydrogenation of cyclynes over palladium on barium carbonate gives *cis*-cyclenes, while hydrogenation with sodium in liquid ammonia yields *trans*-isomers [17].

Acetylenes can also be reduced by chromous salts (see Ex. c.2).

(a) Preparation of a *cis*-olefin (general method). The acetylene, 10 g., in 75–100 ml. of methanol is hydrogenated at room temperature and atmospheric pressure over 200 mg. of 5% palladium on barium sulfate [18] previously deactivated by the addition of 200 mg. (5–6 drops) of pure, synthetic quinoline to the reaction mixture. Hydrogen absorption becomes very slow after 1 equivalent is utilized and the yield of the *cis*-olefin is excellent [1].

(b) Preparation of *threo-trans*-1,3-dihydroxy-2-amino-4-heptadecene (70% by refluxing *threo*-1,3-dihydroxy-2-amino-4-heptadecyne in tetra-

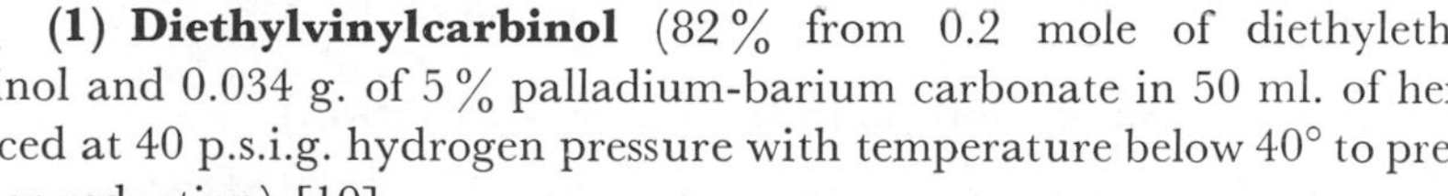

$$\underset{\substack{\|\\ C_{12}H_{25}C \equiv CCHOHCHNH_2CH_2OH}}{} \longrightarrow \underset{\substack{\|\\ HCCHOHCHNH_2CH_2OH}}{C_{12}H_{25}CH}$$

hydrofuran with lithium aluminum hydride for 4 hr.) [15].

(c) Other examples

(1) Diethylvinylcarbinol (82% from 0.2 mole of diethylethynyl carbinol and 0.034 g. of 5% palladium-barium carbonate in 50 ml. of hexane reduced at 40 p.s.i.g. hydrogen pressure with temperature below 40° to prevent further reduction) [10].

(2) *o*-Carboxy-*trans*-stilbene (85% from 0.15 g. of 2-carboxy-diphenylacetylene in 10 ml. of dimethylformamide to which a solution of 0.43 g.

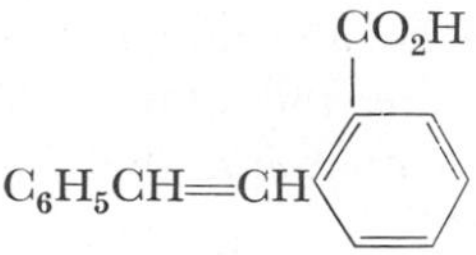

of chromous sulfate pentahydrate in 30 ml. of water and 50 ml. of dimethyl-formamide was added. The solution became red and then green after 3 days. Dilution and extraction from the ethereal extract with aqueous sodium bicarbonate followed by acidification gave the desired acid. The 4-carboxy isomer did not reduce under the same conditions [19].

(3) 3-*trans*-**Hexene** (96% containing about 4% of the *cis*-isomer from 85 mmoles of lithium aluminum hydride in 50 ml. each of THF and diglyme by heating to remove some solvent until internal temperature was 138°, by then adding 50 mmoles of 3-hexyne, and finally by refluxing for 4.5 hr.) [20].

2. From Dienes

Partial reduction of acetylenes is important and widely used because it permits access to *cis*-olefins. On the other hand, partial reduction of dienes is of limited use and rarely employed. It can be carried out by reduction with hydrogen using nickel or zinc chromite [21]. Conjugated dienes can be reduced also with metal hydrides. Potassium hydride is a powerful enough reducing agent to reduce dienes to saturated hydrocarbons, but sodium hydride reduction can be stopped at the olefinic stage. Although it appears that any conjugated, unsaturated compound can be reduced by hydrides, adjustment of conditions can sometimes avoid reduction of the olefinic group [22]:

$$C_6H_5CH=CHCO_2Me + LiAlH_4 \xrightarrow[60°,\ 14.5\ \text{hr.}]{45\ \text{ml.}\ C_6H_6} C_6H_5CH=CHCH_2OH$$

2.9 g. 0.7 g. Cinnamyl alcohol 73%, 96% purity

Ether in place of benzene yielded $C_6H_5CH_2CH_2CH_2OH$.

Hydride reduction may lead to anionic isomerization (see Ex. a).

(a) Preparation of bicyclo[3.3.0]-2-octene. Potassium hydride (0.02

mole) and 1,5-cyclooctadiene (4.7 moles) were stirred in an autoclave for 10–11 hr. at 190°. The mixture was cooled, decanted from potassium hydride, and fractionated to yield 94% of the crude title compound [23].

3. From Aromatic Compounds (Birch Reduction)

Alkali metals in liquid ammonia or amines are high-potential reducing agents [24]. It is important that iron salts, often present in undistilled liquid ammonia, be removed since they inhibit the reducing action of sodium or potassium by promoting metal amide formation [25]. Examples of reduction with lithium and ethylamine are shown in the accompanying formulations. Alcohols added to the liquid amine tend to buffer the solution and furnish more readily available protons to interrupt the reduction at the dihydro stage or to prevent isomerization.

$$C_6H_6 \longrightarrow C_6H_{10} + C_6H_{12}$$

Cyclohexene

51%   17%

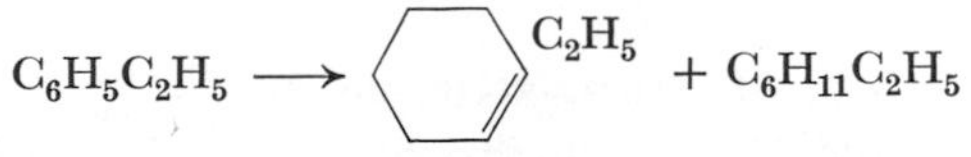

$$C_6H_5C_2H_5 \longrightarrow \langle\text{ring}\rangle C_2H_5 + C_6H_{11}C_2H_5$$

1-Ethylcyclohexene

44%   24%

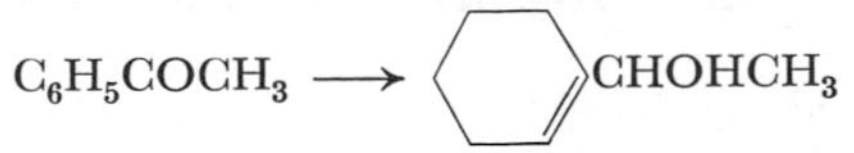

$$C_6H_5COCH_3 \longrightarrow \langle\text{ring}\rangle CHOHCH_3$$

Methyl-1-cyclohexenylcarbinol

65%

Lithium metal has been used most frequently, but potassium is actually more selective in synthesizing 1-alkylcyclohexenes [26]. Ethylenediamine is preferred as the solvent because of its high boiling point and consequent wider selection of reaction temperatures (see Ex. b.1, b.2).

(a) **Preparation of 5,8-dihydro-α-naphthol.** α-Naphthol (0.75 mole) was dissolved in liquid ammonia and 3 g. atoms of lithium was added in small enough pieces to prevent rapid refluxing. The solution turned deep blue and was then treated dropwise with 3 moles of dry ethanol. After evaporation of the ammonia, the product was dissolved in water and cautiously acidified to yield 97–99% of the title compound [27].

(b) **Other examples**

(1) **$\Delta^{9,10}$-Octalin** (97% from 0.2 mole of tetralin in 375 ml. of dry ethylenediamine at 100° to which 1.6 moles of lithium was added slowly; the product probably contained small amounts of other isomers [28, 26].

(2) **1,4-Di-t-butylcyclohexene** (44% containing 12% starting material from 1,4-di-t-butylbenzene in ethylenediamine at 100° under N_2 to which lithium was added) [29].

4. From Benzoins

$$\text{ArCOCHOHAr} \xrightarrow[\text{H}_2]{\text{Zn + HCl}} \text{ArCH}\!=\!\text{CHAr}$$

In a limited number of cases, stilbenes have been obtained by the reduction of benzoins. In the most interesting case, cis-p,p'-diphenylstilbene is produced by the reduction of p,p'-diphenylbenzoin with zinc dust, whereas with amalgamated zinc the reduction product is trans-p,p'-diphenylstilbene [30]; see Ex. a, b).

(a) **Preparation of cis-p,p'-diphenylstilbene.** With stirring and refluxing, hydrogen chloride and hydrogen were passed through a mixture of 1.5 g. of p,p'-diphenylbenzoin and 1 g. of zinc dust in 19 ml. of alcohol for 2.5–3 hr. The yellow precipitate which formed, after diluting the mixture with water, was filtered, washed with water, and dried. Recrystallization from benzene followed by sublimation in a vacuum gave the cis-stilbene (84.6%), m.p. 221–222° [30].

(b) Preparation of *trans-p,p'*-diphenylstilbene (82% from a mixture of *p,p'*-diphenylbenzoin, amalgamated zinc, and hydrochloric acid in alcohol through which hydrogen was passed for 1 hr.) [30].

(c) Preparation of *trans*-stilbene (53–57% from benzoin by reduction with zinc amalgam and hydrochloric acid below 15°) [31].

5. From Ketones (Kishner Reduction-Elimination)

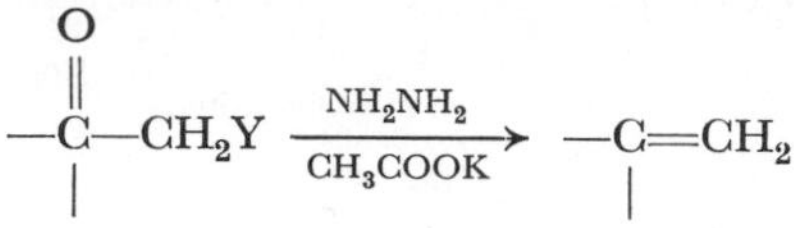

$$\overset{\displaystyle O}{\underset{|}{\overset{||}{C}}}-CH_2Y \xrightarrow[CH_3COOK]{NH_2NH_2} \underset{|}{-C}=CH_2$$

In the normal Kishner reduction the carbonyl group is reduced to a methylene group. A side reaction which sometimes occurs is the formation of the alkene. This elimination reaction has been investigated [32, 33] in cases in which Y is a halogen, a substituted amino, an ether, or a thioether group. The yields in the elimination are not high. The best yield (71%) was obtained from 2-α-fluorocholestanone by adding the α-halo ketone to a two-phase system of boiling cyclohexene or cyclohexane and an excess of hydrazine hydrate and potassium acetate as a buffer [33]. The yield decreased with increasing atomic weight of the halide, the value for the 2-α-iodocholestanone being 54%.

It has been suggested [33] that the reaction proceeds via the hydrazone 2-XXI,

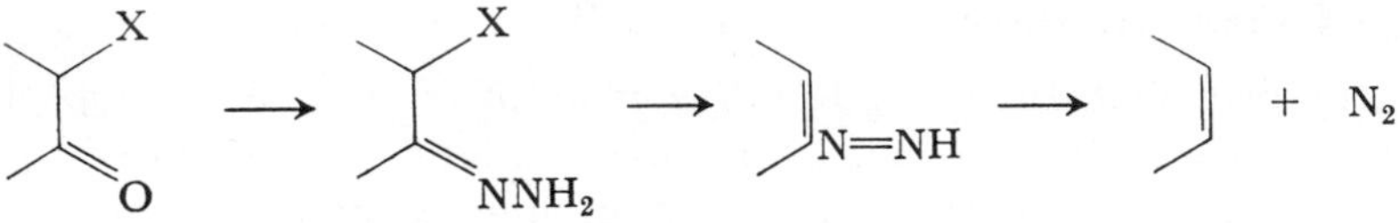

which loses the hydrogen halide to give the alkenyldiimide 2-XXII, which in turn eliminates nitrogen to form the alkene 2-XXIII.

(a) Preparation of 2-cholestene. To 13 ml. (0.4 mole) of hydrazine

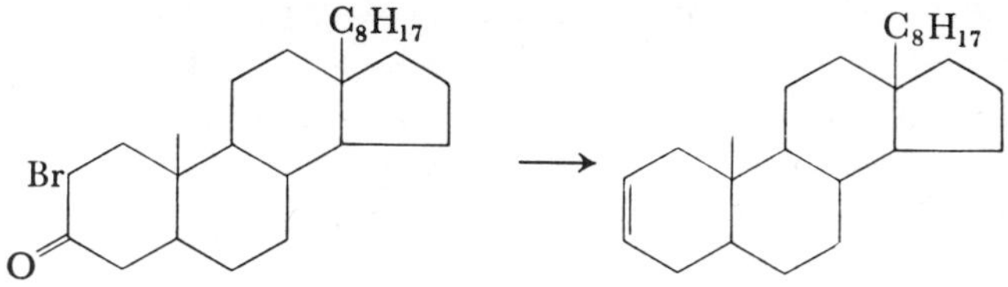

hydrate, 2 g. (20 mmoles) of potassium acetate, and 10 ml. of cyclohexene at the boiling temperature was added 2.004 g. (4.3 mmoles) of 2-α-bromocholestanone in 30 ml. of cyclohexene over 10 min. with stirring at the same temperature. After 30 min. of heating, the mixture was cooled, extracted with ether–water, dried, evaporated, and dissolved in hexane. Percolation through a column of acid-washed alumina (Merck) and evaporation gave 995 mg. (64%) of 2-cholestene, m.p. 72–74° [33].

6. From Enamines

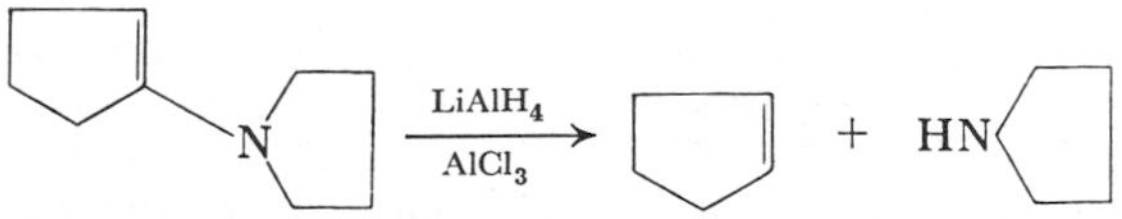

Enamines are reduced to alkenes by lithium aluminum hydride and aluminum chloride [34] or by catalytic hydrogenation over platinum followed by treatment with alcoholic potassium hydroxide and steam [35]. Superior yields (85–98%) are obtained by conversion first into the borane followed by treatment with a carboxylic acid [36]; see Ex. b. Since enamines are produced from carbonyl compounds, the method is of value in synthesizing cyclic alkenes from ketones. In rings containing two carbonyl groups it permits the introduction of one double bond [35].

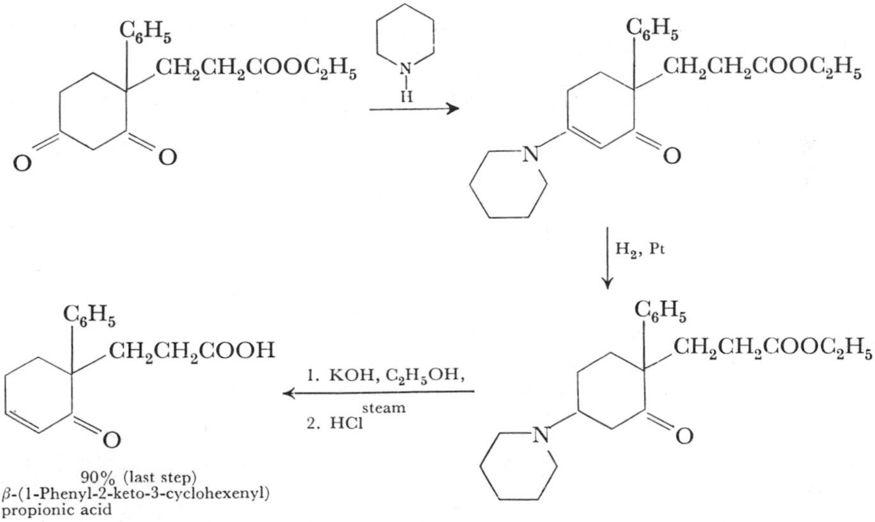

90% (last step)
β-(1-Phenyl-2-keto-3-cyclohexenyl)
propionic acid

(a) Preparation of cyclopentene. The enamine in ether was refluxed with 1 mole of lithium aluminum hydride and 1 mole of aluminum chloride for 5–24 hr. Acidification with cold hydrochloric acid, extraction with ether, and evaporation and distillation of the extract gave cyclopentene (83%) [34].

(b) Preparation of 3-methylcyclohexene. 1-Pyrrolidinocyclopentene methylcyclohexanone and pyrrolidine were treated with sodium borohydride and boron trifluoride in diglyme to form the borohydride derivative:

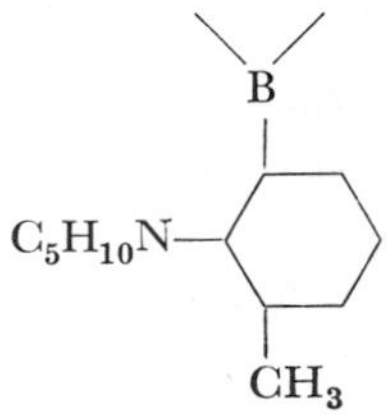

On the addition of acetic acid and refluxing, the title compound was obtained in 95% yield [36].

1. R. L. Augustine, *Catalytic Hydrogenation*, Marcel Dekker, New York, 1965, p. 69.
2. P. N. Rylander, *Catalytic Hydrogenation over Platinum Metals*, Academic Press, New York, 1967, p. 59.
3. H. Lindlar, *Helv. Chim. Acta*, **35**, 446 (1952); H. Lindlar and R. Dubuis, *Org. Syn.*, **46**, 89 (1966).
4. D. J. Cram and N. L. Allinger, *J. Am. Chem. Soc.*, **78**, 2518 (1956).
5. R. J. Tedeschi and G. Clark, Jr., *J. Org. Chem.*, **27**, 4323 (1962).
6. H. C. Brown and C. A. Brown, *J. Am. Chem. Soc.*, **85**, 1005 (1963).
7. K. N. Campbell and L. T. Eby, *J. Am. Chem. Soc.*, **63**, 216 (1941).
8. S. H. Harper and R. J. D. Smith, *J. Chem. Soc.*, 1512 (1955).
9. D. R. Howton and R. H. Davis, *J. Org. Chem.*, **16**, 1405 (1951).
10. G. F. Hennion *et al.*, *J. Org. Chem.*, **21**, 1142 (1956).
11. H. Newman, *Chem. Ind. (London)*, 372 (1963).
12. H. Heusser *et al.*, *Helv. Chim. Acta*, **33**, 370 (1950).
13. H. C. Brown and G. Zweifel, *J. Am. Chem. Soc.*, **81**, 1512 (1959).
14. R. E. Dear and F. L. M. Pattison, *J. Am. Chem. Soc.*, **85**, 622 (1963).
15. E. F. Jenny and J. Druey, *Helv. Chim. Acta*, **42**, 401 (1959).
16. L. H. Slaugh, *Tetrahedron*, **22**, 1741 (1966).
17. A. T. Blomquist *et al.*, *J. Am. Chem. Soc.*, **74**, 3636, 3643 (1952).
18. Ref. 1, App 4B, p. 152.
19. C. E. Castro and R. D. Stephens, *J. Am. Chem. Soc.*, **86**, 4358 (1964).
20. E. F. Magoon and L. H. Slaugh, *Tetrahedron*, **23**, 4509 (1967).
21. G. Natta *et al.*, *Chim. Ind. (Milan)*, **29**, 235 (1947); *C.A.*, **42**, 5839 (1948).
22. E. I. Snyder, *J. Org. Chem.*, **32**, 3531 (1967).
23. L. H. Slaugh, *J. Org. Chem.*, **32**, 108 (1967).
24. A. J. Birch and H. Smith, *Quart. Rev.*, **12**, 17 (1958); H. Smith, *Organic Reactions in Liquid Ammonia*, John Wiley and Sons, New York, 1963, Vol. **1**, Pt. 2.
25. H. O. House, *Modern Synthetic Reactions*, W. A. Benjamin, New York, 1965, p. 70.
26. L. H. Slaugh and J. H. Raley, *J. Org. Chem.*, **32**, 2861 (1967).
27. C. O. Gutsche and H. H. Peter, *Org. Syn.*, Coll. Vol. **4**, 887 (1963).
28. I. Wender *et al.*, *J. Org. Chem.*, **22**, 891 (1957).
29. R. D. Stolow and J. A. Ward, *J. Org. Chem.*, **31**, 965 (1966).
30. E. E. Baroni and K. A. Kovyrzina, *J. Gen. Chem. USSR*, (*Eng. Transl.*), **30**, 1664 (1960).
31. R. L. Shriner and A. Berger, *Org. Syn.*, Coll. Vol. **3**, 786 (1955).
32. N. J. Leonard and S. Gelfand, *J. Am. Chem. Soc.*, **77**, 3269, 3272 (1955).
33. P. S. Wharton *et al.*, *J. Org. Chem.*, **29**, 958 (1964).
34. J. W. Lewis and P. P. Lynch, *Proc. Chem. Soc.*, 19 (1963).
35. P. Kloss, *Chem. Ber.*, **97**, 1723 (1964).
36. J. W. Lewis and A. A. Pearce, *Tetrahedron Letters*, 2039 (1964).

C. Addition and Coupling Reactions

Unsaturated compounds undergo self-addition or addition to other unsaturated compounds. Terminal olefins particularly are prone to undergo addition even on storage, but some nonterminal olefins, such as cyclohexene, form mixtures of unsaturated dimers, trimers, and polymers on standing. The additions are brought about by acid, base, free radical sources, light, or cycloaddition forces. They are discussed more or less on this basis with emphasis on dimerization and trimerization rather than polymerization.

Some additions of unsaturated compounds are found elsewhere (unsaturated acetals from vinyl ethers, Chapter 9, B.2).

References for Section C are on pp. 131–133.

1. From Olefins by Acid or Ziegler-Type Catalysis

$$RCH=CH_2 \xrightarrow{H\oplus} [RCHCH_3] \xrightarrow{RCH=CH_2} \left[RCHCH_2CH \atop CH_3 \right]^{\oplus} \xrightarrow{-H\oplus} RCH=CHCH \atop CH_3$$

Strong Lewis acids such as $AlCl_3$ or BF_3 tend to give polymers, but aqueous acids such as sulfuric or phosphoric give products of lower molecular weight. Since the reaction is of a carbonium ion nature, all the potential reactions of carbonium ions, such as hydride shifts, alkyl migration, cyclization, not to mention further polymerization, are possible. For example, the treatment of tetramethylethylene with aqueous sulfuric acid leads to a fraction of C_{12} olefins containing at least 20 isomers [1]. Nevertheless, the process is simple and economical enough to prove feasible as an occasional method of preparation if good fractionation equipment is available (see Ex. a).

1,5-Dienes with acid give a preponderance of cyclized olefins, either six- or five-membered (see Ex. b).

Of greater interest and of growing importance is the use of Ziegler catalysts in bringing about the formation of dimeric or trimeric olefins. They might be called DNA's of olefin chemistry. The proper choice of catalyst and conditions may lead to amazing results. For example, the catalyst combination which leads to polymerization of butadiene is ideal for the dimerization of ethylene [2]:

$$2\ CH_2=CH_2 \xrightarrow[Al(C_2H_5)_3]{Ti(OC_4H_9)_4} CH_3CH_2CH=CH_2$$

On the other hand, the catalyst which polymerizes ethylene is ideal for the trimerization of butadiene [2]:

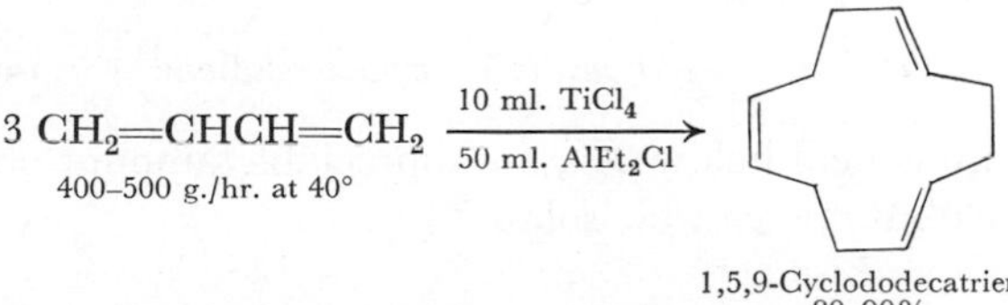

$$3\ CH_2=CHCH=CH_2 \xrightarrow[\substack{400-500\ g./hr.\ at\ 40°}]{10\ ml.\ TiCl_4 \atop 50\ ml.\ AlEt_2Cl}$$

1,5,9-Cyclododecatriene,
80–90%

This compound is now available commercially. Incidentally, it can be converted into $\Delta^{1,6}$-bicyclo[4.6.0]dodecene,

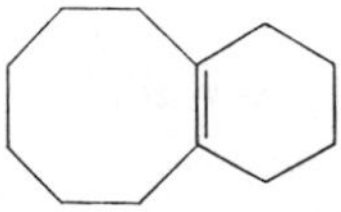

in 20% yield by solution in 80% aqueous sulfuric acid at 15° [3].

Dienes react even with metallic magnesium in tetrahydrofuran [4]:

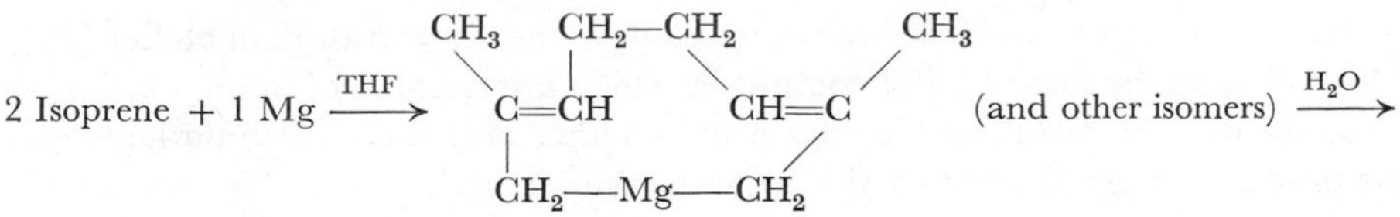

$$2\ \text{Isoprene} + 1\ \text{Mg} \xrightarrow{THF} \begin{array}{c} CH_3 \quad CH_2-CH_2 \quad CH_3 \\ \diagdown \qquad \qquad \diagup \\ C=CH \qquad CH=C \qquad \text{(and other isomers)} \xrightarrow{H_2O} \\ \diagup \qquad \qquad \diagdown \\ CH_2-Mg-CH_2 \end{array}$$

$C_{10}H_{18}$ (mixture of isomeric diolefins)

2,3-Dimethylbutadiene with Ziegler-type catalysts yields an acyclic trimer [2]. Other metal ion catalysts can be used for dimerization, trimerization, or isomerization of olefins [5]. As one example, a rhodium catalyst behaves in the following way [6]:

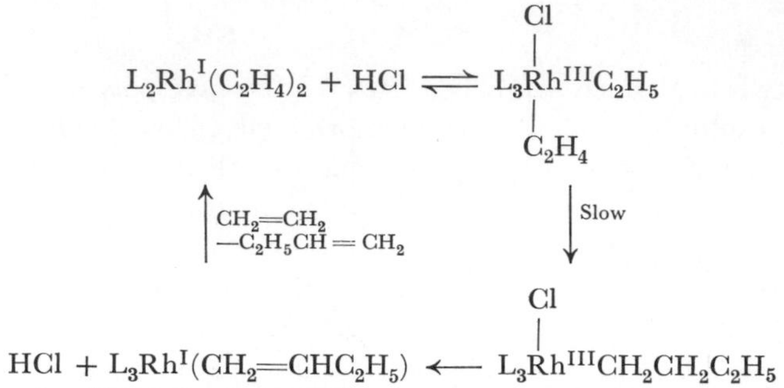

$$L_2Rh^I(C_2H_4)_2 + HCl \rightleftharpoons L_3\overset{\displaystyle Cl}{\underset{\displaystyle C_2H_4}{Rh^{III}C_2H_5}}$$

$$\left| \begin{array}{l} CH_2{=}CH_2 \\ {-}C_2H_5CH{=}CH_2 \end{array} \right. \qquad \Big\downarrow \text{Slow}$$

$$HCl + L_3Rh^I(CH_2{=}CHC_2H_5) \longleftarrow L_3\overset{\displaystyle Cl}{Rh^{III}CH_2CH_2C_2H_5}$$
$$L = (C_6H_5)_3 \text{ P or cyclopentadiene}$$

The formation of butene by this process takes place at temperatures as low as 25°. Substituted olefins of many types also may be dimerized provided a suitable catalyst is found [5, 7]. Such an example is shown:

$$2\ CH_2{=}CHCN + Cu^{2+}\text{acetylacetonate} + Al(iso\text{-}C_4H_9)_3 \xrightarrow[10°]{H_2,\ CHCl_3}$$
8 parts   0.255 part   0.25 part

$$NCCH_2CH{=}CHCH_2CN$$
1,4-Dicyano-2-butene, 2.3 parts

Furthermore, cocyclooligomerizations are now possible using nickel-based coordination compounds [8].

$$CH_2{=}CH_2 + CH_2{=}CHCH{=}CH_2 \rightarrow$$

trans,cis-1,5-cyclodecadiene + 1,5-cyclooctadiene

If the temperature is held below 80°, an appreciable amount of the precursor, cis-1,2-divinylcyclobutane, can be isolated [9]:

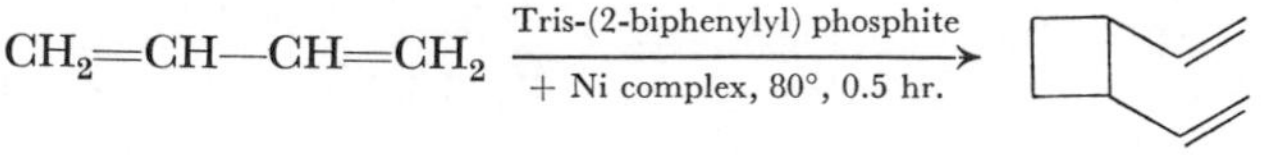

$$CH_2{=}CH{-}CH{=}CH_2 \xrightarrow[\text{+ Ni complex, 80°, 0.5 hr.}]{\text{Tris-(2-biphenylyl) phosphite}}$$

30% conversion, 36% yield

The reaction takes place via the bis-π-allyl complex

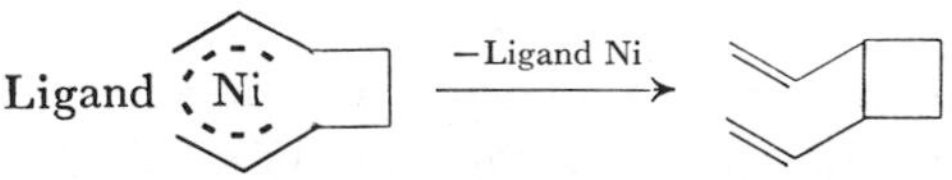

$$\text{Ligand : Ni} \xrightarrow{-\text{Ligand Ni}}$$

Palladium salts as well are known to catalyze the dimerization of olefins [10]. With dienes they form π-allyl complexes which are capable of manipulation to produce dimeric products. For example, another new dimer of butadiene can be produced in good yield by the accompanying reaction [11]. This reaction is simple and free from by-products. The ether can be isolated if desired but, if triphenylphosphine is added to the crude reaction mixture and distillation is

$$C_6H_5OH + CH_2=CHCH=CH_2 \xrightarrow[\text{C}_6\text{H}_5\text{ONa, 1.4} \times 10^{-2} M]{\text{PdCl}_2, 1 \times 10^{-3} M}$$

$$\text{0.4 mole} \text{1.7 mole}$$

$$C_6H_5OCHCH=CH(CH_2)_3CH=CH_2$$

1-Phenoxyoctadiene-2,7
96% conversion based on phenol

carried out at reduced pressure, 1,3,7-octatriene is isolated in 85% yield, 98% pure.

Modified Ziegler catalysts permit the oligomerization of 1,3-dienes with olefins. A nickel chloride-diisobutylaluminum chloride catalyst gives non-stereospecific addition (see Ex. c). On the other hand, an iron acetylacetonate-triethylaluminum catalyst gives cis- addition [12]:

$$\underset{\displaystyle CH_2=\overset{\displaystyle CH_3}{\underset{|}{C}}-CH=CH_2}{} + CH_2=CH_2 \longrightarrow$$

$$CH_2=CHCH_2\overset{\displaystyle CH_3}{\underset{|}{C}}=CHCH_3 + CH_2=CH-CH_2CH=\overset{\displaystyle CH_3}{\underset{|}{C}}-CH_3$$

cis-4-Methyl-1,4-hexadiene,   5-Methyl-1,4-hexadiene,
27%   27%

The orientation pattern of ethylene addition to other dienes is as follows:

$$CH_2=CHCH=CHCH_3 CH_2=\overset{\displaystyle C_6H_5}{\underset{|}{C}}-CH=CH_2$$

$$\uparrow \uparrow \uparrow$$

$$30\% 70\% 96\%$$

Still another catalyst, a complex of cobaltous chloride and triethylaluminum, can be used to prepare C_8 dienes from butadiene and ethylene [13].

To illustrate further the potentialities of various catalysts, an essentially *trans* alkylidenation has been carried out recently with a tungsten modified Ziegler catalyst [14]:

$$\text{2-Pentene} \xrightarrow[\substack{\text{C}_2\text{H}_5\text{AlCl}_2, \\ 25\% \text{ in C}_6\text{H}_6}]{\text{WCl}_6, \text{C}_2\text{H}_5\text{OH}} \quad \begin{array}{l} \text{2-butene, } \text{25 mole }\% \\ \text{2-pentene, 50 mole }\% \\ \text{3-hexene, } \text{25 mole }\% \end{array}$$

The same equilibrium mixture of products can be obtained by starting with any one of the three olefins. The "metatheses" probably occur through formation of a quasi cyclobutane ring complexed to tungsten. The reaction applied to cyclooctene gives macrocyclic polyenes, 9% C_{16}, 5% C_{24}, and even larger rings [15].

Some olefin additions which produce rings proceed stereospecifically in high yield because the intermediate cation has a single preferred conformation which favors the formation of a single product [16]:

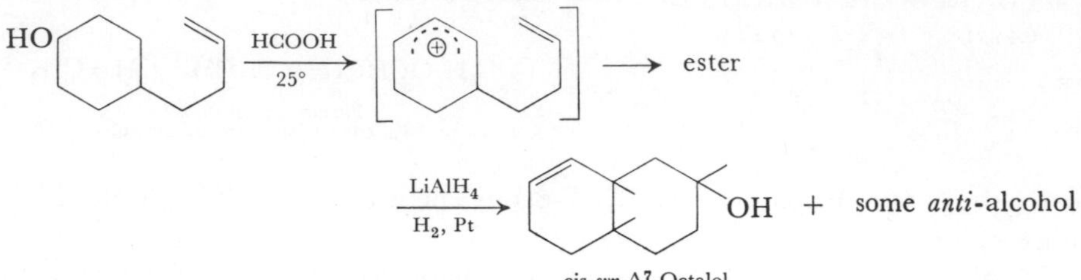

cis-*syn*-Δ⁷-Octalol,
"high yield"

The scope of such interesting and facile addition reactions will probably be broadened considerably in the near future.

(a) Preparation of diisobutylene.   *t*-Butyl alcohol (3.5 l.) was mixed with 2.5 l. of sulfuric acid and 2.5 l. of water and heated on a steam bath for 3 hr. The mixture was cooled, diluted, and the upper layer separated (ca. 2.1 l.). It was then fractionated to yield about 1500 ml. of diisobutylene, b.p. 100–105°. The product consisted of about 4 parts of 2,4,4-trimethyl-1-pentene and 1 part of 2,4,4-trimethyl-2-pentene. About 500 ml. of trimers and polymers was left in the residue. The same mixture was obtained starting with isobutylene rather than *t*-butyl alcohol [17].

(b) Preparation of 1,1-dimethyl-2-isopropenylcyclopentane.   2,7-Dimethyl-2,6-octadiene (20 g.) was refluxed with 10 ml. of 85 % phosphoric acid and the lower-boiling fraction was removed as it formed. The distillate was carefully fractionated to yield 49 % of the titled compound as the first fraction [18].

(c) Preparation of 1,4-hexadiene.   Bis(tributylphosphine)nickel(II) chloride (1 mmole) was dissolved in dry perchloroethylene and 3.7 moles of butadiene distilled into the deep red solution contained in an autoclave. The autoclave was closed, warmed to 65°, and ethylene charged and held at 100 p.s.i.g. on demand. Diisobutylaluminum chloride (6 mmoles) in 4 ml. of perchloroethylene was injected into the autoclave. After 1 hr. at 65–70°, the reaction was terminated by injection of 3 ml. of 2-propanol. The autoclave was then cooled, vented, and the contents fractionated to yield 130 g., 65 %, of a fraction, b.p. 49–83°, consisting of 53 % *trans*- and 21 % *cis*-1,4-hexadiene [19].

2. From 1,3-dienes and Alkenes (Diels-Alder)

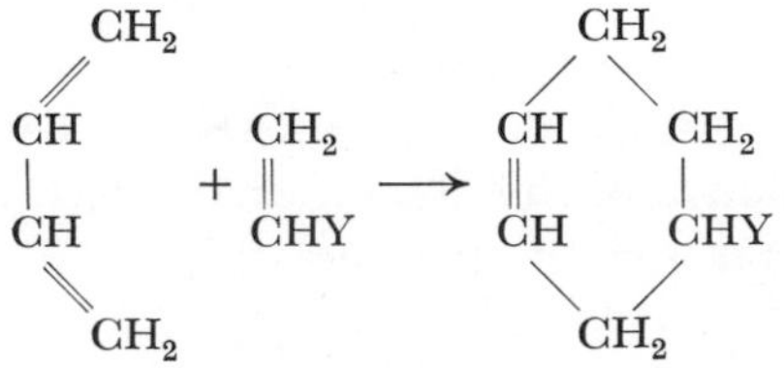

The Diels-Alder reaction has tremendous scope in the preparation of cyclic alkenes usually containing other functional groups. Excellent reviews are available. For theory and brief survey, Sauer and Wassermann are recommended [20]. The most comprehensive review is that by Onishchenko [21]. But there are many others [22–27].

The Diels-Alder is but one of a class of cycloadditions [22, 28]:

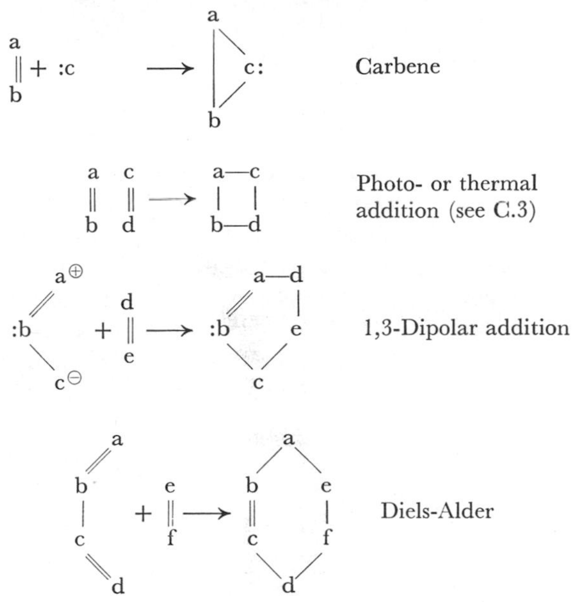

Carbene

Photo- or thermal addition (see C.3)

1,3-Dipolar addition

Diels-Alder

a fifth type may be added:

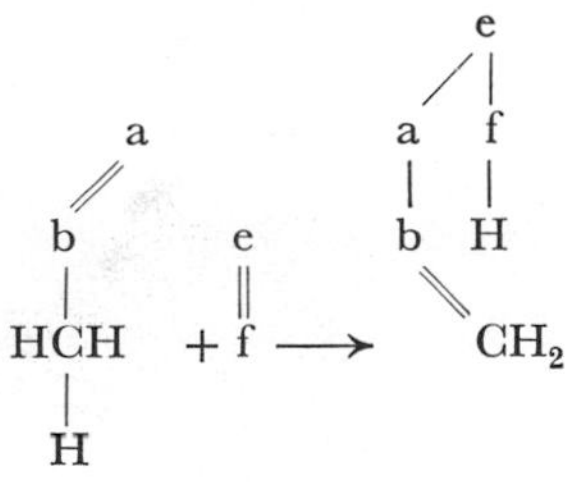

The nature of the hydrogen transfer in the fifth type has not been determined; it may be free radical in nature [29] because of the high temperatures employed. Nevertheless, the reactions of the fifth type are simple to carry out and useful [30]. In its simplest form the Diels-Alder reaction consists of the addition of a conjugated diene to an alkene (the dienophile) to form a substituted cyclohexene. Usually the dienophile contains an electron-withdrawing group Y, such as the carbonyl ($C{=}O$), the cyano (CN), the nitro (NO_2), or the sulfonyl (SO_2). The reaction also occurs if the dienophile contains a triple rather than a double bond or if the double bond is part of a quinone ring. The conjugated systems capable of reacting in general consist of acyclic conjugates (such as 1,3-butadiene), alicyclic conjugates (such as cyclopentadiene), aromatic conjugates

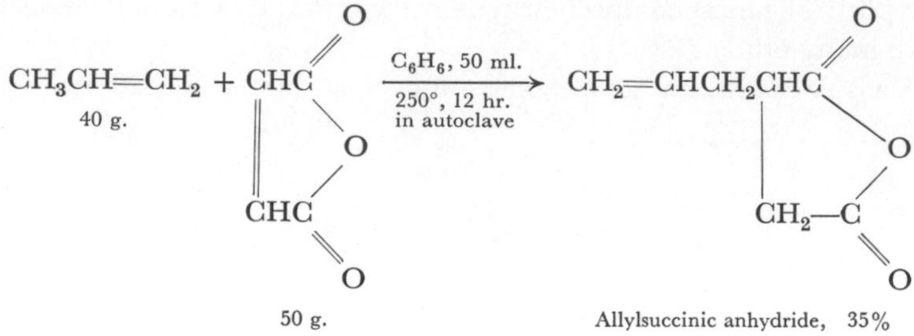

$$CH_3CH{=}CH_2 + \underset{\substack{\text{50 g.}}}{\overset{\substack{40\ \text{g.}}}{\text{(maleic anhydride)}}} \xrightarrow[\substack{250°,\ 12\ \text{hr.}\\ \text{in autoclave}}]{C_6H_6,\ 50\ \text{ml.}} \text{Allylsuccinic anhydride, } 35\%$$

(such as anthracene), and heterocyclic conjugates (such as furan). Thus the synthesis leads to the formation of unsaturated ring systems often not readily available by other methods and, in addition, the reaction possesses remarkable stereospecificity in giving one or, at most, two stereoisomers. It is because of the last-mentioned fact that the reaction has been of such great value in the synthesis of natural products such as cortisone, reserpine, and estrone.

The main facts in connection with the stereospecificity are the following [20]:

1. The diene reacts in the *s-cis* (double bonds in a plane on the same side of the single bond connecting them) and not in the *s-trans* (double bonds in a plane on the opposite sides of the single bond connecting them) configuration:

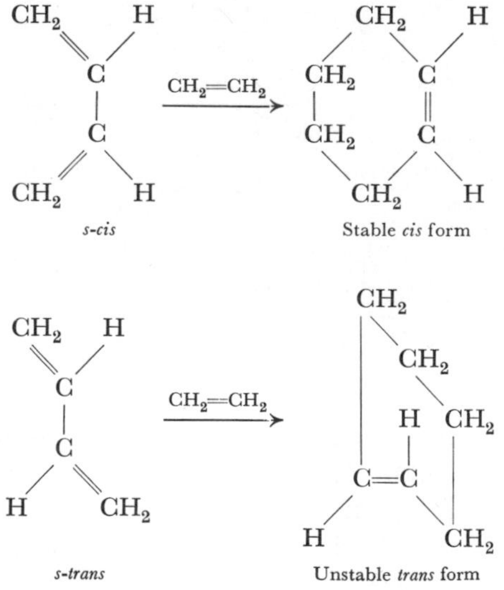

s-cis → (with $CH_2{=}CH_2$) → Stable *cis* form

s-trans → (with $CH_2{=}CH_2$) → Unstable *trans* form

2. The configuration of the diene and the dienophile are unchanged in the adduct. In other words, *cis* addition occurs as is shown in the addition of 1,3-butadiene to dimethyl maleate:

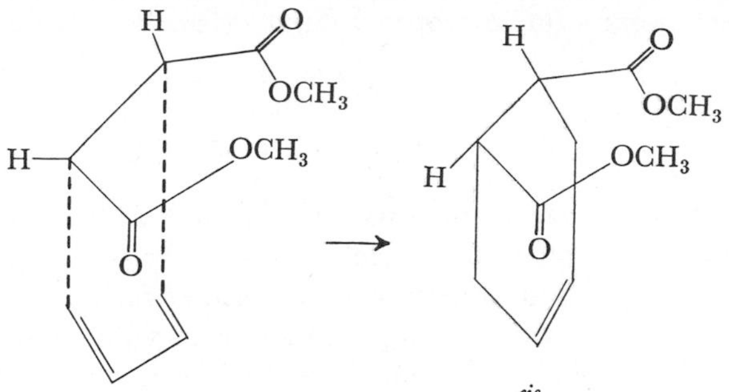

cis

3. In cases in which two ways of addition between the diene and dienophile are possible, the adduct with the *endo* configuration is usually the only or major product, as shown with maleic anhydride and cyclopentadiene:

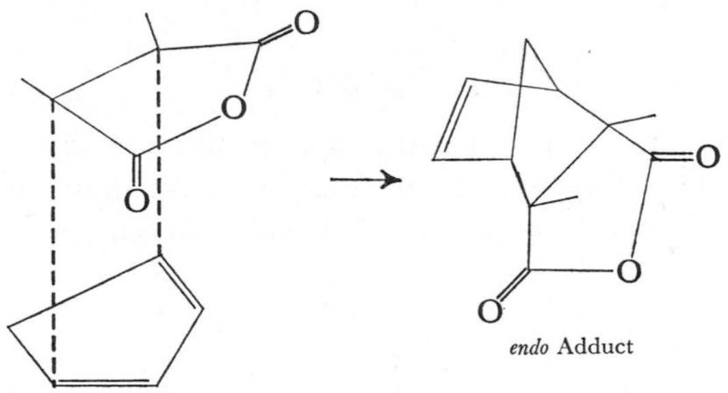

endo Adduct

Geometrical overlap of π-orbitals at primary centers where bonds are forming seems to be less energetic for *endo* isomer formation where components intersect at an angle of 60° than for *exo* formation in which components are parallel [31].

The *endo* rule is most flagrantly violated by the cyclic diene, furan, because its adducts are readily dissociated and thus permit the accumulation of the more thermodynamically stable *exo* isomer. Mixtures of *endo* and *exo* products are obtained with cyclopentadienes and open chain dienophiles, such as methyl acrylate. In these cases the *exo/endo* ratio is solvent-dependent.

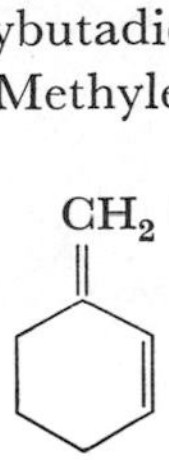

The order of activity of dienes is based to a large extent on electron-release influences of groups attached to the diene, modified by proper orientation of the π-orbitals to overlap with the dienophilic electron-deficient center and modified by steric hindrance. Thus an order of activity of dienes is as follows: cyclopentadiene > 9,10-dimethylanthracene > 1,2-dimethylenecyclohexane > 1,1′-bicyclopentenyl > 1-methoxybutadiene > 2,3-dimethylbutadiene > butadiene > 2-chlorobutadiene. 3-Methylenecyclohexene does not react because of improper orbital orientation. And *s*-*cis*-1-methyl-butadiene

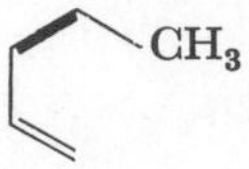

gives only a 4% yield of 1,1-adduct with maleic anhydride. On the other hand, the *trans*-isomer gives a quantitative yield in an exothermic reaction. These facts can be rationalized only on the basis of a steric effect for the *cis*-isomer. Not only dienes but also strained cyclopropanes react with maleic anhydride or other dienophiles, probably forming intermediate diradicals [32] as well as about 5% of the bridgehead adducts.

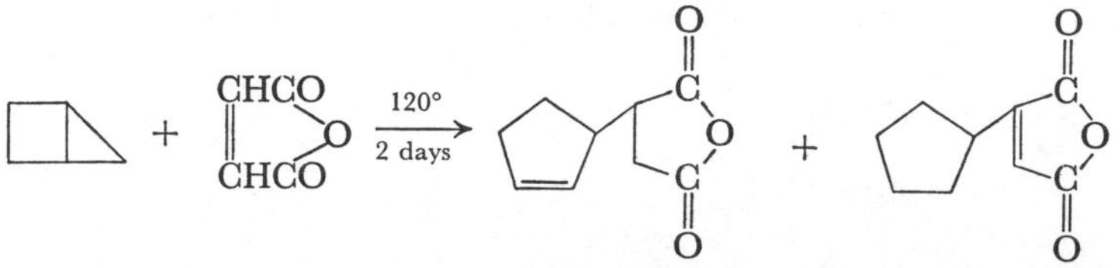

α-Cyclopentenylsuccinic anhydride, 68%   α-Cyclopentylmaleic anhydride, 17.4%

Maleic anhydride is a very good dienophile, but not the best. Dienophiles containing N=N bonds apparently are more reactive than the C=C analogs, and the most reactive of them is N-phenylazodicarboximide:

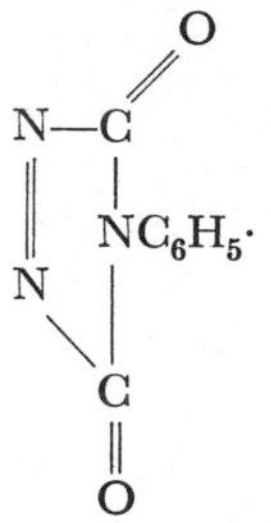

Comprehensive comparative activity tables of dienophiles are not available, but from scattered lists [20] one might conclude that tetracyanoethylene, *p*-benzoquinonedicarboxylic anhydride, 2,3-dicyanomaleic anhydride are highly reactive, while methyl crotonate is feebly reactive. The acid chloride usually is more reactive than the ester or acid. Even the ethylene group may be considered a weak dienophile. For example, butadiene at room temperature usually polymerizes via the *trans* diradical. But, as the temperature of reaction is raised, some cisoid butadiene forms which reacts with a second molecule to form 3-vinyl-1-cyclohexene in a typical Diels-Alder reaction [33]:

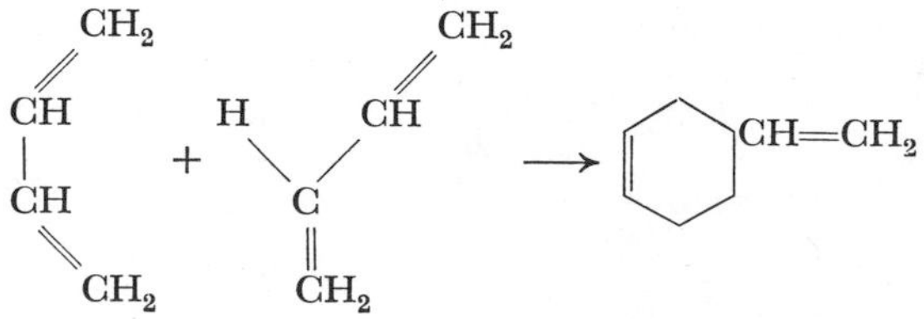

The yields are nearly quantitative when run in the presence of furfural. As the temperature for reaction is raised still higher, from 120 to 270°, where even more cisoid form is available, appreciable amounts of 1,5-cyclooctadiene (2.2 % to 10.6 %, respectively) accompany the vinylcyclohexene product. Usually an inhibitor is added to minimize radical polymerization. Sometimes the complexes formed with the adduct and silver nitrate are of value in purification [34].

For obvious reasons some dienes, such as hexachlorocyclopentadiene, behave like dienophiles. This "inverse electron demand" allows Diels-Alder reactions to proceed under fairly mild conditions with simple olefins [35]:

87%
1,2,3,4,7,7-hexachloro-
5-phenylbicyclo-
[2.2.1]heptene-2

The indifference of the typical Diels-Alder reaction rates to catalysts, solvent polarity, acids, or bases argues for the concerted nature of the electron transfer process, despite the belief of many that orbital overlapping between diene and dienophile occurs in a stepwise fashion. Whatever the mechanism is, the facts are that only three environmental conditions have a pronounced effect on Diels-Alder rates: temperature, high pressure, and the formation of a Lewis acid adduct with the dienophile. With a reactive diene and a dienophile the Diels-Alder reaction is exothermic, and cooling conditions must be observed. With unreactive combinations quite high temperatures and an autoclave to contain the volatiles are necessary. The effect of pressure is truly remarkable [36]:

60% yield, m.p. 118–119°
2,3-benzobicyclo[2,2,2]hexa-
2,7-dien-5,6-dicarboxylic
anhydride

Under normal pressure only traces of the adduct are obtained. But the most important modification for influencing Diels-Alder reactions is the presence of Lewis acid adducts of the dienophile. With reactive dienes no excess Lewis acid should be present to catalyze diene polymerization. The acceleration of the reaction is simply due to the fact that the Lewis acid adduct is more dienophilic than the dienophile itself. The adduct accelerates some Diels-Alder reactions (see Ex. b.3), makes others feasible, and in some cases alters the course of the reaction (see Ex. d.1).

Diels-Alder reactions are reversible, and in cases of relatively unstable adducts approach an equilibrium state between diene, dienophile, and adduct. With some adducts, however, for which some stable product can be formed, the retro Diels-Alder reaction does not return to the original components. Some illustrations follow [37]:

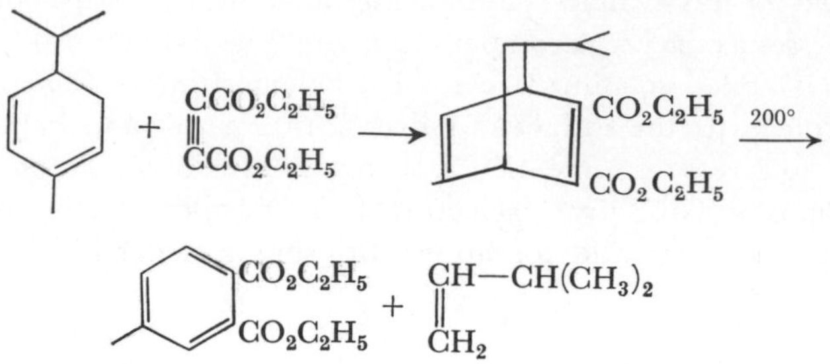

Diethyl-4-methylphthalate   3-Methyl-1-butene

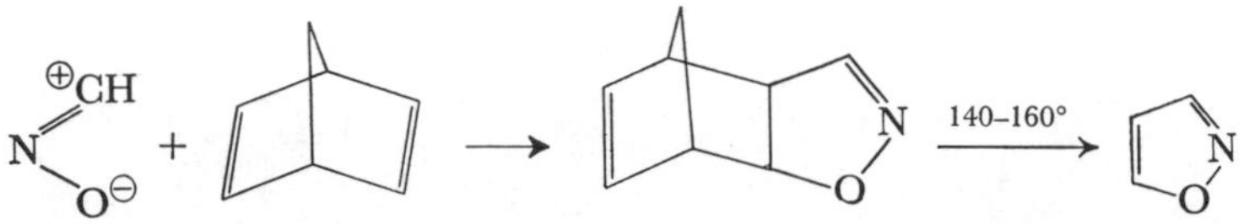

Isoxazole, 90%

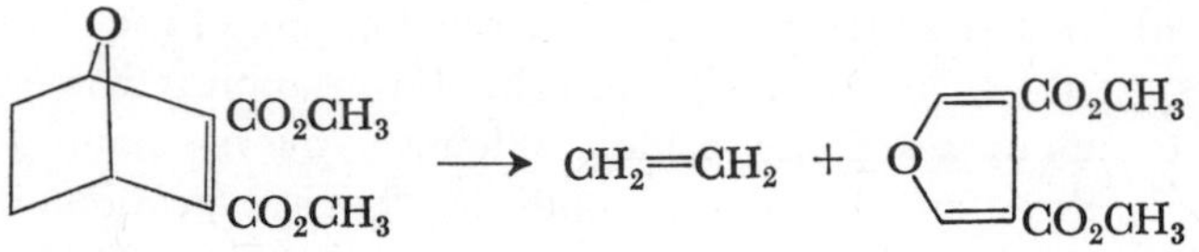

Dihydro derivative of
Diels-Alder adduct

3,4-Dicarbomethoxyfuran

Typical examples of Diels-Alder reactions follow. It will be noted that they always produce a new double bond. In the case of hydrocarbon production, no electron-withdrawing group is present in the dienophile. However, in the more common cases, i.e., formation of adducts of anhydrides, carboxylic acids, esters, aldehydes, quinones, etc., an electron-withdrawing group is present in the dienophile.

(3) Preparation of cyclic hydrocarbons

(1) **Triptycene** (see Alkanes, Chapter 1, H).

(b) Preparation of cyclic acid anhydrides

(1) *cis*-Δ^4-**Tetrahydrophthalic anhydride** (93–97% from maleic anhydride and butadiene in benzene) [38].

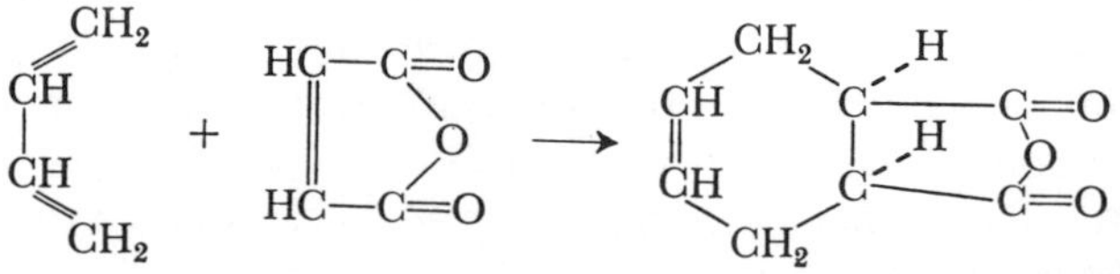

(2) *endo* - 1,4 - **Methylimino - 1,2,3,4 - tetrahydronaphthalene - 2,3-dicarboxylic anhydride** (72% from N-methylisoindole and maleic anhydride in ether at room temperature) [39].

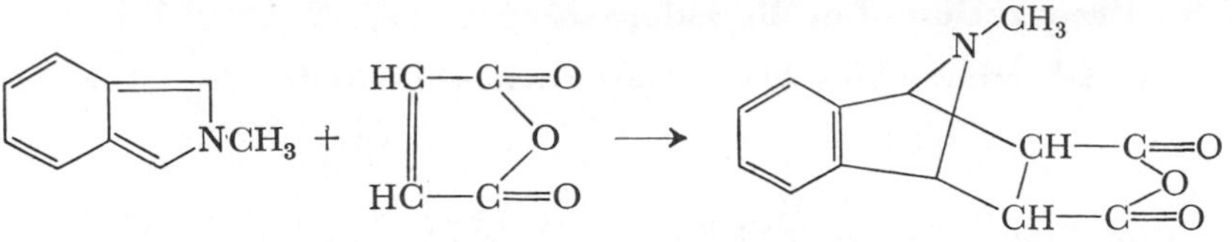

(3) 9,10 - Dihydro - 9,10 - ethanoanthracene - 11,12 - dicarboxylic anhydride (Quantitative yield from equimolar quantities of anthracene,

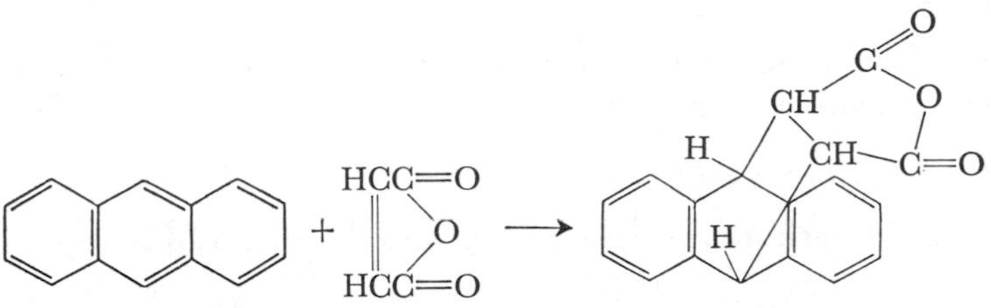

maleic anhydride, and aluminum chloride in methylene chloride within 2 min.) [40].

(c) Preparation of cyclic carboxylic acids

(1) *cis*-Δ^3-**Tetrahydrophthalic acid** (65–70% from *trans*-butadiene-

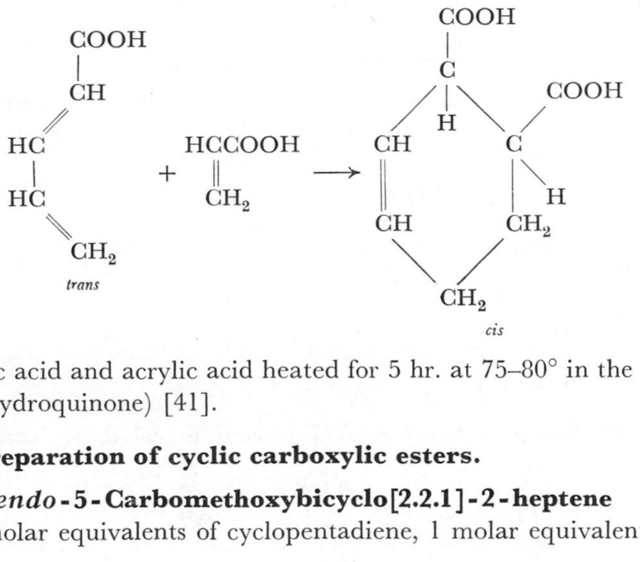

1-carboxylic acid and acrylic acid heated for 5 hr. at 75–80° in the presence of a trace of hydroquinone) [41].

(d) Preparation of cyclic carboxylic esters.

(1) *endo* - **5 - Carbomethoxybicyclo[2.2.1] - 2 - heptene**   (79–91% from 3–4 molar equivalents of cyclopentadiene, 1 molar equivalent of methyl

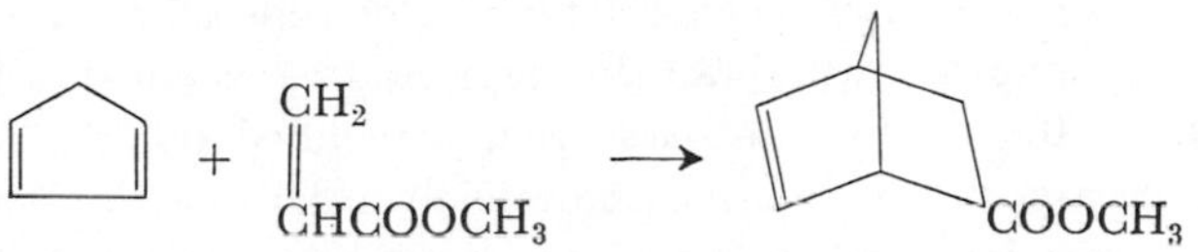

acrylate, and 10% molar equivalent of $AlCl_3(C_2H_5)_2O$ in methylene chloride at 0°. Without $AlCl_3$ as much as 18% of the *exo*-isomer accompanies the product) [42].

(e) Preparation of cyclic aldehydes

(1) 4-Methoxy-3-cyclohexene-1-carboxaldehyde (75% from acro-

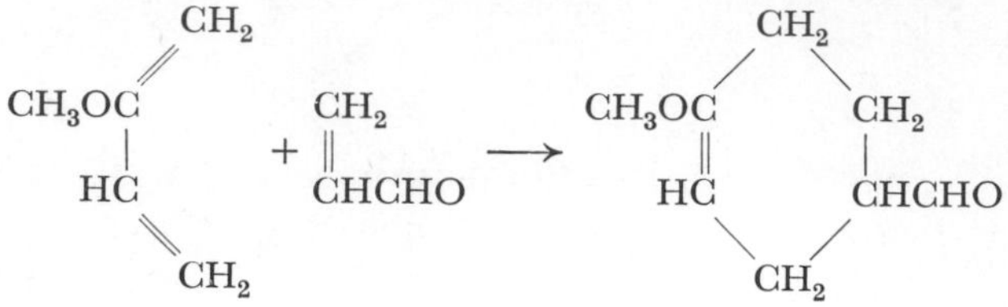

lein, 2-methoxybutadiene, and benzene heated for 30 min. at 160° in a bomb with a trace of hydroquinone) [43].

(f) Preparation of adducts of quinones

(1) 2,3-Dimethylbutadiene-α-naphthoquinone adduct [44].

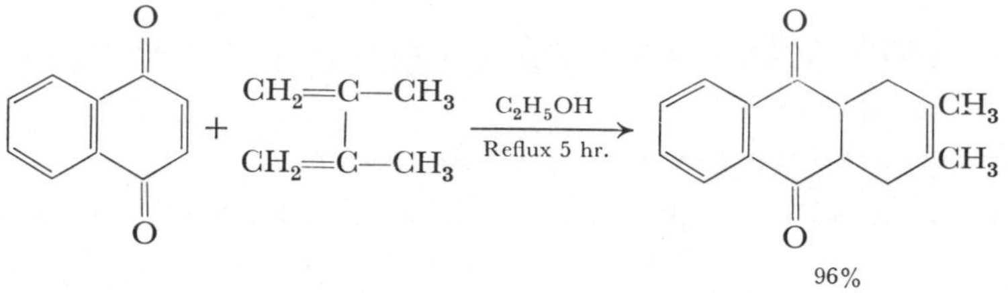

96%

(g) Preparation of cyclic nitro compounds

(1) 4-Nitrocyclohexene (92% from 1.8 moles of butadiene, 1 mole

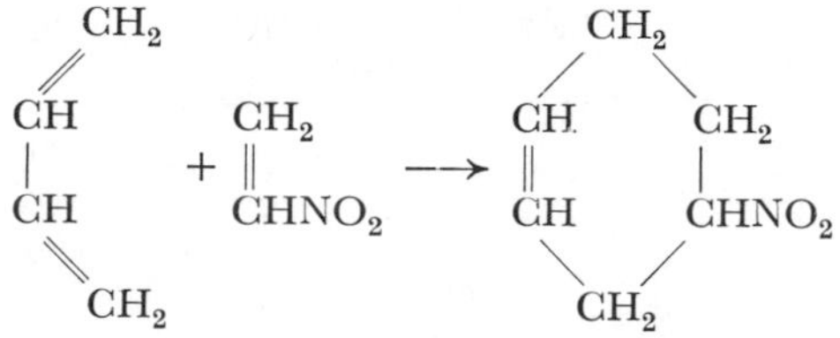

of nitroethylene, and a few drops of acetic acid in benzene heated at 105° for 14 hr. in a sealed tube) [45].

3. From Olefins and Acetylene (Photochemical Reactions)

These addition reactions are supplementary to the Diels-Alder reaction. Whereas the latter involves coalescence of normal π-electrons with electron-deficient centers (assisted by the concerted nature of the coalescence), the photochemical additions involve activated π-electrons (the triplet state). In most cases not much is known about the triplet state, except that the electrons are farther from the nucleus and less well controlled by it. Photochemical reactions, then, are more extensive than Diels-Alder reactions and also less amenable to prediction. For example, benzene and maleic anhydride do not form an adduct under ordinary conditions. But irradiation of this mixture yields an adduct in which 2 molecules of maleic anhydride and 1 of benzene have combined [46, 47].

Tricyclo[4.2.2.0²,⁵]dec-7-
ene-3,4,9,10-tetracarboxylic
acid dianhydride

A large number of parameters can be adjusted to find means to bring about addition. The wavelength of light should be chosen to encompass the absorption band of the olefin or acetylene, and preferably not to encompass the absorption band of the product, for the reason that the substrate in contrast to the product should be excited sufficiently to react. The best procedure is to use the lowest wavelength of light possible to get excitation by choice of proper filters, even though this may extend the reaction time considerably. The use of a sensitizer is another means of bringing about reaction, but occasionally it alters the course of the reaction. The sensitizer essentially is a transfer agent of the light energy. It is activated to the singlet or triplet state and in the latter state activates the substrate by intersystem crossing. The triplet excitation energy should be higher than that of the substrate [48]. A partial list of triplet energies in kilocalories per mole is given: propiophenone, 74.6; benzophenone, 68.5; triphenylene, 66.6; naphthalene, 60.9; pyrene, 48.7. If the triplet energy is lower than that of the substrate, the sensitizer may act as a quencher and prevent reaction. Unfortunately also, in the case of olefins, the ketone sensitizers may react to form oxetanes. Lastly, the choice of solvent may be the difference between success or failure. With all these variables, it is difficult to state in generalities what can and what cannot be done. Rather, specific examples of each type of reaction will be given for the reader to reach his own conclusions. The reactions include *cis-trans* isomerization (see D.1), isomerization of the double bond (see D.1), bridging, and coupling. Unless otherwise stated, they are taken from Kan [49].

Bridging is very common with olefins on irradiation. Cyclobutene formation frequently results from photochemical excitation of 1,3-dienes [50, 51].

1,2-Dimethyl-1-cyclobutene 71%

The greater the dilution of the diene, the higher is the yield. Sensitizers change the course of the reaction, dramatically tending to yield dimers. If two modes of bridging are possible, the cyclobutene usually does not form [52]:

Very small yield

The Woodward-Hoffmann rules [53] would predict, for concerted reactions, a disrotatory process for bridging by thermal means and a conrotatory process for bridging by photochemical excitation, leading in the latter case to the *trans* form, as was observed. A similar preference for larger ring closure is shown by myrcene [54]:

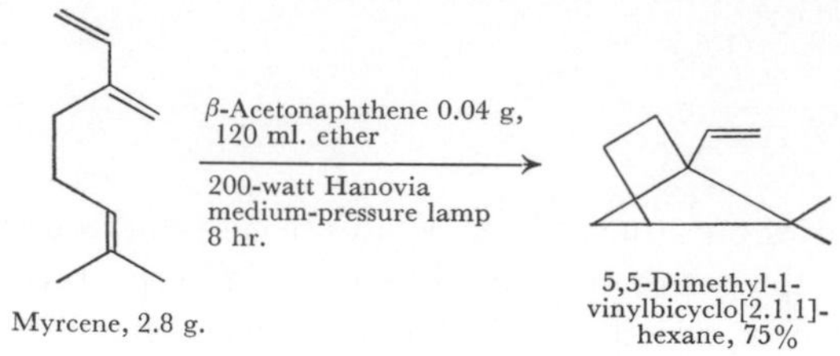

β-Acetonaphthene 0.04 g, 120 ml. ether

200-watt Hanovia medium-pressure lamp 8 hr.

Myrcene, 2.8 g.

5,5-Dimethyl-1-vinylbicyclo[2.1.1]-hexane, 75%

Without a sensitizer the reaction takes a different course [55]:

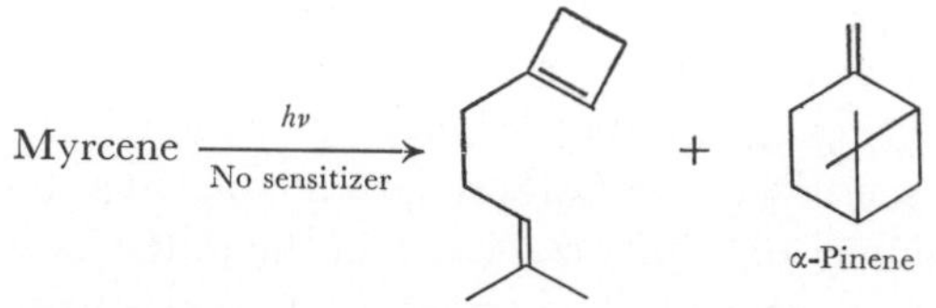

Myrcene $\xrightarrow[\text{No sensitizer}]{h\nu}$ +

α-Pinene

Dewar and Ladenburg benzenes have been produced by photochemical reactions [56]:

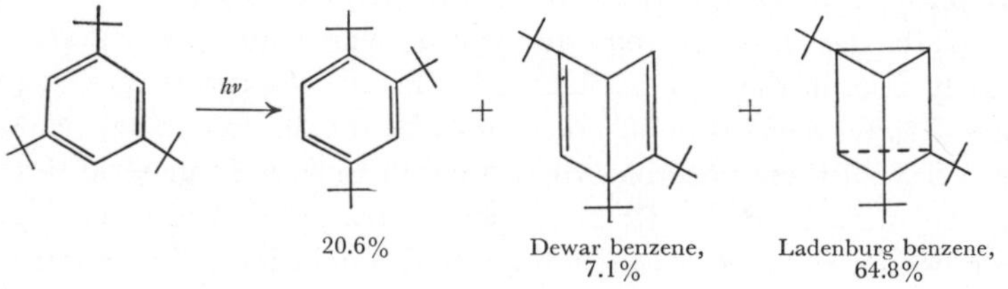

20.6%   Dewar benzene, 7.1%   Ladenburg benzene, 64.8%

Coupling of olefinic derivatives by photochemical means leads in many cases to dimeric cyclobutanes, a process which is not relevant to the preparation of olefins. But, in those cases in which olefins are formed, the course of the reaction is different from the case of thermal dimerization or the Diels-Alder reaction. When butadienes are irradiated in concentrated solution with a sensitizer (as contrasted to dilute with no sensitizer), dimeric olefins are formed [57]:

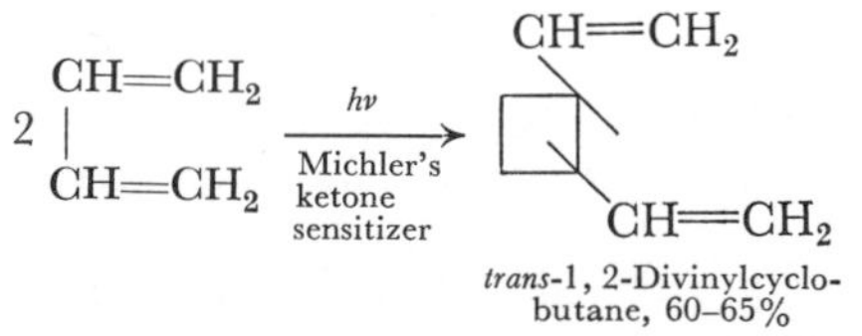

$$2 \begin{array}{c} CH{=}CH_2 \\ | \\ CH{=}CH_2 \end{array} \xrightarrow[\substack{\text{Michler's} \\ \text{ketone} \\ \text{sensitizer}}]{h\nu}$$

trans-1, 2-Divinylcyclo-butane, 60–65%

A small amount of the *cis*-isomer is also formed in the irradiation, but it rearranges to the higher-boiling 1,5-cyclooctadiene during the distillation.

In viewing photochemical reactions from the synthetic standpoint, the problems are many. Choice of proper irradiation, scaling up the size of the reaction, and isolation from a mixture of compounds of very similar physical properties are three of the most important. Yet the rewards are rich in giving access to compounds of most unusual structure.

4. From Conjugated, Unsaturated Systems (Anionic Addition)

Powerful nucleophiles have the ability to add to olefins to give hydrocarbons [58]:

$$ArCH_3 + CH_2{=}CH_2 \xrightarrow[\substack{s\text{-BuCl} \\ \text{(promoter)}}]{\text{Na}} \underset{80\%}{ArCH_2CH_2CH_3}$$

and to conjugated, unsaturated systems to give olefins [59]:

$$ArCH_2CH_3 + CH_2{=}CH{-}CH{=}CH_2 \xrightarrow[\text{CaO}]{\text{K on}} Ar\overset{\displaystyle CH_3}{\overset{|}{C}}HCH_2CH{=}CHCH_3$$

The catalyst employed is sodium or potassium. Yields for the simpler alkylated benzenes vary from 80 to 91 %. By one procedure the 1,3-diene is bubbled through a mixture of the catalyst and the alkylbenzene, after which the mixture is refluxed.

The steps in the synthesis of 5-phenyl-2-pentene are represented as follows [59]:

$$C_6H_5CH_2Na + CH_2{=}CH{-}CH{=}CH_2 \longrightarrow$$

$$C_6H_5CH_2CH_2CH{=}CHCH_2Na \xrightarrow{C_6H_5CH_3} C_6H_5CH_2CH_2CH{=}CHCH_3 + C_6H_5CH_2Na$$

In the addition of the methylsulfinylmethide anion to an alkadiene the reaction proceeds somewhat differently in that methylation occurs [60]:

$$CH_2{=}CH{-}CH{=}CH_2 \xrightarrow{\text{DMSO}} CH_2{=}CHCH{=}CHCH_3 + CH_3SOK$$

cis/trans = 20/80
1-Methyl-1,3-butadiene
50%

The use of tetramethylethylenediamine (TMEDA) in conjunction with alkyllithiums has given more scope to the addition reactions [61]. TMEDA coordinates with the lithium cation yielding a more nucleophilic carbanion. A higher yield with TMEDA is obtained in the following reaction [62]:

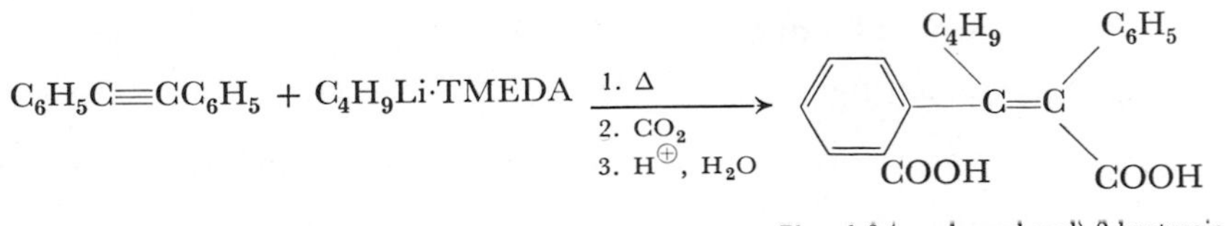

$$C_6H_5C{\equiv}CC_6H_5 + C_4H_9Li{\cdot}TMEDA \xrightarrow[\substack{2.\ CO_2 \\ 3.\ H^{\oplus},\ H_2O}]{1.\ \Delta}$$

α-Phenyl-β-(o-carboxyphenyl)-2-heptenoic acid, 30%

Alkyllithium complexed with TMEDA can also add to some aromatic hydrocarbons (see Ex. b).

Although anionic species rarely add to isolated double bonds, the Grignard reagent has been shown recently to add to diphenylallylcarbinol [63]:

$$(C_6H_5)_2CCH_2CH{=}CH_2 \xrightarrow[\substack{1.\ 36\ hr.,\ 25° \\ 2.\ \overset{\oplus}{H},\ H_2O}]{2CH_2{=}CHCH_2MgBr} (C_6H_5)_2C(CH_2)_4CH{=}CH_2$$
$$\underset{OH}{|} \qquad\qquad\qquad\qquad\qquad\qquad\qquad \underset{OH}{|}$$

1,1-Diphenyl 6-heptenol-1, 70%

(a) **Preparation of 5-phenyl-2-pentene.** Butadiene, 55.6 g. (1.03 moles), was bubbled through a mixture of potassium-on-calcium-oxide catalyst and 347 g. (3.8 moles) of toluene for 6.5 hr. at 91–93°. After the catalyst was destroyed with isopropyl alcohol, the mixture was filtered. The solids were then washed with two 100-ml. portions of toluene and the washings were combined with the filtrate. The filtrate, after being washed with water until neutral and dried, was distilled to give 120 g. (80%) of 5-phenyl-2-pentene [59]; see [64] for use of isoprene as the diene in this reaction.

(b) **Preparation of 1-methyldihydroperylene** [65].

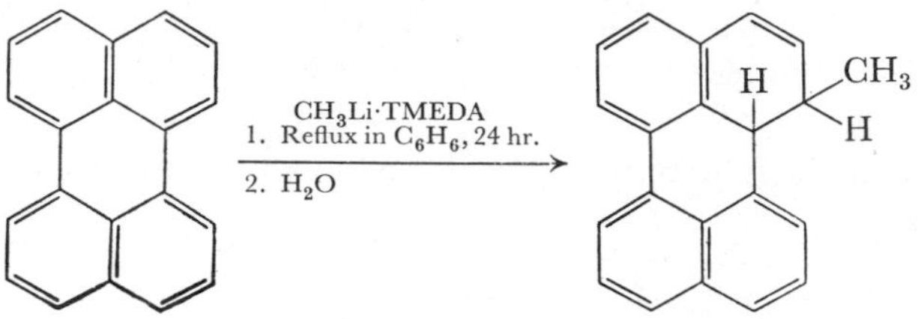

33% (crude)

5. From Organometallics and Carbon Tetrahalides

$$3\ RCH_2M + CF_2Br_2 \longrightarrow RCH{=}CHCH_2R + RCH_2Br + 2\ MF + MBr$$

M = Li or MgBr

This synthesis [66] leads to olefins with an uneven number of carbon atoms and a centrally located double bond. The Grignard reagent or alkyllithium is employed and difluorodibromomethane or trifluoromonobromomethane is utilized as the carbon tetrahalide. Primary halides give alkenes, while secondary ones yield mainly unsaturated halides. It is proposed [66] that a dihalocarbene, as $:CF_2$, is formed and that it is initially inserted between the carbon and metal bond. Details of the mechanism are given in the original. The method is simple in operation and yields with a series of primary halides vary from 37 to 72%.

(a) **Preparation of nonene-4.** To the Grignard reagent prepared in

$$3\ CH_3CH_2CH_2CH_2M + CF_2Br_2 \rightarrow$$
$$CH_3(CH_2)_2CH{=}CH(CH_2)_3CH_3 + CH_3(CH_2)_3Br + 2\ MF + MBr$$

ether from 6.2 g. of magnesium and 35 g. of butyl bromide was added slowly with stirring and cooling to $-70°$, 25 g. of difluorodibromomethane in 100 ml. of ether. The solution was then hydrolyzed with 2 N sulfuric acid; the ether layer was separated, dried, and distilled to give nonene-4 (72%), b.p. 145–146° [66].

This preparation has been checked recently by using trifluorobromomethane [67]. The yield of nonene-4 was 52%. A side product which seems always to accompany the main product was 1-bromo-1-pentene, $C_3H_7CH=CHBr$, about 15%.

6. From Organometallics and a Haloform

$$RCH_2MgBr + CHBr_3 \longrightarrow RCH_2Br + CHBr_2MgBr$$

$$CHBr_2MgBr + RCH_2MgBr \longrightarrow MgBr_2 + (RCH_2CHBrMgBr) \xrightarrow{-MgBr_2}$$

$$(RCH_2CH:) \longrightarrow RCH=CH_2$$

This reaction is supplementary to that in C.5. Yields with a series of alkyl halides range from 43 to 80%.

Recently the Grignard reagent of methylene bromide or iodide has been substituted for the phosphorane [68] (see E.2):

$$CH_2(MgBr)_2 + R_2C=O \longrightarrow \begin{bmatrix} R_2COMgBr \\ | \\ CH_2MgBr \end{bmatrix} \longrightarrow R_2C=CH_2 + \overset{\ominus}{O}MgBr + \overset{\oplus}{M}gBr$$

The synthesis offers a second method for the methylenation of aldehydes and ketones. The reagent is prepared *in situ* by adding an ether solution containing equivalent amounts of methylene bromide or iodide and the carbonyl compound to an excess of magnesium amalgam in ether. Large excesses of the methylene halide should be avoided. Yields with a variety of aldehydes and ketones average 70%.

(a) **Preparation of methylenecyclohexane.**   To 0.2 mole of Grignard reagent from cyclohexyl bromide in 150 ml. of ether, 0.1 mole of bromoform in 30 ml. of ether was added at such a rate as to maintain mild reflux. The solution was decomposed with dilute, cold hydrochloric acid, extracted with ether. The ether extract was dried, concentrated, and the residue distilled to yield 60% of methylenecyclohexane [67]; for preparation (68%) from cyclohexanone, methylene bromide, or iodide, and magnesium amalgam in ether under argon, see [68].

7. From Organometallic Compounds and Unsaturated Halides

$$RMgX + CH_2=CHCH_2X \rightarrow RCH_2CH=CH_2 + MgX_2$$

This synthesis represents a superior method of preparing 1-alkenes, particularly if the halide is sufficiently active. For example, from allyl bromide allylbenzene (82%) [69], allylcyclopentane (70.5%) [70], and 4,4-dimethylpentene-1 (85%) [71] have all been prepared. If ordinary ether is used as the solvent, it may largely be removed before the allyl bromide is added to the Grignard reagent [72]. β-Methallyl halides react normally with the Grignard reagent

$$CH_2=CCH_2X$$
$$|$$
$$CH_3$$

[73], but crotyl halides, $CH_3CH=CHCH_2X$, undergo an allylic rearrangement [74]

$$CH_3CH=CHCH_2X \rightleftharpoons CH_2=CHCHX$$
$$|$$
$$CH_3$$

and thus mixtures are obtained in the reaction of these halides with the Grignard reagent. Vinyl halides also react with Grignard reagents [75], but in these cases very low yields are obtained. It is interesting to note that metallic halides such as cobaltous chloride catalyze this reaction [76]. However, the vinyl halide does form a Grignard reagent in 2-methyltetrahydrofuran or in tetrahydropyran [77] in good yield, and this organometallic compound may be used to form alkenes with an alkyl halide.

This synthesis permits one to produce dimers in one operation from the halide (see Ex. b.1) or from the alcohol (see Ex. b.3).

The mechanism appears to be that of the SN_2 type:

$$CH_2=CHCH_2Cl + \overset{\ominus\oplus}{R MgX} \rightarrow CH_2=CHCH_2R + MgXCl$$

Other types of displacement from unsaturated halides or esters of allylic alcohol may take place with organometallic compounds as shown [78]:

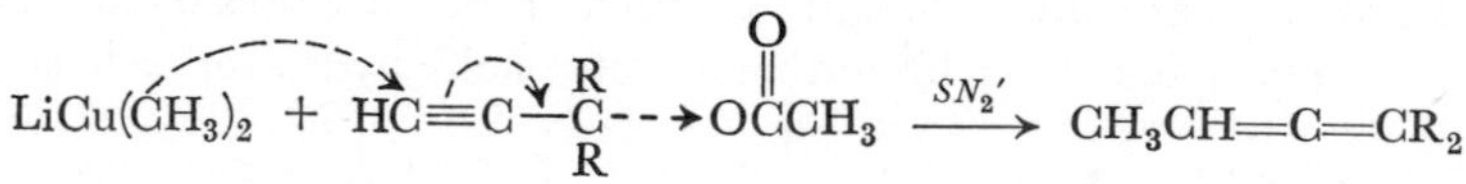

$$LiCu(CH_3)_2 + HC \equiv C\overset{R}{\underset{R}{-C}} \text{-->} OCCH_3 \xrightarrow{SN_2'} CH_3CH=C=CR_2$$

Or the nickel π-complexes formed from the reaction of nickel carbonyl and allyl bromides are readily susceptible to displacement by almost any type of alkyl or aryl iodide or bromide:

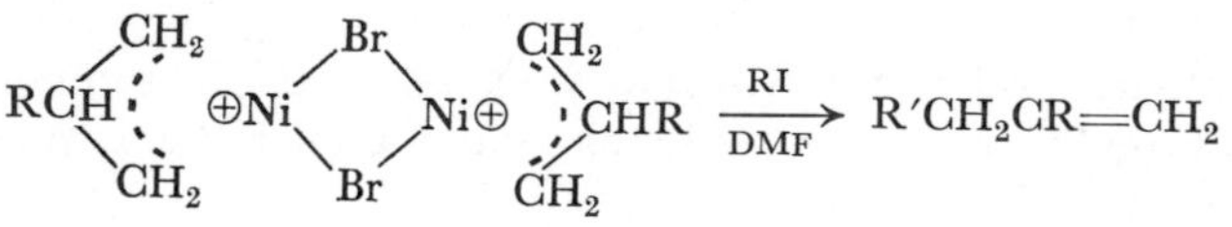

$$RCH \overset{CH_2}{\underset{CH_2}{\langle}} \oplus Ni \overset{Br}{\underset{Br}{\rangle}} Ni \oplus \overset{CH_2}{\underset{CH_2}{\langle}} CHR \xrightarrow[DMF]{RI} R'CH_2CR=CH_2$$

The reaction must be rigorously freed from oxygen for best yields. As one of many examples, 1,4-dimethallylbenzene was made in 97% yield [79]:

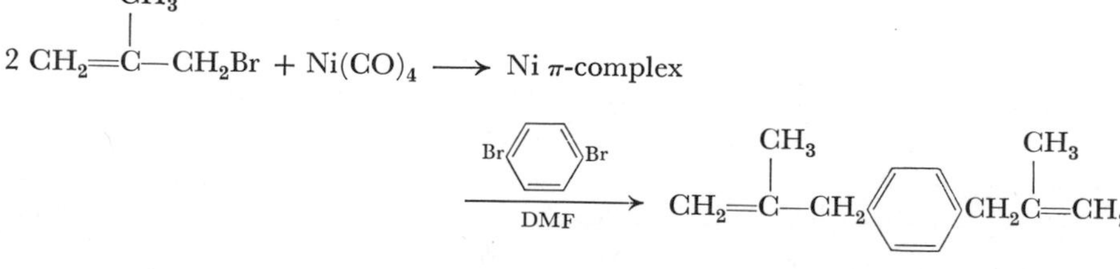

$$\overset{CH_3}{\underset{|}{2\ CH_2=C-CH_2Br}} + Ni(CO)_4 \longrightarrow Ni\ \pi\text{-complex}$$

$$\xrightarrow{DMF} CH_2=\overset{CH_3}{\underset{|}{C}}-CH_2 \langle\rangle CH_2\overset{CH_3}{\underset{|}{C}}=CH_2$$

And 1,12-dibromo-2,10-dodecadiene was converted into 1,5-*trans,trans*-cyclododecadiene in 59% yield [80]:

BrCH$_2$CH=CH(CH$_2$)$_6$CH=CHCH$_2$Br $\longrightarrow$

(a) **Preparation of allylcyclopentane.** To a clear solution of cyclo-pentylmagnesium bromide, prepared from 125 g. (5g. atoms) of magnesium, 745 g. (5 moles) of bromocyclopentane, and 3.1 l. of anhydrous ether was added slowly with stirring 605 g. (5 moles) of allyl bromide at reflux temperature. Stirring for 2 more hr. was followed by cooling and decomposition with cold 6 N hydrochloric acid. From the dried, combined ethereal layer and the ethereal extract of the aqueous layer there was obtained on distillation 387.5 g. (70.5%) of allylcyclopentane, b.p. 121–125° [70].

(b) **Other examples**

(1) **Biallyl** (55–65% from allyl chloride and magnesium in an ice bath for 1 to 1.5 hr.) [81].

(2) **3-Cyclopentyl-1-cyclopentene** (73% from cyclopentylmagnesium bromide and 3-chloro-1-cyclopentene in ether kept overnight and refluxed for 2.5 hr. [82].

(3) **Digeranyl** (71% containing the C-1, C-1′ and C-1, C-3′ dimers

$$\left(\underset{CH_3}{(CH_3)_2C=CHCH_2CH_2C=CHCH_2} \right)_2$$

in a 7:1 ratio, respectively, by mixing 3 moles of methyllithium and 1 mole of titanium trichloride under nitrogen at −78° in ethylene glycol dimethyl ether followed by the addition of 2 moles of geraniol) [83].

8. From Vinyl Grignard Reagents

$$2\,RCH=CHMgBr \xrightarrow[\text{THF}]{\text{2 CuCl}} [2\,RCH=CHCu] \xrightarrow{\text{Warm}} RCH=CHCH=CHR + 2\,Cu$$

Coupling reactions of Grignard reagents via free radicals have been known for a considerable time [84], but high yields were not obtained until the advent of the copper organometallic species. For the preparation of polyenes with the same and different substituents at the ends of the chain see [85].

(a) **Preparation of 2,5-dimethyl 2,4-hexadiene.** One equivalent of 2-methylpropenylmagnesium bromide in THF was added to 2 equivalents of cuprous chloride suspended in THF at −60°. The solution turned deep green and on warming to 20° decolorized. Decomposition with cold, dilute acid, extraction with ether, drying, and distillation gave the diene in 97% yield [86].

1. F. C. Whitmore, *Chem. Eng. News*, **26**, 668 (1948).
2. G. Wilke, *J. Polymer Sci.*, **38**, 45 (1959).
3. E. T. Niles and H. R. Snyder, *J. Org. Chem.*, **26**, 330 (1961).
4. H. Ramsden *et al.*, *Chem. Eng. News*, April 17, 1967, p. 46.

5. C. W. Bird, *Transition Metal Intermediates in Organic Synthesis*, Academic Press, New York, 1967, Chap. 2.
6. R. Cramer, *Accounts Chem. Research*, **1**, 186 (1968).
7. D. A. Cornforth *et al.*, Brit. Patent 1,123,097, August 14, 1968; *C.A.*, **69**, 7148 (1968).
8. G. Wilke, *Angew. Chem. Intern. Ed. Engl.*, **2**, 105 (1963).
9. P. Heimbach and W. Brenner, *Angew. Chem. Intern. Ed. Engl.*, **6**, 800 (1967).
10. A. D. Ketley *et al.*, *Inorg. Chem.*, **6**, 657 (1967).
11. E. J. Smutny, *J. Am. Chem. Soc.*, **89**, 6793 (1967).
12. G. Hata and D. Aoki, *J. Org. Chem.*, **32**, 3754 (1967).
13. M. Iwamoto *et al.*, *J. Org. Chem.*, **32**, 4148 (1967).
14. N. Calderon *et al.*, *Tetrahedron Letters*, 3327 (1967); *J. Am. Chem. Soc.*, **90**, 4133 (1968).
15. E. Wasserman *et al.*, *J. Am. Chem. Soc.*, **90**, 3286 (1968).
16. W. S. Johnson *et al.*, *Accounts Chem. Research*, **1**, 1 (1968); *J. Org. Chem.* **32**, 478 (1967).
17. F. C. Whitmore *et al.*, *J. Am. Chem. Soc.*, **54**, 3706 (1932); **53**, 3136 (1931).
18. P. G. Stevens and S. C. Spalding, Jr., *J. Am. Chem. Soc.*, **71**, 1687 (1949).
19. R. G. Miller *et al.*, *J. Am. Chem. Soc.*, **89**, 3756 (1967).
20. J. Sauer, *Angew. Chem. Intern. Ed. Engl.*, **6**, 16 (1967); **5**, 211 (1966); A. Wassermann, *Diels-Alder Reactions*, Elsevier Pub. Co., New York, 1965.
21. A. S. Onishchenko, *Diene Synthesis*, D. Davey Co., 257 Park Avenue, S., New York, 1964.
22. R. Huisgen *et al.*, in S. Patai *ed.*, *The Chemistry of Alkenes*, Interscience Publishers, New York, 1964, Chap. 11.
23. M. C. Kloetzel, *Org. Reactions*, **4**, 1 (1948); H. L. Holmes, *ibid.*, **4**, 60 (1948); L. W. Butz and A. W. Rytina, *ibid.*, **5**, 136 (1949).
24. L. H. Flett and W. H. Gardner, *Maleic Anhydride Derivatives*, John Wiley and Sons, New York, 1952.
25. K. Alder in *Newer Methods of Preparative Organic Chemistry*, Interscience Publishers, New York, 1948, p. 381.
26. J. G. Martin and R. K. Hill, *Chem. Rev.*, **61**, 537 (1961).
27. L. L. Muller and J. Hamer, *1,2-Cycloaddition Reactions*, Interscience Publishers, New York, 1967.
28. H. Ulrich, *Cycloaddition Reactions of Heterocumulenes*, Academic Press, New York, 1967.
29. H. Shechter and H. C. Barker, *J. Org. Chem.*, **21**, 1473 (1956).
30. Ref. 24, p. 7.
31. W. C. Herndon and L. H. Hall, *Tetrahedron Letters*, 3095 (1967).
32. P. G. Gassman *et al.*, *J. Am. Chem. Soc.*, **90**, 4746 (1968).
33. Ref. 21, Chap. 8.
34. J. K. Stille *et al.*, *J. Am. Chem. Soc.*, **81**, 4273 (1959); **86**, 2188 (1964).
35. Ref. 21, p. 311.
36. W. H. Jones *et al.*, *Tetrahedron*, **18**, 267 (1962).
37. H. Kwart and K. King, *Chem. Rev.*, **68**, 415 (1968).
38. A. C. Cope and E. C. Herrick, *Org. Syn.*, Coll. Vol. **4**, 890 (1963).
39. G. Wittig *et al.*, *Ann. Chem*, **572**, 1 (1951); **584**, 1 (1953).
40. P. Yates and P. Eaton, *J. Am. Chem. Soc.*, **82**, 4436 (1960).
41. K. Alder *et al.*, *Ann. Chem.*, **564**, 79, (1949).
42. J. Sauer and J. Kredel, *Tetrahedron Letters*, 731 (1966).
43. J. I. DeGraw *et al.*, *J. Org. Chem.*, **26**, 1156 (1961).
44. C. F. H. Allen and A. Bell, *Org. Syn.*, Coll. Vol. **3**, 310 (1955).
45. R. B. Kaplan ad H. Shechter, *J. Org. Chem.*, **26**, 982 (1961).
46. D. Bryce-Smith *et al.*, *J. Chem. Soc.*, 4791 (1960); *J. Chem. Soc.*, Pt. C, 390 (1967).
47. E. Grovenstein, Jr., *et al.*, *J. Am. Chem. Soc.*, **83**, 1705 (1961).
48. G. S. Hammond *et al.*, *J. Am. Chem. Soc.*, **86**, 4537 (1964).
49. R. O. Kan, *Organic Photochemistry*, McGraw-Hill Book Co., New York, 1966.
50. K. J. Crowley, *Tetrahedron*, **21**, 1001 (1965).
51. R. Srinivasan, *Advances in Photochemistry*, Interscience Publishers, New York, 1966, Vol. **4**, p. 113.

52. G. J. Fonken, *Tetrahedron Letters*, 549 (1962).
53. R. Hoffmann and R. B. Woodward, *Accounts Chem. Research*, **1**, 17 (1968).
54. R. S. H. Lin and G. S. Hammond, *J. Am. Chem. Soc.*, **89**, 4936 (1967).
55. Ref. 49, p. 35.
56. K. E. Wilzbach and L. Kaplan, *J. Am. Chem. Soc.*, **87**, 4004 (1965).
57. C. D. DeBoer *et al.*, *Org. Syn.*, **47**, 64 (1967).
58. H. Pines *et al.*, *J. Am. Chem. Soc.*, **77**, 554 (1955).
59. G. G. Eberhardt and H. J. Peterson, *J. Org. Chem.*, **30**, 82 (1965).
60. A. Schriesheim *et al.*, *J. Org. Chem.*, **30**, 3233 (1965).
61. G. G. Eberhardt, *Organometallic Chem. Reviews*, **1**, 491 (1966).
62. M. D. Rausch and L. P. Klemann, *J. Am. Chem. Soc.*, **89**, 5732 (1967).
63. J. J. Eisch and G. R. Husk, *J. Am. Chem. Soc.*, **87**, 4194 (1965).
64. H. Pines and N. C. Sih, *J. Org. Chem.*, **30**, 280 (1965).
65. H. E. Zieger and E. M. Laski, *Tetrahedron Letters*, 3801 (1966).
66. V. Franzen and L. Fikentscher, *Chem. Ber.*, **95**, 1958 (1962).
67. J. Villiéras, *Bull. Soc. Chim. France*, 1511 (1967).
68. C. Cainelli *et al.*, *Tetrahedron Letters*, 5153 (1967).
69. E. B. Hershberg, *Helv. Chim. Acta*, **17**, 351 (1934).
70. G. H. Coleman *et al.*, *J. Am. Chem. Soc.*, **68**, 1101 (1946).
71. F. C. Whitmore and A. H. Homeyer, *J. Am. Chem. Soc.*, **55**, 4555 (1933).
72. R. B. Regier and R. W. Blue, *J. Org. Chem.*, **14**, 505 (1949).
73. M. Tamele *et al.*, *Ind. Eng. Chem.*, **33**, 115 (1941).
74. M. S. Kharasch and O. Reinmuth, *Grignard Reactions of Non-metallic Substances*, Prentice-Hall, Englewood Cliffs, N.J., 1954, p. 1148.
75. P. N. Kogerman, *J. Am. Chem. Soc.*, **52**, 5060 (1930).
76. M. S. Kharasch and C. F. Fuchs, *J. Am. Chem. Soc.*, **65**, 504 (1943).
77. H. Normant, *Compt. Rend.*, **239**, 1510 (1954).
78. P. Rona and P. Crabbé, *J. Am. Chem. Soc.*, **90**, 4733 (1968).
79. E. J. Corey and M. F. Semmelhack, *J. Am. Chem. Soc.*, **89**, 2755 (1967).
80. E. J. Corey and E. K. Wat, *J. Am. Chem. Soc.*, **89**, 2757 (1967).
81. A. Turk and H. Chanan, *Org. Syn.*, Coll. Vol. **3**, 121 (1955).
82. G. E. Goheen, *J. Am. Chem. Soc.*, **63**, 744 (1941).
83. E. E. Van Tamelen *et al.*, *J. Am. Chem. Soc.*, **90**, 209 (1968).
84. Ref. 74, Chaps. 5 and 16.
85. L. A. Yanovskaya, *Russ. Chem. Rev.* (*English Transl.*), **36**, 400 (1967).
86. T. Kauffmann and W. Sahm, *Angew. Chem. Intern., Ed. Eng.*, **6**, 85 (1967).

D. Isomerization and Thermal Reactions

1. From Alkenes or Polyalkenes

$$RCH{=}CHCH_2R' \rightarrow RCH_2CH{=}CHR'$$

Under Dehydrohalogenation (A.2) it has been stated that isomerization to a single olefin is rare in basic solution. However, isomerizations, including interconversion of *cis* to *trans* forms, are sometimes possible in good yield in the presence of a base or acid by heating alone or with a catalyst, by irradiation, or via π-complexes of some metal salts [1].

(a) Base isomerization (for base isomerization of ethylenes substituted with groups other than alkyl or aryl see [2]).

References for Section D are on pp. 139–140.

The interconversion by a base may be represented as follows:

$$RCH_2CH=CH_2 + \overset{\ominus}{B} \rightleftharpoons BH + (\overset{\ominus}{RCHCH}=CH_2 \leftrightarrow RCH=CH\overset{\ominus}{CH_2})$$

$$\big\downarrow BH \qquad\qquad\qquad \big\downarrow BH$$

$$RCH_2CH=CH_2 \qquad\qquad RCH=CHCH_3$$

2-Vinylbicyclo[2.2.1]heptane may be transformed quantitatively to the 2-vinylidene isomer with sodium and ethylenediamine at 114° for 45 minutes [3]:

The lithium salt of ethylenediamine in the free diamine isomerizes olefins from terminal to nonterminal ones [4]:

$$CH_3(CH_2)_5CH=CH_2 \rightarrow CH_3(CH_2)_4CH=CHCH_3$$

octene-2
90%
(some 3- and 4-octene)

Not only is the *t*-butoxide anion in dimethyl sulfoxide effective in the rearrangement of acyclic olefins [5], but it is also effective with cyclohexadienes [6], cyclooctadienes, cyclononadienes, and cyclononatrienes [7]:

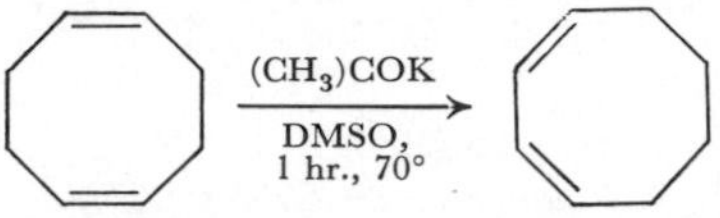

1,5-Cyclooctadiene → 1,3-Cyclooctadiene   (almost quantitative)

The same reagent converts dipentene at 55° into a 5:3:1 mixture of olefins [8]:

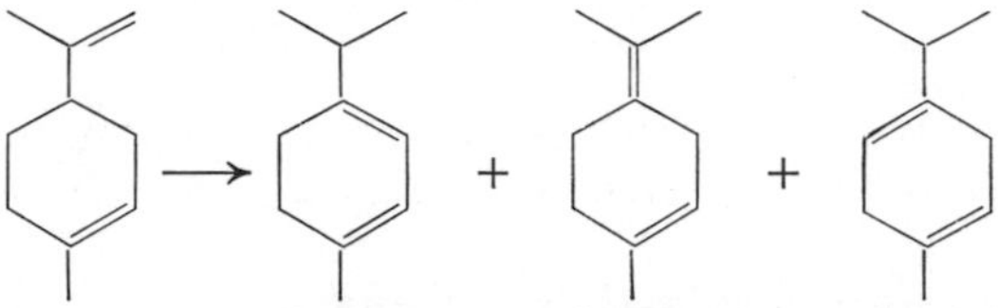

5 parts   3 parts   1 part
α-Terpinene   *p*-Mentha-2,4(8)-diene   γ-Terpinene

The unusual component, *p*-mentha-2,4(8)-diene, which constitutes one-third of the mixture, can be separated easily because of its high boiling point. By recycling the other two components more of the unusual component may be obtained.

(b) Acid isomerization.   Provided no dimerization or polymerization occurs, some olefins tend to isomerize to conjugated systems in the presence of acid. For instance, Δ^5-cholestene-3-one may be isomerized to the Δ^4-isomer in 98% yield by heating in a 95% ethanolic solution of anhydrous oxalic acid [9].

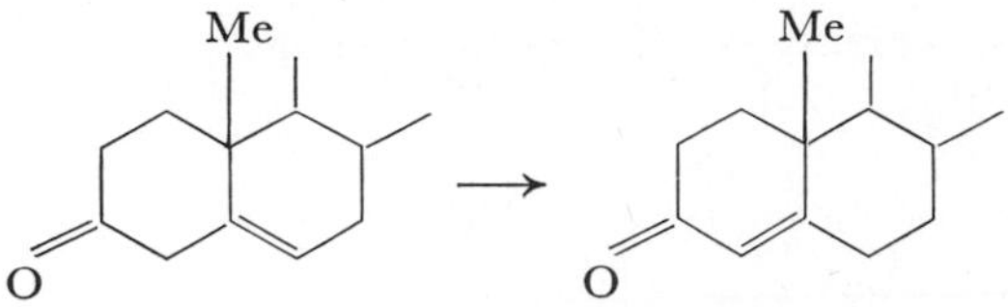

(c) Thermal isomerization.   The reorganization of electronic orbitals occurs to permit rather facile rearrangements, sometimes at relatively low temperatures. These rearrangements, which have been reviewed [10, 11], include the Cope rearrangement. In favorable structures the rearrangements are quantitative, while in others they approach an equilibrium. Examples of diene rearrangements follow:

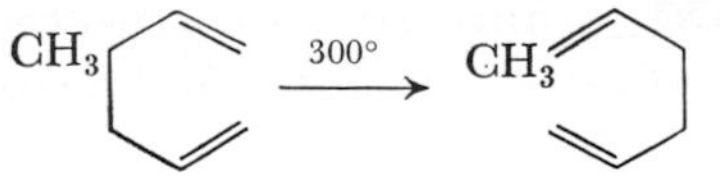

1,5-Heptadiene

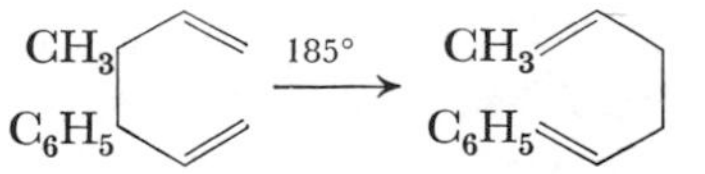

1-Phenyl-1,
5-heptadiene 90%

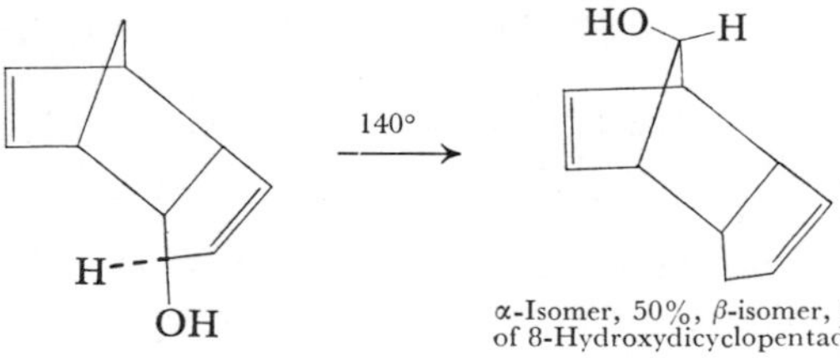

α-Isomer, 50%, β-isomer, 90%,
of 8-Hydroxydicyclopentadiene

An example of a triene rearrangement is

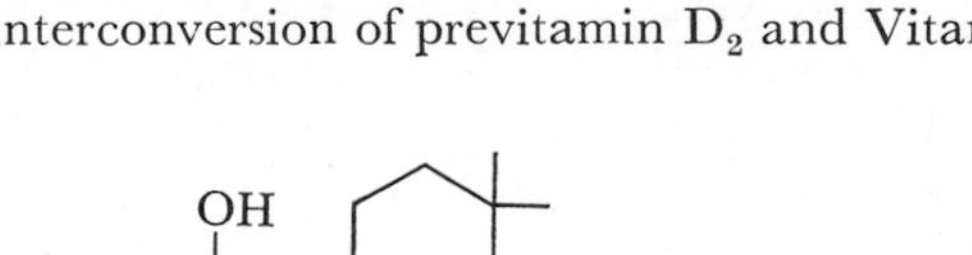

(70% conversion)

$\Delta^{9,14}\,\Delta^{10,11}$-Decahydro-
phenanthrene

Perhaps the most important of these cyclic, concerted rearrangements is the thermal interconversion of previtamin D_2 and Vitamin D_2:

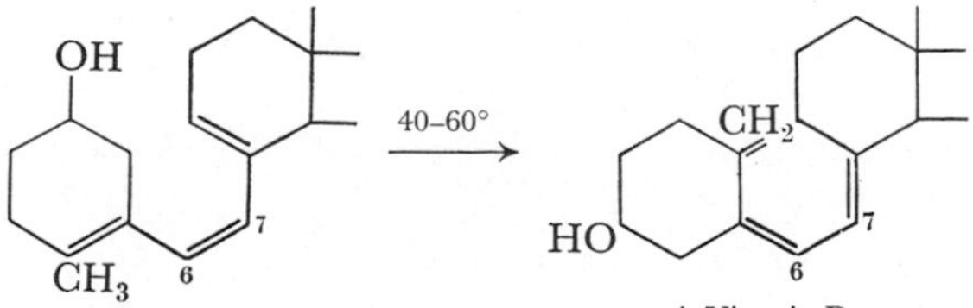

cis-Previtamin D_2

cis-Vitamin D_2

The vitamin exists primarily in the *trans* form about carbons 6 and 7 but, once in the *cis* form, olefin isomerization and tautomerization are favored.

All the preceding thermal transformations should be conducted in the gaseous state or in solutions as dilute as possible to minimize bimolecular reactions.

An example of *cis-trans* isomerization by thermal means is the conversion of oleic to elaidic acid [12]:

$$
\begin{array}{c}
CH_3(CH_2)_7CH \\
\parallel \\
HO_2C(CH_2)_7CH
\end{array}
\xrightarrow[\text{over } N_2 \text{ or } CO_2]{\text{Se, 220-225°, 1 hr.}}
\begin{array}{c}
CH_3(CH_2)_7CH \\
\parallel \\
HC(CH_2)_7CO_2H \\
62\text{-}64\%
\end{array}
$$

(d) Photochemical isomerization. *cis-trans*-Isomerization via irradiation usually leads to a mixture of *cis*- and *trans*-isomers as well as other products [13]. Some examples follow:

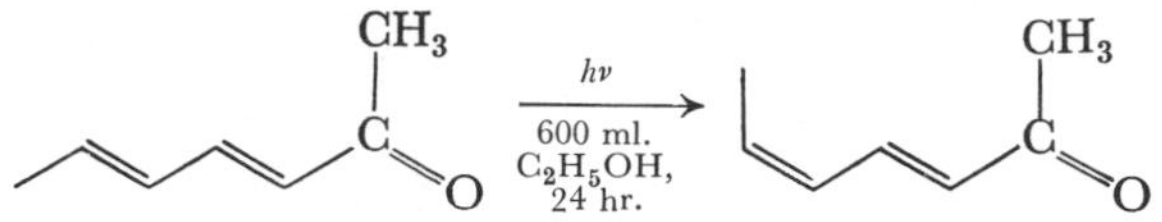

20 g.
trans-trans-Crotonylideneacetone

700 mg., pure
Δ-*cis*-5,6-Isomer   [ref. 14]

cis-2-cyclooctenone ⇌ *trans*-2-cyclooctenone   [ref. 15]
        20%,   80% too reactive to isolate

trans-p-methoxycinnamic acid   1. *hv*, 10 days   *cis-p*-methoxycinnamic acid,   [ref. 16]
as sodium salt in water   2. H⊕   20%, m.p. 64-65°

trans-α-Ionone has been isomerized to *cis*-α-ionone in 23% yield (in ethanol) [17].

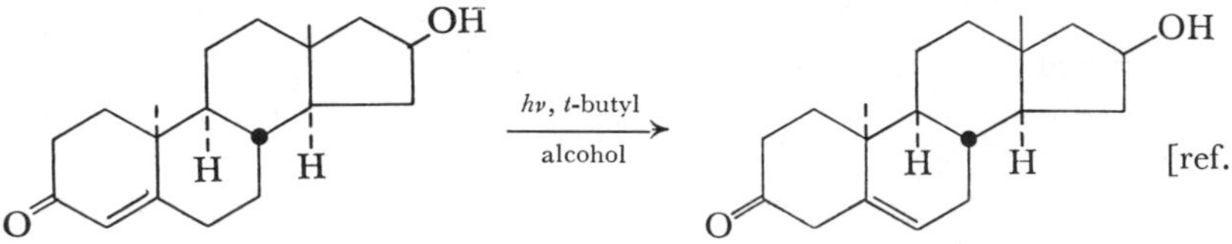

10α-Testosterone, 950 mg.

3-Oxo-17β-hydroxy-Δ⁵-10α-androsten, 197 mg.   [ref. 18]

And endocyclic monolefins tend to give exocyclic olefins, but this reaction is limited to 6- and 7-membered ring compounds [19]:

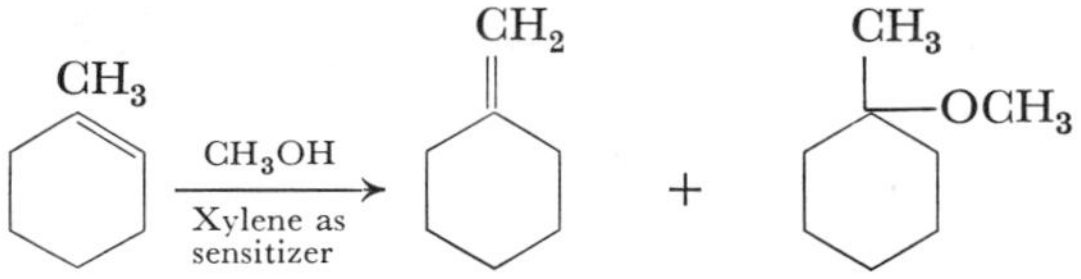

1-Methyl-1-
cyclohexene

33%
Methylene
cyclohexane

44%
1-Methyl-1-methoxy cyclohexane

Traces of acid accelerate isomerization.

α,β-Unsaturated esters isomerize by irradiation in dilute (2–5%) solutions in hydrocarbon, methanol, or ethyl acetate to β,γ-unsaturated esters. Yields are good with the exception of the case of methyl crotonate [20]. This is one of the many cases in which irradiation converts a conjugated into a nonconjugated system.

(e) Isomerization via π-complexes [21].   The unusual results of these rearrangements, carried out under very mild conditions, have begun to attract attention as a means of synthesis. They may be represented as shown:

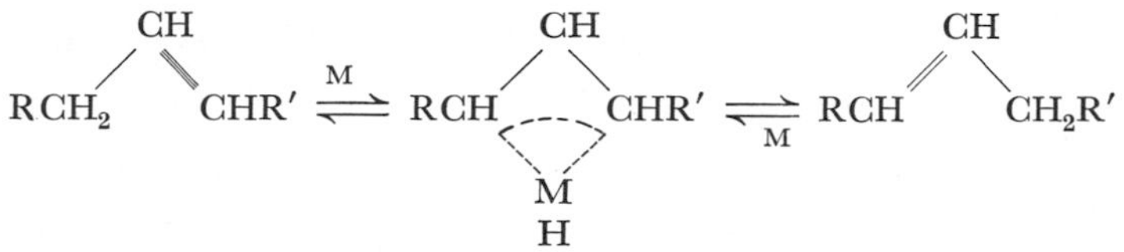

$$
\underset{RCH_2}{\overset{CH}{\diagup}}\ \underset{CHR'}{\overset{\diagdown}{}}\ \overset{M}{\rightleftharpoons}\ RCH \cdots CHR' \underset{M}{\rightleftharpoons}\ \underset{RCH}{\overset{CH}{\diagup}}\ \underset{CH_2R'}{\overset{\diagdown}{}}
$$

A solvent such as ethanol may play a part in the transfer of hydrogen. For example, if deuterated ethanol is used, a deuterium atom may be incorporated into the olefin. An example of the versatility of the rearrangement is shown [22]:

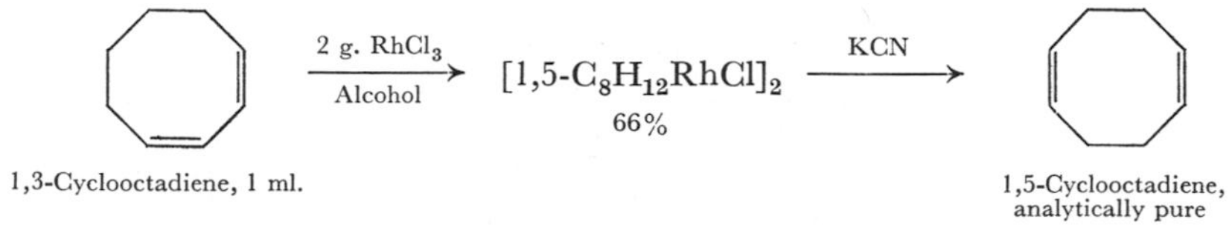

1,3-Cyclooctadiene, 1 ml. $\xrightarrow[\text{Alcohol}]{\text{2 g. RhCl}_3}$ $[1,5\text{-}C_8H_{12}RhCl]_2$ $\xrightarrow{\text{KCN}}$ 1,5-Cyclooctadiene, analytically pure

66%

On the other hand, iron pentacarbonyl isomerizes 1,5-cyclooctadiene to the 1,3-isomer in 98% yield [21]. Indeed, the isomerization of unconjugated to conjugated dienes is the usual result of metal π-complex catalysis. Recently a number of dienes of the type illustrated below has been isomerized with tris(triphenylphosphine) rhodium chloride [23]:

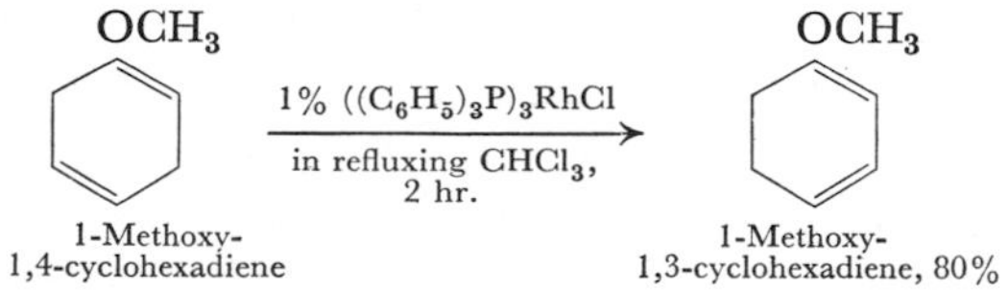

OCH$_3$ $\xrightarrow[\substack{\text{in refluxing CHCl}_3, \\ \text{2 hr.}}]{1\% \ ((C_6H_5)_3P)_3RhCl}$ OCH$_3$

1-Methoxy-1,4-cyclohexadiene   1-Methoxy-1,3-cyclohexadiene, 80%

2. From Alkynes or Diynes

$$RCH_2C\equiv CCH_2R \xrightarrow[\text{Contact time}]{\text{96 sec.}} RCH=CH-CH=CHR$$

The conditions for effecting this rearrangement are similar to those given for alkenes (D.1). For example, at a temperature of 297°, 1,5-hexadiyne mixed with nitrogen may be converted into 3,4-dimethylenecyclobutene, 80% yield [24]:

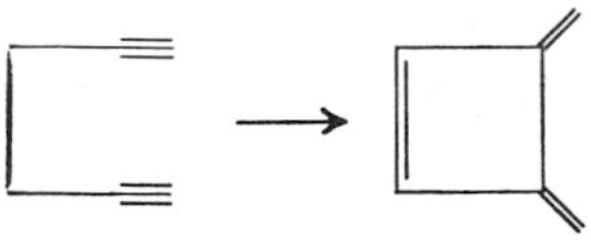

Similarly 1,4-bis-(dimethylamino)-2-butyne may be rearranged to a 1,4-bis-(dimethylamino)-1,3-butadiene in 85% yield with sodium in xylene and hexane at 72° [25]:

$$(CH_3)_2NCH_2C\equiv CCH_2N(CH_3)_2 \rightarrow (CH_3)_2NCH=CHCH=CHN(CH_3)_2$$

The *t*-butoxide anion in dimethyl sulfoxide is an effective reagent [26] as shown:

$$CH_3(CH_2)_3C\equiv CCH=CH_2 \xrightarrow[\substack{DMSO, 1 hr., \\ room temp.}]{(CH_3)_3C\overset{\ominus}{O}} CH_3(CH=CH)_3CH_3$$

1-Octen-3-yne   2,4,6-Octatriene, 87%

3. From Small-Ring Compounds

In these isomerizations [27] a driving force exists to bring about rearrangements at lower temperatures than those of bond dissociation; in this case the relief of strain is accomplished by the opening of small rings. The temperatures, nevertheless, are rather high.

$$\triangle \xrightarrow{400-500°} [\cdot CH_2CH_2CH_2\cdot] \longrightarrow CH_3CH=CH_2$$

(Has never been trapped)   Propene

Bicyclic rings containing a cyclopropane ring are broken down more readily:

Bicyclo[2.1.0]pentane $\xrightarrow{330°}$ Cyclopentene

The norcaradiene compounds, intermediates from the reaction of benzene and carbene, are so unstable that they rearrange under the conditions of the preparation:

Cycloheptatriene
(tropilidene)

Cyclobutane breaks down to ethylene:

$$\square \xrightarrow{420-480°} 2\ CH_2=CH_2 \quad \text{(essentially complete at 125 mm.)}$$

and various cyclobutenes to butadiene:

$$\square \xrightarrow{\text{Above } 200°} CH_2=CH-CH=CH_2$$

Equilibrium between olefin and cyclobutane can be established at various temperatures depending on the type of olefin; for example:

α-Methylmercaptoacrylonitrile   1,2-Dicyano-1,2-dimethylmercaptocyclobutane

4. From Hydrocarbons (Pyrolysis) [28]

Bond dissociation of hydrocarbons occurs in the temperature range 500–700°. Radicals are formed and the final products of these radicals are determined by their relative stabilities. Products most frequently found are hydrogen, methane, ethylene, and particularly 1,3-dienes. For best yield the substrate should be heated as quickly and as uniformly as possible to the temperature of dissociation and then as quickly quenched. In this way, tar formation or carbonization is minimized. Flow methods best fulfill these requirements.

To cite an example, the dry distillation of 5 kg. of rubber gave 250 g. of isoprene, about 2 kg. of dipentene, and 600 g. of mixed triterpene [29].

(a) **Preparation of isoprene.**   Limonene vapor at 20–30 mm. pressure was passed through a tube (35 × 1.6 cm.) which contained a coiled platinum wire (0.5 mm. diameter, 1 meter length) heated to near the glow point by an electric current. Limonene was collected in an ice trap, and isoprene, 60% yield, was collected in a dry ice trap [30].

5. From Hydrocarbons (Dehydrogenation) [31]

The most important dehydrogenation process is the preparation of styrene from ethylbenzene. But alkanes also can be dehydrogenated to alkenes, and alkenes to 1,3-dienes. All these processes are best adapted to industrial manipulation but may be of occasional value in laboratory preparation. The usual dehydrogenation catalyst is a coprecipitated alumina-chromium oxide. A simpler one can be made by adding 100 parts of activated alumina (6–10 mesh) to 50 parts of 10% chromium trioxide in water, filtering, and drying at 220–230°. The specific catalyst for ethylbenzene dehydrogenation is a catalyst containing 72.4% MgO, 18.4% Fe_2O_3, 4.6% CuO, and 4.6% K_2O. The potassium oxide reduces the amount of carbonization to the extent that the catalyst may be used for as long as a year. Dehydrogenation of ethylbenzene is best carried out at about 37% conversion at 600° in which the hydrocarbon and steam are passed over the catalyst bed at about 0.1 atm. The same catalyst and conditions, except that nitrogen rather than steam is the diluent, can be used to dehydrogenate butene to 1,3-butadiene. Recently, high conversion of ethylbenzene to styrene has been brought about by oxidation with sulfur dioxide over a metal phosphate [32].

Other catalyst systems that can be used for dehydrogenation are platinum or palladium metals either with or without a hydrogen acceptor and copper metal; for leading references see [33].

1. A. Maccoll in S. Patai, *ed., The Chemistry of Alkenes*, Interscience Publishers, New York, 1964, Chap. 3.
2. C. D. Broaddus, *Accounts Chem. Research*, **1**, 231 (1968).
3. Union Carbide Neth. Patent 6, 613, 870, April 3, 1967; *C.A.*, **67**, 9369 (1967).
4. I. Wender *et al., J. Org. Chem.*, **23**, 1136 (1958).
5. A. Schriesheim *et al., J. Am. Chem. Soc.*, **83**, 3731 (1961).
6. A. J. Birch *et al., J. Chem. Soc.*, 4234 (1963).
7. P. D. Gardner *et al., J. Am. Chem. Soc.*, **85**, 1553 (1963).
8. A. Schriesheim *et al., J. Org. Chem.*, **33**, 221 (1968).

9. L. F. Fieser, *J. Am. Chem. Soc.*, **75**, 5421 (1953).

10. S. J. Rhoads in P. de Mayo, *Molecular Rearrangement*, John Wiley and Sons, New York, 1963, Vol. **1**, Chap. 11.

11. W. von E. Doering and W. R. Roth, *Angew. Chem. Intern. Ed. Engl.* **2**, 115 (1963).

12. D. Swern and J. T. Scanlon, *Biochem. Prep.*, **3**, 118 (1953).

13. G. M. Wyman, *Chem. Rev.*, **55**, 625 (1955).

14. G. Büchi and N. C. Yang, *J. Am. Chem. Soc.*, **79**, 2318 (1957).

15. P. E. Eaton and K. Lin, *J. Am. Chem. Soc.*, **86**, 2087 (1964).

16. R. Stoermer, *Chem. Ber.*, **44**, 637 (1911).

17. G. Büchi and N. C. Yang, *Helv. Chim. Acta*, **38**, 1338 (1955).

18. H. Wehrli *et al.*, *Helv. Chim. Acta*, **46**, 678 (1963).

19. P. J. Kropp and H. J. Krauss, *J. Am. Chem. Soc.*, **89**, 5199 (1967).

20. R. R. Rando and W. von E. Doering, *J. Org. Chem.*, **33**, 1671 (1968).

21. C. W. Bird, *Transition Metal Intermediates in Organic Synthesis*, Academic Press, New York, 1967, Chap. 3.

22. R. E. Rinehart and J. S. Lasky, *J. Am. Chem. Soc.*, **86**, 2516 (1964).

23. A. J. Birch and G. S. R. Subba Rao, *Tetrahedron Letters*, 3797 (1968).

24. W. D. Huntsman and H. J. Wristers, *J. Am. Chem. Soc.*, **89**, 342 (1967).

25. M. F. Fegley *et al.*, *J. Am. Chem. Soc.*, **79**, 4140 (1957).

26. J. P. C. M. Van Dongen *et al.*, *Rec. Trav. Chim.* **86**, 1077 (1967).

27. R. Breslow in P. de Mayo, *Molecular Rearrangements*, John Wiley and Sons, New York, 1963, Pt. 1, Chap. 4.

28. C. D. Hurd, *The Pyrolysis of Carbon Compounds*, Chemical Catalog Co., New York, 1929, Chap. 3.

29. Ref. 28, p. 119.

30. H. Staudinger and H. W. Klever, *Chem. Ber.*, **44**, 2212 (1911).

31. K. K. Kearby in P. H. Emmett, *Catalysis*, Reinhold Publishing Corp., New York, 1955, Vol. 3, Chap. 10.

32. C. R. Adams, *Chem. Eng. News*, Dec. 25, 1967, p. 58.

33. J. P. Bradbury *et al.*, *Ind. Eng. Chem.*, **51**, 1111 (1959).

E. Condensation Reactions

The aldol or ketol condensation ordinarily yields β-hydroxycarbonyl compounds and therefore is discussed mainly in Chapter 4, G.1. It is true that heat, acid, or a favorable structure may yield instead the unsaturated carbonyl compound in the aldol condensation. Other condensations which yield unsaturated compounds are discussed elsewhere (Claisen-Schmidt, Alcohols Chapter 4, G.1; Perkin, Carboxylic Acids, Chapter 13, D.1, and Knoevenagel, Esters, Chapter 14, C.4). However, an aldol-type reaction expressly designed to form alkenes without other functional groups is discussed in E.1, and the Wittig reaction, of greater range of preparation, in E.2.

1. From Carbonyl Compounds and Cyclopentadienes (Fulvenes)

Fulvenes are the subject of a recent review [1]. Cyclopentadiene tends to

form a stable anion in basic solution

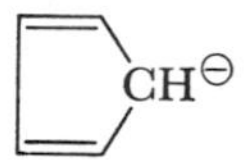

and this with an aldehyde or ketone gives a fulvene

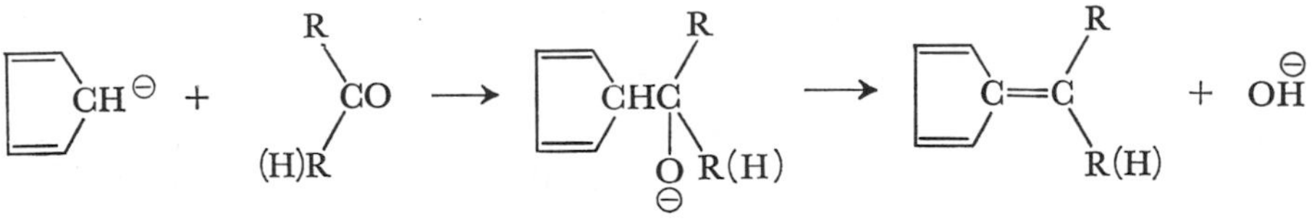

Bases which have been used to effect the condensation are alkali hydroxides or alkoxides, aqueous or alcoholic ammonia, primary or secondary amines, ethylmagnesium bromide, alkyl- or aryllithium compounds and Dowex 1-X10. The fulvenes from straight-chain aldehydes are produced with difficulty. In fact, the parent compound can be produced in a respectable yield only by an indirect method, such as by the treatment of a mixture of 1- and 2-acetoxy-methylcyclopentadienes with triethylamine [2]:

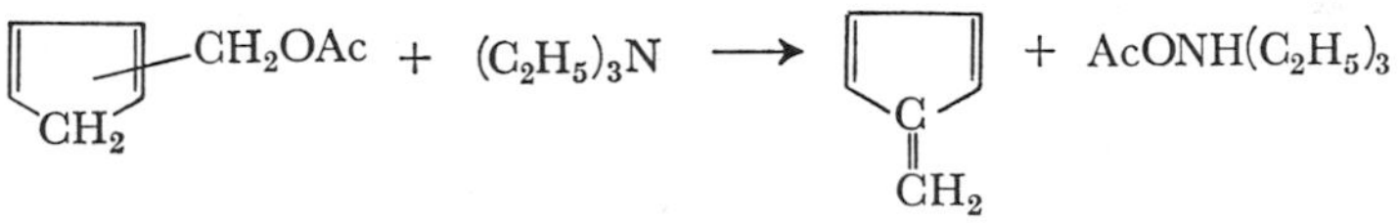

α-Branched aldehydes and ketones react normally to give yields at times as high as 70%.

(a) Preparation of dimethylfulvene. Cyclopentadiene (44 g., 0.67 mole), 23 g. of Dowex 1-X10, and 29 g. of acetone were mixed at 0° and the mixture was allowed to come to room temperature. After 1.5 hr., a vigorous reaction occurred and shaking was begun until the reaction subsided. After standing overnight at room temperature, the ion exchange resin was filtered off and the fulvene, 29.7 g. (46.7%) was recovered from the ether extract by fractionation [3; *see also* 4].

(b) Preparation of benzylidenefluorene. (70% from fluorene, benzaldehyde, potassium hydroxide, and a few drops of piperidine heated in xylene until the calculated amount of water had escaped) [5].

2. From Carbonyl Compounds and Phosphoranes (Wittig)

$$RCHO + (C_6H_5)_3P=CHR' \rightarrow RCH=CHR' + (C_6H_5)_3PO$$

The Wittig reaction is discussed briefly in Chapter 13, D.3, and has been reviewed in some detail [6, 7]. The reaction offers some advantages in the synthesis of alkenes in that: the carbonyl oxygen atom is replaced by the ethylene group without any isomerization; and the mild, alkaline conditions employed permit the preparation of sensitive olefins such as carotenoids, methylene steroids, and other natural products. In fact, it is the only reliable method known for converting a cyclic ketone to the corresponding *exo* cyclic olefin; the Grignard route (see C.6) gives practically only the *endo* cyclic isomer. The disadvantages of the procedure are the expense and the need for carrying along a bulky group which does not appear in the final product.

The Wittig reagent, a phosphorane or ylide, may be prepared from a phosphonium bromide and phenyllithium

$$(Ar)_3\overset{\oplus}{P}{-}CH_3 \; \overset{\ominus}{Br} + C_6H_5Li \rightarrow (Ar)_3P{=}CH_2 + C_6H_6 + LiBr$$

It has also been prepared conveniently by the action of sodium dimethyl sulfoxide on the halide [8] or by means of sodium methoxide in dimethyl formamide. In either case it, like the Grignard reagent, is not isolated from the solution but is utilized in this form.

The reactivity of the phosphorane varies with the nature of the R, R′, and R″ groups present

$$R_3P{=}C\overset{\displaystyle R'}{\underset{\displaystyle R''}{}}$$

Some such as

$$(C_6H_5)_3P{=}CH\overset{\displaystyle O}{\overset{\|}{C}}H$$

do not react with ketones, while $(C_6H_5)_3P = C(C_6H_5)_2$ is even unreactive toward aldehydes. In the synthesis of alkenes a methylenetriphenylphosphorane is often employed since it is a reasonably reactive type and permits some variation through the use of appropriate substituents in the methylene group. For example, 1,2-disubstituted alkenes can be prepared as shown in the accompanying formulation. In such cases, mixtures of *cis*- and *trans*-olefins are

$$RCHO + (C_6H_5)_3P{=}CHR' \rightarrow RCH{=}CHR' + (C_6H_5)_3PO$$

usually formed. To cite one case, benzylidenetriphenylphosphorane and propionaldehyde in benzene give β-ethylstyrene with a *cis/trans* ratio of 26:74 [9]. Of interest to note is the fact that this ratio is dependent on the environment for, when the reaction is carried out in dimethylformamide in the presence of lithium iodide, this *cis/trans* ratio becomes 96:4.

A variety of proton acceptors may be used to effect the Wittig reaction. Examples are butyl- or phenyllithium or an alkali metal alkoxide [6] or sodium dimethyl sulfoxide [8]. Recently it has been shown that improved yields may be obtained by the 1:1 complex of potassium *t*-butoxide and *t*-butanol [10]. A modified Wittig, as shown in the accompanying reaction leads to difluoroalkenes [11]. This reaction has not been too promising with ketones [12], although

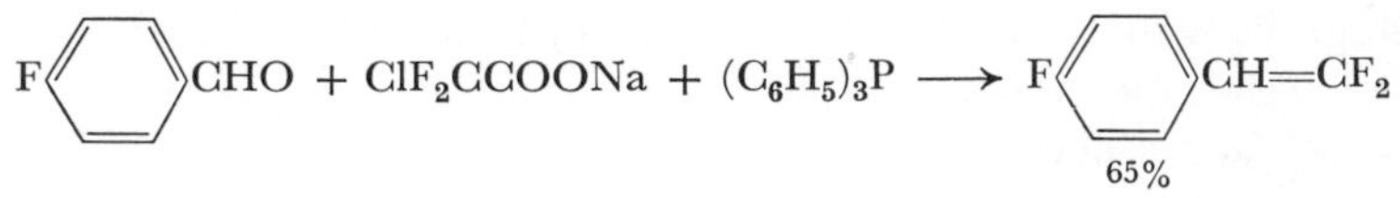

$$F{-}\langle \bigcirc \rangle{-}CHO + ClF_2CCOONa + (C_6H_5)_3P \longrightarrow F{-}\langle \bigcirc \rangle{-}CH{=}CF_2$$

65%

α,α,α-trifluoroacetophenone gives an 80% yield of 2-phenylpentafluoropropene [13].

The synthesis is also applicable to the preparation of 1,3-dienes. For example, triphenylcinnamylphosphonium chloride reacts with benzaldehyde and lithium ethoxide to give 1,4-diphenyl-1,3-butadiene in 60–67% yield [14] (see Ex. b.4).

$$C_6H_5CH{=}CHCH_2Cl \xrightarrow{(C_6H_5)_3P} C_6H_5CH{=}CHCH_2\overset{\oplus}{P}(C_6H_5)_3\overset{\ominus}{Cl} \xrightarrow{C_6H_5CHO,\ LiOC_2H_5}$$

$$C_6H_5CH{=}CHCH{=}CHC_6H_5$$

Phosphonates have also been used with aldehydes or ketones to prepare *trans*-alkenes [15, 16]:

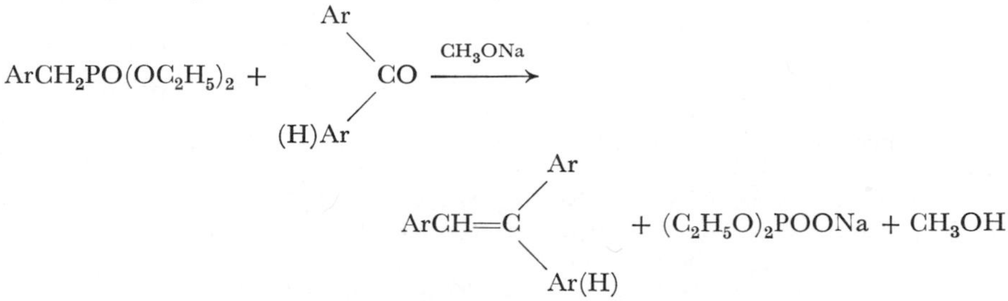

$$ArCH_2PO(OC_2H_5)_2 + \begin{array}{c} Ar \\ \diagdown \\ CO \\ \diagup \\ (H)Ar \end{array} \xrightarrow{CH_3ONa}$$

$$ArCH{=}C\begin{array}{c} Ar \\ \diagup \\ \diagdown \\ Ar(H) \end{array} + (C_2H_5O)_2POONa + CH_3OH$$

Yields with a series of aldehydes and ketones vary from 50 to 85%. The reaction may be carried out by using sodium methoxide in dimethylformamide or sodium hydride in 1,2-dimethoxyethane usually at 30–$40°$. This procedure is preferred for preparing alkenes having an electron-withdrawing group adjacent to the double bond [17].

Phosphonoesters can be used also [18]:

$$(C_2H_5O)_2POCH_2CH{=}CHCO_2C_2H_5 \xrightarrow[\text{2. ArCH}{=}O]{\text{1. NaH in CH}_3OCH_2CH_2OCH_3}$$

$$ArCH{=}CHCH{=}CHCO_2C_2H_5$$

Ethyl 5-arylpent-2,4-dienoates, 54–98%

(a) **Preparation of methylenecyclohexane.** (86.3% from dimethyl sulfoxide, sodium hydride, and methyltriphenylphosphonium bromide to which cyclohexanone was added) [19; *see also* 20].

(b) **Other examples**

(1) **1,1-Diphenyl-1-propene** (97.5% from methyl sulfinyl carbanion and ethyltriphenylphosphonium bromide to which benzophenone was added) [8].

(2) *cis-* and *trans*-**Stilbene** (20.4% *cis* and 62.0% *trans* from benzyl-

$$(C_6H_5)_3P{=}CHC_6H_5 + C_6H_5CHO \rightarrow C_6H_5CH{=}CHC_6H_5 + (C_6H_5)_3PO$$

idene triphenylphosphorane and benzaldehyde; the *cis* form was separated by distillation, the *trans* by crystallization) [21].

(3) **Additional preparations of alkenes.** See Carboxylic Acids, Chapter 13, D.3, Ex. b, c.1, c.2.

(4) 1,4-Diphenyl-1,3-butadiene (60–67%, based on the phosphonium chloride, formed by refluxing (3-chloropropenyl)-benzene and triphenylphosphine in xylene for 12 hr.; in the final step the phosphonium chloride was treated with benzaldehyde and 0.2 M lithium ethoxide in ethanol and allowed to stand for 30 min.) [14].

1. E. D. Bergmann, *Chem. Rev.*, **68,** 41 (1968).
2. H. Schaltegger *et al.*, *Helv. Chim. Acta*, **48,** 955 (1965).
3. G. H. McCain, *J. Org. Chem.*, **23,** 632 (1958).
4. C. E. Boord *et al.*, *J. Am. Chem. Soc.*, **67,** 1237 (1945).
5. D. Lavie and E. D. Bergmann, *Bull. Soc. Chim.* France, 250 (1951).
6. A. Maercker, *Org. Reactions*, **14,** 270 (1965).
7. S. Trippett, *Quart. Rev. (London)*, **17,** 406 (1963).
8. E. J. Corey and M. Chaykovsky, *J. Am. Chem. Soc.*, **84,** 866 (1962); **87,** 1345 (1965).
9. L. D. Bergelson and M. M. Shemyakin, *Tetrahedron*, **19,** 149 (1963) *ibid.*, **23,** 2709 (1967).
10. M. Schlosser and K. F. Christmann, *Angew. Chem. Intern. Ed. Engl.*, **3,** 636 (1964).
11. S. A. Fuqua *et al.*, *Tetrahedron Letters*, 1461 (1964).
12. S. A. Fuqua *et al.*, *Tetrahedron Letters*, 521 (1965); *J. Org. Chem.*, **30,** 2543 (1965).
13. D. J. Burton and F. E. Herkes, *Tetrahedron Letters*, 1883 (1965).
14. R. N. McDonald and T. W. Campbell, *Org. Syn.*, **40,** 36 (1960).
15. E. J. Seus and C. V. Wilson, *J. Org. Chem.*, **26,** 5243 (1961).
16. W. S. Wadsworth, Jr., and W. D. Emmons, *J. Am. Chem. Soc.*, **83,** 1733 (1961).
17. W. S. Wadsworth, Jr., and W. D. Emmons, *Org. Syn.*, **45,** 44 (1965).
18. L. M. Werbel *et al.*, *J. Med. Chem.*, **10,** 366 (1967).
19. E. J. Corey *et al.*, *Org. Reactions*, **14,** 395 (1965).
20. G. Wittig and U. Schoellkopf, *Org. Syn.*, **40,** 66 (1960).
21. G. Wittig and U. Schoellkopf, *Chem. Ber.*, **87,** 1318 (1954).

F. Decarboxylation, Decarbonylation, and Dehydroxylation

In this section an olefinic group is not necessarily created; an unsaturated acid, cyclic ketone, or acyl halide is simply decarboxylated or decarbonylated to yield an olefin. Many other reactions can be envisaged in which a group such as a carbonyl group is eliminated to yield an olefin. But the useful ones seem limited largely to the decarboxylation and decarbonylation methods. In addition to these three types, other related reactions, such as the Meerwein and oxidative decarboxylation, are included in this section.

1. From Unsaturated Acids

$$\text{ArCH}{=}\text{CHCOOH} \xrightarrow[\text{Quinoline}]{\text{Cu}} \text{ArCH}{=}\text{CH}_2 + \text{CO}_2$$

This synthesis, which is similar to the decarboxylation of saturated acids (Chapter 1, F.1), has been employed largely in the preparation of styrenes and stilbenes. It consists usually in heating the unsaturated acid in quinoline with copper powder or a copper salt. In the decarboxylation of cinnamic acids, quinoline and copper powder are preferred [1]. Substituted cinnamic acids give yields of olefins varying from 31 to 82%, and cinnamic acid itself gives an almost quantitative yield of styrene (Ex. b). While these decarboxylations appear to be free radical in nature (depending on the instability of $\text{RCO}_2\cdot$) or to be

References for Section F are on p. 149.

decompositions of complex copper salts, there are indications in the literature that they may proceed via anions or anion radicals as shown [2]:

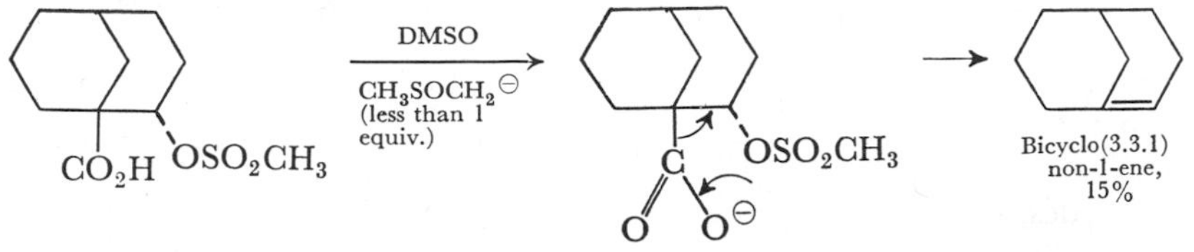

Incidentally, this example shows that Bredt's rule is not infallible. It seems that a double bond can appear at a bridgehead carbon of a bicyclic compound if it is situated in a ring containing eight or more carbon atoms [3].

The decarboxylation of α,β-unsaturated acids proceeds via isomerization to the β,γ-unsaturated isomer [4]:

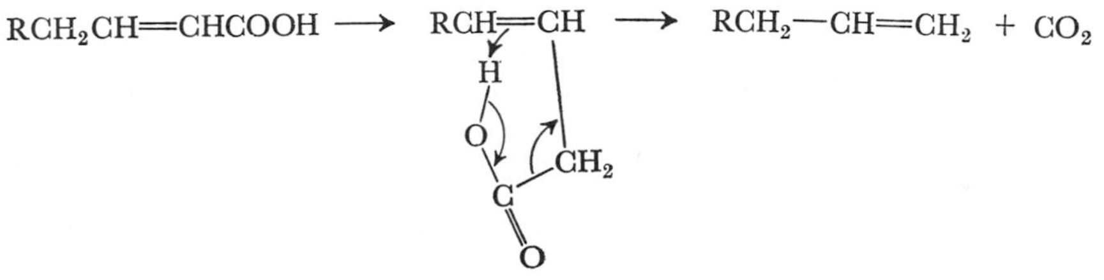

Sometimes it is possible to decarboxylate the unsaturated carboxylic acid without isolation. For example, 1,4-diphenylbutadiene may be prepared in 23–25% yield as shown by the Perkin reaction in the presence of lead monoxide [5]:

$$C_6H_5CH=CHCHO + C_6H_5CH_2COOH \xrightarrow[(CH_3CO)_2O]{PbO}$$

$$C_6H_5CH=CHCH=CHC_6H_5 + CO_2 + H_2O$$

Cyclopropaneacetic acids decarboxylate to give a mixture of olefins. A rather abnormal case is illustrated, however, in which one of the products retains the cyclopropane ring [6]:

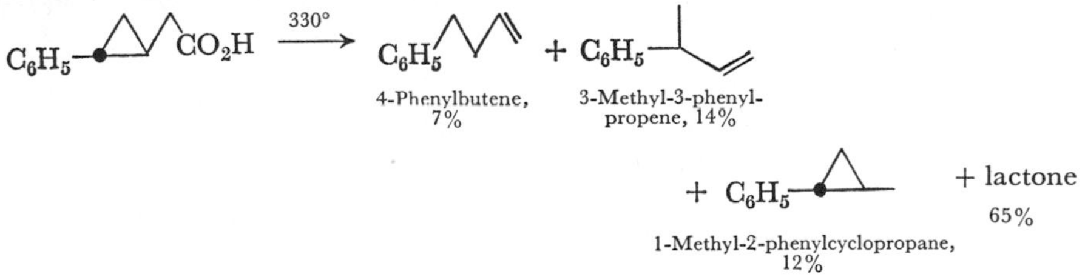

4-Phenylbutene, 7%   3-Methyl-3-phenyl-propene, 14%

1-Methyl-2-phenylcyclopropane, 12%   + lactone 65%

(a) **Preparation of cis-stilbene** (62–65% by heating α-phenyl-cinnamic acid, m.p. 172–173°, in quinoline with copper chromite at 210–220° for 1.25 hr.; the acid isomer, m.p. 137–139°, gave the trans-stilbene) [7].

(b) **Preparation of styrene** (almost 100% by boiling cinnamic acid in quinoline with a trace of copper sulfate) [8].

2. From Unsaturated Acids and Diazonium Salts (Meerwein)

$$ArCH{=}CHCOOH + Ar'\overset{\oplus}{N_2}\overset{\ominus}{X} \rightarrow ArCH{=}CHAr' + N_2 + CO_2 + HX$$

In this synthesis an aryl group is substituted for the carboxyl group via an addition product which loses

$$\underset{ArCH-CHCOOH}{\overset{\overset{X\quad\ Ar'}{|\quad\ \ |}}{}} \longrightarrow ArCH{=}CHAr'$$

carbon dioxide and the hydrogen halide [9]. The phenyl group of the diazonium compound may contain alkyl, halo, ether, ester, or nitro groups [9–11], and aldehydes, ketones, and carboxylic acid derivatives may be substituted for the carboxylic acid [11]. If cinnamalacetic acid is substituted for the monoolefinic acid or derivative, 1,4 dienes are obtained (see Ex. a). The cupric halide

$$C_6H_5CH{=}CHCH{=}CHCOOH \xrightarrow[\text{CuCl}_2]{C_6H_5\overset{\oplus}{N_2}\overset{\ominus}{X}}$$

$$C_6H_5CH{=}CHCH{=}CHC_6H_5 + N_2 + CO_2 + HX$$

and polar solvents such as water, pyridine, or acetone facilitate the reaction, while the diazonium salts of the less basic amines give the best yields. Unfortunately yields are low and the mechanism is not clear [12–14].

(a) Preparation of 1,4-diphenylbutadiene. A mixture of 0.1 mole each of benzenediazonium chloride and cinnamalacetic acid containing 0.2 mole of sodium acetate and 160 ml. of acetone was kept at 10° for 5 hr., during which time 1.4 g. of carbon dioxide was evolved. The acidic product, 0.9 g., was extracted from the ethereal extract with alkali, after which crude 1,4-diphenylbutadiene was obtained. Crystallization from benzene-ligroin gave 5.7 g. (28%), m.p. 149–150° [5].

3. From Succinic Acids

The free radicals of monocarboxylic acids decompose to give coupled products:

$$2\ RCOO{\cdot} \rightarrow RR + 2\ CO_2$$

whereas succinic acid free radicals yield olefins:

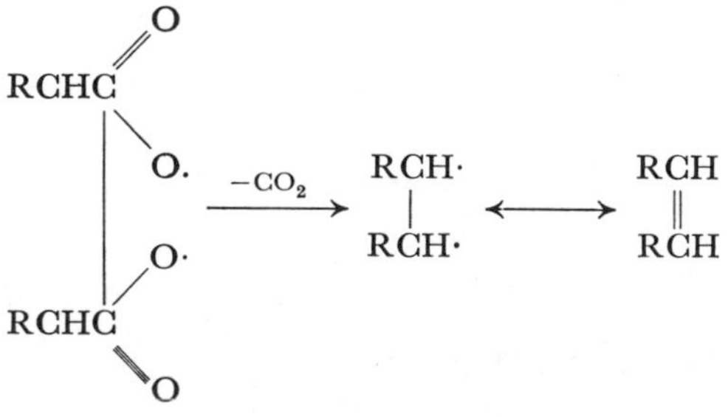

In this oxidative decarboxylation, lead dioxide [16] has been employed, although lead tetraacetate [17] appears to be the preferred reagent. The latter is effective in pyridine or more so in dimethyl sulfoxide containing 1 equivalent of pyridine [18].

Decarboxylation of succinic acids may be brought about also by electrolysis (see Ex. b). Decarboxylation of γ-ketoacids with lead dioxide forms unsaturated ketones in yields varying from 35 to 84% [19].

$$\text{RCOCH}_2\text{CH}_2\text{CO}_2\text{H} \xrightarrow[\substack{\text{Powdered} \\ \text{glass}}]{\text{PbO}_2} \text{RCOCH}=\text{CH}_2$$

(a) Preparation of bicyclo[2.2.2]oct-2-en-5-one. Oxygen was bubbled

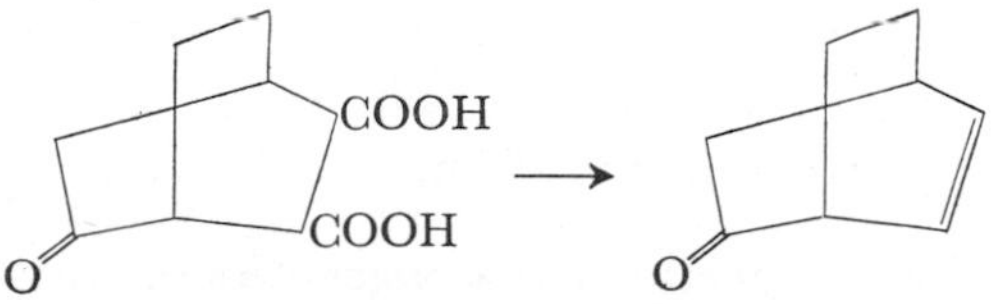

through a solution of 2.83 mmoles of bicyclo[2.2.2]octan-5-one-2,3-dicarboxylic acid in dry pyridine. Lead tetraacetate (50% molar excess) was added and the mixture was immersed in an oil bath at 67°. After carbon dioxide evolution had ceased, the product was poured into dilute nitric acid and extracted with ether. From the ethereal solution, after washing and drying, the ketone was obtained in 84% yield (crude). Sublimation gave a pure product, 52% yield [20].

(b) Preparation of *trans*-bicyclo (6.2.0) decene-9 [21]

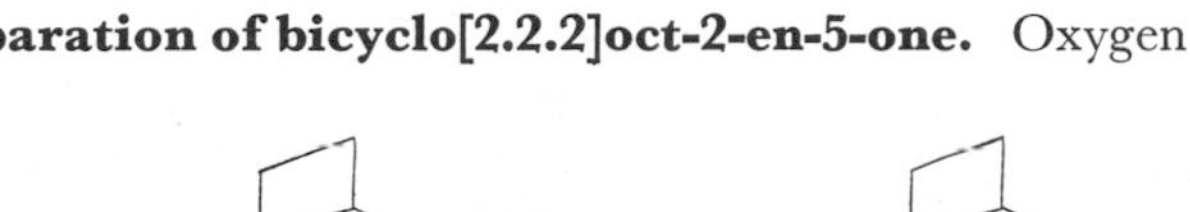

4. From Acid Chlorides

$$\text{Pd} + \text{RCH}_2\text{CH}_2\text{COCl} \rightleftharpoons \text{RCH}_2\text{CH}_2\text{COPdCl} \rightleftharpoons \overset{\overset{\text{CO}}{|}}{\text{RCH}_2\text{CH}_2\text{PdCl}} \rightleftharpoons$$

$$\underset{\substack{\| \\ \text{CH}_2}}{\text{RCH}} + \underset{\substack{| \\ \text{PdCl} \\ | \\ \text{H}}}{\text{CO}} \rightleftharpoons \text{RCH}=\text{CH}_2 + \text{CO} + \text{HCl} + \text{Pd}$$

These equations represent in outline form the ability of palladium (or palladium salts, other noble metals, or rhodium complexes such as $RhCl(Ph_3P)_3$) to decarbonylate acid chlorides or to effect the reverse reaction, the carbonylation of olefins [22]. Catalytic amounts of palladium can be used for the decarbonylation in which the olefin is removed as formed.

(a) **Preparation of nonenes** (58% of a mixture by heating 20 g. of decanoyl chloride and 1 g. of 1% Pd on carbon at 200°) [22].

5. From 1,3-Cyclobutanones

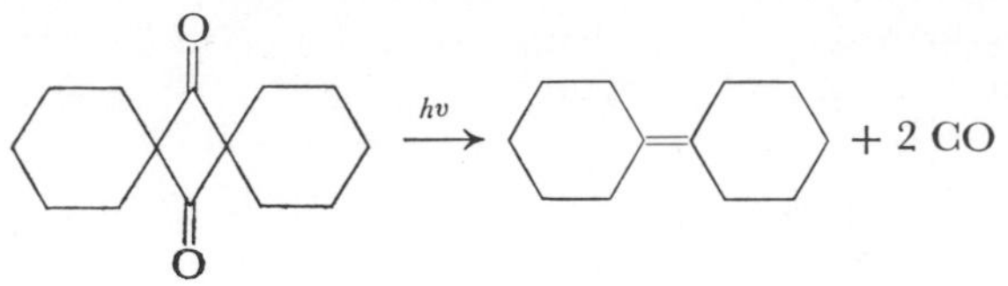

In inert solvents, 1,3-cyclobutanones upon being irradiated decompose to give alkenes. The reaction is general, and the diketone is readily prepared from the cyclohexanecarbonyl chloride via the dimerization of its ketene [23, 24]. Tetramethyl-1,3-cyclobutanedione gives an 80%, and the dispiro dione gives a 49% yield (see Ex. a) of the alkene.

(a) **Preparation of cyclohexylidenecyclohexane** (49% from dispiro [5.1.5.1]tetradecane-7,14-dione by irradiation in methylene chloride for 8–10 hr.) [23].

6. From Diols via Thionocarbonates

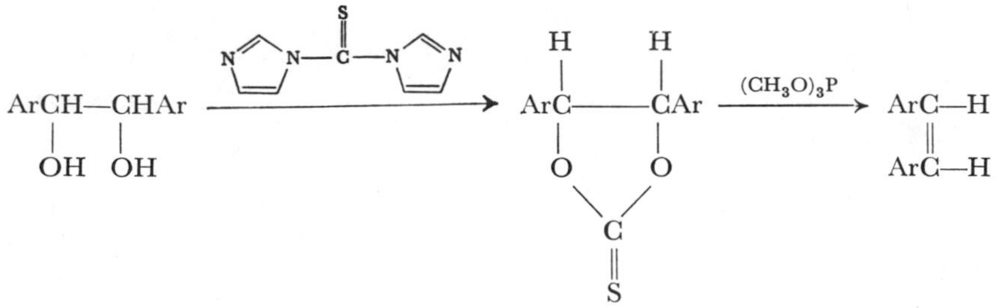

Thiocarbonyldiimidazole, prepared from 2 moles of imidazole and 1 mole of thiophosgene, reacts with 1,2-diols to form a cyclic thionocarbonate, which with trimethylphosphite cleaves to give the alkene [25]. It has been proposed that the trimethylphosphite abstracts sulfur to give the carbene

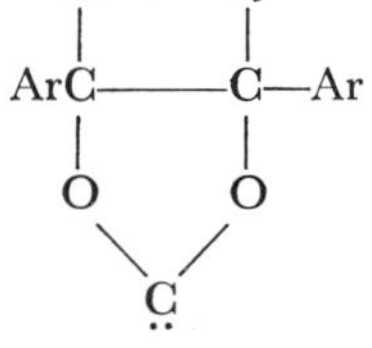

which undergoes *cis* elimination to form the alkene and carbon dioxide. Thus *meso*-hydrobenzoin gives *cis*-stilbene in 92% yield, while the *dl*-isomer forms *trans*-stilbene in 87% yield. Other 1,2-diols give similar yields.

(a) Preparation of *cis*-Stilbene The cyclic thionocarbonate was prepared in high yield by refluxing *meso*-hydrobenzoin with thiocarbonyldiimidazole in toluene or xylene for 30 min. When refluxed for 70–80 hr. under nitrogen in trimethylphosphite, this product gave *cis*-stilbene (92%) [25].

1. C. Walling and K. B. Wolfstirn, *J. Am. Chem. Soc.*, **69,** 852 (1947).
2. J. A. Marshall and H. Faubl, *J. Am. Chem. Soc.*, **89,** 5965 (1967).
3. J. R. Wiseman, *J. Am. Chem. Soc.*, **89,** 5966 (1967).
4. R. T. Arnold *et al.*, *J. Am. Chem. Soc.*, **72,** 4359 (1950).
5. B. B. Corson, *Org. Syn.*, Coll. Vol. **2,** 229 (1943).
6. J. J. Sims and M. Ban, *Tetrahedron Letters*, 5289 (1968).
7. R. E. Buckles and N. G. Wheeler, *Org. Syn.*, Coll. Vol. **4,** 857 (1963).
8. C. Walling and K. B. Wolfstirn, *J. Am. Chem. Soc.*, **69,** 852 (1947).
9. F. Bergmann *et al.*, *J. Org. Chem.*, **9,** 408, 415 (1944); **12,** 57 (1947).
10. R. C. Fuson and H. G. Cooke, Jr., *J. Am. Chem. Soc.*, **62,** 1180 (1940).
11. H. Meerwein *et al.*, *J. prakt. Chem.*, **152,** 237 (1939).
12. W. A. Cowdrey and D. S. Davies, *Quart. Rev.*, (London), **6,** 365 (1952).
13. J. K. Kochi, *J. Am. Chem. Soc.*, **78,** 1228 (1956); **79,** 2942 (1957).
14. S. C. Dickerman *et al.*, *J. Org. Chem.*, **22,** 1070 (1957); *J. Am. Chem. Soc.*, **80,** 1904 (1958).
15. C. F. Koelsch and V. Boekelheide, *J. Am. Chem. Soc.*, **66,** 412 (1944).
16. W. von E. Doering, *J. Am. Chem. Soc.*, **74,** 4370 (1952).
17. C. A. Grob *et al.*, *Helv. Chim. Acta*, **41,** 1191 (1958).
18. L. F. and M. Fieser, *Reagents for Organic Synthesis*, John Wiley and Sons, New York, 1967, p. 555.
19. E. J. Eisenbraun *et al.*, *J. Org. Chem.*, **33,** 2008 (1968).
20. C. M. Cimarusti and J. Wolinsky, *J. Am. Chem. Soc.*, **90,** 113 (1968).
21. P. Radlick, J. J. Sims, E. E. van Tamelen *et al.*, *Tetrahedron Letters*, 5117 (1968).
22. J. Tsuji and K. Ohno, *J. Am. Chem. Soc.*, **90,** 94, 99 (1968).
23. N. J. Turro *et al.*, *Org. Syn.*, **47,** 34 (1967).
24. N. J. Turro *et al.*, *J. Am. Chem. Soc.*, **86,** 955 (1964).
25. E. J. Corey and R. A. E. Winter, *J. Am. Chem. Soc.*, **85,** 2677 (1963).

Chapter 3

ALKYNES
with Brief Comments on
Allenes and Cumulenes

Several reviews on the laboratory synthesis of alkynes are available [1–3]. As is well known, acetylene, the first member of the series, is prepared commercially by the action of water on calcium carbide, a process which has been described in great detail [4]. Brief comments on the syntheses of allenes and cumulenes are made in D.

A. Elimination

It is helpful in this section to relate eliminations to produce olefins with those employed to produce acetylenes. Single eliminations of HX, H_2O, N_2, X_2, Me_3N, Ph_3PO, or other simple molecules are potential pathways to olefins. Similarly the following eliminations exist as possible ways to produce acetylenes: double eliminations of simple molecules; combination of two different single eliminations; and single elimination from a vinyl-bearing substrate.

These possibilities lead to some interesting and sometimes novel conceptions of acetylene syntheses. The first possibility alone suggests that as many as fifteen different eliminations are potential pathways to acetylenes, but some of them are impractical and are simply mentioned under other headings. Allenes as well are produced in some of these eliminations, but fortunately methods of isomerization are available to rid the acetylene of such contamination.

1. From Dihalides and Vinyl Halides (Dehydrohalogenation)

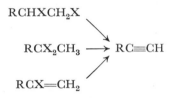

The elimination of a hydrogen halide from dihaloalkanes or haloalkenes gives alkynes. Of the starting materials, the first may be obtained by the addition of the halogen to the appropriate alkene, the second by treatment of the proper ketone with a phosphorus halide, and the third by the partial dehydrohalogenation of either of the first two. Various alkaline reagents are used to effect the elimination, the two most common being potassium hydroxide, either in a hydroxylic solvent such as ethanol, ethylene glycol, or Cellosolve or in the solid state, and sodamide, either suspended in mineral oil or in liquid ammonia [1]. Sometimes a mineral oil is employed with potassium hydroxide in ethanol to maintain a fluid medium [5]. There is some risk of rearrangement in the case of aliphatic acetylenes with the high temperatures (100–200°) usually employed in these reactions. In fact, potassium hydroxide may convert 1-alkynes into 2-alkynes while sodamide, except when used in liquid ammonia, brings about the reverse rearrangement. The sodamide-liquid ammonia reagent exerts a powerful dehydrohalogenation action with mild reaction conditions and is particularly useful in preparing sodium acetylides in solution for use as intermediates in synthesizing other acetylides (see B.1) [6]. Aromatic acetylenes may be synthesized satisfactorily with either potassium hydroxide or sodamide as the dehydrohalogenating agent, although it is interesting to note that a rearrangement occurs here too with 1-halo-2,2-diaryl ethylenes when treated with potassium amide in liquid ammonia [7]. This rearrangement, at least when

$$Ar_2C{=}CHX \xrightarrow{\text{KNH}_2} ArC{\equiv}CAr$$

brought about by butyllithium, does not proceed via a carbene, since the *cis-trans* isomers, when aryl groups are different, do not show equivalent migration of aryl groups [7]. Rearrangements in the alicyclic family have been observed as well [9]. The smaller-ring, more strained acetylenes, such as cyclopentyne,

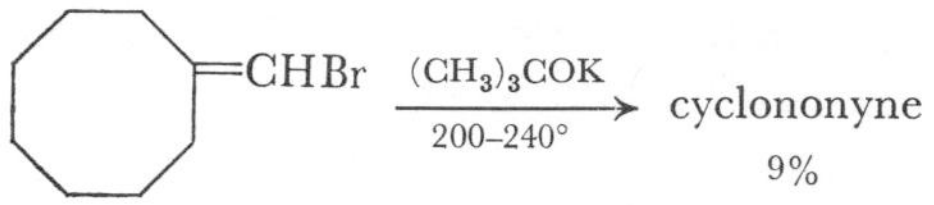

$$\text{(cyclooctylidene)}{=}CHBr \xrightarrow[200-240°]{\text{(CH}_3)_3\text{COK}} \text{cyclononyne} \atop 9\%$$

+ 1,2- (27%) and 1,3-cyclononadiene (25%) + bicyclo[6.1.0]non-1-(2)-ene
40%

trapped as Diels-Alder adducts with 1,3-diphenylisobenzofuran, have been produced in fair yield by the same procedure. The smallest-ring cycloalkyne stable at room temperature is cyclooctyne [10].

It would be expected that dehydrohalogenation of vinyl halides to produce acetylenes would be more difficult than the corresponding process to produce olefins, but exceptions are known [11]. *trans* Eliminations are more rapid than *cis* eliminations if the E_2 mechanism is operable:

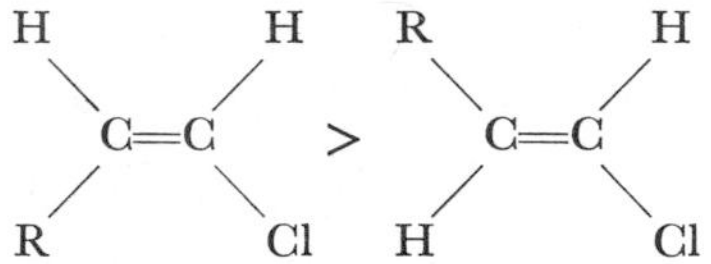

but, if the proton is removed before elimination (when R is strongly electron-withdrawing), the generalization does not hold. Medium effects are large, sometimes puzzling, but encouraging in that, if one medium does not work, another may be found [11].

It should be recalled from Alkenes and Cycloalkenes, Chapter 2, that 1,5-diazabicyclo[4.3.0]nonene-5 is a very effective dehydrohalogenating agent which should find use in the preparation of acetylenes.

(a) **Preparation of phenylpropargyl aldehyde acetal** (80–86 % from α-bromocinnamic aldehyde acetal and potassium hydroxide in absolute ethanol refluxed for 1.5 hr.) [12].

(b) **Other examples**

(1) **3-Isopropylphenylacetylene** [13]

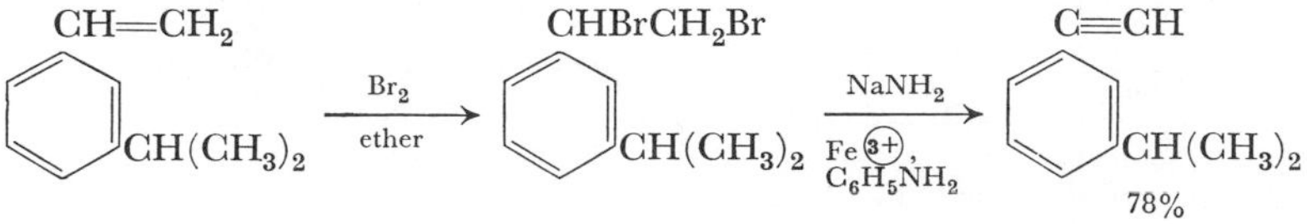

(2) Dibenzoylacetylene (82–95% from *meso*-dibenzoylethylene dibromide by refluxing with purified triethylamine in benzene or acetone) [14].

(3) Dipropargyl (hexa-1,5-diyne) (56% from 1,2,5,6-tetrabromohexane in ether and sodamide in liquid ammonia containing a trace of ferric nitrate) [15].

(4) Isopropylacetylene [16]

$(CH_3)_2CHCH_2CHCl_2$, 141 g., added dropwise to a stirred solution of 320 g.

KOH in 500 g. diethylene glycol at $170° \rightarrow (CH_3)_2CHC{\equiv}CH + (CH_3)_2C{=}C{=}CH_2$

          40%   29%

$$(CH_3)_2C{=}C{=}CH_2 \xrightarrow[\text{Isooctane, } 170°]{\text{NaNH}_2} (CH_3)_2CHC{\equiv}CH$$

          20 g.   75%

(5) 1,3-Butadiyne

$ClCH_2C{\equiv}CCH_2Cl$ added to 40% aqueous NaOH $\xrightarrow[\text{Swept with H}_2]{65-75°}$

  1 part   2.85 parts

$$HC{\equiv}CC{\equiv}CH + CH_2{=}CClC{\equiv}CH + \text{tars}$$

          ca. 60%

The chloroalkyne was removed by preparative gas chromatography. Butadiyne is unstable and should be stored at low temperatures [17]; for numerous examples of dehydrohalogenations *see also* [17].

(6) N-Ethylpropargylamine [18]

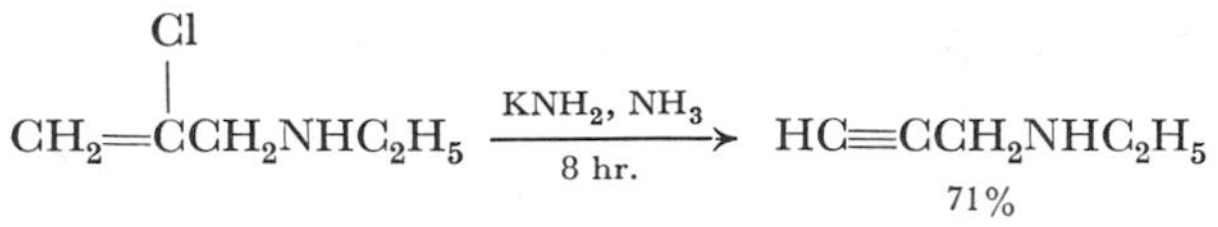

$$\overset{\displaystyle Cl}{\underset{\displaystyle |}{CH_2{=}C}}CH_2NHC_2H_5 \xrightarrow[\text{8 hr.}]{\text{KNH}_2, \text{ NH}_3} HC{\equiv}CCH_2NHC_2H_5$$

                                                      71%

but

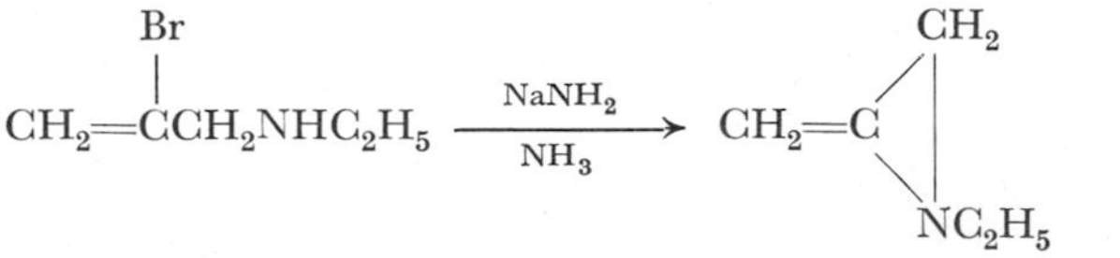

N-Ethylallenimine

(7) *m,m′*-Dimethyltolane

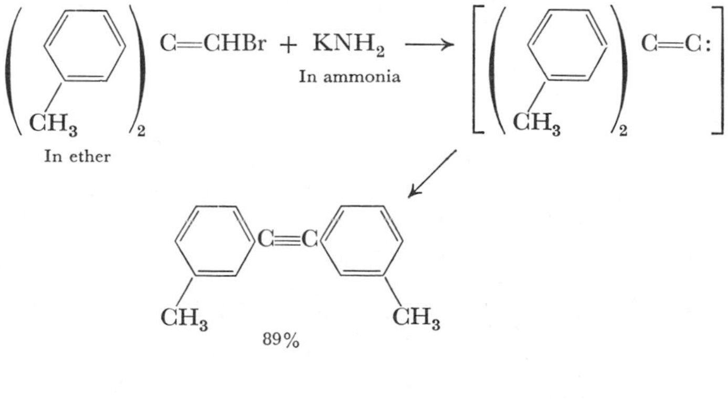

          89%

This reaction is general for unsymmetrical diphenylbromoethenes. The starting material was readily available from the dibromide from which hydrogen bromide is lost spontaneously [7].

2. From β-Chloro Ethers or Vinyl Ethers

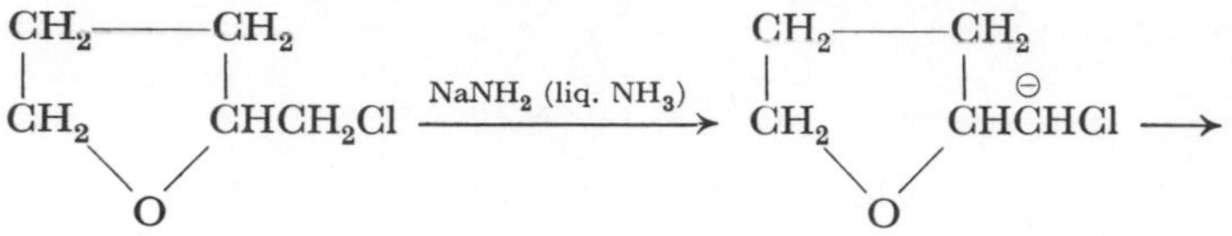

$[^{\ominus}OCH_2CH_2CH_2CH{=}CHCl] \longrightarrow {}^{\ominus}O(CH_2)_3C{\equiv}CH \xrightarrow{NH_4Cl} HO(CH_2)_3C{\equiv}CH$

Tetrahydrofurfuryl chloride with sodium amide in liquid ammonia followed by treatment with ammonium chloride gives 4-pentynol in 75–85% yield [19]; see Ex. a. The mechanism appears to be that shown. A similar cleavage may be obtained with α-chloromethyltetrahydropyran [20]. The former is, of course, much more readily available because it may be produced from furfural, a commercial product. Similarly, o-hydroxyphenylacetylene is obtained from coumarin [21].

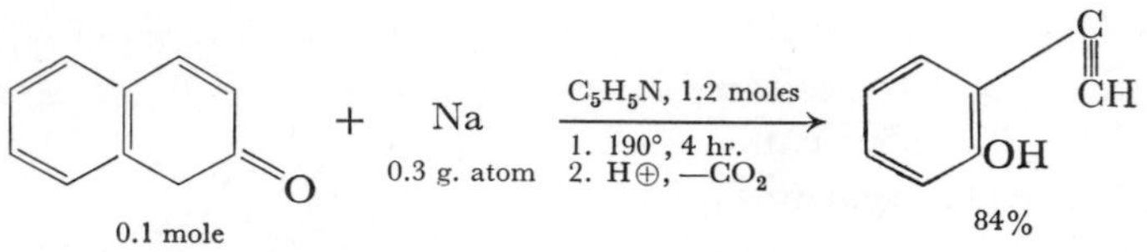

(a) Preparation of 4-pentyn-1-ol (75%–85% from tetrahydrofurfuryl chloride treated with sodamide and a trace of ferric nitrate in liquid ammonia over about 1.5 hr., after which solid ammonium chloride was added) [19].

3. From Tetrahalides or Acetylene Dihalides (Dehalogenation)

$$RCX_2CX_2R \xrightarrow{Zn} RC{\equiv}CR + 2\,ZnX_2$$

At first glance this reaction does not appear to be of much value, as the tetrahalides would seem to be available only from the acetylene, the desired product. But several other sources of tetrahalides are known:

$$2\,ArCCl_3 \xrightarrow{Cu\ powder} ArCCl_2CCl_2Ar \text{ or } ArCCl{=}CClAr$$

(See Ex. a)

$$(Ar)_2CClCCl_3 \xrightarrow[\substack{3.\ Temp.\ rises\ to\ 210° \\ 4.\ Cool\ immediately}]{\substack{1.\ \Delta\ to\ 160° \\ 2.\ Add\ 5\ mg.\ FeCl_3}} ArCCl_2CCl_2Ar$$

85%
(see Ex. b.1)

A wide variety of perfluoroacetylene dihalides is available from the telomerization of tetrafluoroethylene with carbon tetrachloride followed by copper coupling [22]:

$$CF_2{=}CF_2 + CCl_4 \xrightarrow{300°} ClCF_2CF_2CCl_3 \xrightarrow[180°]{4\ Cu} ClCF_2CF_2CCl{=}CClCF_2CF_2Cl$$

Other syntheses of acetylenic dihalides are described as well. These tetrachlorides, acetylene dichlorides, and other perhalides may be dehalogenated with zinc. Moreover, although not exploited as yet, combinations of alkoxyl and halo groups can be visualized which may be potential sources of acetylenes through the loss of the elements of alkoxyl and halo when treated with zinc or magnesium. Debromination of *cis*-2,3-dibromo-2-butene with zinc dust in alcohol at 80° was found to be complete in 1 hour, while to debrominate the *trans*-isomer somewhat more time was required [23].

Some acetylene dihalides may eliminate halogen by treatment with a base to form acetylenes [24]:

$$p\text{-}NO_2C_6H_4CH{=}CHC_6H_4NO_2\text{-}p \xrightarrow[\substack{\text{hot } C_6H_5NO_2, \\ \text{strong lamp}}]{Br_2}$$

$$p\text{-}NO_2C_6H_4\overset{\overset{\displaystyle Br}{|}}{C}{=}\overset{\overset{\displaystyle Br}{|}}{C}C_6H_4NO_2\text{-}p \xrightarrow[\text{in } CH_3OH]{Ca(OH)_2} p\text{-}NO_2C_6H_4C{\equiv}CC_6H_4NO_2\text{-}p$$
$$81\%p,p'\text{-Dinitrotolane, } 61.5\%$$

(a) Preparation of *o,o′*-dichlorodiphenylacetylene, *o,o′*-dichlorotolane [25]

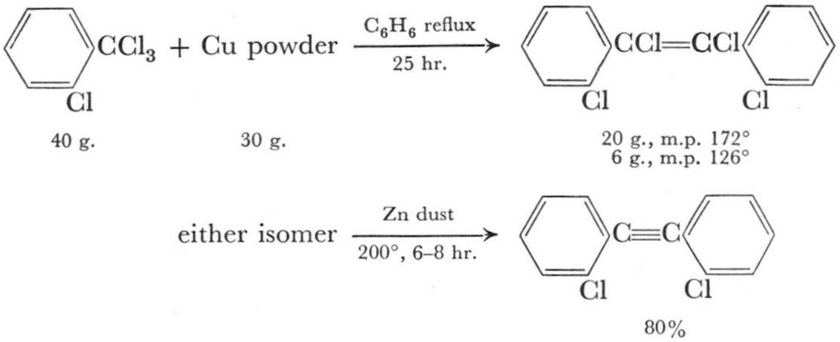

40 g.   30 g.   C$_6$H$_6$ reflux 25 hr.   20 g., m.p. 172° 6 g., m.p. 126°

either isomer $\xrightarrow[200°, 6-8 \text{ hr.}]{Zn \text{ dust}}$ 80%

(b) Other examples

(1) *p,p′*-Dichlorotolane [26]

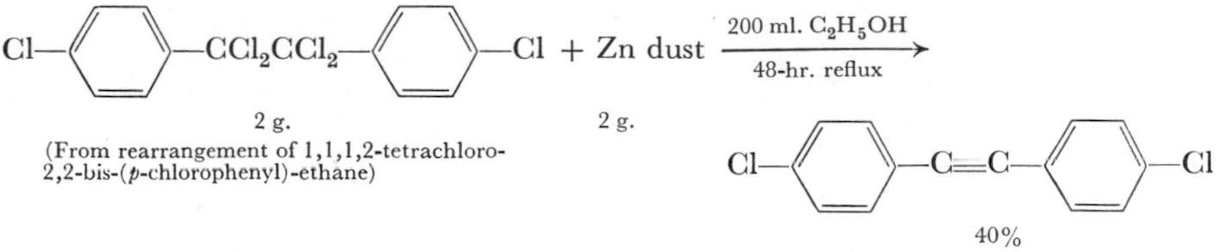

2 g.
(From rearrangement of 1,1,1,2-tetrachloro-2,2-bis-(*p*-chlorophenyl)-ethane)   2 g.   200 ml. C$_2$H$_5$OH 48-hr. reflux   40%

(2) Hexafluoro-2-butyne (63 % by stirring and refluxing 60 g. of zinc dust and 80 ml. of acetic anhydride to which 84 g. of 2,3-dichlorohexafluoro-2-butene in 50 ml. of acetic anhydride was added over a 4-hr. period; head temperature of a 10 in. Vigreux column should not exceed 50°; Zn dust, an additional 30 g., was added, and gentle reflux was continued for another 3 hr., during which all vapors were collected in a dry ice trap; the contents of the trap were distilled by warming to room temperature to yield 34 g. (63 %) of the butyne). Zn dust in acetic anhydride seems to be the reagent of choice [22].

4. From Dihydrazones and Related Compounds

This section deals with double diazo decompositions or with single diazo decompositions combined with other elimination processes.

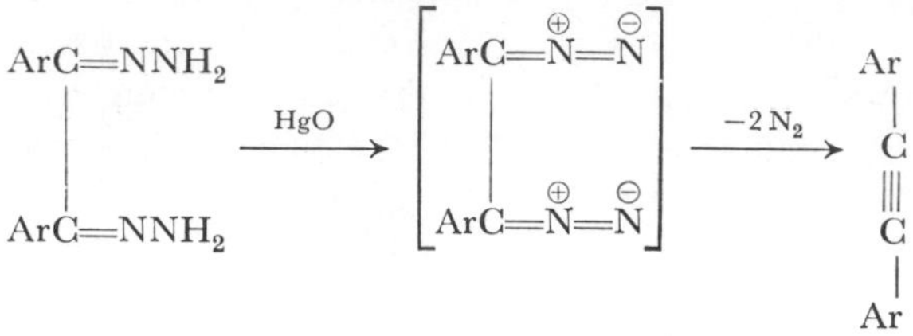

$$
\begin{array}{c}
\text{ArC}=\text{NNH}_2 \\
| \\
\text{ArC}=\text{NNH}_2
\end{array}
\xrightarrow{\text{HgO}}
\left[
\begin{array}{c}
\text{ArC}=\overset{\oplus}{\text{N}}=\overset{\ominus}{\text{N}} \\
| \\
\text{ArC}=\overset{\oplus}{\text{N}}=\overset{\ominus}{\text{N}}
\end{array}
\right]
\xrightarrow{-2\,\text{N}_2}
\begin{array}{c}
\text{Ar} \\
| \\
\text{C} \\
||| \\
\text{C} \\
| \\
\text{Ar}
\end{array}
$$

This synthesis produces acetylenes of good purity, usually in high yield, from the dihydrazones available from 1,2-diketones. The method, although of limited value, is applicable to aliphatic, alicyclic, and aromatic dihydrazones. As oxidizing agents, mercuric oxide alone [27] or with potassium hydroxide and anhydrous sodium sulfate [28], or silver trifluoroacetate in triethylamine [29] are employed.

A useful combined diazo decomposition-dehydrobromination is possible with the readily available pyrazalones [30]:

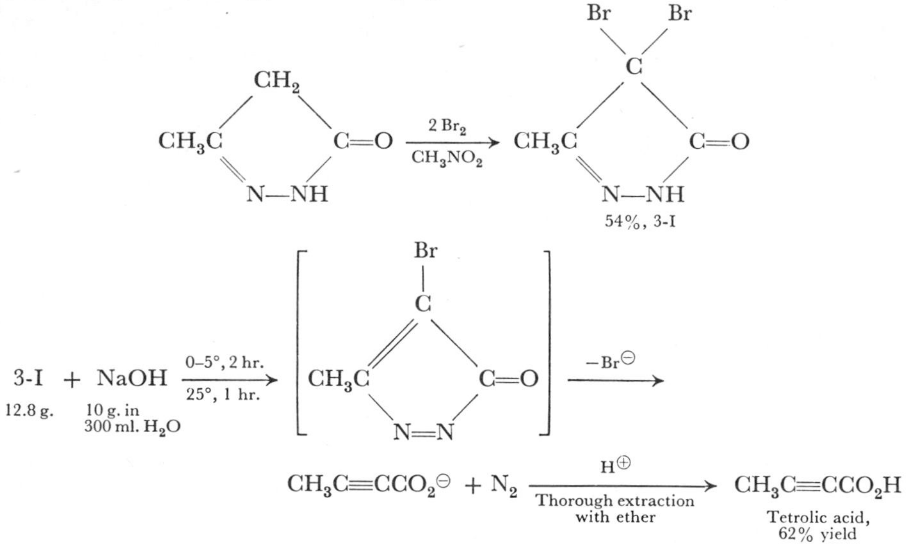

3-I + NaOH 12.8 g. 10 g. in 300 ml. H₂O, 0–5°, 2 hr. 25°, 1 hr.

54%, 3-I

$$\text{CH}_3\text{C}\equiv\text{CCO}_2^{\ominus} + \text{N}_2 \xrightarrow[\text{Thorough extraction with ether}]{\text{H}^{\oplus}} \text{CH}_3\text{C}\equiv\text{CCO}_2\text{H}$$

Tetrolic acid, 62% yield

This is probably the best method of preparing tetrolic acid. The existence of the intermediate above has not been demonstrated and is suggested simply as a rationale for the combination of two different types of elimination.

Another combined elimination involves the decomposition of nitrosooxazolidones [31]:

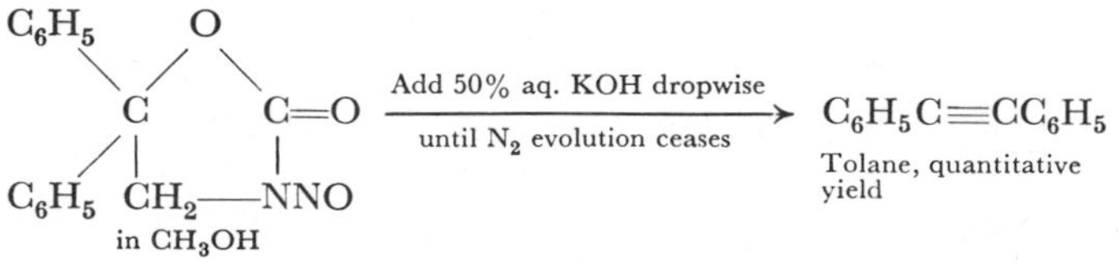

Add 50% aq. KOH dropwise until N₂ evolution ceases → C₆H₅C≡CC₆H₅

Tolane, quantitative yield

5-Phenyl-5-methyl-3-nitroso-2-oxazolidone is converted into 74% phenylpropyne and 16% phenylacetone by the same procedure.

Unsaturated ketones can be converted by a series of steps into an acetylene. A schematic diagram is given here and a specific example is shown in Ex. b.3:

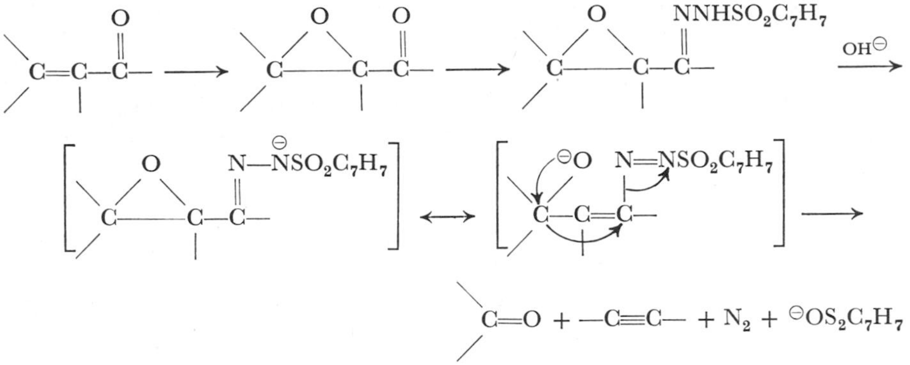

The reaction is reminiscent of a reverse Prins reaction:

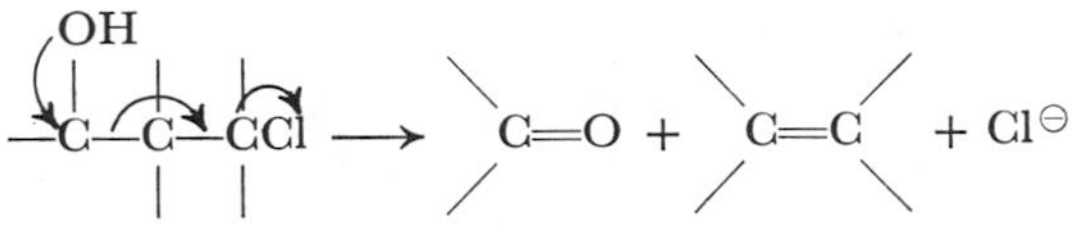

The benzenesulfonate anions of triazoles yield acetylenes on irradiation [32]:

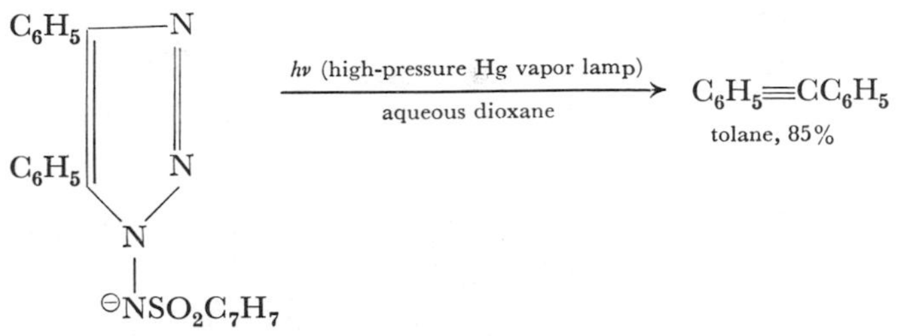

Other substituted triazoles yield other acetylenes, but they were isolated as the Diels-Alder adduct with tetraphenylcyclopentadienone.

In addition, diazo decomposition of vinylamines yields acetylenes. Since vinyl amines are not common, the reaction is limited in scope [33]:

$$(C_6H_5)_2C{=}CHNH_2 \xrightarrow[\substack{C_6H_6, \\ \text{reflux overnight}}]{\text{Isoamyl nitrite}} C_6H_5C{\equiv}CC_6H_5 + \text{some benzophenone}$$

1 g.
(From diphenylacetaldehyde
and ammonia in methanol;
the methanol is essential to
avoid the formation of the
imine)

Tolane, 62%

Finally, when the 1:1 adducts of benzils and triethylphosphite are heated with an excess of the phosphite at 215°, acetylenes are produced in yields varying from 58 to 81% [34]:

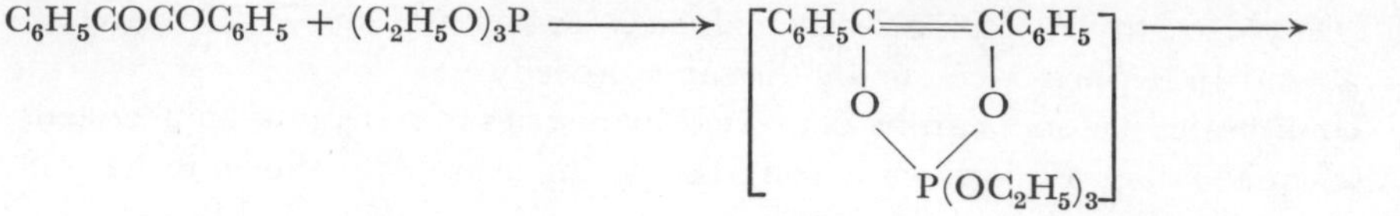

$$C_6H_5COCOC_6H_5 + (C_2H_5O)_3P \longrightarrow \left[\begin{array}{c} C_6H_5C \!=\!= CC_6H_5 \\ | \quad\quad | \\ O \quad\quad O \\ \diagdown \;\; \diagup \\ P(OC_2H_5)_3 \end{array} \right] \longrightarrow$$

$$(C_6H_5)_2C\!=\!C\!=\!O \xrightarrow{(C_2H_5O)_3P} C_6H_5C\!\equiv\!CC_6H_5 + (C_2H_5O)_3PO$$
$$\text{81\%}$$

(a) Preparation of diphenylacetylene, tolane (67–73 % from benzil in *n*-propyl alcohol which was first converted into the dihydrazone by refluxing with 85 % hydrazine hydrate for 60 hr.; the latter in refluxing benzene was oxidized with yellow mercuric oxide) [27]; *see also* [29].

(b) Other examples

(1) ***m*-Chlorophenylphenylacetylene** (80 % from *m*-chlorobenzil dihydrazone, silver trifluoroacetate, and triethylamine in alcohol for 50 min.) [29].

(2) **Cyclodecyne** (67 % from 1,2-cyclodecanedione dihydrazone added in portions with vigorous stirring to a refluxing suspension of mercuric oxide, potassium hydroxide and anhydrous sodium sulfate in toluene for 2.5 hr.) [28].

(3) **4-Cyclopentadecyne-1-one** (60–65 % from bicyclo[10.3.0]-$\Delta^{1,12}$-pentadecene-13-one which was oxidized to the epoxide with 30 % H_2O_2 in dilute

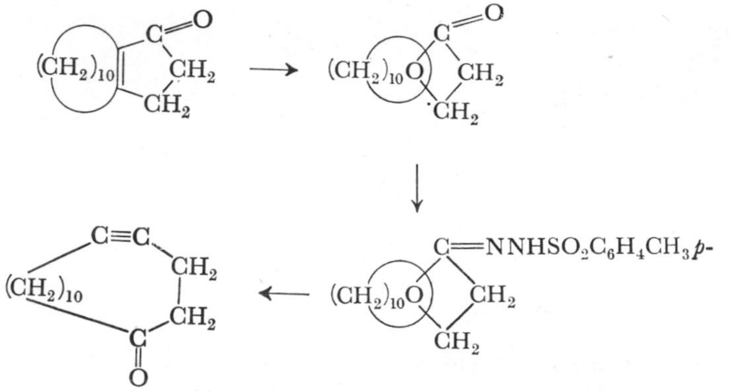

methanolic sodium hydroxide; the oily epoxide was then treated with *p*-toluenesulfonylhydrazine in alcohol, followed by heating with 2 equivalents of sodium methoxide in methanol for 24 hr. Intermediates can be isolated if necessary) [35].

5. From Quaternary Ammonium Bases

$$(CH_3)_3\overset{\oplus}{N}CH_2CH_2\overset{\oplus}{N}(CH_3)_3 \longrightarrow CH\!\equiv\!CH + 2(CH_3)_3N + 2\,H^{\oplus}$$

$$\begin{array}{c} ArC\!=\!CHAr \\ | \\ \underset{\underset{\oplus}{}}{N(CH_3)_3} \end{array} \longrightarrow ArC\!\equiv\!CAr + (CH_3)_3N + H^{\oplus}$$

The decomposition of ethylene bis(trimethylammonium) hydroxides [36] and of the quaternary ammonium salts of enamines [37, 38] gives alkynes. The former are decomposed by heat; for the latter, heat or sodium or potassium amide in liquid ammonia may be employed. This reaction is simply a special case of the β-elimination of ordinary quaternary ammonium ions having β-hydrogens (see Alkenes, Cycloalkenes, and Dienes, Chapter 2, A.8). Alkynes result if two β-hydrogens are eliminated in the bis(trimethylammonium) types or if one is eliminated in the double-bonded trimethylammonium types. Acetylene results in an 80 % yield in the first case, but in the few other cases reported yields are lower.

(a) Preparation of tolane

$$\underset{\substack{| \\ N(CH_3)_2}}{C_6H_5C{=}CHC_6H_5} \xrightarrow{CH_3I} \underset{\substack{| \\ \oplus N(CH_3)_3}}{C_6H_5C{=}CHC_6H_5} \xrightarrow{NaNH_2} C_6H_5C{\equiv}CC_6H_5$$

To α-dimethylaminostilbene in acetonitrile, excess of methyl iodide was added and on addition of ether the quaternary salt precipitated. The freshly prepared salt (25 g., 0.0685 mole) was added to 0.2 mole of sodamide in 400 ml. of liquid ammonia. After the mixture was stirred for 2 hr., solid ammonium chloride was added, and the ammonia was removed on a steam bath as ether was added. Water was then added and the ethereal layer, after being washed with water, then with dilute hydrochloric acid, again with water, was dried with magnesium sulfate and evaporated to give an oil which crystallized on cooling. Recrystallization from methanol-water gave 6.2 g. (51 %) of the acetylene, m.p. 59–60° [37].

(b) Preparation of acetylene (about 80 % by adding 10 % excess of 40 % potassium hydroxide solution to ethylene bis(trimethylammonium dibromide) followed by slow distillation [36].

6. From Olefins

$$RCH{=}CH_2 + 4\ Li \xrightarrow[\text{less}]{100° \text{ or}} RC{\equiv}CLi + 3\ LiH$$

This reaction is limited to terminal olefins [39] from which yields were in the range 15–65 %. A lithium alkene, which eliminates lithium hydride, is probably the intermediate (see Ex. a). In view of the limitation stated above, it is puzzling to note that a nonterminal olefin has been reported to form an acetylene [40]:

$$ClCH_2CH{=}CHCH_2Cl + 2\ LiC_4H_9 \xrightarrow{-15°} C_4H_9CH_2CH{=}CHCH_2C_4H_9 \xrightarrow{C_4H_9Li}$$

$$\underset{\substack{Li \\ | \\ \ }}{C_4H_9CH_2CH{=}CCH_2C_4H_9} \xrightarrow{-LiH} C_4H_9CH_2C{\equiv}CCH_2C_4H_9$$

6-Dodecyne, 7.5%

(a) 1-Trimethylsilyl-1-hexyne, $C_4H_9C{\equiv}CSi(CH_3)_3$. The calculated amount of Li dispersion in paraffin (4 Li to 1 of olefin) was washed with hexane, covered with 1-hexene, and refluxed for 1 hr. Olefins of higher boiling point should be held at 100°. The mixture was cooled and treated with a slight excess of trimethylchlorosilane to yield 65 % of the acetylenic derivative [39].

7. From Acid Chlorides and Certain Phosphoranes

$$\underset{\text{1 equiv.}}{\overset{O}{R\overset{\|}{C}Cl}} + \underset{\text{2 equiv.}}{(C_6H_5)_3P{=}CHY} \longrightarrow \underset{\overset{|}{O^{\ominus}}}{(C_6H_5)_3\overset{\oplus}{P}-\overset{\overset{Y}{|}}{C}{=}CR} + \underset{Cl^{\ominus}}{(C_6H_5)_3\overset{\oplus}{P}CH_2Y}$$

$$\Big\downarrow \overset{\Delta}{\text{under vacuum}} \qquad\qquad \text{3-II}$$

$$(C_6H_5)_3PO + RC{\equiv}CY$$

The synthesis [41] is limited to the formation of acetylenic groups which are conjugated with other unsaturated groups (Y = $CO_2C_2H_5$, $C{\equiv}N$, or C_6H_5). The preparation is a good one except that considerable deadweight, expensive material is carried through to the end of the reaction. At least, the phosphonium salt, 3-II, is recoverable.

(a) Preparation of ethyl hex-2-ynoate. Butyryl chloride, 3.1 g., and 10 g. of the phosphorane $(C_6H_5)_3P{=}CHCO_2C_2H_5$ were allowed to stand overnight in benzene. The phosphonium salt $(C_6H_5)_3\overset{\oplus}{P}CH_2CO_2C_2H_5\overset{\ominus}{Cl}$, 4.8 g., was removed by filtration, the filtrate concentrated, and the residue heated for 1 hr. at 280° (10 mm.) to give a distillate of the acetylenic ester, 80 % based on the phosphorane formed [41].

8. From Some Sulfur Compounds

Desulfurization of suitably substituted sulfur compounds with Raney nickel saturated with hydrogen leads to hydrocarbons (Alkanes, Chapter 1, A.7). However, with Raney nickel, azeotroped with benzene or cyclohexane to remove both water and hydrogen, acetylenes may be produced from thiondisulfides [42]:

$$(p{\text{-}}CH_3OC_6H_4\overset{\overset{S}{\|}}{C}{-}S{-})_2 \xrightarrow[\text{azeotropy with } C_6H_6]{\text{Raney Ni (after}} p{\text{-}}CH_3OC_6H_4C{\equiv}CC_6H_4OCH_3{\text{-}}p'$$

$$p,p'\text{-Dimethoxytolane, 48–65\%}$$

or

$$(p{\text{-}}CH_3OC_6H_4\overset{\overset{S}{\|}}{C}S)_2Pb$$

In a more specific preparation, but one that might have wider application, dicyanoacetylene has been made as follows [43]:

$$NaCN + CS_2 \xrightarrow{\text{Acetone}} \overset{\overset{\displaystyle SNa}{\displaystyle |}}{N}CC{=}\overset{\overset{\displaystyle SNa}{\displaystyle |}}{C}CN \xrightarrow{\text{COCl}_2}$$

$$\overset{\displaystyle O}{\overset{\displaystyle \|}{\underset{\displaystyle S \diagdown \underset{\displaystyle |}{C} \diagup S}{\underset{\displaystyle NCC{=}CCN}{}}}} \xrightarrow[\substack{\text{Quartz tube at} \\ \text{reduced pressure}}]{600-800°} NCC{\equiv}CCN + CS_2 + COS + S$$
$$59\text{-}76\%$$

Dicyanodiazomethane $(N_2C(CN)_2)$ pyrolyzed at 220° yields some dicyano-acetylene.

1. T. L. Jacobs, *Org. Reactions*, **5**, 1 (1949).
2. R. A. Raphael, *Acetylenic Compounds in Organic Synthesis*, Butterworths Scientific Publications, London, 1955.
3. T. F. Rutledge, *Acetylenic Compounds*, Reinhold Book Corporation, New York, 1968.
4. S. A. Miller, *Acetylene—Its Properties, Manufacture and Uses*, Vols. I and II, Academic Press, New York, 1965, 1966.
5. W. H. Puterbaugh and M. S. Newman, *J. Am. Chem. Soc.*, **81**, 1611 (1959).
6. Ref. 2, p. 8.
7. G. H. Coleman et al., *J. Am. Chem. Soc.*, **56**, 132 (1934); **58**, 2310 (1936).
8. D. Y. Curtin et al., *Chem. Ind.* (*London*), 1453 (1957).
9. K. L. Erickson and J. Wolinsky, *J. Am. Chem. Soc.*, **87**, 1142 (1965).
10. L. K. Montgomery and L. E. Applegate, *J. Am. Chem. Soc.*, **89**, 5305 (1967).
11. S. I. Miller, *J. Org. Chem.*, **26**, 2619 (1961).
12. C. F. H. Allen and C. O. Edens, Jr., *Org. Syn.*, Coll. Vol. **3**, 731 (1955).
13. W. E. Parham et al., *J. Am. Chem. Soc.*, **76**, 5380 (1954).
14. R. E. Lutz and W. R. Smithey, Jr., *J. Org. Chem.*, **16**, 51 (1951).
15. R. A. Raphael and F. Sondheimer, *J. Chem. Soc.*, 120 (1950).
16. O. Y. Okhlobystin, *Neftekhimiya*, **1**, 752 (1961); *C.A.*, **57**, 14918 (1962).
17. K. K. Georgieff and Y. Richard, *Can. J .Chem.*, **36**, 1280 (1958); see also P. Pomerantz et al., *J. Res. Nat. Bur. Stand.*, **52**, 51 (1954).
18. A. T. Bottini and J. D. Roberts, *J. Am. Chem. Soc.*, **79**, 1462 (1957).
19. E. R. H. Jones et al., *Org. Syn.*, Coll. Vol. **4**, 755 (1963).
20. G. Eglinton et al., *J. Chem. Soc.*, 2873 (1952).
21. V. Prey and G. Pieh, *Monatsh. Chem.*, **80**, 790 (1949).
22. C. G. Krespan et al., *J. Am. Chem. Soc.*, **83**, 3424 (1961).
23. J. Wislicenus and P. Schmidt, *Ann. Chem.*, **313**, 210 (1900).
24. P. Ruggli and F. Lang, *Helv. Chim. Acta*, **21**, 38 (1938).
25. F. Fox, *Chem. Ber.*, **26**, 653 (1893).
26. W. L. Walton, *J. Am. Chem. Soc.*, **69**, 1544 (1947).
27. A. C. Cope, *Org. Syn.*, Coll. Vol. **4**, 377 (1963).
28. V. Prelog et al., *Helv. Chim. Acta*, **35**, 1598 (1952).
29. M. S. Newman and D. E. Reid, *J. Org. Chem.*, **23**, 665 (1958).
30. L. A. Carpino et al., *J. Org. Chem.*, **31**, 2867 (1966) and earlier papers.
31. M. S. Newman and A. Kutner, *J. Am. Chem. Soc.*, **73**, 4199 (1951).
32. F. G. Willey, *Angew. Chem.*, **76**, 144 (1964).
33. D. Y. Curtin et al., *J. Am. Chem. Soc.*, **87**, 863 (1965).
34. T. Mukaiyama et al., *J. Org. Chem.*, **29**, 2243 (1964).
35. A. Eschenmoser et al., *Helv. Chim. Acta*, **50**, 708 (1967).
36. Y. M. Slobodin and N. A. Selezneva, *Zh. Obshch. Khim.*, **26**, 691 (1956); *C.A.*, **50**, 14502 (1956).

37. C. R. Hauser *et al.*, *J. Am. Chem. Soc.*, **82**, 1786 (1960).
38. F. Kröhnke and M. Meyer-Delius, *Chem. Ber.*, **84**, 941 (1951).
39. D. L. Skinner *et al.*, *J. Org. Chem.*, **32**, 105 (1967).
40. E. Terres and K. Hubbuch, *Erdoel Kohle*, **13**, 940 (1960); *C.A.*, **55**, 11821 (1961).
41. S. T. D. Gough and S. Trippett, *J. Chem. Soc.*, 2333 (1962).
42. K. A. Latif and D. R. Choudhury, *Tetrahedron Letters*, 1735 (1968).
43. E. Ciganek and C. G. Krespan, *J. Org. Chem.*, **33**, 541 (1968).

B. Nucleophilic Reactions

Since terminal acetylenes are weak acids, the anions of their salts are quite basic and show typical nucleophilic character. Syntheses of acetylenes based on this nucleophilic character are mainly alkylation and addition to the carbonyl group.

Although B is devoted mainly to displacement reactions and nucleophilic addition of acetylenic anions, mention should be made here of the fact that the addition of other anions to vinylacetylenes may take place to form substituted acetylenes; for example [1].

$$CH_2{=}CHC{\equiv}CH + NaOCH_3 \xrightarrow[105°, \text{ autoclave}]{CH_3OH} CH_3C{\equiv}CCH_2OCH_3$$
$$\text{1-Methoxy-2-butyne, 61\%}$$

1. From Acetylenic Salts and Alkylating Agents

$$RC{\equiv}CNa + R'X \rightarrow RC{\equiv}CR' + NaX$$
$$RC{\equiv}CMgX + 2\ CH_3SO_2OR' \rightarrow RC{\equiv}CR' + R'X + (CH_3SO_3)_2Mg$$

The early alkylations of alkali acetylides were carried out in liquid ammonia by using organic halides or sulfates as the alkylating agent [2, 3]. Among the halides the bromides are the most satisfactory; however, the reaction is confined to the introduction only of primary alkyl groups possessing no branching on the second carbon atom. In addition, when an alkyl halide is employed the method is not satisfactory for the synthesis of methyl- and ethylacetylenes, and pressure must be employed with the higher alkyl halides. Yields with alkyl bromides from *n*-propyl to *n*-hexyl vary from 40 to 80%. With dimethyl sulfate or diethyl sulfate as the alkylating agent, one alkyl group is substituted and the conversions vary from 50 to 100%. Other esters such as methanesulfonates and *p*-toluenesulfonates as well as the metal acetylides of lithium and potassium have been utilized to a limited extent.

The synthesis of alkyl acetylides may be accomplished directly from the acetylene without isolation of the metallic acetylide. For example, 3-undecyne has been prepared from 1-nonyne in 84% yield by adding the latter to a suspension of sodamide in mineral oil and then adding diethyl sulfate [4]:

$$n\text{-}C_7H_{15}C{\equiv}CH \xrightarrow{NaNH_2} n\text{-}C_7H_{15}C{\equiv}CNa \xrightarrow{(C_2H_5O)_2SO_2} n\text{-}C_7H_{15}C{\equiv}CC_2H_5$$

Another procedure to obtain the higher homologs is to use the lithium acetylide which is not only more reactive, but is also more soluble in liquid ammonia or in dioxane [5]. In this manner, *n*-octyl bromide, lithium acetylide, and

References for Section B are on pp. 167–168.

lithium amide in liquid ammonia gave a 60 % yield of 9-octadecyne and a 41 %

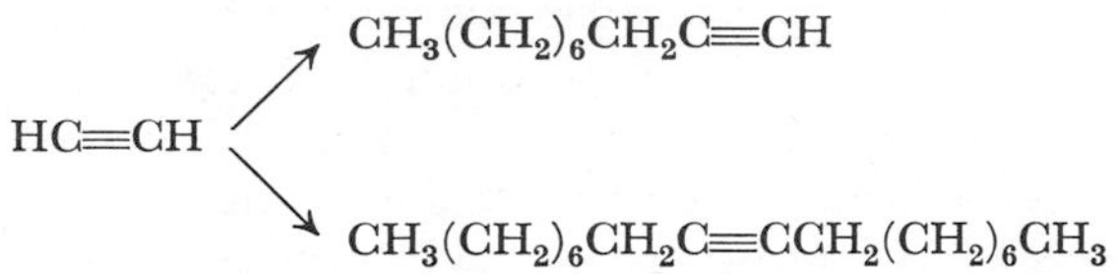

$$HC\equiv CH \Big\langle \begin{array}{l} CH_3(CH_2)_6CH_2C\equiv CH \\[2em] CH_3(CH_2)_6CH_2C\equiv CCH_2(CH_2)_6CH_3 \end{array}$$

yield of 1-decyne. Thus two as well as one of the hydrogens in acetylene are replaceable.

Lithium acetylide stabilized as the ethylenediamine complex may be used effectively in preparing fluoroalkynes from a fluoroalkyl halide [6]. Dimethyl-sulfoxide is employed as the solvent, and the reaction is complete by warming to 60° after the initial reaction has subsided. In this manner, 7-fluoroheptyne may be obtained in 92 % yield:

$$F(CH_2)_5Cl + LiC\equiv CH \cdot EDA \xrightarrow{\text{DMSO}} F(CH_2)_5C\equiv CH$$

7-Fluoro-1-heptyne,
92%

The lithium salts of substituted acetylenes may be prepared from the mercury salts [7]:

$$(C_{15}H_{31}C\equiv C)_2Hg \xrightarrow{\text{2 Li}} 2\,C_{15}H_{31}C\equiv CLi \xrightarrow{(CH_3O)_2SO_2} C_{15}H_{31}C\equiv CCH_3$$

2-Octadecyne, 62%

Some success has also been achieved by using conventional organic solvents [8] at ordinary pressure. With finely divided sodium acetylide prepared in xylene [9] both methyl groups of dimethyl sulfate react to give an 80–85 % yield of propyne:

$$2\,HC\equiv CNa + (CH_3)_2SO_4 \rightarrow 2\,HC\equiv CCH_3 + Na_2SO_4$$

In a mixture of dimethylformamide in xylene (37.5 vol. %), a 2 molar solution (or less) of sodium acetylide with n-butyl bromide gives an 80 % yield of 1-hexyne.

The reaction of a metallic acetylide with an alkylating agent is of the SN_2 type in that the negative acetylide ion replaces, in the case of an alkyl halide, the halide:

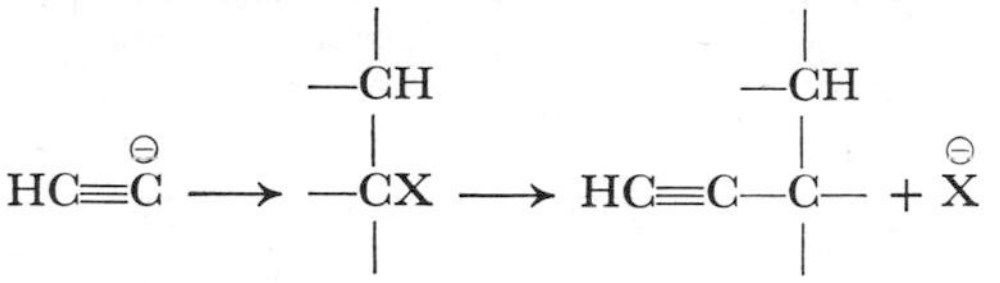

$$HC\equiv \overset{\ominus}{C} \longrightarrow \underset{|}{\overset{|}{-}C}X \longrightarrow HC\equiv C-\underset{|}{\overset{|}{C}}- + \overset{\ominus}{X}$$

Elimination is a competing reaction, and it becomes increasingly prominent in going from primary to secondary to tertiary halides. Thus the alkylation reaction with secondary and tertiary halides gives largely alkenes:

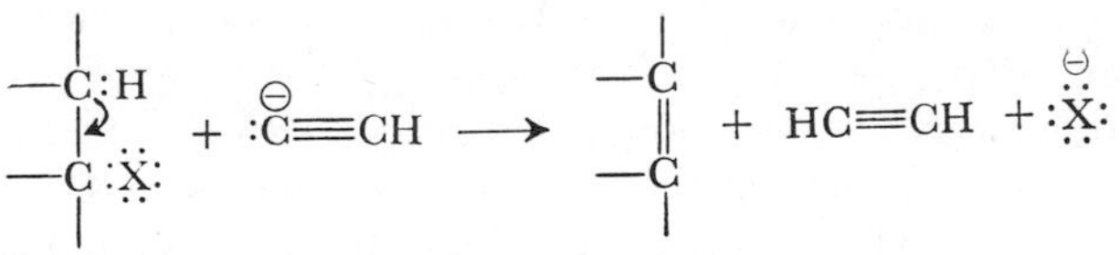

$$\begin{array}{c} -\overset{|}{\underset{|}{C}}:H \\ -\overset{|}{\underset{|}{C}}:\ddot{X}: \end{array} + :\overset{\ominus}{C}\equiv CH \longrightarrow \begin{array}{c} -\overset{|}{C} \\ \| \\ -\overset{|}{C} \end{array} + HC\equiv CH + :\ddot{\overset{\ominus}{X}}:$$

A second method of synthesizing alkylated acetylenes is via the Grignard reagent. Some difficulty is experienced in preparing ethynylmagnesium bromide from acetylene and ethylmagnesium bromide because the bismagnesium bromide is formed almost quantitatively by the usual method of preparation [15]. However, this difficulty may be overcome by saturating the solution with a continuous stream of acetylene gas and by using a solvent such as tetrahydrofuran in which acetylene and the bisbromomagnesium derivative are soluble. In this manner the ethynylmagnesium bromide is present in about an 85% yield. Since this Grignard reagent is less reactive than those containing alkyl or aryl groups, alkyl halides are rather unreactive toward it; in fact, better alkylation results are obtained with alkyl sulfates, p-toluenesulfonates, and methanesulfonates [11]. With alkyl halides and alkynylmagnesium bromides a trace of cuprous chloride or bromide is necessary to effect the reaction.

(a) **Preparation of n-butylacetylene** (70–77% from sodium acetylide in liquid ammonia to which n-butyl bromide was added over a 45–60 min. period, after which the mixture was stirred for 2 more hr.) [3]. For the use of xylene-dimethylformamide solvent see [9].

(b) **Other examples**

(1) **Propyne** (80–85% as determined by infrared from a sodium acetylide slurry in xylene at 90°, to which dimethyl sulfate was added at such a rate as to maintain a temperature of 90–95°; after 1.25 hr. the exothermic reaction ceased and the mixture was refluxed at about 140° for 5–10 min.) [9].

(2) **Benzylphenylacetylene** (72% from phenylethnylmagnesium bromide added to benzyl-p-toluenesulfonate in ether during 2.5 hr. followed by refluxing for 6 hr.) [12].

$$C_6H_5C\equiv CMgBr + 2\text{-}p\text{-}CH_3C_6H_4SO_2OCH_2C_6H_5 \rightarrow$$
$$C_6H_5C\equiv CCH_2C_6H_5 + Mg(OSO_2C_6H_4CH_3\text{-}p)_2 + C_6H_5CH_2Br$$

2. From Acetylenic Salts and Carbonyl Compounds [13]

$$RC\equiv CMgX + \underset{/}{\overset{\backslash}{C}}{=}O \longrightarrow RC\equiv C{-}\underset{|}{\overset{|}{C}}OH$$

or

$$XMgC\equiv CMgX + 2 \underset{/}{\overset{\backslash}{C}}{=}O \longrightarrow {-}\underset{\underset{OH}{|}}{\overset{|}{C}}{-}C\equiv C{-}\underset{\underset{OH}{|}}{\overset{|}{C}}{-}$$

or

$$RC\equiv CMgX + CO_2 \longrightarrow RC\equiv CCOOH$$

The reactions above are quite general and give ready access to a wide variety of acetylenic alcohols, glycols, and acids. In the experience of one of the authors (D.E.P.) the only type that has failed to react is 4-acetoquinoline.

Acetylenic Grignards are used more frequently in carbonyl addition than are sodium or lithium acetylides, probably because of the less basic nature of the former and its tendency to coordinate with carbonyl oxygen atoms [14].

A noteworthy characteristic of acetylenic alcohols or glycols is their ability to cleave to give the original acetylene and carbonyl compounds:

$$\underset{\underset{OH}{|}}{R_2CC\equiv CH} \xrightarrow[\text{NaOH}]{\text{Trace of powdered}} R_2C\!\!=\!\!O + HC\equiv CH$$

Occasionally this reaction is of use in preparing substituted acetylenes (see Ex. b.3).

(a) Preparation of 2-ethynyl-*trans*-2-decalol

Potassium *t*-amylate was prepared from 1.2 g. atoms of potassium and 1 l. of *t*-amyl alcohol. One liter of ether was added and purified acetylene bubbled through the mixture continuously during the dropwise addition of 1.2 moles of *trans*-2-decalone. A precipitate forms in the latter part of the addition. The mixture was made acid and the organic layer washed, dried, and concentrated to yield 109 g. of solid acetylenic alcohol. An additional 73 g. was obtained by distillation of the filtrate under low pressure. Total yield 83% [15].

(b) Other examples

(1) *trans*-2-Decalolcyclohexanolacetylene (99% from 1 mole of 2-ethynyl-*trans*-2-decalol, 1.5 moles of ethylmagnesium bromide in ether to which a slight excess of cyclohexanone was added) [15].

(2) 1,1-Diphenylbut-3-yne-2-ol (75% from ethynylmagnesium bromide prepared from ethylmagnesium bromide in tetrahydrofuran saturated

$$HC\equiv CMgBr \xrightarrow[\text{2. } H_2O]{\text{1. } (C_6H_5)_2CHCHO} (C_6H_5)_2CHC\overset{\displaystyle H}{\underset{\displaystyle C\equiv CH}{\diagup}}\!\!\!-OH$$

with acetylene gas).   This technique overcame the disproportionation:

$$2\ HC\equiv CMgBr \rightarrow BrMgC\equiv CMgBr + HC\equiv CH\cdot$$

Diphenylacetaldehyde was then added dropwise (alkylation of the ethynyl Grignard reagent with benzyl bromide failed) [10].

(3) 1,3-Butadiyne (71–80% from

$$(CH_3)_2COHC\equiv C\!-\!C\equiv CCOH(CH_3)_2$$

0.1 mole, in 100 ml. of xylene to which *no more than* 0.1 g. of powdered sodium hydroxide was added. The mixture was gradually heated to 90°, whereupon 1,3-butadiyne and acetone distilled through a Vigreux column and were collected in a dry ice trap. The temperature of the xylene was gradually raised above 90° until no more gases were evolved. Diacetylene is stable at 0° for 1 hr.,

but can be stored satisfactorily at dry ice temperature [16]. A series of arylacet-
ylenes has been prepared in about 70% yields by a similar alkaline cleavage)
[17].

3. From Acetylenic Salts and Halogens, *ortho* Esters, or Isocyanates

The nucleophilic displacements or additions by acetylenic salts are by no
means limited to alkylating agents or carbonyl compounds. Only brief mention
is made here of a few examples of other types.

$$RC\equiv CH + NaOH + H_2O + Br_2 \quad (10\% \text{ excess}) \rightarrow RC\equiv CBr \quad (\text{almost quantitative})$$

The haloacetylenes must be stored at 0° or lower [18]. On the other hand,
dichloroacetylene is made by the dehydrohalogenation of trichloroethylene. It
is spontaneously inflammable in air, but solutions in ether are manipulative [19].

Acetals are formed from an acetylenic Grignard reagent and orthoformic
ester [20]:

$$C_2H_5C\equiv C(CH_2)_2C\equiv CMgBr + HC(OC_2H_5)_3 \xrightarrow{\text{Ether}}$$

$$C_2H_5C\equiv C(CH_2)_2C\equiv CCH(OC_2H_5)_2$$

2,6-Nonadiynal diethyl acetal, 81%

t-Butyl hypochlorite, on the other hand, substitutes chlorine [21]:

$$CH_3C\equiv CH \xrightarrow{(CH_3)_3COCl} ClCH_2C\equiv CH \quad \text{only}$$

Acetylenic anions react with phenylisocyanate as follows [22]:

$$C_6H_5C\equiv CMgBr + C_6H_5N=C=O \xrightarrow[\text{H}_3O^\oplus]{\text{Ether}} C_6H_5C\equiv CCONHC_6H_5$$

Phenylpropiolicanilide, 90%

4. From Acetylenes, Aldehydes, and Amines (Modified Mannich)

$$R_2NH + R'CH=O + HC\equiv CR'' \xrightarrow{Cu(C_2H_3O_2)_2} R_2NCH\overset{\overset{\displaystyle R'}{|}}{-}C\equiv CR''$$

This reaction is related to the condensation of acetylene with carbonyl
compounds to form acetylenic alcohols. In this case the acetylenic anion probably condenses with the amine-aldehyde adduct:

$$R_2NH + R'CH=O \rightleftharpoons R_2N\overset{\overset{\displaystyle R'}{|}}{\underset{\displaystyle H}{C}}-OH \underset{\rightleftharpoons}{\overset{HC\equiv C^\ominus}{\rightleftharpoons}} R_2N\overset{\overset{\displaystyle R'}{|}}{C}HC\equiv CH + OH^\ominus$$

Moreover, part of the acetylene can act as a source of acetaldehyde to yield
the same product as would be obtained from the addition of acetaldehyde [23]:

$(CH_3)_2NH + CH_3CH{=}O + HCO_2H + CuCl$

500 g., 50%   225 g.   1370 g.   10 g.

$HC{\equiv}CH, N_2$
20 atm.
3–2 parts
25–30°

CH_3
|
$(CH_3)_2N{-}CHC{\equiv}CH$

3-Dimethylamino-1-butyne,
"good yield"

$HC{\equiv}CH, N_2$
15 atm.,
24 hr.,
50–60°

$(CH_3)_2NH + \text{tetralin} + CuCl$

28 g   60 g.   2 g.

This result would lead one to suppose that the same product would be obtained from vinylacetylene; however, different products are obtained [24]:

$$CH_2{=}CHC{\equiv}CH + (CH_3)_2NH \xrightarrow{\text{Autoclave}} (CH_3)_2NCH_2CH{=}C{=}CH_2$$

Equiv. amounts

4-Dimethylamino-1,2-butadiene,
56%

or

$$CH_2{=}CHC{\equiv}CH + (CH_3)_2NH \xrightarrow[100°]{\text{Autoclave}}$$

$$(CH_3)_2NCH_2C{\equiv}CCH_3 + (CH_3)_2NCH_2CH{=}C{=}CH_2$$

4-Dimethylamino-2-butyne,
20%

4-Dimethylamino-1,2-butadiene,
9%

(a) Preparation of 3-dimethylamino-1-propyne. Dimethylamine, 50% aqueous solution, 450 g., 972 g. of 40% formaldehyde, and 40 g. of copper acetate were dissolved in 822 g. of acetic acid in a bomb. The bomb was flushed with nitrogen, closed, and filled with 5 atm. of nitrogen and 10 atm. of acetylene. The bomb was rocked at 35–40° for an unspecified length of time and the contents made basic, extracted, dried, and distilled to yield 580 g. of product [23].

(b) Other examples

(1) 4-(N,N-Dimethylamino)-butyne-2. Vinylacetylene (104 g., 2 moles) and aqueous 25% dimethylamine (720 g., 4 moles) were heated in a nitrogen atmosphere under pressure at 100° for 20 hr. The product was saturated with potassium carbonate and extracted with ether. On drying the ethereal extract and distilling, 39 g. (20%) of the aminobutyne, b.p. 116.5–117°, was obtained [24].

1. R. A. Jacobson et al., J. Am. Chem. Soc., **56**, 1169 (1934).
2. T. H. Vaughn et al., J. Org. Chem., **2**, 1 (1937).
3. K. N. Campbell and B. K. Campbell, Org. Syn., Coll. Vol. **4**, 117 (1963).
4. F. Asinger et al., Chem. Ber., **97**, 1555 (1964).
5. B. B. Elsner and P. F. M. Paul, J. Chem. Soc., 893 (1951).
6. F. L. M. Pattison and R. E. A. Dear, Can. J. Chem., **41**, 2600 (1963).

7. B. B. Elsner and P. F. M. Paul, *J. Chem. Soc.*, 893 (1951).

8. T. F. Rutledge, *J. Org. Chem.*, **22**, 649 (1957).

9. T. F. Rutledge, *J. Org. Chem.*, **24**, 840 (1959).

10. E. R. H. Jones *et al.*, *J. Chem. Soc.*, 4765 (1956).

11. R. A. Raphael, *Acetylenic Compounds in Organic Synthesis*, Butterworths Scientific Publications, London, 1955, p. 17.

12. J. R. Johnston *et al.*, *J. Am. Chem. Soc.*, **60**, 1885 (1938).

13. A. W. Johnson, *The Chemistry of Acetylenic Compounds*, Vol. 1, *Acetylenic Alcohols*, Edward Arnold & Co., London, 1946.

14. W. Kimel *et al.*, *J. Org. Chem.*, **22**, 1611 (1957).

15. C. S. Marvel *et al.*, *J. Am. Chem. Soc.*, **62**, 2659 (1940).

16. R. J. Tedeschi and A. E. Brown, *J. Org. Chem.*, **29**, 2051 (1964).

17. M. S. Shvartsberg *et al.*, *Izv. Akad. Nauk SSSR, Ser Khim.*, 466 (1967); *C.A.*, **67**, 3052 (1967).

18. S. I. Miller *et al.*, *J. Am. Chem. Soc.*, **85**, 1648 (1963).

19. J. H. Wotiz *et al.*, *J. Org. Chem.*, **26**, 1626 (1961).

20. F. Sondheimer, *J. Am. Chem. Soc.*, **74**, 4040 (1952).

21. M. C. Caserio and R. E. Pratt. *Tetrahedron Letters*, 91 (1967).

22. J. R. Johnson and W. L. McEwen, *J. Am. Chem. Soc.*, **48**, 469 (1926).

23. W. Reppe *et al.*, *Ann. Chem.*, **596**, 1 (1955).

24. V. A. Engelhardt, *J. Am. Chem. Soc.*, **78**, 107 (1956).

C. Free Radical and Cyclo Reactions

1. From Acetylenes (Oxidative Coupling, Glaser)

$$RC\equiv CH \xrightarrow{[O]} RC\equiv C-C\equiv CR$$

The oxidation of terminal acetylenes to diacetylenes, known as the Glaser reaction, has become a simple, general, useful method for preparing a wide variety of diacetylenes [1]. This reaction is one of the easiest ways to form carbon-carbon bonds. Yields are usually 90% or higher by bubbling air or oxygen through a mixture of the acetylene and cuprous chloride and an amine such as pyridine or ethylamine. The presence of oxygen shortens the reaction time considerably [2].

Another variation of this procedure is the coupling of a bromoacetylene with another acetylene [3]:

$$C_6H_5C\equiv CBr + HC\equiv C-\underset{\underset{OH}{|}}{C}(CH_3)_2 \xrightarrow{Cu^\oplus, NH_3} C_6H_5C\equiv C-C\equiv C\underset{\underset{OH}{|}}{C}C(CH_3)_2$$

1-Phenyl-5-methyl-1,3-hexadiyn-5-ol, 92%

Still another variation is the oxidation of a terminal acetylene with cupric ion rather than cuprous ion and oxygen.

Oxidative coupling has been useful in the preparation of naturally occurring polyacetylenic compounds, polyenes, and macrocyclic polyacetylenes [4]. For example, a variety of macrocyclic polyacetylenes is obtained from the following single reaction [5]:

$$HC\equiv C(CH_2)_2C\equiv CH \xrightarrow[\substack{1.4 \text{ l. } C_5H_5N, \\ 55°, 3 \text{ hr.}}]{Cu(C_2H_3O_2)_2, 225 \text{ g.}} -[C\equiv C(CH_2)_2C\equiv C]_{\overline{n}}$$

15 g.

$n = 3, 6\%; n = 4, 6\%;$
$n = 5, 6\%; n = 6, 2\%$

References for Section C are on p. 170.

The macrocyclic acetylenes were separated by chromatography. It would seem that a dilution technique with the more efficient oxidation medium ($Cu^{\oplus} + O_2$) might favor the formation of a single species.

It should be mentioned that many of the polyacetylenes are quite unstable in that they either tend to polymerize on standing or to carbonize on heating [4].

(a) Preparation of diphenyldiacetylene (70–80% from phenylacetylene refluxed for 1 hr. in a solution of cupric acetate monohydrate in a 1:1 by volume mixture of pyridine and methanol) [6].

(b) Other examples

(1) **Di-(1-hydroxycyclohexyl)-butadiyne** (93% from 1-ethynyl-cyclohexanol in ethanol added to a solution of ammonium chloride, cuprous chloride, and dilute hydrochloric acid through which air was passed) [7].

(2) **Macrocyclic compound from 1,4-bis(hydroxymethyl) benzene dipropargyl ether** (15% from 1 g. of the ether, 10 g. of copper acetate hydrate in 120 ml. of pyridine under nitrogen at 50° for 20 min.) [8].

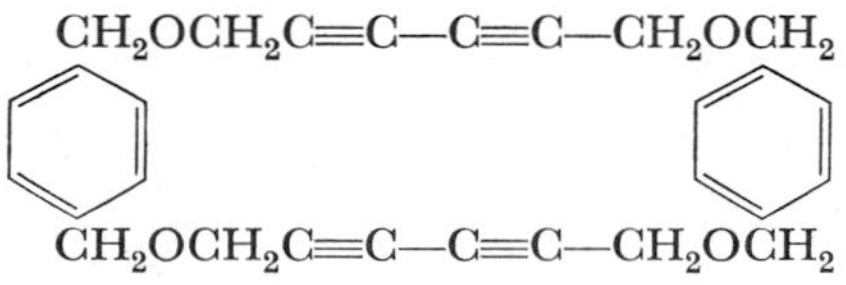

$$CH_2OCH_2C{\equiv}C{-}C{\equiv}C{-}CH_2OCH_2$$

$$CH_2OCH_2C{\equiv}C{-}C{\equiv}C{-}CH_2OCH_2$$

2. From Acetylene and Hydrogen Cyanide

Acetylene at high temperature, about 900°, dissociates into free radicals:

$$HC{\equiv}CH \rightarrow HC{\equiv}C{\cdot} + H{\cdot}$$

In the presence of suitable substrates, coupling products can be obtained. For example, propiolonitrile can be made as follows [9]:

$$HC{\equiv}CH + HCN \xrightarrow[\text{2. Quench}]{\text{1. Hot tube, 900°, fraction of a second}}$$

as mixed gases

$$HC{\equiv}CCN + CH_2{=}CHCN + CH_3CN + CH_4 + CH_2{=}CH_2$$

31% based on acetylene

3. From Cyclooctatetraenes

The conversion of acetylenes into benzenes by cycloaddition has been discussed (see Alkanes, Cycloalkanes, and Arenes, Chapter 1, H). The reverse of this process is more difficult to accomplish because of the resonance stabilization of the benzenoid system. Irradiation would seem to be the energy source of choice and, if it did not succeed, it possibly could with compounds of less aromaticity. Such an example is shown [10]:

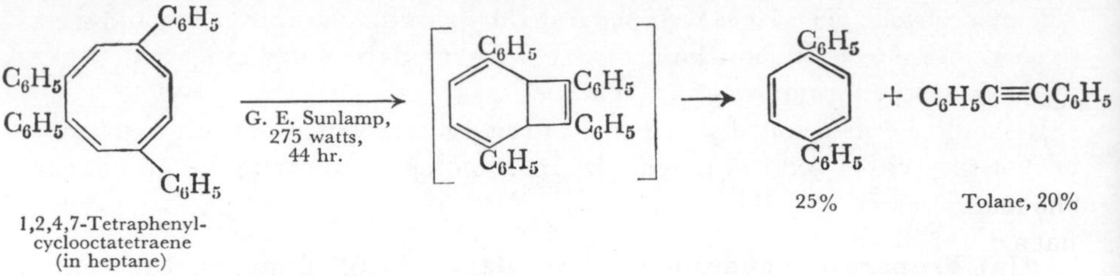

C6H5
C6H5
C6H5
C6H5
G. E. Sunlamp, 275 watts, 44 hr.
C6H5
C6H5
C6H5
C6H5
C6H5
C6H5
+ C6H5C≡CC6H5
25%
Tolane, 20%
1,2,4,7-Tetraphenylcyclooctatetraene (in heptane)

4. From Propargyl Acetoacetates

The acetylenic group does not appear to participate in cycloaddition reactions and, even if it did, it would give rise to allene derivatives. However, the search for cycloaddition reactions showed that the acetylenic group could be carried along in such a reaction [11]:

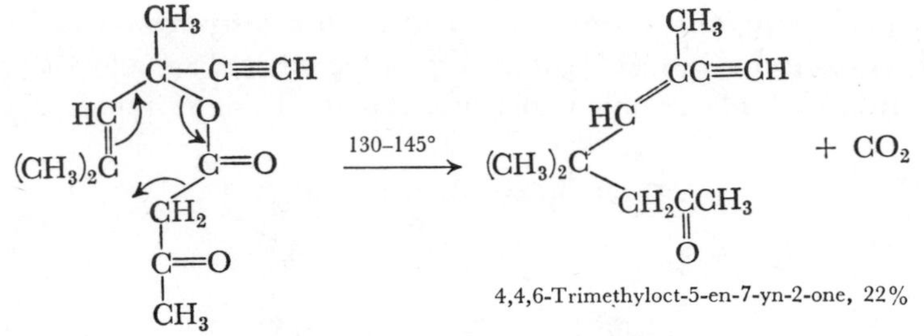

CH3
C—C≡CH
HC
O
(CH3)2C
C=O
CH2
C=O
CH3
130–145°
CH3
CC≡CH
HC
(CH3)2C
CH2CCH3
O
+ CO2
4,4,6-Trimethyloct-5-en-7-yn-2-one, 22%

1. G. Eglington and W. McCrae, *Advances in Organic Chemistry*, Interscience Publishers, New York, Vol. 4, 1963, p. 225; R. A. Raphael, *Acetylenic Compounds in Organic Synthesis*, Butterworths Scientific Publications, 1955, p. 127.
2. R. J. Tedeschi and A. E. Brown, *J. Org. Chem.*, **29,** 2051 (1964).
3. W. Chodkiewicz, *Ann. Chim.* (*Paris*), **2,** 819 (1957).
4. A. M. Sladkov and Y. P. Kudryavtsev, *Russ. Chem. Rev.*, **32,** 229 (1963).
5. F. Sondheimer and R. Wolovsky, *J. Am. Chem. Soc.*, **84,** 260 (1962).
6. I. D. Campbell and G. Eglinton, *Org. Syn.*, **45,** 39 (1965).
7. K. Bowden *et al.*, *J. Chem. Soc.*, 1579 (1947).
8. T. Ando and M. Nakagawa, *Bull. Chem. Soc. Japan*, **40,** 363 (1967).
9. L. H. Krebaum, *J. Org. Chem.*, **31,** 4103 (1966).
10. E. H. White and R. L. Stern, *Tetrahedron Letters*, 193 (1964).
11. I. N. Nazarov and Z. A. Krasnaia, *Bull. Acad. Sci.*, *USSR*, *Div. Chem. Sci.* (*English Transl.*), 844 (1958).

D. Allenes and Cumulenes

Allenes, $R_2C{=}C{=}CHR$, are isomeric with acetylenes, $R_2CHC{\equiv}CR$, and the previous discussion of acetylenes has shown that allenes quite frequently accompany the formation of acetylenes. Extensive literature is available on the preparation of allenes and, fortunately, excellent reviews have been prepared [1–3]. It is the intent neither to repeat nor to abstract these reviews, but rather to reinterpret them to point out the more convenient methods of preparation. Some new methods are described also.

References for Section D are on p. 173.

1. Elimination

As with the preparation of acetylenes, elimination methods are the most important for the preparation of allenes. One of the most important is the elimination of halogen [4]:

$$CH_2\text{=}CBrCH_2Br \xrightarrow[\text{Butyl or isopentyl acetate}]{Zn} CH_2\text{=}C\text{=}CH_2$$

Allene, 95–98%

The higher-boiling esters permit complete debromination to remove the troublesome bromopropene which accompanies the product with lower-boiling solvents. This method is satisfactory provided, of course, that the dibromopropene is available. A more general source of allenes is propargyl halides made from the readily available alcohols. These halides form allenes by a combination reduction-dehydrohalogenation process [5]:

$$(CH_3)_2CC\text{≡}CH \xrightarrow[\substack{\text{Heat gently until} \\ \text{reflux maintains} \\ \text{itself}}]{\text{Zn-Cu couple}}$$
$$\underset{Cl}{|}$$

80% total yield consisting of 98.8% of $(CH_3)_2C\text{=}C\text{=}CH_2$

3-Methyl-1,2-butadiene

This reduction-dehalogenation process is known as the Ginzburg procedure.

Since dibromocyclopropanes are readily available from olefins, a conversion process to allenes from these compounds would be very attractive. Such a process has been developed [6]:

$$\underset{\substack{\text{0.1 mole (cooled to dry ice temp.)}}}{\overset{\text{Br Br}}{\triangle}CH_2\text{—}CH\text{=}CH_2} \underset{\text{25 ml.}}{\text{in dry ether}} \xrightarrow[\substack{\text{Stirred an additional 30 min.,} \\ \text{diluted with water}}]{\substack{\text{CH}_3\text{Li (0.12 mole)} \\ \text{added dropwise in 30 min.}}}$$

$$CH_2\text{=}C\text{=}CHCH_2CH\text{=}CH_2$$

1,2,5-Hexatriene, 46%
(unstable in air)

The advantages of this process, in addition to availability of starting material, are: mild reaction conditions, simplicity, and freedom from isomerization to acetylene. However, some dichloro analogs do not react with methyllithium, but do so with butyllithium [7].

2. Isomerization

Base-catalyzed isomerization is the most common:

$$RCH_2C\text{≡}CH \underset{3\text{-III}}{\overset{B^\ominus}{\rightleftharpoons}} RCH_2C\text{≡}C^\ominus \underset{}{\rightleftharpoons} RCH\text{—}C\text{≡}CH \longleftrightarrow$$

$$RCH\text{=}C\text{=}\overset{\ominus}{CH} \xrightarrow{BH} RCH\text{=}C\text{=}CH_2$$

3-IV

$$\updownarrow$$

$$R\overset{\ominus}{C}\text{=}C\text{=}CH_2$$

$$\updownarrow$$

$$RC\text{≡}C\overset{\ominus}{CH_2} \xrightarrow{BH} RC\text{≡}CCH_3$$

3-V

In practical terms the nonterminal acetylene, 3-V, or the allene, 3-IV, is isomerized by sodamide to the terminal acetylene, 3-III, probably because the anion of 3-III is the stablest and is held in this form by the strong base. A weaker base such as alcoholic sodium hydroxide favors the isomerization of the terminal acetylene, 3-III, first to the allene, 3-IV, and second to the non-terminal acetylene, 3-V. If the rate of isomerization of 3-IV → 3-V is slow, it is possible to prepare allenes by the alkali isomerization of terminal acetylenes, particularly if conjugative groups are attached to the acetylene. The most common method is to pass the acetylene through an alumina column impregnated with potassium hydroxide [1]:

$$ ArC\equiv CCH_2Ar' \xrightarrow[\text{20°, 20 min.}]{Al_2O_3,\ KOH} ArCH\!\!=\!\!C\!\!=\!\!CHAr' $$

About 80% yields

Isomerizations of the SN_2' type are possible occasionally:

$$ X\!\!-\!\!CH_2C\!\!\equiv\!\!CH + B^\ominus \rightarrow CH_2\!\!=\!\!C\!\!=\!\!CHB + X^\ominus $$

For example,

$$ (CH_3)_2CBrC\equiv CH + C_2H_5MgBr \xrightarrow[\text{2. Minimum amount of } H_2O]{\text{1. Ether reflux}} (CH_3)_2C\!\!=\!\!C\!\!=\!\!CHC_2H_5 $$

Dimethylethylallene, 99%

The minimum amount of water prevents isomerization of the product [8]. Cuprous bromide, which probably operates via a cyclo transition state, is helpful in the isomerization [9]:

$$ HC\equiv CCH_2Br + CuBr \xrightarrow{\text{aq. HBr}} CHBr\!\!=\!\!C\!\!=\!\!CH_2 $$

Bromoallene, 60%

3. Recent Methods

Vaporization from a carbon arc into a high vacuum yields C_3, an interesting insertion species [10]:

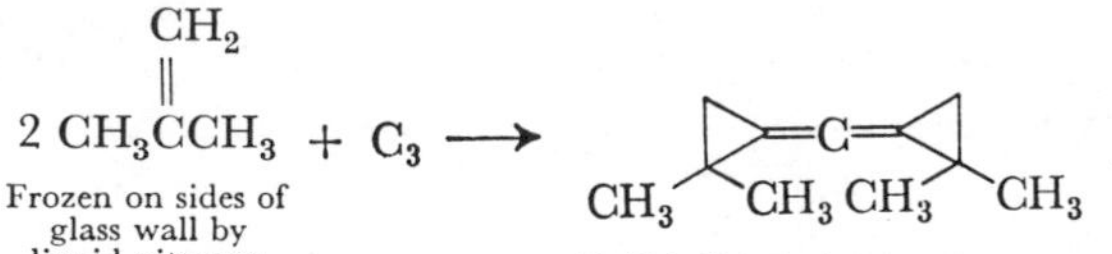

Bis(2,2-dimethylcyclopropylidene) methane, 40% yield based on bulk C vaporized

Special equipment is needed for this method of preparation.

A very specific elimination is that from a pyrazoline hydrazone [11]:

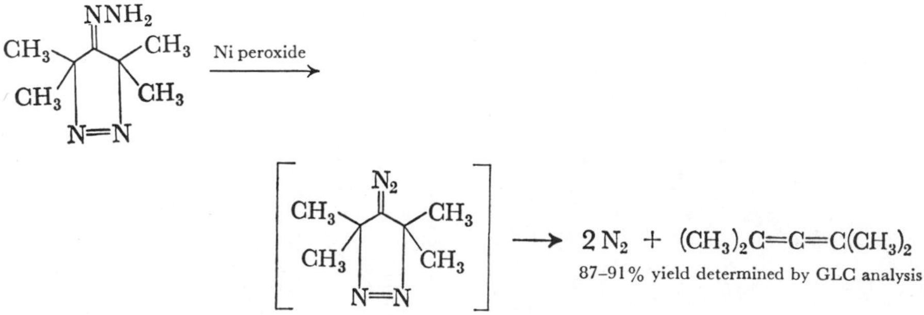

$$ \longrightarrow 2\,N_2 + (CH_3)_2C\!\!=\!\!C\!\!=\!\!C(CH_3)_2 $$

87–91% yield determined by GLC analysis

4. Cumulenes

Cumulenes are accumulated allenes or polyallenes of the general structure, $R_2C[=C=C]_nR_2$. General reviews of their syntheses and properties are available [12]. Many methods of synthesis, based on general allene syntheses, have been conceived. Attention is focused mostly on the acetylenic glycols and alcohols. For instance, a combination of reduction and dehydration is necessary to produce the following cumulene [13]:

$$(C_6H_5)_2CC\equiv C-C\equiv C-C(C_6H_5)_2 \xrightarrow{PI_2,C_5H_5N} (C_6H_5)_2C=C=C=C=C=C(C_6H_5)_2$$

with OH groups below the first and third carbons.

3-VI
1,1,6,6-Tetraphenylhexapentaene,
80% *crude*, m.p. 110–125°

Extensive chromatography yielded a trace of the desired compound as deep red crystals, m.p. ca. 300°. But the most promising method is that of Cadiot [14]:

$$(C_6H_5)_2COHC\equiv CH \xrightarrow[\substack{1.\ \text{Cool to ice temp.}\\2.\ \text{Add }(CH_3CO)_2O,\ 3.5\ \text{ml. dropwise}\\3.\ \text{Dilute with }H_2O,\ \text{filter}}]{0.15\ \text{mole KOH in 10 ml. ether}} 3\text{-VI}$$

0.01 mole

60%

In this transformation the ester from the tertiary carbinol may give the corresponding carbonium ion which then adds to a second molecule:

$$[(C_6H_5)_2\overset{\oplus}{C}C\equiv CH] \longleftrightarrow [(C_6H_5)_2C=C=\overset{\oplus}{C}H] \longrightarrow$$

$$[(C_6H_5)_2C=C=CH-CH=\overset{\oplus}{C}-C(C_6H_5)_2] \xrightarrow[-CH_3COOH]{-H^{\oplus}} 3\text{-VI}$$

with OCCH_3 / ‖ / O (acetate ester group) below.

1. D. R. Taylor, *Chem. Rev.*, **67,** 317 (1967).
2. A. A. Petrov and A. V. Fedorova, *Russ. Chem. Rev.*, **33,** 1 (1964).
3. R. Ya. Levina and E. A. Viktorova, *Uspekhi Khim.*, **27,** 162 (1958).
4. Ya. M. Slobodin and A. P. Khitrov, *J. Gen. Chem.*, *USSR*, (*English Transl.*), **31,** 3680 (1961).
5. T. L. Jacobs and R. D. Wilcox, *J. Am. Chem. Soc.*, **86,** 2240 (1964).
6. L. Skattebøl, *J. Org. Chem.*, **31,** 2789 (1966).
7. L. Skattebøl, *Acta Chem. Scand.*, **17,** 1683 (1963).
8. Y. Pasternak, *Compt. Rend.*, **255,** 3429 (1962).
9. T. L. Jacobs and W. F. Brill, *J. Am. Chem. Soc.*, **75,** 1314 (1953).
10. P. S. Skell and L. D. Wescott, *J. Am. Chem. Soc.*, **85,** 1023 (1963).
11. R. Kalish and W. H. Pirkle, *J. Am. Chem. Soc.*, **89,** 2781 (1967).
12. P. Cadiot *et al.*, *Bull. Soc. Chim. France*, 2176 (1961); H. Fischer in S. Patai, *The Chemistry of Alkenes*, Interscience Publishers, New York, 1964, p. 1025.
13. R. Kuhn and K. Wallenfels, *Chem. Ber.*, **71,** 783 (1938).
14. P. Cadiot, *Ann. Chim.* (Paris), **1** (13), 214 (1956).

Chapter 4

ALCOHOLS

The solvolytic methods of preparation of alcohols (A) are followed by addition mostly to olefins (B). The Brown procedure has made possible the choice of orientation in hydration of olefins, either $RCH{=}CH_2 \rightarrow RCHOHCH_3$ (B.1) or $RCH{=}CH_2 \rightarrow RCH_2CH_2OH$ (B.2), and also the preparation of alcohols with one more carbon atom from olefins, $RCH_2CH_2CH_2OH$ or $(RCH_2CH_2)_3$-COH (B.6). The reduction of various organic oxygen compounds by metal hydrides (C.1) is discussed in enough detail to compare with the alternative and sometimes supplementary catalytic hydrogenation process (C.6). The conversion

$$-\overset{|}{\underset{|}{C}}H \longrightarrow -\overset{|}{\underset{|}{C}}OH$$

is a valuable synthetic process with limitations; it is discussed in D.1, D.2, and D.4 (some biological oxidation processes are included). The addition of various nucleophiles to carbonyl compounds is discussed in E, F, and G. The most important advance in this field is Wittig's directed aldol process (G.1), in which it is possible to select the carbanion and the acceptor without interchange of roles between the two.

A. Solvolysis

1. From Esters

$$RCOOCH_2R' \underset{}{\overset{HOH}{\rightleftharpoons}} RCOOH + R'CH_2OH$$

This synthesis is of limited value because the ester is usually prepared from the alcohol. General conditions for hydrolysis have been described in Houben-Weyl [1]; see also Carboxylic Acids, Chapter 13, A.2. If applicable, the equilibrium may be driven to completion by removing the alcohol as it is formed. Hydrolysis in alkaline solution (saponification), of course, is more widely used than hydrolysis in acid solution. Yields should be good to excellent provided that no interfering factors are present. For instance, a hindered ester may prove difficult to hydrolyze. Although the usual remedy is to resort to alkali in a high-boiling solvent, such as ethylene glycol or diethylene glycol, another rather modern procedure is available. Anhydrous lithium iodide in boiling collidine is a good splitting agent for esters [2]. However, attention is directed to isolation of the acid rather than the alcohol. The alkyl fragment may well end up as an alkyl iodide which would have to be converted into the alcohol as shown in A.2. An unusual but facile splitting is given in Ex. c. Nitrate, acetate, and benzoate esters (but not sulfonate esters) are occasionally split with hydrazine to form the alcohol. This procedure is particularly adaptable to the carbohydrate family [3].

(a) **Preparation of p-nitrobenzyl alcohol** (64–71 % from p-nitrobenzyl acetate and sodium hydroxide in aqueous methanol, and p-iodobenzyl alcohol (81–86 % from p-iodobenzyl acetate)) [4].

(b) **Preparation of 1-acenaphthenol** (about 90 % from 1-acenaphthenol acetate and sodium hydroxide in aqueous methanol) [5].

(c) **Preparation of *trans*-2-phenylcyclopropanol.** Yields by this route are better than by simple hydrolysis, perhaps owing to the tendency of the

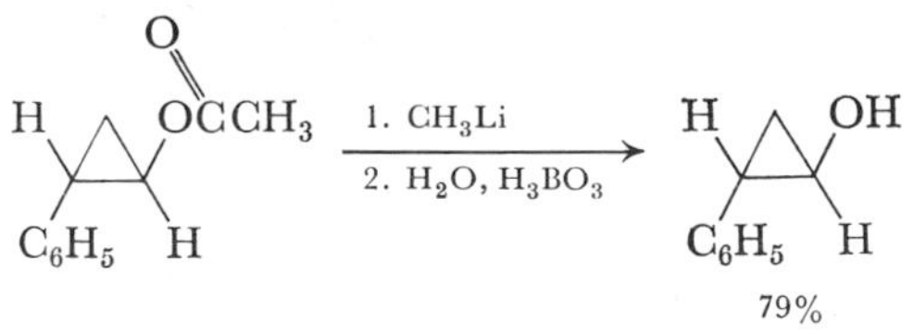

alcohol to rearrange in the former case [6]. Another "anhydrous" route is via reduction of the ester with lithium aluminum hydride [7].

2. From Halides

$$RCH_2X \xrightarrow{HOH} RCH_2OH$$

References for Section A are on p. 184.

This synthesis is not very common but is used commercially in the hydrolysis of the chlorination products of pentane (Sharples process) and of benzyl chloride. Moreover, it is employed in some cases directly or in the better two-step process involving the acetate as an intermediate [8].

Among the three halides the order of reactivity is I > Br > Cl, while among the various types the order is allyl, benzyl, and tertiary > secondary > primary. The allyl and particularly the tertiary halides tend to hydrolyze by the SN_1 mechanism. When this path is taken, the consequences of carbonium ion formation are to be expected and thus olefins are produced as indicated in Table 1 [9]. In addition, in the same SN_1 reaction with ethyl bromide, almost

Table 1 Hydrolysis of Halides in 80% Aqueous Ethanol

	% Olefin
Isopropyl bromide	4.6
sec-Butyl bromide	8.5
t-Butyl chloride	36.3

as much ethyl ether as alcohol is obtained. In the case of an allylic halide hydrolyzing to the alcohol by the SN_1 process, isomerization rather than elimination is the contentious problem. The product again is determined by the

$$C_6H_5CH—CH=CH_2 \xrightarrow[H_2O]{SN_1} C_6H_5CH=CHCH_2OH$$
$$\overset{|}{Cl}$$

relative stabilities of the intermediate carbonium ions and in the specific illustration by the phenyl to vinyl stabilization by resonance. When mechanism SN_1 blends into that of SN_2, such as with solvolysis of secondary halides, mixed products of solvolysis are to be anticipated. If isomerization or elimination is to be avoided in hydrolysis, a strong nucleophilic reagent would seem to be the most effective weapon to confine hydrolysis to the SN_2 mechanism. Aqueous sodium hydroxide is used most frequently, but other stronger nucleophiles, such as sodium peroxide, are worthy of consideration.

Some selectivity is possible in hydrolysis by water or dilute base since the lability of halogen atoms varies greatly. Aromatic, vinyl, and occasionally gem-halogens survive the ordinary conditions of hydrolysis. On the other hand, neighboring groups, particularly sulfur, nitrogen, and oxygen atoms, can so activate halogens that hydrolysis takes place spontaneously in water. These

$$RSCH_2CH_2Cl \xrightarrow{H_2O} RS\overset{\oplus}{\underset{CH_2}{\overset{CH_2}{<}}} \longrightarrow RSCH_2CH_2OH$$

systems should be run in buffered solutions because the reverse reaction is equally fast [10]. The exo-halogen atoms in bicyclic systems show accelerated rates of hydrolysis as well. For example, the rate of hydrolysis of the *exo*-norbornyl chloride is many times faster than that of the *endo*-isomer because of greater stability of the intermediate from the *exo* chloride [11]. The *endo*-

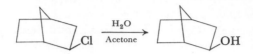

chloride, although much slower in hydrolysis, yields the *exo*-alcohol. All these differences in activity can be utilized on occasion to bring about selectivity in hydrolysis.

Although bases or nucleophilic reagents are used most frequently for hydrolysis of alkyl halides, acids may be useful hydrolytic reagents under certain conditions (see hydrolysis of trihalides, Chapter 13, A.6.).

(a) **Preparation of α-hydroxypalmitic acid.**   Crude α-bromopalmitic acid (105 g.) was added to a solution of 75 g. of potassium hydroxide in 900 ml. of water. The mixture was maintained at the boiling point by superheated steam for 10 hr., whereupon the α-hydroxypalmitic acid was precipitated by addition of excess dilute sulfuric acid. The acid was extracted with ether, washed with water, dried with calcium chloride, and reclaimed by evaporation of the ether. Crystallization from chloroform yielded pure α-hydroxypalmitic acid, 67 g. (80%), m.p. 87° [12].

(b) **Preparation of *trans*-1,2-cyclohexanediol.**   Silver acetate was precipitated by adding excess aqueous potassium acetate to silver nitrate (0.125 mole) dissolved in water. It was filtered and washed three times with glacial acetic acid, diluted with acetic acid, and treated with enough acetic anhydride to make the solution anhydrous. *trans*-1,2-Dibromocyclohexane (0.1 mole) was added to the silver acetate, held at 100–110° for 11 hr., cooled, and filtered to remove silver bromide. The filtrate was concentrated under reduced pressure. The residue consisting mainly of the diacetate was saponified for 2 hr. with 40 ml. of 35% aqueous sodium hydroxide and 40 ml. of ethyl alcohol. The glycol was extracted from the cooled solution by five 25-ml. portions of chloroform, the chloroform extract being dried with potassium carbonate. Concentration of the chloroform gave *trans*-1,2-cyclohexanediol, 70% (or more in larger batches), m.p. 104°, after recrystallization from carbon tetrachloride [13]. An alternative method of extraction of the glycol is to saturate the solution with K_2CO_3 and extract with alcohol.

(c) **Preparation of methyl-bis(β-hydroxyethyl)amine.**   A reaction mixture of 600 ml. of water containing 12 mmoles of 4-I and 60 mmoles of sodium bicarbonate was allowed to stand at 25° for 3 days. The reaction mixture was cooled, acidified with hydrochloric acid, and treated with 11.5 mmoles of picrylsulfonic acid. The picrylsulfonic acid salt of 4-III (24%) precipitated immediately; the acid salt of 4-II was obtained on concentration of the filtrate to 75 ml. under reduced pressure (55%), m.p. 178–180°, after

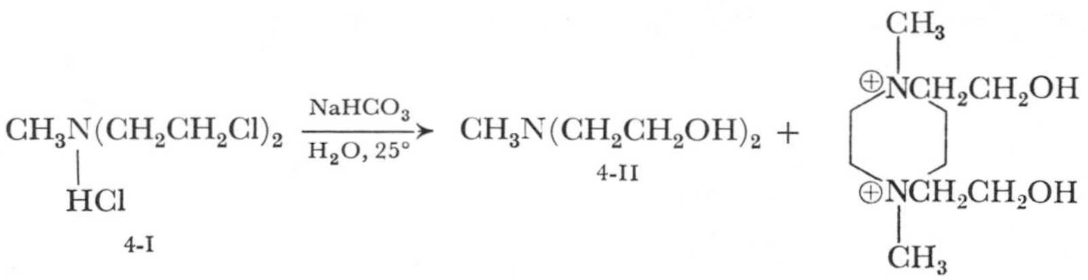

$$CH_3N(CH_2CH_2Cl)_2 \xrightarrow[H_2O,\ 25°]{NaHCO_3} CH_3N(CH_2CH_2OH)_2 +$$

$$\underset{HCl}{|}$$

4-I

4-II

4-III

crystallization from water. Hydrolysis in nonbuffered solutions was incomplete [14].

(d) Other examples

(1) **3,5-Dichloro-2-hydroxybenzyl alcohol** (67% from the benzyl chloride in water at 50° until solution was almost clear) [15].

3. From α-Diazoketones

$$RCOCl \xrightarrow{CH_2N_2} RCOCHN_2 \xrightarrow{H_2O} RCOCH_2OH$$

The Arndt-Eistert reaction, a procedure for converting an acid into the next higher homolog, has been reviewed in detail [16]. The diazoketone is conveniently obtained from the acid chloride or from the α-acylacetaldehyde

$$RCOCH_2CH{=}O + ArSO_2N_3 \longrightarrow RCOCHN_2 + \overset{O}{\overset{||}{HCNHSO_2Ar}}$$

[16a]. In the absence of a catalyst the Wolff rearrangement does not occur, and the diazoketone hydrolyzes to a keto alcohol. The mechanism follows the pattern given in Chapter 13, G.1., except that the acylcarbene does not rearrange but simply adds water to give the keto alcohol. The first step requires anhydrous conditions and involves the preparation of diazomethane as needed. The synthesis has found some use, particularly in the preparation of keto alcohols of ring systems. The yields are satisfactory. Hydrolysis in basic solution, especially with 2-diazo-1,3-diketones, gives cleavage products rather than alcohols [17]:

$$(RCO)_2CN_2(ArCN_2COR) \xrightarrow{OH^\ominus} RCOCHN_2(ArCHN_2) + RCO_2{}^\ominus$$

(a) Preparation of 2-hydroxyacetylfuran

(1) **Diazoacetylfuran.** Furoylchloride, 101 g., in 100 ml. of absolute ether was dropped into an ice-cold ether solution of 65 g. of diazomethane. After the addition was complete, the solution was allowed to come to room temperature and to stand overnight. The removal of the ether by distillation and air jet left a bright yellow oil which crystallized at low temperature [18].

(2) **2-Hydroxyacetylfuran.** Ten and one-half g. (0.077 mole) of diazoacetylfuran was mixed with 150 ml. of 2 N H_2SO_4 and 60 ml. of dioxane while the temperature rose to 45°. The mixture was stirred and held at 45° for

4 hr., after which the temperature was allowed to drop to 25°. After neutralization with solid K_2CO_3 and filtration, the filtrate was distilled from a steam bath at reduced pressure to remove all of the dioxane and some water. The residue obtained was cooled, salted with NaCl, and extracted six times with 30 ml. of ether. From the extract 7.2 g. (74%) of the titled compound was recovered, m.p. 81–82° [19].

4. From Xanthates

$$-CH_2OC\overset{\displaystyle S}{\underset{\displaystyle SCH_2C_6H_5}{\Big\|}} \xrightarrow[CH_3COOH]{Hg(OCOCH_3)_2} -CH_2OC\overset{\displaystyle O}{\underset{\displaystyle SCH_2C_6H_5}{\Big\|}} \xrightarrow[CH_3COOH]{H_2O_2} -CH_2OH$$

This synthesis is a mild method which has been applied recently to the xanthates of carbohydrates. On treatment with mercuric acetate in acetic acid the xanthate produces the thiol ester which is converted into the primary alcohol by hydrogen peroxide, also in acetic acid solution. Methoxy and benzoxy groups are not affected by these reagents. The yields are good.

(a) Preparation of methyl 2,3,4-O-tribenzoyl α-D-glucopyranoside. The 6-S-benzylmonothiolcarbonate, 79%, was obtained from methyl 2,3,4-O-tribenzoyl-D-glucopyranoside 6-S-benzylxanthate in dioxane to which mercuric acetate and acetic acid were added followed by saturation of the mixture with hydrogen sulfide; the monothiolcarbonate in glacial acetic acid containing 30% hydrogen peroxide gave the glucopyranoside, 83% [20].

5. From Amines

$$RCH_2NH_2 \xrightarrow[-N_2]{HONO} RCH_2^{\oplus} \xrightarrow{H_2O} ROH + \text{carbonium ion products}$$

The conversion of an amine into an alcohol is more often a degradative rather than a preparative procedure, but comments about scope are appropriate.

For aliphatic amines the medium should not be too acidic, as the free base is diazotized, not the amine salt. The yields of n-alkanols from n-alkylamines are very poor [21]. For example, n-butylamine yields 25% n-butanol, 13% 2-butanol, and 36% 1-butene. The diazotization procedure for preparing alcohols is useful chiefly in those cases where the carbonium ion cannot rearrange nor form an olefin. However, trans-2-methylcyclohexylamine, similarly to other alicyclic or bicyclic amines, gives a rather good yield of trans-2-methylcyclohexanol on diazotization, while the cis-isomer gives the same product in smaller yield with more isomeric alcohols and some olefin. The conformation is such in the trans-amine (both amino and methyl groups being mainly equatorial) that the carbonium ion formed is protected from intervening rearrangements by the cyclic system to which it is attached. In general, an amino group placed in either an equatorial or an axial position of an alicyclic or bicyclic system yields the equatorial alcohol on diazotization. In fact, the equatorial amine gives a better yield of alcohol than would be anticipated.

The Demjanow rearrangement is a quite specialized method of obtaining alcohols by diazotization of amines. A general discussion is available [22]. The greatest use of the rearrangement is in preparing alcohols with expanded rings.

By-products are unrearranged alcohol and some olefins, which may be difficult to separate (see Ex. c). If the carbon atom bearing the amino substituent is substituted with another alkyl or aryl group, Demjanow ring expansion occurs only to a limited extent. The generalities suggest that the factor allowing acceptable yields in the ring expansion is the favorable energetics of proceeding from a primary to a secondary carbonium ion following diazotization.

(a) **Preparation of 4-diethylaminomethylbenzyl alcohol.** 4-Diethylaminomethylbenzylamine, 100 g., was dissolved in 110 g. of glacial acetic acid, and 20 ml. of water was added. To the cold solution was added gradually a solution of 50 g. of sodium nitrite in a minimum of water. An intense evolution of nitrogen occurred and the hot solution was cooled with water. When the evolution of nitrogen ceased, the solution was heated about 15 min. on a steam bath to complete the reaction. It was then made alkaline and extracted with ether, after which the ethereal layer was dried; the solvent was removed by distillation; and the residue was distilled under reduced pressure. There was obtained 90 g. (89.6%) of 4-diethylaminomethylbenzyl alcohol, b.p. 165° (18 mm.) [23].

(b) **Preparation of cyclohexanol.** Cyclohexylamine, 25 g., was dissolved in 200 ml. of 10% acetic acid and treated with a concentrated aqueous solution of sodium nitrite (at least 0.25 mole). Nitrogen was evolved slowly. After warming for 5 hr. on a steam bath, the nitrogen evolution had ceased. After the usual purification, distillation yielded 4.5 g. of cyclohexene and 17 g. of cyclohexanol (67%). Cyclopentylamine gave equal amounts of cyclopentene and cyclopentanol. Other amino groups attached equatorially to alicyclic rings gave appreciable amounts of the corresponding alcohol, but such groups attached axially gave considerably more olefin [24].

(c) **Preparation of 1-methylcycloheptanol** (25% from 1-methylcyclohexanemethylamine and sodium nitrite in aqueous phosphoric acid at ice temperature and warmed to room temperature; careful fractionation was necessary to separate from 1-ethylcyclohexanol and other products) [25].

6. From Some Quaternary Ammonium Salts

$$\underset{\bigcirc}{\overset{\oplus}{CH_2N(CH_3)_3}Br^{\ominus}} \quad\xrightarrow{\text{NaOH}}\quad \underset{\bigcirc}{CH_2OH}$$

This reaction, like diazotization, is practical only when olefins or isomerization products are structurally prohibitive or at least less likely. It has been applied satisfactorily to the conversion of aminomethyl derivatives of ring systems, through quaternary ammonium salts, into hydroxymethyl derivatives. Sometimes it is desirable to convert the quaternary ammonium salt into the acetate before proceeding to the alcohol. For *o*-methylbenzyl alcohols the method yields purer products than treating the properly substituted Grignard reagent with formaldehyde (Tiffeneau rearrangement) [26]. The synthesis is not difficult and the yields are often high.

(a) Preparation of *o*-methylbenzyl alcohol (77–84 % overall yield from *o*-methylbenzyldimethylamine which was quaternized with ethyl bromide; the quaternary salt was converted into the acetate of the alcohol by refluxing in glacial acetic acid containing anhydrous sodium acetate; the acetate was saponified in 50 % aqueous methanol containing sodium hydroxide) [26].

(b) Other examples

(1) Hydroxymethylferrocene (59–90 % from N,N-dimethylamino-methylferrocene methiodide in refluxing aqueous sodium hydroxide) [27].

(2) 3-Hydroxymethylindole (66 % from gramine methiodide added slowly to a rapidly stirred mixture of ether and 10 % aqueous sodium hydroxide) [28].

7. From Cyclic Ethers

$$\underset{\diagdown\underset{O}{\diagup}}{CH_2(CH_2)_nCH_2} \xrightarrow[\text{H}_2\text{O}]{\text{H}^{\oplus}} HOCH_2(CH_2)_nCH_2OH$$

Acyclic ethers can be cleaved by various reagents, but the reaction is not useful synthetically because strenuous conditions are needed and a derivative of the alcohol is formed in part [29]. However, small-ring ethers give useful products which make even two-step processes look attractive.

$$ROR \xrightarrow{\text{HX}} 2\,RX \quad\text{or}\quad RX + ROH$$
$$\xrightarrow{\text{Na}} RH + RONa$$

Ethylene and trimethylene oxides are hydrolyzed readily to the corresponding glycols in the presence of a catalyst such as trifluoroacetic acid. Transannular participation may be encountered in certain cyclic epoxides. For example, *cis*-cyclooctene oxide with trifluoroacetic acid gives mainly *cis*-1,4-cyclooctane diol on hydrolysis. Even 1-octene oxide shows neighboring group participation in yielding on hydrolysis 90 % 1,2-, 0.7 % 1,3-, 0.5 % 1,4-, 0.3 % 1,5-, 0.2 % 1,6-, and 0.1 % 1,7-octanediol [30]. In the hydrolysis of the epoxides, inversion occurs in that the epoxide from a *cis*-olefin gives the *threo-* and that from a *trans*-olefin gives the *erythro*-glycol [31].

The strong acid perchloric has been of value in the hydrolysis of ethylene oxides (as well as ketals) in the steroid series. To cite one case [32], 3-ethylene-dioxy-5,6-epoxycholestane with aqueous perchloric acid in THF is converted into cholestane-3-one-5α,6β-diol:

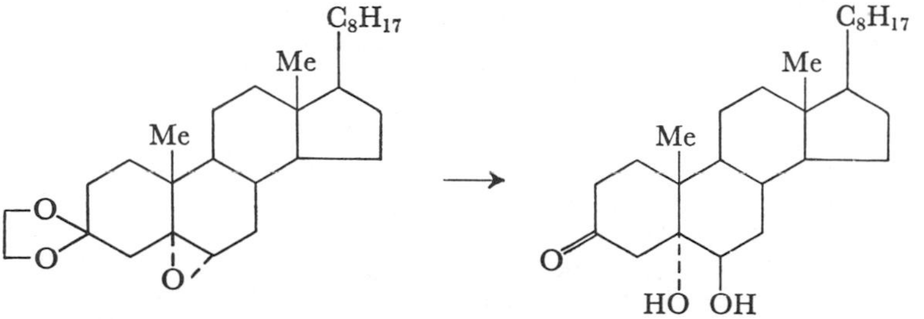

Opening of epoxide rings with organometallic compounds is discussed in E.2.

More strenuous hydrolytic conditions are necessary for the cleavage of tetrahydrofurans and pyrans. In these cases, depending on the reagent, the alcohol group may be converted into the ester or halide, which then must be hydrolyzed.

(a) **Preparation of 1,2,5-trihydroxypentane** (the triacetate was first obtained from tetrahydrofurfuryl alcohol, acetic anhydride and anhydrous zinc chloride in 87–90 % yield; the triol was obtained in 63–71 % yields from the triacetate by hydrolysis with 1 % aqueous sulfuric acid, removal of the acetic acid by steam distillation, and of the sulfuric acid by precipitation with lime) [33].

(b) **Other examples**

(1) **5-Chloropentanediol-1,2** (90 % from 1,2-epoxy-5-chloropentane by heating in dilute sulfuric acid) [34].

(2) **8-Chloro-6-(2-piperidino-1-hydroxyethyl)-2-phenylquinoline** [35]

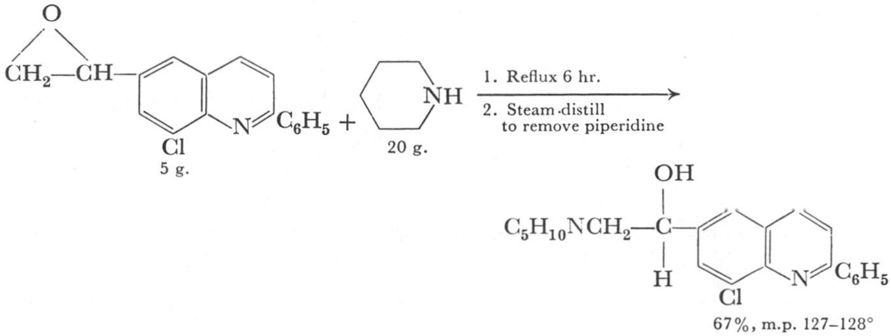

67%, m.p. 127–128°

1. Houben-Weyl, *Methoden der Organischen Chemie*, Vol. 8, Georg Thieme Verlag, Stuttgart, 4th Ed., 1952, p. 418.
2. W. L. Meyer and A. S. Levinson, *J. Org. Chem.*, **28,** 2184 (1963).
3. K. S. Ennor and J. Honeyman, *J. Chem. Soc.*, 2586 (1958).
4. W. W. Hartman and E. J. Rahrs, *Org. Syn.*, Coll. Vol. **3,** 652 (1955).
5. J. Cason, *Org. Syn.*, Coll. Vol. **3,** 3 (1955).
6. C. H. DePuy *et al.*, *J. Org. Chem.*, **29,** 2013 (1964).
7. C. H. DePuy, *Accounts Chem. Research* **1,** 33 (1968).
8. W. W. Hartman and E. J. Rahrs, *Org. Syn.*, Coll. Vol. **3,** 650, 652 (1955).
9. C. K. Ingold, *Structure and Mechanism in Organic Chemistry*, Cornell University Press, Ithaca, N.Y., 1953, p. 442.
10. A. Streitwieser, Jr., *Solvolytic Displacement Reactions*, McGraw-Hill Book Co., New York, 1962, p. 108.
11. H. C. Brown and M.-H. Rei, *J. Am. Chem. Soc.*, **90,** 6216 (1968), for leading references and another explanation.
12. H. R. LeSueur, *J. Chem. Soc.*, **87,** 1895 (1905).
13. S. Winstein and R. E. Buckles, *J. Am. Chem. Soc.*, **64,** 2780 (1942).
14. C. Columbic *et al.*, *J. Org. Chem.*, **11,** 518 (1946).
15. C. A. Buehler *et al.*, *J. Org. Chem.*, **6,** 906 (1941).
16. W. E. Bachmann and W. S. Struve, *Org. Reactions*, **1,** 38 (1942).
16a. M. Regitz and F. Menz, *Chem. Bev.*, **101,** 2622 (1968).
17. J. B. Hendrickson and W. A. Wolf, *J. Org. Chem.*, **33,** 3610 (1968).
18. A. Burger and G. H. Harnest, *J. Am. Chem. Soc.*, **65,** 2382 (1943).
19. F. Kipnis *et al.*, *J. Am. Chem. Soc.*, **70,** 142 (1948).
20. J. J. Willard and E. Pacsu, *J. Am. Chem. Soc.*, **82,** 4349 (1960).
21. H. Zollinger, *Azo and Diazo Chemistry*, Interscience Publishers, New York, 1961, p. 93.
22. P. A. S. Smith and D. R. Baer, *Org. Reactions*, **11,** 157 (1960).
23. A. Funke and O. Rougeaux, *Bull. Soc. Chim. France*, **12,** 1050 (1945).
24. W. Huckel, *Ann. Chem.*, **533,** 1 (1938).
25. R. Kotani, *J. Org. Chem.*, **30,** 350 (1965).
26. W. R. Brasen and C. R. Hauser, *Org. Syn.*, Coll. Vol. **4,** 582 (1963).
27. C. R. Hauser *et al.*, *Org. Syn.*, **40,** 52 (1960).
28. E. Leete and L. Marion, *Can. J. Chem.*, **31,** 781 (1953).
29. Ref. 1, Vol. 1, 1965, Pt. 3, p. 143.
30. A. C. Cope *et al.*, *J. Am. Chem. Soc.*, **85,** 3752 (1963).
31. H. J. Lucas *et al.*, *J. Am. Chem. Soc.*, **63,** 22 (1941).
32. L. H. Sarett *et al.*, *J. Am. Chem. Soc.*, **75,** 422 (1953).
33. O. Grummitt *et al.*, *Org. Syn.*, Coll. Vol. **3,** 833 (1955).
34. R. Paul and H. Normant, *Bull. Soc. Chim. France*, **12,** 388 (1945).
35. J. Wommack and D. E. Pearson, unpublished work.

B. Addition and Substitution (Friedel-Crafts)

Although the number of examples of hydration of olefins is limited, it is nevertheless a most important method of synthesis. H. C. Brown's hydroboration process is included in this section because it is a supplementary tool in broadening the scope of preparing alcohols from olefins. The hydroboration process should be placed more properly in the section on oxidation, but its complementary role to the hydration process suggested collateral discussion. While the simple hydration process yields alcohols in accordance with the Markownikoff rule of addition, the Brown process yields alcohols in violation of this rule.

References for Section B are on pp. 192–193.

$$RCH{=}CH_2 \xrightarrow{\text{Hydration}} RCHOHCH_3$$

$$RCH{=}CH_2 \xrightarrow{\text{Hydroboration}} RCH_2CH_2OH$$

1. From Alkenes

This synthesis is widely employed in manufacturing alcohols from alkenes available in the cracking of petroleum. Ethylene, for example, gives the primary alcohol, ethanol, while unsymmetrical alkenes give either secondary or tertiary alcohols. Sulfuric acid adds in such a manner as to give the stablest carbonium ion:

$$CH_3CH{=}CH_2 \xrightarrow{H_2SO_4} CH_3\overset{\oplus}{C}HCH_3 \longrightarrow \underset{\underset{OSO_2OH}{|}}{CH_3CHCH_3}$$

If the intermediate carbonium ions are of comparable activity, mixtures of alkyl hydrogen sulfates result. It is doubtful whether ethylene forms the primary carbonium ion in sulfuric acid; in fact, it is probable that a concerted addition takes place. The more electron-donating groups (such as alkyl groups) attached to the olefin, the more easily is it solvated by sulfuric acid. For example, iso-butylene can be absorbed in 60–65% aqueous sulfuric acid, whereas ethylene is absorbed efficiently only in concentrated sulfuric acid. Although large excesses of sulfuric acid are used in the normal laboratory hydration proced-ures, there are some indications that as much as 1.5 moles of olefin can be absorbed per mole of sulfuric acid. Tertiary alcohols are extracted rather than distilled from the diluted sulfuric acid mixtures because of their ease of dehy-dration. It is a curious fact that cyclopropane is absorbed about five times as fast as propylene in sulfuric acid [1].

Formic acid mixed with catalytic amounts of a strong acid such as perchloric acid has been used as a means of hydration of terminal olefins. The intermediate formic esters must be hydrolyzed to the alcohols. Isomerization is to be expected as shown in one of the examples which follows. Trifluoroacetic acid is a better hydration agent than formic for branched-chain olefins [2]. Yields of alcohols averaged about 45% for hydration of 2-methyl-2-butene, methylcyclopentene, and methylcyclohexene. In some cases at least, the addition of formic acid in combination with sulfuric acid is stereospecific. For example, trans-7-methyl-2,6-octadienoic acid with a mixture of formic and sulfuric acid cyclizes to give trans-6,6-dimethyl-2-hydroxycyclohexanecarboxylic acid [3], while the cis-dienoic acid gives the cis-hydroxycarboxylic acid [4]. Further discussion is to be found under Chapter 14, B.5.

(a) Preparation of 2-hydroxy-3-nitropropionic acid.   3-Nitroacryl-ic acid, 0.25 g. (0.0017 mole) in 10 ml. of 70% formic acid was heated for 3 hr.

at 85–100°, after which the mixture was diluted with 20 ml. of distilled water and concentrated at reduced pressure. The yellow oil remaining was dissolved in ether, decolorized with charcoal, and filtered. From the filtrate 0.24 g. (83 %) of the hydroxy acid was recovered [5].

(b) Other examples

(1) *trans*-**6,6-Dimethyl-2-hydroxycyclohexanecarboxylic acid** (72 % from *trans*- 7-methyl-2,6-octadienoic acid in formic acid with a small amount of sulfuric acid) [3].

(2) **2- and 3-Hexanol** (55 % from 1-hexene and formic acid, catalyzed by perchloric acid–of which two-thirds was 2- and one-third 3-hexanol) [6].

2. From Alkenes via Boranes and Boric Acid Esters

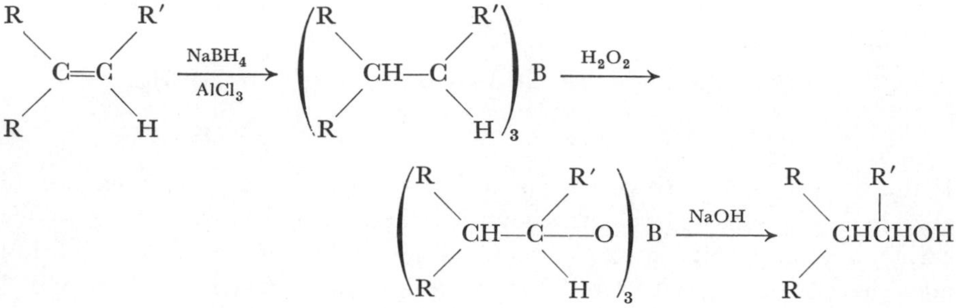

R' = H or alkyl

This synthesis [7] consists of an *anti*-Markownikoff hydration of double bonds and is applicable to the synthesis of both primary and secondary alcohols. Terminal olefins produce primary alcohols in yields of 80–90 %. Several examples are shown to give the extent of orientation control:

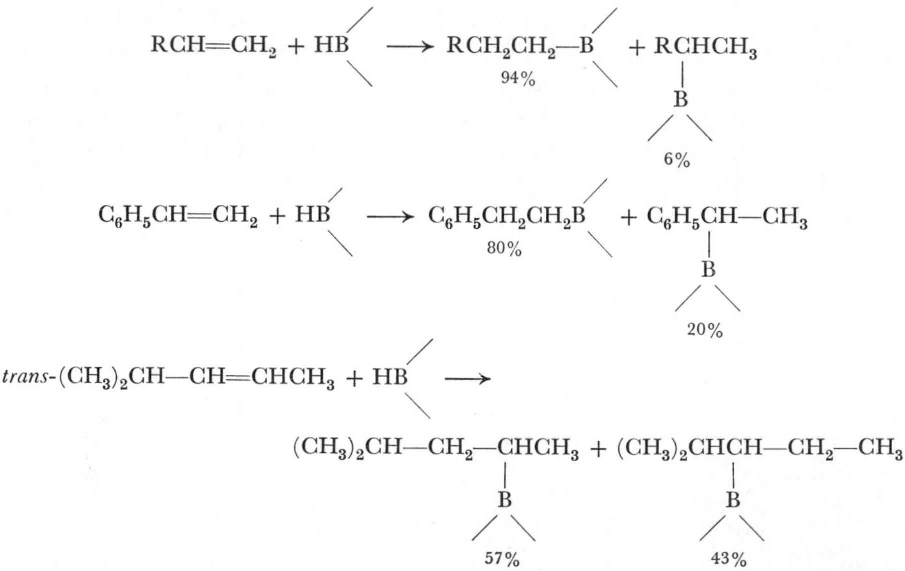

Bis-3-methyl-2-butylborane

$$\overset{\overset{\textstyle CH_3}{|}}{HB(CH-CH(CH_3)_2)_2}$$

is more selective than diborane in the above additions.

Another borane which is quite selective in addition is thexylborane

$$(CH_3)_2CH\overset{\overset{\textstyle CH_3}{|}}{\underset{\underset{\textstyle CH_3}{|}}{C}}BH_2$$

An additional benefit has accrued from the search for boranes with a high degree of selectivity. 9-Borabicyclo [3.3.1] nonane has been found to be stable

in storage and in brief exposure to air [8]. When it is commercially available, it will dispense with elaborate equipment for conversion of terminal olefins into primary alcohols and in other borane conversions where the cyclooctane by-product can be separated from the desired compound.

Internal olefins normally give secondary alcohols, but by heating the re-action mixture to 160° for 1 hr. before the hydrogen peroxide oxidation, primary alcohols are produced. Manipulation is simple and the reaction proceeds rapidly. The active agent, diborane, can be handled with ease, since it is used as generated from sodium borohydride and aluminum chloride or boron tri-fluoride etherate. The reaction is stereospecific in that addition is *cis*, as in the hydroboration (followed by oxidation and hydrolysis) of 1-methylcyclopentene to give *trans*-2-methylcyclopentanol:

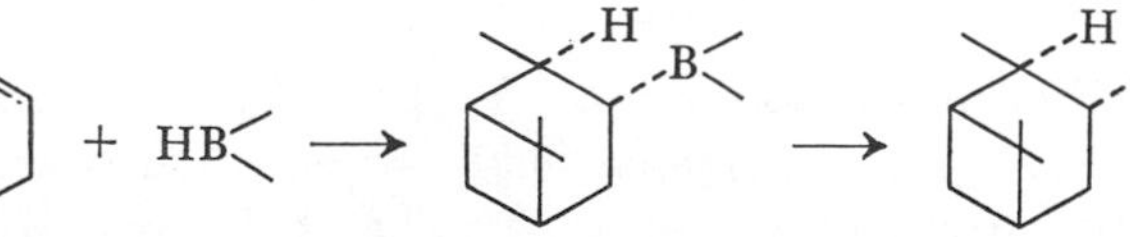

In addition, if hindrance is present, the addition occurs on the less hindered side, as in the case of α-pinene to give isopinocampheol. If (α-pinene)$_2$BH is used

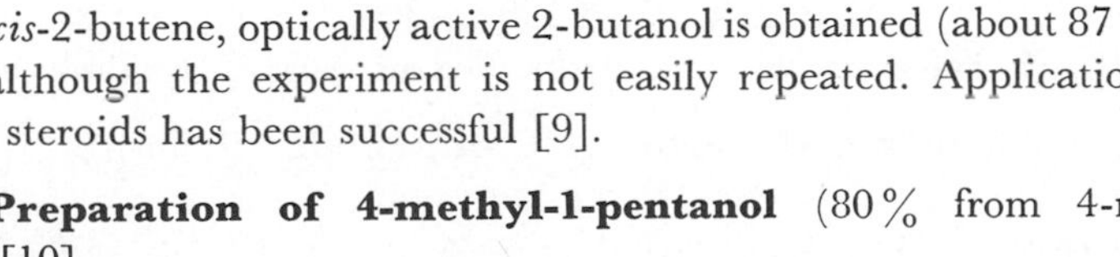

to add to *cis*-2-butene, optically active 2-butanol is obtained (about 87 % optical purity), although the experiment is not easily repeated. Application to un-saturated steroids has been successful [9].

(a) Preparation of 4-methyl-1-pentanol (80 % from 4-methyl-1-pentene) [10].

(b) **Preparation of** *trans*-**2-methylcyclopentanol.** Excess of diborane (from 3.8 g. of sodium borohydride) in diglyme and boron trifluoride
etherate was passed into 16.4 g. (0.2 mole) of 1-methylcyclopentene in 60 ml. of
THF at 0° during 2 hr. After 1 hr. at 25°, several pieces of ice were added to
hydrolyze the excess diborane and the mixture was immersed in an ice bath,
45 ml. of 3 M NaOH was added, and then 25 ml. of 30% hydrogen peroxide
during 1 hr. After 1 hr. more at 25°, the upper layer was separated, the aqueous
layer was extracted with ether, and the combined extracts were dried. Fractionation gave 85% of the *trans* alcohol [11].

(c) **Other examples**

(1) **2,2,5,5-Tetramethyl-3-hexanol** (82.2% from *trans*-2,2,5,5-tetra-
methyl-3-hexene) [12].

(2) **Cholestan-6α-ol** (75% from Δ^5-cholestene) [9].

(3) *cis*-**1,3-Cyclopentanediol** (41% from cyclopentadiene). Other
conjugated dienes may give linear polymers, not useful for preparation of
alcohols [13].

3. From Alkenes (via Mercurials)

$$\text{RCH=CH}_2 + \text{Hg(OAc)}_2 \longrightarrow \underset{\overset{|}{\text{OAc}}}{\text{RCHCH}_2\text{HgOAc}} \xrightarrow{\text{NaBH}_4} \underset{\overset{|}{\text{OH}}}{\text{RCHCH}_3}$$

The reaction steps are simple enough to consider this route favorably with
that of the hydration of olefins. Moreover, in substituted olefins where the
substituent influences the orientation of the mercuriacetate group, the reaction
steps lead to alcohols difficultly accessible by other routes. For example, 3-cyclo-
hexenol may be converted exclusively into *trans*-1,4-cyclohexanediol [14].

84% overall

Sodium borohydride reduction is simpler than hydrazine reduction, however.

Some dihydroxynorbornanes have been prepared recently by the oxymercur-
ation-demercuration process [15].

(a) **Preparation of 2-hexanol.** To 10 mmoles of mercuric acetate in
10 ml. each of water and THF, magnetically stirred, 10 mmoles of 1-hexene was
added slowly and stirred for 10 min. at 25°. Then 10 ml. of 3 M NaOH followed
by 10 ml. of 0.5 M sodium borohydride in 3 M NaOH was added and the
organic layer separated, dried, and distilled to yield 2-hexanol (96%) [16].

4. From Alkenes and Carbonyl Compounds (Prins)

$$\text{R'CH=O} + \text{H}^\oplus \rightleftharpoons \text{R'}\overset{\oplus}{\text{C}}\text{HOH} \xrightarrow[\text{H}_2\text{O}]{\text{RCH=CH}_2} \text{RCHOHCH}_2\text{CHOHR'}$$

This general reaction, which has been reviewed [17, 18], is quite complex, being of a carbonium ion type of mechanism, and yields, in addition to the glycol, a variety of products including the allyl alcohol, $RCH=CHCHOHR'$, rearranged unsaturated alcohol, m-dioxanes, shown below, 4-hydroxytetrahydropyrans,

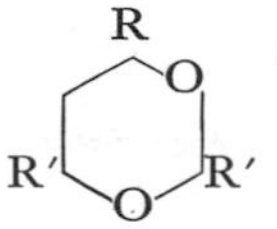

and polymers [19]. Formaldehyde is the most versatile of the aldehydes, tending to yield the m-dioxane as the main product. If one of the R groups is aryl, the m-dioxane can be reductively cleaved to the alcohol (see Ex. a). Yields are quite variable.

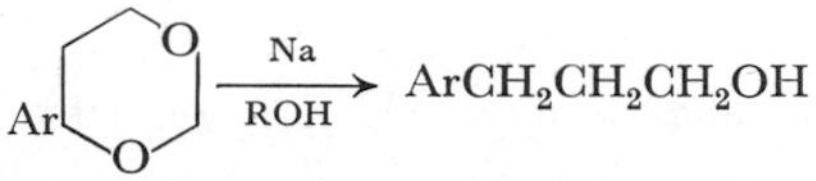

(a) Preparation of 3-phenyl-1-propanol

(1) **4-Phenyl-m-dioxane.** Styrene, 6 moles, was added to a mixture of 16.6 moles of 37 % formaldehyde solution and 96 g. of sulfuric acid, after which the mixture was stirred just below the boiling point (about 93°) for 7 hr. On cooling to room temperature, the product was extracted with benzene from which it was recovered, 805 g. (86%), b.p. 99° (2 mm.) by fractionation under reduced pressure.

(2) **3-Phenyl-1-propanol.** Dry nitrogen was introduced into a mixture of 335 g. of Na (14.56 moles) and 800 g. of dry toluene which was then heated to the boiling point. When the Na had formed small droplets, the heating was ended and a mixture of 1210 g. (8.4 moles) of diisobutylcarbinol and 656 g. (4 moles) of phenyl-m-dioxane was introduced at a rate to produce gentle refluxing. When the metal had disappeared completely (about 2 hr.) during boiling and stirring, the mixture was cooled to 25° and poured into a mixture of ice and the calculated amount of H_2SO_4. The upper layer and the toluene extract of the lower layer were combined, made neutral, and distilled finally under reduced pressure to give diisobutylcarbinol (90%) and 486 g. (89%) of 3-phenyl-1-propanol, b.p. 86° (2 mm.) [20; *see also* 21].

(b) Other examples

(1) **Various substituted alcohols** [20, 22].

(2) **5,5-Dimethyl-3-methylene-1-hexanol** [17]

$$(CH_3)_3CCH_2CCH_3=CH_2 + CH_2O \xrightarrow[18\ hr.]{180°} (CH_3)_3CCH_2C\overset{\displaystyle CH_2CH_2OH}{=}CH_2$$

55%

(3) 1,1,1-Trifluoro-3,4-dimethylpent-4-en-2-ol [23]

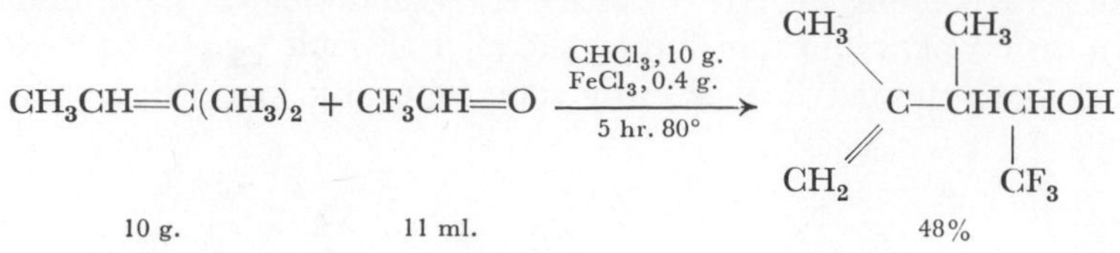

$$CH_3CH{=}C(CH_3)_2 + CF_3CH{=}O \xrightarrow[\text{5 hr. 80°}]{\substack{CHCl_3,\,10\text{ g.}\\FeCl_3,\,0.4\text{ g.}}}$$

$$\begin{array}{cc} CH_3 & CH_3 \\ \diagdown & | \\ C{-}CHCHOH \\ \diagup\diagup & | \\ CH_2 & CF_3 \end{array}$$

10 g.   11 ml.   48%

5. From Alkenes, Carbon Monoxide, and Hydrogen (Hydroformylation)

$$RCH{=}CH_2 + CO + H_2 \longrightarrow RCH_2CH_2CHO + RCHCH_3 \longrightarrow$$
$$\underset{\displaystyle CHO}{|}$$

$$RCH_2CH_2CH_2OH + RCHCH_3$$
$$\underset{\displaystyle CH_2OH}{|}$$

The oxo process, which has been described in some detail [24], is an industrial method for preparing alcohols from olefins. Beginning in 1949, the production of such alcohols has risen hyperbolically each year to the present figure of about one-half billion pounds annually. Almost any cobalt compound or cobalt itself will catalyze the reaction, but the actual catalyst is thought to be dicobalt octacarbonyl, $Co_2(CO)_8$, or cobalt hydrocarbonyl, $HCo(CO)_4$. Normally the gas mixture of carbon monoxide and hydrogen is in a 1 : 1 ratio and temperatures and pressures vary from 70 to 200° and from 100 to 300 atm., respectively. The reaction can be carried out in one stage to produce alcohols, or the aldehydes (see Chapter 10, C.8) may be isolated and then reduced to alcohols.

The exact mechanism of the reaction is unknown, although it appears that cobalt hydrocarbonyl is the essential agent in forming the aldehyde [25]:

$$RCH{=}CH_2 + HCo(CO)_4 + CO \longrightarrow RCH_2CH_2COCo(CO)_4 \xrightarrow{HCo(CO)_4}$$
$$Co(CO)_8 + RCH_2CH_2CHO$$

The ratios of isomeric alcohols obtained in the oxo process are shown in Table 2. Under optimum conditions, the yields of alcohols may approach 90%.

Table 2   Products from the Oxo Process

Olefin	Alcohol	Ratio
Propylene	n-Butanol	60
	2-Methyl-1-propanol	40
1-Butene	n-Pentanol	50
	2-Methyl-1-butanol	50
Isobutylene	3-Methyl-1-butanol	96
	2,2-Dimethylpropanol	4
2,4,4-Trimethyl-1- (or 2-)pentene	3,5,5-Trimethylhexanol	100

6. From Boranes and Carbon Monoxide

$$R_3B + CO \xrightarrow{125°} (R_3CBO) \xrightarrow[NaOH]{H_2O_2} R_3COH$$

A convenient and sometimes superior synthesis of tertiary alcohols is the borane addition to carbon monoxide. The reaction can be carried out at atmospheric pressure at 125° in diglyme solution. In the synthesis of tricyclohexylcarbinol the borane method gives 80% yield, the Grignard method, 7%, and an organosodium method, 19%.

Furthermore, the reaction can be controlled to introduce only one R group into carbon monoxide to make available primary alcohols. The control is brought about by using an excess of lithium borohydride which catalyzes the rate of absorption of carbon monoxide and permits a lower reaction temperature (see Ex. b.1).

(a) Preparation of tri-2-norbornylcarbinol.   Tri-2-norbornylborane

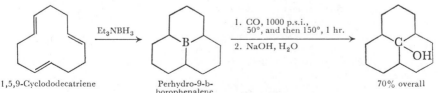

(prepared from 75 mmoles of sodium borohydride and 300 mmoles of norbornene in 150 ml. of diglyme) was treated with 100 mmoles of boron trifluoride diglymate at 0° and the mixture was stirred for 1 hr. at 25°. After 10 ml. of ethylene glycol was added, the system was maintained at 100°, flushed with carbon monoxide, held till absorption was complete, flushed with nitrogen, and heated to 150° for 1 hr. At 0°, 33 ml. of 6 N NaOH and then 33 ml. of hydrogen peroxide (30%) were added and the mixture was warmed to 50° for 3 hr. Addition of water precipitated the carbinol (80%) after recrystallization from pentane [26].

(b) Other examples

(1) *exo*-2-Norbornylmethanol (85%, by analysis, essentially as in Ex. a except the borane was made from the olefin and diborane, LiBH$_4$ was added, and carbon monoxide absorption was carried out at 45°) [27].

(2) Perhydro-9b-phenalenol [28]

| 1,5,9-Cyclododecatriene | → Et$_3$NBH$_3$ → | Perhydro-9-b-borophenalene | 1. CO, 1000 p.s.i., 50°, and then 150°, 1 hr. / 2. NaOH, H$_2$O → | 70% overall |

7. From Cyclic Ethers or Some Aldehydes and Arenes (Friedel-Crafts)

$$ArH + (CH_2)_nO \rightarrow Ar(CH_2)_nOH$$

The creation of an alcoholic group by the Friedel-Crafts reaction is limited largely to the use of an arene with cyclic ethers or some aldehydes [29]. Among the cyclic ethers, best yields are obtained with ethylene or trimethylene oxides and, surprisingly, no F.C. reactions seem to have been run with tetrahydrofuran or tetrahydropyran. With unsymmetrical epoxides, the following orientation takes place:

$$\underset{\displaystyle RCH\text{——}CH_2}{\overset{\displaystyle O}{\triangle}} + ArH \longrightarrow \underset{\displaystyle RCHCH_2OH}{\overset{\displaystyle Ar}{|}}$$

And the main side reactions are formation of either the diarylethane or isomerization.

With aldehydes (or ketones) the main reactions with aromatic hydrocarbons are formation of the diarylmethane (or chloromethylarene) (Chapter 7, D.6). On the other hand, α-ketoaldehydes yield acyloins or benzoins [30]:

$$RCOCHO + ArH \rightarrow RCOCHOHAr$$

Low temperature and a moderating solvent such as carbon bisulfide give the best yields, which are in the range 30–90%. The reaction with aldehydes and aromatic hydrocarbons is similar in some respects to the Prins reaction of aldehydes and olefins (B.4).

(a) Preparation of β-phenylethyl alcohol. Benzene, 1.4 l., was mixed with 400 g. of anhydrous aluminum chloride at 6° and the mixture was stirred while 133 g. of ethylene oxide mixed with dry nitrogen was introduced at such a rate that 7–8 g. of the oxide was consumed per hour. The reaction mixture was decomposed with ice, hydrochloric acid was added, and the hydrocarbon layer was separated and fractionated. The β-phenylethyl alcohol, 220 g. (60% based on ethylene oxide) boiled at 221–222° (743 mm.) [32].

(b) Other examples

(1) 2-(p-Aminophenyl)-hexafluoro-2-propanol (60% from hexafluoroacetone and aniline heated to 170–200°, a very specific synthesis) [31].

(2) 2,4,6-Trimethylbenzoin (63% from mesitylglyoxal) [33].

(3) Phenylpivaloylcarbinol (49% from t-butylglyoxal) [34].

1. C. D. Lawrence and C. F. H. Tipper, *J. Chem. Soc.*, 713 (1955).
2. P. E. Peterson and E. V. P. Tao, *J. Org. Chem.*, **29**, 2322 (1964).
3. R. Helg and H. Schinz, *Helv. Chim. Acta*, **35**, 2406 (1952).
4. A. Eschenmoser et al., *Helv. Chim Acta*, **37**, 964 (1954).
5. H. Shechter et al., *J. Am. Chem. Soc.*, **74**, 3052 (1952).
6. D. Swern et al., *J. Am. Chem. Soc.*, **75**, 6212 (1953).
7. H. C. Brown, *Hydroboration*, W. A. Benjamin, New York, 1962, p. 113; G. Zweifel and H. C. Brown, *Org. Reactions*, **13**, 1 (1963).
8. E. F. Knights and H. C. Brown, *J. Am. Chem. Soc.*, **90**, 5283 (1968).
9. F. Sondheimer et al., *J. Org. Chem.*, **24**, 1034 (1959).
10. H. C. Brown et al., *Org. Reactions*, **13**, 30 (1963).
11. H. C. Brown and G. Zweifel, *J. Am. Chem. Soc.*, **81**, 247 (1959).
12. T. J. Logan and T. J. Flautt, *J. Am. Chem. Soc.*, **82**, 3446 (1960).
13. K. A. Saegebarth, *J. Org. Chem.*, **25**, 2212 (1960).

14. H. B. Henbest and B. Nicholls, *J. Chem. Soc.*, 227 (1959).
15. W. C. Baird, Jr., and M. Buza, *J. Org. Chem.*, **33,** 4105 (1968).
16. H. C. Brown and P. Geoghegan, Jr., *J. Am. Chem. Soc.*, **89,** 1522 (1967).
17. E. Arundale and L. A. Mikeska, *Chem. Rev.*, **51,** 505 (1952).
18. C. W. Roberts in G. A. Olah, *Friedel-Crafts and Related Reactions*, Interscience Publ. Co., New York, Vol. 2, Pt. 2, 1964, Chap. 27.
19. L. J. Delby *et al.*, *J. Org. Chem.*, **33,** 4155 (1968).
20. M. G. J. Beets, *Rec. Trav. Chim.*, **70,** 20 (1951).
21. W. S. Emerson *et al.*, *J. Am. Chem. Soc.*, **72,** 5314 (1950).
22. M. G. J. Beets and H. Van Essen, *Rec. Trav. Chim.*, **70,** 25 (1951); **71,** 343 (1952); E. A. Drukker and M. G. J. Beets, *Rec. Trav. Chim.*, **70,** 29 (1951).
23. R. Pautrat *et al.*, *Bull. Soc. Chim. France*, 1182 (1968).
24. *Higher Oxo Alcohols*, Enjay Co., 15 West 51st Street, New York, 1957; C. W. Bird, *Chem. Rev.*, **62,** 283 (1962).
25. G. L. Karapinka and M. Orchin, *J. Org. Chem.*, **26,** 4187 (1961); J. B. Zachry, *Ann. N.Y. Acad. Sci.*, **125,** 154 (1965).
26. H. C. Brown and M. W. Rathke, *J. Am. Chem. Soc.*, **89,** 2737 (1967).
27. M. W. Rathke and H. C. Brown, *J. Am. Chem. Soc.*, **89,** 2740 (1967).
28. H. C. Brown and E. Negishi, *J. Am. Chem. Soc.*, **89,** 5478 (1967).
29. A. Schriesheim and F. Johnson in G. A. Olah, *Friedel-Crafts and Related Reactions*, Interscience Publishers, New York, Vol. 2, Pt. 1, 1964, Chap. 18 and Vol. 4, 1965, Chap. 47.
30. W. S. Ide and J. S. Buck, *Org. Reactions*, **4,** 286 (1948).
31. H. Hopff and K. Koulen, *Chem. Ber.*, **85,** 897 (1952).
32. E. E. Gilbert *et al.*, *J. Org. Chem.*, **30,** 1001 (1965).
33. R. C. Fuson *et al.*, *Org. Reactions*, **4,** 289 (1948).
34. R. C. Fuson *et al.*, *J. Am. Chem. Soc.*, **61,** 1937 (1939).

C. Reduction

The advent of lithium aluminum hydride on the chemical scene in the late forties brought about a revolutionary change in the methods of making alcohols by reduction. Before its use, hydrogenation with a catalyst under pressure had occupied a position of prominence comparable to that of producing hydrogen with an active metal and an alcohol. Although the metal hydrides tend to eclipse the importance of other methods of preparing alcohols by reduction, the methods are sufficiently supplementary and occasionally unique enough to discuss in separate sections. The metal hydride reduction is followed by the Meerwein-Ponndorf-Verley and Cannizzaro reductions because the mechanisms have a hydride group transfer in common. The active metal-alcohol reduction serves as a bridge leading to the discussion of catalytic hydrogenation. This is followed by a section on bimolecular reduction. The inclusion of the benzoin and acyloin condensations may seem surprising in the latter section and could well have been deferred to the condensation part of the chapter. Nevertheless, reduction is involved and the products are related closely enough to include them with this group. As frequently happens, a miscellaneous collection of reductions concludes this section of the preparation of alcohols.

1. From Organic Oxygen Compounds and Metal Hydrides

Several references on metal hydride reductions are available [1]. Lithium aluminum hydride, a most powerful reducing agent, is the most widely used.

It can reduce carbonyl compounds and even carboxylic acid salts to the corresponding alcohols. It dissolves well in ethers, including diglyme. The mechanism of reduction is unclear in some aspects, particularly the first step. A simple hydride transfer may take place:

$$\begin{array}{ccc} \diagdown & & \diagdown \\ \diagup C{=}O & \longrightarrow & \diagup C{-}OLi \\ H_4AlLi & & H \end{array}$$

or coordination with AlH_3:

$$AlH_3 + \begin{array}{c}\diagdown\\ \diagup\end{array}C{=}O \longrightarrow \begin{array}{c}\diagdown\\ \diagup\end{array}C{=}OAlH_3 \longrightarrow \begin{array}{c}\diagdown\\ \diagup\end{array}C{-}OAlH_2 \longrightarrow \text{etc.}$$

All four hydrogen atoms of the hydride are available for the reduction. The drawbacks of lithium aluminum hydride are that it is expensive and hazardous. Water must be avoided at all costs. Carbon dioxide is known to give a very exothermic reaction, and polyhalogen compounds should not be used in large quantities or as a solvent for the hydride. The hydride is best handled by wrapping tightly in aluminum foil and crushing with a rubber hammer. Grinding with a mortar and pestle has been known to inflame a sample. Sodium borohydride ($NaBH_4$) is a much milder reducing agent which can be handled well in anhydrous ethanol but not very well in methanol. Isopropyl alcohol shows even less decomposition than ethanol, but the solubility of the hydride is limited in this solvent. The decomposition in water is inhibited by adding alkali. Since carbonyl compounds are reduced rather specifically and since operating procedures are so simple (see Ex. e.5), sodium borohydride in hydroxylic solvents is the reagent of choice for reduction of these compounds. With hindered ketones a large excess of sodium borohydride may be needed [2], and with α,β-unsaturated aldehydes the trace of alkali impurities in the reagent should be neutralized with carbon dioxide and the reagent used immediately because of its increased instability [3]; see Ex. e.6 for alternative mode of reduction. Lithium aluminum hydride triturated with t-butyl alcohol, $LiHAl(OC(CH_3)_3)_3$, is a mild reducing agent comparable in some respects to sodium borohydride. A comprehensive study of reduction of aldehydes and ketones with lithium trimethoxyaluminum hydride has been carried out. This reagent is more selective because, in contrast to lithium aluminum hydride, only one type of hydride transfer is involved [4]. To summarize some of the more notable points of the utility of metal hydrides in preparing alcohols: Carbonyl compounds, esters, acid chlorides, and acid salts are all reduced by lithium aluminum hydride, while carbonyl compounds alone of this group are reduced by sodium borohydride. Olefins ordinarily are not reduced; thus unsaturated carbonyl compounds can be reduced selectively to the unsaturated alcohols. However, the double bond in allylic (or triple bond in acetylenic alcohols) is capable of reduction by lithium aluminum hydride. For example, cinnamyl alcohol is reduced to hydrocinnamyl alcohol [5]. Furthermore, terminal double bonds of any sort can be reduced by lithium aluminum hydride at about 100°. The hydride solutions are quite basic in nature, and undesirable (or sometimes desirable) condensation reactions of aldehydes or ketones can occur because of this fact. For example, vigorous refluxing of ethyl cinnamates with lithium aluminum hydride leads to 18–45% of the cyclopropane derivative [6]:

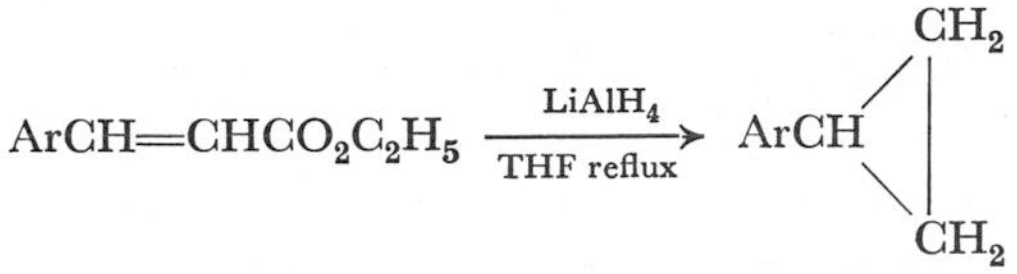

$$ArCH{=}CHCO_2C_2H_5 \xrightarrow[\text{THF reflux}]{\text{LiAlH}_4} ArCH\overset{CH_2}{\underset{CH_2}{\diagup\diagdown}}$$

Coupling can also occur; for example, 9-anthraldehyde and lithium aluminum hydride yield 60% 1,2-dianthrylethane [7].

Both lithium aluminum hydride and sodium borohydride have been employed with aluminum chloride as reducing agents. The former offers an advantage in the reduction of cyclic ketones [8] in that it permits the formation of equatorial alcohols if sufficient steric hindrance exists in the aluminum chloride complex. In this manner *trans*-4-*t*-butylcyclohexanol was prepared in 73–78% yield, as shown in the accompanying formulation. The latter in

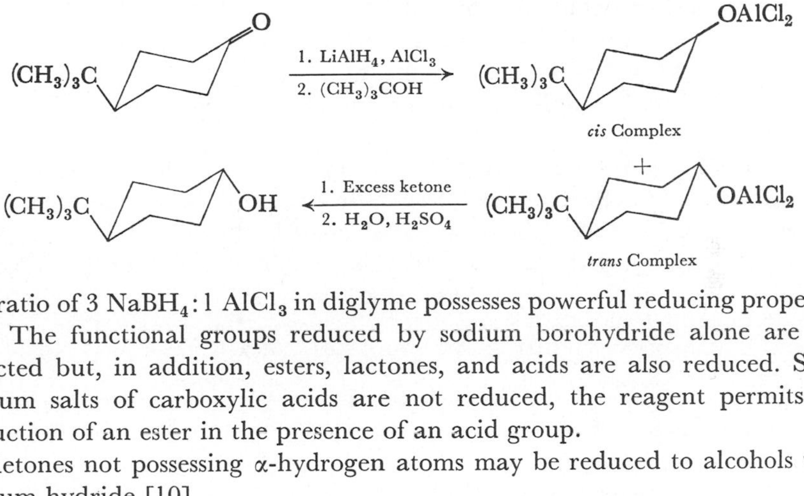

cis Complex

trans Complex

the ratio of 3 $NaBH_4$: 1 $AlCl_3$ in diglyme possesses powerful reducing properties [9]. The functional groups reduced by sodium borohydride alone are still affected but, in addition, esters, lactones, and acids are also reduced. Since sodium salts of carboxylic acids are not reduced, the reagent permits the reduction of an ester in the presence of an acid group.

Ketones not possessing α-hydrogen atoms may be reduced to alcohols with sodium hydride [10].

$$(C_6H_5)_2C{=}O \xrightarrow[\text{Xylene, 145°}]{\text{NaH}} (C_6H_5)_2CHOH$$
Benzhydrol, 83%

(a) Preparation of 2,2-dichloroethanol (64–65% from dichloroacetyl chloride and $LiAlH_4$) [11].

(b) Preparation of 2,2-diphenylethanol. Diphenylacetaldehyde (6.0 g., 0.031 mole) in 50 ml. of ether was reduced with 45 ml. of an ethereal solution of lithium aluminum hydride (0.008 mole + 10% excess) to produce

5.95 g. (97.5%) of the diphenylethanol, b.p. 178–180° (13 mm.), m.p. 59-61° (from petroleum ether) [12].

(c) Preparation of cyclobutanol. Cyclobutanone, 30 g. (0.43 mole) was added to a stirred mixture of 5 g. (0.125 mole) of $LiAlH_4$ in 200 ml. of dry ether, and the whole was refluxed for 1 hr. Enough 10% H_2SO_4 was added with stirring to dissolve the solids and the cyclobutanol was extracted with ether. Distillation gave 8.6 g. (b.p. 121–125°) and 19.1 g. (b.p. 125°) of cyclobutanol (90%) [13].

(d) Preparation of p-chlorobenzyl alcohol. Sodium borohydride, 0.25 mole, in 250 ml. of diglyme was treated with 0.25 mole of finely ground $MgBr_2$, after which the mixture was stirred for 30 min. Then 0.40 mole of ethyl p-chlorobenzoate was added and the mixture was heated on a steam bath for 3 hr. After pouring the contents into a mixture of 500 g. of crushed ice and 50 ml. of concentrated HCl, flakes of the acid appeared. These flakes, when washed with water and dried, weighed 51.7 g. (91%). m.p. 73–74° [14].

(e) Other examples

(1) Benzyl alcohol (72% from benzoyl chloride and lithium aluminum hydride [14a]; (92% from benzoyl chloride and t-butylamine borane) [15].

(2) Bis-(hydroxymethyl)-terphenyl (81% from terphenyl dicarbonyl chloride and lithium aluminum hydride) [16].

(3) 2-(1-Pyrrolidyl)-propanol (80–90% from ethyl α-(1-pyrrolidyl) propionate and lithium aluminum hydride) [17].

(4) 1-Octadecanol (74% from ethyl stearate, sodium borohydride, and $MgCl_2$) [14].

(5) Cinnamyl alcohol (97% from cinnamaldehyde and sodium borohydride, 10% in methanol stabilized with alkali; lithium aluminum hydride reduces the aldehyde to γ-phenylpropyl alcohol) [18].

(6) Cinnamyl alcohol (over 90% on a millimole scale from cinnamaldehyde and aluminum hydride).   The latter is made simply by the reaction

$$3\,LiAlH_4 + AlCl_3 \rightarrow 4\,AlH_3 + 3\,LiCl$$

This reaction appears to be the method of choice for the conversion of α,β-unsaturated carbonyls into unsaturated carbinols and perhaps functional groups other than the carbonyl [19].

(7) p-Hydroxybenzyl alcohol (95% calculated, from the aldehyde and cetyltrimethylammonium borohydride in benzene at 65° for 3 hr., a demonstration of reduction in a nonpolar solvent, the quaternary salt being synthesized readily by metathesis) [20].

2. From Cyclic Ethers and Metal Hydrides

$$R-CH\!-\!\!-CH_2 \xrightarrow{LiAlH_4} RCH_2CH_2OH + RCHOHCH_3$$
$$\diagdown\!\!\diagup$$
$$O$$

The epoxy ring is opened by reduction with lithium borohydride or lithium aluminum hydride. Cleavage may occur to give either a primary or a secondary alcohol. With the hydride alone, the secondary alcohol is usually produced in excess [21, 22]. However, lithium aluminum hydride in the presence of aluminum chloride or bromide gives the primary alcohol in almost quantitative yield [22].

Diborane reduces epoxides to alcohols derived predominantly from anti-Markownikoff ring opening [23]:

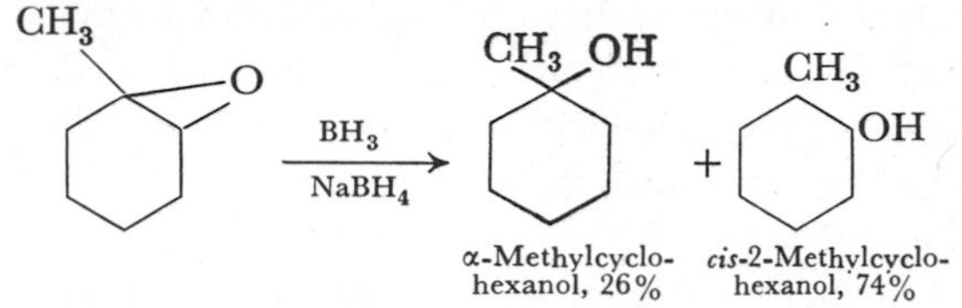

CH3
O
BH3
NaBH4
CH3 OH
CH3
OH
α-Methylcyclo-hexanol, 26%
cis-2-Methylcyclo-hexanol, 74%

Sodium borohydride markedly accelerates the reaction. Another example of the reduction is shown [24]:

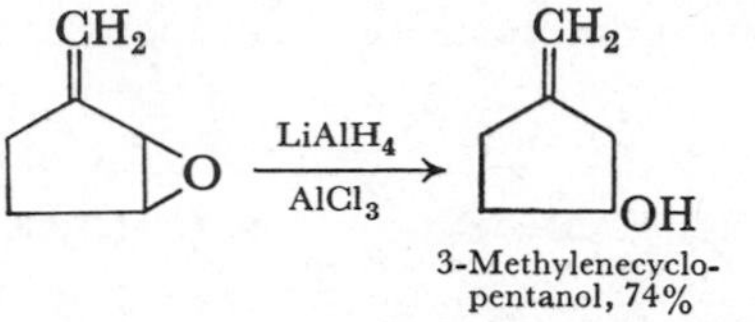

CH2
O
LiAlH4
AlCl3
CH2
OH
3-Methylenecyclo-pentanol, 74%

(a) Preparation of 2-(4-methoxyphenyl)-ethanol. Twenty-three grams of *p*-methoxystyrene oxide was reduced with 6.0 g. (720%) of lithium borohydride in ether. The product weighed 16.4 g. (70%), distilled at 112–115° at 2.5 mm., and melted at 22.6–23.2°. The melting-point curve constructed by the use of authentic 1-(4-methoxyphenyl)-ethanol and 2-(4-methoxyphenyl)-ethanol indicated that the product contained approximately 95% of the primary alcohol [21].

(b) Preparation of 2-butanol. A solution of 14.4 g. of 1,2-epoxybutane in ether was reduced by 3.3 g. (300%) of lithium borohydride. Fractional distillation gave 8.5 g. (57%) of 2-butanol, b.p. 98°, n_D^{20} 1.3957, with no trace of 1-butanol [21].

(c) Other examples

(1) Various *p*-substituted 2-phenylethanols [21].

(2) Various primary and secondary alcohols [25].

3. From Carbonyl Compounds and Other Reducing Agents (Mainly Meerwein-Ponndorf-Verley)

C=O
Al(OCH(CH3)2)2
CHOH + (CH3)2C=O
H
O
C
(CH3)2

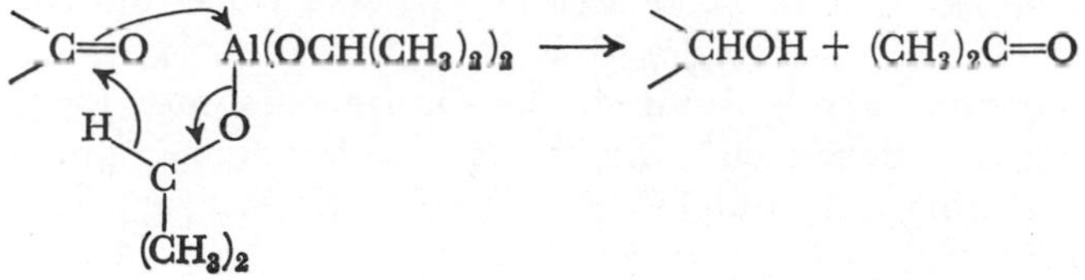

This reaction is similar to the hydride reductions, but the hydride group is supplied from a hydrogen attached to the carbinol group of isopropyl alcohol rather than from a hydrogen attached to a metallic atom. Although the mechanism of reduction is shown above to go through a cyclic transition state, a noncyclic transition state involving 2 moles of alkoxide may compete [26]. The reaction has been discussed in detail [27]. Although the method of reduction has been superseded largely by those in which metal hydrides are employed, it is often quite satisfactory for aliphatic and aromatic aldehydes and ketones. Aluminum isopropoxide (in isopropyl alcohol) is the preferred metal alkoxide, and some degree of selectivity is possible since double bonds, carboxylic esters, nitro groups, and reactive halogens are usually not affected. The success of the reaction depends on driving the equilibrium to completion by removing the low-boiling acetone formed. The yields are usually good. With ketones containing an amino group which can complex with aluminum salts, sodium rather than aluminum isopropoxide is more efficacious (see Ex. c.3).

Unrelated to the MPV reduction, but nevertheless giving the same results, is the biological fermentation of carbonyl compounds [28].

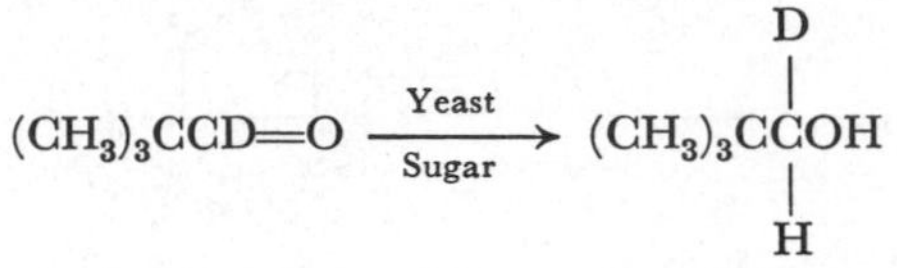

$$(CH_3)_3CCD{=}O \xrightarrow[\text{Sugar}]{\text{Yeast}} (CH_3)_3C\overset{\displaystyle D}{\underset{\displaystyle H}{C}}OH$$

α-d-Neopentyl alcohol,
optically active,
unstated yield

Benzils are reduced readily to benzoins by sodium dithionite in aqueous alcohol [29].

Although many other chemical methods of reducing carbonyl compounds are known (see Ex. c.4), they are rarely used because of the efficiency of sodium borohydride or catalytic reduction.

(a) Preparation of benzyl alcohol (89 % from the aldehyde, aluminum isopropoxide, and isopropyl alcohol using the Hahn condenser which partially condenses the vapor and gives a distillate enriched in acetone) [27].

(b) Preparation of benzhydrol (99 % from the ketone as in Ex. a) [27].

(c) Other examples

(1) Crotyl alcohol (60 % from the aldehyde, isopropyl alcohol, and aluminum isopropoxide prepared *in situ*: a Vigreux column is used to remove acetone) [30].

(2) Trichloroethyl alcohol (84 % from chloral, ethanol, and aluminum ethoxide; because no aldol condensation can take place, the ethoxide is as effective here as the isopropoxide) [31].

(3) Quinine and quinidine, the diastereoisomer (30 % and 60 % yields, respectively, by reduction of quininone with sodium isopropoxide) [32].

(4) n-Heptyl alcohol (75–81 % from the aldehyde, Fe filings, and aqueous acetic acid) [33].

4. From Aldehydes (Cannizzaro)

$$2\,RCHO \xrightarrow{NaOH} RCH_2OH + RCOONa$$

This synthesis has been discussed in detail [34]. It involves the reaction of aldehydes having no α-hydrogen atoms with alkali to give a primary alcohol and the salt of the corresponding acid. Since one molecule of the aldehyde is reduced while another is oxidized, the return of alcohol on the basis of the total amount of aldehyde used would not exceed 50%. This fact often makes the method unattractive. The reaction, however, requires little time, no special reagents, and the apparatus is simple. Finely divided nickel accelerates it [35]. If two different aldehydes are employed, the so-called crossed Cannizzaro results. A common procedure here is to use formaldehyde as one of the components, in which case the other aldehyde is reduced to a primary alcohol in excellent yield [36].

The mechanism of the Cannizzaro reaction may be represented as shown in the accompanying formulations. The most important industrial example of the

$$
\begin{array}{c}
& & & OH \\
& & & | \\
RC{=}O + \overset{\ominus}{O}H & \rightleftharpoons & R{-}C{-}O^{\ominus} \\
| & & | \\
H & & H
\end{array}
$$

$$
\begin{array}{c}
OH \quad\quad H \quad\quad\quad OH \quad\quad H \\
| \quad\quad\quad | \quad\quad\quad\quad | \quad\quad\quad | \\
R{-}C{-}O^{\ominus} + R{-}C{=}O \longrightarrow R{-}C{=}O + R{-}C{-}O^{\ominus} \xrightarrow{Fast} RCO_2{}^{\ominus} + RCH_2OH \\
| \quad\quad\quad\quad\quad\quad\quad\quad\quad\quad\quad\quad\quad | \\
H \quad\quad\quad\quad\quad\quad\quad\quad\quad\quad\quad H
\end{array}
$$

reaction is the preparation of pentaerythritol from acetaldehyde and large amounts of formaldehyde, the first steps being successive Tollens' condensations and the last step being a crossed Cannizzaro between trimethylolacetaldehyde and formaldehyde.

(a) **Preparation of m-hydroxybenzyl alcohol** (normal Cannizzaro) (93.6% from m-hydroxybenzaldehyde together with 94.2% of m-hydroxybenzoic acid) [37].

(b) **Preparation of p-tolylcarbinol** (crossed Cannizzaro) (90% from p-tolualdehyde) [38].

(c) **Other examples**

(1) **4-Quinolylmethanol** (87% from quinoline-4-aldehyde) [39].

(2) **p,p'-Xylylenediol** (68% from terephthalaldehyde and an eightfold excess of formaldehyde) [40].

5. From Esters (Bouveault-Blanc) or Carbonyl Compounds and Active Metal-Alcohols

Carboxylic Acids or Esters (Bouveault-Blanc)

$$RCOOH(R') \rightarrow RCH_2OH$$

One of the older methods of reducing esters to primary alcohols has employed sodium and dry alcohol. In its application to fats and oils, the method gives good yields by using theoretical amounts of both sodium and reducing alcohol, and an inert solvent such as toluene or xylene [41]. An ethanolic solution of phenol improves the yield in the reduction of amino acid esters, while the reduction of aromatic esters is improved by the addition of quinoline or tetrahydroquinoline [42]. A more recent method is the reduction of esters with amalgamated aluminum as shown in Ex. b.3.

In the case shown, sodium-absolute ethanol-liquid ammonia proved to be

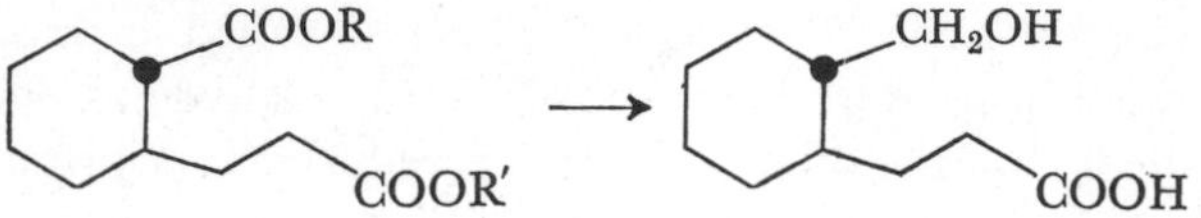

trans-B-(2-Hydroxymethylcyclohexane) propionic acid

superior to the customary sodium and dry ethanol of the Bouveault-Blanc procedure [43].

(a) **Preparation of oleyl alcohol** (49–51 % from ethyl oleate and absolute ethanol to which sodium metal was added) [44].

(b) **Other examples**

(1) **γ-Phenylpropyl alcohol** (90 % from ethyl hydrocinnamate, ethanol, phenol, and a small amount of quinoline to which sodium was added; sodium-alcohol reduction gave only 22 % yield) [42].

(2) **Alcohols from fatty acid esters** [41].

(3) **Phthalyl alcohol** (about 60 % from ethyl phthalate and amalgamated aluminum in alcohol) [45].

Carbonyl Compounds

Although aldehydes and ketones are now usually reduced to primary and secondary alcohols by catalytic hydrogenation or by the use of metallic hydrides, other reducing agents may be employed, particularly for ketones. For example, benzophenone has been reduced to diphenylcarbinol by sodium amalgam, calcium, or magnesium and ethanol, zinc, aluminum, or sodium in strongly alkaline solutions, and photochemically in sodium isopropoxide solution. These reducing agents are effective in that good yields are usually obtained, but they lack the specificity of the modern ones.

(a) **Preparation of benzhydrol, diphenylcarbinol** (95–97 % from the ketone, zinc dust, and aqueous sodium hydroxide) [46].

(b) **Other examples**

(1) **2-Heptanol** (62–65 % from the ketone, sodium, and alcohol) [47].

(2) **Xanthhydrol** (91–95 % from xanthone and sodium amalgam in ethanol at 60–70°) [48].

6. From Organic Oxygen Compounds and Hydrogen with Catalysts

Catalytic hydrogenation is a well-explored field for which general references are available [49]; for high-pressure hydrogenation [50], for low-pressure hydrogenation [51, 52]. The advantages of catalytic hydrogenation in the preparation of alcohols are that it is simple, clean, and frequently quantitative. The disadvantages are that it is nonspecific (olefins are reduced), susceptible to catalyst poisoning, and capable of hydrogenolyzing some alcohols, such as benzyl types and phenyl tosylates, to hydrocarbons [53].

Esters

Catalytic reduction of acids or esters to primary alcohols has been accomplished with catalysts such as Raney nickel or copper chromite at high pressure and high temperature. Methods for the preparation of these catalysts have been published: Raney nickel with different degrees of activity labeled W-1, W-2, W-3, W-4, W-5, W-6 and W-7 [54, 55]; copper-barium-chromium oxide [56, 57]. These methods require special equipment and special techniques and are therefore not so easily carried out as if a metal hydride were employed.

Aldehydes and Ketones

$$\diagdown_{\diagup}\!C{=}O \xrightarrow[\text{Catalyst}]{H_2} \diagdown_{\diagup}\!CHOH$$

A wide variety of catalysts may be used for this facile reduction: platinum from chloroplatinic acid [58, 59] or from ammonium chloroplatinate [60, 61], Raney nickel as already described [54, 55], palladium [62], copper-barium-chromium oxide [56, 57], ruthenium or rhodium on carbon (for ketones in neutral or basic medium) [63]. Triethylamine to which platinic chloride is added has a marked promoter effect on carbonyl reductions with Raney nickel [64]. Also, it is claimed that a more active catalyst than W-7 Raney nickel can be obtained by adding aqueous sodium hydroxide to the nickel alloy rather than adding the alloy to the alkali [65]. It should be noted that the alkali in excess is known to deactivate Raney nickel.

In the low-pressure reduction of aldehydes the platinum black catalyst becomes deactivated during reduction, but can be reactivated by exposure to air [66]. The deactivation can be prevented by introducing a small amount of air with the hydrogen [67] or by the use of platinic oxide and a trace of stannous chloride [52, p. 239].

The rates of platinum-catalyzed hydrogenation of ketones in trifluoroacetic acid at atmospheric pressure are approximately three times those of comparable hydrogenations in acetic acid. A relatively high concentration of the ketone is desirable for a rapid reaction [68]. Palladium, platinum, rhodium, and ruthenium, each on high surface carbon, have been compared as catalysts for rapid and selective hydrogenation of ketones [63]. For cyclopentanone, cyclohexanone, and methyl isobutyl ketone, platinum in aqueous acid and rhodium and ruthenium in neutral or basic solution are the preferred catalysts. For an α,β-unsaturated ketone, such as mesityl oxide, these catalysts lead first to the saturated ketone which is then reduced to the secondary alcohol. For an aromatic ketone, such as acetophenone, there is a tendency for hydrogenation to occur at more than one site but, of the four catalysts, platinum is best for the production of the secondary alcohol.

As a rule the yields are high in these hydrogenations, and some selectivity is possible by the choice of catalysts and other reducing agents. A zinc-copper catalyst has been reported to reduce selectively the carbonyl group of unsaturated ketones [69]. However, the hydride reduction is so outstanding in this selective reduction that a hydrogenation method hardly seems necessary (see C.1, Ex. e.5, and e.6).

The addition of hydrogen by catalytic low-pressure hydrogenation is *cis*, and it occurs from the more accessible side of the double bond [70]. Alkali in Raney nickel or Adams' catalyst can cause isomerization however. The orientation with regard to hydrogenation of cyclohexanones by hydride addition has been discussed rather thoroughly [71]. For 2- and 4-substituted cyclohexanones, the *trans*-alcohols are obtained; for 3-substituted, the *cis* alcohol. This addition is as though the hydrogen atom entered the axial position of the carbon attached to the oxygen atom. The production of cyclohexanols (including fused-ring six-membered cycloalkanols) is carried out by high-pressure hydrogenation of the corresponding phenol. Hydrogenolysis of the higher-molecular-weight alkyl groups of an alkylphenol can be prevented by the addition of pyridine to the hydrogenation mixture [72].

(a) **Preparation of piperonyl alcohol** (3,4-methylenedioxybenzyl alcohol). Piperonal, 90 g., in 70 ml. of purified dioxane was hydrogenated under 300 atm. pressure of hydrogen at 160–175° over 5 g. of copper chromite catalyst (prepared as shown in note 11) [56]. After hydrogenation for $1\frac{1}{4}$ hr. at 160–175°, distillation gave 90 g. (98%) of piperonyl alcohol, b.p. 125–128° (6 mm.) [73; *see also* 74].

(b) **Preparation of α-phenylpropyl alcohol**

(1) **Palladium black.** About 0.2 g. of palladium oxide (PdO) in 100 ml. of water was shaken for 45 min. under a hydrogen pressure of 2.5 atm. The palladium black resulting was washed with water on a glass suction filter and then with methanol in such a way that it was always covered with liquid.

(2) **α-Phenylpropyl alcohol.** Propiophenone, 26.8 g. (0.2 mole), in 50 ml. of methanol in the presence of 0.5 g. of palladium black and 0.0015 mole of nicotinic acid diethylamide, was hydrogenated at 20° and 1 atm. After the absorption of $\frac{1}{5}$ mole of hydrogen in 4 hr., the hydrogenation was terminated. On fractional distillation the filtrate of the mixture gave 26 g. (96%) of α-phenylpropyl alcohol, b.p. 140° (12 mm.) [75].

(c) **Other examples**

(1) **Raney nickel catalyst**

(i) **2,2-Dimethyl-1-pentanol** (94% from 2,2-dimethyl-4-pentenal) [76].

(ii) **2,2-Dimethyl-3-diethylaminopropanol** (80–90% from 2,2-dimethyl-3-diethylaminopropanal) [77].

(iii) **Cyclohexylmethylcarbinol** (96% from cyclohex-3-enyl methyl ketone) [78].

(2) Platinum oxide catalyst

(i) **Anisyl, p-bromobenzyl-, n-butyl, o-chlorobenzyl, p-chlorobenzyl, n-heptyl, p-methoxybenzyl and salicyl alcohols** [74].

(ii) **3β,17β-Diacetoxy-7α-hydroxyandrostane** (78.5% from 3β, 17β-diacetoxy-7-ketoandrostane) [79].

(3) Copper chromite catalyst

(i) **1,3-Difuryl-1-propanol** (91% from furfurylideneacetofuran and copper-chromium oxide catalyst at 100° and 1000 p.s.i. of hydrogen; further hydrogenation with Raney nickel reduced the furan ring system) [80].

(4) Ruthenium on carbon

(i) **Tetramethyl-1,3-cyclobutanediol** (98% from tetramethyl-1,3-cyclobutanedione in methanol by hydrogenation in the presence of ruthenium

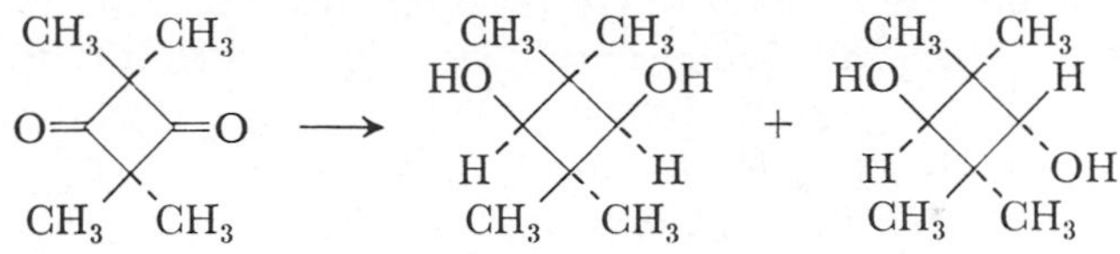

on carbon at 125° and 1000–1500 p.s.i. for 1 hr.; the product was a mixture of about equal parts of *cis*- and *trans*-diol) [81].

7. From O-heterocycles (Hydrogenolysis)

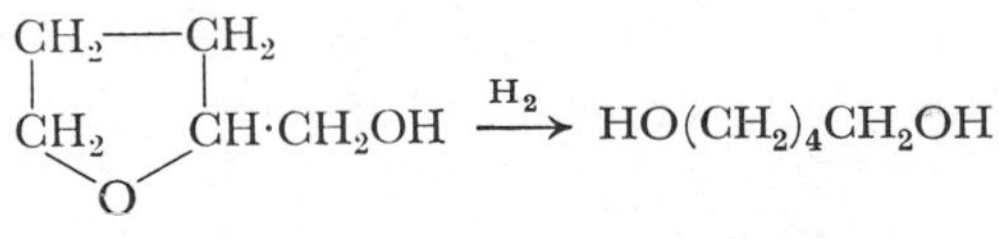

Furans, pyrans, lactols [82, 83], and other oxygen-containing rings may be cleaved under reducing conditions to produce mono- or polyhydric alcohols. Although oxygen-containing substituents are often present in the original heterocycle, such substituents are not necessary, for alkyl furans have been converted into dihydric alcohols by hydrogenation in acid media [84]. The completely reduced furan may be obtained by hydrogenation in the presence of copper chromite [85] at temperatures around 150° but, for ring opening, higher temperatures or the use of a metallic hydride, such as lithium aluminum hydride, at least with thioketals, must be employed. The yields are satisfactory.

(a) **Preparation of 1,5-pentanediol** (40–47% from tetrahydrofurfuryl alcohol, if unreacted alcohol is disregarded, with copper chromite at 250° and 6000 p.s.i. of hydrogen) [86].

(b) **Other examples**

(1) **4,5-Phenanthrenedimethanol** (90% from pseudo ethyl 5-formyl-4-phenanthrenecarboxylate and lithium aluminum hydride) [82].

(2) **4-Hydroxybutyl (or 5-hydroxypentyl) alkyl thioethers** (about 60% from 2-tetrahydrofuranyl (or pyranyl) alkyl thioethers and lithium aluminum hydride and aluminum chloride [87].

$$\text{(tetrahydrofuranyl-SR)} \xrightarrow[\text{AlCl}_3]{\text{LiAlH}_4} \text{HO(CH}_2)_4\text{SR}$$

8. From Carbonyl Compounds or Esters (Mostly Bimolecular Reduction)

The synthesis of pinacols and acyloins (or enediols) is included in this section. The starting materials, carbonyl compounds or esters, rather than being fully reduced to the corresponding alcohol, are partly reduced to an intermediate stage (usually a free radical) from whence a coupling reaction occurs.

Pinacol Reaction [88]

$$2\,R_2C\!\!=\!\!O \longrightarrow R_2C\underset{\underset{OH}{|}}{\quad}\!\!\!-\!\!\!\underset{\underset{OH}{|}}{C}\!\!-\!\!R_2$$

Aliphatic ketones are reduced to glycols by the use of active metals such as sodium, magnesium, or aluminum amalgams of these metals [89]. The yields in these reactions are generally less than 50%. In the reduction of aromatic carbonyl compounds, magnesium and magnesium iodide [90, 91], alkali metals [92], or electrolytic methods [93] are employed. With sodium or magnesium and magnesium iodide, metal ketyls such as

$$\underset{Ar_2\overset{\overset{\displaystyle Na}{|}}{C}ONa}{} \quad or \quad \underset{Ar_2\overset{\overset{\displaystyle MgI}{|}}{C}OMgI}{}$$

appear to be intermediates. These couple (or add to another molecule of ketone) to give pinacolates which hydrolyze to pinacols such as

$$Ar_2C\underset{\underset{OH}{|}}{\quad}\!\!\!-\!\!\!\underset{\underset{OH}{|}}{C}Ar_2$$

The yields are often high. Some benzophenones are best reduced to benzopinacols by exposing them to ultraviolet light in isopropyl alcohol solution:

$$(C_6H_5)_2C\!\!=\!\!O \xrightarrow{h\nu} (C_6H_5)_2C\!\!=\!\!O^* \text{ (activated state)} \xrightarrow{CH_3CHOHCH_3}$$

$$[(C_6H_5)_2\overset{\displaystyle\cdot}{C}\!\!-\!\!OH] + [CH_3\overset{\overset{\displaystyle O}{\|}}{C}CH_3] \xrightarrow[\text{4-IV}]{(C_6H_5)_2C=O} [(C_6H_5)_2\overset{\overset{\displaystyle O}{|}}{C}\!\!-\!\!\overset{\overset{\displaystyle \overset{\cdot}{O}}{|}}{C}(C_6H_5)_2] \xrightarrow{\text{4-IV}}$$

$$CH_3\overset{\overset{\displaystyle O}{\|}}{C}CH_3 + (C_6H_5)_2C\underset{\underset{H}{|}}{\overset{\overset{O}{|}}{\quad}}\!\!\!-\!\!\!\underset{\underset{H}{|}}{\overset{\overset{O}{|}}{C}}(C_6H_5)_2$$

Other sequences may predominate, but the overall stoichiometry is

$$2(C_6H_5)_2C{=}O + CH_3CHOHCH_3 \rightarrow (C_6H_5)_2COHCOH(C_6H_5)_2 + CH_3COCH_3$$

The free radical 4-IV is so reactive that it rarely participates in coupling but rather reduces benzophenone or other species. A variety of ketones undergo the photochemical reduction, but fluorenone and xanthone do not [94].

Mixed pinacols can also be made; for example [95],

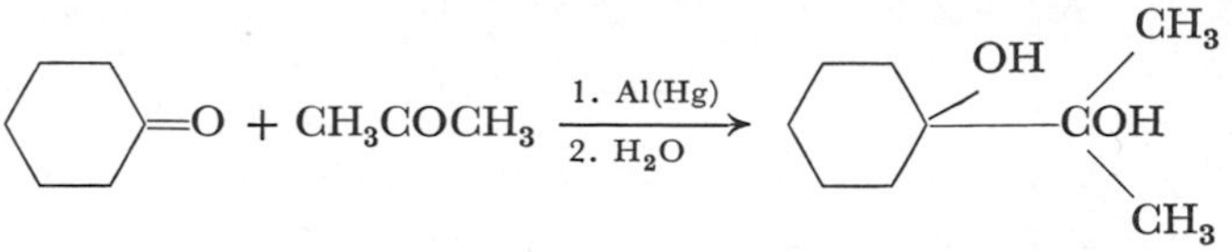

2-(1-Hydroxycyclohexyl)-2-propanol

Related to the pinacol reaction is the Emmert synthesis of heterocyclic carbinols [96]. The yields are quite low, but are compensated for in part by the

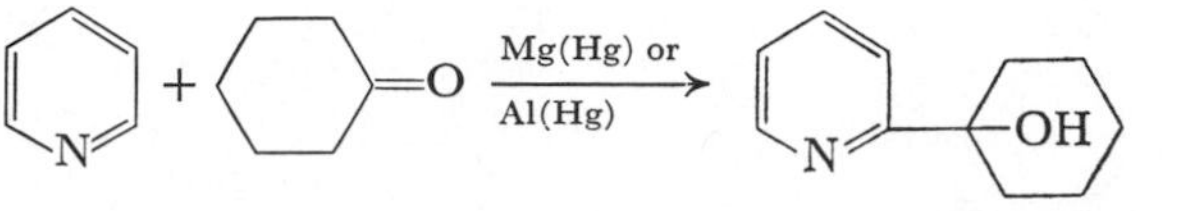

29%
2-(1-Hydroxycyclohexyl)-pyridine

simplicity of the reaction. A variety of carbonyl compounds or heterocycles can be used, but the yields are no better, and usually poorer, than the example given.

Still another coupling type of reaction is the Hammick synthesis as discussed in G.6. Yields are usually better than in the Emmert synthesis.

(a) Preparation of benzopinacol (95% from benzophenone and 2-propanol in the sunlight) [97].

Benzoin Condensation

$$2\ ArCHO \xrightarrow{KCN} ArCOCHOHAr$$

This condensation, in which the cyanide ion is almost a specific catalyst, has been described in some detail [98]. The mechanism may be represented as follows:

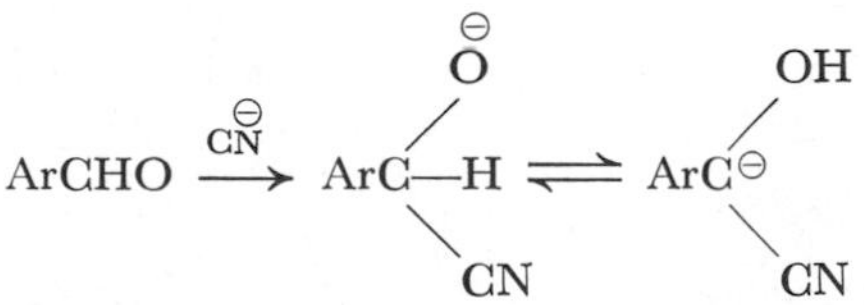

4-V

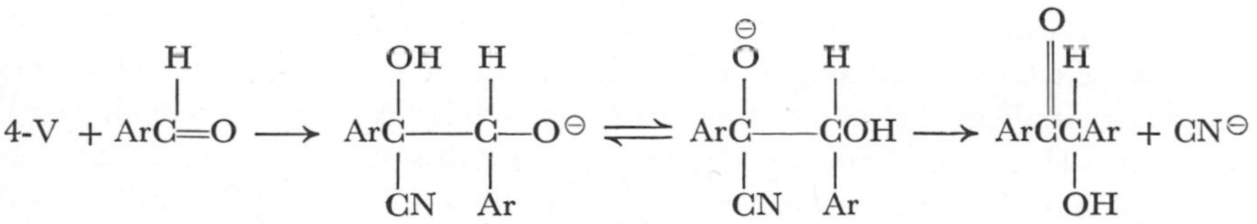

The catalyst must be able to add to the carbonyl group and activate the hydrogen so that the anion 4-V is formed to add to a second molecule of aldehyde. The only other catalysts with this ability seem to be thiazolium salts, the addition product of which then condenses with a second aldehyde group [99].

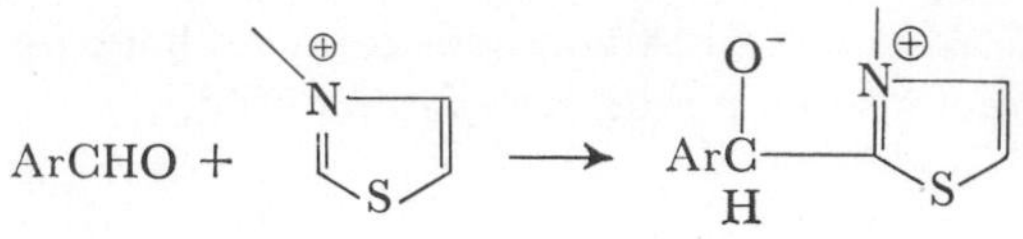

The alkyl-, ethoxy-, halo-, hydroxy-, and aminoaldehydes do not form symmetrical benzoins easily. Among them, greater success is often achieved in forming unsymmetrical benzoins, and of the two benzoins possible, usually only one is obtained. It is the isomer where resonance interaction of an electron-release group occurs with the carbonyl group. Thus

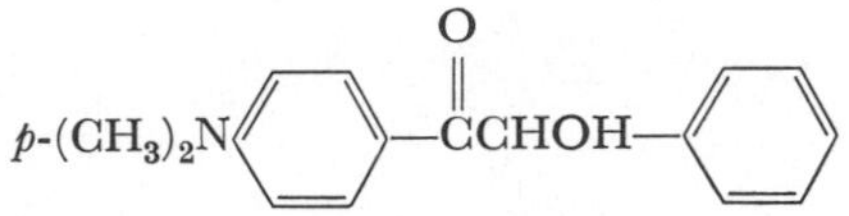

is the product of a mixed condensation. The prediction of which isomer will be obtained is not so easily made in other cases.

The reaction is carried out commonly in aqueous ethanol as the solvent. Sodium or potassium cyanide may be employed; the yields vary from poor to good.

(a) **Preparation of benzoin** (90–92 % from benzaldehyde and sodium cyanide in aqueous alcohol) [100].

(b) **Other examples**

(1) **2′-Chloro-3-ethoxy-4-methoxybenzoin** (81 % from a mixture of o-chloro- and 3-ethoxy-4-methoxy-benzaldehydes and sodium cyanide in aqueous alcohol; benzoins are persistently yellow when crystallized from alcohol, but colorless when crystallized from acetic acid) [101].

(2) **4,4′-Distyrylbenzoin** (85 % from the aldehyde and potassium cyanide in aqueous alcohol) [102].

Benzoin Condensation for Heterocyclic 2-Carboxaldehydes

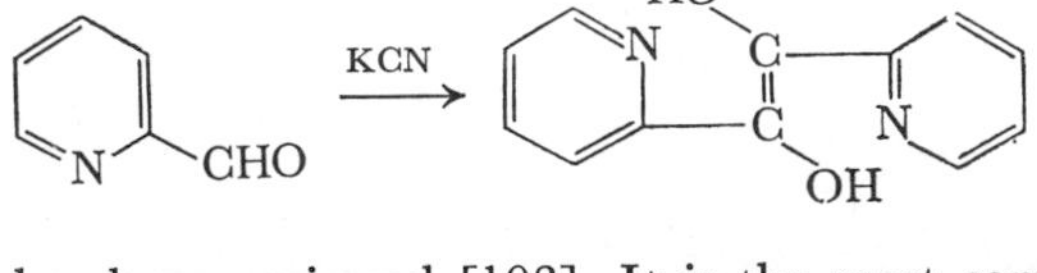

This synthesis has been reviewed [103]. It is the most common method of preparing chelated 1,2-enediols, which are isomeric with benzoins. That the method is applicable only to 2-heterocyclic aldehydes, most of which rings contain a basic nitrogen atom, suggests that the stability of the enediol is due to chelation. Thus the enediol is actually a condensed four-ring system, two of which are chelate rings and all of which contain six members. The reaction may be carried out in alcohol or pyridine as a solvent, and the yields are usually in the range 75–90%. Other reagents such as acetic acid [104], boron trifluoride [105], hydrogen cyanide [106], and methyl heterocycles [107] have been used to effect this condensation.

(a) Preparation of 1,2-di(2-pyridyl)-1,2-ethenediol. A solution of 30 g. of freshly distilled pyridine-2-aldehyde, 50 ml. of pyridine, and 200 ml. of water was heated on a steam bath, and 2 g. of potassium cyanide in 10 ml. of water was added with stirring. The mixture turned dark with the formation of a precipitate, and heating was continued for 30 min. Cooling gave a solid mass which was filtered and washed with about 15 ml. of cold ethanol. Crystallization from pyridine yielded 28.5 g. (95% of yellow to orange needles, m.p. 156°) [108].

(b) Other examples

(1) 1,2-Di-(2-benzothiazolyl)-1,2-ethenediol (90% from 2-benzo-thiazole aldehyde and potassium cyanide in aqueous methanol) [109].

Acyloin Condensation

$$2\ RC\!\!\overset{O}{\underset{}{\diagup}}\!\!OR' \xrightarrow{4\ Na} \begin{matrix} RC-ONa \\ \| \\ RC-ONa \end{matrix} \xrightarrow{H_2O} \begin{bmatrix} RC-OH \\ \| \\ RC-OH \end{bmatrix} \longrightarrow \begin{matrix} RC=O \\ | \\ RCHOH \end{matrix} + 2\ R'ONa$$

This synthesis has been reviewed [110]. It is the principal method of preparing acyloins in which the alkyl groups are identical. Solvents such as ether, benzene, xylene, or an excess of the ester are employed, and the reaction is carried out at the boiling point of the solvent. Sodium in liquid ammonia and ether has given excellent results in the synthesis of steroid-type acyloins [111]. The time required for completion of the reaction varies greatly and it is essential to remove the ammonia before acidification of the acyloin salt [112]. Large-ring acyloins are best prepared by the sodium condensation. Evidently, the ester groups are brought into close proximity on the surface of the sodium, and thus ring formation rather than intermolecular condensation is favored [113].

The most recent advance in acyloin synthesis is the trapping of the enediol as the trimethylsilyl ether, a fact which extends the range of synthesis and perhaps increases the yield [114].

$$CH_3O_2C(CH_3)_2C(CH_3)_2CO_2CH_3 \xrightarrow[2\ (CH_3)_3SiCl]{Na,\ toluene} \begin{matrix} -OSi(CH_3)_3 \\ -OSi(CH_3)_3 \end{matrix} \xrightarrow[THF,\ N_2]{HCl} \begin{matrix} =O \\ -OH \end{matrix}$$

84%, unstable, use immediately

2-Hydroxy-3,3,4,4-tetramethyl-cyclobutanone, 72%

Mixed acyloins from condensation of two different esters give at best 20–25% of the mixed acyloin [115].

(a) Preparation of butyroin (65–70% from ethyl butyrate and finely divided sodium in ether) [116].

(b) Other examples

(1) **Lauroin** (80–90% from methyl laurate and finely divided sodium in xylene) [117].

$$C=O$$

(2) **Sebacoin**, $(CH_2)_8$—CHOH (66% from dimethyl sebacate and sodium in xylene under nitrogen) [118].

Reduction of Acid Chlorides or Benzils

$$ArCOCl \xrightarrow{Mg + MgI_2} \begin{bmatrix} \overset{\displaystyle Cl}{\underset{\displaystyle Cl}{\overset{|}{\underset{|}{Ar-C-OMgI}}}} \\ Ar-C-OMgI \end{bmatrix} \longrightarrow \begin{matrix} ArC=O \\ | \\ ArC=O \end{matrix}$$

$$\xrightarrow[\text{2. } H_2(Pt)]{\text{1. } Mg + MgI_2 + H_2O} \quad \begin{matrix} ArC-OH \\ \| \\ ArC-OH \end{matrix}$$

This synthesis, which is restricted to the formation of hindered enediols, has been reviewed [103]. Magnesium-magnesium iodide is employed as a reducing agent for acid chlorides, and the same reducing agent or hydrogen in the presence of platinum is used for benzils. As indicated in the formulation above, the benzil appears to be an intermediate in the reduction of the acid chloride. Somewhat better yields are obtained from the benzil, but as a rule the acid chloride is more readily available. Both *cis* and *trans* forms may be obtained in these reductions. For example, 2,2′,6,6′-tetramethylbenzil when hydrogenated in methanol containing a drop of acetic acid gave the *cis*-enediol almost quantitatively in 10 minutes. In the absence of acid, the same procedure gave the *trans* form almost quantitatively in 12 hours.

(a) **Preparation of** *cis*-**1,2-(di-2,6-dimethylphenyl)-1,2-ethenediol.** A mixture of 0.5 g. of 2,6-xylil, 50 ml. of methanol, 1 drop of glacial acetic acid, and 0.1 g. of platinum oxide was hydrogenated until the yellow color disappeared (10 min.). The catalyst was removed rapidly by filtration, as was the solvent by distillation under reduced pressure. The white residue was taken up in ether and, after filtration, the filtrate was evaporated almost to dryness. Low-boiling petroleum ether was added, followed by evaporation several times until the enediol crystallized. After cooling in an ice bath, the enediol was filtered and washed several times with low-boiling petroleum ether. The yield of product, m.p. 123–124° (corr.), was almost quantitative [119].

Reductive Coupling

This reaction is relatively new, but does fit into the category of simultaneous reduction and coupling. The product is in reality a mixture of the isomers of the unsaturated alcohol where the double bond is in various positions [120]. Since this preparation is so recent, no comment can be made about the generality

$$CH_3\overset{\overset{\displaystyle O}{\|}}{C}CH_3 + CH_2=CH-CH=CH_2 \xrightarrow[\text{Tetrahydrofuran}]{\text{Na, 0°}} CH_2=CH-\overset{\overset{\displaystyle CH_3}{|}}{CH}-C(CH_3)_2OH$$

1,1,2-Trimethyl-3-butenol-1, 64%

of the reaction, but it does appear promising enough to be included with coupling examples.

9. From Thiol Acid Esters or Hemithioacetals

$$-COSR \xrightarrow[\text{Ni}]{\text{Raney}} -CH_2OH$$

This synthesis has been applied successfully in the field of steroids in going from the carboxylic acid to the primary alcohol via the acid chloride and the thiol acid ester. Although aldehydes are also produced, it was found that the active W-4 Raney catalyst [55] gave a quantitative yield of the alcohol [121]. Certain double bonds and 3-acyl groups (except formyl) were not affected in the reduction [122]. Yields are usually high.

Hemithioacetals also are capable of reduction to alcohols, albeit the yields are low in many cases [123]. The thioacetals of acetylated mono- and disaccharides have been studied in greatest depth, an example of which is shown in Ex. b.2.

(a) **Preparation of 3(β)-acetoxy-22-hydroxy-bisnor-5-cholene.**
Ethyl 3(β)-acetoxy-bisnor-5-thiolcholenate, 0.200 g., in 5 ml. of 95% ethanol

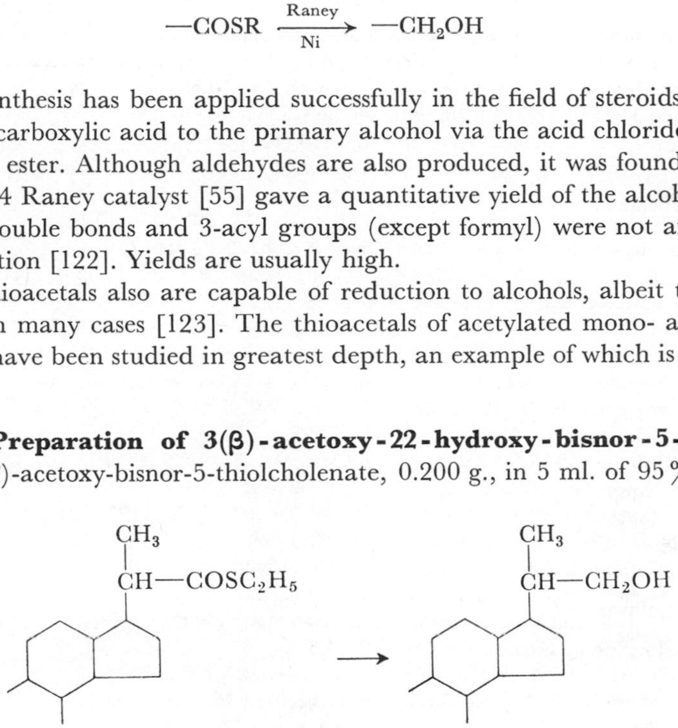

was allowed to stand over W-4 Raney nickel [55] for 1 hr. at 25° with occasional shaking. After separating the nickel by filtration, the filtrate and ethanol washings were combined and diluted with water to give 0.16 g. (93%) of 3(β)-acetoxy-22-hydroxy-bisnor-5-cholene, m.p. 143–148°. Several recrystallizations elevated the m.p. to 152–154.5° [122].

(b) Other examples

(1) **Benzyl and cetyl alcohols, pyridyl-3-carbinol, 3-β-acetoxy-etio-allo-cholanyl-(17)-carbinol, and Δ⁵-3β-acetoxy-etio-cholenyl-(17)-carbinol** [124].

(2) Styracitol (80% from ethyl 1-thio-β-D-manno-pyranoside tetraacetate) [125].

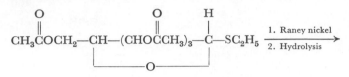

1. H. C. Brown, *Hydroboration*, W. A. Benjamin, New York, 1962; H. O. House, *Modern Synthetic Reactions*, W. A. Benjamin, New York, 1965, p. 23; N. G. Gaylord, *Reduction with Complex Metal Hydrides*, Interscience Publishers, New York, 1956; W. G. Brown, *Org. Reactions*, **6**, 469 (1951): V. M. Micovic and M. L. Mihailovic, *Lithium Aluminum Hydride in Organic Chemistry*, Serbian Academy of Sciences Press, Belgrade, Yugoslavia, 1955; Metal Hydride Technical Bulletin No. 401, October 1963.
2. S. Bernstein *et al.*, *J. Am. Chem. Soc.*, **76**, 6116 (1954).
3. L. F. and M. Fieser, *Reagents for Organic Synthesis*, John Wiley and Sons, New York, 1967, p. 1050.
4. H. C. Brown and P. M. Weissman, *J. Am. Chem. Soc.*, **87**, 5614 (1965).
5. B. Franzus and E. I. Snyder, *J. Am. Chem. Soc.*, **87**, 3423 (1965).
6. M. J. Jorgenson and A. W. Friend, *J. Am. Chem. Soc.*, **87**, 1815 (1965).
7. K. C. Schreiber and W. Emerson, *J. Org. Chem.*, **31**, 95 (1966).
8. E. L. Eliel *et al.*, *J. Am. Chem. Soc.*, **82**, 1367 (1960); *Org. Syn.*, **47**, 16 (1967).
9. H. C. Brown and B. C. Subba Rao, *J. Am. Chem. Soc.*, **78**, 2582 (1956).
10. F. W. Swamer and C. R. Hauser, *J. Am. Chem. Soc.*, **68**, 2647 (1946).
11. C. E. Srogg and H. M. Woodburn, *Org. Syn.*, Coll. Vol. **4**, 271 (1963).
12. V. M. Mićović *et al.*, *Tetrahedron*, **1**, 340 (1957).
13. J. D. Roberts and C. W. Sauer, *J. Am. Chem. Soc.*, **71**, 3925 (1949).
14. H. C. Brown *et al.*, *J. Am. Chem. Soc.*, **77**, 6209 (1955).
14a. R. F. Nystrom and W. G. Brown, *J. Am. Chem. Soc.*, **69**, 1197 (1947).
15. H. Nöth and H. Beyer, *Chem. Ber.*, **93**, 1083 (1960).
16. T. W. Campbell, *J. Am. Chem. Soc.*, **82**, 3126 (1960).
17. R. B. Moffett, *Org. Syn.*, Coll. Vol. **4**, 834 (1963).
18. S. W. Chaikin and W. G. Brown, *J. Am. Chem. Soc.*, **71**, 122 (1949).
19. M. J. Jorgenson, *Tetrahedron Letters*, 559 (1962).
20. E. A. Sullivan and A. A. Hinckley, *J. Org. Chem.*, **27**, 3731 (1962).
21. R. Fuchs and C. A. VanderWerf *J. Am. Chem. Soc.*, **76**, 1631 (1954).
22. E. L. Eliel and D. W. Delmonte, *J. Am. Chem. Soc.*, **78**, 3226 (1956).
23. H. C. Brown and N. M. Yoon, *J. Am. Chem. Soc.*, **90**, 2686 (1968).
24. K. B. Wiberg and J. E. Hiatt, *J. Am. Chem. Soc.*, **90**, 6495 (1968).
25. E. L. Eliel *et al.*, *J. Am. Chem. Soc.*, **78**, 3226 (1956); **80**, 1744 (1958); **82**, 1362 (1960); **84**, 2356 (1962).
26. W. N. Moulton *et al.*, *J. Org. Chem.*, **26**, 290 (1961).
27. A. L. Wilds, *Org. Reactions*, **2**, 178 (1944).
28. H. S. Mosher *et al.*, *J. Am. Chem. Soc.*, **82**, 5938 (1960).
29. T. van Es and O. G. Backeberg, *J. Chem. Soc.*, 1371 (1963).
30. W. G. Young *et al.*, *J. Am. Chem. Soc.*, **58**, 100 (1936).
31. W. Chalmers, *Org. Syn.*, Coll. Vol. **2**, 598 (1943).
32. R. B. Woodward *et al.*, *J. Am. Chem. Soc.*, **67**, 1425 (1945).
33. H. T. Clarke and E. E. Dreger, *Org. Syn.*, Coll. Vol. **1**, 304 (1941).
34. T. A. Geissman, *Org. Reactions*, **2**, 94 (1944).

35. M. Delépine and A. Horeau, *Compt. Rend.*, **204,** 1605 (1937).
36. J. F. Walker, *Formaldehyde*, Reinhold Publishing Corp., New York, 1964, p. 285.
37. G. Lock, *Org. Reactions*, **2,** 111 (1944).
38. D. Davison and M. Weiss, *Org. Syn.*, Coll. Vol. **2,** 590 (1943).
39. B. R. Brown *et al.*, *J. Chem. Soc.*, 1145 (1951).
40. S. E. Hazlet *et al.*, *J. Org. Chem.*, **29,** 2034 (1964).
41. V. L. Hansley, *Ind. Eng. Chem.*, **39,** 55 (1947).
42. W. Enz, *Helv. Chim. Acta*, **44,** 206 (1961).
43. L. A. Paquette and N. A. Nelson, *J. Org. Chem.*, **27,** 2272 (1962).
44. H. Adkins and R. H. Gillespie, *Org. Syn.*, Coll. Vol. **3,** 671 (1955).
45. J. N. Ray *et al.*, *J. Indian Chem. Soc.*, **38,** 705 (1961).
46. F. Y. Wiselogle and H. Sonneborn, III, *Org. Syn.*, Coll. Vol. **1,** 90 (1941).
47. F. C. Whitmore and T. Otterbacher, *Org. Syn.*, Coll. Vol. **2,** 317 (1943).
48. A. F. Holleman, *Org. Syn.*, Coll. Vol. **1,** 554 (1941).
49. H. Adkins, *Reactions of Hydrogen with Organic Compounds*, University of Wisconsin Press, Madison, Wisconsin, 1937.
50. H. Adkins and R. L. Shriner in Gilman's *Organic Chemistry, An Advanced Treatise*, Vol. 1, Chap. 9, John Wiley & Sons, New York, 1945.
51. R. L. Augustine, *Catalytic Hydrogenation*, Marcel Dekker, New York, 1965.
52. P. N. Rylander, *Catalytic Hydrogenation Over Platinum Metals*, Academic Press, New York, 1967.
53. G. W. Kenner and M. A. Murray, *J. Chem. Soc.*, S178 (1949); E. Caspi *et al.*, *J. Chem. Soc.*, 212 (1963).
54. H. R. Billica and H. Adkins, *Org. Syn.*, Coll. Vol. **3,** 180 (1955).
55. Ref. 51, pp. 27–32; 147–149.
56. W. A. Lazier and H. R. Arnold, *Org. Syn.*, Coll. Vol. **2,** 142 (1943).
57. Ref. 51, p. 150.
58. R. Adams *et al.*, *Org. Syn.*, Coll. Vol. **1,** 463 (1941).
59. H. R. Henze *et al.*, *J. Am. Chem. Soc.*, **73,** 4432 (1951).
60. W. F. Bruce, *J. Am. Chem. Soc.*, **58,** 687 (1936).
61. Ref. 58, Note 3.
62. Ref. 51, p. 152.
63. P. N. Rylander *et al.*, *J. Org. Chem.*, **24,** 1855 (1959).
64. D. R. Levering *et al.*, *J. Am. Chem. Soc.*, **72,** 1190 (1950).
65. S. Nishimura and Y. Urushibara, *Bull. Chem. Soc. Japan*, **30,** 199 (1957).
66. W. H. Carothers and R. Adams, *J. Am. Chem. Soc.*, **47,** 1047 (1925).
67. Dr. J. R. Cox, Jr., Private communication, University of Houston, Houston, Texas.
68. P. E. Peterson and C. Casey, *J. Org. Chem.*, **29,** 2325 (1964).
69. L. K. Friedlin *et al.*, *Bull. Acad. Science, USSR Div. Chem. Sci.* (*English Transl.*), 622 (1958).
70. E. G. Peppiatt and R. J. Wicker, *Chem. Ind.* (*London*), 747 (1955).
71. A. V. Kamernitzky and A. A. Akhrem, *Tetrahedron*, **18,** 705 (1962).
72. W. H. Clingman, Jr. and F. T. Wadsworth, *J. Org. Chem.*, **23,** 276 (1958).
73. W. Reeve and J. D. Sterling, Jr., *J. Am. Chem. Soc.*, **71,** 3657 (1959).
74. *See also* W. H. Carothers and R. Adams, *J. Am. Chem. Soc.*, **46,** 1675 (1924).
75. K. Kindler *et al.*, *Ann. Chem.*, **605,** 200 (1957).
76. K. C. Brannock, *J. Am. Chem. Soc.*, **81,** 3379 (1959).
77. W. Wenner, *J. Org. Chem.* **15,** 301 (1950).
78. W. K. Johnson, *J. Org. Chem.*, **24,** 864 (1959).
79. K. Heusler and A. Wettstein, *Helv. Chim. Acta*, **35,** 284 (1952).
80. K. Alexander and G. H. Smith, Jr., *J. Am. Chem. Soc.*, **71,** 735 (1949).
81. R. H. Hasek *et al.*, *J. Org. Chem.*, **26,** 700 (1961).
82. M. S. Newman and H. S. Whitehouse, *J. Am. Chem. Soc.*, **71,** 3664 (1949).
83. R. E. Lutz and C. E. Griffin, *J. Am. Chem. Soc.*, **76,** 4965 (1954).
84. L. E. Schniepp *et al.*, *J. Am. Chem. Soc.*, **69,** 672 (1947).
85. H. E. Burdick and H. Adkins, *J. Am. Chem. Soc.*, **56,** 438 (1934).

86. D. Kaufman and W. Reeve, *Org. Syn.*, Coll. Vol. **3,** 693 (1955).
87. E. L. Eliel *et al.*, *J. Org. Chem.*, **30,** 2448 (1965).
88. R. C. Fuson, *Record Chem. Progress* (Kresge-Hooker Sci. Lib.), **12,** 1 (1951).
89. R. Adams and E. W. Adam, *Org. Syn.*, Coll. Vol. **1,** 459 (1941).
90. M. Gomberg and W. E. Bachman, *J. Am. Chem. Soc.*, **49,** 236 (1927).
91. W. E. Bachman and R. V. Shankland, *J. Am. Chem. Soc.*, **51** 306 (1929).
92. W. E. Bachmann, *J. Am. Chem. Soc.*, **55,** 1179, 2827 (1933).
93. M. J. Allen, *J. Org. Chem.*, **15,** 435 (1950).
94. A. Schönberg and A. Mustafa, *Chem. Rev.*, **40,** 181 (1947).
95. V. R. Skvarchenko *et al.*, *J. Gen. Chem. USSR, Eng. Transl.*, 2117 (1960).
96. H. L. Lochte *et al.*, *J. Am. Chem. Soc.*, **75,** 4477 (1953).
97. W. E. Bachmann, *Org. Syn.*, Coll. Vol. **2,** 71 (1943).
98. W. S. Ide and J. S. Buck, *Org. Reactions*, **4,** 269 (1948).
99. R. Breslow and E. McNelis, *J. Am. Chem. Soc.*, **81,** 3080 (1959).
100. R. Adams and C. S. Marvel, *Org. Syn.*, Coll. Vol. **1,** 94 (1941).
101. J. S. Buck and W. S. Ide, *J. Am. Chem. Soc.*, **54,** 3302 (1932).
102. G. Drefahl and W. Hartrodt, *J. Prakt. Chem.* (4), **4,** 124 (1956).
103. C. A. Buehler, *Chem. Rev.*, **64,** 7 (1964).
104. H. R. Hensel, *Angew. Chem.*, **65,** 491 (1953).
105. C. S. Marvel and J. K. Stille, *J. Org. Chem.*, **21,** 1313 (1956).
106. W. Mathes *et al.*, *Chem. Ber.*, **84,** 452 (1951).
107. H. Andrews *et al.*, *J. Chem. Soc.*, 3827 (1962).
108. C. A. Buehler *et al.*, *J. Org. Chem.*, **20,** 1350 (1955).
109. T. Ukai and S. Kanahara, *J. Pharm. Soc. Japan*, **74,** 45 (1954); *C.A.*, **49,** 1723 (1955).
110. S. M. McElvain, *Org. Reactions*, **4,** 256 (1948).
111. J. C. Sheehan *et al.*, *J. Am. Chem. Soc.*, **75,** 6231 (1953).
112. J. C. Sheehan *et al.*, *J. Am. Chem. Soc.*, **75,** 3997 (1953).
113. K. T. Finley, *Chem. Rev.*, **64,** 573 (1964).
114. G. E. Gream and S. Worthley, *Tetrahedron Letters*, 3319 (1968); J. J. Bloomfield, *ibid.*, 587 (1968).
115. J. W. Lynn and J. English, Jr., *J. Am. Chem. Soc.*, **73,** 4284 (1951).
116. J. M. Snell and S. M. McElvain, *Org. Syn.*, Coll. Vol. **2,** 114 (1943).
117. V. L. Hansley, *Org. Reactions*, **4,** 267 (1948).
118. N. L. Allinger, *Org. Syn.*, Coll. Vol. **4,** 840 (1963).
119. R. C. Fuson *et al.*, *J. Am. Chem. Soc.*, **62,** 2091 (1940).
120. J. K. Kochi, *J. Org. Chem.*, **28,** 1969 (1963).
121. G. B. Spero *et al.*, *J. Am. Chem. Soc.*, **70,** 1907 (1948).
122. A. V. McIntosh, Jr. *et al.*, *J. Am. Chem. Soc.*, **70,** 2955 (1948).
123. G. R. Pettit and E. E. van Tamelen, *Org. Reactions*, **12,** 356 (1962).
124. V. Prelog *et al.*, *Helv. Chim. Acta*, **29,** 684 (1946).
125. J. Fried and D. E. Walz, *J. Am. Chem. Soc.*, **71,** 140 (1949).

D. Oxidation

The direct introduction of hydroxyl into a hydrocarbon would be a most desirable reaction. Any oxidative procedure, however, usually leads to a peroxide or a more highly oxidized substance. Therefore most of the reactions in this section combine not only an oxidative, but also a reductive step. The preparation of glycols also is included together with allylic oxidation and coupling of alcohols to produce glycols.

1. From Hydrocarbons

The oxidation, on which pertinent references [1] are available, is not very specific with any hydrocarbons except those containing a tertiary carbon atom.

References for Section D are on pp. 223–224.

As with all air oxidations, the peroxide seems to accumulate to a certain concentration and further oxidation ceases. Increasing the pressure of oxygen
increases the accumulation of peroxide. Chromium anhydride is a better catalyst
than cobalt naphthenate in some of these oxidations, although sufficient examples are not known to generalize this statement. Water inhibits oxidation
[2]. A relatively obscure, but elegant, example is shown in Ex. (a).

The tertiary hydrogen on a carbon atom adjacent to a carbonyl group may
be oxidized successfully to a hydroxyl group. The specific example given (which
is part of a steroid molecule) has the interesting feature that the intermediate
hydroperoxide is reduced by triethyl phosphite as the former is generated [3].

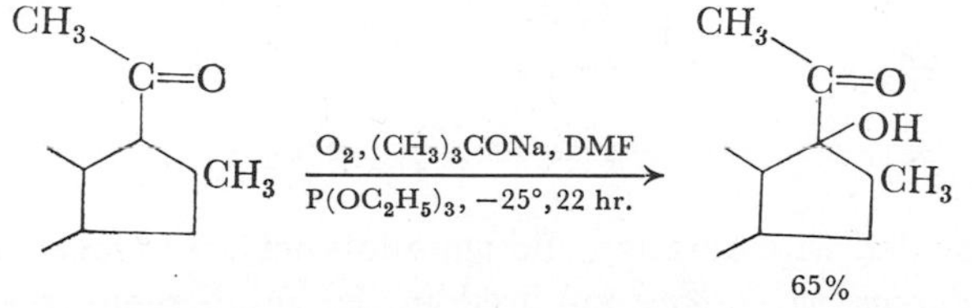

Biological oxidation is also of synthetic value in very specific cases. The
organism, *Fusarium moniliforme*, is capable of bringing about such an oxidation.
The substitution of the hydroxyl group is frequently β to some functional
group, and with another enzyme-substrate system (from *Sporotrichum sulfurescens*
and cyclododecanol) is pinpointed to a distance of about 5.5A from the functional group [4]. However, substitution in the hog liver enzyme-decanoic acid
system yields 10-hydroxydecanoic acid [5]. This area of specific oxidation of
hydrocarbon side chains is a fruitful one and should continue to expand in the
future; see [6] for hydroxylation of amides of azacycloalkanes. An example of
biological β-oxidation is given in (b).

(a) Preparation of cyclodecan-1-ol-6-one from decalin. Oxygen
was passed in a fine stream through decalin (containing tetralin) at 110° for
24 hr. The tetralin hydroperoxide was removed by shaking twice with dilute
aqueous alkali. The unreacted decalin was separated by distillation at 0.4 mm.
(heating no higher than 80°). The oily residue, decalin hydroperoxide, 4-VI,
was crystallized from petroleum ether to give 20 g. of beautiful needles, m.p.
95–96°. Compound 4-VI (8.5 g. in 20 ml. of pyridine) was warmed with 8 g. of
benzoyl chloride in 20 ml. of pyridine at 100° for a short time and the mixture
poured into 500 ml. of cold, dilute sulfuric acid. The ether extract yielded 6.9 g.

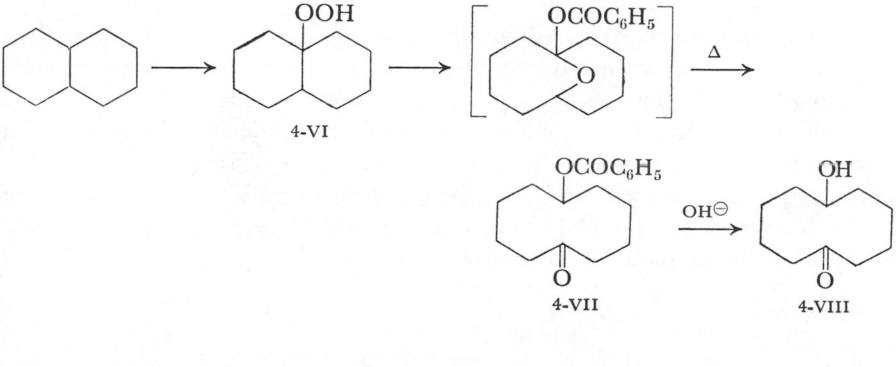

of 4-VII, 58%, m.p. 97–98°. Compound 4-VII (4.8 g.) was saponified with 10 ml. of 1.9 N potassium hydroxide in methanol, 40 ml. of methanol, and 2 ml. of water for 4 hr. The mixture was neutralized with dilute acid and extracted continuously with chloroform. The chloroform residue was recrystallized from benzene to yield 3.7 g. of 4-VIII, 90% [7].

(b) Preparation of 15-α-hydroxyestrone from estrone. Estrone (1 g.) in 25 ml. of dioxane was added to 9 l. of Czapek's broth in a 14-l. fer-

mentor jar provided with aeration and agitation devices. The broth was inoculated with *Fusarium moniliforme* cultured under submerged conditions. After incubation for 18–20 hr. the mixture was extracted thrice with 10 l. of methylene chloride. The crude extract was chromatographed on 50 g. of neutral alumina. Elution with benzene-chloroform (95:5) gave 451 mg. of estrone, and further elution with benzene-chloroform (80:20) gave 340 mg. of crude product, m.p. 223–228°. Recrystallization from acetone-hexane with decolorization gave the pure compound, m.p. 232–233° [8].

2. From Organometallic Compounds [9]

$$RCH_2MgX \xrightarrow{O_2} RCH_2OOMgX \xrightarrow{RCH_2MgX}$$

$$2\,RCH_2OMgX \xrightarrow{H_2O} 2\,RCH_2OH$$

Although air oxidation is not very common, it is sometimes useful in preparing alcohols from halides when the halogen in the original halide is difficult to replace or when the elimination of the hydrogen halide is a problem.

By using *t*-butyl hydroperoxide as the oxidizing agent on several alkyl Grignard reagents, yields of alcohols, based on the hydroperoxide consumed, varied from 90 to 99% [9]; see also Phenols, Chapter 5, B.2. It is proposed that here too the hydroperoxide of the Grignard reagent is first formed; this product then reacts with a second molecule of the Grignard reagent to give 2 molecules of the alkoxymagnesium halide which is readily hydrolyzed to the alcohol.

(a) Preparation of 4,4-dimethylpentanol-1. 1-Bromo-4,4-dimethylpentane was converted into the Grignard reagent in the usual manner; yield 82% in a 0.5-mole run.

A solution of 0.14 mole of the Grignard reagent was cooled and treated with dry oxygen. The product was poured over ice and ammonium chloride; the ether layer was separated and the aqueous layer was extracted with 100 ml. of ether, the extract being added to the main portion. After drying over potassium carbonate, the product was fractionated through a 55 × 1.2 cm. indented column. Fifteen grams (90% based on Grignard reagent) of 4,4-dimethyl-pentanol-1 was obtained: b.p. 158° (737 mm.); n^{20}D 1.4202; d_4^{20} 0.0815 [10].

(b) **Preparation of 1-octanol** (92% based on the hydroperoxide consumed from the Grignard reagent of 1-bromooctane and magnesium, and t-butyl hydroperoxide) [9].

3. From Olefins (Ozonization Followed by Reduction)

Ozone adds to olefins to form several types of ozonides [11] for which the formulas of two follow:

$$\text{RCH=CHR'} \xrightarrow{\text{O}_3}
\left[
\begin{array}{c}
\overset{\ominus}{\text{O}} \\
| \\
\overset{\oplus}{\text{O}} \quad \text{O} \\
| \qquad | \\
\text{R}-\text{C}---\text{C}-\text{R'} \\
| \qquad | \\
\text{H} \quad \text{H}
\end{array}
\right]
\longrightarrow
\begin{array}{c}
\text{R} \quad \text{O} \quad \text{R'} \\
\diagdown \diagup \diagdown \diagup \\
\text{H}-\text{C} \qquad \text{C}-\text{H} \\
| \qquad | \\
\text{O}----\text{O}
\end{array}$$

4-IX   4-X

Compound 4-X is the formula for the common ozonide produced when alkenes are treated with ozone in an inert solvent, while compound 4-IX has been proposed for what is known as the molozonide, which may be produced in ether at temperatures below −110° [12]. Such molozonides decompose with explosive violence above −100°. Although various reducing agents have been employed to reduce the common ozonide, lithium aluminum hydride under controlled conditions is the preferred reagent [13]. With this reagent alcohols are produced in good yield (see Ex. a, b.1, b.2). Another modification is to run the ozonization in methanol-dimethylsulfide to produce the aldehyde directly and, without isolation, to reduce to the alcohol with sodium borohydride in ethanol [14]. Alcohols are also obtained from the molozonides produced from *cis-* and *trans-*alkenes on treatment with an isopropyl Grignard reagent, but in these cases the major product from the *trans* is the 1,2-glycol while the *cis* gives no glycol [15]:

$$\text{trans-2-Pentene} \longrightarrow \text{molozonide} \xrightarrow[\text{reagent}]{\text{Grignard}} \text{threo-dl-pentane-2,3-diol}$$

Other compounds in a high state of oxidation, such as peroxides or hydroperoxides, may be reduced with lithium aluminum hydride [16], or triethyl phosphite [17], to give glycols. The oxidation of the alkene may be conducted in ethanol [18], which offers some advantage over other solvents in that, with 3% ozone in oxygen at temperatures of 0–5°, stable intermediates capable of conversion into various other types are formed. Both monomeric and polymeric peroxides are reducible to give superior yields of alcohols.

(a) **Preparation of *n*-propyl and *n*-amyl alcohols.** Passing ozonized

$$\text{C}_2\text{H}_5\text{CH=CHC}_4\text{H}_9 \xrightarrow{\text{O}_3}
\begin{array}{c}
\text{O} \\
\diagup \diagdown \\
\text{C}_2\text{H}_5\text{CH} \qquad \text{CHC}_4\text{H}_9 \\
| \qquad\qquad | \\
\text{O}----\text{O}
\end{array}
\xrightarrow{\text{LiAlH}_4}$$

$$\text{C}_2\text{H}_5\text{CH}_2\text{OH} + \text{C}_4\text{H}_9\text{CH}_2\text{OH}$$

oxygen (452 mg. ozone/hr.; for ozone preparation see [19]) through a solution of 11.22 g. of 3-octene in 150 ml. of purified *n*-pentane for 10 hr. gave the ozonide. Then dry nitrogen was introduced for 1 hr. at a rate of 11.6 l./hr. and at a temperature of −42 to −38°. Lithium aluminum hydride (149.4 mmoles in 150 ml. of ether) was next added at a rate at which the temperature did not exceed −10° (2 hr.). The mixture was allowed to warm slowly and then refluxed for 15 min., after which it was kept overnight at 25°. Decomposition with a solution of 20 ml. of concentrated H_2SO_4 in 125 ml. of water at −7° produced an organic layer, which was extracted with two 10-ml. portions of water and finally distilled to give 5.25 g. of *n*-propyl alcohol (87.4%) and 7.67 g. of *n*-amyl alcohol (87.0%) [13].

(b) Other Examples

(1) Homophthalyl alcohol

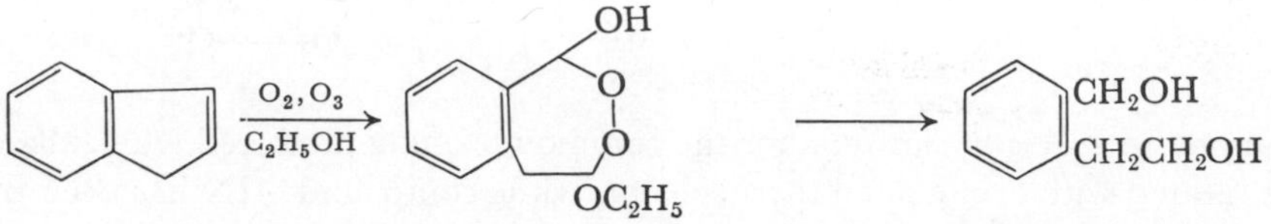

(i) 3-Hydroxy-4,5-benzo-7-ethoxy-1,2-dioxacycloheptene-4.
A solution of 11.6 g. of pure indene in 500 ml. of anhydrous ethanol cooled in ice was treated with a 3% stream of ozone in oxygen (80 l./hr) for 55 min. Evaporation of the ethanolic solution under vacuum at 25° gave 21 g. (99%) of the peroxide, m.p. 105–108° with softening at 95–105°.

(ii) Homophthalyl alcohol.
The peroxide, 21 g., was added to a solution of 20 g. of $LiAlH_4$ in 200 ml. of ether during 30 min. After stirring well for 20 min. under reflux, water was added to decompose the excess metallic hydride and then the mixture was acidified with 25% H_2SO_4. The combined ethereal layer and ethereal extract of the aqueous layer was washed with aqueous Na_2CO_3, dried, and distilled to give 14.5 g. (95%) of the titled alcohol, b.p. 164–165° (2 mm.) [18].

(2) Phenylglycol
(87.5% from styrene peroxide and $LiAlH_4$ in ether-dioxane) [16].

4. From Olefins (Allylic or similar Oxidations)

(a) By oxygen

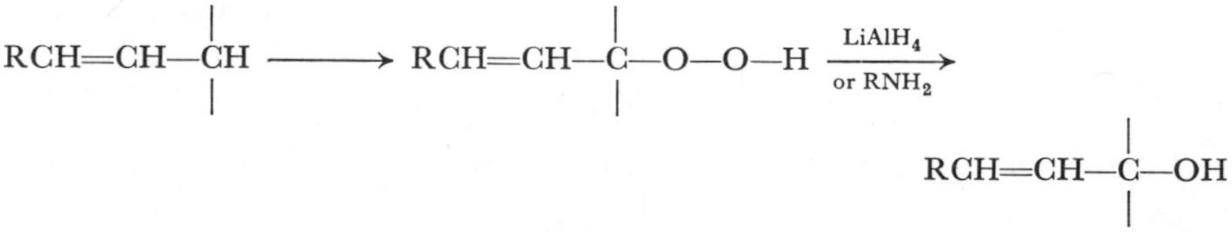

Although some hydroperoxides and peroxides (those of tertiary alcohols) are of sufficient stability to permit distillation under reduced pressure, such compounds are hazardous chemicals. General observations have been recorded on allylic oxidation [20].

For certain hydrocarbons, especially those containing hydrogen atoms α to a double bond or to an aromatic ring system, hydroperoxides may be obtained directly by oxidation with air [21].

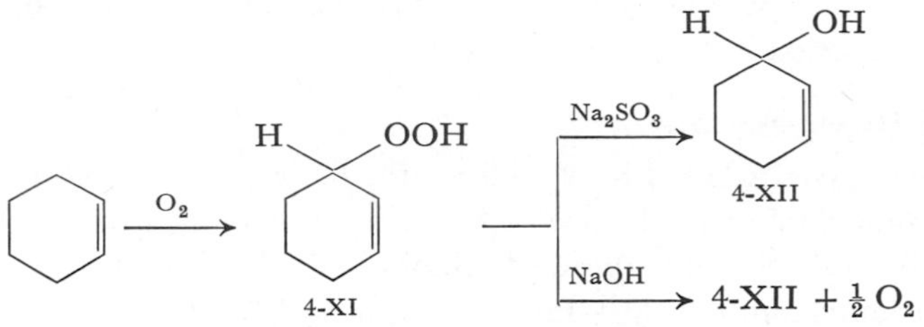

In such cases the synthesis of alcohols by this method becomes of greater importance. A still more efficient oxidation is brought about by dye-sensitized irradiated oxygen for which the general scheme of oxidation is as follows [22].

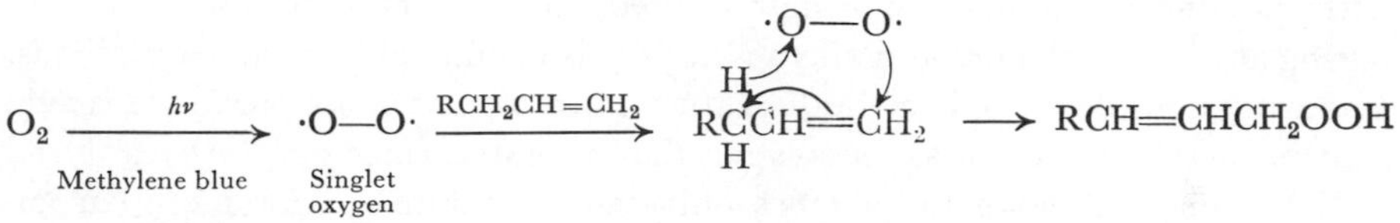

Mixtures of hydroperoxides are obtained if an allylic hydrogen flanks both sides of the double bond in an unsymmetrical olefin. Conjugated dienes yield peroxides. Interestingly, as judged from the identity of the products produced

from either source, singlet oxygen (or whatever the active intermediate is) can be generated chemically by sodium hypochlorite in hydrogen peroxide-methanol solutions or by the dye-sensitized irradiated oxygen procedure [23].

Various reducing agents, such as sodium in alcohol, sodium sulfite, hydrogen in the presence of platinum, lithium aluminum hydride, and amines, have been employed, but a solution of potassium iodide in methanol, ether, and acetic acid appears to be the simplest reducing agent of all [24]. Also, the hydroperoxide can be decomposed by heating with aqueous alkali. The yields are high; in the case of 4-XI, for example, an 80% yield of 4-XII is obtained together with some acidic products [1].

In some cases, air oxidation yields epoxides as the main product, the allylic alcohol being a minor constituent. This seems to be true in the oxidation of cis-cyclooctene with a cobalt salt as a catalyst [25].

(1) Preparation of α-tetralol

(i) **Tetralin hydroperoxide** (44–57% from tetralin and a stream of oxygen at 70° until peroxide content is 25–30%) [26].

(ii) **α-Tetralol.** Tetralin hydroperoxide, 2.5 g., in 20 ml. of dry ether was added with rapid stirring during 30 min. to 0.63 g. of lithium aluminum hydride in 20 ml. of dry ether, and a vigorous reaction occurred. After refluxing for a further 50 min. the complex was decomposed by the addition of cold, dilute sulfuric acid to give 2.14 g. of α-tetralol (95%), which was purified by distillation to give 1.88 g. (83%), b.p. 147° (25 mm.) [27]. (Note Ex. 2.ii for a simpler reduction.)

(2) Other examples

(i) **Hydrindanol-8** (80% from the hydroperoxide by hydrogenation with platinum catalyst in alcohol) [28].

(ii) **α,α-Dimethylbenzyl alcohol** (88% from the hydroperoxide heated in triethylamine at 110–120° for 30 min.) [29].

(b) By selenium dioxide. This synthesis has been reviewed [30].

$$RCH{=}CHCH_2CH_3 \xrightarrow{\text{SeO}_2} RCH{=}CHCHOHCH_3$$

Although the reaction is somewhat unpredictable, it may be useful in preparing alcohols of alkenes and alkynes not readily obtainable by other methods. The oxidations described in the literature have for the most part used acetic acid and acetic anhydride as the solvent. (The acetate formed may be hydrolyzed to the alcohol.) Among the alkenes only those containing 5 or more carbon atoms give alcohols. Yields are better for cycloalkenes than alkenes, some of the former being 50% and over.

(1) Preparation of 2-cyclohexenyl acetate. Freshly distilled cyclo-

hexene, 82.1 g., and 150 ml. of acetic anhydride were heated until the temperature reached 70°, after which 15.2 g. of selenious acid was added. The temperature then rose to 85–90°, where it was kept by heating for 15 min. This process was then repeated twice more, 15 g. of selenious acid being added each time and the temperature being controlled as before, except after the final addition when the temperature was kept at 90–95° for 15 min. and then at 90–100° for 3 hr. The reaction mixture, separated from selenium by decantation, was distilled at ordinary pressure until the temperature reached 140°. Further distillation was accomplished at reduced pressure when 47–51 g. of 2-cyclohexenyl acetate (48–52%), b.p. 57–59° (9 mm.), n^{15}D 1.461 and 8–9 g. of 2-cyclohexenyldiacetate-1,4, b.p. 112–115°/9 mm., n^{15}D 1.471 came over [31].

(c) By lead tetraacetate followed by hydrolysis [32]. Lead tetraacetate substitutes an acetoxy grouping on the carbon atom attached to the doubly bound carbon atom in the allylic group. In addition to allylic substitution, active hydrogen substitution can also be accomplished as shown in Ex. 2i.

(1) Preparation of 7-acenaphthenol (70–74% based on acenaphthene) [33].

(2) Other examples

(i) **Acetyl tartronic ester** ($CH_3CO_2CH(CO_2C_2H_5)_2$, 62% from malonic ester; the acetyltartronic ester can be converted quantitatively into dimethyl hydroxymalonate by cross esterification with methanolic hydrogen chloride) [34].

(d) By N-bromosuccinimide followed by hydrolysis [35]. The mechanism of substitution is probably free radical stemming from a small pool

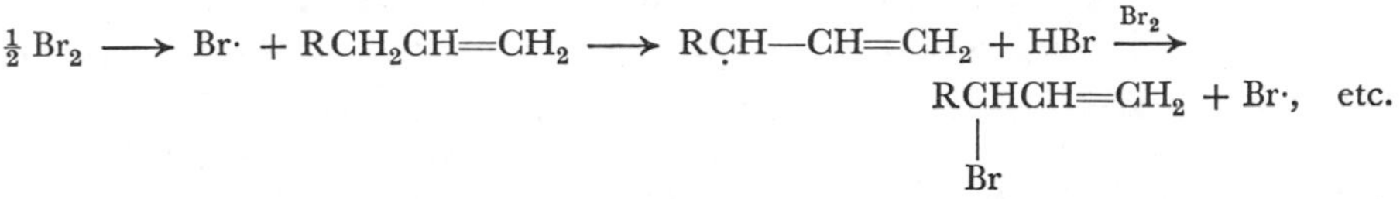

$$\tfrac{1}{2} Br_2 \longrightarrow Br\cdot + RCH_2CH{=}CH_2 \longrightarrow R\overset{.}{C}H{-}CH{=}CH_2 + HBr \xrightarrow{Br_2}$$

$$RCHCH{=}CH_2 + Br\cdot, \quad \text{etc.}$$
$$\underset{Br}{|}$$

of bromine in the N-bromosuccinimide. The hydrogen bromide can cleave N-bromosuccinimide even at −80° to give bromine + succinimide [36]. This maintains the small pool of bromine necessary for allylic substitution. Allylic substitution also occurs at high temperature with olefins and halogen. Isomerization in the allylic radical is possible.

(1) Preparation of 1-bromo-2-octene. 1-Octene (1 mole), 0.35 mole of N-bromosuccinimide, 200 ml. of carbon tetrachloride, and 0.2 g. benzoylperoxide were refluxed for 20 min. After the usual decomposition 63 g. (94%) of a liquid containing 1-bromo-2-octene as the major product and 3-bromo-1-octene as the minor product was obtained. As much as 20% of the latter was present as determined by infrared analysis [37].

(e) By t-butyl perbenzoate followed by hydrolysis. Allylic substi-

$$RH + t\text{-}C_4H_9OOOCC_6H_5 \rightarrow ROOCC_6H_5 + t\text{-}C_4H_9OH$$

tution by t-butyl perbenzoate as catalyzed by cuprous salt is very similar to that by N-bromosuccinimide with respect to orientation. For example, the three isomeric normal butenes are converted into similar mixtures of about 90% α-methylallyl and 10% crotyl benzoates [38].

5. From Alkenes (*cis* or *trans* Addition)

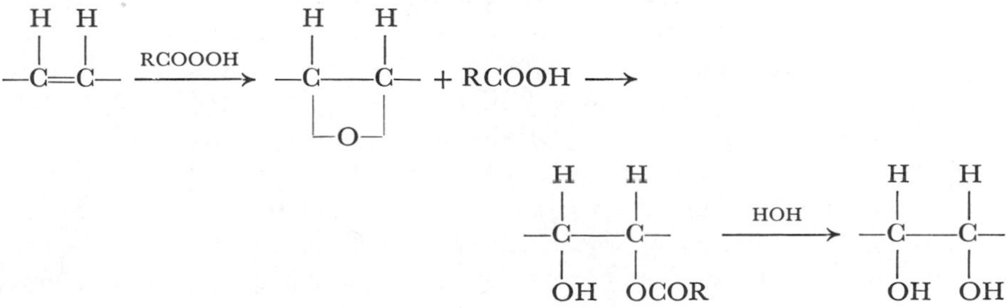

This stereospecific synthesis has been reviewed [39–41]. Various oxidizing agents have been employed. One of the most common is an organic peracid which oxidizes the alkene to an oxirane which may or may not be isolated as an intermediate in the formation of the glycol (see Ethers, Chapter. 6, D.1). Peracetic acid is the usual reagent, although perbenzoic, monoperphthalic, pertrifluoroacetic [42], performic acids and succinyl peroxide [43] have also been employed. In all these reactions, *cis* addition occurs to give the oxirane which leads usually to the *trans* glycol.

The duPont Company has found that polystyrene sulfonic resins (about 10–12 % by weight of the substrate to be oxidized) catalyze the formation of peracetic acid in mixtures of acetic acid and hydrogen peroxide. The resin technique then permits a well-controlled epoxidation reaction with simple removal of the strong acid by filtration. It has been applied mostly to epoxidation of unsaturated fatty oils [40]. Anhydrous peracetic acid is available from acetaldehyde but, since it is not an easy laboratory preparation, a method of preparing it has been developed. In the procedure, water is removed by azeotropic distillation from ethyl acetate [44]. Peracetic acid concentration should be less than 55 % in ethyl acetate at 50° (or less than 30 % at 100°) to be non-detonable! To make longer-chain peracids, methanesulfonic acid seems to be a more useful catalyst than sulfuric acid [45].

Alkaline hydrogen peroxide is effective, particularly in the epoxidation of unsaturated carbonyl compounds as a second method of yielding the *trans*-glycol. The active agent is evidently the peroxide anion, $\overset{\ominus}{O}OH$, which is suitable for addition to olefins containing electron-withdrawing groups. A third method of *trans* hydroxylation is the use of hydrogen peroxide in combination with ferrous sulfate (Fenton's reagent) or with tungstic acid. The active agent in these systems appears to be the hydroxyl free radical:

$$Fe^{(2+)} + H_2O_2 \rightarrow Fe^{(3+)} + OH^{\ominus} + \cdot OH$$

This reagent is more effective in adding to terminal double bonds than is the peracid reagent. The last reagent which results in *trans* hydroxylation is iodine and silver benzoate (Prevost's reagent) [46, 47]. Although this reagent often gives satisfactory yields, it has been used to a limited extent.

A greater variety of reagents is available to accomplish *cis* hydroxylation. One of the oldest of them, dilute alkaline potassium permanganate, gives almost quantitative yields in the hydroxylation of long-chain monoolefinic acids [48]. Although the oxidation procedure is simple, the method becomes tedious with a reagent in low concentration when large quantities are involved.

Perhaps the most widely used method for effecting *cis* hydroxylation is that employing osmium tetroxide alone [49] or with a chlorate [50]. Ether or dioxane is the usual solvent and pyridine is a common catalyst. The osmic ester, which separates at room temperature, usually after several days, is converted into the *cis*-glycol by some agent such as mannitol [51]. Yields are satisfactory. The orientation in the hydrogen peroxide oxidation of cyclohexene can be controlled by the choice of catalyst, the *cis* glycol being obtained in 86 %

yield by the use of osmic acid [49] or inferentially the *trans* glycol by the use of sodium tungstate since it yields the epoxide [52]. *trans*-1,2-Cyclohexanediol has been made also from monopersuccinic acid and cyclohexene emulsified in water in 85 % yield [53].

To avoid the use of the expensive and toxic osmium tetroxide, iodine and silver acetate in wet acetic acid have been employed to effect the *cis* hydroxylation of long-chain olefinic acids [54]. It appears that the iodine and silver acetate first form a *threo*-iodoacetate by *trans* addition and that the latter is inverted by aqueous acetic acid to give the *erythro*-hydroxyacetate or -diacetate, which undergoes no change in configuration on hydrolysis. Thus the net result is *cis* hydroxylation. Yields starting with pure olefins usually run from 80 to 90 %. The method originated by Woodward [55] has been applied successfully to the *cis* hydroxylation of ethylenic bonds in alicyclic systems [56, 57].

(a) **Preparation of *trans*-1,2-cyclohexanediol** (70 % or more from cyclohexene, formic acid and 30 % H_2O_2) [58].

(b) **Preparation of *cis*-5,6-dihydroxy-5,6-dihydro-7,12-dimethylbenz [a] anthracene.** Osmium tetroxide, 2 g., in 40 ml. of dry benzene was

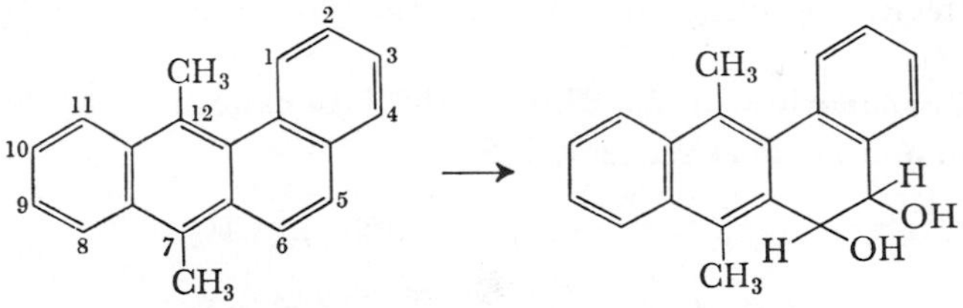

added at 25° under nitrogen to a stirred solution of 2.0146 g. of 7,12-dimethylbenz [a] anthracene in 60 ml. of dry benzene containing 1 ml. of dry pyridine. After standing 2 days the benzene was removed under reduced pressure and the residue produced was dissolved in about 200 ml. of methylene chloride, after which the solution was shaken for 2 hr. with 200 ml. of 5 N NaOH and 60 ml. of 1 M D-mannitol. The latter process was repeated, if necessary, to remove color. The organic layer, after being washed with water, dried, and evaporated, gave a residue which when crystallized from benzene-cyclohexane weighed 1.786 g. (78 %), m.p. 172.5–173.5° [59].

(c) **Other examples**

(1) **3-(8′,9′-Dihydroxypentadecyl)-4-iodoanisole** (59 % from *cis*-3-(pentadecenyl-8′)-anisole from *trans* hydroxylation with iodine and silver benzoate) [60].

(2) ***cis*-2-(2,3-Dimethoxyphenyl)-cyclohexane-1,2-diol** (56 % from 1-(2,3-dimethoxyphenyl)-cyclohexene with iodine and silver acetate) [56].

(3) ***cis*-2-Formyloxy-1-hydroxyindane** (35 % from indene and performic acid) [61].

6. From Alcohols (Coupling or Addition)

These methods which are very specific, do enlarge the scope of the synthesis of glycols and alcohols.

(a) **Coupling.** Acetylenic alcohols, readily made by the condensation of the acetylene alkali metal salts (or Grignard) and carbonyl compounds, can be coupled by air oxidation to diyne glycols. In a similar type of reaction, tertiary alcohols can be coupled to form glycols.

(1) **Preparation of 2,7-dimethylocta-3,5-diyne-2,7-diol.** Compound 4-XIII was dissolved in water with a catalytic amount of cuprous chloride

$$
2\ HC{\equiv}C{-}\underset{\underset{CH_3}{|}}{\overset{\overset{CH_3}{|}}{C}}{-}OH \xrightarrow[O_2]{CuCl} HO\underset{\underset{CH_3}{|}}{\overset{\overset{CH_3}{|}}{C}}C{\equiv}C{-}C{\equiv}C{-}\underset{\underset{CH_3}{|}}{\overset{\overset{CH_3}{|}}{C}}OH
$$

4-XIII   4-XIV

made soluble by an excess of ammonium chloride solution. Air was passed through the solution until maximum precipitation had occurred. The precipitate was filtered and washed with aqueous ammonium chloride to remove the cuprous salt. Recrystallization from xylene gave 4-XIV, 95–100%, m.p. 132–133° [62].

(2) **Preparation of 2,5-dimethyl-2,5-hexanediol** (40–46% from *t*-butyl alcohol and Fenton's reagent) [63].

$$
(CH_3)_3COH \xrightarrow[H_2O_2]{Fe^{2+}} HO\underset{\underset{CH_3}{|}}{\overset{\overset{CH_3}{|}}{C}}CH_2CH_2\underset{\underset{CH_3}{|}}{\overset{\overset{CH_3}{|}}{C}}OH
$$

(b) **Oxidative addition of alcohols to perfluoroalkenes.** This

$$
R_fCF{=}CF_2 + CH_3OH \xrightarrow{Peroxides} R_fCHFCF_2CH_2OH
$$
(R_f = fluorinated alkyl)

synthesis is a new, special one which has been developed for the preparation of fluorinated alcohols. The alcohol adds to the perfluoroolefin in the presence of peroxides by a free radical mechanism. Methyl alcohol produces primary alcohols, other primary alcohols give secondary alcohols, and secondary alcohols yield tertiary alcohols. This is a high-pressure method carried out at temperatures around 100°, and the yields are as high as 90%.

(1) **Preparation of 2,2,3,4,4,4-hexafluorobutanol.** A mixture of 65.5 g. (2.05 moles) of technical grade methanol and 1.5 g. of benzoyl peroxide was charged to the autoclave. The autoclave was cooled with liquid air, evacuated, and 56 g. (0.37 mole) of perfluoropropene ($CF_3CF{=}CF_2$) introduced. The autoclave was then closed and the contents agitated at 110–120° for 3 hr. Only 15 g. of unreacted perfluoropropene was bled from the autoclave at room

temperature. Any undecomposed peroxide was destroyed by the addition of ferrous sulfate or sodium bisulfite to the reaction mixture. Fractionation of the reaction mixture gave 45 g. (90%) of $CF_3CFHCF_2CH_2OH$, b.p. 114.5° (740 mm.), $n^{25}D$ 1.3115 [64].

(2) Other examples

(i) $C_2F_5CFHCF_2CH_2OH$, $CF_3CFHCF(CH_2OH)CF_3$, $C_3H_7CFHCF_2$-CH_2OH, $C_5F_{11}CFHCF_2CH_2OH$, $C_7H_{15}CFHCF_2CH_2OH$, $C_2F_5CFHCF_2$-$CHOHCH_3$, $C_3H_7CFHCF_2CHOHCH_3$, and $C_3H_7CFHCF_2COH(CH_3)_2$ [64].

1. C. E. Frank, *Chem. Rev.*, **46,** 155 (1950); L. F. Marek, *Ind. Eng. Chem.*, **45,** 2000 (1953); G. H. Twigg, *Chem. Ind. (London)*, 4 (1962); and N. M. Emanuel, *Liquid-Phase Oxidation of Hydrocarbons (English Transl.)*, New York, 1967.
2. W. S. Emerson *et al.*, *J. Am. Chem. Soc.*, **70,** 3764 (1948).
3. J. N. Gardner *et al.*, *J. Org. Chem.*, **33,** 3294 (1968).
4. G. S. Fonken *et al.*, *J. Am. Chem. Soc.*, **89,** 672 (1967).
5. S. J. Wakil, *Ann. Rev. Biochem.* **31,** 375 (1962).
6. R. A. Johnson *et al.*, *J. Org. Chem.*, **33,** 3217 (1968).
7. R. Criegee, *Chem. Ber.*, **77,** 22 and 722 (1944); British Patent 963, 945, July 23, 1960; *C.A.*, **61,** 9414 (1964).
8. P. Crabbé and C. Casas-Campillo, *J. Org. Chem.*, **29,** 2731 (1964).
9. S.-O. Lawesson and N. C. Yang, *J. Am. Chem. Soc.*, **81,** 4230 (1959).
10. F. C. Whitmore and A. H. Homeyer, *J. Am. Chem. Soc.*, **55,** 4555 (1933).
11. P. S. Bailey, *Chem. Rev.*, **58,** 925 (1958).
12. F. L. Greenwood, *J. Org. Chem.*, **30,** 3108 (1965).
13. F. L. Greenwood, *J. Org. Chem.*, **20,** 803 (1955).
14. E. J. Corey *et al.*, *J. Am. Chem. Soc.*, **90,** 5618 (1968).
15. F. L. Greenwood, *J. Org. Chem.*, **29,** 1321 (1964).
16. G. A. Russell, *J. Am. Chem. Soc.*, **75,** 5011 (1953).
17. M. S. Kharasch *et al.*, *J. Org. Chem.*, **25,** 1000 (1960).
18. J. L. Warnell and R. L. Shriner, *J. Am. Chem. Soc.*, **79,** 3165 (1957).
19. L. I. Smith *et al.*, *Org. Syn.*, Coll. Vol. **3,** 673 (1955).
20. K. B. Wiberg and S. D. Nielsen, *J. Org. Chem.*, **29,** 3353 (1964).
21. E. H. Farmer and A. Sundralingam, *J. Chem. Soc.*, 121 (1942).
22. A. Schonberg, *Preparative Organische Photochemie*, Springer-Verlag, Berlin, 1958, p. 47.
23. C. S. Foote, *Accounts Chem. Research*, **1,** 104 (1968).
24. A. Nickon and W. L. Mendelson, *J. Am. Chem. Soc.*, **87,** 3921 (1965).
25. I. S. de Roch and J. C. Balaceanu, *Bull. Soc. Chim. France*, 1393 (1964).
26. H. B. Knight and D. Swern, *Org. Syn.*, Coll. Vol. **4,** 895 (1963).
27. D. A. Sutton, *Chem. Ind. (London)*, 272 (1951).
28. R. Criegee and H. Zogel, *Chem. Ber.*, **84,** 215 (1951).
29. C. W. Capp and E. G. E. Hawkins, *J. Chem. Soc.*, 4106 (1953).
30. N. Rabjohn, *Org. Reactions*, **5,** 338 (1949).
31. Y. A. Arbuzov *et al.*, *Bull acad. sci.*, *URSS.*, Classe sci, chim., 163 (1945).
32. L. F. Fieser, *Experiments in Organic Chemstry*, D. C. Heath, New York, 1955, p. 325.
33. J. Cason, *Org. Syn.*, Coll. Vol. **3,** 3 (1955).
34. B. Bak, *Ann. Chem.*, **537,** 286 (1939).
35. C. Djerassi, *Chem. Rev.*, **43,** 271 (1948); L. Horner and E. H. Winkelmann, *Angew. Chem.*, **71,** 349 (1959).
36. R. E. Pearson and J. G. Martin, *J. Am. Chem. Soc.*, **85,** 354 (1963).
37. L. Bateman and J. I. Cunneen, *J. Chem. Soc.*, 941 (1950).
38. H. L. Goering and U. Mayer, *J. Am. Chem. Soc.*, **86,** 3753 (1964).
39. D. Swern, *Org. Reactions*, **7,** 378 (1953).

40. J. G. Wallace, *Hydrogen Peroxide in Organic Chemistry*, E. I. duPont de Nemours Co., 1962.
41. J. O. Edwards, *Peroxide Reaction Mechanisms*, John Wiley and Sons, New York, 1962.
42. W. D. Emmons *et al.*, *J. Am. Chem. Soc.*, **76**, 3472 (1954).
43. J. Blum, *Compt. Rend.*, **248**, 2883 (1959).
44. B. Phillips *et al.*, *J. Org. Chem.*, **23**, 1823 (1958).
45. D. Swern *et al.*, *J. Org. Chem.*, **27**, 1336 (1962).
46. M. Sletzinger and C. R. Dawson, *J. Org. Chem.*, **14**, 670, 849 (1949).
47. C. V. Wilson, *Org. Reactions*, **9**, 332 (1957); F. D. Gunstone, *Advances in Organic Chemistry, Methods and Results*, Interscience Publishers, New York, 1960, Vol. 1, p. 122.
48. A. Lapworth and E. N. Mottram, *J. Chem. Soc.*, **127**, 1628 (1925).
49. R. Criegee *et al.*, *Ann. Chem.*, **522**, 75 (1936).
50. L. Bláha *et al.*, *Collection Czech. Chem. Commun.*, **25**, 237 (1960); Th. Posternak and H. Friedli, *Helv. Chim. Acta*, **36**, 251 (1953).
51. R. Criegee *et al.*, *Ann. Chem.*, **550**, 99 (1942).
52. L. F. and M. Fieser, *Reagents for Organic Synthesis*, John Wiley and Sons, 1967, p. 475.
53. R. Lombard and G. Schroeder, *Bull. Soc. Chim. France*, 2800 (1963).
54. F. D. Gunstone and L. J. Morris, *J. Chem. Soc.*, 487 (1957).
55. R. B. Woodward, U.S. Patent 2,687,435, August 24, 1954; *C.A.*, **49**, 14809 (1955).
56. D. Ginsberg, *J. Am. Chem. Soc.*, **75**, 5746 (1953).
57. L. B. Barkley *et al.*, *J. Am. Chem. Soc.*, **76**, 5014 (1954); P. R. Jefferies and B. Milligan, *J. Chem. Soc.*, 2363 (1956).
58. J. English, Jr., and J. D. Gregory, *Org. Reactions*, **7**, 400 (1953).
59. H. I. Hadler and A. C. Kryger, *J. Org. Chem.*, **25**, 1896 (1960).
60. M. Sletzinger and C. R. Dawson, *J. Org. Chem.*, **14**, 849 (1949).
61. W. E. Rosen *et al.*, *J. Org. Chem.*, **29**, 1723 (1964).
62. R. J. Tedeschi and A. E. Brown, *J. Org. Chem.*, **29**, 2051 (1964).
63. E. L. Jenner, *Org. Syn.*, **40**, 90 (1960).
64. J. D. LaZerte and R. J. Kosher, *J. Am. Chem. Soc.*, **77**, 910 (1955).

E. Organometallic Reactions

A comprehensive reference work on the Grignard reagent [1] is available. Oxidation of Grignard reagents to form alcohols is discussed in D.2 of this chapter.

1. From Carbonyl Compounds, Esters, and Carbonates

$$\underset{R''}{\overset{R'}{\diagdown}}C=O \xrightarrow{RMgX} \underset{R''}{\overset{R'}{\diagdown}}\underset{R}{\overset{OMgX}{C}} \xrightarrow{HOH} \underset{R''}{\overset{R'}{\diagdown}}\underset{R}{\overset{OH}{C}}$$

4-XV

$$\underset{R''}{\overset{R'}{\diagdown}}C=O \xrightarrow{LiR} \underset{R''}{\overset{R'}{\diagdown}}\underset{R}{\overset{OLi}{C}} \xrightarrow{HOH} \underset{R''}{\overset{R'}{\diagdown}}\underset{R}{\overset{OH}{C}}$$

R' = H, alkyl or alkoxy; R'' = alkyl or alkoxy

In the case of the Grignard reagent the reaction proceeds through coordination of the carbonyl compound with the magnesium atom of the reagent

References for Section E are on pp. 230–231.

followed by a slower shift of the alkyl group:

$$\begin{array}{c} R' \\ \diagdown \\ R'' \diagup \end{array} C \!\!=\!\! OMgX \longrightarrow 4\text{—}XV$$

The primary alcohol with one more carbon atom is made by the reaction of the Grignard reagent with formaldehyde. The method can be carried out straightforwardly by vaporizing formaldehyde into the Grignard reagent, but it is inconvenient because of the tendency of formaldehyde to repolymerize and coat the walls of the entrance tube. Three methods are known to avoid this difficulty; one employs paraformaldehyde directly (see Ex. a), another employs methyl formate in which half the Grignard reagent reduces the intermediate aldehyde formed (see Ex. b), and a third generates formaldehyde by heating cyclohexanol hemiformal;

the gas thus formed is swept into the reaction mixture [2]. These methods may not be efficient enough to replace the direct formaldehyde introduction, but they should be considered.

Secondary alcohols are made by the addition of Grignard reagents to aldehydes other than formaldehyde. The reaction is extremely versatile without many exceptions to note. The most important comment to make is that the reaction is quite exothermic in the majority of cases, and precaution must be taken to dissipate heat.

Tertiary alcohols are made by the addition of Grignard reagents to ketones. Here the problem of steric interaction arises and leads to reduction of the ketone or to enolization and condensation. A specific example of reduction is shown. The isopropylmagnesium bromide gives no addition product whatever;

$$((CH_3)_2CH)_2C\!\!=\!\!O + (CH_3)_2CHMgBr \xrightarrow{\;H_2O\;}$$

$$((CH_3)_2CH)_2CHOH + CH_3CH\!\!=\!\!CH_2 + MgBrOH$$

n-propylmagnesium bromide gives about 30% addition (i.e., diisopropyl n-propyl carbinol) and ethylmagnesium bromide about 80% [3]. The corresponding dialkylmagnesium gives no more and sometimes less addition than the Grignard reagent. Another way of making the tertiary alcohols is by the addition of 2 moles of the Grignard reagent to esters or 3 moles to carbonates, if all three alkyl or aryl groups are to be derived from the Grignard reagent. It has been demonstrated in certain cases that anisole as a solvent, which coordinates more weakly than ether with the magnesium atom in the Grignard reagent, gives a higher yield of tertiary alcohols [4]. The stereospecificity of addition can be controlled somewhat by the choice of the proper environment. The addition of the methyl Grignard reagent to 4-t-butylcyclohexanone yields the tertiary alcohol, the cis/trans ratio of which varies from 1.02 to 1.84 in going from methylmagnesium iodide to methylmagnesium bromide in combination with magnesium bromide. Tetrahydrofuran rather than ether as a solvent raises the maximum of the ratio still further to 2.26 [5]. Interestingly, Grignard reagents can be added to cyclic ketals to give alcohols as shown in Ex. c.6.

Addition of Grignard reagents to Mannich bases (and for that matter any aminoketone) generally gives poor yields of alcohols because of complexing of the Grignard reagent with the amino group, but despite this fact the reaction is sometimes valuable. Two examples are shown. Alkyllithiums are superior to Grignard reagents in addition to dialkylaminoketones [8].

$$RMgX + R'COCH_2CH_2NR_2 \longrightarrow \overset{\overset{\displaystyle OMgX}{|}}{\underset{\underset{\displaystyle R}{|}}{R'CCH_2CH_2NR_2}} \qquad \text{[ref. 6]}$$

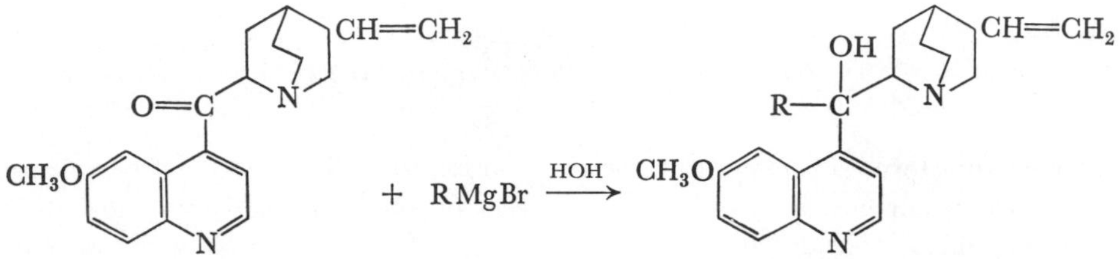

Quininone   + RMgBr $\xrightarrow{\text{HOH}}$ Isobutylquinidine, 40%
R = (CH₃)₂CHCH₂-

[ref. 7]

Organolithium compounds (or other organometallic compounds, particularly organosodium and organopotassium compounds) can be used in place of the Grignard reagent to prepare alcohols; see [9]. They are useful in bringing about addition to a hindered carbonyl rather than reduction as demonstrated in Ex. c.5. Although the sodium acetylide is usually employed as shown (Nef reaction) to prepare ethynylcarbinols, in certain cases lithium acetylide in

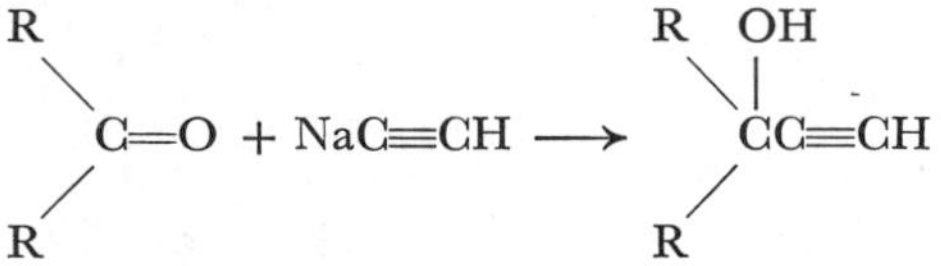

$$\underset{R}{\overset{R}{\diagdown}}C{=}O + NaC{\equiv}CH \longrightarrow \underset{R}{\overset{R}{\diagdown}}\overset{\overset{\displaystyle OH}{|}}{C}C{\equiv}CH$$

liquid ammonia [10] or lithium acetylide in ethylenediamine [11] leads to better yields. For example, β-ionone with lithium acetylide in liquid ammonia gives the alcohol in 95% yield (79% conversion), while with sodium acetylide under similar conditions the yield was 74% (27% conversion).

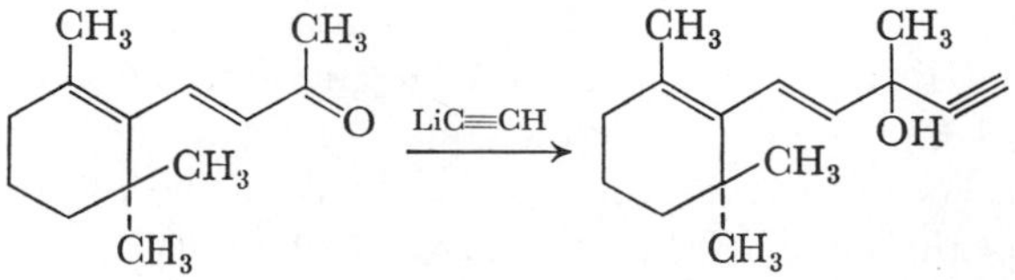

β-Iononeethynlcarbinol   95%

The reference useful in examining the potentialities of organolithium compounds is [12]. These general points can be made: an inert atmosphere must be maintained; organolithium compounds are less complexed with the solvent than Grignard reagents, but exist in polymeric form; and the commercial availability of high-grade butyllithium encourages more general use.

Curiously enough, the commercial metal containing about 0.02–0.2% sodium gives a better yield of alkyllithium than pure lithium metal [13]. A dull coating appears on the pure lithium surface as it is brought into contact with the halide, while the impure lithium surface remains shiny. Simple addition of sodium to lithium does not overcome the difficulties with pure lithium–it must be intimately combined.

(a) Preparation of 1-pentanol from paraformaldehyde. Butylmagnesium bromide from 97 g. of n-butyl bromide was treated with stoichiometric amounts of powdered and dried paraformaldehyde and allowed to stand for 5 days. The mixture was poured onto ice and dilute hydrochloric acid and the organic layer separated, washed with aqueous sodium bisulfite solution, and allowed to stand in contact with this solution for several days to remove $(C_5H_{11}O)_2CH_2$. Distillation of the dried ether solution gave 45.2 g. (92.5% based on paraformaldehyde) of 1-pentanol [14].

(b) Preparation of neopentyl alcohol from methyl formate. To 27 moles of t-butylmagnesium chloride in 10 l. of ether was added 13.5 moles of methyl formate. After the usual isolation procedure 9.75 moles (72%) of neopentyl alcohol was obtained [15].

(c) Other examples

 (1) Cyclohexylcarbinol (64–69% from cyclohexyl chloride, Mg, and formaldehyde vapor) [16].

 (2) Methylisopropylcarbinol (secondary) (53–54% from acetaldehyde, Mg, and isopropyl bromide) [17].

 (3) Triphenylcarbinol (tertiary) (89–93% from bromobenzene, Mg, and ethyl benzoate) [18].

 (4) 1-(α-Pyridyl)-2-propanol (44–50% from α-picoline, phenyllithium, and acetaldehyde) [19].

 (5) Tri-t-Butylcarbinol (81% from t-butyllithium and di-t-butyl ketone) [20].

 (6) 1-Methyl-1-cyclohexyl 2-hydroxyethyl ether [21]

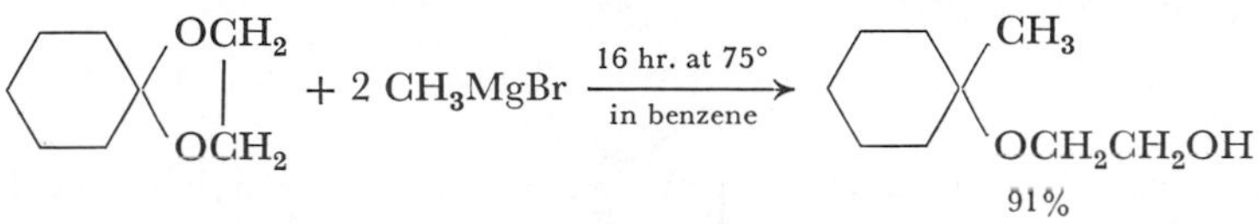

 (7) 1-Methyl-1-cyclobutanol (60% from 1-bromo-4-pentanone, Mg, and a trace of $HgCl_2$ refluxed in THF) [22].

2. From Epoxides and Oxetanes

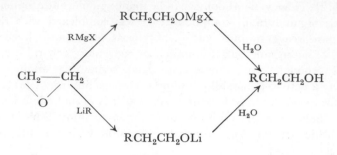

The 3- and 4-membered cyclic ethers open readily on treatment with organolithium compounds or the Grignard reagent, the product on hydrolysis giving an alcohol. The synthesis represents a common procedure for adding carbon atoms to an organic compound. Ethylene oxide and trimethylene oxide give primary alcohols, although with secondary and tertiary Grignard reagents increasing amounts of the halohydrin are produced, presumably because of the reaction of the magnesium halide present on the cyclic ether [23]. In order to avoid the incursion of the halohydrin or products from the rearrangement of the epoxide, i.e.,

$$CH_3CH\overset{O}{\overset{\diagdown}{\underset{}{—}}}CH_2 + MgBr_2 \longrightarrow CH_3CH_2\overset{H}{\underset{|}{C}}{=}O + RMgBr \longrightarrow CH_3CH_2CHOHR$$

aralkylmagnesium or organolithium compounds can be used instead of the Grignard reagent. In the reaction devoid of these complicating factors, the R of the Grignard reagent, being a nucleophilic reagent, should add to the least branched carbon of an unsymmetrical epoxide [24]:

$$CH_3CH\overset{O}{\overset{\diagdown}{\underset{}{—}}}CH_2 \xrightarrow[\text{2. H}_2\text{O}]{\text{1. RMgBr}} CH_3CHOHCH_2R$$

A carbene may intervene as an intermediate with strong bases (Ex. b.3).

(a) Preparation of n-hexyl alcohol (60–62 % from n-butyl bromide and ethylene oxide) [25].

(b) Other examples

(1) Phenylbenzylcarbinol (70–72 % from phenyllithium and styrene oxide) [26].

(2) Propylisopropylcarbinol (63 % from isoamylene oxide and ethylmagnesium bromide) [27].

(3) Nortricyclanol (55 % from exo-2,3-epoxybicyclo[2.2.1]heptane) [28].

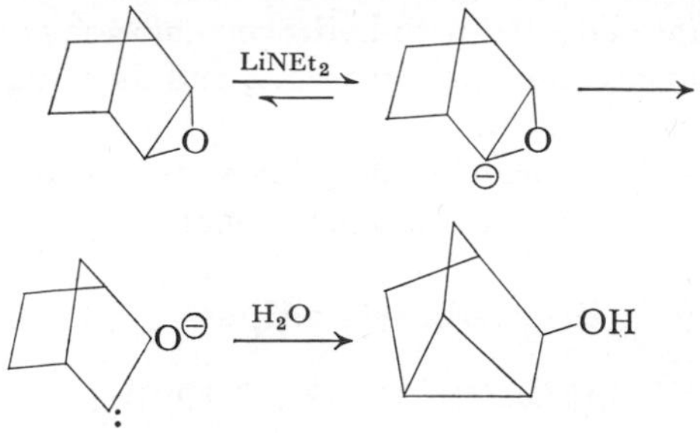

3. From Carbonyl Compounds and α-Halo Esters (Reformatsky)

$$\underset{/}{\overset{\backslash}{C}}{=}O + XZnCH_2COOR \longrightarrow \underset{/}{\overset{\backslash}{C}}\underset{CH_2COOR}{\overset{OZnX}{}} \xrightarrow{H_2O} \underset{/}{\overset{\backslash}{C}}\underset{CH_2COOR}{\overset{OH}{}}$$

This synthesis has been reviewed [29]. It is similar to the Grignard addition except that zinc is commonly used instead of magnesium. However, magnesium is preferred in the case of t-butyl halo esters presumably because self-condensation is retarded (see Ex. b.4). As a rule, zinc gives better yields, perhaps because it has less tendency to add to the carbonyl group of the carboalkoxy function. By this synthesis secondary or tertiary alcohol groups may be located at the β-position or beyond to certain functional groups such as the ester or amino groups. These esters in turn may be hydrolyzed to hydroxy acids or dehydrated to give unsaturated esters. Different forms of zinc have been used, although it is desirable to have the metal pure with a fresh, clean surface [29, 30, 31]. Copper has been used with zinc [32]; in fact, it has been claimed that the addition of copper powder in amounts one-tenth to one-sixth that of zinc dust increases the yield of the hydroxy ester around 10–30% [33]. Iodine, mercuric chloride alone, and with copper(II) ethyl acetoacetate have been used as promoters [34, 35]. Various solvents have been employed, the usual one for ketones being a mixture of equal amounts of benzene and toluene. However, hindered ketones lead to better yields with a mixture of benzene and ether [36].

The yields of the Reformatsky reaction are fair to good. It is the preferred reaction in preparing certain types such as unsaturated acids with branching on the β-carbon atom or the α- and β-carbon atoms [29] and in preparing coumarins [37].

(a) **Preparation of ethyl β-phenyl-β-hydroxypropionate** (61–64% from ethyl bromoacetate, powdered zinc, and benzaldehyde in benzene-ether) [30].

(b) **Other examples**

(1) **Diethyl 2-methyl-3-hydroxy-1,15-pentadecanedioate,**

$$C_2H_5OOC(CH_2)_{11}CHOHCHCOOC_2H_5$$
$$\underset{CH_3}{|}$$

(93.5 % on adding zinc activated with hydrogen chloride and iodine to ethereal ethyl 13-aldehydotridecanoate under nitrogen and then adding ethyl α-bromopropionate to this mixture) [38].

(2) **Ethyl 4-ethyl-2-methyl-3-hydroxyoctanoate** (87 % from zinc foil, 2-ethylhexanal, and ethyl α-bromopropionate in thiophene-free benzene) [31].

(3) **Ethyl α-(α-hydroxybenzyl) adipate,**

$$C_6H_5CHOHCH(CH_2)_3COOC_2H_5$$
$$|$$
$$COOC_2H_5$$

(60 % from a mixture of benzaldehyde, ethyl α-bromoadipate, zinc powder, and some copper(II) ethyl acetoacetate in benzene-toluene to which mercuric chloride was added) [35].

(4) **t-Butyl β-hydroxy-β,β-diphenylisobutyrate** (81 % from t-butyl α-bromopropionate, benzophenone, and magnesium refluxed in ether; zinc is less satisfactory in this reaction) [39].

1. M. S. Kharasch and O. Reinmuth, *Grignard Reactions of Non-Metallic Substances*, Prentice-Hall, Englewood Cliffs, N.J., 1954.
2. C. Michel and T. Tchelitcheff, *Bull. Soc. Chim. France*, 2230 (1964).
3. D. O. Cowan and H. S. Mosher, *J. Org. Chem.*, **27**, 1 (1962).
4. R. N. Lewis and J. R. Wright, *J. Am. Chem. Soc.*, **74**, 1253 (1952).
5. W. J. Houlihan, *J. Org. Chem.*, **27**, 3860 (1962).
6. R. Baltzly and J. W. Billinghurst, *J. Org. Chem.*, **30**, 4330 (1965).
7. R. B. Woodward et al., *J. Am. Chem. Soc.*, **67**, 1425 (1945).
8. J. Wommack and D. E. Pearson, unpublished work.
9. A. Schaap et al., *Rec. Trav. Chim.*, **84**, 1200 (1965).
10. W. Oroshnik and A. D. Mebane, *J. Am. Chem. Soc.*, **71**, 2062 (1949).
11. J. W. Huffman and P. G. Arapakos, *J. Org. Chem.*, **30**, 1604 (1965).
12. *Annotated Bibliography on Organolithium Compounds and Supplements* 1–11, 1949–63, Lithium Corporation of America, Box 428, Bessemer City, North Carolina.
13. C. W. Kamienski and D. L. Esmay, *J. Org. Chem.*, **25**, 1807 (1960).
14. G. V. Medoks and L. E. Ozerskaya, *J. Gen. Chem. USSR (Eng. Transl.)*, **30**, 1641 (1960).
15. L. H. Sommer et al., *J. Am. Chem. Soc.*, **76**, 803 (1954).
16. H. Gilman and W. E. Catlin, *Org. Syn.*, Coll. Vol. **1**, 188 (1941).
17. N. L. Drake and G. B. Cooke, *Org. Syn.*, Coll. Vol. **2**, 406 (1943).
18. W. E. Bachmann and H. P. Hetzner, *Org. Syn.*, Coll. Vol. **3**, 839 (1955).
19. L. A. Walter, *Org. Syn.*, Coll. Vol. **3**, 757 (1955).
20. P. D. Bartlett and E. B. Lefferts, *J. Am. Chem. Soc.*, **77**, 2804 (1955).
21. R. A. Mallory et al., *Proc. Chem. Soc.*, 416 (1964).
22. Y. Leroux, *Bull. Soc. Chim. France*, 359 (1968).
23. R. C. Huston and A. H. Agett, *J. Org. Chem.*, **6**, 123 (1941); S. Searles, *J. Am. Chem. Soc.*, **73**, 124 (1951).
24. R. E. Parker and N. S. Isaacs, *Chem. Rev.*, **59**, 737 (1959).
25. E. E. Dreger, *Org. Syn.*, Coll. Vol. **1**, 306 (1941).
26. S. J. Cristol et al., *J. Am. Chem. Soc.*, **73**, 816 (1951).
27. M. S. Malinovskiĭ and B. N. Konevichev, *Zh. Obshch. Khim.*, **18**, 1833 (1948); *C.A.*, **43**, 3776 (1949).
28. J. K. Crandall, *J. Org. Chem.*, **29**, 2830 (1964).
29. R. L. Shriner, *Org. Reactions*, **1**, 1 (1942).
30. C. R. Hauser and D. S. Breslow, *Org. Syn.*, Coll. Vol. **3**, 408 (1955).
31. K. L. Rinehart, Jr., and E. G. Perkins, *Org. Syn.*, Coll. Vol. **4**, 444 (1963).

32. J. A. Nieuwland and S. F. Daly, *J. Am. Chem. Soc.*, **53,** 1842 (1931).
33. Z. Horii *et al.*, *J. Pharm. Soc., Japan*, **73,** 895 (1953); *C.A.*, **48,** 11329 (1954).
34. D. D. Phillips and D. N. Chatterjee, *J. Am. Chem. Soc.*, **80,** 4364 (1958); R. E. Miller and F. F. Nord, *J. Org. Chem.*, **16,** 728 (1951).
35. R. and S. Gelin, *Compt. Rend.*, **255,** 1400 (1962).
36. J. Cason and R. J. Fessenden, *J. Org. Chem.*, **22,** 1326 (1957).
37. F. Bohlmann, *Chem. Ber.*, **90,** 1512, 1519 (1957).
38. M. Stoll, *Helv. Chim. Acta*, **34,** 678 (1951).
39. T. Moriwake, *J. Org. Chem.*, **31,** 983 (1966).

F. Addition of Simple Anions or Nucleophilic Molecules to Carbonyl Compounds

The Grignard reagent is an example of addition of an anion type of reagent to carbonyl compounds, but it is important enough to have been discussed separately in E. The rest of the anionlike additions are discussed in F and G, those of this section being devoted to the addition of small molecules such as water, bisulfite, alcohol, and hydrogen cyanide. With the exception of cyanohydrins, the products are not very useful other than as vehicles of isolation. But the type of reaction that occurs serves to introduce the more extensive topic of aldol condensation, the products being derived from the addition of a carbanion to a carbonyl compound.

Two types of addition are possible, either that of a nucleophilic anion which needs no catalyst or that of a molecule which depends more or less on an acid catalyst:

$$R_2C{=}O + CN^{\ominus} \longrightarrow R_2C\overset{O^{\ominus}}{\underset{CN}{\diagup}} \xrightarrow{H_2O} R_2C\overset{OH}{\underset{CN}{\diagup}}$$

$$R_2C{=}O + R'OH \underset{}{\overset{H^{\oplus}}{\rightleftharpoons}} R_2\overset{\oplus}{C}OH \rightleftharpoons R_2C\overset{OH}{\underset{\overset{\oplus}{O}R'{-}H}{\diagup}} \underset{}{\overset{-H^{\oplus}}{\rightleftharpoons}} R_2C\overset{OH}{\underset{OR'}{\diagup}}$$

Aldehydes tend to add anions or nucleophilic molecules more readily than ketones. Electron-withdrawing groups attached to the carbonyl group also favor the addition of nucleophiles. For example, chloral, formaldehyde, oxomalonic ester, and perhalo aldehydes and perhalo ketones in general add the elements of water spontaneously to form hydrates:

$$O{=}C(COOC_2H_5)_2 \xrightarrow{H_2O} \overset{HO}{\underset{HO}{\diagup\!\!\!\!\diagdown}}C{-}(COOC_2H_5)_2 + heat$$

Glyoxal   Ethyl dihydroxymalonate

$$O{=}\overset{H}{\underset{}{C}}\overset{H}{\underset{}{C}}{=}O$$

References for Section F are on p. 233.

is probably the most reactive of all aldehydes in this respect. Most aliphatic straight-chain aldehydes form hemihydrates with water,

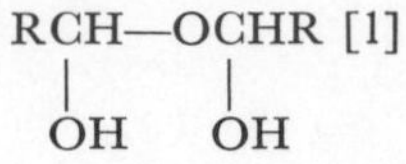

$$\underset{\text{OH}}{\text{RCH}} — \underset{\text{OH}}{\text{OCHR}} \quad [1]$$

On the other hand, some ketones with large groups attached are entirely inert to these addition reagents even with acid catalysts. Such compounds as aceto-mesitylene, diisopropyl ketone, and many benzophenones fall into this group. There is also a series of carbonyl addition compounds which lose the elements of water to form unsaturated compounds, thus mitigating the formation of alcohols. The reagents here have an electron-withdrawing group attached to the atom which adds to the carbonyl group. They are hydroxylamine or hydrazine with all of their substituted derivatives such as phenylhydrazine and semicarbazide. The reason for this loss of water in the rate-determining step is that the electron-withdrawing group has facilitated the ease of removal of the proton on the amino group:

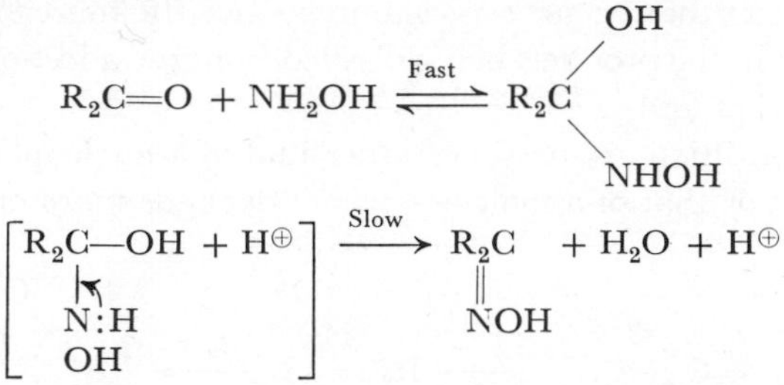

$$\text{R}_2\text{C}=\text{O} + \text{NH}_2\text{OH} \underset{\text{Fast}}{\rightleftarrows} \text{R}_2\text{C} \overset{\text{OH}}{\underset{\text{NHOH}}{\diagdown}}$$

$$\left[\begin{array}{c} \text{R}_2\text{C} — \text{OH} + \text{H}^{\oplus} \\ \text{N:H} \\ \text{OH} \end{array} \right] \xrightarrow{\text{Slow}} \text{R}_2\text{C} \underset{\text{NOH}}{\overset{\parallel}{}} + \text{H}_2\text{O} + \text{H}^{\oplus}$$

Therefore these compounds do not stop at the alcohol stage. Ammonia is a reagent which should form an α-aminoalcohol with a carbonyl compound, but the latter derivative is so reactive that further changes take place [2]. Benzaldehyde yields hydrobenzamide, $C_6H_5CH(N=CHC_6H_5)_2$; carbonyl compounds and secondary amines yield the stable diamino compounds,

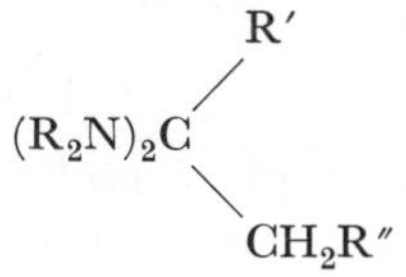

$$(\text{R}_2\text{N})_2\text{C} \overset{\text{R}'}{\underset{\text{CH}_2\text{R}''}{\diagdown}}$$

or the enamines,

$$\text{R}_2\text{NC} \overset{\text{R}'}{\underset{\text{CHR}''}{\diagdown}}$$

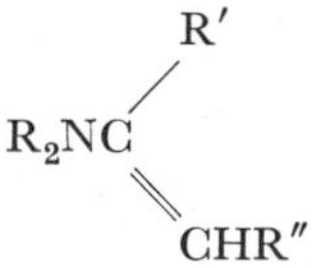

1. From Sodium Bisulfite

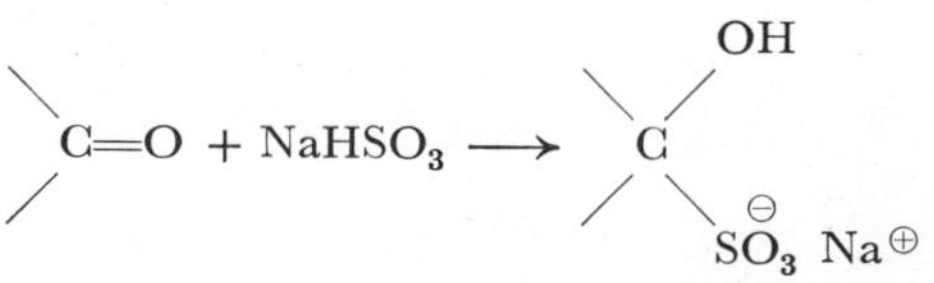

$$\diagdown \text{C}=\text{O} + \text{NaHSO}_3 \longrightarrow \overset{\text{OH}}{\underset{\overset{\ominus}{\text{SO}_3} \ \text{Na}^{\oplus}}{\diagup}} \diagdown \text{C} \diagup$$

These substances are quite frequently sparingly soluble in the reagent solution. Their formation provides a convenient method for the separation of a carbonyl compound from other substances, the carbonyl compound then being regenerated by acid or base. Most aldehydes, methyl aliphatic ketones, cycloheptanone, and smaller cyclic ketones form adducts.

(a) **Preparation of *p*-acetaminobenzaldehyde** (via bisulfite addition compound, about 60% from impure *p*-aminobenzaldehyde) [3].

(b) **Preparation of cycloheptanone** (via bisulfite addition compound, 33–36% from cyclohexanone) [4].

2. From Alkali Cyanides

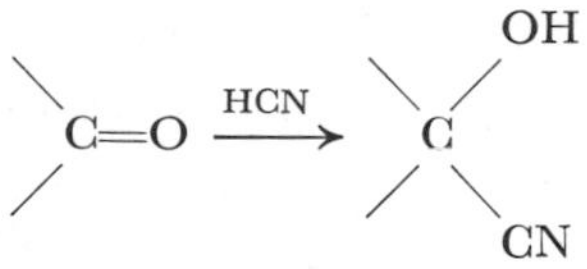

The cyanide ion is more effective in adding to carbonyl compounds than is the bisulfite anion, a fact to be anticipated in view of the ready addition of the cyanide anion to an aromatic ketone, such as acetophenone. If hydrogen cyanide is used as the addition agent, a small amount of sodium cyanide or an aliphatic amine is added to facilitate addition. If hydrogen cyanide is to be avoided, the bisulfite addition compounds can be treated with sodium cyanide.

An unusual addition of cyanide has been reported recently [5].

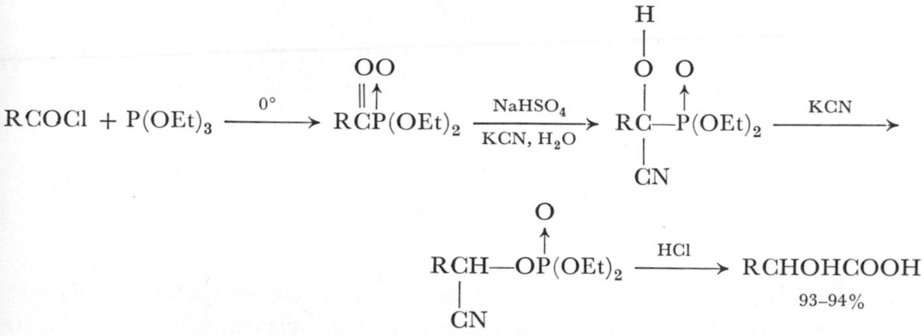

$$ R = C_9H_{19} \text{ or higher} $$

(a) **Preparation of mandelic acid** (50–52% from benzaldehyde sodium bisulfite adduct plus aqueous sodium cyanide) [6].

(b) **Glycolonitrile,** $HOCH_2CN$ (76–80% from formaldehyde, KCN, and dilute H_2SO_4) [7].

1. D. L. Klass et al., J. Org. Chem., **28**, 3029 (1963).
2. R. H. Hasek et al., J. Org. Chem., **26**, 1822 (1961).
3. E. Campaigne et al., Org. Syn., Coll. Vol. **4**, 31 (1963).
4. Th. J. de Boer and H. J. Backer, Org. Syn., Coll. Vol. **4**, 225 (1963).
5. Y. Okamoto et al., Kogyo Kagaku Zasshi, **71**, 187 (1968); C.A., **69**, 3288 (1968).
6. B. B. Corson et al., Org. Syn., Coll. Vol. 1, 336 (1941).
7. R. Gaudry, Org. Syn., Coll. Vol. **3**, 436 (1955).

G. Addition of Carbanions

The principles of carbanion addition to form alcohols are simple. First, a carbanion must be generated from an active hydrogen compound.

$$RONa + CH_3Y \rightleftharpoons ROH + \overset{\ominus}{C}H_2Y$$

The ease of carbanion formation depends on the electron-withdrawing abilities of Y and the stabilization of the anion $\overset{\ominus}{C}H_2Y$ by resonance.

The decreasing order of the ability of Y to form an anion is:

$$NO_2, \quad SO_2R, \quad CH{=}O, \quad \overset{\overset{\displaystyle O}{\|}}{C}{-}R, \quad C{\equiv}N, \quad COOR, \text{ and } C_6H_5$$

With the latter group attached to a methyl, powerful bases must be used to produce a significant concentration of the anion, $C_6H_5CH_2^{\ominus}$. Second, the anion must add to a carbonyl structure:

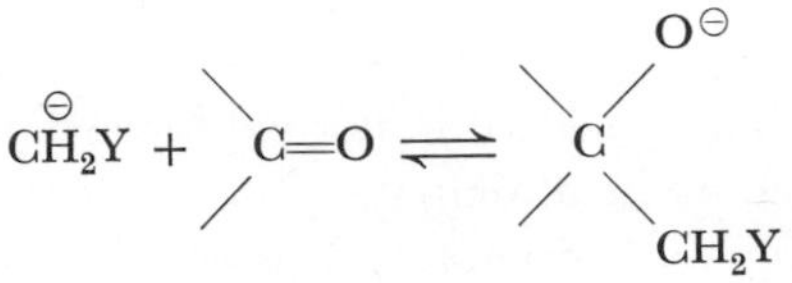

$$\overset{\ominus}{C}H_2Y + \diagup\overset{\diagdown}{C}{=}O \rightleftharpoons \diagup\overset{O^{\ominus}}{\underset{CH_2Y}{\overset{\diagdown}{C}}}$$

To obtain appreciable quantities of the product

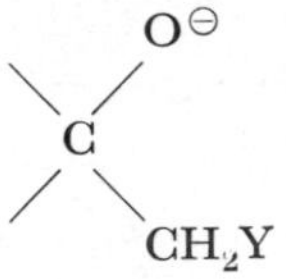

$$\diagup\overset{O^{\ominus}}{\underset{CH_2Y}{\overset{\diagdown}{C}}}$$

its anion must be more stable than $CH_2^{\ominus}Y$. Or another way of stating the same principle is that the acid

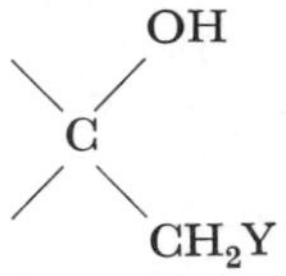

$$\diagup\overset{OH}{\underset{CH_2Y}{\overset{\diagdown}{C}}}$$

must be more acidic than CH_3Y. An unusual situation exists in regard to adducts of benzophenone and similar compounds. Large quantities of base favor the anion, but catalytic quantities favor the cleavage products. Thus precautions must be taken to convert the anion instantaneously into the alcohol in order to prevent cleavage [1]:

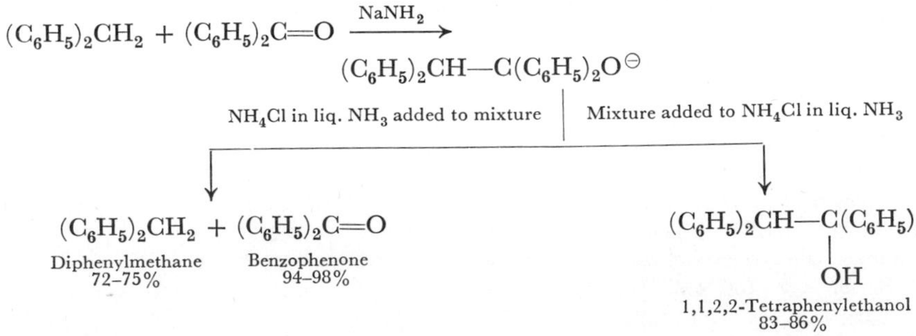

$$(C_6H_5)_2CH_2 + (C_6H_5)_2C{=}O \xrightarrow{\text{NaNH}_2}$$

$$(C_6H_5)_2CH{-}C(C_6H_5)_2O^{\ominus}$$

NH₄Cl in liq. NH₃ added to mixture | Mixture added to NH₄Cl in liq. NH₃

$$(C_6H_5)_2CH_2 + (C_6H_5)_2C{=}O$$

Diphenylmethane   Benzophenone
72–75%   94–98%

$$(C_6H_5)_2CH{-}\underset{OH}{\overset{|}{C}}(C_6H_5)_2$$

1,1,2,2-Tetraphenylethanol
83–86%

References for Section G are on pp. 244–245.

Another factor must be considered: the product

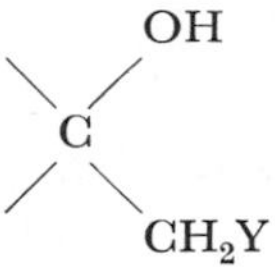

$$\underset{\diagdown}{\overset{\diagup}{C}}\underset{CH_2Y}{\overset{OH}{\diagdown}}$$

has a great tendency to dehydrate to form the unsaturated compound

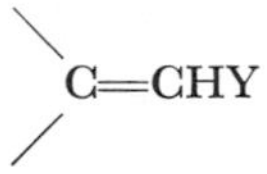

$$\overset{\diagdown}{\underset{\diagup}{C}}\!\!=\!\!CHY$$

particularly if the hydroxyl group is attached to a secondary or tertiary carbon atom. Because the methylene hydrogen is appreciably acidic, removal after the loss of the secondary or tertiary alcohol group in acid solution is quite facile; in basic solution the proton may be removed before or simultaneously with the loss of the hydroxyl group. This chapter, then, is devoted to the preparation of β-hydroxy compounds by taking precautions or finding favorable factors in structure which avoid dehydration.

1. From Carbonyl Compounds (Aldol or Ketol) or from Carbonyl Compounds and Acid Derivatives (Claisen-Schmidt)

$$RCH_2CHO \xrightarrow{\;OH^{\ominus}\;} RCH_2CHOHCHRCH\!=\!O$$

This reaction has been reviewed thoroughly [2] and is discussed in Chapter 10, F.3. Unless otherwise referenced, comments are abstracted from the review. Aldol condensations are quite complex, the choice of catalyst and conditions within narrow limits sometimes determining success or failure. Basic ion exchange resins are the best catalysts for condensations of aldehydes; the yields of aldols, however, decrease as the chain length of the aldehyde increases. Methylanilinomagnesium bromide in ether-benzene is the best catalyst for condensation of ketones. Many side reactions are possible:

1. Dehydration of aldol → $RCH_2CH\!=\!CRCH\!=\!O$ (minimized by avoiding strong acid and high temperatures).

2. Trimerization of aldehyde →

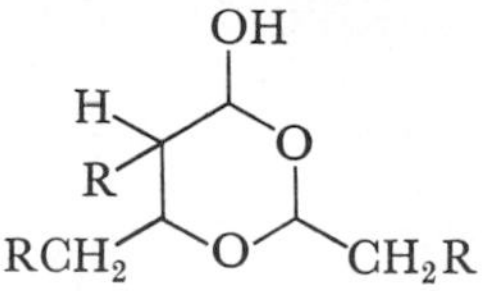

(favored by Na_2CO_3 or other weak base catalysts)

3. Dimerization of aldol →

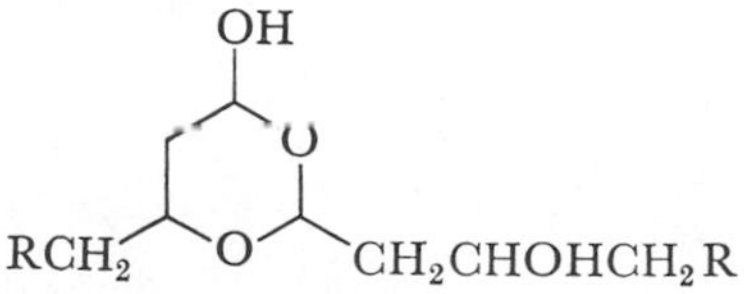

(spontaneously on standing)

4. Retrograde condensation → $RCH_2CH\!=\!O$ (favored by steric effects).

5. Oxidation-reduction $\rightarrow$ $RCH_2CHOHCHRCH_2OH + RCH_2CO_2H$ (or other redox combinations).

6. Michael and other reactions.

Although aldols or ketols may be dehydrated readily to give the corresponding α,β-unsaturated types, in some cases, as in the well-known Claisen-Schmidt reaction, the dehydration occurs, at times in good yield, in the course of the condensation [3]:

$$C_6H_5CHO + CH_3COC_6H_5 \xrightarrow[\substack{\text{Water, alcohol,}\\15-30°}]{\text{NaOH}} C_6H_5CH{=}CHCOC_6H_5$$

Benzalacetophenone 85%

This reaction takes place in the condensation of an aldehyde containing no α-hydrogen atoms with an aliphatic aldehyde or ketone, usually under mild alkaline conditions.

Mixed aldehydes or ketones, or combinations of both, tend to give mixtures of products with the exception of combinations of carbonyl compounds with formaldehyde (G.2). If poor yields are acceptable, other combinations may be employed. However, Wittig's directed aldol synthesis greatly extends the range in using mixed carbonyl compounds [4]:

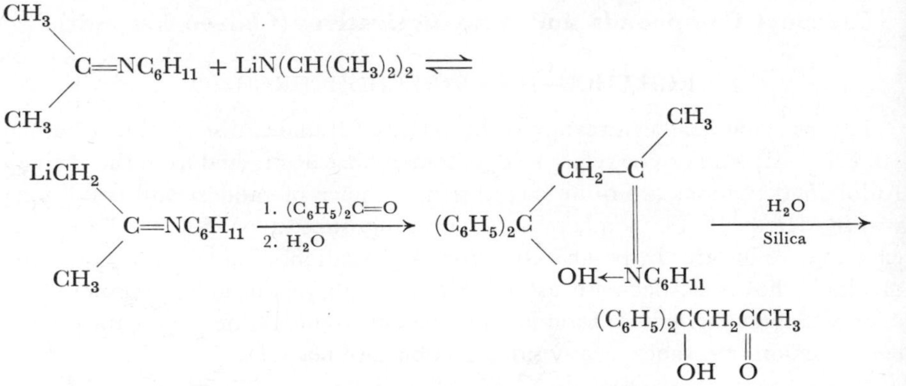

4-4-Diphenyl-4-hydroxy-2-butanone, 68%, m.p. 85°

Use of acid for hydrolysis dehydrates the product, and use of lithium aluminum hydride yields the corresponding β-aminoalcohol.

In certain instances the Michael addition is followed by a ketol condensation as illustrated [5]:

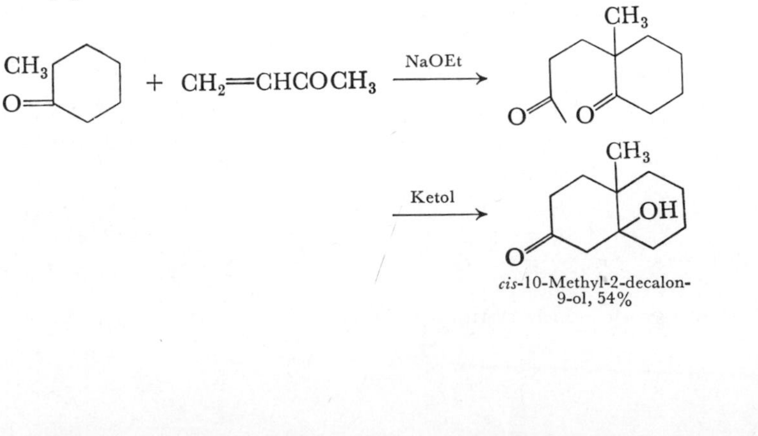

cis-10-Methyl-2-decalon-9-ol, 54%

In a reaction of limited scope, cyclic ketones may be condensed with benzils [6]:

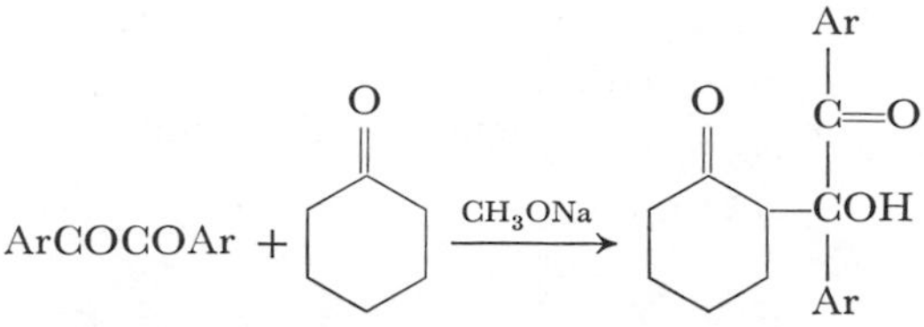

$$ArCOCOAr + \text{(cyclohexanone)} \xrightarrow{CH_3ONa} \text{(product)} \quad 70\text{–}90\%$$

Although not a ketol condensation, one example of the condensation of aldehydes and acid derivatives to form alcohols is given (Ex. c).

(a) Preparation of aldol, $CH_3CHOHCH_2CH{=}O$. Acetaldehyde (100 g.) was added to 200 g. of ice and water and cooled to $-12°$ with mechanical stirring. A solution of 2.5 g. of potassium cyanide in 100 ml. of water was added slowly so that the temperature never rose above $-8°$. The mixture was stored in a refrigerator for 30 hr. and the aldol isolated from it by four extractions with ether after saturating the aqueous solution with salt. Aldol, 40–50 g., b.p. 80–90° at 20 mm., was obtained by distillation of the ether residue [7]; see [8] for an alternative preparation.

(b) Preparation of 1-o-aminophenyl-1-p-methoxyphenyl-2-(2′-pyridyl)-ethanol. 2-Picoline, 0.1 mole, was added in 5 min. to a suspension of

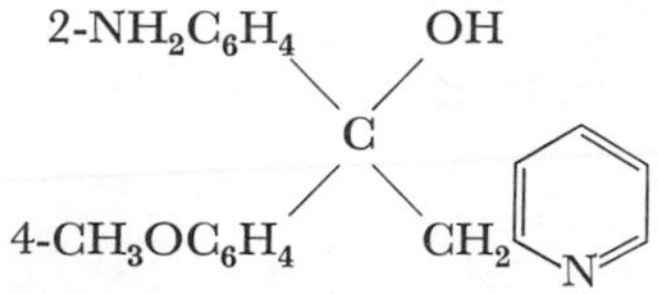

$$
\begin{array}{c}
2\text{-}NH_2C_6H_4 \quad\quad OH \\
\diagdown \quad\quad\quad \diagup \\
C \\
\diagup \quad\quad\quad \diagdown \\
4\text{-}CH_3OC_6H_4 \quad\quad CH_2\text{—(pyridyl)}
\end{array}
$$

sodamide [9] (from 2.5 g. of sodium in 150 ml. of liquid ammonia) and the mixture was stirred for $\frac{1}{4}$ hr. 2-Amino-4′-methoxybenzophenone, 0.05 mole, in 5 ml. of picoline and 40 ml. of dry ether was added during 10 min. and the mixture stirred for 4 hr. After the ammonia was allowed to evaporate, the residue was decomposed with 100 ml. of ether and 30 ml. of water. The yellow titled compound, 10.8 g., m.p. 128–150°, when crystallized from methanol (charcoal) amounted to 8.0 g. (49.6%), m.p. 154–155° [10].

(c) Preparation of methyl 2-α-p-nitrobenzylidencamino-3-hydroxy-3-(p-nitrophenyl)-propionate. p-Nitrobenzaldehyde (7.5 g.) and

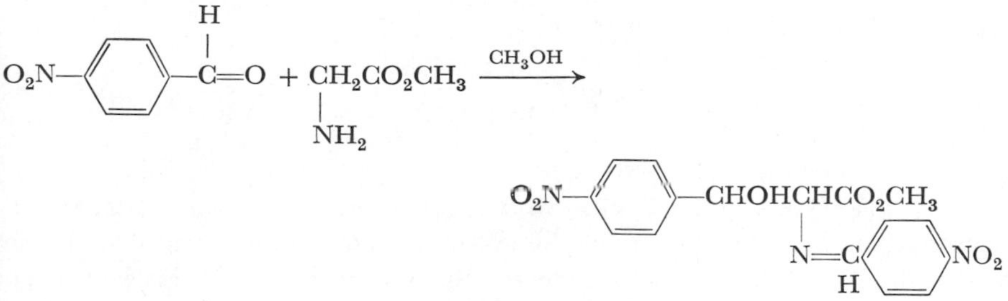

$$O_2N\text{—}\langle\text{—}\rangle\text{—}\overset{H}{\underset{}{C}}{=}O + CH_2CO_2CH_3 \ (\underset{NH_2}{}) \xrightarrow{CH_3OH}$$

$$O_2N\text{—}\langle\text{—}\rangle\text{—}CHOHCHCO_2CH_3 \ \big(\underset{H}{\overset{}{N}}{=}C\text{—}\langle\text{—}\rangle NO_2\big)$$

methyl glycinate (2.3 g.) were allowed to stand in methanol. Nine grams of product deposited from this solution; it melted at 160° after crystallization from isopropyl alcohol. The Schiff base was hydrolysed with methanolic HCl. The active hydrogen on the glycine ester is undoubtedly activated by the intermediate formation of the Schiff base. The *erythro* form

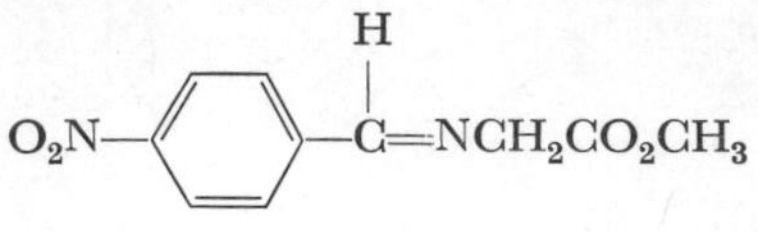

$$O_2N-\langle\bigcirc\rangle-\overset{\overset{\displaystyle H}{|}}{C}{=}NCH_2CO_2CH_3$$

of the phenyl serine was obtained, but with excess methyl glycinate the *threo* form was the predominant isomer [11].

(d) Other examples

(1) **Diacetone alcohol** (71 % by placing barium hydroxide catalyst in a Soxhlet cup, allowing acetone to distill into the catalyst, and extracting the diacetone alcohol from the catalyst; this reaction does not work well with other ketones) [12].

(2) **2,2-Dimethyl-3-hydroxy-3-phenylpropionic acid** (33 % from benzaldehyde, isobutyric anhydride, and sodium isobutyrate; Perkin condensation) [13].

(3) **1,1-Diphenyl-2-(3-pyridyl)-ethanol**

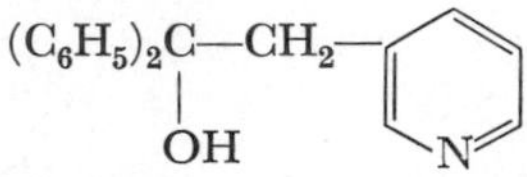

$$(C_6H_5)_2\overset{\overset{\displaystyle}{|}}{\underset{\underset{\displaystyle OH}{|}}{C}}-CH_2-\langle\bigcirc_N\rangle$$

(27 % from 3-picoline, potassium amide, and benzophenone; because the hydrogen atom in 3-picoline is not very active, a powerful base catalyst must be used) [14].

(4) **1-Nitro-2-octanol** (95 % from heptaldehyde, nitromethane, potassium hydroxide, and methanol at 70° for 24 hr.) [15].

(5) **1-Hydroxy-1,1,5-triphenylpentanedione-3,5**

$$(C_6H_5)_2\overset{\overset{\displaystyle}{|}}{\underset{\underset{\displaystyle OH}{|}}{C}}-CH_2COCH_2COC_6H_5$$

(73 % from the dianion of benzoylacetone

$$C_6H_5CO\overset{\ominus}{C}HCO\overset{\ominus}{C}H_2$$

and benzophenone) [16].

(6) **5-Hydroxy-2,4,4-trimethyl-3-octanone** (81 % from butanal, diisopropyl ketone, and N-methylanilino-N-magnesium bromide in benzene; so long as the ketone does not self-condense, this type of reaction works well) [17].

2. From Formaldehyde (Tollen)

This condensation has been reviewed [18]. Almost any active hydrogen compound will add to formaldehyde:

$$\underset{\text{H}}{\overset{\text{H}}{R CH_2 \underset{|}{C}}} = O + H_2C = O \xrightarrow{\text{Ca(OH)}_2} \underset{\underset{\text{CH}_2\text{OH}}{|}}{\overset{\text{CH}_2\text{OH}}{R\underset{|}{C}} - CH = O}$$

4-XVI

If excess formaldehyde is used and prolonged heating, a crossed Cannizzaro reaction will proceed:

$$\text{4-XVI} + CH_2O \xrightarrow{OH^{\ominus}} RC(CH_2OH)_3 + HC\underset{\diagdown O^{\ominus}}{\overset{\diagup\!\!\diagup O}{}}$$

As many hydroxymethyl groups are added as there are active hydrogens, but a single one may be introduced in certain instances by controlling the concentration of the formaldehyde.

(a) **Preparation of pentaerythritol,** $(C(CH_2OH)_4, 73.5\%$ from acetaldehyde) [19].

(b) **Preparation of 1,1,1-trimethylolethane** (94–95% from propionaldehyde) [20].

(c) **Preparation of 2-(2-hydroxyethyl)-pyridine.** An autoclave was charged with 427 g. (4.5 moles) of 2-picoline and 23 g. (0.73 mole) of paraformaldehyde, a molar ratio of six to one, and the mixture was agitated 4 hr. at 165°. From the reaction mixture was obtained 60 g. of 2-(2-hydroxyethyl)-pyridine, a yield of 64% compared with yields of 33–38% in similar experiments which were carried out with methylpyrazine. Less than 5 g. of residue remained in the still pot when the product was distilled, b.p. 110–114° (10 mm.); n^{20}D 1.5374 [21].

(d) **Other examples** with yields are (unless otherwise stated, cited from [22]. This volume is a useful compilation of the linking of one-carbon molecules to other substrates):

(1)

$$C_6H_5CH_2COOH + H_2CO \xrightarrow{OH^{\ominus}} \underset{\underset{CH_2OH}{|}}{C_6H_5 CHCOOH}$$

α-Hydroxymethylphenylacetic
acid, 71–83%, p. 11

(2)

$$CH_3(CH_2)_4C\!\equiv\!CH + H_2CO \xrightarrow{OH^{\ominus}} CH_3(CH_2)_4C\!\equiv\!CCH_2OH$$

Oct-2-yne-1-ol, 57%, p. 47

(3) 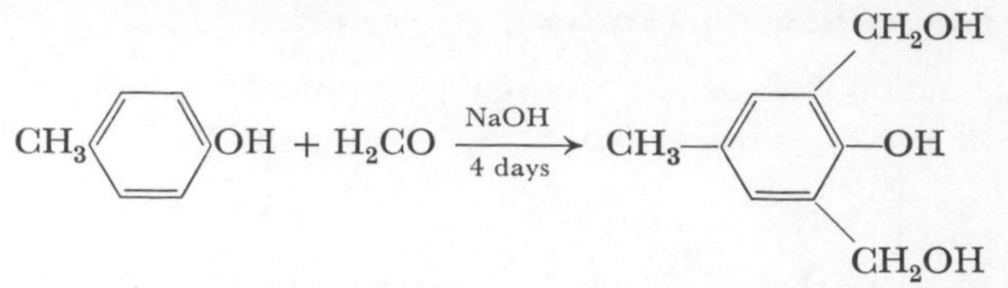 CH_3⟨ ⟩$OH + H_2CO \xrightarrow[\text{4 days}]{\text{NaOH}} CH_3$— —$OH$

with CH_2OH groups

2,6-Di-(hydroxymethyl)-4-cresol,
91%, p. 171

(4) 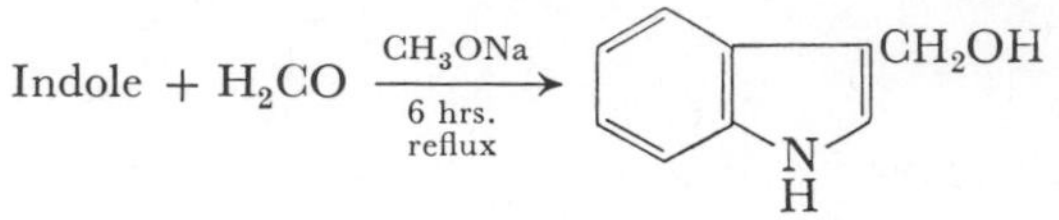   Indole $+ H_2CO \xrightarrow[\substack{\text{6 hrs.}\\\text{reflux}}]{CH_3ONa}$ [indole ring with CH_2OH substituent, N–H]

3-Indolemethanol, 82%, p. 171

(5) 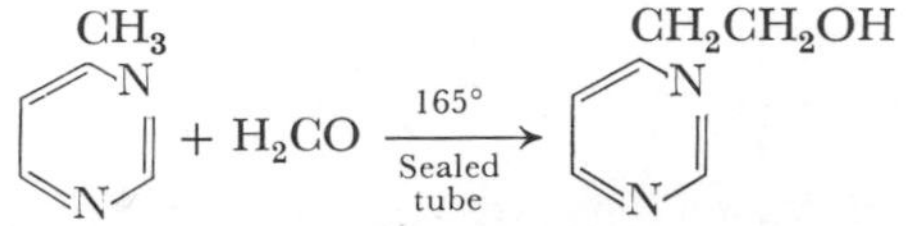   [pyrimidine ring with CH_3, N] $+ H_2CO \xrightarrow[\substack{\text{Sealed}\\\text{tube}}]{165°}$ [pyrimidine ring with CH_2CH_2OH]   [ref. 23]

4-(β-Hydroxyethyl)-
pyrimidine, 87%
conversion

(6) $(CH_3)_3CCOCH_3 + H_2CO \xrightarrow{\text{pH 12}} (CH_3)_3CCOCH(CH_2OH)_2$ [ref. 24]

2-Pivaloyl-1,3-propanediol, 10%

(7) $(C_6H_5)_2CHCN + CH_2O \xrightarrow[\text{C}_5\text{H}_5\text{N, 22 hr.}]{\text{Triton, B}} (C_6H_5)_2CCH_2OH$ [ref. 25]

$|$
CN

94%
α,α-Diphenyl-β-hydroxypropionitrile

3. From Alcohols (Guerbet)

$4\,RCH_2CH_2OH \xrightarrow[\text{Cu bronze}]{\text{Na}}$

$RCH_2CH_2\text{—}CHCH_2OH + RCH_2COOH + RCH_2CH_2OH$
$|$
R

This synthesis occurs at high pressure for primary alcohols which are not branched in the α-position, since an α,β-unsaturated aldehyde, resulting from the corresponding aldol, appears to be an intermediate. However, mixed Guerbet condensations can occur if one of the alcohols lacks branching on the α-carbon atom. The other products obtained in a successful reaction of a straight-chain primary alcohol are the carboxylic acid of the same carbon content and the original alcohol. The synthesis has found limited use because of these restrictions. Small amounts of copper-bronze suppress the oxidation of the alcohol to the corresponding acid in the presence of the sodium alkoxide. The addition of about 0.5 % ferric salt is claimed to more than double the rate of the Guerbet reaction [26]. But dehydrogenation catalysts, such as Raney nickel or palladium, are the most effective in accelerating rates [27]. Yields rarely exceed 70 % when calculated on the basis that 3 moles of the lower produce 1 mole of the higher alcohol [28].

The mechanism is complex and proceeds essentially as follows [27]:

$$RCH_2CH_2OH \xrightarrow{\text{Base}} RCH_2CH{=}O + H_2 \longrightarrow$$

$$RCH_2CH{=}\overset{\overset{\displaystyle R}{|}}{C}CH{=}O \xrightarrow[\text{RCH}_2\text{CH}_2\text{OH}]{H_2 \text{ or}} RCH_2CH_2CHRCH_2OH + RCH_2CH{=}O$$

The condensation as written is partly self-propagating. Addition of the un-saturated aldehyde increases the rate of dehydrogenation but diminishes the amount of hydrogen evolved.

(a) Preparation of 2-hexyldecanol. Octyl alcohol, 130 g., in which 7.6 g. of sodium had been dissolved was placed in an autoclave and heated in the presence of 0.6 g. of copper-bronze. The reaction set in at 210°, as indicated by a sudden rise in pressure. At 295° the pressure was reduced to about 50–60 mm. and heating was continued for 5 hr. Treatment with water gave an organic layer consisting of 45 g. of octyl alcohol, b.p. 195° (34.6%); 42 g. of 2-hexyldecanol, b.p. 170–180° (24 mm.) (52%); and an aqueous layer from which 34 g. of n-caprylic acid, b.p. 120–125° (10 mm.) (70.8%) was isolated [28].

(b) Other examples

(1) 2-Ethylhexanol from n-butyl alcohol (41.5% conversion, 72.6% yield using tripotassium phosphate, CaO, and Cu as a catalyst at 295° for 4 hr.) [29]; (80% conversion, 44% yield by heating butyl alcohol, Na, CaO, and Raney Ni at boiling point, rather than under pressure, for 20 hr.) [30].

4. From Phenylacetic Acid (Ivanov)

The Ivanov reagent is a very specific reagent made by treating sodium phenylacetate with isopropylmagnesium chloride to yield the dianion

$$C_6H_5\overset{\ominus}{C}HCO_2{}^{\ominus}$$

This dianion adds readily to ketones to yield β-hydroxy acids. Recently it has been shown that phenylacetic acid can be converted into the dianion more easily with lithium amide in liquid ammonia.

(a) Preparation of α-(1-hydroxycyclohexyl)-α-phenylacetic acid.

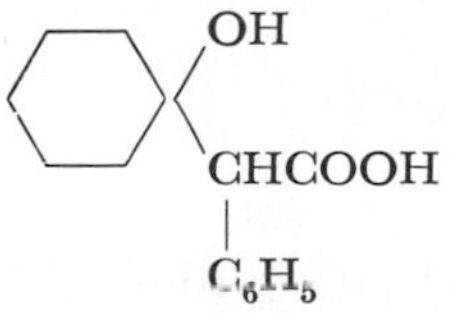

Lithium amide, 0.2 mole, in 250 ml. of liquid ammonia was treated with 0.1 mole of phenylacetic acid in 70 ml. of liquid ammonia. To the green or black solution formed, 0.1 mole of cyclohexanone in 50 ml. of ether was added and a white precipitate formed. After 15 min. the ammonia was removed and 300 ml. of ether was added. The ethereal suspension when shaken with 10% aqueous HCl, dried, and evaporated yielded the crude acid, which upon recrystallization from ethanol gave 21–22 g. (90–93%), m.p. 143–144° [31].

5. From Ketyls and Halides

$$\begin{array}{c} Ar \\ \diagdown \\ C=O \\ \diagup \\ Ar \end{array} \xrightarrow[\text{Liq. NH}_3]{\text{Na}} \begin{array}{c} Ar \quad ONa \\ \diagdown \quad \diagup \\ C \\ \diagup \quad \diagdown \\ Ar \quad Na \end{array} \xrightarrow[\text{2. NH}_4\text{Cl}]{\text{1. (Ar)}_2\text{CHCl}} \begin{array}{c} Ar_2CH \\ | \\ (Ar)_2C\!-\!OH \end{array}$$

1. ArCHO
2. H⊕

ArCHOH
|
(Ar)₂COH

Mono- and dihydric alcohols may be synthesized by treating the alkali metal adduct of an aromatic ketone in liquid ammonia with various reagents [32]. Halides, such as benzyl chloride and benzhydryl chloride, when added to ketyls give monohydric alcohols; aldehydes such as benzaldehyde give 1,2-glycols; and carbon dioxide gives α-hydroxycarboxylic acids [33]. The disodium adducts of the ketone are more readily prepared than those of potassium. Yields of the monohydric and dihydric alcohols vary from 76 to 91%. Benzilic acid was produced from the sodium adduct in a 60–65% yield.

(a) Preparation of 1,1,2,2-tetraphenylethanol

(1) **Disodiobenzophenone.** Benzophenone, 18.2 g., in 50 ml. of dry ether was added to a stirred solution of 4.8 g. of sodium in 250 ml. of liquid ammonia. The solution, which changed in color from blue to purple and finally black, was assumed to contain 0.1 mole of disodiobenzophenone.

(2) **1,1,2,2-tetraphenylethanol.** Benzhydryl chloride, 20.3 g., in 50 ml. of dry ether was added to the disodiobenzophenone solution to form a precipitate. After the ammonia was removed it was replaced by 200 ml. of ether and the suspension formed was poured with stirring into 100 ml. of saturated NH₄Cl solution. The precipitate which formed was crystallized from methylene chloride to give 28.7 g. (82%) of the alcohol, m.p. 238–239° [32].

(b) Other examples

(1) **1,1,2-Triphenylethanediol-1,2** (91% from disodiobenzophenone and benzaldehyde) [32].

(2) **Phenyl 2-pyridyl β-dimethylaminoethylcarbinol** (80% from phenyl 2-pyridyl ketone and β-dimethylaminoethyl chloride) [34].

6. From α (or γ)-Pyridylcarboxylic Acids and Carbonyl Compounds (Hammick)

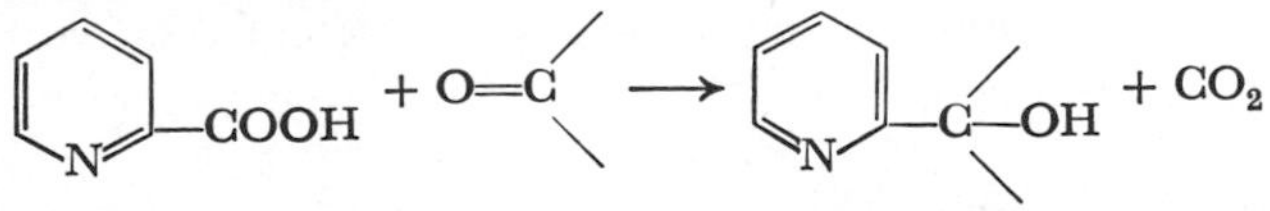

$$\text{(pyridyl)}-\text{COOH} + O=C\diagdown \longrightarrow \text{(pyridyl)}-C-OH + CO_2$$

This synthesis is a special one which has been utilized in the synthesis of carbinols in the N-heterocyclic series. It is applicable only to acids in which the carboxyl group is in position 2 or 4 of the pyridine ring system [35, 36]. Yields are low, but the ease of preparation sometimes makes the method an attractive one. The aldehyde or ketone is used in considerable excess, and the synthesis has been conducted with or without a solvent. In some experiments it has been shown that p-cymene is desirable as a solvent [37]. The aldehyde or ketone probably traps the intermediate α-pyridyl anion

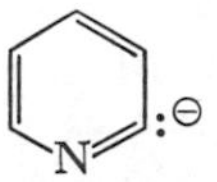

(a) **Preparation of 2-pyridyl-p-tolylcarbinol.** Picolinic acid, 100 g., 600 g. of p-tolualdehyde, and 600 ml. of p-cymene were mixed and refluxed with stirring for 6 hr. under nitrogen. After the evolution of carbon dioxide had ceased, the solution was cooled and extracted with several portions of dilute HCl. The addition of gaseous ammonia to the aqueous acid layer formed an oil which was taken up in ether and recovered in the usual manner to give 79 g. (49%) of 2-pyridyl-p-tolylcarbinol, b.p. 146–152° (1 mm.) [37].

(b) **Preparation of phenyl-α-pyridylcarbinol** (54% from α-picolinic acid and benzaldehyde in refluxing cymene) [38].

7. From Ethers (Wittig Rearrangement)

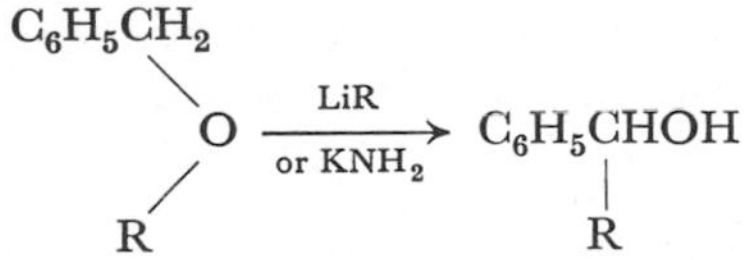

$$C_6H_5CH_2-O-R \xrightarrow[\text{or KNH}_2]{\text{LiR}} C_6H_5\overset{|}{C}HOH-R$$

This synthesis is of limited value, but it has been used recently in the synthesis of secondary and tertiary alcohols of aromatic hydrocarbons. Potassium amide appears to be a better reagent for effecting the rearrangement than the alkyllithium [39]. Yields vary from poor to good.

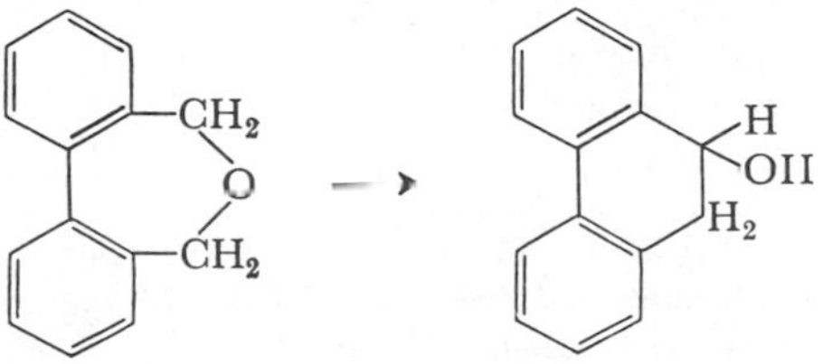

(a) **Preparation of 9-hydroxy-9, 10-dihydrophenanthrene.** Diphenan, 19.6 g., in 75 ml. of ether was added during 5 min. to a stirred suspension of 0.2 mole of potassium amide in 250 ml. of liquid ammonia. At the end of 1 hr. an excess of NH_4Cl was added and the ammonia was allowed to evaporate. From the residue by ether extraction 18.1 g. (92%) of the hydroxydihydrophenanthrene was recovered [40].

(b) Other examples

(1) Benzylphenylcarbinol (61% from dibenzyl ether as in Ex. a) [39].

(2) 9-Benzylfluoren-9-ol (70% from benzyl-9-fluorenyl ether and sodium n-butoxide in butanol at 120° for 2.5 hr.) [41].

8. From Carbonyl (Knoevenagel) or Unsaturated Carbonyl (Michael) Compounds

Occasionally these reactions yield alcohols but, since the Michael condensation yields ketones most frequently and the Knoevenagel, esters, the former is discussed in Chapter 11, G.3, and the latter in Esters, Chapter 14, C.4.

1. P. J. Hamrick, Jr. and C. R. Hauser, *J. Am. Chem. Soc.*, **81**, 3144 (1959).
2. A. T. Nielsen and W. S. Houlihan, *Org. Reactions*, **16**, 1 (1968).
3. E. P. Kohler and H. M. Chadwell, *Org. Syn.*, Coll. Vol. **1**, 78 (1941).
4. G. Wittig, *Record Chem. Progr.* (Kresge-Hooker Sci. Lib.) **28**, 45 (1967).
5. J. A. Marshall and W. I. Fanta, *J. Org. Chem.*, **29**, 2501 (1964); N. C. Ross and R. Levine, *J. Org. Chem.*, **29**, 2341 (1964).
6. C. F. H. Allen and J. A. Van Allan, *J. Org. Chem.*, **16**, 716 (1951).
7. W. J. Hickinbottom, *Reactions of Organic Compounds*, Longmans, Green, and Co., New York, 1948, p. 164.
8. A. T. Nielsen and W. S. Houlihan, *Org. Reactions*, **16**, 80 (1968).
9. L. F. and M. Fieser, *Reagents for Organic Synthesis*, 1967, John Wiley and Sons, New York, p. 1034.
10. A. J. Nunn and K. Schofield, *J. Chem. Soc.*, 716 (1953).
11. E. D. Bergmann et al., *J. Chem. Soc.*, 2564 (1953).
12. J. B. Conant and N. Tuttle, *Org. Syn.*, Coll. Vol. **1**, 199 (1941).
13. C. R. Hauser and D. S. Breslow, *J. Am. Chem. Soc.*, **61**, 793 (1939).
14. A. D. Miller and R. Levine, *J. Org. Chem.*, **24**, 1364 (1959).
15. J. Mathieu and A. Allais, *Cahiers de Synthèse Organique*, Masson, Boulevard Saint-Germain, Paris, 1957, Vol. **1**, p. 58.
16. R. J. Light and C. R. Hauser, *J. Org. Chem.*, **26**, 1716 (1961).
17. A. T. Nielsen et al., *J. Am. Chem. Soc.*, **73**, 4696 (1951).
18. T. A. Geissman, *Org. Reactions*, **2**, 94 (1944).
19. W. Friederich and W. Brün, *Org. Reactions*, **2**, 111 (1944).
20. G. J. Laemmle et al., *Ind. Eng. Chem.*, **52**, 33 (1960).
21. L. J. Kitchen and E. S. Hanson, *J. Am. Chem. Soc.*, **73**, 1838 (1951).
22. Ref. 15, Vol. 1.
23. C. G. Overberger and I. C. Kogon, *J. Am. Chem. Soc.*, **76**, 1879 (1954).
24. D. R. Moore and A. Oroslan, *J. Org. Chem.*, **31**, 2620 (1966).
25. M. Avramoff and Y. Sprinzak, *J. Org. Chem.*, **26**, 1284 (1961).
26. N. L. Gull and J. K. Martzweiller, U.S. Patent 2,829,177, Apr. 1, 1959; *C.A.*, 52, 15564 (1958).
27. S. Veibel and J. I. Nielsen, *Tetrahedron*, **23**, 1723 (1967).
28. Ch. Weizmann et al., *J. Org. Chem.*, **15**, 54 (1950).
29. R. E. Miller and G. E. Bennett, *Ind. Eng. Chem.*, **53**, 33 (1961).
30. J. Bolle and L. Bourgeois, *Compt. Rend.*, **233**, 1466 (1951).

31. P. J. Hamrick, Jr. and C. R. Hauser, *J. Am. Chem. Soc.*, **82,** 1957 (1960).
32. P. J. Hamrick, Jr., and C. R. Hauser, *J. Am. Chem. Soc.*, **81,** 493 (1959).
33. S. Selman and J. F. Eastham, *J. Org. Chem.*, **30,** 3804 (1965).
34. M. Miocque and C. Fauran, *Compt. Rend.*, **259,** 408 (1964).
35. D. LL. Hammick *et al.*, *J. Chem. Soc.*, 1724 (1937); 809 (1939).
36. K. Mislow, *J. Am. Chem. Soc.*, **69,** 2559 (1947).
37. N. Sperber *et al.*, *J. Am. Chem. Soc.*, **71,** 887 (1949).
38. N. H. Cantrell and E. V. Brown, *J. Am. Chem. Soc.*, **75,** 1489 (1953).
39. C. R. Hauser and S. W. Kantor, *J. Am. Chem. Soc.*, **73,** 1437 (1951).
40. C. R. Hauser *et al.*, *J. Org. Chem.*, **18,** 801 (1953).
41. J. Cast *et al.*, *J. Chem. Soc.*, 3521 (1960).

Chapter 5

PHENOLS

A. Solvolysis

1. From Metal Sulfonates

$$ArSO_2ONa \xrightarrow[\text{KOH}]{\text{NaOH}} ArONa \xrightarrow[\text{H}_2\text{O}]{\text{H}^\oplus} ArOH$$

The fusion of alkali metal sulfonates with alkali in the presence of some water is one of the industrial methods for preparing phenols [1]. The synthesis is not always applicable to arylsulfonates which contain substituents such as the chloro, nitro, and carboxy groups, which are sometimes affected by the high temperature and high alkalinity of the reaction mixture. Water is sometimes necessary in the reaction to maintain fluidity [2] and in some cases sodium hydroxide alone is not satisfactory [3]. In such cases either potassium hydroxide alone or mixtures of the two alkali metal hydroxides are employed. The temperatures necessary to achieve maximum return vary usually from 200 to 350°. The alkali phenolate obtained in the fusion is readily converted into the free phenol by acidification with a mineral acid. Yields are in the range 65–85%.

(a) **Preparation of p-cresol** (63–72% from sodium p-toluenesulfonate and a mixture of sodium and potassium hydroxides at 230–330° followed by acidification) [3].

(b) **Other examples**

(1) **1,5-Dihydroxynaphthalene** (83% from sodium 1,5-naphthalene disulfonate and sodium hydroxide at 340°. By use of a drum containing iron spheres, some of which are spiked, the mixture is kept intimately mixed and a low ratio of 6 moles sodium hydroxide to 1 of disulfonate is ample) [4].

(2) **Resorcinol** (quantitative yield by assay from ca. 1 g. of disodium m-benzenedisulfonate and 3 g. of sodium hydroxide at 320° for 2–3 hr. under nitrogen atmosphere. Disulfonation followed by alkali fusion is a potential source of 1,3-dihydroxyarenes) [5]. The benzyne intermediate may occur in this fusion, as any isomer of disodium m-benzenesulfonate yields resorcinol [6].

2. From Halides

$$ArCl \xrightarrow{\text{NaOH}} ArOH$$

Although the halogen atom is difficult to replace from an aryl halide, the reaction has been developed to such an extent (Dow Process) that it is one of the

References for Section A are on pp. 257–258.

commercial methods for preparing phenol. Since the operation requires, among other conditions, a temperature of 400°F., pressure of 4000 p.s.i., and a mechanism for passing the reaction mixture through an autoclave, it is not suitable for use in the laboratory [1].

The main by-product of this reaction is diphenyl ether which can be added initially to suppress its own formation. Among other by-products are 4-hydroxy-biphenyl (and traces of the 2- and 3-isomers), 2,6-diphenylphenol, 2- and 4-phenoxybiphenyl, and hydrocarbons including triphenylene, some of which apparently are derived as a result of a benzyne pathway [7]. Indeed, it is quite likely that at the temperature of about 340° the main pathway for the fusion of any aryl chloride or bromide with alkali is via the benzyne, while at 250° or lower it is largely direct replacement, as witness the behavior of *p*-chloro-toluene in 4 *M* aqueous sodium hydroxide [6]:

		Cresols	
Temp.	Yield	% *m-*	% *p-*
340°	59%	50.4	49.6
250°	12%	14.4	85.6

For an exclusive benzyne pathway, roughly, but not exactly, 50% of *m-* and *p*-cresol would be anticipated.

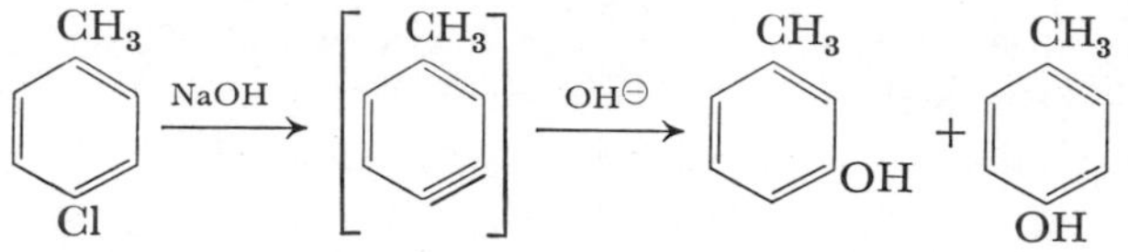

If the solvolysis of the aryl halide is carried out in alcoholic solvents, evidence is available to suggest that the ether is formed first, followed by displacement of the alkoxyl groups [8]:

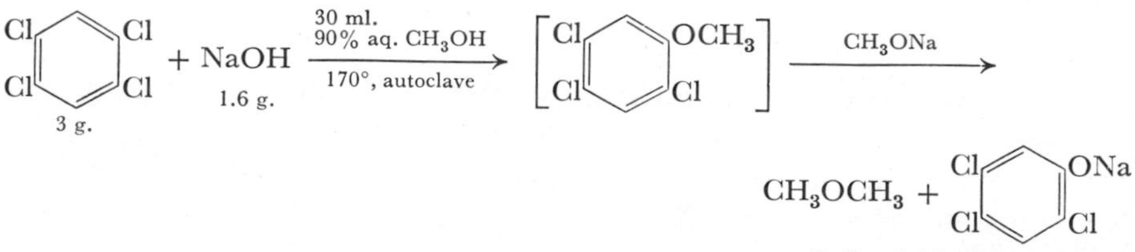

Sodium 2,4,5-trichlorophenolate

With sodium hydroxide in water, no hydrolysis of 1,2,4,5-tetrachlorobenzene takes place under the same conditions; furthermore 2,4,5-trichloroanisole has been shown to cleave to the phenol with sodium methoxide in methanol (see A.4 for other examples of this mode of cleavage). If electron-withdrawing substituents are present in the *o-* and *p*-positions in the ring, the halogen becomes more labile and replacement occurs more readily. For example, *p*-nitrochloro-benzene yields the phenol with 15% sodium hydroxide at 160°, 2,4-dinitrochloro-benzene requires only aqueous sodium carbonate at 130°, and 2,4,6-trinitro-chlorobenzene undergoes the transformation in warm water alone [9]. Although this method finds a very limited use in the laboratory, in the few examples recorded in the literature the yields are good.

With electron-withdrawing groups attached, the mechanism undoubtedly is a nucleophilic displacement:

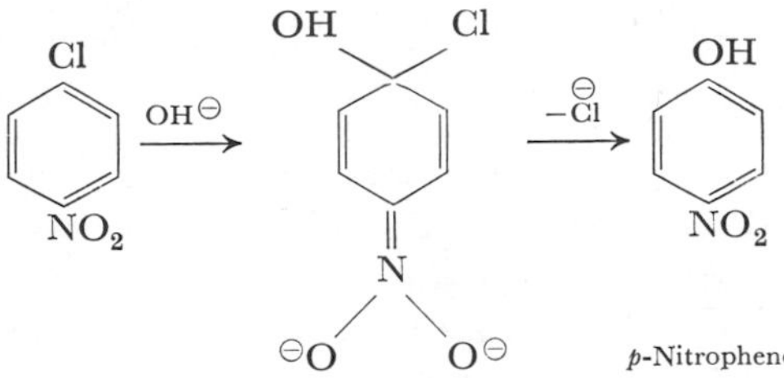

p-Nitrophenol

The ease of halide displacement is then in the order F > Cl > Br > I. Other groups such as nitro, sulfone, or even amino can be displaced also when situated *o*- or *p*- to an electron-withdrawing group [10]:

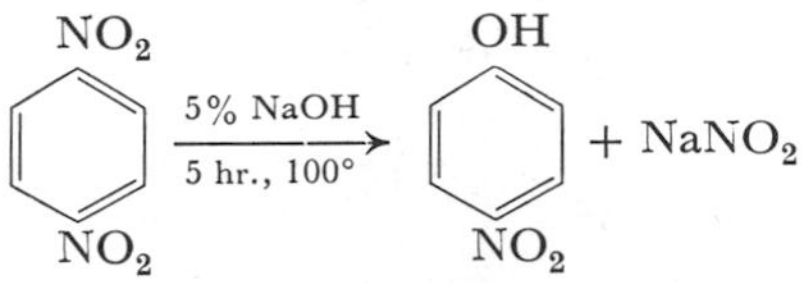

p-Nitrophenol
75–80% after
acidification

Supplementary to the displacement of an *o*- or *p*-activated nitro group, the displacement of a *m*-nitro group can be brought about by visible light [11]:

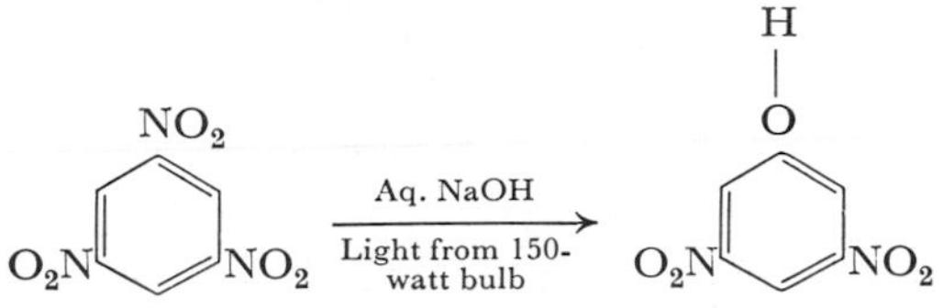

3,5-Dinitrophenol
95%

The deep-red solution fades rapidly. Without light some 3,5-dinitrophenol, picric acid, and other products are formed.

Electron-withdrawing groups which activate the *o*- and *p*-positions include the diazonium, cyano, carboxyl, and carbonyl groups. Positions α or γ to the nitrogen atom in pyridine or quinoline are activated to the extent that many groups can be displaced [12].

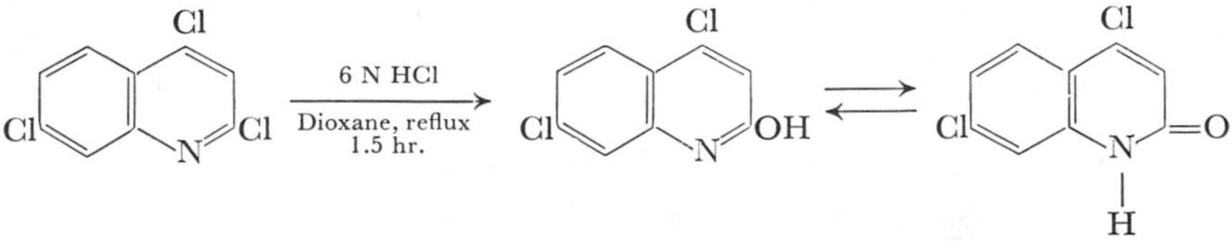

2-Hydroxy-4,7-dichloroquinoline

72%

With nucleophilic agents the 4-chloro substituent seems more active than the 2-chloro. The 2- and 5-positions of benzoquinones are also activated (see Ex. c.2).

Another industrial method of hydrolyzing chlorobenzenes is known as the Prahl process (erroneously called the Raschig process) [13]:

$$C_6H_5Cl + H_2O \xrightarrow[\substack{Ca_3(PO_4)_2 + Cu\ or \\ Mn\ pyrophosphate}]{High\ temp.,\ bed\ of} C_6H_5OH + HCl$$

(a) Preparation of phenol. Commercial potassium *t*-butoxide dissolved in a mixture of DMSO and *t*-butyl alcohol was heated to 125–130° with magnetic stirring, and bromobenzene added all at once. An exothermic reaction ensued which after 1 min. was quenched in ice and water, yielding 42–46 % *t*-butyl phenyl ether. The ether was heated in 6 *N* HCl until dissolved and yielded phenol. The product was isolated as the tribromo derivative in this case but could be isolated by continuous extraction [14].

(b) Preparation of 2,4-dinitrophenol. 2,4-Dinitrochlorobenzene, 10 g., was refluxed with 15 g. of anhydrous sodium carbonate in 150 ml. of water until solution occurred. Acidification gave a precipitate which when washed and dried weighed 8 g. (90 %), m.p. 114° [15].

(c) Other examples

(1) Pseudocumenol-3(2,3,6-trimethylphenol) (82 % from 3-bromo-pseudocumene, cuprous oxide, copper powder, and aqueous sodium hydroxide under pressure at 275° for 3 hr.) [16].

(2) Chloranilic acid (73 % as dihydrate from 10 g. of chloranil and 9 g. of NaOH in 260 ml. of water at 70–80° for 2 hr. followed by acidification

with hydrochloric acid. The crude product, recrystallized from water, came out as a dihydrate, red leaflets, m.p. 283–284°, strongly acidic) [17]. This preparation was repeated by one of us (D.E.P.); while being heated overnight on a steam bath, the reaction mixture turns very dark and deposits a solid. On acidification and filtration the yield of red powder was 80 %. It was dissolved in 15 parts of isopropyl alcohol, filtered hot, and the filtrate treated with 1.5 parts of concentrated hydrochloric acid. On cooling, bright orange-red leaflets were deposited in 60 % yield. Concentration gave a second crop in 15 % yield.

3. From Aromatic Amines (Including Bucherer)

The hydrolysis of aromatic amines is not a general reaction, but still is one that is more prevalent than can be judged from the number of examples in the literature. Electron-withdrawing groups in the *o*- and *p*-positions again facilitate the reaction, but in addition polynuclear aromatic or some heterocyclic amines are subject to hydrolysis because the imino form becomes a greater contributing tautomeric form as annelation increases:

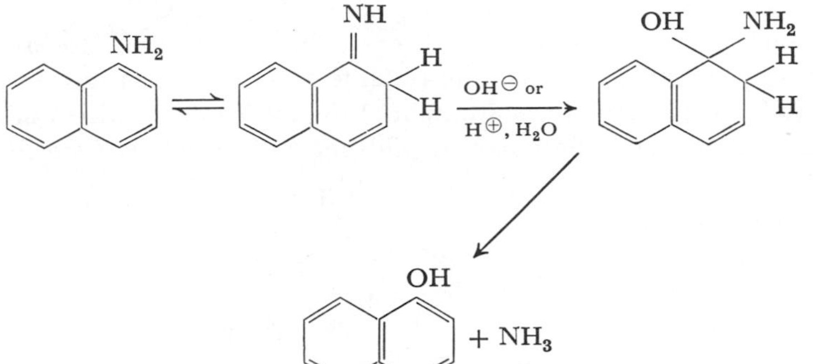

Aniline can not be hydrolyzed under reasonable conditions to phenol, but α-naphthylamine can be converted quantitatively into α-naphthol at 200° with 5% aqueous sulfuric acid. β-Naphthylamine can be hydrolyzed to β-naphthol at 100° with hypophosphorous or sulfurous acid [18]. The nitroanilines are best hydrolyzed via their N-acetyl derivatives [19]:

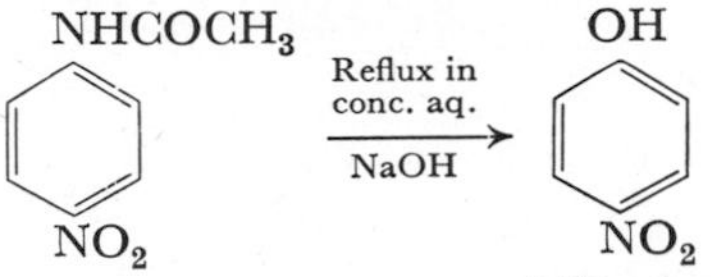

p-Nitrophenol
No yield given

And, when structural features are favorable, the hydrolysis can be accomplished with ease as shown [20]:

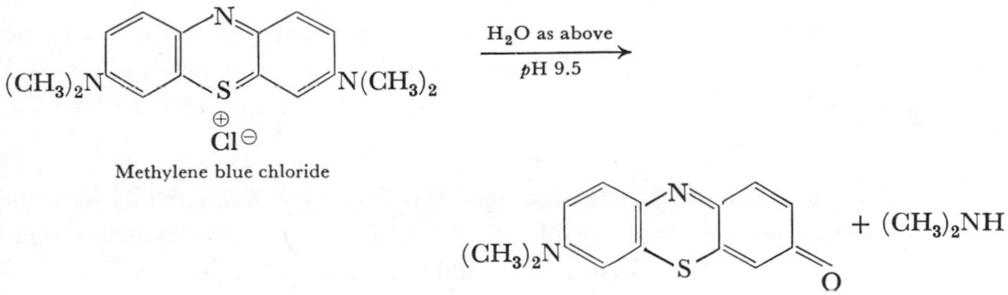

Methylene blue chloride

Methylene violet

The instability of the sulfonium hydroxide serves as part of the driving force or in more general terms $S^{III} \rightarrow S^{II}$ (in basic solution) is an energy-releasing reaction. The decomposition of p-nitrosodialkylanilines can be viewed in a similar light [21]:

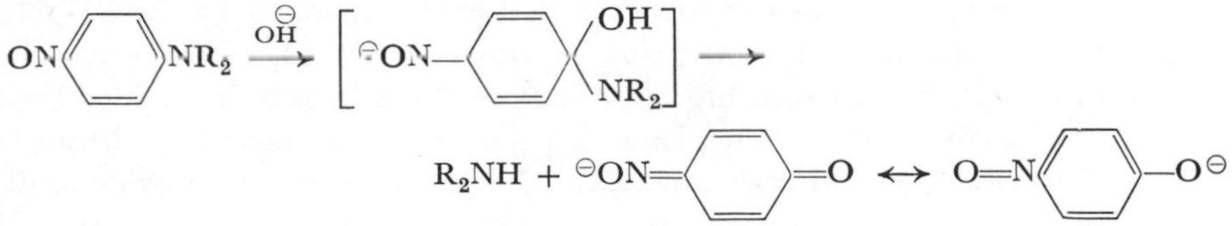

This reaction is a well-known method of preparing secondary amines, but the aromatic product is the anion of a benzoquinone monoxime or, for the purposes here, the anion of a nitrosophenol.

The Bucherer reaction [22] consists of the hydrolysis of an aromatic amine catalyzed by the addition of sodium bisulfite or the reverse of this process. An adduct is formed which favors addition of the nucleophile to the imino form, as in the accompanying formulation. This mechanism differs from that given

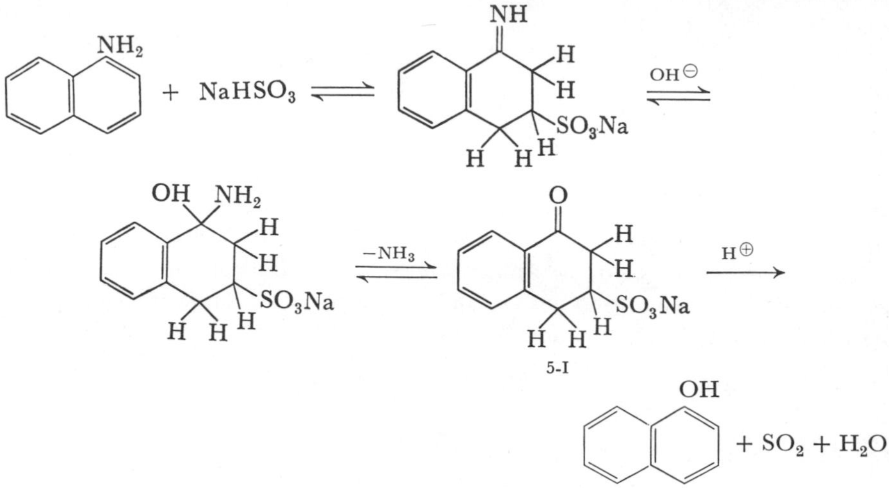

5-I

α-Naphthol

in the review article on the Bucherer reaction [23] in which the sodium bisulfite is shown as adding to the same carbon as the amino group. In view of the fact that the sodium bisulfite adduct of α-naphthol, 5-I, has been demonstrated to have all the chemical characteristics of a ketone [24], the mechanism shown here seems more reasonable. Since the reaction is reversible, amines can be prepared from phenols as well. Only naphthylamines, m-diaminobenzenes, and some heterocyclic amines form sodium bisulfite adducts, a fact which limits the Bucherer reaction considerably.

(a) **Preparation of 1-nitro-2-naphthol** (88–89 % from 0.435 mole of 1-nitro-2-acetaminonaphthalene refluxed for 6–7 hr. in 2.7 l. of water containing 2.8 moles of sodium hydroxide, followed by acidification) [25].

(b) **Preparation of phloroglucinol** (46–53 %, based on 2,4,6-trinitrobenzoic acid, as the dihydrate by refluxing 2,4,6-triaminobenzoic acid in aqueous sodium hydroxide under an inert atmosphere, by concentrating, and then by making slightly acidic) [26].

(c) **Preparation of 1-anthrol.** Crude 1-anthramine, 19.3 g., in 100 ml. of ethanol was precipitated by the slow addition of 200 ml. of hot water. A concentrated technical sodium bisulfite solution, 280 ml., was then added and the mixture was refluxed for 24 hr. Insoluble material was removed by filtration and the filtrate was treated with concentrated aqueous sodium hydroxide until thiazole paper turned permanently red (about 200 ml. of alkali was required). On boiling for an hour, the main quantity of ammonia was removed. After cooling and filtering, the filtrate was acidified immediately with about 100 ml. of glacial acetic acid. Cooling gave a precipitate of 1-anthrol, which when dried weighed 18.6 g. (96%) [27].

(d) **Preparation of 1-naphthol-4,8-disulfonic acid** (quantitative from the amine using a procedure similar to that in Ex. c) [28].

4. From Ethers and Basic Reagents

An example of this fission was shown in A.2:

$$ArOR + R'O^{\ominus} \xrightarrow{150-200°} ArO^{\ominus} + ROR' \qquad (\text{see Ex. a})$$

It could be called alkyl fission, second-order. Grignard reagents can serve the same purpose [29]

$$ArOR + R'MgX \rightarrow ArOMgX + RR' \qquad (\text{see Ex. b})$$

And there is no reason why the hydroxide ion should not cleave aromatic ethers:

$$ArOR + OH^{\ominus} \xrightarrow{\text{Over } 200°} ArO^{\ominus} + ROH \qquad (\text{see Ex. c})$$

In this case the number of examples is very limited because of the higher temperatures necessary to bring about the reaction.

Other basic reagents which have been employed are sodium and butanol (for benzyl phenyl ethers) [30] and sodium or sodium and potassium in liquid ammonia [31]. A mild cleavage agent is the lithium salt of diphenylphosphine [32]:

$$C_6H_5OCH_3 + (C_6H_5)_2\overset{\ominus\;\oplus}{PLi} \xrightarrow[\text{Reflux 4 hr.}]{\text{THF, H}_2\text{O}} C_6H_5OH + (C_6H_5)_2PCH_3$$
$$75\%$$

Benzyl and allyl ethers are cleaved as well by the reagent, but ethyl ethers are affected only slightly.

(a) **Preparation of 2,4,5-trichlorophenol** (58% from the methyl ether and sodium methoxide in methanol at 170°) [33].

(b) **Preparation of cannabidiol** (80% from the dimethyl ether and

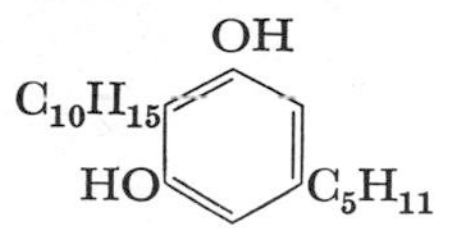

excess methylmagnesium iodide at 155–165° for 15 min.) [34].

(c) **Preparation of protocatechuic acid** [35]

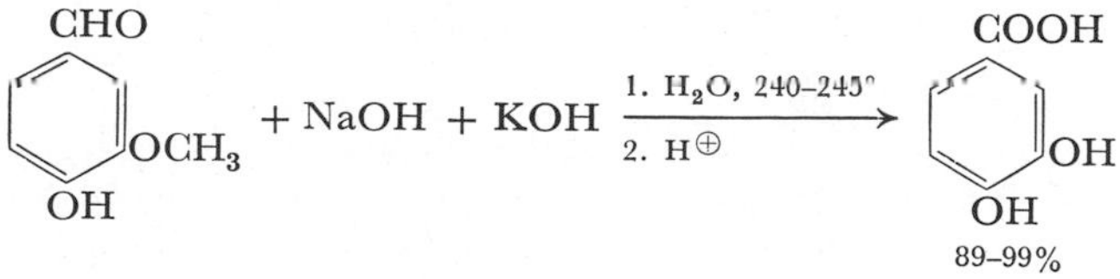

89–99%

(d) Preparation of 4-hydroxy-6-quinaldinecarboxylic acid (88% from 4-methoxy-6-cyanoquinaldine and KOH in glycerol) [36].

5. From Ethers by Acid Cleavage

$$ArOR \xrightarrow{HX} ArOH + RX$$

Many phenols have been synthesized from aryl alkyl ethers by acid cleavage in which the alkyl group is replaced with a hydrogen atom. In fact, phenolic groups are often protected in the form of an alkoxy group, usually methoxy, which can be reconverted into the original group in the last step in the reaction. A great variety of reagents has been used to effect this transformation. Perhaps hydrogen bromide [37, 38] or hydrogen iodide alone or with red phosphorus [39–42], usually with glacial acetic acid as well, are the most common. Other effective reagents are potassium iodide in 95% phosphoric acid [43], pyridine hydrochloride [44] (see Ex. b.3), boron tribromide [45], concentrated hydrochloric acid and acetic acid [46], trifluoroacetic acid (for aryl benzyl ethers [47]), anhydrous aluminum chloride or bromide [48, 49].

Selective cleavage agents are desirable at times. Pyridine hydrochloride has the advantage that it will hydrolyze an ether without affecting a trifluoromethyl group [50]; (see also Ex. b.3):

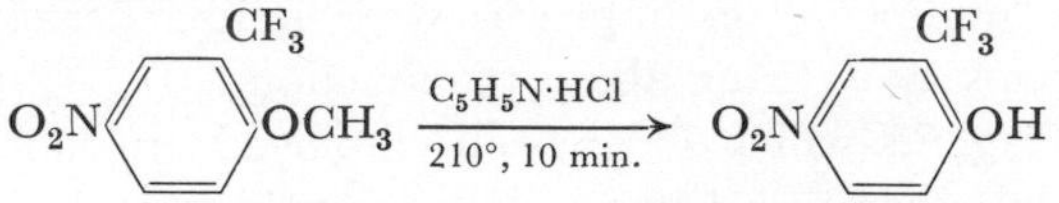

2-Trifluoromethyl-4-nitrophenol

Other hydrolytic reagents give nitrosalicylic acid. Magnesium iodide etherate (prepared from magnesium and iodine in ether) is also selective in cleavage [51]:

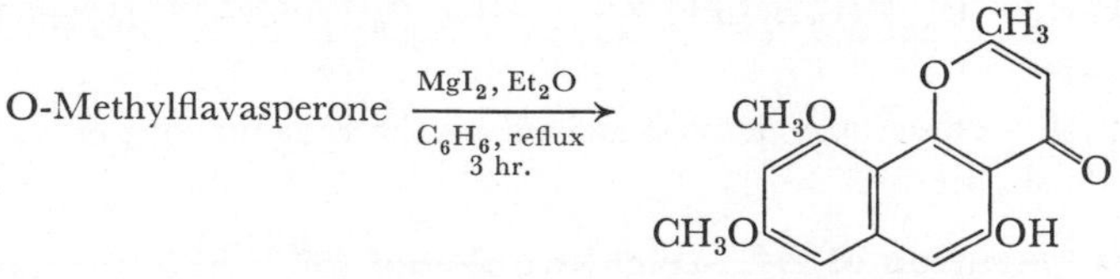

Flavasperone, 63%
(on small scale)

The yields in the cleavage are often in the 80–95% range.

Recently it has been found that the cleavage occurs by using less than an equivalent of a Lewis acid to complex with the more basic site [52]:

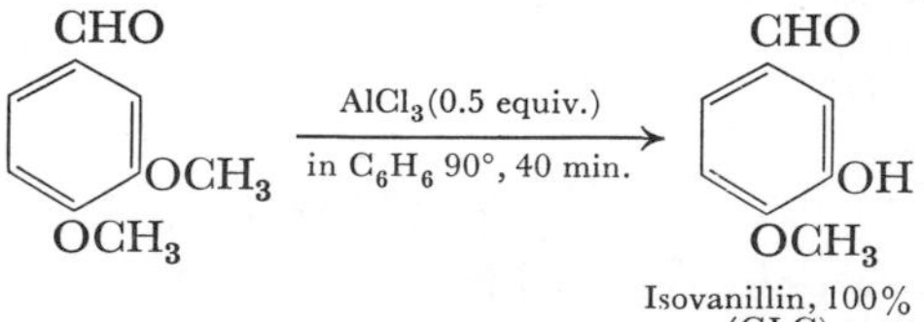

Isovanillin, 100%
(GLC)

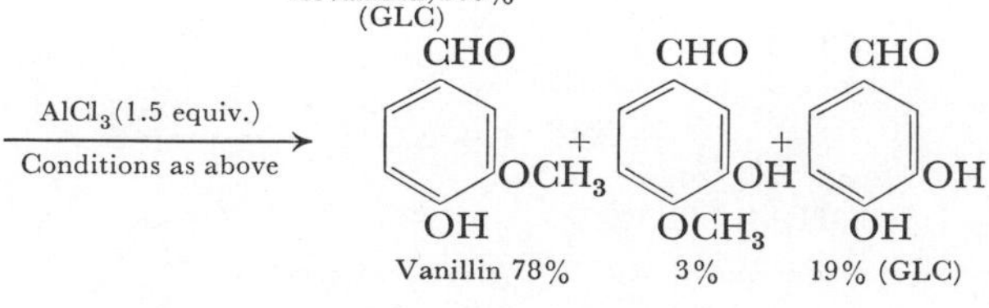

Vanillin 78%   3%   19% (GLC)

The intermediate in this transformation, if a halogen acid is the reagent, appears to be the oxonium ion 5-II, which decomposes to give the phenol 5-III and the carbonium ion 5-IV. The latter unites with the halide ion $X^{\ominus}$ originally formed to produce the alkyl halide.

$$\text{ArOR} + \text{HX} \rightleftharpoons \overset{\oplus}{\underset{\underset{\text{H}}{|}}{\text{ArOR}}} + X^{\ominus}$$

5-II

$$\overset{\oplus}{\underset{\underset{\text{H}}{|}}{\text{ArOR}}} \rightleftharpoons \underset{\text{5-III}}{\text{ArOH}} + \underset{\text{5-IV}}{R^{\oplus}}$$

$$\underset{\text{5-IV}}{R^{\oplus}} + X^{\ominus} \rightleftharpoons \text{RX} \quad \text{or other carbonium ion products}$$

(a) Preparation of N-methyl-3,4-dihydroxyphenylalanine (82% from N-methyl-3-methoxy-4-hydroxyphenylalanine, red phosphorus, hydriodic acid, and acetic anhydride) [41].

(b) Other examples

(1) 2,3-Dihydroxy-5-bromobenzoic acid (96% crude from 2,3-dimethoxy-6-bromobenzoic acid, hydrogen bromide, and glacial acetic acid) [37] (migration of bromine).

(2) 1,4-Dihydroxy-10-methyl-9-anthrone (93% from 1,4-dimethoxy-

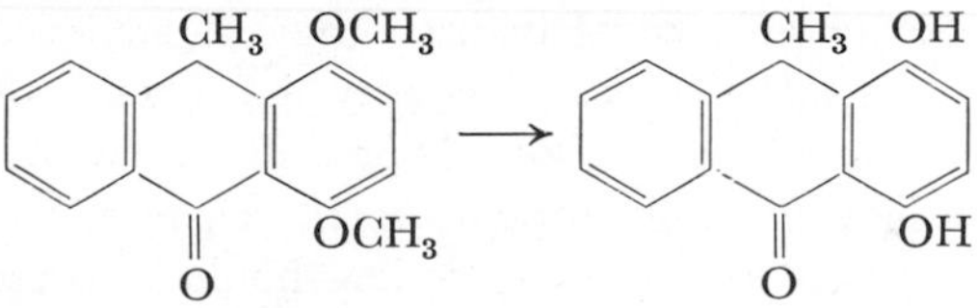

10-methyl-9-anthrone and anhydrous aluminum bromide in dry benzene) [48].

(3) Phenol (82% from anisole heated at 200–220° for 6 hr. with 3 parts pyridine hydrochloride) [53].

6. From Diazonium Salts

$$\text{Ar}\overset{\oplus}{\text{N}_2}X^{\ominus} \xrightarrow{\text{H}_2\text{O}} \text{ArOH} + \text{N}_2 + \text{HX}$$

This synthesis has been widely employed in the preparation of phenols. It is of particular value when a certain orientation is necessary. Since the diazonium salts are produced from amines which are in turn usually derived from nitro compounds, the use of the method assures one of locating the hydroxyl group at the position formerly occupied by the amino or nitro group. To remove any excess nitrous acid from the diazotization of the amine, urea has been employed [54]. For diazonium salts which form with difficulty because of low basicity of the amine, diazonium salt formation can be carried out in concentrated sulfuric acid followed by careful dilution after the diazotization, or the amine can be dissolved in sulfuric acid from which solution the amine may be diluted before diazotization[55]. Nitrosylsulfuric acid is the diazotization reagent in the first of these methods [56].

The synthesis of phenols from diazonium salts is a classic example in which one can recognize many of the factors which detract from the yield and in which one can manipulate in such a way as to combat these factors. The diazonium salt tends to combine with anions. Therefore all such anions of high nucleophilicity should be avoided so that only water or the hydrogen sulfate anion

$$\overset{\oplus}{Ar N} \!\!\equiv\!\! N \xrightarrow{-N_2} Ar^{\oplus} \xrightarrow{X^{\ominus}} ArX\cdot$$

will combine with the aryl carbonium ion or intermediate [57]. The diazonium salt tends to couple with the phenol to produce azo dyes and tarry products. To combat this side reaction the diazonium salt is added in very dilute, cold solution to water boiling sufficiently to remove the phenol as it is formed [58, 59]. Yields may be as high as 80–90 % by this procedure. Since diazonium salts complex with copper ions (in which state less coupling may take place), copper sulfate [60] or a mixture of copper sulfate and urea [61] is dissolved in the boiling water which decomposes the diazonium salt. Control experiments have never been run, however, to prove the efficacy of these additives.

Recently it has been shown that the difficulty in predicting the proper conditions for synthesizing phenols via diazonium salts results from ionic and free radical mechanisms competing [62]. It is suggested that, for best yields, conditions be chosen to promote the radical rather than the ionic mechanism. To accomplish this end, ligand radical transfer may be achieved by treating the diazonium salt with a large excess of copper ion in very dilute sulfuric acid solution at low temperature with no solvent present from which hydrogen atoms can be abstracted. The reaction appears to be triggered by the addition of cuprous oxide, but a large excess of cupric ion should be present (see Ex. b.2).

Purer products and improved yields are sometimes obtained by converting the diazonium salt into the diazonium tetrafluoroborate 5-V which is then converted into the acetyl derivative 5-VI which hydrolyzes readily to the

$$\overset{\oplus}{Ar N_2} Cl^{\ominus} \xrightarrow{HBF_4} \underset{\text{5-V}}{\overset{\oplus}{Ar N_2} \overset{\ominus}{BF_4}} \xrightarrow{CH_3 COOH} \underset{\text{5-VI}}{ArOCOCH_3} \xrightarrow{H_2O} ArOH$$

phenol [63]. For a second way to utilize the tetrafluoroborate see Ex. b.2.

Many substituents such as the chloro [64], the phenoxy [65], the aldehydo [66], the keto [67], the carboxyl [68], the cyano [69], and the nitro [55] remain unaltered in the hydrolysis of the diazonium salt. The synthesis from the amino compound is also applicable to heterocyclic nuclei such as pyridine [70], pyrazine [71], dibenzofuran [72], and thianaphthene [73].

(a) **Preparation of _m_-nitrophenol** (81–86 % from _m_-nitroaniline, sodium nitrite, sulfuric acid, and water at 0–5° followed by hydrolysis in aqueous sulfuric acid) [55].

(b) Other examples

(1) **3,5-Di-*t*-butylphenol** (80% from 3,5-di-*t*-butylaniline hydrogen sulfate, 20% sulfuric acid, and aqueous sodium nitrite at 0° followed by pouring the mixture into 20% sulfuric acid through which steam was passing) [59].

(2) ***p*-Bromophenol** (87% from 0.95 mmole of *p*-bromobenzene-diazonium tetrafluoroborate in 100 ml. of water containing 240 mmoles of $Cu(NO_3)_2 \cdot 3 H_2O$ at room temperature to which 0.9 mmole of cuprous oxide was added to initiate rapid nitrogen evolution; thermal decomposition gave a 53% yield) [62].

(3) **Guaiacol** (80–90% from 220 parts of *o*-anisidine in 240 parts of 50% aqueous sulfuric acid diazotized by 330 parts of 20% sodium nitrite solution, followed by addition of 1 part of urea; the solution was poured into a boiling solution of 200 parts of copper sulfate, 20 parts of urea, and 6 parts of sulfuric acid in 200 parts of water) [61].

1. S. J. Lederman and N. Poffenberger in Kirk and Othmer's *Encyclopedia of Chemical Technology*, The Interscience Publishers, New York, Vol. 15, 2nd Ed., 1968, p. 153.
2. R. N. Shreve and F. R. Lloyd, *Ind. Eng. Chem.*, **42**, 811 (1950).
3. W. W. Hartman, *Org. Syn.*, Coll. Vol. **1**, 175 (1941).
4. H. Sanielevici *et al.*, *Rev. Chim.* (*Bucharest*), **13**, No. 2 88 (1962); *C.A.*, **57**, 15021 (1962).
5. A. P. Shestov and H. A. Osipova, *J. Gen. Chem. USSR*, (*Eng. Transl.*), **26**, 2235 (1956).
6. H. Heaney, *Chem. Rev.*, **62**, 81 (1962).
7. A. Luttringhaus and D. Ambros, *Chem. Ber.*, **89**, 463 (1956).
8. S. M. Shein and V. A. Ignatov, *J. Gen. Chem. USSR*, (*Eng. Transl.*), **32**, 3165 (1962).
9. R. T. Morrison and R. N. Boyd, *Organic Chemistry*, Allyn and Bacon, Boston, 2nd Ed., 1966, p. 834.
10. J. Sauer and R. Huisgen, *Angew. Chem.*, **72**, 294 (1960).
11. V. Gold and C. H. Rochester, *Proc. Chem. Soc.*, 403 (1960).
12. R. C. Elderfield, *Heterocyclic Compounds*, John Wiley and Sons, New York, Vol. **4**, 1952, p. 125.
13. W. H. Prahl, *Chem. Eng. News*, **31**, 4178 (1953); Belgian Patent 620,276, November 14, 1962; *C.A.*, **59**, 511 (1963).
14. M. R. V. Sahyun and D. J. Cram, *Org. Syn.*, **45**, 89 (1965).
15. W. M. Cumming, I. V. Hopper, and T. S. Wheeler, *Systematic Organic Chemistry*, Constable and Company, Ltd., London, 1931, p. 204.
16. L. I. Smith *et al.*, *J. Org. Chem.*, **4**, 318 (1939).
17. C. Graebe, *Ann. Chem.*, **263**, 24 (1891).
18. Elsevier's *Encyclopedia of Org. Chem.*, Series III, Vol. 12B, Elsevier Publishing Co., New York, 1950, p. 1212.
19. P. Wagner, *Chem. Ber.*, **7**, 76 (1874).
20. Ref. 12, Vol. 6, p. 720.
21. C. W. L. Bevan *et al.*, *J. Chem. Soc.*, 4543 (1960).
22. H. Seeboth, *Angew. Chem., Intern. Ed. Engl.*, **6**, 307 (1967).
23. N. L. Drake, *Org. Reactions*, **1**, 105 (1942).
24. S. V. Bogdanov and M. V. Gorelik, *J. Gen. Chem. USSR*, (*Eng. Transl.*), **29**, 140 (1959).
25. W. W. Hartman *et al.*, *Org. Syn.*, Coll. Vol. **2**, 543 (1943).
26. H. T. Clarke and W. W. Hartman, *Org. Syn.*, Coll. Vol. **1**, 455 (1941).
27. H. E. Fierz-David *et al.*, *Helv. Chim. Acta*, **29**, 1718 (1946).
28. H. T. Bucherer, *J. Prakt. Chem.* (2) **69**, 49 (1904); (2) **70**, 345 (1904).
29. H. Meerwein, Houben-Weyl's *Methoden der Organischen Chemie*, Georg Thieme Verlag, Stuttgart, 1965, Vol. 6, Pt. 3, p. 160.
30. B. Loev and C. R. Dawson, *J. Am. Chem. Soc.*, **78**, 6095 (1956).

31. N. A. Nelson and J. C. Wollensak, *J. Am. Chem. Soc.*, **80,** 6626 (1958); K. E. Hamlin and F. E. Fischer, *J. Am. Chem. Soc.*, **75,** 5119 (1953).

32. F. G. Mann and M. J. Pragnell, *Chem. Ind. (London)*, 1386 (1964).

33. S. M. Shein and V. A. Ignatov, *J. Gen. Chem. USSR, (Eng. Transl.)*, **32,** 3165 (1962).

34. R. Mechoulam and Y. Gaoni, *J. Am. Chem. Soc.*, **87,** 3273 (1965).

35. I. A. Pearl, *Org. Syn.*, Coll. Vol. **3,** 745 (1955).

36. C. T. Peng and T. C. Daniels, *J. Am. Chem. Soc.*, **77,** 6682 (1955).

37. G. R. Pettit and D. M. Piatak, *J. Org. Chem.*, **25,** 721 (1960).

38. H. T. Clarke and E. R. Taylor, *Org. Syn.*, Coll. Vol. **1,** 150 (1941); P. N. Craig *et al.*, *J. Am. Chem. Soc.*, **74,** 1316 (1952).

39. C. A. Grob and J. Voltz, *Helv. Chim. Acta*, **33,** 1796 (1950).

40. R. I. Meltzer *et al.*, *J. Org. Chem.*, **22,** 1577 (1957).

41. V. Deulofeu and T. J. Guerrero, *Org. Syn.*, Coll. Vol. **3,** 586 (1955).

42. J. P. Brown *et al.*, *J. Chem. Soc.*, 859 (1949); S. V. Sunthankar and H. Gilman, *J. Org. Chem.* **16,** 8 (1951).

43. A. Furst and C. J. Olsen, *J. Org. Chem.*, **16,** 412 (1951).

44. J. C. Sheehan *et al.*, *J. Am. Chem. Soc.*, **79,** 147 (1957); R. Royer and E. Bisagni, *Bull. Soc. Chim. (France)*, 486 (1954).

45. J. F. W. McOmie and M. L. Watts, *Chem. Ind. (London)*, 1658 (1963).

46. F. E. King *et al.*, *J. Chem. Soc.*, 92 (1952).

47. J. P. Marsh, Jr. and L. Goodman, *J. Org. Chem.*, **30,** 2491 (1965).

48. M. Gates and C. L. Dickinson, Jr., *J. Org. Chem.*, **22,** 1398 (1957).

49. M. Kulka, *J. Am. Chem. Soc.*, **76,** 5469 (1954); H. Mühlemann, *Pharm. Acta Helv.*, **24,** 356 (1949); *C.A.*, **44,** 3481 (1950): H. E. Ungnade and K. T. Zilch, *J. Org. Chem.*, **16,** 64 (1951).

50. R. Filler *et al.*, *J. Org. Chem.*, **27,** 4660 (1962).

51. B. W. Bycroft and J. C. Roberts, *J. Chem. Soc.*, 4868 (1963).

52. R. H. Prager and Y. T. Tan, *Tetrahedron Letters*, 3661 (1967).

53. V. Prey, *Chem. Ber.*, **74,** 1219 (1941).

54. G. F. Grillot and W. T. Gormley, Jr., *J. Am. Chem. Soc.*, **67,** 1968 (1945).

55. R. H. F. Manske, *Org. Syn.*, Coll. Vol. **1,** 404 (1941).

56. L. F. and M. Fieser, *Reagents for Organic Synthesis*, John Wiley and Sons, New York, 1967, p. 755.

57. O. Neunhoffer and H. Kölbel, *Chem. Ber.*, **68,** 255 (1935).

58. J. P. Lambooy, *J. Am. Chem. Soc.*, **72,** 5327 (1950).

59. J. W. Elder and R. P. Mariella, *Can. J. Chem.*, **41,** 1653 (1963).

60. W. Baker, *J. Chem. Soc.*, 476 (1937).

61. R. Bogoczek, Polish Patent 43,278, August 30, 1960; *C.A.*, **57,** 13687 (1962).

62. A. H. Lewin and T. Cohen, *J. Org. Chem.*, **32,** 3844 (1967).

63. L. E. Smith and H. L. Haller, *J. Am. Chem. Soc.*, **61,** 143 (1939).

64. H. H. Hodgson, *J. Am. Chem. Soc.*, **62,** 230 (1940).

65. G. Lock, *Monatsh.*, **55,** 167 (1930).

66. R. B. Woodward, *Org. Syn.*, Coll. Vol. **3,** 453 (1955).

67. L. C. King *et al.*, *J. Am. Chem. Soc.*, **67,** 2089 (1945).

68. P. R. Carter and D. H. Hey, *J. Chem. Soc.*, 150 (1948).

69. M. Silverman and M. T. Bogert, *J. Org. Chem.*, **11,** 34 (1946).

70. W. T. Caldwell *et al.*, *J. Am. Chem. Soc.*, **66,** 1479 (1944).

71. A. E. Erickson and P. E. Spoerri, *J. Am. Chem. Soc.*, **68,** 400 (1946).

72. H. Gilman *et al.*, *J. Am. Chem. Soc.*, **57,** 885 (1935).

73. L. F. Fieser and R. G. Kennelly, *J. Am. Chem. Soc.*, **57,** 1611 (1935).

B. Oxidation

The direct oxidation of an arene to a phenol has long been a reaction of theoretical and practical interest. The reaction is difficult in that the phenol is

References for Section B are on pp. 266–267.

more sensitive to oxidation than the hydrocarbon. Short reaction times partly overcome this difficulty [1]. The irradiation of phenols in the presence of oxygen

$$C_6H_6 + air \xrightarrow[\text{Fluidized bauxite;} \atop \text{rapid quenching}]{300-750°} C_6H_5OH$$

gives a mixture of bisphenols and dihydric phenols [2]. Unfortunately, these methods are not good laboratory ones. Those which are feasible are either limited or designed for oxidation of specific sites on a substituted arene.

1. From Arenes and a Peracid or a Peroxydicarbonate

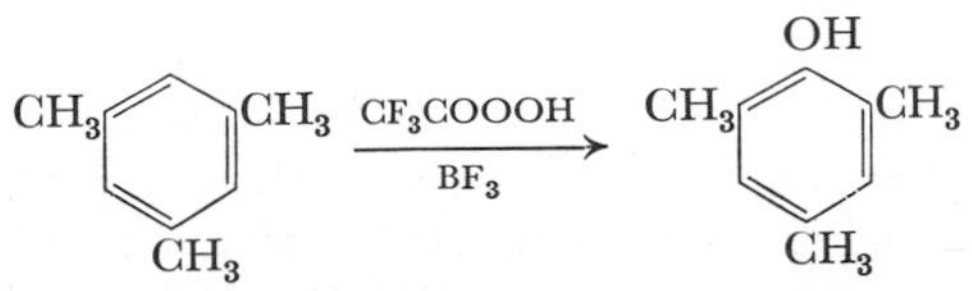

The oxidizing agents employed in this synthesis are hydrogen peroxide in acetic-sulfuric acid [3], hydrogen peroxide and boron trifluoride etherate [4], pertrifluoroacetic acid [5], di-t-butylperoxide [6], and pertrifluoroacetic acid and boron trifluoride [7]. Of these oxidizing agents the latter is the most satisfactory. It gives an 88% yield of mesitol from mesitylene, but in the other oxidations attempted to date the return is less, benzene, for example, giving only a trace of phenol. Common byproducts from benzene ring hydrocarbons are cyclohexadienones and diphenylmethanes. Thus the peracid oxidation seems limited to phenols which are flanked by alkyl groups.

Since the Lewis acid, BF_3, might be expected to produce positive hydroxyl ions from a peroxide, the course of the reaction has been rationalized as an electrophilic attack on the benzene ring [7]:

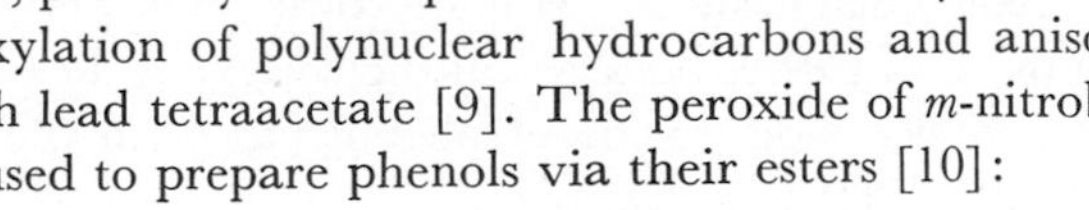

Some success has been achieved in acyloxylation. For example, in nitration with nitric acid and acetic anhydride, some acetoxylation (up to 50% with o-xylene) occurs, probably via the protonated mixed anhydride, CH_3COONO_2 [8], and acetoxylation of polynuclear hydrocarbons and anisole may be accomplished with lead tetraacetate [9]. The peroxide of m-nitrobenzenesulfonic acid has been used to prepare phenols via their esters [10]:

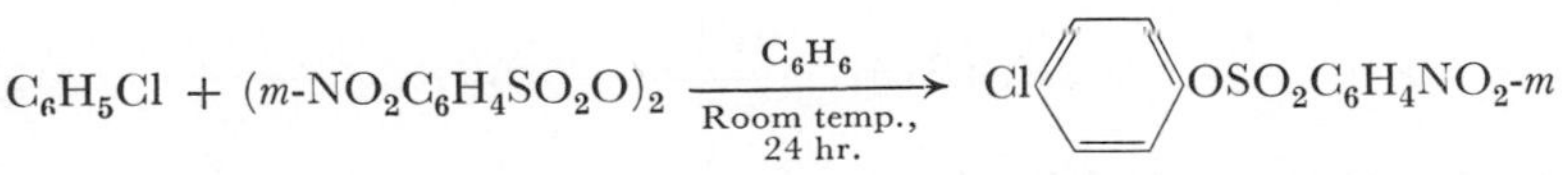

87-91%, 22% o-, 78% p-
p-Chlorophenyl m-nitrobenzenesulfonate

Even toluene gave no *m*-substitution. Perhaps the most general of these procedures is that which utilizes diisopropyl peroxydicarbonate [11]:

$$\text{ArH} + \text{AlCl}_3 + (i\text{-PrOCO}_2)_2 \longrightarrow \underset{\overset{\|}{O}}{\text{ArOC}}\!\!-\!\!\text{OPr-}i + i\text{-PrO}\underset{\overset{\|}{O}}{\text{CO}}\text{AlCl}_2 + \text{HCl}$$

$$\Big\downarrow \begin{array}{l} 1.\ \text{KOH, C}_2\text{H}_5\text{OH} \\ 2.\ \text{H}^{\oplus} \end{array}$$

$$\text{ArOH}$$

Yields of phenols were 66–76% from anisole, mesitylene, and pentamethylbenzene and 48–52% from toluene, durene, and *m*-xylene. However, the phenols were mixtures of isomers: from toluene, 34% *o*-, 11% *m*-, and 55% *p*-cresol; from anisole, 20% *o*- and 80% *p*-methoxyphenol; from *m*-xylene, 91% 2,4- and 9% 2,6-xylenol. Some isopropylation accompanied acyloxylation.

The reaction above can be conducted as well using cupric chloride rather than aluminum chloride in acetonitrile [12] to give with toluene an 85% yield of cresols consisting of 57% *o*-, 15% *m*-, and 28% *p*-.

(a) Preparation of mesitol. Trifluoroacetic anhydride, 35 g., 50 ml. of methylene chloride, and 4 ml. of 90% hydrogen peroxide were mixed at 0° and allowed to warm to room temperature. This solution at 0° was added dropwise to 56.1 g. of mesitylene in 100 ml. of methylene chloride while boron trifluoride was bubbled through the reaction mixture (2.5 hr.) at 7° or less. After warming to room temperature, 100 ml. of water was added and the aqueous layer was separated and washed with three 25-ml. portions of methylene chloride. The combined organic layers were then washed with 10% sodium bisulfite, 10% sodium bicarbonate, and dried over anhydrous magnesium sulfate. After removing the solvent, distillation through a 1-ft. helices-packed column gave 17.7 g. (88.5% based on peracid) of mesitol, b.p. 98° (10 mm.), m.p. 69–70° [7].

2. From Organometallic Compounds

$$2\ \text{ArMgX} + t\text{-BuOOH} \longrightarrow \text{ArOMgX} + t\text{-BuOMgX} + \text{ArH}$$

$$\text{ArOMgX} \xrightarrow{\text{HOH}} \text{ArOH} + \text{MgXOH}$$

Since oxygen with aryl Grignard reagents gives poor yields of phenols [13], various oxidizing agents have been substituted for it. It was found that, by converting the phenylmagnesium bromide into phenylboronic acid and related compounds with methyl borate and then treating the crude product in ether with hydrogen peroxide, yields of phenol based on the bromobenzene utilized were 60–80% [14]: (See Ex. b.2).

$$\text{ArMgX} \xrightarrow{\text{B(OCH}_3)_3} \text{ArB(OCH}_3)_2 \xrightarrow{\text{H}^{\oplus}} \text{ArB(OH)}_2 \xrightarrow{10\%\ \text{H}_2\text{O}_2} \text{ArOH}$$

Later, hydroperoxides, alkali metal hydroperoxides, and tertiary butyl perbenzoates were employed with the Grignard reagent or aryllithium [15–18].

Good yields were obtained in the use of t-butyl hydroperoxide or its magnesium salt [15]. The use of the perbenzoate gives the ether as one of the products,

$$\underset{\substack{\| \\ O}}{ArCOOC(CH_3)_3} + ArMgX \longrightarrow \underset{\substack{\| \\ O}}{ArOC(CH_3)_3} + ArCOMgX$$

and attempts to convert it into the phenol were usually successful when it was pyrolyzed in the presence of p-toluenesulfonic acid [16]. Yields of o-cresol, p-cresol, and p-methoxyphenol from the corresponding ether were quantitative by this procedure.

The synthesis, in which the perbenzoate has been employed, has been formulated as follows with the magnesium dialkyl present in the Grignard reagent:

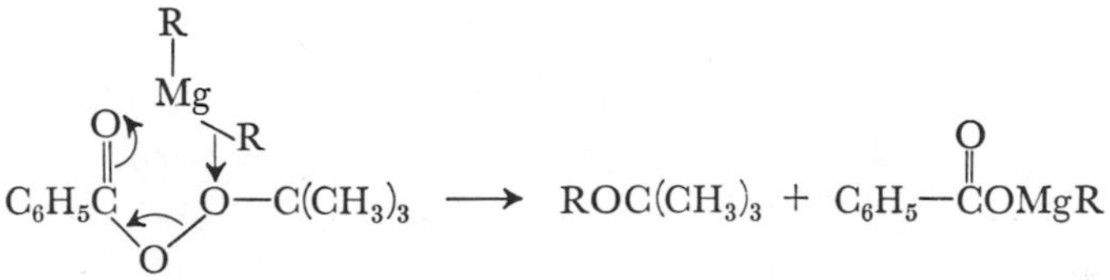

$$C_6H_5\underset{\underset{O}{\|}}{C}\cdots O-C(CH_3)_3 \quad \overset{\substack{R \\ | \\ Mg \\ \diagdown R}}{} \longrightarrow ROC(CH_3)_3 + C_6H_5-\underset{\underset{O}{\|}}{C}OMgR$$

Oxidizing agents other than peroxides can be used. For instance, nitrobenzene at low temperature oxidizes phenyllithium [19]. In this case 1 equivalent of nitrobenzene in tetrahydrofuran to which 2.2 equivalents of phenyllithium is added at $-100°$ gives 2 equivalents of phenol.

(a) **Preparation of 2-hydroxythiophene** (70–76% of 2-t-butoxythiophene from magnesium, bromothiophene, and t-butyl perbenzoate; 89–94% of the phenol from the ether under nitrogen in the presence of a trace of p-toluenesulfonic acid) [17].

(b) **Other examples**

(1) **Phenol** (80% from t-butyl hydroperoxide and ethylmagnesium bromide to form the magnesium salt, and phenylmagnesium bromide in ether at -60 to $-70°$) [15].

(2) **6-Methoxy-2-naphthol** (80% from 6-methoxy-2-bromonaphthalene via the Grignard reagent, treatment of the latter with trimethyl borate, and oxidation of the mixture of naphthylboric acids with 15% hydrogen peroxide in aqueous ammonium chloride solution) [20].

(3) **4-Hydroxy-3,5,2′-trimethoxybiphenyl** (methoxyaucuparin)

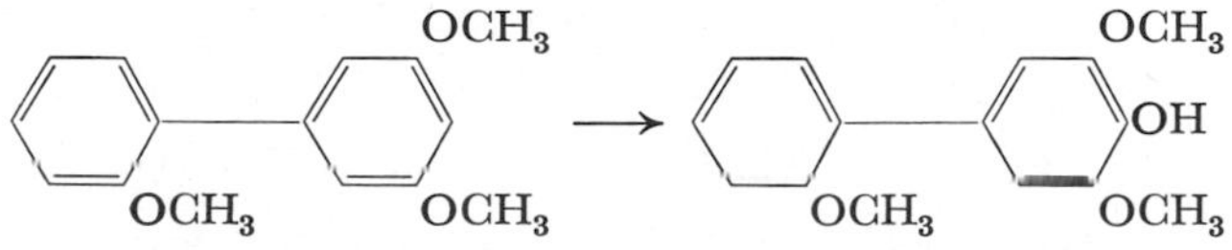

(53% from butyllithium and 3,5,2′-trimethoxybiphenyl to form the aryllithium derivative followed by oxidation with lithium t-butyl hydroperoxide) [21].

3. From Arene Carboxylic Acids

$$(C_6H_5COO)_2Cu \xrightarrow{\Delta} C_6H_5COOC_6H_5 + Cu + CO_2$$

$$C_6H_5COOC_6H_5 + H_2O \longrightarrow C_6H_5OH + C_6H_5COOH$$

This synthesis is accomplished in 1–6 hours at 200–350° by heating the cupric salt alone or in a solvent such as excess carboxylic acid, inert hydrocarbon, or water under pressure, or, if oxygen is introduced, only catalytic amounts of copper are necessary [22, 23]. Anhydrous systems are preferred in producing the phenyl ester and, of the metals investigated, only the cupric salts are of sufficient specificity to be of practical value. The method gives yields as high as 90 % of phenol in the sequence:

Carboxylic acid → cupric salt → ester → phenol

although usually conversions are considerably lower.

Strangely enough, the reaction occurs in such a way that the hydroxyl group appears *ortho* to the position originally occupied by the carboxyl group. For example, *o*- and *p*-substituted benzoic acids give *m*-substituted phenols,

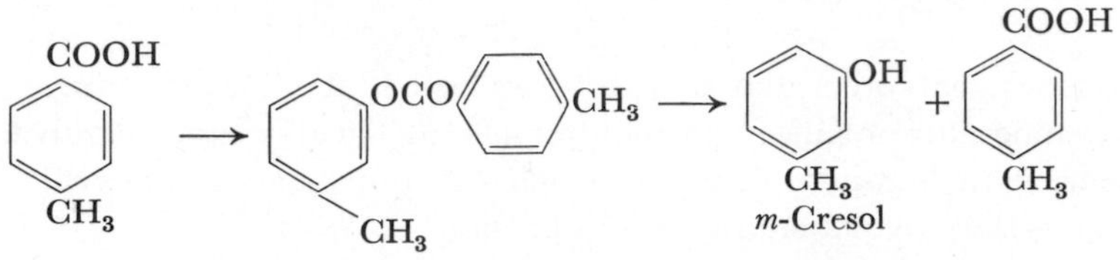

This fascinating reaction not only is unusual in orientation, i.e., insertion of oxygen *ortho* to the carboxyl group, but it also is efficient enough to be a commercial method of production of phenol. It probably proceeds via the formation of an aryl-copper intermediate [24]:

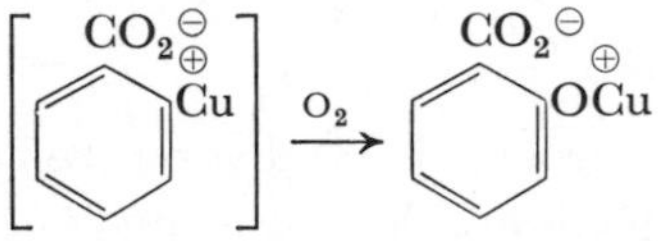

which is somewhat reminiscent of the electrophilic attack of mercuric acetate on aromatic compounds. The orientation is *ortho* because the attack of the copper cation is made from its position in the carboxylate group.

It is also possible to eliminate alkyl groups, *ortho* or *para* to the hydroxyl group, via oxidation to obtain phenols [25]:

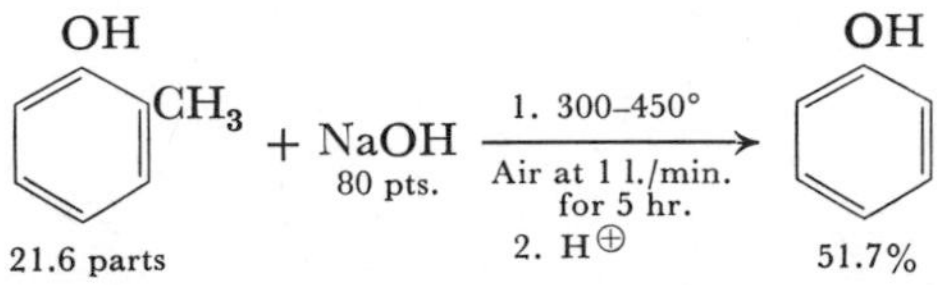

(a) Preparation of phenol. Benzoic acid, 2 moles, 2 moles of $CuSO_4 \cdot 5 H_2O$, and 900 ml. of water were heated in an autoclave with shaking at 315° for 1 hr. The pH was adjusted to 12; the mixture was steam distilled; and 5.1 g.

of phenol was recovered from the distilled water layer. An additional 9.2 g. of 90% pure phenol was recovered from the still residue [22].

4. From Tertiary Hydrocarbons via Hydroperoxides

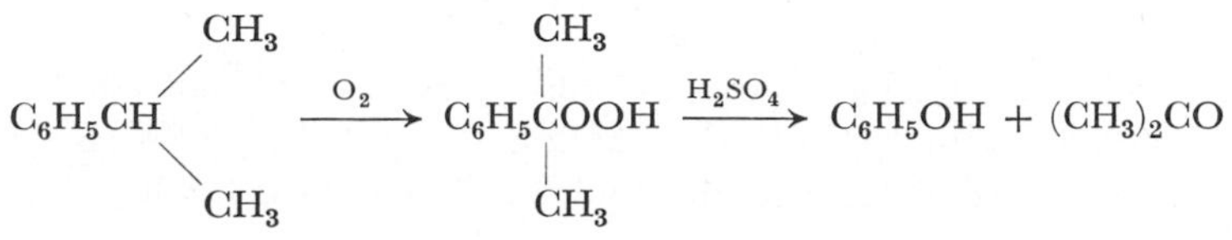

$$C_6H_5CH\begin{smallmatrix}CH_3\\ \\CH_3\end{smallmatrix} \xrightarrow{O_2} C_6H_5\underset{\underset{CH_3}{|}}{\overset{\overset{CH_3}{|}}{C}}OOH \xrightarrow{H_2SO_4} C_6H_5OH + (CH_3)_2CO$$

Although this synthesis is of little value as a laboratory preparation, it is included because it is one of the commercial methods for preparing phenol. Technological developments during World War II made cumene and cumene hydroperoxide available, and it was logical to utilize the latter as a source of phenol and acetone [26]. This synthesis, its mechanism, and examples of it are included under Ketones, Chapter 11, A.7.

5. From Carbonyl Compounds (including Dakin)

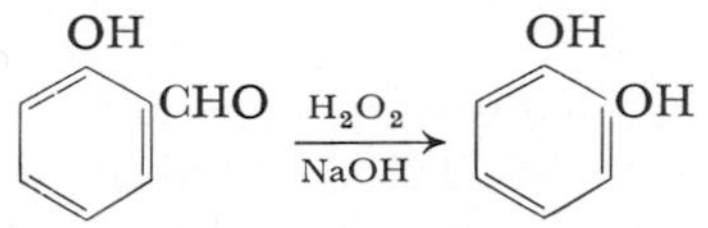

This synthesis is applicable to hydroxy or alkoxy derivatives of benzaldehyde or acetophenone, provided the substituent is *o*- or *p*- to the carbonyl group. Thus the method is very limited in use, but it has been of value in cases in which the appropriate benzaldehydes or acetophenones are available from natural sources. Yields are fair.

The mechanism is unknown but may be rationalized as shown provided that the reaction is ionic:

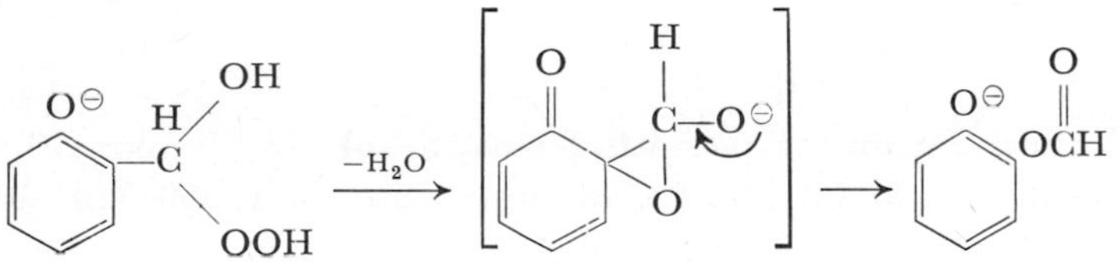

The overall results are the same as the Baeyer-Villiger oxidation of ketones to esters (Carboxylic Esters, Chapter 14, B.4).

(a) **Preparation of pyrogallol 1-monomethyl ether** (68–80% from 2-hydroxy-3-methoxy-benzaldehyde, 6% hydrogen peroxide, and sodium hydroxide) [27].

(b) **Preparation of 2,5-dimethoxyphenol** (50–60% from 2,5-dimethoxy-acetophenone and peracetic acid followed by alkaline hydrolysis; Baeyer-Villiger oxidation) [28].

6. From Cyclic Ketones or Epoxides (Dehydrogenation)

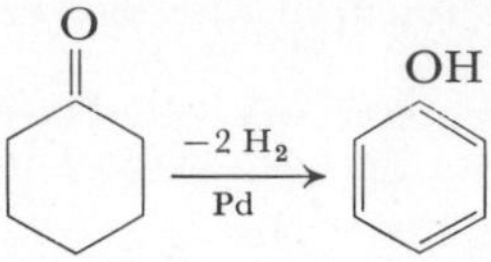

Many cyclic ketones have been dehydrogenated to phenols by the use of catalysts such as sulfur, selenium, palladium, or nickel [29]. A common side reaction is dehydration to yield the hydrocarbon instead of the phenol. The greater the resemblance of the structure to that of the aromatic phenol, the higher is the yield. That is, it would be anticipated that an unsaturated cyclic ketone or, better, a tetralone would give a good yield of the phenol.

The dehydrogenation can be accomplished also by bromination followed by dehydrohalogenation [30]:

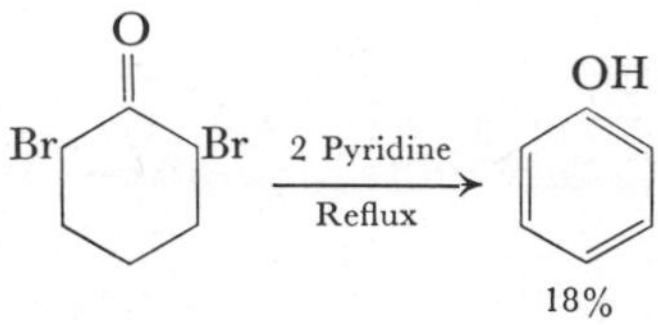

A higher yield is to be expected with more pyridine or a stronger base, such as 1,5-diazabicyclo[3.4.0]nonene-5 (DBN). Similarly the isomeric bromocyclo-epoxides may be converted into phenols by the use of DBN [31]:

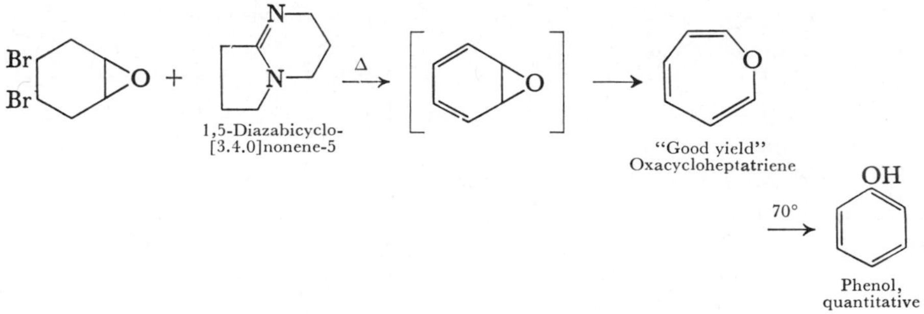

(a) **Preparation of 7-methyl-1-naphthol** (60% from 10 g. of 7-methyl-1-tetralone and 4 g. of palladium on charcoal at 300–320° for 1.5 hr.) [32].

7. From Phenols or Arylamines and Persulfate Salts

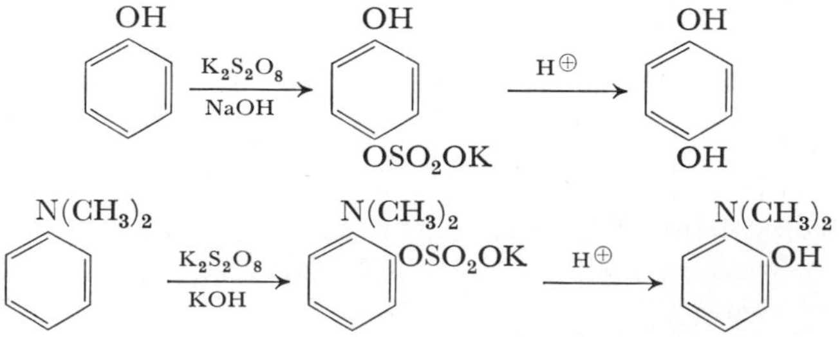

The synthesis of dihydric phenols (Elbs) has been reviewed [33]. The preferred persulfate is that of potassium, although the ammonium salt may also be employed [34]. The sometime addition of ferric chloride offers no advantage, although the yield is sometimes improved by conducting the reaction in solutions saturated with sodium chloride or sodium sulfate. If the *p*-position is occupied, the substituent goes to the *o*-position. Yields are low. The best, for chloroquinol, is 50%. Phenol itself gives an 18% yield of quinol. The method offers a pure product since the intermediate, being a salt, may be separated from organic impurities by ether extraction. The synthesis has found use in the coumarin and flavone series since the hydroxyl group may be introduced into desired positions. A similar reaction occurs for arylamines [35], although in this case only *o*-aminophenols are produced. Unfortunately yields are also low, particularly in the first step.

(a) Preparation of gentisic acid

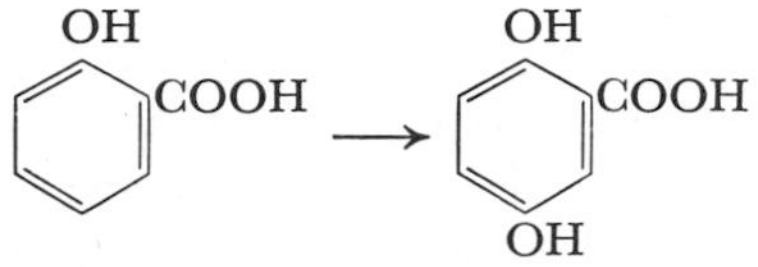

To a solution of 1 mole of salicyclic acid and 4 moles of NaOH in 2 l. of water, 1.2 moles of ammonium persulfate in 2 l. of water was added over 5 hr. at 20–25°. After 15 hr. the mixture was acidified with concentrated H_2SO_4 and then boiled for 1 hr. The filtrate, after decolorization, was extracted with butyl acetate, which extract on evaporation gave 60–78 g. (39–51%) of crude gentisic acid, which when recrystallized melted at 202–204° [36].

(b) Preparation of *o*-dimethylaminophenol (overall yield of 27% from a mixture of dimethylaniline, water, acetone, and KOH to which potassium persulfate was added during 8 hr.; the recovered *o*-dimethylaminophenyl potassium sulfate was heated with concentrated HCl and neutralized) [35].

8. From Nitrobenzenes and the Like

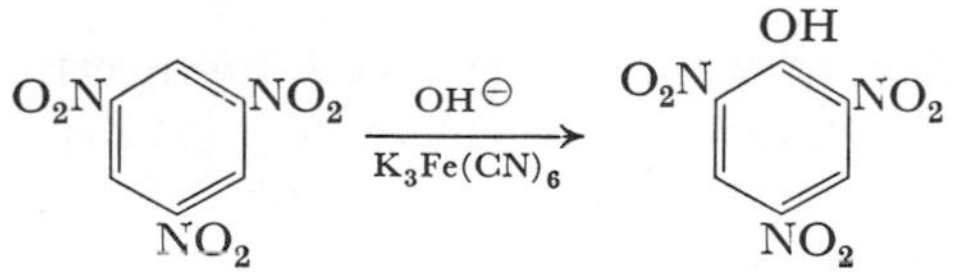

Aromatic compounds with strong electron-withdrawing groups are subject to attack by the hydroxide ion. The intermediate may be oxidized with salts, such as a chlorate, a nitrate, or potassium ferricyanide, or oxygen:

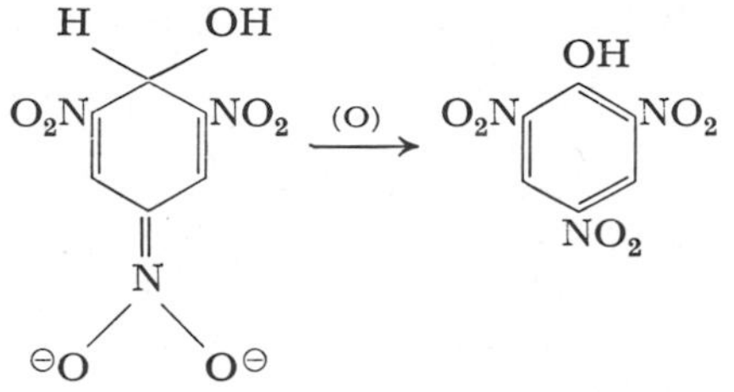

The preparation of alizarin from anthraquinone-β-sulfonic acid, sodium hydroxide, and a nitrate is a well-known industrial example [37]:

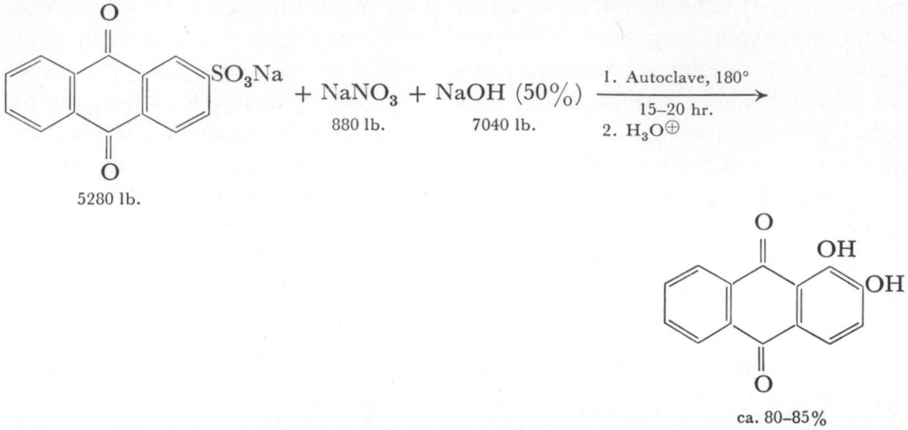

Even nitrobenzene affords some o-nitrophenol, although the experiment was run without the addition of an oxidizing agent, under which circumstances the nitrobenzene serves to oxidize the intermediate [38]. Quinolines yield 2-hydroxy-quinolines by heating with dry potassium hydroxide or hypochlorous acid [39].

9. From Phenols (Oxidative Coupling)

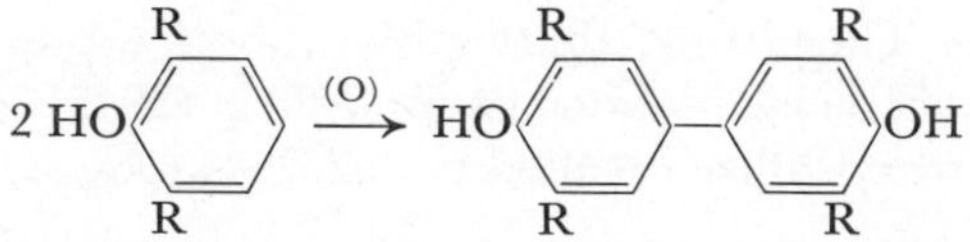

Coupling usually has been carried out with ferric chloride or ferricyanide, but recently manganic tris-(acetylacetonate) has been advocated as a reagent with simple requirements. The reaction is restricted to ortho-substituted phenols with the exception of β-naphthol. The oxidation proceeds by electron transfer, giving rise to an aryloxy radical which couples or dimerizes.

(a) **Preparation of 2,2'-dihydroxy-1,1'-binaphthyl.** Manganic tris-(acetylacetonate) and β-naphthol (1.2:1 ratio) were dissolved in acetonitrile (or carbon bisulfide), and the mixture was gently refluxed under nitrogen for 5 hr. to yield, after workup, 69% of the product [40].

1. Oxidation Process Corp., Belgian Patent 619,935, October 31, 1962; C.A., 59, 511 (1963).
2. H.-I. Joschek and S. I. Miller, J. Am. Chem. Soc., 88, 3273 (1966).
3. D. H. Derbyshire and W. A. Waters, Nature, 165, 401 (1950).
4. J. D. McClure and P. H. Williams, J. Org. Chem., 27, 24 (1962).
5. R. D. Chambers et al., J. Chem. Soc., 1804 (1959).
6. S. Hashimoto et al., Kogyo Kagaku Zasshi, 70 (3), 406 (1967); C.A., 67, 3045 (1967).
7. H. Hart and Charles A. Buehler, J. Org. Chem., 29, 2397 (1964).
8. A. Fischer et al., J. Chem. Soc., 3687 (1964).
9. D. R. Harvey and R. O. C. Norman, J. Chem. Soc., 4860 (1964).
10. R. L. Dannley and G. E. Corbett, J. Org. Chem., 31, 153 (1966).
11. P. Kovacic and M. E. Kurz, J. Am. Chem. Soc., 87, 4811 (1965).
12. M. E. Kurz and P. Kovacic, J. Am. Chem. Soc., 89, 4960 (1967).

13. M. S. Kharasch and O. Reinmuth, *Grignard Reagents of Non-metallic Substances*, Prentice-Hall, Englewood Cliffs, N.J., 1954, p. 1264.
14. M. F. Hawthorne, *J. Org. Chem.*, **22**, 1001 (1957).
15. S.-O. Lawesson and N. C. Yang, *J. Am. Chem. Soc.*, **81**, 4230 (1959).
16. S.-O. Lawesson and C. Frisell, *Arkiv Kemi*, **17**, 393 (1961).
17. C. Frisell and S.-O. Lawesson, *Org. Syn.*, **43**, 55 (1963).
18. M. Nilsson and T. Norin, *Acta Chem. Scand.*, **17**, 1157 (1963).
19. P. Buck and G. Köbrich, *Tetrahedron Letters*, 1563 (1967).
20. R. L. Kidwell and S. D. Darling, *Tetrahedron Letters*, 531 (1966).
21. M. Nilsson and T. Norin, *Acta Chem. Scand.*, **17**, 1157 (1963).
22. W. G. Toland, *J. Am. Chem. Soc.*, **83**, 2507 (1961).
23. W. W. Kaeding, *J. Org. Chem.*, **26**, 3144 (1961).
24. E. J. Strojny, Dow Chemical Company, personal communication.
25. J. S. Mackay and F. J. Vancheri, U.S. Patent 3,071,627, January 1, 1963; *C.A.*, **59**, 511 (1963).
26. S. J. Lederman and N. Poffenberger in Kirk-Othmer's *Encyclopedia of Chemical Technology*, Interscience Publishers, New York, 2nd Ed., 1968, Vol. 15, p. 149.
27. A. R. Surrey, *Org. Syn.*, Coll. Vol. **3**, 759 (1955).
28. C. A. Bartram *et al.*, *J. Chem. Soc.*, 4691 (1963).
29. W. Foerst, *Newer Methods in Preparative Organic Chemistry*, (*Engl. Transl.*), Interscience Publishers, New York, 1948, Vol. 1, p. 27.
30. F. Galinovsky *et al.*, *Monatsh.*, **80**, 288 (1949).
31. E. Vogel *et al.*, *Angew. Chem. Intern. Ed. Engl.*, **3**, 510 (1964).
32. L. Ruzicka and E. Mörgeli, *Helv. Chim. Acta*, **19**, 377 (1936).
33. S. M. Sethna, *Chem. Rev.*, **49**, 91 (1951).
34. W. Baker and N. C. Brown, *J. Chem. Soc.*, 2303 (1948).
35. E. Boyland *et al.*, *J. Chem. Soc.*, 3623 (1953).
36. R. U. Schock, Jr., and D. L. Tabern, *J. Org. Chem.*, **16**, 1772 (1951).
37. A. J. Cofranesco, Ref. 26, Vol. 2, 2nd Ed., 1963, p. 474.
38. A. Wohl, *Chem. Ber.*, **32**, 3486 (1899).
39. R. C. Elderfield, *Heterocyclic Compounds*, John Wiley and Sons, New York, 1952, Vol. 4, p. 136.
40. M. J. S. Dewar and T. Nakaya, *J. Am. Chem. Soc.*, **90**, 7134 (1968).

C. Reduction

1. From Quinones

Since quinones are dehydrogenating agents, a wide variety of substrates, including hydrocarbons, may be used as reducing agents. Quinones with high redox potential, i.e., those with ready reduction to the dihydric phenol, are those with electron-withdrawing groups attached as shown in the accompanying formulas [1].

Various orthodox reducing agents, such as zinc and sodium hydroxide [2], hydrogen in the presence of Raney nickel [3], sodium borohydride in diglyme which has been exposed to air [4], stannous chloride in hydrochloric acid [5], zinc in acetic acid [6], sulfur dioxide in water [7], sodium hydrosulfite [8], and lithium aluminum hydride [9], have been employed to convert quinones into dihydric phenols. These reducing agents usually give a good return of the phenol, although it is sometimes necessary to operate in an inert atmosphere under anhydrous conditions to prevent oxidation of the product. In addition,

References for Section C are on pp. 269–270.

powerful reducing agents such as lithium aluminum hydride may give *trans*-dihydric alcohols from *o*-quinones [10]. The synthesis has had perhaps its greatest value in preparing *o*- and *p*- dihydric phenols in the benzene and naphthalene series.

Relative Rates of Dehydrogenation of 1,2-Dihydronaphthalene with Quinones

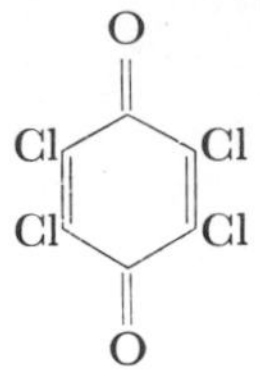

Chloranil

1

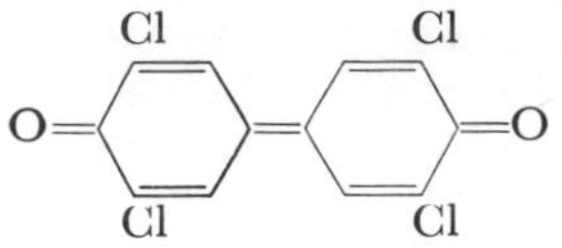

3,3′,5,5′-Tetrachlorodiphenoquinone

1100

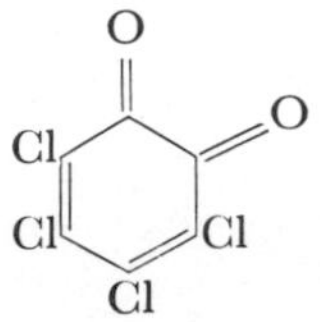

o-Chloranil

4200

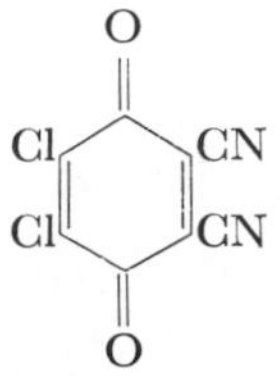

DDQ

5500

Although not a quinone, phloroglucinol is the only trihydric phenol which can be hydrogenolyzed to a dihydric phenol (resorcinol in this case) by sodium borohydride and water at 25° [11].

(a) **Preparation of 1,2,4-trihydroxynaphthalene.** 2-Hydroxy-1,4-naphthoquinone, 15 g., was shaken with ether and hydrosulfite solution (concentration 4 g./20 ml.), and the ethereal solution was washed with brine containing hydrosulfite. It was then introduced from a separatory funnel into a second one containing Drierite supported on glass wool from which it was delivered into a flask heated on a steam bath. The second separatory funnel and the receiver were both flushed with nitrogen and the solvent was removed by flash distillation to give 13.9 g. (93%) of the phenol [8].

(b) **Preparation of 6-benzoyloxy-1,2-naphthohydroquinone** (91% from the 1,2-naphthoquinone, sulfur dioxide, water and methanol) [7].

2. From Unsaturated Polycyclic Diketones and the Grignard Reagent Followed by Dehydration

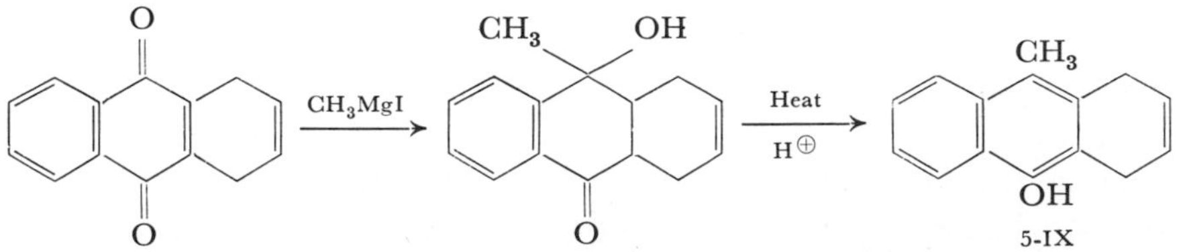

5-VII   5-VIII   5-IX

The first step in this synthesis, the formation of 5-VIII, is of importance in the synthesis of the broad spectrum antibiotics such as terramycin, tetracycline, and aureomycin. The required diketone 5-VII is readily available by the Diels-Alder reaction [12]. To form the ketol 5-VIII it is necessary to add not over a 25% excess of the Grignard reagent to a cold benzene solution of 5-VII. 5-VIII may be dehydrated by heating an alcoholic hydrochloric acid solution in an atmosphere of nitrogen (see Ex. a) to give 5-IX.

(a) Preparation of 10-hydroxy-9-methyl-1,4-dihydroanthracene (5-IX)

(1) **10-Keto-9-hydroxy-9-methyl-1,4,4a,9,9a,10-hexahydroanthracene** (5-VIII). To a stirred solution of 0.05 mole of 5-VII in anhydrous benzene at 5–7° methylmagnesium iodide in anhydrous ether was added. After stirring for 2–3 hr., the mixture was decomposed by pouring it into a mixture of crushed ice and 125 ml. of 1 N hydrochloric acid. Extraction with ether, drying, and evaporation gave 5-VIII, m.p. 135–137° (from alcohol) in a 70% yield.

(2) **10-Hydroxy-9-methyl-1,4-dihydroanthracene** (5-IX). To 5 mmole of 5-VIII in alcohol an equal volume of concentrated hydrochloric acid was added, and the mixture was heated at 60° for 20–30 min. under nitrogen. Cooling gave the phenol 5-IX (83%), m.p. 117–119° (from petroleum ether) [12].

3. From Aromatic Ethers

(See A.4 as well.) The benzyl group is readily removed from benzyl aromatic ethers if hydrogenolysis is stopped after absorption of 1 mole of hydrogen [13]:

$$\text{ArOCH}_2\text{C}_6\text{H}_5 \xrightarrow[\text{Raney Ni}]{\text{H}_2} \text{ArOH} + \text{C}_6\text{H}_5\text{CH}_3$$

This reaction is utilized in protection of the phenolic group, although the most important reductive cleavage is that of diphenyl ethers [14]:

$$\text{C}_6\text{H}_5\text{OC}_6\text{H}_5 + \text{Na} \xrightarrow[\substack{\text{Reflux 4–6 hr.} \\ \text{then dilute and} \\ \text{acidify}}]{\text{Pyridine, 4–6 moles}} \text{C}_6\text{H}_5\text{OH} + \text{C}_6\text{H}_6$$

2–3 g. atoms   90%

Unsymmetrical diphenyl ethers give mixtures of phenols unless substituents, such as the amino group, are present to favor a single type of cleavage:

$$\text{H}_2\text{N}\langle\rangle\text{O}\langle\rangle\text{OCH}_3 \xrightarrow[\text{NH}_3]{\text{Na}} \text{H}_2\text{N}\langle\rangle\text{OH} + \text{HO}\langle\rangle\text{OCH}_3$$

92%   8%
p-Aminophenol   *p*-Methoxyphenol

1. L. M. Jackman, *Advan. in Org. Chem.*, Interscience Publishers, New York, 1960, Vol. 2, p. 329.
2. J. S. Meek *et al.*, *J. Org. Chem.*, **28**, 2572 (1963).
3. A. Marxer, *Helv. Chim. Acta*, **44**, 762 (1961); J. Druey and P. Schmidt, *Helv. Chim. Acta*, **33**, 1080 (1950).

4. G. S. Panson and C. E. Weill, *J. Org. Chem.*, **22**, 120 (1957).
5. H. O. Huisman, *Rec. Trav. Chim.*, **69**, 1133 (1950).
6. H. M. Crawford *et al.*, *J. Am. Chem. Soc.*, **74**, 4087 (1952).
7. M. Gates, *J. Am. Chem. Soc.*, **72**, 228 (1950).
8. L. F. Fieser, *J. Am. Chem. Soc.*, **70**, 3165 (1948).
9. R. F. Nystrom and W. G. Brown, *J. Am. Chem. Soc.*, **70**, 3738 (1948).
10. J. Booth *et al.*, *J. Chem. Soc.*, 1188 (1950).
11. G. I. Fray, *Tetrahedron*, **3**, 316 (1958).
12. M. M. Shemyakin *et al.*, *J. Gen. Chem. USSR*, (*Eng. Transl.*), **29**, 1802 (1959).
13. R. L. Augustine, *Catalytic Hydrogenation*, Marcel Dekker, New York, 1965, p. 135.
14. H. Meerwein, Houben-Weyl's *Methoden der Organischen Chemie*, Georg Thieme Verlag, Stuttgart, 1965, Vol. 6, Pt. 3, p. 168.

D. Electrophilic Reactions

The ready substitution in the aromatic ring of phenols lends itself to the preparation of a great variety of phenols via alkylation, halogenation, nitration, acylation, and similar processes. Exhaustive treatment of these subjects is not given, but the principles are discussed from the viewpoint of isolation of pure *o*-, *m*-, and *p*-isomers and polysubstituted phenols. Several rearrangement reactions of an electrophilic nature are also discussed.

1. From Phenols by Alkylation or Arylation or by Rearrangement

Exhaustive reviews are available [1, 2]. Alkylation can be carried out very readily with a variety of catalysts such as aluminum chloride, sulfuric acid, phosphoric acid, and zinc chloride and with a variety of alkylating agents such as olefins, alcohols, and halides. The substitution is so easy that *sec*-, *t*-, and benzylic halides can be used occasionally without catalysts. Kinetic control (see Alkanes, Cycloalkanes, and Arenes, Chapter 1, D.1) yields mixtures of *o*- and *p*-alkyl phenols, the *para*-isomer often predominating if bulkiness is present in the halide or in the solvent shell of the reacting species. Without a catalyst, the substitution in phenols by the use of an alkyl halide does not appear to take place via ionization of the halide [3]:

$$C_6H_5OH + C_6H_5CHClCH_3 \xrightarrow[\text{rises to } 45°]{\text{Temp.}}$$

2 equiv.   Optically active

33%
p-(α-Phenylethyl)-phenol

41%
o-(α-Phenylethyl)-phenol

para-Substitution occurs with partial inversion of the alkyl halide, a fact which suggests an SN_2 type of replacement. *ortho*-Substitution occurs with partial retention of configuration, a fact which suggests cyclo substitution:

References for Section D are on p. 279.

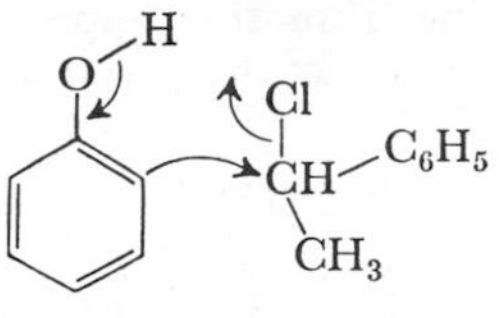

To obtain *ortho*-alkylated phenols exclusively, the attack of the alkylation agent must be made from the position of the hydroxyl group. The procedure of Kolka [4, 5] accomplishes this feat (see Ex. b.1):

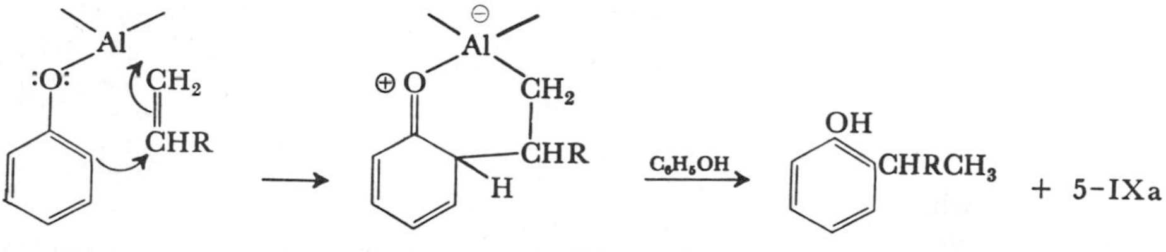

5-IX a

Not only 2-alkylphenols but also 2,6-dialkylphenols can be prepared by this procedure. Where no other alternative is open, a phenol can be *o*-methylated in the following way [6]:

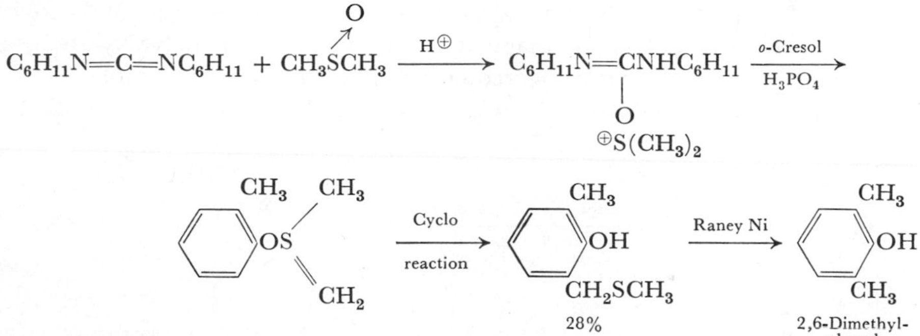

$$C_6H_{11}N\!=\!C\!=\!NC_6H_{11} + CH_3SCH_3 \xrightarrow{H^\oplus} C_6H_{11}N\!=\!\overset{\overset{\displaystyle O}{|}}{C}NHC_6H_{11} \xrightarrow[H_3PO_4]{o\text{-Cresol}}$$

$$\overset{\oplus}{S}(CH_3)_2$$

28%   2,6-Dimethyl-
                                    phenol

To obtain *meta*-alkylated phenols a rearrangement may be employed under thermodynamic control, i.e., with a large excess of Lewis acid and its co-catalyst such as aluminum chloride-hydrogen chloride. Among alkyl groups the methyl migrates the least readily. Because of this fact the rearrangement that occurs with *p*-cresol, for example, must be carried out with liquified hydrogen bromide in an autoclave [7]:

$$p\text{-Cresol, } AlCl_3 + \text{liquefied HBr solvent} \xrightarrow[\text{Autoclave}]{100°} m\text{-Cresol}$$

1 equiv.   1.5 equiv.   90%, no *o*- or *p*-

and disproportionation products

10%

Using aluminum chloride and having hydrogen chloride pass through the *p*-cresol-aluminum chloride complex, about 80% *m*- and 20% *p*-cresol were obtained. All other alkylated phenols, including polymethylated phenols, are more easily rearranged, and a general observation is that the *p*-alkyl group is the first to undergo migration (because the *p*-position is the most susceptible to attack by the proton in the Lewis acid-cocatalyst combination) [8]:

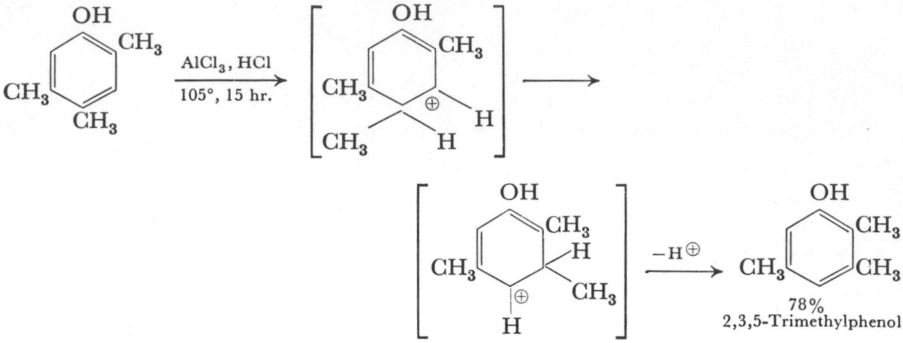

78%
2,3,5-Trimethylphenol

The ease of migration of alkyl groups is in the order: *t*-butyl > isopropyl > ethyl > methyl. Thus conditions milder than those described will bring about migration or disproportionation of groups other than the methyl. Rearrangement of the alkyl group, of course, also can take place in the Friedel-Crafts alkylation of phenol (see Alkanes, Cycloalkanes, and Arenes, Chapter 1, D.1) but to a lesser extent than with arenes, since phenols are so much more reactive. Monoalkylation of phenols is plagued by the incursion of polyalkylation, a condition which can partly be overcome by using an excess of phenol.

Bisphenols are formed from phenols and ketones which are not too hindered as shown [9]:

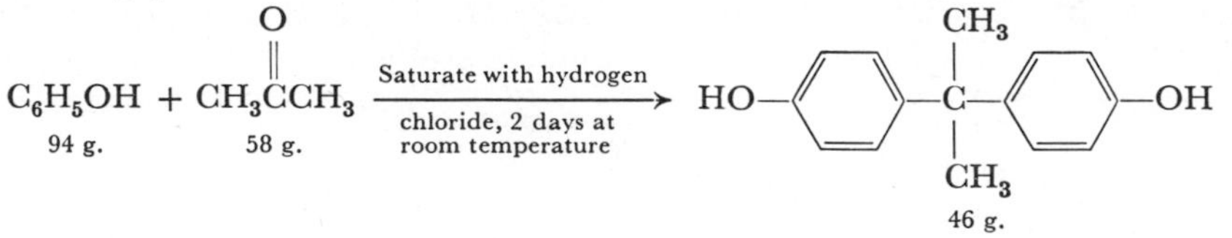

2,2-Di-*p*-hydroxyphenylpropane

The product undergoes an interesting transformation with hydrogen bromide [9]:

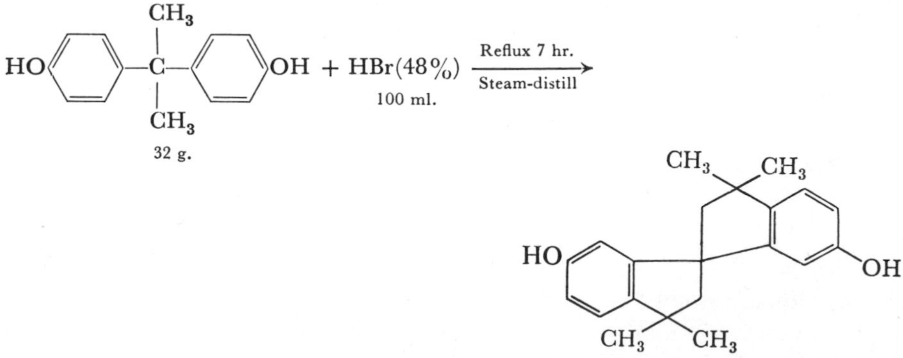

6,6′-Dihydroxy-3,3,3′,3′-tetramethyl-1,1′-spirobiindane, 65%

Benzophenone does not alkylate phenols to form bisphenols except by photochemical means [10]:

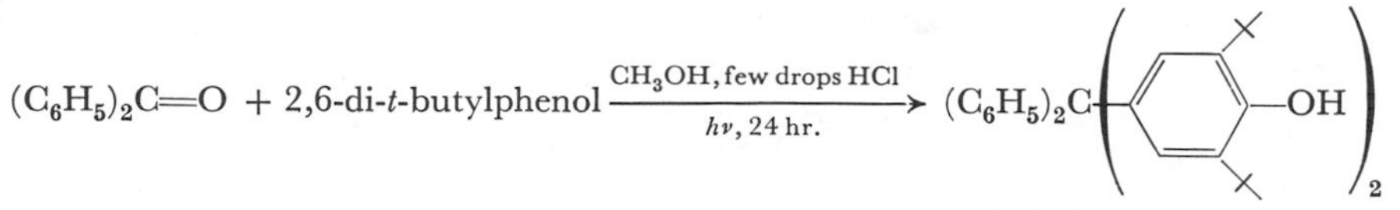

$$(C_6H_5)_2C{=}O + \text{2,6-di-}t\text{-butylphenol} \xrightarrow[h\nu,\, 24\,\text{hr.}]{CH_3OH,\, \text{few drops HCl}} (C_6H_5)_2C{\Big(}{-}\!\!{\bigcirc}\!\!{-}OH{\Big)}_2$$

4,4′-Dihydroxy-3,3′,5,5′-tetra-*t*-butyltetraphenylmethane, 65%

This reaction, of course, is a homolytic rather than electrophilic substitution, but is included here to make the discussion of bisphenol syntheses more complete.

The phenylation of phenol apparently has not been carried out [11] but conceivably could be achieved by reaction of phenol and fluorobenzene in the presence of aluminum chloride. Whether possible or not, easier access to phenylphenols is available from the photolysis of iodophenols [12]:

$$\text{2,4,6-Triodophenol (2 g.)} \xrightarrow[h\nu,\, 24\,\text{hr.}]{285\ \text{ml. of } C_6H_6} \text{2,4,6-triphenylphenol}$$
75%

The reaction is quite general.

(a) Preparation of 2,4,6-triisopropylphenol (95% from 1.5 moles of phenol and 6.75 moles of isopropyl alcohol added dropwise to 800 g. of liquefied hydrogen fluoride at less than 8° and then allowed to evaporate in a hood over a period of 16 hr.) [13]. It is recommended that alkylation of phenols with primary alkyl halides be carried out with a ratio of phenol to aluminum chloride of 1:1.25 and a temperature at which hydrogen halide evolution is moderately fast. These conditions overcome the sluggishness of reaction of phenols and primary alkyl halides and minimize *ortho*-substitution.

(b) Other examples

(1) 2,6-Di-*t*-butylphenol (74% from 4 moles of phenol, 3.6 g. of aluminum turnings, and 100 ml. of toluene to which isobutylene was added in an autoclave at 100° (about 240 p.s.i.). *o-t*-Butylphenol, 9%, and 9% 2,4,6-tri-*t*-butylphenol were also formed. *o*-2-Cyclohexylphenol was made in 42% yield by a similar procedure at 244°) [4].

(2) *m*-Phenylphenol (77% from 1.77 moles of *o*-phenylphenol and 2.24 moles of aluminum chloride in chlorobenzene at steam-bath temperature for 1 hr.) *Note:* The yield would probably be higher if hydrogen chloride were passed in over a longer period of time [14].

2. From Phenols and Halogenating Agents

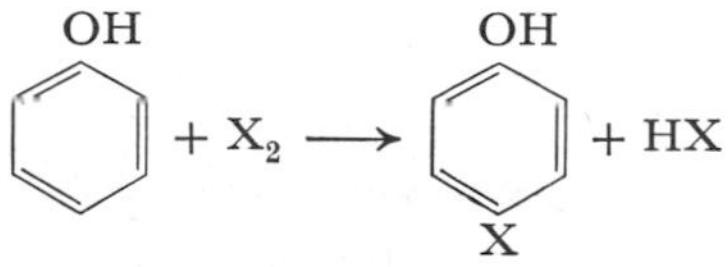

$$\underset{}{\bigcirc}\!\!-OH + X_2 \longrightarrow \underset{X}{\bigcirc}\!\!-OH + HX$$

(See Halides, Chapter 7, D.2.)

3. From Phenols and Nitrating or Sulfonating Agents

The nitration of phenols is discussed under Nitro Compounds, Chapter 20, A.2, and the sulfonation is described here briefly.

The sulfonation of phenol is accomplished with great ease [15]. The proportion of isomers formed in concentrated sulfuric acid is as follows:

T	% o-	% p-
20°	39	61
100°	4	96

As these solutions are allowed to stand, increasing amounts of the *m*-sulfonic acid are formed. An actual preparation, however, would involve separation of the barium salts. A much better method consists of treatment of phenol with the stoichiometric amount of sulfur trioxide at about 50° or the sulfur trioxide-pyridine complex at 170–200°.

4. From Hydrocarbons and Nitrating Agents

Traces of nitrophenols are to be found in nitration of any aromatic hydrocarbon. The phenols probably arise through the steps: nitrosation; reduction to a diazonium salt; decomposition to a phenol; and nitration. Better means, however, are available to bring about nitrohydroxylation. The agent is mercuric nitrate:

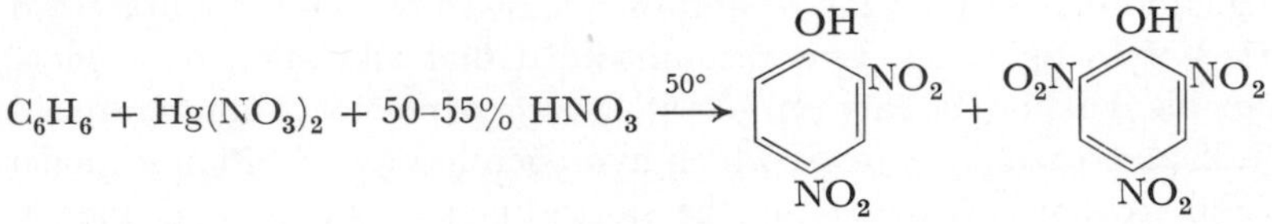

$$C_6H_6 + Hg(NO_3)_2 + 50\text{--}55\% \ HNO_3 \xrightarrow{50°}$$

63–73% total yield of nitrophenols

This reaction was of considerable interest during World War II as a source of picric acid [16].

(a) Preparation of 2,4,6-trinitro-3-hydroxybenzoic acid (11 % from 50 g. of benzoic acid, 5 g. of mercuric nitrate, and 300 g. of nitric acid, $d = 1.35$, heated 20 hr.) [17].

5. From Phenylhydroxylamines (Bamberger Rearrangement)

The general course of this rearrangement is:

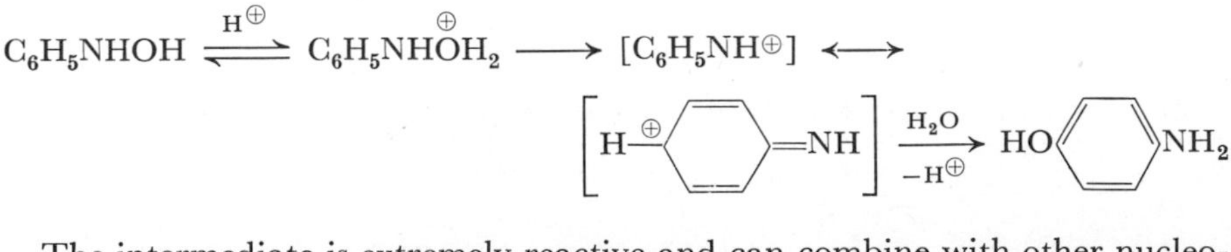

$$C_6H_5NHOH \overset{H^{\oplus}}{\rightleftharpoons} C_6H_5N\overset{\oplus}{H}OH_2 \longrightarrow [C_6H_5NH^{\oplus}] \longleftrightarrow$$

The intermediate is extremely reactive and can combine with other nucleophiles or intermediates. Indeed "brown oils" usually accompany the product. Nevertheless, *p*-aminophenols can be made in reasonable yields (see Ex. a).

Of greater interest is the fact that a phenylhydroxylamine, prepared *in situ* from sulfur sesquioxide (S_2O_6) and a nitro compound, rearranges and oxidizes to a hydroxy quinone [18]:

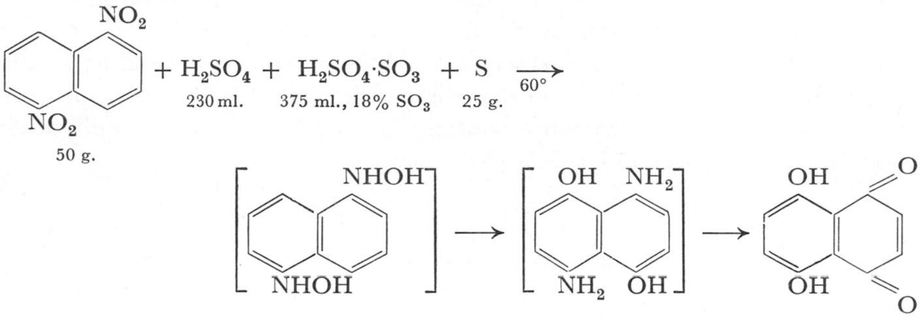

5,8-Dihydroxynaphthoquinone, 60%

The naphthazarin formed probably is quite stabilized; this allows for its survival under such powerful sulfonating conditions. For application to other nitro compounds, sulfonation may be a factor with which to contend.

If phenylhydroxylamines could be formed directly from the amine, the Bamberger rearrangement could be carried out *in situ*. Although yields are rather low because of competing reactions, potassium persulfate seems to be the reagent of choice for this *in situ* preparation, as shown in the following formulation [19]:

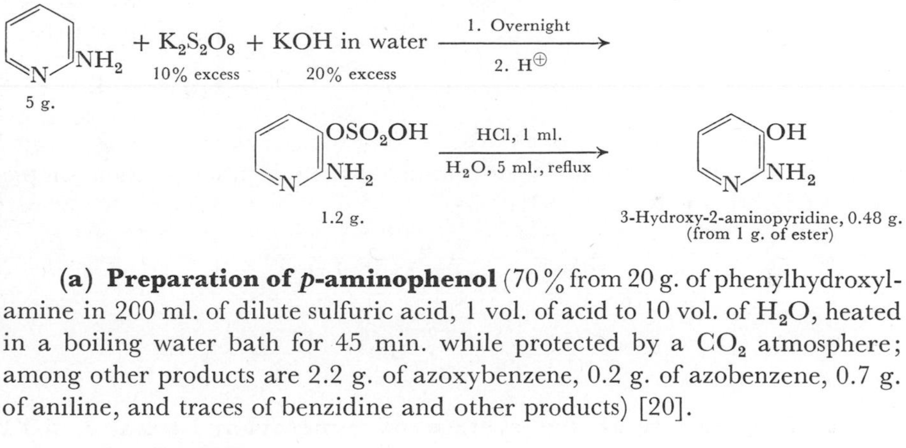

3-Hydroxy-2-aminopyridine, 0.48 g.
(from 1 g. of ester)

(a) **Preparation of p-aminophenol** (70 % from 20 g. of phenylhydroxylamine in 200 ml. of dilute sulfuric acid, 1 vol. of acid to 10 vol. of H_2O, heated in a boiling water bath for 45 min. while protected by a CO_2 atmosphere; among other products are 2.2 g. of azoxybenzene, 0.2 g. of azobenzene, 0.7 g. of aniline, and traces of benzidine and other products) [20].

6. From Cyclodienones

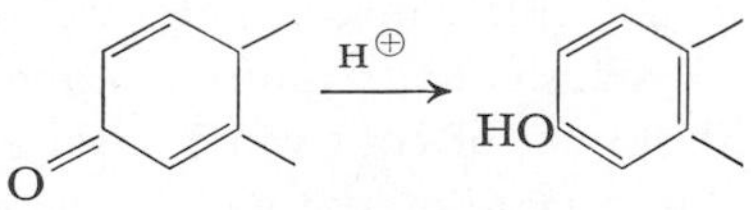

Cyclodienones rearrange to form phenols when treated with mineral acids such as sulfuric [21] and hydrochloric acids [22], with acid anhydrides alone or with an acid catalyst [23–25], with bases [26], with metals such as zinc [27], palladium alone [28], palladium in the presence of hydrogen [29], or with sulfur [30]. The preferred reagent is trifluoroacetic anhydride (see Ex. b.2). The most common of these rearrangements is that of the cyclic dienone into the phenol (dienone-phenol rearrangement) which has been employed with steroids and other natural products for the aromatization of unsaturated keto rings. Yields in these transformations are generally satisfactory.

The dienone-phenol rearrangement takes different courses depending on the reagent used and the structure of the original dienone [31]. For example, 10-methyl-2-keto-$\Delta^{1:9,3:4}$ hexahydronaphthalene (5-X) with acetic anhydride gives 4-methyl-ar-1-tetralol (5-XI) [32]:

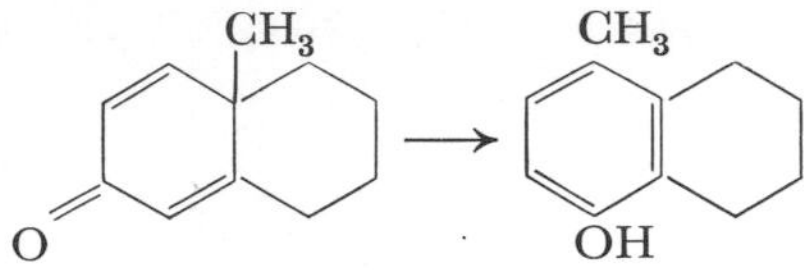

5-X   5-XI

The reaction proceeds via a spiro intermediate:

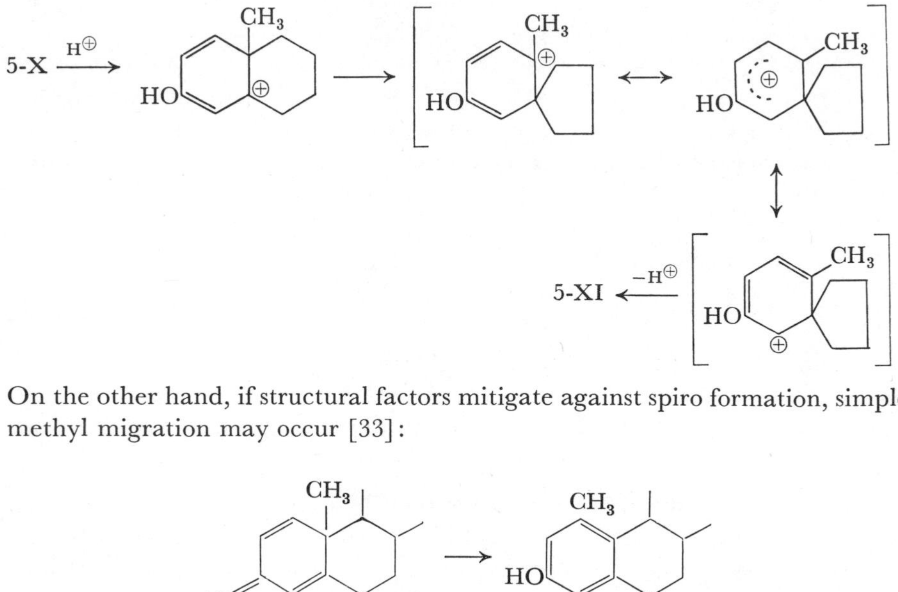

On the other hand, if structural factors mitigate against spiro formation, simple methyl migration may occur [33]:

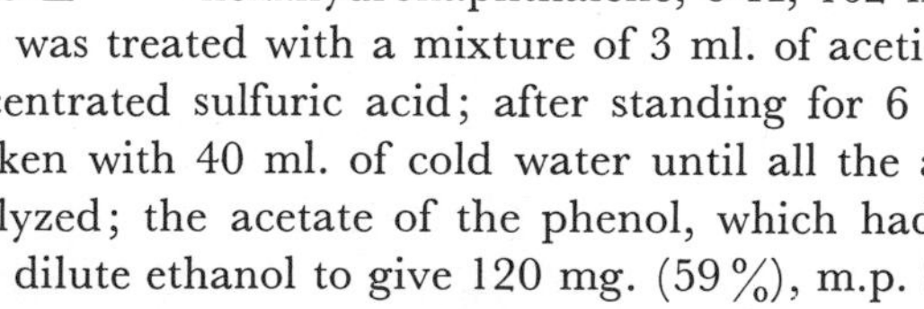

(a) **Preparation of the acetate of 4-methyl-ar-1-tetralol, 5-XI** (10-methyl-2-keto-$\Delta^{1:9,3:4}$-hexahydronaphthalene, 5-X, 162 mg., in 10 ml. of acetic anhydride was treated with a mixture of 3 ml. of acetic anhydride and 100 mg. of concentrated sulfuric acid; after standing for 6 hr. the reaction mixture was shaken with 40 ml. of cold water until all the acetic anhydride had been hydrolyzed; the acetate of the phenol, which had separated, was crystallized from dilute ethanol to give 120 mg. (59%), m.p. 82° [32].

(b) **Other examples**

(1) **3,4,5-Trimethylphenol** (54% from isophorone, oleum, and acetic anhydride) [24].

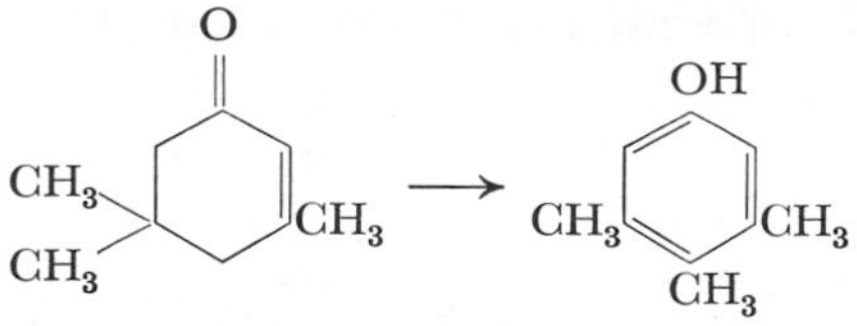

(2) 4-Methoxy-ar-1-tetralol (93% from 10-methoxy-2-keto-$\Delta^{1:9,3:4}$-hexahydronaphthalene and trifluoroacetic anhydride; under similar conditions the 10-benzoxy derivative gave 88% of the 4-benzoxy-ar-2-tetralol) [23].

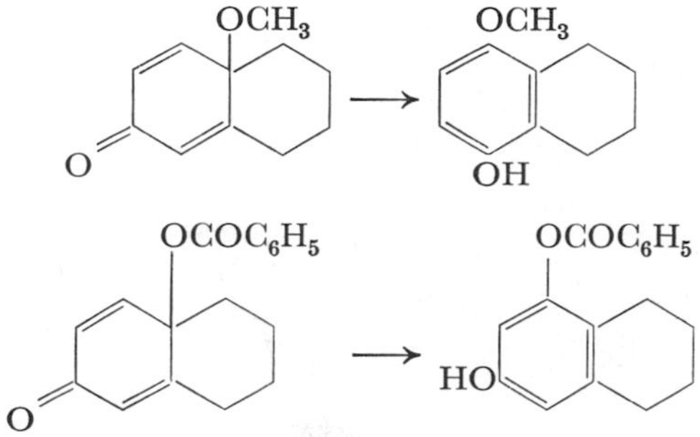

7. From Cyclic Glycols (Pinacol Rearrangement)

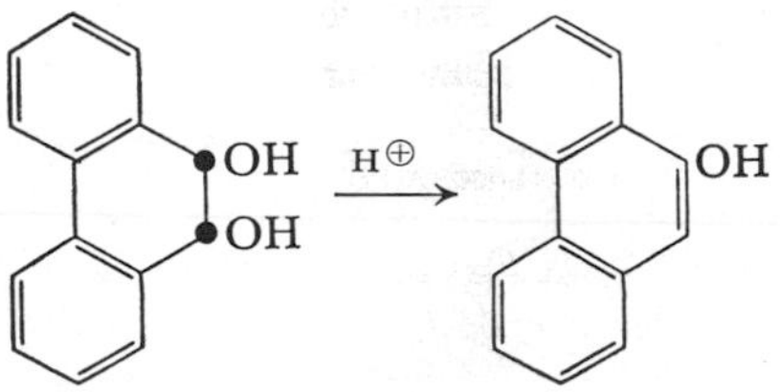

Polycyclic phenols such as 9-phenanthrol are best synthesized by the pinacol rearrangement of the *cis*- or *trans*-9,10-dihydro-9,10-phenanthrenediol [34]. The *cis*-diol is available by treating phenanthrene with osmium tetroxide; the *trans* form may be obtained by the lithium aluminum hydride reduction of 9,10-phenanthraquinone. Either geometrical isomer with a mixture of glacial acetic acid and a trace of sulfuric acid produces 9-phenanthrol in a yield over 90%. In this case the carbonium ion intermediate, 5-XII, achieves stability by the loss of a proton to give the phenol 5-XIII. The pinacol rearrangement leading to

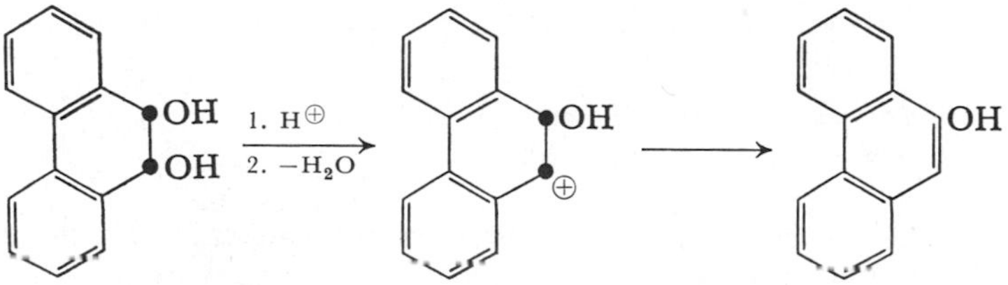

5-XII   5-XIII

carbonyl compounds has been discussed under Aldehydes, Chapter 10, E.1, and Ketones, Chapter 11, E.1.

(a) **Preparation of 9-phenanthrol.** To 0.500 g. of *cis*- or *trans*-9,10-dihydro-9,10-phenanthrenediol in 5 ml. of glacial acetic acid, 0.05 ml. of concentrated sulfuric acid was added. After about 10 min. heating, the solution was cooled to room temperature and the crude phenol separated. One recrystallization of the dry product from methanol gave 0.45 g. (94%) from the *cis*-diol and 0.460 g. (96%) from the *trans*-diol, m.p. 149–150°. A spectral sample prepared by vacuum sublimation melted at 156–157° [34].

8. From Arenes and Some Acylating Agents

A cyclic tetrahydrodiketone is theoretically in the same state of oxidation as a phenol, a fact which suggests that Friedel-Crafts acylation with succinic anhydride or its acid should give rise to phenols, as shown in the accompanying formulation. The reaction is not very successful, probably because the ring

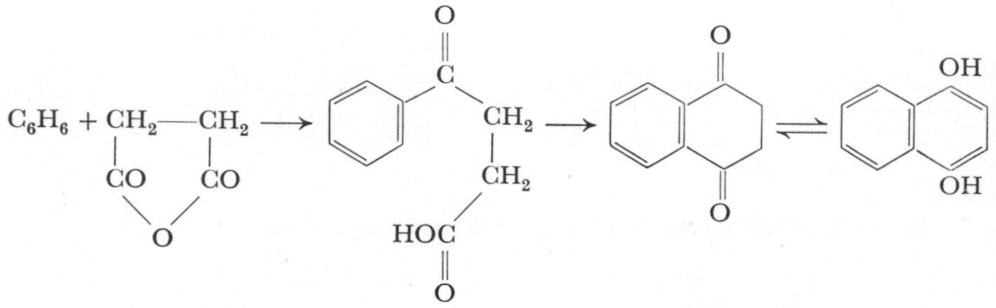

closure of a keto acid is difficult [35]. Such a reaction is reported, but the product is not well-characterized (see Ex. a).

Heterocyclic phenols can be prepared in a wide variety by electrophilic ring closures [36]; see also Amines, Chapter 8, F.3. A versatile, general ring-closing agent is a mixture of phosphorus oxychloride and fused zinc chloride which is claimed to be superior to the individual components [37]. In this particular

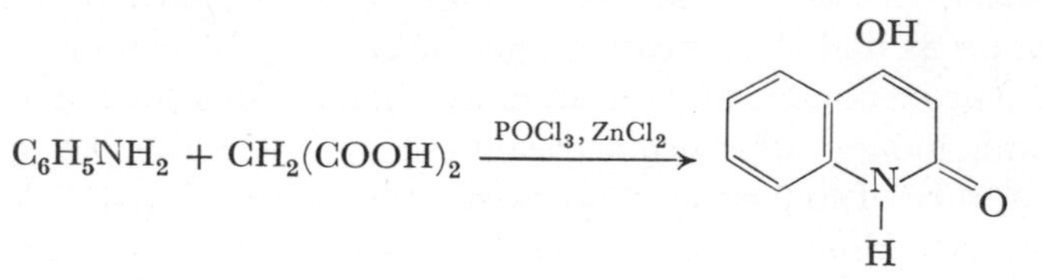

4-Hydroxycarbostyril, 56%

case, the yield was about the same using phosphorus oxychloride alone [38], but the product was more easily purified using the mixed reagent. The mixed reagent has been applied also to the synthesis of a wide variety of hydroxybenzophenones and 4-hydroxycoumarins.

(a) **Preparation of 1,4,5,8-tetrahydroxynaphthalene.** An intimate mixture of hydroquinone and succinic acid was added to a fused mixture of 10 g. of anhydrous aluminum chloride and 2 g. of sodium chloride and heated to 180–200° for 2 min. After cooling, decomposition with water, and extraction, 25% of the phenol was obtained. The phenol may exist as the diketone [39]

OH O

OH O

1. N. I. Shuikin and E. A. Viktorova, *Russian Chem. Rev.*, **29,** 560 (1960).
2. G. A. Olah *Friedel-Crafts and Related Reactions*, Interscience Publishers, New York, 1964, Vol. **2,** Pt. 1.
3. H. Hart *et al.*, *J. Am. Chem. Soc.*, **76,** 4547 (1954).
4. A. J. Kolka *et al.*, *J. Org. Chem.*, **22,** 642 (1957).
5. R. Stroh *et al.* in W. Foerst, *Newer Methods of Preparative Organic Chemistry, Engl. Transl.* Academic Press, New York, 1963, Vol. **2,** p. 337.
6. M. G. Burdon and J. G. Moffatt, *J. Am. Chem. Soc.*, **88,** 5855 (1966).
7. D. E. Pearson and R. D. Wysong, unpublished results.
8. L. A. Fury, Jr., and D. E. Pearson, *J. Org. Chem.*, **30,** 2301 (1965).
9. R. F. Curtis, *J. Chem. Soc.*, 415 (1962).
10. H.-D. Becker, *J. Org. Chem.*, **32,** 2115 (1967).
11. Ref. 2, Vol. 4, comprehensive index.
12. W. Wolf and N. Kharasch, *J. Org. Chem.*, **30,** 2493 (1965).
13. C. C. Price, *Org. Reactions*, **3,** 18 (1946).
14. A. S. Hay, *J. Org. Chem.*, **30,** 3577 (1965).
15. E. E. Gilbert, *Sulfonation and Related Reactions*, Interscience Publishers, New York, 1965, pp. 79, 80.
16. M. Carmack *et al.*, *J. Am. Chem. Soc.*, **69,** 785 (1947).
17. P. B. D. de la Mare and J. H. Ridd, *Aromatic Substitution*, Academic Press, New York, 1959, p. 55.
18. L. F. and M. Fieser, *Reagents for Organic Synthesis*, John Wiley and Sons, New York, 1967, p. 1123.
19. E. Boyland and P. Sims, *J. Chem. Soc.*, 4198 (1958); E. J. Behrman, *J. Am. Chem. Soc.*, **89,** 2424 (1967).
20. E. Bamberger, *Ann. Chem.*, **390,** 131 (1912).
21. Y. Abe *et al.*, *J. Am. Chem. Soc.*, **78,** 1422 (1956).
22. A. S. Dreiding *et al.*, *J. Am. Chem. Soc.*, **75,** 3159 (1953).
23. E. Hecker and E. Meyer, *Chem. Ber.*, **97,** 1926, 1940 (1964); *Angew Chem. Intern. Ed. Engl.*, **3,** 229 (1964).
24. W. von E. Doering and F. M. Beringer, *J. Am. Chem. Soc.*, **71,** 2221 (1949).
25. C. Djerassi and C. R. Scholz, *J. Org. Chem.*, **13,** 697 (1948).
26. E. Zbiral *et al.*, *Monatsh.*, **93,** 15 (1962).
27. K. Tsuda *et al.*, *J. Org. Chem.*, **26,** 2614 (1961).
28. J. W. Cook and R. Schoental, *J. Chem. Soc.*, 288 (1945).
29. F. v. Wessely and F. Sinwel, *Montash.*, **81,** 1055 (1950).
30. F. Bergmann and J. Szmuszkoviz, *J. Am. Chem. Soc.*, **68,** 1662 (1946).
31. N. L. Wendler in P. de Mayo *ed.*, *Molecular Rearrangements*, Interscience Publishers, New York, 1964, Pt. 2, p. 1028.
32. R. B. Woodward and T. Singh, *J. Am. Chem. Soc.*, **72,** 494 (1950).
33. C. Djerassi *et al.*, *J. Am. Chem. Soc.*, **72,** 4540 (1950); **73,** 990 (1951).
34. E. J. Moriconi *et al.*, *J. Org. Chem.*, **24,** 86 (1959).
35. S. M. Sethna, Ref. 2, 1964, Vol. 3, Pt. 2, p. 911.
36. R. M. Acheson, *An Introduction to the Chemistry of Heterocyclic Compounds*, Interscience Publishers, New York, 1966.
37. J. L. Bose and R. C. Shah, *J. Indian Chem. Soc.*, **38,** 701 (1961).
38. E. Ziegler and K. Gelfert, *Monatsh.*, **90,** 822 (1959).
39. D. B. Bruce *et al.*, *J. Chem. Soc.*, 2403 (1953).

E. Nucleophilic Reactions

1. From Carbonyl Compounds

These reactions, which are either acid- or base-catalyzed, consist of the formation of new carbon-carbon bonds leading to phenols. The best aliphatic source of phenols is a malonaldehyde which theoretically could condense with any ketone to form phenols:

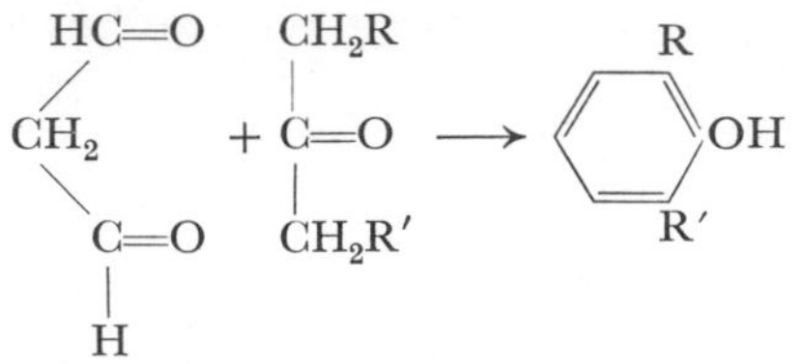

The limitation is the instability of the malonaldehyde. However, the sodium salt of nitromalonaldehyde (*Care!* Can be detonated) is available [1] and it is a potential source of some unusual *p*-nitrophenols (see Ex. a).

A second source of phenols is β,β-triketones of which the simplest is diacetyl-acetone. It cyclizes as shown [2]:

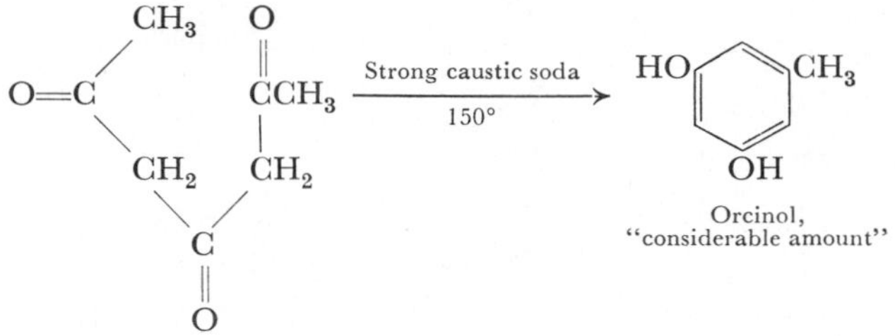

Orcinol,
"considerable amount"

It is quite likely that the biogenesis of many naturally occurring phenols begins with similar ketones [3]. Recently, triketo acids have become available through the elegant process of Harris and Hauser, and they cyclize with ease to resorcylic acid (see Ex. b.1).

Pyrylium salts can be converted into phenols:

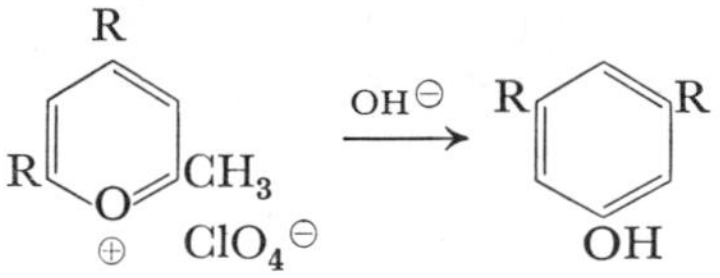

but, since they are prepared in relatively poor yield, this route would not be selected unless an inaccessible phenol could be prepared in this way (see Ex. b.2)

Malonaldehyde and β-triketones are not the only aliphatic sources of phenols. So many combinations and permutations suggested by the preceding synthesis can be visualized theoretically to yield phenols that they are not discussed here. Resonance stabilization of the phenolic anion is a driving force which will be helpful in the realization of significant yields of phenols, and the compendium

References for Section E are on p. 282.

by Dean may suggest other methods [4]. One classic example in the field of anthocyanins is shown [5].

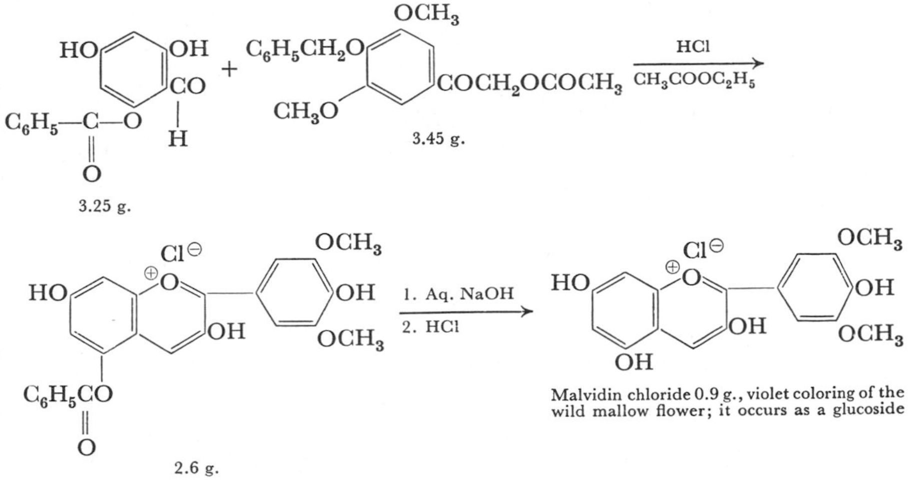

3.25 g.

3.45 g.

2.6 g.

Malvidin chloride 0.9 g., violet coloring of the wild mallow flower; it occurs as a glucoside

(a) Preparation of 2,6-octamethylene-4-nitrophenol (0.3 g. from 2 g.

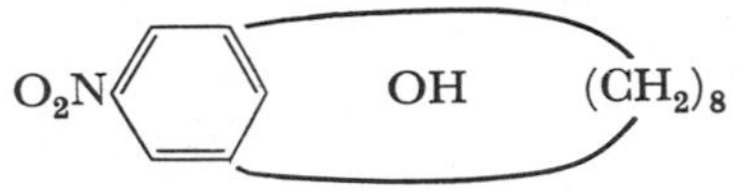

of cycloundecanone and sodium nitromalonaldehyde in 32 ml. of 75% aq. ethanol to which 1 equivalent of a concentrated solution of sodium hydroxide was added, and the mixture shaken from 2-7 days at room temperature; many other bridged nitrophenols were made similarly) [6].

(b) Other examples

(1) 6-Phenyl-β-resorcylic acid [7]

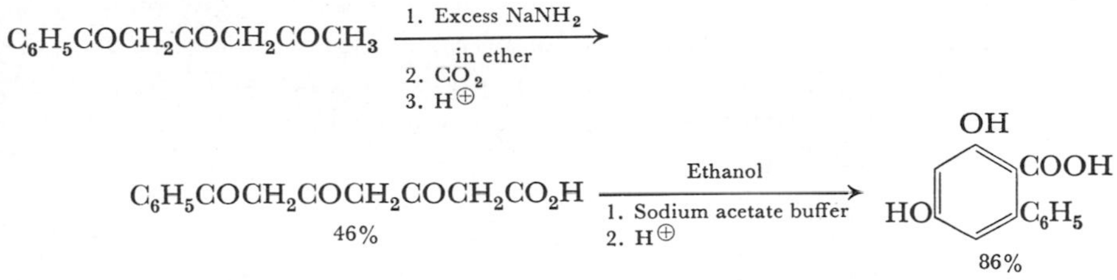

$$C_6H_5COCH_2COCH_2COCH_3 \xrightarrow[\substack{2.\ CO_2 \\ 3.\ H^{\oplus}}]{\substack{1.\ Excess\ NaNH_2 \\ in\ ether}}$$

$$C_6H_5COCH_2COCH_2COCH_2CO_2H \xrightarrow[\substack{1.\ Sodium\ acetate\ buffer \\ 2.\ H^{\oplus}}]{Ethanol}$$

46%

86%

(2) 3-Methyl-5-ethylphenol. The pyrylium salt was made from t-amyl chloride, acetyl chloride, and aluminum chloride, followed by water

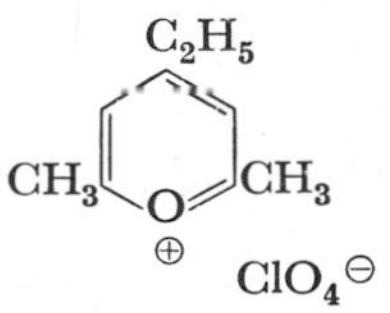

decomposition and treatment with perchloric acid in 27 % yield. The pyrylium salt was refluxed with 4 equiv. of 10 % aq. NaOH for 1 hr. to yield 35–45 % of the phenol on acidification [8].

1. P. E. Fanta, *Org. Syn.*, Coll. Vol. **4,** 844 (1963).
2. J. N. Collie, *J. Chem. Soc.*, **63,** 122 and 329 (1893).
3. A. J. Birch and F. W. Donovan, *Australian J. Chem.*, **6,** 360 (1953); J. F. Snell, *Biosynthesis of Antibiotics*, Academic Press, New York, 1966, Chap. 3.
4. F. M. Dean, *Naturally Occurring Oxygen Ring Compounds*, Butterworths, London, 1963.
5. W. Bradley and R. Robinson, *J. Chem. Soc.*, 1541 (1928).
6. V. Prelog *et al.*, *Helv. Chim. Acta*, **31,** 1325 (1948).
7. T. M. Harris and T. T. Howarth, *Chem. Commum.*, 1253 (1968).
8. A. T. Balaban and C. D. Nenitzescu, *Ann. Chem.*, **625,** 74 (1959).

F. Cyclo Reactions

1. From Allyl Aryl Ethers (Claisen Rearrangement)

The rearrangement of allyl aryl ethers to phenols has been reviewed [1, 2]. If the *o*-position of the ether is unoccupied, the allyl group ordinarily migrates to this position. A small amount of *p*-allylphenol, however, accompanies the *o*-allylphenol [3]. If the *o*-positions are occupied, the allyl group migrates to the *p*-position. It is interesting to note that in the migration to the *o*-position the allyl group is usually inverted, i.e., the original γ-carbon of the ether side chain becomes attached to the carbon of the ring, while in migration to the *p*-position a double inversion occurs, producing essentially no isomerization of the alkyl group. Ring substituents, with the exception of the aldehyde and carboxyl groups, offer little interference to the rearrangement, provided unsubstituted *o*- or *p*-positions are available.

Although much of the publication on the Claisen rearrangement has resulted because of the interest in the mechanism, the method does possess preparative value in the synthesis of allyl phenols. Such phenols, often available in high yield by the rearrangement, have been converted into propyl phenols, substituted phenylacetaldehydes, propenylphenols, etc. [1]. Propenylphenols are made in 80 % yield or better by fusion of the allyl ether with powdered potassium hydroxide at 200° [4]. The storage characteristics of most of these are poor, however. An interesting third type of Claisen rearrangement can be carried out with the allyl ethers of propenylphenols [5]:

References for Section F are on p. 284.

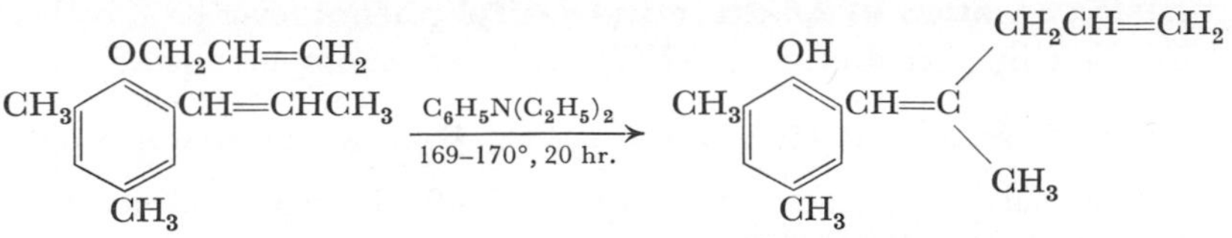

$$\xrightarrow[\text{169–170°, 20 hr.}]{C_6H_5N(C_2H_5)_2}$$

2,4-Dimethyl-6-(β-methylpenta-α-δ-dienyl)-
phenol, 42%

All o- and p-positions must be occupied for this rearrangement to occur.

Since the rearrangement of the o-position is intramolecular and follows first-order kinetics, with large negative entropies of activation, the following mechanism has been proposed:

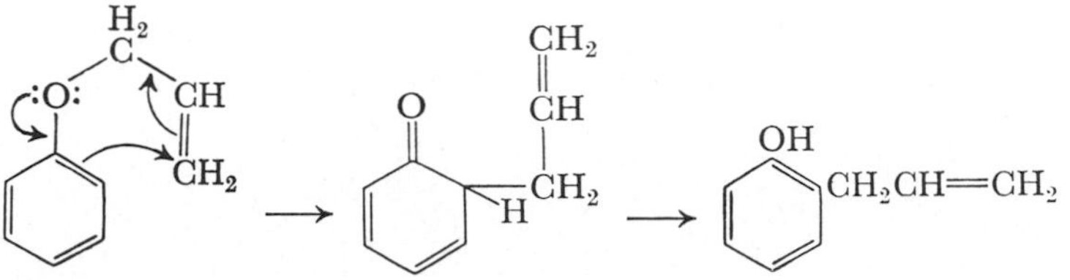

5-XIV

Dienones such as 5-XIV have not only been isolated, but also it has been shown that they exist in equilibrium with the phenol [6].

The rearrangement to the p-position is also intramolecular and first-order, but in this case a full somersault of the allyl group occurs:

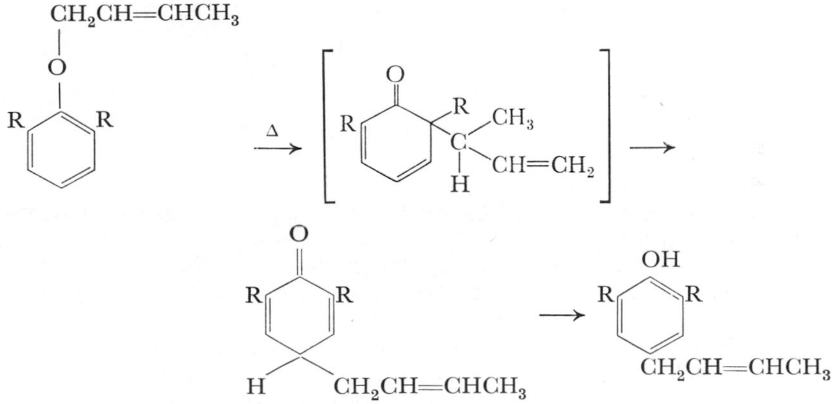

so that no isomerization is found in the crotyl grouping.

(a) **Preparation of o-eugenol** (80–90% by refluxing guaiacol allyl ether) [7].

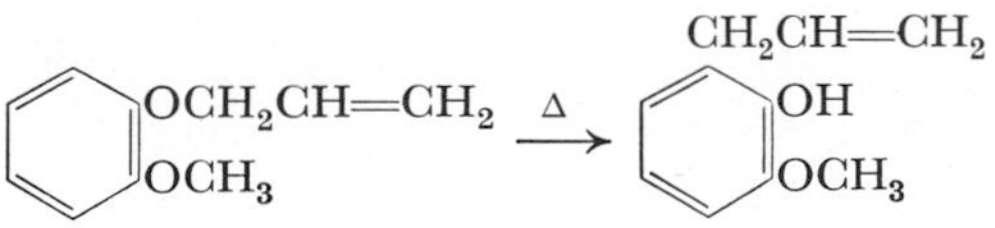

(b) Preparation of 2,6-dimethyl-4-allyl phenol (over 85% by heating allyl 2,6-dimethylphenyl ether at 172° in an inert atmosphere for 7 hr.) [8].

2. From Furans and Hydroxy-Bearing Dienes or Dienophiles

Very few Diels-Alder reactions yield phenols [9]. Benzyne addition to furan eventually leads to a phenol, but this reaction is used more as a means of detection of a benzyne than as a means of synthesis [10]:

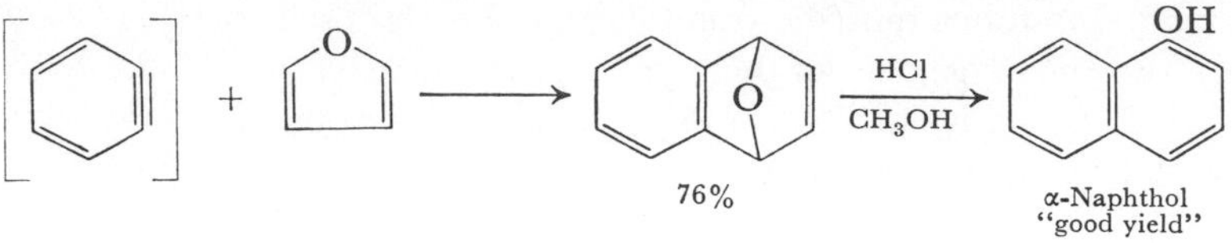

76%

α-Naphthol
"good yield"

Thorough documentation of phenol production by Diels-Alder reactions has not been attempted, but an example is given to illustrate a Diels-Alder synthesis of the type [11]:

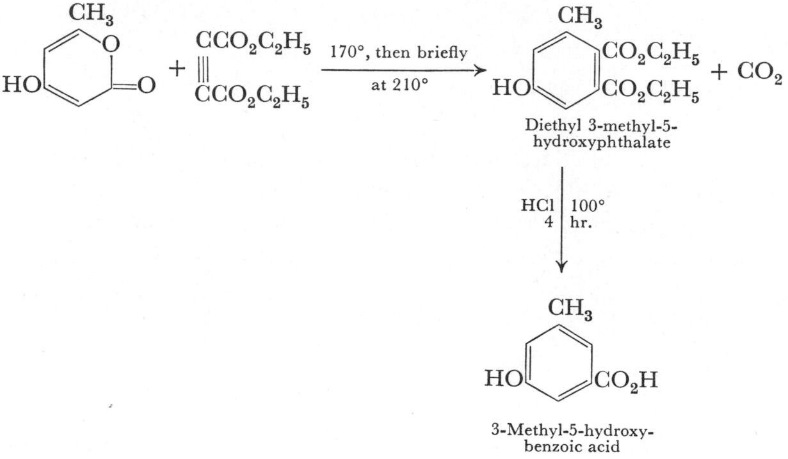

Diethyl 3-methyl-5-
hydroxyphthalate

HCl | 100°
4 | hr.

3-Methyl-5-hydroxy-
benzoic acid

1. D. S. Tarbell, *Org. Reactions*, **2**, 1 (1944).
2. S. J. Rhoads in P. de Mayo's *Molecular Rearrangements*, Interscience Publishers, New York, 1963, Pt. 1, p. 660.
3. E. N. Marvell *et al.*, *J. Org. Chem.*, **30**, 1032 (1965).
4. A. R. Bader, *J. Am. Chem. Soc.*, **78**, 1709 (1956).
5. K. Schmid *et al.*, *Helv. Chim. Acta*, **39**, 708 (1956).
6. D. Y. Curtin and H. W. Johnson, Jr., *J. Am. Chem. Soc.*, **76**, 2276 (1954); **78**, 2611 (1956)
7. C. F. H. Allen and J. W. Gates, Jr., *Org. Syn.*, Coll. Vol. **3**, 418 (1955).
8. D. S. Tarbell and J. F. Kincaid, *J. Am. Chem. Soc.*, **62**, 728 (1940).
9. A. S. Onishchenko, *Diene Synthesis*, D. Davey and Co., New York, 1964.
10. G. Wittig, *Angew. Chem.*, **69**, 245 (1957).
11. K. Alder and H. F. Rickert, *Chem. Ber.*, **70**, 1354 (1937).

Chapter 6

ETHERS

The most comprehensive treatise on the synthesis of ethers is the Houben-Weyl reference [1]. In addition to syntheses applied to ethers in general, it contains sections devoted to specific ether preparations: epoxides, p. 367; oxetanes, p. 489; and tetrahydrofurans, p. 517.

Compared to the syntheses of alcohols or carbonyl compounds, the number of syntheses of ethers is quite limited, and rightfully so. Containing neither active hydrogen nor double bonds, the ether structure is the least reactive of the three types. In the majority of cases it survives acidic or basic hydrolysis and oxidative or reductive attacks. Aliphatic ethers do have the undesirable property of forming peroxides on standing in contact with air. The most flagrant of these are dioxane, tetrahydrofuran, and diisopropyl ether. Many procedures have been recommended to remove peroxides from ether, the latest of which is passage of the ether through a column of Dowex 1, a strongly basic ion exchange resin [2]. But the most efficient procedure used by one of us (D.E.P.) is passage of the ether through an alumina column. The alumina is effective until aliquots of the eluate liberate iodine from equal volumes of glacial acetic acid and concentrated aqueous hydriodic acid, at which point it is replaced.

The synthesis of ethers has been divided into six different categories based on mechanism. Some mechanisms in the categories overlap, and others may seem classified rather superficially or arbitrarily as to the true mechanism. But there is sufficient relationship to bring coherency to the discussion accompanying each section. A most useful development in the synthesis of this type is the conversion of carbonyl compounds to epoxides by dimethylsulfonium methylide (C.4).

1. Houben-Weyl, *Methoden der Organischen Chemie*, Vol. 6, Georg Thieme Verlag, Stuttgart, 1965, Pt. 3, p. 1.
2. R. N. Feinstein, *J. Org. Chem.*, **24**, 1172 (1959).

A. Metathesis

The preparation of ethers by metathesis is a classical one involving the displacement of some leaving group, such as halogen from an alkyl halide, in an SN_2 type of process. The choice of alkoxide and alkyl halide is usually quite obvious,

$$RO^\ominus + R'X \rightarrow ROR' + X^\ominus$$

based as it is on the ease or normalcy of the alkyl halide displacement. For example, inert halides such as bromobenzene, *t*-halides which are prone to give olefins, and some allylic halides which may rearrange are to be avoided, if possible. Ethers containing such groups are preferably prepared by introducing the group through the alkoxide, as

$$ArO^\ominus + RCl \rightarrow ArOR$$

References for Section A are on pp. 299–300.

Relegated to B on electrophilic-type preparations is the group of halides which produce ethers by the SN_1 process. Since these halides, which include such compounds as triphenylmethyl chloride, diphenylmethyl chloride, and allylic types, ionize in the alcohol, no alkoxide anion need be present,

$$Ar_3CCl + ROH \longrightarrow Ar_3C^\oplus + Cl^\ominus \longrightarrow Ar_3\overset{\overset{\displaystyle H}{\vert}}{C}OR^\oplus \longrightarrow Ar_3COR + H^\oplus$$

but a weak base such as collidine may be helpful to prevent the splitting of the ether by acid.

The displacement of halides by alkoxides takes many forms including the typical Williamson synthesis, displacement of sulfate groups, the Ullmann procedure, intramolecular displacement yielding cyclic ethers, including the Darzens synthesis, and displacement of tertiary amines from quaternary bases.

1. From Halides (Williamson)

$$RONa + R'X \rightarrow ROR' + NaX$$

This synthesis, a desirable method particularly for preparing mixed ethers, has been used most widely with the halide. To render the alkoxide soluble an excess of alcohol is used, and thus the procedure requires the separation of the ether formed from a large amount of alcohol. Reasonable yields may be obtained by employing 1 mole of sodium to 8–10 moles of alcohol and 1 mole of alkyl bromide, and then separating the ether by fractional distillation [1]. Traces of alcohol can be removed from the ether by refluxing with sodium metal. Alkenes are formed as a by-product, the amounts for the halide increasing in the order:

$$Primary < secondary < tertiary$$

Of the various alkylating agents, trimethyloxonium 2,4,6-trinitrobenzene sulfonate, appears to be the most reactive [2].

In this displacement reaction various mechanisms compete. For the primary halide the SN_2 mechanism predominates, but for the tertiary halide the principal mechanisms may be SN_1 or E_1 (see discussion in B.2). To complicate the situation further an aryne intermediate appears to be formed in the conversion of bromobenzene to phenyl t-butyl ether on treatment with potassium t-butoxide in dimethyl sulfoxide [3].

Alkyl phenyl ethers may be prepared by the Williamson synthesis in yields varying from 40 to 80% [4]. Acetone has been found to be more satisfactory than alcohol in the synthesis of certain of these ethers [5]. The addition of powdered potassium iodide is effective in increasing the yields probably by an exchange with the alkyl bromide to give the more reactive alkyl iodide.

The potassium iodide accelerator has been used in the preparation of allyl phenyl ethers, 2-substituted 2,2-dinitro-1-alkyl ethers, and aryloxyacetones, the addition in the latter case increasing yields from 20 to more than 90% [6]. The aprotic solvent, dimethylsulfoxide, has not been compared with other solvents in the Williamson synthesis, but it can be anticipated that the nucleophilicity of the alkoxide anion in this solvent will be enhanced and rates of metathesis increased, particularly for hindered phenols [7]:

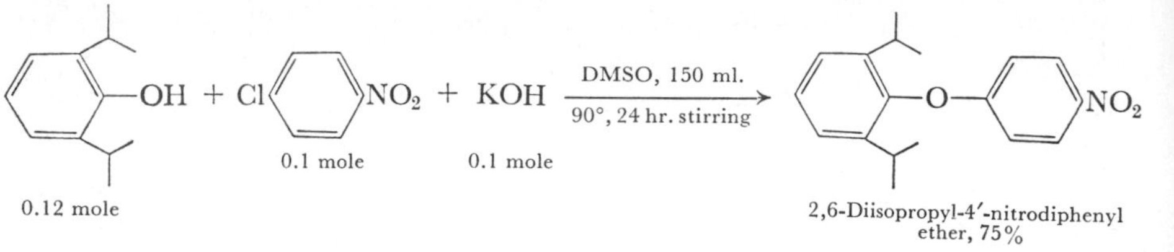
OH + Cl
NO2 + KOH
DMSO, 150 ml.
90°, 24 hr. stirring
O
NO2
0.1 mole
0.1 mole
0.12 mole
2,6-Diisopropyl-4′-nitrodiphenyl ether, 75%

Displacement of halogen from activated aryl halides, proceeding by a different mechanism, is discussed in C.2.

The solvent has a profound effect not only on the rate of metathesis, but also on the nature of the products. In ordinary solvents such as the alcohol of the alkoxide, acetone, or dimethyl sulfoxide, oxygen alkylation is the predominant course of the reaction, but in water, trifluoroethanol, or phenolic solutions, carbon alkylation encroaches and even predominates with such phenols as β-naphthol [8]:

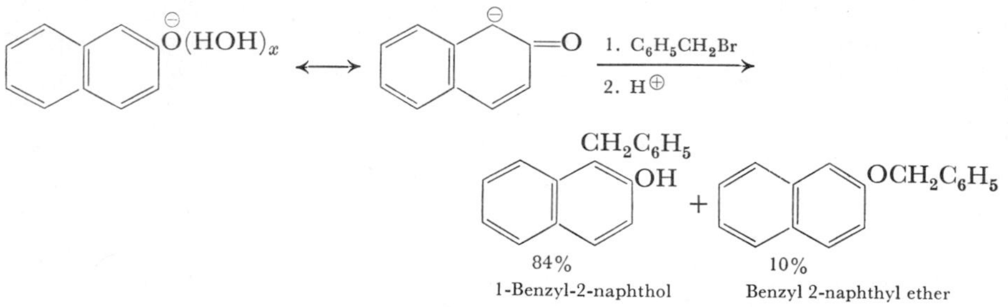
O(HOH)x
=O
1. C6H5CH2Br
2. H⊕
CH2C6H5
OH
OCH2C6H5
84%
1-Benzyl-2-naphthol
10%
Benzyl 2-naphthyl ether

In dimethyl sulfoxide almost 100% oxygen alkylation occurs. The theory is that strong proton donors such as water and trifluoroethanol solvate the anionic oxygen, decreasing its nucleophilic powers to the extent that carbanion alkylation can compete favorably. Another effect of interest is that some carbon alkylation of β-naphthol (but not of benzenoid phenols) occurs in the solvent, dimethyl ether of ethylene glycol. Probably ion pairs exist in this solvent in which the sodium cation of the naphthol shows some electrostatic attraction for the bromide of an alkyl bromide such as methyl bromide:

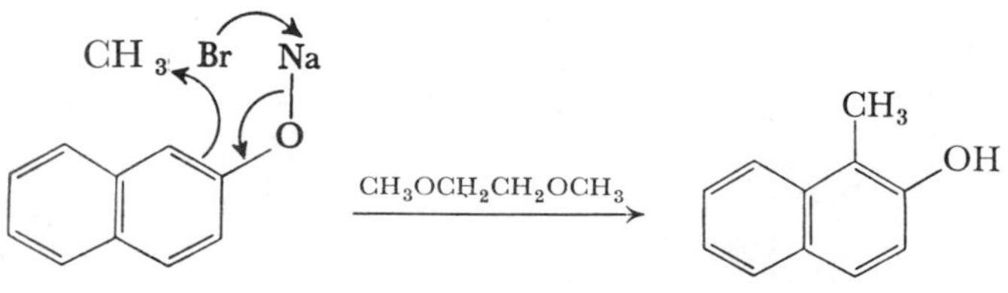
CH3 Br Na
O
CH3OCH2CH2OCH3
CH3
OH

In dimethylsulfoxide the anion is either only loosely associated with the sodium cation or so situated as not to attract the bromide of methyl bromide.

A series of simple and acetylenic ethers has been prepared recently by reaction of the corresponding alcohol with sodamide in liquid ammonia, followed by treatment with an alkyl halide. The yields ranged from 14 to 60%, some dehydrohalogenation accompanying etherification. Lithium amide used in place of sodamide gave no ethers at all [9].

The ordinary preparation of phenoxyacetic acids from sodium salts of phenols in aqueous solution suffers from the competitive hydrolysis of chloroacetic acid to glycolic acid. Two procedures (in Ex. b and c.1) to circumvent the competitive hydrolysis are given.

(a) Preparation of sec-butyl ethyl ether. To 1 l. of anhydrous sec-butyl alcohol was added 10 g. of sodium; after solution 25 ml. of ethyl bromide was added and the mixture was allowed to react for 2 days. This process was repeated until 175 ml. of the bromide and 70 g. of sodium metal had been consumed. The fraction boiling at 80–85° was mixed with 1 l. of water and distilled to give a constant-boiling mixture of ether and water at 67–72°, which again was distilled with water, the final constant-boiling mixture having a b.p. of 71–72°. The ether in the distillate was separated, dried with potassium carbonate, then refluxed with sodium metal, and distilled, b.p. 81° [10].

(b) Preparation of 9-anthroxyacetic acid. Sodium metal (0.25 g. atom) was dissolved in 300 ml. of isopropyl alcohol and anthrone (0.18 mole)

added to the solution, which was warmed to dissolve the ketone. Ethyl chloroacetate (40 ml., 0.38 mole) was added; the mixture allowed to stand for 7 days; and the solution concentrated by distillation until salt appeared. Approximately 100 ml. of 15% sodium hydroxide was added and the rest of the alcohol removed by distillation. Glistening plates of the sodium salt began to deposit from the residue, but they were redissolved by warming. Concentrated hydrochloric acid was added until the solution was acidic, whereupon an oil which solidified separated. The solid was dissolved in 10% aqueous ammonium hydroxide, treated with Norite, and filtered (unreacted anthrone is removed at this stage), and the filtrate reacidified. The yellow crystals, 16.7 g., were recrystallized from a methylcyclohexane-methyl ethyl ketone mixture to give yellow needles, m.p. 187–188°, 11.9 g. (27%), neutralization equivalent 245, calculated 252 [11].

(c) Other examples

(1) 1-Menthoxyacetic acid (78–84% from 1-menthol, sodium, toluene, and monochloroacetic acid) [12].

(2) β-Bromoethyl phenyl ether (56% from sodium phenoxide in water with ethylene bromide) [13].

(3) o-n-Butoxynitrobenzene (75–80% from o-nitrophenol, n-butyl bromide, and potassium carbonate in acetone [14]; dimethylformamide may be used in place of acetone) [15].

(4) α-Methoxydibenzyl ketone.   These conditions avoid the Favorskiĭ

$$C_6H_5CHClCOCH_2C_6H_5 \xrightarrow[CH_3OH]{\text{Lutidine}} C_6H_5CHOCH_3COCH_2C_6H_5$$

90% crude

rearrangement (Esters, Chapter 14, C.8) [16].

(5) β-N-Tosylaminoethyl methyl ether (57% from β-N-tosylaminoethyl chloride and sodium methoxide) [17].

2. From Esters and Related Types

Alkyl sulfates are excellent alkylating agents because of the ease of displacement of the alkyl sulfate anion. The ease of displacement is related to the

$$R \text{ (or Ar)}O^{\ominus} + CH_3OSO_2OCH_3 \rightarrow R(Ar)OCH_3 + \overset{\ominus}{O}SO_2OCH_3$$

6-I

stability of the anion displaced, which in the case of 6-I is resonance-stabilized. The stability of the anion is roughly proportional to the strength of the acid derived from its anion. An approximate order estimated from rate data in hydroxylic solvents is [18]:

$$OSO_2R > I \sim Br > NO_3 \sim Cl > \overset{\oplus}{S}(CH_3)_2 > F > OSO_3^{\ominus} > \overset{\oplus}{N}R_3 > OR > NR_2$$

An approximate order among sulfonates is alkyl sulfate > p-bromobenzenesulfonate > benzenesulfonate > p-toluenesulfonate. Dialkyl sulfates are sufficiently reactive to serve as alkylating agents at room temperature in the presence of phenols or alcohols in alkaline solution. Ethylation has been accomplished also in the absence of alkali [19]. Many functional groups such as the aldehyde [20], the ketone [21], the cyanide [22], the carboxyl [23], the nitro [24], and the ester [25] survive the alkylation. Alkylating agents such as ethylene sulfate and ethylene carbonate [26], methyl-β-naphthalene sulfonate [27], alkyl p-toluenesulfonates [28], dimethylformamide acetal [29], and trimethyl orthoformate [30] have also been employed. For the methyl ethers of allylic [31] and acetylenic [32] alcohols, dimethyl sulfate with sodamide has been used. The yields in these alkylations are generally satisfactory, but the encroachment of the olefin in the treatment of a series of benzenesulfonate esters with alkoxides in dimethylsulfoxide solvent can be seen in Table 3 [33].

The sulfates and tosylates (or their acid chlorides) have been particularly useful in the synthesis of ethers of sugars [34, 35] and steroids [36]. For the former, dimethyl sulfate with barium oxide or barium hydroxide in dimethylformamide or dimethyl sulfoxide or a mixture of these solvents is the preferred reagent [37], while for the latter the free alcohol has proved to be satisfactory.

Table 3

Benzenesulfonate	CH_3ONa		$(CH_3)_3COK$	
	% ether	% olefin	% ether	% olefin
n-Hexyl	90	Trace	69	20
2-Methyl-1-butyl	65	2.3	39	22
2-Octyl	28	47	Trace	79
Cyclopentyl	44	30	None	76
Cyclohexyl	5	85	Trace	83

The interesting example of ether formation from benzoate esters is deserving of comment. Although acyl addition of the anion is by far the predominant reaction, no expected product is isolable because the reaction is reversible. The slower alkyl fission reaction therefore becomes conspicuous [38]:

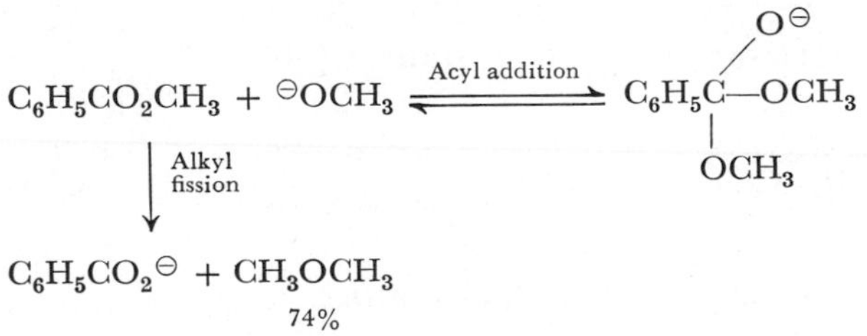

$$C_6H_5CO_2CH_3 + {}^{\ominus}OCH_3 \underset{\text{Acyl addition}}{\rightleftharpoons} C_6H_5\overset{\displaystyle O^{\ominus}}{\underset{\displaystyle OCH_3}{C}}{-}OCH_3$$

$$\Big\downarrow \text{Alkyl fission}$$

$$C_6H_5CO_2{}^{\ominus} + CH_3OCH_3$$
$$74\%$$

Other examples are known (see Ex. c.7, c.8).

Another ester-type reagent which yields ethers with phenols is an O-alkyl-isourea, which can be generated *in situ* from dicyclohexylcarbodiimide and an alcohol [39]:

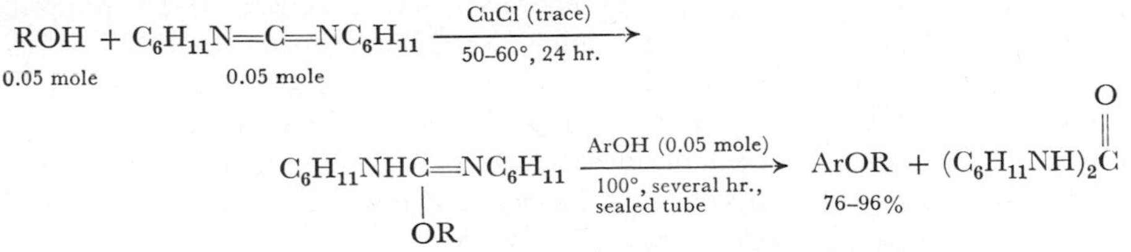

$$ROH + C_6H_{11}N{=}C{=}NC_6H_{11} \xrightarrow[50-60°, 24 \text{ hr.}]{CuCl \text{ (trace)}}$$
0.05 mole   0.05 mole

$$C_6H_{11}NH\underset{\displaystyle OR}{C}{=}NC_6H_{11} \xrightarrow[\substack{100°, \text{ several hr.,} \\ \text{sealed tube}}]{ArOH \text{ (0.05 mole)}} ArOR + (C_6H_{11}NH)_2\overset{\displaystyle O}{\overset{\displaystyle \|}{C}}$$
                                         76-96%

(a) **Preparation of anisole** (72–75% from phenol, sodium hydroxide, and dimethyl sulfate) [40]; this reaction is quantitative on a small scale, a fact which suggests that losses on a large scale are mechanical ones.

(b) **Preparation of n-butyl ethyl ether.** Sodium, 10.8 g., in ether was treated with butyl alcohol, 38 g., and ethyl formate, 45 g., was added dropwise at 0° over a period of 1 hr. The next day the mixture was washed with water, dried, and then refluxed with sodium metal. The distillate of b.p. 91–93° was collected as the product (80 %). This preparation is unusual in that no contamination with dibutyl ether arising from alkoxide exchange with ethyl formate seems to occur [41].

(c) Other examples

(1) **Ethyl propyl ether** (85 % from 1-propanol and ethyl sulfate alone) [42].

(2) *m*-**Methoxybenzaldehyde** (63–72 % from *m*-hydroxybenzaldehyde) [43].

(3) **2,3,4,6-Tetramethyl-*d*-glucose** (46–55 % from glucose) [44].

(4) **1,3-Benzylidene-2-stearyl glyceryl ether** (98.7 % from 1,3-benzylideneglycerol and stearyl *p*-toluenesulfonate in a suspension of potassium in benzene) [28].

(5) **5-Methyl-1,3-benzylidene-L-arabitol** (98 % from 5-tosyl-1,3-benzylidene-L-arabitol, sodium, and methanol) [34].

(6) **22-Phenyl-3-methoxy-22-hydroxy-bisnor-5-cholene** (about 100 % from 22-phenyl-3,22-dihydroxy-bisnor-5-cholene-3-tosylate and methanol) [36].

(7) **Veratrole** (1,2-dimethoxybenzene, 78 % from guaiacol potassium salt and dimethyl phthalate at 200°) [45].

(8) **Ethyl β-methoxy-*cis*-crotonate** [30].

$$CH_3COCH_2COOC_2H_5 + CH(OCH_3)_3 \xrightarrow[\text{Quinoline}]{H_2SO_4} \underset{95\%}{CH_3\overset{OCH_3}{\underset{|}{C}}=CHCOOC_2H_5}$$

3. From Aromatic Halides (Ullmann)

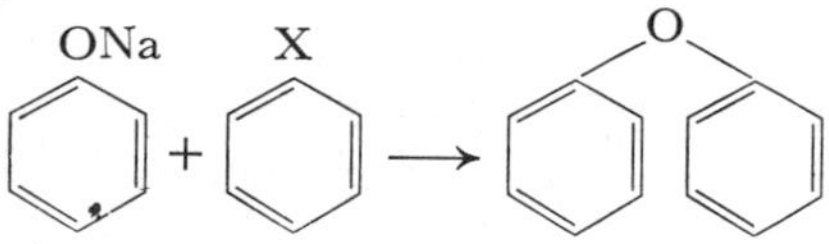

The Ullman reaction is used in the preparation of diaryl ethers. In comparison with aliphatic ether preparations it is much more difficult to accomplish because of the inertness of the aromatic halide. For that reason higher temperatures and the use of powdered copper, cupric or cuprous salts (it does not matter which) are often employed. The mechanism is shown in the accompanying formulation [46]. Ethylene diacetate accelerates the reaction by making the copper salts

$$CuBr + KOAr \longrightarrow CuOAr \xrightarrow{KOAr} \overset{\oplus}{K}\overset{\ominus}{Cu}(OAr)_2 \xrightarrow{Ar'Br}$$
$$ArOAr' + CuOAr + KBr$$

more soluble. By using pyridine and soluble copper salts in the absence of oxygen and water, success has been achieved in preparing diphenoxybenzenes (see Ex. b.1) from dihydric phenols [47].

Special circumstances such as exist in the Smiles rearrangement may bring about ease in displacement of groups other than halogens (see Ex. b.2, b.3). The nitro group facilitates the attack of the anion, and an added driving force is derived from the formation of a carboxyl or sulfinate anion in the products. Also, halogen atoms may be activated as in Ex. b.4 and b.5. In these cases no copper or copper salts are necessary.

(a) **Preparation of 2-methoxydiphenyl ether** (62–67 % from guaiacol, potassium hydroxide, and bromobenzene in the presence of copper) [48].

(b) **Other examples**

(1) *m*-**Diphenoxybenzene** [47].

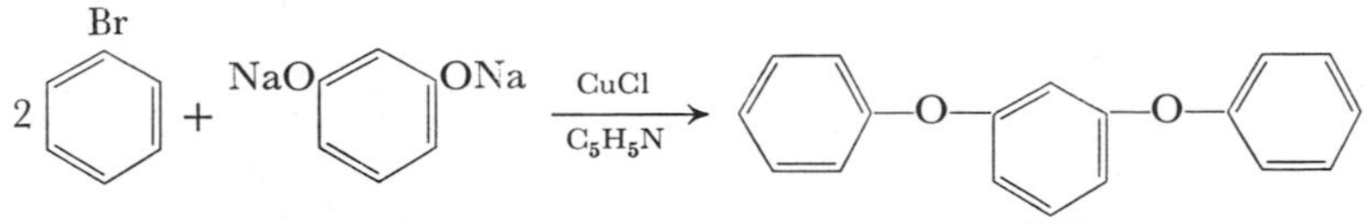

70%

(2) **2-Carboxy-4′-nitrodiphenyl ether** [49].

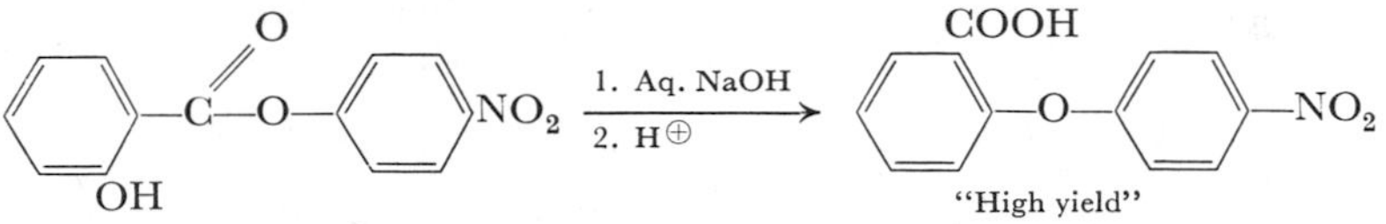

"High yield"

(3) **4-Methyl-2′-nitro-2-sulfinodiphenyl ether** [50].

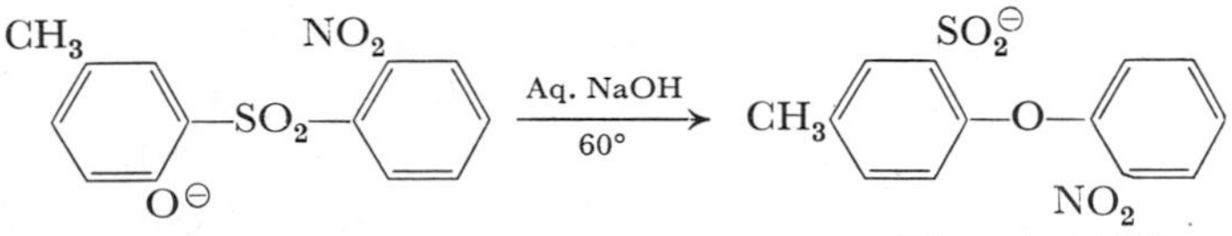

"Theoretical yield"

(4) α-**Naphthoquinonebenzodioxane** [51].

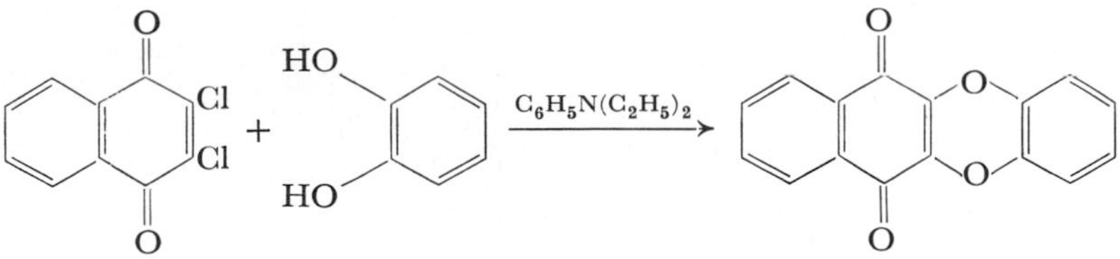

6-II   54%

(5) **2,3-Benzo-1,4-diketophenoxazine-N-tosyl derivative** [51].

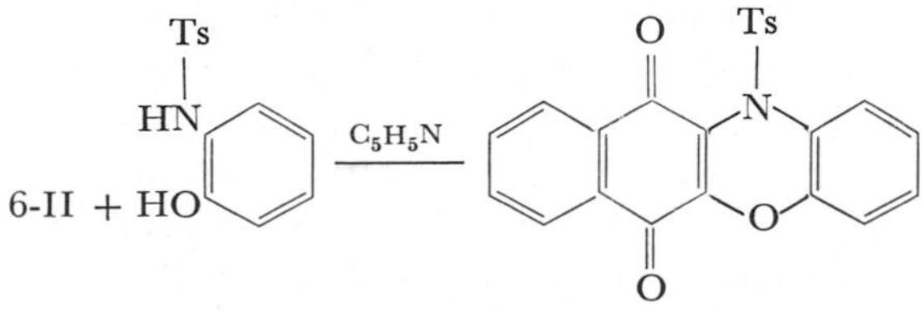

76%

4. From Halohydrins and Related Types (Intramolecular Displacement)

$$RCHXCH_2OH \xrightarrow{OH^\ominus} RCH\underset{\displaystyle O}{\diagdown \diagup} CH_2$$

X = halogen, —OSO$_2$OH, —SO$_2$OH

This synthesis ordinarily starts with halohydrins. In fact one of the earlier industrial methods for preparing ethylene oxide, the most important member, is by treating ethylene chlorohydrin with a base. With some polyunsaturated terpenes, such as squalene, halohydrin formation with N-bromosuccinimide is selective and takes place at the double bond nearest the terminal position. Thus, in this case, squalene-2,3-epoxide becomes available [52]. The oxygen-containing ring formed is called a cyclic ether, or cyclic oxide, an epoxide (if three-membered), or an oxirane. A discussion of the reaction is available [53]. The reaction is a modified Williamson, the ring resulting under alkaline conditions because the halide and hydroxyl groups happen to be part of the same molecule. The mechanism, that of bimolecular nucleophilic displacement, may be represented as in the accompanying formulation. Thus the closure is *trans*.

$$\underset{\displaystyle OH}{\overset{\displaystyle X}{\underset{|}{\overset{|}{CH_2}}}-CH_2} + \overset{\ominus}{OH} \rightleftharpoons H_2O + \underset{\displaystyle O^\ominus}{\overset{\displaystyle \overset{..}{X}}{\underset{|}{CH_2-CH_2}}} \longrightarrow \underset{\displaystyle O}{CH_2CH_2} + \overset{\ominus}{X}$$

A sodium alkoxide, sodium hydroxide, or potassium fluoride has been used to effect the dehydrohalogenation. As a rule the yields are good.

Methyl groups on either carbon atom of ethylene chlorohydrin increase the rate of ring closure [54] and, the ethylene oxide ring is preferred over some larger rings [55]. The ease of displacement to form cyclic ethers is in the order: 5-membered $\cong 6 > 3 > 4$. Specifically in the hydrolysis of chlorohydrins, $Cl(CH_2)_nOH$, the relative rates are as follows:

n	Relative Rate
2	1
3	4
4	1000

Tetrahydrofuran was isolated in 74% yield from the chlorohydrin, $n = 4$, but no cyclic ethers were obtained from the other chlorohydrins [56]. Tetrahydrofuran and tetrahydropyran can be made in 58 and 50% yields, respectively, from tetra- and pentamethylene glycols via the sulfonate ester by adding benzenesulfonyl chloride to a lutidine solution of the glycols [57]. On the other hand, trimethylene oxide is made with difficulty and in poor yield (22.5%) from γ-chloropropyl acetate with strong alkali at about 130° [58]. It has been claimed, however, that trimethylene oxides of all types can be made in yields up to 95%, (30% conversion) by first dissolving the trimethylene glycol in concentrated sulfuric acid and pouring this solution slowly into aqueous sodium hydroxide solution [59].

For ethylene halohydrin ring closure to the epoxide, an inversion of configuration takes place on the carbon atom to which the halogen is attached. The fact is shown in the conversion of *erythro* 3-bromo-2-butanol (6-III) into the *trans* oxide (6-IV) [60]:

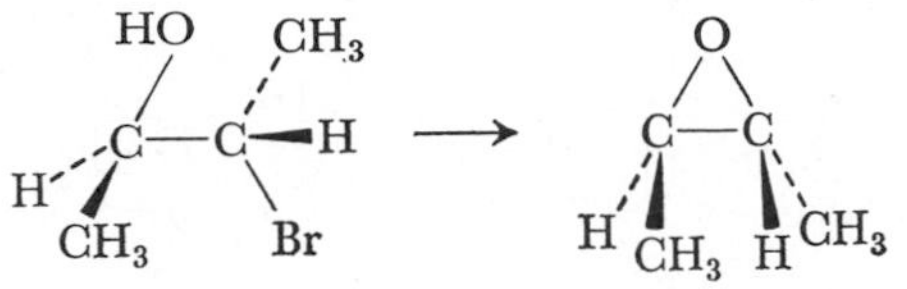

6-III   6-IV

Further support for the rearward nucleophilic attack on the carbon atom holding the halogen is shown by the difference in behavior of the *cis*- and *trans*-halohydrins of cycloalkenes. *trans*-1-Methyl-2-chlorocyclohexanol gives the

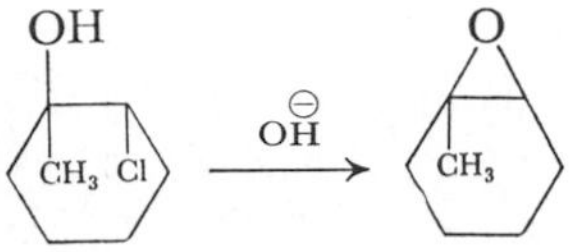

oxide smoothly, while the *cis*-isomer reacts more slowly to give not an oxide, but a carbonyl compound [61].

Besides halide, sulfate and sulfonate groups may be removed from hydroxy compounds to produce cyclic oxides. These eliminations have been employed in complex ring systems, such as the sugars and steroids, in which the oxygen bridge sometimes connects atoms somewhat removed from each other [62].

The Darzens glycidic ester condensation for preparing epoxides is discussed in C.3.

(a) Preparation of 3,4-epoxy-1-butene. 1-Chloro-3-buten-2-ol (1 mole, 106.5 g.) was added to 1.5 moles of 50 % sodium hydroxide at 115–135° during agitation for 1 hr. The 3,4-epoxy-1-butene distilled over as it formed, the vapor temperature being 60–80°. An additional 0.5 mole of 50 % sodium hydroxide was added dropwise as soon as the rate of evolution decreased. Sodium chloride was added to the two-phase distillate and the upper layer was dried and distilled to give 58.9 g. (84 %) of 3,4-epoxy-1-butene, b.p. 65–72° [63].

(b) Other examples

(1) Ethylene oxide (90 % from ethylene chlorohydrin and anhydrous potassium fluoride) [64].

(2) 3-Chloromethyl-3-hydroxymethyl oxetane (76 % from pentaery-

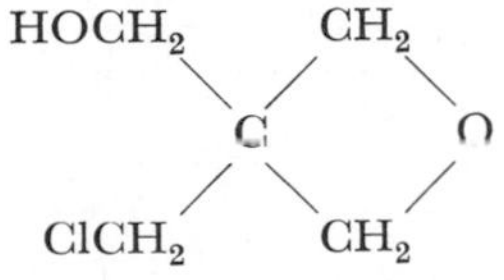

thrityl dichloride and sodium ethoxide) [65].

(3) p-Di(epoxyethyl)benzene (94% from p-di-(α-hydroxy-β-chloro-ethyl)benzene and potassium hydroxide in ethanol) [66].

(4) Scopoline tosylate (86% from teloidine 6,7-di-p-tosylate, sodium, and methanol) [67].

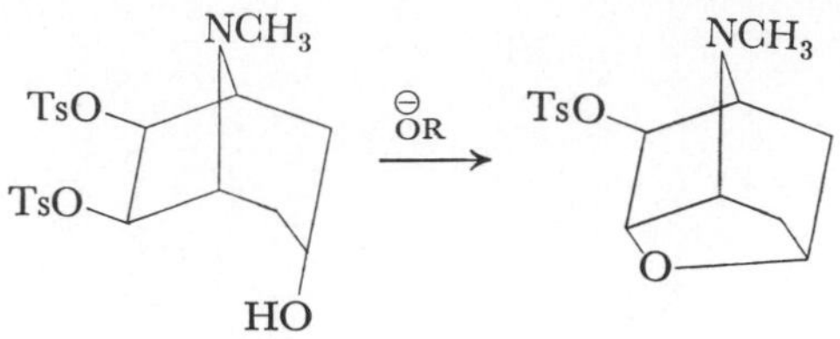

(5) 3,20-Bisethylenedioxy-11β,17α-dihydroxy-5-pregnene-16β,21-epoxide (79% from 21-acetoxy-3,20-bisethylenedioxy-16α-methanesulfonyloxy-5-pregnene-11β,17α-diol and alcoholic potassium hydroxide) [62].

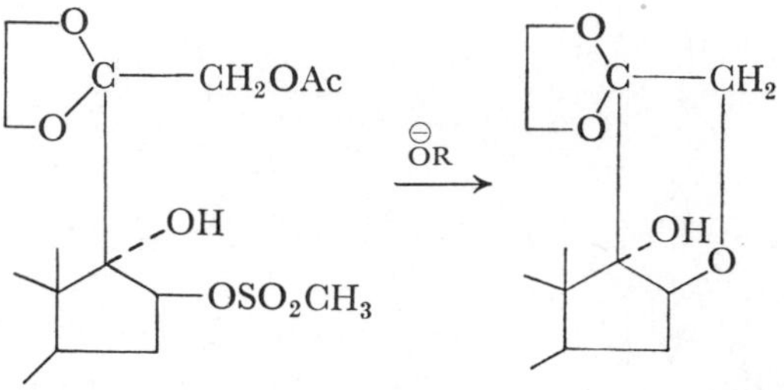

5. From Alcohols or Phenols and Onium Salts or Bases

$$RO^{\ominus} + R'\overset{\overset{\displaystyle CH_3}{|}}{\underset{\underset{\displaystyle CH_3}{|}}{N}}Ar^{\oplus} \longrightarrow ROR' + ArN(CH_3)_2$$

Aryl quaternary bases are used the most frequently and seem to be good etherizing agents for alkaloids containing phenolic or alcoholic groups.

A special reagent, phenyltrimethylammonium ethoxide

$$(C_6H_5\overset{\oplus}{N}(CH_3)_3)\overset{\ominus}{O}C_2H_5$$

is selective in its methylation in that it affects the phenolic hydroxyl but neither the alcoholic hydroxyl nor the tertiary nitrogen [68]. Thus morphine may be converted into codeine which possesses less addiction liability.

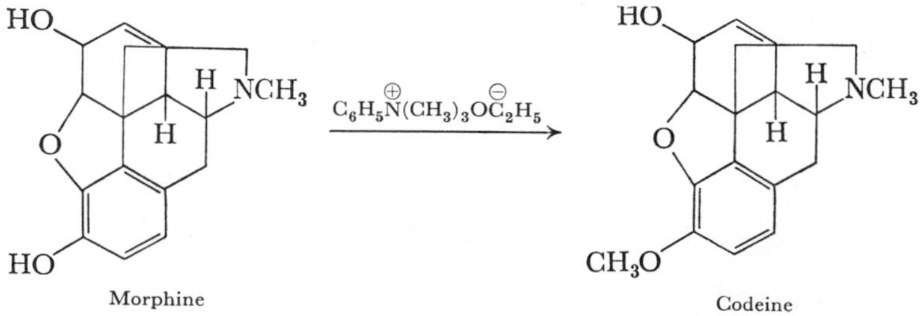

Morphine

Codeine

A second special reagent is trimethylsulfoxonium iodide

$$(CH_3)_3\overset{\oplus}{S} = \overset{\ominus}{O}I$$

which with silver oxide in dimethylformamide converts p-nitrophenol into p-nitroanisole in 78 % yield [69].

A third and perhaps the most powerful alkylating agent is trialkyloxonium borontetrafluoride (preparation: [70]). With alcohols or phenols it reacts exothermically to yield ethers [71]:

$$(C_2H_5)_3\overset{\oplus\ominus}{O}BF_4 + C_6H_5OH \rightarrow C_6H_5OC_2H_5$$

Phenetole, 73%

In the presence of 1 equivalent of sodium hydroxide the yield of phenetole was 91 %.

(a) Preparation of 3-methoxymethylindole. To a solution of 1 g. of sodium in 20 ml. of methanol was added 1 g. of gramine methiodide. After a stream of nitrogen had been passed through the solution for 24 hr. to remove the trimethylamine, the reaction mixture was poured into water and yielded 0.40 g. (79 %) of 3-methoxymethylindole, m.p. 97–98° [72].

(b) Preparation of pilocereine methyl ether.

$$C_{30}H_{41}N_2O_3OH \text{ (OH is phenolic)} + C_6H_5\overset{\oplus}{N}(CH_3)_3\overset{\ominus}{O}SO_2C_6H_5 \rightarrow$$

$$C_{30}H_{41}N_2O_3OCH_3 + C_6H_5SO_2OH + C_6H_5N(CH_3)_2$$

Phenyltrimethylammonium benzenesulfonate, 550 mg., was added to 2.5 ml. of absolute ethanol containing sodium ethoxide (from 50 mg. of sodium). The sodium benzenesulfonate solution was filtered and pilocereine, 420 mg., a cactus alkaloid, added to the filtrate. After warming, 230 mg. of the methyl ether, m.p. 103–105°, was isolated together with 90 mg. of unreacted pilocereine [73].

(c) Preparation of *dl-trans*-1-*p*-chlorophenyl-1,2-diphenylethylene oxide (75 % from dl-α-1-*p*-chlorophenyl-1,2-diphenyl-2-aminoethanol, methyl

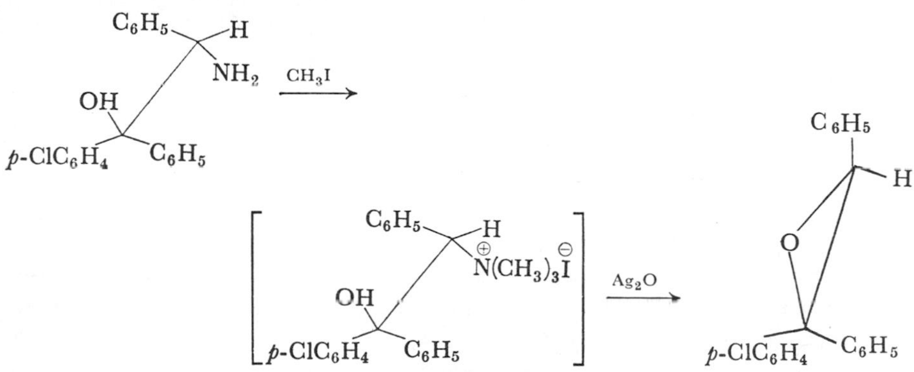

iodide, and silver oxide in a column) [74].

6. From Grignard Reagents and α-Chloroethers and the Like

$$RMgX + CH_2ClOR' \rightarrow RCH_2OR'$$

The reaction of Grignard reagents with α-chloroethers gives yields in the 45–90 % range [75]. Acetals or ketals should react with Grignard reagents to form ethers, but recorded yields are low with the exception of the following [76]:

$$p\text{-ClC}_6\text{H}_4\text{MgBr} + \text{ClCH}_2\text{CH(OEt}_2)_2 \xrightarrow{25°} p\text{-ClC}_6\text{H}_4\overset{\overset{\text{OEt}}{|}}{\text{CH}}\text{CH}_2\text{Cl}$$

p-(α-Ethoxy-β-chloroethyl)-
chlorobenzene, 65%

Ethyl orthoformate should react with two or more equivalents of Grignard reagent to form ethers but apparently it has not been used for this purpose [77].

On the other hand, the reaction of t-butyl peresters with Grignard reagents occasionally is a convenient synthesis of t-butyl ethers (see Ex. b):

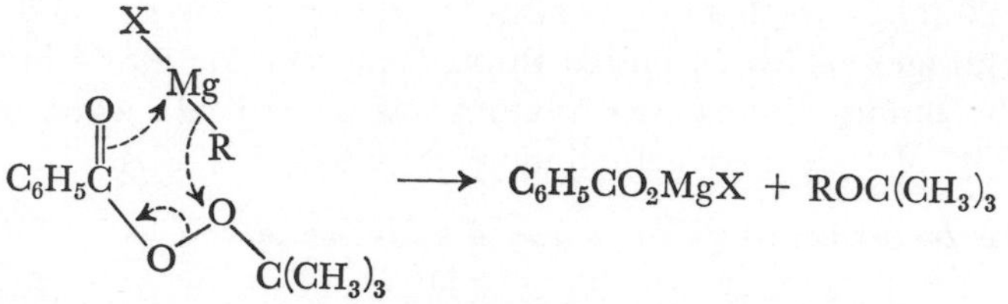

$$\longrightarrow \text{C}_6\text{H}_5\text{CO}_2\text{MgX} + \text{ROC(CH}_3)_3$$

(a) Preparation of alkyl cyclopropyl ethers [78].

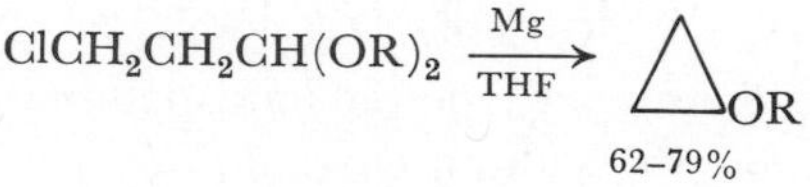

$$\text{ClCH}_2\text{CH}_2\text{CH(OR)}_2 \xrightarrow[\text{THF}]{\text{Mg}} \triangle\text{OR}$$

62–79%

(b) Preparation of t-butyl phenyl ether (78–84%, from t-butyl perbenzoate and phenylmagnesium bromide) [79].

7. From the Disodium Derivative of Benzophenone and Dihaloalkanes

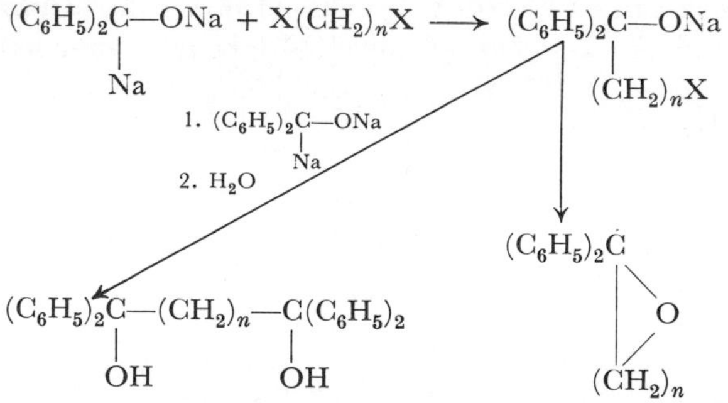

When the disodium derivative of benzophenone is treated with a dihalide, cyclic ethers and glycols are usually produced [80]. However if $n = 1$, only the cyclic ether is obtained (see Ex. a). The yields of ethers when $n = 2$ or more may be increased by using an excess of the dihalide. For example, a tenfold excess of 1,2-dichloroethane gives an 81 % yield of 1,1-diphenyltrimethylene oxide.

(a) Preparation of 1,1-diphenylepoxyethane.   A solution of 1.82 g.
of benzophenone in 15 ml. of dry ether was added to a solution of 0.46 g. of
sodium in 30 ml. of liquid ammonia. The violet solution obtained lost its color
when 0.005 mole of dichloromethane in 10 ml. of dry ether had been added.
Liquid ammonia was now removed and the mixture was decomposed with a
mixture of 10 ml. of water and 10 ml. of ether. The ethereal layer was extracted with 5 % hydrochloric acid, dried, and distilled. The ether, 67 %, came
over at 140–142° (5 mm.) [80].

1. A. I. Vogel, *J. Chem. Soc.*, 616 (1948).
2. H. O. House and B. M. Trost, *J. Org. Chem.*, **30,** 2502 (1965).
3. M. R. V. Sahyun and D. J. Cram, *Org. Syn.*, **45,** 89 (1965).
4. A. I. Vogel, *J. Chem. Soc.*, 616 (1948); W. T. Olson *et al.*, *J. Am. Chem. Soc*, **69,** 2451 (1947).
5. C. M. Brewster and I. J. Putman, Jr., *J. Am. Chem. Soc.*, **61,** 3083 (1939).
6. H. E. Ungnade and L. W. Kissinger, *J. Org. Chem.*, **31,** 369 (1966); C. D. Hurd and P.
   Perletz, *J. Am. Chem. Soc.*, **68,** 38 (1946); L. I. Smith *et al.*, *J. Am. Chem. Soc.*, **62,** 1863 (1940).
7. J. Wright and E. C. Vorgensen, *J. Org. Chem.*, **33,** 1245 (1968).
8. N. Kornblum *et al.*, *J. Am. Chem. Soc.*, **85,** 1148 (1963).
9. M. D. d'Engenières *et al.*, *Bull. Soc. Chim. France*, 2471 (1964).
10. J. F. Norris and G. W. Rigby, *J. Am. Chem. Soc.*, **54,** 2088 (1932).
11. C. V. Breder, M.S. Thesis, Vanderbilt University, Nashville, Tennessee, 1964.
12. M. T. Leffler and A. E. Calkins, *Org. Syn.*, Coll. Vol. **3,** 544 (1955).
13. C. S. Marvel and A. L. Tanenbaum, *Org. Syn.*, Coll. Vol. **1,** 435 (1941).
14. C. F. H. Allen and J. W. Gates, Jr., *Org. Syn.*, Coll. Vol. **3,** 140 (1955).
15. G. Brieger *et al.*, *J. Chem. Eng. Data*, **13,** 581 (1968).
16. A. W. Fort, *J. Am. Chem. Soc.*, **84,** 2620 (1962).
17. C. Grot *et al.*, *Ann. Chem.*, **679,** 42 (1964).
18. J. Hine, *Physical Organic Chemistry*, McGraw-Hill Book Co., New York, 2nd Ed., 1962, p. 184.
19. G. Lagrange *et al.*, *Compt. Rend.*, **254,** 1821 (1962).
20. J. S. Buck, *Org. Syn.*, Coll. Vol. **2,** 619 (1943).
21. S. Bernstein and E. S. Wallis, *J. Am. Chem. Soc.*, **62,** 2871 (1940).
22. J. A. Scarrow and C. F. H. Allen, *Org. Syn.*, Coll. Vol. **2,** 387 (1943).
23. F. Mauthner, *Org. Syn.*, Coll. Vol. **1,** 537 (1941).
24. C. C. Li and R. Adams, *J. Am. Chem. Soc.*, **57,** 1565 (1935).
25. E. D. Amstutz *et al.*, *J. Am. Chem. Soc.*, **68,** 349 (1946).
26. W. W. Carlson and L. H. Cretcher, *J. Am. Chem. Soc.*, **69,** 1952 (1947).
27. W. M. Rodionow, *Bull. Soc. Chim. France*, **45,** 109 (1929).
28. S. C. Gupta and F. A. Kummerow, *J. Org. Chem.*, **24,** 409 (1959).
29. H. Vorbrüggen, *Angew. Chem.*, **75,** 296 (1963).
30. E. E. Smissman and A. N. Voldeng, *J. Org. Chem.*, **29,** 3161 (1964).
31. B. Gredy, *Bull. Soc. Chim. France* (5), **3,** 1093 (1936).
32. B. Gredy, *Ann. Chim.* (Paris), (11), **4,** 42 (1935).
33. C. H. Snyder and A. R. Soto, *J. Org. Chem.*, **30,** 673 (1965).
34. R. Grewe and H. Pachaly, *Chem. Ber.*, **87,** 46 (1954).
35. A. K. Mitra *et al.*, *J. Org. Chem.*, **27,** 160 (1962).
36. F. W. Heyl *et al.*, *J. Am. Chem. Soc.*, **71,** 247 (1949).
37. K. Wallenfeld *et al.*, *Angew. Chem. Intern. Ed. Engl.*, **2,** 515 (1963).
38. J. F. Bunnett *et al.*, *J. Am. Chem. Soc.*, **72,** 2378 (1950).
39. E. Vowinkel, *Chem. Ber.*, **99,** 42, 1479 (1966).
40. G. S. Hiers and F. D. Hager, *Org. Syn.*, Coll. Vol. **1,** 58 (1941).
41. S. B. Sen Gupta and R. Das, *J. Indian Chem. Soc., Ind. and News Ed.*, **13,** 259 (1950); *C.A.*, **46,**
    2481 (1952).
42. G. Lagrange *et al.*, *Compt. Rend.*, **254,** 1821 (1962).
43. R. N. Icke *et al.*, *Org. Syn.*, Coll. Vol. **3,** 564 (1955).

44. E. S. West and R. F. Holden, *Org. Syn.*, Coll. Vol. **3**, 800 (1955).

45. H. King and E. V. Wright, *J. Chem. Soc.*, 1168 (1939).

46. H. Weingarten, *J. Org. Chem.*, **29**, 3624 (1964).

47. A. L. Williams *et al.*, *J. Org. Chem.*, **32**, 2501 (1967).

48. H. E. Ungnade and E. F. Orwoll, *Org. Syn.*, Coll. Vol. **3**, 566 (1955).

49. B. T. Tozer and S. Smiles, *J. Chem. Soc.*, 1897 (1938).

50. A. A. Levy *et al.*, *J. Chem. Soc.*, 3264 (1931).

51. F. Ullmann and M. Ettisch, *Chem. Ber.*, **54**, 259 (1921).

52. E. E. Van Tamelen, *Accounts Chem. Research*, **1**, 111 (1968).

53. S. Winstein and R. B. Henderson in R. C. Elderfield, *ed.*, *Heterocyclic Compounds*, Vol. 1, John Wiley and Sons, New York, 1950, p. 8.

54. H. Nilsson and L. Smith, *Z. physik. Chem.*, **166A**, 136 (1933).

55. W. P. Evans, *Z. physik. Chem.*, **7**, 337 (1891).

56. H. W. Heine *et al.*, *J. Am. Chem. Soc.*, **75**, 4778 (1953).

57. D. D. Reynolds and W. O. Kenyon, *J. Am. Chem. Soc.*, **72**, 1593 (1950).

58. R. Lespieau, *Bull. Soc. Chim. France* (5), **7**, 254 (1940).

59. L. F. Schmoyer and L. C. Case, *Chem. Eng. News*, Aug. 3. 44 (1959).

60. S. Winstein and H. J. Lucas, *J. Am. Chem. Soc.*, **61**, 1576 (1939).

61. P. D. Bartlett and R. H. Rosenwald, *J. Am. Chem. Soc.*, **56**, 1990 (1934).

62. W. S. Allen and S. Bernstein, *J. Am. Chem. Soc.*, **78**, 3223 (1956).

63. R. G. Kadesch, *J. Am. Chem. Soc.*, **68**, 41 (1946).

64. I. L. Knunyants *et al.*, *Zh. Obshch. Khim.*, **19**, 101 (1949); *C.A.*, **43**, 6163 (1949).

65. C. H. Issidorides *et al.*, *J. Org. Chem.*, **21**, 997 (1956).

66. H. Hopff and P. Jaeger, *Helv. Chim. Acta*, **40**, 274 (1957).

67. K. Zeile and A. Heusner, *Chem. Ber.*, **90**, 2809 (1957).

68. W. Rodionow, *Bull. Soc. Chim. France*, **59**, 305 (1926).

69. R. Kuhn and H. Trischmann, *Ann. Chem.*, **611**, 117 (1958).

70. H. Meerwein, *Org. Syn.*, **46**, 120 (1966).

71. H. Meerwein *et al.*, *J. Prakt. Chem.*, **147**, 257 (1937).

72. T. A. Geissman and A. Armen, *J. Am. Chem. Soc.*, **74**, 3916 (1952).

73. C. Djerassi *et al.*, *J. Am. Chem. Soc.*, **75**, 3632 (1953).

74. D. Y. Curtin *et al.*, *J. Am. Chem. Soc.*, **73**, 3453 (1951).

75. M. S. Kharasch and O. Reinmuth, *Grignard Reactions of Nonmetallic Substances*, Prentice-Hall, Englewood Cliffs, N.J., 1954, p. 1071.

76. Ref. 75, p. 1041.

77. Ref. 75, p. 586.

78. C. Feugeas and J.-P. Galy, *Compt. Rend.*, Ser C, **266**, 1175 (1968).

79. C. Frisell and S.-O. Lawesson, *Org. Syn.*, **41**, 91 (1961).

80. D. V. Ioffe, *J. Gen. Chem. USSR* (*Eng. Transl.*), **34**, 3960 (1964).

B. Electrophilic-Type Preparations

In this section, ethers are prepared from the attack of a carbonium ion on another alcohol or from the attack of a species which is well-advanced toward the carbonium ion stage. The tertiary alcohols represent the former:

$$R_3COH \xrightarrow{H^\oplus} [R_3C^\oplus] \xrightarrow{R'OH} R_3COR' + H^\oplus$$

and primary alcohols represent the latter:

$$CH_3CH_2\overset{\displaystyle \overset{H}{|}}{O} \dashrightarrow \overset{\displaystyle \overset{CH_3}{|}}{CH_2^{+\delta}} \dashrightarrow OH_2 \longrightarrow CH_3CH_2O\overset{\displaystyle \overset{CH_3}{|}}{CH_2} + H_2O + H^\oplus$$

References for Section B are on pp. 310–311.

Ether formation also may occur via the alkyl hydrogen sulfate if concentrated sulfuric acid is used. Both types of ether formation are considered here together with other reactions in which electron-deficient species are generated in alcohol solution. They include, in addition to dehydration, the ionization of reactive halides, protonation of olefins and some carbonyls, diazo decompositions, all in the presence of alcohol, and the rather special synthesis of α-haloethers by the addition of alcohols to carbonyl compounds in the presence of hydrogen halide.

Related to the latter synthesis is the preparation of α-haloethers by exchange of acetals with acyl halides to be found in B.8.

1. From Alcohols

$$2\,RCH_2OH \xrightarrow{\ H_2SO_4\ } RCH_2OCH_2R + H_2O$$

This synthesis is the most common method for preparing symmetrical primary ethers. Various dehydrating agents such as sulfuric acid, concentrated hydrochloric acid, p-toluenesulfonic acid, and even heat alone have been employed. For ethers of higher molecular weight, best results are obtained by refluxing the alcohol with acid until the theoretical amount of water has been collected [1]. A convenient way to prepare the symmetrical ethers of diaryl carbinols and α-phenylethanol is to pass the solution of the alcohol in a nonhydroxylic solvent such as benzene through an alumina column at room temperature [2]. Another convenient etherification of benzylic alcohols is by means of iodine catalysis (see Ex. b.5).

For mixed ethers the dehydration synthesis is quite satisfactory provided one of the alcohols can be converted readily into a carbonium ion in dilute acidic solution. For example, ethers containing the t-butyl and primary alkyl radicals may be prepared in good yield by using 15 % sulfuric acid or sodium bisulfate solution [3]. Other mixed ethers prepared by dehydration are triphenylmethyl isoamyl ether (88 %) [4] and benzhydryl β-chloroethyl ether (81–88 %) [5].

On dehydration, glycols may produce cyclic ethers. Dehydrating agents used in such cases have been sulfamide [6], anhydrous hydrogen chloride in benzene [7], p-toluenesulfonic acid monohydrate [8], p-toluenesulfonic in 90 % formic acid [9], phosphoric acid [10], p-toluenesulfonyl chloride in ether (see A.4), [11], and alumina [12]. The yields by these methods are generally good.

(a) **Preparation of isoamyl ether.** A mixture of 1.5 kg. of isoamyl alcohol and 300 g. of p-toluenesulfonyl chloride was refluxed with a condenser having a water separator. As the reaction proceeded, water, isoamyl alcohol, and isoamyl ether collected in the separator. The water was discarded and the alcohol and ether mixture was returned to the reaction flask. Heating was continued until no more water collected (9–10 hr.). The reaction product was dried over potassium carbonate and distilled to give 1036 g. of crude ether boiling above 132°. Since this product contains about 7 % of isoamyl alcohol, an equivalent amount of boric acid and 200 g. of benzene were added. This mixture was distilled to remove first the water as the benzene-water azeotrope, then the benzene, and finally, after cooling, isoamyl ether at 60–61° (10 mm.).

Another distillation over sodium gave 950 g. (70–75%) of pure isoamyl ether [13].

(b) Other examples

(1) **Bis-4,4′-dichlorobenzhydryl ether** (98% from the alcohol and 100% sulfuric acid) [14].

(2) **Benzhydryl β-chloroethyl ether** (81–88% from diphenylcarbinol, ethylenechlorohydrin, and concentrated sulfuric acid) [5].

(3) **Tetrahydrofuran** (92% from tetramethylene glycol and sulfamide) [6].

(4) **t-Butyl isopropyl ether** (82% from t-butyl alcohol added to a mixture of 15% aqueous sodium bisulfate and isopropyl alcohol) [3].

(5) **Methyl trityl ether.**   (Yields decrease regularly as methanol is

$$(C_6H_5)_3COH + I_2 + CH_3OH \xrightarrow[\text{2. } H_2O, Na_2S_2O_3]{\text{1. 2 hr., 25}°} (C_6H_5)_3COCH_3$$
$$\quad\text{10 g.}\qquad\text{10 g.}\quad\text{50 ml.}\qquad\qquad\qquad\qquad\qquad\text{90\%}$$

replaced by ethanol, isopropyl alcohol, or other branched alcohols) [15].

2. From Halides or Esters (*SN*₁ Process)

The halides or esters which tend to ionize by the SN_1 process are the benzyl, allyl, and t-alkyl types. They are not necessarily converted into ethers in good yield because olefin formation may be a competing reaction. If olefin formation is not possible, as in the solvolysis of triphenylmethyl chloride, the yields of ethers are almost quantitative. Pyridine or collidine may be added to prevent cleavage of the ether by the acid liberated. If olefin formation is possible as, for example, by warming t-butyl bromide in aqueous methanol, as much as 10% of olefin is formed together with a preponderance of t-butyl methyl ether and some t-butyl alcohol. n-Butyl bromide warmed in aqueous methanol gives no olefin but, curiously, the ether is formed in direct proportion to the mole percent alcohol present. The bromide, of course, solvolyzes by the SN_2 mechanism [16].

Ethers may be formed from any other sources which generate carbonium ions in alcoholic solution. The synthesis may be started with an olefin as shown (see Ex. c, d):

$$RCH{=}CH_2 \xrightarrow[\text{R′OH}]{H^{\oplus}} \begin{array}{c} RCH{-}CH_3 \\ | \\ OR' \end{array}$$

In fact, alcohols may be added to some olefins by activation with light [17]:

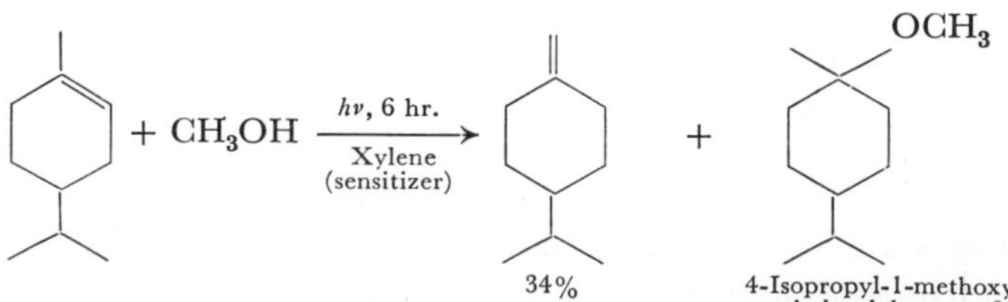

34%   4-Isopropyl-1-methoxy-1-methylcyclohexane, 61% as a *cis-trans* mixture

Methylcyclohexene and methylcycloheptene behave similarly.

The synthesis may also be started with esters or amines (see Ex. h.3) as shown [18]:

$$ROSO_2C_6H_5 \text{ (or other esters)} \xrightarrow[R'OH]{H \oplus} ROR'$$

$$ArNH_2 \text{ (or } RNH_2) \xrightarrow[2.\ R'OH]{1.\ HONO} ArOR' \text{ (or } ROR')$$

The mistaken impression has arisen that aryldiazonium salts are reduced to arenes by alcohols whereas in reality they usually form preponderant amounts of ether [19]. With alkylaryldiazonium salts, yields of ethers may be as high as 90%, but with nitro- or halogen-substituted aryldiazonium salts, yields may be as low as 13%, the reduction products becoming predominant. The more reactive aryl carbonium ion (or similar electron-deficient intermediate) evidently is more prone to hydride removal by the alcohol than to reaction with the oxygen atoms of alcohols [20]. The formation of ethers from aliphatic diazonium salts is not recorded as a good preparative procedure because of side reactions and because diazomethane serves the same purpose, as shown in the next two examples:

$$CH_2N_2 \xrightarrow{C_6H_5OH} CH_3OC_6H_5 \hspace{3cm} \text{(see Ex. h.1, h.2)}$$

$$CH_2N_2 \xrightarrow[R'OH]{HBF_4} [CH_3\overset{\oplus}{N}\equiv N] \xrightarrow{-N_2} CH_3\oplus \xrightarrow{R'OH} CH_3OR' \hspace{1cm} \text{(see Ex. f)}$$

In these reactions, methylation by diazomethane seems to depend on the acidity of the hydroxyl group, proceeding well with nitrophenols but too slowly with alcohols to be practical [21]. However, either fluoroboric acid (Ex. f) or boron trifluoride etherate [22] is acidic enough to produce the methyldiazonium ion which can methylate alcohols via the loss of nitrogen gas. Substituted diazomethane, including diazoketones (see Ex. g), has been employed in alkylation also [23].

(a) **Preparation of triphenylmethyl ethyl ether** (97% from trityl chloride in ethanol) [24].

(b) **Preparation of 6-trityl-β-d-glucose-1,2,3,4-tetraacetate.** Glucose, 120 g., trityl chloride, 193.2 g., and anhydrous pyridine were shaken at room temperature for 5 hr., and then allowed to stand with 360 ml. of acetic anhydride for 12 hr. The mixture was poured into 10 l. of ice water and the precipitate filtered, washed, and air-dried. Ether, 500 ml., was used to extract the α-form, and the residue containing the β-form was recrystallized from alcohol, 175 g., m.p. 166–166.5° [25].

(c) **Preparation of methyl t-butyl ether** (86% from isobutylene and methanol in presence of 3% each of BF_3 and H_2F_2 in an autoclave at 100°) [26].

(d) **Preparation of m-cresyl t-butyl ether** (37.5% yield, 61.2% conversion by bubbling isobutylene through m-cresol and a trace, 0.5 g., of 75% sulfuric acid at room temperature; some nuclear alkylation takes place also, but the simplicity of the method compensates for the low yields) [27].

(e) **Preparation of 9-benzyl-10-α-ethoxybenzylanthracene.** A boiling ethanolic solution of 5 g. of 9,10-dihydroxy-9,10-dibenzyl-9,10-dihydroanthracene was treated with 5 ml. of hydrochloric acid and the mixture was

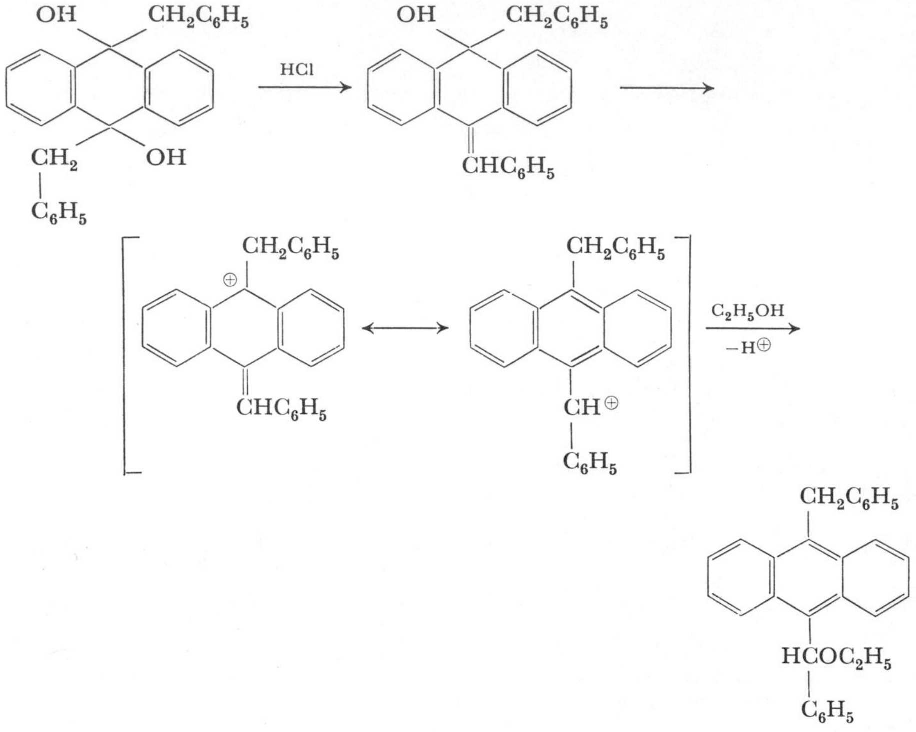

boiled for 6 hr. On cooling the ether was obtained in almost quantitative yield, m.p. 190–191° [28]. The reaction is applicable to other diols of condensed ring systems.

(f) **Preparation of methoxycyclohexane.** The methylation of cyclohexanol was carried out with diazomethane in ether or methylene chloride at 0–25° in the presence of 0.6–0.8 mole percent fluoroboric acid, yield 92% [29].

(g) **Preparation of α-methoxyacetophenone.** Boron trifluoride etherate, 0.3 g., was added to a solution of 7.5 g. of diazoacetophenone in 150 ml. of methanol at 25°. The theoretical amount of nitrogen was evolved in 30 min., after which there was obtained 6.1 g. (79%) of α-methoxyacetophenone, b.p. 124–126° (19 mm.) [30].

(h) **Other examples**

   (1) **Yohimbine methyl ether** (84% from yohimbine in methylene chloride and diazomethane in the presence of aluminum isopropoxide) [31].

   (2) **Ethyl β-(2-bromo-4-nitrophenoxy)propionate** (60% based on phenol consumed from freshly prepared ethyl β-diazopropionate and 2-bromo-4-nitrophenol in ether) [23].

(3) *p*-Ethoxybenzoic acid (50 % from *p*-carboxybenzene diazonium nitrate in ethanol) [32].

3. From Vinyl Esters or Ethers

$$\text{ROH} + \text{AcOCH}{=}\text{CH}_2 \xrightarrow{\text{Hg(OAc)}_2} \begin{array}{c} \text{RO} \\ \diagdown \\ \text{CHCH}_2\text{HgOAc} + \text{AcOH} \longrightarrow \\ \diagup \\ \text{AcO} \end{array}$$

$$\text{ROCH}{=}\text{CH}_2 + \text{Hg(OAc)}_2$$

This synthesis has been utilized in the preparation of a series of vinyl ethers [33]. The ether may be obtained in highest yield (36–98 %) by fractionating it from a solution of the catalyst, a high-boiling vinyl ether, and a lower-boiling alcohol. The mercuric salts of weak acids, such as acetic and benzoic, are specific catalysts and the purity of the catalyst is of great importance [34].

(a) Preparation of ethyl vinyl ether.   Mercuric acetate (5 g., 0.016 mole) in a mixture of dry ethanol (50 g., 1.08 moles) and *n*-butyl vinyl ether (100 g., 1.0 mole) was distilled in a column fitted with a total condensation partial take-off head. By maintaining total reflux until the vapor temperature settled at 36–37° and then adjusting the rate of distillation to keep the boiling point at this value, 71 g. (98 %) of ethyl vinyl ether was obtained [33].

(b) Preparation of β-cyclogeranyl vinyl ether (51 % by refluxing

β-cyclogeraniol, ethyl vinyl ether, and mercuric acetate on a steam bath for 16 hr.) [35].

4. From Cyclic Ethers or Imines by Cleavage

$$\underset{\diagdown\!\!\!_{O}\!\!\!\diagup}{\text{RCH}{-}\!\!-\text{CH}_2} \xrightarrow{\text{R}'\text{OH}} \underset{\underset{\text{OR}'}{|}}{\text{RCHCH}_2\text{OH}}$$

The ring strain in epoxides allows for easy opening of the ring system to produce hydroxyethers. The opening can be catalyzed by both acid and base, each giving a different product. Although practical results do not always agree with predictions, the following orientations would be anticipated:

$$\underset{\diagdown\!\!\!_{O}\!\!\!\diagup}{\text{RCH}{-}\!\!-\text{CH}_2} + {}^{\ominus}\text{OR}' \longrightarrow \left[\underset{\underset{\text{O}^{\ominus}}{|}}{\text{RCHCH}_2\text{OR}'}\right] \xrightarrow{\text{H}^{\oplus}} \underset{\underset{\text{OH}}{|}}{\text{RCH}{-}\text{CH}_2\text{OR}'}$$

$$\underset{\diagdown\!\!\!_{O}\!\!\!\diagup}{\text{RCH}{-}\!\!-\text{CH}_2} \xrightarrow{\text{H}^{\oplus}} \left[\underset{\oplus}{\text{RCHCH}_2\text{OH}}\right] \xrightarrow{\text{R}'\text{OH}} \underset{\underset{\text{OR}'}{|}}{\text{RCHCH}_2\text{OH}} + \text{H}^{\oplus}$$

The acid-catalyzed or noncatalytic cleavage tends to give both primary and secondary alcohols [36, 37]. Results with other α-epoxides are not so clearcut [38].

The ring opening with cyclohexene oxide has been shown to be a *trans* addition, a fact which suggests that the free carbonium ion is not formed [39]:

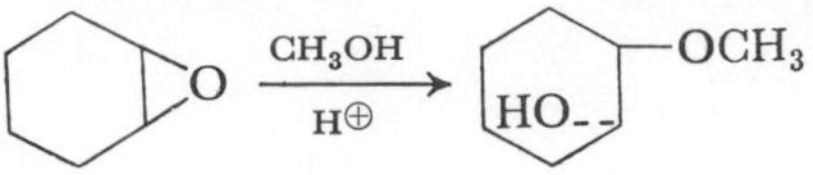

$$\text{cyclohexene oxide} \xrightarrow[\text{H}\oplus]{\text{CH}_3\text{OH}} \text{HO---OCH}_3$$

The mechanism and stereochemistry of such ring openings have been reviewed [40].

Some polymeric products can be expected in epoxide ring opening, as the product can also react as follows:

$$\text{CH}_2\text{---CH}_2 + \text{HOCH}_2\text{CH}_2\text{OH} \longrightarrow \text{HOCH}_2\text{CH}_2\text{OCH}_2\text{CH}_2\text{OH} \longrightarrow$$
$$\text{O}$$

$$\text{HOCH}_2\text{CH}_2\text{OCH}_2\text{CH}_2\text{OCH}_2\text{CH}_2\text{OH}, \quad \text{etc.}$$

The carbowaxes are derived from this type of condensation.

Ethylene imine ring opening leads to amino ethers as shown in Ex. b.

(a) Preparation of *trans*-2 methoxycyclohexanol. A mixture of 49.1 g. of cyclohexene oxide, 202 ml. of anhydrous methanol, and 4 drops of concentrated sulfuric acid was refluxed for 4 hr. After neutralization with barium carbonate the mixture was filtered and the filtrate was distilled first at atmospheric and then at 10 mm. pressure. Methoxycyclohexanol, 53.1 g. (82%), b.p. 72.5–73.2° (10 mm.), n^{25}D 1.4586, was recovered [39].

(b) Preparation of 1-methoxy-2-aminoethane [41]

$$\text{CH}_2\text{---CH}_2 + \text{CH}_3\text{OH} \xrightarrow{\text{BF}_3} \text{H}_2\text{NCH}_2\text{CH}_2\text{OCH}_3$$
$$\text{NH} \qquad\qquad 91\%$$

(c) Other examples

(1) **1-Ethoxy-2-propanol** (81% from propylene oxide, ethanol and sodium hydroxide) [42].

(2) **2-Alloxy-1-phenylethanol** (83% from styrene oxide, allyl alcohol, and sulfuric acid) [43].

5. From protonated ketonizable enols and alcohols

Among the group of electrophilic preparations of ethers is the protonation of ketonizable enols which can then add alcohols to form ethers. This synthesis comprises a rather unusual group of reactions as exemplified for the preparation of methyl β-naphthyl ether:

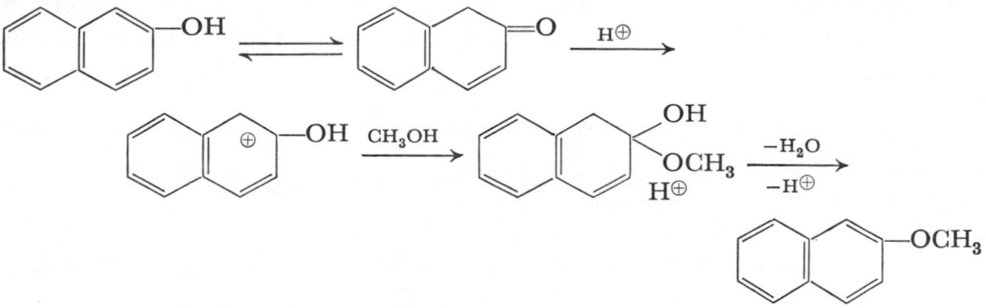

Even β-naphthylamine hydrochloride can be converted into the methyl ether by heating with methanol [44].

Benzenoid phenols, with the exception of resorcinol, do not undergo etherification in this manner. Because of the tendency of this phenol to form ethers, Friedel-Crafts reactions often lead to xanthenes, as is shown in the preparation of fluorescein:

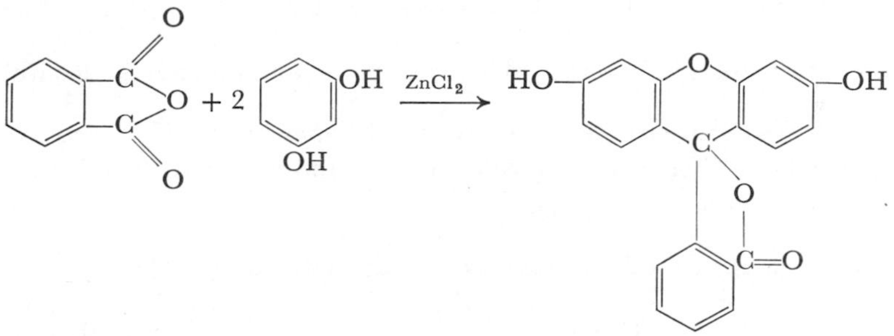

Many examples of similar preparations of benzodioxanes, dioxanes, xanthenes, xanthones, benzofurans, and chromans are to be found in the Elderfield treatise [45].

Benzoin can be transformed to an alkyl ether by a similar process as shown:

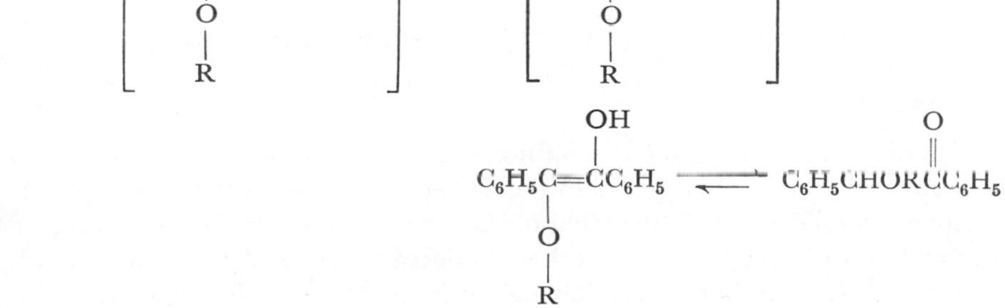

Other like examples are shown, the precise mechanism of which may differ somewhat from those given in the discussion above, but the overall pattern of mechanism shows definite relationships which warrant their collection in this subsection.

(a) **Preparation of n-propyl-β-naphthyl ether.**   n-Propyl alcohol (10 g.) + β-naphthol (10 g.) + $H_2SO_4 \cdot H_2O$ (4 g.) $\xrightarrow[\text{12 hr.}]{100°}$ n-propyl β-naphthyl ether, 94.5%, m.p. 38° [46].

(b) **Preparation of fluorescein** (73–78% from phthalic anhydride, resorcinol, and zinc chloride at 180–190°) [47].

(c) **Preparation of xanthone** (61–63% yield; the stoichiometry suggests

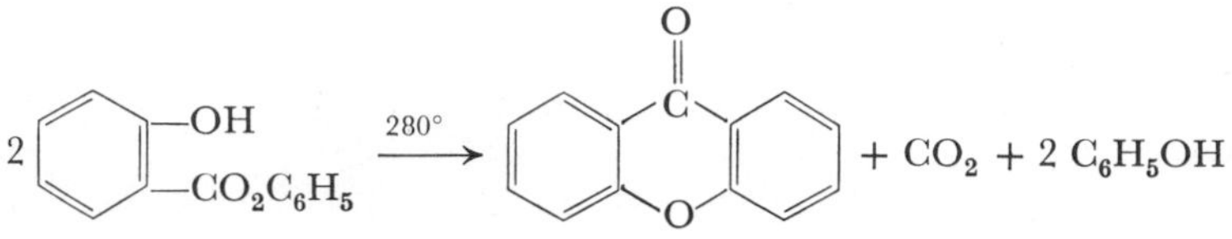

anhydride formation followed by decarboxylation and ether ring closure) [48].

(d) **Preparation of the ethyl ether of benzoin** (50–75% from benzoin, ethyl alcohol, and hydrogen chloride; anisoin and furoin fail to give an ether under the same conditions) [49].

(e) **Preparation of 3-methoxy-2,5-diphenylfuran** [50].

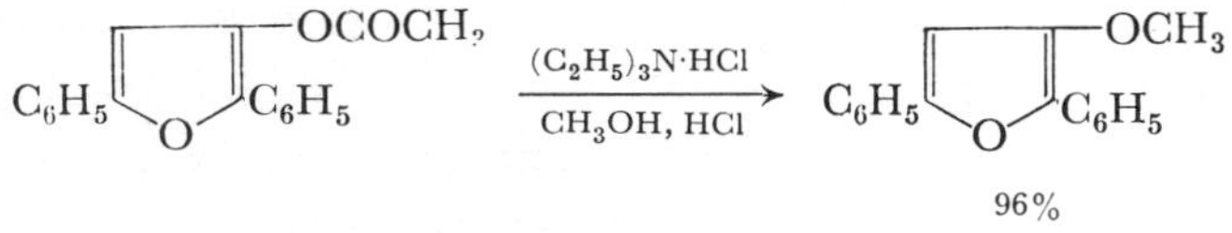

(f) **Preparation of O-methylresorcinol** (30% from 2 moles each of resorcinol, methanol, and potassium hydrogen sulfate in an autoclave at 180°; some dimethyl resorcinol ether is formed also) [51].

(g) **Preparation of 3-ethoxy-2-cyclohexenone** (70–75% from dihydro-resorcinol, ethanol, and p-toluenesulfonic acid) [52].

6. From Acetals and Alkenes and the Like

$$RCH{=}CH_2 + R'CH(OC_2H_5)_2 \xrightarrow{BF_3} RCHCH_2CH_2OC_2H_5$$
$$\underset{OC_2H_5}{|}$$

Oddly enough, this simple modification of the Prins reaction (Alcohols, Chapter 4, B.4) cannot be found in the literature, but so many precedents exist on more complicated structures that it may have been overlooked. For instance, the addition of acetals to vinyl ethers is documented in Acetals and Ketals, Chapter 9, B.2. Addition of aldehydes to some allylic carbinols in the Prins reaction yields ethers [53]:

$$CH_2=CHCH_2CH_2OH + CH_3CH=O + 20\% \text{ aq. } H_2SO_4 \xrightarrow[\text{3 hr.}]{80°, \text{ autoclave}}$$

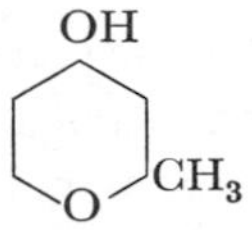

$$\underset{\text{1.44 kg.}}{} \qquad \underset{\text{0.88 kg.}}{} \qquad \underset{\text{2.5 kg.}}{}$$

OH

2-Methyltetrahydro-
pyranol-4, 72%

Addition of orthoformates to esters which can enolize is possible also (see Ex. a).

(a) **Preparation of diethyl ethoxymethylenemalonate** [54].

$$HC(OEt)_3 + CH_2(CO_2Et)_2 \xrightarrow{2 \text{ Ac}_2O}$$

$$C_2H_5OCH=C(CO_2Et)_2 + 2 CH_3CO_2Et + 2 CH_3CO_2H$$
$$\underset{50-60\%}{}$$

7. From Aldehydes by Haloalkylation

$$ROH + R'CHO + HX \longrightarrow R'CH\overset{X}{\underset{OR}{\diagup}} + H_2O$$

This synthesis is a special one employed to prepare haloethers. The method and others for the preparation of α-haloethers have been reviewed [55]. The haloalkylation process in which replaceable hydrogen atoms are removed by an electrophilic attack apparently takes place as shown in the accompanying formulations. The haloethers, being relatively unstable, are freed from the excess

$$\underset{H}{\overset{R'}{\diagdown}}C=O + H^{\oplus} \rightleftharpoons \underset{H}{\overset{R'}{\diagdown}}\overset{\oplus}{C}-OH$$

$$ROH + \underset{H}{\overset{R'}{\diagdown}}\overset{\oplus}{C}-OH \rightleftharpoons RO\underset{R'}{\overset{|}{C}HOH} + H^{\oplus}$$

$$RO\underset{R'}{\overset{|}{C}HOH} + HX \longrightarrow RO\underset{R'}{\overset{|}{C}HX} + H_2O$$

of halogen acid before distillation [56]. The reaction is applicable to both primary and secondary alcohols. Yields as a rule are satisfactory. It should be reemphasized that α-chloroethers are highly reactive substances, usually with poor storage qualities. A review of their synthetic potentialities is available [57].

(a) **Preparation of α-chloroethyl ethyl ether** (87–92%, based on ethanol from ethanol, paraldehyde, and hydrogen chloride) [58].

(b) Preparation of bischloromethyl ether (72–76 % from paraform-aldehyde, concentrated hydrochloric and chlorosulfonic acids) [59].

8. From Acetals

The mechanism is unknown, but the reaction could well proceed by a series of ionization steps in which the products are determined by relative stabilities of carbonium ion intermediates. The copper bronze which occasionally is used

$$RCH(OR')_2 \xrightarrow[-R'OH]{H^\oplus} \left[\begin{array}{c} RCHOR' \\ \oplus \end{array} \right] \underset{R'OH}{\overset{CH_3\overset{O}{\overset{\|}{C}}Cl}{\rightleftarrows}} \underset{Cl}{\overset{|}{RCHOR'}} + \underset{OR' + H^\oplus}{\overset{|}{CH_3C{=}O}}$$

to catalyze the reaction could serve to make the chloride ion more available through complex formation [60]. Yields in this reaction are fair.

(a) Preparation of α-chlorobenzyl methyl ether.   Benzaldehyde dimethylacetal, 76 g., was mixed with 78 ml. of freshly distilled acetyl chloride and 1 ml. of thionyl chloride. The mixture was warmed for ½ hr. on a bath at 55° and then allowed to stand overnight at room temperature. After the methyl acetate and excess acetyl chloride were removed, the residue was distilled at the lowest possible pressure in a bath not above 100°. When 5–10 g. came over, the ether distilled at 71–72°/0.1 mm., yield 60 g. (80 %) [61].

(b) Preparation of chloromethyl-β-chloroisopropyl ether (from di-(β-chloroisopropyl)-formal and benzoyl chloride; yield 65.6 % based on the formal) [62].

1. A. I. Vogel, *J. Chem. Soc.*, 616 (1948).
2. C.-H. Wang, *J. Org. Chem.*, **28**, 2914 (1963).
3. J. F. Norris and G. W. Rigby, *J. Am. Chem. Soc.*, **54**, 2088 (1932).
4. H. A. Smith and R. J. Smith, *J. Am. Chem. Soc.*, **70**, 2400 (1948).
5. S. Sugasawa and K. Fujiwara, *Org. Syn.*, Coll. Vol. **4**, 72 (1963).
6. A. M. Paquin, *Angew. Chem.*, **A60**, 316 (1948).
7. M. S. Newman and H. S. Whitehouse, *J. Am. Chem. Soc.*, **71**, 3664 (1949).
8. H. Wynberg and A. Bantjes, *Org. Syn.*, Coll. Vol. **4**, 534 (1963).
9. R. L. Letsinger and P. T. Lansbury, *J. Am. Chem. Soc.*, **81**, 935 (1959).
10. J. Colonge and R. Marey, *Org. Syn.*, Coll. Vol. **4**, 350 (1963).
11. N. L. Wendler and H. L. Slates, *J. Am. Chem. Soc.*, **80**, 3937 (1958).
12. N. O. Brace, *J. Am. Chem. Soc.*, **77**, 4157 (1955).
13. C. Weygand, *Organic Preparations*, 1945, Interscience Publishers, New York, p. 164.
14. H. A. Smith and R. G. Thompson, *J. Am. Chem. Soc.*, **77**, 1778 (1955).
15. K. G. Rutherford *et al.*, *Can. J. Chem.*, **44**, 2337 (1966).
16. C. K. Ingold, *Structure and Mechanism in Organic Chemistry*, Cornell University Press, Ithaca, New York, 1953, p. 352.
17. J. A. Marshall and R. D. Carroll, *J. Am. Chem. Soc.*, **88**, 4092 (1966).
18. H. Zollinger, *Azo and Diazo Chemistry*, Interscience Publishers, New York, 1961, p. 41.
19. N. Kornblum, *Org. Reactions*, **2**, 262 (1944).
20. D. F. DeTar and T. Kosuge, *J. Am. Chem. Soc.*, **80**, 6072 (1958).
21. A. I. Kosak *et al.*, *J. Am. Chem. Soc.*, **76**, 4481 (1954).
22. E. Müller and W. Runel, *Angew. Chem.*, **70**, 105 (1958).
23. L. L. Braun and J. H. Looker, *J. Am. Chem. Soc.*, **80**, 359 (1958).
24. A. C. Nixon and G. E. K. Branch, *J. Am. Chem. Soc.*, **58**, 492 (1936).

25. D. D. Reynolds and W. L. Evans, *J. Am. Chem. Soc.*, **60,** 2559 (1938).
26. R. D. Morin and A. E. Bearse, *Ind. Eng. Chem.*, **43,** 1596 (1951).
27. D. R. Stevens, *J. Org. Chem.*, **20,** 1232 (1955).
28. G. M. Badger and R. S. Pearce, *J. Chem. Soc.*, 2314 (1950).
29. J. D. Roberts and W. S. Johnson *et al.*, *J. Am. Chem. Soc.*, **80,** 2584 (1958).
30. M. S. Newman and P. F. Beal, III, *J. Am. Chem. Soc.*, **72,** 5161 (1950).
31. A. Popelak and G. Lettenbauer, *Arch. Pharm.*, **295,** 427 (1962).
32. I. Remsen and R. O. Graham, *Am. Chem. J.*, **11,** 319 (1889).
33. W. H. Watanabe and L. E. Conlon, *J. Am. Chem. Soc.*, **79,** 2828 (1957).
34. A. W. Burgstahler and I. C. Nordin, *J. Am. Chem. Soc.*, **83,** 198 (1961).
35. G. Büchi and J. D. White, *J. Am. Chem. Soc.*, **86,** 2884 (1964).
36. H. C. Chitwood and B. T. Freure, *J. Am. Chem. Soc.*, **68,** 680 (1946).
37. A. R. Sexton and E. C. Britton, *J. Am. Chem. Soc.*, **70,** 3606 (1948).
38. P. D. Bartlett and S. D. Ross, *J. Am. Chem. Soc.*, **70,** 926 (1948); W. Reeve and I. Christoffel, *J. Am. Chem. Soc.*, **72,** 1480 (1950).
39. S. Winstein and R. B. Henderson, *J. Am. Chem. Soc.*, **65,** 2196 (1943).
40. S. Winstein and R. B. Henderson in Elderfield *ed.*, *Heterocyclic Compounds*, Vol. 1, John Wiley and Sons, 1950, p. 27.
41. U. Harder *et al.*, *Chem. Ber.*, **97,** 510 (1964).
42. H. C. Chitwood and B. T. Freure, *J. Am. Chem. Soc.*, **68,** 680 (1946).
43. D. Swern *et al.*, *J. Am. Chem. Soc.*, **71,** 1152 (1949).
44. Elsevier's *Encyclopaedia of Organic Chemistry*, Elsevier Publishing Co., New York, 1950, Series III, **12B,** p. 1274.
45. Ref. 40, Vols. 2, 6, 7.
46. W. A. Davis, *J. Chem. Soc.*, **77,** 33 (1900).
47. R. Adams *et al.*, *Laboratory Experiments in Organic Chemistry*, Macmillan Co., New York, 1963, p. 453.
48. A. F. Holleman, *Org. Syn.*, Coll. Vol. **1,** 552 (1941).
49. J. C. Irvine and D. McNicoll, *J. Chem. Soc.*, **93,** 1601 (1908).
50. P. S. Bailey and J. D. Christian, *J. Am. Chem. Soc.*, **71,** 4122 (1949).
51. U. Merz and H. Strasser, *J. Prakt. Chem.*, **61,** 103 (1900).
52. W. F. Gannon and H. O. House, *Org. Syn.*, **40,** 41 (1960).
53. E. Hanschke, *Chem. Ber.*, **88,** 1053 (1955).
54. W. E. Parham and L. J. Reed, *Org. Syn.*, Coll. Vol. **3,** 395 (1955).
55. L. Summers, *Chem. Rev.*, **55,** 301 (1955).
56. J. W. Farren *et al.*, *J. Am. Chem. Soc.*, **47,** 2419 (1925).
57. H. Gross and E. Höft. *Z. Chem.*, **4,** 401 (1964).
58. O. Grummitt *et al.*, *Org. Syn.*, Coll. Vol. **4,** 748 (1963).
59. S. R. Buc, *Org. Syn.*, Coll. Vol. **4,** 101 (1963).
60. F. Straus and H.-J. Weber, *Ann. Chem.*, **498,** 101 (1932).
61. F. Straus and H. Heinze, *Ann. Chem.*, **493,** 191 (1932).
62. J. J. Spurlock and H. R. Henze, *J. Org. Chem.*, **4,** 234 (1939).

C. Nucleophilic-Type Preparations

Anions of alcohols or phenols add readily to double bonds to which electron-withdrawing groups are attached and to acetylenes:

$$RO^{\ominus} + CH_2{=}CHY \text{ (Y = electron-withdrawing group)} \longrightarrow ROCH_2\overset{\ominus}{C}HY \overset{H^{\oplus}}{\longrightarrow}$$
$$ROCH_2CH_2Y$$

$$RO^{\ominus} + HC{\equiv}CR \overset{\overset{\oplus}{H}, 25°}{\longrightarrow} ROCH{=}CHR$$

These additions, having great breadth, are described in C.1.

References for Section C are on pp. 319–320.

Very similar additions of anions take place to aromatic compounds with electron-attracting groups attached. But the addition is followed by the displacement of some anion. These displacements, together with some benzyne

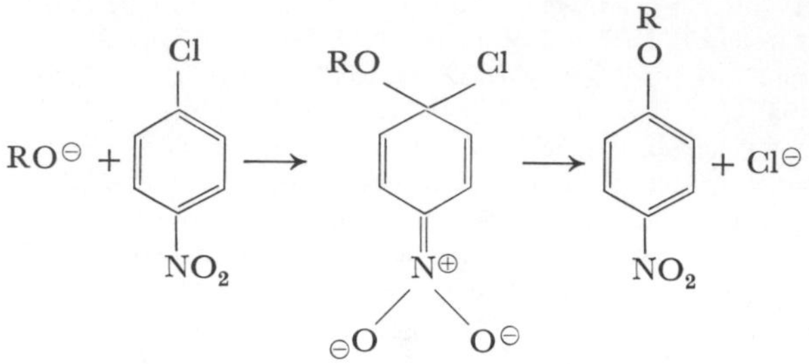

additions, are collected in C.2. The synthesis of epoxides by the Darzens reaction is discussed in C.3. An elegant, new, simple, and extremely useful synthesis of epoxides, the reaction of carbonyl compounds and dimethylsulfonium methylide, is described in C.4.

1. From Ethylenes or Acetylenes

If an electron-withdrawing group is attached to an olefin, the alkoxide anion can attack the carbon with its partially bared nucleus to form an ether:

$$RO^{\ominus} + \overset{+\delta}{CH_2}=\overset{-\delta}{CHY} \longrightarrow ROCH_2-\overset{\ominus}{CHY} \xrightarrow{H^{\oplus}} ROCH_2CH_2Y$$

This addition can take place with great ease when strong electron-withdrawing groups are attached, namely, through delocalization of the negative charges as in:

$$ROCH_2\overset{\ominus}{CHCN} \leftrightarrow ROCH_2CH=C=\overset{\ominus}{N}$$

It has been studied in great detail with acrylonitrile [1]. As a general procedure, acrylonitrile is added slowly with cooling to a solution of the alcohol containing a catalytic amount of sodium ethoxide, and the mixture heated at 80° for several hours. The mixture must be neutralized before isolation of the β-alkoxypropionitrile, because the action is reversible. Substituted acrylonitriles, acrylic esters, some perhaloethylenes, α,β-unsaturated carbonyl or nitro compounds, and maleic and fumaric esters are a few of the unsaturated compounds which add alkoxide anions. The lower limit seems to be reached with ethyl cinnamate, in which case there are indications that about 13 % addition may take place under ordinary circumstances [2]:

$$C_6H_5CH=CHCO_2C_2H_5 + \overset{\ominus}{OC_2H_5} \xrightarrow[\text{2. AcOH}]{\text{1. Reflux in alcohol}} C_6H_5\overset{\overset{OC_2H_5}{|}}{CH}CH_2CO_2C_2H_5$$

In unsymmetrical halo olefins, the orientation is such that the negative charge is situated on the carbon atom with the larger halogen atom which perhaps helps to delocalize the charge by expansion of its shell [3]:

$$CF_2{=}CCl_2 + \overset{\ominus}{O}CH_3 \xrightarrow[CH_3OH]{25°} CH_3OCF_2{-}\overset{\ominus}{C}Cl_2 \longleftrightarrow$$

$$CH_3OCF_2\overset{\overset{\displaystyle Cl}{|}}{C}{=}Cl^{\ominus} \xrightarrow{H_2O} CH_3OCF_2CHCl_2$$

1-Methoxy-1, 1-
difluoro-2,2-
dichloroethane

Symmetrical tetrachloroethylene does not add the methoxide anion under the same conditions.

Acetylenes need no electron-withdrawing groups to bring about reaction with alkoxides. Conditions are rather strenuous, but yields of vinyl ethers are good in the reaction of acetylene under pressure [4]:

$$HC{\equiv}CH + ROH \xrightarrow[\text{Pressure, 150°}]{NaOR} CH_2{=}CHOR$$

Phenoxides also add in alcoholic solution to acetylenes. Vinyl halides with sodium alkoxides form vinyl ethers, probably via the acetylene:

$$CH_2{=}CHCl + C_2H_5O^{\ominus} \xrightarrow{100°} \overset{\ominus}{C}H{=}CHCl \longrightarrow$$

$$HC{\equiv}CH + Cl^{\ominus} \xrightarrow{C_2H_5O^{\ominus}} \overset{\ominus}{C}H{=}CHOC_2H_5 \xrightarrow{H_2O} CH_2{=}CHOC_2H_5$$

85%

(a) Preparation of methyl β-methoxypropionate. One to three grams of sodium in 0.7 mole of methanol was added to a mixture of 0.8 mole of methyl acrylate and 0.1 mole of methanol while the mixture was stirred and kept at 27°. After being refluxed for 16 hr., the mixture was neutralized and distilled at reduced pressure so that the pot temperature did not exceed 100°. The ester, b.p. 55° (23 mm.), n^{20}_D 1.4022, was obtained in 91 % yield [5].

(b) Other examples

(1) Ethyl β-ethoxy-α-methyl crotonate [6].

$$CH_3CCl{=}C(CH_3)CO_2C_6H_5 + NaOC_2H_5 \longrightarrow CH_3\overset{\overset{\displaystyle OC_2H_5}{|}}{C}{=}C(CH_3)CO_2C_2H_5$$

82%

(2) β-Nitro-α-methoxypropionic acid (about 100% from β-nitro-acrylic acid and methanol with no catalyst) [7].

(3) 2-Chloro-1,1,2-trifluoroethyl ethyl ether (88–92% from chloro-trifluoroethylene and ethanol in presence of sodium ethoxide) [8].

(4) 1-Ethoxy-2-trifluoromethylethylene (89% from trifluoromethyl-acetylene and sodium ethoxide in ethanol) [9].

(5) α-Methoxysuccinic acid (good yield from diethyl maleate methanol, and sodium methoxide followed by hydrolysis) [10].

2. From Activated Aromatic Halides and Related Types

The general type of reaction discussed in this subsection is the nucleophilic displacement, as shown for the general example. There can be little doubt of the transitory existence of 6-V [11]. As for the leaving group, Y, the order of ease of displacement is: $C_6H_5SO_2O \simeq F \gg Cl \simeq Br > I > OR^\ominus$ [12]. Indeed, with the 2,4-dinitrophenyl halides the fluoro atom is displaced as an ion about a

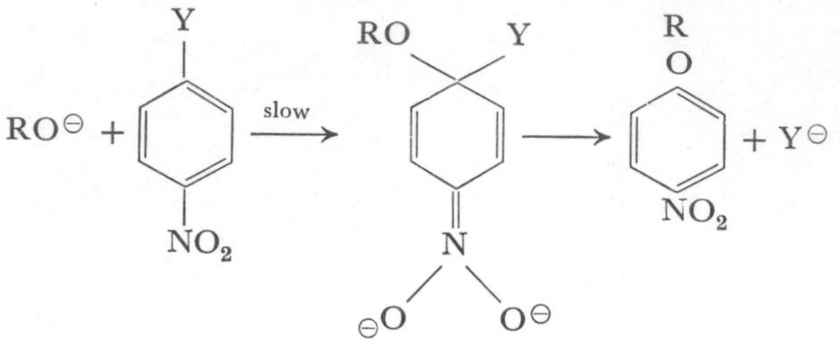

6-V

thousand times faster than the chloro atom, a fact which speaks rather convincingly for the intermediacy of the anion 6-V. The nitro group also is displaced easily from p- and o-dinitrobenzenes as a nitrite anion. As for electron-attracting substituents which facilitate displacement of groups in o- or p-positions to the substituent, the nitro group is by far the best, but others can be used. In a series of substituted o-nitrochlorobenzenes

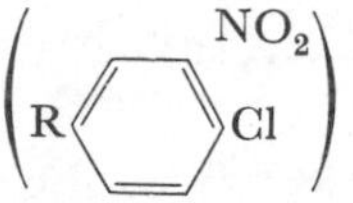

the activation is in the relative order:

$$NO_2 = 170,000, \quad SO_2CH_3 = 18,000, \quad \overset{\oplus}{N}(CH_3)_3 = 5500, \quad COCH_3 = 2100, \quad H = 1$$

Multiple substitution still further increases the facility of displacement, as illustrated by the high reactivity of picryl chloride, 2,4,6-trinitrophenyl chloride. The nitroso group is an exceptionally good activating group, perhaps better than the nitro group, because it forms such a stable anionic species 6-VI. But the

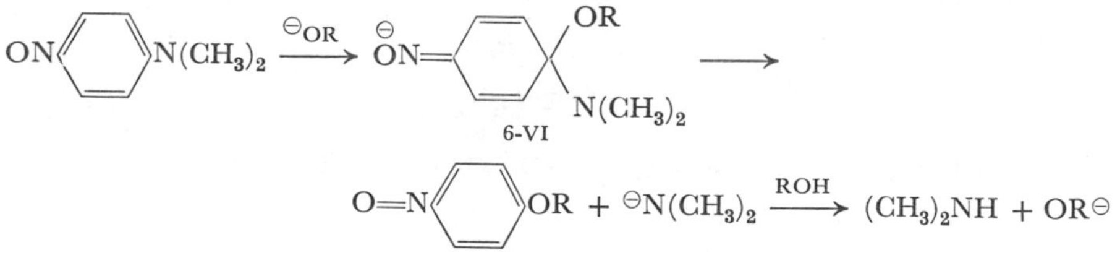

6-VI

reaction has not been used frequently for making ethers. The situation is similar for the diazo, another potent activating group.

Displacements of anions from polynuclear rings is facilitated because these systems are less aromatic and can form the p-quinoid structure (similar to 6-VI) more readily. For instance, 1,4-dinitronaphthalene is converted into the 1-methoxy-4-nitro derivative with sodium methoxide about 100 times faster than 1,4-dinitrobenzene under identical conditions [13].

2- and 4-Halopyridines are also susceptible to attack by nucleophiles. The reaction seems to be somewhat sluggish with the pyridines, since rather high temperatures and copper powder are used with the alkoxide and α-chloropyridine [14]. But 4-chloroquinoline reacts smoothly with sodium methoxide in methanol to form the methyl ether in good yield.

A hazy demarcation exists between the mechanisms which have been classified as metathetical (Ullmann; see A.3), nucleophilic (this subsection), and the benzyne type. The Ullmann reaction may well proceed via the *p*-quinoid intermediate discussed in this subsection, but alternative choices are available for the nucleophilic and benzyne mechanisms, as shown for hypothetical examples:

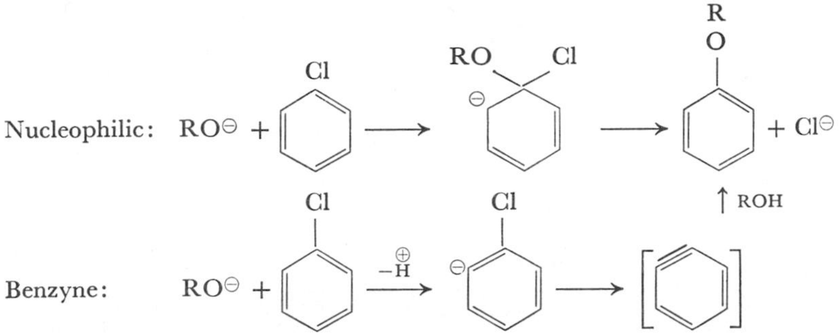

The benzyne intermediate has not been proposed for many ether preparations, but it can be expected when the nucleophilic mechanism is not facilitated by electron-withdrawing groups attached to the substrate. In addition, such powerful bases are needed to abstract the proton from the benzene ring that the benzyne intermediate is rarely encountered in ether synthesis. Two examples are shown:

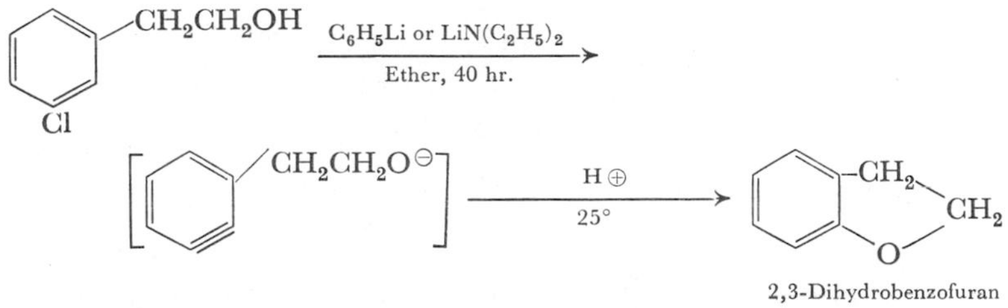

[ref. 15]

$$(CH_3)_3COK + C_6H_5Br \xrightarrow{(CH_3)_2SO} C_6H_5OC(CH_3)_3$$ [ref. 16]

Phenyl *t*-butyl ether
42–46%

Some unusual examples of ether formation are to be found among the following preparations.

(a) Preparation of 2,6-dimethoxybenzonitrile (68–74% from 2-nitro-6-methoxybenzonitrile and methanol containing potassium hydroxide) [17].

(b) 3,5-Dinitroanisole (63–77% from 1,3,5-trinitrobenzene and sodium methoxide; this reaction is most unusual since displacement occurs *meta* to the nitro groups, perhaps by a benzyne intermediate) [18].

(c) 3,5-Dinitroanisole (60–86% from 1,3,5-trinitrobenzene, and alkali metal bicarbonate in aqueous methanol; the methanol and bicarbonate reagents are specific for this reaction) [19].

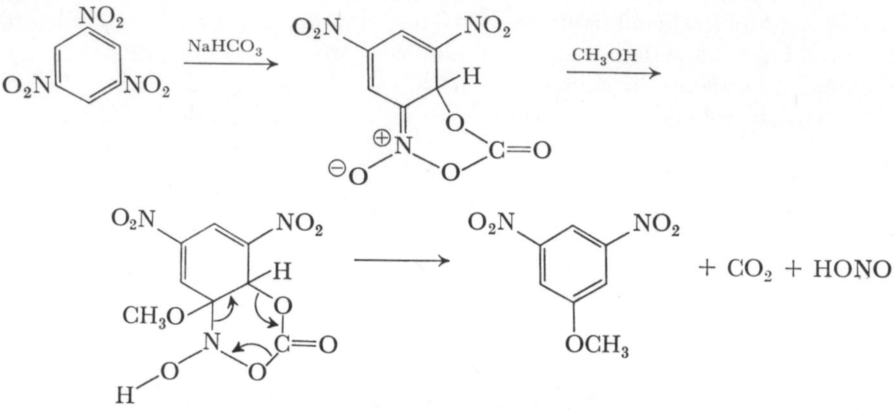

The mechanism is similar to that proposed for the von Richter reaction [20].

(d) Isopropoxypentachlorobenzene (92% from hexachlorobenzene and sodium hydroxide in refluxing isopropyl alcohol; pyridine can also be used to increase the solubility of hexachlorobenzene; the mechanism may be simple displacement rather than nucleophilic addition) [21].

(e) 2-Bromo-2′-nitro-4′-chlorodiphenyl ether (51% from the sodium salt of 2-bromophenol and 2,5-dichloronitrobenzene at 160–170° for 2 hr.) [22].

(f) 2,4-Dinitrophenyl isopropyl ether (75% from 2,4-dinitroanisole and potassium hydroxide in isopropyl alcohol; the *t*-butoxide anion does not displace the methoxyl group under the same conditions) [23].

(g) Dibenzo-*p*-dioxine

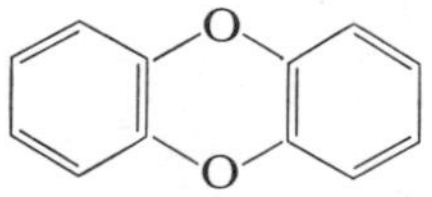

(47–53% from the potassium salt of *o*-bromophenol at 220° in the presence of copper powder) [24].

3. From Carbonyl Compounds and α-Chloroesters or Ketones (Darzens)

A review of this condensation is available [25]:

$$C_6H_5\overset{\displaystyle H}{\overset{|}{C}}{=}O + ClCH_2COOC_2H_5 \xrightarrow{NaOR} C_6H_5CH\underset{\displaystyle O}{\diagdown\!\!\!-\!\!\!\diagup}CHCO_2C_2H_5$$

α-Chloroketones can also be used to obtain epoxides. Potassium *t*-butoxide and sodium hydride [26] seem to be excellent basic catalysts for the Darzens reaction.

(a) Preparation of ethyl methylphenylglycidate (64% from aceto-phenone and ethyl chloroacetate in the presence of sodamide) [27].

(b) Preparation of ethyl β,β-pentamethyleneglycidate, (83–95%

$$\text{[cyclohexylidene]}\text{—CHCO}_2\text{C}_2\text{H}_5$$

from cyclohexanone and ethyl chloroacetate using potassium *t*-butoxide catalyst) [28].

(c) Preparation of 1-phenyl-2-tosyl ethylene oxide ("good yield"

$$\text{C}_6\text{H}_5\text{CH}\underset{\text{O}}{\overset{}{\diagup\!\!\!\diagdown}}\text{CHSO}_2\text{C}_7\text{H}_7$$

from α-chloromethyl *p*-methylphenyl sulfone and benzaldehyde) [25].

4. From Carbonyl Compounds and Dimethylsulfonium Methylide or Dimethyloxosulfonium Methylide

$$\begin{array}{c}\text{R}\\ \diagdown\\ \text{C}=\text{O}\\ \diagup\\ \text{(R)H}\end{array} + (\text{CH}_3)_2\text{S}=\text{CH}_2 \longrightarrow \begin{array}{c}\text{R}\quad\text{O}\\ \diagdown\diagup\diagdown\\ \text{C}\text{—}\text{CH}_2\\ \diagup\\ \text{(R)H}\end{array} + (\text{CH}_3)_2\text{S}$$

$$\begin{array}{c}\text{R}\\ \diagdown\\ \text{C}=\text{O}\\ \diagup\\ \text{(R)H}\end{array} + (\text{CH}_3)_2\!\!\overset{\text{O}}{\underset{}{\overset{\|}{-}\text{S}}}\!=\text{CH}_2 \longrightarrow \begin{array}{c}\text{R}\quad\text{O}\\ \diagdown\diagup\diagdown\\ \text{C}\text{—}\text{CH}_2\\ \diagup\\ \text{(R)H}\end{array} + (\text{CH}_3)_2\text{SO}$$

Epoxides are formed from carbonyl compounds by treatment either with dimethylsulfonium or dimethyloxosulfonium methylide [29, 30]. The dimethyl-sulfonium methylide, the less stable of the two reagents, is prepared from tri-methylsulfonium iodide and a solution of the methylsulfinyl carbanion in dimethylsulfoxide-tetrahydrofuran under nitrogen at 0 to −10°:

$$(\text{CH}_3)_2\overset{\oplus}{\text{S}}\text{CH}_3\overset{\ominus}{\text{I}} + \text{CH}_3\overset{\overset{\text{O}^\ominus}{|}}{\text{S}}=\text{CH}_2\overset{\oplus}{\text{Na}} \longrightarrow (\text{CH}_3)_2\text{S}=\text{CH}_2 + (\text{CH}_3)_2\text{SO} + \text{NaI}$$

On the other hand, the dimethyloxosulfonium methylide is prepared from the oxosulfonium iodide with sodium hydride in dioxane or tetrahydrofuran:

$$(\text{CH}_3)_2\overset{\text{O}}{\overset{\|}{\text{S}}}\!\overset{\oplus}{}\text{CH}_3\overset{\ominus}{\text{I}} + \text{NaH} \longrightarrow (\text{CH}_3)_2\overset{\text{O}}{\overset{\|}{\text{S}}}\!\!=\!\text{CH}_2 + \text{NaI} + \text{H}_2$$

While solutions of the latter are stable for months in an inert atmosphere at 0°, the former, being less stable, should be used immediately.

The sulfonium methylide reacts quickly with ordinary aldehydes and ketones and α,β-unsaturated ketones to give epoxides (oxiranes) with yields from 75 to 97%. The oxosulfonium methylide reacts somewhat more slowly with aldehydes and ketones to also give epoxides with yields from 56 to 90%. With α,β-unsaturated ketones the attack of this reagent, in contrast to that of the sulfonium methylide, is at the carbon-carbon double bond to give a cyclopropane derivative.

$$C_6H_5CH\text{=}CHCOC_6H_5 \xrightarrow{\text{(CH}_3)_2\overset{O}{\overset{\|}{S}}\text{=CH}_2} \underset{\text{1-Phenyl-2-benzoylcyclopropane, 95\%}}{C_6H_5\overset{CH_2}{\overset{\diagup\diagdown}{CH}\text{---}CHCOC_6H_5}} + \text{(CH}_3)_2SO$$

The carbonyl-dimethylsulfonium methylide reaction will find its greatest use in preparing β-aminoethanols (see Ex. b.2):

$$RCH\text{=}O \longrightarrow R\overset{O}{\overset{\diagup\diagdown}{CH}\text{---}CH_2} \xrightarrow{R_2'NH} RCHOHCH_2NR_2'$$

Former methods to produce the intermediate epoxide consisted of sodium borohydride reduction of an α-bromoketone in alcoholic solution.

(a) Preparation of epoxymethanocycloheptane.   A solution of 1.5–

2 M methylsulfinyl carbanion in dimethylsulfoxide-tetrahydrofuran in a salt-ice bath was treated with trimethylsulfonium iodide (1 molar equiv. based on the sodium hydride used to prepare the carbanion) in dimethyl sulfoxide over a 3-min. period. After stirring for 1 min., the cycloheptanone, slightly less than 1 equiv., was added at a moderate rate. Stirring was continued for several minutes in the ice-salt bath and then for 30–60 min. out of the bath. Dilution with 3 volumes of water, extraction with ether, washing, drying, and distilling the ethereal layer gave 2.45 g. (97.2%) of the epoxide [30].

(b) Other examples

   (1) Styrene oxide (56% from benzaldehyde and dimethyloxosulfonium methylide in tetrahydrofuran at 55°) [30].

   (2) 8-Chloro-6-epoxyethyl-2-phenylquinoline   (61.5%, m.p. 99–100°, from 18.7 mmoles of sodium hydride (washed with hexane to remove mineral oil) under nitrogen and 11 ml. of DMSO heated to 65°, cooled, diluted with 11 ml. of THF, cooled to −10°, and treated with 18.7 mmoles of trimethylsulfonium iodide in 20 ml. of THF-DMSO mixture followed by the addition of 9.3 mmoles of 8-chloro-2-phenylquinoline-6-carboxaldehyde in 20 ml. of THF-DMSO; the mixture was stirred at −10° for 10 min., warmed to 25°, and decomposed with cold water. The epoxide was recrystallized from methanol) [31].

5. From 2,5-Dihydrofurans (Rearrangement)

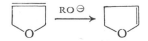

This synthesis represents a general reaction for β,γ-unsaturated cyclic ethers [32, 33] which occurs with potassium t-butoxide in t-butyl alcohol at high temperature under pressure. The 2,5-dihydrofuran may be obtained by the catalytic dehydration of 2-butene-1,4-diol, which in turn is produced by the condensation of acetylene and formaldehyde followed by reduction [34]. 2-Methyl- and 2,3-dimethyl-2,5-dihydrofurans rearrange as well to the corresponding 2,3-isomers. Yields are satisfactory.

(a) **Preparation of 2,3-dihydrofuran.** Potassium, 12 g., was dissolved in 200 ml. of sodium-dried t-butyl alcohol, and 70 g. of the solution and 70 g. (1 mole) of 2,5-dihydrofuran were put in a 0.5-l. stainless steel autoclave which was sealed and heated at 170° for 6 hr. The contents of the cooled autoclave were distilled through a spinning-band column to give 55.3 g. (79%) of 2,3-dihydrofuran, b.p. 53–55° (745 mm.) [33].

1. H. A. Bruson, *Org. Reactions*, **5,** 79 (1949).
2. C. R. Hauser *et al.*, *J. Am. Chem. Soc.*, **85,** 71 (1963).
3. W. T. Miller *et al.*, *J. Am. Chem. Soc.*, **70,** 431 (1948).
4. C. E. Schildknecht *et al.*, *Ind. Eng. Chem.*, **39,** 180 (1947); J. W. Reppe, *Acetylene Chemistry*, C. A. Meyer & Co., 25 Vanderbilt Ave., New York, 1949, pp. 28–48.
5. C. E. Rehberg *et al.*, *J. Am. Chem. Soc.*, **68,** 544 (1946).
6. D. E. Jones *et al.*, *J. Chem. Soc.*, 2349 (1960).
7. H. Shechter *et al.*, *J. Am. Chem. Soc.*, **74,** 3052 (1952). (General reference: V. V. Perekalin, *Unsaturated Nitro Compounds*, Daniel Davey and Co., New York, 1961.)
8. B. Englund, *Org. Syn.*, Coll. Vol. **4,** 184 (1963).
9. A. L. Henne and M. Nager. *J. Am. Chem. Soc.*, **74,** 650 (1952).
10. L. H. Flett and W. H. Gardner, *Maleic Anhydride Derivatives*, John Wiley and Sons, New York, 1952, p. 65.
11. S. D. Ross in S. G. Cohen, A. Streitwieser and R. W. Taft eds., *Progress in Physical Organic Chemistry*, Vol. 1, Interscience Publishers, New York, 1963, p. 31; *see also* W. E. Byrne *et al.*, *J. Org. Chem.*, **32,** 2506 (1967).
12. J. F. Bunnett, *J. Am. Chem. Soc.*, **79,** 5969 (1957); R. E. Parker and T. O. Read, *J. Chem. Soc.*, 9 (1962). C. K. Ingold, *Structure and Mechanism in Organic Chemistry*, Cornell University Press, Ithaca, N.Y., 1953, pp. 797–815.
13. D. H. D. Elias and R. E. Parker, *J. Chem. Soc.*, 2616 (1962).
14. A. J. Hill and W. J. McGraw, *J. Org. Chem.*, **14,** 783 (1949).
15. R. Huisgen *et al.*, *Chem. Ber.*, **93,** 1496 (1960).
16. M. R. V. Sahyun and D. J. Cram, *Org. Syn.*, **45,** 89 (1965).
17. A. Russell and W. G. Tebbens, *Org. Syn.*, Coll. Vol. **3,** 293 (1955).
18. F. Reverdin, *Org. Syn.*, Coll. Vol. **1,** 219 (1941).
19. P. T. Izzo, *J. Org. Chem.*, **24,** 2026 (1959).
20. M. Rosenblum, *J. Am. Chem. Soc.*, **82,** 3796 (1960).
21. A. L. Rocklin, *J. Org. Chem.*, **21,** 1478 (1956).
22. G. E. Bonvicino *et al.*, *J. Org. Chem.*, **26,** 2797 (1961).
23. S. S. Gitis and A. I. Glaz, *J. Gen. Chem. USSR* (*Eng. Transl.*), **30,** 3767 (1960).

24. M. Tomita *et al.*, *J. Pharm. Soc. Japan*, **74,** 934 (1954).
25. M. Ballester, *Chem. Rev.*, **55,** 283 (1955).
26. A. A. Hinckley, *Sodium Hydride Dispersions*, Ventron, Congress Street, Beverly, Massachusetts, May, 1966.
27. C. F. H. Allen and J. Van Allan, *Org. Syn.*, Coll. Vol. **3,** 727 (1955).
28. W. S. Johnson *et al.*, *Org. Syn.*, Coll. Vol. **4,** 459 (1963).
29. E. J. Corey and M. Chaykovsky, *J. Am. Chem. Soc.*, **84,** 867, 3782 (1962); *Tetrahedron Letters*, No. 4, 169 (1963).
30. E. J. Corey and M. Chaykovsky, *J. Am. Chem. Soc.*, **87,** 1353 (1965).
31. J. B. Wommack and D. E. Pearson, unpublished results.
32. M. L. A. Fluchaire and G. Collardeau, U.S. Patent 2,556,325, June 12, 1951; *C.A.*, **46,** 1046 (1952).
33. E. L. Eliel *et al.*, *J. Org. Chem.*, **30,** 2441 (1965).
34. R. Paul *et al.*, *Bull. Soc. Chim. France*, 668 (1950).

D. Oxidation

Oxidative preparations of ethers are not very extensive and seem confined exclusively to the synthesis of epoxides. However, a subsection on miscellaneous oxidations has been added.

1. From Olefins

$$RCH{=}CH_2 \xrightarrow{\quad C_6H_5COOOH \quad} RCH{-}CH_2$$
$$\underset{O}{\diagdown\diagup}$$

An excellent discussion of the methods of synthesizing epoxides is available [1].

This synthesis, one of the most common for preparing cyclic ethers (epoxides or oxiranes), has already been discussed (Alcohols, Chapter 4, D.5) since the type is an intermediate in producing glycols. In addition to peracids [2], hydrogen peroxide in an organic nitrile [3], sodium hypochlorite in pyridine [4] (see Ex. b.5), and succinoyl peroxide in dimethylformamide [5] have been used to synthesize epoxides. The experimental conditions are important to prevent the epoxide ring from opening to give the glycol. In the case of peracetic acid, for example, it is necessary to limit the temperature and reaction time, and to avoid the presence of strong acids, salts, and water [6] With pertrifluoroacetic acid a buffer, such as sodium carbonate, destroys any excess of the peracid after the olefin is consumed, and thus the epoxide remains in solution. In addition, with the use of dilute hydrogen peroxide and an organic nitrile conditions near neutral are maintained. The addition with all these reagents is *cis*.

By the use of a specially prepared solution of peracetic acid in acetone or ethyl acetate [7] a series of glycidic esters has been prepared with yields varying from 22 to 95 % [6] (see Ex. b.4):

$$RCH{=}CHCOOC_2H_5 \xrightarrow{\quad CH_3CO_3H \quad} RCH{-}CHCOOC_2H_5$$
$$\underset{O}{\diagdown\diagup}$$

References for Section D are on p. 323.

Recently a simple procedure for the preparation of volatile epoxides has been reported. The olefin and a stoichiometric amount of m-chloroperbenzoic acid (commercially available) in diglyme were allowed to stand for 24 hr. in a refrigerator. The epoxide was then distilled from the mixture. In this manner, 1-hexene oxide was prepared in 60% yield [8].

The epoxidation appears to be brought about through a convenient 1,3-dipolar transition state as shown in the accompanying formulation [9].

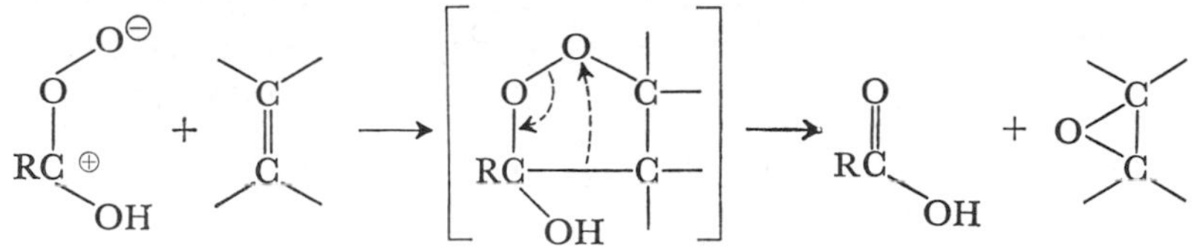

The ease with which epoxidation occurs depends on the groups attached to the olefin. Electron-releasing groups attached to or in close proximity to the doubly bonded carbon atom increase the speed while electron-attracting groups similarly located decrease the speed [10]. In the latter case it may be necessary to employ hydrogen peroxide in a basic medium to produce the oxirane ring [11]. The yields in this synthesis vary from low to high.

(a) Preparation of styrene oxide (69–75% from styrene and perbenzoic acid in chloroform) [12].

(b) Other examples

(1) 1-Octene oxide (80% from 1-octene with 90% hydrogen peroxide, maleic anhydride, and methylene chloride) [13].

(2) 1-Oxaspiro[5.2]octane

(73% from methylenecyclohexane, 50% hydrogen peroxide, benzonitrile, and potassium bicarbonate) [14].

(3) trans-p-Chlorostilbene oxide (88.7% from trans-p-chlorostilbene in chloroform, peracetic acid in acetic acid, and sodium acetate trihydrate) [15].

(4) Ethyl 2,3-epoxybutyrate (75% from 24 moles of ethyl crotonate and 4130 g. of a 22.2% solution of peracetic acid in ethyl acetate at 85° for 5.5 hr.) [6].

(5) 1-Phenyl-2-benzoylethylene oxide (94% from benzalaceto-

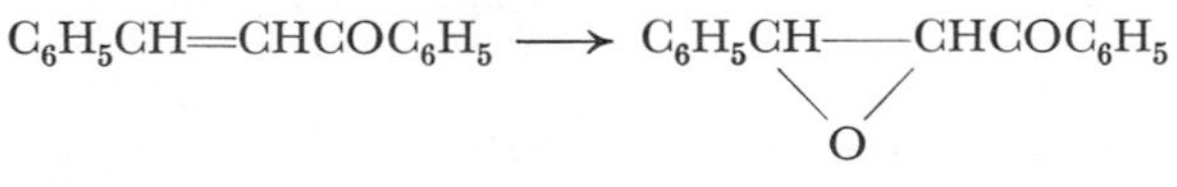

$$C_6H_5CH=CHCOC_6H_5 \longrightarrow C_6H_5CH-CHCOC_6H_5$$

phenone in pyridine to which Clorox (5.25% NaOCl) was added) [4].

(6) 3,4-Epoxycyclooctene (78% from equivalent amounts of 40% peracetic acid and 1,3-cyclooctadiene in methylene chloride and suspended sodium carbonate) [16].

2. Miscellaneous Preparations

One would anticipate the production of ethers by the addition of alkoxy free radicals to olefins (or carbonium ions) or by substitution in alkanes. But addition of alcohols under free radical condition is "of little preparative value to date" [17]. Evidently, the radicals tend to polymerize olefins or to be decomposed themselves by circuitous routes. For example, the major products of the decomposition of di-*t*-butylperoxide are *t*-butyl alcohol, acetone, methane, and isobutylene oxide [18]. Nevertheless, a few isolated examples can be found which might suggest further extensions. They include a rearrangement of triphenylmethyl peroxide, an electrochemical oxidation of an aromatic ether, and an oxidative polymerization of a phenol.

(a) Preparation of diphenyl ether of benzpinacol [19].

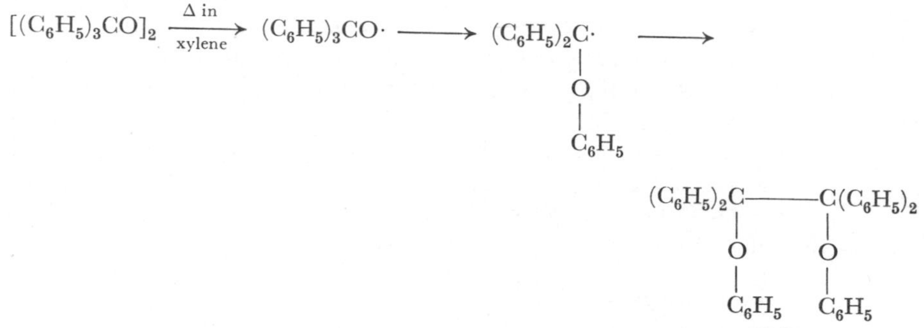

$$[(C_6H_5)_3CO]_2 \xrightarrow[\text{xylene}]{\Delta \text{ in}} (C_6H_5)_3CO\cdot \longrightarrow (C_6H_5)_2\overset{\cdot}{\underset{\underset{C_6H_5}{|}}{\underset{|}{O}}}C\cdot \longrightarrow$$

$$(C_6H_5)_2\overset{}{\underset{\underset{C_6H_5}{|}}{\underset{|}{O}}}C\text{————}\overset{}{\underset{\underset{C_6H_5}{|}}{\underset{|}{O}}}C(C_6H_5)_2$$

60–70%

(b) 3,3,6,6-Tetramethoxy-1,4-cyclohexadiene.   This compound in a

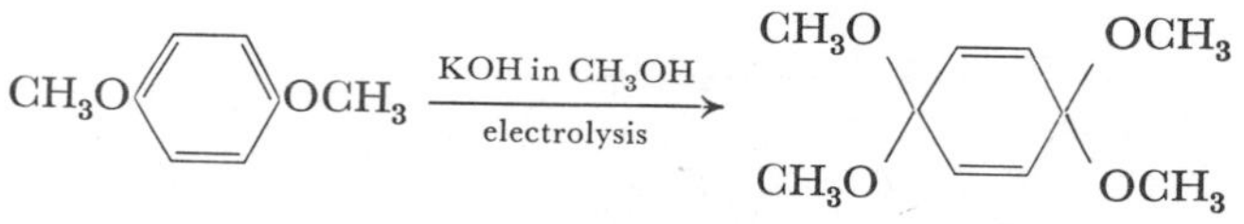

$$\text{CH}_3\text{O}\bigcirc\text{OCH}_3 \xrightarrow[\text{electrolysis}]{\text{KOH in CH}_3\text{OH}}$$

6-VII, 75%

strict sense is not an ether but is included here because of the oxidative nature of the process and the facile transformation of the product to a true ether [20]:

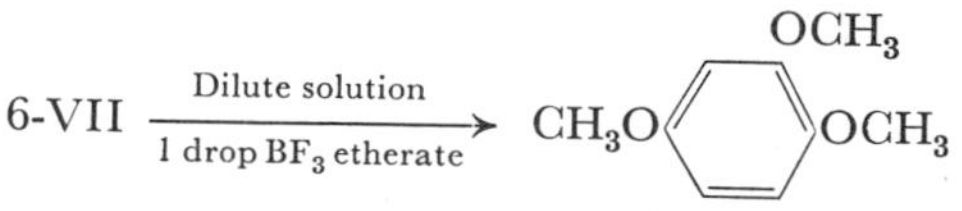

$$\text{6-VII} \xrightarrow[\text{1 drop BF}_3 \text{ etherate}]{\text{Dilute solution}} \text{CH}_3\text{O}\bigcirc\overset{\text{OCH}_3}{\text{OCH}_3}$$

1,2,4-Trimethoxybenzene
81%

(c) N-Methoxymethyl-N-methylaniline [21].

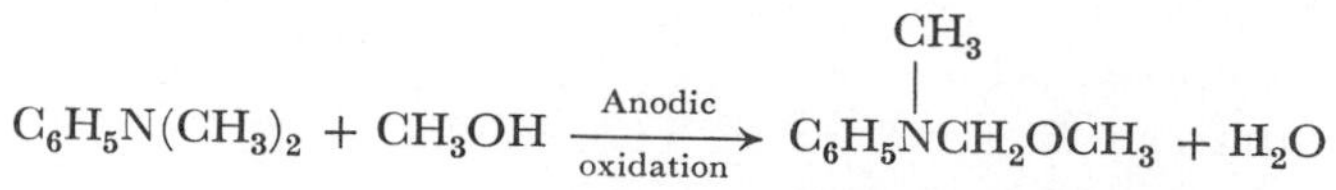

$$C_6H_5N(CH_3)_2 + CH_3OH \xrightarrow[\text{oxidation}]{\text{Anodic}} C_6H_5\overset{\overset{CH_3}{|}}{N}CH_2OCH_3 + H_2O$$

68.7% (Lesser amount of $C_6H_5N(CH_2OCH_3)_2$)

(d) Polyxylenol ether [22].

Nearly quantitative

1. A. Resowsky, *The Chemistry of Heterocyclic Compounds*, Interscience Publishers, 1964, Vol. 19 Pt. 1, p. 1.
2. D. Swern, *Org. Reactions*, **7**, 378 (1953); W. D. Emmons and A. S. Pagano, *J. Am. Chem. Soc.* **77**, 89 (1955).
3. G. B. Payne *et al.*, *J. Org. Chem.*, **26**, 659 (1961).
4. S. Marmor, *J. Org. Chem.*, **28**, 250 (1963).
5. J. Blum, *Compt. Rend.*, **248**, 2883 (1959).
6. P. S. Starcher *et al.*, *J. Am. Chem. Soc.*, **81**, 680 (1959).
7. P. S. Starcher *et al.*, *J. Am. Chem. Soc.*, **79**, 5982 (1957).
8. D. J. Pasto and C. C. Cumbo, *J. Org. Chem.*, **30**, 1271 (1965).
9. H. Kwart and D. M. Hoffman, *J. Org. Chem.*, **31**, 419 (1966).
10. D. Swern, *J. Am. Chem. Soc.*, **69**, 1692 (1947).
11. E. Weitz and A. Scheffer, *Chem. Ber.*, **54**, 2327 (1921).
12. H. Hibbert and P. Burt, *Org. Syn.*, Coll. Vol. **1**, 494 (1941).
13. R. W. White and W. D. Emmons, *Tetrahedron*, **17**, 31 (1962).
14. G. B. Payne, *Tetrahedron*, **18**, 763 (1962).
15. H. O. House, *J. Am. Chem. Soc.*, **77**, 3070 (1955).
16. J. K. Crandall *et al.*, *J. Org. Chem.*, **33**, 423 (1968).
17. G. Sosnovsky, *Free Radical Reactions in Preparative Organic Chemistry*, The Macmillan Co., New York, 1964, p. 122.
18. C. Walling, *Free Radicals in Solution*, John Wiley and Sons, New York, 1957, p. 471.
19. H. Wieland, *Chem. Ber.*, **44**, 2550 (1911).
20. B. Belleau and N. L. Weinberg, *J. Am. Chem. Soc.*, **85**, 2525 (1963).
21. N. L. Weinberg and E. A. Brown, *J. Org. Chem.*, **31**, 4058 (1966).
22. W. A. Butte, Jr., and C. C. Price, *J. Am. Chem. Soc.*, **84**, 3567 (1962).

E. Reduction

The methods of preparation of ethers by reduction are limited. Among the higher oxidation level compounds most likely to be reduced to ethers are the acetals and ketals (E.1) or ketones in the presence of an alcohol (E.2). Less likely candidates are esters and lactones (E.3). Resembling pinacol reduction of carbonyl compounds, a reductive coupling of aldehydes to epoxides should be possible. A reagent, phosphorus hexamethyltriamide, has been found to effect this transformation (E.4), and the variety of α,β-diaryl epoxides available is vastly expanded because of this discovery.

Enol ethers are obvious sources of saturated ethers by reduction. The most important synthesis within this group is the Birch reduction of aryl alkyl ethers to unsaturated cyclohexyl alkyl ethers with alkali metal and liquid ammonia or an amine. This reaction has been discussed previously (Alkenes, Chapter 2, B.3).

References for Section E are on p. 327.

1. From Acetals or Ketals

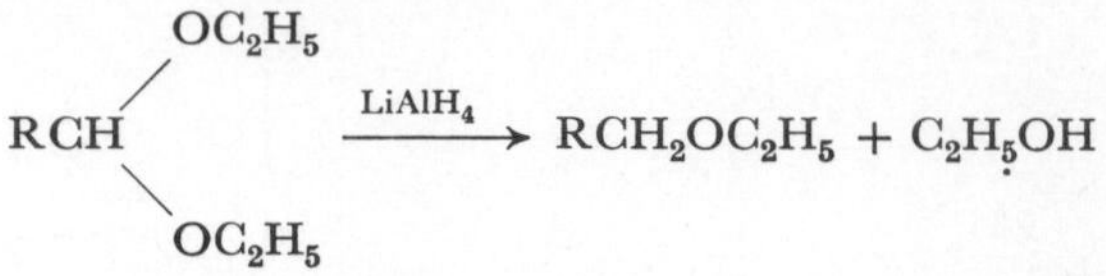

$$\text{RCH} \overset{OC_2H_5}{\underset{OC_2H_5}{<}} \xrightarrow{\text{LiAlH}_4} \text{RCH}_2\text{OC}_2\text{H}_5 + \text{C}_2\text{H}_5\text{OH}$$

A variety of ethers has been synthesized from acetals or ketals with reducing agents such as metal hydrides alone or with aluminum chloride [1–3], trialkylsilanes [4], the Grignard reagent [5], and molecular hydrogen with rhodium as a catalyst [6]. Of these methods the one involving the Grignard reagent is the least satisfactory largely because of the low yields and difficulty of purification. The yields obtained with the other reducing agents are good.

To be more specific, the yields with aluminum chloride-lithium aluminum hydride are generally in the 70–90% range, except when the ether codistills with the by-product alcohol. For best results the ratio of aluminum chloride to lithium aluminum hydride is 4:1 and that of the hydride to the ketal is 1:2 (100% excess). Cyclic ketal 6-VIII with the mixed reducing agent produces the hydroxyether 6-IX in a yield of 94% [7]. This method of synthesis assumes

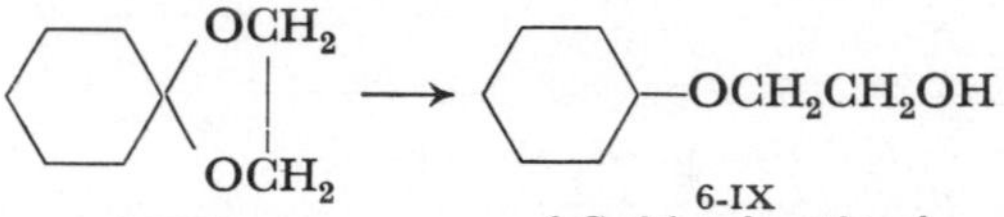

6-VIII

6-IX
2-Cyclohexyloxyethanol

greater importance in view of the recent convenient routes for preparing acetals and ketals [8].

It has been proposed that the reduction with lithium aluminum hydride and aluminum chloride proceeds through the carbonium ion 6-X [9].

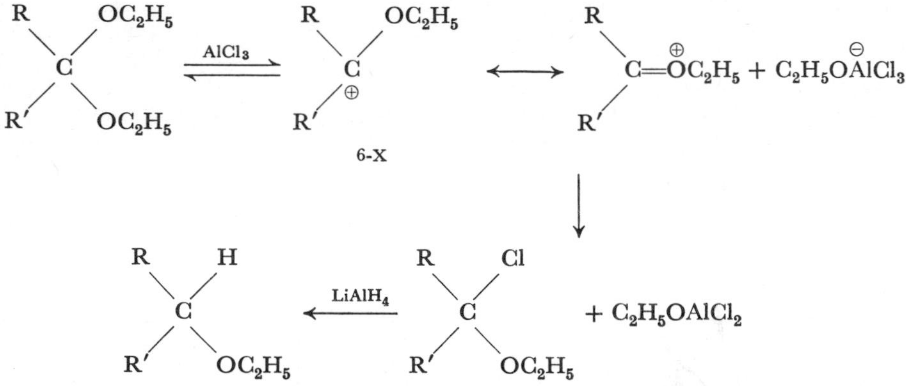

6-X

(a) Preparation of α-phenylethyl ethyl ether. Cold, anhydrous ether, 100 ml., was added cautiously to 13.33 g. of AlCl₃. When the AlCl₃ was dissolved (30 min.), 25 ml. of 1.0 M clear, standardized ethereal lithium aluminum hydride was added with cooling and stirring. After 30 min., 9.7 g. of acetophenone diethyl ketal in 100 ml. of anhydrous ether was added and the mixture was stirred for 2 hr. out of the ice bath. Sulfuric acid, 100 ml. of 10% aqueous, was added, with care at first, to the cool mixture and the aqueous layer was extracted three times with 50 ml. portions of ether. The combined ether layers were dried over anhydrous potassium carbonate, concentrated, and the residue gave 6.14 g. (82%) of ethyl α-phenylethyl ether, b.p. 72–74° (15 mm.), n^{20}D 1.4834 [3].

(b) Other examples

(1) **Butyl ethyl ether** (80% from butyraldehyde diethyl acetal and diisobutyl aluminum hydride) [1].

(2) **Cyclohexyl isopropyl ether** (90% on the basis of the hydrogen consumed from acetone dicyclohexyl ketal and molecular hydrogen in the presence of 5% rhodium on alumina and 1 drop of concentrated hydrochloric acid) [6].

2. From Ketones

$$\text{RCOR}' + \text{R}''\text{OH} \xrightarrow[\text{Pt}]{\text{H}_2, \text{H}^\oplus} \underset{\overset{|}{\text{R}'}}{\text{RCHOR}''}$$

Ethers result when ketones are hydrogenated in the presence of platinum in an acidic alcoholic solution [10]. It appears that the most likely intermediate is the hemiketal. The hydrogenations were conducted under the optimum conditions for cyclohexanone and methanol which were in a molar ratio of 1:15, respectively, and in 2.5 M hydrogen chloride. The alkyl group of the alcohol is attached to the oxygen in the final product. Yields under the conditions chosen for a series of ketones, as determined by gas-liquid chromatography, vary from 39 to 98%.

(a) **Preparation of methyl cycloheptyl ether.** Cycloheptanone, 11.2 g., was hydrogenated with 5% of its weight of platinum oxide and 70 ml. of 2 N methanolic hydrogen chloride at room temperature and atmospheric pressure. The ether recovered amounted to 82% [10].

3. From Esters or Lactones

$$-\text{COOR} \xrightarrow[\text{BF}_3]{\text{NaBH}_4} -\text{CH}_2\text{OR}$$

Esters and lactones in the steroid series have been reduced with boron trifluoride and a metal hydride, sodium borohydride in diglyme-tetrahydrofuran being the most satisfactory [11]. The yields vary greatly in such reductions, the higher ones being obtained with increased branching in the alkyl group of the ester.

(a) **Preparation of 24-*t*-butoxy-5β-cholane.** A solution of 21 g. of boron trifluoride etherate, 2 g. of 5β-cholanic *t*-butyl ester, and tetrahydrofuran-diglyme (35:12.5 ml.) was added dropwise over 15–20 min. to a cooled (ice bath) and stirred mixture (under nitrogen) of 0.37 g. of sodium borohydride

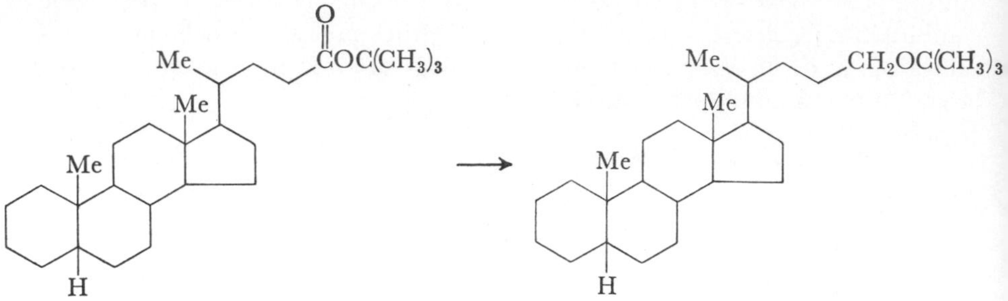

in tetrahydrofuran-diglyme (35:12.5 ml.). The mixture was stirred with cooling for 1 hr. and then heated under reflux for 1 hr. After cooling and cautious addition of 50 ml. of 2 *N* hydrochloric acid and water, the product was extracted with petroleum ether. By chromatographing on activated alumina, the crystalline substance, 1.47 g. (76 %) was eluted with petroleum ether, m.p. 87–89°. Two more crystallizations from ethyl acetate-methanol and two more from acetone gave a m.p. of 95–95.5° [11].

(b) **Preparation of 12α,15-epoxy-12-nor-13β-methyl-11β,14α-abietane** (80 % from dihydroabietic γ-lactone and boron trifluoride etherate-

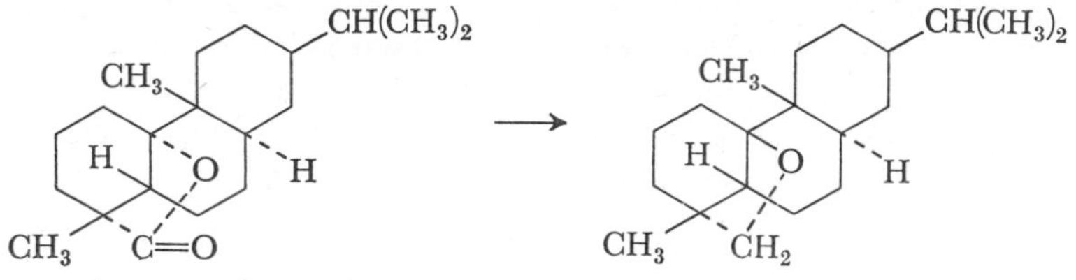

sodium borohydride) [12].

4. From Aldehydes (Semireduction)

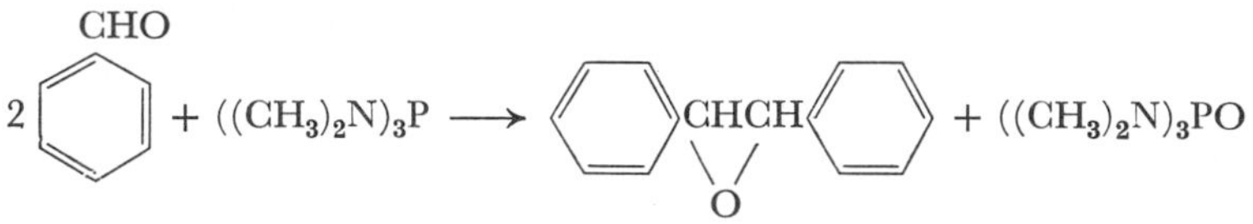

The synthesis of epoxides from aldehydes and hexaalkylphosphorous triamides is a new procedure for forming symmetrical and unsymmetrical epoxides in one step [13]. The reaction has the advantage over the usual epoxide synthesis (from olefins with peroxides or peracids, D.1) in that oxidizable structural features such as thiophene or pyridine rings survive. It is particularly applicable to negatively substituted aromatic and heterocyclic aldehydes. Unfortunately the epoxides obtained consist of mixtures of *cis* and *trans* forms. Yields with a series of aromatic and heterocyclic types vary from 36 to 96 %.

(a) **Preparation of 2,2'-dichloro-α,α'-epoxybibenzyl** (71–81% (50–55% *trans*, 45–50% *cis*) from *o*-chlorobenzaldehyde and hexamethylphosphorous triamide in benzene-ether at 24–36° for 30–50 min. and then at 50° for 15 min.) [14].

1. L. I. Zakharkin and I. M. Khorlina, *Izv. Akad. Nauk SSSR, Otd. Khim. Nauk*, 2255 (1959) *C.A.*, **54,** 10837 (1960).
2. E. L. Eliel and M. Rerick, *J. Org. Chem.*, **23,** 1088 (1958).
3. E. L. Eliel *et al.*, *J. Am. Chem. Soc.*, **84,** 2371 (1962).
4. E. Frainnet and C. Esclamadon, *Compt. Rend.*, **254,** 1814 (1962).
5. I. A. Kaye and I. C. Kogon, *J. Am. Chem. Soc.*, **73,** 4893 (1951).
6. W. L. Howard and J. H. Brown, Jr., *J. Org. Chem.*, **26,** 1026 (1961).
7. R. A. Daignault and E. L. Eliel, *Org. Syn.*, **47,** 37 (1967).
8. N. B. Lorette and W. L. Howard, *J. Org. Chem.*, **25,** 521, 1814 (1960); W. L. Howard and N. B. Lorette, *ibid.*, **25,** 525 (1960).
9. E. L. Eliel, *Record Chem. Progr.* (*Kresge-Hooker Sci. Lib.*), **22,** 129 (1961).
10. M. Verzele *et al.*, *J. Chem. Soc.*, 5598 (1963).
11. G. R. Pettit and D. M. Piatak, *J. Org. Chem.*, **27,** 2127 (1962).
12. G. R. Pettit *et al.*, *J. Org. Chem.*, **26,** 1685 (1961).
13. V. Mark, *J. Am. Chem. Soc.*, **85,** 1884 (1963).
14. V. Mark, *Org. Syn.*, **46,** 31 (1966).

F. Cycloaddition

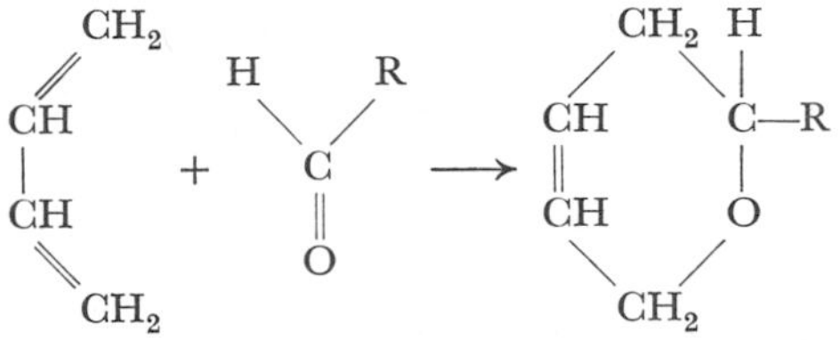

Some of these examples have been mentioned in what seemed to be more appropriate places. The type of syntheses envisioned in this small section is usually of the Diels-Alder nature. For example, furan reacts spontaneously, but mildly, with maleic anhydride to give a quantitative yield of 3,6-endoxo-1,2,3,6-tetrahydrophthalic anhydride [1]. An ether group is not created in this reaction, but certain olefins react with carbonyl compounds to form oxetane derivatives, a reaction which does create an ether group. A number of carbonyl compounds such as formaldehyde [2], chloral [3], carbonyl cyanide and diethyl mesoxalate [4], hexafluoroacetone [5], and hexafluorocyclobutanone [6] have been shown to be dienophiles. The fluoroketones are among the most active of the dienophiles, being excelled only by perfluorothioketones which form the corresponding thioethers [7].

Two other sources of cyclic ethers are 1-(or 2-) methoxybutadienes which react with dienophiles as follows [8]:

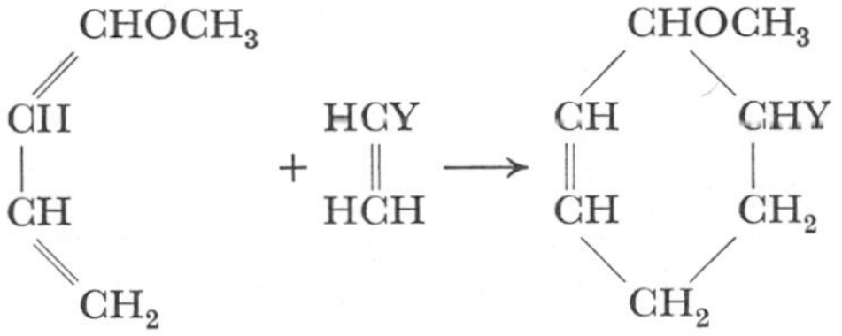

References for Section F are on p. 328.

or vinyl ethers which react with dienes under drastic conditions to form cyclic unsaturated ethers in poor yield [9];

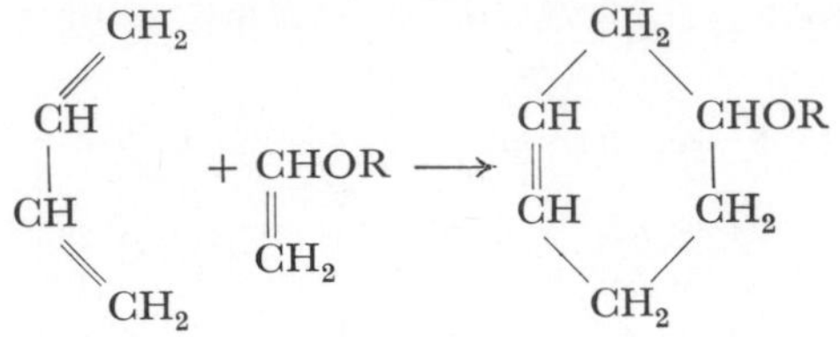

$$
\begin{array}{c}
CH_2 \\
\parallel \\
CH \\
| \\
CH \\
\parallel \\
CH_2
\end{array}
\quad + CHOR \longrightarrow \quad
\begin{array}{c}
CH_2 \\
CH \qquad CHOR \\
\parallel \qquad | \\
CH \qquad CH_2 \\
CH_2
\end{array}
$$

(a) Preparation of 2,2-di(chlorodifluoromethyl)-3-trifluoromethyl perfluorooxetane [10].

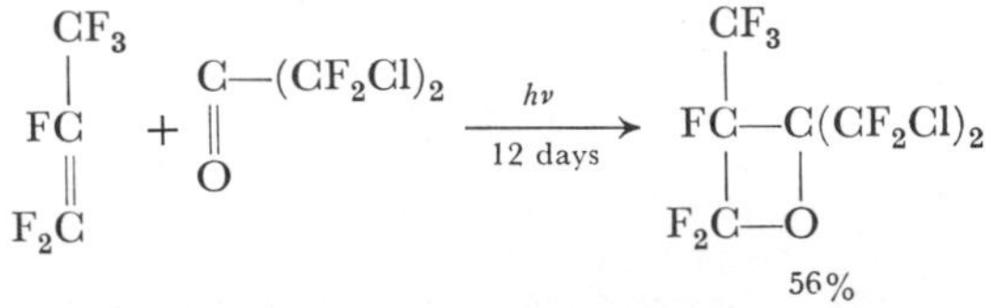

$$
\begin{array}{c}
CF_3 \\
| \\
FC \\
\parallel \\
F_2C
\end{array}
\; + \;
\begin{array}{c}
C-(CF_2Cl)_2 \\
\parallel \\
O
\end{array}
\xrightarrow[\text{12 days}]{h\nu}
\begin{array}{c}
CF_3 \\
| \\
FC-C(CF_2Cl)_2 \\
| \quad | \\
F_2C-O
\end{array}
$$

56%

(b) Preparation of 2,2-dimethyl-3,4-dicyano oxetane [11].

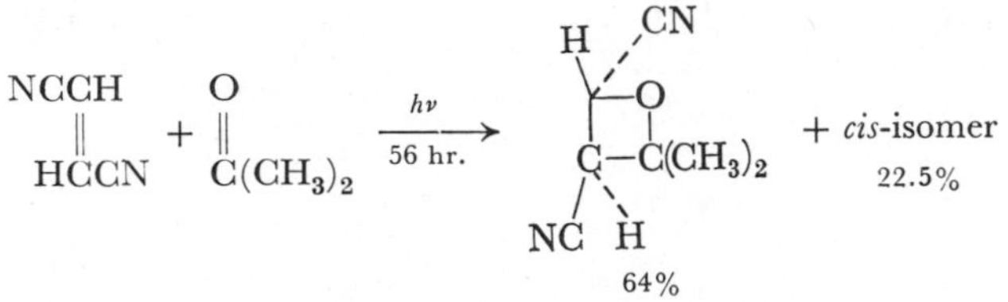

$$
\begin{array}{c}
NCCH \\
\parallel \\
HCCN
\end{array}
\; + \;
\begin{array}{c}
O \\
\parallel \\
C(CH_3)_2
\end{array}
\xrightarrow[\text{56 hr.}]{h\nu}
\begin{array}{c}
H \quad CN \\
\diagdown | \diagup \\
C-O \\
| \quad | \\
C-C(CH_3)_2 \\
\diagup \diagdown \\
NC \quad H
\end{array}
\; + \; cis\text{-isomer}
$$

22.5%

64%

(c) Preparation of 2,4-dimethyl-5,6-dihydropyran (applicable to methylpentadiene but not to butadiene or piperylene) [12].

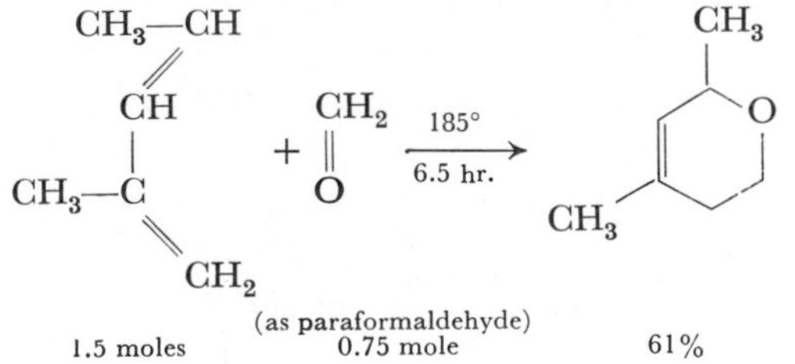

$$
\begin{array}{c}
CH_3-CH \\
\parallel \\
CH \\
| \\
CH_3-C \\
\parallel \\
CH_2
\end{array}
\; + \;
\begin{array}{c}
CH_2 \\
\parallel \\
O
\end{array}
\xrightarrow[\text{6.5 hr.}]{185^\circ}
$$

1.5 moles   (as paraformaldehyde)   61%
                   0.75 mole

1. M. C. Kloetzel, *Org. Reactions*, **4,** 1 (1948).
2. T. L. Gresham and T. R. Steadman, *J. Am. Chem. Soc.*, **71,** 737 (1949).
3. W. J. Dale and A. J. Sisti, *J. Am. Chem. Soc.*, **76,** 81 (1954).
4. O. Achmatowicz and A. Zamojski, *Bull. acad. polon. sci., Classe III*, **5,** 927 (1957); *C.A.*, **52,** 6333 (1958).
5. J. E. Harris, Jr., U.S. Patent 3,136,786, June 9, 1964; *C.A.*, **61,** 4321 (1964).
6. D. C. England, *J. Am. Chem. Soc.*, **83,** 2205 (1961).
7. W. J. Middleton, *J. Org. Chem.*, **30,** 1390 (1965).
8. A. S. Onishchenko, *Diene Synthesis*, D. Davey and Co., New York, 1964, p. 208.
9. Ref. 8, p. 282.
10. J. F. Harris, Jr., and D. D. Coffman, *J. Am. Chem. Soc.*, **84,** 1553 (1962).
11. J. J. Beereboom and M. S. von Wittenau, *J. Org. Chem.*, **30,** 1231 (1965).
12. T. L. Gresham and T. R. Steadman, *J. Am. Chem. Soc.*, **71,** 737 (1949).

Chapter 7

HALIDES

This chapter is subdivided into; A, Displacement; B, Addition to Unsaturated Compounds and Epoxides; C, Aliphatic Substitution; D, Aromatic Substitution; and E, Miscellaneous Reactions. An attempt has been made to cover not only the chlorides and bromides, but also iodides and fluorides. For the latter the most recent review appears to be that of Stephens and Tatlow [1].

A. Displacement

1. From Alcohols and Hydrogen Halides

$$ROH + HX \rightarrow RX + H_2O$$

$$(X = Cl, Br, or I)$$

Although the best way to synthesize primary alkyl chlorides from alcohols is by using thionyl chloride (A.4), the type may be produced in fair yield by the use of concentrated hydrochloric acid and zinc chloride [2]. One disadvantage of the method is that isomeric halides may be produced particularly at high temperatures. Such rearrangements are common in the conversion of branched primary and secondary alcohols into chlorides [3]. Tertiary alcohols are readily transformed into tertiary chlorides by the use of hydrochloric acid without the application of heat [4]. Similarly for benzylic alcohols, such as 1,8-bis-(hydroxymethyl)-naphthalene, concentrated hydrochloric acid alone at low temperature serves well [5]. For reactive alcohols of this nature, including tertiary carbinols, and some bicyclic alcohols that form t-alkyl chlorides, the hydrogenation apparatus of Brown is admirably suited for the introduction of gaseous hydrogen chloride [6]. Concentrated hydrochloric acid (1 ml.) dropped onto stirred sulfuric acid delivers 11 mmoles of hydrogen chloride to the reaction chamber, and the alcohols above are converted into the chlorides in

References for Section A are on pp. 352–356.

74–97% yields in about 100 seconds at 0° with minimum exposure to excess hydrogen chloride.

The use of the hydrogen halide appears to be more common in preparing bromides than chlorides. Hydrogen bromide, 48%, with concentrated sulfuric acid [7] and dry hydrogen bromide [8] seem to be the main reagents which have been employed. The former has been used with success for the lower alcohols but the latter is preferred for higher alcohols [8, 9]. Hydrogen bromide alone has also been used successfully with dihydric alcohols [10, 11]. Phenols are not converted into aryl halides by hydrogen halides. The yields of alkyl halides in these syntheses are usually in the 80–90% range.

Relative to the isomerization problem, it is anticipated that isomerization will occur if the SN_1 mechanism encroaches and if stabler carbonium ions can be formed from the initial species. Table 4 reveals the extent of this isomerization

Table 4

Pentanol	Reagent	% 2-Bromopentane	% 3-Bromopentane
2-	HBr	86	14
3-	HBr	20	80
2-	Aq. HBr-H$_2$SO$_4$	72	28
3-	Aq. HBr-H$_2$SO$_4$	40	60
2-	PBr$_3$	81	19
3-	PBr$_3$	27	73

[12]. 1-Pentanol under the same conditions gives pure 1-bromopentane. The secondary alcohols pass through a transition complex in which bond breaking

$$Br^{\ominus \delta} \dashrightarrow \underset{CH_3}{\overset{R^{\oplus \delta}}{HC}} \dashrightarrow O \overset{H}{\underset{H}{<}}$$

is advanced to the stage where both isomerization and SN_2 displacement take place. Otherwise, if ionization were the sole mechanism (and if hydride shifts are more rapid than the lifetimes of the carbonium ions), identical percentages of products would be expected from both 2- and 3-pentanol. The primary alcohol, 1-pentanol, apparently operates via a pure SN_2 mechanism. However, if some blocking features are incorporated in the structure of the primary alcohol (as well as a driving force to form the more stable t-carbonium ion), the primary alcohol undergoes a Wagner-Meerwein rearrangement:

$$(CH_3)_3CCH_2OH + HBr \rightarrow (CH_3)_3CCH_2^{\oplus} \cdots OH_2 \rightarrow$$

$$(CH_3)_2\overset{\oplus}{C}CH_2CH_3 \rightarrow (CH_3)_2CBrCH_2CH_3$$

To make pure halides from alcohols capable of isomerization or rearrangement, other processes which emphasize the SN_2 or SN_1 mechanism (see A.2) must be employed.

Although hydrogen iodide has been used to convert alcohols into iodides, it is an expensive reagent and its reducing properties tend to affect the iodide formed. Sodium or potassium iodide and 95 % orthophosphoric acid have been shown to be the more satisfactory reagent [13],

$$ROH + KI + H_3PO_4 \rightarrow RI + KH_2PO_4 + H_2O.$$

This reagent converts primary, secondary, and tertiary alcohols into iodides with yields usually around 90 %. The mole ratios employed in the reaction in the case of potassium iodide were 1 of alcohol, 2 of potassium iodide, and 2.9–3.0 of 95 % orthophosphoric acid.

(a) **Preparation of n-butyl chloride** (76–78 % from n-butyl alcohol, concentrated hydrochloric acid, and zinc chloride) [2].

(b) **Other examples**

(1) **n-Dodecyl bromide** (88 % from n-dodecyl alcohol and dry hydrogen bromide) [8].

(2) **1,6-Diiodohexane** (83–85 % from 1,6-hexanediol, 95 % orthophosphoric acid, and potassium iodide) [14].

(3) **Decamethylene bromide** (90 % from decamethylene glycol and dry hydrogen bromide) [10].

2. From Alcohols and Phosphorus Halides

$$ROH + PX_3 \rightarrow ROPX_2 + HX$$

$$X^- \dashrightarrow R^{+\delta} \dashrightarrow \overset{-\delta}{O}PX_2 \rightarrow XR + \overset{\ominus}{O}PX$$

As much as 3 moles of alcohol can be used with 1 mole of the phosphorus trihalide; however, the last mole of alcohol is converted into the halide less easily. The same limitations apply as in the formation of alkyl halides from alcohols and hydrogen halides, namely, that isomerization or rearrangement can take place. Among these displacements recent advances permit some control over either the bond breaking to form SN_1 products or bond forming to yield SN_2 products. One of the mildest of these procedures is that which involves the complex formed between triphenylphosphine and carbon tetrachloride [15].

$$(C_6H_5)_3P + CCl_4 \rightarrow (C_6H_5)_3\overset{\oplus}{P}CCl_3Cl^{\ominus} \xrightarrow{(CH_3)_3CCH_2OH}$$

$$(C_6H_5)_3\overset{\oplus}{P}OCH_2C(CH_3)_3 + (CHCl_3?) + \overset{\ominus}{Cl} \rightarrow (CH_3)_3CCH_2Cl + (C_6H_5)_3PO$$
$$76\%$$

No rearrangement of the neopentyl group occurs. Yields with other alcohols range from 33 to 99 %. Other reagents which will produce neopentyl halides from neopentyl alcohol are phosphorus tribromide in quinoline (47 % yield), [16] and triphenylphosphine dichloride (see Ex. b.3). The latter type of reagent is capable of converting phenols into halides (see Ex. b.7). The results of these methods with neopentyl alcohol and with phenols suggest applicability to other alkyl halide preparations with the promise of less rearrangement or isomerization. Inversion of the carbinol is usually the fate with phosphorus tribromide: the lower the temperature, the cleaner the inversion. For example, it is estimated that the specific rotation of optically active 2-bromobutane should be 39.4° at 25° [17]. The bromide has not been obtained in this state of optical purity by any method of synthesis but the best is as follows:

$$CH_3CHOHCH_2CH_3 \xrightarrow[-15°]{PBr_3} CH_3CHBrCH_2CH_3$$
$$\alpha_D{}^{25} -13.5° \qquad\qquad\qquad \alpha_D{}^{25} +32.09°$$

Other reagents which have been utilized are phosphorus oxyhalide (Ex. b.2), phosphorus pentachloride, triphenyl phosphite halides, $(C_6H_5O)_3PX_2$, [18], triphenyl phosphite alkyl halides [19], diphenyltrichlorophosphorane, $(C_6H_5)_2PCl_3$ (see Ex. b.7) [20], and triphenylphosphine dichloride, $(C_6H_5)_3PCl_2$ [21]. Mixtures such as phosphorus pentachloride and phosphorus oxychloride [22], phosphorus oxychloride and dimethylaniline [23], phosphorus pentachloride and benzoyl chloride [24], and triphenylphosphite and alkyl halides [19] are not uncommon halogenating agents. The latter, in which a trialkyl phosphite is employed, is involved in the Arbusov reaction (see Ex. b.9):

$$\overset{\displaystyle O}{\underset{\displaystyle\uparrow}{}}$$
$$(C_4H_9O)_3P + C_6H_5CH_2Br \longrightarrow C_4H_9Br + C_6H_5CH_2P(OC_4H_9)_2$$
$$\text{Butyl bromide}$$

The phosphorus oxychloride-dimethylaniline mixture appears to be the preferred reagent for converting hydroxypyrimidines and some related nitrogen heterocycles into chlorides (often a difficult conversion).

Phosphorus tribromide alone [25, 26] or in the presence of pyridine [27] has been widely used in the synthesis of bromides. Yields from a series of simple primary and secondary alcohols using only the tribromide vary from 55 to 95%. Again, the newer phosphorus-containing reagents, such as triphenyl phosphite dibromide and triphenylphosphine dibromide, offer some advantage in ease of operation and in yield. The latter with a series of primary, secondary, and tertiary alcohols gives yields of isolated products usually from 75 to 90% [28]. When used with pyridine to neutralize the hydrogen bromide liberated, the former is the preferred reagent for the conversion of primary or secondary acetylenic or allenic alcohols into bromides [29].

Iodides have been produced from a series of primary alcohols by treatment with phosphorus and iodine in yields of over 90% [30]. A less complicated procedure involves the use of triphenyl phosphite-methyl iodide [19], which gives yields of 60 to 95% with a series of primary, secondary, and tertiary alcohols. Glycols and unsaturated alcohols undergo a similar change.

$$(C_6H_5O)_3\overset{\oplus}{P}CH_3I^{\ominus} + ROH \longrightarrow \underset{\underset{\displaystyle R + C_6H_5OH}{\overset{\displaystyle |}{\overset{\displaystyle O}{|}}}}{(C_6H_5O)_2\overset{\oplus}{P}CH_3I^{\ominus}} \longrightarrow RI + \overset{\displaystyle O}{\underset{\displaystyle\uparrow}{}}(C_6H_5O)_2PCH_3$$

For iodides which tend to eliminate to form the olefin, the reagent, *o*-phenylene phosphorochloridite, and subsequent treatment with iodine allow quite mild conditions for displacement and ease in workup [31]: see Ex. b.8:

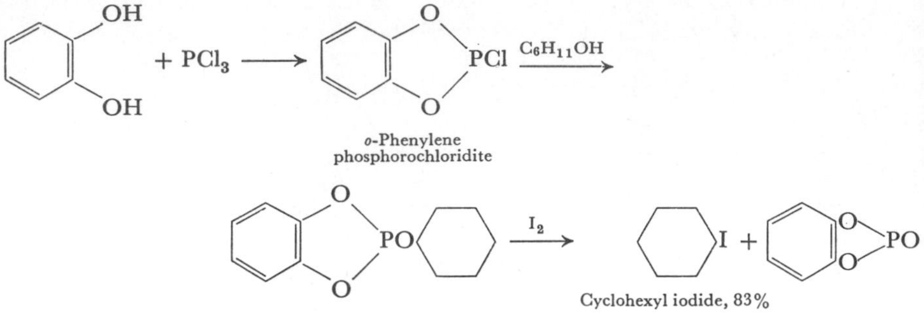

o-Phenylene
phosphorochloridite

Cyclohexyl iodide, 83%

(a) Preparation of *n*-propyl bromide (95 % from *n*-propyl alcohol and phosphorus tribromide) [26].

(b) Other examples

(1) Methyl iodide (95 % from methyl alcohol, iodine, and red and yellow phosphorus) [30].

(2) N-(2-Chloroethyl)aniline hydrochloride (99.5 % from 2-anilino-

$$C_6H_5NHCH_2CH_2OH \cdot HCl \xrightarrow{POCl_3} C_6H_5NHCH_2CH_2Cl \cdot HCl$$

ethanol hydrochloride and phosphorus oxychloride) [32].

(3) Neopentyl chloride (92 % from neopentyl alcohol and triphenylphosphine dichloride in dimethylformamide) [28].

(4) Cyclohexyl bromide (88 % from cyclohexanol and triphenylphosphine dibromide in dimethylformamide [28]; the intermediate

$$(C_6H_5)_3\overset{\oplus}{P}OC_6H_{11}\overset{\ominus}{Br}$$

is too reactive to isolate but one has been isolated in another case) [33].

(5) Allyl iodide (84 % from allyl alcohol and triphenyl phosphite methiodide) [19].

(6) 1,3-Diiodo-2,2-dimethylpropane (75 % from 10 g. of neopentyl glycol, 68 g. of triphenyl phosphite, and 42 g. of methyl iodide heated to 130° and held 36 hr. at this temperature) [34].

(7) *p*-Nitrochlorobenzene (85 % from *p*-nitrophenol and diphenyltrichlorophosphorane in carbon tetrachloride) [20].

(8) Cyclohexyl iodide (83 % based on the alcohol from *o*-phenylene phosphorochloridite and cyclohexanol followed by iodine in methylene chloride at 25° for 6 hr.; the mixture was then washed consecutively with aqueous NaOH, NaHSO$_3$, and NaCl solutions, dried, and distilled) [31].

(9) 1-Chloro-2-iodoethane (85 % from equimolecular amounts of tris-(β-chloroethyl)-phosphite and methyl iodide heated slowly until an exothermic reaction took place and then distilled under reduced pressure) [35].

3. From Carbonyl Compounds

$$\text{ArCOR} \xrightarrow{\text{PCl}_5} \text{ArCCl}_2\text{R}$$

gem-Dihalides may be produced by the action of phosphorus pentachloride on aldehydes or ketones. Other halogenating agents, such as thionyl chloride [36], α,α-dichlorodimethyl ether in the presence of zinc chloride [37], acetyl chloride or acetyl bromide in the presence of the corresponding aluminum halide [38], sulfur tetrafluoride [39], and phenylsulfur trifluoride [40] have also been utilized. The dichlorodimethyl ether-zinc chloride method (see Ex. b.1) is unsatisfactory for aliphatic aldehydes, but the acetyl halide-aluminum halide method has been of value in the synthesis of *gem*-dihalides from halogenated acroleins. The yields vary considerably in these reactions, but values in the 80 and 90 % range are not unusual.

The synthesis has been discussed [41]. Isomeric chloroolefins are common by-products when phosphorus pentachloride is used on carbonyl compounds containing an α-hydrogen. And α-chloroketones can be formed at higher temperatures with phosphorus pentachloride [42]. The *gem*-dihalides are potential intermediates in the synthesis of acetylenes, allenes, xanthene [36], and nor-camphor derivatives [43].

(a) Preparation of norcamphor dichloride.   A solution of 204 g. of

norcamphor in 131 ml. of phosphorus trichloride, cooled in an ice-salt bath, was treated with 435 g. of phosphorus pentachloride in portions over 1 hr. The mixture, allowed to warm slowly to room temperature, was permitted to stand overnight, after which it was poured on 2–3 kg. of ice and extracted with three 1-l. portions of pentane. On washing with water and drying over anhydrous magnesium sulfate, the pentane solution was distilled to remove the solvent and then under reduced pressure until a solid distillate appeared. The solid fraction amounted to 245–260 g. (80–85 %) of the dichloride, b.p. 77–79° (19 mm.); additional material from the forerun increased the yield to 92 % [43].

(b) Other examples

(1) **Dichloromethylbenzene** (97 % from 1 equiv. of benzaldehyde, 1.25 equiv. of α,α-dichlorodimethyl ether, and traces of zinc chloride at 50°. Enolizable ketones yield vinyl chlorides; the dichloromethyl ether is prepared readily) [44].

(2) **1,1-Dibromo-3,3-dichloro-1-propene** (82 % from β,β-dibromo-acrolein, acetyl chloride, and aluminum chloride at room temperature) [38].

(3) **Diphenyldifluoromethane** (97 % from benzophenone and sulfur tetrafluoride in the presence of a trace of hydrogen fluoride at 180° under pressure) [39].

(4) 2,2-Dichloro-3,3-dimethylbutane (48 % crude from equimolecular amounts of phosphorus pentachloride and pinacolone (separated by filtration from 3,3-dimethyl-2-chloro-1-butene) at 0–5°) [45].

4. From Alcohols and Thionyl Chloride

$$ROH + SOCl_2 \xrightarrow{\;C_5H_5N\;} RCl + HCl + SO_2$$

Although this reaction may give a variety of products, particularly in the absence of a hydrogen chloride acceptor [46], it has been used successfully in the syntheses of halides with thionyl chloride alone [47], with thionyl chloride and a trace of a base such as pyridine [48], or with the acid chloride and an equimolar or larger amount of the base. The usual base employed is pyridine, although dimethylaniline [49, 50] has also served in this role. Yields usually run in the 70–90 % range.

With an optically active alcohol, 1 equivalent of pyridine, and thionyl chloride, a Walden inversion occurs. This fact may be accounted for by an SN_2 mechanism [51]:

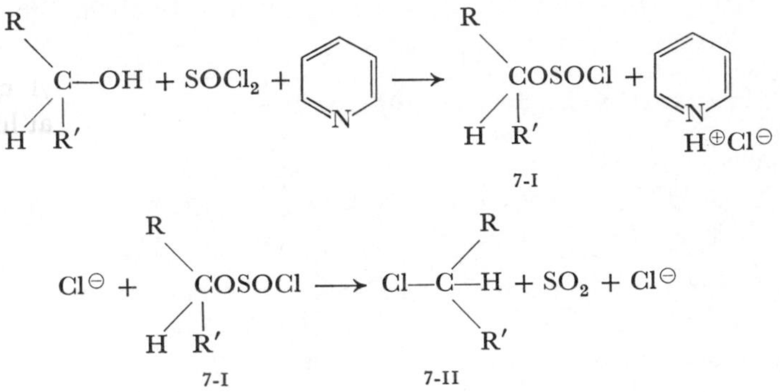

$$\underset{\substack{| \\ H\ R'}}{\overset{\substack{R \\ \diagdown}}{C}}\!\!-OH + SOCl_2 + \bigotimes \longrightarrow \underset{\substack{| \\ H\ R'}}{\overset{\substack{R \\ \diagdown}}{C}OSOCl} + \bigotimes$$

7-I

$$Cl^{\ominus} + \underset{\substack{| \\ H\ R'}}{\overset{\substack{R \\ \diagdown}}{C}OSOCl} \longrightarrow Cl\!-\!\underset{\substack{| \\ R'}}{\overset{\substack{R \\ \diagup}}{C}}\!\!-H + SO_2 + Cl^{\ominus}$$

7-I   7-II

The intermediate chlorosulfite, 7-I, is attacked by the negative chloride ion from the rear to produce the inverted chloride, 7-II.

On the other hand, without a base, retention of configuration is the rule:

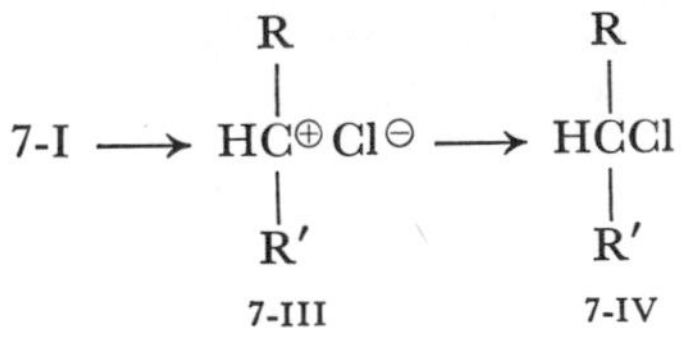

$$7\text{-I} \longrightarrow \underset{\substack{| \\ R'}}{\overset{\substack{R \\ |}}{H}C^{\oplus} Cl^{\ominus}} \longrightarrow \underset{\substack{| \\ R'}}{\overset{\substack{R \\ |}}{H}CCl}$$

7-III   7-IV

proceeding as above or via the SN_i mechanism:

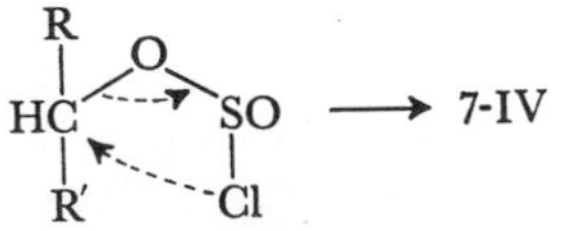

$$\underset{\substack{| \\ R'}}{\overset{\substack{R \\ |}}{H}C}\overset{O}{\diagup\!\!\diagdown}SO \longrightarrow 7\text{-IV}$$

It should be noted that certain halides formed in the use of thionyl chloride may dehydrohalogenate, particularly if pyridine is present, and benzilic acids and like compounds are not converted into halo acids by this reagent. In such cases, phosphorus pentachloride is a better reagent.

The Zollinger reagent, dimethylformamidinium chloride,

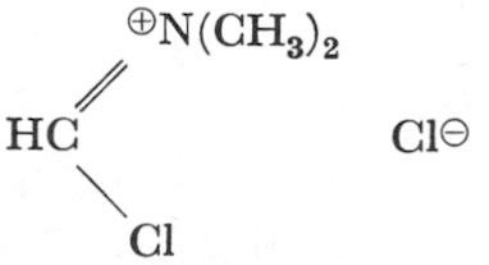

$$\overset{\oplus}{N}(CH_3)_2$$
$$HC\diagdown$$
$$Cl$$ $$Cl^{\ominus}$$

made from thionyl chloride (or phosgene) and dimethylformamide (Acyl Halides, Chapter 15, A.1) has been overlooked in the preparation of alkyl halides but, judging from its reaction in Ex. b.4, it may be of considerable value. To synthesize halides which may react further with dimethylformamide (or with dimethylamine, a hydrolytic product) only catalytic amounts of dimethylformamide should be used. Sometimes very dark reaction mixtures are obtained with thionyl chloride and dimethylformamide, and the difficulty can be traced to impurities in the thionyl chloride which should be purified by the method of L. Friedman [52].

(a) **Preparation of desyl chloride** ($C_6H_5CHClCOC_6H_5$) (74–79 % from benzoin, thionyl chloride, and pyridine) [53].

(b) **Other examples**

(1) **γ-(3,4,5-Trimethoxyphenyl)-propyl chloride** (92 % from γ-(3,4,5-trimethoxyphenyl)-propanol, thionyl chloride, and dimethylaniline) [49].

(2) **β-Dimethylaminoethyl chloride hydrochloride** (87–90 % from β-dimethylaminoethanol and thionyl chloride) [54].

(3) **o-Methylbenzyl chloride** (75–89 % from o-methylbenzyl alcohol and thionyl chloride in benzene containing one drop of pyridine) [55].

(4) **2′,3′,5′-Tribenzoyl-6-azauridine chloride.** If dimethylformamide is used in equimolecular quantities, the chloride atom is replaced by a

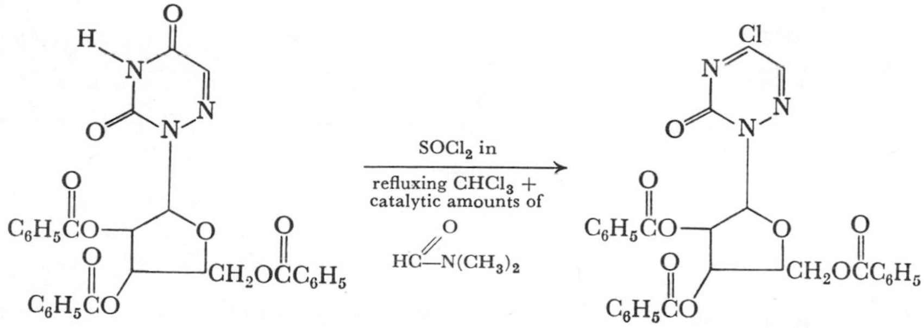

Quantitative yield

dimethylamino grouping [56].

(5) **cis-1,2-Dichlorocyclohexane** (50 % from equimolecular quantities of cyclohexene oxide, pyridine, and thionyl chloride (slight excess) in refluxing chloroform; sulfuryl chloride in place of thionyl chloride gave a 70 % yield of the same product of higher purity; the trans-1,2-dichlorocyclohexane was

prepared in high purity from cyclohexene and phenyliodonium dichloride in refluxing chloroform) [57].

5. From Alcohols via Sulfonates

$$ROH \xrightarrow{R'SO_2Cl} ROSO_2R' \xrightarrow{NaX} RX + NaOSO_2R'$$

In some cases it is desirable to synthesize the halide via the sulfonate. Such is particularly true of some secondary alkyl halides which, when prepared by the action of the phosphorus trihalide, phosphorus pentahalide, or thionyl halide on the secondary alcohol, result not only in low yields, but also in significant amounts of isomeric secondary halides [58]. The route via the sulfonate for primary and secondary alcohols has found favor also in the steroid [59–62] and sugar [63] series. In the first step the alcohol is usually converted into the mesyl or tosyl derivative with methane- or p-toluenesulfonyl chloride, respectively. Various reagents are employed in the second step: lithium chloride in dimethylformamide [63] or ethanol [64] and pyridine hydrochloride in dimethylformamide [62] for the formation of chlorides; hydrogen bromide, 48 % [65], sodium bromide in dimethylformamide or dimethylsulfoxide [58], or in diethylene glycol [68], calcium bromide in dimethylformamide [66], in 2-(2-ethoxyethoxy)ethanol [67], and anhydrous magnesium bromide in ether (for long-chain, unsaturated alkyl mesylates) [69] for bromides; sodium iodide in acetone [59, 70, 71], in butanone [60], in acetonylacetone [72], and potassium iodide in water [73] for iodides; potassium fluoride hydrate in methanol heated under pressure [74] or with anhydrous potassium fluoride in glycol for the conversion of sulfonates of carbohydrates into fluorides [75]. The method is neither applicable to the conversion of optically active alcohols to active halides [58] nor to alcohols sterically hindered to nucleophilic attack [66]. Yields vary from approximately 60 to 90 %.

(a) **Preparation of pentaerythrityl tetrabromide** (68–78 % from 90 % pentaerythritol and benzenesulfonyl chloride followed by treatment with sodium bromide in diethylene glycol) [68].

(b) **Other examples**

(1) **Bicyclo[2.1.1]hexane-1-methyl bromide** (90 % from bicyclo-

[2.1.1]hexane-1-methanol and tosyl chloride in pyridine followed by treatment with lithium bromide in acetone) [76].

(2) **1,1-Bis(iodomethyl)cyclopropane** (79.5 % from 1,1-bis(hydroxymethyl)cyclopropane dibenzenesulfonate and sodium iodide in acetone) [70].

(3) **3-Bromopentane (85 % from 3-amyl tosylate and sodium bromide in DMSO,** dried by partial freezing, and the mixture shaken for 90 hr.; the crude product in ether was washed with cold sulfuric acid; the amyl

tosylate was prepared in 78% yield from 3-pentanol and tosyl chloride in pyridine at 0°; the ester could not be distilled without decomposition) [58].

6. From Halides (Finkelstein Halide Interchange)

$$(Ar)RCl + NaI \xrightarrow{CH_3COCH_3} (Ar)RI + NaCl$$

$$(Ar)RCl + AgF \longrightarrow (Ar)RF + AgCl$$

The interchange of one halogen for another has been accomplished with aliphatic, aromatic, and heterocyclic halides. Perhaps the most common of these is the interchange of chlorine or bromine with iodine by the use of sodium iodide in a solvent such as acetone or methyl ethyl ketone [77–80], ethanol [81], carbon tetrachloride [82], or dimethylformamide [83]. The exchange is more complete if the inorganic halide, such as sodium iodide, is soluble and the exchanged inorganic halide, such as sodium bromide or chloride, is insoluble in the organic solvent. Otherwise, a large excess of the inorganic halide must be used to drive the reaction anywhere near completion. If the exchange is slow, the temperature should be raised by selection of a higher-boiling solvent such as n-butyl alcohol, or resort should be made to the exchange of tosylates with inorganic halides (A.5), or of aromatic halides with cuprous halide (see Ex. b.5):

$$ArBr + Cu_2Cl_2 \xrightarrow[DMSO]{Picoline\ or} ArCl + Cu_2ClBr$$

The Finkelstein interchange in acetone has been considered a reliable SN_2 reaction in which primary are more reactive than secondary halides, which in turn are more reactive than tertiary halides. But it has been pointed out that t-butyl bromide exhibits SN_1 behavior in this solvent—only a 3% yield of t-butyl chloride was obtained in exchange with lithium chloride, the rest being iso-butylene [84].

Activated aromatic halides exchange with sodium halides in dimethylform-amide (see Ex. b.2, b.3), 2- or 4-halopyridines or quinolines with concentrated acid (see Ex. b.6), or other haloheterocycles with traces of acid and sodium iodide (see Ex. b.7). The trace of acid is necessary to give activation:

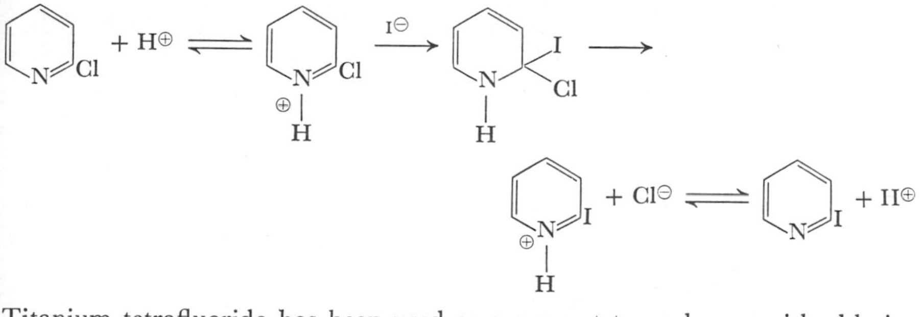

Titanium tetrafluoride has been used as a reagent to exchange with chlorine in a chloroborazine [85]. Aluminum chloride converts alkyl bromides into chlorides at a rate faster than their rearrangement [86]:

$$CH_3CH_2CH_2Br + AlCl_3 \xrightarrow{0°} CH_3CH_2CH_2Cl \xrightarrow{More\ slowly} CH_3CHClCH_3$$

This reaction would be insignificant from the synthetic standpoint were it not for the fact that aluminum chloride (and boron trichloride) has a strong affinity for fluorine and brings about extensive exchange with fluoro and per-fluoro compounds [87]:

$$CF_2{=}CF_2 \xrightarrow[\text{Autoclave}]{\text{AlCl}_3} CFCl{=}CCl_2$$

Of the other interchanges employed, those in which fluorides are produced have been of greatest interest. A review of this reaction is available [88]. In this synthesis, potassium fluoride, zinc fluoride, antimony fluoride, hydrogen fluoride, or bromine trifluoride (see Ex. b.9) have been the principal reagents used. The presence of a small amount of a pentavalent antimony salt to give the so-called Swartz reagent often increases the speed of the reaction and improves the yield. This reagent is usually obtained by adding the free halogen, often chlorine, to antimony trifluoride. Although the reaction may be carried out with the halide and the metallic fluoride at high temperature, if necessary, under pressure, in many cases the procedure may be simplified by employing a solvent at ordinary pressure. In the synthesis of a series of alkyl fluorides from alkyl bromides using potassium fluoride in ethylene glycol, the yields varied from 27 to 46% [89]. In the aromatic series some activation by an electron-withdrawing group, such as the nitro group in the o- or p-position, is necessary for an interchange to occur between the aryl halide and the fluoride ion [90]. With a series of o- or p-monosubstituted chlorobenzenes and potassium fluoride in dimethylformamide or dimethyl sulfoxide, the yields varied from 10 to 81% [91]. 2,4-Dinitrochlorobenzene alone or in a series of solvents gave yields of 2,4-dinitrofluorobenzene varying from 57 to 81%.

Recently it has been shown that dimethyl sulfone is a superior solvent in converting nitrochlorobenzenes or chloropyridines to the fluoro analogs [92]; however, the nitrochlorobenzenes are convertible equally well, if not better, in dimethyl sulfoxide [93]. Silver fluoride is the preferred reagent in converting chloro-s-triazines to fluoro-s-triazines [94].

The solvent of choice in potassium fluoride exchange with perchlorides of three or more carbon atoms is N-methyl-2-pyrrolidone. Extensive fluorine exchange can be brought about with perhalides (see Ex. b.8). Sodium fluoride is rarely used because of its insolubility in solvents of any kind but, even when employed heterogeneously, it yields useful fluorinating agents [95]: For aro-

$$SCl_2 + NaF \xrightarrow[\text{CH}_3\text{CN}]{80°} SF_4$$
$$90\%$$

$$SOCl_2 + NaF \xrightarrow[\text{CH}_3\text{CN}]{80°} SOF_2$$
$$77\% \text{ conversion}$$

matic chlorides and bromides potassium fluoride in sulfolane appears to be the preferred reagent [96].

(a) Preparation of γ-iodobutyronitrile. γ-Bromobutyronitrile, 148 g., was added to a solution of 160 g. of sodium iodide in 825 ml. of dry acetone. After standing for 2 hr. the acetone was distilled off and the product was dissolved in benzene, which solution was washed with sodium bisulfite and vacuum-distilled to give 189 g. (96 %) of γ-iodobutyronitrile, b.p. 73–74° (0.5–1.0 mm.) [78].

(b) Other examples

(1) **n-Hexyl fluoride** (40–45 % from n-hexyl bromide and anhydrous potassium fluoride in dry ethylene glycol) [89].

(2) **2,4-Dinitroiodobenzene** (71 % from 2,4-dinitrochlorobenzene and sodium iodide in dimethylformamide) [97].

(3) **2-(Trifluoromethyl)-4-nitrofluorobenzene** (81 % from 2(trifluoromethyl)-4-nitrochlorobenzene and dry potassium fluoride in dimethylformamide) [91].

(4) **2,4-Dinitrofluorobenzene** (92 % from heating dry potassium fluoride and 2,4-dinitrochlorobenzene at 200°; solvents lower the yield drastically) [98].

(5) **1-Chloro(Cl^{36})-naphthalene** (quantitative yield by refluxing 0.01 mole of 1-bromonaphthalene and 0.011 mole of $CuCl^{36}$ in 25 ml. of DMSO for 1 hr. over N_2) [99]; α-picoline also can be used as a solvent, and a small amount of $CuCl_2$ mixed with Cu_2Cl_2 removes Cu, which tends to produce the hydrocarbon from the halohydrocarbon) [100].

(6) **4-Bromo-7 chloroquinoline** (77 % from 7.14 g. of 4,7-dichloroquinoline refluxed in 65 ml. of freshly distilled hydrobromic acid, concentrated to $\frac{1}{2}$ volume, cooled and made alkaline, filtered, dried, and recrystallized from hexane) [101]. The purity can be checked by nmr: the doublet for the 2-H is at 8.6 and 8.67 ppm for the 4-bromo and at 8.7 and 8.77 ppm for the 4-chloro derivative, both in $CDCl_3$ [101a].

(7) **3,6-Diiodopyridazine**

(74 % from 0.1 mole of 3,6-dibromopyridazine and 0.2 mole of sodium iodide in refluxing acetone to which 6 drops of 50 % hydriodic acid in 15 ml. of acetone was added in three spaced intervals) [79].

(8) **1,1,1,3,3,3-Hexafluoro-2,2-dichloropropane** (69 % from 1 mole of perchloropropane and 9 moles of KF in 1650 ml. of N-methyl-2-pyrrolidone held at 195° while the product distilled out; Hexachlorobutadiene similarly gave 65 % heptafluoro-2-butene:

$$CCl_2{=}CCl—CCl{=}CCl_2 \rightarrow CF_3CH{=}CFCF_3$$

and perchlorocyloalkenes were converted into perfluorocycloalkenes in good yields; dehydrochlorination may take place with compounds such as hexachlorocyclohexane; the reaction conditions were not suitable for exchange of chlorides with only 1 or 2 carbon atoms) [102].

(9) **1,1,1,2-Tetrafluoro-2-bromoethane** (85% from dropwise addition of 1 mole of 1,1,1-trifluoro-2,2-dibromoethane to 0.4 mole of bromine trifluoride in 75 ml. of bromine. The mixture was then warmed to 50° to distill out the product, b.p. 8–9°). Bromine trifluoride probably behaves as follows [103]:

$$2\ BrF_3 \rightleftharpoons \overset{\oplus}{BrF_2}\overset{\ominus}{BrF_4}$$

$$\overset{-\delta}{BrF_4} \rightarrow RBr \rightarrow \overset{+\delta}{BrF_2} \rightarrow FR + BrF_3 + Br_2F_2$$

(10) **Pentaerythrityl tetraiodide** (89–98% from the tetrabromide and NaI in methyl ethyl ketone) [104].

7. From Iminoesters

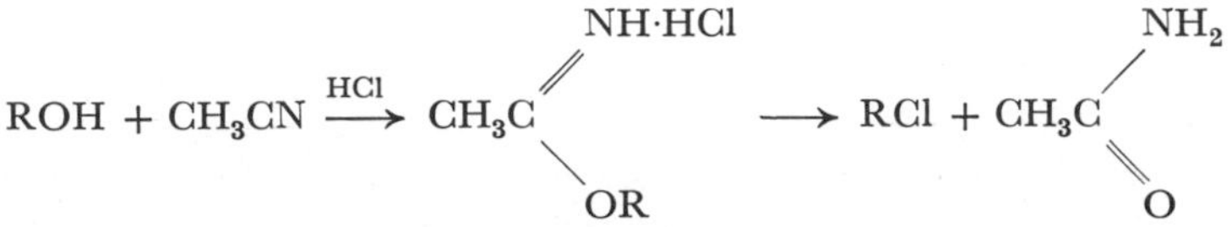

$$ROH + CH_3CN \xrightarrow{HCl} CH_3C\overset{\displaystyle NH\cdot HCl}{\underset{\displaystyle OR}{\big\langle}} \longrightarrow RCl + CH_3C\overset{\displaystyle NH_2}{\underset{\displaystyle O}{\big\langle}}$$

It is well known that iminoesters on pyrolysis give halides and amides. It appears that the mechanism of the reaction [105] is a bimolecular displacement

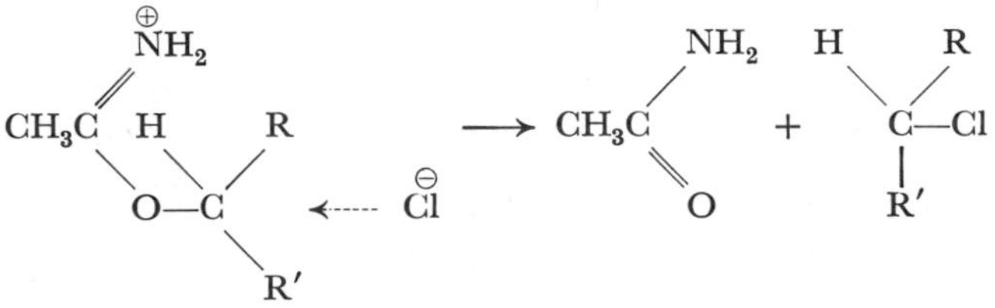

which should produce, from an optically active alcohol, a chloride of the opposite configuration. The experimental results with *sec*-butyl iminoacetate hydrochloride support this view in that *sec*-butyl chloride of the opposite configuration and rotation from that of the corresponding alcohol is obtained. Thus a method is claimed to be available for preparing optically active chlorides in satisfactory yield by some authors and in "disappointing yield" by others [58].

(a) **Preparation of (−)2-butyl chloride.** The dry iminoester hydrochloride, 5.1 g., obtained from (+)2-butanol was pyrolyzed at 130° to give 2.42 g. (78%) of (−)-2-butyl chloride, n^{25}_D 1.3941, $[\alpha]^{25}_D$ −31.2° [105].

8. From Ethers

$$ArOR + HX \rightarrow ArOH + RX$$

Although this synthesis is of more importance in the preparation of phenols (see Chapter 5, A.5), it may be used in the preparation of alkyl halides from aryl alkyl ethers or even from dialkyl ethers. In the case of methyl ethers, the reaction, when hydriodic acid is the reagent, is quantitative and is, in fact, the well-known Zeisel method for the determination of the methoxy group. Advantage is taken of the synthesis in reactions in which halogen atoms would normally be affected. To eliminate the latter possibility the halogen is replaced by the rather unreactive alkoxy group, which in the final step is reconverted into the original halide. In this way polyfunctional types such as halogenated acids [106] and halogenated amines [107] may be prepared.

The usual cleaving reagent for the synthesis of bromides and iodides is the hydrogen halide. Hydrogen bromide may be used alone [108] and in the presence of acetic anhydride [109].

Recently triphenylphosphine dibromide has been employed to cleave dialkyl ethers [100]. To prevent the undesirable reducing action of hydrogen iodide, potassium iodide and phosphoric [111] or polyphosphoric acid [112] are employed to prepare iodides. Acetyl chloride alone [113] or with stannic chloride [114] and α,α-dichlorodimethyl ether with zinc chloride [115] may be used to prepare chlorides. A very practical synthesis of bridgehead halogen compounds utilizes the stannic chloride-acetyl chloride combination (see Ex. b). The yields in the synthesis of these halides are usually in the range 70–90 %.

The synthesis may also serve to prepare dihalides from cyclic ethers such as tetrahydrofurans [116, 117, 111] and tetrahydropyrans [118].

$$\text{(tetrahydrofuran)} \xrightarrow{\text{HX}} XCH_2CH_2CH_2CH_2X$$

Hydrogen chloride alone gives the chlorohydrin [119], but hydrogen chloride with zinc chloride [116], hydrogen bromide with sulfuric acid [117], phosphorus oxychloride with sulfuric acid [120] and potassium iodide with phosphoric acid [111] give the corresponding dihalides. If the cyclic ether is treated with an acid chloride in the presence of zinc chloride, chloroesters [121] are obtained. This type sometimes serves for the preparation of dihalides [118]. The yields

$$\text{(tetrahydropyran)} \xrightarrow[\text{ZnCl}_2]{\text{CH}_3\text{COCl}} CH_3CO_2(CH_2)_5Cl \xrightarrow[\text{H}_2\text{SO}_4]{\text{HBr}} Br(CH_2)_5Cl$$

5-Bromopentyl chloride

in the synthesis of dihalides in one operation usually vary from 75 to 90 %.

The mechanism in these transformations, using the hydrogen halide as the reagent, appears to follow the pattern discussed under Phenols, Chapter 5, A.5.

A relatively new direct method of splitting ethers to halides is via borate esters according to the following stoiochiometry [see Ex. c]:

$$3\,ROR + 2\,X_2 + NaBH_4 \rightarrow 3\,RX + B(OR)_3 + NaX + 2\,H_2$$

However, isomerization may occur in this process.

(a) **Preparation of benzyl chloride.** Benzyl methyl ether, 1 mole, α,α-dichlorodimethyl ether, 1.25 mole, and a trace of zinc chloride were heated at 95–100° for 2 hr. By distillation the chloride was recovered in 88 % yield, b.p. 66–67° (14 mm.) [115].

(b) Preparation of 1-chloro-4-methylbicyclo [2.2.2] octane,

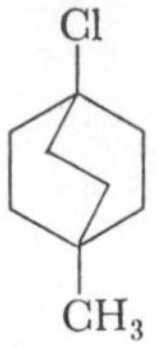

1-Methoxy-4-methylbicyclo [2.2.2] octane, 0.01 mole, was dissolved in 0.02 mole of acetyl chloride, 10 drops of stannic chloride was added with ice cooling, and the mixture was allowed to warm to room temperature. The usual workup to remove acidic material yielded 80% of the desired compound. Thionyl chloride or benzenesulfonyl chloride could be used in place of acetyl chloride. This synthesis is important because the bridgehead ethers are available from Diels-Alder reactions with 2-methoxy-1,3-butadiene [122].

(c) Preparation of 2-iodobutane.

A cold mixture of 11 g. of ICl and 7.8 g. of n-butyl ether was added dropwise to 2 g. of $LiBH_4$ under nitrogen and allowed to warm to 25° for 30–60 min. Dilution with water, extraction, and the usual cleansing of the extract gave 93% of the titled compound. Alcohols also can be converted into halides by this process [123].

(d) Other examples

(1) **Dodecamethylene dibromide** (86% from 1,12-di-o-methoxy-phenoxydodecane and 50% aqueous hydrobromic acid) [109].

(2) **Isopropyl iodide** (90% from diisopropyl ether, potassium iodide, and 95% phosphoric acid) [111].

(3) **1,4-Dibromopentane** (80% from tetrahydrosylvan and 40% hydrobromic acid) [117].

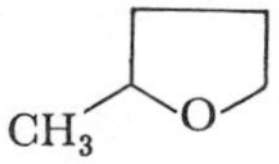

(4) **1,4-Diiodobutane** (96% from THF, potassium iodide, and 95% phosphoric acid) [111].

9. From Diazonium Salts (Sandmeyer)

$$ArN_2X \xrightarrow[HCl]{CuCl} ArCl + N_2 + CuX$$

This synthesis permits the introduction of a halogen for the N_2X group. In fact, since the N_2X group is normally obtained from an amino group which in turn is normally obtained from a nitro group, the method serves for substituting a halogen for a nitro or amino group. Thus the synthesis is of value in obtaining an orientation sometimes not readily achieved by other methods. Chlorides or bromides may be obtained by treating the diazonium salt (Sandmeyer reaction) with the cuprous halide in the presence of the corresponding halogen acid. Yields by this method for a series of toluidines are in the range 70–79% (based on the original amine) [124, 125]. Another method (Gattermann) consists in treating the diazonium salt with copper and the halogen acid. A third related method proceeds via the perbromide obtained by treating the amine hydrobromide with nitrogen trioxide [126] or bromine, sodium nitrite, and hydrobromic acid [127]. By the latter procedure 2-bromopyridine has been obtained in 86–92% yield.

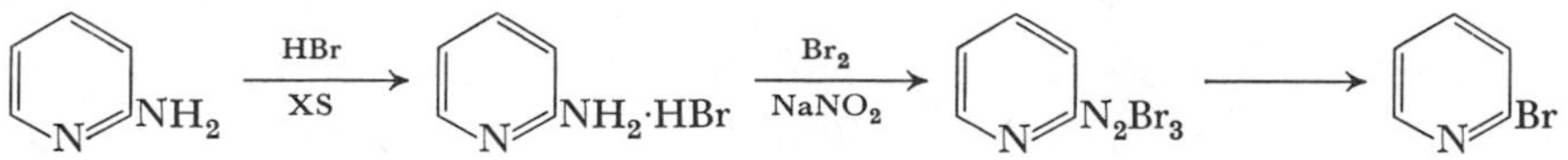

Iodides are obtained from the diazonium salt simply by the addition of an iodide such as potassium iodide [128], in which case the reaction occurs via the triiodide ion, $I_3^{\ominus}$ [129]. The yield in this reaction may be improved by use of the stannic chloride addition product of the diazonium salt and iodine in the presence of ultraviolet light [130].

Still another variation is the transformation of an alkylaminoazobenzene to the halide, thus extending the diazonium reaction to conversion of an aliphatic amine to a halide [131]:

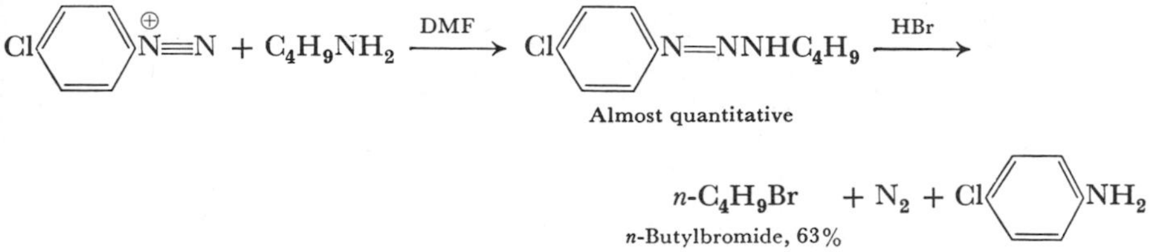

Almost quantitative

$$n\text{-}C_4H_9Br \quad + N_2 + Cl\!\!\left\langle\;\right\rangle\!\!NH_2$$

n-Butylbromide, 63%

Similarly, N-alkylaminoazobenzenes can be converted into esters by treatment with carboxylic acids.

The customary method of synthesizing fluorides is by the Schiemann reaction which has been reviewed [132]. In this method the diazonium tetrafluoroborate is first produced from the amine by diazotizing in the presence of the tetrafluoroborate ion:

$$ArNH_2 + HNO_2 + B\overset{\ominus}{F_4} \rightarrow ArN_2BF_4 + H_2O + \overset{\ominus}{OH}$$

The diazonium tetrafluoroborate, being comparatively stable, may be isolated with safety and, since it usually decomposes at a definite temperature in a clear-cut manner, it serves as a source of the fluoride:

$$ArN_2BF_4 \rightarrow ArF + N_2 + BF_3$$

Decomposition may also be accomplished in an inert solvent, such as toluene or xylene [133]. Yields of fluorides as high as 70% are not uncommon by this method, and it is applicable to the preparation of fluorides in the aromatic and N-heterocyclic series. An improvement over the Schiemann procedure consists in using the aryldiazonium hexafluorophosphate rather than the aryldiazonium tetrafluoroborate as an intermediate [134]. These salts are less soluble than the tetrafluoroborates and can thus be obtained in better yield. In addition, the pyrolysis of the hexafluorophosphate usually gives a better return than the pyrolysis of the tetrafluoroborate. For example, the overall yield in the preparation of o-bromofluorobenzene by the modified procedure is 73–75 % as compared to 40 % by the Schiemann procedure (see Ex. b.4) [135]:

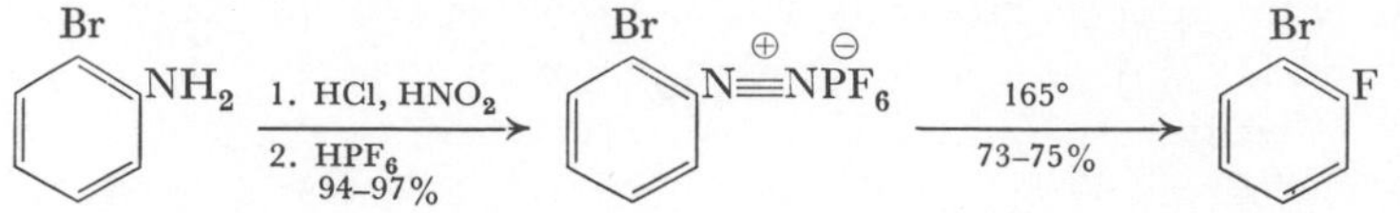

Recently tetramethylurea, a water-miscible solvent, has been shown to serve effectively as a decomposition solvent, particularly in cases in which the aromatic ring of the hexafluorophosphate contains electron-withdrawing substituents [136].

Fluorides may also be prepared from the simple diazonium fluoride by the use of aqueous or anhydrous hydrofluoric acid, sometimes in improved yields, but the method requires special apparatus because of the difficulties involved in handling this acid [132, 137].

It appears that the first step in the Sandmeyer reaction is the formation of $\overset{\ominus}{CuCl_2}$, which reacts with the diazonium ion in the rate-determining step to form

$$\overset{I}{CuCl} + \overset{\ominus}{Cl} \rightarrow \overset{I}{Cu}\overset{\ominus}{Cl_2}$$

$$\overset{\oplus}{ArN_2} + \overset{I}{Cu}\overset{\ominus}{Cl_2} \rightarrow Ar\cdot + N_2 + \overset{II}{CuCl_2}$$

$$\overset{II}{Ar\cdot + CuCl_2} \rightarrow ArCl + \overset{I}{CuCl}$$

an aryl free radical and cupric chloride, which in the third step combine to give the halide and cuprous chloride. Indeed, a major portion of the cuprous halide may be replaced by cupric halide without reducing yields [138]. Thus the copper is oxidized in the second step and reduced in the third. The mechanism of the Schiemann reaction has not been clarified [132].

The scope of diazonium salt preparations has been widened considerably by the discovery that nitroso compounds prepared by nitrosation of hydrocarbons are reduced to diazonium salts by excess nitrous acid or by its anhydride, N_2O_3 [139]:

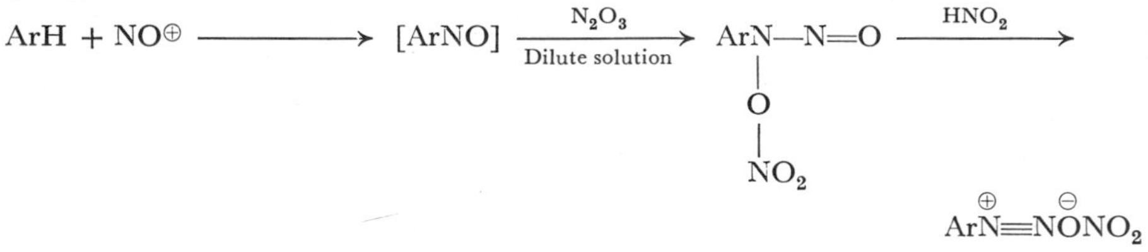

The diazonium salt can be treated with the usual reagents to prepare aryl halides. Three different reagents are employed in the diazotization as shown:

1. For readily substituted aromatic compounds, such as phenols and tertiary aromatic amines, dilute nitrous acid (at least 3 equivalents).

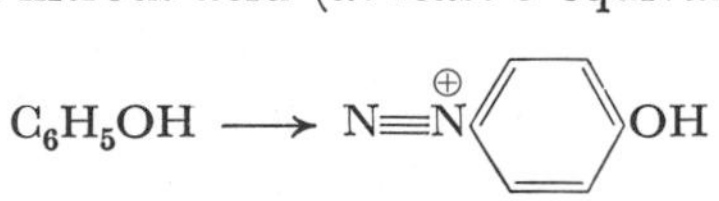

2. For aromatic ethers or polyalkylbenzenes, nitrous acid in strongly acidic media such as concentrated sulfuric acid.

3. For aromatic compounds with electron-withdrawing groups, nitrous acid and catalytic amounts of the mercuric ion in concentrated sulfuric acid.

Since benzynes can be produced by diazotization of anthranilic acids, they serve as intermediates in the preparation of aromatic o-dihalides (see Ex. b.5).

(a) **Preparation of p-chlorotoluene** (70–79% from p-toluidine, sodium nitrite, and hydrochloric acid at 0–5° followed by the addition of cuprous chloride) [125].

(b) **Other examples**

(1) **p-Bromotoluene** (70–73% from p-toluidine, sodium nitrite and sulfuric acid at 15–20° followed by addition to a cuprous bromide-hydrobromic acid solution) [124].

(2) **3-Methoxy-2-nitroiodobenzene** (88% from 3-methoxy-2-nitro-aniline in glacial acetic acid, potassium nitrite, concentrated sulfuric acid, and water at low temperature, followed by addition of potassium iodide) [140].

(3) **p-Fluoroacetanilide** (82% from N-acetyl-p-phenylenediamine, hydrochloric acid, sodium nitrite, and fluoroboric acid at 0° with subsequent stirring of the aqueous solution at room temperature in the presence of copper powder) [141].

(4) **1-Bromo-2-fluorobenzene** (73–75% from o-bromoaniline, sodium nitrite, hydrochloric acid, and hexafluorophosphoric acid at −5 to −10°, after which the hexafluorophosphate salt formed was decomposed by introduction into mineral oil at 165–170°) [135].

(5) **o-Diiodobenzene** (67% from 1 equiv. of the diazonium salt of anthranilic acid and 1 equiv. of iodine in refluxing chloroform; about 2% of o-chloroiodobenzene was formed simultaneously) [142].

10. From Amides (von Braun)

$$\text{RNHCOR}' \xrightarrow{\text{PCl}_5} \text{RCl} + \text{R}'\text{CN} + \text{POCl}_3 + \text{HCl}$$

This synthesis has found limited use in spite of the fact that yields from the N-alkylamides are sometimes in the range 80–90%. As a rule, benzamides rather than nonaromatic amides respond the most satisfactorily [143], and phosphorus pentachloride, phosphorus tribromide [144] or pentabromide [145], or thionyl chloride [143] may be used as the halogenating agent. The reaction requires no solvent, although one such as nitromethane may be employed. When a N-acylpiperidine is subjected to the necessary conditions of the synthesis, ring cleavage occurs and dihalides are produced.

$$\xrightarrow{\text{PBr}_5} \text{Br(CH}_2)_5\text{Br}$$

1,5-Dibromopentane
78%

It is now thought [143] that an imidoyl halide 7-V is the intermediate in the reaction which proceeds as shown. For N-alkylbenzamides the imidoyl halide 7-V forms an imidonium cation 7-VI which dissociates in such a way that an alkyl halide and nitrile are formed.

$$\overset{\overset{\displaystyle X}{\displaystyle |}}{Ar-C}=N-R' \longrightarrow Ar-\overset{\oplus}{C}=N-R'\,\overset{\ominus}{X} \longrightarrow ArC\equiv N + R'X$$

$$\quad\;\; \text{7-V} \qquad\qquad\qquad\quad \text{7-VI}$$

(a) **Preparation of pentamethylene bromide** (65–72 % from benzoyl-piperidine, phosphorus tribromide, and bromine) [144].

(b) **Other examples**

(1) *s*-**Butyl bromide** (61 % from N-benzoyl-*s*-butylamine, phosphorus tribromide, and bromine) [145].

(2) **α-Cyclopentylbenzyl chloride** (80–85 % from N-α-cyclopentyl-benzylbenzamide and thionyl chloride in nitromethane) [143].

11. From Carboxylic Acids or Their Salts (Hunsdiecker, Kochi, and Barton)

$$RCO_2Ag + Br_2 \longrightarrow RBr + CO_2 + AgBr \qquad \text{Hunsdiecker reaction}$$

$$RCO_2Pb(O\overset{\overset{\displaystyle O}{\displaystyle \|}}{C}CH_3)_3 \xrightarrow{\text{LiCl}} RCl + CO_2 + LiPb(O\overset{\overset{\displaystyle O}{\displaystyle \|}}{C}CH_3)_3 \qquad \text{Kochi reaction}$$

$$RCOOH \xrightarrow[I_2,\ h\nu]{Pb(OAc)_4} RI + CO_2 + AcOH + Pb(OAc)_3I \qquad \text{Barton reaction}$$

The use of a silver carboxylate and a halogen to produce a halide containing one less carbon atom is known as the Hunsdiecker reaction. Several fairly recent reviews on this and related reactions are available [146, 147]. The Hunsdiecker reaction gives excellent yields from saturated aliphatic acids containing from two to eighteen carbon atoms. Substituents in any position other than α do not interfere unless they are capable of reacting with the intermediate acyl hypohalite. Silver salts of halogenated esters, such as silver β-bromopropionate, yield dibromides with bromine [148]. ω-Halo esters not readily available by other methods may be prepared from the silver salt of acid esters [149]. In the aromatic series the reaction is less useful, although benzoates

$$ROOC(CH_2)_nCOOAg + X_2 \rightarrow ROOC(CH_2)_nX + CO_2 + AgX$$

possessing electron-attracting substituents give satisfactory yields.

Bromine is the usual halogen employed, but iodine may be substituted if the ratio of components is 1:1. If the salt-to-iodine ratio is 2:1, the Simonini reaction occurs with the result that no organic halide at all is produced.

$$2\,RCOOAg + I_2 \rightarrow RCOOR + CO_2 + 2\,AgI$$

Larger quantities of iodine give both the ester and the organic halide. Of the many solvents employed in the reaction, carbon tetrachloride is probably the

$$3\,RCOOAg + 2\,I_2 \rightarrow RCOOR + RI + 2\,CO_2 + 3\,AgI$$

best. Mercuric and mercurous carboxylates rank next to those of silver in their effectiveness. At least in some cases the difficulty of preparing the needed dry, pure silver carboxylate can be avoided by using the free acid, an excess of red mercuric oxide, and bromine or iodine [150–153].

Although the mechanism of the Hunsdiecker reaction is not fully determined, it would appear that the first step is the formation of the hypohalite 7-VII:

$$RCOOAg + X_2 \rightarrow RCOOX + AgX$$
$$7\text{-VII}$$

In the initiation step the carboxylate radical 7-VIII and the halide radical are

$$RCOOX \rightarrow RCOO\cdot + X\cdot$$
$$7\text{-VIII}$$

formed. In the propagation step the former produces the alkyl radical 7-IX

$$RCOO\cdot \rightarrow R\cdot + CO_2$$
$$7\text{-VIII} \quad 7\text{-IX}$$

which with the halogen gives the alkyl halide 7-X [147].

$$R\cdot + X_2 \rightarrow RX + X\cdot$$
$$7\text{-X}$$

In the Kochi reaction [154, 155] which proceeds via free radicals also, a pre-equilibrium step occurs between acid and lead tetraacetate (see Ex. b.3).

$$RCO_2H + Pb(OAc)_4 \longrightarrow RCO_2Pb(OAc)_3 + AcOH$$

$$\Delta, LiCl$$

$$R\cdot + CO_2 \longrightarrow RCl$$

Fortunately acetic acid does not decarboxylate in high yield under this treatment to methyl chloride. In fact, the Kochi is supplementary to the Hunsdiecker reaction and is most applicable to the preparation of chlorides from *sec*- and *t*-acids.

The Barton reaction [156], like the Kochi reaction, utilizes the free acid which is treated with lead tetraacetate-iodine in an inert solvent under irradiation from a tungsten lamp. The method is most satisfactory for primary and secondary carboxylic acids, in which cases yields as a rule vary from 63 to 100 %. Similar results may be obtained by using *t*-butyl hypoiodite as the reagent (probably the active reagent in the lead tetraacetate-iodine procedure), but the yields in this case are somewhat lower except for some dicarboxylic acids. The mechanism of the Barton reaction appears to be similar to that of the Hunsdiecker reaction.

(a) **Preparation of methyl 5-bromovalerate** (65–68 % from methyl silver adipate and bromine in carbon tetrachloride) [149].

(b) Other examples

(1) **Heptadecyl bromide** (93% of crude from stearic acid, bromine, and mercuric oxide in carbon tetrachloride) [150].

(2) **Ethylene dibromide** (69% from silver β-bromopropionate and bromine in carbon tetrachloride at 0°) [148].

(3) **Chlorocyclobutane** (nearly quantitative yield from 11 meq. of cyclobutanecarboxylic acid, 4.5 meq. of lead tetraacetate, and 6.2 meq. of lithium chloride in 10 ml. of benzene at 81°) [155].

(4) **n-Pentyl iodide** (quantitative from 1 mole each of n-hexanoic acid, lead tetraacetate, and 1.28 moles of iodine in CCl_4 under irradiation) [156].

(5) **Bromocyclopropane** [152].

$$2 \, \triangleright\!\!-\!COOH + HgO + 2\,Br_2 \longrightarrow 2 \, \triangleright\!\!-\!Br + HgBr_2 + 2\,CO_2 + H_2O$$
$$41\text{–}46\%$$

12. From Acid Chlorides and Some Sulfonyl Chlorides

$$Ar\overset{\displaystyle O}{\overset{\|}{C}}Cl \xrightarrow[\Delta]{ClRh(P(C_6H_5)_3)_3} CO + ArCl$$

The ability of carbonyl compounds to form ligands with rhodium salts, which decompose at high temperatures to eliminate carbon monoxide, is well known [157]. With aromatic acid chlorides the product of this reaction is an aryl chloride. The rhodium salt is regenerated in the process so that its action is truly catalytic. This method is preferable to the Hunsdiecker (A.11) and Kochi (A.11) reactions for preparing aryl chlorides. Aroyl iodides are converted into aryl iodides in 53–98% yields by this method [158], but the yields of starting materials prepared from aroyl chlorides are rather low (50–60%).

Although no generally satisfactory procedure for the decomposition of sulfonyl chlorides has been proposed as of now:

$$RSO_2Cl \rightarrow RCl + SO_2$$

it appears that sulfonyl chlorides containing groups with electron-withdrawing substituents are sufficiently labile to produce chlorides [159]. Heating with a

$$C_6H_{13}O\overset{\displaystyle O}{\overset{\|}{C}}NH\!-\!\!\underset{S}{\overset{N-N}{\diagup\diagdown}}\!\!-\!SO_2Cl \xrightarrow[\text{Refluxing } CCl_4]{0.1 \text{ g. Cu bronze}} C_6H_{13}\overset{\displaystyle O}{\overset{\|}{C}}NH\!-\!\!\underset{S}{\overset{N-N}{\diagup\diagdown}}\!\!-\!Cl$$

5 g.   2-Chloro-5-n-hexylcarbonyl-
                                                           amino-1,3,4-thiadiazole, 2 g.

free flame with no catalyst also produces the chloride.

(a) **Preparation of 1-chloronaphthalene.**   α-Naphthoyl chloride (5 g.) and 50–100 mg. of chloro-tris-triphenylphosphine rhodium were heated slowly to the boiling point (the mixture turned from red to yellow and then darkened). Distillation yielded 4.1 g. (96%) of the titled compound. Yields on other aryl chlorides usually varied from 78 to 98% [157]. It is interesting to note that an almost identical preparation was carried out at nearly the same time in another laboratory [160].

13. From Organic Carbonyl (and Related) Compounds and Certain Fluorides

$$RCO_2H + SF_4 \rightarrow RCF_3$$

$$RCN + BrF_3 \rightarrow RCF_3$$

The versatile reagent sulfur tetrafluoride, SF_4, has been shown to convert many different oxygen compounds into mono-, di-, and trifluorides [161]:

$$-COOH \longrightarrow -CF_3 \qquad \diagdown C=O \longrightarrow \overset{|}{C}F_2 \qquad -\overset{|}{C}OH \longrightarrow -\overset{|}{C}F$$

These reactions are conducted in an autoclave (see Ex. a) and perhaps proceed as follows:

$$\diagdown C=O + SF_4 \longrightarrow \diagdown \overset{\oplus}{C} - \overset{\ominus}{O}SF_4 \longrightarrow \overset{F}{\underset{\diagup}{\diagdown}}\overset{|}{C}OSF_3 \overset{SF_4}{\longrightarrow} -\overset{|}{\underset{|}{C}}-F + OSF_2$$

To avoid the use of an autoclave, phenylsulfur trifluoride may be used as the reagent (see Ex. c).

To obtain low-boiling alkyl fluorides, nitriles or ketones are dissolved in liquid hydrogen fluoride, bromine trifluoride is added slowly, and the fluoride is swept out and trapped as it is formed (see Ex. b).

(a) **Preparation of 1,1,1-trifluoro-4-cyclohexylbutane.** γ-Cyclohexylbutyric acid (0.2 mole) and sulfur tetrafluoride (0.6 mole) were heated in a Hastelloy-lined autoclave at 120° for 10 hr. to yield 80% of the titled compound [161]. 1,1,1-Trifluoroheptane has been prepared in 70–80% yield by this procedure [162].

(b) **Preparation of 1,1,1-trifluoro-2-chloroethane.** α-Chloroacetonitrile, 0.0315 mole, in 50 ml. of H_2F_2 in a polypropylene bottle was stirred magnetically at −20° while 0.1 mole of bromine trifluoride in H_2F_2 was added dropwise. The product was swept out with helium as it was formed and trapped in a container cooled with liquid nitrogen; yield 70% [163].

(c) **Preparation of benzal fluoride** (71–80% from 0.1 mole of phenylsulfur trifluoride at 50–70° to which 0.1 mole of benzaldehyde was added dropwise; the mixture then was heated to 100° and evacuated until the product distilled; the difluoride has poor storage properties) [164].

14. From Alcohols and Dialkylaminotetrafluoroethanes

Secondary aliphatic amines add to tetrafluoroethylene or chlorotrifluoroethylene to give reactive fuming liquids which are capable of replacing hydroxyl by fluoride [165]:

$$n\text{-}C_4H_9OH + CFClHCF_2N(C_2H_5)_2 \longrightarrow n\text{-}C_4H_9F + O=C\overset{CHFCl}{\underset{N(C_2H_5)_2}{\diagup}} + HF$$

| 0.2 mole added dropwise | 0.2 mole | 66%, b.p. 31.5° | |

15. From Carboxylic Acids (Halodecarboxylation)

$$CNCH_2COOH \xrightarrow{NBS} Br_2CHCN$$

N-Halosuccinimides convert cyanoacetic acid or its potassium salt into the dihaloacetonitrile. Presumably the elimination of carbon dioxide and the substitution of halogen (halodecarboxylation) is applicable to other acids containing α hydrogen atoms, but no other cases appear to be recorded in the literature. The reaction fails with chloroacetonitrile and with α-cyanopropionic acid. Dibromoacetonitrile and dichloroacetonitrile are produced in yields of 75–87 and 60%, respectively. A mechanism is suggested [166].

(a) **Preparation of dibromoacetonitrile** (75–87%, from cyanoacetic acid and N-bromosuccinimide in water for 26 min., after which the mixture was cooled for 2 hr. in an ice bath) [167].

16. From Sulfonium Salts

This method constitutes the shortest pathway from one alkyl halide to the next higher homologous alkyl iodide:

$$RCH_2X \rightarrow RCH_2CH_2I$$

The path taken is shown:

$$RCH_2X + C_6H_5SCH_2Li \xrightarrow[-70°]{THF} RCH_2CH_2SC_6H_5 \xrightarrow[DMF,\ 75°]{CH_3I,\ NaI}$$

$$RCH_2CH_2\overset{\oplus}{\underset{|\ CH_3}{S}}C_6H_5\ \overset{\ominus}{I} \longrightarrow RCH_2CH_2I + C_6H_5SCH_3$$

For allylic halides it is best to use phenylthiomethyl copper, $C_6H_5SCH_2Cu$, rather than the lithium reagent in order to minimize isomerization problems.

(a) **Preparation of 1-iodoundecane.** 1-Iododecane, 8.2 mmoles, and 10 mmoles of phenylthiomethyllithium were mixed in THF at $-70°$ to give the sulfide as an oil nearly quantitatively. The oil was mixed with 5 ml. of methyl iodide and 1.5 g. of NaI in 10 ml. of DMF and held at 75° under nitrogen. The usual workup gave the iodide, 93% [168].

1. R. Stephens and J. C. Tatlow, *Quart. Rev.*, **16**, 44 (1962).
2. J. E. Copenhaver and A. M. Whaley, *Org. Syn.*, Coll. Vol. **1**, 142 (1941).
3. F. C. Whitmore, *et al.*, *J. Am. Chem. Soc.*, **54**, 3431 (1932); **55**, 1106 (1933); **60**, 2265 (1938); **60**, 2533 (1938).
4. J. F. Norris and A. W. Olmsted, *Org. Syn.*, Coll. Vol. **1**, 144 (1941).
5. V. Boekelheide and G. K. Vick, *J. Am. Chem. Soc.*, **78**, 653 (1956).
6. H. C. Brown and M.-H. Rei, *J. Org. Chem.*, **31**, 1090 (1966).
7. O. Kamm and C. S. Marvel, *Org. Syn.*, Coll. Vol. **1**, 25–35 (1941).
8. E. E. Reid, *et al.*, *Org. Syn.*, Coll. Vol. **2**, 246 (1943).
9. A. I. Vogel, *J. Chem. Soc.*, 636 (1943).
10. W. L. McEwen, *Org. Syn.*, Coll. Vol. **3**, 227 (1955).
11. E. V. Whitehead *et al.*, *J. Am. Chem. Soc.*, **73**, 3632 (1951).
12. H. Pines *et al.*, *J. Am. Chem. Soc.*, **74**, 4063 (1952).
13. H. Stone and H. Shechter, *J. Org. Chem.*, **15**, 491 (1950).

14. H. Stone and H. Shechter, *Org. Syn.*, Coll. Vol. **4,** 323 (1963).
15. I. M. Downie, *et al.*, *Chem. Ind.* (*London*), 900 (1966); *see also* J. Hooz and S. S. H. Gilani, *Can. J. Chem.*, **46,** 86 (1968).
16. L. H. Sommer *et al.*, *J. Am. Chem. Soc.*, **76,** 803 (1954).
17. P. S. Skell *et al.*, *J. Am. Chem. Soc.*, **82,** 410 (1960).
18. D. G. Coe *et al.*, *J. Chem. Soc.*, 2281 (1954).
19. S. R. Landauer and H. N. Rydon, *J. Chem. Soc.*, 2224 (1953).
20. H. Hoffmann *et al.*, *Chem. Ber.*, **95,** 523 (1962).
21. G. A. Wiley *et al.*, *J. Am. Chem. Soc.*, **86,** 964 (1964).
22. A. R. Surrey and R. A. Cutler, *J. Am. Chem. Soc.*, **76,** 1109 (1954).
23. J. Baddiley and A. Topham, *J. Chem. Soc.*, 678 (1944); A. Bendich *et al.*, *J. Am. Chem. Soc.*, **76,** 6073 (1954); N. Yamoaka and K. Aso, *J. Org. Chem.*, **27,** 1462 (1962).
24. A. Albert and J. Clark, *J. Chem. Soc.*, 1666 (1964).
25. C. K. Bradsher *et al.*, *J. Am. Chem. Soc.*, **79,** 1468 (1957); C. E. Boord *et al.*, *J. Am. Chem. Soc.*, **77,** 1751 (1955).
26. C. R. Noller and R. Dinsmore, *Org. Syn.*, Coll. Vol. **2,** 358 (1943).
27. L. H. Smith, *Org. Syn.*, Coll. Vol. **3,** 793 (1955); H. L. Goering *et al.*, *J. Am. Chem. Soc.*, **70,** 3314 (1948); M. S. Newman and J. H. Wotiz *ibid.*, **71,** 1292 (1949).
28. G. A. Wiley, *et al.*, *J. Am. Chem. Soc.*, **86,** 964 (1964).
29. D. K. Black, *et al.*, *Tetrahedron Letters*, 483 (1963).
30. H. S. King, *Org. Syn.*, Coll. Vol. **2,** 399 (1943).
31. E. J. Corey and J. E. Anderson, *J. Org. Chem.*, **32,** 4160 (1967).
32. R. S. Tipton, *J. Org. Chem.*, **27,** 1449 (1962).
33. L. Kaplan, *J. Org. Chem.*, **31,** 3454 (1966).
34. A. Campbell and H. N. Rydon, *J. Chem. Soc.*, 3002 (1953).
35. J. Lorenz and J. Auer, *Angew. Chem.*, **77,** 218 (1965).
36. M. M. Hafez *et al.*, *J. Org. Chem.*, **26,** 3988 (1961).
37. A. Rieche and H. Gross, *Chem. Ber.*, **92,** 83 (1959).
38. C. Raulet and M. Levas, *Compt. Rend.*, **255,** 1406 (1962).
39. W. R. Hasek, *et al.*, *J. Am. Chem. Soc.*, **82,** 543 (1960).
40. W. A. Sheppard, *J. Am. Chem. Soc.*, **84,** 3058 (1962); *Org. Syn.*, **44,** 39, 82 (1964).
41. T. L. Jacobs, *Org. Reactions*, **5,** 20 (1949).
42. M. S. Newman and L. L. Wood, Jr., *J. Am. Chem. Soc.*, **81,** 4300 (1959).
43. K. B. Wiberg, *et al.*, *J. Am. Chem. Soc.*, **83,** 3998 (1961).
44. H. Gross, *et al.*, *Org. Syn.*, **47,** 47 (1967); A. Rieche and H. Gross, *Chem. Ber.*, **92,** 83 (1959).
45. P. D. Bartlett and L. J. Rosen, *J. Am. Chem. Soc.*, **64,** 543 (1942).
46. W. E. Bissinger and F. E. Kung, *J. Am. Chem. Soc.*, **69,** 2158 (1947).
47. H. Gilman and J. E. Kirby, *J. Am. Chem. Soc.*, **51,** 3475 (1929); H. Gilman and A. P. Hewlett, *Rec. Trav. Chim.*, **51,** 93 (1932).
48. M. S. Newman, *J. Am. Chem. Soc.*, **62,** 2295 (1940).
49. H. Rapoport and J. E. Campion, *J. Am. Chem. Soc.*, **73,** 2239 (1951).
50. G. A. Page and D. S. Tarbell, *J. Am. Chem. Soc.*, **75,** 2053 (1953).
51. J. D. Roberts and M. C. Caserio, *Basic Principles of Organic Chemistry*, W. A. Benjamin, New York, 1964, p. 392.
52. L. F. and M. Fieser, *Reagents for Organic Synthesis*, John Wiley and Sons, New York, N.Y., 1967, p. 1158.
53. A. M. Ward, *Org. Syn.*, Coll. Vol. **2,** 159 (1943).
54. L. A. R. Hall *et al.*, *Org. Syn.*, Coll. Vol. **4,** 333 (1963).
55. M. S. Newman, *J. Am. Chem. Soc.*, **62,** 2295 (1940).
56. F. Šorm *et al.*, *Tetrahedron Letters*, 397 (1962).
57. J. R. Campbell *et al.*, *Can. J. Chem.*, **44,** 2339 (1966).
58. J. Cason and J. S. Correia, *J. Org. Chem.*, **26,** 3645 (1961).
59. D. Rosenthal *et al.*, *J. Am. Chem. Soc.*, **85,** 3971 (1963).
60. F. C. Uhle, *J. Am. Chem. Soc.*, **83,** 1460 (1961).
61. F. A. Cutler, Jr., *et al.*, *J. Am. Chem. Soc.*, **80,** 6300 (1958).

62. R. T. Blickinstaff and F. C. Chang, *J. Am. Chem. Soc.*, **80,** 2726 (1958).
63. K. W. Buck and A. B. Foster, *J. Chem. Soc.*, 2217 (1963).
64. M. F. Clarke and L. N. Owen, *J. Chem. Soc.*, 315, 326 (1949).
65. J. H. Looker *et al.*, *J. Org. Chem.*, **27,** 4349 (1962).
66. G. L. Jenkins and J. C. Kellett, Jr., *J. Org. Chem.*, **27,** 624 (1962).
67. G. Eglinton and M. C. Whiting, *J. Chem. Soc.*, 3650 (1950).
68. H. L. Herzog, *Org. Syn.*, Coll. Vol. **4,** 753 (1963).
69. W. J. Baumann and H. K. Mangold, *J. Lipid Res.*, **7,** 568 (1966); *C.A.*, **65,** 10481 (1966).
70. H. O. House *et al.*, *J. Org. Chem.*, **21,** 1487 (1956).
71. A. B. Foster *et al.*, *J. Chem. Soc.*, 2542 (1949).
72. R. M. Hann *et al.*, *J. Am. Chem. Soc.*, **72,** 561 (1950).
73. F. Drahowzal and D. Klamann, *Monatsh. Chem.*, **82,** 970 (1951).
74. E. R. Blakley, *Biochem. Prep.*, **7,** 39 (1960); H. M. Kissman and M. J. Weiss, *J. Am. Chem. Soc.*, **80,** 5559 (1958).
75. N. F. Taylor and P. W. Kent, *J. Chem. Soc.*, 872 (1958).
76. K. B. Wiberg and B. R. Lowry, *J. Am. Chem. Soc.*, **85,** 3188 (1963).
77. H. B. Hass and H. C. Hoffman, *J. Am. Chem. Soc.*, **63,** 1233 (1941).
78. M. S. Newman and R. D. Closson, *J. Am. Chem. Soc.*, **66,** 1553 (1944).
79. P. Coad *et al.*, *J. Org. Chem.*, **28,** 218 (1963).
80. M. Barash and J. M. Osbond, *J. Chem. Soc.*, 2157 (1959).
81. L. C. Swallen and C. E. Boord, *J. Am. Chem. Soc.*, **52,** 651 (1930).
82. F. Boberg and G. Winter, *Ann. Chem.*, **621,** 20 (1959).
83. J. F. Bunnett, and R. M. Conner, *J. Org. Chem.*, **23,** 305 (1958).
84. S. Winstein *et al.*, *Tetrahedron Letters*, No. 16, 24 (1959).
85. K. Niedenzu *et al.*, *Chem. Ber.*, **96,** 2649 (1963).
86. P. J. Trotter, *J. Org. Chem.*, **28,** 2093 (1963).
87. K. R. Raver *et al.*, *J. Gen. Chem. USSR*, (*Eng. Transl.*), **30,** 2347 (1960).
88. A. L. Henne, *Org. Reactions*, **2,** 49 (1944).
89. A. I. Vogel *et al.*, *Org. Syn.*, Coll. Vol. **4,** 525 (1963).
90. J. F. Bunnett and R. E. Zahler, *Chem. Rev.*, **49,** 273 (1951).
91. G. C. Finger and C. W. Kruse, *J. Am. Chem. Soc.*, **78,** 6034 (1956).
92. L. D. Starr and G. C. Finger, *Chem. Ind.* (*London*), 1328 (1962); G. C. Finger *et al.*, *J. Org. Chem.*, **28,** 1666 (1963).
93. Ref. 52, p. 934.
94. E. Kober *et al.*, *J. Org. Chem.*, **27,** 2577 (1962).
95. C. W. Tullock and D. D. Coffman, *J. Org. Chem.*, **25,** 2016 (1960).
96. G. Fuller, *J. Chem. Soc.*, 6264 (1965).
97. J. F. Bunnett and R. M. Conner, *Org. Syn.*, **40,** 34 (1960).
98. N. N. Vorozhtsov, Jr., and G. G. Yakobson, *J. Gen. Chem.*, USSR (*Eng. Transl.*) **27,** 1741 (1957).
99. R. G. R. Bacon and H. A. O. Hill, *J. Chem. Soc.*, 1097 (1964).
100. W. B. Hardy and R. B. Fortenbaugh, *J. Am. Chem. Soc.*, **80,** 1716 (1958).
101. R. M. Peck *et al.*, *J. Am. Chem. Soc.*, **81,** 3984 (1959).
101a. T. G. Barbee, Jr., and D. E. Pearson, unpublished results.
102. J. T. Maynard, *J. Org. Chem.*, **28,** 112 (1963).
103. R. A. Davis and E. R. Larsen, *J. Org. Chem.*, **32,** 3478 (1967).
104. H. B. Schurink, *Org. Syn.*, Coll. Vol. **2,** 476 (1943).
105. C. L. Stevens *et al.*, *J. Am. Chem. Soc.*, **77,** 2341 (1955).
106. F. J. Buckle *et al.*, *J. Chem. Soc.*, 1471 (1949); D. C. Sayles and E. F. Degering, *J. Am. Chem. Soc.*, **71,** 3161 (1949).
107. W. L. Drake, *et al.*, *J. Am. Chem. Soc.*, **68,** 1536 (1946).
108. M. Eckstein, *Dissertationes Pharm.*, **14,** 401 (1962); *C.A.*, **60,** 8029 (1964); N. Sperber *et al.*, *J. Am. Chem. Soc.*, **75,** 1122 (1953).
109. A. W. Nineham, *J. Chem. Soc.*, 2601 (1953).
110. A. G. Anderson, Jr., and F. J. Freenor, *J. Am. Chem. Soc.*, **86,** 5037 (1964).

111. H. Stone and H. Shechter, *J. Org. Chem.*, **15,** 491 (1950).
112. A. C. Cope *et al.*, *J. Am. Chem. Soc.*, **87,** 5452 (1965).
113. K. W. and W. J. Dunning, *J. Chem. Soc.*, 2925 (1950).
114. R. L. Burwell *et al.*, *J. Am. Chem. Soc.*, **73,** 2428 (1951).
115. A. Rieche and H. Gross, *Chem. Ber.*, **92,** 83 (1959).
116. V. I. Lutkova and N. I. Kutsenko, *Zh. Prikl. Khim.*, **32,** 2823 (1959); *C.A.*, **54,** 9717 (1960).
117. N. J. Leonard and J. Figueras, Jr., *J. Am. Chem. Soc.*, **74,** 917 (1952).
118. J. Cason *et al.*, *J. Org. Chem.*, **14,** 37 (1949).
119. D. M. Starr and R. M. Hixon, *Org. Syn.*, Coll. Vol. **2,** 571 (1943).
120. K. Alexander and H. V. Towles, *Org. Syn.*, Coll. Vol. **4,** 266 (1963).
121. M. E. Synerholm, *Org. Syn.*, Coll. Vol. **3,** 187 (1955).
122. Z. Suzuki and K.-I. Morita, *J. Org. Chem.*, **32,** 31 (1967).
123. L. H. Long and G. F. Freeguard, *Chem. Ind.* (*London*), 223 (1965).
124. L. A. Bigelow, *Org. Syn.*, Coll. Vol. **1,** 136 (1941).
125. C. S. Marvel and S. M. McElvain, *Org. Syn.*, Coll. Vol. **1,** 170 (1941).
126. M. S. Newman and W. S. Fones, *J. Am. Chem. Soc.*, **69,** 1221 (1947).
127. C. F. H. Allen and J. R. Thirtle, *Org. Syn.*, Coll. Vol. **3,** 136 (1955).
128. H. J. Lucas and E. R. Kennedy, *Org. Syn.*, Coll. Vol. **2,** 351 (1943).
129. J. G. Carey *et al.*, *Chem. Ind.* (*London*), 1018 (1959).
130. W. E. Lee *et al.*, *J. Am. Chem. Soc.*, **83,** 1928 (1961).
131. E. H. White and H. Scherrer, *Tetrahedron Letters*, 758 (1961).
132. A. Roe, *Org. Reactions*, **5,** 193 (1949).
133. T. L. Fletcher and M. J. Namkung, *Chem. Ind.* (*London*), 179 (1961).
134. K. G. Rutherford *et al.*, *J. Org. Chem.*, **26,** 5149 (1961).
135. K. G. Rutherford and W. A. Redmond, *Org. Syn.*, **43,** 12 (1963).
136. K. G. Rutherford and W. A. Redmond, *J. Org. Chem.*, **28,** 568 (1963).
137. R. L. Ferm and C. A. VanderWerf, *J. Am. Chem. Soc.*, **72,** 4809 (1950).
138. J. K. Kochi, *J. Am. Chem. Soc.*, **79,** 2942 (1957).
139. J. M. Tedder and G. Theaker, *Tetrahedron*, **5,** 288 (1959).
140. L. V. Hankes, *Biochem. Prep.*, **9,** 59 (1962).
141. E. D. Bergmann and M. Bentov, *J. Org. Chem.*, **19,** 1594 (1954).
142. L. Friedman and F. M. Logullo, *Angew. Chem.*, **77,** 217 (1965).
143. W. R. Vaughan and R. D. Carlson, *J. Am. Chem. Soc.*, **84,** 769 (1962).
144. J. von Braun, *Org. Syn.*, Coll. Vol. **1,** 428 (1941).
145. N. J. Leonard and E. W. Nommensen, *J. Am. Chem. Soc.*, **71,** 2808 (1949).
146. C. V. Wilson, *Org. Reactions*, **9,** 332 (1957).
147. R. G. Johnson and R. K. Ingham, *Chem. Rev.*, **56,** 219 (1956).
148. J. C. Conly, *J. Am. Chem. Soc.*, **75,** 1148 (1953).
149. C. F. H. Allen and C. V. Wilson, *Org. Syn.*, Coll. Vol. **3,** 578 (1955).
150. S. J. Cristol and W. C. Firth, Jr., *J. Org. Chem.*, **26,** 280 (1961).
151. S. J. Cristol *et al.*, *J. Org. Chem.*, **27,** 2711 (1962); **29,** 1279 (1964).
152. J. S. Meek and D. T. Osuga, *Org. Syn.*, **43,** 9 (1963).
153. J. A. Davis *et al.*, *J. Org. Chem.*, **30,** 415 (1965).
154. J. K. Kochi, *J. Am. Chem. Soc.*, **87,** 2500 (1965).
155. J. K. Kochi, *J. Org. Chem.*, **30,** 3265 (1965).
156. D. H. R. Barton *et al.*, *Proc. Chem. Soc.*, 309 (1962); *J. Chem. Soc.*, 2438 (1965).
157. J. Blum, *Tetrahedron Letters*, 1605 (1966).
158. J. Blum *et al.*, *J. Org. Chem.*, **33,** 1928 (1968).
159. V. Petrow *et al.*, *J. Chem. Soc.*, 1508 (1958).
160. K. Ohno and J. Tsuji, *J. Am. Chem. Soc.*, **90,** 99 (1968).
161. W. R. Hasek *et al.*, *J. Am. Chem. Soc.*, **82,** 543 (1960).
162. W. R. Hasek, *Org. Syn.*, **41,** 104 (1961).
163. T. E. Stevens, *J. Org. Chem.*, **26,** 1627 (1961).
164. W. A. Sheppard, *Org. Syn.*, **44,** 39 (1964).
165. N. N. Yarovenko and M. A. Raksha, *J. Gen. Chem. USSR*, (*Eng. Transl.*), **29,** 2125 (1959).

166. J. W. Wilt, *J. Org. Chem.*, **21**, 920 (1956).
167. J. W. Wilt and J. L. Diebold, *Org. Syn.*, **38**, 16 (1958).
168. E. J. Corey and M. Jautelat, *Tetrahedron Letters*, 5787 (1968).

B. Addition to Unsaturated Compounds and Epoxides

1. Hydrogen Halides (1)

$$RCH{=}CH_2 \xrightarrow{\text{HX}} RCHXCH_3$$

This synthesis is of little importance in the preparation of the simpler halides since more convenient methods are available. Its greatest preparative value perhaps lies in its use in the synthesis of types such as β-haloaldehydes [2], β-haloesters [3], and β-halonitriles [4]. By these methods, as is shown with an α,β-unsaturated ester, compounds having the halogen in the β- rather than the α-position are obtained. Through such types other β-substituted compounds,

$$CH_2{=}CHCOOCH_3 + HX \rightarrow XCH_2CH_2COOCH_3$$

such as the hydroxy-, the cyano-, the amino-, etc., become readily available. With alkadienes the 1,4-addition compound predominates over the 1,2-adduct to the extent that the addition is of synthetic value [5, 6]. The ease of hydrogen halide addition is in the order of their acidity: HI > HBr > HCl. The hydrogen iodide addition occurs in high yield when potassium iodide and phosphoric acid are employed as the reagent [7, 8]. Addition of hydrogen chloride in aprotic solvents is assisted by stannic chloride [9]. Anhydrous hydrogen fluoride adds well to olefins, but the instability of the monofluoroalkanes in the presence of acids or water tends to reduce yields [10].

The nature of the addition of hydrogen halides to olefins shows variable characteristics reminiscent of the delineations found in SN_1–SN_2 solvolytic reactions. The addition to unsymmetrical olefins is such as to give that intermediate which can best stabilize electron deficiency:

$$RCH{=}CH_2 + HX \longrightarrow [R\overset{\oplus}{CH}{-}CH_3] \xrightarrow{X^{\ominus}} R\overset{\overset{\displaystyle X}{|}}{C}HCH_3$$

$$R_2C{=}CH_2 + HX \longrightarrow [R_2\overset{\oplus}{C}{-}CH_3] \xrightarrow{X^{\ominus}} R_2\overset{\overset{\displaystyle X}{|}}{C}CH_3$$

In the latter case a true carbonium ion may be formed, but in the former case the secondary carbonium ion is of such stability that addition of the halide anion takes place before complete electron deficiency is established. In both cases the majority of additions are a stepwise process. Although it is believed that the halide anion adds in a *trans* fashion to the olefin or acetylene

$$\begin{array}{c} X^{\ominus} \\ \downarrow \\ R\,CH{=}CH_2 \\ \uparrow \\ H^{\oplus} \end{array}$$

References for Section B are on pp. 374–376.

many exceptions are known [11] as shown by the following example [12]:

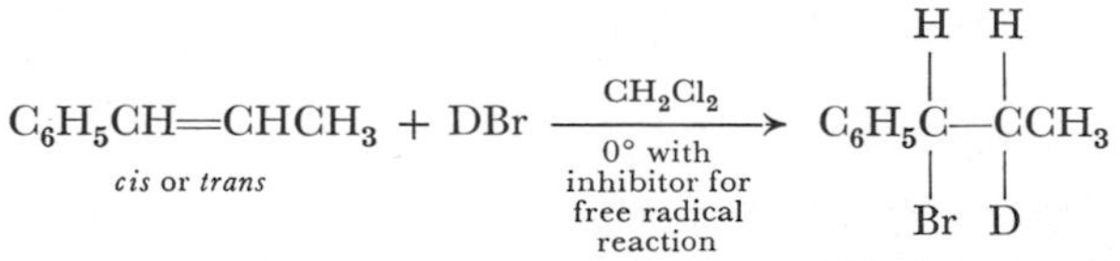

$$C_6H_5CH{=}CHCH_3 + DBr \xrightarrow[\substack{0° \text{ with} \\ \text{inhibitor for} \\ \text{free radical} \\ \text{reaction}}]{CH_2Cl_2}$$

cis or trans

H H
| |
$C_6H_5C{-}CCH_3$
| |
Br D

β-Deutero-α-bromo-
propylbenzene, 88%,
derived from *cis*
addition

Perhaps the addition behavior may be summarized by the statement that the addition is less stepwise and more cleanly *trans* if the intermediate is relatively unstable and the reverse if the intermediate, a tertiary carbonium ion for instance, is more stable. Alkenes which form tertiary or other stable carbonium ions are converted cleanly and rapidly into the chloride by using hydrogen chloride in the apparatus of H. C. Brown [13].

The generalizations above regarding orientation are more satisfactory than the Markownikoff rule since exceptions to the rule, such as the addition of the hydrogen halide to trifluoropropylene [14], are explained readily by the greater

$$CF_3CH{=}CH_2 + HX \rightarrow CF_3CH_2CH_2X$$

stability of $CF_3CH_2CH_2{}^\oplus$ (or the precursor before charge is fully developed) as compared to $CF_3\overset{\oplus}{C}HCH_3$, the latter being destabilized by the strongly electron-withdrawing CF_3 group. The formation of β-haloesters or ketones from the unsaturated compounds is rationalized as shown:

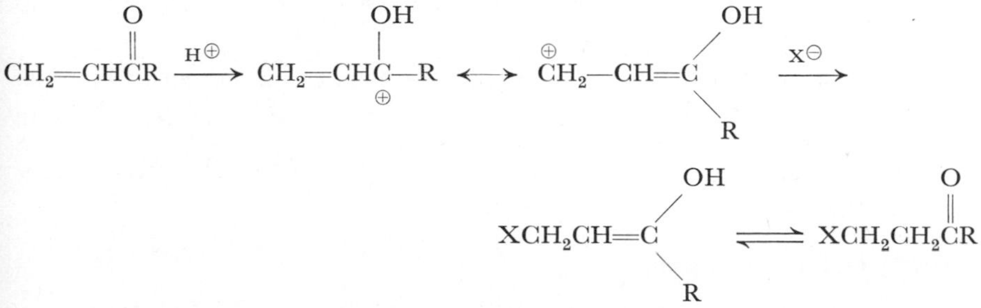

$$CH_2{=}CHCR \xrightarrow{H\oplus} CH_2{=}CHC{-}R \longleftrightarrow CH_2{-}CH{=}C \xrightarrow{X\ominus}$$

$$XCH_2CH{=}C \rightleftharpoons XCH_2CH_2CR$$

7-XI

7-XI is the only stable addition product. Since the transition state has a dispersed charge, addition occurs best in nonpolar solvents.

It is possible to reverse orientation in the addition of hydrogen bromide to olefins by resorting to free radical conditions in which the course of the reaction is as follows:

$$HBr \xrightarrow{RO\cdot} ROH + [Br\cdot] \xrightarrow{RCH{=}CH_2} [R\overset{\cdot}{C}H{-}CH_2Br] \xrightarrow{HBr} RCH_2CH_2Br + [Br\cdot]$$

In this case the stabler free radical which leads to the anti-Markownikoff product is formed. A claim has been made that similar abnormal orientation can be brought about by the use of Linde Molecular Sieve 5A [15]:

$$C_6H_{13}CH{=}CH_2 + HBr \xrightarrow[62°,\ 1\ hr.]{10\ g.\ molecular\ sieve\ 5A} C_6H_{13}CH_2CH_2Br$$

5 g. in 15 g. C_7H_{16} — 91% containing 90% 1-bromooctane

Such addition may take place because of specific orientation on a surface rather than because of a free radical mechanism. For synthetic purposes the alkyl halides prepared by free radical addition are not always free from isomeric bromides [16] and must be purified (see Ex. b.4). Hydrogen chloride does not lend itself well to homolytic addition, although an example is known [17]:

$$CH_3CH{=}CH_2 + HCl \xrightarrow{\text{Di-}t\text{-butyl peroxide}} CH_3CH_2CH_2Cl + CH_3CHClCH_3$$

"Small amounts"

The addition of HX to allenes or acetylenes usually gives the same products:

$$CH_2{=}C{=}CH_2 \text{ (or } CH_3C{\equiv}CH) \xrightarrow{HX} CH_3CX{=}CH_2 \xrightarrow{HX} CH_3CX_2CH_3$$

but the addition of HBr has been shown to give substantial amounts of a cyclobutane derivative [18]. Hydrogen chloride addition to allene is very slow but

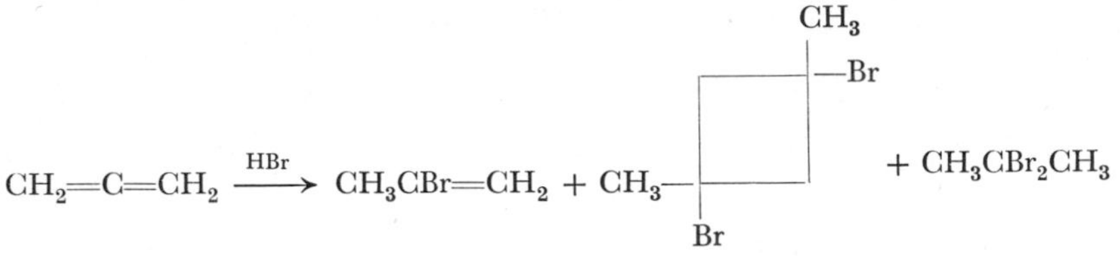

$$CH_2{=}C{=}CH_2 \xrightarrow{HBr} CH_3CBr{=}CH_2 + CH_3{-}\boxed{\quad}{-}Br + CH_3CBr_2CH_3$$

35% cis- and trans-
1,3-Dimethyl-1,3-di-
bromocyclobutane

gives normal addition products.

In addition to the examples which follow, others can be found in Houben-Weyl [19].

(a) Preparation of iodocyclohexane (88–90% from cyclohexene, potassium iodide, and 95% orthophosphoric acid) [7].

(b) Other examples

(1) Methyl-β-bromopropionate (80–84% from methyl acrylate in ether and anhydrous hydrogen bromide at room temperature) [3].

(2) β-Chloropropionitrile (80% from acrylonitrile and anhydrous hydrogen chloride) [4].

(3) 3-Chlorocyclopentene (70–90%, based on cyclopentadiene, from cyclopentadiene and dry hydrogen chloride at a temperature below 0°) [6].

(4) 2-Methyl-3-bromobutane (66% crude, b.p. 47–52°, from 0.1 mole of trimethylethylene and 0.05 mole of lauroyl peroxide in pentane

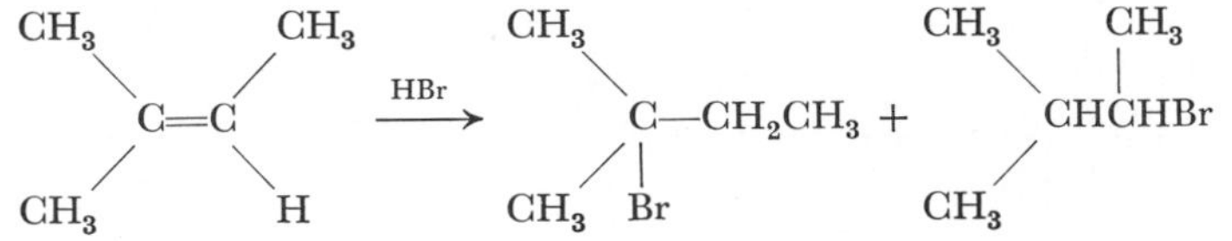

$$\underset{\underset{CH_3}{|}}{\overset{\overset{CH_3}{|}}{C}}{=}\underset{\underset{H}{|}}{\overset{\overset{CH_3}{|}}{C}} \xrightarrow{HBr} \underset{CH_3\ \ Br}{\overset{CH_3}{C{-}CH_2CH_3}} + \underset{CH_3}{\overset{CH_3}{CHCHBr}}$$

through which HBr was passed at room temperature; the accompanying *t*-amyl bromide, consisting of about 40% of the crude mixture, was removed by hydrolysis in 80% aqueous acetone at reflux for 15 min.) [20].

2. Halogens [21]

$$RCH{=}CH_2 + X_2 \rightarrow RCHXCH_2X$$

This synthesis is of interest mostly in preparing intermediates suitable for preparing vinyl halides as shown [22]. It also is of value in the preparation of

$$\underset{\substack{|\\ Br}}{\overset{\substack{CH_3\\|}}{H_2C{-}\underset{|}{\overset{|}{C}}{-}COOH}} \xrightarrow{C_5H_5N} \underset{\substack{|\\ Br}}{CH_3C{=}CH_2}$$

α-Methylvinyl bromide

hexachlorocyclohexane, gammexane (from benzene and chlorine), and similar compounds.

The mechanisms of addition are variable, complex, and similar to the addition of acids to olefins (B.1), although for halogen addition it is plausible to regard the halonium ion as an intermediate. The positive charge on the ion

$$RCH{=}CH_2 + X_2 \longrightarrow RCH\overset{\overset{\displaystyle X}{\diagup\ \diagdown}}{\underset{\oplus}{}}CH_2 \xrightarrow{X^{\ominus}} \underset{\substack{|\\ X}}{RCHCH_2X}$$

need not be symmetrically dispersed before attack by the anion. This mechanism again assumes *trans* addition of the anion, but numerous examples of *cis* addition are now accumulating; such addition seems to depend more on the structure of the substrate than on the type of reagent. Acenaphthylene, *p*-dioxene, cyclooctatetraene, and *cis*-stilbene with various agents, for instance, give more *cis* than *trans* products [21].

Side reactions that may occur in halogen addition are Wagner-Meerwein type rearrangements (see Ex. b.2), substitution, dehydrohalogenation, particularly if one of the halogens resides on a tertiary carbon atom [23], and partial dehalogenation, invariably with *vicinal* iodine atoms, occasionally with *vicinal* bromine atoms.

Addition of Chlorine

Typical examples are to be found in Houben-Weyl [24]. Chlorine additions can be carried out with the free element alone, in a solvent, or with a catalyst [25, 26]. Such addition has also been accomplished with tetrabutylammonium iodotetrachloride [27], sulfuryl chloride [28], phosphorus pentachloride [29], and N-chlorosuccinimide and hydrogen chloride [30]. For deactivated olefins the use of light as a catalyst is recommended, particularly with benzene compounds. A solution of chlorine in benzene exposed to sunlight forms a mixture of *cis-trans* isomers of hexachlorocyclohexane melting at 157°. Similar reactions can be carried out with chlorobenzene, forming mixed heptachlorocyclohexanes, and with alkylbenzenes, giving substitution products of the alkyl group as well as addition products of the benzene ring [31]. For other deactivated olefins a Friedel-Crafts type of catalyst can be used [32]:

$$
\begin{array}{ccc}
\underset{\displaystyle CF_3C=CH_2}{\overset{\displaystyle CH_3}{|}} & \xrightarrow[\text{Trace of FeCl}_3]{Cl_2} & \underset{\displaystyle CF_3CClCH_2Cl}{\overset{\displaystyle CH_3}{|}}
\end{array}
$$

Quantitative yield
2-Methyl-1,2-dichloro-
3,3,3-trifluoropropane

For acrylonitrile, chlorine addition with illumination in carbon tetrachloride or with pyridine alone as a solvent gives comparable yields of around 70% of 2,3-dichloropropionitrile [33]. For allyl alcohol the addition in an equal volume of concentrated hydrochloric acid gives 71% 2,3-dichloropropanol-1, the best yield reported. On the other hand, addition of bromine to allyl alcohol can be carried out more simply in an inert solvent [34]. The addition of halogen (or HX) to fluoroolefins is discussed thoroughly by Knunyants [35].

Addition of Bromine

Bromine addition is usually more satisfactory than that of chlorine, since side reactions are less prominent and mode of addition is more readily controlled. Bromine in carbon tetrachloride is the conventional means of addition even for unsaturated carbonyl compounds, but considerable variations are documented. α,α'-Dibromosuccinic acid is prepared in 72–84% yields from fumaric acid suspended in boiling water to which bromine was added [36]. Dioxane dibromide [37], a crystalline compound, is an excellent mild agent. It gives yields of 100% styrene dibromide or 89% isoprene tetrabromide on slow addition with cooling to styrene and isoprene, respectively [38]. Even bromine formed from magnesium bromide in ether and benzoyl peroxide has been used as an addition agent for cyclohexene to give 83% *trans*-1,2-dibromocyclohexane [39]. However, this reagent is much too elaborate for the addition of bromine to cyclohexene and should therefore be kept in mind for addition to less prosaic olefins. Pyridinium hydrobromide perbromide has the advantage over bromine in being a solid and in being more specific in promoting the addition of bromine. For example [40], with *cis*-stilbene in acetic acid it gives exclusively the *dl*-stilbene dibromide, while with the *trans*-stilbene in the same solvent it gives exclusively the *meso* isomer. Bromine and butadiene form a kinetically controlled

$$
\begin{array}{ccccccc}
C_6H_5CH & & H-\overset{\overset{\displaystyle C_6H_5}{|}}{C}-Br & C_6H_5CH & & H-\overset{\overset{\displaystyle C_6H_5}{|}}{C}-Br \\
\| & \xrightarrow{C_5H_5\overset{\oplus}{N}H\overset{\ominus}{Br_3}} & | & \| & \xrightarrow{C_5H_5\overset{\oplus}{N}H\overset{\ominus}{Br_3}} & | \\
C_6H_5CH & & Br\underset{\underset{\displaystyle C_6H_5}{|}}{C}-H & H-CC_6H_5 & & H-\underset{\underset{\displaystyle C_6H_5}{|}}{C}-Br
\end{array}
$$

mixture of 48% of *trans*-1,4-dibromo-2-butene and 52% of 3,4-dibromo-1-butene [41] but, on heating, the *trans*-1,4-dibromo-2-butene is formed in high yield (see Ex. b.3).

An indirect way to obtain *cis*- and *trans*-vinyl bromides is to proceed via vinyl boranes. They add bromine to form dibromides capable of conversion into *cis*- and *trans*-vinyl bromides [42]; (see Ex. b.7):

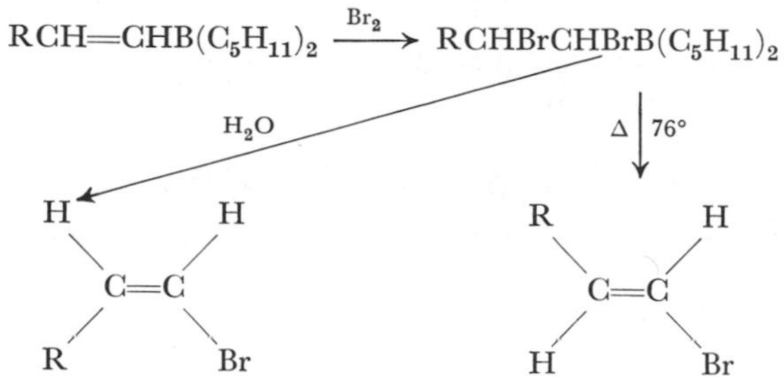

$$RCH{=}CHB(C_5H_{11})_2 \xrightarrow{\ Br_2\ } RCHBrCHBrB(C_5H_{11})_2$$

Addition of Iodine

Since iodine addition is an equilibrium process, the yields of diiodo compounds with simple olefins are not only poor, but the products are also unstable. For example, the position of equilibrium for 0.25 M olefin and 0.02 M iodine in acetic acid at 25° in terms of iodine absorption was found to be as follows [43]: 2-methyl-1-butene, 39%; cyclohexene, 85%; cyclopentene, 51%; 3,3-dimethyl-1-butene, 14%.

Addition of Fluorine

Fluorine addition poses a problem because of the violent nature of the reaction and the instability of fluorides which can readily dehydrofluorinate. Nevertheless, the difficulty has been circumvented to a degree by use of the reagent, xenon difluoride [44]. The origin of B is undoubtedly dehydrofluori-

$$CH_2{=}CH_3 + XeF_2 \xrightarrow{\ 25°\ } CH_2FCH_2F + CHF_2CH_3$$
$$\text{A, 45\%} \qquad\quad \text{B, 35\%}$$

nation of A followed by addition of HF. Lead diacetate difluoride, prepared from lead tetraacetate and hydrogen fluoride, also form *gem*-difluorides from olefins [45].

Addition of Mixed Halogens

Mixed halogens add more rapidly to olefins than the single halogens, the large atom leading the attack, but considerable complications arise from the synthetic

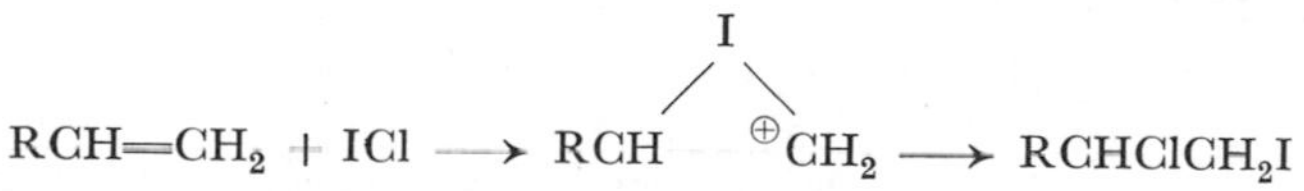

$$RCH{=}CH_2 + ICl \longrightarrow RCH \overset{I}{\underset{}{\diagdown}} \overset{\oplus}{CH_2} \longrightarrow RCHClCH_2I$$

viewpoint. The mixed dihalides tend to dissociate and thus lead to products formed from the single halogens, and furthermore the orientation described in the formulation is not strictly observed [46, 21]. The mixed halogens can be simply mixed in the correct proportion (*Caution:* Usually exothermic reaction) or essentially generated *in situ*. For example, although an intermediate mercurial may intervene, β-iodo-α-chloroethylbenzene and 1-iodo-2-chloro-2-methyl-propane were prepared in 47 and 67% yields, respectively, by mixing the proper olefin with mercuric chloride and iodine in ether [47]. The mixed halogen, IF, prepared according to the equation

$$2 I_2 + IF_5 \xrightarrow{100-175°} 5 IF$$

has been used to add to highly halogenated olefins (the halogen adducts of which are more stable) (see Ex. b.6).

(a) **Preparation of 1,2,3-tribromopropane** (96–98% from allyl bromide and bromine in carbon tetrachloride at 0 to −5°) [48].

(b) **Other examples**

(1) **trans-1,2-Dibromocyclohexane** (83% from cyclohexene, magnesium bromide, and benzoyl peroxide) [39].

(2) **2,4-Dichloro-2,3,5,5-tetramethylhexane** (74% from trans-di-t-

$$
\begin{array}{c}
(CH_3)_3C-CH \\
\parallel \\
H-C-C(CH_3)_3
\end{array}
\xrightarrow{Cl_2}
\begin{array}{c}
\quad\quad\quad Cl \\
\quad\quad\quad | \\
(CH_3)_3C-CHCHC(CH_3)_2 \\
| \quad | \\
Cl \quad CH_3
\end{array}
$$

butylethylene, chlorine, and antimony pentachloride at low temperature by a rearrangement) [25].

(3) **1,4-Dibromo-2-butene** (85–90% from 1,3-butadiene and bromine in carbon disulfide at ice-salt temperature followed by heating the recovered solid at 85–90°) [49].

(4) **3,4,5,6-Tetrachlorocyclohexene** (80 g., 1.3% from 28 moles of benzene, 6 g. of iodine as chain inhibitor, through which 13.5 moles of chlorine was passed at 3 g./min. while two 250-watt lamps irradiated the solution. Considerable fractionation was necessary to remove other chlorinated products, but the desired compound was a crystalline solid, m.p. 32–34°; appreciable amounts of gammexane were formed also) [50].

(5) **1,2,3,4-Naphthalene tetrabromide** (12% yield from 0.1 mole of naphthalene and 1 mole of bromine in 100 ml. of carbon tetrachloride illuminated by an arc lamp for several hr.) [51]; see also [51a] for a 30% yield by a similar reaction.

(6) **1-Iodo-1,1-dichloro-2,2,2-trifluoroethane** (83% conversion, 95%

$$
\begin{array}{c}
Cl \quad\quad F \\
\diagdown \quad \diagup \\
C{=}C \\
\diagup \quad \diagdown \\
Cl \quad\quad F
\end{array}
\xrightarrow[IF_5]{I_2}
\begin{array}{c}
Cl \quad F \\
| \quad | \\
Cl-C-C-F \\
| \quad | \\
I \quad F
\end{array}
$$

purity, from 0.54 mole of IF, 2 g. each of Al shavings and aluminum triiodide, heated to 135° in an autoclave, cooled, and 0.828 mole of 1,1-difluoro-2,2-dichloroethylene added and shaken at 0° for 17 hr.) [52].

(7) **cis- or trans-1-Bromo-1-pentene** (cis-, 65% yield from 50 mmoles each of diisoamylborane and 1-pentyne in CCl_4 to which a slight excess of

bromine in CCl_4 was added, all operations at 0–5°; instantaneous hydrolysis with 2 M NaOH gave the desired compound; for the *trans*-olefin the intermediate dibromoborane was refluxed in CCl_4 under N_2 for 6 hr. to give a 40% yield) [42].

3. Compounds Containing Halogen Attached to a Hetero Atom

The best-known of these reactions is the addition of hypohalous acid to an olefin. The hydroxyhalide formed is a useful intermediate in the preparation of

$$RCH{=}CH_2 + HOX \rightarrow RCHOHCH_2X$$

epoxides, but the reaction is by no means limited to hypohalous acid; in fact, almost any O-halogen or N-halogen compound adds to the unsaturated group. The additions for the most part are ionic and stepwise in nature, the positive halogen initiating the attack:

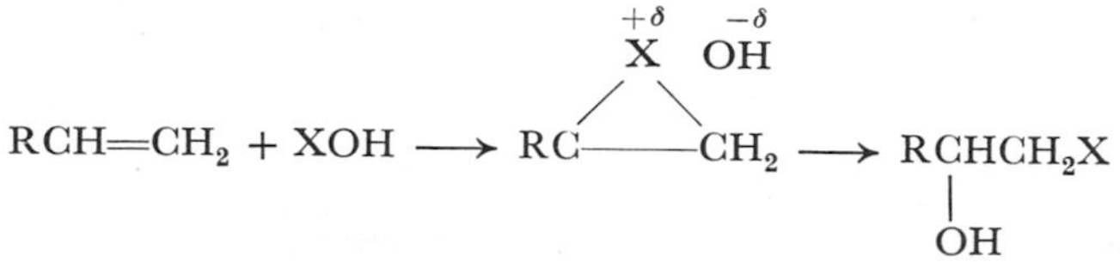

$$RCH{=}CH_2 + XOH \longrightarrow RC\overset{\overset{+\delta}{X}}{\underset{}{\diagup}}\overset{\overset{-\delta}{OH}}{\underset{}{\diagdown}}CH_2 \longrightarrow RCHCH_2X$$
$$\underset{OH}{|}$$

But some reactions are known to be free radical and give anti-Markownikoff products [53]. The reaction as shown is general.

$$(CH_3)_3CCH{=}CH_2 + Cl_2NCO_2C_2H_5 \xrightarrow{\text{25 ml. } C_6H_6}$$

0.05 mole   0.05 mole
added dropwise

$$(CH_3)_3CCHClCH_2\overset{Cl}{\underset{}{\overset{|}{N}}}CO_2C_2H_5 \xrightarrow[H_2O]{NaHSO_3} (CH_3)_3CCHClCH_2NHCO_2C_2H_5$$

Ethyl N-β-chloroneohexylcarbamate, 80%

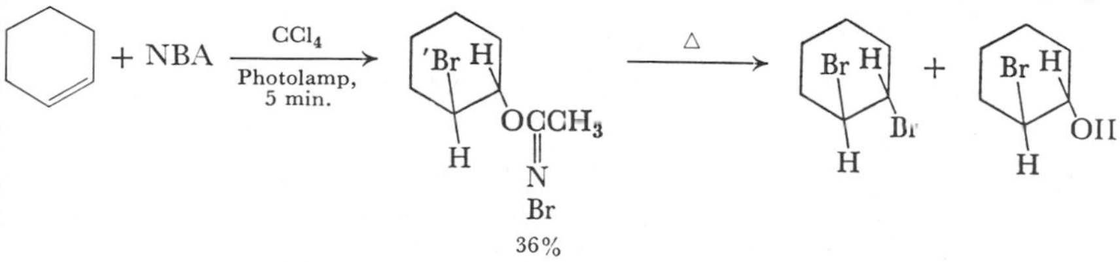

Hypohalites can be generated by adding the halogen to water or aqueous alkali, by passing chlorine into an aqueous sodium hydroxide solution of mercuric chloride [54], by passing chlorine into an aqueous solution containing urea and reprecipitated chalk [55], by employing aqueous calcium hypochlorite and carbon dioxide [56], or *t*-butyl hypochlorite [57]. Emulsifying agents increase the effectiveness of the addition [58, 59]. In the presence of a reactive solvent such as an alcohol or acid, the corresponding halo ether or halo ester is formed [60]. Yields are fair. The halo ether can be formed also from N,N-dibromobenzenesulfonamide and ethanol [61]. In the presence of water, N-bromoacetamide gives bromohydrins [62], but in an inert solvent it gives dibromo adducts [63]. The dibromo adduct is accounted for by the complex series of reactions that occur with olefins and N-bromoacetamide [64]. It appears that the N-bromoacetimidyl radical adds followed by a thermal decomposition to the dibromo adduct. N-Bromoacetamide in the presence of hydrogen chloride or hydrogen fluoride yields bromochloride and bromofluoride adducts, respectively [65]. The Prévost reaction no doubt proceeds via the iodoester [66]:

$$Ag(C_6H_5CO_2)_2I + RCH=CH_2 \longrightarrow \begin{bmatrix} RCHCH_2I \\ | \\ O_2CC_6H_5 \end{bmatrix} \longrightarrow \begin{matrix} RCHCH_2O_2CC_6H_5 \\ | \\ O_2CC_6H_5 \end{matrix}$$

Even ethers can participate in the addition [67].

$$CH_3OCH_2CH_2OCH_3 + NaHCO_3 + Cl_2 + CH_3CH=CH_2 \xrightarrow{40-45°}$$

300 ml.   1 mole   1 mole each added
                                 concurrently

$$\begin{matrix} & CH_3 \\ & | \\ CH_3OCH_2CH_2OCHCH_2Cl & + \text{ other products} \end{matrix}$$

1-Methoxy-2-(1-chloro-2-
propoxy)ethane, 0.32 mole

Iodine azide adds to olefins in the normal orientation pattern [68]:

$$RCH=CH_2 + IN_3 \rightarrow RCHN_3CH_2I$$

but the adducts from unsaturated carbonyl compounds tend to eliminate hydrogen iodide [69]:

$$[N_3CH_2CHI\overset{O}{\overset{\|}{C}}R] \longrightarrow N_3CH=CH\overset{O}{\overset{\|}{C}}R$$

N,N-Dichlorourethane, $Cl_2NCOOC_2H_5$ [70], thiocyanogen chloride, ClSCN [71], thiocyanogen trichloride, $ClSCCl=NCl$ [72], sulfur dichloride, SCl_2 [73], 2-4-dinitrophenylsulfenyl chloride [74],

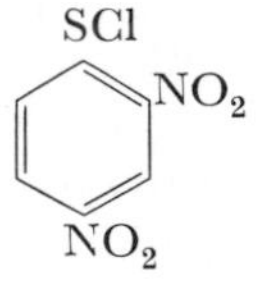

and 2,4-dinitrophenyl selenium trichloride [75],

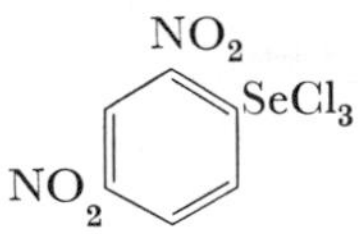

are but a few of the reagents which add to unsaturated compounds. 2,4-Dinitrophenylsulfenyl chloride adds in a manner to suggest a free radical mechanism [76]. The addition of iodine isocyanate to alkenes is a general reaction

$$RCH=CH_2 + ArSCl \longrightarrow \begin{matrix} RCHCH_2SAr \\ | \\ Cl \end{matrix}$$

action and useful in the preparation of many other derivatives such as β-iodoalkylureas, -carbamates, and -amines as well as aziridines [77].

(a) Preparation of 2-chlorocyclohexanol Into a stirred, ice-cooled mixture of 300 ml. of water, 360 g. of urea, and 300 g. of reprecipitated chalk, chlorine was introduced under the surface until the gain in weight was 275 g. (about 1 hr.). To the mixture was added 1 l. of ice-cold water, 1200 g. of ice, 230 ml. of glacial acetic acid, and 395 g. of cyclohexene, after which stirring was started and continued until no top layer of cyclohexene remained and a heavy oil had settled (2–4 hr.). The mixture was then saturated with salt, steam-distilled, and the distillate was extracted with ether. On washing the combined ether extracts with saturated sodium chloride and drying over anhydrous sodium sulfate, the solvent was removed, and distillation of the residue gave 478 g. (74% based on cyclohexene) of 2-chlorocyclohexanol, b.p. 85–86° (17 mm) [55, 54].

(b) Other preparations

(1) **Styrene chlorohydrin** (76% of crude from styrene, carbon dioxide, and calcium hypochlorite in the presence of a trace of alkylbenzene sodium sulfonate) [59].

(2) **1,1-Dichlorohexanone-2** (80% from 0.2 mole of N-chlorosuccinimide in 300 ml. of dry methanol to which 0.1 mole of 1-hexyne was added in 30 min., the slightly exothermic reaction mixture being cooled when the temperature reached 42°; after 6 hr. the mixture was evaporated, extracted with pentane to remove succinimide, the pentane extract evaporated, and the residue distilled; the ketal was hydrolyzed heterogeneously to the dichloroketone in 68% yield by 17% aqueous hydrochloric acid) [78].

(3) **2-Iodocyclohexanol** (70%, m.p. 41–42°, from 0.1 mole of silver perchlorate and the necessary amount of cyclohexene both in ether to which 0.2 mole of iodine was added in portions; the mixture was shaken with water and the product isolated from the ether) [79].

4. Free Radicals

The addition of hydrogen bromide and halogen under free radical conditions has been discussed in preceding sections (B.1 and B.2). There remains the discussion of a series of additions of free radicals derived from halohydrocarbons and like molecules, a discussion which has been reviewed in part [80, 81]. The addition occurs with a variety of halides such as carbon tetrachloride, chloroform, trichlorobromomethane, and many others containing fluorine and iodine as well. Thus it is of interest as a method for synthesizing polyhalides.

$$RCH{=}CH_2 + XCX_3' \longrightarrow R\overset{\overset{\displaystyle H}{|}}{C}XCH_2CX_3'$$

The unsaturated compounds may contain one or more double bonds or a triple bond [82]. As radical producers, acetyl peroxide, benzoyl peroxide, or α,α-azobisisobutyronitrile or related nitriles [82, 83] may be employed. Similar results may usually be obtained photochemically or thermally [see Ex. c.6]. Most of the reactions are conducted in the temperature range 60–100°, sometimes under nitrogen pressure if the olefin is gaseous. In many cases satisfactory yields of the 1 : 1 addition products result in the reaction. The main interference appears to be telomer formation, and sometimes a rearrangement or a dehydro-halogenation of the expected product occurs. In fact, conditions for the formation of telomers have been studied [84]. Telomerization is minimized by the use of large excesses of the halohydrocarbon but, in the case of fluorohalides, can be avoided completely by the use of cuprous chloride as a catalyst. This procedure is an excellent one (see Ex. b).

In the case of chloroform the C—H bond is broken in the addition:

$$RCH{=}CH_2 + CHCl_3 \rightarrow RCH_2CH_2CCl_3$$

With trichloromethyl bromide the C—Br bond is broken:

$$RCH{=}CH_2 + CBrCl_3 \rightarrow RCHBrCH_2CCl_3$$

The latter is a much more satisfactory reaction than the former since the speed is greater and less telomer formation occurs. Both of these factors often operate to give high yields in the trichloromethyl bromide addition. Of several poly-halides containing fluorine, trifluoromethyl iodide appears to have been studied to the greatest extent. Here the C—I bond is usually broken:

$$RCH{=}CH_2 + CF_3I \rightarrow RCHICH_2CF_3$$

These additions are of interest in the synthesis of highly fluorinated products, although in many cases the yields leave something to be desired.

The addition of polyhalides to cyclic olefins also occurs [83]. In such cases the size of the adding group is an important factor. With bulky polar groups, *trans* addition occurs [85, 86] but, with chloroform [87, 86] or ethyl bromo-acetate [88], *cis* addition takes place. These additions are free radical ones which may be represented, in the case of trichloromethyl bromide, as follows. Light or peroxide forms the bromine and trichloromethyl radicals:

$$BrCCl_3 \rightarrow \cdot Br + Cl_3C\cdot$$

and in the first step of the propagation the $Cl_3C\cdot$ radical attacks the olefin to form

$$RCH{=}CH_2 + Cl_3C\cdot \rightarrow \cdot CHRCH_2CCl_3$$
$$\cdot CHRCH_2CCl_3 + BrCCl_3 \rightarrow BrCHRCH_2CCl_3 + \cdot CCl_3$$

the adduct radical which reacts in the second step with trichloromethyl bromide to give the final product and the trichloromethyl radical [89].

(a) Preparation of 1,1,1,3-tetrachloro-3-ethoxypropane.   Dry vinyl

$$CH_2{=}CHOC_2H_5 + CCl_4 \rightarrow CCl_3CH_2CHClOC_2H_5$$

ethyl ether, 288 g., was added under the surface of 1085 g. of refluxing carbon tetrachloride (reagent grade) in the presence of 2.4 g. of α,α-azobisisobutyroni-trile over a 3-hr. period. After an additional hour of refluxing, the excess of carbon tetrachloride was removed under reduced pressure on a steam bath, and on distillation 796 g. (88%), b.p. 64° (7 mm.), of 1,1,1,3-tetrachloro-3-ethoxy-propane was obtained [90].

(b) Preparation of 1,1,2-trifluoro-1,4-dibromo-2-chlorodecane, $CF_2BrCFCl$ CH_2CHBr C_6H_{13}.   1-Octene (0.1 mole),

$$
CF_2Br\!-\!\underset{\underset{Cl}{|}}{\overset{\overset{F}{|}}{C}}\!-\!Br \quad (0.2\ \text{mole}),
$$

CuCl (0.1 g.), and ethanolamine (0.05 mole) were refluxed with stirring in 100 ml. of t-butyl alcohol for 24 hr. The mixture was diluted with water, extracted with ether, and the ether extract dried, evaporated, and fractionated to give a 70% yield of the desired compound, b.p. 88° (0.4 mm.). This reaction is simple, attractive, and applicable to a wide variety of additions including those of internal olefins. It fails if debromination is a more favorable pathway of decomposition than addition to the olefin [91].

(c) Other examples

(1) 1,1,1-Trichloro-3-bromo-3-phenylpropane, $CCl_3CH_2CHBrC_6H_5$ (78% from styrene, trichloromethyl bromide, and a trace of acetyl peroxide at 60–70° for 4 hr. [92]; see Ex. c.6 also).

(2) 1-Iodo-3,3,4,4,5,5,5-heptafluoropentyl acetate (85% from vinyl

$$CH_3COOCH\!=\!CH_2 + CF_3CF_2CF_2I \rightarrow CH_3COOCHICH_2CF_2CF_2CF_3$$

acetate and 1-iodoperfluoropropane in the presence of α,α-azobisisobutyronitrile at 65–80° under nitrogen for 7 hr.) [93].

(3) Dibutyl 3,3,3-trichloro-1-bromopropene-1-boronate (90% from

$$HC\!\equiv\!CB(OC_4H_9)_2 + CCl_3Br \rightarrow CCl_3CH\!=\!CBrB(OC_4H_9)_2$$

dibutyl acetyleneboronate and trichloromethyl bromide under nitrogen in the presence of azobisisobutyronitrile at 95° for 2.5 hr.) [82].

(4) Ethyl γ-bromocaprate (57% from 0.24 mole of 1-octene and 0.96 mole of ethyl bromoacetate to which 0.025 mole of diacetyl peroxide in 0.04 mole of ethyl bromoacetate was added dropwise; the low boiling material was removed by distillation at 20 mm. and the desired material at 0.2 mm., b.p. 93–94°) [94].

(5) 1,1,1-Trichloro-2-m-nitrophenylethane (55% from 13.8 g. of m-nitroaniline diazotized in cold hydrochloric acid to which 16 ml. of 1,1-dichloroethene and cupric chloride in aqueous acetone was added and the mixture allowed to warm until nitrogen evolution was obvious; the addition of diazonium salts to unsymmetrically substituted olefins is general) [95].

(6) 1,1,1-Trichloro-3-bromo-3-phenylpropane 40% from 0.12 mole of styrene and 0.5 mole of bromotrichloromethane under nitrogen heated on a steam bath) [96].

(7) 2,5-Dichloro-4-hexenoic acid (66% from butadiene, dichloro-acetic acid, and catalytic amounts of cuprous chloride in acetonitrile at 110° for 12 hr. in a glass-lined autoclave) [97].

5. Halocarbenes or Halocarbanions

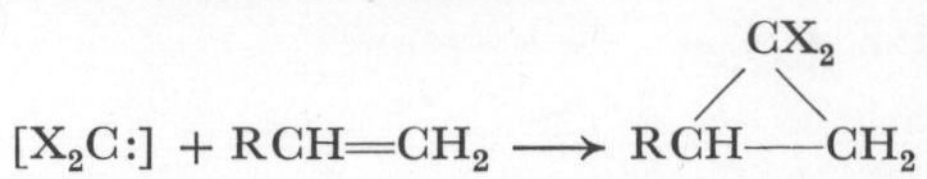

$$[X_2C:] + RCH=CH_2 \longrightarrow RCH\underset{\displaystyle \diagdown \diagup CX_2}{\overline{\hspace{1.2cm}}}CH_2$$

The specific references concerning halocarbenes from J. Hine's treatise [98] are not repeated, but a summary is given together with advances made since this publication. The early procedures generated the halocarbene from haloform and potassium t-butoxide. The presence of alcohol, however, tends to give

$$HCX_3 + KOC(CH_3)_3 \rightleftharpoons \overset{\ominus}{C}X_3 + (CH_3)_3COH \xrightarrow{-X\ominus} [X_2C:]$$

the side reaction $X_2C: + ROH \rightarrow X_2CHOR$, which diminishes the yield of cyclopropane. Better sources of dihalocarbenes, which avoid side reactions, are sodium trihaloacetate in the dimethyl ether of ethylene glycol [99] and t-butyl trihaloacetate [100]:

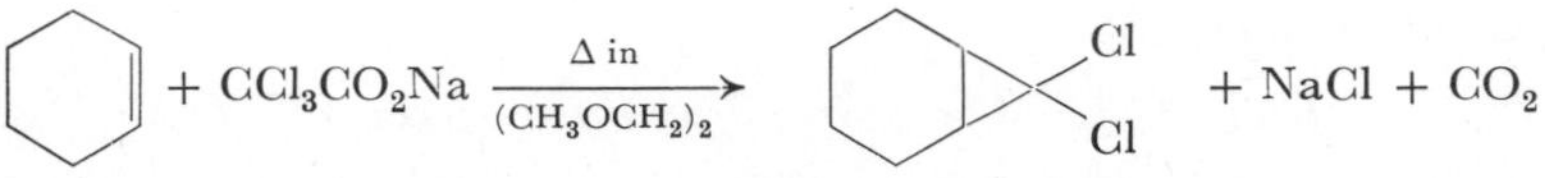

7,7-Dichloronorcarane, 65%

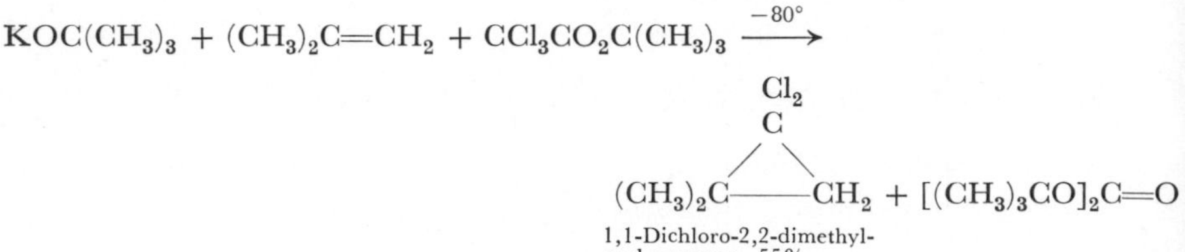

$$KOC(CH_3)_3 + (CH_3)_2C=CH_2 + CCl_3CO_2C(CH_3)_3 \xrightarrow{-80°}$$

$$(CH_3)_2C\underset{\displaystyle \diagup \diagdown \overset{Cl_2}{C}}{\overline{\hspace{1.2cm}}}CH_2 + [(CH_3)_3CO]_2C=O$$

1,1-Dichloro-2,2-dimethyl-
cyclopropane, 55%

Of all halocarbene sources, phenyltrihalomethylmercury invariably gives higher yields, and it appears to be the most versatile, the carbene from this source reacting with such poor acceptors as ethylene, stilbene, and tetrachloroethylene [101]:

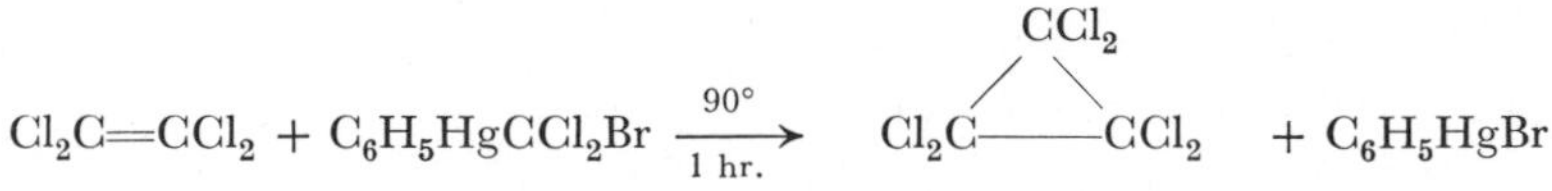

$$Cl_2C=CCl_2 + C_6H_5HgCCl_2Br \xrightarrow[\text{1 hr.}]{90°} Cl_2C\underset{\displaystyle \diagup \diagdown \overset{CCl_2}{}}{\overline{\hspace{1.2cm}}}CCl_2 + C_6H_5HgBr$$

1 mole   0.1 mole   Hexachlorocyclopropane, 74%

Difluorocarbene, generated from trimethyl (trifluoromethyl) tin by the reaction of sodium iodide in 1,2-dimethoxyethane at 80°, adds to cyclohexene to give difluoronorcarane (73%) [102]. Tetramethylethylene under similar conditions

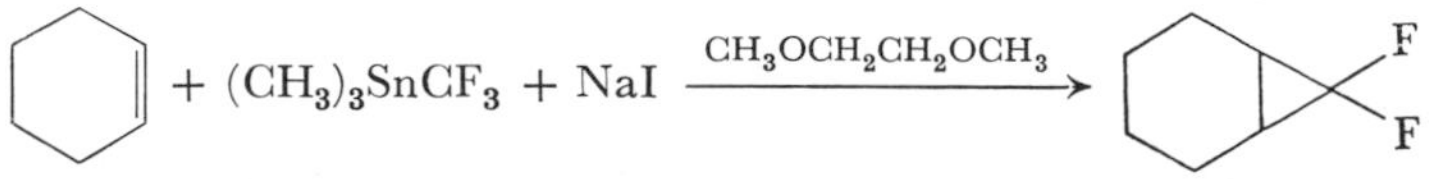

forms 1,1-difluorotetramethylcyclopropane (77%).

Dihalocarbenes have been inserted into an allylic sulfide and benzylic compounds

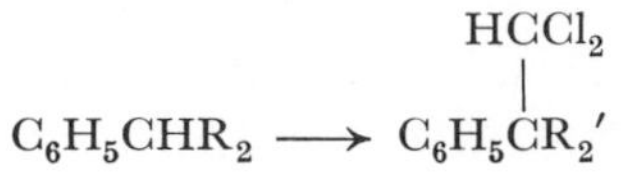

$$C_6H_5CHR_2 \longrightarrow C_6H_5\overset{\displaystyle HCCl_2}{\underset{\displaystyle |}{C}}R_2'$$

and added to acetylenes, allenes, ketene acetals, benzalaniline, indenes, pyrroles, and polynuclear hydrocarbons [98]. In the last three cases, worthwhile rearrangements, one of which is illustrated, can follow addition [103]:

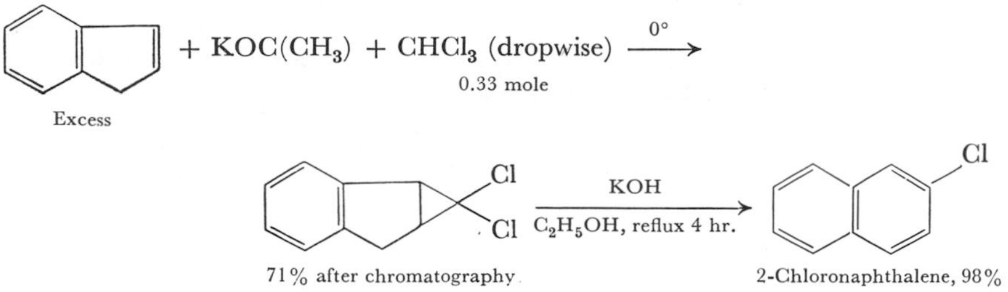

71% after chromatography   2-Chloronaphthalene, 98%

for other transformations, see [104].

Monochlorocarbenes are more difficult to generate, generally give poorer yields, and may involve an organometallic species rather than a carbene. The most recent example is illustrated [105]. Some 7-methylnorcarane is separated as well.

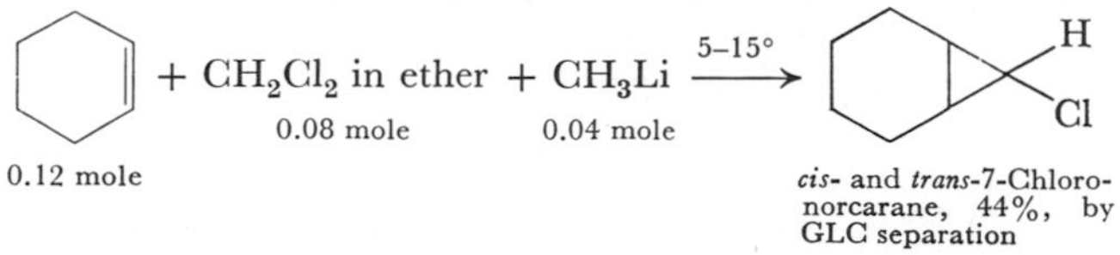

cis- and trans-7-Chloro-norcarane, 44%, by GLC separation

Among reagents which can trap the carbanion, $^{\ominus}CX_3$, before the halocarbene, $:CX_2$, is formed, are carbonyl compounds (see Ex. b.1), acid chlorides, anhydrides (see Ex. b.2), and enamines. Indeed, the latter simply displace the chloride ion from carbon tetrachloride [106]. Yields generally are low, but the reaction is so simple to perform that it has merit.

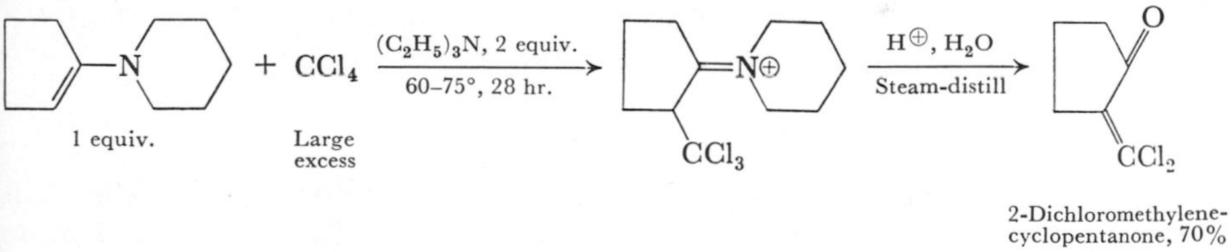

2-Dichloromethylene-cyclopentanone, 70%

If the condensations with haloforms are carried out in hydroxylic solvents, the dihalocarbene intermediate is largely bypassed to form alkoxy- or hydroxy-acids [107]:

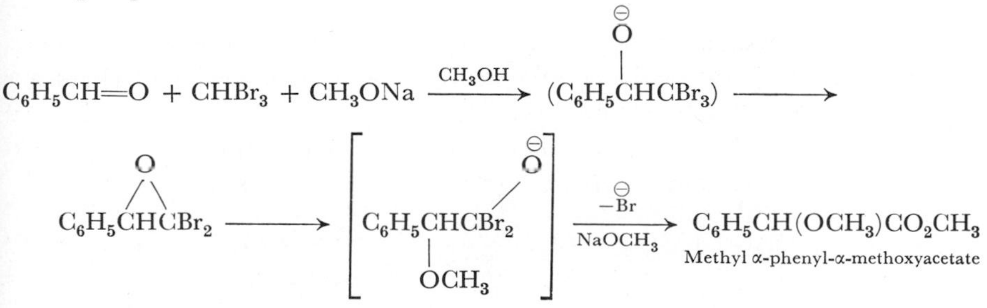

Methyl α-phenyl-α-methoxyacetate

(a) Preparation of 7,7-dibromonorcarane. Phenyl(tribromomethyl)-mercury, 0.105 mole, in 50 ml. of benzene and 0.315 mole of cyclohexene were

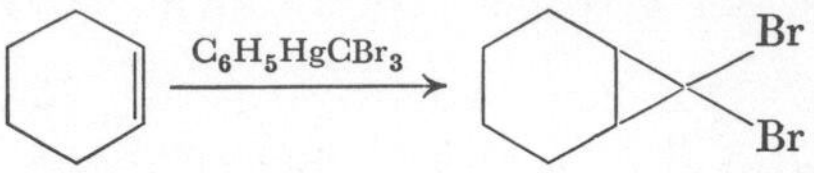

heated with stirring under reflux for 2 hr. The phenylmercuric bromide separated in quantitative yield, and fractional distillation of the filtrate gave 22.5 g. (88%) of 7,7-dibromonorcarane [108].

(b) Other examples

(1) α,α,α-Trichloro-*t*-butyl alcohol (chloretone) (86% from 0.1 mole of sodamide in 250 ml. of liquid ammonia to which a mixture of 5.8 g. of acetone and 11.9 g. of chloroform was added in 4 min.; the mixture was neutralized with 9 g. of NH_4Cl, evaporated, and extracted with ether to yield the desired compound; with other ketones the yields ranged from 18 to 93%) [109].

(2) 3-Hydroxy-3-tribromomethylphthalide

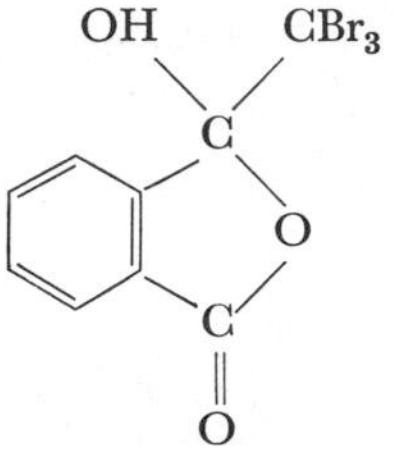

(54% from 1 equiv. each of sodium tribromoacetate and phthalic anhydride heated to 80° in the dimethyl ether of ethylene glycol until CO_2 evolution ceased; the reaction is general for anhydrides, aldehydes, and acid chlorides, but sodium trifluoroacetate does not react) [110].

6. Alkylation or Acylation Reagents

This section comprises a variety of additions of alkyl or acyl halides to unsaturated compounds under Friedel-Crafts conditions. Sometimes elimination

$$R'CH{=}CH_2 + R_3CX \xrightarrow{\text{AlCl}_3} R'CHXCH_2CR_3$$

of HX occurs spontaneously or on treatment with base to give essentially a substitution product. Markownikoff's rule is followed in that the halogen atom adds to the carbon atom containing the smaller number of hydrogen atoms [111]. This method is of little value synthetically except when tertiary halides are employed (see Ex. a). Here yields in the 80 and 90% range have been obtained [112]. Primary and secondary halides lead to large amounts of rearranged products, but polyhalides are quite amenable to the process [113]:

$$CHCl_3 + CCl_2{=}CCl_2 \xrightarrow[\text{Reflux}]{\text{AlCl}_3 \text{ (trace)}} CCl_3CCl_2CHCl_2$$

unsym-Heptachloro-propane, 88–93%

Catalysts such as aluminium chloride, ferric chloride, bismuth chloride, stannic chloride, or zinc chloride are employed, the first two particularly at low temperature.

$$RCOCl + HC\equiv CH \xrightarrow{AlCl_3} RCOCH=CHCl$$

$$RCOCl + CH_2=CHCl \xrightarrow{AlCl_3} \underset{\text{Very unstable}}{RCOCH_2CHCl_2} \longrightarrow RCOCH=CHCl$$

Although aromatic compounds react with alkyl halides to form hydrocarbons containing side chains, by employing vinyl or allyl halides the reaction can be made to occur preferentially at the double bond. Proton donor type of catalysts such as sulfuric acid or hydrogen fluoride [114] are employed. Thus a method

$$ArH + CH_3CH=CHCH_2Cl \xrightarrow{H_2F_2} Ar\overset{\overset{\displaystyle CH_3}{|}}{C}HCH_2CH_2Cl$$

for the synthesis of aromatic compounds having a halogen-substituted side chain is available. Yields here are rarely above the 60–75 % range [115, 116]. The main side reaction appears to be an intermolecular alkylation of certain intermediates followed by hydrogen transfer [114].

But all allylic halides are not this selective toward proton donor acids. For example, 1,3-dichloro-2-methylpropene does not condense with aromatic hydrocarbons in the presence of sulfuric acid, but rather in the presence of aluminum chloride, to give variable products depending on conditions [117]:

$$C_6H_6 + ClCH_2\overset{\overset{\displaystyle CH_3}{|}}{C}=CHCl \xrightarrow{AlCl_3,\ CH_3NO_2} \underset{\substack{\text{1,3-Dichloro-2-phenyl-2-}\\\text{methylpropane, 51%}}}{C_6H_5\overset{\overset{\displaystyle CH_3}{|}}{C}(CH_2Cl)_2}$$

$$\xrightarrow[\text{granules}]{AlCl_3} \underset{\text{γ-Chloro-β-methylallylbenzene, 23%}}{C_6H_5CH_2C(CH_3)=CHCl}$$

The reaction of allyl chloride with benzene is discussed critically in this paper.

It should also be realized that some allylic halides rearrange spontaneously or readily with aluminum chloride [118]:

$$CCl_3CBr=CH_2 \xrightarrow{\text{On standing}} \underset{\text{α,β,β-Trichloroallyl bromide}}{CCl_2=CClCH_2Br}$$

$$CH_2=CHCCl_3 \xrightarrow[\text{Violent reaction}]{AlCl_3\ \text{(trace)}} \underset{\text{β,β-Dichloroallyl chloride}}{Cl_2C=CHCH_2Cl}$$

Even saturated polyhalides undergo rearrangement in the presence of this chloride [119].

$$\underset{\substack{64\%\ dl,\ 26\%\ meso\\\text{2,4-Dichloropentane}}}{CH_3CHClCH_2CHClCH_3} \underset{CS_2,\ 25°}{\overset{AlCl_3}{\rightleftarrows}} \underset{\substack{5\%\ erythro,\ 5\%\ threo\\\text{2,3-Dichloropentane}}}{CH_3CHClCHClCH_2CH_3}$$

(a) Preparation of 1-chloro-3,3-dimethylbutane. *t*-Butyl chloride 3.5 kg., chilled to −20°, was treated with 50 g. of anhydrous aluminum chloride and pure ethylene was introduced as fast as it was absorbed. The temperature was kept at −17 to −20°, while stirring, until the ethylene no longer absorbed readily (2 hr.). The liquid was then decanted into 200 ml. of water and, after shaking, the organic layer was separated, dried over solid potassium hydroxide, and distilled. The fraction boiling at 115–122°, 3.9 kg. (86 %) was 1-chloro-3,3-dimethylbutane [112].

(b) Other examples

(1) **1-Chloro-2-*p*-tolyl-2-methylpropane** (74.5 % from toluene and

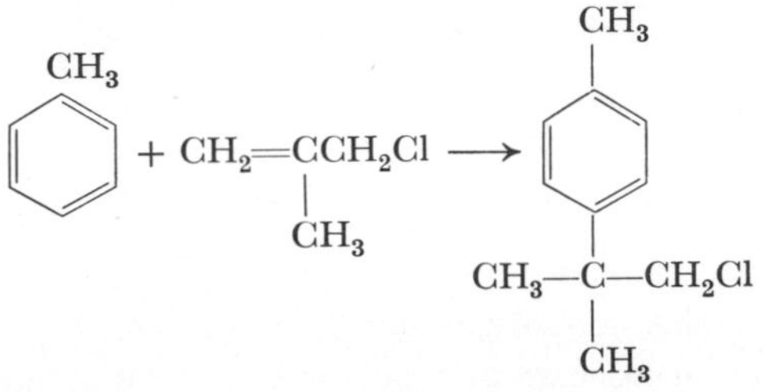

2-methallyl chloride in the presence of sulfuric acid) [116].

(2) **Triphenylchloromethane** (84–86 % based on aluminum chloride, from benzene and carbon tetrachloride) [120].

(3) **3,4-Dichlorocoumarin** (28 % from 0.03 mole each of phenol and

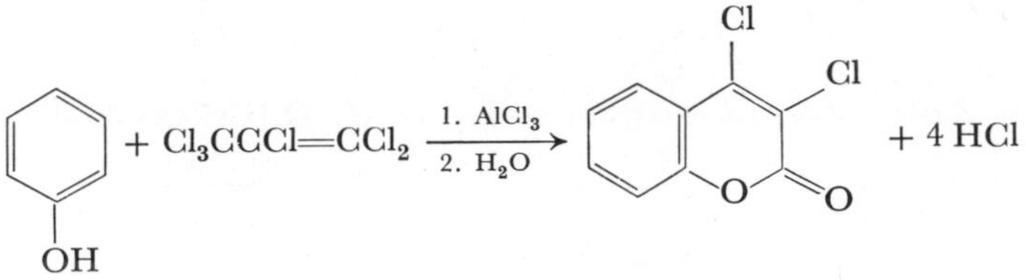

perchloropropylene and 0.06 mole of AlCl$_3$ in CS$_2$; the mixture was hydrolyzed with dilute H$_2$SO$_4$; yields ranged from 2 to 90 % for a series of different phenols) [121].

(4) **1,1,1-Trifluoro-3-phenylpropane** (55 % from 21.6 g. of benzyl

$$C_6H_5CH_2OH + CH_2{=}CCl_2 + 3\,HF \rightarrow F_3CCH_2CH_2C_6H_5$$

alcohol, 26.2 g. of 1,1-dichloroethene in 75 g. of dry hydrogen fluoride at −78° and warmed to 0° for 3 hr.) [122].

7. From Epoxides and Hydrogen Halides

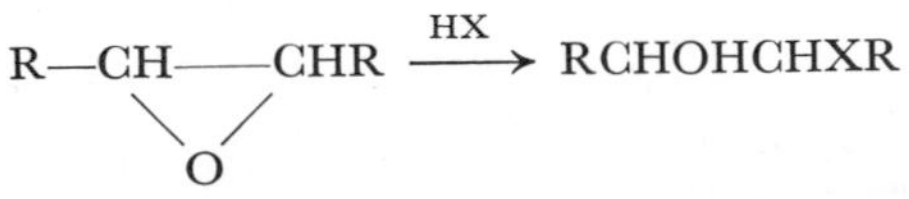

$$R{-}CH{-\!\!-\!\!-}CHR \xrightarrow{\text{HX}} RCHOHCHXR$$

This synthesis is somewhat less strenuous than that described under B.2, since here a halohydrin rather than a dihalide is formed. The reaction has been reviewed [123], and it is of interest since it is usually stereospecific. Cyclohexene oxide, for example, gives the *trans*-bromohydrin with hydrogen bromide [124].

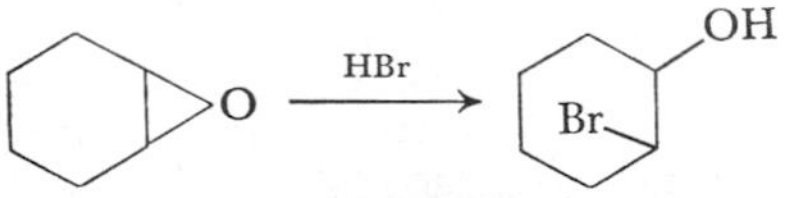

Also, as in acidic hydrolysis to produce glycols, Alcohols, Chapter 4, A.7, the epoxide from a *cis*-olefin such as *cis*-stilbene reacts with hydrogen chloride by inversion to give the *threo*-chlorohydrin while *trans*-stilbene oxide under the same conditions produces the *erythro*-chlorohydrin [125]. There are cases, however, in which no inversion occurs. The *cis*- and *trans*-dypnone oxides, for example, by the action of hydrogen chloride in acetic acid give diastereoiso-meric chlorohydrins which can best be accounted for by a retention of configuration in the ring-opening step [126]. In addition *trans*-N,N-diethyl-3-phenylglycidamide with hydrogen chloride in a nonpolar solvent such as benzene gives the *threo*-chlorohydrin (retention of configuration), while with hydrogen chloride in a polar solvent such as methanol the *erythro*-chlorohydrin (inversion) is formed [127]. The *cis*-glycidamide with either of these solvents gives only the *threo*-chlorohydrin. Some of these results may be explained by neighboring group participation. The yields in these additions are usually high.

(a) Preparation of ethylene bromohydrin (87–92 % from ethylene oxide and 46 % hydrobromic acid) [128].

(b) Other examples

(1) ***trans*-2-Bromocyclohexanol** (73 % from cyclohexene oxide and 48 % hydrobromic acid) [124].

(2) ***erythro*-N,N-Diethyl-3-chloro-2-hydroxy-3-phenylpropion-amide** (inversion) (94 % from *trans*-N,N-diethyl-3-phenylglycidamide and

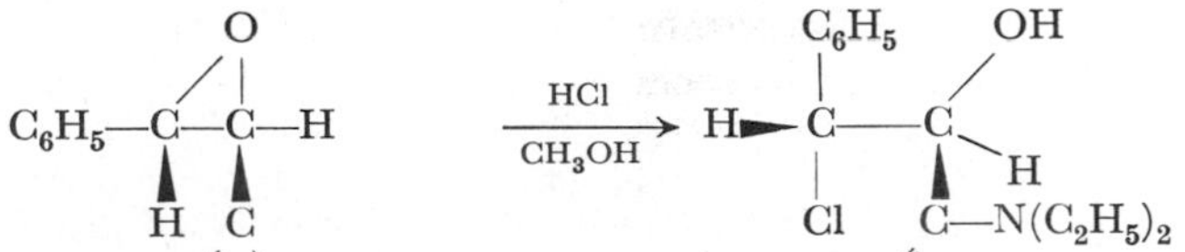

methanolic HCl) [127].

(3) ***threo*-N,N-Diethyl-3-chloro-2-hydroxy-3-phenylpropion-amide** (retention of configuration) (97 % from *trans*-N,N-diethyl-3-phenylgly-

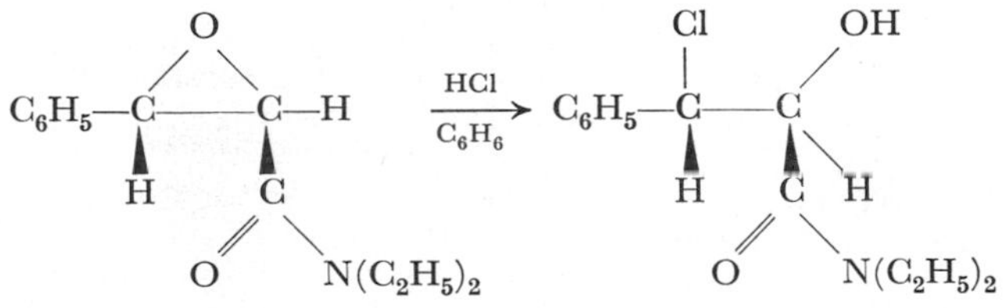

cidamide, C_6H_6, and HCl) [127].

1. P. B. D. de la Mare and R. Bolton, *Electrophilic Additions to Unsaturated Systems*, Elsevier Publishing Co., New York, 1966, Chap. 5.
2. C. Moureu and R. Chaux, *Org. Syn.*, Coll. Vol. **1**, 166 (1941).
3. R. Mozingo and L. A. Patterson, *Org. Syn.*, Coll. Vol. **3**, 576 (1955).
4. R. Stewart and R. H. Clark, *J. Am. Chem. Soc.*, **69**, 713 (1947).
5. Y.-R. Naves *et al.*, *Helv. Chim. Acta*, **30**, 1599 (1947).
6. R. B. Moffet, *Org. Syn.*, Coll. Vol. **4**, 238 (1963).
7. H. Stone and H. Shechter, *Org. Syn.*, Coll. Vol. **4**, 543 (1963).
8. H. Stone and H. Shechter, *J. Org. Chem.*, **15**, 491 (1950).
9. G. Williams, *J. Chem. Soc.*, 1046 (1938).
10. Ref. 1, p. 53.
11. P. E. Peterson and J. E. Duddey, *J. Am. Chem. Soc.*, **88**, 4990 (1966).
12. M. J. S. Dewar and R. C. Fahey, *J. Am. Chem. Soc.*, **85**, 3645 (1963).
13. H. C. Brown and M.-H. Rei, *J. Org. Chem.*, **31**, 1090 (1966).
14. A. L. Henne and S. Kaye, *J. Am. Chem. Soc.*, **72**, 3369 (1950).
15. Shell Intern. Maatschappy N.V., Neth. Appl. 6,410,368; *C.A.*, **63**, 2897 (1965).
16. G. Sosnovsky, *Free Radical Reactions in Preparative Organic Chemistry*, The Macmillan Company, New York, 1964, p. 6; F. R. Mayo and C. Walling, *Chem. Rev.*, **27**, 351 (1940).
17. J. H. Raley *et al.*, *J. Am. Chem. Soc.*, **70**, 2767 (1948).
18. K. Griesbaum *et al.*, *J. Am. Chem. Soc.*, **87**, 3151 (1965).
19. Houben-Weyl, *Methoden der Organischen Chemie*, Georg Thieme Verlag, Stuttgart, 4th Ed., Vol. **5**, Pt. 3, 1962, pp. 99 and 812; Pt. 4, 1960, pp. 102 and 535.
20. C. Walling *et al.*, *J. Am. Chem. Soc.*, **61**, 2693 (1939).
21. Ref. 1, Chaps. 6 and 7.
22. E. A. Braude and E. A. Evans, *J. Chem. Soc.*, 3333 (1956).
23. I. V. Bodrikov and Z. S. Smolyan, *Russ. Chem. Rev. (English Transl.)*, **35**, 374 (1966).
24. Ref. 19, Vol. 5, Pt. 3, p. 529.
25. W. H. Puterbaugh and M. S. Newman, *J. Am. Chem. Soc.*, **81**, 1611 (1959).
26. C. E. Boord *et al.*, *J. Am. Chem. Soc.*, **73**, 3329 (1951).
27. R. E. Buckles and D. F. Knaack, *J. Org. Chem.*, **25**, 20 (1960).
28. M. S. Kharasch and H. C. Brown, *J. Am. Chem. Soc.*, **61**, 3432 (1939).
29. L. Spiegler and J. M. Tinker, *J. Am. Chem. Soc.*, **61**, 940 (1939).
30. C. H. Robinson *et al.*, *J. Am. Chem. Soc.*, **81**, 2191 (1959).
31. M. S. Kharasch and M. G. Berkman, *J. Org. Chem.*, **6**, 810 (1941).
32. A. L. Henne *et al.*, *J. Am. Chem. Soc.*, **72**, 3577 (1950).
33. N. B. Lorette, *J. Org. Chem.*, **26**, 2324 (1961).
34. H. R. Ing, *J. Chem. Soc.*, 1393 (1948).
35. I. L. Knunyants, *Russ. Chem. Rev. (English Transl.)*, **35**, 417 (1966).
36. H. S. Rhinesmith, *Org. Syn.* Coll. Vol., **2**, 177 (1943).
37. L. F. and M. Fieser, *Reagents for Organic Synthesis*, John Wiley and Sons, New York, 1967, p. 333.
38. A. V. Dombrovskiĭ, *Zh. Obshch. Khim.*, **24**, 610 (1954); *C.A.* **49**, 5484 (1955).
39. S.-O. Lawesson and N. C. Yang, *J. Am. Chem. Soc.*, **81**, 4230 (1959).
40. L. F. Fieser, *J. Chem. Ed.*, **31**, 291 (1954).
41. L. F. Hatch *et al.*, *J. Am. Chem. Soc.*, **81**, 5943 (1959).
42. H. C. Brown *et al.*, *J. Am. Chem. Soc.*, **89**, 4531 (1967).
43. P. W. Robertson *et al.*, *J. Chem. Soc.*, 2191 (1950).
44. T.-C. Shieh *et al.*, *J. Am. Chem. Soc.*, **86**, 5021 (1964).
45. J. Bornstein and L. Skarlos, *J. Am. Chem. Soc.*, **90**, 5044 (1968).
46. P. B. D. de la Mare and S. Galandauer, *J. Chem. Soc.*, 36 (1958).
47. S. Winstein and E. Grunwald, *J. Am. Chem. Soc.*, **70**, 828 (1948).
48. J. R. Johnson and W. L. McEwen, *Org. Syn.*, Coll. Vol. **1**, 521 (1941).
49. G. S. Skinner *et al.*, *J. Am. Chem. Soc.*, **72**, 1648 (1950).
50. A. J. Kolka *et al.*, *J. Am. Chem. Soc.*, **73**, 5224 (1951).
51. F. R. Mayo and W. B. Hardy, *J. Am. Chem. Soc.*, **74**, 911 (1952).
51a. J. R. Sampey *et al.*, *J. Am. Chem. Soc.*, **71**, 3697 (1949).

52. M. Hauptschein and M. Braid, *J. Am. Chem. Soc.*, **83,** 2383 (1961).
53. T. A. Foglia and D. Swern, *J. Org. Chem.*, **33,** 766 (1968).
54. G. H. Coleman and H. F. Johnstone, *Org. Syn.*, Coll. Vol. **1,** 158 (1941).
55. M. S. Newman and C. A. VanderWerf, *J. Am. Chem. Soc.*, **67,** 233 (1945).
56. W. S. Emerson, *J. Am. Chem. Soc.*, **67,** 516 (1945).
57. W. E. Hanby and H. N. Rydon, *J. Chem. Soc.*, 114 (1946).
58. C. M. Suter and H. B. Milne, *J. Am. Chem. Soc.*, **62,** 3473 (1940).
59. W. S. Emerson, *J. Am. Chem. Soc.*, **67,** 516 (1945).
60. M. A. Dolliver *et al.*, *J. Am. Chem. Soc.*, **60,** 440 (1938); C. F. Irwin and G. F. Hennion, *J. Am. Chem. Soc.*, **63,** 858 (1941); S. Winstein and R. B. Henderson, *J. Am. Chem. Soc.*, **65,** 2196 (1943).
61. H. L. Holmes and K. M. Mann, *J. Am. Chem. Soc.*, **69,** 2000 (1947).
62. S. Winstein and R. E. Buckles, *J. Am. Chem. Soc.*, **64,** 2780 (1942).
63. R. E. Buckles *et al.*, *J. Am. Chem. Soc.*, **71,** 1157 (1949); *J. Org. Chem.*, **22,** 55 (1957).
64. S. Wolfe and D. V. C. Awang, *J. Am. Chem. Soc.*, **89,** 5287 (1967).
65. Ref. 37, pp. 74, 75.
66. Ref. 37, p. 1007.
67. G. Sumrell *et al.*, *J. Org. Chem.*, **30,** 84 (1965).
68. A. Hassner and F. W. Fowler, *J. Org. Chem.*, **33,** 2686 (1968).
69. F. W. Fowler, A. Hassner and L. A. Levy, *J. Am. Chem. Soc.*, **89,** 2077 (1967).
70. T. A. Foglia and D. Swern, *J. Org. Chem.*, **33,** 766 (1968).
71. R. G. R. Bacon and R. G. Guy, *J. Chem. Soc.*, 318 (1960); R. G. Guy and I. Pearson, *Chem. Ind.* (*London*), 1255 (1967).
72. R. G. R. Bacon, *et al.*, *J. Chem. Soc.* 764 (1958).
73. R. C. Fuson *et al.*, *J. Org. Chem.*, **11,** 469 (1946).
74. N. Kharasch and C. M. Buess, *J. Am. Chem. Soc.*, **71,** 2724 (1949).
75. D. D. Lawson and N. Kharasch, *J. Org. Chem.*, **24,** 857 (1959).
76. D. R. Hogg, *Quart. Rep. Sulfur Chem.*, **2,** 339 (1967).
77. A. Hassner *et al.*, *J. Org. Chem.*, **32,** 540 (1967).
78. S. F. Reed, Jr., *J. Org. Chem.*, **30,** 2195 (1965).
79. L. Birckenbach *et al.*, *Chem. Ber.*, **65,** 1339 (1932).
80. C. Walling, *Free Radicals in Solution*, John Wiley and Sons, New York, 1957, p. 247.
81. Ref. 16, Chap. 2.
82. D. S. Matteson and K. Peacock, *J. Org. Chem.*, **28,** 369 (1963).
83. N. O. Brace, *J. Org. Chem.*, **27,** 3027 (1962).
84. J. Harmon *et al.*, *J. Am. Chem. Soc.*, **72,** 2213 (1950).
85. S. J. Cristol and J. A. Reeder, *J. Org. Chem.*, **26,** 2182 (1961).
86. D. I. Davies, *J. Chem. Soc.*, 3669 (1960).
87. V. A. Rolleri, thesis, University of Delaware (1958), *Dissertation Abstr.* 19, 960 (1958).
88. Ref. 80, p. 267.
89. M. S. Kharasch and M. Sage, *J. Org. Chem.*, **14,** 537 (1949).
90. H. J. Minnemeyer *et al.*, *J. Org. Chem.*, **26,** 4425 (1961).
91. D. J. Burton and L. J. Kehoe, *Tetrahedron Letters*, 5163 (1966).
92. M. S. Kharasch *et al.*, *J. Am. Chem. Soc.*, **69,** 1105 (1947).
93. N. O. Brace, *J. Org. Chem.*, **27,** 3033 (1962).
94. M. S. Kharasch *et al.*, *J. Am. Chem. Soc.*, **70,** 1055 (1948).
95. V. M. Naidan and A. V. Dombrovskii, *J. Gen. Chem. USSR* (*Eng. Transl.*), **34,** 1474 (1964).
96. W. A. Skinner *et al.*, *J. Org. Chem.*, **23,** 1710 (1958).
97. I. L. Mador and J. A. Scheben, U.S. Patent, 3,338,960, August 29, 1967; *C.A.*, **68,** 2055 (1968).
98. J. Hine, *Divalent Carbon*, The Ronald Press Company, New York, 1964, Chap. 3.
99. W. M. Wagner *et al.*, *Rec. Trav. Chim.*, **80,** 740 (1961).
100. W. E. Parham and F. C. Loew, *J. Org. Chem.*, **23,** 1705 (1958).
101. D. Seyferth, *J. Org. Chem.*, **28,** 1163 (1963).
102. D. Seyferth *et al.*, *J. Am. Chem. Soc.*, **87,** 681 (1965).
103. W. E. Parham *et al.*, *J. Am. Chem. Soc.*, **78,** 1437 (1956).

104. W. E. Parham, *Record Chem. Progr.* (Kresge-Hooker Sci. Lib.), **29,** 2 (1968).

105. W. L. Dilling and F. Y. Edamura, *J. Org. Chem.,* **32,** 3492 (1967).

106. J. Wolinsky and D. Chan, *Chem. Commun.,* 567 (1966).

107. W. Reeve and E. L. Compere, Jr., *J. Am. Chem. Soc.,* **83,** 2755 (1961).

108. D. Seyferth *et al., J. Org. Chem.,* **27,** 1491 (1962); *J. Am. Chem. Soc.,* **87,** 4259 (1965); *see also* R. T. Dickerson and H. M. Walborsky, U.S. Patent 3,264,359, August 2, 1966; *C.A.,* **65,** 16882 (1966).

109. H. G. Viehe, U.S. Patent 3,274,227, September 20, 1966; *C.A.,* **65,** 16881 (1966).

110. A. Winston *et al., J. Org. Chem.,* **32,** 2166 (1967).

111. L. Schmerling in G. A. Olah, *ed., Friedel-Crafts and Related Reactions,* Interscience Publishers, New York, 1964, Vol. II, Pt 2, p. 1133.

112. A. Brändström, *Acta Chem. Scand.,* **13,** 610, 611 (1959).

113. M. W. Farlow, *Org. Syn.,* Coll. Vol., **2,** 312 (1943).

114. R. C. Koncos and B. S. Friedman, Ref. 111, Pt. 1, p. 289.

115. W. S. Calcott *et al., J. Am. Chem. Soc.,* **61,** 1010 (1939).

116. V. N. Gramenitskaya *et al., Doklady Akad. Nauk., SSSR,* **118,** 497 (1958); *C.A.,* **52,** 10915 (1958).

117. L. Schmerling *et al., J. Am. Chem. Soc.,* **80,** 576 (1958).

118. A. N. Nesmeyanov, *Selected Works in Organic Chemistry,* The Macmillan Co., New York, 1963, p. 1094 and 1037.

119. W. E. Billups and A. N. Kurtz, *J. Am. Chem. Soc.,* **90,** 1361 (1968).

120. C. R. Hauser and B. E. Hudson, Jr., *Org. Syn.,* Coll. Vol. **3,** 842 (1955).

121. M. S. Newman and S. Schiff, *J. Am. Chem. Soc.,* **81,** 2266 (1959).

122. L. S. German and I. L. Knunyants, *Izv. Akad. Nauk, SSSR, Ser. Khim.,* 685 (1967); *C.A.,* **68,** 2046 (1968).

123. R. E. Parker and N. S. Isaacs, *Chem. Rev.,* **59,** 737 (1959); S. Winstein and R. B. Henderson in Elderfield's *Heterocyclic Compounds,* John Wiley and Sons, New York, Vol. I, 1950, p. 22.

124. S. Winstein, *J. Am. Chem. Soc.,* **64,** 2792 (1942).

125. D. Reulos, *Compt. Rend.,* **216,** 774 (1943); D. Reulos and C. Collin, *Compt. Rend.,* **218,** 795 (1944).

126. H. H. Wasserman and N. E. Aubrey, *J. Am. Chem. Soc.,* **78,** 1726 (1956).

127. C. C. Tung and A. J. Speziale, *J. Org. Chem.,* **28,** 2009 (1963).

128. C. S. Marvel *et al., Org. Syn.,* Coll. Vol. **1,** 117 (1941).

C. Aliphatic Substitution

1. From Aliphatic Hydrocarbons, Alkylarenes, or Alkyl Heterocycles

Halogenation of hydrocarbons is usually a free radical process as shown for chlorination [1]. The chain reaction can be initiated by light or free radicals.

$$Cl_2 \rightarrow 2\ Cl\cdot \qquad (\Delta H = +57\ kcal.)$$

$$Cl\cdot + RH \rightarrow R\cdot + HCl \qquad (\Delta H = 0\ to\ -13\ kcal.)$$

$$R\cdot + Cl_2 \rightarrow RCl + Cl\cdot \qquad (\Delta H = -21\ to\ -27\ kcal.)$$

Sources of halogen other than the elements are N- or O-halocompounds and particularly sulfuryl chloride and peroxides [2]. Halogenation of aliphatic hydrocarbons, which may give more than one substitution product, is generally a poor synthetic method since substitution is largely indiscriminate, only slight preference being shown in the order tertiary H > secondary H > primary H. Moreover, disubstitution rates are nearly the same as monosubstitution rates,

References for Section C are on pp. 390–392.

so that a large ratio of hydrocarbon to halogen must be used to obtain mono-halogenation. The pattern of substitution is illustrated for a gas-phase reaction [3]:

$$X_2 + RH + N_2 \xrightarrow[\text{100 ml./min.}]{\text{2 100-watt lamps}} RX + HX$$

1 part   15 parts   180 parts

Relative selectivity for each H shown for butane and butyl halides, in parts by weight, follows:

CH_3	CH_2	CH_2	CH_3	$ClCH_2$	CH_2	CH_2	CH_3	$BrCH_2$	CH_2	CH_2	CH_3
F_2 1	1.3	1.3	1								
Cl_2 1	3.9	3.9	1	0.7	2.2	4.2	1	0.5	—	4	1
Br_2 1	82	82	1	34	32	82	1				

Selectivity is in the order $Br_2 > Cl_2 > F_2$. The pattern of substitution in liquid-phase chlorination is no more discriminate than in the gas phase [4]:

$$n\text{-}C_4H_9Cl + SO_2Cl_2 + (C_6H_5CO_2)_2 \xrightarrow[80°]{CCl_4}$$

7% 1,1-, 22% 1,2-, 47% 1,3-, and 24% 1,4-dichlorobutane

A remarkable selectivity, however, is shown in the photochemical bromination of alkyl bromides [5]:

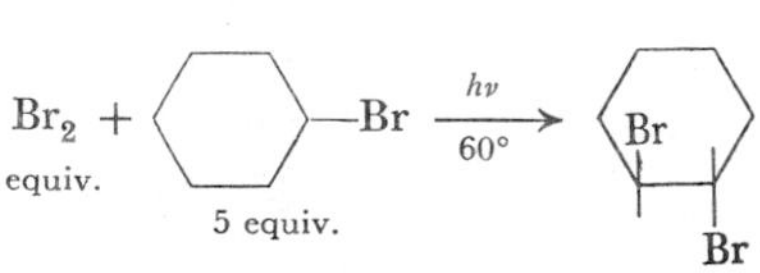

trans-1,2-Dibromo-cyclohexane, 83%, 94% purity

The selectivity is caused by bromine participation in stabilization of the intermediate free radical as shown by the following experiment [6]:

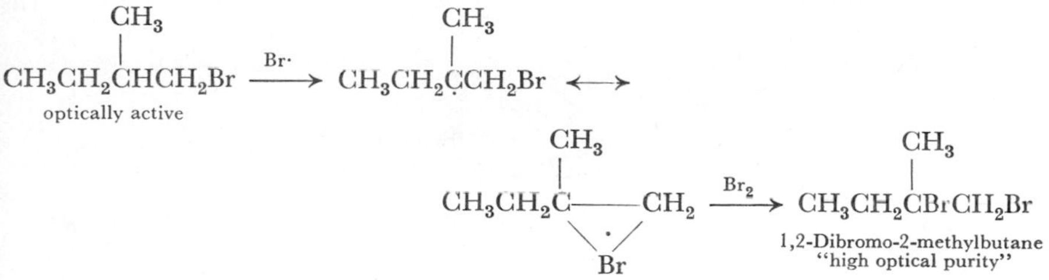

optically active

1,2-Dibromo-2-methylbutane "high optical purity"

Ordinarily, free radicals would be expected to racemize.

In addition to the bromination of alkyl bromides, halogenation is feasible for hydrocarbons which give only one substitution product (see Ex. a) and for side-chain substitution in alkylarenes (see Exs. b, c.1, and c.3). In the latter case, halogenation may be stopped at whatever stage is desired by controlling the amount of halogen introduced. However, the chlorination of ethyl- or

$$ArCH_3 \xrightarrow{X_2} ArCH_2X \xrightarrow{X_2} ArCHX_2 \xrightarrow{X_2} ArCX_3$$

longer-chain alkylbenzenes gives rather indiscriminate substitution in that both α- and β-chlorination take place [7]. This fact, as well as the fact that cyclohexane is about eleven times more reactive than toluene in competitive chlorination, does not speak well for the stabilization of benzylic free radicals, $C_6H_5\dot{C}HCH_3$. The transition state in chlorination may well occur before the formation of free radicals. On the other hand, bromination of long side chains is more selective and yields more α-substitution products.

N-Bromosuccinimide (see C.2), ω-tribromoacetophenone [8], trichloro-methanesufonyl chloride [9], trichloromethanesulfonyl bromide [10], and t-butyl hypochlorite in the presence of light or an initiator, such as azobisiso-butyronitrile [11], are also effective in side-chain halogenation. Indeed, side-chain halogenation of methylarenes with N-bromo- or N-chlorosuccinimide is the method of choice for four reasons: the quantity of halogenating agent may be weighed accurately; the halogen may be supplied in discrete quantities; the reaction conditions, consisting of stirring and refluxing a suspension of the halo compound in carbon tetrachloride subjected to strong light, are simple; and the termination of the reaction may be ascertained by the appearance of the reaction medium, since succinimide floats in carbon tetrachloride whereas the halosuccinimides sink.

The 2- and 4-alkyl groups of pyridines and quinolines are very susceptible to halogenation. For monohalogenation the following heterogenous reaction is superior [12]:

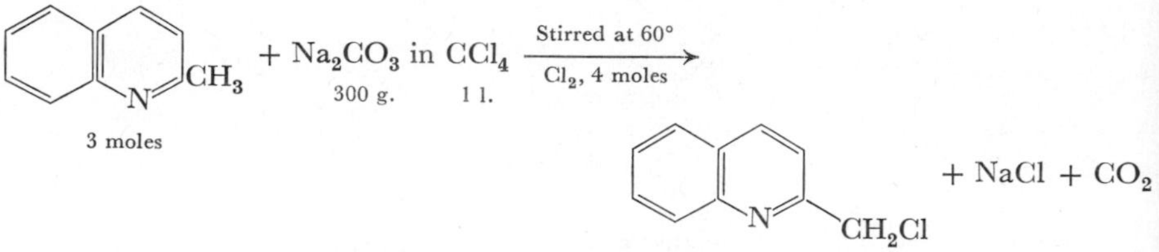

2-Chloromethylquinoline, 86%

These compounds, particularly the 4-halomethyl derivatives, tend to form salts on standing:

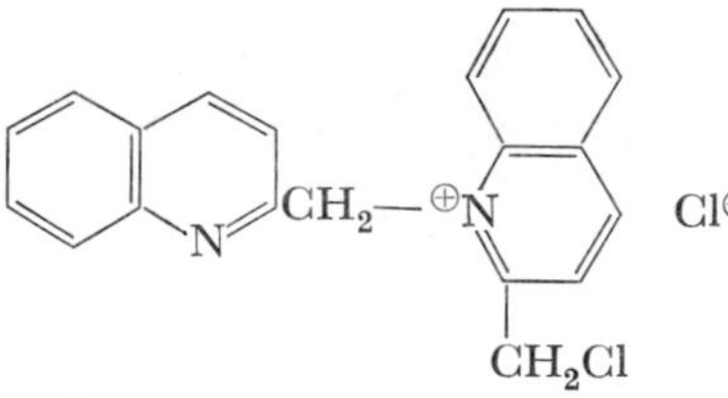

For trihalogenation the best procedure is halogenation in acetic acid containing sodium acetate [13]:

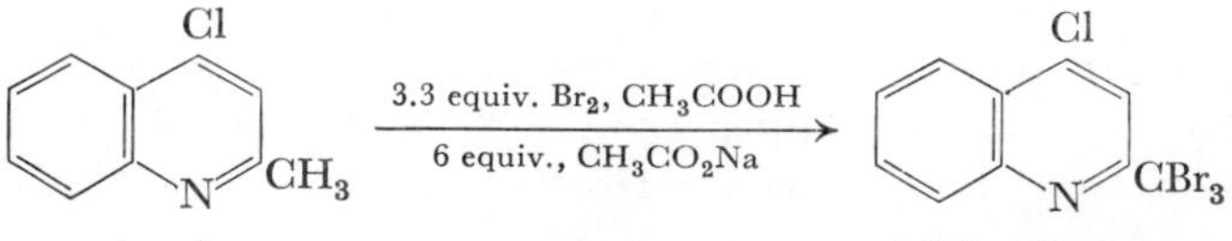

4-Chloro-2-tribromo-methylquinoline 65%, m.p. 121–122°.

The ionic halogenation of hydrocarbons containing a tertiary hydrogen atom (see Ex. c.2) is also possible:

$$R_3CH \xrightarrow{AlCl_3} (R_3C^{\oplus}) \xrightarrow{X_2} R_3CX$$

Moreover, halogenation of olefins at high temperature leads to the formation of unsaturated halides as illustrated for a number of industrial preparations:

$$CH_3CH=CH_2 + Cl_2 \xrightarrow{500–600°} ClCH_2CH=CH_2$$

Allyl chloride. 85–90%

Although fluorination is a violent reaction, three methods are available to moderate the reaction:

1. Passing fluorine and nitrogen over a large heat-contacting surface (this method gives the best results in the preparation of perfluorohydrocarbons) [14].
2. Using cobalt trifluoride or other fluorinating agents as $CoF_3 + RH \rightarrow RF$ (or polyfluorinated compounds) $+ CoF$ [15].
3. Employing the electrolysis of hydrogen fluoride solutions [16] as

$$(CH_3CO)_2O + 10 HF \xrightarrow[\text{Low } T]{\text{Electrolytic,}} 2 CF_3COF + OF_2 + 8 H_2$$

Iodination of butane has been carried out recently [17]:

$$(CH_3)_3COCl + HgI_2 \xrightarrow[0°]{CCl_4} (CH_3)_3COI \xrightarrow{CH_3CH_2CH_2CH_3} CH_3CHICH_2CH_3$$

2 equiv.   1 equiv.   Purple solution   35–39% yield containing 5% 1-iodobutane

(a) Preparation of chlorospiropentane,  Spiropentane and chlorine vapor in the ratio 10:1 were passed slowly through a spiral tube irradiated with two sunlamps and heated with one heat lamp. The condensed product was separated by preparative gas chromatography to give a 32% yield of desired product from 123 g. of spiropentane. Other products were polyhalides and open-chain halides [18].

(b) Preparation of α,α,α′,α′-tetrabromo-o-xylene (74–80% from o-

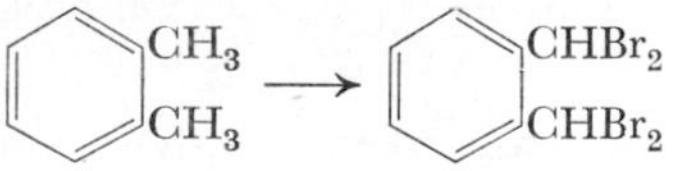

xylene and bromine in the presence of ultraviolet light) [19].

(c) Other examples

(1) 2-Phenyl-2-chloropropane (about 100% from cumene, chlorine, and carbon dioxide in the presence of ultraviolet light) [20].

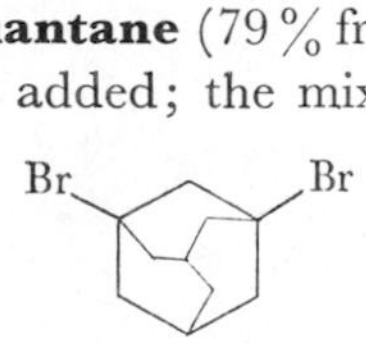

(2) 1,3-Dibromoadamantane (79% from 40 ml. of Br_2, 5 g. of BBr_3 to which 9 g. of adamantane was added; the mixture was heated for 2 hr. on a steam bath and cooled to room temperature overnight; the mono- tri-, and tetrabromo derivatives were prepared also, the latter two with aluminum chloride at higher temperatures) [21].

(3) **2,4,6-Trinitrobenzyl chloride** (85 % from 10 g. of TNT in 100 ml. of THF, a specific solvent, and 50 ml. of methanol to which 100 ml. of 5 % sodium hypochlorite was added, all reagents at 0°; after 1 min. the reaction was quenched by dilution with 1 l. of water containing 10 ml. of concentrated HCl) [22].

2. From Olefins (Allylic Halogenation, Wohl-Ziegler)

$$RCH_2CH{=}CH_2 \xrightarrow{\text{NBS}} RCHBrCH{=}CH_2$$

This synthesis, of theoretical and preparatory interest, has been reviewed [23–25]. Since many other types of reactions may occur in the synthesis, the experimental conditions are very important. For allylic bromination a nonpolar solvent, usually carbon tetrachloride, is employed and, since the N-bromo-succinimide (NBS) is insoluble in the solvent, the reaction is a heterogeneous one. In fact, it has been shown by a study with a series of solvents that a lack of solubility of NBS (zero dipole moment of the solvent) is desirable [24]. Similarly, a study of a series of N-bromoamides and -imides has indicated that N-bromosuccinimide gives yields from 50 to 100 % in allyl substitution in cyclohexene, while the second best, N-bromoacetamide and N-bromophthalimide, give yields around 50 %. In some cases the bromoamide alone leads to a product in satisfactory yield, but in others ultraviolet light, radical generators (azobis-isobutyronitrile, benzoyl peroxides, tetralyl, and cyclohexenyl hydroperoxides), bromine generators with NBS (hydrogen bromide, water, etc.), or t-amines (triethylamine, pyridine) are necessary [26]. Such activators lessen the reaction time and minimize the formation of side products. Substances such as oxygen, bromanil, picric acid, s-trinitrobenzene, and iodine serve as inhibitors. As a rule, the starting material and NBS with or without an activator are heated in carbon tetrachloride at the boiling point, complete reaction having occurred when the NBS is no longer at the bottom of the reaction mixture but is replaced by succinimide floating on the surface.

Although NBS reacts with many types, the main reactions with hydrocarbons are:

1. Straight- or branched-chain alkenes with an end or middle double bond are brominated only once in the allyl position [27]:

$$C_6H_5CH{=}CHCH_3 \rightarrow C_6H_5CH{=}CHCH_2Br$$

Cinnamyl bromide 75%

2. Cycloalkenes react similarly, but often bromine addition occurs and/or elimination of hydrogen bromide from the first-formed product to give an aromatic ring system [24]:

1-Bromo-1,2,3,4,5,6,7,8-octahydronaphthalene
80%

1,2,3,4-Tetrahydronaphthalene
50%

3. In methyl arenes, substitution of one or two hydrogens of the methyl group may occur [24]:

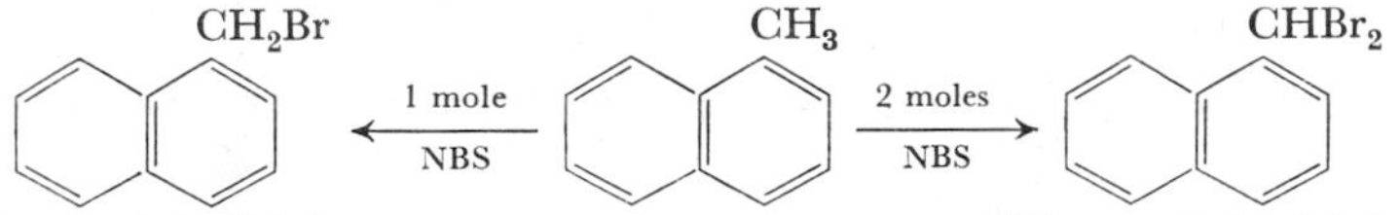

α-Bromomethylnaphthalene
90%

1 mole
NBS

2 moles
NBS

α-Dibromomethylnaphthalene
80%

It has already been indicated that bromine addition products and aromatic hydrocarbons are possible side products in the Wohl-Ziegler reaction of cycloalkenes. Nuclear bromination occurs in the presence of metal chlorides of aluminum, zinc, and ferric iron or sulfuric acid [28]. Alkenes having a double bond at the end of the chain or alkadienes with isolated double bonds may undergo allylic bromination with rearrangement (allylic) to form the unsaturated bromide 7-XII:

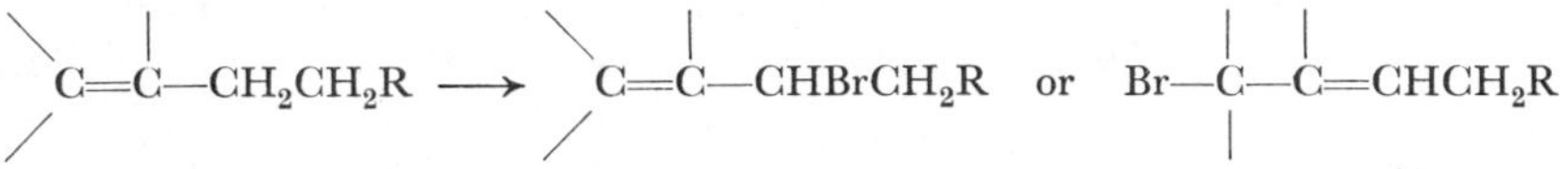

7-XII

One of the few examples of substitution of the halogen for a hydrogen attached to the double-bonded carbon is that of NBS with isolapachol [29]:

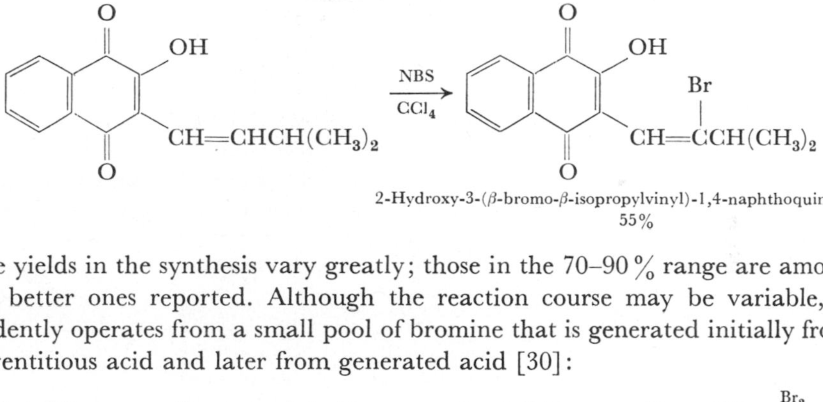

NBS
CCl$_4$

2-Hydroxy-3-(β-bromo-β-isopropylvinyl)-1,4-naphthoquinone
55%

The yields in the synthesis vary greatly; those in the 70–90% range are among the better ones reported. Although the reaction course may be variable, it evidently operates from a small pool of bromine that is generated initially from adventitious acid and later from generated acid [30]:

NBS + HBr ⟶ Br$_2$ + succinimide ⟶ Br· + RH ⟶ R· + HBr $\xrightarrow{\text{Br}_2}$

RBr + Br·

N-Chlorosuccinimide has not been examined in the same depth as NBS, and what results have been communicated (privately) are discouraging with regard to allylic chlorination. It serves other purposes, however (see C.4).

(a) **Preparation of 4-bromo-2-heptene** (58–64% from 2-heptene, NBS,

$$\text{CH}_3\text{CH}_2\text{CH}_2\text{CH}_2\text{CH}{=}\text{CHCH}_3 \xrightarrow{\text{NBS}} \text{CH}_3\text{CH}_2\text{CH}_2\text{CHCH}{=}\text{CHCH}_3$$

Br

and a trace of benzoyl peroxide in carbon tetrachloride at the boiling point) [31].

(b) Other examples

(1) *meso*-α,α'-**Dibromobibenzyl** (91.5 % from bibenzyl, NBS, and a little benzoyl peroxide refluxed in carbon tetrachloride) [32].

(2) **3-Bromocyclohexene** (75 %, based on the bromoimide, from cyclohexene in excess, NBS, and a trace of dimethyl α,α'-azoisobutyrate) [33].

(3) **3-Thenyl bromide** (thiophene, 3-bromomethyl-) (71–79 % from 3-methylthiophene, NBS, and a bit of benzoyl peroxide in benzene) [34].

3. From Acetylenes

$$RC\equiv CH \xrightarrow{\text{NaOX}} RC\equiv CX$$

The preparation of mono- or dihaloacetylenes has been reviewed [35]. A number of methods of synthesis, such as dehydrohalogenation of a dihaloalkene and treatment of metal acetylides with halogen, have been employed, but the simplest and most general one is treatment of the acetylene with hypohalite. For iodination of terminal acetylenes the morpholine-iodine complex with morpholine in excess is effective [36]:

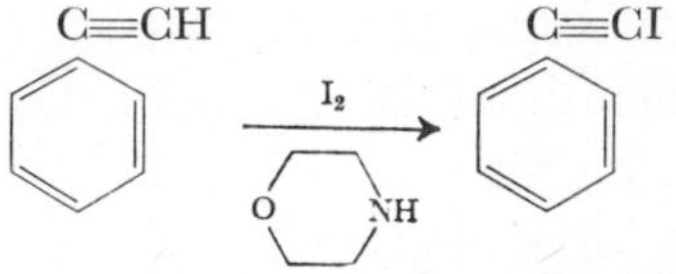

Phenyliodoethyne
75%

The haloacetylenes are highly reactive substances with poor storage capabilities. For example, difluoroacetylene polymerizes spontaneously at room temperature. All preparations and manipulations should be carried out with safety precautions.

(a) Preparation of 1-bromo-3-methyl-1-butyn-3-ol. 3-Methyl-1-

$$\overset{\text{OH}}{\underset{|}{(CH_3)_2CC\equiv CBr}}$$

butyn-3-ol in a large excess of aqueous sodium hydroxide was shaken with a 10 % excess of bromine, extracted with hexane, and the product distilled, b.p. 68° (15 mm.), in practically quantitative yield [37].

(b) Preparation of phenylbromoethyne (73–83 % from NaOH, water, bromine, and phenylacetylene at room temperature for 60 hr.) [38].

4. From Ethers and Sulfides

$$CH_3CH_2OCH_2CH_3 \xrightarrow{Cl_2} CH_3CH_2OCHClCH_3 \xrightarrow{Cl_2} CH_3CHClOCHClCH_3$$

Although α-haloethers can be prepared from an aldehyde, alcohol, and hydrogen halides, the direct halogenation of ethers is readily carried out at low temperatures to give α-halo- or α,α'-dihaloethers (see Ex. a) [39]. Further halogenation does not appear to be useful.

The α-chlorination of sulfides is best carried out with N-chlorosuccinimide, NCS (see Ex. b), probably proceeding via the sulfonium salt [40].

$$RSCH_2R' + \begin{array}{c} CH_2CO \\ \diagdown \\ NCl \\ \diagup \\ CH_2CO \end{array} \longrightarrow \overset{\overset{Cl}{\mid}}{\underset{\overset{\oplus}{C_4H_4O_2N\ominus}}{RSCH_2R'}} \longrightarrow RSCHClR' + C_4H_4O_2NH$$

(a) **Preparation of α-chlorodiethyl ether.** Chlorine was passed through dry diethyl ether at -25 to $-30°$ for 4 hr. until the calculated weight of chlorine had been absorbed. The product was fractionated to give the desired compound (42%). With twice as much chlorine, the α,α'-dichloroethyl ether could be isolated in 57% yield [39].

(b) **Preparation of α-chloroethyl phenyl sulfide.** Equivalent amounts of N-chlorosuccinimide and ethyl phenyl sulfide were mixed in CCl_4 at room temperature; the succinimide was removed by filtration, and the residue from the filtrate was distilled to give the chloro sulfide (75%) [40].

5. From Aliphatic Carbonyl and Nitro Compounds

$$RCH_2CHO \xrightarrow{X_2} RCHXCHO$$

$$RCOCH_3 \xrightarrow{X_2} RCOCH_2X$$

$$RCHNO_2\ominus Na\oplus \xrightarrow{X_2} RCHXNO_2 + NaX$$

The reactivity of the hydrogen α to the carbonyl group permits halogen substitution. The reagent consists of the halogen sometimes in the presence of a solvent such as acetic acid [41, 42], chloroform [43, 44], dimethylformamide [45, 46], or water [47]. Other reagents which may be used are sulfuryl chloride [48–52], selenium oxychloride [53], cupric and potassium bromides [54], or a ω-trihaloacetophenone [55]. Pyridinium hydrobromide perbromide, $C_5H_5NH\cdot Br_3$, and phenyltrimethylammonium perbromide,

$$C_6H_5\overset{\oplus}{N}(CH_3)_3\overset{\ominus}{Br_3}$$

being solids, offer some advantage over liquid bromine [56, 57]. Of the two the latter is now preferred since it is more conveniently prepared and it is more stable. Sometimes catalysts such as aluminum chloride [58], sulfur [59], and bases such as calcium carbonate or sodium hydroxide [60, 61] to neutralize the acid formed are used as well when the free halogen is the reagent.

Although the methods of halogenation of methyl ketones are diverse and easily carried out, the halogenation of a precious ketone can perhaps best be carried out by generation of bromine *in situ* from sodium bromate and hydrobromic acid (see Ex. b).

For the preparation of α-bromoketones with easily substituted ring systems, cupric bromide in refluxing chloroform and ethyl acetate is an excellent reagent for α-bromination (see Ex. c.5) [62]. Indeed, this reagent may have broader applications for α-brominations than have already been demonstrated. Ferric chloride in acetic acid has been used to prepare α,α-dichloroketones [63].

The synthesis is more satisfactory for ketones than for aldehydes since the latter have a greater tendency to polymerize and they also contain a reactive hydrogen attached to the carbonyl group. However, some α-haloaldehydes have been synthesized successfully by this method. Examples are 1-bromo-2-methylcyclohexanecarboxaldehyde (80 %) [60] and α-chloroisobutyraldehyde (60 %) [51]. For the syntheses of α-haloketones the yields are often in the range 80–95 %.

Unsymmetrical ketones which can be halogenated on each side of the carbonyl group tend to give mixtures of α- and α'-bromoketones, particularly if the groups are similar to each other in electronic properties. The amounts of α- and α'-bromoketones vary with the conditions used, as is amply demonstrated by the isolation of four and possibly five isomers of the brominated products of *cis*- and *trans*-α-decalone. Both 2- and 9-bromo isomers are formed [64]. When groups surrounding the carbonyl group are dissimilar, one halogenation product is favored and it will be the one derived from the group which enolizes more readily as shown [65]:

$$CH_3COCH_2Cl \xrightarrow{Cl_2} CH_3COCHCl_2$$

$$Cyclooctanone \xrightarrow{SO_2Cl_2} 2,2\text{-dichlorocyclooctanone}$$
$$80\%$$

Studies on the kinetics of the reaction show that under some circumstances the rate is independent both of the concentration and of the nature of the halogen. These facts indicate that the rate-determining step involves the conversion of the carbonyl compound to another type, perhaps the enol form 7-XIII.

$$RCOCH_2R' \rightleftharpoons \underset{7\text{-XIII}}{RC\!\!=\!\!CHR'} \xrightarrow{X_2} \underset{7\text{-XIV}}{RC\!\!-\!\!CHR'} \longrightarrow \underset{7\text{-XV}}{RCOCHXR'}$$

with OH groups on the carbon atoms and X substituents as shown.

Addition of the halogen to 7-XIII gives the dihalide 7-XIV, which loses the hydrogen halide to give the α-haloderivative 7-XV. In a relatively strong basic solution the anion of the ketone may simply displace bromine from the element or from the sodium hypobromite formed.

The halogenation of aliphatic nitro compounds is carried out by treating the sodium salt of the nitro compound with a halogen (see Ex. c.6) [66]:

$$R_2CHNO_2 \xrightarrow{OH^\ominus} [R_2C\!\!=\!\!NO^\ominus \longleftrightarrow R_2\overset{\ominus}{C}NO_2] \xrightarrow{Br_2} R_2\overset{Br}{\underset{|}{C}}NO_2$$

The fluorination of a nitro compound with perchloryl fluoride, $FClO_3$, is also illustrated (see Ex. c.7) [67]. This reagent seems capable of fluorinating any active methylene compound.

An alternative route to α-bromonitroalkanes is via the NBS bromination of oximes [68]:

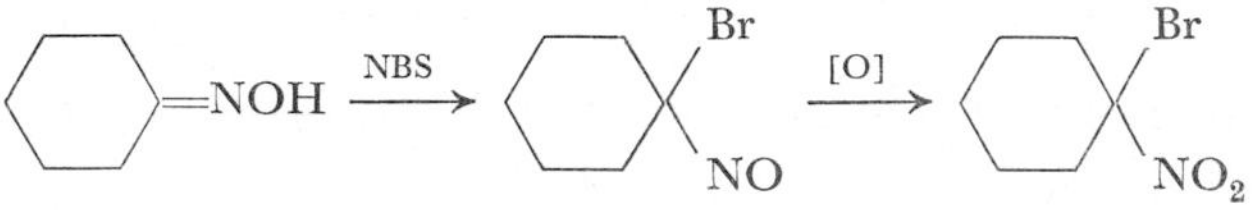

α-Bromonitrocy-
clohexane

(a) **Preparation of α,α-dichloroacetophenone** (80–94% from aceto-phenone and chlorine in glacial acetic acid) [42].

(b) **Preparation of 6,8-dimethyl-2-(4-chlorophenyl)-4-α-bromo-acetylquinoline.** To a stirred mixture of 0.0162 mole of ketone and exactly 0.0054 mole of sodium bromate in 50 ml of acetic acid, 0.064 mole of 47% hydrobromic acid was added dropwise in 15 min. followed by heating at 100° until the red color turned straw yellow. The mixture was cooled, poured over ice, filtered, dried, and recrystallized from ethanol to give 94% of the desired compound in the form of yellow crystals, m.p. 143–145° [69].

(c) **Other examples**

(1) **2-Chloro-2-methylcyclohexanone** (83–85% from 2-methylcyclo-hexanone and sulfuryl chloride in carbon tetrachloride) [49].

(2) **1,4-Bis(chloroacetyl)benzene** (82% from 1,4-diacetylbenzene and chlorine in chloroform) [44].

(3) **6-Methoxy-2-bromoacetylnaphthalene** (80% from 6-methoxy-

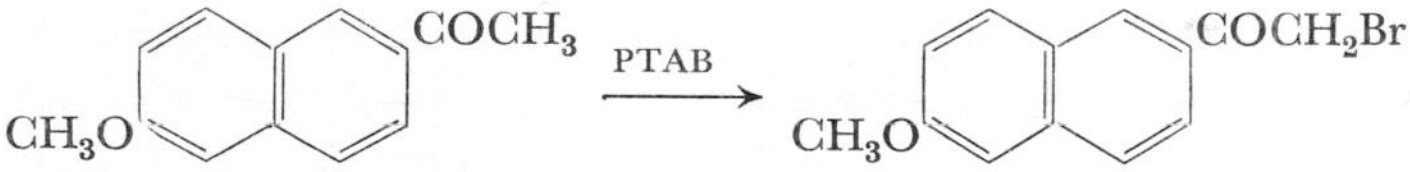

2-acetylnaphthalene and phenyltrimethylammonium perbromide in THF for 1 hr.) [70].

(4) **1-Chloro-1-phenylacetone** (95% from phenylacetone and sulfuryl chloride) [52].

(5) **2,4-Dihydroxyphenacyl bromide** (nearly quantitative yields from 0.03 mole of 2,4-dihydroxyacetophenone and 0.05 mole of commercial, ground cupric bromide refluxed in 50 ml. of a 1:1 or 75 ml. of a 1:2 chloro-form-ethyl acetate mixture; the cuprous bromide was filtered off and the filtrate evaporated) [62].

(6) **1-Bromo-1-nitropropane** (67% from bromine in CS_2 or CCl_4 at 0°, to which the dry sodium salt of 1-nitropropane was added) [66].

(7) **2-Fluoro-2-nitropropane** (32% from 2-nitropropane in basic solution to which perchloryl fluoride was added; acetone was a by-product) [67]; for preparation of 1,1,1-dinitrofluoroalkanes using perchloryl fluoride, see [71].

6. From Acids and Esters

$$RCH_2COOH(R') \xrightarrow{X_2} RCHXCOOH(R') + HX$$

The reactivity of carbonyl compounds toward halogenation decreases in going from aldehydes or ketones to the acid or ester type. As a result the pro-

$$
\begin{array}{c}
RCH_2C=O \\ | \\ (H)R
\end{array}
\quad > \quad
\begin{array}{c}
RCH_2 \\ \diagdown \\ C=O \\ \diagup \\ Cl
\end{array}
\quad > \quad
\begin{array}{c}
RCH_2 \\ \diagdown \\ C=O \\ \diagup \\ HO
\end{array}
\quad > \quad
\begin{array}{c}
RCH_2 \\ \diagdown \\ C=O \\ \diagup \\ R'O
\end{array}
$$

cedure with the last two types is somewhat different from that used with aldehydes and ketones (C.5). The activity of the acid or ester may be increased by conversion first into the acid chloride or anhydride (see Ex. b.5) which is then halogenated. A standard procedure (Hell-Volhard-Zelinsky) is to convert the acid into the acid chloride and then halogenate in one operation by the use of the halogen and phosphorus or the phosphorus halide. By this procedure, yields in the range 80–90 % are often attainable. Chlorination does not always work out satisfactorily in this procedure since, for long-chain esters, mixtures of chlorinated esters are sometimes obtained [72]. Similarly the chlorination of *n*-butyryl chloride with sulfuryl chloride and a peroxide gives 15 % α-, 55 % β-, and 30 % γ-chlorobutyryl chloride [73]. It is more common to employ bromine or sulfuryl chloride without peroxide as the halogenating agent because the results are more specific. Another procedure for conducting the halogenation is to form the acyl chloride with thionyl chloride and then halogenate with bromine or sulfuryl chloride using an excess of the thionyl chloride as a solvent [74, 75]. By this procedure, in the presence of a trace of iodine, a series of long-chain bromo acids, as the methyl esters, was obtained with yields of 59–83 %. Excellent yields (88–92 %) of the ethyl α-bromoesters of several dicarboxylic acids were obtained by starting with the ethyl hydrogen ester, treating with thionyl chloride in excess, bromine, and finally with ethanol [74]. In this

$$
\begin{array}{c}
COOH \\ \diagup \\ (CH_2)_n \\ \diagdown \\ COOC_2H_5
\end{array}
\xrightarrow{SOCl_2}
\begin{array}{c}
COCl \\ \diagup \\ (CH_2)_n \\ \diagdown \\ COOC_2H_5
\end{array}
\xrightarrow{Br_2}
$$

$$
\begin{array}{c}
CHBrCOCl \\ | \\ (CH_2)_{n-1}COOC_2H_5
\end{array}
\xrightarrow{C_2H_5OH}
\begin{array}{c}
CHBrCOOC_2H_5 \\ | \\ (CH_2)_{n-1}COOC_2H_5
\end{array}
$$

manner the formation of α,α'-dibromo dicarboxylic acids so readily produced on direct bromination may be avoided.

α-Bromo acids of the simpler aliphatic acids may be obtained in 61–87 % yields by adding bromine to a solution of the acid in slightly diluted polyphosphoric acid at a temperature near 100° (see Ex. b.5).

The mechanism follows the pattern given (C.5) for aldehydes or ketones. The rate-determining step is the formation of the enol 7-XVI to which the halogen

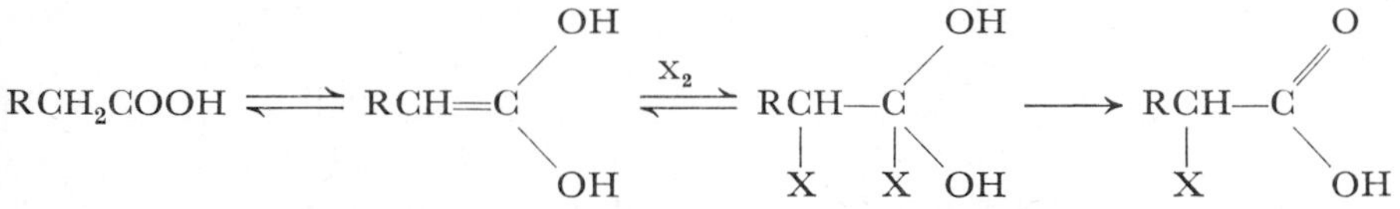

$$RCH_2COOH \rightleftharpoons RCH{=}C\begin{smallmatrix}OH\\ \\OH\end{smallmatrix} \underset{X_2}{\rightleftharpoons} RCH{-}C\begin{smallmatrix}OH\\ \\OH\end{smallmatrix} \longrightarrow RCH{-}C\begin{smallmatrix}O\\ \\OH\end{smallmatrix}$$

adds to form the dihalide 7-XVII, which loses the hydrogen halide to give the α-haloacid 7-XVIII.

(a) Preparation of α-bromo-*n*-caproic acid (83–89% from *n*-caproic acid, bromine, and phosphorus trichloride) [76].

(b) Other examples

(1) Methyl 2-bromododecanoate (90% from dodecanoic acid, bromine, and phosphorus tribromide followed by treatment with methanol) [77].

(2) Methyl α-bromo-*n*-butyrate (76% from *n*-butyric acid, thionyl chloride, a trace of iodine, and bromine followed by methanol) [75].

(3) Ethyl α-chlorophenylacetate (92% from phenylacetic acid, thionyl chloride, and sulfuryl chloride followed by treatment with ethanol) [74].

(4) Ethyl α-bromoadipate (90% from ethyl hydrogen adipate, thionyl chloride, and bromine followed by treatment with ethanol) [74].

(5) α-Bromoacetic acid (67% from 0.1 mole of acetic acid, 10 ml. of PPA, 1 ml. of H_2O, and 0.117 mole of bromine at 80–100°; if volatile, the α-bromoacid may be removed by distillation; if nonvolatile, the mixture is diluted with water and extracted with chloroform) [78].

(6) Chloroethylene carbonate (65–75% from uv irradiation of

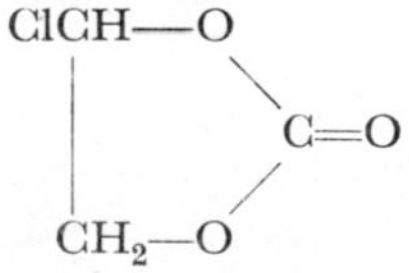

$$\begin{matrix}ClCH{-}O\\ | \quad\quad \diagdown \\ \quad\quad\quad C{=}O\\ | \quad\quad \diagup \\ CH_2{-}O\end{matrix}$$

ethylene carbonate and chlorine at 70°) [79].

7. From α,α,-Dihalogenated Esters

$$R_3B + X_2CHCOOC_2H_5 \xrightarrow{t\text{-}C_4H_9OK} RCHXCOOC_2H_5 + R_2BX$$

α,α-Dihalogenated esters may be monoalkylated by the use of a trialkyl boron in the presence of potassium *t*-butoxide [80]. Equimolecular amounts of the two components are used to prevent two R groups from being introduced into the dihaloester. The organoborane, produced from the olefin and diborane in tetrahydrofuran, is simply treated with the haloester, followed by the addition of the catalyst. The reaction at 0° is practically complete as soon as the base has been added. Yields vary usually from 80 to 98%.

(a) Preparation of ethyl α-bromooctanoate. 1-Hexene, 37.7 ml. (300 mmoles) in 100 ml. of tetrahydrofuran in an ice bath, was hydroborated by the dropwise addition of 50 ml. of a 2.0 M solution of borane (from 300 mmoles of hydride) under nitrogen. Stirring for 1 hr. at 25° was followed by cooling in an ice bath, after which 50 ml. of dry t-butyl alcohol and then 13.0 ml. (100 mmoles) of ethyl dibromoacetate were added. Potassium t-butoxide in t-butyl alcohol, 100 ml. of 1.0 M solution, was added over 30 min. to give an immediate precipitate of potassium bromide. Although glpc analysis indicated a 98% yield of the ester, recovery, through the addition of 30 ml. of 3 N sodium acetate, followed by the dropwise addition of 12 ml. of 30% hydrogen peroxide at 10° or lower, stirring for 30 min., and saturation with salt, gave 21.3 g. (85%) of ethyl α-bromooctanoate [80].

8. From Sulfur Compounds

Carbon tetrachloride is prepared by the chlorination of carbon bisulfide in a two-stage process. High temperatures are necessary, but other sulfur compounds

$$CS_2 + 3\ Cl_2 \rightarrow CCl_4 + S_2Cl_2$$

$$2\ S_2Cl_2 + CS_2 \rightarrow CCl_4 + 6\ S$$

are substituted with more ease (see Ex. a).

(a) Preparation of methylcarbylamine dichloride [81].

$$CH_3N{=}C{=}S + Cl_2 \xrightarrow[\text{bath, 3 hr.}]{\text{Ice-salt}} CH_3N{=}CCl_2$$

in ether   86%, b.p. 78°

9. From Organometallic Compounds

$$(Ar)RM + X_2 \rightarrow (Ar)RX + MX$$

These substitutions involve both aliphatic and aromatic organometallic compounds. They occur readily often in high yield but are valuable only when access to the organometallic compound is easier than to other intermediates. For example, one would rarely synthesize a halide from a Grignard reagent not only because the latter is usually derived from a halide, but also because alcohols are usually readily available as starting materials. Nevertheless, it is a satisfactory reaction and may be used on occasion to exchange halogen or to identify substances. Listed below are some substitutions which are used infrequently:

(1) $RMgX + X_2 \rightarrow RX + MgX_2$

Kharasch suggests that the Grignard reagent be added to a cold solution of the halide [82]:

(2) $ArLi + C_6H_5I \rightarrow ArI + C_6H_5Li$

An example of this reaction is given (Ex. b.2) in which the authors claim that the product is cleaner by this route than by direct iodination. Even less frequently used substitutions from a synthetic viewpoint are the following [83]:

(3) $$ArB(OH)_2 + X_2 + H_2O \rightarrow ArX + B(OH)_3 + HX$$

(4) $$ArSiR_3 + X_2 \rightarrow ArX + R_3SiX$$

(5) $$ArSnR_3 + I_2 \rightarrow ArI + R_3SnI$$

Halodemetalation is of more interest with some mercurials since the mercury atom can be introduced into more unique positions. Ethylene haloethers in wide variety are made by the following process (see Ex. a and b.1):

$$RCH{=}CH_2 + Hg(OCOCH_3)_2 + CH_3OH \longrightarrow$$

$$RCH(OCH_3)CH_2HgOCOCH_3 \xrightarrow[KX]{X_2} RCH(OCH_3)CH_2X + HgX_2$$

Aromatic compounds which are substituted uniquely by the mercuric ion serve as further sources of haloaromatics [84]:

$$C_6H_5OH + Hg(OCOCH_3)_2 \longrightarrow o\text{-}HOC_6H_4HgOCOCH_3 \xrightarrow{NaCl}$$

$$o\text{-}HOC_6H_4HgCl \xrightarrow{I_2} o\text{-}IC_6H_4OH$$

$$\underset{44\%}{} \qquad \underset{63\%}{o\text{-Iodophenol}}$$

or [85]

82–90%

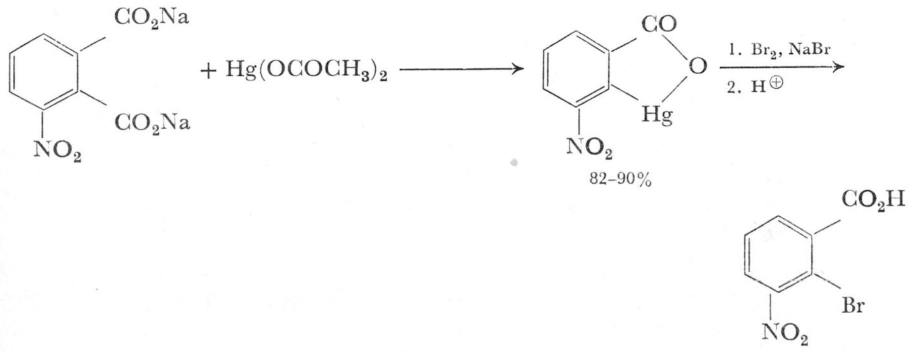

2-Bromo-3-nitrobenzoic
acid, 53–61%

And, finally, organoboranes which ordinarily are quite inert to halogen, may be rapidly displaced by iodine in the presence of 1 equivalent of sodium hydroxide [86]. Only one secondary alkyl group is displaced, in contrast to two

$$R_3B \xrightarrow[NaOH]{2 I_2} 2 RI + RB(OH)_2 + 2 NaI$$

for primary alkyl groups, but resort can be made to disiamylborane $(R_2'BH)$ to produce $R_2'BR$ from olefins, the latter forming RI in yields approaching 100% based on the conversion of the R group to RI.

(a) Preparation of methyl α-bromo-β-methoxypropionate (81–86%, based on mercuric acetate, from methyl acrylate, methanol, mercuric

$$CH_2\!\!=\!\!CHCOOCH_3 \xrightarrow[CH_3OH]{Hg(OAc)_2} CH_3OCH_2CHCOOCH_3 \xrightarrow{KBr}$$
$$\underset{HgOAc}{\mid}$$

$$CH_3OCH_2CHCOOCH_3 \xrightarrow{Br_2} CH_3OCH_2CHBrCOOCH_3$$
$$\underset{HgBr}{\mid}$$

acetate, potassium bromide, and bromine) [87].

(b) Other examples

(1) 2-Methoxy-3-phthalimidopropyl iodide (75% from allylphthalimide, mercuric acetate, methanol, potassium iodide, and iodine) [88].

(2) p-Iododimethylaniline (42–54% yields of recrystallized product from 2 g. of lithium metal in ether to which 0.109 mole of p-bromodimethylaniline was added with stirring; this solution was then added to 0.218 mole of iodobenzene in ether and refluxed for 1 hr.; if iodine was used in place of iodobenzene, the product was colored) [89].

(3) Chloroferrocene (80% from 1 g. of ferrocenylboronic acid and 1.7 g. of $CuCl_2 \cdot 2H_2O$ in 250 ml. of water at 50–60° for 15 min.) [90].

1. G. Chambers and A. R. Ubbelohde, *J. Chem. Soc.*, 285 (1955).
2. M. S. Kharasch and H. C. Brown, *J. Am. Chem. Soc.*, **61,** 2142 (1939).
3. P. S. Fredericks and J. M. Tedder, *J. Chem. Soc.*, 144 (1960).
4. H. C. Brown and A. B. Ash, *J. Am. Chem. Soc.*, **77,** 4019 (1955).
5. W. A. Thaler, *J. Am. Chem. Soc.*, **85,** 2607 (1963).
6. P. S. Skell, *et al., J. Am. Chem Soc.*, **85,** 2849 (1963).
7. H. Goldwhite, *J. Chem. Ed.*, **37,** 295 (1960).
8. F. Kröhnke and K. Ellegast, *Chem. Ber.*, **86,** 1556 (1953).
9. E. S. Huyser *et al., J. Am. Chem. Soc.*, **82,** 5246 (1960); *J. Org. Chem.*, **27,** 3391 (1962).
10. E. S. Huyser, *et al., J. Org. Chem.*, **30,** 38 (1965).
11. C. Walling and B. J. Jacknow, *J. Am. Chem. Soc.*, **82,** 6108 (1960).
12. W. Mathes and H. Schüly, *Angew. Chem.*, **75,** 235 (1963).
13. D. E. Pearson and J. C. Craig, unpublished work; D. L. Hammick, *J. Chem. Soc.*, **123,** 2882 (1923).
14. L. A. Bigelow, *Chem. Rev.*, **40,** 51 (1947).
15. N. I. Gubkina, *et al., Russ. Chem. Rev. (English Transl.)*, **35,** 930 (1966).
16. E. A. Kauck and A. R. Diesslin, *Ind. Eng. Chem.*, **43,** 2332 (1951).
17. D. D. Tanner and G. C. Gidley, *J. Am. Chem. Soc.*, **90,** 808 (1968).
18. D. E. Applequist *et al., J. Am. Chem. Soc.*, **82,** 2368 (1960).
19. J. C. Bill and D. S. Tarbell, *Org. Syn., Coll. Vol.* **4,** 807 (1963).
20. H. Ross and R. Hüttel, *Chem. Ber.*, **89,** 2641 (1956).
21. H. Stetter and C. Wulff, *Chem. Ber.*, **93,** 1366 (1960).
22. K. G. Shipp and L. A. Kaplan, *J. Org. Chem.*, **31,** 857 (1966).
23. C. Djerassi, *Chem. Rev.*, **43,** 271 (1948).
24. L. Horner and E. H. Winkelmann, *Angew. Chem.*, **71,** 349 (1959).
25. R. Filler, *Chem. Rev.*, **63,** 21 (1963).
26. H. J. Dauben, Jr., and L. L. McCoy, *J. Am. Chem. Soc.*, **81,** 4863 (1959).
27. K. Ziegler *et al., Ann. Chem.*, **551,** 80 (1942).
28. H. Schmid, *Helv. Chim. Acta*, **29,** 1144 (1946).
29. K. H. Dudley and H. W. Miller, *Tetrahedron Letters*, 571 (1968).

30. C. Walling *et al.*, *J. Am. Chem. Soc.*, **85**, 3129 (1963).

31. F. L. Greenwood, *et al.*, *Org. Syn.*, Coll. Vol. **4**, 108 (1963).

32. F. D. Greene *et al.*, *J. Am. Chem. Soc.*, **79**, 1416 (1957).

33. M. C. Ford and W. A. Waters, *J. Chem. Soc.*, 2240 (1952).

34. E. Campaigne and B. F. Tullar, *Org. Syn.*, Coll. Vol. **4**, 921 (1963).

35. K. M. Smirnov *et al.*, *Russ. Chem. Rev. (English Transl.)*, **36**, 326 (1967).

36. P. L. Southwick and J. R. Kirchner, *J. Org. Chem.*, **27**, 3305 (1962).

37. F. Straus *et al.*, *Chem. Ber.*, **63**, 1868 (1930); S. I. Miller *et al.*, *J. Am. Chem. Soc.*, **85**, 1648 (1963).

38. S. I. Miller *et al.*, *Org. Syn.*, **45**, 86 (1965).

39. G. E. Hall and F. M. Ubertini, *J. Org. Chem.*, **15**, 715 (1950).

40. D. L. Tuleen and T. B. Stephens, *Chem. Ind. (London)*, 1555 (1966).

41. P. A. Levene, *Org. Syn.*, Coll. Vol. **2**, 88 (1943).

42. J. G. Aston *et al.*, *Org. Syn.*, Coll. Vol. **3**, 538 (1955).

43. L. A. Bigelow and R. S. Hanslick, *Org. Syn.*, Coll. Vol. **2**, 244 (1943).

44. H. Hopff and P. Jaeger, *Helv. Chim. Acta.*, **40**, 274 (1957).

45. E. M. Chamberlin *et al.*, *J. Am. Chem. Soc.*, **79**, 456 (1957).

46. T. Walker *et al.*, *J. Chem. Soc.*, 1277 (1962).

47. M. S. Newman *et al.*, *Org. Syn.*, Coll. Vol. **3**, 188 (1955).

48. P. Delbaere, *Bull. Soc. Chim. Belges*, **51**, 1 (1942); *C.A.*, **37**, 5018 (1943).

49. E. W. Warnhoff *et al.*, *Org. Syn.*, Coll. Vol. **4**, 162 (1963).

50. E. Gudriniece *et al.*, *Izv. Vysshikh. Uchebn. Zavedeniĭ Khim. i Khim. Tekhnol.*, **3**, No. 1, 119 (1960); *C.A.*, **54**, 17352 (1960).

51. C. L. Stevens and B. T. Gillis, *J. Am. Chem. Soc.*, **79**, 3448 (1957).

52. D. P. Wyman and P. R. Kaufman, *J. Org. Chem.*, **29**, 1956 (1964).

53. J. P. Schaefer and F. Sonnenberg, *J. Org. Chem.*, **28**, 1128 (1963).

54. A. W. Fort, *J. Org. Chem.*, **26**, 765 (1961).

55. F. Kröhnke and K. Ellegast, *Chem. Ber.*, **86**, 1556 (1953).

56. C. Djerassi and C. R. Scholz, *J. Am. Chem. Soc.*, **70**, 417 (1948).

57. A. Marquet and J. Jacques, *Tetrahedron Letters*, No. 9, 24 (1959).

58. R. M. Cowper and L. H. Davidson, *Org. Syn.*, Coll. Vol. **2**, 480 (1943).

59. F. A. Long and J. W. Howard, *Org. Syn.*, Coll. Vol. **2**, 87 (1943).

60. I. Heilbron *et al.*, *J. Chem. Soc.*, 737 (1949).

61. K. C. Murdock, *J. Org. Chem.*, **24**, 845 (1959).

62. L. C. King and G. K. Ostrum, *J. Org. Chem.*, **29**, 3459 (1964).

63. Y. Nakatani *et al.*, *Tetrahedron Letters*, 4085 (1967).

64. H. E. Zimmerman and A. Mais, *J. Am. Chem. Soc.*, **81**, 3644 (1959).

65. D. Q. Quan, *Compt. Rend.*, Ser. C, **264**, 320 (1967).

66. S. Trippett and D. M. Walker, *J. Chem. Soc.*, 2976 (1960).

67. H. Shechter and E. B. Roberson, *J. Org. Chem.*, **25**, 175 (1960).

68. D. C. Iffland and G. X. Criner, *J. Am. Chem. Soc.*, **75**, 4047 (1953).

69. J. B. Wommack and D. E. Pearson, unpublished results, adapted from the work of H. King and T. S. Work, *J. Chem. Soc.*, 1307 (1940).

70. A. Marquet and J. Jacques, *Bull. Soc. Chim. France*, 90 (1962).

71. M. J. Kamlet and H. G. Adolph, *J. Org. Chem.*, **33**, 3073 (1968).

72. H. H. Guest and C. M. Goddard, Jr., *J. Am. Chem. Soc.*, **66**, 2074 (1944).

73. M. S. Kharasch and H. C. Brown, *J. Am. Chem. Soc.*, **62**, 925 (1940).

74. E. Schwenk and D. Papa, *J. Am. Chem. Soc.*, **70**, 3626 (1948).

75. H. Reinheckel, *Chem. Ber.*, **93**, 2222 (1960).

76. H. T. Clarke and E. R. Taylor, *Org. Syn.*, Coll. Vol. **1**, 115 (1941).

77. J. Cason *et al.*, *J. Org. Chem.*, **18**, 850 (1953).

78. E. E. Smissman, *J. Am. Chem. Soc.*, **76**, 5805 (1954).

79. M. S. Newman and R. W. Addor, *J. Am. Chem. Soc.*, **77**, 3789 (1955).

80. H. C. Brown *et al.*, *J. Am. Chem. Soc.*, **90**, 1911 (1968).

81. K. A. Petrov and A. A. Neimysheva, *J. Gen. Chem. USSR (Eng. Transl.)*, **29**, 2131 (1959).

82. M. S. Kharasch and O. Reinmuth, *Grignard Reactions of Non-metallic Substances*, Prentice-Hall, Englewood Cliffs, N.J., 1954, p. 1332.

83. R. O. C. Norman and R. Taylor, *Electrophilic Substitution in Benzenoid Compounds*, Elsevier Publishing Co., New York, 1965, Chap. 10.

84. F. C. Whitmore and E. R. Hanson, *Org. Syn.*, Coll. Vol. **1**, 326 (1941).

85. P. J. Culhane, *Org. Syn.*, Coll. Vol. **1**, 125 (1941).

86. H. C. Brown *et al.*, *J. Am. Chem. Soc.*, **90**, 5038 (1968).

87. H. E. Carter and H. D. West, *Org. Syn.*, Coll. Vol. **3**, 774 (1955).

88. B. R. Baker *et al.*, *J. Org. Chem.*, **17**, 68 (1952).

89. H. Gilman and L. Summers, *J. Am. Chem. Soc.*, **72**, 2767 (1950).

90. A. N. Nesmeyanov *et al.*, *Chem. Ber.*, **93**, 2717 (1960).

D. Aromatic Substitution

The literature on the halogenation of hydrocarbons by substitution with halogens is extensive [1–4], largely because of the interest in the theory of the process and the industrial importance of the products. Additional information lies scattered about in every organic laborarory textbook and in *Organic Syntheses*. General theory of orientation, relative to the Friedel-Crafts reaction, is discussed under Alkanes, Cycloalkanes, and Arenes, Chapter 1, Section D.

This section, terminating with a subsection on haloalkylation, discusses halogenation of aromatic hydrocarbons, phenols, anilines, some heterocyclics, and compounds with electron-withdrawing groups.

1. From Aromatic Hydrocarbons

Any alkylbenzene of the order of activity of toluene may be chlorinated or brominated by means of the appropriate halogen in the presence of a catalyst, which tends to develop a positive charge in the complex. The catalyst, which is

$$Cl_2 + FeCl_3 \rightarrow \overset{+\delta}{Cl}\overset{-\delta}{FeCl_4}$$

not always necessary, may be ferric halides, aluminum halides, transition metal halides, iodine, or aluminum amalgam. It is employed in trace amounts. Special reagents such as the Silberrad reagent (sulfur monochloride, aluminum chloride, and sulfuryl chloride), iodine-silver sulfate-sulfuric acid [5], silver trifluoroacetate and bromine or iodine [6], iodine monobromide (for bromination) [7], and iodine monochloride (for iodination) [8–10] may also be used. The most powerful chlorinating agent is claimed to be the Silberrad reagent (see Ex. a) [11].

Chlorination and Bromination of Benzene

Monohalogenation with chlorine or bromine in the presence of a catalyst such as the ferric or aluminum halide occurs at room temperature. As the

References for Section D are on pp. 406–408.

temperature is increased, more halogen atoms are introduced until finally an almost quantitative yield of the hexahalobenzene is produced. The same product is produced in the exhaustive chlorination or bromination of almost any alkylbenzene with the exception of toluene.

Dihalogenation of benzene yields only traces of the *m*-dihalobenzene because of the orientation effects of the first halogen introduced. Conditions for thermodynamic control must be resorted to for increasing the proportion of the *meta* isomer [12]. The equilibrium position for the dichlorobenzenes under the conditions noted is as follows:

$$o\text{-, }m\text{-, or }p\text{-}C_6H_4Cl_2 \xrightarrow[160°,\ 30\ hr.\ or\ more]{AlCl_3\,(0.27\ mole)} 54\%\ m\text{-, }30\%\ p\text{-, and }16\%\ o\text{-}C_6H_4Cl_2$$
$$\text{1 mole}$$

It is stressed that the above is only a single equilibrium position under specified temperature, catalyst, and cocatalyst (HCl) ratio. If the catalyst and cocatalyst ratio is increased, quite likely the equilibrium would favor a greater proportion of the *meta* isomer. The equilibrium for the trichlorobenzenes is as follows [13]:

$$\text{Any }C_6H_3Cl_3 + AlCl_3 + MgSO_4 \xrightarrow[5\ hr.]{200°}$$
$$\begin{matrix} 1 & 0.34 & 0.6 \\ \text{mole} & \text{mole} & \text{mole} \end{matrix}$$

73% 1,2,4-, 24% 1,3,5-, and 3% 1,2,3-trichlorobenzene

The 1,3,5-isomer can be frozen out of this mixture.

Dibromobenzenes behave quite differently in isomerization studies. Whereas chlorine isomerization is mainly intramolecular in nature, bromine isomerization is an intermolecular process. Indeed, bromo- and dibromobenzene have been used as brominating agents of other molecules. The following experiment illustrates the behavior of *o*-dibromobenzene [14]:

$$o\text{-}C_6H_4Br_2 + AlBr_3 \xrightarrow{30°}$$
$$\begin{matrix} 99\ \text{mmoles} & 5\ \text{mmoles} \end{matrix}$$

$$C_6H_5Br + C_6H_3Br_3 + C_6H_4Br_2 \quad (5\%\ m\text{-, }48\%\ p\text{-, and }47\%\ o\text{-})$$

If chlorobenzene is added to the original reactants, the main product is *p*-bromochlorobenzene.

Chlorination and Bromination of Alkyl Benzenes

The preparation of each of the three monochloro isomers of toluene is not carried out by direct chlorination (rather by the Sandmeyer reaction on the corresponding toluidine) because direct monochlorination of toluene yields about an equal mixture of *o*- and *p*-chlorotoluene together with some dichlorotoluene and unreacted toluene. On the other hand, direct dichlorination of toluene yields preparative quantities of 2,4-dichlorotoluene [15]. The orientation is rather surprising since the first chlorine atom introduced, although slightly deactivating, tends to orient the second chlorine atom to a position o or p to

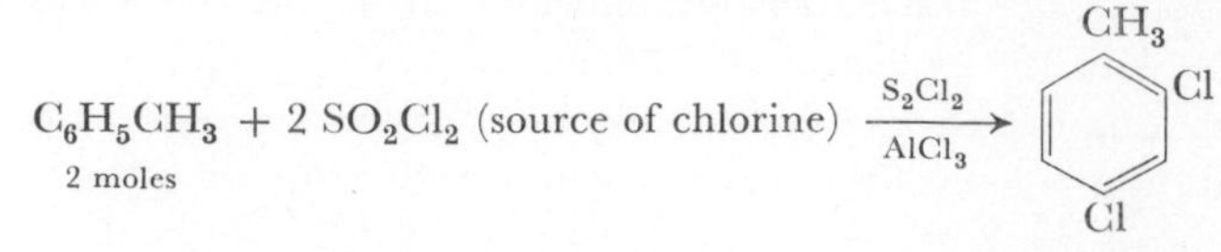

$$C_6H_5CH_3 + 2\ SO_2Cl_2 \text{ (source of chlorine)} \xrightarrow[\text{AlCl}_3]{\text{S}_2\text{Cl}_2}$$

2 moles

2,4-Dichloro-
toluene, 70%

itself, thus anticipating the formation of some 2,5-dichlorotoluene at least. Trichlorination of toluene (regulated by the amount of the chlorine introduced) yields about an equal mixture of 2,4,5- and 2,3,4-trichlorotoluene from which the former may be crystallized. Tetrachlorination of toluene does not seem to be of preparative value, but pentachlorination gives an almost quantitative yield of pentachlorotoluene.

By contrast, mesitylene may be brominated with bromine in carbon tetrachloride without a catalyst (see Ex. b.1) [16]. All tri-, tetra-, or pentaalkylbenzenes are halogenated with equal or greater ease.

Chlorination and Bromination of Polynuclear Aromatic Hydrocarbons

Naphthalene [17], anthracene [18], and phenanthrene [19] are examples of polynuclear hydrocarbons which can be brominated without a catalyst. These compounds form addition products first, from which hydrogen bromide is

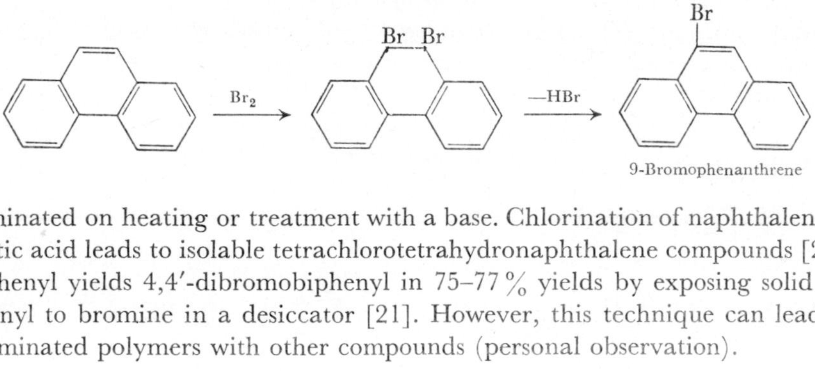

9-Bromophenanthrene

eliminated on heating or treatment with a base. Chlorination of naphthalene in acetic acid leads to isolable tetrachlorotetrahydronaphthalene compounds [20]. Biphenyl yields 4,4'-dibromobiphenyl in 75–77 % yields by exposing solid biphenyl to bromine in a desiccator [21]. However, this technique can lead to brominated polymers with other compounds (personal observation).

Iodination of Aromatic Hydrocarbons

Iodination usually fails either because the iodine molecule is a poor electrophilic attacking agent or because the hydrogen iodide formed reduces the electrophilic attacking species. To obviate these characteristics, silver sulfate and iodine in sulfuric acid (see Ex. b.4) [22], trifluoroacetylhypoiodite, CF_3COOI [6], iodine and peracetic acid in acetic acid [23], or iodine monochloride [8–10] is employed. But contamination of the iodo compound with the chloro compound is to be expected from mixed halogen dissociation. To remove hydrogen iodide as it is formed, it is oxidized with nitric acid (see Ex. b.5), iodic acid (see Ex. b.6), or precipitated as mercuric iodide or other iodide salts. Aryl iodides disproportionate more easily than other aryl halides, and occasionally the reaction may be of synthetic value [24]:

2,5-Diiodo-1,4-xylene, 32%

Fluorination of Aromatic Hydrocarbons

Direct fluorination of aromatic hydrocarbons still remains to be developed [25]. Exchange of the halide or amine group with fluoride (A.6) is the most important method of obtaining aromatic fluorides, with the possible exception of hexafluorobenzene which can be made by the pyrolysis of tribromofluoromethane at 650° [26]. Fluorine reagents such as bromine trifluoride tend to add to benzene derivates rather than substitute, and thus a dehalogenation step must follow to obtain a fluoro aromatic hydrocarbon [27]. These experiments are hazardous.

(a) **Preparation of perchlorotoluene,** $C_6Cl_5CCl_3$. Trichloromethylbenzene was chlorinated with sulfuryl chloride to yield $2,3,4,5\text{-}C_6HCl_4CCl_3$. This compound was further chlorinated with the Silberrad reagent (as modified by the authors): 250 ml. of sulfuryl chloride, 5 g. of S_2Cl_2, and 10 g. of the $C_6HCl_4CCl_3$ were added to a boiling solution of 2.5 g. of $AlCl_3$ in 750 ml. of sulfuryl chloride. The dark solution was concentrated to 50 ml. and refluxed for several hours. If a white solid separated, more sulfuryl chloride was added. The excess of the sulfuryl chloride was removed under reduced pressure, the residue treated with sodium bicarbonate and then strongly acidified. After repeated recrystallizations from benzene, hexane, and acetic acid (to separate hexachlorobenzene), perchlorotoluene (60%), m.p. 71.5–72.5°, was obtained [11].

(b) **Other examples**

(1) **Bromomesitylene** (79–82% from mesitylene and bromine in carbon tetrachloride) [16].

(2) **p-Bromo-t-butylbenzene** (94% from 0.5 mole of t-butylbenzene, 80 g. of bromine, and 1 ml. of pyridine) [28].

(3) **p-Bromotoluene** (60% from 35 ml. of toluene, 12.2 g. of N-bromosuccinimide, and 14 g. of anhydrous ferric chloride held at 60° for 7 hr.) [29].

(4) **Iodobenzene** (78% based on Ag_2SO_4 from 0.3 mole of C_6H_6, 200 ml. of H_2SO_4, 20 ml. of H_2O, 0.1 mole of Ag_2SO_4, and 0.22 mole of iodine stirred into an emulsion for 2 hr.; the solution was diluted, filtered to remove Ag_2SO_4, and the filtrate extracted with ether; chlorobenzene gave 60% p-chloroiodobenzene under the same conditions) [22].

(5) **p-Iodotoluene** (60 % with some *ortho* present from 21 ml. of toluene, 25.4 g. of iodine, and 12 ml. of nitric acid (d = 1.35) under reflux for 2 hr.; the *p*-isomer was frozen out and recrystallized) [30].

(6) **2-Iodo-1,4-xylene** (85 % from *p*-xylene, aqueous acetic and sulfuric acids, iodine, iodic acid, and carbon tetrachloride) [31].

(7) *pseudo* - *gem* - **Bromo** - **carbomethoxy** [2.2] **paracyclophane**

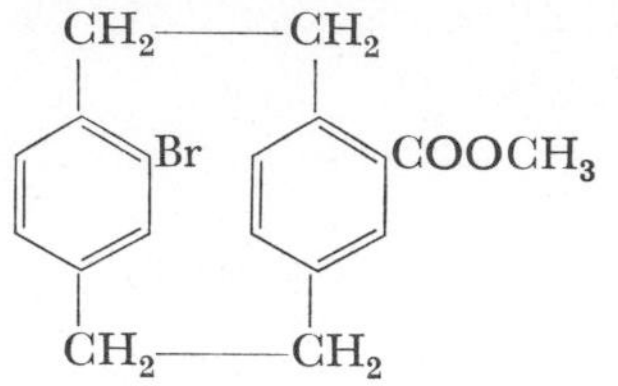

$$CH_2\text{——}CH_2$$

(89 % from carbomethoxyparacyclophane, bromine, and iron powder) [32].

2. From Phenols and Phenol Ethers

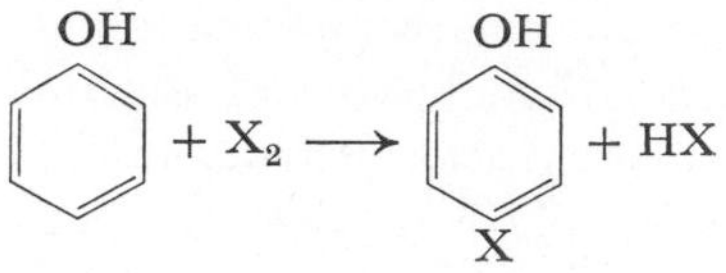

The chlorination or bromination of phenol may be carried out in an inert solvent such as carbon tetrachloride to yield 4-halo-, 2,4-dihalo-, or 2,4,6-trihalophenol by varying the amount of halogen introduced into the system. On the other hand, halogenation of phenol in alkaline, aqueous solution leads mainly to the 2,4,6-trihalophenol regardless of the amount of halogen employed. In this solution the anion of the phenol is attacked, and this ion of the halogenated phenol forms more readily (or to a greater extent) than the phenolic anion. Indeed, further halogenation of trihalophenol in alkaline solution leads only to a tetrahalocyclodienone,

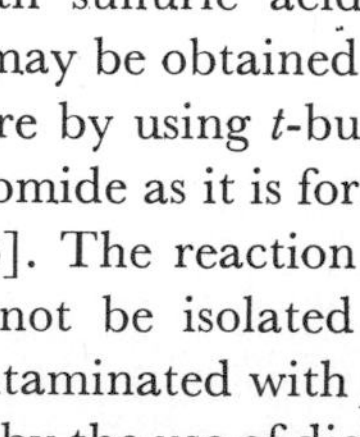

Treatment of the dienone with sulfuric acid gives 2,3,4,6-tetrahalophenol. *o*-Bromophenol of good purity may be obtained in fair yield in the bromination of phenol at dry ice temperature by using *t*-butylamine as a solvent and as a reagent to remove hydrogen bromide as it is formed. This procedure is superior to the *Organic Syntheses* one [33]. The reaction probably occurs via the hypobromite, which, however, cannot be isolated [34]. *o*-Chlorophenol may be prepared similarly, but it is contaminated with *p*-chlorophenol. *p*-Bromophenol (88 %) is available from phenol by the use of dioxane dibromide [35]. *m*-Bromophenol may be made in 77 % yield, containing 12 % *p*-bromophenol, by heating *p*-bromophenol with 2 equivalents of $AlCl_3$ and liquefied HBr as a solvent in an autoclave at 100° [34]. But the best preparation of a *m*-halophenol is still the diazotization of *m*-haloaniline followed by hydrolysis [36].

Polyiodophenols are made readily by adding a solution of iodine and potassium iodide to the phenol dissolved in 60% aqueous ethylenediamine [37]. Even *p*-nitrophenol is diiodinated by this procedure in 34% yield, although veratrole may be monoiodinated in 85–91% yield by iodine in the presence of silver trifluoroacetate (see Ex. b.2). The monoiodophenols are made best by indirect procedures: the *o*-iodo- from *o*-chloromercuriphenol and iodine [38] and the *p*-iodo- from *p*-aminophenol by diazotization and treatment with potassium iodide [39]. For more highly iodinated phenols the method of treatment with morpholine and iodine is worthy of note [40]:

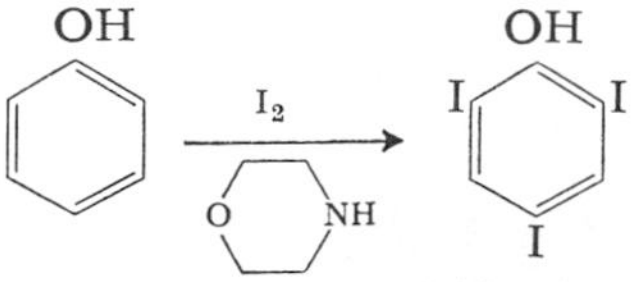

2,4,6-Triiodo-
phenol, 90%

Halogen compounds of unusual orientation may be obtained occasionally via metalation. An example is the preparation of 2-iodoresorcinol dimethyl ether [41]:

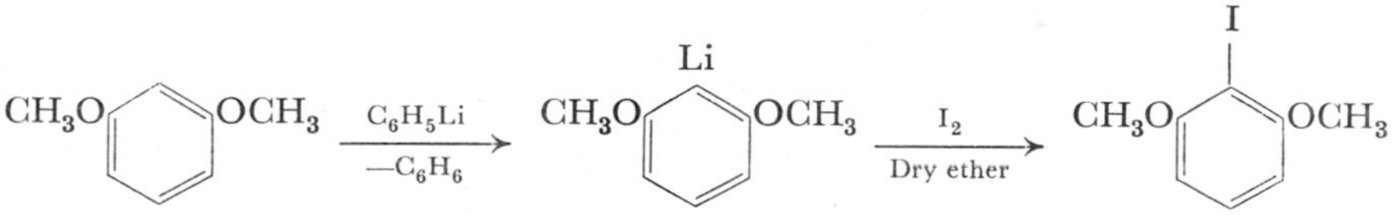

80% overall
1-Iodo-2,6-dimethoxybenzene

Typical procedures for halogenation of phenols are referenced above, but less common procedures are given in the following examples.

(a) **Preparation of 4-chloro-3-methylphenol** (84% from 27 g. of *m*-cresol and 20 ml. of sulfuryl chloride warmed on a steam bath) [42].

(b) **Other examples**

(1) **2,4-Dibromophenol** (87% from 23 g. of phenol in 70 g. of hydrobromic acid to which 80 g. of bromine in 40 g. of hydrobromic acid was added slowly and held at 30° for 1 hr.) [28].

(2) **4-Iodoveratrole** (85–91% from dry silver trifluoroacetate, iodine, and veratrole in chloroform while stirred at room temperature) [43].

(3) **3,5-Dibromo-4-cresol** (83% from 1 equiv. of 2,6-dibromo-4-cresol and 2 equiv. of AlCl$_3$ at 130° for 1 hr.) [44].

(4) **1-Iodo-2-naphthol** (96% from 0.1 mole of β-naphthol in 0.05 mole of 60% aqueous ethylenediamine solution and 10 ml. of alcohol to which 0.1 mole of iodine in aqueous KI solution was added) [37]; dibromination of β-naphthol in inert solvents without using any amine yields 1,6-dibromo-2-naphthol, but dichlorination yields 1,4-dichloro-2-naphthol [45].

(5) 2,4-Dichlorophenol (37% from *p*-chlorophenol and copper chloride hydrate in DMF at 100° for 2 hr.; this is not the best way to prepare 2,4-dichlorophenol but is included to demonstrate a different type of halogenating agent) [46].

(6) 1,4-Dihydroxy-3,5,5-trichloro-2-cyclopentenylcarboxylic acid as ammonium salt (Hantzsch acid, 52% from 0.19 mole of 2,4,6-trichloro-

phenol in 500 ml. of 2.5 M NaOH saturated with chlorine at 0–4°; the chlorine was swept out with N_2 and the mixture was acidified, and after removal of chlorine with nitrogen it was extracted with ether; the ether was then saturated with dry ammonia to precipitate the salt) [47].

3. From Anilines and Anilides

Anilines are very readily halogenated. Because of the tendency of the type to form salts from the hydrogen halide generated in the reaction, anilines are usually acetylated before halogenation. This difficulty is overexaggerated, however, because the salt dissociates, and the free base in equilibrium halogenates. The overall result is that it simply halogenates more slowly since a lower concentration of base is present. In some instances the halosuccinimides as halogenating agents overcome this difficulty (see Ex. b.4). Either the free aniline or acetanilide halogenates to give a mixture of *o*- and *p*-haloanilines which can be separated by crystallization or differences in volatility. If halogenation is carried out in aqueous solvents, the end product contains large amounts of trihaloaniline regardless of the quantity of halogen employed. This result is due to the fact that in aqueous acid solution more of the aniline and less of the bromo- or dibromoaniline is transformed to salts with the hydrogen halide liberated. Thus the haloanilines, being largely present as the free base, halogenate in preference to the aniline.

If the halogenation of acetanilides is carried out in buffered solution (pH 5–7), N-haloacetanilides are obtained. On treatment with strong aqueous acid, they rearrange to a mixture of *o*- and *p*-haloacetanilides, the ratio of products of which differs little from direct halogenation of the acetanilide (Orton rearrangement) [48].

On the other hand, the N-chloroanilines lead to a greater proportion of *ortho* substitution if the solution is not made acidic for rearrangement [49]. The reaction has been shown to proceed via the N-chloroaniline [50]. In fact,

o-and *p*-chloro-N-methylanilines
79% yield, *o/p* ratio = 3.4:1

the fluorination of nitroanilines in liquid hydrogen fluoride or acetonitrile gives N,N-difluoro derivatives [51]. For success the ring must be deactivated considerably.

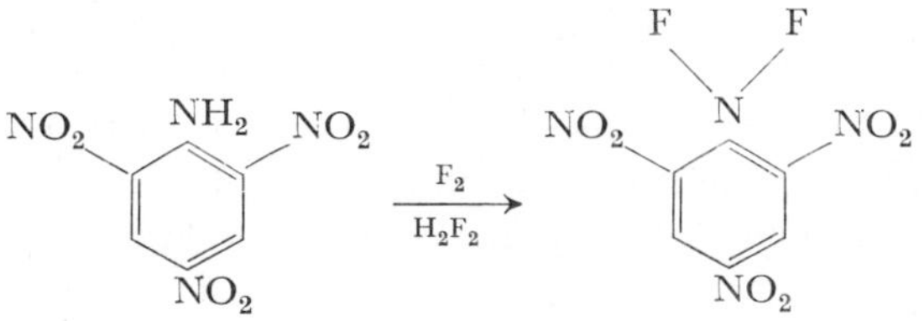

75%
N,N-Difluoro-2,4,6-trinitroaniline

Attempts to change the orientation in the halogenation of anilines has met with only partial success [52]:

$$C_6H_5NH_2 + 2\ AlCl_3 + HCl \longrightarrow C_6H_5\overset{\oplus}{N}H_3\ \overset{\ominus}{AlCl_4} \xrightarrow{Br_2}$$

$$11\%\ o\text{-}BrC_6H_4NH_2 + 50\%\ m\text{-}BrC_6H_4NH_2 + 25\%\ p\text{-}BrC_6H_4NH_2$$

The o,p-bromoanilines were at first considered products of the dissociated free base, but it is now known that a fully developed positive charge on nitrogen is not sufficient to direct substitution entirely to the *meta* position. Even the phenyltrimethylammonium ion, which does not dissociate, gives some o,p-substitution products [53]. However, with the aid of o,p-directing groups in the *para* position of aniline, good yields of 3-halo-substituted anilines can be realized [52]:

$$p\text{-}ClC_6H_4NH_2 + 2\ AlCl_3 + HCl \xrightarrow{Br_2}$$

4-Chloro-3-
bromoaniline,
82%

By contrast, under ordinary conditions of halogenation, 2-bromo-4-chloro-aniline would be obtained.

Monoiodination of anilines to the p-isomer can be carried out readily with iodine itself in the presence of sodium bicarbonate [54]. For a di- or triiodination the ethylenediamine method of Potts should be considered (see Ex. b.2).

(a) **Preparation of 3-chloro-2,6-dimethylacetanilide.** A solution of 15 g. of chlorine in 300 ml. of acetic acid was added rapidly to 33 g. of 2,6-dimethylacetanilide in 400 ml. of acetic acid. After standing overnight, the mixture was diluted with water and filtered. The precipitate was recrystallized from methanol to give the desired compound (80%), m.p. 146–147° [55]. The unusual orientation in this reaction suggests chlorination via the cation $Ar\overset{\oplus}{N}H_2COCH_3$.

(b) **Other examples**

(1) **2-Iodo-4-t-butylaniline** (88% crude yield from 4-t-butylaniline and iodine mixed in equimolecular amounts; to the dark oil, 60 ml. each of

water and ether and an excess of $CaCO_3$ were added and the mixture **was** warmed; the residue from the ether gave the desired compound) [56].

(2) 2-Amino-5-iodobenzoic acid (76% from 0.1 mole of anthranilic acid in 0.05 mole of ethylenediamine (60% aqueous solution) to which 0.1 mole of I_2 in KI was added) [37].

(3) 3,4-Dibromo-N,N-dimethylaniline (57% from dimethylaniline complexed with 2 equiv. of $AlCl_3$; the molten complex was saturated with HCl at 90°, after which slightly more than 2 equiv. of Br_2 was added dropwise with stirring; the cooled mixture was poured into ice, made strongly basic as quickly as possible to convert aluminum hydroxide entirely into sodium aluminate, and steam-distilled) [52].

(4) 4-Amino-3-bromobiphenyl (100% from 0.1 mole of 4-amino-biphenyl in 100 ml. of methylene chloride, cooled to 0°, and treated with a slurry of 0.1 mole of NBS in 100 ml. of methylene chloride followed by stirring for 10 min. at 25°; the solution was washed well with water, dried, and evaporated) [57].

4. From Selected Heterocyclic Compounds

Pertinent references on heterocyclic substitution are available [58–60]. It is the objective of this section to select a few examples to indicate the behavior and problems of heterocyclic halogenation. Halogenation among this type ranges from the most difficult to the most facile. Pyridine is an example of one of the most difficult heterocycles to halogenate under controlled conditions. The reluctance arises from the fact that electrophilic catalysts or the halogen acids formed combine with pyridine to form the salt which is greatly deactivated toward electrophilic attack. For instance, the complex of aluminum chloride and pyridine can be halogenated, but the reaction stops at 50% yield (by analysis) or 30–40% (by isolation) [61]. In this case it is thought that the

$$C_5H_5N \cdot AlCl_3 \xrightarrow[AlCl_3]{Br_2} \quad \overset{Br}{\underset{\underset{AlCl_3}{N}}{\bigcirc}} + C_5H_5\overset{\oplus}{N}H \ \overset{\ominus}{AlCl_3}Br$$

$$\text{7-XIX} \qquad\qquad\qquad\qquad\qquad \text{7-XX}$$

first complex 7-XIX is active enough, but the second complex 7-XX is too inactive to permit substitution. An ingenious method of removing hydrogen bromide is to oxidize it with fuming sulfuric acid, a process which permits the pyridine to be brominated to an extent greater than 50% [62]:

$$2\ C_5H_5N + 2\ SO_3 + Br_2 \xrightarrow{130°,\ \text{sealed tube}} 2\ 3\text{-}BrC_5H_4N + SO_2 + H_2SO_4$$

7.9g.   30 ml. H_2SO_4 4.4 g.   3-Bromopyri-
contg. 65% SO_3 dine, 86%

Another method for forming bromides is to heat the pyridine perbromide hydrobromide complex at 200°. The reaction at this temperature may occur by a free radical mechanism. At 400°, 2-bromopyridine is formed. Nevertheless, the best methods of obtaining 2- and 4-halopyridines are from the corresponding hydroxy or amino compounds (see A.9). At high temperature, exhaustive

$$C_5H_5N \cdot HBr_3 \xrightarrow{200°} 3\text{-}BrC_5H_4N$$
$$30\text{--}40\%$$

chlorination of pyridine with chlorine or phosphorus pentachloride yields pentachloropyridine, which with anhydrous potassium fluoride gives penta-fluoropyridine [63]. Pentachloropyridine is available commercially. Chlori-

$$C_5H_5N + PCl_5 \xrightarrow[\substack{\text{Autoclave, 14 hr.}}]{350°} C_5Cl_5N \xrightarrow[500°]{K_2F_2} C_5F_5N + 3\ ClC_5F_4N$$

0.34 mole   4.0 mole   97% Pentachloro- 18 hr.   Pentafluoro-   7% 3-Chlorotetra-
                              pyridine,   pyridine,   fluoropyridine
                                                          83%

nation of α-picoline at 100° for 50 hr. gives a rather poor yield of perchloro-picoline, $C_5Cl_4NCCl_3$, m.p. 67° [64].

The halogenation of quinoline or isoquinoline may be carried out by the Derbyshire-Waters procedure [65], which employs concentrated sulfuric acid, silver sulfate, and halogen; by the Pearson method, which uses an excess of aluminum chloride and halogen; or by the Eisch procedure, bromination in pyridine [66]. For quinoline the latter method gives an 82% yield of 3-bromo-quinoline. Both the Derbyshire-Waters and Pearson procedures give similar yields and orientation. The orientation follows the pattern:

Quinoline $\xrightarrow[\text{Catalysts}]{X_2}$ 5- + some 8-haloquinoline $\xrightarrow{X_2}$

5,8-dihaloquinoline $\xrightarrow{X_2}$ 5,6,8-trihaloquinoline $\xrightarrow{X_2}$

5,6,7,8-tetrahaloquinoline

These methods combined with the Eisch procedures [66, 67] give access to a variety of haloquinolines except for those substituted in the 2- and 4-positions. Again, the 2- and 4-haloquinolines are made by replacement of hydroxyl or amino groups in these positions. 6-, 7-, and 8-Haloquinolines may be prepared by the Skraup reaction [68].

Furan represents a heterocyclic compound of low aromaticity and high reactivity toward halogenation. Even at −30°, chlorination gives a variety of polychlorinated and some addition products [69]. On the other hand, when an electron-withdrawing group is placed in the 2- position of furan, exemplified by furoic acid, α-acetylfuran, or furfuraldehyde, the ring is stable enough to survive halogenation under most strenuous conditions (see Ex. b.6). 5-Substi-tution occurs in preponderance. Thiophene (see Ex. b.5), imidazole, and pyrazole are halogenated readily, but pyrrole polymerizes when exposed to acidic reagents or acidic byproducts. However, indole, a benzopyrrole, is brominated satisfactorily with pyridinium hydrobromide perbromide [70]:

3-Bromoindole

(a) **Preparation of 5-bromoisoquinoline.** Isoquinoline (0.42 mole) was added in a slow stream to 0.85 mole of stirred, powdered anhydrous aluminum chloride and held at 75° while 0.28 mole of bromine was added as slowly as possible, preferably as vapor. After addition was complete, the black fluid complex was heated for an additional hour, cooled, poured into ice, and made strongly basic. The ether extract was washed, dried, concentrated, and distilled at 0.3 mm. while all fractions boiling above 120° were collected. The solid distillate was recrystallized from pentane to give white needles (78%), m.p. 80° [71].

(b) **Other preparations**

(1) **3-Bromoquinoline** (65% from quinoline, bromine, and sulfur monochloride [72]; 82% by adding dropwise 1 mole of pyridine to a refluxing mixture of 1 mole each of quinoline and Br_2 in CCl_4) [66].

(2) **3-Chloro-8-nitroquinoline** (43% from 0.345 mole of 8-nitroquinoline and 200 g. of sulfur dichloride at 140° for 6 hr.) [73].

(3) **2-Chloro-3,6-dimethylpyrazine** (61% from 20 g. of 2,5-dimethylpyrazine in 500 ml. of carbon tetrachloride saturated with chlorine) [74].

(4) **4-Chloro-2-picolinic acid hydrochloride** (50% from 100 g. of α-picolinic acid hydrochloride in 400 ml. of thionyl chloride which was saturated with sulfur dioxide (vital step) and then heated at 80–90° for 4 days. The thionyl chloride was removed and the residue was poured into water; filtration and concentration of the filtrate gave the derived product) [75].

(5) **2-Iodothiophene** (75% from thiophene, yellow mercuric oxide, and iodine) [76].

(6) **4,5-Dibromo-2-furyl methyl ketone** (62% from 2-acetylfuran, 3 equiv. of $AlCl_3$, and 2 equiv. of Br_2) [77].

(7) **5,5′-Dibromo-4,3′-dimethyl-3,4′-diethylpyrromethene hydrobromide** [78].

5. From Aromatic Compounds with Electron-Withdrawing Groups [79]

$$C_6H_5NO_2 + Fe + Br_2 \xrightarrow{135-145°} m\text{-}NO_2C_6H_4Br + HBr$$
$$60\text{-}75\%$$

Halogenation of aromatic compounds bearing electron-withdrawing groups is sluggish and gives predominantly the *meta*-halo isomer. More strenuous conditions, i.e., higher temperatures and/or stronger electrophilic catalysts, therefore are necessary, for example, silver sulfate with the halogen in sulfuric acid [80]. For substituents that withdraw electrons from the ring by resonance, the minor by-products always contain a larger amount of *ortho*- than *para*-halo isomer, because the *para* position is more deactivated. Considerable scope has been added to *meta* halogenation from the realization that more than 1 equivalent of catalyst must be added to effect halogenation, not for the reason that the group influence is changed, but because the first equivalent of catalyst is consumed in forming the complex. The excess then is free to promote further halogenation. Aromatic esters, acid chlorides, aldehydes, ketones, nitriles, to mention a few types, are halogenated smoothly in this way (see Ex. a). About the only group not amenable to this process is the nitro group—complexing a nitrobenzene with excess aluminum chloride does not improve the yield of *m*-halonitrobenzene and in some cases leads to powerful oxidizing mixtures. The pattern of polyhalogenation for the complexes that can be halogenated is somewhat as follows [81]:

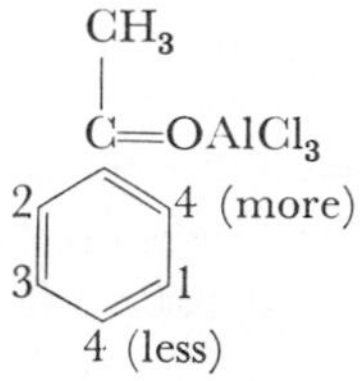

The numbers refer to the sequence of introduction of halogen. But other substituents can change the sequence, as shown for methyl toluate [52]:

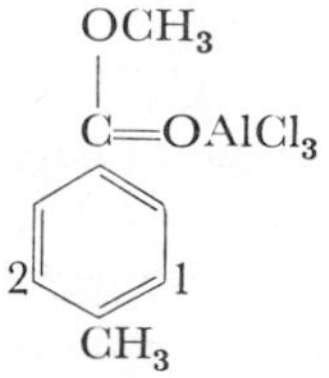

Aromatic compounds with a single deactivating group can be iodinated smoothly with iodine and 20% oleum at room temperature (see Ex. b.1). With more strenuous conditions, polyiodination can be carried out (see Ex. b.2).

If two electron-withdrawing groups are *ortho* or *para* to each other, halogenation may result in the displacement of one of these deactivating groups, particularly the nitro group (see Ex. b.3). Occasionally, mercuration of an aromatic compound can be carried out followed by replacement with halogen. This process generally gives mixtures of isomers which are tedious to separate [83], but the process can be used efficiently to replace carboxyl groups in arene dicarboxylic acids (see C.9).

(a) **2,5-Dibromobenzonitrile.** Benzonitrile, 1.1 mole, was complexed with 3.5 moles of anhydrous aluminum chloride, and 3.1 moles of bromine was added to the stirred, molten mixture at 60° over 3 hr. and heated for an additional 3 hr. The cooled mass was poured into ice, extracted with ether, concentrated, and distilled at 2 mm. up to b.p. 120°. The distillate was recrystallized from benzene to yield 235 g. (79%) of the desired product, m.p. 144–145° [82].

(b) Other examples

(1) 3,5-Diiodo-4-methylbenzoic acid (70% from *p*-toluic acid and 3 equiv. of·I$_2$ in 20% oleum at room temperature for 24 hr.; yields ranged from 28 to 85% for a series of similar compounds; when a methyl group was attached to a ring containing an electron-withdrawing group, the diiodo derivative was obtained, while nitrobenzene, benzoic acid, and other single-substituent compounds gave moniodo derivatives; *m*-dinitrobenzenes did not iodinate) [84].

(2) Tetraiodophthalic anhydride (80–82% from phthalic anhydride and 2.12 equiv. of iodine, added in portions, in 60% fuming sulfuric acid while the temperature was raised gradually to 65° and finally to 170–180°) [85].

(3) 3-Chlorophthalic anhydride (79% from 3-nitrophthalic anhydride at 240° through which chlorine was passed) [86].

(4) 2-Iodofluorenone (58–60% from 0.025 mole each of fluorenone and iodine in 50 ml acetic acid, 1 ml. of nitric acid, and 5 ml. of sulfuric acid at 45–50°) [87].

(5) o-Nitrobromobenzene (92% from nitrobenzene and ferric chloride in carbon tetrachloride to which a mixture of bromine and chlorine in CCl$_4$ was added and then heated to 65–70°; the orientation is most surprising) [88].

6. From Aromatic Compounds by Haloalkylation

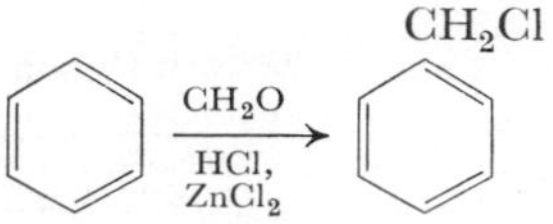

This synthesis, in which one or more haloalkyl groups may be introduced into a molecule, has been reviewed [89, 90]. The reaction has found its greatest application in the aromatic series, although some heterocyclic and aliphatic types respond as well. The alkylating agent is bifunctional, and the conditions employed must be such that only one reactive center is utilized in order that the haloalkyl group may survive. Substituents which are electron-donating promote substitution in the aromatic ring, whereas those which are electron-withdrawing retard substitution. Thus phenols may be chloromethylated with ease; in fact, there is a tendency in this case for the reaction to proceed to the formation of the diphenylmethane. Here indirect halomethylation may be

Catechol + paraformaldehyde + morpholine $\xrightarrow{10°}$

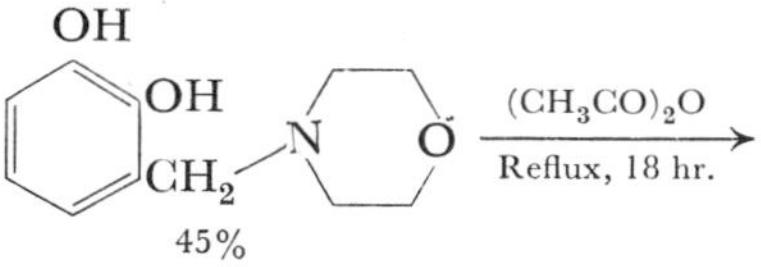

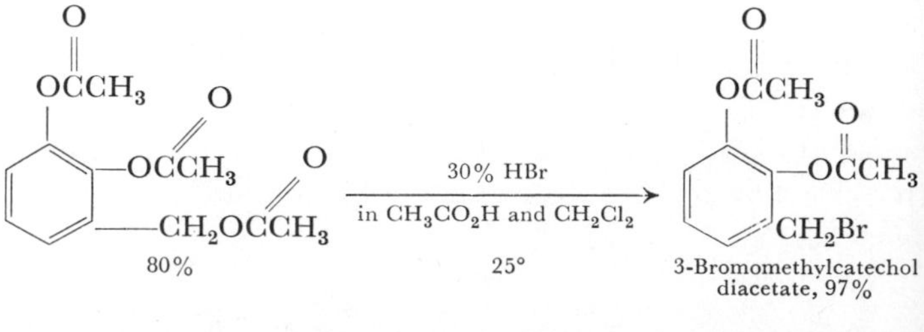

necessary [91]. Nitrobenzenes, on the other hand, chloromethylate with difficulty, if at all.

The reagents, catalysts, and solvents in this synthesis have been employed in great variety. The most common haloalkylating agents are aldehydes and hydrogen halides (formaldehyde or its polymers such as paraformaldehyde in the case of halomethylation), acetals and hydrogen halides, and haloalkyl ethers. The most common catalysts are acidic halides, such as zinc chloride, aluminum chloride, or stannic chloride, or proton acid catalysts such as hydrogen chloride, sulfuric acid, phosphoric acid, or acetic acid. The usual solvents are ether, dioxane, carbon tetrachloride, chloroform, nitrobenzene, or carbon disulfide. In some cases a substance like acetic acid may serve as the catalyst and solvent or a substance like chloromethyl ether may serve as the reagent and catalyst [92]. Mixtures of catalysts are not uncommon. The yields in the synthesis are variable, but are often quite satisfactory.

For a reaction conducted under such varying conditions it is not likely that a single mechanism will suffice. For the chloromethylation of mesitylene with formalin in aqueous acetic acid it has been proposed [93] that protonated formaldehyde 7-XXI is the attacking species. With the hydrocarbon 7-XXII it

$$CH_2O + \overset{\oplus}{H} \rightleftharpoons \overset{\oplus}{C}H_2OH$$
7-XXI

$$ArH + \overset{\oplus}{C}H_2OH \rightleftharpoons ArCH_2OH + \overset{\oplus}{H}$$
7-XXII　　7-XXI　　　　7-XXIII

$$ArCH_2OH + HCl \rightarrow ArCH_2Cl + H_2O$$

produces 7-XXIII, which in the presence of hydrogen chloride gives the benzyl chloride. Other investigators have also regarded the hydroxymethyl cation as the active intermediate [94].

β-Haloethylation also is now possible by taking advantage of the greater activity of the fluoro atom with electrophilic catalysts [95]. Yields ranged from

$$C_6H_6 + FCH_2CH_2Cl \xrightarrow[-10°, 30 \text{ min.}]{BBr_3} C_6H_5CH_2CH_2Cl$$
β-Chloroethylbenzene, 50%

50 to 94% for a number of different hydrocarbons and ethylene fluorohalides.

(a) Preparation of 1-chloromethylnaphthalene (74–77% from naphthalene, paraformaldehyde, glacial acetic acid, 85% phosphoric acid, and concentrated hydrochloric acid [96].

(b) Other examples

(1) 2-Hydroxy-5-nitrobenzyl chloride (69% from p-nitrophenol, methylal, sulfuric and hydrochloric acids with hydrogen chloride passing through the reaction mixture for 4–5 hr.) [97].

(2) 3-Chloromethyl-5-nitrosalicylaldehyde (90% from 5-nitrosalicylaldehyde, chloromethyl methyl ether, and over 4 equiv. of aluminum chloride) [92].

(3) 2-Chloromethyl-5-carboethoxyfuran (83% from ethyl 2-furan-carboxylate, paraformaldehyde, and zinc chloride in chloroform with hydrogen chloride passing through the reaction mixture for 4 hr.) [98].

(4) Bis-(chloromethyl)-durene (67% from durene in a 175–190° boiling petroleum fraction, 40% aqueous formaldehyde, and concentrated hydrochloric acid with hydrogen chloride passing through the mixture for 6 hr.) [99].

1. P. B. D. de la Mare and J. H. Ridd, *Aromatic Substitution, Nitration and Halogenation*, Butterworths Scientific Publications, London, 1959.
2. E. T. McBee and H. B. Hass, *Ind. Eng. Chem.*, **33**, 137 (1941).
3. P. H. Groggins, *Unit Processes in Organic Synthesis*, McGraw-Hill Book Co., New York, 1958, p. 204.
4. H. P. Braendlin and E. T. McBee in Olah ed., *Friedel-Crafts and Related Reactions*, Interscience Publishers, New York, 1964, Vol. III, Pt. 2, p. 1517.
5. I. R. L. Barker and W. A. Waters, *J. Chem. Soc.*, 150 (1952).
6. R. N. Haszeldine and A. G. Sharpe, *J. Chem. Soc.*, 993 (1952).
7. W. Militzer, *J. Am. Chem. Soc.*, **60**, 256 (1938).
8. R. B. Sandin *et al.*, *Org. Syn.*, Coll. Vol. **2**, 196 (1943).
9. G. H. Wollett and W. W. Johnson, *Org. Syn.*, Coll. Vol. **2**, 343 (1943).
10. V. H. Wallingford and P. A. Krueger, *Org. Syn.*, Coll. Vol. **2**, 349 (1943).
11. M. Ballester *et al.*, *J. Am. Chem. Soc.*, **82**, 4254 (1960).
12. V. A. Koptyug, *Isomerization of Aromatic Compounds*, Daniel Davey and Co., 257 Park Ave. S., New York, 1965, Chap. 2.
13. A. A. Spryskov *et al.*, *Zh. Obshch. Khim*, **34**, 237 (1964); *C.A.*, **60**, 10571 (1964).
14. E. C. Kooyman and R. Louw, *Rec. Trav. Chim.*, **81**, 365 (1962).
15. O. Silberrad, *J. Chem. Soc.*, **127**, 2677 (1925).
16. L. I. Smith, *Org. Syn.*, Coll. Vol. **2**, 95 (1943).
17. H. T. Clarke and M. R. Brethen, *Org. Syn.*, Coll. Vol. **1**, 121 (1941).
18. I. M. Heilbron and J. S. Heaton, *Org. Syn.*, Coll. Vol. **1**, 207 (1941).
19. C. A. Dornfield *et al.*, *Org. Syn.*, Coll. Vol. **3**, 134 (1955).
20. P. B. D. de la Mare *et al.*, *Bull. Soc. Chim. France*, 1157 (1966).
21. R. E. Buckles and N. G. Wheeler, *Org. Syn.*, Coll. Vol. **4**, 256 (1963).
22. I. R. L. Barker and W. A. Waters, *J. Chem. Soc.*, 150 (1952).
23. Y. Ogata and K. Nakajima, *Tetrahedron*, **20**, 43 (1964); Y. Ogata and K. Aoki, *J. Am. Chem. Soc.*, **90**, 6187 (1968).
24. H. Suzuki and R. Goto, *Bull. Chem. Soc., Japan*, **36**, 389 (1963).
25. O. R. Pierce and A. M. Lovelace, *Chem. Eng. News*, July 9, 72 (1962).
26. L. A. Wall *et al.*, *J. Am. Chem. Soc.*, **79**, 5654 (1957).
27. L. A. Wall *et al.*, *J. Res. Nat. Bur. Std.*, **62**, 107 (1959).
28. Ref. 1, p. 109.
29. Ref. 1, p. 110.
30. Ref. 1, p. 111.
31. H. O. Wirth *et al.*, *Ann. Chem.*, **634**, 84 (1960).
32. H. J. Reich and D. J. Cram, *J. Am. Chem. Soc.*, **90**, 1365 (1968).
33. R. C. Huston and M. M. Ballard, *Org. Syn.*, Coll. Vol. **2**, 97 (1943).
34. D. E. Pearson *et al.*, *J. Org. Chem.*, **32**, 2358 (1967).
35. L. A. Yanovskaya *et al.*, *Zh. Obshch. Khim.*, **22**, 1594 (1952); *C.A.*, **47**, 8032 (1953).
36. H. E. Ungnade and E. F. Orwoll, *Org. Syn.*, Coll. Vol. **3**, 130 (1955).
37. K. T. Potts, *J. Chem. Soc.*, 3711 (1953).
38. F. C. Whitmore and E. R. Hanson, *Org. Syn.*, Coll. Vol. **1**, 326 (1941).
39. F. B. Dains and F. Eberly, *Org. Syn.*, Coll. Vol. **2**, 355 (1943).
40. P. Chabrier *et al.*, *Compt Rend.*, **245**, 174 (1957).
41. K.-H. Boltze *et al.*, *Ann Chem.*, **709**, 63 (1967).
42. Ref. 1, p. 108.

43. D. E. Janssen and C. V. Wilson, *Org. Syn.*, Coll. Vol. **4**, 547 (1963).
44. Ref. 12, p. 63.
45. P. W. Robertson, *J. Chem. Soc.*, 1883 (1956).
46. E. M. Kosower *et al.*, *J. Org. Chem.*, **28**, 630 (1963).
47. A. W. Burgstahler *et al.*, *J. Org. Chem.*, **31**, 3516 (1966).
48. P. de Mayo, *Molecular Rearrangements*, Interscience Publishers, New York, 1963, Pt. 1, p. 309.
49. R. S. Neale *et al.*, *J. Org. Chem.*, **29**, 3390 (1964).
50. P. Haberfield and D. Paul, *J. Am. Chem. Soc.*, **87**, 5502 (1965).
51. C. L. Coon *et al.*, *J. Org. Chem.*, **33**, 1387 (1968).
52. D. E. Pearson *et al.*, *J. Org. Chem.*, **27**, 447 (1962).
53. J. H. Ridd, *J. Tenn. Acad. Sci.*, **40**, 92 (1965).
54. R. Q. Brewster, *Org. Syn.*, Coll. Vol. **2**, 347 (1943).
55. Ref. 1, p. 106.
56. E. and F. Berliner, *J. Am. Chem. Soc.*, **76**, 6179 (1954); *see also* H. L. Wheeler and L. M. Liddle, *Am. Chem. J.*, **42**, 441 (1909).
57. D. E. Pearson, D. J. Thoennes *et al.*, *J. Heterocycl. Chem.*, **6**, 243 (1969).
58. R. C. Elderfield, *Heterocyclic Compounds*, John Wiley and Sons, New York.
59. A. Weissberger, *The Chemistry of Heterocyclic Compounds*, Interscience Publishers, New York.
60. Ref. 1, Chap. 15.
61. D. E. Pearson *et al.*, *J. Org. Chem.*, **26**, 789 (1961).
62. H. J. Den Hertog *et al.*, *Rec. Trav. Chim.*, **81**, 864 (1962).
63. R. N. Haszeldine *et al.*, *J. Chem. Soc.*, 594 (1965).
64. Y. V. Shcheglov *et al.*, *Agrokhimiya*, 105 (1967); *C.A.*, **68**, 2091 (1968).
65. P. B. D. de la Mare *et al.*, *Chem. Ind.* (*London*), 361 (1958).
66. J. J. Eisch, *J. Org. Chem.*, **27**, 1318 (1962).
67. J. J. Eisch and B. Jaselskis, *J. Org. Chem.*, **28**, 2865 (1963).
68. Ref. 58, 1952, Vol. **4**, p. 17.
69. Ref. 1, p. 204.
70. K. Piers *et al.*, *Can. J. Chem.*, **41**, 2399 (1963).
71. M. Gordon and D. E. Pearson, *J. Org. Chem.*, **29**, 329 (1964).
72. I. M. Hunsberger *et al.*, *J. Am. Chem. Soc.*, **82**, 4430 (1960).
73. R. H. Baker *et al.*, *J. Am. Chem. Soc.*, **68**, 1532 (1946).
74. A. Hirschberg and P. E. Spoerri, *J. Org. Chem.*, **26**, 2356 (1961).
75. H. S. Mosher and M. Look, *J. Org. Chem.*, **20**, 283 (1955).
76. W. Minnis, *Org. Syn.*, Coll. Vol. **2**, 357 (1943).
77. Y. L. Goldfarb and L. D. Tarasova, *Bull. Acad. Sci. USSR, Div. of Chem. Sci.* (*English Transl.*), 1213 (1960).
78. A. Markovac and S. F. MacDonald, *Can. J. Chem.*, **43**, 3364 (1965).
79. J. R. Johnson and C. G. Gauerke, *Org. Syn.*, Coll. Vol. **1**, 123 (1941).
80. D. H. Derbyshire and W. H. Waters, *J. Chem. Soc.*, 573 (1950).
81. D. E. Pearson *et al.*, *J. Org. Chem.*, **23**, 1412 (1958).
82. D. E. Pearson *et al.*, *J. Org. Chem.*, **28**, 3147 (1963).
83. K. A. Kobe and T. F. Doumani, *Ind. Eng. Chem.*, **33**, 170 (1941).
84. J. Arotsky *et al.*, *Chem. Commun.*, 650 (1966).
85. C. F. H. Allen *et al.*, *Org. Syn.*, Coll. Vol. **3**, 796 (1955).
86. M. S. Newman and P. G. Scheurer, *J. Am. Chem. Soc.*, **78**, 5004 (1956).
87. V. T. Slyusarchuk and A. N. Novikov, *Zh. Org. Khim.*, **3**, 1323 (1967); *C.A.*, **67**, 9387 (1967).
88. E. C. Britton and R. M. Tree, Jr., U.S. Patent 2,607,802, August 19, 1952; *C.A.*, **47**, 5437 (1953).
89. R. C. Fuson and C. H. McKeever, *Org. Reactions*, **1**, 63 (1942).
90. G. A. Olah and W. S. Tolgyesi, Ref. 4, 1964, Vol. II, Pt. 2, p. 659.
91. D. L. Fields *et al.*, *J. Org. Chem.*, **29**, 2640 (1964).
92. L. D. Taylor and R. B. Davis, *J. Org. Chem.*, **28**, 1713 (1963).
93. Y. Ogata and M. Okano, *J. Am. Chem. Soc.*, **78**, 5423 (1956).
94. I. N. Nazarov and A. V. Semenovsky, *Bull. Acad. Sci., USSR, Div. Chem. Sci.* (*English Transl.*), 997 (1957).

95. G. A. Olah and S. J. Kuhn, *J. Org. Chem.*, **29,** 2317 (1964).
96. O. Grummitt and A. Buck, *Org. Syn.*, Coll. Vol. **3,** 195 (1955).
97. C. A. Buehler *et al.*, *Org. Syn.*, Coll. Vol. **3,** 468 (1955).
98. O. Moldenhauer *et al.*, *Ann. Chem.*, **580,** 176 (1953).
99. M. J. Rhoad and P. J. Flory, *J. Am. Chem. Soc.*, **72,** 2216 (1950).

E. Miscellaneous Reactions

1. From Polyhalides (Reduction)

$$CHX_3 \xrightarrow{\text{Na}_3\text{AsO}_3} CH_2X_2$$

If a polyhalide is readily available, it may serve as a source of a lower halide of the same carbon content. Reducing agents which have been employed to effect such a change are aluminum amalgam [1], aluminum chloride [2], sodium arsenite [3, 4], copper powder in water [5], mercaptans [6], methylmagnesium bromide in tetrahydrofuran [7], tri-n-butyltin hydride [8], and molecular hydrogen in the presence of platinum [9]. Yields in such reductions are often high. Some tribromomethyl derivatives behave like positive halogen compounds and can be reduced by alcohols (see Ex. b.4), and chloral undergoes a redox reaction with the cyanide ion to form methyl dichloroacetate [10].

(a) **Preparation of methylene iodide** (90–97% from iodoform, sodium arsenite, and sodium hydroxide) [4].

(b) **Other examples**

(1) **1,1-Dichloro-2-(o-chlorophenyl)-2-(p-chlorophenyl)ethane** (65.5% from 1,1,1-trichloro-2-(o-chlorophenyl)-2-(p-chlorophenyl)ethane and aluminum amalgam in 90% ethanol) [1].

(2) **1-Chloronorcamphane** (50–55% from norcamphor dichloride

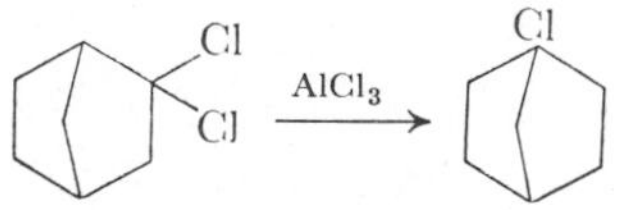

and aluminum chloride in pentane and isopentane) [2].

(3) **Dichloroacetamide** (87% from trichloroacetamide added to 5–20 equiv. of methyl mercaptan and 1.5–5 equiv. of triethylamine at −20° and allowed to warm to room temperature; only trichloroacetic acid derivatives and trichloromethyltriazines could be reduced in this fashion) [6].

(4) **4-Chloro-2-dibromomethylquinoline** (77% from 5 g. of 4-chloro-2-tribromomethylquinoline refluxed for 24 hr. in 65 ml. of 2-propanol) [11].

(5) **1-Bromo-2,2,3-trimethylcyclopropane** (79%, 4 parts *cis* and 1 part *trans*, from 1,1-dibromo-2,2,3-trimethylcyclopropane by reduction with tri-n-butyltin hydride under nitrogen for 1 hr. below 40°) [8].

References for Section E are on p. 410.

2. From α-Haloamides and Hypohalites (Rearrangement)

$$\underset{\underset{Cl}{|}}{\overset{\overset{R}{|}}{R-C}}-CONH_2 \xrightarrow{NaOBr} \underset{\underset{Cl}{|}}{\overset{\overset{R}{|}}{R-C}}-Br$$

$$CF_3(CF_2)_2CONH_2 \xrightarrow{NaOCl(Br)} CF_3(CF_2)_2Cl(Br)$$

Although the Hofmann degradation of an amide usually produces an amine containing one less carbon atom, with α-chloroamides [12] and polyfluorinated amides [13], the reaction takes a different course. For the α-haloamides, *gem*-dihalides, aldehydes or ketones, and α-hydroxycarboxylic acids are produced.

$$\underset{\underset{Cl}{|}}{\overset{\overset{R}{|}}{R-C}}CONH_2 \xrightarrow{NaOBr} \underset{\underset{Cl}{|}}{\overset{\overset{R}{|}}{R-C}}-Br + \underset{\underset{R}{|}}{\overset{\overset{R}{|}}{C}}=O + \underset{\underset{OH}{|}}{\overset{\overset{R}{|}}{R-C}}-\overset{\overset{O}{\|}}{C}-OH$$

The yields of *gem*-dihalides for a series of α-chloro- and α-bromoamides in the aliphatic series varied from 7 to 95%. The reaction appears to be of greater value as a synthetic tool for preparing perfluorinated hydrocarbons in that it avoids the rigorous conditions essential in the silver salt-halogen method (A.11). Although trifluoroacetamide was reported to give hexafluoroethane [14], perfluorobutyramide when treated with sodium hypochlorite or hypobromite gave the corresponding perfluoroalkyl halide, the yield in the case of the bromide being 65–70% (see Ex. a). In attempts to prepare the iodide with sodium hypoiodite, the principal product isolated was heptafluoropropane. Thus it appears that this synthesis may be valuable for preparing perfluoro-bromides containing three or more carbon atoms.

Attempts have been made to account for the products from these halogenated amides on a mechanistic basis [12, 13].

(a) **Preparation of n-perfluoropropyl bromide.** To a solution of 36 g. of sodium hydroxide in 100 ml. of water at 0°, 28.8 g. of bromine was added dropwise at such a rate that the temperature was kept below 5°. On the disappearance of the red color, 32 g. of perfluorobutyramide and 50 ml. of water were added. After stirring for 1 hr., the temperature of the mixture was allowed to rise to 20° over a 1-hr. period and then the solution was refluxed for 2.5 hr. The liquid which collected in a dry ice-acetone cold trap was distilled twice to yield 16 ml. (65–70%) of n-perfluoropropyl bromide, b.p. 15–15.2° (742 mm.) [13].

(b) **Preparation of 2-bromo-2-chloropropane** (95% from α-chloro-isobutyramide, bromine, and sodium hydroxide) [12].

3. From Some N-Haloamines

N-Haloamines on irradiation usually eliminate halogen in a free radical process to produce rearranged products (Hofmann-Löffler reaction) [15]:

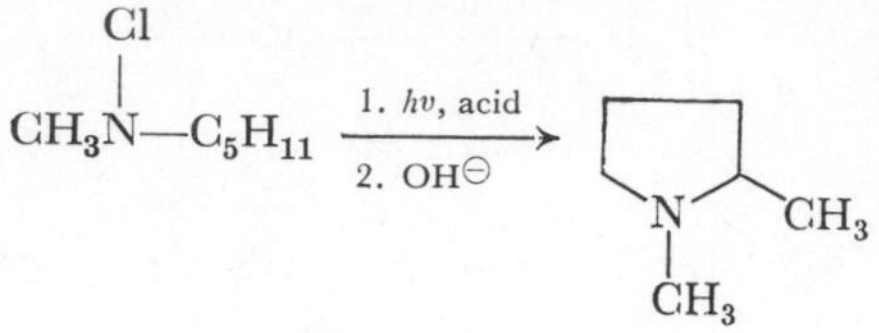

1,2-Dimethylpyrrolidine

In some cases similar rearrangements occur without loss of the halogen. For example, a remarkable reaction has been discovered recently wherein the rearrangement of the N-haloamine is catalyzed by the silver cation as shown [16]:

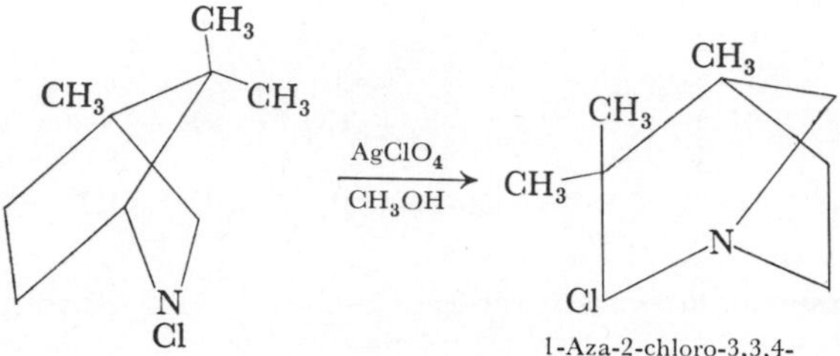

1-Aza-2-chloro-3,3,4-trimethylbicyclo[2.2.1]-heptane, 77%

The reaction may be restricted to bicyclic haloamines.

1. T. Inoi et al., J. Org. Chem., **27**, 4597 (1962).
2. K. B. Wiberg et al., J. Am. Chem. Soc., **83**, 3998 (1961).
3. W. W. Hartman and E. E. Dreger, Org. Syn., Coll. Vol. **1**, 357 (1941).
4. R. Adams and C. S. Marvel, Org. Syn., Coll. Vol. **1**, 358 (1941).
5. H. W. Doughty and G. J. Derge, J. Am. Chem. Soc., **53**, 1594 (1931).
6. E. Kober, J. Org. Chem., **26**, 2270 (1961).
7. D. Seyferth and B. Prokai, J. Org. Chem., **31**, 1702 (1966).
8. D. Seyferth et al., J. Org. Chem., **28**, 703 (1963).
9. J. S. Buck and W. S. Ide, J. Am. Chem. Soc., **54**, 4359 (1932).
10. C. Rosenblum et al., Chem. Ind. (London), 718 (1960).
11. D. E. Pearson and J. C. Craig, unpublished work.
12. C. L. Stevens et al., J. Am. Chem. Soc., **78**, 2264 (1956).
13. D. R. Husted and W. L. Kohlhase, J. Am. Chem. Soc., **76**, 5141 (1954).
14. E. Gryszkiewicz-Trochimowski et al., Rec. Trav. Chim., **66**, 419 (1947).
15. P. de Mayo, Molecular Rearrangements, Interscience Publishers, New York, 1963, Pt. 1, p. 448.
16. P. G. Gassman and R. L. Cryberg, J. Am. Chem. Soc., **90**, 1355 (1968).

Chapter 8

AMINES

The preparation of amines is an inexhaustible subject. Sufficient variety of methods is contained herein, however, to give a general view of the range. For more detailed discussion, Houben-Weyl is recommended [1]. Brief discussions are to be found elsewhere [2–4].

It is helpful in the synthesis of amines to understand the nature of the amino group. The amino group is capable of revealing one of six faces in its reactions to

References for Section A are on pp. 433–436.

prepare other amines as listed:

1. $R_2\overset{H}{N}$:   A weak nucleophile utilizing its free electron pair. Displacement of active halide or other groups (C) or addition to positive centers (D).

2. $R_2\overset{\ominus}{\underset{..}{N}}$:   The anion, a strong nucleophile to use where the amine fails (C.2).

3. $R_2\overset{\oplus}{\underset{..}{N}}$ The nitrenium ion, capable of addition to or substitution in systems with π electrons (F.2) and common as an intermediate in rearrangement reactions (G).

4. $R\overset{..}{N}$:   The nitrene, a highly reactive, uncharged intermediate capable of addition to π systems (F.4), or insertion (F.4), or of inducing rearrangements (G). There is evidence that two types of nitrenes may be formed simultaneously in some cases: an indiscriminate, singlet type and a discriminate, triplet type [5].

5. $R_2\overset{.}{\underset{..}{N}}$ The free radical. A rare species for synthetic work but encountered occasionally in substitution (F.4) or in oxidation reactions.

6. $R_2\overset{.\ \oplus}{N}H$ The ion-free radical. Encountered under the same conditions as above only in strong acid solution. Capable of selective insertion (F.4).

A. Reduction

1. From Nitro Compounds

$$(Ar)RNO_2 \xrightarrow[Pt]{H_2} (Ar)RNH_2$$

This reduction, one of the oldest methods for preparing amines, is still widely used in the laboratory. It involves the use of a reducing agent sufficiently potent to carry the reduction through various intermediates to the amine, which constitutes the final reduction product. At first the method was applied largely to the aromatic series but since nitroalkanes are now readily available satisfactory reductions are possible in this series as well. The synthesis is discussed under the various reducing agents. It is suggested that the reader consult the innumerable examples of Houben-Weyl [6].

Metal and Acid

In the solution of the metal in the acid, the nitro compound is reduced and the amine may be recovered from the reaction mixture, made alkaline, by steam distillation or filtration.

If catalytic methods are eliminated, iron or zinc is the metal of choice for reductive procedures in the elementary laboratory. The advantage of iron is that only a catalytic amount of acid need be added and the iron oxide precipitates out to permit a relatively simple separation. The disadvantage is that the

$$4\,RNO_2 + 9\,Fe + 4\,H_2O \rightarrow 4\,RNH_2 + 3\,Fe_3O_4$$

quality of iron powder from various commercial sources is variable. Some powders give spontaneous, exothermic reduction; others, none. A reliable reagent may be obtained by wetting 200 g. of degreased iron powder with 35 ml. of concentrated hydrochloric acid and redrying. The etched powder may be stored under benzene. Reduction with this reagent may be carried out either in benzene or aqueous alcohol (see Ex. a and Ex. d.1). Furthermore, iron powder in acetic acid is capable of reducing aliphatic, optically active nitro compounds to the amine with retention of configuration [7]. Indeed, iron powder and acetic acid is a favorite combination for the reduction of aromatic nitro compounds. On the other hand, basic reducing agents such as lithium aluminum hydride lead to racemization.

Zinc is a powerful reducing agent, but utilization of its reducing action is wasteful because of rapid reaction with an acid such as hydrochloric. Successful efforts therefore are to be found when an abundance of zinc is supplied during the entire reaction period. If the experimenter is fortunate, the hydrochloride of the amine will separate (see Ex. b) but, if unfortunate, he must make the entire mixture strongly alkaline.

The present authors are impressed with the smooth generation of hydrogen from mossy tin and acid but prejudiced by the difficulty of workup if the amine hydrochloride does not precipitate from the acid solution. The entire mixture should be added slowly to a strong alkaline solution. If tin hydroxide precipitates and remains for any length of time, no reasonable amount of sodium hydroxide dissolves the flocculent solid. Certainly, this method does not lend itself well to large-scale reductions. On the other hand, stannous halide is a useful and selective reducing agent for nitro groups when carbonyl groups are present, such as in the reduction of m-nitrobenzaldehyde [8].

Amalgamated aluminum is another reducing agent best adapted to neutral or basic reducing conditions. For example, 2-nitroacenaphthene suspended in

$$RNO_2 + 2\ Al + H_2O \rightarrow RNH_2 + Al_2O_3$$

boiling ethanol when treated with amalgamated aluminum and then portionwise with boiling water is reduced in 85 % yield to the amino compound [9].

Catalytic Method

This method [10, 11] is so clean and the workup so easy that it should be considered the method of choice in the research laboratory. The only pitfalls are the possibility of poisoning of the catalyst or, for large-scale reductions, the inability to dissipate adequately the heat of hydrogenation [12]. Example (c) illustrates a reasonable concentration of nitro compound to reduce and moreover the surprising effectiveness of used catalyst.

The most common catalysts for the complete hydrogenation of nitro compounds are platinum (low pressure) and Raney nickel (high pressure). Supported catalysts such as platinum on carbon [13, 14], palladium on carbon [14–16], rhodium on alumina [17, 18], and platinum sulfide on carbon, a catalyst less sensitive to poisons and incapable of hydrogenolyzing ring halogen [19], have also been employed.

Metal Hydrides

Nitro groups are customarily reduced with lithium aluminum hydride, one of the most powerful of the metal hydrides. In fact, sodium borohydride in aqueous methanol at 25° does not affect the nitro group [20], but this metal hydride with palladized charcoal in alkaline solution or in aqueous methanol is effective [21]. A complication results with lithium aluminum hydride and tertiary alicyclic nitro compounds in that the intermediate hydroxylamine derivative rearranges to give both primary and secondary amines [22]:

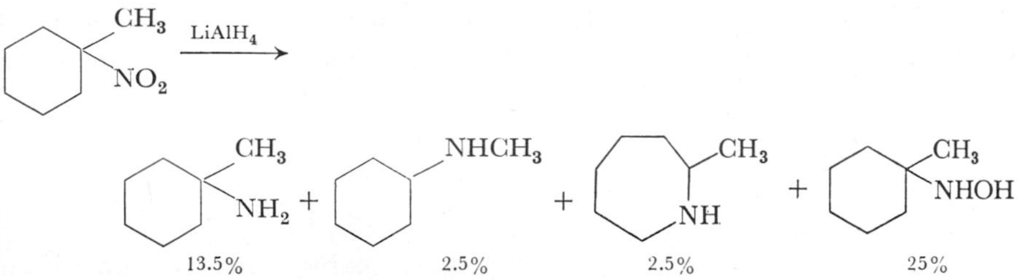

$$\text{13.5\%} \quad\quad \text{2.5\%} \quad\quad \text{2.5\%} \quad\quad \text{25\%}$$

Other Reducing Agents

Reducing agents such as sodium hydrosulfite $(Na_2S_2O_4)$ [23], ferrous sulfate and ammonium hydroxide [24] zinc and water [25], zinc and sodium hydroxide in aqueous alcohol [26], ammonium sulfide [27], hydrazine in the presence of palladized charcoal [28] or in the presence of nickel, platinum, or ruthenium [29], and phenylhydrazine without a catalyst but at high temperature [30] have also been employed to reduce nitro compounds to amines. The Wolff-Kishner method may also be used in producing amines from nitro compounds [31]. Sodium sulfide and ammonium chloride [32], hydrogen sulfide and ammonium hydroxide [33], and sodium sulfide and sulfur [34] have been used to reduce one of two nitro groups in the benzene ring. But, in unsymmetrical dinitrobenzenes, the reduction is not always specific [35, 36]. A most unusual and occasionally helpful reducing agent is cyclohexene in the presence of palladium [37]:

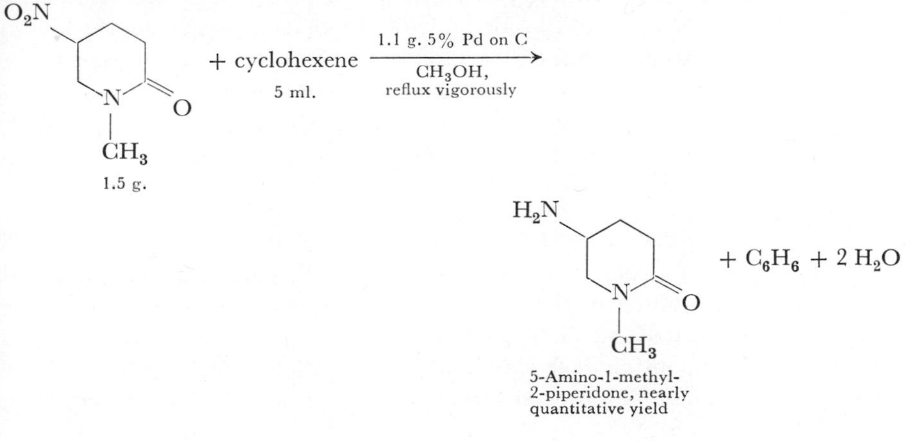

5-Amino-1-methyl-
2-piperidone, nearly
quantitative yield

Anhydrous stannous chloride is an unusual reducing agent for aromatic nitro compounds in that chlorination takes place simultaneously [38]:

$$SnCl_2 \cdot 2\ H_2O + Ac_2O + C_6H_5NO_2 \longrightarrow Cl\langle\!\!\!\bigcirc\!\!\!\rangle NHCOCH_3$$

3 equiv.   9 equiv.   1 equiv.

p-Chloroacetanilide,
quantitative yield

Sodium bisulfite reduces nitro groups to amines but may also form a sulfonic acid or sulfamic acid under the conditions of reduction [39]. Electrolytic reduction is a general method of reduction in that it affords amines in 80–90 % yields [40]. By making the catholytic solution alkaline and by holding the intermediate azoxy compound in solution, electrolytic reduction of nitro compounds leads to hydrazo compounds [41]:

$$ArNO_2 \rightarrow ArNHNHAr$$

The chemical method of making hydrazobenzenes from nitro compounds is by the use of zinc dust and sodium hydroxide. Phenylhydrazine has been used to reduce 4-nitropyridine-N-oxide to the corresponding hydroxylamine [42], although the usual reagent for this reduction is zinc and ammonium chloride [43].

The discussion above by no means exhausts the list of reducing agents for nitro compounds. For example, aniline has been made in 65 % yield by heating nitrobenzene with isopropyl alcohol and solid sodium hydroxide [44].

(a) Preparation of 4-aminobiphenyl hydrochloride (90–92 % from 100 g. of 4-nitrobiphenyl in 100 ml. of benzene and 200 g. of iron powder (pretreated with 35 ml. of conc. hydrochloric acid and then dried) stirred and heated to reflux; then 1 ml. of water was added and the benzene layer was allowed to clear before the next addition of water, which finally totalled 26 ml.; the mixture was centrifuged, the benzene layer separated, and the iron oxide extracted with 100 ml. more of benzene; the combined benzene layers were treated with 220 ml. of conc. hydrochloric acid and the amine hydrochloride filtered) [45].

(b) Preparation of 1,2,3,5-tetraaminobenzene trihydrochloride hydrate (95 % from 20 g. zinc chloride and 100 ml. of conc. hydrochloric acid to which granulated zinc was added followed by 1 g. of picramide; as soon as the yellow color disappeared, another portion of picramide was added until a total of 50 g. was reduced; if the reduction slowed, more granulated zinc was added; the suspended crystals were decanted from the excess zinc and washed with hydrochloric acid) [46].

(c) Preparation of p-t-butylaniline. p-Nitro-t-butylbenzene (0.05 mole) in 125 ml. of alcohol was reduced in a Burgess-Parr apparatus at about 50 lb. pressure of hydrogen with 0.1 g. of platinum oxide. After reduction, the solution was decanted and platinum was left in the reducing bottle. Another 0.05 mole of p-nitro-t-butylbenzene in 125 ml. of alcohol was added and the reduction carried out. This process was repeated 10 times. The time of reduction was 15 min. for the first batch and had increased to 60 min. for the tenth batch. The yield of amino compound was nearly quantitative [47].

(d) Other examples

(1) **2,4-Diaminotoluene** (74% from 2,4-dinitrotoluene, iron, ethyl alcohol, and hydrochloric acid) [48].

(2) **γ-Aminovaleric acid** (98% from γ-nitrovaleric acid in ethanol, Adams' platinum catalyst, and hydrogen (3 moles)) [49].

(3) **m-Aminobenzaldehyde dimethylacetal** (67–78% from m-nitrobenzaldehyde dimethylacetal in anhydrous methanol, Raney nickel, and hydrogen at 1000 lb. pressure and a temperature of 40–70° for about 1.5 hr.) [50].

(4) **4-Hydroxy-3-methoxyphenyl-β-ethylamine** (80% as the picrate

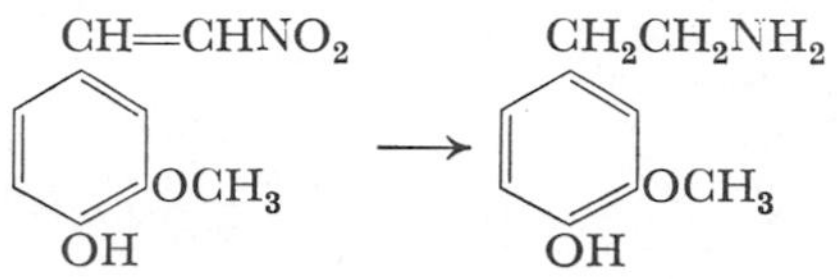

from 4-hydroxy-3-methoxy-β-nitrostyrene in absolute ether and lithium aluminum hydride by the Soxhlet extractor technique) [51].

(5) **2-Aminofluorene** (78–82% from 2-nitrofluorene in aqueous ethanol and zinc dust with a small amount of calcium chloride [25]. Reduction with hydrazine and a trace of palladized charcoal gave a 93–96% yield) [28].

(6) **2-Amino-4-nitrophenol** (58–61% from 2,4-dinitrophenol, sodium sulfide, and ammonium chloride in ammoniacal solution at 80–85°) [32].

(7) **2,5-Dichloroaniline** (99.5% from 2,5-dichloronitrobenzene in methanol, 5% platinum sulfide on carbon, and hydrogen at 34–54 atm. and 85°) [19].

2. From Nitroso, Azo, Hydrazino, Azido, or Related Compounds

Since nitroso and hydroxylamino compounds are intermediates in the reduction of a nitro compound to an amine, identical reducing reagents may be used in these cases. Common reducing agents for the above and other titled compounds are hydrogen in the presence of platinum or Raney nickel, a metal hydride such as lithium aluminum hydride, hydrazine, sodium hydrosulfite, etc. Since these syntheses are of minor importance, they are discussed by simply giving an illustration of each class.

From Nitroso Compounds [52]

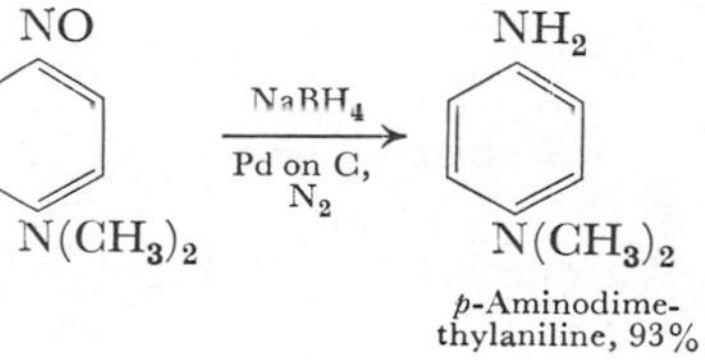

p-Aminodime-
thylaniline, 93%

From Hydroxylamines [53]

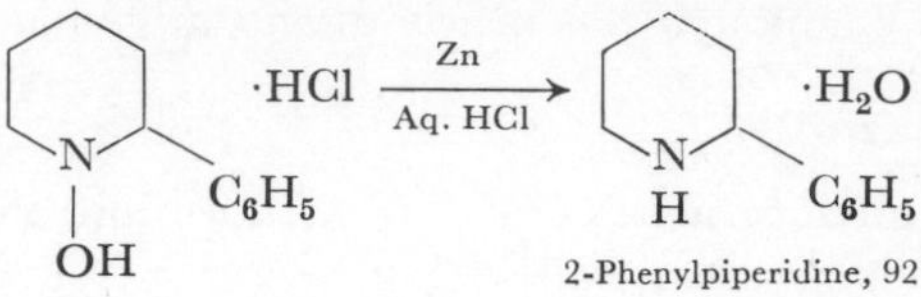

2-Phenylpiperidine, 92%

From N-Oxides

The N-oxides of tertiary amines are readily reduced with palladium and hydrogen, with phosphorus trichloride, and with triphenylphosphine [54], but a remarkable selectivity has been found in the reduction of an aliphatic N-oxide in the presence of an aromatic N-oxide [55]. This synthesis was confirmed with a yield of 86% [56].

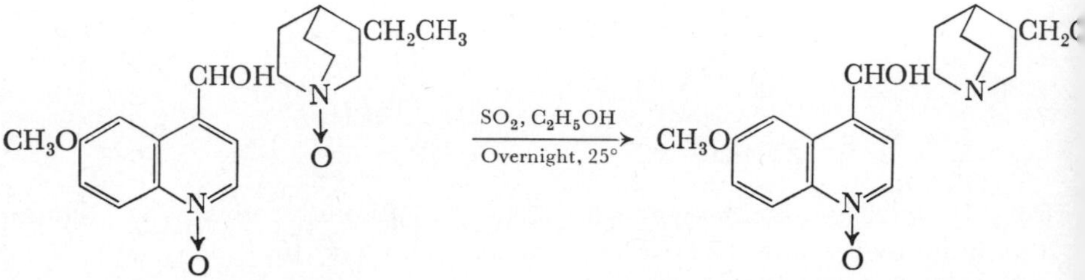

Dihydroquinine-N,N'-dioxide

Dihydroquinine-N-oxide

From Azo Compounds [57]

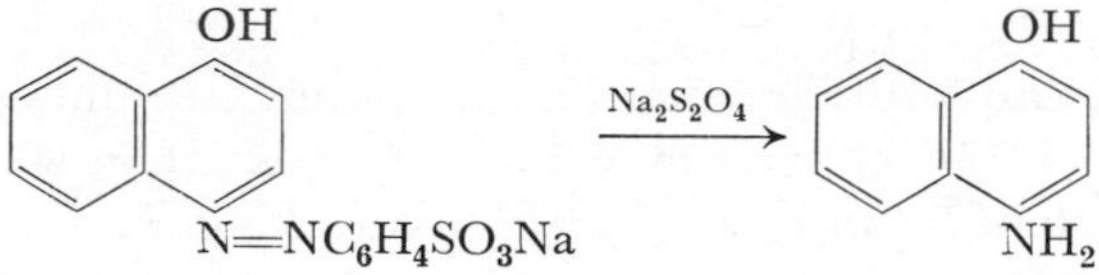

4-Amino-1-naphthol, 72–75% based on the original 1-naphthol

Azo compounds may be reduced to hydrazo compounds by conditions resembling those of the Meerwein-Ponndorf reaction [58]:

$$C_6H_5N{=}NC_6H_5 + NaOH + CH_3OH \xrightarrow[\text{Reflux}]{\substack{\text{9-Fluorenone,} \\ \text{5 g.}}} C_6H_5NHNHC_6H_5$$

(or azoxy), 140 g.   108 g.   192 g.   Hydrazobenzene, 98%

From Hydrazines [59]

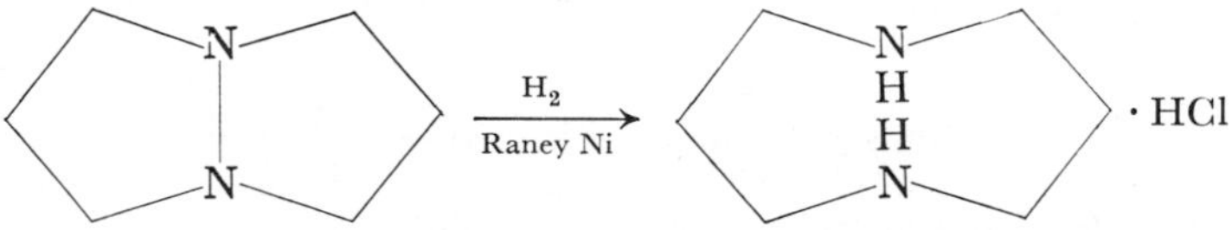

1,5-Diazacyclooctane hydrochloride, 89%

From Azides [60]

$$C_6H_5CH_2CH_2N_3 \xrightarrow{\text{LiAlH}_4} C_6H_5CH_2CH_2NH_2$$

β-Phenylethylamine, 89%

From Nitrosamines [61]

$$m\text{-}CH_3C_6H_4NC_2H_5 \xrightarrow[\text{HCl}]{\text{SnCl}_2} m\text{-}CH_3C_6H_4NHC_2H_5$$

with NO below the nitrogen on the left structure.

N-Ethyl-m-toluidine, 63–66%
from m-toluidine

This method is the last step in a sequence of steps leading to the purification of secondary amines.

3. From Nitriles

$$(R)ArCN \rightarrow (R)ArCH_2NH_2$$

Active metals in an aqueous acid are not satisfactory reducing agents for nitriles since, with such reagents, hydrolysis to the carboxylic acid occurs. The two principal reagents employed in this case are: hydrogen in the presence of a catalyst and metal hydrides. In the catalytic method a complication results because the first-formed imine 8-I reacts with the primary amine to form secondary amines:

$$R\text{—}CN \xrightarrow{H_2} RCH\text{=}NH \xrightarrow{H_2} RCH_2NH_2$$

8-I

$$RCH\text{=}NH + RCH_2NH_2 \rightleftharpoons RCH_2NHCHR \rightleftharpoons RCH_2\text{—}N\text{=}CHR + NH_3$$

with NH_2 below the middle structure, and $\downarrow H_2$ leading to:

$$RCH_2NHCH_2R$$

This difficulty is overcome in the low-pressure catalytic method by using with Raney nickel an acidic solvent such as acetic anhydride, with sodium acetate as a cocatalyst, which removes the primary amine as a salt or acetamide derivative [62]. In the case of a rhodium-on-alumina catalyst, also at low pressure, ammonia is employed to prevent the formation of the secondary amine [63]. Of the two methods the former appears to be the preferred one. In high-pressure hydrogenation an excess of ammonia may also be added to displace the equilibrium, or rhodium with a small amount of lithium hydroxide [64] may be employed.

Metal hydrides such as lithium aluminum hydride or diborane [65] as produced from sodium borohydride and boron trifluoride:

$$3\,NaBH_4 + 4\,BF_3 \rightarrow 3\,NaBF_4 + 2\,B_2H_6$$

are satisfactory reagents for the reduction of nitriles to amines. The use of aluminum chloride with lithium aluminum hydride (1:1) improves the yield [66]. The intermediate with the metal hydride is an imine salt, 8-II, which

$$ArC\text{≡}N \xrightarrow{LiAlH_4} \left[ArCH\text{=}N\overset{\ominus}{\text{—}}AlH_3 \right] \xrightarrow[H_3O^{\oplus}]{LiAlH_4} ArCH_2NH_2$$

8-II

on further reduction gives the amine. The hydrolysis of the imine salt would give an aldehyde, a method which has been realized when lithium triethoxy-aluminohydride (see Aldehydes, Chapter 10, B.4) is the reducing agent. The Stephen reduction with stannous chloride and hydrochloric acid appears to involve a similar intermediate which also gives the aldehyde on hydrolysis (see Aldehydes, Chapter 10, B.6), but which at times is further reduced [67] to give the amine. For low-boiling amines (b.p. less than 117°), lithium aluminum hydride appears to be the reagent of choice (see Ex. a). For high-molecular-weight amines from nitriles, aluminum hydride appears to have some advantage because no tendency exists to form anions of nitriles,

$$RCH_2CN \rightarrow R\overset{\ominus}{C}HCN$$

(see Ex. b), which may produce side reactions.

Diborane reduces nitriles rapidly at room temperature, the intermediate proposed being 8-III [68]:

$$RCN + BH_3 \rightarrow [RCH_2NBH]$$
$$\text{8-III}$$

which probably exists as a trialkyl borazine. This reagent is specific in its action and may be used to reduce nitriles without affecting nitro or ester groups. The intermediate in this case produces no detectable amount of aldehyde.

A new procedure for converting nitriles into secondary amines has been published recently; its chemistry is shown (see Ex. c.6):

$$RC\equiv N + (EtO)_2\overset{\oplus}{C}H\overset{\ominus}{B}F_4 \longrightarrow RC\overset{\oplus}{\equiv}N\overset{\ominus}{E}t\ BF_4 \xrightarrow{CH_3OH}$$

$$\overset{OCH_3}{\underset{|}{RC}}=NEt \xrightarrow{NaBH_4} RCH_2NHEt$$

Homopiperazine and its monomethyl derivatives may be synthesized by catalytic reduction of N-(2-cyanoethyl) ethylenediamines (see Ex. c.8).

(a) Preparation of β,β-dideuterobutylamine (86% from 1 mole of α,α-dideuterobutyronitrile added dropwise to 1 mole of lithium aluminum deuteride in 500 ml. of diethylene glycol diethyl ether held at 5–70°, then solvolyzed with butoxyethanol and distilled) [69].

(b) Preparation of 2,2-diphenylethylamine (91% from 0.266 mole of lithium aluminum hydride in THF treated with 13.03 g. of 100% sulfuric acid to form AlH₃, followed by dropwise addition of a solution of 0.2 mole of diphenylacetonitrile in THF) [70]; see Ex. c.4 for an alternative method.

(c) Other examples

(1) Decamethylenediamine (79–80% from sebaconitrile in 95% ethanol, Raney nickel, liquid ammonia (6–8 moles/mole of dinitrile), and hydrogen (1500 lb. pressure) at 110–125°) [71].

(2) β-Phenylethylamine (97% as the acetyl derivative by hydrogenating phenylacetonitrile in acetic anhydride in the presence of Raney nickel and sodium acetate as a cocatalyst for 45–60 min. at 50° and 50 p.s.i.) [62].

(3) p-(2,3,5,6-Tetramethylbenzoyl) benzylamine hydrochloride (87% from p-cyanobenzoyldurene in chloroform added to a mixture of stannous chloride, hydrogen chloride, and ether at room temperature) [67].

(4) 2,2-Diphenylethylamine (91 % from diphenylacetonitrile, lithium aluminum hydride, and aluminum chloride in ether) [66].

(5) *m*-Nitrobenzylamine hydrochloride (88 % from a tetrahydrofuran solution of *m*-nitrobenzonitrile, diborane (from sodium borohydride and boron trifluoride etherate), and hydrogen chloride) [65].

(6) Amylethylamine (76 % as the hydrochloride from 9 mmoles of valeronitrile and 18 mmoles of diethoxycarbonium tetrafluoroborate, $(EtO)_2\overset{\oplus}{C}H\overset{\ominus}{B}F_4$, made from ethyl orthoformate and boron trifluoride, refluxed 19 hr. in methylene chloride, cooled, evaporated, the residue dissolved in 20 ml. of methanol and treated cautiously with 2 g. of sodium borohydride and stirred 1 hr. at 0°; the mixture was acidified to pH 1, evaporated, made basic, and extracted) [72].

(7) Tryptamine (78 % by the reduction of 3-indoleacetonitrile in

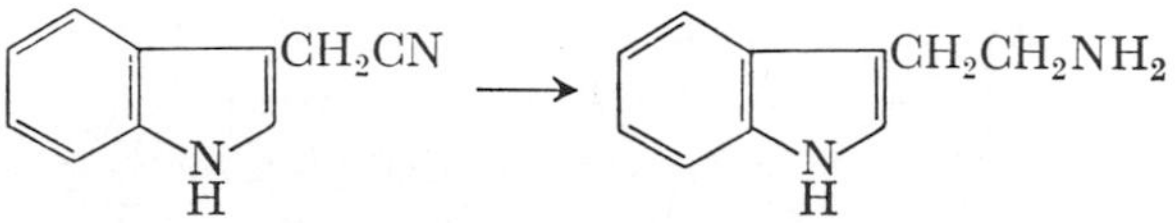

ethanolic ammonia with 2.5 atm. of hydrogen and rhodium on alumina at 25° for 2 hr.; this appears to be the most satisfactory catalytic method for reducing this nitrile) [63].

(8) Homopiperazine.

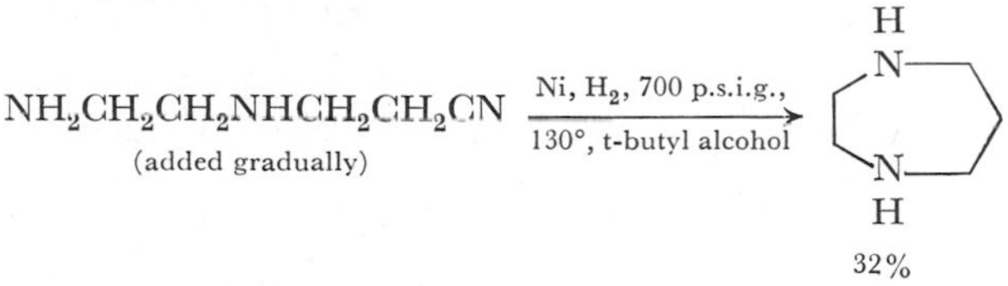

$$NH_2CH_2CH_2NHCH_2CH_2CN \xrightarrow[\text{130°, t-butyl alcohol}]{\text{Ni, H}_2\text{, 700 p.s.i.g.,}}$$

(added gradually)

32%

The remainder is the reduced, higher-boiling, acyclic triamine [73].

4. From Amides, Hydrazides, or Isocyanates

$$(R)ArCONH_2 \xrightarrow{\text{LiAlH}_4} (R)ArCH_2NH_2$$

$$(R)ArCONHNH_2 \xrightarrow{\text{LiAlH}_4} (R)ArCH_2NH_2$$

$$(R)ArN{=}C{=}O \xrightarrow{\text{LiAlH}_4} (R)ArNHCH_3$$

Synthesis from amides is much more common than from acid hydrazides, as might be expected. The principal reducing agent which has been employed in the reduction of amides is a metal hydride such as lithium aluminum hydride [74]. As a rule, the reaction yields an amine of the same number of carbon atoms. However, if limited amounts of lithium aluminum hydride or a less active reducing agent, such as lithium diethoxy-or triethoxyaluminum hydride, is employed, some aldehyde may be obtained. (See Aldehydes, Chapter 10, B.4.) Amides derived from ethylenimine [75], carbazole [76], N-methylaniline [77], and imidazole [78] give substantial yields of aldehyde.

The path of the reaction for substituted amides [79] may be represented as follows:

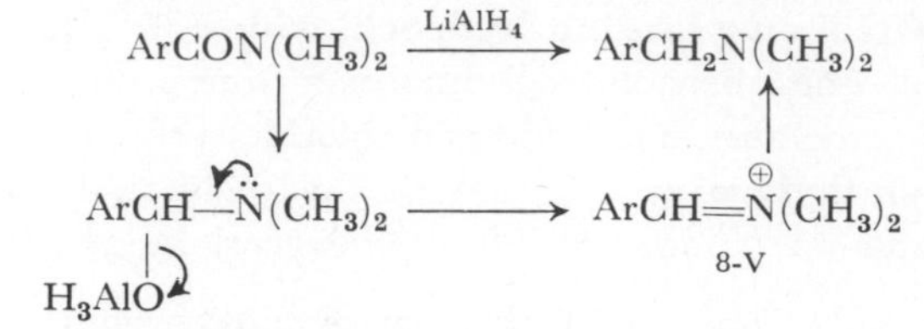

$$ArCON(CH_3)_2 \xrightarrow{\text{LiAlH}_4} ArCH_2N(CH_3)_2$$

$$ArCH-\overset{\curvearrowleft}{\underset{\underset{\text{8-IV}}{\overset{|}{H_3AlO}}}{N}}(CH_3)_2 \longrightarrow ArCH\overset{\oplus}{=}\underset{\text{8-V}}{N}(CH_3)_2$$

The intermediate *gem*-amino alcohol derivative 8-IV may be hydrolyzed to the aldehyde. However, if the electron pair of the nitrogen enters into the elimination process, the arylidenedimethylammonium salt 8-V results and subsequent reduction leads to the amine.

Recently diborane [80] has been shown to be an effective reagent for reducing amides of primary and secondary amines. These reductions in tetrahydrofuran at low temperature require 1 to 8 hours and give yields with a series of amides of 79 to 98% as determined by gas chromatographic analysis, isolation as the picrate, or by titration. And reduction of a series of 2,6-piperazinediones with diborane gave piperazines in yields averaging about 60% [81]. Reduction of amido groups in the presence of ester groups by diborane appears to be selective [82].

Reducing agents less frequently used have been sodium and ethanol and hydrogen in the presence of copper chromite [83].

The reduction of isocyanates with lithium aluminum hydride results in cleavage of the carbon-oxygen bond to give methyl amines [84] as shown:

$$RNCO \xrightarrow{\text{3 LiAlH}_4} LiAl(NR(CH_3))_4 \xrightarrow{\text{4 H}_2O} 4\,RNHCH_3$$

Yields in three cases vary from 78 to 90%.

(a) Preparation of N,N-dimethylcyclohexylmethylamine (88% from N,N-dimethylcyclohexanecarboxamide and lithium aluminum hydride in anhydrous ether under reflux for 16 hr.) [85].

(b) Other examples

(1) Dimethylneopentylamine

$$\underset{\underset{CH_3}{|}}{\overset{\overset{CH_3}{|}}{CH_3CCON(CH_3)_2}} \longrightarrow \underset{\underset{CH_3}{|}}{\overset{\overset{CH_3}{|}}{CH_3CCH_2N(CH_3)_2}}$$

(79% from the addition of N,N-dimethylpivalamide to borane, both in tetrahydrofuran, at 0° followed by 1 hr. reflux) [80].

(2) Decylamine (90% from capric acid amide hydrogenated in the presence of BaO-stabilized copper chromite in liquid ammonia at 350° and 411 atm.; in the absence of liquid ammonia, didecylamine, 73%, was obtained) [83].

(3) Laurylmethylamine (81–95 % from N-methyllauramide and lithium aluminum hydride in ether by the Soxhlet extractor technique) [86].

(4) 3-Amino-2,2-diphenylpropanol-1 (68 % from α,α-diphenyl-β-

$$\underset{\underset{C_6H_5}{|}}{\overset{\overset{C_6H_5}{|}}{HOCH_2C-CONHNH_2}} \longrightarrow \underset{\underset{C_6H_5}{|}}{\overset{\overset{C_6H_5}{|}}{HOCH_2C-CH_2NH_2}}$$

hydroxypropionic acid hydrazide and lithium aluminum hydride, in ethyl-morpholine at 100° for 4 hr.) [87].

5. From Oximes or Hydrazones

$$(R)ArCH{=}NOH \xrightarrow[Pt]{H_2} (R)ArCH_2NH_2$$

$$(R)ArCH{=}NNH_2 \xrightarrow[Pt]{H_2} (R)ArCH_2NH_2$$

Oximes and hydrazones may both be reduced to amines, although the reduction of the former is much more common than that of the latter. Four principal types of reducing agents have been utilized in the reduction: active metal or sodium amalgam and an acid; hydrogen, with a catalyst such as platinum or palladium on carbon or alumina, and Raney nickel; metal hydrides such as lithium aluminum hydride; and sodium and alcohol. The reduction is usually straightforward, the yields commonly being 80 % or better. A few special reducing agents are worthy of mention. Stannous chloride in hydrochloric acid is preferred for the reduction of the difficultly reducible α-oximino carboxylic acids [88], whereas ketoximes often are reduced more satisfactorily with zinc dust and ammonium acetate in aqueous concentrated ammonia and ethanol rather than with the metal in acid solution [89]. Raney nickel in 95 % ethanol has been used with some success in low-pressure hydrogenation, although yields are low on occasion owing to hydrolysis of the oxime to the ketone [90]. Cycloheptanone oxime may be reduced smoothly to the amine by low-pressure hydrogenation in the presence of rhodium-alumina [91]. In the reduction of oximes of phenyl ketones with lithium aluminum hydride both a primary, 8-VI, and a secondary amine, 8-VII, are obtained [92]. The yield of 8-VII was increased with an increase in the electron-releasing ability of the

$$X{-}\underset{\underset{NOH}{||}}{\bigcirc}{-}CCH_3 \longrightarrow X{-}\underset{\underset{NH_2}{|}}{\bigcirc}{-}CHCH_3 + X{-}\bigcirc{-}NHCH_2CH_3$$

$$\text{8-VI} \qquad\qquad \text{8-VII}$$

para substituent, X, an effect similar to that which occurs in the Beckmann rearrangement of the oxime.

(a) Preparation of n-heptylamine (60–73 % from heptaldoxime, sodium, and ethanol at the boiling point) [93].

(b) Other examples

(1) **2,2-Diphenylcyclohexylamine** (80% from 2,2-diphenylcyclo-hexanone oxime in dry isopropyl and ethyl ether dropped into a suspension of lithium aluminum hydride in the same solvent) [94].

(2) **2-Aminopentane** (85% from 2-pentanone oxime in 95% ethanol hydrogenated with Raney nickel at room temperature and 3 atm. original pressure) [90].

(3) **Benzhydrylamine** (91% from benzophenone oxime refluxed with zinc dust and ammonium acetate in conc. ammonium hydroxide and ethanol) [89].

(4) **3-β-Aminoethylpyrazole** (81% from γ-pyrone via the hydrazone

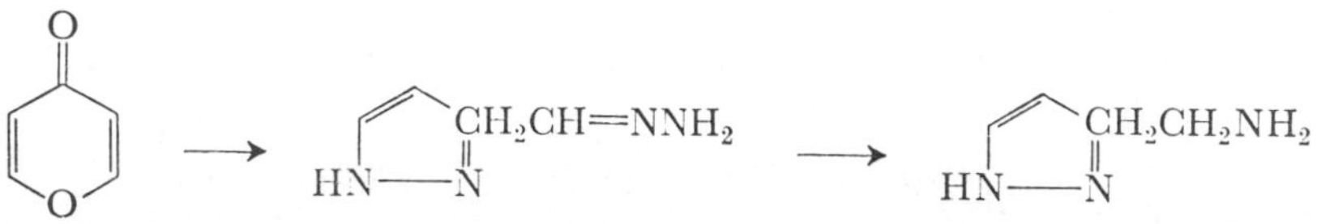

in liquid ammonia, Raney nickel, and hydrogen at 90° and 1500 lb. pressure) [95].

6. From Carbonyl Compounds and Amines (Reductive Alkylation)

$$\diagdown C{=}O + RNH_2 \rightleftharpoons \diagdown C{=}NR \xrightarrow{H_2} \diagdown CHNHR$$

8-VIII

The carbonyl compound and an amine (or ammonia) may be mixed and reduced directly [96], or the azomethine, 8-VIII, if stable, may be isolated and reduced [97]. This synthesis represents one of the best methods to synthesize secondary amines from ketones and primary amines. But the reaction is general in that ammonia, primary and secondary amines (and substances such as a nitro compound which reduces to an amine) may be reductively alkylated with both aldehydes and ketones. With aldehydes and ammonia or primary amines the reaction may proceed beyond the first alkylation stage to give a mixture of primary, secondary, and tertiary amines. The intermediate in alkylating a secondary amine is either the diamine 8-IX or the enamine 8-X:

$$RCH_2CH[NR_2']_2 \quad \text{or} \quad RCH{=}CHNR_2'$$

8-IX   8-X

The most widely used reducing agent is hydrogen with platinum or Raney nickel [98], but sodium, sodium amalgam, or metal hydrides in alcohol [99, 100], dimethylamine borane [101], lead or copper cathodes (electrolytically) or zinc and acid [102–104] may be employed. With aluminum amalgam, benzal butylimine gives considerable amounts of an ethylenediamine derived from a dimerization similar to the pinacol reaction [105].

Yields vary considerably. The most consistently high ones are reported in the use of metal hydrides [99, 100] and dimethylamine borane [101] as reducing agents. For lithium aluminum hydride [100], yields with a series of benzhydryl-ideneimines and related types run from 71 to 94%; for sodium borohydride [99], yields with a series of N-benzylideneaniline types vary from 83 to 98%; for dimethylamine borane [101], yields with N-benzylideneaniline and derivatives cover the range 71–97%.

Raney nickel is capable of producing carbonyl compounds from alcohols [106]. Therefore the Schiff base can be formed and reduced by refluxing an alcohol and amine in the presence of relatively large quantities of Raney nickel (see Ex. b.4).

In the Clemmenson reduction of β-ketoamines, rearrangement, probably occurring through neighboring group participation, is encountered [107]. Examples to illustrate the types of behavior are given:

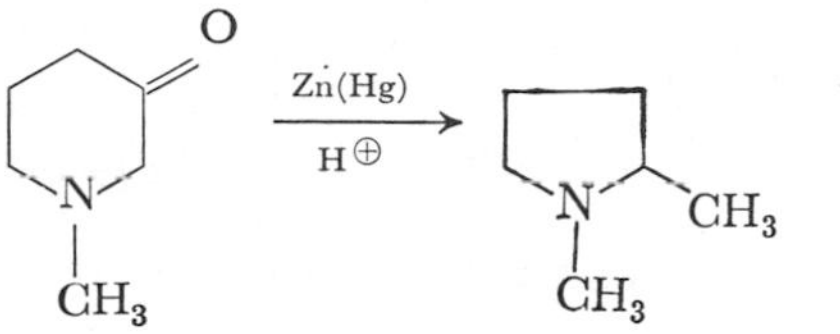

1,2-Dimethylpyrrolidine, 60%

[ref. 108]

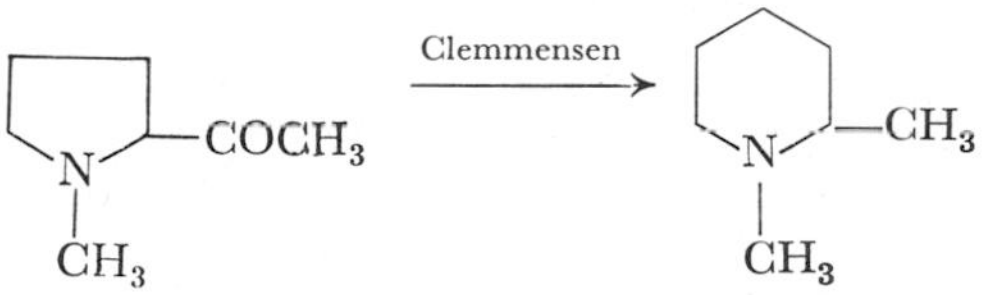

1,2-Dimethylpiperidine, as picrate, 85 mg.

[ref. 109]

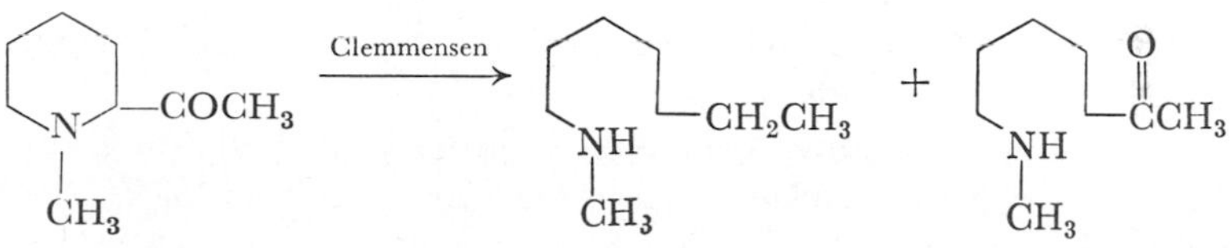

0.52 g. Methyl-*n*-heptyl-amine,

7-Methylaminoheptan-2-one, 0.92 g.

[ref. 109, 110]

On the other hand, 3-oxoquinuclidine does not rearrange on Clemmenson reduction [111].

It is interesting to speculate on the reductive alkylation of an enamine and a carbonyl compound which should give rise to an amine alkylated on a β-carbon atom:

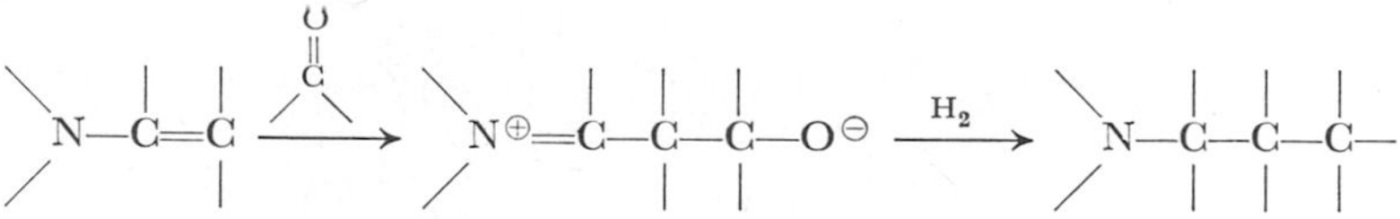

Such an enamine has been generated *in situ* and with ketones has led to the predicted product on hydrogenation [112]:

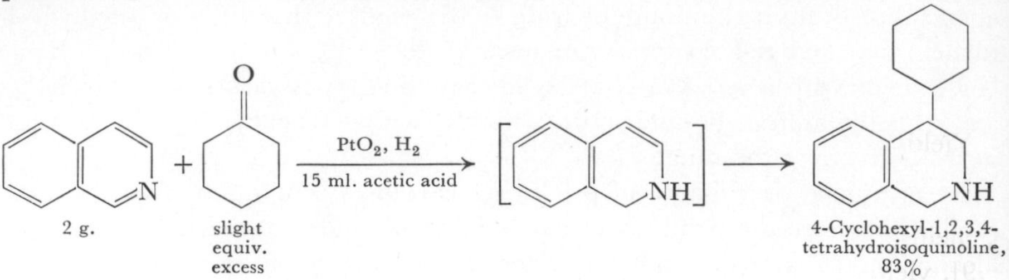

O
+
PtO_2, H_2
15 ml. acetic acid
N
NH
NH
2 g.
slight
equiv.
excess
4-Cyclohexyl-1,2,3,4-
tetrahydroisoquinoline,
83%

But this reaction gave only low yields in another instance [113].

(a) Preparation of benzylamine (89% from benzaldehyde ammonia, and hydrogen at 90 atm. and moderate temperature in the presence of Raney nickel for 30 min.) [114].

(b) Other examples

(1) Phenylisopropylamine (91% from aniline, sodium borohydride, acetone, sodium acetate trihydrate, and acetic acid) [115].

(2) 2-Isopropylaminoethanol (94–95% from ethanolamine and acetone in absolute ethanol hydrogenated at 25 lb. in the presence of platinum) [116].

(3) N,N-Dimethylmesidine (70% from mesidine, 0.075 mole, 40%

NH_2
CH_3
CH_3
CH_3
$N(CH_3)_2$
CH_3
CH_3
CH_3

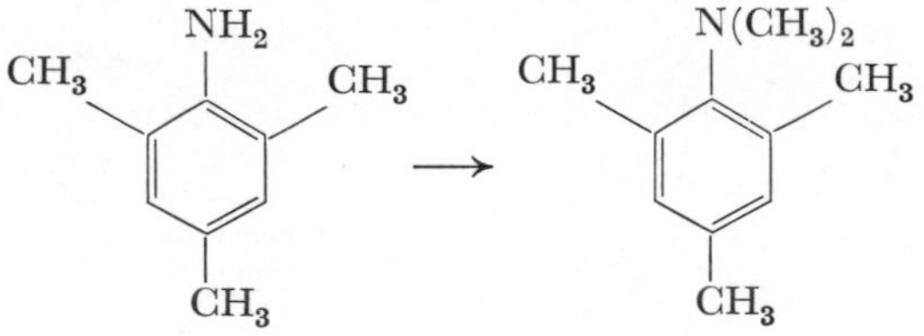

aqueous formaldehyde, 0.23 mole, and amalgamated zinc, 1.53 mole, in glacial acetic and hydrochloric acids) [103].

(4) N-Ethyl-2-naphthylamine (82% from 0.1 mole of 2-naphthylamine and 40 g. of moist Raney nickel refluxed for 4 hr. in 100 ml. of 95% ethanol; isopropyl alcohol and aniline gave only a 50% yield of N-isopropylaniline under the same conditions) [117].

(5) N-Methyl-2,3-dimethoxybenzylamine (86–93% from 0.25 mole of 2,3-dimethoxybenzaldehyde, 0.75 mole of aqueous methylamine in ethanol, and 0.25 mole hydrogen in the presence of W-6 Raney nickel catalyst) [118].

(6) 3-Phenyl-2,3,4,5,6,7-hexahydroindole [119].

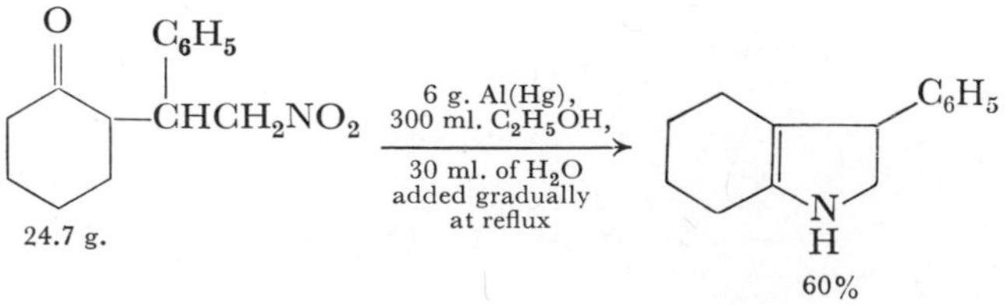

O
C_6H_5
$CHCH_2NO_2$
24.7 g.
6 g. Al(Hg),
300 ml. C_2H_5OH,
30 ml. of H_2O
added gradually
at reflux
C_6H_5
N
H
60%

(7) **Phenylbenzylamine** (97 % from N-benzylideneaniline in a 5–10 % solution of methanol and $NaBH_4$ refluxed 15 min. after careful mixing) [99].

7. From Carbonyl Compounds and Amines (Leuckart-Wallach and Eschweiler-Clarke Reactions)

$$R_2CO \xrightarrow[\text{2. } H_2O]{\text{1. } HCO_2NH_4 \text{ (or } HCONH_2), HCO_2H} R_2CHNH_2 + CO_2$$

Leuckart-Wallach

$$RNH_2 + H_2CO + HCO_2H \longrightarrow RN(CH_3)_2 + CO_2$$

Excess   Eschweiler-Clarke

These reactions, which have been reviewed [120, 121] are excellent methods of preparation for amines. The Leuckart-Wallach reaction can be used to prepare primary, secondary, and tertiary amines depending on the amine employed: ammonia for primary, an alkylamine for secondary, and a dialkylamine for tertiary. It is the method of choice for preparing amines from cyclic ketones and relatively unhindered acyclic ketones, particularly in view of results from the more drastic conditions which have been employed recently (see Ex. b). The Clarke-Eschweiler reaction is an equally elegant method of preparation but restricted to the preparation of tertiary amines with at least one methyl group.

The mechanisms of these reactions are still in dispute but no doubt involve formation of the Schiff base, or *gem*-diamino compounds with dialkylamines, followed by reduction with formic acid:

$$RNH_2 + ^{\diagdown}_{\diagup}C{=}O \; \rightleftharpoons \; RNH\underset{|}{\overset{|}{C}}OH \; \xrightleftharpoons{-H_2O} \; RN{=}C^{\diagup}_{\diagdown}$$

or

$$R_2NH + ^{\diagdown}_{\diagup}C{=}O \; \rightleftharpoons \; R_2N\underset{|}{C}HOH \; \xrightleftharpoons[R_2NH]{-H_2O} \; (R_2N)_2C^{\diagup}_{\diagdown}$$

The first step is heterolytic and affected by catalysts, but the second step, the reduction, is probably homolytic and affected more by heat than by other means.

Many variations are recorded in the literature. For example, formic acid led to better yields in the methylation of amines with formaldehyde [122] and in the formation of pyrenemethylamines from the carboxaldehyde and a formamide [123]. However, ammonia may be methylated to trimethylamine in good yield without the addition of formic acid (see Ex. c.4). In addition, alkaline catalysts such as pyridine, ammonia, or urea in the presence of skeletal nickel accelerate the reaction and lead to the predominant formation of secondary amines [124]. Sometimes magnesium chloride [125] acts as a catalyst.

Of the many reagents used in the Leuckart reaction, it is not possible to name the most effective one. Better yields are obtained with ammonia or an amine and formic acid than with ammonium formate. An ammonium formate-formamide reagent is superior to anhydrous formamide alone [126]. With a series of various types of ketones, yields with this mixed reagent are from 52 to 85 %. In some cases the most satisfactory reagent seems to be formamide or ammonium formate to which sufficient formic acid is added to maintain an acidic medium and to act as well as a reducing agent [127]. On the other hand, as has already been stated, alkaline catalysts in the presence of Raney nickel accelerate the reaction and lead mainly to secondary amine formation [124]. For example, urea, formic acid, cyclohexanone, and skeletal nickel give an 85 % yield of dicyclohexylamine. On the other hand, a formate-formamide mixture, formic acid, and Raney nickel on the same ketone give an 85 % yield of cyclohexylamine (see Ex. c.1) [124].

For the production of primary amines the usual ratio in the Leuckart-Wallach reaction is one of the carbonyl compound to four or five of the ammonium formate or formamide. The excess of ammonium formate tends to reduce the formation of secondary or tertiary amines.

The reaction has been applied to a great variety of aldehydes and ketones, sometimes with satisfactory and, sometimes with unsatisfactory results. A few satisfactory results taken from [120], unless otherwise indicated, follow. Formaldehyde has a tendency to give tertiary amines with ammonia, primary, and secondary amines. Benzaldehyde gives a mixture of primary, secondary, and tertiary amines with ammonium formate, but with piperazine and formic acid it gives an 84 % yield of N,N'-dibenzylpiperazine [128]. Aliphatic ketones produce yields of primary amines varying from 30 to 80 %. From a series of aliphatic-aromatic ketones and ammonia and formic acid, yields ranging from 50 to 85 % are obtained. By substituting a primary or secondary amine for ammonia, satisfactory yields of secondary or tertiary amines may be secured. As has already been indicated, cyclohexanone may be converted satisfactorily into the corresponding primary or secondary amine [124].

(a) **Preparation of α-phenylethylamine** (60–66 % from acetophenone and ammonium formate by heating up to 180–185°) [129].

(b) **Preparation of cyclooctyldimethylamine.** A mixture of 0.79 mole of cyclooctanone and 100 g. of formic acid and 175 g. of dimethylformamide was heated to 190° in a glass-lined autoclave (the glass-lining is essential) for 16 hr., cooled, and poured into mineral acid. After the cyclooctanol was removed from the acid solution by extraction with ether, the acid solution was made basic and the insoluble oil extracted with ether and distilled to give the product in 75 % yield. Benzophenone, 4-heptanone, and α-tetralone, somewhat hindered ketones, gave poor yields of the corresponding amines when treated similarly [130].

(c) **Other examples**

(1) **Cyclohexylamine** (85 % as the hydrochloride from cyclohexanone, a mixture of ammonium formate and formamide, formic acid, and skeletal nickel) [124].

(2) **Dimethyl-3,5,5-trimethylhexylamine** (84% from 3,5,5-trimethyl-hexaldehyde and dimethylamine formate at 60° for 1 hr. and then on the steam bath for 1 hr.) [131].

(3) **N-(1-Pyrenylmethyl)morpholine** (95% as the hydrochloride from morpholine and 90% formic acid heated to 200° followed by the addition of 1-pyrenecarboxaldehyde and more 90% formic acid) [123].

(4) **Trimethylamine** (85–90% from ammonium chloride and para-

$$(CH_2O)_n \xrightarrow{NH_4Cl} (CH_3)_3N \cdot HCl \longrightarrow (CH_3)_3N$$

formaldehyde heated gradually to 160° and held there until carbon dioxide evolution ceased) [132].

8. From Azaaromatic Compounds

Exhaustive catalytic hydrogenation makes available such compounds as piperidine from pyridine and pyrrolidine from pyrrole. Partial hydrogenations make available dihydro and tetrahydro compounds, but it is difficult to generalize on the best reagents or conditions for such reductions. The subject is so inexhaustible that only a few remarks will be made and some references given to sources of information [133–135].

Under a variety of conditions the pyridine ring is reduced more readily than is the benzenoid ring. For example, 1,2,3,4-tetrahydroquinoline is the major product of partial reduction of quinoline whether using chemical or catalytic methods (see Ex. a). A small amount of 5,6,7,8-tetrahydroquinoline can be isolated, however, by acetylating the higher boiling fractions and isolating the basic portion [136]. Tetrahydroquinoline can be obtained also by modifying the Leuckart reaction (A.7) to attain higher temperatures (see Ex. b). Another remarkable example is the specific reduction of the pyridine ring in a series of 4-(2-phenylquinoline)2-pyridyl ketones [137]:

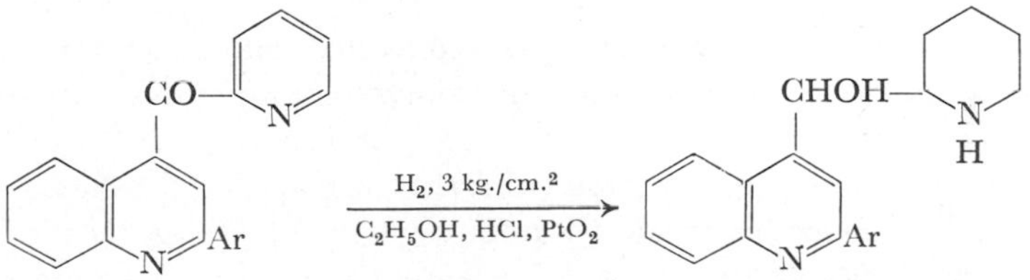

$$\xrightarrow[\text{C}_2\text{H}_5\text{OH, HCl, PtO}_2]{\text{H}_2,\ 3\ \text{kg./cm.}^2}$$

Yields of 12–75% for a series of substituted derivatives

The inertness of the quinoline ring to reduction in this example may be attributed to the presence of the 2-phenyl substituent, as one of us (D. E. P.) has noted no such selectivity when the phenyl substituent is in the 8-position.

Lithium aluminum hydride in ether is the most frequently employed reagent for preparing dihydro derivatives [138]:

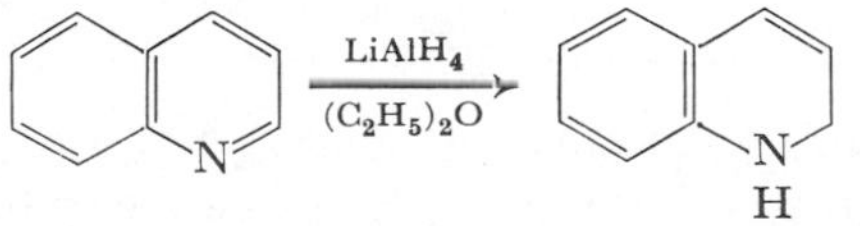

$$\xrightarrow[\text{(C}_2\text{H}_5)_2\text{O}]{\text{LiAlH}_4}$$

1,2-Dihydroquinoline, 91%

Catalytic reduction has been used in certain instances, however [139]. The dihydro compounds tend to disproportionate.

Pyrazines, quinoxaline, and other heterocycles behave similarly to pyridine or quinoline, but heterocyclic compounds with vicinal nitrogen atoms tend to give reductive cleavage products. For example, pyridazine

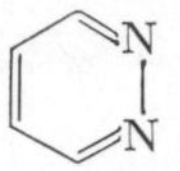

is reduced to tetramethylenediamine by sodium and alcohol [140].

(a) Preparation of 1,2,3,4-tetrahydro-2-(3-hydroxypropyl)-8-ethoxyquinoline (96% from reduction of the corresponding quinoline with

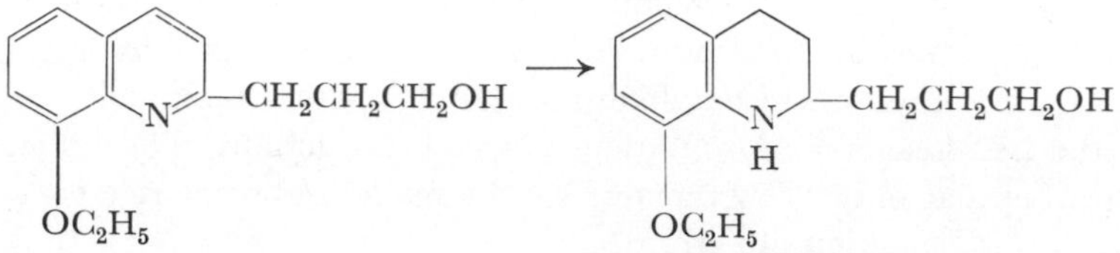

platinum oxide in alcohol at 40 p.s.i. pressure of hydrogen) [141].

(b) Preparation of 1,2,3,4-tetrahydroquinoline (unstated yield from 0.1 mole of quinoline, 86.5 g. of triethylammonium formate, and 1 g. of Raney nickel heated to 160–170°, and the formamide subsequently hydrolyzed) [142].

9. From Benzylamines (Hydrogenolysis)

$$C_6H_5CH_2NR_2 \xrightarrow[\text{Pd on C}]{H_2} C_6H_5CH_3 + R_2NH$$

In this synthesis, tertiary amines are converted into secondary ones and secondary amines give primary ones. The catalyst employed in the hydrogenolysis is usually palladium-on-carbon [143–145] or Raney nickel [146]. Amines containing one benzyl group result in the formation of toluene and the amine usually in average yield. Complications result if two benzyl groups, one or more being substituted, are present in the original amine since the cleavage can occur in two ways, (A) and (B). The effect of various substituents

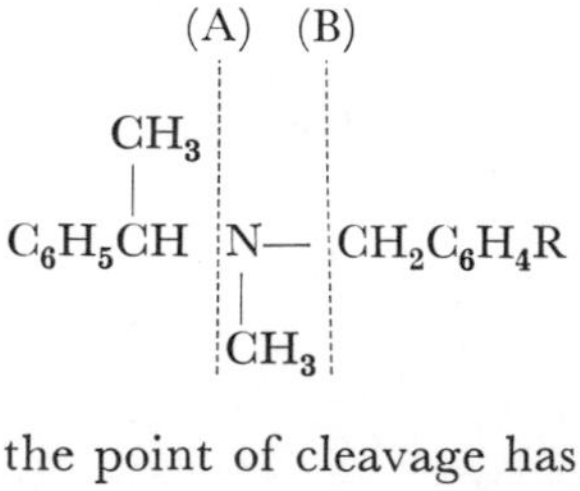

in these benzyl groups on the point of cleavage has been discussed [143–145]. Even the hydrochloride of the original amine cleaves differently from the free amine. For example, the hydrochloride of benzhydrylbenzylmethylamine cleaves at (A) below to give 70 % of benzhydrylmethylamine, while the free base breaks at (B) to give 85 % of benzylmethylamine [145].

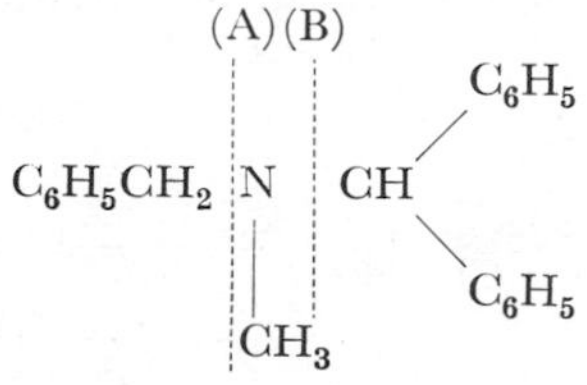

$$\begin{array}{c}(A)(B)\end{array}$$

C₆H₅CH₂ N CH with C₆H₅ / \ C₆H₅, CH₃

A cyclic amine, 2,2-dimethylethylenimine, has also been hydrogenolyzed to *t*-butylamine in the presence of Raney nickel with a yield of 75–82 % [147].

(a) Preparation of α-methylbenzylmethylamine hydrochloride.

$$C_6H_5CH_2N-\overset{CH_3}{\underset{CH_3}{\mid}}CHC_6H_5\cdot HCl \longrightarrow C_6H_5\overset{CH_3}{\underset{CH_3}{\mid}}CHNH\cdot HCl$$

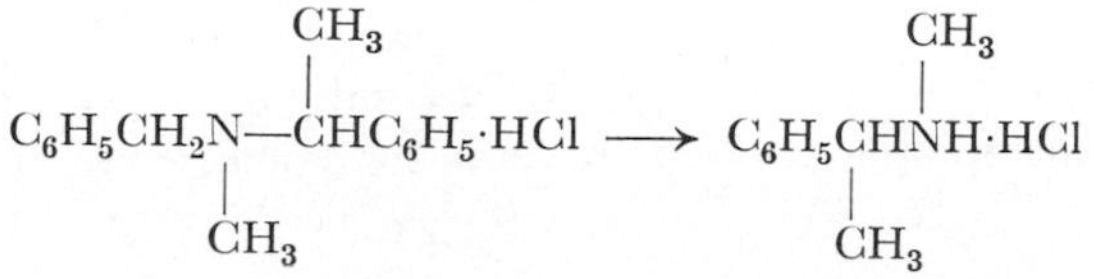

α-Methylbenzylbenzylmethylamine hydrochloride in ethanol was reduced with hydrogen in the presence of palladium-on-carbon. An overpressure of hydrogen (1½ to 3 atm.) was employed and the reduction was regarded as complete when 30 min. or more was required for a pressure drop of 1 lb. Over 90 % of the secondary amine as the hydrochloride was recovered from the reaction mixture [145].

10. From Quaternary Salts (Hydrogenolysis)

$$R_4N^{\oplus} \xrightarrow{[H]} R_3N + RH$$

This synthesis is helpful on occasions when a tertiary amine needs purification via crystallization of the quaternary salts or when the quaternary salt is obtained directly in a synthesis. The most suitable reagent for cleavage of the quaternary salt appears to be lithium aluminum hydride (see Ex. a), but other reagents may serve the same purpose or even alter the course of the cleavage. An example is the cleavage of the methiodide salt of N-methyltetrahydroisoquinoline [148, 149]; see Alkanes, Cycloalkanes, and Arenes, Chapter 1, A.4.

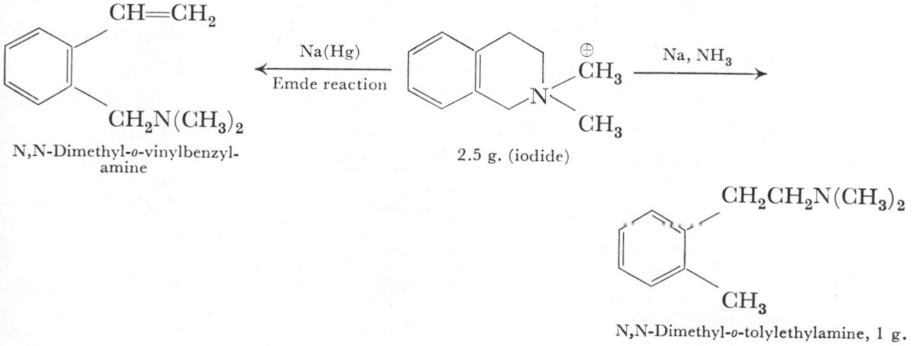

CH=CH₂

CH₂N(CH₃)₂

N,N-Dimethyl-*o*-vinylbenzyl-
amine

Na(Hg) / Emde reaction

2.5 g. (iodide)

Na, NH₃

CH₂CH₂N(CH₃)₂

CH₃

N,N-Dimethyl-*o*-tolylethylamine, 1 g.

A simpler reagent, but one requiring higher temperature, is a mixture of potassium formate and formic acid (see Ex. b).

Sodium thiophenoxide is unique because it is selective in its demethylation in that methoxy groups are not affected [150]:

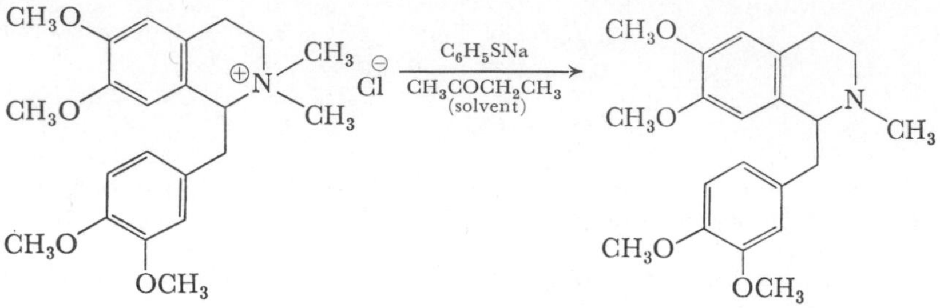

(±)-Laudanosine methochloride   Laudanosine, 85%

It should be noted, however, that, in this demethylation, any ester groups are converted into the thiophenyl derivative of the corresponding carboxylic acid [151].

(a) Preparation of dimethylneomenthylamine. The methiodide

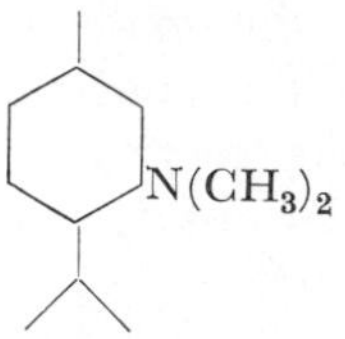

salt of the above compound was refluxed with 5 equiv. of lithium aluminum hydride in THF for 24 hr. and gave the tertiary amine in 74% yield [152].

(b) Preparation of 1,2,3-trimethylpyrrolidine [153].

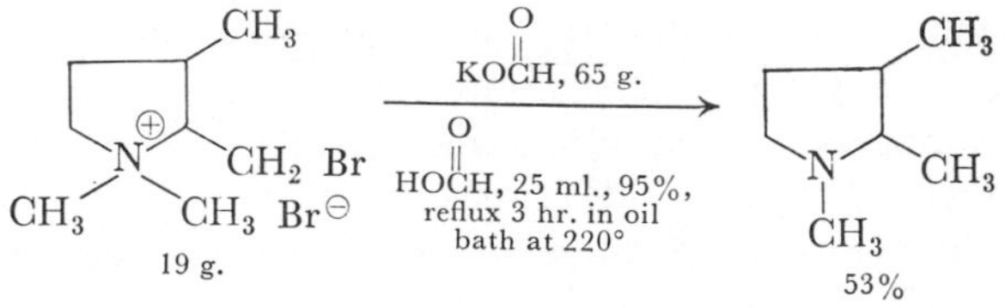

11. From Nitrogen (Fixation)

The fixation of nitrogen remains as a major challenge to organic chemists. The ideal system should contain species which coordinate with nitrogen (and not with oxygen) and reagents which are capable of reduction. Also, the system should be readily regenerated. The closest approach to fulfilling these criteria is the work of van Tamelen [154]:

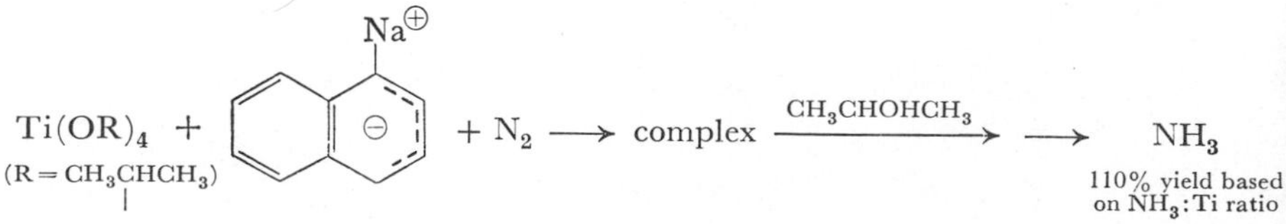

Even air can be used as a source of nitrogen, although there is an appreciable reduction in yield. Further progress is to be expected [155].

1. Houben-Weyl, *Methoden der Organischen Chemie*, G. Thieme Verlag, Stuttgart, 1957, 4th ed., Vol. **11,** Pt. 1.
2. P. A. S. Smith, *The Chemistry of Open-Chain Organic Nitrogen Compounds*, W. A. Benjamin, New York, 1965, Vol. I, pp. 60–78 and 115–122.
3. L. Spialter and J. A. Pappalardo, *The Acyclic Aliphatic Tertiary Amines*, The Macmillan Company, New York, 1965.
4. S. Patai, *Chemistry of the Amino Group*, Interscience Publishers, New York, 1968, pp. 37–77.
5. J. S. McConaghy, Jr., and L. Lwowski, *J. Am. Chem. Soc.*, **89,** 2357 (1967).
6. R. Schröter, Ref. 1, p. 360.
7. H. E. Smith *et al.*, *J. Org. Chem.*, **31,** 684 (1966).
8. J. S. Buck and W. S. Ide, *Org. Syn.*, Coll. Vol. **2,** 130 (1943).
9. G. T. Morgan and H. A. Harrison, *Soc. Chem. Ind.* (*London*), **49,** 413T (1930).
10. R. L. Augustine, *Catalytic Hydrogenation*, Marcel Dekker, New York, 1965, pp. 91–102.
11. P. N. Rylander, *Catalytic Hydrogenation over Platinum Metals*, Academic Press, New York, 1967, pp. 168–203.
12. C. F. H. Allen and J. Van Allan, *Org. Syn.*, Coll. Vol. **3,** 63 (1955).
13. Ref. 10, pp. 35, 152.
14. H. C. Brown and K. Sivasankaran, *J. Am. Chem. Soc.*, **84,** 2828 (1962).
15. Ref. 10, pp. 36, 152.
16. J. A. Berson and T. Cohen, *J. Org. Chem.*, **20,** 1461 (1955).
17. M. Freifelder, *J. Am. Chem. Soc.*, **82,** 2386 (1960).
18. Ref. 10, p. 39.
19. F. S. Dowell and H. Greenfield, *J. Am. Chem. Soc.*, **87,** 2767 (1965).
20. H. Shechter *et al.*, *J. Am. Chem. Soc.*, **74,** 3664 (1952).
21. T. Neilson *et al.*, *J. Chem. Soc.*, 371 (1962).
22. H. J. Barber and E. Lunt, *J. Chem. Soc.*, 1187 (1960).
23. C. T. and C. E. Redemann, *Org. Syn.*, Coll. Vol. **3,** 69 (1965).
24. L. I. Smith and J. W. Opie, *Org. Syn.*, Coll. Vol. **3,** 56 (1955).
25. W. E. Kuhn, *Org. Syn.*, Coll. Vol. **2,** 447 (1943).
26. E. L. Martin, *Org. Syn.*, Coll. Vol. **2,** 501 (1943).
27. G. R. Robertson, *Org. Syn.*, Coll. Vol. **1,** 52 (1941).
28. P. M. G. Bavin, *Org. Syn.*, **40,** 5 (1960).
29. A. Furst *et al.*, *Chem. Rev.*, **65,** 51 (1965).
30. H. Bredereck and H. von Schuh, *Chem. Ber.*, **81,** 215 (1948).
31. Huang-Minlon, *J. Am. Chem. Soc.*, **70,** 2802 (1948).
32. W. W. Hartman and H. L. Silloway, *Org. Syn.*, Coll. Vol. **3,** 82 (1955).
33. K. P. Griffin and W. D. Peterson, *Org. Syn.*, Coll. Vol. **3,** 242 (1955).
34. J. H. Boyer and R. S. Buriks, *Org. Syn.*, **40,** 96 (1960).
35. J. H. and E. K. Weisburger, *J. Org. Chem.*, **21,** 514 (1956).
36. Ref. 1, p. 478 for tables.
37. E. A. Braude *et al.*, *J. Chem. Soc.*, 3586 (1954); Y. Ahmad and D. H. Hey, *J. Chem. Soc.*, 4516 (1954).
38. T. E. de Kiewiet and H. Stephen, *J. Chem. Soc.*, 82 (1931).
39. Ref. 1, p. 457.
40. Ref. 1, p. 472.
41. S. Swann, Jr., in A. Weissberger ed., *Techniques of Organic Chemistry*, Interscience Publishers, New York, 1956, Vol. **2,** p. 486; *Trans. Electrochem. Soc.*, **69,** 307 (1936); **77,** 479 (1940).
42. E. Ochiai and H. Mitarashi, *Chem. Pharm. Bull.*, **11,** 1084 (1963).
43. O. Kamm, *Org. Syn.*, Coll. Vol. **1,** 445 (1941)
44. N. S. Kozlov and M. N. Tovshtein, *J. Org. Chem., USSR* (*Eng. Transl.*), **3,** 132 (1967).
45. R. L. Jenkins *et al.*, *Ind. Chem. Eng.*, **22,** 31 (1930).
46. J. R. E. Hoover and A. R. Day, *J. Am. Chem. Soc.*, **77,** 4324 (1955); R. Nietzki and H. Hagenback, *Chem. Ber.*, **30,** 539 (1897).

47. K. N. Carter, Master's Thesis, Vanderbilt University, 1949.

48. S. A. Mahood and P. V. L. Schaffner, *Org. Syn.*, Coll. Vol. **2**, 160 (1943).

49. W. Theilacker and G. Wendtland, *Ann. Chem.*, **570**, 33 (1950).

50. R. N. Icke *et al.*, *Org. Syn.*, Coll. Vol. **3**, 59 (1955).

51. F. A. Ramirez and A. Burger, *J. Am. Chem. Soc.*, **72**, 2781 (1950).

52. T. Neilson *et al.*, *J. Chem. Soc.*, 371 (1962); *see also* W. M. McLamore, *J. Am. Chem. Soc.*, **73**, 2221 (1951); J. B. Conant and B. B. Corson, *Org. Syn.*, Coll. Vol. **2**, 33 (1943).

53. J. Thesing and H. Mayer, *Chem. Ber.*, **89**, 2159 (1956); *see also* G. E. Utzinger and F. A. Regenass, *Helv. Chim. Acta.*, **37**, 1885 (1954); A. Mustafa and M. Kamel, *J. Am. Chem. Soc.*, **76**, 124 (1954).

54. E. Howard, Jr., and W. F. Olszewski, *J. Am. Chem. Soc.*, **81**, 1482 (1959).

55. G. Kobayashi *et al.*, Japanese Patent 177,997, Feb. 28, 1949; *C.A.*, **45**, 8563 (1951); *J. Pharm. Soc. Japan*, **67**, 101 (1947); *C.A.*, **45**, 9553 (1951).

56. D. J. Thoennes and D. E. Pearson, unpublished work.

57. L. F. Fieser, *Org. Syn.*, Coll. Vol. **2**, 39 (1943); *see also* H. Beyer and G. Wolter, *Chem. Ber.*, **85**, 1077 (1952); J. Lecocq, *Bull. Chim. Soc. France*, 183 (1951).

58. A. A. Sayigh, *J. Org. Chem.*, **25**, 1707 (1960).

59. H. Stetter and H. Spangenberger, *Chem. Ber.*, **91**, 1982 (1958); *see also* G. Losse and J. Müller, *J. Prakt. Chem.*, **12**, 285 (1961); B. Coxon and L. Hough, *J. Chem. Soc.*, 1643 (1961).

60. J. H. Boyer, *J. Am. Chem. Soc.*, **73**, 5865 (1951); *see also* R. Adams and D. C. Blomstrom, *J. Am. Chem. Soc.*, **75**, 3405 (1953); C. A. VanderWerf *et al.*, *J. Am. Chem. Soc.*, **76**, 1231 (1954); A. Streitwieser, Jr., and J. R. Wolfe, Jr., *J. Org. Chem.*, **28**, 3263 (1963).

61. J. S. Buck and C. W. Ferry, *Org. Syn.*, Coll. Vol. **2**, 290 (1943); *see also* F. W. Schueler and C. Hanna, *J. Am. Chem. Soc.*, **73**, 4996 (1951); **74**, 3693 (1952).

62. F. E. Gould *et al.*, *J. Org. Chem.*, **25**, 1658 (1960); **26**, 2602 (1961).

63. M. Freifelder, *J. Am. Chem. Soc.*, **82**, 2386 (1960).

64. Y. Takagi *et al.*, *Sci. Papers Inst. Phys. Chem. Res.* (*Tokyo*), **61**, 114 (1967); *C.A.*, **68**, 9195 (1968).

65. H. C. Brown and B. C. Subba Rao, *J. Am. Chem. Soc.*, **82**, 681 (1960).

66. R. F. Nystrom, *J. Am. Chem. Soc.*, **77**, 2544 (1955).

67. R. C. Fuson *et al.*, *J. Org. Chem.*, **16**, 648 (1951).

68. H. C. Brown and W. Korytnyk, *J. Am. Chem. Soc.*, **82**, 3866 (1960).

69. L. Friedman and A. T. Jurewicz, *J. Org. Chem.*, **33**, 1254 (1968).

70. N. M. Yoon and H. C. Brown, *J. Am. Chem. Soc.*, **90**, 2927 (1968).

71. B. S. Biggs and W. S. Bishop, *Org. Syn.*, Coll. Vol. **3**, 229 (1955).

72. R. F. Borch, *J. Org. Chem.*, **34**, 627 (1969); *Chem. Commun.*, 442 (1968).

73. F. Poppelsdorf and R. C. Myerly, *J. Org. Chem.*, **26**, 131 (1961).

74. N. G. Gaylord, *Reduction with Complex Metal Hydrides*, Interscience Publishers, New York, 1956, p. 544.

75. H. C. Brown and A. Tsukamoto, *J. Am. Chem. Soc.*, **83**, 4549 (1961).

76. G. Wittig and P. Hornberger, *Ann. Chem.*, **577**, 11 (1952).

77. F. Weygand and G. Eberhardt, *Angew. Chem.*, **64**, 458 (1952); F. Weygand *et al.*, *Angew. Chem.*, **65**, 525 (1953); **66**, 174 (1954).

78. H. A. Staub and H. Bräunling, *Ann. Chem.*, **654**, 119 (1962).

79. H. O. House, *Modern Synthetic Reactions*, W. A. Benjamin, New York, 1965, p. 34.

80. H. C. Brown and P. Heim, *J. Am. Chem. Soc.*, **86**, 3566 (1964).

81. D. W. Henry, *J. Heter. Chem.*, **3**, 503 (1966).

82. M. J. Kornet *et al.*, *J. Org. Chem.*, **33**, 3637 (1968).

83. A. Guyer *et al.*, *Helv. Chim. Acta.*, **38**, 1649 (1955).

84. A. E. Finholt *et al.*, *J. Org. Chem.*, **18**, 1338 (1953).

85. A. C. Cope and E. Ciganek, *Org. Syn.*, Coll. Vol. **4**, 339 (1963).

86. C. V. Wilson and J. F. Stenberg, *Org. Syn.*, Coll. Vol. **4**, 564 (1963).

87. B. I. R. Nicolaus *et al.*, *J. Org. Chem.*, **26**, 2253 (1961).

88. R. L. Bixler and C. Niemann, *J. Org. Chem.*, **23**, 575 (1958).

89. J. C. Jochims, *Monatsh. Chem.*, **94**, 677 (1963).

90. D. C. Iffland and T.-F. Yen, *J. Am. Chem. Soc.*, **76,** 4180 (1954).

91. M. Freifelder *et al.*, *J. Org. Chem.*, **27,** 2209 (1962).

92. R. E. Lyle and H. J. Troscianec, *J. Org. Chem.*, **20,** 1757 (1955).

93. C. S. Marvel *et al.*, *Org. Syn.*, Coll. Vol. **2,** 318 (1943).

94. A. Burger and W. B. Bennet, *J. Am. Chem. Soc.*, **72,** 5414 (1950).

95. R. G. Jones and M. J. Mann, *J. Am. Chem. Soc.*, **75,** 4048 (1953).

96. W. S. Emerson, *Org. Reactions*, **4,** 174 (1948).

97. W. S. Emerson, *Org. Reactions*, **4,** 189 (1948).

98. D. G. Norton *et al.*, *J. Org. Chem.*, **19,** 1054 (1954).

99. J. H. Billman *et al.*, *J. Org. Chem.*, **22,** 1068 (1957).

100. J. H. Billman and K. M. Tai, *J. Org. Chem.*, **23,** 535 (1958).

101. J. H. Billman and J. W. McDowell, *J. Org. Chem.*, **26,** 1437 (1961).

102. W. S. Emerson *et al.*, *J. Am. Chem. Soc.*, **62,** 2159 (1940).

103. W. S. Emerson *et al.*, *J. Am. Chem. Soc.*, **63,** 972 (1941).

104. W. S. Emerson *et al.*, *J. Am. Chem. Soc.*, **63,** 2843 (1941).

105. H. Schoenenberger *et al.*, *Arch. Pharm.*, **300,** 258 (1967); *C.A.*, **67,** 1998 (1967).

106. R. G. Rice and E. J. Kohn, *J. Am. Chem. Soc.*, **77,** 4052 (1955).

107. Ref. 1, p. 1002.

108. N. J. Leonard and E. Barthel, Jr., *J. Am. Chem. Soc.*, **72,** 3632 (1950).

109. G. R. Clemo *et al.*, *J. Chem. Soc.*, 2095 (1949).

110. N. J. Leonard and R. C. Sentz, *J. Am. Chem. Soc.*, **74,** 1704 (1952).

111. N. J. Leonard *et al.*, *J. Am. Chem. Soc.*, **75,** 6249 (1953).

112. R. Grewe *et al.*, *Chem. Ber.*, **97,** 119 (1964).

113. W. J. Gensler *et al.*, *J. Org. Chem.*, **33,** 2861 (1968)

114. C. F. Winans, *Org. Reactions*, **4,** 199 (1948).

115. K. A. Schellenberg, *J. Org. Chem.*, **28,** 3259 (1963).

116. E. M. Hancock and A. C. Cope, *Org. Syn.*, Coll. Vol. **3,** 501 (1955).

117. C. Ainsworth, *J. Am. Chem. Soc.*, **78,** 1635 (1956); B. B. Corson and H. Dressler, *J. Org. Chem.*, **21,** 474 (1956).

118. D. M. Balcom and C. R. Noller, *Org. Syn.*, Coll. Vol. **4,** 603 (1963).

119. H. Feuer *et al.*, *Tetrahedron*, **24,** 1187 (1968).

120. M. L. Moore, *Org. Reactions*, **5,** 301 (1949).

121. L. Spialter and J. A. Pappalardo, *The Acyclic Aliphatic Tertiary Amines*, The Macmillan Co., New York, 1965, p. 44.

122. H. T. Clarke *et al.*, *J. Am. Chem. Soc.*, **55,** 4571 (1933).

123. E. Marcus and J. T. Fitzpatrick, *J. Org. Chem.*, **25,** 199 (1960).

124. A. N. Kost and I. I. Grandberg, *Zh Obshch. Khim.*, **25,** 1432 (1955); *C.A.*, **50,** 4800 (1956).

125. J. F. Bunnett and J. L. Marks, *J. Am. Chem. Soc.*, **71,** 1587 (1949).

126. A. W. Ingersoll *et al.*, *J. Am. Chem. Soc.*, **58,** 1808 (1936).

127. F. S. Crossley and M. L. Moore, *J. Org. Chem.*, **9,** 529 (1944).

128. W. T. Forsee and C. B. Pollard, *J. Am. Chem. Soc.*, **57,** 1788 (1935).

129. A. W. Ingersoll, *Org. Syn.*, Coll. Vol. **2,** 503 (1943).

130. R. D. Bach, *J. Org. Chem.*, **33,** 1647 (1968).

131. P. L. de Benneville and J. H. Macartney, *J. Am. Chem. Soc.*, **72,** 3073 (1950).

132. R. Adams and B. K. Brown, *Org. Syn.*, Coll. Vol. **1,** 528 (1941); R. Adams and C. S. Marvel *Org. Syn.*, Coll. Vol. **1,** 531 (1941).

133. Ref. 1, p. 692.

134. R. E. Lyle and P. S. Anderson, *Advances in Heterocyclic Chemistry*, Academic Press, New York, 1966, Vol. **6,** p. 45 and 68.

135. A. Weissberger, *The Chemistry of Heterocyclic Compounds*, Interscience Publishers, New York (numerous volumes).

136. Ref. 1, p. 715.

137. R. E. Lutz, *et al.*, *J. Med. Chem.*, **11,** 273 (1968).

138. K. W. Rosenmund *et al.*, *Chem. Ber.*, **86,** 37 (1953); **87,** 1229 (1954).

139. R. E. Lyle and S. E. Mallett, *Ann. N.Y. Acad. Sci.*, **145,** 83 (1967).

140. Ref. 1, p. 706.
141. J. G. Cannon *et al.*, *J. Heterocyclic Chem.*, **4**, 259 (1967).
142. K. Ito, *Yakugaku Zasshi*, **86**, 1166 (1966); *C.A.*, **66**, 7114 (1967).
143. R. Baltzly and J. S. Buck, *J. Am. Chem. Soc.*, **65**, 1984 (1943).
144. R. Baltzly and P. B. Russell, *J. Am. Chem. Soc.*, **72**, 3410 (1950).
145. R. Baltzly and P. B. Russell, *J. Am. Chem. Soc.*, **75**, 5598 (1953).
146. H. Dahn and U. Solms, *Helv. Chim. Acta.*, **35**, 1162 (1952).
147. K. N. Campbell *et al.*, *Org. Syn.*, Coll. Vol. **3**, 148 (1955).
148. Ref. 1, p. 973.
149. D. B. Clayson, *J. Chem. Soc.*, 2016 (1949).
150. M. Shamma *et al.*, *Tetrahedron Letters*, 1375 (1966).
151. J. C. Sheehan and G. D. Daves, Jr., *J. Org. Chem.*, **29**, 2006 (1964).
152. A. C. Cope *et al.*, *J. Am. Chem. Soc.*, **82**, 4651 (1960).
153. R. Lukes and J. Pliml, *Collection Czech. Chem. Commun.*, **26**, 471 (1961).
154. E. E. van Tamelen *et al.*, *J. Am. Chem. Soc.*, **90**, 1677 (1968).
155. *Chem. Eng. News*, Mar. 24, 1969, p. 48.

B. Hydrolysis or Solvolysis

Any type of acid derivative of an amine is subject to hydrolysis or interchange brought about by a wide variety of acids, including Lewis acids, or bases, including amines (for the latter, see B.2). The subject has been reviewed in some detail [1]. The choice of conditions for splitting are so variable that it depends to a considerable extent on the experimenter's preference. It is suggested, however, that dilute base be used for volatile aliphatic amides and that dilute acid be employed for the aromatic amides. When both the acid and the amine in the amide structure are aromatic, a higher acid concentration, up to 50 % sulfuric acid by volume, may serve to split the amide as well as to give increased solubility. Even these conditions do not work well for some amides because of charring or oxidation. Fortunately, three new procedures are available to use for recalcitrant cases (see B.4). One involves the solvolysis of an amide with methanol using boron trifluoride as a catalyst. The others, designed for sulfonamides, involve splitting with concentrated hydrobromic acid in the presence of phenol or reductive splitting with sodium naphthalene.

Schiff bases or other related carbonyl-amine adducts are best hydrolyzed in acid solution. Other less general, but occasionally useful, hydrolyses are discussed briefly (B.3 and B.6).

1. From Ureas, Urethanes, Isocyanates, and Isothiocyanates

$$RNCO \xrightarrow{H_2O} RNH_2 + CO_2$$

$$\begin{array}{c} RNH \\ \diagdown \\ CO \xrightarrow{H_2O} 2\,RNH_2 + CO_2 \\ \diagup \\ RNH \end{array}$$

$$\begin{array}{c} RNH \\ \diagdown \\ CO \xrightarrow{H_2O} RNH_2 + CO_2 + R'OH \\ \diagup \\ R'O \end{array}$$

References for Section B are on pp. 442–443.

Hydrolysis in acid or alkaline solution produces amines from ureas, urethanes, isocyanates or isothiocyanates. Sodium hydroxide [2, 3], calcium hydroxide [4], and sodium in liquid ammonia [5, 6] have been employed as bases, while hydrofluoric acid [7], formic acid [8], hydrobromic acid [9], and hydrochloric acid [10] have been utilized as acids. A special hydrolytic agent, perbenzoic acid in benzene [11], has been employed to remove the protecting carbothiophenoxy group (C_6H_5SCO-) from substituted amino acids:

$$2 C_6H_5SCONHCH_2COOH \xrightarrow[H_2O]{[O]}$$

$$2 NH_2CH_2COOH + C_6H_5SO_3H + C_6H_5SO_2H + 2 CO_2$$

Isocyanates are intermediates in the Hofmann (G.2) and Curtius (G.3) degradations in which amines are produced. Occasionally, as in the preparation of t-butyl amine from t-butyl urea [12], one-step hydrolytic methods are preferred. The yields on the whole are satisfactory.

(a) **Preparation of allylamine (from the isothiocyanate)** (70–73% from allyl isothiocyanate and 20% hydrochloric acid for 15 hr. under reflux) [10].

(b) **Other examples**

(1) **t-Butylamine (from the urea)** (71–78% from t-butylurea and sodium hydroxide in ethylene glycol after refluxing 4 hr.) [12, 13].
(2) **DL-Phenylalanine (from the urethane)** (65–85% from N-carboallyloxy-DL-phenylalanine in liquid ammonia to which sodium was added until a blue color persisted for 2–3 min.) [5].

2. From Amides or Imides (Including Gabriel)

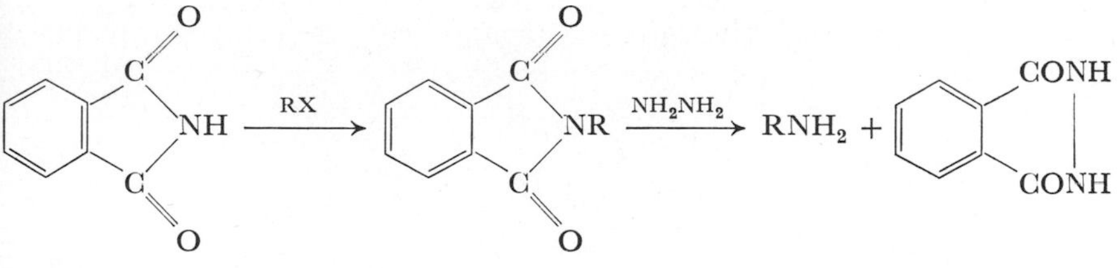

In converting the alkyl halide into the amine with ammonia, not only primary but secondary and tertiary amines are produced as well. To overcome this difficulty one can use potassium phthalimide instead of ammonia. The N-alkylphthalimide formed hydrolyzes exclusively to the primary amine. For a review of the Gabriel reaction, see [15]. The most satisfactory method for preparing N-alkylphthalimides calls for the use of dimethylformamide as the solvent [14]. Since tertiary halides have a tendency to lose hydrogen halide with potassium phthalimide, the *t*-alkylphthalimides have been prepared from *t*-alkylureas and phthalic anhydride [13]; also see C.1. The hydrolysis may be accomplished in acid or alkaline medium but, if milder conditions are desired, the Ing-Manske procedure [13, 14] may be employed. This displacement occurs because a more stable amide is formed:

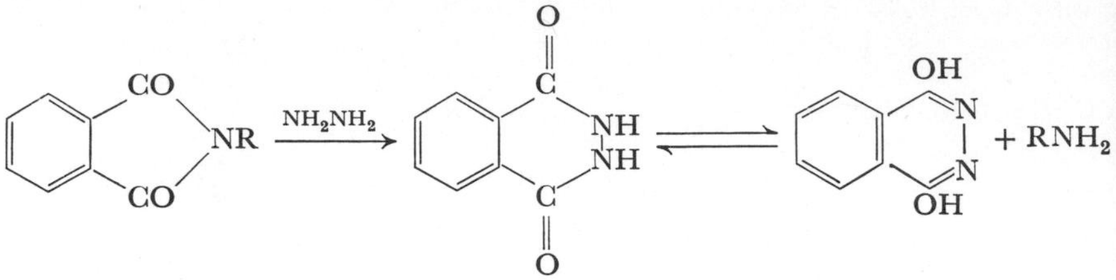

This principle has been applied in a number of isolated instances [16]:

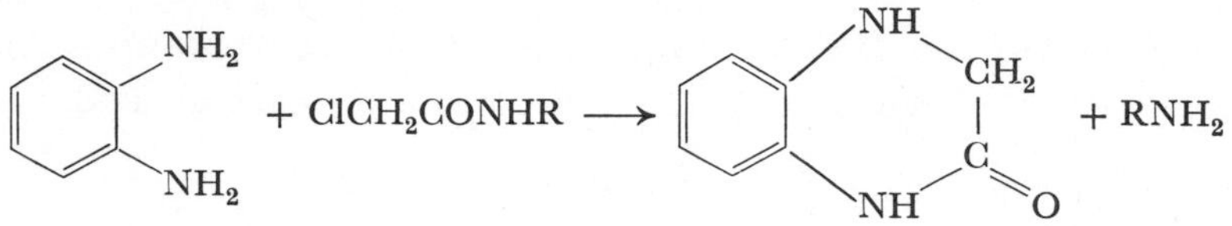

or

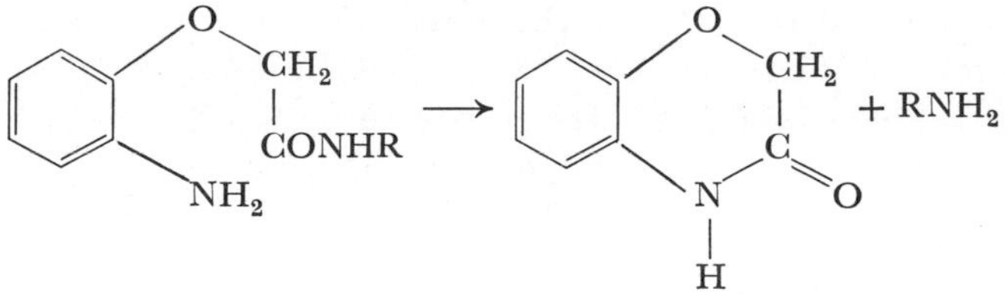

But hydrazine is a strong aminolytic agent even if the type of hydrazide formed is disregarded. Illustratively, an apparently selective hydrazinolysis

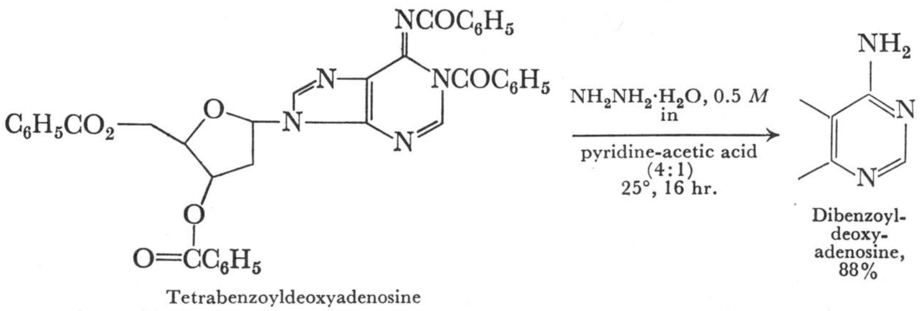

Tetrabenzoyldeoxyadenosine

Dibenzoyl-deoxy-adenosine, 88%

has been demonstrated [17]. The ester groups are unaffected under these conditions.

The α-chloroacetylamides can be solvolyzed to the amine by reaction with thiourea. The solvolysis is specific in that other amide linkages are not affected [18]:

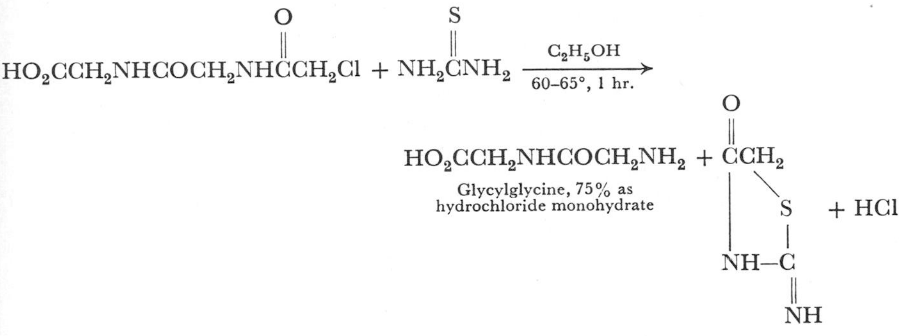

$$HO_2CCH_2NHCOCH_2NHCCH_2Cl + NH_2CNH_2 \xrightarrow[60-65°,\ 1\ hr.]{C_2H_5OH}$$

$$HO_2CCH_2NHCOCH_2NH_2 + CCH_2 + HCl$$

Glycylglycine, 75% as
hydrochloride monohydrate

Pseudothiohydantoin

(a) **Preparation of t-butylamine** (72–88 % of the hydrochloride from t-butylphthalimide and 85 % hydrazine hydrate in 95 % ethanol by refluxing 2 hr.) [13].

(b) **Other examples**

(1) **α,δ-Diaminoadipic acid** (91 % from dimethyl α,δ-diphthalimido-adipate and 48 % hydrogen bromide and glacial acetic acid under reflux for 10 days; aqueous hydrazine hydrate, 85 %, in methanol under reflux for 1 hr. gave a 79.5 % yield) [14].

(2) **α-Amino-γ-butyrolactone** (93 % as the hydrobromide from α-phthalimido-γ-butyrolactone and 24 % hydrobromic acid by refluxing 3 hr.) [19].

3. From Cyanamides

$$R_2NCN \xrightarrow{H_2O} R_2NH + CO_2 + NH_3$$

The hydrolysis of cyanamides may be accomplished in acid or alkaline medium. The synthesis has been of value in preparing secondary amines not attained satisfactorily by alkylation of the primary amine [20]. The dialkyl-cyanamide is readily prepared from sodium or calcium cyanamide and the alkyl bromide [21]. The synthesis has also been of value in the study of N-heterocycles. A methyl group attached to cyclic nitrogen may be converted into hydrogen via the cyanamide [22]:

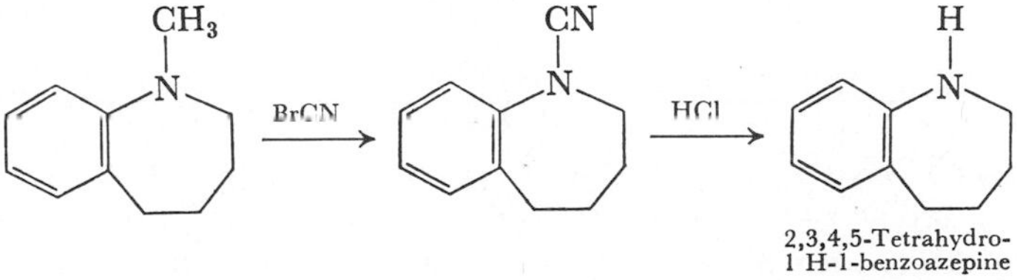

2,3,4,5-Tetrahydro-
1 H-1-benzoazepine

Also, in the degradation of N-heterocycles with cyanogen bromide the cyanogen group which is attached to the nitrogen in the acyclic intermediate may be readily converted into hydrogen [23].

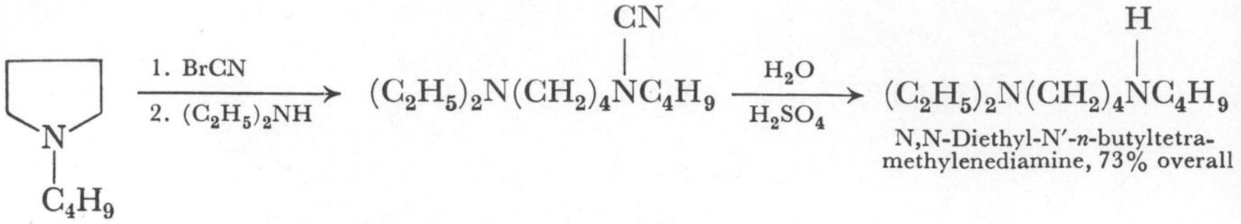

$$\text{(pyrrolidine-}N\text{-C}_4\text{H}_9) \xrightarrow[\text{2. }(C_2H_5)_2NH]{\text{1. BrCN}} (C_2H_5)_2N(CH_2)_4\overset{\overset{\displaystyle CN}{|}}{N}C_4H_9 \xrightarrow[\text{H}_2\text{SO}_4]{\text{H}_2\text{O}} (C_2H_5)_2N(CH_2)_4\overset{\overset{\displaystyle H}{|}}{N}C_4H_9$$

N,N-Diethyl-N'-*n*-butyltetra-methylenediamine, 73% overall

(a) **Preparation of diallylamine** (80–88 % from diallylcyanamide and aqueous sulfuric acid under reflux for 6 hr.) [21].

(b) **Preparation of norpseudotropine** (86.5 % from N-cyanonortropine

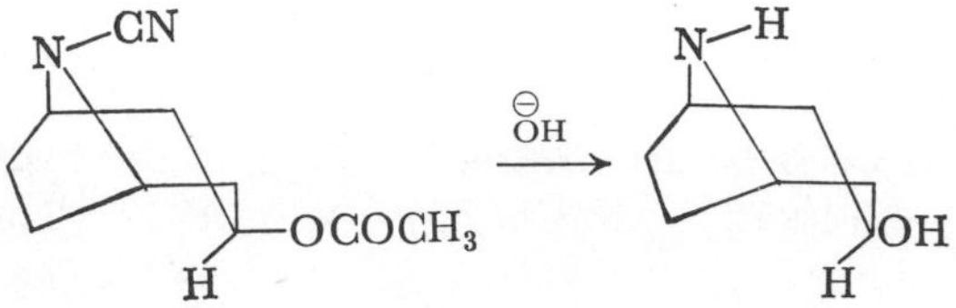

acetate in aqueous sodium hydroxide refluxed 7.5 hr.) [24].

4. From N-Substituted Amides and Sulfonamides

$$RCONHR' \xrightarrow{H_2O} R'NH_2 + RCOOH$$

$$RSO_2NHR' \xrightarrow{H_2O} R'NH_2 + RSO_2OH$$

Primary and secondary amines have been synthesized from N-alkyl-substituted amides and sulfonamides by hydrolysis. Sulfuric acid [25, 26], hydrochloric acid [27], and potassium hydroxide in ethanol [28] have all been employed to promote the hydrolysis. Boron trifluoride in methanol appears to be a most effective agent for amides which are difficult to hydrolyze (see Ex. b), and for acetanilides the method perhaps could be improved by removing the ester as it is formed:

$$ArNHCOCH_3 + CH_3OH \xrightarrow{BF_3} ArNH_2 \cdot BF_3 + CH_3CO_2CH_3$$

Although the sulfonamides are more difficult to hydrolyze than amides, considerable success with the former has been achieved with the use of 48 % hydrogen bromide and phenol [29, 30]. This reaction as shown is actually a reductive cleavage:

$$2\ ArSO_2NR_2 + 5\ HBr + 5\ C_6H_5OH \rightarrow$$
$$ArSSAr + 2\ R_2NH + 5\ p\text{-BrC}_6H_4OH + 4\ H_2O$$

The phenol serves two purposes: it prevents the bromine from reacting with the amine formed, and it increases the solubility of the sulfonamide. In comparing this new reagent with hydrochloric acid, it was found that benzanilide gave a 26 % yield of aniline as the hydrochloride by refluxing for 7 hours with hydrochloric acid, while with the 48 % hydrogen bromide and phenol the yield was

69% after 20 minutes of refluxing. The synthesis has been of value in obtaining primary and secondary amines from the benzenesulfonamides produced in the Hinsberg reaction. Combined with the Beckman rearrangement (Chapter 18, D.5), it permits one to go from the ketoxime via the N-substituted amide or from tosylamides of 1-aminothraquinones (made from 1-chloroanthraquinones and p-toluenesulfonamide) to the amine [26]. The yields as a rule are satisfactory.

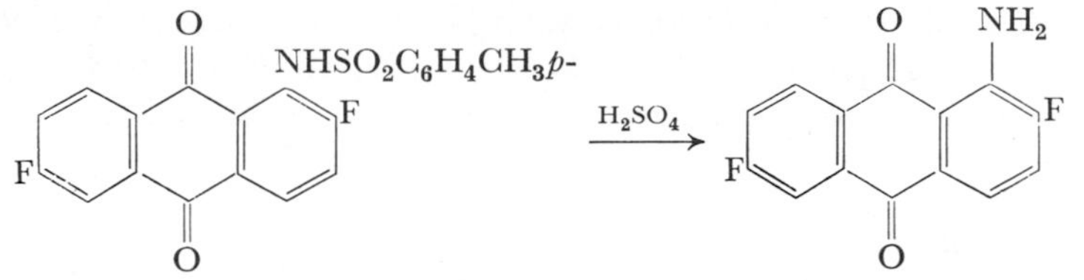

1-Amino-2,6-difluoroanthraquinone

Another reagent which appears to be quite promising, at least for amides of aryl rather than alkylsulfonic acids, is sodium naphthalene in 1,2-dimethoxyethane (see Ex. c.3). In very special cases the enzyme chymotrypsin may be used to remove the benzoyl L-phenylalanyl group, and perhaps others, from their attachment to amine structures [31].

(a) **Preparation of 2-aminobenzophenone** (54%, based on p-toluenesulfonylanthranilic acid, from benzophenone o-(p-tolyl)sulfonamide by warming with concentrated sulfuric acid for 15 min.) [32].

(b) **Preparation of 3,4-dinitro-1-naphthylamine.** 3,4-Dinitro-1-acetaminonaphthalene was added to a methanol solution of 4–8 equiv. of boron trifluoride and refluxed. The exchange was complete in 65 min. at 65°, and the amine was isolated in nearly quantitative yield. Hindered amides took somewhat longer to solvolyze [33].

(c) **Other examples**

(1) **Monomethylethylenediamine** (80%, based on the unmethylated diamide, from N-benzenesulfonyl-N-methyl-N′-acetylethylenediamine and and conc. hydrochloric acid refluxed for 12 hr.) [27].

(2) **Aniline** (93% as the hydrochloride from 2,4,6-trimethylbenzenesulfonanilide, phenol, and 48% hydrobromic acid by refluxing for 20 min.) [29].

(3) **p-Toluidine** (86% from benzenesulfon-p-toluidide added to 3 equiv. of sodium naphthalene in 1,2-dimethoxyethane under nitrogen at room temperature for 1 hr.; yields for various sulfonamides ranged from 62 to 100% with the exception of the poor yields from methanesulfonamides) [34].

5. From Quaternary Imine Salts

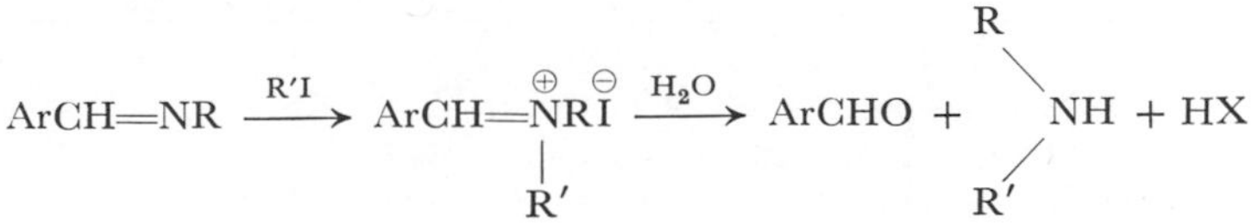

$$\text{ArCH}=\text{NR} \xrightarrow{\text{R'I}} \underset{\underset{\text{R'}}{|}}{\text{ArCH}=\overset{\oplus}{\text{N}}\overset{\ominus}{\text{R}}\text{I}} \xrightarrow{\text{H}_2\text{O}} \text{ArCHO} + \underset{\underset{\text{R'}}{\diagup}}{\overset{\overset{\text{R}}{\diagdown}}{\text{NH}}} + \text{HX}$$

Schiff bases are readily alkylated, and the hydrolysis of the quaternary imine salt formed gives a secondary amine. The method has been used particularly in the preparation of mixed aliphatic secondary amines. In the first step satisfactory results with alkyl halides are obtained only with methyl iodide [35]. By using a dialkyl sulfate in the first step [36] both methyl and ethyl groups have been introduced satisfactorily. This second procedure offers the advantage of simplicity since no pressure vessels are required in the alkylation, but the yields are lower and the products less pure than if the alkyl iodide is employed [37]. Yields vary considerably, but with a series of mixed aliphatic secondary amines in which methyl iodide was the alkylating agent they covered the 52–93% range [37].

(a) Preparation of N-methylethylamine (83–93% from N-benzylideneethylamine and methyl iodide heated under pressure at 100° for 24 hr. followed by hydrolysis first with water and then with sodium hydroxide) [37].

(b) Other examples

(1) N-Methylbutylamine (45–53%, based on *n*-butylamine from N-benzylidenebutylamine and dimethyl sulfate refluxed in benzene followed by steam distillation and alkaline hydrolysis) [36].

(2) β-Phenylisopropylmethylamine (93% from N-benzylidene β-phenylisopropylamine and methyl iodide heated at 100° for some hours followed by hydrolysis in water and in sodium hydroxide) [35].

6. From *p*-Nitrosodialkylanilines

$$ON\langle\bigcirc\rangle NR_2 \xrightarrow{OH^{\ominus},\ H_2O} ON\langle\bigcirc\rangle O^{\ominus} + R_2NH + H_2O$$

The presence of the nitroso group enhances the susceptibility of the *para* position to attack by nucleophiles. But the most important driving force is the resonance stabilization of the anionic product, as shown by the fact that the hydroxide ion is a reagent far superior to the methoxide ion [38]. Sodium bisulfite is another reagent which can be used in place of sodium hydroxide (see Ex. a). The reaction has not been utilized recently for the synthesis of secondary amines.

(a) Preparation of 1-dibutylamino-2-methylaminoethane (64% from *p*-nitroso-N-methyl-N-(dibutylaminoethyl) aniline and about 6 equiv. of sodium bisulfite dissolved in water, stirred 1 hr. at room temperature, and then heated to 76° for 15 min.; the mixture was cooled, made alkaline, and continuously extracted with ether) [39].

1. Houben-Weyl, *Methoden der Organischen Chemie*, G. Thieme Verlag, Stuttgart, 1957, 4th ed., Voll. **11**, Pt. 1, Chap. 8.
2. D. E. Pearson *et al.*, *J. Am. Chem. Soc.*, **70**, 2290 (1948).
3. J. P. Lambooy, *J. Am. Chem. Soc.*, **71**, 3756 (1949).
4. A. Dornow and O. Hahmann, *Arch. Pharm.*, **290**, 20 (1957).
5. C. M. Stevens and R. Watanabe, *J. Am. Chem. Soc.*, **72**, 725 (1950).

6. V. du Vigneaud and G. L. Miller, *Biochem. Prep.*, **2**, 74 (1952).
7. L. A. Carpino, *J. Am. Chem. Soc.*, **79**, 98 (1957).
8. E. Schmidt *et al.*, *Ann. Chem.*, **568**, 192 (1950).
9. N. F. Albertson and F. C. McKay, *J. Am. Chem. Soc.*, **75**, 5323 (1953).
10. M. T. Leffler, *Org. Syn.*, Coll. Vol. **2**, 24 (1963).
11. J. Kollonitsch *et al.*, *Chem. Ber.*, **89**, 2288 (1956).
12. D. E. Pearson *et al.*, *Org. Syn.*, Coll Vol. **3**, 154 (1955).
13. L. I. Smith and O. H. Emerson, *Org. Syn.*, Coll. Vol. **3**, 151 (1955).
14. J. C. Sheehan and W. A. Bolhofer, *J. Am. Chem. Soc.*, **72**, 2786 (1950).
15. M. S. Gibson and R. W. Bradshaw, *Angew. Chem. Intern. Ed. Engl.*, **7**, 919 (1968).
16. R. W. and A. D. Holley, *J. Am. Chem. Soc.*, **74**, 3069 (1952).
17. R. L. Letsinger *et al.*, *Tetrahedron Letters*, 2621 (1968).
18. M. Masaki *et al.*, *J. Am. Chem. Soc.*, **90**, 4508 (1968).
19. Y. Knobler *et al.*, *J. Org. Chem.*, **24**, 1794 (1959).
20. E. B. Vliet, *Org. Syn.*, Coll. Vol. **1**, 201 (1941).
21. E. B. Vliet, *Org. Syn.*, Coll. Vol. **1**, 203 (1941).
22. B. D. Astill and V. Boekelheide, *J. Am. Chem. Soc.*, **77**, 4079 (1955).
23. R. C. Elderfield and H. A. Hageman, *J. Org. Chem.*, **14**, 605 (1949).
24. A. Nickon and L. F. Fieser, *J. Am. Chem. Soc.*, **74**, 5566 (1952).
25. G. W. H. Cheeseman, *J. Chem. Soc.*, 3308 (1955).
26. G. Valkanas and H. Hopff, *J. Chem. Soc.*, 1923 (1963).
27. S. R. Aspinwall, *J. Am. Chem. Soc.*, **63**, 852 (1941).
28. F. F. Blicke and M. U. Tsao, *J. Am. Chem. Soc.*, **68**, 905 (1946).
29. H. R. Snyder *et al.*, *J. Am. Chem. Soc.*, **74**, 2006 (1952).
30. H. R. Snyder *et al.*, *J. Am. Chem. Soc.*, **74**, 4864 (1952).
31. R. W. Holley, *J. Am. Chem. Soc.*, **77**, 2552 (1955).
32. H. J. Scheifele, Jr., and D. F. DeTar, *Org. Syn.*, Coll. Vol. **4**, 34 (1963).
33. L. S. Sihlbom, *Acta. Chem. Scand.*, **8**, 529 (1954).
34. W. D. Clossen, *et al.*, *J. Am. Chem. Soc.*, **89**, 5311 (1967).
35. E. H. Woodruff *et al.*, *J. Am. Chem. Soc.*, **62**, 922 (1940).
36. J. J. Lucier *et al.*, *Org. Syn.*, **44**, 72 (1964).
37. S. Wawzonek *et al.*, *Org. Syn.*, **44**, 75 (1964).
38. C. W. L. Bevan *et al.*, *J. Chem. Soc.*, 4543 (1960).
39. R. Munch *et al.*, *J. Am. Chem. Soc.*, **68**, 1297 (1946).

C. Metathetical or Other Reactions Leading to Metathetical Products

$$R_2NX + \overset{\diagdown}{\underset{\diagup}{C}}{}^{\ominus}\!- \longrightarrow R_2N-\overset{|}{\underset{|}{C}}- + \overset{\cdot\cdot}{X}{}^{\ominus}$$

Alkylation of amines is a classical method of preparation of amines plagued somewhat by a tendency to give di- and trialkylated amines (see C.1 and C.5). However, the reaction may be used when other methods fail or when amines of unusual structure are desired. If the alkyl halides are sluggish in reaction with amines, resort can be made to alkali metal salts of amines (C.2) or sulfonates (C.3).

Other syntheses peculiar to the synthesis of aromatic amines (C.2) and to the displacement of groups other than halide or sulfonate (C.6, C.7, C.8, C.9) are discussed. Numerous examples of metathesis are to be found in the Houben-Weyl reference [1].

References for Section C are on pp. 458–460.

1. From Halides and Amines (Including Ammonia)

$$RX \xrightarrow{NH_3} RNH_2 \cdot HX$$

$$RX \xrightarrow{RNH_2} R_2NH \cdot HX$$

The conversion of halides into amines with ammonia or an amine has been widely used even though the alkylation may not stop at the point desired. To avoid the formation of mixtures, which are particularly common with alkyl halides of low molecular weight, the Gabriel synthesis (B.2) is sometimes used. The use of an excess of ammonia or the amine also retards polyalkylation. Typical results with excess ammonia are: 1 part of octyl bromide to a double volume of liquid ammonia in a bomb at 25° for 1 day gives 45 % of primary and 43 % of secondary amine, or 1 part of benzyl chloride to 8 parts of ammonia gives 53 % of primary and 39 % of secondary amine [2]. However, other combinations give higher yields of primary amine (see Ex. b.3).

Among alkyl halides the best results are obtained with primary halides. Saturated *t*-alkyl halides often undergo dehydrohalogenation, but this side reaction can be circumvented in certain instances by using propargylic halides followed by hydrogenation [3]:

$$HC\equiv CC(CH_3)_2Cl + (CH_3)_2CHNH_2 \text{ in 25 ml. } H_2O \xrightarrow[\text{7 days}]{25°}$$

0.25 mole   0.75 mole

$$HC\equiv CC(CH_3)_2NHCH(CH_3)_2 \xrightarrow[\text{cat.}]{H_2} CH_3CH_2C(CH_3)_2NHCH(CH_3)_2$$

3-Isopropylamino-3-methyl-1-butyne, 58%   Isopropyl-*t*-amylamine, 73%

Special conditions are sometimes necessary to assure success in the synthesis of primary amines. For high-molecular-weight halides, for example, ammonia in ethanol [4] or liquid ammonia [5] may be employed. For aromatic halides of an inactive type, high-pressure ammonolysis at elevated temperature in the presence of a copper catalyst is required [6]. The copper salt should be mixed with potassium acetate to neutralize any halo acid formed [7]. Electron-withdrawing groups in the *o*- and *p*-positions increase the lability of the halogen, and as a result more moderate conditions in the absence of a catalyst may be employed [8]. Six-ring nitrogen heterocyclic compounds with halogens or other groups in the 2- and 4-positions are readily converted into primary amines with or without a catalyst [9, 10]. Many catalysts are known, but the best one appears to be phenol, in which case the intermediate is probably the 2- or 4-phenoxy compound [11].

The aminolysis of halides has been of great value in the synthesis of α-amino acids [12–18]. In such conversions with aqueous or liquid ammonia it has been shown [19, 20] that ammonium salts reduce the formation of secondary and tertiary amines.

Some success has been achieved in the alkylation of primary to secondary amines. By using 4 moles of aniline to 1 of benzyl chloride, N-phenylbenzyl-amine is obtained in a pure form with a yield of 77–78 % [21]. Other N-substituted anilines may be synthesized in a similar manner [22]. In fact, an industrial process for the production of N-methylaniline from chlorobenzene and methylamine in the presence of a copper catalyst has been devised [23]. t-Butyl- or t-octylamines give excellent yields of secondary amines on alkylation without any attempt to prevent tertiary amine formation [24]. However, tertiary amines of t-butyl- or t-octylamine can be formed with the active alkylation agents, methyl iodide and ethylene oxide.

In a similar manner tertiary amines may be obtained from secondary ones [25–27]. Again the synthesis of strictly aromatic types is difficult. However, triphenylamine may be obtained in 82–85 % yield from diphenylamine, phenyl iodide, and potassium carbonate in the presence of a copper catalyst [28].

Neighboring group participation is to be found when the halide atom is incorporated β, γ, or ε to the amino group in the same molecule:

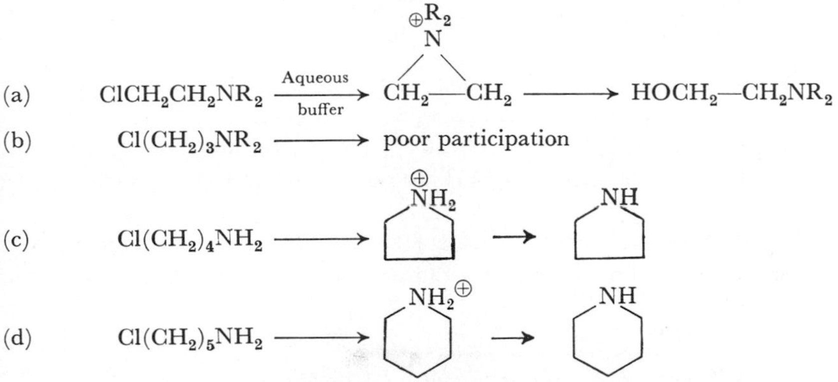

(a) $ClCH_2CH_2NR_2$ $\xrightarrow[\text{buffer}]{\text{Aqueous}}$ $\overset{\overset{\displaystyle \oplus R_2}{\displaystyle N}}{CH_2—CH_2}$ $\longrightarrow$ $HOCH_2—CH_2NR_2$

(b) $Cl(CH_2)_3NR_2$ $\longrightarrow$ poor participation

(c) $Cl(CH_2)_4NH_2$ $\longrightarrow$ (ring with $\overset{\oplus}{N}H_2$) $\rightarrow$ (ring with NH)

(d) $Cl(CH_2)_5NH_2$ $\longrightarrow$ (ring with $NH_2\oplus$) $\rightarrow$ (ring with NH)

The pyrrolidine and piperidine are formed at fantastically fast rates. In the case of equation (a) the neighboring group participation leads to isomerization, the product depending on the type of conditions used:

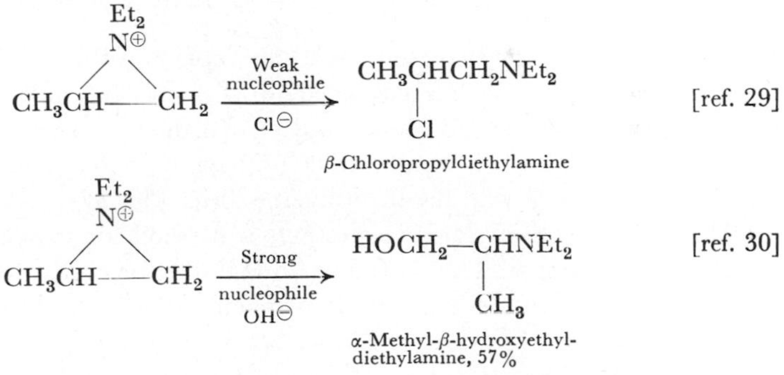

$\overset{\overset{\displaystyle Et_2}{\displaystyle N\oplus}}{CH_3CH—CH_2}$ $\xrightarrow[Cl^\ominus]{\text{Weak}\ \text{nucleophile}}$ $\underset{Cl}{CH_3CHCH_2NEt_2}$ [ref. 29]

β-Chloropropyldiethylamine

$\overset{\overset{\displaystyle Et_2}{\displaystyle N\oplus}}{CH_3CH—CH_2}$ $\xrightarrow[OH^\ominus]{\text{Strong}\ \text{nucleophile}}$ $\underset{CH_3}{HOCH_2—CHNEt_2}$ [ref. 30]

α-Methyl-β-hydroxyethyl-diethylamine, 57%

Interesting ring transformations can be brought about by means of neighboring group participation [31]:

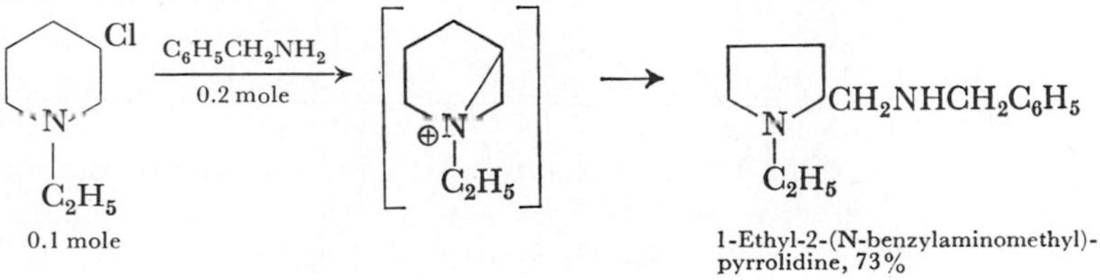

(piperidine ring with Cl, N, C_2H_5) $\xrightarrow[\text{0.2 mole}]{C_6H_5CH_2NH_2}$ [bicyclic intermediate with $\overset{\oplus}{N}$, C_2H_5] $\rightarrow$ (pyrrolidine ring with $CH_2NHCH_2C_6H_5$, N, C_2H_5)

0.1 mole

1-Ethyl-2-(N-benzylaminomethyl)-pyrrolidine, 73%

(a) Preparation of 2,4-dinitroaniline (primary) (68–76 % from 2,4-dinitrochlorobenzene, ammonium acetate, and ammonia at 170° for 6 hr.) [8].

(b) Other examples

(1) Benzylaniline (secondary) (85–87 % from aniline, 4 moles, sodium bicarbonate, water, and benzyl chloride, 1 mole, at 90–95° for 4 hr.) [21].

(2) Triphenylamine (tertiary) (82–85 % from diphenylamine, 1.04 mole, iodobenzene, 1 mole, anhydrous potassium carbonate, 1 mole, and copper powder in nitrobenzene heated for 24 hr.) [28].

(3) Aminoacetal (primary) (71–74 % from chloroacetal, 0.25 mole, and liquid ammonia, about 18 moles, in methanol under pressure at 140° for 10 hr.) [32].

(4) *dl*-Isoleucine (49 % from α-bromo-β-methylvaleric acid and ammonium hydroxide in a closed vessel for 1 week) [16].

(5) N-Ethyl-*m*-toluidine (63–66 % from *m*-toluidine and ethyl bromide at 25° for 24 hr., the secondary amine being purified via the N-nitroso derivative) [33].

(6) N-Phenylanthranilic acid (82–93 % from *o*-chlorobenzoic acid, aniline, potassium carbonate, and a small amount of copper oxide at reflux for 2 hr.) [34].

2. From Halides and Alkali Metal Amides

$$RX + NaNH_2 \xrightarrow[NH_3]{Liquid} RNH_2 + NaX$$

$$RX + LiN(C_2H_5)_2 \longrightarrow RN(C_2H_5)_2 + LiX$$

Two methods have been employed for replacing the halide with the amino group. The older one has consisted in treating the halide with sodium or potassium amide in liquid ammonia [35]. In this synthesis, yields of a series of aliphatic amines varied from 30 to 80 %. The method has found application as well in the pyridine and dibenzothiazine series [36, 37].

The more recent synthesis represents a method for converting a secondary into a tertiary amine. The secondary amine may be converted into the lithium analog by the use of methyl-, butyl-, or phenyllithium, and this product with an alkyl halide gives the tertiary amine. Since the actual isolation of the lithium amide is not necessary, the synthesis may be accomplished by starting with a secondary amine, an alkyl- or aryllithium, and an alkyl halide [38, 39]. Lithium is the preferred alkali metal, while the alkyl bromide or iodide has been commonly employed usually in ether as a solvent. The yields of tertiary amine leave something to be desired, although N,N-diethyl-*n*-octylamine was prepared in 89 % yield (see Ex. b) [38], while for N,N-diisopropyl-*o*-toluidine the yield was 73.5 % [39]. Sodium or potassium salts of imides or sulfonamides are less basic and lend themselves to attack by a wider range of alkyl halides. The reaction of the alkali salt of phthalimide with alkyl halides is, of course, known as the Gabriel synthesis (see B.2). The reaction of the sodium salt of a sulfon-amide with alkyl halides has been applied to the preparation of azetidine [40]:

$$TsSO_2NH_2 + 2\ NaOH + Br(CH_2)_3Br \longrightarrow$$

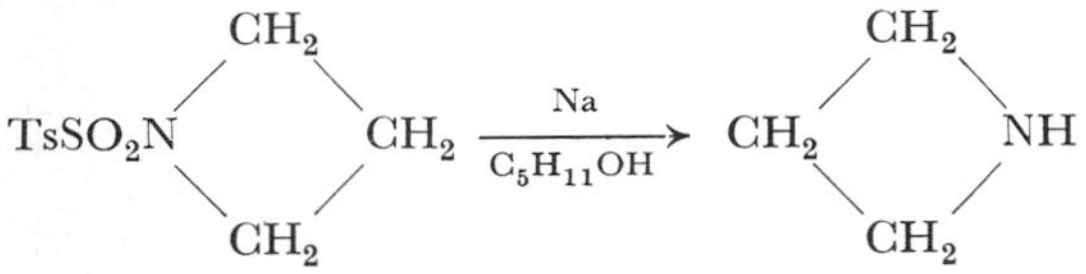

In the case of aromatic halides, displacement with metal amides can take one of two routes, or both simultaneously [41]. The first is normal displacement:

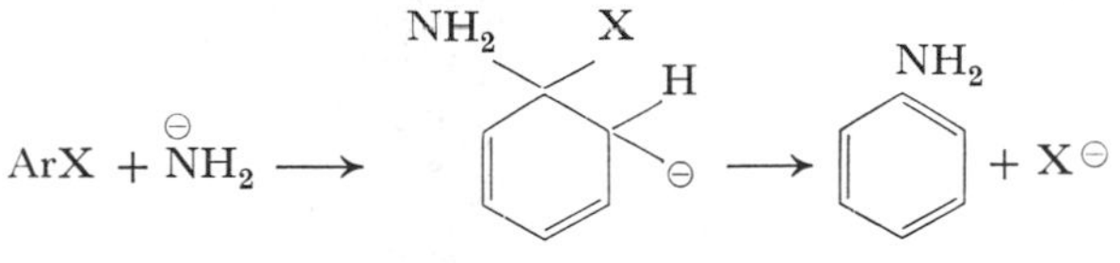

and the second is benzyne formation:

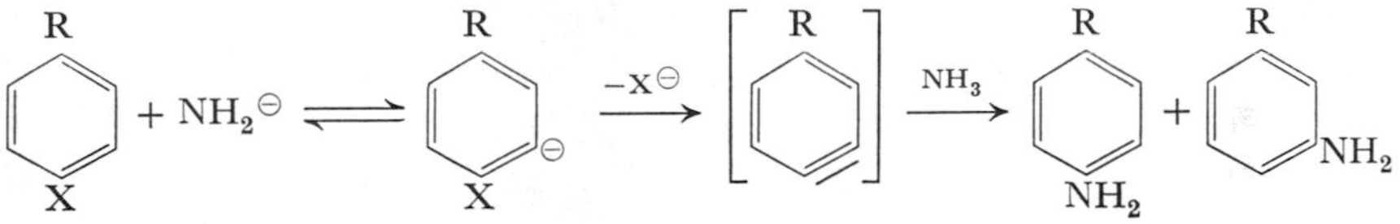

By the benzyne route, a mixture of anilines is obtained with either *p*- or *m*-substituted halides, but only the *meta*-substituted aniline if an *ortho*-substituted halide is used (see Ex. c.1). It is difficult to predict exactly the relative rates of both pathways, but as a rule, the more basic the reagent, the more acidic the *vic*-hydrogen, and the less reactive the halogen, the greater will be the participation of the benzyne intermediate. 1-Halonaphthalene and sodamides apparently react exclusively via the naphthyne to give 33% of 1- and 67% of 2-naphthylamines. Any increase in the amount of 1-naphthylamine is an indication that direct replacement is encroaching.

(a) **Preparation of 2-aminopyridine.** A solution of 2-bromopyridine in an equal volume of anhydrous ether was added to a liquid ammonia sus-pension of sodium amide, after which the mixture was refluxed for 4 hr. Excess of solid ammonium chloride was added; the ammonia was allowed to evaporate; and about 50 ml. of 5% sodium hydroxide solution for each 0.1 mole of the amide ion was added. Saturation of this solution with sodium hydroxide pellets in an ice bath and extraction with ether gave an ethereal solution which, when dried, evaporated, and distilled *in vacuo*, produced a 67% yield of 2-amino-pyridine, b.p. 120–121° (36.5 mm.) [36].

(b) **Preparation of N,N-diethyl-*n*-octylamine.** To 200 ml. (0.2 mole) of a 1 *M* phenyllithium solution there was added dropwise with stirring in 20 min. 16.1 g. (0.22 mole) of diethylamine in 25 ml. of dry ether. Stirring was continued for 10–20 min. more and 0.1 mole of *n*-octyl bromide in an equal volume of ether was added during ½ hr. After the reaction was completed by a

17-hr. reflux, the mixture was cooled and decomposed by the addition of 50 ml. of water. From the ethereal layer and the ether extract of the aqueous layer, the amine was extracted in 5 N hydrochloric acid solution. This solution, after being washed with water and dried, was made strongly basic with 5 N sodium hydroxide solution, salted out, and again extracted with ether. Drying over potassium hydroxide and distillation gave the tertiary amine (89%), b.p. 221–224° (760 mm.) [38].

(c) Other examples

(1) *m*-Anisidine (57.5% from 0.42 mole of *o*-chloroanisole added to 0.85 mole of NaNH$_2$ in 800 ml. of liquid ammonia and stirred for 30 min.) [42].

(2) N-Phenylpiperidine (99% from 0.1 mole of bromobenzene and 0.2 mole of NaNH$_2$ in 30 ml. of piperidine refluxed for about 2 hr.; other secondary amines gave yields varying from 22 to 67%; a benzyne intermediate is involved) [43].

3. From Alkyl Sulfonates or Phosphates and Amines (Including Ammonia)

$$-CH_2OSO_2Ar \xrightarrow{NH_3} -CH_2NH_2$$

$$-CH_2OSO_2Ar \xrightarrow{ArNH_2} -CH_2NHAr$$

Alkyl sulfonates, usually tosylates, may be converted into amines by treatment with ammonia or another amine. For the lower-boiling reagents, such as ammonia and the simpler amines, pressure is necessary. For higher-boiling amines, such as piperidine, a simple refluxing suffices [44]. The method has been applied successfully in the steroid [45, 46] and sugar [47, 48] series. With ammonia and equatorial sulfonate esters in steroidal, decalyl, and cyclohexyl systems, axial amines [46] are produced. The simplicity and high specificity of the method makes this a preferred method for the synthesis of such amines. Indeed, if the halide-amine reaction fails, resort should be made to the sulfonate-amine reaction. If the latter fails, then resort should be made to one of two methods:

Sulfonate-hydrazine method [49]:

$$ROTs + NH_2N(CH_3)_2 \longrightarrow RNHN(CH_3)_2 \xrightarrow[\text{cat.}]{H_2} RNH_2 + (CH_3)_2NH$$

The hydrazine is a better nucleophile than the amine [50].

Sulfonate-azide method [51]. The azide formed from the tosylate may be reduced with lithium aluminum hydride:

$$ROTs + LiN_3 \xrightarrow[\text{Inversion}]{CH_3OH} RN_3 \xrightarrow[\substack{\text{Retention} \\ \text{of configuration}}]{LiAlH_4} RNH_2$$

or with triphenylphosphine [52]:

$$RN_3 + (C_6H_5)_3P \xrightarrow{-N_2} RN{=}P(C_6H_5)_3 \xrightarrow{H_2O} RNH_2 + (C_6H_5)_3PO$$

Interesting quaternary hydrazinium salts can be prepared by a modification of the sulfonate method [53]:

$$C_6H_5N(CH_3)_2 + H_2NOSO_2OH \xrightarrow[\text{reflux 10 min.}]{\substack{\text{75 ml.} \\ \text{absolute alcohol,}}} C_6H_5\overset{\oplus}{\underset{CH_3}{\overset{CH_3}{N}}}NH_2HSO_4^{\ominus}$$

30 g.   5.6 g.

1,1,1-Phenyldimethyl-
hydrazinium hydrogen
sulfate, 38.5%

Aliphatic t-amines give similar hydrazinium salts in an exothermic reaction, and triphenylphosphine with chloramine in a parallel reaction is converted into aminotriphenylphosphine chloride

$$(C_6H_5)_3\overset{\oplus}{P}NH_2Cl^{\ominus}$$

which in turn is transformed to the reactive triphenylphosphine imine, $(C_6H_5)_3$-P=NH [54]. The latter should prove to be an interesting iminating reagent.

By another modification of sulfonate displacement methods, benzyl-type alcohols (or alcohols which form relatively stable carbonium ions) can be converted into amines in some cases by an SN_i mechanism [55]:

$$C_6H_5CHONaCH_3 + (CH_3)_2NSO_2Cl \xrightarrow{\text{Dimethoxyethane}}$$

$$\underset{\underset{C_6H_5CHCH_3}{|}}{OSO_2N(CH_3)_2} \xrightarrow{50-60°} C_6H_5\underset{\underset{N(CH_3)_2}{|}}{CHCH_3} + C_6H_5CH=CH_2$$

α-Dimethylamino-
ethylbenzene, 60%

Alkyl phosphates (99% diethylaniline from aniline and triethyl phosphate) [56], dialkyl phosphorochloridates (94–95% alkylanilines from aniline and the phosphorochloridates) [57], and methyl polyphosphoric acid (see Ex. b.4) are also good alkylating agents for amines.

(a) Preparation of 4-acetaminophenyl-β-piperidinoethyl sulfone.

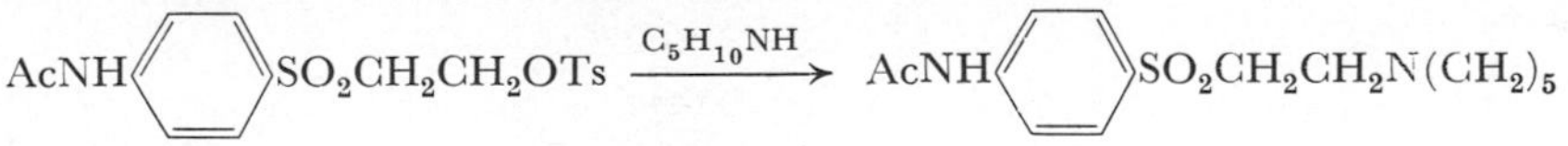

$$AcNH\text{—}\langle\text{—}\rangle\text{—}SO_2CH_2CH_2OTs \xrightarrow{C_5H_{10}NH} AcNH\text{—}\langle\text{—}\rangle\text{—}SO_2CH_2CH_2N(CH_2)_5$$

To 35 g. of 4-acetaminophenyl-β-(p-toluenesulfonyloxy)ethyl sulfone, 22 ml. of piperidine was added. After heating for 2 hr. on a steam bath, the oil which formed solidified when treated with water. It weighed 28.7 g. (99%), m.p. 110° (gas evolution), and was a hydrate which on recrystallization from benzene gave the anhydrous amine, m.p. 121–123° [44].

(b) Other methods

(1) 6-Desoxy-6-amino-methyl-α-D-glucoside (75% from 6-tosyl-methyl-α-D-glucoside and ammonia in methanol under pressure 16 hr. at 120°) [47].

(2) cis-4-t-Butylcyclohexylamine (37% as the hydrochloride from *trans*-4-*t*-butylcyclohexyl tosylate and anhydrous ammonia in a bomb at 95–100° for 24 hr.) [46].

(3) 1,3,3-Trimethylazetidine

(57% from 0.1 mole of 2,2-dimethyl-3-methylamino-1-sulfatopropane and 50 g. of KOH in 200 ml. of water warmed on a steam bath) [58].

(4) Dimethylaniline. A stock solution of 170 g. of PPA and 1 mole of methanol, called methyl polyphosphoric acid, was mixed with 0.2 mole of aniline and heated at 200° for 24 hr., cooled, made basic with KOH, and extracted with ether. The residue from the extract was dimethylaniline, 66% in surprisingly clean and pure condition. The aqueous alkaline layer was concentrated until salt appeared, treated with 100 ml. of 60% aqueous KOH, and distilled again to decompose the quaternary salt, a process which yielded an additional 29% of dimethylaniline in the distillate. The reaction is general for aromatic amines and recommended particularly for weakly basic ones such as diphenylamine [59].

(5) Ethylenimine (34–37% by distilling β-aminoethyl sulfuric acid in aqueous NaOH and collecting the aqueous distillate from which the imine is salted out with KOH pellets) [60].

4. From Phenols and Amines or Ammonia (Including Bucherer)

$$ArOH + RNH_2 \rightarrow ArNH_2 + H_2O$$

$$ArOH + RNH_3{}^{\oplus}\overset{\ominus}{H}SO_3 \rightleftharpoons ArNHR + H_2SO_3 + H_2O \qquad \text{(Bucherer)}$$

Ordinary phenols do not lend themselves well to exchange with amines. Phenols which are less aromatic and show more tautomeric character do exchange with amines probably via ketimine structures. Those best adapted are naphthols, resorcinols, and hydroxyquinolines:

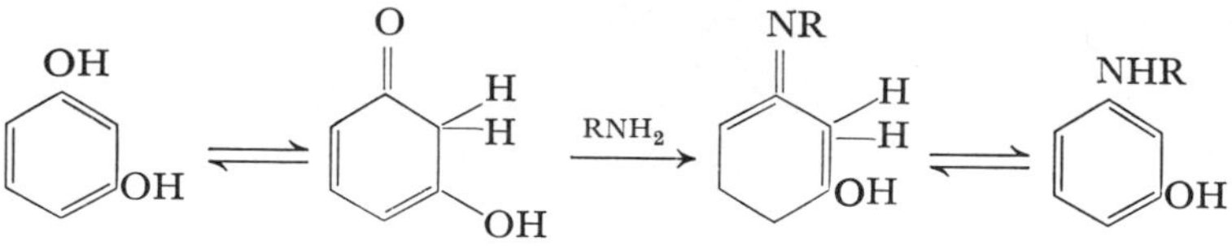

Zinc chloride (see Ex. b.3), iodine [62], calcium chloride [61], and sulfanilic and sulfuric acids [62] have been used as catalysts, but no catalysts are necessary if the phenols are highly tautomeric or if high enough temperatures are used. Several examples are listed to illustrate the range [61, 62]:

5-Aminoresorcinol, 100%, from phloroglucinol and ammonia at 20° for 2 days

3,5-Diaminophenol, from phloroglucinol and ammonia at 20° for 4 weeks.

4-Hydroxydiphenylamine, 79%, from hydroquinone, excess aniline, and sulfanilic acid at 240°

2-Phenylaminonaphthalene 97% from β-naphthol, excess aniline, and 0.1 part sulfuric acid at 170°.

Phenoxazine,

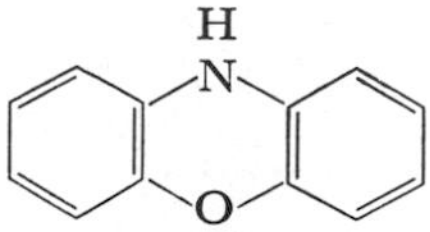

70% from catechol and o-aminophenol with an acid catalyst at 230° [63].

1-Methylphenazine

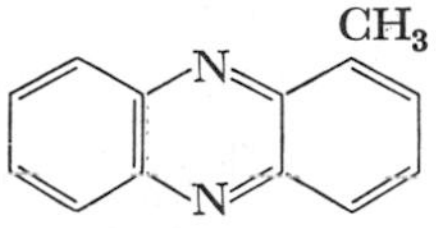

76% from 2,3-diaminotoluene and catechol heated in a sealed tube to 230° for 24 hr. under nitrogen followed by oxidation with a slow stream of oxygen at 230° in an open tube [64].

8-Hydrazinoquinoline, 60–64% from 8-hydroxyquinoline and hydrazine hydrate refluxed for 45 hr. under nitrogen [65–67].

If the less aromatic phenols can add amines, then it stands to reason that the less aromatic amines can add other amines [68]:

$$ArNH_2 + RNH_2 \rightarrow ArNHR + NH_3$$

Several examples are shown:

α-Phenylaminonaphthalene, 90% from α-naphthylamine, excess aniline, and catalytic amounts of p-toluenesulfonic acid at 165–170°.

Di-(4-pyridyl)amine, 72% from 4-aminopyridine refluxed in pyridine containing phosphorus trichloride.

N,N'-Dimethylmelamine, 65–71% from melamine and methylamine hydrochloride at 190–195° for 6 hr.

Indoline

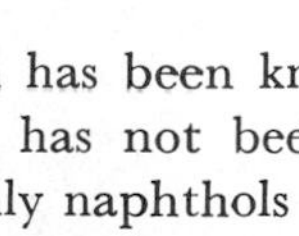

40% from o-aminophenylethylamine dihydrochloride at 300° for 10 min.

Many other heterocyclic and polynuclear amines are made similarly to the examples above.

The Bucherer reaction, which has been known for some 70 years, is used industrially, but its mechanism has not been clarified until quite recently (Phenols, Chapter 5, A.3). Mainly naphthols and resorcinols are employed in this reaction, and the same amines can be reconverted into the phenols since the reaction is entirely reversible. Yields range from fair to excellent [69]. From the discussion in the early part of this section, it is apparent that the Bucherer reaction can be run without a bisulfite catalyst. Example b.4 not only illustrates this point, but also shows that the 2-position of naphthol is less hindered than the 1-position.

(a) **Preparation of β-naphthylamine** (94–96% from β-naphthol and ammonium sulfite at 150° under pressure for 8 hr.) [69].

(b) Other examples

(1) **3-Hydroxy-4-*n*-hexylaniline** (70–80% from *n*-hexylresorcinol, ammonium chloride, sodium bisulfite, and concentrated ammonium hydroxide in a steel bomb at 240–250° for 4 hr.) [70].

(2) **5-Aminoisoquinoline** (66% from 5-hydroxyisoquinoline, conc. ammonium hydroxide, water, and sulfur dioxide under pressure first for 13 hr. at 150° and then for 4 hr. at 160°) [66].

(3) **3-Amino-2-naphthoic acid** (70% from 3-hydroxy-2-naphthoic acid, 28% aqueous ammonia, and zinc chloride heated gradually to 195° in 3 hr. and then maintained at the same pressure, about 400 p.s.i., for 36 hr.) [71].

(4) **7-Methylamino-1-naphthol** (71% from 0.25 mole of 1,7-dihydroxy-naphthalene, 0.5 mole of 40% aqueous methylamine, and 55 ml. of water heated in an autoclave to 140° for 8 hr. Ammonia was not so selective between the 1- and 7-positions as methylamine) [72].

5. From Halomethyl Compounds and Hexamethylenetetramine (Delépine)

$$ArCH_2X \xrightarrow{(CH_2)_6N_4} [ArCH_2(CH_2)_6 \overset{\oplus}{N}_4] \overset{\ominus}{X}$$

$$\downarrow C_2H_5OH, HCl$$

$$ArCH_2NH_2 \xleftarrow{\text{NaOH}} ArCH_2\overset{\oplus}{N}H_3\overset{\ominus}{X}$$

This synthesis consists of the first part of the Sommelet reaction (Aldehydes, Chapter 10, A.9). If the hydrolysis of the complex formed between the halide and hexamethylenetetramine is carried out in a mixture of ethanol and concentrated hydrochloric acid [73], the reaction stops at the primary amine stage. The synthesis thus represents an alternative to that of Gabriel (B.2) in preparation of primary amines. It is successful for primary halides and, since iodides work better than chlorides or bromides, sodium iodide is added when the latter are employed [74]. The method has been utilized successfully in the preparation of simple aliphatic amines [74], certain benzyl amines [75], α-aminoketones [76], aminoalkynes [77], *p*-aminomethylbenzoates [78], α-aminoesters [79], and β-amino acids [80]. Yields vary from 40 to 85%.

(a) **Preparation of β-alanine.** Sodium bicarbonate, 2.74 g., was added to a solution of 5 g. of β-bromopropionic acid in 15 ml. of water and 10 ml. of ethanol. To the neutral solution was added 4.57 g. of hexamethylenetetramine in 10 ml. of water, after which the mixture was allowed to stand for 15 hr. Ethanol, 50 ml., was then added until faint turbidity appeared, at which point scratching led to voluminous crystallization of the betaine complex. Chilling and filtering twice gave a total of 9.5 g. of crystals.

The complex was treated with 120 ml. of ethanol and 15 ml. of conc. hydrochloric acid and refluxed for 15 hr. Concentration to dryness *in vacuo* at 50° followed and then extraction with several portions of ethanol. Filtration and concentration of the filtrate to dryness gave a residue which was boiled under reflux with 50–75 ml. of water for $\frac{1}{2}$ hr. After removing the chloride ion with an excess of silver oxide, the filtrate was saturated with hydrogen sulfide. The colorless filtrate was concentrated *in vacuo* to a few milliliters, after which it was diluted with ethanol. Chilling gave 2.5 g. (85%) of β-alanine, m.p. 199–200° (dec.) [80].

(b) Other examples

(1) **1-Amino-2-heptyne** (63% from the quaternary salt of 1-iodo-2-heptyne and hexamethylenetetramine by stirring for 3 days with alcoholic hydrogen chloride) [77].

(2) **2-Bromoallylamine** (59–72% from hexamethylenetetramine in chloroform to which 2,3-dibromopropene was added slowly and then refluxed. The crude salt was then hydrolyzed) [81].

6. From Ethers, Thioethers, or Lactones and Amines

$$-OR(-SR) \xrightarrow{\text{NH}_3} -NH_2$$

This reaction has been mentioned already as an intermediate step in the reaction of a 2- or 4-halopyridine or haloquinoline with amines in the presence of phenol (C.1). Such a replacement has been accomplished with phenyl-4-pyridylether (Ex. a) and its sulfur analog, which were converted into phenyl 4-pyridylamine with yields of 89 and 70%, respectively, by refluxing with aniline hydrochloride [82]. A similar replacement occurs among certain thioethers in the pyrimidine [83, 84] and purine [85] series. In some of these experiments [85, 83] the reaction was conducted under pressure. The nitroso group in the 5-position in pyrimidines facilitates the displacement of the thiomethyl substituent [86]. Yields are not always satisfactory.

β-Aroxy- or alkoxypropionitriles, on heating, tend to revert to acrylonitriles, which in the presence of an amine yield the aminonitrile [87]:

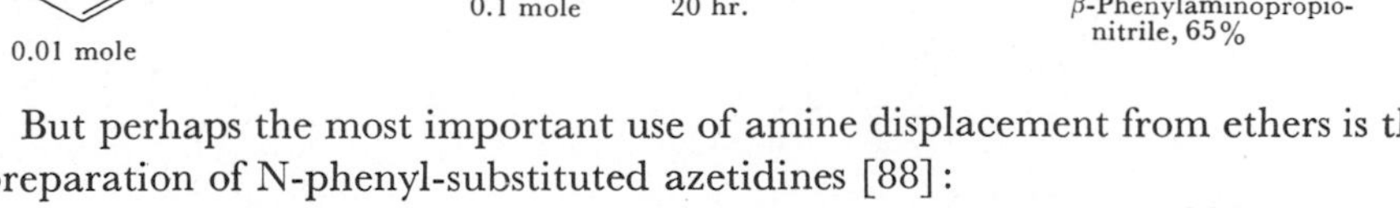

But perhaps the most important use of amine displacement from ethers is the preparation of N-phenyl-substituted azetidines [88]:

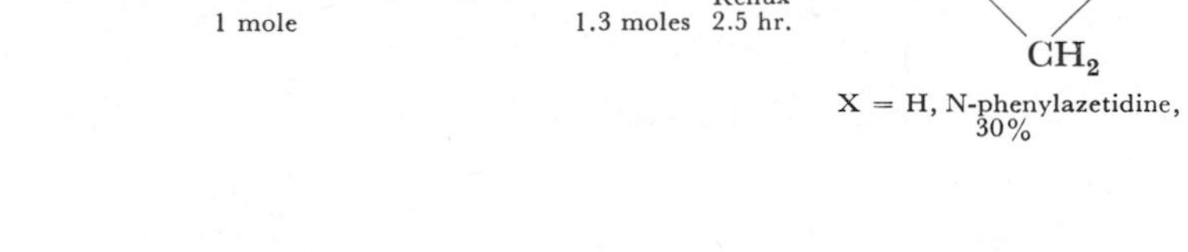

Despite the low yields, this route is the best at present for the preparation of N-arylazetidines.

Lactones and amines, when heated, produce a syrup which is probably a mixture of hydroxyamide and amino acid:

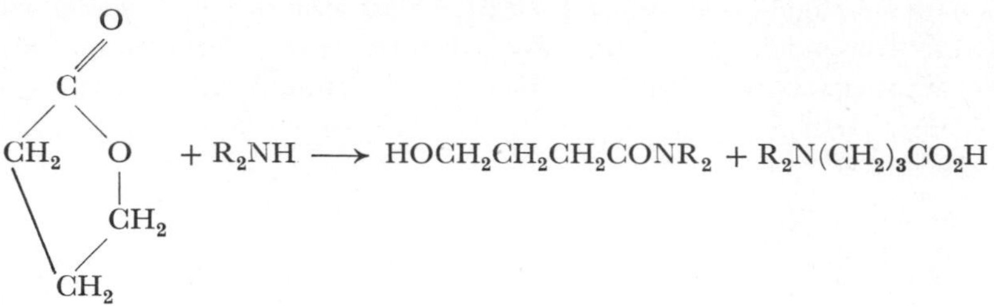

O
C
CH2
O
CH2
CH2
+ R2NH ⟶ HOCH2CH2CH2CONR2 + R2N(CH2)3CO2H

The mixture can be reduced to the aminoalcohol with lithium aluminum hydride (see Ex. b.2).

(a) Preparation of phenyl 4-pyridylamine. Phenyl 4-pyridyl ether, 1.71 g., was heated under reflux with 2.6 g. of aniline hydrochloride for 3 hr. at 180°. After acidification the phenol formed was removed by distillation with steam. On making the solution alkaline, the amine separated (more may be recovered by steam-distilling the aqueous layer). Total yield, 89%; m.p. 172° when crystallized from methanol-water [82].

(b) Other examples

(1) 7-Dimethylamino-*vic*-triazolo(d)pyrimidine (85% from 7-

H
N
N
N
N
N
SCH3
H
N
N
N
N
N
N(CH3)2

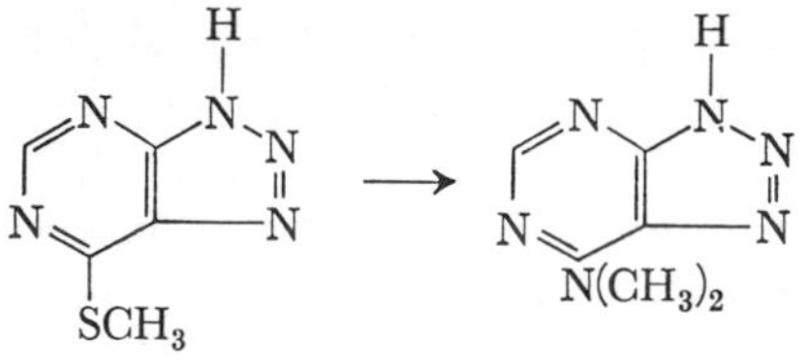

methylthio-*vic*-triazolo(d)pyrimidine and aqueous dimethylamine refluxed for 3 hr.) [84].

(2) 4-Dibutylamino-1-butanol (62% from butyrolactone and dibutylamine at 150° for 4 hr. followed by reduction of the reaction mixture with lithium aluminum hydride) [89].

(3) N-Aminophthalimide (88% from *O*-2,4-dinitrophenylhydroxyl-

O
C
N Na
C
O
+ NO2
NO2
O—NH2
O
C
N NH2
C
O

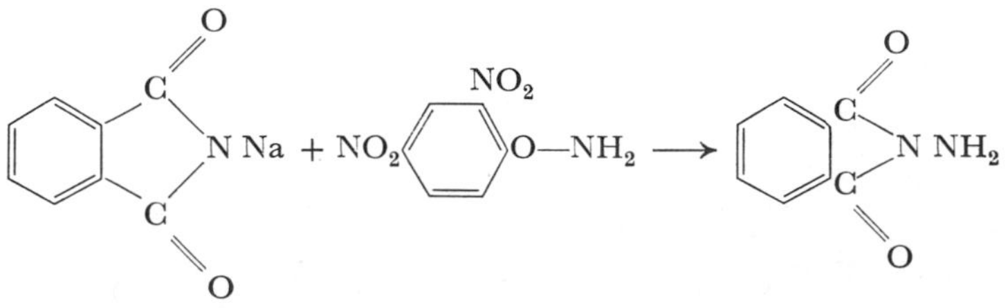

amine and sodium phthalimide heated in DMF) [90].

7. From Alcohols and Ammonia or Amines

$$ROH + NH_3 \xrightarrow{Al_2O_3} RNH_2 + H_2O$$

This synthesis, which is largely of industrial interest, has been summarized [91]. The literature, mostly patents, is somewhat confusing because the results do not always appear to be consistent, perhaps because of the great number of variables involved. With ammonia and an alcohol passed over a catalyst at high temperature or with the reactants and catalyst in an autoclave, primary, secondary, and tertiary amines are obtained. Excess of alcohol leads to a higher percentage of the tertiary amine while an excess of ammonia leads to a higher percentage of the primary amine. The catalysts employed are largely of two types: oxides such as those of aluminum, thorium, silicon, tungsten, magnesium, or chromium; hydrogenation catalysts such as copper, nickel, cobalt, or platinum. Aluminum oxide is the most popular of the oxide catalysts, and the hydrogenation catalysts in the presence of hydrogen are frequently employed. The methylamines have been produced commercially since 1920 from methanol and ammonia using an alumina catalyst under pressure [92].

In the laboratory the synthesis has been of greatest value for the alkylation of liquid amines such as aniline, in that operation at the boiling point of the re-action mixture and at normal pressure is possible. In this manner, using Raney nickel as a catalyst, aniline has been alkylated to give secondary amines with yields varying from 41 to 83% [93] (see Ex. a). The tertiary amine, ethyl-methylheptylamine, has also been produced from methylheptylamine in this manner [94].

Another more specific laboratory preparation is the benzylation of amines with benzyl alcohol and sometimes alcohol in the presence of base, in which case the reaction probably proceeds via the Schiff base [95] (see Ex. c):

$$C_6H_5CH_2OH + OH^\ominus \longrightarrow C_6H_5CHO + H_2O + [H:^\ominus] \xrightarrow{ArNH_2}$$

$$C_6H_5CH=NAr \xrightarrow{[H^\ominus]} C_6H_5CH_2-\overset{\ominus}{N}Ar \xrightarrow{H_2O} C_6H_5CH_2NHAr + OH^\ominus$$

The hydride ion is probably transferred directly to the Schiff base, as indicated by the fact that traces of benzaldehyde, which forms the Schiff base, accelerate the benzylation process.

(a) Preparation of N,N'-diethylbenzidine (60–67% from benzidine and ethanol refluxed for 15 hr. in the presence of Raney nickel) [96].

(b) Preparation of 5-piperidino-4-octanone   C$_3$H$_7$COCHC$_3$H$_7$

(88% from 0.1 mole of butyroin and 0.2 mole of piperidine refluxed in 50 ml. of benzene with azeotropic removal of water, a general reaction for acyloins and secondary amines) [97].

(c) **Preparation of 2-benzylaminopyridine** (98–99 % from 2-amino-pyridine, benzyl alcohol, and 85 % potassium hydroxide heated from 182 to 250° for 30 min. and then for 3 min. more at 250°) [98].

8. From Primary Amines and Trialkyl orthoformates

$$\text{ArNH}_2 \xrightarrow[\text{H}_2\text{SO}_4]{\text{(C}_2\text{H}_5\text{O)}_3\text{CH}} \overset{\overset{\displaystyle \text{C}_2\text{H}_5}{|}}{\text{ArNCHO}} \xrightarrow{\text{H}^\oplus} \text{ArNHC}_2\text{H}_5$$

Primary aromatic amines may be alkylated by treatment with a trialkyl orthoformate in the presence of sulfuric acid followed by hydrolysis [99]. The first product, N-alkylformanilide, results at temperatures above 140°. By this procedure a series of aromatic amines has been alkylated with yields varying from 44 to 78 %.

(a) **Preparation of N-ethyl-p-chloroaniline** (80–86 % of N-ethyl-p-chloroformanilide from p-chloroaniline, triethyl orthoformate, and conc. sulfuric acid heated first to 115–120° and finally to 175–180°; 87–92 % of the amine by acid hydrolysis of N-ethyl-p-chloroformanilide) [100].

9. From Organometallics or Carbanions and Amine Derivatives

$$\text{RMgX} + \text{ClNH}_2 \rightarrow \text{RNH}_2 + \text{MgClX}$$

The average yield of this reaction is about 50–60 % [101 and earlier papers] but it can be improved in isolated instances by using dialkylmagnesium and bringing the reagents together at −60°. An aromatic Grignard reagent tends to cleave in the opposite manner with N-chloroamines [102]:

$$\text{ArMgX} + \text{R}_2\text{NCl} \rightarrow \text{ArCl} + \text{R}_2\text{NMgX}$$

Useful preparations of α-amino acids and α-aminoketones have been devised by using the principle of anionic displacement from N-chloroiminoesters [103] and N,N-dichloroamines [104]:

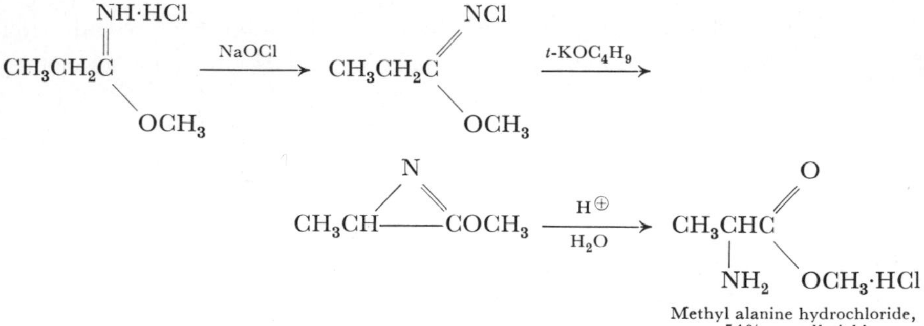

Methyl alanine hydrochloride, 54% overall yield

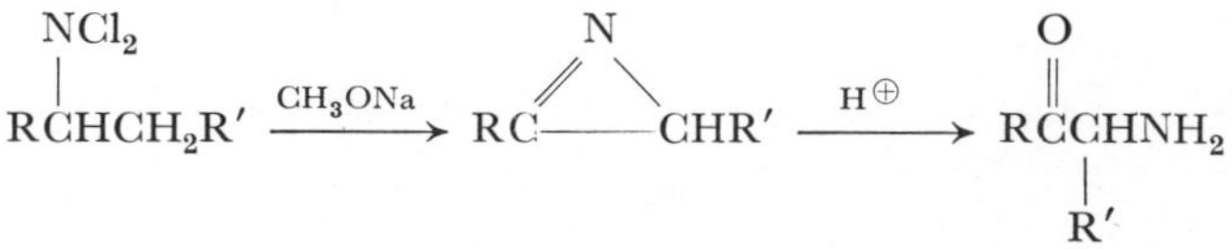

Obviously other amine derivatives can be utilized in the same manner as above:

$$2\,RMgX + CH_3ONH_2 \xrightarrow[\text{2. Water}]{\text{1. Ether}} RNH_2 + MgXOCH_3 \qquad \text{(see Ex. a)}$$

$$RMgX + R_2'NCH_2OR'' \longrightarrow R_2'NCH_2R + MgXOR'' \qquad \text{(see Ex. b.1)}$$

Aromatic aldehydes may be substituted for formaldehyde in the preparation of the aminoether,

$$\overset{\displaystyle Ar}{\underset{\displaystyle R_2'NCHOR''}{\big|}}$$

and these compounds used with Grignard reagents [105]. t-Amines with a vinyl group have been synthesized from a vinyl Grignard reagent according to the equation above [106]. A third amino derivative used is the α-amino-acetonitrile:

$$RMgX + R_2NCR_2'CN \rightarrow R_2NCR_2'R + MgXCN \qquad \text{(see Ex. b.2)}$$

Side reactions which may be troublesome [107] are additions to the nitrile:

$$RMgX + R_2NCR_2'CN \xrightarrow{H_2O} R_2NCR_2'\overset{\displaystyle O}{\overset{\displaystyle \|}{C}}R \;\; + MgXNH_2$$

and coupling:

$$RMgX + R_2NCR_2'CN \xrightarrow{H_2O} \underset{\underset{\displaystyle NR_2}{\big|}}{R_2'C}\text{——}\underset{\underset{\displaystyle NR_2}{\big|}}{CR_2'} + MgXOH + HCN$$

But high yields of the normal displacement product are often obtained; for example, 83% 3-diethylaminohexane or 90% 1-phenyl-2-piperidinobutane [108].

β-Chloroethylamines also react with Grignard reagents to give valuable aminoethyl derivatives, although yields are admittedly low [109]:

$$C_6H_5MgBr + ClCH_2CH_2N(CH_3)_2 \xrightarrow[\text{Reflux}]{C_6H_5CH_3} C_6H_5CH_2CH_2N(CH_3)_2$$

β-Dimethylaminoethylbenzene,
13.4%

Furthermore, boranes may react with chloramines (or hydroxylamine-O-sulfonic acids [110] to form amines (see Ex. b.3):

$$R_3B + 3\,NH_2Cl \;(\text{or}\; NH_2OSO_3H) \rightarrow 3\,RNH_2$$

Essentially this synthesis is a method of conversion of olefins into amines since the boranes are made directly from olefins and diborane.

(a) **Preparation of t-butylamine** (70% from t-butylmagnesium chloride to which methoxyamine in ether was added slowly at -10 to $-15°$ and then refluxed for 2 hr.) [111].

(b) Other examples

(1) **α,β-Diphenyltriethylamine** (92 % from 0.15 mole of benzyl-magnesium chloride to which 0.1 mole of α-diethylaminobenzyl butyl ether in ether was slowly added under nitrogen) [105].

(2) **Tridecyldiethylamine** (41 % from 0.25 mole of dodecylmagnesium chloride to which 0.25 mole of diethylaminoacetonitrile in ether was added slowly and the mixture allowed to stand overnight at room temperature; the subsequent decomposition with cold dilute acid should be carried out in the hood because of hydrogen cyanide evolution) [112].

(3) **2-Phenyl-1-aminopropane** (51.5 % from 0.1 mole of α-methyl-styrene and 33.3 ml. of 1.0 M borane in THF; after standing 1 hr., 3 ml. of

$$C_6H_5\underset{\underset{CH_3}{|}}{C}=CH_2 \xrightarrow{BH_3} (C_6H_5\underset{\underset{CH_3}{|}}{C}HCH_2)_3B \xrightarrow{ClNH_2} C_6H_5\underset{\underset{CH_3}{|}}{C}HCH_2NH_2$$

H$_2$O was added (to destroy excess borane) and then 50 ml. of 3 M sodium hydroxide; chloramine (0.067 mole) solution was added, and the mixture was stirred at 25° for 1 hr., cautiously treated with acid and made strongly basic.) [113].

1. Houben-Weyl, *Methoden der Organischen Chemie*, G. Thieme Verlag, Stuttgart, 1957, 4th ed., Vol. **11**, Pt. 1, Chap. 2.
2. J. v. Braun, *Chem. Ber.*, **70**, 979 (1937).
3. G. F. Hennion and R. S. Hanzel, *J. Am. Chem. Soc.*, **82**, 4908 (1960).
4. O. Westphal and D. Jerchel, *Chem. Ber.*, **73B**, 1002 (1940).
5. J. v. Braun and R. Klar, *Chem. Ber.*, **73B**, 1417 (1940).
6. H. A. Lubs, *The Chemistry of Synthetic Dyes and Pigments*, Reinhold Publishing Corp., New York, 1955, p. 38.
7. T. N. Kurdyumova and L. E. Gordeeva, *J. Gen. Chem.*, USSR (*Eng. Transl.*), **31**, 1456 (1961).
8. F. B. Wells and C. F. H. Allen, *Org. Syn.*, Coll. Vol. **2**, 221 (1943).
9. A. Albert and B. Ritchie, *Org. Syn.*, Coll. Vol. **3**, 53 (1955).
10. O. V. Schickl *et al.*, *Chem. Ber.*, **69**, 2593 (1936).
11. R. C. Elderfield, *Heterocyclic Compounds*, John Wiley and Sons, New York, 1952, Vol. 4, p. 123.
12. J. M. Orten and R. M. Hill, *Org. Syn.*, Coll. Vol. **1**, 300 (1941).
13. W. C. Tobie and G. B. Ayres, *Org. Syn.*, Coll. Vol. **1**, 23 (1941).
14. C. S. Marvel and V. du Vigneaud, *Org. Syn.*, Coll. Vol. **1**, 48 (1941).
15. J. C. Eck and C. S. Marvel, *Org. Syn.*, Coll. Vol. **2**, 374 (1943).
16. C. S. Marvel, *Org. Syn.*, Coll. Vol. **3**, 495 (1955).
17. C. S. Marvel, *Org. Syn.*, Coll. Vol. **3**, 523, 848 (1955).
18. E. N. Safonova and V. M. Belikov, *Russ. Chem. Rev.* (*English Transl.*), **36**, 375 (1967).
19. N. D. Cheronis and K. H. Spitz-Mueller, *J. Org. Chem.*, **6**, 349 (1941).
20. H. H. Sisler and N. D. Cheronis, *J. Org. Chem.*, **6**, 467 (1941).
21. F. G. Willson and T. S. Wheeler, *Org. Syn.*, Coll. Vol. **1**, 102 (1941).
22. W. J. Hickinbottom, *J. Chem. Soc.*, 992 (1930).
23. E. C. Hughes *et al.*, *Ind. Eng. Chem.*, **42**, 787 (1950).
24. N. Bortnick *et al.*, *J. Am. Chem. Soc.*, **78**, 4039 (1956).
25. F. F. Blicke *et al.*, *J. Am. Chem. Soc.*, **61**, 91, 93, 771, 774 (1939).
26. E. T. Borrows *et al.*, *J. Chem. Soc.*, 197 (1947).
27. C. G. Overberger *et al.*, *Org. Syn.*, Coll. Vol. **4**, 336 (1963).
28. F. D. Hager, *Org. Syn.*, Coll. Vol. **1**, 544 (1941).
29. R. C. Fuson *et al.*, *J. Am. Chem. Soc.*, **69**, 2961 (1947).

30. S. R. Ross, *J. Am. Chem. Soc.*, **69,** 2982 (1947).
31. R. H. Reitsema, *J. Am. Chem. Soc.*, **71,** 2041 (1949).
32. R. B. Woodward and W. E. Doering, *Org. Syn.*, Coll. Vol. **3,** 50 (1955).
33. J. S. Buck and C. W. Ferry, *Org. Syn.*, Coll. Vol. **2,** 290 (1943).
34. C. F. H. Allen and G. H. W. McKee, *Org. Syn.*, Coll. Vol. **2,** 15 (1943).
35. R. N. Shreve *et al.*, *Ind. Eng. Chem.*, **29,** 1361 (1937); **33,** 218 (1941).
36. C. R. Hauser and M. J. Weiss, *J. Org. Chem.*, **14,** 310 (1949).
37. H. Gilman and J. F. Nobis, *J. Am. Chem. Soc.*, **67,** 1479 (1945).
38. W. H. Puterbaugh and C. R. Hauser, *J. Org. Chem.*, **24,** 416 (1959).
39. W. G. Young *et al.*, *J. Am. Chem. Soc.*, **82,** 6163 (1960).
40. C. C. Howard and W. Marckwald, *Chem. Ber.*, **32,** 2031 (1899).
41. H. Heaney, *Chem. Rev.*, **62,** 81 (1962); J. F. Bunnett, *Quarterly Rev.*, **12,** 1 (1958).
42. H. Gilman and S. Avakian, *J. Am. Chem. Soc.*, **67,** 349 (1945).
43. J. F. Bunnctt and T. K. Brotherton, *J. Org. Chem.*, **22,** 832 (1957).
44. B. R. Baker and M. V. Querry, *J. Org. Chem.*, **15,** 413 (1950).
45. J. H. Pierce *et al.*, *J. Chem. Soc.*, 694 (1955).
46. J. L. Pinkus *et al.*, *J. Org. Chem.*, **27,** 4356 (1962).
47. F. Cramer *et al.*, *Chem. Ber.*, **92,** 384 (1959).
48. G. S. Skinner *et al.*, *J. Am. Chem. Soc.*, **80,** 3788 (1958).
49. T. Suami and H. Sano, *Tetrahedron Letters*, 2655 (1968).
50. J. O. Edwards and R. G. Pearson, *J. Am. Chem. Soc.*, **84,** 16 (1962).
51. H. E. Smith *et al.*, *J. Org. Chem.*, **31,** 684 (1966).
52. H. Hellmann *et al.*, *Chem. Ber.*, **89,** 2433 (1956).
53. H. H. Sisler *et al.*, *J. Org. Chem.*, **24,** 859 (1959).
54. H. H. Sisler *et al.*, *J. Org. Chem.*, **26,** 1819 (1961).
55. E. H. White and C. A. Elliger, *J. Am. Chem. Soc.*, **87,** 5261 (1965).
56. J. H. Billman *et al.*, *J. Am. Chem. Soc.*, **64,** 2977 (1942).
57. W. Gerrard and G. S. Jeacocke, *Chem. Ind.* (*London*), 1538 (1954).
58. A. G. Anderson, Jr., and M. T. Wills, *J. Org. Chem.*, **33,** 2123 (1968).
59. R. A. Chambers, Ph.D. Thesis, Vanderbilt University, 1962.
60. C. F. H. Allen *et al.*, *Org. Syn.*, Coll. Vol. **4,** 433 (1963).
61. Ref. 1, p. 160.
62. Ref. 1, pp. 164–165.
63. D. E. Pearson *et al.*, Ref. 11, 1957, Vol. 6, p. 685.
64. U. Hollstein, *J. Heterocycl. Chem.*, **5,** 299 (1968).
65. I. A. Krasavin *et al.*, *Metody Polucheniya Khim. Reactivov i Preparatov Gos. Kom. Sov. Min. SSSR po Khim*, No. 7, 5 (1963); *C.A.*, **61,** 3070 (1964); confirmed by D. E. Pearson.
66. R. A. Robinson, *J. Am. Chem. Soc.*, **69,** 1942 (1947).
67. N. N. Woroshtzow and J. M. Kogan, *Chem. Ber.*, **65,** 142 (1932).
68. Ref. 1, p. 250.
69. N. L. Drake, *Org. Reactions*, **1,** 105 (1942).
70. W. H. Hartung *et al.*, *J. Am. Chem. Soc.*, **63,** 507 (1941).
71. C. F. H. Allen and A. Bell, *Org. Syn.*, Coll. Vol. **3,** 78 (1955).
72. D. L. Ross and J. J. Chang, *J. Org. Chem.*, **29,** 1180 (1964).
73. S. J. Angyal, *Org. Reactions*, **8,** 204 (1954).
74. A. Galat and G. B. Elion, *J. Am. Chem. Soc.*, **61,** 3585 (1939).
75. J. Graymore, *J. Chem. Soc.*, 1116 (1947).
76. M. C. Rebstock *et al.*, *J. Am. Chem. Soc.*, **77,** 24 (1955).
77. I. Marszak and M. Koulkes, *Bull. Soc. Chim. France*, 93 (1956).
78. F. F. Blicke and W. M. Lilienfeld, *J. Am. Chem. Soc.*, **65,** 2281 (1943).
79. A. Baniel *et al.*, *J. Org. Chem.*, **13,** 791 (1948).
80. N. L. Wendler, *J. Am. Chem. Soc.*, **71,** 375 (1949).
81. A. T. Bottini *et al.*, *Org. Syn.*, **43,** 6 (1963).
82. D. Jerchel and L. Jakob, *Chem. Ber.*, **91,** 1266 (1958).
83. G. B. Elion *et al.*, *J. Am. Chem. Soc.*, **78,** 217 (1956).

84. R. Weiss *et al.*, *J. Org. Chem.*, **25**, 765 (1960).

85. G. B. Elion *et al.*, *J. Am. Chem. Soc.*, **74**, 411 (1952).

86. R. M. Cresswell and T. Strauss, *J. Org. Chem.*, **28**, 2563 (1963).

87. P. F. Butskus and G. I. Denis, *J. Gen. Chem. USSR (Eng. Transl.)*, **30**, 1350 (1960).

88. L. W. Deady and R. E. J. Hutchinson *et al.*, *Tetrahedron Letters*, 1773 (1968).

89. C. D. Lunsford *et al.*, *J. Org. Chem.*, **22**, 1225 (1957).

90. T. Sheradsky, *Tetrahedron Letters*, 1909 (1968).

91. L. Spialter and J. A. Pappalardo, *The Acyclic Aliphatic Tertiary Amines*, The Macmillan Company, New York, 1965, p. 29; V. A. Nekrasova and N. I. Shuikin, *Russ. Chem. Rev.*, **34**, 843 (1965).

92. R. Williams, Jr., *et al.*, *Chem. Eng. News*, **33**, 3982 (1955).

93. R. G. Rice and E. J. Kohn, *J. Am. Chem. Soc.*, **77**, 4052 (1955).

94. N. J. Leonard and W. K. Musker, *J. Am. Chem. Soc.*, **81**, 5631 (1959).

95. Y. Sprinzak, *J. Am. Chem. Soc.*, **78**, 3207 (1956).

96. R. G. Rice and E. J. Kohn, *Org. Syn.*, Coll. Vol. **4**, 283 (1963).

97. P. Klemmensen *et al.*, *Ark. Kemi*, **28**, 405 (1968).

98. Y. Sprinzak, *Org. Syn.*, Coll. Vol. **4**, 91 (1963).

99. R. M. Roberts and P. J. Vogt, *J. Am. Chem. Soc.*, **78**, 4778 (1956).

100. R. M. Roberts and P. J. Vogt, *Org. Syn.*, Coll. Vol. **4**, 420 (1963).

101. G. H. Coleman *et al.*, *J. Am. Chem. Soc.*, **63**, 1692 (1941).

102. R. J. W. LeFévre, *J. Chem. Soc.*, 1745 (1932).

103. H. E. Baumgarten *et al.*, *J. Am. Chem. Soc.*, **82**, 4422 (1960); **82**, 459 (1960).

104. H. E. Baumgarten and J. M. Peterson, *Org. Syn.*, **41**, 82 (1961).

105. A. T. Stewart, Jr., and C. R. Hauser, *J. Am. Chem. Soc.*, **77**, 1098 (1955).

106. J. Ficini and H. Normant, *Bull. Soc. Chim. France*, 1454 (1957).

107. L. H. Goodson and H. Christopher, *J. Am. Chem. Soc.*, **72**, 358 (1950).

108. P. Bruylants, *Bull. Soc. Chim. Belg.*, **33**, 467 (1924).

109. P. M. G. Bavin *et al.*, *J. Med. Chem.*, **9**, 790 (1966).

110. H. C. Brown *et al.*, *J. Am. Chem. Soc.*, **88**, 2870 (1966).

111. R. Brown and W. E. Jones, *J. Chem. Soc.*, 781 (1946).

112. O. Westphal, *Chem. Ber.* **74B**, 1365 (1941).

113. H. C. Brown *et al.*, *J. Am. Chem. Soc.*, **86**, 3565 (1964).

D. Additions (Mainly Nucleophilic)

This section probes the scope of addition of amines to unsaturated compounds (D.1) and carbonyl compounds (D.2), and includes the fascinating versatility of formaldehyde-amine adducts in Mannich-type reactions (D.3 and D.4); it then proceeds to a consideration of the addition of amines to conjugated combinations of the above types (D.6 and D.7). Additions to epoxides and ethylene imines are discussed also (D.5). Noteworthy developments are the use of tetrakis(dimethylamino)-titanium to prepare *gem*-diamines or enamines (D.2) and capitalization on the reversibility of acrylonitrile-amine additions to prepare pure secondary amines (D.7, Ex. b.1}. Sections E and F also include discussions of additions, the former of the organometallic type and the latter of electrophilic and free-radical types.

1. From Unsaturated Compounds and Amines

$$\underset{/}{\overset{\backslash}{C}}=\underset{\backslash}{\overset{/}{C}} + RNH_2 \longrightarrow \underset{/}{\overset{\backslash}{HC}}-\underset{\backslash}{\overset{/}{C}}-NHR$$

References for Section D are on pp. 471–473.

Addition of amines to unsaturated compounds is difficult unless electron-withdrawing groups are attached to the olefin. Nevertheless, at high temperatures and by the use of various catalysts, simple olefins and amines do add [1] (see Ex. b.2). Styrenes and amines add by means of a sodium metal catalyst which forms the sodamide (see Ex. a). The addition of amines to acetylenes, the catalyst for which is usually the copper acetylide derivative, occurs more readily. The process is called vinylation if the addition stops at the first stage, but a second molecule of acetylene can add (see Ex. b.1):

$$CuC\equiv CCu + R_2NH \longrightarrow CH_2{=}CHNR_2 \xrightarrow{CuC\equiv CCu} \begin{array}{c} C\equiv CH \\ | \\ CH_3{-}CHNR_2 \end{array}$$

(a) **Preparation of N-(β-phenylethyl)-aniline** (65–75% from 3 g. of sodium metal and 2 moles of aniline first heated to reflux (traces of ferrous sulfate hasten the reaction), and then 1.5 moles of styrene were added while the temperature was maintained at 186–196° for 2 hr.) [2].

(b) **Other preparations**

(1) **3-Morpholino-1,4-diphenylbutyne-1** (12% after separation from

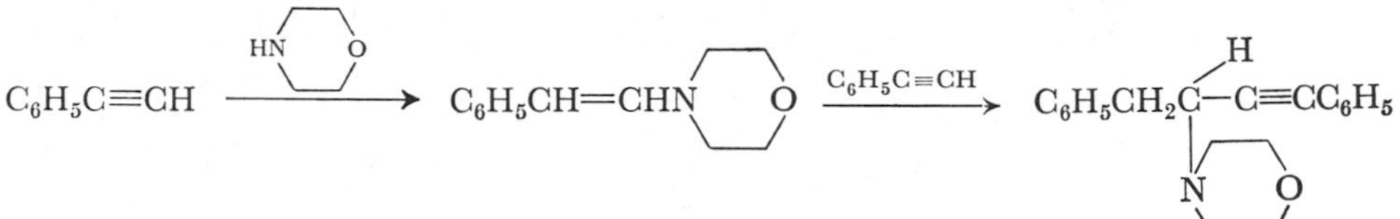

isomeric and other side products from 40 g. of phenylacetylene, 16 g. of morpholine and 2 g. of cuprous chloride heated slowly until an exothermic reaction began at 98° at which point it was maintained for 30 min.) [3].

(2) **N-Ethylpiperidine** (77–83% from 4 moles of piperidine, 5 g. of pyridine, and 4.4 g. of sodium heated briefly with high-speed stirring under nitrogen and then transferred to an autoclave pressurized with ethylene and stirred at 100°) [4].

2. From Carbonyl Compounds or Acid Derivatives and Amines

$$2 R_2NH + CH_2O \rightarrow R_2NCH_2NR_2 + H_2O$$

The condensation of secondary amines with aldehydes alone or with a catalyst such as potassium carbonate occurs readily to give methylene diamines in good yield [5–8]. If higher aliphatic aldehydes containing α-hydrogen atoms and secondary amines are employed, enamines are usually obtained instead. Since certain methylene diamines decompose to give enamines on distillation, it has been inferred that the former are intermediates in the preparation of the latter [9]:

$$(R_2N)_2CHCH_2R' \rightarrow R_2NCH{=}CHR' + R_2NH$$

However, more recent experiments [10] with 1:1 ratios of amine and aldehyde which gave over 50% yield of the enamine suggest that the amino alcohol is also a possible intermediate:

$$C_6H_5CH_2CHO + (CH_3)_2NH \xrightarrow{\text{KOH}}$$

$$[C_6H_5CH_2CHOHN(CH_3)_2] \longrightarrow C_6H_5CH{=}CHN(CH_3)_2$$

β-Dimethylaminostyrene 75%

With primary 1,3-diamines and aldehydes a similar reaction occurs in that a cyclic amine, a hexahydropyrimidine, is formed [11]:

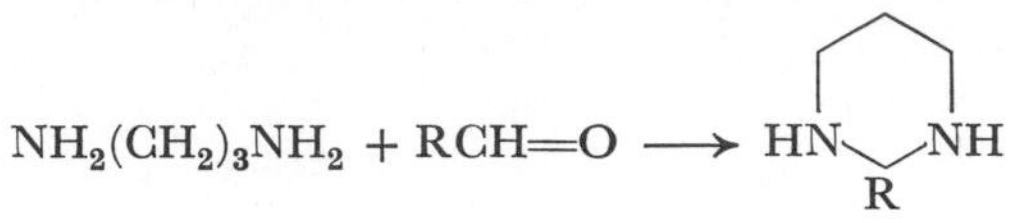

$$NH_2(CH_2)_3NH_2 + RCH{=}O \longrightarrow HN\underset{R}{\qquad}NH$$

The scope of the syntheses above has been broadened recently by means of the reagent tetrakis(dimethylamino)-titanium. Not only diketones react to form enamines [12, 13] but also acid derivatives of any type, but preferably the dialkyl amides, react to form methine triamines (see Ex. b.3):

$$HCON(CH_3)_2 + Ti[N(CH_3)_2]_4 \rightarrow HC[N(CH_3)_2]_3$$

Tris(dimethylamino)
methane

Recently hydrazine and sulfur, which in combination reduce nitro compounds to amines, have been shown to give dihydrotetrazines with nitriles [14]:

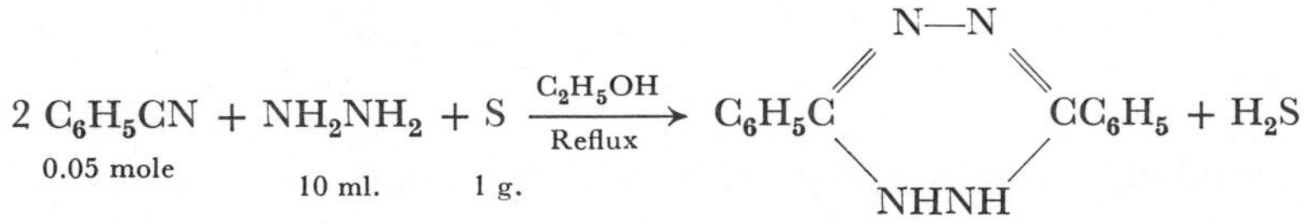

$$2\ C_6H_5CN + NH_2NH_2 + S \xrightarrow[\text{Reflux}]{C_2H_5OH} C_6H_5C\underset{NHNH}{\overset{N-N}{\diagdown\diagup}}CC_6H_5 + H_2S$$

0.05 mole   10 ml.   1 g.

1,2-Dihydro-3,6-diphenyl-1,2,4,5-tetrazine, 82%

Other nitriles give yields varying from 76 to 94%.

(a) Preparation of benzylidene-bis-dimethylamine. Benzaldehyde, 106 g., and 400 g. of 25% aqueous dimethylamine were swirled occasionally while warming the mixture for 10 min. on a steam cone. On cooling, the aqueous layer was saturated with potassium carbonate, after which the upper layer was taken up in 100 ml. of benzene. Drying and distillation of the benzene layer gave 143 g. (80%) of the methylenediamine, b.p. 57–60° (0.9 mm.) [6].

(b) Other examples

(1) 1-Morpholino-1-butene,

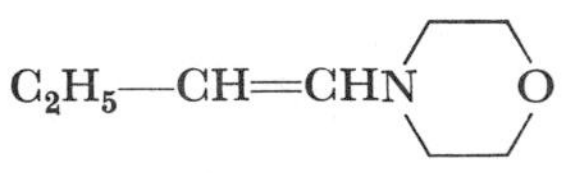

$$C_2H_5{-}CH{=}CHN\underset{\bigcirc}{\qquad}O$$

Morpholine ($\frac{1}{2}$ mole) and n-butyraldehyde ($\frac{1}{4}$ mole) were mixed in the presence of anhydrous potassium carbonate at 5–50°. Distillation of the oil gave a 95% yield of the titled compound [7].

(2) N,N,N',N'-Tetrabenzylmethylenediamine (quantitative by adding 0.05 mole of dibenzylamine to a cold aqueous solution of 0.025 mole of formaldehyde and recrystallizing the precipitate from isopropyl alcohol, m.p. 99–100°) [15].

(3) 1,1,4,4-Tetrakis(dimethylamino)-butadiene (58% from 0.03 mole of tetrakis(dimethylamino)-titanium in ether to which 0.02 mole of

$$\begin{array}{ccc} CH_2CO & & CH{=}C[N(CH_3)_2]_2 \\ | & \diagdown & | \\ & O \xrightarrow{\ Ti[N(CH_3)_2]_4\ } \\ | & \diagup & | \\ CH_2CO & & CH{=}C[N(CH_3)_2]_2 \end{array}$$

succinic anhydride was added and the mixture held at 25° for 60 hr.; tri-(dimethylamino)methane was made similarly from DMF in 83% yield) [13].

3. From Active Hydrogen Compounds, Formaldehyde, and Amines (Mannich)

$$-\overset{|}{\underset{|}{C}}H + CH_2O + HN(CH_3)_2 \xrightarrow{\ HCl\ } -\overset{|}{\underset{|}{C}}-CH_2N(CH_3)_2 \cdot HCl + H_2O$$

This synthesis, reviews of which are available [16, 17], has wide applicability to yield a variety of β-aminoethyl derivatives. The amines in the free base form are not particularly stable, tending to form olefins and polymers:

$$\diagup\hspace{-0.3em}\diagdown\hspace{-1.2em}{}_{CHCH_2N(CH_3)_2} \longrightarrow \diagup\hspace{-0.3em}\diagdown\hspace{-1.2em}{}_{C{=}CH_2} + (CH_3)_2NH$$

The synthesis is applicable to active hydrogen compounds such as aldehydes [18], ketones [19–22], acids [23], esters [24], nitroalkanes [25–28], phenols having *ortho* or *para* positions unoccupied [29], pyrrole compounds (see Ex. b.2), certain heterocyclic compounds with α- or γ-methyl groups [30–32], and acetylene compounds [33] (see Ex. b.5):

$$RC{\equiv}CH + CH_2O + R_2'NH \rightarrow RC{\equiv}CCH_2NR_2'$$

In the synthesis the active hydrogen is replaced by an aminomethyl or substituted aminomethyl group. The reaction may occur with ammonia, primary and secondary amines, aldehydes other than formaldehyde but, obviously, not with tertiary amines. Purer products are obtained from secondary amines since they possess only one replaceable hydrogen atom. To carry out the reaction the components are simply refluxed in an organic solvent, such as methanol, ethanol, isoamyl alcohol [34], nitrobenzene [35], and nitromethane [21], particularly if the aldehyde is not a liquid. By-products are sometimes formed, a fact which is not surprising in view of the reactive groups present in the components. But sometimes the by-products are cyclic types of interesting and useful structure [36]:

$$(CH_3)_2CHCH{=}O + CH_2O + CH_3NH_3Cl \longrightarrow (CH_3)_2CCH{=}O$$

$$\underset{\text{30–35\%}}{\overset{|}{CH_2NHCH_3}}$$

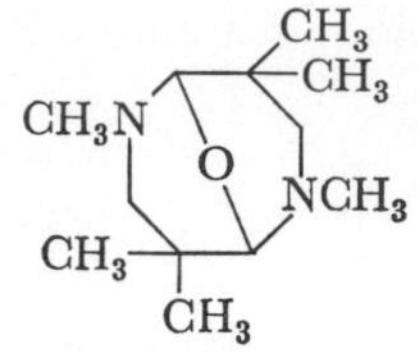

2,4,4,6,8,8-Hexamethyl-9-oxa-2,6-diazabicyclo[3.3.1]nonane, 40%

The latter may be reduced with lithium aluminum hydride to the diazacyclo-octane. Highly branched types do not respond to the Mannich reaction. Yields are variable.

The mechanism of the reaction has not been determined with certainty. It appears that preference is given to the path which involves first the formation of the immonium ion 8-XI [37]:

$$CH_2{=}O + R\overset{\oplus}{N}H_3 \rightarrow CH_2{=}\overset{\oplus}{N}HR + H_2O$$

$$\text{8-XI}$$

which then reacts with the active hydrogen compound (enolic form for alde-hydes or ketones) to give the amino derivative 8-XII:

$$\underset{\underset{O{-}H}{|}}{R{-}C{=}CH_2} + CH_2{=}\overset{\oplus}{N}HR \longrightarrow \underset{\underset{O}{\|}}{RC{-}CH_2CH_2NHR} + H^{\oplus}$$

$$\text{8-XI} \qquad\qquad \text{8-XII}$$

In respect to the intermediate 8-XI, its reactions with phenols and other aromatic nuclei could be classified as electrophilic substitution (Section F), but are discussed in this section because of their obvious relationship to condensation reactions. A further modification, known as the Tscherniac-Einhorn reaction, [38], is a Mannich condensation utilizing hydroxymethyl derivatives of amides or imides for substitution into aromatic rings, such as exist in phenols and anilines or even in hydrocarbons and benzoic acids. Since the positive charge in 8-XI should be as fully developed as possible, N-hydroxymethylamides of strong acids, the most common of which is chloroacetic acid [39] (see Ex. b.6), should be used for the less reactive nuclei:

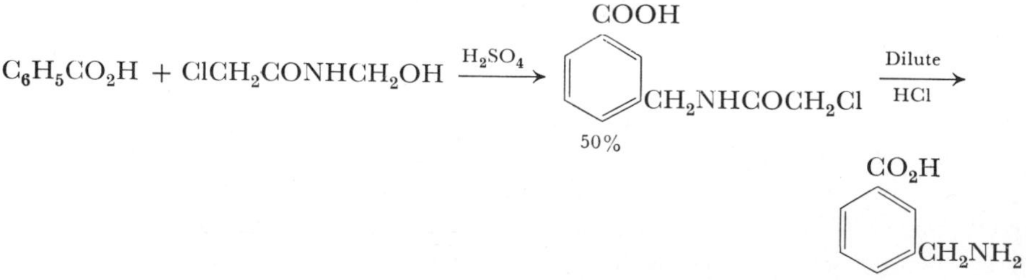

Benzylamine-3-carboxylic acid

But, for N-alkylolamides, stronger acids such as sulfuric acid, PPA, or hydrogen chloride are used. In this form the reaction has versatility and is capable of producing many complex structures as shown for one example [40]:

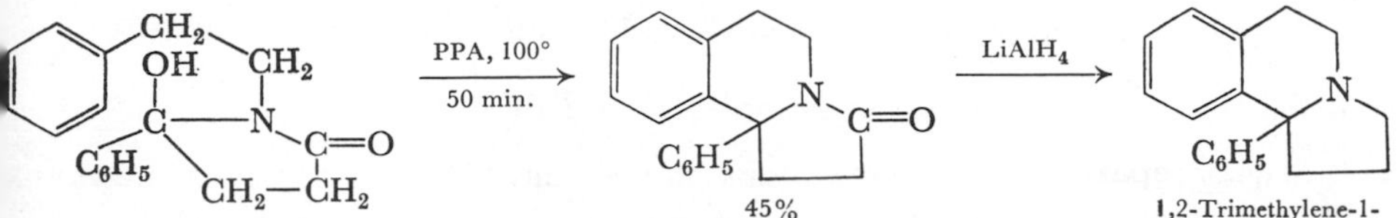

1,2-Trimethylene-1-phenyl-1,2,3,4-tetra-hydroisoquinoline, 89%

(a) Preparation of 2,4,6-tri-(dimethylaminomethyl)-phenol (86% from phenol, dimethylamine, and aqueous formaldehyde) [41].

(b) Other examples

(1) N-(2-Nitroisobutyl)-dimethylamine (74% from formaldehyde, 2-nitropropane, and dimethylamine) [26].

(2) 2-Dimethylaminomethylpyrrole (77% from formalin, pyrrole, and dimethylamine hydrochloride) [42].

(3) 1-Diethylamino-3-butanone (62–70% from diethylamine hydrochloride, paraformaldehyde, acetone, methanol, and a trace of conc. hydrochloric acid) [43].

(4) Diethylaminoacetonitrile (88–90% from diethylamine, sodium cyanide, sodium bisulfite, and 37–40% formaldehyde) [44].

(5) Di-(1-phenylpropargyl)-ethylamine (($C_6H_5C\equiv CCH_2$)$_2NC_2H_5$, 48% from 15 g. of phenylacetylene, 4.8 g. of paraformaldehyde, 6.5 g. of 49% ethylamine, and 2 g. of cuprous chloride in dioxane at 50–60° for 8 hr.) [45].

(6) 1-Aminomethyl-2-methoxynaphthalene. 2-Methoxynaphthalene, 31.6 g., and 30.2 g. of N -methylolbenzamide in 100 ml. of methanol were

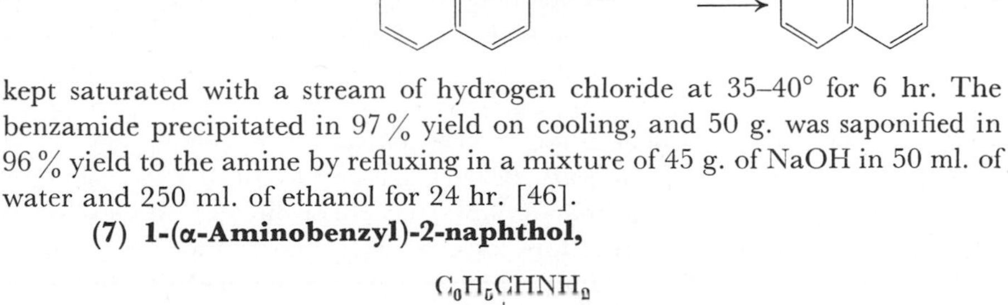

kept saturated with a stream of hydrogen chloride at 35–40° for 6 hr. The benzamide precipitated in 97% yield on cooling, and 50 g. was saponified in 96% yield to the amine by refluxing in a mixture of 45 g. of NaOH in 50 ml. of water and 250 ml. of ethanol for 24 hr. [46].

(7) 1-(α-Aminobenzyl)-2-naphthol,

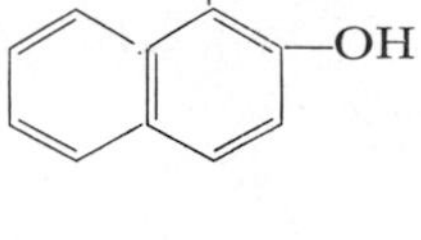

(51–62 % from 2-naphthol, 2 equiv. of benzaldehyde and alcohol saturated with ammonia followed by acid hydrolysis and then neutralization of the N-benzylidene derivative of the titled compound) [47].

4. From Alcohols, Paraformaldehyde, and a Secondary Amine

$$ROH + CH_2O + R_2'NH \rightarrow R_2'NCH_2OR$$

This synthesis is similar to the Mannich reaction (D.3) except that an alcohol is employed as the active hydrogen compound. The method has been utilized in the synthesis of a series of aminoethers [48], the yields varying from 40 to 82 %.

(a) Preparation of diethylaminomethyl ethyl ether (diethylamine, 73 g., 80 g. of 95 % ethanol, and 30 g. of paraformaldehyde were refluxed until the reaction ceased (ice bath was kept in readiness for possible cooling); anhydrous potassium carbonate was then added and the mixture was allowed to stand overnight; the filtrate was fractionated to give a rather constant-boiling portion which was dried over sodium and refractionated to yield 90 g. (69 %) of the aminoether, b.p. 132–134° (756 mm.) [48].

5. From Epoxides or Ethylenimines and Amines (Including Ammonia)

$$\begin{array}{c} CH_2\!\!-\!\!CH_2 \\ \diagdown \;\; \diagup \\ O \end{array} + NH_3 \longrightarrow HOCH_2CH_2NH_2$$

$$\begin{array}{c} CH_2\!\!-\!\!CH_2 \\ \diagdown \;\; \diagup \\ NH \end{array} + RNH_2 \xrightarrow{AlCl_3} NH_2CH_2CH_2NHR$$

Small heterocyclic rings containing oxygen or nitrogen may be opened with the addition of ammonia or an amine. Such additions for epoxides [49] and ethylenimines [50] have been discussed. For epoxides the ammonia addition is usually accomplished under pressure [51–54] but such conditions are less necessary for amines [55, 56]. The yields are much higher, sometimes in the 90 % range, when amines [55, 52, 53] are employed. This method lends itself to the synthesis of 2-dialkylaminoethanols [57, 58] and 1-dialkylamino-2-propanols [55] because of the availability of ethylene and propylene oxides, respectively. Other epoxides such as isobutylene oxide [59], styrene oxide [53], and stilbene oxide [54] respond similarly.

In the case of imines the addition to give diamines is accomplished in the presence of aluminum chloride [60]. Benzene as the solvent at 90° is used for secondary amines, while, for primary amines, tetralin or diphenyl at about 180° is employed. Yields by this procedure for a series of amines added to ethylenimines vary from 77 to 89 %. Similar reactions have been accomplished with ammonia and amines under pressure at 25–120° in the presence of ammonium chloride [61]. Amines give yields as high as 85 % by this procedure, although the maximum for ammonia was 68 %.

Tertiary amine salts also add to ethylene oxides to form an interesting series of quaternary salts [62]:

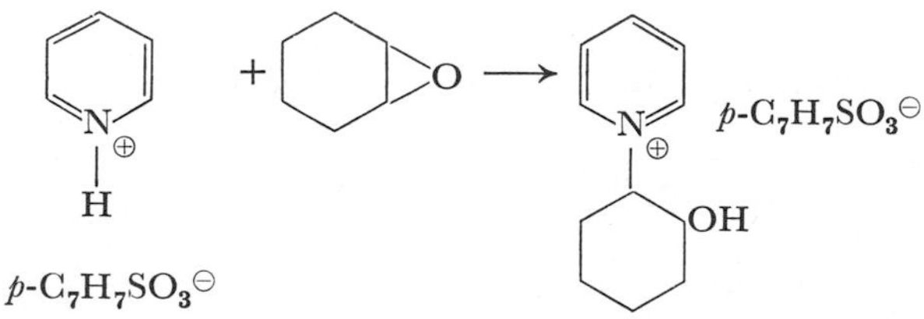

N-(2-Hydroxycyclohexyl)-
pyridinium tosylate,
98–100%

(a) Preparation of β-isopropylaminoethanol. Ethylene oxide, 76.8 g. (1.74 mole) was added to a mixture of 307 g. (5.20 moles) of isopropylamine, 18.0 g. (1.0 mole) of water, and 8.6 g. (about 0.1 mole) of conc. hydrochloric acid over a period of $3\frac{1}{2}$ hr. During this addition the temperature gradually rose to 51°. Refluxing for another 12 hr. gave 137 g. (76%) of the aminoalcohol, b.p. 169–171° [56].

(b) Preparation of N,N-di-*n*-butylethylenediamine (77–89% from a mixture of di-*n*-butylamine (0.9 mole), anhydrous aluminum chloride (0.675 mole), and 100 ml. of dry benzene to which 19.3 g. (0.45 mole) of ethylenimine was introduced over a 30 min. period; stirring for 30 min. followed after which 300 g. of solid KOH was added) [60].

6. From α,β-Unsaturated Aldehydes and Secondary Amines

$$R'CH_2CH{=}CHCHO + R_2NH \xrightarrow{K_2CO_3} R'CH_2\overset{\displaystyle H}{\underset{\displaystyle NR_2}{C}}{-}CH{=}CHNR_2 \; \rightleftharpoons$$

$$R'CH{=}CHCH{=}CHNR_2$$

α,β-Unsaturated aldehydes and secondary amines condense in the presence of a dehydrating agent such as anhydrous potassium carbonate [63] or preferably anhydrous magnesium sulfate [64] to yield ethylenic tertiary diamines and/or dienyl tertiary monoamines [65]. To obtain the diamines, low temperatures [−10 to about 20°] must be employed. In some cases the condensation in which potassium carbonate has been utilized has been modified by carrying out the final distillation in the presence of quinones [66, 67] or polycarboxylic acids [68]. The dienylmonoamines tend to polymerize on standing [69], while the unsaturated diamines darken with time [63]. Yields as a rule are only fair.

(a) Preparation of 1,3-bis(dimethylamino)-1-butene. To anhydrous magnesium sulfate, 120 g. (1 mole), and 115.6 g. (2.56 moles) of dimethylamine in 52 g. of methylcyclopentane at −10° was added, slowly with stirring, 60 g. (0.82 mole) of crotonaldehyde in 67 g. of methylcyclohexane. Then the temperature was allowed to increase to 20–30°, after which stirring was continued for 24 hr. After removing the magnesium sulfate, the solution was distilled to give a 92% yield of 1,3-bis(dimethylamino)-1-butene, b.p. 45–48° (5 mm.) [64].

(b) Preparation of 1-diethylamino-1,3-butadiene,

$$CH_2=CH-CH=CHN(C_2H_5)_2$$

Freshly distilled crotonaldehyde, 105 g., in 150 ml. of benzene was added during 20 min. to a mixture of 225 g. of diethylamine and 60 g. of anhydrous potassium carbonate, while the temperature was kept at −10 to −5°. During 1 hr. the mixture was kept at 0° with frequent swirling and then allowed to come to room temperature and to stand for 4 hr. After decanting the liquid from the potassium carbonate, 0.9 g. of phenanthraquinone was added. Distillation of this mixture under vacuum gave 123 g. of a yellow oil, b.p. 60–70° (12 Torr.). Redistillation produced 114 g. (61%) of 1-diethylamino-1,3-butadiene, b.p. 64–66° (10 Torr.) [66].

7. From α,β-Unsaturated Compounds

$$CH_2=CHY + RNH_2 \rightarrow RNHCH_2CH_2Y$$

$$Y = CN, \overset{O}{\overset{\|}{C}}R', \ CO_2R', \ \text{or} \ CO_2H$$

Addition of amines to unsaturated conjugated compounds takes place readily with perhaps the exception of aromatic amines which require catalysts (see Ex. a). Acrylonitrile is one of the best acceptors, and difficulty may be encountered in stopping at the monoadduct [70]:

$$RNHCH_2CH_2CN + CH_2=CHCN \rightarrow RN(CH_2CH_2CN)_2$$

The formation of the diadduct may be minimized by using an excess of amine. One of the interesting features of acrylonitrile addition is the reversibility of the addition, a fact which permits the β-cyanoethyl group to be used as a protecting group in producing mixed secondary amines (see Ex. b.1):

$$RNHCH_2CH_2CN + R'X \longrightarrow R\overset{R'}{\overset{|}{N}}CH_2CH_2CN \xrightarrow{250-275°}$$

$$RNHR' + CH_2=CHCN$$

The addition to unsaturated ketones has been reviewed [71]. In this case, addition often occurs at room temperature in the absence of pressure or a catalyst. Mesityl oxide, for example, combines with ammonia without the addition of heat to give diacetoneamine [72]:

$$CH_3COCH=C(CH_3)_2 + NH_3 \rightarrow CH_3COCH_2C(CH_3)_2NH_2$$

Diacetoneamine

α,β-Unsaturated aldehydes respond similarly (see D.6). α,β-Unsaturated esters such as those of acrylic provide a means for preparing N-substituted β-amino-propionic esters [73–76].

$$CH_2=CHCOOR + R'NH_2 \rightarrow R'NHCH_2CH_2COOR$$

Excess of the ester must be avoided to prevent the formation of the alkyl di-(carbalkoxyethyl)amine [77]:

$$2\ CH_2=CHCOOC_2H_5 + CH_3NH_2 \rightarrow CH_3N(CH_2CH_2COOC_2H_5)_2$$
Di-β-carbethoxyethylmethylamine

Photochemical addition of amines with secondary alkyl groups to unsaturated esters gives still another product [78]:

$$(CH_3)_2CHNH_2 + CH_3CH=CHCO_2C_2H_5 \xrightarrow[h\nu,\ 24\ hr.]{(C_6H_5)_2C=O}$$

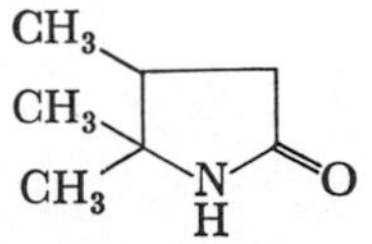

(Reagent and solvent)

2,2,3-Trimethylpyrrolidone, 52%

Additions to unsaturated acids proceed more slowly but nevertheless satisfactorily. By using a basic solvent such as pyridine or α-picoline the more reactive amines react with crotonic acid to give N-alkyl derivatives of β-amino-butyric acid with yields from 65 to 95% [79]. Nucleophilic addition to unsaturated acid esters also occurs smoothly [80] without an attack occurring at the ester group:

$$CH_3OOCCH=CHCOOH \xrightarrow{RNH_2} CH_3OOCCH_2\underset{\underset{NHR}{|}}{C}HCOOH$$

The yields of N-alkyl aspartic acid β-methyl esters obtained from monomethyl maleate varied from 55 to 85%. Although the earlier methods of addition of amines to α,β-unsaturated nitriles required pressure [81], catalysts such as cupric acetate monohydrate [82] have permitted the completion of reactions at ordinary pressure.

Two equivalents of hydroxylamine, one to add, the other to reduce, lead to β-amino acids from unsaturated acids [83]:

$$C_6H_5CH=CHCO_2H \xrightarrow{NH_2OH} C_6H_5\underset{\underset{NHOH}{|}}{C}HCH_2CO_2H \xrightarrow{NH_2OH}$$

$$C_6H_5\underset{\underset{NH_2}{|}}{C}HCH_2CO_2H + C_6H_5\overset{\overset{NOH}{||}}{C}CH_3$$

β-Amino-β-phenylpro-pionic acid, 34%

Acetophenone oxime
14%, side product

The hydroxylamine anion is a powerful nucleophile capable of attacking some aromatic systems to produce amino derivatives, as shown for two examples:

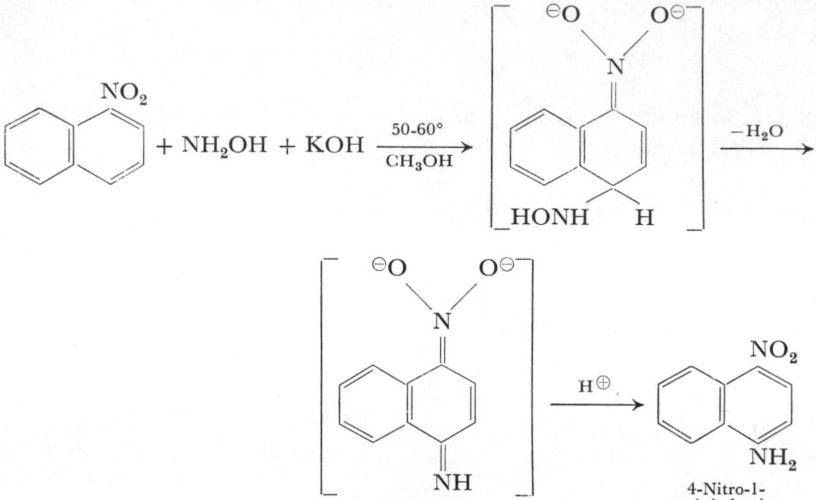

NO_2 + NH_2OH + KOH $\xrightarrow[\text{CH}_3\text{OH}]{50\text{-}60°}$ [... HONH   H] $\xrightarrow{-H_2O}$

$\xrightarrow{H^{\oplus}}$ [ref. 84]

4-Nitro-1-
naphthylamine,
55–60%

NO_2 + NH_2OH + KOH $\xrightarrow{C_2H_5OH}$ [ref. 85]

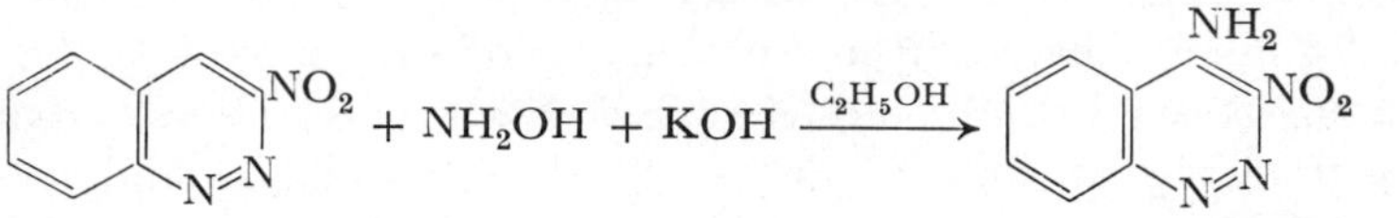

4-Amino-3-nitrocinnoline,
50–55%

(a) Preparation of 3-(o-chloroanilino)propionitrile (90–95 % based

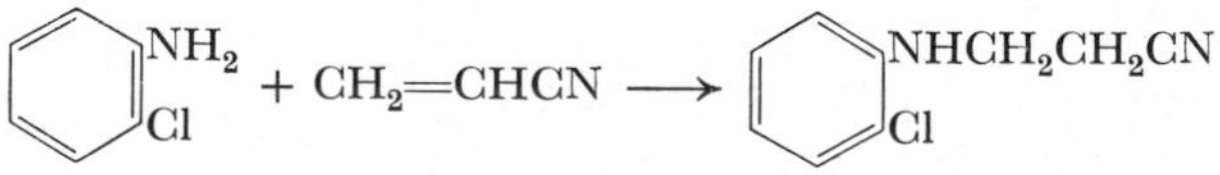

NH_2 + $CH_2{=}CHCN$ $\longrightarrow$ $NHCH_2CH_2CN$

on the amine from 2 moles of o-chloroaniline, 2 moles of acrylonitrile, and cupric acetate monohydrate refluxed for 3 hr. at 95–130°) [82].

(b) Other examples

(1) Dodecyl-β-hydroxyethylamine (0.2 equiv. each of dodecyl-β-

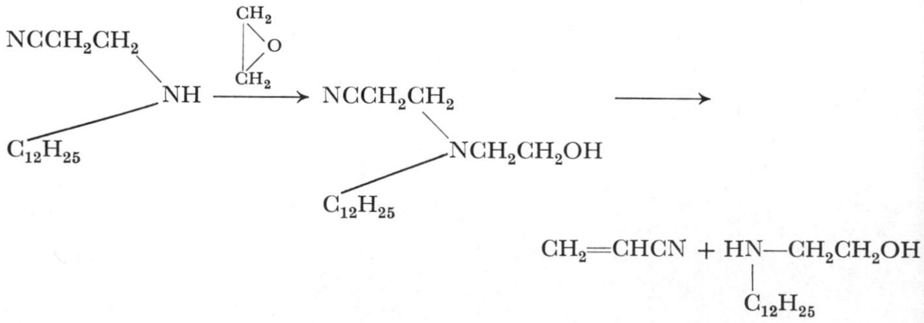

$NCCH_2CH_2$ NH $\xrightarrow{\text{CH}_2\text{-O-CH}_2}$ $NCCH_2CH_2$

$C_{12}H_{25}$ NCH_2CH_2OH

                                      $C_{12}H_{25}$

$\longrightarrow$

$CH_2{=}CHCN$ + $HN{-}CH_2CH_2OH$

                    $C_{12}H_{25}$

cyanoethylamine and ethylene oxide were mixed in 150 ml. of methanol containing 0.2 equiv. of water and held at 40–45° for 4 hr. and 50° for $2\frac{1}{2}$ hr.; dodecyl-β-hydroxyethyl-β-cyanoethylamine was isolated in 90% yield; the latter was heated at 250–275° under slightly reduced pressure to yield acrylonitrile and the desired product in 80% yield) [86].

(2) **N-Benzyl-β-aminobutyric acid** (77–86% from crotonic acid and benzylamine in pyridine heated $1\frac{1}{2}$ hr. at 120–130°) [79].

(3) **Methyl β-anilinopropionate** (75% from aniline, methyl acrylate, and acetic acid refluxed for 8 hr.) [76].

(4) **Di-β-carbethoxyethylmethylamine** (83–86% from 2.71–2.77

$$CH_3NH_2 + 2CH_2 \!\!=\!\! CHCOOC_2H_5 \rightarrow CH_3N(CH_2CH_2COOC_2H_5)_2$$

moles of methylamine and 5.4–5.5 moles of ethyl acrylate in ethanol for 6 days) [77].

(5) **Ethyl β-aminocrotonate** (70% from ethyl acetoacetate in dry ether at 0–5° through which ammonia was passed and the mixture was held at 25° for 24 hr. with Na_2SO_4 as a drying agent) [87].

(6) **3-Dimethylamino-1-propanol** (59–65% from 2 moles of allyl alcohol, 1 mole each of dimethylamine and solid NaOH in an autoclave at 115° for 20 hr.) [88].

1. Houben-Weyl, *Methoden der Organischen Chemie*, Georg Thieme Verlag, Stuttgart, 1957, 4th ed., Vol. **11**, Pt. 1, p. 267.
2. Ref. 1, p. 269.
3. J. D. Rose and R. A. Gale, *J. Chem. Soc.*, 792 (1949).
4. J. Wollensak and R. D. Closson, *Org. Syn.*, **43**, 45 (1963).
5. G. B. Butler, *J. Am. Chem. Soc.*, **78**, 482 (1956).
6. S. V. Lieberman, *J. Am. Chem. Soc.*, **77**, 1114 (1955).
7. P. L. deBenneville and J. H. Macartney, *J. Am. Chem. Soc.*, **72**, 3073 (1950).
8. L. Spialter and J. A. Pappalardo, *The Acyclic Aliphatic Tertiary Amines*, The Macmillan Co., New York, 1965, p. 52.
9. C. Mannich and H. Davidsen, *Chem. Ber.*, **69B**, 2106 (1936).
10. J. R. Geigy, Akt. Ges., British Patent 832,078, April 6, 1960; *C.A.*, **54**, 20877 (1960).
11. J. H. Billman and L. C. Dorman, *J. Pharm. Sci.*, **51**, 1071 (1962).
12. H. Weingarten and M. G. Miles, *J. Org. Chem.*, **33**, 1506 (1968).
13. H. Weingarten and M. G. Miles, *J. Org. Chem.*, **31**, 2874 (1966).
14. M. O. Abdel-Rahman *et al.*, *Tetrahedron Letters*, 3871 (1968).
15. C. V. Breder, Master's Thesis, Vanderbilt University, 1964, p. 16; after the method of J. von Braun and E. Röver, *Chem. Ber.*, **36**, 1196 (1903).
16. F. F. Blicke, *Org. Reactions*, **1**, 303 (1942).
17. B. B. Thompson, *J. Pharm. Sci.*, **57**, 715 (1968).
18. C. Mannich *et al.*, *Chem. Ber.*, **65**, 378 (1932).
19. A. L. Wilds and C. H. Shunk, *J. Am. Chem. Soc.*, **65**, 469 (1943).
20. E. M. Fry, *J. Org. Chem.*, **10**, 259 (1945).
21. S. Winstein *et al.*, *J. Org. Chem.*, **11**, 215 (1946).
22. J. T. Plati *et al.*, *J. Org. Chem.*, **14**, 543 and 873 (1949).
23. C. Mannich and E. Ganz, *Chem. Ber.*, **55**, 3486 (1922).
24. C. Mannich and P. Schumann, *Chem. Ber.*, **69B**, 2299 (1936).
25. M. Senkus, *J. Am. Chem. Soc.*, **68**, 10 (1946).
26. H. G. Johnson, *J. Am. Chem. Soc.*, **68**, 12 and 14 (1946).
27. A. T. Blomquist and T. H. Shelley, Jr., *J. Am. Chem. Soc.*, **70**, 147 (1948).
28. G. B. Butler and F. N. McMillan, *J. Am. Chem. Soc.*, **72**, 2978 (1950).
29. G. F. Grillot and W. T. Gormley, Jr., *J. Am. Chem. Soc.*, **67**, 1968 (1945).

30. G. B. Bachman and L. V. Heisey, *J. Am. Chem. Soc.*, **68,** 2496 (1946).
31. R. F. Holdren and R. M. Hixon, *J. Am. Chem. Soc.*, **68,** 1198 (1946).
32. H. D. Hartough *et al.*, *J. Am. Chem. Soc.*, **70,** 4013 and 4018 (1948).
33. Ref. 8, p. 6.
34. J. van de Kamp and E. Mosettig, *J. Am. Chem. Soc.*, **58,** 1568 (1936).
35. E. M. Fry, *J. Org. Chem.*, **10,** 259 (1945).
36. M. W. Williams, *J. Org. Chem.*, **33,** 3946 (1968).
37. P. Potier *et al.*, *J. Am. Chem. Soc.*, **90,** 5622 (1968).
38. H. E. Zaugg and W. B. Martin, *Org. Reactions*, **14,** 52 (1965).
39. A. Einhorn *et al.*, *Ann. Chem.*, **343,** 207 (1905).
40. M. Winn and H. E. Zaugg, *J. Org. Chem.*, **33,** 3779 (1968).
41. H. A. Bruson and C. W. MacMullen, *Org. Reactions*, **1,** 330 (1942).
42. W. Herz *et al.*, *J. Am. Chem. Soc.*, **69,** 1698 (1947).
43. A. L. Wilds *et al.*, *Org. Syn.*, Coll. Vol. **4,** 281 (1963).
44. C. F. H. Allen and J. A. VanAllan, *Org. Syn.*, Coll. Vol. **3,** 275 (1955).
45. G. R. Kalinina *et al.*, *Probl. Poluch. Poluprod. Prom. Org. Sin. Akad. Nauk SSR, Otd. Obshch. Tekh. Khim.*, 43 (1967); *C.A.*, **68,** 9186 (1968).
46. H. R. Snyder and J. H. Brewster, *J. Am. Chem. Soc.*, **71,** 1058 (1949).
47. M. Betti, *Org. Syn.*, Coll. Vol. **1,** 381 (1941).
48. T. D. Stewart and W. E. Bradley, *J. Am. Chem. Soc.*, **54,** 4172 (1932).
49. R. C. Elderfield, *Heterocyclic Compounds*, John Wiley and Sons, New York, Vol. I, 1950, p. 1.
50. Ref. 49, p. 61.
51. G. E. McCasland and D. A. Smith, *J. Am. Chem. Soc.*, **72,** 2190 (1950).
52. M. T. Leffler and R. Adams, *J. Am. Chem. Soc.*, **59,** 2252 (1937).
53. W. S. Emerson, *J. Am. Chem. Soc.*, **67,** 516 (1945).
54. R. E. Lutz *et al.*, *J. Am. Chem. Soc.*, **70,** 2015 (1948).
55. A. R. Goldfarb, *J. Am. Chem. Soc.*, **63,** 2280 (1941).
56. J. H. Biel, *J. Am. Chem. Soc.*, **71,** 1306 (1949).
57. A. J. W. Headlee *et al.*, *J. Am. Chem. Soc.*, **55,** 1066 (1933).
58. W. H. Horne and R. L. Shriner, *J. Am. Chem. Soc.*, **54,** 2925 (1932).
59. T. L. Cairns and J. H. Fletcher, *J. Am. Chem. Soc.*, **63,** 1034 (1941).
60. G. H. Coleman and J. E. Callen, *J. Am. Chem. Soc.*, **68,** 2006 (1946).
61. L. B. Clapp, *J. Am. Chem. Soc.*, **70,** 184 (1948).
62. L. C. King *et al.*, *J. Am. Chem. Soc.*, **78,** 2527 (1956).
63. C. Mannich *et al.*, *Chem. Ber.*, **69B,** 2113 (1936).
64. R. C. Doss and A. M. Schnitzer, U.S. Patent 2,800,509, July 23, 1957; *C.A.*, **51,** 17979 (1957).
65. Ref. 8, p. 54.
66. S. Hünig and H. Kahanek, *Chem. Ber.*, **90,** 238 (1957).
67. W. Langenbeck *et al.*, *Chem. Ber.*, **75B,** 1483 (1942).
68. W. Lagenbeck *et al.*, German Patent 713,747, October 23, 1941; *C.A.*, **38,** 1532 (1944).
69. K. Bowden *et al.*, *J. Chem. Soc.*, 45 (1946).
70. H. A. Bruson, *Org. Reactions*, **5,** 79 (1949).
71. N. H. Cromwell, *Chem. Rev.*, **38,** 83 (1946).
72. P. R. Haeseler, *Org. Syn.*, Coll. Vol. **1,** 196 (1941).
73. G. Stork and S. M. McElvain, *J. Am. Chem. Soc.*, **69,** 971 (1947).
74. P. L. Southwick and L. L. Seivard, *J. Am. Chem. Soc.*, **71,** 2532 (1949).
75. W. S. Johnson *et al.*, *J. Am. Chem. Soc.*, **71,** 1901 (1949).
76. J. T. Braunholtz and F. G. Mann, *J. Chem. Soc.*, 4166 (1957).
77. R. Mozingo and J. H. McCracken, *Org. Syn.*, Coll. Vol. **3,** 258 (1955).
78. M. Pfau and R. Dulou, *Bull. Soc. Chim. France*, 3336 (1967).
79. A. Zilkha and J. Rivlin, *J. Org. Chem.*, **23,** 94 (1958).
80. A. Zilkha and M. D. Bachi, *J. Org. Chem.*, **24,** 1096 (1959).
81. A. H. Cook and K. J. Reed, *J. Chem. Soc.*, 399 (1945).
82. S. A. Heininger, *Org. Syn.*, Coll. Vol. **4,** 146 (1963).
83. R. E. Steiger, *Org. Syn.*, Coll. Vol. **3,** 91 (1955).

84. C. C. Price and S.-T. Voong, *Org. Syn.*, Coll. Vol. **3,** 664 (1955).
85. H. E. Baumgarten, *J. Am. Chem. Soc.*, **77,** 5109 (1955).
86. P. L. DuBrow and H. J. Harwood, *J. Org. Chem.*, **17,** 1043 (1952).
87. M. C. Mentzer *et al.*, *Bull. Soc. Chim. France*, **12,** 161 (1945).
88. L. P. Kyrides *et al.*, *J. Am. Chem. Soc.*, **72,** 745 (1950).

E. Additions (Mostly Organometallic Methods)

The metathetical reactions of organometallic compounds to produce amines have been discussed in C.9, while the addition reactions of these compounds are discussed in this section. The reactions include obvious methods of addition of which the most interesting, versatile, and useful is the addition of organometallic compounds to azomethines (E.3).

1. From N,N-Dialkylamides (Bouveault) and Other Carbonyl Compounds

$$R_2NCHO \xrightarrow{R'MgX} R'CHO + R_2NCH(R')_2$$

This synthesis for the preparation of aldehydes has been discussed (see Chapter 10, G). If a large excess of Grignard reagent is employed (3 or more moles of 1 of the amide), a fair amount of tertiary amine [1, 2] is sometimes produced. The path which the reaction takes is not entirely clear [3, 4] but comprises a combination of additions and displacement. Although the maximum yield reported appears to be 80% [5], the synthesis has found some value in preparing tertiary amines with bulky alkyl groups and of unusual structure:

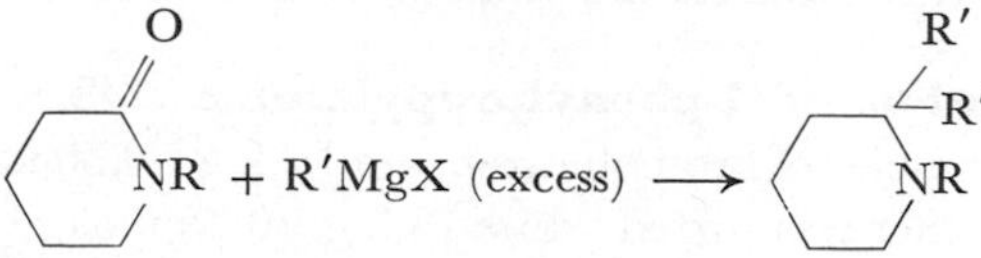

[ref. 6]

With smaller amounts of Grignard reagent, intermediates can be isolated such as:

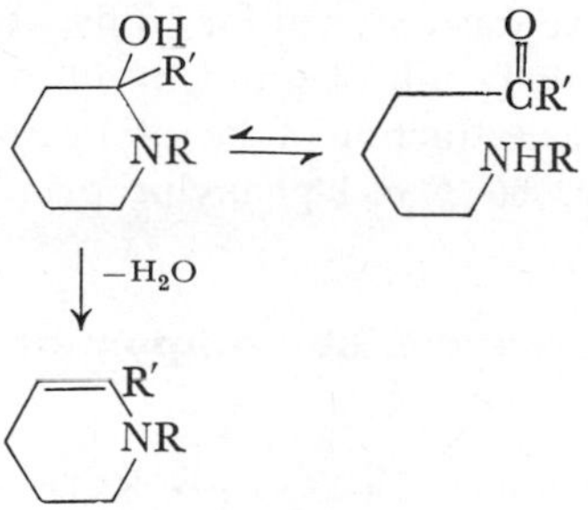

Tetraethyloxamide shows an interesting splitting [7]:

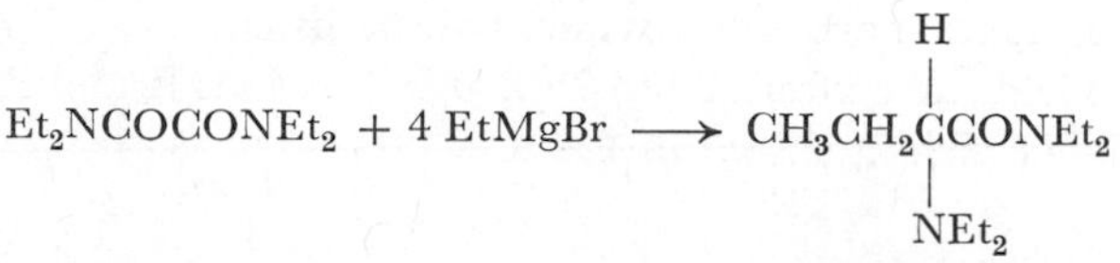

$$Et_2NCOCONEt_2 + 4\ EtMgBr \longrightarrow CH_3CH_2\overset{\overset{\displaystyle H}{|}}{\underset{\underset{\displaystyle NEt_2}{|}}{C}}CONEt_2$$

N,N-Diethyl-α-diethyl-
aminobutyramide, 60%

References for Section E are on pp. 478–479.

Addition of Grignard reagents to *t*-aminoketones forms aminoalcohols of considerable pharmacological interest. However, addition to β-aminoketones, Mannich bases, gives rather low yields because of enolization and consequent salt formation [8].

(a) **Preparation of 5-diisopropylaminononane.** A solution of 10.3 g. (0.08 mole) of diisopropylformamide in 40 ml. of dry ether was added dropwise to the Grignard reagent obtained from 54.8 g. (0.4 mole) of butyl bromide, 10 g. (0.42 g. atom) of magnesium, and 80 ml. of dry ether. The mixture, after standing at 18° overnight, was decomposed in an ice bath with 4 *N* sulfuric acid. After inefficient extractive operations to remove aldehyde and diisopropylamine, the free base was distilled to give 67 % of liquid, b.p. 234° (760 Torr.) [9].

2. From Nitriles Followed by Reduction

$$RMgX + R'CN \longrightarrow R\overset{\overset{\displaystyle NMgX}{\|}}{C}R' \xrightarrow{LiAlH_4} R\overset{\overset{\displaystyle NH_2}{|}}{CH}R'$$

The adduct obtained from the Grignard reagent and a nitrile may be reduced with lithium aluminum hydride to give a disubstituted carbinamine [10]. This procedure eliminates the isolation and subsequent conversion of the ketone which is obtained on hydrolysis of the adduct. Optimum yields were obtained by using 1.2 moles of lithium aluminum hydride per mole of the complex. For five disubstituted carbinamines the yields varied from 23 to 80 %.

(a) **Preparation of 1-phenylpropylamine.** The Grignard reagent from 47.3 g. (0.30 mole) of bromobenzene and 7.2 g. (0.30 mole) of magnesium in 300 ml. of dry ether was stirred while 13.8 g. (0.25 mole) of propionitrile was added dropwise. After the mixture was refluxed for 2 hr., a slurry of 11.4 g. (0.30 mole) of lithium aluminum hydride in 100 ml. of tetrahydrofuran was added slowly. The mixture was refluxed for 18 hr. and then decomposed, with cooling, by the addition of 12 ml. of water, 9 ml. of 20 % sodium hydroxide, and finally 42 ml. of water. Extraction of the solid with ether, followed by drying and distilling, gave 27.1 g. (80 %) of 1-phenylpropylamine, b.p. 78–80° (7 mm.) [10].

3. From Azomethines and Like Compounds

$$RMgX + \overset{\diagdown}{\underset{\diagup}{C}}=N\!- \longrightarrow R\overset{|}{C}N\!-\!MgX$$

This addition takes many forms which will be listed.

Schiff bases. Yields range from 25 to 75 % using various R and R' groups [11], although a higher yield is shown in Ex. a.

$$RMgX + C_6H_5CH\!=\!NR' \longrightarrow C_6H_5\overset{\overset{\displaystyle R}{|}}{CH}NHR'$$

Aromatic aldoximes. The reaction proceeds via the Beckmann rearrangement

$$ArCH{=}NOH + 2\,RMgX \rightarrow (ArNHCH{=}O) \rightarrow ArNHCHR_2$$

and yields are good in some cases [12, 13].

Aromatic ketoximes. The intermediate aziridines, although reactive, can be

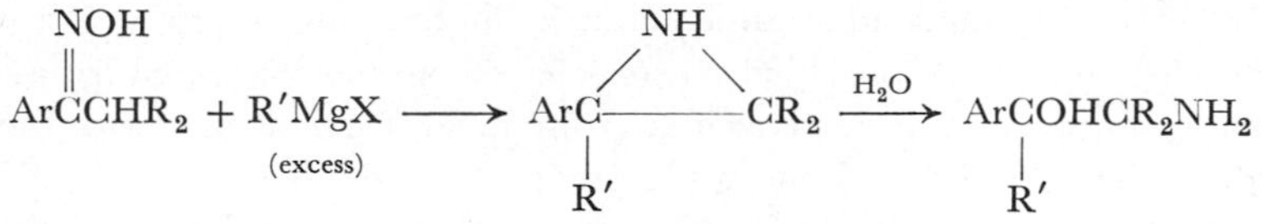

isolated [14–16]. The aminoalcohols have been isolated, after hydrolysis, in yields ranging from 40 to 75 % [17]. Indeed, aziridines can be made simply by simultaneous reduction and condensation with lithium aluminum hydride [18]:

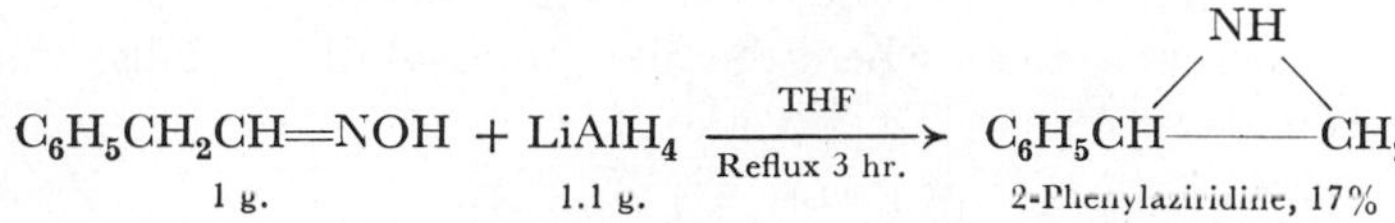

The two reactions above are related to the Neber rearrangement [19]:

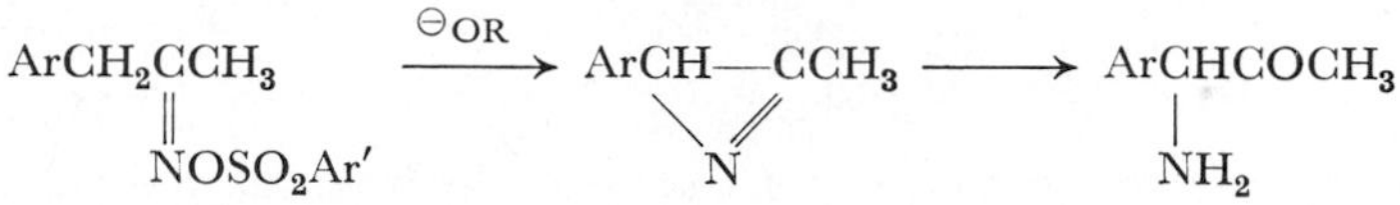

On the other hand, ketoxime ethers add in the expected manner [20]:

$$R_2C{=}NOC_4H_9 + R'Li\ (R' = C_4H_9\ \text{or}\ C_6H_5) \longrightarrow R_2\overset{\overset{\textstyle R'}{\textstyle |}}{C}NHOC_4H_9$$

Hydrazones [20].

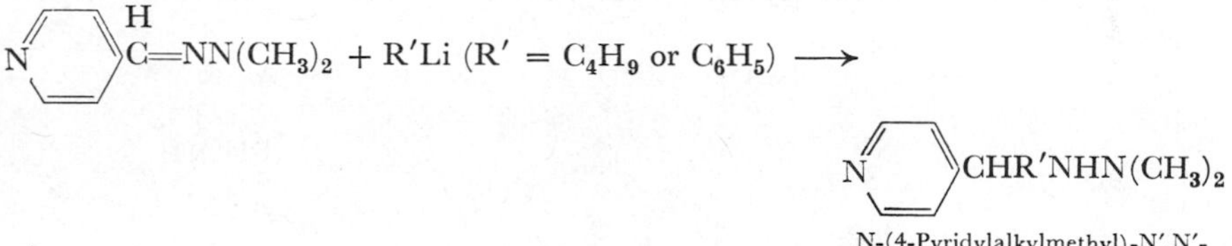

N-(4-Pyridylalkylmethyl)-N′,N′-dimethylhydrazines

Azides and diazoalkanes.

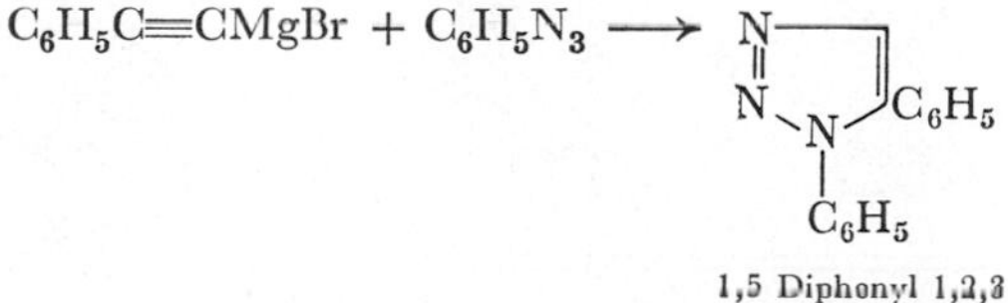

[ref. 21]

1,5 Diphenyl 1,2,3 triazole, 40%

$$(C_6H_5)_2CN_2 + RMgX \longrightarrow (C_6H_5)_2C{=}NNHR \xrightarrow{\overset{\oplus}{H_3O}} RNHNH_2 \quad [\text{ref. 22}]$$

Pyridines and quinolines [23].

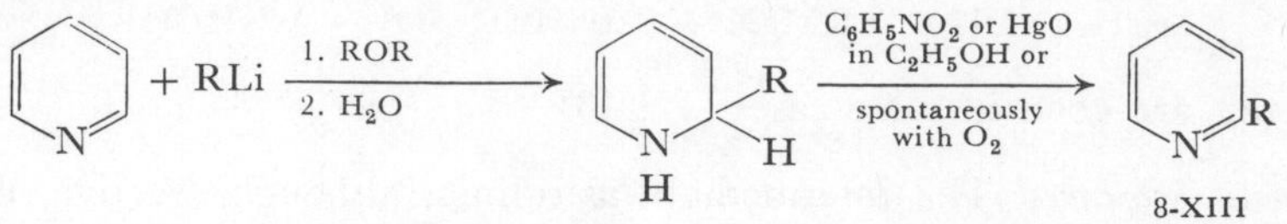

8-XIII

Alkyl- or aryllithiums add rapidly and in high yields to nitrogen aromatic heterocyclic compounds provided no steric effects are encountered from groups adjacent to nitrogen. The more aromatic the heterocyclic compound, the more difficult is the addition and the more facile is rearomatization to 8-XIII. A list in order of ease of addition has been compiled: quinoxaline > acridine ≃ N-benzylidine aniline ≃ phenanthridine > isoquinoline ≃ quinoline > pyridine [24]. The dihydropyridine derivative is rarely isolated because of spontaneous oxidation to the pyridine derivative. Grignard reagents are more sluggish in their addition to the above heterocyclic compounds, but they can be utilized under circumstances where the alkyl- or aryllithium fails, namely, by addition to the N-oxides (see Ex. b.1 and [25]).

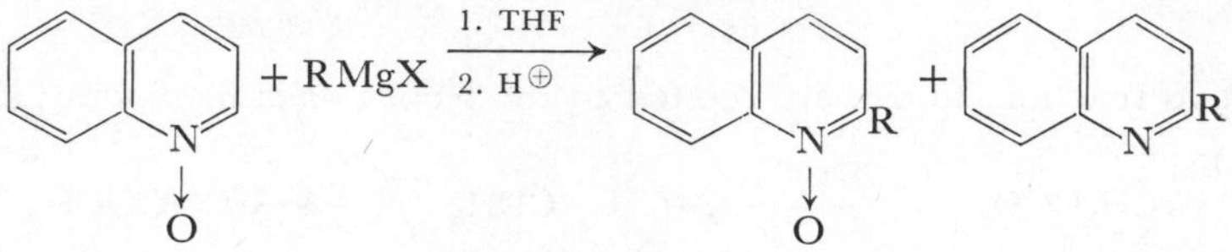

With excess Grignard reagent in refluxing tetrahydrofuran, the amount of 2-aryl(or alkyl)-quinoline-N-oxide is diminished.

(a) Preparation of N-methyl-1,2-diphenylethylamine (91–96 % from benzylmagnesium chloride and N-benzylidenemethylamine) [26].

(b) Other examples

(1) 2-*p*-Chlorophenyldihydroquinine. The di-N-oxide of dihydro-

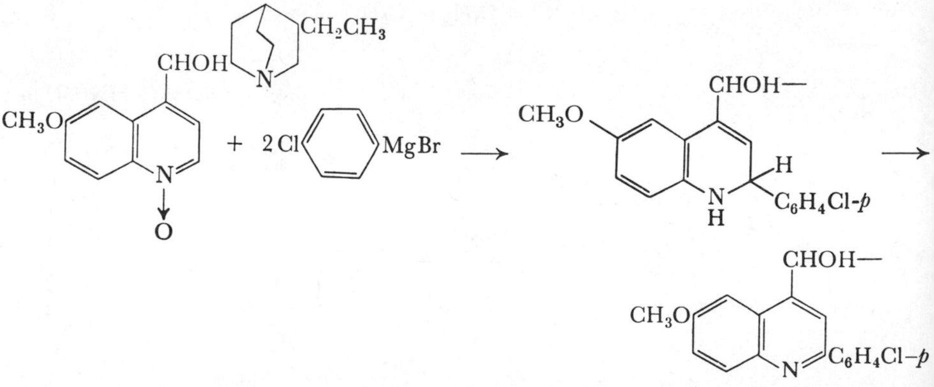

quinine was prepared by the method of Kobayashi [27] and reduced selectively to the quinoline N-oxide by sulfur dioxide in ethanol. This N-oxide was added to an 8-fold excess of *p*-chlorophenylmagnesium bromide in THF, warmed to initiate reaction (a precipitate appeared), and held at reflux for 4 hr. with stirring. The mixture was poured into aqueous ammonium chloride and extracted with ethyl acetate. The ethyl acetate extract was treated with 10% aqueous hydrochloric acid, following which the hydrochloride salt was evaporated to dryness, recrystallized from methanol-ethyl acetate, and made basic with cold aqueous potassium hydroxide. The free base was extracted with methylene chloride and recrystallized from acetonitrile giving 29% of white, fluffy crystals, m.p. 189–191° [28].

(2) **Phenylquinoline** (90% from 0.02 mole of quinoline-N-oxide and 0.1 mole of phenylmagnesium bromide refluxed in THF) [25].

(3) **4-Benzyl-1,4-dihydro-2-methoxyquinoline** (71% from benzyl-

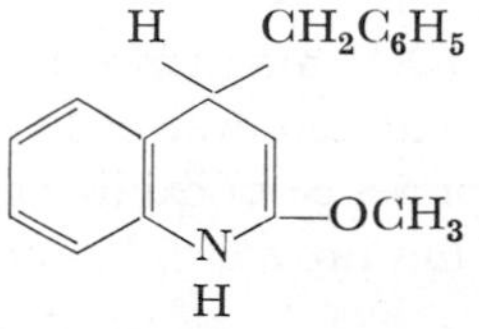

magnesium chloride and 2-methoxyquinoline) [29].

4. From Azomethines and Sodamide

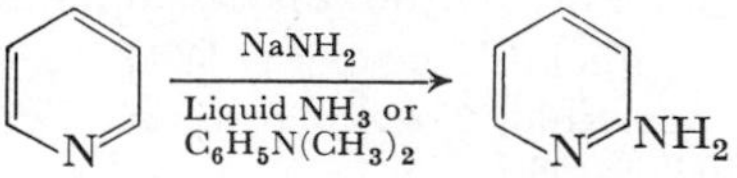

This nucleophilic substitution, which has been reviewed [30, 31], is almost exclusively confined to the heterocyclic series. In fact, the results are most satisfactory for pyridine, quinoline, and their derivatives. The synthesis is of particular value in introducing the amino group into the 2-position of these heterocycles. If the 2-position is occupied, the substituent goes to the 4-position. Some groups such as a carboxyl in position 2 or 4 of quinoline increase the rate of reaction, while others, such as an amino in 2 or a hydroxyl in 2 or 8 of this heterocycle, prevent the reaction. On the other hand, a sulfonic acid or methoxyl group in the 2-position of quinoline is replaced by an amino group. Yields vary from 50 to about 100%.

Sodamide is used in most aminations in the heterocyclic series except when liquid ammonia is used as the solvent. With this solvent, potassium or barium amide is preferred because of greater solubility. Various hydrocarbons, dimethylaniline, diethylaniline, and liquid ammonia have been used as solvents. The use of the dialkylanilines has been of great value in improving yields in the pyridine series. The temperature during reaction is kept as low as possible.

It has already been indicated that the amination of the aromatic nucleus is rare. One such case is that of 1-nitronaphthalene, which yields 1-nitro-4-aminonaphthalene with 55–60% yield on treatment with hydroxylamine and potassium hydroxide [32].

The first step in the amination of pyridine appears to be the formation of the anion 8-XIV, which loses a hydride ion to form aminopyridine 8-XV. In the

presence of the hydride ion, 8-XV may form the anion 8-XVI and hydrogen.

This mechanism suggests that removal of hydride, a high-energy process, should be facilitated by the presence of oxidizing agents which would remove the hydrogen as water. Such has proved to be the case (see Ex. b).

(a) **Preparation of 2-aminopyridine** (66–76 % from pyridine, sodamide, and dimethylaniline at 105–110°) [33].

(b) **Preparation of 4-amino-2-phenylquinoline** (93–98 % from 8.9 mmoles of 2-phenylquinoline, 26.9 mmoles of potassium amide, and 1.61 g. of potassium nitrate at 25° for 4 hr.; without potassium nitrate, the desired product was not formed in any appreciable amount) [34].

1. L. Spialter and J. A. Pappalardo, *The Acyclic Aliphatic Tertiary Amines*, The Macmillan Company, New York, 1965, p. 59.
2. Houben-Weyl, *Methoden der Organischen Chemie*, Georg Thieme Verlag, Stuttgart, 1957, 4th ed., Vol. 11, Pt. 1, p. 820.
3. M. Montagne, *Ann. Chim.*(Paris), **13,** 40 (1930).
4. J. A. Pappalardo and P. O'Brien, Master's Thesis, University of Dayton, 1963.
5. J. Ficini and H. Normant, *Compt. Rend.*, **247,** 1627 (1958).
6. Ref. 2, p. 822.
7. R. Barré, *Compt. Rend.*, **185,** 1051 (1927); *Ann. Chim.* (Paris), **9,** 250 (1928).
8. R. Baltzly and J. W. Billinghurst, *J. Org. Chem.*, **30,** 4330 (1965).
9. F. Kuffner and E. Polke, *Monatsh. Chem.*, **82,** 330 (1951).
10. A. Pohland and H. R. Sullivan, *J. Am. Chem. Soc.*, **75,** 5898 (1953).
11. K. N. Campbell *et al., J. Am. Chem. Soc.*, **70,** 3868 (1948).
12. M. Busch and R. Hobein, *Chem. Ber.*, **40,** 2096 (1907).
13. P. Grammaticakis, *Compt. Rend.*, **210,** 716 (1940).
14. S. Eguchi and Y. Ishii, *Bull. Chem. Soc. Japan*, **36,** 1434 (1963).
15. R. F. Parcell, *Chem. Ind.* (*London*), 1396 (1963).
16. D. S. Tarbell *et al., J. Am. Chem. Soc.*, **75,** 2959 (1953).
17. K. N. Campbell *et al., J. Org. Chem.*, **8,** 99 (1943).
18. K. Kotera *et al., Tetrahedron*, **24,** 3681 (1968).
19. C. O'Brien, *Chem. Rev.*, **64,** 81 (1964).
20. A. Marxer and M. Horvath, *Helv. Chim. Acta.*, **47,** 1101 (1964).
21. G. S. Akimova *et al., Zh. Org. Khim.*, **3,** 968 (1967); *C.A.*, **67,** 9412 (1967).
22. P. A. S. Smith *et al., J. Org. Chem.*, **23,** 1595 (1958).
23. M. S. Kharasch and O. Reinmuth, *Grignard Reactions of Nonmetallic Substances*, Prentice-Hall, Englewood Cliffs, N.J., 1954, pp. 1251–1259.
24. H. Gilman *et al., J. Am. Chem. Soc.*, **79,** 1245 (1957).

25. T. Kato and H. Yamanaka, *J. Org. Chem.*, **30**, 910 (1965).
26. R. B. Moffett, *Org. Syn.*, Coll. Vol. **4**, 605 (1963).
27. G. Kobayashi, *J. Pharm. Soc. Japan*, **70**, 381 (1950); *C.A.*, **45**, 2491 (1951); **67**, 101 (1947); *C.A.*, **45**, 9553 (1951).
28. D. J. Thoennes and D. E. Pearson, unpublished results.
29. R. C. Fuson *et al.*, *J. Org. Chem.*, **16**, 1529 (1951).
30. M. T. Leffler, *Org. Reactions*, **1**, 91 (1942).
31. Ref. 2, p. 9.
32. C. C. Price and S.-T. Voong, *Org. Syn.*, Coll. Vol. **3**, 664 (1955).
33. Ref. 30, p. 99.
34. F. W. Bergstrom. *J. Org. Chem.*, **3**, 424 (1938).

F. Other Additions and Substitutions

Sections D and E were concerned in part with the addition of amines, with or without acid catalysts, or amine anions ($R\overset{\ominus}{N}H$) to unsaturated systems. This section first concentrates on the addition of electron-deficient (or group–deficient) nitrogen atoms, $R_2\overset{\oplus}{N}$, $R_2\overset{..}{N}H$, or $R:\overset{..}{N}:$ or other electron-deficient species. The acceptance of such intermediates is a simplification because re-action may occur before complete formation of the particle has taken place or with some complex of the particle. In contrast to amines and amine anions which add most readily to polarized systems, e.g.,

$$\text{C=C—C}\equiv\text{N} \longleftrightarrow \overset{\oplus}{\text{C}}\text{—C=C=}\overset{\ominus}{\text{N}}$$

the electron or group-deficient nitrogen atoms add most readily to electron-rich systems and indeed in some cases are so reactive that they can insert into C—H bonds. This field is under active investigation at present. Furthermore, this section serves to introduce the later section, G, Molecular Rearrangements, in which are described classical rearrangements, involving similar intermediates, such as the Hofmann degradation. Section F continues with the description of the addition of dipolar substances to unsaturated systems and Diels-Alder reactions which give rise to heterocyclic nitrogen compounds. Most unusual structures are constructed by these syntheses.

1. From Olefins and Nitriles (Ritter Reaction)

$$R_3COH \xrightarrow{H_2SO_4} [R_3\overset{\oplus}{C}] \xrightarrow{R'CN} [R'\overset{\oplus}{C}=NCR_3] \xrightarrow[-H^{\oplus}]{H_2O}$$

$$\underset{\text{R'CNHCR}_3}{\overset{O}{\underset{\|}{}}} \xrightarrow{H_2O} R_3CNH_2$$

The Ritter reaction is discussed in detail under Carboxylic Acid Amides, Chapter 18, D.4. It is an excellent method of making amines with a *t*- or *sec*-alkyl group.

References for Section F are on pp. 492–493.

2. From Hydrocarbons and Haloamines and the Like

$$ArH + R_2NX \xrightarrow[H_2SO_4]{AlCl_3 \text{ or}} ArNR_2 + HX$$

This reaction, which has been reviewed [1], is general but has limited use for synthesis. The limitation is imposed by the formation of appreciable amounts of o-, m-, and p-substituted anilines in a monosubstituted benzene, by modest yields in most cases, and by a rather laborious workup. The formation of o-, m-, p-isomers suggests the incursion of a reactive intermediate, $^{\oplus}NR_2$, or some complex of it. Nevertheless, on occasion the synthesis is quite worthwhile (see Ex. a and Ex. c.1). Among the other nitrogen reagents which have been used are hydroxylamine-O-sulfonic acid and hydroxylamine salts [2] and hydrazoic acid [3]. Orientation is normal with the reagents above, but it is abnormal with nitrogen tri- or dichloride [4]:

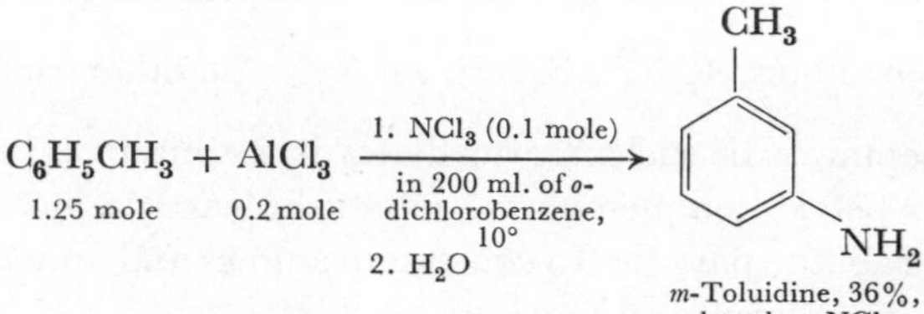

$$C_6H_5CH_3 + AlCl_3 \xrightarrow[\substack{\text{1. NCl}_3 \text{ (0.1 mole)} \\ \text{in 200 ml. of } o\text{-} \\ \text{dichlorobenzene,} \\ 10° \\ \text{2. H}_2O}]{}$$

m-Toluidine, 36%,
based on NCl₃

The orientation is attributed to the reaction of the sigma complex of toluene and NCl₃.

Of course, amino groups which do not participate in the substitution can be carried along in a Friedel-Crafts process to provide amino compounds of considerable variety. Enough aluminum chloride must be added to complex the amino group (see Ex. c.2 and c.3 as well as the following):

$$C_6H_6 + AlCl_3 + CH_2{=}CHCH_2NH_2 \xrightarrow[\substack{\text{2. H}^{\oplus}, H_2O \\ \text{3. HO}^{\ominus}}]{\text{1. Reflux 6 hr.}} C_6H_5CH(CH_3)CH_2NH_2 \quad [\text{ref. 5}]$$

1 mole 0.6 mole   0.2 mole   2-Phenyl-1-aminopropane,
                                                                                   85–94%

$$C_6H_6 + (CH_3)_2COHCH_2NH_2 \xrightarrow[\substack{0.6 \text{ mole}}]{AlCl_3} (CH_3)_2CCH_2NH_2$$

1 mole   0.2 mole $\underset{C_6H_5}{|}$ [ref. 6]

2-Phenyl-2-methyl-1-
aminopropane, 87%

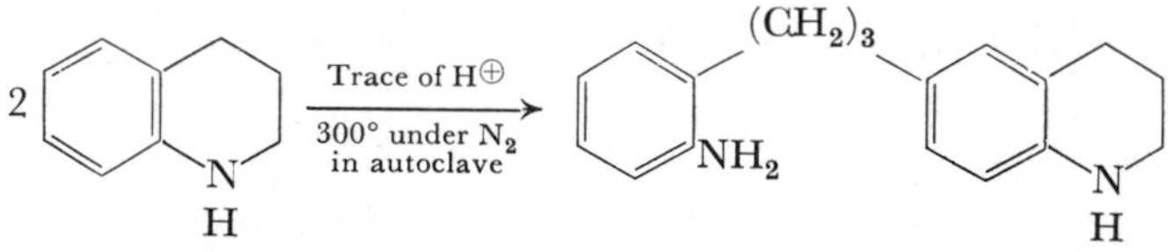

6-(o-Aminophenylpropyl)-1,2,3,4-
tetrahydroquinoline, 40%

Alkylation of aromatic amines is discussed in Olah [8], halogenation under Halides, Chapter 7, D.3, nitration under Nitro Compounds, Chapter 20, A.2, and Acylation in Ketones, Chapter 11, C.1 (see also Ex. c.4). A few typical heterocyclic nitrogen syntheses of an electrophilic nature are discussed elsewhere: Skraup (F.3), benzidine rearrangement (G.6), and Fischer indole synthesis (G.7).

(a) Preparation of p-nitro-p'-dimethylaminodiphenyl (99% from 0.025 mole each of p-nitrodiphenyl and N-chlorodimethylamine in 45 ml. of sulfuric acid at 0° to which 7 g. of ferrous sulfate heptahydrate was added and the mixture warmed to 33° for 30 min.; the reaction may be ion-free radical in nature) [9].

(b) Preparation of 2-methoxy-1-azabicyclo[3.2.1]octane [10].

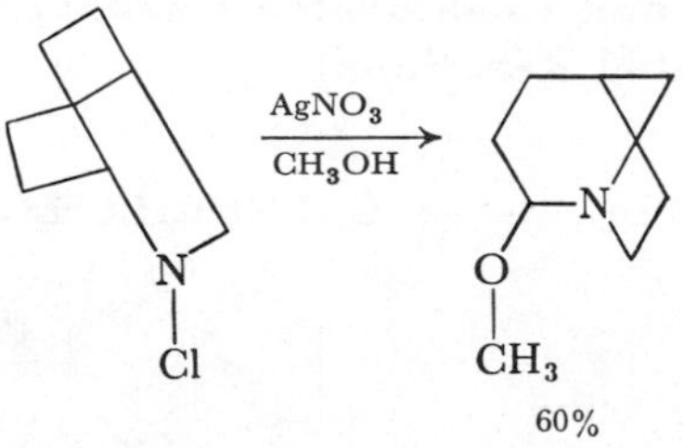

60%

(c) Other preparations

(1) **N-Phenylpiperidine** (60–70% from benzene, N-chloropiperidine, and aluminum chloride from 1–4 hr. at 80–100°) [11].

(2) **p-β-Dimethylaminoethylbenzophenone** (60% from 0.2 mole of

$$C_6H_5CH_2CH_2N(CH_3)_2 + C_6H_5COCl \xrightarrow[\text{2. } H_2O, \text{ OH}^\ominus]{\text{1. AlCl}_3} C_6H_5COC_6H_4CH_2CH_2N(CH_3)_2$$

N,N-dimethylphenylethylamine added dropwise to 0.86 mole of aluminum chloride and the stirred complex kept molten by slight warming while 0.3 mole of benzoyl chloride was added dropwise in 30 min.; the dark green complex was held in a molten condition for 2 hr. and then poured slowly onto ice mixed with 200 g. of sodium hydroxide in water; the oil distilled, b.p. 171–177° (0.35 mm.)) [12].

(3) **o- and p-(β-Aminoethyl)-toluene** (1:1 ratio, 55% from 120 ml. of toluene and 0.38 mole of aluminum chloride to which 0.2 mole of ethylenimine in 80 ml. of toluene was added dropwise at 0° and then refluxed; propylenimine gave a mixture of 2-phenyl-1-propylamine and 1-phenyl-2-propylamine with benzene and aluminum chloride) [13].

(4) **9,10-Diphenyl-9-acridanol**

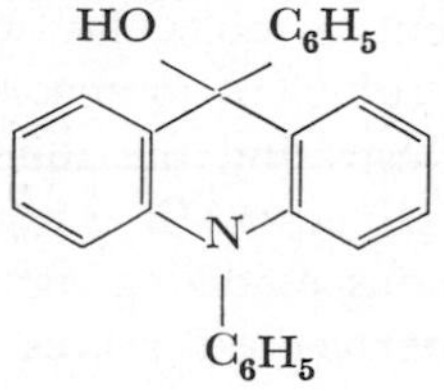

(47% from 1.2 g. of triphenylamine, 0.5 equiv. of benzoic acid, and 10 g. of polyphosphoric acid heated at 190–195° for 30 min.; the ring closure must occur with the precursor, o-benzoyltriphenylamine. Since *para*-acylation is predominant, the relatively high yield of product must arise from deacylation of the *ara*-isomer and reformation of more *ortho*-isomer) [14].

(5) 1-Amino-1-methylcyclohexane (83% from methylcyclohexane, trichloroamine, and AlCl₃) [15].

(6) 1-Aminoadamantane (100% from 1 equiv. of trichloroamine, 2 equiv. of AlCl₃, and 1.5 equiv. of adamantane; other hydrocarbons which rearrange to adamantane may also be used) [16].

3. From Anilines and Unsaturated Carbonyl Compounds (Skraup and Related Reactions)

$$C_6H_5NH_2 + CH_2{=}CHCH{=}O \xrightarrow[130°]{H^{\oplus}} C_6H_5NHCH_2CH_2CH{=}O \longrightarrow$$

The well-known Skraup reaction has been reviewed [17]. Acrolein is usually generated *in situ* from glycerol and sulfuric acid, which is also the ring-closing agent. Oxidizing agents are nitrobenzene or *m*-nitrobenzenesulfonic acid (when a substituted quinoline is made) or arsenic pentoxide. With the latter oxidizing agent, the reaction proceeds more smoothly with less frequent surges of temperature which lead to foaming and spraying. Almost any substituted quinoline can be prepared, the reservations being that the substituent on aniline must survive exposure to hot sulfuric acid and that the aniline should be neither too deactivated nor too activated toward electrophilic attack by a protonated aldehyde group. Acetyl, cyano, and occasionally fluoro are examples of substituents which do not survive, and polyhydroxy and polyamino groups are examples which may induce sulfonation of the ring or more profound oxidation reactions. *meta*-Substituted anilines lead to a mixture of 5- and 7-substituted quinolines which occasionally may be separated. If substituted acroleins are used in place of acrolein, quinolines with substituents in the heterocyclic ring are produced. From the dark, forbidding-looking reaction mixtures obtained it is surprising that such good-to-moderate yields of quinolines can be recovered.

The Doebner-v. Miller reaction is a Skraup reaction and differs only in the milder conditions used for ring closure. Although zinc chloride is sometimes used as the ring-closing reagent, a molecular equivalent of hydrogen chloride (or the salt of the aniline) is sufficient to accomplish this end. Two examples are shown: the first (c.1) illustrating the mild conditions, the second (c.2) revealing the fact that the Doebner v.-Miller reaction may be carried out without the addition of any obvious oxidizing agent; indeed, the yield is better without it. Evidently, an intermediate serves as the oxidizing agent in the reaction suspected (but not proved) to operate as follows (Ex. c.2):

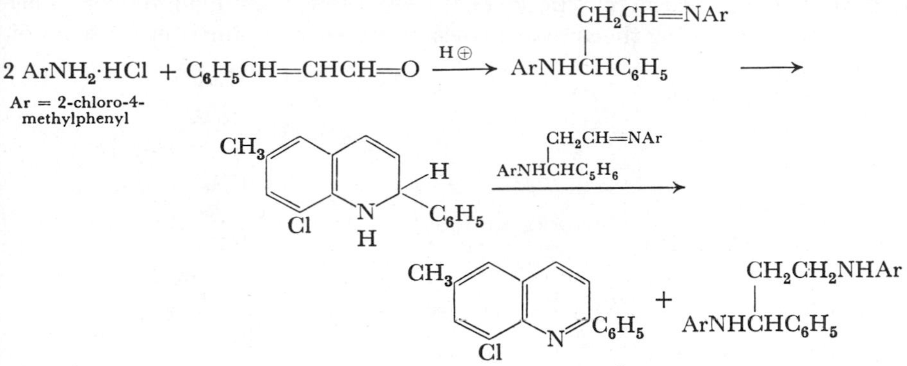

$$2\ ArNH_2 \cdot HCl + C_6H_5CH=CHCH=O \xrightarrow{H\oplus} \overset{\underset{\displaystyle CH_2CH=NAr}{|}}{ArNHCHC_6H_5} \longrightarrow$$

Ar = 2-chloro-4-
methylphenyl

2-Phenyl-6-methyl-8-chloroquinoline

The yield on this premise should be 33% based on aniline. Actual yields cluster around this figure, but they should not be construed as evidence in support of the speculation above. If the Schiff base (ArCH=CHCH=NAr′) is the oxidizing agent, the yield should approach 50%, and reduction products should be detected as reported [18]. Attempts to supply oxidizing agents more efficient than arsenic pentoxide or the Schiff bases have met with some success, as evidenced by the following example [19]:

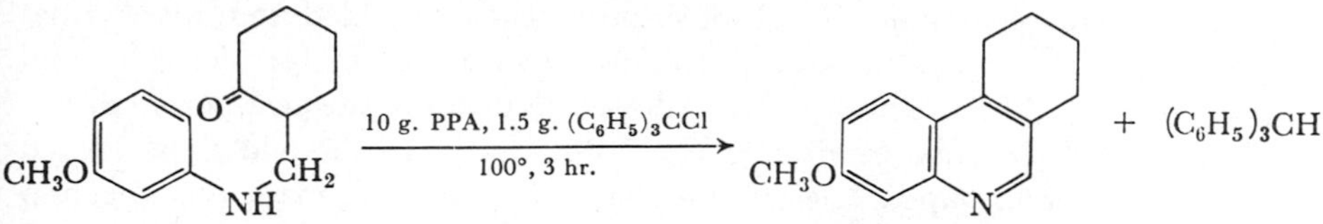

3-Methoxy-7,8,9,10-
tetrahydrophenanthridine,
100%

Trityl chloride as the oxidizing agent, however, does not appear to perform as well in other Doebner v.-Miller reactions.

Countless similar ring closures enumerated in a recent text [20] have been used to synthesize substituted quinolines. Only one more example of recent interest is discussed to illustrate what can be done with unsaturated compounds of higher oxidation state than unsaturated carbonyl compounds. Acetoacetic ester condenses with aniline to form the anilide which with excess strong acid cyclizes to carbostyrils [21]:

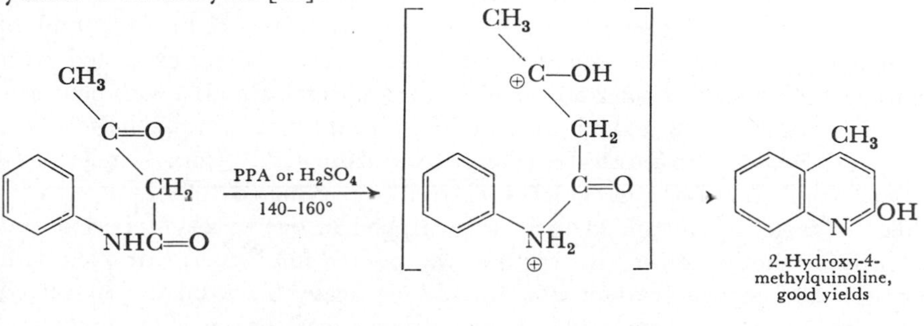

2-Hydroxy-4-
methylquinoline,
good yields

But, if anilide formation can be circumvented, the carbonyl group condenses with the amine and ring closure gives the 4-hydroxy-2-methylquinoline. This can be done by adding the acetoacetic ester last when aniline is in the form of its salt [22]:

$$C_6H_5NH_2 + PPA \text{ (at } 140°) \xrightarrow[\text{dropwise}]{CH_3COCH_2CO_2C_2H_5}$$

4-Hydroxy-2-methylquinoline, 70%

(a) **Preparation of quinoline** (84–91 % yield from ferrous sulfate to moderate the reaction, glycerol, aniline, nitrobenzene, and sulfuric acid heated until a controlled exothermic reaction began) [23].

(b) **Preparation of 6-methoxy-8-nitroquinoline.** To a stirred mixture of 0.3 mole of 2-nitro-4-methoxyacetanilide, 1.45 mole of glycerol, and 0.155 mole of arsenic pentoxide, 50 ml. of sulfuric acid was added and, after initial rise of internal temperature to 75°, the mass was heated to 110° for 1 hr. and then slowly to 125° while water was allowed to distill out over a period of 8–9 hr. A drop of this mixture diluted to 10 ml. with water should show no red coloration on filter paper. The mixture was cooled, diluted, made alkaline, and the precipitate filtered and washed. Recrystallization from toluene gave 67 % of the desired quinoline, m.p. 159–160.5°. The compound was not suitable for catalytic reduction. To remove traces of poisons, 40 g. of the compound was refluxed with 200 ml. of methanol, cooled, and the methanol decanted. This process was repeated and 34–36 g. of crystals, m.p. 159.5–160.5°, was recovered by filtration [24]. The preparation is much less restrictive in operation than the *Organic Syntheses* preparation [25].

(c) **Other examples**

(1) **6-Bromo-4-methyl-8-phenylquinoline** (40–45 % from 0.46 mole of 2-amino-5-bromobiphenyl hydrochloride, 2 moles of commercial methyl vinyl ketone, and 0.152 mole of arsenic pentoxide refluxed 22 hr. in 350 ml. of ethanol; the mixture was concentrated, poured into water, extracted with benzene, and the heavy black oil was eluted from Merck alumina with benzene; the solute was concentrated and recrystallized from hexane, m.p. 63–64°) [26].

(2) **8-Chloro-6-methyl-2-phenylquinoline** (20 % from 1 mole of 2-chloro-4-methylaniline hydrochloride stirred and refluxed in 1 l. of ethanol while 1.5 moles of cinnamaldehyde was dripped in over a 24-hr. period and refluxed for another 24 hr.; the mixture was poured into water, extracted with benzene, and benzene extract evaporated; the heavy black oil was extracted well with 500 ml. of conc. HCl; the acid layer was shaken with methylene

chloride and then dripped into a strongly basic aqueous solution to precipitate the quinoline which was recrystallized from ethanol as slightly yellow needles, m.p. 95–95.5°) [27].

4. From Nitrene Intermediates and the Like

$$R_2\overset{..}{N}{}^{\oplus} + R'H \rightarrow R_2\overset{..}{N}R' + H^{\oplus}$$

(Nitrenium)
ion

$$R_2\overset{.\ \oplus}{NH} + R'H \rightarrow R_2{}^{\oplus}NHR' + H\cdot$$

(Nitrenium
ion-free
radical)

$$R_2\overset{..}{N}\cdot + ArH \rightarrow R_2NAr + H\cdot$$

(Amine
free
radical)

$$R:\overset{..}{N}: + R'H \rightarrow RNHR'$$

(Nitrene)

(See introduction to chapter for a general discussion of these particles.)

The first equation shows an idealized intermediate, the nitrenium ion, similar to that discussed in aromatic substitution in F.2. The use of this intermediate in synthetic work is limited preferably to intramolecular attack on a carbon situated in a position favorable to the nitrogen atom [28]:

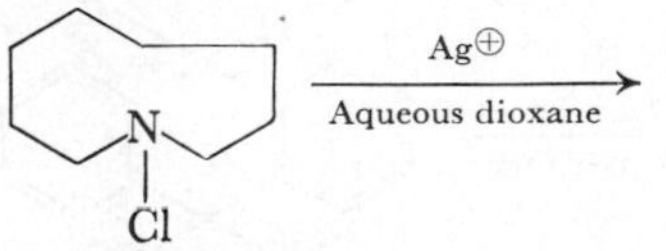

N-Chloroaza-
cyclononane

Indolizidine,
68%

If attack on carbon cannot or does not occur, intermolecular hydride abstraction takes place to give the secondary amine from which the N-chloramine was derived.

On the other hand, some N-chloroamines on irradiation in strongly acid solution have an entirely different pattern of reaction, a reaction general enough to be called the Hofmann-Löffler transformation [29]. It is nitrenium ion-free radical in nature and has the propensity to abstract hydrogen from a δ carbon atom:

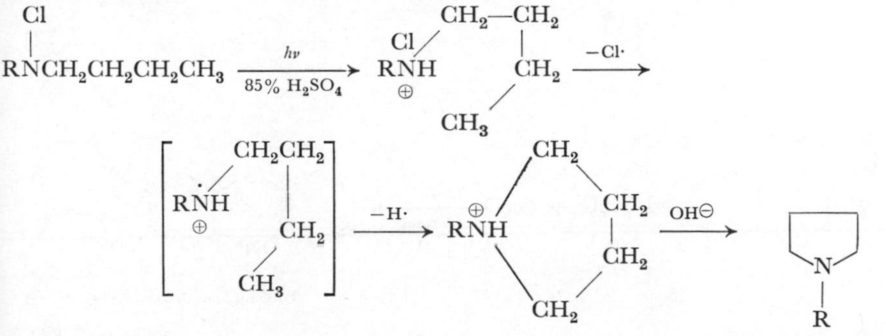

R = C_4H_9, N-Butyl-
pyrrolidine, 90%

The propensity for attack on the δ-carbon has been rationalized in terms of a linear abstraction path, but admittedly, if an epsilon carbon is available, some, but less, attack occurs here to give a piperidine derivative. Ferrous ammonium sulfate or potassium persulfate can be used in place of irradiation to carry out the reaction. It is useful for the preparation of pyrrolidines and azabicyclo-[2.2.1]heptanes:

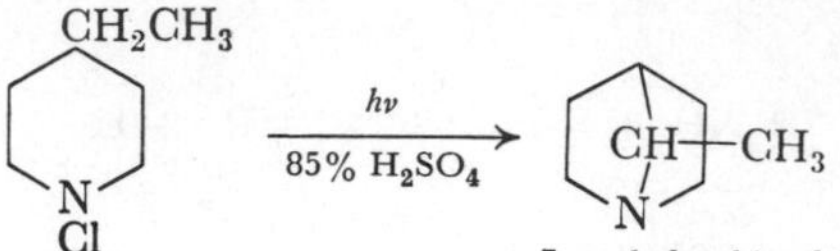

7-methylazabicyclo-
[2.2.1]heptane

The Hofmann-Löffler reaction has been carried out with N-chlorosulfonamides [30]. The ion-free radical also adds to olefins [31]:

$$RNHCl + CH_2{=}CHX \xrightarrow[\substack{CH_3CO_2H \\ h\nu}]{H_2SO_4} RNHCH_2CHXCl$$

50-70%

The amine free radical, $R_2\ddot{N}\cdot$, also can add to olefins or substitute in conjugated, unsaturated systems, the most recent examples of which are shown:

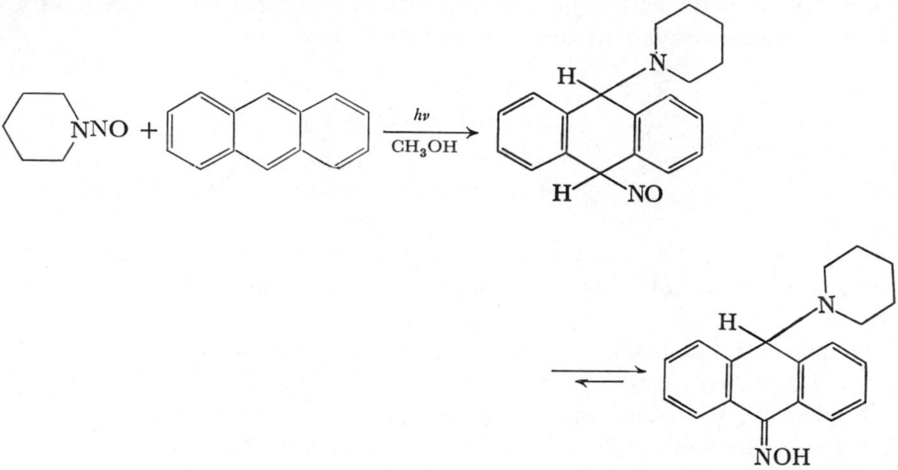

10-N-Piperidinoanthrone
oxime, 52%

[ref. 32]

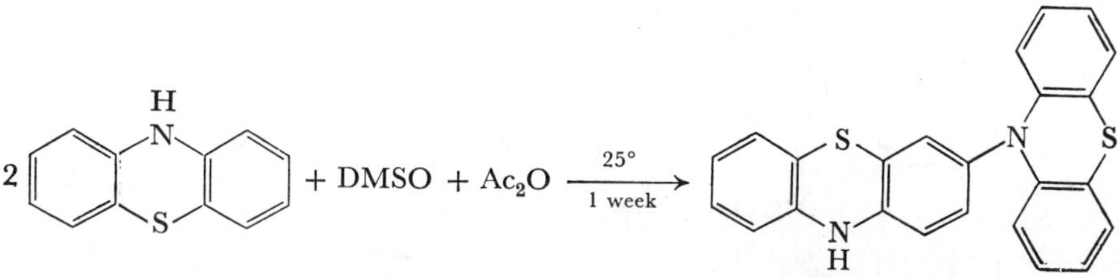

3,10′-Biphenothiazine, 42%

[ref. 33]

The fourth type of intermediate is the nitrene generated from azides by heat or irradiation:

$$RCH_2N_3 \rightarrow RCH_2\ddot{N}: + N_2$$

This intermediate can isomerize to imine $RCH=NH$, abstract intermolecular hydrogen to form an amine RCH_2NH_2, insert intramolecularly to form pyrrolidines (Ex. a) or intermolecularly to form azepines (Ex. b.2), or add to olefins to give aziridines

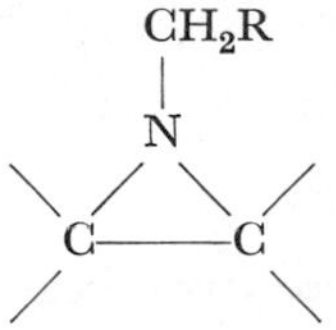

(Ex. b.1) [34, 35]. Aromatic nitrenes give mainly the coupling products, azobenzenes $ArN=NAr$. But, if the nitrene group is in a favorable position for cyclization, ring closure may take place. The best procedure utilizes a relatively new and simple procedure for generating nitrenes *in situ* as shown [36]:

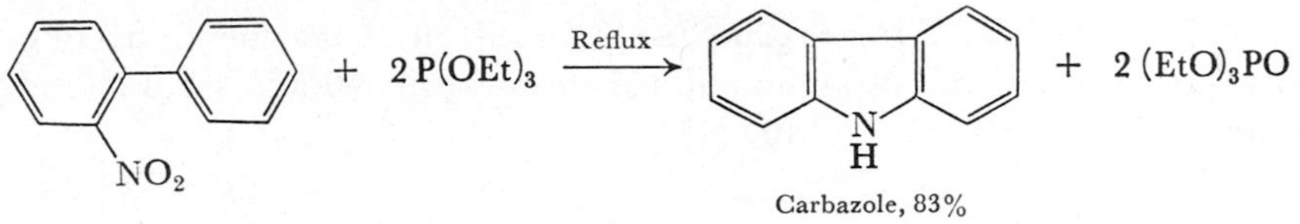

Carbazole, 83%

or

cis

2-Phenylindole, 85%

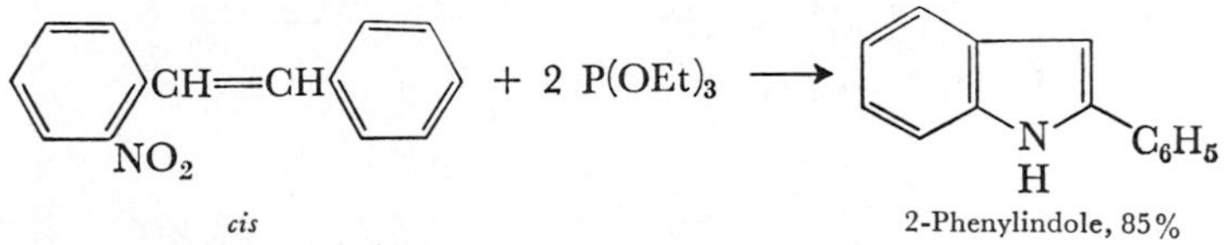

o-Nitrodiphenylsulfide yields phenothiazine.

Whether the nitrene is formed or not, hydroxylamine-*O*-sulfonic acid has been shown to give the product expected from a nitrene [37]:

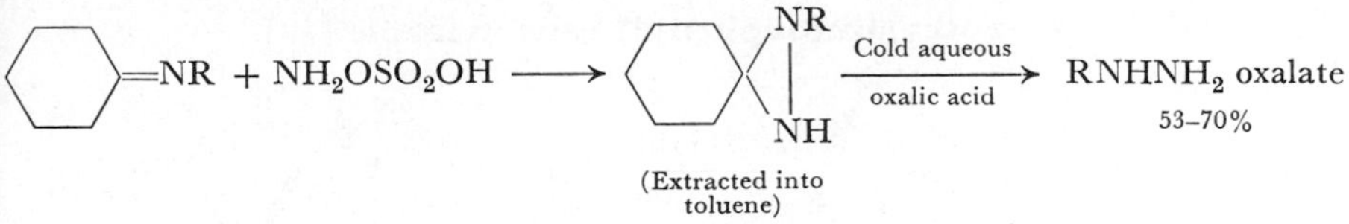

(Extracted into toluene)

RNHNH$_2$ oxalate
53–70%

A recent synthesis of 2H-azirines using the nitrene intermediate has become available [38]:

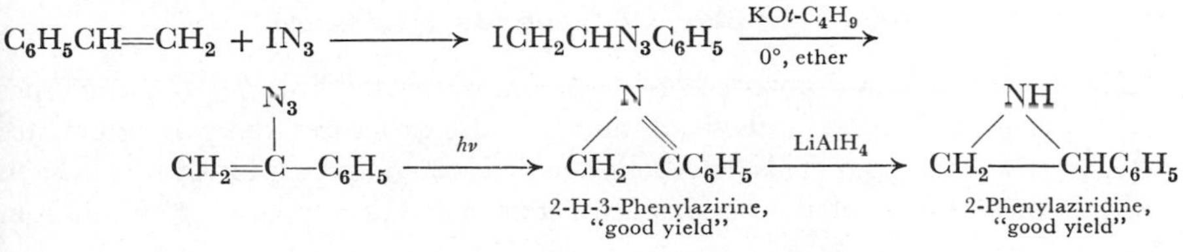

$$C_6H_5CH=CH_2 + IN_3 \longrightarrow ICH_2CHN_3C_6H_5 \xrightarrow[0°, \text{ether}]{KOt\text{-}C_4H_9}$$

2-H-3-Phenylazirine, "good yield"

2-Phenylaziridine, "good yield"

(a) Preparation of 2-ethyl-2-methylindoline (this compound retains

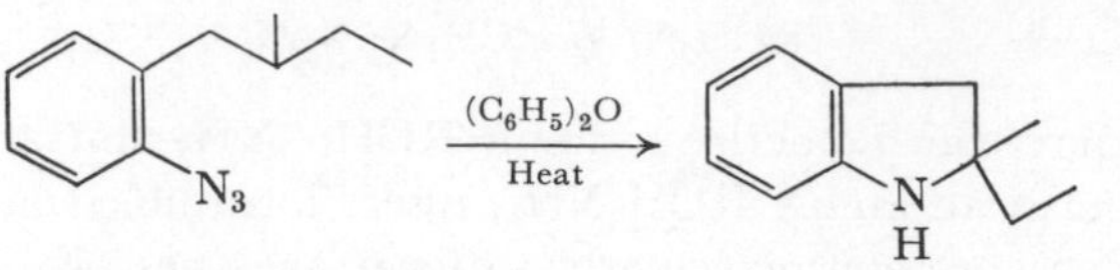

its optical configuration) [39].

(b) Other examples

(1) 1,2-*p*-Methoxyphenyliminoindane

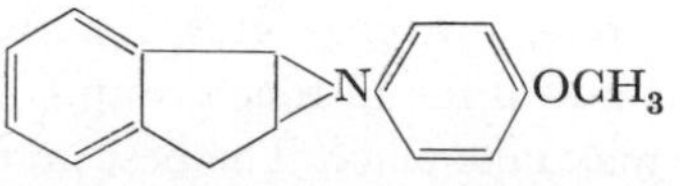

(45 % from indene and *p*-methoxyphenylazide at 130° for 2 hr.) [40].

(2) 2-Anilino-7-H-azepine (41 % based on the azide from 20 g. of phenylazide in 300 ml. of aniline added dropwise in 90 min. to 1200 ml. of aniline at 165° and held for 2 hr. [41]:

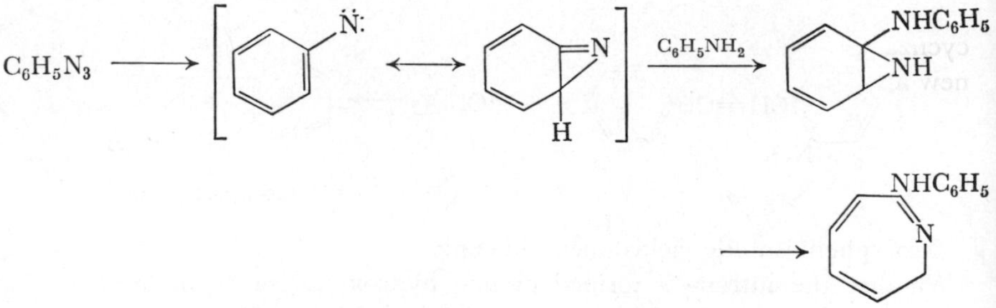

(3) 2-(4-Dimethylaminophenyl)-benzotriazole [42].

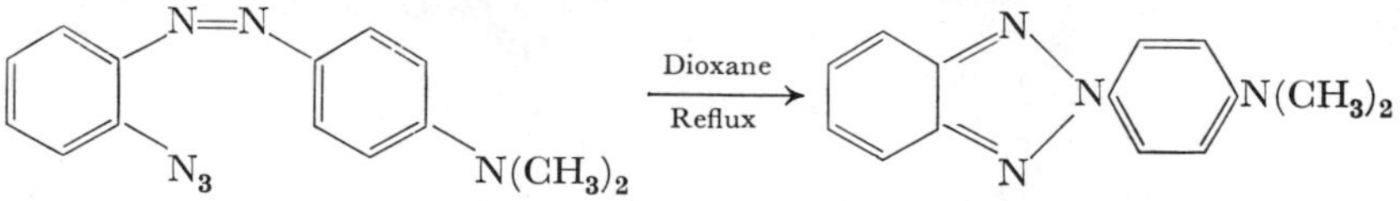

5. From 1,3- or 1,4-dipolar Compounds (Huisgen)

The addition of 1,3-dipolar compounds to unsaturated linkages of many types is a powerful tool in the synthesis of unique cyclic structures, many of which are amino cyclic structures. This field has been delineated and exploited by Huisgen [43]. Typical 1,3-dipolar functions are listed in the accompanying tabulation.

Name	Structure	Structure for Addition
Nitrilium betaine	$-C\equiv\overset{\oplus}{N}-\overset{\ominus}{\underset{..}{C}}\diagdown$	$\longleftrightarrow$	$-\overset{\oplus}{C}=N-\overset{\ominus}{\underset{..}{C}}\diagdown$
Nitrilimines	$-C\equiv\overset{\oplus}{N}-\overset{\ominus}{\underset{..}{N}}-$	$\longleftrightarrow$	$-\overset{\oplus}{C}=N-\overset{\ominus}{\underset{..}{N}}-$
Nitrile oxides or nitrones	$-C\equiv\overset{\oplus}{N}-\overset{\ominus}{\underset{..}{O}}:$	$\longleftrightarrow$	$-\overset{\oplus}{C}=N-\overset{\ominus}{\underset{..}{O}}:$

The nitrilium betaine is derived from imido chlorides in a reaction which proceeds as follows:

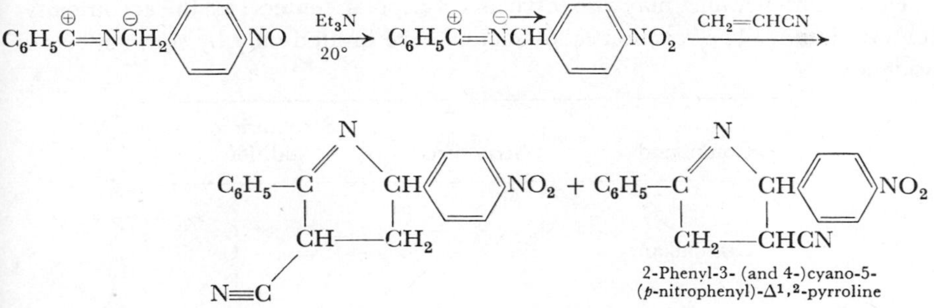

$$C_6H_5\overset{\oplus}{C}=\overset{\ominus}{N}CH_2\!\!\diagup\!\!\diagdown\!\!NO \xrightarrow[20°]{Et_3N} C_6H_5\overset{\oplus}{C}=\overset{\ominus}{N}CH\!\!\diagup\!\!\diagdown\!\!NO_2 \xrightarrow{CH_2=CHCN}$$

2-Phenyl-3- (and 4-)cyano-5-
(p-nitrophenyl)-$\Delta^{1,2}$-pyrroline

Also the nitrilimines are available from tetrazoles, α-chlorobenzylidinephenyl-hydrazones, or α-chloro-β-ketopropylidenephenylhydrazones as shown:

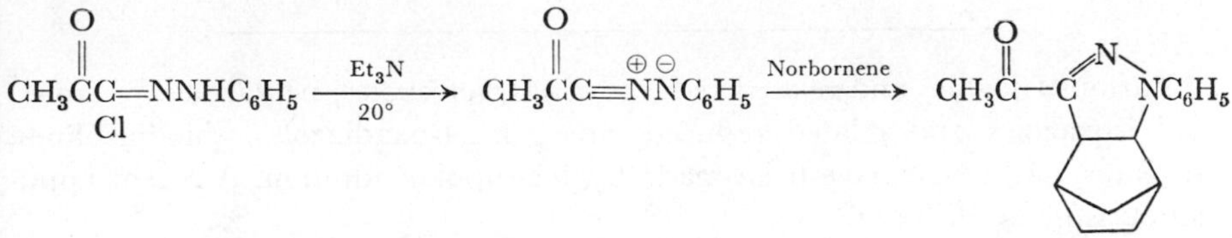

A pyrazoline, 84%

Nitrile oxides (RC≡NO) yield isoxazolines

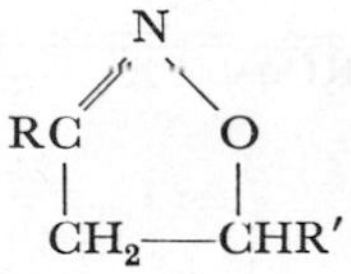

on treatment with various dipolarphiles, while nitrones

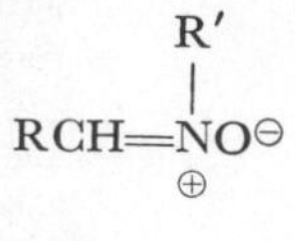

$$\overset{\displaystyle R'}{\underset{\oplus}{RCH=\overset{|}{NO}^{\ominus}}}$$

yield isoxazolidines. The nitrones can be prepared *in situ* as shown [44]:

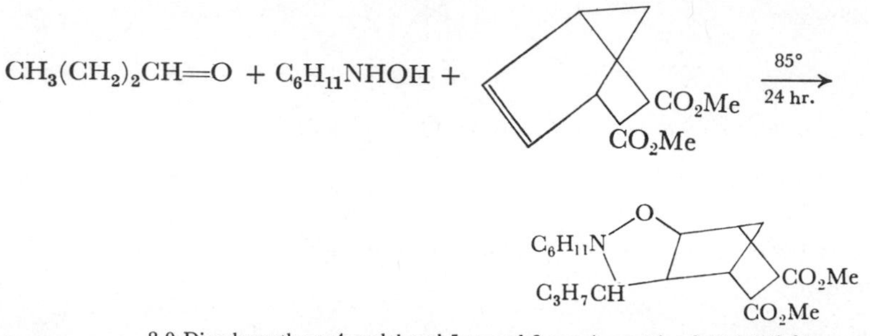

$$CH_3(CH_2)_2CH=O + C_6H_{11}NHOH +$$

$$\xrightarrow[\text{24 hr.}]{85°}$$

8,9-Dicarbomethoxy-4-cyclohexyl-5-propyl-3-oxa-4-aza-tricyclo[5.2.1·0²,⁶]decane

Diazonium betaines may also serve as 1,3-dipolar centers (see the accompanying tabulation.) The reaction scope seems to be limited only by steric effects in addends.

Compound	Structure	Structure for Addition
Diazoalkanes	$\overset{\oplus}{N}{\equiv}\overset{\ominus}{N}{-}\overset{\diagup}{\underset{\diagdown}{C}}$ ↔	$\overset{\oplus}{N}{=}\overset{}{N}{-}\overset{\ominus}{\underset{\diagdown}{C}}\diagup$
Azides	$\overset{\oplus}{N}{\equiv}\overset{}{N}{-}\overset{\ominus}{N}{-}$ ↔	$\overset{\oplus}{N}{=}\overset{}{N}{-}\overset{\ominus}{N}{-}$
Nitrous oxides	$\overset{\oplus}{N}{\equiv}\overset{}{N}{-}\overset{\ominus}{O}{:}$ ↔	$\overset{\oplus}{N}{=}\overset{}{N}{-}\overset{\ominus}{O}{:}$

Various triazoles and related reduced rings, tetrazoles, pentazoles, oxazoles, and isoxazoles and related reduced rings, 1,2,4-oxadiazoles, thiodiazolines, furoxans, and others have been made by 1,3-dipolar addition. A recent application is given in Ex. a.

Similar principles expounded by Huisgen can be applied to 1,4-dipolar compounds in which a carbonyl oxygen acts as the negative end of a dipole [45]:

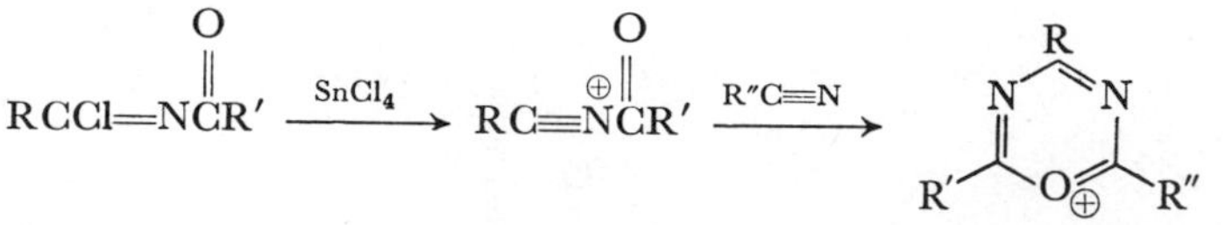

3,5-Diaza-pyrylium salts

(a) Preparation of 1-methyl-4-benzoyl-6-carbethoxy-1,3α-diaza-pentalene (a dehydrogenation is involved in the last step) [46].

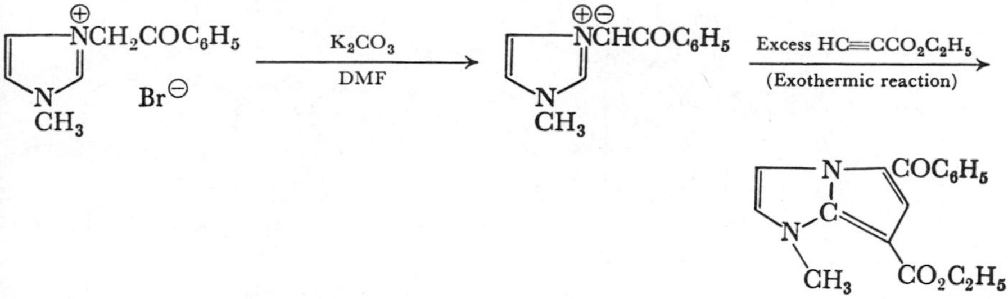

14%

6. From Azo, Azomethine, or Nitroso Compounds (Diels-Alder)

The Diels-Alder reaction has been discussed under Alkenes, Chapter 2, but brief reference is made here to possibilities involving nitrogen atoms. Nitriles are poor dienophiles and give only low yields of dihydropyridines with butadienes [47]. Pyridines and quinolines similarly are poor dienophiles, but their 1,2-dihydro derivatives are good ones [48]:

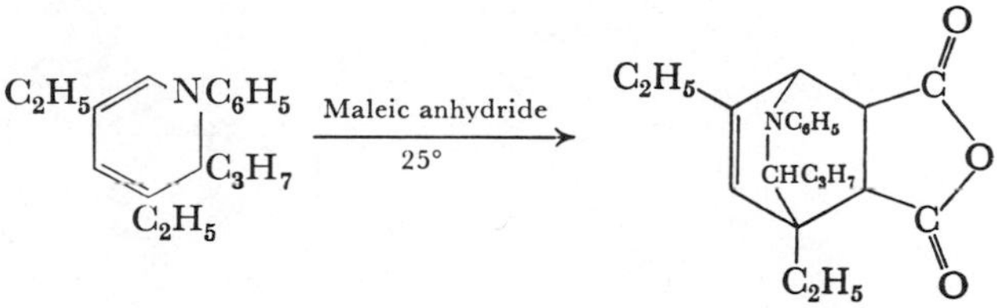

2,4-Diethyl-7-phenyl-8-propyl-7-azabicyclo[2,2,2]oct-2-ene-5,6-dicarboxylic anhydride, 78%

The very versatile ethynylamines (see H.3 for preparation) react directly with the azomethine linkage [49]:

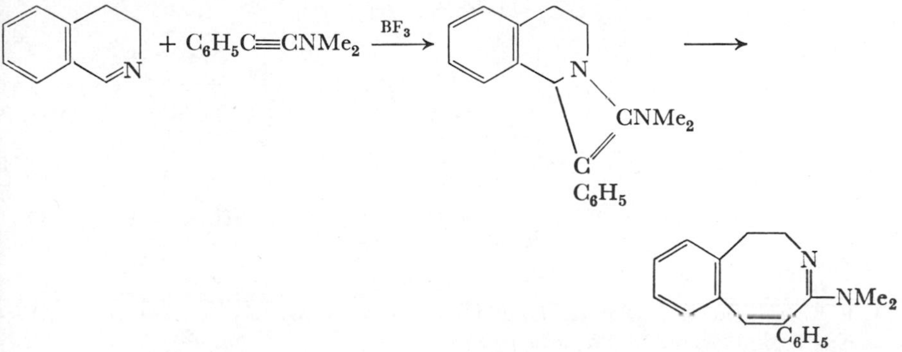

1-Aza-5,6-benzo-2-dimethyl-amino-3-phenylcycloocta-1,3,5-triene (no yield given)

Schiff bases combine with vinyl ethers using boron trifluoride etherate as a catalyst [50],

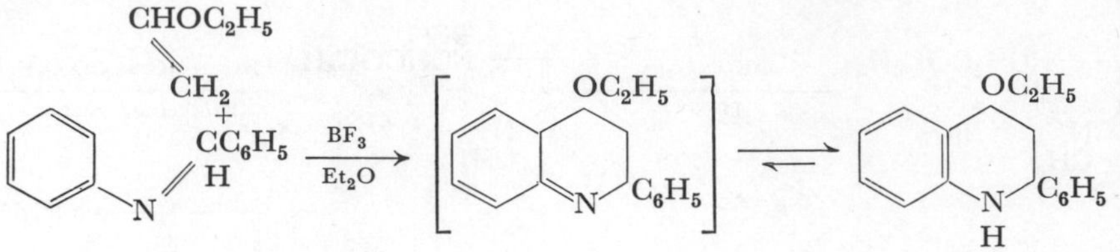

4-Ethoxy-2-phenyl-1,2,3,4-
tetrahydroquinoline

Azodicarboxylic esters, of course, are excellent dienophiles giving access to tetrahydropyridazines [51],

(a) Preparation of N,N′-phenylethyleneindigo [52],

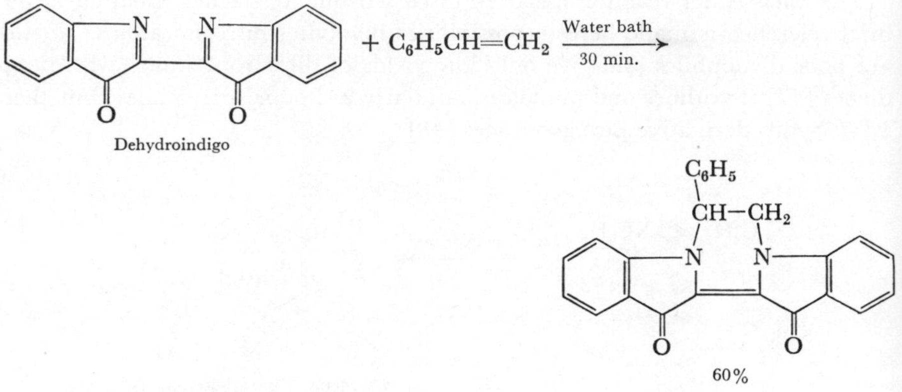

Dehydroindigo

60%

(b) Preparation of α-isopropenyl-α-[1-methyl-1-(N-phenylhydroxylamino)-ethyl]-N-phenylnitrone [53].

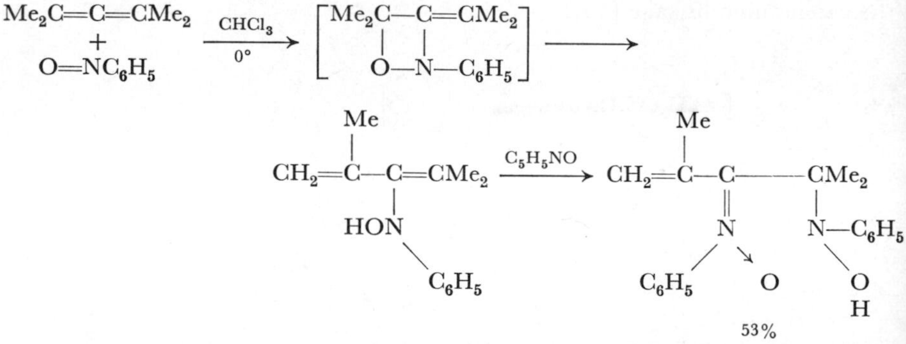

53%

1. P. Kovacic, in G. A. Olah *ed.*, *Friedel-Crafts and Related Reactions*, Interscience Publishers, New York, 1964, Vol. 3, Pt. 2, Chap. 44.
2. P. Kovacic *et al.*, *J. Am. Chem. Soc.*, **83,** 221 (1961); *ibid.* **84,** 759 (1962); *J. Org. Chem.*, **26,** 3013 (1961).

3. G. M. Hoop and J. M. Tedder, *J. Chem. Soc.*, 4685 (1961).
4. P. Kovacic *et al.*, *J. Am. Chem. Soc.*, **88**, 100 (1966).
5. C. M. Suter *et al.*, *J. Am. Chem. Soc.*, **65**, 674 (1943).
6. C. M. Suter and A. W. Ruddy, *J. Am. Chem. Soc.*, **65**, 762 (1943).
7. R. Hausigk, *Tetrahedron Letters*, 2801 (1968).
8. S. H. Patinkin and B. S. Friedman Ref. 1, 1964, Vol. 2, Pt. 1, Chap. 14.
9. F. Minisci and M. Cecere, *Chim. Ind. (Milan)*, **49**, 1333 (1967); *C.A.*, **68**, 9186 (1968); *see also* Netherlands Patent 6,614,947; *C.A.*, **68**, 6614 (1968).
10. P. G. Gassman and B. L. Fox, *J. Am. Chem. Soc.*, **89**, 338 (1967).
11. H. Bock and K.-L. Kompa, *Angew. Chem.*, **77**, 807 (1965).
12. D. E. Pearson and M. Y. Moss, unpublished results.
13. N. Milstein, *J. Heterocycl. Chem.*, **5**, 339 (1968).
14. B. Staskun, *J. Org. Chem.*, **33**, 3031 (1968).
15. P. Kovacic *et al.*, *Chem. Commun.*, 232 (1966); *see* P. Kovacic *et al.*, *J. Org. Chem.*, **33**, 1515 (1968) for other preparations of a similar nature.
16. P. Kovacic and P. D. Roskos, *Tetrahedron Letters*, 5833 (1968).
17. R. H. F. Manske and M. Kulka, *Org. Reactions*, **7**, 59 (1953).
18. W. H. Mills *et al.*, *J. Chem. Soc.*, **119**, 1294 (1921).
19. B. D. Tilak *et al.*, *Tetrahedron Letters*, 1959 (1966).
20. A. Albert, *Heterocyclic Chemistry*, Second Ed., Oxford University Press, New York, 1968, p. 160.
21. B. Staskun, *J. Org. Chem.*, **29**, 1153 (1964); L. Knorr, *Chem. Ann.*, **236**, 69 (1886).
22. K. and C. M. Desai, *Indian J. Chem.*, **5**, 170 (1967).
23. H. T. Clarke and A. W. Davis, *Org. Syn.*, Coll. Vol. **1**, 478 (1941).
24. W. E. Cole, Master's Thesis, Vanderbilt University, Nashville, Tenn., 1949.
25. H. S. Mosher *et al.*, *Org. Syn.*, Coll. Vol. **3**, 568 (1955).
26. D. E. Pearson *et al.*, unpublished work.
27. L. H. Davis, Master's Thesis, Vanderbilt University, Nashville, Tenn., 1969.
28. O. E. Edwards *et al.*, *J. Am. Chem. Soc.*, **87**, 678 (1965).
29. E. J. Corey and W. R. Hertler, *J. Am. Chem. Soc.*, **82**, 1657 (1960).
30. M. Okahara *et al.*, *J. Org. Chem.*, **33**, 3066 (1968).
31. R. S. Neale and N. L. Marcus, *J. Org. Chem.*, **33**, 3457 (1968).
32. Y. L. Chow, *Chem. Commun.*, 330 (1967).
33. Y. Tsujino, *Tetrahedron Letters*, 4111 (1968).
34. L. Horner and A. Christmann, *Angew. Chem. Intern. Ed. Engl.*, **2**, 599 (1963).
35. R. A. Abramovitch and B. A. Davis, *Chem. Rev.*, **64**, 149 (1964).
36. J. I. G. Cadogan, *Quart. Rev.*, **22**, 222 (1968).
37. R. Ohme *et al.*, *J. Prakt. Chem.*, **37**, 257 (1968).
38. A. Hassner and F. W. Fowler, *Tetrahedron Letters*, 1545 (1967); *J. Org. Chem.*, **33**, 2686 (1968).
39. G. Smolinsky and B. I. Feuer, *J. Am. Chem. Soc.*, **86**, 3085 (1964).
40. P. Walker and W. A. Waters, *J. Chem. Soc.*, 1632 (1962).
41. R. Huisgen *et al.*, *Chem. Ber.*, **91**, 1 (1958).
42. J. H. Hall, *J. Org. Chem.*, **33**, 2954 (1968).
43. R. Huisgen, *Proc. Chem. Soc.*, 357 (1961); *J. Org. Chem.*, **33**, 2291 (1968).
44. R. Huisgen *et al.*, *Chem. Ber.*, **101**, 2043 (1968).
45. R. R. Schmidt, *Tetrahedron Letters*, 3443 (1968).
46. V. Boekelheide and N. A. Fedoruk, *J. Am. Chem. Soc.*, **90**, 3830 (1968).
47. A. S. Onishchenko, *Diene Synthesis*, Daniel Davey and Co., 257 Park Ave., S., New York, 1964, p. 126.
48. Ref. 47, p. 579.
49. H. G. Viehe, *Angew. Chem. Intern. Ed. Engl.*, **6**, 767 (1967).
50. Ref. 47, p. 552.
51. Ref. 47, p. 127.
52. Ref. 47, p. 549.
53. R. K. Howe, *J. Org. Chem.*, **33**, 2848 (1968).

G. Molecular Rearrangements

The first five parts of this section are devoted to the rearrangements brought about by electron- or group-deficient nitrogen atoms:

$$-\overset{\oplus}{N}- \quad \text{or} \quad -\overset{..}{\underset{..}{N}}$$

To coordinate all of them, a systematic itemization is given here. The carbonyl oxidation level is the most useful stage at which to establish electron-deficiency.

   Rearrangement.

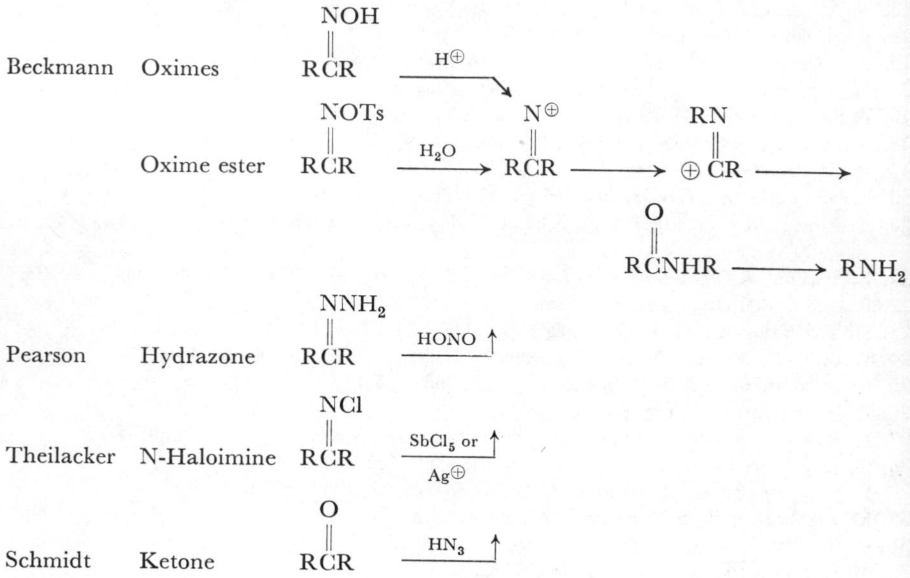

Beckmann   Oximes
$$\overset{\text{NOH}}{\underset{}{\overset{\|}{R C R}}} \xrightarrow{\text{H}\oplus}$$

Oxime ester
$$\overset{\text{NOTs}}{\underset{}{\overset{\|}{R C R}}} \xrightarrow{\text{H}_2\text{O}} \overset{\text{N}\oplus}{\underset{}{\overset{\|}{R C R}}} \longrightarrow \overset{\text{RN}}{\underset{\oplus \text{ C R}}{\overset{\|}{}}} \longrightarrow$$

$$\overset{\text{O}}{\underset{}{\overset{\|}{R C N H R}}} \longrightarrow RNH_2$$

Pearson   Hydrazone
$$\overset{\text{NNH}_2}{\underset{}{\overset{\|}{R C R}}} \xrightarrow{\text{HONO}} \uparrow$$

Theilacker   N-Haloimine
$$\overset{\text{NCl}}{\underset{}{\overset{\|}{R C R}}} \xrightarrow[\text{Ag}\oplus]{\text{SbCl}_5 \text{ or}} \uparrow$$

Schmidt   Ketone
$$\overset{\text{O}}{\underset{}{\overset{\|}{R C R}}} \xrightarrow{\text{HN}_3} \uparrow$$

With the exception of the latter rearrangement and of a special aspect of the Beckmann rearrangement (G.1), they are discussed under Amides and Imides, Chapter 18. A corresponding series of rearrangements, most of which are discussed in this section, are to be found in the acid level of oxidation.

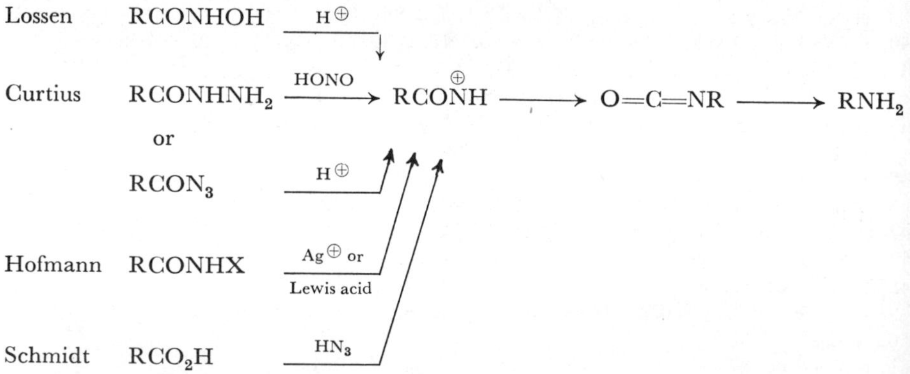

Lossen   RCONHOH $\xrightarrow{\text{H}\oplus}$

Curtius   RCONHNH$_2$ $\xrightarrow{\text{HONO}}$ RCON$\overset{\oplus}{\text{H}}$ $\longrightarrow$ O=C=NR $\longrightarrow$ RNH$_2$

   or

   RCON$_3$ $\xrightarrow{\text{H}\oplus}$

Hofmann   RCONHX $\xrightarrow[\text{Lewis acid}]{\text{Ag}\oplus \text{ or}}$

Schmidt   RCO$_2$H $\xrightarrow{\text{HN}_3}$

References for Section G are on pp. 508–509.

In the case of the Hofmann and Lossen rearrangements, it is more convenient to proceed via the nitrene intermediate rather than via an electron-deficient nitrogen atom, i.e., not

$$RCONHX \xrightarrow[\text{Lewis acid}]{Ag^{\oplus} \text{ or}} R\overset{\overset{\displaystyle O}{\|}}{C}NH^{\oplus}$$

but rather

$$RCONHX \xrightarrow{OH^{\ominus}} RCONX^{\ominus} \xrightarrow{-X^{\ominus}} RCO\ddot{N}:$$

A corresponding series of rearrangements can be listed for the alcohol level, e.g.,

$$R_3CNHOH \xrightarrow{H^{\oplus}} R_2\overset{\oplus}{C}-NHR \longrightarrow R_2C{=}NR$$

but these rearrangements are not useful enough in preparing amines to discuss in any detail [1]. In this section, G.6 and G.7 include rearrangements (benzidine and Fischer indole, respectively) of a different type but nevertheless acid-catalyzed. G.8, on the other hand, discusses an anionic rearrangement (Stevens), and G.9 a collection of miscellaneous rearrangements.

1. From Some Oximes (Semmler-Wolff Aromatization)

For smooth rearrangement of aromatic oximes to acetanilides, the oximino-ethyl grouping

$$\begin{array}{c} NOH \\ \| \\ -CCH_3 \end{array}$$

must twist 90° to a position perpendicular to the plane of the benzene ring to permit coalescence of the electron-deficient nitrogen with the π cloud of the ring [2]. This twisting cannot take place with tetralone oxime because the cyclo oximinoalkyl group is a part of the ring. Therefore a lower-energy transformation of a different type is available via hydride shifts and proton loss [3]:

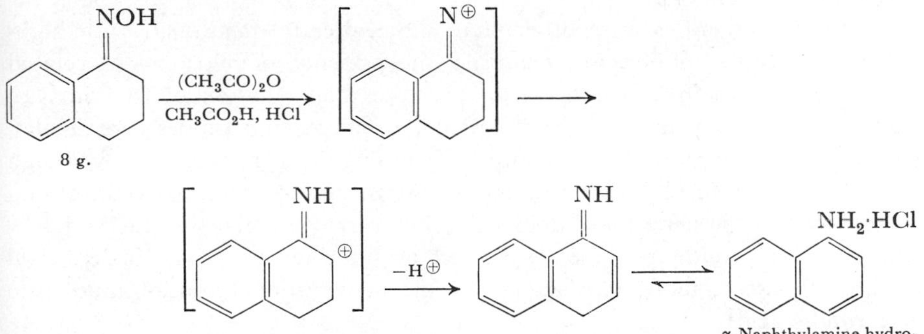

α-Naphthylamine hydrochloride, 5 g. crude

Also, α,β-unsaturated ketoximes may have a lower-energy transformation (than the Beckmann rearrangement) to give aromatization products via similar hydride or alkyl shifts and proton loss [4]:

$$
\underset{\substack{CH_3 \\ CH_3}}{\overset{NOH}{\bigcirc}} CH_3 \xrightarrow[\text{65--100°}]{(CH_3CO)_2O,\ C_5H_5N,\ CH_3COCl}
$$

$$
\underset{\substack{CH_3 \\ CH_3}}{\overset{NHCOCH_3}{\bigcirc}}CH_3 \quad + \quad \underset{\substack{CH_3 \\ CH_3}}{\overset{NHCOCH_3}{\bigcirc}}CH_3
$$

86%

Hydrolysis of the amides and recrystallization from hexane yields 50% of 3,4,5-trimethylaniline and, from the mother liquor, recrystallization of the hydrochlorides yields 20% 2,3,5-trimethylaniline.

2. From Amides via the Bromamide (Hofmann)

$$
RCONH_2 \xrightarrow{NaOX} RCON\overset{\ominus}{X} \xrightarrow{-X^{\ominus}} [RC\overset{..}{O}N:] \longrightarrow RNCO \xrightarrow{H_2O} RNH_2
$$

The Hofmann reaction has been reviewed [5]. The amide is usually dissolved in a slight excess of the cold aqueous hypohalite, after which the mixture is warmed. For higher aliphatic amides in which nitriles are the main product, the procedure is modified to yield amines in that bromine, a methanolic solution of the amide, and sodium methoxide are employed. In this case a urethane is first formed:

$$
RCONH_2 + Br_2 + 2\ NaOCH_3 \rightarrow RNHCOOCH_3 + 2\ NaBr + CH_3OH
$$

but this product may be saponified to yield the amine. Dioxane also may be used as a solvent [6]. Other by-products in the Hofmann degradation are ureas and acylureas, both of which are minimized by homogeneity of the reaction medium and rapid conversion of the amide to amine [7]. For rapid conversion the N-haloamide may be isolated and added to methanolic sodium methoxide. Another way for saponification of the urethane is to distill from 3–4 parts of calcium hydroxide [7].

At times, difficulties arise other than those already mentioned with high-molecular-weight amides. For example, the presence of hydroxyl or related functional groups in a benzene ring promotes halogenation of the ring. In addition, α-hydroxyamides yield aldehydes, α,β-acetylenic amides give nitriles, and highly fluorinated aliphatic amides lead largely to halides [8]. Although sodium or potassium hypobromite is the customary reagent employed, sodium hypochlorite is preferred at times. The latter reagent gives better yields of amines if the amide possesses protected or unprotected aromatic hydroxyl groups. It also is more satisfactory for the conversion of phthalimides into anthranilic acids:

$$
\underset{}{\bigcirc}\overset{C=O}{\underset{C=O}{>}}NH \longrightarrow \underset{}{\bigcirc}\overset{NH_2}{\underset{COOH}{}}
$$

As a rule, yields in the synthesis are satisfactory.

The isocyanate, which can sometimes be isolated, is an intermediate not only in the Hofmann but in the Curtius (G.3) and Lossen (G.4) rearrangements as well.

Some chemistry may be learned from the behavior of perfluoroamides in the Hofmann rearrangement. With ordinary procedures, in the strongly basic aqueous media, no amines are obtained from these halogenated amides but, rather, high yields of bromides or chlorides [9]:

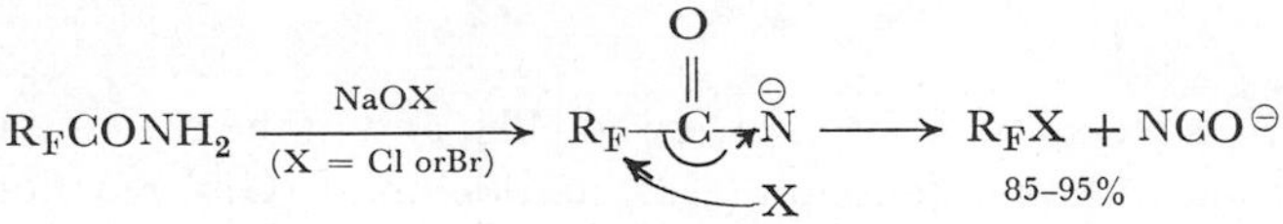

$$R_FCONH_2 \xrightarrow[\text{(X = Cl or Br)}]{\text{NaOX}} R_F-\overset{\overset{\text{O}}{\|}}{C}\overset{\ominus}{\underset{X}{N}} \longrightarrow R_FX + NCO^{\ominus}$$
$$85\text{--}95\%$$

However, the anhydrous sodium salt of the N-bromo derivative may be transformed to the desired isocyanate simply by heating:

$$C_3F_7CO\overset{\overset{\text{Br}}{|}}{N}Na \xrightarrow{170°} C_3F_7N{=}C{=}O + NaBr$$

Perfluoropropyliso-
cyanate, 83%

More chemistry may also be learned from a consideration of the question whether the nitrene or nitrenoid intermediate can be realized from the use of reagents other than hypohalite. Lead tetraacetate turns out to be such a reagent [10]:

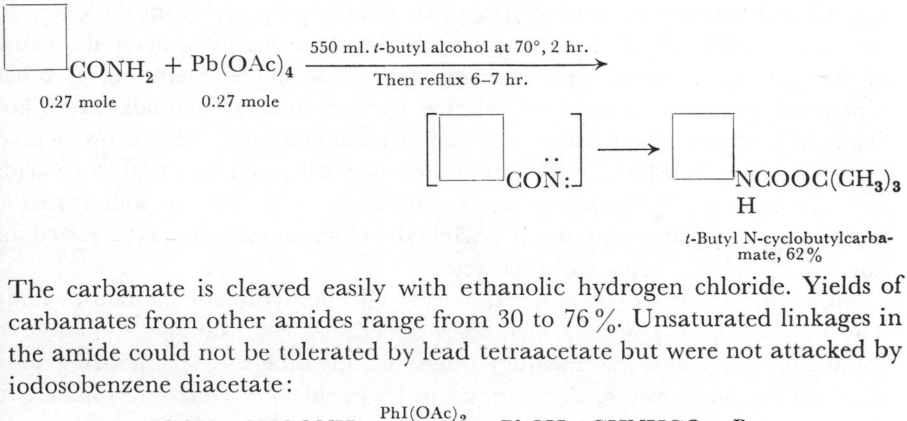

0.27 mole   0.27 mole

t-Butyl N-cyclobutylcarba-
mate, 62%

The carbamate is cleaved easily with ethanolic hydrogen chloride. Yields of carbamates from other amides range from 30 to 76%. Unsaturated linkages in the amide could not be tolerated by lead tetraacetate but were not attacked by iodosobenzene diacetate:

$$PhCH{=}CHCONH_2 \xrightarrow[\text{t-BuOH}]{PhI(OAc)_2} PhCH{=}CHNHCO_2\text{-}t\text{-Bu}$$

t-Butyl N-styrylcarbamate

(a) Preparation of 4-aminoveratrole (80–82% from veratric amide, sodium hypochlorite, and sodium hydroxide at moderate temperature) [11].

(b) Other examples

(1) 3-Aminopyridine (65–71% from nicotinamide, bromine, and sodium hydroxide first at 0° and later at 70–75° for 45 min.) [12].

(2) Tri-n-butylcarbinamine (72% from tributylacetamide, bromine, and sodium hydroxide at 0° for 4 hr., after which the isolated isocyanate was hydrolyzed with hydrochloric acid) [13].

3. From Hydrazides or Acid Chlorides via the Azide (Curtius)

$$RCONHNH_2 \xrightarrow{\text{HONO}}$$

$$RCON_3 \longrightarrow RNCO \xrightarrow{H_2O} RNH_2$$

$$RCOCl \xrightarrow{NaN_3}$$

The Curtius reaction has been reviewed [14], and in the review a comparison of the reaction with the Hofmann (G.2) and Schmidt (G.5) reactions has been made. One may start either with the ester and proceed via the hydrazide or with the acid and proceed via the acid chloride. The azide formed, in either case, in an inert solvent like benzene or chloroform loses nitrogen to give the isocyanate, after which the path taken coincides with the Hofmann reaction. Or the azide may be treated with water or alcohol to yield the urea and urethane, respectively, either of which may be converted into the amine by hydrolysis (see Ex. c). As a rule the yields are satisfactory.

The Curtius degradation has been accomplished successfully on aliphatic, alicyclic, aromatic, and heterocyclic acids. Whether this approach to amines offers any advantage over the Hofmann degradation depends on circumstances. If an ester of the carboxylic acid is available, the Curtius reaction would probably be favored since the hydrazide can be readily prepared from the ester. In starting with the acid, the Hofmann reaction would probably be favored because of the possibility of condensing more steps in a single operation. If other functional groups are present, the choice between the two methods varies. For example, to degrade both carboxyl groups of a malonic acid, the Curtius method must be chosen. In addition, the Curtius reaction is favored for unsaturated acids and aromatic acids containing active halogens (both by the sodium azide method), and acylated amino acids, while the Hofmann reaction is preferred for nonacylated amino acids and keto acids.

Although in many cases either the route via the hydrazide or sodium azide may be chosen, at times one method is preferable. For example, the latter in employing a nonaqueous medium offers an advantage in synthesizing low-molecular-weight amines, the azides and hydrazides of which are difficult to extract from water. Since the higher aliphatic and aromatic esters react poorly with hydrazine, the sodium azide method offers an advantage in these cases. The sodium azide method also is preferred when dealing with unsaturated acids since more side reactions occur in the alternative route.

Side reactions at times interfere with the success of the Curtius reaction. Mono-substituted malonyl azides, for example, give aldehydes via the urethane:

$$RCH \begin{array}{c} CON_3 \\ \\ CON_3 \end{array} \xrightarrow{C_2H_5OH} RCH \begin{array}{c} NHCOOC_2H_5 \\ \\ NHCOOC_2H_5 \end{array} \xrightarrow{H^{\oplus}} RCHO$$

Hydroxy acids sometimes do not form azides properly, or the azide or isocyanate formed may behave abnormally. For example, an azide of an α-hydroxy acid may form an isocyanate which decomposes to give an aldehyde or a ketone:

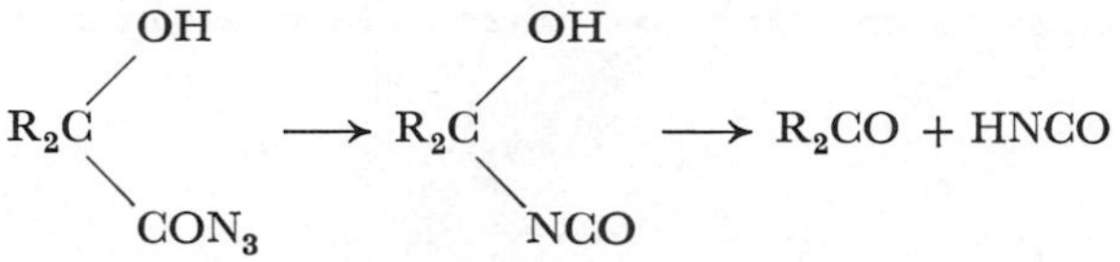

$$R_2C \begin{matrix} OH \\ \\ CON_3 \end{matrix} \longrightarrow R_2C \begin{matrix} OH \\ \\ NCO \end{matrix} \longrightarrow R_2CO + HNCO$$

Similarly, α-halogen azides lead to isocyanates which hydrolyze to aldehydes or ketones.

In the case of *cis*-2-phenylcyclopropanecarboxylic acid the acid chloride route is unsatisfactory because of ready conversion of the *cis*- into the *trans*-acid chloride. To overcome this difficulty the acid was converted into the triethyl-amine salt, which with ethyl chloroformate gave the mixed anhydride, which in turn readily gave the azide with sodium azide [15]:

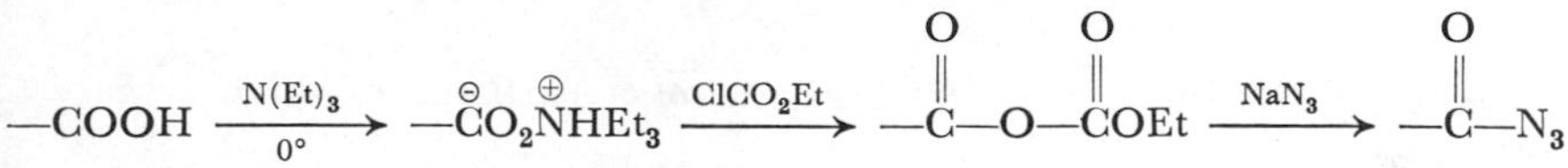

$$-COOH \xrightarrow[0°]{N(Et)_3} -\overset{\ominus}{CO_2}\overset{\oplus}{NHEt_3} \xrightarrow{ClCO_2Et} \overset{O}{\underset{\|}{-C}}-O-\overset{O}{\underset{\|}{C}}OEt \xrightarrow{NaN_3} \overset{O}{\underset{\|}{-C}}-N_3$$

By employing this route, *cis*-2-phenylcyclopropylamine was prepared in 77% yield from the corresponding *cis*-acid.

An interesting conversion which might be considered as related to the Curtius rearrangement occurs in the synthesis of 3-hydroxycinnoline [16]:

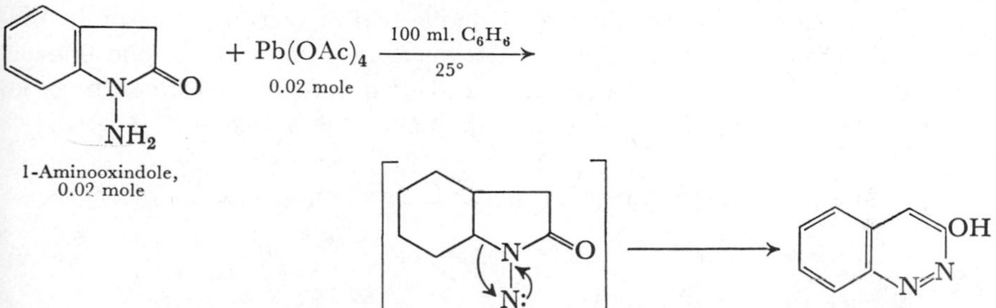

1-Aminooxindole,
0.02 mole

3-Hydroxycinnoline,
78% crude, 63% pure

A similar pathway is followed to generate benzyne [17]:

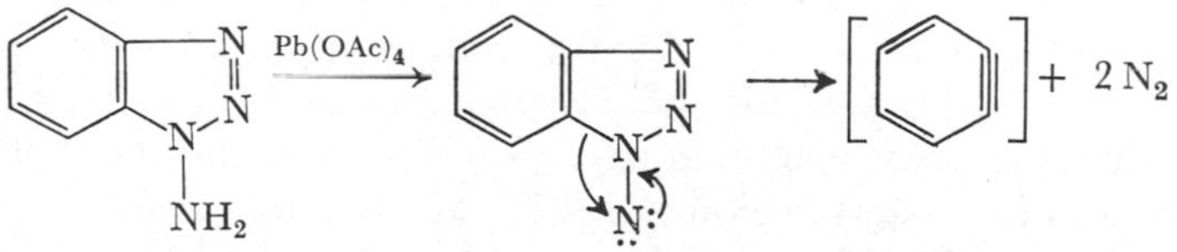

The mechanism of the Curtius reaction is similar to that of the Hofmann reaction (see G.2). The azide also forms the acyl nitrene which rearranges to the isocyanate which in turn hydrolyzes to the amine.

(a) **Preparation of putrescine dihydrochloride** $((CH_2)_4(NH_2 \cdot HCl)_2)$ (73–77% from adipyl dihydrazide, sodium nitrite below 10°, and hydrochloric acid followed by heating) [18].

(b) **Preparation of 3-amino-5-phenyl-2-isoxazoline** (84% based on

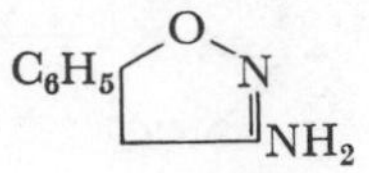

the azide from 5-phenyl-2-isoxazoline-3-carboxylic acid hydrazide treated first with 10% hydrochloric acid and aqueous sodium nitrite in ice and then heated on a steam bath with aqueous trifluoroacetic acid) [19].

(c) **Preparation of** *unsym*-**diphenylhydrazine hydrochloride** [20]:

$$(C_6H_5)_2N\overset{O}{\overset{\|}{C}}N_3 \xrightarrow[\text{Reflux 5 days}]{(CH_3)_3COH} (C_6H_5)_2NNH\overset{O}{\overset{\|}{C}}OC(CH_3)_3 \xrightarrow[\text{Reflux 3 hr.}]{HCl} (C_6H_5)_2NNH_2 \cdot HCl$$
$$78\%$$

(this reaction demonstrates the formation of a carbamate which is easy to hydrolyze.)

4. From Hydroxamic Acids (Lossen)

$$ArCONHOH \xrightarrow{\Delta} ArNCO \xrightarrow{H_2O} ArNH_2$$

The Lossen rearrangement has been reviewed [21]. The reaction occurs by heating hydroxamic acids or their acyl derivatives in inert solvents or in the presence of substances such as thionyl chloride, acetic anhydride, or phosphorus pentoxide. The selection of the derivative and environment determines the type of intermediate, but the final product is the same in either case:

$$RCONHOSO_2Ar \xrightarrow{OH^{\ominus}} RC\overset{\ominus}{ONOSO_2Ar} \longrightarrow [RC\overset{..}{O}N:] \longrightarrow O{=}C{=}NR$$

A nitrene

$$\uparrow -H^{\oplus}$$

$$\text{or}\quad RCONHOH + H^{\oplus} \longrightarrow [RCON\overset{\oplus}{H}] \longrightarrow O{=}C{=}N\overset{H}{\underset{\oplus}{R}}$$

The rearrangement is of most interest from a preparative viewpoint because it permits the direct conversion of acids into amines. By heating aromatic carboxylic acids with hydroxylamine and polyphosphoric acid, usually at 150–170° for 5–10 minutes after the evolution of carbon dioxide begins, the hydroxamic acid is produced *in situ* and rearranged [22]. The synthesis is simpler than the Schmidt reaction (G.5), although it is not so generally applicable as the Curtius (G.3) or Hofmann (G.2) degradation. The yields are as high as 82% although in some cases, particularly in the aliphatic series, only traces of the amine are produced with polyphosphoric acid.

Nitromethane in PPA is a source of hydroxylamine [23]:

$$CH_3NO_2 \xrightarrow{PPA} NH_2OH + CO$$

Therefore acids heated with nitromethane in PPA potentially are capable of conversion into amines. Yields of toluidines or chloroanilines usually are in the 70–80% range (see Ex. b). But aliphatic amines and aromatic amines with electron-withdrawing substituents are better prepared from the ester or acid chloride and hydroxylamine.

(a) **Preparation of β-naphthylamine (use of hydroxylamine).** A mixture of 0.024 mole of hydroxylamine hydrochloride, 0.023 mole of β-naphthoic acid, and 50 g. of polyphosphoric acid was stirred as the temperature was gradually raised. The evolution of carbon dioxide ceased at 160° and the brown mixture was poured over 250 g. of crushed ice. The mixture was filtered and the filtrate, after neutralization with potassium hydroxide, yielded the amine which, when dried, weighed 2.7 g. (82%), m.p. 107–109° [22].

(b) **Preparation of p-chloroaniline (use of nitromethane).** A mixture of 100 g. of polyphosphoric acid, 0.08 mole of p-chlorobenzoic acid, and 0.16 mole of nitromethane was heated with stirring at 115° for 90 min. From the mass poured into 200 g. of crushed ice, stirred, and filtered, 3.1 g. of chlorobenzoic acid was recovered. From the filtrate made strongly alkaline with potassium hydroxide and cooled for 2 hr. at 0°, 6.1 g. (60% conversion, 80% yield) of p-chloroaniline, m.p. 69°, was isolated [23].

5. From Carboxylic Acids and Hydrazoic Acid (Schmidt)

$$RCOOH \xrightarrow{HN_3} RNH_2$$

$$\begin{array}{c} R \\ \diagdown \\ CO \xrightarrow{HN_3} RCONHR \\ \diagup \\ R \end{array}$$

Carboxylic acids and ketones react with hydrazoic acid to produce primary amines and amides, respectively. The amides which are discussed under Carboxylic Acid Amides and Imides, Chapter 18, D.5, may serve as a source of amines. The main concern here, however, is the synthesis of primary amines from carboxylic acids, a subject which has been reviewed [24]. The experimental conditions employed in this case are very similar to those utilized in the case of ketones.

At times, particularly when the carboxylic acid is available, the Schmidt reaction is the preferred method of going to the amine. It possesses the advantage over the Hofmann (G.2) and Curtius (G.3) degradations in being a one step reaction. It cannot be used with acids unstable toward sulfuric acid (usually present to produce hydrazoic acid from sodium azide) or with compounds containing easily sulfonated aromatic rings. In addition, certain functional groups such as the cyano, the oxime, and chloro in the 2-position in pyridine or quinoline are converted into tetrazoles with hydrazoic acid alone or in the presence of sulfuric acid. Such reactions sometimes occur, even though only sufficient hydrazoic acid to react with the carboxylic acid function is employed. It is interesting as well to note that the carboxylic acid function in α-aminoacids is unreactive toward hydrazoic acid. With aromatic acids containing electron-releasing groups in the ring, the conversion can be made satisfactorily in PPA with excess sodium azide at room temperature [25]. In some cases, sodium azide with a mixture of equal volumes of trifluoroacetic anhydride and trifluoroacetic acid has been used to advantage [26]; see Ex. b.3.

Yields are variable but are sometimes in the 80–90 % range. A series of long-chain dicarboxylic acids, with one exception, gave yields of diamines varying from 50 to 87 % [27].

Intermediates identical with those discussed in the Curtius (G.3) and Lossen (G.4) rearrangements are involved in the Schmidt rearrangement of acids, and intermediates in the Schmidt rearrangement of ketones are described under Carboxylic Acid Amides and Imides, Chapter 18, D.6.

(a) Preparation of 2-(m-chlorophenyl)-ethylamine.   m-Chlorohydrocinnamic acid, 80 g., in thiophene-free, absolute benzene was warmed with 300 ml. of conc. sulfuric acid to 50°. With stirring, 40 g. of sodium azide was introduced during 4 hr. and the reaction mixture was held at about 50° without further heating. After an additional hour on a 50° bath, the mixture was cooled and poured on ice. The amine sulfate suspension was then made alkaline with cooling and extracted with ether. Distillation at 111–113° (12 Torr.) gave 57.2 g. (85 %) of the primary amine [28].

(b) Other examples

(1) 2 β-Amino-5,5-dimethylbicyclo[2.1.1]hexane (61 % from 5,5-

dimethylbicyclo[2.1.1]hexane-2β-carboxylic acid, sulfuric acid, chloroform, and sodium azide at 45–55° for 2 hr.) [29].

(2) 4,8-Diamino-2,10-dimethylundecane (75 % as the hydrochloride from the dicarboxylic acid in conc. sulfuric acid and benzene treated with sodium azide at about 30°) [27].

(3) 4-Aminophenanthrene (84 %, based on the isocyanate, from phenanthrene-4-carboxylic acid, 5 g., in 100 ml. of a solution containing equal volumes of trifluoroacetic anhydride and trifluoroacetic acid and an excess of sodium azide at 0–5°, after which the isocyanate, 95.5 %, recovered was treated with KOH) [26].

(4) 5-Amino-6-methyl-3,4-pyridinedicarboxylic acid (demonstra-

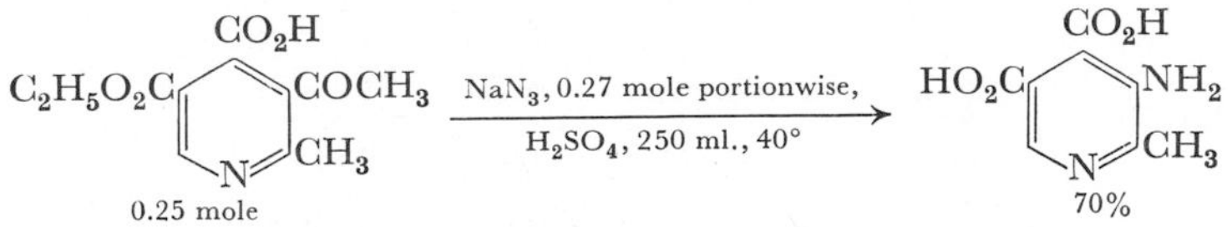

$$C_2H_5O_2C\!\!-\!\!\text{(pyridine)}\!\!-\!\!COCH_3, CH_3 \quad \xrightarrow[\text{H}_2\text{SO}_4, 250 \text{ ml.}, 40°]{\text{NaN}_3, 0.27 \text{ mole portionwise,}} \quad HO_2C\!\!-\!\!\text{(pyridine)}\!\!-\!\!NH_2, CH_3$$

0.25 mole   70%

tion of preference for reaction with the carbonyl rather than the carboxyl group) [30].

6. From Hydrazobenzenes (Benzidine rearrangement)

This rearrangement, which is of some industrial importance, involves the production of a variety of products [31]:

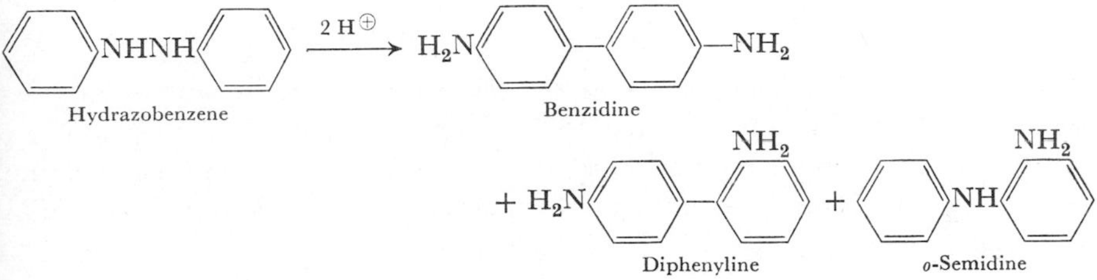

Hydrazobenzene $\xrightarrow{2\,H^{\oplus}}$ Benzidine

+ Diphenyline + o-Semidine

For example, hydrazobenzene itself gives about 70% benzidine and 30% diphenyline. If hydrazobenzene is substituted, the diphenyline is favored, the orientation of groups depending somewhat on the relative stabilities of the π- and σ-complexes to be discussed. With electron-withdrawing substituents, the semidine may be the major rearrangement product. With hydrazonaphthalenes, the o-benzidine is the major product,

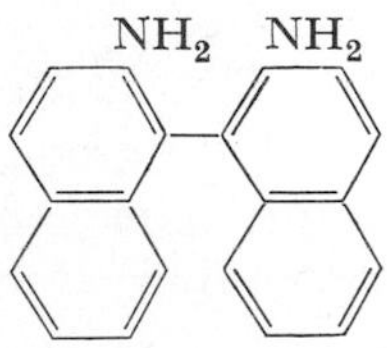

However, the rearrangement of hydrazobenzenes, other than of the parent compound, is not always so clearcut as it should be. Hydrazobenzenes tend to form azobenzenes readily, which products can be detected by the appearance of reddish colors and minimized by the addition of iron powder. On the other hand, aniline is another by-product which is maximized by the incorporation of iron. As an example of the difficulties, p,p'-dimethylhydrazobenzene with acid in alcohol yields only a few percent of the diphenyline, 23% of the azobenzene, and appreciable amounts of p-toluidine [32].

The mechanism of the benzidine rearrangement is one of the most fascinating in the organic field. In limiting cases the rearrangement depends quadratically on the hydrogen ion concentration, a fact which suggests the formation of a dication and consequent weakening of the N,N bond:

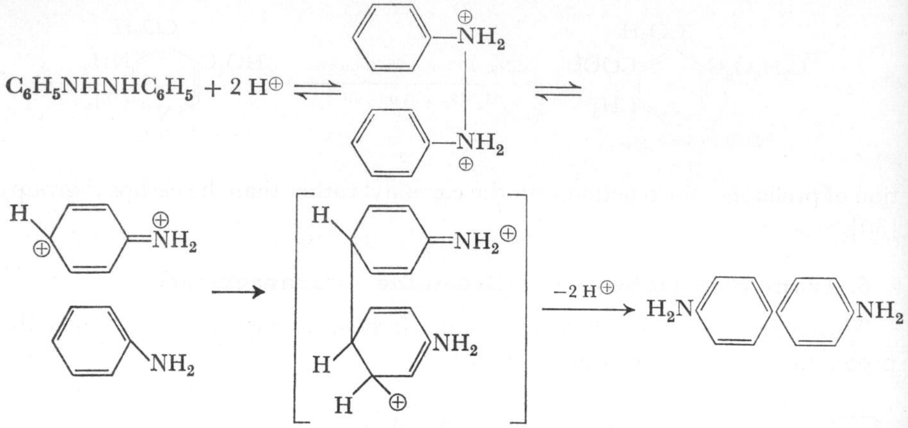

$$C_6H_5NHNHC_6H_5 + 2\,H^\oplus \rightleftharpoons \qquad \rightleftharpoons$$

π-Complex   σ-Complex

The diphenyline would involve an *o*-quinoid form of the π-complex, and the semidine a form which has rotated 180° in the π-complex. This mechanism is limiting in that more easily substituted nuclei, such as the naphthalene ring, do not require complete dication formation before coalescence of orbital electrons from the *para* position begins [33].

(a) **Preparation of benzidine.** Nitrobenzene, 125 g., in 250 ml. of *o*-dichlorobenzene was reduced at 115–125° with 260 g. of zinc dust and 250 g. of 50% sodium hydroxide added alternately in 5–10 g. portions to the stirred solution. The mixture turned red, then colorless, and finally white after 4–10 hr. The potentially inflammable zinc–zinc oxide was filtered off after dilution with water. The organic layer was mixed with an equal volume of ice and 300 ml. of hydrochloric acid and heated to 80°, diluted with 500 ml. more of water. The *o*-dichlorobenzene layer was separated and the aqueous layer treated with 100 g. of anhydrous sodium sulfate. Benzidine sulfate was filtered, suspended in 5 parts of warm water, and treated with sodium carbonate. The precipitated benzidine was filtered, washed, dried; yield 75%. The reduction to the intermediate hydrazobenzene is carried out electrolytically by some industries [34].

7. From Arylhydrazones (Fischer Indole Synthesis)

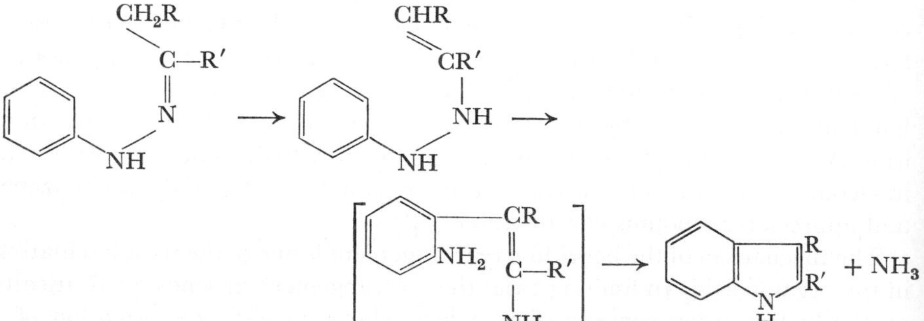

The reaction which has been reviewed [35] is very simple to run. Yields vary, but occasionally are very good. If an *o*-benzidine is formed in the benzidine rearrangement (G.6), the compound may cyclize to a carbazole by a pathway similar to the last step above. Of the cyclizing agents used, concentrated hydrochloric acid, dry hydrogen chloride, boron trifluoride, polyphosphoric acid, cuprous chloride, and zinc chloride, the last was found to be the most effective [36]. Good yields are obtained frequently by simply heating the phenylhydrazone in a neutral, high-boiling solvent such as ethylene glycol [37]. The mechanism of the reaction is not clear but must involve an oxidation in one of the steps. With unsymmetrical ketone phenylhydrazones, the direction of ring closure is determined by the amount and the strength of the acid catalyst [38]:

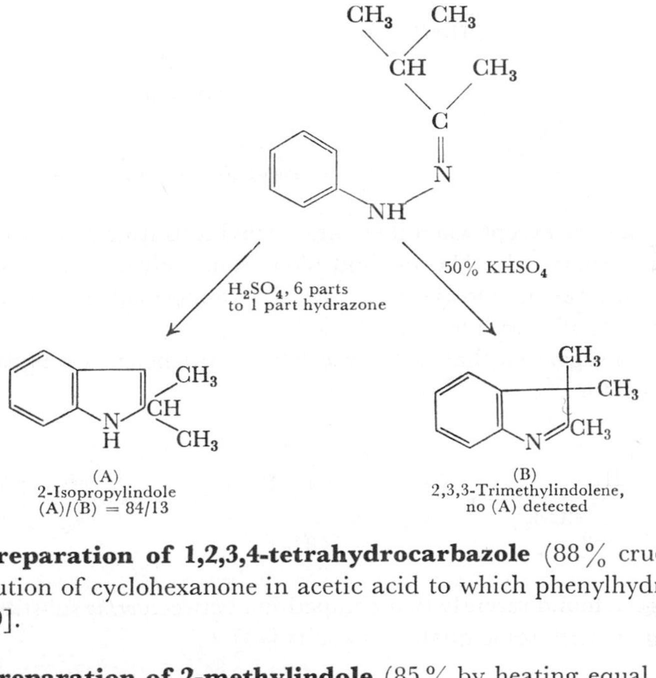

(A)
2-Isopropylindole
(A)/(B) = 84/13

(B)
2,3,3-Trimethylindolene,
no (A) detected

(a) Preparation of 1,2,3,4-tetrahydrocarbazole (88% crude from a boiling solution of cyclohexanone in acetic acid to which phenylhydrazine was added) [39].

(b) Preparation of 2-methylindole (85% by heating equal weights of acetone phenylhydrazone and $ZnCl_2$ in cumene for 1 hr. under nitrogen) [36].

8. From Benzyltrialkylammonium Halides (Stevens, Sommelet-Hauser)

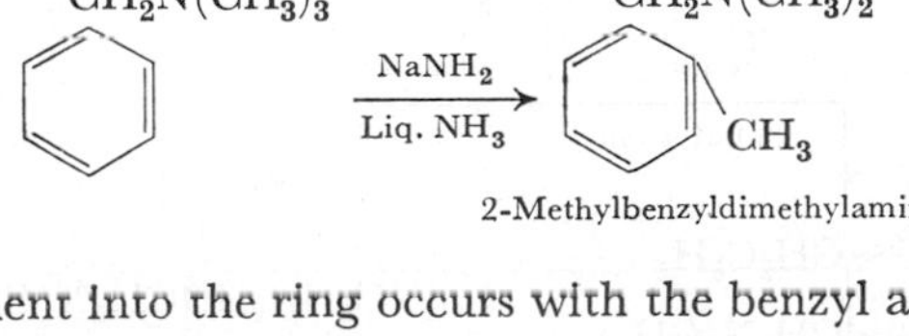

2-Methylbenzyldimethylamine

This rearrangement into the ring occurs with the benzyl and related quaternary ammonium halides [40]. The reagent is sodium amide and the solvent, liquid ammonia. The reaction takes place rapidly in the presence of a small amount of ferric nitrate at the temperature of liquid ammonia. Similar rearrangements which have been effected are:

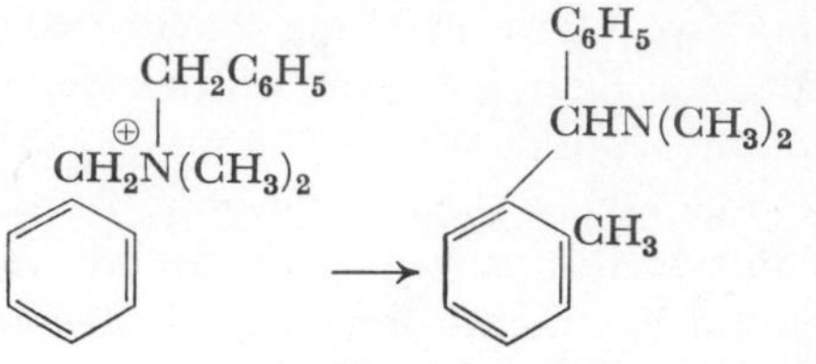

Dimethyl-2-methylbenzhydrylamine

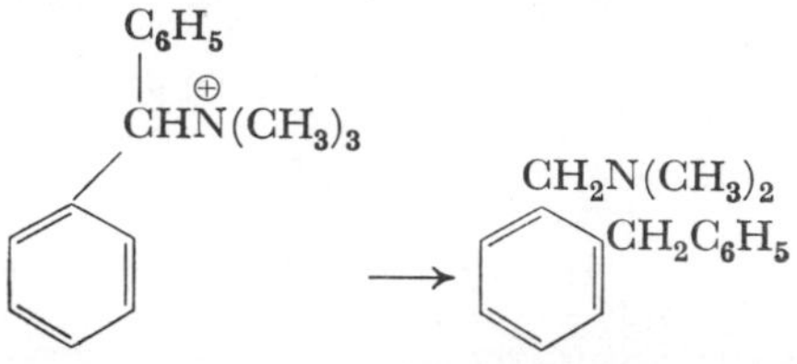

Dimethyl-2-benzylbenzylamine

Yields are superior except when there are methyl substitutents in the ring of the original quaternary salt. The method offers some advantage for introducing *vic*-methyl substituents into the ring over the "abnormal" reaction of benzylmagnesium chlorides with formaldehyde [40].

It has been suggested that in the mechanism the benzene ring serves as an electron acceptor:

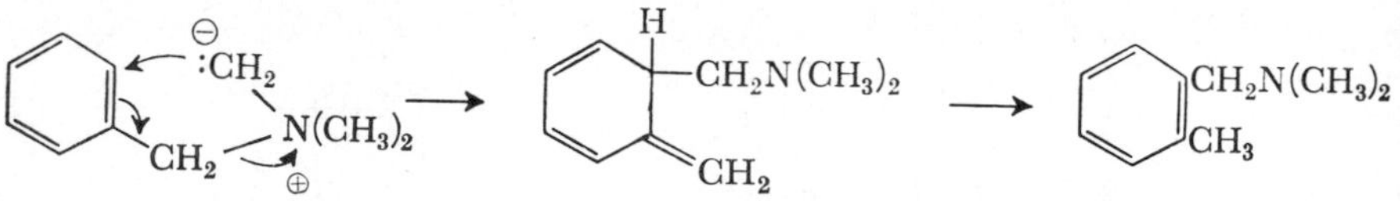

But it has been found recently that competition between *ortho* substitution and a 1,2-shift occurs with some quaternary salts [41]:

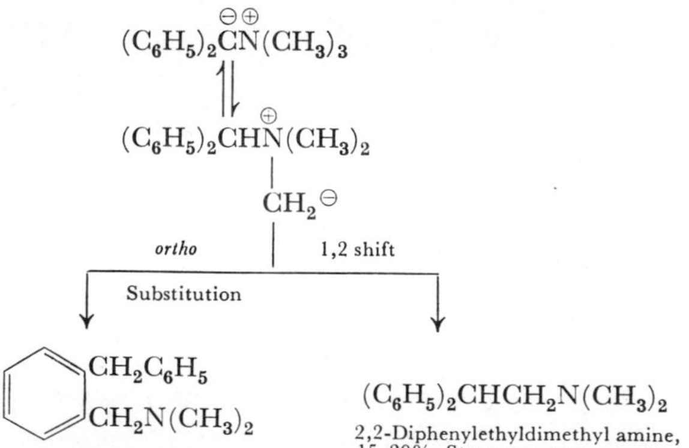

o-Benzylbenzyldimethyl amine, 80–85%, Sommelet-Hauser rearrangement

2,2-Diphenylethyldimethyl amine, 15–20%, Stevens rearrangement

Another manifestation of the Sommelet-Hauser rearrangement is ring enlargement [42]:

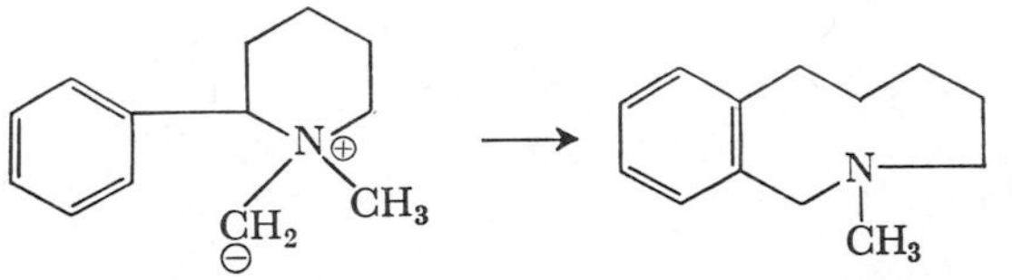

4-Methyl-1,2,3,4,5,6,7,8,9-nonahydro-1,2-benz-4-azacycloundecene-1, 83%

Even tertiary amines, rather than their quaternary salts, form anions [43]:

$$C_6H_5N(CH_3)_2 \xrightarrow[-10^\circ]{BuLi} C_6H_5N(CH_3)CH_2^{\ominus} \xrightarrow[-10 \text{ to } 32^\circ]{C_4H_9I} C_6H_5N(CH_3)CH_2C_4H_9$$

N-Methyl-N-pentylaniline, 40%

The reaction is general but of little preparative value.

(a) Preparation of 2-methylbenzyldimethylamine. Sodium was added to 800 ml. of liquid ammonia until the blue color persisted and then 0.5 g. of granulated ferric nitrate and 1.2 g. atoms of sodium were introduced with stirring. When the blue color disappeared, benzyltrimethylammonium iodide, 1 mole, was added over 10–15 min., again with stirring, and then for 2 hr. more, after which 0.5 mole of NH_4Cl was added. The tertiary amine recovered from the ethereal extract amounted to 134–141.5 g. (90–95%) [44].

9. From Miscellaneous Sources

The rearrangement of N-substituted aromatic amines to nuclear-substituted amines is discussed in more appropriate places:

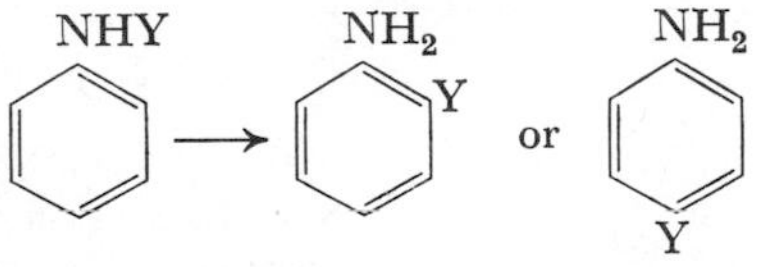

For example, the rearrangement of the aniline with Y = NO_2 is discussed under Nitro Compounds, Chapter 20, A.2. A list of similar rearrangements is appended, however, to recall the possibilities. Rearrangements are known or have been discussed previously where Y = halogen, OH, SO_3H, SO_2Ar, NO, N=NAr, NH_2, NHAr, alkyl, or acyl [45]. In addition, a rearrangement of N-allylamines similar to the Claisen rearrangement has been discovered. N-Allylaniline itself does not rearrange but rather forms aniline and propylene on heating. With less aromatic nuclei, however, rearrangement can comprise the major course of the reaction [46]:

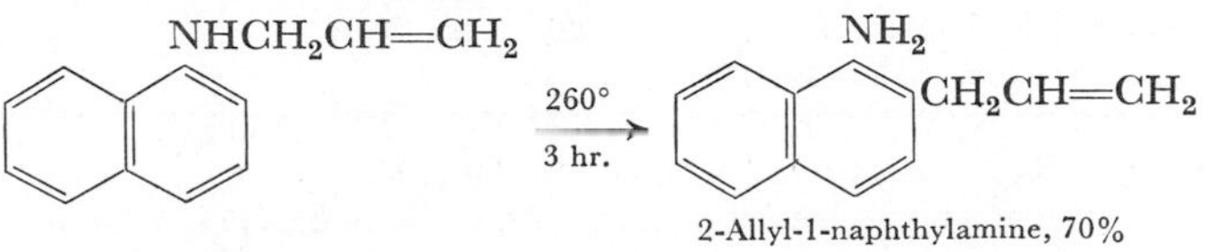

2-Allyl-1-naphthylamine, 70%

Migration of groups from oxygen or sulfur to nitrogen is rather common as envisaged in the following examples:

Smiles rearrangement [47].

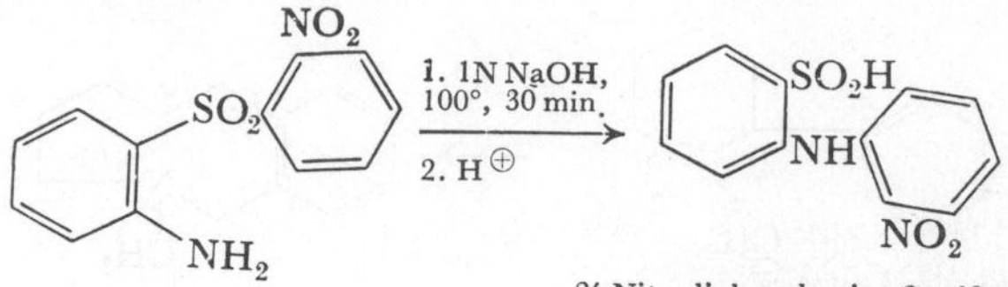

2′-Nitrodiphenylamine-2-sulfenic acid

Cope rearrangement of N-oxides [48]. The benzyl ether was formed in 61 % yield

$$CH_2{=}CHCH_2\overset{\overset{O}{\uparrow}}{N}(CH_3)_2 \xrightarrow{105-110°} (CH_3)_2NOCH_2CH{=}CH_2$$

O-Allyl-N,N-dimethylhydroxylamine,
51%

from the corresponding benzyl N-oxide. Water interferes with the reaction, and groups which have a tendency to eliminate form olefins rather than O-alkylhydroxylamines.

So many rearrangements are to be found in the heterocyclic family that no attempt is made to document them completely. Rather reference is made to a review [49] and a few interesting examples are given:

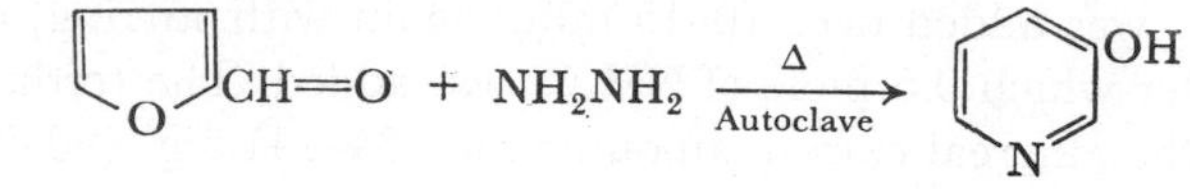

3-Hydroxypyridine

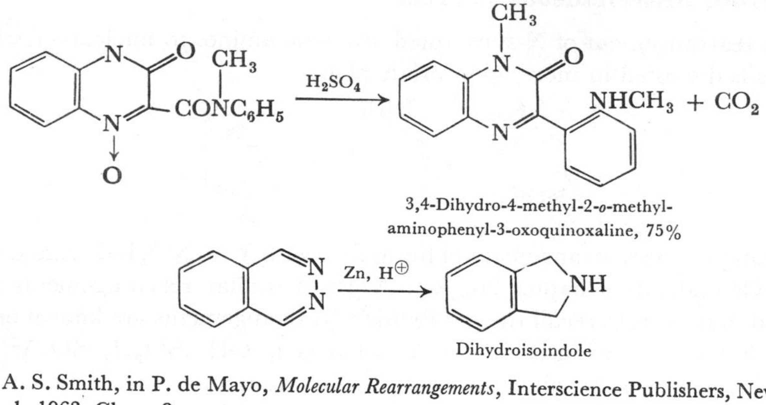

3,4-Dihydro-4-methyl-2-*o*-methyl-
aminophenyl-3-oxoquinoxaline, 75%

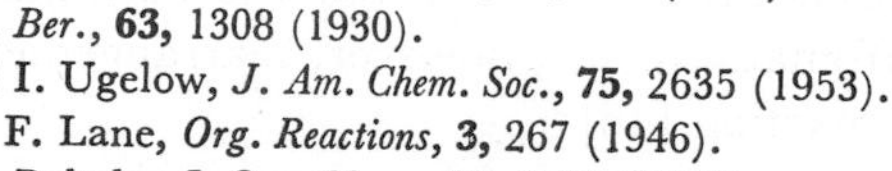

Dihydroisoindole

1. P. A. S. Smith, in P. de Mayo, *Molecular Rearrangements*, Interscience Publishers, New York, Pt. 1, 1963, Chap. 8.
2. D. E. Pearson and W. E. Cole, *J. Org. Chem.*, **20**, 488 (1955).
3. G. Schroeter, *Chem. Ber.*, **63**, 1308 (1930).
4. F. M. Beringer and I. Ugelow, *J. Am. Chem. Soc.*, **75**, 2635 (1953).
5. E. S. Wallis and J. F. Lane, *Org. Reactions*, **3**, 267 (1946).
6. E. Magnien and R. Baltzly, *J. Org. Chem.*, **23**, 2029 (1958).
7. E. Jeffreys, *Am. Chem. J.*, **22**, 14 (1899).
8. D. A. Barr and R. N. Haszeldine, *Chem. Ind.* (*London*), 1050 (1956).
9. D. A. Barr and R. N. Haszeldine, *J. Chem. Soc.*, 30 (1957).
10. H. E. Baumgarten and A. Staklis, *J. Am. Chem. Soc.*, **87**, 1141 (1965); modification suggested by senior author.

11. J. S. Buck and W. S. Ide, *Org. Syn.*, Coll. Vol. **2**, 44 (1943).
12. C. F. H. Allen and C. N. Wolf, *Org. Syn.*, Coll. Vol. **4**, 45 (1963).
13. N. Sperber and R. Fricano, *J. Am. Chem. Soc.*, **71**, 3352 (1949).
14. P. A. S. Smith, *Org. Reactions*, **3**, 337 (1946).
15. J. Weinstock, *J. Org. Chem.*, **26**, 3511 (1961).
16. H. E. Baumgarten *et al.*, *J. Am. Chem. Soc.*, **82**, 3977 (1960).
17. C. D. Campbell and C. W. Rees, *Chem. Commun.*, 192 (1965).
18. P. A. S. Smith, *Org. Syn.*, Coll. Vol. **4**, 819 (1963).
19. W. R. Vaughan and J. L. Spencer, *J. Org. Chem.*, **25**, 1160 (1960).
20. N. Koga and J.-P. Anselme, *J. Org. Chem.*, **33**, 3963 (1968).
21. H. L. Yale, *Chem. Rev.*, **33**, 242 (1943).
22. H. R. Snyder *et al.*, *J. Am. Chem. Soc.*, **75**, 2014 (1953).
23. G. B. Bachman and J. E. Goldmacher, *J. Org. Chem.*, **29**, 2576 (1964).
24. H. Wolff, *Org. Reactions*, **3**, 307 (1946).
25. R. F. Stockel and D. M. Hall, *Nature*, **197**, 787 (1963).
26. K. G. Rutherford and M. S. Newman, *J. Am. Chem. Soc.*, **79**, 213 (1957).
27. D. M. Hall *et al.*, *J. Chem. Soc.*, 1842 (1950).
28. R. Huisgen and H. König, *Chem. Ber.*, **92**, 203 (1959).
29. J. Meinwald and P. G. Gassman, *J. Am. Chem. Soc.*, **82**, 2857 (1960).
30. R. G. Jones, *J. Am. Chem. Soc.*, **73**, 5244 (1951).
31. M. J. S. Dewar, in P. de Mayo, *Molecular Rearrangements*, Interscience Publishers, New York, Pt. 1, 1963, p. 323.
32. R. B. Carlin and G. S. Wich, *J. Am. Chem. Soc.*, **80**, 4023 (1958).
33. C. K. Ingold *et al.*, *J. Chem. Soc.*, 2386 (1962).
34. H. E. Fierz-David and L. Blangey, *Fundamental Processes of Dye Chemistry*, Interscience Publishers, New York, 1949, p. 124.
35. B. Robinson, *Chem. Rev.*, **69**, 227 (1969).
36. N. B. Chapman *et al.*, *J. Chem. Soc.*, 1424 (1965).
37. J. T. Fitzpatrick and R. D. Hiser, *J. Org. Chem.*, **22**, 1703 (1957).
38. H. Illy and L. Funderburk, *J. Org. Chem.*, **33**, 4283 (1968).
39. C. U. Rogers and B. B. Corson, *J. Am. Chem. Soc.*, **69**, 2910 (1947).
40. S. W. Kantor and C. R. Hauser, *J. Am. Chem. Soc.*, **73**, 4122 (1951).
41. K. P. Klein and C. R. Hauser, *J. Org. Chem.*, **31**, 4275 (1966).
42. G. C. Jones and C. R. Hauser, *J. Org. Chem.*, **27**, 3572 (1962).
43. A. R. Lepley and W. A. Khan, *J. Org. Chem.*, **33**, 4362 (1968).
44. W. R. Brasen and C. R. Hauser, *Org. Syn.*, Coll. Vol. **4**, 585 (1963).
45. Houben-Weyl, *Methoden der Organischen Chemie*, Georg Thieme Verlag, Stuttgart, 1957, 4th ed., Vol. 11, Pt. 1, p. 826.
46. S. Marcinkiewicz *et al.*, *Tetrahedron*, **14**, 208 (1961).
47. Ref. 45, p. 913.
48. A. C. Cope and P. H. Towle, *J. Am. Chem. Soc.*, **71**, 3423 (1949); A. H. Wragg *et al.*, *J. Chem. Soc.*, 4057 (1958).
49. G. M. Badger and J. W. Clark-Lewis, in P. de Mayo, *Molecular Rearrangements*, Interscience Publishers, New York, 1963, Pt. 1, Chap. 10.

H. Degradation of Amines

The principal degradation is the dealkylation of amines, used mainly for identification purposes, but on occasion employed in a preparative manner, particularly in the preparation of ethynylamines (H.3). Decarboxylation of α-aminoacids is described briefly (H.2).

References for Section H are on p. 512.

1. From Amines

$$R_3N \rightarrow R_2NH \quad \text{or} \quad R_2NH \rightarrow RNH_2$$

Several examples have been discussed in previous sections:

Hydrogenolysis of quaternary salts, A.10.
Hydrogenolysis of benzylamines, A.9.
Emde reduction of quaternary salts, A.10.
Reverse cyanoethylation, D.7.

Dealkylations from Houben-Weyl [1] will be listed, and those not taken from Houben-Weyl will be documented.

Decomposition of salts.

$$ArNR_2 \cdot HBr \xrightarrow[-RBr]{150°} ArNHR \cdot HBr \xrightarrow[-RBr]{200°} ArNH_2 \cdot HBr$$

Hydrogen bromide was passed through the molten salts at 150° to give high yields of aromatic secondary amines and at 200° to give good yields of aromatic primary amines [2]. Aliphatic amine salts did not dealkylate under these conditions, although at higher temperatures (285°) they are known to undergo dealkylations of no preparative value. Quaternary aromatic salts can be dealkylated with comparative ease:

$$C_6H_5N(CH_3)_3Cl \xrightarrow{\Delta} C_6H_5N(CH_3)_2 + CH_3Cl$$

but are best dealkylated by reduction (A.10), by heating with ethanolamine to yield the amine almost quantitatively [3], or by very mild heating of the quaternary ammonium acetate salt in an aprotic solvent [4]. By the latter method, methyl groups are preferably eliminated (as methyl acetate), and temperatures of 80° are employed for aromatic quaternary ammonium acetates and 110–140° for aliphatic ones using toluene as the solvent with the minimum amount of acetonitrile to bring about miscibility.

Decomposition of amides. An alkyl will split in preference to the aryl group and,

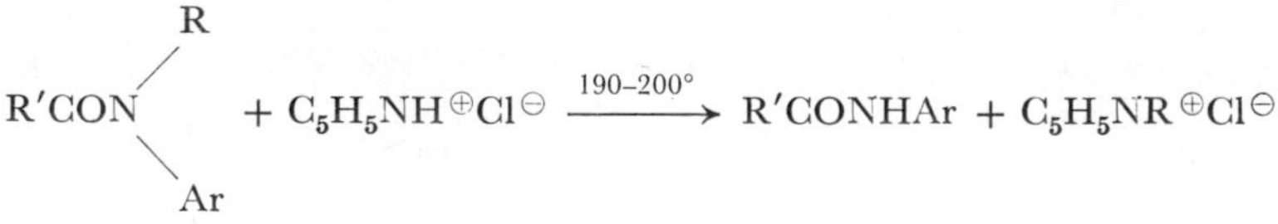

$$R'CON\begin{smallmatrix} R \\ \\ Ar \end{smallmatrix} + C_5H_5NH^{\oplus}Cl^{\ominus} \xrightarrow{190-200°} R'CONHAr + C_5H_5NR^{\oplus}Cl^{\ominus}$$

obviously, alkyls will split in the order tert. > sec. > primary. Yields are good in many cases, but splitting of alkyl groups is incomplete in others [5].

Nitrosation. Contrary to the general impression, nitrous acid reacts with tertiary amines as shown [6]:

$$R_2NCHR_2' \xrightarrow{HONO} R_2NNO + R_2'C{=}O$$

$$\downarrow \text{urea, } H^{\oplus}, \text{ alcohol}$$

$$R_2NH$$

Yields of secondary amines from a series of benzylic tertiary amines range from 53 to 69 %.

Oxidative dealkylation (see Amides, Chapter 18, C.1).

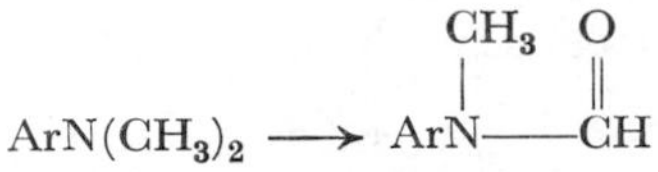

$$ArN(CH_3)_2 \longrightarrow ArN\!\!-\!\!\overset{\overset{\displaystyle CH_3}{|}}{\underset{}{}}\overset{\overset{\displaystyle O}{\|}}{\underset{}{CH}}$$

Halogen dealkylation.

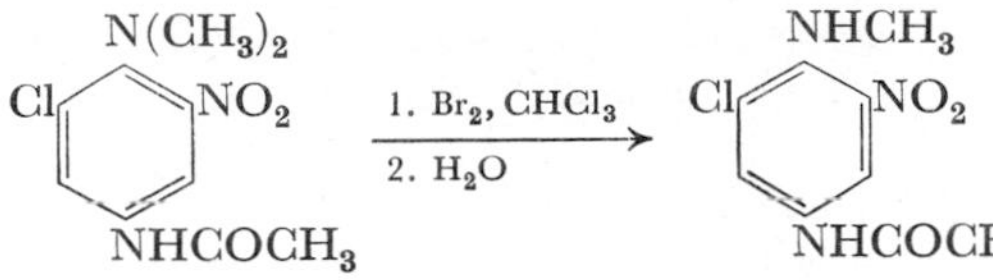

4-Acetamido-2-chloro-6-
nitro-N-methylaniline,
94%

Bromocyanogen dealkylation (von Braun) [7]. This procedure has been used more frequently than the others listed [8]:

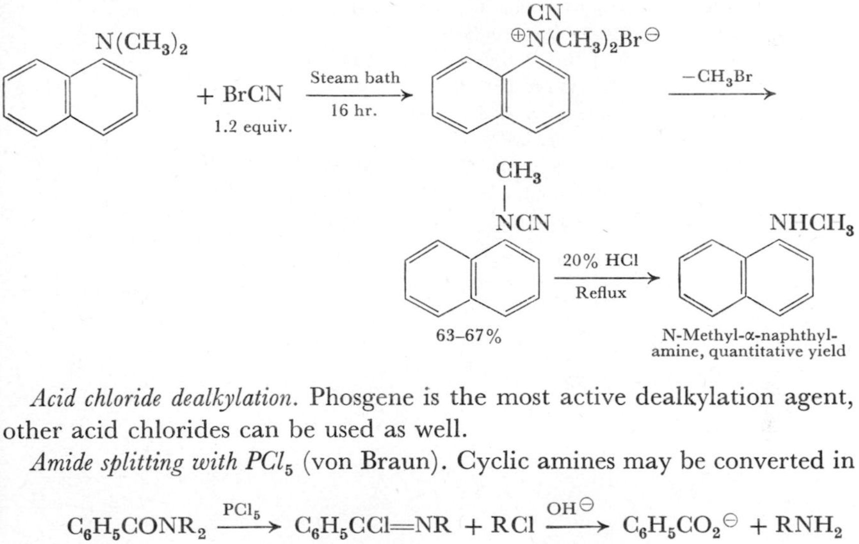

63–67%

N-Methyl-α-naphthyl-
amine, quantitative yield

Acid chloride dealkylation. Phosgene is the most active dealkylation agent, but other acid chlorides can be used as well.

Amide splitting with PCl$_5$ (von Braun). Cyclic amines may be converted in this

$$C_6H_5CONR_2 \xrightarrow{PCl_5} C_6H_5CCl\!\!=\!\!NR + RCl \xrightarrow{OH^{\ominus}} C_6H_5CO_2^{\ominus} + RNH_2$$

manner into ω-chloroamines. For example, piperidine yields 50 % ε-chloro-pentylamine, and tetrahydroquinoline gives *o*-(γ-chloropropyl)-aniline.

Trityl group splitting. Because of its tendency to form a carbonium ion, the trityl group of an amine is readily hydrolyzed and has been used as a protective

$$(C_6H_5)_3CNHR \xrightarrow[H_2O]{H^{\oplus}} (C_6H_5)_3COH + RNH_3^{\oplus}$$

group in peptide synthesis [9].

2. From α-Amino Acids

$$RCH(NH_2)COOH \xrightarrow{\Delta} RCH_2NH_2 + CO_2$$

(see Alkanes, Chapter 1, F.1). In addition α-acylamino ketones may be obtained by heating α-amino acids and acid anhydrides in pyridine (see Ketones, Chapter 11, F.1).

3. From Ethynylquaternary Salts

This preparation could well be included in H.1 but is given prominence here because of its importance in producing a new class of compounds, the ethynylamines, which are proving extremely versatile in their reactions. For example, it is claimed that they are better dehydrating agents than dicyclohexylcarbodiimide. This method involves the degradation of a quaternary salt [10]:

$$C_6H_5C\equiv CBr \xrightarrow[\text{40 hr., 55°}]{(CH_3)_3N} C_6H_5C\equiv \overset{\oplus}{C}N(CH_3)_3 \xrightarrow{-CH_3Br} C_6H_5C\equiv CN(CH_3)_2$$
$$Br^\ominus$$

The phenylethynyldimethylamine has been produced in 100-g. lots by this procedure and is now available commercially (Fluka).

1. Houben-Weyl, *Methoden der Organischen Chemie*, Georg Thieme Verlag, Stuttgart, 1957, 4th ed., vol. 11, Pt. 1, p. 961.
2. R. A. Chambers and D. E. Pearson, *J. Org. Chem.*, **28,** 3144 (1963).
3. S. Hünig and W. Baron, *Chem. Ber.*, **90,** 395 (1957).
4. N. D. V. Wilson and J. A. Joule, *Tetrahedron*, **24,** 5493 (1968).
5. D. Klamann and E. Schaffer, *Chem. Ber.*, **87,** 1294 (1954).
6. P. A. S. Smith and R. N. Loeppky, *J. Am. Chem. Soc.*, **89,** 1147 (1967).
7. H. A. Hageman, *Org. Reactions*, **7,** 198 (1953).
8. H. W. J. Cressman, *Org. Syn.*, Coll. Vol. **3,** 608 (1955); Ref. 1, p. 983.
9. L. Zervas and D. M. Theodoropoulos, *J. Am. Chem. Soc.*, **78,** 1359 (1956).
10. H. G. Viehe, *Angew. Chem. Intern. Ed. Engl.*, **6,** 767 (1967).

ACETALS
AND KETALS

Although acetals and ketals are of lesser importance among the types of organic compounds, twenty-six methods of synthesis are given. An extensive review also is available [1], and another is devoted exclusively to the synthesis of ketals derived from tetrahydrofuran [2].

Actually, this reaction is often used to protect the carbonyl group from attack in a later reaction [3, 4]. The alcoholic group may be protected as well by addition to dihydropyran to form the ketal, an alkoxytetrahydropyran [5]. The acetals or ketals occasionally are useful intermediates in the synthesis of polyfunctional compounds or heterocycles.

The simplest methods are those in which the acetals or ketals are made directly from the carbonyl compound and an alcohol, preferably a glycol as ring formation seems to be an added driving force (see A). The water is removed by azeotropy using a Dean-Stark tube. However, many other ingenious methods of making acetals and ketals are to be found in other sections, as for example the electrolytic alkoxylation procedures (E). The preparation of mixed acetals

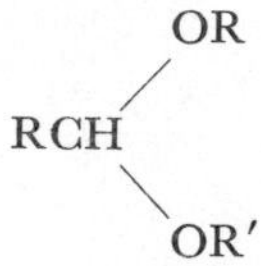

$$RCH\begin{matrix}OR\\\\OR'\end{matrix}$$

or ketals as compared to that of simple acetals

$$(RCH(OR)_2)$$

is mentioned in several sections (Exs: A.8, c.3; B.1, b.3, and E.1.b), and the inference is that mixed acetals disproportionate most readily to the simple acetals. However, if precautions are taken to avoid the presence of both free alcohol and traces of acid, the mixed acetals are stable for indefinite periods [6].

A word of caution must be given to those who handle acetals or ketals. Some of them are peroxide formers, and due care should be taken with pot residues from distillation of acetals or ketals which have been stored for any length of time.

A. Nucleophilic Addition of Alcohols to Carbonyl Groups or Exchange

1. From Carbonyl Compounds and Alcohols

Carbonyl compounds and alcohols form equilibrium mixtures with the hemiacetals or hemiketals:

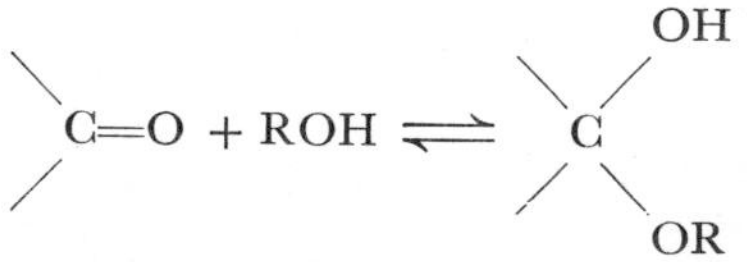

$$\begin{matrix}\diagdown\\\diagup\end{matrix}C=O + ROH \rightleftharpoons \begin{matrix}\diagdown\\\diagup\end{matrix}C\begin{matrix}OH\\\\OR\end{matrix}$$

References for Section A are on pp. 526–528.

The equilibrium lies far to the left if the carbonyl compound is a ketone or a sterically hindered aldehyde, but farther to the right in favor of the hemiacetal or hemiketal if electron-withdrawing groups are attached to the α-carbon of the carbonyl compounds, or if favorable cyclization conformations are existent. For example, monosaccharides exist mainly in the hemiacetal form, and paraldol of the following structure forms spontaneously from freshly distilled aldol [7]:

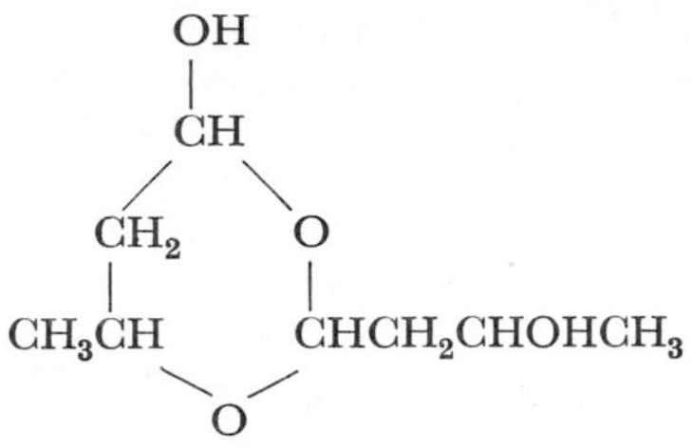

$$
\begin{array}{c}
\text{OH} \\
| \\
\text{CH} \\
\diagup \quad \diagdown \\
\text{CH}_2 \qquad \text{O} \\
| \qquad\quad | \\
\text{CH}_3\text{CH} \qquad \text{CHCH}_2\text{CHOHCH}_3 \\
\diagdown \qquad \diagup \\
\text{O}
\end{array}
$$

The equilibrium reaction leading to the hemi type is catalyzed by both acid and base. Acetal or ketal formation, on the other hand, is not catalyzed by base because of the poor leaving properties of the hydroxide ion, but is catalyzed by acid leading to the handy formation of the resonance-stabilized (and solvated) carbonium ion 9-I, which permits easy passage to the acetal or ketal in the

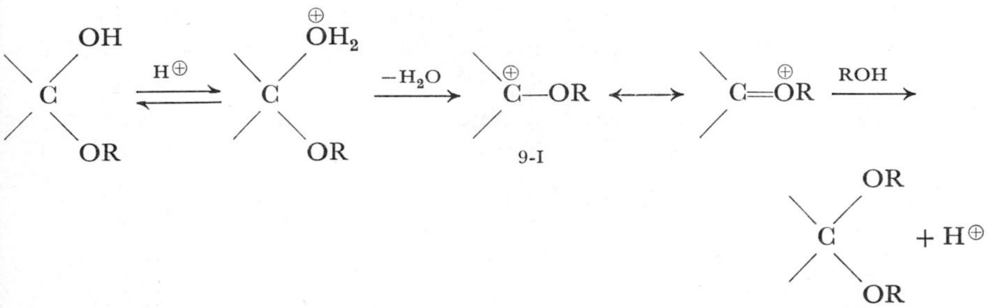

$$
\begin{array}{ccc}
\diagdown \;\;\; \text{OH} & \diagdown \;\;\; \overset{\oplus}{\text{OH}_2} & \\
\underset{\diagup}{\text{C}} \quad \overset{\text{H}^{\oplus}}{\rightleftharpoons} \quad \underset{\diagup}{\text{C}} \quad \overset{-\text{H}_2\text{O}}{\longrightarrow} \quad \overset{\oplus}{\underset{\diagup}{\text{C}}}\text{—OR} \;\longleftrightarrow\; \underset{\diagup}{\text{C}}{=}\overset{\oplus}{\text{OR}} \quad \overset{\text{ROH}}{\longrightarrow} \\
\diagup \;\;\; \text{OR} & \diagup \;\;\; \text{OR} & \text{9-I}
\end{array}
$$

$$
\begin{array}{c}
\text{OR} \\
\diagdown \quad \diagup \\
\underset{\diagup}{\text{C}} \qquad +\,\text{H}^{\oplus} \\
\diagup \quad \diagdown \\
\text{OR}
\end{array}
$$

presence of alcohol or return to the carbonyl compound in the presence of water. Conditions or devices to improve yields of acetals or ketals are based on the best methods of generating 9-I and exposing it to the maximum concentration of alcohol and minimum concentration of water. Indeed, other syntheses, to be discussed later, also depend on the generation of 9-I.

Aldehydes and primary alcohols give high yields of acetals by simply mixing the reactants in the presence of an acidic reagent such as hydrogen chloride. Aldehydes and secondary alcohols generally give lower yields under these conditions [8]. A great variety of catalysts, such as hydrogen chloride [9–11], calcium chloride [12], ammonium chloride [13], hydrogen bromide [14], p-toluenesulfonic acid [15–17], boron trifluoride etherate [18], barium oxide and an alkyl iodide [19], and malonic acid, oxalic acid, or selenium dioxide [20] have been employed to effect the reaction.

To prevent the reverse reaction from occurring, it is customary to make the reaction mixture alkaline, and then for separation distillation is often employed.

Here advantage may be taken of the azeotrope formed with benzene or petroleum ether, and various types of water separators [15, 21, 22], as discussed in more detail under Carboxylic Esters, Chapter 14, A.1, may be employed.

Glycols, glycerol [15], pentaerythritol [11], and many other polyhydric alcohols condense with carbonyl compounds, including ketones, to give cyclic acetals or ketals. Neopentyl glycol has been proposed as a good reagent for protecting the carbonyl function [23]:

$$\text{(cyclopentanone with } CH_3) + HOCH_2\overset{\overset{\displaystyle CH_3}{|}}{\underset{\underset{\displaystyle CH_3}{|}}{C}}CH_2OH \xrightarrow[\text{Dean-Stark apparatus}]{p\text{-Toluenesulfonic acid}} \text{(neopentyl glycol ketal)}$$

2-Methylcyclopentanone neopentyl glycol ketal, 82%

Although rates of subsequent hydrolysis are somewhat lower than those of ketals of straight-chain alcohols [24], relative rates of ketal formation with ethylene glycols under these conditions show that most ketal formations are complete in 2 hours. However, benzophenone ketalization (5 hours) and that of mesityl oxide or methyl isobutyl ketone are much slower [25].

Ketones and alcohols exposed to acid without water removal have unfavorable equilibrium constants from the standpoint of synthesis of ketals [26]. For instance, the following ketones (1 part) and alcohols (4 parts) passed slowly through Dowex 50 (strong acid-exchange resin) gave the conversions in the accompanying tabulation [27]. The situation is not good, therefore, for the

	% Conversion	
	At 24°	At −28°
Acetone-methanol	11	32
Acetone-ethanol	2	17
4-Methyl-2-pentanone-methanol	4	9
Cyclohexanone-methanol	46	86

preparation of ketals from volatile alcohols and ketones. (Azeotropy can be used on the relatively non-volatile types.) However, good success in preparing ketals from volatile alcohols and ketones has been achieved by using 2,2-dimethoxypropane with the ketone and alcohol in acid solution [28, 29]. It appears that the purpose of the 2,2-dimethoxypropane is to increase the concentration of the alcohol by reaction with water, $(CH_3)_2C(OCH_3)_2 + H_2O \rightleftharpoons (CH_3)_2CO + 2 CH_3OH$. By removing the acetone as it is formed by distillation, this equilibrium is driven toward the alcohol side. Apparently a series of equilibria are involved in this reaction since 2,2-dimethoxypropane facilitates the formation of ketals of ketones other than acetone and alcohols other than methanol. The effect of the ketal can be illustrated in the case of cyclohexanone dimethyl ketal, a ketal which can be produced with 78% conversion from equal volumes of cyclohexanone and methanol in the presence of a trace of hydrogen chloride [30]. With 2,2-dimethoxypropane and p-toluenesulfonic acid as the catalyst, the yield is 95% [28].

Carbonyl compounds with multiple or strong electron-withdrawing groups, such as hexachloroacetone, readily form the hemiketal but proceed on to the ketal with great difficulty. An ingenious circumvention of this difficulty is the application of a Williamson type of synthesis for ethers [31]: (see Ex. b.6 also):

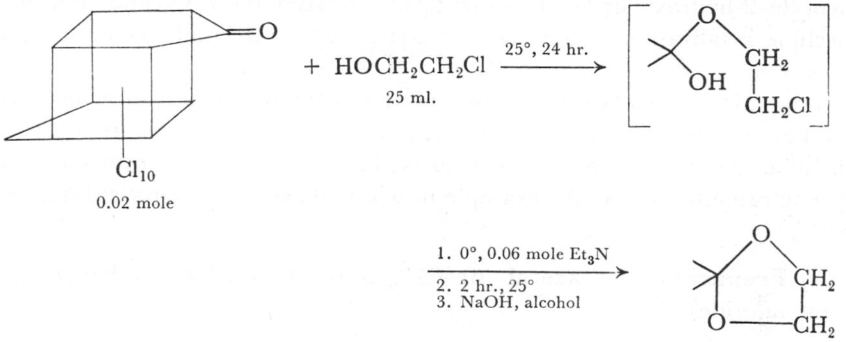

Cl_{10}
0.02 mole

$+ HOCH_2CH_2Cl \xrightarrow{25°, 24 \text{ hr.}}$
25 ml.

$\xrightarrow[\substack{2. \ 2 \text{ hr., } 25° \\ 3. \ NaOH, \text{ alcohol}}]{1. \ 0°, \ 0.06 \text{ mole } Et_3N}$

Decachloropentacyclo-
$[5.3.0.0^{2,6}.0^{3,9}.0^{4,8}]$-decan-5-one
cyclic ethylene ketal, 93%

The same principle has been applied in making some ketene ketals [32]:

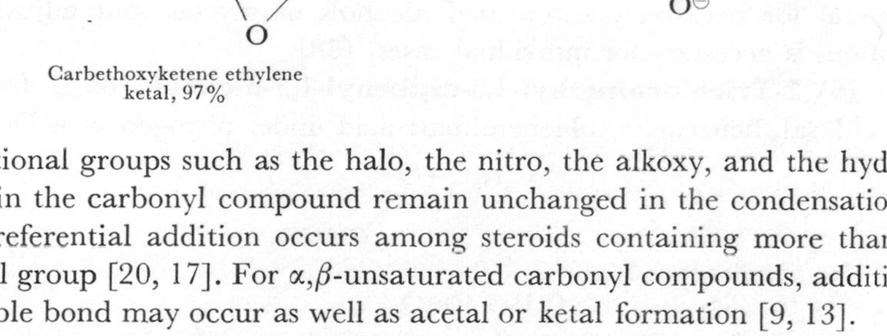

$C_2H_5O_2CCH_2CO_2CH_2CH_2Br \xrightarrow[THF]{NaH} C_2H_5O_2C\overset{\ominus}{C}HCO_2CH_2CH_2Br$

$\xleftarrow{-Br^{\ominus}} C_2H_5O_2CCH=COCH_2CH_2Br$
$\overset{|}{O^{\ominus}}$

Carbethoxyketene ethylene
ketal, 97%

Functional groups such as the halo, the nitro, the alkoxy, and the hydroxymethyl in the carbonyl compound remain unchanged in the condensation. At times preferential addition occurs among steroids containing more than one carbonyl group [20, 17]. For α,β-unsaturated carbonyl compounds, addition to the double bond may occur as well as acetal or ketal formation [9, 13].

Glycosides are readily formed from monosaccharides and alcohols in the presence of hydrogen chloride. Of further interest to sugar chemists is the preparation of isopropylidene acetals or ketals from the hydroxyl groups of monosaccharides [33]. Usually, the five-membered ring is formed in preference to the six-membered ring when both possibilities are present. For example, glycerol and acetone give almost theoretical yields of 1,2-isopropylidene glycerol [34]. On the other hand, benzaldehyde and glycerol form both five- and six-membered acetals from which the 1,3-isomer

$$
\begin{array}{l}
\text{CH}_2\text{——O} \\
\quad| \qquad\quad \diagdown\text{H} \\
\text{CHOH} \qquad \text{CC}_6\text{H}_5 \\
\quad| \qquad\quad \diagup \\
\text{CH}_2\text{——O}
\end{array}
$$

can be separated by crystallization in 22 % yield [35, 36]. 1,3-Isopropylidene glycerol can be made indirectly from the 1,3-benzylidene glycerol by benzoylation of the 2-hydroxyl group, followed by hydrolysis of the 1,3-acetal, and then by acetone ketalization of the 1,3-hydroxyl groups and finally saponification [37].

Heterocyclic compounds are sources of aldehydes or ketones since they can sometimes be cleaved by acids to form acetals. For example, furan with methanolic hydrogen chloride is converted in rather low yield into succinaldehyde tetramethyl acetal. An example in which there is a better yield is given in b.8.

(a) **Preparation of acetal** (61–64 % from acetaldehyde, ethanol, and calcium chloride) (12).

(b) **Other examples**

(1) *m*-**Nitrobenzaldehyde dimethyl acetal** (76–85 % from *m*-nitrobenzaldehyde, methanol, and hydrochloric acid) [10].

(2) *dl*-**Isopropylideneglycerol** (87–90 % from acetone, glycerol, and *p*-toluenesulfonic acid monohydrate) [15].

(3) **3,3-Dimethoxy-20-oxo-5α-pregnane** (87 % from allopregnanedione, methanol, and *p*-toluenesulfonic acid) [17].

(4) **Acetophenone dipropyl ketal** (63 % from acetophenone, propyl alcohol, 2,2-dimethoxypropane, and a trace of *p*-toluenesulfonic acid) [28].

(5) **Bis-(2,2,2-tribromoethyl)-formal** (87 % from paraformaldehyde added to a solution of 2,2,2-tribromoethanol in 90 % sulfuric acid. The method is general for negatively substituted alcohols or glycols, but adjustment of conditions is necessary for individual cases) [38].

(6) **2-Trichloromethyl-4,5-diphenyl-1,3-dioxole** (85 % from benzoin, chloral, benzene, *p*-toluenesulfonic acid under nitrogen in a Dean-Stark

$$
\begin{array}{l}
\text{C}_6\text{H}_5\text{C——O} \\
\quad\ \ \|\qquad\quad \diagdown \\
\quad\ \ \|\qquad\qquad \text{CHCCl}_3 \\
\quad\ \ \|\qquad\quad \diagup \\
\text{C}_6\text{H}_5\text{—C—O}
\end{array}
$$

apparatus; the reaction is not general for α-hydroxy ketones but is noteworthy because of the normally difficult acetalization of chloral) [39].

(7) **2(1-Ethyl-1-pentenyl)-1,3-dioxolane** (71 % from 3 moles each of 2-ethyl-3-propylacrolein and ethylene glycol in 1 l. of benzene containing 2 g. of *p*-toluenesulfonic acid, the water (55 ml.) being removed as it formed by means of the Dean-Stark apparatus) [40].

(8) 4-Oxopentanal dimethyl acetal (80% from α-methylfuran, 500 g., in 1500 ml. of methanol and 20 ml. of 80% methanolic hydrogen chloride refluxed for 34 hr.) [41].

2. From Carbonyl Derivatives and Alcohols

Carbonyl derivatives, subject to hydrolysis, can also be solvolyzed with alcohol to yield the corresponding acetals or ketals. Occasionally this method is of benefit when the aldehyde is unstable or when the derivative is made directly and solvolyzed before isolation (see Ex. b.4).

(a) Preparation of glutardialdehyde tetraethyl acetal. Glutaraldehyde dioxime (180 g.) and ethyl nitrite (240 g.) in 50 ml. of 95% ethanol and 5 ml. of acetic acid were held at 0° until no more nitrous oxide was evolved. Ethanolic hydrogen chloride (100 g. 3%) and calcium chloride (40 g.) were added, and the mixture held at 0° for 5 hr. and then 48 hr. at room temperature. The usual workup yielded 175 g. (51%) of the acetal, b.p. 97–100° (3 mm.) [42].

(b) Other examples

(1) Benzaldehyde diethyl acetal (78% from benzalaniline (36 g.), sulfuric acid (9.8 g.) in 75 ml. of ether to which ethanol (37 g.) was added; the mixture was allowed to stand for 4 days) [43].

(2) Palmitaldehyde dimethyl acetal (84% from palmitaldehyde 2,4-dinitrophenylhydrazone, boron trifluoride-methanol or 10% methanolic hydrogen chloride and levulinic acid; apparently the free aldehyde is an intermediate) [44].

(3) 2-Hexene-4,5-dione-1-al diethyl acetal (80% from the tosylate

$$\underset{\displaystyle CH_3CCCH=CHC(OC_2H_5)_2}{\overset{\displaystyle \overset{O\ O}{\underset{\| \|}{}}\quad \overset{H}{\underset{|}{}}}{}}$$

ester of methyl furyl ketoxime warmed in alcohol; an example of extremely easy nucleophilic attack of alcohol) [45].

(4) Phenylglyoxal diethyl acetal (53% from 1.0 mole of acetophenone in 1500 ml. of ethanol and 2.61 mole of nitrosyl chloride at 24–30°, after which the mixture was held at 63° for 2.5 hr. and then finally refluxed with 2 N aqueous NaOH) [46].

3. From Carbonyl Compounds and Epoxides

$$RCHO + CH_2\overset{\displaystyle O}{\overset{\diagup\ \diagdown}{}}CH_2 \xrightarrow{SnCl_4} RCH\begin{matrix} O-CH_2 \\ | \\ O-CH_2 \end{matrix}$$

Epoxides and aldehydes or ketones in the presence of stannic chloride form acetals or ketals readily [47, 48]. It is common to use dry carbon tetrachloride as a solvent and to hold the temperature of the exothermic reaction down to

20–30°. Yields from a variety of aliphatic and aromatic ketones vary from 45 to 83%.

The mechanism probably involves the addition of the ethylene oxide-stannic chloride complex to the aldehyde

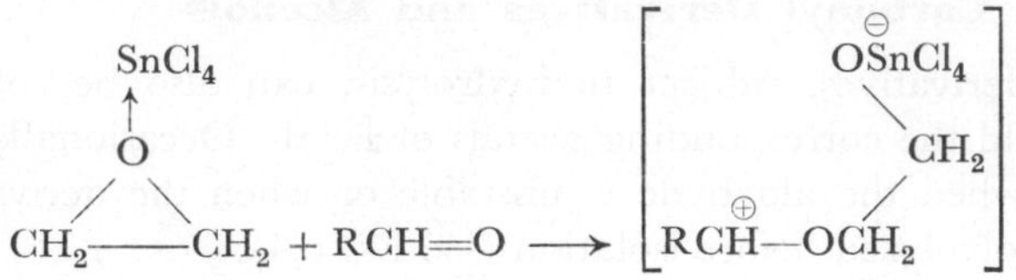

followed by rapid ring closure:

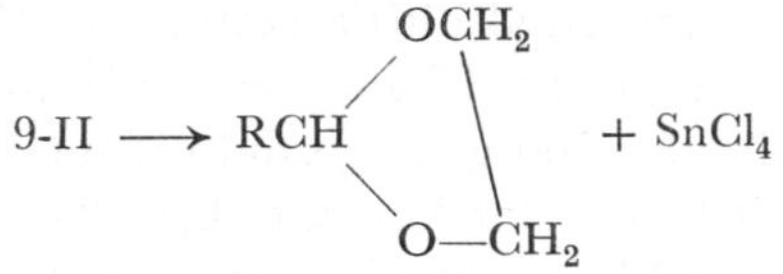

(a) Preparation of crotonaldehyde γ-chloropropylene acetal.

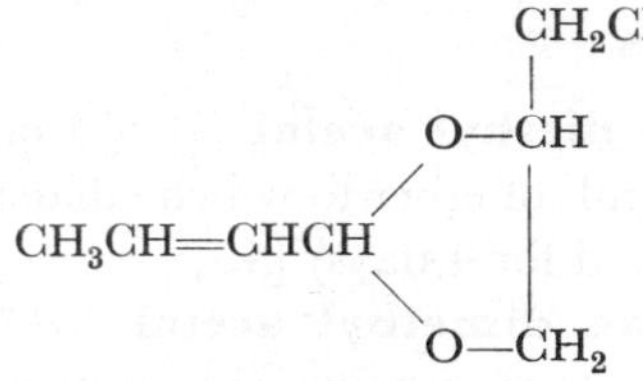

Crotonaldehyde, 31.5 g., 44 g. of epichlorohydrin, and 10 g. of stannic chloride in carbon tetrachloride were permitted to react. There was obtained 53 g. of an oil, b.p. 65–72° (2 mm.) (75% yield based on the aldehyde). Another distillation gave 45 g. of colorless oil, b.p. 68–70° (1.5 mm.) [48].

(b) Preparation of diethyl ketone γ-bromopropylene ketal (68.5% from diethylketone, epibromohydrin, and stannic chloride) [48].

4. From Carbonyl Compounds and Orthoesters or Orthosilicates

$$\diagup_{\diagdown} C{=}O \xrightarrow{\text{HC(OR)}_3} \diagup_{\diagdown} C(OR)_2 + HCOOR$$

A fourth method for synthesizing acetals or ketals from carbonyl compounds is by the use of orthoesters. Typical catalysts used in alcoholic solution are hydrochloric acid [49], hydrogen chloride [50, 51], ferric chloride [52], ammonium chloride [53, 54], ammonium nitrate [55], and *p*-toluenesulfonic acid [56].

It is claimed that in some cases ethyl orthosilicate is superior to ethyl orthoformate in this synthesis [57]. This reagent with hydrogen chloride as catalyst gave a 76% yield of β-ethoxypropionaldehyde diethyl acetal from acrolein in ethanol [58]. Thus, as in A.1, addition as well as acetal formation may occur with α,β-unsaturated aldehydes.

In a study of ketal formation using ethyl orthoformate and alcohols higher than ethyl, yields of ketals were similar to those of acetals, usually in the 70–90% range [56]. A series of equilibria are involved here, again stemming from the relatively facile formation of the key intermediate carbonium ion 9-III [28].

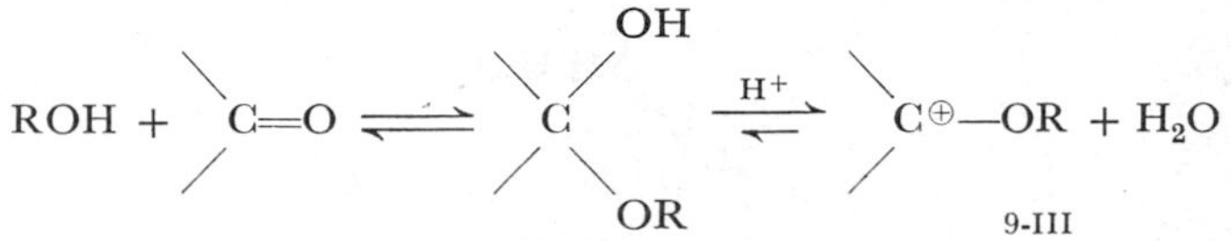

$$ROH + \underset{/}{\overset{\backslash}{C}}{=}O \rightleftharpoons \underset{/}{\overset{\backslash}{C}}\overset{OH}{\underset{OR}{\Big\langle}} \underset{}{\overset{H^+}{\rightleftharpoons}} \underset{/}{\overset{\backslash}{C^{\oplus}}}{-}OR + H_2O$$

9-III

In a mixture of alcohols (of equal activity) it is expected that the ratio of respective acetals or ketals derived from the various alcohols will be formed in proportion to the ratios of the concentrations of the alcohols. The ethyl orthoformate plays the role of dehydrating agent as it reacts with acid and water to form ethyl formate:

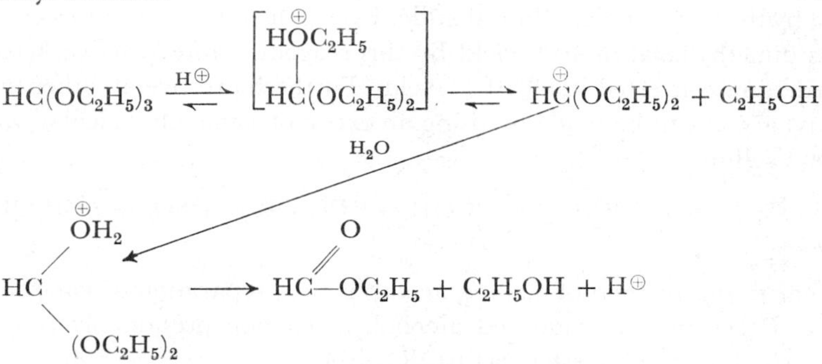

$$HC(OC_2H_5)_3 \underset{}{\overset{H^{\oplus}}{\rightleftharpoons}} \left[\begin{array}{c} \overset{\oplus}{HOC_2H_5} \\ | \\ HC(OC_2H_5)_2 \end{array} \right] \rightleftharpoons \overset{\oplus}{HC}(OC_2H_5)_2 + C_2H_5OH$$

$$H_2O$$

$$\underset{\backslash}{\overset{/}{HC}}\overset{\overset{\oplus}{OH_2}}{\underset{(OC_2H_5)_2}{\Big|}} \longrightarrow HC{-}OC_2H_5 + C_2H_5OH + H^{\oplus}$$

(a) **Preparation of α-bromocinnamaldehyde acetal** (82–86% from α-bromocinnamic aldehyde, ethyl orthoformate, and ammonium chloride in absolute ethanol) [53].

(b) **Other examples**

(1) **Acrolein diethyl acetal** (72–80% from acrolein, ethyl orthoformate, and ammonium nitrate in absolute ethanol) [55].

(2) **Ethyl p-acetylbenzoate diethyl ketal** (81% from ethyl p-acetylbenzoate, ethyl orthoformate, and a trace of conc. hydrochloric acid in absolute ethanol) [51].

(3) **Acetone di-n-butyl ketal** (81% from acetone, n-butyl alcohol, and ethyl orthoformate in the presence of p-toluenesulfonic acid) [56].

(4) **5,5-Dimethoxy-2,2,4,4-tetrakis-(trifluoromethyl)-1,3-dioxo-lane** (79% based on the ketone in the reaction:

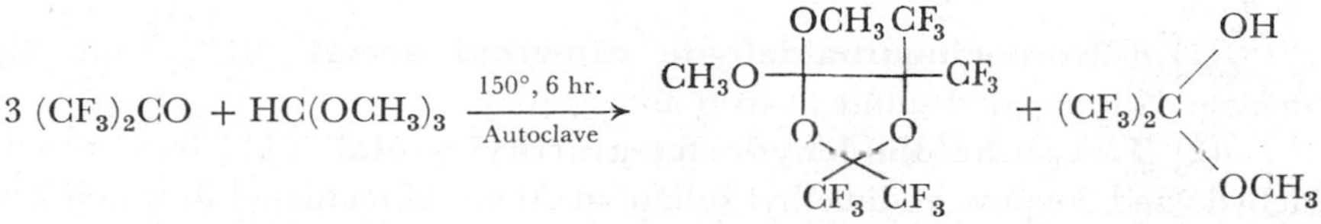

$$3\,(CF_3)_2CO + HC(OCH_3)_3 \xrightarrow[\text{Autoclave}]{150°,\,6\,hr.} CH_3O\overset{OCH_3\ CF_3}{\underset{O\diagdown_{\overset{|}{\underset{CF_3\ CF_3}{\diagup}}}O}{\Big|{-}\Big|}}CF_3 + (CF_3)_2C\overset{OH}{\underset{OCH_3}{\Big\langle}}$$

this reaction is general for orthoformates and the highly reactive hexafluoroacetone) [59].

5. From Carbonyl Compounds and Esters

Ordinary esters, of course, are not efficient in forming acetals or ketals from carbonyl compounds, but some types with special properties are good acetalizing agents. Formimidoester hydrochloride

is one example. It is sometimes called nascent orthoformic ester and as such can acetalize or ketalize carbonyl compounds.

Another interesting ester which can perform acetalization or ketalization is dimethyl sulfite; it probably owes its unique properties to the driving force offered by the formation of sulfur dioxide. Even benzophenone can be converted into its dimethylketal in 40 % yield by this reagent. More reactive ketones or aldehydes are converted in 63–91 % yields. Furthermore, acetals other than the dimethyl acetal can be made by using an excess of some other alcohol with the dimethyl sulfite:

$$RCH{=}O + SO(OCH_3)_2 + 2\ R'OH \rightarrow RCH(OR')_2 + SO_2 + 2\ CH_3OH$$
(Excess)

Similar to this method of making acetals is the preparation of some steroidal ketals from selenium dioxide and alcohol, a reaction presumably proceeding via the selenium dioxide ester, $SeO(OR)_2$ [60].

A notably active ester is dimethyl sulfate which has been used to convert a number of aldehydes into the dimethylacetals in 61–98 % yields (Ex. c.2 and Ex. c.3).

(a) Preparation of acetophenone dimethyl ketal (quantitative yield from acetophenone, 120 g., and methyl imidoformate hydrochloride, 200 g., in 250 g. absolute methanol after allowing the mixture to stand with occasional shaking for 8 days) [61].

(b) Preparation of cycohlexanone dimethyl ketal (79 % from cyclohexanone, 49 g., and dimethyl sulfite, 60 g., in 60 ml. of absolute methanol containing 1 ml. of 11 % hydrogen chloride in methanol; the evolution of sulfur dioxide is regulated by adjusting the temperature of the mixture) [62].

(c) Other examples

(1) α-Bromocinnamaldehyde dimethyl acetal (91 % from the aldehyde and dimethyl sulfite as in b above) [63].
(2) Terephthaldialdehyde tetramethyl acetal (61 % from the dialdehyde and 2 equiv. of dimethyl sulfate in 20 ml. of methanol to which 2.6–3 equiv. of 2 N sodium hydroxide were added dropwise with ice-cooling; the mixture was then heated for 10 min. with a water bath; the ethereal solution

was shaken with an alkaline solution of hydroxylamine to remove any free aldehyde, and the ether dried and evaporated) [64].

(3) 1,3-Dimethoxyphthalan

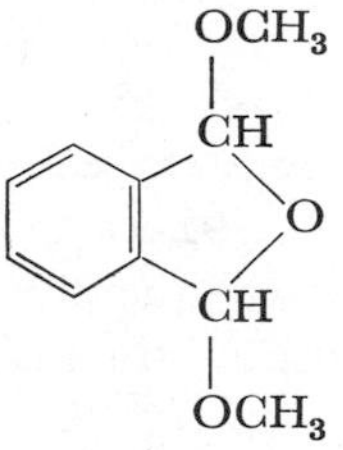

$$
\begin{array}{c}
OCH_3 \\
| \\
CH \\
\diagdown O \\
CH \\
| \\
OCH_3
\end{array}
$$

(89 % from *o*-phthaldialdehyde, dimethyl sulfate, and aqueous sodium hydroxide treated as in c.2 above) [64].

6. From Acetals or Ketals and Alcohols
(Transacetalization or Transketalization)

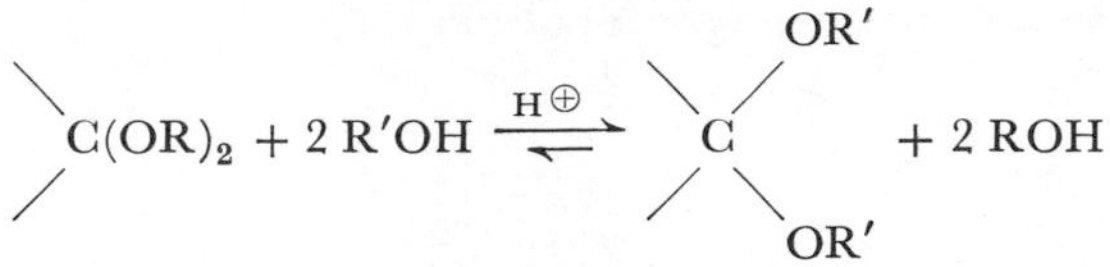

$$
\diagdown C(OR)_2 + 2\,R'OH \underset{\longleftarrow}{\overset{H^{\oplus}}{\rightleftharpoons}} \diagup C \diagdown_{OR'}^{OR'} + 2\,ROH
$$

All equilibria feed through the carbonium ion

$$
\diagdown \overset{\oplus}{C}OR \diagup
$$

and in this case equilibrium is more easily obtained than with the free carbonyl group, since neither reagents nor products above are resonance-stabilized (as in the carbonyl group).

Satisfactory, if not high, yields, using such varied catalysts as a mineral acid [65, 66], chloroacetic acid [67], arenesulfonic acids [68, 28], sulfosalicylic acid [69], boron trifluoride [70], potassium acid sulfate [71], or Dowex 50-H$^{\oplus}$ resin [72], have been obtained. With ethylene chlorohydrin as the alcohol, no catalyst is necessary [73].

The recent availability of acetone dimethyl ketal has made this method of synthesis more attractive since the by-product methanol can readily be removed by distillation to drive the equilibrium toward the ketal side [28]. The use of this ketal in synthesizing others avoids the need for less accessible ortho-esters (see A.4) or acetylenes (see B.4). Primary and secondary monohydric alcohols [29] and polyhydric alcohols [67–70] may be utilized in the reaction, although the yields for the secondary monohydric alcohols are not so satisfactory as for the others. Mixed ketals were obtainable by using 1 mole of the alcohol for each mole of the ketal.

In the discussion above the alkoxy group or groups of the acetal or ketal have been replaced by treatment with an alcohol; less common, but also possible, is the replacement of the remainder of the molecule by treatment with an aldehyde or ketone [74, 75]:

$$
\underset{\substack{\text{H} \quad\quad \text{OR}}}{\overset{\substack{\text{Ar} \quad\quad \text{OR}}}{C}} \xrightarrow{\text{Ar'CHO}} \underset{\substack{\text{H} \quad\quad \text{OR}}}{\overset{\substack{\text{Ar'} \quad\quad \text{OR}}}{C}} + \text{ArCHO}
$$

(a) Preparation of chloroacetaldehyde dibutyl acetal. Chloro-acetaldehyde dimethyl acetal 77.5 g., 92.5 g. of *n*-butanol, and 0.1 ml. of sulfuric acid were distilled until the theoretical quantity of methanol (50 ml.) came over. Neutralization of the residue with potassium carbonate followed by distillation gave the acetal, 109.2 g. (84%), b.p. 71° (0.4 mm.) [65].

(b) Other examples

(1) **Acetone dipropyl ketal** (63%, see A.1, Ex. b.4).

(2) **Bromoacetaldehyde o-xylylene acetal** (80% from bromoacet-aldehyde diethyl acetal, phthalyl alcohol, and a trace of *p*-toluenesulfonic acid) [68].

(3) **1,2-Hexadecylideneglycerol acetal.**

$$
\text{C}_{15}\text{H}_{31}\text{CH(OCH}_3)_2 + \text{CH}_2\text{OHCHOHCH}_2\text{OH} \longrightarrow \underset{\substack{\text{O—CH}_2}}{\overset{\substack{\text{O—CH}}}{\text{C}_{15}\text{H}_{31}\text{CH}}}\underset{\substack{\\ \\ \\ \text{CH}_2}}{\overset{\substack{\text{HOCH}_2\\ \\ \text{CH}}}{}}
$$

(93% from palmital dimethyl acetal, glycerol, and a trace of sulfosalicyclic acid) [69].

(4) **Butyraldehyde dibutyl acetal** (80% from butyraldehyde and acetaldehyde dibutyl acetal with 0.5–1% conc. hydrochloric acid) [75].

7. From Mercaptals and Alcohols

$$
\underset{|}{\text{CH(SR)}_2} \xrightarrow{\text{2 R'OH}} \underset{|}{\text{CH(OR')}_2} + 2\,\text{RSH}
$$

Mercaptals are readily produced from carbonyl compounds by treatment with an alkyl mercaptan under acidic conditions. In the presence of an alcohol and a catalyst an interchange occurs whereby the mercaptal is converted into an acetal. The usual catalysts are mixtures such as mercuric chloride with calcium sulfate or mercuric oxide [76, 77] and mercuric chloride and cadmium carbonate [78]. The reaction has found its greatest use in the sugar series where it is possible to prepare chain acetals from mercaptals in which ring closure is prevented by substituent groups such as the acetyl [79].

$$
\underset{\substack{\text{(CHOAc)}_4 \\ \text{CH}_2\text{OAc}}}{\text{HC(SC}_2\text{H}_5)_2} \xrightarrow[\substack{\text{CaCO}_3, \\ \text{HgCl}_2}]{\text{CH}_3\text{OH}} \underset{\substack{\text{(CHOAc)}_4 \\ \text{CH}_2\text{OAc}}}{\text{HC(OCH}_3)_2} \xrightarrow{\text{Ba(OC}_2\text{H}_5)_2} \underset{\substack{\text{(CHOH)}_4 \\ \text{CH}_2\text{OH}}}{\text{HC(OCH}_3)_2}
$$

Glucose dimethyl acetal

Yields vary from fair to good.

(a) Preparation of 3,4-di-O-benzoyl-2-deoxy-5-O-triphenylmethyl-D-ribose dimethyl acetal. Yellow mercuric oxide, 16.1 g., 74 mmoles, was

$$
\begin{array}{ccc}
HC(SC_3H_7)_2 & & HC(OCH_3)_2 \\
| & & | \\
CH_2 & & CH_2 \\
| & & | \\
CHOCOC_6H_5 & \longrightarrow & CHOCOC_6H_5 \\
| & & | \\
CHOCOC_6H_5 & & CHOCOC_6H_5 \\
| & & | \\
CH_2OC(C_6H_5)_3 & & CH_2OC(C_6H_5)_3
\end{array}
$$

added to a solution of 13.36 g., 18.6 mmoles, of 3,4-di-O-benzoyl-2-deoxy-5-O-triphenylmethyl-D-ribose diisopropyl mercaptal in 500 ml. of dry, boiling methanol. While stirring, a solution of 15.1 g., 56 mmoles, of mercuric chloride in 60 ml. of methanol was added in 1 min. and then while still stirring the suspension was refluxed for 15 min. The hot solution was filtered, the precipitate washed with methanol, and the filtrate concentrated to a syrup and extracted with methylene chloride. The methylene chloride extract after washing, drying, and concentrating gave a residue which when recystallized twice from methanol yielded 88% of the acetal [77].

8. From *gem*-Dihalides, *gem*-Disulfates, or α-Chloroethers

$$
RCHX_2 \xrightarrow{\text{2 R'ONa}} RCH(OR')_2 + 2\,NaX
$$

Just as *gem*-dihalides can be converted into aldehydes in alkaline or acidic medium (Aldehydes, Chapter 10, D.4), so these dihalides produce acetals or ketals when treated with an alkali metal alkoxide. Yields are fair to good. Similarly, the disulfate of glyoxal is easily converted into the diacetal, in this case without using the alkoxide reagent (Ex. c.1):

$$
\begin{array}{c}
CH \\
\diagup\ |\ \diagdown \\
O \quad\ |\ \quad O \\
|\ \quad |\ \quad | \\
SO_2 \quad |\quad SO_2 \xrightarrow{\ C_2H_5OH\ } \quad
\begin{array}{l}
CH(OC_2H_5)_2 \\
| \\
CH(OC_2H_5)_2
\end{array}\\
|\ \quad |\ \quad | \\
O\!-\!CH\!-\!O
\end{array}
$$

Vinyl halides probably first add alkoxy groups to form α-haloethers and then react further to form the acetals. (See B.1, Ex. b.4 and Ex. b.5.)

Formals are made also from alcohols and the commercially available methylene disulfate or from the corresponding α-chloromethyl ethers (see Ethers, Chapter 6, B.7 and B.8.) in the following way:

$$
ROCH_2Cl +
\underset{\underset{^\ominus Cl^-}{\overset{\oplus}{N}CH_2OR}}{\bigcirc\!\!\!N}
\longrightarrow
\underset{\text{(Quantitative)}}{\bigcirc\!\!\!N}
\xrightarrow{\ R'OH\ }
CH_2\!\!\diagup^{OR}_{\diagdown OR'}
+
\bigcirc\!\!\!\underset{H^{\oplus}\ {}^{\ominus}Cl}{N}
$$

(See Ex. c.3).

(a) **Preparation of ethyl diethoxyacetate** (45–50% from dichloro-acetic acid, sodium ethoxide, ethanol, and hydrogen chloride) [80].

(b) **4,6-Dimethoxy-2-dimethoxymethyl-s-triazine** (81% from 2-dibromomethyl-4,6-dimethoxy-s-triazine and sodium methoxide) [81].

(c) **Other examples**

(1) **Glyoxal tetramethyl acetal** (79% from glyoxal disulfate in absolute methanol) [82].

(2) **p-Nitrobenzophenone diethyl ketal** (nearly quantitative crude yield from 4.7 g. of *gem* dichloride in a well stirred mixture of 35 g. of powdered Na_2CO_3 in 200 ml. of absolute ethanol refluxed for 3 hr.) [83].

(3) **Formaldehyde butyloctyl acetal** (49% from 0.1 mole each of N-(octyloxymethyl)-pyridinium chloride and *n*-butyl alcohol heated for 5 hr. at 120–130°; two fractionations are necessary to separate the simple symmetrical formals which form; the b.p. of the mixed formal is 119–120° (3 mm.), n^{20}D 1.4240; aliphatic, symmetrical formals can be made in 82–91% yield by this method) [84].

1. H. Meerwein, in Houben-Weyl's, *Methoden der Organischen Chemie*, Vol. 6, G. Thieme Verlag, Stuttgart, Pt. 3, 1965, p. 199; literature reviewed up to 1963.
2. H. Kröper, Ref. 1, p. 688.
3. L. Schmid *et al.*, *Monatsh.*, **83**, 185 (1952).
4. H. Schinz and G. Schäppi, *Helv. Chim. Acta*, **30**, 1483 (1947).
5. W. E. Parham and E. L. Anderson, *J. Am. Chem. Soc.*, **70**, 4187 (1948).
6. Dr. W. Stevens, University of Leiden, Leiden, Netherlands; private communication.
7. M. Vogel and D. Rhum, *J. Org. Chem.*, **31**, 1775 (1966).
8. R. E. Dunbar and H. Adkins, *J. Am. Chem. Soc.*, **56**, 442 (1934).
9. W. L. Evans *et al.*, *Org. Syn.*, Coll. Vol. **2**, 137 (1943).
10. R. N. Icke *et al.*, *Org. Syn.*, Coll. Vol. **3**, 644 (1955).
11. C. H. Issidorides and R. Gulen, *Org. Syn.*, Coll. Vol. **4**, 679 (1963).
12. H. Adkins and B. H. Nissen, *Org. Syn.*, Coll. Vol. **1**, 1 (1941).
13. C. G. Alberti and R. Sollazzo, *Org. Syn.*, Coll. Vol. **3**, 371 (1955).
14. S. M. McElvain and D. Kundiger, *Org. Syn.*, Coll. Vol. **3**, 123 (1955).
15. M. Renoll and M. S. Newman, *Org. Syn.*, Coll. Vol. **3**, 502 (1955).
16. W. S. Allen *et al. J. Am. Chem. Soc.*, **76**, 6116 (1954).
17. M.-M. Janot *et al.*, *Bull. Soc. Chim. France*, 2109 (1961).
18. C. R. Engel and S. Rakhit, *Can. J. Chem.*, **40**, 2153 (1962).
19. R. Kuhn and H. Trischmann, *Chem. Ber.*, **94**, 2258 (1961).
20. H. Uberwasser *et al.*, *Helv. Chim. Acta*, **46**, 344 (1963).
21. C. E. Rehberg, *Org. Syn.*, Coll. Vol. **3**, 46 (1955).
22. S. Natelson and S. Gottfried, *Org. Syn.*, Coll. Vol. **3**, 381 (1955).
23. M. S. Newman and R. J. Harper, Jr., *J. Am. Chem. Soc.*, **80**, 6350 (1958).
24. T. H. Fife and L. Hagopian, *J. Org. Chem.*, **31**, 1772 (1966).
25. M. Sulzbacher *et al.*, *J. Am. Chem. Soc.*, **70**, 2827 (1948).
26. R. Garrett and D. G. Kubler, *J. Org. Chem.*, **31**, 2665 (1966).
27. N. B. Lorette *et al.*, *J. Org. Chem.*, **24**, 1731 (1959).
28. N. B. Lorette and W. L. Howard, *J. Org. Chem.*, **25**, 521 (1960).
29. W. L. Howard and N. B. Lorette, *ibid.*, **25**, 525 (1960).
30. R. E. McCoy *et al.*, *J. Org. Chem.*, **22**, 1175 (1957).
31. R. J. Stedman *et al.*, *J. Org. Chem.*, **33**, 1280 (1968).
32. C. O. Parker, *J. Am. Chem. Soc.*, **78**, 4944 (1956).

33. M. L. Wolfrom and A. Thompson, in W. Pigman, *The Carbohydrates*, Academic Press, New York, 1957, Chap. 4, p. 188, for full discussion.
34. L. Smith and J. Lindbergh, *Chem. Ber.*, **64,** 505 (1931).
35. H. Hibbert and N. M. Carter, *J. Am. Chem. Soc.*, **51,** 1601 (1929).
36. B. F. Stimmel and C. G. King, *J. Am. Chem. Soc.*, **56,** 1724 (1934).
37. N. M. Carter, *Chem. Ber.*, **63,** 2399 (1930).
38. K. G. Shipp and M. E. Hill, *J. Org. Chem.*, **31,** 853 (1966).
39. R. Webb and A. Duke, *J. Chem. Soc.*, 4320 (1962); H. J. Dietrich and J. V. Karabinos, *J. Org. Chem.*, **31,** 1127 (1966).
40. D. L. Heywood and B. Phillips, *J. Org. Chem.*, **25,** 1699 (1960).
41. Ref. 1, p. 269.
42. P. Baudart, *Bull. Soc. Chim. France*, **11,** 336 (1944); R. H. Hall and B. K. Howe, *J. Chem. Soc.*, 2480 (1951).
43. Ref. 1, p. 220.
44. V. Mahadevan *et al.*, *J. Lipid Res.*, **6**(3), 434 (1965); *C.A.*, **63,** 8184 (1965).
45. A. P. Dunlop and F. N. Peters, *The Furans*, Reinhold Publishing Corp., New York, 1953, p. 659.
46. D. T. Manning and H. A. Stansbury, Jr., *J. Org. Chem.*, **26,** 3755 (1961).
47. T. Bersin and G. Willfang, *Chem. Ber.*, **70B,** 2167 (1937).
48. G. Willfang, *Chem. Ber.*, **74B,** 145 (1941).
49. L. Ruzicka *et al.*, *Helv. Chim. Acta*, **31,** 422 (1948).
50. E. Vogel and H. Schinz, *Helv. Chim. Acta*, **33,** 116 (1950).
51. L. Schmid *et al.*, *Monatsh.*, **83,** 185 (1952).
52. J. Bornstein *et al.*, *J. Am. Chem. Soc.*, **78,** 83 (1956).
53. C. F. H. Allen and C. O. Edens, Jr., *Org. Syn.*, Coll. Vol. **3,** 731 (1955).
54. C. E. Ballou, *Biochem. Prepn.*, **7,** 45 (1960).
55. J. A. VanAllan, *Org. Syn.*, Coll. Vol., **4,** 21 (1963).
56. C. A. MacKenzie and J. H. Stocker, *J. Org. Chem.*, **20,** 1695 (1955).
57. B. Helferich and J. Hausen, *Chem. Ber.*, **57,** 795 (1924); C. Weygand, *Organic Preparations*, Interscience Publishers, New York, 1945, p. 187.
58. C. E. Feazel and W. G. Berl, *J. Am. Chem. Soc.*, **72,** 2278 (1950).
59. R. A. Braun, *J. Org. Chem.*, **31,** 1147 (1966).
60. E. P. Oliveto *et al.*, *J. Am. Chem. Soc.*, **76,** 6113 (1954).
61. K. Alder and H. Niklas, *Ann. Chem.*, **585,** 97 (1954).
62. W. Voss, *Ann. Chem.*, **485,** 283 (1931).
63. F. Wille and F. Knörr, *Chem. Ber.*, **85,** 841 (1952).
64. E. Schmitz, *Chem. Ber.*, **91,** 410 (1958).
65. G. Eglinton *et al.*, *J. Chem. Soc.*, 1860 (1954).
66. W. E. Parham and J. D. Jones, *J. Am. Chem. Soc.*, **76,** 1068 (1954).
67. T. Matsuda and M. Sugishita, *Bull. Chem. Soc. Japan*, **35,** 1446 (1962).
68. R. Grewe and A. Struve, *Chem. Ber.*, **96,** 2819 (1963).
69. C. Piantadosi *et al.*, *J. Am. Chem. Soc.*, **80,** 6613 (1958).
70. U. Faass and H. Hilgert, *Chem. Ber.*, **87,** 1343 (1954).
71. E. H. Pryde *et al.*, *J. Org. Chem.*, **29,** 2083 (1964).
72. G. E. Ham, *J. Chem. Eng. Data*, **8,** 280 (1963).
73. S. M. McElvain and A. N. Bolstad, *J. Am. Chem. Soc.*, **73,** 1988 (1951).
74. E. Bograchov, *J. Am. Chem. Soc.*, **72,** 2268 (1950).
75. M. F. Shostakovskiĭ *et al.*, *Izv. Akad. Nauk SSSR, Otd. Khim. Nauk*, 378 (1956); *C.A.*, **50,** 15504 (1956).
76. J. H. Jordaan and W. J. Serfontein, *J. Org. Chem.*, **28,** 1395 (1963).
77. D. L. MacDonald and H. C. Fletcher, Jr., *J. Am. Chem. Soc.*, **81,** 3719 (1959).
78. H. R. Bolliger, *Helv. Chim. Acta*, **34,** 989 (1951).
79. Ref. 33, p. 199.
80. R. B. Moffett, *Org. Syn.*, Coll. Vol. **4,** 327 (1963).
81. E. Kober and C. Grundmann, *J. Am. Chem. Soc.*, **80,** 5547 (1958).

82. D. H. Grangaard and C. B. Purves, *J. Am. Chem. Soc.*, **61**, 428 (1939).

83. W. W. Kaeding and L. J. Andrews, *J. Am. Chem. Soc.*, **74**, 6189 (1952).

84. D. N. Kurssanow *et al.*, in A. N. Nesmeyanov and P. A. Bobrow, *Synthesen Organischer Verbindungen*, Veb Verlag Technik, Berlin, 1959, Vol. 1, p. 90.

B. Addition to Unsaturated Linkages

1. From Vinyl Ethers, Halides, or Acetylenes and an Alcohol or Phenol

$$CH_2=CHOR \underset{\longleftarrow}{\overset{H^\oplus}{\longrightarrow}} (CH_3\overset{\oplus}{CHOR}) \underset{\longleftarrow}{\overset{R'OH}{\longrightarrow}} \underset{OR'}{\overset{CH_3CHOR}{\underset{|}{}}} + H^\oplus$$

The same intermediate carbonium ion as discussed in A.1 may be generated from vinyl ethers and acid, and naturally it then becomes a source of acetals or ketals.

Vinyl ethers, halides, and acetylenes also have the property of adding nucleophilic reagents, such as alkoxide bases, which permit the formation of acetals as shown:

$$CH_2=CHOR + \overset{\ominus}{O}C_2H_5 \longrightarrow \left[\underset{OC_2H_5}{\overset{\ominus}{\underset{|}{CH_2CHOR}}}\right] \underset{H_2O}{\overset{HOC_2H_5}{\underset{or}{\longrightarrow}}} \underset{OC_2H_5}{\overset{CH_3CHOR}{\underset{|}{}}} \quad [ref. 1]$$

$$CH_2=CHX + \overset{\ominus}{O}C_2H_5 \longrightarrow CH_2^\ominus{-}CHXOC_2H_5 \overset{-\overset{\ominus}{X}}{\longrightarrow}$$

$$CH_2=CHOC_2H_5 \underset{2.\ C_2H_5OH}{\overset{1.\ \overset{\ominus}{O}C_2H_5}{\longrightarrow}} CH_3CH(OC_2H_5)_2$$

(See Ex. b.4 and Ex. b.5.) Acetylenes add the alkoxide anion to form first the vinyl ether and then the acetal or ketal (see Ex. b.6). No catalysts are needed in some cases if a high enough temperature is employed (see Ex. b.3).

In electrophilic reactions the addition is catalyzed by acid [2, 3] and phosphorus oxychloride [4]. Mercaptans also add to give O,S-acetals [5]. Yields are sometimes in the 70–90 % range.

(a) Preparation of 2-(3-chloropropoxy)-tetrahydropyran. A mixture of equimolar quantities of dihydropyran and 3-chloropropyl alcohol with a few drops of conc. hydrochloric acid, was allowed to stand for 3 hr. with occasional shaking. After ether had been added, the solution was shaken with 10 % sodium hydroxide solution to remove all acid. Drying followed by distillation gave a 78 % yield of the ketal, b.p. 103° (14 mm.) [2].

(b) Other examples

(1) Acetoacetaldehyde dimethyl acetal (83–85 % from 1-methoxy-1-buten-3-yne, methanol, and dilute sulfuric acid) [3].

$$HC\equiv CCH=CHOCH_3 \overset{CH_3OH}{\underset{H^\oplus}{\longrightarrow}} CH_3COCH_2CH(OCH_3)_2$$

References for Section B are on p. 534

(2) **2-Phenoxytetrahydropyran** (77% from dihydropyran, phenol, and a trace of concentrated hydrochloric acid) [2].

(3) **Acetaldehyde ethyl n-butyl acetal** (46% from 200 g. of vinyl n-butyl ether and 92 g. of absolute ethyl alcohol in an autoclave at 100–120° for 8–10 hr.; yields are of course higher for simple, symmetrical acetals; lower yields are obtained in an ordinary flask at 98–99° reflux temperature if 1 drop of concentrated hydrochloric acid is added; about 30 different acetals are described) [6].

(4) **3-Oxo-6-methylheptanal dimethyl acetal** (90% from 2-chlorovinyl 3-methylbutyl ketone and absolute methanol containing 1 equiv. of NaOH) [7].

(5) **α-Benzoylacetaldehyde dimethyl acetal** (60% from β-benzoyl-vinyl chloride and sodium methoxide in methanol for 1 day at room temperature [8].

(6) **Benzoylacetaldehyde diethyl acetal** (71% from 6.5 g. of phenyl ethynyl ketone in 25 ml. of absolute ethanol added to a solution of sodium ethoxide, from 1.15 g. of Na, in 50 ml. of absolute ethanol held at 0–5° for 1 hr.) [9].

2. From Vinyl Ethers

$$CH_2{=}CHOR \xrightarrow[\text{Hg(OAc)}_2]{\text{BF}_3} CH_2{=}CHCH_2CH(OR)_2$$

Vinyl ethers polymerize with extreme ease in the presence of acid catalysts, but moderation in polymerization can be brought about by the use of boron trifluoride and mercuric acetate as catalysts [10]. The path of the reaction is thought to be:

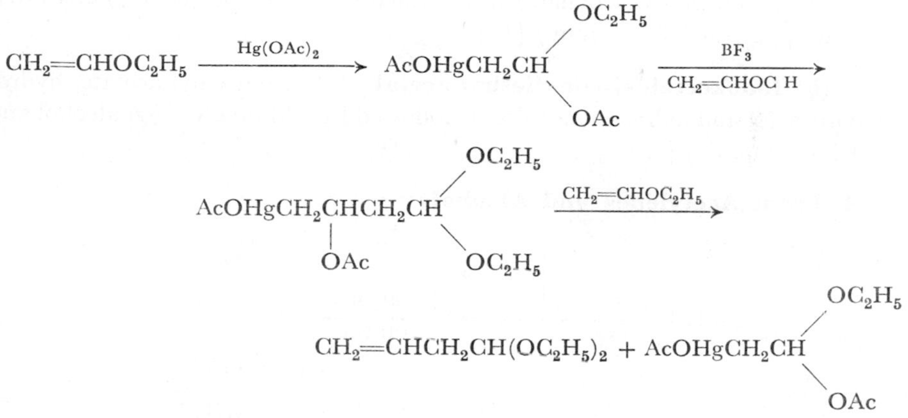

$$CH_2{=}CHOC_2H_5 \xrightarrow{\text{Hg(OAc)}_2} AcOHgCH_2CH\underset{OAc}{\overset{OC_2H_5}{{<}}} \xrightarrow[CH_2{=}CHOC\ H]{\text{BF}_3}$$

$$AcOHgCH_2CHCH_2CH\underset{OC_2H_5}{\overset{OC_2H_5}{{<}}} \quad \underset{OAc}{|} \xrightarrow{CH_2{=}CHOC_2H_5}$$

$$CH_2{=}CHCH_2CH(OC_2H_5)_2 + AcOHgCH_2CH\underset{OAc}{\overset{OC_2H_5}{{<}}}$$

Hydrolysis of the unsaturated acetal yields crotonaldehyde rather than vinylacetaldehyde, and pyrolysis leads to good yields of 1-ethoxybutadiene.

Mixtures of vinyl ether and acetals without the use of mercuric acetate give analogous ether acetals (see Ex. b):

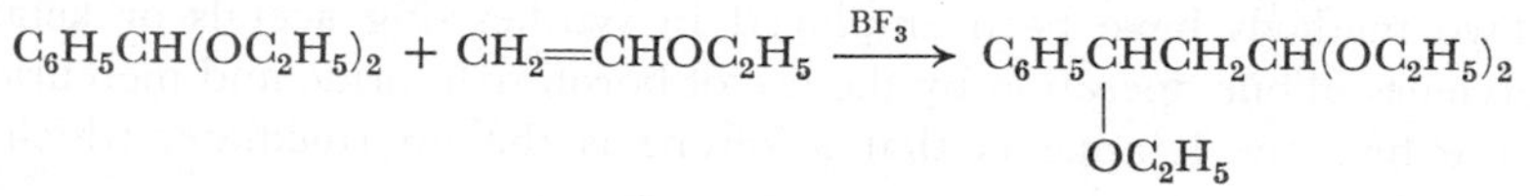

$$C_6H_5CH(OC_2H_5)_2 + CH_2{=}CHOC_2H_5 \xrightarrow{\text{BF}_3} C_6H_5CHCH_2CH(OC_2H_5)_2 \atop |\ OC_2H_5$$

(a) **Preparation of 1,1-diethoxy-3-butene** (72 % from 10 moles of vinyl ethyl ether added to a mixture of 21 g. of 32 % boron trifluoride in ether, 0.15 moles of mercuric acetate, and 100 ml. of ether in 14 min. at 42 ± 3°) [10].

(b) **Preparation of β-ethoxyhydrocinnamaldehyde diethyl acetal** (72 % from the ethyl acetal of benzaldehyde, vinyl ethyl ether, and a catalytic amount of boron trifluoride at 49° for several hours) [11].

3. From Vinyl Esters and Alcohols

$$CH_2{=}CHOCOR + R'OH \xrightarrow[BF_3]{HgO} CH_3CH(OR')_2$$

This synthesis appears to take place in two steps: (1) Electrophilic addition to the carbon-carbon double bond as shown in B.1 followed by (2) an interchange of the acyl for an alkoxy group [12, 13]. Mercuric oxide-boron trifluoride or iodine monochloride has been used as a catalyst. The reaction proceeds best under anhydrous conditions and is suitable for the preparation of acetals or ketals, the yields of acetals being in the 80–90 % range with primary alcohols and 38 % with isopropyl alcohol. Tertiary alcohols do not give acetals by this method [12].

(a) **Preparation of n-butyl acetal.** To n-butyl alcohol, 148 g. (2.0 moles), 1 g. of red mercuric oxide, and 1 ml. of boron trifluoride diethyl etherate was added 86 g. (1.0 mole) of vinyl acetate over a 10 min. period while the temperature was kept below 55°. Stirring was continued for 1 hr., after which the mixture was poured into a suspension of 56 g. (0.5 mole) of sodium carbonate in 250 ml. of water. The upper layer was separated, dried over anhydrous potassium carbonate, and distilled to give 153 g. (88.5 %) of n-butyl acetal, b.p. 74.5–76° (14 mm.) [12].

(b) **Iodoacetaldehyde diethyl acetal** (84 % from vinyl acetate, hydrochloric acid, and iodine monochloride followed by addition to ethyl alcohol and calcium chloride) [13].

4. From Acetylenes and Alcohols

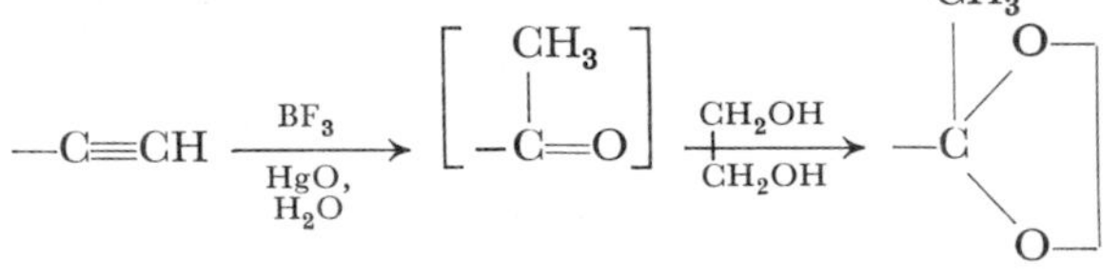

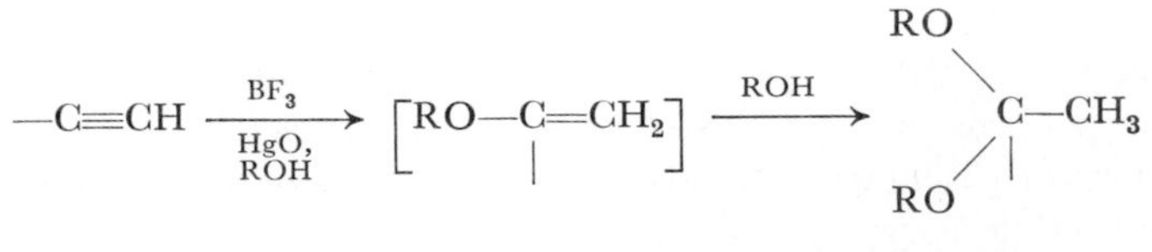

Two methods have been employed in synthesizing acetals or ketals from acetylenes in one operation by the use of boron trifluoride and mercuric oxide. In the first one it appears that a ketone is the intermediate, which in the presence of an alcohol gives the acetal or ketal [14]. By the second method it has been suggested that a vinyl ether is the intermediate which with the alcohol gives the final product [15]. The yields by these methods are in the 70–80% range.

(a) **Preparation of 2-(1'-hydroxycylcopentyl)-2-methyl-1,3-dioxo-lane.**   Mercuric oxide, 4.5 g., was covered with 2 ml. of the freshly distilled

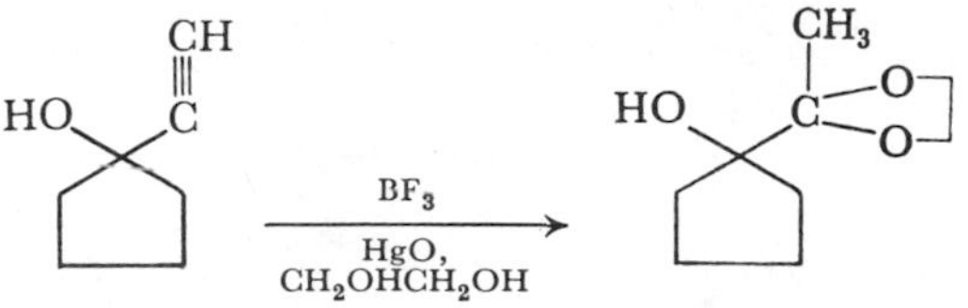

boron trifluoride-ether and 5 g. of dry ethylene glycol. After warming for a few minutes and then cooling, a mixture of 20 g. of ethylene glycol and 35 g. of 1-ethynylcyclopentanol was introduced dropwise with stirring. Stirring was continued for an additional 1.5–2 hr. and then on the next day 5 g. of anhydrous potassium carbonate was stirred into the solution for a few minutes. Centrifugation and distillation gave 34 g. (74%) of the ketal, b.p. 69–70° (1 mm.) [14].

(b) **Preparation of acetaldehyde di(carbomethoxymethyl) acetal** (81% from methyl hydroxyacetate and acetylene in the presence of boron trifluoride-methanol (1:1) and mercuric oxide) [15].

5. From Acetylene, Acyl Halides, and Alcohols

$$HC\equiv CH \xrightarrow[AlCl_3]{RCOCl} RCOCH=CHCl \xrightarrow[NaOH]{R'OH} RCOCH_2CH\begin{smallmatrix}OR'\\ \\OR'\end{smallmatrix}$$

This synthesis involves two steps: (1) addition of the acyl halide to acetylene in the presence of aluminum chloride to form a β-chlorovinyl ketone, and (2) reaction of the ketone formed with an alcohol and sodium hydroxide to form a β-ketoacetal [16, 17]. Both steps are low-temperature reactions carried out under anhydrous conditions, the first being in a solvent such as carbon tetrachloride or trichlorethylene. Overall yields vary from 50 to 70%.

In this reaction the first step is that of electrophilic addition to form the β-chlorovinyl ketone which presumably forms the acyl acetylene 9-IV in the presence of sodium methoxide:

$$RCOCH=CHCl \longrightarrow [RCOC\equiv CH] \xrightarrow{R'OH} RCOCH_2CH(OR')_2$$
$$9\text{-IV}\phantom{] \xrightarrow{R'OH} RCOCH_2CH(O}9\text{-V}$$

9-IV then adds the alcohol as indicated in B.4 to give the acetal 9-V.

(a) **Preparation of acetoacetaldehyde dimethyl acetal.**   A solution of 1.05 moles of sodium hydroxide in 350 ml. of absolute methanol was added with stirring to 1.0 mole of β-chlorovinyl methyl ketone in 150 ml. of absolute methanol at $-15°$ to $-10°$ (2 hr.). The mixture was poured into a liter of ice-cold, saturated salt solution and extracted with four 100-ml. portions of ether. Distillation of the dried extracts gave an 81 % yield (50 % based on the original acyl halide) of the acetal, b.p. $38°$ (2 mm.) [16].

(b) **Preparation of ethyl 8,8-dimethoxy-6-oxooctanoate** (62–70 % from adipic acid ethyl ester chloride, acetylene, and aluminum chloride followed by treatment with sodium hydroxide in methanol) [17].

6. From Acetylenes and Ethyl Orthoformate

$$RC\equiv CH + HC(OC_2H_5)_3 \rightarrow RC\equiv CCH(OC_2H_5)_2 + C_2H_5OH$$

Although acetylenic acetals may be prepared via the Grignard reagent (C.2) it is also possible to prepare them directly from acetylenes in the presence of a catalyst such as zinc chloride, zinc iodide, zinc nitrate, or cadmium iodide [18]. The method offers the advantage of requiring readily available starting ma-terials, although for low-boiling acetylenes a bomb must be employed. By using orthoesters higher in the series than orthoformic, ketals are obtained. Yields vary from 15 to 80 %.

The reaction appears to be an ionic one [18] in which the carbonium ion 9-VI formed by the catalyst from ethyl orthoformate adds to the acetylene

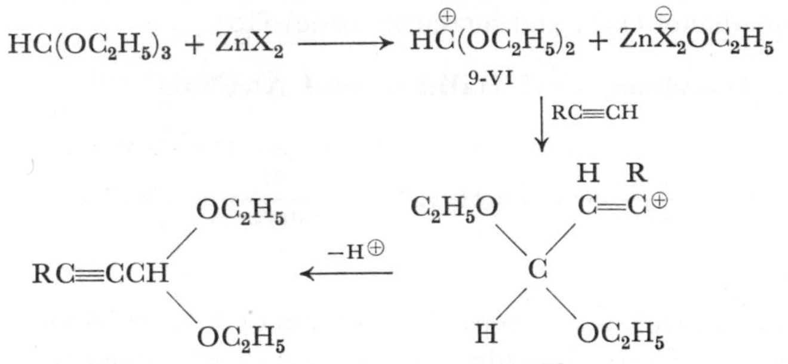

$$HC(OC_2H_5)_3 + ZnX_2 \longrightarrow H\overset{\oplus}{C}(OC_2H_5)_2 + Zn\overset{\ominus}{X_2}OC_2H_5$$

forming the carbonium ion 9-VII which by the loss of a proton gives the acetal 9-VIII.

(a) **Preparation of phenylpropargylaldehyde diethyl acetal** (72–78 % from phenylacetylene and ethyl orthoformate in the presence of zinc iodide) [19].

(b) **Methyl phenylethynyl diethyl ketal** (34 % from phenylacetylene,

$$C_6H_5C\equiv C-C(OC_2H_5)_2$$
$$|$$
$$CH_3$$

ethyl orthoacetate, and zinc chloride) [18].

7. From Alkenes and Formaldehyde (Prins)

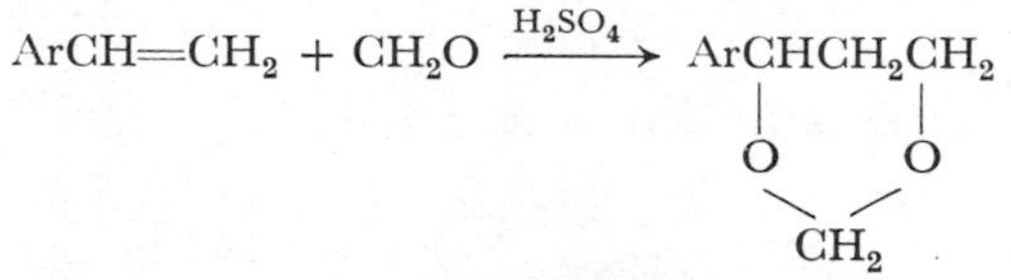
ArCH=CH2 + CH2O
H2SO4
ArCHCH2CH2
O O
CH2

This synthesis has been reviewed [20] and discussed under Alcohols, Chapter 4, b.4. In the Prins reaction the use of formaldehyde more than any other aldehyde leads to *m*-dioxanes as the product. The condensation may be carried out with acid catalysts to give highly variable yields, sometimes excellent, other times poor, accompanied by a wide variety of products typical of carbonium ion reactions. A palladium chloride-copper chloride catalyst has been recommended when the yield with an acid catalyst is poor [21]:

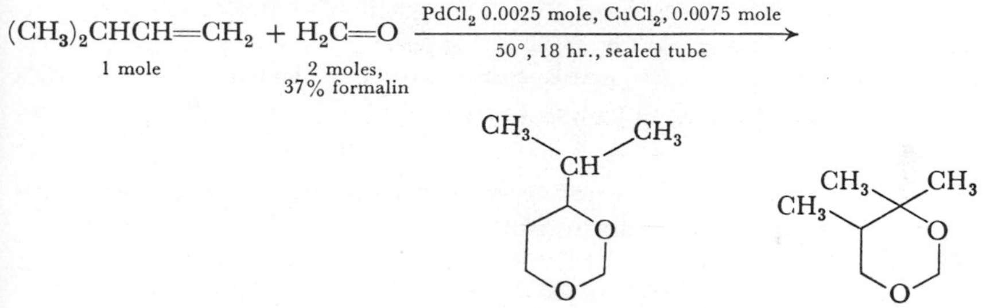
(CH3)2CHCH=CH2 + H2C=O
PdCl2 0.0025 mole, CuCl2, 0.0075 mole
50°, 18 hr., sealed tube
1 mole
2 moles, 37% formalin
CH3 CH3
CH
O
O
CH3 CH3
CH3
O
O

4-Isopropyl-1,3-dioxane, 52% 4,4,5-Trimethyl-1,3-dioxane, 8%

Paraformaldehyde may be used in place of formalin.

Olefins also may be coupled with dioxolane or trioxane by a photochemical process [22]:

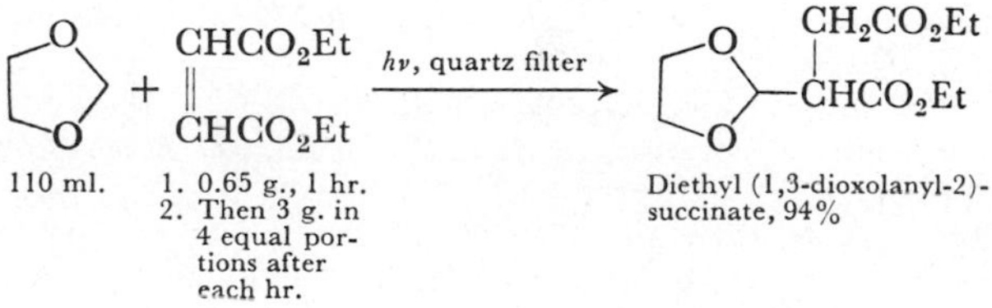
O
O
CHCO2Et
CHCO2Et
hν, quartz filter
CH2CO2Et
CHCO2Et
110 ml.
1. 0.65 g., 1 hr.
2. Then 3 g. in 4 equal portions after each hr.
Diethyl (1,3-dioxolanyl-2)-succinate, 94%

Simple, terminal olefins give yields of 20–50% by this process. A small amount of 2,2′-dioxolane

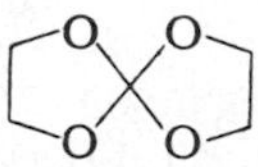

accompanies the products. Yields of products from trioxane and olefins are somewhat lower than those from dioxolane.

(a) Preparation of 4-phenyl-1,3-dioxane (72–88% from styrene, formalin, and sulfuric acid) [23].

(b) Preparation of 4,4-dimethyl-*m*-dioxane (60% based on formaldehyde treated with isobutylene and 25% sulfuric acid) [20].

8. From Tetracyanoethylene and Alcohols

Tetracyanoethylene has recently become available through a synthesis from malonitrile [24]. With alcohols, this alkene, in the presence of a catalyst such as urea or zinc acetate, forms dicyanoketene ketals which may be converted into a variety of related types [25]. Yields vary from 50 to 94%.

(a) Preparation of dicyanoketene ethylene ketal (77–85% from tetracyanoethylene, ethylene glycol, and urea) [26].

(b) Preparation of dicyanoketene diethyl ketal (72% from tetracyanoethylene, ethyl alcohol, and urea) [25].

1. A. Dornow and F. Ische, *Chem. Ber.*, **89,** 876 (1956).
2. W. E. Parham and E. L. Anderson, *J. Am. Chem. Soc.*, **70,** 4187 (1948).
3. W. Franke *et al.*, *Chem. Ber.*, **86,** 793 (1953).
4. H. B. Henbest *et al.*, *J. Chem. Soc.*, 3646 (1950).
5. F. Kipnis *et al.*, *J. Am. Chem. Soc.*, **73,** 1783 (1951).
6. M. F. Shostakovskiĭ and N. A. Gerschtein, in A. N. Nesmeyanov and P. A. Bobrow, *Synthesen Organischer Verbindungen*, Veb Berlag Technik, Berlin, Vol. 2, 1956, pp. 150.
7. C. C. Price and J. A. Pappalardo, *Org. Syn.*, **32,** 79 (1952).
8. N. K. Kochetkov *et al.*, *J. Gen. Chem. USSR (Eng. Transl.)*, **29,** 2533 (1959).
9. K. Bowden *et al.*, *J. Chem. Soc.*, 945 (1946).
10. R. I. Hoaglin *et al.*, *J. Am. Chem. Soc.*, **80,** 5460 (1958).
11. B. M. Mikhailov and L. S. Povarov, *Izv. Akad. Nauk SSSR, Otd. Khim. Nauk*, 1239 (1957); *C.A.*, **52,** 6253 (1958).
12. W. J. Croxall *et al.*, *J. Am. Chem. Soc.*, **70,** 2805 (1948).
13. S. Akiyoshi and K. Okuno, *J. Am. Chem. Soc.*, **74,** 5759 (1952).
14. J. D. Billimoria and N. F. Maclagan, *J. Chem. Soc.*, 3257 (1954).
15. D. D. Coffman *et al.*, *J. Org. Chem.*, **13,** 223 (1948).
16. C. C. Price and J. A. Pappalardo, *J. Am. Chem. Soc.*, **72,** 2613 (1950).
17. U. Schmidt and P. Grafen, *Chem. Ber.*, **92,** 1177 (1959).
18. B. W. Howk and J. C. Sauer, *J. Am. Chem. Soc.*, **80,** 4607 (1958).
19. B. W. Howk and J. C. Sauer, *Org. Syn.*, Coll. Vol. **4,** 801 (1963).
20. E. Arundale and L. A. Mikeska, *Chem. Rev.*, **51,** 505 (1952).
21. S. Sakai *et al.*, *Chem. Commun.*, 1073 (1967).
22. I. Rosenthal and D. Elad, *J. Org. Chem.*, **33,** 805 (1968).
23. R. L. Shriner and P. R. Ruby, *Org. Syn.*, Coll. Vol. **4,** 786 (1963); *see also* R. W. Shortridge, *J. Am. Chem. Soc.*, **70,** 873 (1948).
24. R. A. Carboni, *Org. Syn.*, Coll. Vol. **4,** 877 (1963).
25. W. J. Middleton and V. A. Engelhardt, *J. Am. Chem. Soc.*, **80,** 2788 (1958).
26. C. L. Dickinson and L. R. Melby, *Org. Syn.*, Coll. Vol. **4,** 276 (1963).

C. Displacement by Alkoxides or Complexes

1. From Ketones, Alkoxides, and Alkylating Agents

The question arises whether advantage can be taken of the product 9-IX derived from the addition of the alkoxide ion to ketones to form ketals:

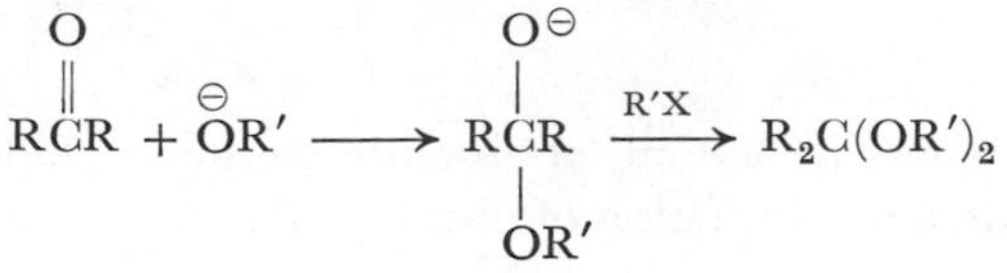

$$
\underset{\text{RCR}}{\overset{\text{O}}{\|}} + \overset{\ominus}{\text{OR}'} \longrightarrow \underset{\underset{\text{OR}'}{|}}{\overset{\overset{\text{O}\ominus}{|}}{\text{RCR}}} \xrightarrow{\text{R}'\text{X}} R_2C(OR')_2
$$

9-IX

Only 1,2-diketones such as benzil and 9,10-phenanthraquinone seem to ketalize in this manner (see Ex. a).

Formaldehyde, however, has been acetalized with acrylonitrile [1]:

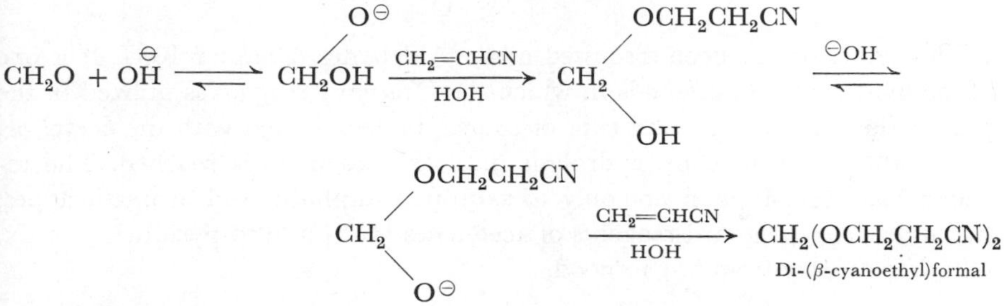

$$
CH_2O + \overset{\ominus}{OH} \rightleftharpoons \overset{\overset{O\ominus}{\diagup}}{CH_2OH} \xrightarrow[\text{HOH}]{CH_2=\text{CHCN}} \underset{\underset{OH}{\diagdown}}{\overset{\overset{OCH_2CH_2CN}{\diagup}}{CH_2}} \xrightarrow{\overset{\ominus}{OH}} \rightleftharpoons
$$

$$
\underset{\underset{O\ominus}{\diagdown}}{\overset{\overset{OCH_2CH_2CN}{\diagup}}{CH_2}} \xrightarrow[\text{HOH}]{CH_2=\text{CHCN}} CH_2(OCH_2CH_2CN)_2
$$
Di-(β-cyanoethyl)formal

Similar to the above processes is the conversion of α-haloketones into α-hydroxyketals (see Ex. b).

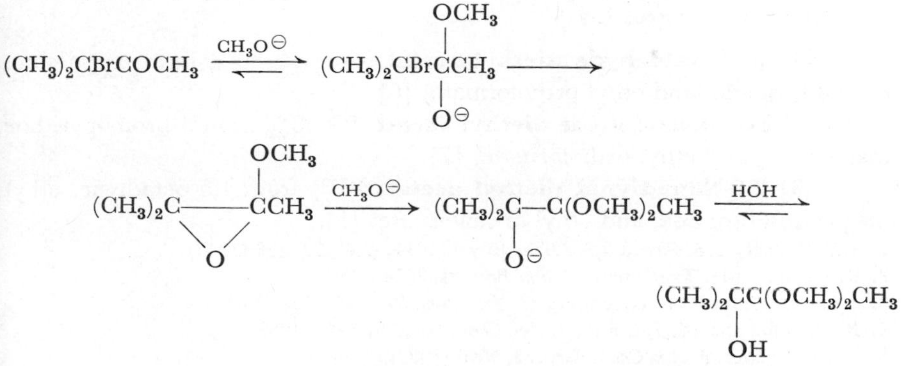

$$
(CH_3)_2CBrCOCH_3 \underset{\longleftarrow}{\overset{CH_3O\ominus}{\longrightarrow}} \underset{\underset{O\ominus}{|}}{\overset{\overset{OCH_3}{|}}{(CH_3)_2CBrCCH_3}} \longrightarrow
$$

$$
\underset{\diagdown O \diagup}{(CH_3)_2C{-}\!\!\!{-}CCH_3} \xrightarrow{CH_3O\ominus} \underset{\underset{O\ominus}{|}}{\overset{\overset{OCH_3}{|}}{(CH_3)_2C{-}C(OCH_3)_2CH_3}} \xrightarrow{\text{HOH}} \rightleftharpoons
$$

$$
\underset{\underset{OH}{|}}{(CH_3)_2CC(OCH_3)_2CH_3}
$$
3-Hydroxy-3-methyl-2-butanone
dimethyl ketal

(a) **Preparation of 9,10-phenanthraquinone-9,9-dimethyl ketal** (nearly quantitative yield on a microscale from the quinone, methyl iodide, and barium oxide with a small amount of barium hydroxide hydrate, the latter being essential, in DMF at 20° for 14 hr.) [2].

References for Section C are on p. 536.

(b) Preparation of 3-hydroxy-3-methyl-2-butanone dimethyl ketal
(76 % from 3-bromo-3-methyl-2-butanone, 0.875 mole, added slowly to sodium

$$CH_3\overset{\overset{\displaystyle Br}{|}}{\underset{\underset{\displaystyle CH_3}{|}}{C}}-COCH_3 \longrightarrow CH_3\overset{\overset{\displaystyle OH}{|}}{\underset{\underset{\displaystyle CH_3}{|}}{C}}-\overset{\overset{\displaystyle OCH_3}{}}{\underset{\underset{\displaystyle CH_3}{}}{C}}{\overset{}{\diagdown}}_{OCH_3}$$

methylate, 0.875 mole, in 300 ml. of absolute methanol, followed by filtration
of sodium bromide and distillation of filtrate) [3].

2. From Grignard Reagents and Ethyl Orthoformate

$$RMgX \xrightarrow{HC(OC_2H_5)_3} RCH\underset{OC_2H_5}{\overset{OC_2H_5}{}} + C_2H_5OMgX$$

This synthesis has been discussed under Aldehydes, Chapter 10.G. It is one
of the most common methods in which the Grignard reagent is utilized in the
preparation of aldehyde and can, of course, be terminated with the acetal or,
if desirable, continued by hydrolysis until the aldehyde is reached. The reaction has been applied not only to saturated aliphatic and aromatic types,
but also to the Grignard reagents of acetylenes [4, 5] and to pyridines (see Ex.
b.2). Yields vary from fair to good.

(a) Preparation of p-tolualdehyde diethyl acetal (over 74 %; see
Aldehydes, Chapter 10.G, Ex. a.2).

(b) Other examples

(1) n-Hexaldehyde diethyl acetal (over 45–50 % from magnesium,
n-amyl bromide, and ethyl orthoformate) [6].

(2) Nicotinaldehyde diethyl acetal (50–58 % from 3-bromopyridine,
magnesium, and ethyl orthoformate) [7].

(3) 2,6-Nonadiynal diethyl acetal (81 % from 1,5-octadiyne, ethylmagnesium bromide, and ethyl orthoformate) [5].

1. J. F. Walker, U.S. Patent 2,352,671, July 4, 1944; C.A., **39,** 223 (1945).
2. R. Kuhn and H. Trischmann, Chem. Ber., **94,** 2258 (1961).
3. J. G. Aston and R. B. Greenburg, J. Am. Chem. Soc., **62,** 2590 (1940).
4. R. G. Jones and M. J. Mann, J. Am. Chem. Soc., **75,** 4048 (1953).
5. F. Sondheimer, J. Am. Chem. Soc., **74,** 4040 (1952).
6. G. B. Bachman, Org. Syn., Coll. Vol. **2,** 323 (1943).
7. J. P. Wibaut et al., Rec. Trav. Chim., **71,** 1021, 798 (1952).

D. Oxidation or Reduction

1. From Ethylene Derivatives by Oxidation

$$\langle\rangle CH{=}CH_2 \xrightarrow[Tl(OCOCH_3)_3]{CH_3OH} \langle\rangle CH_2CH(OCH_3)_2$$

References for Section D are on p. 539.

This oxidation has been carried out with several ethylene derivatives which give a variety of products [1]. Some, however, such as styrene and *p*-methoxystyrene, gave yields of the acetals of 64% and 93% (based on the starting material consumed), respectively. Cyclohexene gave a 62% yield of cyclopentylcarboxaldehyde dimethyl acetal. It is suggested that the oxidation takes the following path:

$$Tl(OAc)_3 + \overset{\oplus}{H} \longrightarrow Tl(\overset{\oplus}{OAc})_2 + AcOH$$

$$C_6H_5CH{=}CH_2 \xrightarrow{Tl(\overset{\oplus}{OAc})_2} C_6H_5\overset{\oplus}{C}H{-}CH_2{-}Tl\overset{\displaystyle OAc}{\underset{\displaystyle OAc}{\big\langle}}$$

$$\Big\downarrow {CH_3OH}$$

$$C_6H_5\overset{\oplus}{C}H{-}CH_2 \xleftarrow[\substack{-TlOAc \\ -AcOH,}]{\overset{\oplus}{H}} C_6H_5CH{-}CH_2{-}Tl\overset{\displaystyle OAc}{\underset{\displaystyle OAc}{\big\langle}}$$
$$\;\;\;\underset{\displaystyle OCH_3}{\big|} \qquad\qquad\qquad \underset{\displaystyle OCH_3}{\big|}$$

$$\Big\downarrow {C_6H_5\ migration}$$

$$CH_3O\overset{\oplus}{C}H{-}CH_2C_6H_5 \xrightarrow[H^{\oplus}]{CH_3OH} \overset{\displaystyle CH_3O}{\underset{\displaystyle CH_3O}{\big\rangle}}CHCH_2C_6H_5$$

(a) Preparation of homoanisaldehyde dimethyl acetal. Thallic acetate, 20 g., 7 g. of *p*-methoxystyrene, and 50 ml. of methanol were heated for 2 days under reflux. Distillation at 0.1 Torr. gave 7.5 g. (93% based on the starting material consumed) of the acetal, b.p. 80–87°) [1].

2. From Ethers and *t*-Butyl Peracetate (or Perbenzoate)

In this free radical reaction, both the acetate and the *t*-butyl acetal are formed, the latter probably being derived from the acetate:

$$\underset{O}{\boxed{}} + CH_3CO_3C(CH_3)_3 \xrightarrow[C_6H_6]{CuBr} \underset{O}{\boxed{}}\overset{\displaystyle O}{\overset{\|}{OCCH_3}} \xrightarrow{-CH_3CO_2H}$$

$$\underset{O}{\boxed{}} \xrightarrow{(CH_3)_3COH} \underset{O}{\boxed{}}OC(CH_3)_3$$

(a) Preparation of 2-*t*-butoxytetrahydrofuran. A mixture of tetrahydrofuran (133 g.), benzene (150 ml. containing 75% *t*-butyl peracetate) and cuprous bromide (0.1 g.) was heated under reflux for 14 hr., cooled, diluted with 50 ml. of ether, and washed with dilute sodium carbonate solution. The ether-benzene extract was dried and concentrated and the residue distilled to give 2-acetoxytetrahydrofuran (4 g., b.p. 34–40° (0.1 mm.) and 2-*t*-butoxytetrahydrofuran (46 g., 45%, b.p. 40–42° (13 mm.) [2].

(b) Preparation of 1-ethoxyethyl acetate (75 % from ethyl ether, *t*-butyl peracetate and traces of cuprous bromide under uv irradiation) [3].

(c) Preparation of *t*-butyl methyl acetal of *p*-chlorobenzaldehyde (36 % from *p*-chlorobenzyl methyl ether and *t*-butyl hydroperoxide at 110°) [4].

3. From Diazoalkanes

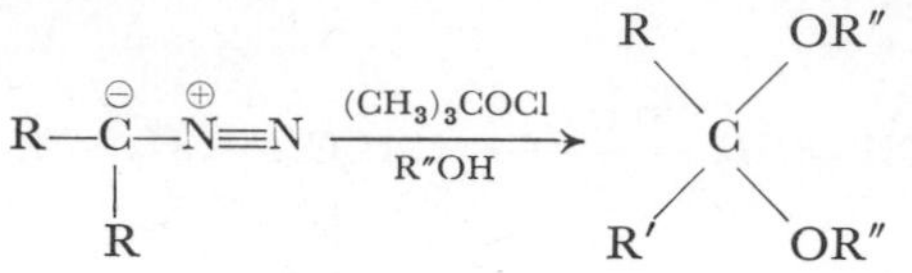

$$R-\overset{\ominus}{C}-\overset{\oplus}{N}\equiv N \xrightarrow[R''OH]{(CH_3)_3COCl} \begin{array}{c} R \\ R' \end{array}\!\!C\!\!\begin{array}{c} OR'' \\ OR'' \end{array}$$

Hydrolysis of diazoalkanes in the absence of catalysts leads to alcohols [5]. If the hydrolysis is carried out in an oxidative environment under the proper conditions, a ketal is obtained, as shown in Ex. a.

(a) Preparation of benzophenone diethyl ketal (43 % as the 2,4-dinitrophenylhydrazone from diazodiphenylmethane in ethanol treated with *t*-butyl hypochlorite dropwise at −10°) [6].

4. From Alcohols by Oxidation

The oxidation of alcohols in acidic, nearly anhydrous solution may well give acetals or ketals as intermediates or by-products. The method apparently is impractical for acetal preparation and rarely used for ketal preparation. A new approach which apparently gives poor yields at present involves an oxidation by oxygen [7]:

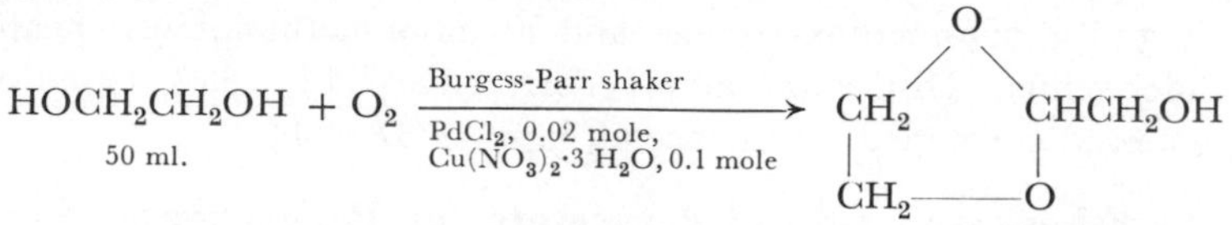

$$HOCH_2CH_2OH + O_2 \xrightarrow[\substack{PdCl_2, 0.02\ mole, \\ Cu(NO_3)_2\cdot 3\ H_2O, 0.1\ mole}]{Burgess\text{-}Parr\ shaker} \begin{array}{c} O \\ CH_2\ \ \ \ CHCH_2OH \\ | \ \ \ \ \ \ \ | \\ CH_2\text{——}O \end{array}$$

50 ml.

2-Hydroxymethyldioxolane,
ca. 3 g.

5. From Orthoesters by Reduction

$$HC(OC_2H_5)_3 \xrightarrow{LiAlH_4} CH_2(OC_2H_5)_2 + C_2H_5OH$$

Several orthoesters have been reduced to acetals by the use of metallic hydrides such as lithium aluminum hydride [8, 9] and diisobutyl aluminum hydride [10]. Yields are good.

(a) Preparation of β-methylmercaptopropionaldehyde dimethyl acetal.   Lithium aluminum hydride, 0.25 *M*, in 1 *M* ether solution, was added to a boiling 0.33 *M* solution of methyl ortho-β-methylmercaptopropionate in benzene. After refluxing for 4 hr., the complex was decomposed with a 30 % Rochelle salt solution and the benzene extract was dried and distilled to give a 97 % yield of the acetal, b.p. 73° (0.9 mm.) [8].

(b) Preparation of benzaldehyde diethyl acetal (95 % from triethyl orthobenzoate and diisobutylaluminum hydride) [10].

1. H.-J. Kabbe, *Ann. Chem.*, **656**, 204 (1962).
2. G. Sosnovsky, *J. Org. Chem.*, **25**, 874 (1960); *Tetrahedron*, **13**, 241 (1961).
3. G. Sosnovsky, *J. Org. Chem.*, **28**, 2934 (1963).
4. R. L. Huang *et al.*, *Chem. Commun.*, 1251 (1968).
5. H. Zollinger, *Azo and Diazo Chemistry*, Interscience Publishers, New York, 1961, p. 102.
6. H. Baganz and H.-J. May, *Angew. Chem., Intern. Ed. Engl.*, **5**, 420 (1966).
7. W. C. Lloyd, *J. Org. Chem.*, **32**, 2816 (1967).
8. C. J. Claus and J. L. Morgenthau, Jr., *J. Am. Chem. Soc.*, **73**, 5005 (1951).
9. C. J. Claus and J. L. Morgenthau, Jr., U.S. Patents 2,786,872, March 26, 1957; 2,830,092 April 8, 1958.
10. L. I. Zakharkin and I. M. Khorlina, *Izv. Akad. Nauk SSSR, Otd. Khim. Nauk*, 2255 (1959); *C.A.*, **54**, 10837 (1960).

E. Electrolytic Alkoxylation

A variety of rather specific electrolytic methods is available for the preparation of ketals.

1. Electrolysis of α-Alkoxycarboxylic Acids in Alcohol

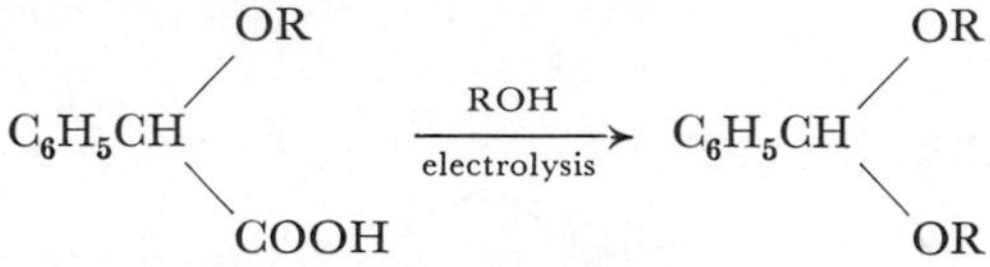

Electrolysis of α-alkoxy phenyl- and diphenylacetic acids results in the formation of the corresponding acetal or ketal as the case may be [1]. If the R groups in the acid and alcohol are identical, simple acetals or ketals are produced; if they are different, mixed types are obtained. This method appears to be one of the most reliable for forming mixed acetals or ketals. Yields are in the 60–75 % range. α-Arylthiodiphenyl acetic acids on electrolysis give ketals of benzophenone [2].

It has been suggested that the free radical 9-X under electrolytic conditions

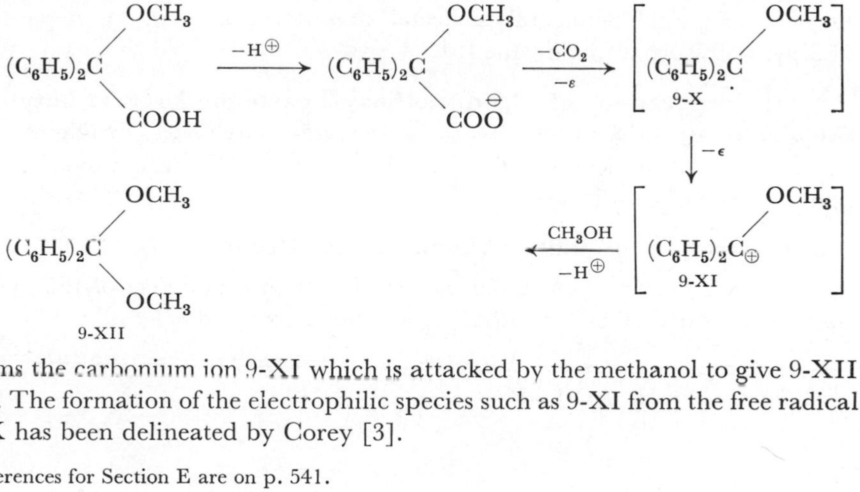

forms the carbonium ion 9-XI which is attacked by the methanol to give 9-XII [2]. The formation of the electrophilic species such as 9-XI from the free radical 9-X has been delineated by Corey [3].

References for Section E are on p. 541.

(a) Preparation of benzaldehyde dimethyl acetal. A solution of 8.1 g. of α-methoxyphenylacetic acid in 180 ml. of absolute methanol, to which sufficient sodium had been added to neutralize about 3% of the acid, was electrolyzed until the electrolyte became slightly alkaline. After evaporation of the solvent under reduced pressure, the residue was distilled to give 4.5 g. (61.6%) of the acetal, b.p. 196° [1].

(b) Preparation of benzaldehyde methyl ethyl acetal (74% by electrolysis of α-ethoxydiphenylacetic acid in methanol) [1].

2. Electrolysis of Furans in Alcohol

Furans and substituted furans can be electrolyzed in alcoholic solutions containing ammonium bromide as the electrolyte to form ketals (or o-esters) [4]:

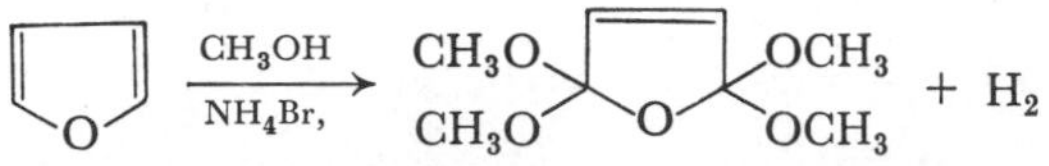

45.6 amp.-hr.   2,2,5,5-Tetramethoxy-2,5-dihydrofuran

Bromine is generated above and adds to the furan to give a product which is solvolyzed:

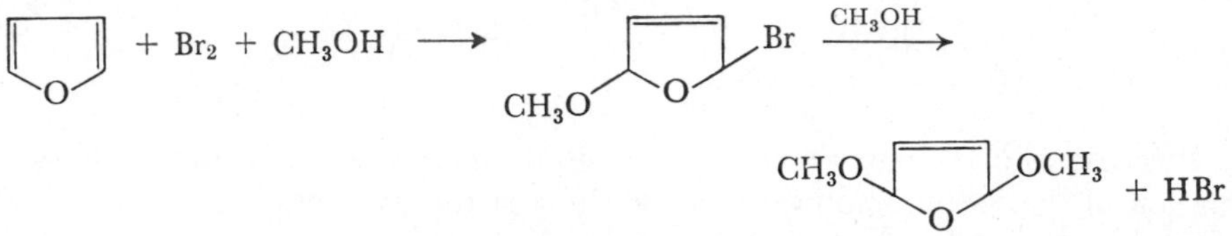

2,5-Dimethoxy-2,5-dihydrofuran

This reaction can be carried out without electrolysis, but the product is usually less stable because of bromine-containing impurities.

The electrolytic method is adaptable to large-scale production and has considerable scope in the synthesis of furan derivatives; even methyl furoate in methanol can be electrolyzed to 2,5-dimethoxy-2-carbomethoxy-2,5-dihydro-furan in 68% yield. Some of these acetals are intermediates in the preparation of 3-pyridols, such as pyridoxine [5].

(a) Preparation of 2,5-dimethoxy-2-carbomethoxy-5-t-butyl-2,5-dihydrofuran (97% from 2-carbomethoxy-5-t-butylfuran, methanol, and conc. sulfuric acid electrolyzed in a cell containing a nickel cathode and a platinum anode at −22°) [6].

3. Electrolysis of Dialkoxybenzenes in Alcohol

In a similar manner, but using basic rather than acidic electrolytes, hydroquinone dimethyl ether (and isomers) can be alkoxylated [7]:

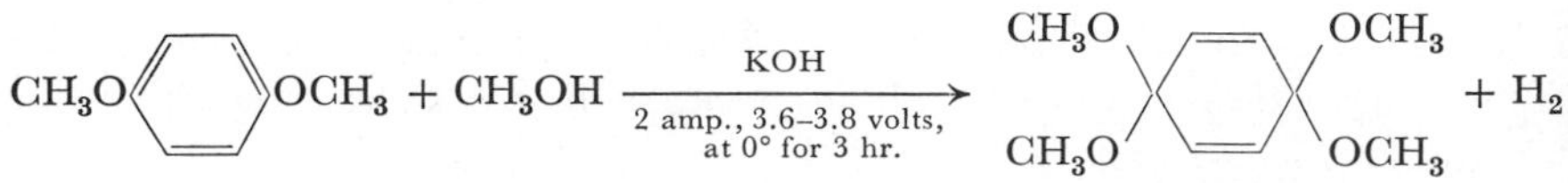

75%
p-Benzoquinone tetramethyl ketal

The product from resorcinol dimethyl ether indicates that the attack takes place *para* to one of the methoxy groups:

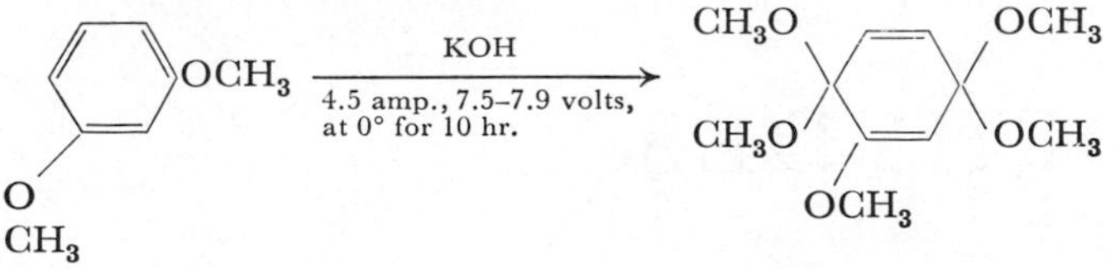

61%
2,3,3,6-Pentamethoxy-1,
4-cyclohexadiene

The mechanism of this and other electrolytic alkoxylations is uncertain and may involve the insertion of methoxy free radicals or the attack of a methoxy anion on an electron-deficient substrate [7]:

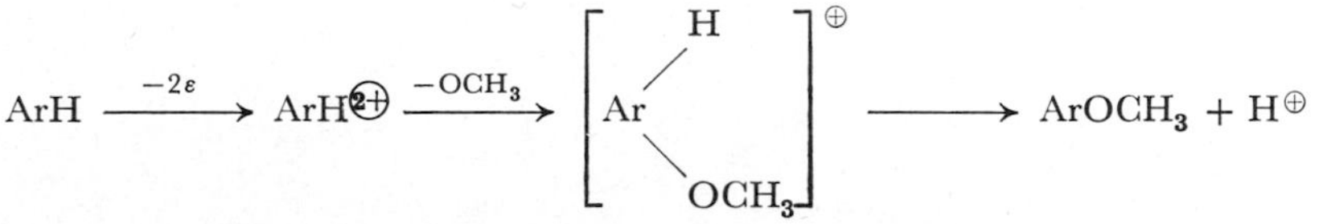

1. B. Wladislaw and A. M. J. Ayres, *J. Org. Chem.*, **27**, 281 (1962).
2. B. Wladislaw, *Chem. Ind. (London)*, 1868 (1962).
3. E. J. Corey *et al.*, *J. Am. Chem. Soc.*, **82**, 2645 (1960).
4. N. Elming, *Advan. Org. Chem.*, **2**, 67 (1960).
5. N. Elming and N. Clauson-Kaas, *Acta Chem. Scand.*, **9**, 23 (1955).
6. N. Elming *et al.*, *Acta Chem. Scand.*, **9**, 17 (1955).
7. B. Belleau and N. L. Weinberg, *J. Am. Chem. Soc.*, **85**, 2525 (1963).

Chapter 10

ALDEHYDES

The syntheses of aldehydes are more diverse and extensive than the syntheses of any other class with the possible exception of amines and ketones. Four reasons are apparent. First, the aldehyde, in the middle of the different oxidation levels of organic compounds, may be prepared both by oxidation and by reduction. Second, the aldehyde structure, being a reactive functional group—liquid members often can be air-oxidized or polymerized by traces of acid—has served as a synthetic challenge to the ingenuity of the chemist. Third, aldehydes are valuable intermediates. Fourth, since few truly *general* syntheses of aldehydes are available, the discovery of numerous good, but less general, methods has been desirable.

The methods which follow start with oxidation, of which there are twenty different sections. Here a reagent is needed which will not carry the oxidation beyond the aldehyde stage. Five potential reagents as well as active manganese dioxide are discussed in A.1 while others are found in A.5. Of the methods of reduction, the Brown procedure using lithium trialkoxyaluminohydrides is coming to the forefront as a method of reducing acid chlorides (B.3) and nitriles (B.4). The Friedel-Crafts syntheses (C) in which the aldehyde group may be attached to aromatic rings or introduced into ethylenic groups follow. The numerous hydrolytic methods (D) are discussed, and they become extensive since many heterocyclic ring compounds may be hydrolyzed to aldehydes, a fact which is important when it is realized that 1,3-dithiacyclohexanes (D.3) after alkylation and dihydrooxazines

after alkylation and reduction yield a variety of aldehydes on hydrolysis. Some of the rearrangements (E) described are ingenious and, if they are viewed collectively, it is not difficult to visualize other rearrangements which may lead to aldehydes. Although the condensation (F) and organometallic methods (G) are well known classical procedures, recent progress has been made among these methods in the alkylation of aldehydes and aldehyde derivatives. Even the time-honored Reimer-Tiemann reaction has been subject to recent study. The mechanism has been clarified, and an understanding of it should lead to more ideal conditions for improving yields. Finally the electrocyclic reactions (H) lead to aldehydes of complicated structure which may be difficult to obtain by other methods.

Three general sources of information are available on the synthesis of aldehydes. Of these the Houben-Weyl reference [1] is the most comprehensive. The discussion in this chapter is arranged differently and is much more abbreviated than the Houben-Weyl treatment. Three other references [2–4] are worthy of note.

A. Oxidation

Compounds in a lower oxidation level, such as hydrocarbons, primary alcohols, halides, amines, olefins, are capable potentially of being oxidized to the corresponding aldehyde. As mentioned in the introduction, however, difficulties arise because of the general instability of aldehydes, particularly those in the liquid state. They are susceptible to further oxidation not only by the oxidizing agent, but also by the air. Furthermore, the aliphatic aldehydes have poor storage characteristics and polymerize to paraldehyde-type molecules

$$RCHO \rightarrow RCH—[O—CHR—]_nO—$$

probably catalytically by the acid ($RCOOH$) produced by air oxidation. Liquid aromatic aldehydes are also unstable to air but usually form the acid rather than the polymeric aldehyde. Thus the synthesis of an aldehyde in high yield may not depend as much on the ideality of a reagent as on the efficacy with which the product is handled. Specific means of overcoming these problems are described in the following oxidation sections. But it should be remembered that the instability problem is inherent in all syntheses of aldehydes.

1. From Primary Alcohols

$$RCH_2OH \rightarrow RCHO$$

The yields in general are only fair to good because of the tendency, already mentioned, of the aldehyde to oxidize further to the acid. The ester is quite often the first product of this overoxidation if carried out in acid solution [5]:

$$RCH_2OH \longrightarrow RCO_2H \xrightarrow{RCH_2OH} RCOOCH_2R$$

Four methods are available to avoid overoxidation:

1. Removal of the aldehyde as it is formed (see Ex. a). The aldehyde is the most volatile of the possible components: alcohol, aldehyde, acid, and ester. If the temperature of the reaction mixture is held above the boiling point of the aldehyde, it will distill as it is formed. The limitation, of course, is that the method is restricted to aldehydes which boil below a reasonable temperature (see Ex. b for extending temperature range). It is possible that the method could be further exploited by sweeping with an inert gas, such as carbon dioxide, below the boiling point of the aldehyde.

2. Rapid conversion of the aldehyde into a stable derivative. If the reaction is conducted in acetic anhydride, for example, the diacetate, which seems to be somewhat more resistant to oxidation than the aldehyde, is formed:

$$RCHO \xrightarrow{(CH_3CO)_2O} RCH(OCOCH_3)_2$$

The diacetate, of course, is hydrolyzed readily by aqueous acid [6].

3. Protection by an inert solvent. The oxidation medium in this case is heterogeneous, and success depends on the preferential diffusion of the alcohol

References for Section A are on pp. 567–570.

to the aqueous oxidation phase. A case in point is the use of "active" manganese dioxide in an ether solvent. The manganese dioxide is not only insoluble in the medium but also is a mild oxidizing agent. It has been particularly useful in the preparation of aldehydes from benzylic and allylic alcohols (see Ex. d). The active manganese dioxide may be prepared from aqueous solutions of equivalent amounts of manganese sulfate and potassium permanganate [7], in the presence of alkali [8], from concentrated aqueous potassium permanganate added to aqueous manganese sulfate [9], and by the air pyrolysis of manganese carbonate or oxalate [10]. A commercial grade has also been employed [11]. It is difficult to generalize on the relative activity of the various oxide prep- arations because of the great number of variables involved and the incon- sistencies which appear in the literature. The Attenburrow oxide seems to have been used the most widely, usually in a satisfactory manner. The Harfenist oxide oxidizes benzyl alcohols to the aldehyde in good yield but does not affect allyl alcohols appreciably. However, if the oxide is washed with nitric acid, the reverse is true [10]. In a later investigation [12] in which various types of manganese dioxide were employed it was reported that the Attenburrow and Morton oxides are superior to that of Mancera in the oxidation of benzyl and allyl alcohols (see Ex. d). Fortunately the commercial manganese dioxide now available is regarded as more active than the Attenburrow product [13].

4. Use of selective oxidizing agents. Five such types of reagents are described.

Dimethyl sulfoxide. Oxidation with dimethyl sulfoxide has broad applicability to primary alcohols [14]. With inexpensive or common alcohols, it is perhaps best simply to heat the alcohol in DMSO while air is bubbled through to give yields of 25–85 % on selected examples (see Ex. g.7). With more precious alco- hols, it is preferable to utilize a combination of reagents which aid in the formation of intermediate sulfoxonium salts, 10-III via the cation 10-II, as shown with dicyclohexylcarbodiimide (see Ex. g.6). Reagents other than 10-I

$$C_6H_{11}N{=}C{=}NC_6H_{11} + CH_3SOCH_3 \xrightarrow[\text{pyridinium trifluoroacetate}]{H_3PO_4 \text{ or}}$$

10-I

$$\underset{\substack{\overset{|}{N}H \\ \overset{|}{C_6H_{11}} \\ \text{10-II}}}{C_6H_{11}N{=}\overset{\overset{\displaystyle CH_3}{|}}{C}{-}O\overset{\oplus}{S}\underset{\overset{|}{CH_3}}{}} \xrightarrow{RCH_2OH} (C_6H_{11}NH)_2CO + \underset{\substack{\overset{|}{CH_3} \\ \text{10-III}}}{\overset{\overset{\displaystyle CH_3}{|}}{\oplus}SOCH_2R} \underset{\longleftarrow}{\xrightarrow{\text{Weak base}}}$$

$$\left[\underset{\substack{\overset{|}{CH_3} \\ \text{10-IV}}}{\overset{\overset{\displaystyle \overset{\ominus}{CH_2}}{|}}{\oplus SOCH_2R}} \right] \longrightarrow RCH{=}O + CH_3SCH_3$$

which aid in the formation of the sulfoxonium salt are acetic anhydride or phosphorus pentoxide. And alkyl chloroformates form a sulfoxonium salt, 10-V, similar to 10-II, which can abstract a proton as in 10-IV [15]:

$$RCH_2OCOCl + CH_3SOCH_3 \longrightarrow RCH_2O\overset{\displaystyle O}{\overset{\displaystyle \|}{C}}O\overset{\oplus}{\underset{\displaystyle CH_3}{\overset{\displaystyle CH_3}{S}}} + \xrightarrow{Et_3N}$$

10-V

$$RCHO + CH_3SCH_3 + CO_2$$

It is reasonable to anticipate that sulfoxonium salts formed in any other way will be subsequently oxidized to the aldehyde. The salts can be formed from alkyl iodides, tosylates, or reactive halides (see A.10) or by diazotization of primary amines which form relatively stable carbonium ions [16]:

$$C_6H_5CH_2NH_2 \xrightarrow[\text{DMSO, } 100°, \text{ 2 hr.}]{2 \text{ NaNO}_2, \text{ 3 CF}_3CO_2H} [C_6H_5CH_2^{\oplus}] \longrightarrow$$

$$\left[C_6H_5CH_2O\overset{\oplus}{\underset{\displaystyle CH_3}{\overset{\displaystyle CH_3}{S}}} \right] \xrightarrow{CF_3CO_2^{\ominus}} C_6H_5CH{=}O + CH_3SCH_3$$

Benzaldehyde, 82%

Ceric ammonium nitrate [17]. The procedure with this reagent is simplicity itself. The aqueous reagent (2.1 equivalents) is added to a stirred aqueous or dilute acetic acid solution (or heterogeneous mixture) of the alcohol heated at 50–100° until the orange-red color fades. The aldehyde is extracted in the usual manner with ether or methylene chloride. Yields are excellent with benzylic alcohols and good with cyclopropanemethanol, but the reaction may not be general.

Chromium trioxide in dry pyridine [18], *t-butyl chromate* (Ex. f), *or dipyridine* Cr^{VI} *oxide*. The three reagents capitalize on the fact that anhydrous conditions mitigate against further oxidation to the acid. The latter reagent appears to be the most promising: yields are high, operation is simple, the reagent may be stored, and the scope seems wide, although it is true that only one aliphatic and three aromatic aldehydes had been synthesized at the time of writing [19]:

$$C_6H_{13}CH_2OH + CrO_3 \cdot 2 \text{ } C_5H_5N \xrightarrow[25°, \text{ 5–15 min.}]{CH_2Cl_2} C_6H_{13}CH{=}O$$

1 equiv.,   6 equiv., red   Heptaldehyde, 93%

Potassium hypochlorite. Since this reagent is relatively sluggish in oxidizing benzaldehydes to the corresponding acids, it has been used in oxidizing benzylic alcohols to benzaldehydes (Ex. c). N-Chlorosuccinimide and N-bromoacetamide, similar reagents, are used occasionally for the same purpose [20].

Iodosobenzene [21]:

$$C_6H_5CH_2OH + C_6H_5IO + \text{dioxane} \xrightarrow[25°]{N_2} C_6H_5CH{=}O$$

10 mmoles   11 mmoles   20 ml.   Benzaldehyde, 85%

It was demonstrated that iodosobenzene does not oxidize benzaldehyde under these conditions. Unfortunately, the reaction has not been investigated in depth.

Less selective oxidizing agents may be employed if limiting quantities are adhered to:

Chromic anhydride in dilute acetic acid (Ex. g.4);
Selenium dioxide (for long-chain fatty aldehydes) [22];
Lead tetraacetate (for pyridinecarboxaldehydes) [23];
1-Chlorobenzotriazole (for aromatic aldehydes) [24]

(a) Preparation of propionaldehyde (removal of aldehyde by distillation as it is formed; 45–49% from *n*-propyl alcohol, potassium dichromate, and sulfuric acid) [25].

(b) Preparation of 2,2,3-trimethyl-3-butenal (28% from 100 g. of the alcohol, 35 ml. of acetic acid, and 15 ml. of propionic acid brought to distillation through a Podbielniak column to which 85 g. of CrO_3 in 15 ml. of water, 700 ml. of acetic acid, and 300 ml. of propionic acid were added at the same rate as distillation; the distillate was neutralized, extracted, and the extract distilled) [26].

(c) Preparation of *o*-methoxybenzaldehyde. Enough bleaching powder to give 0.1 mole of potassium hypochlorite (such as H.T.H., Monsanto, 35% active chlorine) was treated with potassium carbonate until the pH was 9–11. The calcium carbonate was filtered and washed with water. The filtrate was mixed with 0.1 mole of *o*-methoxybenzyl alcohol, dissolved in 15 ml. of methanol and 100 ml. of water. The mixture was shaken overnight at room temperature, extracted with ether, and washed with aqueous sodium bisulfite solution. The bisulfite solution was acidified to a pH of 2 to liberate *o*-methoxybenzaldehyde, 51%; 2,4-dinitrophenylhydrazone m.p. 249–252° [27].

(d) Preparation of acrolein. Allyl alcohol (2 g.) in 48 g. of petroleum ether was refluxed over 5 g. of activated manganese dioxide for 19 hr. A 99% yield of acrolein, as the 2,4-dinitrophenylhydrazone, was isolated from the petroleum ether. This yield is higher than that obtained in the usual preparations because of the small quantities of substrate used [12].

(e) Preparation of *p*-nitrobenzalaniline. *p*-Nitrobenzylaniline (0.05 mole), activated manganese dioxide, and 500 ml. of benzene were refluxed in an apparatus with Dean-Stark tube for water removal. After slightly more than the theoretical amount of water was collected (several hours), the benzene was filtered and concentrated to yield 82% *p*-nitrobenzalaniline. This compound could be hydrolyzed readily to *p*-nitrobenzaldehyde [28].

(f) Preparation of benzaldehyde. *t*-Butyl chromate obtained from the reaction of 20 g. of chromium trioxide and 44.4 g. of *t*-butyl alcohol was dried by azeotroping with benzene. The ester in 70 ml. of benzene was added gradually to a well-cooled solution of 16.2 g. of benzyl alcohol and 70 ml. of benzene. After 7 days at room temperature, 19.5 g. of hydrazine hydrate (85%) in 50 ml. of water was added gradually with cooling and stirring, followed by the addition of 250 ml. of 20% sulfuric acid. The benzene layer, after washing and drying, yielded 15 g. of benzaldehyde and 0.8 g. of benzoic acid [29].

(g) Other examples

(1) **Cinnamaldehyde** (80% from the alcohol using silver trifluoro-acetate and iodine) [30].

(2) **Propiolaldehyde** (HC≡CCHO, 35–41% using chromium tri-oxide and aqueous sulfuric acid) [31].

(3) **o-(-o-Dimethylaminomethylphenyl) benzaldehyde** (90% from the alcohol and *t*-butyl chromate in benzene at room temperature for 4 days, the excess reagent being decomposed with oxalic acid) [32].

(4) **3,6-Dimethoxy-2,4,5-trimethylbenzaldehyde** (83% from the alcohol in aqueous acetic acid using only a slight excess of the required amount of chromic anhydride; the groups surrounding the aldehyde protect it from further oxidation) [33].

(5) **4-Pyridinecarboxaldehyde** (68% from the corresponding alcohol and manganese dioxide in benzene refluxed for several hr.) [34].

(6) **Cholane-24-al** (84% from cholane-24-ol, in dry DMSO to which dry pyridine, trifluoroacetic acid, and DCC were added in the order given) [35].

(7) **Benzaldehyde** (80% from 0.1 mole of benzyl alcohol and 0.7 mole of DMSO heated at reflux for 14 hr. with air passing through the solution) [36].

2. From Primary Alcohols by Dehydrogenation

$$\text{RCH}_2\text{OH} \xrightarrow{\text{Catalyst}} \text{RCHO} + \text{H}_2$$

This synthesis, largely of industrial interest, has found little use as a laboratory method because the apparatus is somewhat involved and special catalysts are often required. Various catalysts such as copper, silver, mixtures of these metals, and copper chromite have been employed. The latter on Celite gave yields from 53 to 67% with eight primary alcohols [37]. In comparing in the presence of some air, copper on kieselguhr, copper-silver on kieselguhr, silver on copper gauze, and copper-silver on pumice, it was found that the latter was the most effective [38]. More recently ethyl alcohol passed over a supported copper catalyst containing 5% cobalt and 2% chromium as promotors gave acetaldehyde in a conversion as high as 95% with a yield of 88% [39].

(a) **Preparation of acetaldehyde** (88% as the bisulfite from ethanol using a copper nitrate-cobalt oxide-chromic oxide catalyst impregnated on short-fibered asbestos at 275°) [39].

(b) **Preparation of other aliphatic aldehydes** (67% from 1-propanol, 62% from 1-butanol, 53% from 1-hexanol, each passed through a column containing copper-chromium oxide on Celite heated at 300–345°) [37].

3. From Primary Alcohols and Aluminum *t*-Butoxide (Oppenauer)

This reaction is discussed more fully under Ketones, Chapter 11, A.2, but mention is made here of its application to aldehyde synthesis [40]:

$$RCH_2OH \xrightarrow{\text{Al } t\text{-butoxide}} RCHO$$

Cinnamaldehyde, anisaldehyde, benzophenone, fluorenone [41] with aluminum triphenoxide, or the aluminum salt of the alcohol to be oxidized, can be used as the hydrogen acceptor. Of these a high-boiling one is preferred since the aldehyde formed can be removed by distillation. Cinnamaldehyde is superior to benzaldehydes as a hydrogen acceptor (Ex. b).

(a) **Preparation of α-cyclocitral** (66 % by conversion of α-cyclogeraniol into the aluminum salt and oxidation with anisaldehyde as the temperature is raised from 122 to 170° under 12 mm. pressure) [40].

(b) **Preparation of benzaldehyde** (94 % from benzyl alcohol, cinnamaldehyde, and aluminum benzyloxide) [29].

4. From 1,3-Dioxanes

$$
\begin{array}{ccc}
\begin{array}{c} R_1 \\ \diagdown \\ \diagup \\ R_2 \end{array} C \begin{array}{c} O-CH_2 \\ \diagup \\ \diagdown \\ O-CH_2 \end{array} C \begin{array}{c} R \\ \diagup \\ \diagdown \\ R \end{array} & \xrightarrow[\text{or}\atop SiO_2]{\text{Pumice}} & \begin{array}{c} R_1 \\ \diagdown \\ \diagup \\ R_2 \end{array} CH \begin{array}{c} O-CH_2 \\ \diagup \\ \diagdown \\ OHC \end{array} C \begin{array}{c} R \\ \diagup \\ \diagdown \\ R \end{array}
\end{array}
$$

This reaction logically follows the Oppenauer reaction because a hydride shift is involved in the rate-determining step of both reactions, although it is true that each reaction is run under entirely different conditions. The mechanism may be represented as follows [42]:

$$
\begin{array}{c} R_1 \\ \diagdown \\ \diagup \\ R_2 \end{array} C \begin{array}{c} O-CH_2 \\ \diagup \\ \diagdown \\ O-CH_2 \end{array} CR_2 \longrightarrow
\left[\begin{array}{c} R_1 \\ \diagdown \\ \diagup \\ R_2 \end{array} \overset{\delta+}{C} \begin{array}{c} O-CH_2 \\ \diagup \\ \diagdown \\ HCH \end{array} CR_2 \right] \longrightarrow
$$

$$
\overset{\delta-}{O}
$$

$$
\begin{array}{c} R_1 \\ \diagdown \\ \diagup \\ R_2 \end{array} CH\ OCH_2CR_2CHO
$$

The rearrangement occurs at high temperature to give β-alkoxy aldehydes when the 1,3-dioxanes are passed over pumice or certain forms of silica. The 1,3-dioxanes of simple aliphatic aldehydes and ketones rearrange well at about 400° on pumice; silica is active at 250–350°, but the yields are lower. Formals respond less satisfactorily than other acetals and ketals. A benzal ($R_1 = C_6H_5$, $R_2 = H$) is isomerized readily, but ring substituents generally retard the rearrangement. If the R's are H, rearrangement occurs, but the product is cleaved to acrolein and an alcohol. Yields are generally good.

(a) **Preparation of benzyloxypivalaldehyde.** The original article [43] must be consulted for details on the equipment and procedure for the vapor-phase isomerization of *m*-dioxanes. The vapor from 2-phenyl-5,5-dimethyl-dioxane-1,3 passed over pumice at 340–360° yielded, on distillation through a 100-mm. micro Vigreux column, a 75 % yield of the substituted pivalaldehyde, b.p. 139–140° (21 mm.).

5. From Benzylic (or Allylic) Alcohols and Halides with Nitrogen-Oxygen Reagents

Nitro or nitroso compounds are mild and quite suitable oxidizing agents for the conversion of hydroxybenzyl alcohols (or halides) into hydroxyaldehydes [29, 44]. The alcohols are obtained easily by formylation or the chlorides by chloromethylation of phenols [45].

The sodium salt of 2-nitropropane also is used for the oxidation of benzyl halides, but nitrobenzenes and alkali or a nitroso compound and hydrogen chloride are employed for benzyl alcohols. The following products are formed with the various oxidizing agents:

$$\text{Benzyl alcohol} \xrightarrow[\text{OH}^\ominus]{\text{2-Nitropropane}} \text{benzaldehyde} + \text{acetone oxime}$$

$$\xrightarrow[\overset{\ominus}{\text{OH}}]{\text{Nitrobenzene}} \text{benzaldehyde} + \text{azobenzene}$$

$$\xrightarrow[\overset{\oplus}{\text{H}}]{p\text{-Nitrosodimethylaniline}} \text{benzaldehyde} + p\text{-aminodimethylaniline}$$

In pyridine and alkali the *p*-nitrosodimethylaniline oxidation takes a different course:

$$\text{ArCH}_2\text{Cl} \xrightarrow{\text{C}_5\text{H}_5\text{N}} \text{ArCH}_2\overset{\oplus}{\text{N}}\text{C}_5\text{H}_5\text{Cl}^\ominus + \text{ONC}_6\text{H}_4\text{N}(\text{CH}_3)_2 \xrightarrow{\text{OH}^\ominus}$$

$$\text{ArCH}\overset{\overset{\text{O}}{\uparrow}}{=}\text{NC}_6\text{H}_4\text{N}(\text{CH}_3)_2 \xrightarrow[\text{H}_2\text{O}]{\text{H}^\oplus} \text{ArCHO}$$
Nitrone

The latter procedure, known as the Kröhnke reaction, is quite mild and adaptable not only to benzyl alcohols but also to allylic alcohols. In this reaction the pyridinium quaternary salt can be formed from the allylic or benzylic halide and then treated with *p*-nitrosodimethylaniline in base to yield a nitrone as shown in the previous equation. The nitrone is then decomposed with acid to form the aldehyde.

Another excellent procedure for the preparation of benzaldehydes from the alcohols uses the reagent, dinitrogen tetroxide, N_2O_4, at low temperatures. The yields reported are in the range of 91–98 % for 12 benzaldehydes, a fact which makes this synthetic route appear to be an unusually attractive one. (See Ex. c.)

Similarly phenylhydroxylamine has been used to oxidize benzyl halides. The yields reported [46] are low but possibly could be raised by azeotroping.

(a) Preparation of o-nitrobenzaldehyde (Kröhnke procedure, 47–53% yield (based on N-bromosuccinimide) from o-nitrobenzylpyridinium bromide, made from o-nitrotoluene and N-bromosuccinimide followed by reaction with pyridine, and p-nitrosodimethylaniline hydrochloride in ethanol to which aqueous sodium hydroxide was added at 0–5°; the nitrone was filtered and treated with hot 6 N sulfuric acid to precipitate the aldehyde) [47].

(b) Preparation of 2-p-chlorophenyl-6,8-dichloro-7-quinolinecarboxaldehyde. To 0.08 mole each of sodium ethoxide and 2-nitropropane in 350 ml. of absolute ethanol, 0.08 mole of 2-p-chlorophenyl-6,8-dichloro-7-bromomethylquinoline, m.p. 177–180.5°, was added and the mixture was refluxed for 7 hr. The solvent was removed and the residue in methylene chloride was washed with water, aqueous 10% NaOH, dried, and evaporated. Recrystallization from ethyl acetate gave pale yellow crystals, 16.3 g. (60%), m.p. 199–201.5° [48].

(c) Preparation of benzaldehydes using dinitrogen tetroxide. The benzyl alcohol (0.1 mole) was dissolved in 2–3 times its volume of chloroform and carbon tetrachloride and cooled to 0°. An ice-cold solution of dinitrogen tetroxide (0.13 mole) in 30 ml. of chloroform or carbon tetrachloride was added, the mixture held at 0° for 15 min., and stored overnight at room temperature. The solvent was removed at water aspirator pressure and the residue in ether washed with aqueous sodium bicarbonate and dried. Distillation at reduced pressure yielded the benzaldehydes in 91–98% yields (12 aldehydes). The reaction probably occurs through the attack of the $\cdot NO_2$ free radical, although the overall equation is simply as follows [49]:

$$ArCH_2OH + N_2O_4 \rightarrow ArCH{=}O + N_2O_3 + H_2O$$

(d) Other examples of 2-nitropropane oxidations

(1) o-Tolualdehyde (68–73% from o-xylyl bromide and 2-nitropropane with sodium ethoxide) [50].

(2) 9,10-Anthracenedicarboxaldehyde (70% from 9,10-bis(chloromethyl)anthracene, 2-nitropropane, sodium ethoxide, and dimethyl sulfoxide at room temperature) [51].

(e) Other examples of Kröhnke oxidation

(1) Terephthalaldehyde (77% from the pyridinium salt) [52].

(2) 3-Acetoxypregna-5,7-diene-21-al (about 30% on a milligram scale from the corresponding bromide) [53].

(3) Phytenal

$$(CH_3)_2CHCH_2CH_2(CH_2\overset{\overset{\displaystyle CH_3}{|}}{C}HCH_2CH_2)_2CH\overset{\overset{\displaystyle CH_3}{|}}{C}{=}CH{-}CH{=}O,$$

(about 34% from phytyl bromide) [54].

6. From Ethylenic Compounds via the Ozonide and Related Reactions

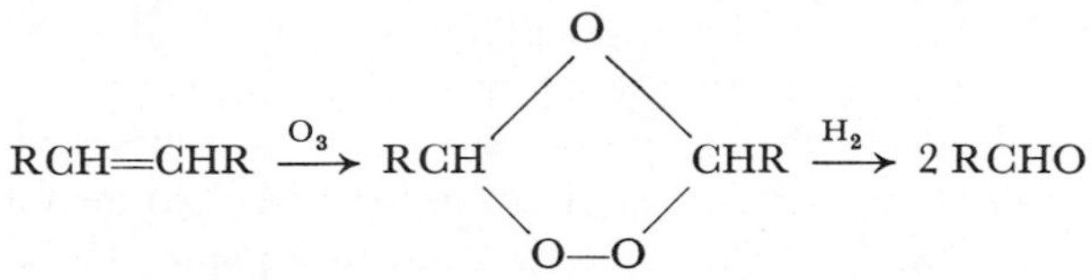

$$RCH{=}CHR \xrightarrow{O_3} RCH\diagdown\!\!\!\overset{\displaystyle O}{\underset{\displaystyle O{-}O}{\diagup}}\!\!\!\diagup CHR \xrightarrow{H_2} 2\,RCHO$$

Various oxidizing agents have been employed to cleave ethylenes to produce aldehydes. A common procedure is to use ozone alone or mixed with oxygen and to reduce the ozonide with hydrogen in the presence of a catalyst such as palladium-on-charcoal. In some cases the ozonide is converted into the aldehyde in the normal course of the reaction [55]; in others, it can be reduced in the reaction mixture by palladium-on-charcoal [56], by a metal such as zinc in acetic acid [57], by triphenylphosphine [58], or by sodium or potassium iodide in acetic acid [59]. In the presence of groups such as the oxygen in pyran, an excess of ozone must be avoided [56]. It should be realized that ozonization of aliphatic olefins yields some by-products. For example, ozone at $-70°$ with 2-pentene yields mainly 2-pentene ozonide, but also some butene and hexene ozonides [60], a disturbing fact when it is realized that the reaction is a means of locating double bonds. Other oxidizing agents such as sodium dichromate and sulfuric acid (with sulfanilic acid to remove the aldehyde formed and to act as a dispersing agent [61], (see Ex. c.3)) nitrobenzene [44], and sodium metaperiodate in the presence of osmium tetroxide [62] have been used. In the latter case the osmium tetroxide adds to the double bond to form an osmate ester which is oxidized by the periodate to give the carbonyl compound with the regeneration of osmium tetroxide (see Ex. c.1):

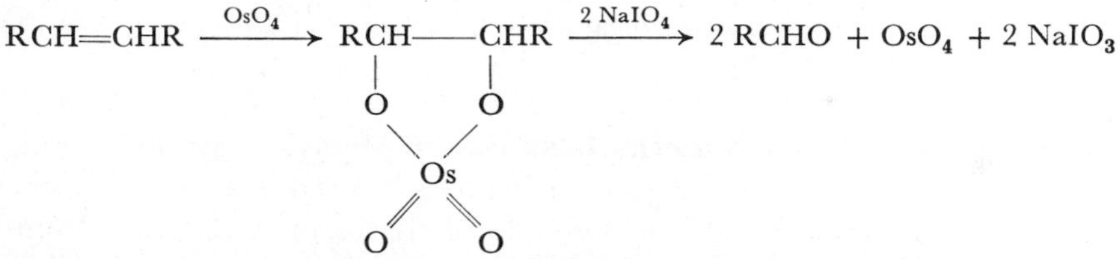

$$RCH{=}CHR \xrightarrow{OsO_4} \underset{\underset{\displaystyle Os}{\diagup\diagdown}}{RCH{-\!\!-}CHR} \xrightarrow{2\,NaIO_4} 2\,RCHO + OsO_4 + 2\,NaIO_3$$

Thus the result is identical with that obtained on ozonization followed by reductive cleavage.

In some aromatic ring systems the double-bond character is often sufficiently pronounced to permit ozonide formation. Such ozonides may be cleaved with acetic acid-potassium hydroxide-potassium hypochlorite [63] or with sodium iodide-acetic acid [59, 64] to produce aldehydes. In the ozonization of acenaphthylene it was found that ozone with nitrogen gives a better yield of the dialdehyde hydrate (73.5%) than ozone-oxygen (16.5%) [64]:

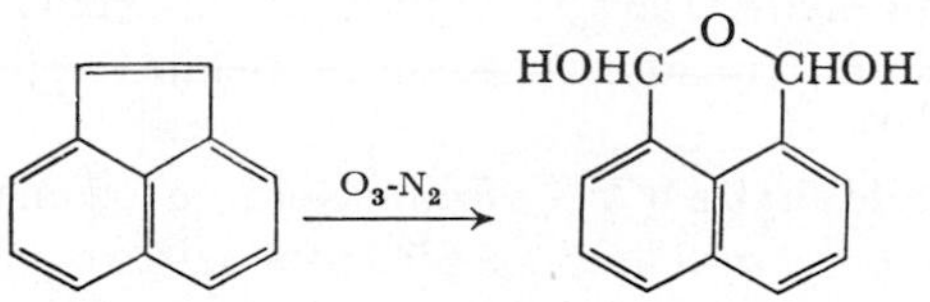

If the ozonolysis is effected at low temperature ($-40°$ or lower) in a hydroxylic solvent, such as methanol, the hydroperoxide formed may be reduced with trimethyl phosphite [65] or dimethyl sulfide (see Ex. c.2). This method appears to be preferred, particularly among natural product chemists.

Palladium chloride-olefin complexes are potential sources of aldehyde via oxidation. Ethylene and palladium chloride have been studied as a system for the industrial production of acetaldehyde; with aqueous sodium acetate the complex yields vinyl acetate [66].

(a) **Preparation of vanillin.**   Oil of clove (86% eugenol) 100 g. was

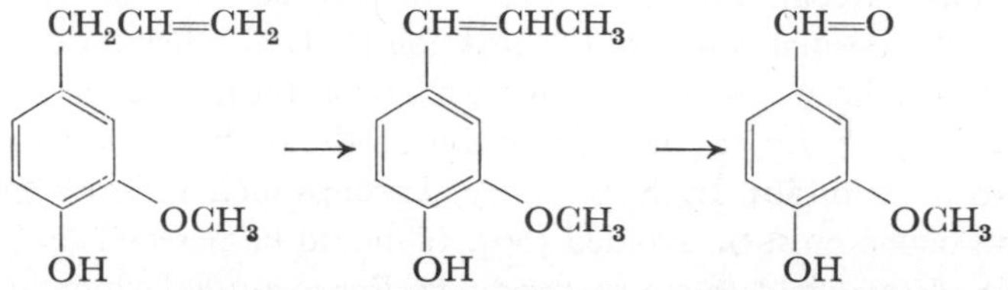

10-VI

added to 50 g. of potassium hydroxide and 200 ml. of water. At 125°, 15 ml. of terpenes were removed by distillation (with water). Aniline (400 g.) was added directly and the mixture distilled to give about 100 g. of aniline. The eugenol had rearranged to propenylguaiacol, 10-VI, in this process. After cooling, 300 g. of nitrobenzene and 100 g. of 5% aqueous sodium hydroxide were added and the mixture held at 105° for 2 hr. with agitation, followed by steam distillation. In the residue the sodium salt of vanillin and azobenzene remained. The azobenzene was filtered and the filtrate acidified to obtain vanillin in 79% yield [44]. Syringaldehyde, 3,5-dimethoxy-4-hydroxybenzaldehyde, was made similarly in 80% yield [67].

(b) **Preparation of 5-methylhexanal.**   6-Methyl-1-heptene (0.5 mole) in 200 ml. of methylene chloride was cooled to $-78°$ and treated with a stream of 6% ozonized oxygen at 20 l./hr. for 12 hr. This mixture was added dropwise to 32.5 g. of zinc dust in 300 ml. of 50% aqueous acetic acid with stirring. Considerable heat was evolved. After refluxing for 1 hr., the mixture was cooled and extracted with ether. The ether was washed with aqueous potassium iodide solution until all peroxides were eliminated, then with aqueous NaOH, HCl, and finally with water saturated with sodium chloride. The aldehyde, b.p. 144°, n^{20}D 1.4114, was obtained in 62% yield [68].

(c) **Other examples**

(1) **Benzaldehyde** (85% as the 2,4-dinitrophenylhydrazone from *trans*-stilbene, osmium tetroxide, and sodium metaperiodate in dioxane at 24–26° for 95 min.) [62].

(2) **1-Heptaldehyde** (75% from 1-octene ozonized in methanol at $-60°$ to which dimethyl sulfide was added; the latter compound is apparently

an excellent reducing agent for this system) [69]:

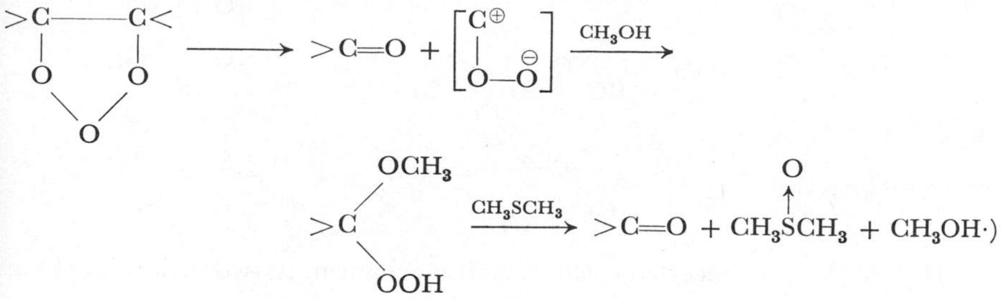

$$>\underset{\underset{O}{|}}{C}—\underset{\underset{O}{|}}{C}< \longrightarrow >C=O + \begin{bmatrix} \overset{\oplus}{\underset{|}{C}} \\ \underset{O—O}{|} \ominus \end{bmatrix} \xrightarrow{CH_3OH}$$

$$>\underset{\overset{|}{OOH}}{\overset{\overset{OCH_3}{|}}{C}} \xrightarrow{CH_3SCH_3} >C=O + CH_3\overset{\overset{O}{\uparrow}}{S}CH_3 + CH_3OH\cdot)$$

(3) Piperonal (86.5 % from isosafrole and sodium dichromate, sulfuric acid, and sulfanilic acid; potassium permanganate was too vigorous an oxidizing agent) [61].

(4) 2,4-Diacetyl-D-erythrose (94.5 % from 1,4,6-triacetyl pseudoglucal and ozone) [70].

(5) 5-Formyl-4-phenanthroic acid (32–38 % from pyrene and ozone) [63].

(6) Diphenaldehyde (80–90 % from phenanthrene and ozone) [59].

7. From Glycols

$$RCHOHCHOHR \xrightarrow[\underset{Pb(OAc)_4}{or}]{HIO_4} 2\,RCHO$$

The oxidation with periodic acid has been reviewed in detail [71], and the oxidizing properties of lead tetraacetate have also been described [72]. The main types which may be oxidized by these reagents are 1,2-glycols, 1,2-hydroxyamines, α-hydroxyaldehydes, and α-hydroxyketones. For the most part the degradation with either reagent proceeds at room temperature and the yields are so high when simpler aldehydes are produced that the method sometimes serves as an analytical procedure. For simpler 1,2-glycols, both reagents oxidize the *cis* more rapidly than the *trans* form although, with the higher glycols, this situation is reversed [73]. In the case of lead tetraacetate the cleavage proceeds without a catalyst or is catalyzed by trichloroacetic acid or base [74]. The reaction has been applied widely in the carbohydrate series for synthesis and determination of structure.

One of the sources of a starting glycol is a series of reactions beginning with an acid chloride [75]:

$$RCOCl \xrightarrow{CH_2N_2} RCOCHN_2 \xrightarrow{CH_3CO_2H}$$
$$RCOCH_2OCOCH_3 \xrightarrow{Al[OCH(CH_3)_2]_3} RCHOHCH_2OH$$

Yields are satisfactory except for α,β-unsaturated acid chlorides.

The mechanism for the oxidation with periodic acid has been studied [76, 77]. In the first step a cyclic periodate ester, 10-VII, is formed between the ion,

$IO_4^{\ominus}$ (or its octahedral dihydrate, $H_4IO_6^{\ominus}$) and the glycol:

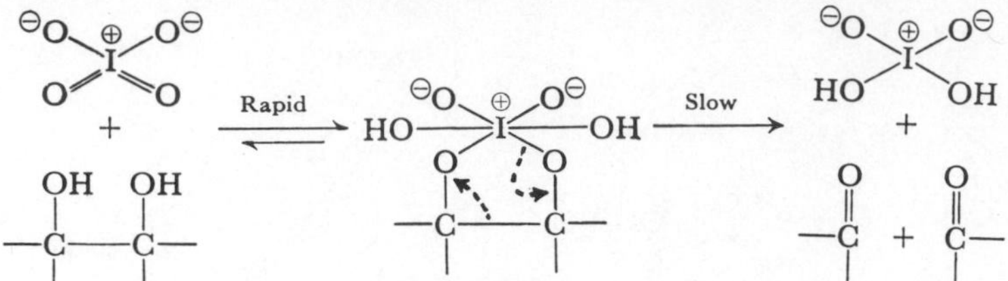

10-VII cleaves by a concerted electron-shift mechanism to two molecules of the carbonyl compound and the hydrated iodate ion, $H_2IO_4(IO_3 + H_2O)$.

A concerted mechanism involving a base has also been proposed for the lead tetraacetate oxidation [78]:

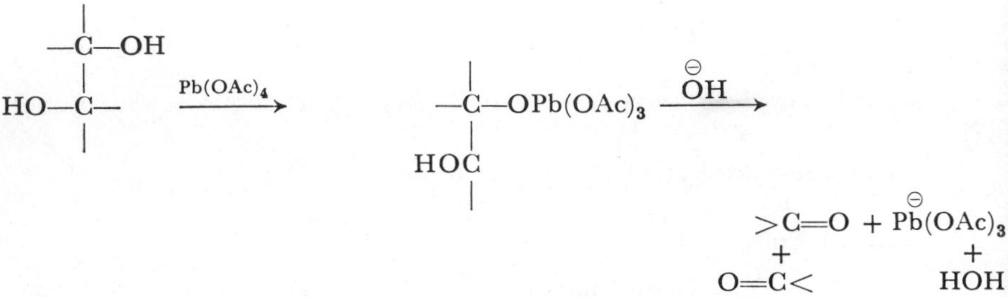

The uncatalyzed reaction probably proceeds via a cyclo mechanism.

Oxygenation of 1,2-glycols, although not a preparative method, has been demonstrated to proceed via aldehyde formation [79]:

trans-1,2-Cyclohexanediol + Co(OAc)$_2$ + O$_2$ $\xrightarrow[\substack{\text{Shaken } 3\frac{1}{2} \text{ hr.} \\ \text{at } 100°}]{C_6H_5CN, \text{ 50 ml.}}$

45 mmoles   1 mmole

12.8 meq. of aldehyde consisting of adipodialdehyde and 1-cyclopentenecarboxaldehyde

(a) **Preparation of pelargonaldehyde** (89% from 9,10-dihydroxy-stearic acid and periodic acid in alcohol at 40°) [80].

$$CH_3(CH_2)_7CHOHCHOH(CH_2)_7COOH \rightarrow CH_3(CH_2)_7CHO + OCH(CH_2)_7COOH$$

Pelargonaldehyde

(b) **Other examples**

(1) *n*-**Butyl glyoxylate** (77–87% by the oxidation of di-*n*-butyl d-

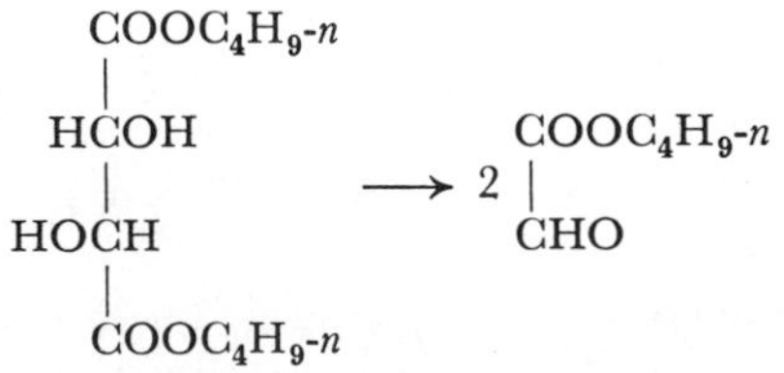

tartrate with lead tetraacetate) [81].

(2) **Mesitylphenylacetaldehyde** (almost quantitative from the oxidation of 3-mesityl-3-phenyl-1,2-propanediol with periodic acid) [82].

(3) L-**Xylose** (80 % from 2,4-benzal-D-sorbitol and periodic acid) [83].

8. From Aldoses (Degradation)

$$\begin{matrix} \text{CH}{=}\text{O} \\ | \\ \text{CHOH} \\ | \end{matrix} \xrightarrow[\text{pH 11}]{\text{NaOCl}} \begin{matrix} \text{CO}_2{}^{\ominus} \\ | \\ \text{CHOH} \\ | \end{matrix} \xrightarrow[\text{pH 5}]{\text{NaOCl}} \begin{matrix} \text{CH}{=}\text{O} + \text{CO}_2 \\ | \end{matrix}$$

A recent development in carbohydrate chemistry has been an improved method of decarboxylating the aldonic acid without isolation of the acid (or lactone). Formerly, this decarboxylation was carried out by hydrogen peroxide and ferrous ion catalysis [84]. The development applies to both mono- and di-saccharides which have free aldehyde groups.

(a) **Preparation of 3-O-α-D-glucopyranosyl-α-D-arabinose.** β-Maltose monohydrate, 10 g. in water buffered with NaOH to pH 11, was treated with 500 ml. of 0.334 N NaOCl adjusted to pH 11 with NaOH and Na_2CO_3. The mixture was held at 25° for 22 hr. and the pH was adjusted to 5 with HCl and 300 ml. of 0.266 N NaOCl was added. The solution was concentrated and the residue was chromatographed on a carbon-Celite column with 5 % aqueous ethanol. The eluate on concentration gave the crude product as a syrup (32.6 %). Another chromatographing was necessary to obtain a crystalline product [85].

9. From Halomethyl Compounds and Hexamethylenetetramine (Sommelet)

$$\text{ArCH}_2\text{X} \xrightarrow{(\text{CH}_2)_6\text{N}_4} [\text{ArCH}_2(\text{CH}_2)_6\text{N}_4]^{\oplus}\text{X}^{\ominus} \xrightarrow[\text{acetic acid}]{50\%} \text{ArCH}{=}\text{O}$$

The Sommelet reaction is an excellent method of obtaining aromatic aldehydes. Indeed, if the aldehyde must be derived from the corresponding methyl compound, the two methods of choice are the Sommelet reaction or selenium dioxide oxidation. If the methyl group is activated such as with o- or p-electron-withdrawing groups or in α- or λ-methylpyridines or quinolines (A.13), the SeO_2 method is preferred; if not activated, the Sommelet method is preferred. In this case the methyl group is brominated, and without isolation the benzyl bromide is converted into the hexamethylenetetramine quaternary salt as described [86].

The mechanism may be debatable [87] but undoubtedly involves a shift from a benzylamine to a protonated iminoformaldehyde species:

$$\text{ArCH}_2\text{NH}_2 + \text{CH}_2{=}\overset{\oplus}{\text{NH}_2} \longrightarrow [\text{ArCH}{=}\text{NH}_2]^{\oplus} + \text{CH}_3\text{NH}_2 \xrightarrow{\text{H}_2\text{O}}$$
$$\text{ArCH}{=}\text{O} + \text{NH}_3$$

Whatever the mode of transfer of hydride ion, the important point in regard to synthesis is that a large excess of hexamethylenetetramine should be present to prevent the Schiff base of the benzylamine, $ArCH_2N{=}CH_2$, from acting as the oxidizing agent and forming N-methylbenzylamine. The synthesis is employed in producing aromatic (50–80% yields), heterocyclic, and aliphatic (50% yields if aldehyde is removed as it is formed) aldehydes. The reaction may lead to failure if *ortho* positions in benzyl alcohol are occupied or if strongly electron-attracting substituents are present. Phenolic aldehydes produce condensation products with the formaldehyde present in the mixture. There are other inexplicable cases in which the Sommelet reaction fails. In such circumstances it is suggested that an attempt be made to brominate the methylarene with N-bromosuccinimide (Halides, Chapter 7, C.1), after which oxidation with hexamine, dimethylsulfoxide, or trimethylamine-N-oxide (A.10) may lead to the desired aldehyde.

As the equation above suggests, aldehydes may be produced from primary amines of the type $ArCH_2NH_2$ and hexamethylenetetramine. In fact, diazotization in the presence of dimethyl sulfoxide accomplishes this end [88].

(a) Preparation of 1-naphthaldehyde (75–82% from 1-chloromethyl-naphthalene) [89].

(b) Other examples

(1) 2-Thiophenecarboxaldehyde (45–52% from 2-chloromethylthio-phene) [90].

(2) Pyridine-3-carboxaldehyde (57% from 3-aminomethylpyridine) [91].

(3) 2-Carbethoxyindole-3-aldehyde (68–72% from 2-carbethoxy-3-dimethylaminomethylindole) [92].

(4) Heptaldehyde (51% from an aqueous solution of N-heptyl-hexammonium iodide added to a boiling solution of 50% aqueous acetic acid while the aldehyde was distilled out with steam) [93].

10. From Halides and Miscellaneous Oxidizing Agents (Dimethyl Sulfoxide, Potassium t-Butyl Hydroperoxide, Sodium Dichromate or Trimethylamine Oxide)

$$RCH_2X \xrightarrow{\text{TsONa}} RCH_2OTs \xrightarrow[\text{Na}_2\text{CO}_3]{\text{DMSO}} RCHO$$

$$RCH_2X \xrightarrow{\text{DMSO,}} RCHO$$

$$(CH_3)_3COOK,$$
$$Na_2Cr_2O_7 + Na_2CO_3,$$
$$\text{or}$$
$$(CH_3)_3NO$$

Halomethyl compounds via esters such as the tosylate may be oxidized smoothly to the corresponding aldehyde [94]. With benzyl tosylate, sodium bicarbonate, and dimethyl sulfoxide, the oxidation is complete in less than 5 minutes at 100°, but with alkyl tosylates the oxidation is carried out best at 150°. Yields are in the range 65–84%.

Direct oxidation with the sulfoxide may be accomplished with iodomethyl compounds in 3–4 minutes at around 150° [95]. In a similar manner, phenacyl halides respond to dimethyl sulfoxide [96] or preferably to the first-formed nitrate (from $AgNO_3$) treated with dimethyl sulfoxide in the presence of a catalytic amount of sodium acetate [97]. Yields with a series of straight-chain aliphatic iodides vary from 25 to 86%. Yields with secondary iodides are less satisfactory. The method is also applicable to the oxidation of ethyl bromoacetate in which case ethyl glyoxalate is obtained in 70% yield [98].

$$BrCH_2COOC_2H_5 + (CH_3)_2SO \rightarrow OHCCOOC_2H_5 + HBr + (CH_3)_2S$$

In this case, 1,2-epoxy-3-phenoxypropane is used as a scavenger for hydrogen bromide while methyl bromide is added to convert the dimethyl sulfide into trimethylsulfonium bromide and thus repress side reactions.

Potassium t-butyl hydroperoxide (see Ex. b.2) and alkali metal dichromates (see Ex. b.3) are also mild oxidizing agents which may be used directly to produce the aldehyde.

Tertiary amine oxides with halides form salts which decompose readily to give aldehydes [99]:

$$RCH_2Br + (CH_3)_3NO \rightarrow RCH_2O\overset{\oplus}{N}(CH_3)_3Br^{\ominus} \rightarrow RCHO + (CH_3)_3N \cdot HBr$$

Yields by this method usually vary from 25 to 66%; indeed, this method appears to be the only way to obtain 3-nitroquinoline-4-carboxaldehyde (35%).

(a) Preparation of heptaldehyde (using dimethylsulfoxide). 1-Iodoheptane, 7.0 g., was added to a solution of 11 g. of silver tosylate (prepared by mixing equivalent amounts of silver oxide and p-toluenesulfonic acid monohydrate) in 100 ml. of acetonitrile at 0–5° (protected from light). After standing over night, the product was added to ice water and the mixture was extracted with ether. The oil resulting by concentrating the dried ether solution under reduced pressure was added to a mixture of 20 g. of sodium bicarbonate and about 150 ml. of dimethyl sulfoxide through which nitrogen was bubbling at 150° (foaming occurs). After 3 min. at this temperature the mixture was cooled rapidly to room temperature, and the aldehyde was recovered as the 2,4-dinitrophenylhydrazone, m.p. 106–107°. Yield 6.9 g. (70%) [94].

(b) Other examples

(1) Hexaldehyde (86% as the 2,4-dinitrophenylhydrazone by oxidizing n-hexyl iodide in a solution of DMSO at 150° for 3 min.) [95].

(2) 2,4,6-Trimethylbenzaldehyde from the halide and potassium t-butyl peroxide. To 16.9 g. of 2,4,6-trimethylbenzyl chloride dissolved in 100 ml. of methanol was added 12.8 g. of potassium t-butyl peroxide. The reaction mixture was refluxed for 5 hr.; the methanol was removed by distillation; the residue in ether was washed with water and fractionated after concentration to yield 58% of the aldehyde [100].

(3) p-Tolualdehyde from the halide and sodium dichromate [100].

$$3\,p\text{-}CH_3C_6H_4CH_2Cl + Na_2Cr_2O_7 + \tfrac{1}{2}\,Na_2CO_3 \xrightarrow[\text{20 hr. reflux}]{\text{150 ml. } H_2O}$$

41 g.   30 g.   6 g.

$$3\,p\text{-}CH_3C_6H_4CH{=}O + 3\,NaCl + Cr_2O_3 + \tfrac{1}{2}\,CO_2 + 1\tfrac{1}{2}\,H_2O$$

90% (by steam distillation)

11. From Methylarenes

The oxidation of the aromatic hydrocarbons poses a more difficult problem than oxidation of the alcohols, but nevertheless it can be done. This subject is discussed in A.11 and A.12, the first being devoted to the more general reagents and the second to a specific reagent. The most useful general reagent is chromic anhydride in acetic anhydride:

$$ArCH_3 \xrightarrow[\text{(CH}_3\text{CO)}_2\text{O}]{\text{CrO}_3} ArCH(OCOCH_3)_2 \xrightarrow[\text{H}_2\text{O}]{\text{H}^{\oplus}} ArCH{=}O$$

The overall yields are rather low.

Manganese dioxide in sulfuric acid is another general oxidizing agent for methylarenes. In order to prevent the oxide from converting the aldehyde into the acid, theoretical amounts, added portionwise, must be employed. Efficient agitation and a large excess of sulfuric acid are helpful.

Lead dioxide, nickel oxide, and cerium dioxide have also been used to oxidize methylarenes, but they offer no advantages over manganese dioxide [101].

(a) Preparation of p-nitrobenzaldehyde (43–51 % from p-nitrotoluene and chromic and acetic anhydrides in sulfuric acid) [102].

(b) Preparation of 2,5-dichloroterephthalaldehyde (32 % from 2,5-dichloro-p-xylene and chromic and acetic anhydrides) [103].

12. From Methylarenes and Chromyl Chloride (Etard Oxidation)

This reagent is powerful enough to ignite volatile alcohols and other organic compounds. However, both carbon tetrachloride and carbon bisulfide are suitable as solvents, but must be added with external cooling and stirring. The reaction apparently is an electrophilic attack on the methyl group to form the diester [104]:

$$ArCH_3 + 2\,CrO_2Cl_2 \longrightarrow ArCH(OCrCl_2OH)_2 \xrightarrow[\text{H}_2\text{O}]{\text{H}^{\oplus}} ArCH{=}O$$

Attempts have been made to form aldehydes from alicyclic hydrocarbons such as methylcyclohexane. A trace of an olefin was necessary to initiate the reaction, and the yield of hexahydrobenzaldehyde was poor [105].

(a) Preparation of p-iodobenzaldehyde. To p-iodotoluene (0.2 mole in 150 ml. of carbon tetrachloride), chromyl chloride (0.42 mole) in an equal volume of carbon tetrachloride was added slowly with stirring for 1 hr. After refluxing for about 20 hr., the mixture was poured into ice and water containing sodium sulfite. Dilute hydrochloric acid was added to dissolve the chromium salts. The aldehyde was removed by three extractions with carbon tetrachloride; after drying and concentrating it distilled at 145–150° (25 mm.), 27–30 g. (58–64 %). Recrystallization from ethanol gave a melting point of 75° [106].

(b) Other examples

(1) **2-Nitroanisaldehyde** (50 % from 2-nitro-4-methoxytoluene) [107].

(2) **2-Ethoxy-4-nitrobenzaldehyde** (40 % as the thiosemicarbazone from 2-ethoxy-4-nitrotoluene) [108].

13. From Methyl Heterocycles and Selenium Dioxide

The oxidation of methyl heterocycles to yield aldehydes has been reviewed [109]. It appears to be important to use freshly prepared selenium dioxide [110] in such oxidations. If the original hydrocarbon is a liquid, no solvent may be necessary, but solvents such as ethyl alcohol, ethyl acetate, dioxane, and xylene have been employed particularly for solids. The heterocyclic aldehyde may be produced either exclusively or mixed with the acid. Yields have been reported as high as 90 %, but 50 % or even lower is much more common. In spite of this fact, the synthesis is important in the series owing to the availability of many methyl heterocycles. It will be noted in the examples that lower temperatures are used with 2-methylquinolines, and they apply as well to 4-methylquinolines.

(a) Preparation of 3,8-dimethylquinoline-2-carboxaldehyde (82 % from 2,3,8-trimethylquinoline and SeO_2 refluxed in alcohol) [111].

(b) Other examples

(1) **Quinoline-8-carboxaldehyde** (70 % from SeO_2 and 8-methyl-quinoline heated to 150° and finally at 250°) [112].

(2) **Quinaldehyde-N-oxide** (approximately 54 % from quinaldine-N-oxide and SeO_2 refluxed in pyridine) [113].

(3) **Isoquinoline-3-carboxaldehyde** (48 % by adding SeO_2 in small portions to 3-methylisoquinoline stirred at 180–220°) [114].

14. From Methyl Ketones or Methylene Aldehydes and Selenium Dioxide

$$RCOCH_3 \xrightarrow{SeO_2} RCOCHO$$
$$RCH_2CHO \longrightarrow RCOCHO$$

The versatility of selenium dioxide as an oxidizing agent has been reviewed [109]. Active methyl or methylene groups are often oxidized by this reagent. In the first case, an aldehyde is obtained and, in the second, a ketone. Although glyoxal has been obtained in a 90 % yield from acetaldehyde [115], the method appears to have been employed more for oxidizing methyl ketones than for aldehydes containing α-hydrogen atoms. The oxidation of the methyl in a benzene ring to the aldehyde group is unusual, but the conversion is common if the methyl is attached to a heterocyclic ring (see A.13). For methyl ketones the reaction involves a simple reflux of the compound alone or in the presence of some solvent such as dioxane, ethanol, or acetic acid. Yields as a rule are not high.

(a) **Preparation of phenylglyoxal** (69–72 % from acetophenone) [116].

(b) **Preparation of glyoxal** (72–74 % as the bisulfite from paraldehyde and selenious acid) [117].

15. From Methyl Ketones or Arenes by Nitrosation

Nitrosation raises the level of oxidation to the aldehyde stage:

$$\underset{RCCH_3}{\overset{O}{\|}} \xrightarrow[\text{NOCl}]{\cdot\text{HONO or}} \underset{RCCH=NOH}{\overset{O}{\|}}$$

The α-oximino ketones can be hydrolyzed to α-ketoaldehydes, and have the added versatility of reacting with Grignard reagents to form eventually α-hydroxyaldehydes:

$$\underset{RCCH=NOH}{\overset{O}{\|}} \xrightarrow[\text{2. H}_2\text{O}]{\text{1. R}'\text{MgX (excess)}} \underset{RC(OH)CH=O}{\overset{R'}{|}}$$

At least 2 equivalents of the Grignard reagent are necessary to bring about the addition reaction, as shown in Ex. a.

Methylarenes in the few instances reported [118] give excellent yields of the oximes, as for example:

$$C_6H_5CH_3 + NOCl \rightarrow C_6H_5CH=NOH$$
(nearly quantitative)

Picolines have resisted ordinary methods of nitrosation, but capitalization has been made of the fact that the anions of the picolines can attack alkyl nitrites to yield oximes according to the following equations [119]:

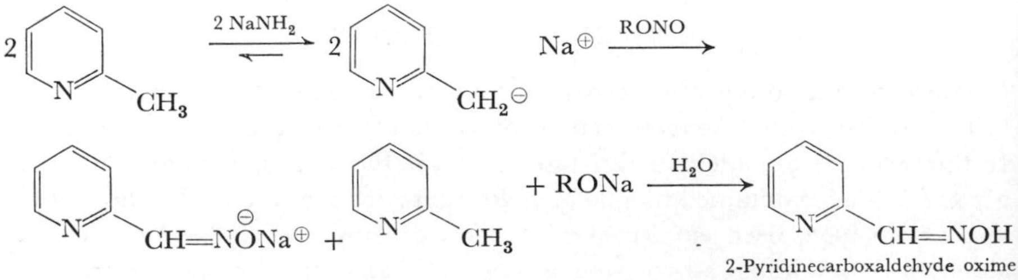

2-Pyridinecarboxaldehyde oxime

This type of attack had been known as early as 1901 when Lapworth nitrosated the anion of *o*- and *p*-nitrotoluene [120].

(a) Preparation of methyl phenyl glycolic aldehyde

(1) Isonitrosoacetone. A mixture of 100 g. of acetoacetic ester and 50 g. of potassium hydroxide in 1800 ml. of water was treated, after 24 hr. (or after 4 hr. or more), with 62 g. of sodium nitrite in 200 ml. of water and, after cooling to 5–6°, with 430 g. of 20% sulfuric acid. The reaction product was neutralized with 140 g. of 30% sodium hydroxide and cooled with 80 g. of ice, after which the mixture was extracted twice with ether to eliminate insoluble organic substances. The aqueous layer was then acidified with 20% sulfuric acid at 5° and extracted with ether to yield 56 g. (80% of isonitrosoacetone, m.p. 67–68°.

(2) Methyl phenyl glycolic aldoxime. Isonitrosoacetone, 29 g., was condensed with phenylmagnesium bromide (from 32 g. of magnesium and 210 g. of phenyl bromide) in the usual manner. On extraction with ether and distilling, 26 g. of a liquid boiling at 150–170° (7 mm.) was obtained. Purification by further distillation gave 20 g. (36%) of methyl phenyl glycolic aldehyde oxime, b.p. 155–156° (5 mm.).

(3) Methyl phenyl glycolic aldehyde. The oxime, 16 g., was treated with 15 ml. of conc. hydrochloric acid and 15 ml. of 35% formaldehyde. While agitating mechanically for 2 hr., the temperature was increased by heating to about 50–60°. The red-violet solution was allowed to stand overnight, after which it was extracted with ether. The dark red extract turned to yellow when neutralized with sodium bicarbonate. Drying over sodium sulfate and distilling gave 8 g. (54%) boiling between 108 and 112° (6 mm.). The pure aldehyde boils at 101° (4 mm.) [121].

(b) Preparation of α-pyridylcarboxaldehyde oxime.

Sodium (2 g. atoms) was added to 400 ml. of liquid ammonia containing 0.5 g. of ferric nitrate monohydrate, and α-picoline (3 moles) was then added to the sodamide solution during 30 min. Butyl nitrite (1 mole) was added dropwise to the intensely red-colored solution, after which the mixture was stirred for 1 hr. It was decomposed with 1.5 moles of ammonium sulfate in 300 ml. of water and the oxime extracted with ether, concentrated, and distilled, b.p. 110° (0.9 mm.), m.p. 113–113.5°, 75% based on the requirement of 2 g. atoms of sodium (numerous other examples are given) [119].

16. From Ethers by Peroxidation

n-Alkyl ethers react with *t*-butyl perbenzoate to form mixed acetals which can be hydrolyzed to an aldehyde. The reaction has not been studied in depth but seems to have merit. (See Acetals, Chapter 9, D.2.)

(a) Preparation of *n*-butyl *t*-butyl butyral [122]:

$$C_3H_7CH_2OC_4H_9 + C_6H_5CO_3C(CH_3)_3 \xrightarrow[\text{48 hr., 90°}]{\text{Cu}_2\text{Br}_2 \text{ (0.03 mole)}}$$

0.35 mole   0.1 mole

$$\begin{bmatrix} C_3H_7CH-OC_4H_9 \\ | \\ OCOC_6H_5 \end{bmatrix} \longrightarrow [CH_3CH_2CH=CH-OC_4H_9] \xrightarrow{(CH_3)_3COH}$$

$$CH_3CH_2CH_2-\overset{\displaystyle H \quad O\text{-}t\text{-}C_4H_9}{\underset{\displaystyle OC_4H_9}{C}}$$

48%, b.p. 100° (17 mm.), n^{25}D 1.4148

17. From Phenylcarbinol Oxidative Splitting

Although aldehyde preparations from Grignard reagents are discussed under G, the Grignard preparation in this reaction takes place before a cleavage step which does involve an oxidation [123]. The diazonium salt essentially is the oxidizing agent, but the driving force is the re-formation of a resonance-stabilized system from 10-IX. The first step is:

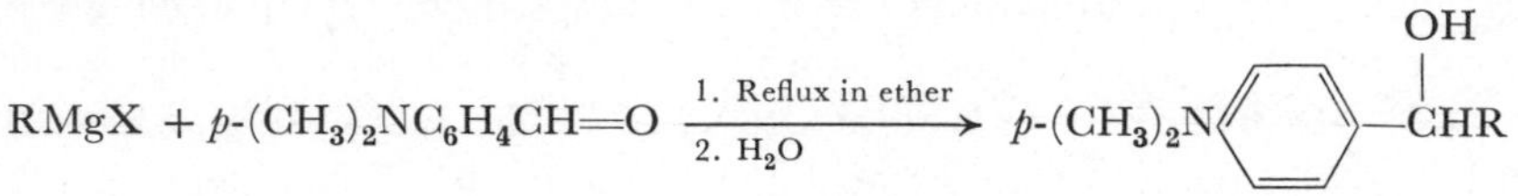

$$RMgX + p\text{-}(CH_3)_2NC_6H_4CH=O \xrightarrow[\text{2. }H_2O]{\text{1. Reflux in ether}} p\text{-}(CH_3)_2N\langle\ \rangle\overset{\displaystyle OH}{\underset{}{-CHR}}$$

10-VIII

Yields with either aliphatic or aromatic Grignard reagents range from 50–75 %.

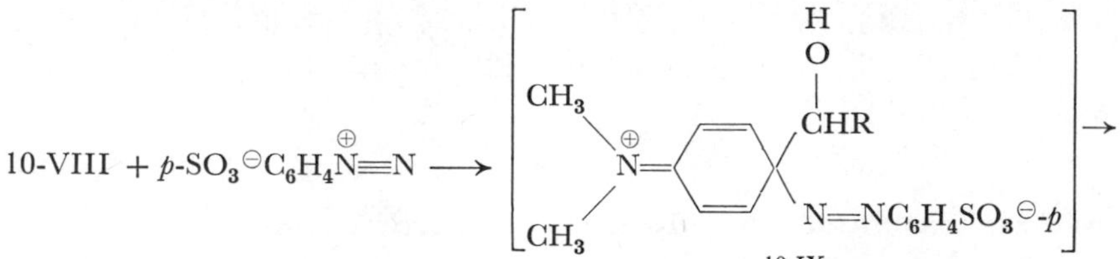

$$10\text{-VIII} + p\text{-}SO_3^{\ominus}C_6H_4\overset{\oplus}{N}{\equiv}N \longrightarrow \begin{bmatrix} & H \\ & O \\ CH_3 & | \\ \diagdown & CHR \\ \overset{\oplus}{N}= \langle\ \rangle \\ \diagup & N=NC_6H_4SO_3^{\ominus}\text{-}p \\ CH_3 & \end{bmatrix} \rightarrow$$

10-IX

$$RCH=O + (CH_3)_2NC_6H_4N=NC_6H_4SO_3H\text{-}p$$

Yields in the second step range from 45 to 82 %.

(a) Preparation of 4-methylvaleraldehyde. Sulfanilic acid, 60 g., 0.3 mole, in 200 ml. of water containing 18.4 g. of sodium carbonate was treated with 64 ml. of hydrochloric acid, and diazotized at 0–5° with 24.4 g. of sodium nitrite in 75 ml. of water. The solution was buffered by the addition of 70 g. of sodium acetate in 200 ml. water to about pH 6. A solution of 0.2 mole of 10-VIII, R = $(CH_3)_2CHCH_2CH_2$—, in 750 ml. of acetone was added to the stirred diazonium solution at 0–5° under nitrogen for 30-min. and the mixture was held for an additional 30 min. with removal of the ice bath. The mixture was then diluted with water, extracted with ether, and the extract dried and distilled under nitrogen, yielding 60 % of 4-methylvaleraldehyde [123].

18. From Nitroalkanes, Sodium Salt (Nef), and Other Redox Systems

$$[RCHNO_2]Na \xrightarrow[H^{\oplus}]{H_2O} RCHO + N_2O$$

The acidification of the salts of primary nitroparaffins proceeds in two ways. Dilute acid gives the aldehyde, while more concentrated acid produces the hydroxamic acid. In either case it seems definitely established that the *aci* form of the nitroparaffin is an intermediate [124–126]. Although the other steps in the formation of the aldehyde are less definite, it has been suggested [126] that the path followed is:

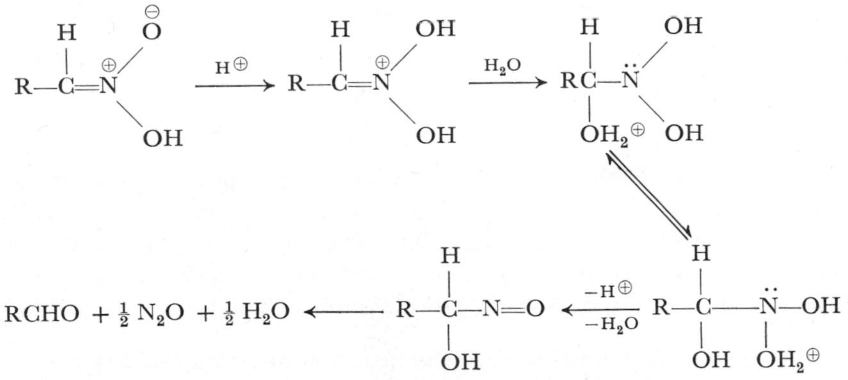

The synthesis, which has been reviewed [124] is applicable to primary nitroalkanes which give aldehydes and secondary nitroalkanes which give ketones (see Ketones, Chapter 11, A.9). The method has been applied in the sugar series, in which case the sugar and nitromethane are put into solution in methanol containing sodium methoxide to form the epimeric aci-nitroalcohols which are recovered by acidification with aqueous sulfuric acid. Yields from nitroalkanes vary from 32 to 86% [127, 126], while from nitro polyhydric alcohols they run from 70 to 80% [128].

A superior modification of the Nef reaction consists of titration of the potassium nitronate in the presence of magnesium sulfate, which apparently precipitates the hydroxyl ion and thus prevents formation of the potassium isonitrosonate. In this reaction [129]

$$3\,RCH{=}NO_2K + 2\,KMnO_4 + H_2O \rightarrow 3\,RCHO + 2\,MnO_2 + 3\,KNO_2 + 2\,KOH$$

it is sometimes necessary to use 70–90% of the theory of the permanganate to prevent oxidation of the aldehyde to the carboxylic acid. Yields of aldehydes as 2,4-dinitrophenylhydrazones vary from 68 to 97%.

(a) Preparation of p-nitrobenzaldehyde. Ethyl nitronic ester of *p*-nitrophenylmethane, 3 g., was added to 60 ml. of 3.12 N sulfuric acid (about 10% water, about 81% ethanol, and about 9% acid, all by volume) with stirring and cooling with ice. The stirred mixture was kept at 0° under nitrogen for 1.3 hr. and then at 25° for 18 hr. After cooling to 0°, 90 g. of ice was added and the product was extracted with ether, the ether layer being washed with aqueous sodium bicarbonate. Further washing with water, drying, and evaporation gave 2.60 g. of yellow oil. Treatment with 2,4-dinitrophenylhydrazine gave 4.0 g. (82%) of the 2,4-dinitrophenylhydrazone of *p*-nitrobenzaldehyde, m.p. 308–310° after the usual purification [126].

(b) Other examples

(1) **D-Manno-D-*gala*-heptose** (70% from 1-nitro-1-desoxy-D-manno-D-*gala*-heptitol, alkali, and acid) [128]:

$$
\begin{array}{ccc}
\text{CHO} & \text{CH}_2\text{NO}_2 & \text{CHO} \\
| & | & | \\
\text{HOC—H} & \text{HC—OH} & \text{H—C—OH} \\
| & | & | \\
& \text{HO—C—H} & \text{HO—C—H} \\
& | & | \\
\text{Mannose} & & \\
\end{array}
$$

CHO $\xrightarrow[\text{CH}_3\text{ONa}]{\text{CH}_3\text{NO}_2}$ ... $\xrightarrow[\text{2. H}^{\oplus}]{\text{1. NaOH}}$ ...

Mannose   1-Nitro-1-desoxy-D-manno-D-*gala*-heptitol   D-manno-D-*gala*-heptose

(2) **Benzaldehyde.** The potassium salt of phenylnitromethane, 4.5 mmoles, was dissolved in water containing 0.01 mole of potassium hydroxide and 0.04 mole of magnesium sulfate. The solution, volume 500 ml., was titrated with 3.1 mmoles of potassium permanganate, and the aldehyde was removed by steam distillation and recovered as the 2,4-dinitrophenylhydrazone, 97% [129].

19. From β-Ketosulfoxides (Pummerer Rearrangement)

This reaction is similar to the Nef reaction in overall results in that an internal oxidation-reduction occurs [130]:

$$
\text{C}_6\text{H}_5\text{CO}_2\text{C}_2\text{H}_5 + \text{CH}_3\overset{\uparrow \text{O}}{\text{S}}\text{CH}_3 \xrightarrow{\ominus \text{OC(CH}_3)_3} \text{C}_6\text{H}_5\text{CO}\overset{\ominus}{\text{C}}\text{H}\overset{\uparrow \text{O}}{\text{S}}\text{CH}_3 \xrightarrow{\text{H}^{\oplus}}
$$

$$
\text{C}_6\text{H}_5\text{COCH}_2\overset{\uparrow \text{O}}{\text{S}}\text{CH}_3 \longrightarrow \text{C}_6\text{H}_5\text{COC}\overset{\diagup \text{SCH}_3}{\underset{\diagdown \text{OH}}{\text{H}}}
$$

82% overall
Methyl hemimercaptal of phenylglyoxal

(See Ex. a.) The exact mechanism, however, is not known. The advantage of this reaction is that the ketosulfoxide need not be isolated, but can be converted into the methyl hemimercaptal by acidification.

(a) Preparation of the methyl hemimercaptal of phenylglyoxal.

Potassium, 4 g., in 100 ml. of *t*-butyl alcohol was cooled to room temperature, and 100 ml. of dimethyl sulfoxide was added. The solution was distilled at a pressure of about 2 mm. (bath temperature 65–70°) until the sulfoxide began to come over (b.p. 43°). The excess *t*-butyl alcohol was removed in this way and 15 g. of ethyl benzoate was added dropwise to the somewhat solid mixture which was stirred at room temperature with a flow of nitrogen. After being kept for 60 min. more at room temperature, the mixture was heated under vacuum of about 3 mm. for 60 min. at 60°, and about 20 ml. of the solvent came over. The addition of 100 ml. of water and extraction of the aqueous solution resulting with 100 ml. of ether produced an aqueous layer which was acidified with a mixture of 30 ml. of conc. hydrochloric acid and 30 ml. of water. Colorless crystals began to form in 1 hr. and after 2 days they were removed, washed with water, and dried to give 15 g. of the methyl hemimercaptal (82%), m.p. 99–100°. One recrystallization elevated the m.p. to 101° [130].

20. From Carboxylic Acids or Acid Anhydrides

$$ArCH_2COOH \xrightarrow[(CH_3CO)_2O]{2\ C_5H_5NO} ArCHO + 2\ C_5H_5N + CO_2 + 2\ CH_3COOH$$

Carboxylic acids (or acid anhydrides) containing two α-hydrogens may be oxidized to aldehydes, while those containing one, under the same conditions, give ketones [131]. The oxidizing agent is pyridine-N-oxide and, if the acid is employed, acetic anhydride is also used. Yields both with phenylacetic acid, which gives an aldehyde, and diphenylacetic acid, which gives a ketone, are in the 60–70% range. The reaction is of greater theoretical than preparative interest, involving both a decarboxylation and oxidation step.

Of more general preparative value is the similar oxidation and decarboxylation of an α-bromoacid (Ex. a.)

(a) Preparation of butyraldehyde (67% as 2,4-dinitrophenylhydrazone from 50 mmoles of α-bromovaleric acid and 200 mmoles of pyridine-N-oxide refluxed for 24 hr. in xylene, while the aldehyde was swept out with N_2) [132].

1. O. Bayer, in Houben-Weyl's *Methoden der Organischen Chemie*, 4th ed., Vol. 7, G. Thieme Verlag, Stuttgart, 1954, Pt. 1.
2. J. Carnduff, *Quart. Rev. (London)*, **20,** 169 (1966).
3. K. Kulka, *Am. Perfumer Aromat.*, **69,** 31, Feb. 1957.
4. L. N. Ferguson, *Chem. Rev.*, **38,** 227 (1946).
5. W. J. Hickinbottom, *Reactions of Organic Compounds*, Longmans, Green and Co., New York, 1948, p. 106.
6. S. V. Lieberman and R. Connor, *Org. Syn.*, Coll. Vol. **2,** 441 (1943).
7. R. A. Morton *et al.*, *Biochem. J.*, **42,** 516 (1948).
8. J. Attenburrow *et al.*, *J. Chem. Soc.*, 1094 (1952).
9. O. Mancera *et al.*, *J. Chem. Soc.*, 2189 (1953); *J. Am. Chem. Soc.*, **77,** 4145 (1955).
10. M. Harfenist *et al.*, *J. Org. Chem.*, **19,** 1608 (1954).
11. M. Z. Barakat *et al.*, *J. Chem. Soc.*, 4685 (1956).
12. R. J. Gritter and T. J. Wallace, *J. Org. Chem.*, **24,** 1051 (1959).
13. L. F. and M. Fieser, *Reagents for Organic Synthesis*, John Wiley and Sons, New York, 1967, p. 638.
14. W. W. Epstein and F. W. Sweat, *Chem. Rev.*, **67,** 247 (1967).
15. D. H. R. Barton *et al.*, *J. Chem. Soc.*, 1855 (1964).
16. K. H. Scheit and W. Kampe, *Angew. Chem.*, **77,** 811 (1965).
17. W. S. Trahanovsky *et al.*, *J. Org. Chem.*, **32,** 2349, 3865 (1967).
18. H. O. House, *Modern Synthetic Reactions*, W. A. Benjamin, Inc., New York, 1965, p. 88.
19. J. C. Collins *et al.*, *Tetrahedron Letters*, 3363 (1968).
20. R. Filler, *Chem. Rev.*, **63,** 21 (1963).
21. T. Takaya *et al.*, *Bull. Chem. Soc. Japan*, **41,** 1032 (1968).
22. H. P. Kaufmann and D. B. Spannuth, *Chem. Ber.*, **91,** 2127 (1958).
23. V. M. Mićović and M. L. Mihailović, *Rec. Trav. Chim.*, **71,** 970 (1952).

24. C. W. Rees and R. C. Storr, *Chem. Commun.*, 1305 (1968).
25. C. D. Hurd and R. N. Meinert, *Org. Syn.*, Coll. Vol. **2**, 541 (1943).
26. R. A. Schneider and J. Meinwald, *J. Am. Chem. Soc.*, **89**, 2023 (1967).
27. C. Y. Meyers, *J. Org. Chem.*, **26**, 1046 (1961).
28. E. F. Pratt and T. P. McGovern, *J. Org. Chem.*, **29**, 1540 (1964).
29. Ref. 3, **70**, 45, Aug. 1957.
30. E. D. Bergmann and I. Shahak, *J. Chem. Soc.*, 1418 (1959).
31. J. C. Sauer, *Org. Syn.*, Coll. Vol. **4**, 813 (1963).
32. H. W. Bersch and A. V. Mletzko, *Arch. Pharm.*, **291**, 91 (1958).
33. L. I. Smith *et al.*, *J. Org. Chem.*, **4**, 323 (1939).
34. E. P. Papadopoulos *et al.*, *J. Org. Chem.*, **31**, 615 (1966).
35. J. G. Moffatt, *Org. Syn.*, **47**, 25 (1967).
36. V. J. Traynelis and W. L. Hergenrother, *J. Am. Chem. Soc.*, **86**, 298 (1964).
37. R. E. Dunbar and M. R. Arnold, *J. Org. Chem.*, **10**, 501 (1945).
38. R. R. Davies and H. H. Hodgson, *J. Chem. Soc.*, 282 (1943).
39. J. M. Church and H. K. Joshi, *Ind. Eng. Chem.*, **43**, 1804 (1951).
40. C. Djerassi, *Org. Reactions*, **6**, 207 (1951).
41. E. W. Warnhoff and P. Reynolds-Warnhoff, *J. Org. Chem.*, **28**, 1431 (1963).
42. C. S. Rondestvedt, Jr., *J. Am. Chem. Soc.*, **84**, 3319 (1962).
43. C. S. Rondestvedt and G. J. Mantell, *J. Am. Chem. Soc.*, **84**, 3307 (1962).
44. Ref. 3, **71**, 51, Jan. 1958.
45. R. C. Fuson and C. H. McKeever, *Org. Reactions*, **1**, 63 (1942).
46. G. E. Utzinger, *Ann. Chem.*, **556**, 50 (1944).
47. A. Kalir, *Org. Syn.*, **46**, 81 (1966).
48. L. C. Washburn and D. E. Pearson, unpublished results.
49. B. O. Field and J. Grundy, *J. Chem. Soc.*, 1110 (1955).
50. H. B. Hass and M. L. Bender, *Org. Syn.*, Coll. Vol. **4**, 932 (1963).
51. B. H. Klanderman, *J. Org. Chem.*, **31**, 2618 (1966).
52. F. Kröhnke, *Chem. Ber.*, **71**, 2583 (1938).
53. H. Reich, *Helv. Chim. Acta*, **23**, 219 (1940).
54. P. Karrer and A. Epprecht, *Helv. Chim. Acta*, **24**, 1039 (1941).
55. F. M. Dean *et al.*, *J. Chem. Soc.*, 792 (1961).
56. R. Aneja *et al.*, *Tetrahedron*, **2**, 203 (1938); **3**, 230 (1958).
57. D. A. Shepherd *et al.*, *J. Am. Chem. Soc.*, **77**, 1212 (1955).
58. O. Lorenz and C. R. Park, *J. Org. Chem.*, **30**, 1976 (1965).
59. P. S. Bailey and R. E. Erickson, *Org. Syn.*, **41**, 41 (1961).
60. L. D. Loan *et al.*, *J. Am. Chem. Soc.*, **87**, 737 (1965).
61. R. R. Davies and H. H. Hodgson, *Soc. Chem. Ind.* (*London*), **62**, 90 (1943).
62. R. Pappo *et al.*, *J. Org. Chem.*, **21**, 478 (1956).
63. R. E. Dessy and M. S. Newman, *Org. Syn.*, Coll. Vol. **4**, 484 (1963).
64. J. K. Stille and R. T. Foster, *J. Org. Chem.*, **28**, 2703 (1963).
65. W. S. Knowles and Q. E. Thompson, *J. Org. Chem.*, **25**, 1031 (1960).
66. J. Tsuji *et al.*, *J. Am. Chem. Soc.*, **86**, 4851 (1964); *Tetrahedron Letters*, 1061 (1963).
67. I. A. Pearl, *J. Am. Chem. Soc.*, **70**, 1746 (1948).
68. A. L. Henne and P. Hill, *J. Am. Chem. Soc.*, **65**, 752 (1943).
69. J. J. Pappas *et al.*, *Tetrahedron Letters*, 4273 (1966).
70. W. G. Overend *et al.*, *J. Chem. Soc.*, 1358 (1949).
71. E. L. Jackson, *Org. Reactions*, **2**, 341 (1944).
72. R. Criegee, *Newer Methods of Preparative Organic Chemistry*, Vol. 2, Academic Press, New York, 1963, p. 367.
73. L. F. and M. Fieser, *Reagents for Organic Synthesis*, John Wiley and Sons, New York, 1967. p. 546.
74. R. P. Bell *et al.*, *J. Chem. Soc.*, 1696 (1958).
75. E. Mosettig, Grundman reaction, *Org. Reactions*, **8**, 225 (1954).
76. C. J. Buist and C. A. Bunton, *J. Chem. Soc.*, 1406 (1954).

77. F. R. Duke and V. C. Bulgrin, *J. Am. Chem. Soc.*, **76,** 3803 (1954).

78. L. P. Kuhn, *J. Am. Chem. Soc.*, **76,** 4323 (1954).

79. G. de Vries and A. Schors, *Tetrahedron Letters*, 5689 (1968).

80. G. King, *Org. Reactions*, **2,** 363 (1944).

81. F. J. Wolf and J. Weijlard, *Org. Syn.*, Coll. Vol. **4,** 124 (1963).

82. R. C. Fuson and T.-L. Tan, *J. Am. Chem. Soc.*, **70,** 602 (1948).

83. E. Dimant and M. Banay, *J. Org. Chem.*, **25,** 475 (1960).

84. E. A. Davidson, *Carbohydrate Chemistry*, Holt, Rinehart, and Winston Co., New York, 1967, p. 136.

85. R. L. Whistler and K. Yagi, *J. Org. Chem.*, **26,** 1050 (1961).

86. S. J. Angyal, *Org. Reactions*, **8,** 197 (1954).

87. H. R. Snyder and J. R. Demuth, *J. Am. Chem. Soc.*, **78,** 1981 (1956).

88. K. H. Scheit and W. Kampe, *Angew Chem.*, **77,** 811 (1965).

89. S. J. Angyal *et al.*, *Org. Syn.*, Coll. Vol. **4,** 690 (1963).

90. K. B. Wiberg, *Org. Syn.*, Coll. Vol. **3,** 811 (1955).

91. S. J. Angyal *et al.*, *J. Chem. Soc.*, 1740 (1953).

92. H. R. Snyder *et al.*, *J. Am. Chem. Soc.*, **74,** 5110 (1952).

93. S. J. Angyal *et al.*, *J. Chem. Soc.*, 1737 (1953).

94. N. Kornblum *et al.*, *J. Am. Chem. Soc.*, **81,** 4113 (1959).

95. A. P. Johnson and A. Pelter, *J. Chem. Soc.*, 520 (1964).

96. N. Kornblum *et al.*, *J. Am. Chem. Soc.*, **79,** 6562 (1957).

97. N. Kornblum and H. W. Frazier, *J. Am. Chem. Soc.*, **88,** 865 (1966).

98. I. M. Hunsberger and J. M. Tien, *Chem. Ind.* (*London*), 88 (1959).

99. V. Franzen and S. Otto, *Chem. Ber.*, **94,** 1360 (1961); V. Franzen, *Org. Syn.*, **47,** 96 (1967).

100. Ref. 3, **70,** 37, Nov. 1957.

101. Ref. 3, **70,** 39, Dec. 1957.

102. S. V. Lieberman and R. Connor, *Org. Syn.*, Coll. Vol. **2,** 441 (1943); *see also* T. Nishimura, *Org. Syn.*, Coll. Vol. **4,** 713 (1963).

103. J. R. Naylor, *J. Chem. Soc.*, 4085 (1952).

104. R. A. Stairs, *Can. J. Chem.*, **42,** 550 (1964); O. H. Wheeler, *Can. J. Chem.*, **38,** 2137 (1960).

105. A. Tillotson and B. Houston, *J. Am. Chem. Soc.*, **73,** 221 (1951).

106. O. H. Wheeler, *Can. J. Chem.*, **36,** 667 (1958).

107. W. R. Boon, *J. Chem. Soc.*, S230 (1949).

108. L. Katz and W. E. Hamlin, *J. Am. Chem. Soc.*, **73,** 2801 (1951).

109. N. Rabjohn, *Org. Reactions*, **5,** 331 (1949).

110. H. Kaplan, *J. Am. Chem. Soc.*, **63,** 2654 (1941).

111. A. Burger and L. R. Modlin, Jr., *Org. Reactions*, **5,** 347 (1949).

112. V. M. Rodionov and M. A. Berkengeim, *J. Gen. Chem. USSR*, **14,** 330 (1944); *C.A.*, **39,** 4076 (1945).

113. C. A. Buehler *et al.*, *J. Org. Chem.*, **26,** 1410 (1961).

114. C. E. Teague, Jr., and A. Roe, *J. Am. Chem. Soc.*, **73,** 688 (1951); Cf. H. E. Baumgarten and J. E. Dirks, *J. Org. Chem.*, **23,** 900 (1958).

115. H. L. Riley *et al.*, *J. Chem. Soc.*, 1875 (1932).

116. H. A. Riley and A. R. Gray, *Org. Syn.*, Coll. Vol. **2,** 509 (1943).

117. A. R. Ronzio and T. D. Waugh, *Org. Syn.*, Coll. Vol. **3,** 438 (1955).

118. O. Touster, *Org. Reactions*, **7,** 327 (1953).

119. S. E. Forman, *J. Org. Chem.*, **29,** 3323 (1964).

120. A. Lapworth, *J. Chem. Soc.*, **79,** 1284 (1901).

121. P. Fréon, *Ann. Chim.* (*Paris*), (11), **11,** 453 (1939).

122. G. Sosnovsky, *J. Org. Chem.*, **25,** 874 (1960).

123. M. Stiles and A. J. Sisti, *J. Org. Chem.*, **25,** 1691 (1960).

124. W. E. Noland, *Chem. Rev.*, **55,** 137 (1955).

125. M. F. Hawthorne, *J. Am. Chem. Soc.*, **79,** 2510 (1957).

126. N. Kornblum and R. A. Brown, *J. Am. Chem. Soc.*, **87,** 1742 (1965).

127. K. Johnson and E. F. Degering, *J. Org. Chem.*, **8,** 10 (1943).

128. J. C. Sowden and R. Schaffer, *J. Am. Chem. Soc.*, **73,** 4662 (1951).

129. H. Shechter and F. T. Williams, Jr., *J. Org. Chem.*, **27,** 3699 (1962).

130. H.-D. Becker *et al.*, *J. Am. Chem. Soc.*, **85,** 3410 (1963).

131. T. Cohen *et al.*, *Tetrahedron Letters*, 237 (1965); C. Rüchart *et al.*, *ibid.*, 233 (1965).

132. T. Cohen and I. H. Song, *J. Org. Chem.*, **31,** 3058 (1966).

B. Reduction

For the preparation of aldehydes by reduction, attention is focused on the various acid derivatives, such as the acid chlorides, amides, nitriles, and esters. Within each group the choice of a wide variety of reducing agents, ranging from hydrogen gas to lithium aluminum hydride to hydrazine, is available with the limitation that reduction must stop at the aldehyde stage. For this reason, special conditions, specific catalysts, or means of derivatization are to be found in those reductions which give noteworthy yields of aldehydes. The twelve methods of reduction are not in the order of their worthiness; indeed, some of them are characteristic of specific types of aldehydes and thus eliminate a general comparison. However, for the general types, attention should be directed to the Brown reduction of acid chlorides (B.3) and to the reduction of nitriles (B.4 and B.7), which methods may supplant the classical procedures described. Nitrile reduction with Raney nickel and formic acid seems to be a particularly attractive method because of its simplicity (B.7).

1. From Acid Chlorides via the Reissert's Compound

The re-formation of the aromatic ring is the driving force which brings about reduction of the acid chloride. Detailed discussions of the aldehyde synthesis [1] and of the chemistry of Reissert's compounds [2] are available. This method of reduction is of limited value because the results are erratic. The reaction can be carried out in aqueous or in anhydrous media, or in liquid sulfur dioxide. 2-Methyl-, 5-nitro-, 5-amino-, 8-dimethylamino-, and 8-acetoxyquinolines have failed to lend themselves well to the preparation of Reissert's compounds.

(a) **Preparation of o-nitrobenzaldehyde** (58% from o-nitrobenzoyl chloride) [3].

(b) **Other examples**

(1) **Acetaldehyde** ((73% from acetyl chloride), **propionaldehyde** (36% from propionyl chloride), **n-Butyraldehyde** (62% from n-butyryl chloride), all being recovered as p-nitrophenylhydrazones) [4].

References for Section B are on pp. 582–583.

(2) **Cinnamaldehyde** (83% of the *p*-nitrophenylhydrazone from cinnamoyl chloride) [5].

2. From Acid Chlorides via the Thiol Ester

$$RCOCl \xrightarrow{RSH} RCOSR \xrightarrow[H_2]{Ni} RCHO$$

A discussion of this synthesis in some detail has been published [6]. The acid chlorides may be converted into the thiol esters by the use of a mercaptan or the lead alkyl mercaptide. Ordinarily, standard Raney nickel catalyst, W-1, or the more active W-4, leads to the alcohol but, if the catalyst is partially deactivated by refluxing in acetone for an hour or two, aldehydes are obtained in satisfactory yields. This synthesis has been utilized in the carbohydrate [7] and steroid [8] fields. The large amount of Raney nickel (about 10 parts to 1) necessary for the reduction detracts from the usefulness of the method.

(a) **Preparation of 3β-acetoxy-5-cholen-24-al** (63% from the thiol ester) [9].

(b) **Preparation of propionaldehyde** ((73% as the sodium bisulfite complex from the thiol ester) **and of aldehydo-D-ribose tetraacetate** (22% from the thiol ester)) [7].

3. From Acid Chlorides (Rosenmund and Brown)

$$RCOCl \xrightarrow[Pd(BaSO_4)]{H_2} RCHO \qquad \text{(Rosenmund)}$$

This reduction is attractive because of the simplicity of operation [10]. The acid chloride (10 parts) is dissolved in a solvent such as toluene, and hydrogen is passed through the heated solution containing 1 part of 5% palladium on barium sulfate until the theoretical amount of hydrogen chloride has been swept out of the solution. Further reduction of the aldehyde to the alcohol may be prevented by using "regulators" to deactivate the palladium. Although many regulators have been advocated, the most recent one recommended is tetramethylthiourea (about 2 mg./g. of catalyst) [11]. Apparently, some type of palladium sulfide is formed. However, it appears that a regulator is not always necessary and that overreduction can be prevented by maintaining the solution at the lowest temperature at which hydrogen chloride is evolved. Aliphatic dibasic acid chlorides do not give good yields of aldehydes, nor do those acid chlorides which tend to lose carbon monoxide, such as triphenylacetyl chloride.

The method of Brown and Subba Rao, although differing in reagents, appears to be similar to the Rosenmund reduction in its applicability [12]:

$$LiAlH_4 + 3 (CH_3)_3COH \xrightarrow{Diglyme} \underset{10\text{-}X}{LiAlH[OC(CH_3)_3]_3} + 3 H_2$$

$$10\text{-}X + RCOCl \xrightarrow{-78°} RCH{=}O$$

The operation is simple and gives about 60–90% yields with aromatic, unsaturated, or heterocyclic acid chlorides containing such varied substituents as nitro, cyano, and carboethoxy groups. Aliphatic acid chlorides give yields in the range 40–60%. Although three hydrogens of lithium aluminum hydride should be displaced by *t*-butyl alcohol for best results, the yields do not seem to depend crucially on whether the stoichiometry is exact.

The two preceding methods of reduction, the Rosenmund and Brown-Subba Rao reactions, are without doubt superior methods of making aldehydes if one takes into consideration all aspects of synthesis, particularly the isolation.

Recently it has been shown that tri-*n*-butyltin hydride reduces acyl halides to aldehydes and esters at room temperature; the former are sometimes produced in high yield [13].

(a) Preparation of β-naphthaldehyde (Rosenmund, 74–81% from the acid chloride, hydrogen, palladium on barium sulfate, and a quinoline sulfur "regulator") [14]; for preparations of catalyst see [15].

(b) Preparation of terephthalaldehyde (Brown-Subba Rao reaction). *t*-Butyl alcohol (1.2 moles) was added dropwise through a pressure-equilibrated dropping funnel over a period of 1 hr. to a stirred solution of 0.4 mole of LiAlH$_4$ in 150 ml. of diglyme. The resulting lithium tributoxyaluminohydride was cooled to −78° and 0.2 mole of terephthalyl chloride in 200 ml. of diglyme added over a period of 2 hr. On dilution and recrystallization from water, 77% of terephthalaldehyde, m.p. 114–115°, was obtained [12].

(c) Other examples of Rosenmund reduction

(1) 2,4,6-Trimethylbenzaldehyde (70–80% from the acid chloride) [16].

(2) Benzaldehyde (almost quantitative from the acid chloride) [17].

4. From Nitriles, Amides, or Acylpyrazoles or Imidazoles

$$R\text{—}CN \ \text{or} \ RCONR_2' \xrightarrow[\text{2. H}^{\oplus}]{\text{1. LiAlH(OC}_2\text{H}_5)_3} RCHO$$

So much work has been done on the reduction of nitriles that it is difficult to state with conviction which method is the best. Perhaps the elegant procedure of H. C. Brown and co-workers is the method of choice (see Ex. a). Yields for aromatic and heterocyclic aldehydes are equal to or superior to those of other methods, and moreover yields of aliphatic aldehydes are good contrasted to the low or nil yields from other types of reductions. The only criticism of the Brown procedure at present is that it is so new that it has not been seasoned enough to test its limitations.

As is true on reduction of an acid chloride with lithium tri-*t*-butoxyaluminohydride (B.3), a somewhat less bulky and perhaps better reducing agent is needed for nitriles. Thus lithium triethoxyaluminohydride, which is made *in situ* by addition of ethyl acetate to lithium aluminum hydride, has been selected:

$$2 \ LiAlH_4 + 3 \ CH_3CO_2C_2H_5 \xrightarrow{\text{Ether}} 2 \ LiAlH(OC_2H_5)_3$$

$$RC{\equiv}N + LiAlH(OC_2H_5)_3 \xrightarrow[]{} RC{=}\overset{\overset{\displaystyle Li}{\displaystyle \nearrow \quad (\text{solvent})}}{\underset{\displaystyle H}{N}}Al(OC_2H_5)_3$$

$$\text{10-XI}$$

The presence of some lithium diethoxyaluminodihydride from incomplete reaction with ethyl acetate does not affect the yields adversely. The lithium atom attached to the nitrogen evidently inhibits further addition of a hydride ion so that the intermediate 10-XI is not reduced to the alcohol. Not only is the lithium ion essential to the success of the reduction, but also the solvent, ether, plays a role, for yields are superior in this solvent as compared to tetrahydrofuran or diglyme [18].

Reducing agents other than hydrides have been employed for the reduction of nitriles in the sugar family. The following equation illustrates a method of proceeding to the next higher member [19]:

$$\underset{|}{CH{=}O} \xrightarrow{\text{HCN}} \underset{|}{\overset{CN}{\underset{|}{CHOH}}} \xrightarrow[\text{Pd(OH)}_2\text{—BaSO}_4]{H_2} \underset{|}{\overset{CH{=}O}{\underset{|}{CHOH}}}$$

The isolation technique is involved enough to consult the original literature for details. The method above is applicable also to the reduction of aminocyanohexoses [20].

The success of lithium triethoxyaluminohydride with nitriles carries over to the preparation of aldehydes from N,N-dimethylamides and indeed suggests consideration as replacement for other methods of reduction of amides as described by Mosettig [21]. It should be noted in this reference, however, that notable syntheses have been achieved in making highly conjugated aldehydes, such as $C_6H_5(CH{=}CH)_nCH{=}O$ where $n = 2$, 4, and 6, from the corresponding amides and stoichiometric amounts of $LiAlH_4$.

Lithium aluminum hydride without deactivation by alcohols may be used for the reduction of acylpyrazoles (Ex. c.1) or acylimidazoles (c.2). The acylpyrazoles are prepared by simple acylation of pyrazoles or from ketene intermediates and pyrazoles [22]:

$$RCOCHN_2 \xrightarrow{h\nu} [RCH{=}C{=}O] \xrightarrow{\text{3,5-Dimethylpyrazole}} RCH_2CON{\underset{\diagdown}{\overset{\diagup}{}}}\begin{array}{l} N{=}CCH_3 \\ \ \ | \\ C{=}CH \\ \ | \\ CH_3 \end{array}$$

The reduction of acylimidazoles is worthy of consideration as a means of conversion of an expensive acid into an aldehyde

(a) Preparation of caproaldehyde from capronitrile. Lithium aluminum hydride (0.3 mole) was dissolved in 300 ml. of dry ether in a nitrogen atmosphere. To the stirred solution, ethyl acetate (0.45 mole) was added over a period of 75 min. while the temperature was maintained at 3–7° and stirring was continued an additional 30 min. Capronitrile (0.3 mole) was added to the stirred solution maintained at −10° in 5 min. The mixture became viscous as the temperature rose to 12° and was stirred an additional 50 min. at 3°. The mixture was decomposed with 300 ml. of 5 N sulfuric acid and extracted with ether. The purified and dried ether extract was distilled through a 12-in. Vigreux column to give caproaldehyde, 16.6 g. (55%), b.p. 51–55° (53–55 mm.), n^{20}D 1.4042 [18].

(b) Preparation of cyclohexanecarboxaldehyde from N,N-dimethylcyclohexanecarboxamide. The lithium triethoxyaluminohydride (0.375 mole) was made as described in Ex. a. To the stirred slurry at ice-bath temperature, N,N-dimethylcyclohexanecarboxamide (0.375 mole) was added as rapidly as possible without permitting too vigorous a refluxing of the ether. The reaction mixture was stirred for an additional hour and worked up as in Ex. a; b.p. of the aldehyde 74–78° (20 mm.), n^{20}D 1.4499, 78% [23].

(c) Other examples

(1) 4-Methylphenylacetaldehyde. 1-(4-Methylphenylacetyl)-3,5-dimethylpyrazole, 0.01 mole, in 80 ml. of absolute ether was treated with stirring at 0° with 0.0033 mole of lithium aluminum hydride in portions; after 10 hr. stirring at 0°, the mixture was hydrolyzed with 20 ml. of 2 N sulfuric acid. The ethereal phase was separated and the aqueous phase was extracted twice more with 30 ml. of ether. The purified ether extract yielded 1 g. (76%) of an oil which produced a 2,4-dinitrophenylhydrazone of m.p. 131° with decomposition [22].

(2) Benzaldehyde (77% from 1 mole of benzoyl chloride and 2 moles

$$C_6H_5COCl + 2 HN\overset{=N}{\diagup} \longrightarrow C_6H_5CON\overset{=N}{\diagup} \xrightarrow{LiAlH_4} C_6H_5CHO$$

of imidazole in THF at 25°, and reduction of the amide formed with 0.25 mole of $LiAlH_4$ in ether at −20° for 30–60 min.) [24].

5. From Acid Anilides via the Imidochloride and the Anil (Sonn and Müller)

$$ArCONHC_6H_5 \xrightarrow{PCl_5} ArCCl{=}NC_6H_5 \xrightarrow[HCl]{SnCl_2}$$

$$ArCH{=}NC_6H_5 \cdot HCl \xrightarrow{H_2O} ArCHO$$

This method has been reviewed in some detail [25]. It is not applicable to simple aliphatic anilides since the imido chlorides obtained are transformed readily into the corresponding enamines. α,β-Unsaturated anilides produce aldehydes satisfactorily with yields as high as 92%. Although stannous chloride is the usual reducing agent employed, it is unsatisfactory for purely aliphatic α,β-unsaturated anilides. Chromous chloride has been substituted in such cases although it is not always satisfactory. In fact, it appears that some clarification is needed on the preparation and use of this reagent.

The synthesis has been applied most widely to the aromatic series in which substituents unaffected by phosphorus pentachloride or stannous chloride are present. In such cases the yields are generally good. In the heterocyclic series the addition of hydrogen chloride to the imidochloride sometimes gives an amine rather than the desired anil.

Anilides may be obtained from arylmagnesium bromides and phenyliso-cyanate, a method which utilizes an aryl bromide as starting material (Ex. b.1). Also, since imidochlorides are intermediates in the Beckmann rearrangement of benzophenone oximes with phosphorus pentachloride, the oximes can be used as starting materials for the synthesis of aldehydes (see Ex. b.3).

(a) Preparation of cinnamaldehyde (92 % from the anilide) [26].

(b) Other examples

(1) *o*-Tolualdehyde (62–70 % from the anilide which was made from *o*-tolylmagnesium bromide and phenylisocyanate) [27].

(2) 3-Phenanthraldehyde (85 % from the anilide) [28].

(3) 1,2,3,4-Tetrahydrophenanthrene-9-carboxaldehyde   (68 % from the oxime of 9-benzoyl-1,2,3,4-tetrahydrophenanthrene and phosphorus pentachloride followed by reduction with stannous chloride) [29].

6. From Nitriles via the Imidochloride and Anil (Stephen)

This reaction is much more versatile than the Sonn-Müller reaction (B.5), but it does resemble the latter in certain aspects:

$$2\ RC\!\!\equiv\!\!N + SnCl_2 + 6\ HCl \xrightarrow{\text{Ether}} (RCH\!\!=\!\!\overset{\oplus}{N}H_2)_2SnCl_6{}^{\ominus} \xrightarrow{H_2O} RCH\!\!=\!\!O$$

It has been discussed in detail [30], but some experimental manipulations are still a matter of controversy. The aromatic nitriles give excellent yields, particularly if the solvent ethyl formate or ethyl acetate is substituted for ether. The esters solubilize the starting materials and permit fairly efficient precipitation of the imino-hydrochloride-stannic chloride complex which serves to drive the reaction to completion [31]. The yields for aliphatic aldehydes are reported to be low, but it is claimed that, if the reduction is carried out in an anhydrous medium, using a large excess of stannous chloride (7:1 molar ratio) and a prolonged reaction time (7 days), yields of 55–67 % are achieved [32]. It is generally agreed that anhydrous conditions are best probably because the imidoester, likely to form from ether splitting or ester interchange in the presence of water, is inert to reduction under Stephen reaction conditions. But the Russian investigators take exception to the statement that anhydrous stannous chloride prepared by dehydration of the hydrate with acetic anhydride is of dependable quality. Rather, they conduct the dehydration in a vacuum desiccator over concentrated sulfuric acid.

(a) **Preparation of β-naphthaldehyde** (73–80% from the nitrile and stannous chloride dehydrated with acetic anhydride) [33].

(b) **Preparation of isovaleraldehyde.** Anhydrous stannous chloride (7 mmoles, dehydrated in a vacuum desiccator over sulfuric acid) in 7 ml. of absolute ether was saturated with hydrogen chloride until two layers were formed. Isovaleronitrile (1 mmole) was added and hydrogen chloride again passed through the solution until saturated. After 7 days the ether and acid were removed from the mixture by evaporation at water aspirator pressure. The residue was dissolved in 100 ml. of water and steam-distilled, the aqueous distillate yielding 61–64% isovaleraldehyde [32].

(c) **Other examples**

(1) **4-Methyl-5-thiazolecarboxaldehyde** (40% from the cyanide) [34].

(2) **α-Naphthaldehyde** (20% from the nitrile using ethyl acetate as a solvent; the reaction seems somewhat more sensitive to steric effects than the Sonn-Müller reduction, at least as judged from this yield) [31].

7. From Nitriles, Raney Nickel, and Reducing Agents

$$RCN \xrightarrow[\substack{NaH_2PO_2 \\ CH_3COOH}]{Raney\ Ni} RCHO$$

This synthesis is applicable largely to preparation of aromatic aldehydes. The Raney nickel appears to play two roles in that it liberates hydrogen from aqueous sodium hypophosphite solutions and acts as a catalyst as well. The procedure is simple because no special apparatus nor heating is required. No reduction beyond the aldehyde occurs. Yields are usually satisfactory although some types such as nitrocyanides are not affected [35]. Recently, Raney nickel in formic acid has been used to effect this reduction (see Ex. b.2). The method is extremely simple. A third method of reducing nitriles with Raney nickel involves the use of hydrazine, but the intermediate azine is isolated and must be hydrolyzed to the aldehyde in a second step (see B.8).

(a) **Preparation of benzaldehyde.** Benzonitrile, 1 g., and 2 g. of hydrated sodium hypophosphite in 29 ml. of 1:1:2 water-acetic acid-pyridine (more acetic acid was added to effect solution, if necessary) were treated with 0.3–0.4 g. of Raney nickel and stirred at 40–45° for about 1 hr. The catalyst was filtered and washed with warm aqueous alcohol, and the combined filtrates were treated with 2,4-dinitrophenylhydrazine in the usual manner to give the 2,4-dinitrophenylhydrazone of benzaldehyde in 90% yield [35].

(b) **Other examples**

(1) **β-Naphthaldehyde** (90% from the nitrile) [35].

(2) **p-Chlorobenzaldehyde** (quantitative crude yield from the nitrile and equal weight of Raney nickel alloy refluxed in 75% aqueous formic acid for 1 hr.) [36].

8. From Nitrile Derivatives (Imidoesters, Imidazolines, Semicarbazides and Hydrazides)

Although the intermediates listed in the accompanying scheme can be used directly, it is more convenient to prepare these compounds *in situ* from the nitriles. Under these conditions the nitrile may be reduced, and the imino-aldehyde or derivative trapped by the derivatizing agent. Originally the

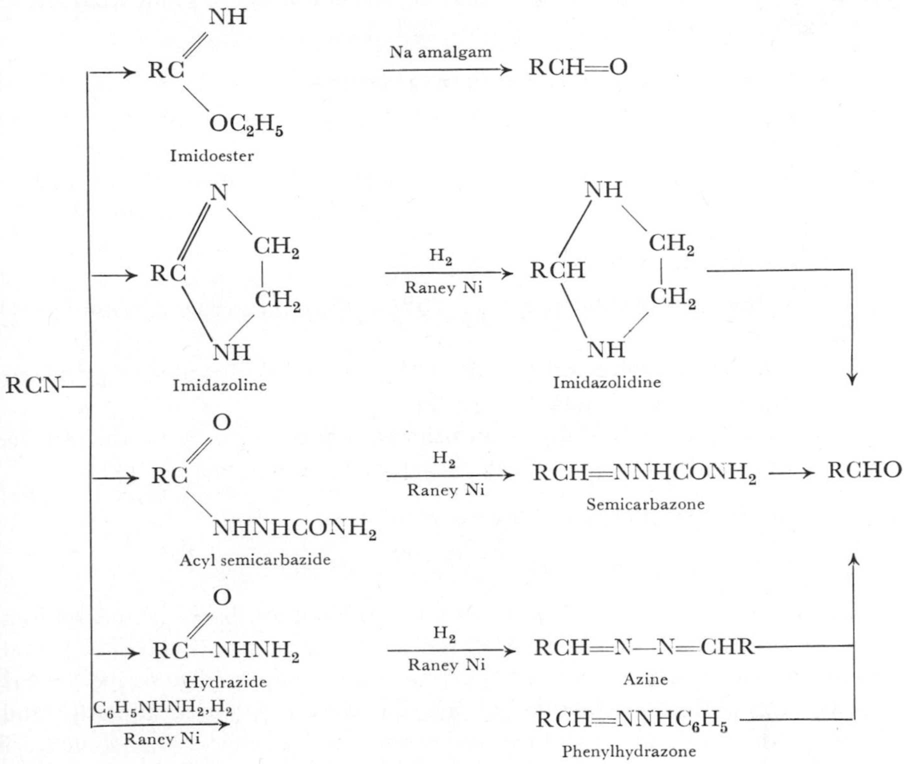

$$
RCN- \begin{cases}
\overset{NH}{\underset{OC_2H_5}{RC}} & \xrightarrow{\text{Na amalgam}} RCH{=}O \\[2ex]
\text{Imidoester} \\[2ex]
\text{Imidazoline} & \xrightarrow[\text{Raney Ni}]{H_2} \text{Imidazolidine} \\[2ex]
\overset{O}{\underset{NHNHCONH_2}{RC}} & \xrightarrow[\text{Raney Ni}]{H_2} RCH{=}NNHCONH_2 \longrightarrow RCHO \\
\text{Acyl semicarbazide} & \text{Semicarbazone} \\[2ex]
\overset{O}{RC{-}NHNH_2} & \xrightarrow[\text{Raney Ni}]{H_2} RCH{=}N{-}N{=}CHR \\
\text{Hydrazide} & \text{Azine} \\
\underset{\text{Raney Ni}}{C_6H_5NHNH_2,H_2} & RCH{=}NNHC_6H_5 \\
& \text{Phenylhydrazone}
\end{cases}
$$

cyanide was converted into the imidoester which was reduced with sodium amalgam in the presence of phenylhydrazine, and the phenylhydrazone formed was hydrolyzed to the aldehyde [37]. Semicarbazide, but not aniline, may be substituted for phenylhydrazine. More recently the cyanide was treated with semicarbazide hydrochloride or dianilinoethane and the mixture was reduced with hydrogen in the presence of Raney nickel [38]. The intermediate amidines and imidazolines have also been reduced using sodium and ethanol in liquid ammonia [39]. The azine intermediate method seems particularly attractive because of the simplicity of operation, but the semicarbazide intermediate method may be superior in certain instances owing to the ease of isolation of the semicarbazone. Yields of around 90% of aldehyde are claimed by converting the aliphatic or aromatic nitrile, 1 mole, and phenylhydrazine, 4 moles, first into the phenylhydrazone (by hydrogenation in the presence of Raney nickel) which is hydrolyzed with concentrated hydrochloric acid [40].

Direct preparation of aldehydes by reduction of nitriles with Raney nickel and various reducing agent is given under B.7.

(a) Preparation of benzaldehyde.   2-Phenylimidazoline [40a], 7.3 g., in 20 ml. of ethanol and 100 ml. of liquid ammonia was reduced by the addition of 2.3 g. of sodium. Water, 100 ml., was added, and crystals, m.p. 53–54°, separated after standing overnight. These were dissolved in 25 ml. of 2 N hydrochloric acid and an oil separated. Conversion into the semicarbazone of benzaldehyde gave 6.12 g. (75%), m.p. 221° [41].

(b) Preparation of o-benzylbenzaldehyde azine.   One gram of o-benzylbenzonitrile, 0.5 g. of Raney Ni, 10 ml. of 85% hydrazine hydrate, and 50 ml. of ethanol were heated at 50–55° until the evolution of ammonia had ceased. The catalyst was filtered, and the filtrate concentrated until crystals of the azine began to appear. Yields were 88–91% for four different nitriles [42].

(c) Other examples

(1) **Phenylacetaldehyde** (over 70% as the semicarbazone from benzyl cyanide) [43].

(2) **m-Methoxybenzaldehyde** (94% as 2,4-dinitrophenylhydrazone from 2-m-methoxyphenylimidazoline) [39].

(3) **3-Indoleacetaldehyde semicarbazone** (68% from the nitrile, semicarbazide hydrochloride, Raney nickel, and sodium acetate) [44].

9. From Esters, Lactones, or ortho-Esters

$$\text{RCOOR}' \xrightarrow{(i\text{-}C_4H_9)_2AlH} \text{RCHO}$$

Various reducing agents, such as diisobutylaluminum hydride and sodium diisobutylaluminum dihydride [45], sodium aluminum hydride [46], and lithium tri-t-butoxyaluminohydride [47] have been employed to convert esters into aldehydes. Diisobutylaluminum hydride with a series of aliphatic and aromatic esters, preferably in toluene or hexane at $-70°$, gave yields, sometimes as the 2,4-dinitrophenylhydrazone, of 48–90%. In the case of three esters reduced with sodium diisobutylaluminum dihydride in ether solution at $-70°$, the yields varied from 60 to 80%. Sodium aluminum hydride with a series of aliphatic and aromatic esters in tetrahydrofuran or in a mixture of tetrahydrofuran and pyridine at -45 to $-65°$ gave yields for the 2,4-dinitrophenylhydrazones of 25–88%. The yields were somewhat better for aliphatic than for aromatic aldehydes. Lithium tri-t-butoxyaluminohydride is so mild as a reducing agent that it does not affect appreciably alkyl esters and many other groups readily reduced by lithium aluminum hydride. However, in addition to acid chlorides, it does affect many phenyl esters to yield aldehydes when solutions of the ester in tetrahydrofuran at 0° are treated with the metal hydride. Yields with the phenyl esters varied from 33 to 77%, except in the case of phenyl cyclopropylcarboxylate and phenyl benzoate which remained unchanged. Thus selective reduction of the carboxylic acid group is possible by proceeding via the acid chloride or phenyl ester. Although the former path would ordinarily be preferred, cases might arise in which it would not be feasible to continue by way of the acid chloride group.

Lithium aluminum hydride may be employed for reducing perfluoroesters to perfluoroaldehydes at $-70°$ [48]. See Ex. c.1. Apparently the perfluoroaldehydes are not reduced under these conditions because of their tendency to solvate.

Lactones (Ex. c.2) and *ortho*-esters (c.3) may be reduced by stoichiometric amounts of lithium aluminum hydride but, in the sugar family, lactones have been reduced more frequently with sodium amalgam (c.4).

(a) Preparation of acetaldehyde.   To 40 mmoles of phenyl acetate in 10 ml. of THF was added, with stirring at $0°$, 30 ml. of a THF solution containing 40 mmoles of lithium tri-*t*-butoxyaluminohydride. All reagents were dry and the reaction was carried out under nitrogen. After 4 hr. no hydride remained, and there was recovered a 70% yield of acetaldehyde 2,4-dinitrophenylhydrazone, m.p. 145° [47].

(b) Preparation of β-phenylpropionaldehyde (88% as the 2,4-dinitrophenylhydrazone from methyl hydrocinnamate and sodium aluminum hydride in THF at -45 to $-65°$) [46].

(c) Other preparations

(1) Trifluoroacetaldehyde (71% from ethyl trifluoroacetate and lithium aluminum hydride in ether at $-70°$) [48].

(2) α-Methyl-β-hydroxycaproaldehyde (64% from 0.1 mole of the corresponding lactone and slightly over 0.025 mole of $LiAlH_4$ in THF at $-10°$) [49].

(3) Dimethyl acetal of β-methylmercaptopropionaldehyde (97% from the corresponding *ortho*-ester and exactly one-fourth the molar equivalent of $LiAlH_4$ in benzene under reflux) [50].

(4) Arabinose (56% from arabinolactone in water at a pH of 3–3.5 with 2.5–3 equiv. of 2.5% sodium amalgam at 5–10°) [51].

10. From Esters via Acyl Arylsulfonylhydrazides (McFadyen and Stevens)

$$ArCOOR \xrightarrow{NH_2NH_2} ArCONHNH_2 \xrightarrow{Ar'SO_2Cl}$$

$$ArCONHNHSO_2Ar' \xrightarrow[Na_2CO_3]{\text{Ethylene glycol}} ArCH{=}O + N_2 + Ar'SO_2Na$$

10-XII

In this synthesis the sulfonylation is brought about to make the hydrazide 10-XII sufficiently acidic to form an anion subject to ready decomposition to nitrogen gas, the sulfinate salt, and the corresponding aldehyde. The synthesis has been reviewed in some detail [52]. Good yields are obtained from benzoic acid esters, fair yields from heterocyclic esters, and poor or no yields from aliphatic esters. It is now known that the poor yield from aliphatic esters is caused by aldol condensation of the aliphatic aldehyde in the alkaline medium [53]. Pivalaldehyde, for example, which has no active methylene group, can be formed in 40% yield on short exposure (30 seconds) of the corresponding tosylacylhydrazide to the sodium carbonate-ethylene glycol medium. More recently (see Ex. b.4) [54], this yield has been increased to 50% by removing the aldehyde as it is formed. Solids such as powdered soft glass may increase the yield slightly by lowering the temperature of pyrolysis in ethylene glycol [55].

In a few cases it has been possible to eliminate the sulfonylation step in the sequence above by converting the acylhydrazide directly into the aldehyde by alkaline oxidation [56, 57].

(a) Preparation of benzaldehyde (70% from the acid hydrazide) [58]; *see also* [55].

(b) Other examples

(1) 4-Amino-2-methylpyrimidine-5-carboxaldehyde (44% from the acid hydrazide) [59].

(2) Nicotinaldehyde (60–70% from the acid hydrazide) [57, 55].

(3) Apocamphane-1-carboxaldehyde (60% from the tosyl acylhydrazide on heating for 30 sec. in ethylene glycol and sodium carbonate) [53].

(4) Pivalaldehyde (50% from *p*-toluenesulfonylpivalhydrazide added slowly to a refluxing solution of ethylene glycol containing 1 equiv. of water and 2 equiv. of sodium hydroxide with an arrangement to collect the volatile aldehyde as formed. About 16% neopentyl alcohol and 14% pivalic acid were formed concurrently. The tosylhydrazide of butyric acid gave 10% butyraldehyde under similar conditions, the lower yield being due in all probability to the presence of α-hydrogen atoms in the aldehyde) [54].

11. From Acids

$$\text{ArCO}_2\text{H} \rightarrow \text{ArCH}{=}\text{O}$$

Although not used often, this reduction has been achieved by the use of sodium amalgam (Ex. a) and by high-temperature reaction with formic acid or its salts (Ex. b.1, b.2, and b.3).

(a) Preparation of salicylaldehyde. Salicylic acid, 15 g., *p*-toluidine, 18 g. (which subsequently forms the Schiff base of the aldehyde), and enough sodium carbonate to neutralize the acid were dissolved in 1 l. of hot water. Boric acid, 15 g., and 250 g. of sodium chloride were added to the solution at room temperature. To the agitated mixture 2% sodium amalgam (330–430 g.) was added gradually while keeping the solution on the acidic side by concurrent addition of boric acid. The reaction was complete when no salicylic acid precipitated on acidification of an aliquot. The mixture was acidified and steam-distilled to yield 7.5 g. of salicylaldehyde [60]; for the use of sodium bisulfite in place of *p*-toluidine see [61].

(b) Other examples

(1) *p*-Isopropyl-α-methylhydrocinnamaldehyde. A tube of good Pyrex glass, 1 m. long and 50 mm. wide, was filled with granulated pumice containing manganous oxide. A vapor mixture of 600 g. of formic acid and 400 g. of p-isopropyl-α-methylhydrocinnamic acid was passed through the tube in the course of 1 hr. at 360°. The vapor was condensed at the outlet and fractionated to give about 80 % of the aldehyde, b.p. 119° (6 mm.) [60].

(2) **Lauric aldehyde** (31 % conversion, 90 % yield (on acid consumed) from lauric acid and formic acid with titanium oxide in a Carius tube) [62].

(3) **Stearic aldehyde** (19 % from 3 parts of manganous formate and 1 part of manganous stearate; yields vary from 19 to 60 % in the aliphatic series) [63].

12. From Diazoalkanes

This synthesis, which involves a special reductive procedure of a diazomethane derivative, has been successful for obtaining the aldehydes of s-triazines which could not be prepared by conventional methods. The diazo compound is obtained by treating cyanuric chloride with diazomethane [64]. In alkaline or neutral medium, preferably with a lower alkyl thiol in alcohol, the diazo compound may be reduced with sodium at 0° to the hydrazone which may be converted into the aldehyde with 2,4-dinitrobenzaldehyde [65]. Yields are good for the first, but poor for the second, step in the sequence.

(a) Preparation of 4,6-bis(thiomethyl)-s-triazine-2-carboxaldehyde

(1) **4,6-Bis-thiomethyl-s-triazine-2-carboxaldehyde hydrazone.** 2-Diazomethyl-4,6-dichloro-s-triazine, 47.5 g., was stirred at 0° into a solution of 11.5 g. of sodium in 500 ml. of absolute ethanol and 100 g. of methanethiol. After stirring the mixture for 1 day at room temperature, the precipitate was filtered off and extracted 5 times with hot ethanol. After standing for a while at −20°, the hydrazone, 35 g., crystallized from the extracts. More, 11.5 g., was obtained on concentrating the mother liquor and cooling. Total yield, 87.5 %. Recrystallized from aqueous acetone, yellow needles, m.p. 179–181°, were obtained.

(2) **4,6-Bis-thiomethyl-s-triazine-2-carboxaldehyde.** The hydrazone, 13 g., and 24 g. of 2,4-dinitrobenzaldehyde in 200 ml. of ethanol and 10 ml. of water were refluxed for 15 hr. After cooling and filtering from the dinitrobenzalazine, the solvents were removed under reduced pressure to give a dark, oily residue which was extracted with ligroin. On keeping the extracts at −25° for a few days, white crystals of the hemi-ethyl acetal, 5.05 g. (33.4 %), precipitated. Sublimation twice between 70 and 120° at 0.5 mm. gave 3.4 g. of the yellow, free aldehyde (28 %), m.p. 102–102.5° [65].

1. E. Mosettig, *Org. Reactions*, **8**, 220 (1954).
2. W. E. McEwen and R. L. Cobb, *Chem. Rev.*, **55**, 511 (1955).
3. G. L. Buchanan *et al.*, *Org. Reactions*, **8**, 223 (1954).
4. J. M. Grosheintz and H. O. L. Fischer, *J. Am. Chem. Soc.*, **63**, 2021 (1941).
5. G. Wittig *et al.*, *Ann. Chem.*, **577**, 1 (1952).
6. E. Mosettig, *Org. Reactions*, **8**, 229 (1954).
7. M. L. Wolfrom and J. V. Karabinos, *J. Am. Chem. Soc.*, **68**, 1455 (1946).
8. A. V. McIntosh, Jr., *et al.*, *J. Am. Chem. Soc.*, **70**, 2955 (1948).
9. A. V. McIntosh, Jr., *et al.*, *Org. Reactions*, **8**, 231 (1954).
10. E. Mosettig and R. Mozingo, *Org. Reactions*, **4**, 362 (1948).
11. S. Affrossman and S. J. Thomson, *J. Chem. Soc.*, 2024 (1962).
12. H. C. Brown and B. C. Subba Rao, *J. Am. Chem. Soc.*, **80**, 5377 (1958).
13. H. G. Kuivila and E. J. Walsh, Jr., *J. Am. Chem. Soc.*, **88**, 571 (1966).
14. E. B. Hershberg and J. Cason, *Org. Syn.*, Coll. Vol. **3**, 627 (1955).
15. R. Mozingo, *Org. Syn.*, Coll. Vol. **3**, 685 (1955).
16. R. P. Barnes, *Org. Syn.*, **21**, 110 (1941).
17. C. Weygand and W. Meusel, *Chem. Ber.*, **76**, 503 (1943).
18. H. C. Brown and C. P. Garg, *J. Am. Chem. Soc.*, **86**, 1085 (1964).
19. R. Kuhn and P. Klesse, *Chem. Ber.*, **91**, 1989 (1958).
20. R. Kuhn and W. Kirschenlohr, *Ann. Chem.*, **600**, 115, 126 (1956).
21. E. Mosettig, *Org. Reactions*, **8**, 252 (1954).
22. W. Ried *et al.*, *Ann. Chem.*, **642**, 121 (1961).
23. H. C. Brown and A. Tsukamoto, *J. Am. Chem. Soc.*, **86**, 1089 (1964).
24. H. A. Staab, *Angew. Chem.*, *Intern. Ed. Engl.*, **1**, 351 (1962).
25. E. Mosettig, *Org. Reactions*, **8**, 240 (1954).
26. A. Sonn and E. Müller, *Org. Reactions*, **8**, 243 (1954).
27. J. W. Williams *et al.*, *Org. Syn.*, Coll. Vol. **3**, 818 (1955).
28. W. E. Bachmann, *Org. Reactions*, **8**, 244 (1954).
29. G. H. Coleman and R. E. Pyle, *J. Am. Chem. Soc.*, **68**, 2007 (1946).
30. E. Mosettig, *Org. Reactions*, **8**, 246 (1954).
31. T. Stephen and H. Stephen, *J. Chem. Soc.*, 4695 (1956).
32. M. M. Shemiakin *et al.*, *J. Gen. Chem. USSR* (*Eng. Transl.*), **28**, 952 (1958).
33. J. W. Williams, *Org. Syn.*, Coll. Vol. **3**, 626 (1955).
34. C. R. Harington and R. C. G. Moggridge, *Org. Reactions*, **8**, 252 (1954).
35. O. G. Backeberg and B. Staskun, *J. Chem. Soc.*, 3961 (1962).
36. T. Van Es and B. Staskun, *J. Chem. Soc.*, 5775 (1965).
37. F. Henle, *Chem. Ber.*, **35**, 3039 (1902); **38**, 1362 (1905).
38. H. Plieninger and G. Werst, *Chem. Ber.*, **88**, 1956 (1955).
39. A. J. Birch *et al.*, *Chem. Ind.* (*London*), 1559 (1954).
40. A. Gaiffe and R. Pollaud, *Compt. Rend.*, **252**, 1339 (1961); **254**, 496 (1962).
40a. P. Oxley and W. F. Short, *J. Chem. Soc.*, 497 (1947).
41. A. J. Birch *et al.*, *Austr. J. Chem.*, **7**, 256 (1954).
42. W. W. Zajac, Jr. and R. H. Denk, *J. Org. Chem.*, **27**, 3716 (1962).
43. H. Plieninger and G. Werst, *Angew. Chem.*, **67**, 156 (1955).
44. J. N. Coker, *Chem. Eng. News*, **39**, 18, 54, Sept. 1961.
45. L. I. Zakharkin and I. M. Khorlina, *Tetrahedron Letters*, No. **14**, 619 (1962).
46. L. I. Zakharkin *et al.*, *Tetrahedron Letters*, No. **29**, 2087 (1963).
47. P. M. Weissman and H. C. Brown, *J. Org. Chem.*, **31**, 283 (1966).
48. O. R. Pierce and T. G. Kane, *J. Am. Chem. Soc.*, **76**, 300 (1954).
49. G. E. Arth, *J. Am. Chem. Soc.*, **75**, 2413 (1953).
50. C. J. Claus and J. L. Morgenthau, Jr., *J. Am. Chem. Soc.*, **73**, 5005 (1951).
51. W. M. Sandstorm *et al.*, *J. Am. Chem. Soc.*, **69**, 915 (1947).
52. E. Mosettig, *Org. Reactions*, **8**, 232 (1954).
53. M. Sprecher *et al.*, *J. Org. Chem.*, **26**, 3664 (1961).
54. H. Babad *et al.*, *Tetrahedron Letters*, 2927 (1966).

55. M. S. Newman and E. G. Caflisch, Jr., *J. Am. Chem. Soc.*, **80**, 862 (1958).
56. L. Kalb and O. Gross, *Chem. Ber.*, **59**, 727 (1926); C. Niemann and J. T. Hays, *J. Am. Chem. Soc.*, **65**, 482 (1943).
57. H. N. Wingfield *et al.*, *J. Am. Chem. Soc.*, **74**, 5796 (1952).
58. J. S. McFadyen and T. S. Stevens, *Org. Reactions*, **8**, 235 (1954).
59. D. Price *et al.*, *Org. Reactions*, **8**, 236 (1954).
60. K. Kulka, *Am. Perfumer Aromat.*, **70**, 47, Sept. 1957.
61. H. Weil and H. Ostermeier, *Chem. Ber.*, **54**, 3217 (1921).
62. R. R. Davies and H. H. Hodgson, *J. Chem. Soc.*, **84** (1943).
63. P. Mastagli *et al.*, *Compt. Rend.*, **248**, 1830 (1959).
64. C. Grundmann and E. Kober, *J. Am. Chem. Soc.*, **79**, 944 (1957).
65. E. Kober and C. Grundmann, *J. Am. Chem. Soc.*, **80**, 5547 (1958).

C. Friedel-Crafts Reactions

The direct introduction of the aldehyde group into the aromatic nucleus is a most useful and important reaction, a complete discussion of which has been published recently [1]. The Gattermann (C.1) and Gatterman-Koch (C.2) methods are classical, but more modern procedures such as the dimethyl-formamide-phosphorus oxychloride (C.6) and dichloromethyl ether (C.4) processes are simpler and seem to excel the older methods. Moreover, the formyl fluoride-boron trifluoride system is also proving quite successful as a formylating agent (C.3). Liberty is taken in this section to consider as Friedel-Crafts-type reactions not only aromatic substitution, but also any reaction in which a positive (electron-seeking) reagent attacks an unsaturated center to form an aldehyde derivative. Thus acylation or substitution of olefins or vinyl ethers is to be found here. Arylations via diazonium salts are also included although their mechanism is less certain (C.11).

1. From Arenes, Zinc Cyanide, and Hydrogen Chloride (Gattermann)

$$ArH + Zn(CN)_2 + HCl \xrightarrow{ZnCl_2} ArCH{=}NH{\cdot}HCl$$

$$\downarrow H_2O$$

$$ArCHO$$

This synthesis has been reviewed [2, 1].

Originally hydrogen cyanide was employed in the Gattermann synthesis, but the use of zinc cyanide is more convenient and the yields are as satisfactory [3]. In contrast to the Gattermann-Koch synthesis, this method has been applied successfully to the synthesis of aldehydes of phenols and ethers. The yields vary from poor to good. Because of the development of more convenient methods little use has been made of this synthesis in recent literature, but one interesting modification has been employed for aliphatic substitution (see Ex. b).

References for Section C are on pp. 591–592.

(a) Preparation of mesitaldehyde (75–81 % from mesitylene) [4].

(b) Preparation of β-salicyloyl-α-arylacetaldehydes [5]:

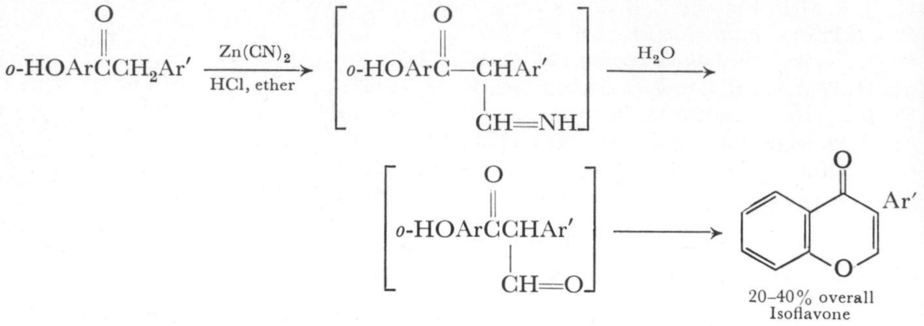

$$o\text{-HOArCCH}_2\text{Ar}' \xrightarrow[\text{HCl, ether}]{\text{Zn(CN)}_2} \left[o\text{-HOArC—CHAr}' \atop \text{CH=NH} \right] \xrightarrow{\text{H}_2\text{O}}$$

$$\left[o\text{-HOArCCHAr}' \atop \text{CH=O} \right] \longrightarrow$$

20–40 % overall
Isoflavone

(c) Preparation of 3,5-dimethylpyrrole-2-carboxaldehyde (92 % from the pyrrole) [6].

2. From Arenes, Carbon Monoxide, and Hydrogen Chloride (Gattermann-Koch)

$$\text{C}_6\text{H}_6 \xrightarrow[\text{AlCl}_3]{\text{CO + HCl}} \text{C}_6\text{H}_5\text{CHO}$$

This synthesis has been reviewed [7]. Mostly of industrial interest, it has been applied to the introduction of the formyl group into benzene and its homologs. An alkyl group directs the substituent almost exclusively to the *para* position. Side reactions, such as alkylation, dealkylation, and migration of alkyl groups, sometimes occur with benzene homologs. The usual catalyst employed is aluminum chloride which at atmospheric pressure is mixed with a carrier such as cuprous chloride. (Only a mixture of cuprous chloride and aluminum chloride forms a complex with carbon monoxide at atmospheric pressure.) As a rule, 1 mole of catalyst is used with each mole of hydrocarbon. Benzene is added to homologs to be formylated to prevent the formation of dialkyl benzenes. Ordinarily, atmospheric pressure and temperatures from 35 to 40° have been employed. Chlorosulfonic acid added to formic acid is a convenient source of carbon monoxide and hydrogen chloride. Yields vary greatly, the maximum of 90 % being possible in the formylation of benzene.

(a) Preparation of *p*-tolualdehyde (46–51 % from toluene, carbon monoxide, hydrogen chloride, AlCl$_3$, and Cu$_2$Cl$_2$) [8].

(b) Preparation of *p*-phenylbenzaldehyde (73 % from biphenyl in benzene and carbon monoxide, hydrogen chloride, AlCl$_3$, and Cu$_2$Cl$_2$ at 35–40°) [9].

3. From Arenes and Formyl Fluoride

$$\text{ArH} + \text{FCHO} \xrightarrow{\text{BF}_3} \text{ArCHO} + \text{HF}$$

This synthesis is similar to the Gattermann-Koch, but in this case the formyl halide is isolated before being used. Formyl fluoride formylates aromatic hydrocarbons (and, incidentally, alcohols, phenols, carboxylic acid salts, thiols, and primary and secondary amines) [10]. The reagent may be prepared from formic acid and potassium hydrogen fluoride or from acetic formic anhydride and anhydrous hydrogen fluoride. Boron fluoride is the catalyst of choice, and yields with aromatic hydrocarbons vary from 56 to 78%.

(a) **Preparation of mesitaldehyde.**   Mesitylene, 0.5 mole, in CS_2 was treated at 0–10° with a slow stream of 1:1 formyl fluoride and boron trifluoride with stirring. After 3 hr., when 0.5 mole of formyl fluoride had been absorbed, the reaction was stopped and stirring was continued for another 0.5 hr. Washing with cold water, drying, and fractionating gave 70% of mesitaldehyde [10].

4. From Arenes and Dichloromethyl Alkyl Ethers

$$ArH + Cl_2CHOCH_3 \xrightarrow{AlCl_3} ArCHO + CH_3Cl + HCl$$

This synthesis follows the pattern of the usual Friedel-Crafts reaction in which the reaction mixture is hydrolyzed in water or with alkali. Yields for a series of about sixteen aldehydes, mostly aromatic, vary from 37 to 92% [11]. Dichloromethyl methyl ether is made conveniently from methyl formate and phosphorus pentachloride in about 60% yield. Chloromethylene dibenzoate also has been used as a formylating agent [12].

(a) **Preparation of mesitaldehyde.**   Mesitylene, about 0.12 mole, was dissolved in methylene chloride; the solution was cooled to 0° and treated with 0.2 mole of titanium tetrachloride. With stirring there was added 0.1 mole of dichloromethyl methyl ether dropwise rapidly. The reaction begins usually immediately with vigorous hydrogen chloride evolution, and a dark, oily, or solid product separates. Stirring was continued without cooling until the hydrogen chloride evolution let up noticeably (about 5–15 min.) and then the mixture was poured on ice and the organic layer was washed with water, then with aqueous bicarbonate and once again with water. On evaporation of the solvent, the residue was steam-distilled and the aldehyde was extracted from the distillate with ether, after which the aldehyde, b.p. 115–116° (12 mm.) in 85% yield was recovered from the dried ethereal solution by distillation [11].

(b) **Preparation of thiophenecarboxaldehyde** (90% from thiophene) [11].

5. From Phenols and Triethyl Orthoformate Followed by Hydrolysis (Formylation)

Phenols may be converted into acetals with ethyl orthoformate in the presence of aluminum chloride. The acetals on hydrolysis under acidic conditions give aldehydes [13]. The method is simple and requires no external heating. Best yields (89% or better) were obtained with trihydric phenols.

(a) Preparation of 1-hydroxy-2-naphthaldehyde. α-Naphthol, 10 mmoles, in 20 ml. of benzene and 10 ml. of ethyl orthoformate were treated with stirring with 2 g. of powdered $AlCl_3$. Stirring was continued for 10 min., after which the mixture was cooled and treated with 30 ml. of cold 5% HCl. The ethereal extract was then washed with aqueous Na_2CO_3, concentrated to 50 ml., and steam-distilled. Extraction of the distillate residue with ether gave the aldehyde (97%) [13].

6. From Arenes or Olefins and Formamides (Vilsmeier)

$$ArH \xrightarrow[\substack{\text{or} \\ C_6H_5N(CH_3)CHO}]{HCON(CH_3)_2} ArCHO$$

This synthesis has been applied to aromatic and nitrogen- and sulfur-containing ring systems, sometimes in the presence of hydroxy, alkoxy, dimethylamino, or acetal groups. It succeeds with some acetals to form malonic aldehydes (Ex. b.4), ketones to form chlorovinyl aldehydes (b.6 and b.8), and some olefins to form unsaturated aldehydes (b.5). It apparently fails with naphthalene and the simple benzenoid hydrocarbons. A comprehensive bibliography of the reaction is available [14]. Phosphorus oxychloride is the usual condensing agent and, if the mixture is not homogeneous, solvents such as ethylene dichloride or o-dichlorobenzene are employed. Yields are good to excellent. The most noteworthy preparation is that of pyrrolecarboxaldehyde. Pyrrole is very susceptible to polymerization in acidic media; yet the aldehyde forms and is thereafter stable.

(a) Preparation of indole-3-carboxaldehyde (97% from indole) [15].

(b) Other examples

(1) 9-Anthraldehyde (74–84% from anthracene) [16].

(2) 2-Pyrrolecarboxaldehyde (78–79% from pyrrole) [17]. When applied to pyrrole the synthesis is called the Vilsmeier-Haack reaction [18].

(3) p-Dimethylaminobenzaldehyde (80–84% from dimethylaniline) [19].

(4) Methylmalonic aldehyde sodium salt [20].

$$CH_3CH_2CH(OC_2H_5)_2 + HCON(CH_3)_2 + COCl_2 \xrightarrow{ClCH_2CH_2Cl}$$

$$\underset{\substack{\text{α-Methyl-β-dimethyl-} \\ \text{aminoacrolein}}}{(CH_3)_2N\overset{H}{C}{=}\overset{\overset{\displaystyle CH_3}{|}}{C}CHO} \xrightarrow[70°]{\text{aq. NaOH}} \underset{90.5\%}{[CH_3\overset{\ominus}{C}(CHO)_2]\overset{\oplus}{N}a}$$

(5) 2,4-Dimethoxycinnamaldehyde (the anilinoaldehyde was made by addition of N-methylaniline to propargylic aldehyde) [21].

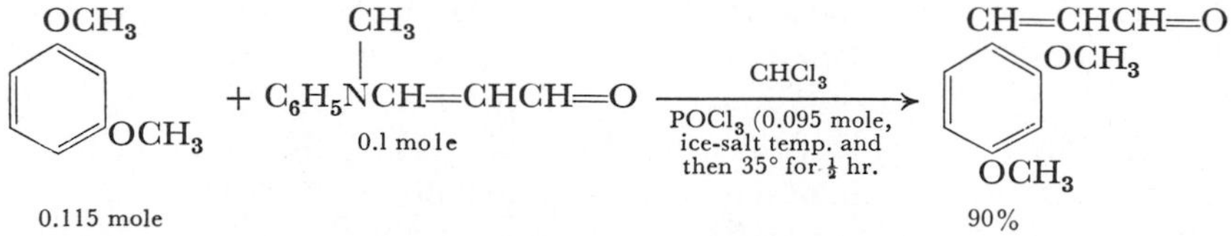

(6) 2-Chlorocyclopentenecarboxaldehyde (54% from cyclopentanone, dimethylformamide, and phosphorus oxychloride) [22].

(7) p-Methoxycinnamaldehyde (68% from p-methoxystyrene, dimethylformamide, and phosphorus oxychloride at moderate temperature) [23].

(8) β-p-Chlorophenyl-β-chloroacrolein (30% from 400 mmoles of

$$\underset{\substack{\\}}{p\text{-ClC}_6\text{H}_4\overset{\overset{\text{Cl}}{|}}{\text{C}}=\text{CHCHO}}$$

DMF and 100 ml. of trichloroethylene to which 300 mmoles of $POCl_3$ and then 200 mmoles of p-chloroacetophenone in 100 ml. of trichloroethylene were added dropwise and stirred for 3 hr.) [24].

7. From Tertiary Anilines or Phenols and Formaldehyde Derivatives Followed by Oxidation

$$\text{C}_6\text{H}_5\text{N(CH}_3)_2 + 2\text{ H}_2\text{CO} + p\text{-(CH}_3)_2\text{NC}_6\text{H}_4\text{NO} \xrightarrow{\text{HCl}}$$

$$p\text{-(CH}_3)_2\text{NC}_6\text{H}_4\text{N}=\text{CHC}_6\text{H}_4\text{N(CH}_3)_2$$
10-XIII

$$\text{10-XIII} + \text{H}_2\text{CO} \xrightarrow{\text{CH}_3\text{CO}_2\text{H}} p\text{-(CH}_3)_2\text{NC}_6\text{H}_4\text{CH}=\text{O}$$
p-Dimethylaminobenzaldehyde

Another modification, known as the Duff reaction, utilizes hexamethylenetetramine as both the aminomethylating and oxidizing agent (see Exs. b.1 and b.2):

$$\text{C}_6\text{H}_5\text{OH} \xrightarrow{\text{(CH}_2)_6\text{N}_4} o\text{-(HOC}_6\text{H}_4\text{CH}_2)_2\text{NH} \xrightarrow{\overset{\oplus}{\text{CH}_2=\text{NH}_2}}$$

$$o\text{-HOC}_6\text{H}_4\text{CH}=\text{NCH}_2\text{C}_6\text{H}_4\text{OH-}o + \text{CH}_3\text{NH}_2$$

The aldehydes from phenols are predominantly *ortho*-substituted and those from N,N-dialkylanilines *para*-substituted. The yields in the Duff reaction leave something to be desired, about 15–20% for phenols [25] and 35–45% for anilines [26], but are compensated in part by the simplicity of operation. The reaction is unsuccessful with nitro- or thiophenols and with 2-hydroxypyridine.

(a) Preparation of p-dimethylaminobenzaldehyde (59% from dimethylaniline, formalin, and conc. HCl heated for 10 min. and treated with p-nitrosodimethylaniline to form 10-XIII; the latter was treated with formaldehyde in 50% acetic acid to obtain the aldehyde) [27].

(b) Other examples

(1) **Eugenol-5-carboxaldehyde** (eugenol, 10 ml., and 40 g. of hexamethylenetetramine were dissolved in 75 ml. of acetic acid and held at 100° for 6 hr. The darkened, hot solution was then treated with 50 ml. of conc. hydrochloric acid in water, cooled, and extracted with ether. A small amount of 20% sodium hydroxide was used to neutralize and extract the acetic acid, and a larger amount to precipitate the bright yellow sodium salt of the product. The salt was dissolved in water and 3 g. of eugenol-5-carboxaldehyde precipitated with acid, another gram being obtained from the mother liquor by extraction [28].

(2) **p-Dimethylaminobenzaldehyde** (38% from dimethylaniline and hexamine in ethanol heated in acetic-formic acids [26].

8. From Ethylenic Compounds by Hydroformylation (Oxo Process)

$$RCH{=}CH_2 + CO + H_2 \xrightarrow[\text{catalyst}]{\text{Co}} RCH_2CH_2CHO$$

This synthesis has been discussed under Alcohols, Chapter 4, B.5, and reviewed [29]. It permits a variety of unsaturated compounds to be converted into saturated aldehydes containing an additional carbon atom [30, 31]. Recently it has been found that yields are increased by the addition of benzonitrile, which apparently stabilizes the acyl cobalt complex [32]. Yields in the reactions average about 50%.

Hydroformylation has been accomplished recently at atmospheric pressure by capitalizing on the observation that lithium trimethoxyaluminum hydride in combination with trialkylboranes fixes carbon monoxide [33]:

$$R_3B + CO \rightleftarrows R_3\overset{\ominus}{B}\overset{\oplus}{C}O \rightleftarrows R_2\overset{\overset{\text{O}}{\|}}{B}CR \xrightarrow{\text{LiHAl(OCH}_3)_3} \underset{\underset{\text{OLiAl(OCH}_3)_3}{|}}{R_2BCHR}$$

Oxidation with hydrogen peroxide then gives aldehydes (Ex. b).

(a) Preparation of γ-acetoxybutyraldehyde. Allyl acetate, 50 g.,

$$CH_2{=}CHCH_2OAc + CO + H_2 \rightarrow O{=}\overset{\overset{\text{H}}{|}}{C}CH_2CH_2CH_2OAc$$

was placed in 40 ml. of the ether solution containing 2.2 g. of dicobalt octacarbonyl (for preparation see [34]) and 40 ml. of ether in a steel bomb. Carbon monoxide, 3200 p.s.i., and hydrogen, 1600 p.s.i., were added and the mixture was shaken and heated at 115° but less than 125° until no more pressure drop occurred. After removing the ether, the crude product 46 g. (69%) which distilled at 60–90° (10 mm.), was 92% pure. On fractionation it boiled at 59–60° (1 mm.) and was 94% pure [30].

(b) **Preparation of heptaldehyde.** Boron hydride, 52 mmoles calcd. as BH_3, in THF was added to 150 mmoles of 1-hexene and stirred for 0.5 hr. Lithium trimethoxyaluminum hydride (55 mmoles) was added and then carbon monoxide until absorption ceased (30 min.). The buffer, 2.7 M $NaHPO_4$-Na_2HPO_4 was added to minimize hydrolysis before the addition of 18 ml. of 30% H_2O_2, a process which held the temperature at 25°. The yield of aldehyde, analyzed by conversion into the alcohol, was 98% disregarding a 3% impurity [33].

9. From Vinyl Ethers and Acetals (or *ortho*-Esters)

$$CH_2=CHOR + ArCH=CHCH(OC_2H_5)_2 \xrightarrow[\text{2. } CH_3COONa + CH_3COOH]{\text{1. } ZnCl_2 + CH_3COOH}$$

$$Ar(CH=CH)_2CHO$$

The mechanism of this synthesis, if one starts with an orthoformate, is probably as follows:

$$HC(OC_2H_5)_3 + ZnCl_2(BF)_3 \longrightarrow \left[H\overset{\oplus}{C}(OC_2H_5)_2 \right] \xrightarrow{CH_2=CHOC_2H_5}$$

$$\left[(C_2H_5O)_2CHCH_2\overset{\oplus}{C}HOC_2H_5 \right] \xrightarrow{C_2H_5OH} (C_2H_5O)_2CHCH_2CH(OC_2H_5)_2$$

Malonaldehyde tetraethyl acetal

The diacetal is readily hydrolyzed to the dialdehyde, preferably in sodium acetate and acetic acid [35].

Similar results have been obtained from aldehydes and acetals by a free radical reaction [36]:

$$CH_2=CHCH(OC_2H_5)_2 + CH_3CHO \xrightarrow[80°]{(C_6H_5CO)_2O_2}$$

$$CH_3COCH_2CH_2CH(OC_2H_5)_2 \xrightarrow[H_2O]{H^{\oplus}} CH_3COCH_2CH_2CHO$$

β-Acetylpropionaldehyde,
53% overall yield

(a) **Preparation of 5-phenyl-2,4-pentadienal.** A mixture of 41.2 g of

$$C_6H_5CH=CHCH(OC_2H_5)_2 + CH_2=CHOC_2H_5$$

$$\downarrow \quad OC_2H_5$$

$$C_6H_5(CH=CH)_2CHO \longleftarrow C_6H_5CH=CHCHCH_2CH(OC_2H_5)_2$$

the diethyl acetal of cinnamaldehyde and 2 ml. of a 10% solution of zinc chloride in acetic acid was heated to 50°, after which 14.8 g. of vinyl ether was added at a rate at which the temperature did not exceed this value. The mixture was held at this temperature for 1 hr., cooled to room temperature, and shaken with 40 ml. of a 10% solution of sodium hydroxide and 40 ml. of diethyl ether. Distillation of the ether layer gave 48.5 g. (87%) of 5-phenyl-1,1,3-triethoxypentene-4, b.p. 157–159° (3 mm.).

The triethoxypentene, 25 g., mixed with 38 ml. of glacial acetic acid and 3.8 g. of sodium acetate, was boiled for 3 hr. in a stream of nitrogen. Cooling to room temperature was followed by the addition of 75 ml. of water and 40 g. of sodium carbonate, after which the aldehyde was extracted with ether. Distillation and fractionation of the residue in vacuum gave 10.8 g. (76%) of the aldehyde, b.p. 134–135° (2 mm.), m.p. 37–38° [35].

10. From Vinyl Esters via the Chloromercurialdehyde or via an Alkylpalladium Adduct

$$\underset{RCH=\overset{\overset{\displaystyle H}{|}}{C}OCOCH_3}{} \xrightarrow[KCl]{Hg(OCOCH_3)_2} \underset{ClHg\overset{\overset{\displaystyle R}{|}}{C}H\overset{\overset{\displaystyle H}{|}}{C}=O}{} \xrightarrow{(C_6H_5)_3CCl} \underset{(C_6H_5)_3C\overset{\overset{\displaystyle R}{|}}{C}H\overset{\overset{\displaystyle H}{|}}{C}=O}{}$$

This synthesis is applicable to the preparation of aldehydes or ketones (see Ketones, Chapter 11, C.8). The enol acetates with mercuric acetate and potassium chloride are readily converted into chloromercurialdehydes or ketones which with highly substituted methyl chlorides give β-substituted aldehydes or ketones [37]. The yields are fair, and the reaction is limited in scope because of its restriction to utilization of substituted methyl halides which form carbonium ions readily.

Vinyl acetates also react with arylpalladium chloride to form aldehydes in part [38]:

$$ArPdCl + CH_3CH=CHOCOCH_3 \longrightarrow (CH_3\overset{\overset{\displaystyle Ar}{|}}{C}HCHOCOCH_3) \longrightarrow$$
$$\underset{PdCl}{|}$$

$$CH_3\overset{\overset{\displaystyle Ar}{|}}{C}=CHOCOCH_3 + HPdCl \longrightarrow CH_3\overset{\overset{\displaystyle Ar}{|}}{C}HCH=O$$

Ar = p-anisyl, 21%

The reaction is not of preparative value because stilbenes and enol acetates accompany the aldehyde. A similar reaction produces aldehydes from allylic alcohols [39]:

$$C_6H_5PdCl + CH_2=CHCH_2OH \longrightarrow (C_6H_5CH_2\overset{\overset{\displaystyle PdCl}{|}}{C}HCH_2OH) \xrightarrow{-HPdCl}$$
$$C_6H_5CH_2CH_2CH=O$$

β-Phenylpropionaldehyde, 35%

(a) Preparation of β,β,β-triphenylpropionaldehyde

(1) α-Chloromercuripropionaldehyde. To 250 ml. of water was added 96 g. (0.3 mole) of mercuric acetate, and the mixture was shaken to promote solution. To this mixture was added 30 g. (0.3 mole) of 1-propenyl acetate and the whole was chilled, after which 22.8 g. (0.3 mole) of potassium chloride in 200 ml. of cold water was added until a trace of oil appeared. This was removed and triturated until crystalline and the crystals were used for later seeding. The product was dried for several days under vacuum at room temperature. The aldehyde recovered, m.p. 60–70° (dec.), was 63 g. (71%) based on the 1-propenyl acetate.

(2) β,β,β-Triphenylpropionaldehyde. Under anhydrous conditions, 0.3 mole of trityl chloride was dissolved in 600 ml. of dry benzene. To this solution, with stirring, was added 0.3 mole of the chloromercurialdehyde. After stirring overnight at room temperature, the mixture was refluxed for 2 hr. and cooled. The insoluble mercuric salts were removed by filtration and the filtrate was washed repeatedly with 10 % sodium carbonate solution until all the mercuric salts, which had been precipitated by the washing, were removed. After another washing with water and filtration, the benzene was distilled to leave a residue which, when crystallized from hexane and then methanol, weighed 60 g. (65 %). m.p. 99.5–102° [40].

11. From Diazonium Salts and Formaldoxime

$$ArN_2X + CH_2{=}NOH \longrightarrow ArCH{=}NOH \xrightarrow{H_2O} ArCHO$$

This reaction, for purposes of classification, may be looked on as an attack of an aryl carbonium ion-like particle in a copper complex on formaldoxime:

$$ArN_2^{\oplus} \longrightarrow [Ar]^{\oplus} \xrightarrow{CH_2{=}NOH} [ArCH_2{-}NOH]^{\oplus}$$

$$\downarrow{-H^{\oplus}}$$

$$ArCHO \xleftarrow[H^{\oplus}]{H_2O} ArCH{=}NOH$$

The true mechanism, however, is unknown and may be of an SN_2 or free radical type. With oximes of aldehydes other than formaldehyde, ketones are formed (see Ketones, Chapter 11, C.9).

The synthesis of the oxime from the diazonium salt is most satisfactory in an acid medium of pH 5.5–6.0 (sodium acetate as a buffer) in the presence of cupric sulfate-sodium sulfite mixture as a catalyst [41]. The aldehyde is obtained from the oxime by acid hydrolysis or by treatment with aqueous ferric ammonium sulfate. Yields rarely exceed 40–50 %, although 2-nitroanisaldehyde was obtained in a 63 % yield [42]; see Ex. b). The method has been employed to prepare a variety of substituted benzaldehydes [43], but no aldehyde was obtained when *o*-cyano- or *o*-ethoxycarbonyl groups were present in the diazonium salt.

(a) Preparation of 2-bromo-4-methylbenzaldehyde (35–45 % from diazotized 2-bromo-4-methylaniline and hydrochloric acid at −5 to +5° which was treated with an aqueous mixture of paraformaldehyde, hydroxylamine hydrochloride, and hydrated sodium acetate to which hydrated cupric sulfate and sodium sulfite had been added) [43].

(b) Preparation of 2-nitroanisaldehyde (63 % from 2-nitroanisidine and formaldoxime) [42].

1. G. A. Olah and S. J. Kuhn, *Friedel-Crafts and Related Reactions*, Vol. 3, John Wiley and Sons, New York, 1964, Pt. 2, Chap. 38.
2. W. E. Truce, *Org. Reactions*, **9,** 37 (1957).

3. R. Adams and I. Levine, *J. Am. Chem. Soc.*, **45**, 2373 (1923); R. Adams and E. Montgomery, *ibid.*, **46**, 1518 (1924).

4. R. C. Fuson *et al.*, *Org. Syn.*, Coll. Vol. **3**, 549 (1955).

5. L. Farkas *et al.*, *Chem. Ber.*, **91**, 2858 (1958).

6. H. Fischer and W. Zerweck, *Org. Reactions*, **9**, 57 (1957).

7. N. N. Crounse, *Org. Reactions*, **5**, 290 (1949).

8. G. H. Coleman and D. Craig, *Org. Syn.*, Coll. Vol. **2**, 583 (1943).

9. D. H. Hey, *Org. Reactions*, **5**, 298 (1949).

10. G. A. Olah and S. J. Kuhn, *J. Am. Chem. Soc.*, **82**, 2380 (1960).

11. A. Rieche *et al.*, *Chem. Ber.*, **93**, 88 (1960).

12. Ref. 1, p. 1189.

13. H. Gross *et al.*, *Chem. Ber.*, **96**, 308 (1963).

14. Ref. 1, p. 1211.

15. P. N. James and H. R. Snyder, *Org. Syn.*, Coll. Vol. **4**, 539 (1963).

16. L. F. Fieser *et al.*, *Org. Syn.*, Coll. Vol. **3**, 98 (1955).

17. R. M. Silverstein *et al.*, *Org. Syn.*, Coll. Vol. **4**, 831 (1963).

18. A. Ermili *et al.*, *J. Org. Chem.*, **30**, 339 (1965).

19. E. Campaigne and W. L. Archer, *Org. Syn.*, Coll. Vol. **4**, 331 (1963).

20. Z. Arnold and F. Sorm, *Collection Czech. Chem. Commun.*, **23**, 452 (1958).

21. C. Jutz, *Chem. Ber.*, **91**, 850 (1958).

22. W. R. Benson and A. E. Pohland, *J. Org. Chem.*, **30**, 1126 (1965).

23. C. J. Schmidle and P. G. Barnett, *J. Am. Chem. Soc.*, **78**, 3209 (1956).

24. M. Weissenfels *et al.*, *Z. Chem.*, **6**, 471 (1966); *C.A.*, **66**, 5200 (1967).

25. J. C. Duff, *J. Chem. Soc.*, 547 (1941).

26. J. C. Duff, *J. Chem. Soc.*, 276 (1945).

27. R. Adams and G. H. Coleman, *Org. Syn.*, Coll. Vol. **1**, 214 (1941).

28. K. V. Rao *et al.*, *Proc. Indian Acad. Sci.*, *Sect. A*, **30**, 114 (1949).

29. C. W. Bird, *Chem. Rev.*, **62**, 283 (1962).

30. H. Adkins and G. Krsek, *J. Am. Chem. Soc.*, **70**, 383 (1948); **71**, 3051 (1949).

31. L. A. Wetzel *et al.*, *J. Am. Chem. Soc.*, **72**, 4939 (1950).

32. L. Roos and M. Orchin, *J. Org. Chem.*, **31**, 3015 (1966).

33. H. C. Brown *et al.*, *J. Am. Chem. Soc.*, **90**, 499 (1968).

34. I. Wender *et al.*, *Inorg. Syn.*, **5**, 190 (1957).

35. B. M. Mikhailov and G. S. Ter-Sarkisyan, *J. Gen. Chem. USSR* (*Eng. Transl.*), **29**, 2524 (1959).

36. A. Mondon, *Angew. Chem.*, **64**, 224 (1952).

37. A. N. Nesmeyanov *et al.*, *Izv. Akad. Nauk, SSSR, Otd. Khim. Nauk*, 601 (1949); *C.A.*, **44**, 7225 (1950).

38. R. F. Heck, *J. Am. Chem. Soc.*, **90**, 5535 (1968).

39. R. F. Heck, *J. Am. Chem. Soc.*, **90**, 5526 (1968).

40. D. Y. Curtin and M. J. Hurwitz, *J. Am. Chem. Soc.*, **74**, 5381 (1952).

41. W. F. Beech, *J. Chem. Soc.*, 1297 (1954).

42. R. B. Woodward *et al.*, *Tetrahedron*, **2**, 1 (1958).

43. S. D. Jolad and S. Rajagopal, *Org. Syn.*, **46**, 13 (1966).

D. Hydrolysis or Hydration

The obvious hydrolytic methods have been included in more appropriate places throughout this chapter and in the Ketone chapter (Chapter 11, D.5). One hydrolytic method particularly suitable for aldehyde derivatives is given in this section (D.1). There remains a series of compounds in which the hydrolytic step assumes a more important role because the starting compounds are not produced

References for Section D are on pp. 599–600.

from aldehydes. Outstanding among these compounds are 1,3-dithianes and dihydro-1,3-oxazines, for they may be alkylated or alkylated and reduced to give a variety of aldehydes on hydrolysis (D.3):

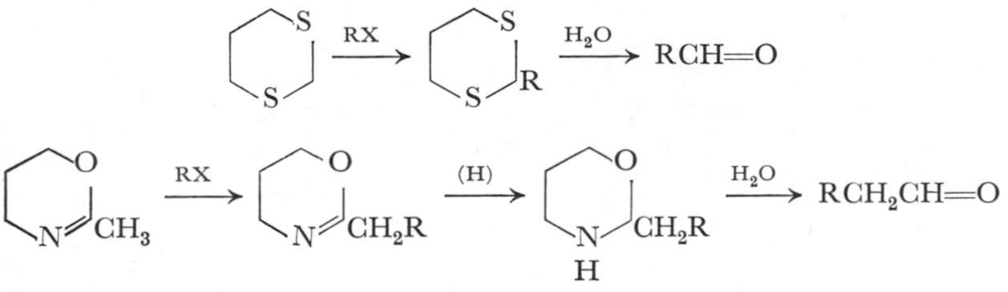

Many other heterocyclic compounds may be hydrolyzed to specific aldehydes (D.2 and D.3) and, among open-chain compounds, vinyl ethers, divinyl ethers, their nitrogen and sulfur analogs, and *gem*-disubstituted compounds (D.4 and D.5) are potential sources of aldehydes. (Unless otherwise referenced, examples are taken from [1].) Hydration of acetylene is described in D.6.

1. From Aldehyde 2,4-Dinitrophenylhydrazones

Although the compounds above can be hydrolyzed under acidic conditions (see Ketones, Chapter 11, D.5), the free aldehydes may undergo side reactions precluding their isolation. A simple, mildly alkaline hydrolysis (Ex. a) has been devised which permits isolation of a larger number of aldehydes, albeit in yields around 60%. Here acid hydrolysis cannot be used without seriously affecting the aldehyde.

(a) **Preparation of citronellal** (60% from 3 g. of citronellal 2,4-dinitrophenylhydrazone and 7.5 g. of $KHCO_3$ in 75 ml. each of water and ethylene glycol refluxed to give a product, which after two steam distillations, is of 98% purity) [2].

2. From Five-membered Heterocyclic Rings

Furan, pyrrole, and thiophene are potential sources of succinaldehyde by means of hydrolytic splitting. An alkaline cleavage of pyrrole to succinaldehyde dioxime is shown in Ex. a. Bromine can be added to furan, the product of which may be converted into 2,5-dimethoxydihydrofuran. This compound by hydrolysis is a source of maleic dialdehyde as shown in Ex. b. Other manipulations of the furan nuclei lead to many different aldehydes of unusual structure. Isoxazole is a potential source of cyanoacetaldehydes, while oxazole derivatives can be converted into aminoacetaldehydes. Again, many different substituents can be attached to yield substituted derivatives of the aldehydes above. The list is by no means exhausted, but enough variety is given in the examples to visualize the potentialities. It must be added that most of the aldehydes of this group are highly unstable and must be isolated as derivatives or used immediately in some ensuing process.

(a) Succinic dialdehyde oxime [3].

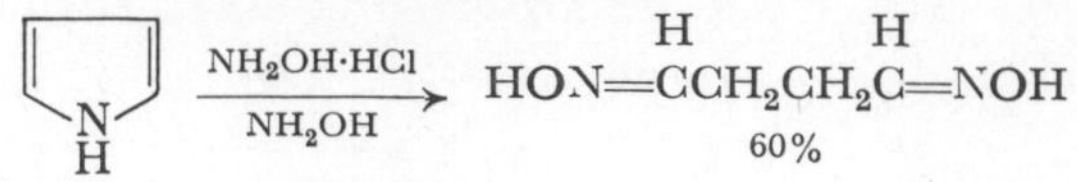

$$\underset{\underset{H}{N}}{\boxed{}} \xrightarrow[\text{NH}_2\text{OH}]{\text{NH}_2\text{OH·HCl}} \underset{60\%}{\text{HON}=\overset{H}{C}\text{CH}_2\text{CH}_2\overset{H}{C}=\text{NOH}}$$

(b) Maleic dialdehyde [4].

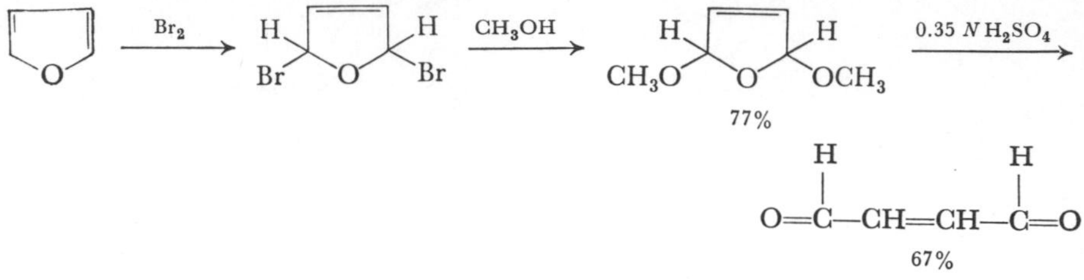

$$\underset{77\%}{} \qquad \underset{67\%}{O=\overset{H}{\underset{|}{C}}-CH=CH-\overset{H}{\underset{|}{C}}=O}$$

(c) Mucobromic acid [5].

$$O=\overset{H}{C}CBr=CBrCO_2H + CO$$

64–67%, a stable aldehyde

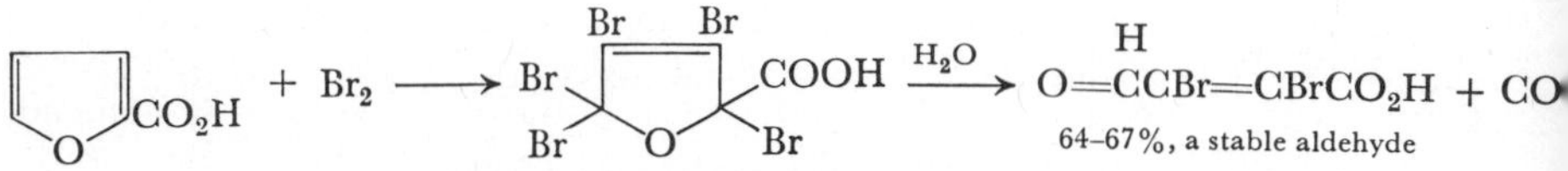

(d) Phenylacetaminomalondialdehyde:

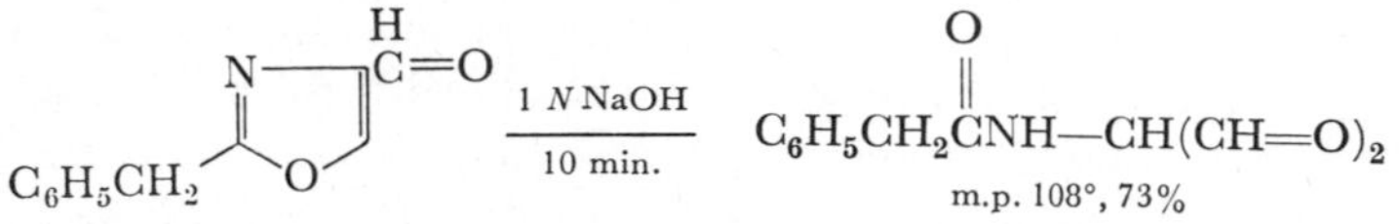

$$C_6H_5CH_2\overset{O}{\overset{\|}{C}}NH-CH(CH=O)_2$$

m.p. 108°, 73%

Oxazole itself is quite resistant to the attack of alkali [6].

(e) Diethylacetaldehyde [7].

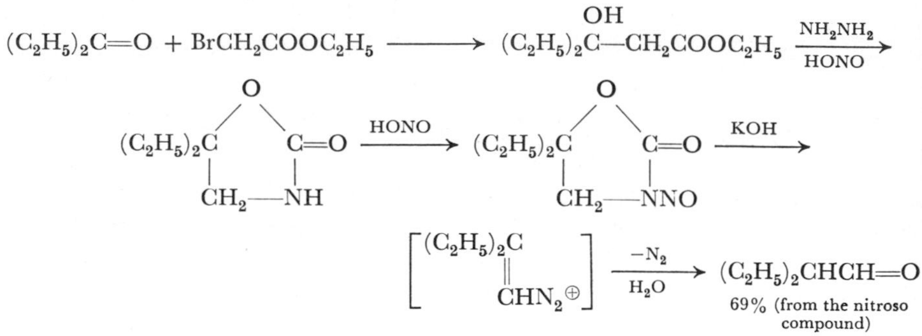

$$(C_2H_5)_2C=O + BrCH_2COOC_2H_5 \longrightarrow (C_2H_5)_2\overset{OH}{\underset{|}{C}}-CH_2COOC_2H_5 \xrightarrow[\text{HONO}]{\text{NH}_2\text{NH}_2}$$

$$(C_2H_5)_2\overset{O}{\underset{\underset{CH_2-NH}{|}}{C}}\diagup\diagdown\overset{}{C}=O \xrightarrow{\text{HONO}} (C_2H_5)_2\overset{O}{\underset{\underset{CH_2-NNO}{|}}{C}}\diagup\diagdown\overset{}{C}=O \xrightarrow{\text{KOH}}$$

$$\left[(C_2H_5)_2\overset{}{\underset{\underset{CHN_2}{\|}}{C}}\oplus\right] \xrightarrow[\text{H}_2\text{O}]{-\text{N}_2} (C_2H_5)_2CHCH=O$$

69% (from the nitroso compound)

3. From Six-membered Heterocyclic Rings

Two important general heterocyclic sources of aldehydes are 1,3-dithiane and dihydro-1,3-oxazines. The generality of synthesis arises from the fact that both these compounds may be converted into anions and the anions treated in the usual manner with alkyl halides, polymethylene halides, carbonyl compounds, epoxides or others to produce the aldehyde desired after hydrolysis. 2-Alkyl-1,3-dithianes have been used for making ketones (see Ketones, Chapter

11, D.1), but 1,3-dithiane itself is a source of aldehydes [8]:

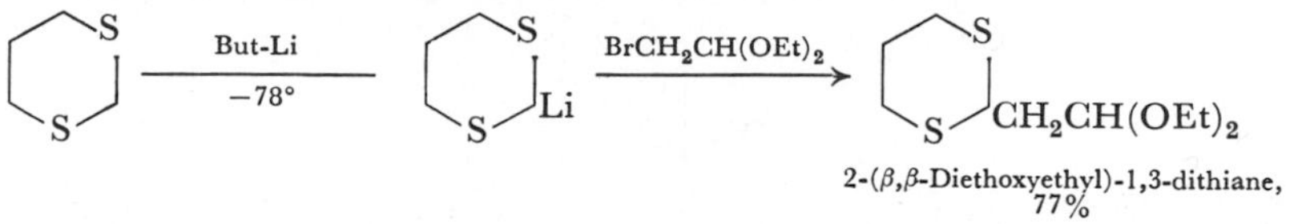
But-Li
−78°
Li
BrCH2CH(OEt)2
CH2CH(OEt)2
2-(β,β-Diethoxyethyl)-1,3-dithiane, 77%

The product is a potential source of a malonic aldehyde by mild hydrolysis, although the aldehyde cannot be isolated as such.

On the other hand, the dihydro-1,3-oxazines have been studied in detail in regard to the synthesis of aldehydes [9]:

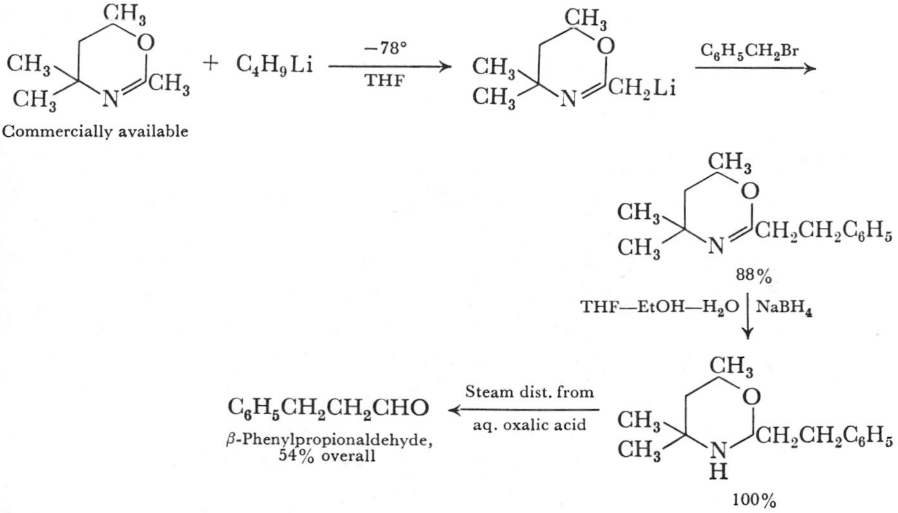
Commercially available
+ C4H9Li
−78°
THF
CH2Li
C6H5CH2Br
CH2CH2C6H5
88%
THF—EtOH—H2O
NaBH4
100%
Steam dist. from
aq. oxalic acid
C6H5CH2CH2CHO
β-Phenylpropionaldehyde, 54% overall

Cyclopropanecarboxaldehyde was made in 69% yield from the anion above and 1-bromo-2-chloroethane, while many unsaturated aldehydes were produced from the anion to which a carbonyl compound had been added (see G for similar use of a dihydroquinazoline).

Other six-membered heterocyclic compounds are sources of specific aldehydes.

2,3-Dihydropyran, readily available from tetrahydrofurfuryl alcohol, is a source of 5-hydroxypentanal, as shown in Ex. a. Manipulation of the dihydropyran structure can lead to substituted hydroxypentanals. Furthermore, the dihydropyran structure can be formed by the Diels-Alder reaction between unsaturated carbonyl compounds and vinyl alkyl ethers and, by subsequent hydrolysis, glutaraldehydes or ketoaldehydes are formed:

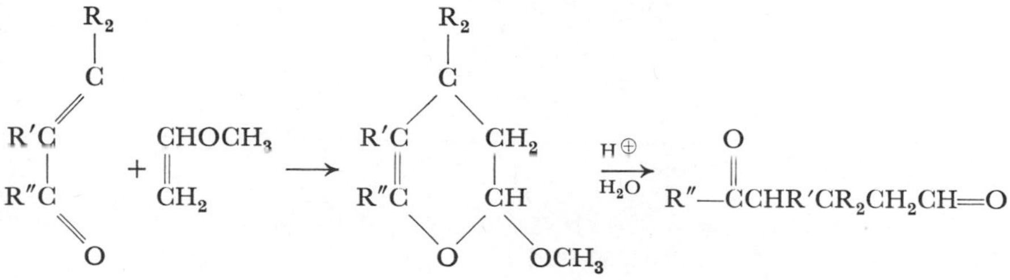
CHOCH3
CH2
H⊕
H2O
R″—CCHR′CR2CH2CH=O

Pyridine compounds are potential sources of glutacondialdehyde compounds:

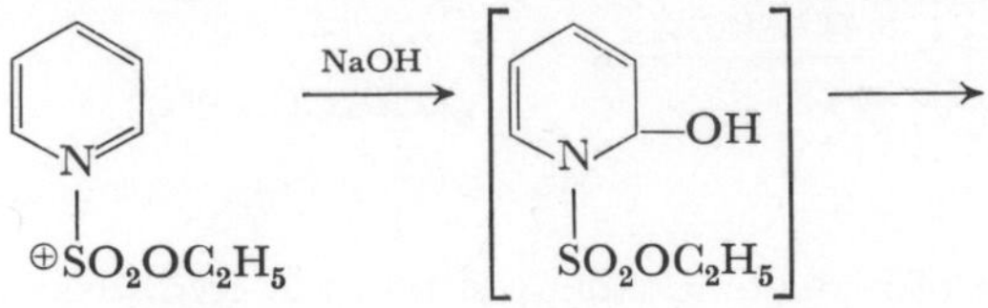

$$NaOCH\!=\!CHCH\!=\!CHCH\!=\!O \xrightarrow{\text{H}\oplus} [O\!=\!CHCH_2CH\!=\!CHCH\!=\!O] \longrightarrow$$

10-XIV

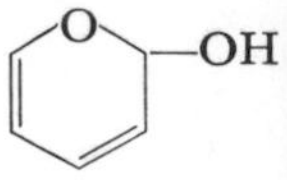

10-XV
2-Hydroxy-
2 H-pyran

Compound 10-XIV is not very stable and cyclizes to 10-XV spontaneously. Quaternization of the nitrogen frequently is carried out with 2,4-dinitrochlorobenzene, in which case the 2,4-dinitroanil of glutacondialdehyde is isolated.

A similar series of reactions applies to the quinoline and isoquinoline family, although considerable work remains to be done to clarify the course of the reactions. Evidently, with the dihydrohydroxyquinolines, the cyclic form exists in equilibrium with the aminocinnamaldehydes (see Ex. c.4):

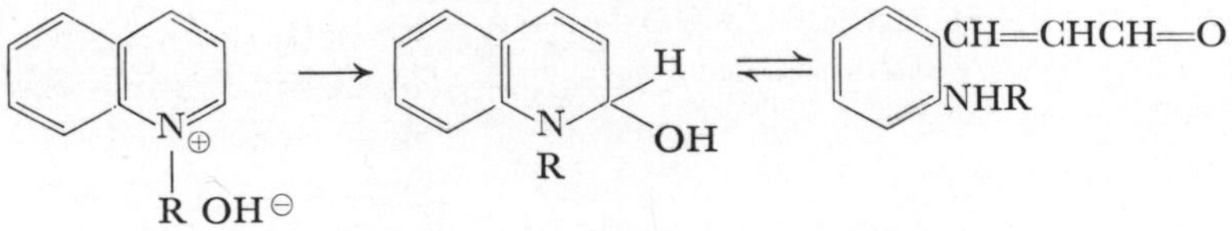

(a) **Preparation of 5-hydroxypentanal** (74–79% from 2,3-dihydropyran) [10].

(b) **Preparation of 5-keto-3-phenylhexanal** [11]:

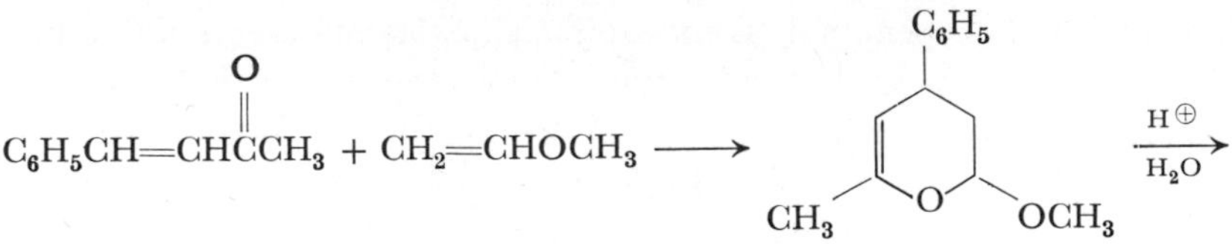

$$CH_3\overset{\overset{\displaystyle O}{\|}}{C}CH_2CH(C_6H_5)CH_2CH\!=\!O$$

85%

(c) **Other examples**

(1) **O-Ethylmalic dialdehyde** (86% as the *p*-nitrophenylhydrazone from 2,3,5-triethoxytetrahydrofuran) [12].

(2) 5-Hydroxypent-2-enal (55 % from 2-ethoxy-5,6-dihydropyran):

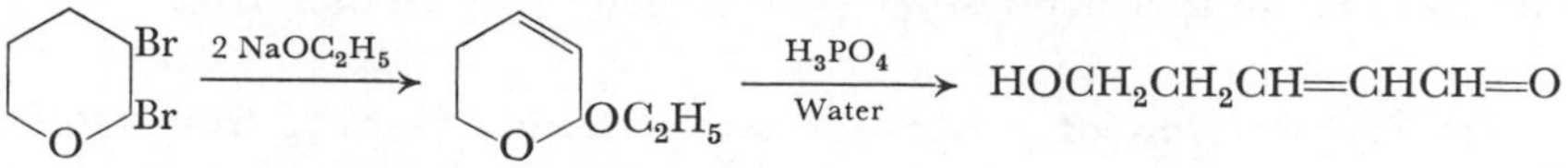

$$\text{(dibromopyran)} \xrightarrow{\text{2 NaOC}_2\text{H}_5} \text{(ethoxypyran)} \xrightarrow[\text{Water}]{\text{H}_3\text{PO}_4} \text{HOCH}_2\text{CH}_2\text{CH=CHCH=O}$$

(3) Glutacondialdehyde as sodium salt (32 % from pyridine, sodium hydroxide, and ethyl chlorosulfonate) [13].

$$\begin{array}{c} \text{CH=CHCH=O} \\ | \\ \text{CH=CHO}^{\ominus}\text{Na}^{\oplus} \end{array}$$

(4) o-Benzamidocinnamaldehyde

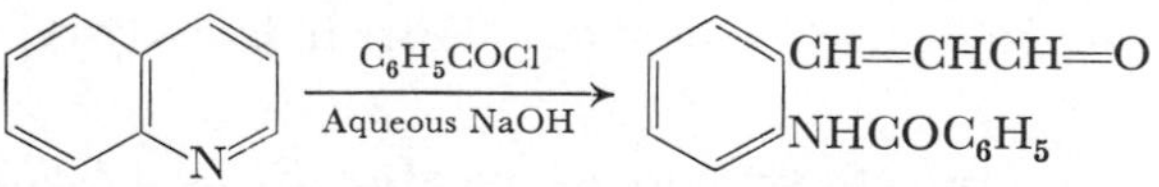

$$\text{(quinoline)} \xrightarrow[\text{Aqueous NaOH}]{\text{C}_6\text{H}_5\text{COCl}} \begin{array}{c} \text{CH=CHCH=O} \\ \text{NHCOC}_6\text{H}_5 \end{array}$$

13%, m.p. 186°

4. From *gem*-Dihalides

$$\text{ArCH}_3 \xrightarrow[\text{NBS}]{\text{Br}_2 \atop \text{or}} \text{ArCHBr}_2 \xrightarrow{\text{H}_2\text{O}} \text{ArCHO}$$

This synthesis (for application to ketone synthesis see Chapter 11, D.2) may be applied largely to methylarenes, in which case the halogen usually introduced has been chlorine or bromine with or without irradiation. Recently, however, N-bromosuccinimide has been used successfully in the halogenation. As a rule, alkali or silver nitrate may be employed to effect the hydrolysis, although the dihalides may also be converted into the aldehyde in yields from 60 to 90 % via the morpholine derivative [15]:

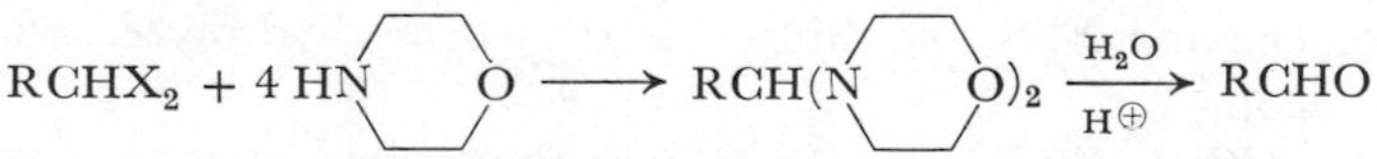

$$\text{RCHX}_2 + 4\ \text{HN} \overbrace{}\text{O} \longrightarrow \text{RCH(N} \overbrace{}\text{O)}_2 \xrightarrow[\text{H}^{\oplus}]{\text{H}_2\text{O}} \text{RCHO}$$

The method has been applied to *m*-cresol by first acetylating the hydroxyl group [16]. The synthesis carries over into the aliphatic family, where the appropriate dichlorinated alkanes can be obtained by chlorination, Friedel-Crafts reactions with vinyl chloride, free radical additions, or reduction of trihalo compounds. Recently, alkyl-substituted aromatic hydrocarbons, such as ethylbenzene, cumene, have been converted into *gem*-dihalides by the action of dichlorocarbene on the hydrocarbon [17]:

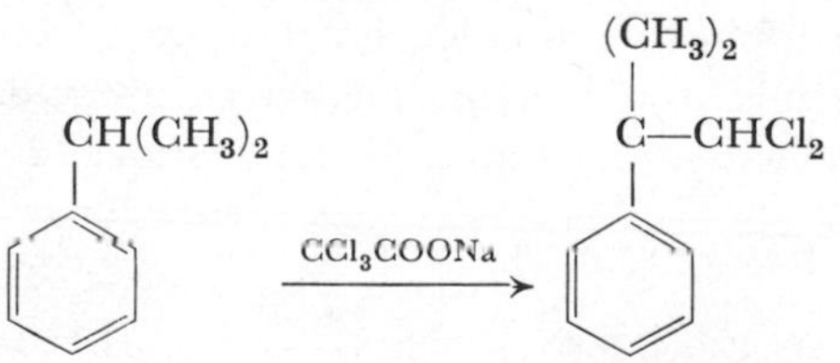

$$\begin{array}{ccc} \text{CH(CH}_3)_2 & & \text{(CH}_3)_2 \\ | & & | \\ \text{(benzene ring)} & \xrightarrow{\text{CCl}_3\text{COONa}} & \text{C—CHCl}_2 \\ & & \text{(benzene ring)} \end{array}$$

β,β-Dichloro-*t*-
butylbenzene

When these compounds are capable of hydrolysis, they serve as good sources of aldehydes of unique structure even though yields may be only fair.

(a) Preparation of p-bromobenzaldehyde (60–69% from p-bromotoluene) [18].

(b) Preparation of neohexaldehyde, $(CH_3)_3CCH_2CH=O$. t-Butyl chloride, 100 g., vinyl chloride, 55 g., and ferric chloride, 10 g., were held overnight in an autoclave or sealed tube at room temperature. After the usual workup, neohexylidine dichloride, $((CH_3)_3CCH_2CHCl_2$, b.p. 57° (31 mm.)), was obtained in 77% yield. Neohexylidine dichloride, 14 g., and 30 ml. of water were heated in a sealed tube to 300° to yield 60% neohexaldehyde, b.p. 102–103°. Magnesium oxide may be used as a buffer in hydrolysis [19].

(c) Preparation of trichloroacrolein, $CCl_2=CClCH=O$. Pentachloropropene (from the dehydrohalogenation of hexachloropropane), 1 mole, was stirred at room temperature with 200 ml. of conc. sulfuric acid until hydrogen chloride ceased evolving. Starting material, 120 g., was removed and the sulfuric acid was diluted with crushed ice and water. Extraction with ether and usual purification gave trichloroacrolein, 48 g., b.p. 57–58° (12 mm.) [20].

(d) Other examples

(1) 3,5-Di-t-butyl-4-hydroxybenzaldehyde (77–85% from the di-t-butylcresol in 80% aqueous acid to which 2 equiv. of Br_2 were added) [21].
(2) p-(Triphenylsilyl) benzaldehyde (67% from the dibromide and 3 equiv. of $AgNO_3$ in aqueous methyl Cellosolve) [22].
(3) 2,4-Bis-(acetamino)benzaldehyde (79% from 2,4-bis(diacetamino)toluene and NBS, the dibromide then being hydrolyzed with aqueous Na_2CO_3) [23].
(4) o-Phthalaldehyde (59–64% from tetrabromination of o-xylene and hydrolysis with potassium oxalate in aqueous alcohol) [24].

5. From Halogenated Methyl Esters and Related Types

$$ArCHBrOOCC_6H_5 \xrightarrow{\text{HOH}} ArCHO + C_6H_5COOH + HBr$$

This synthesis is applicable to halogenated methyl esters in which an aryl or acyl group is also substituted in the methyl group. Quaternary salts of α,α-dibromoacetophenone hydrolyze in a similar manner [25]:

$$C_6H_5COCHBrN \overset{\oplus}{\underset{}{}} \overset{\ominus}{}Br \xrightarrow{\text{HOH}} C_6H_5COCHO$$

Phenylglyoxal

(a) Preparation of phthalaldehydic acid (78–83 % by heating

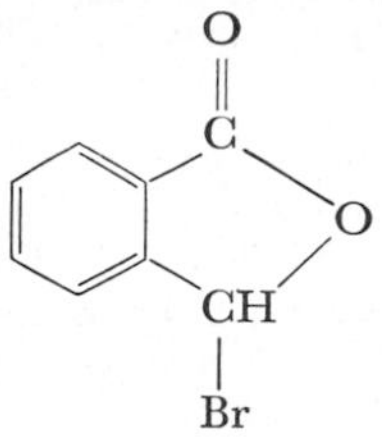

in H_2O) [26].

(b) Preparation of phenylglyoxal (82 % from the acetate of benzoyl carbinol) [27].

6. Hydration of Some Acetylenes

$$HC\equiv CH \xrightarrow[H_2SO_4]{H_2O} CH_3CHO$$

Acetylene itself is the only homolog of the series which yields an aldehyde. The hydration reaction, a commercial method for producing acetaldehyde, is catalyzed by the presence of the mercuric ion which evidently forms a bis-acetylene mercuric ion complex before hydration [28]. Acetylenes, substituted with strongly electron-withdrawing groups, hydrate to yield at least some aldehyde [29]:

$$CF_3C\equiv CH \xrightarrow[\substack{H_2SO_4}]{Hg^{\oplus\oplus}} CF_3CH_2CHO + CF_3COCH_3$$

1 part   2 parts
β,β,β-Trifluoro- α,α,α-Trifluoroacetone
propionaldehyde

The best way to obtain aldehydes from acetylenes is by the nucleophilic addition of an alkoxide ion to form the vinyl ether which subsequently can be hydrolyzed to the aldehyde:

$$RC\equiv CH \xrightarrow[NaOC_2H_5]{HOC_2H_5} RCH=CHOC_2H_5 \xrightarrow[HOH]{H^{\oplus}} RCH_2CHO$$

It would be expected that Brown's hydroboration of acetylenes would lead to aldehydes. In fact, the vinyl organoboranes obtained from terminal acetylenes undergo the usual oxidation with alkaline hydrogen peroxide. For example, 1-hexaldehyde has been made in 88 % yield from 1-hexyne [30]. A useful isomerization of acetylenic acetates to unsaturated aldehydes is discussed in E.2.

(a) Preparation of trifluoromethylvinyl ethyl ether. Trifluoromethylacetylene (10 g.) was bubbled through a solution of 100 ml. of ethyl alcohol in which 2 g. of sodium had been dissolved. The alcohol refluxed from the heat of reaction. The ether was isolated in the usual way, b.p. 102–103°, 89 % yield (acid hydrolysis would undoubtedly give the aldehyde) [31].

1. Houben-Weyl, *Methoden der Organischen Chemie*, 4th ed., Vol. 7, G. Thieme Verlag, Stuttgart, 1954, Pt. 1, pp. 255–268.
2. G. W. O'Donnell, *Australian J. Chem.*, **21**, 271 (1968).
3. S. P. Findley, *J. Org. Chem.*, **21**, 644 (1956).

4. D. L. Hufford *et al.*, *J. Am. Chem. Soc.*, **74**, 3014 (1952).

5. C. F. H. Allen and F. W. Spangler, *Org. Syn.*, Coll. Vol. **3**, 621 (1955).

6. J. W. Cornforth *et al.*, *J. Chem. Soc.*, 1549 (1949).

7. M. S. Newman and A. Kutner, *J. Am. Chem. Soc.*, **73**, 4199 (1951).

8. E. J. Corey and D. Seebach, *Angew. Chem.*, *Intern. Ed. Engl.*, **4**, 1075, 1077 (1965).

9. A. I. Meyers *et al.*, *J. Am. Chem. Soc.*, **91**, 764, 765 (1969) and preceding paper.

10. G. F. Woods, Jr., *Org. Syn.*, Coll. Vol. **3**, 470 (1955).

11. R. I. Longley, Jr., and W. S. Emerson, *J. Am. Chem. Soc.*, **72**, 3079 (1950).

12. A. Stoll *et al.*, *Helv. Chim. Acta*, **36**, 1500 (1953).

13. P. Baumgarten, *Chem. Ber.*, **57**, 1622 (1924).

14. I. W. Elliott, *J. Org. Chem.*, **29**, 305 (1964).

15. M. Kerfanto, *Compt. Rend.*, **252**, 3457 (1961); **254**, 493 (1962); *Angew. Chem.*, *Intern. Ed. Engl.*, **1**, 459 (1962).

16. E. L. Eliel and K. W. Nelson, *J. Chem. Soc.*, 1628 (1955).

17. E. K. Fields, *J. Am. Chem. Soc.*, **84**, 1744 (1962).

18. G. H. Coleman and G. E. Honeywell, *Org. Syn.*, Coll. Vol. **2**, 89 (1943).

19. L. Schmerling, *J. Am. Chem. Soc.*, **68**, 1650 (1946).

20. A. Roedig and E. Degener, *Chem. Ber.*, **86**, 1469 (1953).

21. L. A. Cohen, *J. Org. Chem.*, **22**, 1333 (1957).

22. H. Gilman *et al.*, *J. Am. Chem. Soc.*, **78**, 1689 (1956).

23. J. J. and R. K. Brown, *Can. J. Chem.*, **33**, 1819 (1955).

24. J. C. Bill and D. S. Tarbell, *Org. Syn.*, Coll. Vol. **4**, 807 (1963).

25. F. Kröhnke, *Chem. Ber.*, **66**, 1386 (1933).

26. R. L. Shriner and F. J. Wolf, *Org. Syn.*, Coll. Vol. **3**, 737 (1955).

27. W. Madelung and M. E. Oberwegner, *Chem. Ber.*, **65**, 931 (1932).

28. W. L. Budde and R. E. Dessy, *J. Am. Chem. Soc.*, **85**, 3964 (1963).

29. R. N. Haszeldine and K. Leedham, *J. Chem. Soc.*, 3483 (1952).

30. H. C. Brown, *Hydroboration*, W. A. Benjamin, Inc., New York, 1962, p. 233.

31. A. L. Henne and M. Nager, *J. Am. Chem. Soc.*, **74**, 650 (1952).

E. Rearrangements by Acid Catalysis

Rearrangements in this section are confined to hydrogen, alkyl, or aryl shifts to electron-deficient carbon atoms. They comprise the pinacol, Tiffeneau, ethylene oxide, and related types and make a rather consistent group in which to discuss the rules governing rearrangement. It cannot be said that any of the rearrangements are general methods of making aldehydes. Rather, they are limited methods which on occasion are helpful.

1. From Pinacols

All pinacols do not give aldehydes (see Ketones, Chapter 11, E.1), and therefore the factors governing rearrangement must be understood in order to realize the limitations. The first rule is that the hydroxyl group is lost from that carbon which forms the most stable carbonium ion:

$$R_2COHCH_2OH \xrightarrow[\text{}]{H^{\oplus}} R_2\overset{\overset{\displaystyle\oplus}{OH_2}}{\underset{|}{C}}-CH_2OH \xrightarrow{-H_2O} \left[R_2\overset{\oplus}{C}CH_2OH\right] \longrightarrow$$

$$R_2CHCH{=}O + H^{\oplus}$$

References for Section E are on pp. 607–608.

The second rule is that the group which is attacked by the carbonium ion is one that can best satisfy the electron deficiency. Usually, the order is aryl > alkyl > hydrogen. In the example shown above, no doubt can exist about which group is attacked because both groups are hydrogen. In general therefore *vic*-glycols with one primary alcohol group are sources of aldehydes. On the other hand, when both hydroxyl groups are secondary, the possibility exists that a ketone 10-XVII will form:

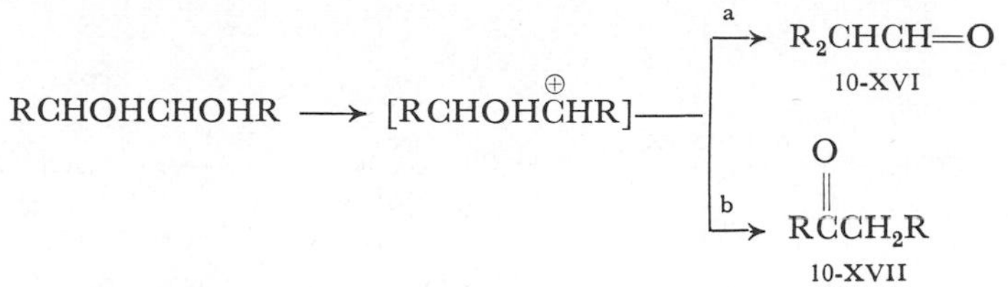

$$RCHOHCHOHR \longrightarrow [RCHOH\overset{\oplus}{C}HR] \longrightarrow \begin{cases} \overset{a}{\longrightarrow} R_2CHCH{=}O \\ \quad\quad \text{10-XVI} \\ \\ \overset{b}{\longrightarrow} R\overset{O}{\overset{\|}{C}}CH_2R \\ \quad\quad \text{10-XVII} \end{cases}$$

According to the sequence of migratory powers, step a should take preference over step b. It usually does, but a complication arises. The expected aldehyde 10-XVI may undergo another rearrangement to form 10-XVII:

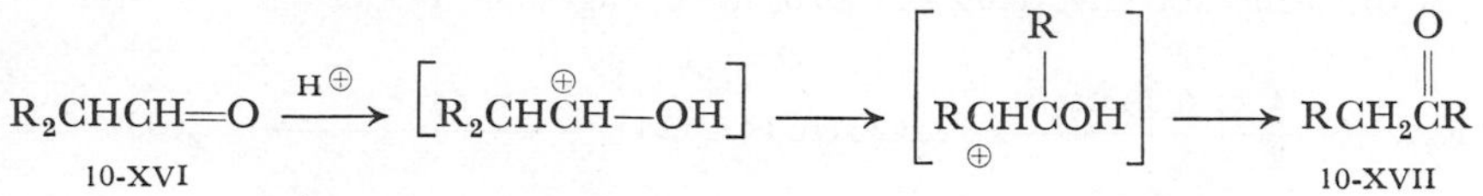

$$R_2CHCH{=}O \xrightarrow{H\oplus} \left[R_2CH\overset{\oplus}{C}H{-}OH \right] \longrightarrow \left[\begin{matrix} R \\ | \\ RCH\overset{}{\underset{\oplus}{C}}OH \end{matrix} \right] \longrightarrow RCH_2\overset{O}{\overset{\|}{C}}R$$

$$\text{10-XVI} \quad\quad\quad\quad\quad\quad\quad\quad\quad\quad\quad\quad\quad\quad\quad\quad\quad \text{10-XVII}$$

The latter rearrangement takes place under stronger acid conditions than those for the pinacol rearrangement [1]; therefore dilute acid rearrangement favors aldehyde formation. Many examples are known where aldehydes are obtained in 20 % aqueous sulfuric acid and ketones from the same pinacol in concentrated sulfuric acid [2]. On the other hand, one example is known in which the hydride shift leading to a ketone is favored in very weak acid solution [3]:

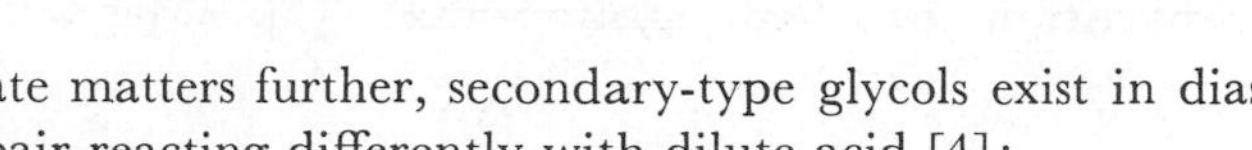

$$(C_6H_5)_2COHCHOHC_6H_5 \xrightarrow[\text{conc.}]{H_2SO_4} (C_6H_5)_3CCH{=}O$$

$$\Big\downarrow \xrightarrow[\text{HCl}]{\text{Dioxane-water}} (C_6H_5)_2CHC\overset{O}{\overset{\|}{C}}C_6H_5$$

To complicate matters further, secondary-type glycols exist in diastereomeric pairs, each pair reacting differently with dilute acid [4]:

$$CH_3CHOHCHOHCH_3 \xrightarrow{\text{Dilute } H_3PO_4} CH_3COC_2H_5 + (CH_3)_2CHCH{=}O$$

| *meso* | 42% | 18% |
| *dl* | 59% | 1% |

It is difficult to predict from these results when aldehyde formation is favored.

Various derivatives of *vic*-glycols also can be used to prepare aldehydes as, for example, chlorohydrins, amino alcohols (by deamination of ethanolamines with nitrous acid), and alkoxy alcohols. The latter, 10-XVIII and 10-XX, containing one tertiary and one alkylated primary alcoholic group, may be produced by the Grignard reaction:

$$C_2H_5OCH_2COOC_2H_5 \xrightarrow[\text{2. H}_2\text{O}]{\text{1. 2 RMgX}} \underset{\underset{OH}{|}}{C_2H_5OCH_2CR_2} \xrightarrow{\overset{\oplus}{H}} R_2CHCHO$$

$$10\text{-XVIII} \qquad 10\text{-XIX}$$

or

$$C_2H_5OCH_2COR \xrightarrow[\text{2. H}_2\text{O}]{\text{1. RMgX}} \underset{\underset{OH}{|}}{\overset{\overset{R}{|}}{C_2H_5OCH_2CR'}} \xrightarrow{\overset{\oplus}{H}} \overset{\overset{R'}{|}}{RCCHCHO}$$

$$10\text{-XX} \qquad 10\text{-XXI}$$

and on treatment with acids give the aldehydes 10-XIX and 10-XXI. Both reactions have been accomplished giving overall yields of aldehydes from 32 to 59% [5, 6].

A third method involving the Grignard reagent is as follows:

$$R_2C{=}O \xrightarrow[\text{2. H}_2\text{O}]{\text{1. ClMgCH}_2\text{OC}_2\text{H}_5} R_2COHCH_2OC_2H_5 \xrightarrow{H^{\oplus}} R_2CHCH{=}O$$

Yields reported in a single paper average 70% [7].

(a) Preparation of isobutyraldehyde. The yield is quantitative if the isobutylene glycol is distilled from 12% aqueous sulfuric acid as the aldehyde is formed. Under reflux, the isobutylene glycol acetal of isobutyraldehyde is formed in substantial quantities [8].

(b) Preparation of 2,3-dimethylbutanal [5].

$$(CH_3)_2CH\overset{\overset{CH_3}{|}}{C}OHCH_2OCH_3 \xrightarrow[\text{4 hr., 100}°]{\substack{\text{Anhydrous} \\ \text{oxalic acid}}} (CH_3)_2CH\overset{\overset{CH_3}{|}}{C}{-}CH{=}O$$

61% (isolated via bisulfite adduct)

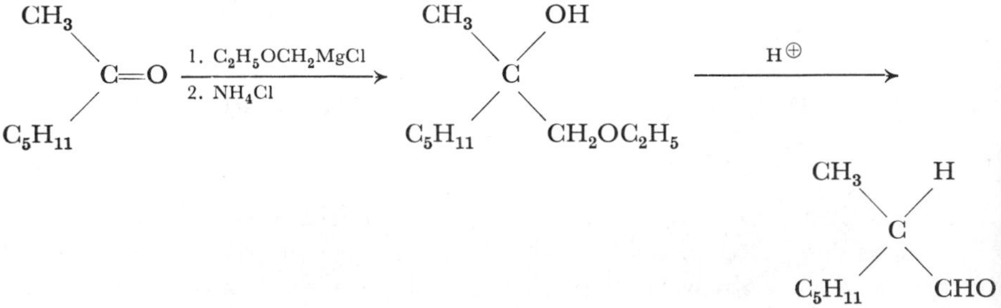

(c) Preparation of 2-methylheptanal [7]. Magnesium, 9.7 g., covered with dry tetrahydrofuran was treated with 0.4 g. of mercuric chloride and about 8 g. of chloromethyl ethyl ether in tetrahydrofuran. When reaction began, 30 g. more of the chloromethylethyl ether and 35 g. of methyl n-amyl ketone in tetrahydrofuran were added simultaneously to the solution at 0° over a period of 3 hr. After warming to room temperature and decomposing the Grignard reagent with aqueous ammonium chloride, and recovery, 1-ethoxy-2-methyl-2-heptanol, b.p. 87° (12 mm.) was obtained in 93 % yield. The ethoxy-alcohol was refluxed with an equal weight of formic acid and the mixture then stirred with 0.1 N sulfuric acid for several hours at 0°. The aldehyde was recovered in the usual manner, 71 % yield, b.p. 50° (11 mm.), n^{22}D 1.4142.

(d) Other examples

(1) **Diphenylacetaldehyde** (60 % from hydrobenzoin and solid oxalic acid at 155°) [9].

(2) **Acrolein** (33–48 % from glycerol and potassium acid sulfate) [10].

(3) **Phenylacetaldehyde** (52 % as the bisulfite from β-iodo-α-hydroxy-ethylbenzene and aqueous silver nitrate) [11].

(4) **Phenylacetaldehyde** (58 % from styrene glycol passed over a bed of phosphoric acid-pumice at 200–225°) [12].

(5) **α-Phenylpropionaldehyde** [13]

$$\underset{\substack{\text{Norephedrin}\\\text{40 g.}}}{C_6H_5CHOH\overset{\overset{\displaystyle CH_3}{|}}{C}HNH_2} \xrightarrow{\text{HONO}} \underset{\substack{|\\C_6H_5\\\text{4.4 g.}}}{CH_3CHCHO} \quad \text{(as the optically active semicarbazone)}$$

2. From Unsaturated Alcohols

It is tempting to state that the rearrangement mechanism of allyl alcohols is identical with that of the pinacol rearrangement, except that protonation of the allyl alcohol replaces the dehydration step of the pinacol:

$$\underset{\overset{|}{CH_2=CCH_2OH}}{\overset{CH_3}{}} \xrightarrow{H^{\oplus}} \left[\underset{\overset{|}{CH_3\overset{\oplus}{C}CH_2OH}}{\overset{CH_3}{}}\right] \longrightarrow (CH_3)_2CHCH=O$$

The preceding statement is quite valid for allyl alcohols with α-alkyl or α-aryl substituents such as

$$\underset{CH_2=CCH_2OH}{\overset{R}{}}$$

the corresponding glycol having been isolated in one case under rearrangement conditions [14]. On the other hand, the rearrangement mechanism of allyl alcohols with β-substituents, RCH=CHCH_2OH, is less predictable, the protonation step being less strongly directed. As a result, both aldehydes and ketones (see Chapter 11, E.2, for ketone discussion) may be obtained [15]:

$$CH_3CH{=}\overset{\overset{\displaystyle CH_3}{|}}{C}CH_2OH$$

$$\longrightarrow [CH_3CH_2\overset{\overset{\displaystyle CH_3}{|}}{\underset{\oplus}{C}}CH_2OH] \longrightarrow CH_3CH_2\overset{\overset{\displaystyle CH_3}{|}}{C}HCH{=}O$$

25%
2-Methylbutanal

$$H^{\oplus}$$

$$\longrightarrow [CH_3CH{=}\overset{\overset{\displaystyle CH_3}{|}}{C}CH_2{}^{\oplus}] \longleftrightarrow [CH_3\overset{\overset{\displaystyle CH_3}{|}}{C}H\overset{\oplus}{C}{=}CH_2] \xrightarrow[-H^{\oplus}]{H_2O} [CH_3\overset{\overset{\displaystyle OH}{|}}{C}H\overset{\diagup{}^{CH_3}}{C}{=}CH_2]$$

$$\xrightarrow{H^{\oplus}} [CH_3\overset{\overset{\displaystyle OH}{|}}{C}H\overset{\oplus}{C}(CH_3)_2] \longrightarrow CH_3\overset{\overset{\displaystyle O}{\|}}{C}CH(CH_3)_2$$

63%
Isopropyl methyl ketone

Allyl alcohols of the type $R_2COHCH{=}CH_2$ have not been converted into aldehydes, but instead have been isomerized to the primary alcohol [16]:

$$R_2COHCH{=}CH_2 \xrightarrow{H^{\oplus}} R_2C{=}CH{-}CH_2OH$$

Acetylenic alcohols, quite readily available from the condensation of acetylenic salts with aldehydes or ketones, give small amounts of aldehydes on acidic hydrolysis [17], but the method of choice is the isomerization of the acetylene acetates with silver ion to allenic acetates followed by hydrolysis to α,β-unsaturated aldehydes (Ex. b). The method is particularly suitable for synthesizing aldehydes in the steroid family.

Unsaturated alcohols of any type are isomerized to aldehydes by means of iron pentacarbonyl and irradiation [18]:

$$CH_2{=}CH(CH_2)_7CH_2OH + Fe(CO)_5 \xrightarrow[\text{Pentane, }20°,\ 1-6\ \text{hr.}]{200\ \text{w. quartz lamp}} C_9C_{19}CH{=}O$$

5 g.   0.3 g.   Decanal, 54%

Yields from other unsaturated alcohols may be quite low, however.

(a) **Preparation of 2-methylhexanal.** β-Butylallyl alcohol, 20 g.,

$$CH_2{=}\overset{\overset{\displaystyle C_4H_9}{|}}{C}{-}CH_2{-}OH$$

was refluxed under nitrogen in 0.4 N solution of sulfuric acid containing 1 vol. of dioxane to 3 vol. of water for 48 hr. (If the alcohol is water-soluble, the dioxane can be omitted.) The mixture was cooled and extracted with ether; the ether extract was dried and distilled to yield 90% of the aldehyde, semicarbazone, m.p. 92–93° [15].

(b) Preparation of *trans*-3-β-acetoxypregna-5,17(20)-diene-21-al [19].

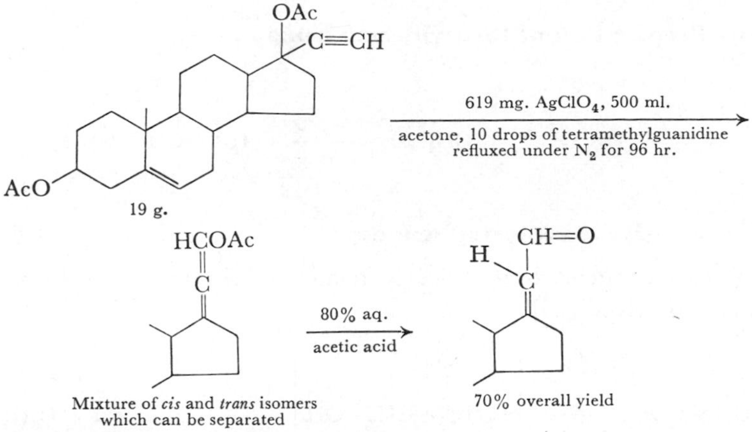

619 mg. AgClO$_4$, 500 ml.

acetone, 10 drops of tetramethylguanidine refluxed under N$_2$ for 96 hr.

19 g.

80% aq.

acetic acid

Mixture of *cis* and *trans* isomers which can be separated

70% overall yield

3. From Ethylene Oxides

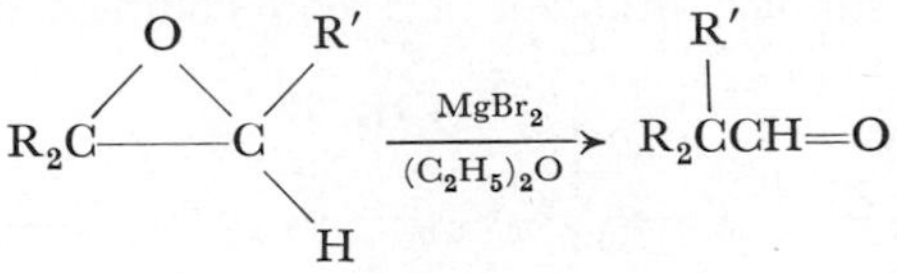

This rearrangement has been reviewed [20], and influences concerning the direction of ring opening and relative migratory powers of the substituents have been discussed [21]. The influences and migratory powers are similar to those discussed for pinacols, E.1, but operating conditions differ from those of the pinacol in utilizing anhydrous solvents and catalysts such as boron trifluoride or magnesium bromide etherate. A typical rearrangement is illustrated:

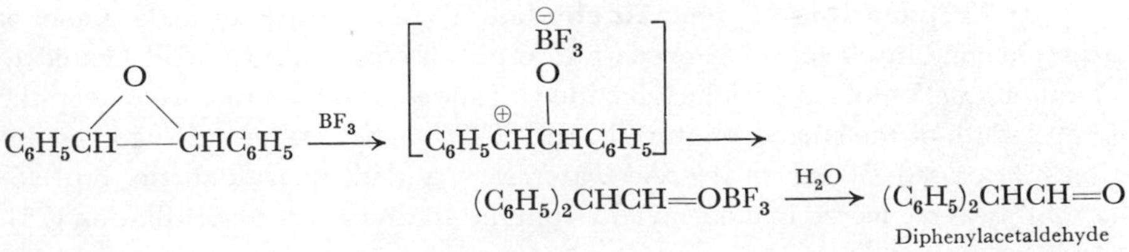

$(C_6H_5)_2CHCH{=}OBF_3 \xrightarrow{H_2O} (C_6H_5)_2CHCH{=}O$

Diphenylacetaldehyde

Yields with various catalysts and isomers have been presented as shown in the accompanying tabulation [22]. The epoxide rearrangement is applicable to

	Diphenylacetaldehyde, % yield		
	BF$_3$	MgBr$_2$ etherate	LiN(C$_2$H$_5$)$_2$
cis-Stilbene oxide	64	65	0
trans-Stilbene oxide	79	35	66

substituted oxides in which one of the carbon atoms of the oxirane is bonded to two alkyl groups or one unsaturated group (including aryl).

(a) Preparation of diphenylacetaldehyde (74–82% from *trans*-stilbene oxide and boron trifluoride etherate) [23].

(b) Preparation of formyldesoxybenzoin [24].

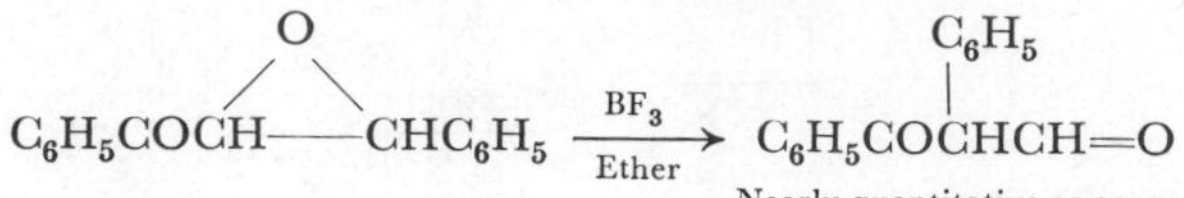

$$C_6H_5COCH{-}\!\!\!\overset{O}{\triangle}\!\!\!{-}CHC_6H_5 \xrightarrow[\text{Ether}]{BF_3} \overset{\displaystyle C_6H_5}{\underset{}{C_6H_5COCHCH{=}O}}$$

Nearly quantitative as copper salt

4. From α-Hydroxyacetophenones

This rearrangement is not clearly related to the pinacol type because a cleavage is involved [25]:

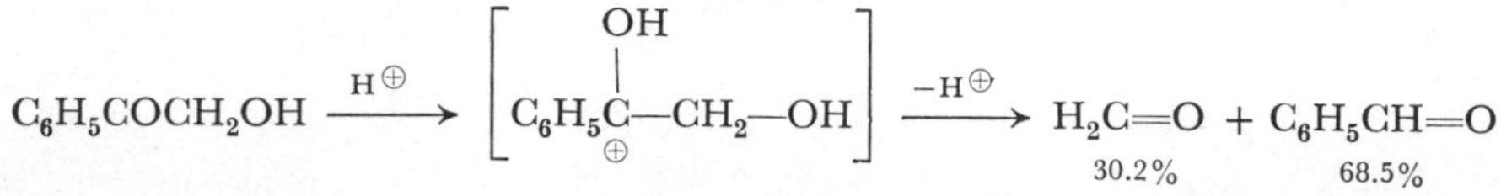

$$C_6H_5COCH_2OH \xrightarrow{H^{\oplus}} \left[\underset{\oplus}{\overset{OH}{C_6H_5C{-}CH_2{-}OH}}\right] \xrightarrow{-H^{\oplus}} H_2C{=}O + C_6H_5CH{=}O$$

$$\qquad\qquad\qquad\qquad\qquad\qquad\qquad 30.2\% \qquad 68.5\%$$

Azides behave in a similar, although in a more complex, way (see Ex. a):

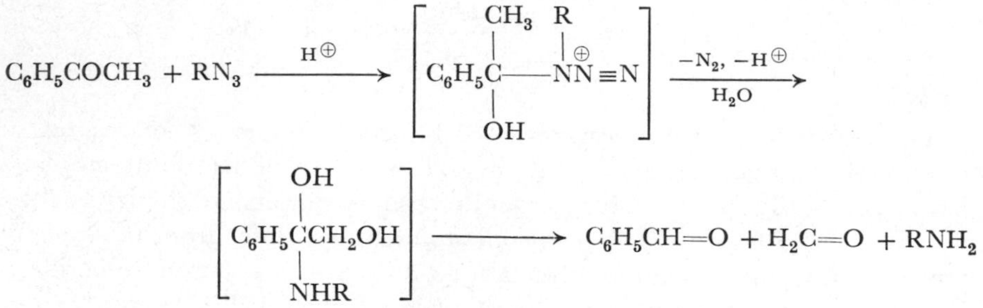

$$C_6H_5COCH_3 + RN_3 \xrightarrow{H^{\oplus}} \left[\overset{CH_3\ R}{\underset{OH}{C_6H_5C{-}NN{\equiv}N}}\right] \xrightarrow[H_2O]{-N_2,\ -H^{\oplus}}$$

$$\left[\overset{OH}{\underset{NHR}{C_6H_5CCH_2OH}}\right] \longrightarrow C_6H_5CH{=}O + H_2C{=}O + RNH_2$$

(a) Preparation of benzaldehyde. To a mixture of 0.05 mole of acetophenone in 50 ml. of benzene and 6 ml. of conc. sulfuric acid heated to about 60°, 0.05 mole of cyclohexyl azide was added at such a rate as to keep the temperature of the stirred mixture in the 70–75° range. When nitrogen evolution had ceased, 50 ml. of ice and water were added. Benzaldehyde, b.p. 35° (2 mm.) was recovered in a 85% yield from the organic layer by distillation [25].

5. From Dihydrazides

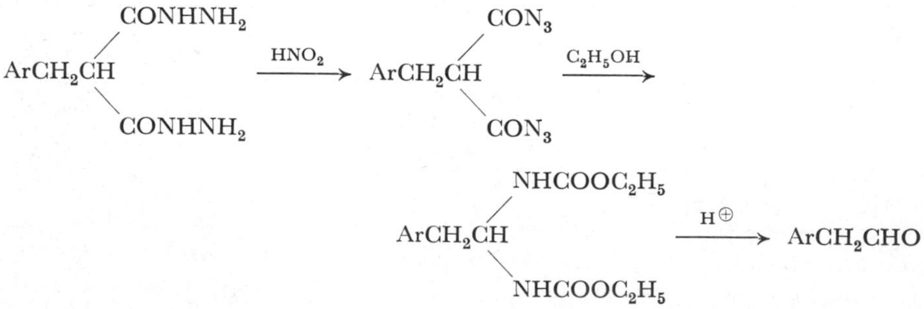

$$ArCH_2CH\overset{\displaystyle CONHNH_2}{\underset{\displaystyle CONHNH_2}{{<}}} \xrightarrow{HNO_2} ArCH_2CH\overset{\displaystyle CON_3}{\underset{\displaystyle CON_3}{{<}}} \xrightarrow{C_2H_5OH}$$

$$ArCH_2CH\overset{\displaystyle NHCOOC_2H_5}{\underset{\displaystyle NHCOOC_2H_5}{{<}}} \xrightarrow{H^{\oplus}} ArCH_2CHO$$

Aldehydes are one of the side products of the Curtius reaction, which has been reviewed [26]. The dihydrazide, which may be obtained from the substituted diethylmalonate, is readily converted into the diazide, which normally rearranges to give the diisocyanate, which in turn in alkali produces the diamine. However, the diazide in the presence of alcohol rearranges to the urethane of a *gem*-diamine which is hydrolyzed rapidly by mineral acids to the aldehyde. The overall yield is often quite satisfactory.

A similar reaction may occur if a single rearrangeable group is present (Weerman rearrangement):

$$RCHOH\overset{\overset{\textstyle O}{\|}}{C}NH_2 \xrightarrow{\ NaOCl\ } RCHOHNCO \xrightarrow{\ H_2O\ } RCH{=}O + CO_2 + NH_3$$

α-Bromo, amino, or unsaturated amides can be used to produce aldehydes in this manner.

(a) Preparation of phenylacetaldehyde (98 % from benzylmalonic ester isolated as the benzoyl hydrazone) [27].

(b) Other examples

(1) Butyraldehyde (3.76 g. as the diisocyanate,

$$CH_3CH_2CH_2CH(NCO)_2$$

by heating the diazide from 5 g. of *n*-propylmalonic acid dihydrazide until nitrogen evolution ceased) [28].

(2) Tetramethylarabinose. Pentamethylgluconamide (1 g. in 12 ml. of water) was treated with 4.9 ml. of sodium hypochlorite solution (58.2 g. per l.) at 0° for 48 hr. The solution was acidified, treated with barium carbonate, filtered, and the filtrate evaporated to dryness, giving 0.7 g. of syrup, b.p. 85° (0.01 mm.), $\alpha_D^{17.5°}$ 16.6° (in water) [29].

1. S. Danilov and E. Venus-Danilova, *Chem. Ber.*, **60,** 1050 (1927).
2. O. Bayer, Houben-Weyl's *Methoden der Organischen Chemie*, Vol. 7, G. Thieme Verlag, Stuttgart, 1954, Pt. 1, p. 239.
3. C. J. Collins, *J. Am. Chem. Soc.*, **77,** 5517 (1955).
4. E. R. Alexander and D. C. Dittmer, *J. Am. Chem. Soc.*, **73,** 1665 (1951).
5. R. A. Barnes and W. M. Budde, *J. Am. Chem. Soc.*, **68,** 2339 (1946).
6. L. F. Fieser *et al.*, *J. Am. Chem. Soc.*, **61,** 2134 (1939).
7. H. Normant and C. Crisan, *Bull. Soc. Chim. France*, (5), 459 (1959).
8. G. Hearne *et al.*, *Ind. Eng. Chem.*, **33,** 805 (1941).
9. S. Danilov and E. Venus-Danilova, *Chem. Ber.*, **59,** 1032 (1926); S. Danilov, *ibid.*, **60,** 2390 (1927).
10. H. Adkins and W. H. Hartung, *Org. Syn.*, Coll. Vol. **1,** 15 (1941).
11. S. Winstein and L. L. Ingraham, *J. Am. Chem. Soc.*, **77,** 1738 (1955).
12. W. S. Emerson, U.S. Patent 2,444,400, June 29, 1948; *C.A.*, **43,** 3461 (1949).
13. A. McKenzie *et al.*, *Chem. Ber.*, **65,** 798 (1932).
14. G. Hearne *et al.*, *Ind. Eng. Chem.*, **33,** 805 (1941).
15. M. B. Green and W. J. Hickinbottom, *J. Chem. Soc.*, 3262 (1957).
16. Ref. 2, p. 228.
17. E. D. Bergmann, *J. Am. Chem. Soc.*, **73,** 1218 (1951).

18. R. Damico and T. J. Logan, *J. Org. Chem.*, **32,** 2356 (1967).

19. W. R. Benn, *J. Org. Chem.*, **33,** 3113 (1968).

20. R. E. Parker and N. S. Isaacs, *Chem. Rev.*, **59,** 737 (1959).

21. S. Winstein and R. B. Henderson, *Heterocyclic Compounds*, Vol. 1, John Wiley and Sons, New York, 1950, p. 1.

22. H. O. House, *J. Am. Chem. Soc.*, **77,** 3070 (1955); A. C. Cope *et al.*, *ibid.*, **80,** 2844 (1958).

23. D. J. Reif and H. O. House, *Org. Syn.*, Coll. Vol. **4,** 375 (1963).

24. H. O. House, *J. Am. Chem. Soc.*, **76,** 1235 (1954).

25. J. H. Boyer and L. R. Morgan, Jr., *J. Am. Chem. Soc.*, **81,** 3369 (1959).

26. P. A. S. Smith, *Org. Reactions*, **3,** 337 (1946).

27. T. Curtius and O. E. Mott, *Org. Reactions*, **3,** 384 (1946).

28. T. Curtius and W. Lehmann, *J. Prakt. Chem.*, **125,** 211 (1930).

29. W. N. Haworth *et al.*, *J. Chem. Soc.*, 1975 (1938).

F. Condensations

The first reaction to be discussed is the condensation of ethyl formate with active methylene compounds. The second is the Reimer-Tiemann reaction, which creates hydroxybenzaldehydes. Although the remainder of the condensations do not produce a new aldehyde group, they are included because of the similarity of principles involved. These include the aldol, Mannich, and Michael condensations. Among these condensations, attention is drawn to the development of directed aldol syntheses (F.3). Previously, little headway had been made in controlling the condensation products from two different aldehydes, $RCH_2CH=O + R'CH_2CH=O$, but now the problem has been solved. Lastly, the problem of alkylation of aldehydes, including the interesting developments of recent times, is discussed.

1. Formylation with Ethyl Formate (Claisen)

$$CH_3COCH_2CH_3 \xrightarrow[\substack{\text{Na or} \\ \text{NaOC}_2\text{H}_5}]{\text{HCOOC}_2\text{H}_5} CH_3CO\overset{\overset{\displaystyle CH_3}{|}}{C}HCHO + O\overset{\overset{\displaystyle H}{|}}{C}CH_2COCH_2CH_3$$

$$\text{10-XXII} \qquad\qquad \text{10-XXIII}$$

Ethyl formate reacts with sufficient variation in the aldol condensation to deserve a special section (for other Claisen condensations see Ketones, Chapter 11, F.2 and Carboxylic Acid Esters, Chapter 14, C.1). A list of such reactions is available [1]. If the ketone has two different active methylene groups as in methyl ethyl ketone, condensation can take place with either alkyl group to give 10-XXII and 10-XXIII; and indeed does to a significant extent [2]. According to Roch, formylation of the methyl carbon is preferred even in the case of phenylacetone, although this product is easily reverted to the starting materials, thus permitting some methylene condensation [3]. Indeed, methyl condensation seems to predominate, if alcohol-free solvents are used in the condensation. If the problem of formylation at two different sites in a ketone is eliminated, the condensation reaction is straightforward, giving yields of 50–80 %, but the product must be handled with extreme care because of its tendency to polymerize or condense to some more complicated structure [4]. The sodium salt is isolated usually [5, 6], but the free acid may be obtained by rapid extraction from a cold, acidified solution of the sodium salt. An example of an undesirable condensation from the viewpoint of obtaining aldehydes is as follows [7]:

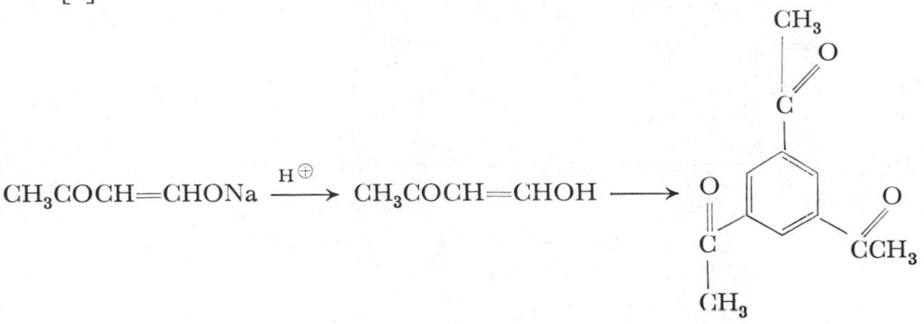

$$CH_3COCH{=}CHONa \xrightarrow{H^{\oplus}} CH_3COCH{=}CHOH \longrightarrow$$

1,3,5-Triacetylbenzene

On the other hand, ketoaldehydes of the structure RCOCHRCH=O are more stable.

Ethyl formate also can be condensed with esters having α-hydrogen atoms to yield β-aldehydoesters, whereas ethyl orthoformates condense with vinyl ethers to give acetals as shown:

$$(C_2H_5O)_3CH + CH_2{=}CHOC_2H_5 \xrightarrow{BF_3} (C_2H_5O)_2\overset{H}{\underset{}{C}}CH_2\overset{H}{\underset{}{C}}(OC_2H_5)_2$$

Malonaldehyde tetraethyl diacetal

The mechanism is that of the Friedel-Crafts type and is discussed in C.9, which is devoted to reactions of this nature.

(a) Preparation of 4-chlorobenzoylacetaldehyde, sodium salt. A mixture of 6 g. of methyl formate and 15.5 g. of 4-chloroacetophenone was dropped into an ice-cooled suspension of 5.7 g. of 95 % commercial sodium methoxide in 100 ml. of dry toluene. On removing the ice bath, the sodium salt of the product separated as a thick, white precipitate in about 10 min. Absolute methanol, 30 ml., was added to facilitate stirring and the reaction mixture was stirred overnight at room temperature. Filtration and drying at 60° gave 12.9 g. (56 %) of the sodium salt of the aldehyde [6].

(b) Preparation of 2-hydroxymethylenecyclohexanone (70–74 % from cyclohexanone, ethyl formate, and sodium in ether, the reaction being initiated by anhydrous ethanol) [8].

2. Reimer-Tiemann Condensation

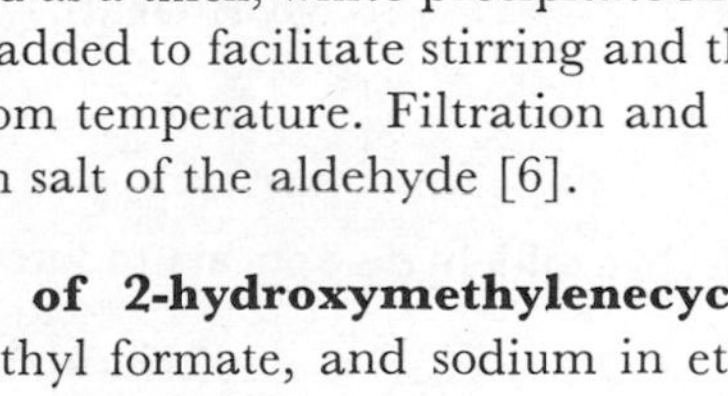

$$C_6H_5OH \xrightarrow[\text{NaOH}]{CHCl_3}$$

Although this reaction differs in several respects from formylation with ethyl formate, it bears a resemblance in consisting of the attack of a nucleophile on a

formic acid derivative, in this case the dichlorocarbene derived from chloroform:

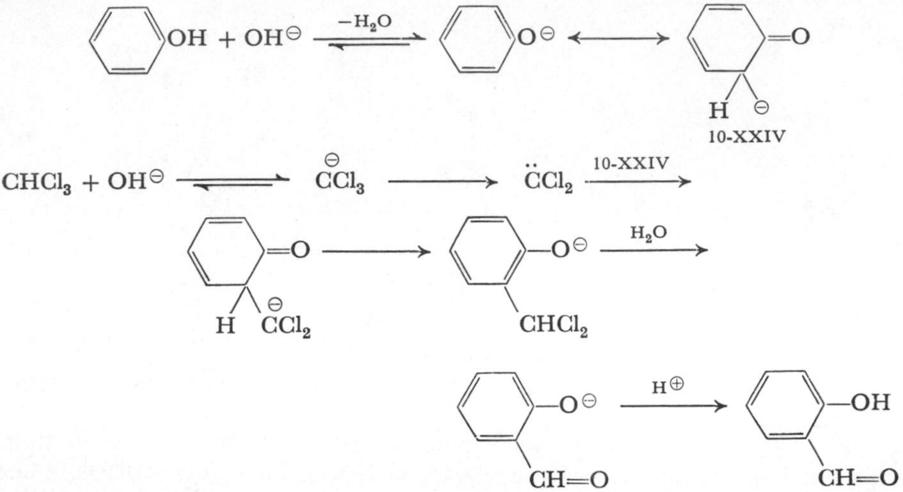

Form 10-XXIV combines with the dichlorocarbene with subsequent hydrolysis
of the ensuing *gem*-dihalide to give salicylaldehyde [9]. The ratio of *ortho/para*
substitution is about 2.2 in concentrated reaction mixtures and smaller in dilute
solutions. Furthermore, the mechanism above implies that an excess of alkali
is needed to generate the dichlorocarbene, not just an equivalent amount to
form sodium phenoxide.

Salicylaldehydes may also be prepared from the aryloxymagnesium salt and
ethyl orthoformate as shown (see Ex. b):

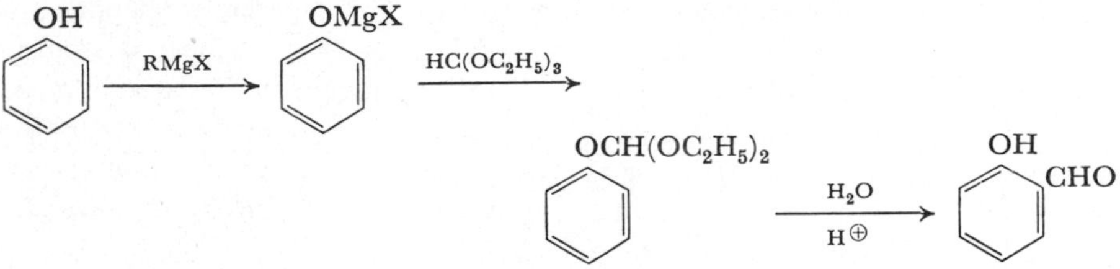

None of the *p*-hydroxybenzaldehyde appears to form by this method.

(a) **Preparation of 2-hydroxy-1-naphthaldehyde** (38–48% from β-
naphthol and NaOH in aqueous alcohol to which chloroform was added at
70–80°) [10].

(b) **Preparation of 2-hydroxy-3-methylbenzaldehyde** (42% from
0.1 mole each of *o*-cresol and ethylmagnesium bromide in ether to which 20–
30 ml. of ethyl orthoformate was added and the mixture distilled until b.p.
reached 100°; the residue was decomposed with dilute acid and steam-distilled;
the oil in the steam distillate was converted into the 2,4-dinitrophenylhydrazone;
yields of other salicylaldehydes ranged from 7 to 55%; those from phenols
with electron-withdrawing groups were very poor) [11].

3. Aldol and Related Mannich and Michael Condensations

The principles involved in making unsaturated aldehydes or β-hydroxy-aldehydes have been discussed under Alcohols, Chapter 4, G.1, but brief recapitulation is given here:

$$RCH_2CH{=}O + OH^{\ominus} \rightleftharpoons H_2O + R\overset{\ominus}{C}HCH{=}O \xrightarrow{RCH_2CH{=}O}$$

$$\underset{R}{\underset{|}{RCH_2\overset{O^{\ominus}}{C}HCHCH{=}O}} \xrightarrow[H_2O]{} \underset{\underset{R}{|}}{RCH_2\overset{OH}{\underset{|}{C}}HCHCH{=}O} \xrightarrow{\text{Slow}} \underset{\underset{R}{|}}{RCH_2CH{=}CCH{=}O}$$

All steps, except the last, are rapid and reversible [12]. Aqueous bases are the more common catalysts, but an extremely wide variety of weak bases and acidic catalysts are capable of effecting condensation. Among the most important, because of their ease of separation from the reaction mixture, are the ion exchange resins. They are discussed briefly under the specific example (b) devoted to the use of such a catalyst.

When it is realized that two different aldehydes can be condensed, the possibilities become greater. Formaldehyde is by far the most commonly used aldehyde in a mixture because it acts solely as the acceptor of the anion without forming any carbanions itself. Thus, many aldehydes containing hydroxymethyl groups are available. Also, mixed unsaturated aldehydes have been condensed [13]:

$$C_6H_5CH{=}CHCH{=}O + 3\ CH_3CH{=}CHCH{=}O \rightarrow C_6H_5(CH{=}CH)_7CH{=}O$$
$$\omega\text{-Phenylpentadecaheptaenal}$$

But, other than the two types above, the condensation of two different aldehydes (or an aldehyde and a ketone) has led to mixtures of products unworthy of consideration for synthesis. The problem of directed aldol condensation, however, has been resolved. The elegant method of Wittig permits the selection of the active methylene component as the anion and the selection of the carbonyl group as the recipient of the anion as illustrated [14]:

$$CH_3CH{=}NC_6H_{11} + LiN[CH(CH_3)_2]_2 \xrightarrow{0°,\ 10\ \text{min.}}$$
10-XXV

$$LiCH_2CH{=}NC_6H_{11} \xrightarrow[\substack{2.\ H_2O}]{\substack{1.\ (C_6H_5)_2C{=}O,\\ \text{ether},\ {-}70°}} (C_6H_5)_2\overset{CH_2\cdot CH}{\underset{OH}{C}}NC_6H_{11} \xrightarrow{H_3O^{\oplus}}$$
92%

$$(C_6H_5)_2C{=}CHCH{=}O$$
β,β-Diphenylacrolein, 100%

The Schiff base, 10-XXV, serves as the anion and can be added to a variety of other aldehydes or ketones without the difficulty of obtaining mixed aldols.

Indeed, the method above is superior to the phosphorus ylide method for extending aldehyde chains, since the ylide

$$\overset{\ominus}{C}RCH{=}O$$
$$|$$
$$\overset{\oplus}{P}(C_6H_5)_3$$

does not add to ketones.

Two more specific but related types of condensation are the Mannich and Michael condensations. The Mannich condensation (see Amines, Chapter 8, D.3) has been used infrequently to make substituted aldehydes for which examples are listed in *Organic Reactions* [15]; (one is shown in Ex. c.2):

$$RCH_2CH{=}O + H_2C{=}O + R_2'NH \longrightarrow \overset{\displaystyle CH{=}O}{\underset{10\text{-XXVI}}{R\overset{|}{C}HCH_2NR_2'}}$$

The latter can be used to make unsaturated aldehydes:

$$10\text{-XXVI} \overset{\Delta}{\longrightarrow} \overset{\displaystyle CH{=}O}{R\overset{|}{C}{=}CH_2} + R_2'NH$$

On the other hand, the Michael condensation (see Ketones, Chapter 11, G.3) has wide-ranging prospects for making aldehydes [16]. Either an aldehyde can be condensed with innumerable types of α,β-unsaturated carbonyl compounds as, for example,

$$RCH_2CH{=}O + CH_2{=}CHC{\equiv}N \overset{Base}{\longrightarrow} \overset{\displaystyle CH{=}O}{R\overset{|}{C}HCH_2CH_2CN}$$

or an α,β-unsaturated aldehyde can be condensed with an active methylene compound:

$$RCH{=}CHCH{=}O + CH_3CO\overset{\ominus}{C}HCO_2C_2H_5 \overset{\text{1. Warm}}{\underset{\text{2. } H_3O^{\oplus}}{\longrightarrow}} \overset{\displaystyle RCHCH_2CH{=}O}{\underset{\displaystyle CH_3COCHCO_2C_2H_5}{\diagdown}}$$

Further condensation may occur as shown in Ex. c.4.

Lastly, methyl groups, such as those in 2,4-dinitrotoluene and α- and γ-picolines, are reactive enough to form anions capable of attacking nitroso compounds. The resulting Schiff bases then can be hydrolyzed to the corresponding aldehydes as shown in Ex. c.5.

(a) Preparation of aldol, $CH_3CHOHCH_2CH{=}O$ (Alcohols, Chapter 4, G.1, Ex. a.).

(b) Preparation of α-ethyl-β-propylacrolein. Butyraldehyde, 100 g., and Deacidite (a weak base-ion exchange resin washed with acetic acid, water, and then dried), 30 g., were heated for 36 hr. on a steam bath. The resin was filtered, and the filtrate distilled to yield 85 % of the desired aldehyde, b.p.

70–73° (10 mm.). Strong quaternary base resins react exothermically with aliphatic aldehydes and lose their strength from reaction with the acidic by-products of the condensation. Neither weak acid nor base ion-exchange resins were as effective as the weak base resin washed with acid [17].

(c) Other examples

(1) β-(β-Naphthylamino)-β-phenylpropionaldehyde (as intermediate) [18].

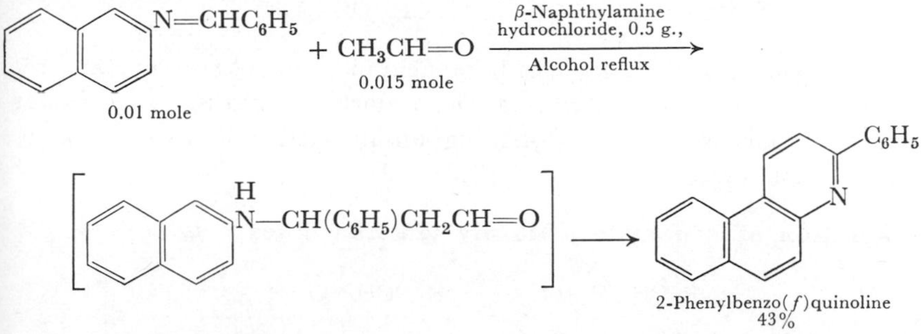

2-Phenylbenzo(f)quinoline
43%

(2) Bis-(dimethylaminomethyl)-hydroxymethylacetaldehyde (Mannich) [15].

$$CH_3CH=O + CH_2O + (CH_3)_2NH \longrightarrow [(CH_3)_2NCH_2]_2\overset{\displaystyle CH_2OH}{\underset{}{C}}CH=O$$

Nearly quantitative

(3) α-(β-Cyanoethyl)isobutyraldehyde (Michael) [16].

$$(CH_3)_2CHCH=O + CH_2=CHCN \xrightarrow[\text{2. Aqueous KCN}]{\text{1. Ion exchange resin or}} (CH_3)_2\overset{\displaystyle CH_2CH_2CN}{\underset{}{C}}CH=O$$

40% from 1,
79% from 2

(4) 5-Carbethoxy-5-cyano-1-cyclohexene-1-carboxaldehyde (Michael) [16].

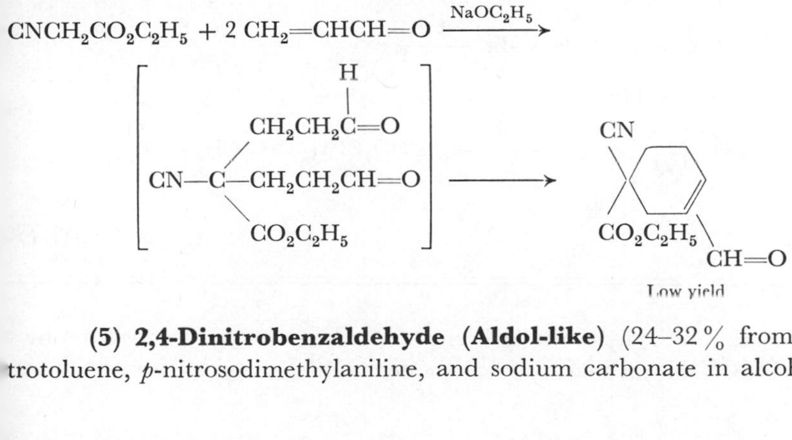

$$CNCH_2CO_2C_2H_5 + 2\ CH_2=CHCH=O \xrightarrow{NaOC_2H_5}$$

Low yield

(5) 2,4-Dinitrobenzaldehyde (Aldol-like) (24–32% from 2,4-dinitrotoluene, p-nitrosodimethylaniline, and sodium carbonate in alcohol) [19].

(6) 3,5-Diketo-5-phenylpentanal (Aldol) [20].

$$CH_3COCH_2CH{=}O \xrightarrow[NH_3]{2\,NH_2^{\ominus}} \overset{\ominus}{C}H_2CO\overset{\ominus}{C}HCH{=}O \xrightarrow[2.\ H_2O]{1.\ C_5H_5CO_2CH_3} C_6H_5COCH_2COCH_2CH{=}O$$

52%, m.p. 85–87°

(7) 2-Pentyl-2-nonenal (Aldol) (Nearly quantitative from refluxing

$$2\ CH_3(CH_2)_5CHO \longrightarrow CH_3(CH_2)_5CH{=}\underset{\displaystyle CH_3(CH_2)_3CH_2}{\underset{|}{C}}CHO$$

0.33 mole of heptanal and 0.2 mole of boric acid in 220 g. of *m*-xylene and removing the water of reaction with a Dean-Stark apparatus, an excellent general method for preparing high-boiling unsaturated aldehydes, probably involving an enol borate intermediate) [21].

4. Alkylation of Aldehydes, Mainly via Enamines

$$R_2CHCH{=}O \xrightarrow{RX} R_3CCH{=}O$$

Direct alkylation of aldehydes is rarely encountered, although some have been described, such as the ethylation of hydrocinnamaldehyde [22]. But the alkylation of aldehydes containing an *alpha* tertiary hydrogen atom can be accomplished via enamines by the method of Opitz and Mildenberger [23]:

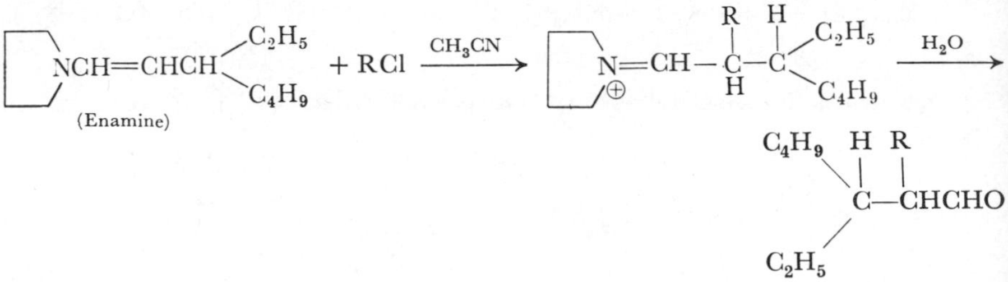

C2H5
NCH=CHCH
C4H9
(Enamine)
+ RCl
CH3CN
R H
N=CH—C—C
H
C2H5
C4H9
H2O
C4H9 H R
C—CHCHO
C2H5

Yields with allyl, crotyl, and benzyl halides vary from 19 to 78%.

The method of Stork and Dowd extends the alkylation to less active halides [24]:

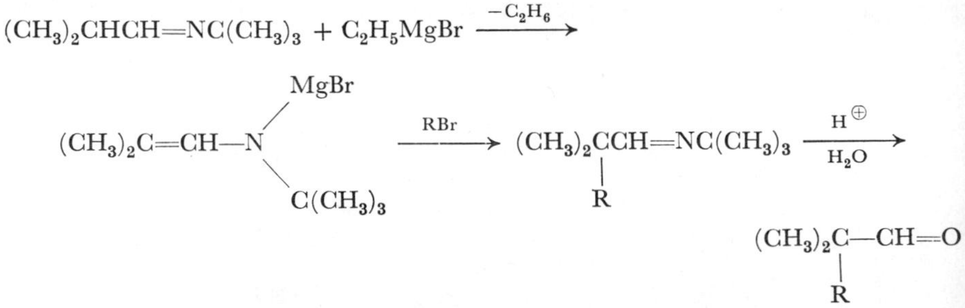

(CH3)2CHCH=NC(CH3)3 + C2H5MgBr
−C2H6
MgBr
(CH3)2C=CH—N
C(CH3)3
RBr
(CH3)2CCH=NC(CH3)3
R
H⊕
H2O
(CH3)2C—CH=O
R

A difference is noted in the alkylation products using crotyl bromide ($CH_3CH{=}CHCH_2Br$) in that a rearrangement of the crotyl group occurs by the Opitz procedure [25]:

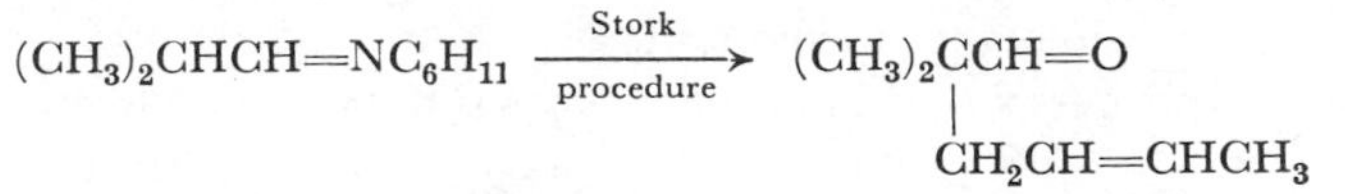

$$(CH_3)_2CHCH=NC_6H_{11} \xrightarrow[\text{procedure}]{\text{Stork}} (CH_3)_2CCH=O$$

$$\overset{|}{CH_2CH=CHCH_3}$$

2,2-Dimethyl-4-hexanal 71%

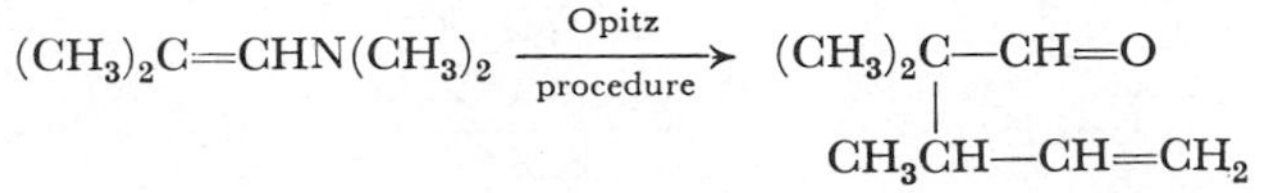

$$(CH_3)_2C=CHN(CH_3)_2 \xrightarrow[\text{procedure}]{\text{Opitz}} (CH_3)_2C-CH=O$$

$$\overset{|}{CH_3CH-CH=CH_2}$$

2,2,3-Trimethyl-4-pentanal 67%

The Opitz procedure probably involves alkylation of the nitrogen atom followed by a cyclic electron-shift mechanism giving isomerization of the crotyl group. In general, alkylation of aldehydes via enamines is less satisfactory than in the case of ketones (Chapter 11, G.2). For the enamines of aldehydes and alkyl halides, only allyl halides give a satisfactory return; among the α,β-unsaturated ketones, only vinyl ketones with an unsubstituted double bond respond well [26].

Condensations can be carried out in acidic as well as basic solutions. In the former case, an electron-deficient particle attacks the enol form of the carbonyl group. An illustration is shown in Ex. c whereby propargyl alcohol alkylates isobutyraldehyde in an acidic medium.

(a) Preparation of allyl-*n*-butylethylacetaldehyde. The pyrrolidine enamine was heated with allyl bromide in acetonitrile for a brief period

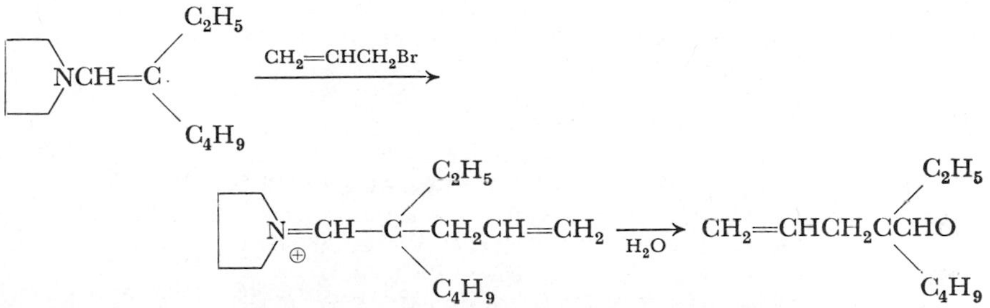

and poured into water. The trisubstituted acetaldehyde was recovered by distillation, b.p. 83–85° (10 mm.), 75% yield [23].

(b) Preparation of 2-butylheptanal (50% from the *t*-butylimine of heptaldehyde and ethylmagnesium bromide, followed by treatment with *n*-butyl iodide) [24].

(c) Preparation of 2,2-dimethylpenta-3,4-dienal [27].

$$HC\equiv CCH_2OH + (CH_3)_2CHCH=O \xrightarrow[\text{Diisopropylbenzene,}]{p\text{-}CH_3C_6H_4SO_3H, 1 \text{ g.}} $$
$$200 \text{ g. at b.p.}$$

5 moles   7 moles

$$CH_2=C=CHC(CH_3)_2CH=O$$

2 moles, 40%, b.p. 131°

5. From Butadienes and Amines

Butadienes are subject to nucleophilic attack and, in the presence of a basic catalyst and a primary amine, can yield an aldimine or alkylated aldimine:

$$RNH_2 + CH_2{=}CHCH{=}CH_2 \xrightarrow{\text{NaH}} CH_3CH{=}CHCH{=}NR \xrightarrow{CH_2{=}CH{-}CH{=}CH_2}$$

$$CH_3CH_2CHCH{=}NR \xrightarrow{CH_2{=}CH{-}CH{=}CH_2} CH_3CH_2\overset{\displaystyle |}{\underset{\displaystyle |}{C}}{-}CH{=}NR$$

with side groups $CH_2CH{=}CHCH_3$ (left) and $CH_2CH{=}CHCH_3$ (both, right).

On hydrolysis, mixtures of aldehydes can be anticipated, and adjustment of butadiene-amine ratios gives some control of the products.

(a) Preparation of α,α-dibutenylbutyraldehyde (78 % from 2 parts of butadiene to one of t-butylamine and catalytic amounts of sodium hydride in an autoclave at about 80° followed by hydrolysis) [28].

1. C. R. Hauser et al., Org. Reactions, **8,** 59 (1954).
2. R. P. Mariella and E. Godar, J. Org. Chem., **22,** 566 (1957).
3. L.-M. Roch, Ann. Chim. (Paris), (13), **6,** 105 (1961).
4. C. R. Hauser et al., Org. Reactions, **8,** 87 (1954).
5. R. P. Mariella, J. Am. Chem. Soc., **69,** 2670 (1947).
6. R. S. Long, J. Am. Chem. Soc., **69,** 990 (1947).
7. R. L. Frank and R. H. Vorland, Org. Syn., Coll. Vol. **3,** 829 (1955).
8. C. Ainsworth, Org. Syn., Coll. Vol. **4,** 536 (1963).
9. J. Hine and J. M. van der Veen, J. Am. Chem. Soc., **81,** 6446 (1959).
10. A. Russell and L. B. Lockhart, Org. Syn., Coll. Vol. **3,** 463 (1955).
11. G. Casnati et al., Tetrahedron Letters, 243 (1965).
12. D. S. Noyce and W. L. Reed, J. Am. Chem. Soc., **81,** 624 (1959).
13. J. Schmitt, Ann. Chem., **547,** 270 (1941).
14. G. Wittig, Record Chem. Progr. (Kresge-Hooker Sci. Lib.), **28,** 45 (1967).
15. F. F. Blicke, Org. Reactions, **1,** 303 (1942).
16. E. D. Bergmann et al., Org. Reactions, **10,** 179 (1959).
17. M. J. Astle and J. A. Zaslowsky, Ind. Eng. Chem., **44,** 2867 (1952).
18. N. S. Kozlov and I. A. Shur, J. Gen. Chem. USSR (Eng. Transl.), **29,** 3739 (1959).
19. G. M. Bennett and E. V. Bell, Org. Syn., Coll. Vol. **2,** 223 (1943).
20. T. M. Harris et al., J. Am. Chem. Soc., **87,** 3186 (1965).
21. R. D. Offenhauer and S. F. Nelsen, J. Org. Chem., **33,** 775 (1968).
22. O. Bayer, in Houben-Weyl's Methoden der Organischen Chemie, Vol. 7, 1954, G. Thieme Verlag, Stuttgart, Pt. 1, p. 100.
23. G. Opitz and H. Mildenberger, Angew. Chem., **72,** 169 (1960).
24. G. Stork and S. R. Dowd, J. Am. Chem. Soc., **85,** 2178 (1963).
25. K. C. Brannock and R. D. Burpitt, J. Org. Chem., **26,** 3576 (1961).
26. G. Stork et al., J. Am. Chem. Soc., **85,** 207 (1963).
27. B. Thompson, Brit. Patent 971,751; C.A., **62,** 446 (1965).
28. E. A. Zuech et al., J. Org. Chem., **31,** 3713 (1966).

G. Organometallic Methods

Grignard reactions involve an anionic attack on some substrate, a process resembling to a certain extent the condensation reactions just discussed. For the preparation of aldehydes, the anionic attack takes place on some formic acid

References for Section G are on p. 619.

derivative, coordination first occurring through the magnesium atom as, for example:

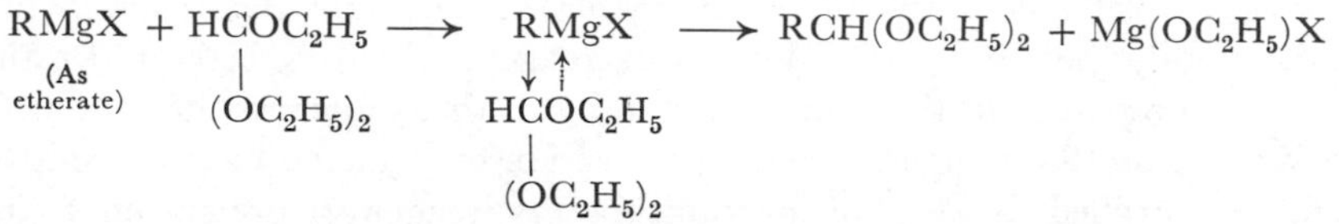
RMgX + HCOC2H5 ⟶ RMgX ⟶ RCH(OC2H5)2 + Mg(OC2H5)X
(As etherate)
(OC2H5)2
HCOC2H5
(OC2H5)2

Although Grignard reagents have been mentioned in other sections (such as A.15, E.1, F.2, and F.4) they are confined in this section to attack on formic acid derivatives and carbon bisulfide, a subsequent reduction step being included with the latter reagent. The reactions, other than that with ethyl orthoformate as shown above and *p*-dimethylaminobenzaldehyde (see A.17), are listed below. Numerous examples of each type are to be found in the Kharasch and Reinmuth treatise [1]:

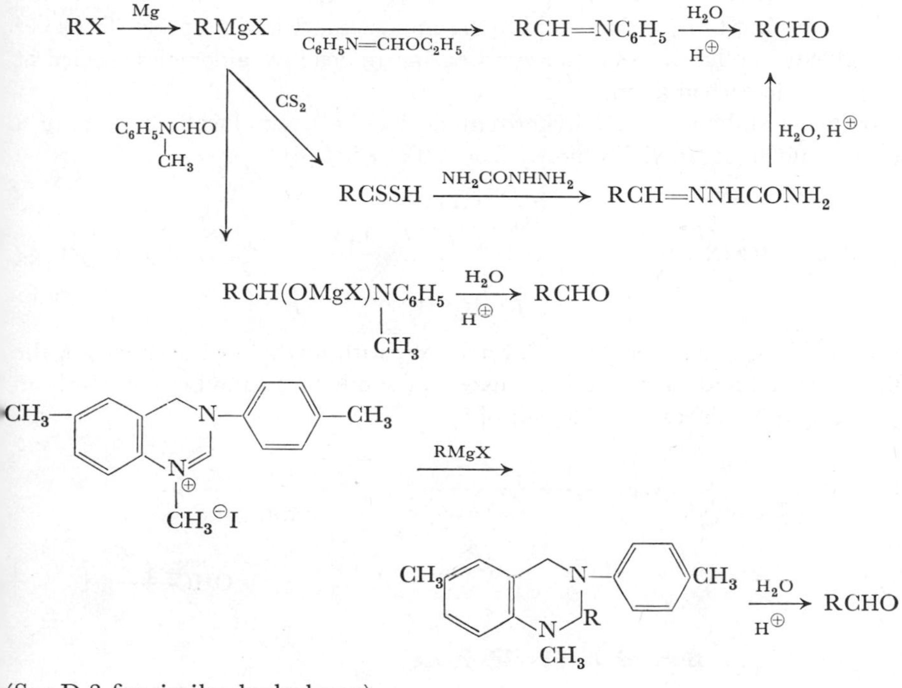
RX
Mg
RMgX
C6H5N=CHOC2H5
RCH=NC6H5
H2O
H⊕
RCHO
CS2
C6H5NCHO
CH3
H2O, H⊕
RCSSH
NH2CONHNH2
RCH=NNHCONH2
RCH(OMgX)NC6H5
CH3
RCHO
CH3
N
CH3
N
CH3⊖I
RMgX
R
RCHO

(See D.3 for similar hydrolyses.)

Various reagents, such as ethyl orthoformate, ethoxymethyleneaniline, carbon disulfide-semicarbazide, methyl formanilide, the methiodide of 6-methyl-3-*p*-tolyl-3,4-dihydroquinazoline and *p*-dimethylaminobenzaldehyde (A.17) have been employed in the synthesis of aldehydes from the Grignard reagent. Ethyl orthoformate has been shown to be superior to methylformanilide in starting with aromatic halogen compounds [2]. In a study of three of these reagents [3] it has been shown that, from a series of bromomethylbenzenes, ethoxymethyleneaniline gave yields of the aldehydes from 60 to 82%, ethyl orthoformate from 43 to 74%, and carbon disulfide-semicarbazide from 0 to

60%. Thus, at least from among these halides, ethoxymethyleneaniline and ethyl orthoformate are the preferred reagents. With both of these reagents the synthesis is relatively simple. For ethoxymethyleneaniline the reaction proceeds smoothly, but the method involves a reagent which is costly and rather difficult to prepare. On the other hand, ethyl orthoformate is cheap, but for success in its use the temperature of the reaction on removal of ether must be carefully controlled. Because of this fact, the first reagent is preferable if large amounts of material are involved. In a later study [4] with the methiodide of 6-methyl-3-p-tolyl-3,4-dihydroquinazoline, which is readily prepared in a one-step process from p-toluidine, formaldehyde, and formic acid [5], it is claimed that this reagent offers some advantage: over ethyl orthoformate in that there is no need for the application of heat over long periods; and over ethoxy-methyleneaniline in that it is less costly and easier to prepare. The yields in using the latter reagent with a series of aliphatic and aromatic halides run from 34 to 95 % when the aldehyde is recovered as the 2,4-dinitrophenylhydrazone. It is interesting to note that the dihydroquinazoline when prepared from C^{14} formaldehyde offers a means for synthesizing radioactive aldehydes labeled at the aldehyde carbon atom.

With alkyllithiums, dimethylformamide has been employed to prepare a series of aldehydes in yields mostly from 50 to 85 % [6]:

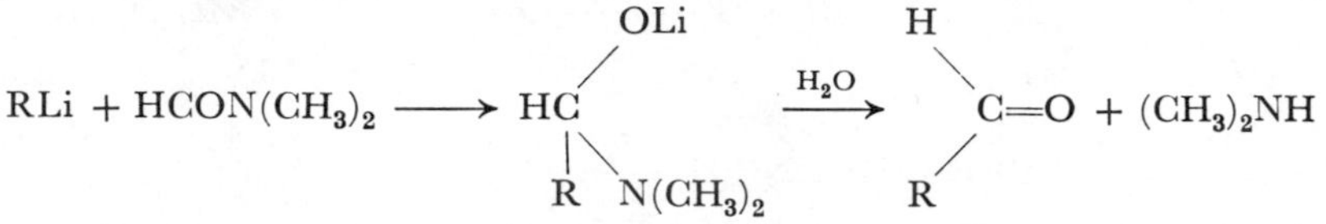

Isopropyllithium did not react. Nevertheless, with methyl or butyllithium, the dimethylformamide addition is an excellent synthetic method, particularly in preparing quinolinecarboxaldehydes [7]:

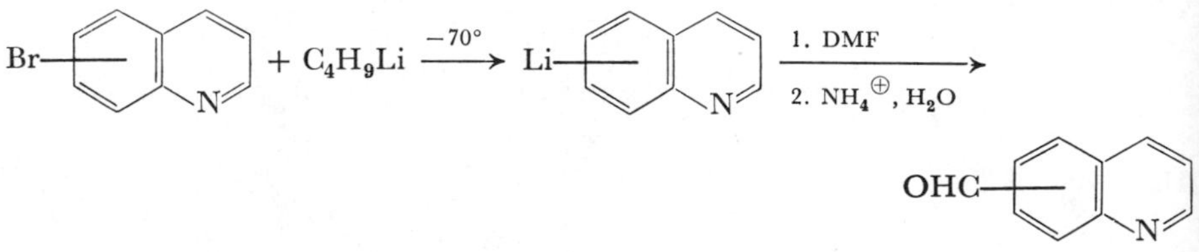

(a) Preparation of p-tolualdehyde

(1) **Ethoxymethyleneaniline procedure.** p-Bromotoluene, 15 g. in 100 ml. of ether, was converted into the Grignard reagent which was treated with 13.4 g. of ethoxymethyleneaniline in 30 ml. of ether dropwise at room temperature. After refluxing the mixture for 30 min., it was decomposed with ice and hydrochloric acid and refluxed for 30 min. to hydrolyze the anil. Steam distillation in an atmosphere of carbon dioxide gave the aldehyde which was extracted into ether and purified in the usual manner. Yield, 16.2 g. (82 %) [3].

(2) **Orthoformate procedure.**   The Grignard reagent was prepared in the usual manner under nitrogen from 20.8 g. of p-bromotoluene and 3.3 g

of magnesium in ether. Ethyl orthoformate, 22 g., in ether was added and the mixture was refluxed for 5 hr. The ether was distilled off on a steam bath, and near the end a vigorous reaction occurred. Here the flask was quickly immersed in an ice bath until the reaction ceased. After standing overnight, 50 g. of ice and 125 ml. of cold 5 N hydrochloric acid were added, the ether was evaporated, and the mixture was refluxed for 30 min. under carbon dioxide. The aldehyde which was recovered as the bisulfite weighed 20.3 g. (74%) [3].

(b) Other examples

(1) **p-Chlorobenzaldehyde** (methiodide of 6-methyl-3-p-tolyl-3,4-dihydroquinazoline procedure; 62% from p-chlorobromobenzene) [4].

(2) **n-Hexaldehyde** (45–50% from n-amyl bromide by the orthoformate procedure) [8].

(3) **m-Formylstyrene** (71% from m-bromostyrene by the dimethylformamide method) [9].

(4) **Heptanal** (85% as the 2,4-dinitrophenylhydrazone from n-hexyllithium and dimethylformamide in ether followed by the addition of saturated aqueous ammonium chloride) [6].

1. M. S. Kharasch and O. Reinmuth, *Grignard Reactions of Nonmetallic Substances*, Prentice-Hall, Englewood Cliffs, N.J., 1954.
2. L. I. Smith and M. Bayliss, *J. Org. Chem.*, **6**, 437 (1941).
3. L. I. Smith and J. Nichols, *J. Org. Chem.*, **6**, 489 (1941).
4. H. M. Fales, *J. Am. Chem. Soc.*, **77**, 5118 (1955).
5. E. C. Wagner, *J. Org. Chem.*, **2**, 157 (1937).
6. E. A. Evans, *J. Chem. Soc.*, 4691 (1956).
7. D. E. Pearson et al., *J. Heterocycl. Chem.*, **6**, 243 (1969).
8. G. B. Bachman, *Org. Syn.*, Coll. Vol. **2**, 323 (1943).
9. W. J. Dale et al., *J. Org. Chem.*, **26**, 2225 (1961).

H. Electrocyclic and Decarboxylative Reactions

General electrocyclic reactions, such as the Diels-Alder reaction, have been discussed previously (Alkenes, Chapter 2, C.2). Those discussed in H.1 and H.2 relate to specific methods for the preparation of aldehydes. Decarboxylation of various types and mechanisms are discussed in H.3.

1. Allyl Vinyl Ether Rearrangement and Other Selected Pyrolytic Reactions

The first reaction, elucidated in 1938, would seem so specific that extensions would not be possible [1]:

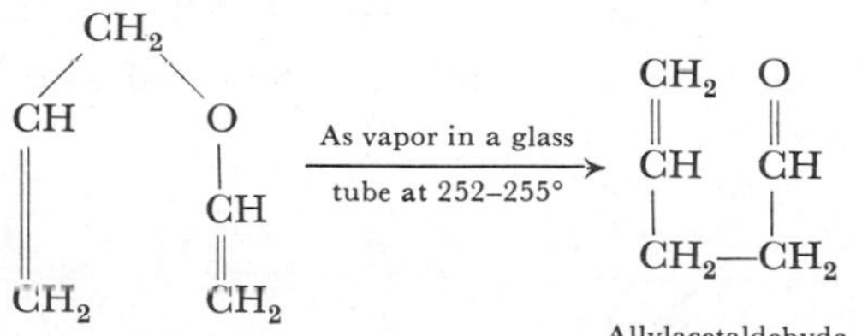

Allylacetaldehyde, 50%

References for Section H are on p. 622.

However, many substituted allyl vinyl ethers are now available by the acid decomposition of diallylacetals. Furthermore, the acetals can be generated from an aldehyde and allyl alcohol and converted into the ether in one operation [2].

$$R_2CHCH{=}O \xrightarrow{\ CH_2=CHCH_2OH\ } R_2CHCH(OCH_2CH{=}CH_2)_2 \xrightarrow{\ H^\oplus\ }$$
$$R_2C{=}CHOCH_2CH{=}CH_2$$

Other pyrolytic reactions, which are less dependent on concerted movement of electrons, may be brought about by higher temperatures of conversion. Two such examples are shown:

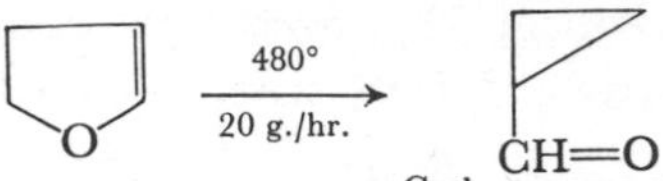

Cyclopropanecarbox-
aldehyde, 24% conversion

[ref. 3]

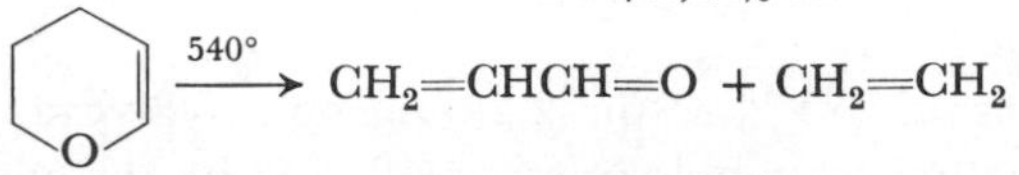

$$CH_2{=}CHCH{=}O + CH_2{=}CH_2$$

[ref. 4]

Acrolein, 85%

(a) **Preparation of 2,2-dimethyl-4-pentenal.** Isobutyraldehyde diallyl acetal (2.3 moles) with 0.2 ml. of phosphoric acid was distilled through a 1-ft. Vigreux column and the crude allyl alcohol was collected between 95 and 117°. Then, by distillation over a 3-hr. period, the product was collected between 130 and 140°, washed with water, dried, and redistilled, b.p. 124–125°, 77% [2].

(b) **Preparation of 2-allyl-2-ethylpent-4-enal.** Butyraldehyde, 1 mole, allyl alcohol, 2 moles, 25 ml. of benzene, and 0.25 g. of p-toluenesulfonic acid were refluxed with a Dean-Stark water removal trap until 20 ml. of water was obtained. Diphenyl ether, 200 g., was added to the residue and refluxing was continued for 6 days when 18 ml. more of water collected. The product was removed at 50–90° at 4–5 mm. pressure and redistilled to give the unsaturated aldehyde, b.p. 49–53° (5 mm.), 36% yield [2].

(c) **Cyclopentenylacetaldehyde** [5].

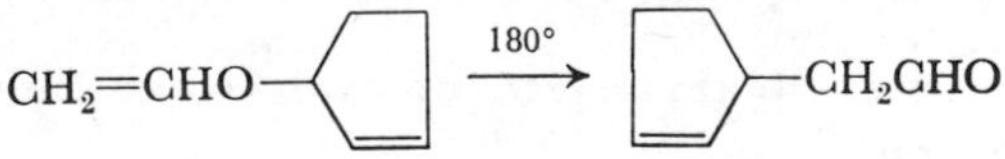

81%

2. β-Hydroxyolefin Rearrangement

$$RCHOHCH_2CH{=}CHCH_2COOH \xrightarrow{\ 450-500°\ } RCHO + CH_2{=}CH(CH_2)_2COOH$$

This synthesis has been accomplished with several β-hydroxyolefins and the mechanism suggested is as follows [6, 7]:

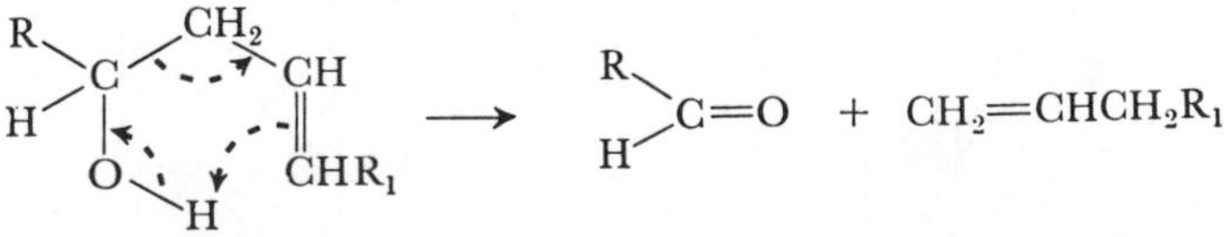

The yields are satisfactory.

(a) Preparation of heptaldehyde. 9-Octadecen-1,12-diol, 40.5 g., was introduced (30 g./hr.) into a Pyrex tube filled with glass helices heated at 500°, and swept with nitrogen. On fractionation of the condensate there was obtained 9.8 g. (60%) of heptaldehyde, b.p. 153–155°, and 14.5 g. (60%) of Δ^{10}-undecenol-1, b.p. 124–130° (23 mm.), m.p. −5° [6].

(b) Preparation of heptaldehyde (28.9% of the weight of castor oil as compared to theory of 34%) [8].

3. From α-Keto- and α-Hydroxyacids

$$RCOCO_2H \xrightarrow[\text{2. } Fe^{\oplus\oplus\oplus}, H_2O_2]{\text{1. } R_3N \text{ or } ArNH_2} RCOCO_2^{\ominus} \longrightarrow$$

$$CO_2 + [R\overset{\ominus}{C}=O] \xrightarrow{R_3\overset{\oplus}{N}H} RCH=O$$

The various methods for making α-ketoacids are recounted under Ketones, Chapter 11, D.1, D.3, D.5, and A.6, and Carboxylic Acids, Chapter 13, B.10. The decarboxylation can be accomplished smoothly by heating in quinoline, N,N-dimethyl-p-toluidine, or in aniline. With the latter reagent the Schiff base is formed, and it must be hydrolyzed to the aldehyde (Ex. c.5). The possibility exists that, if primary aromatic amines catalyze the decomposition, decarboxylation may occur by a cyclic electron-shift mechanism. Since α-aminoacids can be oxidized to α-ketoacids, they are potential sources of aldehydes, as illustrated in the preparation of 3-indoleacetaldehyde in 90% yield (as the bisulfite adduct) from tryptophane [9], or acetaldehyde in 25–35% yield from alanine [10].

α-Hydroxyacids can be oxidized and decarboxylated to aldehydes in a single operation using iron salts and hydrogen peroxide. This reaction is most useful in preparing tetroses and pentoses, the yields ranging from 20 to 80% [11]. The degradation of α-hydroxyacids can take still another form. Simple heating of the pure acids in a carbon dioxide atmosphere brings about loss of carbon monoxide and formation of the aldehyde with yields ranging from 57 to 96% [12]. The reaction is in reality an internal oxidation-reduction:

$$RCHOHCO_2H \xrightarrow{\Delta} RCH=O + CO + H_2O$$

A modification is known in which the α-methoxycarboxylic acid is heated with copper to give high yields of the aldehyde [13].

Lastly, glycidic esters can be decarboxylated in acidic or basic solution [14]:

$$R_2C{\overset{\diagdown}{\underset{O}{\diagup}}}CHCO_2C_2H_5 \xrightarrow[H_2O]{NaOH} R_2C{\overset{\diagdown}{\underset{O}{\diagup}}}CH-CO_2^{\ominus} \longrightarrow CO_2 + R_2CHCH=O$$

Yields in decarboxylation vary from 25 to 82% [15].

(a) Preparation of p-hydroxybenzaldehyde from a ketoacid. p-Hydroxybenzoylformic acid (50 g.) and p-toluidine (250 g.) were heated at 85–90° for 40 min. and then briefly at 130°. The mixture was cooled, diluted with 500 ml. of benzene, filtered and the filtrate cake refluxed with 5% aqueous sulfuric acid, the liberated aldehyde then being extracted with benzene (80%) [16].

(b) Preparation of undecaldehyde from α-hydroxylauric acid.
α-Hydroxylauric acid (20 g.) was heated gradually in an atmosphere of carbon dioxide to a temperature of 190°, and refluxing at 190–200° was continued for 15 min. On distillation, the aldehyde-water mixture was removed in about 5 min. and the aldehyde, purified through the bisulfite compound, was obtained as a water-white oil which polymerized overnight to a waxy solid, m.p. 40–48°, which, on analysis, contained 96%, 15.1 g., of undecaldehyde [12].

(c) Other examples

(1) α-Phenylpropionaldehyde (65–70% from phenylmethylglycidic ester) [17].

(2) 2-Naphthaldehyde (86% from 2-naphthylglyoxylic acid in N,N-dimethyl-*p*-toluidine at 120° until carbon dioxide evolution ceased) [18].

(3) Phthalaldehydic acid (40–41% from naphthalene oxidized to the glyoxylic acid and then decarboxylated to the titled compound by heating with hydrochloric acid and sodium bisulfite) [19].

(4) 9,10-Diphenylanthracene-2-carboxaldehyde (35% based on the hydrocarbon by conversion of ethyl 9,10-diphenylanthracene glyoxylate into the anilide and then into the aldehyde with 25% H_2SO_4) [20].

(5) Indole-3-aldehyde (93% from ethyl indole-3-glyoxylic acid anil which in turn was made from the acid and aniline) [21].

(6) Indole-3-aldehyde (70% from tryptophan and ferric chloride) [22].

(7) *d*-Arabinose (80% from calcium gluconate) [23].

(8) Benzaldehyde (75% from phenylglyoxylic acid, benzoic anhydride, and pyridine refluxed in benzene) [24].

1. C. D. Hurd and M. A. Pollack, *J. Am. Chem. Soc.*, **60**, 1905 (1938).
2. K. C. Brannock, *J. Am. Chem. Soc.*, **81**, 3379 (1959).
3. C. L. Wilson, *J. Am. Chem. Soc.*, **69**, 3002 (1947).
4. J. G. M. Bremner *et al.*, *J. Chem. Soc.*, 1018 (1946).
5. R. K. Hill and A. G. Edwards, *Tetrahedron Letters*, 3239 (1964).
6. R. T. Arnold and G. Smolinsky, *J. Am. Chem. Soc.*, **81**, 6443 (1959).
7. R. T. Arnold and G. Smolinsky, *J. Org. Chem.*, **25**, 129 (1960).
8. A. A. Vernon and H. K. Ross, *J. Am. Chem. Soc.*, **58**, 2430 (1936).
9. R. A. Gray, *Arch. Biochem. Biophys.*, **81**, 480 (1959).
10. A. Schönberg *et al.*, *J. Chem. Soc.*, 2504 (1951).
11. W. G. Overend *et al.*, *J. Chem. Soc.*, 1358 (1949).
12. R. R. Davies and H. H. Hodgson, *Soc. Chem. Ind.* (*London*), **62**, 128 (1943).
13. G. Darzens and A. Levy, *Compt. Rend.*, **196**, 348 (1933).
14. M. E. Dullaghan and F. F. Nord, *J. Org. Chem.*, **18**, 878 (1953).
15. O. Bayer, in Houben-Weyl's *Methoden der Organischen Chemie*, 4th ed., Vol. 7, G. Thieme Verlag, Stuttgart, 1954, Pt. 1, p. 328.
16. K. Kulka, *Am. Perfumer Aromat.*, **70**, 47, Sept. 1957.
17. C. F. H. Allen and J. van Allan, *Org. Syn.*, Coll. Vol. **3**, 733 (1955).
18. J. Cymerman-Craig *et al.*, *Australian J. Chem.*, **9**, 222 (1956).
19. J. H. Gardner and C. A. Naylor, Jr., *Org. Syn.*, Coll. Vol. **2**, 523 (1943).
20. R.-G. Douris, *Compt. Rend.*, **229**, 224 (1949).
21. J. Elks *et al.*, *J. Chem. Soc.*, 629 (1944).
22. M. E. Rafelson, Jr., *et al.*, *J. Biol. Chem.*, **211**, 725 (1954).
23. R. C. Hockett and C. S. Hudson, *J. Am. Chem. Soc.*, **56**, 1632 (1934).
24. T. Cohen and I. H. Song, *J. Am. Chem. Soc.*, **87**, 3780 (1965).

Chapter 11

KETONES

A. Oxidation

1. From Alcohols (Secondary)

$$\overset{\displaystyle\diagdown}{\underset{\displaystyle\diagup}{\text{CHOH}}} \longrightarrow \overset{\displaystyle\diagdown}{\underset{\displaystyle\diagup}{\text{C}}}\text{=O}$$

This synthesis has been widely employed in the preparation of ketones, often in high yields. The oxidizing agents used are essentially the same as those utilized for converting primary alcohols into aldehydes (Chapter 10, A.1). Mechanisms have been discussed previously (Chapter 10, A.1), are mentioned again under Carboxylic Acids (Chapter 13, B.1), and are described in detail in the literature [1].

Oxidizing mixtures such as sodium or potassium dichromate and sulfuric acid, chromium trioxide in acetic or sulfuric acids or in pyridine are the most common. The Jones reagent (chromium trioxide in aqueous sulfuric acid) added to the alcohol in acetone (see Exs. a and b) offers the advantage of rapid oxidation with a high yield under mild conditions [2]. This method is preferred if other easily oxidized functions are not present, or for small-scale preparations. A modification of the Jones procedure [3], in which a slight excess of sodium dichromate, a stoichiometric amount of sulfuric acid, and the solvent water were used, led to superior yields of ketones from cyclanols (see Ex. c).

In the field of steroids where specificity is desired, milder oxidizing agents such as N-bromoacetamide or N-bromosuccinimide [4] are utilized. For example, N-bromosuccinimide converts the 6β-hydroxyl group into the keto group in cholestane-3β-5α-6β-triol, but does not affect the hydroxyl groups in positions 3 and 5 [5]. Similarly the 11α-hydroxyl group in pregnan-3α-11α-diol-20-one is relatively inert toward N-bromoacetamide, while the 3α group is affected [6]. For oxidation procedures at room temperature the order of increasing intensity of attack is: N-bromosuccinimide in aqueous acetone, aqueous potassium chromate added to the steroid in acetic acid heavily buffered with sodium acetate, aqueous potassium chromate-acetic acid, and chromium trioxide in aqueous acetic acid [7].

Of value as well in the field of steroids is dimethyl sulfoxide with N,N-dicyclohexylcarbodiimide and an acid such as phosphoric, phosphorous, or cyanoacetic, or pyridinium phosphate, or pyridinium trifluoroacetate [8, 9]; see Ex. f.7. Testosterone with these reagents gives a 92 % yield of Δ^4-androstene-3,17-dione on standing overnight at 25°:

References for Section A are on pp. 643–646.

It is interesting to note that these reagents oxidize an equatorial 11α-hydroxy-steroid, whereas the 11β-epimer is unaffected.

The Pfitzner-Moffatt reagent has proved to be superior as an oxidizing agent for converting other complex secondary alcohols into ketones. Examples are carbohydrate mesylates having a free hydroxyl group [10] and certain sub-stituted α-D-altro- and α-D-glucopyranosides, each with a free hydroxyl group [11]. In these cases the molecules are unaffected except at the secondary alcohol group. Dimethyl sulfoxide-acetic anhydride also has proved to be an effective oxidizing agent, particularly for sterically hindered hydroxyl groups [12].

A few comments on various oxidizing agents follow. The use of ethyl ether with chromic acid proved advantageous, particularly in the case of ketones capable of undergoing epimerization [13]; see Ex. f.8. In the case of indole alkaloids, in which the indole moiety is particularly sensitive to oxidation, an oxidation mixture of N,N'-dicyclohexylcarbodiimide, orthophosphoric acid, and dimethylsulfoxide has proved to be superior [14]. N-Chlorosuccinimide, being a stronger oxidizing agent than N-bromosuccinimide, will convert a greater variety of alcohols into the corresponding ketones [4]. Manganese dioxide is capable of oxidizing α-phenylcarbinols to ketones in good yields. This oxidation is best carried out in a Dean-Stark apparatus to remove the water formed [15]. Aliphatic secondary alcohols are not oxidized so readily. Another mild oxidizing agent, which has the advantage of utilization in a nonpolar medium, benzene at 25°, is 4-phenyl-1,2,4-triazoline-3,5-dione:

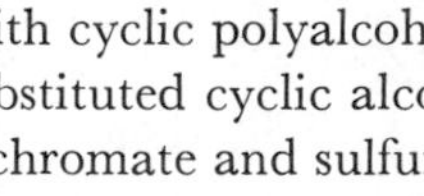

$$C_6H_5-N \overset{\quad}{\underset{\quad}{\diagdown}} = O$$

Yields of ketones with this reagent range from 62 to 90% [16].

Of some selectivity is the air oxidation of secondary alcohols using a platinum catalyst [17]. For instance, with cyclic polyalcohols only axial hydroxyl groups are oxidized (see Ex. f.4). Substituted cyclic alcohols can be oxidized to open-chain ketoacids by sodium dichromate and sulfuric acid (see Ex. f.5), probably via the olefin [18]:

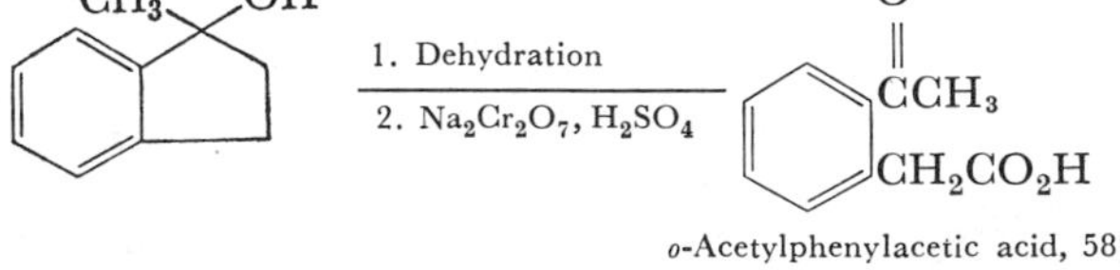

CH₃   OH → 1. Dehydration, 2. Na₂Cr₂O₇, H₂SO₄ →

o-Acetylphenylacetic acid, 58%

(a) Preparation of cyclooctanone (92–96% from cyclooctanol in acetone added to chromium trioxide dissolved in aqueous sulfuric acid; the Jones oxidation procedure is rapid and specific) [19].

(b) Preparation of nortricyclanone

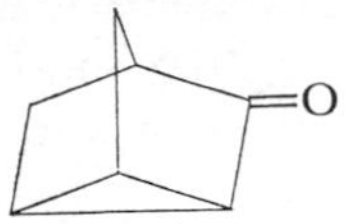

(79–88% from the alcohol by the Jones oxidation procedure) [20].

(c) Preparation of 4-ethylcyclohexanone (90% from 4-ethylcyclohexanol, 1 mole, slurried vigorously in water to which 0.40 mole of sodium dichromate and 1.33 mole of sulfuric acid in water were added dropwise) [3].

(d) Preparation of 1-menthone (83–85% from menthol oxidized with sodium dichromate and sulfuric acid) [21]; *see also* [13].

(e) Preparation of cholestanone (83–84% from dihydrocholesterol oxidized with sodium dichromate and acetic-sulfuric acids) [22].

(f) Other examples

(1) Cholestane-3β-5α-diol-6-one (97% from cholestane-3β-5α-6βtriol oxidized with N-bromosuccinimide in aqueous methanol-ether) [5].

(2) Yohimbinone (80% from yohimbine, N,N′-dicyclohexylcarbodiimide, orthophosphoric acid, and dimethyl sulfoxide) [14].

(3) Isobutyrophenone (77% from the carbinol, manganese dioxide, and benzene in a Dean-Stark apparatus) [15].

(4) *myo*-Inosose-2

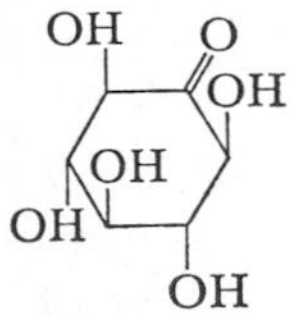

(30% from *myo*-inositol, platinum reduced from Adams catalyst, and oxygen; product purified via the phenylhydrazone) [17].

(5) 6-Oxoheptanonic acid (55% from 2-methylcyclohexanol, chromic oxide, and sulfuric acid) [23].

(6) 3-β-Benzyloxy-5-keto-5,6-*seco*-cholestanic-6-acid [24]:

(7) Cholestanone (80% from cholestanol) [9]:

$$\rangle HCOH + DCC + DMSO \xrightarrow{H^{\oplus}} \rangle C{=}O + C_6H_{11}NHC(=O)NHC_6H_{11} + CH_3SCH_3$$

80%

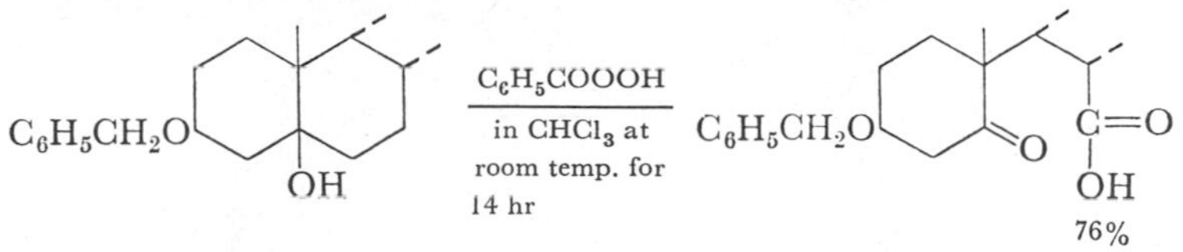

(8) 6-*m*-Methoxyphenylhex-1-ene-3-one (65% from the carbinol in a two-phase system of ether and water to which 30–40% excess of 8 N chromic acid was added dropwise over 8 hr.; the unsaturated carbinols are prepared by Grignard addition to acraldehyde or by addition of vinylmagnesium bromide to aldehydes) [25].

2. From Alcohols (Secondary) and Aluminum *t*-butoxide (Oppenauer)

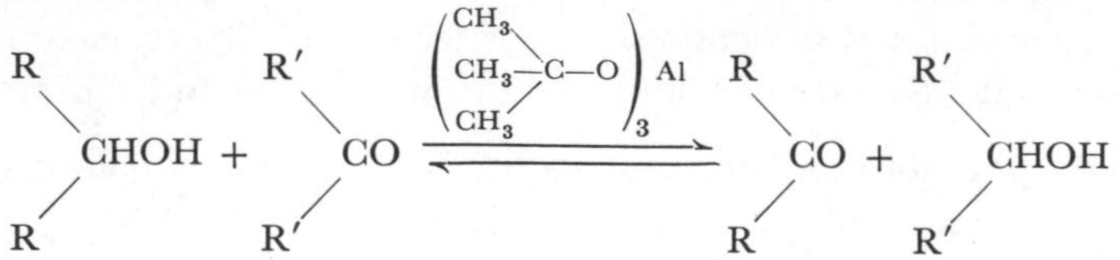

This synthesis, which is the reverse of the Meerwein-Pondorff-Verley Reduction, has been reviewed [26]. Although it is applicable to both aldehydes (Chapter 10, A.3) and ketones, its greatest value has been in the preparation of the latter, particularly in the field of steroids where the mild conditions associated with the reaction are so desirable. For the synthesis of aldehydes it is preferable to use the aluminate of the alcohol to be oxidized and an aldehyde boiling about 50° higher to serve as the hydrogen acceptor. The product may be distilled from this mixture under reduced pressure [27]. In the usual reaction, aluminum *t*-butoxide, isopropoxide, or phenoxide may be used with acetone-benzene or cyclohexanone-toluene as the hydrogen acceptor. Time is reduced by refluxing with the higher-boiling cyclohexanone-toluene mixture, although, with sensitive compounds such as steroids, the reaction is carried out at lower temperatures. Some specificity is exhibited in the oxidation since substituents such as allyl, vinyl, ethynyl, benzal, and some other unsaturated side chains are not attacked. Yields are variable but may be high.

In a modification, quinine has been oxidized to quininone in 95% yield by the use of potassium *t*-butoxide with benzene as the solvent and benzophenone as the oxidant [28]:

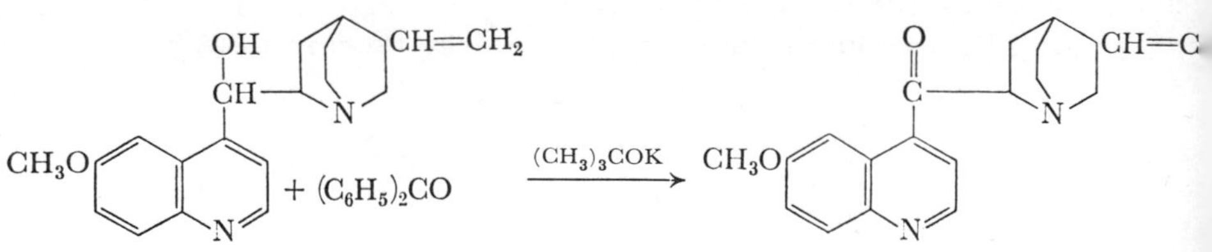

For aminoalcohols which tend to complex with aluminum alkoxides, the system, potassium *t*-butoxide, fluorenone, and benzene, is far superior as an oxidizing medium. Quininone has also been prepared in high yield from quinine by means of this medium [29], although the product is cleaner when the oxidation is conducted at 25° for 12 hours rather than at 80° for 10 minutes (personal observation by one of us).

Some remarks concerning the association of aluminum alkoxides relative to their use in the Tishchenko reaction may be pertinent to Oppenauer oxidation (see Esters, Chapter 14, D.1).

(a) **Preparation of 4-cholesten-3-one** (70–81% and 73–83% from cholesterol) [30].

(b) **Other examples**

(1) $\Delta^{20,23}$-**24,24-Diphenylcholadiene-3,11-dione** (86% from $\Delta^{20,23}$-24,24-diphenylcholadien-3-ol-11-one) [31].

(2) **Methyl 12-oxooleate** (76% from methyl ricinoleate, acetone, and aluminum triphenoxide after 25 hr.) [32].

3. From Alcohols (Secondary) by Dehydrogenation

$$\diagdown\!\!\text{CHOH} \xrightarrow{\text{Catalyst}} \diagdown\!\!\text{CO} + H_2$$

The disadvantages of this synthesis and the catalysts employed have been discussed under Aldehydes, Chapter 10, A.2. Copper chromite on Celite gives yields of ketones from 20 to 80% with eight secondary alcohols [33]. A liquid-phase dehydrogenation at or below the boiling point using Raney nickel catalyst produces aliphatic ketones with yields from 29 to 95% [34]. By the use of cyclohexanone as a hydrogen acceptor, yields of 30–80% are obtained with a variety of secondary alcohols [35].

Industrial dehydrogenations are carried out with such catalysts as copper-zinc alloy or copper at about 325° or Raney nickel at a lower temperature; (see [36] for some examples using these catalysts).

Primary alcohols can be converted into ketones by a combined condensation-dehydrogenation reaction:

$$2\ RCH_2CH_2OH \xrightarrow[375-475°]{\text{Chromia}} RCH_2\overset{\overset{\displaystyle O}{\|}}{C}CH_2R + CO + 3\ H_2$$

Mixed primary alcohols can be used; for example, a mixture of ethanol and n-octyl alcohol gives 42% methyl n-heptyl ketone [37]

(a) **Preparation of cholestenone.** Toluene, 150 ml., cyclohexanone, 50 ml., Raney nickel, 10–15 g., and dihydrocholesterol, 5 g., were refluxed for 24 hr. with stirring. After filtering off the catalyst, the toluene and cyclohexanone were removed by distillation under reduced pressure. The residue was taken up in ether, filtered, and the product was recovered by evaporation. Crystallization from ethanol gave an 80% yield of cholestenone, m.p. 127–129° [35].

(b) **Other examples**

(1) **2-Octanone** (95% from 2.6 g. of 2-octanol distilled 4 hr. from 10 g. of nickel as hydrogen is evolved) [34].

(2) **Di-*n*-amyl ketone** (47% from *n*-hexyl alcohol passed over a chromia catalyst at 425°; see original article [37] for preparation of catalyst and other details).

4. From Acyloins or Benzoins

$$(Ar)RCOCHOHR(Ar) \xrightarrow[\text{CH}_3\text{COOH}]{\text{CuSO}_4} (Ar)RCOCOR(Ar)$$

Benzoins or acyloins, available from the benzoin or sodium-ester condensations, are oxidized with remarkable ease to the 1,2-diketones. A variety of mild oxidizing agents can be used. For benzoins, nitric acid [38], air and copper sulfate in pyridine [39], lead tetraacetate in glacial acetic acid [40], potassium bromate and potassium carbonate [41], Fehling's solution [42], catalytic amounts of copper, iron, nickel, or cobalt salts with ammonium nitrate in 80% acetic acid [43], and an excess of thallous ethoxide in benzene [44] have been employed as oxidizing agents. Although thallous ethoxide is not satisfactory for converting acyloins into diketones, cupric acetate in acetic acid works well in this oxidation [45, 46]. The yields in these conversions are usually high.

For benzoins which exist largely in the enediol form, oxidation to the benzil occurs by bubbling air through a solution of the compound in some organic solvent. The two principal types which exist in such tautomeric forms are those which are sterically hindered [47] as,

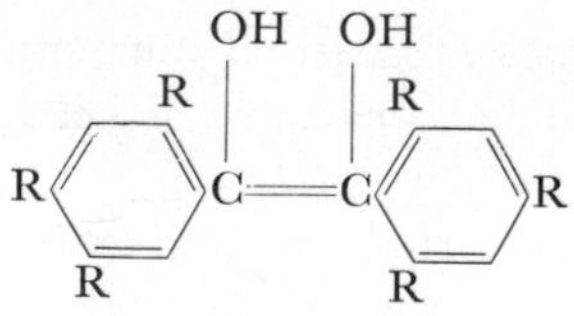

and those in which chelation is possible [47], as

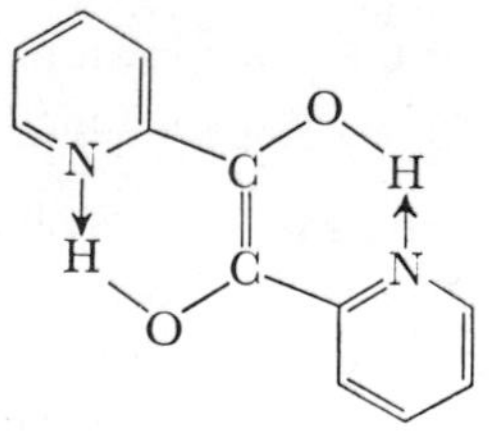

(a) **Preparation of benzil** (86% from benzoin, air, copper sulfate, and pyridine) [39].

(b) **Other examples**

(1) **Furil** (83% from furoin, thallous ethoxide in ethanol, and nitrobenzene) [48].

(2) Sebacil

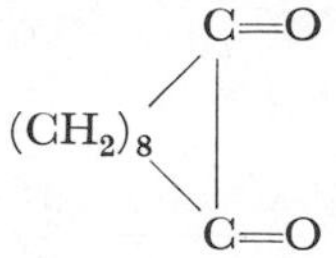

$$\begin{array}{c} C{=}O \\ (CH_2)_8 \Big\langle \\ C{=}O \end{array}$$

(88–89 % from sebacoin, cupric acetate, and 50 % aqueous acetic acid) [45].

5. From Olefins via the Ozonide and Related Reactions

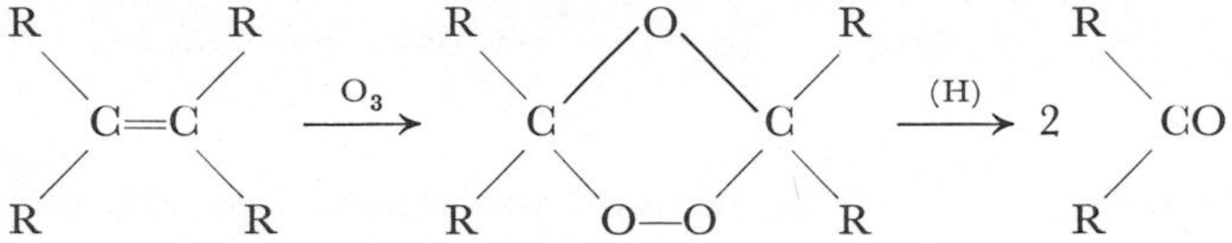

$$\underset{R}{\overset{R}{\diagdown}}C{=}C\underset{R}{\overset{R}{\diagup}} \xrightarrow{O_3} \underset{R}{\overset{R}{\diagdown}}C\underset{O-O}{\overset{O}{\diagup}}C\underset{R}{\overset{R}{\diagdown}} \xrightarrow{(H)} 2 \ \underset{R}{\overset{R}{\diagdown}}CO$$

This synthesis is applicable both to aldehydes (Chapter 10, A.6) and ketones, but it has been employed much less for the latter than for the former. Usually oxygen containing some ozone is passed through a suspension or solution of the alkene and the ozonide is decomposed by some reducing agent such as zinc and acetic acid [49] or Raney nickel [50]. If formic acid is used as the solvent, the ketone may be recovered from the solution on neutralization with alkali [51].

The use of sodium periodate in the presence of a trace of osmium tetroxide in converting alkenes into carbonyl compounds has been discussed under Aldehydes, Chapter 10, A.6. Among the successes of the method in producing ketones is the oxidation of cyclic compounds containing doubly bound methylene groups attached to the ring, as in the case of F-dihydrogarryfoline diacetate [52]:

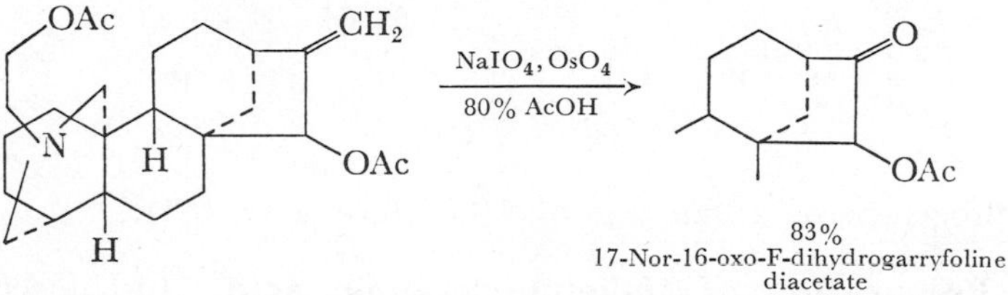

83%
17-Nor-16-oxo-F-dihydrogarryfoline
diacetate

Potassium periodate and small amounts of potassium permanganate accomplish the same type of oxidation (see Ex. d.5).

If a milder oxidizing agent, such as a peracid, is used, olefins are converted into epoxides, which in the acidic medium are transformed into ketones (see E.3):

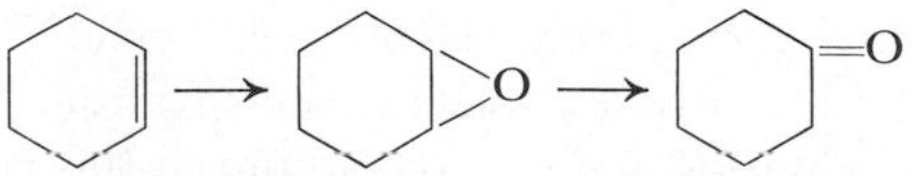

Cyclohexanone

Intermediates, other than the epoxide, such as the glycol [53] and the glycol ester [54], have been isolated in the reaction. The best method is as shown [55]:

$$(CH_3)_2C{-}C(CH_3)_2 + CF_3CO_3H \text{ in } CH_2Cl_2 \longrightarrow$$
0.06 mole   0.066 mole

$$\text{epoxide} \xrightarrow[\text{0.066 mole}]{BF_3 \cdot Et_2O} (CH_3)_3CCOCH_3$$
(not isolated)   Pinacolone 75%

Even some conjugated systems may be epoxidized in this manner:

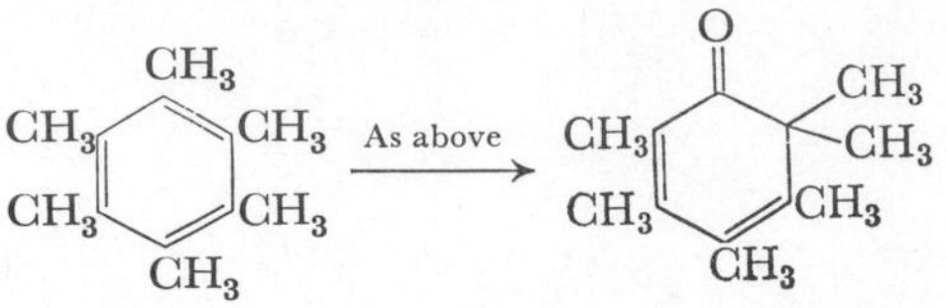

88%
Hexamethylcyclohexadienone

Another elegant epoxidation with oxygen and subsequent rearrangement is given in Ex. c.

(a) Preparation of 4,4-Bis-(chloromethyl)-2-pentanone. A suspension of 9 g. of the dimer of methallyl chloride (2-methyl-4,4-bis-(chloromethyl)-1-pentene) in 100 ml. of 90% formic acid was cooled to 0°; oxygen containing 6% ozone was passed through the solution at a rate of 125 ml./min. until the exit gases colored an acidic solution of potassium iodide; after the addition of 50 ml. of water, the solution was neutralized at 25° with 10 M sodium hydroxide and extracted with ether; from the extract dried over sodium sulfate and fractionated, there was obtained on distillation 8.6 g. (95%), b.p. 74–75° (3 mm.), of the ketone [51].

(b) Preparation of 2-indanone (69–81% from indene, 88% formic acid,

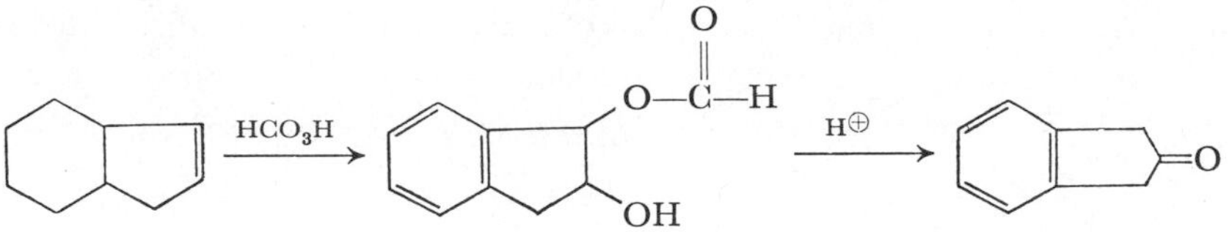

30% hydrogen peroxide followed by 7% sulfuric acid) [54].

(c) Preparation of 10-ketoundecanoic acid. 10-Undecenoic acid (0.15 mole) was added dropwise in 2.5 hr. to a solution of 0.02 mole each of palladium chloride and cupric chloride dihydrate in 50 ml. of dimethylformamide and 4 ml. of water at 60–70° while oxygen was passed through in a fine stream at 3.3 l./hr. The solution was poured into cold, dilute hydrochloric acid and the precipitated acid filtered, washed, dried, and recrystallized from pentane and then ethyl acetate, m.p. 57–59°, 83% yield. The olefin was fed in slowly to prevent build-up of the olefin concentration, which retarded the reaction. The cupric chloride kept the palladium in the Pd(2+)state [56].

(d) Other examples

(1) 3(β),11(α)-Pregnanediol-20-one-3-acetate (71% crude by ozonization of (3(β)-11(α)-hydroxyetiocholanyl)-methyldiphenylethylene) [49].
(2) 4-Methyl-3-tetralone (84% from 4-methyl-1,2-dihydronaphthalene and perbenzoic acid in chloroform at ice-bath temperature) [57].

(3) 2-(2′,3′,4′-Trimethoxyphenyl)-cyclohexanone (78% from 1-

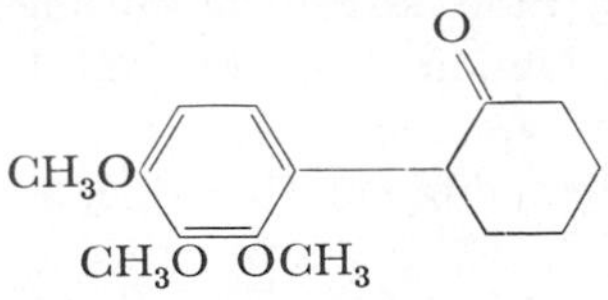

CH₃O   CH₃O OCH₃   O

(2′,3′,4′- trimethoxyphenyl)-cyclohexene and perbenzoic acid in ethyl acetate at 3°) [58].

(4) Methyl ursan-3β-ol-12-one-28-oate acetate

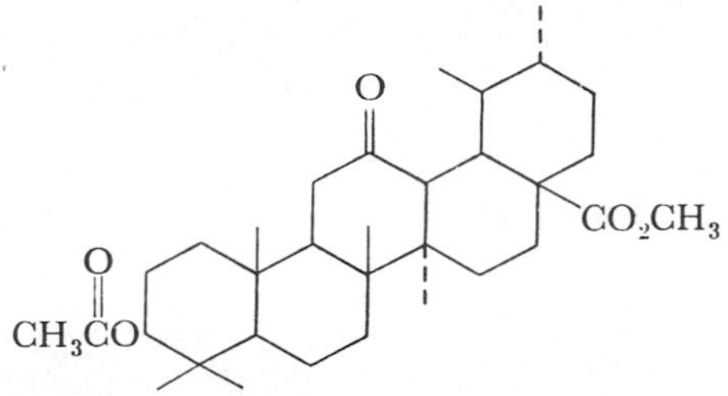

O   CO₂CH₃   CH₃CO

(79% from acetyl methyl ursolate

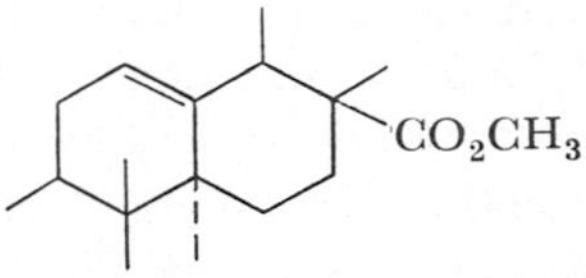

CO₂CH₃

peroxytrifluoroacetic acid and Na₂CO₃ in dichloromethane) [59].

(5) 5,5-Dimethylbicyclo[2.1.1.]hexane-2-one (59% from the 2-methylene-5,5-dimethylbicyclo[2.1.1]hexane, potassium periodate, a small amount of potassium permanganate, and potassium carbonate in water after shaking vigorously for 16 hr.) [60].

6. From Methylene Derivatives

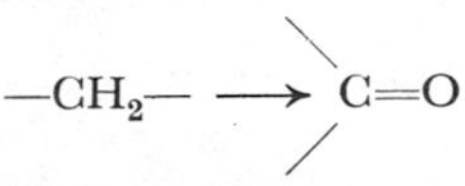

$$-CH_2- \longrightarrow C=O$$

Aliphatic hydrocarbons can be oxidized in the vapor state to ketones in fair yields using oxygen and hydrogen bromide, the latter serving as a free radical source (Ex. a). Cyclohexane oxidation has been studied in considerable detail because of obvious industrial applications. All the industrial oxidative methods give a spread of products including the hydroperoxide, alcohol, ketone, and cleavage compounds and do not seem well adapted to the laboratory. Perhaps the best laboratory route from hydrocarbon to ketone is that employing nitrosation in the presence of ultraviolet light. Under this influence cyclohexanone oxime, for example, has been obtained in 45–65% conversion from cyclohexane, nitrosyl chloride, and concentrated hydrochloric acid at −5 to 5° [61].

On the other hand, methylene groups in aromatic hydrocarbons are oxidized in the liquid phase by a variety of reagents. Air oxidations are shown in examples under b. Although conversions are low, the syntheses are appealing because of the simplicity of operation. Chromium trioxide seems to be of most general use as a catalyst. Etard's reagent, chromyl chloride (CrO_2Cl_2), is not selective, at least in the oxidation of decalin [62]. Examples b.3 and b.4 indicate that compounds in their anionic form are oxidized with remarkable ease, particularly with the system, dimethylsulfoxide, and potassium t-butoxide [63]. Chemical oxidations are shown in examples under c.

Ketones are oxidized readily to diketones as shown in the examples under d. Selenium dioxide seems to be the reagent of choice [64], although in one instance 2-methylcyclohexanone was dehydrogenated as well as oxidized to yield Δ^3-cyclohexene-1,2-dione [65], and in another, 3-benzylcamphor was dehydrogenated without forming a diketone [66]. The oxidation with selenium dioxide seems to proceed via the enol ester of the ketone and selenious acid [67]. With α-bromoketones, oxidation gives α-ketoesters in satisfactory yields [68]:

$$C_6H_5COCH_2Br \xrightarrow{SeO_2} (C_6H_5COCOBr) \xrightarrow{CH_3OH} C_6H_5COCO_2CH_3$$

Methyl phenylglyoxylate, 80%

(See also Ex. b.5.)

Other oxidizing agents, such as ferric chloride in 50% aqueous acetic acid [69], air in the presence of cobalt benzoate [70], air in alkaline medium [71], and silver trifluoroacetate and iodine [72] have been used to synthesize α-diketones. Nitrous acid also is a good reagent for obtaining α,β-diketones (see Ex. d.3). The intermediate α-oximinoketone is hydrolyzed readily with cold, concentrated hydrochloric acid. One limitation is that cycloalkanones seem to have a tendency to dinitrosate to yield an α,α'-dioximinoketone (see Ex. d.4). Some cyclic ketones may cleave to yield ketoesters [73]:

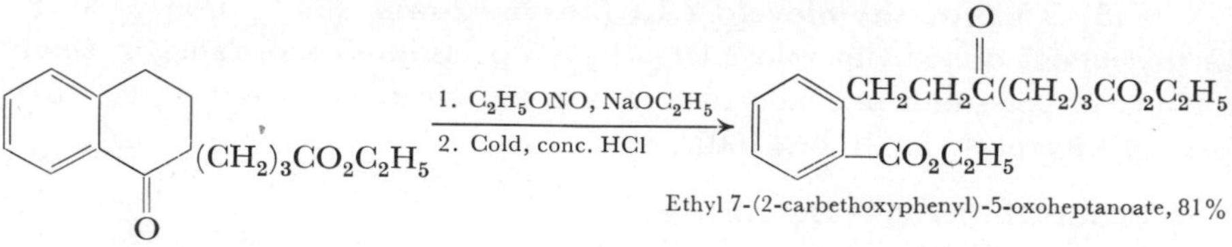

Ethyl 7-(2-carbethoxyphenyl)-5-oxoheptanoate, 81%

The Japp-Klingemann reaction is related to the nitrosation reaction and has some interesting possibilities in the preparation of α,β-diketones or α-ketoacids:

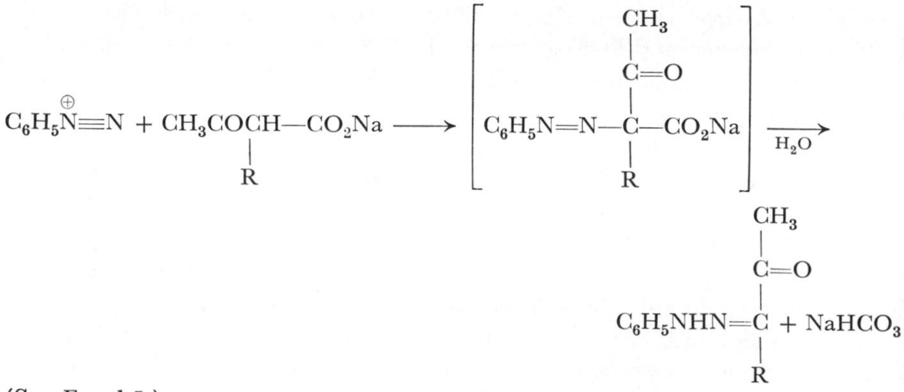

(See Ex. d.5.)

This reaction has been reviewed [74]. The acetoacetic ester is saponified before addition of the diazonium salt. Cyclic ketoesters undergo ring opening if the ketoester is not presaponified:

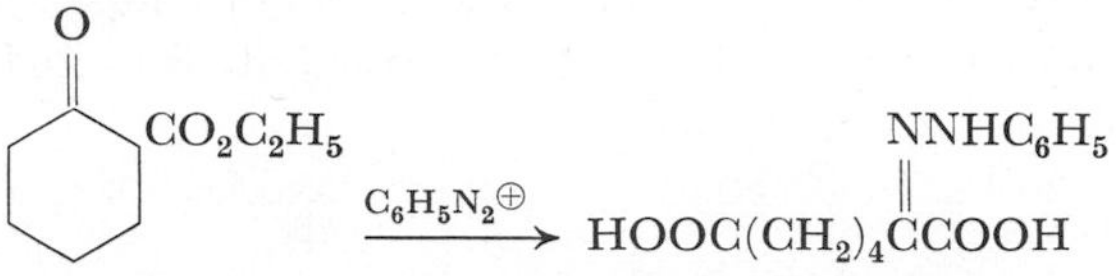

$$\text{C}_6\text{H}_5\text{N}_2^{\oplus} \longrightarrow \text{HOOC(CH}_2)_4\overset{\overset{\displaystyle\text{NNHC}_6\text{H}_5}{\|}}{\text{C}}\text{COOH}$$

α-Ketopimelic acid phenylhydrazone

(a) **Preparation of acetone** (75% based on hydrocarbon consumed in a 2:2:1 mixture of propane, oxygen, and hydrogen bromide at 190°) [75].

(b) **Air oxidation of methylene groups**

(1) **Methyl *p*-acetobenzoate** (40–42% conversion, 60% yield from methyl *p*-ethylbenzoate, air, chromium oxide, and calcium carbonate at 150°) [76]. (Similar oxidations have been effected to prepare *o*-nitroacetophenone [77] and the isomeric ethyl acetophenones [78]. The presence of *p*-nitrophenol in the air oxidation of *p*-nitroethylbenzene appears to reduce the accumulation of hydroperoxide and to increase the amount of *p*-nitroacetophenone formed to 75%, the highest yield recorded for this preparation) [79].

(2) **α-Tetralone** (23–36% conversion, 44–56% yield from tetrahydro-naphthalene, air, and no catalyst at 75°) [80].

(3) **2-Acetylfluorenone** (73% from 2-acetylfluorene, pyridine, and oxygen at room temperature; a variety of fluorenes gave yields of 45–91%; the anion of the fluorene

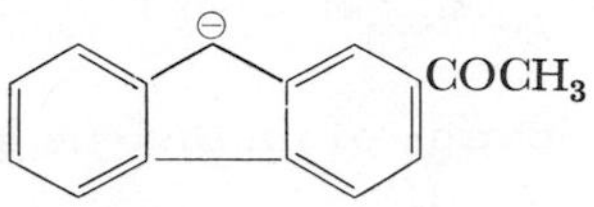

formed to a slight extent in pyridine solution is susceptible to oxidation, a fact which accounts for the strikingly mild conditions; see b.4) [81].

(4) **2,2′-Carbonyl-bis-(5-acetylthiophene)**(88% from 2,2′-methylene-bis-(5-acetylthiophene) in alcoholic potassium hydroxide and oxygen at room temperature) [82].

(5) **Methyl phenylglyoxylate** (86% based on the ester consumed from methyl phenylacetate, air, and cobalt benzoate at 110–115° for 36 hr.) [70].

(c) **Chemical oxidation of methylene groups**

(1) ***p*-Diacetylbenzene** (88% from *p*-ethylacetophenone and aqueous potassium permanganate and magnesium nitrate at 60°) [83].

(2) **2-Carboxyfluorenone** (67–74% from 2-acetylfluorene and sodium dichromate in acetic acid and acetic anhydride) [84].

(3) **4-Benzoylpyridine** (81% from 4-benzylpyridine and selenium dioxide in acetic acid) [85].

(4) **2-Benzoylpyridine** (86% from 2-benzylpyridine and aqueous potassium permanganate) [86].

(5) 6-Benzoylphenanthridine (72% from 6-benzyl-5,6-dihydrophenanthridine and sodium dichromate in acetic acid) [87].

(6) Δ²-Cyclohexenone (37% conversion from cyclohexene and chromic anhydride in aqueous acetic acid at room temperature; 25% adipic acid accompanies the ketone) [88]; the intermediate is probably an allylic free radical [89]:

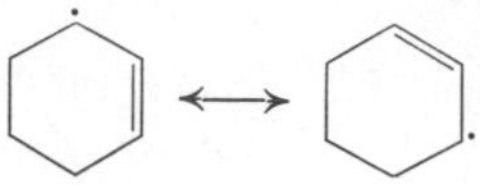

(7) Ethyl cyclohexylglyoxalate (76% based on the oximino ester from ethyl cyclohexylacetate, sodium nitrite, and sodium ethoxide at −10°; the intermediate oximino ester is hydrolyzed with formalin and hydrochloric acid) [90].

(8) Ethyl oxomalonate (74–76% from diethyl malonate and nitrogen trioxide) [91].

(9) Phenyl-9-oxodehydroabietinol (70% from phenyl dehydroabietinol acetate and CrO_3 in 80% aqueous acetic acid at 27°; the methylene group adjacent to aromatic ring is oxidized) [92].

(10) 2,2-Dimethyl-4-acetoxy-4-acetyltetrahydropyran (90% from the 4-ethyl derivative) [93]:

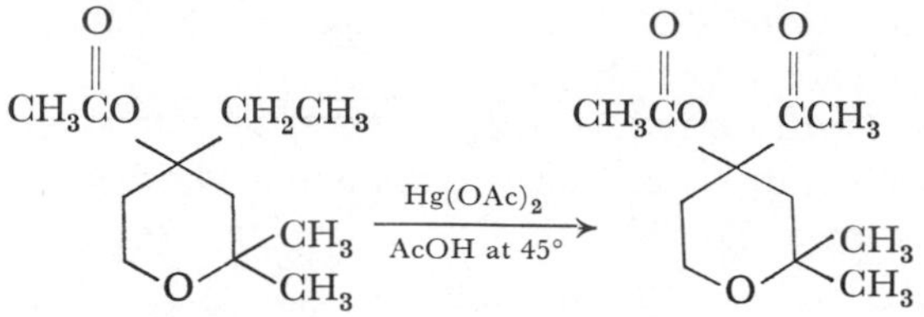

(d) Chemical oxidation of methylene groups in ketones

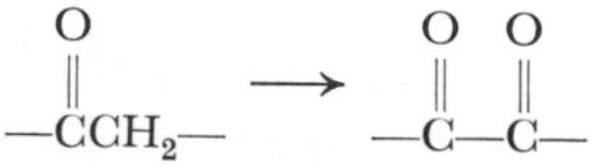

(1) 1,2-Cyclohexanedione (60% from cyclohexanone and selenium dioxide in aqueous dioxane) [94].

(2) Benzil (86% from phenyl benzyl ketone and selenium dioxide in acetic anhydride; many other examples are shown) [95].

(3) Ethyl α-oximinoacetoacetate (63% from ethyl acetoacetate, sodium nitrite, and glacial acetic acid) [96].

(4) 5-Methyl-1,2,3-cyclohexanetrione-1,3-dioxime (70% from 4-methylcyclohexanone, isoamyl nitrite, and conc. hydrochloric acid at −5°) [97].

(5) 1,2-Cyclohexanedione monophenylhydrazone (almost quantitative from ethyl cyclohexanone-2-carboxylate in cold aqueous potassium hydroxide and benzenediazonium salt) [74].

(6) Ethyl pyruvate o-nitrophenylhydrazone (83% from ethyl 2-methylacetoacetate in cold aqueous potassium hydroxide to which o-nitrobenzenediazonium salt was added immediately; no saponification took place and the product was deacetylated) [74]:

$$\text{CH}_3\text{COCH}(\text{CH}_3)\text{CO}_2\text{C}_2\text{H}_5 + \text{ArN}_2{}^{\oplus} \longrightarrow \underset{\underset{\text{R}}{|}}{\overset{\overset{\text{N}=\text{NAr}}{|}}{\text{CH}_3\text{COC}(\text{CH}_3)\text{CO}_2\text{C}_2\text{H}_5}} \xrightarrow{\text{OH}\ominus}$$

$$\overset{\text{NNHAr}}{\underset{\|}{}}$$

$$\text{CH}_3\text{CO}_2{}^{\ominus} + \text{CH}_3\overset{\|}{\text{C}}\text{CO}_2\text{C}_2\text{H}_5$$

7. From Tertiary Hydrocarbons via Hydroperoxides

$$\underset{\underset{\text{R}}{|}}{\overset{\overset{\text{R}}{|}}{\text{Ar}-\text{C}-\text{H}}} \xrightarrow{\text{O}_2} \underset{\underset{\text{R}}{|}}{\overset{\overset{\text{R}}{|}}{\text{Ar}-\text{C}-\text{OOH}}} \xrightarrow{\text{HClO}_4} \text{ArOH} + \text{RCOR}$$

This oxidation has been reviewed [98]. As has been indicated under Alcohols, Chapter 4, D.4a, hydrocarbons containing hydrogen atoms α to a double bond or to an aromatic ring system may be converted into hydroperoxides by air oxidation [99]. Such hydroperoxides have also been produced from triaryl-methyl chlorides by treatment with hydrogen peroxide and the catalyst stannic chloride [100]. It should be mentioned again that hydroperoxides, with the possible exception of those of tertiary alcohols, are explosive in nature. Although in acid medium (acetic with a trace of perchloric acid, sulfuric acid, or ferrous sulfate) good yields of the ketone may be obtained, the method has found limited use.

The decomposition of the hydroperoxide gives products which vary greatly with the nature of the medium. Fortunately, under acidic conditions (acetic acid in the presence of 0.1 mole percent of perchloric), the yield of the ketone and hydroxy compound may be quantitative as it is in the case of α,α-dimethyl benzyl hydroperoxide [101]. The reaction proceeds via the oxonium ion as shown:

$$\underset{\underset{\text{CH}_3}{|}}{\overset{\overset{\text{CH}_3}{|}}{\text{C}_6\text{H}_5\text{COOH}}} \xrightarrow[-\text{H}_2\text{O}]{\text{H}\oplus} \underset{\underset{\text{CH}_3}{|}}{\overset{\overset{\text{CH}_3}{|}}{\text{C}_6\text{H}_5\text{CO}\oplus}} \longrightarrow \underset{\underset{\text{CH}_3}{|}}{\overset{\overset{\text{CH}_3}{|}}{\oplus\text{C}-\text{OC}_6\text{H}_5}} \xrightarrow{\underset{\underset{\text{CH}_3}{|}}{\overset{\overset{\text{CH}_3}{|}}{\text{C}_6\text{H}_5\text{COOH}}}}$$

$$(\text{CH}_3)_2\text{CO} + \text{C}_6\text{H}_5\text{OH} + \underset{\underset{\text{CH}_3}{|}}{\overset{\overset{\text{CH}_3}{|}}{\text{C}_6\text{H}_5\text{CO}\oplus}}$$

(a) Preparation of p-nitrobenzophenone.

(1) **p-Nitrophenylmethylhydroperoxide.** To 1 g. of p-nitrotri-phenylmethyl chloride in 15 ml. of anhydrous ether at 0°, there was added 2 ml. of 90% hydrogen peroxide and the mixture was recooled in an ice bath. Stannic chloride, 0.5 ml., was added slowly with stirring after 5 min. and then, after 2 hr. or more in the ice bath, the ether solution was extracted three times with ice water, once with a saturated ice-cold solution of ferrous ammonium

sulfate, and finally with ice water to remove the ferrous ion. The ether, after drying with sodium sulfate, was evaporated and the residue when crystallized from alcohol-water gave crystals of the hydroperoxide monohydrate (61–80%), m.p. 99–101°.

(2) *p*-Nitrobenzophenone.   To *p*-nitrotriphenylmethylhydroperoxide 0.422 g., in 5 ml. of glacial acetic acid, there was added with cooling 8 ml. of perchloric acid. Addition of water gave the ketone which when washed and dried weighed 0.268 g. (94%), m.p. 136–137°. The phenol recovered as tribromophenol weighed 0.331 g. (81%), m.p. 92–93.5° [100].

(b) **Preparation of acetone** (74% from isopropylbenzene hydroperoxide and aqueous sulfuric acid) [102].

8. From Glycols

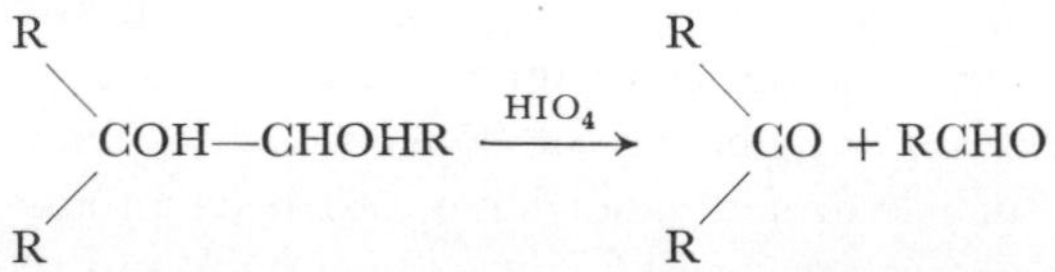

$$\underset{R}{\overset{R}{\diagdown}}\text{COH—CHOHR} \xrightarrow{\text{HIO}_4} \underset{R}{\overset{R}{\diagdown}}\text{CO} + \text{RCHO}$$

This degradation has been discussed under Aldehydes, Chapter 10, A.7. Whereas primary and secondary alcoholic groups lead to formaldehyde and a higher aldehyde, respectively, the tertiary groups produce ketones. The two principal oxidizing agents are periodic acid and lead tetraacetate [103]. Although this synthesis has found wider application in the aldehyde series, it has been used as well in the ketone series for preparatory or structural determination purposes, particularly among cyclic types such as the carbohydrates. Yields are sometimes high.

(a) **Preparation of Δ⁴-androstene-11-ol-3,17-dione** (11-II) (60% from Δ⁴-pregnene-11,17,20,21-tetrol-3-one, 11-I, and periodic acid) [104]:

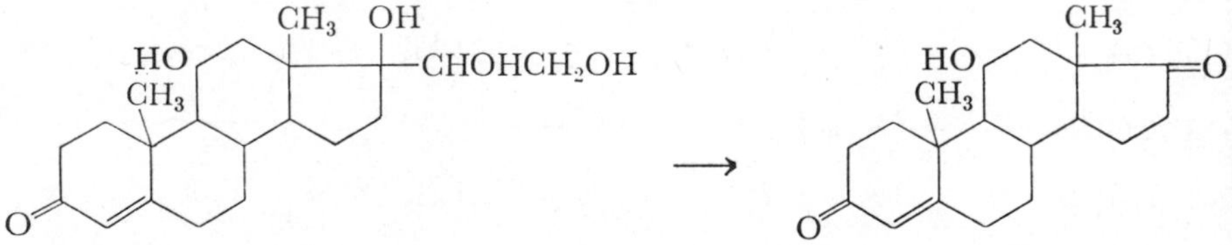

11-I   11-II

(b) **Preparation of etiocholan-3α-ol-11,17-dione** (85% from pregnan-3α,17α-diol-11,20-dione by reduction followed by oxidation with lead tetraacetate in acetic acid) [105]:

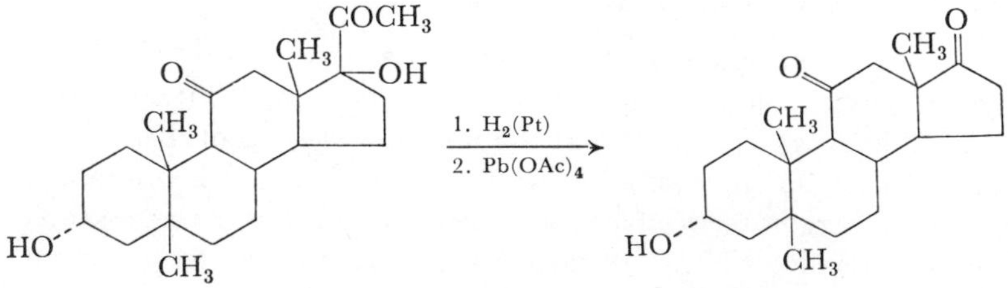

9. From Nitroalkanes, Sodium Salt (Nef), and Other Internal Redox Systems

The oxidations in this section comprise those in which one part of the molecule serves as an oxidizing agent for another part. Some similarity in

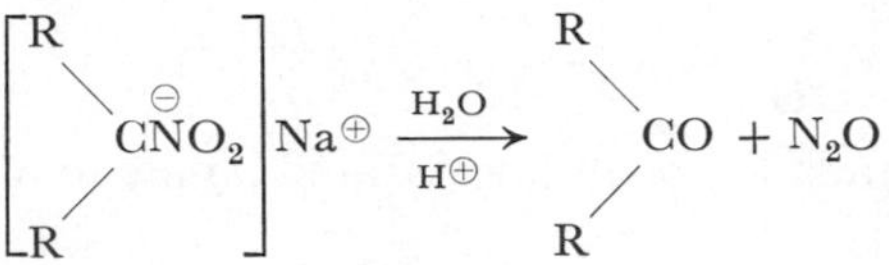

$$\left[\begin{array}{c} R \\ \diagdown \\ \overset{\ominus}{C}NO_2 \\ \diagup \\ R \end{array}\right] Na^{\oplus} \xrightarrow[H^{\oplus}]{H_2O} \begin{array}{c} R \\ \diagdown \\ CO + N_2O \\ \diagup \\ R \end{array}$$

overall results may be found in the internal Cannizzaro in which pyruvaldehyde

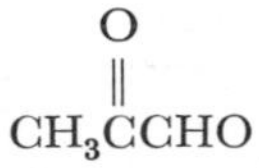

$$\underset{\underset{CH_3CCHO}{\parallel}}{O}$$

is converted into lactic acid ($CH_3CHOHCO_2H$) by alkali.

This synthesis has been reviewed [106] and discussed under Aldehydes (Chapter 10, A.18), to which type it is also applicable. The method gives good yields of ketones from simple nitroalkanes and from nitrocycloalkanes like nitrocyclohexane. In some cases, such as the adducts of cyclopentadiene and 1-nitroolefins, the reaction fails.

An elegant modification of the Nef reaction consists of titration of the potassium nitronate with aqueous potassium permanganate for which the reaction proceeds rapidly as follows:

$$3\ R_2C=NO_2K + 2\ KMnO_4 + H_2O \rightarrow 3\ R_2C=O + 2\ MnO_2 + 3\ KNO_2 + 2\ KOH$$

Not only ketones but also aldehydes can be prepared in this way, and furthermore the reaction is of greater scope than the Nef reaction (see Ex. b).

An interesting application of the Nef reaction is its use in the conversion of a ketone into an isomeric one, the specific example being the conversion of cholestane-3-one into cholestane-2-one [107]:

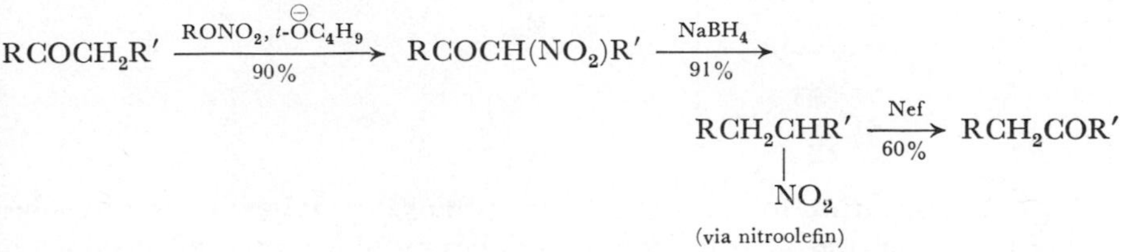

$$RCOCH_2R' \xrightarrow[90\%]{RONO_2,\ t\text{-}\overset{\ominus}{O}C_4H_9} RCOCH(NO_2)R' \xrightarrow[91\%]{NaBH_4}$$

$$RCH_2\underset{\underset{NO_2}{|}}{CHR'} \xrightarrow[60\%]{Nef} RCH_2COR'$$

(via nitroolefin)

Other types of internal oxidation-reduction can be visualized. If the oxidation state of a carbon atom containing a hydroxyl, alkoxyl, or amino group is raised by insertion of another amino group via rearrangement, an intermediate is obtained which can be hydrolyzed to a ketone:

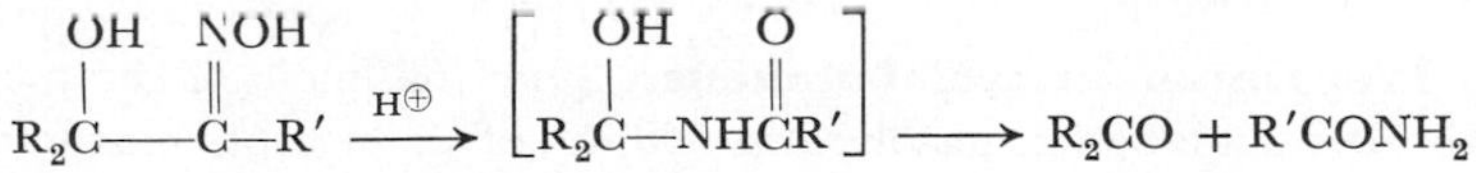

$$\underset{\underset{R_2C\text{---}C\text{---}R'}{|\qquad\ |}}{\overset{OH\ \ NOH}{\ \ |\quad\ \ \parallel}} \xrightarrow{H^{\oplus}} \left[\underset{\underset{R_2C\text{---}NHCR'}{|\qquad\quad}}{\overset{OH\quad O}{\ |\qquad \parallel}}\right] \longrightarrow R_2CO + R'CONH_2$$

For example, the Beckmann rearrangement of an oxime of a cyclic β-ketoether yields a series of ketones [108]:

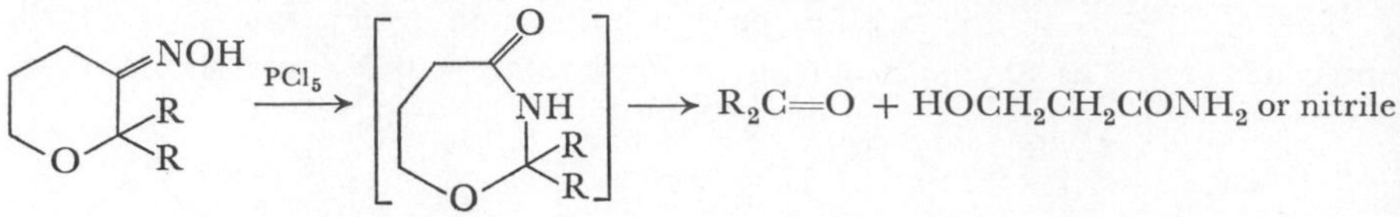

NOH
PCl5
R
R
O
O
NH
R
R
O
R2C=O + HOCH2CH2CONH2 or nitrile

Or the Hofmann degradation of an unsaturated amide to a vinylamine should lead to a ketone [109]:

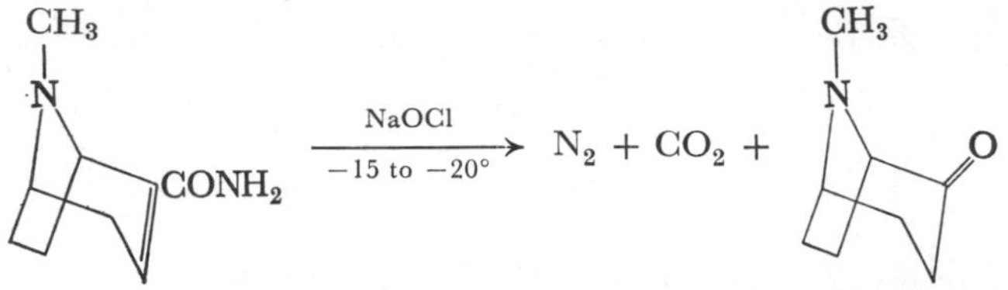

CH3
N
CONH2
NaOCl
−15 to −20°
N2 + CO2 +
CH3
N
O

L-(+)-2-Tropinone, 64%

And, as a final example, but not one that exhausts the field, the preparation of ketoacids from amino acids by the Weygand technique can be construed as another example of internal oxidation-reduction [110]:

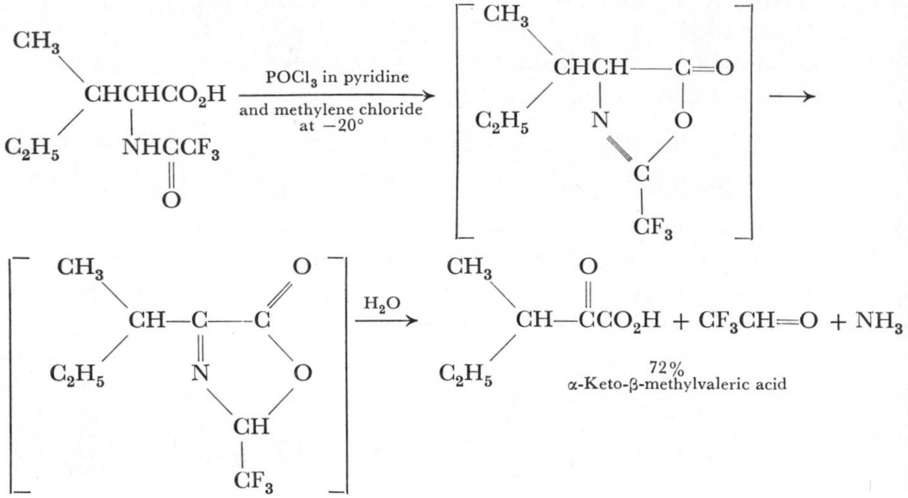

CH3
CHCHCO2H
C2H5
NHCCF3
O
POCl3 in pyridine
and methylene chloride
at −20°
CHCH—C=O
N
O
C
CF3
CH—C—C
H2O
CH—CCO2H + CF3CH=O + NH3
72%
α-Keto-β-methylvaleric acid

(a) Preparation of 2(2,3-dimethoxyphenyl)cyclohexanone. A solution of 1-(2,3-dimethoxyphenyl)-2-nitrocyclohexane, 4.8 g. in 30 ml. of ethanol, was treated with 5 g. of sodium hydroxide in 20 ml. of ethanol and the mixture was added slowly with stirring to 400 ml. of 10% sulfuric acid at 0°. The total was kept in ice for 15 min., warmed to 60°, cooled, extracted with ether, and distilled. The ketone, 2.4 g., b.p. 180–190° (bath) (0.1 mm.) crystallized from ether-light petroleum, m.p. 67–69° [111].

(b) Preparation of cyclobutanone. One millimole of the potassium salt of nitrocyclobutane was dissolved in 500 ml. of water containing 0.01 mole of potassium hydroxide and 0.04 mole of magnesium sulfate. The solution at 0–5° was titrated with aqueous permanganate until the purple color persisted. The ketone was removed by steam distillation and was converted into the 2,4-dinitrophenylhydrazone, 94% yield [112].

(c) **Preparation of 9,10-dihydro-9,10-(11-ketoethano)-anthracene** (87% from 9,10-dihydro-9,10-(11-nitroethano)-anthracene made alternately basic (pH 10) with potassium hydroxide and acidic (pH 2) with hydrochloric acid) [113].

10. From Alkenes via Hydroboration

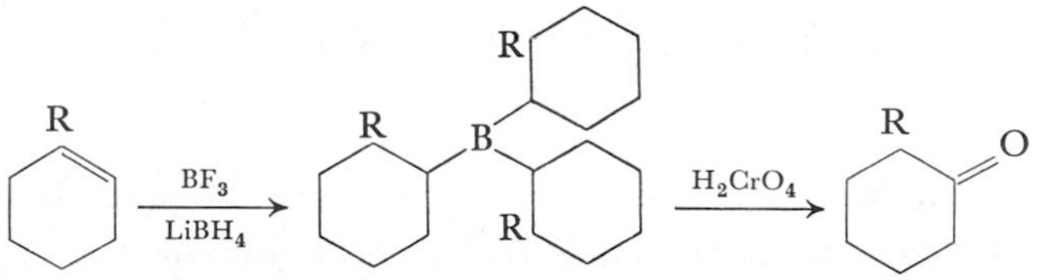

It has already been shown that organoboranes when oxidized with alkaline hydrogen peroxide yield a product which, when subjected to alkaline hydrolysis, yields alcohols (Chapter 4, B.2). If chromic acid is used as the oxidizing agent, ketones instead of alcohols are obtained [114]. The procedure is simple. The olefin is hydroborated in ethyl ether; a slight excess of aqueous chromic acid is added; and the mixture is refluxed for 2 hours. The method is a convenient one for synthesizing 2-alkyl or 2-arylcycloalkanones from the unsubstituted ketone:

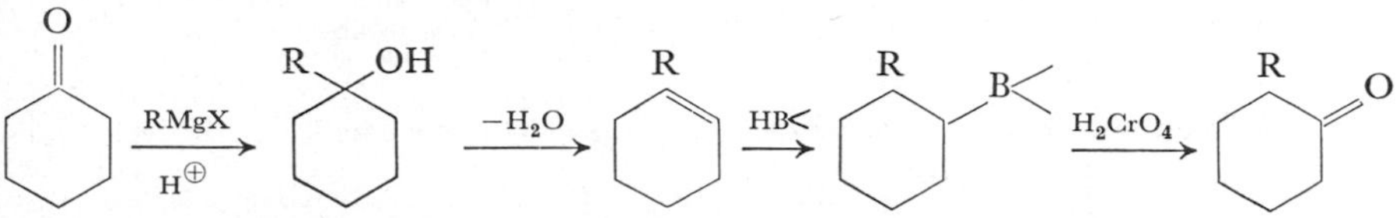

Yields in the oxidation of 1-alkyl- and 1-arylcycloalkenes vary from 63 to 87%.

(a) **Preparation of 2-methylcyclohexanone** (87% from 1-methyl-cyclohexene, lithium borohydride, and boron trifluoride etherate in ether followed by the addition of sodium dichromate dihydrate, sulfuric acid and water) [114].

11. From Alkenes with Cyanogen Azide

Cyanogen azide, a highly reactive reagent, which can be handled safely in dry acetonitrile, is made easily from cyanogen bromide and sodium azide. It reacts with olefins to form cyanoiminoketones which can be hydrolyzed to ketones [115]:

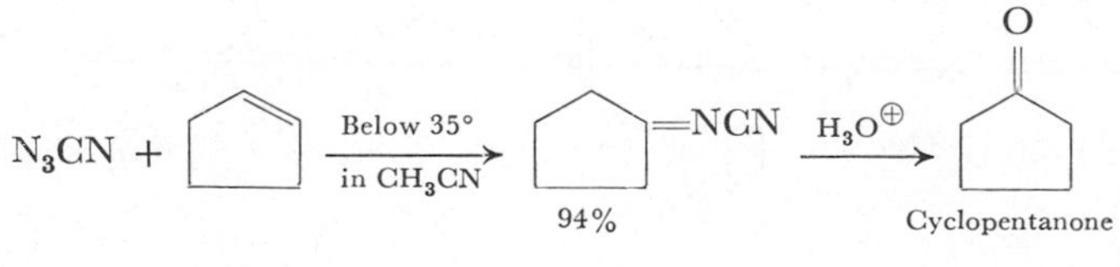

12. From Carboxylic Acids or Acid Anhydrides

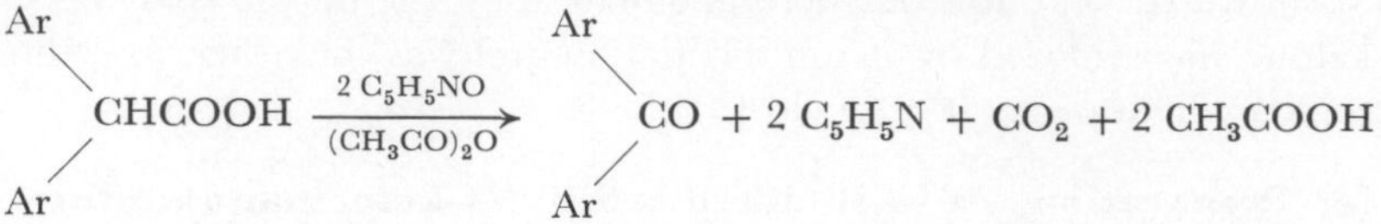

$$\begin{array}{c} Ar \\ \diagdown \\ CHCOOH \\ \diagup \\ Ar \end{array} \xrightarrow[\text{(CH}_3\text{CO)}_2\text{O}]{2\,C_5H_5NO} \begin{array}{c} Ar \\ \diagdown \\ CO \\ \diagup \\ Ar \end{array} + 2\,C_5H_5N + CO_2 + 2\,CH_3COOH$$

This synthesis, which gives either aldehydes or ketones, is discussed under Aldehydes, Chapter 10, A.20.

13. From Amines

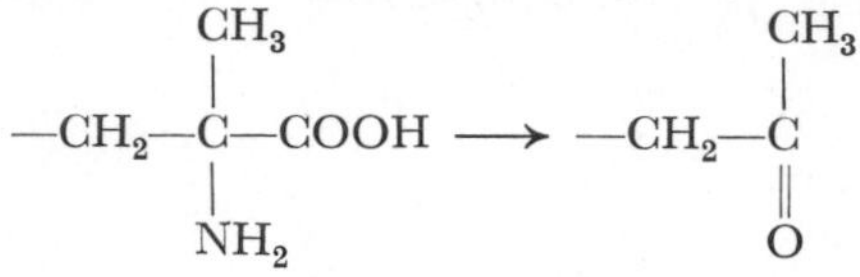

$$\begin{array}{c} CH_3 \\ | \\ -CH_2-C-COOH \\ | \\ NH_2 \end{array} \longrightarrow \begin{array}{c} CH_3 \\ | \\ -CH_2-C \\ \| \\ O \end{array}$$

Oxidation of amines to iminoketones or enamines represents a potential pathway to ketones. The reaction is not general, but has some specialized applications as shown in the examples. Example a is similar to the oxidation of an amino acid to an aldehyde (Aldehydes, Chapter 10, H.3).

In addition, dienamines may be oxidized and hydrolyzed to unsaturated 1,4-diketones [116]:

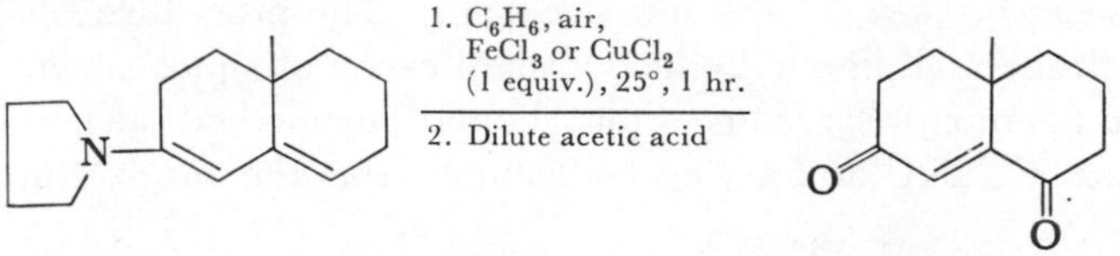

1. C_6H_6, air, FeCl$_3$ or CuCl$_2$ (1 equiv.), 25°, 1 hr.

2. Dilute acetic acid

10-Methyl-$\Delta^{1(9)}$-octalin-2,8-dione, 80–85%

(a) Preparation of 3,4-dimethoxyphenylacetone. α-Methyl 3,4-

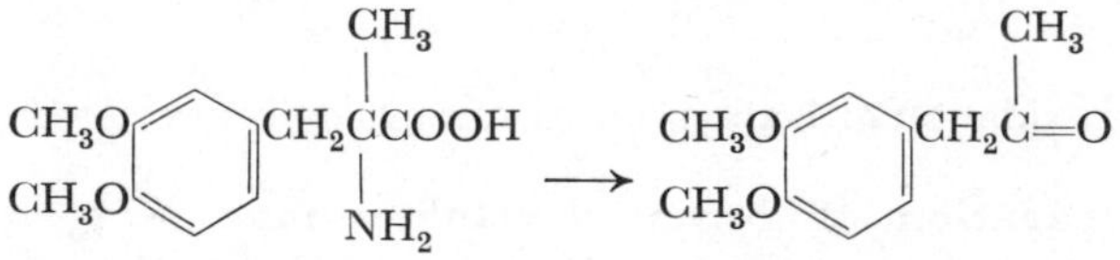

dimethoxyphenylalanine (4 mmoles) in 25 ml. of water with 10 ml. of benzene was titrated with 14 ml. of sodium hypochlorite (0.3 N active chlorine) until the starch-iodide test was positive. The ketone was extracted from the aqueous layer with benzene-ether, dried, and concentrated to yield 92 % of 3,4-dimethoxy-phenylacetone [117].

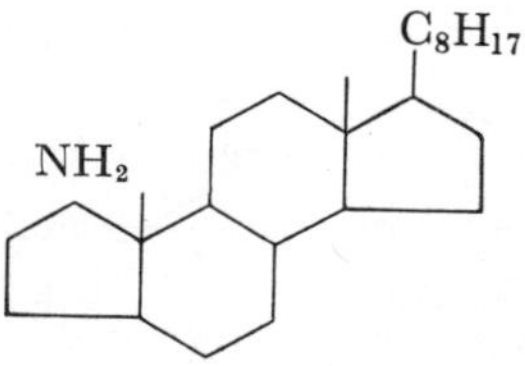

(b) Preparation of A-norcholestane-1-one (73 % from noramine and *t*-butyl hypochlorite, followed by dehydrohalogenation to the iminoketone with sodium ethoxide and by hydrolysis with dilute sulfuric acid) [118].

14. From Halides

The hypothetical route is a combined solvolytic-oxidative procedure. In the majority of cases it is best simply to oxidize the alcohol corresponding to the halide, but in specialized cases the combined solvolytic-oxidative technique offers unique advantages as shown in Ex. a.

Aryl-substituted phenacyl bromides are solvolyzed and oxidized to diketones by a combination of sodium acetate and dimethyl sulfoxide (see Aldehydes, Chapter 10, A.10).

(a) Preparation of tetrachlorocyclopentene-1,2-dione (90% from

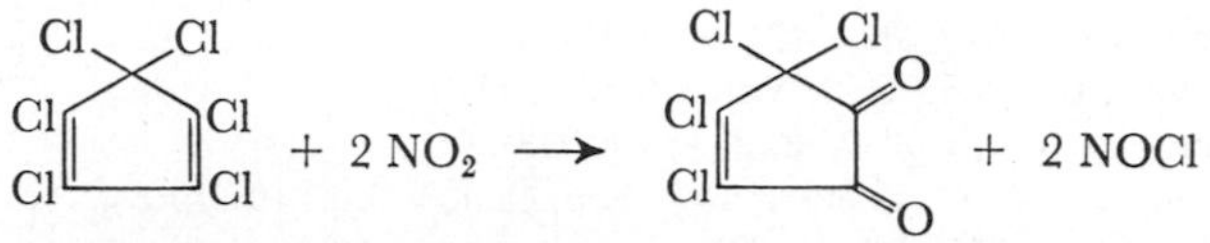

hexachlorocyclopentadiene and NO_2 at 60° for 6 hr. in an autoclave; the product is somewhat unstable in moist air) [119].

15. From Epoxides

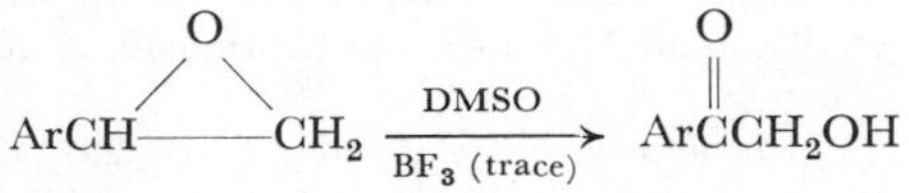

Oxidation of epoxides with dimethyl sulfoxide in the presence of catalytic amounts of boron trifluoride produces α-hydroxyketones in yields from 45 to 76% [120].

(a) Preparation of 2-hydroxycyclohexanone. Cyclohexene oxide,

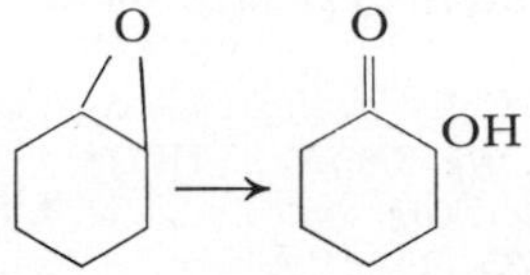

10 g. (0.102 mole), and 0.12 ml. of boron trifluoride etherate in 40 ml. of dimethyl sulfoxide were heated on a steam bath for 22 hr. More catalyst (0.06 and 0.04 ml.) was added at the end of 15 and 20 hr., respectively. After pouring the mixture in ice water and extracting with chloroform, the dried extract was distilled to give 8.8 g. (76%) of 2-hydroxycyclohexanone [120].

1. W. A. Waters, *Mechanisms of Oxidation of Organic Compounds*, John Wiley and Sons, New York, 1964; R. Stewart, *Oxidation Mechanisms. Applications to Organic Chemistry*, W. A. Benjamin, New York, 1964.
2. K. Bowden *et al.*, *J. Chem. Soc.*, 39 (1946); E. R. H. Jones *et al.*, *ibid.*, 157, 2548, 3019 (1953); see also C. Djerassi *et al.*, *J. Org. Chem.*, **21**, 1547 (1956).
3. A. S. Hussey and R. H. Baker, *J. Org. Chem.*, **25**, 1434 (1960).
4. R. Filler, *Chem. Rev.*, **63**, 21 (1963).

5. L. F. Fieser and S. Rajagopalan, *J. Am. Chem. Soc.*, **71**, 3935, 3938 (1949).

6. E. P. Oliveto *et al.*, *J. Am. Chem. Soc.*, **75**, 1505 (1953).

7. L. F. Fieser and S. Rajagopalan, *J. Am. Chem. Soc.*, **72**, 5530 (1950).

8. K. E. Pfitzner and J. G. Moffatt, *J. Am. Chem. Soc.*, **85**, 3027 (1963).

9. K. E. Pfitzner and J. G. Moffatt, *J. Am. Chem. Soc.*, **87**, 5661, 5670 (1965).

10. B. R. Baker and D. H. Buss, *J. Org. Chem.*, **30**, 2304 (1965).

11. B. R. Baker and D. H. Buss, *J. Org. Chem.*, **30**, 2308 (1965).

12. J. D. Albright and L. Goldman, *J. Am. Chem. Soc.*, **89**, 2416 (1967).

13. H. C. Brown and C. P. Garg, *J. Am. Chem. Soc.*, **83**, 2952 (1961).

14. J. D. Albright and L. Goldman, *J. Org. Chem.*, **30**, 1107 (1965).

15. E. F. Pratt and J. F. Van de Castle, *J. Org. Chem.*, **26**, 2973 (1961).

16. R. C. Cookson *et al.*, *Chem. Commun.*, 744 (1966).

17. K. Heyns and H. Paulsen, in W. Foerst, *Newer Methods of Preparative Organic Chemistry*, Vol. 2, Academic Press, New York, 1963, p. 303.

18. J. O. Halford and B. Weissman. *J. Org. Chem;* **18**, 30 (1953).

19. E. J. Eisenbraun, *Org. Syn.*, **45**, 28 (1965).

20. J. Meinwald *et. al.*, *Org. Syn.*, **45**, 77 (1965).

21. L. T. Sandborn, *Org. Syn.*, Coll. Vol. **1**, 340 (1941).

22. W. F. Bruce, *Org. Syn.*, Coll. Vol. **2**, 139 (1943).

23. J. R. Schaeffer and A. O. Snoddy, *Org. Syn.*, Coll. Vol. **4**, 19 (1963).

24. L. Knof, *Ann. Chem.*, **656**, 183 (1962).

25. A. E. Vanstone and J. S. Whitehurst, *J. Chem. Soc.*, (C) 1972 (1966).

26. C. Djerassi, *Org. Reactions*, **6**, 207 (1951).

27. H. Schinz *et al.*, *Helv. Chim. Acta*, **31**, 2235 (1948); R. Rüegg and O. Jeger, *ibid.*, **31**, 1753 (1948); A. Lauchenauer and H. Schinz, *ibid.*, **32**, 1265 (1949).

28. R. B. Woodward *et al.*, *J. Am. Chem. Soc.*, **67**, 1425 (1945); *see also* H. Rapoport *et al.*, *J. Org. Chem.*, **15**, 1103 (1950) and M. Gates and G. Tschudi, *J. Am. Chem. Soc.*, **78**, 1380 (1956).

29. E. W. Warnhoff and P. Reynolds-Warnhoff, *J. Org. Chem.*, **28**, 1431 (1963).

30. R. V. Oppenauer, *Org. Syn.*, Coll. Vol. **3**, 207 (1955); J. F. Eastham and R. Teranishi, *Org. Syn.*, Coll. Vol. **4**, 192 (1963).

31. A. Wettstein and Ch. Meystre, *Org. Reactions*, **6**, 235 (1951).

32. K. Mihara and K. Takaoka, *Nippon. Kagaku. Zasshi*, **79**, 282 (1958); *C.A.*, **54**, 4366 (1960).

33. R. E. Dunbar and M. R. Arnold, *J. Org. Chem.*, **10**, 501 (1945).

34. R. Paul, *Bull. Soc. Chim.*, *France*, **8**, 507 (1941).

35. E. C. Kleiderer and E. C. Kornfeld, *J. Org. Chem.*, **13**, 455 (1948).

36. Houben-Weyl's *Methoden der Organischen Chemie*, 4th ed., Vol. 4, G. Thieme Verlag, Stuttgart, 1955, Pt. 2, p. 339.

37. V. I. Komarewsky and J. R. Coley, *J. Am. Chem. Soc.*, **63**, 700, 3269 (1941).

38. R. Adams and C. S. Marvel, *Org. Syn.*, **1**, 25 (1921).

39. H. T. Clarke and E. E. Dreger, *Org. Syn.*, Coll. Vol. **1**, 87 (1941).

40. E. Baer, *J. Am. Chem. Soc.*, **62**, 1597 (1940).

41. A. Lespagnol and P. Ponthieu, *Bull. Soc. Chim.*, *France*, **11**, 541 (1944).

42. R. Kuhn *et al.*, *Chem. Ber.*, **76**, 900 (1943).

43. M. Weiss and M. Appel, *J. Am. Chem. Soc.*, **70**, 3666 (1948).

44. L. P. McHatton and M. J. Soulal, *J. Chem. Soc.*, 2771 (1952).

45. A. T. Blomquist and A. Goldstein, *Org. Syn.*, Coll. Vol. **4**, 838 (1963).

46. P. Rüggli and P. Zeller, *Helv. Chim. Acta*, **28**, 741 (1945).

47. C. A. Buehler, *Chem. Rev.*, **64**, 7 (1964).

48. L. P. McHatton and M. J. Soulal, *J. Chem. Soc.*, 4095 (1953).

49. T. Reichstein *et al.*, *Helv. Chim. Acta*, **27**, 821 (1944).

50. N. C. Cook and F. C. Whitmore, *J. Am. Chem. Soc.*, **63**, 3540 (1941).

51. K. E. Wilzbach *et al.*, *J. Am. Chem. Soc.*, **70**, 4069 (1948).

52. H. Vorbrüggen and C. Djerassi, *Tetrahedron Letters*, 119 (1961); *J. Am. Chem. Soc.*, **84**, 2990 (1962).

53. E. D. Bergmann et al., J. Chem. Soc., 1369 (1950).

54. J. E. Horan and R. W. Schiessler, Org. Syn., **41**, 53 (1961).

55. H. Hart and L. R. Lerner, J. Org. Chem., **32**, 2669 (1967).

56. W. H. Clement and C. M. Selvitz, J. Org. Chem., **29**, 241 (1964).

57. H. Hock et al., Chem. Ber., **83**, 227 (1950).

58. C. D. Gutsche and F. A. Fleming, J. Am. Chem. Soc., **76**, 1771 (1954).

59. R. A. Micheli, J. Org. Chem., **27**, 666 (1962).

60. J. Meinwald and P. G. Gassman, J. Am. Chem. Soc., **82**, 2857 (1960).

61. V. Martello et al., Chim. Ind. (Milan), **38**, 932 (1956); C.A., **51**, 5714 (1957); see also O. Touster, Org. Reactions, **7**, 327 (1953).

62. C. D. Nenitzescu et al., Tetrahedron, **24**, 4667 (1968).

63. G. A. Russell and A. G. Bemis, J. Am. Chem. Soc., **88**, 5491 (1966).

64. N. Rabjohn, Org. Reactions, **5**, 331 (1949).

65. M. Godchot and G. Cauquil, Compt. Rend., **202**, 326 (1936).

66. J. Vene, Compt. Rend., **216**, 772 (1943).

67. E. J. Corey and J. P. Schaefer, J. Am. Chem. Soc., **82**, 918 (1960).

68. J. P. Schaefer and E. J. Corey, J. Org. Chem., **24**, 1825 (1959).

69. E. B. Reid and J. F. Yost, J. Am. Chem. Soc., **72**, 5232 (1950).

70. P. G. Sergeev and A. M. Sladkov, J. Gen. Chem. USSR (Eng. Transl.), **27**, 895 (1957).

71. B. Camerino et al., Tetrahedron Letters, 554 (1961).

72. E. D. Bergmann and I. Shahak, J. Chem. Soc., 1418 (1959).

73. R. Huisgen et al., Ann. Chem., **641**, 63, 71 (1961).

74. R. R. Phillips, Org. Reactions, **10**, 143 (1959).

75. F. F. Rust and W. E. Vaughan, Ind. Eng. Chem., **41**, 2595 (1949).

76. W. S. Emerson and G. F. Deebel, Org. Syn., Coll. Vol. **4**, 579 (1963).

77. W. S. Emerson et al., J. Am. Chem. Soc., **69**, 706 (1947).

78. E. D. Bergmann and C. Resnik, J. Org. Chem., **17**, 1291 (1952).

79. M. Schulz et al., German Patent 1,276,675, May 9, 1968; C.A., **69**, 7184 (1968).

80. R. B. Thompson, Org. Syn., **20**, 94 (1940).

81. Y. Sprinzak, J. Am. Chem. Soc., **80**, 5449 (1958).

82. T. L. Cairns et al., J. Am. Chem. Soc., **73**, 1270 (1951).

83. J. Holsten and E. H. Pitts, Jr., J. Org. Chem., **26**, 4151 (1961).

84. G. Rieveschl, Jr., and F. E. Ray, Org. Syn., Coll. Vol. **3**, 420 (1955).

85. R. E. Lyle et al., J. Org. Chem., **24**, 330 (1959).

86. E. H. Huntress and H. C. Walter, J. Am. Chem. Soc., **70**, 3702 (1948).

87. H. Gilman and J. J. Eisch, J. Am. Chem. Soc., **79**, 4423 (1957).

88. F. C. Whitmore and G. W. Pedlow, Jr., J. Am. Chem. Soc., **63**, 758 (1941).

89. H. O. House, Modern Synthetic Reactions, W. A. Benjamin, New York, 1965, p. 101.

90. R. Fischer and T. Wieland, Chem. Ber., **93**, 1387 (1960).

91. A. W. Dox, Org. Syn., Coll. Vol. **1**, 266 (1941).

92. R. P. Jacobsen, J. Am. Chem. Soc., **73**, 3463 (1951).

93. S. A. Vartanyan and V. N. Zhamagortsyan, Izv. Akad. Nauk Arm., SSR, Khim. Nauk. **15**, 353 (1962); C.A., **59**, 3874 (1963).

94. C. C. Hach et al., Org. Syn., Coll. Vol. **4**, 229 (1963).

95. H. H. Hatt et al., J. Chem. Soc., 93 (1936).

96. O. Touster, Org. Reactions, **7**, 327 (1953).

97. D. C. Batesky and N. S. Moon, J. Org. Chem., **24**, 1694 (1959).

98. C. E. Frank, Chem. Rev., **46**, 155 (1950).

99. E. H. Farmer and A. Sundralingam, J. Chem. Soc., 121 (1942).

100. P. D. Bartlett and J. D. Cotman, Jr., J. Am. Chem. Soc., **72**, 3095 (1950).

101. M. S. Kharasch et al., J. Org. Chem., **15**, 748, 753, 763 (1950).

102. G. P. Armstrong et al., J. Chem. Soc., 666 (1950).

103. E. L. Jackson, Org. Reactions, **2**, 341 (1944).

104. T. Reichstein, Org. Reactions, **2**, 364 (1944).

105. H. L. Herzog et al., J. Am. Chem. Soc., **75**, 266 (1953).

106. W. E. Noland, *Chem. Rev.*, **55,** 137 (1955).
107. A. Hassner *et al.*, *J. Org. Chem.*, **33,** 1733 (1968).
108. R. K. Hill, *J. Org. Chem.*, **27,** 29 (1962).
109. M. R. Bell and S. Archer, *J. Am. Chem. Soc.*, **82,** 4642 (1960).
110. F. Weygand *et al.*, *Ann. Chem.*, **658,** 128 (1962).
111. J. A. Barltrop and J. S. Micholson, *J. Chem. Soc.*, 2524 (1951).
112. H. Shechter and F. T. Williams, Jr., *J. Org. Chem.*, **27,** 3699 (1962).
113. W. E. Noland *et al.*, *J. Am. Chem. Soc.*, **78,** 2233 (1956).
114. H. C. Brown and C. P. Garg, *J. Am. Chem. Soc.*, **83,** 2951 (1961).
115. F. D. Marsh *et al.*, *Chem. Eng. News* (Dec. 27, 1965), p. 29.
116. S. K. Malhotra *et al.*, *J. Am. Chem. Soc.*, **90,** 6565 (1968).
117. H. L. Slates *et al.*, *J. Org. Chem.*, **29,** 1424 (1964).
118. M. P. Cava and B. R. Vogt, *Tetrahedron Letters*, 2813 (1964).
119. R. M. Scribner, *J. Org. Chem.*, **30,** 3657 (1965).
120. T. Cohen and T. Tsuji, *J. Org. Chem.*, **26,** 1681 (1961).

B. Reduction

1. From Nitroalkenes

$$\underset{\underset{NO_2}{|}}{ArCH{=}CR} \xrightarrow[CH_3COOH]{Zn} ArCH_2COR$$

This synthesis occurs for nitroalkenes or nitrocycloalkenes in which the nitro group is attached to a doubly bound carbon atom. The steps involved in the reduction are [1]:·

$$\underset{\underset{NO_2}{|}}{ArCH{=}CR} \xrightarrow[HCl]{Fe} \left[\underset{\underset{NHOH}{|}}{ArCH{=}CR}\right] \longrightarrow \underset{\underset{NOH}{\|}}{ArCH_2{-}CR} \xrightarrow[HCl]{H_2O} \underset{\underset{O}{\|}}{ArCH_2CR}$$

If sufficient acid is present, the reduction and hydrolysis to the ketone may be carried out in one operation. The method has been of value in the steroid series in that nitro groups may be substituted on doubly bound carbon atoms by nitration and these groups in turn may be replaced by a carbonyl oxygen atom [2, 3]. Common reducing agents are zinc and acetic acid or iron and hydrochloric acid alone or in the presence of ferric chloride. The yields are fair.

(a) **Preparation of o-methoxyphenylacetone** (63–71% from o-methoxybenzaldehyde, nitroethane, *n*-butylamine, and iron, ferric chloride, and hydrochloric acid) [4]:

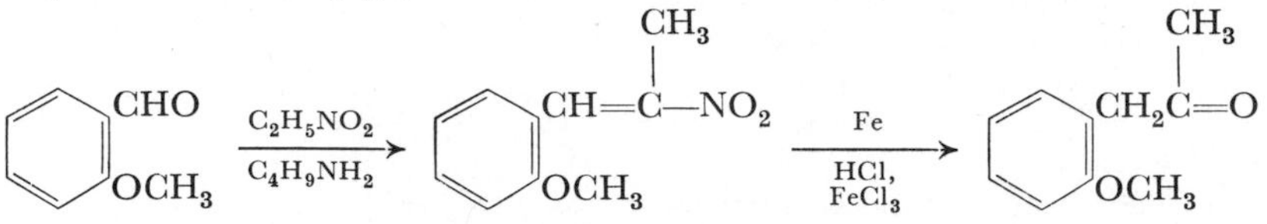

(b) **Other examples**

(1) **Cholestane-3β-ol-6-one acetate** (73% from 6-nitrocholesteryl acetate, zinc dust, and acetic acid) [3].

References for Section B are on pp. 650–651.

(2) 1-Anisylbutan-2-one (86% as the 2,4-dinitrophenylhydrazone from 1-anisyl-2-nitro-1-butene in glacial acetic acid and hydrogen over 10% palladized charcoal) [5].

2. From Phenols

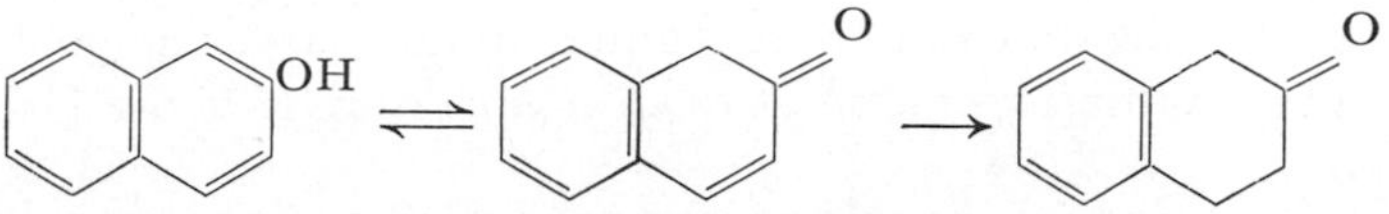

Phenols such as resorcinol and β-naphthol may be partially reduced to cyclo-alkanones with reducing agents such as hydrogen in the presence of catalysts, Raney nickel or reduced Universal Oil Products hydrogenation catalyst [6, 7], copper chromite [8], and palladium with N-ethylmorpholine [9]. It is obvious that the synthesis would have limited application although in some cases the yields are high.

(a) Preparation of dihydroresorcinol (1,3-cyclohexanedione) (85–95% from resorcinol, sodium hydroxide, reduced Universal Oil Products hydrogenation or Raney nickel catalyst, and hydrogen at 1000–1500 p.s.i. at 50°) [6]; for low-pressure hydrogenation with rhodium on alumina catalyst (85%) see [10].

(b) Preparation of β-tetralone (80% from β-naphthol, copper chromite catalyst, and hydrogen at 1500–2500 p.s.i. at 200°) [8].

3. From Phenols or Phenol Ethers (Birch Reduction)

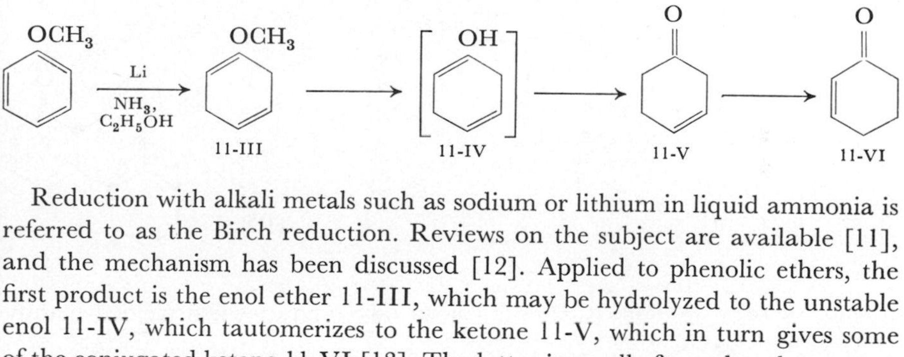

Reduction with alkali metals such as sodium or lithium in liquid ammonia is referred to as the Birch reduction. Reviews on the subject are available [11], and the mechanism has been discussed [12]. Applied to phenolic ethers, the first product is the enol ether 11-III, which may be hydrolyzed to the unstable enol 11-IV, which tautomerizes to the ketone 11-V, which in turn gives some of the conjugated ketone 11-VI [13]. The latter is usually formed to the greatest extent. However, hydrolysis of the enol ether with oxalic acid or 95% acetic acid favors the unconjugated ketone, while with a mineral acid the product is the conjugated ketone [14]. Reduction of either leads to the saturated ketone.

Since enol ethers hydrolyze readily to ketones (D.3), this synthesis offers a method for converting aromatic ethers into ketones in simple or complex ring systems, such as exist in the steroids [15]. In these cases experimental conditions, such as the metal used, the method of its addition, and the amount of ethanol employed, are important.

The phenols, in contrast to the ethers, may be reduced by increasing the lithium metal substrate level from 1.5 g. atoms/mole of phenol to 4 g. atoms/mole [16]. The liquid ammonia should be free from colloidal iron salts which accelerate the rate of consumption of the metal [17].

An aliphatic secondary amine-lithium-alcohol combination serves as a comparable reducing medium to the liquid ammonia system and has the advantage of permitting operation at room temperature or above [18].

(a) Preparation of 2-cyclohexenone and 3-cyclohexenone as 2,4-dinitrophenylhydrazones. Anisole, 15 g., in 50 ml. of ether and 200 ml. of liquid ammonia was treated with 4.5 g. of lithium (see original article [13] for details) and then with 35 g. of absolute ethanol over a 30-min. period. Extraction with ether and evaporation gave 12.8 g. (84%) of 2,5-dihydroanisole, 80% pure, b.p. 148–149° (745 mm.). Hydrolysis with aqueous hydrochloric acid in the presence of 2,4-dinitrophenylhydrazine in refluxing solution gave the conjugated ketone derivative (95%); in the cold the unconjugated ketone derivative was obtained (88%) [13].

(b) Preparation of *dl*-13,14-dehydro-18-nor-D-homoepiandrosterone (11-VIII) **and *dl*-16,17-dehydro-18-nor-D-homoepiandrosterone** (II-IX). From 1 g. of 11-VII treated with lithium, liquid ammonia, and

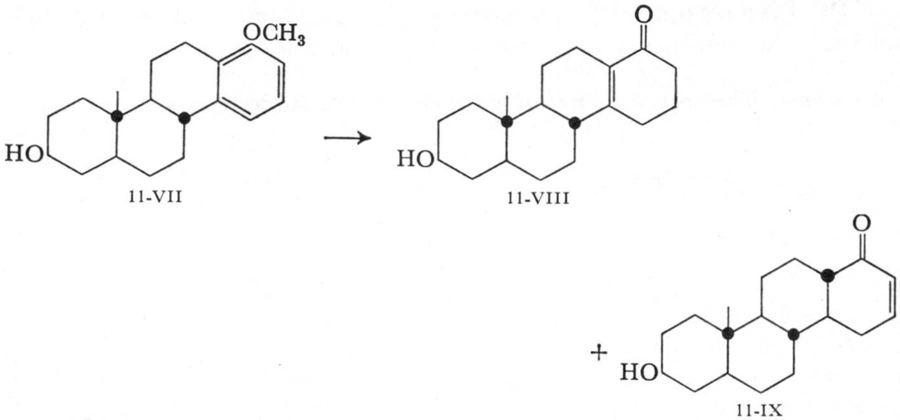

11-VII   →   11-VIII

+   11-IX

ethanol there was obtained 0.980 g. of an oily residue which yielded mainly 11-VIII and 11-IX [15].

4. From Furans

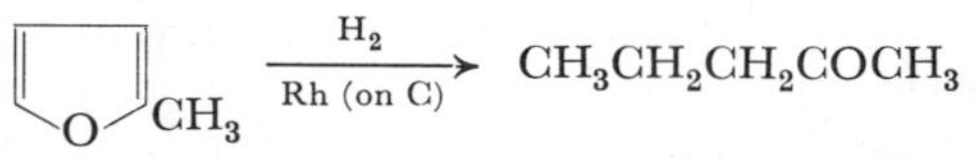

$$\text{furan-CH}_3 \xrightarrow[\text{Rh (on C)}]{\text{H}_2} CH_3CH_2CH_2COCH_3$$

The products obtained on reduction of furans vary greatly with the nature of the catalyst, the temperature, and the pressure. As a rule, at lower temperatures hydrogen in the presence of catalysts such as osmium and palladium saturate the ring only. Other catalysts such as platinum on carbon or rhodium on carbon lead to ring cleavage as well, the extent with the latter catalyst being almost quantitative at temperatures around 300° [19].

(a) Preparation of pentanone-2. 2-Methylfuran, 30 g., was hydrogenated in the vapor phase over rhodium on carbon at a volume rate of 0.1 hr.$^{-1}$. The product, after being dried and distilled through a column of 40 theoretical plates, gave a return of approximately 100% of pentanone-2 [19]. This ketone has also been prepared in approximately 95% yield by hydrogenation in the presence of Raney Pt or Raney Cu (prepared by NaOH leaching of 10% Pt-Al or 30% Cu-Al, respectively, at 275°) [20].

5. From α-Diketones and α-Ketols

$$C_6H_5COCOC_6H_5 \xrightarrow{\text{HI}} C_6H_5COCH_2C_6H_5$$

$$C_2H_5COC(OH)(C_2H_5)_2 \xrightarrow{\text{HI}} C_2H_5COCH(C_2H_5)_2$$

α-Diketones and α-ketols, with the exception of benzoins, can be reduced to saturated ketones with hydriodic acid in refluxing acetic acid [21]. This synthesis offers some advantage over the zinc dust-hydrochloric acid reduction of α-ketols such as sebacoin [22] in that temperature control is not so important and yields are comparable.

The steps in the reduction of α-diketones appear to be two: reduction to the α-ketol; and reduction of the α-ketol to the ketone. Since benzil is difficult to reduce to desoxybenzoin, different mechanisms have been advanced for enolizable and nonenolizable ketones. The original reference [21] should be consulted for the possible mechanistic paths.

(a) Preparation of cyclodecanone. Sebacil, over 92% pure, 3.2 g. in 40 ml. of acetic acid, and 35 g. of 47% hydriodic acid were refluxed for 2.5 hr. The cooled mixture was poured into a two- or three-fold quantity (by volume) of cold sodium hydroxide solution containing enough sodium bisulfite to remove the iodine color. Three or four extractions with ether and recovery from the ethereal solutions in the usual manner gave 2.5 g. of a yellow oil which proved to be 94% cyclodecanone by vapor-phase chromatography (80%) [21].

(b) Preparation of cholestan-4-one (70% from 4-acetoxycholestan-3-

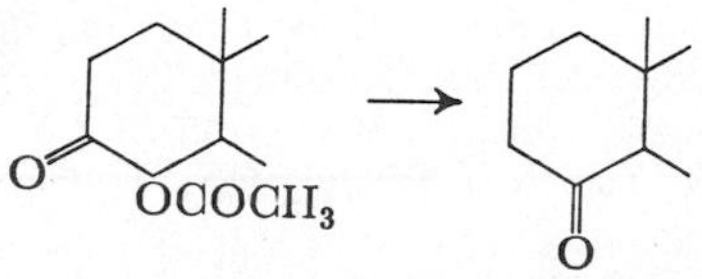

one, hydriodic and acetic acids) [21].

6. From Unsaturated Ketones

Under conditions similar to the pinacol reduction, conjugated ketones frequently couple to give diketones:

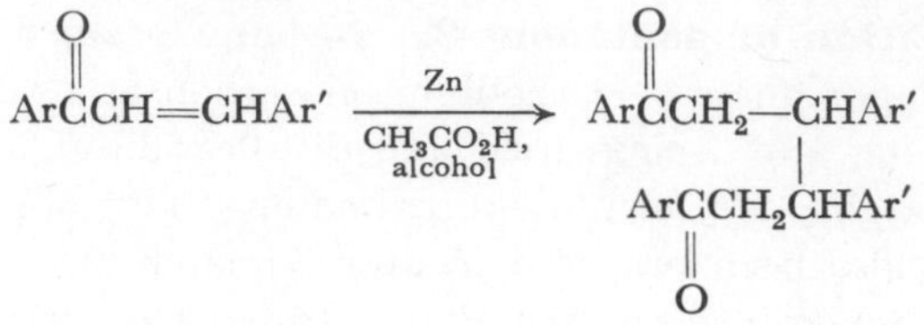

$$\underset{O}{\overset{O}{\underset{\parallel}{ArCCH}}} =CHAr' \xrightarrow[\substack{CH_3CO_2H, \\ alcohol}]{Zn} \underset{O}{\overset{O}{\underset{\parallel}{ArCCH_2}}} -CHAr'$$

$$\underset{\parallel}{\overset{}{ArCCH_2CHAr'}}$$

Amalgamated aluminum and sodium also have been used as reducing agents. Aliphatic unsaturated ketones are apt to cyclize after reductive coupling: [23]

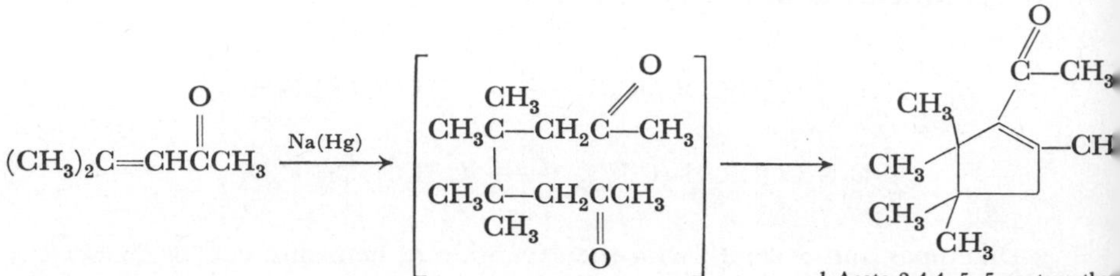

1-Aceto-2,4,4, 5, 5-pentamethyl cyclopentene-1, 60–70%

(a) Preparation of 1,6-di-*o*-methoxyphenyl-3,4-diphenylhexane-1,6-dione (67% of a separable mixture of *meso* and *dl* forms from benzal *o*-methoxyacetophenone and zinc dust in refluxing alcohol containing acetic acid) [24].

7. From Acid Chlorides

Reduction of acid chlorides ordinarily leads to aldehydes. Reductive coupling, on the other hand, should yield an α,β-diketone. In a reaction restricted to hindered acid chlorides, the coupling catalyst, magnesium-magnesium iodide does lead to the desired product [25]:

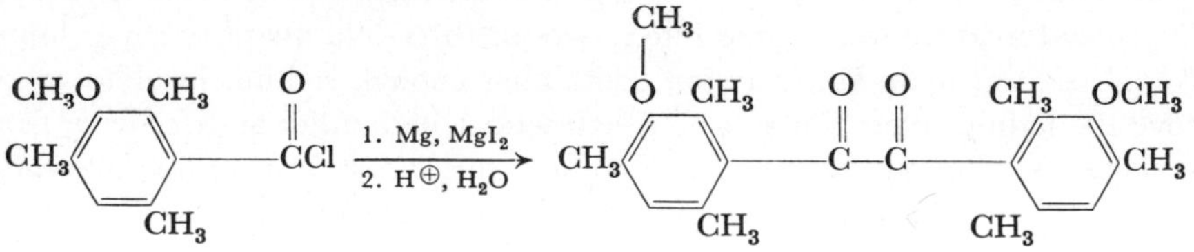

3,3'-Dimethoxymesitil, 62%

1. H. B. Hass *et al.*, *J. Org. Chem.*, **15,** 8 (1950).
2. L. Blunschy *et al.*, *Helv. Chim. Acta*, **29,** 199 (1946).
3. C. E. Anagnostopoulos and L. F. Fieser, *J. Am. Chem. Soc.*, **76,** 532 (1954).
4. R. V. Heinzelman, *Org. Syn.*, Coll. Vol. **4,** 573 (1963).
5. T. C. Myers *et al.*, *J. Am. Chem. Soc.*, **77,** 5655 (1955).
6. R. B. Thompson, *Org. Syn.*, Coll. Vol. **3,** 278 (1955); D. Papa *et al.*, *J. Org. Chem.*, **14,** 366 (1949).
7. D. Papa, *J. Am. Chem. Soc.*, **71,** 3246 (1949).
8. H. Adkins *et al.*, *J. Am. Chem. Soc.*, **70,** 4247 (1948).

9. G. Stork and E. L. Foreman, *J. Am. Chem. Soc.*, **68,** 2172 (1946).

10. J. C. Sircar and A. I. Meyers, *J. Org. Chem.*, **30,** 3206 (1965).

11. A. J. Birch *et al.*, *Quart. Rev.*, **4,** 69 (1950); **12,** 17 (1958).

12. J. K. Brown *et al.*, *Tetrahedron Letters*, 2621 (1966).

13. A. L. Wilds and N. A. Nelson, *J. Am. Chem. Soc.*, **75,** 5360 (1953).

14. D. Burn and V. Petrow, *J. Chem. Soc.*, 364 (1962); P. Radlick, *J. Org. Chem.*, **30,** 3208 (1965); B. Weinstein and A. H. Fenselau, *ibid.*, **30,** 3209 (1965).

15. W. S. Johnson *et al.*, *J. Am. Chem. Soc.*, **78,** 6331 (1956).

16. J. Fried *et al.*, *J. Am. Chem. Soc.*, **89,** 1044 (1967).

17. L. F. and M. Fieser, *Reagents for Organic Synthesis*, John Wiley and Sons, New York, 1967. p. 54.

18. R. A. Benkeser *et al.*, *J. Org. Chem.*, **28,** 1094 (1963).

19. N. I. Shuĭkin and I. F. Bel'skii, *J. Gen. Chem. USSR (Eng. Transl.)*, **29,** 1063 (1959).

20. N. I. Shuĭkin and I. F. Bel'skii, *Dokl. Akad. Nauk SSSR*, **125,** 345 (1959); *C.A.*, **53,** 20015 (1959).

21. W. Reusch and R. LeMahieu, *J. Am. Chem. Soc.*, **86,** 3068 (1964).

22. A. C. Cope *et al.*, *Org. Syn.*, Coll. Vol. **4,** 218 (1963).

23. C. Harries and F. Hübner, *Ann. Chem.*, **296,** 295 (1897).

24. J. Jack *et al.*, *J. Chem. Soc.*, 3684 (1954).

25. R. C. Fuson *et al.*, *J. Am. Chem. Soc.*, **63,** 2645 (1941).

C. Friedel-Crafts and Related Acylations

The vast scope of acylation reactions in ketone synthesis has been reviewed recently [1]. It is not the intent to repeat what is to be found in that text but rather to outline the general principles and to make several comments which will be of help to the synthetic chemist. The most common Friedel-Crafts acylation reaction involves the attack of a complexed acyl halide on an aromatic ring:

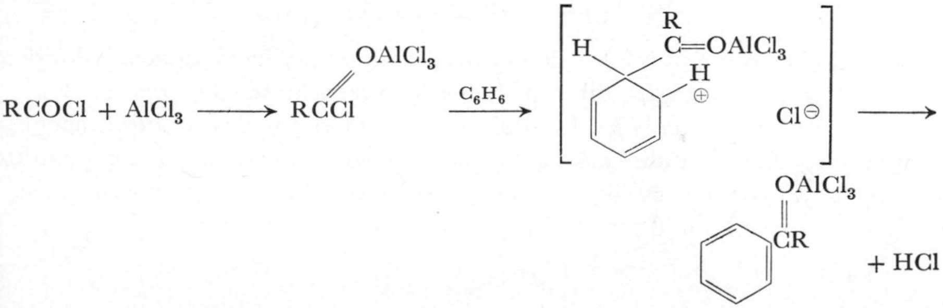

The catalyst. The purpose of the catalyst is to develop the maximum positive charge on the carbon atom of the acyl group so that it can attack the aromatic nucleus. Under the most favorable circumstances, it is doubtful that the free acylium ion

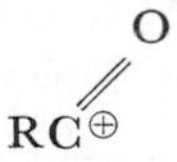

or the complexed acylium ion

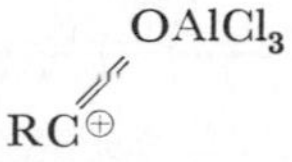

References for Section C are on pp. 665–667.

is the attacking agent. It appears more likely that these ions are complexed with anions or solvent molecules. Aluminum chloride (or bromide) without solvent is probably the most reactive reagent, and its action can be moderated by the use of such solvents as carbon disulfide, nitrobenzene, or nitromethane. The solvents have an additional advantage in bringing the aluminum chloride complexes into solution. On the other hand, iodine, or occasionally no catalyst at all, can be used in the acylation of such active nuclei as exist in anisole, thiophene, or polynuclear hydrocarbons [2]. A great variety of catalysts of intermediate activity (see [3] for a detailed discussion) have been described. In recent years the use of polyphosphoric acid has become more popular.

The acyl group. Any acid derivative can be used, but the acyl halide or anhydride is the most suitable. At least 1 equivalent of aluminum chloride for acyl halides, or 2 equivalents for anhydrides is necessary for maximum yields of ketones. The acylating agent is bulky, weakly electrophilic, and therefore highly selective in its orienting ability. Thus *para* substitution is favored, and deactivated nuclei, such as exist in acetophenones, aromatic acid derivatives, benzonitriles, nitrobenzenes, quinolines, pyridines, and the like, cannot be acylated. Acid halides tend to eliminate carbon monoxide in the presence of aluminum chloride if the remaining portion of the alkyl group is stable as a cation:

$$R_3CCOCl \xrightarrow{-CO} R_3C^{\oplus}$$

$$HCOCl \xrightarrow{-CO} H^{\oplus}$$

$$ClCOCOCl \longrightarrow {}^{\oplus}\overset{\displaystyle O}{\overset{\parallel}{C}}Cl$$

Thus oxalyl chloride in the Friedel-Crafts reaction may yield an acid chloride, a ketone, or a diketone, the diketone being favored with reactive nuclei.

The aromatic nucleus or olefin. In addition to aromatic hydrocarbons, phenols, aryl ethers, halides, anilines, and the like, the following types of compounds can be acylated: trialkylsilylbenzenes, ferrocenes, furans and furoic esters, thiophenes, pyrroles and pyrrole esters, carbazoles, indolizines

pyrazoles, imidazolones (but not imidazoles), hydroxythiazoles (but not thiazoles), phenothiazines (but not phenazine). In addition, olefins, hydrocarbons which can be dehydrogenated to olefins, and acetylenes can be acylated (the Kondakov reaction, known also as the Darzens reaction).

1. From Aromatic or Heterocyclic Compounds and Acylating Agents

$$ArH + RCOCl \xrightarrow{AlCl_3} Ar\overset{\displaystyle OAlCl_3}{\overset{\parallel}{C}}R \quad + HCl$$

This synthesis has been reviewed thoroughly [4].

An important point of interest is that acylation is usually irreversible because of resonance stabilization existing between the acyl group and the aromatic nucleus. However, if the acyl group is tilted out of the plane of the aromatic nucleus by *ortho*-substituents, several interesting consequences result:

The acylation becomes reversible. Thus acylation can be adjusted to give a kinetically controlled ketone or a thermodynamically controlled ketone. As an example, acetylation of naphthalene in nitrobenzene yields the kinetically controlled product, α-acetonaphthalene, but, in ethylene chloride or in no solvent, the thermodynamically controlled product, β-acetonaphthalene results. The Hayashi rearrangement is another example of the reversible nature of some acylations [5]:

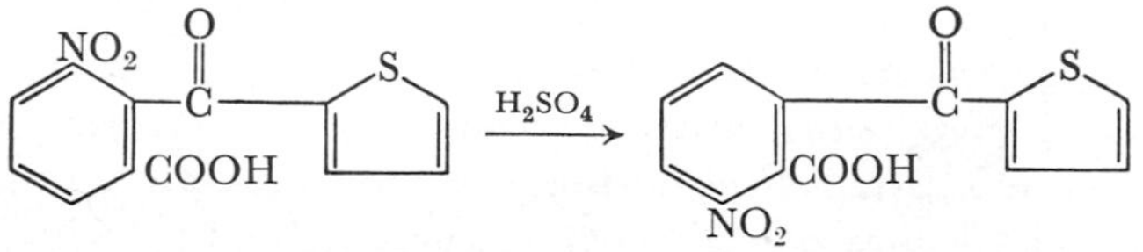

6-Nitro-2(2-thenoyl)-benzoic acid

Diacylation can be accomplished. Since the acetyl group is tilted out of the plane of the ring, the aromatic nucleus is not so deactivated as would be expected. Mesitylene, for example, can be diacetylated.

Rearrangement of alkyl groups can occur. Alkylbenzenes are readily rearranged or disproportionated. If an equilibrium exists between the complexed ketone and hydrocarbon, the alkyl groups can migrate to positions where the acylation

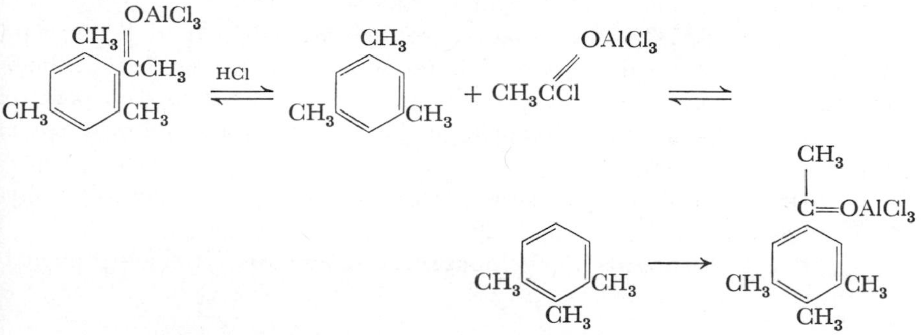

product is formed irreversibly as shown. Fortunately, this reaction does not occur at temperatures below 120°. Nevertheless, to avoid alkyl position isomerization and disproportionation, it is best to use the Perrier modification of the Friedel-Crafts reaction in which the acylation agent and aluminum chloride are first complexed and the aryl hydrocarbon is added last. An example of disproportionation is shown when acetyl chloride is added slowly to a slurry of ethylbenzene and aluminum chloride in carbon disulfide in that the main product is 2,4-diethylacetophenone (unpublished observation by D.E.P); see [6] for isomerization in acetylation of *p-t*-butyltoluene.

With polyphosphoric acid (PPA) as catalyst and solvent, the yields are highest when the reactants are soluble in PPA or when intramolecular acylation is involved [7, 8]. Thus phenols and their ethers are the most suitable for inter-molecular and ω-arylalkanoic acids for intramolecular acylations (see Exs. c.6 and c.9). The acylation in PPA proceeds via the mixed anhydride [9]

(a) **Preparation of 3-benzoylpyridine** (90–96% from nicotinic acid, thionyl chloride, benzene, and anhydrous aluminum chloride) [10].

(b) **Preparation of α-tetralone** (91–96% from benzene, γ-butyrolactone, and anhydrous aluminum chloride [11]; 75–86% from γ-phenylbutyric acid and polyphosphoric acid heated to 90° [12].

(c) **Other examples**

(1) **3-Acetyl-2,5-dimethylfuran** (87% from 2,5-dimethylfuran, acetic anhydride, and stannic chloride) [13].

(2) **2,2′,4,4′-Tetrahydroxybenzophenone** (75% from β-resorcylic acid, resorcinol, phosphorus oxychloride, and fused zinc chloride) [14].

(3) **Methyl β-(2,3,4-trimethoxybenzoyl) propionate** (79% from trimethylpyrogallol, β-carbomethoxypropionic acid, and polyphosphoric acid) [15].

(4) **3-Chloro-4-fluoroacetophenone** (80% from o-chlorofluoroben-zene, acetyl chloride, aluminum chloride in carbon disulfide; orientation control by the fluorine atom) [16].

(5) **4,4′-Bis-(dimethylamino)-benzil** (38–42% from dimethylaniline, oxalyl chloride, and aluminum chloride in carbon disulfide) [17].

(6) **6-Acetyl-2-methoxynaphthalene** (82% from 2-methoxynaphtha-lene, acetic acid in PPA at 80–85° for several hours) [18].

(7) **2-Methyl-3,4-dimethoxyacetophenone** (70% from 2,3-dimeth-oxytoluene, acetic anhydride, and PPA, the substitution taking place in the 6-position of the starting compound; if aluminum chloride is used in place of PPA or if the acylating agent is bulkier, appreciable 5-substitution takes place) [19].

(8) **4-Aminobenzophenone** (84% from benzanilide, benzoic anhy-dride, and PPA at 150° for 3 hr.) [20].

(9) **1,2,5,6-Dibenz-1,5-cyclooctadiene-3-one** (93% from o-(β-phenyl-

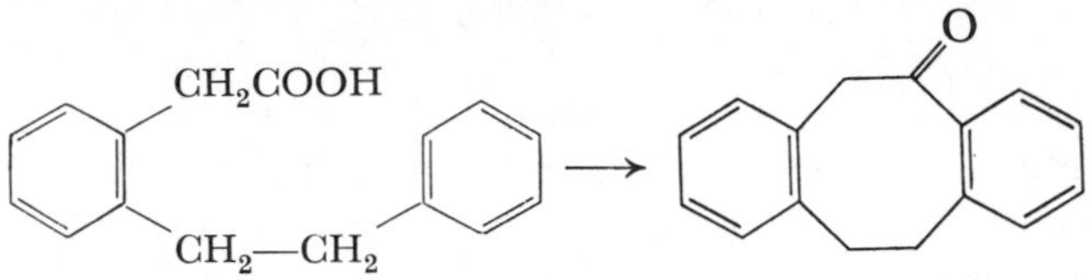

ethyl)-phenylacetic acid in hot PPA for 2 hr.) [21].

2. From Phenols and Nitriles (Hoesch)

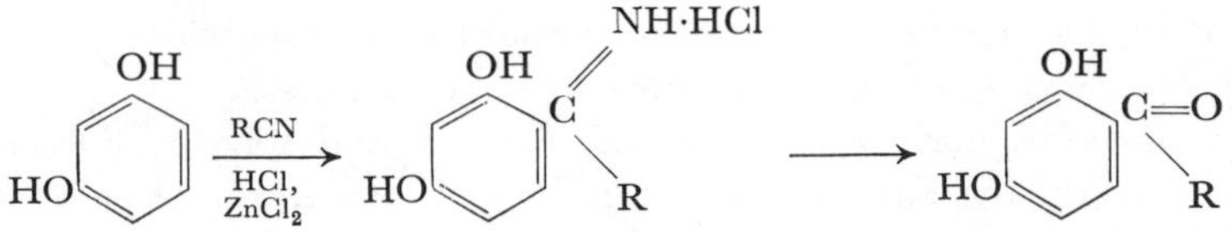

Excellent treatments of this synthesis are available [22]. The four components in the reaction are: an aromatic compound such as a phenol, a phenol ether, a pyrrole or furan derivative; a nitrile; an acid such as hydrogen chloride or sulfuric acid; and a catalyst such as zinc chloride, zinc bromide, aluminum chloride, or ferric chloride, but in some cases no catalyst is needed.

In this synthesis no shifts of substituents occur. Yields for aromatic compounds are improved if more than one substituent of the donor type (OH, OR, R) is present, particularly if they occupy the *m*-position, for in this way activation occurs on the same carbon atoms. A halogen atom *ortho* to a hydroxyl reduces the activity, whereas a nitro group in the same position inhibits the reaction. The reaction can be applied to hydrocarbons and their halogen derivatives, particularly if aluminum chloride is used as the catalyst and if an excess of a hydrocarbon solvent is employed. Yields vary somewhat, but in most favorable cases 75–90 % is achieved.

It is now generally thought that the attacking particle is a complexed carbonium ion

$$R-\overset{\oplus}{C}=NH$$

formed by the addition of a proton to the nitrile. This ion attacks the phenol to give the stablest σ-complex, 11-X:

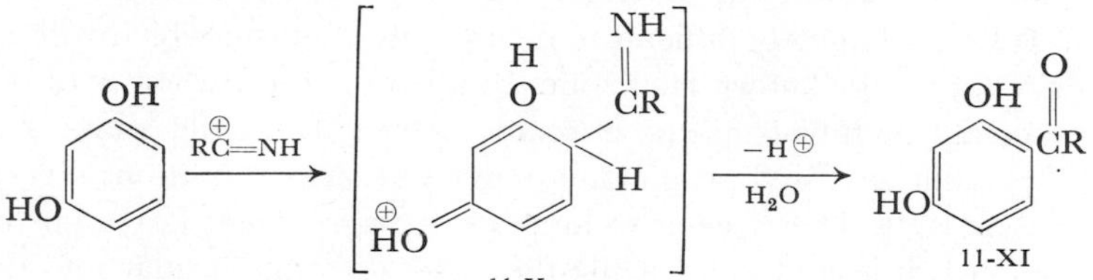

which loses a proton and hydrolyzes to form the ketone 11-XI.

(a) **Preparation of phloroacetophenone** (74–87 % from phloroglucinol, anhydrous acetonitrile, HCl, and fused zinc chloride) [23].

(b) **Other examples**

(1) **2,4-Dimethoxy-6-hydroxyphenyl benzoylhydroxymethyl ketone** (62 % from phloroglucinol dimethyl ether, benzoylhydroxyaceto-

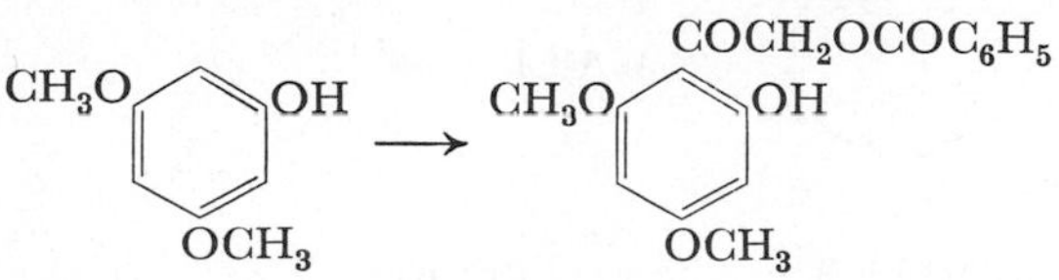

nitrile, and HCl) [24].

3. **From Phenolic Esters (Fries Rearrangement)**

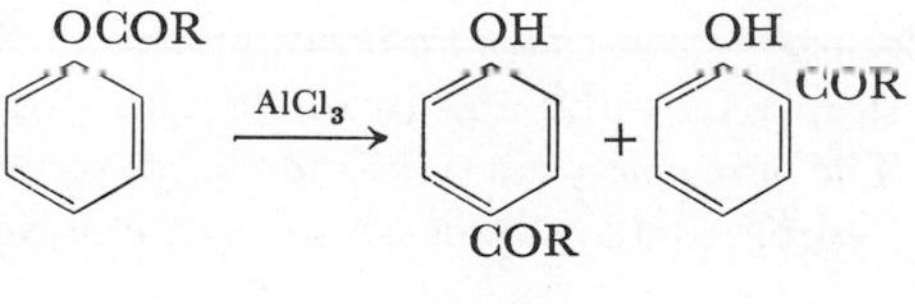

Several reviews are available on this synthesis [25, 26]. The synthesis may be regarded as a variant of the Friedel-Crafts reaction since the catalysts are identical; the phenol and acid chloride of the F.-C. can produce the ester utilized in the Fries; and phenolic ketones may be prepared by both methods. However, in many cases the Fries method is the better of the two for the preparation of the ketones of phenols, in that it avoids the complication resulting from the formation of aluminum phenoxides. It suffers, of course, from the fact that o- and p-hydroxyketones are usually both produced.

The catalysts, mostly metallic halides, employed in the reaction present great variety. The most useful halides appear to be aluminum chloride, stannic chloride, titanium tetrachloride, and zinc chloride. Other types of service have been p-toluenesulfonic acid, hydrogen fluoride, boron trifluoride, and particularly PPA; sometimes no catalyst is employed. In the latter case, irradiation is used in a polar solvent such as methanol, and in such cases the reaction is called the photo-Fries. For example, 28% 2,2'- and 32% 2,4'-dihydroxybenzophenone are obtained from the irradiation of phenyl salicylate [27]. The thermal reaction is ordinarily carried out by heating a mixture of the ester and catalyst to 80–180° or by using a solvent at a lower temperature. The usual solvents are nitrobenzene, tetrachloroethane, carbon disulfide, or chlorobenzene.

In spite of many studies of the Fries reaction, the mechanism is not as yet clear. It has been extremely difficult to devise a mechanism which will account satisfactorily for all the known facts; one troublesome illustration may be cited. m-Cresyl acetate in nitrobenzene at room temperature gives the 4-hydroxyketone [28]; while m-cresyl propionate under the same conditions gives the 2-hydroxyketone [29]. There seems to be tentative agreement [25] that the *para* rearrangement is intermolecular while the *ortho* is intramolecular and that the attacking acylium ion is $\overset{\oplus}{R}CO$

11-XII

Thus in the complex 11-XII, the acylium ion

$$CH_3\overset{\oplus}{C}O$$

may leave and attack the electron-rich *para* position, in all probability of another molecule, or it may simply move by an intramolecular process to the electron-rich *ortho* position. The *ortho-para* ratio is increased somewhat by increased temperature, excess hydrogen chloride or aluminum chloride.

(a) **Preparation of o- and p-hydroxypropiophenone** (32–35 % *ortho* and 45–50 % *para* from phenyl propionate and anhydrous aluminum chloride in carbon disulfide; the *para* is recovered as a solid and the *ortho* as an oil) [30].

(b) **Other examples**

(1) **2-Hydroxy-6-methylpropiophenone** (93 % from 3-methylphenyl-propionate and anhydrous aluminum chloride at 120–140°) [29].

(2) **Hydroxyacetophenones** (*ortho*-substituted, 20 %, and *para*-substituted, *ca.* 53–65 %, starting either from phenylacetate or from phenol and acetic acid with hot PPA; phenyl benzoate gives very poor yields of the hydroxybenzophenone under the same conditions) [31].

(3) **2,4-Dihydroxyisobutyrophenone** (69 % from isobutyric acid, resorcinol, and dried IR-120 Amberlite strong acid resin at 160° for several hours; pyrogallol can be acylated well similarly, phloroglucinol moderately well, and hydroquinone very poorly) [32].

(4) **2,5-Diacetylquinol** (60 % from quinol diacetate, 2 g., added at 140°

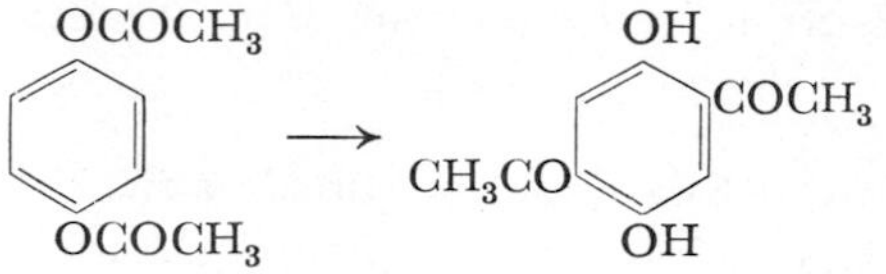

to a mixture of 10 g. of anhydrous $AlCl_3$ and 2 g. of NaCl, after which the temperature was raised rapidly to 180° and kept at 180–200° for 2 min.) [33].

4. From Aromatic or Heterocyclic Compounds and Bifunctional Acylating Agents

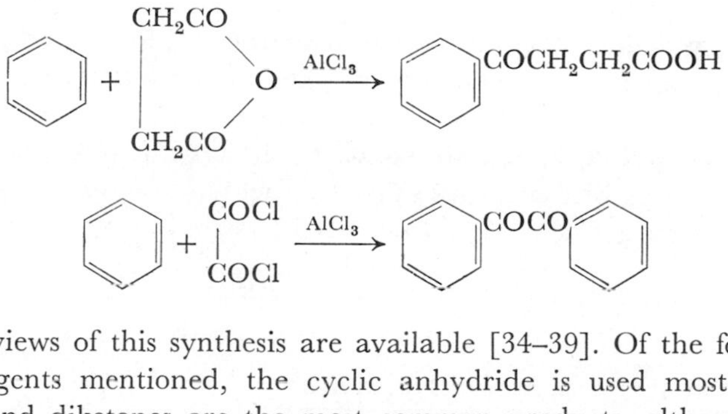

Many reviews of this synthesis are available [34–39]. Of the four types of acylating agents mentioned, the cyclic anhydride is used most frequently. Ketoacids and diketones are the most common products, although it is not surprising that there are others in view of the number of reactive groups present in the reaction mixture.

The most common catalyst, except perhaps when an acid is the acylating agent, is aluminum chloride. Others which have been employed are stannic chloride, zinc chloride, boron trifluoride, hydrogen fluoride, perchloric acid, phosphoric acid, and polyphosphoric acid. The number of moles of aluminum chloride recommended, if no other complexing group is present, are given in the accompanying tabulation [40]. Usually a solvent is employed, but an excess of

Acylating Agent	Product Required	
	Keto Acid	Diketone
Anhydride	2–2.2	3–3.5
Diacid	(3–3.6)	4–5.5
Ester acid chloride	2–2.2	—
Dichloride	(1–2)	2–2.2

one of the components may serve as a diluent. Typical solvents are nitrobenzene, nitromethane, symmetrical tetrachloroethane, carbon disulfide, and benzene. The literature should be consulted on experimental conditions which are often of great importance [34, 41].

One of the most important condensations of cyclic anhydrides is the preparation of *o*-benzoylbenzoic acid from phthalic anhydride and benzene, followed by ring closure to anthraquinone. This synthesis has been reviewed [42].

(a) **Preparation of β-benzoylpropionic acid** (77–82 % from benzene, succinic anhydride, and anhydrous aluminum chloride, using excess benzene as the solvent) [43].

(b) **Preparation of 1,4-dibenzoylbutane** (75–81 %, based on the adipic acid, from adipyl chloride, benzene, and anhydrous aluminum chloride, using excess benzene as the solvent) [44].

(c) **Other examples**

(1) **2,5-Diphenyl 3,4-dimesitoylfuran** (74 % from 2,5-diphenylfuran-3,4-dicarbonyl chloride, excess mesitylene, and anhydrous aluminum chloride) [45].

(2) **ω-Benzoylpelargonic acid,** $C_6H_5CO(CH_2)_8COOH$ (80 % from sebacic acid, excess benzene, and silicon tetrachloride to form a mixed anhydride, which is treated with anhydrous $AlCl_3$) [46].

5. From Arenes and N-Substituted Amides

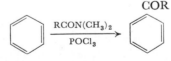

This synthesis has been discussed under Aldehydes, Chapter 10, C.6, in which substituted formamides are the attacking reagents (Vilsmeir reaction). By the use of homologous amides, ketones may be obtained in the presence of phosphorus oxychloride. In this manner, dimethylaniline [47], pyrrole and a variety of indoles [48, 49] have been acylated. Attempts to acylate β-ethoxynaphthalene, thiophene, and fluorene were unsuccessful [49]. On the other hand, pyrroles condense very well to give ketones (see Ex. b.2). The yields are variable.

(a) Preparation of 5-benzyloxy-3-acetylindole. To N,N-dimethylacetamide, 18 ml., at 5°, 7 ml. of phosphorus oxychloride was added at a rate to keep the temperature below 20°. Then a solution of 12.5 g. of 5-benzyloxyindole and 9 ml. of N,N-dimethylacetamide was added at a rate to keep the temperature below 40°. After heating the mixture to 87° for 2 hr., the mass was cooled, dissolved in water, and extracted with ether. The aqueous solution was made basic with sodium hydroxide and the solid produced was washed with water, refluxed in alcohol containing Darco 60, and filtered. The solution on cooling yielded 10.5 g. (71%) of the ketone, m.p. 189–190° [49].

(b) Other examples

(1) p-Dimethylaminobenzophenone (85% from dimethylaniline, phosphorus oxychloride, and benzanilide) [47].

(2) 2-Chloroacetylpyrrole (54% from N,N-diethylchloroacetamide, phosphorus oxychloride, and pyrrole at room temperature overnight) [50].

6. From Arenes and Unsaturated Ketones and α-Haloketones

Unsaturated ketones are potential alkylating agents in the Friedel-Crafts reaction [51]; see Sec. H.6 for arylation of unsaturated ketones with Grignard reagents.

$$\text{ArCH}\!-\!\text{CHCC}_6\text{H}_5 \xrightarrow[\text{AlCl}_3]{\text{Ar'H}} \text{ArCH}\!-\!\text{CH}_2\text{CC}_6\text{H}_5$$

The possibility exists that one aryl group can replace another in the alkylation:

$$\text{ArCH}\!=\!\text{CHCC}_6\text{H}_5 \xrightarrow[\text{AlCl}_3]{\text{2 Ar'H}} \text{Ar'}^2\text{CHCH}_2\text{CC}_6\text{H}_5$$

α-Haloketones can also be used as alkylating agents, but give rather poor yields.

Although unrelated to Friedel-Crafts reactions, the Meerwein arylation procedure can be applied to unsaturated ketones to obtain products similar to those described above:

$$\text{C}_6\text{H}_5\text{CH}\!=\!\text{CHCCH}_3 + \text{ArN}\!\equiv\!\text{NCl} \xrightarrow{\text{Cu}} $$

$$\text{C}_6\text{H}_5\text{CHClCHCCH}_3 \quad \text{or} \quad \text{C}_6\text{H}_5\text{CH}\!=\!\text{CCCH}_3$$

Yields range from 20 to 45% and orientation of aryl groups is based on radical stability of the intermediate [52].

(a) **Preparation of α,α-diphenylacetone** (53–57 % from α-bromo-α-phenylacetone, benzene, and aluminum chloride) [53].

(b) **Preparation of 3,3-diphenylpropiophenone** (90 % from benzal-acetophenone, benzene, and aluminum chloride) [54].

(c) **Preparation of 1,1,3-triphenylpropanone** (17 % from α-bromo-dibenzyl ketone, benzene, and aluminum chloride) [55].

7. From Unsaturated Compounds and Acylating Agents

For the most part, this section discusses the addition of acyl groups to olefins. If electrophilic in nature, i.e., involving an acylium ion or complex thereof, the reaction is known as the Kondakof:

$$RCH{=}CH_2 \xrightarrow[\text{SnCl}_4]{\text{R'COCl}} \underset{\underset{\text{Cl}}{|}}{RCH}{-}CH_2COR' \longrightarrow RCH{=}CHCOR'$$

But, since complexes of carbon monoxide with boranes or with nickel[IV] give similar additions to unsaturated compounds, these reactions are discussed also (see F.7 for acyl free radical addition to olefins). The section concludes with a brief discussion of the addition of carbonium ions to α,β-unsaturated carbonyl compounds and a longer discussion of the dimerization of ketenes.

The Kondakof reaction, being extremely versatile, makes available a wide variety of ketones which would be difficult to prepare otherwise. An excellent review is available by investigators, who have studied the reaction in depth [56]. The yields are not very high because of so many potential side reactions involving the olefinic group. Possible products are the β-chloroketones, isomeric unsaturated ketones, diacylation compounds including pyrylium salts [56a] alkyl halides, and polymers. The best catalyst is stannic chloride, but zinc chloride as well as aluminum chloride has been shown to be effective, if care is taken to use only equivalent amounts of acyl halide and aluminum chloride. This can be done by dissolving aluminum chloride and the acyl halide in methylene chloride and decanting the solution from excess aluminum chloride. The olefin should be added last. Yields by this procedure range from 40 to 57 % [57].

If unsaturated or aromatic acids are used, the unsaturated ketone may undergo ring closure [58]:

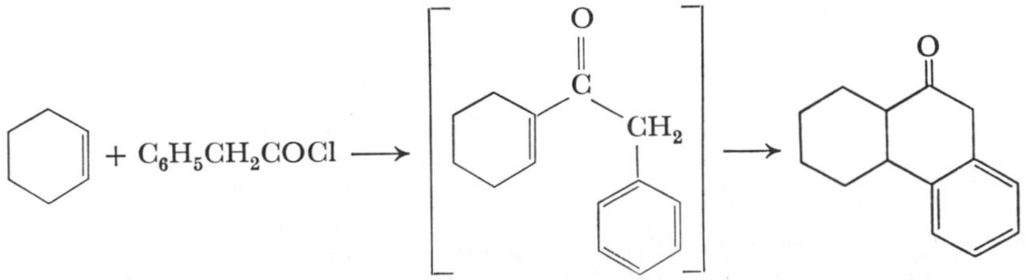

9-keto-4b,5,6,7,8,8a,9,10-octahydrophenanthrene

Indeed, intramolecular ring closure is a means of producing steroid-type compounds [59]:

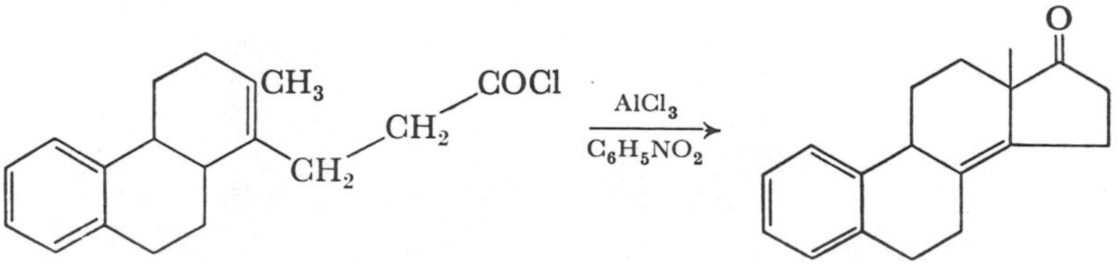

β-(2-Methyl-3,4,4a,9,10,10a-hexahydrophenanthryl) propionyl chloride

18-Norandrost-1,3,5(10),8(14)-tetraen-17-one

Olefinic acids of many different types have been cyclized as shown [60]:

Oct-5-enoic acid $\xrightarrow{\text{PPA}}$ 50% lactones + 50% of a mixture of 2-n-propylcyclopentenone and 2-ethylcyclohexenone

If a solvent which is a source of hydride ions is employed, saturated ketones are formed [61]:

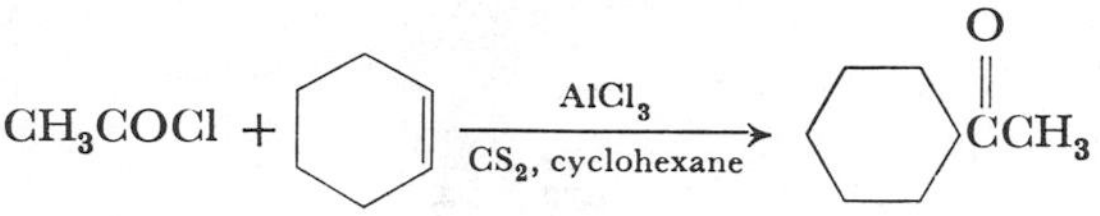

53%
Cyclohexyl methyl ketone

Saturated hydrocarbons can also be acylated, but they must first be dehydrogenated *in situ* to the olefin. Rather more strenuous conditions are needed, and skeletal carbon rearrangments can occur [62]:

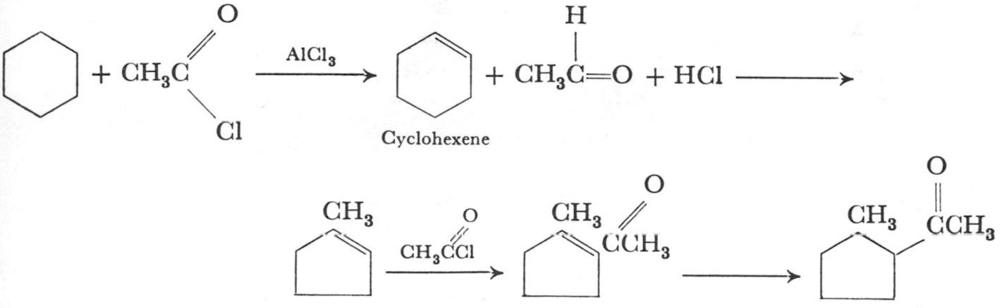

Cyclohexene

2-Methylacetyl-cyclopentane

Tetralin leads to a diketone [63]:

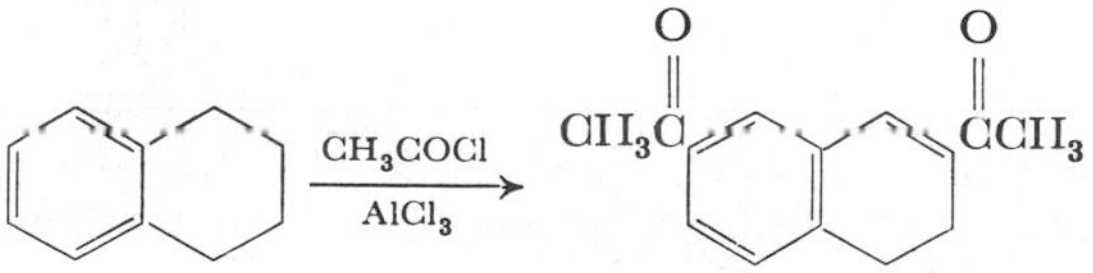

3,4-Dihydro-2,7-diacetylnaphthalene

Acids are acylated probably via their enolic forms [64]:

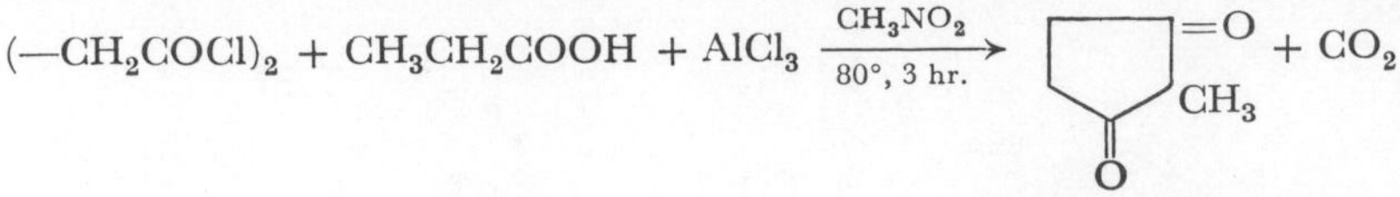
(—CH2COCl)2 + CH3CH2COOH + AlCl3
CH3NO2
80°, 3 hr.
=O
+ CO2
CH3
O

2-Methyl-1,3-cyclopentadione, 45%

Acetylenes are acylated in poor to fair yields to give β-halovinyl ketones [65]:

$$RC{\equiv}CH \xrightarrow[AlCl_3,\ CCl_4,\ \text{low temp.}]{R'COCl} RC(Cl){=}CHC({=}O)R'$$

A recent modification extends the range of addition to acetylenes [66]:

$$\underset{Ar\,=\,p\text{-}CH_3C_6H_4\text{-}}{ArLi} + Ni(CO)_4 \xrightarrow{\text{Ether}} \underset{0.05\ \text{mole}}{Li(ArCONi(CO)_3)} \xrightarrow[1.\ -70°,\ 5\ \text{hr.}\quad 2.\ 4\ N\ HCl]{CH_3C{\equiv}CH,\ 0.05\ \text{mole}} ArCOCH(CH_3)CH_2COAr$$

1,2-Di-*p*-toluylpropane, 69%

No doubt, other additions will be carried out with this remarkable acylation reagent.

In addition ketones can be formed from alkanes and carbon monoxide [67] in the presence of an electrophilic catalyst (see similar reaction for preparation of acids, Chapter 13, F.4):

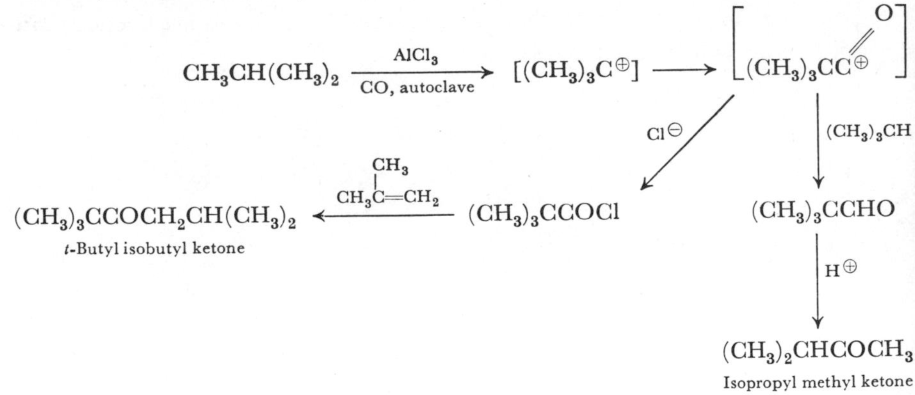
CH3CH(CH3)2
AlCl3
CO, autoclave
[(CH3)3C⊕]
(CH3)3CC⊕ O
Cl⊖
(CH3)3CH
(CH3)3CCOCl
(CH3)3CCHO
CH3
CH3C=CH2
(CH3)3CCOCH2CH(CH3)2
t-Butyl isobutyl ketone
H⊕
(CH3)2CHCOCH3
Isopropyl methyl ketone

Both ketones have been isolated from this mixture together with pivalic acid. In a similar manner, 3-pentanone has been obtained in over 50% yield from ethylene, carbon monoxide, hydrogen, and a cobalt catalyst [68]; see also Ex. b.4.

Another method of fixation of carbon monoxide is addition to boranes. This reaction is versatile in permitting selection of each radical group flanking the carbonyl group [69]:

$$\text{H}\!\!-\!\!\text{BH}_2 + \text{CH}_3\overset{\overset{\displaystyle \text{CH}_3}{|}}{\text{C}}\!\!=\!\!\text{CH}_2 \xrightarrow[0°]{\text{THF}} \text{H}\!\!-\!\!\text{BHCH}_2\text{CH(CH}_3)_2 \xrightarrow{\text{CH}_2\!=\!\text{CHCH}_2\text{CO}_2\text{C}_2\text{H}_5}$$

(2,3-Dimethyl-2-butyl)borane

$$\underset{\underset{\displaystyle (\text{CH}_2)_3\text{CO}_2\text{C}_2\text{H}_5}{\diagdown}}{\overset{\overset{\displaystyle \text{CH}_2\text{CH(CH}_3)_2}{\diagup}}{\text{H}\!\!-\!\!\text{B}}} \xrightarrow[\text{2. H}_2\text{O}_2,\ \text{CH}_3\text{CO}_2\text{Na}]{\text{1. CO, H}_2\text{O (1000 p.s.i.)}} (\text{CH}_3)_2\text{CHCH}_2\overset{\overset{\displaystyle \text{O}}{\|}}{\text{C}}\!\!-\!\!(\text{CH}_2)_3\text{CO}_2\text{C}_2\text{H}_5$$

Ethyl 7-methyl-5-oxooctanoate, 84%

1,3-Dienes, such as isoprene, yield cyclopentanones, such as 3-methylcyclopentanone, 46%.

A fourth method of fixation of carbon monoxide is via palladium salts [70]:

$$\text{ArHgCl} + \text{PdCl}_2 \longrightarrow \text{ArPdCl} \xrightarrow{\text{CO}} \text{ArCOPdCl} \xrightarrow{\text{ArPdCl}} \text{Ar}_2\text{CO} + \text{PdCl}_2 + \text{Pd}$$

Yields of benzophenones are low.

Ketones via their enol forms can be acylated to β-diketones in fair to good yields [71]:

$$\text{CH}_3\overset{\overset{\displaystyle \text{O}}{\|}}{\text{C}}\text{CH}_3 + \text{CH}_3\text{COCl} \xrightarrow{\text{BF}_3} \text{CH}_3\overset{\overset{\displaystyle \text{O}}{\|}}{\text{C}}\text{CH}_2\overset{\overset{\displaystyle \text{O}}{\|}}{\text{C}}\text{CH}_3 + \text{HCl}$$

80–85%
2,4-Pentadione

Enol esters are particularly good acylating agents; isopropenyl acetate, for example, yields ethyl diacetoacetate with acetoacetic ester and sulfuric acid as a catalyst. Pyrones are available by diacylation of ketones followed by ring closure [72]:

$$\text{C}_6\text{H}_5\text{CH}_2\overset{\overset{\displaystyle \text{O}}{\|}}{\text{C}}\text{CH}_2\text{C}_6\text{H}_5 \xrightarrow[\text{PPA}]{\text{CH}_3\text{CO}_2\text{H}} \text{CH}_3\overset{\overset{\displaystyle \text{O}}{\|}}{\text{C}}\!\!-\!\!\text{CH(C}_6\text{H}_5)\overset{\overset{\displaystyle \text{O}}{\|}}{\text{C}}\text{CH(C}_6\text{H}_5)\overset{\overset{\displaystyle \text{O}}{\|}}{\text{C}}\text{CH}_3 \xrightarrow{-\text{H}_2\text{O}}$$

2,6-Dimethyl-3,5-diphenyl-
4-pyrone, 48%

The enol form of an amide also can be acylated (see Ex. b.3). Examples not covered by Nenitzescu [56] are given.

Not only acylium but also carbonium ions add to unsaturated compounds as shown by the following two examples:

$$\text{CH}_3\text{C}\!\!\equiv\!\!\text{CCH}_2\text{CH}_2\text{OSO}_2\text{C}_6\text{H}_4\text{NO}_2\!\!-\!\!m \xrightarrow[\text{CF}_3\text{COONa}]{\text{CF}_3\text{COOH}}$$ [ref. 73]

2-Methylcyclobutanone,
52%

$$\text{C}_6\text{H}_5\text{CHOHCHC}_6\text{H}_5\text{CH}_2\text{CCl}\!\!=\!\!\text{CH}_2 \xrightarrow{\text{HCOOH}}$$ [ref. 74]

4-Chloro-1,2-diphenylpent-4-enol-1

trans-3,4-Diphenylcyclopen-
tanone, 30%

(a) Preparation of ethyl 4-acetyl-3-methylbut-2-enoate [75]. A

$$(CH_3)_2C{=}CHCO_2C_2H_5 \xrightarrow[\text{AlCl}_3]{\text{CH}_3\text{COCl}} CH_3\overset{\overset{\text{O}}{\|}}{C}CH_2\overset{\overset{\text{CH}_3}{|}}{C}{=}CHC\overset{\overset{\text{O}}{\|}}{O}C_2H_5$$
72%

later report indicates that the product is probably a mixture of Δ^2 and Δ^3 unsaturated esters [76].

(b) Other examples

(1) **2-Acetyl-1,3-cyclopentanedione** (50% from succinic anhydride, isopropenylacetate, 2 equiv. of aluminum chloride in refluxing ethylene dichloride; similarly 2-acetyl-1,3-cyclohexanedione (40%) was prepared) [77].

(2) **Piperitenone** [78.]

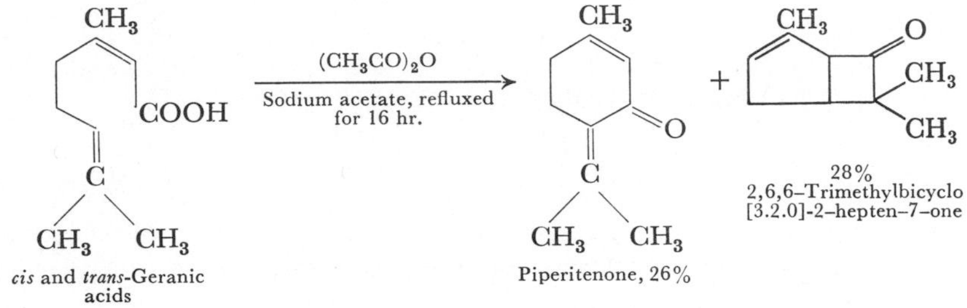

cis and trans-Geranic acids

Piperitenone, 26%

28%
2,6,6–Trimethylbicyclo
[3.2.0]–2–hepten–7–one

(3) 5-Diethylamino-2,2-dichloro-4-methyl-3-(2H)-furanone [79].

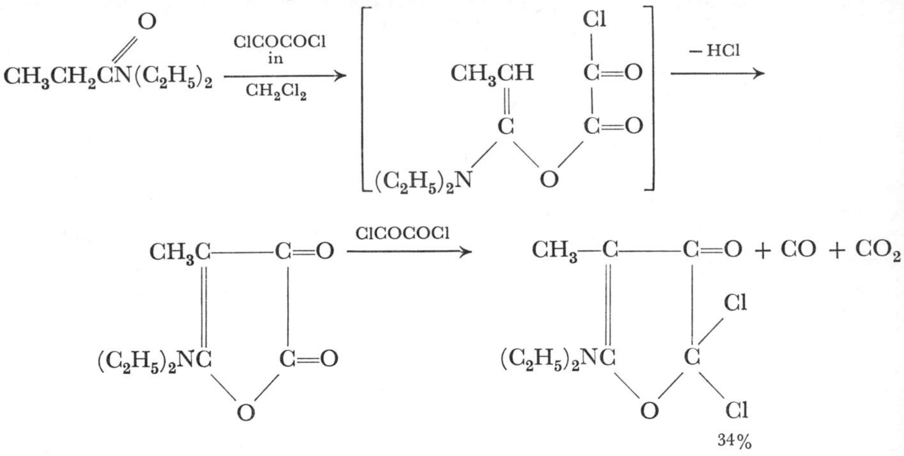

(4) 2-Methyl-2-phenyl-1-indanone [80].

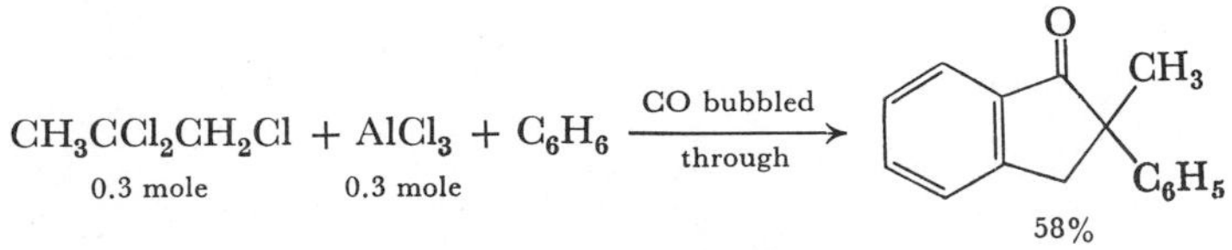

$$CH_3CCl_2CH_2Cl + AlCl_3 + C_6H_6 \xrightarrow[\text{through}]{\text{CO bubbled}}$$
0.3 mole   0.3 mole

58%

8. From Vinyl Esters via the Chloromercuriketone

$$RCH{=}\overset{\overset{\displaystyle R'}{|}}{C}OCOCH_3 \xrightarrow[KCl]{Hg(OCOCH_3)_2} ClHg\overset{\overset{\displaystyle R}{|}}{C}H\overset{\overset{\displaystyle R'}{|}}{C}{=}O \xrightarrow{(C_6H_5)_3CCl} (C_6H_5)_3C\overset{\overset{\displaystyle R}{|}}{C}H\overset{\overset{\displaystyle R'}{|}}{C}{=}O$$

This synthesis has been discussed under Aldehydes, Chapter 10, C.10.

9. From Aldoximes and Diazonium Salts

$$ArCH{=}NOH + ArN_2Cl \longrightarrow \overset{\displaystyle Ar}{\underset{\displaystyle Ar}{>}}C{=}NOH \xrightarrow[H^{\oplus}]{H_2O} \overset{\displaystyle Ar}{\underset{\displaystyle Ar}{>}}C{=}O$$

This method, applicable to aromatic ketones, has been discussed under Aldehydes, Chapter 10, C.11. If an oxime rather than formaldoxime is employed, ketones rather than aldehydes are obtained. Yields are low, lower, in fact, than for the synthesis of aldehydes by the same procedure. Thus the method would not be attractive except in special cases, such as in the preparation of 4-acetylphthalic acid from dimethyl 4-aminophthalate [81].

1. G. Olah, *Friedel-Crafts and Related Reactions*, Vol. 3, Pt. 1 and 2, Interscience Publishers, New York, 1964.
2. X. A. Dominguez et al., *J. Am. Chem. Soc.*, **76**, 5150 (1954).
3. Ref. 1, Vol. 1, 1964, p. 201; see also K. L. Rinehart, Jr. et al., *J. Am. Chem. Soc.*, **84**, 3263 (1962).
4. P. H. Gore, Ref. 1, Pt. 1, pp. 1–382.
5. M. S. Newman and K. G. Ihrman, *J. Am. Chem. Soc.*, **80**, 3652 (1958).
6. W. J. Heintzelman and B. B. Corson, *J. Org. Chem.*, **22**, 25 (1957).
7. F. Uhlig and H. R. Snyder, *Advan. Org. Chem.*, **1**, 35 (1960).
8. J. P. Marthe and S. Munavalli, *Bull. Soc. Chim. France*, 2679 (1963).
9. R. G. Downing and D. E. Pearson, *J. Am. Chem. Soc.*, **84**, 4956 (1962).
10. F. J. Villani and M. S. King, *Org. Syn.*, Coll. Vol. **4**, 88 (1963).
11. C. E. Olson and A. R. Bader, *Org. Syn.*, Coll. Vol. **4**, 898 (1963).
12. H. R. Snyder and F. X. Werber, *Org. Syn.*, Coll. Vol. **3**, 798 (1955).
13. P. H. Williams et al., *J. Am. Chem. Soc.*, **82**, 4883 (1960).
14. P. K. Grover et al., *J. Chem. Soc.*, 3982 (1955).
15. P. D. Gardner, *J. Am. Chem. Soc.*, **76**, 4550 (1954).
16. B. K. Diep et al., *J. Chem. Soc.*, 2784 (1963).
17. C. Tüzün et al., *Org. Syn.*, **41**, 1 (1961).
18. B. P. Gupta and C. A. Haksar, *Agra. Univ. J. Res. Sci.*, Pt. 2, **11**, 165 (1962); *C.A.*, **59**, 1546 (1963).
19. J. D. Edwards et al., *J. Org. Chem.*, **29**, 3028 (1964).
20. D. A. Denton and H. Suschitzky, *J. Chem. Soc.*, 4741 (1963).
21. A. C. Cope and R. D. Smith, *J. Am. Chem. Soc.*, **77**, 4596 (1955).
22. W. Ruske, Ref. 4, p. 383; P. E. Spoerri and A. S. DuBois, *Org. Reactions*, **5**, 387 (1949).
23. K. C. Gulati et al., *Org. Syn.*, Coll. Vol. **2**, 522 (1943).
24. R. Kuhn and I. Low, *Chem. Ber.*, **77**, 202 (1944).
25. A. Gerecs, Ref. 4, p. 499.
26. A. H. Blatt, *Chem. Rev.*, **27**, 413 (1940); *Org. Reactions*, **1**, 342 (1942).
27. D. V. Rao and V. Lamberti, *J. Org. Chem.*, **32**, 2896 (1967).
28. K. W. Rosenmund and W. Schnurr, *Ann. Chem.*, **460**, 56 (1928).
29. R. Baltzly and A. Bass, *J. Am. Chem. Soc.*, **55**, 4292 (1933).

30. E. Miller and W. H. Hartung, *Org. Syn.*, Coll. Vol. **2**, 543 (1943).

31. Ref. 7, p. 62.

32. P. Price and S. S. Israelstam, *J. Org. Chem.*, **29**, 2800 (1964).

33. R. H. Thomson et al., *J. Chem. Soc.*, 2403 (1953).

34. A. G. Peto, Ref. 1, Pt. 1, p. 535.

35. G. Baddeley, *Quart. Rev. (London)*, **8**, 355 (1954).

36. E. Berliner, *Org. Reactions*, **5**, 229 (1949).

37. P. H. Gore, *Chem. Rev.*, **55**, 229 (1955).

38. R. E. Kirk and D. F. Othmer, *Encyclopedia of Chemical Technology*, 2nd ed., Vol. 10, Interscience Publishers, New York, 1966.

39. C. A. Thomas, *Anhydrous Aluminum Chloride in Organic Chemistry*, Reinhold Publishing Corp, New York, 1941.

40. M. F. Carrol, *Mfg. Chemist*, **26**, 97 (1955).

41. Ref. 36, p. 254.

42. M. A. Perkins in H. A. Lubs, *The Chemistry of Synthetic Dyes and Pigments*, Reinhold Publishing Corp., New York, 1955, Chap. 7.

43. L. F. Somerville and C. F. H. Allen, *Org. Syn.*, Coll. Vol. **2**, 81 (1943).

44. R. C. Fuson and J. T. Walker, *Org. Syn.*, Coll. Vol. **2**, 169 (1943).

45. D. V. Nightingale and B. Sukornick, *J. Org. Chem.*, **24**, 497 (1959).

46. Y. K. Yur'ev et al., *Zh. Obshch. Khim.*, **24**, 1568 (1954); *C.A.*, **49**, 11589 (1955).

47. H. H. Bosshard and H. Zollinger, *Helv. Chim. Acta*, **42**, 1659 (1959).

48. P. Wolff and W. Werner, Ger. Patent 614,326, June 6, 1935; *C.A.*, **29**, 5861 (1935); E. Ghigi and A. Drusiani, *Atti accad. sci. ist. Bologna, Classe sci. fis., Rend.* (XI), **5**, No. 2, 56 (1957–58); *C.A.*, **54**, 5613 (1960).

49. W. C. Anthony, *J. Org. Chem.*, **25**, 2049 (1960).

50. A. Ermili et al., *J. Org. Chem.*, **30**, 339 (1965).

51. R. Koncos and B. S. Friedman, Ref. 1, Vol. 2, Pt. 1, 1964, p. 332.

52. C. S. Rondestvedt, Jr., *Org. Reactions*, **11**, 189 (1960).

53. E. M. Schultz and S. Mickey, *Org. Syn.*, Coll. Vol. **3**, 343 (1955).

54. D. Vorlander and A. Friedberg, *Chem. Ber.*, **56**, 1144 (1923).

55. A. C. B. Smith and W. Wilson, *J. Chem. Soc.*, 1342 (1955).

56. C. D. Nenitzescu and A. T. Balaban, Ref. 1, Vol. 3, Pt. 2, 1964, p. 1033.

56a. A. T. Balaban, Tetrahedron Letters, 4643 (1968).

57. N. Jones et al., *J. Chem. Soc.*, 2354 (1963).

58. Ref. 56, p. 1050.

59. Ref. 56, p. 1056.

60. M. F. Ansell and M. H. Palmer, *Quart. Rev. (London)*, **18**, 211 (1964).

61. Ref. 56, p. 1069.

62. Ref. 56, p. 1073.

63. Ref. 56, p. 1080.

64. H. Schick et al., *Angew. Chem., Intern, Ed. Engl.*, **6**, 80 (1967).

65. Ref. 56, p. 1081.

66. Y. Sawa et al., *J. Org. Chem.*, **33**, 2159 (1968).

67. Ref. 56, p. 1085.

68. E. A. Naragon et al., U.S. Patent 2,699,453, January 11, 1955; *C.A.*, **50**, 1893 (1956).

69. H. C. Brown and E. Negishi, *J. Am. Chem. Soc.*, **89**, 5285, 5477 (1967).

70. R. F. Heck, *J. Am. Chem. Soc.*, **90**, 5546 (1968).

71. Ref. 56, p. 1095.

72. Ref. 56, p. 1098.

73. M. Hanack and I. Herterich, *Tetrahedron Letters*, 3847 (1966).

74. P. T. Lansbury and E. J. Nienhouse, *J. Am. Chem. Soc.*, **88**, 4290 (1966).

75. I. Alkonyi, *Chem. Ber.*, **98**, 3099 (1965).

76. G. Lohaus et al., *Chem. Ber.*, **100**, 658 (1967).

77. F. Merenyi and M. Nilsson, *Acta Chem. Scand.*, **17**, 1801 (1963).

78. J. J. Beereboom, *J. Org. Chem.*, **30**, 4230 (1965).

79. A. J. Speziale *et al.*, *J. Org. Chem.*, **30**, 4303 (1965).
80. H. Bruson and H. L. Plant, *J. Org. Chem.*, **32**, 3356 (1967).
81. W. F. Beech, *J. Chem. Soc.*, 1297 (1954).

D. Hydrolysis or Hydration

1. From Heterocyclic Rings

As is true in the synthesis of aldehydes, Chapter 10, D.2 and D.3, a variety of heterocyclic rings may be cleaved in alkaline or acidic solution to give ketones. Examples are furans [1, 2], tetrahydrofurans [3], lactones [4, 5], azlactones [6], hydantoins [7], 5-alkylidene-2-thiooxazolid-4-ones [8], pyrrolones [9], rhodanines [10], and, of wider applicability, the 1,3-dithianes. The latter may be alkylated and hydrolyzed as shown [11] (see also Aldehydes, Chapter 10, D.3) to give a variety of ketones:

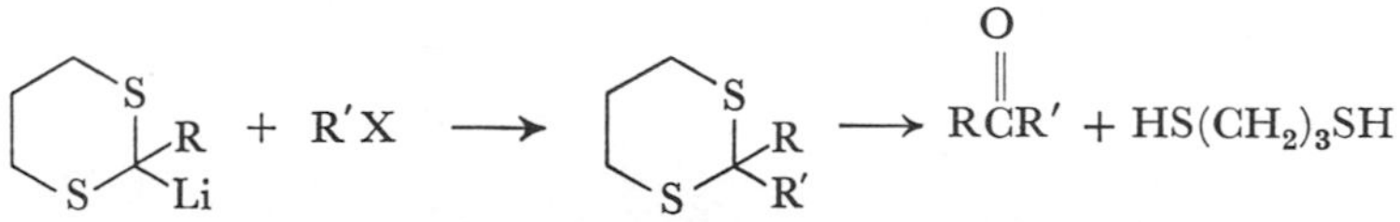

(a) Preparation of acetonylacetone (from a furan).   2,5-Dimethyl

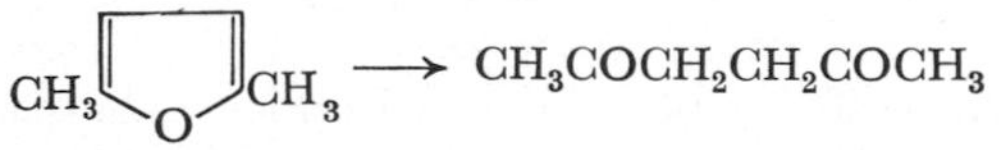

furan was hydrolyzed with a mixture of glacial acetic acid, sulfuric acid, and water to give an 86–90 % yield of acetonylacetone [2].

(b) Preparation of 4,5-diketoheneicosaldehyde (from a tetrahydrofuran).   2-Heptadecanoyl-2,5-dimethoxytetrahydrofuran was refluxed with

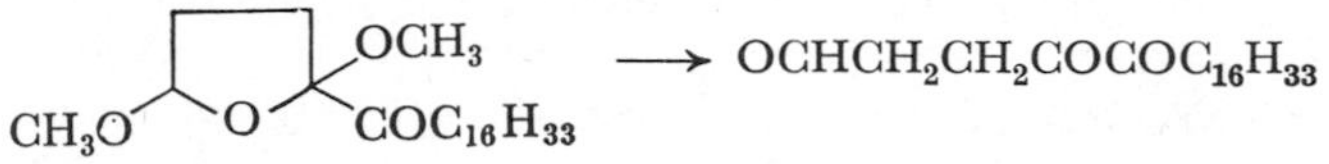

hydrogen chloride in dioxane-water for 2 hr. to give a 75 % yield of 4,5-diketoheneicosaldehyde [3].

(c) Preparation of 5-chloro-2-pentanone (from a lactone).   α-Acetyl-γ-butyrolactone in dilute hydrochloric acid in an ice bath was heated

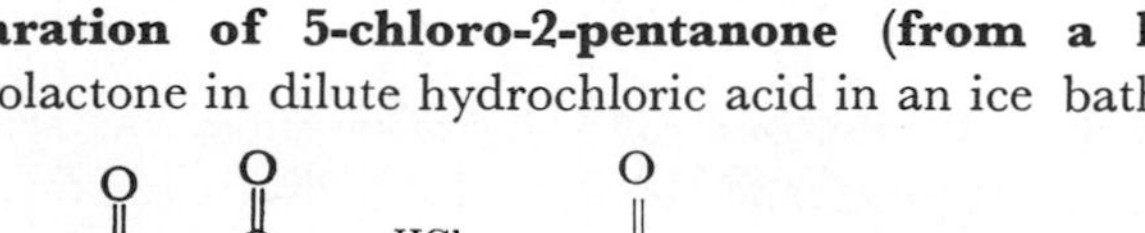

slowly to give the ketone, b.p. 70–72° (20 mm.) (70–82 %) [4].

(d) Preparation of 4-methoxyphenylpyruvic acid (from a hydantoin).   5-(4'-Methoxybenzal)hydantoin was refluxed with 20 % aqueous

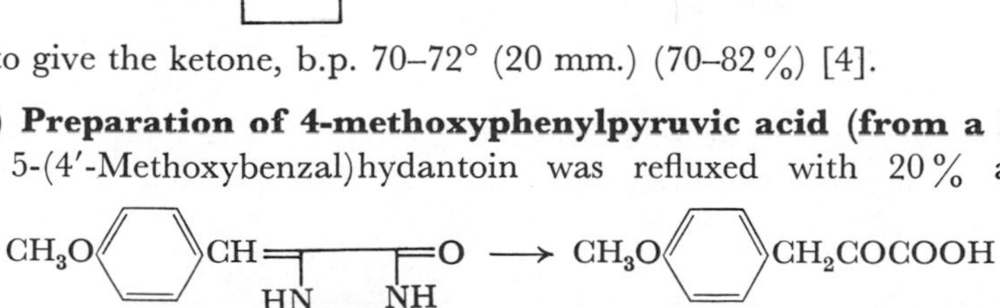

References for Section D are on pp. 675–676.

sodium hydroxide for 3 hr. at a bath temperature of 170–180° while nitrogen was bubbled through the mixture. The acid was recovered (70%) [7].

(e) **Preparation of 2-nitro-4-hydroxy-3-methoxyphenylpyruvic acid (from an azlactone).** The azlactone was refluxed for $2\frac{1}{2}$ hr. with dilute

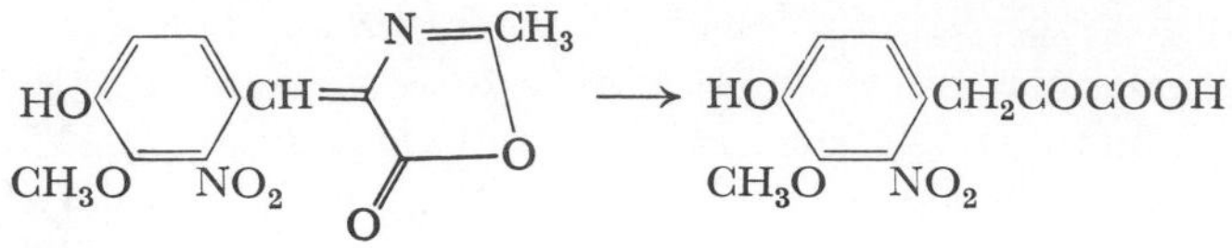

hydrochloric acid to give the acid (88%) [6].

2. From *gem*-Dihalides

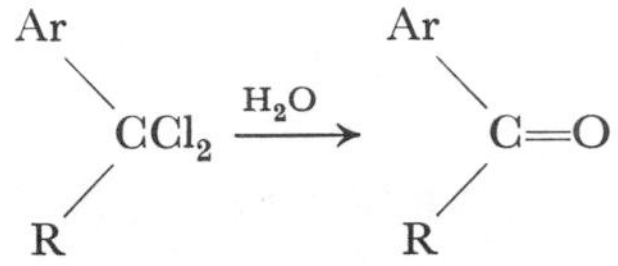

Again, this synthesis has been discussed under Aldehydes, Chapter 10, D.4. It has found less use in the preparation of ketones, although in some cases the yields are high. An interesting application has been in the conversion of dibenzoylmethane into diphenyl triketone (—COCH$_2$CO— → —COCOCO—) [12]; see Ex. b.1. Water alone, or a small amount of water and cupric chloride in nitrobenzene as a solvent [13], or silver trifluoroacetate (see Ex. b.3), or sodium acetate in acetic acid has been used to effect the hydrolysis.

(a) **Preparation of benzophenone** (80–89% based on benzene converted into diphenyldichloromethane (carbon tetrachloride and aluminum chloride) and finally by hydrolysis with steam into benzophenone) [14].

(b) Other examples

(1) **Diphenyl triketone** (59% based on dibenzoylmethane converted into dibenzoyldibromomethane (bromine in chloroform), then into diphenyl triketone hydrate (fused sodium acetate, glacial acetic acid followed by water), and finally distillation) [12].

(2) **Fluorenone** (90% from 9,9-dibromofluorene and sodium acetate in acetic acid) [15].

(3) **Benzocyclobutenedione** (89% from 1,1,2,2-tetrabromobenzocyclobutene refluxed for 12 hr. with silver trifluoroacetate in acetonitrile and a

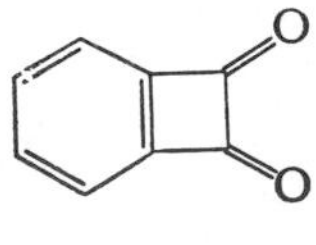

small amount of water) [16].

3. From Vinyl Halides, Ethers, and Amines

Substituted vinyl chlorides, ethers, or amines are subject to hydrolysis, the former with difficulty, the latter two with ease:

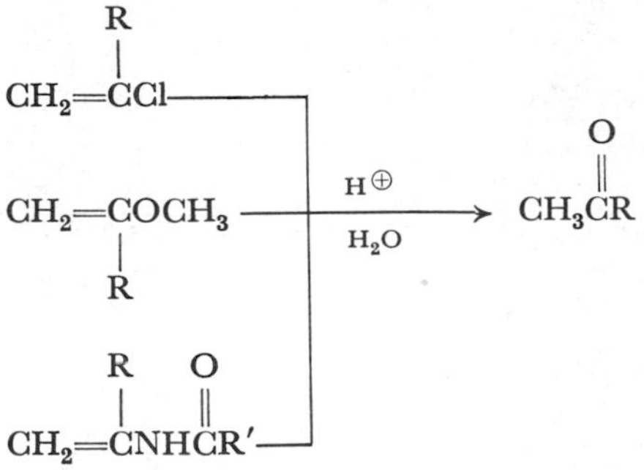

Another reaction of the vinyl ether leading to a hydrolytic product is that with the Grignard reagent (see Ex. e.1):

$$CH_2{=}COCH_3 + R'MgX \xrightarrow{H_3O^\oplus} CH_3COR + Mg(OCH_3)X + R'H$$
(with R above the first carbon)

The interest or, better, curiosity in this reaction stems from the source of the active hydrogen which yields the hydrocarbon. Vinyl ethers are readily available from the reaction of acetylenes with alcohols or of ketones with ethyl orthoformate, the latter being a common way to protect keto groups [17]. They are hydrolyzed almost instantaneously to ketones by acid and in some instances with boiling water and no catalyst.

(a) **Preparation of 3-ketoindane-1-carboxylic acid.** 3-Chloroindene-1-carboxylic acid, 0.3 g., was refluxed with 10 ml. of hydrochloric acid for 4 hr. After cooling, the solution was neutralized and extracted with ether. From the ethereal solution there was recovered 0.27 g., m.p. 114–116° (hydrate m.p. 83.5°), of the ketone [18].

(b) **Preparation of (butan-3-onyl) dehydroindolizidinium iodide** (90% from (3-chloro-2-butenyl)dehydroindolizidinium iodide, potassium hydroxide in chloroform and sulfuric acid followed by isolation as the iodide [19]:

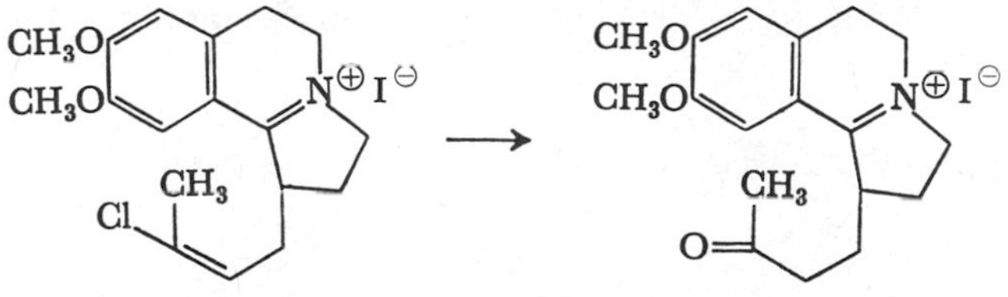

(c) **Preparation of α-butyl-γ-acetylbutyric acid** (86% from butyl-(3-chloro-2-butenyl) acetic acid and sulfuric acid) [20].

(d) **Preparation of phenylpyruvic acid** (84–95% from prolonged boiling of ethyl α-methoxycinnamate with dilute sulfuric acid; the ester was made by condensation of benzaldehyde with ethyl α-methoxyacetate; the hydrolysis was more difficult than that of the other vinyl ethers) [21].

(e) Other examples

(1) 5-Hendecanone (butyl hexyl ketone, 76% from 5-butoxy-5-hendecene and butylmagnesium bromide) [22].

(2) Ethyl 3-(*o*-benzamidophenyl)-pyruvate [23].

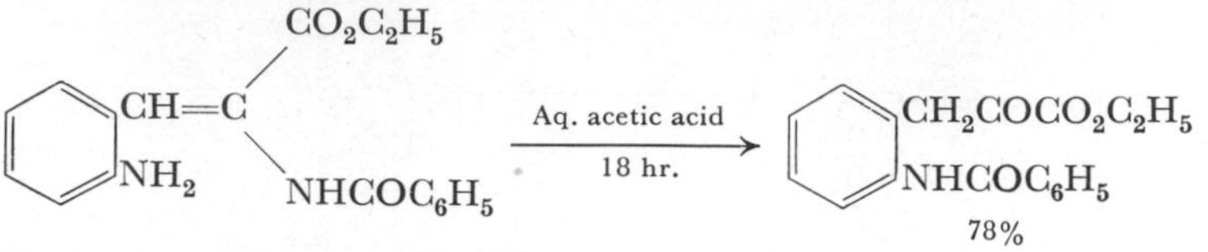

4. From Acetylenes and Allenes

$$RC{\equiv}CH \xrightarrow[HgSO_4,\ H_2SO_4]{H_2O} RCOCH_3$$

Although acetylene on hydration gives acetaldehyde, all other homologs of the series produce ketones. The reaction to yield acetaldehyde from acetylene has been discussed under Aldehydes, Chapter 10, D.6 (see also [24]).

The hydration of acetylenic alcohols, readily available from the condensation of acetylene salts with carbonyl compounds, is a most handy source of ketones. The hydration can be manipulated to obtain either the hydroxy or unsaturated ketone, as for example [25]:

1-Acetylcyclohexene   1-Acetylcyclohexanol

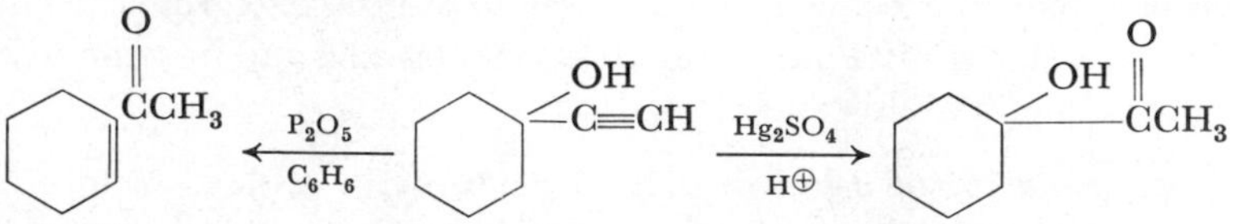

Many steroid-type ketones have been synthesized in this manner (see Ex. c.3).

The Meyer-Schuster rearrangement may occur with either acetylenic alcohols or glycols. For example, butyne-1,4-diol yields the following ketol [26]:

$$HOCH_2C{\equiv}CCH_2OH \xrightarrow[\substack{10\%\ H_2SO_4 \\ \text{in } H_2O,\ 6\text{–}8\ \text{hr.} \\ 25^\circ}]{5\%\ HgCl_2} HOCH_2\overset{O}{\overset{\|}{C}}CH{=}CH_2$$

1-Hydroxy-but-3-en-2-one

Another example illustrates the control of product by appropriate conditions.

11-XIII   11-XIV

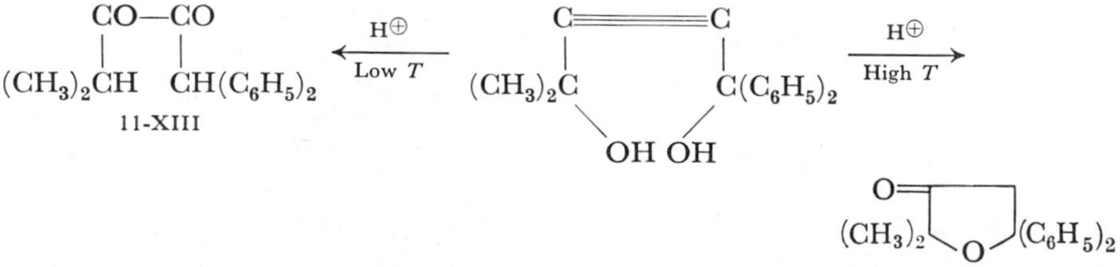

With 7.6% sulfuric acid and 92.4% ethanol at 30–35° for 6 hours, 10.5 g. gives a 90% yield of 2-methyl-5,5-diphenyl-3,4-pentanedione, 11-XIII while with

7 ml. of sulfuric acid and 80 ml. of ethanol at 60 to 70° for 8 hours 2,2-dimethyl-5,5-diphenyltetrahydro-3-furanone, 11-XIV, is produced with an 86% yield [27]. Allenes also yield ketones on hydration (see Ex. c.4).

(a) **Preparation of 1-acetylcyclohexanol** (65–67% from 1-ethynyl-cyclohexanol, mercuric oxide, sulfuric acid, and water [28]; with a mercury resin (Dowex 50 stirred into a solution containing mercuric sulfate in dilute sulfuric acid), methanol, and water the yield of crude 1-acetylcyclohexanol was 84%) [29].

(b) **Preparation of 1-acetylcyclohexene** (56–70% from 1-ethylnyl-cyclohexanol, phosphorus pentoxide, and benzene) [25].

(c) **Other examples**

(1) *o*-**Chlorodibenzoylmethane** (about 100% from *o*-chlorophenyl-benzoylacetylene in conc. sulfuric acid at room temperature) [30].

(2) **4-Octanone** (89% on basis of alkyne consumed from 4-octyne, acetic acid, water, and mercury resin) [29].

(3) **5-Methoxy-17α-acetyl-17β-acetoxy-des-A-estratriene-5,7,9**

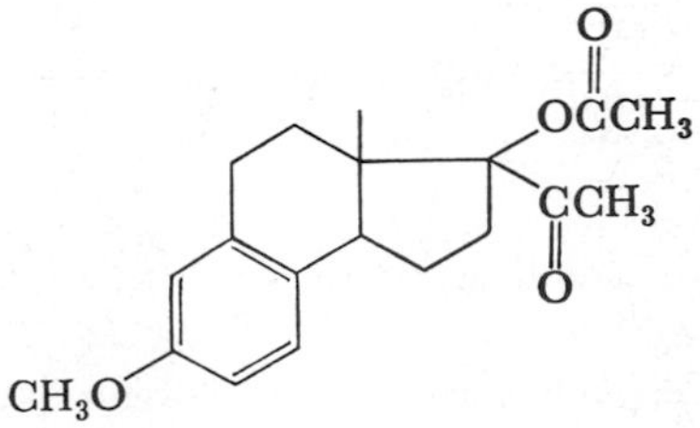

(96% from the corresponding acetylenic acetate refluxed for 7 hr. in ethanol with Dowex 50, acid form, impregnated beforehand with mercuric acetate) [31].

(4) **Cyclotridecanone** (76% from 0.05 mole of the corresponding allene and 30 ml. of 88% sulfuric acid containing 0.2 g. of hydroquinone held

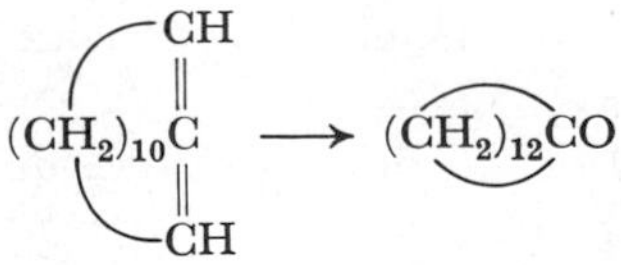

at 25° for 20 min; the allene was prepared by the addition of dichlorocarbene to cyclododecene, the cyclopropane derivative then being dehydrochlorinated) [32].

5. **From Ketone Derivatives**

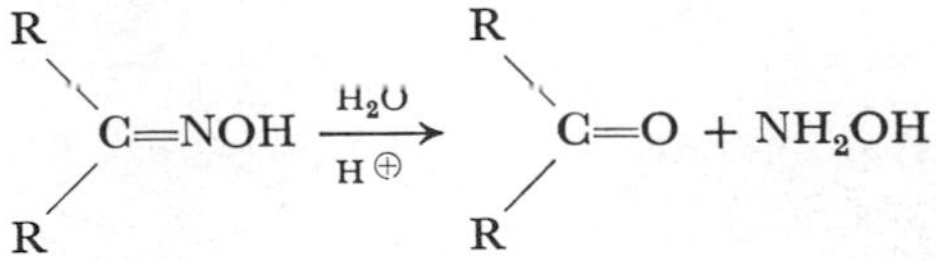

This synthesis may appear to be of little value since the ketone derivatives are usually prepared from ketones. However, such is not always the case as will be shown.

1. Oximes may be prepared in three ways not involving a ketone group:

$$\text{ArCOCH}_2\text{R} + \text{HONO} \longrightarrow \underset{\overset{\|}{\text{NOH}}}{\text{ArCOCR}} \qquad \text{[ref. 33]}$$

$$\text{RCHXCOOC}_2\text{H}_5 + \text{NaONO} \longrightarrow \underset{\overset{\|}{\text{NOH}}}{\text{RCCOOC}_2\text{H}_5} \qquad \text{[ref. 34]}$$

$$\text{ArCHO} + \text{RCH}_2\text{NO}_2 \longrightarrow \text{ArCH}{=}\text{C(R)NO}_2 \xrightarrow[\text{H}_2\text{O}]{\text{Fe, H}^{\oplus}} \text{ArCH}_2\text{C(R)}{=}\text{NOH}$$

$$\text{[ref. 35]}$$

If no other groups present are hydrolyzable, the oximes are usually best hydrolyzed by dissolving in concentrated hydrochloric acid at room temperature.

2. Enol esters may be produced from acetylene derivatives [36]:

$$\text{CF}_3\text{C}{\equiv}\text{CCF}_3 \xrightarrow[\substack{\text{(CH}_3\text{CO)}_2\text{O,}\\ \text{CH}_3\text{COONa}}]{\text{CH}_3\text{COOH}} \underset{\overset{|}{\text{OCOCH}_3}}{\text{CF}_3\text{CH}{=}\text{CCF}_3} \longrightarrow \text{CF}_3\text{CH}_2\text{COCF}_3$$

and on refluxing with a small amount of sulfuric acid in n-butanol, ketones are produced in excellent yields.

3. Other derivatives, which may or may not be isolated, are intermediates in the Nef (A.9), Hoesch (C.2), and Grignard (H) reactions.

Ketone derivatives may be converted into ketones, exclusive of the method previously mentioned for oximes, by a variety of methods, as shown:

1. Treatment of the semicarbazone with formaldehyde in ethanol under nonacidic conditions or with pyruvic acid under mildly acidic conditions [37].

2. Steam distillation of the semicarbazone with phthalic acid until all the ketone has been carried over [38].

3. Heating the oxime or 2,4-dinitrophenylhydrazone with levulinic acid-hydrochloric acid (see Ex. b.3).

4. Treatment of the ketal with aqueous perchloric acid in tetrahydrofuran [39]; (see also Alcohols, Chapter 4, A.7).

5. Cleavage of 2,4-dinitrophenylhydrazones with stannous chloride and hydrochloric acid [40].

6. Refluxing the oxime in dibutyl ether with 5% (by weight of the oxime) of boron trifluoride etherate and an equivalent amount of iron pentacarbonyl [41].

(a) Preparation of acetylbenzoyl (66–70% from isonitrosopropiophenone and 10% H_2SO_4) [42].

(b) Other examples

(1) **Phenylpyruvic acid** (88–94% from α-acetaminocinnamic acid and hydrochloric acid) [43]:

$$C_6H_5CH=CCOOH \xrightarrow[\text{HCl}]{H_2O} C_6H_5CH_2CCOOH$$

$$\underset{\text{NHCOCH}_3}{|} \qquad\qquad \underset{\text{O}}{\|}$$

(2) **1,1,1,4,4,4-Hexafluoro-2-butanone** (95% from 1,1,1,4,4,4-hexa-fluoro-2-buten-2-ol acetate and a trace of sulfuric acid in n-butanol) [36].

(3) **1,8-Diketo-4,7-methano-3α,4,7,7α-tetrahydroindene** (70% from

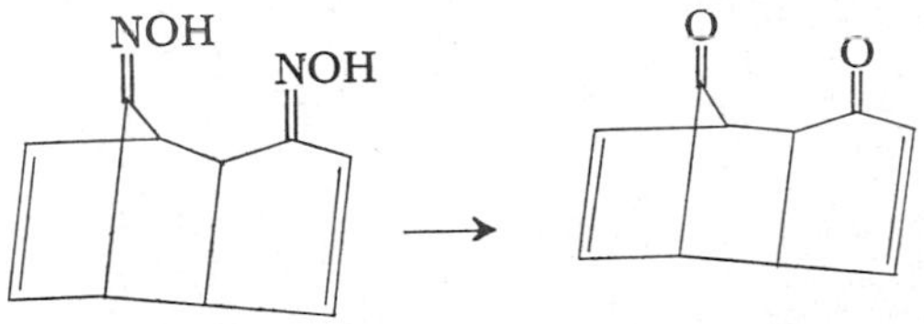

1,8-dioximino-4,7-methano-3α,4,7,7α-tetrahydroindene by heating for 3 hr. with levulinic acid to which 10 volume % of 1 N HCl had been added) [44].

6. From β-ketoesters (Acetoacetic Ester Synthesis)

$$CH_3COCH_2COOC_2H_5 \longrightarrow CH_3COCHCOOC_2H_5 \longrightarrow$$
$$\underset{R}{|}$$

$$CH_3COCHCOOH \longrightarrow CH_3COCH_2R + CO_2$$
$$\underset{R}{|}$$

The use of β-ketoesters in the synthesis of carboxylic acids is discussed under Carboxylic Acids, Chapter 13, E.1. Concentrated alkali favors cleavage of β-ketoesters, while dilute alkali favors saponification of the ester group to the sodium salt, which on acidification readily decarboxylates to the ketone. Methods for the synthesis of β-keto esters are given under Carboxylic Esters, Chapter 14, C.1, and methods for the alkylation of acetoacetic esters are similar to those employed for malonic esters (Carboxylic Esters, Chapter (14, C.6)). At times these esters are not isolated, but are converted directly into ketones. Although carboxylic acids and carboxylic esters are other possible products of this hydrolysis, by the proper experimental conditions satisfactory yields of ketones may often be obtained.

A great variety of substances has been employed to effect the hydrolysis. Among these are sulfuric acid [45], phosphoric acid [46], acetic-sulfuric acid [47], and a cation exchange resin such as Dowex 50 W × 2 [48].

Two modifications have been introduced to overcome the difficulty of hydrolyses of some acetoacetic esters. The first capitalizes on the ready pyrolysis or hydrolysis of t-butyl esters. Example (a) demonstrates that simply heating the t-butyl acetoacetate brings about isobutylene formation and decarboxylation of the ester, but it is more common to add traces of p-toluenesulfonic acid (see F.3, Ex. c.6). The second deals with the pyrolysis of ethyl acetoacetates at about 525° (see Ex. b).

Not only acetoacetic esters, but also β-ketoesters in general and acylmalonic esters (F.3) are hydrolyzed and decarboxylated to ketones. These methods lend themselves well to the preparation of ketones of the cyclic type via Dieckmann condensations or Thorpe ring closures (see Ex. c.1 and c.2). In addition, β-ketosulfoxides, sometimes more readily available than the corresponding β-ketoesters, can be alkylated and reductively cleaved to produce ketones (see Ex. c.5):

$$\text{RCOCH}_2\text{SOCH}_3 \longrightarrow \text{RCOCH(R}')\text{SOCH}_3 \xrightarrow{\text{Al(Hg)}} \text{RCOCH}_2\text{R}'$$

Lastly, α-aminoacids react with acetic anhydride in pyridine to form β-ketoacid-type intermediates which decarboxylate to ketones (Dakin-West reaction). Thus α-aminoacids are sources of α-acetamidoketones (see Ex. c.4).

(a) Preparation of 2,4-dinitrophenylacetone. 2,4-Dinitroiodobenzene was mixed with the ethoxymagnesio salt of t-butyl acetoacetate. The crude product from this reaction was heated at 100° for 4 hr., until gas evolution had ceased, to give the title compound (88%) [49].

(b) Preparation of methyl n-amyl ketone. By saponification of the corresponding acetoacetic ester, 25–61% yield based on the ethyl acetoacetate used [45]. By pyrolysis of the corresponding acetoacetic ester, at 525°, 71% yield, based on unrecovered ester; no reagents are needed and the workup is simplified; the pyrolysis of ethyl isopropylacetoacetate to yield 82% of methylisobutyl ketone is superior to the saponification method [50].

(c) Other examples

(1) 4-Cyano-4-(3',4'-dimethoxyphenyl)-cyclohexanone (78% by

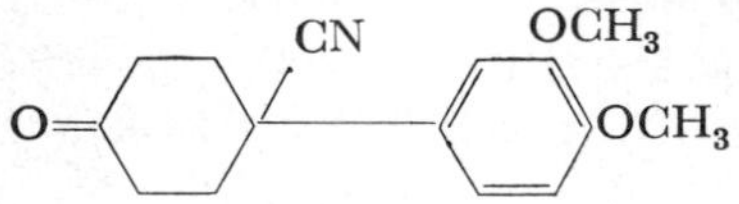

heating 2-carbomethoxy-4-cyano-4-(3',4'-dimethoxyphenyl)-cyclohexanone in acetic acid-conc. hydrochloric acid mixture) [51].

(2) 2,2-Diphenylcyclopentanone (81% from 5-cyano-2,2-diphenyl-cyclopentanoneimine

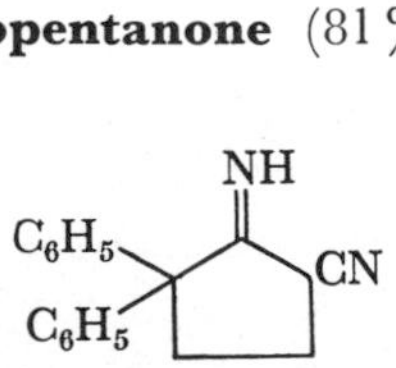

by refluxing in dilute sulfuric acid) [52].

(3) 1,4-Cyclohexanedione (2,5-dicarbethoxy-1,4-cyclohexanedione was obtained in 64–68% yields by self-condensation of diethyl succinate with sodium ethoxide; the ester was hydrolyzed and decarboxylated by heating with water in an autoclave at 195° for 15 min. to yield 81–89% of 1,4-cyclo-hexanedione) [53].

(4) Acetamidoacetone (70–78% from glycine, acetic anhydride, and pyridine refluxed for 6 hr. (Dakin-West reaction)) [54].

(5) 4-Nonanone (69% from 1-methylsulfinyl-2-heptanone, ethyl iodide, and sodium hydride in dimethyl sulfoxide followed by reductive cleavage of the crude product with aluminum amalgam in aqueous tetrahydrofuran) [55].

7. From β-Hydroxyesters, Acetylenic Alcohols, and Nitro Compounds

β-Hydroxyesters [56], acetylenic alcohols [57], and nitro alcohols [58] are cleaved by base to ketones, a reverse reaction of the aldol condensation. The process is of greater importance as a degradative rather than a preparative reaction.

(a) Preparation of 4-keto-1,2,3,4-tetrahydrophenanthrene [59]:

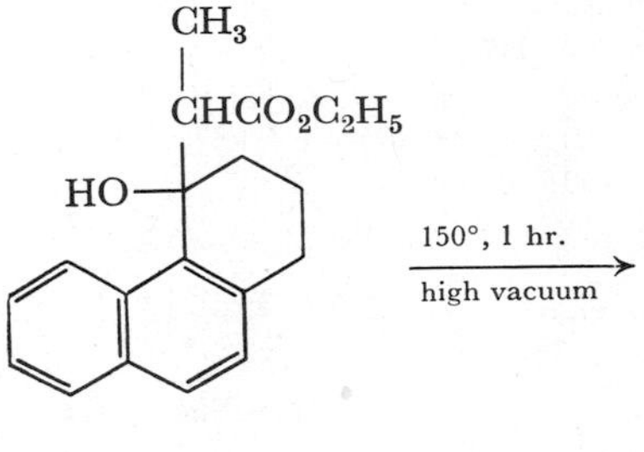

$$\text{150°, 1 hr.} \quad \text{high vacuum}$$

100%

(b) Preparation of methyl neopentyl ketone [58]:

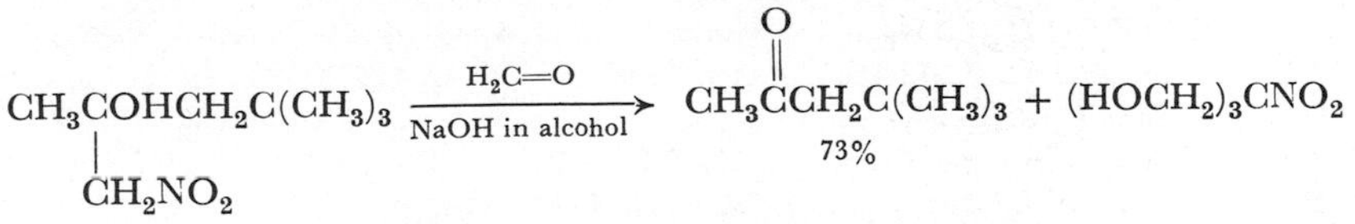

$$\underset{\underset{CH_2NO_2}{|}}{CH_3COHCH_2C(CH_3)_3} \xrightarrow[\text{NaOH in alcohol}]{H_2C=O} \underset{73\%}{CH_3\overset{O}{\overset{\|}{C}}CH_2C(CH_3)_3} + (HOCH_2)_3CNO_2$$

1. C. Grundmann and W. Ruske, *Chem. Ber.*, **86,** 939 (1953); F. Micheel and W. Flitsch, *ibid.*, **88,** 509 (1955); H. Hunsdiecker, *ibid.*, **75,** 447 (1942).
2. D. M. Young and C. F. H. Allen, *Org. Syn.*, Coll. Vol. **2,** 219 (1943).
3. W. R. Boehme, *J. Am. Chem. Soc.*, **82,** 498 (1960).
4. G. W. Cannon *et al.*, *Org. Syn.*, Coll. Vol. **4,** 597 (1963).
5. R. Crewe *et al.*, *Chem. Ber.*, **89,** 1978 (1956).
6. S. F. MacDonald, *J. Chem. Soc.*, 376 (1948).
7. G. Billek, *Monatsh.*, **92,** 343 (1961).
8. T. E. Gorizdra and S. N. Baranov, *Zh. Obshch., Khim.,* **26,** 3092 (1956); *C.A.,* **51,** 8069 (1957).

9. R. Lukeš and A. Zobáčová, *Collection Czech. Chem. Commun.*, **24**, 3189 (1959); *C.A.*, **54**, 4507 (1960).
10. R. Gaudry and R. A. McIvor, *Canad. J. Chem.*, **29**, 427 (1951).
11. E. J. Corey and D. Seebach, *Angew. Chem., Intern. Ed. Engl.*, **4**, 1075, 1077 (1965).
12. L. A. Bigelow and R. S. Hanslick, *Org. Syn.*, Coll. Vol. **2**, 244 (1943).
13. W. Deuschel, *Helv. Chim. Acta*, **34**, 2403 (1951).
14. C. S. Marvel and W. M. Sperry, *Org. Syn.*, Coll. Vol. **1**, 95 (1941).
15. G. Wittig and F. Vidal, *Chem. Ber.*, **81**, 368 (1948).
16. M. P. Cava *et al.*, *J. Am. Chem. Soc.*, **85**, 2076 (1963).
17. P. L. Julian *et al.*, *J. Am. Chem. Soc.*, **73**, 1982 (1951).
18. M. Donbrow, *J. Chem. Soc.*, 1963 (1959).
19. V. Prelog *et al.*, *Helv. Chim. Acta*, **42**, 1301 (1959).
20. G. T. Tatevosyan and M. O. Melikyan, *J. Gen. Chem. USSR (Eng. Transl.)*, **17**, 975 (1947); *C.A.* **42**, 1566 (1948).
21. Th. Gröger and E. Waldmann, *Monatsh.*, **89**, 370 (1958).
22. C. M. Hill *et al.*, *J. Am. Chem. Soc.*, **77**, 352 (1955).
23. A. P. Martinez *et al.*, *J. Org. Chem.*, **26**, 860 (1961).
24. V. Migrdichian, *Organic Synthesis*, Vol. 2, Reinhold Publishing Corp., New York, 1957, p. 983.
25. J. H. Saunders, *Org. Syn.*, Coll. Vol. **3**, 22 (1955).
26. Ref. 24, p. 1002.
27. E. D. Venus-Danilova *et al.*, *Zh. Obshch. Khim.*, **20**, 2222 (1950); *C.A.*, **45**, 7075 (1951).
28. G. W. Stacy and R. A. Mikulec, *Org. Syn.*, Coll. Vol. **4**, 13 (1963).
29. M. S. Newman, *J. Am. Chem. Soc.*, **75**, 4740 (1953).
30. C. L. Bickel, *J. Am. Chem. Soc.*, **69**, 73 (1947).
31. R. Bucourt *et al.*, *Bull. Soc. Chim., France*, 1923 (1963).
32. M. Mühlstädt and J. Graefe, *Chem. Ber.*, **100**, 223 (1967).
33. W. H. Hartung and F. Crossley, *Org. Syn.*, Coll. Vol. **2**, 363 (1943).
34. R. H. Barry and W. H. Hartung, *J. Org. Chem.*, **12**, 460 (1957).
35. H. B. Hass *et al.*, *J. Org. Chem.*, **15**, 8 (1950).
36. A. L. Henne *et al.*, *J. Am. Chem. Soc.*, **72**, 4195 (1950).
37. E. B. Hershberg, *J. Org. Chem.*, **13**, 542 (1948).
38. N. Jones *et al.*, *J. Chem. Soc.*, 2354 (1963).
39. G. I. Poos *et al.*, *J. Am. Chem. Soc.*, **75**, 422 (1953).
40. N. H. Cullinane and B. F. R. Edwards, *J. Chem. Soc.*, 1311 (1958); J. Demaecker and R. H. Martin, *Bull. Soc. Chim. Belges.*, **68**, 365 (1959).
41. H. Alper and J. T. Edward, *J. Org. Chem.*, **32**, 2938 (1967).
42. W. W. Hartmann and L. J. Roll, *Org. Syn.*, Coll. Vol. **3**, 20 (1955).
43. R. M. Herbst and D. Shemin, *Org. Syn.*, Coll. Vol. **2**, 519 (1943).
44. C. H. DePuy and B. W. Ponder, *J. Am. Chem. Soc.*, **81**, 4629 (1959).
45. J. R. Johnson and F. D. Hager, *Org. Syn.*, Coll. Vol. **1**, 351 (1941).
46. W. M. Dehn and K. E. Jackson, *J. Am. Chem. Soc.*, **55**, 4284 (1933); F. Langer and F. v. Wessely, *Monatsh. Chem.*, **88**, 298 (1957).
47. C. R. Hauser *et al.*, *J. Am. Chem. Soc.*, **69**, 119 (1947).
48. M. J. Astle and J. A. Oscar, *J. Org. Chem.*, **26**, 1713 (1961).
49. D. S. Deorha and S. P. Sareen, *J. Indian Chem. Soc.*, **41**, 735 (1964).
50. W. S. Bailey and J. J. Daly, Jr., *J. Org. Chem.*, **22**, 1189 (1957).
51. E. C. and M. G. Horning *et al.*, *J. Am. Chem. Soc.*, **74**, 773 (1952).
52. N. R. Easton and S. J. Nelson, *J. Am. Chem. Soc.*, **75**, 640 (1953).
53. A. T. Nielsen and W. R. Carpenter, *Org. Syn.*, **45**, 25 (1965).
54. J. D. Hepworth, *Org. Syn.*, **45**, 1 (1965).
55. P. G. Gassman and G. D. Richmond, *J. Org. Chem.*, **31**, 2355 (1966).
56. C. R. Hauser and W. R. Dunnavant, *J. Org. Chem.*, **25**, 1296 (1960).
57. D. Craig *et al.*, *J. Org. Chem.*, **24**, 240 (1959).
58. H. Baldock *et al.*, *J. Chem. Soc.*, 2627 (1949).
59. H. Dannenberg and H. Brachert, *Chem. Ber.*, **84**, 504 (1951).

E. Rearrangements

1. From Pinacols

$$\text{RCHOHCHOHR} \xrightarrow{\text{H}_2\text{SO}_4} \text{RCOCH}_2\text{R}$$

This rearrangement has been discussed under Aldehydes, Chapter 10, E.1, where it was noted that aldehydes or ketones or both are sometimes obtained. The mechanism has been reviewed recently [1]. The fact that both aldehydes and ketones are sometimes obtained makes the rearrangement unattractive as a preparatory method. If it should be used, it is desirable to employ concentrated acid as the catalyst since it has been shown that there is more probability that ketones rather than aldehydes, when both are possible, will be formed under such conditions [2]. The use to which the rearrangement has been put in the synthesis of ketones appears to have been confined largely to cases such as symmetrical tertiary-tertiary 1,2-glycols, in which the rearrangement is clear-cut:

$$\underset{\underset{\text{Ar}}{|} \; \underset{\text{Ar}}{|}}{\text{ArCOHCOHAr}} \longrightarrow \left[\underset{\underset{\text{Ar}}{|} \; \underset{\text{Ar}}{|}}{\text{ArC}^{\oplus}\text{COHAr}} \right] \longrightarrow \underset{\underset{\text{Ar}}{/}\;\underset{\text{Ar}}{|}}{\overset{\overset{\text{Ar}}{\diagdown}}{\text{CCOAr}}}$$

or in the case of symmetrical secondary-secondary 1,2-glycols which give ketones, particularly at higher acid concentrations:

$$\text{ArCHOHCHOHAr} \rightarrow [\text{Ar}\overset{\oplus}{\text{C}}\text{HCHOHAr}] \rightarrow \text{ArCH}_2\text{COAr}$$

In addition such glycols give ready access to spiro ketones even though yields are low [3]:

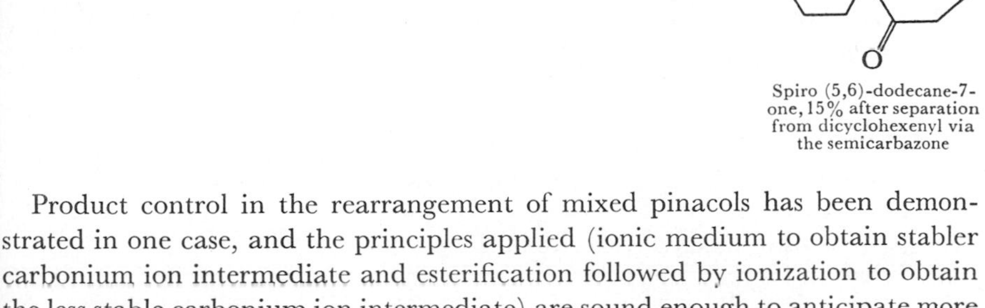

Spiro (5,6)-dodecane-7-one, 15% after separation from dicyclohexenyl via the semicarbazone

Product control in the rearrangement of mixed pinacols has been demonstrated in one case, and the principles applied (ionic medium to obtain stabler carbonium ion intermediate and esterification followed by ionization to obtain the less stable carbonium ion intermediate) are sound enough to anticipate more general applicability [4].

References for Section E are on pp. 685–686.

Mixed pinacols can be made handily, albeit in only fair yields, by the reaction of a mixture of two carbonyl compounds and amalgamated magnesium. Although attrition also is to be expected in rearrangement, a wide variety of ketones is available. A recent study exemplifies the possibilities [5]:

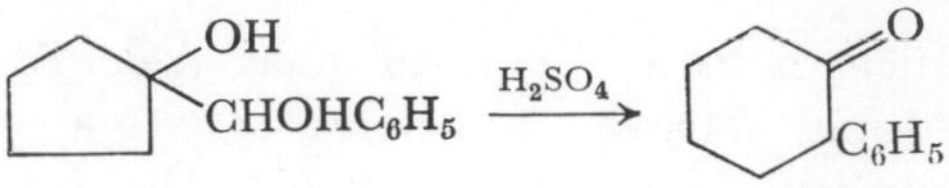

93% product
distribution
(not yield)
2-Phenylcyclohexanone

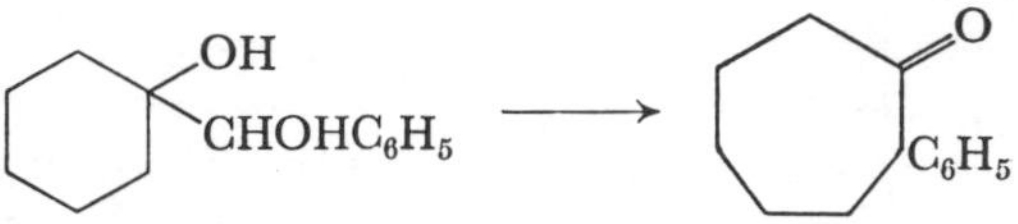

65%
2-Phenylcycloheptanone

(a) β-Benzopinacolone (95–96% from benzopinacol and iodine in glacial acetic acid) [6].

(b) Other examples

(1) α-Phenylisobutyrophenone (85% from 2-methyl-1,1-diphenyl-propane-1,2-diol and fused zinc chloride in acetic anhydride) [7].

(2) 4,4'-Bis(dimethylamino) desoxybenzoin (96% from 4,4'-bis-(dimethylamino) hydrobenzoin and aqueous hydrochloric acid) [8].

(3) Cyclopentenone (61% by heating a mixture of 3,4- and 3,5-cyclopentenediol with a trace of p-toluenesulfonic acid; this synthesis gives ready access to a cyclic α,β-unsaturated ketone) [9].

2. From Allyl Alcohols

$$CH_2=C-CHOH \xrightarrow{H_2SO_4} CH_3CHCOR$$
$$\underset{R\ \ R}{\mid\ \ \mid} \qquad\qquad \underset{R}{\mid}$$

This rearrangement has been covered under Aldehydes, Chapter 10. E.2. Whereas allyl alcohols of the type

$$CH_2=CCH_2OH$$
$$\underset{R}{\mid}$$

give high yields of aldehydes, secondary ones of the type

$$CH_2=C-CHOH$$
$$\underset{R\ \ R}{\mid\ \ \mid}$$

give almost quantitative yields of the ketone [10]. Allyl alcohols with R groups on the two doubly bound carbon atoms such as

$$RCH{=}CCH_2OH$$
$$|$$
$$R$$

occupy an intermediate position in that mixtures of aldehydes and ketones are obtained. The 2-methylenecycloalkanols 11-XV and the 1-cycloalkenemethanols

11-XV
(x = 1 or 2)

11-XVI
(x = 1 or 2)

11-XVI both give the ketones in yields of 86 and 74 %, respectively as the 2,4-dinitrophenylhydrazones [11]. These rearrangements occur in the presence of sulfuric acid and have limited preparative value because of the rare nature of the starting materials.

(a) **Preparation of methyl isopropyl ketone.** α,β-Dimethylallyl alcohol, 20 g., was refluxed with 1 l. of 0.4 N sulfuric acid under nitrogen for 48 hr. After saturating the solution with salt, the product was extracted with ether and finally recovered as the semicarbazone. Yield 92 % [10].

(b) **Preparation of 2-methylcyclohexanone** (86 % as the 2,4-dinitro-phenylhydrazone from 2-methylenecyclohexanol and 50 % sulfuric acid) [11].

3. From Ethylene Oxides

$$\overset{\displaystyle O}{\underset{\displaystyle ArCH{-}{-}{-}CHAr}{\diagup\diagdown}} \xrightarrow{\text{LiN}(C_2H_5)_2} ArCOCH_2Ar$$

Both aldehydes (Chapter 10, E.3) and ketones may be obtained in the rearrangement of ethylene oxides, a review of which is available [12]. It is obvious that, if the oxide has no hydrogen atoms attached to the carbon atoms, only ketones would be produced. Otherwise aldehydes are also a possibility. With acid catalysts such as boron trifluoride etherate at moderate temperatures, it has been shown, Aldehydes, Chapter 10, E.3, that diphenylacetaldehyde is produced from both the cis and trans forms of stilbene oxide. Magnesium bromide etherate also gives the aldehyde from the cis-oxide, but from the trans isomer both diphenylacetaldehyde and desoxybenzoin are obtained [13]. In contrast, a basic catalyst such as lithium diethylamide gives desoxybenzoin with cis-stilbene oxide and diphenylacetaldehyde with the trans isomer. Other ethylene oxides such as triphenylethylene and 1,1-diphenyl-2-p-tolyethylene oxides give the corresponding ketones with this basic catalyst. The basic catalyst, tri-n-butylphosphine oxide, gives both desoxybenzoin and diphenylacetaldehyde from cis- and trans-stilbene oxides in the ratio 9:1, respectively [14]. Thus it appears that acid catalysts of the weaker type at moderate temperatures, as a rule, produce aldehydes with such cyclic oxides, while basic ones generally produce ketones [15]. A two-path mechanism has been proposed for this rearrangement:

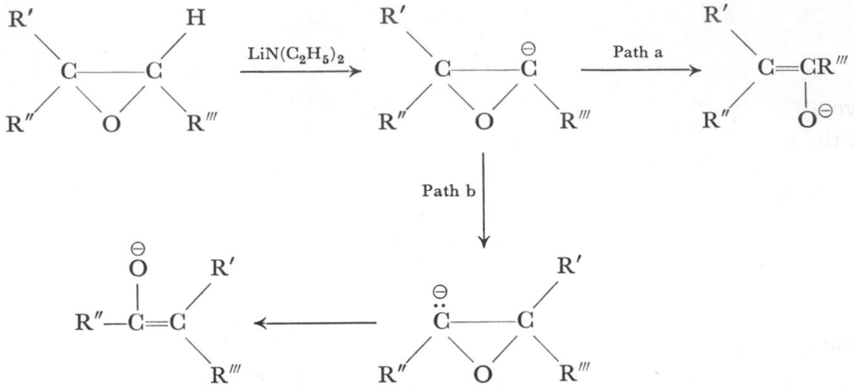

In *trans*-stilbene oxides, path b is followed in that a phenyl group (R') has migrated to give the aldehyde (R" = H). In *cis*-stilbene oxides, path a is followed and the product is desoxybenzoin (R" = H).

Recently it has been shown that α-olefin oxides, which ordinarily give aldehydes, yield ketones when treated with dicobalt octacarbonyl (see Ex. b) [16].

And still more recently two new reagents, lithium perchlorate, and propyl iodide and sodium iodide in dimethyl sulfoxide, as shown, have been utilized in epoxide rearrangements:

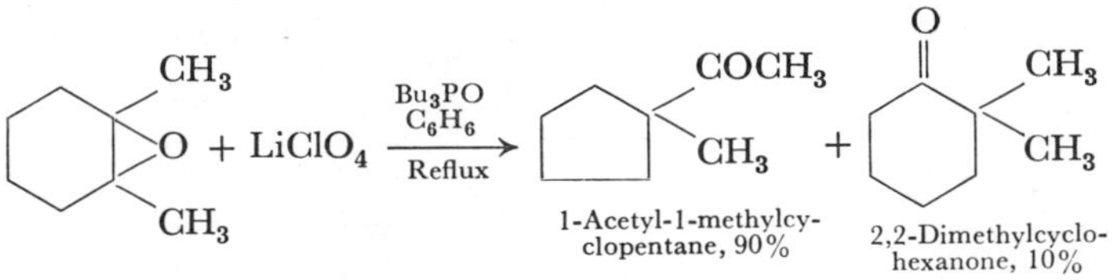

1-Acetyl-1-methylcy-
clopentane, 90%

2,2-Dimethylcyclo-
hexanone, 10%

The lithium salt may be solubilized by the addition of 1 equivalent of tributyl-phosphine oxide [17].

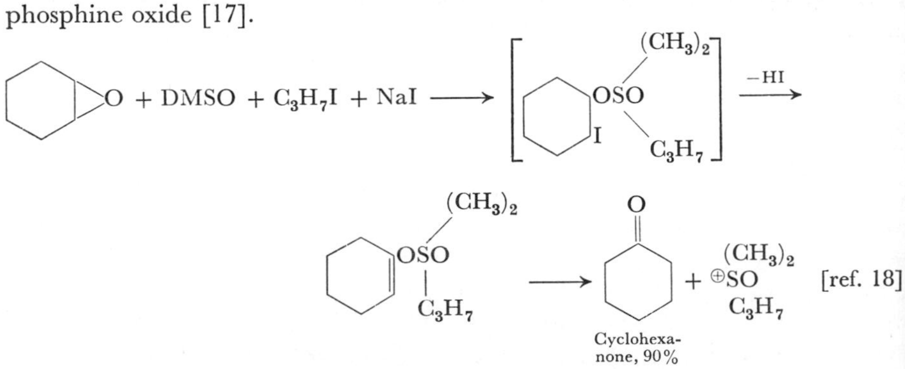

Cyclohexa-
none, 90%

[ref. 18]

In the terpene series, epoxide rearrangements are more complicated and the course cannot be predicted as readily as shown:

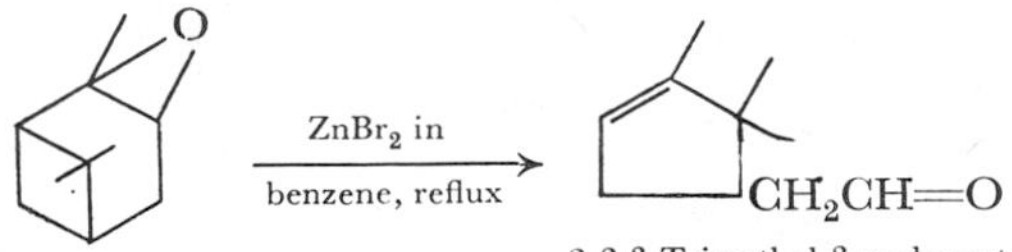

$$\text{α-Pinene oxide} \xrightarrow[\text{benzene, reflux}]{\text{ZnBr}_2 \text{ in}} \text{2,2,3-Trimethyl-3-cyclopentene-1-acetaldehyde, 70\%}$$

[ref. 19]

Glycidic esters yield pyruvic esters [20]:

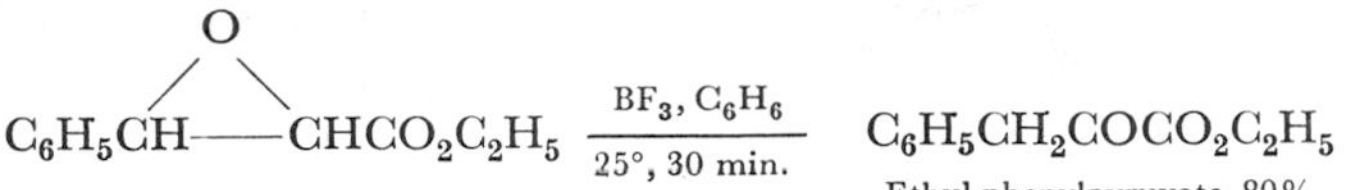

$$C_6H_5CH\text{----}CHCO_2C_2H_5 \xrightarrow[25°, 30 \text{ min.}]{BF_3, C_6H_6} C_6H_5CH_2COCO_2C_2H_5$$

Ethyl phenylpyruvate, 80%
as the 2,4-dinitrophenyl-
hydrazone

(a) Preparation of desoxybenzoin. To 60 ml. of a solution containing 0.0544 g. of butyllithium per ml. at −15° under dry nitrogen there was added a solution of 5.2 ml. of dry ethylamine in 25 ml. of anhydrous ether. After stirring for 15 min. at −15°, a solution of 5.02 g. of *cis*-stilbene oxide in 25 ml. of anhydrous ether was added dropwise. When the temperature of the reaction mixture rose to that of the room, the mixture was refluxed for 1 hr. After pouring into 100 ml. of ice water, there was recovered from the ether layer 5.04 g. of impure desoxybenzoin. Recrystallization from ethanol gave a 70% yield of a pure product [15].

(b) Acetone (75% from propylene oxide and dicobalt octacarbonyl) [16].

4. From Amino Alcohols (or Diamines)

$$(C_6H_5)_2C \begin{matrix} OH \\ \\ CH_2NH_2 \end{matrix} \xrightarrow[H^{\oplus}]{NaNO_2} C_6H_5COCH_2C_6H_5$$

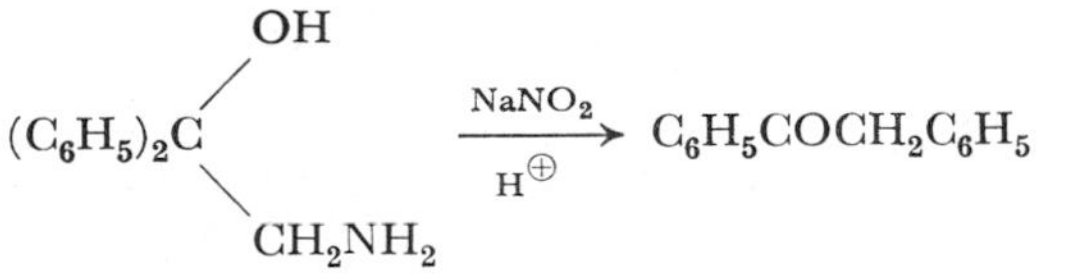

This rearrangement of amino alcohols or diamines occurs with sodium nitrite in acid, usually acetic, solution to produce a ketone. The method is part of the overall scheme to produce electron deficiency from a number of different sources and thus gives rise to the same products from a common intermediate as discussed under Aldehydes, Chapter 10, E.1:

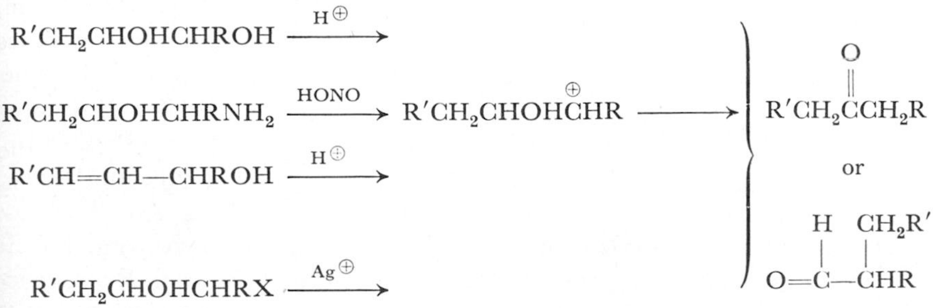

$$R'CH_2CHOHCHROH \xrightarrow{H^{\oplus}}$$

$$R'CH_2CHOHCHRNH_2 \xrightarrow{HONO} R'CH_2CHOHCHR \xrightarrow{}$$

$$R'CH=CH\text{---}CHROH \xrightarrow{H^{\oplus}}$$

$$R'CH_2CHOHCHRX \xrightarrow{Ag^{\oplus}}$$

$$\begin{matrix} O \\ \parallel \\ R'CH_2CCH_2R \end{matrix}$$

or

$$\begin{matrix} H & CH_2R' \\ | & | \\ O=C\text{---}CHR \end{matrix}$$

Ring enlargement results in the case of amino alcohols such as 1-(amino-methyl) cyclohexanol to give cycloheptanone [21].

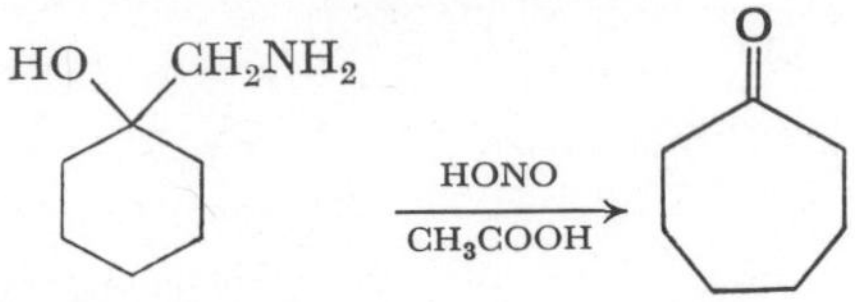

This reaction is sometimes called the Tiffeneau rearrangement.

In some instances the aminohydrin is not easily prepared from the addition of hydrogen cyanide to the ketone and subsequent reduction. The preferred route then may be via the enolate ether as shown [22]:

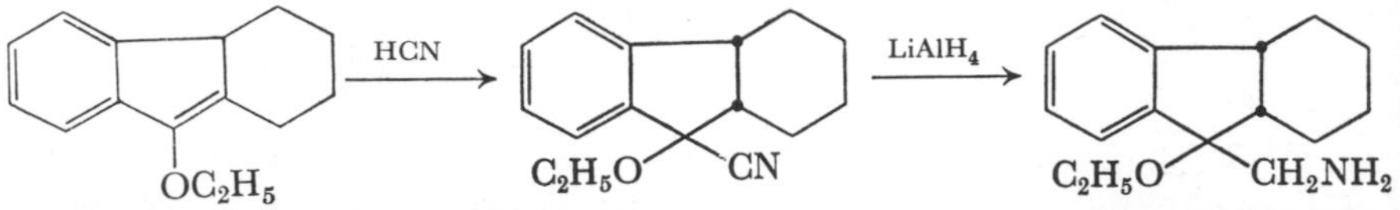

9-Ethoxy-9-aminomethyl-
1,2,3,4,4a-*cis*-9a-*cis*-hexa-
hydrofluorene, 55% overall

(a) **Preparation of cycloheptanone** (40–42% from cyclohexanone via 1-(nitromethyl)-cyclohexanol and 1-(aminomethyl)-cyclohexanol) [23].

(b) **Other examples**

(1) *p*-**Anisyl benzhydryl ketone** (97% from *dl*-α-1,2-diphenyl-1-(*p*-anisyl)-2-aminoethanol, sodium nitrite, and 50% acetic acid) [24].

(c) **Benzyl phenyl ketone** (90% from 1,1-diphenylethylenediamine, sodium nitrite, and aqueous acetic acid) [25].

5. **From Aldehydes or Ketones (Homologization)**

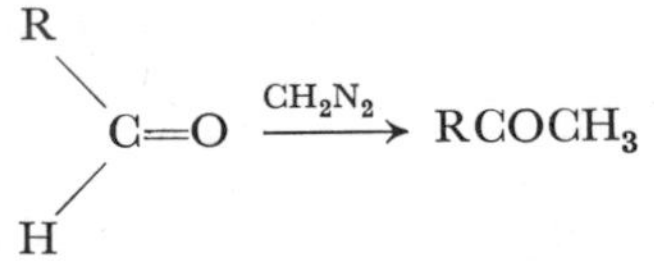

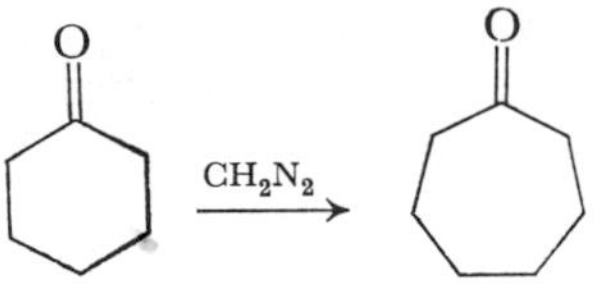

This synthesis has been reviewed [26, 27]. Unfortunately it is of limited value in preparative work because of the variety of products formed. For example, simple aliphatic aldehydes with diazomethane usually give homologous ketones with the isomeric epoxides as well. The yield of ketone varies with the complexity of the aldehyde, with the nature of the solvent, with the substituents present on the α-carbon of the aldehyde, etc. For aromatic aldehydes the products depend largely on the solvent and the substituents in the ring. To cite one example, benzaldehyde gives a 97% yield of acetophenone in the absence of methanol [28], but only a 40–50% yield if the alcohol is present [29].

Fortunately aromatic (and heterocyclic) aldehydes give superior yields of the ketone with diazoethane and other higher diazoalkanes [30, 31].

Acyclic ketones have a greater tendency to produce oxides than do aldehydes, but the simplest ones give appreciable amounts of the homolog. In the case of carbocyclic ketones (and ketene) the ring-enlarged ketone and usually some oxide are also produced. The method, in fact, is best suited to the preparation of cycloheptanones and cyclooctanones. Cycloheptanone is obtained in a 63% yield from cyclohexanone [32]; cycloöctanone in about a 45% yield from cycloheptanone [32]. Ethanol accelerates the rate of addition of diazoalkanes to cyclohexanones and increases the yield of product, as shown in one case [33]:

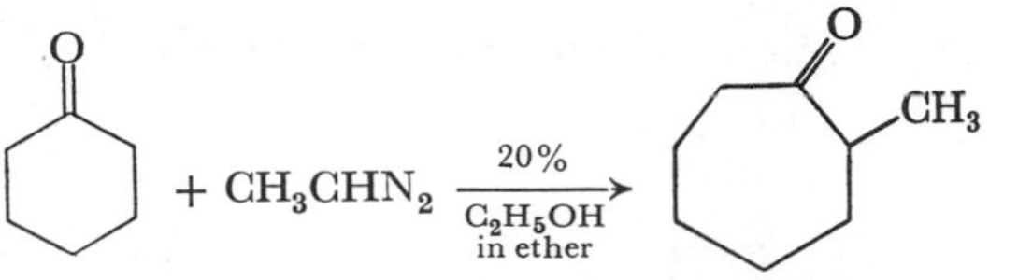

2-Methylcycloheptanone,
92% containing 9% of
the epoxide

Heterocyclic ketones, α,β-unsaturated ketones, and quinones yield more complicated products, some of which are ketones, with diazomethane [26].

The reaction is carried out in an inert solvent such as ether or, if a catalytic effect is desired, in water or alcohols, usually at a low temperature. Sometimes salts such as lithium chloride or aluminum chloride are employed to promote the reaction. As a source of diazomethane, the comparatively stable p-tolyl sulfonylmethylnitrosamide [34] may be employed.

Thus this synthesis for preparing ketones is the most satisfactory one if the reactants are an aromatic or heterocyclic aldehyde, with a diazoalkane higher than diazomethane, or a cycloalkanone containing about six or seven carbon atoms.

The course of the reaction is regarded to be as indicated below [35]. The aldehyde (R' = H) or ketone (R = CH_3, C_2H_5, etc.) accepts the diazomethane to form the zwitterion 11-XVII which on the loss of N_2 forms the zwitterion 11-XVIII:

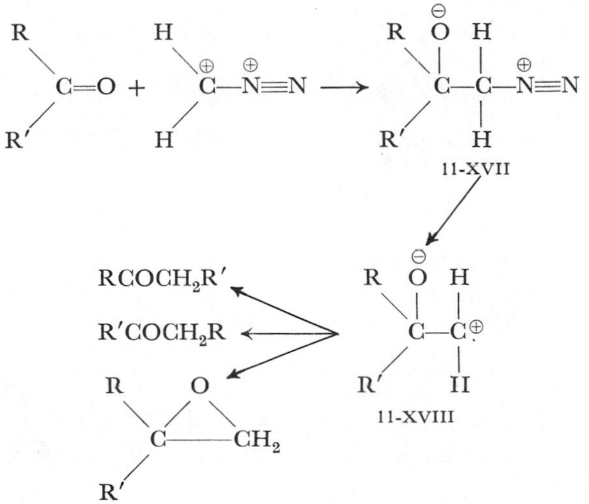

The latter becomes stabilized by forming the ketone, the aldehyde, or the epoxide.

(a) **Preparation of cycloheptanone** (63 % from cyclohexanone, nitroso-methylurethane, and sodium carbonate in methanol at 20–25°) [36].

(b) **Other examples**

(1) **2-Phenylcycloheptanone** (76 % from cyclohexanone, phenyl-diazomethane in methanol at room temperature) [37].

(2) **Ethyl 2-furyl ketone** (almost quantitative from furfural and diazoethane in ether) [31].

(4) **Methyl β-(2-ketocycloheptyl)propionate** [38]:

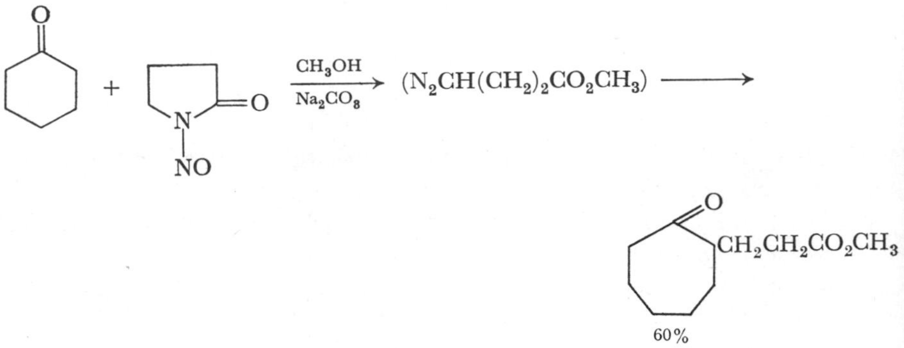

6. **From Aldehydes or Ketones**

$$\begin{array}{c} Ar \\ \diagdown \\ CHCHO \longrightarrow ArCH_2COAr \\ \diagup \\ Ar \end{array}$$

Early experiments in this rearrangement consisted in heating the aldehyde with aluminum sulfate as a catalyst. By this procedure, trimethylacetaldehyde was converted quantitatively into isopropyl methyl ketone:

$$(CH_3)_3CCHO \xrightarrow{350°} (CH_3)_2CHCOCH_3$$

and triphenylacetaldehyde gave α,α-diphenylacetophenone at 25 mm. [39]:

$$(C_6H_5)_3CCHO \xrightarrow[340°]{320-} (C_6H_5)_2CHCOC_6H_5$$

Catalysts which are hydration catalysts for acetylenes, sulfuric acid, or mercury salts and sulfuric acid, have also been employed [40]. Aryl groups facilitate the rearrangement to ketones, i.e., $(C_6H_5)_3CCHO$ gives a better yield of the ketone than does $(C_6H_5)_2CHCHO$. The latter gives a 60–65 % yield of desoxybenzoin on heating with 50–75 % sulfuric acid, whereas the oxime gives almost a quantitative yield. The aldehydes may be added at $-10°$ or $-20°$ to concentrated sulfuric acid, after which the mixture is stirred and poured on ice to yield the ketone. The work of E. D. Venus-Danilova should be consulted for more examples [41].

Until recently, rearrangements were believed to be confined to highly substituted aldehydes and ketones, in which an intermediate carbonium ion of favorable stability was formed. But it has now been shown that the rearrangement is much more general using the strong acid, perchloric acid, as the rearrangement catalyst. No oxygen migration takes place in the rearrangement of 3-pentanone but rather a series of alkyl group shifts [42]:

$$CH_3CH_2CCH_2CH_3 \underset{}{\overset{HClO_4}{\rightleftharpoons}} CH_3CH_2\overset{\oplus}{C}CH_2CH_3 \longrightarrow \overset{\oplus}{C}H_2\overset{OH}{\underset{CH_3}{C}}CH_2CH_3 \overset{C_2H_5}{\underset{Shift}{\longrightarrow}}$$

$$CH_3CH_2CH_2\overset{OH}{\underset{\oplus}{C}}CH_3 \overset{-H\oplus}{\longrightarrow} CH_3CH_2CH_2\overset{O}{\overset{\|}{C}}CH_3$$

63%, Methyl propyl ketone

However, in cases of more highly branched ketones, oxygen migration (via an epoxide) does occur [43] (see also Ex. b). Another promising catalyst for this rearrangement is phosphoric acid on Celite, over which vapors of the ketone are passed several times [44].

(a) **Preparation of 3,4-dimethyl-2-pentanone.** Diisopropyl ketone was heated with about 20 times its weight of 70 % aqueous perchloric acid on a steam bath to give about 70 % of the desired product mixed with the starting ketone [42].

(b) **Preparation of 3-methyl-3-phenyl-2-butanone** [45].

$$C_6H_5\overset{O}{\overset{\|}{C_{14}}}C(CH_3)_3 \overset{72\% HClO_4}{\underset{Room\ temp.,\ 25\ hr.}{\longrightarrow}} CH_3\overset{C_6H_5}{\underset{CH_3}{C}}-\overset{O}{\overset{\nearrow}{C_{14}}}CH_3 + CH_3-\overset{C_6H_5}{\underset{CH_3}{C}}-\overset{O}{\overset{\nearrow}{C_{14}}}CH_3$$

94%

1. Y. Pocker, in P. de Mayo, *Molecular Rearrangements*, Vol. 1, Interscience Publishers, New York, 1963, Chap. 1.
2. S. N. Danilov and E. Venus-Danilova, *Chem. Ber.*, **60**, 1050 (1927).
3. N. V. Elagina and B. A. Kazanskiĭ, *Dokl. Akad. Nauk. SSSR*, **124**, 1053 (1959); *C.A.*, **53**, 14969 (1959).
4. T. E. Zalesskaya and I. K. Lavrova, *J. Org. Chem. USSR, Eng. Transl.*, **1**, 1226 (1965).
5. D. G. Botteron and G. Wood, *J. Org. Chem.*, **30**, 3871 (1965).
6. W. E. Bachmann, *Org. Syn.*, Coll. Vol. **2**, 73 (1943).
7. R. E. and G. G. Lyle, *J. Org. Chem.*, **18**, 1058 (1953).
8. M. J. Allen, *J. Am. Chem. Soc.*, **73**, 1841 (1951).
9. C. H. DePuy and K. L. Eilers, *J. Org. Chem.*, **24**, 1380 (1959).
10. M. B. Green and W. J. Hickinbottom, *J. Chem. Soc.*, 3262 (1957).
11. A. S. Dreiding and J. A. Hartman, *J. Am. Chem. Soc.*, **78**, 1216 (1956).
12. R. E. Parker and N. S. Isaacs, *Chem. Rev.*, **59**, 737 (1959).

13. H. O. House, *J. Am. Chem. Soc.*, **77**, 3070 (1955).

14. D. E. Bissing and A. J. Speziale, *J. Am. Chem. Soc.*, **87**, 1405 (1965).

15. A. C. Cope *et al.*, *J. Am. Chem. Soc.*, **80**, 2844 (1958).

16. J. L. Eisenman, *J. Org. Chem.*, **27**, 2706 (1962).

17. B. Rickborn and R. M. Gerkin, *J. Am. Chem. Soc.*, **90**, 4193 (1968).

18. G. W. Kenner *et al.*, *Chem. Commun.*, 227 (1968).

19. J. B. Lewis and G. W. Hedrick, *J. Org. Chem.*, **30**, 4271 (1965).

20. H. O. House *et al.*, *J. Am. Chem. Soc.*, **80**, 6386 (1958).

21. C. D. Gutsche and D. Redmore, *Carbocyclic Ring Expansion Reactions*, Academic Press, New York, 1968, p. 74.

22. W. E. Parham and L. J. Czuba, *J. Am. Chem. Soc.*, **90**, 4030 (1968).

23. H. J. Dauben, Jr., *et al.*, *Org. Syn.*, Coll. Vol. **4**, 221 (1963).

24. P. I. Pollak and D. Y. Curtin, *J. Am. Chem. Soc.*, **72**, 961 (1950).

25. R. Granger and H. Técher, *Compt. Rend.*, **250**, 2581 (1960).

26. C. D. Gutsche, *Org. Reactions*, **8**, 364 (1954).

27. Ref. 21, Chapt. 4.

28. F. Schlotterbeck, *Chem. Ber.*, **40**, 479 (1907).

29. E. Mosettig and K. Czadek, *Monatsh. Chem.*, **57**, 291 (1931).

30. D. W. Adamson and J. Kenner, *J. Chem. Soc.*, 181 (1939); E. Mosettig and K. Czadek, *Monatsh. Chem.*, **57**, 291 (1931).

31. J. Ramonczai and L. Vargha, *J. Am. Chem. Soc.*, **72**, 2737 (1950).

32. E. P. Kohler *et al.*, *J. Am. Chem. Soc.*, **61**, 1057 (1939).

33. J. A. Marshall and J. J. Partridge, *J. Org. Chem.*, **33**, 4090 (1968).

34. Th. J. de Boer and H. J. Backer, *Org. Syn.*, Coll. Vol. **4**, 943 (1963).

35. H. O. House *et al.*, *J. Am. Chem. Soc.*, **82**, 4099 (1960).

36. E. P. Kohler *et al.*, *Org. Reactions*, **8**, 401 (1954); *see also* Th. J. de Boer and H. J. Backer, *Org. Syn.*, Coll. Vol. **4**, 225 (1963).

37. C. D. Gutsche and E. F. Jason, *Org. Reactions*, **8**, 401 (1954).

38. C. D. Gutsche and I. Y. C. Tao, *J. Org. Chem.*, **32**, 1778 (1967).

39. S. N. Danilov, *Zh. Fiz. Khim.*, **61**, 723 (1929); *C.A.*, **23**, 4670 (1929).

40. S. N. Danilov, *Zh. Obshch. Khim.*, **18**, 2000 (1948); *C.A.*, **43**, 4632 (1949).

41. V. F. Kazimirova and L. A. Pavlova, *J. Gen. Chem., USSR, Eng. Transl.*, **29**, 3817 (1959).

42. A. Fry *et al.*, *J. Org. Chem.*, **25**, 1252 (1960).

43. K. Bhatia and A. Fry, *J. Org. Chem.*, **34**, 806 (1969).

44. A. Fry, Univ. of Arkansas, Fayetteville, Ark., private communication.

45. T. B. Remizova and T. E. Zalesskaya, *Zh. Obshch. Khim.*, **34**, 1395 (1964); *C.A.*, **61**, 5473 (1964).

F. Acylations

Acylation as a method of preparation for ketones has been summarized [1] and is discussed in a number of other chapters (Carboxylic Esters, Chapter 14,C and Aldehydes, Chapter 10,F). Some repetition is necessary for coherency, but it will be minimized. Most of the acylations, with the exception of those in F.7, are concerned with the addition of carbanions to esters or other acid derivatives. Other condensations of a nucleophilic nature to produce ketones are to be found elsewhere (acyloin and benzoin condensations, Alcohols, Chapter 4, C.8).

1. From Acids (Decarboxylation)

(See Aldehydes, Chapter 10, H.3, for corresponding preparation of aldehydes.) Although this reaction may not be a true acylation, it accomplishes the

References for Section F are on pp. 695–697.

same purpose as the Claisen condensation, namely, the eventual formation of a symmetrical ketone:

$$(RCO_2)_2Ba \longrightarrow R\overset{\overset{\displaystyle O}{\|}}{C}R + BaCO_3$$

The mechanism is under dispute. It may be simple decarboxylation to an anion which attacks the carboxyl group:

$$R\overset{\overset{\displaystyle O}{\|}}{C}\!-\!\overset{\ominus}{O} \longrightarrow CO_2 + \left[\overset{\ominus}{R}\right] \xrightarrow{RCO\overset{\ominus}{O}} \left[\begin{array}{c} \overset{\ominus}{O} \\ R \diagup \\ RC \\ \diagdown O \\ \ominus \end{array}\right] \xrightarrow{2\overset{\oplus}{H}} R_2C\!=\!O + H_2O$$

Or it may be that a β-ketoacid is formed by a Claisen-type condensation, after which the keto acid decarboxylates. However, the latter explanation could hardly apply to the synthesis of 2,2,5,5-tetramethylpentanone in 72 % yield by distillation of the barium salt of 2,2,5,5-tetramethyladipic acid [2].

The simpler carboxylic acids give excellent yields when their vapors are passed over a metallic oxide such as manganese or thorium dioxide at 400–500° [3]. More recently it has been found that a thoria "aerogel" catalyst is superior to other forms of thorium dioxide [4].

Various modifications have been tried in an attempt to improve the yield. For higher acids such as lauric and 9-undecenoic, it was found that the esters over thoria "aerogel" gave yields superior to the free acids, ethyl laurate giving 93 % and ethyl 9-undecenoate giving 86 % of the ketone [5]. For mixed acids over thoria at 400° an excess of the short-chain acid minimizes the formation of the long-chain symmetrical ketone with the result that the yields of mixed ketones are improved. In this manner, methyl benzyl and other alkyl aryl ketones have been prepared with yields of 55–65 % [6]; *see also* [7]. Phenylacetic acids and acetic anhydride are readily condensed and decarboxylated to phenylacetones in refluxing pyridine [8]. Other catalysts such as manganous oxide [9], potassium fluoride [2], fluorosulfonic acid [10], and a mixture of thorium and aluminum oxides [11] are also effective. Sometimes the acids are first converted into salts [12, 13] which are then decomposed by heat, although the vapor-phase method is of greater industrial importance.

Iron salts give ketones in fair to excellent yields from straight-chain acids (see Ex. b.4). Lead salts decompose in two distinct phases [14]:

$$2\ (RCO_2)_2Pb \xrightarrow[-CO_2]{300°} R_2CO + (RCO_2Pb)_2O + CO_2$$

$$\downarrow 370°$$

$$R_2CO + 2\ PbO + CO_2$$

The Dakin-West reaction is a decarboxylation procedure applied to amino acids (see Ex. b.5). γ-Ketoacids apparently decarboxylate to yield α,β-unsaturated ketones [15]:

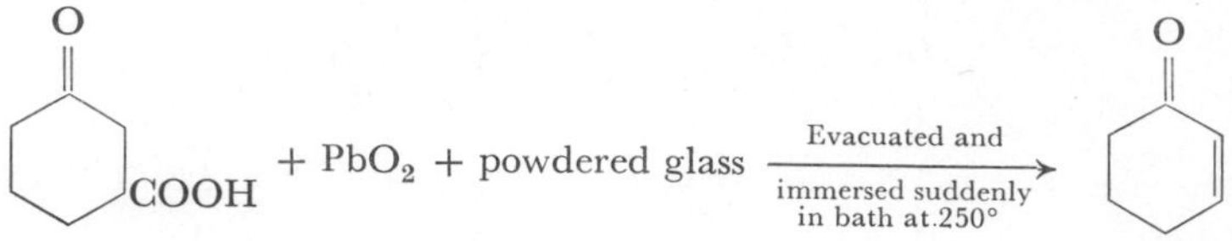

2-Cyclohexenone,
92%

(a) **Preparation of benzyl methyl ketone** (55–65% by passing the vapors of phenylacetic (1 mole) and acetic (2 moles) acids over thorium dioxide at 430–450°) [6].

(b) **Other examples**

(1) **Acetophenone** (88% by passing a 1:5 mixture of benzoic and acetic acids over an aluminum oxide-thorium oxide catalyst at 420°) [11].

(2) **Cyclopentanone** (75–80% by heating a mixture of adipic acid and barium hydroxide at 285–295°) [12]; *see also* [2].

(3) **Stearone** (81–87% by heating a mixture of stearic acid and magnesium oxide at 335–340°) [13].

(4) **10-Nonadecanone** (96% from 1 equiv. of decanoic acid and 1.1 equiv. of hydrogen-reduced iron powder by refluxing until carbon dioxide evolution ceased) [16].

(5) **3-Acetamido-2-butanone** (81–88% from alanine and acetic anhydride in refluxing pyridine; with a variety of amino acids the yields vary from 27 to 88%) [17].   (For hydrolysis to the amine hydrochloride, see [18].)

(6) **Laurone** (46–55% from lauroyl chloride and triethylamine followed by hydrolysis and decarboxylation) [19].

2. From Esters (Claisen, Dieckmann)

The Claisen and Dieckmann ester condensations are discussed under Carboxylic Acid Esters, Chapter 14, C.1 and C.2, and hydrolysis and decarboxylation of the β-ketoesters in this chapter, D.6.

3. From Malonic and Acetoacetic Esters and β-Diketones

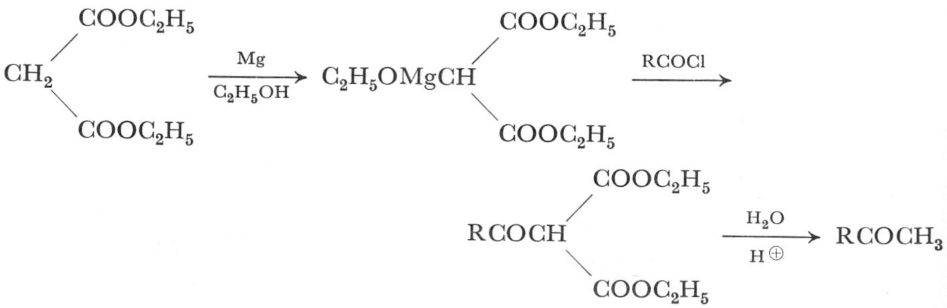

Malonic ester may be acylated, preferably via the ethoxymagnesium derivative, and the diethyl acylmalonate may be hydrolyzed and decarboxylated completely to give the ketone [20]. The hydrolysis and subsequent decarboxylation occur best under acidic conditions such as those supplied by aqueous acetic and sulfuric acids [20], propionic and sulfuric, followed by treatment with 10 *N* sulfuric [21], or glacial acetic acid containing acetic anhydride and some *p*-toluenesulfonic acid [22]. This synthesis is particularly valuable in the synthesis of *o*- and *p*-nitroacetophenone and *o*-chloroacetophenone [23]. Monoalkyl substituted malonic esters can be acylated also, but the saponification of the ethyl ester, $RCOCR'(COOC_2H_5)_2$, offers difficulty. The corresponding *t*-butyl ester, on the other hand, can be hydrolyzed without difficulty (see Ex. a).

β-Diketones may be prepared by acylation of acetoacetic ester and subsequent hydrolysis, but the synthesis is complicated by O-acylation accompanying or replacing carbon acylation. O-Acylation is favored in pyridine solution and C-acylation in alkoxide or hydroxide solution. Or C-acylation may be favored if the metal cation-to-oxygen bond is more covalent [24]. The O-ester can be converted to the carbon acylation product by heating with potassium carbonate [25]. An interesting application of this migration is the Baker-Venkataraman rearrangement [26]:

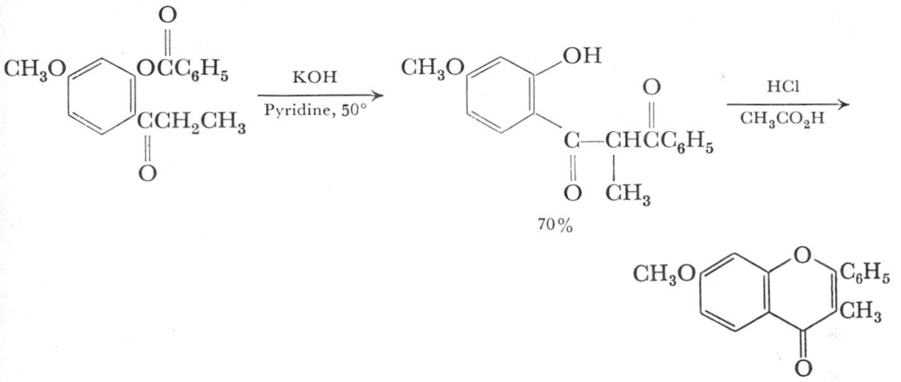

7-Methoxy-3-methylflavone, 85%

Another complication is that basic conditions may give ketone cleavage rather than saponification:

$$CH_3COCH(COR)CO_2C_2H_5 \xrightarrow[2.\ H_3O^{\oplus}]{1.\ \text{Saponification}} CH_3COCH_2COR + CO_2 + C_2H_5OH$$

$$\downarrow \text{Ketone cleavage}$$

$$CH_3COCH_3 + RCO_2H + CO_2 + C_2H_5OH$$

The ketone cleavage can be avoided by using benzyl acetoacetate for acylation followed by hydrogenolysis of the benzyl ester (see Ex. c.4) or by using *t*-butyl acetoacetate for acylation followed by acidic hydrolysis of the *t*-butyl ester (see Ex. c.6).

Many variations and permutations of these procedures can be conceived involving, for example, acylation of a diketone (see Ex. c.5).

(a) Preparation of styryl β-phenylethyl ketone (79% from di-*t*-butyl benzylmalonate, sodium hydride in benzene, and cinnamoyl chloride, followed by heating with *p*-toluenesulfonic acid) [27].

(b) Preparation of *o*-nitroacetophenone (82–83% from malonic ester, magnesium ethoxide, and *o*-nitrobenzoyl chloride followed by hydrolysis with sulfuric acid in acetic acid) [23].

(c) Other examples

(1) Ethyl benzoylacetate (68–71% from ethyl acetoacetate in naphtha and simultaneous addition of aqueous sodium hydroxide and benzoyl chloride from two separate funnels, followed by hydrolysis of the aqueous layer with ammonium chloride) [28].

(2) Diethyl benzoylmalonate (an acylation of diethyl malonate, 68–75% from the ester, magnesium, absolute alcohol, and mixed benzoic-carbonic anhydride [29].

(3) β-(2-Nitro-3,4,5-trimethoxyphenyl)-3′-methoxypropiophenone (80–85% from di-*t*-butyl 2-nitro-3,4,5-trimethoxybenzylmalonate, lithium amide, ferric nitrate (trace), *m*-methoxybenzoyl chloride, and acetic acid containing acetic anhydride and *p*-toluenesulfonic acid) [22].

(4) 2,4-Heptanedione (49% from benzyl acetoacetate, magnesium methoxide, and butyryl chloride followed by hydrogenolysis and decarboxylation) [30].

(5) α,α-Di-(phenylacetyl)-acetophenone (54% from phenylacetyl-acetophenone and phenylacetyl chloride with sodium metal in ether) [31].

(6) Stearoylacetone (quantitative yield from *t*-butyl stearoylacetoacetate by heating with a trace of *p*-toluenesulfonic acid at 160–170° until gas evolution had ceased) [32].

4. From Ketones or Enamines

$$CH_3COCH_2R' + RCOOC_2H_5 \xrightarrow{\text{Base}} RCOCH_2COCH_2R' + C_2H_5OH$$

Type I

or

$$CH_3COCH_2R' + (RCO)_2O \xrightarrow{\text{BF}_3} RCOCHCOCH_3 + RCOOH$$
$$\underset{R'}{|}$$

Type II

Two methods have been used in the acylation of ketones to produce β-diketones. By Type I the ketone is treated with an ester in the presence of a base, preferably sodium amide or sodium hydride [33], and in Type II the ketone is treated with an anhydride in the presence of boron trifluoride. In the case of a symmetrical ketone or an unsymmetrical one with only one reactive group attached to the carbonyl group, the product is the same by Type I or II. How-ever, if there are two different reactive groups, CH_3 and CH_2R as in Type I and II, two different products generally result. This synthesis, employing a variety of ethyl esters and anhydrides as acylating agents, gives yields mostly in the 30–60% range.

The path taken in Type I has already been discussed since it is a Claisen condensation (see Carboxylic Acid Esters, Chapter 14, C.1), but an example (b.3) is given which speaks well for the advantages of sodium hydride in di-methyl sulfoxide as the condensing agent. Type II is an acid-catalyzed acylation brought about by the Lewis acid, boron trifluoride. This appears to first form the acylium ion [34]

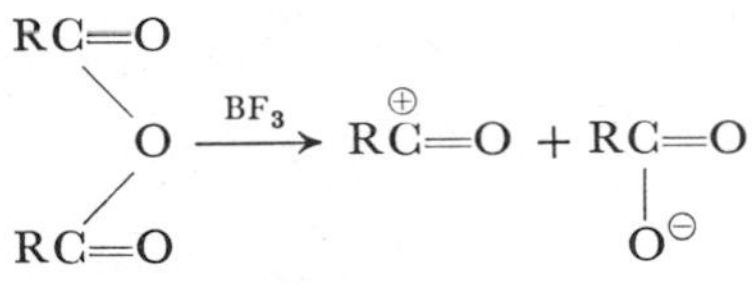

$$
\begin{array}{c}
RC{=}O \\
\diagdown \\
O \xrightarrow{BF_3} \overset{\oplus}{RC}{=}O + RC{=}O \\
\diagup \qquad\qquad\qquad | \\
RC{=}O \qquad\qquad O^{\ominus}
\end{array}
$$

which attacks the enolic form of the ketone

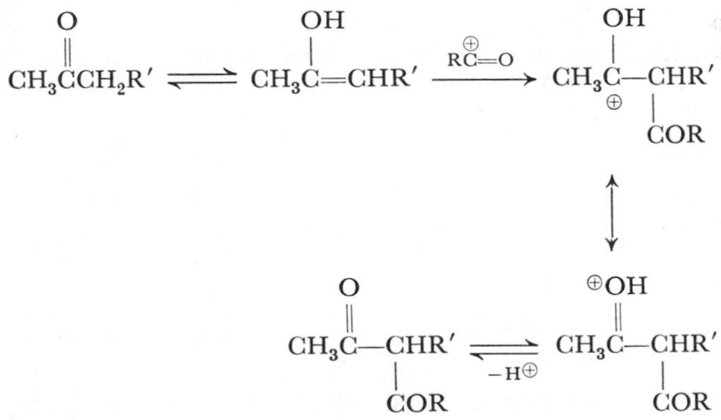

$$
\underset{\substack{\| \\ CH_3CCH_2R'}}{O} \rightleftharpoons CH_3\underset{|}{\overset{OH}{C}}{=}CHR' \xrightarrow{\overset{\oplus}{RC}{=}O} CH_3\underset{\substack{| \\ \underset{COR}{\overset{\oplus}{}}}}{\overset{OH}{C}}{-}CHR'
$$

$$
\underset{\substack{\| \\ CH_3C{-}CHR' \\ | \\ COR}}{O} \underset{-H^{\oplus}}{\rightleftharpoons} \underset{\substack{\| \\ CH_3C{-}CHR' \\ | \\ COR}}{\overset{\oplus}{OH}}
$$

Enamines of ketones have been diacylated in one example [36]:

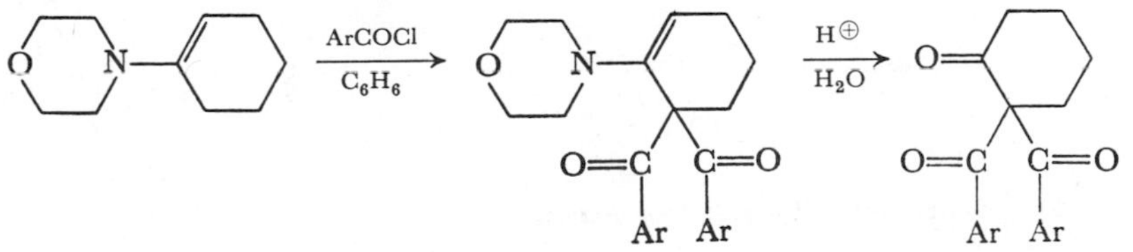

2,2-Di(*o*-chlorobenzoyl)-
cyclohexanone, 40%,
Ar = *o*-ClC₆H₄-

However, monoacylation occurs in other examples [35, 37].

Cyclohexanone may be monoacylatėd in about 50% yield using sodium *t*-amylate in benzene as the catalyst [38]; in 69% yield with sodamide in ether [39]; but usually in better yields with boron trifluoride [40].

The thallium salts of β-diketones may be isolated as crystalline, nonhygroscopic salts which lend themselves well to acylation and alkylation [41]:

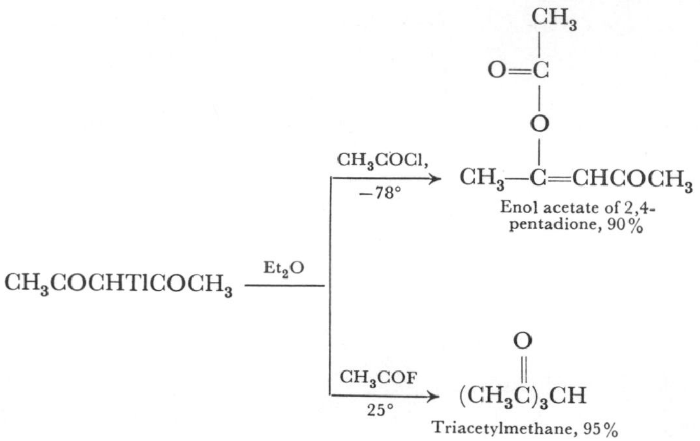

$$
CH_3COCHTlCOCH_3 \xrightarrow{Et_2O}
\begin{cases}
\xrightarrow[-78°]{CH_3COCl,} & CH_3-C=CHCOCH_3 \quad \overset{\displaystyle CH_3 \atop \displaystyle | \atop \displaystyle O=C \atop \displaystyle | \atop \displaystyle O \atop \displaystyle |}{} \\
& \text{Enol acetate of 2,4-pentadione, 90\%} \\[2em]
\xrightarrow[25°]{CH_3COF} & (CH_3\overset{O}{\overset{\|}{C}})_3CH \\
& \text{Triacetylmethane, 95\%}
\end{cases}
$$

The thallium halide is removed by filtration and the filtrate simply concentrated.

(a) Preparation of acetylacetone (80–85% based on acetone treated with acetic anhydride and boron trifluoride [41a]; 38–45% based on acetone treated with ethyl acetate and sodium ethoxide) [42].

(b) Other examples

   (1) Dibenzoylmethane (62–71% based on acetophenone treated with ethyl benzoate and sodium ethoxide) [43].

   (2) 3-Propionyl-2-heptanone (47% based on methyl *n*-amyl ketone treated with propionic anhydride and boron trifluoride) [44].

   (3) 1,3-Di(cyclohexenyl)-1,3-propanedione (72% from 4-carbomethoxycyclohexene, 1 mole, 0.5 mole of 4-acetylcyclohexane, and 1 mole of NaH in DMSO for 4 hr. at 60°) [45].

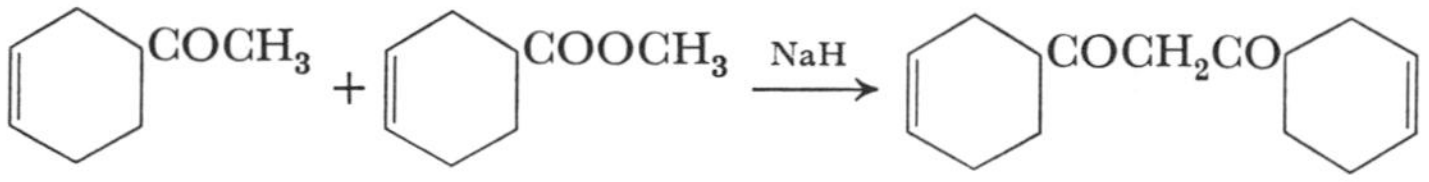

(4) 3-Isopropyl-2,4-pentanedione.

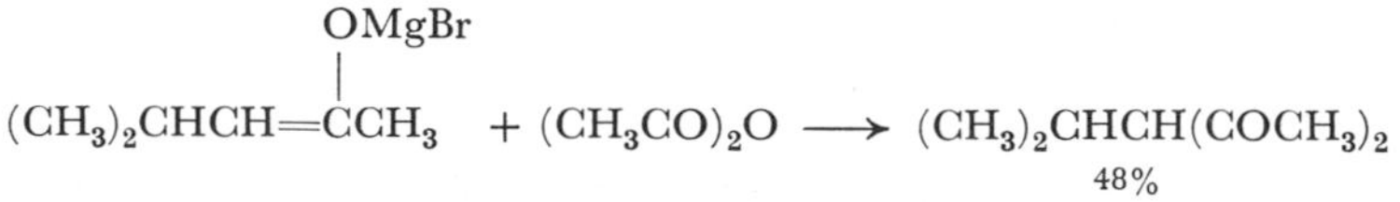

$$(CH_3)_2CHCH=\overset{\displaystyle OMgBr \atop \displaystyle |}{C}CH_3 \; + \; (CH_3CO)_2O \longrightarrow (CH_3)_2CHCH(COCH_3)_2$$
$$48\%$$

The remainder is the product of O-acylation [46].

5. From Nitriles, Diazoalkanes, and 2-(or 4-)Alkylpyridines or -quinolines

Nitriles are acylated using very strong base catalysts (see Ex. a). On the other hand, diazoalkanes are acylated without a catalyst [47]:

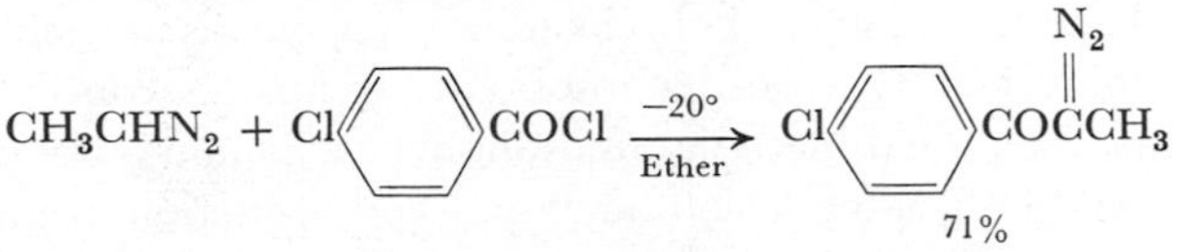

$$CH_3CHN_2 + Cl\!-\!\langle\bigcirc\rangle\!-\!COCl \xrightarrow[\text{Ether}]{-20°} Cl\!-\!\langle\bigcirc\rangle\!-\!\overset{\overset{\displaystyle N_2}{\|}}{C}OCCH_3$$

71%

Although the diazoketone is usually rearranged to the ketene:

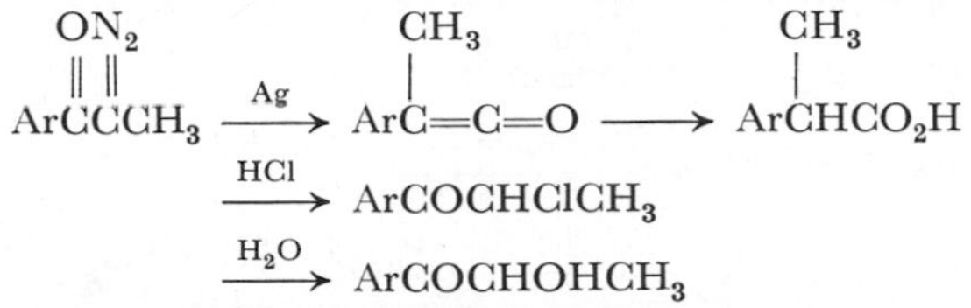

$$\underset{ArCCCH_3}{\overset{ON_2}{\underset{\|\ \|}{}}} \xrightarrow{Ag} \underset{ArC=C=O}{\overset{CH_3}{\underset{|}{}}} \longrightarrow \underset{ArCHCO_2H}{\overset{CH_3}{\underset{|}{}}}$$

$$\xrightarrow{HCl} ArCOCHClCH_3$$

$$\xrightarrow{H_2O} ArCOCHOHCH_3$$

in the absence of silver catalyst, the diazoketone can be hydrolyzed either to the chloride or alcohol (see Ex. b). The 2- and 4-alkyl groups of pyridine or quinoline compounds have active methylene groups and can be acylated. Phenyllithium (Ex. c) and the more readily available sodium diisopropylamide (Ex. d) have been used as condensing agents.

(a) **Preparation of α-phenylacetoacetonitrile** (66–73% from benzyl cyanide, ethyl acetate, and sodium ethoxide) [48].

(b) **Preparation of benzyl chloromethyl ketone** (83–85% from phenylacetyl chloride, diazomethane, and hydrogen chloride) [49].

(c) **Preparation of 3-(2-pyridyl)-2-butanone** (52% from 2-ethyl-pyridine, phenyllithium, and methyl acetate) [50].

(d) **Preparation of 3-phenacylpyridine** (78% from 2 equiv. of 3-methylpyridine, 2 equiv. of sodium diisopropylamide, and 1 equiv. of ethyl benzoate, all three in benzene at 5°) [51].

6. From Dimethyl sulfoxide, Dimethyl Sulfone, or N,N-Dimethyl-methanesulfonamide

$$CH_3SOCH_3 + RCOOR' \xrightarrow{NaH} RCOCH_2SOCH_3 \xrightarrow{AlHg} RCOCH_3$$

$$CH_3SO_2CH_3 + RCOOR' \xrightarrow{NaH} RCOCH_2SO_2CH_3 \xrightarrow{AlHg} RCOCH_3$$

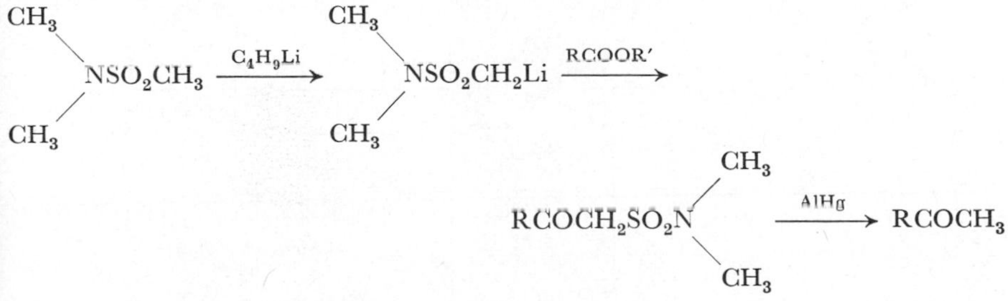

$$\underset{CH_3}{\overset{CH_3}{\diagdown}}N\!-\!SO_2CH_3 \xrightarrow{C_4H_9Li} \underset{CH_3}{\overset{CH_3}{\diagdown}}N\!-\!SO_2CH_2Li \xrightarrow{RCOOR'}$$

$$RCOCH_2SO_2N\underset{CH_3}{\overset{CH_3}{\diagup}} \xrightarrow{AlHg} RCOCH_3$$

Dimethyl sulfoxide and dimethyl sulfone may be converted into carbanions ($\overline{\text{CH}}_2\text{SOCH}_3$ or $\overline{\text{CH}}_2\text{SO}_2\text{CH}_3$, respectively) with sodium hydride or potassium t-butoxide, and on treatment with an ester the β-ketosulfoxide or β-ketosulfone is formed [52, 53]. Reduction of the latter with aluminum amalgam or zinc dust in ethanol-acetic acid gives methyl ketones. The yields in each of these steps vary from 70 to over 98%. In order to obtain the ketone from N,N-dimethyl-methanesulfonamide, the dimethylaminosulfonyl carbanion is obtained through the use of butyllithium, after which acylation and reduction are accomplished as before [52]. To avoid the reduction step, sulfinamides may be acylated to give ketones [54]:

$$\text{CH}_3\text{SONHC}_6\text{H}_4\text{CH}_3\text{-}p\text{-} + 2\ \text{C}_4\text{H}_9\text{Li} \xrightleftharpoons[]{\text{THF}}$$

$$\overset{\ominus}{\text{CH}}_2\text{SO}\overset{\ominus}{\text{N}}\text{C}_6\text{H}_4\text{CH}_3\text{-}p\text{-} \xrightarrow[\text{2. H}^{\oplus}, \text{H}_2\text{O}]{\text{1. 1 equiv. C}_{17}\text{H}_{35}\text{COOCH}_3, -78°, 2\ \text{hr.}}$$

$$\text{C}_{17}\text{H}_{35}\text{COCH}_3 + \text{SO}_2 + p\text{-CH}_3\text{C}_6\text{H}_4\text{NH}_2$$
2-Nonadecanone, 50%

This reaction is general.

(a) Preparation of p-methoxyacetophenone (87% by reduction of ω-(methylsulfinyl)-p-methoxyacetophenone (obtained from dimethyl sulfoxide, potassium t-butoxide, and methyl p-anisate) with zinc dust in ethanol-acetic acid; the original articles should be consulted for manipulative details) [55, 53].

7. From Aldehydes and Olefins or Biacetyl and Cyclohexane

Aldehydes react with olefins in a free radical reaction to form ketones:

$$\text{RCH}{=}\text{O} + \text{R}'\text{CH}{=}\text{CH}_2 \longrightarrow \overset{\text{O}}{\overset{\|}{\text{R}\text{C}}}\text{CH}_2\text{CH}_2\text{R}'$$

For a summary see [56].

The aldehyde should be used in large excess to suppress telomerization. Conversions are quite low, particularly with low-molecular-weight aldehydes, but yields occasionally are good, especially with α,β-unsaturated esters (see Exs. b.1, b.2, and b.3). Tetrahydrofuran has also been combined with olefins under free radical conditions (see Ex. b.4). Peroxides, ultraviolet light, or air (see Ex. b.5) has been employed as free radical sources or initiators.

Advantage also may be taken in free radical reactions of the ease of splitting of biacetyl to form methyl ketones [57]:

$$\overset{\text{O}\quad\text{O}}{\overset{\|\quad\|}{\text{C}_6\text{H}_5\text{COOCC}_6\text{H}_5}} + \text{C}_6\text{H}_{12} \longrightarrow [\text{C}_6\text{H}_{11}\cdot] \xrightarrow{\text{CH}_3\text{COCOCH}_3}$$

$$\begin{bmatrix} \overset{\dot{\text{O}}\text{O}}{\overset{|\ \|}{\text{CH}_3\text{CCCH}_3}} \\ | \\ \text{C}_6\text{H}_{11} \end{bmatrix} \longrightarrow \overset{\text{O}}{\overset{\|}{\text{CH}_3\text{CC}_6\text{H}_{11}}} + [\text{CH}_3\text{CO}\cdot]$$

Methyl cyclohexyl ketone, 66%
(by analysis)

(a) Preparation of 7-pentadecanone (75% from 1-octene and heptaldehyde to which acetyl peroxide was added) [58].

(b) Other examples

(1) Ethyl enanthylsuccinate (71–76% from enanthaldehyde and ethyl maleate using benzoyl peroxide at 80–90°) [59].

(2) Ethyl acetylsuccinate (78% from acetaldehyde, ethyl maleate, and benzoylperoxide at about 76° in an autoclave) [60].

(3) Ethyl butyrylsuccinate (84% from butyraldehyde and ethyl maleate γ-irradiated from cobalt-60) [61].

(4) 4-Dodecanone (41% from 1-octene and tetrahydrofuran with di-t-butyl peroxide in an autoclave at 150° for 2 hr.; the THF free radical probably rearranges to $CH_3CH_2CH_2\overset{.}{C}=O$) [62].

(5) Nonanone-2 (70% containing 10% 3-methyloctanone-2 from acetaldehyde, 1-heptene, and a small amount of cobalt acetate under reflux for 12 hr. while air was passed through the mixture) [63].

8. From Some Ketenes

Some highly substituted ketenes, notably dimethylketene, spontaneously dimerize to 1,3-cyclobutanediones:

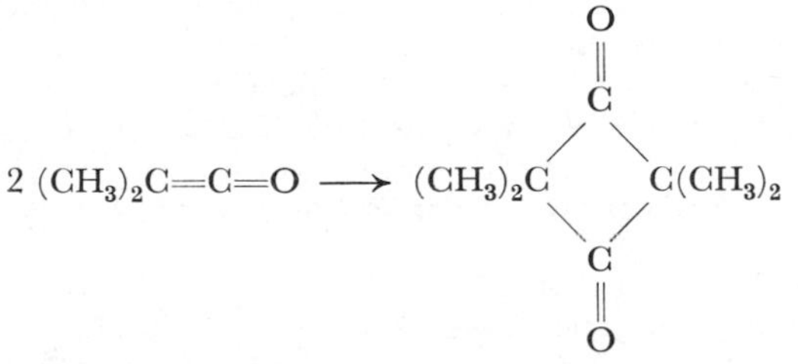

$$2 (CH_3)_2C{=}C{=}O \longrightarrow$$

Tetramethylcyclobutane-1,3-dione,
essentially quantitative

This compound reacts with alcohols or phenols under base catalysis to yield β-ketoesters [64]:

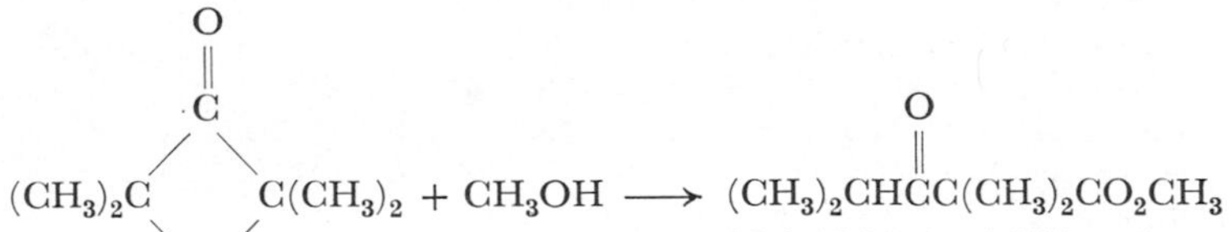

$$+ CH_3OH \longrightarrow (CH_3)_2CHCC(CH_3)_2CO_2CH_3$$

Methyl 2,2,4-trimethyl-3-ketovalerate

1. J. Mathieu and A. Allais, *Cahiers de Synthèse Organique*, Vol. 4, Libraires de l'Acádemie de Médecine, Paris, France, 1958.
2. L. Rand *et al.*, *J. Org. Chem.*, **27**, 1034 (1962).
3. S. Swann, Jr., *Ind. Eng. Chem.*, **26**, 388 (1934) and reference quoted.
4. S. Swann, Jr., *et al.*, *Ind. Eng. Chem.*, **26**, 388 (1934).
5. S. Swann, Jr., *et al.*, *Ind. Eng. Chem.*, **26**, 1014 (1934).
6. R. M. Herbst and R. H. Manske, *Org. Syn.*, Coll. Vol. **2**, 389 (1943).

7. W. Winkler, *Chem. Ber.*, **81**, 256 (1948).

8. J. A. King and F. H. McMillan, *J. Am. Chem. Soc.*, **73**, 4911 (1951).

9. D. M. Cowan *et al.*, *J. Chem. Soc.*, 171 (1940); V. M. Potapov and A. P. Teren'tev, *Zh. Obshch. Khim.*, **28**, 3323 (1958); *C.A.*, **53**, 14028 (1959).

10. W. Baker *et al.*, *J. Chem. Soc.*, 1376 (1951).

11. V. Martello and S. Ceccotti, *Chim. Ind. (Milan)*, **38**, 289 (1956); *C.A.*, **50**, 15454 (1956).

12. J. F. Thorpe and G. A. R. Kon, *Org. Syn.*, Coll. Vol. **1**, 192 (1941).

13. A. G. Dobson and H. H. Hatt, *Org. Syn.*, Coll. Vol. **4**, 854 (1963).

14. S. Gerchakov and H. P. Schultz, *J. Org. Chem.*, **32**, 1656 (1967).

15. D. V. Hertzler *et al.*, *J. Org. Chem.*, **33**, 2008 (1968).

16. R. Davis and H. P. Schultz, *J. Org. Chem.*, **27**, 854 (1962).

17. R. H. Wiley and O. H. Borum, *Org. Syn.*, Coll. Vol. **4**, 5 (1963).

18. J. D. Hepworth, *Org. Syn.*, **45**, 1 (1965).

19. J. C. Sauer, *Org. Syn.*, Coll. Vol. **4**, 560 (1963).

20. H. G. Walker and C. R. Hauser, *J. Am. Chem. Soc.*, **68**, 1386 (1946).

21. R. E. Bowman, *J. Chem. Soc.*, 322 (1950).

22. D. S. Tarbell *et al.*, *J. Am. Chem. Soc.*, **82**, 3982 (1960).

23. G. A. Reynolds and C. R. Hauser, *Org. Syn.*, Coll. Vol. **4**, 708 (1963).

24. H. O. House, *Record Chem. Progr. (Kresge-Hooker Sci. Libr.)*, **28**, 99 (1967).

25. Ref. 1, p. 40.

26. W. D. Ollis and D. Weight, *J. Chem. Soc.*, 3826 (1952).

27. G. S. Fonken and W. S. Johnson, *J. Am. Chem. Soc.*, **74**, 831 (1952).

28. J. M. Straley and A. C. Adams, *Org. Syn.*, Coll. Vol. **4**, 415 (1963).

29. J. A. Price and D. S. Tarbell, *Org. Syn.*, Coll. Vol. **4**, 285 (1963).

30. B. R. Baker *et al.*, *J. Org. Chem.*, **17**, 77 (1952).

31. A. Becker, *Helv. Chim. Acta*, **32**, 1114 (1949).

32. A. Treibs and K. Hintermeier, *Chem. Ber.*, **87**, 1163 (1954).

33. R. Levine *et al.*, *J. Am. Chem. Soc.*, **67**, 1510 (1945); F. W. Swamer and C. R. Hauser, *ibid.*, **72**, 1352 (1950); N. Green and F. B. LaForge, *ibid.*, **70**, 2287 (1948).

34. D. J. Cram and G. S. Hammond, *Organic Chemistry*, 2nd ed., McGraw-Hill Book Co., New York, 1964, p. 376.

35. J. Szmuszkovicz, *Advan. Org. Chem.*, **4**, 1 (1963).

36. R. D. Campbell and J. A. Jung, *J. Org. Chem.*, **30**, 3711 (1965).

37. K. Blaha and O. Cervinka, *Advances in Heterocyclic Chemistry*, Vol. 6, Academic Press, New York, 1966, p. 197.

38. G. Vavon and J.-M. Conia, *Compt. Rend.*, **233**, 876 (1951).

39. C. R. Hauser *et al.*, *J. Am. Chem. Soc.*, **69**, 2649 (1947).

40. C. R. Hauser *et al.*, *J. Am. Chem. Soc.*, **75**, 5030 (1953).

41. E. C. Taylor *et al.*, *J. Am. Chem. Soc.*, **90**, 2421 (1968).

41a. C. E. Denoon, Jr., *Org. Syn.*, Coll. Vol. **3**, 16 (1955).

42. H. Adkins and J. L. Rainey, *Org. Syn.*, Coll. Vol. **3**, 17 (1955).

43. A. Magnani and S. M. McElvain, *Org. Syn.*, Coll. Vol. **3**, 251 (1955).

44. J. T. Adams and C. R. Hauser, *J. Am. Chem. Soc.*, **67**, 284 (1945).

45. J. J. Bloomfield, *J. Org. Chem.*, **27**, 2742 (1962).

46. Ref. 24, p. 109.

47. A. L. Wilds and A. L. Meader, Jr., *J. Org. Chem.*, **13**, 763 (1948).

48. P. L. Julian *et al.*, *Org. Syn.*, Coll. Vol. **2**, 487 (1943).

49. W. D. McPhee and E. Klingsberg, *Org. Syn.*, Coll. Vol. **3**, 119 (1955).

50. C. Osuch and R. Levine, *J. Org. Chem.*, **21**, 1099 (1956).

51. S. Raynolds and R. Levine, *J. Am. Chem. Soc.*, **82**, 472 (1960).

52. E. J. Corey and M. Chaykovsky, *J. Am. Chem. Soc.*, **87**, 1345 (1965).

53. G. A. Russell and G. J. Mikol, *J. Am. Chem. Soc.*, **88**, 5498 (1966).

54. E. J. Corey and T. Durst, *J. Am. Chem. Soc.*, **90**, 5548 (1968).

55. G. A. Russell *et al.*, *J. Am. Chem. Soc.*, **85**, 3410 (1963).

56. G. Sosnovsky, *Free Radical Reactions in Preparative Organic Chemistry*, The Macmillan Co., New York, 1964, p. 125.
57. W. G. Bentrude and K. R. Darnall, *J. Am. Chem. Soc.*, **90**, 3588 (1968).
58. M. S. Kharasch *et al.*, *J. Org. Chem.*, **14**, 248 (1949).
59. T. M. Patrick, Jr., and F. B. Erickson, *Org. Syn.*, Coll. Vol. **4**, 430 (1963).
60. T. M. Patrick, Jr., *J. Org. Chem.*, **17**, 1009 (1952).
61. R. H. Wiley and J. R. Harrell, *J. Org. Chem.*, **25**, 903 (1960).
62. T. J. Wallace and R. J. Gritter, *J. Org. Chem.*, **26**, 5256 (1961).
63. G. I. Nikitin *et al.*, *Bull. Acad. Sci. USSR, Div. Chem. Sci. (English Transl.)*, 1083 (1966).
64. R. H. Hasek *et al.*, *J. Org. Chem.*, **26**, 700 (1961).

G. Alkylations

Alkylations may be discussed in the same manner as acylations but are confined to ketones, since the alkyl group does not introduce a carbonyl group. Alkylations of acetoacetic ester have been documented [1]. In addition to alkylation of ketones (G.1) and enamines (G.2) and to carbanion reactions with unsaturated carbonyl compounds (G.3, Michael), similar nucleophilic reactions which produce ketones are discussed elsewhere: aldol or ketol (Alcohols, Chapter 4, G.1), Knoevenagel (Carboxylic Esters, Chapter 14, C.4), and Claisen-Schmidt (Alcohols, Chapter 4, G.1).

1. From Ketones

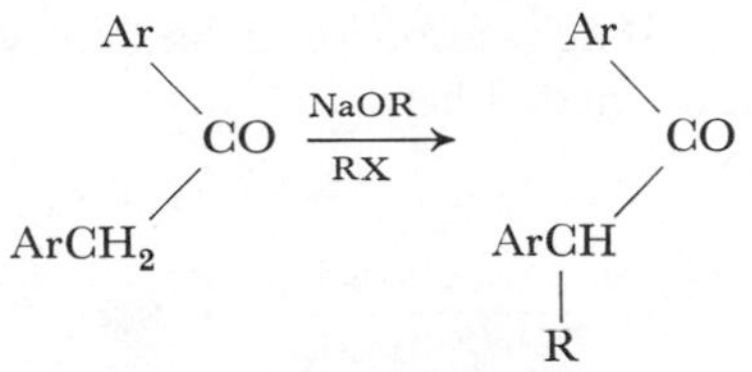

The alkylation of β-ketoesters is discussed under D.6 and of ketones here. Such ketones containing α-hydrogen atoms have been alkylated by first forming the anion ArCOCHAr with sodium or potassium alkoxide, sodium or potassium, potassium carbonate, sodamide, or sodium hydride, which then attacks an alkyl halide such as RX to give the alkylated ketone

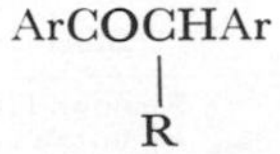

$$\begin{matrix} \text{ArCOCHAr} \\ | \\ \text{R} \end{matrix}$$

The Haller-Bauer reaction, as it is called, is not entirely satsifactory largely for two reasons [2]: a strong base is necessary to convert the carbonyl compound into its anion; and, since many carbonyl compounds contain more than one α-hydrogen atom, the reaction does not stop after monoalkylation has been accomplished. Thus undesirable mixtures of alkylation products are often obtained with accompanying low yields for any specific product. An improved method for carrying out such a reaction is through the use of enamines of ketones, a method discussed under G.2.

References for Section G are on pp. 712–713.

To be more specific, unsymmetrical alkyl ketones are alkylated predominantly on the least substituted carbon atom, with the exception of alkyl cycloalkanones as shown [3]:

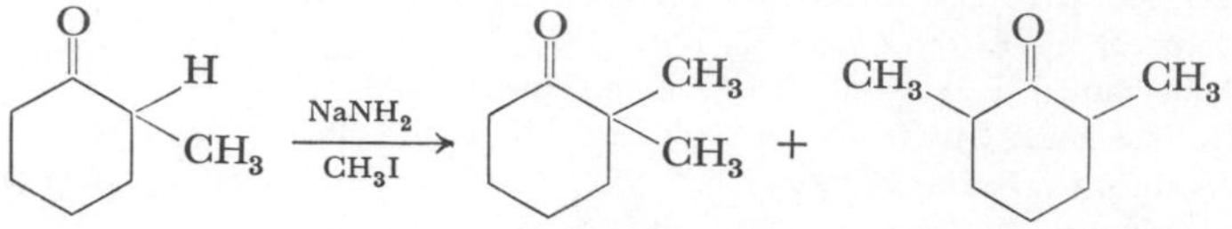

85–90%
2,2-Dimethyl-
cyclohexanone

15–10%
2,6-Dimethyl-
cyclohexanone

A good method of separation of the two ketones, via formylation of the mixture, has been reported (see Ex. b.6).

A very simple procedure for obtaining methyl ketones from 2,4-pentanedione has been demonstrated [4]:

$$CH_3\overset{O}{\overset{\|}{C}}CH_2\overset{O}{\overset{\|}{C}}CH_3 + RX \xrightarrow[K_2CO_3]{Alcohol} CH_3\overset{O}{\overset{\|}{C}}CHR + CH_3CO_2K + KX$$

$$(R = ArCH_2{}^-, \ 60\text{–}78\% \ \text{yields})$$

Alkylation can also be brought about in a limited way by refluxing benzyl alcohol with a high-boiling methyl ketone:

$$C_6H_5\overset{O}{\overset{\|}{C}}CH_3 + C_6H_5CH_2OH \xrightarrow[\substack{LiOCH_2C_6H_5, \\ \text{Dean-Stark apparatus}}]{Xylene} C_6H_5\overset{O}{\overset{\|}{C}}CH_2CH_2C_6H_5 + H_2O$$

70%
Phenyl β-phenylethyl ketone

The reaction proceeds via the formation of benzalacetophenone and its subsequent reduction by benzyl alcohol [5].

To introduce one alkyl group into a dione, the crystalline thallium salt of the dione is the preferred intermediate in that it gives an almost quantitative yield of product [6]:

$$(CH_3CO)_2CHTl + RI \xrightarrow[\substack{\text{Remove TlI} \\ \text{by filtration}}]{Ether} (CH_3CO)_2CHR$$

By a repetition of this process a second alkyl group may be introduced, also in almost quantitative yield. To introduce two alkyl groups into the dione in one step, the alkyl halide and sodium hydride in dimethyl sulfoxide constitute the preferred reagent [7]:

3,3-Dimethyl-2,4-pentanedione
64%

Occasionally diketones, particularily enolic types, may be alkylated with allyl alcohols as exemplified by the following useful reaction [8]:

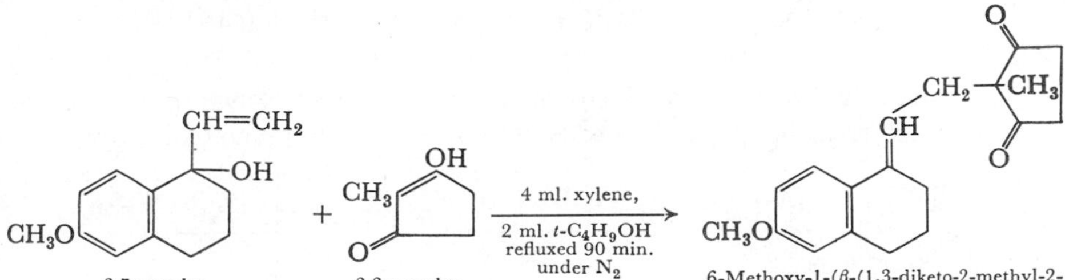

6-Methoxy-1-(β-(1,3-diketo-2-methyl-2-cyclopentyl)-ethylidene)-1,2,3,4-tetra-hydronaphthalene, 70%

Alkylation of the dianion of diketones, 11-XIX, can be accomplished as shown [9]:

$$C_6H_5COCH_2COCH_3 \xrightarrow[\text{Ether}]{\text{KNH}_2} C_6H_5\overset{\ominus}{C}OCHCO\overset{\ominus}{C}H_2 \xrightarrow[\text{2. H}_2\text{O}]{\text{1. } C_6H_5CH_2Cl}$$

11-XIX

$$C_6H_5COCH_2COCH_2CH_2C_6H_5$$

77%

α-(β-Phenylpropionyl)acetophenone

Similarly the trianions of triketones can be alkylated [10]:

$$C_6H_5COCHNaCOCHNaCOCH_2Na \xrightarrow[\text{2. H}^{\oplus}]{\text{1. } C_6H_5CH_2Cl}$$

$$C_6H_5COCH_2COCH_2COCH_2CH_2C_6H_5$$

1,7-Diphenylheptanetrione-1,3,5, 63%

These anions can be acylated as well. Alkylation or acylation occurs at the site of the most nucleophilic carbon atom.

Arylation of the dianion can be accomplished in the same manner using diaryliodonium salts:

$$\text{11-XIX} \xrightarrow{(C_6H_5)_2\text{ICl}} C_6H_5COCH_2COCH_2C_6H_5$$

1,4-Diphenyl-1,3-butanedione

The yields range from 21 to 92% among a series of different diketones [11]. Other arylations of ketones are discussed in the Friedel-Crafts Section C.6. In addition, anions of β-diketones or ketoesters can be coupled with iodine (see Ex. b.8).

Any olefin with an electron-attracting group Y attached as in CH_2=CHY is a potential alkylating agent of the methylene group of a ketone (see Ex. b.3). Indeed, acrylonitrile is considered to be the most searching of all in this respect and can cyanoethylate (or polycyanoethylate) almost any ketone, using base catalysts [12]. If the vinyl alkylating agent is unstable, the corresponding t-amine or quaternary salt can be used to generate the olefin *in situ*:

$$\underset{\text{(Mannich base)}}{C_6H_5\overset{\overset{\text{O}}{\|}}{C}CH_2CH_2N(CH_3)_2} \longrightarrow \underset{\text{Phenyl vinyl ketone}}{C_6H_5\overset{\overset{\text{O}}{\|}}{C}CH=CH_2} + (CH_3)_2NH$$

or

$$C_6H_5\overset{O}{\overset{\|}{C}}CH_2CH_2\overset{\oplus}{N}(CH_3)_3 \longrightarrow C_6H_5\overset{O}{\overset{\|}{C}}CH{=}CH_2 + (CH_3)_3\overset{\oplus}{N}H$$

Phenyl vinyl ketone

And similarly a carbene capable of attacking a ketone methylene group can be generated from the quaternary salts of benzylic-like compounds [13].

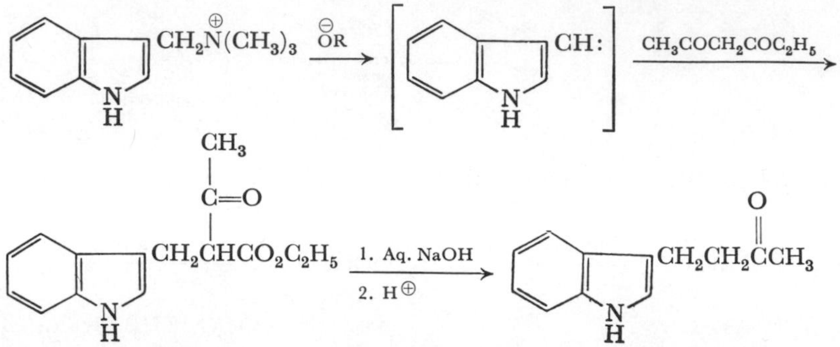
CH2N(CH3)3
OR
CH:
CH3COCH2COC2H5
CH3
C=O
CH2CHCO2C2H5
1. Aq. NaOH
2. H⊕
CH2CH2CCH3
N
H

95%

ca. 70%
2-(3-Indolyl) ethyl methyl ketone

The Mannich reaction itself is a source of aminoketones:

$$R\overset{O}{\overset{\|}{C}}CH_3 + CH_2O + R_2'NH \longrightarrow R\overset{O}{\overset{\|}{C}}CH_2CH_2NR_2'$$

This reaction has been reviewed [14], and an example is given in b.2. Since Mannich base salts are readily reduced to the saturated ketone, this reaction has provided a means of methylating ketones [15]:

$$(C_6H_5)_2CH\overset{O}{\overset{\|}{C}}CH_2CH_2N(CH_3)_2{\cdot}HCl \xrightarrow[\text{H}_2\text{, Raney Ni, 1200 p.s.i. at 80°}]{\text{Alcohol}} (C_6H_5)_2CH\overset{O}{\overset{\|}{C}}CH_2CH_3$$

72%
Benzhydryl ethyl ketone

α,β-Unsaturated ketones may be alkylated *alpha* to the carbonyl group by a unique process of reduction and trapping of the intermediate enolate by reaction with methyl iodide, as illustrated for one of the intermediates in the synthesis of *dl*-D-homotestosterone [16]:

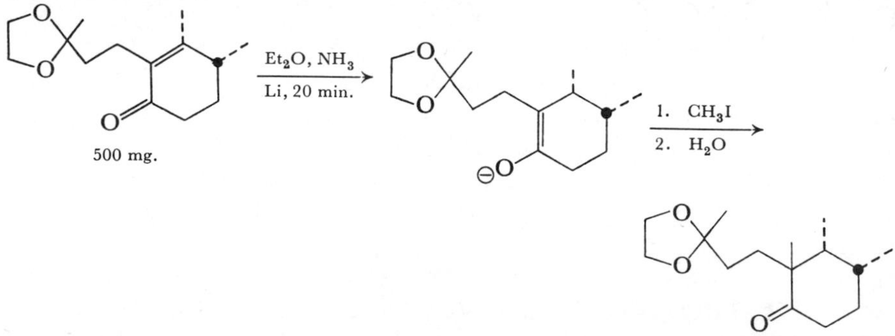
Et2O, NH3
Li, 20 min.
1. CH3I
2. H2O

500 mg.

460 mg.

Furthermore, the methyl group is introduced stereospecifically to give only the β-methyl isomer.

A second method of trapping the enolate begins with the α-bromo ketone [17]:

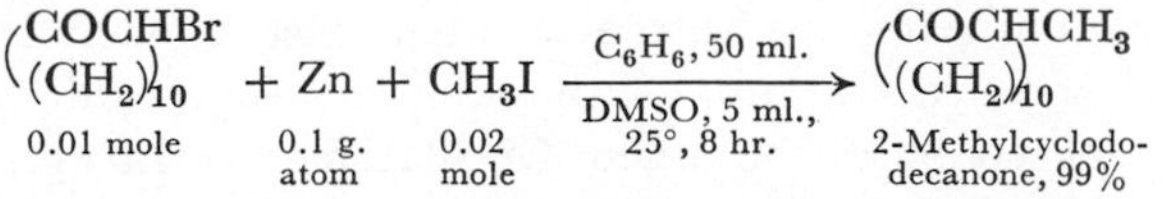

$$\left(\begin{matrix}COCHBr\\(CH_2)_{10}\end{matrix}\right) + Zn + CH_3I \xrightarrow[\substack{DMSO,\ 5\ ml.,\\25°,\ 8\ hr.}]{C_6H_6,\ 50\ ml.} \left(\begin{matrix}COCHCH_3\\(CH_2)_{10}\end{matrix}\right)$$

0.01 mole   0.1 g.   0.02   2-Methylcyclododo-
              atom   mole   canone, 99%

α-Bromoketones may also be alkylated by using trialkylboranes [18]:

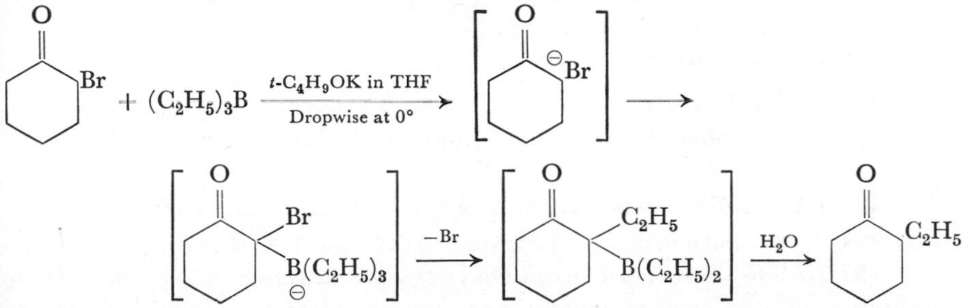

2-Ethylcyclo-
hexanone, 68%

Diazoketones may be alkylated by trialkylboranes in yields ranging from 36 to 89% [19]:

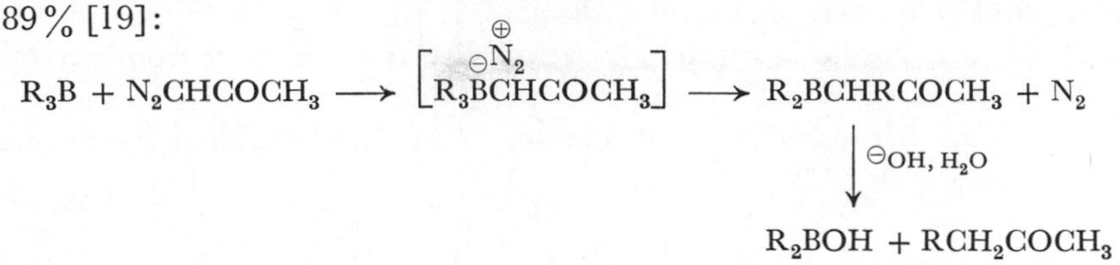

$$R_3B + N_2CHCOCH_3 \longrightarrow \left[R_3\overset{\ominus}{B}CH\overset{\overset{\oplus}{N_2}}{C}OCH_3\right] \longrightarrow R_2BCHRCOCH_3 + N_2$$

$$\Big\downarrow \ominus OH, H_2O$$

$$R_2BOH + RCH_2COCH_3$$

Alkylation of ketones may also be brought about by free radical means:

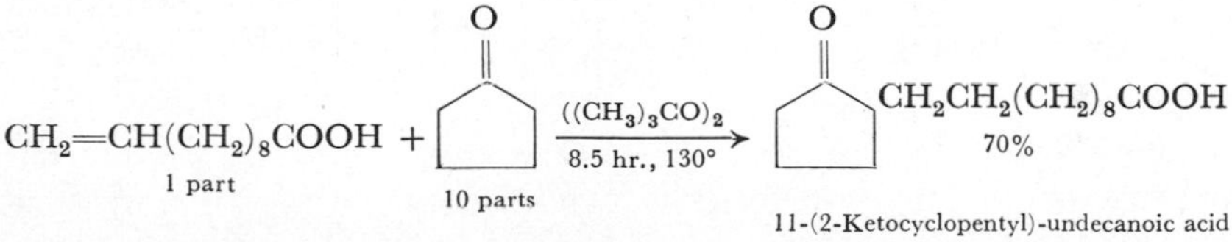

$CH_2{=}CH(CH_2)_8COOH +$ $((CH_3)_3CO)_2$ $CH_2CH_2(CH_2)_8COOH$
1 part   8.5 hr., 130°   70%
                10 parts

11-(2-Ketocyclopentyl)-undecanoic acid

See [20]. The substitution is not always terminal. Ultraviolet light has also been used for alkylations by olefins.

(a) Preparation of 1-methyl-5,8-dimethoxy-2-tetralone.   Potassium,

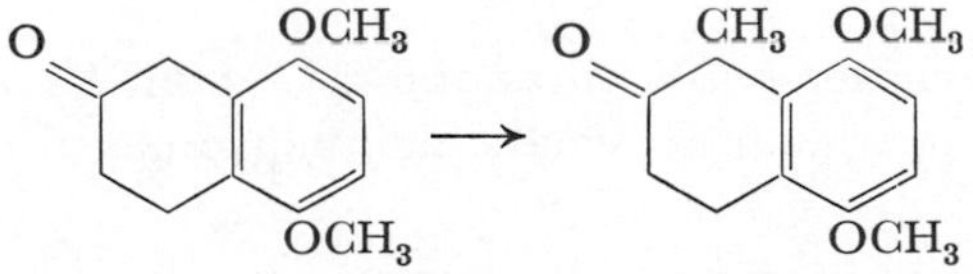

390 mg., was pulverized in xylene, after which 20 ml. of anhydrous benzene and 1.92 g. of 5,8-dimethoxy-2-tetralone was added. The mixture, in an anhydrous, nitrogen atmosphere, evolved hydrogen as solution occurred. After

boiling under reflux for 40 min., a precipitate formed and 2 ml. of methyl iodide was added. Heating again for 30 min. and allowing to stand overnight produced a brown precipitate; a little 2 N sulfuric acid was added; the benzene was removed by distillation under reduced pressure; and the red-brown oily mass was taken up in water and extracted three times in ether. Impurities were removed by treatment several times with sodium hydrogen sulfite, after which a residue of 2.2 g. of red oil was obtained. Distillation gave 1.63 g., b.p. 127–128° (0.2 mm.). A molecular distillation gave 0.2 g. more to make the yield of the methyl ketone 90% [21].

(b) Other examples

(1) **1,2-Dimethylcyclopentyl phenyl ketone** (49% from 2-methylcyclopentyl phenyl ketone, sodamide, and methyl iodide in benzene) (22).

(2) **1-Diethylamino-3-butanone** (62–70% from diethylamine hydrochloride, acetone, paraformaldehyde, and hydrochloric acid [23].

(3) **2,2,6,6-Tetra-(-β-cyanoethyl)-cyclohexanone** (88% from cyclohexanone in *t*-butyl alcohol containing potassium hydroxide to which acrylonitrile was added at 45°) [24].

(4) **3-Methylpentane-2,4-dione** (75–77% from the dione, methyl iodide, and anhydrous potassium carbonate) [25].

(5) **1,1-Di-(carbomethoxymethyl)-2-tetralone** (94% from tetralone, methyl bromoacetate, and sodium hydride) [26].

(6) **2,2-Dimethylcyclohexanone.** 2-Methyl-1-decalone has been obtained similarly in 77% yield [27].

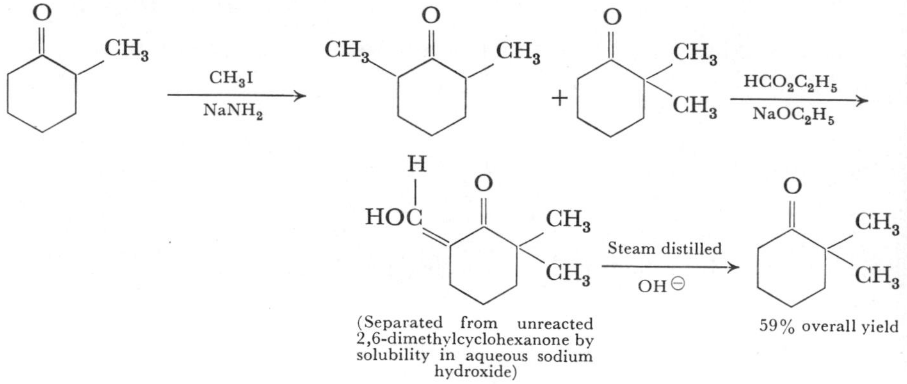

(Separated from unreacted 2,6-dimethylcyclohexanone by solubility in aqueous sodium hydroxide)

59% overall yield

(7) **4,4′-Phenylene-di-2-butanone** (50% from 2,4-pentanedione, α,α′-dichloro-*p*-xylene, potassium carbonate, and anhydrous alcohol under reflux) [4].

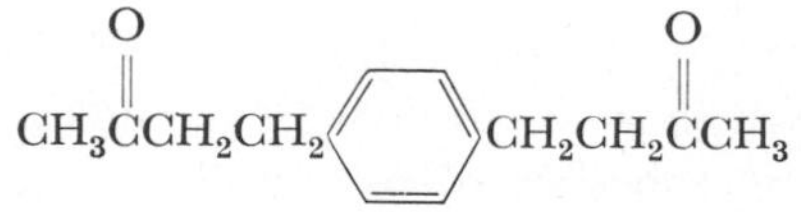

(8) Tetraacetylethane (41–59% from 2,4-pentanedione, aq. sodium hydroxide, and iodine) [28].

(9) Geranylacetone, (74% from linalol, acetoacetic ester, and

$$(CH_3)_2C{=}CH(CH_2)_2C(CH_3){=}CH(CH_2)_2COCH_3$$

aluminum isopropoxide by heating for 9 hr. up to 196°) [29].

(10) 1-(1-Phenyl-3-ketobutyl)-1,3-cyclohexanedione (50% from

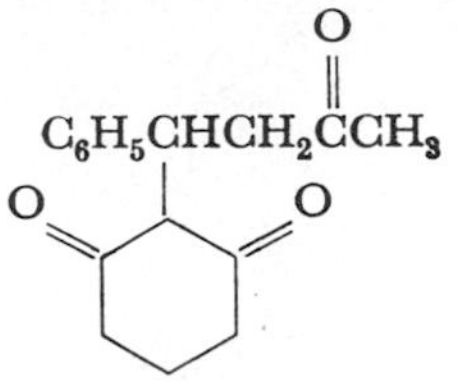

benzalacetone, 1,3-cyclohexanedione, and potassium ethoxide in alcohol) [30].

(11) 2-(2-Pyrazylethyl)-cyclohexanone (68% from 2-vinylpyrazine,

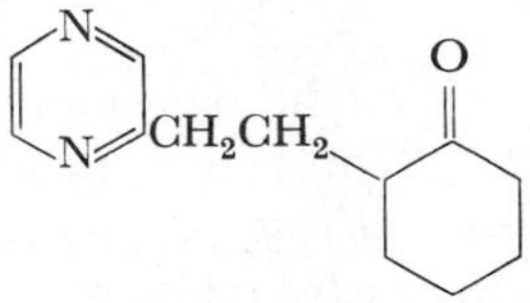

cyclohexanone, and small pieces of sodium at 90–160°) [31].

2. From Enamines

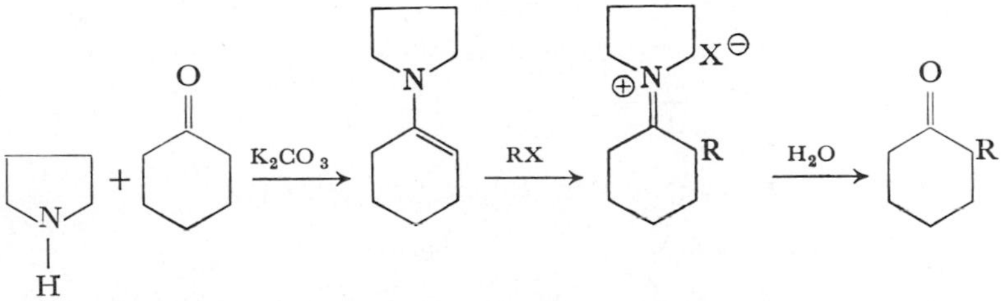

Excellent treatments of this synthesis are available [32]. Enamines (α,β-unsaturated amines) are prepared by the reaction of a secondary amine on an aldehyde or ketone in the presence of a dehydrating agent such as anhydrous potassium carbonate or p-toluenesulfonic acid [33]. However, superior yields of N-pyrrolidyl enamines of certain steroidal ketones may be obtained by azeotropic distillation with benzene [34]. The amine usually employed is pyrrolidine, morpholine, or piperidine. The procedure is satisfactory for ketones with the exception of some monosubstituted acetones and very unreactive ones, although it is less so for aldehydes (see Aldehydes, Chapter 10, F.4).

Recently titanium tetrachloride and dimethylamine have been shown to be an effective combination for converting hindered ketones into enamines [35]. A side reaction in the preparation of enamines in general is the oxidation of the secondary amine to the imine 11-XX (which may polymerize) by the enamine [36]:

$$\text{HNCH}_2\text{—} \longrightarrow \text{N}\text{=}\text{CH—}$$

11-XX

A variety of substituents may be introduced into the enamine. Readily ionizable halides, such as those of allyl, benzyl, propargyl, and the α-halo-substituted ethers, ketones, esters, and nitriles, give good yields of the alkylated ketone.

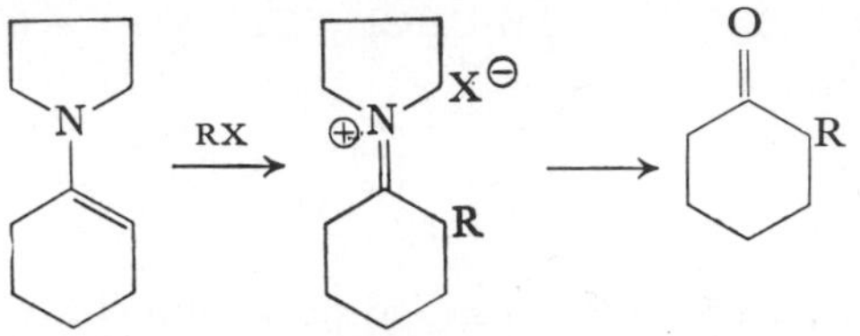

(R = CH₂=CHCH₂—, C₆H₅CH₂—, CH≡CCH₂—, CH₃CH₂OCH₂CH₂—,
CH₃CH₂COCH₂—, CH₃CHCOOC₂H₅, CH₃CHCN)

As it happens, these products are the ones most difficult to prepare via the conventional β-ketoester path. Simple alkyl groups give fair yields of the 2-alkyl ketones except in the case of β-tetralone; then the yields are high. The enamine from pyrollidine is the most, while that from morpholine is the least, nucleophilic toward attack of halides. If N-alkylation is troublesome, the enamine of N,N-dicyclohexylamine may divert the main course to C-alkylation. Polar solvents such as acetonitrile are more effective in promoting alkylation than are nonpolar ones.

The mechanism may be represented as follows [37]:

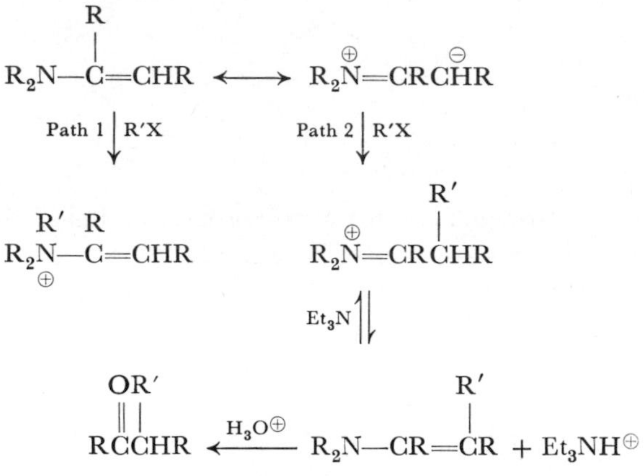

As a rule, path 2 prevails and often 1 equivalent of triethylamine and the enamine, in place of 2 equivalents of enamine, are used.

The most important alkylations of enamines are those with electrophilic olefins, such as α,β-unsaturated ketones, nitriles, or esters, since they yield valuable intermediates for further reactions, particularly annelation. The behavior of methyl vinyl ketone is illustrated [38]:

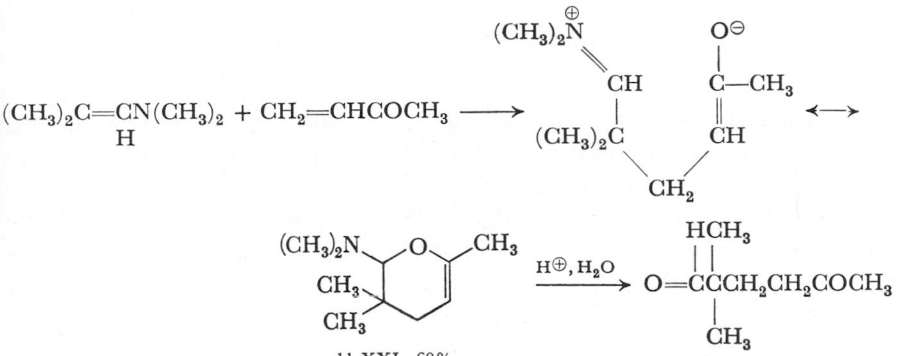

(CH3)2C=CN(CH3)2 + CH2=CHCOCH3
H
(CH3)2N
CH
(CH3)2C
CH2
C—CH3
CH
H⊕, H2O
O=CCCH2CH2COCH3
HCH3
CH3

11-XXI 60%

2,2-Dimethyl-5-ketohexanal, 72%

The intermediate dihydropyran, 11-XXI, may be condensed to a cyclobutylamine by phenyllithium. A final reaction is that with aldehydes in which α-monoalkylidene types are produced (see Ex. b.2).

(a) Preparation of 1-methyl-2-tetralone

(1) 2-Tetralone pyrrolidine enamine. 2-Tetralone, 5 g., with 4 g. of pyrrolidine in 100 ml. of benzene was refluxed under nitrogen for 3 hr. The enamine was obtained (93 %) on removing the solvent. Recrystallization from petroleum ether gave a m.p. of 81–82°.

(2) 1-Methyl-2-tetralone. The enamine from 10 g. of β-tetralone and 7 g. of pyrrolidine was refluxed for 10 hr. with 20 ml. of methyl iodide in 50 ml. of dioxane. After adding 25 ml. of water and 1 ml. of acetic acid followed by further heating for 4 hr., the solvent was removed under reduced pressure. Purification in the usual manner gave 9 g. (81 %) of 1-methyl-2-tetralone, b.p. 138–142° (20 mm.) [2].

(b) Other examples

(1) Methyl 2-methyl-3-(2-oxocyclohexyl)-propionate (80 % from methyl methacrylate and the pyrrolidine enamine of cyclohexanone) [2].

(2) 2-Benzylidenecyclopentanone (84 % from benzaldehyde and the morpholine enamine of cyclopentanone) [39].

(3) α-Dichloromethylenecyclopentanone (70 % from the enamine

O
Cl
CCl

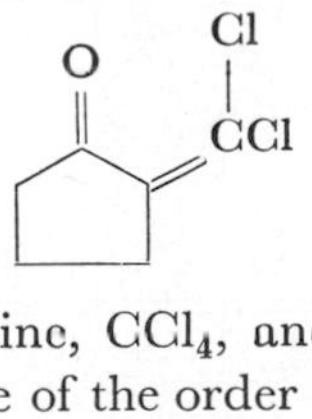

of cyclopentanone and morpholine, CCl_4, and 2 equiv. of triethylamine; the reaction is general but yields are of the order of 30 % from enamines of acyclic ketones) [40].

3. From Unsaturated Carbonyl Compounds and Carbanions (Michael)

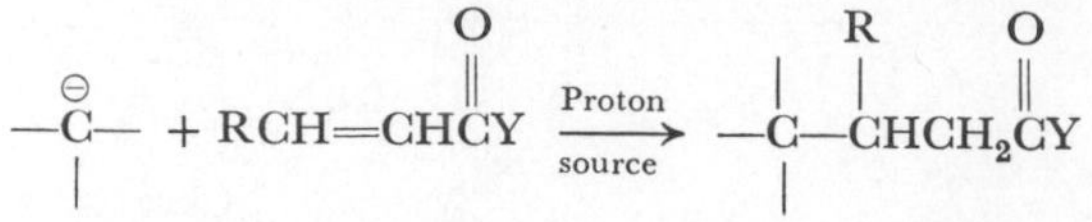

$$\underset{|}{\overset{|}{-}}\overset{\ominus}{\underset{|}{C}}- + RCH{=}CHCY \xrightarrow[\text{source}]{\text{Proton}} -\overset{O}{\underset{|}{C}}-\overset{R}{\underset{|}{CH}}CH_2\overset{O}{\overset{\|}{C}}Y$$

The Michael condensation has been reviewed [41]. Relative to the synthesis of ketones, it consists of the 1,4 addition of an active methylene compound to α,β-olefinic or acetylenic ketones.

The course of some of these condensations is complex and it varies with the experimental conditions employed. For instance, the use of triethylamine catalyst has been shown to give the simple Michael adduct, whereas sodium ethoxide leads to a ring-closed adduct [42]:

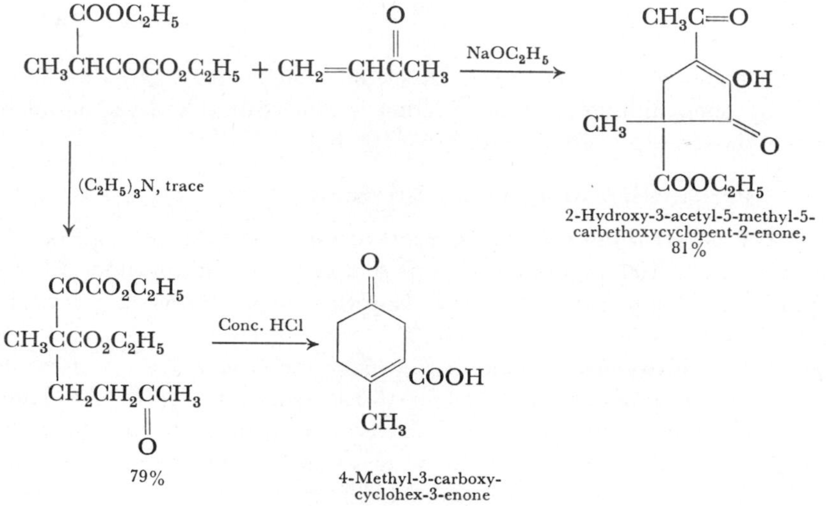

2-Hydroxy-3-acetyl-5-methyl-5-carbethoxycyclopent-2-enone, 81%

4-Methyl-3-carboxy-cyclohex-3-enone

The orientation of the Michael condensation is changed also by the structure of the addend to which the anion adds:

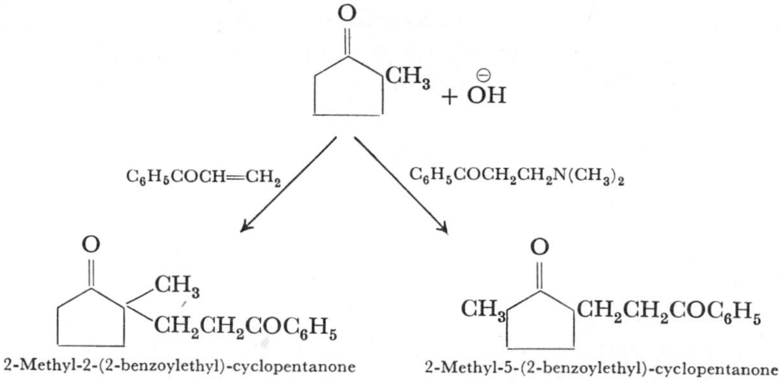

2-Methyl-2-(2-benzoylethyl)-cyclopentanone

2-Methyl-5-(2-benzoylethyl)-cyclopentanone

It would be pleasing to find that this orientation control is general [43].

The Mannich base may serve not only as the source of the unsaturated ketone but also as the catalyst in what is known as a thermal Michael reaction [44].

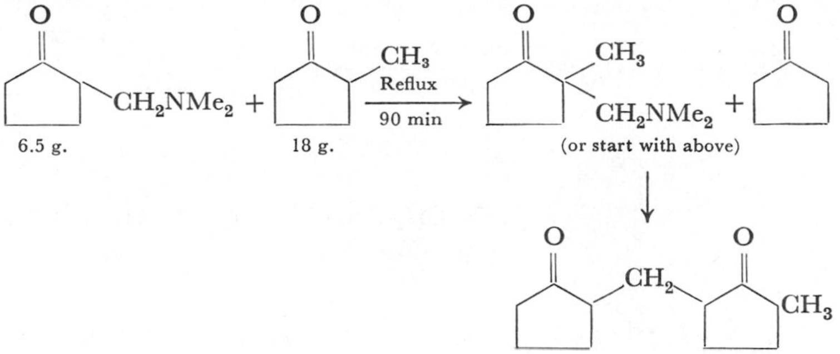

2-Methyl-5-(2'-oxocyclopentylmethyl)-
cyclopentanone, 40–50%

Acetoethylation of ketones in low yield may be accomplished with methyl vinyl ketone [45]:

$$ArCOCH_3 + CH_2{=}CHCOCH_3 \xrightarrow{OH^{\ominus}} ArCOCH_2CH_2CH_2COCH_3$$

A similar reaction occurs with esters, β-ketoesters, malonic esters, and amines, sometimes in good yield [46].

In a more recent example illustrating the use of the Grignard reagent as a base catalyst, 1,3-diphenyl-4,4,6-trimethyl-1,5-heptanedione,

$$((CH_3)_2CHCOC(CH_3)_2CH(C_6H_5)CH_2COC_6H_5)$$

has been obtained in 90 % yield from diisopropylketone and isopropylmagnesium chloride, the enolate formed having added to benzalacetophenone [47]. The Michael condensation has been extended to enamines (see Ex. a).

(a) Preparation of tricyclo-[6.2.2.0²·⁷]-dodecane-3,9-dione (45 % overall yield from cyclohexenone and pyrrolidine perchlorate with benzene

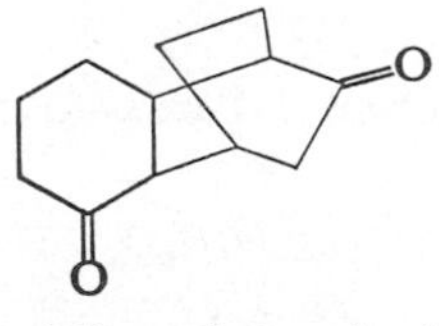

reflux and Dean-Stark tube followed by hydrolysis with aqueous sodium hydroxide) [48].

4. From Allylic Rearrangements

Ketones can be prepared by Claisen-type rearrangements of allylic aceto-acetic esters [49]; see Ex. a.

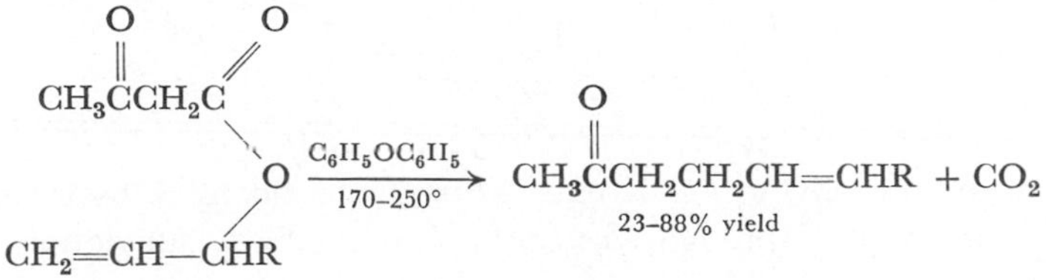

This reaction occurs also with allyl enolethers but is not so selective [50]:

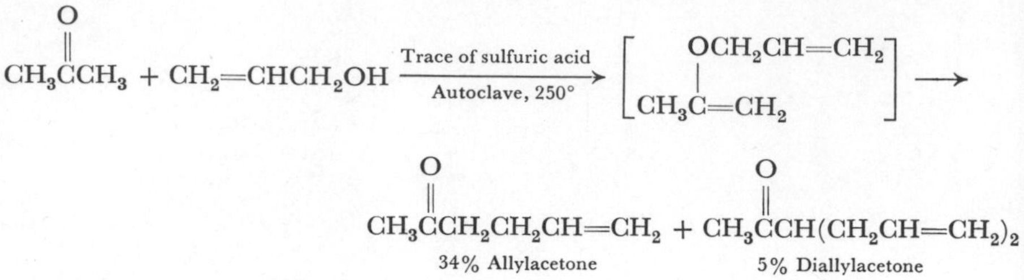

$$CH_3\overset{\displaystyle O}{\overset{\|}{C}}CH_3 + CH_2{=}CHCH_2OH \xrightarrow[\text{Autoclave, 250°}]{\text{Trace of sulfuric acid}} \left[\begin{array}{c} OCH_2CH{=}CH_2 \\ | \\ CH_3C{=}CH_2 \end{array} \right] \longrightarrow$$

$$CH_3\overset{\displaystyle O}{\overset{\|}{C}}CH_2CH_2CH{=}CH_2 + CH_3\overset{\displaystyle O}{\overset{\|}{C}}CH(CH_2CH{=}CH_2)_2$$

34% Allylacetone   5% Diallylacetone

The yield of diallylated ketone (both unsymmetrical and symmetrical) can be increased by increasing the alcohol/ketone ratio. Not only allyl but also alkyl enol ethers rearrange to ketones. Because the conditions are much more vigorous, the reaction is not so general and it is hampered by side reactions [51]:

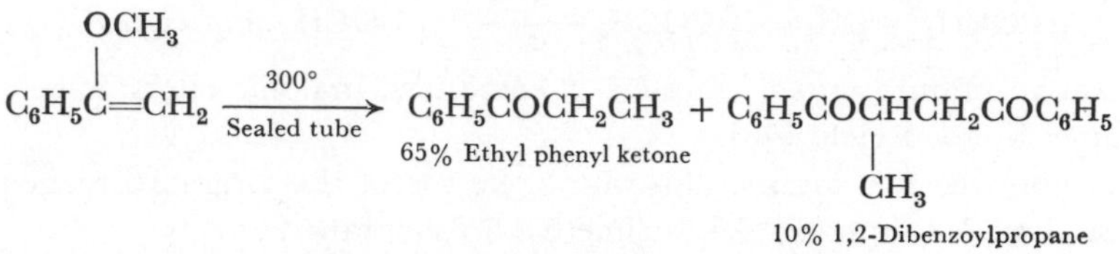

$$C_6H_5\overset{\displaystyle OCH_3}{\overset{|}{C}}{=}CH_2 \xrightarrow[\text{Sealed tube}]{300°} C_6H_5COCH_2CH_3 + C_6H_5COCHCH_2COC_6H_5$$

65% Ethyl phenyl ketone

$$\overset{|}{\underset{CH_3}{}}$$

10% 1,2-Dibenzoylpropane

The enol ethers may be alkylated without a catalyst:

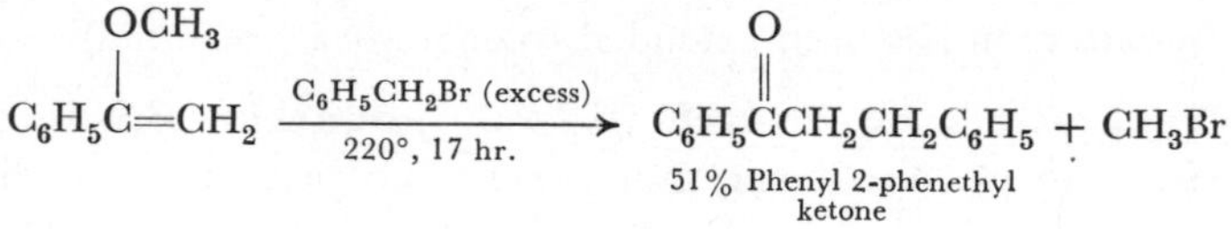

$$C_6H_5\overset{\displaystyle OCH_3}{\overset{|}{C}}{=}CH_2 \xrightarrow[\text{220°, 17 hr.}]{C_6H_5CH_2Br \text{ (excess)}} C_6H_5\overset{\displaystyle O}{\overset{\|}{C}}CH_2CH_2C_6H_5 + CH_3Br$$

51% Phenyl 2-phenethyl
ketone

(a) Preparation of 3-phenyl-1-hexene-5-one, (74% from cinnamyl

$$CH_3COCH_2CH(C_6H_5)CH{=}CH_2$$

acetoacetate in diphenyl ether at 250°) [49].

5. From Aldehydes or Their Derivatives

Alkylation of an aldehyde to prepare a ketone has never been carried out as follows:

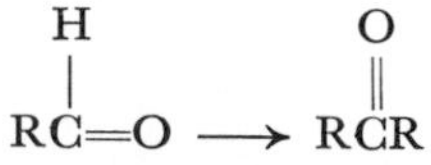

$$R\overset{\displaystyle H}{\overset{|}{C}}{=}O \longrightarrow R\overset{\displaystyle O}{\overset{\|}{C}}R$$

However, if the carbonyl group is masked or if the aldehyde hydrogen atom is made more acidic through the use of derivatives, alkylation can be accomplished. Two methods are known, the first of which is recent [52]:

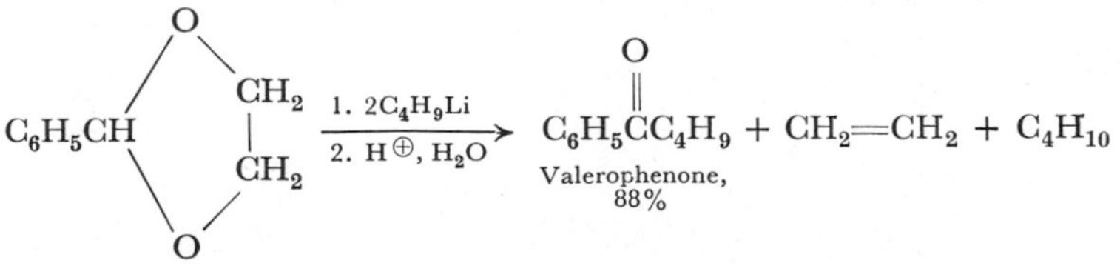

$$C_6H_5CH \underset{O}{\overset{O}{<}} \underset{CH_2}{\overset{CH_2}{|}} \xrightarrow[\text{2. H}^{\oplus}, \text{H}_2\text{O}]{\text{1. 2C}_4\text{H}_9\text{Li}} C_6H_5\overset{O}{\overset{\|}{C}}C_4H_9 + CH_2{=}CH_2 + C_4H_{10}$$

Valerophenone, 88%

This reaction proceeds via lithium benzoate.

The second, being more classic, consists of the alkylation of aminonitrile derivatives of aldehydes:

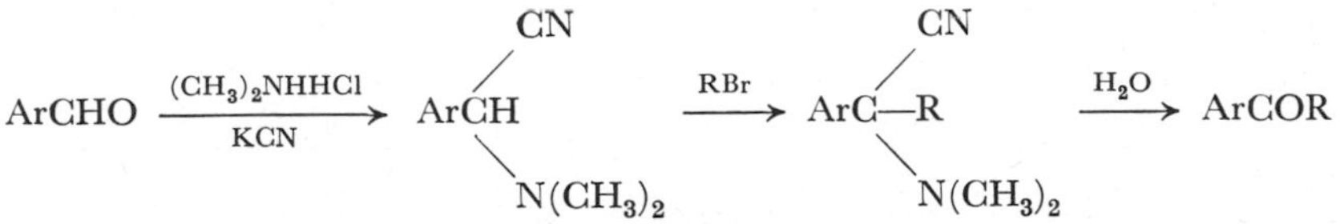

$$ArCHO \xrightarrow[\text{KCN}]{(CH_3)_2NHHCl} ArCH\overset{CN}{\underset{N(CH_3)_2}{<}} \xrightarrow{RBr} ArC{-}R \overset{CN}{\underset{N(CH_3)_2}{|}} \xrightarrow{H_2O} ArCOR$$

In the first step in this synthesis the aminonitrile is obtained from the aldehyde by the use of a secondary amine hydrochloride and potassium cyanide [53]. This product, with potassium amide and an alkyl halide in liquid ammonia, yields the alkylated aminonitrile [54], which on hydrolysis in acid solution gives the ketone. It appears that the unsubstituted aminonitrile forms the anion

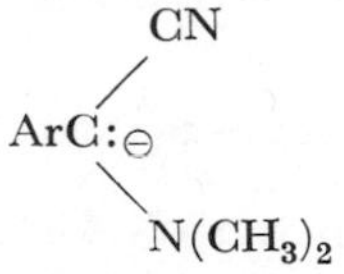

$$ArC\overset{CN}{\underset{N(CH_3)_2}{<}}{:}_{\ominus}$$

which displaces the halide from an alkyl halide. In the process, side reactions of the active carbanion may be eliminated by the simultaneous addition of the reactants [55]. The method is applicable to both aliphatic and aromatic aldehydes [56, 54], although in some cases difficulties are experienced in producing the original aminonitriles [53]. The most satisfactory method for preparing this type is to proceed via the sodium bisulfite addition compound of the aldehyde [57]. The best yields have been obtained with aromatic aldehydes, in which case the benzylated aminonitrile has been produced in yields of 80–100% and the hydrolysis to the ketone in yields of 90–92% [54]. Thus the synthesis offers particular promise in the preparation of substituted desoxybenzoins.

Alkali metal cyanides also convert benzalanilines into benzildianils in a reaction reminiscent of a benzoin condensation followed by oxidation [58]:

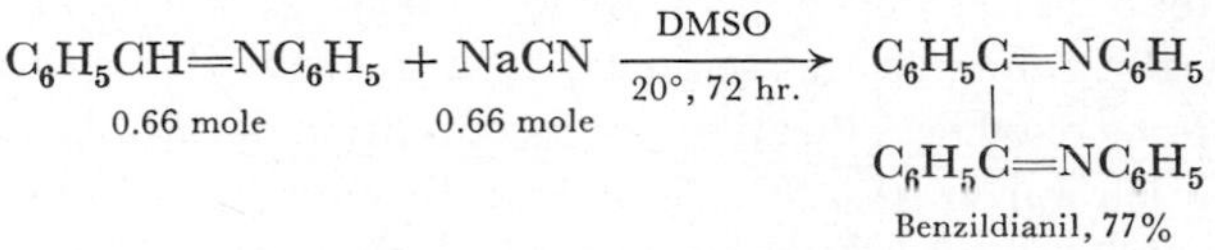

$$C_6H_5CH{=}NC_6H_5 + NaCN \xrightarrow[\text{20}^\circ, \text{72 hr.}]{\text{DMSO}} \begin{array}{c} C_6H_5C{=}NC_6H_5 \\ | \\ C_6H_5C{=}NC_6H_5 \end{array}$$

0.66 mole   0.66 mole

Benzildianil, 77%

The reaction is general, but concentration and temperature dependent.

Lastly, tetraaminoethylenes yield ketones with aldehydes [59] (for the mechanism see [60]):

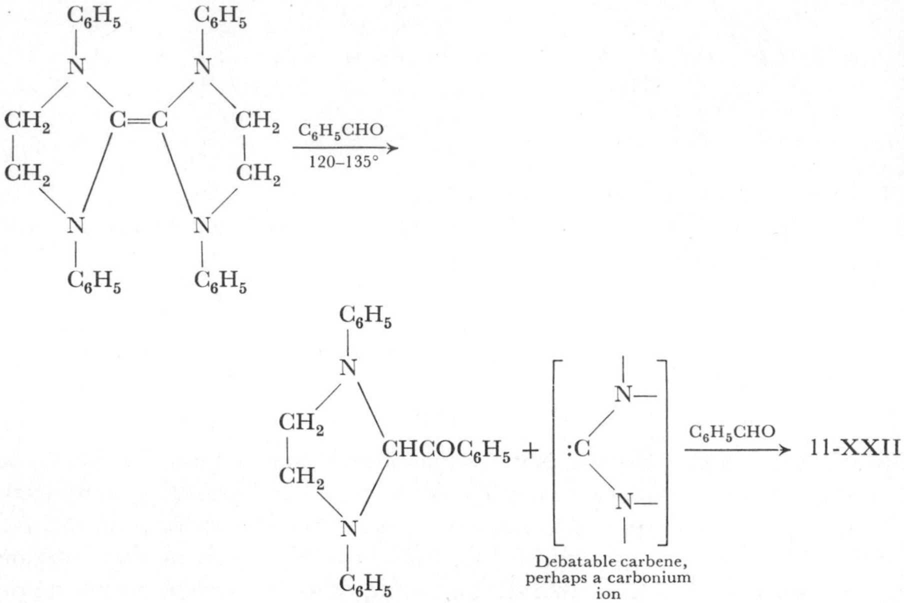

11-XXII, 1,3-Diphenyl-
2-benzoylimidazolidine, 90%

(a) Preparation of 4-chloro-α-phenylacetophenone.

(1) 2-(N,N-Dimethylamino)-2-(4-chlorophenyl)-acetonitrile. To a stirred slurry of 0.71 mole of sodium bisulfite in 150 ml. of water was added 0.71 mole of p-chlorobenzaldehyde followed after 15 min. by 128 g. of 25% aqueous dimethylamine. The mixture was stirred for 30 min., then cooled in an ice bath, after which 0.71 mole of sodium cyanide in 10 ml. of water was added dropwise. On removing the ice bath, stirring was continued for 3 hr. and the mixture was filtered to give a precipitate which was washed with water and dried under reduced pressure. Recrystallization from absolute ethanol produced 114 g. (84%), m.p. 40–43° (43.5–44.5° after two recrystallizations from hexane) of the aminochlorophenylacetonitrile [57].

(2) 2-(N,N-Dimethylamino)-2-(4-chlorophenyl)-2-benzylacetoni- trile. Potassium amide, 0.05 mole, in 250 ml. of commercial anhydrous liquid ammonia was stirred and there was added a solution of 9.75 g. (0.05 mole) of the aminochlorophenylacetonitrile in 100 ml. of anhydrous ether, followed, after 5 min., by 6.3 g. (0.05 mole) of benzyl chloride in 50 ml. of ether. After 5 min. the ammonia was removed on a steam bath as it was replaced by an equal volume of ether. Filtration and evaporation of the ethereal filtrate gave a residue which on recrystallization from hexane gave 13.6 g. (96%) of 2-(N,N-dimethylamino)-2-(4-chlorophenyl)-3-phenylpropionitrile, m.p. 104– 107.5° (105–107° after several recrystallizations from hexane [54]).

(3) 4-Chloro-α-phenylacetophenone. The substituted propionitrile, 1.0 g., was hydrolyzed by heating a solution in 25 ml. of ethanol and 50 ml. of 6 *N* hydrochloric acid on a steam bath for 20 hr. Dilution of the cooled solution with water gave a white solid which, after four crystallizations from ethanol, produced 0.74 g. (92%), m.p. 106–106.5°, of the ketone [54].

(b) Ethyl butyl ketone (63% from α-dimethylaminobutyronitrile, *n*-butyl bromide, and potassium amide in liquid ammonia followed by refluxing with dilute hydrochloric acid) [61].

6. From Certain Amides

The N-benzoyl group in Reissert compounds, such as 1-cyano-1,2-dihydro-isoquinoline, is susceptible to alkylation by readily ionized alkyl groups. Thc re-formation of the resonance-stabilized aromatic ring serves as a driving force [62]:

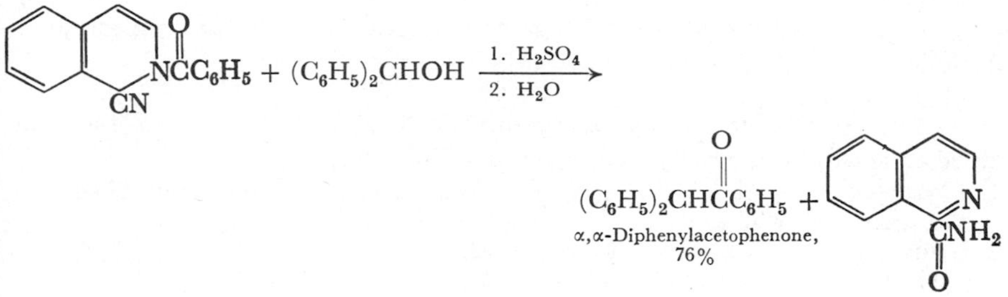

α,α-Diphenylacetophenone, 76%

7. From Phenols

The question arises whether a Reimer-Tiemann synthesis is available for ketones as it is for aldehydes. If it is to be found, a substituted chlorocarbene,

$$R\overset{\displaystyle Cl}{\overset{|}{C}}:$$

must be generated from one of two sources:

$$R\underset{\displaystyle Cl}{\underset{|}{\overset{\displaystyle H}{\overset{|}{C}}}}Cl \xrightarrow{OH^{\ominus}} R\overset{\displaystyle Cl}{\overset{|}{C}}: + H_2O + \overset{\ominus}{Cl}$$

$$RCCl_3 \xrightarrow{OH^{\ominus}} R\overset{\displaystyle Cl}{\overset{|}{C}}: + HOCl + Cl^{\ominus}$$

No examples could be found for the first equation, but one for the second equation was available [63]:

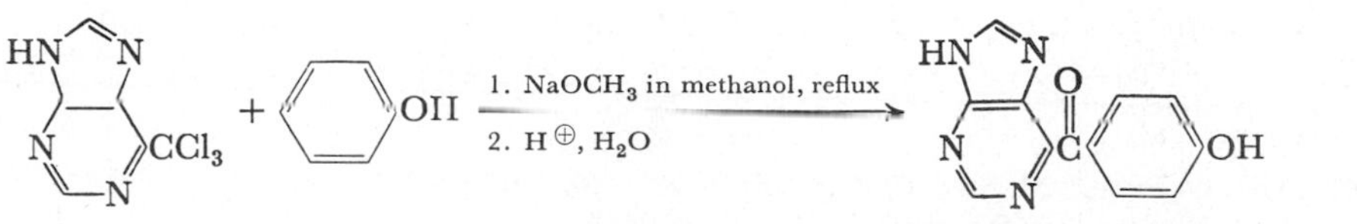

6-(*p*-Hydroxybenzoyl)-purine, 79%

Alternatively, the reaction may proceed by simple displacement with the phenoxide anion, although one must then accept carbon rather than oxygen alkylation as the course of the reaction.

Entry into the cyclodienone family of compounds may be brought about by alkylation of phenols, usually 2,6-di- or more highly substituted ones. Conditions are most important to favor C- over O-alkylation [64].

8. From Ketene and Diazomethane

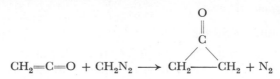

This reaction occurs at low temperature to form cyclopropanone which is so unstable that it can be used only in solution (such as in methylene chloride) at about −78°. It polymerizes on raising the temperature or it reacts rapidly with alcohols at −78° to form the hemiketal.

(a) Preparation of cyclopropanone (50–60% from diazomethane added to ketene in methylene chloride at −78°) [65].

1. J. Mathieu and A. Allais, *Cahiers de Synthèse Organique*, Vol. 2, Libraires de l'Académie de Médecine, Paris, France, 1957.
2. G. Stork *et al.*, *J. Am. Chem. Soc.*, **85,** 207 (1963).
3. J.-M. Conia, *Bull. Soc. Chim. France*, 1040 (1956).
4. S. Boatman *et al.*, *J. Org. Chem.*, **30,** 3321 (1965).
5. E. F. Pratt and A. P. Evans, *J. Am. Chem. Soc.*, **78,** 4950 (1956).
6. E. C. Taylor *et al.*, *J. Am. Chem. Soc.*, **90,** 2421 (1968).
7. J. J. Bloomfield, *J. Org. Chem.*, **26,** 4112 (1961).
8. N. L. Wendler *et al.*, *J. Org. Chem.*, **33,** 3126 (1968).
9. C. R. Hauser and T. M. Harris, *J. Am. Chem Soc.*, **80,** 6360 (1958); T. M. Harris and C. M. Harris, *J. Org. Chem.*, **31,** 1032 (1966).
10. C. R. Hauser *et al.*, *J. Org. Chem.*, **30,** 4263 (1965).
11. K. G. Hampton *et al.*, *J. Org. Chem.*, **29,** 3511 (1964).
12. H. A. Bruson, *Org. Reactions*, **5,** 79 (1949).
13. D. O. Holland and J. H. C. Nayler, *J. Chem. Soc.*, 280 (1953).
14. F. F. Blicke, *Org. Reactions*, **1,** 303 (1942).
15. E. M. Schultz and J. B. Bicking, *J. Am. Chem. Soc.*, **75,** 1128 (1953).
16. G. Stork and J. E. McMurry, *J. Am. Chem. Soc.*, **89,** 5464 (1967).
17. T. A. Spencer *et al.*, *J. Am. Chem. Soc.*, **89,** 5727 (1967).
18. H. C. Brown *et al.*, *J. Am. Chem. Soc.*, **90,** 6218 (1968).
19. J. Hooz and S. Linke, *J. Am. Chem. Soc.*, **90,** 5936 (1968).
20. G. Sosnovsky, *Free Radical Reactions in Preparative Organic Chemistry*, The Macmillan Co., New York, 1964, p. 133.
21. C. A. Grob and W. Jundt, *Helv. Chim. Acta*, **31,** 1691 (1948).
22. G. Wash *et al.*, *J. Am. Chem. Soc.*, **63,** 2975 (1941).
23. A. L. Wilds *et al.*, *Org. Syn.*, Coll. Vol. **4,** 281 (1963).
24. H. A. Bruson and T. W. Riener, *J. Am. Chem. Soc.*, **64,** 2850 (1942).
25. A. W. Johnson *et al.*, *Org. Syn.*, **42,** 75 (1962).
26. M. D. Soffer *et al.*, *J. Am. Chem. Soc.*, **72,** 3704 (1950).
27. W. J. Bailey and M. Madoff, *J. Am. Chem. Soc.*, **76,** 2707 (1954).
28. R. G. Charles, *Org. Syn.*, Coll. Vol. **4,** 869 (1963).

29. J. Dreux and J. Colonge, *Bull. Soc. Chim. France*, 1312 (1955).
30. H. Stetter and M. Coenen, *Chem. Ber.*, **87**, 869 (1954).
31. G. M. Singerman and R. Levine, *J. Org. Chem.*, **30**, 4379 (1965).
32. J. Szmuszkovicz, *Advan. Org. Chem.*, **4**, 1 (1963); K. Blaha and O. Cervinka, *Advances in Heterocyclic Chemistry*, Vol. 6, Academic Press, New York, 1966, p. 147.
33. S. Hünig *et al.*, *Org. Syn.*, **41**, 65 (1961).
34. F. E. Heyl and M. E. Herr, *J. Am. Chem. Soc.*, **75**, 1918 (1953).
35. W. A. White and H. Weingarten *J. Org. Chem.*, **32**, 213 (1967).
36. A. G. Cook and C. R. Schulz, *J. Org. Chem.*, **32**, 473 (1967).
37. G. Stork *et al.*, *J. Am. Chem. Soc.*, **76**, 2029 (1954).
38. I. Fleming and M. H. Karger, *J. Chem. Soc.* (C), 226 (1967).
39. L. Birkofer *et al.*, *Chem. Ber.*, **95**, 1495 (1962).
40. J. Wolinsky and D. Chan, *Chem. Commun.*, 567 (1966).
41. E. D. Bergmann *et al.*, *Org. Reactions*, **10**, 179 (1959).
42. D. J. Goldsmith and J. A. Hartman, *J. Org. Chem.*, **29**, 3520, 3524 (1964).
43. G. L. Buchanan and G. W. McLoy, *Chem. Commun.*, 504 (1965).
44. G. L. Buchanan, A. C. W. Curran *et al.*, *Tetrahedron*, **24**, 4565 (1968).
45. N. C. Ross and R. Levine, *J. Org. Chem.*, **29**, 2341 (1964).
46. N. C. Ross and R. Levine, *J. Org. Chem.*, **29**, 2346 (1964).
47. Y. Maroni-Barnaud *et al.*, *Tetrahedron Letters*, 2243 (1966).
48. N. J. Leonard and W. J. Musliner, *J. Org. Chem.*, **31**, 639 (1966).
49. W. Kimel and A. C. Cope, *J. Am. Chem. Soc.*, **65**, 1992 (1943).
50. N. B. Lorette, *J. Org. Chem.*, **26**, 4855 (1961).
51. C. W. Mortenson and M. A. Spielman, *J. Am. Chem. Soc.*, **62**, 1609 (1940).
52. K. D. Berlin *et al.*, *J. Org. Chem.*, **30**, 226 (1965).
53. D. B. Luten, Jr., *J. Org. Chem.*, **3**, 588 (1939).
54. C. R. Hauser and G. F. Morris, *J. Org. Chem.*, **26**, 4740 (1961).
55. K. Ziegler and H. Ohlinger, *Ann. Chem.*, **495**, 84 (1932).
56. Z. Welvart, *Bull. Soc. Chim. France*, 1653 (1961).
57. G. F. Morris and C. R. Hauser, *J. Org. Chem.*, **26**, 4741 (1961).
58. J. S. Walia *et al.*, *Tetrahedron Letters*, 195 (1969).
59. H.-W. Wanzlick and E. Schikora, *Chem. Ber.*, **94**, 2389 (1961).
60. D. M. Lemal *et al.*, *J. Am. Chem. Soc.*, **86**, 2518 (1964).
61. Z. Welvart, *Bull. Soc., Chim. France*, 1653 (1961).
62. T.-K. Liao and W. E. McEwen, *J. Org. Chem.*, **26**, 5257 (1961).
63. S. Cohen *et al.*, *J. Org. Chem.*, **28**, 1379 (1963).
64. A. J. Waring, *Advances in Alicyclic Chemistry*, Vol. 1, Academic Press, New York, 1966, p. 129.
65. N. J. Turro and W. B. Hammond, *J. Am. Chem. Soc.*, **88**, 3673 (1966).

H. Organometallic Methods

The Grignard reagents react with a great number of carboxylic acid derivatives to form ketones. Among them are esters, lactones, acid chlorides, nitriles, acid anhydrides, acid amides, and sometimes even carboxylic acids. The most complete treatment of these reactions is that of Kharasch and Reinmuth [1]. The alkyl- or aryllithiums may also be employed and in some cases offer an advantage, as discussed under Carboxylic Acids, Chapter 13, C.1. Of particular interest is the use of methyllithium in preparing methyl ketones directly from carboxylic acids [2]. Other metal alkyls or aryls are rarely used but the cadmium analog has proved to be of particular value in the synthesis of ketones from acid chlorides as is shown in H.2.

References for Section H are on p. 723.

The mechanism of these addition reactions has not been determined with certainty. Not only is $RMgX(RMgX \cdot 2(C_2H_5)_2O$ in ether) an oversimplification for the formula of the reagent, but also intermediates are difficult to identify, and the variations in the nature of the products formed are too numerous to permit a simple explanation [3]. It is common to regard coordination as occurring between the magnesium atom and the oxygen atom of the carbonyl group, after which a nucleophilic attack of the R group on the carbon atom of the carbonyl group occurs:

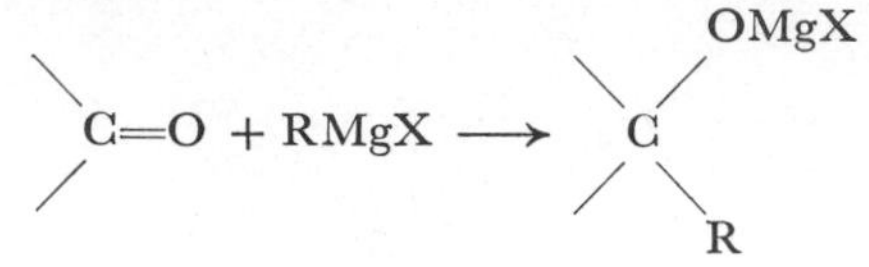

$$\underset{/}{\overset{\backslash}{C}}=O + RMgX \longrightarrow \underset{/\ \ \backslash}{\overset{\backslash\ \ OMgX}{C}}{\ \ R}$$

In the case of esters or lactones, a second molecule of RMgX normally replaces the OR group with R:

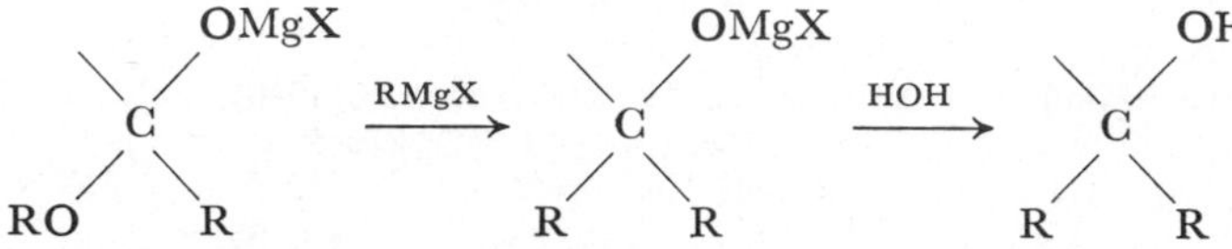

$$\underset{RO\ \ \ \ R}{\overset{\backslash\ /\ OMgX}{C}} \xrightarrow{RMgX} \underset{R\ \ \ R}{\overset{\backslash\ /\ OMgX}{C}} \xrightarrow{HOH} \underset{R\ \ \ R}{\overset{\backslash\ /\ OH}{C}}$$

with the result that tertiary alcohols are formed on hydrolysis. However, in some cases the elimination of ROMgX occurs by an unknown mechanism to give the ketone:

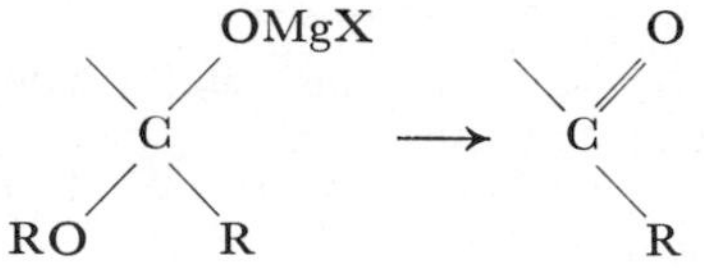

$$\underset{RO\ \ \ R}{\overset{\backslash\ /\ OMgX}{C}} \longrightarrow \underset{R}{\overset{\backslash\ //\ O}{C}}$$

In the case of acid chlorides the complex with the Grignard also forms, but it reorganizes to the ketone-magnesium halide complex:

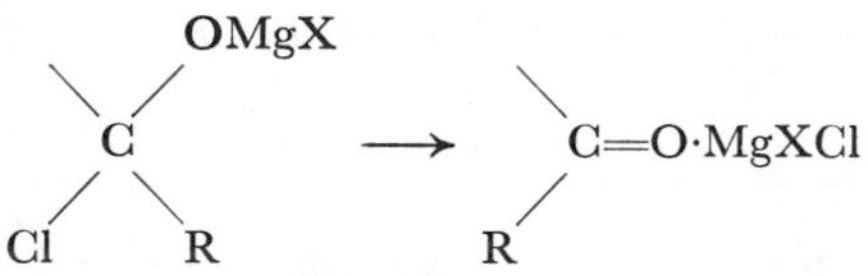

$$\underset{Cl\ \ \ R}{\overset{\backslash\ /\ OMgX}{C}} \longrightarrow \underset{R}{\overset{\backslash}{C}}=O \cdot MgXCl$$

With nitriles the mechanism is somewhat more firmly established. It appears that a complex is also first formed which in the rate determining step rearranges:

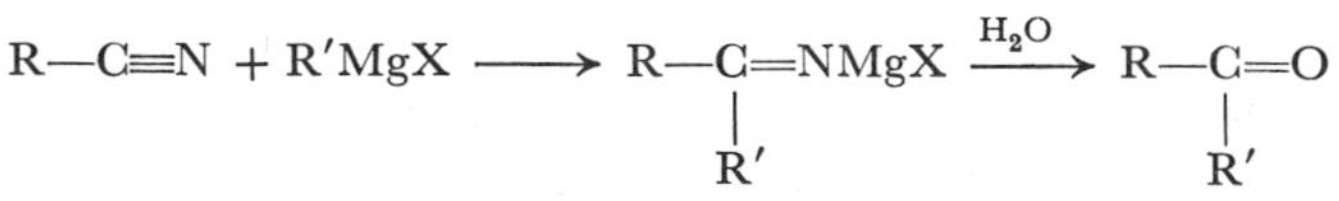

$$R-C\equiv N + R'MgX \longrightarrow R-\underset{R'}{\overset{\displaystyle |}{C}}=NMgX \xrightarrow{H_2O} R-\underset{R'}{\overset{\displaystyle |}{C}}=O$$

the R' with its pair of electrons migrating to the nitrile carbon.

Except in a limited manner, no attempt has been made to deal with the numerous side reactions of the Grignard reagent. For greater detail on this subject Kharasch and Reinmuth [1] should be consulted.

1. From Esters, Lactones, and Polyfunctional Ketones

$$\text{RCOOR}' + \text{R}''\text{MgX} \longrightarrow \underset{\overset{\displaystyle |}{\text{R}'\text{O}} \quad \overset{\displaystyle |}{\text{R}''}}{\overset{\text{R} \quad \text{OMgX}}{\overset{\diagdown \diagup}{\text{C}}}} \xrightarrow{\text{H}_2\text{O}} \underset{\text{R}''}{\overset{\text{R}}{\overset{\diagdown}{\diagup}\text{CO}}}$$

Since the ketone is susceptible to further addition to the Grignard reagent to form tertiary alcohols, steric effects favor ketone formation. Thus, branched-chain alkylmagnesium halides, pyrrylmagnesium halides, and di-ortho-substituted arylmagnesium halides tend to stop at the ketone stage (see Ex. c.3). However, in the case of α,β-unsaturated esters, the nature of the Grignard reagent appears to be unimportant since ketones are invariably the main products of such reactions. Alkyllithiums may be substituted for the Grignard reagent, and lactones, being inner esters, may be converted into ketones by these reagents. The yields vary from fair to good.

(a) Preparation of methyl 8-benzhydryl-1-naphthyl ketone. Methyl

$$(\text{C}_6\text{H}_5)_2\text{CH} \quad \text{COOCH}_3 \xrightarrow{\text{CH}_3\text{Li}} (\text{C}_6\text{H}_5)_2\text{CH} \quad \text{COCH}_3$$

8-benzhydryl-1-naphthoate, 1.10 g., m.p. 167–167.5°, was added cautiously to a solution of methyllithium prepared from 0.76 g. of lithium and 14.2 g. of methyl iodide in 50 ml. of ether. After refluxing the mixture for 5 hr., it was hydrolyzed with ice water and the organic matter was extracted with ether. From the residue obtained from the ethereal solution, there was obtained by trituration with petroleum ether 0.96 g. (92%) of the ketone, m.p. 164–165° after recrystallization from ethanol [4].

(b) Tetradecafluoro-4-heptanone (61.5% with 13% recovered starting material from ethyl heptafluorobutyrate, phenylmagnesium bromide, and heptafluoro-1-iodopropane) [5].

(c) Other examples

(1) β,β-Diphenylbutyrophenone (17% from ethyl acetoacetate with excess phenylmagnesium bromide; not a good method of preparation but it shows the course of addition to β-ketoesters under forcing conditions) [6].

(2) 5-Allyl-5-hydroxy-2,4,4-trimethyl-7-octen-3-one (61% from

$$(\text{CH}_2\!\!=\!\!\text{CHCH}_2)_2\text{C}\underset{\overset{\displaystyle |}{\text{CH}_3}}{\overset{\overset{\displaystyle \text{OH}}{|}}{\underset{}{-}}}\underset{}{\overset{\overset{\displaystyle \text{CH}_3}{|}}{\text{C}}}\underset{}{-}\overset{\overset{\displaystyle \text{O}}{\|}}{\text{C}}\text{CH}(\text{CH}_3)_2$$

tetramethylcyclobutanedione-1,3 and magnesium in ether to which allyl bromide was added dropwise; this reaction illustrates that the Grignard acceptor can be present during formation of the Grignard reagent, at least with allyl-magnesium bromide) [7].

(3) **Hexamethylacetone** (80% from t-butyllithium and ethyl pivalate; 70% from t-butyllithium and diethyl carbonate; 71% from t-butylmagnesium chloride and pivaloyl chloride in THF) [8].

2. From Acid Chlorides and Anhydrides

$$\underset{RCCl}{\overset{O}{\parallel}} + R'MgX \xrightarrow{FeCl_3} \underset{RCR'}{\overset{OMgXCl}{\parallel}} \xrightarrow{H_2O} \underset{RCR'}{\overset{O}{\parallel}}$$

For many years this reaction was thought to be quite inferior for preparing ketones because of further addition of the Grignard reagent to produce tertiary alcohols. Since the publication of a review [9] and two definitive papers [10, 11], the reaction has taken on respectability and assumes an important role in making methyl and perhaps other ketones. The improvements suggested are the use of low temperature (-60 to $-70°$) and the addition of about 2 equivalents of anhydrous ferric chloride and a mixed toluene-ether solvent to avoid the un-workable material which occasionally is obtained at low temperatures with ether alone. For ω-carbethoxyacid chlorides, $C_2H_5OOC(CH_2)_nCOCl$, the alkyl-cadmiums [9] or ferric chloride-catalyzed Grignard reagents [11] give improved yields of ketones as compared to those obtained with ordinary Grignard reagents.

Another method of obtaining ketones is the addition of the alkylcadmium to the acid chloride:

$$2\,RMgX \xrightarrow{CdCl_2} \underset{\text{or } RCdCl}{R_2Cd} \xrightarrow{R'COCl} \underset{RCR'}{\overset{O}{\parallel}}$$

Alkylcadmiums have the advantage that they do not add to ketones readily, so that the encroachment of the tertiary alcohol is not a problem. They are limited in use, however, to primary alkyl groups; secondary alkylcadmiums are un-stable [12]. Pure dialkylcadmiums do not react with acid chlorides but do react in the presence of magnesium bromide or, preferably, lithium bromide. This reaction promises to enlarge the scope of dialkylcadmium additions (see Ex. b.6).

The organozinc compounds have found some use, but as a rule they are not so satisfactory as the cadmium analogs. The preparation of the zinc types is more difficult; the use of ether as a solvent is undesirable because of ether cleavage and ester formation; and the greater reactivity of the zinc compounds sometimes leads to a reaction with the carbonyl group of the ketone.

(a) **Preparation of 2-hexanone** (72% from addition of 2 moles of butyl-magnesium chloride to a solution of 6 moles of acetyl chloride and 3 g. of ferric chloride in ether at $-70°$ under nitrogen; the reaction is complete within 5 min; the yield is only ca. 30% without ferric chloride) [10]; for an improved yield but with excess of acetic anhydride see [13].

(b) Other examples

(1) γ-Ketocaproic acid (53% from succinyl chloride and ethyl-magnesium bromide at 0°) [14].

(2) o-Acetylbenzoic acid (62% from dimethylcadmium and o-phthalic anhydride) [15].

(3) Propiophenone (84% from diethylcadmium in benzene and benzoyl chloride) [16].

(4) Hexamethylacetone (72% from t-butylmagnesium chloride and trimethylacetyl chloride in the presence of cuprous chloride) [17]; see also H.1, Ex. c.3.

(5) Ethyl 9-keto-10-methyloctadecanoate (93% from 2-decyl-magnesium bromide, anhydrous zinc chloride, and ω-carbethoxyoctanoyl-chloride) [18].

(6) Decane-5,6-dione (37% from oxalyl chloride, and dibutylcadmium in tetrahydrofuran at −20° to which lithium bromide was added to activate dibutylcadmium; dialkylcadmiums are inert in attacking acid chlorides, but magnesium salts and, particularly, lithium bromide activate them) [19].

3. From Nitriles

$$RC{\equiv}N + R'MgX \longrightarrow \underset{\underset{R'}{|}}{RC{=}NMgX} \xrightarrow{\text{H}_2\text{O}} \underset{\text{Iminoketone}}{R\overset{\overset{\text{NH}}{||}}{C}R'} \longrightarrow R\overset{\overset{\text{O}}{||}}{C}R'$$

The nitrile synthesis is the best general method of making ketones. No tendency exists for the addition to go beyond the iminoketone stage, and the reaction between nitrile and Grignard reagent proceeds nicely in the range of 35 to 100°. Yields with benzonitriles are very good, with higher molecular weight aliphatic nitriles good, and with acetonitrile fair, provided that for the latter the ether is replaced with benzene, as shown in Table 5 [20].

The intermediate iminoketone which usually is spontaneously hydrolyzed to the ketone during workup can be isolated in good yield by decomposing the

Table 5 Influence of the Solvent on the Reaction of the Grignard Reagent and Acetonitrile

Product[a]	% Yield	
	In Benzene	In Ether
Acetophenone	68	42
3,5-Dibromoacetophenone	52	21
2-Heptanone	44	14
Methyl benzyl ketone	34	16
2-Decanone	49	—

[a] The side reaction gives a triazine-type polymer.

Grignard-nitrile complex with methanol. The magnesium methoxybromide formed is easily filtered [21].

(a) Preparation of ω-methoxyacetophenone (71–78% based on methoxyacetonitrile when treated with phenylmagnesium bromide) [22].

(b) Other examples

(1) 9-Acetylphenanthrene (52–59% from 9-cyanophenanthrene and methylmagnesium iodide) [23].

(2) o-Acetovanillone (2-hydroxy-3-methoxyacetophenone, 75% from o-veratronitrile and 2 equiv. of methylmagnesium iodide) [24].

(3) 5-Benzoylpentanoic acid (85% from α-cyanocyclopentanone which was added to more than 2 equiv. of phenylmagnesium bromide; not an addition to nitrile but rather to carbonyl) [25].

4. From Amides and Imidazolides

$$RCONH_2 + 2\ R'MgX \longrightarrow \underset{\underset{R'}{|}}{RC}\overset{\overset{NHMgX}{/}}{{-}}OMgX \overset{H_2O}{\longrightarrow} \underset{R'}{\overset{R}{\diagdown}}CO$$

This synthesis is the least attractive of all those involving the Grignard reagent. It has been employed, however, to prepare t-butyl n-alkyl ketones from amides with yields varying from 42 to 78% [26] and desoxybenzoins from benzyl magnesium bromides and amides with yields of 42–85% [27]. Similarly, from mandelamide or p-methoxyphenylacetamide, the hydroxyketones were obtained [28]. The synthesis appears to be the most attractive if alkyllithiums and amides (Exs. b and c.1) or Grignard reagents and imidazolides (Ex. a) are used.

(a) Preparation of benzophenone. To 8.2 g. (48 mmoles) of N-benzoylimidazole in 300 ml. of THF at −50°, a solution of phenylmagnesium bromide (53 mmoles) (from 1.27 g. of magnesium and 8.23 g. of bromobenzene in 300 ml. of THF) was added during 3 hr. with stirring. After another 4 hr., the mixture was treated with 200 ml. of 2 N HCl and then the main part of the THF was removed by vacuum distillation. Extraction with ether and washing of the ethereal layer with Na_2CO_3 until neutral and then with water, followed by evaporation, gave the ketone, which after two crystallizations from ethanol weighed 6.3 g. (72%) [29].

(b) Preparation of ethyl cyclohexyl ketone. Lithium ribbon, 1.6 g. atoms, in 800 ml. of anhydrous ether at −10° under nitrogen was treated with 0.83 mole of ethyl bromide in 200 ml. of anhydrous ether over a 2-hr. period. After stirring for 1 hr. more at −10°, the temperature was lowered to −20° and a solution of 0.8 mole of N,N-dimethylcyclohexanecarboxamide in 200 ml. of ether was added slowly over a 1.5-hr. period. Stirring was continued under nitrogen for 3 hr. while the temperature was allowed to rise to 25°, and then the solution was cooled to −10° after which 500 ml. of a cold, saturated ammonium chloride solution was added slowly. After stirring for 30 min., the ether layer was separated, washed with 1 N hydrochloric acid and water and then dried. Distillation gave a 70% yield of the ketone, b.p. 73–77° (8 mm.) [30].

(c) Other examples

(1) **2-Phenyl-5(ω-phenacyl) pyrrocoline** (66% from 2-phenyl-5-pyrrocolymethyllithium and N,N-dimethylbenzamide) [31].

(2) **3,4-Dimethoxydesoxybenzoin** (71% from benzylmagnesium chloride and veratramide) [32].

5. From Haloketones

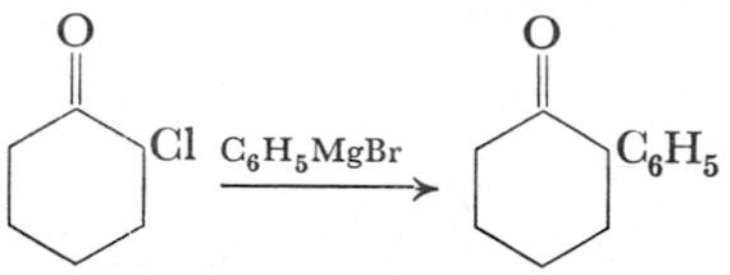

This synthesis is of theoretical interest since the aryl group, at least in part, replaces the halogen atom in the ring [33]. The path to produce such a product is then apparently:

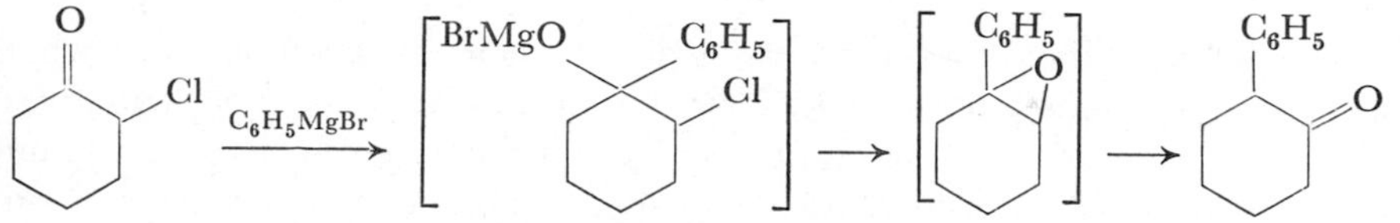

But phenyl, rather than hydride, migration may occur. Unfortunately, if there are other substituents in the ring, mixtures are obtained, a fact which detracts from the synthetic value of the method [33]. However, in unsubstituted rings, ketones of value as intermediates for polynuclear syntheses may be obtained [34].

Phenacyl halides, when treated with a Grignard reagent, form adducts which undergo a similar rearrangement [35]:

$$C_6H_5COCH_2Cl \xrightarrow{RMgX} \underset{\underset{OMgX}{|}}{\overset{\overset{R}{|}}{C_6H_5CCH_2Cl}} \longrightarrow RCOCH_2C_6H_5$$

The phenyl migrates in preference to the ethyl, as is to be expected from the results in analogous systems.

In the reactions of cyclic chloroketones inverse addition is preferred and the intermediate chlorohydrin, which has the *cis* configuration, may be isolated [34]. Yields of ketones in both types of reactions are usually in the range 50–70%.

(a) **Preparation of 2-phenylcyclohexanone.** 2-Chlorocyclohexanone, 280 g. in 750 ml. of dry ether, was added with stirring to 750 ml. of 3.22 M phenylmagnesium bromide over 1 hr. The ether was removed by distillation until the foaming residue almost filled the flask, after which 700 ml. of dry benzene was added and the mixture was refluxed for 8 hr. Hydrolysis with cold water and distillation of the benzene layer gave a 58% yield of the ketone, f.p. 53–55°, b.p. 136–137° (5–6 mm.) [36].

(b) Other examples

(1) **2-Phenylcyclopentanone** (50% from 2-chlorocyclopentanone and phenylmagnesium bromide) [37].

(2) **Desoxybenzoin** (62% from phenacyl chloride and phenylmagnesium bromide by replacing ether with benzene and refluxing 4 hr.) [35].

6. From Unsaturated Ketones

The Grignard reagent must add in a 1,4-manner to an unsaturated ketone in order to yield a ketone [38]:

$$\underset{1\ 2\quad\ 3}{R'MgX + R\overset{O}{\overset{\|}{C}}CH{=}CHR} \longrightarrow R\overset{O}{\overset{\|}{C}}CH_2CH\overset{\displaystyle R}{\underset{\displaystyle R'}{\diagup\!\!\!\diagdown}}$$

To judge from the confusing results available, generalizations are somewhat hazardous. But some remarks, based for the most part on common sense, will be made. 1,4-Addition is favored by relatively strong electron withdrawal and no steric hindrance at carbon-3 above, by appreciable steric hindrance in R on carbon-1, by bulky and less reactive Grignard reagents, and by the introduction of a small amount of cuprous bromide or iodide (the latter being more of an empirical observation). When R on carbon-1 is phenyl,

$$ArCH{=}CH{-}\overset{O}{\overset{\|}{C}}R$$

the 1,4-addition product is invariably predominant (except if R is H on carbon-3 as in $ArCH{=}CHC\overset{O}{\overset{\|}{}}H$). Recently it has been shown that two active agents [39] which bring about 1,4-addition are $Li\overset{\oplus}{}Me_2\overset{\ominus}{Cu}$ and $MeCuP(n\text{-}Bu)_3$.

(a) Preparation of *cis*-9-methyl-2-decalone (60% from 2-keto-$\Delta^{1,9}$-

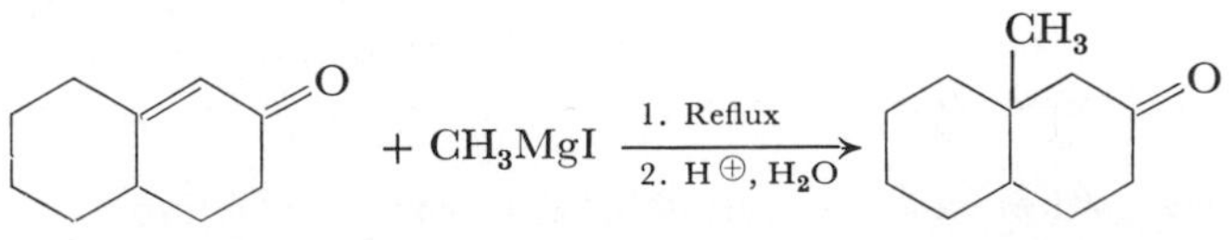

octalin, methylmagnesium iodide, and a trace of cuprous bromide) [40].

(b) Preparation of 3-isopropenylcyclohexanone (68 % from 2-cyclo-hexene-1-one, isopropenylmagnesium bromide, and 5 mole % of cuprous iodide;

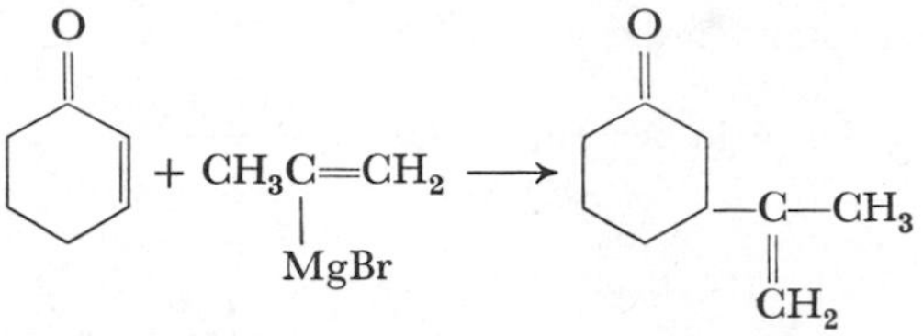

another example in which cuprous iodide promotes 1,4-addition) [41].

7. From Acid Derivatives and Alkylidene Phosphoranes (Wittig)

$$RCOCl \xrightarrow{2 \ R'CH=P(C_6H_5)_3} RCO\overset{\overset{R'}{|}}{C}=P(C_6H_5)_3 \xrightarrow{H_2O} RCOCH_2R' + (C_6H_5)_3PO$$

The alkylidene phosphoranes (via $R'\overset{\ominus}{C}H\overset{\oplus}{P}(C_6H_5)_3$) add to acid derivatives in a fashion reminiscent of the addition of Grignard reagents just discussed. This synthesis has been reviewed [42]; see also Alkenes, Chapter 2, E.2, and Carboxylic Acids, Chapter 13, D.3. Esters [43], thioesters [44], acid halides [43, 44], and N-acylimidazoles [45] form β-ketoalkylphosphonium salts 11-XXIV with alkylidene phosphoranes 11-XXIII which may be synthesized by a variety of methods [42]:

$$(C_6H_5)_3P=CHR + R'COX \longrightarrow (C_6H_5)_3\overset{\oplus}{P}CHR \quad \overset{\ominus}{X}$$

11-XXIII   11-XXIV $\underset{COR'}{|}$

$$(C_6H_5)_3P=CR$$
$$\underset{COR'}{|}$$
11-XXV

In the case of esters or thioesters (X = OR or SR), HX is eliminated to give a β-ketoalkylidene phosphorane 11-XXV. In the case in which X = halogen or imidazole, 11-XXIV reacts with another molecule of the alkylidene phosphorane 11-XXIII to give 11-XXV:

$$(C_6H_5)_3\overset{\oplus}{P}CHR \underset{COR'}{\overset{|}{\underset{\ominus}{X}}} + \text{11-XXIII} \longrightarrow (C_6H_5)_3P=CR \underset{COR'}{|} + (C_6H_5)_3\overset{\oplus}{P}CH_2RX^{\ominus}$$

11-XXIV   11-XXV

The hydrolysis or reduction of 11-XXV by zinc and acetic acid or at the mercury cathode gives the ketone:

$$\text{11-XXV} + H_2O \rightarrow RCH_2COR' + (C_6H_5)_3PO$$

The yields of ketones from a series of β-keto alkylidene phosphoranes containing alkyl, aryl, and cycloalkyl groups vary from 74 to 95% [44].

Ylides react with aromatic nitriles to form ketones. Lithium iodide as a catalyst is essential to give good yields [46]:

$$(C_6H_5)_3P\!\!=\!\!CHC_6H_5 + LiI \text{ (in benzene refluxed)} \xrightarrow{p\text{-ClC}_6\text{H}_4\text{CN}}$$

0.02 mole   0.02 mole

$$(C_6H_5)_3\overset{\oplus}{P}\!\!-\!\!CHC_6H_5 \xrightarrow[\text{Steam distill}]{H_3O^\oplus} p\text{-ClC}_6H_4COCH_2C_6H_5$$

$$\underset{\ominus}{}N\!\!=\!\!CC_6H_4Cl\text{-}p$$

4'-Chlorodesoxybenzoin, 81%

The Wittig reaction also lends itself to the synthesis of α-ketoaldehydes and β-ketoesters. For example, a diazoketone reacts with triphenylphosphine in ether to form an α-ketotriphenylphosphazine which with nitrous acid gives the α-ketoaldehyde [47]:

$$\underset{\displaystyle RCCH\!\!=\!\!\overset{\oplus}{N}\!\!=\!\!\overset{\ominus}{N}}{\overset{O}{\|}} + P(C_6H_5)_3 \longrightarrow \underset{\displaystyle RCCH\!\!=\!\!N\!\!-\!\!N\!\!=\!\!P(C_6H_5)_3}{\overset{O}{\|}} \xrightarrow{HNO_2}$$

$$\underset{\displaystyle RCCHO + N_2 + (C_6H_5)_3PO}{\overset{O}{\|}}$$

Yields of crude α-ketoaldehydes, based on the phosphazine, are 62–90%. The synthesis of β-ketoesters from acid chlorides is as shown [48]:

$$\underset{\displaystyle RCCl}{\overset{O}{\|}} + 2\ \overset{\ominus}{N}\!\!=\!\!\overset{\oplus}{N}\!\!=\!\!CHCOOC_2H_5 \longrightarrow \underset{\displaystyle \underset{\|}{\overset{\displaystyle RCCCOOC_2H_5}{}}}{\overset{O}{\|}} +$$

$$ClCH_2CO_2C_2H_5 + N_2$$

$$\underset{\displaystyle \overset{\ominus}{N}}{\overset{\displaystyle \overset{\oplus}{N}}{\|}}$$

$$\downarrow P(C_6H_5)_3$$

$$\underset{\displaystyle RCCH_2COOC_2H_5}{\overset{O}{\|}} \xleftarrow[\text{Wolff-Kishner}]{\text{Base}} \underset{\displaystyle \underset{NNH_2}{\overset{\|}{}}}{\underset{\displaystyle RCCCOOC_2H_5}{\overset{O}{\|}}} \xleftarrow[\text{ZnCl}_2]{H_2O} \underset{\displaystyle \underset{N\!\!-\!\!N\!\!=\!\!P(C_6H_5)_3}{}}{\underset{\displaystyle RCCCOOC_2H_5}{\overset{O}{\|}}}$$

Treatment of the first-formed α-diazo-β-ketoester with triphenylphosphine gives in good yield the phosphazine, which hydrolyzes to the hydrazone, readily convertible into the β-ketoester with a base. Yields of β-ketoesters based on the phosphazine vary from 64 to 77%.

(a) Preparation of ethyl pentyl ketone. (α-Hexanoylethylidene)-triphenylphosphine from 0.022 mole of the alkylidene phosphorane in 50–80 ml. of 80% methanol was treated with 1-2 ml. of 2 N sodium hydroxide solution and boiled under reflux for 12 hr.; glacial acetic acid, 3 ml., was added and the ketone (95%) formed was separated by steam [44].

1. M. S. Kharasch and O. Reinmuth, *Grignard Reactions of Nonmetallic Substances*, Prentice-Hall, Englewood Cliffs, N.J., 1954.
2. H. Gilman and P. Van Ess, *J. Am. Chem. Soc.*, **55,** 1258 (1933): C. Tegner, *Acta Chem. Scand.*, **6,** 782 (1952); J. F. Ahrens and D. A. Van Dorp, *Rec. Trav. Chim.*, **65,** 338 (1946); **66,** 759 (1947); C. H. DePuy *et al.*, *J. Org. Chem.*, **29,** 2813 (1964).
3. M. S. Newman, *Steric Effects in Organic Chemistry*, John Wiley and Sons, New York, 1956, p. 395.
4. R. L. Letsinger and P. T. Lansbury, *J. Am. Chem. Soc.*, **81,** 935 (1959).
5. E. T. McBee *et al.*, *J. Am. Chem. Soc.*, **77,** 6387 (1955).
6. J. P. Freeman, *J. Am. Chem. Soc.*, **80,** 1926 (1958).
7. M. P. Dreyfuss, *J. Org. Chem.*, **28,** 3269 (1963).
8. J.-E. Dubois *et al.*, *Bull. Soc. Chim. France*, 1150 (1967).
9. D. A. Shirley, *Org. Reactions*, **8,** 28 (1954).
10. W. C. Percival *et al.*, *J. Am. Chem. Soc.*, **75,** 3731 (1953).
11. J. Cason and K. W. Kraus, *J. Org. Chem.*, **26,** 1768 (1961).
12. J. Cason and R. Fessenden, *J. Org. Chem.*, **25,** 477 (1960).
13. M. S. Newman and W. T. Booth, Jr., *J. Am. Chem. Soc.*, **67,** 154 (1945).
14. J. Cason and E. J. Reist, *J. Org. Chem.*, **23,** 1668 (1958).
15. P. L. deBenneville, *J. Org. Chem.*, **6,** 462 (1941).
16. J. Cason, *J. Am. Chem. Soc.*, **68,** 2078 (1946).
17. N. C. Cook and W. C. Percival, *J. Am. Chem. Soc.*, **71,** 4141 (1949).
18. G. A. Schmidt and D. A. Shirley, *J. Am. Chem. Soc.*, **71,** 3804 (1949).
19. J. Kollonitsch, *J. Chem. Soc.*, A, 456 (1966).
20. D. E. Pearson and M. Gordon, unpublished work.
21. P. L. Pickard and T. L. Tolbert, *J. Org. Chem.*, **26,** 4886 (1961).
22. R. B. Moffett and R. L. Shriner, *Org. Syn.*, Coll. Vol. **3,** 562 (1955).
23. J. E. Callen *et al.*, *Org. Syn.*, Coll. Vol., **3,** 26 (1955).
24. E. D. Amstutz, *J. Am. Chem. Soc.*, **71,** 3836 (1949).
25. M. Lamant and R. Mornet, *Bull. Soc. Chim. France*, 3041 (1965).
26. F. C. Whitmore *et al.*, *J. Am. Chem. Soc.*, **61,** 683 (1939).
27. S. S. Jenkins, *J. Am. Chem. Soc.*, **55,** 703, 2896 (1933); S. S. Sanford and E. M. Richardson, *ibid.*, **55,** 1618 (1933).
28. S. S. Sanford, *J. Am. Chem. Soc.*, **56,** 682 (1934).
29. H. A. Staab *et al.*, *Ann. Chem.*, **655,** 90 (1962); *Angew. Chem. Intern. Ed. Engl.*, **1,** 351 (1962).
30. P. T. Izzo and S. R. Safir, *J. Org. Chem.*, **24,** 701 (1959).
31. R. J. Windgassen, Jr. *et al.*, *J. Am. Chem. Soc.*, **81,** 1459 (1959).
32. M. D. Farooq *et al.*, *Chem. Ber.*, **92,** 2555 (1959).
33. M. S. Newman and W. T. Booth, Jr., *J. Org. Chem.*, **12,** 737 (1947).
34. A. S. Hussey and R. R. Herr, *J. Org. Chem.*, **24,** 843 (1959).
35. R. L. Huang, *J. Chem. Soc.*, 4089 (1957).
36. M. S. Newman and M. D. Farbman, *J. Am. Chem. Soc.*, **66,** 1550 (1944).
37. R. T. Arnold *et al.*, *J. Am. Chem. Soc.*, **72,** 3153 (1950).
38. Ref. 1, p. 196.
39. H. O. House *et al.*, *J. Org. Chem.*, **31,** 3128 (1966).
40. A. J. Birch and R. Robinson, *J. Chem. Soc.*, 501 (1943).
41. H. O. House *et al.*, *J. Org. Chem.*, **31,** 2667 (1966).
42. S. Trippett, *Quart. Rev. (London)*, **17,** 406 (1963); A. Maercker, *Org. Reactions*, **14,** 270 (1965).
43. S. Trippett, *J. Chem. Soc.*, 1266 (1961).
44. H. J. Bestmann and B. Arnason, *Chem. Ber.*, **95,** 1513 (1962).
45. H. J. Bestmann *et al.*, *Angew. Chem.*, **74,** 243 (1962); H. A. Staab and N. Sommer, *ibid.*, **74,** 294 (1962).
46. R. G. Barnhardt, Jr., and W. E. McEwen, *J. Am. Chem. Soc.*, **89,** 7009 (1967).
47. H. J. Bestmann *et al.*, *Chem. Ber.*, **92,** 1345 (1959); **96,** 2259 (1963).
48. H. J. Bestmann and H. Kolm, *Chem. Ber.*, **96,** 1948 (1963).

Chapter 12

QUINONES AND RELATED SUBSTANCES

Although the preparation of quinones is limited in scope, quinones are important members of the organic family. Their structure is observed in some synthetic dyes and in many naturally occurring substances such as pigments, vitamins, and enzymes [1–3].

A. Oxidation

Oxidation is the principal mode of preparation of quinones. The process can be carried out on the corresponding hydrocarbon, the phenol or aniline, or the o- or p-dihydroxy-, diamino- or hydroxyaminoaromatic derivatives. It is reasonable

References for Section A are on pp. 735–736.

to assume that conditions will be milder and yields higher, the nearer the substrate material is to the quinone in oxidation state, i.e., the ease of preparation of benzoquinone from the following sources should be in the order: hydroquinone > phenol ≫ benzene. On the other hand, many dihydroxy- or diaminoaromatic derivatives are unstable and undergo oxidative polymerization. Those which are unstable should be protected from air and used as soon as possible.

This section discusses preparation from the various substrates at different oxidation levels, as described above, as well as some unique oxidative coupling reactions of phenols.

1. From Hydrocarbons

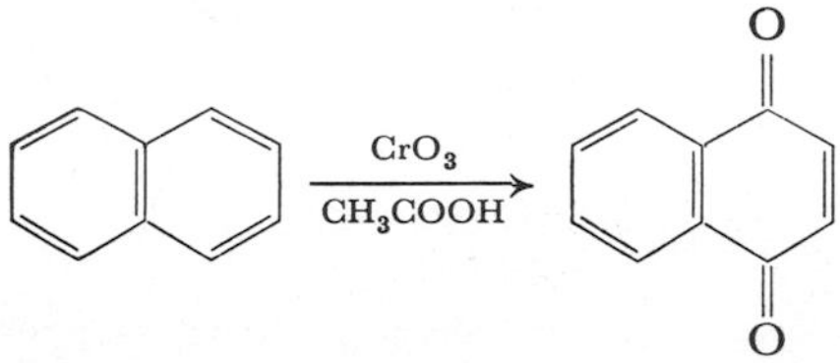

This synthesis often has the advantage of starting with readily available materials, but the yields of quinones obtained are often less than 50 percent. Various oxidizing agents such as chromium trioxide in acetic acid [4], sodium dichromate in acetic acid [5, 6], chromium trioxide in sulfuric acid [7], sodium dichromate in carbon tetrachloride-aqueous sulfuric acid [8], hydrogen peroxide and acetic acid [9], sodium chlorate in acetic-sulfuric acids and a trace of vanadium pentoxide [10], and periodic acid in dimethyl formamide [11] may be employed.

Because of the generally poor yields, quinones perhaps should be prepared from hydrocarbons only if the quinone is relatively stable and the hydrocarbons readily oxidizable. One of the best examples which fits these criteria is the oxidation of anthracene to anthraquinone (see Ex. a). Periodic acid at times gives superior yields (see Ex. b.3), but it gives no reaction with bi- or terphenyl, perylene, coronene, triphenylene, chrysene, or picene, and with pyrene it gives bipyrenyl.

(a) **Preparation of anthraquinone.** Finely powdered anthracene, 90 g., 0.5 g. of vanadium pentoxide, and 76 g. of sodium chlorate in 1 l. of acetic acid and 200 ml. of 2% sulfuric acid were warmed until a vigorous reaction set in. After heat evolution had subsided, the mixture was refluxed for an hour, cooled, and filtered, to yield 88–91% of the anthraquinone [10]; *see also* [12].

(b) **Other examples**

(1) **9,10-Phenanthraquinone** (44–48% from phenanthrene and chromium trioxide in sulfuric acid) [7].

(2) 2-Methyl-5,8-diacetyl-1,4-naphthoquinone (75 % from 2-methyl-1,4-dihydro-5,8-diacetylnaphthalene and chromium trioxide in acetic acid) [13].

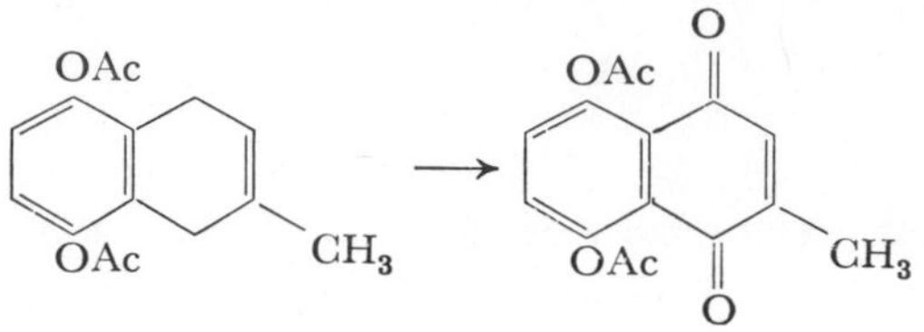

(3) Naphthacenequinone (80–85 % from 0.005 mole of naphthacene

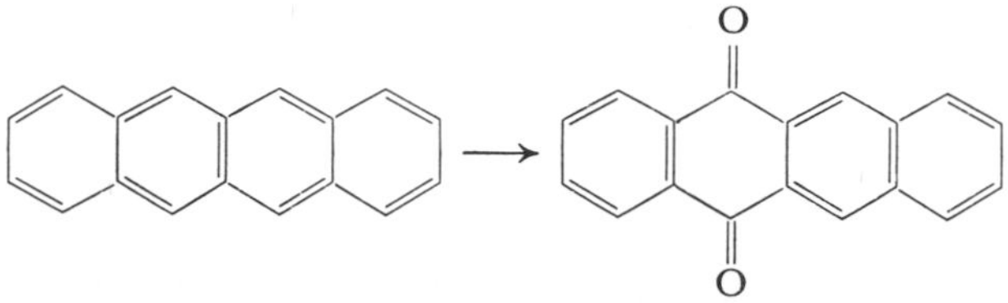

and 0.02 mole of periodic acid in 100 ml. of DMF at 120° for 5 min. and then at 70° for 30 min., followed by dilution with water) [11].

2. From Arylamines or Phenols

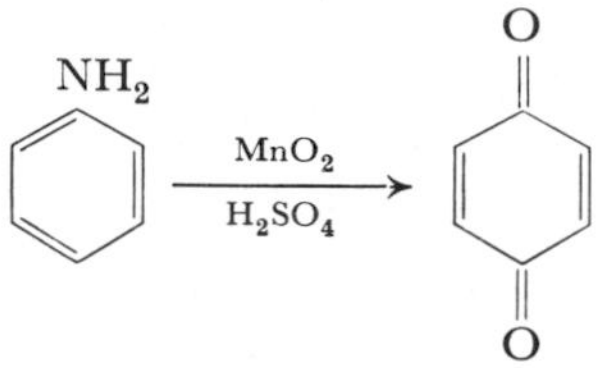

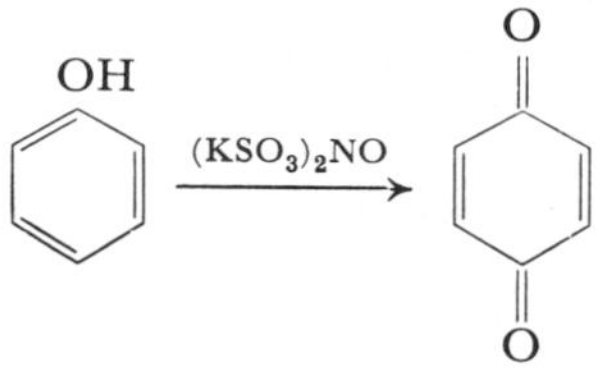

Both arylamines and phenols may be oxidized to quinones. A commercial process for the manufacture of p-benzoquinone utilizes manganese dioxide and sulfuric acid as the oxidizing agent [14]. Another older, commercial process employs sodium dichromate and sulfuric acid in oxidizing either aniline or phenol to quinone. A widely used laboratory oxidizing agent, which is effective both for amines and for phenols, is potassium nitrosodisulfonate, $ON(SO_3K)_2$ (Fremy's Salt), a stable nitrosyldisulfonate radical. For a series of phenols [15] this oxidizing agent gives yields of quinones varying from 50 to 99%, while from amines [16] the yields usually vary from 49 to 96%. For sensitive quinones, such as o-benzoquinone and stilbenequinone, silver oxide is the preferred oxidizing agent [17]. Other laboratory reagents are hydrogen peroxide and acetic acid [18], and potassium ferricyanide [19].

The manganese dioxide-sulfuric acid oxidation of aniline is thought to follow a free radical course [20]. The free radical 12-I combines with aniline to form

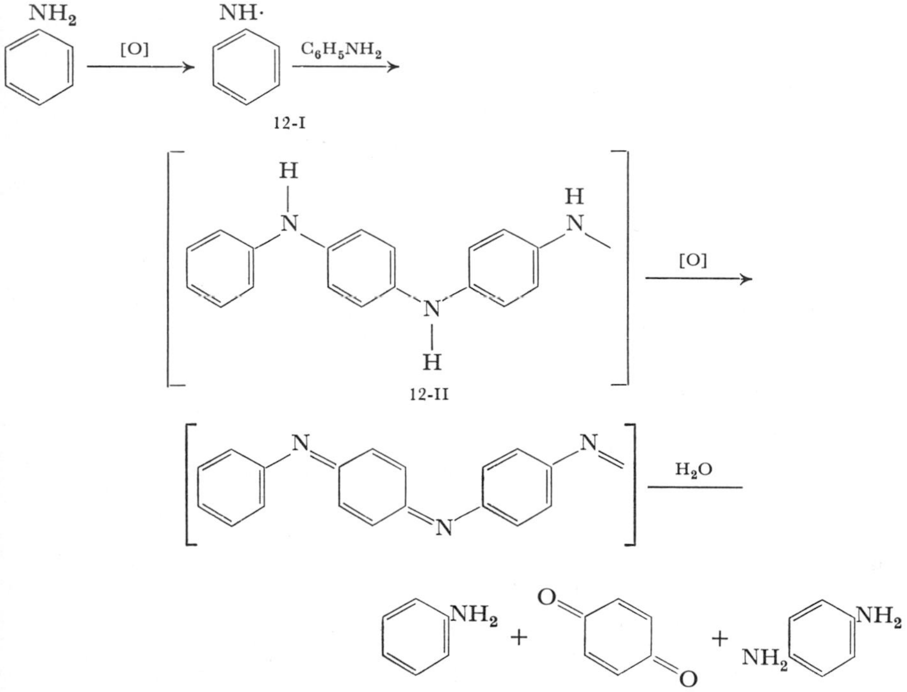

12-I

12-II

12-III

the complex amine 12-II, which by oxidation and hydrolysis gives the quinone 12-III and aniline and p-phenylenediamine available for further oxidation.

A free radical mechanism has also been suggested for the potassium nitrosodisulfonate oxidation of a phenol [15]. The free radical, being a hybrid of many canonical forms including 12-IV and 12-V, couples with potassium nitrosodisulfonate to form 12-VI which achieves stability by the elimination of $HN(SO_3K)_2$:

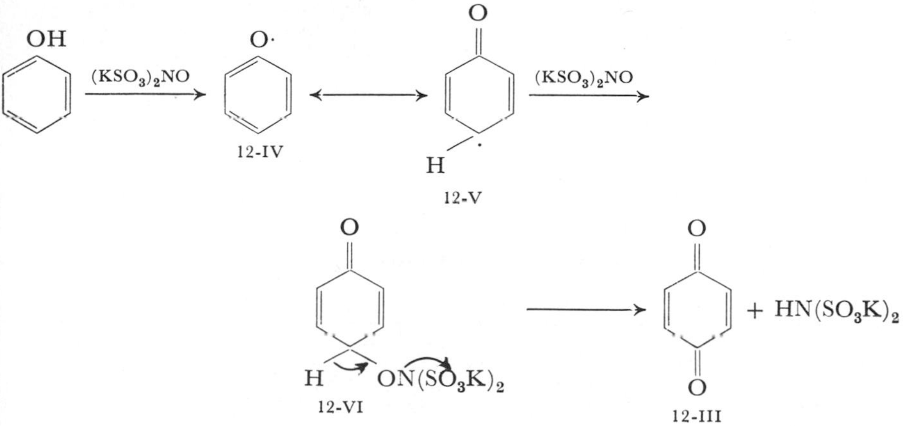

12-IV

12-V

12-VI

12-III

Because of its free radical nature, oxidation of phenols can take unusual courses including, among others, coupling reactions which are discussed in A.3.

(a) Preparation of 2-methyl-6-methoxy-1,4-benzoquinone. 2-Methyl-6-methoxyaniline, 1.37 g., in 30 ml. of acetone was oxidized by the addition of a solution of 6 g. of potassium nitrosodisulfonate in 100 ml. of *M*/6 sodium dihydrogen phosphate and 150 ml. of water. After 1 hr., 250 mg. of yellow crystals, m.p. 148–149°, separated. By chloroform extraction 1.21 g. more, m.p. 136–137° (dec.), was obtained. Total yield 96%. Recrystallization three times from benzine (80–110°) gave a m.p. of 152° [16].

(b) Other examples

(1) *o*-Toluquinone (82% from *o*-cresol, potassium nitrosodisulfonate, and sodium acetate in methanol-water) [15].

(2) Chloranil (60% from 0.5 mole of phenol in 1 l. of conc. hydrochloric acid through which chlorine was passed until saturated; the saturated,

O
Cl
Cl
Cl
Cl
O

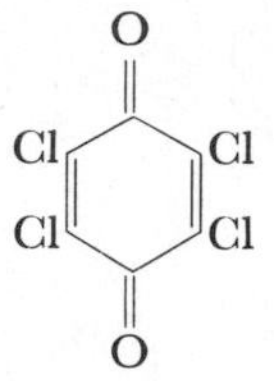

stirred solution at 80–85° was then treated with 250 ml. of nitric acid added dropwise over 3 hr. and held an additional 20 hr. at 80–85°, when crystals had separated) [21].

(3) Stilbenequinone (40–45% by refluxing 4,4′-dihydroxystilbene in acetone with silver oxide for 1 hr.) [17].

HO
CH=CH
OH
O=
=CH—CH=
=O

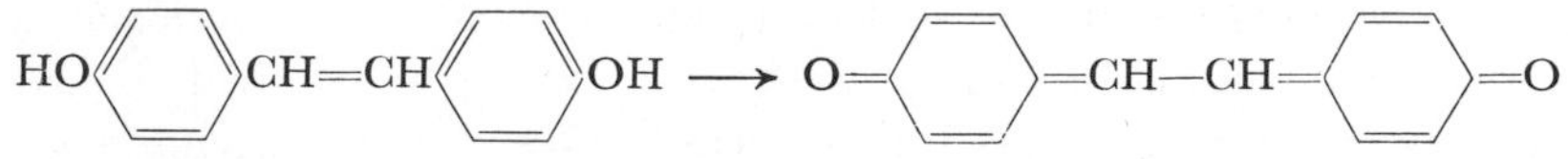

(4) 4,5-Dimorpholine-1,2-benzoquinone [22]:

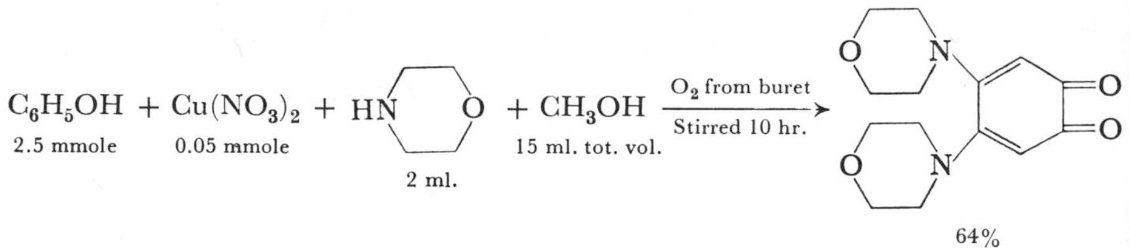

C6H5OH + Cu(NO3)2 + HN O + CH3OH
2.5 mmole
0.05 mmole
2 ml.
15 ml. tot. vol.
O2 from buret
Stirred 10 hr.
O N
O N
=O
=O
64%

3. From Phenols (by Oxidative Coupling)

R
HO
R
[O]
R
R
O=
=O
R
R

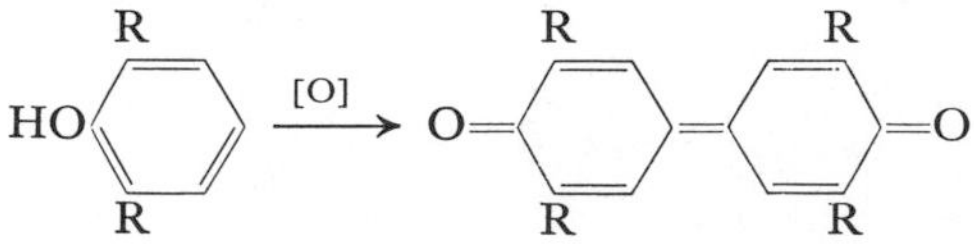

Oxidative coupling has been reviewed [23]. Most of the oxidations are restricted to phenols with at least *ortho* substituents. They occur in alkaline solution with mild oxidizing agents and probably proceed via the relatively stable free radical shown:

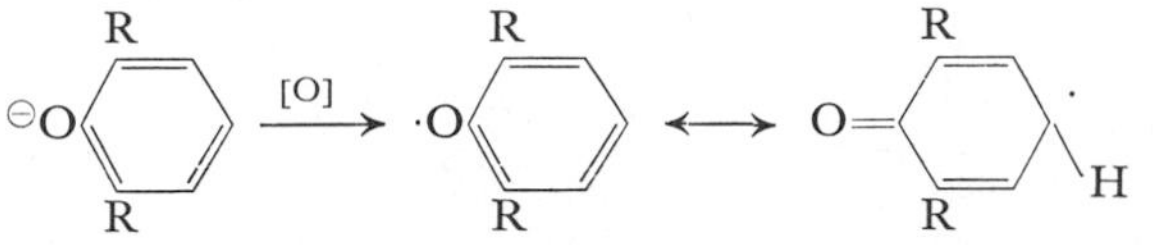

The free radical either couples with another or attacks the anion of the initial phenol. The progress of free radical oxidation of phenols is not always straight-forward. For example, Pummerer's ketone, the result of coupling followed by ring closure, is one of the products frequently encountered in the oxidation of *p*-cresol [23]:

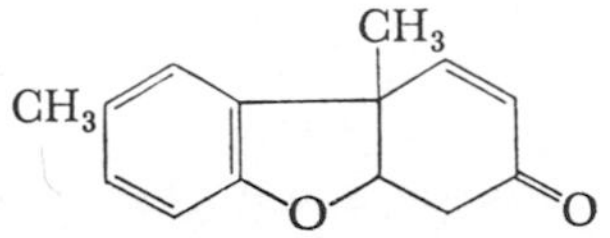

(a) **Preparation of 3,3′,5,5′-tetramethoxydiphenoquinone.** 2,6-Dimethoxyphenol was dissolved in a mixture of sodium hydroxide in alcohol and water and treated with an aqueous solution of potassium ferricyanide, $K_3Fe(CN)_6$. The purple crystals were filtered and washed to give a 96% yield of quinone. 2,6-Dimethylphenol gave only a 50% yield of the diphenoquinone [24].

(b) **Other examples**

(1) **3,3′,5,5′-Tetra-*t*-butylstilbene-4,4′-quinone** (40% from a solution of 0.09 mole of 2,6-di-*t*-butyl-4-cresol and 32 g. of potassium hydroxide in

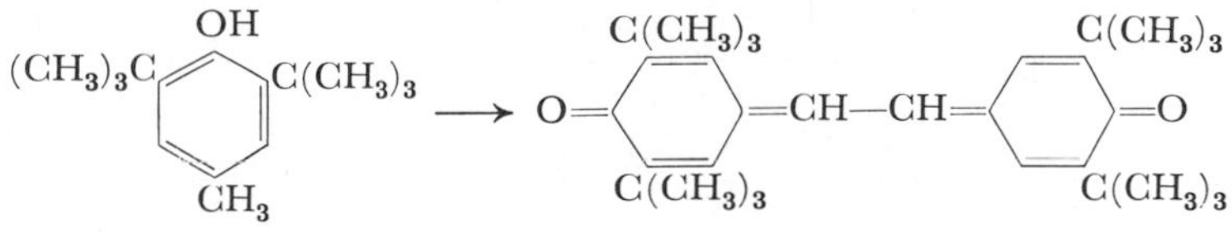

300 ml. of water treated with 0.55 mole of $K_3Fe(CN)_6$ and allowed to stand for 24 hr.) [23, 25].

(2) **3,3′,5,5′-Tetra-*t*-butyldiphenoquinone** (92% from 10 mmoles of 2,6-di-*t*-butyl-4-bromophenol stirred with 20 g. of active MnO_2 in 175 ml. of benzene for 35 min.) [26].

(3) **4,5-Diacetyl-3-aminophenoxazone** (80% from the aminophenol and the theoretical amount of ferricyanide) [27]:

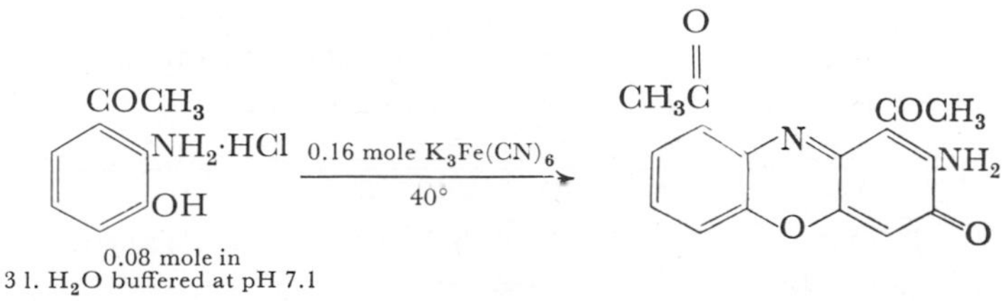

4. From o- or p-Dihydroxybenzenes, Diaminobenzenes, and Related Types

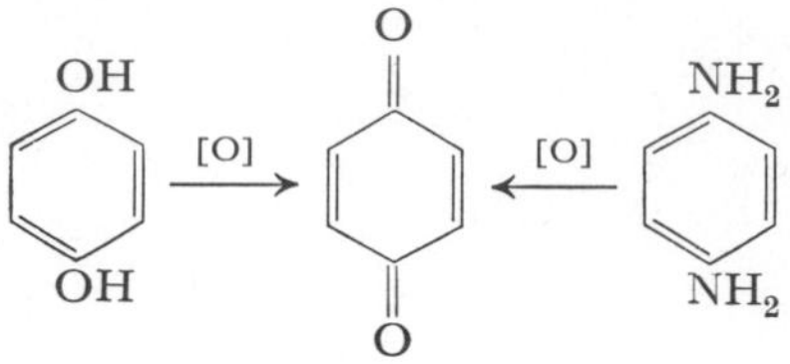

A great variety of oxidizing agents has been employed in converting the o- or p-dihydroxybenzenes or diaminobenzenes into quinones [28]. For the p-dihydroxy type, ferric chloride [29, 30], sodium hypobromite in aqueous alcoholic sulfuric acid [31], sodium bromate in acetic acid [32], lead dioxide in aqueous alcoholic hydrogen chloride [33], nitric acid in ether [34], lead tetraacetate in glacial acetic acid [35], hydrogen peroxide in aqueous sodium hydroxide [36], lead dioxide [37], silver oxide, sodium sulfate, and anhydrous ether [38], mercuric oxide [39], nitrogen oxides [40], ferric sulfate and hydrochloric acid [41], sodium nitrite in aqueous acetic acid [42], sodium chlorate, aqueous sulfuric acid, and a trace of vanadium pentoxide [9], and sodium dichromate and sulfuric acid [43] have all been employed. For the conversion of α-tocopherol into α-tocopherylquinone, an aqueous solution of gold chloride is a specific oxidizing agent [44]:

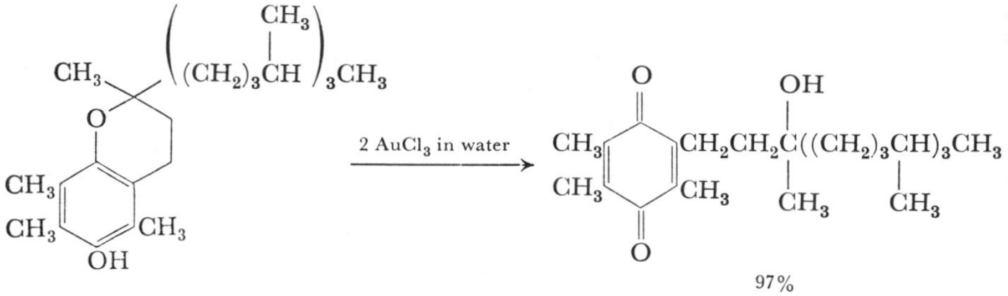

97%

Of all the agents listed for the oxidation of hydroquinones, perhaps the most desirable from the standpoint of product isolation is the nitrogen oxide, N_2O_4 (see Ex. b.3). For the o-dihydroxy type [45], sodium iodate in aqueous alcohol [46] and silver oxide and sodium sulfate in dry ether [47] have been utilized. For the o-methoxyphenolic types, sodium periodate has been employed [48].

p-Aminophenols or their N-acetyl derivatives, other starting materials for the synthesis of p-quinones, have been oxidized with sodium or potassium dichromate in sulfuric acid [49–51], with a mixture of 70% nitric and glacial acetic acids [52], with nitrous acid [53], with ferric sulfate [54], and with sodium dichromate in acetic acid [55]. The o-aminophenols have been converted into quinones by nitric acid [56, 57], by ferric chloride in aqueous hydrochloric acid

[58], and by sodium nitrite and sulfuric acid in a sodium acetate solution [59]. Other disubstituted benzenes such as p-bromophenols [60], p-nitrosophenols [61], and p-diaminobenzenes [62] have also been oxidized to quinones.

The great variety of oxidizing agents employed suggests that no one is satisfactory in all cases. The one selected will be determined largely by the structure of the compound to be oxidized. It is difficult to generalize, but perhaps a few statements may be in order. Sensitive types, such as some of the o-quinones, are obtained best in an anhydrous solvent such as ether or benzene, with an oxidizing agent such as silver oxide and a drying agent to absorb any water formed in the reaction [63]. p-Aminophenols are usually oxidized by sodium or potassium dichromate or manganese dioxide in dilute sulfuric acid [34]. However, ferric chloride and the sulfate have been used to oxidize aminophenols as well as quinols. Tri- and tetrasubstituted quinones can be prepared from the appropriate quinol by oxidation with concentrated nitric acid in ether at low temperature. Under the proper experimental conditions most quinones may be obtained in a satisfactory yield.

The oxidation of a p-dihydroxybenzene appears to occur through the semiquinone 12-VII, a radical anion of moderate stability in basic medium because

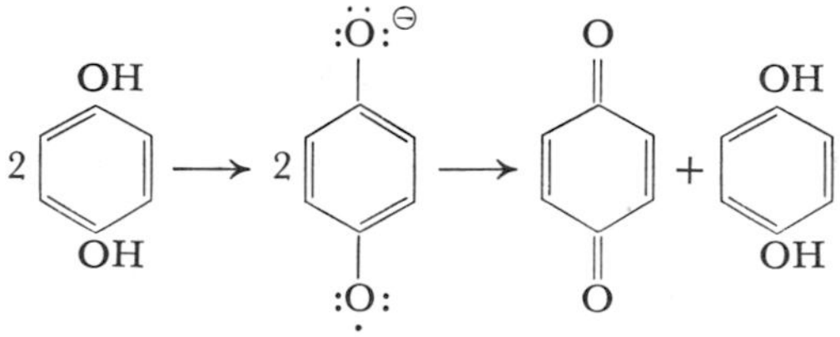

12-VII

of contributions to the resonance hybrid of many canonical forms. On acidification, quinone and p-dihydroxybenzene are formed sometimes as a π-molecular complex [64].

Quinone diimines

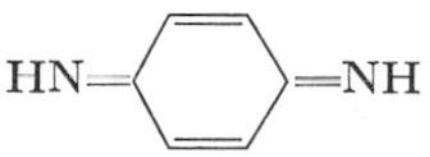

are relatively unstable in comparison with the corresponding quinones, but they can be prepared in nonaqueous solvents from diamines or hydroxyamines using silver oxide oxidation [65]. On the other hand, quinone disulfonimides

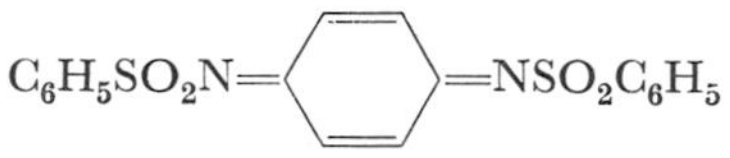

are quite stable and readily prepared by lead tetraacetate oxidation (see Ex. b.7). Simple quinone methides are unstable but may be intermediates in further conversions [66]:

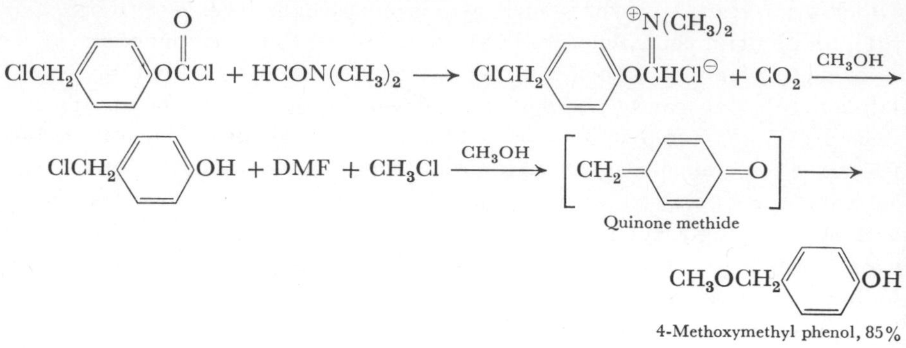

The *p*-chloromethylphenol may be isolated or carried on without isolation by using an excess of methanol. More highly substituted quinone methides are stable (see Ex. b.8).

(a) Preparation of *p*-benzoquinone (86–92 % from hydroquinone, sodium dichromate and sulfuric acid) [43]; *see also* [10].

(b) Other examples

(1) **Ammonium 1,2-naphthoquinone-4-sulfonate** (94–98 % from 1-amino-2-hydroxy-4-naphthalene sulfonic acid and aqueous nitric acid) [57].

(2) **Duroquinone (2,3,5,6-tetramethyl-*p*-benzoquinone)** (90 % from the addition compound of stannic chloride and 2,3,5,6-tetramethyl *p*-phenylene-diamine hydrochloride, $(C_6(CH_3)_4(NH_2\cdot HCl)_2)_2\cdot SnCl_4$, and ferric chloride in aqueous hydrochloric acid) [62].

(3) **2,5-Dimethylbenzoquinone** (97 % from 20 g. of the hydroquinone in 300 ml. of CCl_4 held and stirred at 25° while 6 ml. of N_2O_4 was added drop-wise from a pipet in 5 min. and stirred for 5 min. longer; the quinone was filtered off and recrystallized from a mixture of chloroform and benzene; the N_2O_4, b.p. 20–30°, was prepared by distillation of a mixture of 33 ml. of sulfuric acid, 83 ml. of 98 % nitric acid, and 100 g. of arsenious oxide; yields of quinone, with the exception of benzoquinone, were in general very good) [67].

(4) **1,4-Naphthoquinone** (78–81 % from 1,4-aminonaphthol hydro-chloride and potassium dichromate in sulfuric acid) [49].

(5) **1,4-Naphthoquinone** (ca. 75 % overall yield from butadiene and

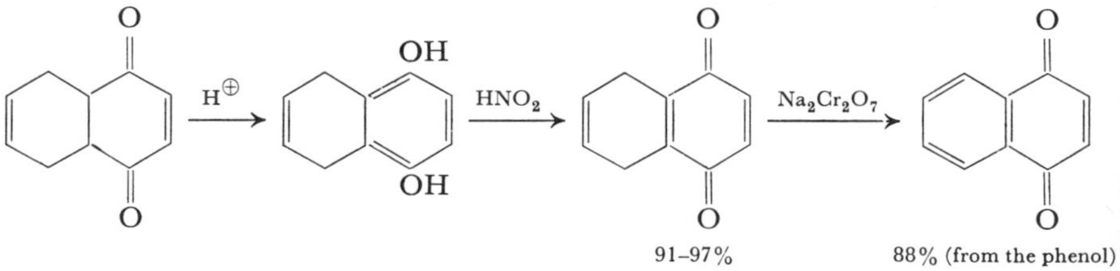

91–97 %   88% (from the phenol)

benzoquinone followed by oxidation of the Diels-Alder adduct); this prepar-ation is included to illustrate methods of proceeding from one benzoquinone to a polyclic quinone via the Diels-Alder adduct; the reference given is quite detailed in describing the purity and irritating properties of the product [68].

(6) **2,3',5',6-Tetra-*t*-butylindophenol** (60–75 % from 4-amino-2,6-di-*t*-butylphenol heated to form the diphenylamine which was then oxidized by

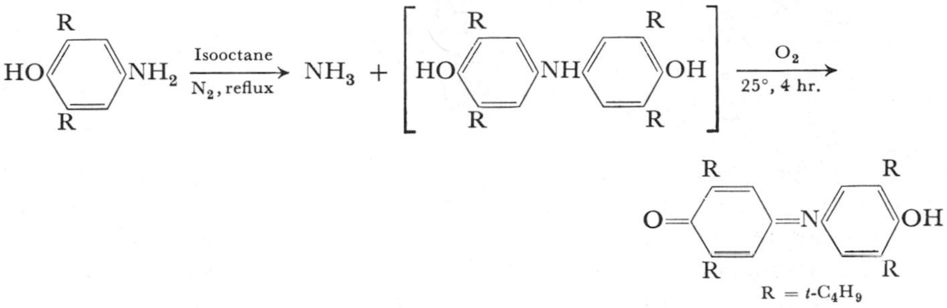

$$R = t\text{-}C_4H_9$$

air; elimination of ammonia from *p*- or *o*-hydroxyanilines is much easier than from anilines) [69].

(7) **4-Methyl-*o*-quinonedibenzenesulfonimide** (67 % from 9 g. of the disulfonamide and 10 g. of lead tetraacetate in acetic acid for 30 min. at 25°;

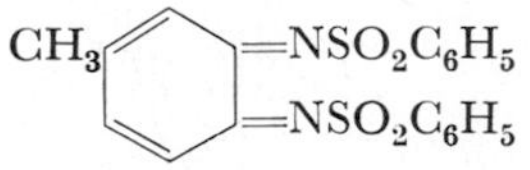

a few drops of glycerol were added to destroy the excess of oxidizing agent, and the product separated as orange-yellow crystals essentially in a pure state) [70].

(8) **2,6-Dibromo-4-cycloheptatrienylidinebenzoquinone** (37 % overall yield by the accompanying reactions; the quinonemethide is deep purple in color) [71].

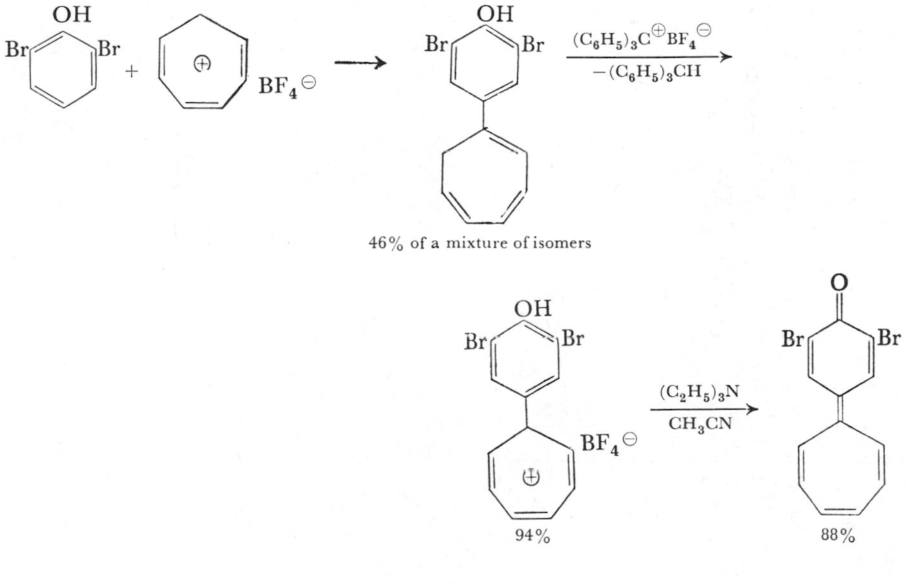

46% of a mixture of isomers

94%   88%

5. From o-Quinones

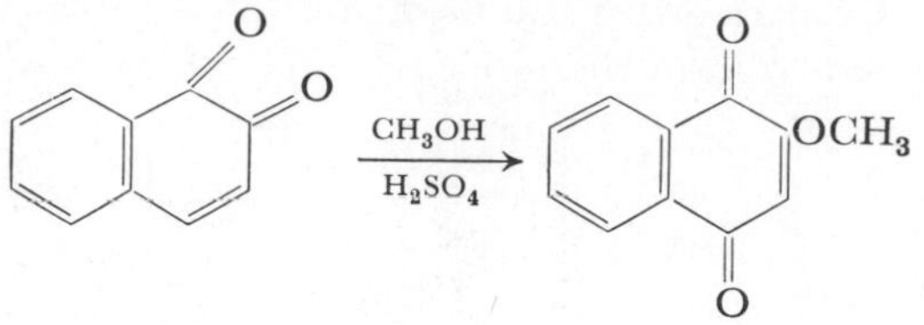

The oxidation of o-benzoquinones to hydroxy (or methoxy) p-benzoquinones is seldom used as a method of synthesis. It is sufficiently common, however, to mention several examples. 2-Hydroxy-3-methylnaphthoquinone-1,4 results from 3-methylnaphthoquinone-1,2 by treatment with selenium dioxide [72]. In fact, both of these quinones are obtained in low yield from 3-methyltetralone-1 by oxidation with selenium dioxide:

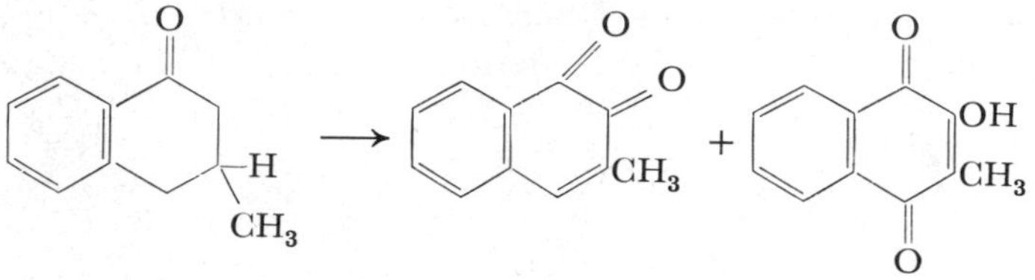

Similarly ammonium 1,2-naphthoquinone-4-sulfonate may be converted into 2-hydroxy-1,4-naphthoquinone (see Ex. a).

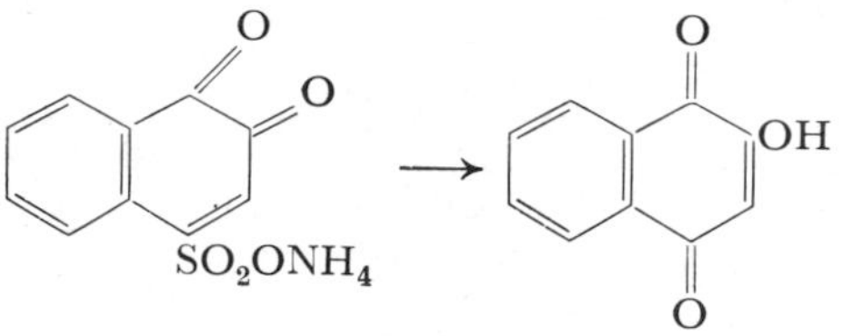

Finally, o-quinones, available from catechols or 1,2-dihydroxynaphthalenes and chloranil, yield p-quinones in alcoholic solution by treatment with a trace of concentrated sulfuric acid [73]. In this manner, 3-methoxycatechol gives a 71 % yield of 2,6-dimethoxy-p-quinone:

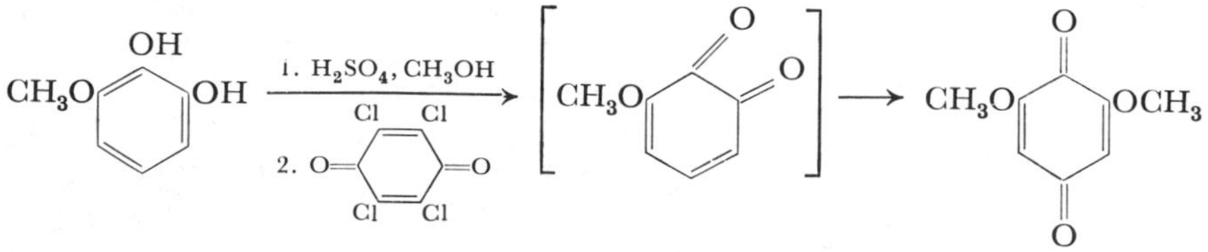

The o-quinone is a likely intermediate since 3-methoxy-o-quinone in methanol and a trace of concentrated sulfuric acid gave about a 33% yield of the 2,6-dimethoxy-p-quinone. Evidently the transformation is favored by the greater stability of the p- as compared to the o-quinone. A possible pathway is suggested:

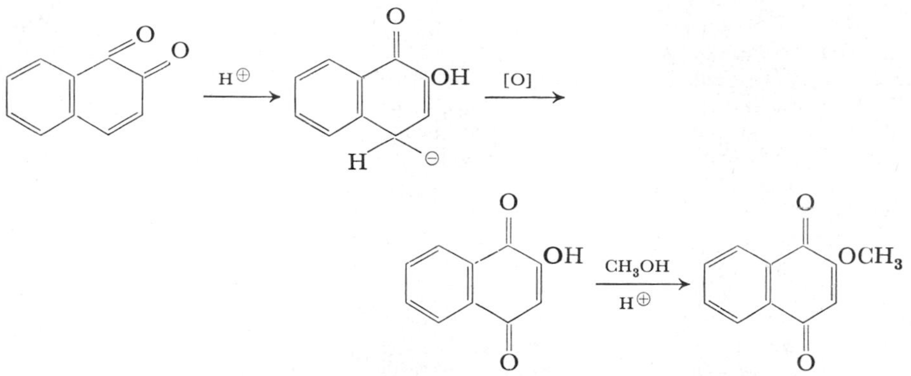

(a) Preparation of 2-hydroxy-1,4-napthoquinone (58–65% from ammonium 1,2-naphthoquinone-4-sulfonate and conc. sulfuric acid in methanol) [74].

1. R. H. Thomson, *Naturally Occurring Quinones*, Butterworths Scientific Publications, London, 1957.
2. L. F. and M. Fieser, *Advanced Organic Chemistry*, Reinhold Publishing Corp., New York, 1961, Chap. 26.
3. L. F. and M. Fieser, *Topics in Organic Chemistry*, Reinhold Publishing Corp., New York, 1963, Chaps. 1 and 9.
4. L. F. Fieser, *J. Biol. Chem.*, **133**, 391 (1940).
5. P. Lambert and R. H. Martin, *Bull. Soc. Chim. Belges*, **61**, 124 (1952).
6. C. F. H. Allen and J. A. Van Allan, *Org. Syn.*, Coll. Vol. **3**, 1 (1955).
7. R. Wendland and J. LaLonde *Org. Syn.*, Coll. Vol. **4**, 757 (1963).
8. J. Hyman and C. F. Peter, U.S. Patent, 2,402,226, June 18, 1946; *C.A.*, **40**, 610 (1946); L. Li and W. H. Elliott, *J. Am. Chem. Soc.*, **74**, 4089 (1952).
9. R. T. Arnold and R. Lawson, *J. Org. Chem.*, **5**, 250 (1940).
10. H. W. Underwood, Jr., and W. L. Walsh, *Org. Syn.*, Coll. Vol. **2**, 553 (1943).
11. A. J. Fatiadi, *Chem. Commun.* 1087 (1967).
12. H. E. Fierz-David and L. Blangey, *Fundamental Processes of Dye Chemistry*, Interscience Publishers, New York, 1949, p. 224.
13. F. Farina *et al.*, *Tetrahedron Letters*, **No. 19**, 9 (1959).
14. J. R. Thirtle, in R. E. Kirk and D. F. Othmer, *Encyclopedia of Chemical Technology*, Vol. 16, 1968, p. 907.
15. H.-J. Teuber and W. Rau, *Chem. Ber.*, **86**, 1036 (1953).
16. H.-J. Teuber and M. Hasselbach, *Chem. Ber.*, **92**, 674 (1959).
17. K.-H. König *et al.*, *Chem. Ber.*, **93**, 554 (1960).
18. D. Bryce-Smith and A. Gilbert, *J. Chem. Soc.*, 873 (1964).
19. D. Schulte-Frohlinde and F. Erhardt, *Ann. Chem.*, **671**, 92 (1964).
20. Ref. 2, p. 846.
21. Ref. 12, p. 146.
22. W. Brackman and E. Havinga, *Rec. Trav. Chim.*, **74**, 937 (1955).
23. H. Musso, *Angew. Chem.*, **75**, 965 (1963).

24. C. G. Haynes *et al.*, *J. Chem. Soc.*, 2823 (1956).
25. C. D. Cook *et al.*, *J. Am. Chem. Soc.*, **77,** 1783 (1955).
26. H.-D. Becker, *J. Org. Chem.*, **29,** 3068 (1964).
27. A. Butenandt *et al.*, *Ann. Chem.*, **588,** 106 (1954).
28. J. Cason, *Org. Reactions*, **4,** 305 (1948).
29. M. M. Shemyakin *et al.*, *J. Gen. Chem. USSR*, **29,** 1802 (1959).
30. W. Alcalay, *Helv. Chim. Acta*, **30,** 578 (1947).
31. V. A. Zagorevskiĭ and D. A. Zykov, *Zh. Prikl. Khim.*, **32,** 2815 (1959); *C.A.*, **54,** 9818 (1960).
32. P. D. Bartlett *et al.*, *J. Am. Chem. Soc.*, **72,** 1003 (1950).
33. P. W. D. Mitchell, *Canad. J. Chem.*, **41,** 550 (1963).
34. M. F. Ansell *et al.*, *J. Chem. Soc.*, 3028 (1963).
35. H. Brockmann and W. Müller, *Chem. Ber.*, **91,** 1920 (1958).
36. T. R. Seshadri and G. B. Venkatasubramanian, *J. Chem. Soc.*, 1660 (1959).
37. R. Kuhn and I. Hammer, *Chem. Ber.*, **83,** 413 (1950).
38. R. J. S. Beer *et al.*, *J. Chem. Soc.*, 2029 (1951).
39. G. Jacini and T. Bacchetti, *Gazz. Chim. Ital.*, **80,** 757 (1951); *C.A.*, **46,** 4500 (1952).
40. R. Pummerer *et al.*, *Chem. Ber.*, **75,** 1976 (1942).
41. L. I. Smith and P. F. Wiley, *J. Am. Chem. Soc.*, **68,** 894 (1946).
42. L. F. Fieser, *J. Am. Chem. Soc.*, **70,** 3165 (1948).
43. E. B. Vliet, *Org. Syn.*, Coll. Vol. **1,** 482 (1941).
44. P. Karrer *et al.*, *Helv. Chim. Acta*, **21,** 939 (1938); **23,** 455 (1940).
45. W. M. Horspool, *Quart. Rev.* **23,** 204 (1969).
46. J. M. Bruce, *J. Chem. Soc.*, 2366 (1959).
47. R. Willstätter and A. Pfannenstiel, *Chem. Ber.*, **37,** 4744 (1904).
48. H.-D. Becker, *J. Org. Chem.*, **34,** 2026 (1969).
49. L. F. Fieser, *Org. Syn.*, Coll. Vol. **1,** 383 (1941).
50. J. Cason *et al. J. Org. Chem.*, **13,** 403 (1948).
51. R. E. Harman, *Org. Syn.*, Coll. Vol. **4,** 148 (1963).
52. D. J. Cram, *J. Am. Chem. Soc.*, **71,** 3953 (1949); M. G. Ettlinger, *J. Am. Chem. Soc.*, **72,** 3666 (1950).
53. E. Kremers *et al.*, *Org. Syn.*, Coll. Vol. **1,** 511 (1941).
54. L. I. Smith and F. L. Austin, *J. Am. Chem. Soc.*, **64,** 528 (1942).
55. W. M. McLamore, *J. Am. Chem. Soc.*, **73,** 2221 (1951).
56. W. Langenbeck *et al.*, *Chem. Ber.*, **87,** 496 (1954).
57. E. L. Martin and L. F. Fieser, *Org. Syn.*, Coll. Vol. **3,** 633 (1955).
58. L. F. Fieser, *Org. Syn.*, Coll. Vol. **2,** 430 (1943); M. Gates, *J. Am. Chem. Soc.*, **72,** 228 (1950).
59. H. H. Hodgson and S. Birtwell, *J. Chem. Soc.*, 539 (1944).
60. J. M. Lyons and R. H. Thomson, *J. Chem. Soc.*, 2910 (1953).
61. W. M. McLamore, *J. Am. Chem. Soc.*, **73,** 2225 (1951).
62. L. I. Smith, *Org. Syn.*, Coll. Vol. **2,** 254 (1943).
63. J. Cason, *Org. Reactions*, **4,** 305 (1948).
64. D. J. Cram and G. S. Hammond, *Organic Chemistry*, 2nd ed., McGraw-Hill Book Co., New York, 1964, p. 519.
65. E. H. Rodd, *Chemistry of Carbon Compounds*, Vol. 3, Elsevier Publishing Co., New York, 1956, Pt. B, p. 715.
66. V. A. Pattison *et al.*, *J. Org. Chem.*, **33,** 1084 (1968).
67. A. G. Brook, *J. Chem. Soc.*, 5040 (1952).
68. L. F. and M. Fieser, *Reagents for Organic Synthesis*, John Wiley and Sons, New York, 1967, p. 714.
69. G. M. Coppinger, *Tetrahedron*, **18,** 61 (1962).
70. R. Adams and C. N. Winnick, *J. Am. Chem. Soc.*, **73,** 5687 (1951).
71. J. J. Looker, *J. Org. Chem.*, **32,** 2941 (1967).
72. F. Weygand and K. Schröder, *Chem. Ber.*, **74,** 1844 (1941).
73. L. Horner and S. Göwecke, *Chem. Ber.*, **94,** 1291 (1961).
74. L. F. Fieser and E. L. Martin, *Org. Syn.*, Coll. Vol. **3,** 465 (1955).

B. Electrophilic Reactions

1. From o-Aroylbenzoic Acids

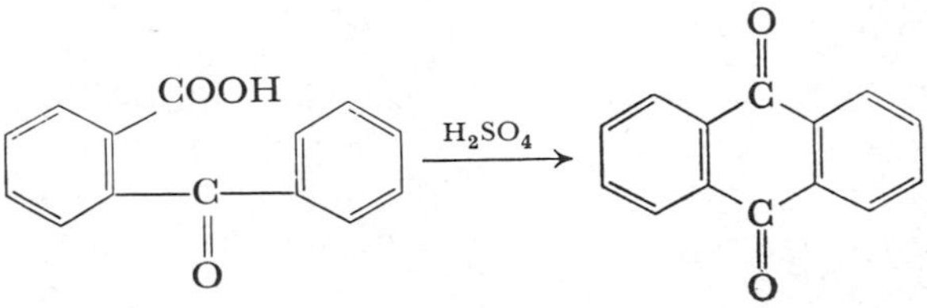

The intramolecular condensation of *o*-aroylbenzoic acids produces quinones. Common dehydrating agents are fuming sulfuric acid [1], benzoyl chloride and concentrated sulfuric acid [2], benzoyl chloride and zinc chloride [3], and polyphosphoric acid [4]. Yields with PPA are sufficiently quantitative to carry out kinetics studies; the half-life for ring closure of *o*-benzoylbenzoic acid in PPA (81.6% P_2O_5 content) at 80° was found to be 85 minutes, and that for *p*-chloro-*o*-benzoylbenzoic acid to be 8500 minutes [5]. These results are given to emphasize the point that lower temperatures (and longer reaction times) than usually recommended for PPA ring closures are desirable. Sulfonation occurs in some ring closures with sulfuric acid, and the addition of boric acid has been recommended as an inhibitor of this side reaction [6]. Since the quinones are readily reduced to hydrocarbons, the synthesis has been of value in preparing polycyclic hydrocarbons [2, 3].

Occasionally an *o*-benzoylbenzoic acid resists ring closure; in this case the corresponding *o*-benzylbenzoic acid may be used [7]:

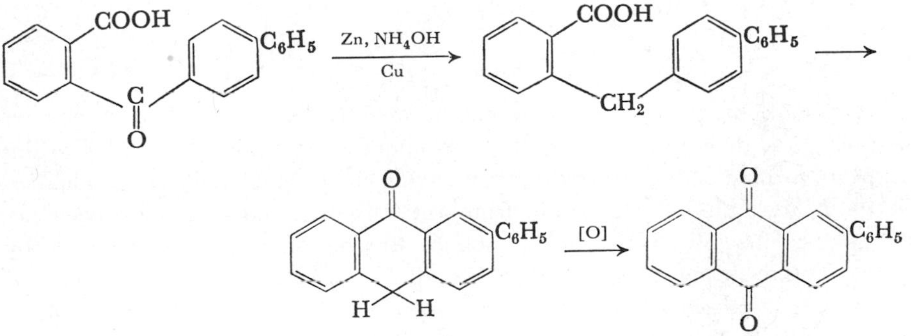

2-Phenyl-9,10-anthraquinone

Perhaps, in these cases, polymerization of the keto acid via intermolecular acylation of the 4-phenyl group competes with ring closure to make the more devious path as shown the more attractive one.

References for Section B are on p. 739.

(a) Preparation of β-methylanthraquinone (81–90% from *p*-toluyl-*o*-benzoic acid and fuming sulfuric acid) [1].

(b) Other preparations

(1) 2′,3′,3,4-Naphthopentaphene-1′,4′,8,13-diquinone, (12-IX) (65%

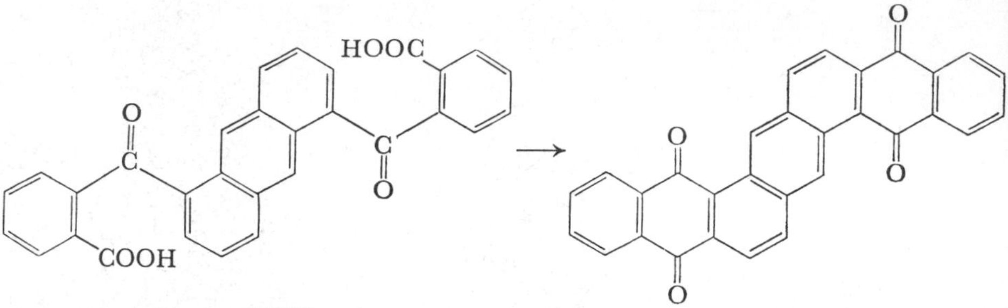

12-VIII   12-IX

from 1,5-bis-(*o*-carboxybenzoyl) anthracene (12-VIII), benzoyl chloride and zinc chloride in nitrobenzene) [3].

(2) Anthraquinone (quantitative from *o*-benzoylbenzoic acid in PPA at 140–150° for 40 min.) [8].

2. From Phenols and Nitrous Acid

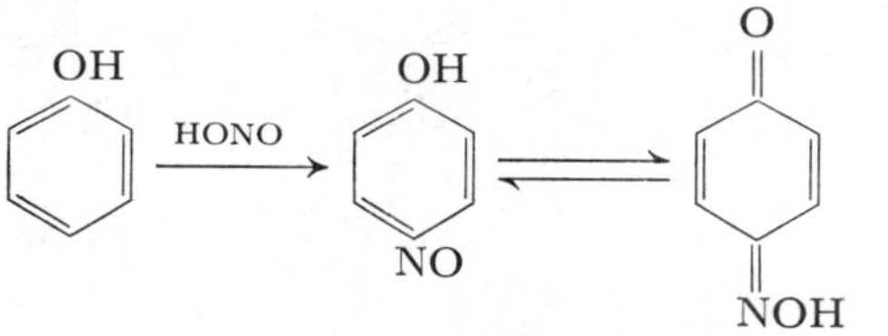

It appears that the nitrosophenol, first formed by the action of nitrous acid on a phenol, is in equilibrium with the quinone monoxime. This equilibrium appears to exist in transformations involving nitrosophenol [9, 10]. The isolation of the quinone monoxime from the mixture seems to have been accomplished first by Hodgson [11]. Actually the monoxime is best prepared by nitrosation in strong sulfuric acid solution [12]. The type may also be prepared in low yield (but with a simple isolation procedure) by oxidation of a mixture of an aromatic hydrocarbon and hydroxylamine (see Ex. a).

The nitrosophenols are useful in preparing indophenols which are substituted quinone monoximes [13] (see Ex. b):

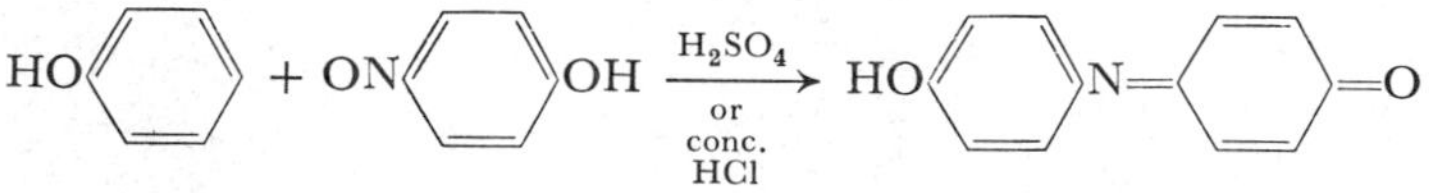

This reaction is the basis of the Liebermann nitroso test for a phenol or a nitrite, the deep blue color of the indophenol being capable of detection at very low concentration.

(a) Preparation of o-nitrosophenol (o-benzoquinone monoxime). To a stirred, heterogeneous mixture of 2 g. of sodium pentacyano-ammine-ferroate in 100 ml. of water, 25 ml. of benzene, and 50 ml. of ligroin there was added 2 g. of hydroxylamine hydrochloride followed by the addition of 4 ml. of Merck Superoxol. The nitrosophenol in low yield dissolved with a deep green color in the ligroin layer from which it was separated and dried [14]. Yields with a wide variety of substrates were higher if the hydrocarbon was omitted and replaced by a stoichiometric amount of the corresponding phenol [15].

(b) Preparation of the indophenol of carbazole and nitrosophenol. Carbazole, 440 lb. in 5330 lb. of sulfuric acid at 0°, was added carefully to 330

lb. of p-nitrosophenol (88%) in 2360 lb. of sulfuric acid held at −20 to −23° and then agitated for 1 hr. The mixture was poured into 12 tons of water and 5 tons of ice containing 293 lb. of iron powder, and the diluted mass was allowed to come to room temperature, at which time the precipitate was filtered, washed, and dried to yield 700 lb. of the desired product [10].

1. L. F. Fieser, *Org. Syn.*, Coll. Vol. **1**, 353 (1941).
2. E. Clar, *Chem. Ber.*, **81**, 63 (1948); *J. Chem. Soc.*, 2440 (1949).
3. E. Clar, *Chem. Ber.*, **81**, 169 (1948).
4. F. Uhlig and H. R. Snyder, *Advan. Org. Chem.*, **1**, 51 (1960).
5. R. G. Downing and D. E. Pearson, *J. Am. Chem. Soc.*, **84**, 4956 (1962).
6. P. H. Groggins and H. P. Newton, *Ind. Eng. Chem.*, **22**, 157 (1930).
7. H. A. Lubs, *The Chemistry of Synthetic Dyes and Pigments*, Reinhold Publishing Corp., New York, 1955, p. 342.
8. II. R. Snyder and F. X. Werber, *J. Am. Chem. Soc.*, **72**, 2965 (1950).
9. L. F. and M. Fieser, *Advanced Organic Chemistry*, Reinhold Publishing Corp., New York, 1961, p. 854.
10. R. E. Kirk and D. F. Othmer, *Encyclopedia of Chemical Technology*, Vol. 13, Interscience Encyclopedia, New York, 1954, p. 456.
11. H. H. Hodgson and F. H. Moore, *J. Chem. Soc.*, **123**, 2499 (1923).
12. H. Kraaijeveld and E. Havinga, *Rec. Trav. Chim.*, **73**, 537 (1954).
13. E. H. Rodd, *Chemistry of Carbon Compounds*, Vol. 3, Elsevier Publishing Co., New York, 1956, Pt. B, p. 720.
14. O. Baudisch, *J. Am. Chem. Soc.*, **63**, 622 (1941).
15. G. Cronheim, *J. Org. Chem.*, **12**, 1 (1947).

C. Condensation Reactions

1. From Aromatic o-Dialdehydes and 1,4-Cyclohexanedione

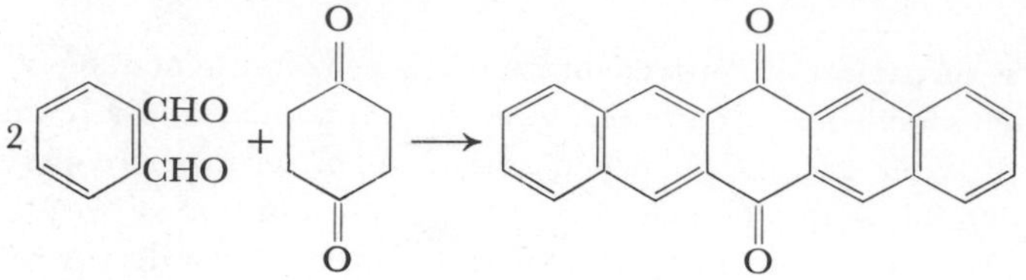

This synthesis proceeds in good yield when 2 moles of the dialdehyde is added to 1 mole of 1,4-cyclohexanedione in alkaline solution. The reaction appears to have been attempted only with o-phthalaldehyde [1] and naphthalene-2,3-dialdehyde [2].

(a) **Preparation of heptacene-7, 16-quinone.** Naphthalene-2,3-dialdehyde and 1,4-cyclohexanedione in alcoholic solution in the presence of a

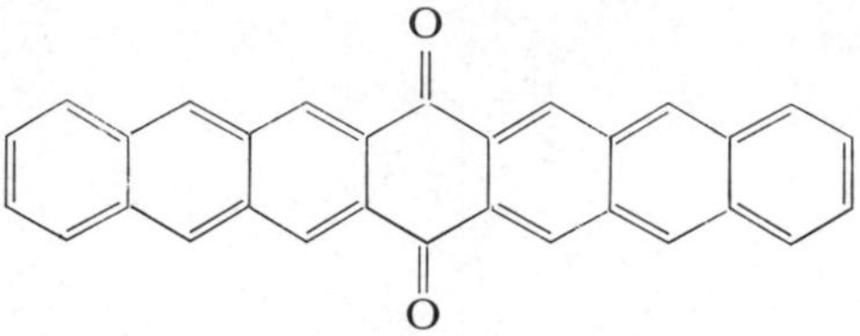

few drops of 5% aqueous potassium hydroxide reacted almost immediately to give a quantitative yield of the quinone. After warming for a short time at 50°, the crude product was removed by filtration. The product is difficult to purify; perhaps the least loss was obtained by continuous extraction with dimethylformamide in a Soxhlet extractor. The melting point of the quinone is indefinite; it sublimes by heating over 400° with considerable decomposition [2].

2. From Aromatic o-Dialdehydes and Glyoxal

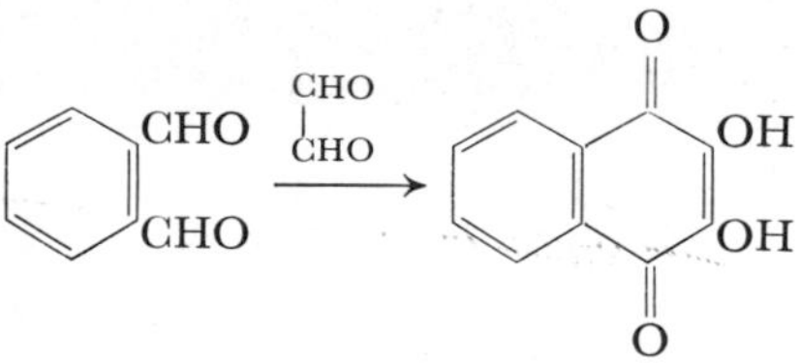

Dihydroxy-p-quinones have been synthesized from aromatic o-dialdehydes and glyoxal in the presence of the cyanide ion and air in a weakly alkaline medium [3, 4]. Yields are usually low, but sometimes they are as high as 60% or above. The reaction appears to involve a double benzoin condensation in which

References for Section C are on p. 743.

the diketodihydroxytetrahydronaphthalene 12-X rearranges to the tetrahydroxy-naphthalene 12-XI which in the presence of air oxidizes to the dihydroxy-quinone 12-XII.

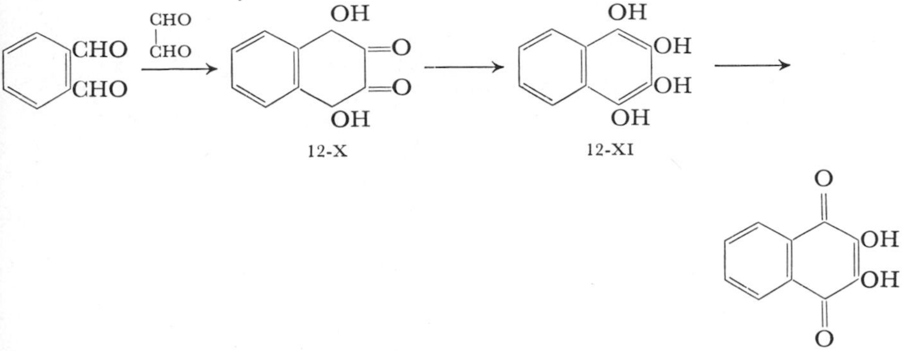

12-X   12-XI

12-XII,
Isonaphthazarin

(a) **Preparation of isonaphthazarin (2,3-dihydroxynaphthoqui-none-1,4).** o-Phthalaldehyde, 1.4 g. in 10 ml. of dioxane, was treated with 3 g. of glyoxal-sodium bisulfite, 0.5 g. of potassium cyanide, and 50 ml. of 2 N sodium carbonate solution. Stirring at 20° in the air produced a deep violet color, and after 15 min. the solution was acidified with conc. hydrochloric acid to produce 1.2 g. of isonaphthazarin (60 %) as a red powder which after washing and drying melted at 287° [3].

3. From 1,4-Dihydroxynaphthalene-3-carboxaldehydes

The construction of naphthoquinones with unsaturated side chains is of interest because of their relationship in structure to vitamin K_1. The unsaturated side chain may be introduced by three methods.

(a) Friedel-Crafts type of alkylation with allylic alcohols [5]:

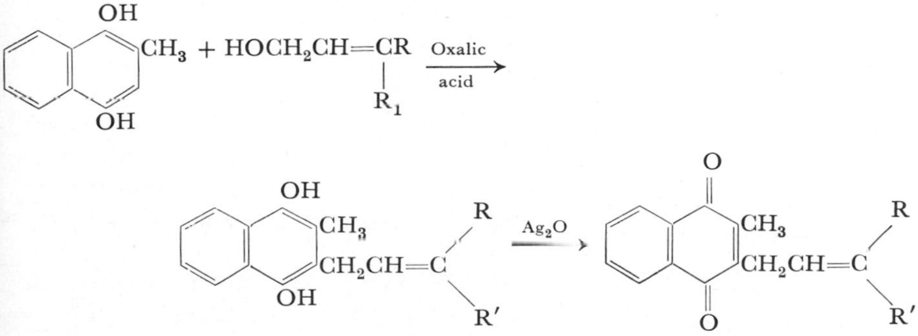

(b) Vinylation of 2-hydroxy-1,4-naphthoquinones [6].   This reaction does not succeed, however, without the presence of the 2-hydroxy group. A

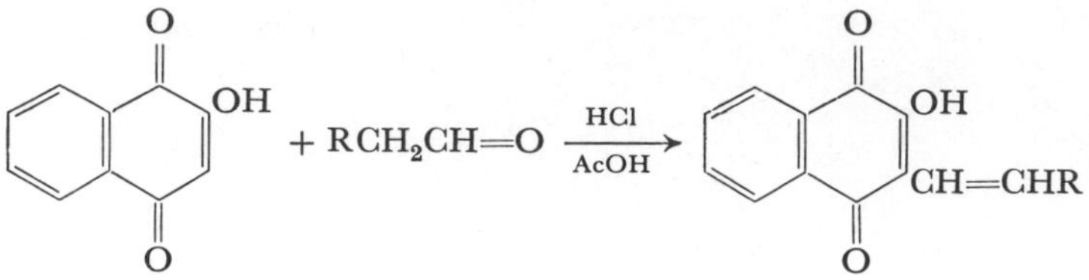

more general route involving a novel protecting group is now available as shown in method c.

(c) From   carboxaldehydes   and   phosphoranes   (Wittig)   [7].

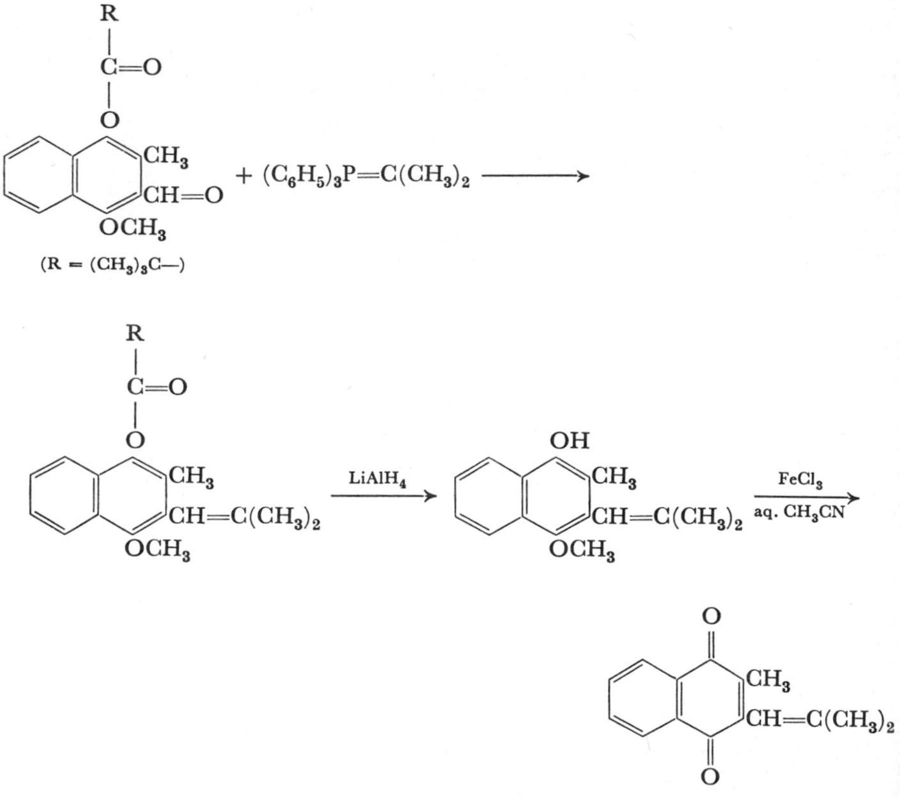

$(R = (CH_3)_3C-)$

2-Methyl-3-(2-methyl-1-propenyl)-
1,4-naphthoquinone,
88% overall from the aldehyde

Furthermore, the naphthalene phosphorane may be used if the carbonyl compound is available:

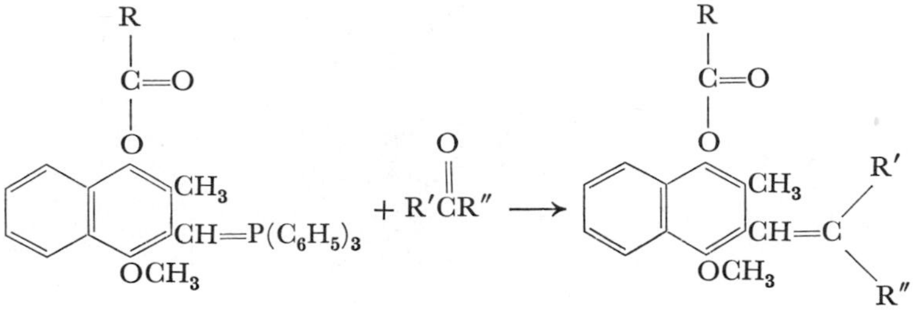

1. W. Ried and F. Anthöfer, *Angew Chem.*, **65,** 601 (1953).
2. W. Ried and F. Anthöfer, *Angew. Chem.*, **66,** 604 (1954).
3. F. Weygand et al., *Chem. Ber.*, **75B,** 625 (1942).
4. F. Weygand et al., *Chem. Ber.*, **76,** 818 (1943); **80,** 391 (1947); *Chem. Ind. (London),* 106 (1954).
5. L. F. and M. Fieser, *Reagents for Organic Synthesis,* John Wiley and Sons, New York, 1967, p. 765.
6. S. C. Hooker, *J. Am. Chem. Soc.*, **58,** 1163 (1936).
7. H. Rapoport et al., *J. Org. Chem.*, **33,** 4351 (1968).

Chapter 13

CARBOXYLIC ACIDS

Since the carboxylic acids, aside from carbonic acid, occupy the highest oxidation level, a great many syntheses involve oxidation (B). However, the hydrolysis of acid derivatives in the same oxidation level (A) is sometimes preferred. To complement the preparation of acids by oxidation, reduction of carbon dioxide under the title, carbonation of organometallic compounds, is discussed under (C). Although all types of carboxylic acid syntheses could be classified under these three headings, it is more helpful to separate methods involving profound change into more descriptive titles. Thus the classifications of condensation (D), alkali cleavage (E), electrophilic substitution and additions (F) and rearrangements (G) have been added.

A. Hydrolysis

The acid derivatives may be arranged in a decreasing order of reactivity as shown:

acid halide > acid anhydride > ester > amide (nitrile) > salt

Only the first two are hydrolyzed spontaneously and completely in the majority of cases, depending on the solubility in the aqueous medium. The rest of the

References for Section A are on pp. 758–760.

derivatives react slowly with water and approach an equilibrium state. For complete reaction it is best to saponify to the sodium salts and recover the free acid by acidification. Perhaps the most powerful reagent to split recalcitrant esters and other acid derivatives is potassium *t*-butoxide (A.2). In converting one derivative into another it is more feasible to proceed from the more reactive to the less reactive.

Occasionally devious means, such as diazotization of amides (A.3) or oxidation of thiolesters (A.2), are necessary to accomplish hydrolyses but, since the overall objective is conversion of the acid derivative into the acid, these methods are included under hydrolysis.

Indirect methods of obtaining acids are not completely listed in the chapter. For example, many unusual acids are prepared by the Diels-Alder reaction in which the carboxyl group is carried in the dienophile (Alkenes, Chapter 2, C.2). In addition, acids may be obtained by hydrolysis or saponification of some acid derivatives prepared in an unusual way. For these syntheses, Chapters 14 through 19 should be consulted.

1. From Acid Halides or Anhydrides

$$RCOX \xrightarrow{H_2O} RCOOH$$

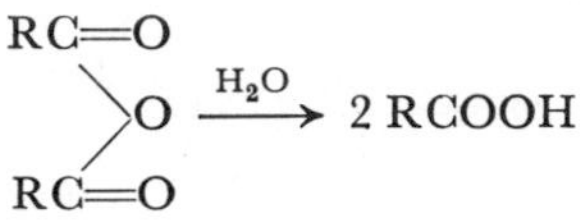

$$\begin{matrix} RC{=}O \\ \diagdown \\ O \xrightarrow{H_2O} 2\ RCOOH \\ \diagup \\ RC{=}O \end{matrix}$$

Although these methods are very satisfactory, they are not widely used because the acid halide or anhydride is usually prepared from the acid. The mechanism in both cases usually consists of an SN_2 attack on the positively polarized carbon of the carbonyl group:

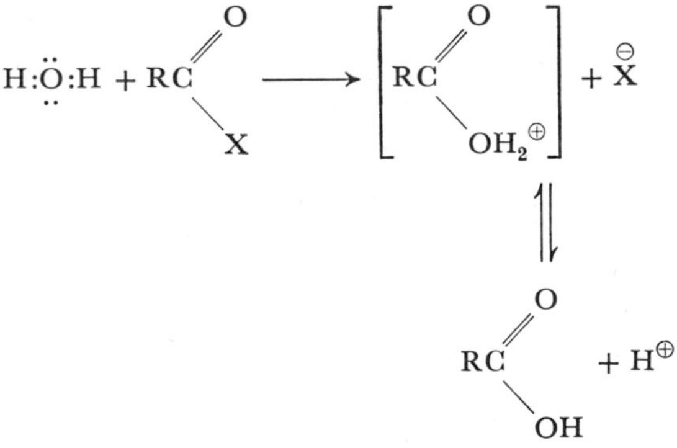

In the SN_2 mechanism it is implied that the order of activity in hydrolysis or other nucleophilic attack is RCOF > RCOCl > RCOBr > RCOI, but this conjecture has not been tested among a wide enough variety of acids for one to have complete trust in its validity. It is implied further that electron-withdrawing groups attached to the acid chloride or anhydride will increase the rate of hydrolysis, but the matter is complex as SN_1 character may dominate SN_2 at a certain point when substituent influences will be reversed [1]. Superimposed on these electrical effects are steric influences. Newman's "rule of six" is helpful in this regard [2]. A last consideration in determining the ease of hydrolysis of acid chlorides or anhydrides is water solubility. For instance, acetyl chloride reacts violently with water, but hexadecanoyl chloride reacts so slowly that its exothermicity with water is barely detectable. Both acyl groups are of about comparable activity so that the hydrolysis of hexadecanoyl chloride is probably controlled by diffusion transport into the water phase rather than by activity. To be more specific about activity, several classes of activities are given:

1. Active acid chlorides or anhydrides. Aliphatic types, oxalyl chloride, acid chlorides with electron-withdrawing groups attached, such as 3,5-dinitrobenzoyl chloride, phthalic anhydride, or other 5-membered (or less) cyclic anhydrides, and some mixed anhydrides, such as acetic formic anhydride or acetic trifluoroacetic anhydride.

2. Acid chlorides or anhydrides of intermediate activity. Aromatic, α,β-unsaturated types, sulfonyl halides.

3. Unreactive acid halides or anhydrides. Hindered acid derivatives such as pentachlorobenzoyl chloride, whose reaction with methanol at reflux temperature takes about 30 hours for completion.

To have some anchoring point to compare the hydrolysis of these acid derivatives on a practical scale, the half-lives of several anhydrides of the active class are shown [3]:

Anhydride	Half-life at 25°
Acetic	4.5 min.
Succinic	4.4 min.
Glutaric	4.3 min.
3,3-Dimethylglutaric	63.0 min.

It should be recalled that the acid chloride corresponding to the anhydride is the more reactive and that the tabulation refers to kinetic results where activities are close to 1. In more concentrated solutions where activities are considerably lower than 1, the rates of hydrolyses will be slower.

Variations in mechanism are encountered. For example, some acid chlorides may be induced to undergo an SN_1 hydrolysis by means of powerful electrophilic reagents such as mercuric perchlorate [4]. Electron-releasing groups will favor the change of mechanism. In special cases certain acids do not undergo simple hydrolysis but rather lose carbon monoxide:

$$RC\overset{O}{\underset{X}{\diagdown}} + H_2O \xrightarrow{SN_1} R^{\oplus} + CO + HX$$

When R can form a stable carbonium ion or a proton, as when R is t-butyl or hydrogen, respectively, decarbonylation is to be expected in an SN_1 mechanism [5].

The acid chloride hydrolysis is involved in the Hell-Volhard-Zelinsky synthesis in which the acid chlorides of α-halogenated acids are hydrolyzed to the free acids. Anhydrides are sometimes found in nature and are thus likely starting materials for the preparation of acids. It is of interest to note that the hydrolysis of the anhydrides is catalyzed by tertiary amine bases [6].

(a) **Preparation of β,β-diphenylacrylic acid.** Crude β,β-diphenyl-acrylyl chloride, 1 mole (from the reaction of 1,1-diphenylethylene and oxalyl chloride, see F.4) was stirred in an ice-cold Na_2CO_3 solution for 2 hr. The mixture was then boiled with about 3.6 l. of water to dissolve the sodium salt, and charcoal was added. Recovery of the free acid from the salt by acidification amounted to 95%, m.p. 167° [7].

(b) **Other examples from anhydrides**

(1) **Citraconic acid** (94% from the anhydride) [8].
(2) **Itaconic acid** (24–39% from the anhydride) [9].
(3) **Mesaconic acid** (43–52% from citraconic anhydride) [10].

2. From Esters

$$RCOOR' \xrightarrow{\text{HOH}} RCOOH$$

Esters may be hydrolyzed in the presence of a base or an acid, although the use of the former is the more common procedure. In acidic hydrolysis, protonation of the carbonyl oxygen makes the carbonyl carbon more susceptible to nucleophilic attack by water:

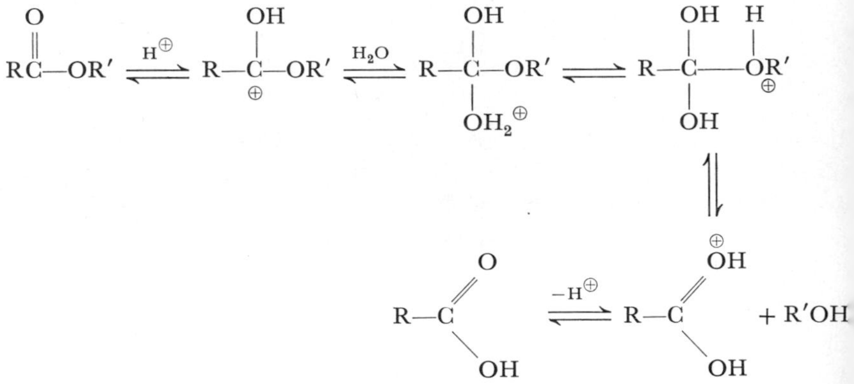

Because the hydrogen ion plays such a predominant role or because compensating factors cancel each other (i.e., electron-releasing groups increase the extent of protonation but decrease the tendency of the carbonyl group to be attacked by water), substituent effects in acid hydrolysis by the SN_2 mechanism are negligible. Steric effects of course still play a very important role in specific cases. One mole each of an aliphatic straight-chain monoester and water form an equilibrium mixture of about two-thirds ester and one-third acid.

The basic hydrolysis mechanism is simpler in that the attack is made by the strong nucleophilic reagent $^{\ominus}OH$:

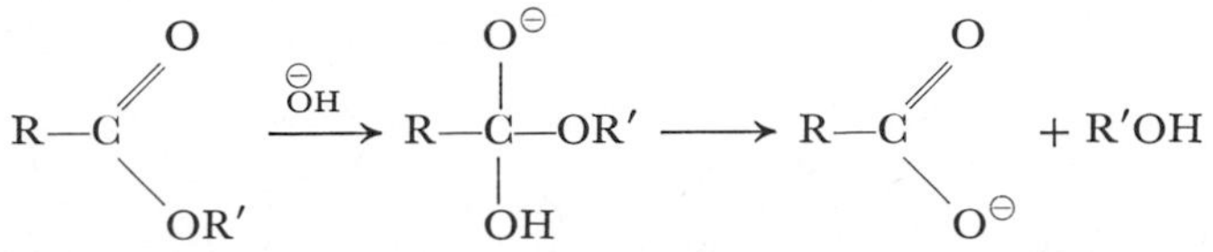

This reaction appears to be largely irreversible because of the formation of the resonance-stabilized carboxylate anion. Substituent effects are important, i.e., clcctron-withdrawing groups accelerate saponification, and the steric effect is still a factor to be contended with as it was in acid hydrolysis.

Acids with groups sensitive to water may be prepared from the corresponding pyrone and lithium aluminum hydride [11]:

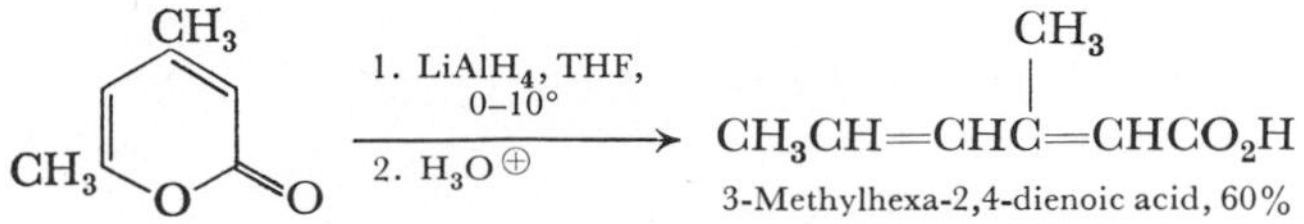

The ring oxygen of the pyrone in this reaction is displaced by hydride, perhaps via an alkyl fission, second-order reaction.

As natural products, esters are sometimes available as starting materials and in some cases substituted benzoic acids are best obtained through the use of the ester [12]. Special methods of hydrolysis requiring 100% sulfuric acid are necessary for sterically hindered esters, such as those of the 2,4,6-trialkyl benzoic acids. These hydrolyses proceed through the SN_1 mechanism [13]. Esters of sterically hindered acids and alcohols, such as t-butyl 2,4,6-trimethyl-benzoate, have been hydrolyzed with 18% hydrochloric acid [14]. Axial carbethoxy esters in the diterpene field such as methyl desoxypodocarpate

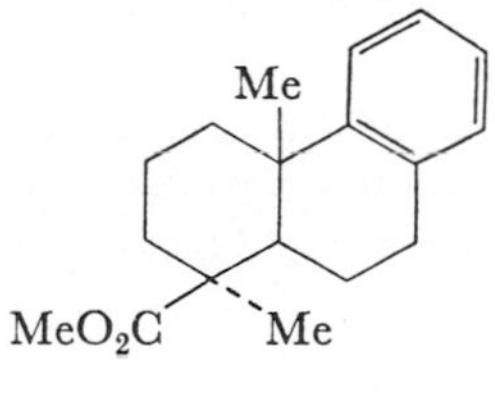

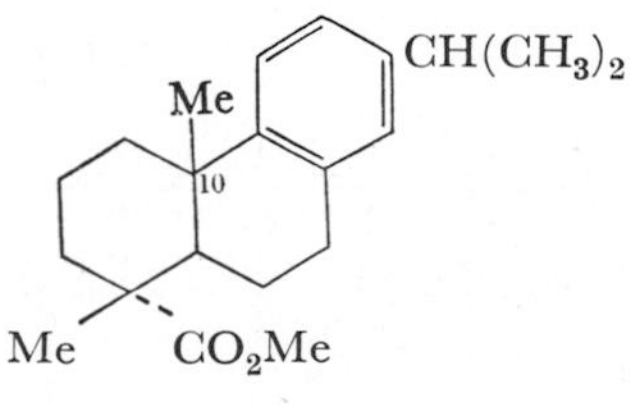

have been hydrolyzed in a tetrahydrofuran solution by lithium and liquid ammonia [15], while similar ring systems such as methyl dehydroabietate

and methyl O-methyl podocarpate

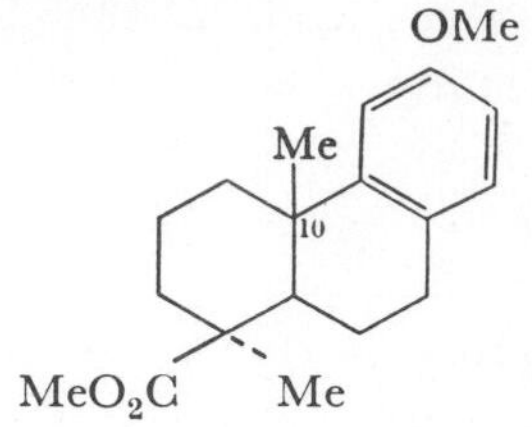

have been hydrolyzed with potassium *t*-butoxide in dimethyl sulfoxide. The former, since it possesses a tertiary ester group which is equatorial and not hindered by the axial methyl group at 10, hydrolyzes in less than an hour, while the latter in which the two groups in question are axial requires 2 hours at 56°. On the other hand, methyl triisopropylacetate

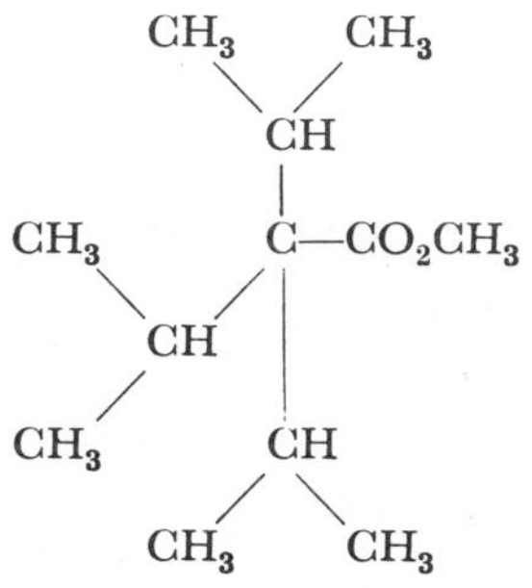

requires about 4 hours at 100° for complete hydrolysis [16].

Other special methods for the hydrolysis of esters which contain sensitive groups are available. For example, phenacyl esters are hydrolyzed with considerable ease by sodium thiophenoxide (Ex. b.5). Thiolesters are cleaved readily by oxidation with mercuric oxide or boron trifluoride and benzaldehyde (Ex. b.6). Anhydrous acrylic acid in 74–78% yield may be obtained from the methyl ester by heating with 98% formic acid in the presence of a catalytic amount of sulfuric acid and hydroquinone as a polymerization inhibitor [17].

Esters of acids sensitive to both acids and bases or esters resistant to basic hydrolysis may be hydrolyzed by lithium iodide in pyridine, 2,6-lutidine, or 2,4,6-collidine [18]. For example, 10-carbethoxy-1,1-dimethyl-*trans*-decalin may be hydrolyzed in 94% yield to the carboxylic acid with lithium iodide in *sym*-collidine [19]:

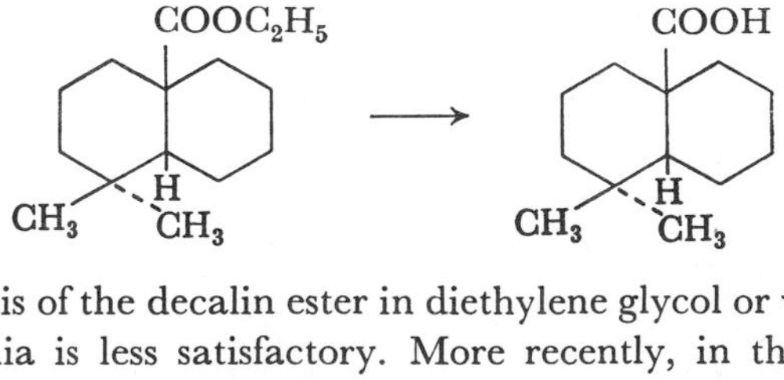

Alkaline hydrolysis of the decalin ester in diethylene glycol or with alkali metals in liquid ammonia is less satisfactory. More recently, in the case of methyl glycyrrhetate, a hindered triterpenoid ester, dimethyl formamide, substituted for the heterocyclic base, permitted the hydrolysis to occur in less time [20].

For rather efficient conversion into the acid, carried out by a transesterification, a procedure using methanesulfonic acid in 90% formic acid has been recommended (Ex. b.7).

If saponification fails or if acid esters are desired from diesters, pyrolytic methods are available [21]:

$$C_2H_5O_2C(CH_2)_nCO_2C_2H_5 \xrightarrow{\Delta} C_2H_5O_2C(CH_2)_nCOOH + CH_2{=}CH_2$$

$$RCH(CO_2C_2H_5)_2 \xrightarrow{\Delta} RCH_2COOH$$

By this method, half-acids are obtained in 69–80% yields, and substituted acetic acids from substituted malonic acid esters in 38–84% yields [22].

(a) **Preparation of *m*-nitrobenzoic acid** (90–96% from the methyl ester) [12].

(b) **Other examples**

(1) **Myristic acid** (89–95% from trimyristin) [23].
(2) **Linoleic acid** (90% from the ethyl ester) [24].
(3) **Cyclopentan-2-oneoxalic acid** (81–92% from the ethyl ester and potassium hydroxide in alcohol-water below 5° followed by acidification) [25].
(4) ***o*-Hydroxyhydrocinnamic acid** (90% from dihydrocoumarin and 40% aqueous sodium hydroxide) [26].
(5) **N-Phthaloylglycine** (99% from phenacyl N-phthaloylglycinate and 2 equiv. of sodium thiophenoxide in dimethylformamide at room temperature) [27].
(6) **Trimethylacrylic acid** (81–86% from ethyl trimethylthiolacrylate and the complex of boron trifluoride and benzaldehyde or from oxidation with a mixture of mercuric oxide and chloride; the thiolester is obtained from 1-ethylthiolpropyne, $CH_3C{\equiv}CSC_2H_5$) [28].
(7) **Nicotinic acid** (76% from 0.1 mole of methyl nicotinate, 0.1 mole of methanesulfonic acid, and 100 ml. of formic acid by refluxing for 5 hr.) [29].

(c) **Preparation of ethyl hydrogen sebacate.** With a nitrogen flow of 60 bubbles per min., 233 g. of diethyl sebacate was added dropwise to a Vycor pyrolysis tube filled with Pyrex helices and heated externally at 440° for 11 hr. After half the material had been added, the pyrolysis tube was changed to avoid charring. On fractionation the pyrolysate, 218.7 g., gave 59.1 g. (28%) of ethyl hydrogen sebacate, b.p. 152° (0.6 mm.); 137.5 g. (59%) of diethyl sebacate; and 12.3 g. of sebacic acid, m.p. 135°. Yield of ethyl hydrogen sebacate, based on the diester consumed, was 69% [21].

(d) **Preparation of phenylacetic acid** (64% from ethyl phenylmalonate (as in Ex. c)) [22].

3. From Amides

$$RCONH_2 \xrightarrow[\text{or } OH^{\ominus}]{H^{\oplus}} RCOOH$$

This synthesis occurs in good yield in acidic or alkaline medium, although the latter appears to be the more widely used. The mechanisms of these reactions are essentially the same as in the hydrolysis of esters. In acidic medium, hydrolysis again involves the attack by water on the protonated amide whereas, in basic medium, the powerful nucleophilic hydroxide ion attacks the free amide [30].

When the carboxyl group is produced from the cyanide group, the amide group is an intermediate which may or may not be isolated. Difficultly hydrolyzable amides have been converted into acids successfully by the use of 100% phosphoric acid [31] or nitrous acid [32]. Hindered amides have been converted into the acid by 75% sulfuric acid and solid sodium nitrite [33] and by conversion into the N-methyl derivative (with sodium hydride and methyl iodide), then into the N-nitroso derivative (with nitrosyl chloride), and finally into the acid (with methanolic potassium hydroxide followed by acidification) [34]. In addition nitrosonium tetrafluoroborate, $\overset{\oplus}{N}O\overset{\ominus}{BF_4}$, in acetonitrile [35] and nitrosyl chloride in chloroform (in the sugar series) [36] have been used for the same purpose.

The phenylhydrazide group can be removed by mild oxidation to yield acids which are sensitive to normal hydrolytic processes (see Ex. b.4).

(a) Preparation of benzocyclobutene-1-carboxylic acid. Benzocyclobutene-1-carboxamide, 1 g. in 15 ml. of hot 20% aqueous NaOH was heated for 5 hr. The cooled solution was acidified with concentrated HCl and extracted twice with 100 ml. portions of a 5:1 petroleum ether (30–60°)-ether mixture. From the extract there was recovered 0.975 g. (97.5%) of the acid, m.p. 74.5–76% [37].

(b) Other examples

(1) **Benzoylglycyl-DL-leucine** (68% from the amide and 1 N NaOH at 100°) [38].

(2) **Diisopropylacetic acid** (83% from the amide, solid $NaNO_2$ and 75% H_2SO_4 [33].

(3) **2-Methyl-5-sulfonamido-3-furoic acid** (70% from the diamide and barium hydroxide in water) [39].

(4) **Ethyl N-carbobenzoxy-α-L-glutamylglycinate** (82% from the

$$HO_2C(CH_2)_2CHCONHCH_2COOC_2H_5$$
$$|$$
$$NHCOOCH_2C_6H_5$$

phenylhydrazide and activated manganese dioxide) [40].

4. From Nitriles

$$RCN \rightarrow RCOOH$$

This method is widely used in the preparation of carboxylic acids because the cyano group is readily introduced into organic molecules. Even benzonitriles can now be produced in high yields from the reaction of aryl bromides and cuprous cyanide in dimethylformamide or N-methylpyrrolidone [41].

The hydrolyses may be carried out in acidic or alkaline medium. Either a proton and water or a hydroxide ion and water produce the intermediate acid amide, the mechanism of which has been discussed in A.3. It has been found that alkali either in ethylene glycol [42] or in glycerol [43] is more effective in promoting hydrolysis than are other reagents.

The hydrolysis of cyanides has found rather wide use in the synthesis of α-hydroxy- and α-aminocarboxylic acids since the appropriate cyanide may be obtained from the aldehyde or ketone by the addition of hydrogen cyanide alone or in the presence of ammonia:

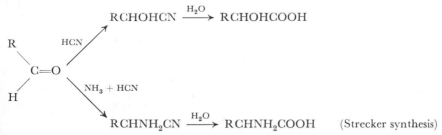

In the Strecker synthesis the nitrile anion probably adds to the iminoaldehyde (RCH=NH) which is in equilibrium with the aldehyde and ammonia. The two essential reagents may be produced from ammonium cyanide generated from a mixture of sodium cyanide and ammonium chloride. As a rule the yields are more satisfactory for the simple than for the substituted carboxylic acids.

(a) Preparation of 3-benzyl-3-methylpentanoic acid (91–93 % from the nitrile and a base in ethylene glycol) [44].

(b) Other examples

  (1) Mesitylacetic acid (87 % from the nitrile and sulfuric acid) [45].

  (2) o-Toluic acid (80–89 % from the nitrile and 75 % aqueous sulfuric acid) [46].

  (3) 4-Hydroxyquinaldine-6-carboxylic acid (88 % from the nitrile and a base in glycerol) [43].

  (4) Nicotinic acid [47]. Sodium peroxide seems to be a superior reagent for the hydrolysis of nitriles to amides (87 % yield of nicotinamide). The amides can be converted into acids by procedures already described. Perhaps a single-step hydrolysis of nitriles with sodium peroxide could be devised.

  (5) Atrolactic acid (29–30 % from acetophenone via the cyanohydrin which without isolation was hydrolyzed to the amide with dilute acid and then to the acid with dilute base) [48].

  (6) dl-α-Aminophenylacetic acid (33–37 % from benzaldehyde via the aminonitrile which was then hydrolyzed with dilute acid) [49].

  (7) General preparation of diacids from glycols via the dinitriles (the latter being hydrolyzed in 94–99 % yields by means of conc. hydrochloric acid) [50].

5. From Aldoximes

$$RCH_2CH=\!\!=\!NOH \xrightarrow{\text{KOH}} RCH_2COOH$$

This reaction was first reported in 1936 [51]. It is applicable to aliphatic and aromatic aldoximes, the most frequently employed solvent being diethylene glycol at 170–190° [52]. The reaction appears to proceed through the cyanide which under the conditions employed is hydrolyzed to the amide and finally to the acid. It offers another method for going from the aldehyde to the acid.

Alternatively, the oxime can be converted quantitatively into the nitrile by means of acetic anhydride and the nitrile hydrolyzed as in A.4.

(a) Preparation of phenylacetic acid. β-Phenylacetaldoxime (10 mmoles) was heated in 100 ml. of diethylene glycol at 190° with about 50 mmoles of potassium hydroxide for 3 hr. under a nitrogen atmosphere. The cooled mixture was diluted with water and extracted four times with half volumes of methylene chloride, these combined extracts being washed with 100 ml. of aqueous sodium chloride solution. Drying the organic extracts over sodium sulfate, filtering, and evaporating the solvent left as a residue the neutral fraction of the reaction mixture. The original aqueous alkaline solution then was acidified with hydrochloric acid to pH 7, and again extracted with methylene chloride. A similar treatment of the organic phase yielded the weakly acidic products. Finally, acidification of the aqueous solution to pH 2 and similar extraction with methylene chloride yielded the strongly acidic products. From these extracts there was obtained an 80% yield of phenylacetic acid [52].

(b) Other examples

(1) Benzoic (95%), **cinnamic** (68%), **isobutyric** (89%), **enanthalic** (63%), and **pivalic** (38%) **acids** [52].

6. From Trihalides

$$RCX_3 \xrightarrow[\text{H}_2\text{O}]{\text{H}_2\text{SO}_4} RCO_2H$$

Aliphatic 1,1,1-trichloroalkanes are inert toward nucleophilic agents but can be hydrolyzed readily with 95% sulfuric acid. Some charring occurs, and nitric acid (sp. gr. 1.50–1.51) has been shown to give better yields than sulfuric acid in some cases [53]. The reaction would seem to be of limited value, but actually trichloroalkanes are obtained readily from the telomerization of carbon tetrachloride or bromotrichloromethane with olefins [54]:

$$CCl_4 + CH_2=\!\!=\!CH_2 \rightarrow CCl_3(CH_2CH_2)_nCl$$

The tetrachloroalkanes can be separated by fractionation and hydrolyzed to ω-chloroacids. Benzotrichlorides of course are readily available and can be hydrolyzed similarly. Chloral may be used to introduce the trichloroethyl group into compounds [55], which can be hydrolyzed to acids. Trichloroethanols from the condensation of haloforms with aldehydes are capable of hydrolysis to α-hydroxy acids [56]. The trifluoromethyl group would seem to be quite inert but can be hydrolyzed by sulfuric acid [57]. Indeed, the trifluoromethyl group in o- and p- (but not m-) trifluorophenols is extremely labile to base [58].

(a) **Preparation of 3,5-dichlorobenzoic acid.**   Chlorine was passed into 50 g. of irradiated 3,5-dichlorotoluene at 185–190° until the weight increase amounted to 28 g. (about 8 hr.). To the product, 78 g., was added dropwise with stirring 200 g. of 8 % fuming H_2SO_4. After stirring for 30 hr., the acid was precipitated by pouring the mixture onto cracked ice. Purification in the usual manner gave 53 g. (90%), m.p. 181° [59].

(b) **Other examples**

(1) **4,6-Dimethoxyphthalide-3-carboxylic acid** (85–90% from the trichloromethylphthalide) [60].

(2) **β-(2-Pyrimidyl) acrylic acid** (92% from the chloral adduct of 2-methylpyrimidine) [61].

(3) **γ,γ,γ-Trifluorocrotonic acid** (51% from 1,1,1-trichloro-4,4,4-trifluoro-3-iodobutane treated with ethanolic KOH and then H_2SO_4 at 80° for 6 hr.) [62].

(4) **p-Nitrophenylacetic acid** (98% from 1-(p-nitrophenyl)-2,2,2-trichloroethane and sulfuric acid) [63].

7. From Dihalides and the Like

$$-CH=CCl_2 \xrightarrow[\text{2. HOH}]{\text{1. } H_2SO_4} -CH_2CO_2H$$

The reaction resembles the hydrolysis of benzotrichlorides, the first chlorine being difficult to remove, but rapid carbonyl formation ensues thereafter. In this case, however, prior addition of sulfuric acid may take place followed by rapid ketolization and hydrolysis of the acid chloride:

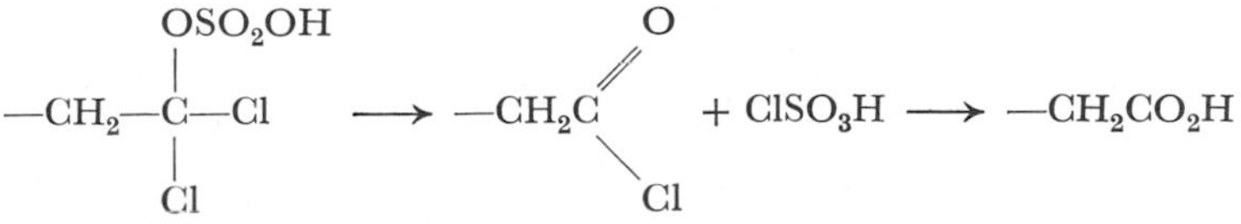

$$-CH_2-\underset{\underset{Cl}{|}}{\overset{\overset{OSO_2OH}{|}}{C}}-Cl \longrightarrow -CH_2\overset{O}{\underset{Cl}{\diagdown}}C + ClSO_3H \longrightarrow -CH_2CO_2H$$

The yields are good.

A source of γ-keto acids is alkylidene succinic acids which may be converted as follows [64]:

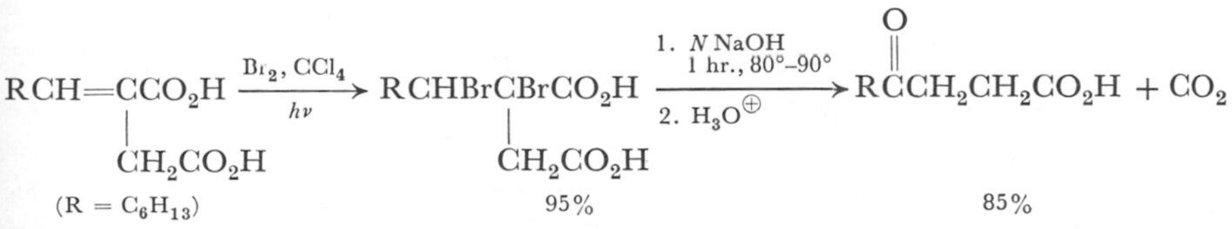

$$\underset{\underset{(R = C_6H_{13})}{\overset{|}{CH_2CO_2H}}}{RCH=CCO_2H} \xrightarrow[h\nu]{Br_2, CCl_4} \underset{\underset{95\%}{\overset{|}{CH_2CO_2H}}}{RCHBrCBrCO_2H} \xrightarrow[\text{2. } H_3O^{\oplus}]{\overset{\text{1. } N \text{ NaOH}}{\text{1 hr., 80°–90°}}} \underset{85\%}{R\overset{O}{\overset{\|}{C}}CH_2CH_2CO_2H + CO_2}$$

This reaction creates a carbonyl rather a carboxyl group but is included here because of its potential in obtaining acids. For the same reason the conversion of pyrazolones into acetylenic or olefinic acids is described [65]:

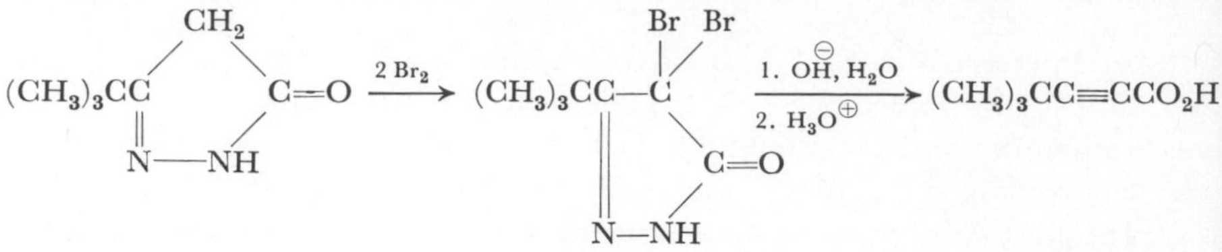

$$(CH_3)_3CC \underset{N-NH}{\overset{CH_2}{\diagdown}} C=O \xrightarrow{2\,Br_2} (CH_3)_3CC-\underset{\underset{N-NH}{C=O}}{\overset{Br\ Br}{|\ |}}C \xrightarrow[2.\ H_3O^{\oplus}]{1.\ OH,\ H_2O} (CH_3)_3CC{\equiv}CCO_2H$$

(From β-ketoesters)   94%   t-Butylpropargylic acid, 57%

Olefinic acids may be prepared from the same source by mono- rather than dibromination.

(a) Preparation of 2-(p-bromophenyl)-propanoic acid. 3-(p-Bromophenyl)-1,1-dichloro-1-propene, 20 g., and 50 g. of 93% sulfuric acid were heated for 1 hr. at 120°, after which the mixture was quenched in water to give a 93% yield of p-bromohydrocinnamic acid, m.p. 135° [66].

(b) Other examples

(1) **10-Aminodecanoic acid** (84% from 1,1-dichloro-10-amino-1-decene) [67].

(2) **α,δ-Dichlorovaleric acid** (78% from 1,1,5-trichloropentene-1, chlorine, and 93% H_2SO_4 at 15–20°)[68].

8. From α-Ethynylcarbinols

The α-ethynylcarbinols are readily available from the condensation of acetylene with ketones and are potentially good sources of some acids. The best method seems to proceed via an α-bromoacetylene as shown:

$$(CH_3)_2COHC{\equiv}CH \xrightarrow[h\nu]{Br_2} (CH_3)_2COHCBr{=}CHBr \xrightarrow{P_2O_5}$$

$$CH_2{=}C(CH_3)CBr{=}CHBr \xrightarrow[CH_3OH]{KOH} (CH_2{=}C(CH_3)C{\equiv}CBr) \xrightarrow[H_2O]{H^{\oplus}}$$

$$(CH_3)_2C{=}CHCO_2H$$

(a) Preparation of β,β-dimethylacrylic acid. Dimethylethynylcarbinol, 21 g., was brominated in 150 ml. of petroleum ether with 41 g. of bromine using a 200-watt light bulb (2 hr. at 35°). The dibromide in hexane was dehydrated by gradual addition of 40 g. of phosphorus pentoxide to the vigorously stirred solution followed by brief refluxing. The petroleum ether layer was washed, dried, and concentrated. The residue was mixed with 90 g. of potassium hydroxide in 300 ml. of methanol and refluxed for 30 hr. The methanol was removed by distillation and replaced by 50 ml. of water. Careful acidification with concentrated hydrochloric acid and filtration gave some crystalline dimethylacrylic acid. The filtrate extracted with ether gave more of the acid. The combined crops, recrystallized from water, gave colorless crystals, m.p. 69–70° (56%) [69].

9. From Diethyl Malonate

$$\underset{\substack{\diagup\; COOC_2H_5 \\ CH_2 \\ \diagdown\; COOC_2H_5}}{}\xrightarrow[\text{2. } C_2H_5ONa]{\text{1. RX}}\underset{\substack{\diagup\; COOC_2H_5 \\ RCH \\ \diagdown\; COOC_2H_5}}{}\xrightarrow[\text{3. Heat}]{\substack{\text{1. NaOH} \\ \text{2. } H\oplus}} RCH_2COOH$$

$$\underset{\substack{\diagup\; COOC_2H_5 \\ RCH \\ \diagdown\; COOC_2H_5}}{}\xrightarrow[\text{2. } C_2H_5ONa]{\text{1. R'X}}\underset{\substack{R\diagdown\quad\diagup\; COOC_2H_5 \\ C \\ R'\diagup\quad\diagdown\; COOC_2H_5}}{}\xrightarrow[\text{3. Heat}]{\substack{\text{1. NaOH} \\ \text{2. } H\oplus}} \underset{\substack{R\diagdown \\ CHCOOH \\ R'\diagup}}{}$$

This synthesis has been used widely in the preparation of mono- and di-substituted acetic acids if the molecular weight is high and if the alkyl groups are straight-chained. Occasionally, secondary halides have been used as alkylating agents, but never tertiary halides. To widen the scope of alkylation, sodium hydride in the solvent dimethylformamide has been used, for example, to ethylate diethyl t-butylmalonate [70], but sodium hydride in tetramethylurea seems to be the best environment for difficult alkylations of malonic esters [71]. The α-hydrogens in the malonic ester are quite acidic, attached as they are to a carbon atom located between two carboethoxy groups. This effect is generally attributed to the resonance stabilization of the carbanion. The first step of the reaction leads to the formation of

$$\underset{\substack{\overset{\oplus}{N}a\overset{\ominus}{C}H \\ \diagdown\; COOC_2H_5}}{\diagup\; COOC_2H_5}$$

the carbanion of which by a nucleophilic attack on the alkyl halide produces

$$\underset{\substack{RCH \\ \diagdown\; COOC_2H_5}}{\diagup\; COOC_2H_5}$$

This ester hydrolyzes by the usual mechanism to give an unstable dicarboxylic acid which loses carbon dioxide spontaneously or by heating to produce the monocarboxylic acid. As might be expected, the substitution of two branched groups may be difficult [72]. The yields are satisfactory, if branching in R and R' is not excessive.

By a modification of this method, α-aminocarboxylic acids may be synthesized. The alkyl halide with diethyl formamidomalonate in the presence of sodium hydride gives the alkylated malonate, which on hydrolysis with dilute acid produces the aminocarboxylic acid [73, 74]:

$$\underset{\substack{\diagup\; COOC_2H_5 \\ HC-NHCHO \\ \diagdown\; COOC_2H_5}}{}\xrightarrow[\text{NaH}]{\text{RX}}\underset{\substack{\diagup\; COOC_2H_5 \\ RC-NHCHO \\ \diagdown\; COOC_2H_5}}{}\xrightarrow{H\oplus}\underset{\substack{RCHCOOH \\ | \\ NH_2}}{}$$

R may be alkyl or allyl, or gramine

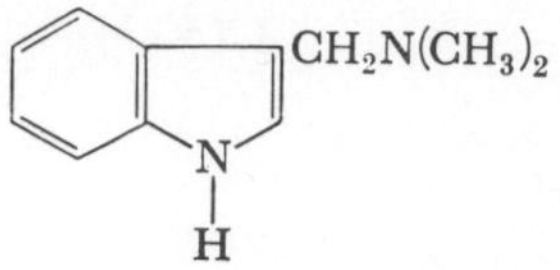

may be used as the alkylating agent [75]. α-Hydroxycarboxylic acids may be prepared in a similar manner by utilizing diethyl acetoxymalonate in place of the formamidomalonate [76].

(a) Preparation of 3-methylpentanoic acid (62–65 % from the malonate) [77].

(b) Other examples

(1) Pelargonic acids (66–75 % from ethyl malonate) [78].

(2) α-Bromoisovaleric acid (55–66 % from isopropylmalonic ester) [79].

(3) Propargylglycine (69 % from propargyl bromide) [73].

10. From Hydantoins

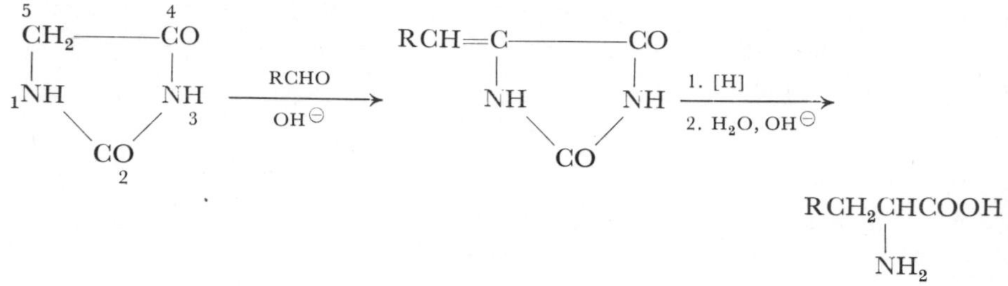

Aliphatic, aromatic, and heterocyclic aldehydes condense with hydantoins, the attack being an aldol condensation at carbon atom 5, in the presence of sodium acetate in a mixture of acetic acid and acetic anhydride or pyridine containing traces of piperidine. The resulting product may be reduced with phosphorus and hydriodic acid, ammonium sulfide, or stannous chloride to the alkylated hydantoin which may be hydrolyzed in acid or alkaline, usually barium hydroxide, medium to the α-amino acid. The mechanism of the hydrolysis is similar to that of amides. This synthesis has been reviewed [80]. The substituted hydantoin may also be obtained from the aldehyde or ketone cyanohydrins and ammonium carbonate. Yields are generally good.

(a) Preparation of N-methyl-3,4-dihydroxyphenylalanine (about 42 % from the condensation product of vanillin and creatinine) [81].

(b) Preparation of dl-tryptophan (90 % from 5-(3-indolymethyl) hydantoin) [82].

1. J. Hine, *Physical Organic Chemistry*, McGraw-Hill Book Co., New York, 1962, p. 294.
2. M. S. Newman, *Steric Effects in Organic Chemistry*, John Wiley and Sons, 1956, p. 206.
3. L. Eberson, *Acta Chem. Scand.*, **18**, 534 (1964).
4. H. K. Hall, Jr., and C. H. Lueck, *J. Org. Chem.*, **28**, 2818 (1963).

5. M. E. Grundy *et al.*, *J. Chem. Soc.*, 581 (1958).

6. V. Gold and E. G. Jefferson, *J. Chem. Soc.*, 1409 (1953).

7. F. Bergmann *et al.*, *J. Am. Chem. Soc.*, **70**, 1612 (1948).

8. R. L. Shriner *et al.*, *Org. Syn.*, Coll. Vol. **2**, 140 (1943).

9. R. L. Shriner *et al.*, *Org. Syn.*, Coll. Vol. **2**, 368 (1943).

10. R. L. Shriner *et al.*, *Org. Syn.*, Coll. Vol. **2**, 382 (1943).

11. K. Yamada *et al.*, *Bull. Chem. Soc., Japan*, **34**, 1873 (1961); **35**, 1329 (1962).

12. O. Kamm, and J. B. Segur, *Org. Syn.*, Coll. Vol. **1**, 392 (1941).

13. M. S. Newman, *J. Am. Chem. Soc.*, **63**, 2431 (1941).

14. S. G. Cohen and A. Schneider, *J. Am. Chem. Soc.*, **63**, 3382 (1941).

15. E. Wenkert and B. G. Jackson, *J. Am. Chem. Soc.*, **80**, 217 (1958).

16. F. C. Chang and N. F. Wood, *Tetrahedron Letters*, 2969 (1964).

17. C. E. Rehberg, *Org. Syn.*, Coll. Vol. **3**, 33 (1955).

18. A. Eschenmoser *et al.*, *Helv. Chim. Acta*, **43**, 113 (1960).

19. W. L. Meyer and A. S. Levinson, *J. Org. Chem.*, **28**, 2184 (1963).

20. P. D. G. Dean, *J. Chem. Soc.*, 6655 (1965).

21. W. J. Bailey and W. G. Carpenter, Jr., *J. Org. Chem.*, **29**, 1252 (1964).

22. W. J. Bailey and J. J. Daly, Jr., *J. Org. Chem.*, **29**, 1249 (1964).

23. G. D. Beal, *Org. Syn.*, Coll. Vol. **1**, 379 (1941).

24. J. W. McCutcheon, *Org. Syn.*, Coll. Vol. **3**, 526 (1955).

25. R. Mayer *et al.*, *J. Prakt. Chem.*, **14**, 261 (1961).

26. T. Matsuura and A. Nishinaga, *J. Org. Chem.*, **29**, 3168 (1964).

27. J. C. Sheehan and G. D. Daves, Jr., *J. Org. Chem.*, **29**, 2006 (1964).

28. L. B. Bos and J. F. Arens, *Rec. Trav. Chim.*, **82**, 339 (1963).

29. B. Loev, *Chem. Ind.* (*London*), 193 (1964).

30. A. Bruylants and F. Kezdy, *Record Chem. Progr.* (*Kresge-Hooker Sci. Lib.*), **21**, 213 (1960).

31. G. Berger and S. C. J. Olivier, *Rec. Trav. Chim.*, **46**, 600 (1927).

32. F. C. Whitmore and D. P. Langlois, *J. Am. Chem. Soc.*, **54**, 3438 (1932); N. Sperber *et al.*, *ibid.*, **70**, 3091 (1948).

33. S. Sarel and M. S. Newman, *J. Am. Chem. Soc.*, **78**, 5416 (1956).

34. M. E. Kuehne, *J. Am. Chem. Soc.*, **83**, 1492 (1961).

35. G. A. and J. A. Olah, *J. Org. Chem.*, **30**, 2386 (1965).

36. M. L. Wolfrom and H. B. Wood, *J. Am. Chem. Soc.*, **73**, 730 (1951).

37. M. P. Cava *et al.*, *J. Am. Chem. Soc.*, **80**, 2257 (1958).

38. T. Wieland and H. Fritz, *Chem. Ber.*, **86**, 1186 (1953).

39. J. F. Scully and E. V. Brown, *J. Org. Chem.*, **19**, 894 (1954).

40. R. B. Kelly, *J. Org. Chem.*, **28**, 453 (1963).

41. L. Friedman and H. Shechter, *J. Org. Chem.*, **26**, 2522 (1961); M. S. Newman and H. Boden, *ibid.*, **26**, 2525 (1961).

42. M. S. Newman and R. M. Wise, *J. Am. Chem. Soc.*, **78**, 450 (1956).

43. C. T. Peng and T. C. Daniels, *J. Am. Chem. Soc.*, **77**, 6682 (1955).

44. F. S. Prout *et al.*, *Org. Syn.*, Coll. Vol. **4**, 93 (1963).

45. R. C. Fuson and N. Rabjohn, *Org. Syn.*, Coll. Vol. **3**, 557 (1955).

46. H. T. Clarke and E. R. Taylor, *Org. Syn.*, Coll. Vol. **2**, 588 (1943).

47. J. G. Wallace, *Hydrogen Peroxide in Organic Chemistry*, E. I. duPont de Nemours Co., 1962, p. 42.

48. E. L. Eliel and J. P. Freeman, *Org. Syn.*, Coll. Vol. **4**, 58 (1963).

49. R. E. Steiger, *Org. Syn.*, Coll. Vol. **3**, 84 (1955).

50. K. E. Miller *et al.*, *J. Chem. Eng. Data*, **9**, Pt. 2, 227 (1964).

51. E. Jordan and C. R. Hauser, *J. Am. Chem. Soc.*, **58**, 1304 (1936).

52. H. Rapoport and W. Nilsson, *J. Org. Chem.*, **27**, 629 (1962).

53. A. N. Nesmeyanov, *Selected Works in Organic Chemistry*, The Macmillan Co., New York, 1963, p. 1023.

54. G. Sosnovsky, *Free Radical Reactions in Preparative Organic Chemistry*, The Macmillan Co., New York, 1964, Chap. 2.

55. A. McKenzie and W. S. Dennler, *J. Chem. Soc.*, 1596 (1926); H. Brockmann *et al.*, *Chem. Ber.*, **90**, 2302 (1957).
56. W. Reeve and E. L. Compere, Jr., *J. Am. Chem. Soc.*, **83**, 2755 (1961).
57. G. M. LeFave, *J. Am. Chem. Soc.*, **71**, 4148 (1949).
58. R. G. Jones, *J. Am. Chem. Soc.*, **69**, 2346 (1947).
59. G. H. Stempel, Jr. *et al.*, *J. Am. Chem. Soc.*, **73**, 455 (1951).
60. P. Fritsch, *Ann. Chem.*, **296**, 344 (1897).
61. R. G. Jones *et al.*, *J. Am. Chem. Soc.*, **72**, 3539 (1950).
62. R. N. Haszeldine, *J. Chem. Soc.*, 922 (1953).
63. V. M. Naidan and A. V. Dombrovskii, *J. Gen. Chem. USSR (Eng. Transl.)*, **34**, 1474 (1964).
64. A. Takeda *et al.*, *J. Org. Chem.*, **31**, 616 (1966).
65. L. A. Carpino *et al.*, *J. Org. Chem.*, **31**, 2867 (1966); *J. Am. Chem. Soc.*, **80**, 5796 (1958).
66. A. N. Nesmeyanov *et al.*, *Izv. Akad. Nauk, SSSR, Otd. Khim. Nauk*, 993 (1955); *C.A.*, **50**, 11278 (1956).
67. A. N. Nesmeyanov *et al.*, *J. Gen. Chem. USSR (Eng. Transl.)*, **27**, 2481 (1957).
68. A. N. Nesmeyanov *et al.*, *Dokl. Akad. Nauk, SSSR*, **103**, 1029 (1955); *C.A.*, **50**, 9998 (1956).
69. I. N. Nazarov *et al.*, *Bull. Acad. Sci. USSR, Div. Chem. Sci. (Engl. Transl.)*, 1306 (1958).
70. D. E. Pearson *et al.*, *J. Med. Chem.*, **10**, 1078 (1967).
71. A. Lüttringhaus and H. W. Dirksen, *Angew. Chem. Intern. Ed. Engl.*, **3**, 260 (1964).
72. C. R. Hauser *et al.*, *J. Am. Chem. Soc.*, **66**, 309 (1944).
73. H. Gershon *et al.*, *J. Am. Chem. Soc.*, **76**, 3484 (1954).
74. N. A. Dobson and R. A. Raphael, *J. Chem. Soc.*, 3642 (1958).
75. H. M. Kissman and B. Witkop, Ref. 10, *J. Am. Chem. Soc.*, **75**, 1967 (1953).
76. M. J. Gortatowski and M. D. Armstrong, *J. Org. Chem.*, **22**, 1217 (1957).
77. C. S. Marvel *et al.*, *Org. Syn.*, Coll. Vol. **2**, 416 (1943).
78. E. E. Reid and J. R. Ruhoff, *Org. Syn.*, Coll. Vol. **2**, 474 (1943).
79. C. S. Marvel and V. duVigneaud, *Org. Syn.*, Coll. Vol. **2**, 93 (1943).
80. R. J. Block, *Chem. Rev.*, **38**, 501 (1946); E. Ware, *ibid.*, **46**, 403 (1950).
81. V. Deulofeu and T. J. Guerrero, *Org. Syn.*, Coll. Vol. **3**, 586 (1955).
82. J. Elks *et al.*, *J. Chem. Soc.*, 629 (1944).

B. Oxidation

The oxidizing agents most commonly employed are potassium permanganate, potassium dichromate (or chromic anhydride), and dilute nitric acid. The advantages of potassium permanganate are that it is powerful as an oxidizer, and forms the insoluble manganese dioxide which is removed easily from the potassium salt of the acid dissolved in the aqueous medium. Its disadvantages are that it cannot be dissolved well in nonaqueous media and it is unstable and releases oxygen in boiling water or refluxing pyridine-water solutions. Alkalinity increases this tendency [1]. Although the examples to be cited do not follow this practice, it seems best to bring about intimate contact between the alcohol and the aqueous permanganate solution by vigorous stirring and to hold the temperature at the lowest feasible point.

Chromic acid oxidations invariably are conducted in acetic acid solution which occasionally creates isolation problems. There is need for the development of a procedure which omits the acetic acid. Such a procedure involving intimate contact between the oxidizing solution and an ether solution of the alcohol is available for the preparation of ketones and might be adapted to the preparation of acids [2].

References for Section B are on pp. 774–776.

N-Halosuccinimides give low yields of esters in reaction with alcohols [3]. Other oxidizing agents are mentioned in the following examples, among them being oxygen which has some attractive features as described in B.1 and B.5.

1. From Alcohols or Carbonyl Compounds

$$RCH_2OH \xrightarrow{[O]} RCHO \xrightarrow{[O]} RCO_2H$$

This synthesis is widely employed in the preparation of carboxylic acids even though the yields are often not high. The usual chemical oxidizing agents are employed as well as oxygen in the presence of a catalyst, such as platinum on carbon or platinum oxide [4]. The latter has some unusual applications, as shown in Ex. c.9 and c.10.

In some cases the intermediate aldehyde may be isolated, and under acidic conditions the ester is sometimes obtained (see Ex. c.3). The mechanism of the oxidation, depending on the oxidizing agent used and the nature of the medium, takes such a variety of forms that no attempt will be made to give a complete discussion here. To cite one case [5], a primary alcohol in acidic solution with potassium dichromate forms the alkyl chromate which on acid catalysis yields the aldehyde

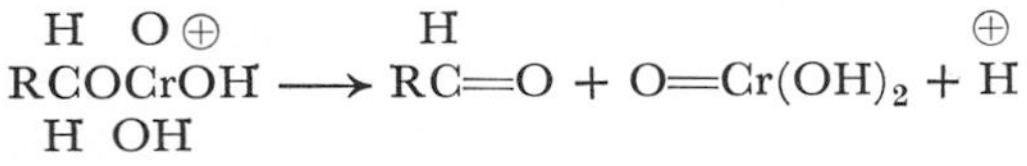

$$\underset{\substack{H \\ H}}{\overset{O\oplus}{RCOCrOH}} \longrightarrow \underset{H}{\overset{H}{RC}}=O + O=Cr(OH)_2 + \overset{\oplus}{H}$$

The aldehyde is then oxidized further in a similar manner perhaps via the hydrate. In the case of benzaldehyde in neutral or acidic solution of potassium permanganate, the steps appear to be:

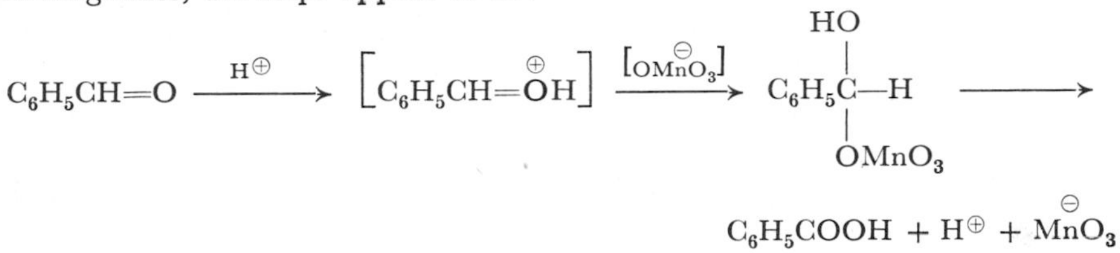

$$C_6H_5CH=O \xrightarrow{H\oplus} \left[C_6H_5CH=\overset{\oplus}{O}H\right] \xrightarrow{[O\overset{\ominus}{M}nO_3]} \underset{OMnO_3}{\overset{HO}{\underset{|}{C_6H_5\overset{|}{C}-H}}} \longrightarrow$$

$$C_6H_5COOH + H^\oplus + \overset{\ominus}{M}nO_3$$

More comprehensive discussion of oxidation mechanisms can be found in recent references [6].

Although cleavage of aliphatic ketones by oxidation is not a usual method for the synthesis of acids, this process has been used with some success in converting cyclic ketones into dicarboxylic acids.

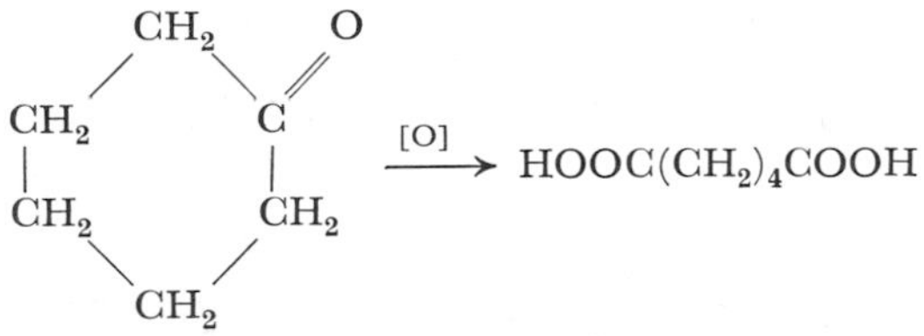

$$\xrightarrow{[O]} HOOC(CH_2)_4COOH$$

Nitric acid in the presence of vanadium pentoxide as catalyst is customarily used as the oxidizing agent for these ketones or the corresponding alcohols. The yields are fair. Cyclic ketones can also be converted into dibasic acids in moderate to excellent yield by air oxidation in the presence of potassium hydroxide dissolved in hexamethylphosphoramide [7]. Acetophenone yields benzoic acid under the same conditions.

Argentic oxide is a versatile oxidizing agent for the conversion of alcohols, aldehydes, or α-amino acids into acids [8]. An advantage of argentic oxide is that it oxidizes unsaturated aldehydes to unsaturated acids without affecting the olefinic double bond [9]:

3-Cyclohexenyl-
carboxylic acid, 97%

For oxidation of conjugated unsaturated aldehydes it is best to carry out the oxidation with an alkali cyanide, under which conditions the reaction proceeds as shown [9]:

$$RCH{=}CHCH{=}O \xrightarrow{\text{NaCN}} RCH{=}CHCHOHCN \xrightarrow{\text{MnO}_2}$$

$$RCH{=}CHC\overset{\overset{\displaystyle O}{\|}}{C}CN \xrightarrow{\text{OH}^{\ominus}} RCH{=}CHCO_2H + CN^{\ominus}$$

If aldehydes are available from specific syntheses, they may be oxidized with hydrogen peroxide (see Exs. b and c.8).

(a) Preparation of di-*t*-butylacetic acid.   To a stirred solution of 86.5 g. of 2,2-di-*t*-butylethanol in 270 ml. of sulfuric-acetic acid solution (50 ml. of concentrated H_2SO_4, 100 ml. of water, and 500 ml. of acetic acid) was added 240 ml. of chromic acid solution (125 g. of chromic acid and 125 ml. of water diluted to 500 ml. with acetic acid) during $1\frac{1}{2}$ hr. The mixture, after standing overnight, was heated on a steam bath for 1 hr. and 300 ml. of water was added. The organic product extracted in ether-benzene and recovered in the usual manner weighed 35.5 g. On reoxidation of the neutral fraction 41.1 g. more was obtained to make a total of 76.6 g. (82%), m.p. 72–74°. Additional purification by low temperature crystallization and vacuum sublimation elevated the m.p. to 80.5–81.5° [10].

(b) Preparation of glutaric acid.   3,4-Dihydro-1,2-pyran-2-carboxaldehyde (267 g.) was hydrolyzed with 560 g. of 0.0216 N sulfuric acid for $2\frac{1}{2}$ hr. at room temperature. The sulfuric acid was neutralized with excess calcium carbonate and the calcium sulfate filtered. The filtrate contained α-hydroxyadipaldehyde (ca. 90%). To this filtrate 840 g. of 30% hydrogen peroxide was added in 6 portions at regular intervals over a 48-hr. period. The mixture was then heated at 80° for an additional 8 hr. and then evaporated under water aspirator pressure. Glutaric acid was obtained as a white solid in 89% yield [11].

(c) Other examples from the alcohol

(1) Lauric acid (96% from dodecyl alcohol on a small scale using platinum oxide and air) [12].

(2) 10-Fluorodecanoic acid (93% from the alcohol with chromium trioxide in acetic acid) [13].

(3) n-Butyl n-butyrate (41–47% from n-butyl alcohol, $Na_2Cr_2O_7$, and H_2SO_4 [14]; the ester is encountered frequently in oxidation of alcohols).

(4) Adipic acid (72% from cyclohexanol and HNO_3) [15].

(5) Isobutyric or other fatty acids (use of potassium permanganate at room temperature) [16].

(6) δ-Benzoylvaleric acid (81% from 1-phenyl-1-cyclohexanol [17]; general oxidation of cyclic tertiary alcohols to keto acids by chromium trioxide in acetic acid).

(7) β-Chloropropionic acid (78–79% from trimethylene chlorohydrin by means of nitric acid) [18].

(8) Hydroxybenzoic acids from hydroxybenzaldehydes (a quantitative conversion using hydrogen peroxide in acetone [19]; the hydroxyaldehydes are obtainable by the Reimer-Tiemann reaction).

(9) Trimethylolacetic acid (50% from pentaerythritol, 30 g., 10% platinum on charcoal, 20 g., oxygen and gradual addition of sodium bicarbonate in water at 35°) [20].

(10) 2-Keto-L-gulonic acid (62% from ʟ-sorbose, 5% platinum on

$$CH_2(CHOH)_3COHCO_2H$$
$$\underset{\textstyle O}{\rule{1.2cm}{0.4pt}}$$

charcoal, $NaHCO_3$, water, and oxygen; the product can be rearranged readily to ascorbic acid) [21].

(11) Palmitic acid [22]:

$$C_{15}H_{31}CH_2OH \xrightarrow[\substack{-10° \\ \text{2. 50 hr., 18°}}]{\text{1. } N_2O_4} C_{15}H_{31}COOH$$

Cetyl alcohol   palmitic acid
90%

(d) Other examples from carbonyl compounds

(1) β-Chloropropionic acid (60–65% from acrolein via β-chloropropionaldehyde; nitric acid oxidizing agent) [23].

(2) n-Heptanoic acid (85–90% from the aldehyde and acidic potassium permanganate) [24].

(3) Piperonylic acid (78–84% from piperonal and alkaline potassium permanganate) [25].

(4) Δ⁹,¹²-Stearadiynoic acid (78% crude by oxidation of the aldehyde

$$CH_3(CH_2)_4C\equiv CCH_2C\equiv C(CH_2)_7CHO \rightarrow CH_3(CH_2)_4C\equiv CCH_2C\equiv C(CH_2)_7COOH$$

with $AgNO_3$ under nitrogen followed by the addition of NaOH) [26].

(5) 3-Thenoic acid (95–97% from the aldehyde added in small portions to alkaline silver nitrate cooled in an ice bath) [27].

2. From Carbonyl Peroxides

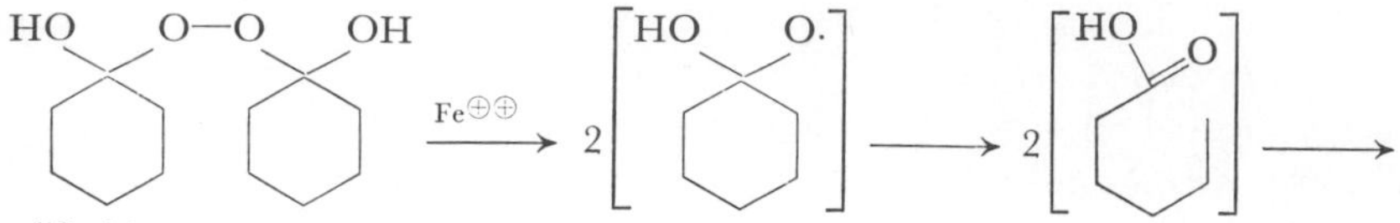

"Cyclohexanone peroxide"

$$HOOC(CH_2)_{10}COOH$$

1,12-Dodecanedioic acid

Carbonyl peroxides are made by the oxidation of alcohols or the hydrogen peroxide treatment of ketones. Sometimes they are available commercially. A promising dimerization process of these peroxides to form acids has been recorded as shown in the following example. The example is limited in application but thought-provoking.

(a) Preparation of 1,12-dodecanedioic acid. Solid ferrous sulfate

Cyclohexanone peroxide (commercially available) $\xrightarrow{\text{FeSO}_4}$ 1,12-dodecanedioic acid

heptahydrate, 72.3 g., was dissolved under nitrogen in 700 ml. of deaerated methanol. To the stirred solution at room temperature 24.6 g. of cyclohexanone peroxide in 300 ml. of methanol was added dropwise in 5 min. The mixture was brought to pH 11, the methanol removed by distillation, and the residue saponified for 1 hr. The cooled mixture was filtered, and the filtrate acidified to yield crude 1,12-dodecanedioic acid, m.p. 125–126.5° (52%) [28].

3. From Carbonyl Compounds by Treatment with Alkali (Cannizzaro)

$$2\,RCHO \xrightarrow{\text{NaOH}} RCOONa + RCH_2OH$$

This synthesis has been discussed previously under Alcohols, Chapter 4, C.4.

(a) Preparation of m-bromobenzoic acid (96% from m-bromobenzaldehyde) [29].

4. From Benzils by Treatment with Alkali (Benzilic Acid Rearrangement)

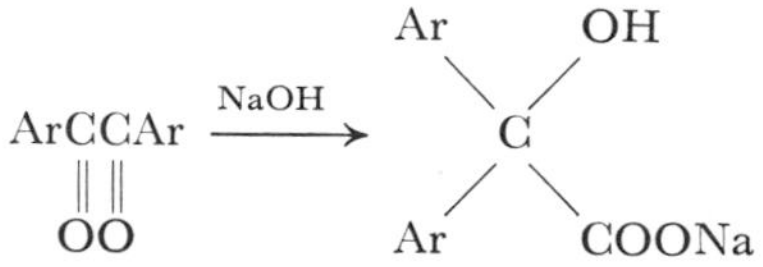

This synthesis is the most common method for preparing benzilic acids [30]. It has been applied as well to a limited extent to aliphatic [31] and alicyclic diketones [32]. The reaction is applicable to benzoins which may be converted into the benzilic acid via the benzil in one operation [33], as well as to α-ketoaldehydes and related types [34]. The mechanism may be represented as follows:

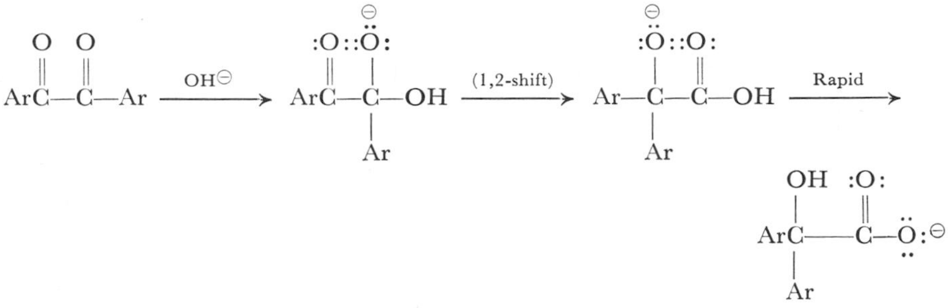

The yields are usually good.

A similar rearrangement occurs to give benzilic ester, 93%, when benzil in dry benzene is treated with potassium t-butoxide-t-butyl alcohol [35]. The yield of methyl benzilate when sodium methoxide is used is much less.

(a) **Preparation of benzilic acid** (84–90% from benzaldehyde) [33].

(b) **Preparation of 2,2'-dimethoxybenzilic acid** (89% from o-anisil) [36].

5. From Cyclic Ketones and from Selected Phenols by Photochemical Oxidation

A recent review [37] describes the conversion of many cyclic ketones into open-chain acids by photochemical activation. The conversion occurs via the ketene [38]:

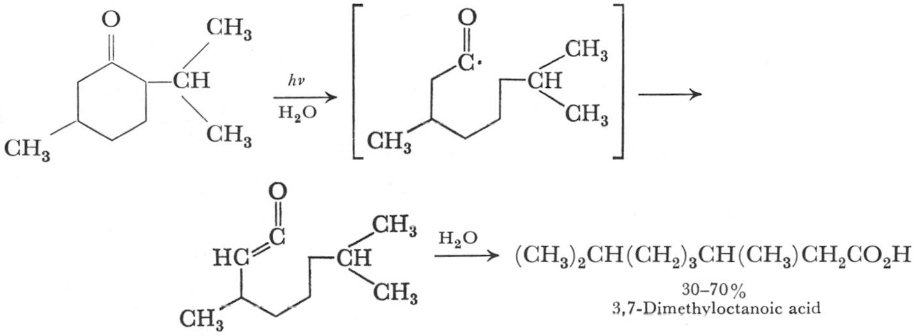

Not only cycloalkanones but also terpene and steroid ketones have yielded acids by this method. The reaction is remindful of the Cannizzaro synthesis insofar as an internal reduction-oxidation step is involved.

Some selected phenols have been oxidized photochemically, as shown in Ex. a.

(a) Preparation of 1-carboxyfluorene-9-acetic acid (82% from 3-hydroxyfluoranthrene in 1 N sodium hydroxide by ultraviolet light exposure for 14 days) [39].

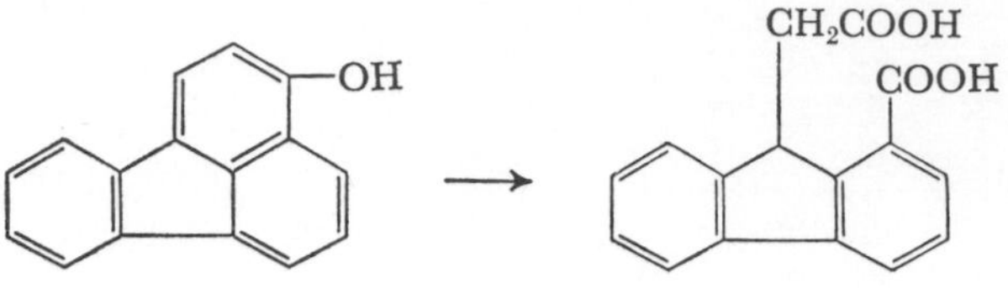

6. From Alkenes

$$RCH{=}CH_2 \xrightarrow{\text{KMnO}_4} RCOOH$$

Although this method has been employed rather frequently for the determination of structure, it is not very commonly used for the preparation of carboxylic acids. A few cases may be found in the literature with alkaline potassium permanganate the most common oxidizing agent. As is true with many oxidation processes, the yields are not always high. The reaction produces a sequence of products, most of which do not survive in the presence of a powerful oxidizing agent. First to appear is the epoxide or glycol which is cleaved to the aldehyde, the latter finally being oxidized to the carboxylic acid.

Isotopic assay frequently involves degradation of one acid to a lower homolog. At least a half-dozen methods are available [40]. The classical Barbier-Wieland degradation is the best known.

$$RCH_2CO_2Me \xrightarrow{C_6H_5MgBr} RCH_2\overset{\overset{\displaystyle H}{\overset{\displaystyle |}{\underset{\displaystyle |}{O}}}}{C}(C_6H_5)_2 \xrightarrow{-H_2O}$$

$$RCH{=}C(C_6H_5)_2 \xrightarrow{[O]} RCO_2H + (C_6H_5)_2CO$$

The method has been used recently to synthesize a series of monofluoroalkanoic acids [41]; see Ex. b.3 also. However, other methods of degradation for isotope assay are equally efficient, if not superior: Beckmann second-order rearrangement of α-ketoxime essentially converting RCH_2CO_2H into RCO_2H in 50–75% overall yield (see G.3) and Varrentrapp cleavage of an unsaturated acid, essentially converting RCH_2CH_2COOH into $RCOOH$ in about 50% overall yields (see E.2).

The following reagent is quite efficient for oxidation of olefins and may well be an improvement on the Barbier-Wieland degradation. Sodium metaperiodate with potassium permanganate as a catalyst cleaves the olefinic double bond rapidly at 25° [42–44]. Disubstituted or trisubstituted olefins give carboxylic acids, probably via the glycol and ketol:

$$\begin{array}{ccccc}
RCH & RCHOH & RCHOH & RCOOH \\
\| & \longrightarrow & | & \longrightarrow & | & \longrightarrow & + \\
R'CH & R'CHOH & R'CO & R'COOH
\end{array}$$

Terminal olefins can be made to give formaldehyde in high yield:

$$RCH{=}CH_2 \rightarrow RCHO + CH_2O$$

while those with an isopropylidine group may give a quantitative yield of acetone.

The method has been useful in locating the double bond in unsaturated natural products. The desired pH may be obtained by the use of potassium carbonate, and insoluble materials may be dissolved in *t*-butanol, pyridine, or dioxane. To cite one example, the structure of the sex attractant of the gypsy moth was elucidated by its degradation to 3-acetoxy-1-nonanoic acid and 7-hydroxyheptanoic acid [45]:

$$CH_3(CH_2)_5\underset{\underset{OCOCH_3}{|}}{CH}CH_2CH{=}CH(CH_2)_5CH_2OH \longrightarrow$$

$$CH_3(CH_2)_5\underset{\underset{OCOCH_3}{|}}{CH}CH_2COOH + HOOC(CH_2)_5CH_2OH$$

92%   (71% as pimelic acid)

An interesting variant of the periodate-permanganate reagent is the periodate-ruthenium tetroxide reagent which has been used in the Barbier-Wieland degradation [46]; see Ex. b.3 also.

A combination of dinitrogen tetroxide, N_2O_4, and nitric acid has been shown to convert *t*-butyl alcohol, presumably via the olefin, into α-hydroxyisobutyric acid (see Ex. b.4). Whether the reaction is general for terminal alkenes is not known.

(a) Preparation of 4-acetamidoquinaldinic acid.

(1) **4-Acetamido-2-styrylquinoline.** 4-Aminoquinaldine, 1 mole, 3 moles of acetic anhydride, and 3 moles of benzaldehyde were heated for 3 hr. at 155–160°. The excess of aldehyde was removed by steam distillation and the solid remaining was ground under methanol and crystallized from 67 % ethanol to give 91 % of product, m.p. 199.5°.

(2) **4-Acetamidoquinaldinic acid.** The styrylquinoline, 4.6 g. in 32 ml. of pyridine and 6.5 ml. of water, was stirred while 6.7 g. of $KMnO_4$ was added during 1 hr. at an internal temperature of 18–20°. Water (total, 20 ml.) was added from time to time to keep the mixture fluid and stirring was continued for 30 min. After acidification (to Congo paper) and decolorization with $NaHSO_3$, the precipitated acid which formed was purified in the usual manner to give a product (82 %) m.p. 237° [47].

(b) Other examples

(1) **Azelaic acid** (32–36 % from castor oil and potassium permanganate) [48].

(2) **α,β,β-Trichlorodifluoropropionic acid** (74 % from 1,3,4,4-tetrachlorotetrafluoro-1-butene and potassium permanganate) [49].

(3) Norlithocolic acid (80% by Barbier-Wieland degradation of lithocolic acid (3-hydroxycholanic acid) using sodium metaperiodate with catalytic amounts of ruthenium dioxide) [50].

(4) α-Hydroxyisobutyric acid (52% from t-butyl alcohol, perhaps via the olefin, dinitrogen tetroxide, and nitric acid at 65°) [51].

(5) Azelaic and pelargonic acids (quantitative yields from methyl

$$CH_3(CH_2)_7CH=CH(CH_2)_7COOCH_3 \rightarrow CH_3(CH_2)_7COOH + HOOC(CH_2)_7COOH$$

Pelargonic acid   Azelaic acid

oleate and potassium carbonate in 30% t-butanol oxidized by a solution of sodium metaperiodate containing a trace of potassium permanganate) [44].

7. From Alkenes and Alkynes via the Ozonide

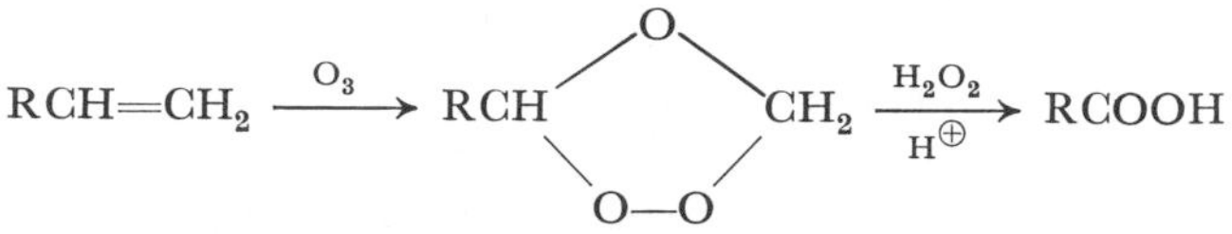

Although this reaction has long been employed in locating the site of unsaturation in alkenes and alkynes, it also has been used as a method of synthesis. A review of the reaction appeared in 1940 [52]. Ozonides are reasonably stable unless considerable unsaturation is present. For the preparation of acids it is now common practice to employ hydrogen peroxide after ozonization has occurred in order to assure complete oxidation. Yields are often very satisfactory.

Recently cycloolefins in emulsion with aqueous, alkaline hydrogen peroxide have been converted in one step into α,ω-alkanedicarboxylic acids usually in 60–80% yield [53]. (the ozonide is not involved in this reaction):

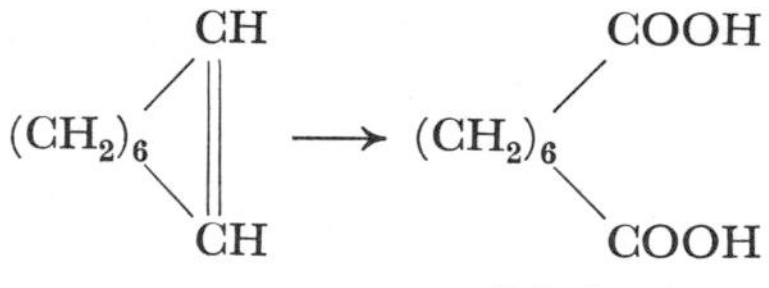

Suberic acid

(a) Preparation of 3-Pyridylacetic acid. 3-Allylpyridine, 0.01 mole, in 3 ml. of water dissolved on the addition of 4.5 ml. of approx. 3 N HCl. A slow stream of ozone was passed through the solution, kept cool with water, and after 45 min. the theoretical amount of ozone had been absorbed. Then 2 ml. of 30% hydrogen peroxide was added to complete the oxidation of the aldehydes formed to carboxylic acids. The reaction product, dried in a vacuum desiccator, was obtained in nearly quantitative yield as calculated for 3-pyridylacetic acid hydrochloride. The recrystallized product melted at 153–154° [54].

(b) Other examples

(1) 5-Formyl-4-phenanthroic acid (32–38% from pyrene in dimethylformamide and ozone) [55].

(2) Benzoylglycolic acid (73% from the dibenzoate ester of but-2-yne-1,4-diol and ozone) [56].

(3) **Azelaic acid** (81% from 9-octadecynedioic acid) [57].

(4) **Suberic acid** (63% from a mixture of cyclooctene, an emulsifier (polyoxyethylated lauryl alcohol), sodium hydroxide, and 30% hydrogen peroxide through which ozone at 5° was passed) [53].

8. From Alkylarenes, Heterocycles, or Phenols

$$-CH_3 \xrightarrow{KMnO_4} -COOH$$

The method of oxidizing alkyl groups in carbocyclic systems to carboxyl groups is a common one. A variety of oxidizing agents has been employed, although potassium permanganate or sodium dichromate in acid solution is most frequently used. In some heterocyclic systems, as, for example, the 2- and 4-alkylpyridines or quinolines, selenium dioxide in sufficient quantity to override aldehyde formation or oxygen in the presence of potassium *t*-butoxide in dimethylformamide may be employed [58]. For alkylarenes which are difficult to oxidize, and there are many, conversion into the acid may be more efficient by initial bromination to the benzyl bromide followed by oxidation with alkaline potassium permanganate.

Aqueous sodium dichromate at high temperatures and autogenous pressures has proved to be an unusual oxidizing agent, yielding terminal acids from alkylarenes (Ex. c.4), and it is an exceptionally fine agent for oxidizing alkyl groups attached to polynuclear hydrocarbons (Ex. c.5) [59]. Mention should be made also of the specific side-chain oxidation of fluorotoluenes or xylenes with dilute nitric acid (Ex. c.6).

Oxidation of an alkylphenol to a hydroxybenzoic acid is not feasible under ordinary circumstances because of the sensitivity of the ring to oxidation. However, a specific reagent, lead dioxide, appears to accomplish this oxidation at least in one case [60]:

$$\text{(Mixed with some water)} + KOH + PbO_2 \xrightarrow[\text{2. } H_3O^\oplus]{\text{1. 240–250°}} \text{2-Hydroxyisophthalic acid, 46–61\%}$$

(a) **Preparation of o-chlorobenzoic acid** (76–78% from the *o*-chlorotoluene consumed) [61].

(b) **Preparation of p-acetylbenzoic acid.** Air at 0.8 l./min. was passed through a column of 60 g. of *p*-methylacetophenone and 0.01 g. of nickel benzoate at 115–120° for 56 hr. The acid, 28.5 g., m.p. 205–206°, was isolated in 70% yield based on the ketone consumed. If oxygen were more soluble in organic media, air oxidation would be used more frequently. Increase in pressure is an assistance in increasing solubility [62]

(c) **Other examples**

(1) **Isophthalic acid** (89% from *m*-xylene and aqueous ammonium sulfate at 325° in an autoclave) [63].

(2) **p-(Triphenylsilyl)benzoic acid** (81% from triphenyl-(p-tolyl)-silane and chromium trioxide in acetic anhydride-acetic acid and a bit of H_2SO_4) [64].

(3) **2-(β,β,β-Trichloroethyl)-4-chlorobenzoic acid** (93% from 2-trichloroethyl-4-chlorotoluene, sodium dichromate, and sulfuric acid in acetic acid) [65].

(4) **Phenylacetic acid** (96% from ethylbenzene, sodium dichromate, and water at 275° in an autoclave [66]; this reaction is surprising in that a terminal oxidation product is obtained; the attack may take place at the α-carbon to form a chromate ester which then isomerizes).

(5) **2,3-Naphthalenedicarboxylic acid** (87–93% from 2,3-dimethyl-naphthalene, sodium dichromate, and water at 250° in an autoclave) [67].

(6) **4-Fluoro-2-nitrobenzoic acid** (69% from 4-fluoro-2-nitrotoluene and 15% aq. nitric acid at 190° for 5 hr. under pressure) [68].

(7) **Isonicotinic acid** (61% from 4-picoline, potassium t-butoxide in dimethylformamide under 4 atm. pressure of oxygen in a Parr apparatus at room temperature for 10 min.; t-butyl alcohol is added to suppress oxidation of dimethylformamide) [58]; for operating at atmospheric pressure and elevated temperatures, hexamethylphosphoramide has been recommended as a solvent stable to oxygen) [69].

9. From Furans (Substituted)

There are a few cases in the literature in which carboxylic acids have been produced in satisfactory yield by the destruction of a furan ring with potassium permanganate. Thus dimethylethylacetic acid and 1-methyl-1-cyclohexane-carboxylic acid have been prepared from the appropriate furan [70]. More recently 1-benzyl-3-methylpyrazole-5-carboxylic acid has been obtained by the oxidation of 5-furyl-1-benzyl-3-methylpyrazole [71].

(a) **Preparation of dimethylethylacetic acid.** 5-t-Amylfuran-2-carboxylic acid, 9 g., was dissolved in a solution of 18 g. of potassium hydroxide in 180 g. of water, and a solution of 35 g. of potassium permanganate in 600 g. of water was dropped slowly into the cooled and agitated solution. A permanent color was obtained after the stated quantity of permanganate was added. Now sulfuric acid (1:2) was added with shaking and cooling to 10° to permit a strong Congo acid reaction and, for the destruction of the oxalic acid present, stirring was continued for ¼ hr. Then sodium bisulfite was dropped in until the manganese dioxide was dissolved, and the excess sulfur dioxide was removed by the addition of permanganate solution until a brown coloration was obtained. The dimethylethylacetic acid was distilled off with steam, extracted with ether, and distilled in vacuum. Yield 4 g. (79%), b.p. 79–81° (11 mm.) [70].

(b) Preparation of pyrazolecarboxylic acids (usually 70–80% from furylpyrazoles) [71].

10. From Aryl Methyl Ketones (to α-Keto Acids)

$$\text{ArCOCH}_3 \xrightarrow{\text{KMnO}_4} \text{ArCOCOOH}$$

This synthesis must be carried out under mild conditions to prevent further oxidation to a benzoic acid. The method is particularly applicable to high-melting ketones in a homogeneous solution of pyridine, potassium permanganate being the oxidizing agent. The yields are only fair [72].

(a) Preparation of phthalonic acid. A mixture of potassium hydroxide, 1 g., in 100 ml. of water and 1.75 g. of 2-acetylbenzoic acid in 150 ml. of pyridine was treated gradually over 2 hr. with 4.75 g. of potassium permanganate in 400 ml. of water at 10–15° with stirring and cooling. After short standing, excess permanganate was destroyed by addition of a minimum of sodium sulfite, and the manganese dioxide was filtered. The filtrate was acidified with hydrochloric acid and extracted continuously with ether for 6 hr. Distillation of the ether extract gave an oil which crystallized. Extraction of the crystals with chloroform gave phthalonic acid in the soluble portion, 0.94 g., m.p. 142°, 45%. The chloroform-insoluble portion was o-phthalic acid [72].

11. From α-Keto acids

$$\text{RCOCOOH} \xrightarrow{\text{H}_2\text{O}_2} \text{RCOOH}$$

This reaction has been employed in a limited number of cases for oxidizing α-keto acids in the presence of alkali to carboxylic acids by the use of 30% hydrogen peroxide. In many instances the α-ketocarboxylic acid is obtained as the ester by replacing the active hydrogen through treatment with diethyl oxalate and sodium methoxide.

(a) Preparation of 3,4-dimethoxyphenylacetic acid [73]:

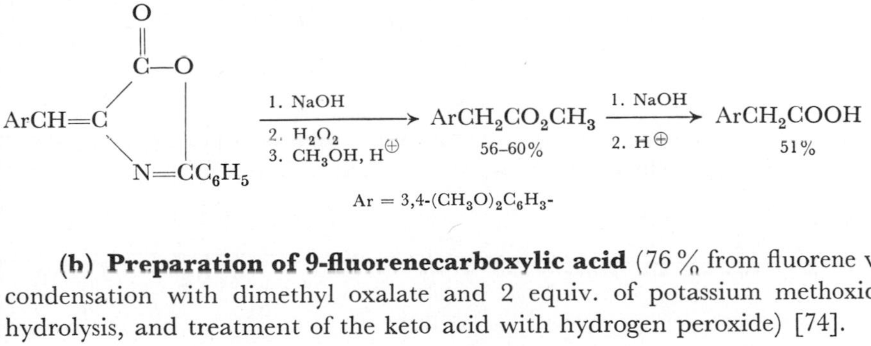

Ar = 3,4-(CH₃O)₂C₆H₃-

(b) Preparation of 9-fluorenecarboxylic acid (76% from fluorene via condensation with dimethyl oxalate and 2 equiv. of potassium methoxide, hydrolysis, and treatment of the keto acid with hydrogen peroxide) [74].

12. From Methyl Ketones and β-Diketones

(a) Treatment with halogen and alkali (haloform reaction).   In

$$RCOCH_3 \xrightarrow[\text{NaOH}]{\text{X}_2} RCOCX_3 \xrightarrow{\text{NaOH}} RCOONa + CHX_3$$

many cases this synthesis gives a satisfactory conversion of the acetyl into the carboxyl group. Various combinations of chlorine or bromine and sodium or potassium hydroxide, the corresponding hypohalites, or even commercial bleaching agents, are used as the reagent. If the methyl ketone is water-insoluble, vigorous stirring and the addition of a small amount of detergent facilitates the reaction. Excess of chlorine in the hypohalite reagent is to be avoided [75]. Alkyl ketones higher than methyl may be used if there are two α-hydrogen atoms present [76]. Methylene and methyl groups attached to an aromatic ring are affected if the nucleus contains an acetyl group which is also oxidized to carboxyl [77]. Thus p-methylacetophenone gives terephthalic acid:

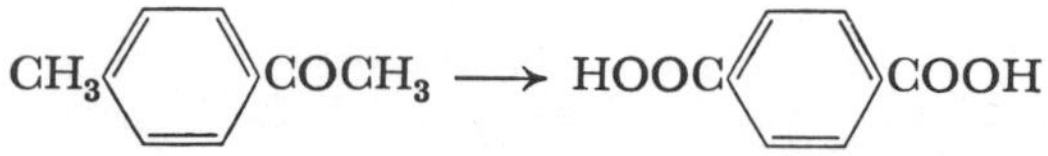

47%

and 2-acetyl-9,10-dihydrophenanthrene gives 2,2',4-tricarboxybiphenyl at pH 12–13:

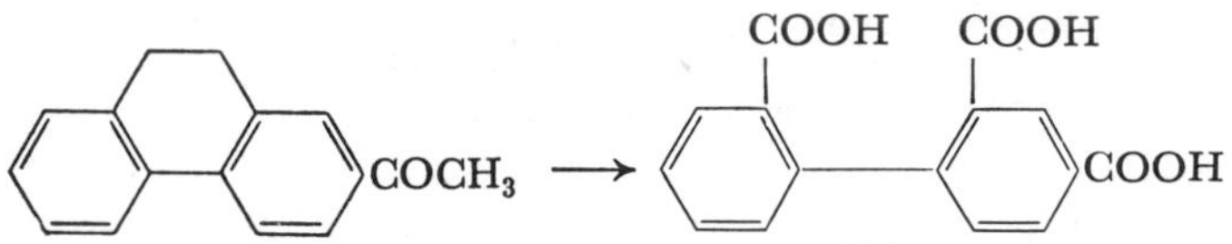

49%

The mechanism of this multistep reaction consists first of the removal of an α-hydrogen atom by the hydroxyl ion followed by an electrophilic attack of the positive halogen ion of the hypohalite:

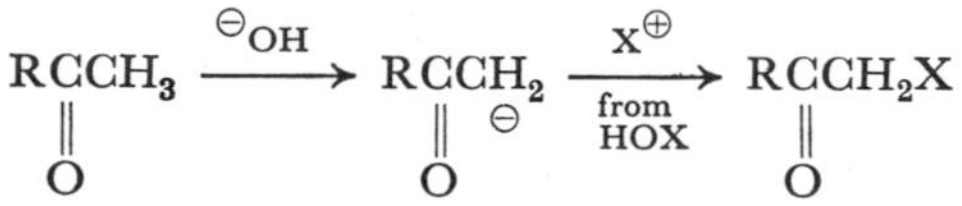

The electronegative character of the halogen makes the α-hydrogens which remain more readily replaceable, and thus substitution of halogen on the same carbon atom occurs until

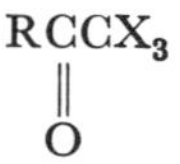

$$\underset{\text{O}}{\overset{\|}{R}CCX_3}$$

is obtained. The product is then subject to a nucleophilic attack of the hydroxy ion to form an anion which cleaves to give the anion of the acid and the haloform:

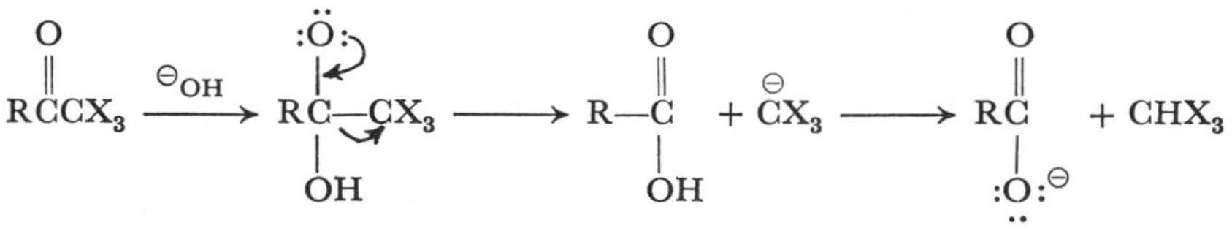

Not only are methyl ketones converted into acids by hypohalites, but also those compounds which split to form methyl ketones. β-Diketones form such a class of compounds. Under typical hypohalite conditions they react as follows:

$$\underset{O}{\overset{O}{\underset{\parallel}{R\overset{\parallel}{C}CH_2\overset{\parallel}{C}R}}} \xrightarrow[\text{NaOCl}]{\text{NaOH}} 2\,RCO_2Na + CHCl_3$$

Halogenation may precede splitting.

This synthesis is of greatest value with cyclic diketones when diacids are the product, as shown in Ex. a.2. An alternative reagent for the conversion of 1,3-diketones into diacids is aqueous sodium periodate [78]:

$$\longrightarrow HOOC(CH_2)_3COOH$$
Glutaric acid, 86.5%

(1) Preparation of β-naphthoic acid (88% from methyl-β-naphthyl ketone) [75].

(2) Preparation of β,β-dimethylglutaric acid (91–96% from methone, 5,5-dimethyl-1,3-cyclohexanedione) [79].

(3) Other examples

    (i) β,β-Dimethylacrylic acid (49–53% from mesityl oxide) [80].

    (ii) 5-n-Butylpyridine-2-carboxylic acid (90% from the ethyl ketone) [81].

(b) By treatment with iodine-pyridine and alkali. This method,

$$ArCOCH_3 \xrightarrow[C_5H_5N]{I_2} (ArCOCH_2\overset{\oplus}{N}C_5H_5)\overset{\ominus}{I} \xrightarrow{\text{NaOH}} ArCOONa$$

devised by King [82], represents a second way for converting the acetyl into the carboxyl group. It has been employed, giving a high yield, to a limited extent in cases in which the haloform reaction is unsatisfactory [83, 84].

The mechanism of this synthesis appears to bear some resemblance to that of the haloform reaction. Iodine in the presence of the base pyridine substitutes for an α-hydrogen atom to produce $ArCOCH_2\overset{\oplus}{N}C_5H_5I^{\ominus}$ which is attacked by the hydroxyl ion to give an anion which cleaves to produce the carboxylic acid:

$$\underset{O}{\overset{O}{\underset{\parallel}{Ar\overset{\parallel}{C}CH_2\overset{\oplus}{N}C_5H_5I^{\ominus}}}} \xrightarrow{\overset{\ominus}{O}H} \underset{OH}{\overset{:\overset{\ominus}{\overset{..}{O}}:}{\underset{|}{ArC-CH_2\overset{\oplus}{N}C_5H_5I}}} \longrightarrow$$

$$\underset{OH}{\overset{:O:}{\underset{|}{Ar\overset{..}{C}}}} + CH_2\overset{\ominus}{\overset{\oplus}{N}}C_5H_5I^{\ominus} \longrightarrow \underset{:\underset{\ominus}{\overset{..}{O}}:}{\overset{:O:}{\underset{|}{Ar\overset{\parallel}{C}}}} + CH_2\overset{\oplus}{N}C_5H_5I^{\ominus}$$

(1) Preparation of 1-naphthoic acid.

(i) 1-Naphthoylmethyl pyridinium iodide. Iodine, 0.1 mole, was added to 0.1 mole of 1-acetylnaphthalene in 30–50 ml. of pyridine. After heating for 30 min. on the steam bath, the solution was allowed to stand over-night, after which the excess of pyridine was removed and the mixture of pyridine hydroiodide and substituted β-ketoalkylpyridinium iodide was separated by taking advantage of the insolubility of the latter in water, methyl, or ethyl alcohol. Purification by crystallization gave a 92 % yield, m.p. 219–220°.

(ii) 1-Naphthoic acid. To a suspension of 1–5 g. of the substituted β-ketoalkylpyridinium iodide in 50–100 ml. of water or 50 % ethyl alcohol, 1–3 g. of sodium hydroxide was added. After heating for 1 hr. on the steam bath, the acidic fractions were isolated in the usual manner. Yield 90 %; m.p. 160–161° [82].

(2) Other examples

(i) Various aromatic acids (80–90 % from the pyridinium iodides) [82].

(ii) Various hydroxybenzoic acids (40–98 % from the pyridinium iodides) [85].

(iii) 5-Indanecarboxylic acid (75 % from the pyridinium iodide) [83].

(iv) 6-Carboxydehydroabietic acid (70–80 % from the pyridinium iodide) [84].

1. J. W. Ladbury and C. F. Cullis, *Chem. Rev.*, **58,** 403 (1958).
2. H. C. Brown and C. P. Garg, *J. Am. Chem. Soc.*, **83,** 2952 (1961).
3. R. Filler, *Chem. Rev.*, **63,** 21 (1963).
4. K. Heyns and H. Paulsen in W. Foerst, *Newer Methods of Preparative Organic Chemistry*, Vol. 2, Academic Press, New York, 1963, p. 303.
5. C. R. Noller, *Chemistry of Organic Compounds*, 3rd ed., W. B. Saunders Co., Philadelphia, 1965, p. 151.
6. R. Stewart, *Oxidation Mechanisms. Applications to Organic Chemistry*, W. A. Benjamin, Inc., New York, 1964; W. A. Waters, *Mechanisms of Oxidation of Organic Compounds*, John Wiley and Sons, New York, 1964.
7. A. Schriesheim *et al.*, *J. Org. Chem.*, **30,** 3768 (1965).
8. T. G. Clarke *et al.*, *Tetrahedron Letters*, 5685 (1968).
9. E. J. Corey *et al.*, *J. Am. Chem. Soc.*, **90,** 5616 (1968).
10. M. S. Newman *et al.*, *J. Am. Chem. Soc.*, **82,** 2498 (1960).
11. J. G. Wallace, *Hydrogen Peroxide in Organic Chemistry*, E. I. duPont de Nemours Co., 1962, p. 57.
12. K. Heyns and L. Blazejewicz, *Tetrahedron*, **9,** 67 (1960).
13. F. L. M. Pattison *et al.*, *J. Am. Chem. Soc.*, **78,** 2255 (1956).
14. G. R. Robertson, *Org. Syn.*, Coll. Vol. **1,** 138 (1941).
15. B. A. Ellis, *Org. Syn.*, Coll. Vol. **1,** 18 (1941).
16. W. J. Hickinbottom, *Reactions of Organic Compounds*, Longmans, Green and Co., New York 1948, p. 108.
17. L. F. Fieser and J. Szmuskovicz, *J. Am. Chem. Soc.*, **70,** 3352 (1948).
18. S. G. Powell *et al.*, *Org. Syn.*, Coll. Vol. **1,** 168 (1941).
19. Ref. 11, p. 56.
20. K. Heyns and M. Beck, *Chem. Ber.*, **89,** 1648 (1956).
21. K. Heyns, *Ann. Chem.*, **558,** 171 (1947).
22. W. Langenbeck and M. Richter, *Chem. Ber.*, **89,** 202 (1956).
23. C. Moureu and R. Chaux, *Org. Syn.*, Coll. Vol. **1,** 166 (1941).

24. J. R. Ruhoff, *Org. Syn.*, Coll. Vol. **2**, 315 (1943).
25. R. L. Shriner and E. C. Kleiderer, *Org. Syn.*, Coll. Vol. **2**, 538 (1943).
26. H. M. Walborsky *et al.*, *J. Am. Chem. Soc.*, **73**, 2590 (1951).
27. E. Campaigne and W. M. LeSuer, *Org. Syn.*, Coll. Vol. **4**, 919 (1963).
28. N. Brown *et al.*, *J. Am. Chem. Soc.*, **77**, 1756 (1955).
29. G. Lock, *Org. Reactions*, **2**, 112 (1944).
30. C. D. Shacklett and H. A. Smith, *J. Am. Chem. Soc.*, **75**, 2654 (1953).
31. B. H. Nicolet and A. E. Jurist, *J. Am. Chem. Soc.*, **44**, 1136 (1922).
32. O. Wallach, *Ann. Chem.*, **414**, 296 (1918); **437**, 148 (1924).
33. D. A. Ballard and W. M. Dehn, *Org. Syn.*, Coll. Vol. **1**, 89 (1941).
34. J. G. Aston *et al.*, *Org. Syn.*, Coll. Vol. **3**, 538 (1955).
35. W. von E. Doering and R. S. Urban, *J. Am. Chem. Soc.*, **78**, 5938 (1956).
36. A. H. Ford-Moore, *J. Chem. Soc.*, 952 (1947).
37. G. Quinkert, *Angew. Chem.*, **77**, 229 (1965).
38. G. O. Schenck and F. Schaller, *Chem. Ber.*, **98**, 2056 (1965).
39. A. Sieglitz *et al.*, *Chem. Ber.*, **95**, 3013 (1962).
40. A. Murray, III, and D. L. Williams, *Organic Synthesis with Isotopes*, Interscience Publishers, New York, 1958, Pt. I, p. 42.
41. F. L. M. Pattison and R. L. Buchanan, *Biochem J.*, **92**, 100 (1964).
42. R. U. Lemieux and E. Von Rudloff, *Canad. J. Chem.*, **33**, 1701, 1710 (1955).
43. E. Von Rudloff, *Canad. J. Chem.*, **33**, 1714 (1955).
44. E. Von Rudloff, *Canad. J. Chem.*, **34**, 1413 (1956).
45. M. Jacobson *et al.*, *J. Am. Chem. Soc.*, **83**, 4819 (1961).
46. G. Stork *et al.*, *J. Am. Chem. Soc.*, **85**, 3419 (1963).
47. R. Royer, *J. Chem. Soc.*, 1803 (1949).
48. J. W. Hill and W. L. McEwen, *Org. Syn.*, Coll. Vol. **2**, 53 (1943).
49. R. N. Haszeldine and J. E. Osborne, *J. Chem. Soc.*, 3880 (1955).
50. Y. Yanuka and S. Sarel, *Bull. Res. Council Israel*, **6A**, 286 (1957); *C.A.*, **52**, 408 (1958).
51. SNAM, French Patent 1,373,229, Oct. 9, 1964; *C.A.*, **62**, 9016 (1965).
52. L. Long, Jr., *Chem. Rev.*, **27**, 437 (1940).
53. M. I. Fremery and E. K. Fields, *J. Org. Chem.*, **28**, 2537 (1963).
54. J. P. Wibaut and H. C. Beyerman, *Rec. Trav. Chim.*, **70**, 977 (1951).
55. R. E. Dessy and M. S. Newman, *Org. Syn.*, Coll. Vol. **4**, 484 (1963).
56. R. Criegee and M. Lederer, *Ann. Chem.*, **583**, 29 (1953).
57. W. J. Gensler and H. N. Schlein, *J. Am. Chem. Soc.*, **77**, 4846 (1955).
58. H. Schriesheim *et al.*, *J. Org. Chem.*, **28**, 410 (1963).
59. H. Shechter *et al.*, *J. Org. Chem.*, **30**, 1453 (1965).
60. C. Graebe and H. Kraft, *Chem. Ber.*, **39**, 794 (1906); D. Todd and A. E. Martell, *Org. Syn.*, **40**, 48 (1960).
61. H. T. Clarke and E. R. Taylor, *Org. Syn.*, Coll. Vol. **2**, 135 (1943).
62. P. G. Sergeev and A. M. Sladkov, *J. Gen. Chem., USSR (Eng. Transl.)*, **27**, 893 (1957).
63. W. G. Toland, *J. Am. Chem. Soc.*, **82**, 1911 (1960).
64. H. Gilman *et al.*, *J. Am. Chem. Soc.*, **78**, 1689 (1956).
65. M. S. Newman and L. L. Wood, Jr., *J. Org. Chem.*, **23**, 1236 (1958).
66. R. H. Reitsema and N. L. Allphin, *J. Org. Chem.*, **27**, 27 (1962).
67. L. Friedman, *Org. Syn.*, **43**, 80 (1963).
68. G. Valkanas and H. Hopff, *J. Chem. Soc.*, 1925, 3475 (1963).
69. T. J. Wallace and F. A. Baron, *J. Org. Chem.*, **30**, 3520 (1965).
70. T. Reichstein *et al.*, *Helv. Chim. Acta*, **18**, 721 (1935).
71. A. P. Terent'ev *et al.*, *J. Gen. Chem. USSR (Eng. Transl.)*, **30**, 2901 (1960).
72. J. Cymerman-Craig *et al.*, *Australian J. Chem.*, **9**, 222 (1956).
73. H. R. Snyder *et al.*, *Org. Syn.*, Coll. Vol. **2**, 333 (1943).
74. A. Campbell and S. H. Tucker, *J. Chem. Soc.* 2623 (1949).
75. M. S. Newman and H. L. Holmes, *Org. Syn.*, Coll. Vol. **2**, 428 (1943).
76. R. Levine and J. R. Stephens, *J. Am. Chem. Soc.*, **72**, 1642 (1950).

77. D. D. Neiswender, Jr. *et al.*, *J. Am. Chem. Soc.*, **82**, 2876 (1960).
78. M. L. Wolfrom and J. M. Bobbitt, *J. Am. Chem. Soc.*, **78**, 2489 (1956).
79. W. T. Smith and G. L. McLeod, *Org. Syn.*, Coll. Vol. **4**, 345 (1963).
80. L. I. Smith *et al.*, *Org. Syn.*, Coll. Vol. **3**, 302 (1955).
81. E. Hardegger and E. Nikles, *Helv. Chim. Acta*, **40**, 1016 (1957).
82. L. C. King, *J. Am. Chem. Soc.*, **66**, 894 (1944).
83. R. T. Arnold *et al.*, *J. Am. Chem. Soc.*, **72**, 4193 (1950).
84. Y. T. Pratt, *J. Am. Chem. Soc.*, **73**, 3803 (1951).
85. L. C. King *et al.*, *J. Am. Chem. Soc.*, **67**, 2089 (1945).

C. Carbonation and Carboxymethylation of Organometallic Compounds

1. Carbonation

$$RMgX \xrightarrow{CO_2} RCOOMgX \xrightarrow{H_3\overset{\oplus}{O}} RCOOH$$

$$RLi \xrightarrow{CO_2} RCOOLi \xrightarrow{H_3\overset{\oplus}{O}} RCOOH$$

Organometallic compounds are widely used to give high yields in the synthesis of carboxylic acids. The Grignard reagent is the most common, but in recent years the use of organolithium compounds has been on the increase, probably in part because of the availability from suppliers of the simpler ones used in the metalation. The mechanism of the reaction, as shown with the Grignard reagent, probably consists first of rapid coordination of the magnesium atom with the oxygen atom of carbon dioxide followed by a rate-determining nucleophilic attack of the R group:

$$
\begin{array}{ccc}
C_2H_5\ddot{O}C_2H_5 & C_2H_5OC_2H_5 & O \\
| & | & \| \\
RMgX \quad + CO_2 \xrightarrow{Rapid} & XMgO{=}C{=}O \xrightarrow{Slow} & XMgOCR + C_2H_5OC_2H_5 \\
| & | & \\
C_2H_5\ddot{O}C_2H_5 & R &
\end{array}
$$

This mechanism is surmised from the behavior of ketones and nitriles with Grignard reagents.

Discussions of the halogen-metal interconversion [1] and the metalation reaction with organolithium compounds [2] are available. The lithium compound is more reactive than the Grignard, can be prepared and used in a hydrocarbon solvent, and is less likely to give reduction or 1,4-addition products. The only disadvantages of organolithium compounds are: they are more expensive than the Grignard reagent; the environment must be more rigorously dry; and a nitrogen atmosphere is almost imperative. For both reagents, the organolithium or the Grignard reagent, carbonation may proceed beyond the acid salt stage to the ketone, particularly if the salt is soluble in the medium [3].

From a practical standpoint for a crucial synthesis, the solvent, ether or tetrahydrofuran, should never be exposed to the laboratory atmosphere. Rather, they should be distilled from lithium aluminum hydride into the reaction vessel or percolated through a column containing Dow Desiccant 812 or

References for Section C are on p. 780.

alumina directly into the vessel. As long as the temperature is constant or increasing, the solvent vapor blankets the reagents but, if the temperature is decreasing, air will diffuse into the system, a fault which can be prevented by passing a slow stream of nitrogen under a head of 5–10 cm. of mineral oil into the closed vessel.

The carbonation of a haloalkyllithium derived from a *gem*-dihalide yields a hydroxyacid (Ex. c.6). Similar treatment of a symmetrical dichloroethylene yields a dichloroacrylic acid (Ex. c.7). Olefins can be treated with amylsodium and then carbon dioxide to yield β,γ-unsaturated acids (Ex. c.8). Ketones such as 4-*t*-butylcyclohexanone may be carboxylated to the β-keto acid with trityl-potassium and dry ice (Ex. c.10), but the more common reagent for preparing acids which tend to decarboxylate is magnesium methyl carbonate; here chelation prevents the reverse reaction [4]; (see example c.5 also):

$$Mg(OCH_3)_2 \xrightarrow[DMF]{CO_2} (CH_3OCO)_2Mg \xrightarrow[50°, 4 \text{ hr.}]{CH_3NO_2} O=C \begin{array}{c} \overset{H \quad O}{\underset{\diagup}{C=N}} \\ \diagup \quad \diagdown \\ \diagdown \quad O \\ O-Mg \end{array} \xrightarrow[Ice]{H_3O^\oplus}$$

$$NO_2CH_2CO_2H$$
α-Nitroacetic acid,
63%

(a) **Preparation of mesitoic acid** (86–87 % from bromomesitylene) [5].

(b) **Preparation of 9-phenylfluorene-9-carboxylic acid** (80 % from 9-phenylfluorene and *n*-butyllithium) [6].

(c) **Other examples**

(1) **Acids through the use of the Grignard reagent** (68–86 % from the halide) [7].

(2) **Acids through the use of organolithium compounds** [8].

(3) **Pentachlorobenzoic acid** (77 % crude from pentachlorophenyl-magnesium chloride) [9].

(4) **2,5-Diphenyladipic acid** (*meso*-racemic mixture, 40 % from dimerization of styrene with metallic sodium followed by carbonation) [10].

(5) **3-Carboxy-5-methoxy-2-tetralone** (45 % from 5-methoxy-2-tetralone and magnesium methyl carbonate in DMF) [11].

(6) **Diphenylglycolic acid.** The intermediate chloroalkyllithium is stable at −100° but decomposes rapidly above −65° [12].

$$(C_6H_5)_2CCl_2 \xrightarrow[THF \text{ at } -100°]{LiC_4H_9} (C_6H_5)_2\overset{Li}{CCl} \xrightarrow[H_2O]{CO_2} (C_6H_5)_2COHCOOH$$
40%

(7) *trans*-**1,2-Dichloroacrylic acid** (99 % from symmetrical *trans*-dichloroethylene and *n*-butyllithium in THF at −110° followed by carbonation) [13].

(8) $\Delta^{2,3}$**-Cyclohexenyl carboxylic acid** (rather poor yield from cyclohexene and amylsodium followed by carbonation; cyclopentene yields the cyclopentenyl carboxylic acid). For references quoted see [14].

(9)   5-Hydroxydibenzo[a,d]cycloheptadiene-5-carboxylic   acid

(49% from the potassium ketyl of dibenzocyclo[a,d]heptadiene-5-one and carbon dioxide [15].

(10) 2-Keto-5-*t*-butylcyclohexanecarboxylic   acid (89.5% from 4-*t*-butylcyclohexanone and triphenylmethylpotassium in dry ether poured over dry ice under nitrogen) [16].

(11) β-*t*-Butylacrylic acid (86% from 1-chloro-3,3-dimethyl-1-butene, prepared from *t*-butyl chloride and vinyl chloride, and sodium in kerosene followed by addition of carbon dioxide) [17].

2. Carboxymethylation

$$\underset{RCCH_2R'}{\overset{O}{\overset{\|}{}}} \xrightarrow{LiNH_2} \underset{RCCHLi}{\overset{OR'}{\overset{\|\,|}{}}} \xrightarrow[\text{2. H}^{\oplus},\,H_2O]{\text{1. BrCH}_2\text{COOLi}} \underset{RCCHCH_2COOH}{\overset{OR'}{\overset{\|\,|}{}}}$$

Ketones with active hydrogen atoms may be carboxymethylated by treating the lithium enolate with the lithium salt of haloacetic acid. The first step is the usual metal-for-active-hydrogen replacement, while the second step follows the SN_2 mechanism. The lithium salt of the acid is prepared *in situ* by adding 2 equivalents of the ketone to more than 3 equivalents of lithium amide in liquid ammonia and, after replacing the ammonia by ether, 1 equivalent of the halo acid in ether is added. Yields in five syntheses vary from 48 to 76%.

(a) Preparation   of   1-keto-1,2,3,4-tetrahydronaphthalene-2-acetic acid.   α-Tetralone, 2 moles, was added to 3 moles plus a 5% excess of lithium amide in liquid ammonia, which was then replaced by ether, after which 1 mole of bromoacetic acid in ether was added. The mixture was refluxed for 10 hr. and the acid was recovered by addition of water, separation of the basic aqueous layer, clarification with charcoal and acidification; yield 76% [18].

(b) Preparation of β-benzoylvaleric acid (60% from butyrophenone) [18].

3. Carbonation of Phenol Salts (Kolbe)

The Kolbe synthesis involves the reaction of an alkali salt of a phenol with carbon dioxide to form a hydroxybenzoic acid. Schmitt's modification using relatively low temperature and longer reaction times allows for the complete formation of sodium phenyl carbonate which then can be converted more or less completely into the hydroxyacid [19]:

$$C_6H_5ONa + CO_2 \longrightarrow C_6H_5OCO_2Na \xrightarrow{\Delta}$$

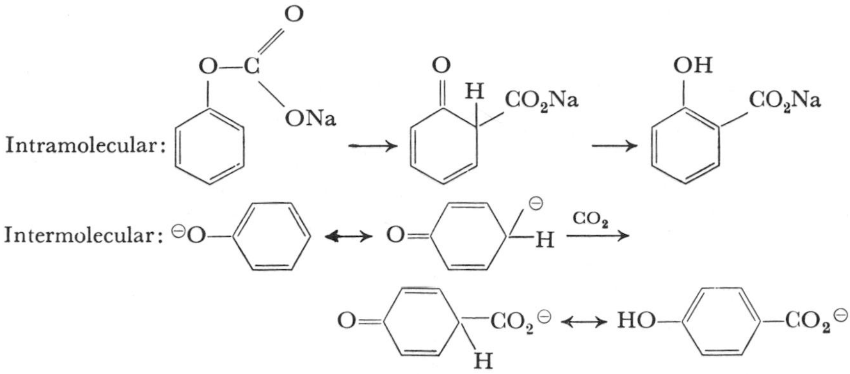

Intramolecular:

Intermolecular:

At low temperature the intramolecular route seems predominant, although it is the more reversible route. At high temperature (and usually with potassium salts) the *para* isomer can be formed almost quantitatively. In this case the intermolecular route is favored. Another example is the carbonation of the sodium salt of β-naphthol. At 150° or less, the sodium salt of 1-carboxy-2-naphthol is formed but, at 200° or more, the salt of 3-carboxy-2-naphthol is formed.

(a) **Preparation of 2,4-dihydroxybenzoic (β-resorcylic) acid** (57–60% from resorcinol) [20]; polyhydroxybenzenes, such as resorcinol, are converted into the corresponding acids, such as β-resorcyclic acid, with remarkable ease merely by passing carbon dioxide through an alkaline solution of the phenol.

4. Rearrangement of Dicarboxylic Acid Salts (Henkel)

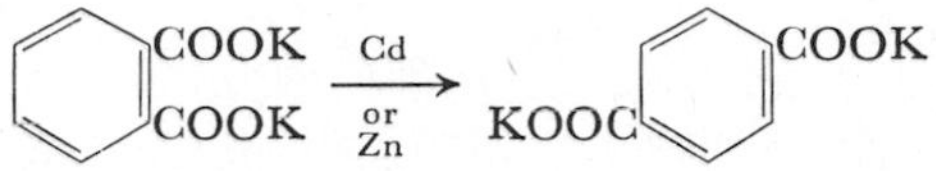

Like the alkali salts of salicyclic acids, which may be rearranged to those of *p*-hydroxybenzoic acids, the salts of dicarboxylic acids may be rearranged to more symmetrical isomers; [21, 22] and earlier papers. The tautomeric transformation of this reaction is obscure but, from the possible course of the reaction, one can see why high temperatures are necessary:

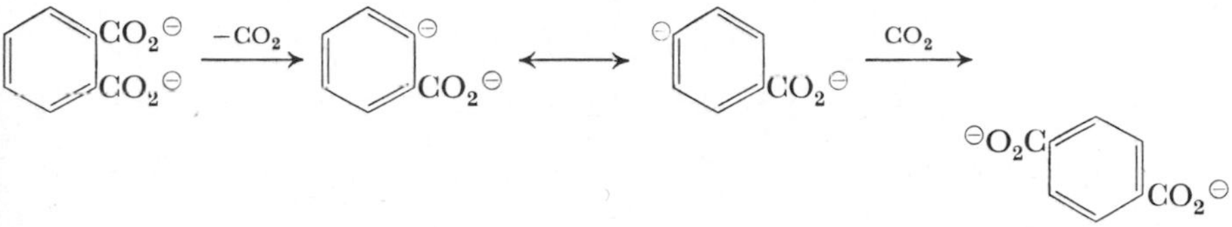

Best results are obtained at a temperature of 400–420° in the absence of moisture at 10 atm. pressure (the reaction also occurs at atmospheric pressure) in the presence of cadmium or zinc or their compounds. Among the rearrangements effected by this procedure are phthalic to terephthalic acid, 1,8-naphthalene-dicarboxylic acid to the 2,6-isomer, 2,2'-biphenyldicarboxylic acid to the 4,4'-acid, and 2,3-pyridinedicarboxylic acid to the 2,5-dicarboxylic acid.

(a) **Preparation of terephthalic acid.** An intimate mixture of 1.6 g. of finely powdered, dry dipotassium phthalate and 0.08 g. of cadmium iodide was heated in a metal bath at $415 \pm 2°$ for 2.5 hr. The product was dissolved in 100 ml. of hot water, filtered, and to the boiling filtrate an excess of hydrochloric acid was added. The precipitate resulting was filtered while hot, washed thoroughly with hot water, and dried. Terephthalic acid (0.93 g., 85%), the dimethyl ester of which melted at 140°, was recovered [23].

(b) **Preparation of naphthalene-2,6-dicarboxylic acid** (57–61% from heating 100 g. of potassium naphthalene-1,8-dicarboxylate and 4 g. anhydrous cadmium chloride, protected by a CO_2 atmosphere, in a stirred or shaken autoclave at 400–430° for about 3 hr. followed by acidification) [24].

1. R. G. Jones and H. Gilman, *Org. Reactions*, **6**, 339 (1951).
2. H. Gilman and J. W. Morton, Jr., *Org. Reactions*, **8**, 258 (1954).
3. G. A. Holmberg, *Acta Chem. Scand.*, **6**, 1137 (1952).
4. M. Stiles and H. L. Finkbeiner, *J. Am. Chem. Soc.*, **81**, 505 (1959); **85**, 616 (1963).
5. D. M. Bowen, *Org. Syn.*, Coll. Vol. **3**, 553 (1955).
6. H. Gilman and J. W. Morton, Jr., *Org. Reactions*, **8**, 287 (1954).
7. *Org. Syn.*, Coll. Vol. **1**, 361, 363, 524 (1941); **2**, 425 (1943).
8. *Org. Reactions*, **8**, 290, 291, 292 (1954).
9. D. E. Pearson and D. Cowan, *Org. Syn.*, **44**, 78 (1964).
10. C. E. Frank *et al.*, *J. Org. Chem.*, **26**, 307 (1961).
11. S. W. Pelletier *et al.*, *J. Org. Chem.*, **31**, 1747 (1966).
12. D. F. Hoeg *et al.*, *J. Am. Chem. Soc.*, **87**, 4147 (1965).
13. G. Köbrich and K. Flory, *Tetrahedron Letters*, 1137 (1964).
14. R. A. Finnegan and R. S. McNees, *J. Org. Chem.*, **29**, 3234 (1964).
15. M. A. Davis *et al.*, *J. Med. Chem.*, **6**, 513 (1963).
16. J. Sicher *et al.*, *Collection Czech. Chem. Commun.*, **26**, 847 (1961).
17. A. Brändström, *Acta Chem. Scand.*, **13**, 610 (1959).
18. W. H. Puterbaugh and R. L. Readshaw, *Chem. Ind. (London)*, 255 (1959).
19. A. S. Lindsey and H. Jeskey, *Chem. Rev.*, **57**, 583 (1957).
20. M. Nierenstein and D. A. Clibbens, *Org. Syn.*, Coll. Vol. **2**, 557 (1943).
21. B. Raecke, *Angew. Chem.*, **70**, 1 (1958).
22. Y. Ogata *et al.*, *J. Org. Chem.*, **27**, 3373 (1962).
23. Y. Ogata *et al.*, *J. Am. Chem. Soc.*, **79**, 6005 (1957).
24. B. Raecke and H. Schirp, *Org. Syn.*, **40**, 71 (1960).

D. Condensation

α,β-Unsaturated acids, β-hydroxy acids, and β-keto acids are available from aldol-type and Claisen-type condensation reactions.

1. From Aromatic Aldehydes and Anhydrides (Perkin)

$$\text{ArCHO} + (\text{CH}_3\text{CO})_2\text{O} \xrightarrow[\text{2. H}^{\oplus}]{\text{1. CH}_3\text{COOK}} \text{ArCH}{=}\text{CHCOOH} + \text{CH}_3\text{COOK}$$

This synthesis, which has been reviewed [1], is one of the oldest methods for preparing α,β-unsaturated acids. The carboxylate anion of the salt reacts with

References for Section D are on pp. 783–784.

the anhydride to form the anion of the anhydride, i.e.,

$$\overset{\ominus}{C}H_2\overset{\overset{\displaystyle O}{\|}}{C}O\overset{\overset{\displaystyle O}{\|}}{C}CH_3$$

which attacks the carbon of the aldehyde to produce an aldol anion, i.e.,

$$Ar\overset{\overset{\displaystyle O^{\ominus}}{|}}{C}HCH_2\overset{\overset{\displaystyle O}{\|}}{C}O\overset{\overset{\displaystyle O}{\|}}{C}CH_3$$

The latter in the presence of acid is converted into the aldol which in turn loses water and hydrolyzes to produce the α,β-unsaturated acid. The synthesis is essentially applicable only to aromatic aldehydes, aliphatic aldehydes with no active hydrogen atoms, and to acetic or monosubstituted acetic acids. In some cases, triethylamine serves as a basic catalyst (see Ex. b). With phthalides and phenylacetic acid, benzalphthalides are produced [2]. Yields are fair.

(a) **Preparation of 2-furylacrylic acid** (65–70% from furfural, acetic anhydride, and freshly fused potassium acetate at 150° for 4 hr.) [3].

(b) **Preparation of *trans-o-nitro-α-phenylcinnamic* acid** (71–72% from *o*-nitrobenzaldehyde, phenylacetic acid, acetic anhydride, and triethylamine refluxed for 15 min.) [4].

2. From Aromatic Aldehydes and Malonic Acid (Doebner)

$$ArCHO + CH_2(COOH)_2 \xrightarrow{\text{Pyridine}} ArCH=CHCOOH$$

This synthesis, which also occurs by the aldol mechanism (for an alternative mechanism see [5]), offers a second method for preparing α,β-unsaturated acids. A discussion of the synthesis is available [1, 6, 5]. The synthesis resembles the Knoevenagel reaction (Carboxylic Esters, Chapter 14, C.4). Pyridine alone or with piperidine is the preferred catalyst for preparing arylidenemalonic acids (see, however, [7]). In some cases the modification offers advantages over the Perkin condensation. Although confined largely to aldehydes, it is convenient, works well for hydroxyaldehydes, and usually gives good results in a reasonable amount of time. Morover, any type of aliphatic aldehyde can be used. Recently the procedure has been modified by using a small amount of pyridine as a catalyst and by warming the reactants on a steam bath in the absence of alcohol [6]. The yields for a series of substituted cinnamic acids vary from 31 to 82%.

(a) **Preparation of 2-furylacrylic acid** (91–92% from furfural, malonic acid, and pyridine) [8].

(b) **Other examples**

(1) *m*-**Nitrocinnamic acid** (75–80% from *m*-nitrobenzaldehyde, malonic acid, and pyridine refluxed in 95% ethanol for 6–8 hr.) [6].

(2) **Cyclohexylidenecyanoacetic acid** (65–76% from cyclohexanone, cyanoacetic acid, and ammonium acetate in benzene refluxed for 3 hr.) [9].

3. From Ylides (Wittig)

The Wittig reaction has been discussed previously (Alkenes, Chapter 2, E.2), but examples of its application to the synthesis of acids are described here.

Ylide + an alkyl halide (see Ex. a):

$$2 \begin{array}{c} HCCO_2C_2H_5 \\ \| \\ P(C_6H_5)_3 \end{array} + C_6H_5CH_2Br \longrightarrow \begin{array}{c} C_6H_5CH_2CCO_2C_2H_5 \\ \| \\ P(C_6H_5)_3 \end{array} + \begin{array}{c} CH_2CO_2C_2H_5 \\ | \\ {}^{\oplus}P(C_6H_5)_3Br^{\ominus} \end{array} \xrightarrow{\overset{\ominus}{OH}}$$

$$C_6H_5CH_2CH_2CO_2C_2H_5 \xrightarrow[H_2O]{H^{\oplus}} C_6H_5CH_2CH_2COOH$$
$$\text{Hydrocinnamic acid}$$

Ylide + an aldehyde:

$$\begin{array}{c} C_6H_5CH_2CCO_2C_2H_5 \\ \| \\ P(C_6H_5)_3 \end{array} + RCHO \longrightarrow \begin{array}{c} C_6H_5CH_2CCO_2C_2H_5 \\ \| \\ RCH \end{array}$$

The reaction of (carbalkoxyalkylidene)triphenylphosphorane with aldehydes is a powerful tool for the synthesis of both *cis* and *trans* unsaturated fatty acids. The *trans* acid is usually favored if the condensation is carried out in a non-polar solvent. The *cis* acid is favored in a strongly polar solvent [10]:

$$(C_6H_5)_3P{=}CH(CH_2)_6CO_2C_2H_5 + CH_3(CH_2)_6CHO \xrightarrow[\substack{2.\ OH^{\ominus} \\ 3.\ H_3O^{\oplus}}]{1.\ DMF}$$

$$\overset{H\quad H}{CH_3(CH_2)_6C{=}C(CH_2)_6CO_2H}$$
$$\textit{cis}\text{-Hexadec-8-enoic acid, } 56\%$$

(a) Preparation of hydrocinnamic acid

(1) **(Benzylcarboethoxymethylene)-triphenylphosphorane.** (Carboethoxymethylene)-triphenylphosphorane, 5.0 g. in absolute ethyl acetate, was treated with 1.5 ml. of benzyl bromide and the mixture was refluxed for 5 hr. The precipitated hydrobromide of the starting material, 2.69 g. (87%), was removed by filtration. Evaporation of the filtrate in vacuum gave 2.80 g. (75%) of (benzylcarboethoxymethylene)-triphenylphosphorane, m.p. 186–187°, after digesting several times with ethyl acetate.

(2) **Hydrocinnamic Acid.** The substituted triphenylphosphorane, 13.6 g., was heated with 10% aq. methanolic potassium hydroxide (H_2O to CH_3OH ratio was 1:1) for 1 hr. under reflux. After cooling, the contents of the flask were treated with ten times its quantity of water, after which the triphenylphosphine oxide precipitated. Filtration and evaporation of the filtrate to 50–100 ml. was followed by acidification with sulfuric acid to pH of 1–3.

Several extractions with ether, drying over magnesium sulfate, and evaporation gave 4.56 g. (95%) of the acid, m.p. and mixed m.p. 48° [11].

(b) Preparation of *o*-chloro-α-benzylcinnamic acid. (Benzylcarboethoxymethylene)-triphenylphosphorane, 10 g., and 3.4 g. of *o*-chlorobenzaldehyde in 400 ml. of absolute ethyl acetate were boiled for 24 hr. After saponification and acidification the acid was extracted with ether to give 4.8 g. (75%), m.p. 122–124° [11].

(c) Other examples

(1) **Butyric, allylacetic, succinic, and γ-benzylidenebutyric acids** (66, 52, 99, 78%, respectively, from the properly substituted phosphorane) [11].

(2) **Benzylidenesuccinic, α-benzylcinnamic, and α-cinnamylcinnamic acids** (40, 42, 60% respectively, from the appropriate phosphorane and benzaldehyde) [11].

4. From Ketones via the Enamine and Morpholide (Carboxyethylation)

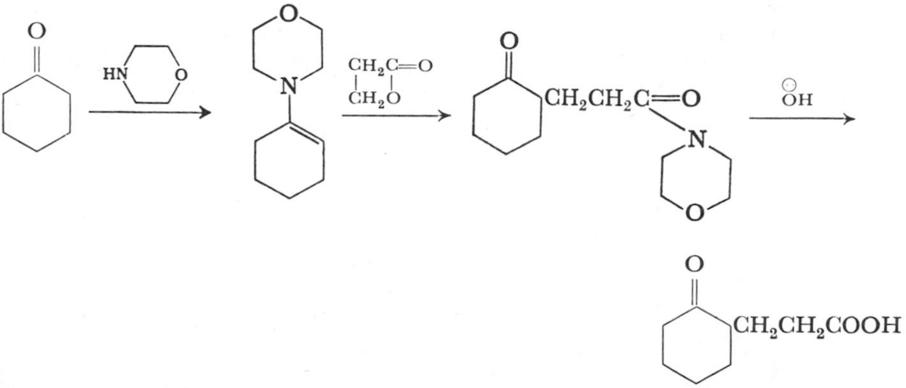

In a preliminary paper [12] the synthesis of (2-ketocyclohexane) 3-propionic acid is described. The enamine is treated with β-propiolactone, readily available from ketene and formaldehyde, to give the morpholide which may be hydrolyzed to yield the keto acid. The synthesis represents a new method of carboxyethylation.

(a) Preparation of (2-ketocyclohexane) 3-propionic acid. 1-Morpholine-1-cyclohexene [13], 0.6 mole, and 0.4 mole of β-propiolactone in 100 ml. of chlorobenzene were refluxed for 4 hr. at 154–155° [12]; distillation gave 78 g., 83%, of the morpholide, b.p. 187–188° (1 mm.); basic hydrolysis produced the acid, 77%, [12].

1. J. R. Johnson, *Org. Reactions*, **1**, 210 (1942).
2. R. Weiss, *Org. Syn.*, Coll. Vol. **2**, 61 (1943).
3. J. R. Johnson, *Org. Syn.*, Coll. Vol. **3**, 426 (1955).

4. D. F. DeTar, *Org. Syn.*, Coll. Vol. **4,** 730 (1963).

5. G. Jones, *Org. Reactions*, **15,** 204 (1967).

6. R. H. Wiley and N. R. Smith, *Org. Syn.*, Coll. Vol. **4,** 731 (1963).

7. T. Boehm and M. Grohnwald, *Arch. Pharm.*, **274,** 329 (1936); *C.A.*, **30,** 6752 (1936).

8. S. Rajagopalan and P. V. A. Raman, *Org. Syn.*, Coll. Vol. **3,** 425 (1955).

9. A. C. Cope *et al.*, *Org. Syn.*, Coll. Vol. **4,** 234 (1963).

10. L. D. Bergelson and M. M. Shemyakin, *Angew. Chem. Intern. Ed. Eng.*, **3,** 250 (1964).

11. H. J. Bestman and H. Schulz, *Chem. Ber.*, **95,** 2921 (1962).

12. S.-O. Lawesson, *Acta Chem. Scand.*, **18,** 2201 (1964).

13. S. Hünig *et al.*, *Org. Syn.*, **41,** 65 (1961).

E. Alkali Cleavage

Some of these methods are relatively new and useful routes to the preparation of acids of unique structure, particularly those cited in E.3. Essentially they are reverse Claisen or aldol condensations.

1. From β-Keto esters (Acetoacetic Ester Synthesis)

$$CH_3COCH_2COOC_2H_5 \xrightarrow[NaOC_2H_5]{RX} CH_3COCHCOOC_2H_5 \xrightarrow[H^{\oplus}]{NaOH} RCH_2COOH$$
$$\underset{R}{|}$$

In the first step of this synthesis the anion of the sodium salt of acetoacetic ester, a resonance hybrid, and the alkyl halide produces C-alkylation. In the last step the mechanism is essentially the reverse of that of the Claisen ester condensation in that cleavage occurs to give two acid anions (retrograde Claisen):

$$CH_3COCHCOOC_2H_5 \xrightarrow{[2\overset{\ominus}{O}H]} \left[\begin{array}{c} :\overset{..}{O}:{}^{\ominus} \quad :\overset{..}{O}:{}^{\ominus} \\ | \qquad | \\ CH_3C\!\!-\!\!-\!\!CH\!\!-\!\!C\!\!-\!\!OC_2H_5 \\ | \qquad | \qquad | \\ OH \quad R \quad OH \end{array} \right] \longrightarrow$$

$$CH_3COOH + \left[R\overset{\ominus}{C}HCOOH \right] + \left[\overset{\ominus}{O}C_2H_5 \right] \longrightarrow$$

$$CH_3C\overset{O}{O} + RCH_2C\overset{\ominus}{O}O + HOC_2H_5$$

The final alkaline hydrolysis gives ketones as well even though strong alkali is used, and thus the yields are often unsatisfactory. However, a procedure is now available which mitigates ketone formation by using only traces of sodium ethoxide as the catalyst in an excess of absolute ethanol and by removing ethyl acetate continuously as it is formed [1]:

$$CH_3COCHCOOC_2H_5 \xrightarrow[NaOC_2H_5]{C_2H_5OH} RCH_2COOC_2H_5 + CH_3COOC_2H_5$$
$$\underset{R}{|}$$

References for Section E are on p. 789.

(a) Preparation of 6-chloro-2,4-dinitrophenylacetic acid [2]:

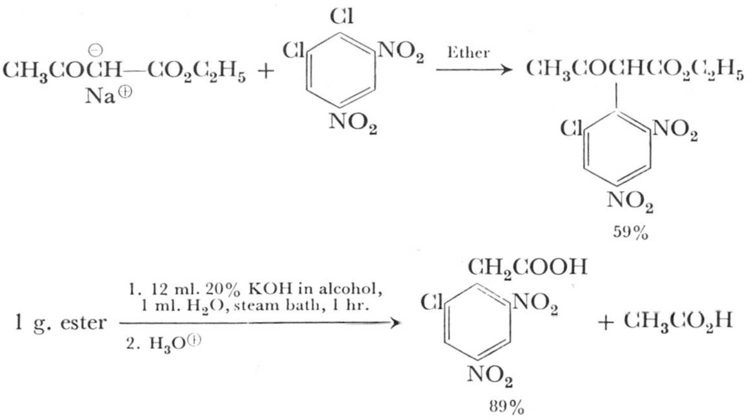

$$CH_3COCH-CO_2C_2H_5 \overset{\ominus}{\underset{Na^\oplus}{}} + \quad [\text{Cl, Cl, NO}_2, NO_2 \text{ ring}] \quad \xrightarrow{\text{Ether}} \quad CH_3COCHCO_2C_2H_5$$

[ring with Cl, NO$_2$, NO$_2$]

59%

1 g. ester $\xrightarrow[\substack{2.\ H_3O^\oplus}]{\substack{1.\ 12\ \text{ml. } 20\% \text{ KOH in alcohol,} \\ 1\ \text{ml. } H_2O, \text{ steam bath, 1 hr.}}}$ [ring with CH$_2$COOH, Cl, NO$_2$, NO$_2$] $+ CH_3CO_2H$

89%

(b) Preparation of methylethylacetic acid (60% from methylethylacetoacetic ester) [3].

2. From Unsaturated Acids (Varentrapp)

$$CH_3(CH_2)_nCH=CHCOOH \xrightarrow{KOH} CH_3(CH_2)_nCOOH + CH_3COOH$$

In this synthesis, yields are quite satisfactory in an inert atmosphere despite the high-temperature conditions. In fact, the method is employed as an isotopic assay method in which an acid is degraded by two carbon atoms [4]:

$$C_6H_{13}CH_2CO_2H \xrightarrow[Br_2]{SOCl_2} C_6H_{13}CHBrCOCl$$

↓ Neopentyl alcohol

$$C_5H_{11}CH=CHCO_2CH_2C(CH_3)_3 \xleftarrow{C_6H_5N(C_2H_5)_2} C_6H_{13}CHBrCO_2CH_2C(CH_3)_3$$

↓ 1. KOH, H$_2$O, evaporate
  2. KOH, 300–350°
  3. H$_3$O$^\oplus$

$$CH_3COOH + C_5H_{11}COOH$$

27%   50%
overall   overall
           Hexanoic acid

This cleavage occurs not only with α,β-unsaturated acids but with almost any unsaturated acid. The phenomenon is due to the fact that the double bond migrates reversibly in both directions along the chain, but the irreversible fission occurs with the α,β-unsaturated isomer [5].

The mechanism is complex [6] but in its simplest form may be viewed as a reverse aldol combined with a bimolecular redox reaction:

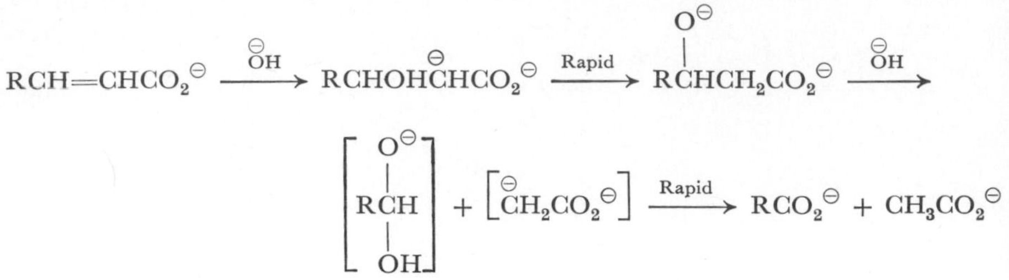

$$RCH{=}CHCO_2^{\ominus} \xrightarrow{\overset{\ominus}{OH}} RCHOH\overset{\ominus}{C}HCO_2^{\ominus} \xrightarrow{Rapid} R\overset{\overset{\displaystyle O^{\ominus}}{|}}{C}HCH_2CO_2^{\ominus} \xrightarrow{\overset{\ominus}{OH}}$$

$$\left[\overset{\overset{\displaystyle O^{\ominus}}{|}}{\underset{\underset{\displaystyle OH}{|}}{RCH}} \right] + \left[\overset{\ominus}{C}H_2CO_2^{\ominus} \right] \xrightarrow{Rapid} RCO_2^{\ominus} + CH_3CO_2^{\ominus}$$

Unsaturated nitriles would be expected to undergo the same reaction [7]:

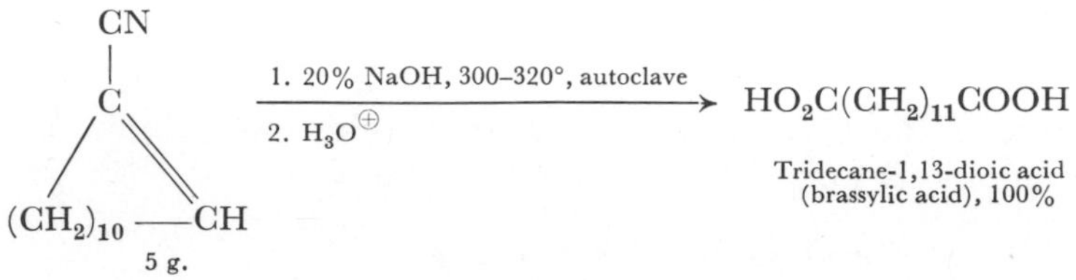

$$\xrightarrow[\text{2. } H_3O^{\oplus}]{\text{1. 20\% NaOH, 300–320°, autoclave}} HO_2C(CH_2)_{11}COOH$$

Tridecane-1,13-dioic acid
(brassylic acid), 100%

(a) **Preparation of pivalic acid.** β-t-Butylacrylic acid (see C.1, Ex. c. 11), 1100 g., was mixed carefully with a solution of 1150 g. of KOH (85–87%) in 1150 ml. of water in a stainless steel vessel. The vessel was provided with a cover and was heated electrically up to 300°. The cooled reaction mixture in water was acidified and the organic layer (plus the benzene extract of the aqueous layer) was dried and distilled to give 760 g. (87%) of the acid, b.p. 162–165°, m.p. 34–35° [8].

(b) **Preparation of nonanoic acid** (80% from 10-undecenoic acid) [5].

3. From Ketones

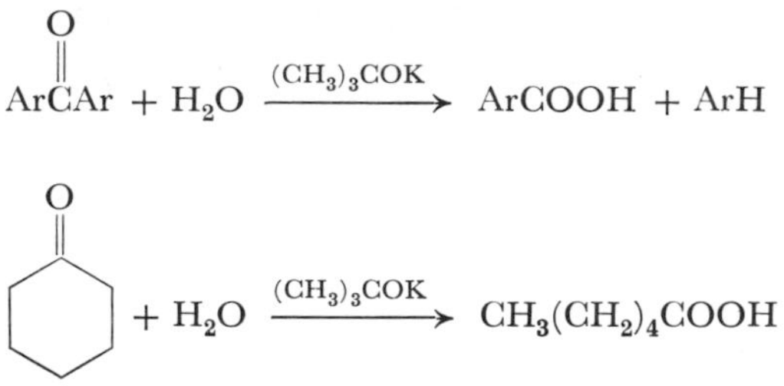

$$Ar\overset{\overset{\displaystyle O}{\|}}{C}Ar + H_2O \xrightarrow{(CH_3)_3COK} ArCOOH + ArH$$

$$+ H_2O \xrightarrow{(CH_3)_3COK} CH_3(CH_2)_4COOH$$

This synthesis involves the cleavage of unenolizable aromatic or cyclic ketones with potassium hydroxide [9], with barium hydroxide or constant-boiling hydrochloric acid [10], with sulfuric acid [11], or with potassium t-butoxide-dimethyl sulfoxide [12, 13]. The potassium t-butoxide-dimethyl sulfoxide is the preferred reagent because it is effective at room temperature. It is thought that the reaction in alkaline solution involves an attack of the negative hydroxyl (or t-butoxide) on the positive carbonyl carbon to produce an anion capable of undergoing cleavage to the carboxylic acid:

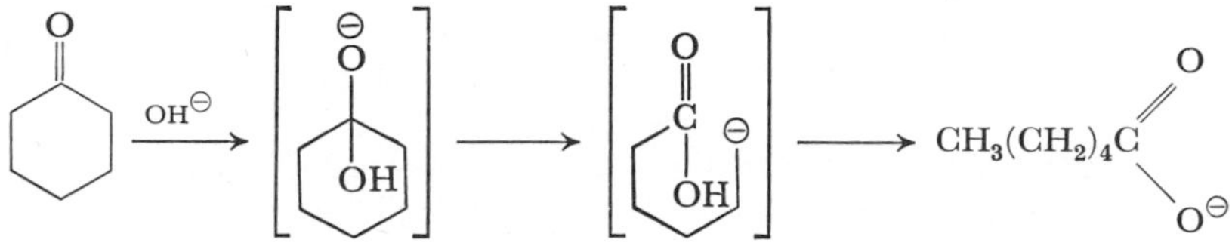

However, this mechanism has been questioned because nortricyclanone undergoes cleavage with complete retention of configuration [13].

This synthesis has found its greatest value among cyclic ketones, as is shown by the exploitation of Stetter in splitting the dihydroresorcinols [10]. Dihydroresorcinol is a good source of acids because it can be alkylated, dimerized with aldehydes, or condensed with unsaturated compounds:

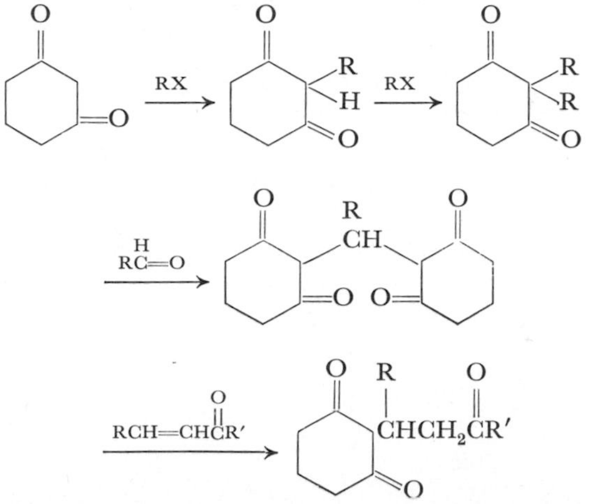

Each of these products is capable of being cleaved to keto acids either by barium hydroxide or, preferably, by constant-boiling hydrochloric acid. Alternatively, the products may be simultaneously split and reduced to acids by using Wolff-Kishner conditions (potassium hydroxide, hydrazine, diethylene glycol); for example,

$$\text{2-Ethyldihydroresorcinol} \xrightarrow[\text{Kishner}]{\text{Wolff-}} \text{octanoic acid}$$

Indeed, the single-step gives better yields than the two-step process.

To minimize O-alkylation of dihydroresorcinol, the process should be carried out in as ionic a medium as possible (see Ethers, Chapter 6, A.1).

(a) **Preparation of γ-(6-methyl-3-keto-1-cyclohexen-1-yl)-butyric acid.**   A mixture of 0.5 g. of 1,6-dioxo-8a-methyl-1,2,3,4,6,7,8,8a-octahydro-naphthalene and 5 ml. of 10% KOH was allowed to stand for 20 min. at 27°,

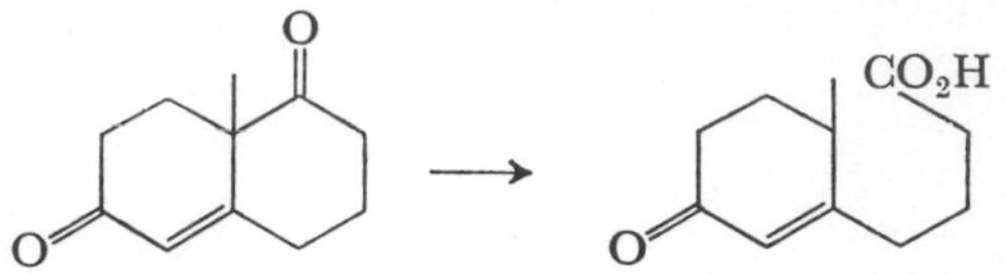

after which it was heated on a steam bath for 5 min., cooled, diluted with 5 ml, of water, and acidified. After extraction with ether, the solution was worked up. as usual, to yield the titled acid (75%), m.p. 77.5–78.0° [14].

(b) **Other examples**

(1) **6-(3,5-Dimethylphenyl) hexanoic acid** (94% from 1,2(1′,3′-dimethylbenzo) cyclooceten-1-one-3 and sulfuric acid) [11].

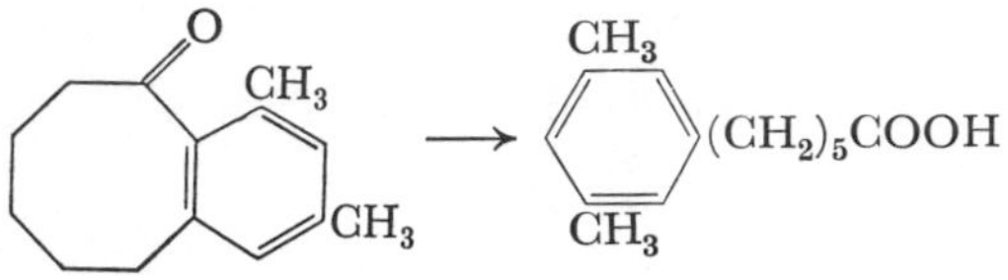

(2) **Benzoic acid** (100% from benzopinacolone, $(C_6H_5)_3CCOC_6H_5$, potassium *t*-butoxide-dimethyl sulfoxide, and a trace of water at 25° for 4 min.) [12].

(3) **Δ³-Cyclopentenylacetic acid** (80% from dehydronorcamphor and potassium *t*-butoxide in dimethyl sulfoxide at room temperature) [15].

(4) **2-Carboxydiphenyl** (97% from fluorenone and finely divided potassium hydroxide in boiling toluene; sodium hydroxide gave a very poor yield; a

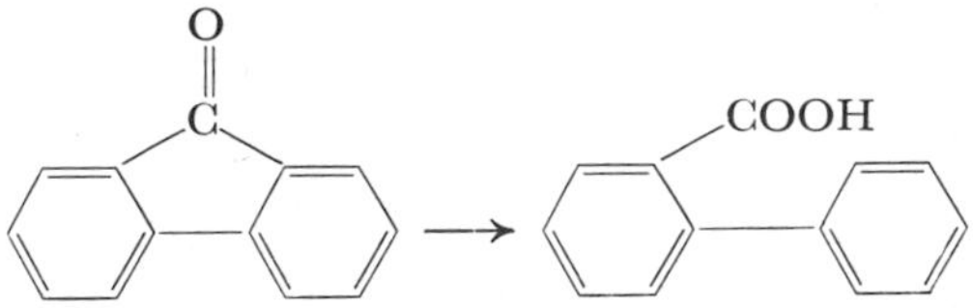

small amount of 2′-(9-hydroxy-9-fluorenyl)-biphenyl-2-carboxylic acid accompanies the product) [16].

(5) **5-Oxononanoic acid** (81% from propyldihydroresorcinol and aq. barium hydroxide; the keto group can be reduced to yield nonanoic acid) [17].

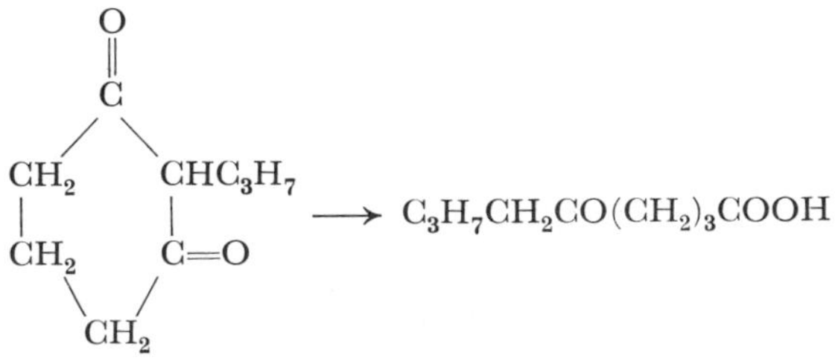

(6) Brassylic acid (Nearly 100% from methylene-2,2'-dihydro-resorcinol in diethylene glycol containing hydrazine and powdered sodium hydroxide at 195°) [18].

$$
\text{(cyclic diketone structure)} \longrightarrow HOOC(CH_2)_{11}COOH
$$

1. J. J. Ritter and T. J. Kaniecki, *J. Org. Chem.*, **27**, 622 (1962).
2. A. B. Sen and P. M. Bhargava, *J. Indian Chem. Soc.*, **25**, 403 (1948).
3. W. M. Cumming *et al.*, *Systematic Organic Chemistry*, 4th ed., D. Van Nostrand Co., Princeton, N.J., 1950, p. 200.
4. G. D. Hunter and G. Popjàk, *Biochem. J.*, **50**, 163 (1951).
5. P. Linstead *et al.*, *Tetrahedron*, **8**, 221 (1960).
6. W. A. Bonner and R. T. Rewick, *J. Am. Chem. Soc.*, **84**, 2334 (1962).
7. L. I. Zakharkin and V. V. Korneva, *J. Org. Chem. USSR* (*Eng. Transl.*), **2**, 740 (1966).
8. A. Brändstrom, *Acta Chem. Scand.*, **13**, 610 (1959).
9. W. E. Bachmann and E. J.-H. Chu, *J. Am. Chem. Soc.*, **58**, 1118 (1936).
10. H. Stetter in W. Foerst, *New Methods of Preparative Organic Chemistry*, Vol. **2**, Academic Press, New York, 1963, p. 51.
11. W. M. Schubert *et al.*, *J. Am. Chem. Soc.*, **76**, 5462 (1954).
12. P. G. Gassman and F. V. Zalar, *Tetrahedron Letters*, 3031 (1964).
13. P. G. Gassman and F. V. Zalar, *Tetrahedron Letters*, 3251 (1964).
14. S. Swaminathan and M. S. Newman, *Tetrahedron*, **2**, 88 (1958).
15. P. G. Gassman and F. V. Zalar, *Chem. Eng. News*, **42** (No. 16), 44 (1964).
16. G. W. Kenner *et al.*, *J. Chem. Soc.*, 1756 (1962).
17. H. Stetter and W. Dierichs, *Chem. Ber.*, **85**, 61 (1952).
18. H. Stetter and W Dierichs, *Chem. Ber.*, **85**, 290 (1952).

F. Substitution and Addition (Mainly Electrophilic)

This section includes reactions of the Friedel-Crafts type (F.1 and F.2) in which substitution occurs in arenes. Also covered are addition reactions of alkenes (or other aliphatic types) in which carbon monoxide (F.3) or formic acid (F.4 and F.5) is involved. The reactions are discussed more or less in the order of their importance. The Reppe method (F.3) is largely of industrial interest, while the Koch-Haaf procedures, although simple in operation, suffer in that mixtures are produced.

1. From Aromatic Compounds by Acylation or Carboxylation (Friedel-Crafts)

$$
C_6H_6 \xrightarrow[AlCl_3]{COCl_2} C_6H_5COCl \xrightarrow{H_2O} C_6H_5COOH
$$

$$
C_6H_6 \xrightarrow[AlCl_3]{\underset{CO}{\overset{CO}{(CH_2)_n}}O} C_6H_5CO(CH_2)_nCOOH
$$

References for Section F are on pp. 795–796.

Carbonyl chloride (phosgene) or oxalyl chloride in the presence of a catalyst often introduces into aromatic ring systems an acid chloride group which is hydrolyzed readily to the acid. One must be aware of the possibility of the acid chloride further reacting to form a symmetrical ketone. The catalyst, such as aluminum chloride, forms the acylium ion ($\overset{\oplus}{C}OCl$) or a precursor where the carbon of the carbonyl is partially charged. In any form the acylium ion or its precursor is a weak attacking agent since resonance stabilization can delocalize the positive charge:

$$:\overset{\oplus}{\underset{\displaystyle \overset{\|}{\text{O}}}{C}}Cl \;\longleftrightarrow\; :\overset{\overset{\oplus}{\text{O}}}{\underset{..}{C}}\!-\!Cl \;\longleftrightarrow\; \overset{\overset{\|}{\text{O}}}{C}\!=\!Cl^{\oplus}$$

Therefore substitution is usually restricted to benzenoid compounds having substituents which increase the electron density of the ring or to other aromatic compounds in which substitution occurs more easily than in benzene.

Dibasic anhydrides or even dibasic acids may be employed to introduce acyl-carboxyl groups into ring systems. A discussion of the reaction when anhydrides are involved is available [1]. It usually gives yields from 50 to 100 % with sufficiently activated aromatic and heterocyclic compounds.

It is interesting to note that acids in small amounts have been obtained by the carbonation of aliphatic hydrocarbons in the presence of aluminum chloride at temperatures of 90–100° and at 30–40 atm. (initial pressure) [2]; (see F.5 for alternative procedure).

(a) Preparation of 1-azulenecarboxylic acid. Azulene, 1 mmole, was dissolved in 3 ml. of dry toluene, and 6 ml. of a 20 % solution of phosgene in toluene was introduced with cooling. The reaction mixture, which possesses at the beginning the color of the hydrocarbon, became red-violet to red at room temperature. At the end of 60 min. the excess of phosgene was removed either by careful warming under reflux or by applying a slight vacuum. Since the azulene-carbonyl chloride is so easily hydrolyzable, the corresponding acid was obtained by adding the reaction mixture with stirring to 150 ml. of ice water. The solution was then made alkaline with dilute sodium carbonate, and it was extracted several times with ether. The aqueous solution containing the sodium salt was then acidified, extracted with benzene from which dark red crystals (95 %), m.p. 188–190°, separated on removing the solvent [3].

(b) Other examples

(1) Anthracene-9-carboxylic acid (67 % from anthracene and oxalyl chloride) [4].

(2) Substituted azulenecarboxylic acids (73–98 %) [3].

(3) β,β-Di(p-chlorophenyl)-acrylic acid (low yield from the dichloro-phenylethylene and ethoxalyl chloride at b.p. [5]; good yield from 1,1-di-(p-chlorophenyl)-ethylene and oxalyl chloride at room temperature; similarly, oxalyl bromide formed acid bromides from aliphatic and alicyclic alkenes and some aromatic hydrocarbons) [6].

(4) Cyclohexanecarbonyl chloride (60% from cyclohexane and oxalyl chloride exposed to ultraviolet light at the b.p. of the latter) [7].

(5) 3-Methyl-1,2-phthalic acid from *o*-toluic acid [8].   Ingenious method for introducing an *ortho* carboxyl group. *o*-Toluyl chloride was treated with lead isothiocyanate to yield

$$o\text{-CH}_3\text{—C}_6\text{H}_4\overset{\displaystyle \overset{\text{O}}{\|}}{\text{C}}\text{—NCS}$$

Friedel-Crafts acylation of the isothiocyanate yields 3-methylthiophthalimide in 45% yield which can be hydrolyzed to 3-methylphthalic acid in 80% yield.

(6) β-(*p*-Methoxybenzoyl)propionic acid (85% from anisole and succinic anhydride [9]; general procedure for succinolylation of arenes and aryl ethers).

(7) ω-Benzoylpelargonic acid (80% from sebacic acid [10]; use of a dicarboxylic acid).

2. From Aromatic Compounds by Alkylation (Friedel-Crafts)

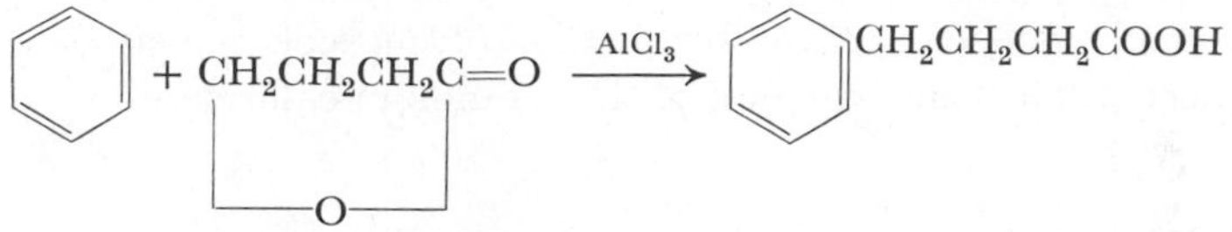

Although the yields in this reaction are not high owing to the formation of by-products, some acids have been prepared by this method. The mechanism is the usual electrophilic attack of Friedel-Crafts reactions in which alkylation takes precedence over acylation. No isomerization of the side chain occurs with the use of aluminum chloride, and the phenyl radical is attached at the position of the alcohol function in the parent hydroxy acid [11]. In the case of indoles and pyrroles, no catalyst is needed and the substituent is attached to positions 3 and 2, respectively [12]. An alternative, and perhaps superior, preparation of 3-indolebutyric acid (82%) is the reaction of indole, butyrolactone, and potassium hydroxide at 250–290° under autogenous pressure [13]. The reaction proceeds by migration of the carboxypropyl group from nitrogen to the 3-position; it is reminiscent in some respects of enamine alkylation.

Although good procedures have not been worked out for all reactions, practically any lactone, hydroxy acid, or unsaturated acid may be used to introduce groups into a ring system. In the case of unsaturated acids, migration of the double bond may occur to produce a mixture of acids [14]:

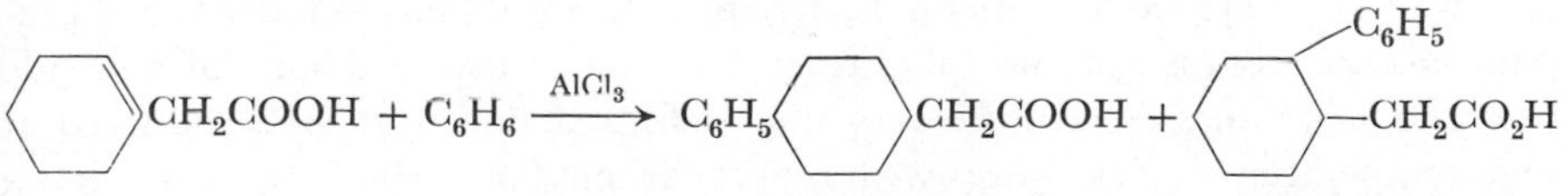

4-(Phenylcyclohexyl)-acetic acid   2-(Phenylcyclohexyl)-acetic acid

The usual catalyst does not always appear to be necessary. For example, α-naphthylacetic acid has been produced from naphthalene and chlorine-containing glycolic acid polymers alone [15].

(a) Preparation of γ-phenylbutyric acid. Powdered AlCl₃, 0.7 mole, was added in small portions to a stirred solution of 0.5 mole of γ-butyrolactone in 100 g. of benzene, and after 4 hr., 100 ml. of benzene was added and the whole was allowed to stand overnight at 25°. On being heated for 1 hr. on a water bath, the mixture was hydrolyzed with ice and HCl, and distillation of the benzene layer gave 36 g. (44%) of the acid, b.p. 120–125° (1 mm.); m.p. 48–49° after recrystallization [11]. It has been stated that the ω-chloroacid (i.e., ω-chlorovaleric acid rather than valerolactone) gives a better yield in the Friedel-Crafts reaction [16].

(b) Diethyl vinylmalonate. A mixture of 0.2 mole each of the reagents in the equation was treated with 0.25 mole of AlCl₃ while being stirred and

$$CH_2{=}CH{-}OC_4H_9 + CH_2(CO_2C_2H_5)_2 \xrightarrow[25°]{AlCl_3} CH_2{=}CH{-}CH(CO_2C_2H_5)_2$$
$$39\%$$

held, after an initial temperature surge, at 28–30° for 4–5 hr. The product was isolated in the normal way, b.p. 172–174° at 75 mm., n^{20}D 1.4378 [17]. The yields probably can be improved by use of a solvent such as methylene chloride and by more prudent introduction of the aluminum chloride into the reaction mixture.

(c) Other examples

(1) **γ- and δ-Phenylvaleric acids** (61% and 51%, respectively, from the lactones) [11].

(2) **3-Indolepropionic acid** (56% from acrylic acid, indole and acetic anhydride) [18]; it is claimed to be the best method for the preparation of the compound).

3. Hydrocarboxylation of Olefins and Acetylenes (Reppe)

$$RCH{=}CH_2 \xrightarrow{Ni(CO)_4} RCHCH_3$$
$$\underset{COOH}{|}$$

$$RC{\equiv}CH \xrightarrow{Ni(CO)_4} RC{=}CH_2$$
$$\underset{COOH}{|}$$

This method of hydrocarboxylation of unsaturated compounds has been reviewed [19–21]; see also Alcohols, Chapter 4, B.2. It has found limited preparative use because of the laboratory techniques involved and of the side products sometimes formed. Usually the olefin reaction must be conducted at high temperature and pressures with nickel carbonyl or carbon monoxide and a nickel salt. Other carbonyls such as that of cobalt may be employed, although under these conditions more side products are formed. Alcohols may be substituted for olefins, although in this case the temperature must be increased and a nickel halide added. Yields with alkenes vary from 20 to 90%.

For a discussion of the mechanism, see [20]. More suitable methods for producing carboxylic acids from alkenes in the laboratory are given under F.4.

The hydrocarboxylation of acetylenes proceeds similarly to that of alkenes except that the conditions are less strenuous. Typical reaction conditions are 150° and 30 atm. In this case the addition of formic acid is *cis*. Yields from acetylenes rarely exceed 60%.

(a) **Preparation of propionic acid.**   A solution of 30 g. of $NiCl_2$ in 30 g. of water under a total of 100 atm. pressure (50 of ethylene and 50 of CO) was heated to 285°. The pressure then increased to about 270 atm. After a decline in the pressure due to CO absorption, the pressure was 200 atm. Total time required was 16 hr. From the filtrate there was obtained on distillation 109.5 g. (89%) of propionic acid [22].

(b) **Preparation of decene-5-carboxylic acid-5** (44% from decyne-5 in ethanol, water, acetic acid, and nickel carbonyl at 75°; 8% of the ethyl ester was also obtained) [23].

4. From Alcohols, Alkenes, Alkyl Halides, or Esters and Formic-Sulfuric acid, or 1,1-Dichloroethylene in the Presence of Boron Trifluoride-Sulfuric Acid (Koch-Haaf)

These syntheses [24, 25] resemble the oxo process but are more suitable for the laboratory. In the first case a carboxylic acid with one more carbon atom, and in the second, a carboxylic acid with two more carbon atoms, is produced. Alkenes are probably always intermediates, and under the acid conditions prevailing they probably form carbonium ions first. Isomerization of the carbonium ion tends to give the more stable tertiary carbonium ion, and thus the methods are of more value in preparing branched rather than straight-chain acids. The procedures are not difficult, but separation from the other acids formed is essential. Yields may be as high as 90%.

(a) **Preparation of 1-methylcyclohexanecarboxylic acid (addition of one carbon atom)** (89–94% from 2-methylcyclohexanol, 96% sulfuric acid, and 98–100% formic acid at 15–20° for 2 hr.) [26].

(b) **Other examples (addition of one carbon atom)**

(1) **2-Methylbutyric acid** (approx. 40% yield from 1-butanol and 2-butanol); **2,2-dimethylbutyric acid** (83% from 2,2-dimethylpropanol-1) [24].

(c) **Preparation of β,β-dimethylbutyric acid (addition of two carbon atoms).** A mixture of *t*-butyl alcohol, 74 g., and 145 g. of 1,1-dichloroethylene was added to 200 ml. of sulfuric acid containing 8 % of boron trifluoride at 5–7° over a 2-hr. period with vigorous stirring. The mixture was allowed to stand for 2 hr., after which it was hydrolyzed with ice. The separated product was taken up in ether, extracted with dilute sodium hydroxide, and finally acidified. Yield 93 g. (80 %), b.p., 80–82° (12 mm.). The same acid was prepared in 79 % yield from *t*-butyl chloride and in 75 % yield from isobutylene [27].

(d) **Preparation of β-methylhydrocinnamic acid (addition of two carbon atoms)** (63 % from α-phenylethyl chloride, sulfuric acid, and 1,1-dichloroethylene) [28].

5. From Tertiary, Saturated Hydrocarbons, Formic Acid, *t*-Butyl Alcohol, and Sulfuric Acid

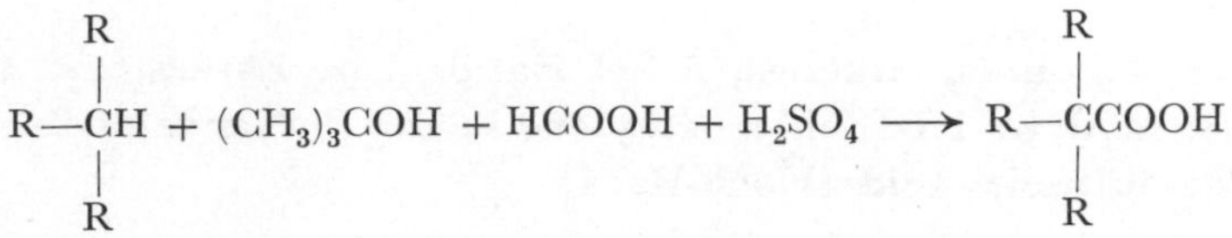

$$\begin{array}{ccc}
R & & R \\
| & & | \\
R\!-\!CH + (CH_3)_3COH + HCOOH + H_2SO_4 \longrightarrow & R\!-\!CCOOH \\
| & & | \\
R & & R
\end{array}$$

This synthesis [29] is related to the preparation of carboxylic acids from alkenes (F.4). In this case it appears that the tertiary butyl carbonium ion extracts the tertiary hydrogen from the hydrocarbon to produce a carbonium ion which reacts with carbon monoxide (from the formic acid) and water to give the carboxylic acid [30]. Simple in operation, the method has been used to convert isopentane into 2,2-dimethylbutanoic acid, 2,3-dimethylbutane into 2,2,3-trimethylbutanoic acid, and methylcyclohexane into 1-methylcyclohexanecarboxylic acid. However, mixtures of acids are often obtained with the result that the best yields of any one are about 70 %.

(a) **Preparation of 1-adamantanecarboxylic acid** (56–61 % from

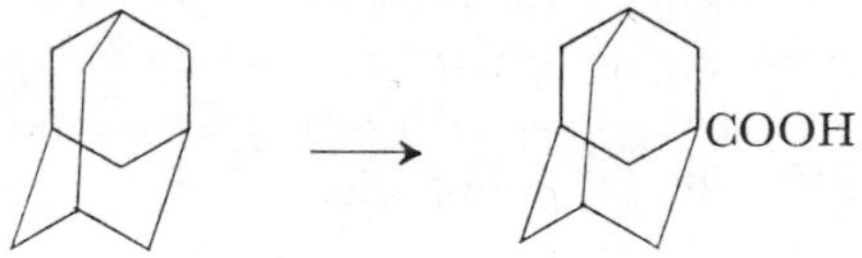

adamantane, *t*-butyl alcohol, 98–100 % formic acid, and 96 % sulfuric acid in carbon tetrachloride at 17–25°) [31].

6. From Xanthhydrols and Malonic Acid

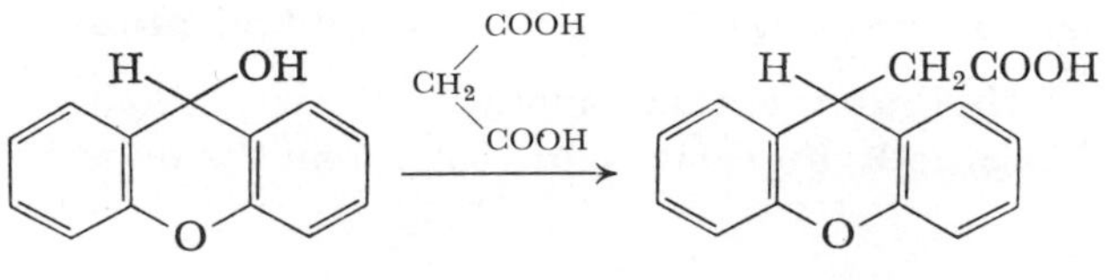

A few cases of this synthesis are on record. The procedure is extremely simple and the yields are satisfactory. The carbonium ion is formed readily from xanthhydrol, which may then attack an anion or enolic form of the malonic acid-pyridine salt. This mechanism, however, is simply conjecture.

(a) Preparation of 9-thioxanthenacetic acid. Thioxanthhydrol, 42 g., 30 g. of malonic acid, and 80 ml. of pyridine were mixed and heated for 2 hr. at 60–70°, then for 2 hr. at 90–95°. Acidification with 600 ml. of 2 N HCl yielded an oil which soon crystallized. Purification in the usual manner gave 45 g. (90%) of the acid, m.p. 167–168° (from 50% alcohol) [32].

(b) Preparation of 9-xanthenacetic acid (80% from xanthhydrol) [33].

7. From Alkenes and α-Halocarboxylic Acids

$$RCH{=}CH_2 + BrCH_2CO_2H \xrightarrow{h\nu} RCH_2CH_2CH_2CO_2H$$

The scope of this reaction has not been delineated but may well be general for symmetrical olefins at least. If α-iodoacetic acid is used, both the saturated and unsaturated acids are obtained.

(a) Preparation of α-cyclohexylacetic acid [34]:

1. E. Berliner, *Org. Reactions*, **5,** 229 (1949).
2. H. Hopff and T. Zimmerman, *Helv. Chim. Acta*, **47,** 1293 (1964).
3. W. Treibs *et al.*, *Chem. Ber.*, **92,** 1216 (1959).
4. H. G. Latham, Jr. *et al.*, *J. Am. Chem. Soc.*, **70,** 1079 (1948).
5. F. Bergmann and A. Kalmus, *J. Chem. Soc.*, 4521 (1952).
6. W. Treibs and H. Orttmann, *Chem. Ber.*, **93,** 545 (1960).
7. F. Runge, *Z. Elektrochem.*, **56,** 779 (1952).
8. P. A. S. Smith and R. O. Kan, *J. Org. Chem.*, **29,** 2261 (1964).
9. L. F. Fieser and E. B. Hershberg, *Org. Reactions*, **5,** 262 (1949).
10. Y. K. Yur'ev *et al.*, *Zh. Obshch. Khim.*, **24,** 1568 (1954); *C.A.*, **49,** 11589 (1955).
11. R. V. Christian, Jr., *J. Am. Chem. Soc.*, **74,** 1591 (1952).
12. J. Harley-Mason, *Chem. Ind.* (*London*), 886 (1951); *J. Chem. Soc.*, 2433 (1952).
13. H. E. Fritz, *J. Org. Chem.*, **28,** 1384 (1963).
14. W. S. Johnson, *Rec. Chem. Progr.* (*Kresge-Hooker Sci. Lib.*), **10,** 52 (1949).
15. P. L. Southwick *et al.*, *J. Am. Chem. Soc.*, **83,** 1358 (1961).
16. A. N. Nesmeyanov, *Selected Works in Organic Chemistry*, The Macmillan Co., New York, 1963, p. 1056.
17. Z. I. Torgashina and A. L. Naumchuk, *J. Gen. Chem. USSR* (*Eng. Transl.*), **26,** 3731 (1956).
18. H. E. Johnson and D. G. Crosby, *J. Org. Chem.,*, **25,** 569 (1960).
19. W. Reppe, *Ann. Chem.*, **582,** 1 (1953).
20. C. W. Bird, *Chem. Rev.*, **62,** 283 (1962).

21. C. W. Bird, *Transition Metal Intermediates in Organic Synthesis*, Academic Press, New York, 1967, Chap. 7; G. N. Schrauzer, *Advances in Organometallic Chemistry*, Vol. 2, Academic Press, New York, 1964, p. 1.
22. W. Reppe and H. Kröper, *Ann. Chem.*, **582**, 38 (1953).
23. E. R. H. Jones *et al.*, *J. Chem. Soc.*, 48 (1951).
24. H. Koch and W. Haaf, *Ann. Chem.*, **618**, 251 (1958).
25. K. Bott, *Chem. Ber.*, **100**, 978, 2791 (1967).
26. W. Haaf, *Org. Syn.*, **46**, 72 (1966).
27. K. Bott, *Chem. Ber.*, **100**, 978 (1967).
28. K. Bott, *Chem. Ber.*, **100**, 2791 (1967).
29. W. Haaf and H. Koch, *Ann. Chem.* **638**, 122 (1960).
30. H. Koch and W. Haaf, *Angew. Chem.*, **72**, 628 (1960).
31. H. Koch and W. Haaf, *Org. Syn.*, **44**, 1 (1964).
32. R. G. Jones *et al.*, *J. Am. Chem. Soc.*, **70**, 2843 (1948).
33. K. Ziegler, *Ann. Chem.*, **434**, 34 (1923).
34. N. Kharasch *et al.*, *Chem. Commun.*, 435 (1967).

G. Rearrangements

The preparation of acids by rearrangement comprises a rather heterogeneous collection of reactions including the Arndt-Eistert, Favorskiĭ, second-order Beckmann, and the acid hydrolysis of nitroalkanes.

1. From Acid Chlorides via Diazoketones (Arndt-Eistert and Wolff Rearrangement)

$$RCOCl \xrightarrow{CH_2N_2} RCOCHN_2 \xrightarrow[H_2O]{Ag_2O} RCH_2COOH$$

This synthesis, reviewed in detail [1], transforms a carboxylic acid to the next higher homolog via the diazoketone. The three steps involved generally can be accomplished in one day with overall yields of 50–80%. The reaction requires anhydrous materials, and the usual precautions must be observed in working with diazomethane. Agents such as silver oxide, sodium thiosulfate, potassium hydroxide in methanol-water, a tertiary amine in benzyl alcohol, or simply irradiation have been used to catalyze the decomposition of the diazoketone:

$$RCOCH{=}\overset{\oplus}{N}{=}\overset{\ominus}{N} \xrightarrow{-N_2} (RCOCH{:}) \longrightarrow O{=}C{=}CHR \longrightarrow RCH_2CO_2H$$

but the most reliable, homogeneous catalyst is silver benzoate in triethylamine and *t*-butyl alcohol [2]. Optically active diazoketones retain their configuration through the Wolff rearrangement [3].

This synthesis has been applied to cyclic diazoketones which give carboxylic acids with ring contraction. In these cases fair yields are obtained by irradiation of the acetic acid solution [4]:

References for Section G are on p. 800.

(a) **Preparation of α-naphthylacetic acid** (79–88% from the diazo-ketone) [5].

(b) **Preparation of indole-3-carboxylic acid** (ring contraction). Quinoline-3,4-quinone-3-diazide, 2 g., was dissolved in 40 ml. of acetic acid and, after the addition of 160 ml. of water, the solution was placed in sunlight or exposed to a continuous arc lamp. The vessel was cooled by surrounding it with ice. The weak yellow color of the solution changed generally to brown and the product produced separated as a light brown, crystalline precipitate. After a test with phloroglucinol in alkaline solution was negative, the mixture was filtered. The residue was dissolved in bicarbonate solution, treated with charcoal, and the reaction product was precipitated by the addition of hydrochloric acid. By recrystallization from aqueous acetone, there was obtained 0.9 g. of indole-3-carboxylic acid in white needles, m.p. 218° [6].

(c) **Other examples**

(1) **Various acids** [7].

(2) **Substituted α-propionic acids** (44–70% from the acid chloride) [8].

(3) **Various indole- and pyrrolecarboxylic acids** [9] (ring contractions).

(4) **α,α,α-Phenyldimethylacetic acid from α-bromoisobutyro-phenone** [10]:

$$C_6H_5COC(CH_3)_2Br \xrightarrow{Ag^{\oplus}} C_6H_5C(CH_3)_2COOH$$
$$40\%$$

This example at first glance seems quite different from the Wolff rearrangement. But in the diazoketone, group deficiency is established on loss of nitrogen. Therefore, if electron deficiency is established by other means, rearrangement should also occur. In this example, $C_6H_5CO\overset{\oplus}{C}(CH_3)_2$ is the intermediate. Other ketones with electron deficiency on the α-carbon should also undergo rearrangement.

2. From Cyclic α-Haloketones (Favorskiĭ)

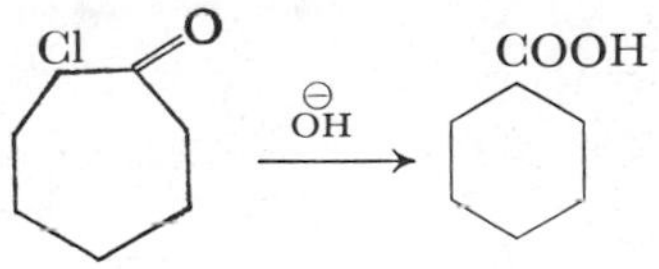

This synthesis has been reviewed [11]. It is applicable to both acyclic and cyclic types, although with the former the products of the reaction are not so predictable as in the case of the latter. For this reason the synthesis has found wider application in the cyclic series. In the alicyclic series in which the halogen is α and in which the keto group is a part of the ring system, rearrangement occurs in the presence of alkali to give carboxylic acids with ring contraction. Thus α-halocyclanones containing six to ten carbon atoms give acids with yields varying usually from 40 to 75%. Other nucleophilic bases such as alkoxides or amines may be employed to effect the rearrangement to give esters (see Carboxylic Esters, Chapter 14, C.8) or amides, respectively. A mild base, sodium bicarbonate, has been employed in one example [12]:

$$\underset{\text{46 g.}}{BrCH_2\overset{O}{\overset{\|}{C}}CHBrCH_3} + \underset{\text{1 mole}}{NaHCO_3} \xrightarrow[\text{Stirred 1–2 hr. at 25°}]{\text{1 l. }H_2O} \underset{cis\text{-Crotonic acid, 9.3 g.}}{CH_3CH{=}CHCOOH}$$

This acid (neat) undergoes slow isomerization to the *trans* isomer at 25°.

The mechanism of this reaction is variable, but in most reactions of α-chloroketones with bases the unstable cyclopropanone derivative 13-III is accepted as the intermediate [13]. According to this interpretation the ketone 13-I loses an α-hydrogen to form the anion 13-II which loses a negative chlorine ion to form the cyclopropane intermediate 13-III. The latter, on being attacked by the hydroxyl ion of the base, cleaves with ring contraction to give the carboxylic acid 13-IV:

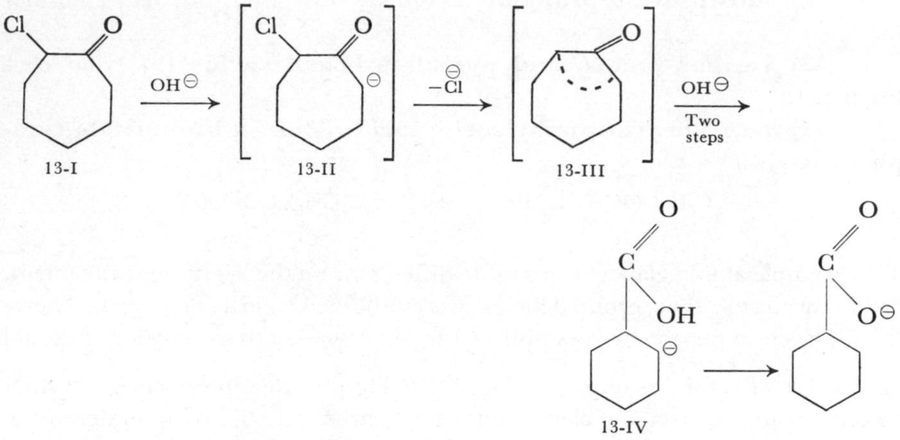

Higher temperatures favor the formation of the cyclopropanone intermediate to form esters and reduce the tendency to form α-alkoxyketones as shown [13]:

$$RCH_2\overset{O}{\overset{\|}{C}}CH_2Cl \xrightarrow[\text{Low temp.}]{OR^{\ominus}} RCH_2\underset{OR}{\overset{O^{\ominus}}{C}}CH_2Cl \xrightarrow{-Cl^{\ominus}} RCH_2\underset{OR}{C}\overset{O}{\frown}CH_2 \xrightarrow{OR^{\ominus}}$$

$$RCH_2\underset{OR}{\overset{O^{\ominus}}{C}}{-}CH_2OR \xrightarrow{-OR^{\ominus}} RCH_2\overset{O}{\overset{\|}{C}}{-}CH_2OR$$

(a) Preparation of cyclohexanecarboxylic acid (69% from 2-chlorocycloheptanone) [14] (ring contraction).

(b) Other examples

(1) **Diphenyl-4-carboxylic acid** (97% from 2-chloro-5-phenyl-tropone) [15] (ring contraction).

(2) **5-Nonenoic acid from 2-chlorocyclopentanone** [16].   Although

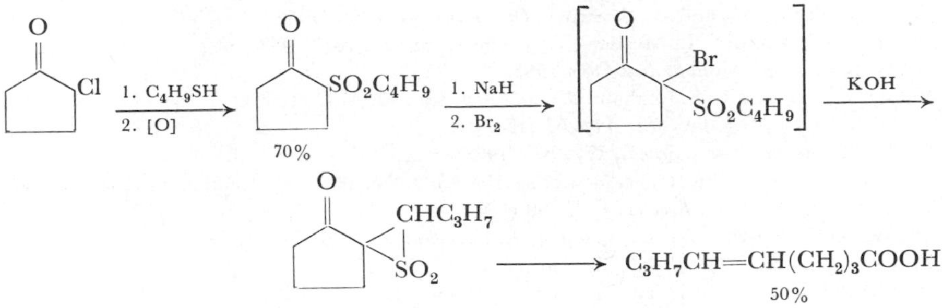

70%

50%

this discovery is too recent to assess fully, the potentiality of this reaction to prepare unsaturated acids appears promising.

3. From Oximes (Beckmann Rearrangement, Second-Order)

$$RC{-}C{-}C_6H_5 \longrightarrow RCN + C_6H_5COOH$$

$$\downarrow$$

$$RCOOH$$

This reaction is discussed under Nitriles, Chapter 19, C.2. It is employed as a method of degradation of an acid to the next lower homolog.

4. From Nitro Compounds (Primary)

$$RCH_2NO_2 \xrightarrow[H^\oplus]{H_2O} RCONHOH \xrightarrow[H^\oplus]{H_2O} RCOOH + H_3\overset{\oplus}{N}OH$$

This internal redox reaction is of limited value in the synthesis of carboxylic acids. In fact, no references to the method have been found in the recent literature. Of the few examples available, the best yields appear to have been obtained from simple nitroalkanes [17].

(a) **Preparation of butyric acid.** Equimolecular quantities of 1-nitrobutane and 65% H_2SO_4 were mixed and heated to 140° with stirring. The mixture was then heated at reflux for 2 hr., during which the temperature rose to 158°. Vacuum distillation of the crude fatty acid layer at a pressure not exceeding 10 mm. gave a product which on a second distillation led to a 94% yield of the acid [17].

(b) Other examples

(1) **Acetic, propionic, and isobutyric acids** (90, 96, and 90%, respectively, from the nitroalkane) [17].

1. W. E. Bachmann and W. S. Struve, *Org. Reactions*, **1**, 38 (1942).
2. M. S. Newman and P. F. Beal, III, *J. Am. Chem. Soc.*, **72**, 5163 (1950).
3. K. J. Sax and W. Bergmann, *J. Am. Chem. Soc.*, **77**, 1910 (1955).
4. O. Süs *et al.*, *Ann. Chem.*, **579**, 133 (1953); **583**, 150 (1953).
5. F. Arndt and B. Eistert, *Org. Reactions*, **1**, 50 (1942).
6. O. Süs *et al.*, *Ann. Chem.*, **583**, 150 (1953).
7. W. E. Bachmann and W. S. Struve, *Org. Reactions*, **1**, 55 (1942).
8. A. L. Wilds and A. L. Meader, Jr., *J. Org. Chem.*, **13**, 763 (1948).
9. O. Süs and K. Möller, *Ann. Chem.*, **593**, 91 (1955).
10. A. C. Cope and E. S. Graham, *J. Am. Chem. Soc.*, **73**, 4702 (1951).
11. A. S. Kende, *Org. Reactions*, **11**, 261 (1960).
12. C. Rappe, *Acta Chem. Scand.*, **17**, 2766 (1963).
13. F. G. Bordwell and R. G. Scamehorn, *J. Am. Chem. Soc.*, **90**, 6751 (1968) and references cited.
14. C. D. Gutsche, *Org. Reactions*, **11**, 290 (1960).
15. W. von E. Doering and L. H. Knox, *J. Am. Chem. Soc.*, **75**, 297 (1953).
16. J. Ficini and G. Stork, *Bull. Soc. Chim. France*, 723 (1964).
17. S. B. Lippincott and H. B. Hass, *Ind. Eng. Chem.*, **31**, 118 (1939).

CARBOXYLIC ESTERS

A. Solvolysis

Solvolytic methods comprise the largest group of ester preparations because all acid derivatives are interconvertible by one means or another. The relative, decreasing order of activity of these derivatives is:

Acid chloride > acid anhydride > ester > amide, nitrile > salt

The ester falling in the middle can be expected to be made easily from acid chlorides or acid anhydrides, to exist in equilibrium with other esters (under acid catalytic conditions), and to be made with some difficulty from amides and salts. However, the preparation from salts receives an additional driving force when the elimination or precipitation of some insoluble inorganic salt occurs. In short, then, esters may be prepared from acids (esterification) as well as from the five other types mentioned above. In addition, other starting materials, such as the ketene, closely related to the anhydride, and 1,1,1-trihalides or α-dihaloethers, in the same oxidation level as esters, may be employed. A unique preparation to be discussed is the pyrolysis of copper salts (A.12) in which an unusual orientation feature is found.

1. From Carboxylic Acids (Esterification)

$$RCOOH + R'OH \underset{}{\overset{H^{\oplus}}{\rightleftharpoons}} RCOOR' + H_2O$$

One of the most common methods of synthesis employed for esters is that of esterification, since the desired acid and alcohol are often available. The method is most satisfactory for primary, less so for secondary, and still less so for tertiary alcohols. To arrive at the point of equilibrium as quickly as possible, catalysts such as sulfuric acid, hydrogen chloride, an aryl sulfonic acid, or boron trifluoride etherate are employed, although some esters of benzyl alcohol [1] and formic acid [2] may be prepared without a catalyst. Boron trifluoride etherate is preferred if the alcohol or acid is of such a nature as to be affected by acids [3]. Two equivalents of boron trifluoride are desirable for the esterification

References for Section A are on pp. 822–825.

of aromatic acids [4]. When it is desirable to have no acid in the medium, strong acid ion exchange resins can be used as a catalyst [5]. The rate of ester formation is dependent on the surface area of the resin, and in the case of large-molecular-weight acids this surface area may be further limited by poor diffusion of the acid into the resin. An example of the esterification of furfuryl alcohol, an alcohol which polymerizes with acid in the medium, by means of a strong acid ion exchanger is shown (see Ex. a).

Recently conditioned Linde 3A molecular sieve pellets have been employed to form methyl nonanoate from the acid, alcohol, and sulfuric acid in 96% yield [6].

To drive the equilibrium as far as possible toward the ester side, one of the starting materials, usually the alcohol, is employed in excess, or one of the products, water, is removed by azeotropic distillation and the solvent, benzene or toluene, returned to the reaction mixture by means of the Dean-Stark apparatus [7, 8]. Other methods of removing water are to distill the azeotrope into a Soxhlet cup containing a drying agent such as magnesium sulfate [9] or chemically by means of the reaction with the dimethyl acetal of acetone to form acetone and methanol [10]. The azeotropic distillation using the Dean-Stark apparatus is the best method of preparing esters, particularly esters made from high-boiling alcohols. Methyl alcohol, however, poses a problem in this procedure because of its volatility. In this case a special bubble column is used to remove the intermediate fractions containing the water [11]. But, when large amounts of sulfuric acid are not harmful to the acid, methanol, the carboxylic acid, and sulfuric acid may be simply refluxed and the methyl ester extracted into toluene following the procedure of Klostergaard for preparation of ethyl esters such as triethyl citrate [12]. Indeed, a simple semimicro procedure similar to the method above has been worked out whereby methyl esters are formed and separated as efficiently as they would be by reaction of the acid with diazomethane (see Ex. b). Lastly, methyl esters of aliphatic and aromatic acids may be prepared conveniently in yields of 87–98% by refluxing the acid (1 mole), methanol (3 moles), and sulfuric acid in ethylene or methylene chloride for 6–15 hours [13]. This method does not require the elaborate equipment of that of Weissberger and Kibler and is more convenient for large-scale preparations than that of Baker [14].

The path taken in esterification may be represented as follows:

$$
\underset{\text{O}}{\overset{\text{O}}{R-\overset{\|}{C}-OH}} \ \underset{\longleftarrow}{\overset{H^{\oplus}}{\longrightarrow}} \ \left[\underset{14\text{-}I}{R-\overset{OH}{\underset{\oplus}{\overset{|}{C}}}-OH}\right] \ \underset{\longleftarrow}{\overset{R'OH}{\longrightarrow}} \ \left[\underset{14\text{-}II}{R-\overset{OH}{\underset{\overset{|}{R'OH}}{\overset{|}{C}}-OH}}\right] \ \rightleftharpoons
$$

$$
\left[R-\overset{OH}{\underset{\overset{|}{OR'}}{\overset{|}{C}}-\overset{\oplus}{OH_2}}\right] \ \underset{\longleftarrow}{\overset{-H_2O}{\longrightarrow}} \ \left[R-\overset{OH}{\underset{\overset{|}{OR'}}{\overset{|}{C}{\oplus}}}\right] \ \underset{\longleftarrow}{\overset{-H^{\oplus}}{\longrightarrow}} \ \underset{\underset{\text{OR'}}{\overset{O}{R-\overset{\|}{C}}}}{}
$$

14-III

The electrophilic attack of the proton from the acid forms the carbonium ion 14-I which undergoes a nucleophilic attack by the alcohol to give 14-II which by rearrangement and elimination of water and a proton gives the ester 14-III [15, 16]. Of course, allyl and benzyl alcohols and tertiary alcohols tend to react by a carbonium ion mechanism. Because of this tendency, tertiary alcohols are converted mainly into olefins rather than esters, and other means of synthesis of the tertiary alkyl esters must be sought as described in the paragraphs to follow.

Steric effects play an important role in esterification since the carbon atom in the carbonyl group of the acid is transformed from a planar trigonal structure (sp^2 hybridized orbital) to the tetrahedral structure (sp^3 hybridized orbital) in the transition complex, 14-II. Newman's rule of six is helpful in determining the influence of steric effects on the rate of esterification of aliphatic acids, but examination of models is the best way for estimating steric factors [17]. Substituents in the *ortho* positions of aromatic acids also interfere with esterification by ordinary methods. In the case of *o,o'*-dialkyl substitution, esterification may be accomplished by preparing a solution of the acid in 100 percent sulfuric acid and pouring it into the alcohol [18]. Success appears to depend on the formation of a planar carbonium ion

$$\overset{\oplus}{C}=O$$

R — (benzene ring) — R
R

(the usual protonated carboxyl group

$$-C\overset{\overset{\oplus}{O}H}{\diagdown_{OH}}$$

in this case cannot become coplanar with the ring) which reacts with the alcohol to give the ester. The yields vary from fair to good by this procedure.

Ethyl esters may be prepared (84–94 % yields) by heating the acid and an excess of ethyl orthoformate until solution results [19]. In two cases *p*-toluene-sulfonic acid or dimethylformamide is added. Hindered acids such as 2,4,6-trinitro- and 2,4,6-trimethylbenzoic acids are esterified in moderate yield.

Recently a new method for esterifying hindered acids has been reported [20]. In this case the reaction of the alcohol and acid is catalyzed with trifluoracetic anhydride. Excellent yields are usually obtained with hindered acids such as 9-anthroic and mesitoic and simple or hindered alcohols or phenols. If both the phenol and acid are hindered and an alternative path, such as carbon acylation of the phenol, is available, the esterification may fail. The mechanism suggested follows two paths, both involving the protonated anhydride 14-IV:

$$\underset{\text{14-IV}}{\overset{\overset{\displaystyle O\quad\;\; O}{\displaystyle \|\quad\;\; \|}}{ArC-O-CCF_3} + H^{\oplus} \rightleftharpoons \overset{\overset{\displaystyle OHO}{\displaystyle |\;\;|}}{ArCOC-CF_3}}$$

The latter may react directly with the alcohol to give the ester 14-V:

$$\underset{\text{14-V}}{14\text{-IV} + ROH \rightarrow ArCOOR} + CF_3COOH + H^{\oplus}$$

or an acylium ion 14-VI may be involved in forming 14-V:

$$14\text{-IV} \rightleftharpoons CF_3COOH + \underset{\text{14-VI}}{Ar\overset{\oplus}{C}{=}O} \xrightarrow{ROH} \underset{\text{14-V}}{ArCOOR} + \overset{\oplus}{H}$$

A similar procedure is the reaction of the acid in 20 parts of pyridine with 2 equivalents of benzenesulfonyl- or p-toluenesulfonyl chloride to form the acid anhydride. To this cold solution 1 equivalent of alcohol (or phenol) is added and the mixture is diluted with ice water. The procedure is recommended for the preparation of t-alkyl esters [21] and esters of tertiary acetylenic alcohols [22, 23].

For esterification of acids under mild conditions, dimethylformamide diethyl acetal is satisfactory [24]. For example, with benzoic acid and 2 equivalents of the formamide in benzene at 80°, in 1 hour the yield of ester is over 90%:

$$C_6H_5COOH + \underset{\displaystyle C_2H_5O}{\overset{\displaystyle C_2H_5O}{\diagdown}}\!\!\!H{-}C{-}N(CH_3)_2 \longrightarrow$$

$$\underset{\text{Ethyl benzoate}}{C_6H_5COOC_2H_5 + HCON(CH_3)_2 + C_2H_5OH}$$

Amino acids may be esterified satisfactorily with the alcohol and thionyl chloride as a catalyst [25, 26] and in some cases with p-toluenesulfonic acid and a large excess of dimethyl sulfite [27]:

$$\underset{\displaystyle NH_2}{C_6H_5CH_2CHCOOH} + (CH_3O)_2SO + TsOH \longrightarrow$$

$$\underset{\displaystyle NH_3^{\oplus}(TsO)^{\ominus}}{C_6H_5CH_2CHCOOCH_3 + SO_2 + CH_3OH}$$

Methyl α-amino-β-phenyl-
alanine tosylate, 99%

Steroid alcohols may often be acetylated best with acetic acid and acetic anhydride in the presence of a trace of perchloric acid [28].

Aryl esters may be prepared by treatment of the acid with the phenol in the presence of phosphorus oxychloride [29] or polyphosphoric acid [30].

If both primary and secondary alcoholic groups are present in a molecule such as in the antibiotic, chloramphenicol, partial esterification ordinarily yields primary monoesters. The ester of the secondary alcoholic group has been produced by the ingenious scheme shown [31].

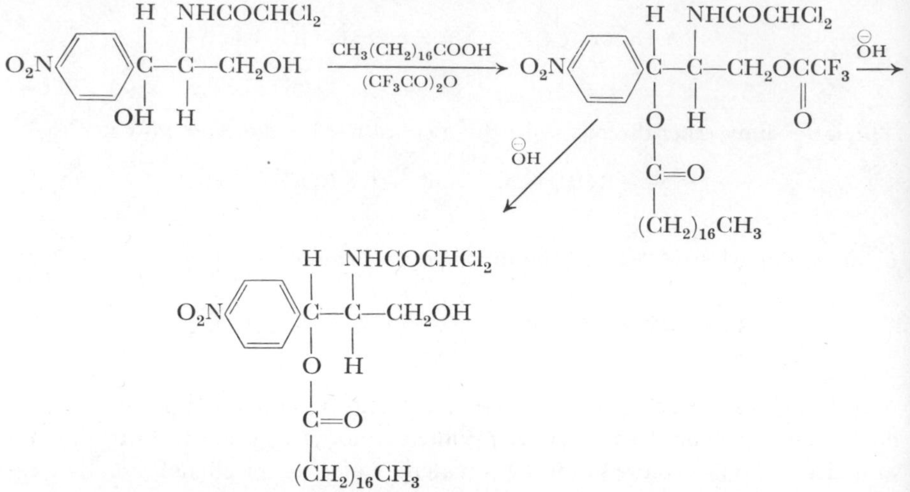

Chloramphenicol stearic ester

Diesters from ketones or aldehydes and acids can be formed probably via the anhydride (see Ex. d) and from ester interchange (see A.7).

(a) **Preparation of furfuryl acetate** (21 % by refluxing 0.2 mole of furfuryl alcohol, 0.4 mole of acetic acid, 74 ml. of benzene, and 20 g. of Zeo-Karb H (strong acid ion exchange resin) with Dean-Stark tube attachment until water ceases to form) [32].

(b) **Preparation of methyl esters for gas chromatography.** A mixture of 0.5 ml. of fatty acid in 2 ml. of methanol and 0.5 ml. of sulfuric acid was heated to boiling and cooled. Water, 2 ml., was added and then 2 ml. of hexane with shaking. An aliquot of the hexane layer containing all the methyl ester in about a 20 % solution can be injected directly into the gas chromatograph [33].

(c) **Preparation of ethyl mandelate** (82–86 % from ethanol and mandelic acid in the presence of hydrogen chloride) [34].

(d) **Preparation of isopropylidene malonate (Meldrum's acid).**

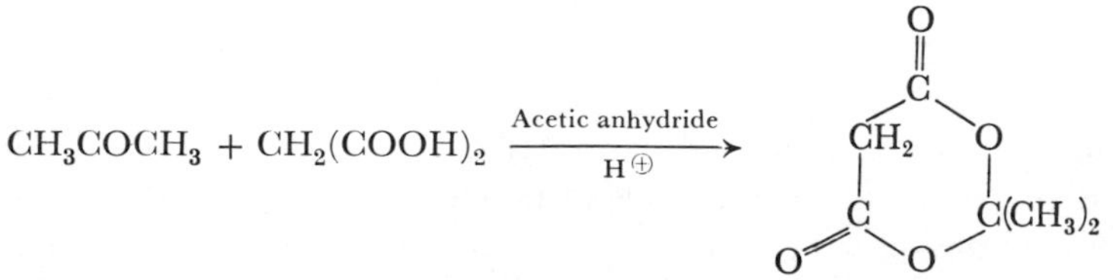

$$CH_3COCH_3 + CH_2(COOH)_2 \xrightarrow[H^{\oplus}]{\text{Acetic anhydride}}$$

To a stirred suspension of 0.5 mole of malonic acid in 60 ml. of acetic anhydride, 1.5 ml. of sulfuric acid was added dropwise. To the resulting solution, 40 ml.

of acetone was added while cooling to maintain room temperature. After standing overnight in the refrigerator, the crystals were filtered and washed with cold water (35 g., 49% yield). They can be recrystallized in about 70% yield from acetone to which water is added, m.p. 94–95° (dec.) [35].

(e) Other examples

(1) **Ethyl adipate** (95–97% based on the acid, from the acid and ethanol; water removed as toluene azeotrope) [36].

(2) **Ethyl bromoacetate** (65–70% based on bromoacetic acid; water removed as benzene azeotrope) [8].

(3) **Mesityl mesitoate** (94% from mesitoic acid, mesitol, and trifluoroacetic anhydride by stirring at room temperature for 20 min.) [20]; see also [37] for preparation of 1- or 2-monoacylglycerides by this procedure.

(4) **α-Acetoxy-N-cyclohexylisovaleramide** (69% from acetic acid,

$$CH_3COOH + \underset{/}{\overset{\backslash}{C}}{=}O + {-}C{\equiv}NR \longrightarrow CH_3CO_2\overset{O}{\overset{\|}{\underset{|}{C}}}CNHR$$

isobutyraldehyde, and cyclohexylisonitrile, $C_6H_{11}NC$ (Passerini reaction)) [38].

(5) **p-Cresyl hexanoate** (quantitative from p-cresol, hexanoic acid, and a trace of sulfuric acid in toluene with Dean-Stark apparatus; contrary to reports in textbooks, phenols can be esterified directly) [39].

(6) **L-Glutamic dibenzyl ester hydrochloride** (69% from the acid, alcohol, and thionyl chloride, the product being precipitated by dry ether) [26].

2. From Acid Chlorides

$$ROH + R'COCl \rightarrow R'COOR + HCl$$

This synthesis has been widely used because it avoids the reversibility associated with esterification. The hydrogen chloride may escape as such, or it may be absorbed in a base such as sodium hydroxide, dimethylaniline, or pyridine, or by magnesium. Sometimes the use of the sodium alcoholate is desirable. Dimethylaniline [40] and magnesium [41] have been employed in preparing the esters of tertiary alcohols, which have a tendency to form halides in the absence of a substance which consumes the hydrogen chloride liberated. In the esterification of phenols, magnesium [42] has been used frequently, a reaction for which a quantitative study is available [43]. The highest yields of phenyl esters are obtained from the thallium salts of phenols and the acid chloride (see Ex. d.6).

The simplest procedure is the Schotten-Baumann method which consists in shaking the acid chloride with aqueous sodium hydroxide containing the appropriate alcohol. For best yields the acid chloride (and ester formed) should be insoluble in water so that reaction occurs at the interface of the organic and aqueous layers. The insolubility of the ester in the aqueous phase retards its saponification. At the end of the reaction the ester is free from acid chloride and hydrogen chloride and can be extracted and dried immediately. A rather old and neglected study of the Schotten-Baumann reaction states that the yields of esters are increased by lowering the temperature of the reaction, by increasing the concentration of alcohol (which can be done by adding alkali gradually), and by using potassium hydroxide rather than sodium hydroxide. Furthermore, acetic anhydride gives a higher yield of ester than acetyl chloride [43]. The pyridine method is the most comprehensive and perhaps would be used exclusively if it were not for the nuisance of ridding the ester of traces of pyridine, which are usually washed out with dilute acid.

The effectiveness of the acid halide reaction with alcohols may be enhanced by prior reaction of the acid halide with a Lewis acid to form an acylium salt. The simplest of these appears to be methyl oxocarbonium hexafluoroantimonate, $CH_3\overset{\oplus}{C}OSbF_6{}^{\ominus}$, which alone or in nitromethane reacts rapidly with alcohols to form acetates in yields from 62 to 87 % (see Ex. d.7).

It appears that the reaction occurs because of the nucleophilic attack of the alcohol on the carbonyl carbon of the acid chloride. The complex thus formed

$$R'-\overset{\overset{\displaystyle \overset{\ominus}{O}}{|}}{\underset{\underset{\displaystyle ROH}{\uparrow}}{\overset{\oplus}{C}}}-Cl$$

loses hydrogen chloride rapidly to give the ester, R'COOR.

In the case of the formation of the reactive and unstable benzyl tosylates, it is desirable to form the alkoxide first with sodium hydride and then treat with the acid chloride [44]. A similar method, but without isolation of the sodium alkoxide, was applied to the formation of the tosylate of 2-nitro-3,4,5-trimethoxybenzyl alcohol, in which the hydroxyl of the alcohol is strongly bonded to the nitro group [45]:

$$\underset{\underset{\displaystyle OCH_3}{CH_3O\diagdown\diagup OCH_3}}{\overset{\overset{\displaystyle CH_2OH}{\diagup NO_2}}{}} + NaH + TsCl \longrightarrow \underset{\underset{\underset{\displaystyle 90\text{–}94\%}{\displaystyle OCH_3}}{CH_3O\diagdown\diagup OCH_3}}{\overset{\overset{\displaystyle CH_2OTs}{\diagup NO_2}}{}}$$

Many functional groups are not affected in the synthesis. In acyl halides having double bonds [46], acetoxy [47], amino alkyl [48], and 2-furanacrylyl [49] groups the groups remain unaffected. Hydroxyesters may be prepared from a diol and the acid chloride [50]; haloesters may be obtained similarly from halohydrins [51]. The yields are often high.

(a) **Preparation of phenyl benzoate** (unstated yield from 1 g. of phenol in 10 % aqueous sodium hydroxide to which 1 ml. of benzoyl chloride was added dropwise with vigorous shaking between drops; after complete addition, the mixture was shaken further until the odor of benzoyl chloride had disappeared; the ester was filtered, washed, and recrystallized from ethanol) [52].

(b) **Preparation of benzoates, *p*-nitrobenzoates, and 3,5-dinitro-benzoates** (from the acid chloride and alcohol in pyridine) [53].

(c) **Preparation of *t*-butyl acetate** (63–68% with dimethylaniline as the base [40]; 45–55% with magnesium [41]).

(d) **Other examples**

(1) **Dimethyl α-butyl-α-ethyl glutarate** (about 100% when the monoester acid chloride was introduced into methanol vapor at 170–180°) [54].

(2) **α-Methyl glucoside tetrastearate** (92% from the glucoside using reduced pressure to remove the hydrogen chloride formed) [55].

(3) **Phenyl cinnamate** (63–75% from the acid, phenol, and $SOCl_2$) [56].

(4) **3-Methylcyclopentyl tosylate** (over 64% from the alcohol and *p*-toluenesulfonyl chloride; this esterification is useful in that the ester can be converted into the bromide without isomerization) [57].

(5) **Ethyl benzylsulfonate** (quantitative yield from benzylsulfonyl chloride, triethylamine, and ethyl alcohol proceeding via the sulfene, $C_6H_5CH{=}SO_2$) [58].

(6) **Phenyl pivalate** (96% from thallium phenolate suspended and stirred at 25° in dry ether to which pivalyl chloride was added; thallium chloride was removed by filtration and the filtrate was concentrated) [59].

(7) **Ethyl acetate.** Methyl oxocarbonium hexafluoroantimonate, 0.3 mole, in nitromethane or in small portions was added with stirring to 0.6 mole of ethanol. The mixture was washed with water, dried over Na_2SO_4, and fractionated to give a 70% yield of ethyl acetate [60].

3. From Acid Anhydrides

$$ROH + (R'CO)_2O \rightarrow R'COOR + R'COOH$$

or

$$\underset{\underset{H(R)}{|}}{RCH_2C}{=}O + (CH_3CO)_2O \longrightarrow \underset{\underset{H(R)}{|}}{RCH}{=}\overset{\overset{O}{\|}}{C}OCCH_3$$

Like acid chlorides, acid anhydrides have been widely used in forming esters with hydroxy compounds. The reaction has been employed as such or catalysts such as sulfuric acid [61], zinc chloride [62, 63], chlorosulfonic acid [64], acetyl chloride [65], sodium acetate [66], boric acid [67], ferric sulfate [68], sodium methoxide [69], and *p*-toluenesulfonic acid monohydrate [70] have been used. A variety of phenols may be acetylated in aqueous alkaline solution [71], while cyclic anhydrides may be partially or completely esterified [69, 70]. The mechanism of the synthesis follows the same pattern as when an acid chloride is used, the leaving group being a carboxylate anion rather than a halide anion. Since the carboxylate anion is not so thermodynamically stable in solution as the halide anion, the anhydride reaction with alcohols can be expected to be somewhat less exothermic than the acid halide reaction. Contrary to some textbook reports, *t*-alcohols can be esterified with anhydrides and acidic catalysts (see Ex. c.4) or triphenylmethyl sodium (see Ex. c.5).

Enolizable aldehydes may be converted into the enol acetates by heating with acetic anhydride and potassium acetate [72]. Yields with three aldehydes vary from 35 to 61 %. Enolizable ketones do not respond so readily, but success may be achieved with acetic anhydride and *p*-toluenesulfonic acid [73]. Yields with three ketones vary from 32 to 68%. It is interesting to note that under the conditions employed with ketones only the methylenic hydrogen atoms are involved. The enol acetates have been utilized as intermediates in the synthesis of α-bromoaldehydes and ketones as shown:

$$\underset{\displaystyle\text{RCH}=\overset{\displaystyle|}{\underset{\displaystyle}{\text{C}}}\text{—H(R}')}{\overset{\displaystyle\text{OCOCH}_3}{}} \xrightarrow{\text{Br}_2} \underset{\displaystyle\text{RCHBr}\overset{\displaystyle|}{\underset{\displaystyle|}{\text{C}}}\text{—H(R}')}{\overset{\displaystyle\text{OCOCH}_3}{}}\underset{\text{Br}}{} \xrightarrow{\text{CH}_3\text{OH}}$$

$$\overset{\displaystyle\text{O}}{\overset{\displaystyle\|}{\text{RCHBrC}}}\text{—H(R}') + \text{CH}_3\text{COOCH}_3 + \text{HBr}$$

Mixed anhydrides derived from the alkyl chloroformate decompose on heating to form esters as follows:

$$\overset{\text{O O}}{\overset{\|\ \|}{\text{RCOCOR}'}} \xrightarrow{\text{A}} \text{RCO}_2\text{R}' + \text{CO}_2$$

$$2\,\overset{\text{O O}}{\overset{\|\ \|}{\text{RCOCOR}'}} \xrightarrow{\text{B}} (\text{RCO})_2\text{O} + \text{O}=\text{C(OR}')_2 + \text{CO}_2$$

$$\Big\downarrow \text{C}$$

$$2\,\text{RCO}_2\text{R}' + \text{CO}_2$$

A number of mixed anhydrides from aliphatic acids decompose to give esters by path A, while those from aromatic acids decompose by both paths A and B. At 250°, step C occurs as well [74]. Trifluoroacetic acid and phenyl chloroformate on mixing gave phenyl trifluoroacetate in 75–80% yield [75]. The decarboxylations above are of a cyclic electron-shift nature.

(a) Preparation of benzoin acetate (86–90 % by heating benzoin with acetic anhydride, glacial acetic acid, and H$_2$SO$_4$) [76].

(b) Preparation of cyclohexanone enol acetate (68 % by heating 1 mole of the ketone, 2 moles of acetic anhydride, and 1 mole of *p*-toluenesulfonic acid for 4 hr.) [73].

(c) Other examples

(1) Diethyl *cis*-Δ⁴-tetrahydrophthalate (83–86 % from the anhydride, ethanol, and *p*-toluenesulfonic acid monohydrate) [70].

(2) **Hydroquinone diacetate** (96–98% from hydroquinone, acetic anhydride, and sulfuric acid) [61].

(3) **Methyl hydrogen succinate** (95–96% from the anhydride and methanol) [77].

(4) *t*-**Amyl acetate** (80% of high purity from 110 ml. of the alcohol, 100 ml. of the anhydride, and 4 drops of conc. hydrochloric acid, the mixture being held at 100–112° for 3 hr.) [78].

(5) *t*-**Butyl acid phthalate** (65% from phthalic anhydride, *t*-butyl alcohol, and triphenylmethyl sodium) [79].

4. From Ketenes, Ketene Acetals, and Isocyanates

$$R_2C{=}C{=}O + R'OH \rightarrow R_2CHCOOR'$$

Although the synthesis often gives high yields, it is not a common method because the ketene is usually not available or is troublesome to prepare [80]. Catalysts such as sulfuric acid or *p*-toluenesulfonic acid promote the reaction [81]. The method has been used successfully in the acetylation of lactic esters [82] and of tertiary alcohols and phenols [81, 83]. Some functional groups such as those of aldehydes [81] and ketones [84] are attacked by ketenes. In some cases these carbonyl compounds are converted into enol acetates:

$$\underset{}{RCOCH_3} \longrightarrow \underset{OH}{RC{=}CH_2} \xrightarrow{CH_2{=}C{=}O} \underset{OCOCH_3}{RC{=}CH_2}$$

The addition of the alcohol to the ketene appears to follow the pattern associated with carbonyl groups. The first formed enol

$$R_2C{=}C\overset{OH}{\underset{OR'}{<}}$$

tautomerizes to give the ester

$$R_2CHC\overset{O}{\underset{OR'}{<}}$$

Since ketenes are intermediates in the irradiation of most ketones, the ketones become potential sources of esters if irradiated in the presence of alcohols [85] (see Carboxylic Acids, Chapter 13, B.5):

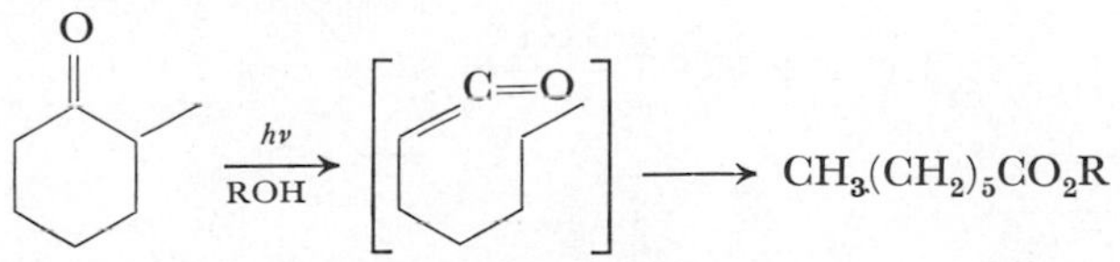

$$\xrightarrow[ROH]{h\nu} \left[\quad \right] \longrightarrow CH_3(CH_2)_5CO_2R$$

Alcohols add to diketenes to give acetoacetates [86] as follows:

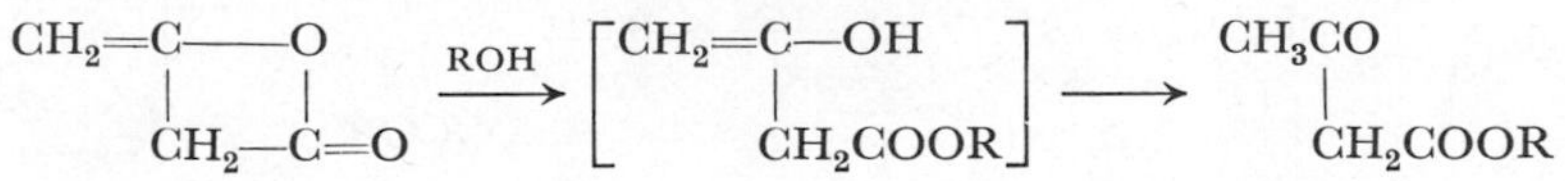

$$\underset{CH_2-C{=}O}{\overset{CH_2{=}C-O}{|}} \xrightarrow{ROH} \left[\underset{CH_2COOR}{\overset{CH_2{=}C-OH}{|}} \right] \longrightarrow \underset{CH_2COOR}{\overset{CH_3CO}{|}}$$

This method has been used to advantage in preparing acetoacetates with unsaturated R groups [87]. In these cases, sodium alkoxide proved to be the preferable catalyst. These esters may also be obtained from the acid chloride via the diketene in one operation [88].

Ketene acetals also can be converted spontaneously and quantitatively into esters as illustrated [89]:

$$(C_6H_5)_2C{=}C(OCH_3)_2 \xrightarrow[\text{Trace of acid}]{\text{H}_2\text{O in dioxane}} (C_6H_5)_2CHCO_2CH_3$$

Methyl diphenylacetate

$$\xrightarrow{\text{Br}_2} (C_6H_5)_2CBrCO_2CH_3$$

Methyl α-bromodiphenylacetate

The synthesis of carbamates from isocyanates and alcohols resembles the formation of esters from ketenes:

$$RN{=}C{=}O + R'OH \rightarrow RNHCOOR'$$

The reaction is spontaneous proceeding without a catalyst and giving high yields of carbamates. Tertiary alcohols, however, tend to yield olefins. For the preparation of phenyl carbamates from isocyanates and phenols, a few drops of triethylamine or pyridine are used as a catalyst. The isocyanate ester synthesis is useful as a means of identifying alcohols [90].

(a) Preparation of *t*-butyl acetate. To 0.1 mole of *t*-butyl alcohol was added 1 drop of conc. sulfuric acid or 0.01 g. of *p*-toluenesulfonic acid, after which ketene was passed through the solution in an ice bath for 70 min. The solution was treated with 10 ml. of 6 *N* sodium hydroxide and then washed with 10 ml. of water. The ether extraction of these two aqueous solutions was added to the ester layer and the total was dried with potassium carbonate and distilled. *t*-Butyl acetate, b.p. 94–95°, was obtained in a yield of 86–89% [81].

(b) Other examples

(1) Ethyl α-acetoxypropionate (98% from ethyl lactate, ketene, and sulfuric acid) [82].

(2) Crotyl acetoacetate (83% from diketene, crotyl alcohol, and sodium) [87].

5. *gem*-Diesters or α-Chloroesters from Carbonyl Compounds

$$RCHO + R'COCl \xrightarrow{\text{ZnCl}_2} RCH\overset{\displaystyle OCOR'}{\underset{\displaystyle Cl}{<}}$$

or

$$RCHO + (R'CO)_2O \xrightarrow{\text{BF}_3} RCH\overset{\displaystyle OCOR'}{\underset{\displaystyle OCOR'}{<}}$$

These addition reactions have been used to a limited extent to produce α-haloalkyl esters or acylals [91]. The former are produced by mixing the components alone or in the presence of a trace of zinc chloride [92], the yields being in the 50% range. The latter are produced best with acetic anhydride when a small amount of boron trifluoride etherate is used, but with propionic, and generally with *n*-butyric, anhydride the best results are obtained by employing 10% more than an equivalent of the catalyst. With both amounts of catalysts, mixtures are obtained with acetic anhydride and formaldehyde, and the reaction failed with succinic anhydride and propionaldehyde [93]. Yields in these additions are in the 60–80% range. Acids such as sulfuric may also be used in the anhydride additions but the yields are usually lower [94], except when sodium acetate is added [95].

Examples of the preparation of isopropylidene malonate and of α-acetoxy-N-cyclohexylisovaleramide have already been given (A.1, Ex. d, and Ex. e.4, respectively).

(a) **Preparation of α-chlorocrotyl acetate.** Freshly distilled, pure crotonaldehyde, 52.5 g., was mixed with 58.9 g. of pure acetyl chloride. The temperature of the mixture rose to 64° but was held thereafter within the 35–40° range by cooling. After standing overnight at 18°, the mixture was distilled under vacuum to give 79 g. (71%) boiling at 64–68° (9 mm.). From this there was obtained by distillation a purer fraction, b.p. 64–65° (8.5 mm.) [96].

(b) **Furfural diacetate** (65–70% from acetic anhydride, furfural, sulfuric acid, and sodium acetate) [95].

6. From Nitriles

$$RCN \xrightarrow[H^\oplus]{R'OH} RC\begin{smallmatrix}\overset{\oplus}{NH_2}\\\diagup\\\diagdown\\OR'\end{smallmatrix} \xrightarrow{H_2O} RC\begin{smallmatrix}O\\\diagup\\\diagdown\\OR'\end{smallmatrix}$$

Although nitriles may be converted into esters via the amide [97], the usual method is to proceed via the imino ester which may or may not be isolated as an intermediate. Various catalysts such as hydrogen chloride [98], hydrogen bromide and hydrogen iodide [99], sulfuric acid [100], and benzene- and *p*-toluenesulfonic acid monohydrates [101] have been employed. The arene sulfonic acids are preferred over hydrogen chloride or sulfuric acid. Yields are more satisfactory with hydrogen bromide and hydrogen iodide than with the chloride [99]. As the equation indicates, imino esters may be obtained if water is excluded from the reaction [99]. In some cases no water is necessary in the reaction leading to the iminoester [102], while in others no reaction occurs in its absence [103].

This synthesis has found wide application in preparing polyfunctional types. Malonic esters, for example, are usually obtained from α-cyanocarboxylic acid esters [104,105]. From cyanohydrins, hydroxy esters [106], from keto cyanides, keto esters [107–109], and from amino cyanides, amino esters [110–112] are obtained. The yields are generally satisfactory.

Difficultly solvolyzed amides may be esterified by the procedure given under esterification of amides (see A.14, Ex. a).

(a) Preparation of glycine ethyl ester hydrochloride (87–90 % from methyleneaminoacetonitrile, ethanol, water, and hydrogen chloride) [110].

(b) Other examples

(1) Ethyl phenylacetate (90 % from phenylacetonitrile, ethanol, and hydrogen bromide, followed by hydrolysis) [99].

(2) Ethyl α-phenylacetoacetate (50–81 % from α-phenylacetoaceto-nitrile, hydrogen chloride, and ethanol, followed by hydrolysis) [107].

(3) n-Heptyl phenylacetate (70 % by refluxing 0.40 mole each of phenylacetonitrile, n-heptyl alcohol, and p-toluenesulfonic acid monohydrate for 6 hr.) [101].

7. From Esters and Alcohols (Transesterification)

$$RCOOR' + R''OH \underset{\underset{OH^{\ominus}}{or}}{\overset{H^{\oplus}}{\rightleftharpoons}} RCOOR'' + R'OH$$

Transesterification involves an equilibrium which may be shifted in the forward direction by using an excess of one of the reactants or, preferably, by removing one of the components, $RCOOR''$ or $R'OH$. If $R'OH$ is lower-boiling than $R''OH$, separation may be possible by fractionation or in some cases advantage is taken of azeotropes for the removal of products [113]. This interchange is catalyzed by acids such as sulfuric [114, 115] or p-toluenesulfonic [113] or by bases such as aluminum alkoxides [116, 113]. As a rule, the conversion occurs best with primary alcohols, methyl alcohol being the most responsive, although secondary alcohols may at times be used successfully [117, 113].

The synthesis lends itself well to the preparation of higher fatty acid alkyl esters from fats, oils, and waxes [114, 115]. Other esters such as the acrylates [113] and γ-diethylamino-α-phenylbutyrates [112] have been produced to advantage in large numbers by this interchange. The most important enol ester to be used for the synthesis of esters is isopropenyl acetate. This reagent is readily available and produces esters in high yield by exchange with an alcohol in the presence of acid catalysts:

$$\underset{\underset{CH_3COCOCH_3}{||}}{CH_2} + ROH \xrightarrow{H^{\oplus}} \underset{CH_3}{\overset{CH_3}{\underset{|}{C}}}{=}O + \underset{\overset{O}{||}}{ROCCH_3}$$

The reaction may proceed via the ketene, a reagent from which isopropenyl acetate is made, but the possibility exists also that an acylium ion, which attacks the alcohol, is formed on acid catalysis. An example of ester formation is given (b) and also one of enol ester formation (c).

The mechanism of transesterification of ordinary esters follows the pattern of acid- and base-catalyzed ester hydrolysis as shown under Carboxylic Acids,

Chapter 13, A.2. β-Ketoesters need no catalysts for transesterification, and the exchange of alcohol groups proceeds quantitatively at steam-bath temperature. *t*-Butyl alcohol, however, is very sluggish in exchange [118].

(a) **Preparation of *n*-butyl acrylate** (78–94 % from methyl acrylate, *n*-butyl alcohol, and *p*-toluenesulfonic acid) [113].

(b) **Preparation of glycerol triacetate.** Isopropenyl acetate (100 g.), glycerol (30 g.), and 3 drops of sulfuric acid were distilled slowly until 57 g. of acetone was recovered. The residue was neutralized with magnesium carbonate, and glycerol triacetate distilled, b.p. 113–115° (2 mm.), quantitative yield [119].

(c) **Preparation of 1-acetoxy-1,3-butadiene.**   A mixture of isopropenyl acetate (2 kg.), *p*-toluenesulfonic acid (20 g.), and cupric acetate (5 g.) was added to a 1 in. × 6 ft. packed column held at a temperature above the boiling point of acetone. Simultaneously crotonaldehyde (1050 g.) was added to the column at the same rate. The crude enol ester, collected in the flask attached to the bottom of the column, was flashed off at 2–5 mm. pressure and redistilled, b.p. 58° (40 mm.), n^{20}D 1.4460 (90 %) [119].

(d) **Other examples**

   (1) **Butyl oleate** (70 % from olive oil, butyl alcohol, and conc. sulfuric acid) [115].
   (2) **Benzyl γ-diethylamino-α-phenylbutyrate** (88 % from benzyl alcohol, methyl γ-diethylamino-α-phenylbutyrate, and sodium ethoxide) [112].

8. From Lactones and Alcohols (Transesterification)

$$CH_2CH_2CH_2C\!=\!O \xrightarrow[\text{HCl}]{C_2H_5OH} HOCH_2CH_2CH_2COOC_2H_5$$
$$\llcorner\!\!-\!\!O\!\!-\!\!\lrcorner$$

This synthesis has found limited use in the preparation of esters. Hydrogen bromide in alcoholic solution gives the bromoester [120, 121], while hydrogen chloride produces the hydroxy ester [122]. Other halides such as phosphorus pentabromide [120] and thionyl chloride [123] have been used instead of the hydrogen halide. β-Propiolactone seems to have been studied the most thoroughly [124]. In this case with the alcohol, two products are formed depending on the catalyst. With bases the cleavage occurs at the oxygen-carbonyl carbon bond to give hydroxy esters:

$$\begin{array}{c} CH_2\!\!-\!\!CH_2 \\ |\qquad | \\ O\!\!-\!\!-\!\!C\!\!=\!\!O \end{array} + ROH \xrightarrow{OH^{\ominus}} HOCH_2CH_2COOR$$

On the other hand, without a catalyst or with an acidic catalyst, cleavage occurs at the oxygen-methylene carbon bond to give β-alkoxy acids:

$$\begin{array}{c} CH_2\!\!-\!\!CH_2 \\ |\qquad | \\ O\!\!-\!\!-\!\!C\!\!=\!\!O \end{array} + ROH \xrightarrow{H^{\oplus}} ROCH_2CH_2COOH$$

The mechanism of the cleavage follows the pattern covered for esters (Carboxylic Acids, Chapter 13, A.2) since lactones may be regarded as inner esters. The yields are satisfactory.

(a) **Preparation of ethyl α-ethyl-γ-hydroxybutyrate.** α-Ethylbutyrolactone, 150 g., in 500 ml. of absolute ethanol was saturated with dry hydrogen chloride and allowed to stand for 3 days. The alcohol was removed by distillation and the residue was poured into water. The ester layer was added to ethereal extracts of the aqueous layer and the total was dried and distilled to give 175 g. (84%) of the ester, b.p. 78–80° (3 mm.), d_4^{25} 0.9609 [122]; for a similar preparation of ethyl γ-bromobutyrate see [125].

(b) Other examples

(1) **Ethyl-4-methylheptanoate** (86.5% from γ-methyl-γ-propylbutyrolactone, thionyl chloride, ethanol, and hydrogen chloride) [123].

(2) **Methyl hydracrylate** (85% from β-propiolactone, sodium hydroxide, and methanol followed by acidification) [124].

9. From Esters and Acids (Transesterification)

$$RCH_2COOH + R'COOR'' \rightarrow RCH_2COOR'' + R'COOH$$

This interchange has been accomplished, often with good yields, in the case of N-acylamino acids and in producing vinyl esters of a variety of aliphatic and aromatic carboxylic acids. The N-acylamino acids, such as phthaloyl glycine or tosyl glycine, were converted into methyl or ethyl esters by the use of the alkyl formate or acetate [126]. The most satisfactory catalysts were 96% sulfuric acid or p-toluenesulfonic acid monohydrate. Nonacylated amino acids, with the exception of phenylalanine, could not be converted into esters successfully with sulfuric acid as the catalyst.

The "vinyl interchange" requires as a catalyst the mercuric salt of a strong acid [127]. The temperature range necessary is 20–80°; higher temperatures produce ethylidene diesters and acid anhydrides. The synthesis gives monomers of greater purity than those obtained by the addition of acetylene to the acid (B.5). It is suggested that in this case the mercury salt forms the acetylene complex 14-VII:

$$CH_3COOCH{=}CH_2 + HgSO_4 \longrightarrow CH_3COOH + \underset{\underset{\text{14-VII}}{HgSO_4}}{HC{\equiv}CH}$$

which reacts with the acid to give the ester 14-VIII:

$$RCOOH + \underset{\underset{\text{14-VII}}{HgSO_4}}{HC{\equiv}CH} \longrightarrow RCOOCH{=}CH_2 + HgSO_4 \quad \text{14-VIII}$$

For a mechanism of vinyl ether formation by this method see [128].

(a) Preparation of vinyl laurate (53–59% from lauric acid, vinyl acetate, mercuric acetate, and a trace of 100% sulfuric acid) [129].

(b) Preparation of methyl N-tosylglycinate (90% from N-tosyl-glycine, methyl acetate, and 96% sulfuric or p-toluenesulfonic acid monohydrate at 20° for 4–5 days) [126].

10. From Salts with Alkyl Halides

$$RCO_2Na + R'X \xrightarrow[\text{CH}_3\text{CON(CH}_3)_2]{\text{DMF or}} RCO_2R' + NaX$$

This metathesis is a high yield reaction when run with halides amenable to SN_2 displacement in aprotic solvents. Methyl esters are formed in high yield and good purity with methyl iodide, sodium bicarbonate and the acid in dimethyl-acetamide at room temperature [130]. A number of other esters has been prepared with other alkyl halides (see Ex. a). The sodium or silver salt may be employed with triethylamine as a catalyst (see Ex. c.2) or the free acid alone with triethylamine [131]. Phenacyl esters, prepared by this method, are used for the characterization of carboxylic acids.

In the preparation of trityl esters, trityl bromide is preferred to the chloride because it is more reactive and less hygroscopic and it can be used with the dry silver, sodium, or potassium salt [132].

The reaction of the salt, tetraethylammonium acetate, with an optically active halide gives the ester with less racemization than if acetic acid is employed [133]:

$$\underset{(-)}{\overset{\overset{\displaystyle CH_3}{|}}{C_6H_5CHCl}} + (C_2H_5)_4\overset{\ominus}{N}(O\overset{\ominus}{C}OCH_3) \longrightarrow \underset{\underset{\text{α-Phenylethyl acetate}}{(+)}}{\overset{\overset{\displaystyle CH_3}{|}}{C_6H_5CHOCOCH_3}} + (C_2H_5)_4\overset{\oplus}{N}Cl^{\ominus}$$

The reaction of the cuprous salt of an acid with an aryl halide in a high-boiling solvent is the only feasible way of preparing aryl esters from an aryl halide. An organocopper compound is the intermediate [134]:

$$p\text{-}CH_3OC_6H_4Br + CuO_2CC_6H_5 \xrightarrow[140-160°]{\text{Xylene}} \underset{p\text{-Anisyl benzoate, 85\%}}{p\text{-}CH_3OC_6H_4O_2CC_6H_5}$$

Vinyl halides may also be used in this reaction to prepare vinyl esters.

(a) Preparation of butyl acetate (95–98% from butyl bromide and potassium acetate in DMF at 90–100° for 2 hr.; DMSO may also be used as the solvent) [135].

(b) Preparation of p-nitrobenzyl acetate (78–82% from sodium acetate and p-nitrobenzyl chloride in glacial acetic acid) [136].

(c) Other examples

(1) p-Bromophenacyl 5-hexynoate (91.5% from the sodium salt and p-bromophenacyl bromide) [137].

(2) **Benzyl benzoate** (95% from the sodium salt and benzyl chloride using triethylamine catalyst) [138].

(3) **Trityl propionate** (86% from 0.03 mole of trityl bromide, 0.04 mole of silver propionate in 100 ml. of anhydrous cyclohexane by refluxing under nitrogen for 3 hr.) [132].

11. From Salts or Acids and Other Alkylating Agents

$$RCOONa \xrightarrow{(CH_3)_2SO_4} RCOOCH_3 + NaOSO_2OCH_3$$

$$\xrightarrow{CH_3OSOCl} RCOOCH_3 + SO_2 + NaCl$$

In this synthesis, dimethyl sulfate is the most common alkylating agent, but alkyl pyrocarbonates [139], alkyl silicates [140], alkyl phosphonates [141], acetals [142, 143], and even alkyl chlorosulfites [144] have been employed as well. In the last case, pyrolysis of the reaction mixture liberates sulfur dioxide to give the ester. As a rule, the alkylation is carried out in alkaline medium under which conditions both hindered and unhindered carboxyl groups are alkylated. By operating under acidic conditions the unhindered group alone is alkylated [145]. The yields are often high.

In a recent method, the free acid is alkylated by the use of an alkyl sulfate and dicyclohexylethylamine as the proton acceptor [146]:

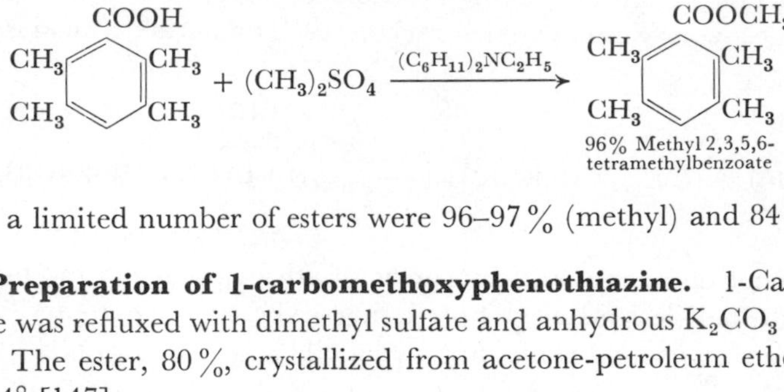

96% Methyl 2,3,5,6-tetramethylbenzoate

Yields for a limited number of esters were 96–97% (methyl) and 84% (ethyl).

(a) Preparation of 1-carbomethoxyphenothiazine. 1-Carboxyphenothiazine was refluxed with dimethyl sulfate and anhydrous K_2CO_3 in acetone for 18 hr. The ester, 80%, crystallized from acetone-petroleum ether, melted at 113–114° [147].

(b) Other examples

(1) **Ethyl 4-formylphenoxyacetate** (92.5% by the use of ethyl pyrocarbonate and the acid at 80° for 3 hr.) [139].

(2) **Various esters of aromatic carboxylic acids** (28–90% using methyl sulfate and sodium bicarbonate in acetone or methanol) [148].

(3) **Butyl 2,4,6-trimethylbenzoate** (71% using butyl chlorosulfite and the sodium salt with or without benzene followed by pyrolysis) [144].

12. From Copper Salts

A most unusual preparation of aryl benzoates has become available. When the copper salt of an aromatic acid is pyrolyzed, the ester is formed in good yield. The unusual feature is that the aryl group becomes attached *ortho* to the carboxylic group originally present but lost during reaction, as shown:

$$(CH_3)_3C—\underset{\text{0.56 mole}}{\boxed{}}—CO_2H + \underset{\substack{\text{0.25}\\\text{mole}}}{CuO} \xrightarrow[\Delta \text{ to } 288°]{\text{15 ml. xylene}}$$

$$(CH_3)_3C\overset{}{\boxed{}}—\overset{\overset{O}{\|}}{C}—O—\boxed{} + CO_2 + H_2O$$
$$ C(CH_3)_3$$

m-t-Butylphenyl p-t-butylbenzoate,
58%, m.p. 41°

The mechanism may be of the cyclo substitution type, but is still open to investigation [149]. Air can be substituted for part of the copper oxide.

13. From Tetraalkylammonium Salts

$$\overset{\ominus\oplus}{ArCOON(CH_3)_4} \xrightarrow{\text{Heat}} ArCOOCH_3$$

This synthesis was originally applied to the preparation of esters of sterically hindered benzoic acids. The methyl esters of 2,4,6-trimethyl- and 2,4,6-triethyl-benzoic acids were produced in yields varying from 63 to 90% [150]. The ester may also be obtained directly from the quaternary ammonium halide by heating with sodium acetate and acetic acid [151].

(a) **Preparation of o-methylbenzyl acetate** (88–91% from 2-methyl-benzylethyldimethylammonium bromide, sodium acetate, and glacial acetic acid) [151].

14. From Amides

$$RCONH_2 \xrightarrow{\text{NaOBr}} RCONHBr \longrightarrow RNCO \xrightarrow{\text{R'OH}} RNHCOOR'$$

The conversion of amides into esters is not a thermodynamically favored reaction; it must be forced as shown in Ex. a. In fact, the reaction takes place with ease only under unusual circumstances. One such case is the smooth conversion of o- or p-nitroacetanilides into esters by alcohols in the presence of sodium alkoxide [152]. The alcoholysis takes place evidently via the anion of the nitroacetanilides, $CH_3CON\overset{\ominus}{C_6H_4NO_2}$.

Amides can be converted into isocyanates by the Hofmann hypobromite reaction, the isocyanates then being converted into carbamates by alcohol (see Ex. b). Since urea is converted into isocyanic acid, $HN{=}C{=}O$, by heating, it can react with alcohols via the isocyanic acid to produce alkylcarbamates (urethanes) as derivatives. For instance, benzyl carbamate is made readily by refluxing urea and benzyl alcohol until ammonia ceases to be evolved:

$$CO(NH_2)_2 + C_6H_5CH_2OH \rightarrow C_6H_5CH_2OCONH_2 + NH_3$$

Any source of an isocyanate is a potential source of a carbamate if the reaction is run in an alcohol as a solvent. For example, the irradiation of an acid azide in ethanol produces the urethane as shown [153] (see Ex. b also):

Ethyl urethane of 2-amino-1,6,6-trimethyl-
bicyclo[1.1.2]hexane, ca. 70%

Allophanates may be made from isocyanic acid which probably dimerizes first in acid solution:

$$HN=C=O \longrightarrow NH_2CON=C=O \xrightarrow{ROH} NH_2CONHCO_2R$$

Many of the Cellosolve allophanates are crystalline solids suitable for derivatives.

A most useful, although restricted, method for converting an amide into an ester is the rearrangement of a β-hydroxyethylamide with concentrated hydrochloric acid:

$$C_6H_5CONHCH_2CH_2OH \xrightarrow{HCl} C_6H_5CO_2CH_2CH_2NH_2 \cdot HCl$$

2-Aminoethyl benzoate hydrochloride

Example d illustrates this reaction, which is in agreement with the preferential cleavage of an amide in dioxane-hydrogen chloride solution [154]:

4-Benzoyldecahydroquinoline
hydrochloride

(a) Preparation of methyl benzoate from benzamide. Methanol (1 mole) was added slowly to commerical polyphosphoric acid (162 g.). Benzamide (0.2 mole) was added to the methyl polyphosphate and the mixture heated for 40 hr. at 190°. After pouring into water, a few grams of black polymeric material were removed by filtration and the filtrate extracted with benzene. The benzene extract was concentrated and the residue distilled at 100–105° at water aspirator pressure to yield 22 g. (80%) of methyl benzoate. The residue from the distillation yielded 3 g. of benzoic acid. Similar treatment of benzanilide yielded 75% methyl benzoate [155].

(b) Preparation of methyl pentadecyl carbamate. Palmitic amide (0.1 mole) in 69 g. of methanol was mixed with sodium methoxide (0.2 mole) in 115 g. of methanol. Bromine (0.1 mole) was added and the mixture heated on the steam bath for 10 min. The solution was neutralized with a few drops of acetic acid and the alcohol removed by distillation. The residue was washed with water to remove sodium bromide, dissolved in hexane to remove the insoluble palmitic amide, and recrystallized from alcohol (83–94%), m.p. 61–62° [156].

(c) Preparation of 2-methoxyethyl allophanate. 2-Methoxyethanol (15.2 g.) in 150 ml. of dioxane was treated with dry sodium cyanate (39 g.) over a period of 2 hr. while dry hydrogen chloride was passed through the mixture. The mixture was then evaporated under water aspirator pressure, and the allophanate extracted with ether, 21 g. (65%). It was recrystallized from ether, yielding colorless crystals, m.p. 163° [157].

(d) Preparation of β-N-n-heptylaminoethyl p-nitrobenzoate hydrochloride (ca. 75% from N-p-nitrobenzoyl-N-heptyl β-aminoethanol and

$$p\text{-}NO_2C_6H_4CO \diagdown \atop C_7H_{15} \diagup N CH_2CH_2OH \xrightarrow{\text{HCl}} p\text{-}NO_2C_6H_4COOCH_2CH_2NHC_7H_{15}\cdot HCl$$

1.1 equiv. of conc. hydrochloric acid) [158].

15. From Trihalides

$$C_6H_5CCl_3 \xrightarrow[\text{2. CH}_3\text{CH}_2\text{CH}_2\text{OH}]{\text{1. 100\% H}_2\text{SO}_4} C_6H_5COOC_3H_7$$

This synthesis has been accomplished with various substituted benzotrifluorides and benzotrichlorides in yields ranging from 42 to 90% [159]. For more labile trihalides, as 3,3,3-trichloro-1,1-di-(p-chlorophenyl)-1-propene, refluxing with methanol alone yields a 95% return of the ester [160]:

$$(p\text{-}ClC_6H_4)_2C{=}CHCCl_3 \xrightarrow{\text{CH}_3\text{OH}} (p\text{-}ClC_6H_4)_2C{=}CHCOOCH_3$$
Methyl β,β-di-(p-chlorophenyl)-acrylate

(a) Preparation of n-propyl benzoate. Benzotrichloride, 48.9 g., and 35 g. of 100% sulfuric acid were heated until a reaction occurred, and then for 1 hr. during agitation. The cool mixture was then carefully poured with stirring into 60 g. of propanol. Refluxing for 30 min. and pouring the mixture in cold water gave an ester layer which was purified in the usual manner to give the ester, b.p. 229–231° (81%) [159].

(b) Preparation of methyl β,β-di-(p-chlorophenyl) acrylate (95% from 3,3,3-trichloro-1,1-di-(p-chlorophenyl)-1-propene and methanol) [160].

16. From α,α-Dihaloethers

$$ROCCl_2R \xrightarrow{\text{H}_2\text{O}} RO\overset{\displaystyle O}{\overset{\|}{C}}R$$

α,α-Dichloroalkyl ethers may be hydrolyzed to esters by water [161]. For α,α-difluoroalkyl ethers more powerful reagents such as 96% sulfuric acid must be employed [162, 163]. The yields in these hydrolyses run from 55 to 83%. The mechanism of the cleavage has not been determined with certainty [163], but it can be assumed that the hydrolysis occurs with greater ease than that of trihalides because of the influence of the ether grouping:

$$ROCCl_2R \longrightarrow R\overset{\oplus}{O}CClR \longleftrightarrow RO=\overset{\oplus}{C}ClR \xrightarrow[-H^{\oplus}]{H_2O} RO\overset{\overset{\displaystyle OH}{\displaystyle |}}{C}ClR \longrightarrow$$

$$RO\overset{\overset{\displaystyle O}{\displaystyle \|}}{C}R + Cl^{\ominus} + H^{\oplus}$$

(a) **Preparation of ethyl chlorofluoroacetate** (65–68% from 2-chloro-1,1,2-trifluoroethyl ethyl ether and 96% sulfuric acid) [162].

(b) **Preparation of ethyl acetate** (85% from α,α-dichlorodiethyl ether and water) [161].

17. From Carboxylic Acids and Divinyl Mercury

$$(Ar)RCOOH + (CH_2=CH)_2Hg \rightarrow (Ar)RCOOCH=CH_2 + CH_2=CH_2 + Hg$$

This synthesis is applicable only in the preparation of vinyl esters, vinyl ethers (from the phenol), and vinyl sulfides (from the thiophenol or alkyl thiol) [164]. To carry out the reaction one simply warms the aromatic or aliphatic carboxylic acid with divinylmercury, prepared from mercuric chloride and vinyl magnesium bromide in tetrahydrofuran [165] either alone or in a solvent. In the absence of a solvent the reaction is usually over 50% complete in less than 5 minutes on warming on a steam bath. For safety the reaction must be conducted in a well-ventilated hood since divinylmercury is extremely toxic. The intermediate vinylmercuric ester, $RCOOHgCH=CH_2$, may be isolated if the reaction is carried out in an inert solvent. Yields of the vinyl esters vary from 38 to 74%.

(a) **Preparation of vinyl acetate.**   Divinyl mercury, 25.5 g. (0.1 mole), and 13.2 g. (0.22 mole) of acetic acid were heated on a steam bath for 30 min. Extraction with 100 ml. of ether, washing with 5% NaHCO$_3$ and water, drying, and distilling at atmospheric pressure gave 74% of the ester [164].

1. T. J. Thompson and G. J. Leuck, *J. Am. Chem. Soc.*, **44,** 2894 (1922).
2. A. I. Vogel, *J. Chem. Soc.*, 624 (1948).
3. A. V. Topchiev et al., *Boron Fluoride and Its Compounds as Catalysts in Organic Chemistry*, Pergamon Press, New York, 1959, p. 269.
4. G. Hallas, *J. Chem. Soc.*, 5770 (1965).
5. C. Calmon and T. R. E. Kressman, *Ion Exchangers in Organic and Biochemistry*, Interscience Publishers, New York, 1957, p. 668; M. B. Bochner et al., *Ind. Eng. Chem. Fundamentals*, **4,** 314 (1965).
6. L. F. Fieser and M. Fieser, *Reagents for Organic Synthesis*, John Wiley and Sons, New York, 1967, p. 705.
7. C. E. Rehberg, *Org. Syn.*, Coll. Vol. **3,** 46 (1955).
8. S. Natelson and S. Gottfried, *Org. Syn.*, Coll. Vol. **3,** 381 (1955).
9. B. R. Baker et al., *J. Org. Chem.*, **12,** 138 (1947).
10. N. B. Lorette and J. H. Brown, Jr., *J. Org. Chem.*, **24,** 261 (1959).
11. A. Weissberger and C. J. Kibler, *Org. Syn.*, Coll. Vol. **3,** 610 (1955).

12. H. Klostergaard, *J. Org. Chem.*, **23,** 108 (1958).
13. R. O. Clinton and S. C. Laskowski, *J. Am. Chem. Soc.*, **70,** 3135 (1948).
14. Ref. 6, p. 678.
15. M. L. Bender, *Chem. Rev.*, **60,** 53 (1960).
16. H. Zimmermann and J. Rudolph, *Angew. Chem.*, **77,** 65 (1965).
17. M. S. Newman, *Steric Effects in Organic Chemistry*, John Wiley and Sons, New York, 1956, p. 203.
18. M. S. Newman, *J. Am. Chem. Soc.*, **63,** 2431 (1941).
19. H. Cohen and D. J. Mier, *Chem. Ind. (London)*, 349 (1965).
20. R. C. Parish and L. M. Stock, *J. Org. Chem.*, **30,** 927 (1965).
21. J. H. Brewster and C. J. Ciotti, Jr., *J. Am. Chem. Soc.*, **77,** 6214 (1955).
22. G. F. Hennion and S. O. Barrett, *J. Am. Chem. Soc.*, **79,** 2146 (1957).
23. J. Klosa, *Angew. Chem.*, **69,** 135 (1957).
24. H. Vorbrüggen, *Angew. Chem., Intern. Ed. Engl.*, **2,** 211 (1963); H. Brechbühler *et al.*, *ibid.*, **2,** 212 (1963); *Helv. Chim. Acta*, **48,** 1746 (1965).
25. M. Brenner and W. Huber, *Helv. Chim. Acta*, **36,** 1109 (1953).
26. R. P. Patel and S. Price, *J. Org. Chem.*, **30,** 3575 (1965).
27. J. M. Theobald *et al.*, *J. Chem. Soc.*, 1927 (1963).
28. T. Reichstein *et al.*, *Helv. Chim. Acta.*, **36,** 1295 (1953).
29. G. H. Daub and W. S. Johnson, *Org. Syn.*, Coll. Vol. **4,** 390 (1963).
30. A. R. Bader and A. D. Kontowicz, *J. Am. Chem. Soc.*, **75,** 5416 (1953).
31. L. Almirante and G. Tosolini, *J. Org. Chem.*, **26,** 177 (1961).
32. S. Sussman, *Ind. Eng. Chem.*, **38,** 1228 (1946).
33. M. Rogozinski, *J. Gas Chromatog.*, **2,** 328 (1964).
34. E. L. Eliel *et al.*, *Org. Syn.*, Coll. Vol. **4,** 169 (1963).
35. D. Davidson and S. A. Bernhard, *J. Am. Chem. Soc.*, **70,** 3426 (1948).
36. V. M. Mićović, *Org. Syn.*, Coll. Vol. **2,** 264 (1943).
37. P. F. E. Cook and A. J. Showler, *J. Chem. Soc.*, 4594 (1965).
38. J. W. McFarland, *J. Org. Chem.*, **28,** 2179 (1963).
39. R. D. Offenhauer, *J. Chem. Ed.*, **41,** 39 (1964).
40. C. R. Hauser *et al.*, *Org. Syn.*, Coll. Vol. **3,** 142 (1955).
41. A. Spassow, *Org. Syn.*, Coll. Vol. **3,** 144 (1955).
42. A. Spassow, *Chem. Ber.*, **75,** 779 (1942).
43. F. A. Menalda, *Rec. Trav. Chim.*, **49,** 967 (1930).
44. J. K. Kochi and G. S. Hammond, *J. Am. Chem. Soc.*, **75,** 3443 (1953).
45. D. S. Tarbell *et al.*, *J. Am. Chem. Soc.*, **82,** 3982 (1960).
46. E. B. Womack and J. McWhirter, *Org. Syn.*, Coll. Vol. **3,** 714 (1955).
47. T. L. Gresham *et al.*, *J. Am. Chem. Soc.*, **70,** 1003 (1948).
48. F. F. Blicke and W. M. Lilienfeld, *J. Am. Chem. Soc.*, **65,** 2281 (1943).
49. P. D. Bartlett and S. D. Ross, *J. Am. Chem. Soc.*, **69,** 460 (1947).
50. S. M. McElvain and T. P. Carney, *J. Am. Chem. Soc.*, **68,** 2592 (1946).
51. A. H. Ford-Moore, *Org. Syn.*, Coll. Vol. **4,** 84 (1963).
52. W. J. Hickinbottom, *Reactions of Organic Compounds*, Longmans, Green and Co., New York, 1948, p. 98.
53. R. L. Shriner *et al.*, *The Systematic Identification of Organic Compounds*, John Wiley and Sons, New York, 5th ed., 1964, pp. 246–247.
54. J. Cason and K. W. Kraus, *J. Org. Chem.*, **26,** 2624 (1961).
55. C. G. Youngs, *J. Am. Oil Chemists' Soc.*, **35,** 416 (1958).
56. E. B. Womack and J. McWhirter, *Org. Syn.*, Coll. Vol. **3,** 714 (1955).
57. G. L. Jenkins and J. C. Kellett, Jr., *J. Org. Chem.*, **27,** 624 (1962).
58. J. F. King and T. Durst, *J. Am. Chem. Soc.*, **86,** 287 (1964).
59. E. C. Taylor *et al.*, *J. Am. Chem. Soc.*, **90,** 2422 (1968).
60. G. A. Olah *et al.*, *J. Am. Chem. Soc.*, **84,** 2733 (1962).
61. W. W. Prichard, *Org. Syn.*, Coll. Vol. **3,** 452 (1955).
62. R. H. Baker and F. G. Bordwell, *Org. Syn.*, Coll. Vol. **3,** 141 (1955).

63. C. E. Braun and C. D. Cook, *Org. Syn.*, **41,** 79 (1961).
64. J. Erdos, *Angew. Chem.*, **63,** 329 (1951).
65. F. Adickes, *J. Prakt. Chem.*, **161,** 271 (1943).
66. The Miner Laboratories, *Org. Syn.*, Coll. Vol. **1,** 285 (1941).
67. V. K. Kuskov and V. A. Zhukova, *Izv. Akad. Nauk SSSR, Otd. Khim. Nauk,* 733 (1956); *C.A.*, **51,** 1877 (1967).
68. S. Niraz, *Roczniki Chem.*, **31,** 1047 (1957).
69. S. F. Birch *et al.*, *J. Chem. Soc.*, 1363 (1952).
70. A. C. Cope and E. C. Herrick, *Org. Syn.*, Coll. Vol. **4,** 304 (1963).
71. F. D. Chattaway, *J. Chem. Soc.*, 2495 (1931).
72. P. Z. Bedoukian, *J. Am. Chem. Soc.*, **66,** 1325 (1944).
73. P. Z. Bedoukian, *J. Am. Chem. Soc.*, **67,** 1430 (1945).
74. T. B. Windholz, *J. Org. Chem.*, **25,** 1703 (1960).
75. M. Green, *Chem. Ind. (London)*, 435 (1961).
76. B. B. Corson and N. A. Saliani, *Org. Syn.*, Coll. Vol. **2,** 69 (1943).
77. J. Cason, *Org. Syn.*, Coll. Vol. **3,** 169 (1955).
78. K. S. Mazdiyasni *et al.*, *J. Chem. Eng. Data*, **11,** 277 (1966).
79. K. G. Rutherford *et al.*, *J. Org. Chem.*, **28,** 582 (1963).
80. R. N. Lacey, *Advan. Org. Chem.*, 213 (1960).
81. C. D. Hurd and A. S. Roe, *J. Am. Chem. Soc.*, **61,** 3355 (1939).
82. H. V. Claborn and L. T. Smith, *J. Am. Chem. Soc.*, **61,** 2727 (1939).
83. C. D. Hurd and W. A. Hoffman, *J. Org. Chem.*, **5,** 212 (1940).
84. B. H. Gwynn and E. F. Degering, *J. Am. Chem. Soc.*, **64,** 2216 (1942).
85. G. Quinkert, *Angew. Chem., Intern. Ed. Engl.*, **4,** 211 (1965).
86. A. B. Boese, Jr., *Ind. Eng. Chem.*, **32,** 16 (1940).
87. W. Kimel and A. C. Cope, *J. Am. Chem. Soc.*, **65,** 1992 (1943).
88. J. C. Sauer, *J. Am. Chem. Soc.*, **69,** 2444 (1947).
89. S. M. McElvain *et al.*, *J. Am. Chem. Soc.*, **73,** 3807 (1951).
90. Ref. 53, p. 241.
91. C. D. Hurd and F. O. Green, *J. Am. Chem. Soc.*, **63,** 2201 (1941).
92. L. H. Ulrich and R. Adams, *J. Am. Chem. Soc.*, **43,** 660 (1921).
93. C. R. Hauser *et al.*, *J. Am. Chem. Soc.*, **72,** 847 (1950).
94. R. Wegscheider and E. Späth, *Monatsh.*, **30,** 825 (1909).
95. R. T. Bertz, *Org. Syn.*, Coll. Vol. **4,** 489 (1963).
96. E. Späth and H. Schmid, *Chem. Ber.*, **73,** 243 (1940).
97. J. M. Bobbitt and D. A. Scola, *J. Org. Chem.*, **25,** 560 (1960).
98. W. Davey and D. J. Tivey, *J. Chem. Soc.*, 1230 (1958).
99. D. J. Morgan, *Chem. Ind. (London)*, 854 (1959).
100. L. P. Kyrides *et al.*, *J. Org. Chem.*, **12,** 577 (1947).
101. F. L. James and W. H. Bryan, *J. Org. Chem.*, **23,** 1225 (1958).
102. H.-F. Piepenbrink, *Ann. Chem.*, **572,** 83 (1951).
103. A. C. Cope and S. W. Fenton, *J. Am. Chem. Soc.*, **73,** 1668 (1951).
104. M. M. Rising and T.-W. Zee, *J. Am. Chem. Soc.*, **49,** 541 (1927).
105. H. Dahn and H. Hauth, *Helv. Chim. Acta*, **42,** 1214 (1959).
106. J. W. E. Glattfeld and E. C. Lee, *J. Am. Chem. Soc.*, **62,** 354 (1940).
107. R. H. Kimball *et al.*, *Org. Syn.*, Coll. Vol. **2,** 284 (1943).
108. H. A. Smith *et al.*, *J. Am. Chem. Soc.*, **71,** 3772 (1949).
109. B. Abramovitch and C. R. Hauser, *J. Am. Chem. Soc.*, **64,** 2720 (1942).
110. C. S. Marvel, *Org. Syn.*, Coll. Vol. **2,** 310 (1943).
111. F. C. Whitmore *et al.*, *J. Am. Chem. Soc.*, **67,** 1071 (1945).
112. J. H. Billman *et al.*, *J. Am. Chem. Soc.*, **69,** 2058 (1947).
113. C. E. Rehberg, *Org. Syn.*, Coll. Vol. **3,** 146 (1955).
114. J. C. Sauer *et al.*, *Org. Syn.*, Coll. Vol. **3,** 605 (1955).
115. E. E. Reid *et al.*, *Org. Syn.*, Coll. Vol. **2,** 468 (1943).
116. M. Reimer and H. R. Downes, *J. Am. Chem. Soc.*, **43,** 945 (1921).

117. G. B. Hatch and H. Adkins, *J. Am. Chem. Soc.*, **59**, 1694 (1937).

118. A. R. Bader *et al.*, *J. Am. Chem. Soc.*, **73**, 4195 (1951).

119. H. J. Hagemeyer, Jr., and D. C. Hull, *Ind. Eng. Chem.*, **41**, 2920 (1949).

120. R. P. Linstead and E. M. Meade, *J. Chem. Soc.*, 935 (1934).

121. H. N. Rydon, *J. Chem. Soc.*, 1340 (1937); G. B. Brown and C. W. H. Partridge, *J. Am. Chem. Soc.*, **66**, 839 (1944).

122. E. R. Meincke and S. M. McElvain, *J. Am. Chem. Soc.*, **57**, 1443 (1935).

123. J. Cason *et al.*, *J. Am. Chem. Soc.*, **66**, 1764 (1944).

124. T. L. Gresham *et al.*, *J. Am. Chem. Soc.*, **70**, 1004 (1948).

125. J. Lavety and G. R. Proctor, *Org. Syn.*, **45**, 42 (1965).

126. E. Taschner and C. Wasielewski, *Ann. Chem.*, **640**, 142 (1961).

127. R. L. Adelman, *J. Org. Chem.*, **14**, 1057 (1949).

128. W. H. Watanabe and L. E. Conlon, *J. Am. Chem. Soc.*, **79**, 2828 (1957).

129. D. Swern and E. F. Jordan, Jr., *Org. Syn.*, Coll. Vol. **4**, 977 (1963).

130. A. J. Parker, *Advan. Org. Chem.*, **5**, 37 (1965).

131. W. T. Moreland, Jr., *J. Org. Chem.*, **21**, 820 (1956); R. H. Mills *et al.*, *Chem. Ind. (London)*, 2144 (1962).

132. K. D. Berlin *et al.*, *J. Org. Chem.*, **27**, 3595 (1962).

133. J. Steigman and L. P. Hammett, *J. Am. Chem. Soc.*, **59**, 2536 (1937).

134. T. Cohen and A. H. Lewin, *J. Am. Chem. Soc.*, **88**, 4521 (1966).

135. S. Yoneda *et al.*, *Kogyo Kagaku Zasshi*, **69**, 641 (1966); *C.A.*, **66**, 1002 (1967).

136. W. W. Hartman and E. J. Rahrs, *Org. Syn.*, Coll. Vol. **4**, 650 (1955).

137. K. E. Schulte and K. P. Reiss, *Chem. Ber.*, **86**, 777 (1953).

138. F. C. Whitmore *et al.*, *Ind. Eng. Chem.*, **39**, 1300 (1947).

139. W. Thoma and H. Rinke, *Ann. Chem.*, **624**, 30 (1959).

140. G. Sumrell and G. E. Ham, *J. Am. Chem. Soc.*, **78**, 5573 (1956).

141. F. W. Hoffman and H. D. Weiss, *J. Am. Chem. Soc.*, **79**, 4759 (1957).

142. H. Brechbühler *et al.*, *Angew. Chem.*, **75**, 296 (1963).

143. E. Vorbrüggen, *Angew. Chem.*, **75**, 296 (1963).

144. M. S. Newman and W. S. Fones, *J. Am. Chem. Soc.*, **69**, 1046 (1947).

145. M. Cachia and H. Wahl, *Compt. Rend.*, **245**, 1249 (1957).

146. F. H. Stodola, *J. Org. Chem.*, **29**, 2490 (1964).

147. S. P. Massie and P. K. Kadaba, *J. Org. Chem.*, **21**, 347 (1956).

148. U. N. Hirwe *et al.*, *J. Univ. Bombay, Sect. A*, **22**, Pt. 5 (Science No. 35), 14 (1954); *C.A.*, **49**, 11594 (1955).

149. W. G. Toland, *J. Am. Chem. Soc.*, **83**, 2507 (1961).

150. R. C. Fuson *et al.*, *J. Am. Chem. Soc.* **61**, 1290 (1939).

151. W. R. Brasen and C. R. Hauser, *Org. Syn.*, Coll. Vol. **4**, 582 (1963).

152. E. Marvell *et al.*, *J. Am. Chem. Soc.*, **79**, 1076 (1957).

153. L. Horner and E. Spietschka, *Chem. Ber.*, **88**, 934 (1955).

154. E. A. Mistryukov and V. F. Kucherov, *Izv. Akad. Nauk, SSR, Otd. Khim. Nauk*, 1345 (1961); *C.A.*, **56**, 2423 (1962).

155. R. A. Chambers, *Kinetic and Mechanistic Study on Dealkylation of Alkylarylamines and the Methylation of Aromatic Amines with Methanol in Polyphosphoric Acid*, Ph.D. Thesis, Vanderbilt University, 1962.

156. E. Jeffreys, *Am. Chem. J.*, **22**, 14 (1899).

157. E. S. Lane, *J. Chem. Soc.*, 2764 (1951).

158. J. R. Reasenberg and S. D. Goldberg, *J. Am. Chem. Soc.*, **67**, 933 (1945).

159. G. M. LeFave and P. G. Scheurer, *J. Am. Chem. Soc.*, **72**, 2464 (1950).

160. A. Roedig and H. Niedenbruck, *Chem. Ber.*, **90**, 673 (1957).

161. L. Heslinga *et al.*, *Rec. Trav. Chim.*, **76**, 969 (1957).

162. B. Englund, *Org. Syn.*, Coll. Vol. **4**, 423 (1963).

163. J. A. Young and P. Tarrant, *J. Am. Chem. Soc.*, **71**, 2432 (1949).

164. D. J. Foster and E. Tobler, *J. Am. Chem. Soc.*, **83**, 851 (1961).

165. G. F. Reynolds *et al.*, *J. Org. Chem.*, **23**, 1217 (1958).

B. Electrophilic-Type Syntheses

In this section, preparations are considered in which an electrophilic species derived from a functional group bearing an alkyl substituent attacks an acid or an acylium-type ion attacks an alcohol. The section is not clearly separable from Section A, as obviously some esters are formed by an SN_1 cleavage of an alcohol. The latter was included in Section A because the dividing line between SN_1 and SN_2 mechanism is illusory and one mechanism blends into the other. On the other hand, the preparations in this section have no clear-cut counterpart. They involve electron-deficient or octet-deficient intermediates.

1. From Amines

Diazotization of amines in carboxylic acid or anhydride solutions would be expected to yield esters, but few experiments directed toward this synthesis seem to be recorded [1]; see Ex. b. Competing reactions such as olefin formation must be taken into account in such esterifications. A more elaborate method has been devised by White in which the amide is nitrosated and the N-nitrosoamide is decomposed in acetic acid and acetic anhydride (see Ex. a).

Still another method is the decomposition of triazene, 14-IX, as shown:

$$RNH_2 + C_6H_5\overset{\oplus}{N}\equiv N \xrightarrow{HCON(CH_3)_2} \underset{\text{14-IX}}{RNHN=NC_6H_5} \xrightarrow{R'COOH}$$

$$R'CO_2R + N_2 + C_6H_5NH_2$$

Yields of esters are usually on the order of 40–60% with olefin formation encroaching [2].

(a) Preparation of butyl acetate.   Butylamine (0.01 mole) and acetic acid (0.015 mole) were chilled to $-80°$ and treated with acetic anhydride (0.01 mole). The excess acid then was removed by distillation. The crude amide in the residue was treated with a mixture of 10 ml. of acetic acid and 50 ml. of acetic anhydride, cooled to $0°$, and treated portionwise with 15 g. of sodium nitrite. The mixture was then maintained at a temperature at which nitrogen evolved steadily. After evolution had ceased, the mixture was poured into water, extracted with ether, dried, and distilled. Yield of butyl acetate was 56% [3].

(b) Preparation of D-α-Acetoxyisovaleric acid (59% from D-valine and isoamyl nitrite in glacial acetic acid containing sodium acetate) [4].

2. Alkylation of Acids by Diazoalkanes and of Olefins by Diazoesters

$$RCOOH + CH_2N_2 \rightarrow RCOOCH_3 + N_2$$

This synthesis has been discussed under Ethers (Chapter 6, B.2). Since the carboxylic acid is one of the most active hydrogen types, superior yields of the methyl esters are usually obtained by the use of diazomethane, a good preparation of which is from the commercially available bis-(N-methyl-N-nitroso)-terephthalamide [5]. This procedure offers some advantage over other methods

References for Section B are on pp. 834–835.

of producing esters with complex peptides in which the free amino group is protected by the 2,4-dinitrophenyl group [6].

The method is also applicable to substituted diazoalkanes such as the diazoketones readily available from acid chlorides and diazomethane:

$$RCOCl \xrightarrow{CH_2N_2} RCOCHN_2 \xrightarrow{R'COOH} RCOCH_2OCOR'$$

Ester groups when a part of the diazo compound may also be introduced by addition to unsaturated compounds [7]:

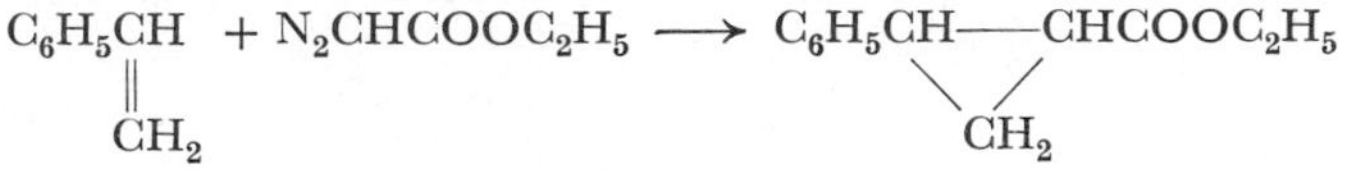

$$C_6H_5CH + N_2CHCOOC_2H_5 \longrightarrow C_6H_5CH\!-\!\!-\!\!CHCOOC_2H_5$$

$$\underset{CH_2}{\|} \qquad\qquad\qquad \underset{CH_2}{\diagdown\diagup}$$

1-Phenyl-2-carbethoxycyclopropane

or by alkylation of heterocyclic rings [8]:

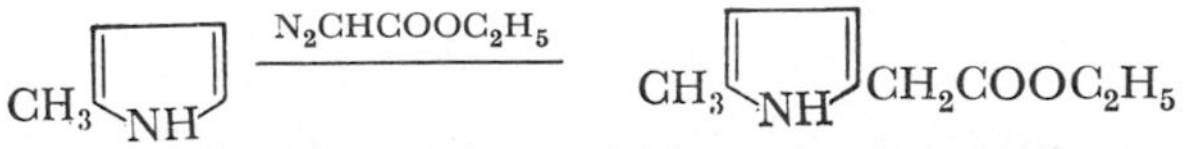

$$CH_3 \underset{NH}{\boxed{}} \xrightarrow{N_2CHCOOC_2H_5} CH_3 \underset{NH}{\boxed{}} CH_2COOC_2H_5$$

Ethyl 2-methylpyrrole-5-acetate

The yields when diazoesters are employed are not so satisfactory as in the other cases.

(a) Preparation of dimethyl 2-methyl-3-amino-4,5-pyridinedicarboxylate. A suspension of 17 g. of 2-methyl-3-amino-4,5-pyridinedicarboxylic acid monohydrate in 200 ml. of methanol was treated in a 500-ml. solution of methylene chloride containing the diazomethane produced from 42 g. of nitrosomethylurea. When the evolution of nitrogen ceased, the solution was filtered and evaporated under reduced pressure to a syrup. This was dissolved in ethyl acetate, filtered, and the filtrate was again evaporated to a syrup. Crystallization from petroleum ether gave 16 g. (90%) of the dimethyl ester. Recrystallization from ethyl acetate-ether-petroleum ether gave a melting point of 94–95° [9].

(b) Other examples

(1) **Benzyl mandelate** (70% from mandelic acid and phenyldiazomethane which was obtained from azibenzil) [10].

(2) **1-(−)-Hydroxy-N-acetylproline methyl ester** (95% from the free acid) [11].

(3) **β-Naphthyl acetoxymethyl ketone** (72% from β-naphthoyl chloride, diazomethane, and acetic acid) [12].

(4) **Benzhydryl benzoate** (100% from benzoic acid and diphenyldiazomethane) [13].

3. Rearrangement of Diazoketones (Arndt-Eistert and Wolff Rearrangement)

$$RCOCl \xrightarrow{CH_2N_2} RCOCHN_2 \xrightarrow[ROH]{Ag_2O} RCH_2COOR'$$

This synthesis has been discussed in some detail under Carboxylic Acids, Chapter 13, G.1. The present synthesis differs from that method only in that an alcohol rather than water is used in the last step. The agents employed for this step are very much the same as in the acid preparation, although some success in producing esters has also been achieved by irradiation (see Ex. b.2), by the use of cuprous iodide in acetonitrile [14], and by silver benzoate in triethylamine [15]; see Ex. b.1. In the use of the latter, it has been shown that a hydrogen atom α to both the carbonyl and the diazo groups is necessary for success.

(a) **Preparation of ethyl α-naphthylacetate** (73–82 % from the diazo-ketone, silver oxide, and ethanol) [16].

(b) **Other examples**

(1) **Methyl p-methoxyphenylacetate** (84 % from the diazoketone, silver benzoate, triethylamine, and methanol) [15].

(2) **Methyl β,γ,γ,γ-tetrachlorocrotonate** (71 % from the diazoketone in methanol by irradiation with ultraviolet light) [17].

4. Oxidation of Carbonyl Compounds with a Peracid (Baeyer-Villiger)

$$\begin{array}{c} R \\ \diagdown \\ CO \\ \diagup \\ R' \end{array} \xrightarrow{CH_3CO_2OH} \begin{array}{c} RO \\ \diagdown \\ C{=}O \\ \diagup \\ R' \end{array}$$

A review of this synthesis is available [18]. It is applicable to both aldehydes and ketones, but its use has been confined largely to ketones. Cyclanones in ethanol with Caro's acid produce only fair yields of ethyl esters of ω-hydroxy aliphatic acids of the same carbon content [19], but they give good yields of lactones with performic acid (see Ex. b.5). Hydroxybenzaldehydes yield hydroxyphenyl formates [20]. Peracetic acid [20, 21], perbenzoic acid [22]. trifluoroperacetic acid [23], hydrogen peroxide with boron trifluoride [24], Caro's acid [19], permaleic acid [25], and the commercially available m-chloroperbenzoic acid have been employed as oxidizing agents. Of these the trifluoroperacetic acid is particularly attractive. With this reagent, transesterification of the ester with the acid may be avoided by the use of a buffer such as disodium hydrogen phosphate. The yields vary but may be as high as 80–90 %.

The use of peracetic acid in an inert solvent in the oxidation of cyclopentanone and various cyclohexanones produces monomeric lactones in high yield [26] (see Ex. b.5):

ε-Caprolactone

In contrast, cycloheptanone and cyclooctanone give considerable amounts of the corresponding dibasic acids.

However, higher ketones, such as cyclododecanone, may be converted into lactones (51%) by the use of a great excess of peracetic acid in acetone and concentrated sulfuric acid [27]. The principal other product (25%) is dodecanedioic acid.

The mechanism of the Baeyer-Villiger reaction has been the subject of much study [28]. It appears that a complex 14-X is first formed.

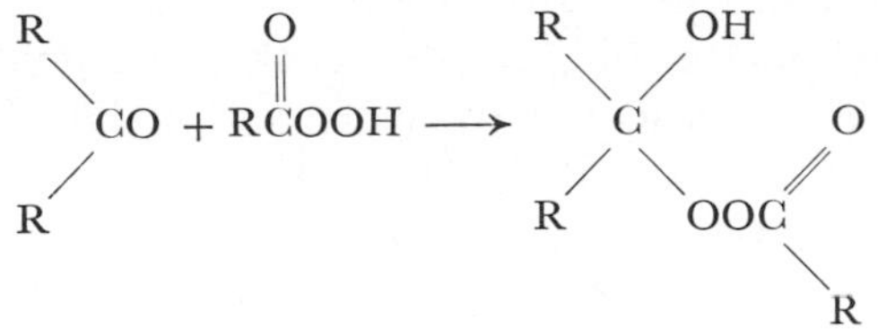

14-X

This by the loss of

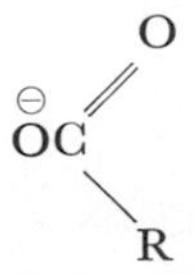

and the concerted migration of an R group in the remaining cation 14-XI gives 14-XII, which on the loss of a proton produces the ester 14-XIII [24]. The

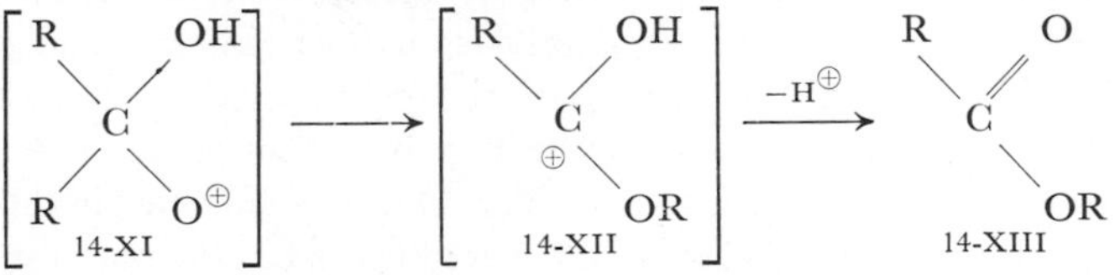

R group which becomes attached to the electron-deficient oxygen atom is that which is the more electron-releasing, i.e., the order $Ar > (CH_3)_2CH > CH_3CH_2 > CH_3$ is the same as the order of migration in the Beckmann rearrangement.

(a) Preparation of isobutyl acetate. A pertrifluoroacetic acid solution was obtained by adding dropwise 50.8 ml. of trifluoroacetic anhydride to a suspension of 8.2 ml. of 90% hydrogen peroxide in 50 ml. of cold methylene chloride. Addition of this solution was carried out over 20 min. to a stirred suspension of 130 g. of dry, finely ground disodium hydrogen phosphate in 150 ml. of methylene chloride and 20 g. of methyl isobutyl ketone. Boiling occurred during the process and at the end the solution was refluxed for 30 min. The filtrate and washings from 100 ml. of methylene chloride were washed with 50 ml. of 10% sodium carbonate solution and dried over magnesium sulfate. After removing most of the solvent by distilling at atmospheric pressure, the liquid remaining was fractionated in a Todd column packed with a Monel metal spiral and equipped with a variable reflux ratio head. The ester, 19.5 g. (84%), was obtained, b.p. 114–115.5° [23].

(b) Other examples

(1) Ethyl benzoate (90 % from benzaldehyde diethyl acetal, peracetic acid, and a few drops of sulfuric acid) [21].

(2) o-Hydroxyphenyl formate (88 % from salicylaldehyde, peracetic acid, and acetic anhydride) [20].

(3) Phenyl p-nitrobenzoate (95 % from p-nitrobenzophenone, peracetic, glacial acetic and sulfuric acids) [29].

(4) Butyl crotonate (73 % from crotonaldehyde dibutylacetal at 60° to which peracetic acid in ethyl acetate was added, the products being fractionated with ethylbenzene as azeotroping solvent) [30].

(5) ε-Caprolactone (90 % by heating 5 moles of cyclohexanone and 1790 g. of a 25.5 % solution of peracetic acid in ethyl acetate, 6 moles, at 40° for 6.5 hr.) [26]; *see also* [31].

5. Addition of Carboxylic Acids to Alkenes or Alkynes

$$\underset{\underset{\text{CH}_3}{}}{\overset{\text{CH}_3}{>}}\text{C}=\text{CH}_2 + \text{RCOOH} \xrightarrow{\text{H}_2\text{SO}_4} \underset{\underset{\text{CH}_3 \quad \text{OCOR}}{}}{\overset{\text{CH}_3}{>}}\text{C}-\text{CH}_3$$

This addition has been discussed previously under Alcohols, Chapter 4, B.1.

Carboxylic acids add readily to alkenes in accordance with Markownikoff's rule to gives esters. Sometimes catalysts such as sulfuric acid [32] or boron trifluoride [33] are added. Although the synthesis has been employed to a limited extent, it may offer some advantage in the preparation of tertiary esters, particularly if the appropriate compounds are not available for a more conventional ester synthesis. When sulfuric acid is used as the catalyst, the optimum conditions for success are: low reaction temperature; liberal excess of one of the reactants; relatively large quantity of catalyst; and anhydrous conditions [32].

Formic acid addition to alkenes has been accomplished effectively with 90 % perchloric acid as a catalyst. This catalyst may also be used in acetic acid additions, although the reaction is not so complete [34].

The reaction proceeds in a similar manner with alkynes:

$$\text{R}-\text{C}\equiv\text{CH} + \text{R}'\text{COOH} \longrightarrow \underset{\underset{\text{OCOR}'}{|}}{\text{R}-\text{C}}=\text{CH}_2$$

In this case, boron trifluoride and mercuric oxide have been used as catalysts [35]. Since the reaction is reversible in the presence of mercuric salts of strong acids, vinyl esters may be prepared from vinyl acetate and high-molecular-weight acids (see A.9, Ex. a) [36]. The yields in all these additions are variable.

The internal addition of the olefinic to the acid group in unsaturated acids to form lactones is an important phase of the chemistry of unsaturated acids. Protonic acid catalysis leading to a carbonium ion of the olefinic group tends to give lactones, whereas Lewis acid catalysis or other reagents leading to the acylium ion tend to yield unsaturated ketones [37]. All other factors being equal, γ-lactones form in preference to δ-lactones:

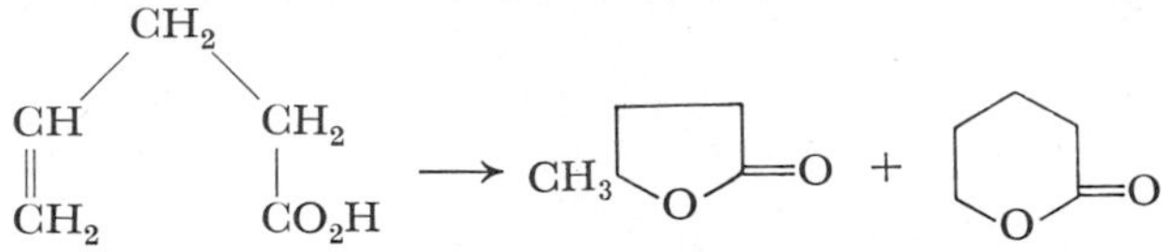

Preferred
γ- and δ-Valerolactones

The double bond may be situated in any position of the acid, since isomerization will take place followed by lactonization, the least reversible of the reactions involved. Nevertheless, a δ-lactone can be isomerized under acidic catalysis to a γ-lactone, or vice versa, depending on relative stabilities. Interesting examples of the preceding transformations have been described [37].

(a) **Preparation of di-*t*-butyl malonate** (58–60 % from malonic acid, isobutylene, and sulfuric acid) [38].

(b) **Other examples**

(1) **N-Carbobenzoxy-O-*t*-butyl-L-threonine *t*-butyl ester** (about 90 % from N-carbobenzoxy-L-threonine, an excess of liquid isobutylene, and

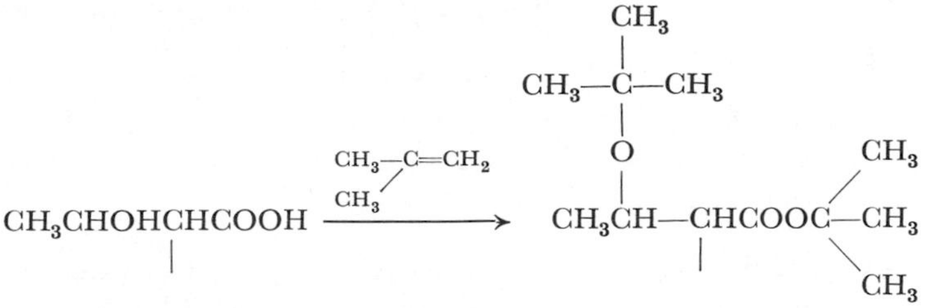

sulfuric acid in a pressure vessel) [39].

(2) **Vinyl chloroacetate** (42–49 % from monochloroacetic acid, acetylene, mercuric oxide, and a trace of hydroquinone) [40].

(3) **Cyclohexyl formate** (90 % from cyclohexene and 90 % formic acid containing a trace of perchloric acid) [41].

(4) **1-Phenyltrimethylene diacetate-1,3** (92 % from phenylcyclopropane and thallium triacetate; cyclopropane ring opening) [42].

6. Carbalkoxylation with Carbon Monoxide and a Base

$$RI + CO + CH_3OH + (C_6H_{11})_2NCH_2CH_3 \xrightarrow{NaCo(CO)_4}$$

$$RCOOCH_3 + (C_6H_{11})_2\overset{\oplus}{N}CH_2CH_3I^{\ominus}$$
$$\underset{H}{|}$$

This synthesis is a modification of hydrocarboxylation as described under Carboxylic Acids, Chapter 13, F.3. The reaction may be accomplished with carbon monoxide under pressure during electrolysis [43]:

$$RCH=CH_2 + CO + \overset{\ominus}{C}H_3O \xrightarrow[\text{(Pt}_x\text{(CO)}_y)^n]{\text{Anode}} RCH=CHCOOCH_3$$

and in the presence of hydrogen chloride in ethanol [44]:

$$C_2H_5OOCC\equiv CCOOC_2H_5 + C_2H_5OH + CO \xrightarrow[PdCl_2]{HCl}$$

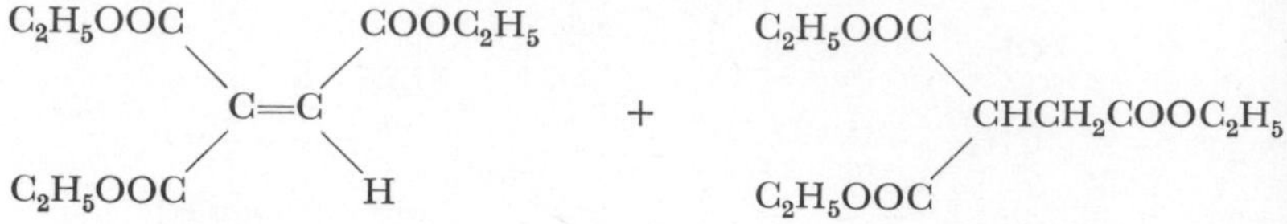

1,1,2–Tricarboethoxyethylene   (Main products)   1,1,2–Tricarboethoxyethane

or in the presence of a hindered amine and methanol as shown in the titled equation above [45]. The catalyst in the first case is a platinum carbonyl complex, in the second, metallic palladium or palladium chloride, and in the third, sodium cobalt carbonylate. Alkenes or acetylenes are starting materials in the first and second cases, while halides, sulfates, and sulfonates are employed in the third case. These are low yield reactions, the maximum (third case) being 80 %. Of the three methods, the one utilizing the sodium cobalt carbonylate catalyst is the most promising; it possesses an additional advantage over the other two in being effective at atmospheric pressure.

In a somewhat similar process, aryl iodides, but not bromides, react with nickel carbonyl and ethanol at reflux temperature to produce esters (see Ex. b):

$$ArI + Ni(CO)_4 + C_2H_5OH \rightarrow ArCOOC_2H_5$$

(a) **Preparation of methyl nonanoate** (56 % from 1-iodooctane, carbon monoxide, methanol, dicyclohexylethylamine, and a small amount of sodium cobalt carbonylate at 50° for 20 hr.) [45].

(b) **Preparation of ethyl benzoate** (60 % by refluxing equimolar quantities of nickel carbonyl and iodobenzene in ethanol) [46].

7. Cleavage of Ethers

$$ROR \xrightarrow{R'COCl} R'COOR + RCl$$

This synthesis was originally carried out with boron trifluoride and an aliphatic acid, in which case the aliphatic ethers gave yields not exceeding 55 % [47]. More recently, yields as high as 85 % were obtained from an ether of a diglyceride by employing an acid chloride and silver perchlorate [48]. Apparently the ether adds to the positive carbon of the acid chloride with subsequent elimination of RCl:

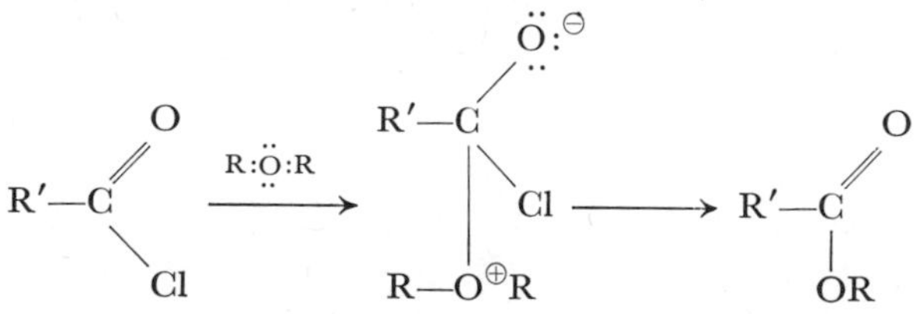

The method is also applicable to cyclic ethers. For example, tetrahydrofuran and benzoyl chloride in the presence of zinc chloride gave 4-chlorobutyl benzoate [49] (see Ex. b):

$$C_6H_5COCl + \underset{O}{\boxed{}} \xrightarrow{ZnCl_2} C_6H_5COO(CH_2)_4Cl$$

Similarly the reaction of acetic anhydride and zinc chloride leads to ring cleavage [50].

(a) Preparation of glycerol 1-myristate 2,3-bis(4-nitrobenzoate). Dry silver perchlorate, 1.03 g., was dissolved in 35 ml. of absolute nitromethane, 2 g. Drierite was added, and then, after shaking and allowing the mixture to stand for a short time, 3.16 g. of glycerol 1-trityl ether 2,3-bis-(4-nitrobenzoate) was added. After another short period, 1.23 g. of myristoyl chloride was added first and then pyridine until the color disappeared. On treatment with 50 ml. of ethyl acetate followed by slight warming, the Drierite and silver chloride were filtered off and the solvent was removed by distillation. The residue was then heated slightly in 75 ml. of petroleum ether, after which the insoluble N-trityl pyridinium perchlorate was filtered off. On distilling the filtrate under reduced pressure to remove the petroleum ether, there was obtained a final residue which when crystallized from 30 ml. of 99% ethanol gave 2.55 g. (85%) of glycerol 1-myristate-2,3-bis-(4-nitrobenzoate), m.p. 66–68° [48].

(b) Preparation of 4-chlorobutyl benzoate (78–83% from refluxing tetrahydrofuran, benzoyl chloride, and freshly fused $ZnCl_2$) [49].

8. Alkylation of Esters and Related Processes

Alkylation of esters by electron-deficient species may be carried out in two ways:

$$R_3B + BrCH_2CO_2R' \rightarrow RCH_2CO_2R' + R_2BBr$$

or

$$R^\oplus + \underset{OEt}{\overset{OH}{C=C}} \longrightarrow RCCO_2Et + H^\oplus$$

In regard to the organoborane process above, the procedure, which is simple and rapid, is applicable to α-mono- and α,α-dihalocarboxylic acid esters. The organoborane (prepared from the olefin and diborane in tetrahydrofuran) is treated with the chloro- or bromoester followed by the addition of potassium t-butoxide in t-butyl alcohol. The reaction is essentially complete after the addition, yields usually being in the 80–100% range (see Ex. a).

One of the most useful alkylations is the t-butylation of malonic esters in which the t-butyl carbonium ion adds to the enol form of malonic ester (see Ex. b).

Carbalkoxylation of alkenes or arenes would seem to be a general reaction, and one such reaction has been found [51]:

$$C_6H_5O\overset{\overset{\displaystyle O}{\|}}{C}Cl + C_6H_6 + AlCl_3 \longrightarrow C_6H_5O_2CC_6H_5$$

Phenyl benzoate, 64%

However, no similar reactions are recorded for chloroformates, a failure which has been attributed to the instability of chloroformates in the presence of aluminum chloride [52]. Perhaps milder catalysts and conditions would widen the scope of carbalkoxylation.

Carbalkoxymethylation has been accomplished recently by a novel reagent, manganese(III) triacetate. It presumably reacts with olefins via some electron-deficient species of the acetate moiety such as $CH_2=C(OMn(OAc)_2)_2$. A specific example is shown [53]:

$$CH_3 \atop \underset{45\ g.}{C_6H_5C{=}CH_2} + \underset{80\ g.}{Mn(OAc)_3{\cdot}2\ H_2O} \xrightarrow[\text{180 ml. Ac}_2O]{\text{360 ml. CH}_3CO_2H}$$

γ-Phenyl-γ-valerolactone, 72%

Yields with other alkenes range from 10 to 39% but were not optimized.

(a) Preparation of ethyl α-bromooctanoate (85% from trihexylboron and ethyl α,α-dibromoacetate in THF followed by addition of 1 equiv. of potassium *t*-butoxide in *t*-butyl alcohol; the second bromine atom can be replaced by repeating the process) [54].

(b) Preparation of diethyl *t*-butylmalonate. Malonic ester (1 mole) was cooled and stirred while 0.5 mole of anhydrous aluminum chloride was added as rapidly as possible. *t*-Butyl chloride (1 mole) was then added dropwise to the amber-colored liquid maintained at 75° and the mixture heated for several hours after end of addition. G.L.C. analysis indicated a 100% yield of the *t*-butyl ester *based on the AlCl₃*. No more AlCl₃ could be added, however, because the mixture became too pasty and difficult to stir. Pouring into water, extraction of the organic phase, and fractionation with a spinning-band column gave 55 g. (51% based on AlCl₃) of the ester, b.p. 55–57° (2 mm.), analytically pure [55]. This preparation does not darken as that prepared using nitromethane as a solvent [56].

(c) Preparation of diethyl vinylmalonate. Two-tenths mole of butyl vinyl ether and diethyl malonate and 0.05 mole of aluminum chloride were warmed to 28–30° for 4–5 hr. to give a 39% yield of the titled ester [57].

1. H. Zollinger, *Azo and Diazo Chemistry*, Interscience Publishers, New York, 1961, p. 93.
2. E. H. White and H. Scherrer, *Tetrahedron Letters*, 758 (1961).
3. E. H. White, *J. Am. Chem. Soc.*, **77**, 6011 (1955).
4. P. A. Plattner *et al.*, *Helv. Chim. Acta*, **46**, 927 (1963).
5. J. A. Moore and D. E. Reed, *Org. Syn.*, **41**, 16 (1961).
6. H. Hörmann *et al.*, *Chem. Ber.*, **89**, 933 (1956).
7. A. Burger and W. L. Yost, *J. Am. Chem. Soc.*, **70**, 2198 (1948).
8. G. R. Clemo and T. P. Metcalfe, *J. Chem. Soc.*, 606 (1936).
9. R. G. Jones and E. C. Kornfeld, *J. Am. Chem. Soc.*, **73**, 107 (1951).
10. P. Yates and B. L. Shapiro. *J. Org. Chem.*, **23**, 739 (1958).

11. A. Neuberger, *J. Chem. Soc.*, 429 (1945).
12. R. G. Linville and R. C. Elderfield, *J. Org. Chem.*, **6,** 270 (1941).
13. L. I. Smith and K. L. Howard, *Org. Syn.*, Coll. Vol. **3,** 351 (1955).
14. P. Yates and J. Fugger, *Chem. Ind. (London)*, 1511 (1957).
15. M. S. Newman and P. F. Beal, III, *J. Am. Chem. Soc.*, **72,** 5163 (1950).
16. F. Arndt and B. Eistert, *Org. Reactions*, **1,** 52 (1942).
17. A. Roedig and H. Lunk, *Chem. Ber.*, **87,** 971 (1954).
18. C. H. Hassall, *Org. Reactions*, **9,** 73 (1957).
19. R. Robinson and L. H. Smith, *J. Chem. Soc.*, 371 (1937).
20. A. von Wacek and A. von Bezard, *Chem. Ber.*, **74,** 845 (1941).
21. D. L. Heywood and B. Phillips, *J. Org. Chem.*, **25,** 1699 (1960).
22. S. L. Friess, *J. Am. Chem. Soc.*, **71,** 14 (1949); L. H. Sarett, *ibid.*, **69,** 2899 (1947).
23. W. D. Emmons and G. B. Lucas, *J. Am. Chem. Soc.*, **77,** 2287 (1955).
24. J. D. McClure and P. H. Williams, *J. Org. Chem.*, **27,** 24 (1962).
25. R. W. White and W. D. Emmons, *Tetrahedron*, **17,** 31 (1962).
26. P. S. Starcher and B. Phillips, *J. Am. Chem. Soc.*, **80,** 4079 (1958).
27. K. Kosswig *et al.*, *Ann. Chem.*, **681,** 28 (1965).
28. M. F. Hawthorne *et al.*, *J. Am. Chem. Soc.*, **80,** 6393 (1958).
29. W. von E. Doering and L. Speers, *Org. Reactions*, **9,** 93 (1957).
30. D. L. Heywood and B. Phillips, U.S. Patent 3,240,798, March 15, 1966; *C.A.*, **64,** 17429 (1966).
31. F. Weiss, Belg. Patent 646,938, Aug. 17, 1964; *C.A.*, **64,** 603 (1966).
32. R. Altschul, *J. Am. Chem. Soc.*, **68,** 2605 (1946).
33. J. A. Nieuwland *et al.*, *J. Am. Chem. Soc.*, **56,** 2689 (1934).
34. D. Swern *et al.*, *J. Am. Chem. Soc.*, **75,** 6212 (1953).
35. G. F. Hennion and J. A. Nieuwland, *J. Am. Chem. Soc.*, **56,** 1802 (1934).
36. R. L. Adelman, *J. Org. Chem.*, **14,** 1057 (1949).
37. M. F. Ansell and M. H. Palmer, *Quart. Rev.*, **18,** 211 (1964).
38. W. S. Johnson *et al.*, *Org. Syn.*, Coll. Vol. **4,** 261 (1963).
39. H. C. Beyerman and J. S. Bontekoe, *Proc. Chem. Soc.*, 249 (1961).
40. R. H. Wiley, *Org. Syn.*, Coll. Vol. **3,** 853 (1955).
41. G. F. Bloomfield, *J. Chem. Soc.*, 3329 (1953).
42. R. J. Ouellette *et al.*, *J. Am. Chem. Soc.*, **86,** 2744 (1964).
43. T. Inoue and S. Tsutsumi, *J. Am. Chem. Soc.*, **87,** 3525 (1965).
44. J. Tsuji and T. Nogi, *J. Org. Chem.*, **31,** 2641 (1966).
45. R. F. Heck and D. S. Breslow, *J. Am. Chem. Soc.*, **85,** 2779 (1963).
46. N. L. Bauld, *Tetrahedron Letters*, 1841 (1963).
47. J. A. Nieuwland *et al.*, *J. Am. Chem. Soc.*, **55,** 2857 (1933).
48. H. Bredereck *et al.*, *Chem. Ber.*, **94,** 812 (1961).
49. M. E. Synerholm, *Org. Syn.*, Coll. Vol. **3,** 187 (1955).
50. O. Grummitt *et al.*, *Org. Syn.*, Coll. Vol. **3,** 833 (1955).
51. W. H. Coppock, *J. Org. Chem.*, **22,** 325 (1957).
52. C. D. Nenitzescu and A. T. Balaban, in Olah, *Friedel-Crafts and Related Reactions*, Vol. 3, Interscience Publishers, New York, 1964, Pt. 2, p. 1053.
53. J. B. Bush, Jr., and H. L. Finkbeiner, *J. Am. Chem. Soc.*, **90,** 5903 (1968).
54. H. C. Brown *et al.*, *J. Am. Chem. Soc.*, **90,** 1911 (1968).
55. D. E. Pearson, unpublished work.
56. P. Boldt and H. Militzer, *Tetrahedron Letters*, 3599 (1966).
57. Z. I. Torgashina and A. L. Naumchuk, *J. Gen. Chem. USSR (Eng. Transl.)* **26,** 3731 (1956).

C. Nucleophilic-Type Syntheses

Two large classes of reactions are discussed in this section: condensation reactions of the Claisen type with all types of ramifications and cleavage of

References for Section C are on pp. 851–853.

carbonyl compounds with alkoxides. A brief discussion of the principles involved are given in C.1. Considerable progress has been made of late in condensation reactions, particularly with regard to choice of catalysts.

1. From Esters and Other Active Hydrogen Types with Esters (Claisen)

$$CH_3CO_2C_2H_5 \underset{}{\overset{CH_3CH_2ONa}{\rightleftharpoons}} \overset{\ominus}{CH_2}C\overset{\textstyle O}{\underset{\textstyle OC_2H_5}{\big\langle}} \underset{}{\overset{CH_3CO_2C_2H_5}{\rightleftharpoons}}$$

$$CH_3\underset{\underset{\textstyle C_2H_5}{\overset{\textstyle |}{\underset{\textstyle |}{O}}}}{\overset{\textstyle O^{\ominus}}{C}}\!-\!CH_2CO_2C_2H_5 \rightleftharpoons CH_3\overset{\overset{\textstyle O^{\ominus}}{\textstyle |}}{C}\!=\!CHCO_2C_2H_5 \longleftrightarrow$$

$$CH_3\overset{\textstyle O}{\overset{\textstyle \|}{C}}\!-\!\overset{\ominus}{C}HCO_2C_2H_5 + C_2H_5OH$$

This reaction has been reviewed [1] and only the essential points of these reviews are given in addition to significant advances made more recently. (For other applications of the Claisen reaction, see Aldehydes, Chapter 10, F.1, and Ketones, Chapter 11, F.2.) The equilibrium part in the first step depends on the basicity of the catalyst, and the acidity of the hydrogen atom on the α-carbon of the ester. The second step is rate-controlling and depends on the reactivity of the ester group (paralleling its saponification rate) and the concentration of the ester anion; the second step is also quite dependent on steric effects. The third step occurs rapidly, but the equilibrium favors the formation of the acetoacetate anion because it is the weakest of all bases present and therefore the stablest. If no anion can be formed by the β-keto ester, such as in ethyl isobutyrylisobutyrate, a powerful base catalyst, sodium triphenylmethide, rather than sodium ethoxide, the original catalyst, must be employed. Here the ethoxide ion, a product of the condensation, is a weaker base than the triphenylmethide ion, thus favoring ethoxide, triphenylmethane, and β-keto ester formation.

The condensation of mixed esters, usually feasible with a large excess of ethyl oxalate, ethyl trifluoroacetate, methyl picolinate, ethyl benzoate, or ethyl isobutyrate, and some other ester with an active methylene group such as ethyl propionate, has been improved of late by use of the base catalyst, diisopropylaminomagnesium bromide. By means of this catalyst, the two esters can be used in equal quantities and the product, obtained in 47–51 % yields, is virtually free of self-condensed ester [2]. Evidently, the catalyst converts the ester with the active methylene group completely into its anion. This catalyst has proved effective in the self-condensation of various α-aryloxyacetic esters (22–65% yields), although other catalysts could be used in this reaction [3]. More modern practice would utilize lithium diisopropylamide, prepared from butyllithium and diisopropylamine in tetrahydrofuran or other aprotic solvents, rather than the magnesium salt [4]. Methyl α,α-dimethoxyacetate is so reactive on addition of anions that it can be used in mixed ester condensations (50–85 % yields) with ordinary base catalysts [5]. Biphthalide

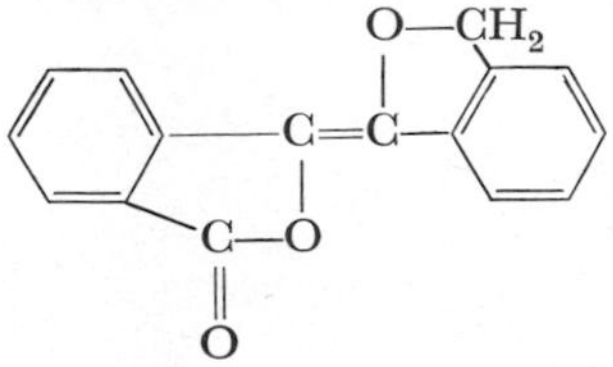

has been made recently in 75 % yield from self-condensation of phthalide by the use of the increasingly popular base catalyst system, potassium t-butoxide in dimethylsulfoxide [6]. Biphthalyl

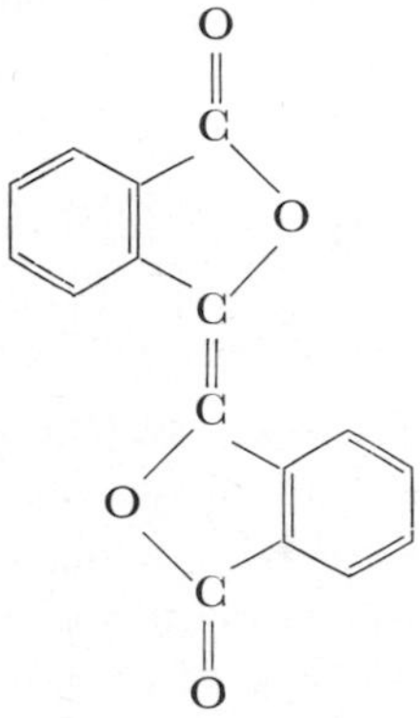

is made conveniently by heating phthalic anhydride and triethyl phosphite, a carbene perhaps being the intermediate [7]. Acyloins have been condensed with ethyl cyanoacetate to yield unsaturated lactones [8]:

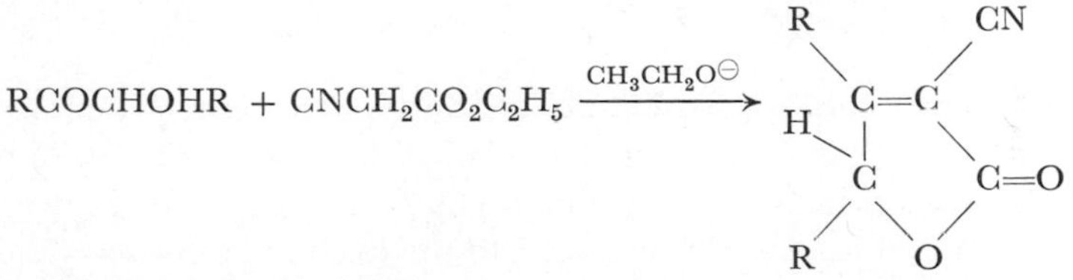

$$RCOCHOHR + CNCH_2CO_2C_2H_5 \xrightarrow{CH_3CH_2O^\ominus}$$

32–80% yields

Sodamide has been employed widely in the Claisen condensation [9], and another condensing agent which has been quite popular is sodium hydride [10–13]. In some cases in which esters containing two active hydrogen atoms are employed, diethyl carbonate in excess replaces the second hydrogen atom by an ethyl group [14].

When diethyl oxalate is employed, the intermediate α-keto ethyl ester produced when heated alone or preferably with powdered glass or boric acid [15, 16] loses carbon monoxide to give the diester.

Ketones [17] and heterocyclic compounds containing methyl groups in the 1- or 4-position [18] may also be acylated by base condensation. Recently, ethyl diethoxyphosphinyl formate, several ketones containing α-hydrogen atoms, and sodium hydride gave yields of β-keto esters, after treatment with ethanol-sulfuric acid, varying from 67 to 81 % [19]:

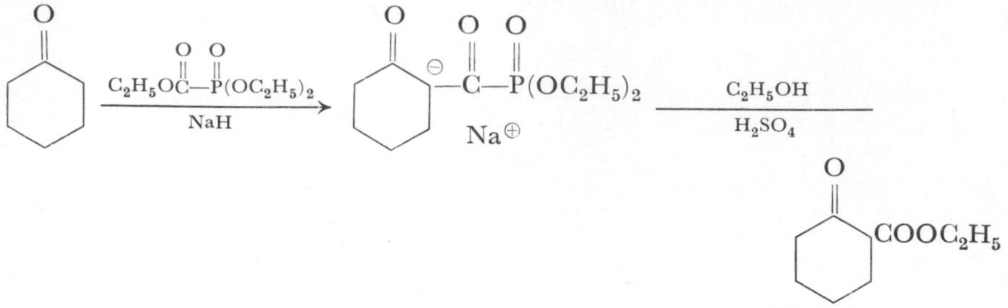

2-Carbethoxycyclohexanone

Modifications occur if the ester containing active hydrogen is condensed with ethyl orthoformate in the presence of acetic anhydride and zinc chloride [20]. In this case an ethoxymethylene malonate is obtained.

$$CH_2(COOC_2H_5)_2 + HC(OC_2H_5)_3 \xrightarrow[ZnCl_2]{(CH_3CO)_2O} C_2H_5OCH{=}C(COOC_2H_5)_2$$

The yields in these syntheses are usually satisfactory.

(a) Preparation of methyl α-2-furoylpropionate. Diisopropyl-aminomagnesium bromide in 250 ml. of ether was made by adding diiso-propylamine (0.4 mole) dropwise to ethylmagnesium bromide (0.4 mole). A mixture of 0.2 mole each of methyl furoate and methyl propionate was added to the stirred base catalyst over a period of 30 min., after which stirring was continued for an additional 2 hr. The mixture was poured onto 300 g. of cracked ice mixed with 25 ml. of sulfuric acid, and the ether layer separated and washed with water and aqueous sodium bicarbonate solution. The ether was dried and concentrated, and the residue distilled at 2 mm. pressure, b.p. 114–116°, 17 g. (47 %) [2].

(b) Preparation of 3-phthalanylidenephthalide. Commercial potassium butoxide (2.8 g., 23 mmoles) was added to a solution of phthalide (2.7 g., 20 mmoles) in 100 ml. of dimethylsulfoxide. After 1 hr. at room temperature, 50 ml. of ice water was added to the dark red solution. Acidification with 50 ml. of 0.5 N hydrochloric acid gave a fine, crystalline, colorless precipitate, m.p. 155°, 2 g. (75 %). The compound on melting lost the elements of water to form the titled compound (90 %), m.p. 225–227°, yellow with greenish fluorescence [6].

(c) Preparation of ethyl phenylcyanoacetate (70–78 % from phenyl-acetonitrile, diethylcarbonate, and sodium ethoxide) [21].

(d) Other examples

(1) Cetylmalonic ester (68.5–71 % from ethyl stearate, diethyl oxalate, and sodium ethoxide; decarbonylation by heating under reduced pressure) [22].

(2) Ethyl α-tetralone-β-carboxylate (87–92 % from α-tetralone, diethyl oxalate, and sodium ethoxide; decarbonylation by heating with glass powder) [17].

(3) Ethyl _n_-butyryl acetate (60 % from methyl _n_-propyl ketone, ethyl carbonate, and sodium hydride) [12].

(4) Ethyl benzoylacetate (68–71 % from ethyl acetoacetate, benzoyl chloride, aqueous sodium hydroxide, stirred in technical naphtha) [23].

2. From Diesters (Dieckmann)

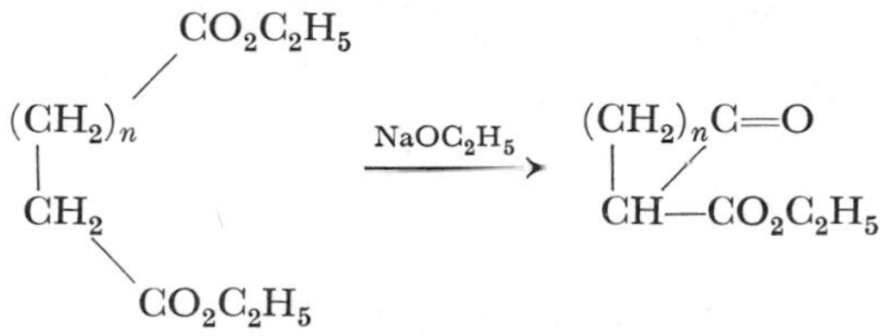

$$
\begin{array}{c}
\quad\quad CO_2C_2H_5 \\
\quad\quad \diagup \\
(CH_2)_n \\
\quad| \\
CH_2 \\
\quad\diagdown \\
\quad\quad CO_2C_2H_5
\end{array}
\xrightarrow{\text{NaOC}_2\text{H}_5}
\begin{array}{c}
(CH_2)_n\,C=O \\
\quad| \diagup \\
CH—CO_2C_2H_5
\end{array}
$$

The condensation, which has been reviewed [24], ordinarily is restricted to the formation of five- and six-membered rings, but occasionally can be made to succeed for larger rings using high dilution techniques [25]. Ethyl succinate, which might be thought to be the source of a three-membered ring, first dimerizes under Dieckmann conditions and then cyclizes to form 2,5-dicarbethoxy-1,4-cyclohexanedione (see Ex. b.1). Different diesters also can be used, diethyl glutarate and diethyl oxalate yielding 65 % 3,5-dicarbethoxy-1,2-cyclopentanedione [26].

The reversibility of the Dieckmann condensation is nicely illustrated by the isomerization of the following ketoester:

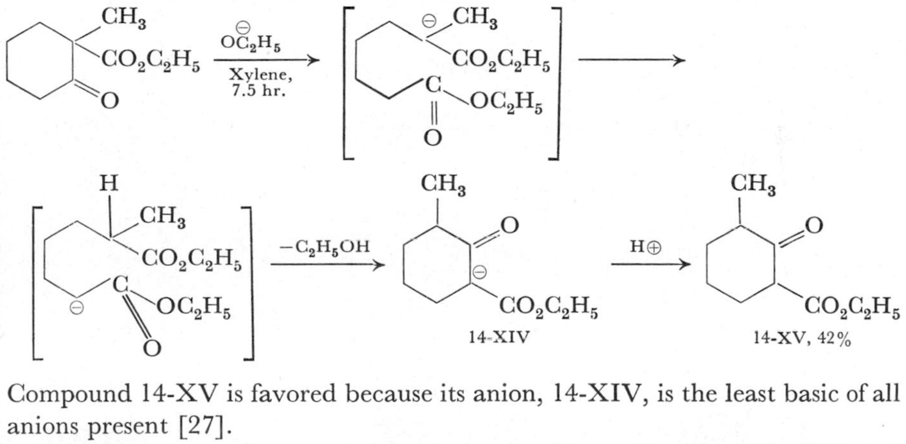

14-XIV   14-XV, 42%

Compound 14-XV is favored because its anion, 14-XIV, is the least basic of all anions present [27].

In addition to sodium and potassium ethoxide, sodium hydride [28] and potassium _t_-butoxide [29] have also been used to effect the Dieckmann condensation. In the condensation of ethyl γ-phenylbutyrate with diethyl oxalate the use of potassium rather than sodium ethoxide reduced the time from 24 to 12 hours [30]. The potassium _t_-butoxide in refluxing xylene under nitrogen with high-speed stirring followed by hydrolysis and decarboxylation has permitted the synthesis of C_{13}–C_{15} monoketones in 24–48% yields and C_{18}-, C_{20}-, C_{22}-, and C_{24}-diketones usually in somewhat lower yields:

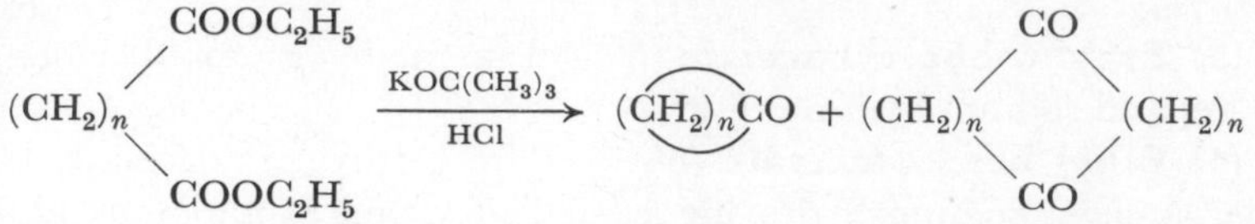

$$(CH_2)_n \overset{COOC_2H_5}{\underset{COOC_2H_5}{\Big\langle}} \xrightarrow[\text{HCl}]{KOC(CH_3)_3} (CH_2)_nCO + (CH_2)_n \overset{CO}{\underset{CO}{\Big\rangle}} (CH_2)_n$$

The use of sodium hydride dispersion has permitted the time required for completion of the condensation to be lessened (see Ex. a).

The acyloin condensation of diesters (Alcohols, Chapter 4, C.8) may produce a Dieckmann product if the sodium ethoxide formed is not scavenged by trimethyl chlorosilane [31]. The Thorpe-Ziegler condensation of dinitriles, which bears some resemblance to the Dieckmann, is discussed elsewhere (Nitriles, Chapter 19, B.2).

(a) **Preparation of 2-carbethoxycyclopentanone** (79–82% from diethyl adipate, excess sodium in benzene, and traces of ethanol at refluxing temperature overnight) [32]; with the sodium hydride dispersion in toluene at 40°, the reaction may be completed (65–80%) in 1.5 hr. [33].

(b) **Other examples**

(1) **2,5-Dicarbethoxy-1,4-cyclohexanedione** (64–68% from sodium

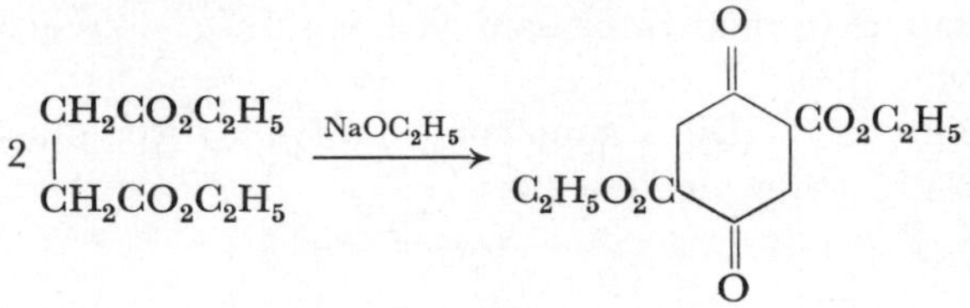

$$2 \; \overset{CH_2CO_2C_2H_5}{\underset{CH_2CO_2C_2H_5}{|}} \xrightarrow{NaOC_2H_5}$$

ethoxide and diethyl succinate in refluxing ethanol for 24 hr.) [34].

3. From Diethyl Succinate and Carbonyl Compounds (Stobbe)

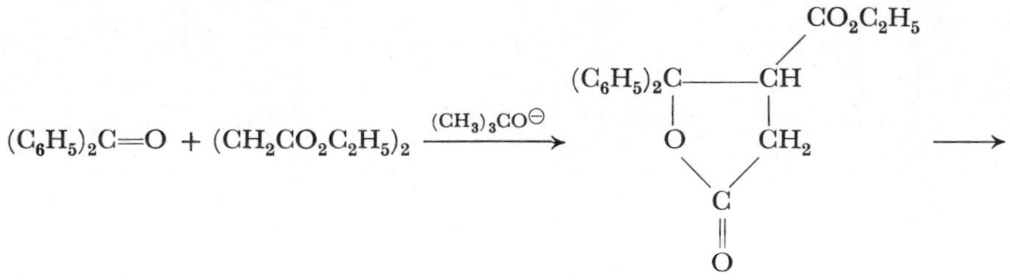

$$(C_6H_5)_2C{=}O + (CH_2CO_2C_2H_5)_2 \xrightarrow{(CH_3)_3CO^{\ominus}}$$

14-XVI

$$(C_6H_5)_2C{=}C{-}CO_2C_2H_5 \xrightarrow{H^{\oplus}} (C_6H_5)_2C{=}C$$
$$\overset{|}{CH_2}$$
$$\overset{|}{CO_2^{\ominus}}$$

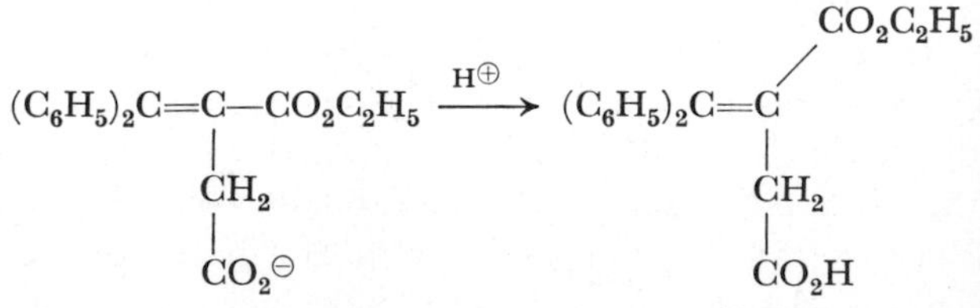

14-XVII

Benzophenone condenses difficultly with carbanions; yet, in the Stobbe condensation with diethyl succinate, it condenses smoothly and in high yield. The facility with which the reaction occurs suggests the intermediacy of some species such as 14-XVI which adds driving force. Diethyl succinate, therefore, is particularly facile in carbonyl condensations. This reaction has been reviewed [35]. The synthetic utility of the reaction lies in the value of the decarboxylation product, i.e., unsaturated acids or lactones [36]:

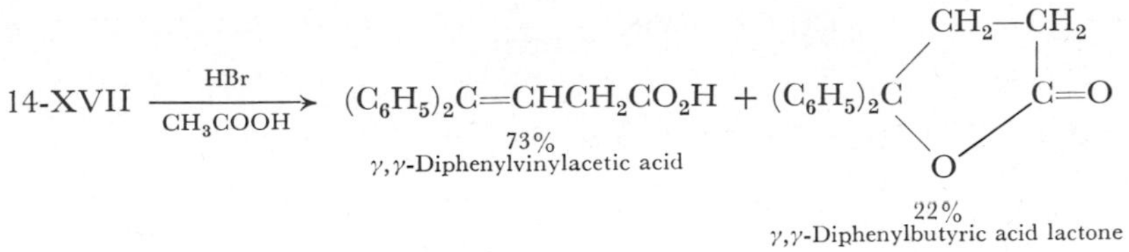

$$14\text{-XVII} \xrightarrow[\text{CH}_3\text{COOH}]{\text{HBr}} (\text{C}_6\text{H}_5)_2\text{C}{=}\text{CHCH}_2\text{CO}_2\text{H} + (\text{C}_6\text{H}_5)_2\text{C}$$

73%
γ,γ-Diphenylvinylacetic acid

22%
γ,γ-Diphenylbutyric acid lactone

Of late, the term Stobbe condensation has been used more freely to describe condensations of dimethyl homophthalate [37], although the analog, diethyl glutarate, has no particular propensity to participate in the reaction. Although the original base employed in the condensation was sodium ethoxide, potassium *t*-butoxide is to be preferred [38]. It has now been shown that sodium hydride is an even better base [39]. A dispersion of this hydride in hexane at 22° with benzophenone and diethyl succinate led to a 95.7% yield of sodium ethyl diphenylmethylenesuccinate in 1 hour [40]:

$$(\text{C}_6\text{H}_5)_2\text{CO} + \overset{\overset{\displaystyle \text{CH}_2\text{COOC}_2\text{H}_5}{|}}{\text{CH}_2\text{COOC}_2\text{H}_5} + 2\,\text{NaH} \longrightarrow$$

$$(\text{C}_6\text{H}_5)_2\text{C}{=}\overset{\overset{\displaystyle \text{CH}_2\text{COONa}}{|}}{\text{C}}\text{COOC}_2\text{H}_5 + \text{C}_2\text{H}_5\text{ONa} + 2\,\text{H}_2$$

(a) Preparation of β-carbethoxy-γ,γ-diphenylvinylacetic acid (14-XVII, 92–94% from benzophenone, diethyl succinate, and potassium *t*-butoxide) [41].

(b) Preparation of methyl *o*-carboxyphenyl-3-methylbut-2-enoate (88% from dimethyl homophthalate, *o*-C$_6$H$_4$(CO$_2$CH$_3$)CH$_2$CO$_2$CH$_3$, acetone, and sodium hydride in benzene) [37].

4. From Carbonyl Compounds and Esters (Aldol, Knoevenagel or Doebner Condensation)

The aldol condensation (see Alcohols, Chapter 4, G.1) of carbonyl compounds with monoesters is not a very common method, perhaps because self-condensation products encroach and because diesters condense with themselves so readily. Nevertheless instances are known:

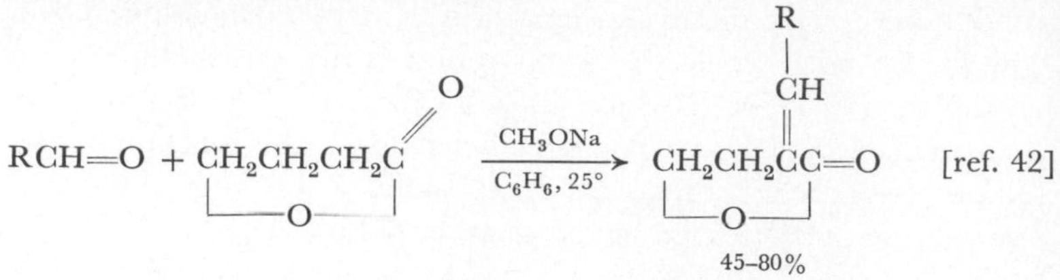

$$RCH{=}O + CH_2CH_2CH_2\overset{O}{\underset{\underset{O}{\rule{1.2em}{0.4pt}}}{C}} \xrightarrow[C_6H_6,\,25°]{CH_3ONa} \overset{R}{\underset{\underset{O}{\rule{1.2em}{0.4pt}}}{CH}} CH_2CH_2\overset{\|}{C}C{=}O \quad [ref.\ 42]$$

45–80%

$$C_6H_5CH{=}O + CH_2ORCOOR \xrightarrow[\substack{C_6H_5CH_3,\\C_2H_5OH}]{Na} C_6H_5CH{=}CORCOOR \quad [ref.\ 43]$$

ca. 50%

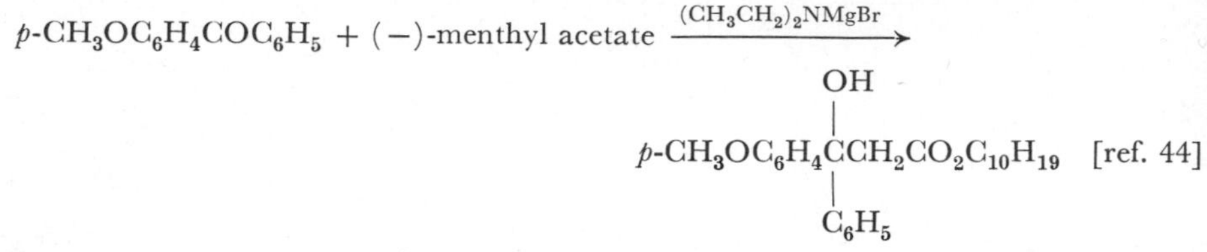

$$p\text{-}CH_3OC_6H_4COC_6H_5 + (-)\text{-menthyl acetate} \xrightarrow{(CH_3CH_2)_2NMgBr}$$

$$p\text{-}CH_3OC_6H_4\overset{OH}{\underset{C_6H_5}{\underset{|}{\overset{|}{C}}}}CH_2CO_2C_{10}H_{19} \quad [ref.\ 44]$$

67%
Menthyl 3-(p-anisyl)-3-phenyl-
hydracrylate
16% optical purity

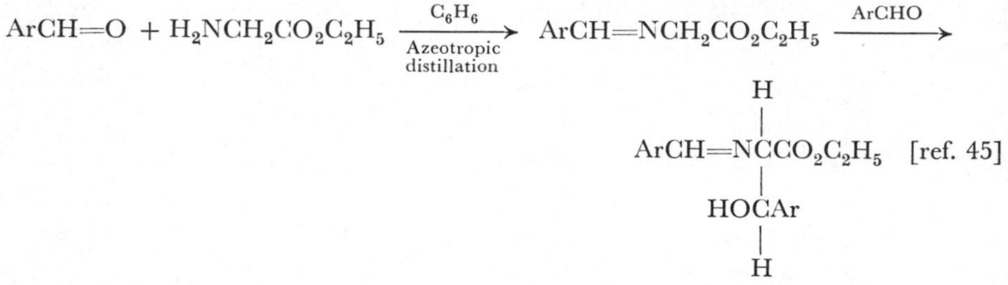

$$ArCH{=}O + H_2NCH_2CO_2C_2H_5 \xrightarrow[\substack{Azeotropic\\distillation}]{C_6H_6} ArCH{=}NCH_2CO_2C_2H_5 \xrightarrow{ArCHO}$$

$$ArCH{=}N\overset{H}{\underset{\underset{H}{\underset{|}{HOCAr}}}{\underset{|}{C}}}CO_2C_2H_5 \quad [ref.\ 45]$$

Ethyl N-arylidenearyl serine

The carbanion of bromoform as well as that of esters adds to aldehydes as shown:

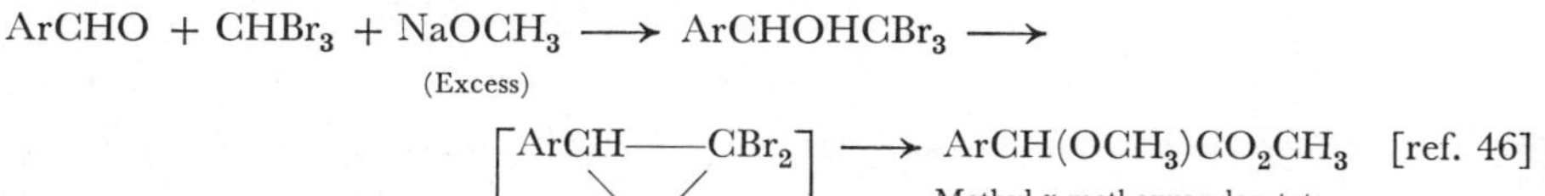

$$ArCHO + CHBr_3 + NaOCH_3 \longrightarrow ArCHOHCBr_3 \longrightarrow$$

(Excess)

$$\left[\overset{ArCH\rule{1.5em}{0.4pt}CBr_2}{\underset{O}{\diagdown\ \diagup}}\right] \longrightarrow ArCH(OCH_3)CO_2CH_3 \quad [ref.\ 46]$$

Methyl α-methoxyarylacetates,
0–79%

Of much greater versatility is the condensation of carbonyl compounds with diesters or ketoesters:

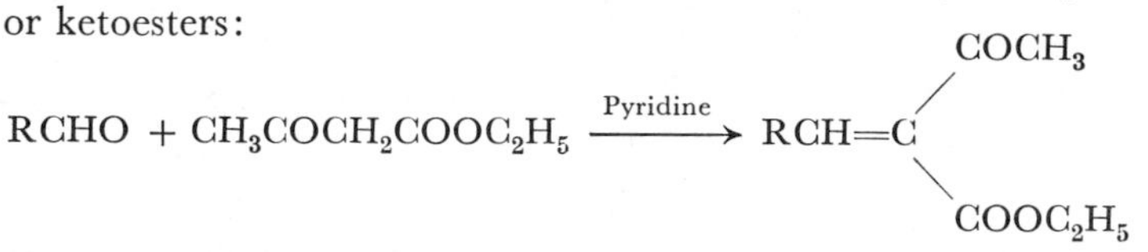

$$RCHO + CH_3COCH_2COOC_2H_5 \xrightarrow{Pyridine} RCH{=}C\overset{COCH_3}{\underset{COOC_2H_5}{<}}$$

This condensation has been discussed for dicarboxylic acids as the Doebner modification of the Perkin reaction, under Carboxylic Acids, Chapter 13, D.2, where the mechanism is given. It is also known as the Knoevenagel reaction, although some investigators make an arbitrary distinction by confining the Knoevenagel to cases in which ammonia and a primary or secondary amine are used, and the Doebner to cases in which pyridine, usually with a little piperidine, is the base [47]. For a recent review of the reaction see [48]. In place of the diester or keto ester any compound having an active methylene group may be substituted. Such activation is due to the direct attachment of a group, such as nitro, cyano, acyl, carboxy, carbalkoxy, etc., and in most cases two such groups are necessary to furnish sufficient activation.

A great variety of catalysts has been employed in the reaction. The most general one appears to be the already mentioned pyridine with or without piperidine. However, piperidine or other secondary amines are suitable for condensations involving malonic esters, β-ketonic esters, and related types. Ammonium acetate appears to be a preferred catalyst for condensations of ethyl cyanoacetate with hindered ketones, while primary amines, such as benzylamine, are preferred with unhindered ketones [49]. In fact, in some cases a trace of benzylamine in piperidine increases the yield of ester substantially [49]. The Cope catalyst, ammonium or an amine acetate with an inert solvent such as benzene, toluene, or chloroform and a smaller amount of acetic acid, is important for condensations involving cyanoacetic esters [50]. The same catalyst has been employed successfully with other esters such as malonic and acetoacetic esters. For acetoacetic ester, piperidine was found to be more satisfactory as a catalyst than piperidine acetate or acetamide and acetic acid [51]. Amino acids, such as β-alanine and ε-aminocaproic acid in the presence of acetic acid, are more effective than piperidine acetate in the condensation of acetone with ethyl cyanoacetate [52]. It has been shown that a small amount of benzoic acid in the piperidine catalyst accelerates the reaction, but a large amount retards it [53]. Amyl amine acetate is claimed to be a superior catalyst for the condensation of acetophenone with ethyl cyanoacetate [54]. Basic resins, such as Amberlite IR-4B [55] and Dowex 3 [56], preferably as the acetate, have been used successfully in the condensation. Fluorides of caesium, rubidium, and potassium have also been employed [57].

The amount of catalyst varies greatly in the preparations described in the literature, except in the case of the Cope catalyst. Here 0.2 mole of catalyst is commonly used with each mole of the active methylene compound. In the most active methylene compounds, such as malononitrile, a catalyst is not always necessary.

Although unsaturated esters may also be prepared by the Claisen and Reformatsky reactions, these methods, in contrast to the Knoevenagel, are not satisfactory for preparing cinnamic esters which contain halogen, nitro, or phenolic groups in the ring [58]. The yields in the Knoevenagel condensation are satisfactory, particularly with aldehydes.

(a) **Preparation of ethyl benzalmalonate** (89–91 % from benzaldehyde, diethyl malonate, and piperidine benzoate) [59].

(b) **Other examples**

(1) **Ethyl n-butylcyanoacetate** (94–96 % from butyraldehyde, ethyl cyanoacetate, and acetic acid under the reducing conditions produced by palladium and hydrogen) [60].

(2) Ethyl cyclohexylidinecyanoacetate (quantitative yield from cyclohexanone, ethyl cyanoacetate, and Dowex 3-acetate refluxed in benzene) [56].

(3) Ethyl α-cyano-β-methyl-2-pentenoate (80 % from methyl ethyl ketone and ethyl cyanoacetate in benzene boiled with a small amount of ε-aminocaproic acid and acetic acid for 7 hr.) [52].

5. From Esters and Unsaturated Esters (Michael)

Since the principles of the Michael condensation have been discussed elsewhere (Ketones, Chapter 11, G.3), this discussion is restricted to examples which produce esters [61] for the most part such as:

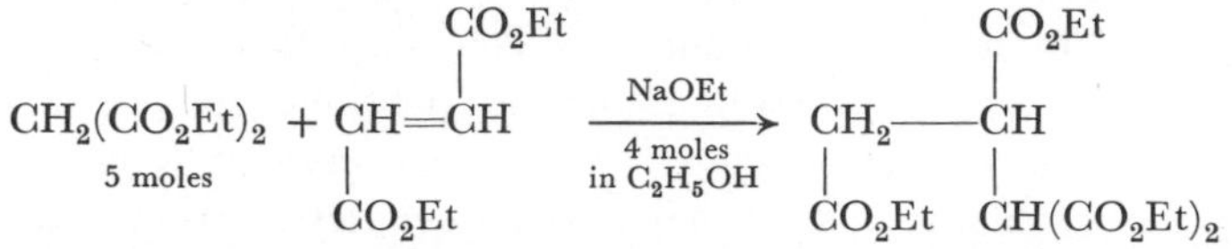

$$CH_2(CO_2Et)_2 + \begin{matrix} CO_2Et \\ | \\ CH{=}CH \\ | \\ CO_2Et \end{matrix} \xrightarrow[\substack{4\ moles \\ in\ C_2H_5OH}]{NaOEt} \begin{matrix} CO_2Et \\ | \\ CH_2{-}CH \\ | \quad | \\ CO_2Et \ CH(CO_2Et)_2 \end{matrix}$$

$$\text{5 moles} \qquad\qquad \text{4.1 moles} \qquad\qquad \substack{\text{1,1,2,3-Tetracarbethoxypropane,} \\ \text{93%}}$$

Abnormal products may be formed on occasion depending on the quantity or the type of catalyst. For instance, normal condensation occurs in the following reaction with small quantities of alkoxide to give 14-XVIII. However, in the

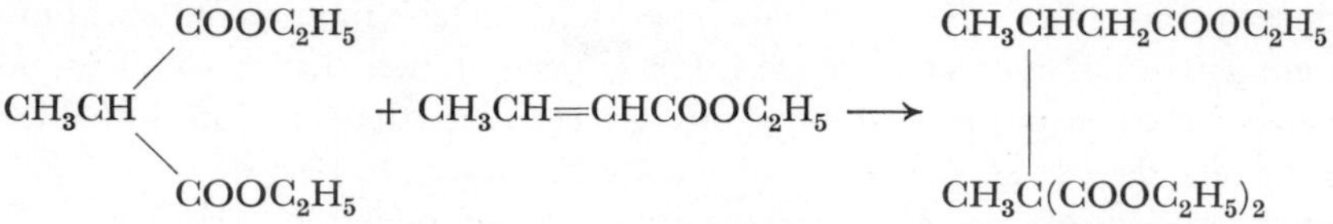

$$\begin{matrix} COOC_2H_5 \\ \diagup \\ CH_3CH \\ \diagdown \\ COOC_2H_5 \end{matrix} + CH_3CH{=}CHCOOC_2H_5 \longrightarrow \begin{matrix} CH_3CHCH_2COOC_2H_5 \\ | \\ CH_3C(COOC_2H_5)_2 \end{matrix}$$

$$\substack{\text{14-XVIII} \\ \text{1,3,3-Tricarboethoxy-2-methylbutane}}$$

presence of larger quantities of catalyst the product is 14-XIX. The most widely

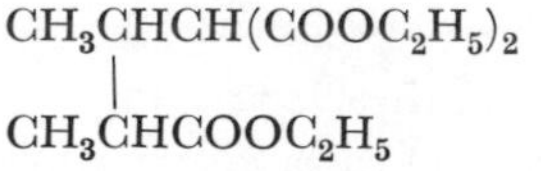

$$\begin{matrix} CH_3CHCH(COOC_2H_5)_2 \\ | \\ CH_3CHCOOC_2H_5 \end{matrix}$$

$$\substack{\text{14-XIX} \\ \text{1,1,3-Tricarboethoxy-2-methylbutane}}$$

accepted explanation [62] assumes that a Dieckmann condensation occurs in 14-XVIII to give 14-XX, which in turn cleaves to give 14-XIX, the abnormal

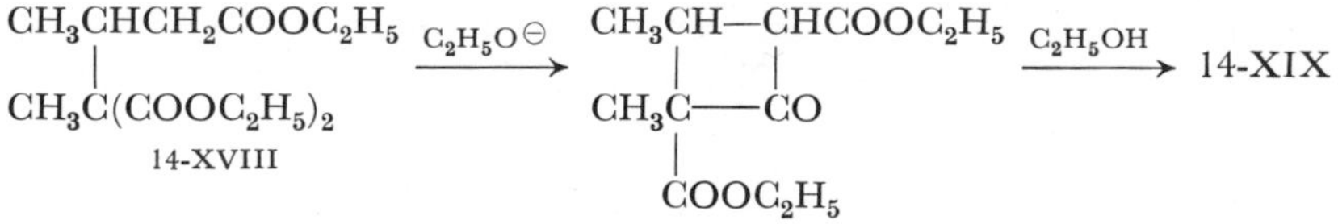

$$\begin{matrix} CH_3CHCH_2COOC_2H_5 \\ | \\ CH_3C(COOC_2H_5)_2 \end{matrix} \xrightarrow{C_2H_5O^{\ominus}} \begin{matrix} CH_3CH{-}CHCOOC_2H_5 \\ | \quad\quad | \\ CH_3C{-}\!-\!CO \\ | \\ COOC_2H_5 \end{matrix} \xrightarrow{C_2H_5OH} \text{14-XIX}$$

$$\text{14-XVIII} \qquad\qquad\qquad\qquad \text{14-XX}$$

condensation product.

Another abnormality results with sodium hydride dispersed in oil as the condensing agent [63] if there is produced in the addition: a three-carbon chain with a carbanion at one end and an appropriate group, usually halogen, at the other; and the presence at the carbanion end of the three-carbon chain of some group, such as carboalkoxy, which can stabilize the anion. Under these conditions cyclopropane derivatives are obtained as shown:

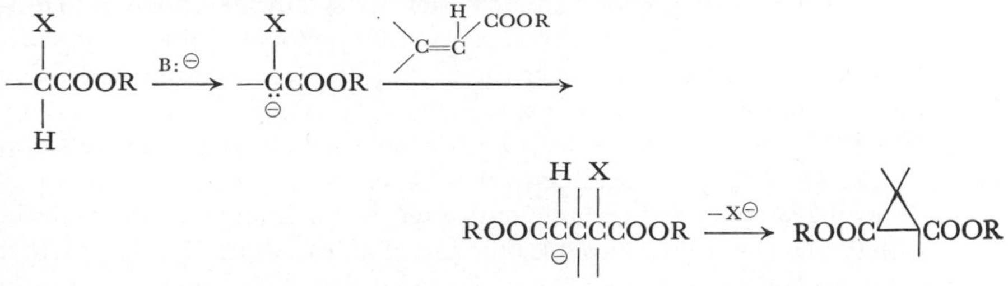

Such derivatives may be produced in *cis* and *trans* forms in total yields as high as 74%.

(a) Preparation of diethyl 6-keto-4-methyl-2-heptene-1,5-dicarboxylate (75% from ethyl acetoacetate, ethyl sorbate, and potassium *t*-butoxide) [64].

(b) Preparation of ethyl 2,3,5-triphenyl-5-ketopentanoate (90% from benzalacetophenone, ethyl α-phenylacetate, and $\frac{1}{3}$ equiv. of $NaOC_2H_5$ in alcohol at 25° for 20 hr.) [65].

(c) Preparation of warfarin. The marked accelerating effect of

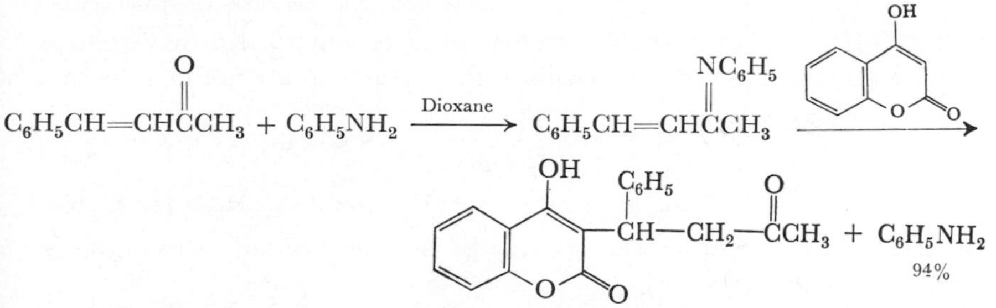

aniline as compared to tertiary amines suggested that condensation was occurring via the Schiff base. This effect should be examined in other Michael condensations. As the senior author states, write to him if comments are favorable, but write elsewhere if they are not [66].

6. From Alkyl or Acyl Halides with Esters

$$RCH_2COOR' \xrightarrow[NaOC_2H_5]{R''X} RCH(R'')COOR'$$

$$CH_2(COOR)_2 \xrightarrow[C_2H_5OH]{Mg} C_2H_5OMgCH(COOR)_2 \xrightarrow[\text{or } ArCOOCOOR]{ArCOCl} ArCOCH(COOR)_2$$

Esters containing α-hydrogen atoms may be converted into substituted esters via the attack of the carbanion on the alkyl halide (see Carboxylic Acids, Chapter 13, A.9, and the comprehensive review [67]). Agents employed to produce carbanions from the esters have been sodium ethoxide [68], sodium triphenylmethide [69], potassium triphenylmethide [70], and potassium hydroxide in acetal-type solvents like 1,1-diethoxyethane [71]. Simple esters of the type RCH_2COOR' give poor yields because of the Claisen competing reaction between 2 moles of the ester leading to the β-keto ester.

Better results are obtained with malonic ester in which single alkyl groups in great variety have been introduced by the use of alkyl halides [72, 68], diethyl sulfate [73], and ethyl p-toluenesulfonate [74]. Yields in these cases vary greatly [75], the better ones reported being for the preparation of the methyl (79–83 %) [68], the n-butyl (80–90 %) [72], the n-amyl (80 %) [76], and the α-naphthylmethyl (82 %) [77] malonates. Two alkyl groups [75] may also be introduced into malonic ester, although the yields are low if the two groups are branched [78].

When dihalides are used in equimolar amounts with malonic ester, cycloalkanedicarboxylic acid esters are obtained. The ester and trimethylene bromide, for example, react as follows [79, 80]:

$$CH_2(COOC_2H_5)_2 + BrCH_2CH_2CH_2Br \xrightarrow{NaOC_2H_5} \begin{array}{c} CH_2-CH_2 \\ | \quad\quad | \\ CH_2-C(COOC_2H_5)_2 \end{array}$$

Diethyl cyclobutane-1,1-dicarboxylate

Smaller [81] and larger [82] ring esters of dicarboxylic acids have also been produced by this method. The yields in these reactions are not high because of the competing reaction:

$$(C_2H_5OOC)_2CH_2 + X(CH_2)_nX + CH_2(COOC_2H_5)_2 \xrightarrow{NaOC_2H_5}$$

$$(C_2H_5OOC)_2CH(CH_2)_nCH(COOC_2H_5)_2$$

In fact, the long-chain tetra esters may be produced in fair yield by using an excess of malonic ester [83].

Many other modified alkyl groups have been introduced into malonic ester. Tricarboxylic esters have been produced with haloesters [84]; halomalonic esters with ethylene halides [85]; alkoxymalonic esters with haloethers [86]; and ketomalonic esters with haloketones [87].

Acyl malonates are prepared by treating the alkoxymagnesium malonic ester with a mixed carboxylic-carbonic anhydride [88] or an acyl chloride [89].

Alkylations of monoalkyl-substituted malonic esters are best carried out in the nonprotonic solvent, dimethylformamide. The rate of alkylation of the sodium salt of diethyl n-butylmalonate is many times faster in this solvent than in benzene, probably because the anion is less solvated and therefore more nucleophilic [90]. The sodium salt of the monoalkyl-substituted malonic esters can be made from sodium hydroxide and the water of reaction removed by azeotropic distillation with benzene without saponifying the ester (or by sodium hydride dispersed in mineral oil). Moreover, monoesters such as ethyl diphenylacetate can be alkylated by means of a dimethylformamide-sodium hydride system [91]; see Ex. c.2 also. It is interesting to note that ethyl aceto-acetate in 50% aqueous alcohol solutions can be alkylated in about 40% yield [92].

Diethyl t-butylmalonate has been a difficult ester to prepare. It can be made by prolonged exposure of t-butyl chloride and the sodium salt of diethyl malo-nate in absolute alcohol in 29% yield [93] or by the addition of methylmagnesium iodide to diethyl isopropylidenemalonate in 60% yield [94]; the 1,4-addition of the Grignard reagent to these unsaturated esters has been improved by the addition of cuprous cyanide [95]. A third method consisting of an electrophilic attack of t-butyl chloride has been described in B.8, Ex. b.

Free radical methods of alkylation and acylation of esters have been made available recently (see Ex. c.5 for alkylation and Ex. b for acylation).

(a) Preparation of ethyl n-butylmalonate (80–90% from diethyl malonate, n-butyl bromide, and sodium ethoxide) [72].

(b) Preparation of ethyl enanthylsuccinate [96]:

$$CH_3(CH_2)_5CH{=}O + \begin{matrix} CHCO_2C_2H_5 \\ \| \\ CHCO_2C_2H_5 \end{matrix} \xrightarrow{(C_6H_5CO_2)_2} CH_3(CH_2)_5\overset{\overset{\displaystyle O}{\|}}{C}{-}\begin{matrix} CHCO_2C_2H_5 \\ | \\ CH_2CO_2C_2H_5 \end{matrix}$$

71–76%

(c) Other examples

(1) Ethyl 1,1-cyclobutanedicarboxylate (21–23% recovered as the free acid from ethyl malonate, trimethylene bromide, and sodium ethoxide) [79].

(2) Ethyl α,α-dimethylbutyrate (58% from ethyl isobutyrate, ethyl iodide, and sodium triphenylmethide) [69].

(3) Triethyl 1-phenylpropane-2,2,3-tricarboxylate (95% from ethyl benzylmalonate, ethyl bromoacetate, and sodium ethoxide) [84].

(4) Diethyl benzoylmalonate (68–75% from ethoxymagnesium-malonic ester and mixed benzoic carbonic anhydride) [88].

(5) Diethyl n-octylmalonate [97]:

$$C_6H_{13}CH{=}CH_2 + \text{diethyl malonate} + \text{di-}t\text{-butyl peroxide} \xrightarrow[80 \text{ hr.}]{105°}$$

1 mole   30 moles   0.05 mole

$$CH_3(CH_2)_7CH\begin{matrix} \diagup COOC_2H_5 \\ \diagdown COOC_2H_5 \end{matrix}$$

79%

(6) Triethyl methanetricarboxylate (88–93% from Mg, absolute ethanol, ethyl malonate, ether, and ethyl chloroformate) [98].

7. From Grignard Reagents and Carbonate Esters

This reaction is best carried out with the less reactive Grignard reagents and by addition of the reagent to the ester (inverse addition), both factors mitigating against the possibility of further reaction to form ketones or carbinols. Either diethyl carbonate or ethyl chloroformate can be used.

(a) Preparation of ethyl α-naphthoate (68–73% from α-naphthyl-magnesium bromide and diethyl carbonate) [99].

(b) Preparation of diethyl p-tolylsulfonylmalonate [100].   The

$$p\text{-}CH_3C_6H_4SO_2CH_2MgBr + ClCO_2C_2H_5 \rightarrow p\text{-}CH_3C_6H_4SO_2CH(CO_2C_2H_5)_2$$

22% conversion, 74% yield

intermediate, $p\text{-}CH_3C_6H_4SO_2CH_2CO_2C_2H_5$, is converted rapidly to its anion, which reacts with a second chloroformate molecule to form the product.

(c) Preparation of 2-methylnaphthoic acid [101]:

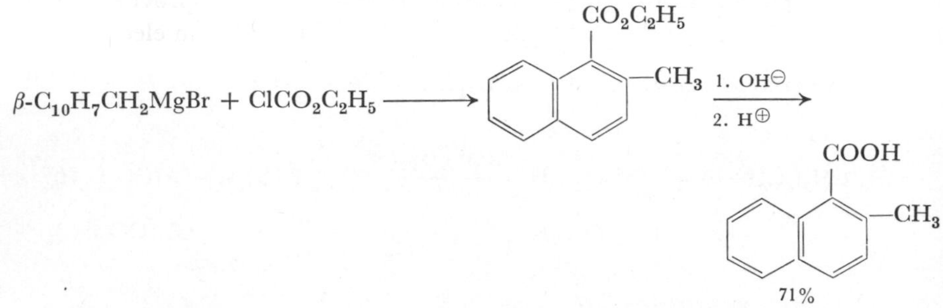

$$\beta\text{-}C_{10}H_7CH_2MgBr + ClCO_2C_2H_5 \longrightarrow$$

71%

8. Reformatsky, Darzens and Favorskiĭ Reactions

Reformatsky reaction (see Alcohols, Chapter 4, E.3).

$$RCH{=}O + \underset{\underset{Br}{|}}{CH_2CO_2C_2H_5} \xrightarrow{Zn} \underset{\underset{OH}{|}}{RCH}{-}CH_2CO_2C_2H_5$$

Darzens reaction (see Ethers, Chapter 6, C.3).

$$R_2C{=}O + ClCH_2CO_2R \xrightarrow{base} R_2C\underset{O}{\diagdown\diagup}CHCO_2R$$

Favorskiĭ reaction (see Carboxylic Acids, Chapter 13, G.2).

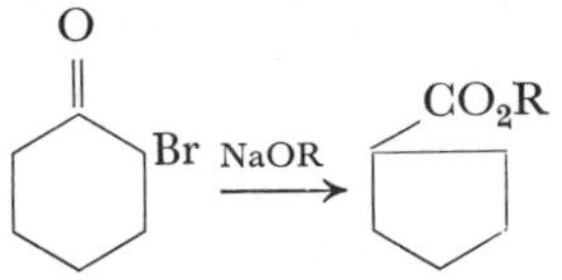

9. Cleavage of Carbonyl Compounds

Since the Claisen condensation is reversible, cleavage of β-keto esters can be brought about readily to give esters. Dialkyl β-keto esters cleave easily without concomitant ketone formation:

$$CH_3COCR_2CO_2C_2H_5 \xrightarrow{NaOC_2H_5} CH_3CO_2C_2H_5 + R_2CHCO_2C_2H_5$$

Monoalkyl β-keto esters, usually difficult to cleave without ketone formation, have been found to cleave nicely in absolute alcohol with catalytic amounts of sodium ethoxide to yield the substituted ester (see Ex. a).

β-Diketones can be expected to cleave also, and this reaction has been utilized in the preparation of long-chain acids [102]:

$$\text{(cyclohexanone-dione with R)} \xrightarrow{\text{OH}^\ominus} RCH_2\overset{\overset{\text{O}}{\|}}{C}(CH_2)_3\overset{\overset{\text{O}}{\nearrow}}{C}\diagdown_{ONa}$$

Although salts are produced by treatment with sodium hydroxide, esters can be produced by means of sodium ethoxide.

Benzils undergo cleavage by sodium cyanide in alcohol:

$$C_6H_5COCOC_6H_5 \xrightarrow[C_2H_5OH]{CN^\ominus} C_6H_5\overset{\overset{\text{O}^\ominus}{|}}{\underset{\underset{\text{CN}}{|}}{C}}-COC_6H_5 \longrightarrow$$

$$C_6H_5\overset{\overset{\text{O}}{\diagup\diagdown}}{\underset{\underset{\text{CN}}{|}}{C}}-\overset{}{\underset{\underset{\text{O}^\ominus}{|}}{C}}C_6H_5 \longrightarrow C_6H_5\overset{\overset{\text{OH}}{|}}{\underset{\underset{\text{CN}}{|}}{C}}-\overset{\overset{\text{OC}_2H_5}{|}}{\underset{\underset{\text{O}^\ominus}{|}}{C}}-C_6H_5 \longrightarrow$$

$$C_6H_5\overset{\overset{\text{O}}{\nearrow}}{C}\diagdown_{OC_2H_5} + C_6H_5\overset{\overset{\text{OH}}{|}}{\underset{\underset{\text{CN}}{|}}{C}}{}^\ominus \longrightarrow C_6H_5\overset{\overset{\text{O}^\ominus}{|}}{\underset{\underset{\text{CN}}{|}}{C}}H \rightleftharpoons C_6H_5CH{=}O + CN^\ominus$$

The mechanism is more complex than illustrated, but the equation is sufficient to predict that, in unsymmetrical benzils, the more positive carbon of the two carbonyls will be attacked by cyanide to yield the aldehyde of that fragment and the ester of the other [103]. This is a degradative rather than a preparative procedure.

(a) **Preparation of ethyl 4-pentenoate.** To 1.5 g. of sodium dissolved in 81 g. of absolute ethanol, ethyl α-allylacetoacetate (122 g., 0.72 mole) was added and the mixture was slowly distilled through a 5-ft. column packed with glass helices, which procedure removed the ethanol-ethyl acetate azeotrope at such a rate that the vapor temperature was at 72–73°. After about 22 hr. the remaining alcohol was distilled and further distillation of the residue yielded the titled ester, b.p. 142–144°, 81 g. (88%) [104].

(b) **Other examples**

(1) **Preparation of ethyl benzoylacetate** (77–79% from ethyl benzoylacetoacetate, ammonia, and ammonium chloride) [105].

(2) Preparation of ethyl α-butyrylglutarate (70% from butyryl acetoacetic ester, β-iodopropionic acid ester, and sodium ethoxide) [106].

10. Cleavage of Trihaloketones (Haloform)

$$\text{ArCOCX}_3 \xrightarrow[\text{C}_2\text{H}_5\text{OH}]{\text{NaOC}_2\text{H}_5} \text{ArCOOC}_2\text{H}_5$$

Although the hydrolysis of trihaloketones in alkaline solution gives the salts of carboxylic acids (Chapter 13, B.12a, Ex. 1), in the presence of an alcohol the corresponding esters are recovered [107]. Very high yields have been obtained with polyhydroxy-ω-trichloroacetophenones, 2-trichloroacetyl coumarones, and 2-methyl-3-trichloroacetylindole by treating them with methanol or ethanol containing a trace of potassium hydroxide [108]. Strangely enough, the procedure failed with 4-hydroxy-2-methoxy- and 4-hydroxy-ω-trichloro-2,6-dimethoxyacetophenone, but was successful with every ω-trichloro-*o*-hydroxyacetophenone tested. The mechanism of this conversion appears to be similar to that given in Chapter 13, except that $OR^{\ominus}$ attacks the trihaloketone instead of $OH^{\ominus}$.

(a) Preparation of methyl 2-methylindole-3-carboxylate. A solution of 0.5 g. of 2-methyl-3-trichloroacetylindole in 17 ml. of methanol containing 1 drop of aq. 60% potassium hydroxide was refluxed for 5 min. On cooling and diluting with water, methyl 2-methylindole-3-carboxylate separated in quantitative yield. Crystallization from methanol gave m.p. 165° [108].

(b) Other examples

(1) 5-Carbomethoxy-8-methoxytetralin (80% from 5-acetyl-8-methoxytetralin, calcium hypochlorite, potassium carbonate, potassium hydroxide, and aqueous methanol) [107].

(2) Ethyl benzoate (85% from trichloroacetophenone, absolute ethanol, and sodium ethoxide) [109].

11. Cleavage of Substituted Malonic Esters

$$\text{R}_2\text{C(COOC}_2\text{H}_5)_2 \xrightarrow{\text{NaOC}_2\text{H}_5} \text{R}_2\text{CHCOOC}_2\text{H}_5$$

The synthesis of monocarboxylic acids from substituted malonic esters is covered under Acids, Chapter 13, A.9. In the present method the monoester is recovered and thus the procedure differs from that used to get the monocarboxylic acid. The diester is cleaved in three different ways: by sodium ethoxide (Ex. a), by sodium or potassium (Ex. b), or by sodium cyanide (Ex. c). Although occasionally valuable as a method of synthesis, the cleavage is more often encountered as a side reaction in the preparation of barbiturates [110]. Yields are variable, but may be as high as 81%.

(a) Preparation of ethyl diphenylacetate. Diphenylmalonic ester, 6.24 g., in a solution of 0.46 g. of sodium in 50 ml. of absolute ethanol was rerefluxed for 30 min. After removing the alcohol by distillation under reduced pressure, water was added to the residue and the mixture was extracted with ether. From the dried extract there was recovered by distillation 4.33 g. (81 %) of ethyl diphenylacetate, b.p. 186–189° (20 mm.), m.p. 60–61° [111].

(b) Preparation of ethyl di-*n*-propylacetate (61 % from di-*n*-propyl malonic ester and sodium) [112].

(c) Preparation of carboethoxycyclobutane (75 % by heating 0.5 mole of 1,1-dicarboethoxycyclobutane and 1 mole of NaCN in DMSO for 4 hr. and then removing the ester by distillation) [113].

1. C. R. Hauser and B. E. Hudson, Jr., *Org. Reactions*, **1**, 266 (1942); H. O. House, *Modern Synthetic Reactions*, W. H. Benjamin, New York, 1965, Chap. 9.
2. E. E. Royals and D. G. Turpin, *J. Am. Chem. Soc.*, **76**, 5452 (1954).
3. J. Munch-Petersen, *Acta Chem. Scand.*, **5**, 519 (1951).
4. T. M. Harris *et al.*, *J. Am. Chem. Soc.*, **91**, 517 (1969); P. L. Creger, *ibid.*, **89**, 2500 (1967).
5. E. E. Royals and A. G. Robinson, III, *J. Am. Chem. Soc.*, **78**, 4161 (1956).
6. H.-D. Becker, *J. Org. Chem.*, **29**, 3070 (1964).
7. F. A. Ramirez *et al.*, *J. Am. Chem. Soc.*, **83**, 173 (1961).
8. I. Hori, *Sci. Papers Inst. Phys. Chem. Res. (Tokyo)*, **56**, No. 2, 178 (1962); *C.A.*, **58**, 5508 (1963).
9. R. Levine and C. R. Hauser, *J. Am. Chem. Soc.*, **66**, 1768 (1944).
10. S. B. Soloway and F. B. LaForge, *J. Am. Chem. Soc.*, **69**, 2677 (1947).
11. N. Green and F. B. LaForge, *J. Am. Chem. Soc.*, **70**, 2287 (1948).
12. M. Jackman *et al.*, *J. Am. Chem. Soc.*, **70**, 2884 (1948).
13. F. W. Swamer and C. R. Hauser, *J. Am. Chem. Soc.*, **72**, 1352 (1950).
14. V. H. Wallingford and D. M. Jones, *J. Am. Chem. Soc.*, **64**, 578 (1942).
15. V. Prelog and W. Hinden, *Helv. Chim. Acta*, **27**, 1854 (1944).
16. Pl. A. Plattner *et al.*, *Helv. Chim. Acta*, **29**, 730 (1946).
17. J. Mathieu and A. Allais, *Cahiers de Synthèse Organique*, Masson et Cie, 1958, Vol. 4, p. 57.
18. M. J. Weiss and C. R. Hauser, *J. Am. Chem. Soc.*, **71**, 2023 (1949).
19. I. Shahak, *Tetrahedron Letters*, 2201 (1966).
20. W. E. Parham and L. J. Reed, *Org. Syn.*, Coll. Vol. **3**, 395 (1955).
21. E. C. Horning and A. F. Finelli, *Org. Syn.*, Coll. Vol. **4**, 461 (1963).
22. D. E. Floyd and S. E. Miller, *Org. Syn.*, Coll. Vol. **4**, 141 (1963).
23. J. M. Straley and A. C. Adams, *Org. Syn.*, Coll. Vol. **4**, 415 (1963).
24. J. P. Schaefer and J. J. Bloomfield, *Org. Reactions*, **15**, 1 (1967).
25. N. S. Leonard and R. C. Sentz, *J. Am. Chem. Soc.*, **74**, 1704 (1952).
26. R. C. Fuson, *Reactions of Organic Compounds*, John Wiley and Sons, New York, 1962, p. 484.
27. N. W. Vul'fson and V. I. Zaretskii, *J. Gen. Chem. USSR (Eng. Transl.)*, **29**, 2704 (1959).
28. F. F. Blicke *et al.*, *J. Am. Chem. Soc.*, **75**, 5418 (1953).
29. N. J. Leonard and C. W. Schimelplenig, Jr., *J. Org. Chem.*, **23**, 1708 (1958).
30. E. B. Hershberg and L. F. Fieser, *Org. Syn.*, Coll. Vol. **2**, 194 (1943).
31. J. J. Bloomfield, *Tetrahedron Letters*, 591 (1968).
32. P. S. Pinkney, *Org. Syn.*, Coll. Vol. **2**, 116 (1943).
33. A. A. Hinckley *et al.*, *Sodium Hydride Dispersions*, Ventron, Beverly, Mass., May 1966, p. 8.
34. A. T. Nielsen and W. R. Carpenter, *Org. Syn.*, **45**, 25 (1965).
35. W. S. Johnson and G. H. Daub, *Org. Reactions*, **6**, 1 (1951).
36. W. S. Johnson *et al.*, *J. Am. Chem. Soc.*, **69**, 74 (1947).
37. H. J. E. Loewenthal and R. Pappo, *J. Chem. Soc.*, 4799 (1952).
38. W. S. Johnson *et al.*, *J. Am. Chem. Soc.*, **67**, 1357, 1360, 1366 (1945); **69**, 74 (1947).
39. G. H. Daub and W. S. Johnson, *J. Am. Chem. Soc.*, **70**, 418 (1948); **72**, 501 (1950).
40. Ref. 33, p. 7.
41. W. S. Johnson and W. P. Schneider, *Org. Syn.*, Coll. Vol. **4**, 132 (1963).
42. H. Zimmer and J. Rothe, *J. Org. Chem.*, **24**, 28 (1959).
43. Th. Groger and E. Waldmann, *Monatsh.*, **89**, 370 (1958).

44. K. Sisido *et al.*, *J. Am. Chem. Soc.*, **82,** 125 (1960).
45. E. D. Bergmann *et al.*, *J. Chem. Soc.*, 2564 (1953).
46. W. Reeve and E. L. Compere, Jr., *J. Am. Chem. Soc.*, **83,** 2755 (1961).
47. J. R. Johnson, *Org. Reactions*, **1,** 210 (1942).
48. G. Jones, *Org. Reactions*, **15,** 204 (1967).
49. S. Dev, *J. Indian Chem. Soc.*, **30,** 665 (1953).
50. A. C. Cope *et al.*, *J. Am. Chem. Soc.*, **59,** 2327 (1937); **63,** 3452 (1941).
51. A. C. Cope and C. M. Hofmann, *J. Am. Chem. Soc.*, **63,** 3456 (1941).
52. F. S. Prout, *J. Org. Chem.*, **18,** 928 (1953).
53. Y. Agata and M. Tsuchida, *J. Am. Chem. Soc.*, **81,** 2092 (1959).
54. R. W. Hein *et al.*, *J. Org. Chem.*, **26,** 4874 (1961).
55. M. J. Astle and J. A. Zaslowsky, *Ind. Eng. Chem.*, **44,** 2867 (1952); P. Mastagli *et al.*, *Bull. Soc. Chim. France*, 796 (1956).
56. R. W. Hein *et al.*, *J. Org. Chem.*, **26,** 4874 (1961).
57. L. Rand *et al.*, *J. Org. Chem.*, **27,** 3505 (1962).
58. A. Galat, *J. Am. Chem. Soc.*, **68,** 376 (1946).
59. C. F. H. Allen and F. W. Spangler, *Org. Syn.*, Coll. Vol. **3,** 377 (1955).
60. E. R. Alexander and A. C. Cope, *Org. Syn.*, Coll. Vol. **3,** 385 (1955).
61. H. T. Clarke and T. F. Murray, *Org. Syn.*, Coll. Vol. **1,** 272 (1944).
62. N. E. Holden and A. Lapworth, *J. Chem. Soc.*, 2368 (1931).
63. L. L. McCoy, *J. Am. Chem. Soc.*, **80,** 6568 (1958); **84,** 2246 (1962); *J. Org. Chem.*, **25,** 2078 (1960).
64. D. E. Ames and R. E. Bowman, *Org. Reactions*, **10,** 269 (1959).
65. R. Connor and W. R. McClellan, *J. Org. Chem.*, **3,** 570 (1939).
66. K. P. Link *et al.*, *Tetrahedron Letters*, No. **13,** 23 (1960).
67. A. C. Cope *et al.*, *Org. Reactions*, **9,** 107 (1957).
68. N. Weiner, *Org. Syn.*, Coll. Vol. **2,** 279 (1943).
69. B. E. Hudson, Jr., and C. R. Hauser, *J. Am. Chem. Soc.*, **62,** 2457 (1940).
70. C. R. Hauser *et al.*, *J. Am. Chem. Soc.*, **66,** 1230 (1944).
71. Ch. Weizmann *et al.*, *J. Org. Chem.*, **15,** 918 (1950).
72. R. Adams and R. M. Kamm, *Org. Syn.*, Coll. Vol. **1,** 250 (1941).
73. C. D. Hurd *et al.*, *J. Am. Chem. Soc.*, **57,** 2033 (1935).
74. D. H. Peacock and P. Tha, *J. Chem. Soc.*, 2303 (1928).
75. H. A. Shonle *et al.*, *J. Am. Chem. Soc.*, **52,** 2440 (1930).
76. P. Karrer *et al.*, *Helv. Chim. Acta*, **13,** 1292 (1930).
77. L. F. Fieser and M. D. Gates, Jr., *J. Am. Chem. Soc.*, **62,** 2335 (1940).
78. C. R. Hauser *et al.*, *J. Am. Chem. Soc.*, **66,** 309 (1944).
79. G. B. Heisig and F. H. Stodola, *Org. Syn.*, Coll. Vol. **3,** 213 (1955).
80. J. Cason and C. F. Allen, *J. Org. Chem.*, **14,** 1036 (1949).
81. G. H. Jeffery and A. I. Vogel, *J. Chem. Soc.*, 1804 (1948).
82. T. L. Jacobs and W. H. Florsheim, *J. Am. Chem. Soc.*, **72,** 256, 261 (1950).
83. P. C. Guha and N. K. Seshadriengar, *Chem. Ber.*, **69,** 1212 (1936); B. Arbusow and O. Schapschinskaja, *ibid.*, **68,** 437 (1935).
84. A. Weizmann, *J. Org. Chem.*, **8,** 285 (1943).
85. E. F. Rosenberg *et al.*, *J. Am. Chem. Soc.*, **56,** 1339 (1934).
86. R. B. Wagner, *J. Am. Chem. Soc.*, **71,** 3214 (1949).
87. C. D. Hurd and M. L. McAuley, *J. Am. Chem. Soc.*, **70,** 1650 (1948).
88. J. A. Price and D. S. Tarbell, *Org. Syn.*, Coll. Vol. **4,** 285 (1963).
89. G. A. Reynolds and C. R. Hauser, *Org. Syn.*, Coll. Vol. **4,** 708 (1963).
90. H. E. Zaugg *et al.*, *J. Am. Chem. Soc.*, **82,** 2895 (1960); *J. Org. Chem.*, **26,** 644 (1961).
91. H. E. Zaugg *et al.*, *J. Org. Chem.*, **26,** 647 (1961).
92. R. Brettle, *Chem. Ind.* (*London*), 1700 (1965).
93. H. F. Van Woerden, *Rec. Trav. Chim.*, **82,** 920 (1963).
94. M. T. Bush and W. D. Beauchamp, *J. Am. Chem. Soc.*, **75,** 2949 (1953).
95. F. S. Prout and M. M. E. Abdulslam, *J. Chem. Eng. Data*, **11,** 616 (1966).

96. T. M. Patrick, Jr., and F. B. Erickson, *Org. Syn.*, Coll. Vol. **4,** 430 (1963).
97. J. C. Allen *et al.*, *Chem. Ind. (London)*, 830 (1961).
98. H. Lund and A. Voigt, *Org. Syn.*, Coll. Vol. **2,** 594 (1943).
99. F. C. Whitmore and D. J. Loder, *Org. Syn.*, Coll. Vol. **2,** 282 (1943).
100. L. Field *et al.*, *J. Am. Chem. Soc.*, **78,** 4389 (1956).
101. S.-O. Lawesson, *Acta Chem. Scand.*, **12,** 1 (1958).
102. H. Stetter, in Foerst's, *Newer Methods of Preparative Organic Chemistry*, Vol. **2,** Academic Press, New York, 1963, p. 51.
103. H. Kwart and M. M. Baevsky, *J. Am. Chem. Soc.*, **80,** 580 (1958).
104. J. J. Ritter and T. J. Kaniecki, *J. Org. Chem.*, **27,** 622 (1962).
105. R. L. Shriner *et al.*, *Org. Syn.*, Coll. Vol. **2,** 266 (1943).
106. A. Franke and A. Kroupa, *Monatsh. Chem.*, **69,** 167 (1936).
107. R. T. Arnold *et al.*, *J. Am. Chem. Soc.*, **66,** 208 (1944).
108. W. B. Whalley, *J. Chem. Soc.*, 3229 (1951).
109. J. Houben and W. Fischer, *Chem. Ber.*, **64,** 240 (1931).
110. D. E. Pearson *et al.*, *J. Med. Chem.*, **10,** 1078 (1967).
111. A. C. Cope and S. M. McElvain, *J. Am. Chem. Soc.*, **54,** 4319 (1932).
112. F. Krollpfeiffer and A. Rosenberg, *Chem. Ber.*, **69B,** 465 (1936).
113. A. P. Krapcho *et al.*, *Tetrahedron Letters*, 215 (1967).

D. Oxidative- and Reductive-Type Syntheses

Only oxidations or reductions of a nature peculiar to the formation of esters are considered in this section. Obviously the discussion would be quite extensive if the reduction of unsaturated or keto esters was included. Actually the latter reductions are quite straightforward and are conducted largely in a manner similar to that used for converting ethylenes or ketones into alkanes. In dealing with syntheses in which the ester group is actually produced, five involving oxidation and three involving reduction are included.

1. Intermolecular Oxidation-Reduction of Aldehydes (Tishchenko)

$$2\ RCHO \xrightarrow{Al(OC_2H_5)_3} RCOOCH_2R$$

This synthesis is similar to the Cannizzaro reaction except in this case the ester is produced rather than the corresponding alcohol and acid. It also differs from the Cannizzaro reaction in being applicable to aldehydes containing α-hydrogen atoms. The most common catalysts are the aluminum alkoxides. However, alkoxides of metals such as sodium [1], magnesium, and calcium [2] have also been employed. With mildly basic catalysts such as magnesium ethoxide, calcium ethoxide, magnesium-aluminum ethoxide $(Mg(Al(OC_2H_5)_4)_2)$ and sodium-magnesium ethoxide $(Na_2Mg(OC_2H_5)_4)$, the aldol is first formed from saturated aldehydes containing two α-hydrogen atoms, and this in turn gives the ester of the 1,3-diol [3]:

$$2\ RCH_2CHO \longrightarrow \underset{\overset{|}{R}}{RCH_2CHOHCHCHO} \longrightarrow \underset{\overset{|}{R}}{RCH_2CHOHCHCH_2OCOCH_2R}$$

Strangely enough, the highly basic sodium ethoxide gives only an aldol condensation with such aldehydes. In mixed condensations the ester from the acid

References for Section D are on pp. 857–858.

of the aldehyde which undergoes the Tishchenko reaction most rapidly is formed in the larger amounts [4]. The yields of esters for the simpler aliphatic aldehydes are variable, but may reach quantitative values [2, 3, 5].

The first step in the mechanism [6] appears to be the formation of the co-ordination compound 14-XXI, which reacts with the oxygen of the carbonyl

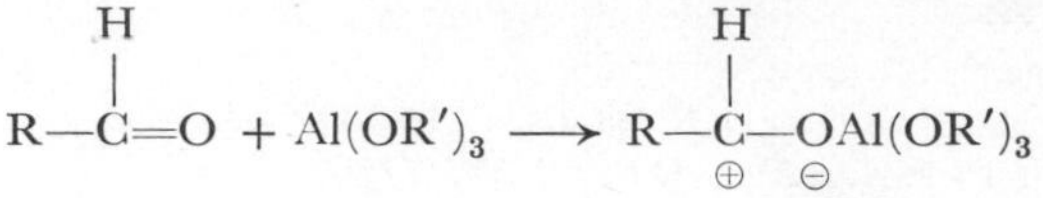

$$
\underset{}{\overset{\overset{\displaystyle H}{|}}{R-C=O}} + Al(OR')_3 \longrightarrow \underset{\underset{14\text{-}XXI}{\overset{\oplus\ \ominus}{}}}{\overset{\overset{\displaystyle H}{|}}{R-C-OAl(OR')_3}}
$$

group of another aldehyde molecule to give 14-XXII. The loss of the aluminum

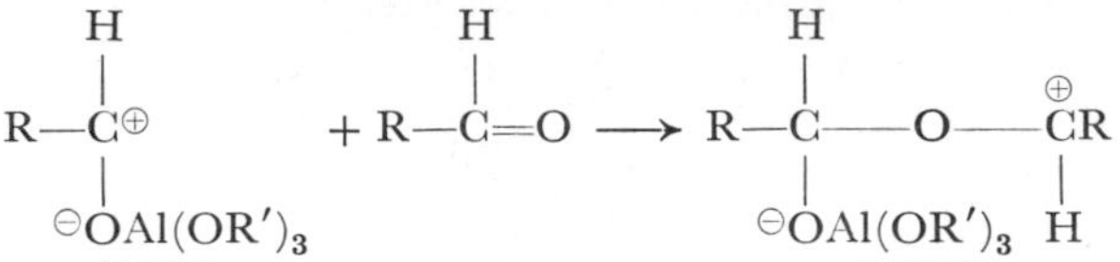

$$
\underset{\underset{14\text{-}XXI}{\ominus OAl(OR')_3}}{\overset{\overset{\displaystyle H}{|}}{R-C^{\oplus}}} + \underset{}{\overset{\overset{\displaystyle H}{|}}{R-C=O}} \longrightarrow \underset{\underset{14\text{-}XXII}{\ominus OAl(OR')_3\ \ H}}{\overset{\overset{\displaystyle H}{|}}{R-C-O-\overset{\oplus}{C}R}}
$$

alkoxide and the transfer of a hydride ion leads to the ester 14-XXIII. Aluminum

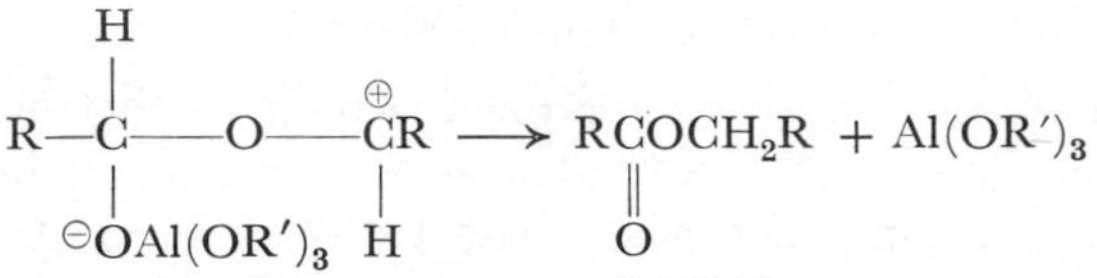

$$
\underset{\underset{14\text{-}XXII}{\ominus OAl(OR')_3\ \ H}}{\overset{\overset{\displaystyle H}{|}}{R-C-O-\overset{\oplus}{C}R}} \longrightarrow \underset{14\text{-}XXIII}{\overset{\overset{\displaystyle}{}}{RCOCH_2R}} + Al(OR')_3
$$

alkoxides are quite associated and do not complex well with aldehydes substituted with electron-withdrawing groups. For example, chloral gives at most 2% of the Tishchenko product with aluminum ethoxide. The solution to this problem is to use an alkoxide of a more acidic alcohol where less association occurs as is shown [7]:

$$
CCl_3CH=O + Al(OCH_2CCl_3)_3 \xrightarrow[\substack{1\ hr.,\\30°}]{C_6H_6} \overset{\overset{\displaystyle O}{\|}}{CCl_3COCH_2CCl_3}
$$

0.05 mole   0.0015 mole   2,2,2-Trichloroethyl
                                                        trichloroacetate, 99%

In a mechanism unrelated to that of the Tishchenko reaction, but nevertheless involving a redox series of transformations, benzil has been converted into trans-α,α'-stibenediol dibenzoate [8]:

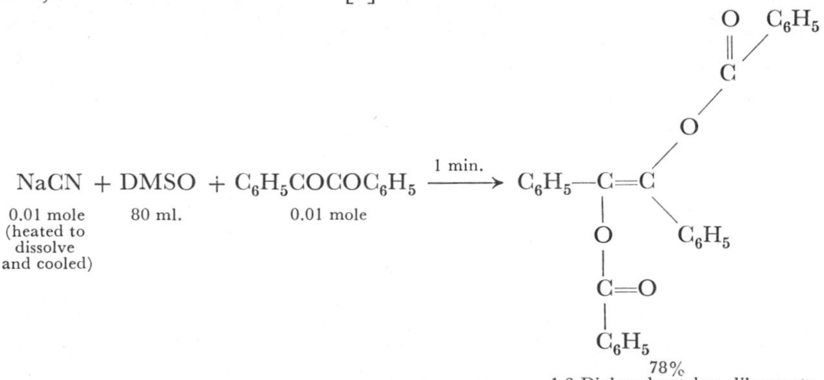

$$
NaCN + DMSO + C_6H_5COCOC_6H_5 \xrightarrow{1\ min.} C_6H_5-C=C
$$

0.01 mole   80 ml.   0.01 mole
(heated to
dissolve
and cooled)

78%
1,2-Diphenylacetylene dibenzoate

(a) Preparation of benzyl benzoate (90–93 % from benzaldehyde and sodium benzoxide) [1].

(b) Other examples

(1) Butyl butyrate (81.6 % from butyraldehyde and aluminum ethoxide) [2].

(2) Benzyl acetate (56.8 % from benzaldehyde and acetaldehyde with aluminum isopropoxide) [6].

2. Oxidation of Primary Alcohols Followed by Esterification

$$RCH_2OH \xrightarrow[H_2SO_4]{Na_2Cr_2O_7} RCOOCH_2R$$

Primary alcohols may be oxidized to carboxylic acids and, if an excess of the alcohol is present, esters may be produced in the same operation. Oxidizing agents employed have been sodium dichromate and sulfuric acid [9], N-chlorosuccinimide, and t-butyl hypochlorite [10]. Of the last two oxidizing agents, the t-butyl hypochlorite is the more satisfactory. This reagent gives yields as high as 89 % with some primary alcohols although for others, such as benzyl alcohol, the aldehyde is the main product.

Oxidation of carbonyl compounds to esters (Baeyer-Villiger) has been discussed in B.4. Caro's acid, H_2SO_5, has the unique property of oxidizing a mixture of an alcohol and an aldehyde to an ester [11]:

$$RCH{=}O + CH_3OH \longrightarrow \underset{OCH_3}{R\overset{H}{\underset{|}{C}}OH} \xrightarrow{H_2SO_5} \underset{OCH_3}{R\overset{H}{\underset{|}{C}}OOSO_2OH} \longrightarrow R\overset{O}{\overset{\|}{C}}OCH_3 + H_2SO_4$$

(a) Preparation of *n*-butyl *n*-butyrate (41–47 % from technical *n*-butyl alcohol, sodium dichromate, and commercial sulfuric acid) [9].

(b) Preparation of *n*-hexyl caproate (83 % from *n*-hexyl alcohol and t-butyl hypochlorite in pyridine-carbon tetrachloride) [10].

3. Oxidation of Ethers

$$RCH_2OCH_2R \rightarrow RCOOCH_2R$$

No doubt, many reagents can effect this seldom used conversion, but the most recent one suggested is trichloroisocyanuric acid [12]:

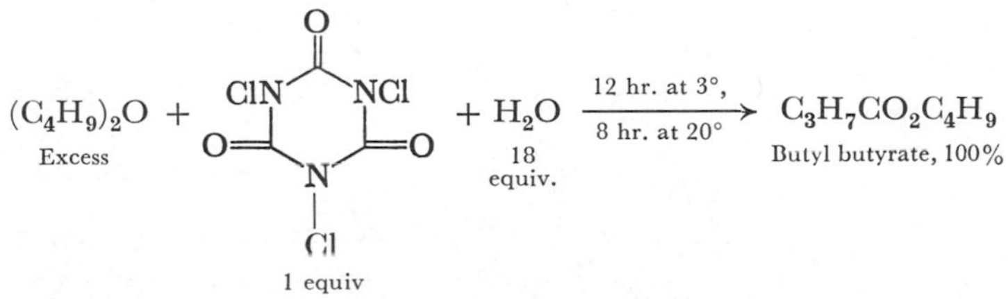

$$(C_4H_9)_2O + \text{[trichloroisocyanuric acid]} + H_2O \xrightarrow[\text{8 hr. at 20°}]{\text{12 hr. at 3°,}} C_3H_7CO_2C_4H_9$$

Excess   18 equiv.   Butyl butyrate, 100%

1 equiv

Benzyl ethers yield benzaldehyde rather than benzoate esters.

4. Oxidation of Methyl Arenes in Acid Anhydrides

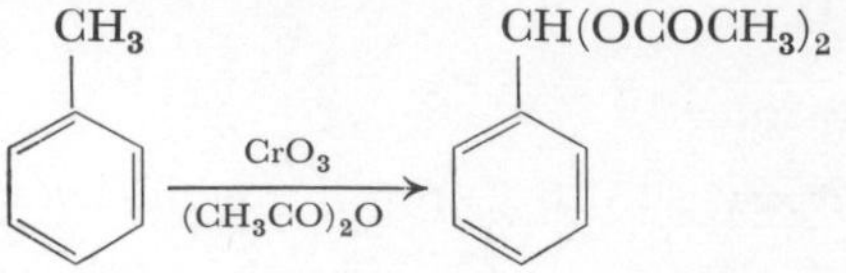

This synthesis has already been discussed under Aldehydes, Chapter 10, A.11, since the diacetates may be hydrolyzed to produce aldehydes. Although different oxidizing agents may be employed, chromium trioxide is the most common. Yields are usually low.

(a) Preparation of o- and p-nitrobenzaldiacetates (65–66 % p- and 36–37 % o-, from the nitrotoluene, chromium trioxide, acetic anhydride, and sulfuric acid) [13].

5. Electrolysis of Acid Ester Salts (Kolbe)

$$2\ C_2H_5OOC(CH_2)_nCOOK \xrightarrow{\text{Electrolysis}} C_2H_5OOC(CH_2)_{2n}COOC_2H_5$$

(See Alkanes, Cycloalkanes, and Arenes, Chapter 1, G.7.)

6. Acyloxylation

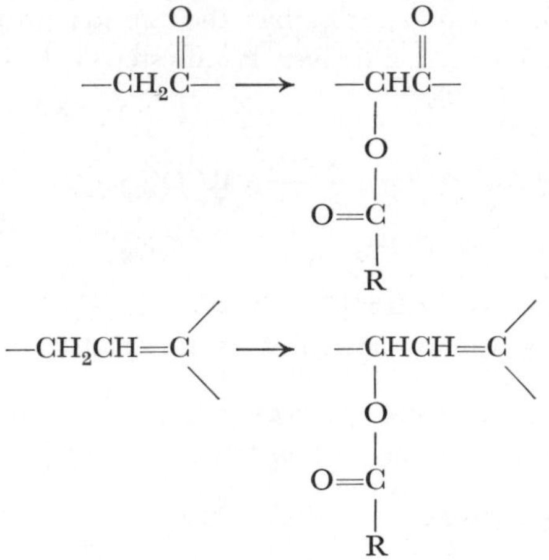

Two methods are available for the acyloxylation of ketones. The first employs lead tetraacetate in acetic acid alone (see Ex. a) or catalyzed by boron trifluoride [14] to introduce the α-acetoxy group. The second brings the activity of thallium triacetate to bear on the enamine of the ketone [15]:

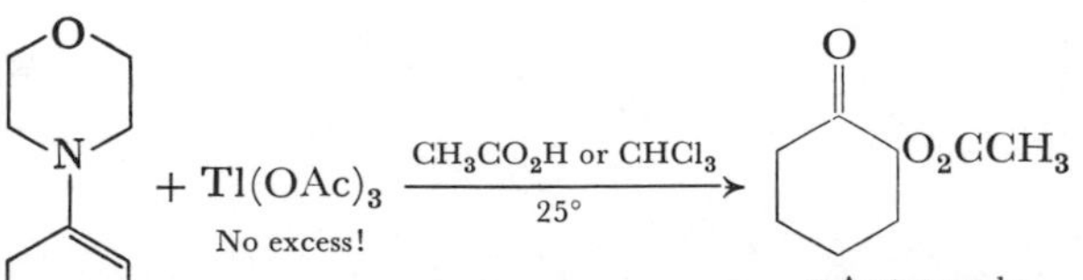

α-Acetoxycyclo-
hexanone, 73%

The reaction is of greater interest because a mixture of cyclohexanone and morpholine may be used in place of the enamine. But yields are variable and usually lower with other enamines.

The best method for acyloxylation of alkenes is by means of t-butyl hydroperoxide and a copper salt of an organic acid in the proportions given [16]:

$$CH_2{=}CHCH_2R + \underset{\substack{\text{1 mole}}}{HOOC(CH_3)_3} + \underset{\substack{\text{0.1 mole}}}{Cu(O\overset{\displaystyle O}{\overset{\displaystyle \|}{C}}R)_2} \xrightarrow[\substack{\text{Or slight warming}\\\text{until color changes}}]{25°} CH_2{=}CH\overset{\displaystyle R}{\overset{\displaystyle |}{C}}HO\overset{\displaystyle O}{\overset{\displaystyle \|}{C}}R$$

10 moles

If an optically active acid is used as the copper salt, a partially optically active ester is formed. Yields based on the copper salt are good.

(a) Preparation of Δ⁵-cholestene-4α-ol-3-one acetate. Δ^5-Cholestene-3-one, 50 g., in 300 ml. of benzene at 32° was diluted with 600–950 ml.

of acetic acid, and 1.3–1.5 moles of lead tetraacetate was added. After standing overnight, water was added and the mixture was extracted with ether. From the ethereal layer there was recovered a 45–55 % yield of the acetate [17].

7. Reductive Acylation of Carbonyl Compounds

$$\underset{\displaystyle -\overset{\displaystyle O}{\overset{\displaystyle \|}{C}}-}{} + RCOCl \text{ (or anhydride)} \xrightarrow{[H]} -\underset{\displaystyle |}{\overset{\displaystyle |}{\underset{\displaystyle H}{C}}}{-}\overset{\displaystyle O}{\overset{\displaystyle \|}{OCR}}$$

Reductive acylation has been accomplished recently with triphenyltin hydride and acetyl chloride [18]:

$$\underset{\substack{\text{0.4 mmole}}}{CH_3COCl} + \underset{\substack{\text{0.41 mmole}}}{C_6H_5COCH_3} + \underset{\substack{\text{0.53 mmole}}}{(C_6H_5)_3SnH} \xrightarrow[\text{0.5 hr. at 25°}]{C_6H_6}$$

$$C_6H_5\overset{\displaystyle CH_3}{\overset{\displaystyle |}{\underset{\displaystyle |}{\underset{\displaystyle O}{\underset{\displaystyle |}{\overset{\displaystyle |}{\underset{}{}}}}}}CHCH_3$$

α-Phenylethyl acetate, 100%

1. O. Kamm and W. F. Kamm, *Org. Syn.*, Coll. Vol. **1**, 104 (1941).
2. F. J. Villani and F. F. Nord, *J. Am. Chem. Soc.*, **69**, 2605 (1947).
3. M. S. Kulpinski and F. F. Nord, *J. Org. Chem.*, **8**, 256 (1943).
4. I. Lin and A. R. Day, *J. Am. Chem. Soc.*, **74**, 5133 (1952).
5. W. C. Child and H. Adkins, *J. Am. Chem. Soc.*, **45**, 3013 (1923); **47**, 798 (1925).
6. I. Lin and A. R. Day, *J. Am. Chem. Soc.*, **74**, 5133 (1952).
7. T. Saegusa and T. Ueshima, *J. Org. Chem.*, **33**, 3310 (1968).
8. J. C. Trisler and J. L. Frye, *J. Org. Chem.*, **30**, 306 (1965).

9. G. R. Robertson, *Org. Syn.*, Coll. Vol. **1**, 138 (1941).

10. C. A. Grob and H. J. Schmid, *Helv. Chim. Acta*, **36**, 1763 (1953).

11. A. Nishihara and I. Kubota, *J. Org. Chem.*, **33**, 2525 (1968).

12. E. C. Juenge and D. A. Beal, *Tetrahedron Letters*, 5819 (1968).

13. T. Nishimura, *Org. Syn.*, Coll. Vol. **4**, 713 (1963).

14. H. B. Henbest *et al.*, *J. Chem. Soc.*, 4472 (1961); 6 (1965).

15. M. E. Kuehne and T. J. Giacobbe, *J. Org. Chem.*, **33**, 3359 (1968).

16. D. B. Denney *et al.*, *J. Org. Chem.*, **30**, 3151 (1965) and references quoted.

17. L. F. Fieser and R. Stevenson, *J. Am. Chem. Soc.*, **76**, 1728 (1954).

18. L. Kaplan, *J. Am. Chem. Soc.*, **88**, 4970 (1966).

Chapter 15

ACYL HALIDES

The acid halides are the most reactive of the acid derivatives, acetyl chloride, for example, hydrolyzing rapidly and exothermally in water. Less water-soluble acid halides can be handled in a moist atmosphere over short periods of time or even washed with water, but they must be protected from moisture on storage in order to prevent hydrolysis from taking place.

Being quite reactive, the acid halides must be made from active reagents. The most common are inorganic acid halides, but less reactive reagents can be used if the acid halide is removed as it is formed. For example, acetyl chloride can be distilled almost quantitatively from a mixture of benzoyl chloride and acetic acid. This is the method of choice for low-boiling acid halides (A.2). The method of second choice, but one of wider applicability, is the use of a mixture of thionyl chloride and dimethylformamide, which forms as an intermediate, dimethylformamidinium chloride, a most reactive reagent (A.1).

Other acid derivatives can be converted into the corresponding acid halide: anhydrides (A.4), esters or salts (A.3). Amides, however, are not suitable as they are converted into nitriles or imidochlorides. But a specific type of amide, N-acylimidazole, is most useful for conversion under relatively mild conditions with dry hydrogen chloride (see A.6).

Compounds other than acid derivatives which can be converted into acid halides are small in number because of the activity of the latter and their high oxidative state. No reductive procedures are available obviously, and only a few oxidative ones are to be found (B). But some of the latter are interesting, if for no other reason, because oxidative procedures for the synthesis of acid halides are used sparingly.

A. Metathesis

1. From Carboxylic Acids and Inorganic Halides

$$RCOOH \xrightarrow{PX_3} RCOX$$

The most common method for the synthesis of acyl halides is from the corresponding acid. Usually phosphorus trichloride or pentachloride or thionyl chloride is the inorganic reagent used to prepare the acyl chlorides. The phosphorus halides have the advantage of high activity but the disadvantage of forming volatile phosphorus oxychlorides which may contaminate the distilled acyl chlorides. Thionyl chloride, on the other hand, forms gaseous by-products:

$$RCOOH + SOCl_2—RCOCl + HCl + SO_2$$

and leaves only the problem of separating excess thionyl chloride (b.p. 79°) from the acid chloride. To prepare acyl bromides, phosphorus tribromide is the usual reagent [1]; see Ex. e.4 also.

Several variations of procedures have been recommended without clear-cut proof that they are advantageous. The mixed reagent, phosphorus pentachloride and acetyl chloride, has been used to convert acids other than acetic into their acyl chlorides [2]; or thionyl chloride has been employed with iodine [3] or with a trace of pyridine [4]. Zinc chloride is claimed to be a useful catalyst [5]. Triphenylphosphine in carbon tetrachloride gave "good" yields of acid chlorides and no acid is produced as a by-product [6]. On the other hand, thionyl chloride itself, without any so-called catalyst added, showed no latent period and converted the following into their respective acyl chlorides at 20° in the times indicated: acetic acid, 1 hour; isobutyric acid, 5 hours; benzoic acid, 50 hours; diphenylacetic acid, 38% in 5 days; trichloroacetic acid, 0% in 9 days [7].

References for Section A are on pp. 867–868.

Prolonged heating of an acid containing α-hydrogen atoms in excess thionyl chloride with pyridine as a catalyst may lead to serious side reactions [8]:

$$C_6H_5CH_2CH_2CO_2H \xrightarrow[\text{14 hr.}]{SOCl_2,\ C_5H_5N} C_6H_5CH_2\overset{\overset{\displaystyle Cl}{|}}{\underset{\underset{\displaystyle SCl}{|}}{C}}COCl$$

α-Chloro-α-chlorothio-
β-phenylpropionyl
chloride, 61%

The mechanism involves the formation of the mixed anhydride:

$$RCOOH + SOCl_2 \longrightarrow R\overset{\overset{\displaystyle O}{||}}{C}\ \overset{\overset{\displaystyle O}{|}}{\underset{\underset{\displaystyle Cl}{|}}{S}}O + HCl$$

which decomposes by an internal electron shift (i) or by SN_2 displacement (ii).

(i) $R\overset{O}{C}\curvearrowright SO \longrightarrow RCOCl + SO_2 \xleftarrow{-Cl^{\ominus}} Cl\overset{O}{S}O\overset{O}{C}R$ (ii)
         $\underset{Cl}{\ }$ $Cl^{\ominus}$

The role of the pyridine catalyst seems pointless to discuss since the much superior reagent, dimethylformamidinium chloride, undoubtedly will replace the thionyl chloride-pyridine method. It is formed from dimethylformamide and thionyl chloride:

$$HCON(CH_3)_2 + SOCl_2 \longrightarrow HC\overset{\overset{\displaystyle \oplus}{N(CH_3)_2}}{\underset{\underset{\displaystyle Cl}{\ }}{\diagup}}\ \overset{\ominus}{Cl}$$

It is important to use thionyl chloride free from sulfur monochloride and dichloride. Their presence may cause considerable darkening of the reaction mixture. Some commercial grades are satisfactory but, if not, they may be purified by the method of Friedman and Wetter using distillation from triphenyl phosphite [9]. The reagent can be used as such after isolation (m.p. 138–142°), or formed *in situ* simply by using dimethylformamide as the solvent for thionyl chloride, or used in amounts of about 10% of an equivalent of dimethylformamide to thionyl chloride [10]. Sulfonic acids or trichloroacetic

$$\overset{\overset{\displaystyle \oplus}{N(CH_3)_2}}{\underset{\underset{\displaystyle Cl\ Cl^{\ominus}}{\ }}{CH}} \leftrightarrow \overset{\overset{\displaystyle N(CH_3)_2}{\ }}{\underset{\underset{\displaystyle Cl^{\ominus}}{\ }}{\overset{\oplus}{H}CCl}} \xrightarrow{RCOOH} \left[\overset{\overset{\displaystyle N(CH_3)_2}{|}}{\underset{\underset{\underset{\underset{\displaystyle R}{|}}{\underset{\displaystyle C=O}{|}}}{\underset{\displaystyle HO^{\oplus}}{|}}}{\overset{\ominus}{Cl}HCCl}} \right] \longrightarrow RCCl + H\overset{O}{C}N(CH_3)_2 + HCl$$

acid which are not converted into acid chlorides by other reagents succumb to the effectiveness of this reagent and give high yields of acid chlorides. The effectiveness of the reagent stems from its very superior acylative ability and the driving force of forming the carbon-oxygen double bond.

Recently acyl bromides have been prepared in yields of 50–80 % by treatment of the carboxylic acid with triphenylphosphine dibromide (see Ex. e.4):

$$RCOOH + (C_6H_5)_3PBr_2 \rightarrow RCOBr + (C_6H_5)_3PO + HBr$$

Diacyl halides may be prepared by the use of phosphorus pentachloride [11] or thionyl chloride [12]. Double bonds or substituents such as the halogen, alkoxy, phenoxy, nitro, or cyano group remain unchanged if present in the carboxylic acid undergoing conversion.

Caesium chloride appears to be an efficient catalyst for acid chloride preparations through the use of thionyl chloride [13].

(a) Preparation of trifluoroacetyl chloride (89 % from 0.4 mole of acid, 0.04 mole of dimethylformamide, and 0.44 mole of thionyl chloride at 80–85° for $2\frac{1}{2}$ hr; without dimethylformamide, no acid chloride forms) [10].

(b) Preparation of benzenesulfonyl chloride (6.2 g. of dimethyl-formamidinium chloride and 5 g. of benzenesulfonic acid were dissolved in 120 ml. of dimethylformamide and held for 24 hr. at 20°; the mixture was poured into ice water and extracted with benzene; the benzene extract was dried, evaporated, and distilled at 11 torr., b.p. 115°, to give 75 % of the acid chloride [10]; see A.3, Ex. a, for preparation from sodium salt of a sulfonic acid).

(c) Preparation of 2,3-0-isopropylideneglyceric acid chloride. A "good" yield is claimed according to the equation:

$$(C_6H_5)_3P + CCl_4 \longrightarrow (C_6H_5)_3\overset{\oplus}{P}CCl_3 \xrightarrow{RCOOH}$$
$$(C_6H_5)_3\overset{\oplus}{P}O_2CR + CHCl_3 \xrightarrow{Cl^{\ominus}} RCOCl + (C_6H_5)_3PO$$

The advantage of this reaction is that no acid is generated, and acid-reactive groups such as the ketal can survive the conversion [6].

(d) Preparation of mesitoyl chloride (90–97 % from mesitoic acid and thionyl chloride) [14].

(e) Other examples

(1) p-Nitrobenzoyl chloride (90–96 % from p-nitrobenzoic acid and phosphorus pentachloride) [15]; *see also* [3]. It should also be noted that phosphorus pentachloride in methylene chloride is monomeric [16], a fact which suggests that methylene chloride may be an effective solvent for some acyl chloride preparations with phosphorus pentachloride.

(2) Oleoyl chloride (97–99 % crude from oleic acid by a continuous process in which the acid is added to a column containing glass helices through which thionyl chloride is distilled) [17].

(3) **DL-Valyl chloride hydrochloride** (86% from DL-valine, phosphorus pentachloride and acetyl chloride) [2].

(4) **Benzoyl bromide** (72% from 0.1 mole of triphenylphosphine in chlorobenzene to which 0.1 mole of bromine was added at 0–5°, followed by the addition of 0.1 mole of benzoic acid and refluxing for 30 min.; anhydrides may also be used) [18].

2. From Carboxylic Acids and Acyl Halides or Aliphatic α,α-Dihaloethers

$$RCOOH \xrightarrow[\substack{\text{or} \\ X_2CHOCH_3}]{ArCOX} RCOX$$

Two organic reagents, the acyl halide and the aliphatic α,α-dihaloether, have been employed rather widely in converting carboxylic acids into acyl halides. Benzoyl chloride [19] and fluoride [20] for example serve well in the synthesis of the volatile acyl halides, while benzenesulfonyl chloride in hydrocarbon solvents serves in a similar manner since the by-product, benzenesulfonic acid, is insoluble [21]. Phthaloyl chloride has also been employed [22, 23], as has been the oxalyl chloride and bromide [24, 25]. It is claimed that unsaturated acyl chlorides are prepared best by the use of oxalyl chloride [26]. The use of this reagent in benzene at low temperature prevents racemization [27] and rearrangement [28]. The yields using acyl halides are variable, but often quite satisfactory.

More recent reagents are α,α-dichlorodimethyl- or α,α-dichlorodiethyl ether [29–32] which sometimes give superior yields of the acyl chloride. In some cases, catalysts such as zinc chloride are helpful [31]. In a somewhat similar manner, α,α-difluorotrimethylamine trihydrofluoride gives a 90% yield of benzoyl fluoride from benzoic acid [33], and the like reagent, 2-chloro-1,1,2-trifluorotriethylamine, prepared readily from the tetrahaloethylene and diethylamine, gives creditable yields of the acid fluoride by the reaction (see Ex. b.5):

$$RCOOH + \underset{\underset{N(C_2H_5)_2}{|}}{F_2CCHFCl} \longrightarrow RCOF + (C_2H_5)_2N\overset{\overset{\displaystyle O}{\|}}{C}CHFCl + HF$$

(a) **Preparation of butyryl chloride.** Phthaloyl chloride, 340 g., heated to about 140°, was treated slowly with 132 g. of *n*-butyric acid; as the reaction proceeded, butyryl chloride began to distill over, the last portion of which was recovered under vacuum. Redistillation gave 146.5 g. (91.5%) of the acyl chloride [22].

(b) **Other examples**

(1) **Oleoyl chloride** (90% or better from oleic acid and oxalyl chloride) [25]; care should be taken with the reagent, oxalyl chloride, as it dissociates photochemically to free radicals capable of attacking hydrocarbon structures and with the reagent, oxalyl bromide, which dissociates thermally; see C.1.

(2) **Pentaacetylgluconic acid chloride** (about 100% from penta-acetylgluconic acid hydrate and α,α-dichlorodimethyl ether) [32].

(3) **Phthalylglycyl chloride** (96% from phthalylglycine and α,α-dichlorodiethyl ether) [30].

(4) **N-Benzylaspartic acid monochloride,** $HOOCCH_2CH(COCl)$-$NHCH_2C_6H_5$ (84% from N-benzyl aspartic acid, oxalyl chloride in benzene and a drop of pyridine) [34].

(5) **Acetyl fluoride.** Acetic acid, 12.0 g. (0.2 mole), was added drop-wise to 37.8 g. (0.2 mole) of 2-chloro-1,1,2-trifluorotriethylamine. The gases liberated were passed through calcined KF and trapped in a receiver at $-70°$. Distillation gave 7.3 g. (59%) of the fluoride [35].

3. From Esters or Salts

$$RCOOC_2H_5(Na) \xrightarrow{SOCl_2} RCOCl$$

At times it is desirable to prepare the acyl halides from the ester or salt rather than from the free carboxylic acid. Or one may start with the ester and proceed to the acyl halide via the salt [36]. The reagent of choice is thionyl chloride and dimethyl formamide (see Ex. a). Other reagents (thionyl chloride, phosphorus pentachloride, phosphorus oxychloride, and α,α-dichloromethyl ether) and the catalyst are similar to those employed when the carboxylic acid is the starting material. This procedure has been used widely in synthesizing fluoroacyl chlorides, presumably because of the availability and nonvolatility of the salt, the starting material [37, 38]. It also serves well in the preparation of ester-acyl halides, such as ethoxalyl chloride which has been obtained from the diethyl ester [39] and from the potassium ethyl ester [40]. It is interesting to note that β-propiolactone, an inner ester, gives an 87% yield of β-chloropropionyl chloride with thionyl chloride, while with acetyl chloride in the presence of a trace of sulfuric acid a 67% yield of β-acetoxypropionyl chloride is obtained [41]. The yields in these conversions are variable, but are sometimes in the 80–90% range.

(a) **β-Naphthalenesulfonyl chloride** (practically quantitative from 0.2 mole of the sodium salt of the sulfonic acid, and 0.22 mole of thionyl chloride in 75 ml. of dimethylformamide at 10–12°; the mixture was poured into ice water, and crystalline sulfonyl chloride filtered and washed; also a practically quantitative yield was obtained from the sodium salt (0.2 mole) and 50 ml. of thionyl chloride and 0.02 mole of dimethylformamide at reflux temperature; no sulfonyl chloride is formed without dimethylformamide; see discussion in A.1.) [42].

(b) **Preparation of N-trifluoroacetyl-L-glutamic acid 5-chloride 1-ethyl ester.** The dicyclohexylammonium 1-ethyl N-trifluoroacetyl-L-glutamate, 1.375 g., was heated for 45 min. with 3 ml. of thionyl chloride in

$$C_2H_5OOCCHCH_2CH_2COONH_2(C_6H_{11})_2 \longrightarrow C_2H_5OOCCHCH_2CH_2COCl$$
$$\qquad | \qquad\qquad\qquad\qquad\qquad\qquad\qquad\qquad\qquad\qquad |$$
$$\quad NHCOCF_3 \qquad\qquad\qquad\qquad\qquad\qquad\qquad\qquad NHCOCF_3$$

30 ml. of benzene. On cooling to 0° and standing at this temperature for 1 hr., the mixture was filtered to separate the dicyclohexyl ammonium salt. The filtrate was evaporated to dryness under reduced pressure and the residue was dissolved twice in benzene and evaporated under reduced pressure to give 0.865 g. (99%) of the crude acyl chloride. Purification by sublimation in high vacuum yielded 0.760 g. (87%), m.p. 88–89° [43].

(c) Other examples

(1) **Trifluoroacetyl chloride** (90% from sodium trifluoroacetate and phosphorus oxychloride [38].

(2) **Methoxalyl chloride** (65% from potassium methyl oxalate and thionyl chloride) [40].

(3) **Benzoyl chloride** (86% from sodium benzoate, α,α-dichloromethyl ether, and a trace of zinc chloride) [31].

(4) **γ-Chlorobutyryl chloride** (83% from 258 parts of γ-butyrolactone, 393 parts of thionyl chloride, and 10 parts of zinc chloride at 50–60° for several hr.) [44].

4. From Acid Anhydrides

$$\begin{matrix} RCO \\ \diagdown \\ O \xrightarrow{\ SOCl_2\ } 2\,RCOCl + SO_2 \\ \diagup \\ RCO \end{matrix}$$

This synthesis possesses limited value because many acid anhydrides are obtained from acyl halides. However, it is useful, particularly when applied to cyclic anhydrides which are ordinarily not obtained from acyl halides. To prepare the acyl chlorides, thionyl chloride, phthaloyl chloride, phosphorus pentachloride, or α,α-dichloromethyl ether is the usual reagent. The presence of an inorganic halide such as zinc chloride acts as a catalyst in the reaction [22, 5, 31]. Acyl fluorides may be obtained by treating the acid anhydride with hydrogen fluoride [45] or with potassium acid fluoride or potassium fluoride [46, 47]. Yields are generally good.

(a) **Preparation of symmetrical o-phthaloyl chloride** (92% from phthalic anhydride and phosphorus pentachloride) [48].

(b) Other examples

(1) **Fumaryl chloride** (82–95% from maleic anhydride, phthaloyl chloride, and a trace of anhydrous zinc chloride) [49].

(2) **Succinyl chloride** (95.5% from succinic anhydride, α,α-dichlorodimethyl ether, and a trace of zinc chloride) [31].

(3) **Propionyl fluoride** (91.5% from propionic anhydride and anhydrous hydrogen fluoride at −10°) [45].

(4) **1,8-Naphthalenedicarboxylic dichloride** (80% from the corresponding anhydride, phosphorus pentachloride at 160° and then refluxing for 5 hr.) [50].

(5) **Benzoyl bromide** (71 % from the anhydride by the procedure given in A.1, Ex. e.4).

5. From Acyl Halides (Halide Exchange)

$$RCOCl \xrightarrow{\text{HX}} RCOX$$

This synthesis has been employed in some cases in which other methods failed. The free acid or the salt in the presence of acetic acid [46], acetic anhydride [51], or tetramethylene sulfone (see Ex. b.4) has been used as the reagent. Acyl bromides, iodides, and fluorides have been synthesized by this method, although in recent years it appears to have been used largely for the latter. At temperatures between -5 and $0°$ at atmospheric pressure, anhydrous hydrogen fluoride in slight excess gives yields of 79–94 % from a series of aliphatic and aromatic acyl chlorides [45].

(a) **Preparation of benzoyl fluoride** (75–80 % from benzoyl chloride to which anhydrous, liquefied hydrogen fluoride was added in a polyolefin bottle; general method for many acid fluoride preparations) [52].

(b) **Other examples**

(1) **Acetyl fluoride** (76 % from acetyl chloride, potassium fluoride, and acetic acid) [46].

(2) **Oxalyl bromide** (85 % from oxalyl chloride and hydrogen bromide [53]; this acyl bromide could not be prepared from oxalic acid and phosphorus pentabromide).

(3) **Benzoyl iodide** (rather poor yield from benzoyl chloride and sodium iodide warmed to 50°, where spontaneous reaction begins, followed by heating at 60–70° for 4 hr.; the product was extracted with carbon tetrachloride and distilled, b.p. 115–118° (15 mm.). The brown color of the distillate is discharged with a drop of mercury. Stearoyl iodide prepared similarly decomposes on storage even in the dark) [54].

(4) **Cyclopropylcarbonyl fluoride** [55]:

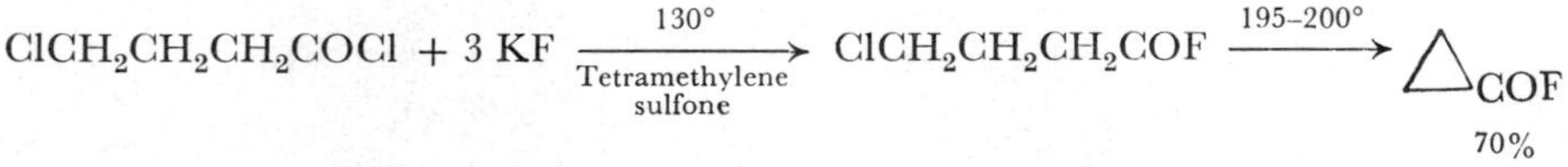

$$ClCH_2CH_2CH_2COCl + 3\ KF \xrightarrow[\text{Tetramethylene sulfone}]{130°} ClCH_2CH_2CH_2COF \xrightarrow{195–200°} \triangle COF$$

70%

6. From N-Acylimidazoles

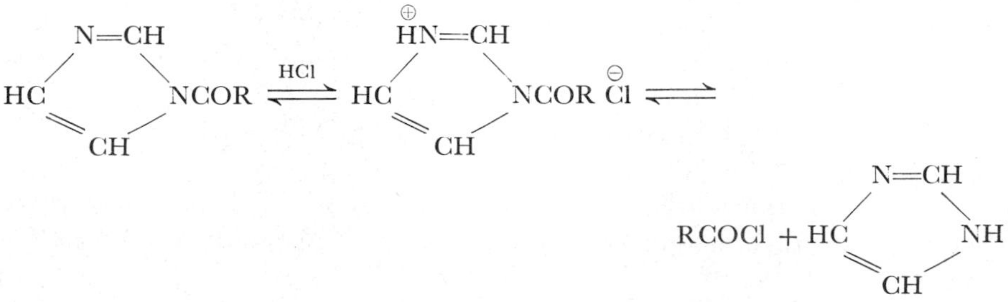

The N-acylimidazoles are cleaved by hydrogen chloride with remarkable ease to imidazole hydrochloride, the acyl fragment being converted into the acyl halide. The reaction appears to be general and is recommended particularly for the conversion of semimicro quantities of an acid or for an acid which is labile to acidic conditions stronger than dry hydrogen chloride. The acid is converted readily into the N-acylimidazole by reaction with 1,1'-carbonyldiimidazole:

$$\left(\begin{array}{c} H \\ C \\ N \diagdown\diagup N \\ HC{=}C \\ H \end{array}\right)_2 C{=}O \ + \ RCOOH \ \xrightarrow{THF} \ \begin{array}{c} H \\ C \\ N \diagdown\diagup N \end{array} \overset{O}{\overset{\|}{CR}} \ + \ CO_2 \ + \ imidazole$$

(a) Preparation of p-methoxybenzoyl chloride (91 % from a solution of N-p-methoxybenzoylimidazole in boiling ethylene chloride to which hydrogen chloride was added for 15 min.; the imidazole hydrochloride was separated by filtration, and the filtrate concentrated and then fractionated) [56].

7. From 1,1,1-Trihalides

$$C_6H_5CCl_3 \ \xrightarrow{ZnO} \ C_6H_5COCl$$

These preparations are of more interest to industrial than to research chemists. A large number of patents describe the formation of benzoyl chloride from equivalent quantities of benzotrichloride and water with reagents such as trace amounts of ferric chloride at 100°, or with an equivalent amount of ethanol at 130–140°, or with chloroacetic acid and zinc chloride at 80–90° [57].

The conversion of benzotrichloride to acid chlorides by means of acids is not described well in the literature but obviously can be brought about as claimed in a patent [58].

(a) Preparation of benzoyl chloride (58% from benzotrichloride and pure zinc oxide at 100° for 1 hr.) [59].

(b) Preparation of benzoyl fluoride (65% from benzotrichloride and zinc fluoride (1.5 parts) at 120° for 6 hr.; benzoyl chloride probably forms first from oxides in the zinc fluoride and then is converted into the fluoride) [59].

1. E. A. Caulson, *J. Chem. Soc.*, 1406 (1934); T. M. Burton and E. F. Degering, *J. Am. Chem. Soc.*, **62,** 227 (1940).
2. H. Zinner and G. Brossmann, *J. Prakt. Chem.*, (4), **5,** 91 (1957).
3. V. I. Zaionts, *Zh. Prikl. Khim.*, **33,** 711 (1960); *C.A.*, **54,** 20965 (1960).
4. J. W. Ralls and B. Riegel, *J. Am. Chem. Soc.*, **77,** 6073 (1955).
5. P. Ruggli and A. Maeder, *Helv. Chim. Acta*, **26,** 1476 (1943).
6. J. B. Lee, *J. Am. Chem. Soc.*, **88,** 3440 (1966).
7. W. Gerrard and A. M. Thrush, *J. Chem. Soc.*, 2117 (1953).
8. A. J. Krubsack and T. Higa, *Tetrahedron Letters*, 5149 (1968).
9. L. F. and M. Fieser, *Reagents for Organic Synthesis*, John Wiley and Sons, New York, 1967, p. 1158.
10. H. Zollinger *et al.*, *Helv. Chim. Acta*, **42,** 1653 (1959).

11. H. Feuer and S. M. Pier, *Org. Syn.*, Coll. Vol. **4**, 554 (1963).

12. Ref. 9, p. 1159.

13. M. Beg and H. Singh, *Fette Seifen Anstrichm.*, **71**, 367 (1969); *C.A.*, **71**, 221 (Sept. 1, 1969).

14. R. P. Barnes, *Org. Syn.*, Coll. Vol. **3**, 555 (1955).

15. R. Adams and R. L. Jenkins, *Org. Syn.*, Coll. Vol. **1**, 394 (1941).

16. M. S. Newman, Abstracts, Organic Division, American Chemical Society Meeting, April 5, 1965.

17. C. F. H. Allen *et al.*, *Org. Syn.*, Coll. Vol. **4**, 739 (1963).

18. H.-J. Bestmann and L. Mott, *Ann. Chem.*, **693**, 132 (1966).

19. H. C. Brown, *J. Am. Chem. Soc.*, **60**, 1325 (1938).

20. A. I. Mashentsev, *J. Gen. Chem. USSR (Eng. Transl.)*, **16**, 203 (1946); *C.A.*, **41**, 706 (1947).

21. D. J. Cram and G. S. Hammond, *Organic Chemistry*, 2nd ed., McGraw-Hill Book Co., New York, 1964, p. 353.

22. L. P. Kyrides, *J. Am. Chem. Soc.*, **59**, 206 (1937).

23. M. Bubner and H. L. Schmidt, *Pharmazie*, **18**, 668 (1963); *C.A.*, **60**, 5629 (1964).

24. R. Adams and L. H. Ulich, *J. Am. Chem. Soc.*, **42**, 599 (1920).

25. B. F. Daubert *et al.*, *J. Am. Chem. Soc.*, **65**, 2142 (1943).

26. S. T. Bauer, *Oil Soap*, **23**, 1 (1946); *C.A.*, **40**, 1330 (1946).

27. S. Ställberg-Stenhagen, *J. Am. Chem. Soc.*, **69**, 2568 (1947).

28. J. E. H. Hancock and R. P. Linstead, *J. Chem. Soc.*, 3490 (1953).

29. L. Heslinga *et al.*, *Rec. Trav. Chim.*, **76**, 969 (1957).

30. L. Heslinga and J. F. Arens, *Rec. Trav. Chim.*, **76**, 982 (1957).

31. A. Rieche and H. Gross, *Chem. Ber.*, **92**, 83 (1959).

32. H. Gross and I. Farkas, *Chem. Ber.*, **93**, 95 (1960).

33. Z. Arnold, *Collection Czech. Chem. Commun.*, **28**, 2047 (1963).

34. L. and V. Arsenijevic and A. F. Damanski, *Compt. Rend.*, **256**, 4039 (1963).

35. N. N. Yarovenko and M. A. Raksha, *J. Gen. Chem. USSR (Eng. Transl.)*, **29**, 2125 (1959).

36. E. Hecker, *Chem. Ber.*, **88**, 1666 (1955).

37. J. H. Simons and E. O. Ramler, *J. Am. Chem. Soc.*, **65**, 389 (1943); W. E. Truce, *J. Am. Chem. Soc.*, **70**, 2828 (1948).

38. S. G. Cohen *et al.*, *J. Am. Chem. Soc.*, **71**, 3439 (1949).

39. K. Kindler *et al.*, *Chem. Ber.*, **76B**, 308 (1943).

40. P. L. Southwick and L. L. Seivard, *J. Am. Chem. Soc.*, **71**, 2532 (1949).

41. T. L. Gresham *et al.*, *J. Am. Chem. Soc.*, **72**, 72 (1950).

42. H. H. Bosshard *et al.*, *Helv. Chim. Acta*, **42**, 1653 (1959).

43. F. Weygand and R. Geiger, *Chem. Ber.*, **90**, 634 (1957).

44. Badische Anilin- and Soda Fabrik, German Patent 804,567, April 26, 1951; *C.A.*, **45**, 8031 (1951).

45. G. A. Olah and S. J. Kuhn, *J. Org. Chem.*, **26**, 237 (1961).

46. A. N. Nesmeyanov and E. J. Kahn, *Chem. Ber.*, **67B**, 370 (1934).

47. A. I. Mashentsev, *Zh. Obshch. Khim. USSR*, **15**, 915 (1945); *C.A.*, **40**, 6443 (1946).

48. E. Ott, *Org. Syn.*, Coll. Vol. **2**, 528 (1943).

49. L. P. Kyrides, *Org. Syn.*, Coll. Vol. **3**, 422 (1955).

50. J. Arient and J. Marhan, *Collection Czech. Chem. Commun.*, **28**, 1292 (1963).

51. N. O. Calloway, *J. Am. Chem. Soc.*, **59**, 1474 (1937).

52. G. A. Olah and S. J. Kuhn, *Org. Syn.*, **45**, 3 (1965).

53. H. Staudinger and E. Anthes, *Chem. Ber.*, **46**, 1426 (1913).

54. D. W. Theobald and J. C. Smith, *Chem. Ind. (London)*, 1007 (1958).

55. R. E. A. Dear and E. E. Gilbert, *J. Org. Chem.*, **33**, 1690 (1968).

56. H. A. Staab *et al.*, *Ann. Chem.*, **694**, 78 (1966).

57. Beilstein, *Handbuch der Organischen Chemie*, 4th ed., Second Supplement, Vol. 9, Springer-Verlag, Berlin, 1949, p. 159.

58. H. Kaltschmitt and A. Tartter, German Patent 804,567, April 26, 1951; *C.A.*, **45**, 8031 (1951).

59. W. Davies and J. H. Dick, *J. Chem. Soc.*, 2808 (1932).

B. Oxidation

Some of the methods described are true oxidation procedures in that the oxidation state of an alcohol, alkyl chloride, or aldehyde is increased to the acid oxidation state. But included here also are other oxidation procedures in which an acid derivative is oxidized to some intermediate which is easier to cleave to the acid halide.

1. From Aldehydes

$$RCH{=}O \xrightarrow{Cl_2} RCOCl + HCl$$

Advantage is taken in this reaction of the ease of abstraction of the hydrogen atom of an aldehyde by free radicals which can be generated by irradiation [1] or by thermal decomposition of some free radical source [2].

$$\tfrac{1}{2} Cl_2 \xrightarrow{h\nu} Cl\cdot \xrightarrow{RCH{=}O} R\overset{\cdot}{C}{=}O + HCl \xrightarrow{Cl_2} RCOCl + Cl\cdot, \quad \text{etc.}$$

This mechanism suggests that sulfuryl chloride, a good source of chlorine free radicals, should be an excellent reagent for converting benzaldehyde into benzoyl chloride. The reaction has been run but yields are not given [3].

(a) **Preparation of o-chlorobenzoyl chloride** (70–72% from o-chlorobenzaldehyde and chlorine at 140–160°; no irradiation) [4].

(b) **Preparation of benzoyl chloride** (80–90% from benzaldehyde and t-butyl hypochlorite in carbon tetrachloride) [5].

(c) **Preparation of benzoyl chloride** (65% from benzaldehyde and N-chlorosuccinimide in benzene) [6].

2. From Perhalides

Where oxidation sites are not too extensive, as in perhalides, acyl chlorides can be formed sometimes by irradiation in the presence of oxygen. For example, trichloroethylene irradiated for 152 hours as oxygen was passed through gave a 16% yield of dichloroacetyl chloride and pentachloroethane gave a 6% yield of trichloroacetyl chloride [7]. Side reactions in oxidations of this sort include carbon-carbon bond splitting.

The best oxidation example is the preparation of trifluoroacetyl chloride from 2,2,2-trifluorodichloroethane as given in Ex. a. The mechanism is complicated but possibly occurs via abstraction of hydrogen or chlorine followed by formation of an intermediate peroxy free radical as shown [8]:

$$CF_3CHCl_2 \xrightarrow{Cl\cdot} CF_3\overset{\cdot}{C}Cl_2 \xrightarrow{O_2} CF_3CCl_2O_2\cdot \xrightarrow{\cdot Cl} CF_3CCl_2O_2Cl \xrightarrow{-\cdot OCl}$$

$$CF_3CCl_2O\cdot \longrightarrow CF_3\overset{O}{\overset{\|}{C}}Cl + Cl\cdot$$

A small amount of chlorine to serve as the initiator increases the yield of the acid chloride considerably.

References for Section B are on p. 870.

(a) **Preparation of trifluoroacetyl chloride** (90% from vapor-phase irradiation of 3.35 mmoles of 2,2,2-trifluorodichloroethane, 9.5 mmoles of chlorine, and 16.4 mmoles of oxygen in a 260-ml. quartz flask for 30 min.; product was shaken with mercury to remove chlorine and distilled) [8].

3. From Acid Hydrazides

$$RCONHNH_2 \xrightarrow[HCl]{Cl_2} RCOCl + N_2 + 3 HCl$$

Although the acyl portion is not oxidized (and therefore cannot be categorized by definition as an oxidation), the hydrazide portion is oxidized to a convenient leaving group, nitrogen. The method is simple and has particular advantages for preparation of some types of acyl halides which cannot be prepared by normal methods, as, for example, α-nitroacyl halides.

(a) **Preparation of o-chlorobenzoyl chloride.** A solution of the hydrazide of o-chlorobenzoic acid in nitromethane was saturated with hydrogen chloride, and chlorine was passed through the mixture until the precipitate which formed dissolved. There was recovered by distillation a 74% yield of the acyl chloride, b.p. 119–120° (20 mm.) [9].

(b) **Preparation of α-nitrobutyryl chloride** (79–96% from α-nitrobutyryl hydrazide in methylene chloride at 0° through which excess chlorine was passed; no concomitant hydrogen chloride was added; the product was distilled under reduced pressure directly from the reaction mixture) [10].

4. From Thiolesters

Direct halogenation of thiolesters leads to acid chlorides:

$$C_6H_5COSCH_3 \xrightarrow[\substack{Liquid\ butane, \\ dry\ ice\ temp.}]{Cl_2} C_6H_5COCl + CH_3SCl_3$$

The crystalline methylsulfur trichloride can be removed by filtration or centrifugation.

(a) **Preparation of benzoyl chloride** (86% from methyl thiolbenzoate and chlorine in liquid butane at dry ice temperature; mixture was centrifuged and the supernatant liquid distilled) [11].

1. E. and E. Levas, *Compt. Rend.*, **235**, 61 (1952).
2. S. Winstein and F. H. Seubold, Jr., *J. Am. Chem. Soc.*, **69**, 2916 (1947).
3. T. H. Durrans, *J. Chem. Soc.*, **121**, 44 (1922).
4. H. T. Clarke and E. R. Taylor, *Org. Syn.*, Coll. Vol. **1**, 155 (1941).
5. D. Ginsburg, *J. Am. Chem. Soc.*, **73**, 702 (1951).
6. M. F. Hebbelynck and R. H. Martin, *Bull. Soc. Chim. Belges.*, **60**, 54 (1951).
7. E. Muller and K. Ehrmann, *Chem. Ber.*, **69B**, 2207 (1936).
8. R. N. Haszeldine and F. Nyman, *J. Chem. Soc.*, 387 (1959).
9. L. A. Carpino, *Chem. Ind.* (*London*), 123 (1956).
10. L. W. Kissinger and H. E. Ungnade, *J. Org. Chem.*, **24**, 1244 (1959).
11. I. B. Douglass and C. E. Osborne, *J. Am. Chem. Soc.*, **75**, 4582 (1953).

C. Halocarbonylation

Oxalyl chloride or bromide is the most common reagent in the group, as discussed in C.1, but carbon monoxide and chlorine sources may be used as well, as shown in C.2.

1. By the Use of Oxalyl Chloride or Bromide

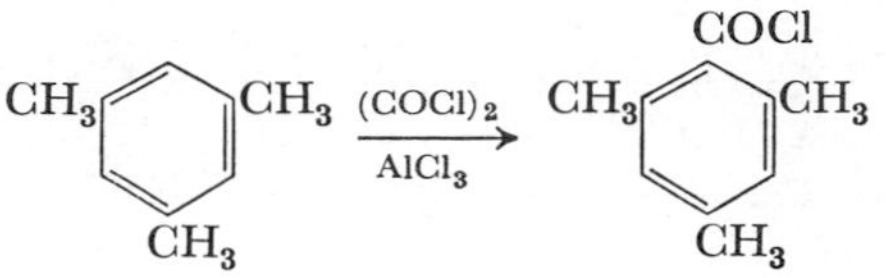

$$RCH_2COBr \xrightarrow{(COBr)_2} RCH(COBr)_2$$

In this synthesis the chlorocarbonyl or bromocarbonyl group is substituted for a hydrogen atom in alkanes and cycloalkanes [1], arenes [2–4], olefins [5, 3], ketones [6], or even acyl halides [7]. The mechanism follows different patterns. It appears to be free radical at temperatures at which acyl bromides are bromocarbonylated:

$$BrCOCOBr \xrightarrow[100°]{CCl_4} 2\ Br\overset{\cdot}{C}{=}O \xrightarrow{RCH_2COBr}$$

$$(HCOBr) + R\overset{\cdot}{C}HCOBr \xrightarrow{\cdot COBr} RCH(COBr)_2$$

$$\downarrow$$

$$CO + HBr$$

A like mechanism prevails in the chlorocarbonylation of cyclohexane in the presence of light or a peroxide [1]. In the chlorocarbonylation of diarylethylenes using oxalyl chloride, substituents in the aryl groups influence the rate of the reaction in such a way as to suggest an ionic mechanism [5]:

$$(C_6H_5)_2C{=}CH_2 \xrightarrow{(COCl)_2} (C_6H_5)_2C{=}CHCOCl$$

β,β-Diphenylacrylyl chloride

Arenes have been chlorocarbonylated without catalysts [2]:

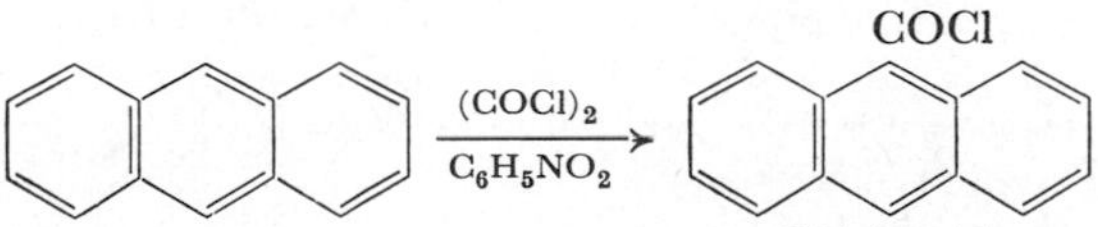

67% (of acid)
Anthracene-9-carboxylic
acid chloride

or under Friedel-Crafts conditions as well [4]:

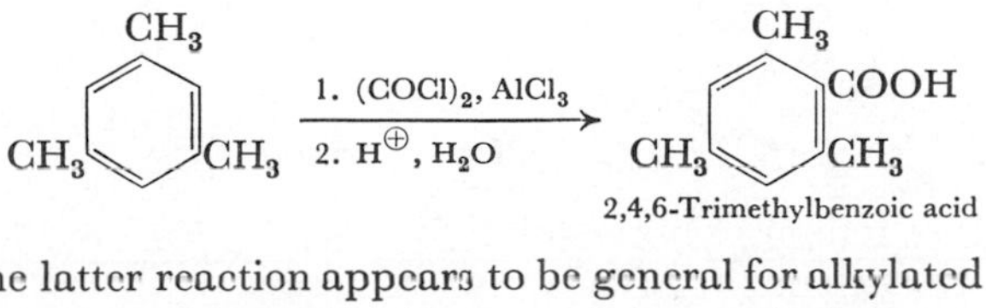

2,4,6-Trimethylbenzoic acid

(see Ex. a). The latter reaction appears to be general for alkylated benzenes and polynuclear arenes, readily oxidizable types being excluded.

References for Section C are on p. 872.

(a) **Preparation of 2,4,6-trimethylbenzoyl chloride** (65–76 % as the acid from 1 mole of mesitylene, 1.10 moles of oxalyl chloride, and 1.10 moles of anhydrous aluminum chloride in dry CS_2 at 10–15°) [4].

(b) **Preparation of dimethylmalonyl bromide** (76 % from 0.8 mole of isobutyryl bromide and 0.4 mole of oxalyl bromide, the limiting reagent, in CCl_4 at 100–110° for 18 hr.) [7].

2. By the Use of Carbon Monoxide and a Chlorine Source

$$(R\cdot) + CO \longrightarrow (R\overset{\cdot}{C}O) \xrightarrow{CCl_4} RCOCl + \cdot CCl_3$$

This equation points out that a competitive abstraction of chlorine occurs between the alkyl free radicals to form RCl and the acyl free radical to form RCOCl. Evidently, acid chloride formation is favored at high carbon monoxide pressures since reasonable yields of acid chloride are obtained under these conditions. The abstraction of chlorine by the acyl free radical also seems to be strongly inhibited by a combination of metal salts and the acyl halide, and for this reason the reaction must be carried out in a glass-lined autoclave [8]:

$$C_6H_{12} + CCl_4 + \text{di-}t\text{-butylperoxide} + CO \xrightarrow[130°, 16 \text{ hr.}]{\text{Autoclave,}}$$

C_6H_{12}	CCl_4	di-t-butylperoxide	CO
2 moles	1 mole	5%	6000 p.s.i.

Cyclohexylcarbonyl chloride,
44–46% based on CCl_4

The ratio of acid chloride to cyclohexyl chloride was 5:1 in this experiment.

Another similar free radical reaction can be brought about at atmospheric pressure by palladium chloride [9]:

$$C_6H_5HgCl + PdCl_2 + CO \xrightarrow[25°]{CH_3CN} (C_6H_5PdCl) \longrightarrow C_6H_5COCl + Pd + HCl$$

2 mmoles   2 mmoles

Benzoyl chloride,
10%

Some benzophenone and biphenyl were formed concurrently.

The reactions above are not at present well-adapted to laboratory preparations.

1. M. S. Kharasch and H. C. Brown, *J. Am. Chem. Soc.*, **64,** 329 (1942).
2. E. Mosettig *et al.*, *J. Am. Chem. Soc.*, **70,** 1079 (1948).
3. W. Treibs and H. Orttmann, *Chem. Ber.*, **93,** 545 (1960).
4. P. E. Sokol, *Org. Syn.*, **44,** 69 (1964).
5. F. Bergmann *et al.*, *J. Am. Chem. Soc.*, **70,** 1612 (1948).
6. W. Treibs *et al.*, *Chem. Ber.*, **93,** 551 (1960).
7. W. Treibs and H. Orttmann, *Chem. Ber.*, **91,** 297 (1958).
8. W. A. Thaler, *J. Am. Chem. Soc.*, **88,** 4278 (1966).
9. P. M. Henry, *Tetrahedron Letters*, 2285 (1968).

D. From Ketenes

$$CH_2=C=O + (Ar)_3CCl \rightarrow (Ar)_3C-CH_2COCl$$

Active chlorine-containing compounds add to ketene to produce acyl chlorides [1]. Sometimes a catalyst such as aluminum chloride is necessary. Although

References for Section D are on p. 873.

α,β-dichloroethyl ethyl ether, α-chlorobenzyl methyl ether, and triphenyl-chloromethane react under the conditions employed (room temperature in benzene or nitrobenzene), benzyl chloride, benzyl bromide, benzotrichloride, benzoyl chloride, chloroacetone, and 2-chloro-2-nitropropane do not. However, trichloroacetyl chloride has been added to dimethylketene (see Ex. b). Yields are usually low.

(a) **Preparation of β-methoxy-β-phenylpropionyl chloride.** Gaseous ketene (10 equiv.) was bubbled through α-chlorobenzyl methyl ether and the mixture was then cooled to room temperature. Distillation under reduced pressure gave a 70% yield of the acyl chloride, b.p. 91–105° (3 mm.) (major portion boiled at 100–101° (3 mm.)) [1].

(b) **Preparation of 4,4,4-trichloro-2,2-dimethyl-3-ketobutanoyl chloride** [2]:

$$CCl_3COCl + (CH_3)_2C{=}C{=}O \xrightarrow[\text{25° overnight}]{C_6H_{14}} CCl_3CO\underset{\underset{CH_3}{|}}{\overset{\overset{CH_3}{|}}{C}}COCl$$

0.0912 mole   0.037 mole   61%

1. A. T. Blomquist *et al.*, *J. Am. Chem. Soc.*, **69**, 2356 (1947).
2. W. T. Brady and L. Smith, *J. Org. Chem.*, **33**, 4550 (1968).

Chapter 16

CARBOXYLIC ACID ANHYDRIDES

A. Solvolysis

1. From Carboxylic Acids

Heating carboxylic acids to remove water is successful only with dibasic ones that yield cyclic anhydrides. If certain dibasic esters are the starting materials, a mineral acid such as sulfuric is employed [1] (see also Ex. a). Cyclic anhydrides also can be polymerized to form cyclic polymers [2].

References for Section A are on pp. 881–882.

Heating monocarboxylic acids alone produces some anhydride, but it is feasible to drive the reaction to completion only if an acid such as benzoic, which has no α-hydrogen atoms, is employed. Actually the process is very slow. For example, the following acids were held at their boiling points for varying lengths of time and produced the following percentages of anhydride [3], cinnamic acid, 5 % (4 min.); hydrocinnamic acid, 30 % (60 min.); myristic acid, 30 % (12 min.); benzoic acid, 50 % (48 hr.); adipic acid, 6 % (4 min.).

Various derivatives of imidazole such as N,N'-carbonyldiimidazole, N-trifluoro- or N-trichloroacetyl imidazole may be used to form anhydrides of acids such as maleic, palmitic, phthalic, and benzoic [4].

Aliphatic monocarboxylic acids may be converted into anhydrides in 77–81 % yields by heating with a mixture of a tertiary phosphine and *trans*-dibenzoylethylene (see also Ex. c.3) [5]:

$$2 RCOOH + C_6H_5COCH{=}CHCOC_6H_5 + (n\text{-}C_4H_9)_3P \rightarrow$$
trans
$$(RCO)_2O + C_6H_5COCH_2CH_2COC_6H_5 + n\text{-}(C_4H_9)_3PO$$

The best, but also most expensive, method of producing anhydrides from aliphatic aminocarboxylic acids is by the use of dicyclohexylcarbodiimide (see Ex. b):

$$2 RNHCH_2COOH + C_6H_5N{=}C{=}NC_6H_5 \rightarrow$$
$$(RNHCH_2CO)_2O + C_6H_5NHCONHC_6H_5$$

An economical process using phosphorus pentoxide dispersed on kieselguhr and asbestos has been worked out for the synthesis of sulfonic anhydrides [6]; the granular form of phosphorus pentoxide, Granusic, may overcome the dispersion problem (see also Ex. A.2, b.6 for alternative preparation of sulfonic anhydrides), but it does not seem to have been applied to carboxylic acids. Separation of the anhydride from the phosphoric acid may be accomplished with hot benzene.

(a) Preparation of 3,4-dihydro-1,2-naphthalic anhydride (73–81 % from ethyl-γ-phenylbutyrate employed to prepare ethyl α-ethoxalyl-γ-phenylbutyrate which was cyclized by conc. sulfuric acid) [7]:

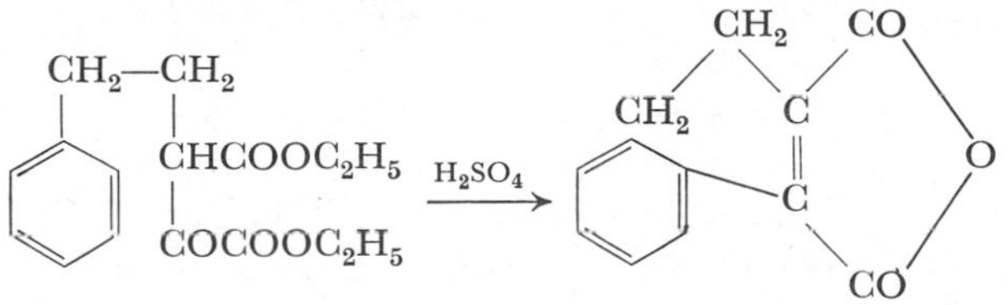

(b) Preparation of carbobenzyloxy-DL-phenylalanine anhydride (90 % from 1.1 mmoles of the amino acid and 0.6 mmole of DCC in ether, the precipitated anhydride being extracted from the dicyclohexylurea with dry ethyl acetate) [8].

(c) Other examples

(1) 3-Hydroxyphthalic anhydride (77–91 % from 3-aminophthalic anhydride via 3-hydroxyphthalic acid which was sublimed at 160–180° (0.2 mm.)) [9].

(2) Hydrastic anhydride

(83 % from hydrastic acid at 190°) [10].

(3) Propionic anhydride (78 % from propionic acid by refluxing for 2 hr. with a mixture of tri-*n*-butylphosphine and *trans*-dibenzoylethylene in benzene) [5].

2. From Carboxylic Acids or Anhydrides and Acylating Agents, and Related Reactions

$$\text{RCOOH} + \text{R'COCl} \longrightarrow \overset{\overset{\text{O}}{\|}}{\text{RC}}\text{O}\overset{\overset{\text{O}}{\|}}{\text{CR'}} + \text{HCl}$$

Acid anhydrides are most commonly prepared by this method in which the acylating agent may be an acid chloride, anhydride, or similar reagent. Since many of the anhydrides are reactive substances, the best methods are those which involve a minimum of separation steps and, preferably, no operation except filtration or distillation under reduced pressure. The best methods are to be found in A.3. Mixed acid anhydrides tend to disproportionate in the presence of acid or base catalysts or by heat [11, 12]:

$$2\ \overset{\overset{\text{O}}{\|}}{\text{RC}}\text{O}\overset{\overset{\text{O}}{\|}}{\text{CR'}} \xrightarrow{\text{Heat}} (\overset{\overset{\text{O}}{\|}}{\text{RC}})_2\text{O} + (\overset{\overset{\text{O}}{\|}}{\text{R'C}})_2\text{O}$$

The acid anhydride, if stable to water, may be prepared by simply adding the right amount of water to the acid chloride complexed with pyridine [13]; *see also* [14]:

$$\text{C}_6\text{H}_5\text{COCl} + \text{C}_5\text{H}_5\text{N} + \text{H}_2\text{O} \xrightarrow[\substack{\text{2. Pour into ice and} \\ \text{dilute HCl}}]{\substack{\text{1. 40 ml. dioxane} \\ \text{at 10°}}} (\text{C}_6\text{H}_5\text{CO})_2\text{O}$$

0.1 mole   10 ml.   1 ml.   Benzoic anhydride 82%

Or, it may be prepared by adding the acid in place of water (see Ex. b.3).

Among the mixed anhydrides which are stable are acetyl formate [15], acylcarbonates [16], and acyltrifluoroacetates. The latter, which are made simply by mixing trifluoroacetic anhydride and the proper amount of carboxylic acid:

$$\text{RCOOH} + (\text{CF}_3\text{CO})_2\text{O} \longrightarrow \text{CF}_3\overset{\overset{\text{O}}{\|}}{\text{C}}\text{—O—}\overset{\overset{\text{O}}{\|}}{\text{CR}} + \text{CF}_3\text{COOH}$$

are powerful acylating agents.

The acylation of oxygen (or carbon) with this reagent takes place with the acyl group of the weaker acid of the mixed anhydride, but acylation of nitrogen may occur with the acyl group of the stronger acid:

$$RNH_2 + CF_3\overset{O}{\overset{\|}{C}}O\overset{O}{\overset{\|}{C}}R \longrightarrow RNHCOCF_3$$

For dicarboxylic acids such as malonic, trifluoroacetic anhydride forms anhydrides (Ketenes, Chapter 17, A.2). In the references cited for the following compounds, either the mixed anhydride or, more frequently, the symmetrical anhydride is formed. A summary of the acylating agents used includes acid chlorides [17–20], carboxylic acid anhydrides [21–27], a mixture of acetyl chloride and acetic acid [28], ethyl chlorocarbonate [16, 29], carbonyl chloride [30], thionyl chloride [31, 32], benzenesulfonyl chloride [33], and phosphorus oxychloride [34]. Of the substances sometimes used with the acylating or related agent, pyridine is the most common [33, 21, 18, 35, 36, 32]. Instead of pyridine, triethylamine [30], neutral salts [17], and sulfuric acid [22, 37] have also been employed.

The method is applicable not only to types readily converted into acids under the experimental conditions followed, such as cyanides and esters [37], but also to dicarboxylic acids, many of which form anhydrides readily [22, 28, 23, 37, 24, 19, 38, 39, 34, 26, 27]. Yields are usually high, particularly when pyridine is present.

An unusual combination of dehydration and oxidation occurs in the preparation of arylmaleic anhydride [40]:

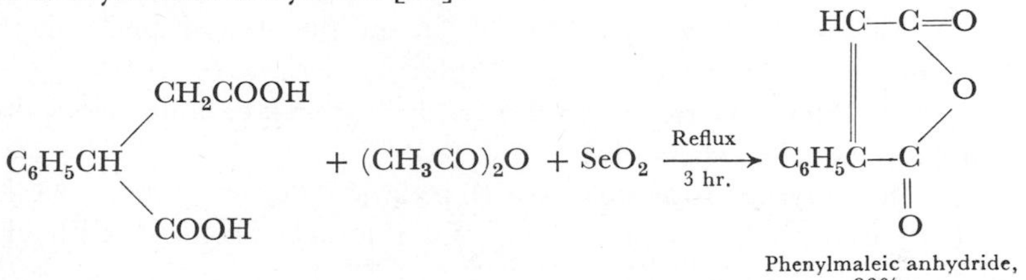

$$C_6H_5CH\begin{array}{c} CH_2COOH \\ \\ COOH \end{array} + (CH_3CO)_2O + SeO_2 \xrightarrow[\text{3 hr.}]{\text{Reflux}} C_6H_5C\overset{HC-C=O}{\underset{O}{\overset{\|}{C}}}\diagdown O$$

Phenylmaleic anhydride, 86%

In the case of an acylating agent like acetyl chloride when employed alone, the mechanistic path appears to be similar to that followed when an alcohol reacts with the acyl halide (Carboxylic Esters, Chapter 14, A.2). The complex formed by the nucleophilic attack of the carboxylic acid on the acid chloride

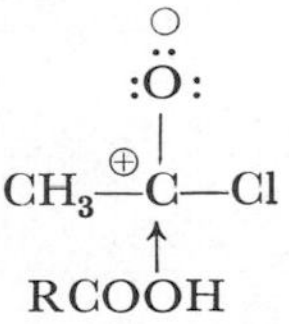

$$CH_3\overset{\oplus}{\underset{\underset{RCOOH}{\uparrow}}{C}}-Cl \quad \overset{\overset{O}{\ddot{\cdot}\ddot{O}\colon}}{}$$

loses hydrogen chloride readily to give the acid anhydride

$$CH_3\overset{O}{\overset{\|}{C}}-O-\overset{O}{\overset{\|}{C}}R$$

If RCOOH is higher-boiling than CH_3COOH, an interchange may occur to give the anhydride of the higher boiling acid rather than the mixed anhydride:

$$
\underset{\substack{\uparrow \\ RCOOH}}{\begin{array}{c} R-C=O \\ \diagdown \\ O \\ \diagup \\ CH_3C=O \end{array}} \xrightarrow{\overset{O}{RC-OH}} \begin{array}{c} R-C=O \\ \diagdown \\ O \\ \diagup \\ CH_3C=O \end{array} \longrightarrow \begin{array}{c} R-C=O \\ \diagdown \\ O \\ \diagup \\ R-C=O \end{array} + CH_3COOH
$$

With pyridine, displacement takes place from the quaternary salt [41], but see [42]:

$$
\overset{O}{\underset{\|}{RCOH}} + \overset{O}{\underset{\|}{RCC_5H_5\overset{\oplus\ominus}{N}Cl}} \longrightarrow \overset{O\ \ O}{\underset{\|\ \ \|}{RCOCR}} + C_5H_5N \cdot HCl
$$

With stronger bases one must be aware of the possibility of ketene formation as an intermediate.

(a) **Preparation of benzoic anhydride** (72–74 % from benzoic acid and acetic anhydride) [43]; *see also* [33].

(b) **Other examples**

(1) **Succinic anhydride** (93–95 % from succinic acid and acetyl chloride) [20].

(2) **Ricinoleic anhydride** (95 % from ricinoleic acid, ethyl chloroformate, and triethylamine) [29].

(3) **n-Heptanoic anhydride** (97 % from n-heptanoic acid, n-heptanoyl chloride, and pyridine) [18]; *see also* [44].

(4) **p-Chlorobenzoic anhydride** (90 % from p-chlorobenzoyl chloride, pyridine, and aqueous HCl) [44].

(5) **Benzoyl trifluoroacetate** (b.p. 56–57° at 3.5 mm., ca. 60 % from 4.65 g. of benzoic anhydride and 4.33 g. of trifluoroacetic anhydride held at 40° for 90 min. and then distilled) [12].

(6) **Benzenesulfonic anhydride** (46 % by heating benzenesulfonyl chloride and oxalic acid at 200° [45]; see also A.1 for dehydration of a sulfonic acid).

(7) **Nicotinic anhydride** (87–93 % from nicotinic acid, triethylamine, and phosgene in dry benzene at 7°, after which the temperature was raised to the boiling point) [46].

3. From the Salts of Carboxylic Acids and Acylating Agents

$$RCOONa + RCOCl \rightarrow (RCO)_2O + NaCl$$

This synthesis represents perhaps the best general method for preparing anhydrides. Salts of alkali metals, silver, or tertiary amines [30] and acyl chlorides, including carbonyl chloride or oxalyl chloride, phosphorus oxychloride [47], sulfur halides (see Ex. c.3), or even carbon bisulfide (see Ex. b) may be employed. When it is important to obtain high conversions or where other methods fail, the procedure as shown is recommended [48]:

$$RCOOH + TlOEt \longrightarrow \underset{\substack{\text{Ether} \\ \text{suspension}}}{RCO_2Tl} \xrightarrow{R'COCl} R\overset{O}{\overset{\|}{C}}-O-\overset{O}{\overset{\|}{C}}R' + TlCl$$

Here the cost and toxicity of thallium salts must be weighed against the high yields (95–100%). However, recognizing that acid catalyzes disproportionation and, in the case shown, decarbonylation, Schijf and Stevens have described the conditions necessary for preparing mixed anhydrides [49]:

$$\underset{\substack{\text{1 mole}}}{CH_3COCl} + \underset{\substack{\text{Dry,} \\ \text{1 mole}}}{HCO_2Na} \xrightarrow[\substack{\text{Stir 24 hr., 0°,} \\ \text{filter and distill} \\ \text{under reduced} \\ \text{pressure}}]{\text{100 ml. THF}} \underset{\substack{\text{Acetic-formic} \\ \text{anhydride, 60\%}}}{H\overset{O}{\overset{\|}{C}}O\overset{O}{\overset{\|}{C}}CH_3}$$

The mechanism follows the pattern indicated when the free acid is employed with the acyl halides (see A.2) [30].

(a) Preparation of nicotinic anhydride.   Potassium nicotinate, 32.24 g., dry and finely divided, in a suspension of anhydrous benzene was treated with 12.69 g. of oxalyl chloride in 45 ml. of anhydrous benzene, with stirring in an ice bath for 20 min. The mixture was allowed to stand for another 15 min. with the bath removed; it was then stirred for 1 hr. and refluxed for 1 hr. After filtering while hot and washing with hot benzene, the combined filtrate and washings were concentrated to about 100 ml. and cooled to give 17.23 g. of the anhydride, m.p. 123–126°. The total yield was 85% when 2.09 g. more was recovered from the mother liquor by concentrating followed by the addition of hexane [50]; *see also* [51, 30].

(b) Preparation of benzoic anhydride.   Silver benzoate, 2.90 g., in 8 ml. of dry carbon disulfide was heated in a sealed tube at 100° for 8 hr. After

$$4 RCO_2Ag + CS_2 \rightarrow ((RCO_2)_4C) \rightarrow 2 (RCO)_2O + CO_2$$

cooling, the tube was opened and on warming carbon dioxide was evolved. Extraction with dry ether and evaporation gave 1.40 g. (98%) of benzoic anhydride, m.p. 42–43°, without additional purification [52].

(c) Other examples

(1) **Phenylacetic anhydride** (87% from phenylacetyl chloride and sodium phenylacetate) [53].

(2) **Benzoyl trifluoroacetate** (80% from benzoyl chloride and silver trifluoroacetate by flash distillation of an ethereal solution at 1.0 mm.) [54].

(3) **Propionic anhydride** (85% of 90% purity from an exothermic reaction of the salt, sulfur, and bromine by holding the temperature at 50° according to the equation: $8 RCO_2Na + S + 3 Br_2 \rightarrow 4 (RCO)_2O + 6 NaBr + Na_2SO_4$) [55].

4. From Acyl Chlorides and Acetic Anhydride

$$2\,RCOCl + (CH_3CO)_2O \rightarrow (RCO)_2O + 2\,CH_3COCl$$

This synthesis was first applied to the preparation of benzoic anhydride [56]. Later a thorough study, including a comparison with other methods of synthesis, was made of its use in the preparation of fatty acid anhydrides [57]. Yields of 90–95% were obtained for these higher aliphatic acid anhydrides, including some unsaturated ones, by using 100–300% excess of acetic anhydride. This method, at least for the anhydrides prepared by its use, gives yields which compare favorably with those obtained with the use of pyridine in A.2, which is one of the better methods of preparing acid anhydrides.

(a) **Preparation of palmitic anhydride.**   Palmitoyl chloride, 9.02 g., and 6.69 g. (300% excess) of acetic anhydride were heated in a flask arranged for fractionation through a Vigreux column. When the boiling point of the liquid rose to 60° (flask temperature 130–150°), the pressure was reduced gradually to 100 mm. and finally to 10 mm. for 15 min., while the flask temperature was 160°. At this point the excess acetic anhydride had been removed and the contents of the flask solidified on cooling. Recrystallization from Skellysolve F using charcoal gave 7.45 g. (91.6%) of palmitic anhydride, m.p. 63.8–64° [57].

(b) **Preparation of phenylacetic anhydride** (88.4% from phenylacetyl chloride and acetic anhydride) [58].

5. From Carboxylic Acids and Ketenes

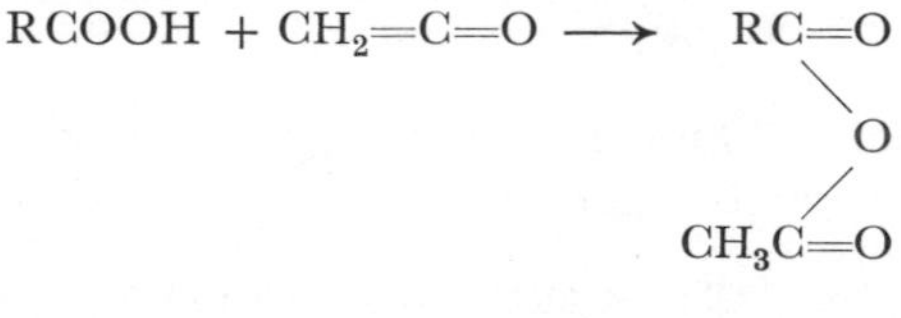

$$RCOOH + CH_2{=}C{=}O \longrightarrow \begin{array}{c} RC{=}O \\ \diagdown \\ O \\ \diagup \\ CH_3C{=}O \end{array}$$

This synthesis is a convenient method for preparing simple and mixed anhydrides, although it is dependent on the not always available ketene. The reaction proceeds at low temperature and the product may be recovered by distillation. Mixed anhydrides are produced with ordinary ketene and the homologs of acetic acid and aromatic acids [59], but these may be converted into simple anhydrides by distillation at ordinary pressure or by interchange with the higher-boiling acid as discussed under A.2. Yields are good.

A more convenient, but less reactive, acylating agent than ketene is isopropenyl acetate [60]:

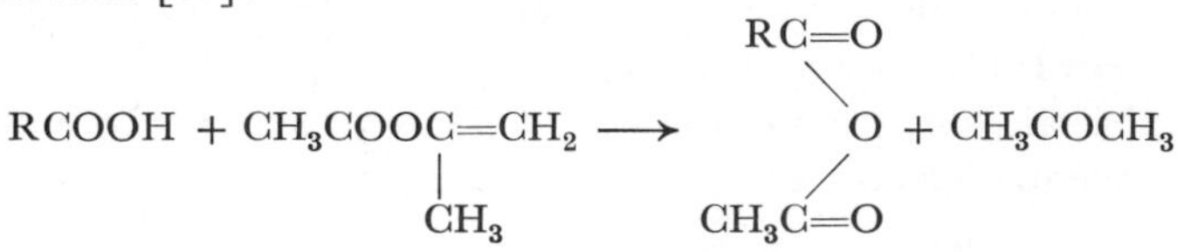

$$RCOOH + \underset{\underset{CH_3}{|}}{CH_3COOC{=}CH_2} \longrightarrow \begin{array}{c} RC{=}O \\ \diagdown \\ O \\ \diagup \\ CH_3C{=}O \end{array} + CH_3COCH_3$$

The course of the reaction follows the normal nucleophilic addition to ketene, the unstable enol formed tautomerizing to the anhydride:

$$\underset{\underset{O}{\overset{\|}{RCOH}}}{\overset{\uparrow}{CH_2=C=O}} \longrightarrow \left[\underset{\underset{O}{\overset{\|}{RCO}}}{\overset{|}{CH_2=C-OH}}\right] \longrightarrow \underset{RC=O}{\overset{CH_3C=O}{\diagdown O \diagup}}$$

(a) Preparation of *n*-caproic anhydride (80–87% from ketene and *n*-caproic acid) [61].

(b) Preparation of benzoic acetic anhydride (quantitative yield from ketene and benzoic acid) [59].

6. From Mercuric Carboxylates and a Phosphine

$$(RCOO)_2Hg + (R')_3P \rightarrow (RCO)_2O + Hg + (R')_3PO$$

$$2\,RCOOH + (R')_3P + (C_6H_5)_2Hg \rightarrow (RCO)_2O + Hg + (R')_3PO + 2\,C_6H_6$$

These syntheses proceed under comparatively mild conditions. In the first method a phosphite may be used instead of a phosphine, and a mercurous carboxylate may be substituted for the mercuric salt. Of the limited number of anhydrides prepared by these methods the yields usually vary from 60 to 80% [62, 63]. Intimate details of the mechanism may be found in the references given.

(a) Preparation of acetic anhydride. Diphenylmercury, 5.3 g., tri-*n*-butylphosphine, 3.0 g., and glacial acetic acid, 1.8 g., in 10 ml. of anhydrous benzene were refluxed for 2 hr. After filtering off the mercury, the filtrate was concentrated and then distilled to give 1.38 g. (80%) of acetic anhydride, b.p. 134–136° [63].

(b) Preparation of propionic anhydride (59% from triethyl phosphite and mercurous propionate) [62].

1. L. Eberson, *Acta Chem. Scand.*, **13**, 40 (1959).
2. J. W. Hill and W. H. Carothers, *J. Am. Chem. Soc.*, **55**, 5023 (1933).
3. D. Davidson and P. Newman, *J. Am. Chem. Soc.*, **74**, 1515 (1952).
4. H. A. Staab *et al.*, *Chem. Ber.*, **95**, 2073 (1962).
5. I. Kuwajima and T. Mukaiyama, *J. Org. Chem.*, **29**, 1385 (1964).
6. L. Field and J. W. McFarland, *Org. Syn.*, Coll. Vol. **4**, 940 (1963).
7. E. B. Hershberg and L. F. Fieser, *Org. Syn.*, Coll. Vol. **2**, 194 (1943).
8. D. H. Rammler and H. G. Khorana, *J. Am. Chem. Soc.*, **85**, 1997 (1963).
9. E. L. Eliel *et al.*, *J. Am. Chem. Soc.*, **77**, 5092 (1955).
10. W. Reeve and H. Myers, *J. Am. Chem. Soc.*, **73**, 1371 (1951).
11. J. M. Tedder, *Chem. Rev.*, **55**, 787 (1955).
12. E. J. Bourne *et al.*, *J. Chem. Soc.*, 2006 (1954).
13. H. Adkins and Q. E. Thompson, *J. Am. Chem. Soc.*, **71**, 2242 (1949).
14. W. Gerrard and A. M. Thrush, *J. Chem. Soc.*, 741 (1952); 2117 (1953).
15. C. D. Hurd *et al.*, *J. Am. Chem. Soc.*, **68**, 789 (1946).
16. J. A. Price and D. S. Tarbell, *Org. Syn.*, Coll. Vol. **4**, 285 (1963).
17. V. V. Katyshkina and M. Y. Kraft, *J. Gen. Chem. USSR (Eng. Transl.)*, **29**, 65 (1959).
18. W. Gerrard and A. M. Thrush, *J. Chem. Soc.*, 2117 (1953).
19. J. A. McRae *et al.*, *Canad. J. Res.*, **28B**, 73 (1950).
20. L. F. Fieser and E. L. Martin, *Org. Syn.*, Coll. Vol. **2**, 560 (1943).

21. G. N. Walker, *J. Am. Chem. Soc.*, **79**, 3508 (1957).

22. R. L. Shriner and C. L. Furrow, Jr., *Org. Syn.*, Coll. Vol. **4**, 242 (1963).

23. S. F. Birch *et al.*, *J. Org. Chem.*, **20**, 1178 (1955).

24. O. Grummitt *et al.*, *Org. Syn.*, Coll. Vol. **3**, 449 (1955).

25. R. Adams and J. L. Anderson, *J. Am. Chem. Soc.*, **73**, 136 (1951).

26. J. Cason, *Org. Syn.*, Coll. Vol. **4**, 630 (1963).

27. E. C. Horning and A. F. Finelli, *Org. Syn.*, Coll. Vol. **4**, 790 (1963).

28. A. Ziekha and Y. Liwschitz, *J. Chem. Soc.*, 4397 (1957).

29. J. S. Nelson *et al.*, *J. Org. Chem.*, **28**, 1905 (1963).

30. H. Rinderknecht and V. Ma, *Helv. Chim. Acta*, **47**, 162 (1964).

31. R. P. Linstead and M. Whalley, *J. Chem. Soc.*, 3722 (1954).

32. F. H. Carpenter, *J. Am. Chem. Soc.*, **70**, 2964 (1948).

33. J. H. Brewster and C. J. Ciotti, Jr., *J. Am. Chem. Soc.*, **77**, 6214 (1955).

34. M. Hauptschein *et al.*, *J. Am. Chem. Soc.*, **74**, 4005 (1952).

35. R. Kuhn and I. Löw, *Chem. Ber.*, **77**, 211 (1944).

36. K. Miescher *et al.*, *Helv. Chim. Acta*, **28**, 991 (1945).

37. N. N. Saha *et al.*, *J. Am. Chem. Soc.*, **81**, 3670 (1959).

38. M. N. Donin *et al.*, *J. Am. Chem. Soc.*, **73**, 4286 (1951).

39. V. du Vigneaud and G. L. Miller, *Biochem. Prepn.*, **2**, 79 (1952).

40. R. K. Hill, *J. Org. Chem.*, **26**, 4745 (1961).

41. S. I. Lur'e, *Zh. Obshch. Khim.*, **18**, 1517 (1948); *C.A.*, **43**, 4240 (1949).

42. W. Gerrard and A. M. Thrush, *J. Chem. Soc.*, 741 (1952).

43. H. T. Clarke and E. J. Rahrs, *Org. Syn.*, Coll. Vol. **1**, 91 (1941).

44. C. F. H. Allen *et al.*, *Org. Syn.*, Coll. Vol. **3**, 28 (1955).

45. R. G. Shepherd, *J. Org. Chem.*, **12**, 275 (1947).

46. H. Rinderknecht and M. Gutenstein, *Org. Syn.*, **47**, 89 (1967).

47. E. H. Rodd, *Chemistry of Carbon Compounds*, 2nd ed., Vol. Ic, Elsevier Publishing Co., New York, 1965, p. 157.

48. E. C. Taylor *et al.*, *J. Am. Chem. Soc.*, **90**, 2422 (1968).

49. R. Schijf and W. Stevens, *Rec. Trav. Chim.*, **85**, 627 (1966).

50. A. W. Schrecker and P. B. Maury, *J. Am. Chem. Soc.*, **76**, 5803 (1954).

51. C. O. Badgett, *J. Am. Chem. Soc.*, **69**, 2231 (1947).

52. D. Bryce-Smith, *Proc. Chem. Soc.*, 20 (1957).

53. J. A. King and F. H. McMillan, *J. Am. Chem. Soc.*, **73**, 4911 (1951).

54. A. F. Ferris and W. D. Emmons, *J. Am. Chem. Soc.*, **75**, 232 (1953).

55. J. Orshansky and E. Bograchov, *Chem. Ind.*, (*London*), 382 (1944).

56. F. Zetzsche *et al.*, *Helv. Chim. Acta*, **9**, 181 (1926).

57. N. O. V. Sonntag *et al.*, *J. Am. Oil Chemists' Soc.*, **31**, 151 (1954).

58. C. D. Hurd and A. G. Prapas, *J. Org. Chem.*, **24**, 388 (1959).

59. C. D. Hurd and M. F. Dull, *J. Am. Chem. Soc.*, **54**, 3427 (1932).

60. H. V. Hagemeyer, Jr., and D. C. Hull, *Ind. Eng. Chem.*, **41**, 2920 (1949).

61. J. W. Williams and J. A. Krynitsky, *Org. Syn.*, Coll. Vol. **3**, 164 (1955).

62. T. Mukaiyama *et al.*, *J. Org. Chem.*, **28**, 917 (1963).

63. T. Mukaiyama *et al.*, *J. Org. Chem.*, **28**, 2024 (1963).

B. Oxidation

1. From Acenaphthenes

1,8-Naphthalic anhydride

References for Section B are on p. 884.

This oxidation occurs with a dichromate or chromic anhydride [1] in an acid solution such as acetic or hydrochloric acid [2]. Yields are often nearly quantitative [3]. Some variation in time was observed in oxidizing 2-acyl-4-halo- and 2-acyl-5-haloacenaphthenes with sodium dichromate in glacial acetic acid, although in no case was more than $1\frac{1}{2}$ hours required.

The reaction, of course, is capable of extension to other hydrocarbons provided the anhydride is stable under the oxidizing conditions. In the case of acenaphthenes, the oxidation probably proceeds via the diketone, which type is discussed in B.2.

(a) Preparation of 2-acetyl-4-chloro-1,8-naphthalic anhydride.

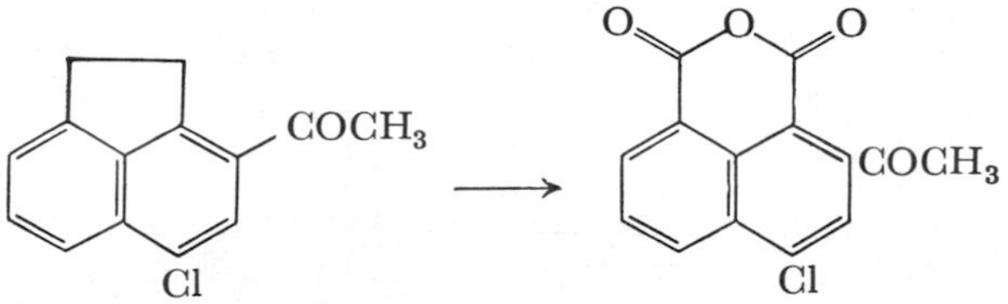

3-Acetyl-5-chloroacenaphthene, 10 g., in 120 ml. of glacial acetic acid was treated with 50 g. of sodium dichromate. When the reaction had subsided, the solution was refluxed for 10 min., and then poured into 400 ml. of hot water. The anhydride which separated was filtered off and washed with water. Yield was practically quantitative. Recrystallization from benzene or ethyl acetate gave a product, m.p. 283.5–284° (decomposition and sublimation) [3].

(b) Preparation of 2,5-dibenzoyl-1,8-naphthalic anhydride (58% from 3,6-dibenzoylacenaphthene, glacial acetic acid, and chromic anhydride) [1].

2. From α-Diketones and o-Quinones

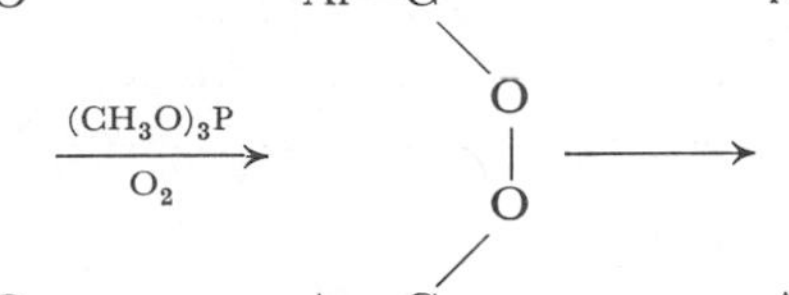

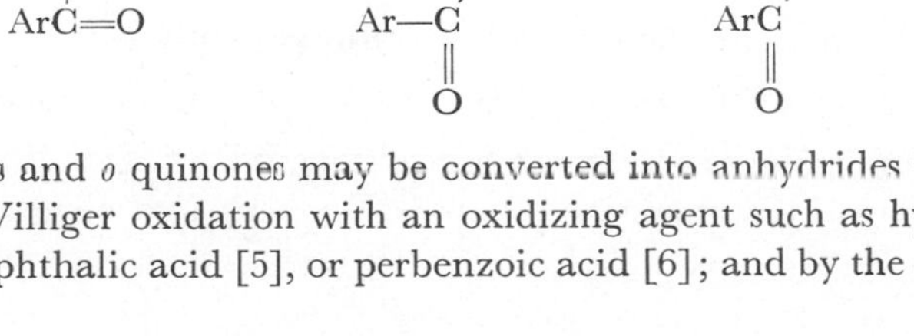

α-Diketones and o-quinones may be converted into anhydrides in two ways: by a Baeyer-Villiger oxidation with an oxidizing agent such as hydrogen peroxide [4], perphthalic acid [5], or perbenzoic acid [6]; and by the air oxidation of the adduct formed between the diketone and a trialkyl phosphite [7]. In the first method, in which α-diketones are cleaved with hydrogen peroxide, perchloric acid is an effective catalyst [8]. In the second method, triphenyl phosphine may be substituted for the trialkyl phosphite and ozone for oxygen [9]. A mechanism for the oxygen or ozone oxidation has been advanced [10]. Yields are sometimes fair by the first method, but poor by the second.

The implication of this synthesis is that acyl peroxides in general can be reduced to anhydrides by phosphites (or phosphines) [11] but, since acyl peroxides are made from anhydrides, the process would have few applications.

(a) Preparation of diphenylmaleic anhydride. To 250 mg. of diphenylcyclobutenedione in 20 ml. of carbon tetrachloride, three 10-ml. portions of 30–35% hydrogen peroxide were added at 2-hr. intervals with shaking. After a total of 28 hr. of shaking, the carbon tetrachloride layer was separated and washed first with 30 ml. of 10% ferrous sulfate solution and then twice with 250 ml. of water. After drying, there was recovered from this solution 210 mg. (79%) of the anhydride, m.p. 154–155° (155–156° when recrystallized from 95% ethanol) [4].

(b) Preparation of benzoic anhydride (about 30% from the adduct of benzil and trimethyl phosphite and oxygen) [7].

3. From Aldehydes

Air oxidation of aldehydes to anhydrides has its complications because water is produced simultaneously with the anhydride. Such a reaction would appear to be more readily adapted to industrial than to laboratory methods.

(a) Preparation of butyric anhydride (20% from butyraldehyde and traces of cobalt metal at 40° through which oxygen was passed for 150 min.) [12].

1. H. J. Richter and F. B. Stocker, *J. Org. Chem.*, **24**, 214 (1959).
2. M. M. Dashevskiĭ, *J. Gen. Chem. USSR (Eng. Transl.)*, **30**, 1296 (1960).
3. D. V. Nightingale and W. S. Wagner, *J. Org. Chem.*, **25**, 32 (1960).
4. A. T. Blomquist and E. A. LaLancette, *J. Am. Chem. Soc.*, **83**, 1387 (1961).
5. P. Karrer et al., *Helv. Chim. Acta*, **31**, 1210 (1948); **32**, 950, 1932 (1949).
6. P. Karrer and L. Schneider, *Helv. Chim. Acta*, **30**, 859 (1947).
7. F. A. Ramirez et al., *J. Am. Chem. Soc.*, **82**, 2651 (1960).
8. J. E. Leffler, *J. Org. Chem.*, **16**, 1785 (1951).
9. F. A. Ramirez et al., *J. Am. Chem. Soc.*, **83**, 492 (1961).
10. F. A. Ramirez et al., *J. Am. Chem. Soc.*, **82**, 5763 (1960).
11. D. B. Denney and M. A. Greenbaum, *J. Am. Chem. Soc.*, **79**, 979 (1957).
12. T. Vrbaški and I. Brihta, *Arkiv. Kem.*, **24**, 111 (1952); *C.A.*, **49**, 163 (1955).

C. Electrophilic Reactions

1. From Carboxylic Acids and Methoxyacetylene

$$2\,RCOOH + CH_3OC\equiv CH \rightarrow (RCO)_2O + CH_3COOCH_3$$

This synthesis permits the conversion of many carboxylic acids into their anhydrides at or below room temperature. Crotonic and pivalic acids require

References for Section C are on p. 885.

mercuric oxide and phosphoric acid, respectively, as catalysts at room temperature. Cinnamic and palmitic acids remain inert [1]. Yields are fair to good.

It is suggested that an adduct 16-I is formed between the acetylene and the carboxylic acid:

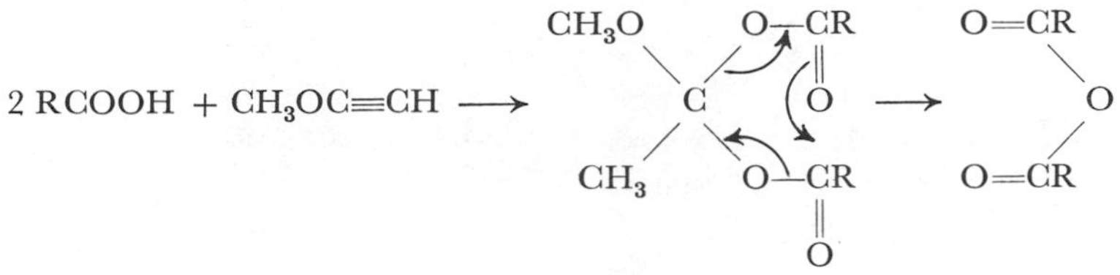

$$2\ RCOOH + CH_3OC\equiv CH \longrightarrow$$

16-I   16-II

and its decomposition via a cyclic transition state gives the anhydride 16-II [1, 2].

(a) **Preparation of phenylacetic anhydride.** Methoxyacetylene, 0.5 ml., was added to 1.4 g. of phenylacetic acid in 2 ml. of methylene dichloride. The mixture was allowed to stand at 20° for 18 hr., after which the solvent was removed by distillation to give 1.24 g. (96%) of the anhydride, m.p. 72–72.5° after recrystallization from light petroleum (b.p. 60–80°) [1].

2. Acid Anhydrides from the Oxo Process

$$CH_3COOCH_3 + CO \xrightarrow[H_2]{CoI_2} (CH_3CO)_2O$$

The oxo process has been discussed under Alcohols, Chapter 4, B.5, and Aldehydes, Chapter 10, C.8. Methyl acetate, carbon monoxide, and hydrogen in N-methylpyrrolidine in the presence of cobalt iodide have been heated to 190° under 700 atm. pressure for 17 hours to give a 60% yield of acetic anhydride [3].

1. G. Eglinton *et al.*, *J. Chem. Soc.*, 1860 (1954).
2. H. H. Wasserman and P. S. Wharton, *J. Am. Chem. Soc.*, **82**, 1411 (1960).
3. W. Reppe and H. Friederich, U.S. Patent 2,789,137, April 16, 1957; *C.A.*, **51,** 15553 (1957).

D. Cycloadditions

Since the Diels-Alder reaction utilizes maleic anhydride as a dienophile, many substituted anhydrides may be produced by this procedure (see Alkenes, Chapter 2, C.2).

Chapter 17

KETENES AND
KETENE DIMERS

A thorough discussion of the methods of preparing ketenes and ketene dimers is available [1], and the synthesis of ketene itself has been discussed more recently [2]. The methods bear some resemblance to those used in the preparation of olefins in that pyrolysis or elimination is often involved. The compounds bear a close resemblance to the structurally related ketenimines, $R_2C{=}C{=}NR$ [3], or carbodiimides, $RN{=}C{=}NR$ [4]. The type, being extremely reactive, may form a dimer immediately on liberation, or it may react with any solvent present if it contains active hydrogen. In fact, it is not unusual for the ketene to be passed directly into a reaction medium and thus to be isolated not in its original form, but as a derivative.

A. Pyrolysis or Decomposition

1. From Acids, Anhydrides, Ketones, and Esters

$$CH_3COCH_3 \rightarrow CH_2{=}C{=}O + CH_4$$

References for Section A are on p. 890.

It appears that any organic compound containing an acetyl group will give some ketene on pyrolysis. This synthesis is applicable to higher members of the series, but its use has been confined largely to the first member of the series which can be prepared best both commercially and in the laboratory by this method. Of the various laboratory procedures attempted [5], the best yield (90–95%) is obtained by passing acetone over Chromel A wire at 700–750°. If pyrolysis is carried out in a chamber, lower temperatures, around 500°, are preferable to avoid further decomposition to carbon monoxide and olefinic gases. The ketene obtained from acetone is mixed with methane, which may create an absorption problem. Pyrolysis of acetic anhydride [6] or of diketene (see A.4), on the other hand, does not yield adventitious gases.

Photolytic decompositions are potential pathways to ketenes. However, too many side reactions are possible for this process to be of practical value at the present time. For example, the photolysis of tetramethyl-1,3-cyclobutanedione yields about 20% dimethylketene (by spectral analysis) [7].

The pyrolysis of acetone, for example, is a free radical chain reaction in which the methyl radical is formed in the initiation step:

$$CH_3COCH_3 \rightarrow 2\ CH_3\cdot + CO$$

In the propagation step the acetonyl radical is produced:

$$CH_3COCH_3 + CH_3\cdot \rightarrow CH_3COCH_2\cdot + CH_4$$

and this decomposes to give ketene and the methyl free radical, available for further reaction:

$$CH_3COCH_2\cdot \rightarrow CH_2{=}C{=}O + CH_3\cdot$$

Ketene is comparable to phosgene both in the degree of its toxicity and in the type of physiological symptoms produced (pulmonary edema).

(a) **Preparation of ketene** (approaching 95% from acetone passed over Chromel A wire at 700–750°) [8].

(b) **Preparation of methylketene** (90% of fairly pure product containing a high percentage of dimer from propionic anhydride vapor passed through a carbon-coated quartz tube at about 550° (5 mm.)) [9].

2. From Malonic Acid Derivatives

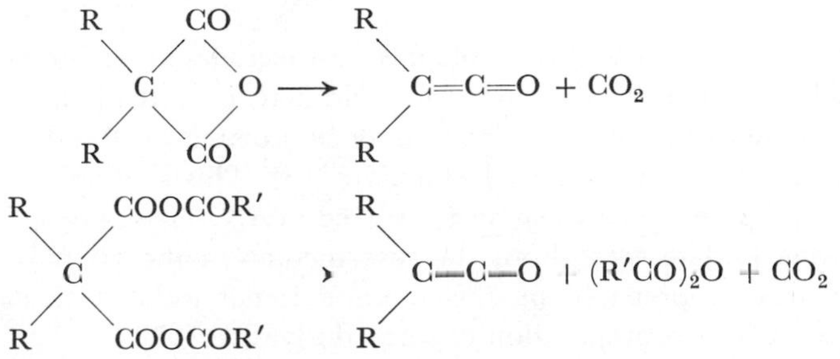

Ketenes have been synthesized by the decomposition of malonic anhydrides or mixed malonic acid anhydrides, provided there were two alkyl or aryl substituents in the malonic acid derivative. The anhydrides can be prepared from the disubstituted malonic acid by treatment with acetic anhydride in the presence of a small amount of sulfuric acid, followed by neutralization of the acid and evaporation to dryness. The anhydride residue is decomposed at low pressure to give the ketene. Dialkylketenes, low in the series, have been prepared in 50–80% yields by this method [1].

Originally the mixed anhydrides were obtained from disubstituted malonic acids and diphenylketene [10]. To avoid the use of this not easily prepared ketene, trifluoroacetic anhydride has been substituted for it [11]. This anhydride with a dialkylmalonic acid in equimolar amounts gives the acid anhydride 17-I, which with pyridine gives the malonic anhydride 17-II, capable of

$$R_2C(COOH)_2 + (CF_3CO)_2O \longrightarrow \underset{\text{17-I}}{R_2C(COOH)(COOCOCF_3)} \xrightarrow{C_5H_5N}$$

$$\underset{\text{17-II}}{R_2C(CO)_2O} + C_5H_5\overset{\oplus}{N}HCF_3C\overset{\ominus}{O}O$$

conversion into the ketene by pyrolysis. By this procedure, dibutylketene was obtained in 85% yield and diethylketene in 56% yield.

(a) Preparation of dimethylketene (65% from dimethylmalonic acid and acetic anhydride containing a trace of conc. sulfuric acid) [12].

(b) Preparation of dibutylketene (85% based on dibutylmalonic anhydride obtained from dibutylmalonic acid and trifluoroacetic anhydride followed by treatment with pyridine) [11].

3. From Ketohydrazones or Diazoketones

$$\underset{\underset{NNH_2}{\|}}{C_6H_5COCC_6H_5} \xrightarrow{HgO} \underset{\underset{\underset{N\ominus}{\|}}{N\oplus}}{C_6H_5COCC_6H_5} \longrightarrow \underset{C_6H_5}{\overset{C_6H_5}{\diagdown}}C=C=O + N_2$$

Ketohydrazones which are available from α-diketones and hydrazine may be decomposed with mercuric oxide [13] to yield ketenes as the final product. Or the intermediate in the reaction above may be isolated [14] and subjected to irradiation [15] or treated with silver oxide [16]. This synthesis is ordinarily employed to prepare diphenylketene as utilized in A.2. Yields are 58% from the ketohydrazone [13] or 92% from the intermediate azibenzil [15]. In many cases the ketene as prepared by this method is not isolated as such, but is employed *in situ* in the preparation of some derivative [17].

This synthesis has also been applied to the formation of β-ketoesters through the sequence

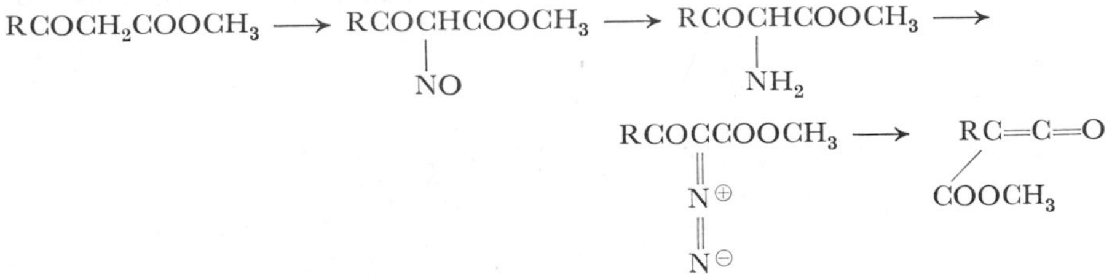

$$RCOCH_2COOCH_3 \longrightarrow \underset{\underset{NO}{|}}{RCOCHCOOCH_3} \longrightarrow \underset{\underset{NH_2}{|}}{RCOCHCOOCH_3} \longrightarrow$$

$$\underset{\underset{\underset{N\ominus}{\|}}{\underset{N\oplus}{\|}}}{RCOCCOOCH_3} \longrightarrow \underset{\underset{COOCH_3}{\diagup}}{RC{=}C{=}O}$$

By this scheme, for example, a 47% yield of the dimer of methylcarbomethoxy-ketene was obtained [18]. Similarly the diazo compound from methyl benzoyl-acetate gave phenylcarbomethoxyketene in 70% yield [19].

(a) **Preparation of diphenylketene** (58% from benzil monohydrazone, mercuric oxide, and anhydrous calcium sulfate in benzene) [13]; *see also* [15].

4. Depolymerization of Diketenes

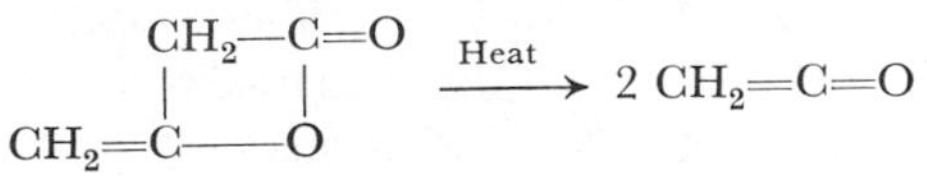

$$\underset{\underset{CH_2{=}C{-}O}{\overset{\quad|\quad\diagup}{}}}{\overset{CH_2{-}C{=}O}{\overset{|\quad|}{}}} \xrightarrow{\text{Heat}} 2\ CH_2{=}C{=}O$$

Diketene, the simplest dimer, is now available commercially and is perhaps best represented as the lactone of an enolic form of acetoacetic acid [20]. Pyrolysis of this product with a hot platinum filament or a hot tube at 550–600° gives ketene in quantitative yield [21]; see Ex. b. Other diketenes which have been studied respond similarly. For example, the dimer of dimethylketene gives an 86% yield of the monomer by decomposition over hot Nichrome wire [22], while the polymer of phenylcarbomethoxyketene when distilled in vacuum produces the monomer in 70–72% yield, based on the methyl benzoyldiazo-acetate used in producing the polymer [19]. This method is of limited value, particularly for ketenes other than the first member of the series. However, it is well to remember that some of the methods of synthesis (see B.2) yield diketenes rather than ketenes. In these cases the method may possess merit.

(a) **Preparation of dimethylketene** (86% from dimethylketene dimer by vaporization in a nitrogen atmosphere in the presence of hot Nichrome wire) [23].

(b) **Preparation of ketene** (46–55% from ketene dimer by heating to 550° in a nitrogen atmosphere) [24].

5. From Acid Chlorides

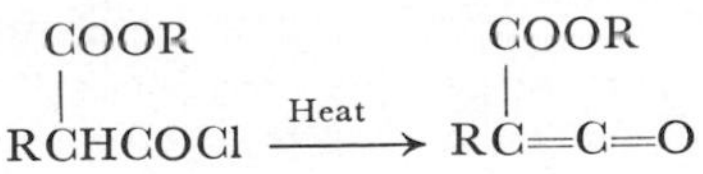

$$\underset{\underset{RCHCOCl}{\overset{|}{}}}{\overset{COOR}{}} \xrightarrow{\text{Heat}} \underset{\underset{RC{=}C{=}O}{\overset{|}{}}}{\overset{COOR}{}}$$

One example of the pyrolysis of an acid chloride to a ketene is recorded in the literature [25]. α-Carbethoxy-β-phenylisovaleroyl chloride 17-III, on being distilled gives (α,α-dimethylbenzyl)-carbethoxyketene 17-IV in high yield. The

$$
\underset{\text{17-III}}{C_6H_5C(CH_3)_2\overset{\overset{\displaystyle COOC_2H_5}{|}}{C}HCOCl} \longrightarrow \underset{\text{7-IV}}{C_6H_5C(CH_3)_2\overset{\overset{\displaystyle COOC_2H_5}{|}}{C}{=}C{=}O}
$$

method failed for the synthesis of t-butylcarbethoxyketene. It is interesting to note that these two ketenes, in contrast to ethylcarbethoxyketene [26] and phenylcarbethoxyketene [27], do not undergo dimerization. This inactivity has been attributed to the bulkiness of the group containing the tertiary carbon atom.

(a) Preparation of (α,α-dimethylbenzyl)-carbethoxyketene.

Thionyl chloride, 18 g. (0.15 mole) with 25 g. (0.1 mole) of α-carbethoxy-β-phenyl-isovaleric acid was allowed to stand overnight, after which the mixture was heated on a steam bath for 1 hr. The excess thionyl chloride was removed under reduced pressure and the yellow liquid remaining was distilled twice under reduced pressure to give 21.3 g. (92%) of the ketene, b.p. 151–154° 15 mm. [25].

1. W. E. Hanford and J. C. Sauer, *Org. Reactions*, **3**, 108 (1946).
2. R. N. Lacey, *Advan. Org. Chem.*, **2**, 213 (1960); G. Quadbeck, in Foerst, *Newer Methods of Preparative Organic Chemistry*, Academic Press, Vol. 2, New York, 1963, p. 133.
3. C. L. Stevens and G. H. Singhal, *J. Org. Chem.*, **29**, 34 (1964).
4. J. R. Schaeffer, *Org. Chem. Bull.*, **33**, No. 2 (1961).
5. J. W. Williams and C. D. Hurd, *J. Org. Chem.*, **5**, 122 (1940).
6. G. J. Fisher *et al.*, *J. Org. Chem.*, **18**, 1055 (1953).
7. I. Haller and R. Srinivasan, *Canad. J. Chem.*, **43**, 3165 (1965).
8. J. W. Williams and C. D. Hurd, *J. Org. Chem.*, **5**, 122 (1940); *Org. Reactions*, **3**, 132 (1946).
9. A. D. Jenkins, *J. Chem. Soc.*, 2563 (1952).
10. H. Staudinger *et al.*, *Chem. Ber.*, **46**, 3539 (1913); *Helv. Chim. Acta*, **6**, 291 (1923).
11. A. C. Duckworth, *J. Org. Chem.*, **27**, 3146 (1962).
12. H. Staudinger, *Org. Reactions*, **3**, 135 (1946).
13. L. I. Smith and H. H. Hoehn, *Org. Syn.*, Coll. Vol. **3**, 356 (1955).
14. C. D. Nenitzescu and E. Solomonica, *Org. Syn.*, Coll. Vol. **2**, 496 (1943).
15. L. Horner *et al.*, *Ann. Chem.*, **573**, 17 (1951).
16. R. Pfleger and A. Jäger, *Chem. Ber.*, **90**, 2460 (1957).
17. H. J. Bestman and H. Fritzsche, *Chem. Ber.*, **94**, 2477 (1961); W. Kirmse and L. Horner, *ibid.*, **89**, 2759 (1956).
18. G. Schroeter, *Chem. Ber.*, **49**, 2697 (1916).
19. H. Staudinger and H. Hirzel, *Chem. Ber.*, **49**, 2522 (1916).
20. V. V. Perekalin and T. A. Sokolova, *Russian Chem. Rev. (English Transl.)*, **25**, 1351 (1956).
21. A. B. Boese, Jr., *Ind. Eng. Chem.*, **32**, 16 (1940).
22. Ref. 1, p. 120.
23. Ref. 1, p. 136.
24. S. Andreades and H. D. Carlson, *Org. Syn.*, **45**, 50 (1965).
25. M. S. Newman and E. A. Zuech, *J. Org. Chem.*, **27**, 1436 (1962).
26. H. Staudinger and St. Bereza, *Chem. Ber.*, **42**, 4908 (1909).
27. H. Staudinger and H. Hirzel, *Chem. Ber.*, **50**, 1024 (1917).

B. Elimination

1. From α-Haloacid Halides

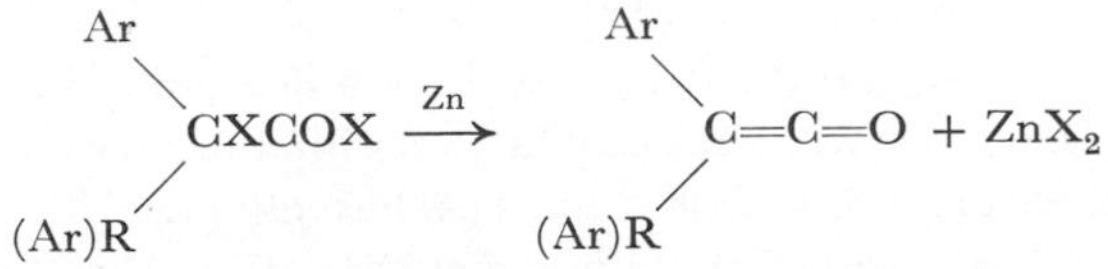

$$\underset{(Ar)R}{\overset{Ar}{\diagdown}}CXCOX \xrightarrow{Zn} \underset{(Ar)R}{\overset{Ar}{\diagdown}}C{=}C{=}O + ZnX_2$$

This is an older method of synthesis which has been employed particularly for the preparation of aromatic ketoketenes such as methylphenylketene (70%) [1], diphenylketene (unstated yield) [2], and di-p-xenylketene (over 60%) [3]. The procedure is simple in that the α-haloacid halide in an organic solvent under carbon dioxide or nitrogen is dehalogenated with zinc to give the ketene which may be used in solution or recovered by distillation. The reaction is not attractive for the synthesis of ketene or monosubstituted ketenes [4]. This limitation and the unavailability of α-haloacid halides accounts for the rare use of the method in recent years. Perhaps the use of triphenylphosphine as the dehalogenating agent will make this method more attractive (see Ex. c).

(a) **Preparation of dimethylketene** (46–54% from α-bromoisobutyryl bromide and zinc in ethyl acetate under a nitrogen atmosphere) [5].

(b) **Preparation of dichloroketene** (60–70% from trichloroacetyl bromide stirred in ether with Zn dust activated with Cu) [6].

(c) **Preparation of diphenylketene** [7]:

$$(C_6H_5)_2CHCOOH \xrightarrow[\text{2. Ice, water}]{\text{1. Br}_2 \text{ (excess), P, CCl}_4} \underset{89\%}{(C_6H_5)_2CBrCOBr} \xrightarrow[\text{Benzene (under N}_2)]{(C_6H_5)_3P, 5-10°}$$

$$(C_6H_5)_2C{=}C{=}O + (C_6H_5)_3PBr_2\downarrow$$

Concentrated and distilled
with hydroquinone under
reduced pressure, 81.5%

2. From Acid Chlorides

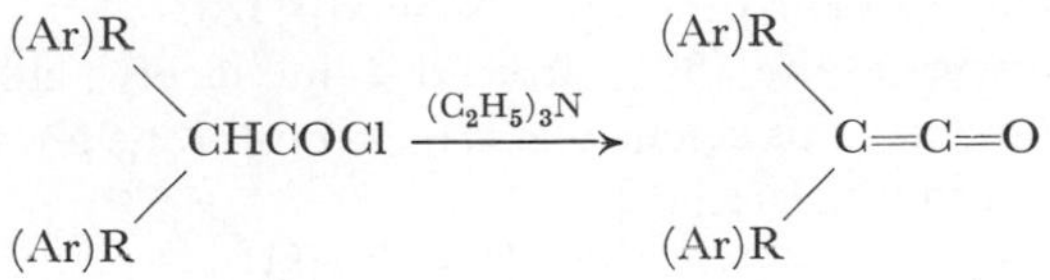

$$\underset{(Ar)R}{\overset{(Ar)R}{\diagdown}}CHCOCl \xrightarrow{(C_2H_5)_3N} \underset{(Ar)R}{\overset{(Ar)R}{\diagdown}}C{=}C{=}O$$

The extreme ease of ketene formation in this reaction suggests that acylations with acid chlorides in the presence of tertiary amines may proceed via the ketene. Both mono- and disubstituted acetyl chlorides dehydrohalogenate in the presence of a base to give the ketene. Although bases such as pyridine [8], quinoline [9], sodamide in liquid ammonia (see Ex. b.4), and low-molecular-weight aliphatic tertiary amines [10–13] have been employed, trialkylamines

References for Section B are on pp. 892–893.

are the preferred bases [12, 13]. Dimethylaniline and pyridine are unsatisfactory in the dehydrohalogenation of alkyl-substituted acetyl halides [12]. By one procedure using preferred bases the reaction was sometimes complete at room temperature in 1–3 hours [12].

One limitation of this synthesis is that dimers are obtained with the more reactive ketenes (see Ex. b.3). Thus aliphatic acid halides with one alkyl group attached to the α-carbon atom yield dimers exclusively [12, 13]. It is interesting to note, however, that none of the other methods of synthesizing ketenes described gives aldoketenes, ((R)ArCH=C=O). By this synthesis, dimers of this type at least are obtained. Yields vary greatly, although with the higher aliphatic acid halides 90% or above may be achieved for the dimers. Aromatic ketenes containing two aryl groups have been obtained as monomers with yields varying from 78% to an almost quantitative yield [14, 8, 15]. For further discussion of ketene dimerization, which results also in the formation of 1,3-butanediones, see Ketones, Chapter 11, F.8.

(a) Hexadecylketene dimer.   A solution of 15 g. of stearoyl chloride, b.p. 150° (1 mm.), in 225 ml. of anhydrous benzene was seeded with 0.5 g. of triethylamine hydrochloride. After addition of 6 g. of triethylamine at room temperature, the mixture was set aside for 16 hr. The amine salt was separated by inverted filtration [16], and on concentrating the filtrate on a steam bath under reduced pressure, the residue was taken up in ligroin. Cooling gave 12 g. (90.3%) of the dimer, m.p. 62–63° [12].

(b) Other examples

(1) Mesitylphenylketene (78% of the monomer from mesitylphenylacetic acid, thionyl chloride, and pyridine) [8].

(2) 4-Chlorophenoxyketene (77.5% from 4-chlorophenoxyacetyl chloride and a small excess of triethylamine) [17].

(3) Pentamethyleneketene (32% from hexahydrobenzoyl chloride and 1.1 equiv. of triethylamine in ether, the hydrochloride being removed without exposure to moisture; the ether was evaporated and the residue distilled, b.p. 40–41° (3 mm.); if the ether solution was allowed to stand for about 4 days, the crystalline ketene dimer was isolated) [18].

(4) Di-*t*-butylketene (57% from di-*t*-butylacetyl chloride and sodamide in liquid ammonia; this ketene seems to be one of the few aliphatic ketenes which is stable as a monomer) [19].

1. H. Staudinger and L. Ruzicka, *Ann. Chem.*, **380,** 278 (1911).
2. H. Staudinger, *Ann. Chem.*, **356,** 51 (1907); *Chem. Ber.*, **38,** 1735 (1905).
3. E. Schilow and S. Burmistrow, *Chem. Ber.*, **68,** 582 (1935).
4. C. D. Hurd *et al.*, *J. Org. Chem.*, **8,** 367 (1943).
5. C. W. Smith and D. G. Norton, *Org. Syn.*, Coll. Vol. **4,** 348 (1963).
6. W. T. Brady *et al.*, *J. Org. Chem.*, **31,** 626 (1966).
7. S. D. Darling and R. L. Kidwell, *J. Org. Chem.*, **33,** 3974 (1968).
8. R. C. Fuson *et al.*, *J. Am. Chem. Soc.*, **66,** 1464 (1944).
9. H. Staudinger and S. Schotz, *Chem. Ber.*, **53,** 1105 (1920).
10. J. C. Sauer, U.S. Patent 2,268,169; *C.A.*, **36,** 2737 (1942).
11. L. L. Miller and J. R. Johnson, *J. Org. Chem.*, **1,** 135 (1936).

12. J. C. Sauer, *J. Am. Chem. Soc.*, **69,** 2444 (1947); *Org. Syn.*, **31,** 68 (1951).
13. A. Sturzenegger *et al.*, *J. Org. Chem.*, **28,** 920 (1963).
14. H. Staudinger, *Chem. Ber.*, **44,** 1619 (1911).
15. E. C. Taylor *et al.*, in L. F. and M. Fieser, *Reagents for Organic Synthesis*, John Wiley and Sons, New York, p. 344.
16. R. W. Bost and E. W. Constable, *Org. Syn.*, Coll. Vol. **2,** 610 (1943).
17. C. M. Hill *et al.*, *J. Am. Chem. Soc.*, **72,** 2286 (1950).
18. C. M. Hill and M. E. Hill, *J. Am. Chem. Soc.*, **75,** 2765 (1953).
19. M. S. Newman *et al.*, *J. Am. Chem. Soc.*, **82,** 2498 (1960).

Chapter 18

CARBOXYLIC ACID AMIDES AND IMIDES

The scope of this chapter is limited almost entirely to the titled compounds, although some references are given to ureas and carbamates. However, the diverse heteroatom amides, such as guanidines, thioamides, hydrazides, and hydroxamic acids, are omitted. Their syntheses will be obvious from the methods described in this chapter, but specific detail can be found in the literature [1].

A. Solvolysis

A general discussion of solvolytic methods is available [2]. The wide range of activity among acids and their derivatives limits the number of *general* methods, although two stand out: from acid chlorides and ammonia or amines (A.2); and from acids and amines in the presence of a carbodiimide (A.1). An attractive third procedure is the reaction of a substituted sodamide (RNHM) with an ester (A.4).

As a protective group, t-butyloxycarbonyl, $(CH_3)_3COC-$, shows some advantages over the more traditionally used benzyloxycarbonyl group (see A.12).

1. From Carboxylic Acids and Their Ammonium Salts

$$RCOOH + R'NH_2 \rightleftharpoons RCOONH_3R' \rightarrow RCONHR' + H_2O$$

Since an equilibrium is involved in this reaction, an excess of either amine or acid is desirable, the choice depending on which component is less costly. Furthermore, the reaction is not general in that it succeeds well with straight-chain, aliphatic acids, but poorly with aromatic or hindered acids or with less basic amines.

References for Section A are on pp. 911–913.

The most elegant synthesis of amides from acids and amines is the carbodi-imide method:

$$RCOOH \xrightarrow[\text{THF, low temp.}]{C_6H_{11}N=C=NC_6H_{11}} \left[RC\!\!\overset{O}{\diagup}\!\!-O-C\!\!\overset{NC_6H_{11}}{\diagdown_{NHC_6H_{11}}} \right] \xrightarrow{R'NH_2}$$

$$RC\!\!\overset{O}{\diagup}\!\!-NHR' + (C_6H_{11}NH)_2C=O$$

(See Ex. a.) Yields of amides are almost quantitative. The method is particu-larly useful for the preparation of peptides [3] and suffers only from the expense of the reagent. Water soluble or dispersed acids, including peptides, may be converted into amides in water by means of water-soluble carbodiimides, such as N-ethyl-N'-3-dimethylaminopropylcarbodiimide, at *ca.* pH 4.7 [4].

The reaction has been carried out in a number of ways. With less volatile substances the components can be heated together [5]; nitrogen may be passed through to remove water [6]; or the water of reaction may be removed as an azeotrope with toluene (or other solvent) [7]. In some cases, anhydrides [8], acetone condensates such as

$$C_6H_5CH\!\!-\!\!-\!\!C=O$$
$$\quad | \qquad \quad |$$
$$\quad O \qquad \quad O$$
$$\qquad \diagdown \quad \diagup$$
$$\qquad \quad C$$
$$\qquad \diagup \quad \diagdown$$
$$\quad CH_3 \qquad CH_3$$

or phosphazoanilides, R—N=P—NHR (from phosphorus trichloride) [9] can be prepared *in situ* and amidated without isolation. One of the best mixed anhydride systems to prepare amides is the lithium salt of the acid and sulfur trioxide in dimethyl formamide:

$$RCOOLi + SO_3{\cdot}DMF \longrightarrow RCOOSO_2OLi \xrightarrow{R_2'NH} RCONR_2'$$

In this manner lysergic acid dimethylamide [10] and peptides [11] have been prepared. It has also been possible to prepare amides by subjection of acid and amine vapors to silica gel surface at 280° [12] and by azeotroping the water with xylene after using catalytic amounts of Amberlite IR 120 (H⁺) [13]. But even in aqueous solution, some tendency exists for the amide to form an equilibrium mixture with the amine and acid, particularly in the case of dibasic acids [14].

It is tempting to suggest that amide formation from the salt takes place via the anhydride, but the anhydride has not been detected in the reaction mixture [15]. The mechanism evidently is complex.

Five-membered cyclic imides form readily from dicarboxylic acids with 1 equivalent [16] or with 2 equivalents of amine [17]:

$$\begin{array}{ccc} CH_2COOH & & CH_2CO \\ | & \xrightarrow[\text{2 RNH}_2]{\text{RNH}_2\ \text{or}} & | \quad \diagdown \\ | & & \quad\quad NR \\ CH_2COOH & & CH_2CO \diagup \end{array}$$

or from diammonium salts by heating [18]:

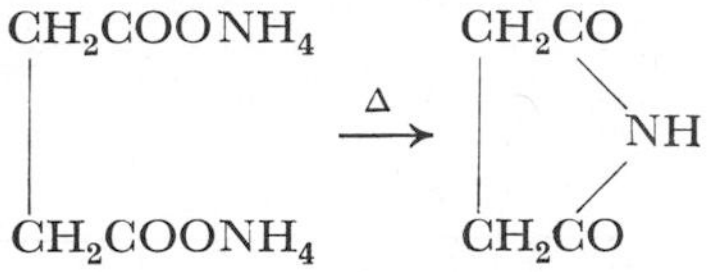

$$CH_2COONH_4 \atop | \atop CH_2COONH_4 \quad \xrightarrow{\Delta} \quad {CH_2CO \atop | \atop CH_2CO} {\Large >} NH$$

Most frequently, however, imides are prepared by heating the dicarboxylic acid monoamide (see Ex. c.7)

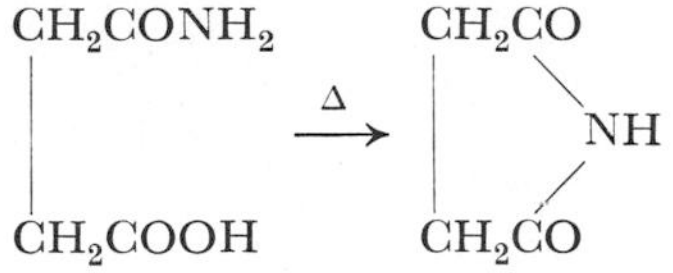

$$CH_2CONH_2 \atop | \atop CH_2COOH \quad \xrightarrow{\Delta} \quad {CH_2CO \atop | \atop CH_2CO} {\Large >} NH$$

These acid amides can also be converted into imides via the isoimide [19]:

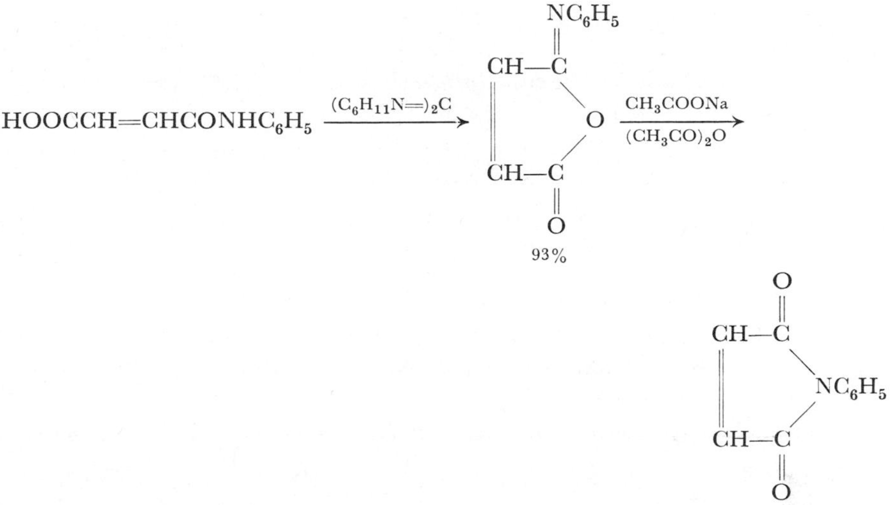

$$HOOCCH{=}CHCONHC_6H_5 \xrightarrow{(C_6H_{11}N=)_2C}$$

NC$_6$H$_5$

CH—C

CH—C

O

93%

$$\xrightarrow[(CH_3CO)_2O]{CH_3COONa}$$

O

CH—C

CH—C

O

NC$_6$H$_5$

N-Phenylmaleimide, 90%

Diamides are converted into cyclic imides by heating alone [20] or in α-methylnaphthalene [21]. Ester amides yield cyclic imides when heated under reduced pressure [22].

Six-membered cyclic imides form somewhat less readily than five-membered ones [16], and seven-membered ones, such as adipimide, form with difficulty (about 1 % yield from pyrolysis of adipamide) [23]. Imides of monocarboxylic acids are made usually from the amide and corresponding anhydride (see A.3).

It is interesting to note that heating of specific mixtures of α-aminoacids gives peptide-like molecules, called protenoids, of molecular weight about 5000 and of non-random distribution, glutamic and aspartic acids, lysine, and alanine being incorporated more readily than other amino acids [24].

Dipeptides are converted into 2,5-piperazinediones by heating the formate salt of the dipeptide ester [25] or by heating the dipeptide itself in phenol just below the latter's boiling point [26]:

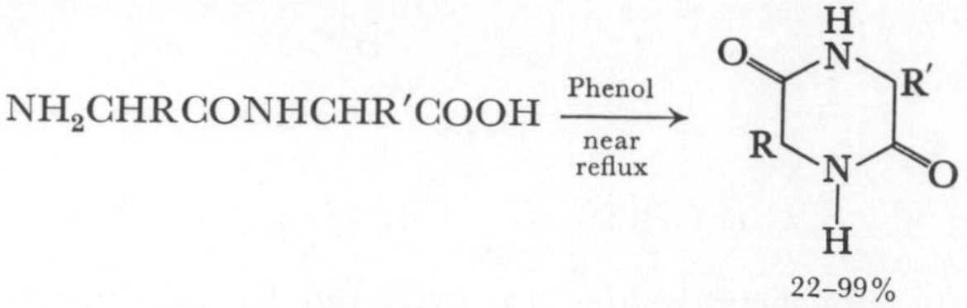

$$NH_2CHRCONHCHR'COOH \xrightarrow[\substack{near \\ reflux}]{Phenol}$$

22–99%

A formic acid-triethylamine complex of high boiling point is capable of partial dealkylation of aromatic tertiary amines to form formamides [27]:

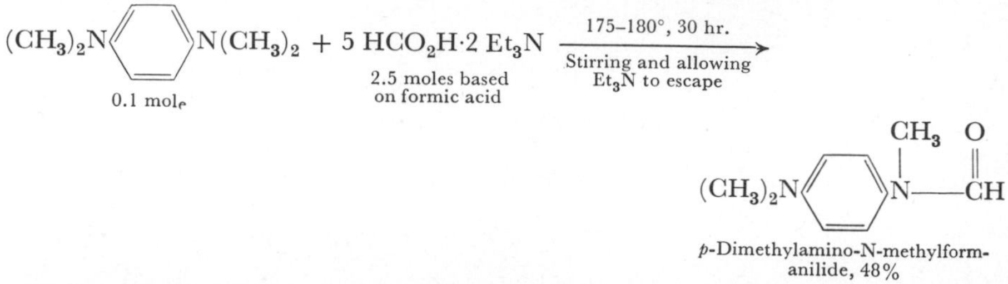

$$(CH_3)_2N\langle\!\!\!\!\!\bigcirc\!\!\!\!\!\rangle N(CH_3)_2 + 5\ HCO_2H \cdot 2\ Et_3N \xrightarrow[\substack{Stirring\ and\ allowing \\ Et_3N\ to\ escape}]{175-180°,\ 30\ hr.}$$

0.1 mole   2.5 moles based
                  on formic acid

$$(CH_3)_2N\langle\!\!\!\!\!\bigcirc\!\!\!\!\!\rangle N\overset{CH_3}{\underset{}{|}}\!-\!\overset{O}{\overset{\|}{C}}H$$

p-Dimethylamino-N-methylform-
anilide, 48%

(a) Preparation of N-acetylphenethylamine (75 % from phenethyl-amine, acetic acid, and DCC in THF at 0° for 3 hr., the pH being adjusted to 7.5 with trifluoro- or trichloroacetic acid; with aromatic amines which are less basic, it is not necessary to adjust the pH) [28].

(b) Preparation of N-methylformanilide (93–97 % from formic acid and methylaniline in toluene) [7].

(c) Other examples

(1) Acetamide (87–90 % from ammonium acetate) [29].

(2) N-Methyl-α-phenylsuccinimide (83 % from phenylsuccinic acid and aq. methylamine heated) [17].

(3) Ethyl 4-carboxamidobicyclo[2.2.2]octane-1-carboxylate (87 %

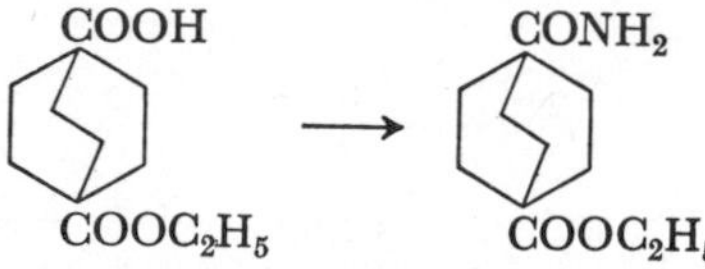

COOH   CONH₂

$\longrightarrow$

COOC₂H₅   COOC₂H₅

from ethyl hydrogen bicyclo[2.2.2]octane-1,4-dicarboxylate, triethylamine, ethyl chlorocarbonate, and ammonia) [30].

(4) N,N²-Dibenzyl-*dl*-aspartimide hydrochloride. N,N²-Dibenzyl-α-asparagine, 4 g., was dissolved in 110 ml. of acetyl chloride and

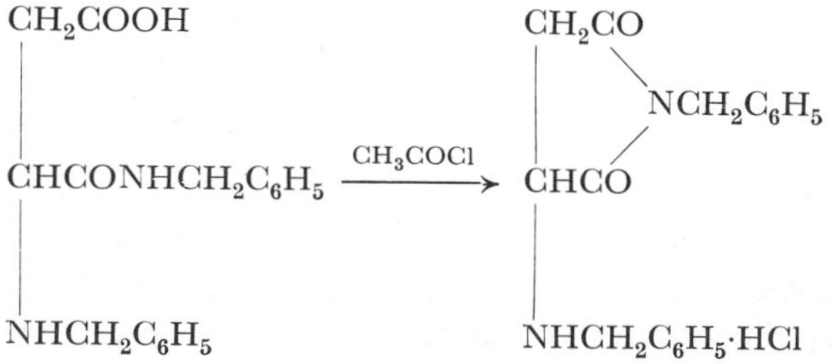

CH₂COOH   CH₂CO

                                    ＼
                                     NCH₂C₆H₅
                                    ／
CHCONHCH₂C₆H₅ $\xrightarrow{CH_3COCl}$ CHCO

NHCH₂C₆H₅   NHCH₂C₆H₅·HCl

left for 6 hr., after which the solid formed was removed and washed with acetyl chloride to give 2.6 g. (59%), m.p. 191°. Recrystallization from ethanol elevated the melting point to 195° [31].

(5) **N-Ethylphthalimide** (quantitative by shaking N-ethylphthalamide in water) [20].

(6) **N,N-Diethylvaleramide** (over 80% from the reaction shown;

$$SiCl_4 + 4 C_4H_9CO_2H \xrightarrow{C_6H_6} (C_4H_9CO_2)_4Si + 4 HCl \xrightarrow{4 (C_2H_5)_2NH}$$

$$4 C_4H_9CON(C_2H_5)_2$$

a 50% excess of acid and silicon tetrachloride relative to the amine gave maximum yields; the method suffers from the formation of gelatinous silicic acid) [32].

(7) **Glutarimide** (58–65% by heating the monamide in a bath at 220–225° for 3–4 hr.) [33].

2. From Acid Chlorides

$$RCOCl \xrightarrow{NH_3} RCONH_2$$

This synthesis represents perhaps the best method for preparing amides. Yields are usually in the 80–90% range, and purification of the product is rarely difficult. Ammonia, ammonium salts [34], and primary or secondary amines are the customary amidating agents. The reaction with concentrated aqueous ammonia usually is quite exothermic [35]. One equivalent of the amine is lost as hydrochloride:

$$RCOCl + 2 R'NH_2 \rightarrow RCONHR' + R'NH_3Cl$$

a circumstance which is of no concern in most cases, but which can be mitigated for water-insoluble (or less reactive) acid chlorides by shaking the acid chloride with amine in aqueous base solution. Another method of utilizing a valuable amine completely is to add the acid chloride to a mixture of the amine and a tertiary amine such as triethylamine. Organic solvents such as ethylene dichloride [36, 37], ether [38, 39], benzene [40], carbon tetrachloride [41], chloroform [42], and toluene [43] have been used to mitigate the exothermicity of amidation. Superior yields of benzoylated aminoacids are obtained by treating 1 mole of the acid with 1 mole of benzoyl chloride in the presence of 2 moles of aqueous sodium hydroxide at 1° [44].

Two unusual amidating reagents are N-phenyl-N',N'-dimethylformamidine [38]:

$$RCOCl + C_6H_5N=CHN(CH_3)_2 \longrightarrow \underset{\oplus}{C_6H_5N} = CHN(CH_3)_2 \xrightarrow{NaOH}$$

where the group on the amidine nitrogen is $\overset{O}{\overset{\|}{CR}}$

$$C_6H_5NHCOR$$

and dimethylformamide [45]:

$$RCOCl + HCON(CH_3)_2 \rightarrow RCON(CH_3)_2 + CO + HCl$$

(See Ex. b.4.) The latter reagent avoids the use of gaseous dimethylamine or aqueous solutions.

(a) Preparation of isobutyramide (78–83% from isobutyryl chloride and 28% cold aqueous ammonia) [46].

(b) Other examples

(1) Benzamide (92% from benzoyl chloride and ammonium acetate in acetone) [34].

(2) N-Ethyl-α-bromoacetamide (82% from bromoacetyl bromide, ethylamine hydrochloride, and aq. sodium hydroxide in ethylene dichloride at −10°) [37].

(3) N-(2-Pyridyl)-2′-thiophenecarboxamide (85% from the sodium derivative of 2-aminopyridine and 2-thenoyl chloride in toluene) [43].

(4) N,N-Dimethylbenzamide (97% from benzoyl chloride and dimethylformamide heated to 150° for 4 hr.; if anhydrides are used, a drop of sulfuric acid accelerates the amidation) [45].

3. From Acid Anhydrides

$$
\begin{array}{c}
RC{=}O \\
\diagdown \\
O \xrightarrow{\ NH_3\ } RCONH_2 + RCOOH \\
\diagup \\
RC{=}O
\end{array}
$$

$$
\begin{array}{c}
RCHC{=}O \\
| \quad \diagdown \\
O \xrightarrow{\ NH_3\ } \begin{array}{c} RCHC{=}O \\ | \qquad \diagdown \\ NH \\ | \qquad \diagup \\ RCHC{=}O \end{array} \\
| \quad \diagup \\
RCHC{=}O
\end{array}
$$

Ammonia, amines, urea, and urethanes are the usual amidating agents in this synthesis, which appears to have been employed more widely in preparing imides rather than amides. For acyclic anhydrides, amides are produced while cyclic anhydrides give imides, acid amides, or diamides depending on the reagent and the experimental conditions. Recently, superior yields of amides have been obtained by the use of mixed anhydrides produced from the free acid, diethylamine, and ethyl chloroformate (see examples under A.1 for mixed anhydrides prepared *in situ*). In most cases, mixed anhydrides give acid derivatives derived from the weaker acid portion of anhydrides with the exception of those from trifluoroacetic acid. Here mixtures of amides are obtained [47]. A large number of trifluoroacetamides have been described recently for identification of amines by gas chromatography [48]. Acylated amino acids may be obtained in good yield from the free acid and the acid anhydride [49], while the corresponding imino acids are produced from the free acid and the cyclic anhydride, best in the presence of 0.1 equivalent of triethylamine [50]. The yields in these transformations are usually 80% or higher.

Acyclic imides are produced by treating the amide with the corresponding anhydride in the presence of an acid catalyst (see Ex. b.4).

For less reactive amines such as diphenylamine or 2,4-dinitroaniline, acetylation with acetic anhydride should be carried out with an acid catalyst, such as a drop of sulfuric or other strong acid, in such a way that charring does not take place.

(a) Preparation of phthalimide (95–97% from phthalic anhydride and ammonium hydroxide) [51]; *see also* [52] for preparation with urea.

(b) Other examples

(1) **Acetylglycine** (89–92% from glycine, water, and acetic anhydride) [49].

(2) ***t*-Butylphthalimide** (72–76% from phthalic anhydride and *t*-butylurea) [53].

(3) **4-Chlorophthalamide** (75% from 4-chlorophthalimide and aq. ammonia) [20].

(4) **Dipropionamide (propionimide)** (74% from propionamide, propionic anhydride, and 0.1 molar equiv. of 100% sulfuric acid at 100° for 1 hr.; without the sulfuric acid, the main product was propionitrile) [54].

(5) **N,N-Dimethylbenzamide** (from benzoic anhydride, dimethylformamide, and sulfuric acid catalyst; see A.2, Ex. b.4).

(6) **Maleimide** (55% overall yield from the Diels-Alder adduct of furan and maleic anhydride, subjected to bromination, imidation, and finally a retrograde Diels-Alder) [55].

4. From Esters, Lactones, or Phthalides

$$RCOOR' \xrightarrow{NH_3} RCONH_2 + R'OH$$

This synthesis has been accomplished with ammonia or unhindered amines as the amidating agent, although the use of the former is much more common. The ammonia has been employed in aqueous solution [56–59], in alcoholic solution [60], and as liquid ammonia [61–64].

With ethyl formate and primary and secondary amines, amides may be produced in good yield by refluxing for 2 hours or more [65]:

$$HCOOC_2H_5 + \underset{\text{H} \quad \text{NH}_2}{\bigcirc} \xrightarrow[\text{2 hr.}]{\text{Reflux}} \underset{\text{H} \quad \text{NHCHO}}{\bigcirc} + C_2H_5OH$$

N-Cyclohexylformamide, 90%

One of the best leaving groups in esters is the *p*-nitrophenyl. *p*-Nitrophenyl esters are used frequently in making peptides, and in this regard it has been shown that the inclusion of 10 equivalents of imidazole decreases the time of peptide formation from 48 hours to 1 hour, a process no doubt occurring via the acylimidazole [66]. Lesser amounts of imidazole decrease the time of completion to intermediate values.

The reaction of esters with ammonia is catalyzed by water, glycols, and related compounds [67], while salts, such as ammonium chloride, promote a similar reaction with amines [68]. Isopropenyl acetate is a reactive ester quite suitable for preparing acetamides [69]. In some cases, such as in the conversion of a triethyl ester into a triamide [62], pressure is applied to promote the reaction. In other cases, unreactive esters, such as diethyl diethylmalonate, react satisfactorily with sodioanilide [70], a reaction which bears some resemblance to the use of an aminomagnesium halide (Bodroux reaction) for the conversion of esters into amides (see Ex. c.3). Indeed, the use of amine-metal salts (RNHNa) is the most searching method of converting esters into amides (see Ex. b). α,α-Disubstituted malonic esters may also be converted into diamides with yields from 50 to 76% by heating with formamide in the presence of sodium methoxide [71].

Chelated peptide esters may be produced in 1 minute at 20° by treating the ester, $Co(en)_2(GlyOCH_3)(ClO_4)_3$, with the amino acid or peptide ester in acetone, dimethylsulfoxide, or sulfolane [72]:

$$Co(en)_2(GlyOCH_3)(ClO_4)_3 + \text{peptide-OR} \rightarrow$$

$$(Co(en)_2(Gly\text{-peptide-OR}))(ClO_4)_3 + CH_3OH$$

(en = ethylenediamine, GlyOCH$_3$ = methyl glycinate, R = CH$_3$)

Lactones or phthalides also are converted into imides with amines [73] or by opening the ring with potassium cyanide, treating the salt formed with a mineral acid, and heating the acid [33]:

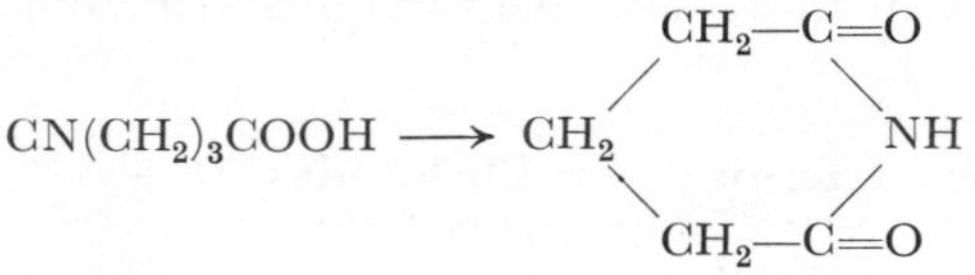

$$CN(CH_2)_3COOH \longrightarrow$$

Glutarimide,
58–65%

Imide formation also can be carried out simultaneously with condensation [74]:

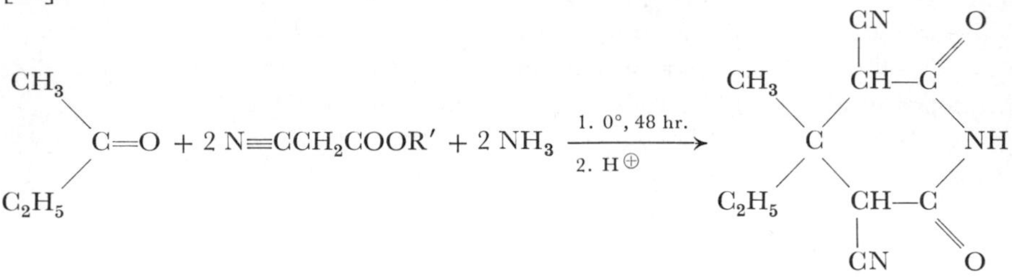

α,α'-Dicyano-β-ethyl-β-methyl-glutarimide, 65–70%

(a) Preparation of cyanoacetamide (86–88% from ethyl cyanoacetate and conc. aqueous ammonia) [59].

(b) Preparation of butyro-p-toluidide. p-Toluidine, 1 g., dissolved in 10 ml. of dimethylsulfoxide, was treated with 0.5 g. of sodium hydride (50% dispersion in mineral oil) and warmed until no further hydrogen evolution was noted. The deep purple solution was cooled to room temperature and treated with *n*-propyl butyrate. After 1 hr. standing, the mixture was poured into water, extracted with ether, and the ether extract washed with dilute acid and base and dried. Evaporation gave 1.1 g. of reddish crystals which recrystallized from hexane, using Norite, deposited colorless, fluffy needles, 0.5 g., m.p. 72–73°. This procedure is a general method of derivatizing the acid portion of an ester [75].

(c) Other examples

(1) **Fumaramide** (80–88 % from diethyl fumarate, conc. ammonium hydroxide, and ammonium chloride) [57].

(2) **Lactamide** (70–74 % from ethyl lactate and liquid ammonia) [64].

(3) **Hexanoanilide** (87 % from aniline, methylmagnesium iodide, and

$$2\ C_6H_5NH_2 \xrightarrow{\ 2\ CH_3MgI\ } 2\ [C_6H_5NHMgI] \xrightarrow{\ CH_3OOCC_5H_{11}\ } C_6H_5NHCOC_5H_{11}$$

methyl hexanoate; 2 equiv. of amine metal salt needed for maximum yields) [76].

(4) **2-Ketomorpholine** (34 % from aziridinium borotetrafluoride and

butyl glycolate at 100° followed by neutralization of the acidic medium) [77].

(5) **N,N-Dimethylbenzohydrazide** (56 % from 0.05 mole of ethyl benzoate, 0.2 mole of 1,1-dimethylhydrazine, and 0.1 mole of sodium methoxide refluxed in 50 ml. of methanol for 24 hr.; the reaction is general but does not proceed without the methoxide base catalyst) [78].

5. From Nitriles

$$RCN \xrightarrow{\ H_2O\ } RCONH_2$$

The reagent of choice in this synthesis is aqueous sodium hydroxide containing about 6–12 % hydrogen peroxide (see Exs. a and c.4). The hydroperoxide anion (HOO^-) is several thousand times more reactive than the hydroxyl anion. Aromatic nitriles give almost quantitative yields, although 30 % hydrogen peroxide must be used with *o*-substituted ones. Alkyl cyanides do not always give good results.

Acids which have been employed as reagents are concentrated hydrochloric [79], concentrated sulfuric [80], polyphosphoric [81], hydrogen halides in glacial acetic [82, 83], boron trifluoride in acetic and water [84], and a mixture of concentrated hydrochloric and concentrated sulfuric [85]. Aqueous sodium hydroxide [86], potassium hydroxide in aqueous ethylene glycol monoethyl ether [87], and resins Amberlite IRA 400-OH and Permutit ESB [88, 89] are some of the alkaline reagents which have been utilized in the hydrolysis. In the hydrolysis of dicyanides [86, 90, 91], imides are often obtained. The strongly basic anion exchange resins such as Amberlite IRA-400-OH and Permutit

ESB permit the formation of cyanocarboxylic acid amides both from aliphatic dinitriles having more than one methylene group between the functional groups and from aromatic dinitriles, the *o*-dinitriles being an exception [89]. The highly substituted nitrile, tributylacetonitrile, when treated with potassium hydroxide or sulfuric acid at varying concentrations, gave the best yield of amide with 80% sulfuric [92]. Hydrolysis to amides under anhydrous conditions at room temperature may be achieved in high yield by the use of sodium methylsulfinylmethide [93]. Yields usually vary from 75 to 95%.

Amidines

or their salts are best prepared from nitriles, ammonium chloride, and ammonia at 125–150° under autoclave conditions [94].

(a) Preparation of nicotinamide. Nicotinonitrile (2 g.), ethyl alcohol (10 ml.), and 30% hydrogen peroxide (7.5 g., 3.5:1 molar ratio with nitrile) were mixed and chilled. Aqueous sodium hydroxide (6.3 N, 1 ml.) was added slowly so that the temperature of the reaction did not exceed 50°. When the exotherm subsided, the mixture was heated for an additional 6 hr. at 50°, neutralized with 5% sulfuric acid, and evaporated under reduced pressure. The dry residue was extracted 15 times with 10 ml. portions of hot ethyl acetate, the combined portions being evaporated to yield 87.4% of nicotinamide [95].

(b) Preparation of β-cyclopentylglutarimide (97% from β-cyclopentylglutaronitrile and conc. hydrochloric acid by the procedure described in [96, 90].

(c) Other examples

(1) *p*-Toluamide (97% from *p*-tolunitrile, water, and the boron trifluoride-acetic acid complex) [84].

(2) 4-Methyl-3-pyridinecarboxamide (89% from 3-cyano-4-methylpyridine, Amberlite IRA 400-OH and water at reflux temperature) [88].

(3) 2,3-Diphenyltartarodiamide (82% from 2,3-diphenyltartaronitrile and hydrogen bromide-glacial acetic acid at room temperature) [82].

(4) Veratric amide (87–92% from veratronitrile, 3% H_2O_2, and 25% aq. KOH at 45°) [97].

(5) *o*-Toluamide (95% from *o*-tolunitrile and PPA at 115° for 1.5 hr.) [81].

6. From Keto Nitriles

Keto nitriles are versatile acylating agents, and with amines they give high yields of amide.

$$RCOCN + R'NH_2 \rightarrow RCONHR' + HCN$$

The cyano group in reactions of this nature is called a pseudo halogen.

(a) **Preparation of N-benzoylephedrine.** Ephedrine and a slight excess of benzoyl cyanide in ether deposited the amide in 75 % yield after standing for a short time; O,N-dibenzoylephedrine was made in 65 % yield from the same components in refluxing dimethylformamide [98].

7. From 1,1,1-Trihalides

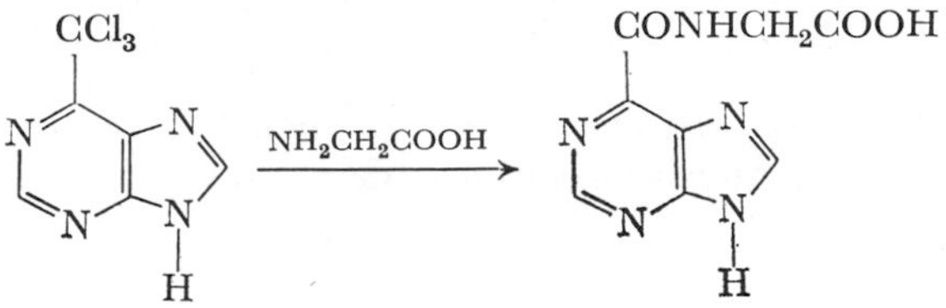

6-Trichloromethylpurine reacts, under mild conditions, with the amino groups of amino acids, peptides, and primary and secondary amines to yield amides containing the purine group [99]. The synthesis offers a simple way to introduce the purinoyl group into systems of likely biological interest. Yields vary from 18 to 83 %. Some speculation about the path taken in acidic or neutral media or in the pH range 7–9 has been advanced [99].

(a) **Preparation of (N-purinoylglycyl) glycylglycine.** A mixture of 1 g. of 6-trichloromethylpurine, 1 g. of glycylglycylglycine, and 3 g. of sodium bicarbonate in 100 ml. of water was stirred at room temperature for 4–5 hr. After acidification to pH 1 with concentrated hydrochloric acid, 1 g. (71 %) of (N-purinoylglycyl) glycylglycine precipitated. Purification was accomplished by dissolving the product in dilute aqueous ammonia, treatment with charcoal, and reacidification. Decomposition point was 283–285° [99].

8. From Ketenes, Diketenes, or Carbodiimides

$$CH_2{=}C{=}O \xrightarrow{\ NH_3\ } CH_3CONH_2$$

Many active hydrogen compounds such as water (see Carboxylic Acids, Chapter 13, G.1), alcohols (see Carboxylic Esters, Chapter 14, A.4), and ammonia and amines add to ketenes to produce carboxylic acids, carboxylic esters, and amides, respectively. The acylation of amines by ketenes can be accomplished in the presence of hydroxy compounds since the acylation of the hydroxyl groups does not proceed well unless catalyzed by acid [100]. In a similar manner, amines add to diketene, a commercial product, to give amides of acetoacetic acid [101]. The isolation of the ketene may be avoided by starting with the diazodiketone [102] which, when irradiated with the amine by ultraviolet light in a nitrogen atmosphere, gives the amide:

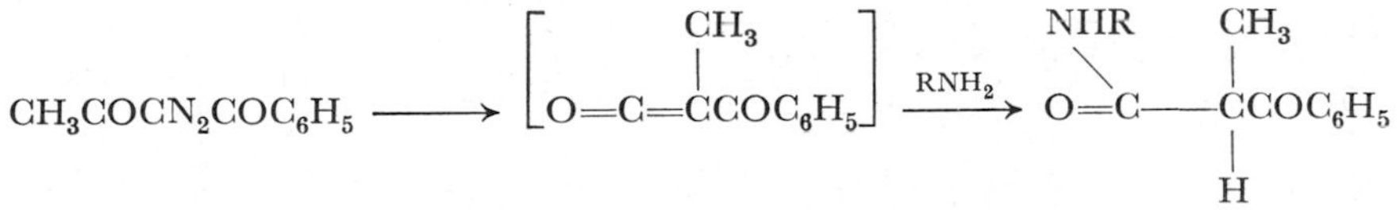

Yields in this synthesis are fair. The mechanism is discussed under Carboxylic Acids, Chapter 13, G.1.

Disubstituted ketenes react with Schiff bases to form β-lactams [103]:

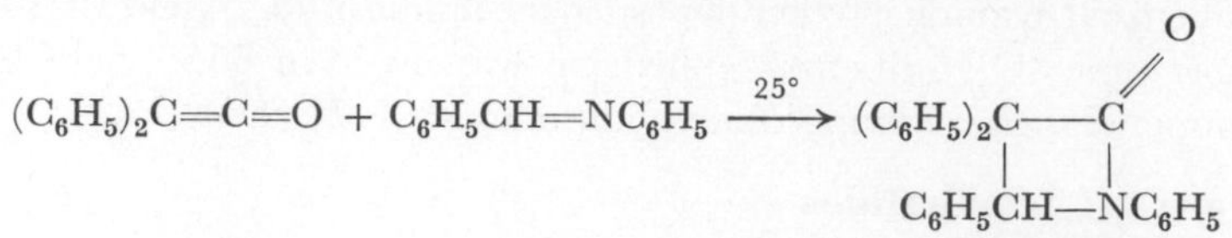

$$(C_6H_5)_2C=C=O + C_6H_5CH=NC_6H_5 \xrightarrow{25°} (C_6H_5)_2C-\overset{\displaystyle O}{\overset{\|}{C}}$$

$$\underset{C_6H_5CH-NC_6H_5}{}$$

1,3,3,4-Tetraphenyl-2-
azetidinone, 72%

In a relatively unexplored field, dicyclohexylcarbodiimide has been found to react with electronegatively substituted phenols to form ureas [104]:

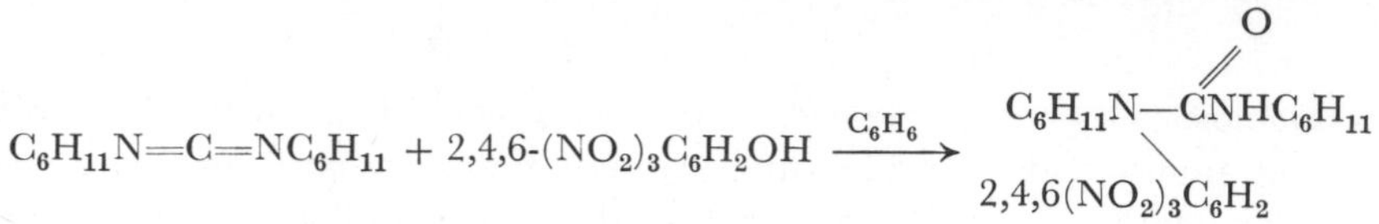

$$C_6H_{11}N=C=NC_6H_{11} + 2,4,6\text{-}(NO_2)_3C_6H_2OH \xrightarrow{C_6H_6}$$

$$\underset{2,4,6(NO_2)_3C_6H_2}{C_6H_{11}N-\overset{\displaystyle O}{\overset{\|}{C}}NHC_6H_{11}}$$

"Good yield"
N,N′-Dicyclohexyl-N(2,4,6-
trinitrophenyl) urea

Other phenols, specifically catechols, give the isoimide,

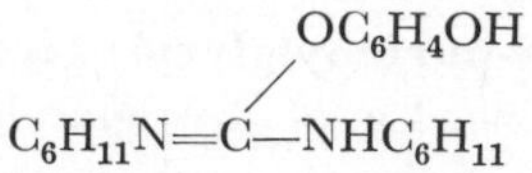

$$C_6H_{11}N=\overset{OC_6H_4OH}{\overset{\diagup}{C}}-NHC_6H_{11}$$

N,N′-Dicyclohexyl-0-(2-hydroxyphenyl)-isourea

(a) Preparation of acetoacetanilide (74% from the ketene dimer and aniline) [101].

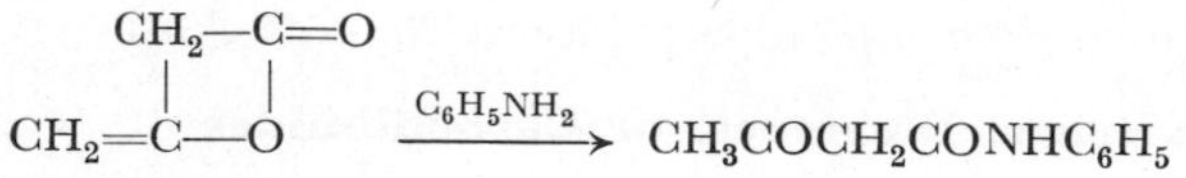

$$\underset{CH_2=\overset{|}{C}-O}{\overset{\displaystyle CH_2-C=O}{\overset{|}{}}} \xrightarrow{C_6H_5NH_2} CH_3COCH_2CONHC_6H_5$$

(b) Other examples

(1) 4-Chlorophenoxyacetamide (73% from 4-chlorophenoxyketene and ammonia) [105].

(2) α-Methylbenzoylacetanilide [102]:

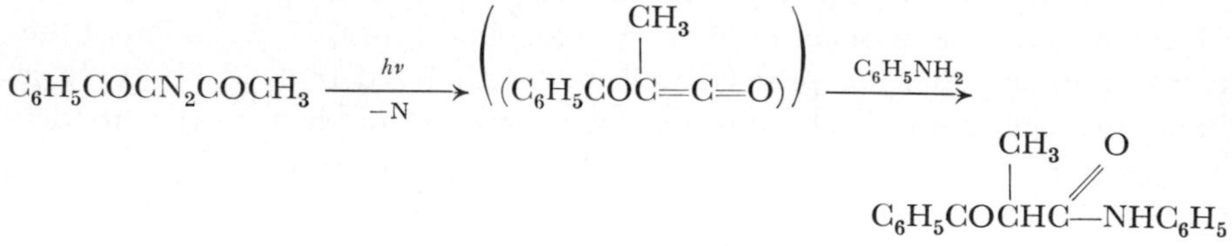

$$C_6H_5COCN_2COCH_3 \xrightarrow[-N]{hv} \left(\underset{(C_6H_5CO\overset{|}{C}=C=O)}{\overset{\displaystyle CH_3}{\overset{|}{}}} \right) \xrightarrow{C_6H_5NH_2}$$

$$C_6H_5CO\overset{CH_3}{\underset{|}{C}}H\overset{O}{\overset{\|}{C}}-NHC_6H_5$$

α-Methyl-α-benzoylacetanilide,
44%

9. From Thioamides

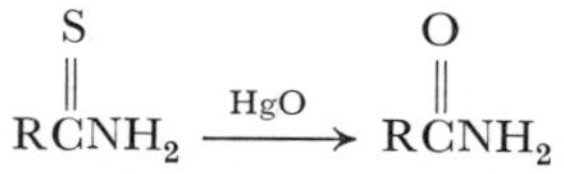

$$\underset{RCNH_2}{\overset{\displaystyle S}{\overset{\|}{}}} \xrightarrow{HgO} \underset{RCNH_2}{\overset{\displaystyle O}{\overset{\|}{}}}$$

Various reagents, such as silver nitrate in alkaline solution or mercuric oxide [106], litharge in boiling aqueous propanol [107], and potassium hydroxide in 95% ethanol [108], have been employed to convert thioamides into amides. Yields vary from fair to good.

(a) Preparation of 4-pyridineacetomorpholide hydrochloride. 4-Pyridinethioacetomorpholide, 250 g., in 1200 ml. of 95% ethanol containing 68 g. of potassium hydroxide was refluxed for 72 hr., after which the mixture was poured into 2 volumes of water and concentrated to $\frac{1}{3}$ volume. The amide, 207 g. (89%), when crystallized from ethanol, melted at 204–205° [108]; *see also* [107].

10. From Thiolacids

$$\text{ArCOSH} \xrightarrow{\text{NH}_2\text{CH}_2\text{COOH}} \text{ArCONHCH}_2\text{COOH}$$

The use of the thiolacid offers some advantage over the simple carboxylic acid in reaction with optically active amino acids containing several functional groups: the reaction occurs at a sufficiently low temperature to prevent racemization [109]. The *o*-carboethoxythiolbenzoic acid employed was obtained from phthaloyl sulfide [110]. It reacted readily with the amino acid at 80° to give the amide 18-I, which with hydrobromic-acetic acid produced the imide 18-II. Yields are fair.

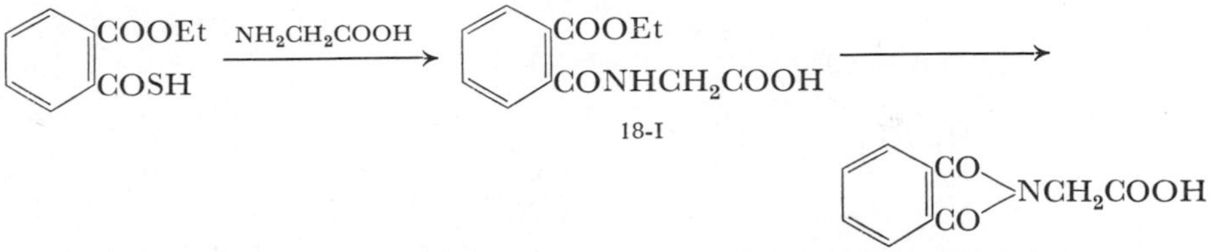

18-I

18-II,
N-Carboxymethylphthalimide

(a) Preparation of S-benzyl-N-phthaloyl-L-cysteine. *o*-Carboethoxythiolbenzoic acid in dimethylformamide was heated with benzylcysteine

$$\begin{array}{c}\text{NH}_2\text{CHCOOH}\\ |\\ \text{CH}_2\text{SCH}_2\text{C}_6\text{H}_5\end{array}$$

at 80°. Hydrogen sulfide evolved and there was produced N-(O-carboethoxybenzoyl)-S-benzylcysteine. The crude compound in a mixture of glacial acetic acid and 48% hydrobromic acid (2:1) was heated at 50° for 1 hr. to give S-benzyl-N-phthaloyl-L-cysteine, $(\alpha)_D$ −167° in 65% yield [109].

11. From Amides

$$\text{RCONH}_2 + \text{R}'\text{NH}_2 \rightarrow \text{RCONHR}' + \text{NH}_3$$

This synthesis of N-substituted amides has been accomplished by heating the amide with the amine, a process which may be selective (see Ex. d), with the amine hydrochloride [111], or with the amine in the presence of boron trifluoride [112]. Formylation of aromatic amines has been successful with dimethylformamide in the presence of sodium methoxide [113]. The yields of formanilides vary from 35 to 97%. In the case of 2,5-dimethoxyaniline, sod-amide or sodium hydride gives an improved yield over sodium methoxide.

Carbon dioxide has a catalytic effect, probably operating via a cyclo addition intermediate, in promoting an exchange between dimethylformamide and amines [114]:

$$\text{HCON(CH}_3)_2 + \text{C}_4\text{H}_9\text{NH}_2 + \text{CO}_2 \xrightarrow[\text{5 hr.}]{60°} \text{HCONHC}_4\text{H}_9$$

N-Butylformamide,
46%

The reaction appears to be restricted to the amide dimethylformamide and to aliphatic amines.

Exchange is feasible between amides, such as N-methylacetanilide, and acids if the acid from the amide can be removed by distillation [15], or in certain cases with acids and such diverse compounds as urea (see A.13), dicyandiamide [115], diphenylphosphinic amides [116], sulfamide [117], or a combination of phosphorus pentoxide and dimethylformamide (see Ex. c).

β-Lactams are best prepared by an amide exchange [103]:

$$
\begin{array}{l}
(\text{CH}_3)_2\text{C—COOH} \\
\quad | \\
\text{C}_6\text{H}_5\text{CHNCOCH(CH}_3)_2 \\
\quad | \\
\text{CH}_2\text{C}_6\text{H}_5
\end{array}
\longrightarrow
\begin{array}{l}
\quad\quad\quad\quad\text{O} \\
\quad\quad\quad\quad\| \\
(\text{CH}_3)_2\text{C—C} \\
\quad | \quad\quad \backslash \\
\text{C}_6\text{H}_5\text{CH——N} \\
\quad\quad\quad\quad | \\
\quad\quad\quad\text{CH}_2\text{C}_6\text{H}_5
\end{array}
+ (\text{CH}_3)_2\text{CHCOOH}
$$

1-Benzyl-3,3-dimethyl-4-phenyl-
2-azetidinone, 50–60%

O-Aryldialkylthiocarbamates are converted by heat into S-aryl isomers in good yield [118]:

$$
\underset{\text{ArOCNR}_2}{\overset{\text{S}}{\overset{\|}{}}} \longrightarrow \underset{\text{ArSCNR}_2}{\overset{\text{O}}{\overset{\|}{}}}
$$

(a) **Preparation of N-acetyl-α-naphthylamine.** α-Naphthylamine hydrochloride, 17.95 g., was heated with 10 g. of acetamide for several minutes, when the precipitation of ammonium chloride was complete. Treatment with hot water, filtration, and washing with water gave a product which, when crystallized from alcohol, gave 14.8 g. (80%) of the acetylated amine, m.p. 132° [111].

(b) **Preparation of o-ethylformanilide** (97% from 0.15 mole of o-ethylaniline and 0.3 mole of sodium methoxide in 150 ml. of dimethylformamide refluxed for 30 min.) [113].

(c) **Preparation of N,N-dimethylacylamides** (40–95% from various carboxylic acids, 0.3 mole, and P$_2$O$_5$, 0.15 mole in 200 ml. of DMF, held at 50–70° and then refluxed for 10 hr., after which the excess DMF was removed under reduced pressure; N,N-dimethylacetamide does not exchange with acids so readily as does DMF) [119].

(d) Preparation of 3′,5′-O-dibenzoyldeoxyadenosine [120]:

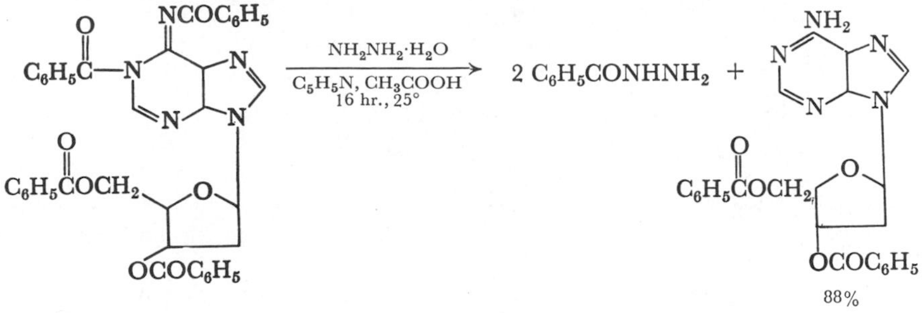

12. From Acylazides

$$RCON_3 + R'NH_2 \xrightarrow{MgO} RCONHR' + Mg\begin{smallmatrix}N_3\\ \diagup\\ \diagdown\\ OH\end{smallmatrix}$$

The reaction is general [121] but finds its most elegant use in derivatizing amino acids using *t*-butyl azidoformate [122]:

$$(CH_3)_3COCN_3 + NH_2CHRCOOH \xrightarrow[\text{Aq. dioxane}]{MgO} (CH_3)_3COCNHCHRCOOH$$

The reagent is stable and can be used in aqueous solutions to derivatize the amino acid, and the derivative can be cleaved instantly under mild acidic conditions (trifluoroacetic or hydrofluoric acid).

(a) Preparation of N-*t*-butyloxycarbonyl-L-methionine (90 % from 40 mmoles of L-methionine and 80 mmoles of magnesium oxide suspended in 100 ml. of 50 % aq. dioxane to which 80 mmoles of *t*-butylazidoformate was added, after which the mixture was held at 40–50° for 20 hr.) [123].

13. From Isocyanates or Isothiocyanates and Precursors

$$NH_2CONH_2 \xrightarrow{\Delta} HNCO \xrightarrow{RCOOH} NH_2\overset{O}{\overset{\|}{C}}-O\overset{O}{\overset{\|}{C}}R \longrightarrow RCONH_2 + CO_2$$

Ureas heated with acids give a fair yield of the amide, a process which probably proceeds via the acyl carbamate. If alkylated ureas are used, N-substituted amides are obtained. Since the isocyanate is an intermediate, an obvious advantage is to start with it rather than the urea.

Urea heated with an amine furnishes the substituted urea, again via the isocyanate:

$$NH_2CONH_2 + RNH_2 \rightarrow RNHCONH_2 + NH_3$$

An interesting application of this reaction is the removal of *o*-phenylenediamine as a derivative from a mixture with the *p*-diamine by adding urea in an amount equivalent to the *o*-diamine content [124]:

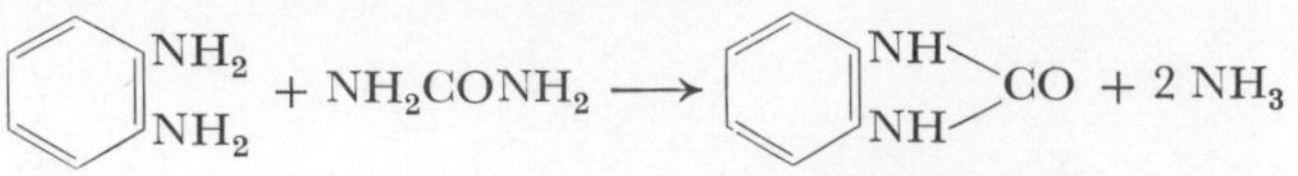

$$\text{(benzene ring)}\begin{array}{c}NH_2\\NH_2\end{array} + NH_2CONH_2 \longrightarrow \text{(benzene ring)}\begin{array}{c}NH\\NH\end{array}{>}CO + 2\,NH_3$$

Benzimidazolinone-2

With alcohol the isocyanates give carbamates, an improved synthesis of difficultly accessible ones being now available (see Ex. c.3).

(a) N-Ethylacetamide (almost quantitative yield by heating acetic acid and N,N'-diethylurea at 160° and then at 180° for several hours) [125].

(b) *n*-Heptamide (73–80 % by heating 1 equiv. of heptanoic acid with 2 equiv. of urea gradually to 140° and then holding at 170–180° for 4 hr.) [126]; *see also* [127].

(c) Other examples

(1) N-Phenylsuccinimide (74 % from succinic anhydride and diphenylthiourea at 180–185° for 2 hr.) [128].

(2) N,N'-Dimyristylhexamethylenediamine (ca. 60 % by heating hexamethylenediisocyanate and 2 equiv. of myristic acid gradually to 150° for 90 min.) [129].

(3) N-*t*-Butylcarbamate (69 % from 2 parts of sodium cyanate, and 1 part of *t*-butyl alcohol in benzene to which 2 parts of trifluoroacetic anhydride were added dropwise; the mixture was then allowed to stand overnight) [130].

(4) β,β-Thiodipropionic acid diamide (85 % from the acid and urea heated at 160–170° for 6 hr.) [127].

14. From Isocyanides

Isocyanides derive their reputation from the nauseous odor, toxicity, and complex reaction patterns. To a certain extent they resemble carbodiimides in forming imine-anhydrides with acids:

$$RN{=}C + R'COOH \longrightarrow RN{=}\overset{H}{\underset{|}{C}}{-}O{-}\overset{O}{\overset{\|}{C}}R'$$

The latter break down to formamides:

$$RN{=}\overset{H}{\underset{|}{C}}{-}O{-}\overset{O}{\overset{\|}{C}}R' \xrightarrow{R'COOH} RNH\overset{O}{\overset{\|}{C}}H + (R'CO)_2O$$

Isocyanides also show tendencies to attack electron-deficient centers such as the carbon of carbonyl groups [131]:

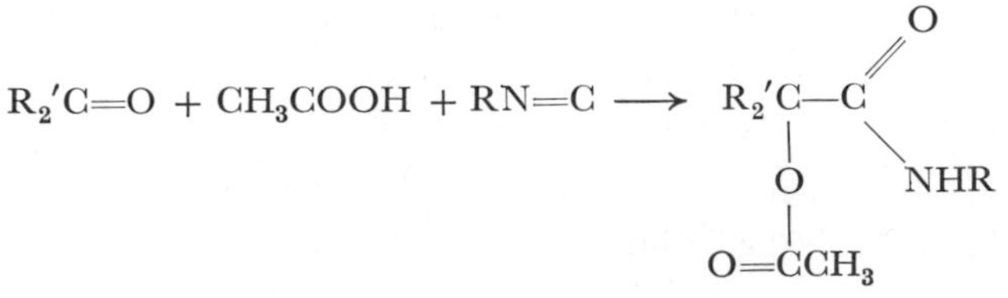

$$R_2'C{=}O + CH_3COOH + RN{=}C \longrightarrow R_2'C{-}\overset{O}{\overset{\diagup}{C}}{\diagdown}_{NHR} \quad \text{(Passerini reaction)}$$
$$\underset{\underset{O{=}CCH_3}{|}}{O}$$

Similarly to the Strecker reaction, the inclusion of an amine yields α-amino acid amides (see Ex. a).

(a) N-Cyclohexyl-α-dimethylaminoisovaleramide (94% from iso-butyraldehyde, cyclohexylisocyanide, and dimethylamine in acetic acid) [132].

1. P. A. S. Smith, *Open-Chain Nitrogen Compounds*, Vols. 1 and 2, W. A. Benjamin, Inc., New York, 1965, 1966; H. Henecka and P. Kurtz, Houben-Weyl, *Methoden der Organischen Chemie*, Vol. 8, G. Thieme Verlag, Stuttgart, 1952, Pt. 3, pp. 653–708.
2. V. Migrdichian, *Organic Synthesis*, Reinhold Publishing Corp., Vol. 1, New York, 1957, p. 367.
3. B. V. Bocharov, *Russ. Chem. Rev. (English Transl.)*, **34**, 212 (1965).
4. D. G. Hoare and D. E. Koshland, Jr., *J. Biol. Chem.*, **242**, 2447 (1967).
5. C. N. Webb, *Org. Syn.*, Coll. Vol. **1**, 82 (1941).
6. J. Weijlard et al., *J. Am. Chem. Soc.*, **71**, 1889 (1949).
7. L. F. Fieser and J. E. Jones, *Org. Syn.*, Coll. Vol. **3**, 590 (1955).
8. J. S. Nelson et al., *J. Org. Chem.*, **28**, 1905 (1963).
9. H. W. Grimmel et al., *J. Am. Chem. Soc.*, **68**, 539 (1946).
10. W. L. Garbrecht, *J. Org. Chem.*, **24**, 368 (1959); G. W. Kenner, *Chem. Ind. (London)*, 15 (1951).
11. N. F. Albertson, *Org. Reactions*, **12**, 255 (1962).
12. Y. I. Leitman and M. S. Pevzner, *Zh. Prikl. Khim.*, **36**, 632 (1963); *C.A.*, **59**, 7422 (1963).
13. M. Walter et al., *Helv. Chim. Acta*, **44**, 1546 (1961).
14. T. Higuchi and T. Miki, *J. Am. Chem. Soc.*, **83**, 3899 (1961).
15. R. N. Ring et al., *J. Org. Chem.*, **27**, 2428 (1962).
16. G. B. Hoey and C. T. Lester, *J. Am. Chem. Soc.*, **73**, 4473 (1951).
17. C. A. Miller and L. M. Long, *J. Am. Chem. Soc.*, **73**, 4895 (1951); W. Schneider and H. Goetz, *Arch. Pharm.*, **294**, 506 (1961); *C.A.*, **56**, 3449 (1962).
18. H. T. Clarke and L. D. Behr, *Org. Syn.*, Coll. Vol. **2**, 562 (1943).
19. R. J. Cotter et al., *J. Org. Chem.*, **26**, 10 (1961).
20. F. S. Spring and J. C. Woods, *J. Chem. Soc.*, 625 (1945).
21. H. Stetter and H. Hennig, *Chem. Ber.*, **88**, 789 (1955).
22. R. A. Barnes and J. C. Godfrey, *J. Org. Chem.*, **22**, 1043 (1957).
23. H. K. Hall, Jr., and A. K. Schneider, *J. Am. Chem. Soc.*, **80**, 6409 (1958).
24. S. W. Fox et al., *Arch. Biochem. Biophys.*, **102**, 439 (1963); *C.A.*, **59**, 14217 (1963).
25. D. E. Nitecki et al., *J. Org. Chem.*, **33**, 864 (1968).
26. K. D. Kopple and H. G. Ghazarian, *J. Org. Chem.*, **33**, 862 (1968).
27. N. J. Leonard et al., *J. Org. Chem.*, **33**, 318 (1968).
28. A. Buzas et al., *Compt. Rend.*, **260**, 2249 (1965).
29. G. H. Coleman and A. M. Alvarado, *Org. Syn.*, Coll. Vol. **1**, 3 (1941).
30. J. D. Roberts et al., *J. Am. Chem. Soc.*, **75**, 637 (1953).
31. Y. Liwschitz et al., *J. Am. Chem. Soc.*, **78**, 3067 (1956).
32. H. V. Eynde, *Ind. Chim. Belges*, **28**, 1363 (1963).
33. G. Paris et al., *Org. Syn.*, Coll. Vol. **4**, 496 (1963).
34. P. A. Finan and G. A. Fothergill, *J. Chem. Soc.*, 2824 (1962).
35. T. Fujisawa and S. Sugasawa, *Tetrahedron*, **7**, 185 (1959).
36. A. P. Martinez et al., *J. Org. Chem.*, **26**, 4501 (1961); K. C. Murdock and R. B. Angier, *ibid.*, **27**, 3317 (1962).
37. W. B. Weaver and W. M. Whaley, *J. Am. Chem. Soc.*, **69**, 515, 1144 (1947).
38. W. Falk, *J. Prakt. Chem.*, **15**, 228 (1962).
39. J. M. Bobbitt and T.-T. Chou, *J. Org. Chem.*, **24**, 1106 (1959); W. Treibs et al., *Chem. Ber.*, **92**, 1216 (1959); H. Goldstein and G. Preitner, *Helv. Chim. Acta*, **27**, 612 (1944).
40. G. E. Philbrook, *J. Org. Chem.*, **19**, 623 (1954); H. Erlenmeyer and Ch. J. Morel, *Helv. Chim. Acta*, **28**, 362 (1945).
41. A. H. Schlesinger and E. J. Prill, *J. Am. Chem. Soc.*, **78**, 6123 (1956); M. Heintzeler, *Ann. Chem.*, **569**, 97 (1950).

42. G. T. Newbold *et al.*, *J. Chem. Soc.*, 1855 (1948).
43. L. P. Kyrides *et al.*, *J. Am. Chem. Soc.*, **69,** 2239 (1947).
44. R. E. Steiger, *J. Org. Chem.*, **9,** 396 (1944).
45. G. M. Coppinger, *J. Am. Chem. Soc.*, **76,** 1372 (1954).
46. R. E. Kent and S. M. McElvain, *Org. Syn.*, Coll. Vol. **3,** 490 (1955).
47. J. M. Tedder, *Chem. Rev.*, **55,** 787 (1955).
48. M. Pailer and W. J. Huebsch, *Monatsh. Chem.*, **97,** 1541 (1966).
49. R. M. Herbst and D. Shemin, *Org. Syn.*, Coll. Vol. **2,** 11 (1943).
50. A. K. Bose *et al.*, *J. Org. Chem.*, **23,** 1335 (1958).
51. W. A. Noyes and P. K. Porter, *Org. Syn.*, Coll. Vol. **1,** 457 (1941).
52. I. Sausa, *Chem. Zvesti*, **16,** 574 (1962); *C.A.*, **59,** 1527 (1963).
53. L. I. Smith and O. H. Emerson, *Org. Syn.*, Coll. Vol. **3,** 151 (1955).
54. D. Davidson and H. Skovronek, *J. Am. Chem. Soc.*, **80,** 376 (1958).
55. J. A. Berson and R. Swidler, *J. Am. Chem. Soc.*, **76,** 2835 (1954).
56. E. G. Howard and R. V. Lindsey, Jr., *J. Am. Chem. Soc.*, **82,** 158 (1960).
57. D. T. Mowry and J. M. Butler, *Org. Syn.*, Coll. Vol. **4,** 486 (1963).
58. J. H. Gorvin, *J. Chem. Soc.*, 732 (1945); W. A. Jacobs and M. Heidelberger, *Org. Syn.*, Coll. Vol. **1,** 153 (1941).
59. B. B. Corson *et al.*, *Org. Syn.*, Coll. Vol. **1,** 179 (1941).
60. F. Micheel and R. Habendorff, *Chem. Ber.*, **90,** 1590 (1957); P. B. Russell, *J. Am. Chem. Soc.*, **72,** 1853 (1950); M. Viscontini *et al.*, *Helv. Chim. Acta*, **35,** 451 (1952); R. A. Baxter and F. S. Spring, *J. Chem. Soc.*, 229 (1945); A. H. Gowenlock *et al.*, *ibid.*, 622 (1945).
61. E. C. Taylor, Jr., and A. J. Crovetti, *J. Am. Chem. Soc.*, **78,** 214 (1956).
62. J. Rudinger and Z. Pravda, *Collection Czech. Chem. Commun.*, **23,** 1947 (1958).
63. A. O. Zoss and G. F. Hennion, *J. Am. Chem. Soc.*, **63,** 1151 (1941).
64. J. Kleinberg and L. F. Audrieth, *Org. Syn.*, Coll. Vol. **3,** 516 (1955).
65. I. Ugi *et al.*, *Org. Syn.*, **41,** 13 (1961); J. Moffat *et al.*, *J. Org. Chem.*, **27,** 4058 (1962).
66. R. H. Mazur, *J. Org. Chem.*, **28,** 2498 (1963).
67. M. Gordon *et al.*, *J. Am. Chem. Soc.*, **71,** 1245 (1949).
68. O. C. Dermer and J. King, *J. Org. Chem.*, **8,** 168 (1943).
69. H. J. Hagemeyer, Jr., and D. C. Hull, *Ind. Eng. Chem.*, **41,** 2920 (1949).
70. E. S. Stern, *Chem. Ind.* (*London*), 277 (1956).
71. W. Hackbart and M. Hartmann, *J. Prakt. Chem.*, **14,** 1 (1961).
72. D. A. Buckingham *et al.*, *J. Am. Chem. Soc.*, **89,** 4539 (1967).
73. J. Blair *et al.*, *J. Chem. Soc.*, 708 (1955).
74. H. H. Farmer and N. Rabjohn, *Org. Syn.*, Coll. Vol. **4,** 441 (1963).
75. D. E. Pearson, unpublished results.
76. H. L. Bassett and C. R. Thomas, *J. Chem. Soc.*, 1188 (1954).
77. E. Pfeil and U. Harder, *Angew. Chem.*, *Intern. Ed. Engl.*, **6,** 178 (1967).
78. R. F. Smith *et al.*, *J. Org. Chem.*, **33,** 851 (1968).
79. W. Wenner, *Org. Syn.*, Coll. Vol. **4,** 760 (1963).
80. R. A. Turner and C. Djerassi, *J. Am. Chem. Soc.*, **72,** 3081 (1950); R. G. Jones, *ibid.*, **73,** 5610 (1951); R. H. Wiley and W. E. Waddey, *Org. Syn.*, Coll. Vol. **3,** 560 (1955).
81. H. R. Snyder and C. T. Elston, *J. Am. Chem. Soc.*, **76,** 3039 (1954).
82. F. Micheel *et al.*, *Chem. Ber.*, **94,** 132 (1961).
83. E. C. Horning and R. U. Schock, Jr., *J. Am. Chem. Soc.*, **70,** 2945 (1948).
84. C. R. Hauser and D. S. Hoffenberg, *J. Org. Chem.*, **20,** 1448 (1955).
85. F. Nerdel and H. Rachel, *Chem. Ber.*, **89,** 671 (1956).
86. C. L. Dickinson, *J. Am. Chem. Soc.*, **82,** 4367 (1960).
87. M. Orchin and L. Reggel, *J. Am. Chem. Soc.*, **73,** 436 (1951).
88. J. M. Bobbitt and D. A. Scola, *J. Org. Chem.*, **25,** 560 (1960).
89. C. Berther, *Chem. Ber.*, **92,** 2616 (1959).
90. H. L. Lochte and E. N. Wheeler, *J. Am. Chem. Soc.*, **76,** 5548 (1954).
91. K. Hoffmann *et al.*, *Helv. Chim. Acta*, **40,** 387 (1957).
92. N. Sperber *et al.*, *J. Am. Chem. Soc.*, **70,** 3091 (1948).

93. W. Roberts and M. C. Whiting, *J. Chem. Soc.*, 1290 (1965).

94. F. C. Schaefer and A. P. Krapcho, *J. Org. Chem.*, **27**, 1255 (1962).

95. J. G. Wallace, *Hydrogen Peroxide in Organic Chemistry*, E. I. duPont Co., 1962, p. 42.

96. C. S. Marvel and W. F. Tuley, *Org. Syn.*, Coll. Vol. **1**, 289 (1941).

97. J. S. Buck and W. S. Ide, *Org. Syn.*, Coll. Vol. **2**, 44 (1943).

98. A. Dornow and H. Theidel, *Chem. Ber.*, **88**, 1267 (1955).

99. S. Cohen *et al.*, *Biochemistry*, **2**, 176 (1963).

100. G. Quadbeck, in W. Foerst, ed., *Newer Methods of Preparative Organic Chemistry*, Vol. 2, translated by F. K. Kirchner, Academic Press, New York, 1963, p. 138.

101. J. W. Williams and J. A. Krynitsky, *Org. Syn.*, Coll. Vol. **3**, 10 (1955).

102. L. Horner and E. Spietschka, *Chem. Ber.*, **85**, 225 (1952).

103. J. C. Sheehan and E. J. Corey, *Org. Reactions*, **9**, 388 (1957).

104. A. C. Piñol and M. M. Manas, *Chem. Commun.*, 229 (1967).

105. C. W. Hill *et al.*, *J. Am. Chem. Soc.*, **72**, 2286 (1950).

106. P. Chabrier and S. H. Renard, *Bull. Soc. Chim. France*, D272 (1949).

107. A. R. Katritzky, *J. Chem. Soc.*, 2586 (1955).

108. T. S. Gardner *et al.*, *J. Org. Chem.*, **19**, 753 (1954).

109. K. Balenović and B. Gašpert, *Chem. Ind. (London)*, 115 (1957).

110. A. Reissert and H. Holle, *Chem. Ber.*, **44**, 3027 (1911).

111. A. Galat and G. Elion, *J. Am. Chem. Soc.*, **65**, 1566 (1943).

112. F. J. Sowa and J. A. Nieuwland, *J. Am. Chem. Soc.*, **59**, 1202 (1937).

113. G. R. Pettit *et al.*, *J. Org. Chem.*, **24**, 895 (1959); **26**, 2563 (1961).

114. Y. Otsuji *et al.*, *Bull. Chem. Soc. Japan*, **41**, 1485 (1968).

115. M. T. Dangyan *et al.*, *Proc. Acad. Sci.*, *Armenian SSR*, II, **4**, 107 (1945); *C.A.*, **40**, 3399 (1946).

116. I. N. Zhmurova *et al.*, *Zh. Obshch. Khim.*, **29**, 2083 (1959); *C.A.*, **54**, 8681 (1960).

117. A. V. Kirsanov and Y. M. Zolotov, *Zh. Obshch. Khim.*, **19**, 2201 (1949); *C.A.*, **44**, 4446 (1950).

118. M. S. Newman and H. A. Karnes, *J. Org. Chem.*, **31**, 3980 (1966).

119. H. Schindbauer, *Monatsh. Chem.*, **99**, 1799 (1968).

120. R. L. Letsinger *et al.*, *Tetrahedron Letters*, 2621 (1968).

121. W. L. Garbrecht, *J. Org. Chem.*, **24**, 368 (1959).

122. L. F. and M. Fieser, *Reagents for Organic Synthesis*, John Wiley and Sons, New York, 1967, p. 84.

123. R. Schwyzer *et al.*, *Helv. Chim. Acta*, **42**, 2622 (1959).

124. J. M. Cross, French Patent 1,477,922; *C.A.*, **68**, 6614 (1968).

125. E. Cherbuliez and F. Landolt, *Helv. Chim. Acta*, **29**, 1438 (1946).

126. J. L. Guthrie and N. Rabjohn, *Org. Syn.*, Coll. Vol. **4**, 513 (1963).

127. J. H. MacGregor and C. Pugh, *J. Chem. Soc.*, 736 (1950).

128. A.-U. Rahman *et al.*, *J. Org. Chem.*, **27**, 3315 (1962).

129. C. L. Agre *et al.*, *J. Org. Chem.*, **20**, 695 (1955).

130. B. Loev and M. F. Kormendy, *J. Org. Chem.*, **28**, 3421 (1963).

131. R. H. Baker and D. Stanonis, *J. Am. Chem. Soc.*, **73**, 699 (1951).

132. J. W. McFarland, *J. Org. Chem.*, **28**, 2179 (1963).

B. Reduction

1. From Hydroxamic Acids

$$\text{RCONHOH} \xrightarrow[\substack{\text{Raney} \\ \text{Ni}}]{\text{H}_2} \text{RCONH}_2$$

Although hydroxamic acids would not ordinarily be considered as starting materials for the synthesis of amides, it has been found that they may be reduced in yields of 76–97% by hydrogen in the presence of Raney nickel [1].

References for Section B are on p. 915.

The hydroxamic acids were prepared in the usual manner from the ester by treatment with hydroxylamine.

(a) Preparation of lauramide. Laurohydroxamic acid, 4 g., and about 1 g. of Raney nickel in 75 ml. of ethanol were shaken in a Parr hydrogenation apparatus under hydrogen at 50 p.s.i. for about 3 hr. At this point the ferric chloride test for the hydroxamic acid was negative, and the catalyst was filtered and the filtrate was reduced under reduced pressure to about one-third volume. The addition of water precipitated the amide which when dried, gave 3.24 g. (97%) of product, m.p. 101–102° [1].

2. From Mono- and Diacyl Hydrazides and Related Types

$$RCONHNH_2 \xrightarrow[\substack{Raney \\ Ni \\ alcohol}]{} RCONH_2 + NH_3$$

This synthesis, which occurs for alkyl, aryl, and heterocyclic carboxylic acid hydrazides, is of limited value, but it does permit the conversion of heterocyclic hydrazides of interest as possible medicinals into amides [2, 3]. Unsubstituted 1,2-diacylhydrazides are also cleaved readily, although some difficulty is experienced if there are substituents other than hydrogen on the nitrogen atoms. The 1-acyl-2-alkylidenehydrazides undergo smooth hydrogenolysis [4]. Substantial quantities of Raney nickel in alcohol is the usual catalyst, but potassium ferricyanide and aqueous ammonia have also been employed [2]. The yields are poor to fair.

(a) Preparation of cyclohexanecarboxamide. A mixture of 2 g. of cyclohexanecarboxylic acid hydrazide, 100 ml. of 95% ethanol, and about 10 g. of moist Raney nickel was heated under reflux with stirring for about 3 hr. After removing the nickel by filtration, the filtrate was evaporated to dryness, after which the residue was extracted with benzene. Cooling gave 1 g. (56%) of cyclohexanecarboxamide, m.p. 188–189° [3].

(b) Preparation of benzamide (60% from 1,2-dibenzoylhydrazine and Raney nickel W-2 (see Alcohols, Chapter 4, C.6) or commercial Raney nickel in absolute ethanol) [4].

3. From Schiff Bases (Reductive Acylation)

$$3 \, ArCH{=}NAr + (CH_3)_3NBH_3 + 3 \, CH_3COOH \longrightarrow$$

$$\begin{array}{c} 3 \, ArCH_2NAr \\ | \\ COCH_3 + (CH_3)_3N + B(OH)_3 \end{array}$$

In this synthesis the Schiff base is reduced to the secondary amine which in the presence of acetic acid is acetylated to give the amide [5]. Substituents in the aryl groups not affected are:

$$Cl, \, OCH_3, \, NO_2, \, OH, \, COOC_2H_5, \, SO_2NH_2$$

By using an excess of the reducing agent with a series of Schiff bases, yields varying from 25 to 88% are obtained.

(a) **N-p-Chlorophenyl-N-p-chlorobenzylacetamide.** To a suspension of 0.1 mole of N-p-chlorobenzylidene-p-chloroaniline in 50 ml. of glacial acetic acid in an ice bath a solution of 0.113 mole of trimethylamine borane in 30 ml. of glacial acetic acid was added slowly with stirring until no increase in reaction temperature occurred. The remainder of the reducing solution was then added rapidly and the mixture was refluxed for 12 hr. On making the mixture alkaline and extraction of the solid formed, there was recovered in the usual manner 19.52 g. (66.4%) of the acetamide [5].

1. R. M. Gipson *et al.*, *J. Org. Chem.*, **28**, 1425 (1963).
2. A. Giner-Sorolla and A. Bendich, *J. Am. Chem. Soc.*, **80**, 3932 (1958).
3. C. Ainsworth, *J. Am. Chem. Soc.*, **76**, 5774 (1954).
4. R. L. Hinman, *J. Org. Chem.*, **22**, 148 (1957).
5. J. H. Billman and J. W. McDowell, *J. Org. Chem.*, **27**, 2640 (1962).

C. Oxidation

Oxidation of amines, particularly tertiary amines, would appear to be a feasible process. Difficulty is encountered because of possibilities of so many side reactions such as N-oxide or Schiff base formation, dehydrogenation, or coupling. Nevertheless, although limited, oxidation procedures are available for special cases. Furthermore, they are available wherein an organic substrate is oxidized to an acid under amidating conditions. Perhaps the most useful of these is the Willgerodt reaction.

1. From Tertiary Amines

Manganese dioxide, in very large excess to utilize the few active oxygen sites, is the reagent of choice [1]:

$$C_6H_5N(CH_3)_2 \xrightarrow[\text{18 hr. at 20°}]{\text{MnO}_2 \text{ (150 g), CHCl}_3} C_6H_5N-\overset{CH_3}{\underset{}{|}}\overset{H}{\underset{}{|}}C=O$$

N-Methylformanilide, 78%

N-Methylaniline yields 83% formanilide. Diethylaniline yields the same product, a fact which suggests that dehydrogenation of the ethyl chain first occurs followed by oxidative splitting. p-Nitro- and o-alkyldimethylanilines are not oxidized by this procedure. Aliphatic amines, such as diethylcyclohexylamine, produce carbonyl compounds such as cyclohexanone (85%). The ring in piperazine and triethylenediamine can be opened by oxidation (see Ex. a).

Recently a few steroidal tertiary amines have been oxidized with chromic anhydride in pyridine to form N-formyl derivatives in high yield [2] (see Ex. b.2), and secondary aliphatic amines have been oxidized to the corresponding amides in low yields [3]:

$$(C_4H_9)_2NH + 2\,HgO \xrightarrow[\text{210°, 6 hr.}]{\text{Autoclave,}} CH_3(CH_2)_2CONHC_4H_9 + 2\,Hg + H_2O$$

Butylbutyramide, 40% based on HgO

References for Section C are on pp. 918–919.

Also aromatic tertiary amines have been oxidized with oxygen and platinum black; for example, benzyldimethylamine to N-benzyl-N-methylformamide in 85% yield on a small scale [4].

(a) Preparation of N,N′-dibenzyl-N,N′-diformylethylenediamine

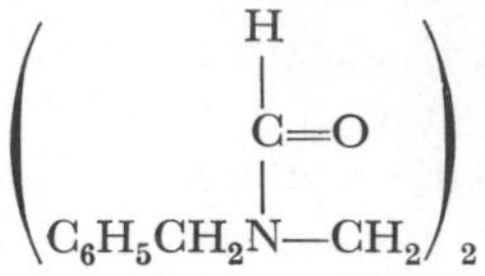

$$\left(\begin{array}{c} H \\ | \\ C{=}O \\ | \\ C_6H_5CH_2N{-}CH_2 \end{array} \right)_2$$

(80% from 1,4-dibenzylpiperazine and a large excess of manganese dioxide in chloroform stirred for 18 hr.) [5].

(b) Other examples

(1) **1-Methyl-2-pyridone** (65–70% from the quaternary salt of pyridine and dimethyl sulfate treated with potassium ferricyanide in aq. base solution) [6].

(2) **N-Formyl paravallarine** (97% from N-methylparavallarine by

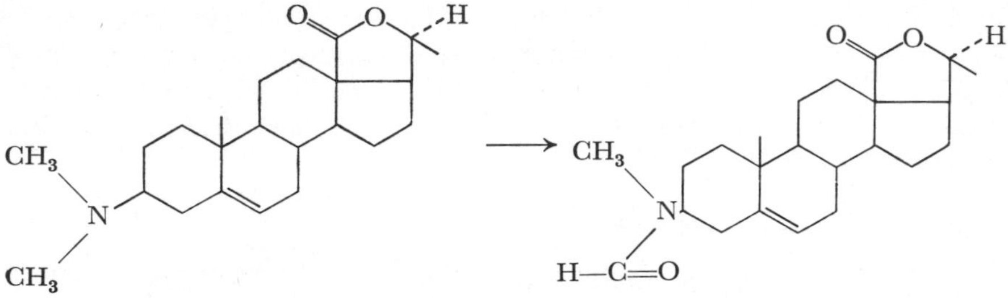

oxidation with chromic anhydride in pyridine at 45°) [2].

2. From Ketones (Willgerodt)

$$ArCOCH_3 \xrightarrow{(NH_4)_2S_x} ArCH_2CONH_2$$

$$ArCOCH_2CH_3 \xrightarrow{(NH_4)_2S_x} ArCH_2CH_2CONH_2$$

This remarkable rearrangement has been reviewed [7]. Strangely enough, the carbon skeleton remains unchanged in the reaction and the terminal methyl is always converted into the carbamide group. As a synthetic method, the rearrangement has enjoyed its greatest success with aromatic-aliphatic ketones. Strictly aliphatic ketones also respond, but usually with lower yields [8]. Other types which form amides with ammonium polysulfide are alkenes, alkynes [9], mercaptans [10], and cyclic oxides [11], but in these cases the return is also usually low.

In the original method the compound to be rearranged was heated under pressure with aqueous ammonium polysulfide at 160–200°. Later it was found that better yields and purer products were obtained at lower temperature by employing dioxane as a solvent [12] or by using a mixture of sulfur, ammonium hydroxide, and pyridine as the reagent [13, 9]. A final modification, that of Kindler, has the advantage of avoiding pressure vessels. It consists in heating the ketone with sulfur and a dry amine such as morpholine [14]. Although Kindler's procedure gives the thioamide, it does not interfere with the usefulness of this method of synthesis in the preparation of intermediates for conversion into carboxylic acids.

In the most favorable cases in which acetyl groups are present in heterocyclic or fused ring systems, amides may be produced by the Willgerodt reactions in yields as high as 90% or above, but in most cases the yields are considerably lower. Alkyl, alkoxy, and inert halogen groups do not interfere, but easily oxidizable or reducible groups such as the amino, nitro, or formyl do [7].

From what has already been said it is apparent that the usefulness of the Willgerodt reaction is limited. Readily available acetyl derivatives in the aromatic or heterocyclic series would suggest its use in the synthesis of the corresponding amide. For example, the diamide of 1,4-diphenylbutane *p,p'*-diacetic acid could be prepared to advantage from the hydrocarbon by the sequence [15]:

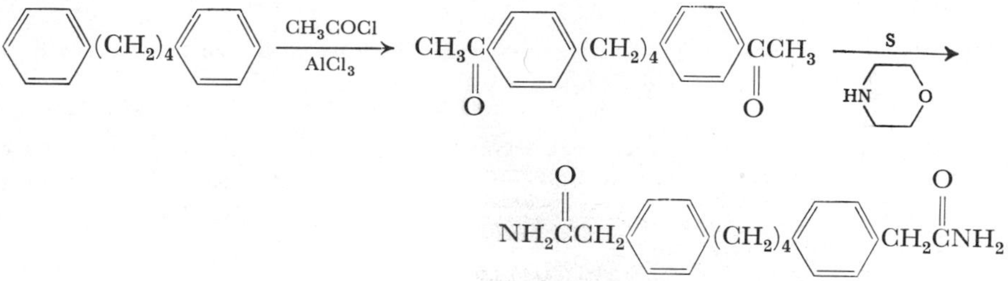

The mechanism of the reaction is not entirely clear [16]. Suggested steps, if one starts with a carbonyl compound containing an alkyl group larger than the methyl, are:

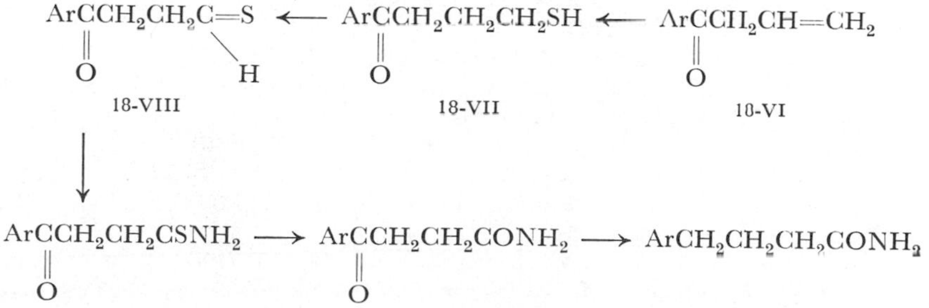

18-III 18-IV 18-V 18-VIII 18-VII 18-VI 18-IX 18-X 18-XI

In steps 18-III to 18-IV the activating influence of the carbonyl group produces the alkene which in steps 18-IV through 18-VII leads to the terminal mercapto group. The latter is then converted in turn into the keto thioaldehyde, 18-VIII, to the keto thioamide 18-IX, and finally to the keto amide 18-X, which by reduction gives the unsubstituted amide 18-XI. With such a maze of possible steps, it is not likely that a single mechanism prevails.

(a) **Preparation of phenylacetamide** (82 % from acetophenone, conc. aqueous ammonia, sulfur and pyridine under pressure) [17].

(b) **Preparation of 1-pyrenylacetamide** (92 % from 1-acetylpyrene, sulfur, conc. aqueous ammonia, hydrogen sulfide, and dioxane under pressure) [18].

3. From Trihaloacroleins

In a very specific reaction, trihaloacroleins with secondary amines yield dihaloacrylamides:

$$CCl_2{=}CClCH{=}O + R_2NH \rightarrow CHCl{=}CClCONR_2$$

An internal oxidation-reduction is involved, the possible mechanisms of which have been discussed extensively [19].

(a) **Preparation of α,β-dichloroacryl-N,N-diethylamide.** Freshly distilled α,β,β-trichloroacrolein, 0.1 mole, in about 200 ml. of anhydrous ether was added in small portions to 0.2 mole of dry diethylamine in twice its volume of anhydrous ether, while the mixture was cooled in a stream of water. After standing for 2–3 hr., the precipitate of the diethylamine hydrochloride which formed was filtered off. The filtrate, after removal of the ether, was distilled through a Vigreux column to give an 80 % yield of the diethylamide, b.p. 77–79° (0.6 mm.) [19].

1. H. B. Henbest and A. Thomas, *J. Chem. Soc.*, 3032 (1957).
2. A. Cavé *et al.*, *Tetrahedron*, **23,** 4691 (1967).
3. D. M. Fenton, U.S. Patent 3,385,891; *C.A.*, **70,** 295 (Jan. 6, 1969).
4. G. T. Davis and D. H. Rosenblatt, *Tetrahedron Letters*, 4085 (1968).
5. H. B. Henbest *et al.*, *J. Chem. Soc.*, 3559 (1960).
6. E. A. Prill and S. M. McElvain, *Org. Syn.*, Coll. Vol. **2,** 419 (1943).
7. M. Carmack and M. A. Spielman, *Org. Reactions*, **3,** 83 (1946).
8. L. Cavalieri *et al.*, *J. Am. Chem. Soc.*, **67,** 1783 (1945).
9. M. Carmack and D. F. DeTar, *J. Am. Chem. Soc.*, **68,** 2029 (1946).
10. J. A. King and F. H. McMillan, *J. Am. Chem. Soc.*, **68,** 1369 (1946).
11. R. T. Gerry and E. V. Brown, *J. Am. Chem. Soc.*, **75,** 740 (1953).
12. L. F. Fieser and G. W. Kilmer, *J. Am. Chem. Soc.*, **62,** 1354 (1940); W. E. Bachmann and J. C. Sheehan, *ibid.*, **62,** 2687 (1940).
13. D. F. DeTar and M. Carmack, *J. Am. Chem. Soc.*, **68,** 2025 (1946).
14. E. Schwenk and E. Bloch, *J. Am. Chem. Soc.*, **64,** 3051 (1942).
15. D. J. Cram and G. S. Hammond, *Organic Chemistry*, 2nd ed., McGraw-Hill Book Co., New York, 1964, p. 567.
16. W. G. Dauben and J. B. Rogan, *J. Am. Chem. Soc.*, **78,** 4135 (1956); J. M. Harkin, translation of H. Krauch and W. Kunz, *Organic Name Reactions*, John Wiley and Sons, New York, 1964, p. 488; F. Asinger *et al.*, *Angew. Chem.*, **75,** 1050 (1963).

17. M. Carmack and D. F. DeTar, *Org. Reactions*, **3**, 97 (1946).
18. W. E. Bachman and M. Carmack, *Org. Reactions*, **3**, 96 (1946).
19. C. Raulet and E. Levas, *Bull. Soc. Chim. France*, 2139 (1963).

D. Electrophilic-Type Syntheses

1. From Hydrocarbons and Carbamic Acid Chlorides (Friedel-Crafts)

This synthesis has been reviewed [1]. It has been applied mostly to the preparation of monoamides or diamides of polyaromatic hydrocarbons [2, 3]. Yields vary from 37 to 95 % in the aromatic series. The amides serve as intermediates in the synthesis of carboxylic acids and aldehydes.

Carbamoyl chloride is somewhat unstable, but its complex with anhydrous aluminum chloride can be stored indefinitely and used for acylation.

(a) Preparation of diphenyl-4-carboxylic acid amide. Carbamic acid chloride, 220 g., and 370 g. of aluminum chloride were mixed in 800 ml. of benzene with cooling, and 308 g. of diphenyl was introduced. With stirring the mixture was heated to 50°, after which hydrogen chloride evolution occurred. The temperature was elevated to 70° in 2–3 hr. and then to 80° for 2 hr. On the addition of ice water the amide precipitated. The benzene and unchanged diphenyl were removed by steam, after which washing and drying gave 390 g. of the amide, m.p. 202–210°. Further purification by boiling with 5–6 l. of water, washing, and drying yielded 330 g. (90 % based on the diphenyl consumed) of the amide, m.p. 222–223° [2].

(b) Other examples

(1) **2,5-Dimethylbenzoic acid-N-methylanilide** (62 % from *p*-xylene, N-methyl-N-phenylcarbamoyl chloride, and aluminum chloride) [3].

(2) **N,N-Diphenyl-3,4-dimethoxybenzamide** (65 % from veratrole, diphenylcarbamoyl chloride, and aluminum chloride in ethylene chloride) [4].

2. From Hydrocarbons and Phenylisocyanates

In a neglected but promising area of derivatization, it has been found that aromatic hydrocarbons or ethers react with phenylisocyanate in the presence of aluminum chloride to form amides [5]; this method has been used successfully in the identification of aromatic hydrocarbons by one (D.E.P.) of us:

$$ \text{ArH} + \text{C}_6\text{H}_5\text{N=C=O} \xrightarrow{\text{AlCl}_3} \left[\text{C}_6\text{H}_5\text{NHCCl} \atop O \right] \longrightarrow \text{ArC—NHC}_6\text{H}_5 $$

Phenols are converted into carbamates [6], but phenol ethers yield the alkoxybenzanilide.

References for Section D are on pp. 927–928.

Similar to this conversion is the Friedel-Crafts treatment of an aroyliso-thiocyanate to yield a thionimide (Carboxylic Acids, Chapter 13, F.1, Ex. b.5):

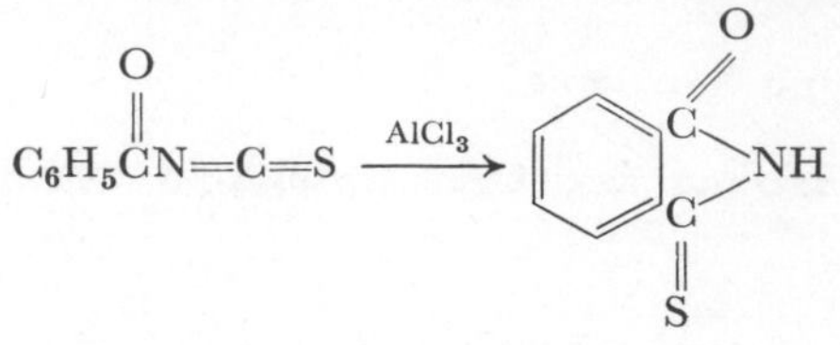

$$C_6H_5\overset{\overset{\displaystyle O}{\|}}{C}N{=}C{=}S \xrightarrow{AlCl_3}$$

ca. 45%

3. From Amido Derivatives of Aldehydes

$$RCH(NH\overset{\overset{\displaystyle O}{\|}}{C}CH_3)_2 + ArH \longrightarrow RCHAr$$

$$\underset{\underset{\displaystyle O}{\underset{\displaystyle \|}{NHCCH_3}}}{}$$

NCH₂OH + ArH ⟶ ArCH₂N

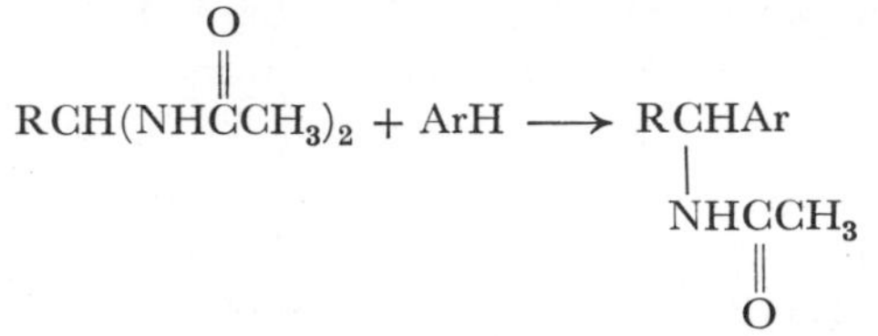

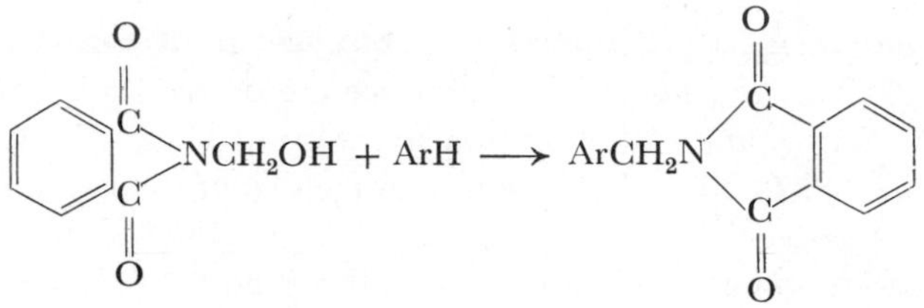

(see E.2, Ex. a, for preparation).

These reactions and the preparations of the reagents have been reviewed [7]; see Amines, Chapter 8, D.3 including Ex. b.6. Since the electrophilic reagent is a weak attacking agent, the aromatic nuclei, ArH, should be activated by hydroxy or amino groups for best yields, or the aromatic nuclei may be replaced by aliphatic enolic compounds (see Ex. b).

The amido derivatives of aldehydes are by no means restricted to the types listed above but are quite diverse in nature [7].

(a) Preparation of N-(2-hydroxy-3,5-dinitrobenzyl)-phthalimide [7]:

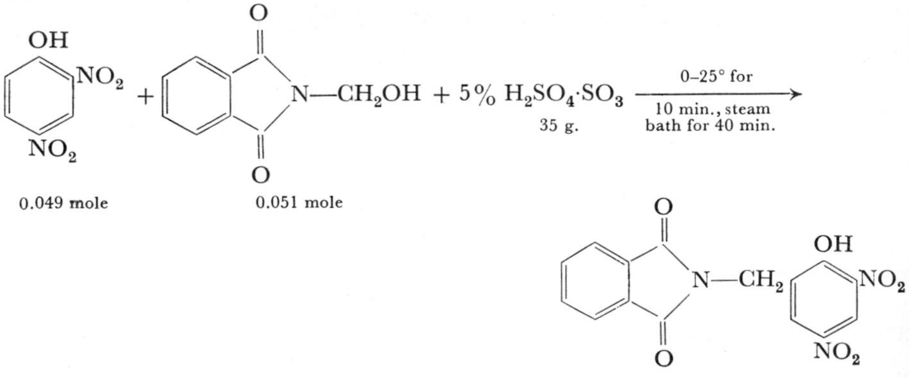

0.049 mole   0.051 mole

95%

(b) Preparation of diethyl α-acetamidobenzylmalonate [7]:

$$C_6H_5CH(NHCOCH_3)_2 + CH_2(CO_2C_2H_5)_2 + (CH_3CO)_2O \xrightarrow[\text{3 hr.}]{150-155°}$$

0.05 mole   0.05 mole   25 ml.

$$C_6H_5CHCH(CO_2C_2H_5)_2$$
$$|$$
$$NHCOCH_3$$
62%

4. From Alkenes and Nitriles (Ritter)

Olefins which can form carbonium ions in sulfuric acid are capable of alkylating nitriles:

$$RCH{=}CH_2 + R'CN \xrightarrow[H_2SO_4]{H_2O} R'\overset{O}{\overset{\|}{C}}NH\overset{CH_3}{\overset{|}{C}H}\diagdown_R$$

In the reaction which has been reviewed [7a], it is important to generate a carbonium ion which is fairly stable since the cyanide is a poor carbonium ion trap. Thus alcohols (especially tertiary ones, some secondary ones, benzyl and allyl alcohols), nonconjugated, unsaturated acids, haloolefins, bicyclic olefins, and unsaturated ketones (in low yield) have been added to nitriles [8, 9]. Hydrogen cyanide, generated *in situ*, has been converted into N-alkyl-formamides by olefins or other sources of carbonium ions. Among the alkenes, methyl crotonate and maleic acid and, among the nitriles, cyanogen and phenylacetonitrile do not appear to react. With the proper polyfunctional alcohols or nitriles, the Ritter reaction is capable of producing heterocyclic compounds [10].

Although the alkylation of amides by the Ritter reaction has not been examined critically, an example of an atypical alkylation has been found (see Ex. b.2). Lastly the Ritter reaction and methylolation may be run simultaneously (see Ex. b.3). Yields are usually good.

Other alkylations of a nucleophilic nature are described in E.

(a) Preparation of N-(1,1,3,3-tetramethylbutyl)-acetamide. To a stirred solution of 100 g. of conc. sulfuric acid and 45 g. of acetonitrile in 500 ml. of acetic acid, 112 g. of diisobutylene was added dropwise while the temperature was held at or below 50°. After standing overnight, the mixture was diluted and the solid filtered and washed thoroughly to yield 80% of product [11].

(b) Other examples

(1) N-Formyl-α,α-dimethyl-β-phenethylamine (65-70% from sodium cyanide, α,α-dimethyl-β-phenethyl alcohol, and sulfuric acid in acetic acid) [8].

(2) N-(2-Tetrahydropyranyl)-benzamide (73% from an equimolecular mixture of benzamide and 2,3-dihydropyran in benzene to which a small amount of ethyl ether saturated with hydrogen chloride was added) [12].

(3) N,N′-Diacetyl-2,5-dimethyl-1,4-diaminomethylbenzene (55%
from 144 g. of paraformaldehyde, 1750 ml. of acetic acid, 360 ml. of sulfuric

$$CH_3CONHCH_2\underset{CH_3}{\overset{CH_3}{\bigcirc}}CH_2NHCOCH_3$$

acid, and 214 g. of acetonitrile to which, after the initial reaction had subsided,
212 g. of *p*-xylene was added; from filtrate of workup, 21% of N-(2,5-dimethyl-
benzyl)-acetamide was recovered) [13].

5. From Oximes (Beckmann Rearrangement) and Related Types

$$\underset{NOH}{\overset{ArCR}{\|}}\xrightarrow{PCl_5}\underset{NHAr}{\overset{O=CR}{|}}$$

Few rearrangements have been studied more widely than the Beckmann,
which has recently been reviewed [14, 15]. Although most of the publications
deal with the mechanism of the reaction and the stereochemical configuration of
the oximes employed, there are cases in which the rearrangement offers a method
of preparative value, such as the synthesis of ε-caprolactam (see Ex. a and b.1) or
certain aromatic amines such as 1-, 2-, 3-, and 9-acetaminophenanthrenes [16].

The most common reagents used in the rearrangement of ketoximes are
concentrated sulfuric acid, phosphorus pentachloride in ether and hydrogen
chloride in a mixture of acetic acid and acetic anhydride (Beckmann's mixture)
[17], although recently it has been shown that polyphosphoric acid gives
superior yields and more definitive products [18–20]. In addition, trifluoroacetic anhydride offers advantages in cases in which water-soluble amides are
formed [21]. The benzenesulfonyl esters (and other esters) of oximes rearrange
in neutral or aqueous alkaline solutions, conditions which occasionally offer
advantages [22, 23].

The first step in the rearrangement appears to be reversible protonation of
the oxime by the acid reagent to give 18-XII, which on the loss of water brings
about the rearrangement as follows:

$$\underset{NOH}{\overset{ArCR}{\|}}\overset{H^{\oplus}}{\rightleftharpoons}\underset{\overset{\oplus}{NOH_2}}{\overset{ArCR}{\|}}\xrightarrow{-H_2O}\left[\underset{\oplus N}{\overset{ArCR}{\|}}\right]$$

18-XII

$$\underset{ArNH}{\overset{O}{\diagdown}}\underset{}{\overset{}{CR}}\overset{-H^{\oplus}}{\rightleftharpoons}\left[\underset{ArN}{\overset{\oplus OH}{\underset{\|}{H\diagup\diagdown CR}}}\right]\overset{H_2O}{\longleftarrow}\left[\underset{ArN}{\overset{\oplus CR}{\|}}\right]$$

As is well-known, the electron-deficient nitrogen atom attacks the group *anti*
to the hydroxyl group originally present. Thus the *syn* and *anti* forms of a
specific oxime potentially are capable of producing two different amides.

Experimentally, however, these results are rarely attainable. Acylphenones (ArCOR) almost always give only one oxime (in which the aryl group is *anti* to hydroxy), and most aliphatic, unsymmetrical ketones give mixtures of *syn* and *anti* oximes which cannot even be separated by gas chromatography. Moreover, in the unsymmetrical benzophenone oxime family, where it is most practical to separate *syn* and *anti* forms, rearrangement conditions may bring about an equilibration of *syn* and *anti* forms, a process which gives mixtures of amides or the amide corresponding to the more stable oxime. Phosphorus pentachloride in ether at low temperature appears to be the most reliable reagent to prevent isomerization [24].

The Beckmann rearrangement also occurs for aldoximes, in which case either the nitrile or the amide is formed:

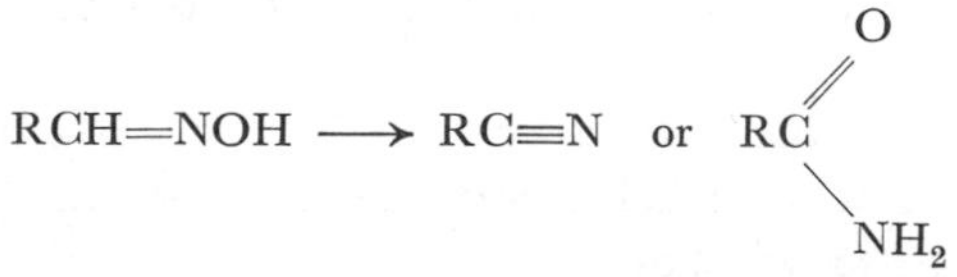

$$RCH{=}NOH \longrightarrow RC{\equiv}N \quad or \quad RC\overset{\displaystyle O}{\underset{\displaystyle NH_2}{\diagup\diagdown}}$$

The catalysts employed here have been copper, Raney nickel, boron trifluoride, trifluoroacetic acid, phosphorus pentachloride, and sulfuric acid [25]. A recent one, nickel acetate in a homogeneous mixture, appears to be the best [26].

Ketimines [27], hydrazones, and semicarbazones [28] likewise undergo a type of Beckmann rearrangement. For the former, the reagent employed was peracetic acid and, for the last two types, 90% sulfuric acid containing 1.5 equivalents of sodium nitrite was the preferred reagent. Yields from these various derivatives of aldehydes and ketones are at times high.

Some rather unusual rearrangements, related to the Beckmann transformation or at least to those involving group-deficient nitrogen atoms, have been described in the literature. For example, N-nitrosoacetamides are capable of forming oxamides (see Ex. b.5):

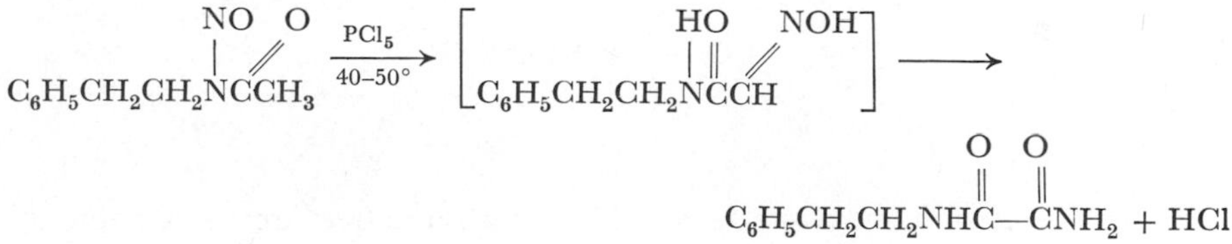

$$C_6H_5CH_2CH_2\overset{\displaystyle NO}{N}\overset{\displaystyle O}{C}CH_3 \xrightarrow[40-50°]{PCl_5} \left[C_6H_5CH_2CH_2\overset{\displaystyle HO}{N}\overset{\displaystyle NOH}{C}CH \right] \longrightarrow$$

$$C_6H_5CH_2CH_2NH\overset{\displaystyle O}{C}-\overset{\displaystyle O}{C}NH_2 + HCl$$

N-Phenethyloxamide 90%

And nitrene (:NH) may be inserted in specific phenols to yield amides (see Ex. b.6):

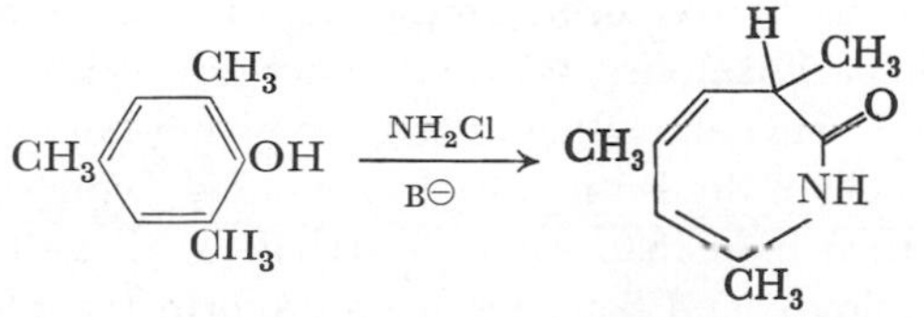

3,5,7-Trimethyl-1,3-dihydro-
2H-azepin-2-one

(a) **Preparation of ε-caprolactam** (89 % from 250 ml. of 95 % sulfuric acid to which 226 g. of cyclohexanone oxime was added very cautiously in 20–30 g. portions, the mixture being held at 60° for 90 hr. (*Caution:* a violent, exothermic reaction occurs at about 70°); the mixture was then cooled and neutralized with ammonium hydroxide) [29, 18].

(b) **Other examples**

(1) **ε-Caprolactam.** This method is faster than that in Ex. a, but yields a product of lower quality; 88 % by addition of cyclohexanone oxime to hot sulfuric acid dispersed with a mixture of benzene and chlorobenzene, the heat of rearrangement being dissipated by the refluxing of the solvent [30].

(2) **Acetanilide** (97 % from acetophenone oxime and polyphosphoric acid) [18].

(3) **p-Dimethylaminobenzamide** (95 % from p-dimethylaminobenzaldoxime isomerized with nickel acetate tetrahydrate in xylene containing a small amount of piperidine) [26].

(4) **Benzamide** (75–100 % from benzaldoxime and Raney nickel) [31].

(5) **N-Phenethyloxamide** (90 % from N-nitroso-N-acetyl-2-phenylethylamine and phosphorus pentachloride at 40–50° until hydrogen chloride evolution ceased) [32].

(6) **3,5,7-Trimethyl-1,3-dihydro-2H-azepin-2-one** (50–55 % from dropwise addition of a cold (−70°) ethereal solution of chloramine to a well-stirred solution of sodium 2,4,6-trimethylphenoxide in excess 2,4,6-trimethylphenol held at 120–150°) [33].

6. From Carbonyl Compounds and Hydrazoic Acid or from Alkenes (Schmidt)

$$\begin{array}{c} R \\ \diagdown \\ CO \xrightarrow{\ N_3H\ } RCONHR \\ \diagup \\ R \end{array}$$

By treatment with hydrazoic acid, acids give amines (see Amines, Chapter 8, G.5), aldehydes give nitriles and formyl derivatives of the amine, and ketones give amides. The synthesis known as the Schmidt reaction has been reviewed [34]. Although the greatest use of the method lies in the preparation of amines from carboxylic acids, it has been employed to some extent with carbonyl compounds, particularly ketones, to prepare amides. With cyclic ketones, such as the 4-piperidones, ring enlargement occurs to give 5-homopiperazinones [35].

The synthesis involves the ketone, hydrazoic acid, which because of its high toxicity must be used only in a well-ventilated hood, and a catalyst. A low temperature is employed and benzene or chloroform is the customary solvent. Of the great variety of catalysts employed, sulfuric and hydrochloric acids appear to be the most common, although recently it has been shown that the use of polyphosphoric acid as a solvent and catalyst gives superior yields [36]. The hydrazoic acid may be used as such or it may be generated *in situ* from sodium azide in the acidic medium. Various methods of addition have been employed [34].

For aliphatic ketones, concentrated aqueous hydrochloric acid is a useful solvent and the preferred catalyst. In the case of insoluble aliphatic ketones, hydrogen chloride in alcohol or dioxane serves well. Trichloroacetic acid as the catalyst gives excellent results with the less basic aryl alkyl ketones. For the still less basic diaryl ketones, sulfuric acid is preferred; 2 moles of this catalyst with 1 of the ketone in trichloroacetic acid produces a homogeneous reaction medium and reduces the possibility of sulfonation [37]. Yields in the synthesis are variable.

With unsymmetrical ketones, if the mechanism which follows is correct, two diazoketimine ions, the *syn* and the *anti*, are possible. These should lead to two different amides if the *anti* shift associated with the Beckmann rearrangement is valid. Attempts to determine the migratory aptitude of groups by the isolation of the relative amounts of isomeric amides have shown that, as a rule, the group which preferentially migrates from carbonyl to nitrogen is the one which has the greatest bulk [38, 39], although *o*-substituted aryl ketones appear to be an exception [40].

The pathway of reaction involves the electron-deficient nitrogen atom common to this type of rearrangement [37, 41]

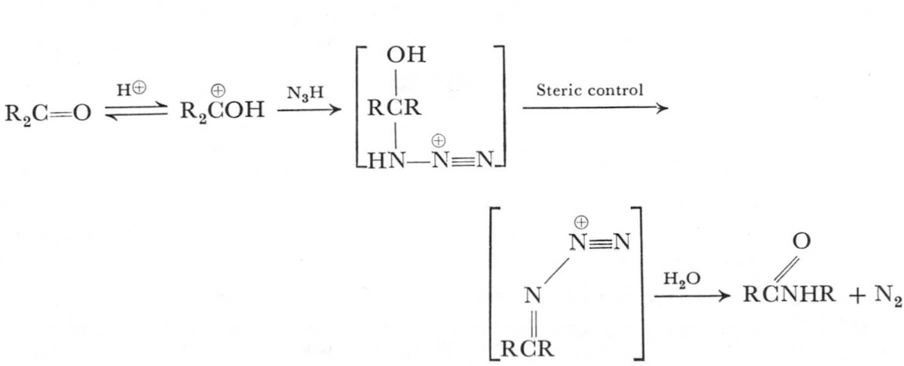

Acyl azides are potential sources of amides. They react with olefins to form pyrazolines which spontaneously decompose to aziridines [42]:

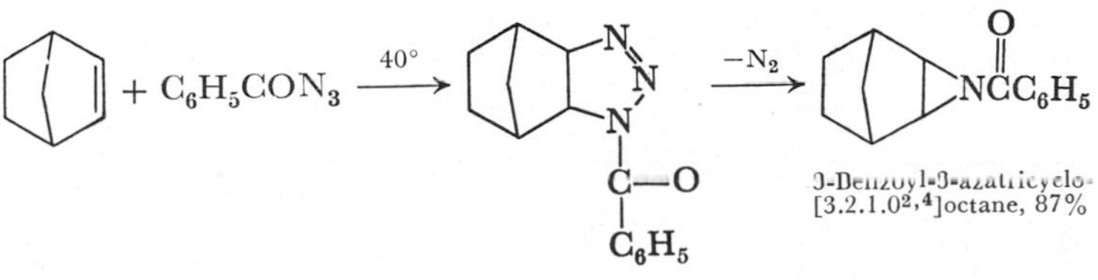

3-Benzoyl-3-azatricyclo-
[3.2.1.0²,⁴]octane, 87%

And, similarly, the unusual, reactive nitrile imines can be converted into amides [42]:

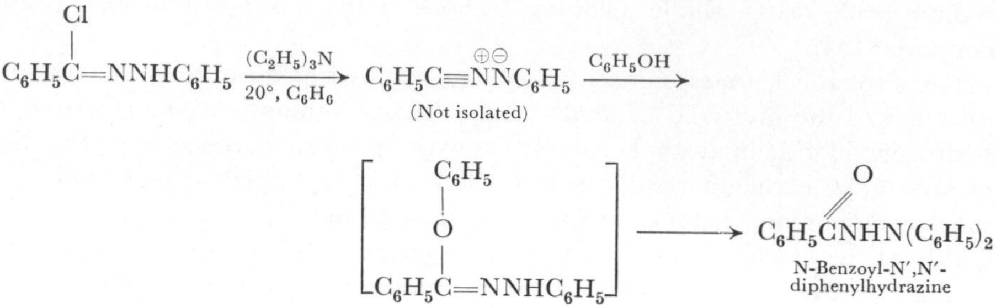

$$
\underset{|}{\overset{Cl}{C_6H_5C}}=NNHC_6H_5 \xrightarrow[20°,\ C_6H_6]{(C_2H_5)_3N} \underset{\text{(Not isolated)}}{C_6H_5C\equiv\overset{\oplus}{N}\overset{\ominus}{N}C_6H_5} \xrightarrow{C_6H_5OH}
$$

$$
\left[
\begin{array}{c}
C_6H_5 \\
| \\
O \\
| \\
C_6H_5C=NNHC_6H_5
\end{array}
\right]
\longrightarrow
\underset{\substack{\text{N-Benzoyl-N',N'-}\\ \text{diphenylhydrazine}}}{C_6H_5\overset{O}{\overset{\|}{C}}NHN(C_6H_5)_2}
$$

(a) **Preparation of β-acetonaphthalide.** To 1.70 g. of β-acetonaphthone in 15 g. of trichloroacetic acid at 60°, 1.0 g. of powdered sodium azide

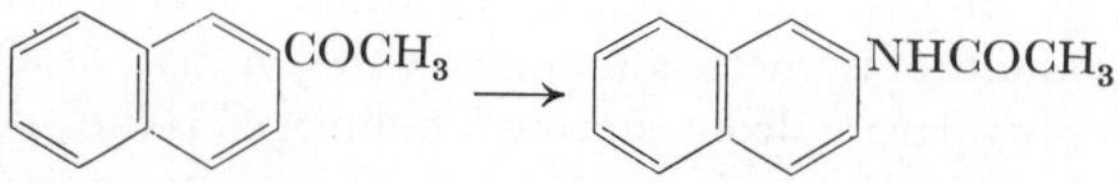

was added. After stirring occasionally during 4 hr. at the same temperature, the mixture was treated with 75 ml. of water and made alkaline with 10 ml. of ammonium hydroxide. The white product formed was recovered and washed with water and warm petroleum ether to give 1.75 g. (95%), m.p. 125–130°. Recrystallization from aqueous alcohol gave a melting point of 132–134° [37].

(b) **Other examples**

   (1) **Acetanilide** (90% from acetophenone and hydrogen azide in benzene in the presence of sulfuric acid) [39]; *see also* [36].
   (2) **Benzanilide** (99% from benzophenone, sodium azide, and polyphosphoric acid) [36].
   (3) **2,2,7,7-Tetramethyl-5-homopiperazinone** (88% from 2,2,6,6-tetramethyl-4-piperidone monohydrate, sodium azide, and conc. sulfuric acid

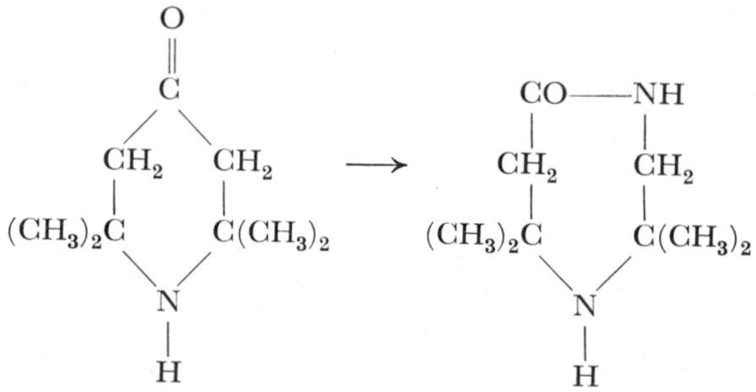

in chloroform) [35].

7. From Carboxylic Acids or Aralkyl Ketones and Nitrosylsulfuric Acid

This reaction takes a different course from that of the Schmidt reaction (D.6) or the Beckmann rearrangement (D.5) [43]:

$$R_2CHCOOH + NO_2SO_2OH \longrightarrow \underset{\underset{NO}{|}}{R_2CCOOH} \longrightarrow$$

$$CO_2 + (R_2C\!\!=\!\!NOH) \xrightarrow[\text{rearrangement}]{\text{Beckmann}} RC\!\!-\!\!NHR$$

or

$$ArCOCHR_2 \xrightarrow{NO_2SO_2OH} \underset{\underset{NO}{|}}{ArCOCR_2} \longrightarrow$$

$$ArCOOH + R_2C\!\!=\!\!NOH \longrightarrow RC\!\!-\!\!NHR$$

Actually the synthesis is a means of degradation of an acid or a ketone. It finds its greatest use in preparing cyclic amides (see Ex. a). A similar reaction is the formation of ε-caprolactam (60%) from cyclohexanecarboxylic acid, nitric and sulfuric acids in liquid sulfur dioxide [44].

(a) **Preparation of laurolactam** (88.5% from cyclododecanecarboxylic acid which was nitrosated with equimolecular amounts of nitrosylsulfuric acid in 15–30% oleum at 65–70° in chloroform) [45].

(b) **Preparation of N-butylvaleramide** (50% from 1 mole of dibutyl-acetic acid in 100 ml. of cyclohexane at 80° to which a mixture of 1 mole of nitrosylhydrogen sulfate, 1 mole of H_2SO_4, and 1 mole of SO_3 was added dropwise over 1 hr. and the mixture heated an additional 15 min.) [46].

8. From Metal Carbonyls

Metal carbonyls such as $Mn_2(CO)_{10}$ or $Co_2(CO)_8$ react with amines under pressure to yield a mixture of formamides and ureas [47].

Palladium chloride, carbon monoxide, and amines also form a mixture of urea and oxamide [48]:

$$C_{10}H_{21}NH_2 \xrightarrow[\substack{PdCl_2, 0.5\ g. \\ 180°, 20\ hr.}]{CO,\ 100\ kg./cm.^2} (C_{10}H_{21}NH)_2C\!\!=\!\!O + (C_{10}H_{21}NHC)_2 + H_2$$

$$\underset{7.9\ g.}{} \qquad \underset{\substack{2.2\ g. \\ \text{1,3-Didecylurea}}}{} \qquad \underset{\substack{3.8\ g. \\ \text{N,N'-Didecyloxamide}}}{}$$

(a) **Preparation of N,N-diethylacrylamide** (66% from aqueous diethylamine, nickel cyanide, hydroquinone, carbon monoxide, and acetylene at 25 atm. for 25 hr.) [49].

1. G. A. Olah and J. A. Olah, in G. A. Olah, *Friedel Crafts and Related Reactions*, Vol. 3, Interscience Publishers, New York, 1964, Pt. 2, p. 1262.
2. H. Hopff and H. Ohlinger, *Angew. Chem.*, **61**, 183 (1949).

3. F. Weygand and R. Mitgau, *Chem. Ber.*, **88,** 301 (1955).
4. J. F. K. Wilshire, *Australian J. Chem.*, **20,** 575 (1967).
5. R. Leuckart, *J. Prakt. Chem.* (2), **41,** 301 (1890).
6. J. B. McKinley *et al.*, *Ind. Eng. Chem., Anal. Ed.*, **16,** 304 (1944); H. E. French and A. F. Wirtel, *J. Am. Chem. Soc.*, **48,** 1736 (1926).
7. H. E. Zaugg and W. B. Martin, *Org. Reactions*, **14,** 52 (1966).
7a. L. J. Krimen and D. J. Cota, *Org. Reactions*, **11,** 213 (1969).
8. J. J. Ritter and J. Kalish, *Org. Syn.*, **44,** 44 (1964).
9. P. J. Scheuer *et al.*, *J. Org. Chem.*, **22,** 674 (1957).
10. F. Johnson and R. Madroñero, in A. R. Katritzky and A. J. Boulton, *Advances in Heterocyclic Chemistry*, Vol. 6, Academic Press, New York, 1966, p. 95.
11. J. J. Ritter and P. P. Minieri, *J. Am. Chem. Soc.*, **70,** 4045 (1948).
12. A. J. Speziale *et al.*, *J. Org. Chem.*, **26,** 4311 (1961).
13. C. L. Parris and R. M. Christenson, *J. Org. Chem.*, **25,** 1888 (1960).
14. L. G. Donaruma and W. Z. Heldt, *Org. Reactions*, **11,** 1 (1960).
15. P. A. S. Smith, in P. DeMayo, *Molecular Rearrangements*, Interscience Publishers, New York, 1963, Pt. 1, Chap. 8.
16. W. E. Bachmann and C. H. Boatner, *J. Am. Chem. Soc.*, **58,** 2097 (1936); E. Mosettig and J. W. Krueger, *J. Org. Chem.*, **3,** 317 (1938).
17. L. G. Donaruma and W. Z. Heldt, *Org. Reactions*, **11,** 1 (1960).
18. E. C. Horning and V. L. Stromberg, *J. Am. Chem. Soc.*, **74,** 2680 (1952).
19. R. T. Conley and M. C. Annis, *J. Org. Chem.*, **27,** 1961 (1962).
20. R. K. Hill and O. T. Chortyk, *J. Am. Chem. Soc.*, **84,** 1064 (1962).
21. W. D. Emmons, *J. Am. Chem. Soc.*, **79,** 6522 (1957).
22. Ref. 14, p. 45.
23. W. Z. Heldt, *J. Org. Chem.*, **26,** 1695 (1961).
24. Ref. 14, p. 55.
25. Ref. 14, p. 41.
26. L. Field *et al.*, *J. Am. Chem. Soc.*, **83,** 1983 (1961).
27. C. R. Hauser and D. S. Hoffenberg, *J. Am. Chem. Soc.*, **77,** 4885 (1955).
28. D. E. Pearson *et al.*, *J. Am. Chem. Soc.*, **75,** 5905 (1953).
29. D. E. Pearson *et al.*, *J. Org. Chem.*, **19,** 1815 (1954); *see also* C. S. Marvel and J. C. Eck, *Org. Syn.*, Coll. Vol. **2,** 371 (1943).
30. D. E. Pearson *et al.*, *J. Org. Chem.*, **19,** 1815 (1954).
31. R. Paul, *Compt. Rend.*, **204,** 363 (1937); *Bull. Soc. Chim. France*, 1115 (1937).
32. M. Murakami *et al.*, *J. Am. Chem. Soc.*, **83,** 2002 (1961).
33. L. A. Paquette, *J. Am. Chem. Soc.*, **84,** 4987 (1962); W. Theilacker and K. Ebke, *Angew. Chem.*, **75,** 208 (1963).
34. H. Wolff, *Org. Reactions*, **3,** 307 (1946).
35. S. C. Dickerman and H. G. Lindwall, *J. Org. Chem.*, **14,** 530 (1949).
36. R. T. Conley, *J. Org. Chem.*, **23,** 1330 (1958).
37. P. A. S. Smith, *J. Am. Chem. Soc.*, **70,** 320 (1948).
38. P. A. S. Smith and J. P. Horowitz, *J. Am. Chem. Soc.*, **72,** 3718 (1950).
39. J. K. Sanford *et al.*, *J. Am. Chem. Soc.*, **67,** 1941 (1945).
40. P. A. S. Smith, *J. Am. Chem. Soc.*, **76,** 431 (1954).
41. M. S. Newman and H. Gildenhorn, *J. Am. Chem. Soc.*, **70,** 317 (1948).
42. R. Huisgen, in P. A. S. Smith, *Open-Chain Nitrogen Compounds*, Vol. 2, W. A. Benjamin, Inc., New York, 1966, Chap. 10.
43. Y. Ogata *et al.*, *J. Am. Chem. Soc.*, **85,** 3649 (1963).
44. N. Tokura *et al.*, *Bull. Soc. Chem. Japan*, **38,** 849 (1965).
45. W. Ziegenbein and W. Lang, *Angew. Chem., Intern. Ed. Engl.*, **2,** 149 (1963).
46. K. Smeykal *et al.*, *J. Prakt Chem.*, **30,** 126 (1965).
47. F. Calderazzo, *Inorg. Chem.*, **4,** 293 (1965).
48. J. Tsuji and N. Iwamoto, *Chem. Commun.*, 380 (1966).
49. W. Reppe, *Ann. Chem.*, **582,** 33 (1953).

E. Nucleophilic-Type Syntheses

These methods consist of alkylation, E.1, or hydroxyalkylation, E.2, of amides or imides, addition of amines to electronegatively substituted alkenes, E.3, Grignard addition to isocyanates and unsaturated amides, E.4, and ketone cleavage by amides, E.5.

1. From Amides or Imides by Alkylation

$$RCONHAr \xrightarrow[R'I]{Na} RCONAr$$
$$\underset{R'}{|}$$

Amides are obtained from N-monosubstituted amides by treatment of the sodium salt with methyl iodide or dimethyl sulfate [1, 2]. The sodium salt may be obtained with the use of sodium metal or sodium hydride. Yields for a series of N-monosubstituted amides vary from 53 to 89%.

For alkylation of imides, see Gabriel synthesis of amines, Chapter 8, B.2.

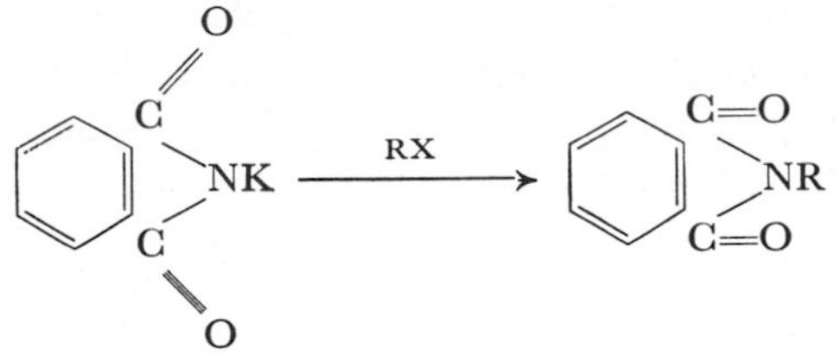

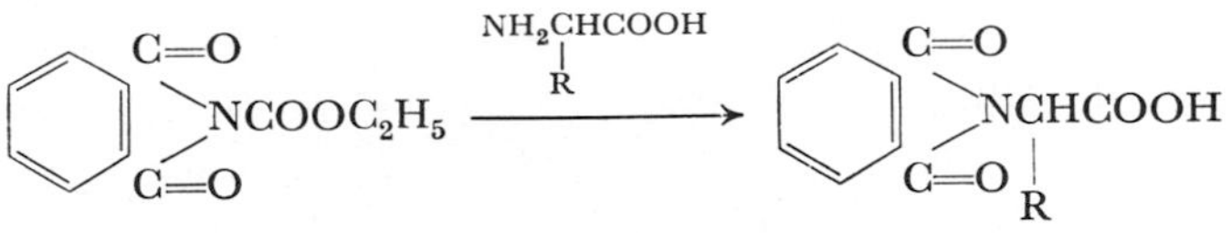

Three ways, methods A, B, and C, have been developed for synthesizing substituted phthalimides. Method A is the Gabriel synthesis in which the potassium salt may be alkylated by an alkyl halide. Recently yields of 89% and better have been obtained [3] by this method when a solvent such as dimethylformamide, in which the potassium phthalimide is appreciably soluble, is utilized. Method B permits the synthesis of phthaloyl amino acids under mild conditions in yields varying from 65 to 95% [4]. In this method the potassium phthalimide is converted into N-carbethoxyphthalimide by the use of ethyl

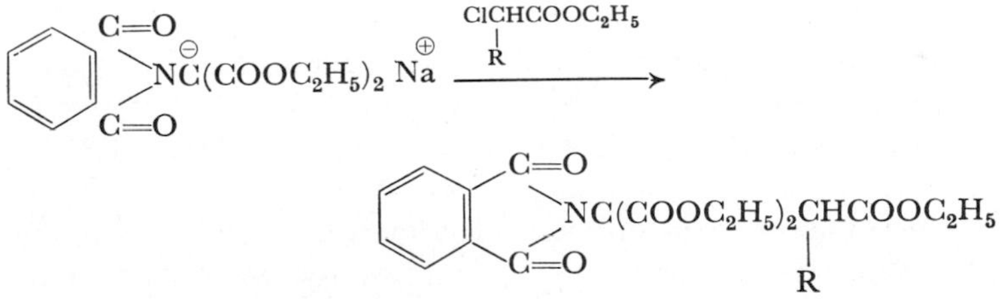

References for Section E are on pp. 934–935.

chlorocarbonate in dimethylformamide; phthalimide gives the same product on treatment with triethylamine and ethyl chlorocarbonate in the same solvent. In the final step of this method the optical activity of the amino acid is not destroyed. By method C the potassium phthalimide is converted into diethyl sodium phthalimidomalonate by the use of ethyl bromomalonate [5] and sodium [6], and the sodium salt is then treated with the chloroester [7].

Recently, the dianion of acetanilide has been made from butyllithium in ether-hexane and thus the alkylation at carbon rather than at nitrogen is possible [8]:

$$C_6H_5\overset{\ominus}{N}CO\overset{\ominus}{C}H_2 \xrightarrow[\text{2. H}_2\text{O}]{\text{1. C}_6\text{H}_5\text{CH}_2\text{Cl}} C_6H_5NHCOCH_2CH_2C_6H_5$$

β-Phenylpropionanilide, 69%

(a) Preparation of N-acetyl-N-methyl-*p*-anisidine. To 1.4 g. of sodium hydride in 50 ml. of dry xylene was added 8.3 g. of N-acetyl-*p*-anisidine in 200 ml. of boiling xylene and the mixture was refluxed with stirring for 20 hr. under an atmosphere of nitrogen. On cooling, the reflux condenser was replaced by a dry ice-acetone condenser and 20 g. of methyl iodide was added, after which the mixture was refluxed for 8 hr. more. From the filtrate plus the washings from 50 ml. of dry benzene there was obtained on distillation 8.0 g. of a product, b.p. 137–140° (3.5 mm.), m.p. 50–56°. One crystallization from ligroin (b.p. 90–120°) gave 7.3 g. (80%) of the substituted amide, m.p. 57–59° [2].

(b) Preparation of phthaloyl glycine. Glycine, 1.5 g., 5.75 g. of sodium carbonate decahydrate, and 4.5 g. of N-carbethoxyphthalimide were added to 30 ml. of water and the mixture was stirred for about 15 min. Phthaloyl-glycine separated after filtration followed by acidification of the filtrate. Crystallization from water and drying gave 3.72 g. (90.5%), m.p. 191° [4].

(c) Other examples

(1) Phthalimidoacetophenone (92% from potassium phthalimide and phenacyl bromide in dimethylformamide) [3].

(2) Triethyl α-phthalimidoethane-α,α,β-tricarboxylate (95–99% from diethyl sodium phthalimidomalonate and ethyl chloroacetate) [7].

(3) N-(β-Phenylpropionyl)-α-benzoylacetamide (69% from the tripotassio salt of N-acetyl-α-benzoylacetamide and benzyl chloride in liquid ammonia) [9]:

$$C_6H_5CO\overset{\overset{\text{K}}{|}}{C}HCO\overset{\overset{\text{K}}{|}}{N}CO\overset{}{C}H_2K \xrightarrow[\text{2. HOH}]{\text{1. C}_6\text{H}_5\text{CH}_2\text{Cl}} C_6H_5COCH_2CONHCOCH_2CH_2C_6H_5$$

2. From Amides or Imides by Hydroxyalkylation

$$ArCONH_2 + CH_2O \xrightarrow{\text{K}_2\text{CO}_3} ArCONHCH_2OH$$

Under mild conditions in neutral or alkaline medium, formaldehyde condenses with amides or imides to give N-hydroxymethyl derivatives [10, 11] in good yields but most higher aldehydes condense further to yield alkylidene-diacylamides, $RCH(NHCOR')_2$ [12]. A considerable number of hydroxy-methyl derivatives have been described by Einhorn [13], one of which has been oxidized to a formamide.

Alkylation of amides with acetaldehyde is a feasible process (see Ex. c) and, in a similar manner, alkylation has been carried out with N-vinylamides [14]:

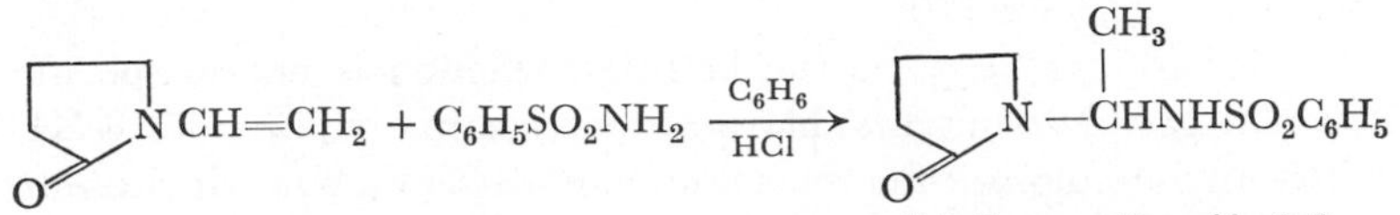

N-(α-Benzenesulfonamidoethyl)-pyrrolidinone-2, 95%

The alkylation is fairly general for compounds with active hydrogen atoms.

(a) **Preparation of N-hydroxymethylphthalimide.** Phthalimide, 511 g., 260 ml. of 40% formalin, and 1750 ml. of water were refluxed until a clear solution was obtained (5 min. at the boiling point). After cooling over-night, there was recovered 594 g. (96%), m.p. 137–141°, of N-hydroxymethyl phthalimide. Recrystallization from alcohol gave 94% of the original at the same melting point [10].

(b) **Preparation of N-hydroxymethylphenylacetamide** (51 g. from 50 g. of phenylacetamide and formaldehyde in the presence of potassium carbonate) [11].

(c) **Preparation of ethylidenebisacetamide** (60% from acetaldehyde, acetamide, and catalytic amounts of 60% perchloric acid heated for 1 hr.) [15].

3. From Imides or Amines and Electronegatively Substituted Alkenes

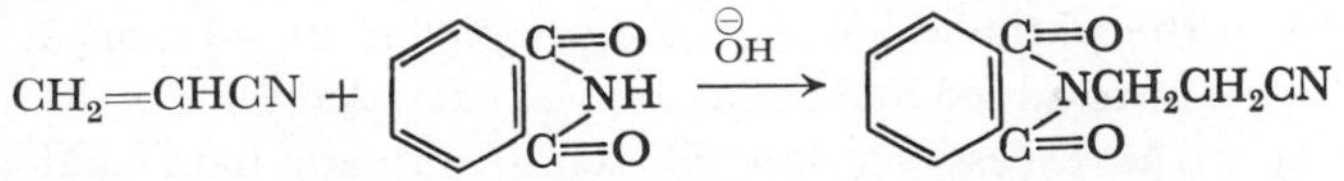

Substituted phthalimides have been synthesized by the addition of alkenes such as acrylonitrile or 2-vinylpyridine to phthalimide in presence of trimethyl-benzylammonium hydroxide (Triton B) [16, 17]. Yields with acrylonitrile were quantitative, and with 2-vinylpyridine 75%. The β-phthalimidopropionitrile is a useful intermediate in the synthesis of β-alanine.

Enamines alkylate unsaturated acids to form amides [18]:

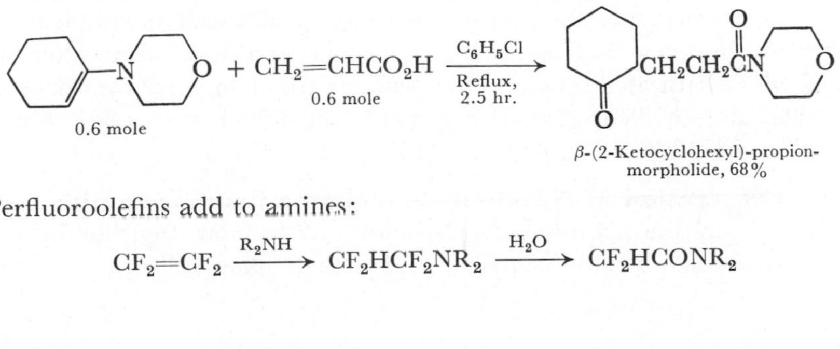

β-(2-Ketocyclohexyl)-propion-morpholide, 68%

Perfluoroolefins add to amines:

$$CF_2{=}CF_2 \xrightarrow{R_2NH} CF_2HCF_2NR_2 \xrightarrow{H_2O} CF_2HCONR_2$$

This synthesis, which is of interest in connection with the preparation of fluorinated amides, may be accomplished in one or two steps. In the one-step process the amine, borax, and tetrafluoroethylene are heated under pressure at 130°, the yields with four amines (aniline, N-methylaniline, butylamine, and dibutylamine) varying from 51 to 90% [19]. Without the use of borax and with diethylamine the intermediate N-(1,1,2,2-tetrafluoreothyl)-diethylamine was obtained in 80.5% yield [20].

With chlorotrifluoroethylene the addition reaction is not so specific [21]. *n*-Butylamine gave N-*n*-butyl-α-chloro-α-fluoroacetimidyl fluoride 18-XIII and N,N'-di-*n*-butyl-α-chloro-α-fluoroacetamidine 18-XIV, but with aniline only the amidine was obtained. Either of these products may be hydrolyzed to the halogenated amide. For secondary amines, such as diethylamine, the simple

$$ClCF{=}CF_2 \xrightarrow{C_4H_9NH_2} [CHClFCF_2NHC_4H_9]$$

$$\text{CHClFC} \underset{\substack{\diagup NC_4H_9 \\ \diagdown NHC_4H_9}}{} \xleftarrow{C_4H_9NH_2} \text{CHClFCF}{=}NC_4H_9$$

18-XIV   18-XIII

addition product was isolated from a mixture of the two components by working in a pressure vessel at room temperature. It hydrolyzed quantitatively to the amide. Diaryl amines did not add to chlorotrifluoroethylene.

(a) **Preparation of β-phthalimidopropionitrile.** Phthalimide, 29.4 g., was refluxed with 100 ml. of acrylonitrile while 2.5 ml. of a 40% solution of trimethylbenzylammonium hydroxide (Triton B) was added directly into the reaction mixture over a 10-min. period. Refluxing was continued for another 10 min. and, if the phthalimide was not completely in solution, a few more drops of catalyst were added and refluxing was extended for 5 or 10 min. more. After removing the excess acrylonitrile on the steam bath under reduced pressure, the residue of β-phthalimidopropionitrile amounted to 40 g. (quantitative). Recrystallization from alcohol or water gave a melting point of 154–155.5° [16].

(b) **Preparation of difluoroacetanilide.** In a stainless steel-lined shaker tube were placed 93 g. of freshly distilled aniline and 15 g. of borax decahydrate. After sweeping with nitrogen, the tube was closed, cooled in carbon dioxide-acetone, and charged with 50 g. of tetrafluoroethylene. The mixture was heated at 130° with shaking for 8 hr. and the product, after being washed with a saturated solution of potassium carbonate, dried over potassium carbonate, and distilled, gave 61.6 g. (71%) of difluoroacetanilide, b.p. 90° (1 mm.) and 114° (5 mm.), m.p. 58° [19].

(c) **Preparation of N,N-diethyl-α-chloro-α-fluoroacetamide** (quantitative yield by the hydrolysis N-(2-chloro-1,1,2-trifluoroethyl)-diethylamine prepared from chlorotrifluoroethylene and diethylamine) [21].

4. From Isocyanates and Like Compounds by the Addition of Grignard Reagents or Carbanions

$$\text{RMgX} + C_6H_5N{=}C{=}O \xrightarrow{H_2O} C_6H_5NHCOR$$

$$(\text{or } C_6H_5N{=}C{=}S) \qquad\qquad (\text{or } C_6H_5NHC\underset{\underset{S}{\|}}{R})$$

$$R\overset{\overset{O}{\diagup}}{C}{-}\overset{\ominus}{C}HY + C_6H_5N{=}C{=}O \xrightarrow{H_2O} C_6H_5NHCOC\overset{\overset{Y}{\diagup}}{H}{-}\overset{\overset{O}{\|}}{C}R$$

The first method has been used as a method of identification of alkyl or aryl halides. Yields are very good [22]. Triethylaluminum may be substituted for the Grignard reagent to give N-substituted propionic acid amides from isocyanates, usually in yields of 81–99 % [23]. The scope of the second method, the addition of carbanions, has not been delineated completely, but it appears that, when Y is electron-withdrawing, the reaction is feasible [24].

Comparable to these reactions is the addition of Grignard reagents to carbodiimides to yield amidines [25]:

$$C_6H_5N{=}C{=}NC_6H_5 + C_6H_5MgBr \longrightarrow C_6H_5\overset{\overset{NC_6H_5}{\diagup}}{C}{-}NHC_6H_5$$

N-Phenyl-N′-phenyl-
benzamidine, 70%

Conjugated unsaturated amides react with Grignard reagents as follows [26]:

$$C_6H_5MgBr + C_6H_5CH{=}CHCON(CH_3)_2 \xrightarrow{H_2O} (C_6H_5)_2CHCH_2CON(CH_3)_2$$

N,N-Dimethyl-β,β-diphenylpropionamide,
93%

(a) Preparation of N-(α-furyl)-propionamide (89 % from α-furylisocyanate and ethylmagnesium bromide) [27].

(b) Preparation of α-benzoyl-α-nitroacetanilide (83 % from α-nitroacetophenone, phenylisocyanate, and trimethylamine in benzene and ether for 2 hr.) [24].

5. From Ketones by Cleavage with Sodamide (Haller-Bauer)

$$C_6H_5COCR_3 \xrightarrow[\text{2. } H_2O]{\text{1. NaNH}_2} C_6H_6 + R_3CCONH_2$$

This synthesis, which has been reviewed [28], is a general reaction for diaryl ketones, hexaalkyl acetones, and t-alkyl aryl ketones. Although of limited value, the synthesis has been of service in preparing trialkylacetic acids via trialkylacetophenones and trialkylacetamides [29]. Cleavage of the ketone occurs by refluxing with sodamide in a hydrocarbon solvent such as benzene, toluene, or xylene. Among the ketones high in the series, results are not satisfactory unless two of the alkyl groups are methyl groups [29, 30]. For lower ketones, yields are sometimes in the 80–90 % range.

Since the Haller-Bauer reaction is restricted to nonenolizable ketones, the cleavage path is divorced from the anion formation of an active hydrogen compound. Instead, a carbanion is formed by the addition of the amide anion to the ketone:

$$C_6H_5COCR_3 \underset{NH_2^{\ominus}}{\rightleftharpoons} C_6H_5\overset{\overset{O^{\ominus}}{|}}{\underset{\underset{NH_2}{|}}{C}}-CR_3 \longrightarrow R_3CCONH_2 + \left[C_6H_5^{\ominus}\right] \overset{H_2O}{\longrightarrow} C_6H_6 + O\overset{\ominus}{H}$$

and the carbanion dissociates to give the amide and the stablest anion, which in the case shown is phenyl.

(a) Preparation of α,α-dimethyl-β-phenylpropionamide (69% from

$$C_6H_5CH_2\overset{\overset{CH_3}{|}}{\underset{\underset{CH_3}{|}}{C}}COC_6H_5 \xrightarrow[\text{2. H}_2\text{O}]{\text{1. NaNH}_2} C_6H_5CH_2\overset{\overset{CH_3}{|}}{\underset{\underset{CH_3}{|}}{C}}CONH_2 + C_6H_6 + NaOH$$

2,2-dimethyl-1,3-diphenylpropan-1-one by heating with sodamide in toluene on a steam bath for 5 hr.) [31].

1. E. Thielepape, *Chem. Ber.*, **68**, 751 (1935).
2. W. S. Fones, *J. Org. Chem.*, **14**, 1099 (1949).
3. J. C. Sheehan and W. A. Bolhofer, *J. Am. Chem. Soc.*, **72**, 2786 (1950).
4. G. H. L. Nefkens, *Nature*, **185**, 309 (1960).
5. A. E. Osterberg, *Org. Syn.*, Coll. Vol. **1**, 271 (1941).
6. G. Barger and T. E. Weichselbaum, *Org. Syn.*, Coll. Vol. **2**, 384 (1943).
7. M. S. Dunn and B. W. Smart, *Org. Syn.*, Coll. Vol. **4**, 55 (1963).
8. R. L. Gray and C. R. Hauser, *J. Am. Chem. Soc.*, **89**, 1647 (1967).
9. J. J. Wolfe and C.-L. Mao, *J. Org. Chem.*, **32**, 1977 (1967).
10. S. R. Buc, *J. Am. Chem. Soc.*, **69**, 254 (1947).
11. R. D. Haworth et al., *J. Chem. Soc.*, 1493 (1950).
12. H. E. Zaugg and W. B. Martin, *Org. Reactions*, **14**, 52 (1966).
13. A. Einhorn, *Ann. Chem.*, **361**, 113 (1908).
14. R. A. Hickner et al., *J. Org. Chem.*, **32**, 729 (1967).
15. N. Yanaihara and M. Saito, *Chem. Pharm. Bull.* (*Tokyo*), **15**, 128 (1967); *C.A.*, **66**, 9778 (1967).
16. A. Galat, *J. Am. Chem. Soc.*, **67**, 1414 (1945).
17. F. K. Kirchner et al., *J. Org. Chem.*, **14**, 388 (1949).
18. P. Klemmensen and S.-O. Lawesson, *Ark. Kemi*, **26**, 317 (1967).
19. D. D. Coffman et al., *J. Org. Chem.*, **14**, 747 (1949).
20. N. N. Yarovenko et al., *J. Gen. Chem. USSR* (*Eng. Transl.*), **27**, 2305 (1957).
21. R. L. Pruett et al., *J. Am. Chem. Soc.*, **72**, 3646 (1950).
22. A. M. Schwartz and J. R. Johnson, *J. Am. Chem. Soc.*, **53**, 1063 (1931); R. B. Carlin and L. O. Smith, Jr., *ibid.*, **69**, 2007 (1947).
23. H. Reinheckel and D. Jahnke, *Chem. Ber.*, **97**, 2661 (1964).
24. A. Dornow et al., *Ann. Chem.*, **594**, 191 (1955).
25. M. Busch and R. Hobein, *Chem. Ber.*, **40**, 4296 (1907).
26. G. Gilbert, *J. Am. Chem. Soc.*, **77**, 4413 (1955).
27. H. M. Singleton and W. R. Edwards, Jr., *J. Am. Chem. Soc.*, **60**, 540 (1938).
28. F. W. Bergstrom and W. C. Fernelius, *Chem. Rev.*, **12**, 43 (1933); **20**, 413 (1937); K. E. Hamlin and A. W. Weston, *Org. Reactions*, **9**, 1 (1957).

29. C. L. Carter and S. N. Slater, *J. Chem. Soc.*, 130 (1946).

30. N. P. Buu-Hoï, *Rec. Trav. Chim.*, **65**, 246 (1946).

31. L. L. Abell et al., *Org. Reactions*, **9**, 16 (1957).

F. Free Radical Reactions

The discovery that free radicals of the nature of ·CONH$_2$ or ·CH$_2$CONH$_2$ can be generated photochemically or from free radical sources has led to some interesting syntheses of amides. These free radicals add to olefins in particular, but can be induced to substitute in aromatic hydrocarbons. With di-*t*-butyl peroxide as catalyst, the reaction is complicated by the intrusion of more than one free radical [1]:

$$RCH{=}CH_2 + HC\overset{\displaystyle O}{\|}{-}N(CH_3)_2 \xrightarrow[\text{Di-}t\text{-butyl peroxide}]{115\text{–}150°}$$

$$RCH_2CH_2CON(CH_3)_2 + RCH_2CH_2CH_2\overset{\displaystyle O}{\overset{\|}{N}}CH \quad \text{and higher telomers}$$

52%

|
CH$_3$

35%

N-*t*-Butylformamide yields only the product, RCH$_2$CH$_2$CONHC(CH$_3$)$_3$, while N,N-dimethylacetamide yields RCH$_2$CH$_2$CH$_2$N(CH$_3$)COCH$_3$. Thus, either amidation or aminoalkylation can be carried out selectively by choice of N-*t*-butylformamide or N,N-dimethylacetamide, respectively. However, amidation has been accomplished with acetamide [2]:

$$C_6H_{13}CH{=}CH_2 + CH_3CONH_2 \xrightarrow[150°]{\text{Di-}t\text{-butyl peroxide}} C_6H_{13}CH_2CH_2CH_2CONH_2$$

Decanamide, 31%

If no olefin is present, a variety of products is formed from dimethylformamide [1]:

$$HCON(CH_3)_2 \xrightarrow[\text{peroxide}]{\text{Di-}t\text{-butyl}}$$

$$HC\overset{\displaystyle O}{\|}NCH_2CH_2\overset{CH_3}{\overset{/}{N}}CH\overset{\displaystyle O}{\|} \; + \; HC\overset{\displaystyle O}{\|}{-}N\overset{CH_3}{\overset{|}{C}}H_2CN(CH_3)_2\overset{\displaystyle O}{\|} \; + \; (CH_3)_2N\overset{OO}{\overset{\|\,\|}{CC}}N(CH_3)_2$$

N,N'-Diformyl-N, N'-dimethyl-
ethylenediamine, 72%

N-Formyl-N-methylaminoacetic
acid dimethylamide, 22%

N,N,N',N'-Tetramethyl-
oxamide, 6%

Mesityloxide peroxide apparently is superior to di-*t*-butylperoxide in excitation of formamide [3].

Formamide free radicals are generated by abstraction of hydrogen by photochemically active acetone, and the radicals then add to olefins and arenes (see Exs. a, b.1, and b.2):

$$[\cdot CONH_2] + RCH{=}CH_2 \longrightarrow [RCHCH_2CONH_2] \xrightarrow{HCONH_2}$$

$$[RCHCH_2CONH_2] + [\cdot CONH_2], \quad \text{etc.}$$

References for Section F are on p. 936.

(a) **Preparation of octanamide.** A mixture of 1-heptene (0.5 g.), formamide (40 g.), *t*-butyl alcohol (35 ml.), and acetone (5 ml.) was irradiated for 45 min. with a high-pressure mercury lamp. Then a solution of heptene (4.4 g.), *t*-butyl alcohol (10 ml.), and acetone (7 ml.) was added in 10 equal portions at 45-min. intervals and irradiation was continued for another 6 hr. The solvents were removed by distillation at atmospheric pressure and the formamide at 0.2 mm. pressure. The residue was dissolved in acetone, filtered to remove traces of oxamide; removal of solvent and addition of water gave 44% of octanamide [4].

(b) Other examples

(1) **Norbornane-2-*exo*-carboxamide** (87% from norbornene, formamide, *t*-butyl alcohol, and acetone irradiated as in Ex. a) [5].

(2) **1-Naphthamide** (20% from irradiation of naphthalene, formamide, and acetone; benzene yielded 15% benzamide; toluene, 23% β-phenylpropionamide) [6].

1. L. Friedman and H. Shechter, *Tetrahedron Letters*, 238 (1961).
2. R. J. Gritter and R. S. Woosley, *J. Chem. Soc.*, 5544 (1963).
3. A. Rieche *et al.*, *Angew. Chem.*, **73**, 621 (1961); *Z. Chem.*, **4**, 177 (1964).
4. D. Elad and J. Rokach, *J. Org. Chem.*, **29**, 1855 (1964).
5. D. Elad and J. Rokach, *J. Chem. Soc.*, 800 (1965).
6. D. Elad, *Tetrahedron Letters*, 77 (1963).

G. Cycloadditions

The Diels-Alder reaction has been conducted with N-alkylmaleimides [1]. 1,3- or 1,4-Dipolar additions also have given amides in certain cases [2]. An interesting example of 1,4-dipolar addition is shown [3]:

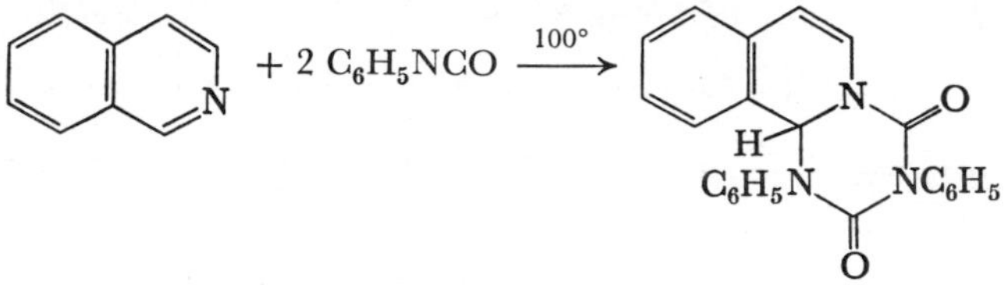

2,4-Dioxo-1,3-diphenyl-1,3,4,11b-tetrahydro-2H-*s*-triazino-[2.1-a]-isoquinoline, 70%

1. M. C. Kloetzel, *Org. Reactions*, **4**, 1 (1948).
2. R. Huisgen, *Proc. Chem. Soc.*, 357 (1961); *Chem. Ber.*, **100**, 1107 (1967).
3. R. Huisgen, *Chem. Ber.*, **100**, 1107 (1967).

Chapter 19

NITRILES (CYANIDES)

General references on the synthesis of nitriles are available [1]. Both the classical alkyl halide-sodium cyanide metathetical reaction and the exchange reaction between aryl halides and cuprous cyanide have been vastly improved by the use of aprotic solvents (Section A.1). These methods together with the dehydration of amides (C.1) and of oximes (C.4) still are the most common, general, and reliable routes to nitriles. Nitrile-carboxylic acid interchange is a relatively new method particularly useful for the preparation of low-boiling nitriles (C.5). Addition reactions perhaps should first be considered for preparation of cyanide groups attached to a *tertiary* carbon atom (D). Most of the other methods are not so general, but may be appropriate and even vital for the preparation of a specific nitrile from a single available source. For example, in a hypothetical case, if adamantyl cyanide were needed with only adamantane as the starting material, halogenation followed by metathetical exchange with cyanide might be considered but also direct cyanylation or carboxylation followed by amidation and dehydration (C.1) would offer alternative routes.

A. Metathesis

1. From Halides

$$RCH_2X \xrightarrow{\text{NaCN}} RCH_2CN$$

$$ArX \xrightarrow{\text{CuCN}} ArCN + CuX$$

Aliphatic Halides

This synthesis is of value only for primary or secondary halides, since tertiary halides which can dehydrohalogenate give little or no nitrile. As was stated under Alcohols, Chapter 4, A.2, the reactivity of the halides in decreasing order is $-I > -Br > -Cl$. Thus the yield of nitrile obtained from a primary halide may be increased by the introduction of a promoter, such as sodium iodide, which by exchange converts the chloride first into the iodide, which then reacts with the metallic cyanide [2] to form the nitrile. The metallic cyanide employed with the alkyl halides is usually sodium cyanide, although sometimes the potassium or cuprous salt [3, 4] is used. Basic anion exchange resins as Amberlite

References for Section A are on pp. 945–946.

IRA 400 or Dowex 21K after conversion into the cyanide form have also been employed [5].

Many substituted primary alkyl halides may be converted, with yields usually varying from 75 to 90%, into nitriles. The equations which follow are typical:

$$CH_2OHCH_2Cl \xrightarrow{NaCN} CH_2OHCH_2CN \qquad [ref. 6]$$
$$79-80\%$$

$$Br(CH_2)_3Br \xrightarrow{NaCN} NC(CH_2)_3CN \qquad [ref. 7]$$
$$77-86\%$$

$$C_6H_5CH_2Cl \xrightarrow{NaCN} C_6H_5CH_2CN \qquad [ref. 8]$$
$$80-90\%$$

$$CH_2{=}CHCH_2Cl \xrightarrow[KI]{CuCN} CH_2{=}CHCH_2CN \qquad [ref. 3]$$
$$79-84\%$$

$$ClCH_2COONa \xrightarrow{NaCN} NCCH_2COONa \xrightarrow[C_2H_5OH]{HCl} NCCH_2COOC_2H_5 \quad [ref. 9]$$
$$77-80\%$$

The solvent to be used in the synthesis is important. Originally ethanol was widely employed, but the reaction is slow. Higher-boiling solvents such as ethylene glycol [10], methyl Cellosolve [11], tetrahydrofurfuryl alcohol [12], and dimethylformamide [11] lead to some improvement, although these solvents are not entirely satisfactory, particularly because the yields with secondary halides are low and tertiary halides give little or no nitrile. Dimethyl sulfoxide [13, 14] is now the reagent of choice. In this solvent, primary alkyl chlorides react rapidly with sodium cyanide to give the nitriles in excellent yield. Secondary chlorides such as 2-chlorobutane and chlorocyclopentane react in about 3 hours to give the corresponding nitriles in moderate yields (65–70%). This solvent may also be employed to advantage with primary and secondary alkyl bromides.

Interferences occasionally develop in the synthesis. Alcohol as a solvent sometimes, as in the case of certain benzyl halides, leads to extensive alcoholysis. Tar formation is also common when benzyl halides are treated with a metallic cyanide. The higher allylic halides undergo the allylic rearrangement. Sometimes the halogen is not replaceable, as is often the case with vinyl halides and haloethers having a halogen on the carbon atom β to the ether linkage.

To overcome these difficulties as much as possible, cuprous cyanide in an anhydrous solvent such as pyridine, ether, or benzene is used. Alcoholysis is prevented by the use of solvents such as acetone [15], acetonitrile [16], and phenylacetonitrile [17]. Acylnitriles (RCOCN) also are best prepared from the acyl halides and cuprous cyanide [18].

Aromatic Halides

The conversion of aromatic halides into nitriles is accomplished best with cuprous cyanide. The reagent has been employed with pyridine, quinoline, dimethylformamide [19], and N-methylpyrrolidone [20] as solvents or with no solvent at 250–260° (Rosenmund-von Braun). The induction period by the latter method may be shortened by the addition of a small amount of a nitrile, and traces of copper sulfate exert a catalytic effect [21]. Of these methods, those using dimethylformamide and N-methylpyrrolidone as solvents are preferred. The former has been studied in some depth and gives yields varying from 75 to 100% with a series of aryl chlorides and bromides. An improvement in the procedure for decomposing the complexes of the nitrile and cuprous halide by the use of ferric chloride or ethylenediamine has been achieved as well. N-Methylpyrrolidone [20], a good solvent for cuprous cyanide, usually permits the reaction to occur in a short period of time. With a limited number of aryl halides, mostly bromides, the yields usually range from 82 to 92%.

Coupling sometimes occurs with sodium cyanide in dimethyl sulfoxide if α-hydrogen atoms are present in the nitrile first formed [22]:

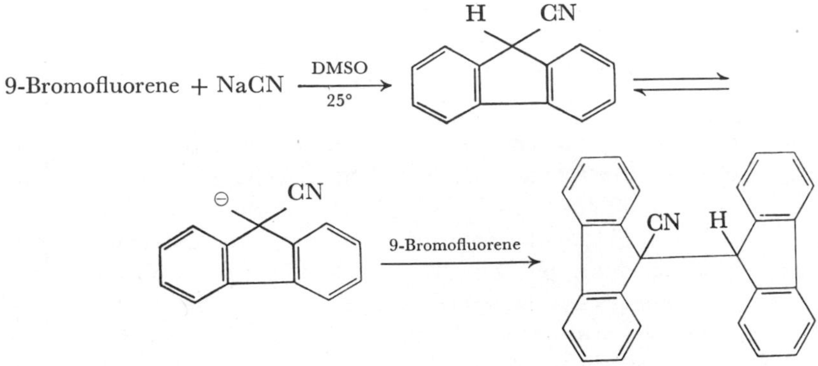

9-Cyano-9,9′-bifluorene, 77%

Heterocyclic Halides

The halides of pyridine [23], quinoline [24], and isoquinoline [25] yield nitriles satisfactorily with cuprous cyanide alone [20].

(a) **Preparation of 4-methoxyphenylacetonitrile** (74–81% based on the corresponding alcohol from 4-methoxybenzyl chloride (anisyl chloride), sodium cyanide, sodium iodide, and dry acetone) [26].

(b) Other examples

(1) **α-Naphthonitrile** (82–90% from α-bromonaphthalene, dry cuprous cyanide, and pyridine at 215–225° for 15 hr.) [27]; for shorter heating period, see [19].

(2) **9-Cyanophenanthrene** (87% from 9-bromophenanthrene and cuprous cyanide at 260°) [28].

(3) **Valeronitrile** (93% from 1-chlorobutane, 1 mole, sodium cyanide, 1.1 moles, and dimethyl sulfoxide (250 ml.)) [13].

(4) **3-Cyanoquinoline** (78–92% from 1 equiv. of 3-bromoquinoline and $1\frac{1}{2}$ equiv. of cuprous cyanide melted together and then distilled at about 50 mm. pressure) [24].

2. From Esters (Sulfates or Sulfonates)

$$R_2SO_4 + NaCN \rightarrow RCN + ROSO_3Na$$

$$ArOSO_2CH_3 + NaCN \rightarrow ArCN + CH_3SO_3Na$$

In the older synthesis of nitriles it was common to start with alkyl sulfates and convert them into nitriles by the use of sodium or potassium cyanide. This method works well, particularly for lower aliphatic nitriles. Sulfonates have also been used as starting materials or as intermediates in converting the alcohol into the nitrile. The most common esters employed appear to be the methane- or p-toluenesulfonates. In this synthesis various solvents such as methanol [29], ethanol [30], dimethylformamide [31], N-methyl-2-pyrrolidone [32], and dimethyl sulfoxide [33] have been utilized. From the limited number of experiments conducted in these solvents, it appears that aprotic dipolar solvents such as the last three mentioned offer some advantage in nucleophilic substitution reactions of this type. Yields of nitriles in such solvents are usually in the 80–90% range.

(a) **Preparation of acetonitrile, $CH_3C^{14}N$** (90% from dimethyl sulfate and $NaC^{14}N$) [34].

(b) **Other examples**

(1) **3-(1-Anthryl)butyronitrile**

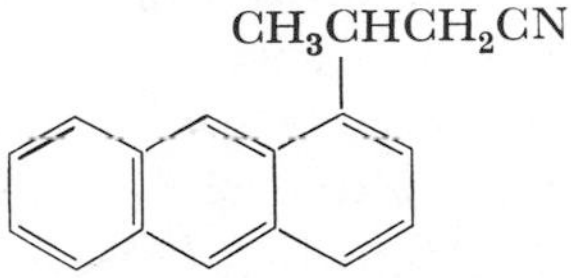

$$CH_3CHCH_2CN$$

(85% based on the alcohol, from the methanesulfonate and sodium cyanide, threefold excess, in dimethylformamide at 40–60° for 3 hr.) [31].

(2) **3α-Cyano-5α-cholestane**

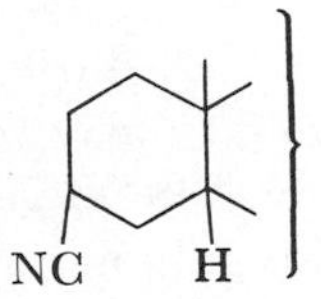

(81% from the tosylate and calcium cyanide in N-methyl-2-pyrrolidone containing a small amount of t-butyl alcohol at 90° for 20 hr.) [32].

(3) *trans*-**4,5-Bis-(cyanomethyl)-cyclohexene**

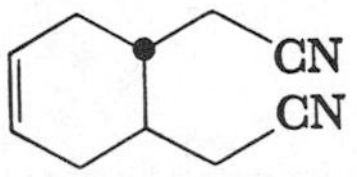

(92.5% from the ditosylate and sodium cyanide in dry dimethylsulfoxide at 90–95°) [33].

3. From Metal Sulfonates

$$ArSO_2ONa + NaCN \rightarrow ArCN + Na_2SO_3$$

Just as phenols may be obtained from sodium aryl sulfonates (Chapter 5, A.1) by heating with sodium hydroxide, so nitriles may be obtained by similar treatment with sodium cyanide. A review of this reaction is available [35]. Although the synthesis offers a method for preparing nitriles and carboxylic acids, which are readily obtained from nitriles, it appears to have been used to a minor extent in recent years. Factors which have contributed to the unattractiveness of the method are low yields, and rearrangement or replacement of ring substituents other than the sulfonate group. The synthesis has been of major interest until quite recently in the synthesis of 3-cyanopyridine, the precursor of 3-nicotinamide (niacinamide) [36]:

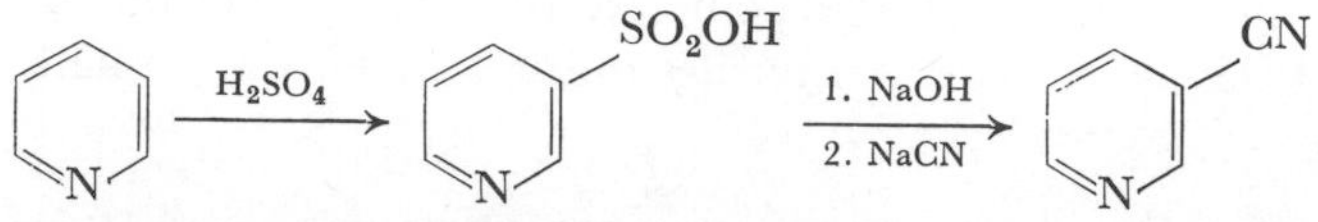

For some preparations, potassium ferrocyanide has been used in place of sodium cyanide [37].

(a) **Preparation of 1-naphthonitrile.** Equimolecular weights of sodium 1-naphthalenesulfonate and sodium cyanide were intimately mixed, placed in an iron vessel, and heated strongly up to 285–300° until vapors ensued. The vapors were trapped in a cooled flask and in a second vessel containing benzene through which the noncondensable material was passed. Although yields were not specifically mentioned in this preparation, which is the latest one found, they usually range from 60 to 70% [38].

4. From Quaternary Salts and Some Tertiary Amines

$$-CH_2N(CH_3)_3I \xrightarrow{NaCN} -CH_2CN$$

Quaternary salts are converted into nitriles by treatment with sodium or potassium cyanide [39]. The cyanide may be formed in aqueous, aqueous-alcoholic [40], or dimethylformamide [41] solution. The method has been used in the conversion of Mannich bases into nitriles [39]:

$$C_6H_5COCH_2CH_2N(CH_3)_2 \cdot HCl \xrightarrow[H_2O]{KCN} C_6H_5COCH_2CH_2CN$$

β-Benzoylpropionitrile

The displacement is limited mainly to Mannich bases or those which eliminate the amine readily:

$$RCOCH_2CH_2NR_2' \longrightarrow RCOCH=CH_2 \xrightarrow[\text{2. } H_2O]{\text{1. } CN^\ominus} RCOCH_2CH_2CN$$

Yields are usually in the 65–90% range.

(a) Preparation of phthalimidoacetonitrile. The methyl iodide of

N-dimethylaminomethylphthalimide, 3.46 g., and 0.5 g. of sodium cyanide in 25 ml. of dimethylformamide were heated at the boiling point until the evolution of trimethylamine subsided (3–4 hr.). The solution was distilled under reduced pressure and the residue was stirred with water, after which the nitrile separated. Recrystallization from water gave 1.4 g. (76%), m.p. 123° [41].

(b) Preparation of β-indolylacetonitrile (94% from trimethylskatyl-

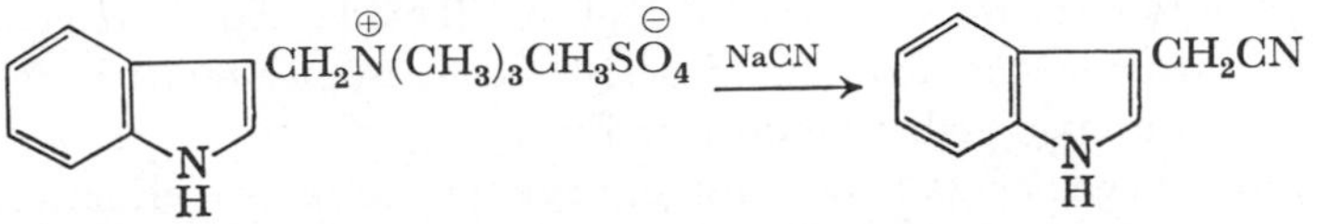

ammonium methyl sulfate and aq. sodium cyanide) [42].

(c) Preparation of 2,3-dicyano-2-heptene. Both abnormal dis-

$$C_4H_9C{\equiv}CCH_2\overset{\oplus}{N}(CH_3)_3I^{\ominus} + KCN \xrightarrow[\text{Reflux 2 hr.}]{\text{70 ml. DMF}} C_4H_9CCN{=}CCNCH_3 + (CH_3)_3N$$

2.8 g. (made similarly to a Mannich base) 4 g. 1.2 g.

placement (SN_2') and addition must have occurred in this reaction [43].

5. From Diazonium Salts (Sandmeyer)

$$\overset{\oplus}{\text{Ar}}\overset{\ominus}{N_2X} \xrightarrow{\text{CuCN}} \text{ArCN} + N_2 + \text{CuCl}$$

This synthesis has been discussed in more detail under Halides, Chapter 7, A.9, to which type it is also applicable. Methods for the preparation of the cuprous cyanide are available [44, 45]. Although the double salt of potassium cyanide and nickel cyanide may be substituted for the cuprous cyanide [46, 47], in the limited extent to which the synthesis has been used, particularly in recent years, cuprous cyanide or double or complex salts thereof such as $Na_3Cu(CN)_5NH_3$ [48] have been employed to the greatest extent. It is important in carrying out the reaction to neutralize the diazonium salt before addition to the cyanide solution in order to avoid the escape of hydrogen cyanide. Yields are usually in the 65–85% range.

(a) Preparation of o-tolunitrile (64–70% from o-toluidine, hydrochloric acid, sodium nitrite, and cuprous cyanide at 0–5°) [44].

(b) Preparation of 4,4′-dicyanodiphenyl (66% from benzidine diazotized in the usual manner; the neutralized solution was then added to a solution of nickel chloride and potassium cyanide) [47].

6. From Anions and Cyanogen Compounds or Cyanogen Itself

$$RO^{\ominus} + ClCN \longrightarrow ROCN$$

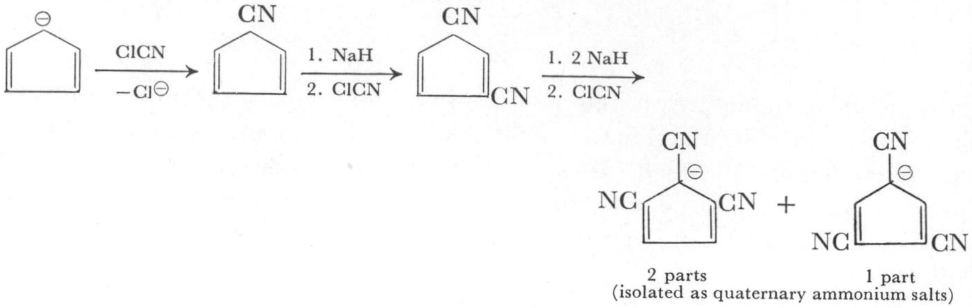

2 parts   1 part
(isolated as quaternary ammonium salts)

The cyanates formed by the reaction of an alkoxide and cyanogen chloride are rather unstable and tend to trimerize or to isomerize with acid. Bulky alkoxides tend to give higher yields (see Ex. a).

The anion of cyclopentadiene is readily cyanylated to give either mono-, di-, or tricyanocyclopentadienes [49]. The monocyano compound is the most unstable and tends to dimerize. The tricyano compound is obtained as a mixture of two isomers which can be separated by fractional crystallization of the quaternary ammonium salts. To obtain tetra- and pentacyanocyclopentadienes, the potassium salts of the tricyanocyclopentadienes are treated with aluminum chloride and chlorocyanogen.

The anion from the Grignard reagent reacts as follows with cyanogen:

$$RMgX + (CN)_2 \rightarrow RCN + MgCNX$$

This reaction has no practical applications but it does possess one unique aspect. In displacement of benzylmagnesium halides with cyanogen (for a recent synthesis see [50]), *ortho* substitution occurs exclusively (see Ex. b):

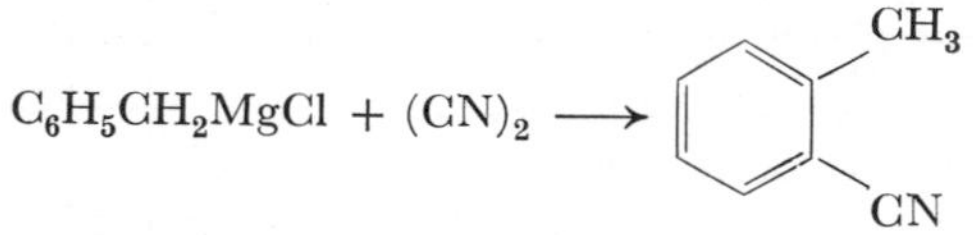

Reagents such as carbon dioxide give some *ortho* substitution or addition products, but mainly normal products, i.e., $C_6H_5CH_2COOH$.

(a) Preparation of bicyclo[2.2.2]octane-1,4-dicyanate. 1,4-Dihydroxybicyclo[2.2.2]octane was treated with sodium hydride (or butyllithium) to yield the dialkoxide which by the addition of cyanogen chloride gave the dicyanate in 41 % yield. The dicyanate rearranged to diisocyanate when passed through acid-washed alumina [51].

(b) Preparation of o-tolunitrile. Benzylmagnesium chloride in ether from 26 g. of benzyl chloride and 10 g. of magnesium was added to 11 g. of cyanogen in ether at 0°, followed by brief refluxing. The mixture was decomposed with water and extracted with ether, after which the extract yielded 52 %

crude *o*-tolunitrile. Cyanogen bromide gave only benzyl bromide with the same Grignard reagent [52].

1. D. T. Mowry, *Chem. Rev.*, **42,** 189 (1948); V. Migrdichian, *Organic Synthesis*, Vol. 1, Reinhold Publishing Corp., New York, 1957, p. 420.
2. H. B. Hass and J. R. Marshall, *Ind. Eng. Chem.*, **23,** 352 (1931).
3. C. W. Smith and H. R. Snyder, *Org. Syn.*, Coll. Vol. **3,** 852 (1955).
4. J. V. Supniewski and P. L. Salzberg, *Org. Syn.*, Coll. Vol. **1,** 46 (1941).
5. M. Gordon *et al.*, *J. Org. Chem.*, **28,** 698 (1963).
6. E. C. Kendall and B. McKenzie, *Org. Syn.*, Coll. Vol. **1,** 256 (1941).
7. C. S. Marvel and E. M. McColm, *Org. Syn.*, Coll. Vol. **1,** 536 (1941).
8. R. Adams and A. F. Thal, *Org. Syn.*, Coll. Vol. **1,** 107 (1941).
9. J. K. H. Inglis, *Org. Syn.*, Coll. Vol. **1,** 254 (1941).
10. R. N. Lewis and P. V. Susi, *J. Am. Chem. Soc.*, **74,** 840 (1952).
11. H. B. Copelin, U.S. Patent 2,715,137, August 9, 1955; *C.A.*, **50,** 7126 (1956).
12. A. O. Rogers, U.S. Patent 2,415,261, Feb. 4, 1947; *C.A.*, **41,** 3119 (1947).
13. L. Friedman and H. Shechter, *J. Org. Chem.*, **25,** 877 (1960).
14. R. A. Smiley and C. Arnold, *J. Org. Chem.*, **25,** 257 (1960).
15. C. L. Hewett and R. H. Martin, *J. Chem. Soc.*, 1396 (1940).
16. G. Wittig and H. Petri, *Ann. Chem.*, **513,** 26 (1934).
17. C. L. Hewett, *J. Chem. Soc.*, 293 (1940).
18. J. Thesing and D. Witzel, *Angew. Chem.*, **68,** 425 (1956).
19. L. Friedman and H. Shechter, *J. Org. Chem.*, **26,** 2522 (1961).
20. M. S. Newman and H. Boden, *J. Org. Chem.*, **26,** 2525 (1961).
21. C. F. Koelsch and A. G. Whitney, *J. Org. Chem.*, **6,** 795 (1941).
22. J. F. Cavalla *et al.*, *Chem. Ind.* (*London*), 1961 (1967).
23. S. M. McElvain and M. A. Goese, *J. Am. Chem. Soc.*, **63,** 2283 (1941).
24. H. Gilman and S. M. Spatz, *J. Am. Chem. Soc.*, **63,** 1553 (1941).
25. F. T. Tyson, *J. Am. Chem. Soc.*, **61,** 183 (1939).
26. K. Rorig *et al.*, *Org. Syn.*, Coll. Vol. **4,** 576 (1963).
27. M. S. Newman, *Org. Syn.*, Coll. Vol. **3,** 631 (1955).
28. J. E. Callen *et al.*, *Org. Syn.*, Coll. Vol. **3,** 212 (1955).
29. R. Grewe and H. Pachaly, *Chem. Ber.*, **87,** 46 (1954).
30. R. Grewe and E. Nolte, *Ann. Chem.*, **575,** 1 (1952).
31. M. S. Newman and S. Otsuka, *J. Org. Chem.*, **23,** 797 (1958).
32. H. B. Henbest and W. R. Jackson, *J. Chem. Soc.*, 954 (1962).
33. J. J. Bloomfield and P. V. Fennessey, *Tetrahedron Letters*, 2273 (1964).
34. E. V. Brown *et al.*, *J. Am. Chem. Soc.*, **73,** 3735 (1951).
35. C. M. Suter, *The Organic Chemistry of Sulfur*, John Wiley and Sons, New York, 1944, p. 438.
36. A. P. Sachs and J. F. Couch, in R. E. Kirk and D. F. Othmer, *Encyclopedia of Chemical Technology*, Vol. 9, The Interscience Encyclopedia, Inc., New York, 1952, p. 305.
37. *Elsevier's Encyclopedia of Organic Chemistry*, Ser. 3, Vol. 12B, Elsevier Publishing Co., New York, 1953, p. 4010.
38. F. C. Whitmore and A. L. Fox, *J. Am. Chem. Soc.*, **51,** 3363 (1929).
39. E. B. Knott, *J. Chem. Soc.*, 1190 (1947).
40. T. A. Geissman and A. Armen, *J. Am. Chem. Soc.*, **74,** 3916 (1952).
41. H. Hellman, *Chem. Ber.*, **87,** 1690 (1954).
42. J. Thesing and F. Schülde, *Chem. Ber.*, **85,** 324 (1952).
43. K. Schlögl and H. Pelousek, *Monatsh. Chem.*, **92,** 51 (1961).
44. H. T. Clarke and R. R. Read, *Org. Syn.*, Coll. Vol. **1,** 514 (1941).
45. H. J. Barber, *J. Chem. Soc.*, 79 (1943).
46. A. Korczynski *et al.*, *Compt. Rend.*, **171,** 182 (1920); A. Korczynski and B. Fandrich, *ibid.*, **183,** 421 (1926).
47. C. DeMilt and M. Sartor, *J. Am. Chem. Soc.*, **62,** 1954 (1940).
48. H. Hagenest and F. W. Stauf, U.S. Patents 1,879,209 (1932); *C.A.*, **27,** 997 (1933); 1,962,559 (1934); *C.A.*, **28,** 4848 (1934).

49. O. W. Webster, *J. Am. Chem. Soc.*, **88**, 3046 (1966).
50. G. B. Kauffman *et al.*, *J. Chem. Ed.*, **45**, 141 (1968).
51. J. C. Kauer and W. W. Henderson, *J. Am. Chem. Soc.*, **86**, 4732 (1964).
52. V. F. Raaen and J. F. Eastham, *J. Am. Chem. Soc.*, **82**, 1349 (1960).

B. Nucleophilic Reactions

1. From Nitriles by Alkylation

$$(Ar)_2CHCN \xrightarrow[\text{Liq. NH}_3]{\text{KNH}_2} [(Ar)_2\overset{\ominus}{C}CN]K^{\oplus} \xrightarrow{\text{ArCH}_2\text{Cl}} (Ar)_2\underset{|}{C}CN$$
$$CH_2Ar$$

Alkylation of nitriles having α-hydrogens may be accomplished by the reaction of an alkyl halide on the nitrile treated with sodium ethoxide in ethanol [1], with methylmagnesium iodide in ether [2], with lithium diethylamide [3], and with sodamide in ether [4]. More recently, better yields (89–99%) [5] were obtained by the use of potassium amide and the alkyl halide in liquid ammonia. Apparently the potassium salt of the nitrile forms first, and it reacts with the halide to give the alkylated nitrile. It was found that, if the β-hydrogen in these nitriles is activated by at least one phenyl group, dehydrocyanation will occur with potassium amide in liquid ammonia. Thus α,α,β-triphenylpropionitrile yields triphenylethylene in 94% yield under these conditions:

$$C_6H_5$$
$$C_6H_5CH_2\underset{|}{\overset{|}{C}}CN \xrightarrow[\text{Liq. NH}_3]{\text{KNH}_2} C_6H_5CH{=}C(C_6H_5)_2$$
$$C_6H_5$$

A similar process in which naphthalene-sodium first α-metalates the nitrile has been reported [6]:

$$(C_6H_5)_2CHCN + C_2H_5Br \xrightarrow{\text{C}_{10}\text{H}_8\text{-Na}} (C_2H_5)_2\underset{|}{C}{-}CN$$
$$C_2H_5$$
$$77\%$$
Triethylacetonitrile

In fact, sodamide or butyllithium is capable of metalating acetonitrile [7]:

$$C_4H_9Li + CH_3CN \xrightarrow[\text{THF, C}_6\text{H}_{14}]{-80°} \left[\overset{\ominus}{C}H_2CN \right] \xrightarrow[\substack{\text{No cooling for}\\\text{10 min.}}]{(C_6H_5)_2CO} (C_6H_5)_2\overset{\text{OH}}{\underset{CH_2CN}{C}}$$

β,β-Diphenyl-β-hydroxy-
propionitrile, 89%

References for Section B are on p. 950.

For dinitriles dialkylation is accomplished best with the alkyl halide and sodium hydride in dimethyl sulfoxide [8]:

$$CH_2(CN)_2 + 2\ CH_3I \xrightarrow[\text{NaH}]{\text{DMSO}} (CH_3)_2C(CN)_2$$

2,2-Dicyanopropane,
60%

The range of alkylation may always be enlarged by the attachment of other leaving groups to facilitate the process. The most obvious one is the attachment of a carbethoxy group:

$$NCCH_2CO_2C_2H_5 \xrightarrow[\text{RX}]{\text{NaOC}_2\text{H}_5} NCCHRCO_2C_2H_5 \xrightarrow{\text{H}_3\text{O}^{\oplus}} RCH_2CN + CO_2$$

A less obvious one is the attachment of a diazo group [9]:

$$R_3B + N_2CHCN \xrightarrow{\text{THF, 0° then 25°, 2 hr.}} R_3\overset{\ominus}{B}CHCN \xrightarrow{-N_2} \underset{N_2^{\oplus}}{\overset{|}{}}$$

63 mmoles   30 mmoles

$$\underset{R_2BCHCN}{\overset{R}{\overset{|}{}}} \xrightarrow{\text{Cold 3 }N\text{ KOH}} RCH_2CN + R_2BOH$$

When R was cyclopentyl, the yield of cyclopentylacetonitrile was 81%, and the reaction is general for any terminal olefin.

(a) **Preparation of α,α,β-triphenylpropionitrile** (95–99% from diphenylacetonitrile, potassium, benzyl chloride, and a few crystals of ferric nitrate hydrate in liquid ammonia) [10].

(b) **Preparation of α-cyclohexylphenylacetonitrile** (65–77% from phenylacetonitrile, sodium, liquid ammonia, a crystal of ferric nitrate hydrate, and bromocyclohexane) [11].

2. From Nitriles by Acylation or Aldolization

$$RCOOC_2H_5 + NCCH_2COOR' \xrightarrow{\text{NaOC}_2\text{H}_5} \underset{\overset{|}{CN}}{RCOCHCOOR'} + C_2H_5OH$$

The Claisen condensation and its course have been discussed in some detail under Carboxylic Acid Esters, Chapter 14, C.1, and Ketones, Chapter 11, F.2. The nitrile group serves as a better activator than a carbethoxy group for forming anions from adjacent methylene groups and as a poorer acceptor of carbanions. The reaction has been used widely in synthesizing a variety of substituted nitriles.

If two molecules of a nitrile are used, a Claisen reaction also occurs as shown:

$$RCH_2CN + RCH_2CN \longrightarrow RCH_2\overset{\overset{\displaystyle NH}{\|}}{C}CHCN \xrightarrow{H_2O} RCH_2\overset{\overset{\displaystyle O}{\|}}{C}CHCN$$

R   R

By using high dilution techniques an intramolecular condensation similar to the Dieckmann condensation (see Carboxylic Acid Esters, Chapter 14, C.2) occurs. Known as the Thorpe-Ziegler condensation [12], this cyclization employs an ether-soluble metal alkyl anilide, usually sodium N-methylanilide. It produces cyclic cyanoenamines as illustrated in Ex. c, the equation for which follows:

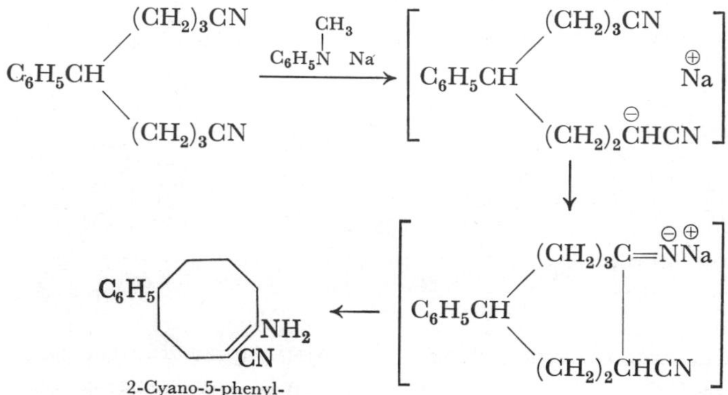

2-Cyano-5-phenyl-cyclooctene-1-ylamine

The products from the condensation of carbonyl compounds and ethyl cyanoacetate may be hydrogenated, hydrolyzed, and decarboxylated (or pyrolyzed) [13] to nitriles, as for example:

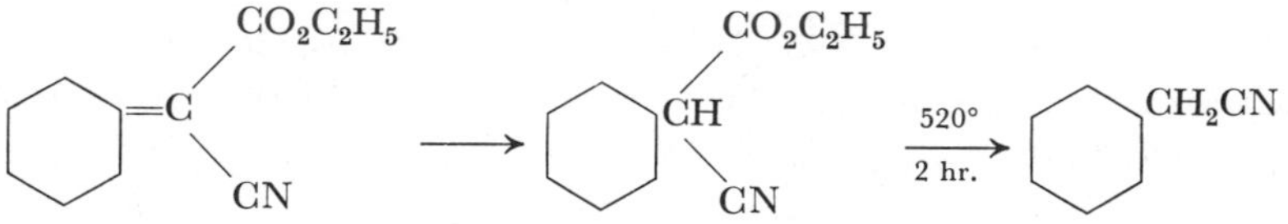

80 g.

Cyclohexylacetonitrile, 77%

(a) Preparation of α-(4-chlorophenyl)-γ-phenylacetoacetonitrile

$$C_6H_5CH_2COOC_2H_5 \xrightarrow[NaOC_2H_5]{p\text{-}ClC_6H_4CH_2CN} (p\text{-}ClC_6H_4\overset{\ominus}{\underset{CN}{C}}COCH_2C_6H_5)Na^{\oplus} \xrightarrow{HCl}$$

$$p\text{-}ClC_6H_4\underset{CN}{CH}COCH_2C_6H_5$$

(74–82 % from ethyl phenylacetate, 4-chlorophenylacetonitrile, sodium, and absolute ethanol) [14].

(b) Preparation of 2,3-diphenylsuccinonitrile (68–72 % from benzyl

$$C_6H_5CHO \xrightarrow[\text{NaCN}]{C_6H_5CH_2CN} C_6H_5CH-CHC_6H_5$$
$$\underset{CN}{|} \quad \underset{CN}{|}$$

cyanide, benzaldehyde, and sodium cyanide; an aldol followed by addition of the cyanide ion to $C_6H_5C\!=\!CHC_6H_5$) [16].
$$\underset{CN}{|}$$

(c) Preparation of 2-cyano-5-phenylcyclooctene-1-ylamine (or 2-cyano-5-phenylcyclooctane-1-ylimine). To a well-stirred, boiling solution of sodium methylanilide (prepared from 12 g. of sodium, 40.4 g. of naphthalene, 70 g. of methylaniline, and 785 ml. of anhydrous ether under nitrogen), a solution of 11.8 g. of 5-phenylazelanitrile ($C_6H_5CH[(CH_2)_3CN]_2$) in 250 ml. of anhydrous ether was added in 8 hr. The reaction mixture was then refluxed for 1 hr., cooled, and treated with water cautiously. From the ethereal layer and an ether extract of the aqueous layer there was recovered 6.4 g. (54 %) of crude iminonitrile, m.p. 87.4–96.6° [17].

3. From Carbonyl Compounds and Malonitrile or Cyanoacetic Esters

$$\underset{(R)H}{\overset{R}{\diagdown}}C\!=\!O + CH_2(CN)_2 \longrightarrow \underset{(R)H}{\overset{R}{\diagdown}}C\!=\!\underset{}{\overset{CN}{\diagup}}C\!-\!CN + H_2O$$

Malonitrile condenses with aldehydes or ketones (Knoevenagel reaction, see Carboxylic Acid Esters, Chapter 14, C.4) in the presence of ammonium acetate-acetic acid to give olefinic dicyanides which are valuable intermediates for subsequent reactions (see D.7 for similar reactions of an olefinic mononitrile) [18, 19]. Other catalysts effective in the condensation are amino acids and amino phenols [15, 20–22]. Yields with a series of ketones, mostly containing at least one aryl group, vary from 60 to 95 %.

Perhaps a more common procedure is to substitute a cyanoacetic ester for malonitrile. In the presence of ammonium acetate and acetic acid in benzene, from which mixture the water may be removed azeotropically, aliphatic ketones unbranched at the α-carbon atom give yields of alkylidene esters of 75–87 %; for aromatic ketones, yields are 60–80 % [23, 24].

(a) Preparation of α-tetralidenemalonitrile. α-Tetralone, 0.5 mole,

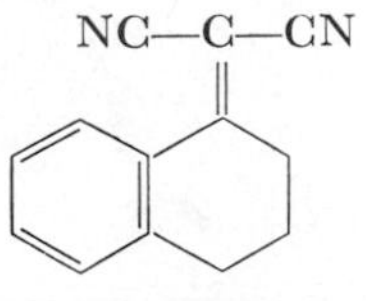

0.6 mole of malonitrile, 4 g. of anhydrous ammonium acetate, and 12 ml. of glacial acetic acid were refluxed in benzene for 6 hr. while the water formed was

removed in a Dean-Stark trap. From the benzene solution after washing and drying, there was recovered by evaporation 81 % of the dinitrile [19].

(b) **Preparation of benzalmalonitrile** (100 % from saturated, 70 % aqueous-alcoholic solution of benzaldehyde and 1 equiv. of malonitrile and a trace of glycine) [22].

(c) **Ethyl (1-methylbutylidene) cyanoacetate** (80 % from 0.5 mole of ethyl cyanoacetate, 0.55–0.6 mole of methyl propyl ketone, 0.05 mole of

$$\begin{array}{ccc} CH_3 & & CN \\ & \diagdown & \diagup \\ & C{=}C & \\ & \diagup & \diagdown \\ CH_3CH_2CH_2 & & COOC_2H_5 \end{array}$$

NH$_4$OAc, 0.1 mole of glacial acetic acid, and 50 ml. of benzene at 130–160° with the use of a Dean-Stark separator) [24].

(d) **Preparation of ethyl** sec-**butylidenecyanoacetate** (81–87.5 %

$$C_2H_5COCH_3 \xrightarrow[\beta\text{-alanine}]{NCCH_2COOC_2H_5} \quad C_2H_5{-}\overset{\displaystyle CH_3}{\underset{\displaystyle CN}{\overset{\displaystyle |}{\underset{\displaystyle |}{C}}}}{=}CCOOC_2H_5$$

from butanone, ethyl cyanoacetate, β-alanine, and glacial acetic acid (Knoevenagel reaction)) [15].

1. K. Neure, Ann. Chem., **250**, 140 (1889).
2. K. Sisido et al., J. Am. Chem. Soc., **72**, 2270 (1950).
3. J. Cason et al., J. Org. Chem., **15**, 850 (1950).
4. P. Ramart, Bull. Soc. Chim. France, **35**, 196 (1924).
5. C. R. Hauser and W. R. Brasen, J. Am. Chem. Soc., **78**, 82 (1956).
6. L. Horner and H. Güsten, Ann. Chem., **652**, 99 (1962).
7. E. M. Kaiser and C. R. Hauser, J. Org. Chem., **33**, 3402 (1968).
8. J. J. Bloomfield, J. Org. Chem., **26**, 4112 (1961).
9. J. Hooz and S. Linke, J. Am. Chem. Soc., **90**, 6891 (1968).
10. C. R. Hauser and W. R. Dunnavant, Org. Syn., Coll. Vol. **4**, 962 (1963).
11. E. M. Hancock and A. C. Cope, Org. Syn., Coll. Vol. **3**, 219 (1955).
12. J. P. Schaefer and J. J. Bloomfield, Org. Reactions, **15**, 1 (1967).
13. W. J. Bailey and J. J. Daly, Jr., J. Am. Chem. Soc., **81**, 5397 (1959).
14. S. B. Coan and E. I. Becker, Org. Syn., Coll. Vol. **4**, 174 (1963).
15. F. S. Prout et al., Org. Syn., Coll. Vol. **4**, 93 (1963).
16. R. B. Davis and J. A. Ward, Jr., Org. Syn., Coll. Vol. **4**, 392 (1963).
17. A. C. Cope and R. J. Cotter, J. Org. Chem., **29**, 3467 (1964).
18. D. T. Mowry, J. Am. Chem. Soc., **67**, 1050 (1945).
19. E. Campaigne et al., J. Org. Chem., **27**, 4428 (1962).
20. H. D. Dakin, J. Biol. Chem., **7**, 49 (1909).
21. F. S. Prout, J. Org. Chem., **18**, 928 (1953).
22. J. B. Bastus, Tetrahedron Letters, 955 (1963).
23. A. C. Cope et al., J. Am. Chem. Soc., **59**, 2327 (1937).
24. A. C. Cope et al., J. Am. Chem. Soc., **63**, 3452 (1941).

C. Elimination

The following sections are not in order of their importance. Dehydration of amides (C.1) and of oximes (C.4) are the most common methods of preparation, and nitrile exchange (C.5) is becoming more popular. The rest of the eliminations are for more specialized cases and are placed in an order to relate to the preceding or succeeding sections.

1. From Amides

$$RCONH_2 \xrightarrow{\quad P_2O_5 \quad} RCN$$

One of the most widely used methods of synthesizing nitriles starts with the amide, which can be readily obtained if the corresponding acid is available. The dehydration may be accomplished with chemical reagents or catalysts. Examples of chemical reagents used in the synthesis are phosphorus pentoxide [1–3], phosphorus pentachloride [4], phosphorus oxychloride [5], thionyl chloride [6, 7], acetic anhydride [8], carbonyl chloride [9], benzene- and p-toluenesulfonyl chloride [10], sulfamic acid [11], ammonium sulfamate [12] (see Ex. b.6), and newer reagents such as dicyclohexylcarbodiimide [13] and pyrocatechyl phosphorotrichloridate [14]. Reagents such as carbonyl chloride, benzene- or p-toluene sulfonyl chloride, phosphorus oxychloride, and dicyclohexylcarbodiimide are sometimes used in a nucleophilic solvent such as pyridine or other tertiary amines. Sometimes a salt adds to the effectiveness of the reagent, as in the case of the addition of sodium chloride [15] or sodium metabisulfite [16] to phosphorus oxychloride, or of lithium chloride to phosphorus pentoxide [17]. Mixtures such as phosphorus pentachloride and oxychloride or thionyl chloride and dimethylformamide [18] have also been used to advantage in these dehydrations. To convert acid amides into β-cyanoesters, ethyl chloroformate (or an acid chloride) and triethylamine have been used effectively [19]:

$$\underset{\underset{HCCONH_2}{\|}}{HCCOOH} + 2 \; ClCOOC_2H_5 + 2 \; (C_2H_5)_3N \longrightarrow$$

$$\underset{\underset{HCCN}{\|}}{HC-COOC_2H_5} + C_2H_5OH + 2 \; CO_2 + 2 \; (C_2H_5)_3\overset{\oplus}{N}H\overset{\ominus}{Cl}$$

91%

Ethyl β-cyanoacrylate

Certain advantages may be mentioned in the use of these reagents. With thionyl chloride, the by-products, sulfur dioxide and hydrogen chloride, are eliminated as gases during the course of the reaction. Elimination is also promoted with carbonyl chloride, benzenesulfonyl chloride, and phosphorus oxychloride in the presence of a basic solvent such as pyridine. Pyrocatechyl phosphorotrichloridate is soluble in organic solvents and it appears to lead to no side reactions. Yields under the most favorable conditions are often in the 80–95 % range.

References for Section C are on pp. 960–962.

It has long been known that amides may be dehydrated by heat alone [20]. In this process the yield may be improved by the use of certain catalysts such as alumina, lamp black, pumice, silica, aluminum phosphate, or manganese dioxide [21]. Such methods, because of the special catalysts and complicated equipment requirements, are seldom used in the laboratory, but they are of importance commercially. In fact, in industry, fatty acid nitriles [22] are prepared from the acid or ester and ammonia, rather than from the amide, in the presence of a contact catalyst. For example, capronitrile, $CH_3(CH_2)_4CN$, is available from a mixture of caproic acid and ammonia passed over red-hot pumice, while oleonitrile, $CH_3(CH_2)_7CH=CH(CH_2)_7CN$, results from passing methyl oleate and ammonia over heated alumina. In the laboratory an elegant method to proceed from the acid to the nitrile under mild conditions is by means of the reagent chlorosulfonylisocyanate [23] (see E.1 for other uses of this reagent):

$$HO_2C(CH_2)_4COOH + 2\ ClSO_2N=C=O \longrightarrow$$

$$\underset{\displaystyle}{ClSO_2NH\overset{\overset{\textstyle O}{\|}}{C}(CH_2)_4\overset{\overset{\textstyle O}{\|}}{C}NHSO_2Cl} \xrightarrow[CH_2Cl_2]{(C_2H_5)_3N} N\equiv C(CH_2)_4CN + 2\ ClSO_2\overset{\ominus}{O}$$

Adiponitrile, 81.5%

Yields are good to excellent for a number of preparations.

The feasibility of rearranging isonitriles to nitriles makes the dehydration procedure a potential pathway for lengthening the chain of amines [24]:

$$C_6H_5NH\overset{\overset{\textstyle O}{\|}}{C}H \xrightarrow[\substack{Silica\ gel \\ or\ silicates}]{460-560°} (C_6H_5NC) \longrightarrow C_6H_5CN$$

Benzonitrile, 83.5%

The nitrile can be reduced to the amine, the amine converted into the formamide, and the dehydration-isomerization repeated to give phenylacetonitrile [25].

No attempt will be made to suggest mechanisms for all of the preceding reagents but a typical example is shown, the behavior of which can be applied to other reagents. Dehydration of an amide with an anhydride is an equilibrium process [26]:

$$C_6H_5CONH_2 \xrightarrow[\Delta,\ neat]{(C_6H_5CO)_2O} [C_6H_5\overset{\overset{\textstyle NH}{\|}}{C}OCOC_6H_5]$$

19-I

$$C_6H_5\overset{\overset{\textstyle O}{\|}}{C}NH\overset{\overset{\textstyle O}{\|}}{C}C_6H_5 \qquad\qquad C_6H_5CN + C_6H_5COOH$$

19-II, 2.7%   90%

The nitrile arises by elimination from the isoimide anhydride intermediate, 19-I, and the imide, 19-II, by rearrangement. Since the latter reaction is effected by strong mineral acid, one can increase the amount of imide formed by using hydrogen chloride or benzoyl chloride. But, for nitrile formation, the strong acid should be avoided.

(a) Preparation of 2-ethylhexanonitrile (86–94% from 2-ethyl-hexanamide and thionyl chloride in dry benzene at 75–80°) [7].

(b) Other examples

(1) **Nicotinonitrile** (83–84% from a mixture of nicotinamide and phosphorus pentoxide heated under reduced pressure) [2].

(2) **4-Cyanofluorene** (82% from a mixture of fluorene-4-carbonamide, sodium chloride, and phosphorus oxychloride heated on a sand bath) [27].

(3) **Carbobenzoxy-β-cyano-L-alanine**

$$\begin{array}{c} CH_2CN \\ | \\ CHCOOH \\ | \\ NHCO_2CH_2C_6H_5 \end{array}$$

(78% from carbobenzoxy-L-asparagine and N,N'-dicyclohexylcarbodiimide in pyridine at 16–20°) [13].

(4) **1,1-Bis(*p*-chlorophenyl)-acetonitrile** (89% from 1,1-bis-(*p*-chlorophenyl)-acetamide and acetic anhydride by refluxing for 4 hr.) [8].

(5) **Isobutyronitrile** (90% by passing isobutyramide over alumina-silica gel at 425°) [28].

(6) **Stearonitrile** (92% from 0.15 mole of ammonium sulfamate, 0.1 mole of stearamide heated to 150°, held, and then slowly heated to 200°; the nitrile was removed by vacuum distillation; the reaction was held at 150° to permit the formation of $RCONHSO_3NH_4$ which then eliminates ammonium bisulfate at higher temperatures; yields were good for aliphatic amides but poor for benzamides) [29].

(7) *cis*-**Hexahydrophthalonitrile** (95% by heating the *cis*-diamide in pyridine at 60–65° while bubbling $COCl_2$ through the suspension) [30].

2. From Some Substituted Amides (von Braun) and Some Oximes (Second-Order Beckmann Rearrangement)

$$RNHCOR' \xrightarrow{PCl_5} RCl + R'CN + POCl_3 + HCl$$

This synthesis has been discussed under Halides, Chapter 7, A.10. As the equation above shows, it produces a nitrile as well as a halide. The choice of the method for the synthesis of nitriles would be rare indeed, although such nitriles have been recovered. For example, N-α-cyclopentylbenzylbenzamide in nitrobenzene when treated with thionyl chloride gave an 80.5% yield of benzonitrile as well as an 80–85% yield of α-cyclopentylbenzyl chloride (see Chapter 7, A.10, Ex. b.2) [31].

The second-order Beckmann rearrangement also leads to nitriles. The abnormal course of this rearrangement is restricted to those oximes in which one of the groups is capable of forming a relatively stable carbonium ion:

$$\overset{\overset{\displaystyle NOH}{\displaystyle \|}}{R\overset{}{C}R'} \longrightarrow [R\overset{\overset{\displaystyle N^{\oplus}}{\displaystyle \|}}{C}R'] \longrightarrow [R^{\oplus}] + R'C{\equiv}N$$

R must be t-butyl, acyl, aminomethyl, hydrogen, or a group of similar characteristics. Two examples are shown:

$$C_6H_5\overset{\overset{\displaystyle HON}{\displaystyle \|}}{C}\!-\!\overset{\overset{\displaystyle O}{\displaystyle \|}}{C}\, Mes \xrightarrow[NaOH]{C_6H_5SO_2Cl} C_6H_5CN + Mes\ COONa \qquad [ref.\ 32]$$

(Mes = 2,4,6-trimethylphenyl)   56%   80% as acid

$$C_6H_5\overset{\overset{\displaystyle NOH}{\displaystyle \|}}{C}CH_2N\!\!\bigcirc \xrightarrow[KOH]{p\text{-}CH_3C_6H_4SO_2Cl} C_6H_5CN + \left(\bigcirc\!\!\overset{\oplus}{N}{=}CH_2 \right) \xrightarrow[-CH_2O]{H_2O}$$

87%

$$\bigcirc\!\!NSO_2C_6H_4CH_3\text{-}p \quad [ref.\ 33]$$

96%

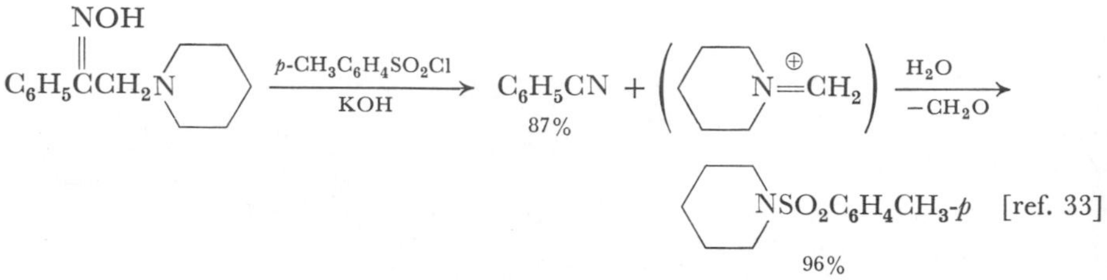

Although the above are not desirable means of synthesis, an adaptation of this method is used to degrade an acid by one carbon atom (see Ex. a):

$$RCH_2COOH \longrightarrow RCH_2COC_6H_5 \xrightarrow[H^{\oplus}]{RONO} R\overset{\overset{\displaystyle HON}{\displaystyle \|}}{C}COC_6H_5 \xrightarrow[NaOH]{C_6H_5SO_2Cl}$$

$$RCN \xrightarrow{H^{\oplus}} RCOOH$$

(a) **Preparation of tetradecyl cyanide.** Palmitophenone was made in 89% yield from labeled palmitoyl chloride, benzene, and aluminum chloride in a small-scale reaction. The ketone in a mixture of hydrochloric acid and dioxane was treated with isoamyl nitrite, the solution made basic, and then treated with p-toluenesulfonyl chloride in the cold followed by warming. The basic solution was extracted with pentane, which gave the nitrile on evaporation in a yield of at least 70% [34].

3. From Aldehydes (or Quinones) and Hydrazoic Acid (Schmidt) or from Diazides

$$RCHO \xrightarrow[H_2SO_4]{HN_3} RC{\equiv}N + RNHCHO$$

As discussed in C.2, a nitrile is the product from the second order Beckmann rearrangement of oximes. Since the intermediate in the Schmidt reaction is quite similar to, if not identical with, the intermediate in the Beckmann rearrangement, the nitrile formed might be considered the product of a second-order Schmidt reaction. The formamide is the product of the true rearrangement. The ratio of nitrile to formamide probably is controlled by the orientation of the leaving group with respect to R:

$$
\text{RCHO} \xrightarrow[\text{H}_2\text{SO}_4]{\text{HN}_3}
\underset{\substack{\downarrow \text{ } -\text{N}_2, \\ \text{H}_2\text{O}, -\text{H}^{\oplus}}}{\overset{\overset{\displaystyle \overset{\oplus}{\text{N}}\!\!\equiv\!\!\text{N}}{\diagup}}{\underset{\|}{\text{N}}}{\text{RCH}}}
\quad + \quad
\underset{\downarrow -\text{N}_2, -\text{H}^{\oplus}}{\overset{\overset{\displaystyle \text{N}\!\equiv\!\text{N}^{\oplus}}{\diagdown}}{\underset{\|}{\text{N}}}{\text{RCH}}}
$$

$$
\underset{\text{RNHC}=\text{O}}{\overset{\text{H}}{|}} \qquad\qquad \text{RC}\!\equiv\!\text{N}
$$

Very little can be done to alter this ratio, although more concentrated and larger amounts of sulfuric acid seem to increase the nitrile/formamide ratio slightly. But yields of nitriles may be as high as 70 %. The Schmidt reaction has been discussed in more detail under Amines, Chapter 8, G.5, and references to nitrile formation are to be found in the review article on this reaction [35].

The Schmidt reaction suggests that nitriles could be obtained from hydrazones of aldehydes by diazotization:

$$
\text{RCH}\!=\!\text{NNH}_2 \xrightarrow[\text{H}^{\oplus}, -\text{N}_2]{\text{RONO}} [\text{RCH}\!=\!\text{N}]^{\oplus} \xrightarrow{-\text{H}^{\oplus}} \text{RC}\!\equiv\!\text{N}
$$

but this reaction does not appear to be of preparative value.

As indicated above, azides in acid solution form electron-deficient nitrogen intermediates which can eliminate one proton to form nitriles. This fact suggests the possibility of decomposing azides in neutral medium to form intermediate nitrenes (the latter part of F.1 describes nitrile formation via a nitrene made by oxidation):

$$
\text{RCH}_2\text{N}_3 \xrightarrow{-\text{N}_2} [\text{RCH}_2\overset{..}{\text{N}}:] \longrightarrow \text{RCH}\!=\!\text{NH}
$$

A nitrene

The monoazides do not give nitriles, however, but decompose with explosive violence (unless highly diluted) to give the imine and hydrogen abstraction or insertion products. On the other hand, some diazides decompose to give a dinitrene which spontaneously forms the dinitrile [36]:

In decalin added
dropwise to
refluxing decalin

Via dinitrene

cis,cis-1,4-Dicyano-
1,3-butadiene, 79%

Quinones when added to excess sodium azide in trichloroacetic acid may give rise to nitriles via consecutive addition and decomposition of 2 equivalents of the azide ion [37]:

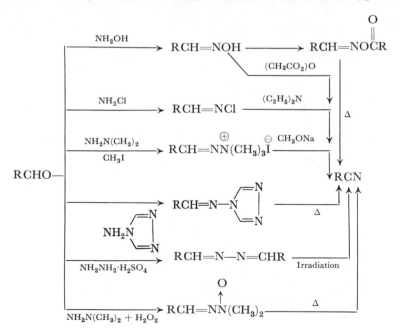

32%
γ-Lactone of 5-cyano-4-hydroxy-
3-amino-2-methylhexadien-2,4-
oic acid

(a) Preparation of vanillonitrile. Vanillin (100 g., 0.658 mole) was dissolved in 375 ml. of conc. sulfuric acid at 0–10°, and powdered sodium azide (45 g., 0.69 mole) was added with stirring over $1\frac{1}{2}$ hr. at the same temperature. Stirring was continued for $\frac{1}{2}$ hr. more with the cooling bath removed, after which the mixture was chilled and 900 ml. of water was added at such a rate that the temperature did not rise above 18°. The crude nitrile, which separated (yield about 70%), melted at 87.5–88.3° [38].

4. From Aldehydes via Oximes, Azines, and Related Types

Of the many N-containing derivatives of aldehydes which have been utilized in preparing nitriles, the oxime is the most common. It is readily prepared and may be dehydrated under mild conditions by a variety of reagents such as acetic anhydride alone or with sodium acetate [39–41], thionyl chloride [42], α,β-dibromopropionitrile [43], phosphorus pentoxide-ethanol (3:4) [44], phenyl isocyanate in the presence of a trace of triethylamine [45], polyphosphoric acid [46], and benzoyl chloride in pyridine [47]. In some cases the nitrile may be prepared without isolating the intermediate aldoxime. For example, anisaldehyde with fused sodium acetate and hydroxylamine hydrochloride in acetic acid gave a 67% yield of 4-methoxybenzonitrile [48]. Also a series of aldehydes has been converted into nitriles either by treatment with O,N-bis(trifluoroacetyl)-hydroxylamine, mostly in yields from 70–90% [49], by mild pyrolysis of the benzoate esters of the oxime

$$ArCH{=}NO\overset{\overset{\displaystyle O}{\|}}{C}C_6H_5 \quad [50]$$

or by refluxing the aldehyde with diammonium hydrogen phosphate and 1-nitropropane in glacial acetic acid [51]; see Ex. b.6. Finally, both aliphatic and aromatic aldehydes may be converted directly into nitriles by refluxing a formic acid solution with hydroxylamine hydrochloride and sodium formate, the yields for the aromatic types varying from 81 to 97% [52].

The synthesis of nitriles via the oxime is applicable not only to aliphatic and aromatic types but also to sugars [40, 47] and heterocyclic aldehydes such as those of indoles [42], quinolines [53], pyrroles [54], and furans [55]. It has been applied as well to α-ketocarboxylic acids whose oximes may be degraded to the nitrile by heating in the absence of a solvent [56], or the original acid may be converted directly into the nitrile by heating with hydroxylamine hydrochloride, pyridine, and ethanol [57] or with hydroxylamine hydrochloride alone in aqueous solution [58]. Yields in these degradations are often 90% or better.

Derivatives related to the aldoximes which have also served as intermediates en route to nitriles are N-chlorimines [59], N,N,N-trimethylhydrazonium iodides [60], N,N-dimethylhydrazone oxides (see Ex. b.5) [61], 4-alkylideneamino-1,2,4-triazoles [62], and azines [63, 64]. These require reagents somewhat rarer than hydroxylamine, although the azines are readily prepared in good yield from hydrazine sulfate [65]. Heat or various reagents, as already indicated, are necessary to convert the intermediate into the nitrile. It is interesting to note that irradiation of a series of aromatic azines in nonpolar solvents gives nitriles with yields from 80 to 95% [64]. For the most part these methods are more satisfactory when applied to aromatic than to aliphatic aldehydes. The method involving 4-alkylideneamino-1,2,4-triazoles has been applied to a greater extent than the other four to the heterocyclic and polynuclear series of compounds.

(a) Preparation of veratronitrile (70–76% from veratraldehyde, hydroxylamine hydrochloride, and sodium hydroxide followed by dehydration of the aldoxime formed with acetic anhydride) [41].

(b) Other examples

(1) Benzonitrile (89% from benzaldehyde, hydroxylamine hydrochloride, and sodium formate in formic acid) [52].

(2) Cinnamonitrile (82%, based on the quaternary hydrazonium salt, from cinnamaldehyde, N,N-dimethylhydrazine, and methyl iodide followed by treatment of the N,N,N-trimethylhydrazonium iodide with sodium methoxide) [60].

(3) 2-Cyanothiophene (70% from 2-thiophenecarboxaldehyde and 4-amino-1,2,4-triazole heated in C_6H_6 with a trace of p-toluenesulfonic acid followed by heating the azomethine formed above its melting point) [62].

(4) p-Nitrobenzonitrile (95% from p-nitrobenzaldehyde, hydrazine sulfate, and aq. ammonia and irradiation of the azine formed until the golden color disappeared) [64].

(5) Benzonitrile (50% from 0.25 mole each of benzaldehyde and

$$C_6H_5CH{=}NN(CH_3)_2 \longrightarrow C_6H_5CN + (CH_3)_2NOH$$
$$\downarrow$$
$$O$$

$unsym$-dimethylhydrazine refluxed in 170 ml. of methanol; more, 70 ml., methanol was added at room temperature followed by dropwise addition of 148 ml. of 30% hydrogen peroxide. The oxidation to the N-oxide was exothermic. The reaction mixture was diluted with water and the nitrile extracted with ether) [61].

(6) Indole-3-carbonitrile (48–63% by refluxing indole-3-carboxaldehyde, diammonium hydrogen phosphate, and 1-nitropropane in glacial acetic acid for 12.5 hr.) [66].

5. From Carboxylic Acids and Nitriles (Exchange)

$$ArCOOH \xrightarrow{m\text{-}C_6H_4(CN)_2} ArCN$$

An exchange reaction between aromatic acids and aromatic nitriles [67] has led to the production of nitriles from carboxylic acids in high yield. The method depends on using a high-boiling nitrile with a lower-boiling acid to form an equilibrium mixture containing as well a low-boiling nitrile and a higher-boiling acid. By distillation the low-boiling nitrile may be removed to drive the equilibrium to completion.

It has been proposed that the reaction occurs through the isoimide 19-III as an intermediate. This then rearranges, perhaps simultaneously, to the imide

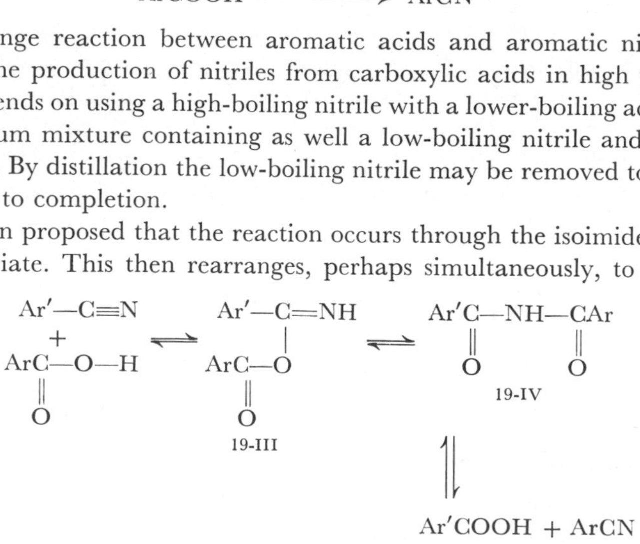

$$
\begin{array}{ccccc}
Ar'{-}C{\equiv}N & & Ar'{-}C{=}NH & & Ar'C{-}NH{-}CAr \\
+ & \rightleftharpoons & | & \rightleftharpoons & \| \qquad \| \\
ArC{-}O{-}H & & ArC{-}O & & O \qquad O \\
\| & & \| & & \text{19-IV} \\
O & & O & & \\
& & \text{19-III} & & \\
\end{array}
$$

$$\Updownarrow$$

$$Ar'COOH + ArCN$$

19-IV which in turn is in equilibrium with the nitrile and carboxylic acid [67]; *see also* [68].

More recently, lower aliphatic nitriles have been obtained by heating the acid with adiponitrile in the presence of an acid catalyst [69]. Valeronitrile was obtained in this way in a 91 % yield. Similarly valeric acid has been converted into the nitrile (72 %) by distillation with dicyandiamide [70].

(a) **Preparation of p-chlorobenzonitrile.** Isophthalonitrile, 128 g., and 78.2 g. of *p*-chlorobenzoic acid were heated to a flask temperature of 259° when distillation began at 220° overhead temperature. While the pot temperature rose to 294° in 40 min., 64 g. (93 %) of crude nitrile was collected at 220–223°, m.p. 88–91° [67].

(b) **Other examples**

(1) **Valeronitrile** (91 % from valeric acid, adiponitrile, and *p*-toluenesulfonic acid by refluxing) [69].

(2) **Pivalonitrile** (56 % from 1 mole of pivalic acid, 3 moles of phenylacetonitrile, and 0.01 mole of sulfuric acid refluxed for 30 min.; the nitrile slowly distilled through a 6-in. Vigreux column; redistillation with 0.1 volume of *o*-toluidine was conducted to remove acid) [71].

6. From Carboxylic Acids, Sulfonamides, and Phosphorus Pentachloride (Exchange)

$$ArCOOH \xrightarrow[\text{2 PCl}_5]{\text{ArSO}_2\text{NH}_2} ArCN + ArSO_2Cl + 2\ POCl_3 + 3\ HCl$$

A series of aromatic nitriles have been synthesized by this method, the yields of purified products varying from 63 to 79 % [72]. Aliphatic acids give low yields.

A similar exchange has been accomplished by using a sulfonic acid for the phosphorus pentachloride [73]. Yields in this case are variable, although sometimes values as high as 90 % are obtained. A series of equilibria similar to those described for nitrile-acid exchange (C.5) are proposed to account for these results.

(a) **Preparation of p-nitrobenzonitrile** (77–82 % from *p*-nitrobenzoic acid, *p*-toluenesulfonamide, and phosphorus pentachloride) [72].

(b) **Preparation of benzonitrile** (72 % from benzoic acid, *p*-toluenesulfonamide, and *p*-toluenesulfonic acid monohydrate) [73].

7. From Acid Chlorides and Trichlorophosphazosulfonyl Arenes

$$ArCOCl \xrightarrow{\text{Ar}'\text{SO}_2\text{N}=\text{PCl}_3} ArCN + Ar'SO_2Cl + POCl_3$$

A series of aromatic nitriles have been synthesized by heating the acid chloride with *o*-(trichlorophosphazosulfonyl)toluene [74]. Yields are usually in the 80–90 % range. It has been suggested that the reaction proceeds as follows:

$$ArCOCl \xrightarrow{\text{Ar}'\text{SO}_2\text{N}=\text{PCl}_3} Ar'SO_2Cl + ArCON=PCl_3 \longrightarrow ArCN + POCl_3$$

(a) **Preparation of benzonitrile.** An equimolecular mixture of benzoyl chloride and *o*-(trichlorophosphazosulfonyl)toluene was heated for 20 min. at

200–205°. On distillation, phosphoryl chloride and o-toluenesulfonyl chloride came over first, after which benzonitrile (80.5%) followed [74].

8. From α-Halonitroalkanes or N-Haloamides and Phosphines

The reactions as shown are not fully explored, but they show interesting characteristics of triphenylphosphine [75]:

$$n\text{-}C_7H_{15}\overset{\overset{\displaystyle Br}{|}}{C}HNO_2 \xrightarrow{(C_6H_5)_3P} n\text{-}C_7H_{15}C\equiv\overset{\oplus}{N}\text{—}\overset{\ominus}{O} + (C_6H_5)_3PO + HBr$$

$$\downarrow {\scriptstyle (C_6H_5)_3P \text{ (or (RO)}_3P)}$$

$$n\text{-}C_7H_{15}CN + (C_6H_5)_3PO \text{ (or (RO)}_3PO)$$

Octanoyl nitrile,
72% overall
with $(C_6H_5)_3P$

The reducing agents in the last step offer an advantage over zinc dust and acetic acid or tin and hydrochloric acid in that the reduction gives superior yields under mild conditions [76]. With five nitrile oxides the yields vary from 80 to 98% when trimethyl or triethyl phosphite is the reducing agent.

Similarly, some N-haloamides may be deoxodehydrohalogenated to nitriles:

$$C_6H_5CH_2CONHBr \xrightarrow{(C_6H_5)_3P} C_6H_5CH_2\overset{\overset{\displaystyle OP(C_6H_5)_3}{|}\;\;Br}{C}=NH \xrightarrow{\hspace{1cm}} C_6H_5CH_2CN + (C_6H_5)_3PO\cdot HBr$$

Phenylacetonitrile,
60%

The latter sequence should have broader future applications.

1. W. S. Bishop, *Org. Syn.*, Coll. Vol. **3**, 584 (1955); E. T. McBee *et al.*, *Ind. Eng. Chem.*, **39**, 391 (1947); D. B. Reisner and E. C. Horning, *Org. Syn.*, Coll. Vol. **4**, 144 (1963).
2. P. C. Teague and W. A. Short, *Org. Syn.*, Coll. Vol. **4**, 706 (1963).
3. R. E. Kent and S. M. McElvain, *Org. Syn.*, Coll. Vol. **3**, 493 (1955).
4. H. Richtzenhain and P. Nippus, *Chem. Ber.*, **82**, 408 (1949); B. B. Corson *et al.*, *Org. Syn.* Coll. Vol. **2**, 379 (1943).
5. C. F. Koelsch, *J. Org. Chem.*, **26**, 1003 (1961); B. Liberek, *Chem. Ind.* (*London*), 987 (1961); J. Budĕšinský and J. Kopecký, *Collection Czech. Chem. Commun.*, **20**, 52 (1955); R. Delaby. *et al.*, *Compt. Rend.*, **242**, 2644 (1956); *Bull. Soc. Chim. France*, 1294 (1956); D. Price and F. D. Pickel, *J. Am. Chem. Soc.*, **63**, 1067 (1941); J. P. English *et al.*, *J. Am. Chem. Soc.*, **68**, 1039 (1946); A. R. Surrey, *Org. Syn.*, Coll. Vol. **3**, 535 (1955).
6. J. C. Thurman, *Chem. Ind.* (*London*), 752 (1964); H. Goldstein and R. Voegeli, *Helv. Chim. Acta*, **26**, 1125 (1943).
7. J. A. Krynitsky and H. W. Carhart, *Org. Syn.*, Coll. Vol. **4**, 436 (1963).
8. P. Weiss *et al.*, *J. Am. Chem. Soc.*, **71**, 2650 (1949).
9. P. M. Brown *et al.*, *J. Chem. Soc.*, 2882 (1957).
10. C. R. Stephens *et al.*, *J. Am. Chem. Soc.*, **77**, 1701 (1955).
11. A. V. Kirsanov and Y. M. Zolotov, *Zh. Obshch. Khim.*, **20**, 284 (1950); *C.A.*, **44**, 6384 (1950).
12. J. L. Boivin, *Canad. J. Res.*, **28B**, 671 (1950).
13. C. Ressler and H. Ratzkin, *J. Org. Chem.*, **26**, 3356 (1961).
14. H. Gross and J. Gloede, *Chem. Ber.*, **96**, 1387 (1963).
15. A. R. Surrey, *Org. Syn.*, Coll. Vol. **3**, 535 (1955).

16. R. M. Herbst and K. R. Wilson, *J. Org. Chem.*, **22**, 1142 (1957).

17. E. M. Gal and A. T. Shulgin, *J. Am. Chem. Soc.*, **73**, 2938 (1951).

18. J. C. Thurman, *Chem. Ind. (London)*, 752 (1964).

19. C. K. Sauers and R. J. Cotter, *J. Org. Chem.*, **26**, 6 (1961).

20. T. Mukaiyama *et al.*, *J. Org. Chem.*, **27**, 803 (1962).

21. D. T. Mowry, *Chem. Rev.*, **42**, 262 (1948).

22. K. S. Markley *et al.*, in R. E. Kirk and D. F. Othmer, *Encyclopedia of Chemical Technology*. Vol. 6, The Interscience Encyclopedia, Inc., New York, 1951, p. 222.

23. H. Vorbrüggen, *Tetrahedron Letters*, 1631 (1968).

24. F. Becke and O. P. Swoboda, Ger. Patent 1,117,121, Jan. 8, 1960; *C.A.*, **56**, 10051 (1962).

25. F. Becke, *Allg. Prakt. Chem.*, 618 (1966); *C.A.*, **66**, 2688 (1967); F. Becke *et al.*, *Chemiker Zeit.*, **89**, 807 (1965).

26. D. Davidson and H. Skovronek, *J. Am. Chem. Soc.*, **80**, 376 (1958).

27. W. E. Bachmann and C. E. Brockway, *J. Org. Chem.*, **13**, 384 (1948).

28. M. A. Naylor and A. W. Anderson, *J. Am. Chem Soc.*, **75**, 5392 (1953).

29. P. E. Gagnon *et al.*, *Can. J. Chem.*, **34**, 1662 (1956).

30. R. P. Linstead *et al.*, *J. Chem. Soc.*, 3730 (1954).

31. W. R. Vaughan and R. D. Carlson, *J. Am. Chem. Soc.*, **84**, 769 (1962).

32. J. P. Freeman, *J. Org. Chem.*, **26**, 3507 (1961).

33. H. P. Fischer *et al.*, *Helv. Chim. Acta*, **45**, 2539 (1962).

34. W. G. Dauben *et al.*, *J. Am. Chem. Soc.*, **75**, 2347 (1953).

35. H. Wolff, *Org. Reactions*, **3**, 307 (1946).

36. J. H. Hall and E. Patterson, *J. Am. Chem. Soc.*, **89**, 5856 (1967).

37. H. W. Moore and H. R. Shelden, *J. Org. Chem.*, **33**, 4019 (1968).

38. C. Schuerch, Jr., *J. Am. Chem. Soc.*, **70**, 2293 (1948).

39. M. F. Browne and R. L. Shriner, *J. Org. Chem.*, **22**, 1320 (1957).

40. H. T. Clarke and S. M. Nagy, *Org. Syn.*, Coll. Vol. **3**, 690 (1955).

41. J. S. Buck and W. S. Ide, *Org. Syn.*, Coll. Vol. **2**, 622 (1943).

42. F. P. Doyle *et al.*, *J. Chem. Soc.*, 2853 (1956).

43. T. Mukaiyama and T. Hata, *Bull. Chem. Soc. Japan*, **33**, 1712 (1960).

44. T. Mukaiyama and T. Hata, *Bull. Chem. Soc. Japan*, **34**, 99 (1961).

45. T. Mukaiyama and H. Nohira, *J. Org. Chem.*, **26**, 782 (1961).

46. R. T. Conley and F. A. Mikulski, *J. Org. Chem.*, **24**, 97 (1959).

47. E. R. de Labriola and V. Deulofeu, *J. Org. Chem.*, **12**, 726 (1947).

48. J. H. Hunt, *Chem. Ind. (London)*, 1873 (1961).

49. J. H. Pomeroy and C. A. Craig, *J. Am. Chem. Soc.*, **81**, 6340 (1959).

50. J. H. M. Hill and L. D. Schmookler, *J. Org. Chem.*, **32**, 4025 (1967).

51. H. M. Blatter *et al.*, *J. Am. Chem. Soc.*, **83**, 2203 (1961).

52. T. van Es, *J. Chem. Soc.*, 1564 (1965).

53. W. Borsche and W. Ried, *Ann. Chem.*, **554**, 269 (1943).

54. H. Fischer and W. Zerweck, *Chem. Ber.*, **56**, 519 (1923).

55. N. Williams, *Chem. Ber.*, **60**, 2509 (1927).

56. R. Adams and A. W. Schrecker, *J. Am. Chem. Soc.*, **71**, 1186 (1949).

57. K. N. F. Shaw *et al.*, *J. Org. Chem.*, **23**, 1171 (1958).

58. A. Ahmad and I. D. Spenser, *Can. J. Chem.*, **39**, 1340 (1961).

59. E. J. Poziomek *et al.*, *J. Org. Chem.*, **25**, 2135 (1960).

60. R. F. Smith and L. E. Walker, *J. Org. Chem.*, **27**, 4372 (1962).

61. R. F. Smith *et al.*, *J. Org. Chem.*, **31**, 4100 (1966).

62. H. G. O. Becker and H. J. Timpe, *Z. Chem.*, **4**, 304 (1964).

63. E. Klingsberg, *J. Org. Chem.*, **25**, 572 (1960).

64. J. E. Hodgkins and J. A. King, *J. Am. Chem. Soc.*, **85**, 2679 (1963).

65. H. H. Hatt, *Org. Syn.*, Coll. Vol. **2**, 395 (1943).

66. H. M. Blatter *et al.*, *Org. Syn.*, **43**, 58 (1963).

67. W. G. Toland and L. L. Ferstandig, *J. Org. Chem.*, **23**, 1350 (1958).

68. R. H. Wiley and W. B. Guerrant, *J. Am. Chem. Soc.*, **71**, 981 (1949).

69. P. S. Pyryalova and E. N. Zil'berman, *Tr. po Khim. i. Khim. Tekhnol.*, (2) 353 (1963); *C.A.*, **61**, 4211 (1964).

70. M. T. Dangyan and M. A. Oganisyan, *Proc. Acad. Sci. Armenian SSR*, **2**, No. 2, 41 (1945); *C.A.*, **40**, 2783 (1946).

71. L. Turner, *J. Chem. Soc.*, 1686 (1956).

72. C. S. Miller, *Org. Syn.*, Coll. Vol. **3**, 646 (1955).

73. P. Oxley *et al.*, *J. Chem. Soc.*, 763 (1946).

74. A. V. Kirsanov, *Zh. Obshch. Khim.*, **22**, 274 (1952); *C.A.*, **46**, 11135 (1952).

75. S. Trippett and D. M. Walker, *J. Chem. Soc.*, 2976 (1960).

76. C. Grundmann and H.-D. Frommeld, *J. Org. Chem.*, **30**, 2077 (1965).

D. Addition Reactions

1. From Unsaturated Compounds and Hydrogen Cyanide (or R$_2$AlCN)

$$HC{\equiv}CH \xrightarrow{HCN} CH_2{=}CHCN \xrightarrow{HCN} CNCH_2CH_2CN$$

The addition of hydrogen cyanide to alkenes and alkynes is of industrial interest and therefore has been the subject of many patents [1]. In the presence of cuprous chloride and ammonium chloride adjusted to a pH of 1 with hydrochloric acid, hydrogen cyanide has been combined with acetylene to produce about two-thirds of the acrylonitrile in the U.S. [2] or the reaction occurs by vapor-phase addition at 300–700° [3]:

$$HC{\equiv}CH + HCN \xrightarrow[\text{700°–900°}]{\text{300–700°}} \begin{array}{l} CH_2{=}CHCN \\ HC{\equiv}CCN + \text{various other products} \end{array}$$

Hydrogen cyanide also adds to alkenes, but again a catalyst is essential to effect the union. Catalysts which have been employed are alumina, cobalt cyanide-on-alumina, cobalt-on-alumina, and dicobalt octacarbonyl [4]. By introducing alkoxy or acyloxy substituents in the alkene, addition occurs more readily. With the α,β unsaturated type such as RCH=CHX, where X = COOR, CN, COR, NO$_2$, or C$_6$H$_5$, a catalyst is not always necessary, if the customary excess of the cyanide plays this role. A second negative substituent of this type in the α-position leads to further ease of addition. For α,β unsaturated aldehydes, such as RCH=CHCHO, cyanohydrin formation predominates over 1,4- addition to the carbon-carbon double bond, a reaction which also occurs to some extent in the addition to some ketones [1]. The 1,4-addition which resembles the Michael addition (see Alcohols, Chapter 4, G.8) has been widely used in the preparation of mono- and dicarboxylic acids and related types with excellent results.

A reagent effecting hydrocyanation much more efficiently is diethylaluminum cyanide [5] prepared as shown (and being checked by *Organic Synthesis*):

$$(C_2H_5)_3Al + HCN \xrightarrow[0°]{C_6H_6} (C_2H_5)_2AlCN + C_2H_6$$

This reagent adds rapidly to unsaturated ketones and can be expected to give different stereoselectivity than hydrogen cyanide. The *in situ* prepared reagent,

References for Section D are on pp. 973–974.

which is not so active as the isolated reagent, has been shown to give a different ratio of adducts than hydrogen cyanide [6]:

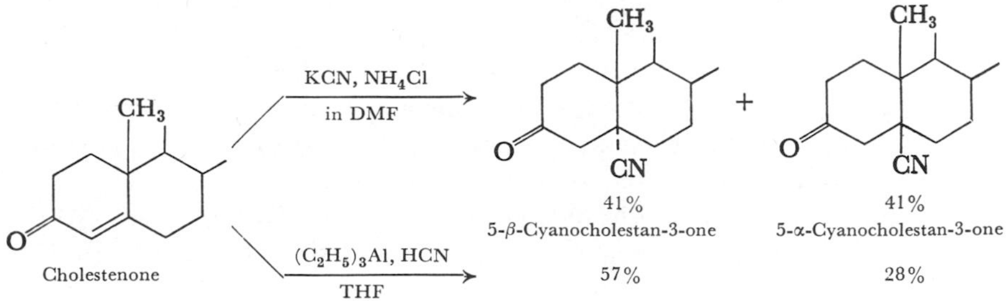

41%	41%
5-β-Cyanocholestan-3-one	5-α-Cyanocholestan-3-one
57%	28%

With some ketones, as in the case shown, the stereoselectivity is even higher [6–8].

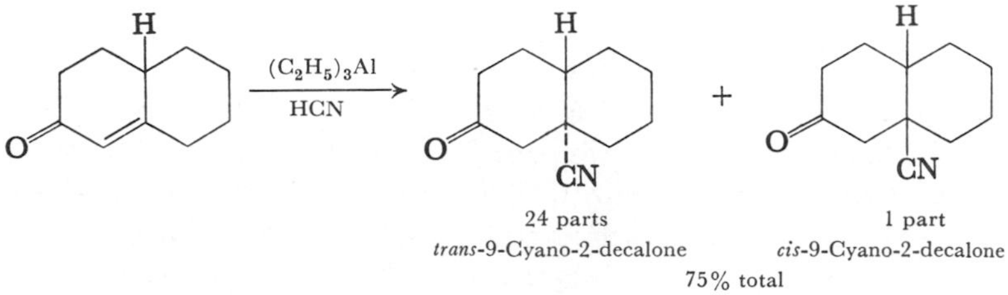

24 parts / 1 part
trans-9-Cyano-2-decalone / cis-9-Cyano-2-decalone
75% total

(a) **Preparation of α-phenyl-β-benzoylpropionitrile** (93–96% from benzalacetophenone (1 mole), glacial acetic acid (1 mole), and potassium cyanide (2 moles) in aq. ethanol at 35°) [9].

(b) **Other examples**

(1) **β-(2-Pyridyl)-propionitrile** (67% from 2-vinylpyridine (1 mole), acetic anhydride (2 moles), and potassium cyanide (2 moles) in water) [10].

(2) **Isobutyronitrile** (75% conversion, based on cyanide, from hydrogen cyanide and propylene in the presence of dicobalt octacarbonyl and triphenylphosphine at 130° and 100 atm.) [4].

(3) **Succinonitrile** (95% from acrylonitrile and anhydrous hydrogen cyanide in the presence of 1–3% of sodium cyanide at 30–50°) [11].

2. From Carbonyl Compounds and Hydrogen Cyanide

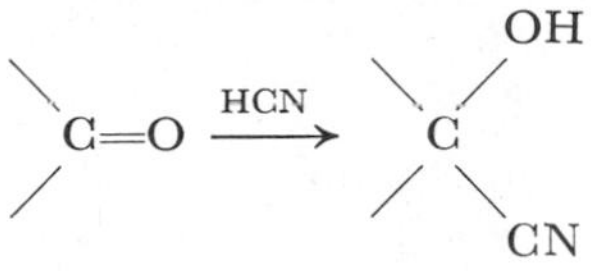

Three principal methods have been utilized to effect this addition [1]: (i) anhydrous hydrogen cyanide may be added to the aldehyde or ketone containing a trace of an alkali cyanide at about 0°, (ii) the sodium bisulfite addition product may be formed first and then treated with sodium cyanide:

$$RCHO \xrightarrow{NaHSO_3} RCHOHSO_3Na \xrightarrow{NaCN} RCHOHCN + Na_2SO_3$$

and (iii) an interchange between an aldehyde and a ketone cyanohydrin may be effected:

$$RCHO \xrightarrow{\quad R_2'C\overset{OH}{\underset{CN}{\diagup}}\quad} RCHOHCN + R_2'CO$$

Method (i) is satisfactory for most aldehydes and ketones. It has the advantage of ease of isolation and purification since water and inorganic salts are present in minimum amounts. The return by this method is dependent on the degree of dissociation of the cyanohydrin at equilibrium. These values vary considerably with the structure of the carbonyl compound and the temperature. Method (ii) avoids the presence of hydrogen cyanide fumes and minimizes the formation of benzoins from aromatic aldehydes. Method (iii), of value in the synthesis of lower water-soluble cyanohydrins, also avoids the handling of anhydrous hydrogen cyanide and simplifies isolation since aqueous solutions of inorganic salts are avoided.

A few limitations to the addition of hydrogen cyanide may be mentioned. Although the addition is satisfactory for aliphatic aldehydes and ketones and aromatic aldehydes, the yields are poor for aryl alkyl ketones, and no addition at all occurs to diaryl ketones. α,β-Unsaturated ketones and quinones (see Exs. a and b) tend to add hydrogen cyanide in a 1,4 manner, while highly enolizable ketones usually do not form cyanohydrins satisfactorily. As has already been indicated, aromatic aldehydes undergo the benzoin condensation.

The synthesis is of value in carbohydrate chemistry in that it is one step in the well-known Kiliani-Fischer method of going from one aldose to the next higher homolog. It has also been useful in synthesizing other nitriles, particularly by the replacement of the hydroxyl group with other groups.

For a discussion of the mechanism see Alcohols, Chapter 4, F; examples other than shown here are given in Chapter 4, F.2.

(a) Preparation of 2,3-dicyanohydroquinone

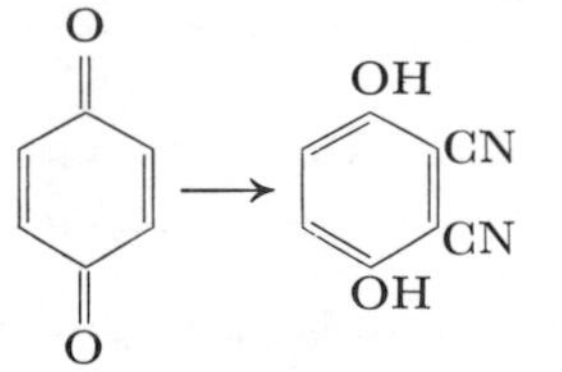

Sulfuric acid, 25 ml., and 100 ml. of ethanol were added to 20 g. of benzoquinone suspended in 500 ml. of ethanol. Then 40 g. of potassium cyanide in 100 ml. of water was added in 30 min. while the temperature was held at 24–28°. The brown mixture was acidified and the sulfate salt filtered and washed with alcohol. The filtrate was reduced in volume to 150 ml. under reduced pressure and minimum heating. The crystalline product was recrystallized from water with charcoal and gave a 64% yield of the desired compound [12].

(b) Preparation of 3-cyano-1-amino-2-carboxy-4-butylaminoanthraquinone (unstated yield from 10 parts of 1-amino-2-carboxy-4-butyl-aminoanthraquinone, 20 parts of sodium cyanide, 2 parts of sodium bicarbonate

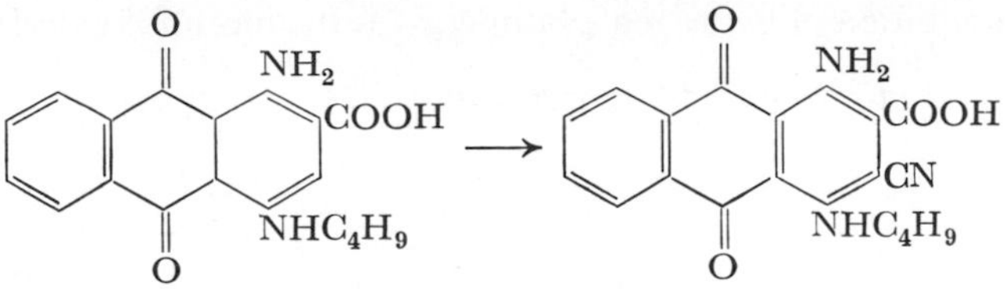

in 350 ml. of water at 75–80° for 5 hr.; the overall consequence is substitution, but the intermediate steps possibly involve oxidation and 1,4 addition of cyanide) [13].

3. From Carbonyl Compounds, Hydrogen Cyanide, and Ammonia (Strecker) or like Compounds

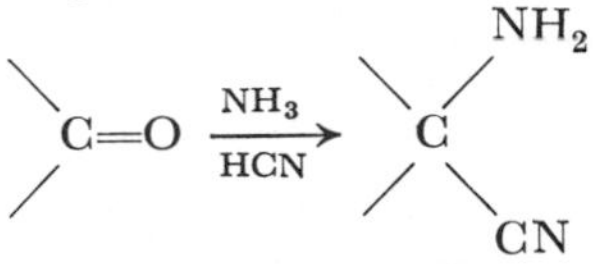

This synthesis is a modification of that in D.2. Instead of simply adding hydrogen cyanide to produce cyanohydrins, ammonia and hydrogen cyanide are added to yield aminonitriles. The method is of importance in that α-amino acids, the hydrolytic products of α-amino nitriles, may be synthesized by this route [14].

A variety of methods has been employed in carrying out this reaction. The most common procedure is to add a solution, alcoholic if necessary, of the carbonyl compound to an aqueous solution of sodium cyanide and ammonium chloride. If N-alkyl- or N-arylamino nitriles are desired, the ammonia is replaced by an amine. A second procedure consists in using ammonium cyanide for ammonium chloride and the alkali cyanide and allowing the single salt and the carbonyl compound in alcoholic solution to stand for several days [15]. A third procedure [16] consists in forming the sodium bisulfite addition product of the carbonyl compound first and then letting it react with ammonia or the amine and the cyanide. This method is only suitable for preparing amino nitriles from formaldehyde and the simpler amines. Finally, the cyano-hydrin may be treated with ammonia or the amine [17], a procedure which gives amino nitriles from acetone cyanohydrin and ammonia, alkyl amines or aryl amines, usually with yields in the 60–90 % range.

Hydantoins are possible side products, particularly if ammonium carbonate is used as a reagent [18]:

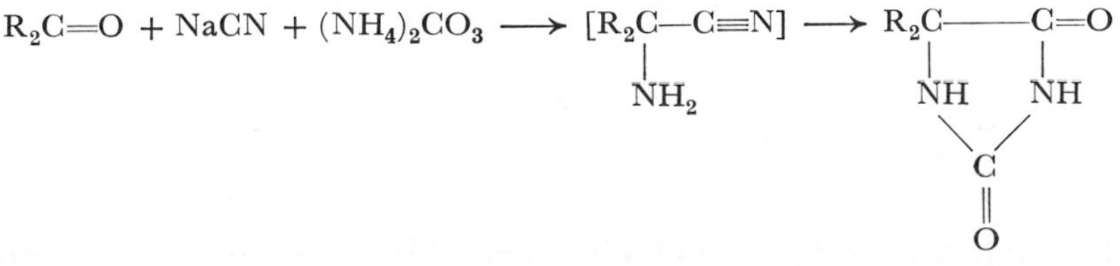

Although the steps in the Strecker reaction have been the subject of con-troversy, as it is usually conducted, it is thought that the imine 19-V is first formed and addition of hydrogen cyanide gives the amino nitrile 19-VI:

$$RCHO \xrightarrow{NH_3} [RCH{=}NH] \xrightarrow{HCN} \underset{\underset{CN}{|}}{RCHNH_2}$$

$$\phantom{RCHO \xrightarrow{NH_3}}\text{19-V}\phantom{[RCH{=}NH] \xrightarrow{HCN}}\text{19-VI}$$

(a) Preparation of α-(diethylamino)phenylacetonitrile. Diethyl-amine (0.50 mole) was neutralized with conc. hydrochloric acid and benzaldehyde (0.50 mole) was added. The mixture was stirred while an aqueous solution of potassium cyanide (0.55 mole) was added dropwise. After heating on a steam bath from 1–2 hr., the mixture was cooled. If the product did not crystallize, it was separated and distilled; yield 80%, b.p. 78–80° (0.05 mm.) [19].

(b) Other examples

(1) Diethylaminoacetonitrile (88–90% from the sodium bisulfite addition product of formaldehyde, diethylamine, and sodium cyanide in water) [20].

(2) α-(1-Pyrrolidyl)isobutyronitrile (88.7% from acetone cyano-

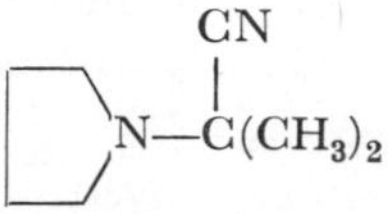

$$\underset{}{\text{N}}-\overset{\overset{\text{CN}}{|}}{\text{C}}(CH_3)_2$$

hydrin and pyrrolidine below boiling point of the mixture) [21].

(3) 1,2-Di-1-(1-cyano)cyclohexylhydrazine (66–70% from cyclohexanone, hydrazine sulfate, and sodium cyanide) [22]; this compound may be oxidized to the azo derivative and pyrolyzed to yield 1,1′-dicyano-1,1′-bicyclohexyl in 65–69% yield [23].

4. From Azomethines and Hydrogen Cyanide (and Related Reactions)

$$RCH{=}NH \xrightarrow{HCN} \underset{\underset{CN}{|}}{RCHNH_2}$$

$$RCH{=}NOH \xrightarrow{HCN} \underset{\underset{CN}{|}}{RCHNHOH}$$

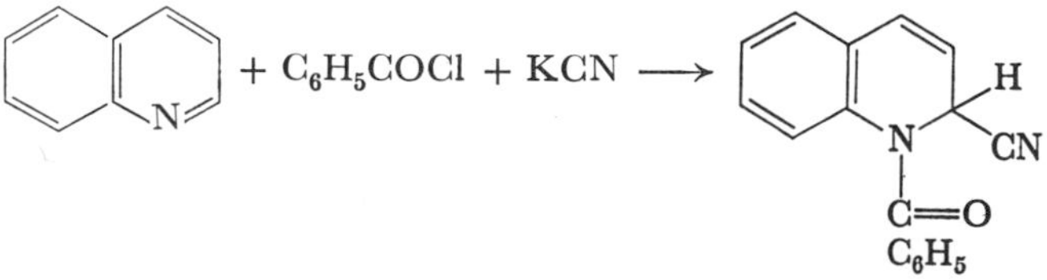

Reissert compound

Hydrogen cyanide adds readily to imines [24] and oximes [25] to form α-aminonitriles and α-hydroxyaminonitriles, respectively (see Ex. a). Actually, better yields are obtained by employing a sodium cyanide-phosphate buffer [26] although aqueous hydrogen cyanide with pyridine [27] has also been used successfully. Yields in such additions are satisfactory.

Considerable progress in the preparation of Reissert compounds [28] has been made since publication of the preceding review. The most general method involves a heterogeneous system of potassium cyanide in water and quinoline (or isoquinoline) in methylene chloride to which an acyl chloride is added dropwise [29]; see Ex. b. The method is applicable to all substituted quinolines and isoquinolines except those with substituents in the 2- or 8-positions of quinoline and in the 1- or 3-positions of isoquinoline. The Reissert compounds may be converted into quinoline carbonitriles by treatment with phosphorus pentachloride (see Ex. c.1) or by treatment of the sulfonyl Reissert compound with strong base (see Ex. c.2):

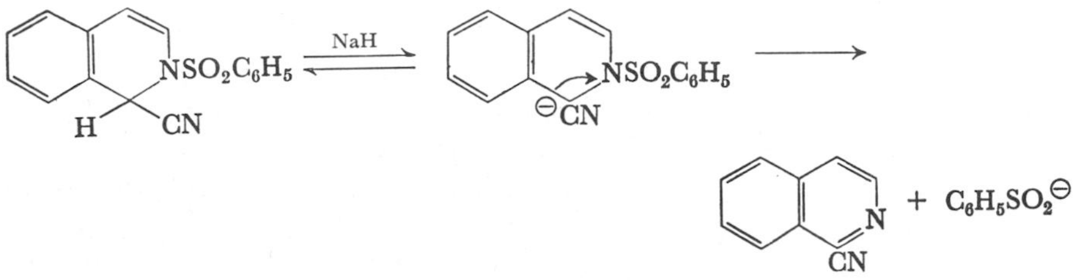

Unfortunately, yields are not uniformly high for the latter method; for example, treatment of quinoline and cyanide, with benzenesulfonyl chloride gives quinaldonitrile directly but in yields of only 24%. Under these circumstances, on an untested heterocycle, it would be best to attempt the benzenesulfonyl reaction first because of its simplicity and, if that does not succeed, retreat to the preparation of the Reissert compound with benzoyl chloride followed by the reaction with phosphorus pentachloride.

Nonaqueous systems such as benzene also have been used in the preparation of Reissert compounds [28], but they have no advantage over the heterogeneous water-methylene chloride system.

Cyanopyridines and quinolines also are prepared in a reaction which has some of the characteristics of the Reissert reaction [30]:

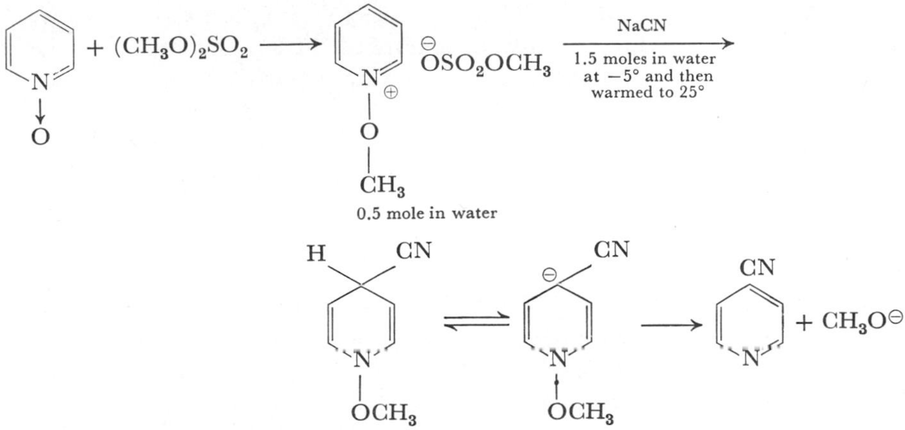

The mechanism is shown for the 4-cyanopyridine, but both 4- and 2-cyano-pyridine, with yields of 32 and 49%, respectively, are formed and must be separated. Quinoline-N-oxide gives 2-cyanoquinoline in 93% yield similarly: yields of the 2-cyano are in general higher than those of the 4-cyano derivative.

(a) **Preparation of α-hydroxyaminovaleronitrile** (Butyraldehyde oxime, 20 g., and 10 ml. of hydrogen cyanide were stirred in a pentane emulsion and then a second 10-ml. portion of HCN was added after 24 hr.; total time of stirring was 60 hr.; the reaction was first held at 0° and then at room temperature; the solid was filtered and precipitated from ether-pentane to give a 75% yield) [31].

(b) **Preparation of 4-hydroxy-1-benzoyl-2-cyano-1,2-dihydroquino-line (Reissert compound of 4-hydroxyquinoline).** Benzoyl chloride, 0.032 mole, was added dropwise in 2 hr. to a stirred mixture of 0.016 mole of 4-hydroxyquinoline in 20 ml. of methylene chloride and 0.048 mole of potassium cyanide in 8 ml. of water and the mixture was stirred an additional 6–8 hr. The methylene chloride layer was separated, after which the aqueous layer was extracted once with methylene chloride and the combined organic layers were washed with acid, base, and water, and dried. On evaporation, the desired compound was obtained in 98% yield. Yields ranged from 9 to 99% for 38 quinolines and isoquinolines [29]. This method is more general than the ordinary procedure [32].

(c) **Other examples**

(1) **1-Cyanoisoquinoline** (1-cyano-2-benzoyl-1,2-dihydroisoquinoline, 130 g., and 130 g. of phosphorus pentachloride were rubbed together and warmed at 125–130° until the evolution of hydrogen chloride began; the warming was continued under reflux until a deep red liquid resulted on cooling; neutralization with sodium carbonate and steam distillation gave 62 g. of the cyanide, m.p. 67–85°; two recrystallizations from ligroin-benzene left 41 g. (53%), m.p. 88–89°) [33].

(2) **Isoquinaldonitrile.** The benzenesulfonyl Reissert compound was made in 90% yield from a stirred mixture of 77.5 g. of potassium cyanide and 0.356 mole of isoquinoline in 625 ml. of water to which 95 ml. of benzene-sulfonyl chloride was added in 2 hr. followed by 4 hr. of stirring. The Reissert compound, 0.012 mole, and 0.58 g. of sodium hydride, 50% oil dispersion, were refluxed in xylene to give the nitrile in quantitative yield. Quinoline treated similarly without formation of the Reissert intermediate gave quinaldonitrile in 24% yield [34].

5. From O-Heterocyclic Compounds and Hydrogen Cyanide

$$CH_2\!\!-\!\!CH_2 \xrightarrow{\text{HCN}} HOCH_2CH_2CN$$

Cleavage of the rings in cyclic ethers, lactones, and phthalides with hydrogen cyanide results in the formation of nitriles. Ethylene oxide, for example, gives ethylene cyanohydrin in almost quantitative yield at low temperature [35]. Phthalide yields the potassium salt of o-carboxyphenylacetonitrile by simply heating the solid with potassium cyanide (see Ex. a). The γ-lactone of γ-anisylbutyric acid and potassium cyanide at 210° gives a product which on acidification leads to a 30% yield of γ-anisyl-β-cyanobutyric acid [36]:

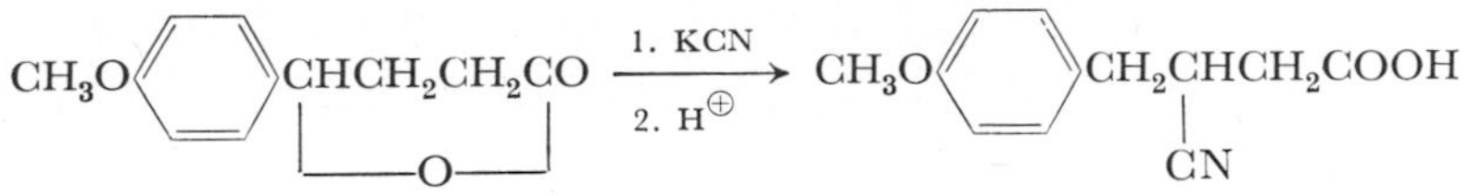

$$\text{CH}_3\text{O}\diagdown\diagup\text{CHCH}_2\text{CH}_2\text{CO} \xrightarrow[\text{2. H}^{\oplus}]{\text{1. KCN}} \text{CH}_3\text{O}\diagdown\diagup\text{CH}_2\text{CHCH}_2\text{COOH}$$

It will be noted that addition occurs in a manner similar to that of an α,β-unsaturated acid.

The ring of a pyrylium salt is opened [37] by a cyanide to give unsaturated ketonitriles (for general reactions of pyrylium salts, see [38]):

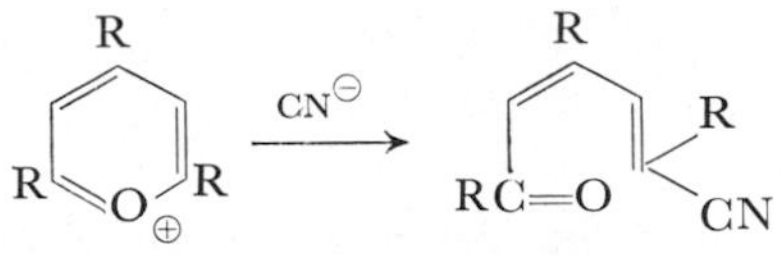

$$\text{R} \diagdown\diagup \text{O}_{\oplus} \text{R} \quad \xrightarrow{\text{CN}^{\ominus}} \quad \text{RC}=\text{O} \diagdown\diagup \text{R}_{\text{CN}}$$

(a) Preparation of o-carboxyphenylacetonitrile (67–83% from phthalide and potassium cyanide at 180–190° for 4–5 hr., followed by acidification) [39].

6. From Cyanogen Compounds

Cyanogen and cyanogen halide additions to unsaturated compounds have not been studied in detail. The one recent example indicates that addition is facilitated by means of electrophilic catalysts (see Ex. a).

The von Braun reaction [40] consists of an initial addition of cyanogen bromide to a tertiary amine followed by a thermal decomposition of the quaternary salt (see Ex. b.1):

$$\text{R}_3\text{N} + \text{CNBr} \rightarrow \text{R}_3\overset{\oplus}{\text{N}}\text{CNBr}^{\ominus} \rightarrow \text{R}_2\text{NCN} + \text{RBr}$$
$$\text{Dialkylcyanamide}$$

A similar reaction is the cleavage of sulfides [41]:

$$\text{R}_2\text{S} + \text{CNBr} \longrightarrow \text{R}_2\overset{\oplus}{\text{S}}\text{CNBr}^{\ominus} \xrightarrow{\Delta} \text{RSCN} + \text{RBr}$$

Cyanoazides usually add to olefins to form triazoles which spontaneously decompose to a mixture of alkylidene cyanamide and N-cyanoaziridine [42]:

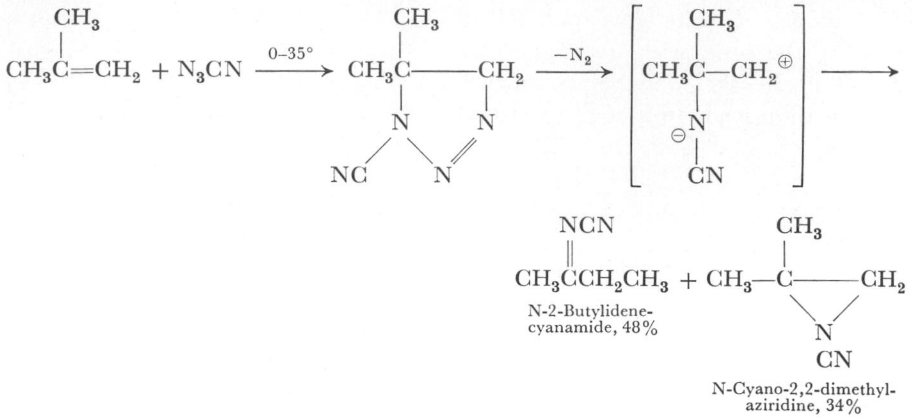

$$CH_3C{=}CH_2 + N_3CN \xrightarrow{0-35°} \text{(triazole)} \xrightarrow{-N_2} \text{[cation]} \longrightarrow$$

N-2-Butylidene-cyanamide, 48%

N-Cyano-2,2-dimethyl-aziridine, 34%

Cyclopentene yields 94% cyclopentylidene cyanamide and no aziridine. Another rearrangement is shown in Ex. b.2. Care must be taken in handling of cyanogen azide because of its explosive nature. Usually, the reagent is prepared *in situ* in acetonitrile solvent at concentrations of no more than 5%.

(a) Preparation of *trans*-β-bromo-β-phenylacrylonitrile

($C_6H_5CBr{=}CHCN$).   Cyanogen bromide, 15 g., and aluminum bromide, 25 g., were dissolved in 200 ml. of tetrachloroethane and heated to 40–45° for 2 hr. Phenylacetylene in tetrachloroethane was then added dropwise and the mixture poured onto ice and water to give the desired compound (32%) from the organic phase [43].

(b) Other examples

(1) N-Cyano-2-(4-bromobutyl)-3-methylpiperidine [44]:

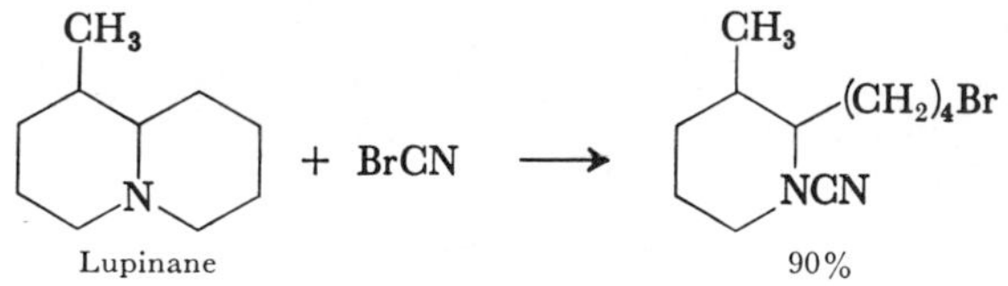

Lupinane   + BrCN ⟶   90%

(2) N-Cyano-2-azabicyclo[3.2.1]octa-3,6-diene [45]:

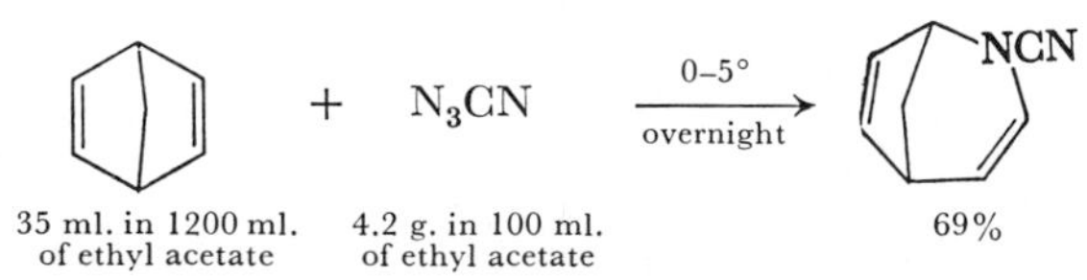

+ N_3CN $\xrightarrow[\text{overnight}]{0-5°}$

35 ml. in 1200 ml.   4.2 g. in 100 ml.   69%
of ethyl acetate   of ethyl acetate

7. From Acrylonitriles (Cyanoethylation)

$$CH_2{=}CHCN \xrightarrow{ROH} ROCH_2CH_2CN$$

$$\underset{NC}{\overset{NC}{\diagdown}}C{=}C\underset{CN}{\overset{CN}{\diagup}} \xrightarrow{ROH} \underset{NC}{\overset{NC}{\diagdown}}C{=}C\underset{OR}{\overset{OR}{\diagup}}$$

Organic and inorganic compounds containing active hydrogen atoms add to acrylonitrile and related unsaturated nitriles to form compounds having a cyanoethyl ($-CH_2CH_2CN$) or related group. The reaction, which is similar to the Michael type of addition, has been reviewed [46]. Although no cyanide group is created in the addition, the reaction has such wide application in synthetic organic chemistry that some discussion of it seems to be in order.

The types which add to the unsaturated nitrile are in the main:

(i) Compounds with one or more NH groups as ammonia, primary and secondary amines, imides, amides.

(ii) Compounds with one or more $-OH$ or $-SH$ groups as water, alcohols, phenols, oximes, hydrogen sulfide, mercaptans.

(iii) Acidic compounds, other than carboxylic acids, as hydrogen cyanide, hydrogen chloride, hydrogen bromide.

(iv) Compounds with the formula CHX_3 in which X equals chlorine or bromine.

(v) Carbanions of all types.

Usually a basic catalyst in small quantity is essential in the reaction. For this purpose, sodium or potassium or compounds thereof such as the oxide, hydroxide, alkoxide, carbonate, or cyanide have been employed. Quaternary ammonium hydroxides such as benzyltrimethylammonium hydroxide (Triton B) are particularly effective since they are soluble in organic solvents. Triethylamine is effective with active methylene compounds (see Ex. b.3) and alkylpyridine methiodides [47, 48]. In the former case the rate increases with the solvating ability of the solvent. Cupric acetate has been found to be of advantage in the addition of anilines [49].

The addition reaction is carried out satisfactorily if the acrylonitrile is added slowly to the second component and if adequate cooling facilities are available. Low temperature is essential if polymerization is to be avoided. Solvents such as benzene, dioxane, or pyridine serve to make the reaction a homogeneous one as well as to moderate the reaction. Yields are usually in the 70–90 % range.

Tetracyanoethylene is an α,β-unsaturated nitrile of high reactivity which has been prepared from malonitrile [50]. It is also an example of a cyanocarbon, a class of compounds receiving some attention of late [51]. This cyanide reacts with butadiene, anthracene, 2-vinyl naphthalene, and other 1,3-dienes to give tetracyanocyclohexenes 19-VII, rapidly at room temperature [52]. Such a

$$\underset{\underset{C(CN)_2}{\parallel}}{C(CN)_2} \xrightarrow{\text{CH}_2=\text{CH}-\text{CH}=\text{CH}_2} \quad \text{(CN)}_2 \text{(CN)}_2$$

19-VII
4,4,5,5-Tetracyano-
1-cyclohexene

compound has a tendency with nucleophilic reagents to yield compounds in which either one or two cyano groups attached to the same carbon are replaced with groups such as alkoxy, hydroxy, or aminoaryl. For example, with an alcohol it gives a dicyanoketene ketal, 19-VIII, while with a glycol the cyclic

$$\underset{\substack{\text{O}\underline{\quad}\text{CH}_2 \\ 19\text{-IX}}}{\overset{C(CN)_2}{\underset{\displaystyle |}{\overset{\parallel}{\underset{\displaystyle |}{C}}-\text{O}-\text{CH}_2}}} \xleftarrow{\underset{\text{CH}_2\text{OH}}{\text{CH}_2\text{OH}}} \underset{C(CN)_2}{\overset{C(CN)_2}{\parallel}} \xrightarrow{\text{ROH}} \underset{19\text{-VIII}}{\overset{C(CN)_2}{\underset{C(OR)_2}{\parallel}}}$$

ketal 19-IX is produced (see Ex. b.1). These reactions illustrate only a few of the many unusual nitriles available through the use of tetracyanoolefins. For other reactions see [52, 53].

(a) Preparation of β-(2-hydroxyethylmercapto)propionitrile, HOCH₂CH₂SCH₂CH₂CN (92–94 % from acrylonitrile and 2-mercaptoethanol at moderate temperature) [54].

(b) Other examples

(1) Dicyanoketene ethylene ketal (77–85 % from tetracyanoethyl-

$$\underset{\text{NC}}{\overset{\text{NC}}{}}\diagdown \underset{\diagup}{\overset{}{}} C=C \underset{\diagup}{\overset{\diagup}{}} \underset{\text{O}-\text{CH}_2}{\overset{\text{O}-\text{CH}_2}{}} \mid$$

ene and ethylene glycol in the presence of urea at 70–75°) [55].

(2) 2,5-Diamino-3,4-dicyanothiophene (79–85 % from tetracyano-

$$\underset{\text{H}_2\text{N}}{\overset{\text{NC}}{}}\diagdown \underset{\text{S}}{\overset{}{}} \diagup \underset{\text{NH}_2}{\overset{\text{CN}}{}}$$

ethylene and hydrogen sulfide in acetone, carbon disulfide, and pyridine at 0–5°) [56].

(3) γ-Acetyl-γ-carbethoxypimelonitrile (77 % from 0.1 mole each

$$\text{CH}_3\text{COCH}_2\text{COOC}_2\text{H}_5 + \text{CH}_2=\text{CHCN} \longrightarrow \underset{\underset{\text{CH}_2\text{CH}_2\text{CN}}{\displaystyle |}}{\overset{\overset{\text{CH}_2\text{CH}_2\text{CN}}{\displaystyle |}}{\text{CH}_3\text{COCCOOC}_2\text{H}_5}}$$

of ethyl acetoacetate and acrylonitrile, 30 ml. each of isopropyl alcohol and water, and 10 ml. of triethylamine at 25° for 1 day) [47].

(4) 2-β-Cyanoethyl-2-carbethoxycyclohexanone (85% from 0.6

$$\text{(cyclohexanone with } COOC_2H_5\text{)} + CH_2{=}CHCN \longrightarrow \text{(cyclohexanone with } COOC_2H_5 \text{ and } CH_2CH_2CN\text{)}$$

mole of 2-carbethoxycyclohexanone, 0.6 mole of acrylonitrile, and 3 ml. of 40% Triton B in 100 ml. of dioxane for 3 hr. at 25°) [57].

1. D. T. Mowry, *Chem. Rev.*, **42**, 189 (1948).
2. W. O. Fugate, in R. E. Kirk and D. F. Othmer, *Encyclopedia of Chemical Technology*, Vol. 1, The Interscience Publishers, New York, 1963, p. 344.
3. L. J. Krebaum, *J. Org. Chem.*, **31**, 4103 (1966).
4. P. Arthur, Jr., *et al.*, *J. Am. Chem. Soc.*, **76**, 5364 (1954).
5. W. Nagata and M. Yoshioka, *Tetrahedron Letters*, 1913 (1966).
6. W. Nagata *et al.*, *Tetrahedron Letters*, 461 (1962), and earlier papers.
7. W. Nagata *et al.*, *Tetrahedron Letters*, 865 (1963).
8. W. Nagata *et al.*, *Chem. Pharm. Bull.* (*Tokyo*), **11**, 226 (1963); *C.A.*, **59**, 7388 (1963).
9. C. F. H. Allen and R. K. Kimball, *Org. Syn.*, Coll. Vol. **2**, 498 (1943).
10. V. Boekelheide *et al.*, *J. Am. Chem. Soc.*, **75**, 3243 (1953).
11. Ref. 1, p. 229.
12. A. M. Creighton and L. M. Jackman, *J. Chem. Soc.*, 3138 (1960).
13. British Celanese, Ltd., British Patent 593,485, Oct. 17, 1947; *C.A.*, **42**, 1743 (1948).
14. R. E. Steiger, *Org. Syn.*, Coll. Vol. **3**, 66, 84, 88 (1955).
15. R. M. Herbst and T. B. Johnson, *J. Am. Chem. Soc.*, **54**, 2463 (1932).
16. D. B. Luten, Jr., *J. Org. Chem.*, **3**, 588 (1939).
17. R. A. Jacobson, *J. Am. Chem. Soc.*, **67**, 1996 (1945); **68**, 2628 (1946).
18. J. Mathieu and A. Allais, *Cahiers de Synthèse Organique*, Vol. 1, Masson et Cie., Paris, 1957, p. 112.
19. L. H. Goodson and H. Christopher, *J. Am. Chem. Soc.*, **72**, 358 (1950).
20. C. F. H. Allen and J. A. Van Allan, *Org. Syn.*, Coll. Vol. **3**, 275 (1955).
21. R. B. Moffett, *J. Org. Chem.*, **14**, 862 (1949).
22. C. G. Overberger *et al.*, *Org. Syn.*, Coll. Vol. **4**, 274 (1963).
23. C. G. Overberger and M. B. Berenbaum, *Org. Syn.*, Coll. Vol., **4**, 273 (1963).
24. R. Tiollais, *Bull. Soc. Chim. France*, **14**, 966 (1947).
25. C. C. Porter and L. Hellerman, *J. Am. Chem. Soc.*, **66**, 1652 (1944).
26. H. A. Lillevik *et al.*, *J. Org. Chem.*, **7**, 164 (1942).
27. F. Adickes, *J. Prakt. Chem.*, **161**, 271 (1943).
28. W. E. McEwen and R. L. Cobb, *Chem. Rev.*, **55**, 511 (1955).
29. F. D. Popp *et al.*, *J. Org. Chem.*, **26**, 4930 (1961); **27**, 297 (1962).
30. W. E. Feely and E. M. Beavers, *J. Am. Chem. Soc.*, **81**, 4004 (1959).
31. C. D. Hurd and J. M. Longfellow, *J. Org. Chem.*, **16**, 761 (1951).
32. J. Weinstock and V. Boekelheide, *Org. Syn.*, Coll. Vol. **4**, 641 (1963).
33. J. J. Padbury and H. G. Lindwall, *J. Am. Chem. Soc.*, **67**, 1268 (1945).
34. F. D. Popp *et al.*, *Chem. Ind.* (*London*), 140 (1965).
35. Ref. 1, p. 244.
36. Ref. 1, p. 246.
37. C. R. Noller, *Chemistry of Organic Compounds*, 3rd ed., W. B. Saunders Co., Philadelphia, 1965, p. 684.
38. K. Dimroth, *Angew. Chem.*, **72**, 331 (1960).
39. C. C. Price and R. G. Rogers, *Org. Syn.*, Coll. Vol. **3**, 174 (1955).
40. H. A. Hageman, *Org. Reactions*, **7**, 198 (1953); K. W. Bentley, *Technique of Organic Chemistry*, Vol. 11, Interscience Publishers, New York, 1963, Pt. 2, p. 773.
41. D. S. Tarbell and D. P. Harnish, *Chem. Rev.*, **49**, 15 (1951).

42. F. D. Marsh and M. E. Hermes, *J. Am. Chem. Soc.*, **86,** 4506 (1964).

43. I. Iwai *et al., Chem. Pharm. Bull. (Tokyo)*, **12,** 1446 (1964); *C.A.*, **62,** 10366 (1965).

44. K. Winterfeld and F. W. Holschneider, *Chem. Ber.*, **64B,** 137 (1931).

45. A. G. Anastassiou, *J. Org. Chem.*, **31,** 1131 (1966).

46. H. A. Bruson, *Org. Reactions*, **5,** 79 (1949); N. M. Bikales, Kirk-Othmer, *Encyclopedia of Chemical Technology*, 2nd ed., Vol. 6, Interscience Publishers, New York, 1965, p. 634.

47. J. A. Adamcik *et al., J. Org. Chem.*, **28,** 336 (1963).

48. J. A. Adamcik *et al., J. Org. Chem.*, **29,** 572 (1964).

49. S. A. Heininger, *Org. Syn.*, Coll. Vol. **4,** 146 (1963).

50. R. A. Carboni, *Org. Syn.*, Coll. Vol. **4,** 877 (1963).

51. B. C. McKusick and T. L. Cairns, Kirk-Othmer, *Encyclopedia of Chemical Technology*, 2nd ed., Vol. 6, Interscience Publishers, New York, 1965, p. 625.

52. T. L. Cairns *et al., J. Am. Chem. Soc.*, **80,** 2775 (1958).

53. R. E. Merrifield and W. D. Phillips, *J. Am. Chem. Soc.*, **80,** 2778 (1958); B. C. McKusick *et al., J. Am. Chem. Soc.*, **80,** 2806 (1958); W. J. Middleton *et al., ibid.*, **80,** 2832 (1958), and previous papers.

54. L. L. Gershbein and C. D. Hurd, *Org. Syn.*, Coll. Vol. **3,** 458 (1955).

55. C. L. Dickinson and L. R. Melby, *Org. Syn.*, Coll. Vol. **4,** 276 (1963).

56. W. J. Middleton, *Org. Syn.*, Coll. Vol. **4,** 243 (1963).

57. J. C. Sheehan and C. E. Mumaw, *J. Am. Chem. Soc.*, **72,** 2127 (1950).

E. Substitution Reactions

1. From an Arene and a Cyanylation Source (including Friedel-Crafts)

$$ArH + ArCHBrCN \xrightarrow{AlCl_3} (Ar)_2CHCN + HBr$$

The Friedel-Crafts reaction has found little use in the synthesis of nitriles. The first successful attempts as a preparative method were those of Karrer [1], who synthesized benzonitrile from benzene and cyanogen bromide with a 69 % yield based on the cyanogen bromide. More recently a series of diarylacetonitriles have been synthesized with yields usually in the 65–80 % range [2].

A patent is recorded on the direct introduction of the nitrile group into benzene, probably via a free radical mechanism [3]:

$$C_6H_6 + CuCN \xrightarrow[\substack{2.5 \text{ hr.} \\ \text{autoclave}}]{330°} C_6H_5CN + \text{polycyanobenzenes}$$

780 g.   180 g.

More recently, chlorosulfonylisocyanate has been used for introducing the nitrile group into arenes [4]:

86%
9-Cyanoanthracene

References for Section E are on p. 977.

The procedure is attractive because of its simplicity. Similar substitution occurs in olefins [5].

(a) Preparation of diphenylacetonitrile. Bromine, 17 ml. at 105–110°, was added to 35 g. of pure benzyl cyanide (chlorine-free) with stirring for 30 min. After another 30 min. at the same temperature, the crude bromobenzyl cyanide was dissolved in 150 ml. of benzene and 42 g. of powdered aluminum chloride was added during 20–30 min. in portions of a size sufficient to maintain a temperature of 45–50°. The reaction mixture was then heated for 1 hr. at 60–65°, after which most of the benzene was distilled off and the hot residue was treated with a mixture of 400 ml. of water, 400 g. of ice, and 20 ml. of hydrochloric acid. A steam distillation for 20 min. gave a reddish oil which soon solidified. Purification by solution in benzene, washing with water, drying, and distillation gave 49 g. of the nitrile, b.p. 121–125° (0.2 mm.). Crystallization from 1.2 parts of ethanol left 43 g. (74%), m.p. 75° [2]; *see also* [6].

2. From Aliphatic Compounds and a Cyanylation Source

$$RH + XCN \xrightarrow{h\nu} RCN + HX$$

Substitution of cyanide into aliphatic compounds is not an ordinary preparative procedure, but on occasion it is a direct pathway to inaccessible nitriles [7]. The general behavior is as follows: Cyanogen chloride can be excited directly by ultraviolet light or indirectly by means of a sensitizer, acetyl chloride or acetone. Yields are poor but conversions good, as is true of many photochemical reactions. Substitution into hydrocarbons is indiscriminate, except perhaps for partial preferential substitution of a tertiary hydrogen atom. Irradiation of heptane and cyanogen chloride gives a mixture of at least four octanonitriles and of tetrahydrofuran and cyanogen chloride a mixture of 2- and 3-cyanotetrahydrofurans (see Ex. a for experimental details).

Cyanamides may be prepared by thermal decomposition of cyanogen azide (see Ex. b.1) in the presence of an alkane

$$N_3CN + RH \xrightarrow{-N_2} [:NCN] \longrightarrow RNHCN$$

And, in a third type of substitution, acrylonitriles may be prepared from olefins and palladium cyanide (see Ex. b.2):

$$CH_2{=}CH_2 + Pd(CN)_2 \xrightarrow{autoclave} CH_2{=}CHCN$$

Unfortunately, higher homologs of ethylene or cyclic olefins give poorer yields of nitriles than ethylene itself because of an increased tendency for hydrogen disproportionation.

(a) Preparation of α-ethoxypropionitrile. One mole each of diethyl

$$\overset{\displaystyle CH_3}{\underset{\displaystyle |}{C_2H_5OC_2H_5 + XCN \longrightarrow C_2H_5OCHCN + HX}}$$

ether and cyanogen chloride, 0.05 mole of acetone (sensitizer), and 18 g. of finely divided sodium bicarbonate were stirred in a water-cooled vessel flushed with nitrogen. The mixture was irradiated for 1 hr. with a high-pressure mercury lamp, the temperature of the mixure being held at 15-20°. The excess cyanogen chloride was swept out with nitrogen, the sodium bicarbonate and sodium chloride filtered, and the filtrate dried and fractionated, yielding 17 g. of the desired compound, b.p. 129–131°. The yield was low, but the conversion high, on the basis of recovered cyanogen chloride [8].

(b) Other examples

(1) **N-(3-Methyl-3-pentyl)-cyanamide** (41 % from cyanogen chloride

$$
\underset{\underset{CH_3CH_2CHCH_2CH_3}{\overset{|}{\underset{}{}}}}{CH_3} + ClCN + NaN_3 \longrightarrow \underset{\underset{CH_2CH_3}{\overset{|}{\underset{}{}}}}{\overset{CH_3}{\overset{|}{NCNHCCH_2CH_3}}} + N_2 + NaCl
$$

and sodium azide in 3-methylpentane stirred at 0° (the concentration of cyanogen azide should be no more than 5 % in the hydrocarbon!); the excess cyanogen chloride was flushed out; the sodium chloride was removed by filtration; and the filtrate warmed to 50° or to a temperature where nitrogen gas evolution was evident) [9].

(2) **Acrylonitrile** (51 % from 0.03 mole of palladium(II) cyanide, 30 ml. of benzonitrile and ethylene at 55 kg./cm.^{2} for 5 hr. in an autoclave) [10].

3. From Triphenylcarbinols and Cyanoacetic Acid

$$
Ar_3COH + NCCH_2CO_2H \xrightarrow[CH_3COOH]{ZnCl_2} \left[\underset{19\text{-}X}{(Ar)_3C\overset{\overset{CN}{\overset{|}{}}}{CHCO_2H}} \right] \longrightarrow Ar_3CCH_2CN + CO_2
$$

This reaction is of limited scope but of interest in illustrating the attack of a carbonium ion on the enolic form, $NCCH{=}C\overset{\diagup OH}{\diagdown OH}$, of an active methylene compound. The type of reaction is not without precedent (Esters, Chapter 14, B.8). If zinc chloride is omitted, the intermediate acid, 19-X, can be isolated together with a product from the Ritter reaction involving attack of the carbonium ion on nitrogen of the nitrile group.

(a) **Preparation of β,β,β-tri(4-chlorophenyl)propionitrile.** A mixture of 728 g. of tri(4-chlorophenyl)carbinol, 340 g. of cyanoacetic acid, 720 g. of glacial acetic acid, and 137 g. of anhydrous zinc chloride was heated under reflux with stirring for 3 hr.; from the clear solution obtained, crystals of the nitrile separated on cooling; when filtered and washed with methanol, they weighed 678 g. (88 %), m.p. 170–172° [11].

1. P. Karrer and E. Zeller, *Helv. Chim. Acta*, **2**, 482 (1919); P. Karrer *et al.*, *ibid.*, **3**, 261 (1920).
2. D. Shapiro, *J. Org. Chem.*, **14**, 839 (1949).
3. W. G. Toland, Jr., U.S. Patent 2,780,637, Feb. 5, 1957; *C.A.*, **51**, 13922 (1957).
4. G. Lohaus, German Patent 1,259,893, Feb. 1, 1968; *C.A.*, **68**, 6557 (1968).
5. K. Matterstock and G. Lohaus, German Patent 1,253,704, Nov. 9, 1967; *C.A.*, **68**, 6596 (1968).
6. C. M. Robb and E. M. Schultz, *Org. Syn.*, Coll. Vol. **3**, 347 (1955).
7. A. Schönberg, *Preparative Organic Photochemistry*, Springer Verlag, New York, 1968, p. 260.
8. E. Muller and H. Huber, *Chem. Ber.*, **96**, 2319 (1963), and previous papers.
9. A. G. Anastassiou and H. E. Simmons, *J. Am. Chem. Soc.*, **89**, 3177 (1967).
10. Y. Odaira *et al.*, *J. Am. Chem. Soc.*, **88**, 4105 (1966).
11. M. Schorr, *Ann. Chem.*, **661**, 157 (1963).

F. Oxidation

1. From Amines (Dehydrogenation)

$$(Ar)RCH_2NH_2 \xrightarrow{NiO_2} (Ar)RCN$$

Various attempts have been made to dehydrogenate amines containing α-hydrogen atoms to nitriles [1]. One of the first observations in this connection was the formation of nitriles from aliphatic amides containing more than five carbon atoms in the Hofmann degradation, presumably via the primary amine first formed. This method, in which chloramine-T in 0.2 N sodium hydroxide or sodium hypochlorite is the reagent, has been employed successfully in converting α-amino acids to nitriles (see Ex. a) [2].

The use of nickel peroxide as the reagent appears to offer one of the most promising methods of dehydrogenation [3]. In this case the amine in benzene solution is simply stirred at room temperature or under reflux with the peroxide. A series of benzyl amines gave yields from 55.5 to 87.5%, while three long, straight-chain aliphatic amines gave yields from 73 to 95.8%.

A third method, applied to aliphatic amines containing α-methylene groups, is oxidation with lead tetraacetate (1 mole of amine to 2 of the acetate) in benzene [4]. This reaction probably proceeds as shown:

$$RCH_2NH_2 \xrightarrow{Pb(OAc)_4} [RCH{=}NH] \longrightarrow RCN$$

Yields do not exceed 61%. The same oxidizing agent converts both 2-amino-benzotriazole [5] and *o*-phenylenediamine [6] into *cis,cis*-muconitrile:

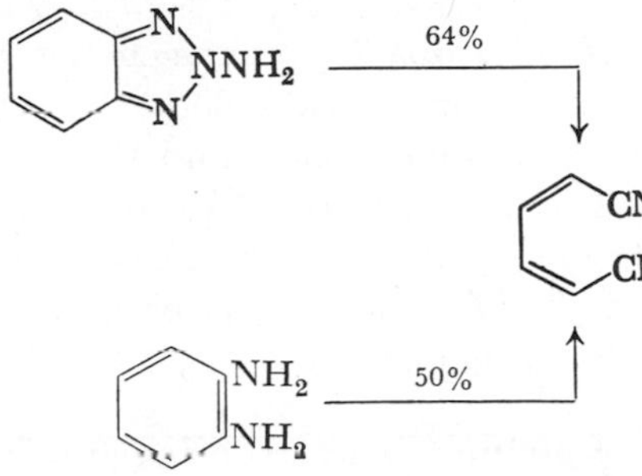

(a) Preparation of cyanomethylimidazole (73–76% of crude from

$$
\begin{array}{ccc}
\text{HC}\!\!=\!\!\text{C—CH}_2\text{CHCOOH} & & \text{HC}\!\!=\!\!\text{C—CH}_2\text{CN} \\
|\quad\quad | \qquad | & \xrightarrow{\text{NaOCl}} & |\quad\quad | \\
\text{HN}\quad\text{N}\quad\text{NH}_2\cdot\text{HCl} & & \text{HN}\quad\text{N} \\
\diagdown\;\diagup & & \diagdown\;\diagup \\
\text{C} & & \text{C} \\
\text{H} & & \text{H}
\end{array}
$$

histidine monohydrochloride monohydrate and sodium hypochlorite solution at 10–20°) [2].

(b) Preparation of p-methoxybenzonitrile (87.5% from p-methoxy-benzylamine and nickel peroxide in benzene) [3].

2. From Primary Alcohols or Aldehydes and Ammonia

$$
\text{RCH}_2\text{OH} \longrightarrow \text{RCHO} \xrightarrow[\text{oxid. agent}]{\text{NH}_3} \text{RCN}
$$

Several methods have been reported for a one-stage conversion of primary alcohols or aldehydes into nitriles. By the first method in the aliphatic series, butyl alcohol and ammonia passed over a catalyst containing 3% reduced nickel on activated aluminum oxide at 300° gives 81.5% butyronitrile [7]. In a second method aromatic aldehydes give nitriles with yields usually in the 70% range by interaction with ammonium dibasic phosphate, nitropropane, and glacial acetic acid [8]. By a third method primary aromatic alcohols or aldehydes may be converted into nitriles by oxidation in a methanolic solution of ammonia containing a strong base such as sodium methoxide and a copper complex [9]. With a series of aromatic aldehydes, which usually give better yields than the corresponding primary alcohols, the yields varied from 40 to 90%. A fourth method [10] consists in treating the aldehyde in benzene with ammonia and lead tetraacetate. By this procedure better results were also obtained with aromatic aldehydes, which gave yields varying from 64 to 90%.

In the last three methods it has been suggested that the Schiff base, ArCH=NH, is first formed and this product is then oxidized to the nitrile, ArCN.

(a) Preparation of benzonitrile. Oxygen was bubbled through 100 ml. of a solution of 4 mmoles of cupric chloride dihydrate, 400 mmoles of ammonia, 30 mmoles of sodium methoxide, and 100 mmoles of benzaldehyde in methanol at 30°. After 6 hr., the mixture was diluted with water, acidified, and extracted with ether. After the ether had been removed by evaporation and the unused aldehyde by formation of the bisulfite addition compound, a 79% yield of benzonitrile was obtained by distillation [9].

(b) Preparation of p-chlorobenzonitrile (90% from p-chlorobenzalde-hyde, ammonia, and lead tetraacetate in benzene) [10].

3. From Oxazoles

Oxazoles, made from acyloins and formamide, may be converted into cyanocarboxylic acids by light-activated oxygen [11]:

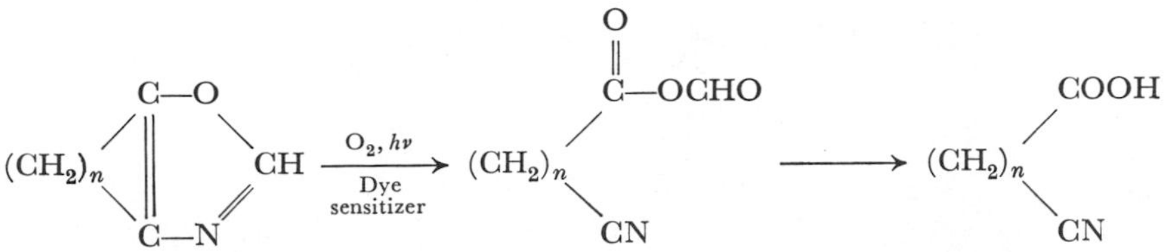

$$
(CH_2)_n \overset{C-O}{\underset{C-N}{\big|\big|}} CH \xrightarrow[\substack{Dye \\ sensitizer}]{O_2,\ h\nu} (CH_2)_n \overset{\overset{O}{\parallel}}{\underset{CN}{C-OCHO}} \longrightarrow (CH_2)_n \overset{COOH}{\underset{CN}{}}
$$

(a) **Preparation of δ-cyanovaleric acid** (4,5-tetramethylene oxazole ($n = 4$) was irradiated in CH_2Cl_2 with a 275-watt sunlamp for 12–14 hr. as dry oxygen was passed through the stirred solution containing the sensitizer methylene blue; the anhydride obtained on evaporation gave during workup the desired cyanocarboxylic acid; yield based on the oxazole was 80–90%) [11].

1. D. T. Mowry, *Chem. Rev.*, **42**, 247 (1948).
2. H. Bauer and H. Tabor, *Biochem. Prep.*, **5**, 97 (1957).
3. K. Nakagawa and T. Tsuji, *Chem. Pharm. Bull.* (*Tokyo*), **11**, 296 (1963); *C.A.*, **59**, 3827 (1963).
4. N. Lj. Mihailović *et al.*, *Tetrahedron Letters*, 461 (1965).
5. C. D. Campbell and C. W. Rees, *Chem. Commun.*, 192 (1965).
6. K. Nakagawa and H. Onoue, *Chem. Commun.*, 396 (1965).
7. M. A. Popov and N. I. Shuĭkin, *Izv. Akad. Nauk SSSR, Otd. Khim. Nauk*, 713 (1958); *C.A.*, **52**, 19924 (1958).
8. H. M. Blatter *et al.*, *J. Am. Chem. Soc.*, **83**, 2203 (1961); *Org. Syn.*, **43**, 58 (1963).
9. W. Brackman and P. J. Smit, *Rec. Trav. Chim.*, **82**, 757 (1963).
10. K. N. Parameswaran and O. M. Friedman, *Chem. Ind.* (*London*) 988 (1965).
11. H. H. Wasserman and E. Druckrey, *J. Am. Chem. Soc.*, **90**, 2440 (1968).

Chapter 20

NITRO COMPOUNDS

A. Electrophilic Reactions

1. From Aliphatic Compounds (Substitution)

$$RH + HNO_3 \rightarrow RNO_2 + H_2O$$

This synthesis has been widely employed and as a result the literature on the subject is extensive. Reviews have been published [1]. It appears that the

References for Section A are on pp. 990–991.

reaction proceeds by a free radical mechanism, frequently via nitrogen dioxide [2]. Some interesting, seemingly dormant, work is to be found in the Houben-Weyl series [3] and will be included in the following discussion.

Alkanes

Alkanes may be nitrated by nitric acid in the liquid or vapor phase, although the latter is the more important of the two processes. For rapid reaction the temperature is usually in the 375–450° range. Under these conditions an alkane such as n-butane produces all the possible mononitrobutanes as well as some 1-nitropropane, nitroethane, and nitromethane. The formation of such mixtures detracts from the method in laboratory preparative work, although it is the most important technical method for production of nitroalkanes. Contrary to present views which imply that dilute nitric acid is simply an oxidizing agent, the older German literature demonstrates that substantial quantities of nitroalkanes may be obtained by simply heating the hydrocarbon with dilute nitric acid, $d = 1.05$, in a closed flask at 105–108° (see Ex. a). Both aliphatic and araliphatic hydrocarbons may be nitrated by this procedure. It is true, however, that nitrogen pentoxide, which must be diluted in carbon tetrachloride because of its high activity, may be an equally satisfactory, if not superior, reagent [4]:

$$C_6H_{12} + N_2O_5 \xrightarrow[\text{Reflux}]{\text{CCl}_4} C_6H_{11}NO_2 + C_6H_{11}ONO_2$$

$$\text{Nitrocyclo-} \quad \text{Cyclohexyl}$$
$$\text{hexane, 39\%} \quad \text{nitrate, 41\%}$$

Nitryl chloride in cyclohexane with light activation does not give nitrocyclohexane, but rather cyclohexanone oxime; chlorocyclohexane is a by-product [5].

Polynitroalkanes are produced by more indirect methods, as indicated below. Tetranitromethane may be prepared by the nitration of acetic anhydride [6]:

$$4\,(CH_3CO)_2O + 4\,HNO_3 \xrightarrow[\text{25° for 7 days}]{\text{10°, then}} C(NO_2)_4 + 7\,CH_3COOH + CO_2$$

$$d = 1.52 \qquad\qquad 57\text{--}65\%$$

although a higher yield (90%) is obtained from the reaction of ketene with nitric acid [7]. Chlorotrinitromethane is made by the nitration of acetylene with fuming nitric acid and catalytic amounts of mercuric nitrate [8] followed by the addition of concentrated hydrochloric acid.

$$HC{\equiv}CH + HNO_3 \xrightarrow[\text{30°. 8 hr.}]{\text{Hg(NO}_3)_2} HC(NO_2)_3 \xrightarrow[\text{HNO}_3]{\text{HCl}} ClC(NO_2)_3$$

Hexanitroethane may be made in 90% yield by nitration of the intriguing compound, dipotassium salt of tetranitroethane [9, 10]:

$$\text{Picric acid} \xrightarrow{\text{Ca(OBr)}_2} CBr_3NO_2 \xrightarrow[\text{KNO}_2, \, 30°]{\text{KCN}}$$

$$\xrightarrow[\text{H}_2\text{SO}_4]{\text{HNO}_3} (NO_2)_3C{-}C(NO_2)_3$$

$$90\%, \text{ last step}$$

These polynitro compounds were handled with little concern when originally prepared, but they should be treated with respect. Hexanitroethane has been considered a propellant component of rocket fuel, and modern laboratories in which these compounds are investigated use hoods in isolated rooms with fall-out walls.

Active Methylene Compounds

A specific reaction is obtained in the nitration of active methylene compounds. For example, in the method of choice ethyl nitromalonate may be obtained in 92% yield from ethyl malonate and fuming nitric acid [11], while the nitration of diethyl isobutylmalonate with 100% nitric and polyphosphoric acids gives the isobutylmalonate ester in 78% yield [12]. A very convenient reagent for nitration of active methylene compounds in basic solution is acetone cyanohydrin nitrate (see Ex. c.1). In reality this reaction can be viewed as an SN_2 displacement:

$$(RCO)_2CH^{\ominus} + NO_2O\underset{\underset{CH_3}{|}}{\overset{\overset{CH_3}{|}}{C}}CN \longrightarrow (RCO)_2CHNO_2 + O=\underset{\underset{CH_3}{|}}{\overset{\overset{CH_3}{|}}{C}} + CN^{\ominus}$$

Ethyl nitrate behaves in somewhat the same way but perhaps with not the same efficiency (see B.3).

Amines and Amides

Acetone cyanohydrin nitrate may be used to make nitroamines, $RNHNO_2$ (see Ex. c.2), but dinitrogen pentoxide is an equally efficient reagent [13]. The preparation of dinitrogen pentoxide is documented [14]. Some amides can be nitrated with nitric acid [15]:

$$NO_2NHCONHCONH_2 + HNO_3 \ (d = 1.52) \xrightarrow[\substack{\text{in the dark under vacuum} \\ \text{over soda lime and } H_2SO_4}]{\text{Evaporate in desiccator}}$$

$$NO_2NHCONHCONHNO_2 \text{ as residue}$$

Dinitrobiuret, 80% yield after
recrystallization from methanol

Alcohols

Alcohols may be nitrated to nitrate esters. It is such a well-known reaction that only one of the lastest references dealing with this subject will be given [16].

(a) Preparation of α-nitroethylbenzene. Ethylbenzene and 5–6 parts of dilute nitric acid, $d = 1.075$, were placed in a closed flask and heated to 105–108° for several hours. The product after extraction and fractionation was obtained in 44% yield [17]; *see also* [18] for extensions.

(b) Preparation of diethyl isobutylnitromalonate (78% from diethyl isobutylmalonate and 100% nitric and polyphosphoric acids at 60°) [12].

(c) Other examples

(1) Diethyl nitromalonate (45 % from 0.3 mole of diethyl malonate, 0.3 mole of sodium hydride in THF to which 0.1 mole of acetone cyanohydrin nitrate was added and the mixture stirred under reflux for 2 hr.; unreacted NaH was destroyed with alcohol and the mixture poured into acid and ice (*Caution!* HCN), and extracted with ether; the ether was then extracted with 10 % sodium hydroxide, the alkaline layer acidified, and the oil distilled; sodium ethoxide cannot be used because it reacts with the nitro ester) [19].

(2) N-Nitromorpholine (57–64 % from 0.4 mole of morpholine and 0.2 mole of acetone cyanohydrin nitrate heated carefully to 60° whereupon the temperature rose to 110° and then was held at 80° for 1 hr.; the mixture was poured into 10 % acid (*Caution!* HCN) and extracted with methylene chloride; the solid in the concentrate was recrystallized from ethanol to give a product, m.p. 52–54°; for larger-scale or for other amines, acetonitrile may be used as the moderating solvent) [20].

2. From Aromatic Compounds

$$ArH + HNO_3 \rightarrow ArNO_2 + H_2O$$

Barring the Friedel-Crafts reaction, nitration is the most versatile and productive of all aromatic substitution processes. General references are [21, 22].

Nitric acid, density 1.42, 68 %, and fuming nitric acid, density 1.49, 98 %, may be used alone with easily nitrated nuclei. Colorless fuming nitric acid may be obtained by distillation below 0° under reduced pressure, but at 25° it turns red from nitrogen oxides. Anhydrous nitric acid ($d = 1.51$) is made by distillation of ordinary fuming nitric acid from an equal volume of sulfuric acid [6]. The most commonly used reagent is a mixed acid (nitric and sulfuric using ordinary acids for simple nitrations and fuming acids for difficult or multinitrations). The choice is sometimes difficult to make. One of the best ways to control the amount of nitrating agent and to minimize water accumulation is to use a mixture of potassium nitrate and sulfuric acid. Water, of course, does not affect the nitration process directly because nitration is an irreversible reaction, but it does affect the type of nitrating agent. The most powerful nitrating agent should produce the nitronium ion, $\overset{\oplus}{NO_2}$, in a medium as devoid of anions as possible and the few anions that are present should be stable ones, such as $\overset{\ominus}{BF_4}$, $\overset{\ominus}{AlCl_4}$ and $\overset{\ominus}{HSO_4}$; of these agents the most important is nitronium tetrafluoroborate [23, 24]. In aprotic solvents such as sulfolane such agents behave as though the aryl group was displacing the $BF_4^{\ominus}$ group from the ion pair:

$$ArH + NO_2BF_4 \rightarrow ArNO_2 + HBF_4$$

In strongly acidic media, such as 100 % sulfuric acid, nitronium tetrafluoroborate is a powerful nitrating agent, a fact which suggests a higher concentration of nitronium ion or, less likely but not impossible, the intervention of the protonated nitronium ion, $\overset{(2+)}{NO_2H}$. (For a free radical mechanism of nitration see [2].)

Nitronium tetrafluoroborate is preferred for nitrating aromatic compounds, such as nitriles, which are susceptible to acid-catalyzed hydrolysis. Although nitriles may be mononitrated by the usual reagents, they cannot be dinitrated by them. On the other hand, nitronium tetrafluoroborate in nonaqueous, acid-free systems permits dinitration to occur [23, 25]; see also Ex. b.8. Nitronium pyrosulfate $(NO_2HS_2O_7)$ in 100% sulfuric acid is capable of forming picryl fluoride in 40% yield from 2,4-dinitrofluorobenzene (see Ex. b.6). Another method of forcing nitration is to raise the temperature, but limitation is placed on heating because of the powerful oxidizing components present. In general, any nitration process carried on above a temperature of 130–150° should be considered hazardous.

On the other end of the activity scale, nitric acid in acetic acid is a mild nitrating agent useful for preparing nitropolyalkylbenzenes. Nitric acid in carbon tetrachloride is another mild nitrating agent which has the unusual property of nitrating mesitylene about 300 times faster at 0° than at 40° [26]. This behavior is explained by the assumption that more highly associated forms of nitric acid (at lower temperatures) may be able to generate a relatively larger concentration of nitronium ion:

$$(HNO_3)_x H_2 \overset{\oplus}{N}O_3 \rightleftarrows NO_2{}^{\oplus} + H_2O + (HNO_3)_x$$

Even milder methods, but somewhat specific for phenols and anilines, involve the use of tetranitromethane (see Ex. b.14 and a recent discussion [27]) and dilute nitrous acid to give nitrosation preponderantly in the *para* position followed by oxidation to the nitro compound (see Ex. b.16 and also Phenols, Chapter 5, D.3 and D.4).

Although orientation rules and mechanisms have been discussed for other aromatic substitutions (Alkanes, Chapter 1, D.1, and Halides, Chapter 7, D.1 and D.5), discussion in this chapter is limited to characteristics of nitration. Nitration of alkylbenzenes yields mainly *o*- and *p*-nitroalkylbenzenes. The amount of *meta*-nitrotoluene is about 2%, but can be raised to 4.3% at 60°. The second nitro group orients *meta* to the first, if such a position is available, but exceptions are known [28]:

| | 60–80% | 40–20% |
| | 1,4-Dimethyl-2,3-
dinitrobenzene | 1,4-Dimethyl-2,6-
dinitrobenzene |

Although in normal cases the *meta*-dinitro derivative is in great predominance, among the minor substitutents the *o/p* ratio of dinitro derivatives is always greater than 1 because the position *para* to the first nitro group is the most deactivated position. For instance, the nitration of nitrobenzene yields 93% *m*-, *ca.* 6% *o*-, and less than 1% *p*-dinitrobenzene (see [29] for a more extensive list).

Phenols and anilines are nitrated with great ease and, moreover, are subject to manipulation in the orientation of the entering nitro group. Acetanilide with mixed acid yields about 90 % p-nitroacetanilide. The nitration of aniline in 100 % sulfuric acid yields 64 % m- and 36 % p-nitroanilines. Pure m-nitroaniline is more conveniently prepared by reduction of m-dinitrobenzene. Nevertheless, the strength of acid employed can be a useful means of controlling orientation in nitration [30]:

Weak acid method

$$
CH_3O\text{—}C_6H_4\text{—}NHCOCH_3 \text{ in } CH_3COOH\text{—}(CH_3CO)_2O \xrightarrow[\substack{2.\ H^\oplus, H_2O \\ 3.\ Base}]{\substack{1.\ H_2SO_4 + KNO_3 \\ \text{added to solution}}} CH_3O\text{—}C_6H_3(NO_2)\text{—}NH_2
$$

2-Nitro-4-anisidine, 70%

Strong acid method:

$$
CH_3O\text{—}C_6H_4\text{—}NHCOCH_3 \text{ in } CH_3COOH \xrightarrow[\substack{3.\ Base}]{\substack{2.\ H^\oplus, H_2O}} CH_3O\text{—}C_6H_3(NO_2)\text{—}NH_2
$$

added to $H_2SO_4 + KNO_3$

3-Nitro-4-anisidine, unstated yield

Employing the principle of nuclear attack from the position of the substituent, one should obtain high proportions of o-nitroanilines from nitramines [31]:

$$
C_6H_5NHNO_2 \xrightarrow[\text{at freezing point}]{H_2SO_4\cdot H_2O} o\text{-} + p\text{-}NH_2C_6H_4NO_2
$$

93%   7%
Total yield, 90%

Nitramines may be prepared by the acetone cyanohydrin nitrate method (A.1) or in limited cases by nitrosation of secondary amines followed by simultaneous oxidation and nuclear attack (see Ex. b.15). It is claimed also that o- and p-nitroacetanilides in ratios of 4.0 to 9.1 are obtained in fair yields by nitration of the appropriate acetanilide with acetyl nitrate (nitric acid in acetic anhydride) [32]. This report is partially substantiated by the 49–55 % yield of 2-amino-3-nitrotoluene obtained by nitration of o-toluidine [33]. Acetyl nitrate may form a hydrogen-bonded species with the acetamido grouping to guide the nitrating agent to the *ortho* position.

Nitration of phenols may be carried out with dilute nitric acid to yield a high proportion (*ca.* 60%) of o-nitrophenol [34]. The *ortho* isomer may be easily separated from the *para* by its much greater volatility. m-Nitrophenol is best prepared from m-nitroaniline. The oxynitration of hydrocarbons to nitrophenols has been discussed under Phenols, Chapter 5, D.4.

Naphthalene is readily nitrated to give about 10 parts of α- to 1 part of β-nitronaphthalene [35]. The second nitro group enters the unsubstituted ring to yield about 2 parts of 1,8-dinitro- to 1 part of 1,5-dinitronaphthalene [36]. The orientation pattern is similar for other polynuclear compounds, the first nitro group enters very readily at the most active site and the second nitro group enters a different ring of the polynuclear compound.

The nitration of heterocyclic compounds covers such a vast field that only brief discussion is given here. The heterocyclic treatises [37, 38] and the review on nitration of heterocyclic nitrogen compounds [39] should be consulted for specific examples. Nitration conditions range from the very mildest with heterocycles of low aromaticity such as thiophene [40] or pyrrole [39] to the most vigorous with nitrogen heterocycles of high aromaticity and deactivation by salt formation. For example, pyridine in fuming sulfuric acid with potassium nitrate at 300° yields about 20 % 3-nitropyridine [41]. No doubt, improvements could be made on this nitration. On the other hand, the nitration of pyridine N-oxide proceeds readily to give 4-nitropyridine-1-oxide (see Ex. b.2).

Some of the nitration products obtained in *Organic Syntheses*, Collective Volumes, are listed: *m*-nitroacetophenone, **2,** 434; *m*-nitrobenzaldehyde, **3,** 644; methyl *m*-nitrobenzoate, **1,** 372; 3,5-dinitrobenzoic acid, **3,** 337; 4-nitrophthalimide, **2,** 459; 2-nitrofluorene, **2,** 447; *p*-nitrophenylacetonitrile, **1,** 396; dinitrodurene, **2,** 254; or, from the single volumes, *o*-amino-*p*′-nitrobiphenyl from *o*-aminobiphenyl, **46,** 85 (1966); 2,4,5,7-tetranitrofluorenone, **42,** 95 (1962); 1,3-dinitro-4,6-dichlorobenzene, **40,** 96 (1960). Some unusual nitrations are to be found in Exs. b.12 through b.16.

(a) Preparation of 2,6-dinitroaniline (30–36 % based on chlorobenzene converted first into 1-chloro-2,6-dinitro-4-benzenesulfonic acid potassium salt with potassium nitrate and sulfuric acid, then into the corresponding amino derivative with ammonium hydroxide, and finally into 2,6-dinitroaniline with aqueous sulfuric acid) [42].

(b) Other examples

    (1) 6-Nitroveratraldehyde (73–79 % from veratraldehyde and conc. nitric acid at 18–22°) [43].

    (2) 3-Methyl-4-nitropyridine-1-oxide (70–73 % from 3-methylpyridine-1-oxide, fuming nitric acid, and conc. sulfuric acid) [44].

    (3) 2-Nitro-4-cymene (78–82 % from *p*-cymene, glacial acetic, and conc. sulfuric-nitric acids at −10 to −15°) [45].

    (4) Picric acid (90 % from 1 mole of phenol disulfonated with 300 g. of 100 % sulfuric acid at 100–110°, cooled to 0°, after which 3.5 moles of nitric acid in the form of 50 % mixed acid was added dropwise; a 55 ml. portion was heated to 110–125° until foaming ceased and then the rest added dropwise to the 55-ml. portion held at the same temperature) [46].

    (5) 4-Nitro-1,3-xylene (98 % from 1 mole of *m*-xylene, 1.1 mole each of 81 % sulfuric acid and conc. nitric acid at 30° for 1 hr.) [47].

    (6) Picryl fluoride (40 % from 0.05 mole of 2,4-dinitrofluorobenzene, 60 g. of 100 % sulfuric acid, 0.15 mole of nitronium tetrafluoroborate (or pyrosulfate) heated to 110–120° for 12 hr.) [23].

    (7) *p*-Chloronitrobenzene (89 % from 0.1 mole of titanium tetrachloride and 0.4 mole of chlorobenzene at 0° to which 0.2 mole of nitryl chloride was added and the mixture allowed to warm to room temperature) [23].

(8) 3,5-Dinitro-2-tolunitrile (48–53% from *o*-tolunitrile and nitronium tetrafluoroborate in tetramethylenesulfone) [48].

(9) *m*-Dinitrobenzene (87% from 0.25 mole of nitrobenzene and 0.3 mole of fuming nitric acid saturated with 17 g. of boron trifluoride and heated to 80°) [49].

(10) Trinitrobenzene. "Good yield" from *m*-dinitrobenzene in fuming nitric and 60% fuming sulfuric acid at 110° [50]. This compound ordinarily is prepared from the oxidation of 2,4,6-trinitrotoluene followed by decarboxylation of the acid formed [51].

(11) 5-Bromo-8-nitroisoquinoline (96% from 5-bromoisoquinoline and potassium nitrate in sulfuric acid) [52].

(12) *m*-Nitrobenzoyl chloride (90% from benzoyl chloride and nitrogen pentoxide in carbon tetrachloride containing phosphorus pentoxide in suspension at −10°; no hydrolysis takes place; nitrogen pentoxide may act violently with aromatic compounds if not diluted with carbon tetrachloride) [53].

(13) *o*-Dinitrobenzene (91.7% by treatment of 1 g. of *o*-nitrophenylmercuric chloride with 30 ml. of HNO_3, *d* 1.386, at 95° for 10 min.; the mercuration of nitrobenzene with mercuric acetate yields a mixture of 37% *o*- and 57% *m*-derivatives from which the former may be separated by extracting with cold acetone) [54].

(14) 2-Nitro-4-methyldimethylaniline-nitroform salt (71% from 1 equiv. each of N,N-dimethyl-*p*-toluidine and boric acid in alcohol to which 1 equiv. of tetranitromethane was added; this reagent nitrates phenols in pyridine) [55].

(15) *p'*-(*o*-Nitrophenylamino)-azobenzene-*p*-sulfonic acid (Azo yellow). (Methylene green is made similarly from methylene blue [56].)

$$HO_3SC_6H_4N{=}NC_6H_4NHC_6H_5 \xrightarrow[H_2O, 5°]{NaNO_2}$$

$$HO_3SC_6H_4N{=}NC_6H_4\overset{NO}{\underset{|}{N}}C_6H_5 \xrightarrow[\text{Raise temp. slowly to 68°}]{60\% HNO_3} HO_3SC_6H_4N{=}NC_6H_4NH{-}\langle\text{(NO}_2)\rangle$$

"Good yield"

(16) 2-Chloro-6-nitrothymol (from 2 g. of 6-chlorothymol in hexane floated for a week on a considerable volume of water containing sodium nitrite and excess hydrochloric acid; this procedure is useful for phenols or anilines which oxidize readily) [57].

(17) 5,7-Dinitro-8-hydroxyquinoline (54% from 0.6 g. of 5-nitro-8-hydroxyquinoline, 0.67 g. of sodium nitrite refluxed for 15 min. in 200 ml. of 1% nitric acid) [58].

(18) 3-Hydroxy-2-nitropyridine (74% from 1 mole of 3-hydroxypyridine in 650 ml. of conc. H_2SO_4 to which a mixture of 48 ml. of HNO_3, *d* 1.5, and 92 ml. of conc. H_2SO_4 was added at such a rate so as to hold the temperature at 40–45°; if an alkyl group is in the 2-position of 3-hydroxypyridine, the nitration occurs largely at the 4-position) [59].

3. From Olefins (Addition) and Their Adducts (Elimination)

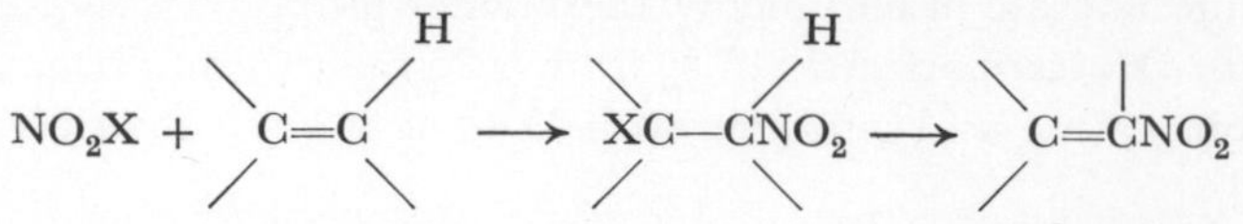

$$NO_2X + \quad \overset{\diagdown}{\underset{\diagup}{C}}=\overset{H}{\underset{\diagdown}{C}} \quad \longrightarrow \quad XC-\overset{H}{\underset{\diagdown}{C}}NO_2 \longrightarrow \quad \overset{\diagdown}{\underset{\diagup}{C}}=\overset{|}{C}NO_2$$

Nitryl reagents which add well to olefins are nitryl chloride [60]; *see also* [61], dinitrogen tetroxide, N_2O_4 [62], and dinitrogen trioxide [63]. Fuming nitric acid, acting probably via the nitrogen oxides, has been used to nitrate unsaturated steroids (see Ex. b.6).

Nitryl chloride, having a dipole $NO_2{}^{+\delta}Cl^{-\delta}$, adds in a straightforward manner to unsaturated compounds [64].

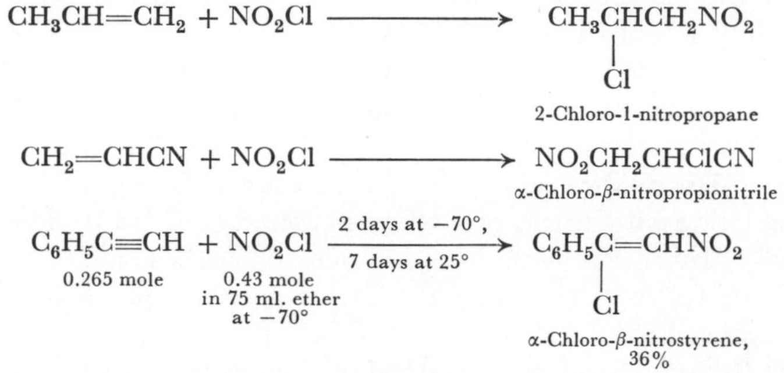

$$CH_3CH{=}CH_2 + NO_2Cl \longrightarrow \underset{\underset{Cl}{|}}{CH_3CHCH_2NO_2}$$

2-Chloro-1-nitropropane

$$CH_2{=}CHCN + NO_2Cl \longrightarrow NO_2CH_2CHClCN$$

α-Chloro-β-nitropropionitrile

$$C_6H_5C{\equiv}CH + NO_2Cl \xrightarrow[\text{7 days at } 25°]{\text{2 days at } -70°,} \underset{\underset{Cl}{|}}{C_6H_5C{=}CHNO_2}$$

0.265 mole   0.43 mole
in 75 ml. ether
at −70°

α-Chloro-β-nitrostyrene,
36%

But certain aspects of dinitrogen tetroxide or trioxide addition suggest free radical characteristics (see Exs. a and b.2; also [65]):

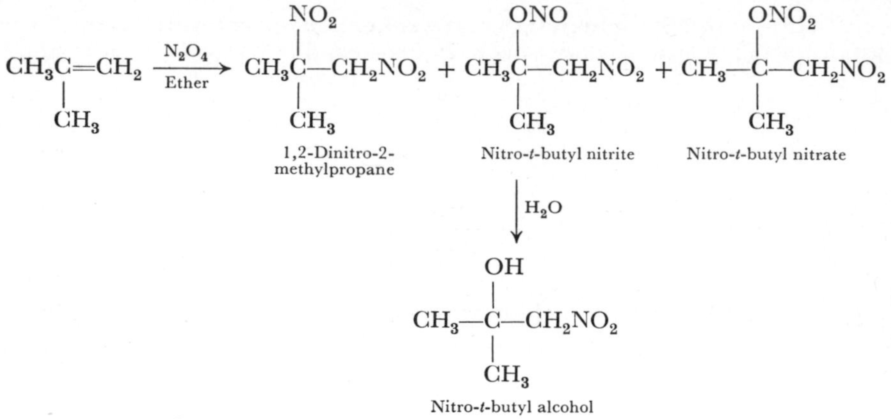

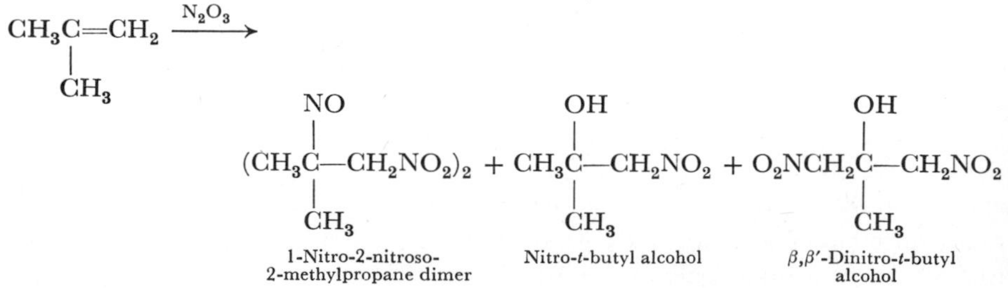

$$\underset{\underset{CH_3}{|}}{CH_3C{=}CH_2} \xrightarrow[\text{Ether}]{N_2O_4} \underset{\underset{CH_3}{|}}{CH_3\overset{\overset{NO_2}{|}}{C}-CH_2NO_2} + \underset{\underset{CH_3}{|}}{CH_3\overset{\overset{ONO}{|}}{C}-CH_2NO_2} + \underset{\underset{CH_3}{|}}{CH_3-\overset{\overset{ONO_2}{|}}{C}-CH_2NO_2}$$

1,2-Dinitro-2-
methylpropane   Nitro-*t*-butyl nitrite   Nitro-*t*-butyl nitrate

$$\downarrow H_2O$$

$$\underset{\underset{CH_3}{|}}{CH_3-\overset{\overset{OH}{|}}{C}-CH_2NO_2}$$

Nitro-*t*-butyl alcohol

$$\underset{\underset{CH_3}{|}}{CH_3C{=}CH_2} \xrightarrow{N_2O_3}$$

$$\left(\underset{\underset{CH_3}{|}}{CH_3\overset{\overset{NO}{|}}{C}-CH_2NO_2}\right)_2 + \underset{\underset{CH_3}{|}}{CH_3\overset{\overset{OH}{|}}{C}-CH_2NO_2} + \underset{\underset{CH_3}{|}}{O_2NCH_2\overset{\overset{OH}{|}}{C}-CH_2NO_2}$$

1-Nitro-2-nitroso-
2-methylpropane dimer   Nitro-*t*-butyl alcohol   β,β′-Dinitro-*t*-butyl
alcohol

The dinitroalkane is the predominant product in dinitrogen tetroxide additions. Nitroolefins may be prepared from the above adducts with or without isolation of the adducts. Basic reagents such as anhydrous sodium acetate [60], pyridine [60, 66], N,N-dimethylaniline [60], potassium hydroxide [67], and, best of all, triethylamine (see Ex. b.3) have been employed for nitrohalides; anhydrous sodium acetate and acetic anhydride [68], sodium carbonate [69], and sodium bicarbonate [70] for nitroacetates (for preparation see Ex. b.7); and sodium hydroxide [63] for dinitro compounds. Yields of the nitroolefins are in the 70–95 % range. In the case of 1-chloro-1-nitropropane with potassium hydroxide, 3,4-dinitro-3-hexene was obtained in small yield [67].

Elimination from Mannich bases prepared from nitroalkanes also is a means of preparing nitroolefins [71]:

$$CH_3CH_2NO_2 + R_2NH + H_2C{=}O \longrightarrow CH_3\overset{\overset{\displaystyle NO_2}{|}}{C}HCH_2NR_2 \xrightarrow[\substack{105° \text{ under reduced} \\ \text{pressure}}]{BF_3 \cdot (C_2H_5)_2O}$$

$$CH_3CNO_2{=}CH_2$$
2-Nitropropene, 70–77%

(a) **Preparation of methyl 2-chloro-3-nitropropionate** (70% from methyl acrylate in ether at 0° through which nitryl chloride was bubbled; impurities which must be separated by fractionation are methyl 2,3-dichloro- and perhaps 2,3-dinitropropionate) [60].

(b) **Other preparations**

(1) **2-Chloro-1-nitropropane** (40% from propylene in CCl$_4$ at 0° through which NO$_2$Cl was passed) [72].

(2) **1,2-Dinitro-2-methylpropane** (42% from isobutylene in ether at 0° through which dinitrogen tetroxide was passed; other products, after hydrolysis, were 1-nitro-2-methyl-2-propanol, 28%, and small amounts of 1,3-dinitro-2-methyl-2-propanol; dinitrogen trioxide with 29 g. of isobutylene treated in the same manner gave 16.6 g. of the dimer of 1-nitro-2-nitrosoiso-butane, 8.9 g. of nitro-*t*-butyl alcohol, and 3 g. of 1,3-dinitro-2-methyl-2-propanol) [63].

(3) **1-Nitrocyclooctene** (96% from 0.43 mole of N$_2$O$_4$ in 150 ml. of ether at 9–12° to which 0.4 mole of cyclooctene was added in 24 min. while 13 mmoles of O$_2$ was bubbled through the mixture (more O$_2$ increases the amount of the side product, nitroketone; no O$_2$ gives a lower yield of product); triethylamine, 1.2 mole, was then added at 4–12°) [73].

(4) **1-Chloro-1-bromo-2-nitroethane** (85% from vinyl bromide and nitryl chloride at low temperature) [74].

(5) **Methyl 3-nitroacrylate** (92% from methyl 2-chloro-3-nitro-propionate (see Ex. a) and anhydrous sodium acetate stirred in ether) [60].

(6) **6-Nitrocholesteryl acetate** (79% from cholesteryl acetate and fuming nitric acid in ether followed by treatment with aqueous sodium hydroxide) [75].

(7) *trans*-1-Acetoxy-2-nitro-1-phenylcyclohexane (49% from 0.05 mole of 70% of HNO$_3$ in 35 ml. of acetic anhydride at $-20°$ to which 0.025 mole of 1-phenylcyclohexene in 10 ml. of acetic anhydride was added in 5 min.; chromatographing on a silica gel column gave in addition nitroolefins and 19% of the titled *cis* isomer) [76].

1. H. B. Hass and E. F. Riley, *Chem. Rev.*, **32**, 373 (1943); N. Levy and J. D. Rose, *Quart. Rev. (London)*, **1**, 358 (1947); J. L. Martin and P. J. Baker, Jr., Kirk-Othmer, *Encyclopedia of Chemical Technology*, 2nd ed., Vol. 13, Interscience Publishers, New York, 1967, p. 864; A. V. Topchiev, *Nitration of Hydrocarbons and Other Organic Compounds*, Pergamon Press, New York, 1959; N. Kornblum, *Org. Reactions*, **12**, 101 (1962); P. Noble, Jr., *et al.*, *Chem. Rev.*, **64**, 19 (1964).

2. A. I. Titov, *Tetrahedron*, **19**, 557 (1963).

3. J. Houben, *Die Methoden der Organischen Chemie*, G. Thieme Verlag, Leipzig, 3rd ed., Vol. 4, 1941, pp. 131–341.

4. A. I. Titov and N. V. Shchitov, *Dokl. Akad. Nauk, SSSR*, **81**, 1085 (1951); *C.A.*, **46**, 7992 (1952).

5. E. Müller and H.-G. Padeken, *Chem. Ber.*, **99**, 2971 (1966).

6. P. Liang, *Org. Syn.*, Coll. Vol. **3**, 803 (1955).

7. G. Darzens and G. Lévy, *Compt. Rend.*, **229**, 1081 (1949).

8. A. K. Macbeth and D. D. Pratt, *J. Chem. Soc.*, **119**, 354 (1921).

9. F. Allsop and J. Kenner, *J. Chem. Soc.*, **123**, 2296 (1923); L. Hunter, *ibid.*, **123**, 543 (1923).

10. Ref. 3, p. 179.

11. D. I. Weisblat and D. A. Lyttle, *J. Am. Chem. Soc.*, **71**, 3079 (1949).

12. J. P. Kispersky and K. Klager, *J. Am. Chem. Soc.*, **77**, 5433 (1955).

13. T. E. Stevens *et al.*, *J. Org. Chem.*, **23**, 311 (1958).

14. N. S. Gruenhut *et al.*, *Inorg. Syn.*, **3**, 78 (1950); J. H. Robson, *J. Am. Chem. Soc.*, **77**, 107 (1955).

15. J. Thiele and E. Uhlfelder, *Ann. Chem.*, **303**, 93 (1898).

16. L. T. Eremenko, *Tetrahedron*, **20**, Suppl. 1, 37 (1964).

17. Ref. 3, 159.

18. M. Konowalow, *Chem. Ber.*, **28**, 1850 (1895).

19. W. D. Emmons and J. P. Freeman, *J. Am. Chem. Soc.*, **77**, 4391 (1955).

20. J. P. Freeman and I. G. Shepard, *Org. Syn.*, **43**, 83 (1963).

21. P. B. D. de la Mare and J. H. Ridd, *Aromatic Substitution, Nitration and Halogenation*, Academic Press, New York, 1959.

22. A. V. Topchiev, *Nitration of Hydrocarbons and Other Organic Compounds*, Pergamon Press, New York, 1959; L. F. Albright, in Kirk-Othmer, *Encyclopedia of Chemical Technology*, 2nd ed., Vol. 13, The Interscience Encyclopedia, Inc., New York, 1967, p. 784; H. A. Lubs, *Chemistry of Synthetic Dyes and Pigments*, Reinhold Publishing Corp., New York, 1955, pp. 12, 71, 350.

23. S. J. Kuhn and G. A. Olah, *J. Am. Chem. Soc.*, **83**, 4564 (1961).

24. G. A. Olah *et al.*, *J. Am. Chem. Soc.*, **84**, 3687 (1962).

25. G. A. Olah *et al.*, *J. Chem. Soc.*, 4257 (1956).

26. T. G. Bonner *et al.*, *Tetrahedron Letters*, 1665 (1968).

27. T. C. Bruice *et al.*, *J. Am. Chem. Soc.*, **90**, 1612 (1968).

28. K. A. Kobe and T. B. Hudson, *Ind. Eng. Chem.*, **42**, 356 (1950).

29. Ref. 21, p. 83.

30. A. H. Blatt and N. Gross, *J. Am. Chem. Soc.*, **75**, 1245 (1953).

31. E. D. Hughes *et al.*, *J. Chem. Soc.*, 2678 (1950); 4354 (1958).

32. B. M. Lynch *et al.*, *Canad. J. Chem.*, **46**, 1141 (1968).

33. J. C. Howard, *Org. Syn.*, Coll. Vol. **4**, 42 (1963).

34. F. Arnall, *J. Chem. Soc.*, **125**, 811 (1924).

35. M. J. S. Dewar and T. Mole, *J. Chem. Soc.*, 1441 (1956).

36. R. C. Fuson, *Reactions of Organic Compounds*, John Wiley & Sons, 1962, p. 36.

37. R. C. Elderfield, *Heterocyclic Compounds*, Vols. 1–9, John Wiley and Sons, New York.

38. A. Weissberger, *The Chemistry of Heterocyclic Compounds*, Interscience Publishers, New York.
39. K. Schofield, *Quart. Rev.*, **4**, 382 (1950).
40. V. S. Babasinian, *Org. Syn.*, Coll. Vol. **2**, 466 (1943).
41. Ref. 37, Vol. 1, p. 409.
42. H. P. Schultz, *Org. Syn.*, Coll. Vol. **4**, 364 (1963).
43. C. A. Fetscher, *Org. Syn.*, Coll. Vol. **4**, 735 (1963).
44. E. C. Taylor, Jr., and A. J. Crovetti, *Org. Syn.*, Coll. Vol. **4**, 654 (1963).
45. K. A. Kobe and T. F. Doumani, *Org. Syn.*, Coll. Vol. **3**, 653 (1955); *see also* K. A. Kobe and E. M. Langworthy, *Ind. Eng. Chem.*, **49**, 801 (1957).
46. H. E. Fierz-David and L. Blangey, *Fundamental Processes of Dye Chemistry*, Interscience Publishers, New York, 1949, p. 150.
47. K. A. Kobe and H. M. Brennecke, *Ind. Eng. Chem.*, **46**, 728 (1954).
48. G. A. Olah and S. J. Kuhn, *Org. Syn.*, **47**, 56 (1967).
49. A. V. Topchiev *et al.*, *Boron Fluoride and Its Compounds as Catalysts in Organic Chemistry*, Pergamon Press, New York, 1959, p. 254.
50. Ref. 21, p. 50.
51. H. T. Clarke and W. W. Hartman, *Org. Syn.*, Coll. Vol. **1**, 541 (1944).
52. M. Gordon and D. E. Pearson, *J. Org. Chem.*, **29**, 329 (1964).
53. K. E. Cooper and C. K. Ingold, *J. Chem. Soc.*, 836 (1927).
54. Y. Ogata and M. Tsuchida, *J. Org. Chem.*, **21**, 1065 (1956).
55. Ref. 3, p. 232.
56. H. E. Fierz-David and L. Blangey, *Fundamental Processes of Dye Chemistry*, Interscience Publishers, New York, 1949, p. 275.
57. Ref. 21, p. 55.
58. T. Urbanski and W. Kutkiewicz, *Tetrahedron*, **20**, Suppl. 1, 97 (1964).
59. R. C. DeSelms, *J. Org. Chem.*, **33**, 478 (1968).
60. H. Shechter *et al.*, *J. Am. Chem. Soc.*, **74**, 3052 (1952).
61. F. W. Stacey and J. F. Harris, *Org. Reactions*, **13**, 150 (1963).
62. N. Levy and C. W. Scaife, *J. Chem. Soc.*, 1093 (1946).
63. N. Levy *et al.*, *J. Chem. Soc.*, 52 (1948).
64. R. D. Campbell and F. J. Schultz, *J. Org. Chem.*, **25**, 1877 (1960).
65. H. Shechter, *Record Chem. Progr.* (*Kresge-Hooker Sci. Libr.*), **25**, 55 (1964).
66. W. E. Parham and J. L. Bleasdale, *J. Am. Chem. Soc.*, **73**, 4664 (1951).
67. D. E. Bisgrove *et al.*, *Org. Syn.*, Coll. Vol. **4**, 372 (1963).
68. H. Burton *et al.*, *J. Chem. Soc.*, 1062 (1950).
69. H. Shechter and F. Conrad, *J. Am. Chem. Soc.*, **76**, 2716 (1954).
70. J. C. Sowden and H. O. L. Fischer, *J. Am. Chem. Soc.*, **69**, 1963 (1947).
71. W. D. Emmons *et al.*, *J. Am. Chem. Soc.*, **75**, 1993 (1953).
72. J. Ville and G. Dupont, *Bull. Soc. Chim. France*, 804 (1956).
73. W. K. Seifert, *J. Org. Chem.*, **28**, 125 (1963).
74. W. Steinkopf and M. Kühnel, *Chem. Ber.*, **75B**, 1323 (1942).
75. C. E. Anagnostopoulos and L. F. Fieser, *J. Am. Chem. Soc.*, **76**, 532 (1954).
76. F. G. Bordwell and E. W. Garbisch, Jr., *J. Org. Chem.*, **28**, 1765 (1963).

B. Metathesis

1. From Halides

$$RCH_2Br(I) + AgNO_2 \rightarrow RCH_2NO_2 + AgBr(I)$$

$$RCHXR' + NaNO_2 \rightarrow RCH(NO_2)R' + NaX$$

The ambident character of the nitrite anion imposes certain restrictions on the choice of reagents for preparation of nitro compounds [1]. If the transition complex has considerable carbonium ion character, it will tend to coalesce with that atom on which the negative charge mainly resides:

$$R^{\oplus} (or^{+\delta}) + \overset{\ominus}{O}NO \rightarrow RONO$$

But, if not, the nucleophilic character of the nitrogen atom will be utilized:

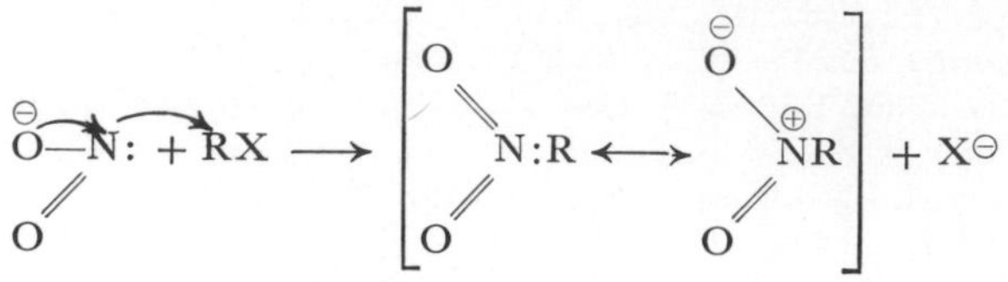

Thus silver nitrite should be used only with substrates, such as primary halides or particularly primary halides subjected to electron-withdrawing substituents, which are difficult to convert into carbonium ions. For the same reason, sodium nitrite is far superior to silver nitrite for the synthesis of all types of secondary nitro compounds. Specific details of synthesis, reviewed by Kornblum [2], will be outlined in which the principles discussed above will be seen to apply. The experimental conditions are important in order to minimize the formation of by-products such as nitrites, nitrates, alcohols. By using silver nitrite and a primary bromide or iodide at 0° and by allowing the temperature to rise to that of the room, good yields of nitro compounds may be obtained. Similarly, good yields are obtained with α-iodo esters and with branched-chain primary bromides and iodides if the branching is β to the carbon atom to which the halogen is attached. In fact, the use of silver nitrite is listed as a preferred procedure in the synthesis of primary nitroalkanes and α-nitro esters [3]. Primary chlorides are not affected by silver nitrite at room temperature; secondary halides give poor yields; and tertiary halides give little or none of the nitro compound. For best results it is desirable to keep the reaction mixture at 0° for the first 24 hours and to avoid exposure to light. Anhydrous diethyl ether is the best solvent, although petroleum ether, cyclohexane, and benzene have also been used.

With sodium nitrite, solvents such as dimethylformamide and dimethyl sulfoxide are employed in order that: the speed of the reaction may be fast; and sufficient amounts of the halide and alkali nitrite are dissolved for the reaction to occur. In this case the formation of a nitroso derivative of the product by reaction with the nitrite ester by-product:

$$RR'CHNO_2 + RR'CHONO \rightarrow RR'C(NO)NO_2 + RR'CHOH$$

may be troublesome unless the workup is conducted promptly or unless, when dimethylformamide is the solvent, urea alone or urea and a phenol, such as phloroglucinol, catechol, or resorcinol, are added to the reaction mixture. Only the phenol is necessary with dimethyl sulfoxide. The urea increases the solubility of the sodium nitrite in dimethylformamide while the phenol acts as a nitrite ester scavenger. This method, usually conducted at room temperature, gives yields with primary alkyl bromides and iodides somewhat lower than when silver nitrite is used, 60% vs. 80% in round numbers. However, it is a preferred method in synthesizing secondary nitro alkanes, α-nitro esters, and α-nitronitriles [3].

(a) **Preparation of 1-nitrooctane** (75–80% from 1-bromooctane and silver nitrite in dry ether immersed in an ice bath for 24 hr. and then at room temperature for about 40 hr.) [4].

(b) **Other preparations**

(1) **Ethyl α-nitrobutyrate** (68–75% from ethyl α-bromobutyrate, sodium nitrite, and phloroglucinol in dimethylformamide at room temperature) [5].

(2) **1,4-Dinitrobutane** (41–46% from 1,4-diiodobutane added to a stirred suspension of silver nitrite in absolute ether at 0°) [6].

2. From Nitro Compounds and Alkyl or Aryliodonium Halides

$$\text{RX} + \overset{\ominus}{\text{CH}_3\text{CHNO}_2} \longrightarrow \text{CH}_3\text{CHNO}_2 + \text{X}^{\ominus}$$
$$\underset{\text{R}}{|}$$

This synthesis is of limited value, but it has led to the preparation of 1-*p*-nitrobenzylnitroethane with an 83% yield (see Ex. a) [7]. Alkylation on oxygen or carbon is possible, but the latter appears to occur if the stability of *aci* form of the nitro compound is high. In applying the method to halides of secondary nitro compounds, low yields of dinitroalkanes are obtained [8]:

$$
\begin{array}{c}
\text{R} \\
| \\
\text{NO}_2\text{—C—X}\\
| \\
\text{R}
\end{array}
+
\begin{array}{c}
\text{R}\\
|\ominus\\
\text{R—CNO}_2\text{Na}^{\oplus}
\end{array}
\longrightarrow
\begin{array}{c}
\text{R R}\\
|\ |\\
\text{NO}_2\text{—C—C—NO}_2 + \text{NaX}\\
|\ |\\
\text{R R}
\end{array}
$$

(a) **Preparation of 1-*p*-nitrobenzylnitroethane.** Sodium nitroethane, 24.45 g., prepared from 5.75 g. of sodium, a small amount of ethanol, and 131.2 g. of nitroethane, was treated with 42.9 g. of *p*-nitrobenzyl chloride in 500 ml. of absolute ethanol and the mixture was refluxed for 11 hr. on a steam bath. The filtrate was evaporated to dryness and the residue was dissolved in benzene, from which solution the *p*-nitrobenzaldehyde present was extracted with sodium bisulfite solution. On removing the benzene, the residue was distilled to give 28 g. (83%), b.p. 145–160° (0.5–0.6 mm.) of the dinitroalkane. Crystallized from alcohol, the solid melted at 51° [7].

(b) **Preparation of 2-nitro-2-phenyloctane** [9]:

$$
\begin{array}{c}
\text{C}_6\text{H}_{13}\\
|\\
\text{CH}_3\text{C}\\
\|\\
\text{NONa}\\
\swarrow\\
\text{O}\\
\text{20 mmoles}
\end{array}
+ (\text{C}_6\text{H}_5)_2\overset{\oplus\ominus}{\text{IOSO}_2\text{C}_7\text{H}_7}
\xrightarrow[\text{DMF}]{\text{15 ml.}}
\begin{array}{c}
\text{C}_6\text{H}_5\\
|\\
\text{CH}_3\text{—C—C}_6\text{H}_{13}\\
|\\
\text{NO}_2\\
54\%
\end{array}
$$

20 mmoles

3. From Carbanions and Nitrate Esters

$$ArCH_2CN + CH_3ONO_2 \xrightarrow{C_2H_5ONa} ArC{=}NO_2Na$$
$$\underset{CN}{|}$$

$$\Big\downarrow \substack{NaOH \\ H_2O}$$

$$ArCH_2NO_2 \xleftarrow{HCl} ArC{=}NO_2Na$$
$$\underset{COONa}{|}$$

This synthesis has been reviewed in some detail [2]. It has been applied to a variety of active methylene types such as nitriles [10], arylacetic esters [11], dinitriles [12], cyclic ketones [13], malonic and acetoacetic esters [14], and N,N-dialkylamides (see Ex. c.1). The original catalysts employed were sodium or potassium alkoxides, but potassium amide (see Ex. c.2) or potassium *t*-butoxide is preferred [13]. With the latter catalyst, α,α'-dinitrocycloalkanones may be formed in high yield [15]:

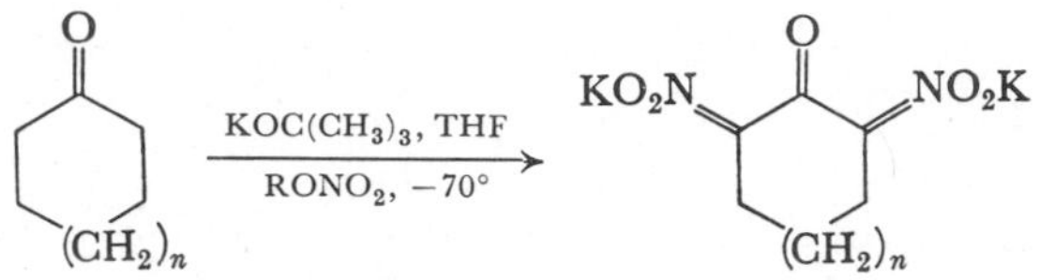

20-I

Furthermore, the monosalts can be readily cleaved to α,ω-dinitroalkanes:

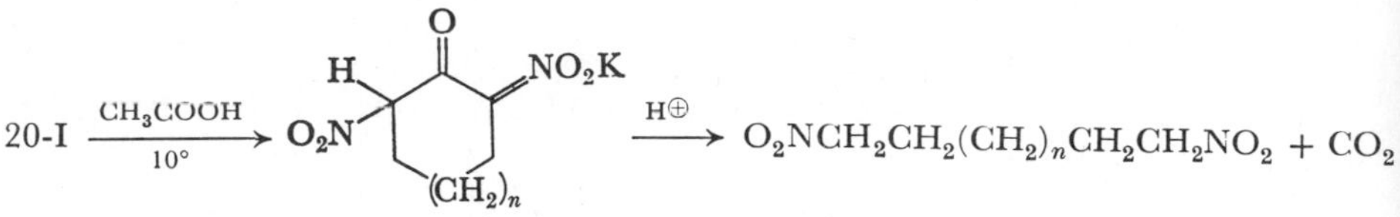

$$20\text{-}I \xrightarrow[10°]{CH_3COOH} O_2N\ \cdots \xrightarrow{H^\oplus} O_2NCH_2CH_2(CH_2)_nCH_2CH_2NO_2 + CO_2$$

Although this synthesis gives satisfactory yields in some cases, its evaluation is difficult. The nitro salts formed first are often hygroscopic and difficult to purify. In fact, yields are sometimes based on the bromine addition products formed practically quantitatively. In addition, the conversion of the salt into the nitro compound may lead to decomposition to an aldehyde or ketone [16]. To add to these complications, in some cases, such as those involving ethyl malonate and acetoacetic ester, alkylation rather than nitration occurs [17]. However, it has been found that the use of acetone cyanhydrin nitrate [14] permits the nitration of these compounds (see A.1 for other nitrations). For α-nitroesters this synthesis is listed as a preferred method [2], although it is inferior to the synthesis from α-halo esters and sodium nitrite (B.1).

The mechanism of the reaction is regarded to be a nucleophilic displacement in which the carbanion 20-II attacks the nitrate ester to yield the nitro compound 20-III:

$$>\!CH\!: + \overset{\ominus}{N}O_2OR \longrightarrow\ >\!CHNO_2 + O\overset{\ominus}{R}$$

20-II   20-III

The older literature states that the anion of pyrrole yields 3-nitropyrrole with nitrate esters:

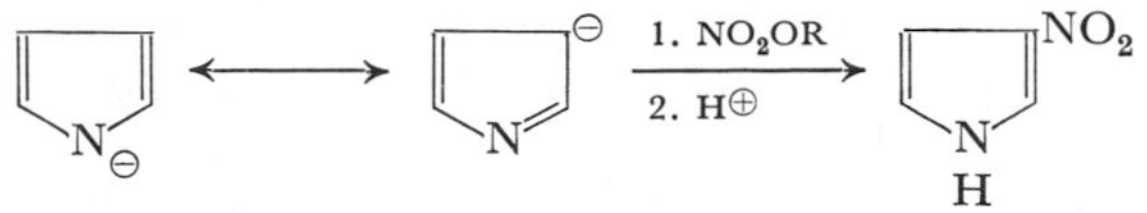

But it has been shown lately that only small amounts of 3-nitropyrrole are obtained by this method and that the compound is prepared best by condensation of nitromalonaldehyde and ethyl glycinate followed by decarboxylation (see E.3, Ex. b.2).

(a) Preparation of phenylnitromethane (50–55% from benzyl cyanide and methyl nitrate in the presence of sodium ethoxide) [10].

(b) Preparation of dipotassium 2,5-dinitrocyclopentanone (55% of analytically pure salt from cyclopentanone, amyl nitrate, and sublimed potassium t-butoxide in THF at $-30°$) [13].

(c) Other examples

(1) N,N-Dimethyl-α-bromo-α-nitrobutyramide [18]:

$$CH_3CH_2CH_2CON(CH_3)_2 + (CH_3)_3COK + (C_2H_5)_2O \xrightarrow[0.11 \text{ mole } C_5H_{11}ONO_2]{-70°,}$$

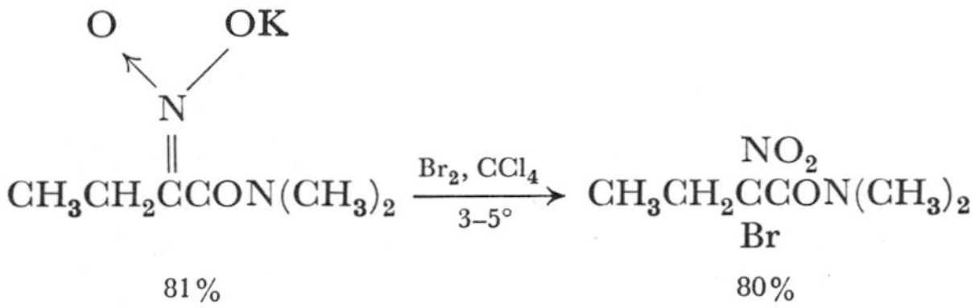

$$\underset{0.1 \text{ mole}}{} \quad \underset{0.165 \text{ mole}}{} \quad \underset{130 \text{ ml.}}{}$$

(2) 2-Nitrocycloheptanone (79% from 0.115 mole each of cycloheptanone and KNH_2 in 300 ml. of liquid ammonia for 2 hr., followed by the addition of 0.21 mole of amyl nitrate in 10 min., evaporation of NH_3, and addition of acetic acid at $0°$) [19].

4. From Diazonium Compounds

$$\overset{\oplus}{Ar}\overset{\ominus}{N_2X} \xrightarrow[\text{Catalyst}]{NaNO_2} ArNO_2 + NaX + N_2$$

Amino groups may be replaced by nitro groups by proceeding via the diazonium salt. Although the synthesis has limited application, it serves perhaps best when a certain orientation of substituents is desired. For example, p-dinitrobenzene, not available by direct nitration, may be prepared in 76% yield from p-nitroaniline by conversion first into the diazonium salt which with sodium nitrite gives the dinitro compound [20]. By proceeding via the diazonium fluoborate, the yield is 67–82% [21]. Various diazonium salts, such as the chloride, sulfate [22], fluoborate [21], and cobaltinitrite [23], have been employed. Catalysts used in the conversion of these salts into the nitro compounds have been cuprous oxide and copper sulfate, sodium sulfite and copper sulfate [20], and copper powder [21]. Yields for conversion of one amino group may run as high as 80%, although much lower values are obtained in the conversion of two.

(a) Preparation of *p*-dinitrobenzene (67–82% from *p*-nitroaniline by treatment first with nitrous and fluoboric acid, and finally with sodium nitrite in the presence of copper powder) [21].

(b) Preparation of 1,4-dinitronaphthalene (52–60% from 1-amino-4-nitronaphthalene by treatment first with sodium nitrite and sulfuric acid and finally with sodium nitrite in the presence of copper sulfate and sodium sulfite) [24].

1. N. Kornblum et al., J. Am. Chem. Soc., **83**, 2779 (1961); **77**, 6269 (1955).
2. N. Kornblum, Org. Reactions, **12**, 101 (1962).
3. Ref. 2, p. 129.
4. N. Kornblum and H. E. Ungnade, Org. Syn., Coll. Vol. **4**, 724 (1963).
5. N. Kornblum and R. K. Blackwood, Org. Syn., Coll. Vol. **4**, 454 (1963).
6. H. Feuer and G. Leston, Org. Syn., Coll. Vol. **4**, 368 (1963).
7. F. W. Hoover and H. B. Hass, J. Org. Chem., **12**, 501 (1947).
8. L. W. Seigle and H. B. Hass, J. Org. Chem., **5**, 100 (1940).
9. N. Kornblum and H. J. Taylor, J. Org. Chem., **28**, 1424 (1963).
10. A. P. Black and F. H. Babers, Org. Syn., Coll. Vol. **2**, 512 (1943).
11. W. Wislicenus and R. Grützner, Chem. Ber., **42**, 1930 (1909).
12. H. Feuer and C. Savides, J. Am. Chem. Soc., **81**, 5830 (1959).
13. H. Feuer et al., J. Am. Chem. Soc., **78**, 4364 (1956).
14. W. D. Emmons and J. P. Freeman., J. Am. Chem. Soc., **77**, 4391 (1955).
15. H. Feuer and R. S. Anderson, J. Am. Chem. Soc., **83**, 2960 (1961).
16. N. Kornblum and G. E. Graham, J. Am. Chem. Soc., **73**, 4041 (1951).
17. J. W. Nef, Ann. Chem., **309**, 172 (1899).
18. H. Feuer and B. F. Vincent, Jr., J. Org. Chem., **29**, 939 (1964).
19. H. Feuer et al., J. Org. Chem., **33**, 3622 (1968).
20. H. Hodgson et al., J. Chem. Soc., 1512 (1948).
21. E. B. Starkey, Org. Syn., Coll. Vol. **2**, 225 (1943).
22. H. Zollinger, Azo and Diazo Chemistry, Interscience Publishers, New York, 1961, pp. 14–20.
23. H. H. Hodgson and E. R. Ward, J. Chem. Soc., 127 (1947).
24. H. H. Hodgson et al., Org. Syn., Coll. Vol. **3**, 341 (1955).

C. Condensation Reactions

Anions of nitromethane and its homologs add readily to carbonyl and unsaturated carbonyl compounds, and azomethines. These additions are discussed in C.1 and C.2. Supplementary to these reactions is the addition of nucleophiles to α-nitroolefins, as discussed in C.2. It is interesting to note that no comments could be found in the literature on the Claisen condensation of nitroalkanes with esters, although implications were found that the reaction may be abnormal [1]:

$$CH_3NO_2 + C_6H_5COCl \xrightarrow{C_5H_5N} C_6H_5CONHCONHCOC_6H_5$$

N,N'-Dibenzoylurea

References for Section C are on pp. 1002–1003.

1. From Aldehydes, Ketones, or Schiff Bases

$$RCH=O + R'CH_2NO_2 \xrightarrow{\text{Base}} RCHOH\overset{R'}{\underset{|}{C}}HNO_2 \longrightarrow RCH=\overset{R'}{\underset{|}{C}}HNO_2$$

This reaction has been reviewed as a section of the Knoevenagel reaction [2]. Nitroalcohols or nitroalkenes may be synthesized by the action of nitroalkanes containing an α-hydrogen atom with carbonyl compounds or azomethines (Schiff bases) under alkaline conditions, an aldol-type reaction (Alcohols, Chapter 4, G.1). The nitroalcohol may readily dehydrate if derived from an aromatic aldehyde, but otherwise is dehydrated via the acetate ester [3]:

$$RCHOH\overset{R'}{\underset{|}{C}}HNO_2 \xrightarrow{Ac_2O} RCH\overset{\overset{\overset{O}{\parallel}}{O\overset{}{C}CH_3}\ R'}{\underset{}{C}}HNO_2 \xrightarrow{Na_2CO_3} RCH=CR'NO_2$$

They may be converted not only into nitroolefins but also to α-nitroketones by oxidation [4].

Two important characteristics of the condensation reaction must be taken into account. First, nitromethane (and less probably its homologs) undergoes self-condensation in strong base solution [5]:

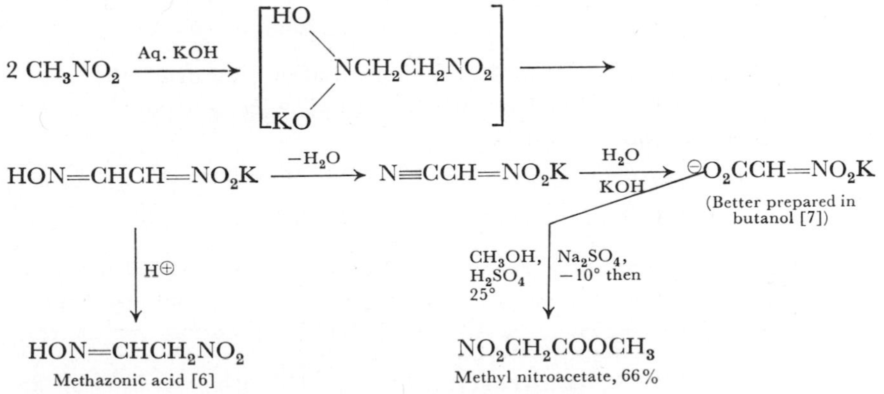

$$2\ CH_3NO_2 \xrightarrow{\text{Aq. KOH}} \left[\begin{array}{c} HO \\ \diagdown \\ NCH_2CH_2NO_2 \\ \diagup \\ KO \end{array} \right] \longrightarrow$$

$$HON=CHCH=NO_2K \xrightarrow{-H_2O} N\equiv CCH=NO_2K \xrightarrow[KOH]{H_2O} {}^{\ominus}O_2CCH=NO_2K$$

(Better prepared in butanol [7])

$$\downarrow H^{\oplus} \qquad\qquad\qquad\qquad \begin{array}{c} CH_3OH, \\ H_2SO_4 \\ 25° \end{array} \Big| \begin{array}{c} Na_2SO_4, \\ -10° \text{ then} \end{array}$$

$$HON=CHCH_2NO_2 \qquad\qquad NO_2CH_2COOCH_3$$

Methazonic acid [6]   Methyl nitroacetate, 66%

For the same reason, better yields of nitroalcohols are obtained by short exposure times of the nitroalkane to base catalysts [8]:

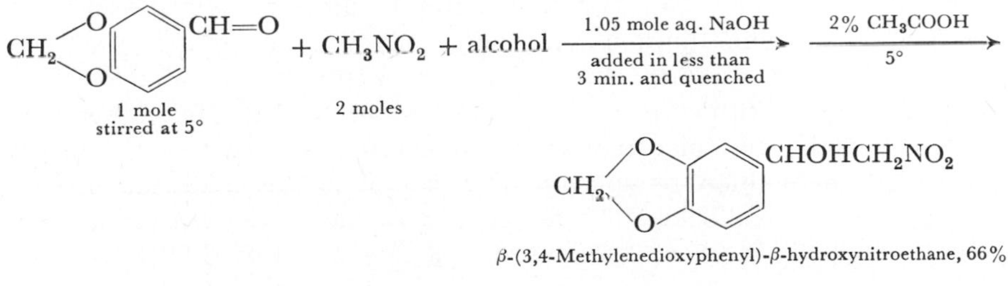

1 mole
stirred at 5°   2 moles

1.05 mole aq. NaOH
added in less than
3 min. and quenched

2% CH₃COOH
5°

$β$-(3,4-Methylenedioxyphenyl)-$β$-hydroxynitroethane, 66%

The preceding reaction conditions are general except that a *p*-hydroxy group attached to the aromatic aldehyde inhibits the reaction. Second, two or even three aldehyde molecules, particularly in the case of formaldehyde with nitromethane, may condense with the nitroalkane:

$$3 \ CH_2O \ + \ CH_3NO_2 \ \rightarrow \ (HOCH_2)_3CNO_2$$

Trihydroxymethyl-
nitromethane

In this case, potassium fluoride in isopropyl alcohol is an effective catalyst to stop at the mono condensation stage and indeed is a good general catalyst [9]:

$$(CH_2O)_n \ + \ KF \ + \ (CH_3)_2CHOH \ \text{at } 30\text{--}35° \ \xrightarrow[\text{2. Filter off catalyst}]{\substack{\text{1. 2 moles } CH_3NO_2 \\ \text{in 2 hr. and hold} \\ \text{4 hr. more}}} \ HOCH_2CH_2NO_2$$

1 mole   0.05 mole   300 ml.   β-Nitroethanol, 73%

Other aliphatic aldehydes and nitroalkanes give yields of condensation products in the range 36–86%.

Condensation with ketones is more sluggish. For instance, only nitromethane, and not its homologs, is capable of addition to cyclic ketones [2]:

Nitromethyl-
cyclohexene-1

Other very complex products arise in this condensation, however [10].

The Mannich reaction succeeds well with nitroalkanes; nitromethane and nitroethane tend to yield diadducts

$$CH_3NO_2 \ + \ 2 \ H_2CO \ + \ 2 \ R_2NH \ \longrightarrow \ R_2NCH_2\overset{\displaystyle NO_2}{\underset{\displaystyle |}{C}}HCH_2NR_2 \ + \ 2 \ H_2O$$

but 1-nitropropane gives a monoadduct (see Ex. c.3).

Although salts of dinitromethane are not too stable, they are stable enough to undergo condensation (see Ex. b) and to form Mannich products (see Ex. c.4).

(a) Preparation of β-nitrostyrene (80–83% from benzaldehyde, nitromethane, sodium hydroxide, and hydrochloric acid) [11].

(b) Preparation of ethyl α-hydroxy-β,β-dinitropropionate. Ethyl ethoxyhydroxyacetate, 5.92 g., was added drop by drop to a stirred suspension of 5.76 g. of potassium dinitromethane in 20 ml. of water. Solution was completed by warming on a steam bath and then the mixture was allowed to cool to room temperature with stirring. The potassium salt, 8.85 g. (90%), separated and, after washing with methanol and recrystallization from water, it melted at 144° with decomposition. Acidification of 28.4 g. of the salt in 200 ml. of water with 20% sulfuric acid below 10° followed by extraction with ether yielded 23.6 g. (94%) of ethyl α-hydroxy-β,β-dinitropropionate, $n^{25}D$ 1.4573 [12].

(c) Other examples

(1) **2-(2,3-Diethoxyphenyl)-2-hydroxy-1-nitroethane** (93% from 2,3-diethoxybenzaldehyde, nitromethane, potassium hydroxide, and 50% acetic acid) [13].

(2) **2′-Bromo-α-nitrostilbene** (86% from 2-bromobenzylidene-*n*-butylamine and phenylnitromethane in glacial acetic acid) [14].

(3) **1-Morpholino-2-nitrobutane** (68% from 0.2 mole of 1-nitropropane added to a mixture of 17 ml. of 37% formaldehyde and 17 ml. of morpholine held at 0°, allowed to warm and then held at 25° for 2 days) [15].

(4) **1,1-Dinitro-2-piperidinoethane** (94% from potassium dinitro-

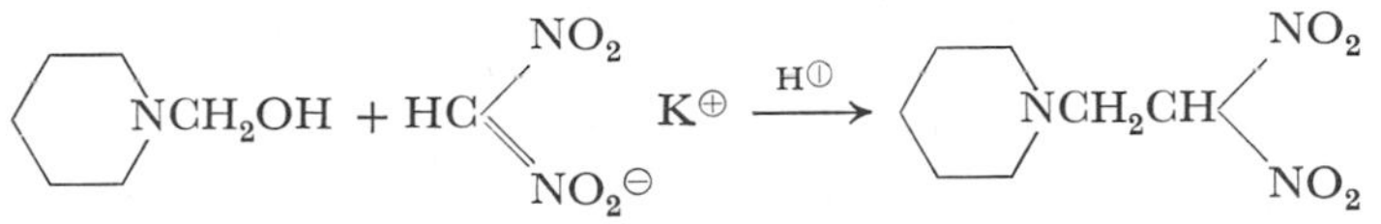

methane acidified with 0.50 N sulfuric acid at 0° and added to N-hydroxymethylpiperidine) [16].

2. From Unsaturated Carbonyl Compounds and the Like (Michael)

$$CH_2\!\!=\!\!CHCOOC_2H_5 \xrightarrow[\text{NaOC}_2\text{H}_5]{\text{CH}_3\text{NO}_2} NO_2CH_2CH_2CH_2COOC_2H_5$$

The Michael condensation pertaining in part to the addition of nitroalkanes to α,β-unsaturated carbonyl compounds has been reviewed [17]. Nitroalkanes which contain active hydrogen atoms respond in a manner to give nitro compounds in which the anion containing the nitro groups adds to the doubly bound carbon atom farthest removed from the activating group in the alkene (aldehyde, ketone, ester, or nitrile).

As in other Michael condensations, an alkaline catalyst is usually employed. Sodium alkoxides [18], a catalytic amount of sodium hydroxide in an alcohol [19], diethylamine [20], benzyltrimethylammonium butoxide [21, 22], calcium hydride in methanol [23], Triton B (benzyltrimethylammonium hydroxide) [24], and liquid ammonia alone [25] have been used in these condensations. Yields are in the range 75–92% except when a nitroolefin is employed. Choice of catalyst is important in obtaining such high yields [21]. A further example is the superior performance of potassium fluoride as a catalyst [26]:

$$C_2H_5O_2CCH\!\!=\!\!CHCO_2C_2H_5 + CH_3CH_2NO_2 \xrightarrow[\substack{0.25\ \text{mole,}\\ 04\text{-}70°,\ 10\ \text{hr.}}]{K_2F_2} C_2H_5O_2C\overset{\displaystyle \overset{CH_3}{|}\ \overset{CHNO_2}{|}}{C}HCH_2CO_2C_2H_5$$

0.2 mole   0.2 mole   20-IV

Diethyl 3-nitrobutane-1,2 dicarboxylate, 45.5%

Compound 20-IV is unstable in the presence of dibutylamine, a potential catalyst:

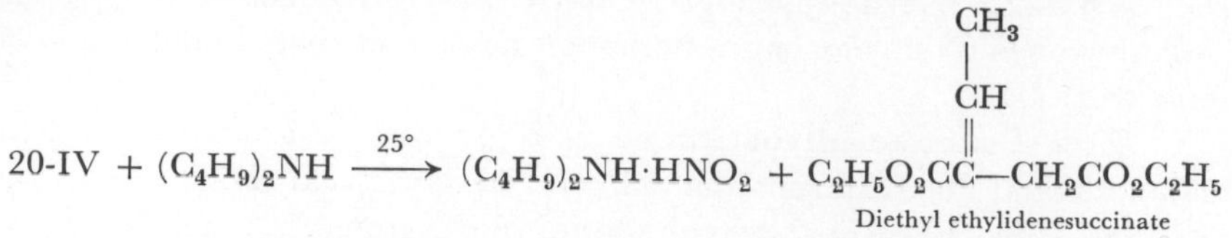

$$20\text{-IV} + (C_4H_9)_2NH \xrightarrow{25°} (C_4H_9)_2NH \cdot HNO_2 + C_2H_5O_2C\underset{\|}{\overset{\overset{\displaystyle CH_3}{|}\overset{\displaystyle CH}{\|}}{C}}\!\!-CH_2CO_2C_2H_5$$

Diethyl ethylidenesuccinate

In the Michael addition to nitroolefins, potassium fluoride (see Ex. b.3 for use of the fluoride with acrylonitrile) or traces of aqueous sodium hydroxide have been used as catalysts. Sodium acetate or the salt of the nitro compound is employed if the nitroolefin is generated *in situ* from the acetate ester [27]:

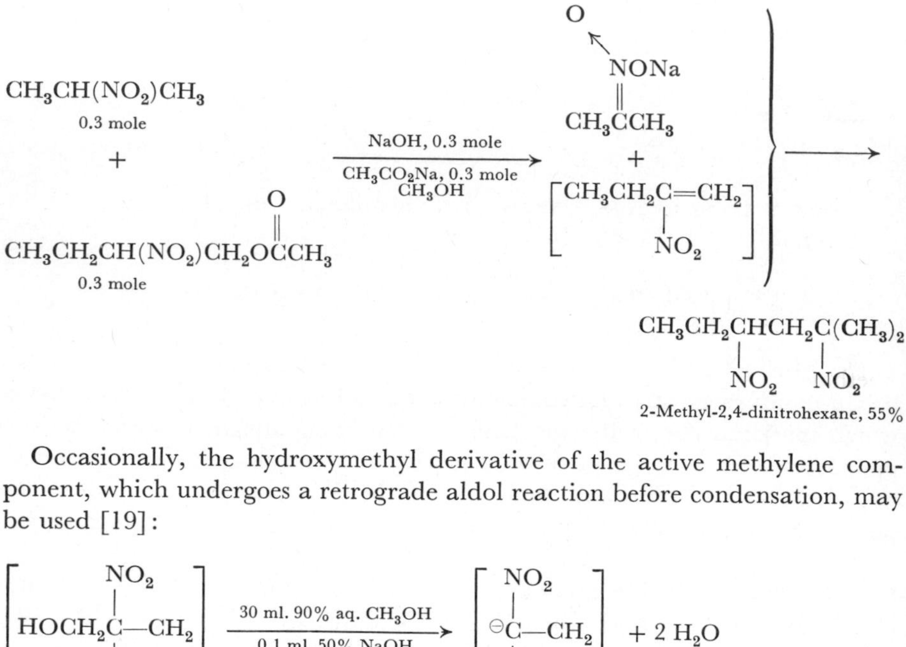

$$
\begin{array}{c}
CH_3CH(NO_2)CH_3 \\
\text{0.3 mole} \\
+ \\
CH_3CH_2CH(NO_2)CH_2O\overset{\overset{\displaystyle O}{\|}}{C}CH_3 \\
\text{0.3 mole}
\end{array}
\xrightarrow[\substack{CH_3CO_2Na,\ 0.3\ mole \\ CH_3OH}]{NaOH,\ 0.3\ mole}
\left\{
\begin{array}{c}
CH_3\overset{\overset{\displaystyle O}{\nwarrow}}{\underset{\|}{\overset{\displaystyle NONa}{C}}}CH_3 \\
+ \\
\left[CH_3CH_2\underset{NO_2}{\overset{|}{C}}\!=\!CH_2\right]
\end{array}
\right\}
\longrightarrow
$$

$$CH_3CH_2\underset{NO_2}{\overset{|}{C}}HCH_2\underset{NO_2}{\overset{|}{C}}(CH_3)_2$$

2-Methyl-2,4-dinitrohexane, 55%

Occasionally, the hydroxymethyl derivative of the active methylene component, which undergoes a retrograde aldol reaction before condensation, may be used [19]:

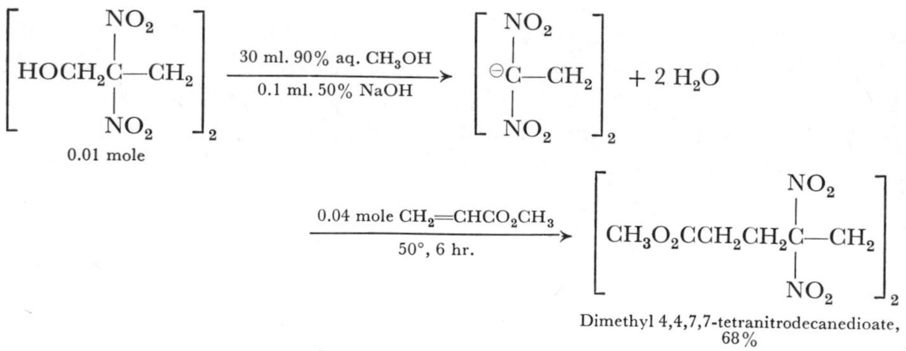

$$
\left[HOCH_2\underset{NO_2}{\overset{\overset{\displaystyle NO_2}{|}}{C}}\!-CH_2\right]_2
\xrightarrow[\substack{0.1\ ml.\ 50\%\ NaOH}]{30\ ml.\ 90\%\ aq.\ CH_3OH}
\left[{}^{\ominus}\underset{NO_2}{\overset{\overset{\displaystyle NO_2}{|}}{C}}\!-CH_2\right]_2 + 2\ H_2O
$$

0.01 mole

$$
\xrightarrow[50°,\ 6\ hr.]{0.04\ mole\ CH_2\!=\!CHCO_2CH_3}
\left[CH_3O_2CCH_2CH_2\underset{NO_2}{\overset{\overset{\displaystyle NO_2}{|}}{C}}\!-CH_2\right]_2
$$

Dimethyl 4,4,7,7-tetranitrodecanedioate, 68%

The addition of hydride anion from sodium borohydride provides another pathway for the preparation of nitroalkanes [28]:

$$RCH\!=\!CHNO_2 + NaBH_4 \xrightarrow{CH_3CN} RCH_2\!-\!\overset{\ominus}{C}HNO_2 \xrightarrow{H^{\oplus}} RCH_2CH_2NO_2$$

However, not all Michael additions of nitroethylenes are normal, as is suggested by the following reaction [29]:

$$C_6H_5CH=CHNO_2 + CN^\ominus \xrightarrow[\text{2. H}^\oplus, \text{ reflux}]{\text{1. Heat}} \begin{array}{l} C_6H_5CHCOOH \\ | \\ C_6H_5CHCOOH \end{array}$$

Retroaldol and perhaps oxidative dimerizations may be involved.

The ylide of trimethylsulfoxide has been added to nitro aromatics, albeit in poor yields [30]:

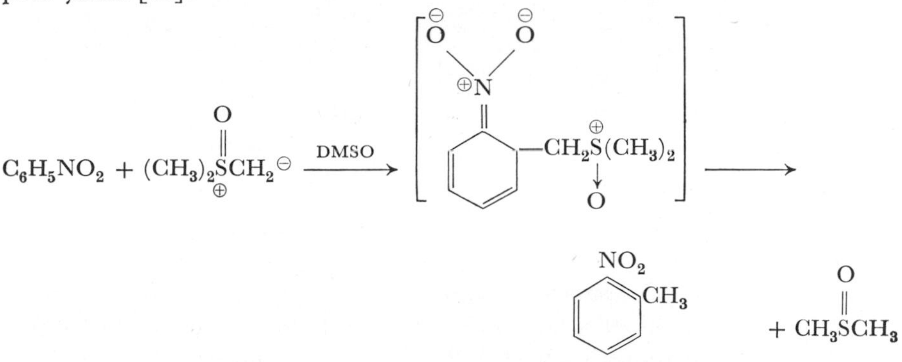

$$C_6H_5NO_2 + (CH_3)_2\overset{\oplus}{S}CH_2^\ominus \xrightarrow{\text{DMSO}}$$

o-Nitrotoluene, total yield 35%
in which the ratio of *o*- to
p-nitrotoluene was about 10–15/1

The Grignard addition also may be 1,4 like the Michael (see Ex. b.5). It occurs even with aromatic compounds [31]:

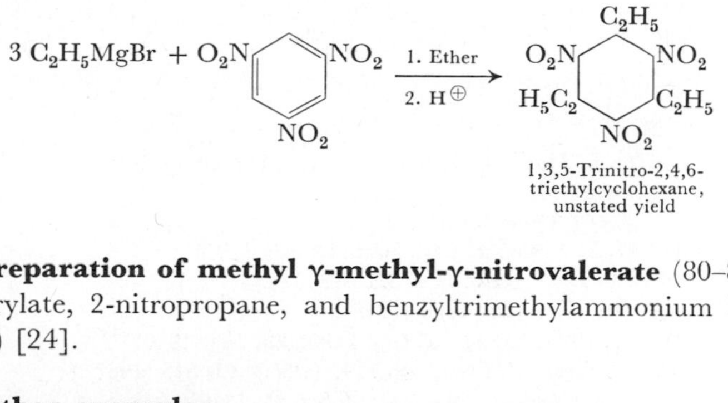

1,3,5-Trinitro-2,4,6-
triethylcyclohexane,
unstated yield

(a) Preparation of methyl γ-methyl-γ-nitrovalerate (80–86 % from methyl acrylate, 2-nitropropane, and benzyltrimethylammonium hydroxide (Triton B)) [24].

(b) Other examples

(1) 4-Methyl-4-nitro-1,3-diphenyl-1-pentanone   (85–92 % from benzalacetophenone, 2-nitropropane, and calcium hydride in dry methanol) [23].

(2) N-Cyclohexyl-2-methyl-3-nitropiperidine [32]:

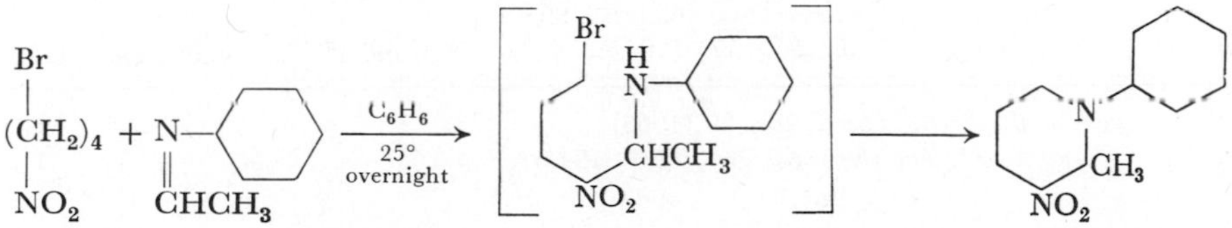

60% crude

(3) γ-Methyl-γ-nitropimelonitrile [33]:

$$CH_3CH_2NO_2 + K_2F_2 + 2\ CH_2{=}CHCN \xrightarrow[40-40^\circ,\ 24\ hr.]{Absolute\ C_2H_5OH} NCCH_2CH_2\overset{\overset{\displaystyle CH_3}{|}}{\underset{\underset{\displaystyle NO_2}{|}}{C}}CH_2CH_2CN$$

0.1 mole · 5 g. · 0.2 mole

64%

(4) 2,2-(Bis-β-nitroethyl)-5,5-dimethylcyclohexane-1,3-dione

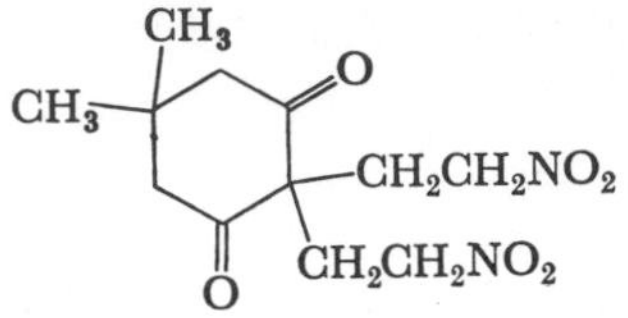

(52% from 5,5-dimethylcyclohexane-1,3-dione in sodium methoxide and methanol to which nitroethylene was added at 25°) [34].

(5) 1-Nitro-2,2-dimethylbutane (60% from 1 equiv. of ethylmagnesium bromide in ether added to a stirred solution of 1 equiv. of 1-nitro-2-methyl-1-propene in dry ether at 0–10° followed by reflux for 1 hr.) [35].

(6) 2,4-Diphenyl-6-isopropylnitrobenzene. A general reaction for

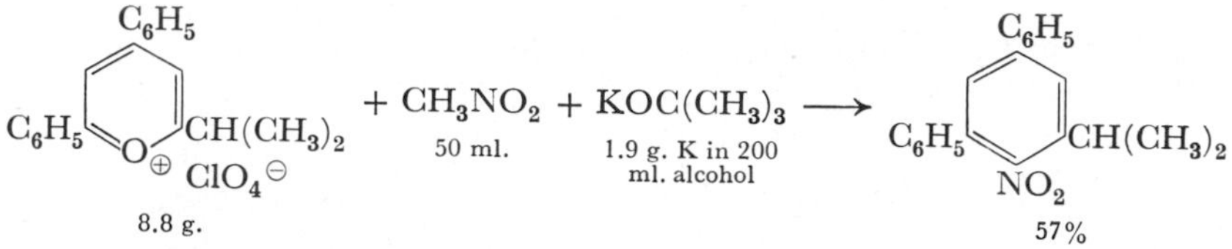

8.8 g.

+ CH₃NO₂ + KOC(CH₃)₃ ⟶

50 ml. · 1.9 g. K in 200 ml. alcohol

57%

pyrylium salts [36].

1. R. H. Terss and W. E. McEwen, *J. Am. Chem. Soc.*, **76,** 580 (1954).
2. G. Jones, *Org. Reactions*, **15,** 204 (1967).
3. H. B. Hass *et al.*, *J. Org. Chem.*, **15,** 8 (1950).
4. C. D. Hurd and M. E. Nilson, *J. Org. Chem.*, **20,** 927 (1955).
5. H. B. Hass and E. F. Riley, *Chem. Rev.*, **32,** 373 (1943).
6. D. J. Morgan, *J. Org. Chem.*, **23,** 1069 (1958).
7. V. E. Matthews and D. G. Kubler, *J. Org. Chem.*, **25,** 266 (1960).
8. R. A. Heacock *et al.*, *Canad. J. Chem.*, **39,** 1143 (1961); **41,** 543 (1963).
9. S. Kambe and H. Yasuda, *Bull. Chem. Soc., Japan*, **41,** 1444 (1968).
10. H. O. House and R. W. Magin, *J. Org. Chem.*, **28,** 647 (1963).
11. D. E. Worrall, *Org. Syn.*, Coll. Vol. **1,** 413 (1941).
12. L. W. Kissinger *et al.*, *J. Org. Chem.*, **22,** 1658 (1957).
13. G. Tsatsus, *Bull. Soc. Chim. France*, 884 (1949).
14. D. N. Robertson, *J. Org. Chem.*, **25,** 47 (1960).
15. M. Zief and J. P. Mason, *J. Org. Chem.*, **8,** 1 (1943).
16. L. Zeldin and H. Shechter, *J. Am. Chem. Soc.*, **79,** 4708 (1957).
17. E. D. Bergmann *et al.*, *Org. Reactions*, **10,** 179 (1959).
18. M. A. T. Rogers, *J. Chem. Soc.*, 590 (1943); A. Lambert and H. A. Piggott, *ibid.*, 1489 (1947).
19. H. Feuer *et al.*, *J. Org. Chem.*, **28,** 339 (1963).
20. M. C. Kloetzel, *J. Am. Chem. Soc.* **69,** 2271 (1947).

21. N. J. Leonard and D. L. Felley, *J. Am. Chem. Soc.*, **71**, 1758 (1949).
22. N. J. Leonard *et al.*, *J. Am. Chem. Soc.*, **72**, 2537 (1950); **73**, 857 (1951).
23. N. Fishman and S. Zuffanti, *J. Am. Chem. Soc.*, **73**, 4466 (1951).
24. R. B. Moffett, *Org. Syn.*, Coll. Vol. **4**, 652 (1963).
25. S. Wakamatsu and K. Shimo, *J. Org. Chem.*, **27**, 1609 (1962).
26. T. Urbanski *et al.*, *Tetrahedron*, **20**, Suppl. 1, 285 (1964).
27. H. Feuer and R. Miller, *J. Org. Chem.*, **26**, 1348 (1961).
28. A. Hassner *et al.*, *J. Org. Chem.*, **33**, 1733 (1968); A. I. Meyers and J. C. Sircar, *ibid.*, **32**, 4134 (1967).
29. M. Holleman, *Rec. Trav. Chim.*, **23**, 283 (1904).
30. V. J. Traynelis and J. V. McSweeney, *J. Org. Chem.*, **31**, 243 (1966).
31. T. Severin, *Angew. Chem.*, **70**, 164 (1958).
32. J. E. Dolfini and E. J. Swain, *J. Org. Chem.*, **33**, 2079 (1968).
33. J. M. Patterson and M. W. Barnes, *Bull. Chem. Soc. Japan*, **40**, 2715 (1967).
34. V. V. Perekalin and K. S. Parfenova, *Dokl. Akad. Nauk SSSR*, **124**, 592 (1959); *C.A.*, **53**, 13142 (1959).
35. G. D. Buckley and E. Ellery, *J. Chem. Soc.*, 1497 (1947).
36. K. Dimroth *et al.*, *Tetrahedron*, **20**, Suppl. 1, 261 (1964).

D. Oxidation

1. From Amines

If the amino compound is more readily available than the nitro compound, oxidation of the former to the latter may be possible. For example, *tertiary* nitroalkanes cannot be prepared from the halide and silver nitrite, and the liquid- and vapor-phase nitration methods can hardly be classified as laboratory syntheses. However, these nitro compounds may be prepared in superior yields by the oxidation of primary amines in which the amino group is attached to a tertiary carbon atom [1]. Similarly, amino compounds are more readily available than nitro compounds in the pyridine and quinoline series because of the success of direct amination. Oxidation of these with hydrogen peroxide in sulfuric acid has been accomplished in fair yield [2]. Furthermore, this method of synthesis is sometimes of value when a certain orientation is desired in an aromatic ring. For example, an attractive method for synthesizing 2,4,6-tribromonitrobenzene (90%) is by the oxidation of the readily available 2,4,6-tribromoaniline with hydrogen peroxide and maleic anhydride [3]. The permaleic acid formed here is undoubtedly a powerful oxidizing agent, exceeded only by pertrifluoroacetic acid, for amines (see Ex. a). Only the peracids are useful in oxidizing aromatic amines.

References for Section D are on pp. 1007–1008.

Other oxidation agents which have been used in the synthesis are potassium permanganate [4] and peracetic acid [5]. Potassium permanganate and peracetic acid have been employed satisfactorily with *tertiary* alkylamines; peracetic acid with *secondary* alkylamines [5]; peracetic acid, hydrogen peroxide and sulfuric acid, hydrogen peroxide and trifluoroacetic acid, and hydrogen peroxide and maleic anhydride with aromatic amines; and hydrogen peroxide and sulfuric acid with aminopyridines. The method is not of value for the synthesis of primary nitroalkanes.

It has been suggested that the path taken in the oxidation [4, 5] is as follows:

$$R_3C{-}NH_2 \longrightarrow \underset{\underset{\text{20-V}}{O}}{R_3C{-}\overset{\downarrow}{NH_2}} \longrightarrow \underset{\underset{\text{20-VI}}{OH}}{R_3C{-}NH} \longrightarrow \underset{\text{20-VII}}{R_3C{-}NO} \longrightarrow \underset{\text{20-VIII}}{R_3C{-}NO_2}$$

The attack of oxygen on the nitrogen gives the amine oxide 20-V which rearranges to the hydroxylamine 20-VI, which is oxidized first to the nitroso compound 20-VII and finally to the nitro compound 20-VIII. The mechanism implies that intermediates related to the structures mentioned above should also be oxidized to nitro compounds. Examples bear out this implication; for instance, nitrosamines may be oxidized to nitroamines [6] (see also D.3):

$$R_2NNO \xrightarrow{\text{CF}_3\text{CO}_2\text{OH}} R_2NNO_2$$

(a) Preparation of nitrobenzene. Hydrogen peroxide 90%, 0.2 mole, was added to 40 ml. of trifluoroacetic acid. Aniline, 0.05 mole, was added all at once to the stirred peracid. The internal temperature rose steadily but was held at 50° by intermittent cooling. The yield of nitrobenzene was 79% [6].

(b) Preparation of 4-nitro-2,2,4-trimethylpentane. *t*-Octylamine, 4-amino-2,2,4-trimethylpentane, 0.2 mole, was dissolved in 500 ml. of acetone with magnesium sulfate in 125 ml. of water. To this well-stirred solution, 1.2 moles of solid $KMnO_4$ was added in small portions. The product was separated by steam distillation after removal of acetone; yield 69–82%. Oxidation of *t*-butylamine may be carried out without adding acetone [7].

(c) Other examples

(1) **2,4,6-Tribromonitrobenzene** (90% from 2,4,6-tribromoaniline, 90% hydrogen peroxide, and maleic anhydride) [3].

(2) **2-Nitro-4-methylpyridine** (68% from 2-amino-4-methylpyridine, 30% hydrogen peroxide, and fuming sulfuric acid) [2].

2. From Nitro Compounds and Silver Nitrite

$$\underset{}{RC}{\overset{H}{=}}NO_2^{\ominus} + 2\,Ag^{\oplus} + NO_2^{\ominus} \longrightarrow RCH(NO_2)_2 + 2Ag$$

This method is general for the preparation of *gem*-dinitro compounds, the synthesis of which has been reviewed recently [8].

The classical method is that of ter Meer:

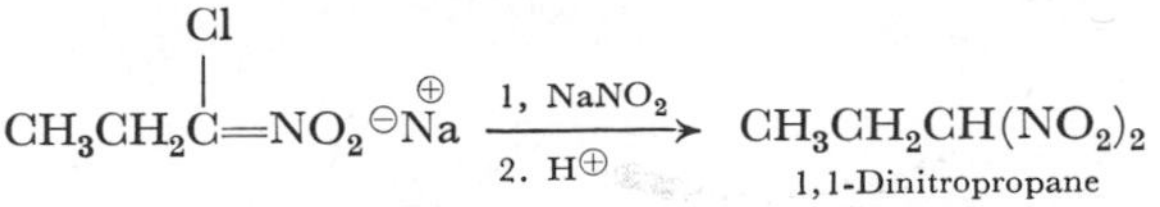

$$CH_3CH_2\overset{\overset{\displaystyle Cl}{\displaystyle |}}{C}{=}NO_2{}^\ominus\overset{\oplus}{Na} \quad \xrightarrow[\text{2. } H^\oplus]{\text{1, NaNO}_2} \quad CH_3CH_2CH(NO_2)_2$$
1,1-Dinitropropane

But the Kaplan-Shechter method [9] shown in the first equation is more direct. One limitation of the latter reaction is that it fails when applied to 1,3-dinitropropane and 1,4-dinitrobutane. To overcome this difficulty in part, the nitro compound may be hydroxymethylated, then treated with silver nitrate and sodium nitrite and dehydroxymethylated [10]:

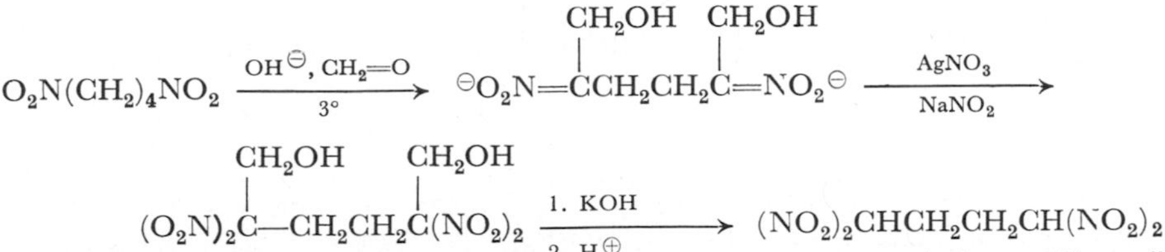

$$O_2N(CH_2)_4NO_2 \xrightarrow[3°]{OH^\ominus, CH_2{=}O} {}^\ominus O_2N{=}\overset{\overset{\displaystyle CH_2OH}{\displaystyle |}}{C}CH_2CH_2\overset{\overset{\displaystyle CH_2OH}{\displaystyle |}}{C}{=}NO_2{}^\ominus \xrightarrow[NaNO_2]{AgNO_3}$$

$$(O_2N)_2\overset{\overset{\displaystyle CH_2OH}{\displaystyle |}}{C}{-}CH_2CH_2\overset{\overset{\displaystyle CH_2OH}{\displaystyle |}}{C}(NO_2)_2 \xrightarrow[\text{2. } H^\oplus]{\text{1. KOH}} (NO_2)_2CHCH_2CH_2CH(NO_2)_2$$
1,1,4,4-Tetranitrobutane, 49% overall

The accompanying tabulation shows a comparison of the yields of $\alpha,\alpha,\omega,\omega$-tetranitroalkanes by the direct Kaplan-Shechter procedure and the indirect methylol procedure.

$(O_2N)_2HC(CH_2)_nCH(NO_2)_2$		
	Percent Yield	
n	Direct	Indirect
1	0	0
2	0	49
3	10	25
4	84	70

The *aci* forms of nitroalkanes have been oxidized to *gem*-dinitroalkanes with dinitrogen tetroxide [11]:

$$C_6H_5CH{=}NO_2H \xrightarrow[\text{Ether}]{N_2O_4} C_6H_5CH(NO_2)_2$$
α,α-Dinitrotoluene, 44.5%

(3) Preparation of 1,1-dinitroethane. A solution of 0.2 mole of nitroethane, 4 g. of sodium nitrite, and 8.5 g. of sodium hydroxide in 80 ml. of water was poured into a stirred solution of 0.41 mole of silver nitrate in 120 ml. of water made slightly alkaline with a few drops of NaOH over which 150 ml. of ether floated. The temperature was originally held at 0–5° but soon rose to 20° and was held at 25° for 30 min. The silver was filtered, washed with ether, and the combined ether extracts were dried and concentrated. On distilling, the residue gave the product in 78% yield [9].

3. From Oximes

$$ArCH{=}NOH \xrightarrow{\text{CF}_3\text{COOOH}} ArCH_2NO_2$$

$$\underset{\text{NOH}}{\overset{\displaystyle RCCOOR'}{\|}} \xrightarrow[\text{NH}_4\text{NO}_3]{\text{HNO}_3} \underset{\text{NO}}{\overset{\displaystyle \overset{\text{NO}_2}{|}}{\underset{|}{RCCOOR'}}} \xrightarrow[\text{Illumination}]{\text{O}_2} \underset{\text{NO}_2}{\overset{\displaystyle \overset{\text{NO}_2}{|}}{\underset{|}{RCCOOR'}}}$$

Oximes have been converted directly into nitro compounds by the use of a variety of oxidizing agents [12, 13]. Perhaps the most common of these are pertrifluoroacetic acid [14] and dinitrogen tetroxide [15]. In the indirect procedure the oxime with N-bromosuccinimide may be converted first into the nitrosobromo derivative which with nitric acid and 30% hydrogen peroxide is oxidized to the corresponding α-bromo nitro derivative. The latter is finally reduced to the unsubstituted nitro compound with sodium borohydride [16, 17].

Dinitro esters may be obtained satisfactorily from some α-oximino esters by oxidation to the α-nitro α-nitroso ester with ammonium nitrate and 100% nitric acid followed by oxidation with oxygen under the influence of illumination (see Ex. b.2). Dinitrogen tetroxide also produces *gem*-dinitro compounds but with serious side reactions (see Ex. b.3):

$$4\,C_6H_5CH{=}NOH + 7\,N_2O_4 \rightarrow 4\,C_6H_5CH(NO_2)_2 + 4\,NO + N_2 + 4\,HNO_3$$

$$4\,C_6H_5CH{=}NOH + 5\,N_2O_4 \rightarrow 4\,C_6H_5CHO + 4\,HNO_3 + 8\,NO + N_2$$

Of these methods of preparing mononitro compounds, the pertrifluoroacetic acid method is of wider application and usually gives better yields. When it is applied to aldoximes and ketoximes, yields varying from 40 to 77% are obtained [14]. For monofunctional oximes the oxidation procedure is modified in that acetonitrile is used as the solvent and a buffer, such as sodium bicarbonate for aliphatic and sodium hydrogen phosphate for alicyclic and aromatic oximes, is employed. The presence of urea improves the yield by serving as a scavenger for any nitrogen oxides formed.

The three-step method has been applied to the oxidation of cycloketoximes [16] and aliphatic ketoximes [17] with overall yields varying from 10 to 50%. It is not satisfactory for aliphatic aldoximes and aromatic ketoximes. However, in contrast to the pertrifluoroacetic acid oxidation procedure, which is sensitive to steric hindrance, it can be used to convert a branched oxime, such as pinacolone oxime, into the pure nitro compound:

$$
\underset{\substack{\big/ \\ CH_3}}{\overset{\substack{CH_3 \\ \big\backslash}}{CH_3-C-CCH_3}} \overset{NOH}{\parallel} \longrightarrow \underset{\substack{\big/ \\ CH_3}}{\overset{\substack{CH_3 \\ \big\backslash}}{CH_3-C-C-CH_3}} \underset{H}{\overset{NO_2}{\mid}}
$$

The *aci* form of the nitro compound 20-IX is generally regarded as the intermediate in the oxidation of oximes. Because of resonance stabilization,

$$
R_2C{=}NOH \to R_2C{=}NO_2H \to R_2CHNO_2
$$
$$
\text{20-IX} \qquad\qquad \text{20-X}
$$

this form is not destroyed but isomerizes to the nitro form 20-X.

(a) **Preparation of phenylnitromethane.** A solution of pertrifluoro-acetic acid (5.5 ml. of 90% hydrogen peroxide, 34.0 ml. of trifluoroacetic anhydride, and 50 ml. of acetonitrile) was added during 75 min. to a stirred refluxing mixture of 2.0 g. of urea, 78 g. of dibasic sodium phosphate, and 12.1 g. of benzaldehyde oxime in 200 ml. of acetonitrile. After the refluxing was continued for an additional hour, the mixture was cooled and added to 400 ml. of water. The combined extracts from four 100-ml. portions of methylene chloride were then washed with three 100-ml. portions of 10% sodium bi-carbonate solution, after which they were dried and distilled under reduced pressure to give 10.6 g. (77%) of phenylnitromethane, b.p. 97–99° (4 mm.) [14].

(b) **Other examples**

(1) **2-Nitropentane** (38% from 2-pentanone oxime first by conversion into the bromonitroso derivative with N-bromosuccinimide, then to the bromo-nitro derivative with 30% hydrogen peroxide and conc. nitric acid, and finally reduction with sodium borohydride) [17].

(2) **Ethyl α,α-dinitrobutyrate** (ethyl α-oximinobutyrate, 0.05 mole, and 0.5 mole each of 100% HNO_3 and ammonium nitrate at 20°; the blue α-nitroso-α-nitro ester which separated was oxidized to the titled compound (78%) by passing oxygen through the liquid illuminated with a 135-watt light) [12].

(3) **α,α-Dinitrotoluene** (43% by passing dinitrogen tetroxide through a cold ethereal solution of benzaldoxime: benzaldehyde always accompanied the product) [18]; for other examples see [15].

1. N. Kornblum, *Org. Reactions*, **12**, 115 (1962).
2. R. H. Wiley and J. L. Hartman, *J. Am. Chem. Soc.*, **73**, 494 (1951).
3. R. W. White and W. D. Emmons, *Tetrahedron*, **17**, 31 (1962).
4. N. Kornblum *et al.*, *J. Am. Chem. Soc.*, **78**, 4003 (1956).
5. W. D. Emmons, *J. Am. Chem. Soc.*, **79**, 5528 (1957).
6. W. D. Emmons and A. F. Ferris, *J. Am. Chem. Soc.*, **75**, 4623 (1953).
7. N. Kornblum and W. J. Jones, *Org. Syn.*, **43**, 87 (1963).
8. H. Feuer, *Tetrahedron*, **20**, Suppl. 1, 103 (1964).
9. R. B. Kaplan and H. Shechter, *J. Am. Chem. Soc.*, **83**, 3535 (1961).
10. H. Feuer *et al.*, *J. Org. Chem.*, **27**, 3598 (1962).
11. S. S. Novikov *et al.*, *J. Gen. Chem. USSR (Eng. Transl.)*, **28**, 2340 (1958).

12. H. E. Ungnade and L. W. Kissinger, *J. Org. Chem*., **24**, 666 (1959).
13. L. Canonica, *Gazz. Chim. Ital.*, **77**, 92 (1947); *C.A.*, **42**, 1885 (1948); E. M. Cherkasova and N. N. Mel'nikov, *Zh. Obshch. Khim.*, **19**, 321 (1949); *C.A.*, **43**, 6569 (1949).
14. W. D. Emmons and A. S. Pagano, *J. Am. Chem. Soc.*, **77**, 4557 (1955).
15. L. F. Fieser and W. von E. Doering, *J. Am. Chem. Soc.*, **68**, 2252 (1946).
16. D. C. Iffland and G. X. Criner, *J. Am. Chem. Soc.*, **75**, 4047 (1953).
17. D. C. Iffland and T.-F. Yen, *J. Am. Chem. Soc.*, **76**, 4083 (1954).
18. Ref. 11, p. 2333.

E. Miscellaneous Reactions

Several unrelated methods of preparation of nitro compounds, too indirect or too narrow in scope to warrant separate sections, are described briefly in this section.

1. From Dienes and Nitroethylenes (Diels-Alder)

Nitroethylenes are excellent dienophiles. A list of their additions to butadiene is recorded [1]. For a general discussion and an example (g.1) see Alkenes, Cycloalkenes, and Dienes, Chapter 2, C.2.

2. From Polynitro Aromatics and Diazomethane

A curious combination addition and insertion process occurs in the reaction of trinitrobenzene and diazomethane [2]:

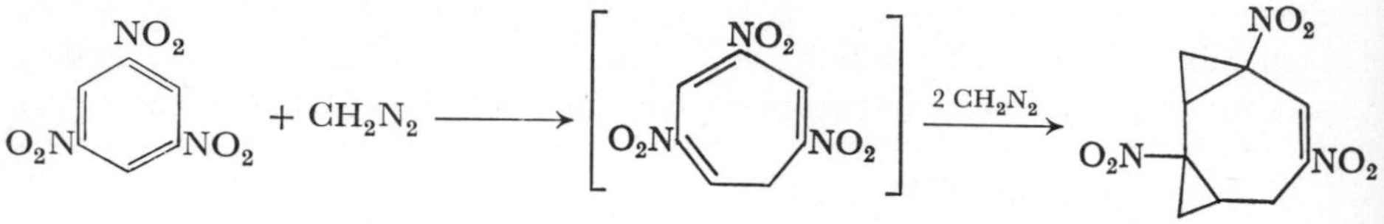

1,4,6-Trinitrotricyclo-
[6.1.0.0^{2,4}]-non-5-ene

Other trinitro aromatics such as picric acid undergo similar reactions.

3. From Nitrocarboxylic Acids (Decarboxylation)

$$NO_2CHCOOH \longrightarrow NO_2CH_2R$$
$$\overset{|}{R}$$

Decarboxylation has been employed to a limited extent in preparing nitro compounds from nitrocarboxylic acids (see Alkanes, Cycloalkanes, and Arenes, Chapter 1, F.1, for decarboxylation procedures). Examples are the synthesis of 1,3,5-trinitrobenzene [3] and 2- (see Ex. b.1) and 3-nitrobenzofurans [4]. In the oxidation of compounds such as TNT to obtain the carboxylic acid for decarboxylation, the presence of nitric acid or acetic acid should be avoided, as the hydroxy group may be introduced in place of the carboxyl group [5]:

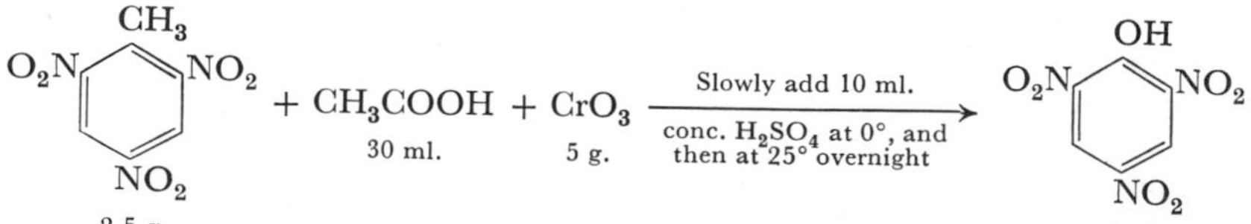

References for Section E are on p. 1009.

The elimination of carbon dioxide is also one step in sequences which have been utilized for the synthesis of nitro compounds. Cases of this sort are the synthesis of nitromethane from nitroacetic acid, which is decarboxylated with extreme ease [6]:

$$\text{ClCH}_2\text{COOH} \xrightarrow{\text{NaNO}_2} \text{NO}_2\text{CH}_2\text{COOH} \longrightarrow \text{CH}_3\text{NO}_2 + \text{H}_2\text{O}$$

phenylnitromethane from benzyl cyanide (see B.3), and *o*-nitroacetophenone from diethyl malonate [7] (see Ex. a):

$$\text{CH}_2\begin{array}{l}\diagup \text{COOC}_2\text{H}_5 \\ \diagdown \text{COOC}_2\text{H}_5 \end{array} \xrightarrow[\text{C}_2\text{H}_5\text{OH}]{\text{Mg}} \text{C}_2\text{H}_5\text{OMgCH(COOC}_2\text{H}_5)_2 \xrightarrow{\textit{o}\text{-NO}_2\text{C}_6\text{H}_4\text{COCl}}$$

$$\textit{o}\text{-NO}_2\text{C}_6\text{H}_4\text{COCH(COOC}_2\text{H}_5)_2 \xrightarrow[\text{H}\oplus]{\text{H}_2\text{O}} \textit{o}\text{-NO}_2\text{C}_6\text{H}_4\text{COCH}_3$$

Yields are variable although they are sometimes in the 80% range.

(a) **Preparation of *o*-nitroacetophenone** (82–83% from diethyl malonate, magnesium, ethyl alcohol, *o*-nitrobenzoyl chloride, and dilute sulfuric acid) [7].

(b) **Other examples**

(1) **2-Nitrodibenzofuran** (72% from 2-nitro-6-dibenzofurancarboxylic acid by heating with powdered copper in quinoline) [4].

(2) **3-Nitropyrrole** (41% from 1 g. of 2-carboxy-4-nitropyrrole, made from α-nitromalonaldehyde and ethyl glycinate, 0.5 g. of copper chromite heated in 5 ml. of quinoline at 210–222° for 1 hr.) [8].

1. A. S. Onishchenko, *Diene Synthesis*, Daniel Davey Co., New York, 1964, Table VII, p. 91.
2. Th. J. DeBoer, *Tetrahedron*, **20**, Suppl. 1, 339 (1964).
3. H. T. Clarke and W. W. Hartman, *Org. Syn.*, Coll. Vol. **1**, 541 (1941).
4. H. Gilman *et al.*, *J. Am. Chem. Soc.*, **61**, 643 (1939).
5. I. Dunstan *et al.*, *Tetrahedron*, **20**, Suppl. 1, 431 (1964).
6. F. C. Whitmore and M. G. Whitmore, *Org. Syn.*, Coll. Vol. **1**, 401 (1941).
7. G. A. Reynolds and C. R. Hauser, *Org. Syn.*, Coll. Vol. **4**, 708 (1963).
8. K. J. Morgan and D. P. Morrey, *Tetrahedron*, **22**, 57 (1966).

AUTHOR INDEX

Carnduff, J., 567
Carney, T. P., 823
Carothers, W. H., 211(2), 881
Carpenter, F. H., 882
Carpenter, L. L., 48
Carpenter, W. G., 759
Carpenter, W. R., 676, 851
Carpino, L. A., 161, 443, 760, 870(2)
Carrol, M. F., 666
Carroll, R. D., 310
Carter, C. L., 935
Carter, H. E., 392
Carter, K. N., 434
Carter, N. M., 527(2)
Carter, P. R., 258
Casas-Campillo, C., 223
Case, L. C., 300
Caserio, M. C., 102, 168, 353
Casey, C., 211
Cash, D. J., 48
Casnati, G., 616
Cason, J., 28, 103, 184, 223, 231, 353, 355, 391, 582, 723(4), 736(3), 823, 824, 825, 852, 881, 950
Caspi, E., 20, 211
Cast, J., 245
Castro, C. E., 104, 112
Catlin, W. E., 28, 230
Caulson, E. A., 867
Cauquil, G., 645
Cava, M. P., 22, 646, 676, 759
Cavalieri, L., 918
Cavalla, J. F., 945
Cavé, A., 918
Ceccotti, S., 696
Cecere, M., 493
Cervinka, O., 696, 713
Chabrier, P., 406, 913
Chadwell, H. M., 244
Chaikin, S. W., 210
Chalmers, W., 210
Chamberlin, E. M., 391
Chambers, G., 390
Chambers, R. A., 459, 512, 825
Chambers, R. D., 266
Chan, D., 376, 713
Chanan, H., 133
Chang, F. C., 104, 354, 759
Chang, J. J., 459
Chapman, N. B., 509
Charles, R. G., 712
Chattaway, F. D., 824
Chatterjee, D. N., 231
Chaux, R., 374, 774

Chaykovsky, M., 33, 144(2), 320(2), 696
Cheeseman, G. W. H., 443
Chenier, P. J., 105
Cherbuliez, E., 913
Cherkasova, E. M., 1008
Cheronis, N. D., 458(2)
Child, W. C., 857
Chitwood, H. C., 311(2)
Chodkiewicz, W., 170
Chortyk, O. T., 928
Chou, T. T., 911
Choudhury, D. R., 162
Chow, Y. L., 493
Christenson, R. M., 928
Christian, J. D., 311
Christian, R. V., Jr., 795
Christmann, A., 493
Christmann, K. F., 144
Christoffel, I., 311
Christopher, H., 460, 973
Chu, E. J. H., 789
Church, J. M., 568
Ciganek, E., 105, 162, 434
Cimarusti, C. M., 149
Ciotti, C. J., Jr., 823, 882
Claborn, H. V., 824
Clapp, L. B., 472
Clar, E., 49, 739(2)
Clark, G., Jr., 112
Clark, J., 353
Clark, R. H., 374
Clarke, H. T., 210, 257, 258, 391, 406, 435, 493, 644, 759, 775, 852, 870, 882, 911, 945, 961, 991, 1009
Clarke, M. F., 354
Clarke, T. G., 774
Clark-Lewis, J. W., 509
Claus, C. J., 539(2), 582
Clauson-Kaas, N., 541
Clayson, D. B., 436
Clement, W. H., 645
Clemo, G. R., 435, 834
Cleveland, E. A., 29
Clibbens, D. A., 780
Clingman, W. H., Jr., 211
Clinton, R. O., 823
Clossen, W. D., 443
Closson, R. D., 33, 354, 471
Coad, P., 354
Coan, S. B., 950
Cobb, R. L., 582, 973
Coburn, E. R., 75, 103
Cocker, W., 52
Coe, D. G., 353

Mao, C. L., 934
Marchand, B., 20(2)
Marcinkiewicz, S., 509
Marckwald, W., 459
Marcus, E., 435
Marcus, N. L., 493
Marek, L. F., 217
Marey, R., 310
Marhan, J., 868
Mariella, R. P., 258, 616(2)
Marion, L., 184
Mark, V., 33, 82, 104, 327(2)
Markley, K. S., 961
Markovac, A., 407
Marks, J. L., 435
Marmor, S., 323
Maroni-Barnaud, Y., 713
Marquet, A., 391(2)
Marsden, E., 66
Marsh, F. D., 646, 974
Marsh, J. P., Jr., 258
Marshall, J. A., 149, 244, 310,
  686
Marshall, J. R., 945
Marsi, K. L., 47
Marszak, I., 459
Martell, A. E., 775
Martello, V., 645, 696
Marthe, J. P., 665
Martin, E. L., 19(2), 433, 736(2),
  881
Martin, J. G., 132, 223
Martin, J. L., 990
Martin, R. H., 676, 735, 870, 945
Martin, W. B., 472, 928, 934
Martinez, A. P., 676, 911
Martzweiller, J. K., 244
Marvel, C. S., 20, 168, 212(2),
  299, 352, 355, 376, 410, 435(2),
  458(4), 644, 676, 760(2), 824,
  913, 928, 945
Marvell, E., 825
Marvell, E. N., 48, 284
Marxer, A., 269, 478
Masaki, M., 443
Mashentsev, A. I., 867, 868
Maslova, M. A., 48(2)
Mason, J. P., 1002
Mason, S. F., 20
Massie, S. P., 825
Mastagli, P., 583, 852
Mathes, W., 212, 390
Mathieu, J., 244(2), 695, 696,
  712, 851, 973
Matlack, G. M., 104
Matsuda, K., 55

Matsuda, T., 527
Matsuura, T., 759
Matterstock, K., 977
Matteson, D. S., 375
Matthews, V. E., 1002
Maury, P. B., 882
Mauthner, F., 299
May, J., 539
Mayer, H., 434
Mayer, J., 20
Mayer, R., 759
Mayer, U., 223
Maynard, J. T., 354
Mayo, F. R., 374(2)
Mazdiyasni, K. S., 824
Mazur, R. H., 912
McAuley, M. L., 852
McBee, E. T., 104, 406(2), 723,
  960
McCain, G. H., 144
McCasland, G. E., 472
McClellan, W. R., 852
McClure, J. D., 266, 835
McColm, E. M., 945
McConaghy, J. S., Jr., 433
McCoy, L. L., 390, 852
McCoy, R. E., 526
McCracken, J. H., 66, 472
McCrae, W., 170
McCutcheon, J. W., 759
McDonald, R. N., 144(2)
McDowell, J. W., 435, 915(2)
McElvain, S. M., 55, 212(2), 355,
  472, 526, 527, 696, 823, 824,
  825, 853, 912, 918, 960
McEwen, W. E., 582, 713, 723,
  973, 1002
McEwen, W. L., 168, 352, 374,
  775
McFadyen, J. S., 583
McFarland, J. W., 823, 881, 913
McGovern, T. P., 568
McGraw, W. J., 319
McHatton, L. P., 644(2)
McIntosh, A. V., Jr., 212, 582(2)
McIvor, R. A., 676
McKay, F. C., 443
McKee, G. H. W., 459
McKee, R. A., 54
McKeever, C. H., 407, 568
McKenzie, A., 607, 760
McKenzie, B., 945
McKinley, J. B., 928
McKusick, B. C., 104, 974(2)
McLamore, W. M., 434, 736(2)
McLeod, G. L., 776

SUBJECT INDEX

The Table of Contents in front of each chapter serves as a general index and this Section as a specific index. Symbols occasionally to be found in this Section are (O) for oxidation, (H) for reduction, (-H) for dehydrogenation and (H → NO_2) for nitration.

nitriles, 946
phenols, 270
Alkylation, reductive, 424
Alkylation, reductive, of an
    enamine and a carbonyl
    compound, 426
Alkylation, side-chain, 30
  experimental conditions
  mechanism, 31
Alkylbenzenes
 by rearrangement (Jacobsen), 43
N-Alkylbenzenesulfonamides (N-
    vinylamides), 931
Alkylcyclohexanes (arenes), 14
Alkylidenation, trans, 115
Alkylidenediacylamides (RCHO +
    amide), 931
N-Alkylphthalimides, 437
Alkynes (see also Acetylenes)
 from acetylene dihalides or
    tetrahalides (OH⁻ or Zn),
    154
 from acid chlorides (phospho-
    ranes), 160
 from benzils and triethylphos-
    phite (Δ), 157
 from alkali metal salts (RX or
    R₂SO₄), 162
 from lower alkynes (NaNH₂ and
    R₂SO₄), 162
 from the benzenesulfonate anions
    of triazoles (hν), 157
 from β-chlorocyclic ethers
    (NaNH₂-liquid NH₃), 154
 from cyclooctatetraenes (hν),
    169–170
 from gem dihalides, vicinal
    dihalides, and vinyl
    halides (NaNH₂, KNH₂,
    KOC(CH₃)₃, or KOH), 151,
    152
  rearrangement, 151, 152
 from dihydrazones (CF₃COOAg-
    (C₂H₅)₃N, HgO), 156
 from ethylene-bis-(trimethyl-
    ammonium), hydroxides (Δ),
    159
 from ethynyl Grignard reagents
    (alkylating agents), 164
 from Li acetylides (RBr and
    LiNH₂-liquid NH₃), 162, 163
 from nitrosooxazolidones (KOH),
    156
 from propargyl acetoacetates (Δ),
    170
    ⁻ pyrazolones (Br₂ and NaOH),

    156
 from quaternary ammonium salts
    of enamines (Δ, NaNH₂-
    liquid NH₃, KNH₂-liquid
    NH₃), 159
 from thiondisulfides (Raney Ni),
    160
 from vinyl amines (isoamyl
    nitrite), 157
 from unsaturated ketones via
    the p-toluenesulfonylhy-
    drazone, 157
Alkynes, aminomethyl
 from acetylenes (R₂NH and RCHO),
    166
Alkynes, aryl
 from acetylenic alcohols (NaOH),
    165
Alkynes, chloromethyl
 from methylalkynes [(CH₃)₃COCl)],
    166
Alkynes, halo-
 from the alkyne (X₂ + NaOH), 166
Alkynyl acetals
 from acetylenic Grignard rea-
    gents [HC(OC₂H₅)₃], 166
Alkynyl alcohols
 from >C=O compounds (RC≡CMgX),
    164
Alkynyl carboxylic acids
 from CO₂ (RC≡CMgX), 164
Alkynyl glycols
 from >C=O (XMgC≡CMgX), 164
Allene
 from 2,3-dibromo-1-propene, 171
 from 2,3-dichloropropene, 82
Allenes
 from acetylenes (base-catalyzed
    isomerization), 171
 from acetylenes [LiCu(CH₃)₂],
    130
 from allyldibromocyclopropanes
    (CH₃Li), 171
 from bromoacetylenes (RMgX,
    CuBr), 172
 from dihaloethylenes (Zn-ester),
    171
 from propargyl halides (Zn-Cu),
    171
 from pyrazoline hydrazones (Ni
    peroxide), 172
Allenic bromides (the alcohol),
    333
Allophanates (HNCO), 820
2-Alloxy-1-phenylethanol
 from styrene oxide, 306

anhydride adducts, 53
by dehydrogenation of hydro-
    aromatic hydrocarbons, 49,
    50
from diazonium borofluorides
    (NaBH$_4$), 16
from diazonium salts (H$_3$PO$_2$ pre-
    ferred), 16
from esters (alkali metal car-
    bonates, LiI, KOH), 53
from the Grignard reagent (HOH),
    22
from hexafluorophosphonium
    salts [(CH$_3$)$_2$NCON(CH$_3$)$_2$],
    17
from phenols
  via diethyl phosphates, 7
  via tetrazole ethers, 7
  via p-toluenesulfonates, 7
from quinones (H), 14
Arenes, alkyl
from alkyl arenes (olefin with
    alkali metals, C$_6$H$_5$CH$_2$Na,
    RNa, or K graphite), 30
from anions (RX), 25
from the arene [(CH$_3$)$_3$COK-DMSO],
    32
from arenes (RCH=CH$_2$, ROH, RX,
    carbonyl compounds, sul-
    fonates) (Friedel-Crafts),
    37, 38
  catalysts, 37
from ArI[LiCu(CH$_3$)$_2$], 26
via the benzyne, 25
from Grignard reagent or alkyls
    of metals (RX), 23
Arenes, methyl
from benzyltrimethyl ammonium
    salts (H$_2$-Pd-C-NaOH,
    LiAlH$_4$, Na-Hg-H$_2$O, hν), 10
Arenes, polynuclear
from o-acyldiphenylmethane
    (HBr-CH$_3$COOH or HBr-H$_2$SO$_4$),
    39, 40
from arenes (AlCl$_3$, AlCl$_3$-HCl,
    AlCl$_3$-CuCl$_2$), 46
from aromatic ketones (Elbs),
    39
Arndt-Eistert reaction, 179
Aromatic acids
from the pyridinium iodides,
    774
Aromatic ring activation, 249
Aroylaminoacids
from thiol [NH$_2$CH(R)COOH], 907
Arylation of phenols or dianions,

270, 699
Arylcyclopropanes
from ethyl cinnamate (LiAlH$_4$),
    195
Arylidenemalonic acids
from the aldehyde (Doebner), 781
9-Arylmethylanthracene
from 10-arylmethylene-9-anthrone,
    14
Arylmethylformamide
from the aryldimethylamine, 511
Arylpropyl alcohols
from aryl m-dioxanes (Na-ROH),
    189
Arylpyridines
by coupling of pyridine with
    the diazonium salt, 56
by coupling of pyridine with
    the N-nitrosoacetylamine,
    56
Arylthiophenes
by coupling of thiophene with
    the diazonium salt, 56
by coupling of thiophene with
    the N-nitrosoacetylamine,
    56
Aspartic acid, β-methyl esters
from monomethyl maleate (RNH$_2$),
    469
Atrolactic acid
from acetophenone via the
    cyanhydrin, 753
1-Aza-5,6-benzo-2-dimethylamino-
    3-phenylcyclooocta-1,3,5-
    triene
from 3,4-dihydroisoquinoline,
    491
1-Aza-2-chloro-3,3,4-trimethyl-
    bicyclo (2·2·1) heptane
from 4,7,7-trimethyl-2-chloro-
    2-azabicyclo (2·2·1)
    heptane, 410
Azelaic acid
from castor oil, 767
from methyl oleate, 768
from 9-octadecynedioic acid, 769
Azetidine
from trimethylene bromide, 447
Azetidines, N-phenyl
from γ-phenylaminopropyl ethers,
    453
Azide structure, 490
Aziridines
from acyl azides (olefins), 925
from aromatic ketoximes RMgX,
    LiAlH$_4$), 475

by the Reformatsky reaction,
   230
t-Butyl hypochlorite, 166, 378,
   855(2), 869
N-2-Butylidenecyanamide
 from isobutylene, 970
t-Butyl isobutyl ketone
 from isobutane, 662
t-Butyl isopropyl ether
 from the two alcohols, 302
t-Butyl methyl acetal of p-
   chlorobenzaldehyde
 from p-chlorobenzyl methyl
   ether, 538
1-n-Butylnaphthalene
 from 1-bromonaphthalene via the
   Li derivative, 26
Butyl oleate
 from olive oil, 815
N-t-Butyloxycarbonylamino acids
 from t-butyl azidoformate
   (NH$_2$CHRCOOH-MgO), 909
N-t-Butyloxycarbonyl-L-methionine
 from t-butylazidoformate, 909
t-Butyl peracetate, 537(3), 538
t-Butyl perbenzoate, 260, 261(2),
   537, 563, 564
m-t-Butylphenyl p-t-butylbenzoate
 from p-t-butylbenzoic acid, 819
t-Butyl phenyl ether
 from t-butyl perbenzoate, 298
t-Butylphthalimide
 from anhydride, 901
t-Butylpropargylic acid
 from 3-t-butylpyrazolin-5-one,
   756
5-n-Butylpyridine-2-carboxylic
   acid
 from the ethyl ketone, 773
N-Butylpyrrolidine
 from N-chlorodi-n-butylamine,
   485
t-Butyl N-styrylcarbamate
 from cinnamamide, 497
Butyl 2,4,6-trimethylbenzoate
 from the Na salt, 818
N-Butylvaleramide
 from dibutylacetic acid, 927
Butyraldehyde
 from α-bromovaleric acid, 567
 from 1-butanol, 549
 from n-butyryl chloride, 570
 from n-propylmalonic acid
   diazide, 607
Butyraldehyde dibutyl acetal
 from acetaldehyde dibutyl acetal,

   524
Butyric acid
 from 1-nitrobutane, 799
 by Wittig reaction, 783
Butyric anhydride
 from butyraldehyde, 884
Butyroin
 from ethyl butyrate, 207
Butyro-p-toluidide
 from n-propyl butyrate, 902
Butyryl chloride
 from n-butyric acid, 863

Camphane
 from 4-benzenesulfonycamphane,
   16
 from 4-phenylmercaptocamphane,
   16
Cannabidiol
 from the dimethyl ether, 253
Cannizzaro reaction, 764
Cannizzaro (crossed), 199, 239
Caproaldehyde
 from capronitrile, 573
n-Caproic anhydride
 from n-caproic acid, 881
ε-Caprolactam
 from cyclohexanecarboxylic acid,
   927
 from cyclohexanone oxime, 922,
   924(2)
ε-Caprolactone
 from cyclohexanone, 828, 830
Capronitrile
 from caproic acid, 952
cis-Carane
 from car-3-ene, 18
Carbalkoxylation, 831
 of arenes, 833
 to esters, 831
Carbalkoxymethylation
 to lactones, 833
Carbalkoxymethylation of alkenes
 by manganese III triacetate, 834
Carbamates
 from amides via Hofmann rearrangement, 819
 from amides (NaOBr and ROH), 819
 from amides [Pb(OCOCH$_3$)$_4$], 497
 from isocyanates (ROH), 909
Carbamates, aryl
 from isocyanates [ArOH-
   (C$_2$H$_5$)$_3$N or C$_5$H$_5$N], 812
 from phenols (C$_6$H$_5$NCO-AlCl$_3$),
   919
Carbamates, unsaturated

from phenacylchloride, 720
from cis- and trans-stilbene
    oxide, 679, 680, 681
Desoxybenzoins
from aldehydes via α-dimethyl-
    aminophenylacetonitriles,
    709
from the amide (RLi or RMgX),
    718
Desulfonation
to alkanes, 45
to arenes, 44
Desulfurization
to alkynes, 160
mechanism, 15
Desyl chloride
from benzoin, 337
Deuterium
incorporation in benzene ring,
    22
β-Deutero-α-bromopropylbenzene
from cis- or trans-1-phenylpro-
    pene, 357
Diacetone alcohol
from acetone, 238
Diacetonamine
from mesityl oxide, 468
3β, 17β-Diacetoxy-7α-hydroxyan-
    drostane
from the ketone, 203
4,5-Diacetyl-3-aminophenoxazone
from the hydrochloride of 2-
    amino-3-acetylphenol, 729
p-Diacetylbenzene
from p-ethylacetophenone, 635
N,N'-Diacetyl-2,5-dimethyl-1,4-
    diaminomethylbenzene
from p-xylene, 922
2,4-Diacetyl-D-erythrose
from 1,4,6-triacetyl pseudoglucal,
    555
2,5-Diacetylquinol
from quinol diacetate, 657
Diacids
from glycols via the dinitriles,
    753
Diacyl halides
from the dicarboxylic acid
    (PCl$_5$ or SOCl$_2$), 862
Dialkylhydrazines
from hydrazones (RLi), 475
Dialkylthiocarbamates, S-aryl
from O-aryl dialkylthiocarba-
    mates, 908
Dialkynes
from acetylenes (O$_2$ and

CuCl-C$_5$H$_5$N) (Glaser), 168
from acetylenes (RC≡CBr-Cu$^+$-NH$_3$),
    168
from acetylenes, terminal (Cu^{++}),
    168
Diallylamine
from the cyanamide, 440
Diamides
from cyclic anhydrides, 900
from α,α-disubstituted malonic
    acids, 902
Diamines
from dicarboxylic acids (Schmidt),
    502
from ethylenimine (NH$_3$ or amines
    with AlCl$_3$ or NH$_4$Cl), 466
cyclic 1,3-, 462
t-Diamines, ethylenic
from unsaturated aldehydes
    (R$_2$NH-K$_2$CO$_3$ or R$_2$NH-MgSO$_4$),
    467
α,δ-Diaminoadipic acid
from the diphthalimidoadipate,
    439
Diaminobiphenyls
from hydrazobenzenes (H$^+$), 503
2,5-Diamino-3,4-dicyanothiophene
from tetracyanoethylene, 972
4,8-Diamino-2,10-dimethylundecane
from the dicarboxylic acid, 502
3,5-Diaminophenol
from phloroglucinol, 450
2,4-Diaminotoluene
from 2,4-dinitro-, 417
Di-n-amyl ketone
from n-hexyl alcohol, 630
1,2-Di-9-anthrylethane
from 9-anthraldehyde, 6, 195
1,5-Diazabicyclo(3.4.0)nonene-5,
    DBN, 78(3), 79, 264(2)
1,5-Diazabicyclo(4.5.0)undecene-5,
    78, 79
1,5-Diazacyclooctane hydro-
    chloride
from 1,5-diazabicyclo(3.3.0)
    octane, 418
3,5-Diazapyrylium salts
from the N-acylnitrilium salt,
    490
Diazoalkane structure, 490
Diazo ketones
from acid chlorides, 179
from α-acylacetaldehydes, 179
Diazomethane
from bis-(N-methyl-N-nitroso)-
    terephthalamide, 826

from o-diazidobenzene, 955

1,4-Dicyano-2-butene
  from acrylonitrile, 114

1,2-Di-1-(1-cyano)cyclohexylhy-
    drazine
  from cyclohexanone, 966

Dicyandiamide, 908, 959

1,2-Dicyano-1,2-dimethylmercapto-
    cyclobutane
  from α-methylmercaptoacryloni-
    trile, 138

4,4'-Dicyanodiphenyl
  from the tetrazonium salt, 943

Di-(β-cyanoethyl)-formal
  from H₂CO, 535

α,α-Dicyano-β-ethyl-β-methylglu-
    tarimide
  from a cyanoacetic ester, 902

2,3-Dicyano-2-heptene
  from 2-heptynyltrimethyl-
    ammonium iodide, 943

2,3-Dicyanohydroquinone
  from benzoquinone, 964

Dicyanoketene diethyl or cyclic
    ketals
  from [CN₂C=C(CN)₂], 534, 972

2,2-Dicyanopropane
  from methylene cyanide, 947

1,3-Di-(cyclohexenyl)-1,3-
    propanedione
  from 4-carbomethoxycyclohexene,
    692

Dicyclohexylcarbodiimide, DCC,
    875(2), 896, 897, 906,
    933, 951, 953

N,N'-Dicyclohexyl-0-(2-
    hydroxyphenyl)-isourea
  from DCC, 906

N,N'-Dicyclohexyl-N(2,4,6-
    trinitrophenyl)urea
  from DCC, 906

N,N'-Didecyloxamide
  from decylamine, 927

1,3-Didecylurea
  from decylamine, 927

β,β-Dideuterobutylamine
  from the nitrile, 420

cis- or trans-1,2-(2,6-Dimethyl-
    phenyl)-1,2-ethenediol
  from 2,6-xylil, 208

Dieckmann condensation, 674, 688,
    839, 844

Diels-Alder reaction (see also
    Cycloaddition), 116, 269,
    491, 939, 1008
  inverse electron demand, 121

Lewis acid adducts, 121
  stereospecificity and facts on,
    118-122

Dienes
  from alkynes or diynes
    [(CH₃)₃CO⁻, Na], 137
  from pinacols [Al₂O₃, Ca₃(PO₄)₂,
    HBr], 73
  by pyrolysis, 139

Dienes, C₈
  from butadiene and ethylene, 115

1,3-Dienes
  from alkenes (alumina-chromium
    oxide), 139
  from 4-halo-2-butenes (KOH), 79
  from 3-halocyclohexene
    [C₆H₅N(CH₃)₂], 79
  from β,γ-unsaturated halides
    [(C₆H₅)₃P], 143
  from unsaturated phosphono-
    esters (>C=O compounds
    and NaH), 143
  from vinyl Grignard reagents
    (CuCl), 131

1,4 Dienes
  from 2,4-diolefinic acids
    (ArN₂⁺X⁻) (Meerwein), 146

Dienone-phenol rearrangement,
    275, 276

Dienophiles, carbonyl, 327

p-Di(epoxyethyl) benzene
  from p-di-(α-hydroxy-β-
    chloroethyl) benzene, 296

Diesters
  from acid ester salts (Kolbe),
    65, 856

gem-Diesters
  from the aldehyde, 813

Diesters of aldehydes or ketones
  from the acid [(CH₃CO)₂O-H⁺],
    806

1,1-Diethoxy-3-butene
  from vinyl ethyl ether, 530

Diethoxycarbonium tetrafluoro-
    borate, 420, 421

2-(β,β-Diethoxyethyl)-1,3-
    dithiane
  from 1,3-dithiane, 595

2-(2,3-Diethoxyphenyl)-2-hydroxy-
    1-nitroethane
  from nitromethane, 999

Diethylacetaldehyde
  from 5,5-diethyloxazolidone, 594

Diethyl α-acetamidobenzylmalonate
  from N,N'-benzylidenebisacetamide,
    921

Ethyl pentyl ketone
  from (α-hexanoylethylidene)-
    triphenylphosphine, 722
Ethyl phenylacetate
  from phenylacetonitrile, 814
Ethyl α-phenylacetoacetate
  from α-phenylacetoacetonitrile,
    814
Ethyl phenylcyanoacetate
  from phenylacetonitrile, 838
Ethyl β-phenyl-β-hydroxypro-
    pionate
  by the Reformatsky, 229
Ethyl phenyl ketone
  from methyl α-phenylvinyl ether,
    708
Ethyl phenylpyruvate
  from ethyl β-phenylglycidate,
    681
N-Ethylphthalimide
  from the amide, 899
N-Ethylpiperidine
  from ethylene, 461
N-Ethylpropargylamine
  from 2-chloro-3-ethylamino-1-
    propene, 153
α-Ethyl-β-propylacrolein
  from butyraldehyde, 612
Ethyl propyl ether
  from ethyl sulfate, 292
Ethyl pyruvate o-nitrophenylhydra-
    zone
  from ethyl 2-methylacetoacetate,
    636
β-Ethylstyrene, cis and trans
  from benzylidenetriphenyl-
    phosphorane, 142
Ethyl α-tetralone-β-carboxylate
  from α-tetralone, 839
N-Ethyl-m-toluidine
  from ethyl bromide, 446
  from N-nitroso-N-ethyl-m-
    toluidine, 419
Ethyl 2,3,5-triphenyl-5-
    ketopentanoate
  from benzalacetophenone, 845
Ethyl urethane of 2-amino-1,6,6-
    trimethylbicyclo (1.1.2)
    hexane
  from the acid azide, 820
Ethyl vinyl ether
  from n-butyl vinyl ether, 305
α-Ethynyl carbinols
  from acetylene (RCOR), 756
Ethynylcarbinols
  via lithium acetylide, 226

  via sodium acetylide, 226
2-Ethynyl-trans-2-decalol
  from trans-2-decalone, 165
Ethynylmagnesium bromide,
    preparation, 164
Etiocholan-3α-ol-11,17-dione
  from pregnan-3α,17α-diol-11,20-
    dione, 638
o-Eugenol
  from guaiacol allyl ether, 283
Eugenol-5-carboxaldehyde
  from eugenol, 588

Favorskii rearrangement, 797,
    798, 848
Fenton's reagent, 220
Ferric chloride
  Friedel-Crafts catalyst, 37
  inhibitor in Birch reduction, 18
  oxidizing agent in quinone
    synthesis, 770
Finkelstein halide interchange,
    339
Fischer indole synthesis, 504
Fission, alkyl, in ether forma-
    tion, 291
Flavasperone
  from the methyl ether, 254
Fluoralkynes
  from Li acetylide
    $-NH_2CH_2CH_2NH_2$ (fluoralkyl
    halide), 163
Fluoranthene
  by dehydrogenation of 6b,7,10,
    10a-tetrahydrofluoranthene,
    51
9-Fluorenecarboxylic acid
  from fluorene, 771
Fluorenes, 9-alkyl
  from fluorene (RONa), 31
Fluorenone
  from 9,9-dibromofluorene, 668
Fluorescein
  from resorcinol, 307, 308
Fluorides
  from alcohols [$CFClHCF_2N(C_2H_5)_2$],
    351
  from alcohols (SF_4), 351
  from alkanes (F_2-N_2, or COF_3),
    379
  from anhydrides (HF-electrolysis),
    379
Fluorides, alkyl
  from halides (BrF_3, HF, or a
    metal fluoride), 340
  from ketones or nitriles

2,2,5,5-Tetramethoxy-2,5-
dihydrofuran
from furan, 540
3,3',5,5'-Tetramethoxydipheno-
quinone
from 2,6-dimethoxyphenol, 729
Tetramethylallene
from 4-keto-3,3,5,5-tetramethyl-
pyrazoline hydrazone, 172
Tetramethylarabinose
from pentamethylgluconamide, 607
p-(2,3,5,6-Tetramethylbenzoyl)-
benzylamine hydrochloride
from p-cyanobenzoyldurene, 420
N-(1,1,3,3-Tetramethylbutyl)-
acetamide
from diisobutylene, 921
Tetramethyl-1,3-cyclobutanediol
from the diketone, 203
Tetramethylcyclobutane-1,3-dione
from dimethylketene, 695
Tetramethylenediamine
from pyridazine, 430
Tetramethylene sulfone, 866(2)
Tetramethylethylene
from tetramethyl-1,3-
cyclobutanedione, 148
2,3,4,6-Tetramethyl-d-glucose
from methyl sulfate, 292
2,2,5,5-Tetramethyl-3-hexanol
from the trans-3-hexene, 188
2,2,7,7-Tetramethyl-5-
homopiperazinone
by Schmidt, 926
1,1,2,3-Tetramethylindene
from 1,1-dibromotetramethylcy-
clopropane, 39
2,2,5,5-Tetramethylpentanone
from the adipic acid, 687
Tetramethylthiourea, 571
Tetramethylurea, 17, 757
1,2,3,5-Tetraminobenzene trihydrochloride
from picramide, 416
1,1,4,4-Tetranitrobutane
from 1,4-dinitrotetramethylene,
1005
Tetranitroesters
from α,ω-bismethylol derivatives
of $\alpha,\alpha,\omega,\omega$-tetranitroal-
kanes (CH$_3$OH-NaOH-CH$_2$=
CHCOOR), 1000
2,4,5,7-Tetranitrofluorenone
from 9-fluorenone, 986
Tetranitromethane, 981, 984
1,3,3,4-Tetraphenyl-2-azetidinone

from diphenyl ketene, 906
1,1,4,4-Tetraphenylbutane
from the alkene, 30
1,1,2,2-Tetraphenylethanol
from benzophenone, 234
from disodiobenzophenone, 242
Tetraphenylethylene
from diazodiphenylmethane, 99(2)
from diphenyldichloromethane, 81
from 1,1,2,2-tetraphenylethanol,
74
from tetraphenylethylene
sulfide, 98
1,1,6,6-Tetraphenylhexapentaene,
173
1,2,3,4-Tetraphenylnaphthalene
from diphenyliodonium-2-
carboxylate, 68
Tetrazole
from 5-aminotetrazole, 16
Tetrolic acid
from 3-methyl-2-pyrazolin-5-one
via the dibromide, 156
Thallic acetate, 536, 537(2), 856
Thallium bromide, 28
Thallium phenoxide, 807
Thallium ethoxide, 879
Theilacker rearrangement, 494
3-Thenoic acid
from the aldehyde, 764
3-Thenyl bromide
from 3-methylthiophene, 382
Thiocarbonyldiimidazole, 148, 149
Thiocyanides
from sulfides (CNBr and Δ), 969
Thiocyanogen chloride and
trichloride, 364
β,β-Thiodipropionic acid
diamide
from urea, 910
Thionimides
from aroylisothiocyanates
(AlCl$_3$), 920
Thionyl fluoride
from the chloride, 340
2-Thiophenealdehyde
from 2-chloromethylthiophene,
558
Thiophenecarboxaldehyde
from thiophene, 585
Thiophthalimide
from benzoylisothiocyanate,
920
9-Thioxantheneacetic acid
from thioxanthhydrol, 795
Thorium dioxide aerogel, 687